Handbook of
Heat Transfer

OTHER McGRAW-HILL HANDBOOKS OF INTEREST

AMERICAN SOCIETY OF MECHANICAL ENGINEERS · ASME Handbooks:

    Engineering Tables   Metals Engineering—Processes

    Metals Engineering—Design   Metals Properties

BAUMEISTER AND MARKS · Standard Handbook for Mechanical Engineers

BEEMAN · Industrial Power Systems Handbook

BRADY · Materials Handbook

BURINGTON AND MAY · Handbook of Probability and Statistics with Tables

CALLENDER · Time-Saver Standards

CARRIER AIR CONDITIONING COMPANY · Handbook of Air Conditioning System Design

CARROLL · Industrial Instrument Servicing Handbook

CONSIDINE · Process Instruments and Controls Handbook

CONSIDINE AND ROSS · Handbook of Applied Instrumentation

CROCKER AND KING · Piping Handbook

DUDLEY · Gear Handbook

EMERICK · Handbook of Mechanical Specifications for Buildings and Plants

EMERICK · Heating Handbook

EMERICK · Troubleshooters' Handbook for Mechanical Systems

FACTORY MUTUAL ENGINEERING DIVISION · Handbook of Industrial Loss Prevention

FINK AND CARROLL · Standard Handbook for Electrical Engineers

FLÜGGE · Handbook of Engineering Mechanics

GARTMANN · De Laval Engineering Handbook

HARRIS · Handbook of Noise Control

HARRIS AND CREDE · Shock and Vibration Handbook

HEYEL · The Foreman's Handbook

KALLEN · Handbook of Instrumentation and Controls

KING AND BRATER · Handbook of Hydraulics

KLERER AND KORN · Digital Computer User's Handbook

KOELLE · Handbook of Astronautical Engineering

KORN AND KORN · Mathematical Handbook for Scientists and Engineers

LEGRAND · The New American Machinists' Handbook

MACHOL · System Engineering Handbook

MAGILL, HOLDEN, AND ACKLEY · Air Pollution Handbook

MANAS · National Plumbing Code Handbook

MANTELL · Engineering Materials Handbook

MAYNARD · Industrial Engineering Handbook

MERRITT · Building Construction Handbook

MORROW · Maintenance Engineering Handbook

PERRY · Chemical Engineers' Handbook

PERRY · Engineering Manual

ROSSNAGEL · Handbook of Rigging

ROTHBART · Mechanical Design and Systems Handbook

SHAND · Glass Engineering Handbook

SOCIETY OF MANUFACTURING ENGINEERS:

    Die Design Handbook   Manufacturing Planning and Estimating Handbook

    Handbook of Fixture Design   Tool Engineers Handbook

STANIAR · Plant Engineering Handbook

STREETER · Handbook of Fluid Dynamics

TOULOUKIAN · Retrieval Guide to Thermophysical Properties Research Literature

TRUXAL · Control Engineers' Handbook

Handbook of Heat Transfer

Edited by

WARREN M. ROHSENOW
Professor of Mechanical Engineering
Massachusetts Institute of Technology

JAMES P. HARTNETT
Head, Department of Energy Engineering
University of Illinois at Chicago Circle

McGRAW-HILL BOOK COMPANY

New York   St. Louis   San Francisco   Düsseldorf   Johannesburg
Kuala Lumpur   London   Mexico   Montreal   New Delhi   Panama
Rio de Janeiro   Singapore   Sydney   Toronto

Library of Congress Cataloging in Publication Data

Rohsenow, Warren M
    Handbook of heat transfer.

    1. Heat—Transmission—Handbooks, manuals, etc.
I. Hartnett, James P., joint author. II. Title.
QC320.R528   536'.2   72-11529
ISBN 0-07-053576-0

1234567890 KPKP 76543

*The editors for this book were Harold B. Crawford, Daniel N.
Fischel, and Don A. Douglas, the designer was Naomi Auerbach,
and its production was supervised by Stephen J. Boldish. The
text was edited by William Begell and it was set in Press Roman
by Scripta Technica, Inc.*

It was printed and bound by The Kingsport Press.

Contents

vi   Contents

Index follows Section 19.

Contributors

ARTHUR E. BERGLES, *Chairman, Department of Mechanical Engineering, Iowa State University.* (**Section 10.** Techniques to Augment Heat Transfer)

ROBERT D. CESS, *State University of New York, Stony Brook, New York.* (**Section 6.** Free-convection Boundary-layer Heat Transfer)

R. V. DUNKLE, *Division of Mechanical Engineering, C.S.I.R.O. Highett, Victoria, Australia.* (**Section 15.** Radiation, *Part C,* Radiation Exchange in an Enclosure with a Participating Gas)

E. R. G. ECKERT, *Regent's Professor of Mechanical Engineering, University of Minnesota.* (**Section 15.** Radiation, *Part A,* Relations and Properties)

WARREN H. GIEDT, *Professor of Mechanical Engineering, University of California, Davis.* (**Section 9.** Rarefied Gases)

PETER GRIFFITH, *Professor of Mechanical Engineering, Massachusetts Institute of Technology.* (**Section 12.** Condensation, *Part B,* Dropwise Condensation. **Section 14.** Two-phase Flow)

J. P. HARTNETT, *Head, Department of Energy Engineering, University of Illinois at Chicago Circle.* (**Section 1.** Introduction, *Part A,* Basic Concepts of Heat Transfer. **Section 17.** Mass Transfer Cooling)

HENRYK HURWICZ, *Chief Advance Technology Engineer, Aero/Thermodynamics and Nuclear Effects, Research and Development, Advance Systems and Technology, McDonnell Douglas Astronautics Company–West.* (**Section 16.** Ablation. **Section 19.** High-temperature Thermal Protection Systems)

WARREN IBELE, *Professor of Mechanical Engineering, University of Minnesota.* (**Section 2.** Thermophysical Properties)

MAMORU INOUYE, *Ames Research Center, NASA, Moffett Field, California.* (**Section 8.** Forced Convection, External Flows)

WILLIAM M. KAYS, *Professor of Mechanical Engineering, Stanford University.* (**Section 7.** Forced Convection, Internal Flow in Ducts)

A. C. MUELLER, *Engineering Department, E. I. du Pont de Nemours & Co., Inc.* (**Section 18.** Heat Exchangers)

VICTOR PASCHKIS, *Consultant, Professor Emeritus of Mechanical Engineering and Formerly Director of the Heat and Mass Flow Analyzer Laboratory, Columbia University.* (**Section 5.** Analog Methods)

S. S. PENNER, *Department of Applied Mechanics and Engineering Sciences, University of California, San Diego, La Jolla.* (**Section 15.** Radiation, ***Part D,*** Equilibrium Radiation Properties of Gases)

H. C. PERKINS, *Aerospace and Mechanical Engineering Department, The University of Arizona.* (**Section 7.** Forced Convection, Internal Flow in Ducts)

P. RAZELOS, *Associate Professor of Engineering Science, Richmond College of the City University of New York.* (**Section 4.** Methods of Obtaining Approximate Solutions)

J. E. ROGAN, *Branch Chief, Aero/Thermodynamics and Nuclear Effects, Research and Development, Advance Systems and Technology, McDonnell Douglas Astronautics Company–West.* (**Section 16.** Ablation. **Section 19.** High-temperature Thermal Protection Systems)

WARREN M. ROHSENOW, *Professor of Mechanical Engineering, Massachusetts Institute of Technology.* (**Section 1.** Introduction, ***Part A,*** Basic Concepts of Heat Transfer. **Section 12.** Condensation, ***Part A,*** Film Condensation. **Section 13.** Boiling)

MARY F. ROMIG, *Lecturer in Astronomy, University of Southern California, Los Angeles.* (**Section 11.** Electric and Magnetic Fields)

MORRIS W. RUBESIN, *Ames Research Center, NASA, Moffett Field, California.* (**Section 8.** Forced Convection, External Flows)

P. J. SCHNEIDER, *Lockheed Missiles & Space Company, Sunnyvale, California.* (**Section 1.** Introduction, ***Part B,*** Mathematical Methods. **Section 3.** Conduction)

E. M. SPARROW, *Heat Transfer Laboratory, Department of Mechanical Engineering, University of Minnesota.* (**Section 15.** Radiation, ***Part B,*** Radiant Interchange between Surfaces Separated by Nonabsorbing and Nonemitting Media)

D. ROGER WILLIS, *Professor of Engineering Science, University of California, Berkeley.* (**Section 9.** Rarefied Gases)

Preface

The groundwork for this Handbook was laid nearly a decade ago when the editors, a group of the authors, and a representative of the publisher met at a lunch in Philadelphia to discuss the project. We agreed at that time that the Handbook should present up-to-date information on heat transfer in a distilled and readily usable form intended for practicing engineers, research specialists, educators, students, and technicians: in brief, for anyone dealing with the solution of heat transfer problems. To achieve this objective, the overall subject of heat transfer was subdivided into approximately 20 sub-areas and outstanding specialists accepted the responsibility for presenting the heat transfer information available in their respective sub-areas.

The Handbook includes a large collection of thermophysical properties. Solutions and methods of solution of heat conduction problems by formal mathematics, by various approximate methods, and by analog methods are presented. Current information on convection, both forced and free convection, including internal and external flows is assembled. The influence of magnetic fields on heat transfer is covered. Heat transfer between surfaces and rarefied gases is treated. The current information on heat transfer in film and dropwise condensation, boiling, and two-phase flow is reported. A large section on radiation and radiation properties is included. Mass transfer in stationary and convection systems, including a special chapter on ablation, is covered. Special topics include high temperature protection systems, techniques for augmenting heat transfer, and a detailed presentation of heat exchanger design procedures.

We acknowledge with appreciation the cooperation of William Begell, who edited the book, as well as the excellent workmanship of Joseph Egerton, Gail Parker, Keith Wilkinson, and Vic Enfield, all of Scripta Technica, in the preparation and composition of the text. We also wish to thank Daniel Fischel, Harold Crawford, and Don Douglas of McGraw-Hill for their continued support and encouragement over these years. Finally, we wish to extend special appreciation to our authors for their hard work in the preparation of their chapters and for their patience with the editors in awaiting the final production of the Handbook. As a result of their efforts, we believe that we have a publication which will be valuable to the engineering community. We would appreciate receiving suggestions from readers as to how the Handbook's usefulness might be improved in future editions.

W. M. Rohsenow

J. P. Hartnett

Section 1

Introduction

Part A

Basic Concepts of Heat Transfer

J. P. HARTNETT

University of Illinois at Chicago Circle, Chicago, Illinois

W. M. ROHSENOW

Massachusetts Institute of Technology, Cambridge, Massachusetts

1.   Types of Heat Transfer Mechanisms

Heat is defined as energy transferred by virtue of a temperature difference or gradient and is vectorial in the sense that it flows from regions of higher temperature to regions of lower temperature. The basic modes of heat transfer are *conduction* and *radiation*.

Conduction is the transfer of heat from one part of the body at a higher temperature to another part of the same body at a lower temperature, or from one body at a higher temperature to another body at a lower temperature in physical contact with it. The conduction process takes place at the molecular level and involves the transfer of energy from the more energetic molecules to those with a lower energy level. This can be easily visualized within gases where we note that the average kinetic energy of molecules in the higher-temperature regions is greater than those in the lower temperature regions. The more energetic molecules, being in constant and random motion, periodically collide with molecules of a lower energy level and exchange energy and momentum. In this manner there is a continuous transport of energy from the high temperature regions to those of lower temperature. In liquids the molecules are more closely spaced than in gases, but the molecular energy exchange process is qualitatively similar to that in gases. In solids which are non-conductors of electricity (dielectrics), heat is conducted by lattice waves caused by atomic motion. In solids which are good conductors of electricity this lattice vibration mechanism is only a small contribution to the energy transfer process, with the principal contribution being that due to the motion of free electrons which move about in the same way as molecules in a gas.

At the macroscopic level we state that the heat flux is proportional to the temperature gradient with the proportionality factor being identified as the thermal conductivity k:

$$\frac{q}{A} = -k \left(\frac{\partial T}{\partial y} \right) \tag{1}$$

This relationship is used for the conduction process in solids, liquids, and gases. From the foregoing, as one might expect, the magnitude of the thermal conductivity of electrically conducting solids is higher than for dielectrics and solids in general have higher conductivity than liquids.

In treating conduction problems it is often convenient to introduce another property which is related to the thermal conductivity, namely the thermal diffusivity α

$$\alpha = \frac{k}{\rho c} \tag{2}$$

Here ρ is the density and c is the specific heat.

Radiation, or more correctly thermal radiation, is electromagnetic radiation emitted by a body by virtue of its temperature. Thus thermal radiation is of the same nature as visible light, X-rays, and radio waves, the difference between them being in their wave lengths. The eye is sensitive to electromagnetic radiation in the region from 35 to 75 microns; this is identified as the visible regions of the spectrum. Radio waves have a wave length of 10^4 microns and above, X-rays have wave lengths of 0.01 to 1, while thermal radiation occurs in rays from 0.1 to 100 microns. All heated solids and liquids as well as some gases emit thermal radiation. On the macroscopic level, the calculation of thermal radiation is based on the Stefan-Boltzmann law which relates the energy flux emitted by an ideal radiator to the fourth power of the absolute temperature

$$e_b = \sigma T^4 \tag{3}$$

Here σ is the Stefan-Boltzmann constant. Engineering surfaces in general do not perform as ideal radiators and for real surfaces the above law is modified to read

$$e = \epsilon \sigma T^4 \tag{4}$$

The term ϵ is called the emissivity of the surface with a value between 0 and 1.

Convection, sometimes identified as a separate mode of heat transfer, relates to the transfer of heat from a bounding surface to a fluid in motion, or to the heat transfer across a flow plane within the interior of the flowing fluid. If the fluid motion is induced by a pump, blower, fan, or some similar device, the process is called forced convection. If the fluid motion occurs as a result of the density differences produced by the heat transfer itself, the process is called free or natural convection. Detailed inspection of the heat transfer process in these cases reveals that the basic heat transfer mechanisms are conduction and radiation, both of which are generally influenced by the fluid motion. In convective processes involving heat transfer to or from a boundary surface exposed to a low velocity fluid stream, it is convenient to introduce a heat transfer coefficient h defined by Eq. (5), which is known as Newton's law of cooling

$$\frac{q}{A} = h(T_f - T_s) \tag{5}$$

Here T_s is the surface temperature and T_f is a characteristic fluid temperature.

For surfaces in unbounded convection, such as plates, tubes, bodies of revolution, etc., immersed in a large body of fluid, it is customary to define h in Eq. (5) with T_f being the temperature of the fluid far away from the surface often identified as $T_{f\infty}$. For bounded convection, such as fluids flowing in tubes, channels, across tubes in bundles, etc., T_f is usually taken as the enthalpy-mixed-mean temperature, customarily identified as T_m.

The heat transfer coefficient so defined may include both radiation and conductive contributions. If the radiation contribution is negligible, then the total transfer is due to conduction. In this case we may note

$$h = \frac{q/A}{T_f - T_s} = \frac{-k(\partial T/\partial y)_s}{T_f - T_s} \tag{6}$$

The heat transfer coefficient is then recognizable as the gradient of dimensionless temperature at the surface. It is sensitive to the geometry, to the physical properties of the fluid, and to the fluid velocity.

For convective processes involving high velocity gas flows (high subsonic or supersonic), a more meaningful and useful definition of the heat transfer coefficient is given by

$$\frac{q}{A} = h(T_r - T_s) \tag{5a}$$

Here T_r, commonly called the adiabatic wall temperature or the recovery temperature, is the equilibrium temperature the surface would attain in the absence of any heat transfer to or from the surface and in the absence of radiation exchange between the surroundings and the surface. In general the adiabatic wall temperature is dependent on the fluid properties and the properties of the bounding wall. Generally, the adiabatic wall temperature is reported in terms of a dimensionless recovery factor r defined as

$$T_r = T_f + r\frac{V^2}{2C_p}$$

The value of r for gases normally lies between 0.8 and 1.0. It can be seen that for low-velocity flows the recovery temperature is equal to the free stream temperature T_f. In this case, Eq. (5a) reduces to Eq. (5). From this point of view, Eq. (5a) can be taken as the generalized definition of the heat transfer coefficient.

In some physical situations, it is possible to determine analytically the details of the flow field and the temperature distribution and thereby to evaluate the heat transfer coefficient h. In these cases which are amenable to analysis, if the heat transfer process involves both radiation and conduction it is convenient to determine the magnitude of each mode and to define the heat transfer coefficient in terms of the conductive contribution alone as given by Eq. (6). Needless to say, in determining the total heat transfer the radiation contribution must be added. Unfortunately, in many engineering problems the convective heat transfer cannot be determined analytically but must be evaluated by experiment. In such cases, if radiation is an important mechanism it is usually impossible to separate the conductive and radiative modes and the heat transfer coefficient may not be interpreted as the dimensionless temperature gradient; rather it is defined by Eq. (5). The chapters on forced and free convection will concentrate on the determination of the heat transfer coefficient.

2. Rate Equations

Equation (1) for heat conduction relates the heat flux to the temperature gradient. It is an example of the type of phenomenological equation which is used for the prediction of other rate processes. Of particular interest here are the rate equations for transfer of momentum and mass.

a. Momentum Transfer. Consider a fluid confined between two parallel plates separated by a distance S with the bottom plate at rest and the top one moving with a velocity V. Under steady-state conditions the velocity distribution will be linear as shown in Figure 1. To sustain the motion, a force F must be exerted on the top plate

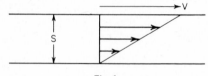

Fig. 1.

and an equal and opposite force is transmitted to the plate at rest. The ratio of this force to the area of the plate F/A is called the shearing stress τ and for many fluids of engineering importance its magnitude is directly proportional to the velocity V and inversely proportional to the plate spacing S

$$\tau = \mu \frac{V}{S} \tag{7}$$

The proportionality factor μ is called the dynamic viscosity and is a property of the fluid. Fluids such as water and air which conform to this relationship are called Newtonian fluids. There are other more complex fluids called non-Newtonian which follow more complex laws and these will be treated in later chapters. Extending the basic Newtonian law to a more general fluid motion we may write

$$\tau = \mu \left(\frac{dV_x}{dy} \right) \tag{8}$$

This shearing stress τ may be interpreted in terms of the transport of momentum ρV_x. The faster moving molecules transfer momentum to their slower moving neighbors and consequently there is a net transfer of x momentum in the y direction. This may be seen from the Newtonian relationship restated as follows

$$\tau = \left(\frac{\mu}{\rho}\right)\frac{d}{dy}(\rho V_x) \tag{9}$$

The ratio (μ/ρ) has been given its own identity, and is called the kinematic viscosity,

$$\nu = \frac{\mu}{\rho} \tag{10}$$

The dimensions of kinematic viscosity are L^2/T which in the English system of units becomes ft²/hr. For liquids, both the dynamic viscosity μ and the kinematic viscosity ν are primarily dependent on temperature and are relatively insensitive to pressure except in the neighborhood of the critical point. For gases the dynamic viscosity is also temperature dependent, showing little sensitivity to pressure whereas the kinematic viscosity is strongly dependent on both temperature and pressure, being inversely proportional to pressure. Generally, the dynamic viscosity is higher for liquids than for gases, while the kinematic viscosity of gases tends to be higher than for liquids. As an example, at atmosphere pressure and 70°F the dynamic viscosity of water is approximately 50 times higher than that of air, while the kinematic viscosity of air is approximately 10 times that of water.

   b. **Mass Transfer.** Consider a stagnant pure dry gas positioned between two parallel surfaces separated by a distance S. The lower surface is wetted, which can be accomplished by using a porous wick for this surface or by maintaining a thin layer of liquid on a solid surface. The upper surface is so selected that it is capable of absorbing any vapor which may be transferred from the lower surface. The partial density of the vapor immediately above the wetted surface is maintained at C_s, while the partial density at the ideal absorbing surface is found to be negligible. If these conditions are maintained for a sufficiently long period, a steady state will ensue in which it will be found that the partial density profile is a linear one. Under these conditions, the amount of vapor transported from the lower surface to the upper one is found to be directly proportional to the value of C_s and inversely proportional to the plate spacing

$$\frac{W}{A} = D\frac{C_s}{S} \tag{11}$$

   Here W is the mass of vapor transported and A is the surface area. The proportionality factor D is called the mass diffusivity or the ordinary coefficient of diffusion, having units of L^2/T or ft²/hr in the English system. If this relationship is extended to more complex systems we have

$$\frac{W}{A} = -D\frac{\partial C}{\partial y} \tag{12}$$

This relation is called Fick's Law of Diffusion.

3. Basic Fluid Mechanics

In dealing with convection problems it is important to have an understanding of the behavior of fluids in motion over external surfaces or through enclosed channels. For liquids or gases flowing over external surfaces under continuum conditions, it will be found that the relative velocity between the surface and the fluid goes to zero at the surface. Moving away from the surface, the velocity increases rapidly toward the free-stream value, effectively reaching the free-stream value at a distance S not far from the surface. The thin region where the velocity is varying is called the boundary layer, a term suggested by Prandtl who first recognized this basic phenomenon. Since the shearing stress is proportional to the product of the viscosity and the velocity gradient, it is clear that substantial shearing stresses will occur only in the boundary layer where a velocity gradient exists, whereas outside the boundary the shearing stress will be vanishingly small. Accordingly, it may be stated that the effects of viscosity are confined to the boundary layer, whereas outside the boundary layer the flow may be considered to be inviscid. Thus in analyzing the flow field over external surfaces, the inviscid flow equations may be used to predict the free-stream flow field. The resulting velocity distribution may then be used in conjunction with the boundary layer equations which include the influence of viscosity for the prediction of the flow field in the immediate vicinity of the wall. In this manner the drag on external surfaces can be determined.

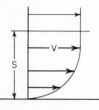

Fig. 2.

In the case of heat transfer (or mass transfer) to or from external surfaces placed in a flow field, it will be found that there is a thermal boundary layer (or concentration boundary layer) analogous to the velocity boundary layer, within which the influence of thermal conductivity (or diffusivity) is confined. Outside this region the flow is essentially nonconducting and nondiffusing.

In the case of a fluid flow through an enclosed channel, a boundary layer begins at the channel entrance. In this entrance region there is an inviscid core flow and a

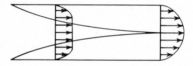

Fig. 3.

viscous boundary layer. Some distance downstream the boundary layers grow together and the velocity is at a maximum in the central region of the duct, decreasing to a value of zero at the bounding surfaces.

a. Laminar and Turbulent Flows. Osborne Reynolds in 1883 reported that there are two basically different types of fluid motion which he identified as laminar flow and turbulent flow. For example, in the case of flow over a flat plate geometry the boundary layer motion near the leading edge of the plate is smooth or streamlined. Locally within the boundary layer the velocity is constant and invariant with time.

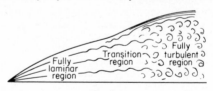

Fig. 4.

In this region momentum and energy transfer occur by a diffusion process as described by the Newtonian shearing stress law and by the Fourier conduction relationship. This is the region of laminar flow. If the plate is long enough or the velocity sufficiently high and we proceed far downstream, the nature of the flow is markedly changed. At any point in the boundary layer the velocity varies with time about some mean value as shown in Fig. 5. The exchange of momentum and energy is now no longer controlled by diffusional processes. Rather macroscopic eddies randomly move from one fluid

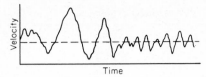

Fig. 5.

layer to another, and in the process momentum and energy are transferred. The analysis of transport processes in turbulent flows is inherently more difficult than in the laminar cases and, in general, the treatment is semi-empirical in nature.

The flow does not change abruptly from laminar to turbulent motion, but rather there is an intermediate region connecting the well-defined laminar and the well-defined turbulent motion. This is the transition region. It has been found that the laminar boundary layer begins to experience transition where the dimensionless quality $(u_e\,x/\nu)$, called the critical Reynolds number for flow over external surfaces, is of the order of 500,000, but this is dependent on the level of turbulence in the free stream.

For flow in circular tubes, it has been found that the flow is generally laminar if the Reynolds number $\bar{u}\,d/\nu$, where $\bar{u}$ is mean velocity, d is pipe diameter, and ν is kinematic viscosity, is lower than 2,300. If this Reynolds number is greater than 10,000, the flow is considered to be fully turbulent. In the 2,300 to 10,000 region, the flow is described as transition flow. It is possible to shift these Reynolds values by minimizing the disturbances in the inlet flow, but for general engineering applications the numbers cited are representative.

   b.   **Flow Separation.**   In region of adverse pressure gradient such as encountered in flow over curved bodies, the boundary layer, in effect, separates from the surface. At this location the shear stress goes to zero and beyond this point there is a reversal of flow in the vicinity of the wall as shown in Fig. 6. In this separated region, the boundary layer equations are no longer valid and the analysis of the flow is generally very difficult.

Fig. 6.

4. Units

Generally, the English system of engineering units is used throughout the handbook, although in a few chapters the authors have selected other systems. To assist the handbook user, a conversion table (Table 2) is given to aid rapid calculation in any system of units. Furthermore, when possible, engineering results are presented in a dimensionless fashion, independent of the unit system, and it is a relatively straightforward matter to proceed from the dimensionless number to the desired dimensioned quantity. A listing of dimensionless groups frequently encountered in heat transfer is given in Table 1.

5. General Equations

The following are some of the general equations encountered in heat transfer. They have been collected in this section for use as a ready reference.

   a.   **Continuity Equations.**
Vector form

$$\frac{\partial \rho}{\partial t} + \nabla \cdot (\rho \mathbf{V}) = 0$$

or

$$\frac{D\rho}{Dt} + \rho \nabla \cdot \mathbf{V} = 0$$

Incompressible

$$\nabla \cdot \mathbf{V} = 0$$

TABLE 1.   Dimensionless Groups

Group*	Symbol	Name
$\Delta p/\rho V^2$	Eu	Euler number
$\alpha t/r_0^2$	Fo	Fourier number
$(L/d)(k/Vd\rho c_p)$	Gz [= (L/d)/RePr]	Graetz number
$g\beta (\Delta T) L^3\rho^2/\mu^2$	Gr	Grashof number
λ/L	Kn	Knudsen number
α/D	Le	Lewis number
V/V_{sound}	Ma	Mach number
$hL/k, hd/k$	Nu	Nusselt number
$Vd\rho c_p/k$	Pe (= RePr)	Peclet number
$c_p\mu/k$	Pr	Prandtl number
$g\beta (\Delta T)L^3\rho^2 c_p/\mu k$	Ra (= GrPr)	Rayleigh number
$\rho VD/u, \rho VL/\mu$	Re	Reynolds number
$\mu/\rho D$	Sc	Schmidt number
$h_p d/D$	Sh	Sherwood number
$h/c_p G$	St (= Nu/RePr)	Stanton number
$V_\infty^2/C_p (\Delta T)_0$	E	Eckert number
V^2/gL	Fr	Froude number
$f_r d/V$	St	Strouhal number
$\rho V^2 L/\sigma$	We	Weber number

*f_r = frequency of oscillation
σ = surface tension.

Cartesian: $\quad \dfrac{D}{Dt} = \dfrac{\partial}{\partial t} + u\dfrac{\partial}{\partial x} + v\dfrac{\partial}{\partial y} + w\dfrac{\partial}{\partial z}$

Cylindrical: $\quad \dfrac{D}{Dt} = \dfrac{\partial}{\partial t} + v_r\dfrac{\partial}{\partial r} + \dfrac{v_\theta}{r}\dfrac{\partial}{\partial \theta} + v_z\dfrac{\partial}{\partial z}$

Spherical: $\quad \dfrac{D}{Dt} = \dfrac{\partial}{\partial t} + v_r\dfrac{\partial}{\partial r} + \dfrac{v_\theta}{r}\dfrac{\partial}{\partial \theta} + \dfrac{v_\phi}{r\sin\theta}\dfrac{\partial}{\partial \phi}$

Cartesian: u, v, and w are the velocities in the x, y, and z directions respectively.

$$\frac{\partial \rho}{\partial t} + \frac{\partial}{\partial x}(\rho u) + \frac{\partial}{\partial y}(\rho v) + \frac{\partial}{\partial z}(\rho w) = 0$$

Incompressible

$$\frac{\partial u}{\partial x} + \frac{\partial v}{\partial y} + \frac{\partial w}{\partial z} = 0$$

Cylindrical: v_r, v_θ, and v_z are the velocities in the r, θ, and z directions respectively.

$$\frac{\partial \rho}{\partial t} + \frac{1}{r}\frac{\partial}{\partial r}(r\rho v_r) + \frac{1}{r}\frac{\partial}{\partial \theta}(\rho v_\theta) + \frac{\partial}{\partial z}(\rho v_z) = 0$$

Incompressible

$$\frac{1}{r}\frac{\partial}{\partial r}(rv_r) + \frac{1}{r}\frac{\partial v_\theta}{\partial \theta} + \frac{\partial v_z}{\partial z} = 0$$

Spherical: v_r, v_θ, and v_ϕ are the velocities in the r, θ, and ϕ directions respectively

$$\frac{\partial \rho}{\partial t} + \frac{1}{r^2}\frac{\partial}{\partial r}(r^2 \rho v_r) + \frac{1}{r \sin\theta}\frac{\partial}{\partial \theta}(\rho v_\theta \sin\theta) + \frac{1}{r \sin\theta}\frac{\partial}{\partial \phi}(\rho v_\phi) = 0$$

Incompressible

$$\frac{1}{r^2}\frac{\partial}{\partial r}(r^2 v_r) + \frac{1}{r \sin\theta}\frac{\partial}{\partial \theta}(v_\theta \sin\theta) + \frac{1}{r \sin\theta}\frac{\partial v_\phi}{\partial \phi} = 0$$

b. Momentum Equations (Navier-Stokes).

P = pressure
F = body force per unit volume
μ = viscosity
λ = second coefficient of viscosity ($\lambda = -\frac{2}{3}\mu$, monatomic gas)
$\zeta = \lambda + \frac{2}{3}\mu$ (= zero for monatomic gas)

Vector form

$$\rho \frac{D\mathbf{V}}{Dt} = \rho\left[\frac{\partial \mathbf{V}}{\partial t} + (\mathbf{V}\cdot\nabla)\mathbf{V}\right] = \rho\left[\frac{\partial \mathbf{V}}{\partial t} + \nabla\left(\frac{V^2}{2}\right) - \mathbf{V}\times(\nabla\times\mathbf{V})\right]$$

$$= -\nabla P + \mathbf{F} - \nabla\times[\mu(\nabla\times\mathbf{V})] + \nabla\left[\left(\zeta + \frac{4}{3}\mu\right)\nabla\cdot\mathbf{V}\right]$$

or in terms of λ

$$\rho\left[\frac{\partial \mathbf{V}}{\partial t} + (\mathbf{V}\cdot\nabla)\mathbf{V}\right] = \rho\left[\frac{\partial \mathbf{V}}{\partial t} + \nabla\left(\frac{V^2}{2}\right) - \mathbf{V}\times(\nabla\times\mathbf{V})\right]$$

$$= -\nabla P + \mathbf{F} - \nabla\times[\mu(\nabla\times\mathbf{V})] + \nabla[(\lambda + 2\mu)\nabla\cdot\mathbf{V}]$$

ρ, μ constant

$$\rho\frac{D\mathbf{V}}{Dt} = \rho\left[\frac{\partial \mathbf{V}}{\partial t} + (\mathbf{V}\cdot\nabla)\mathbf{V}\right] = -\nabla P + \mathbf{F} + \mu\nabla^2\mathbf{V}$$

Cartesian

$$\nabla\cdot\mathbf{V} = \frac{\partial u}{\partial x} + \frac{\partial v}{\partial y} + \frac{\partial w}{\partial z}$$

$$\rho\frac{Du}{Dt} = F_z - \frac{\partial P}{\partial x} + \frac{\partial}{\partial x}\left[2\mu\frac{\partial u}{\partial x} + \lambda\nabla\cdot\mathbf{V}\right] + \frac{\partial}{\partial y}\left[\mu\left(\frac{\partial u}{\partial y} + \frac{\partial v}{\partial x}\right)\right] + \frac{\partial}{\partial z}\left[\mu\left(\frac{\partial w}{\partial x} + \frac{\partial u}{\partial z}\right)\right]$$

$$\rho\frac{Dv}{Dt} = F_y - \frac{\partial P}{\partial y} + \frac{\partial}{\partial y}\left[2\mu\frac{\partial v}{\partial y} + \lambda\nabla\cdot\mathbf{V}\right] + \frac{\partial}{\partial z}\left[\mu\left(\frac{\partial v}{\partial z} + \frac{\partial w}{\partial y}\right)\right] + \frac{\partial}{\partial x}\left[\mu\left(\frac{\partial u}{\partial y} + \frac{\partial v}{\partial x}\right)\right]$$

$$\rho\frac{Dw}{Dt} = F_z - \frac{\partial P}{\partial z} + \frac{\partial}{\partial z}\left[2\mu\frac{\partial w}{\partial z} + \lambda\nabla\cdot\mathbf{V}\right] + \frac{\partial}{\partial x}\left[\mu\left(\frac{\partial w}{\partial x} + \frac{\partial u}{\partial z}\right)\right] + \frac{\partial}{\partial y}\left[\mu\left(\frac{\partial v}{\partial z} + \frac{\partial w}{\partial y}\right)\right]$$

ρ, μ constant

$$\rho\left(\frac{\partial u}{\partial t} + u\frac{\partial u}{\partial x} + v\frac{\partial u}{\partial y} + w\frac{\partial u}{\partial z}\right) = -\frac{\partial P}{\partial x} + F_x + \mu\left(\frac{\partial^2 u}{\partial x^2} + \frac{\partial^2 u}{\partial y^2} + \frac{\partial^2 u}{\partial z^2}\right)$$

$$\rho\left(\frac{\partial v}{\partial t} + u\frac{\partial v}{\partial x} + v\frac{\partial v}{\partial y} + w\frac{\partial v}{\partial z}\right) = -\frac{\partial P}{\partial y} + F_y + \mu\left(\frac{\partial^2 v}{\partial x^2} + \frac{\partial^2 v}{\partial y^2} + \frac{\partial^2 v}{\partial z^2}\right)$$

$$\rho\left(\frac{\partial w}{\partial t} + u\frac{\partial w}{\partial x} + v\frac{\partial w}{\partial y} + w\frac{\partial w}{\partial z}\right) = -\frac{\partial P}{\partial z} + F_z + \mu\left(\frac{\partial^2 w}{\partial x^2} + \frac{\partial^2 w}{\partial y^2} + \frac{\partial^2 w}{\partial z^2}\right)$$

Cylindrical

$$\nabla\cdot\mathbf{V} = \frac{1}{r}\frac{\partial}{\partial r}(rv_r) + \frac{1}{r}\frac{\partial v_\theta}{\partial\theta} + \frac{\partial v_z}{\partial z} = \frac{\partial v_r}{\partial r} + \frac{v_r}{r} + \frac{1}{r}\frac{\partial v_\theta}{\partial\theta} + \frac{\partial v_z}{\partial z}$$

$$\rho\left[\frac{Dv_r}{Dt} - \frac{v_\theta^2}{r}\right] = F_r - \frac{\partial P}{\partial r} + \frac{\partial}{\partial r}\left[2\mu\frac{\partial v_r}{\partial r} + \lambda\nabla\cdot\mathbf{V}\right]$$

$$+ \frac{1}{r}\frac{\partial}{\partial\theta}\left[\mu\left(\frac{1}{r}\frac{\partial v_r}{\partial\theta} + \frac{\partial v_\theta}{\partial r} - \frac{v_\theta}{r}\right)\right] + \frac{\partial}{\partial z}\left[\mu\left(\frac{\partial v_r}{\partial z} + \frac{\partial v_z}{\partial r}\right)\right] + \frac{2\mu}{r}\left(\frac{\partial v_r}{\partial r} - \frac{1}{r}\frac{\partial v_\theta}{\partial\theta} - \frac{v_r}{r}\right)$$

$$\rho\left[\frac{Dv_\theta}{Dt} + \frac{v_r v_\theta}{r}\right] = F_\theta - \frac{1}{r}\frac{\partial P}{\partial r} + \frac{1}{r}\frac{\partial}{\partial\theta}\left[\frac{2\mu}{r}\frac{\partial v_\theta}{\partial\theta} + \lambda\nabla\cdot\mathbf{V}\right]$$

$$+ \frac{\partial}{\partial z}\left[\mu\left(\frac{1}{r}\frac{\partial v_z}{\partial\theta} + \frac{\partial v_\theta}{\partial z}\right)\right] + \frac{\partial}{\partial r}\left[\mu\left(\frac{1}{r}\frac{\partial v_r}{\partial\theta} + \frac{\partial v_\theta}{\partial r} - \frac{v_\theta}{r}\right)\right] + \frac{2\mu}{r}\left[\frac{1}{r}\frac{\partial v_r}{\partial\theta} + \frac{\partial v_\theta}{\partial r} - \frac{v_\theta}{r}\right]$$

$$\rho\frac{Dv_z}{Dt} = F_z - \frac{\partial P}{\partial z} + \frac{\partial}{\partial z}\left[2\mu\frac{\partial v_z}{\partial z} + \lambda\nabla\cdot\mathbf{V}\right] + \frac{1}{r}\frac{\partial}{\partial r}\left[\mu r\left(\frac{\partial v_r}{\partial z} + \frac{\partial v_z}{\partial r}\right)\right]$$

$$+ \frac{1}{r}\frac{\partial}{\partial\theta}\left[\mu\left(\frac{1}{r}\frac{\partial v_z}{\partial\theta} + \frac{\partial v_\theta}{\partial z}\right)\right]$$

ρ, μ constant

$$\rho\left[\frac{\partial v_r}{\partial t} + v_r\frac{\partial v_r}{\partial r} + \frac{v_\theta}{r}\frac{\partial v_r}{\partial\theta} + v_z\frac{\partial v_r}{\partial z} - \frac{v_\theta^2}{r}\right] = F_r - \frac{\partial P}{\partial r}$$

$$+ \mu\left[\frac{\partial^2 v_r}{\partial r^2} + \frac{1}{r}\frac{\partial v_r}{\partial r} + \frac{1}{r^2}\frac{\partial^2 v_r}{\partial\theta^2} + \frac{\partial^2 v_r}{\partial z^2} - \frac{v_r}{r} - \frac{2}{r^2}\frac{\partial v_\theta}{\partial\theta}\right]$$

$$\rho\left[\frac{\partial v_\theta}{\partial t} + v_r\frac{\partial v_\theta}{\partial r} + \frac{v_\theta}{r}\frac{\partial v_\theta}{\partial\theta} + v_z\frac{\partial v_\theta}{\partial z} + \frac{v_r v_\theta}{r}\right] = F_\theta - \frac{1}{r}\frac{\partial P}{\partial\theta}$$

$$+ \mu\left[\frac{\partial^2 v_\theta}{\partial r^2} + \frac{1}{r}\frac{\partial v_\theta}{\partial r} + \frac{1}{r^2}\frac{\partial^2 v_\theta}{\partial\theta^2} + \frac{\partial^2 v_\theta}{\partial z^2} + \frac{2}{r^2}\frac{\partial v_r}{\partial\theta} - \frac{v_\theta}{r^2}\right]$$

$$\rho\left[\frac{\partial v_z}{\partial t} + v_r\frac{\partial v_z}{\partial r} + \frac{v_\theta}{r}\frac{\partial v_z}{\partial\theta} + v_z\frac{\partial v_z}{\partial z}\right] = F_z - \frac{\partial P}{\partial z} + \mu\left[\frac{\partial^2 v_z}{\partial r^2} + \frac{1}{r}\frac{\partial v_z}{\partial r} + \frac{1}{r^2}\frac{\partial^2 v_z}{\partial\theta^2} + \frac{\partial^2 v_z}{\partial z^2}\right]$$

Spherical

$$\rho\left[\frac{Dv_r}{Dt} - \frac{v_\theta^2 + v_\phi^2}{r}\right] = F_r - \frac{\partial P}{\partial r} + \frac{\partial}{\partial r}\left[2\mu\frac{\partial v_r}{\partial r} + \lambda\nabla\cdot\mathbf{V}\right]$$

$$+ \frac{1}{r}\frac{\partial}{\partial\theta}\left[\mu\left\{r\frac{\partial}{\partial r}\left(\frac{v_\theta}{r}\right) + \frac{1}{r}\frac{\partial v_r}{\partial\theta}\right\}\right] + \frac{1}{r\sin\theta}\frac{\partial}{\partial\phi}\left[\mu\left\{\frac{1}{r\sin\theta}\frac{\partial v_r}{\partial\phi} + r\frac{\partial}{\partial r}\left(\frac{v_\phi}{r}\right)\right\}\right]$$

$$+ \frac{\mu}{r}\left[4\frac{\partial v_r}{\partial r} - \frac{2}{r}\frac{\partial v_\theta}{\partial\theta} - \frac{4v_r}{r} - \frac{2}{r\sin\theta}\frac{\partial v_\phi}{\partial\phi} - \frac{2v_\theta\cot\theta}{r} + r\cot\theta\frac{\partial}{\partial r}\left(\frac{v_\theta}{r}\right) + \frac{\cot\theta}{r}\frac{\partial v_r}{\partial\theta}\right]$$

$$\rho\left[\frac{Dv_\theta}{Dt} + \frac{v_r v_\theta}{r} - \frac{v_\phi^2\cot\theta}{r}\right] = F_\theta - \frac{1}{r}\frac{\partial P}{\partial\theta} + \frac{1}{r}\frac{\partial}{\partial\theta}\left[\frac{2\mu}{r}\left(\frac{\partial v_\theta}{\partial\theta} + v_r\right) + \lambda\nabla\cdot\mathbf{V}\right]$$

$$+ \frac{1}{r\sin\theta}\frac{\partial}{\partial\phi}\left[\mu\left\{\frac{\sin\theta}{r}\frac{\partial}{\partial\theta}\left(\frac{v_\phi}{\sin\theta}\right) + \frac{1}{r\sin\theta}\frac{\partial v_\theta}{\partial\phi}\right\}\right] + \frac{\partial}{\partial r}\left[\mu\left\{r\frac{\partial}{\partial r}\left(\frac{v_\theta}{r}\right) + \frac{1}{r}\frac{\partial v_r}{\partial\theta}\right\}\right]$$

$$+ \frac{\mu}{r}\left[2\left(\frac{1}{r}\frac{\partial v_\theta}{\partial\theta} - \frac{1}{r\sin\theta}\frac{\partial v_\phi}{\partial\phi} - \frac{v_\theta\cot\theta}{r}\right)\cdot\cot\theta + 3\left\{r\frac{\partial}{\partial r}\left(\frac{v_\theta}{r}\right) + \frac{1}{r}\frac{\partial v_r}{\partial\theta}\right\}\right]$$

$$\rho\left[\frac{Dv_\phi}{Dt} + \frac{v_\phi v_r}{r} + \frac{v_\theta v_\phi\cot\theta}{r}\right] = F_\phi - \frac{1}{r\sin\theta}\frac{\partial P}{\partial\phi}$$

$$+ \frac{1}{r\sin\theta}\frac{\partial}{\partial\phi}\left[\frac{2\mu}{r}\left(\frac{1}{\sin\theta}\frac{\partial v_\phi}{\partial\phi} + v_r + v_\theta\cot\theta\right) + \lambda\nabla\cdot\mathbf{V}\right]$$

$$+ \frac{\partial}{\partial r}\left[\mu\left\{\frac{1}{r\sin\theta}\frac{\partial v_r}{\partial\phi} + r\frac{\partial}{\partial r}\left(\frac{v_\phi}{r}\right)\right\}\right] + \frac{1}{r}\frac{\partial}{\partial\theta}\left[\mu\left\{\frac{\sin\theta}{r}\frac{\partial}{\partial\theta}\left(\frac{v_\phi}{\sin\theta}\right) + \frac{1}{r\sin\theta}\frac{\partial v_\theta}{\partial\phi}\right\}\right]$$

$$+ \frac{\mu}{r}\left[3\left\{\frac{1}{r\sin\theta}\frac{\partial v_r}{\partial\phi} + r\frac{\partial}{\partial r}\left(\frac{v_\phi}{r}\right)\right\} + 2\cot\theta\left\{\frac{\sin\theta}{r}\frac{\partial}{\partial\theta}\left(\frac{v_\phi}{\sin\theta}\right) + \frac{1}{r\sin\theta}\frac{\partial v_\theta}{\partial\phi}\right\}\right]$$

ρ, μ constant

$$\rho\left[\frac{\partial v_r}{\partial t} + v_r\frac{\partial v_r}{\partial r} + \frac{v_\theta}{r}\frac{\partial v_r}{\partial\theta} + \frac{v_\phi}{r\sin\theta}\frac{\partial v_r}{\partial\phi} - \frac{v_\theta^2 + v_\phi^2}{r}\right]$$

$$= F_r - \frac{\partial P}{\partial r} + \mu\left[\frac{1}{r^2}\frac{\partial}{\partial r}\left(r^2\frac{\partial v_r}{\partial r}\right) + \frac{1}{r^2\sin\theta}\frac{\partial}{\partial\theta}\left(\sin\theta\frac{\partial v_r}{\partial\theta}\right)\right.$$

$$\left. + \frac{1}{r^2\sin^2\theta}\frac{\partial^2 v_r}{\partial\phi^2} - \frac{2v_r}{r^2} - \frac{2}{r^2}\frac{\partial v_\theta}{\partial\theta} - \frac{2v_\theta\cot\theta}{r^2} - \frac{2}{r^2\sin\theta}\frac{\partial v_\phi}{\partial\phi}\right]$$

$$\rho \left[\frac{\partial v_\theta}{\partial t} + v_r \frac{\partial v_\theta}{\partial r} + \frac{v_\theta}{r} \frac{\partial v_\theta}{\partial \theta} + \frac{v_\phi}{r \sin\theta} \frac{\partial v_\theta}{\partial \phi} + \frac{v_r v_\theta}{r} - \frac{v_\phi^2 \cot\theta}{r} \right]$$

$$= F_\theta - \frac{1}{r} \frac{\partial P}{\partial \theta} + \mu \left[\frac{1}{r^2} \frac{\partial}{\partial r} \left(r^2 \frac{\partial v_\theta}{\partial r} \right) + \frac{1}{r^2 \sin\theta} \frac{\partial}{\partial \theta} \left(\sin\theta \frac{\partial v_\theta}{\partial \theta} \right) \right.$$

$$\left. + \frac{1}{r^2 \sin^2\theta} \frac{\partial^2 v_\theta}{\partial \phi^2} + \frac{2}{r^2} \frac{\partial v_r}{\partial \theta} - \frac{v_\theta}{r^2 \sin^2\theta} - \frac{2\cos\theta}{r^2 \sin^2\theta} \frac{\partial v_\phi}{\partial \phi} \right]$$

$$\rho \left[\frac{\partial v_\phi}{\partial t} + v_r \frac{\partial v_\phi}{\partial r} + \frac{v_\theta}{r} \frac{\partial v_\phi}{\partial \theta} + \frac{v_\phi}{r \sin\theta} \frac{\partial v_\phi}{\partial \phi} + \frac{v_\phi v_r}{r} + \frac{v_\theta v_\phi \cot\theta}{r} \right]$$

$$= F_\phi - \frac{1}{r \sin\theta} \frac{\partial P}{\partial \phi} + \mu \left[\frac{1}{r^2} \frac{\partial}{\partial r} \left(r^2 \frac{\partial v_\phi}{\partial r} \right) + \frac{1}{r^2 \sin\theta} \frac{\partial}{\partial \theta} \left(\sin\theta \frac{\partial v_\phi}{\partial \theta} \right) \right.$$

$$\left. + \frac{1}{r^2 \sin^2\theta} \frac{\partial^2 v_\phi}{\partial \phi^2} - \frac{v_\phi}{r^2 \sin^2\theta} + \frac{2}{r^2 \sin^2\theta} \frac{\partial v_r}{\partial \phi} + \frac{2\cos\theta}{r^2 \sin^2\theta} \frac{\partial v_\theta}{\partial \phi} \right]$$

c.   Energy Equations.

e = internal energy (per unit mass).
P = pressure.
Q = internal heat generation.
q_r = radiation heat flux vector.
T = temperature.
k = thermal conductivity.
ρ = mass density
Φ = mechanical or viscous dissipation function.
i = enthalpy (per unit mass).

Vector Form

$$\frac{\partial Q}{\partial t} + \Phi + \nabla \cdot (k \nabla T) - \nabla \cdot \mathbf{q}_r = \rho \frac{De}{Dt} + P \nabla \cdot \mathbf{V} = \rho \left[\frac{De}{Dt} + P \frac{D}{Dt} \left(\frac{1}{\rho} \right) \right]$$

$$\rho \frac{Di}{Dt} = \frac{DP}{Dt} + \frac{\partial Q}{\partial t} + \Phi + \nabla \cdot (k \nabla T) - \nabla \cdot \mathbf{q}_r$$

ρ, k constant

$$\frac{\partial Q}{\partial t} + \Phi + \kappa \nabla^2 T - \nabla \cdot \mathbf{q}_r = \rho \frac{De}{Dt}$$

$$\rho \frac{Di}{Dt} = \frac{DP}{Dt} + \frac{\partial Q}{\partial t} + \Phi + k \nabla^2 T - \nabla \cdot \mathbf{q}_r$$

For perfect gases

$$\frac{Di}{Dt} = c_p \frac{DT}{Dt}$$

$$\frac{De}{Dt} = c_v \frac{DT}{Dt}$$

Cartesian

$$\nabla \cdot \mathbf{V} = \frac{\partial u}{\partial x} + \frac{\partial v}{\partial y} + \frac{\partial w}{\partial z}$$

$$\nabla^2 = \frac{\partial^2}{\partial x^2} + \frac{\partial^2}{\partial y^2} + \frac{\partial^2}{\partial z^2}$$

$$\frac{\partial Q}{\partial t} + \Phi + \frac{\partial}{\partial x}\left(k\frac{\partial T}{\partial x}\right) + \frac{\partial}{\partial y}\left(k\frac{\partial T}{\partial y}\right) + \frac{\partial}{\partial z}\left(k\frac{\partial T}{\partial z}\right) - \nabla \cdot \mathbf{q}_r$$

$$= \rho\left[\frac{De}{Dt} + P\frac{D}{Dt}\left(\frac{1}{\rho}\right)\right] = \rho\frac{De}{Dt} + P\nabla \cdot \mathbf{V}$$

$$\rho\frac{Di}{Dt} = \frac{DP}{Dt} + \frac{\partial Q}{\partial t} + \Phi + \frac{\partial}{\partial x}\left(k\frac{\partial T}{\partial x}\right) + \frac{\partial}{\partial y}\left(k\frac{\partial T}{\partial y}\right) + \frac{\partial}{\partial z}\left(k\frac{\partial T}{\partial z}\right) - \nabla \cdot \mathbf{q}_r$$

ρ, k constant

$$\frac{\partial Q}{\partial t} + \Phi + k\nabla^2 T - \nabla \cdot \mathbf{q}_r = \rho\frac{De}{Dt}$$

$$\rho\frac{Di}{Dt} = \frac{DP}{Dt} + \frac{\partial Q}{\partial t} + \Phi + k\nabla^2 T - \nabla \cdot \mathbf{q}_r$$

$$\Phi = 2\mu\left[\left(\frac{\partial u}{\partial x}\right)^2 + \left(\frac{\partial v}{\partial y}\right)^2 + \left(\frac{\partial w}{\partial z}\right)^2 + \frac{1}{2}\left(\frac{\partial u}{\partial y} + \frac{\partial v}{\partial x}\right)^2 + \frac{1}{2}\left(\frac{\partial v}{\partial z} + \frac{\partial w}{\partial y}\right)^2\right.$$

$$\left. + \frac{1}{2}\left(\frac{\partial w}{\partial x} + \frac{\partial u}{\partial z}\right)^2\right] + \lambda\left[\frac{\partial u}{\partial x} + \frac{\partial v}{\partial y} + \frac{\partial w}{\partial z}\right]^2$$

Cylindrical

$$\nabla \cdot \mathbf{V} = \frac{1}{r}\frac{\partial}{\partial r}(rv_r) + \frac{1}{r}\frac{\partial v_\theta}{\partial \theta} + \frac{\partial v_z}{\partial z}$$

$$\nabla^2 = \frac{\partial^2}{\partial r^2} + \frac{1}{r}\frac{\partial}{\partial r} + \frac{1}{r^2}\frac{\partial^2}{\partial \theta^2} + \frac{\partial^2}{\partial z^2}$$

$$\frac{\partial Q}{\partial t} + \Phi + \frac{1}{r}\frac{\partial}{\partial r}\left(rk\frac{\partial T}{\partial r}\right) + \frac{1}{r^2}\frac{\partial}{\partial \theta}\left(k\frac{\partial T}{\partial \theta}\right) + \frac{\partial}{\partial z}\left(k\frac{\partial T}{\partial z}\right) - \nabla \cdot \mathbf{q}_r$$

$$= \rho\frac{De}{Dt} + P\nabla \cdot \mathbf{V} = \rho\left[\frac{De}{Dt} + P\frac{D}{Dt}\left(\frac{1}{\rho}\right)\right]$$

$$\rho \frac{Di}{Dt} = \frac{DP}{Dt} + \frac{\partial Q}{\partial t} + \Phi + \frac{1}{r} \frac{\partial}{\partial r}\left(rk\frac{\partial T}{\partial r}\right) + \frac{1}{r^2}\frac{\partial}{\partial \theta}\left(k\frac{\partial T}{\partial \theta}\right) + \frac{\partial}{\partial z}\left(k\frac{\partial T}{\partial z}\right) - \nabla \cdot \mathbf{q}_r$$

ρ, k constant

$$\frac{\partial Q}{\partial t} + \Phi + k\nabla^2 T - \nabla \cdot \mathbf{q}_r = \rho \frac{De}{Dt}$$

$$\rho \frac{Di}{Dt} = \frac{DP}{Dt} + \frac{\partial Q}{\partial t} + \Phi + k\nabla^2 T - \nabla \cdot \mathbf{q}_r$$

$$\Phi = \mu\left[2\left\{\left(\frac{\partial v_r}{\partial r}\right)^2 + \left(\frac{1}{r}\frac{\partial v_\theta}{\partial \theta} + \frac{v_r}{r}\right)^2 + \left(\frac{\partial v_z}{\partial z}\right)^2\right\} + \left(\frac{\partial v_z}{\partial \theta} + \frac{\partial v_\theta}{\partial z}\right)^2 \right.$$

$$\left. + \left(\frac{\partial v_r}{\partial z} + \frac{\partial v_z}{\partial r}\right)^2 + \left(\frac{1}{r}\frac{\partial v_r}{\partial \theta} + \frac{\partial v_\theta}{\partial r} - \frac{v_\theta}{r}\right)^2\right] + \lambda\left[\frac{\partial v_r}{\partial r} + \frac{1}{r}\frac{\partial v_\theta}{\partial \theta} + \frac{v_r}{r} + \frac{\partial v_z}{\partial z}\right]^2$$

Spherical

$$\nabla \cdot \mathbf{V} = \frac{1}{r^2}\frac{\partial}{\partial r}(r^2 v_r) + \frac{1}{r\sin\theta}\frac{\partial}{\partial \theta}(v_\theta \sin\theta) + \frac{1}{r\sin\theta}\frac{\partial v_\phi}{\partial \phi}$$

$$\nabla^2 = \frac{1}{r^2}\frac{\partial}{\partial r}\left(r^2\frac{\partial}{\partial r}\right) + \frac{1}{r^2\sin\theta}\frac{\partial}{\partial \theta}\left(\sin\theta\frac{\partial}{\partial \theta}\right) + \frac{1}{r^2\sin^2\theta}\frac{\partial^2}{\partial \phi^2}$$

$$\frac{\partial Q}{\partial t} + \Phi + \frac{1}{r^2}\frac{\partial}{\partial r}\left(r^2 k\frac{\partial T}{\partial r}\right) + \frac{1}{r^2\sin\theta}\frac{\partial}{\partial \theta}\left(k\sin\theta\frac{\partial T}{\partial \theta}\right)$$

$$+ \frac{1}{r^2\sin^2\theta}\frac{\partial}{\partial \phi}\left(k\frac{\partial T}{\partial \phi}\right) - \nabla \cdot \mathbf{q}_r = \rho\frac{De}{Dt} + P\nabla \cdot \mathbf{V} = \rho\left[\frac{De}{Dt} + P\frac{D}{Dt}\left(\frac{1}{\rho}\right)\right]$$

$$\rho\frac{Di}{Dt} = \frac{DP}{Dt} + \frac{\partial Q}{\partial t} + \Phi + \frac{1}{r^2}\frac{\partial}{\partial r}\left(r^2 k\frac{\partial T}{\partial r}\right) + \frac{1}{r^2\sin\theta}\frac{\partial}{\partial \theta}\left(k\sin\theta\frac{\partial T}{\partial \theta}\right)$$

$$+ \frac{1}{r^2\sin^2\theta}\frac{\partial}{\partial \phi}\left(k\frac{\partial T}{\partial \phi}\right) - \nabla \cdot \mathbf{q}_r$$

ρ, k constant

$$\frac{\partial Q}{\partial t} + \Phi + k\nabla^2 T - \nabla \cdot \mathbf{q}_r = \rho\frac{De}{Dt}$$

$$\rho\frac{Di}{Dt} = \frac{DP}{Dt} + \frac{\partial Q}{\partial t} + \Phi + k\nabla^2 T - \nabla \cdot \mathbf{q}_r$$

$$\Phi = \mu\left[2\left\{\left(\frac{\partial v_r}{\partial r}\right)^2 + \left(\frac{1}{r}\frac{\partial v_\theta}{\partial \theta} + \frac{v_r}{r}\right)^2 + \left(\frac{1}{r\sin\theta}\frac{\partial v_\phi}{\partial \phi} + \frac{v_r}{r} + \frac{v_\theta\cot\theta}{r}\right)^2\right\}\right.$$

$$+ \left\{ \frac{1}{r \sin \theta} \frac{\partial v_\theta}{\partial \phi} + \frac{\sin \theta}{r} \frac{\partial}{\partial \theta} \left(\frac{v_\phi}{\sin \theta} \right) \right\}^2$$

$$+ \left\{ \frac{1}{r \sin \theta} \frac{\partial v_r}{\partial \phi} + r \frac{\partial}{\partial r} \left(\frac{v_\phi}{r} \right) \right\}^2 + \left\{ r \frac{\partial}{\partial r} \left(\frac{v_\theta}{r} \right) + \frac{1}{r} \frac{\partial v_r}{\partial \theta} \right\}^2 \Bigg]$$

$$+ \lambda \left[\frac{\partial v_r}{\partial r} + \frac{1}{r} \frac{\partial v_\theta}{\partial \theta} + \frac{2 v_r}{r} + \frac{1}{r \sin \theta} \frac{\partial v_\phi}{\partial \phi} + \frac{v_\theta \cot \theta}{r} \right]^2$$

6. Conversion Tables

TABLE 2

	To convert number of	To	Multiply by
Length	inch	cm	2.540
	ft	m	0.3048
Area	ft^2	m^2	0.0929
Volume	ft^3	m^3	0.02832
Mass	lbm	kg	0.45359
	Slugs	kg	14.594
Force	lbf	Newtons	4.4482
Density	lbm/ft^3	kg/m^3	16.02
Work	ft-lbf	mkg	0.1383
	hp-hr	mkg	273,700
Heat	Btu	kcal	0.2520
	Chu	Btu	1.800
	Btu	Joules	1054.35
	Btu	ft-lbf	778.26
	kw-hr	Btu	3412.75
Specific heat	Btu/lbm-°F	cal/gC	1.000
	Btu/lbm-°F	Wsec/kgmC	4184.0
Pressure	lbf/in^2, psi	kgf/cm^2	0.070309
	psi	atm	0.068046
	psi	bars	0.068948
	psi	dynes/cm^2	68947.0
Surface tension	lbf/ft	dynes/cm	6.8519 x 10^{-5}

TABLE 3. Heat Flux, q/A

Multiply number of by ⟶ To obtain ↓	$\dfrac{Btu}{ft^2 \text{-} hr}$	$\dfrac{W}{cm^2}$	$\dfrac{kcal}{hr\text{-}m^2}$	$\dfrac{cal}{sec\text{-}cm^2}$
Btu/ft^2-hr	1	3,170.75	0.36865	13,277.26
W/cm^2	3.154 x 10^{-4}	1	1.163 x 10^{-4}	4.1868
kcal/hr-m^2	2.7126	8,600	1	2.778 x 10^{-5}
cal/sec-cm^2	7.536 x 10^{-5}	0.2389	36,000	1

TABLE 4.   Heat Transfer Coefficient, h

Multiply number of → / To obtain ↓	$\dfrac{Btu}{hr\text{-}ft^2\text{-}^\circ F}$	$\dfrac{W}{cm^2\text{-}^\circ C}$	$\dfrac{cal}{sec\text{-}cm^2\text{-}^\circ C}$	$\dfrac{kcal}{hr\text{-}m^2\text{-}^\circ C}$
Btu/hr-ft²-°F	1	1761	7376	0.20489
W/cm²-°C	5.6785 x 10⁻⁴	1	4.186	1.163 x 10⁻⁴
cal/sec-cm²-°C	1.356 x 10⁻⁴	0.2391	1	2.778 x 10⁻⁵
kcal/hr-m²-°C	4.8826	8600	36000	1

TABLE 5.   Thermal Conductivity, k

Multiply number of → / To obtain ↓	$\dfrac{Btu}{hr\text{-}ft\text{-}^\circ F}$	$\dfrac{W}{cm\text{-}^\circ C}$	$\dfrac{cal}{sec\text{-}cm\text{-}^\circ C}$	$\dfrac{kcal}{hr\text{-}m\text{-}^\circ C}$	$\dfrac{Btu\ in}{hr\text{-}ft^2\text{-}^\circ F}$
Btu/hr-ft-°F	1	57.793	241.9	0.6722	0.08333
W/cm-°C	0.01730	1	4.186	0.01171	1.442 x 10⁻³
cal/sec-cm-°C	4.134 x 10⁻³	0.2389	1	2.778 x 10⁻³	3.445 x 10⁻⁴
kcal/hr-m-°C	1.488	86.01	360	1	0.1240
Btu in./hr-ft²-°F	12	693.5	2903	8.064	1

TABLE 6.   Viscosity, μ

Multiply number of → / To obtain ↓	$\dfrac{lbm}{ft\text{-}hr}$	$\dfrac{lbf\text{-}sec}{ft^2}$	Centipoise	$\dfrac{kgm}{m\text{-}hr}$	$\dfrac{kgf\text{-}sec}{m^2}$
lbm/ft-hr	1	116,000	2.42	0.672	23733
lbf-sec/ft²	0.00000862	1	0.00002086	0.00000579	0.2048
Centipoise	0.413	47,880	1	0.278	9807
kgm/m-hr	1.49	172,000	3.60	1	35305
kgf-sec/m²	0.0000421	4.882	0.0001020	0.0000284	1

100 centipoise = 1 Poise = 1g/sec-cm = 1 dyne sec/cm².

TABLE 7.   Kinematic Viscosity, ν

Multiply number of → / by → To obtain ↓	$\dfrac{ft^2}{hr}$	Stokes	$\dfrac{m^2}{hr}$	$\dfrac{m^2}{sec}$
ft^2/hr	1	3.875	10.764	38,751
Stokes	0.25806	1	2.778	10^4
m^2/hr	0.092903	0.3599	1	3600
m^2/sec	0.00002581	10^{-4}	0.0002778	1

Part **B**

Mathematical Methods

P. J. SCHNEIDER
Lockheed Missiles & Space Company, Sunnyvale, California

1. Fundamentals

A condensation is given of those basic mathematical operations, functions, and elements of calculus that are frequently drawn upon in heat transfer analyses. Handbooks devoted exclusively to general mathematical formulas and theorems [1] and to mathematical functions and their tabulation [2] are available for supporting reference.

a. Elementary Operations

Real and Complex Numbers. If *real* numbers are described by a, b, then *complex* (*imaginary*) numbers and their *conjugates* are represented by $a + bi$ and $a - bi$, where the *unit imaginary number* i obeys the relations

$$\left.\begin{array}{ll} i^{4n+1} = i \quad, & i^{4n+2} = -1 \\ i^{4n+3} = -i \quad, & i^{4n+4} = 1 \end{array}\right\} \, n = 0, 1, 2, \ldots \tag{13}$$

Fundamental operations with complex numbers include

$$(a + bi) + (c + di) = (a + c) + (b + d)i$$
$$(a + bi)(c + di) = (ac - bd) + (ad + bc)i$$
$$\frac{(a + bi)}{(c + di)} = \left(\frac{ac + bd}{c^2 + d^2}\right) + \left(\frac{bc - ad}{c^2 + d^2}\right)i$$

If r is the *absolute value* of $a + bi$ and β is its *amplitude*, then the *trigonometric (polar) form* of $a + bi$ in the Argand plane with real x and imaginary y (Fig. 7) is

$$a + bi = r(\cos\beta + i \sin\beta) \tag{14}$$

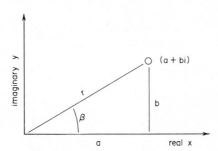

Fig. 7. Graphics of a complex number.

De Moivre's theorem for power expansions of the trigonometric form is

$$(\cos\beta + i \sin\beta)^n = \cos n\beta + i \sin n\beta \tag{15}$$

where the n are positive integers.

Polynomials. An nth degree *polynomial* $f(x)$ with real or complex coefficients a_n

$$f(x) = a_0 x^n + a_1 x^{n-1} + \cdots + a_n$$

is factorable to the form

$$f(x) = a_0(x - x_1)(x - x_2) \cdots (x - x_n) \tag{16}$$

where the x_n are roots of $f(x) = 0$. The roots x_n are related to the coefficients a_n by [3]

$$x_1 + x_2 + \cdots + x_n = -\frac{a_1}{a_0}$$

$$x_1 x_2 + x_1 x_3 + x_2 x_3 + \cdots + x_{n-1} x_n = \frac{a_2}{a_0} \tag{17}$$

$$x_1 x_2 x_3 + x_1 x_2 x_4 + \cdots + x_{n-2} x_{n-1} x_n = -\frac{a_3}{a_0}$$

$$\vdots$$

$$x_1 x_2 x_3 x_4 \cdots x_{n-3} x_{n-2} x_{n-1} x_n = (-1)^n \frac{a_n}{a_0}$$

(17)
(cont.)

Cubic Equations. The general *cubic* $(n = 3)$, $y^3 + by^2 + cy + d = 0$, is reducible (substitute $y = x - b/3$) to the form $x^3 + vx + w = 0$, where $v = (3c - b^2)/3$ and $w = (2b^3 - 9bc + 27d)/27$. The three roots of the reduced cubic (*Cardan's solution*) are

$$x_1 = A + B \ , \quad x_{2,3} = -\frac{1}{2}(A + B) \pm \frac{\sqrt{3}}{2}(A - B)i$$

(18)

where

$$A = \left[-\frac{w}{2} + \left(\frac{w^2}{4} + \frac{v^3}{27} \right)^{1/2} \right]^{1/3} \ , \quad B = \left[-\frac{w}{2} - \left(\frac{w^2}{4} + \frac{v^3}{27} \right)^{1/2} \right]^{1/3}$$

If $(w^2/4 + v^3/27) > 0$, there is one real root and two conjugate complex roots.

Quartic Equations. The general *quartic* $(n = 4)$, $y^4 + ay^3 + by^2 + cy + d = 0$, is reducible (substitute $y = x - a/4$) to the form $x^4 + ux^2 + vx + w = 0$. The four roots of the reduced quartic for positive v (*Descartes-Euler solution*) are

$$x_1 = -\sqrt{z_1} - \sqrt{z_2} - \sqrt{z_3} \ , \quad x_2 = -\sqrt{z_1} + \sqrt{z_2} + \sqrt{z_3}$$

$$x_3 = \sqrt{z_1} - \sqrt{z_2} + \sqrt{z_3} \ , \quad x_4 = \sqrt{z_1} + \sqrt{z_2} - \sqrt{z_3}$$

(19)

where z_1, z_2, and z_3 are roots of the unreduced cubic equation

$$z^3 + \left(\frac{u}{2} \right) z^2 + \left(\frac{u^2 - 4w}{16} \right) z - \frac{v^2}{64} = 0$$

Partial Fractions. A proper fraction (numerator of lower degree than denominator) can be reduced to *partial fractions* whose individual denominators are prime factors of the denominator in the original proper fraction. Several examples arise in heat conduction with radiation and combined radiation/convection boundary conditions:
- Proper fraction $1/(x^4 - w^4)$ with real linear factors in denominator. Express denominator as linear factors and use method of *undetermined coefficients*

$$\frac{1}{x^4 - w^4} = \frac{1}{(x^2 + w^2)(x^2 - w^2)} = \frac{A}{x^2 + w^2} + \frac{B}{x^2 - w^2} = \frac{1}{2w^2(x^2 - w^2)} - \frac{1}{2w^2(x^2 + w^2)}$$

- Proper fraction $1/(x^4 + vx - w^4)$ without linear factors in denominator. Express as the partial fractions

$$\frac{1}{x^4 + vx - w^4} = \frac{A}{x - x_1} + \frac{B}{x - x_2} + \frac{C}{x - x_3} + \frac{D}{x - x_4}$$

where x_1, x_2, x_3, and x_4 are roots of the reduced quartic $x^4 + vx - w^4 = 0$ given by Eqs. (19). Solve for A, B, C, and D using the polynomial relations in Eqs. (17).

Power Series. The infinite series of terms

$$\sum_{n=0}^{\infty} a_n(x - x_0)^n = a_0 + a_1(x - x_0) + a_2(x - x_0)^2 + \cdots$$

is called a *power series* in x. For the series to converge, its sum must approach a finite value. By the *Cauchy ratio test,* if the absolute value of the $(n + 1)$ term divided by the nth term approaches a limit as $n \to \infty$ which is less than unity, the series converges.

The power series is useful in expanding an arbitrary function $f(x)$ near $x = x_0$. By successive differentiation

$$f(x) = \sum_{n=0}^{\infty} \frac{(x - x_0)^n}{n!} \frac{d^n}{dx^n} f(x_0) = f(x_0) + (x - x_0) \frac{d}{dx} f(x_0)$$

$$+ \frac{(x - x_0)^2}{2!} \frac{d^2}{dx^2} f(x_0) + \cdots$$

(20)

where $n! = 1 \cdot 2 \cdot 3 \cdots n$. This is the *Taylor* series. For valid representation of $f(x)$, all derivatives of $f(x_0)$ must exist and be continuous and finite. If $x_0 = 0$, Eq. (20) is a *Maclaurin* series. Example expansions are the binomial and geometric series

$$(a + x)^n = a^n + na^{n-1}x + \frac{n(n - 1)}{2!} a^{n-2}x^2$$

$$+ \frac{n(n - 1)(n - 2)}{3!} a^{n-3}x^3 + \cdots \quad ; \quad x^2 < a^2$$

$$(a - bx)^{-1} = \frac{1}{a}\left(1 + \frac{b}{a} x + \frac{b^2}{a^2} x^2 + \cdots\right) \qquad ; \quad b^2x^2 < a^2$$

Power series are used to expand elementary functions for solving differential equations (term-by-term differentiation permissible), for formulating finite-difference approximations to differential equations, for performing formal integrations (term-by-term integration permissible), and for calculating approximate values of $f(x)$ for small x. Approximations to arbitrary functions are obtained by dropping all but the first two or so terms. The *remainder* dropped after n terms in the Taylor series (*Lagrange formula*) is

$$R_n = \frac{(x - x_0)^n}{n!} \frac{d^n}{dx^n} f(X) \quad , \quad X = x_0 + c(x - x_0) ; \quad 0 < c < 1 \quad (21)$$

b. Elementary Functions. A number of integrals recurrent in heat transfer theory have solutions in terms of both elementary and higher transcendental functions. Both types of functions are reviewed and sources given for available tabulations of their numerical values.

Logarithmic Function. If $N = a^b$, then b is the *logarithm* of N to the *base a*, expressed as $b = \log_a N$. *Common (Briggsian)* logarithms have a base of 10 (abbreviated log), while *natural(Napierian)* logarithms used in the calculus are to base e (abbreviated ln), where $e = \lim_{n \to \infty} (1 + 1/n)^n = 2.71828$. For a change of base between the two systems, $\ln N = \log N / \log e = \log N / 0.43429$. The transcendental *logarithmic functions* $y = \log x$ and $y = \ln x$ are shown in Fig. 8.

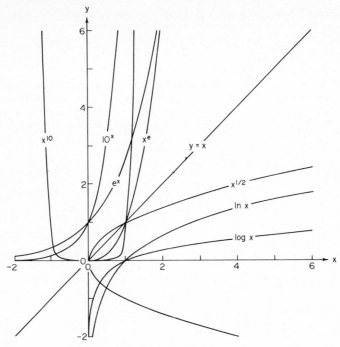

Fig. 8. Logarithmic, exponential, and power functions.

Basic properties of logarithms are

$$\ln xz = \ln x + \ln z \quad , \quad \ln \frac{x}{z} = \ln x - \ln z$$

$$\ln x^z = z \ln x \qquad , \quad \ln x^{1/z} = \frac{1}{z} \ln x$$

(22)

Useful series expansions include

$$\ln x = 2\left[\left(\frac{x-1}{x+1}\right) + \frac{1}{3}\left(\frac{x-1}{x+1}\right)^3 + \frac{1}{5}\left(\frac{x-1}{x+1}\right)^5 + \cdots\right] \qquad ; \quad 0 < x < \infty$$

$$\ln(1 \pm x) = \pm x - \frac{1}{2}x^2 \pm \frac{1}{3}x^3 - \frac{1}{4}x^4 \pm \cdots \qquad ; \quad -1 < x < 1$$

$$\ln\left(\frac{1+x}{x}\right) = 2\left[\frac{1}{(2x+1)} + \frac{1}{3(2x+1)^3} + \frac{1}{5(2x+1)^5} + \cdots\right] \qquad ; \quad x > 0$$

$$\ln\left(\frac{1+x}{1-x}\right) = 2\left(x + \frac{1}{3}x^3 + \frac{1}{5}x^5 + \cdots\right) \qquad ; \quad -1 < x < 1$$

$$\ln\left(\frac{x+1}{x-1}\right) = 2\left(\frac{1}{x} + \frac{1}{3x^3} + \frac{1}{5x^5} + \cdots\right) \qquad ; \quad x^2 > 1$$

$$\ln(x + z) = \ln x + 2\left[\left(\frac{z}{2x + z}\right) + \frac{1}{3}\left(\frac{z}{2x + z}\right)^3 + \frac{1}{5}\left(\frac{z}{2x + z}\right)^5 + \cdots\right] \quad ; \quad \begin{array}{l} x > 0 \\ -x < z < \infty \end{array}$$

Exponential and Power Functions. Considering x variable, an *exponential function* is $y = a^x$ (image of logarithmic function) and a *power function* is $y = x^a$ (Fig. 8). Both functions obey the laws of exponents

$$e^x e^y = e^{x+y} \quad , \quad (e^x)^y = e^{xy} , \tag{23}$$
$$\frac{e^x}{e^y} = e^{x-y} \quad , \quad (e^x)^{1/y} = e^{x/y} .$$

Useful series expansions for exponential functions include

$$a^x = 1 + (x \ln a) + \frac{1}{2!}(x \ln a)^2 + \frac{1}{3!}(x \ln a)^3 + \cdots \quad ; \quad -\infty < x < \infty$$

$$e^x = 1 + x + \frac{1}{2!}x^2 + \frac{1}{3!}x^3 + \cdots \quad ; \quad -\infty < x < \infty$$

$$e^{-x^2} = 1 - x^2 + \frac{1}{2!}x^4 - \frac{1}{3!}x^6 \pm \cdots \quad ; \quad -\infty < x < \infty$$

Hyperbolic Functions. The derived *hyperbolic functions* (Fig. 9) represent linear combinations of exponential functions according to

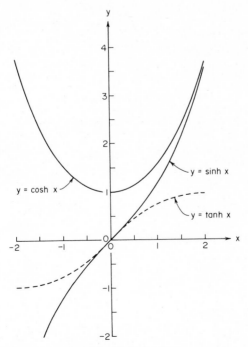

Fig. 9. Hyperbolic functions.

$$\sinh x = \frac{1}{2}(e^x - e^{-x}) = -i \sin ix$$

$$\cosh x = \frac{1}{2}(e^x + e^{-x}) = \cos ix \qquad (24)$$

$$\tanh x = \frac{\sinh x}{\cosh x} = -i \tan ix$$

These transcendental functions are interrelated by the formulas

$$\cosh^2 x - \sinh^2 x = 1 \quad , \quad \sinh x = \frac{\tanh x}{\sqrt{1 - \tanh^2 x}} \qquad (25)$$

The *inverse* hyperbolic functions are denoted $\sinh^{-1}x$, $\cosh^{-1}x$, and $\tanh^{-1}x$.
Some of the more frequently used identities for the three basic functions include

$$\sinh 2x = 2 \sinh x \cosh x = \frac{2 \tanh x}{1 - \tanh^2 x}$$

$$\cosh 2x = \sinh^2 x + \cosh^2 x = \frac{1 + \tanh^2 x}{1 - \tanh^2 x}$$

$$\tanh 2x = \frac{2 \tanh x}{1 + \tanh^2 x}$$

$$\sinh(x \pm y) = \sinh x \cosh y \pm \cosh x \sinh y$$

$$\cosh(x \pm y) = \cosh x \cosh y \pm \sinh x \sinh y$$

$$\tanh(x \pm y) = \frac{\tanh x \pm \tanh y}{1 \pm \tanh x \tanh y}$$

$$\sinh x \pm \sinh y = 2 \sinh\frac{1}{2}(x \pm y) \cosh\frac{1}{2}(x \mp y)$$

$$\cosh x + \cosh y = 2 \cosh\frac{1}{2}(x + y) \cosh\frac{1}{2}(x - y)$$

$$\cosh x - \cosh y = 2 \sinh\frac{1}{2}(x + y) \sinh\frac{1}{2}(x - y)$$

$$\tanh x \pm \tanh y = \frac{\sinh(x \pm y)}{\cosh x \cosh y}$$

The basic series expansions are

$$\sinh x = x + \frac{1}{3!}x^3 + \frac{1}{5!}x^5 + \cdots$$

$$\cosh x = 1 + \frac{1}{2!}x^2 + \frac{1}{4!}x^4 + \cdots$$

$$\tanh x = x - \frac{1}{3}x^3 + \frac{2}{15}x^5 - \frac{17}{315}x^7 \pm \cdots$$

Many other identities and expansions are available [1, 2, 4]. Numerical values of the hyperbolic functions themselves are widely tabulated (Table 9).

Circular Functions.   The periodic *circular* (*trigonometric*) *functions* are defined in terms of one of the acute angles β of a right triangle (Fig. 7, real y) as

$$\sin \beta = \frac{b}{r} = -i \sinh i\beta$$

$$\cos \beta = \frac{a}{r} = \cosh i\beta \qquad (26)$$

$$\tan \beta = \frac{\sin \beta}{\cos \beta} = -i \tanh i\beta$$

These transcendental functions (Fig. 10) are interrelated as

$$\cos^2 \beta + \sin^2 \beta = 1 \quad , \quad \sin \beta = \frac{\tan \beta}{\sqrt{1 + \tan^2 \beta}} \qquad (27)$$

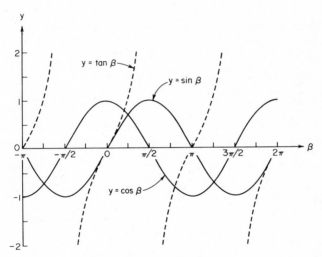

Fig. 10. Circular functions.

The inverse circular functions are denoted $\sin^{-1}\beta$, $\cos^{-1}\beta$, and $\tan^{-1}\beta$. Some useful identities for the three basic functions include

$$\sin 2\beta = 2 \sin \beta \cos \beta$$

$$\cos 2\beta = \cos^2 \beta - \sin^2 \beta = 1 - 2 \sin^2 \beta$$

$$\tan 2\beta = \frac{2 \tan \beta}{1 - \tan^2 \beta}$$

$$\sin(\beta \pm y) = \sin \beta \cos y \pm \cos \beta \sin y$$

$$\cos(\beta \pm y) = \cos \beta \cos y \mp \sin \beta \sin y$$

$$\tan(\beta \pm y) = \frac{\tan \beta \pm \tan y}{1 \mp \tan \beta \tan y}$$

$$\sin\beta + \sin\gamma = 2 \sin\frac{1}{2}(\beta + \gamma) \cos\frac{1}{2}(\beta - \gamma)$$

$$\sin\beta - \sin\gamma = 2 \cos\frac{1}{2}(\beta + \gamma) \sin\frac{1}{2}(\beta - \gamma)$$

$$\cos\beta + \cos\gamma = 2 \cos\frac{1}{2}(\beta + \gamma) \cos\frac{1}{2}(\beta - \gamma)$$

$$\cos\beta - \cos\gamma = -2 \sin\frac{1}{2}(\beta + \gamma) \sin\frac{1}{2}(\beta - \gamma)$$

$$\tan\beta + \tan\gamma = \frac{\sin(\beta + \gamma)}{\cos\beta \cos\gamma}$$

$$\tan\beta - \tan\gamma = \frac{\sin(\beta - \gamma)}{\cos\beta \cos\gamma}$$

$$a \sin\beta - b \cos\beta = \sqrt{a^2 + b^2} \sin\left(\beta - \tan^{-1}\frac{b}{a}\right)$$

$$\sin^2\beta - \sin^2\gamma = \sin(\beta + \gamma) \sin(\beta - \gamma)$$

$$\cos^2\beta - \sin^2\gamma = \cos(\beta + \gamma) \cos(\beta - \gamma)$$

$$\cos^2\beta - \cos^2\gamma = -\sin(\beta + \gamma) \sin(\beta - \gamma)$$

$$\sin\beta \sin\gamma = \frac{1}{2} \cos(\beta - \gamma) - \frac{1}{2} \cos(\beta + \gamma)$$

$$\sin\beta \cos\gamma = \frac{1}{2} \sin(\beta - \gamma) + \frac{1}{2} \sin(\beta + \gamma)$$

$$\cos\beta \cos\gamma = \frac{1}{2} \cos(\beta - \gamma) + \frac{1}{2} \cos(\beta + \gamma)$$

Series expansions of the basic functions are

$$\sin\beta = \beta - \frac{1}{3!}\beta^3 + \frac{1}{5!}\beta^5 \mp \cdots$$

$$\cos\beta = 1 - \frac{1}{2!}\beta^2 + \frac{1}{4!}\beta^4 \mp \cdots$$

$$\tan\beta = \beta + \frac{1}{3}\beta^3 + \frac{2}{15}\beta^5 + \frac{17}{315}\beta^7 + \cdots \quad ; \quad \beta^2 < \frac{\pi^2}{4}$$

Many other identities and expansions are known [1, 2, 4], and numerical values of the circular functions are accurately tabulated (Table 9).

The following identities relate the exponential functions of complex exponent or base and the power function of complex exponent to the logarithmic and circular functions

$$a^{ix} = e^{ix \ln a} = \cos(x \ln a) + i \sin(x \ln a)$$

$$i^x = e^{x \ln i} = \cos\frac{\pi}{2}x + i \sin\frac{\pi}{2}x$$

$$x^i = e^{i \ln x} = \cos(\ln x) + i \sin(\ln x)$$

The first of these leads to the *Euler identity*

$$e^{\pm ix} = \cos x \pm i \sin x \qquad (28)$$

whereby $\sin x = (e^{ix} - e^{-ix})/2i$ and $\cos x = (e^{ix} + e^{-ix})/2$.

Gamma and Beta Functions. An important integral related to the factorial function $n! = 1 \cdot 2 \cdot 3 \cdots n$ is

$$\Gamma(n) = \int_0^\infty e^{-x} x^{n-1} dx \quad ; \quad n > 0 \qquad (29)$$

where the n are positive integers. This definite integral is called the *gamma function* (*Euler's integral of the second kind*). By integrating Eq. (29)

$$\Gamma(n) = \frac{1}{n} \Gamma(n + 1) = (n - 1)\Gamma(n - 1)$$

$$= \frac{\Gamma(n + m)}{n(n + 1)(n + 2) \cdots (n + m - 1)} \quad ; \quad m = 1, 2, 3, \ldots \qquad (30)$$

In general, $\Gamma(n + 1) = n!$ where $n = 0, 1, 2, \ldots$. Equation (30) permits calculation of $\Gamma(n)$ for negative nonintegral values of n (Fig. 11). Suppose $\Gamma\left(-\frac{1}{2}\right)$ is required.

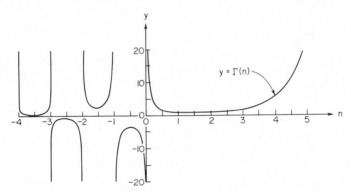

Fig. 11. Gamma function.

From Eq. (29), $\Gamma\left(\frac{1}{2}\right) = \sqrt{\pi}$. Then by Eq. (30), $\Gamma\left(-\frac{1}{2}\right) = \Gamma\left(\frac{1}{2}\right) / \left(-\frac{1}{2}\right) = -2\sqrt{\pi}$. An *incomplete* gamma function is

$$\Gamma_y(n) = \int_0^y e^{-x} x^{n-1} dx \quad ; \quad n > 0 \qquad (31)$$

The *beta function* (*Euler's integral of the first kind*) is related to the gamma function as

$$B(m, n) = \int_0^1 x^{m-1}(1 - x)^{n-1} dx = \frac{\Gamma(m)\Gamma(n)}{\Gamma(m + n)} \quad ; \quad m, n > 0 \qquad (32)$$

An incomplete beta function is

$$B_y(m, n) = \int_0^y x^{m-1}(1 - x)^{n-1} dx \quad ; \quad 0 \le y \le 1 \qquad (33)$$

Since a number of important integrals have solutions in terms of gamma and beta functions, these functions have been accurately calculated and tabulated (Table 9).
Error Function.   The simple integral (Fig. 12)

$$\mathrm{erf}\, x \;=\; \frac{2}{\sqrt{\pi}} \int_{0}^{x} e^{-\lambda^2}\, d\lambda \tag{34}$$

does not have a solution in terms of elementary functions. It arises frequently in heat

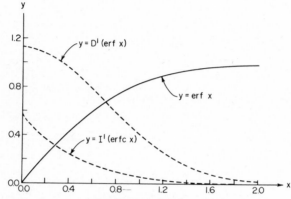

Fig. 12. Error function.

transfer analyses and is called the *error function* (*probability integral*). A useful series expansion is

$$\mathrm{erf}\, x \;=\; \frac{2}{\sqrt{\pi}} \left(x - \frac{1}{3} x^3 + \frac{1}{5\cdot 2!} x^5 - \frac{1}{7\cdot 3!} x^7 \pm \cdots \right)$$

The first derivative (Fig. 12) is

$$\frac{d}{dx}(\mathrm{erf}\, x) \;=\; D^1(\mathrm{erf}\, x) \;=\; \frac{2}{\sqrt{\pi}} e^{-x^2}$$

A *complementary* error function is defined as

$$\mathrm{erfc}\, x \;=\; 1 - \mathrm{erf}\, x \;=\; \frac{2}{\sqrt{\pi}} \int_{x}^{\infty} e^{-\lambda^2}\, d\lambda \tag{35}$$

The repeated integrals of this complementary function are important. The first integral (Fig. 12) is

$$\int_{x}^{\infty} \mathrm{erfc}\, \lambda\, d\lambda \;=\; I^1(\mathrm{erfc}\, x) \;=\; \frac{1}{\sqrt{\pi}} e^{-x^2} - x\, \mathrm{erfc}\, x$$

and a recurrence formula for the remaining integrals [5] is

$$I^n(\mathrm{erfc}\, x) \;=\; \frac{1}{2n}[I^{n-2}(\mathrm{erfc}\, x) - 2x I^{n-1}(\mathrm{erfc}\, x)] \quad ; \quad n = 1, 2, 3, \ldots \tag{36}$$

Numerical values of the error function and its derivatives are tabulated (Table 9).

Sine, Cosine, and Exponential Integral Functions. A number of other simple integrals arise in heat transfer predictions that are not integrable with elementary functions. These include (Fig. 13) the *sine integral function*

$$\text{Si}\, x \;=\; \int_0^x \frac{1}{x}\sin x\,dx \;=\; x - \frac{1}{3!}\frac{x^3}{3} + \frac{1}{5!}\frac{x^5}{5} \mp \cdots \tag{37}$$

where $\text{Si}(\infty) = \pi/2$, the *cosine integral function*

$$\text{Ci}\, x \;=\; -\int_x^\infty \frac{1}{x}\cos x\,dx \;=\; \gamma + \ln x - \frac{1}{2!}\frac{x^2}{2} + \frac{1}{4!}\frac{x^4}{4} \mp \cdots \quad ; \quad x > 0 \tag{38}$$

where $\gamma = $ *Euler's constant* $= \lim_{n \to \infty}(1 + 1/2 + 1/3 + \cdots + 1/n - \log n) = 0.57722$ and $\text{Ci}(0) = -\infty$, and the *exponential integral function*

$$\text{Ei}(-x) \;=\; -\int_x^\infty \frac{1}{x}e^{-x}\,dx \quad ; \quad x > 0 \tag{39}$$

where $\text{Ei}(0) = -\infty$. Numerical values of these functions are also available (Table 9).

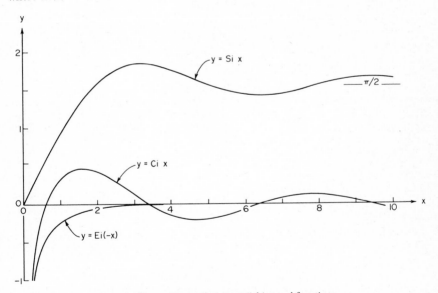

Fig. 13. Sine, cosine, and exponential integral functions.

Combined Functions. A number of elementary functions reoccur in characteristic combined forms in heat transfer studies:
- The combined exponential and power relation

$$y \;=\; (\pi z)^{-1/2}e^{-x^2/z}$$

is the *plane source function* of heat conduction (Fig. 14). This function, composed of two independent variables x and z, describes the temperature y in an infinite solid at

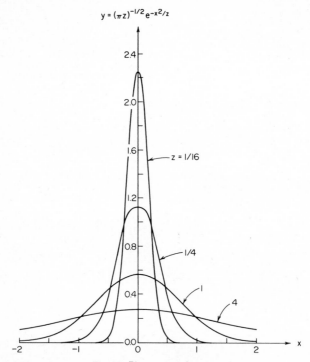

$$y = (\pi z)^{-1/2} e^{-x^2/z}$$

Fig. 14. Plane source function.

time z and linear distance x from a plane instantaneous heat source of unit strength at $x = 0$. The area under the curves (proportional to heat liberated by source) is constant.

- The similar combination

$$y = (\pi z)^{-3/2} e^{-x^2/z}$$

is the *spherical source function* of heat conduction [6] (Fig. 15). Here an instantaneous point source at radius $x = 0$ liberates heat of unit strength into an infinite solid.

- The combined exponential and power relation of one independent variable

$$y = \frac{1}{x^5 (e^{1/x} - 1)}$$

is the *Planck function* for thermal radiation (Fig. 16). The peak value $y = 21.20144$ (proportional to maximum radiation intensity) occurs at $x = 0.20141$ (wave length for maximum intensity). Integral of the curve (proportional to total radiated energy) equals $\pi^4/15$.

- The combined exponential and circular relation

$$y = e^{-x/2} \sin x$$

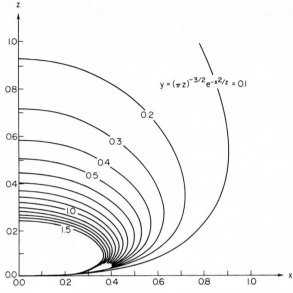

Fig. 15. Spherical source function.

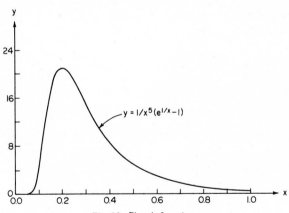

Fig. 16. Planck function.

is the *damped wave function* for conduction with harmonic temperature boundary conditions (Fig. 17). The functions $y = \pm e^{-x/2}$ form an envelope of the damped

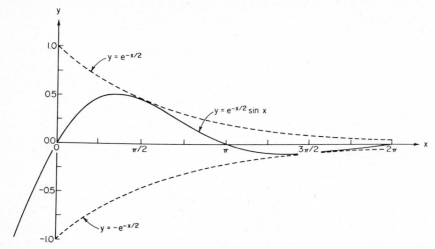

Fig. 17. Damped wave function.

harmonic in the solid, but do not represent the locus of maxima and minima in the sine wave; these occur at $x = \tan^{-1}2$.

Sources for Tabulated Functions. Short numerical tabulations of the elementary logarithmic, exponential, power, hyperbolic, and circular functions are listed in numerous and readily available handbooks of mathematics [1, 2, 4], physics [7], and engineering [8, 9]. However, these abbreviated tables are generally unsuited for detailed numerical calculations, and less available sources must be consulted. An index to the more elaborate mathematical tables suitable for accurate computational work (tabulating numerical values over small intervals of the independent variable, each with a large number of significant digits) has been compiled with references through 1944 [10]. A recent handbook [11] contains an index to 115 such references.

Table 9 gives selected references for accurate numerical tabulations that are generally available in mathematics, physics, and engineering libraries of universities in the United States.

TABLE 9. References for Tabulated Mathematical Functions

$\ln x$	[2,* 12,* 13]	$\sinh^{-1}x$	[2, 12, 18]	$\sin^{-1}\beta$	[2, 12, 20]	$B_y(m, n)$	[24]
$e^{\pm x}$	[2, 12, 14]	$\cosh^{-1}x$	[12, 18]	$\cos^{-1}\beta$	[12]	$\operatorname{erf} x$	[2, 12, 25]
x^a	[12,† 15]	$\tanh^{-1}x$	[2, 12, 18]	$\tan^{-1}\beta$	[2, 12, 21]	$D^n(\operatorname{erf}x)$	[2, 6, 25]
$\sinh x$	[2, 12, 16]	$\sin\beta$	[2, 12, 19]	$\Gamma(n)$	[2, 12, 22]	$\operatorname{Si} x$	[6, 26]
$\cosh x$	[2, 12, 16]	$\cos\beta$	[2, 12, 19]	$\Gamma_y(n)$	[23]	$\operatorname{Ci} x$	[6, 26]
$\tanh x$	[2, 12, 17]	$\tan\beta$	[2, 12, 17]	$B(m, n)$	[24]	$\operatorname{Ei}(-x)$	[6, 26]

*These two references tabulate a relatively large number of functions under a single cover and are therefore more condensed than the alternate (specialized) references cited.
†$a = 2, 3, 4, 5, 1/2, 1/3, 1/4, 1/5$.

Although all data referenced above are for real arguments, numerical tabulations that are occasionally required for complex arguments $x + iy$ are also available [2, 10, 11], but to a lesser extent.

Constants. Some mathematical constants that reoccur in computations with the elementary functions are listed in Table 10.

TABLE 10. Selected Mathematical Constants

$2^{1/2}$ = 1.41421	π = 3.14159	e = 2.71828	$\ln 2$ = 0.69315
$2^{-1/2}$ = 0.70711	$\pi^{1/2}$ = 1.77245	$e^{1/2}$ = 1.64872	$\ln 3$ = 1.09861
$3^{1/2}$ = 1.73205	$\pi^{-1/2}$ = 0.56419	$e^{1/3}$ = 1.39561	$\ln \pi$ = 1.14473
$3^{-1/2}$ = 0.57735	$(2\pi)^{1/2}$ = 2.50663	e^{π} = 23.1407	$\ln \pi^{1/2}$ = 0.57238
$10^{1/2}$ = 3.16228	$(2/\pi)^{1/2}$ = 0.79789	$e^{\pi/2}$ = 4.81048	$\ln (2\pi)^{1/2}$ = 0.91894
$2^{1/3}$ = 1.25992	$\pi^{1/3}$ = 1.46459	$e^{\pi/4}$ = 2.19328	$\ln \gamma$ = -0.54954
$3^{1/3}$ = 1.44225	π^{e} = 22.4592	e^{e} = 15.1543	$\log 10$ = 2.30259
$10^{1/3}$ = 2.15443	$180/\pi$ = 57.2958	e^{γ} = 1.78107	$\log e$ = 0.43429

c. Elements of Calculus

Total Derivative. Given a continuous, single-variable function $y = f(x)$, *increments* (finite-difference values) of y and x are denoted Δy and Δx, while *differentials* (limit-difference values) are denoted dy and dx The first *total derivative* of y with respect to x (rate-of-change of y with respect to x) is denoted dy/dx and has the value (Fig. 18)

$$D^1 = \frac{dy}{dx} = \lim_{\Delta x \to 0} \frac{\Delta y}{\Delta x} = \lim_{\Delta x \to 0} \frac{1}{\Delta x}[f(x + \Delta x) - f(x)] = \tan\beta \tag{40}$$

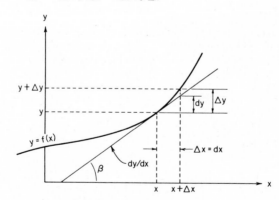

Fig. 18. Graphics of a derivative.

The second derivative of $y = f(x)$ has the value

$$\frac{d^2 y}{dx^2} = \frac{d}{dx}\left(\frac{dy}{dx}\right)$$

For *inverse* functions $x = f(y)$ the corresponding derivatives are

$$\frac{dy}{dx} = \frac{1}{dx/dy} \quad , \quad \frac{d^2 y}{dx^2} = -\frac{d^2 x/dy^2}{(dx/dy)^3}$$

Total derivatives of the elementary functions are tabulated [4].

The following two rules apply to differentiation of *composite* functions $y = f(u, v)$, where $u = f_1(x)$ and $v = f_2(x)$

$$y = uv \quad ; \quad \frac{dy}{dx} = \frac{d}{dx}(uv) = u\frac{dv}{dx} + v\frac{du}{dx}$$

$$y = \frac{u}{v} \quad ; \quad \frac{dy}{dx} = \frac{d}{dx}\left(\frac{u}{v}\right) = \frac{v\,du/dx - u\,dv/dx}{v^2}$$

$$(41)$$

These rules are useful for *parametric differentiation*. If $y = f_1(z)$ with $z = f_2(x)$

$$\frac{dy}{dx} = \frac{dy}{dz}\frac{dz}{dx}$$

$$\frac{d^2y}{dx^2} = \frac{d}{dx}\left(\frac{dy}{dz}\frac{dz}{dx}\right) = \frac{dy}{dz}\frac{d^2z}{dx^2} + \left(\frac{dz}{dx}\right)^2\frac{d^2y}{dz^2}$$

If $y = f_1(z)$ with $x = f_2(z)$

$$\frac{dy}{dx} = \frac{dy/dz}{dx/dz}$$

$$\frac{d^2y}{dx^2} = \frac{d}{dx}\left(\frac{dy/dz}{dx/dz}\right) = \frac{(d/dz)\,[(dy/dz)/(dx/dz)]}{dx/dz} = \frac{(dx/dz)(d^2y/dz^2) - (dy/dz)(d^2x/dz^2)}{(dx/dz)^3}$$

Partial Derivative. Given a continuous, double-variable function $y = f(x_1, x_2)$, *partial derivatives* of y with respect to x_1 or x_2 are written $\partial y/\partial x_1$ and $\partial y/\partial x_2$ to denote one variable being held constant (x_2 in former, x_1 in latter). The first partial of y with respect to x_1 has the value

$$\frac{\partial y}{\partial x_1} = \lim_{\Delta x_1 \to 0} \frac{1}{\Delta x_1}[f(x_1 + \Delta x_1, x_2) - f(x_1, x_2)]$$

$$(42)$$

and a limit of similar form holds for $\partial y/\partial x_2$. The three possible second partial derivatives are

$$\frac{\partial^2 y}{\partial x_1^2} = \frac{\partial}{\partial x_1}\left(\frac{\partial y}{\partial x_1}\right), \quad \frac{\partial^2 y}{\partial x_1 \partial x_2} = \frac{\partial}{\partial x_1}\left(\frac{\partial y}{\partial x_2}\right) = \frac{\partial}{\partial x_2}\left(\frac{\partial y}{\partial x_1}\right), \quad \frac{\partial^2 y}{\partial x_2^2} = \frac{\partial}{\partial x_2}\left(\frac{\partial y}{\partial x_2}\right)$$

Equations (41) hold similarly for partial differentiation of composite functions. Parametric partial differentiation is also similar. If $y = f(x_1, x_2)$ with $x_1 = f_1(u, v)$ and $x_2 = f_2(u, v)$

$$\frac{\partial y}{\partial u} = \frac{\partial y}{\partial x_1}\frac{\partial x_1}{\partial u} + \frac{\partial y}{\partial x_2}\frac{\partial x_2}{\partial u}$$

$$\frac{\partial y}{\partial v} = \frac{\partial y}{\partial x_1}\frac{\partial x_1}{\partial v} + \frac{\partial y}{\partial x_2}\frac{\partial x_2}{\partial v}$$

In the general multivariable case where $x_1, x_2, \ldots$ depend on z alone as

$y = f(x_1, x_2, \ldots)$ with $x_1 = f_1(z)$, $x_2 = f_2(z)$, ..., the *total* derivative is

$$\frac{dy}{dz} = \frac{\partial y}{\partial x_1} \frac{dx_1}{dz} + \frac{\partial y}{\partial x_2} \frac{dx_2}{dz} + \cdots$$

For the *implicit* function $f(x, y) = 0$ with x independent and y dependent, the total first and second derivatives are

$$\frac{dy}{dx} = -\frac{\partial f/\partial x}{\partial f/\partial y}$$

$$\frac{d^2y}{dx^2} = -\frac{(\partial f/\partial y)^2(\partial^2 f/\partial x^2) - 2(\partial f/\partial x)(\partial f/\partial y)(\partial^2 f/\partial x\partial y) + (\partial f/\partial x)^2(\partial^2 f/\partial y^2)}{(\partial f/\partial y)^3}$$

If $f(x, y, z) = 0$ with x and y independent and z dependent

$$\frac{\partial z}{\partial x} = -\frac{\partial f/\partial x}{\partial f/\partial z} \quad , \quad \frac{\partial z}{\partial y} = -\frac{\partial f/\partial y}{\partial f/\partial z}$$

Integration. An *integral* of $f(x)$ is a function $F(x)$ whose derivative is $f(x)$. Thus, if $f(x) = (d/dx)F(x)$ is continuous over a bounded interval

$$F(x) = \int f(x)\, dx + c \tag{43}$$

where $f(x)$ is the *integrand* and c is an *integration constant.* The fundamental theorem for *indefinite integrals* is

$$\int_a^x f(x)\, dx = F(x)\Big]_a^x = F(x) - F(a) \tag{44}$$

Here a and x on the integral are *integration limits,* the latter indefinite.

A *definite (Riemann) integral* is defined as the limit (Fig. 19)

$$I^1 = \lim_{\substack{n \to \infty \\ \Delta x_i \to 0}} \sum_{i=1}^{n} f(\lambda_i)\Delta x_i = \int_a^b f(x)\, dx = F(b) - F(a) \tag{45}$$

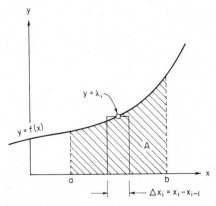

Fig. 19. Graphics of an integral.

Geometrically, I^1 is the shaded area A bounded by the x axis, the ordinates $x = a$ and $x = b$, and the curve $y = f(x)$. The *mean value* of $f(x)$ between a and b is therefore

$$\bar{f} = \frac{1}{b - a} \int_a^b f(x)\,dx = \frac{F(b) - F(a)}{b - a} \tag{46}$$

An *improper* definite integral has either an infinite integration limit(s) or a discontinuous integrand. For the former

$$\int_a^\infty f(x)\,dx = \lim_{\epsilon \to 0} \int_a^{\infty - \epsilon} f(x)\,dx \quad ; \quad 0 \le a \le \infty$$

This integral converges if the limit exists. If the limits of integration are both infinite

$$\int_{-\infty}^\infty f(x)\,dx = \lim_{\epsilon_1 \to 0} \int_{-\infty + \epsilon_1}^b f(x)\,dx + \lim_{\epsilon_2 \to 0} \int_b^{\infty - \epsilon_2} f(x)\,dx$$

For the latter, if both limits are finite but the integrand has a discontinuity $f(x_0) = \infty$, $a < x_0 < b$, then

$$\int_a^b f(x)\,dx = \lim_{\epsilon_1 \to 0} \int_a^{x_0 - \epsilon_1} f(x)\,dx + \lim_{\epsilon_2 \to 0} \int_{x_0 + \epsilon_2}^b f(x)\,dx$$

A large number of indefinite and definite (proper and improper) integrals have been evaluated and tabulated [4, 27, 28].

Parametric integration is performed with the rule of *integration by parts*. If $u = u(x)$ and $v = v(x)$ for $a \le x \le b$

$$\int_a^b u\,dv = uv\Big]_a^b - \int_a^b v\,du \tag{47}$$

If $u = u(x)$ and its derivable inverse function $x = x(u)$ are single-valued, the rule for *change-of-variable* gives

$$\int_a^b f(x)\,dx = \int_{u(a)}^{u(b)} f[x(u)]\frac{dx}{du}\,du \tag{48}$$

If a continuous function $f(x, y)$ is integrated with respect to x within $a \le x \le b$ and with y fixed as a *parameter*

$$\int_a^b f(x, y)\,dx = F(y)$$

The total derivative of this function is

$$\frac{dF}{dy} = \frac{\partial}{\partial y} \int_a^b f(x, y)\, dx = \int_a^b \frac{\partial}{\partial y} f(x, y)\, dx$$

If, additionally, $a = a(y)$ and $b = b(y)$, then by the *Liebnitz rule*

$$\frac{dF}{dy} = \int_a^b \frac{\partial}{\partial y} f(x, y)\, dx + f(b, y)\frac{db}{dy} - f(a, y)\frac{da}{dy}$$

Stieltjes Integral.   Given two bounded functions $f(x)$ and $g(x)$, the *Stieltjes integral* of $f(x)$ with respect to $g(x)$ is

$$\int_a^b f(x)\, dg(x) = \lim_{\substack{n \to \infty \\ \Delta x_i \to 0}} \sum_{i=1}^n f(\lambda_i)[g(x_i) - g(x_{i-1})] \tag{49}$$

Although $f(x)$ must be continuous, it is not necessary that $g(x)$ in the above integral be continuous from a to b [29]. This is advantageous when $g(x)$ has *discontinuities* as, for example, when calculating heat transfer to a surface with discontinuous surface temperature. If $f(x)$ is a convective unit surface conductance and $g(x)$ a surface temperature function, then the integral in Eq. (49) represents the surface heat input which cannot be calculated in the Riemann sense unless both $f(x)$ and $g(x)$ are continuous. If $g(x)$ has discontinuities at $x = x_1, x = x_2, \ldots, x = x_n$ (Fig. 20), the Stieltjes integral may be interpreted as

$$\int_a^b f(x)\, dg(x) = \int_a^b f(x)\frac{dg(x)}{dx}\, dx + \sum_{i=1}^n f(x_i)[g(x_i^+) - g(x_i^-)] \tag{50}$$

that is, a Riemann integral plus the sum of individual contributions at all x_i evaluated on each side $(+, -)$ of the n discontinuities.

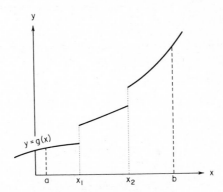

Fig. 20. Discontinuous function.

Suppose it is required to calculate the convective heat transfer to a surface whose temperature varies in the flow direction. Let the prescribed wall temperature t_w vary as a continuous function of x. Then the local heat flux q'' depends on local unit surface conductance h as

$$q'' = \int_0^x h(x, \xi) \, dt_w(\xi)$$

If t_w is not continuous, but broken by n finite steps occurring at ξ_i (Fig. 21), the heat flux is [30]

$$q'' = \int_0^x h(x, \xi) \, \frac{dt_w(\xi)}{d\xi} \, d\xi + \sum_{i=1}^n h(x, \xi_i) \, \Delta t_w(\xi_i)$$

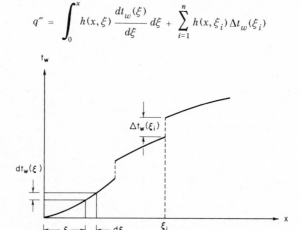

Fig. 21. Discontinuous wall temperature.

This is equivalent to the Stieltjes interpretation in Eq. (50).

2. Solution of Ordinary Differential Equations

Certain *ordinary* (*total*) differential equations characteristically reoccur in steady-state flow and heat transfer processes and, under certain conditions, in transient heat conduction. Methods of solving (integrating) these equations depend on their reduction to a recognized standard form [31].

a. First-order Equations. Ordinary differential equations of *first order* and of the *first degree* in dy/dx are expressed implicitly as

$$f\left(\frac{dy}{dx}, x, y\right) = 0 \tag{51}$$

or explicitly as

$$M(x, y) \, dx + N(x, y) \, dy = 0 \tag{52}$$

Separable Form. If the *coefficients* $M(x, y)$ and $N(x, y)$ are constant, or if $M = M(x)$ and $N = N(y)$ only, then Eq. (52) is *separable* and has the general solution

$$\int M(x) \, dx + \int N(y) \, dy = c \tag{53}$$

If separation is not evident, try variable transformations (for example, $u = xy$) to reduce original equation to a separable form.

Homogeneous Form. If $M(x, y)$ and $N(x, y)$ are both *homogeneous* in that $M(ax, ay) = a^n M(x, y)$, and if each term in M and N is of the same degree (n) in x and y, then Eq. (52) can be written as $dy + f(y/x)\,dx = 0$. Substituting $u = y/x$ reduces this to a separable form whose general solution is

$$x = c \exp\left[-\int \frac{1}{f(u) + u}\,du\right] \quad , \quad f(u) = \frac{M(1, u)}{N(1, u)} \tag{54}$$

Exact Form. If $\partial M(x, y)/\partial y = \partial N(x, y)/\partial x$, then the left side of Eq. (52) is an *exact differential* and the general solution is

$$\int M(x, y)\,dx + \int\left[N(x, y) - \frac{\partial}{\partial y}\int M(x, y)\,dx\right]dy = c \tag{55}$$

where y is constant when integrating with respect to x. If $\partial M/\partial y \neq \partial N/\partial x$, find a multiplying *integrating factor* $u(x, y)$ such that $u(M dx + N dy)$ is an exact differential solvable by Eq. (55). This integrating factor must satisfy the differential equation $N \partial u/\partial x - M \partial u/\partial y + (\partial N/\partial x - \partial M/\partial y)u = 0$.

Linear Form. The first-order differential equation

$$\frac{dy}{dx} + M(x)y = N(x) \tag{56}$$

is *linear* because it is of the first degree in the dependent variable and its derivative. Equation (56) can be reduced to exact form on multiplying by the integrating factor $e^{\int M dx}$. Its general solution is then

$$y = e^{-\int M(x)dx}\left[\int N(x)\,e^{\int M(x)dx}\,dx + c\right] \tag{57}$$

The *Bernoulli equation*

$$\frac{dy}{dx} + M(x)y = N(x)y^n \quad ; \quad n \neq 1 \tag{58}$$

is reducible to linear form on dividing by y^n and substituting $u = y^{1-n}$. Its general solution is then

$$y = e^{-\int M(x)dx}\left[(1 - n)\int N(x)\,e^{(1 - n)\int M(x)dx}\,dx + c\right]^{\frac{1}{1-n}} \tag{59}$$

The indefinite integrations in Eqs. (53) to (55), (57), and (59) can generally be performed by reference to tabulated integrals [4, 27, 28].

b. Second-order Linear Equations. Ordinary linear differential equations of *second order* and of the *(first degree)* in d^2y/dx^2 and dy/dx, and having *constant coefficients,* are expressed implicitly as

$$f\left(\frac{d^2y}{dx^2}, \frac{dy}{dx}, x, y\right) = 0 \tag{60}$$

Certain characteristic forms of Eq. (60) reoccur with one or more terms missing. Examples and general solutions are:

- $\dfrac{d^2 y}{dx^2} = a$; $y = \dfrac{a}{2} x^2 + c_1 x + c_2$

- $\dfrac{d^2 y}{dx^2} = f(x)$; $y = x \displaystyle\int f(x)\,dx - \int x f(x)\,dx + c_1 x + c_2$

- $\dfrac{d^2 y}{dx^2} = f(y)$; $x = \displaystyle\int \dfrac{1}{\sqrt{2\int f(y)dy + c_1}}\,dy + c_2$

Exs. $\dfrac{d^2 y}{dx^2} - m^2 y = a$; $y = c_1 e^{mx} + c_2 e^{-mx} - \dfrac{a}{m^2}$

   $\dfrac{d^2 y}{dx^2} + m^2 y = a$; $y = c_1 \sin mx + c_2 \cos mx + \dfrac{a}{m^2}$

- $\dfrac{d^2 y}{dx^2} = f(x, y)$

Ex. $\dfrac{d^2 y}{dx^2} + m^2 y = a \sin nx + b \cos nx$;

   $n \neq m, \quad y = \dfrac{a \sin nx + b \cos nx}{m^2 - n^2} + c_1 \sin mx + c_2 \cos mx$

   $n = m, \quad y = \dfrac{(b \sin mx - a \cos mx)x}{2m} + c_1 \sin mx + c_2 \cos mx$

- $\dfrac{d^2 y}{dx^2} = f\left(\dfrac{dy}{dx}\right)$; $x = \displaystyle\int \dfrac{1}{f(p)}\,dp + c$, $p = \dfrac{dy}{dx}$

Solve for p (note $dp/dx = p\,dp/dy$), resubstitute $p = dy/dx$, and solve resulting first-order equation.

- $\dfrac{d^2 y}{dx^2} = f\left(\dfrac{dy}{dx}, x\right)$; $y = \displaystyle\int p(x)\,dx + c$, $p = \dfrac{dy}{dx}$

- $\dfrac{d^2 y}{dx^2} = f\left(\dfrac{dy}{dx}, y\right)$; $y = \displaystyle\int \dfrac{1}{p(y)}\,dy + c$, $p = \dfrac{dy}{dx}$

Ex. $\dfrac{d^2 y}{dx^2} + 2b \dfrac{dy}{dx} + m^2 y = a$;

   $b^2 < m^2$, $y = e^{-bx}(c_1 \sin \sqrt{m^2 - b^2}\, x + c_2 \cos \sqrt{m^2 - b^2}\, x) + \dfrac{a}{m^2}$

   $b^2 = m^2$, $y = e^{-bx}(c_1 x + c_2) + \dfrac{a}{m^2}$

   $b^2 > m^2$, $y = e^{-bx}[c_1 \exp(\sqrt{b^2 - m^2}\, x) + c_2 \exp(-\sqrt{b^2 - m^2}\, x)] + \dfrac{a}{m^2}$

c. General Linear Equations. The general nth-order *nonhomogeneous* linear differential equation with *variable coefficients*

$$\frac{d^n y}{dx^n} + f_1(x) \frac{d^{n-1}y}{dx^{n-1}} + \cdots + f_{n-1}(x) \frac{dy}{dx} + f_n(x)y = f(x) \tag{61}$$

is homogeneous if $f(x) = 0$. The general solution of Eq. (61) is the sum of a *complementary function* y_c (solution of homogeneous equation) and a *particular integral* y_p (one solution of nonhomogeneous equation) as

$$y = y_c + y_p \tag{62}$$

A large number of particular integrals y_p are tabulated [22].

Homogeneous with Constant Coefficients. The homogeneous form of Eq. (61) with constant coefficients is

$$\frac{d^n y}{dx^n} + b_1 \frac{d^{n-1}y}{dx^{n-1}} + \cdots + b_{n-1}\frac{dy}{dx} + b_n y = 0 \tag{63}$$

Let r_n be the n roots of the *auxiliary equation* $r^n + b_1 r^{n-1} + \cdots + b_{n-1}r + b_n = 0$:
- For all roots real and different

$$y = y_c = c_1 e^{r_1 x} + c_2 e^{r_2 x} + \cdots + c_n e^{r_n x}$$

- For m equal roots, $r_1 = r_2 = \cdots = r_m$, and the rest real

$$y = y_c = (c_1 + c_2 x + c_3 x^2 + \cdots + c_m x^{m-1})e^{r_1 x} + c_{m+1}e^{r_{m+1}x} + \cdots + c_n e^{r_n x}$$

- For two conjugate complex roots, $r_1 = \psi + i\omega$ and $r_2 = \psi - i\omega$, and the rest real

$$y = y_c = (c_1 \sin \omega x + c_2 \cos \omega x)e^{\psi x} + c_3 e^{r_3 x} + \cdots + c_n e^{r_n x}$$

- For two double conjugate complex roots, $r_1 = r_2 = \psi + i\omega$ and $r_3 = r_4 = \psi - i\omega$, and the rest real

$$y = y_c = [(c_1 + c_2 x)\sin \omega x + (c_3 + c_4 x)\cos \omega x]e^{\psi x} + c_5 e^{r_5 x} + \cdots + c_n e^{r_n x}$$

Euler Form. The homogeneous equation with variable coefficients

$$x^n \frac{d^n y}{dx^n} + b_1 x^{n-1}\frac{d^{n-1}y}{dx^{n-1}} + \cdots + b_{n-1}x \frac{dy}{dx} + b_n y = 0 \tag{64}$$

is known as *Euler's equation*. This can be transformed to a linear homogeneous form with constant coefficients by the substitution $x = e^u$, giving Eq. (63) with independent variable x replaced by u. Examples are:

- $x^2 \dfrac{d^2 y}{dx^2} + x \dfrac{dy}{dx} - m^2 y = 0$; $\quad y = c_1 x^m + c_2 x^{-m}$

- $x^2 \dfrac{d^2 y}{dx^2} + 2x \dfrac{dy}{dx} - my = 0$; $\quad y = c_1 x^{\sqrt{m+\frac{1}{4}} - \frac{1}{2}} + c_2 x^{-\sqrt{m+\frac{1}{4}} - \frac{1}{2}}$

Nonhomogeneous with Constant Coefficients. The nonhomogeneous form of Eq. (61) with constant coefficients is

$$\frac{d^n y}{dx^n} + b_1 \frac{d^{n-1} y}{dx^{n-1}} + \cdots + b_{n-1} \frac{dy}{dx} + b_n y = f(x) \tag{65}$$

Let $Dy = dy/dx$ (in general $D^n y = d^n y/dx^n$) be a *differential operator* (D operating on y). Then Eq. (65) is

$$F(D)y = (D^n + b_1 D^{n-1} + \cdots + b_{n-1} D + b_n)y = f(x) \tag{66}$$

Particular integrals are found by substituting assumed functions for y_p in Eq. (66) and then expanding, equating coefficients of like terms, and solving for the constant coefficients in y_p (method of *undetermined coefficients*) [31]:

- For $f(x) = x^n + a_1 x^{n-1} + \cdots + a_{n-1} x + a_n$

$$y_p = x^n + c_1 x^{n-1} + \cdots + c_{n-1} x + c_n$$

If D^s appears as a *factor* of $F(D)$, multiply right side of y_p by x^s.

- For $f(x) = a e^{mx}$

$$y_p = c e^{mx}$$

If $(D - m)^s$ appears as a factor of $F(D)$, multiply right side of y_p by x^s.

- For $f(x) = a \sin mx$ (or $a \cos mx$)

$$y_p = c_1 \sin mx + c_2 \cos mx$$

If $(D^2 + m^2)^s$ appears as a factor of $F(D)$, multiply right side of y_p by x^s.

A general solution of the nonhomogeneous second-order equation

$$\frac{d^2 y}{dx^2} + 2b \frac{dy}{dx} + m^2 y = f(x)$$

is given by Eq. (62) with y_c from the indicated solutions of Eq. (60) and y_p given by

$$b^2 < m^2 \; ; \; y_p = \frac{\left[\sin\sqrt{m^2 - b^2}\,x \int f(x) e^{bx} \cos\sqrt{m^2 - b^2}\,x\,dx - \cos\sqrt{m^2 - b^2}\,x \int f(x) e^{bx} \sin\sqrt{m^2 - b^2}\,x\,dx \right] e^{-bx}}{\sqrt{m^2 - b^2}}$$

$$b^2 = m^2 \; ; \; y_p = \left[x \int f(x) e^{bx} dx - \int x f(x) e^{bx} dx \right] e^{-bx}$$

$$b^2 > m^2 \; ; \; y_p = \frac{\left[e^{-(b - \sqrt{b^2 - m^2})x} \int f(x) e^{(b - \sqrt{b^2 - m^2})x} dx - e^{-(b + \sqrt{b^2 - m^2})x} \int f(x) e^{(b + \sqrt{b^2 - m^2})x} dx \right]}{2\sqrt{b^2 - m^2}}$$

Solutions of ordinary differential equations not obtainable by the methods given here can generally be found by use of the *Laplace transform* (Sec. B.5).

3. Special Ordinary Differential Equations of Heat Conduction

Certain characteristic first- and second-order ordinary differential equations arise in analyses of heat conduction in bodies of different fundamental shape [32]. Although solutions of the differential equations for bodies with *plane* boundaries appear in terms of the elementary functions reviewed in Sec. B.1, general solutions for *cylindrical, spherical,* and *elliptical* boundaries are generated in terms of higher mathematical functions not yet considered.

a. Radiation Equation

Pure Radiation Environment. The temperature response of a plate with internal heat sources and of infinite internal thermal conductance, suddenly exposed to pure surface *radiative* heating or cooling, satisfies the ordinary first-order differential equation (y = temperature, x = time)

$$\frac{dy}{dx} + a_1(y^4 - m_1^4) = 0 \tag{67}$$

Equation (67) is separable; its general solution, using partial fractions, is

$$x = \frac{1}{2a_1 m_1^3}\left[\frac{1}{2}\ln\left(\frac{y/m_1 + 1}{y/m_1 - 1}\right) + \tan^{-1}\left(\frac{y}{m_1}\right)\right] + c \tag{68}$$

in which the dependent variable y cannot be solved for explicitly.

Combined Convection/Radiation Environments. If the aforementioned plate is suddenly exposed to a combination of *radiative* and *convective* environments, the separable differential equation is

$$\frac{dy}{dx} + a_2(y^4 + by - m_2^4) = 0 \tag{69}$$

The general solution of Eq. (69), using partial fractions and Eqs. (17), is

$$x = \frac{1}{a_2}\left\{A_1 \ln(y - r_1) + A_2 \ln(y - r_2) + \frac{A_3}{2}\ln[(y - \eta)^2 + \phi^2] + \frac{A_4}{\phi}\tan^{-1}\left(\frac{y - \eta}{\phi}\right)\right\} + c \tag{70}$$

where

$$A_1 = \frac{1}{8\sqrt{\tau - \eta^2}[\eta\sqrt{\tau - \eta^2} + (\tau/2 + \eta^2)]} \quad , \quad A_2 = \frac{1}{8\sqrt{\tau - \eta^2}[\eta\sqrt{\tau - \eta^2} - (\tau/2 + \eta^2)]}$$

$$A_3 = \frac{\eta}{\tau^2 + 8\eta^4} \quad , \quad A_4 = \frac{\tau/2 - \eta^2}{\tau^2 + 8\eta^4}$$

$$r_1 = -\eta - \sqrt{\tau - \eta^2} \quad , \quad r_2 = -\eta + \sqrt{\tau - \eta^2}$$

$$\eta = \frac{\sqrt{\kappa}}{2} \quad , \quad \phi = \sqrt{\tau + \frac{\kappa}{4}} \quad , \quad \tau = \sqrt{m_2^4 + \frac{\kappa}{4}}$$

$$\kappa = \left(\frac{b^2}{2} + \sqrt{\frac{b^4}{4} + \frac{64}{27}m_2^{12}}\right)^{1/3} + \left(\frac{b^2}{2} - \sqrt{\frac{b^4}{4} + \frac{64}{27}m_2^{12}}\right)^{1/3}$$

b. Bessel Equation.

A linear second-order differential equation occurring in problems of potential with *cylindrical* geometry is *Bessel's equation*

$$x^2\frac{d^2y}{dx^2} + x\frac{dy}{dx} + (x^2 - \nu^2)y = 0 \tag{71}$$

The general solution of Eq. (71) is a linear combination of two independent series solutions (*cylindrical harmonics*) that depend on ν and are called νth-order Bessel functions of the first and second kind [33].

Bessel Functions. Let n denote zero or a positive integer. If $\nu = n$, the nth-order Bessel functions of the *first kind* (Bessel's first solution) are represented by the series

$$J_n(x) = \sum_{p=0}^{\infty} \frac{(-1)^p}{p!(n+p)!} \left(\frac{x}{2}\right)^{n+2p} \quad ; \quad n = 0, 1, 2, \ldots \tag{72}$$

and those of the *second kind* (Weber's form of Neumann's second solution) are given by the series

$$Y_n(x) = \frac{2}{\pi} \left\{ \left[\ln\left(\frac{x}{2}\right) + \gamma\right] J_n(x) - \frac{1}{2} \sum_{p=0}^{n-1} \frac{(n-p-1)!}{p!} \left(\frac{x}{2}\right)^{-(n-2p)} \right.$$

$$\left. + \frac{1}{2} \sum_{p=0}^{\infty} \frac{(-1)^{p+1}[f(p) + f(n+p)]}{p!(n+p)!} \left(\frac{x}{2}\right)^{n+2p} \right\} \tag{73}$$

where $\gamma = 0.57722$ is Euler's constant, $f(0) = 0$, and $f(p) = \sum_{m=1}^{p} 1/m \ (m \geq 1)$. The general solution of Eq. (71) for *zero or positive integral n* is then

$$y = c_1 J_n(x) + c_2 Y_n(x) \quad ; \quad n = 0, 1, 2, \ldots \tag{74}$$

The first three Bessel functions of the first kind and integral n are given in Fig. 22.

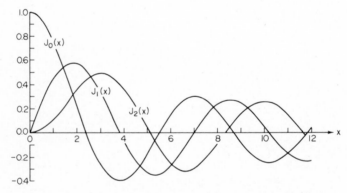

Fig. 22. Zero-, first-, and second-order Bessel functions of the first kind.

Intersections of these curves with the x axis represent the important *roots* (*zeros*) of $J_n(x) = 0$. Figure 23 shows the corresponding first three Bessel functions of the second kind. Both $J_n(x)$ and $Y_n(x)$ behave like damped sinusoidal functions.

If $\pm\nu$ is *nonintegral*, the functions $J_\nu(x)$ and $J_{-\nu}(x)$ are independent solutions calculated as

$$J_\nu(x) = \sum_{p=0}^{\infty} \frac{(-1)^p}{p!\,\Gamma(\nu+p+1)} \left(\frac{x}{2}\right)^{\nu+2p} \tag{75}$$

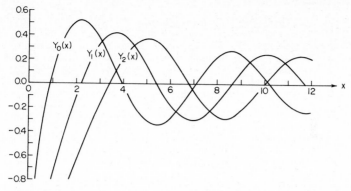

Fig. 23. Zero-, first-, and second-order Bessel functions of the second kind.

where Γ is gamma function (Sec. B.1). The general solution of Eq. (71) is then

$$y = c_1 J_\nu(x) + c_2 J_{-\nu}(x) \quad ; \quad \nu \neq 0 \tag{76}$$

When ν is half an odd integer ($\nu = n + \frac{1}{2}$), the $J_\nu(x)$ are expressible in closed form as, for example,

$$J_{1/2}(x) = \sqrt{\frac{2}{\pi x}} \sin x \quad , \quad J_{3/2}(x) = \sqrt{\frac{2}{\pi x}} \left(\frac{1}{x} \sin x - \cos x \right)$$

$$J_{-1/2}(x) = \sqrt{\frac{2}{\pi x}} \cos x \quad , \quad J_{-3/2}(x) = -\sqrt{\frac{2}{\pi x}} \left(\sin x + \frac{1}{x} \cos x \right)$$

or in general

$$J_{n+1/2}(x) = \frac{2n-1}{x} J_{n-1/2}(x) - J_{n-3/2}(x) \tag{77}$$

Useful derivatives are

$$\frac{d}{dx} J_\nu(ax) = \frac{\nu}{x} J_\nu(ax) - a J_{\nu+1}(ax) = -\frac{\nu}{x} J_\nu(ax) + a J_{\nu-1}(ax)$$

$$\frac{d}{dx} x^\nu J_\nu(ax) = ax^\nu J_{\nu-1}(ax) \quad , \quad \frac{d}{dx} x^{-\nu} J_\nu(ax) = -ax^{-\nu} J_{\nu+1}(ax)$$

The complex functions

$$H_\nu^{(1)}(x) = J_\nu(x) + iY_\nu(x) \quad , \quad H_\nu^{(2)}(x) = J_\nu(x) - iY_\nu(x)$$

are referred to as Bessel functions of the *third kind* or, in particular, as νth-order *Hankel functions of the first and second kind* [34].

Modified Bessel Functions. The linear second-order equation

$$x^2 \frac{d^2 y}{dx^2} + x \frac{dy}{dx} - (x^2 + \nu^2)y = 0 \tag{78}$$

is reducible to Bessel's standard form, Eq. (71), by substituting a new independent variable ix for x. The Bessel functions $J_\nu(ix)$ and $Y_\nu(ix)$ with imaginary arguments are therefore solutions to Eq. (78). In order to retain real variables, the following real functions $I_\nu(x)$ and $K_\nu(x)$ are defined as *νth-order modified Bessel functions of the first and second kind*

$$I_\nu(x) = i^{-\nu}J_\nu(ix) = \sum_{p=0}^{\infty} \frac{1}{p!(\nu + p)!}\left(\frac{x}{2}\right)^{\nu+2p} = I_n(x)$$

$$K_\nu(x) = \frac{\pi}{2\sin\nu\pi}[I_{-\nu}(x) - I_\nu(x)]$$

$$K_n(x) = \frac{\pi}{2}i^{n+1}[J_n(ix) + iY_n(ix)] = \frac{\pi}{2}i^{n+1}H_n^{(1)}(ix)$$

If ν is *zero or a positive integer*, the general solution of Eq. (78) is

$$y = c_1I_n(x) + c_2K_n(x) \quad ; \quad n = 0, 1, 2, \ldots \tag{79}$$

In contrast to the $J_n(x)$ and $Y_n(x)$ functions that have an infinity of real roots, the $I_0(x)$ has no real roots, the $I_n(x)$ for $n > 0$ have only one real root $(x = 0)$, and the $K_n(x)$ have no real roots (Fig. 24).

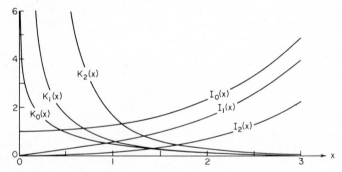

Fig. 24. Zero-, first-, and second-order modified Bessel functions of the first and second kind.

If $\pm\nu$ is *nonintegral*, the general solution of Eq. (78) is

$$y = c_1I_\nu(x) + c_2I_{-\nu}(x) \quad ; \quad \nu \neq 0 \tag{80}$$

For the modified functions of order equal to half an odd integer

$$I_{1/2}(x) = \sqrt{\frac{2}{\pi x}}\sinh x \quad , \quad I_{3/2}(x) = \sqrt{\frac{2}{\pi x}}\left(-\frac{1}{x}\sinh x + \cosh x\right)$$

$$I_{-1/2}(x) = \sqrt{\frac{2}{\pi x}}\cosh x \quad , \quad I_{-3/2}(x) = \sqrt{\frac{2}{\pi x}}\left(\sinh x - \frac{1}{x}\cosh x\right)$$

or, in general,

$$I_{n+1/2}(x) = \frac{1 - 2n}{x}I_{n-1/2}(x) + I_{n-3/2}(x) \tag{81}$$

Useful derivatives are

$$\frac{d}{dx} I_\nu(ax) = \frac{\nu}{x} I_\nu(ax) + a I_{\nu+1}(ax) = -\frac{\nu}{x} I_\nu(ax) + a I_{\nu-1}(ax)$$

$$\frac{d}{dx} K_\nu(ax) = \frac{\nu}{x} K_\nu(ax) - a K_{\nu+1}(ax) = -\frac{\nu}{x} K_\nu(ax) - a K_{\nu-1}(ax)$$

$$\frac{d}{dx} x^\nu I_\nu(ax) = a x^\nu I_{\nu-1}(ax) \quad , \quad \frac{d}{dx} x^{-\nu} I_\nu(ax) = a x^{-\nu} I_{\nu+1}(ax)$$

$$\frac{d}{dx} x^\nu K_\nu(ax) = -a x^\nu K_{\nu-1}(ax) \quad , \quad \frac{d}{dx} x^{-\nu} K_\nu(ax) = -a x^{-\nu} K_{\nu+1}(ax)$$

Abbreviated numerical tabulations of the Bessel functions and roots of various transcendental Bessel equations are available [2, 6, 32, 34], as are more elaborate tables for accurate computational work [35–38].

Bessel/Kelvin Functions.   The differential equation

$$x^2 \frac{d^2 y}{dx^2} + x \frac{dy}{dx} - (ix^2 + n^2)y = 0 \tag{82}$$

is also similar to Bessel forms, Eqs. (71) and (78), for integral n. Solutions of Eq. (82) are generally taken as

$$y = c_1 J_n(i^{3/2}x) + c_2 K_n(i^{1/2}x) \quad ; \quad n = 0, 1, 2, \ldots$$

in which the $J_n(i^{3/2}x)$ and $K_n(i^{1/2}x)$ are series consisting of alternating real and imaginary terms. Individual sums of the real and imaginary parts are defined as separate functions such that

$$y = c_1(\mathrm{ber}_n x + i\,\mathrm{bei}_n x) + c_2(\mathrm{ker}_n x + i\,\mathrm{kei}_n x) \tag{83}$$

The new functions $\mathrm{ber}_n x$, $\mathrm{bei}_n x$, $\mathrm{ker}_n x$, and $\mathrm{kei}_n x$ are referred to respectively as: *Bessel-real, Bessel-imaginary, Kelvin-real,* and *Kelvin-imaginary.* The Bessel-real and Bessel-imaginary functions of zero order (Fig. 25) are of special practical importance.

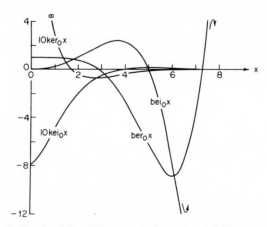

Fig. 25. Zero-order Bessel-real, Bessel-imaginary, Kelvin-real, and Kelvin-imaginary functions.

Their series representations are given by

$$\text{ber}_0 x = \sum_{p=0}^{\infty} \frac{(-1)^p}{[(2p)!]^2} \left(\frac{x}{2}\right)^{4p} \tag{84}$$

$$\text{bei}_0 x = \sum_{p=0}^{\infty} \frac{(-1)^p}{[(2p+1)!]^2} \left(\frac{x}{2}\right)^{4p+2} \tag{85}$$

Numerical values of $\text{ber}_0 x$, $\text{bei}_0 x$, $\text{ker}_0 x$, and $\text{kei}_0 x$ are tabulated [2, 34, 39].

Generalized Bessel Form. A linear second-order differential equation, if reducible to Bessel's form by a change of variable, can be solved directly by inspection on comparing it to the *generalized Bessel equation*

$$x^2 \frac{d^2 y}{dx^2} + [(1 - 2A)x - 2Bx^2]\frac{dy}{dx}$$
$$+ [C^2 D^2 x^{2C} + B^2 x^2 - B(1 - 2A)x + A^2 - C^2 \nu^2]y = 0 \tag{86}$$

where ν (or n) is the order of Bessel's equation. Having evaluated by inspection the constants A, B, C, D, and ν, the *generalized solution* of Eq. (86) is

$$y = x^A e^{Bx}[c_1 J_\nu(Dx^C) + c_2 Y_\nu(Dx^C)] \tag{87}$$

Given, for example, $d^2 y/dx^2 - axy = 0$. Comparing this with Eq. (86); $A = 1/2$, $B = 0$, $C = 3/2$, $D = \frac{2}{3} i\sqrt{a}$, and $\nu = 1/3$. Then by Eq. (87)

$$y = \sqrt{x}\left[c_1 J_{1/3}\left(\frac{2}{3} i\sqrt{a}\, x^{3/2}\right) + c_2 Y_{1/3}\left(\frac{2}{3} i\sqrt{a}\, x^{3/2}\right)\right]$$

c. Legendre Equation. A linear second-order differential equation arising in problems of potential with *spherical* geometry is *Legendre's equation*

$$(1 - x^2)\frac{d^2 y}{dx^2} - 2x\frac{dy}{dx} + \nu(\nu + 1)y = 0 \tag{88}$$

The general solution of Eq. (88) is a linear combination of two independent solutions (*zonal harmonics*) that depend on ν as

$$y = c_1 P_\nu(x) + c_2 Q_\nu(x) \tag{89}$$

The functions $P_\nu(x)$ and $Q_\nu(x)$ are infinite series referred to as ν*th-degree Legendre functions of the first and second kind* [40]. Generally only the first-kind solutions are needed.

When $\nu = n$ (zero or a positive integer) the $P_\nu(x)$ reduce to *Legendre polynomials*

$$P_n(x) = \sum_{p=0}^{N} \frac{(-1)^p(2n - 2p)!}{2^n p!(n - p)!(n - 2p)!} x^{n-2p} \quad \begin{cases} N = \dfrac{n}{2} & \text{for } n \text{ even} \\[2mm] N = \dfrac{1}{2}(n - 1) & \text{for } n \text{ odd} \end{cases}$$

The first six polynomials are

$$P_0(x) = 1 \qquad , \qquad P_1(x) = x$$

$$P_2(x) = \frac{1}{2}(3x^2 - 1) \qquad , \qquad P_3(x) = \frac{1}{2}(5x^3 - 3x) \tag{90}$$

$$P_4(x) = \frac{1}{8}(35x^4 - 30x^2 + 3) \qquad , \qquad P_5(x) = \frac{1}{8}(63x^5 - 70x^3 + 15x)$$

With the substitution of $x = \cos\psi$, the polynomials $P_n(\cos\psi)$ are called *Legendre coefficients* and are written as

$$P_0(\cos\psi) = 1 \qquad\qquad , \quad P_1(\cos\psi) = \cos\psi$$

$$P_2(\cos\psi) = \frac{1}{4}(3\cos2\psi + 1) \qquad , \quad P_3(\cos\psi) = \frac{1}{8}(5\cos3\psi + 3\cos\psi)$$

$$P_4(\cos\psi) = \frac{1}{64}(35\cos4\psi + 20\cos2\psi + 9) , \quad P_5(\cos\psi) = \frac{1}{128}(63\cos5\psi + 35\cos3\psi$$

$$+ 30\cos\psi)$$

These functions resemble $\cos n\psi$ (Fig. 26).

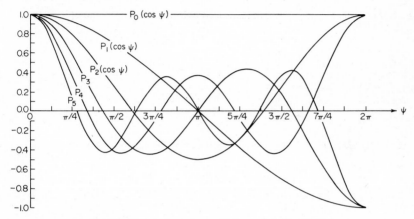

Fig. 26. First six Legendre coefficients of the first kind.

The Legendre functions of the *second kind* are infinite series for integral n but expressible in closed form $(-1 < x < 1)$ as

$$Q_0(x) = \tanh^{-1}x \qquad\qquad , \quad Q_1(x) = xQ_0(x) - 1$$

$$Q_2(x) = P_2(x)Q_0(x) - \frac{3}{2}x \qquad , \quad Q_3(x) = P_3(x)Q_0(x) - \frac{5}{2}x^2 + \frac{2}{3} \qquad (91)$$

$$Q_4(x) = P_4(x)Q_0(x) - \frac{35}{8}x^3 + \frac{55}{24}x \quad , \quad Q_5(x) = P_5(x)Q_0(x) - \frac{63}{8}x^4 + \frac{49}{8}x^2 - \frac{8}{15}$$

Both $P_n(x)$ and $Q_n(x)$ satisfy the recurrence formula

$$P_n(x) = \frac{1}{n}[(2n-1)xP_{n-1}(x) - (n-1)P_{n-2}(x)] \qquad (92)$$

Numerical tabulations are available for both the Legendre polynomials and coefficients [2, 6, 32, 40].

d. Mathieu Equation. A linear second-order differential equation occurring in problems of potential with *elliptic* geometry is *Mathieu's equation*

$$\frac{d^2y}{dx^2} + (a - 2p\cos2x)y = 0 \qquad (93)$$

Two general solutions of Eq. (93) are possible $(a = \pm n)$, each representing linear combinations of two independent solutions that depend on both x and p as

$$y = c_1 ce_n(x, p) + c_2 fe_n(x, p)$$
$$y = c_3 se_n(x, p) + c_4 ge_n(x, p)$$

(94)

where n is an integer [41]. The general functions $ce_n(x, p)$ and $se_n(x, p)$ are periodic and called *nth-order Mathieu functions of the first kind,* "*ce*" and "*se*" referring to cosine-elliptic and sine-elliptic (the type of functions when $p = 0$). The second solutions $fe_n(x, p)$ and $ge_n(x, p)$ are nonperiodic *nth-order Mathieu functions of the second kind.* Generally only the first-kind solutions are needed.

The $ce_n(x, p)$ and $se_n(x, p)$ are of period 2π when n is odd, and of period π when n is even. These functions are therefore expressible as Fourier series (Sec. B.4) for $n = 0, 1, 2, \ldots$

$$ce_{2n}(x, p) = \sum_{r=0}^{\infty} A_{2r}^{2n} \cos 2rx$$

$$ce_{2n+1}(x, p) = \sum_{r=0}^{\infty} A_{2r+1}^{2n+1} \cos(2r + 1)x$$

(95)

$$se_{2n}(x, p) = \sum_{s=0}^{\infty} B_{2s}^{2n} \sin 2sx$$

$$se_{2n+1}(x, p) = \sum_{s=0}^{\infty} B_{2s+1}^{2n+1} \sin(2s + 1)x$$

where the A's and B's depend on p. The first cosine-elliptic Mathieu function of the first kind (Fig. 27) is

$$ce_0(x, p) = 1 - \frac{p}{2} \cos 2x + \frac{p^2}{32} \cos 4x - \frac{p^3}{128} \left(\frac{1}{9} \cos 6x - 7 \cos 2x \right) + \cdots$$

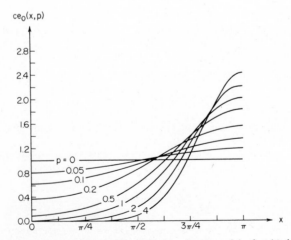

Fig. 27. Zero-order cosine-elliptic Mathieu function of the first kind.

Substituting ix for x reduces Eq. (93) to the *modified Mathieu equation*

$$\frac{d^2 y}{dx^2} - (a - 2p\cosh 2x)y = 0 \tag{96}$$

whose *nth-order modified Mathieu functions of the first kind* are

$$Ce_n(x, p) = ce_n(ix, p)$$
$$Se_n(x, p) = -ise_n(ix, p)$$

General solutions can occur as combined functions of $ce_n(x, p)$ and $Ce_n(x, p)$. Only a limited number of Mathieu functions and associated roots have been calculated [6, 42].

4. Expansion of Arbitrary Functions

Determination of the unknown coefficients in a series representation of an arbitrary function is a necessary step in deriving exact solutions to boundary-value problems by the classical separation-of-variables method (Sec. B.5). Particular solutions, for example, of the Fourier, Poisson, and Laplace differential equations in rectangular, cylindrical, spherical, and elliptical coordinates (Sec. 3.A) require expansions of periodic functions into infinite, convergent series of circular, Bessel, Legendre, and Mathieu functions. Expansions in circular functions are the familiar Fourier series.

The basis of expanding $f(x)$ in terms of characteristic functions is that these functions shall necessarily be *orthogonal* [43]. Two functions $\zeta_m(x)$ and $\zeta_n(x)$ are said to be orthogonal with respect to a *weighting function* $\xi(x)$ over the interval $a < x < b$ if

$$\int_a^b \xi(x)\,\zeta_m(x)\,\zeta_n(x)\,dx = 0 \quad ; \quad m \neq n \tag{97}$$

Let $f(x)$ be expanded in an infinite series of orthogonal functions $\zeta_n(x)$, $a < x < b$, as

$$f(x) = \sum_{n=1}^{\infty} a_n\,\zeta_n(x)$$

Multiplying both sides of this by $\xi(x)\,\zeta_m(x)$ and integrating over limits $x = a, b$ gives

$$\int_a^b \xi(x)\,\zeta_m(x)\,f(x)\,dx = \int_a^b \xi(x)\,\zeta_m(x)\left[\sum_{n=1}^{\infty} a_n\,\zeta_n(x)\right]dx = \sum_{n=1}^{\infty} a_n \int_a^b \xi(x)\,\zeta_m(x)\,\zeta_n(x)\,dx$$

By the orthogonality condition, Eq. (97), the right-hand sum vanishes if $m \neq n$. If $m = n$, the a_n are equal to

$$a_n = \frac{\int_a^b \xi(x)\,\zeta_n(x)\,f(x)\,dx}{\int_a^b \xi(x)\,\zeta_n^2(x)\,dx}$$

a. Fourier Series. Many arbitrary *periodic* functions $f(x)$ of period 2π may be expanded in a *Fourier series* consisting of an infinite sum of the *circular* sine and cosine functions according to

$$f(x) = \frac{a_0}{2} + \sum_{n=1}^{\infty} (a_n \cos nx + b_n \sin nx) \quad ; \quad -\pi < x < \pi \tag{98}$$

in which the constant *coefficients* (*amplitudes*) a_n and b_n are to be determined for positive integral n [43, 44, 45].

Fourier Coefficients. Owing to the orthogonality property of the circular functions (weighting function of unity) in the symmetrical 2π interval $x = -\pi, \pi$

$$\int_{-\pi}^{\pi} \sin mx \cos nx dx = 0$$

$$\int_{-\pi}^{\pi} \sin mx \sin nx dx = 0 \quad ; \quad m \neq n$$
$$= \pi \quad ; \quad m = n$$

$$\int_{-\pi}^{\pi} \cos mx \cos nx dx = 0 \quad ; \quad m \neq n$$
$$\pi \quad ; \quad m = n$$

The *Fourier coefficients* a_n and b_n of the expansion in Eq. (98) are derivable from the above orthogonality integrals as

$$a_n = \frac{1}{\pi} \int_{-\pi}^{\pi} f(x) \cos nx dx \quad ; \quad n = 0, 1, 2, \ldots \tag{99}$$

$$b_n = \frac{1}{\pi} \int_{-\pi}^{\pi} f(x) \sin nx dx \quad ; \quad n = 1, 2, 3, \ldots \tag{100}$$

If $f(x)$ is an *even* function, $f(-x) = f(x)$, then $b_n = 0$ and

$$a_n = \frac{2}{\pi} \int_{0}^{\pi} f(x) \cos nx dx \quad ; \quad n = 0, 1, 2, \ldots$$

Similarly if $f(x)$ is an *odd* function, $f(-x) = -f(x)$, then $a_n = 0$ and

$$b_n = \frac{2}{\pi} \int_{0}^{\pi} f(x) \sin nx dx \quad ; \quad n = 1, 2, 3, \ldots$$

For functions neither even nor odd, both the cosine and sine terms are needed to represent the given function.

In order that a function be Fourier expandable, it is sufficient that $f(x)$ be periodic, single-valued, finite function having a finite number of maxima, minima, and finite discontinuities (*Dirichlet's conditions*). The derivative of $f(x)$ cannot, in general, be obtained by termwise differentiation of its Fourier series (as can a power series, Sec. B.1), while the integral of $f(x)$ can be obtained by integrating the series term by term. For termwise differentiation, $f(x)$ must also be continuous over $-\pi < x < \pi$ and satisfy the condition $f(-\pi) = f(\pi)$.

Consider the periodic "sawtooth" function given by $f(x) = x$ over $-\pi < x < \pi$ and $f(x) = 0$ at $x = -\pi, \pi$. This function is odd and thus expandable in terms of sines alone. With $a_n = 0$ and $b_n = -(2/n) \cos n\pi$

$$f(x) = -2 \sum_{n=1}^{\infty} \frac{1}{n} \cos n\pi \sin nx = 2 \left(\sin x - \frac{1}{2} \sin 2x + \frac{1}{3} \sin 3x \mp \cdots \right)$$

Figure 28 shows how the sine series builds up to approximate the given function $y = f(x) = x$ by

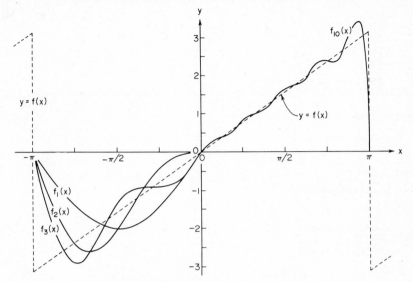

Fig. 28. Fourier sine-series approximation to $f(x) = x$.

successive addition of terms. The solid curves shown over the range $-\pi < x < 0$ are $f_1(x) = 2(\sin x)$, $f_2(x) = 2(\sin x - \frac{1}{2} \sin 2x)$ and $f_3(x) = 2(\sin x - \frac{1}{2} \sin 2x + \frac{1}{3} \sin 3x)$. The approximation slowly improves as more terms are added. The single solid curve $f_{10}(x)$ over the range $0 < x < \pi$ is the corresponding ten-term approximation. At the discontinuities $(x = \mp\pi)$, the series converges to an *average* of the function on either side of the discontinuity.

For expansion of periodic functions of period 2π over *nonsymmetrical* intervals $p < x < p + 2\pi$, the Fourier coefficients are

$$a_n = \frac{1}{\pi} \int_p^{p+2\pi} f(x) \cos nx \, dx \tag{101}$$

$$b_n = \frac{1}{\pi} \int_p^{p+2\pi} f(x) \sin nx \, dx \tag{102}$$

For expansions over the symmetrical but *arbitrary* interval $-L < x < L$, the Fourier series is

$$f(x) = \frac{a_0}{2} + \sum_{n=1}^{\infty} \left(a_n \cos \frac{n\pi}{L} x + b_n \sin \frac{n\pi}{L} x \right) \tag{103}$$

with coefficients

$$a_n = \frac{1}{L} \int_{-L}^{L} f(x) \cos \frac{n\pi}{L} x \, dx \tag{104}$$

$$b_n = \frac{1}{L} \int_{-L}^{L} f(x) \sin \frac{n\pi}{L} x \, dx \tag{105}$$

and similarly as in Eqs. (101) and (102) for nonsymmetrical and arbitrary intervals $p < x < p + 2L$.

Periodic functions of two variables $f(x, y)$ are expandable as a *double Fourier series* if $f(x, y)$ is of period 2π in both x and y. Considering a sine expansion for odd functions, the double series is

$$f(x, y) = \sum_{m=1}^{\infty} \sum_{n=1}^{\infty} a_{mn} \sin \frac{m\pi}{L} x \sin \frac{n\pi}{l} y \quad ; \quad \begin{array}{c} -L < x < L \\ -l < y < l \end{array} \tag{106}$$

and its coefficients are

$$a_{mn} = \frac{4}{Ll} \int_0^L \int_0^l f(x, y) \sin \frac{m\pi}{L} x \sin \frac{n\pi}{l} y \, dx \, dy \quad ; \quad m, n = 1, 2, 3, \ldots \tag{107}$$

Frequently the n in Eqs. (98) and (103) are not simple positive integers but consecutive positive *roots* (*zeros*) μ_n of a trigonometric equation that depends on boundary conditions. These characteristic numbers μ_n are called *eigenvalues*. The development of Fourier coefficients in this instance is accomplished in the usual way. Consider the series $f(x) = \sum_{n=1}^{\infty} a_n \cos(\mu_n/L) x$, where, for example, the μ_n are roots of $\mu_n \tan \mu_n = c$, $n = 1, 2, 3, \ldots$. The coefficients a_n are obtained by multiplying both sides of this equation by $\cos(\mu_m/L)x \, dx$ with $m \neq n$, integrating over the interval $0 < x < L$ (assuming convergence of the integral), collecting the surviving (nonzero) term that results from application of the orthogonality properties, and finally using the trigonometric equation for μ_n to simplify the expression for a_n. Roots μ_n of various trigonometric equations are tabulated in a number of sources [2, 23].

In certain applications it is convenient to represent an arbitrary function as a series of complex exponential terms. Using Euler's identity, Eq. (28), for $e^{\pm ix}$, the *exponential Fourier series* form corresponding to Eq. (103) is

$$f(x) = \frac{1}{2} \sum_{n=-\infty}^{\infty} a_n \exp\left(i \frac{n\pi}{L} x\right) \tag{108}$$

with coefficients for all even and odd functions given by

$$a_n = \frac{1}{L} \int_{-L}^{L} f(x) \exp\left(-i \frac{n\pi}{L} x\right) dx \quad ; \quad n = 0, \pm 1, \pm 2, \ldots \tag{109}$$

Fourier Integral. The Fourier series for finite intervals $-L < x < L$ in Eq. (103) becomes an integral for infinite intervals $-\infty < x < \infty$. This *Fourier integral* representation, valid for all x and all integrable, nonperiodic functions satisfying Dirichlet's (*piecewise regular*) conditions, is

$$f(x) = \frac{1}{\pi} \int_0^\infty [a(\lambda) \cos\lambda x + b(\lambda) \sin\lambda x] d\lambda$$

with coefficients

$$a(\lambda) = \int_{-\infty}^\infty f(x') \cos\lambda x' dx'$$

$$b(\lambda) = \int_{-\infty}^\infty f(x') \sin\lambda x' dx'$$

These relations, analogous to the Fourier series, can be combined as the double integral

$$f(x) = \frac{1}{\pi} \int_0^\infty \int_{-\infty}^\infty f(x') \cos\lambda(x' - x) dx' d\lambda \quad ; \quad -\infty < x < \infty \tag{110}$$

where x' is the variable of integration for the inner integral (considering x a dummy variable and λ held constant). The Fourier integral, like the series, converges to $f(x)$ wherever the function is continuous and to the mean value of $f(x)$ at points of discontinuity.

If $f(x)$ is an *even* function, $b(\lambda) = 0$ and Eq. (110) reduces to

$$f(x) = \frac{2}{\pi} \int_0^\infty \cos\lambda x \int_0^\infty f(x') \cos\lambda x' dx' d\lambda \tag{111}$$

Similarly, if $f(x)$ is *odd*, $a(\lambda) = 0$ and Eq. (110) becomes

$$f(x) = \frac{2}{\pi} \int_0^\infty \sin\lambda x \int_0^\infty f(x') \sin\lambda x' dx' d\lambda \tag{112}$$

Equations (111) and (112) are known respectively as the *Fourier cosine and sine integrals* and hold generally for any function (even or odd) in the range $x > 0$. The *exponential Fourier integral* form

$$f(x) = \frac{1}{2\pi} \int_{-\infty}^\infty e^{i\lambda x} \int_{-\infty}^\infty f(x') e^{-i\lambda x'} dx' d\lambda$$

leads to *Fourier transforms* (similar to the *Laplace transform,* Sec. B.5) with useful operational properties for solving differential equations [45].

b. Bessel Series. Expansions of an arbitrary function in an infinite series of νth-order *Bessel functions* (Sec. B.3) are required in obtaining boundary-value solutions to differential equations in cylindrical coordinates. Expansions in nth-order Bessel functions of the first kind are of special interest.

Assume that a piecewise regular function $f(x)$, defined over the interval $0 < x < L$, can be expanded in an infinite *Bessel (Fourier-Bessel) series* as

$$f(x) = \sum_{j=1}^\infty a_j J_n(\mu_j x) \quad ; \quad n = 0, 1, 2, \ldots \tag{113}$$

Here the μ_j are consecutive roots of a Bessel equation derived from boundary conditions. Two important examples are

$$J_n(\mu_j x)\Big]_L = 0 \tag{114}$$

$$cJ_n(\mu_j x)\Big]_L + \mu_j \frac{d}{dx} J_n(\mu_j x)\Big]_L = 0 \tag{115}$$

The orthogonality property of the first-kind Bessel functions in the interval $0 < x < L$ (weighting function x) gives

$$\int_0^L xJ_n(\mu_i x) J_n(\mu_j x)dx = 0 \quad ; \quad i \neq j$$

If Eq. (114) applies, the above orthogonality integral with $i = j$ reduces to

$$\int_0^L xJ_n^2(\mu_j x)dx = \frac{L^2}{2} J_{n+1}^2(\mu_j L)$$

while if Eq. (115) applies

$$\int_0^L xJ_n^2(\mu_j x)dx = \frac{L^2(\mu_j^2 + c^2) - n^2}{2\mu_j^2} J_n^2(\mu_j L)$$

By multiplying both sides of the assumed expansion by $xJ_n(\mu_i x)dx$ with $i \neq j$ and integrating over $0 < x < L$, the respective a_j for the two examples, Eqs. (114) and (115), are

$$a_j = \frac{2}{L^2 J_{n+1}^2(\mu_j L)} \int_0^L xf(x) J_n(\mu_j x)dx \tag{116}$$

$$a_j = \frac{2\mu_j^2}{\left[L^2(\mu_j^2 + c^2) - n^2\right]J_n^2(\mu_j L)} \int_0^L xf(x) J_n(\mu_j x)dx \tag{117}$$

Roots μ_j of various Bessel equations are tabulated in a number of sources [2, 6, 32].

c. Legendre Series. Expansions of an arbitrary function in a series of νth-degree *Legendre functions* (Sec. B.3) arise in boundary-value problems in spherical coordinates. Expansions in nth-degree Legendre polynomials of the first kind are of particular interest.

Consider that $f(x)$, satisfying Dirichlet's conditions, can be represented by the infinite *Legendre series*

$$f(x) = \sum_{n=0}^{\infty} a_n P_n(x) \tag{118}$$

where $P_n(x)$ are the *Legendre polynomials* defined over the range $-1 < x < 1$. Like the Fourier and Bessel series, the Legendre polynomials comprise an orthogonal system with the properties

$$\int_{-1}^{1} P_m(x) P_n(x) dx = 0 \qquad ; \quad m \neq n$$

$$= \frac{2}{2n + 1} \qquad ; \quad m = n$$

The series coefficients a_n, obtained by multiplying both sides of Eq. (118) by $P_m(x)$ with $m \neq n$ and integrating over the interval $-1 < x < 1$, are

$$a_n = \frac{2n + 1}{2} \int_{-1}^{1} f(x) P_n(x) dx \tag{119}$$

If $f(x) = f(\cos\psi)$ is expanded in a series of Legendre coefficients $P_n(\cos\psi)$, $0 < \psi < \pi$, the corresponding a_n in Eq. (118) are*

$$a_n = \frac{2n + 1}{2} \int_{0}^{\pi} f(x) \sin\psi \, P_n(\cos\psi) d\psi \tag{120}$$

5. Solution of Partial Differential Equations

In contrast to ordinary differential equations whose *general solutions* are independent of boundary conditions (Sec. B.2), the solution of a partial differential equation requires satisfying certain boundary conditions in the course of combining *particular integrals*.

The general class of *linear* and *homogeneous* second-order partial differential equations in $u(x, \theta)$ is

$$f_1(x) \frac{\partial^2 u}{\partial x^2} + g_1(\theta) \frac{\partial^2 u}{\partial \theta^2} + f_2(x) \frac{\partial u}{\partial x} + g_2(\theta) \frac{\partial u}{\partial \theta} + f_3(x)u + g_3(\theta)u = 0 \tag{121}$$

where the coefficients f and g are constants or functions only of x and θ respectively. Examples are the elliptic *Laplace equation* $\partial^2 u/\partial x^2 + \partial^2 u/\partial y^2 = 0$ and the parabolic *Fourier equation* $\partial^2 u/\partial x^2 - (1/\alpha)\partial u/\partial \theta = 0$. These *linear* equations have the property that if $u_1, u_2, \ldots, u_n$ are independent particular solutions, then the sum $u = \sum_{n=1}^{\infty} u_n$ is also a solution (providing the series converges and is termwise differentiable). This property allows a solution to be constructed in terms of *eigenfunctions* (characteristic functions) that represent particular integrals associated with sets of *eigenvalues* (characteristic values) generated to satisfy required boundary conditions [32].

a. Separation of Variables

Homogeneous Form. Linear partial differential equations that are *homogeneous* can be rendered solvable for a restricted class of boundary conditions by applying the elementary *separation-of-variables method*. This method serves to reduce the original nth-order equation in m independent variables to an equivalent system of m separate ordinary differential equations each of order n.

To separate the variables in Eq. (121), assume a solution for $u(x, \theta)$ of the form

$$u(x, \theta) = X(x) \Theta(\theta) \tag{122}$$

Mathieu series (Sec. B.3) are developed similarly by virtue of the orthogonality of Mathieu functions over the range $x = 0, 2\pi$ [41, 44].

where X is a function of x alone and Θ a function of θ alone. Substituting this primitive product solution in Eq. (121) gives the equality

$$\frac{1}{X}\left(f_1\frac{\partial^2 X}{\partial x^2} + f_2\frac{\partial X}{\partial x} + f_3 X\right) = -\frac{1}{\Theta}\left(g_1\frac{\partial^2\Theta}{\partial\theta^2} + g_2\frac{\partial\Theta}{\partial\theta} + g_3\Theta\right)$$

Since, by hypothesis, the left side of this equation is independent of θ and the right side is independent of x, each side must be independently equal to a common *separation constant,* say μ. Equation (121) is thus reduced, by the assumption of Eq. (122), to the following system of two ordinary differential equations

$$f_1\frac{d^2 X}{dx^2} + f_2\frac{dX}{dx} + (f_3 + \mu)X = 0 \tag{123}$$

$$g_1\frac{d^2\Theta}{d\theta^2} + g_2\frac{d\Theta}{d\theta} + (g_3 - \mu)\Theta = 0 \tag{124}$$

The separation constant μ may be chosen as positive, zero, negative, or imaginary. Each choice leads to different solutions (particular integrals) of Eq. (121), only one of which generally has a form suitable to fit required boundary conditions.

Suppose $\partial^2 u/\partial x^2 + (1/x)\partial u/\partial x = (1/\alpha)\partial u/\partial\theta$. Assuming $u = X\Theta$ with a negative separation constant $-\mu$ leads to $d^2 X/dx^2 + (1/x)dX/dx + \mu X = 0$ and $d\Theta/d\theta + \alpha\mu\Theta = 0$ whose general solutions are given respectively by Eqs. (87) and (57). Thus $u_\mu = [c_1 J_0(\sqrt{\mu}\,x) + c_2 Y_0(\sqrt{\mu}\,x)]e^{-\alpha\mu\theta}$

Having determined the general solutions $X(x)$ and $\Theta(\theta)$ of Eqs. (123) and (124), their product $u = X\Theta$ must be forced to satisfy boundary conditions to complete the solution. This latter operation requires the separation constant μ to assume sets of eigenvalues μ_n (Sec. B.4). Since each eigenvalue represents a possible particular solution, the complete solution of Eq. (121) must be built up as an infinite series (sum) of orthogonal eigenfunctions.

Nonhomogeneous Form. Certain linear but *nonhomogeneous* partial differential equations can be solved by assuming a solution of the form

$$u(x,\theta) = v(x,\theta) + w(x) \tag{125}$$

where $w(x)$, a function of x alone, is determined such that $v(x,\theta)$ will satisfy the homogeneous form of the original partial differential equation.

The equation $\partial^2 u/\partial x^2 - (1/\alpha)\partial u/\partial\theta = f(x)$ is not homogeneous because the right side is nonzero. If a solution of the form in Eq. (125) is assumed, then v and w must satisfy $\partial^2 v/\partial x^2 + \partial^2 w/\partial x^2 - (1/\alpha)\partial v/\partial\theta = f(x)$. If $w(x)$ is determined such that $d^2 w/dx^2 = f(x)$, then $v(x,\theta)$ will satisfy the homogeneous equation $\partial^2 v/\partial x^2 - (1/\alpha)\partial v/\partial\theta = 0$ whose solution can be obtained by the separation-of-variables method.

b. Particular Integrals. Some example *particular integrals* (primitive solutions) for characteristic partial differential equations of heat transfer, derived generally by the separation-of-variables method, are as follows (t = temperature function, c = integration constant, μ = real separation constant, and λ = dummy variable):

Conduction, steady state
- Two-dimensional rectangular

$$\frac{\partial^2 t}{\partial x^2} + \frac{\partial^2 t}{\partial y^2} = 0 \; ; \; t_\mu = (c_1 \sin \mu x + c_2 \cos \mu x)(c_3 e^{\mu y} + c_4 e^{-\mu y})$$

- Two-dimensional cylindrical

$$\frac{\partial^2 t}{\partial r^2} + \frac{1}{r}\frac{\partial t}{\partial r} + \frac{1}{r^2}\frac{\partial^2 t}{\partial \phi^2} = 0 \; ; \; t_\mu = (c_1 \sin \mu\phi + c_2 \cos \mu\phi)(c_3 r^\mu + c_4 r^{-\mu})$$

- Two-dimensional spherical

$$r\frac{\partial^2}{\partial r^2}(rt) + \frac{1}{\sin\psi}\frac{\partial}{\partial\psi}\left(\sin\psi\frac{\partial t}{\partial\psi}\right) = 0 \; ; \; t_\mu = (c_1 r^\mu + c_2 r^{-\mu-1})P_\mu(\cos\psi)$$

Conduction, transient state
- One-dimensional rectangular

$$\frac{\partial^2 t}{\partial x^2} = \frac{1}{\alpha}\frac{\partial t}{\partial\theta} \; ; \quad t_\mu = c\exp(-\mu^2\alpha\theta \pm i\mu x)$$

$$t_\mu = c\exp(-\mu^2\alpha\theta)\sin\mu x$$

$$t_\mu = c\exp(-\mu^2\alpha\theta)\cos\mu x$$

$$t_\mu = c\exp\left(\pm\sqrt{\frac{\mu}{2\alpha}}\,x\right)\left[\cos\left(\mu\theta \pm \sqrt{\frac{\mu}{2\alpha}}\,x\right) \pm i\sin\left(\mu\theta \pm \sqrt{\frac{\mu}{2\alpha}}\,x\right)\right]^*$$

$$t_\mu = c\frac{1}{2\sqrt{\pi\alpha\theta}}\exp\left[-\frac{(x-\mu)^2}{4\alpha\theta}\right]^\dagger$$

$$t_\mu = c\frac{2}{\sqrt{\pi}}\int_0^{x/2\sqrt{\alpha\theta}} e^{-\lambda^2}d\lambda = c\operatorname{erf}\frac{x}{2\sqrt{\alpha\theta}}^\dagger$$

Convection, steady state
- Laminar flow between parallel plates

$$\frac{\partial^2 t}{\partial x^2} + \frac{\partial^2 t}{\partial y^2} - a\frac{\partial t}{\partial x} = 0 \; ; \quad t_\mu = c\exp\left[-\left(\sqrt{\frac{a^2}{4}+\mu^2}-\frac{a}{2}\right)x\right]\cos\mu y$$

- Laminar flow in a tube

$$\frac{\partial^2 t}{\partial x^2} + \frac{\partial^2 t}{\partial r^2} + \frac{1}{r}\frac{\partial t}{\partial r} - a\frac{\partial t}{\partial x} = 0 \; ; \quad t_\mu = c\exp\left[-\left(\sqrt{\frac{a^2}{4}+\mu^2}-\frac{a}{2}\right)x\right]J_0(\mu y)$$

*Based on imaginary separation constants $\pm i\mu$.
†Not obtainable by separation of variables.

The availability of particular integrals permits simplification of a given partial differential equation. Suppose $\partial^2 u/\partial x^2 - (b^2/\alpha)u = (1/\alpha)\partial u/\partial \theta$. If a solution is assumed as $u(x, \theta) = e^{-b^2\theta}v(x, \theta)$, then $v(x, \theta)$ must satisfy the simpler equation $\partial^2 v/\partial x^2 = (1/\alpha)\partial v/\partial \theta$. This, in turn, can be reduced to an ordinary differential equation by letting $v(x, \theta) = e^{-\alpha\theta}w(x)$, for then $w(x)$ satisfies $d^2 w/dx^2 + w = 0$.

c. Laplace Transform. The operational technique of solving differential equations known as the *Laplace-transform method* is generally more powerful than integration by separation of variables in that the technique is manipulatively simpler for linear ordinary and partial differential equations, and because certain classes of nonlinear and nonhomogeneous equations and boundary conditions can be handled that are otherwise intractable by the latter method. In addition, the mechanics of the Laplace-transform method are independent of whether finite or infinite domains are considered, and frequently solution forms are obtained that converge more rapidly (early response times) than the corresponding forms derived by separation of variables. Another notable feature of the method is its success in problems where the dependent variable or its derivative is discontinuous.

The *direct Laplace transformation* of a real piecewise continuous function $f(\theta)$, denoted $\mathcal{L}\{f(\theta)\}$, is defined for positive θ in terms of a new variable s (positive real or complex) as the integral

$$\mathcal{L}\{f(\theta)\} = \bar{f}(s) = \int_0^\infty e^{-s\theta}f(\theta)d\theta \tag{126}$$

The new function $\bar{f}(s)$ is called the *Laplace transform* of $f(\theta)$ with respect to θ. A transform is thus obtained simply by multiplying the known function $f(\theta)$ by $e^{-s\theta}$ and integrating with respect to θ from 0 to ∞. For example, if $f(\theta) = 1$ then $\bar{f}(s) = 1/s$. The known function $f(\theta)$ is conversely called the *inverse transform* of $\bar{f}(s)$ and is denoted as

$$f(\theta) = \mathcal{L}^{-1}\{\bar{f}(s)\} \tag{127}$$

Thus, 1 is the inverse transform of $1/s$.

Although the transform in Eq. (126) exists and the integral converges only under certain limitations as to the character of $f(\theta)$ and the range of s [45], most engineering functions behave within these limitations. The inverse transform in Eq. (127) is *unique* (in the applied sense) in that one and only one $f(\theta)$ can be found to correspond to a given transform $\bar{f}(s)$.

Operational Properties. The following properties are useful consequences of the Laplace transformation [Eq. (126)]:

$$\mathcal{L}\{c_1 f_1(\theta) + c_2 f_2(\theta)\} = c_1\bar{f}_1(s) + c_2\bar{f}_2(s) \tag{128}$$

$$\mathcal{L}\{e^{\pm a\theta}f(\theta)\} = \bar{f}(s \mp a) \tag{129}$$

$$\mathcal{L}\{\theta^n f(\theta)\} = (-1)^n \frac{d^n}{ds^n}\bar{f}(s) \quad ; \quad n = 1, 2, 3, \ldots \tag{130}$$

$$\mathcal{L}\left\{\frac{\partial}{\partial\theta}f(\theta)\right\} = s\bar{f}(s) - f(0) \tag{131}$$

$$\mathcal{L}\left\{\frac{\partial^2}{\partial\theta^2}f(\theta)\right\} = s^2\bar{f}(s) - sf(0) - \frac{\partial}{\partial\theta}f(0) \tag{132}$$

$$\mathcal{L}\left\{\frac{\partial^n}{\partial x^n} f(\theta)\right\} = \frac{\partial^n}{\partial x^n} \bar{f}(s) \quad ; \quad n = 1, 2, 3, \ldots \tag{133}$$

$$\mathcal{L}\left\{\int_a^\theta f(\theta)d\theta\right\} = \frac{1}{s}\bar{f}(s) - \frac{1}{s}\int_0^a f(\theta)d\theta \tag{134}$$

$$\mathcal{L}\left\{\int_0^\theta f_1(\tau) f_2(\theta - \tau)d\tau\right\} = \bar{f}_1(s)\bar{f}_2(s) \tag{135}$$

Equation (128) is the *linearity* characteristic of the transform; Eq. (129) is the *translation* (*shifting*) *theorem;* Eq. (130) gives the *n*th derivative of $\bar{f}(s)$; Eqs. (131) and (132) are transforms of the first and second derivatives of $f(\theta)$, where $f(0) = f_{\theta=0}$; Eq. (133) is the transform for derivatives with respect to variables other than θ; Eq. (134) gives the transform for the integral of $f(\theta)$; and Eq. (135) is the *convolution* (*Faltung* or *Borel*) *theorem* (τ a dummy variable).

Method. Consider a partial differential equation for the response function $u(x,\theta)$ where u is dependent variable and x and θ are independent spatial and time variables. The transform of $u(x,\theta)$ with respect to θ is

$$\bar{u}(x,s) = \int_0^\infty e^{-s\theta} u(x,\theta)d\theta \tag{136}$$

Solution of the differential equation by Laplace transform consists of: an integral transformation of each term in the equation (and associated boundary conditions) by Eq. (136) into a *subsidiary equation* in $\bar{u}(x,s)$; solving of the subsidiary equation subject to boundary conditions; and performing an inverse Laplace transformation to obtain $u(x,\theta)$ from $\bar{u}(x,s)$. Derivation of the subsidiary equation is assisted by the properties in Eqs. (128) to (135).

Equations (131) and (132) show that transformation of derivatives has two important effects, namely the reduction of order of the original derivative and the introduction of boundary (initial) conditions. Transformation to a subsidiary equation has the effect of reducing the number of original independent variables by one. Thus, in the case of an ordinary differential equation the subsidiary equation is algebraic, while a partial differential equation reduces to an ordinary differential equation; in both cases the transformed problem is simpler than the original.

The technique of performing an inverse Laplace transformation, i.e., the mechanics of determining the $u(x,\theta)$ that transforms to $\bar{u}(x,s)$, is formally that of contour integration involving solution of complex integrals in the *inversion theorem*

$$u(x,\theta) = \frac{1}{2\pi i} \lim_{\beta \to \infty} \int_{\gamma-i\beta}^{\gamma+i\beta} e^{z\theta}\bar{u}(x,z)dz \tag{137}$$

A large number of inverse transforms have been worked out in this way and tabulated [46, 47, 48]. Useful Laplace-transform pairs are given in Table 11.

In the process of inverse transformation, the solved subsidiary equation must generally be resolved into partial fractions (Sec. B.1) in order to expose the direct transforms $\bar{f}(s)$ tabulated in Table 11 in the form of polynomials [45]. If the direct transform is factorable as a product of known inverse transforms, then the convolution theorem [Eq. (135)] can be used to obtain the direct transform $f(\theta)$ from

TABLE 11. Laplace-transform Pairs $(\lambda = \sqrt{s/\alpha})$*

	Direct transform $= \bar{f}(s)$	Inverse transform $= f(\theta)$
1	$\dfrac{1}{s}$	1
2	$\dfrac{1}{s}\ln s$	$-\ln(c\theta)$; $\ln c = \gamma = 0.57722$
3	$\dfrac{1}{s+a}$	$e^{-a\theta}$
4	$\dfrac{1}{s^{a+1}}$; $a > -1$	$\dfrac{\theta^a}{\Gamma(a+1)}$
5	$\dfrac{a}{s^2+a^2}$	$\sin a\theta$
6	$\dfrac{s}{s^2+a^2}$	$\cos a\theta$
7	$e^{-\lambda x}$	$\dfrac{x}{2\sqrt{\pi\alpha\theta^3}}\, e^{-x^2/4\alpha\theta}$
8	$\dfrac{e^{-\lambda x}}{\lambda}$	$\sqrt{\alpha/\pi\theta}\, e^{-x^2/4\alpha\theta}$
9	$\dfrac{e^{-\lambda x}}{s}$	$\operatorname{erfc}\dfrac{x}{2\sqrt{\alpha\theta}}$
10	$\dfrac{e^{-\lambda x}}{s^{1+n/2}}$; $n = 0,1,2,\ldots$	$(4\theta)^{n/2} i^n \operatorname{erfc}\dfrac{x}{2\sqrt{\alpha\theta}}$
11	$\dfrac{e^{-\lambda x}}{\lambda s}$	$2\sqrt{\alpha\theta/\pi}\, e^{-x^2/4\alpha\theta} - x \operatorname{erfc}\dfrac{x}{2\sqrt{\alpha\theta}}$

Table of transforms (continued). Entries 14–20.

No.	Transform	Inverse function
14	$\dfrac{e^{-\lambda x}}{s^2}$	$\left(\dfrac{x^2}{2\alpha}+\theta\right)\operatorname{erfc}\dfrac{x}{2\sqrt{\alpha\theta}}-x\sqrt{\dfrac{\theta}{\pi\alpha}}\,e^{-x^2/4\alpha\theta}$
15	$\dfrac{e^{-\lambda x}}{\lambda + a}$	$\sqrt{\dfrac{\alpha}{\pi\theta}}\,e^{-x^2/4\alpha\theta}-a\alpha\,e^{ax+a^2\alpha\theta}\operatorname{erfc}\left(\dfrac{x}{2\sqrt{\alpha\theta}}+a\sqrt{\alpha\theta}\right)$
16	$\dfrac{e^{-\lambda x}}{\lambda(\lambda + a)}$	$\alpha\,e^{ax+a^2\alpha\theta}\operatorname{erfc}\left(\dfrac{x}{2\sqrt{\alpha\theta}}+a\sqrt{\alpha\theta}\right)$
17	$\dfrac{e^{-\lambda x}}{\lambda s(\lambda + a)}$	$\dfrac{1}{a}\left[\operatorname{erfc}\dfrac{x}{2\sqrt{\alpha\theta}}-e^{ax+a^2\alpha\theta}\operatorname{erfc}\left(\dfrac{x}{2\sqrt{\alpha\theta}}+a\sqrt{\alpha\theta}\right)\right]$
		$\dfrac{2\sqrt{\alpha\theta/\pi}}{a}\,e^{-x^2/4\alpha\theta}-\dfrac{1}{a^2}\left[(1+ax)\operatorname{erfc}\dfrac{x}{2\sqrt{\alpha\theta}}-e^{ax+a^2\alpha\theta}\operatorname{erfc}\left(\dfrac{x}{2\sqrt{\alpha\theta}}+a\sqrt{\alpha\theta}\right)\right]$
18	$\dfrac{e^{-\lambda x}}{\lambda^{n+1}(\lambda + a)}$	$\dfrac{\alpha}{(-a)^n}\left[e^{ax+a^2\alpha\theta}\operatorname{erfc}\left(\dfrac{x}{2\sqrt{\alpha\theta}}+a\sqrt{\alpha\theta}\right)-\sum_{m=0}^{n-1}(-2a\sqrt{\alpha\theta})^{m}\,i^{m}\operatorname{erfc}\dfrac{x}{2\sqrt{\alpha\theta}}\right]$
19	$\dfrac{e^{-\lambda x}}{(\lambda + a)^2}$	$\alpha(1+ax+2a^2\alpha\theta)\,e^{ax+a^2\alpha\theta}\operatorname{erfc}\left(\dfrac{x}{2\sqrt{\alpha\theta}}+a\sqrt{\alpha\theta}\right)-2a\sqrt{\alpha^3\theta/\pi}\,e^{-x^2/4\alpha\theta}$
20	$\dfrac{e^{-\lambda x}}{s(\lambda + a)^2}$	$\dfrac{1}{a^2}\left[\operatorname{erfc}\dfrac{x}{2\sqrt{\alpha\theta}}-2a\sqrt{\alpha\theta/\pi}\,e^{-x^2/4\alpha\theta}-(1-ax-2a^2\alpha\theta)\,e^{ax+a^2\alpha\theta}\operatorname{erfc}\left(\dfrac{x}{2\sqrt{\alpha\theta}}+a\sqrt{\alpha\theta}\right)\right]$
	$\dfrac{e^{-\lambda x}}{s - a}$	$\dfrac{e^{a\theta}}{2}\left[e^{-x\sqrt{a/\alpha}}\operatorname{erfc}\left(\dfrac{x}{2\sqrt{\alpha\theta}}-\sqrt{a\theta}\right)+e^{x\sqrt{a/\alpha}}\operatorname{erfc}\left(\dfrac{x}{2\sqrt{\alpha\theta}}+\sqrt{a\theta}\right)\right]$

TABLE 11. Laplace-transform Pairs ($\lambda = \sqrt{s/\alpha}$)* (Continued)

	Direct transform $= \bar{f}(s)$	Inverse transform $= f(\theta)$
21	$\dfrac{e^{-\lambda x}}{(s-a)^2}$	$\dfrac{e^{a\theta}}{2}\left[\left(\theta - \dfrac{x}{2\sqrt{a\alpha}}\right)e^{-x\sqrt{a/\alpha}}\,\mathrm{erfc}\left(\dfrac{x}{2\sqrt{\alpha\theta}} - \sqrt{a\theta}\right) + \left(\theta + \dfrac{x}{2\sqrt{a\alpha}}\right)e^{x\sqrt{a/\alpha}}\,\mathrm{erfc}\left(\dfrac{x}{2\sqrt{\alpha\theta}} + \sqrt{a\theta}\right)\right]$
22	$\dfrac{e^{-\lambda x}}{\lambda(s-a)}$	$\dfrac{e^{a\theta}}{2}\sqrt{\alpha/a}\left[e^{-x\sqrt{a/\alpha}}\,\mathrm{erfc}\left(\dfrac{x}{2\sqrt{\alpha\theta}} - \sqrt{a\theta}\right) - e^{x\sqrt{a/\alpha}}\,\mathrm{erfc}\left(\dfrac{x}{2\sqrt{\alpha\theta}} + \sqrt{a\theta}\right)\right]$
23	$\dfrac{e^{-\lambda x}}{(s-a)(\lambda+b)}$; $a \neq ab^2$	$\dfrac{e^{a\theta}}{2}\left[\dfrac{\sqrt{\alpha}}{\sqrt{a}+b\sqrt{\alpha}}\,e^{-x\sqrt{a/\alpha}}\,\mathrm{erfc}\left(\dfrac{x}{2\sqrt{\alpha\theta}} - \sqrt{a\theta}\right) - \dfrac{\sqrt{\alpha}}{\sqrt{a}-b\sqrt{\alpha}}\,e^{x\sqrt{a/\alpha}}\,\mathrm{erfc}\left(\dfrac{x}{2\sqrt{\alpha\theta}} + \sqrt{a\theta}\right)\right]$ $+\dfrac{ba}{a-b^2\alpha}\,e^{bx+b^2\alpha\theta}\,\mathrm{erfc}\left(\dfrac{x}{2\sqrt{\alpha\theta}} + b\sqrt{\alpha\theta}\right)$
24	$\dfrac{e^{-\lambda x}}{s^{3/4}}$	$\dfrac{1}{\pi}\sqrt{x/2\alpha^2\theta}\;e^{-x^2/8\alpha\theta}\,K_{\frac{1}{4}}(x^2/8\alpha\theta)$
25	$K_0(\lambda x)$	$\dfrac{1}{2\theta}\,e^{-x^2/4\alpha\theta}$
26	$s^{\nu/2}\,K_\nu(x\sqrt{s})$	$\dfrac{x^\nu}{(2\theta)^{\nu+1}}\,e^{-x^2/4\theta}$
27	$s^{\nu/2-1}\,K_\nu(x\sqrt{s})$	$\dfrac{2^{\nu-1}}{x^\nu}\displaystyle\int_{x^2/4\theta}^{\infty} e^{-\mu}\mu^{\nu-1}d\mu$

28	$\dfrac{1}{\sqrt{s}} K_{2\nu}(\lambda x)$	$\dfrac{1}{2\sqrt{\pi\theta}} e^{-x^2/8\alpha\theta} K_\nu(x^2/8\alpha\theta)$
29	$\dfrac{e^{x/s}}{s}$	$I_0(2\sqrt{x\theta})$
30	$\dfrac{e^{xs - x\sqrt{(s+a)(s+b)}}}{\sqrt{(s+a)(s+b)}}$	$e^{\frac{1}{2}(a+b)(x+\theta)} I_0[\tfrac{1}{2}(a-b)\sqrt{(2x+\theta)\theta}]$
31	$[s - \sqrt{s^2 - x^2}]^\nu$; $\nu > 0$	$\dfrac{\nu x^\nu}{\theta} I_\nu(x\theta)$
32	$\dfrac{e^{x(\sqrt{s+a} - \sqrt{s+b})^2}}{\sqrt{s+a}\sqrt{s+b}(\sqrt{s+a}+\sqrt{s+b})^{2\nu}}$; $\nu \geq 0$	$\dfrac{\theta^{\nu/2} e^{-\frac{1}{2}(a+b)\theta} I_\nu[\tfrac{1}{2}(a-b)\sqrt{(4x+\theta)\theta}]}{(a-b)(4x+\theta)^{\nu/2}}$
33	$I_\nu(\lambda x')K_\nu(\lambda x)$; $x > x'$ $I_\nu(\lambda x)K_\nu(\lambda x')$; $x < x'$	$\dfrac{1}{2\theta} e^{-(x^2 + x'^2)/4\alpha\theta} I_\nu(xx'/2\alpha\theta)$; $\nu \geq 0$

*Adapted from Ref. 5 by permission of Clarendon Press, Oxford.

the original inverse transform $\bar{f}(s)$.* In cases where an inverse transformation cannot be effected, the integrals must be evaluated numerically.

Consider the boundary-value problem

$$\frac{\partial^2 u}{\partial x^2} = \frac{1}{\alpha}\frac{\partial u}{\partial \theta} \quad \left\{ \begin{array}{ll} u = 0 & ; \quad \theta = 0, \ x \geq 0 \\[2mm] \dfrac{\partial u}{\partial x} = a(u - u_1) & ; \quad x = 0, \ \theta > 0 \\[2mm] u \neq \infty & ; \quad x \to \infty, \ \theta > 0 \end{array} \right.$$

This is to be solved using Laplace transforms with respect to θ. By Eq. (133) $\mathcal{L}\{\partial^2 u/\partial x^2\} = \partial^2 \bar{u}/\partial x^2$, and by Eq. (131) along with the initial condition $\mathcal{L}\{\partial u/\partial \theta\} = s\bar{u} - u(x,0) = s\bar{u}$. Treating s in $\bar{u}(x,s)$ as a parameter and using Eq. (128), the subsidiary equation is

$$\frac{d^2\bar{u}}{dx^2} = \frac{s}{\alpha}\bar{u}$$

By Eqs. (128) and (133) and the tabulated transform-pair No. 1, the two boundary conditions transform for all θ to

$$\left\{ \begin{array}{ll} \dfrac{d\bar{u}}{dx} = a\left(\bar{u} - \dfrac{u_1}{s}\right) & ; \quad x = 0 \\[3mm] \bar{u} \neq \infty & ; \quad x \to \infty \end{array} \right.$$

The general solution of the homogeneous subsidiary equation (Sec. B.2) is

$$\bar{u} = c_1 e^{\sqrt{s/\alpha}\,x} + c_2 e^{-\sqrt{s/\alpha}\,x}$$

where, by application of the transformed boundary conditions, $c_1 = 0$ and $c_2 = au_1/s(\sqrt{s/\alpha} + a)$. Thus

$$\bar{u} = au_1 \frac{e^{-\sqrt{s/\alpha}\,x}}{s(\sqrt{s/\alpha} + a)}$$

The inverse transform of this (given by the tabulated transform-pair No. 15) is the problem solution as

$$u = u_1\left[\operatorname{erfc}\frac{x}{2\sqrt{\alpha\theta}} - e^{ax + a^2\alpha\theta}\operatorname{erfc}\left(\frac{x}{2\sqrt{\alpha\theta}} + a\sqrt{\alpha\theta}\right)\right]$$

---

*Suppose $\bar{f}(s) = 1/(s^2 - a^2)$. This can be expressed as a product $\bar{f}_1(s)\bar{f}_2(s)$, where $\bar{f}_1(s) = 1/(s + a)$ and $\bar{f}_2(s) = 1/(s - a)$. The tabulated transform-pair No. 3 gives the inverse of $\bar{f}_1(s)$ as $e^{-a\theta}$ and of $\bar{f}_2(s)$ as $e^{a\theta}$. Using Eq. (135) in its inverse form

$$f(\theta) = \mathcal{L}^{-1}\{\bar{f}_1(s)\bar{f}_2(s)\} = \int_0^\theta f_1(\tau)f_2(\theta - \tau)d\tau$$

the inverse transform of $\bar{f}(s)$ becomes

$$f(\theta) = \int_0^\theta e^{-a\tau}e^{a(\theta - \tau)}d\tau = e^{a\theta}\int_0^\theta e^{-2a\tau}d\tau = \frac{1}{a}\sinh a\theta$$

d. Combined Solutions. Under certain conditions, individual solutions of partial differential equations for simple problems can be combined to yield solutions of a more complex problem. Generally the characteristic that permits these combinations is the *linearity property* of the governing differential equations.

Product Solutions. Solutions for response functions in multidimensional regions can be obtained simply as the product of simpler one-dimensional solutions providing the differential equation is linear and boundary conditions are homogeneous. For example, *product solutions* for conduction temperature response of a cube can be formulated directly by multiplying together three solutions representing one-dimensional response in the three mutually perpendicular directions of the cube.

Consider the response function $u(x, y, \theta)$ in a two-dimensional region $-x_1 \leq x \leq x_1$ and $-y_1 \leq y \leq y_1$ with initial and boundary conditions

$$\frac{\partial^2 u}{\partial x^2} + \frac{\partial^2 u}{\partial y^2} = \frac{1}{\alpha} \frac{\partial u}{\partial \theta} \quad \begin{cases} u = X_1(x)\, Y_1(y) & ; \quad \theta = 0 \\[2mm] \dfrac{\partial u}{\partial x} = au & ; \quad x = \pm x_1 \\[2mm] \dfrac{\partial u}{\partial y} = bu & ; \quad y = \pm y_1 \end{cases}$$

Since the differential equation is linear, the product $u = X(x, \theta)\, Y(y, \theta)$ is the desired two-dimensional solution if $X(x, \theta)$ and $Y(y, \theta)$ satisfy the one-dimensional problems

$$\frac{\partial^2 X}{\partial x^2} = \frac{1}{\alpha} \frac{\partial X}{\partial \theta} \quad \begin{cases} X = X_1(x) & ; \quad \theta = 0 \\[2mm] \dfrac{\partial X}{\partial x} = aX & ; \quad x = \pm x_1 \end{cases}$$

and

$$\frac{\partial^2 Y}{\partial y^2} = \frac{1}{\alpha} \frac{\partial Y}{\partial \theta} \quad \begin{cases} Y = Y_1(y) & ; \quad \theta = 0 \\[2mm] \dfrac{\partial Y}{\partial y} = bY & ; \quad y = \pm y_1 \end{cases}$$

The solution $u = XY$ clearly satisfies the original initial condition and the two boundary conditions; it also satisfies the original differential equation since

$$Y \frac{\partial^2 X}{\partial x^2} + X \frac{\partial^2 Y}{\partial y^2} = \frac{1}{\alpha}\left(Y \frac{\partial X}{\partial \theta} + X \frac{\partial Y}{\partial \theta} \right) = \frac{1}{\alpha}\left(\alpha Y \frac{\partial^2 X}{\partial x^2} + \alpha X \frac{\partial^2 Y}{\partial y^2} \right)$$

Superposition Solutions. The solution of a linear partial differential equation satisfying multiple nonhomogeneous boundary conditions is obtained by adding together (*superimposing*) individual solutions for which only one initial or boundary condition is nonhomogeneous.

Suppose $u(x, \theta)$ is to be obtained in the region $0 \leq x \leq x_1$ subject to

$$\frac{\partial^2 u}{\partial x^2} = \frac{1}{\alpha} \frac{\partial u}{\partial \theta} \quad \begin{cases} u = F(x) & ; \quad \theta = 0 \\[2mm] u = f(\theta) & ; \quad x = 0 \\[2mm] u = g(\theta) & ; \quad x = x_1 \end{cases}$$

The linear sum $u = u_1 + u_2 + u_3$ is the desired solution providing u_1, u_2, and u_3 individually satisfy the differential equation and the following *decomposed* initial and boundary conditions:

$\theta = 0$	$x = 0$	$x = x_1$
$u_1 = F(x)$	$u_1 = 0$	$u_1 = 0$
$u_2 = 0$	$u_2 = f(\theta)$	$u_2 = 0$
$u_3 = 0$	$u_3 = 0$	$u_3 = g(\theta)$

e. Superposition Integral (*Duhamel's Theorem*). If a time-dependent input (*excitation*) function $y = f(\theta)$ is resolved into a set of consecutive step functions of equal duration $\Delta \tau$ (Fig. 29), each step input will stimulate a response at time θ equal to

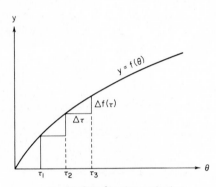

Fig. 29. Step-function excitation.

$\Delta f(\tau) u_1(x, \theta - \tau)$. If the system described by $u(x, \theta)$ is *linear*, then the system response to all step inputs $\tau_1, \tau_2, \ldots, \tau_n$ is represented by the sum (*superposition*)

$$u(x, \theta) = \sum_{i=1}^{n} \frac{\Delta f(\tau_i)}{\Delta \tau_i} u_1(x, \theta - \tau_i) \Delta \tau_i$$

From the definition in Eq. (45), this becomes (in the limit as $n \to \infty$ and $\Delta \tau_i \to 0$) the *superposition integral*

$$u(x, \theta) = \int_0^\theta \frac{df(\tau)}{d\tau} u_1(x, \theta - \tau) d\tau \tag{138}$$

or, using Eq. (48),

$$u(x, \theta) = \int_0^\theta f(\tau) \frac{\partial}{\partial \theta} u_1(x, \theta - \tau) d\tau \tag{139}$$

The integrals in Eqs. (138) and (139) are useful in solving linear partial differential equations subject to *transient* boundary conditions (*Duhamel's theorem*). Use of Duhamel's method in heat conduction, for example, serves to reduce the problem of temperature response in bodies of zero initial temperature exposed to time-varying ambient (or surface) temperature to that of exposure to a constant unit ambient (or surface) temperature.*

Suppose it is required to solve the Fourier temperature-response equation

$$\frac{\partial^2 t}{\partial x^2} = \frac{1}{\alpha}\frac{\partial t}{\partial \theta} \quad \left\{ \begin{array}{lll} t = 0 & ; & \theta = 0 \\[2mm] \dfrac{\partial t}{\partial x} = b[t - t_a(\theta)] & ; & x = 0 \end{array} \right.$$

where $t_a(\theta)$ is the transient ambient temperature. Using Eq. (139) the solution is

$$t(x, \theta) = \int_0^\theta t_a(\tau)\frac{\partial}{\partial \theta}t_1(x, \theta - \tau)d\tau$$

where τ is a dummy variable and $t_1(x, \theta)$ is a solution of the same (but simpler) problem of zero initial temperature and exposure to a *constant* ambient temperature of unity; $t_a(\theta) = 1$. The choice between Eqs. (138) and (139) leads to different solution forms, one of which may have advantages of more rapid convergence.

f. Variable Transformations. Multivariable transformations are frequently employed to reduce the complexity of a partial differential equation (and/or its boundary conditions) and to combine several partial differential equations into a single governing equation. The basis of such transformations generally stems from physical considerations of the problem, such as coordinate substitutions for moving sources and boundaries and similarity concepts for boundary-layer flows and heat transfer.

Moving Source. Consider that an excitation source s (Fig. 30) moves through a

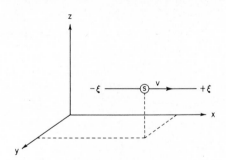

Fig. 30. Moving coordinate.

region (x, y, z) with constant unidirectional velocity v (parallel to x), and suppose that the field-response function $u(x, y, z, \theta)$ must satisfy

$$\frac{\partial^2 u}{\partial x^2} + \frac{\partial^2 u}{\partial y^2} + \frac{\partial^2 u}{\partial z^2} = \frac{1}{\alpha}\frac{\partial u}{\partial \theta}$$

where x, y, and z are fixed coordinates. For example, s may be a point, line, or plane *heat source* and $u(x, y, z, \theta)$ the temperature response of its surroundings.

*For broader applications in heat conduction, see Ref. 5.

A solution for u is simplified by transforming to a *moving coordinate* ξ whose origin coincides with s and moves at the source velocity v. Then the surrounding temperature field with respect to the moving coordinate depends only on location ξ and is independent of time θ (*quasi steady-state condition*). The transformation is effected by defining a new set of position and time variables as

$$\xi = x - v\theta \quad , \quad \theta' = \theta$$

Then $\partial u/\partial x = \partial u/\partial \xi$, $\partial^2 u/\partial x^2 = \partial^2 u/\partial \xi^2$, and $\partial u/\partial\theta = -v\partial u/\partial\xi + \partial u/\partial\theta'$. For the steady state $\partial u/\partial\theta' = 0$, whereby the foregoing partial differential equation becomes

$$\frac{\partial^2 u}{\partial \xi^2} + \frac{\partial^2 u}{\partial y^2} + \frac{\partial^2 u}{\partial z^2} = \frac{1}{\alpha}\frac{\partial u}{\partial \xi}$$

Solution of the simplified problem $u(\xi, y, z)$ is returned to the original variables $u(x, y, z, \theta)$ by substituting $x - v\theta$ for ξ.

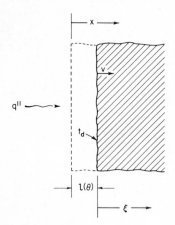

Fig. 31. Receding boundary.

Moving Boundary. If a solid $x \geq 0$ is suddenly exposed to a heat flux q'' at $x = 0$ (Fig. 31), its surface may attain the decomposition temperature t_d of the material and absorb a portion of the heat input due to endothermic phase change (melting, vaporization, sublimation, etc.). Considering that the decomposition products are immediately removed on formation, the surface will recede into the solid at some variable rate v until the instantaneous *ablation depth* is

$$l(\theta) = \int_0^\theta v\,d\theta$$

If the solid is initially at a uniform temperature t_0, and assuming a constant heat of decomposition H at the receding boundary, the nonlinear differential system describing the temperature response of the surviving solid (ignoring the starting transient) is

$$\frac{\partial^2 t}{\partial x^2} = \frac{1}{\alpha}\frac{\partial t}{\partial \theta} \left\{ \begin{array}{llll} t = t_0 & ; & \theta = 0 & , \ x \geq 0 \\ t = t_d & ; & x = l & , \ \theta > 0 \\ q'' = -k\dfrac{\partial t}{\partial x} + \rho H v & ; & x = l & , \ \theta > 0 \end{array} \right.$$

A solution for $t(x, \theta)$ is simplified by eliminating the moving boundary $x = l(\theta)$ through transformation to a moving coordinate ξ whose origin coincides with the instantaneous ablation surface as

$$\xi = x - l(\theta) \quad , \quad \theta' = \theta$$

Then the transformed boundary-value problem becomes

$$\frac{\partial^2 t}{\partial \xi^2} + \frac{v}{\alpha}\frac{\partial t}{\partial \xi} = \frac{1}{\alpha}\frac{\partial t}{\partial \theta'} \quad \left\{ \begin{array}{l} t = t_0 \qquad\qquad ; \quad \theta = 0 \;, \; \xi \geq 0 \\ t = t_d \qquad\qquad ; \quad \xi = 0 \;, \; \theta > 0 \\ q'' = -k\dfrac{\partial t}{\partial \xi} + \rho H v \;; \quad \xi = 0 \;, \; \theta > 0 \end{array} \right.$$

If q'' is constant, then v is constant and the temperature history is quasi-steady ($\partial t/\partial \theta' = 0$) in the moving frame of reference.

Similarity. A number of technically important boundary-value problems in partial differential equations possess so-called *similarity solutions* derived through similarity transformations of the independent variables. In the case of certain boundary-layer flows, for instance, the longitudinal velocity profile $u(y)$ behaves such that it can be *scaled* between two arbitrary flow stations x_1 and x_2. The velocity profiles are said, in this sense, to be *similar*; that is, the profiles at x_1 and x_2 differ only by constant (but different) scaling factors on u and y. A large number of potential flows exhibit this property [49]. Some problems in heat conduction also admit to similarity solutions, reflecting the fact that instantaneous temperature profiles in solids can also be similar (*affine*) in certain cases.

Although no general rules are known for distinguishing boundary-value problems as to whether similarity solutions do or do not exist, it is a characteristic of those systems that do possess such solutions to have lacking a physical length with which to nondimensionalize the independent variables. Examples are given below:

• *Boundary-layer Flow.* An infinite plate, initially at rest, is suddenly accelerated to a constant velocity u_0 (Fig. 32). The transient velocity profile $u(y, \theta)$ that develops in the surrounding viscous

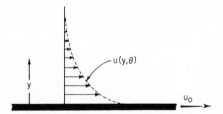

Fig. 32. Boundary layer on moving plate.

fluid satisfies the differential system (*momentum equation*)

$$\frac{\partial^2 U}{\partial y^2} = \frac{1}{\nu}\frac{\partial U}{\partial \theta} \quad \left\{ \begin{array}{llll} 1) & U = 0 \quad ; & \theta = 0 \;, & y \geq 0 \\ 2) & U = 0 \quad ; & y = \infty \;, & \theta > 0 \\ 3) & U = 1 \quad ; & y = 0 \;, & \theta > 0 \end{array} \right.$$

where $U = u/u_0$.

A *similarity variable* $\eta(y, \theta)$ is to be found such that substitution of $U(y, \theta) = f(\eta)$ will reduce the partial differential system to an ordinary one while simultaneously satisfying the single initial condition and the double boundary conditions. Since the resulting ordinary differential equation will require only one initial and one boundary condition, it is necessary that $\eta(y, 0) = \eta(\infty, \theta)$. A similarity variable that achieves this is $\eta = y/\sqrt{2\nu\theta}$, and with this the system to be solved becomes

$$\frac{d^2 f}{d\eta^2} + \eta\frac{df}{d\eta} = 0 \quad \left\{ \begin{array}{lll} 1), 2) & f(\eta) = 0 \quad ; & \eta = \infty \\ 3) & f(\eta) = 1 \quad ; & \eta = 0 \end{array} \right.$$

• *Heat Convection.* A fluid of uniform approach temperature t_a flows over an infinite plate of uniform temperature t_w (Fig. 33). If the boundary-layer velocity profile is prescribed as

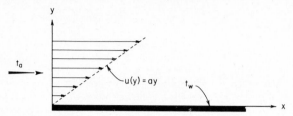

Fig. 33. Heat convection in flow along stationary plate.

$u(y) = ay$, then the temperature profile $t(y, x)$ that develops in the fluid adjacent to the plate satisfies the differential system (*energy equation*)

$$\frac{\partial^2 T}{\partial y^2} - \frac{a}{\alpha} y \frac{\partial T}{\partial x} = 0 \qquad \begin{cases} 1) \ T = 1 & ; \quad x = 0 \ , \ y \geq 0 \\ 2) \ T = 1 & ; \quad y = \infty \ , \ x > 0 \\ 3) \ T = 0 & ; \quad y = 0 \ , \ x > 0 \end{cases}$$

where $T = (t - t_w)/(t_a - t_w)$.

Substitution of $T(y, x) = f(\eta)$ with the similarity variable $\eta = y(a/3\alpha x)^{1/3}$ reduces the problem to solution of the ordinary differential system

$$\frac{d^2 f}{d\eta^2} + \eta^2 \frac{df}{d\eta} = 0 \qquad \begin{cases} 1), 2) \ f(\eta) = 1 & ; \quad \eta = \infty \\ \ \ \ \ 3) \ f(\eta) = 0 & ; \quad \eta = 0 \end{cases}$$

• *Heat Conduction.* A semi-infinite solid $x \geq 0$, initially at a uniform temperature t_0, is suddenly exposed at its surface to a constant heat flux q'' (Fig. 34). The resulting temperature

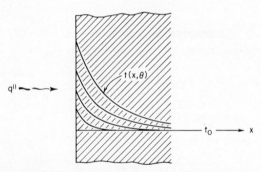

Fig. 34. Heat conduction in semi-infinite solid.

history $t(x, \theta)$ in the solid satisfies the differential system (*heat conduction equation*)

$$\frac{\partial^2 T}{\partial x^2} = \frac{1}{\alpha} \frac{\partial T}{\partial \theta} \qquad \begin{cases} 1) \ T = 0 & ; \quad \theta = 0 \ , \ x \geq 0 \\ 2) \ T = 0 & ; \quad x = \infty \ , \ \theta > 0 \\ 3) \ q'' = -k \dfrac{\partial T}{\partial x} & ; \quad x = 0 \ , \ \theta > 0 \end{cases}$$

where $T = t - t_0$.

Assuming a function of the form $T(x, \theta) = -(q''/k)\sqrt{2\alpha\theta}\, f(\eta)$ along with the similarity variable $\eta = x/\sqrt{2\alpha\theta}$ reduces the problem to solution of the reduced differential system

$$\frac{d^2f}{d\eta^2} + \eta\frac{df}{d\eta} - f = 0 \qquad \left\{ \begin{array}{ll} 1),\ 2) & f(\eta) = 0 \quad ; \quad \eta = \infty \\[2ex] 3) & \dfrac{df(\eta)}{d\eta} = 1 \quad ; \quad \eta = 0 \end{array} \right.$$

It is noted that in all three of the above examples the similarity variable is of the form $\eta = Au/v^n$, where u and v are independent variables.

REFERENCES

1. G. A. Korn and T. M. Korn, "Mathematical Handbook for Scientists and Engineers," McGraw-Hill Book Company, Inc., New York, 1961.
2. "Handbook of Mathematical Functions," AMS-55, National Bureau of Standards, Washington, D.C., March 1965.
3. L. E. Dickson, "First Course in the Theory of Equations," John Wiley & Sons, Inc., New York, 1922.
4. R. S. Burington, "Handbook of Mathematical Tables and Formulas," 3d ed., McGraw-Hill Book Company, New York, 1949.
5. H. S. Carslaw and J. C. Jaeger, "Conduction of Heat in Solids," 2d ed., Oxford University Press, London, 1959.
6. E. Jahnke and F. Emde, "Tables of Functions," 4th ed., Dover Publications, Inc., New York, 1945.
7. E. U. Condon and H. Odishaw, "Handbook of Physics," McGraw-Hill Book Company, New York, 1958.
8. T. Baumeister and L. S. Marks, "Mechanical Engineers' Handbook," 6th ed., McGraw-Hill Book Company, New York, 1958.
9. J. H. Perry, "Chemical Engineers Handbook," 3d ed., McGraw-Hill Book Company, New York, 1952.
10. A. Fletcher et al., "An Index of Mathematical Tables," McGraw-Hill Book Company, New York, 1946.
11. "American Institute of Physics Handbook," 2d ed., McGraw-Hill Book Company, New York, 1963.
12. L. J. Comrie, "Chambers's Six-Figure Mathematical Tables," vol. I—Logarithmic Values, vol. II—Natural Values, D. Van Nostrand Co., Inc., New Jersey, 1949.
13. "Table of Natural Logarithms," AMS-31, 53, National Bureau of Standards, Washington, D.C., 1953 and 1958.
14. "Tables of Exponential Function $e^{\pm x}$," AMS-14, National Bureau of Standards, Washington, D.C., 1951.
15. L. M. Milne-Thomson, "Standard Table of Square Roots," Bell and Sons, London, 1929.
16. "Tables of Circular and Hyperbolic Sines and Cosines for Radian Arguments," AMS-36, 3d ed., National Bureau of Standards, Washington, D.C., 1953.
17. "Tables of Circular and Hyperbolic Tangents and Cotangents for Radian Arguments," 2d ed., Columbia University Press, New York, 1947.
18. "Tables of the Inverse Hyperbolic Functions," Computation Laboratory, Harvard University Press, Cambridge, Mass., 1949.
19. "Tables of Sines and Cosines for Radian Arguments," AMS-43, National Bureau of Standards, Washington, D.C., 1955.
20. "Table of Arcsin x," National Bureau of Standards, Columbia University Press, New York, 1945.
21. "Table of Arctan x," AMS-26, National Bureau of Standards, Washington, D.C., 1953.
22. C. D. Hodgman et al., "Handbook of Chemistry and Physics," 41st ed., Chemical Rubber Co., Inc., Cleveland, 1959–1960.
23. K. Pearson, "Tables of the Incomplete Γ-Function," Cambridge University Press, London, 1951.
24. K. Pearson, "Tables of the Incomplete Beta-Function," University of London, London, 1934.
25. "Tables of the Error Function and Its Derivative," AMS-41, National Bureau of Standards, Washington, D.C., 1954.

26. "Tables of Sine, Cosine, and Exponential Integrals," vols. I & II, Columbia University Press, New York, 1940.
27. B. O. Pierce and R. M. Foster, "A Short Table of Integrals," 4th ed., Ginn and Co., Inc., Boston, 1956.
28. W. Gröbner and N. Hofreiter, "Integraltafel," 2d ed., vol. 1—Indefinite Integrals, vol. 2—Definite Integrals, Springer-Verlag, Vienna, 1957–1958.
29. V. Widder, "Advanced Calculus," Prentice-Hall, Inc., New York, 1947.
30. E. R. G. Eckert and R. M. Drake, "Heat and Mass Transfer," 2d ed., McGraw-Hill Book Company, New York, 1959.
31. H. W. Reddick, "Differential Equations," John Wiley & Sons, Inc., New York, 1943.
32. P. J. Schneider, "Conduction Heat Transfer," Addison-Wesley Publishing Co., Inc., Reading, Mass., 1957.
33. F. E. Relton, "Applied Bessel Functions," Blackie, London, 1946.
34. N. W. McLachlan, "Bessel Functions for Engineers," 2d ed., Oxford University Press, London, 1955.
35. "Tables of Bessel Functions of the First Kind," National Bureau of Standards, Columbia University Press, New York, 1947 and 1951.
36. "Tables of Ordinary Bessel Functions of the Second Kind of Orders 0 through 9," five vols., *Ballistic Res. Lab. Rept.* 1197, Aberdeen Proving Ground, Maryland, 1963.
37. "Tables of Bessel Functions of Fractional Order," vols. I & II, National Bureau of Standards, Columbia University Press, New York, 1948 and 1949.
38. "Table of the Bessel Functions $J_0(z)$ and $J_1(z)$ for Complex Arguments," 2d ed., National Bureau of Standards, Columbia University Press, New York, 1947.
39. "Tables of the Bessel-Kelvin Functions ber, bei, ker, kei and Their Derivatives for the Argument Range 0(0.01)107.50," *NASA TR* R-32, Washington, D.C., 1959.
40. W. E. Byerly, "Fourier Series and Spherical, Cylindrical and Ellipsoidal Harmonics," Ginn and Co., Inc., Boston, 1893.
41. N. W. McLachlan, "Theory and Application of Mathieu Functions," Oxford University Press, London, 1947.
42. E. T. Kirkpatrick and W. F. Stokey, "Transient Heat Conduction in Elliptic Plates and Cylinders," *J. Heat Transfer, ASME,* 81, No. 1, February 1959.
43. R. V. Churchill, "Fourier Series and Boundary Value Problems," McGraw-Hill Book Company, New York, 1941.
44. K. E. Miller, "Partial Differential Equations in Engineering Problems," Prentice-Hall, Inc., New York, 1953.
45. R. V. Churchill, "Operational Mathematics," 2d ed., McGraw-Hill Book Company, New York, 1958.
46. H. S. Mickley, T. K. Sherwood, and C. E. Reed, "Applied Mathematics in Chemical Engineering," 2d ed., McGraw-Hill Book Company, New York, 1957.
47. A. Erdélyi et al., "Tables of Integral Transforms," vol. I, McGraw-Hill Book Company, New York, 1954.
48. G. A. Campbell and R. M. Foster, "Fourier Integrals for Practical Applications," D. Van Nostrand Co., Inc., New York, 1948.
49. H. Schlichting, "Boundary Layer Theory," McGraw-Hill Book Company, New York, 1955.

NOMENCLATURE

General Constants

$$A, a, B, b, C, c, D, M, m, N, n, p, s, u, v, w, \eta, \kappa, \lambda, \nu, \tau, \phi, \psi, \omega$$

General Coefficients

$$a, b, M, N$$

General Variables

$$s, t, u, v, x, y, z, \theta$$

General Functions

$$F, f, g, u, X, Y, \zeta, \Theta, \xi$$

English Symbols

a	base of logarithm
$B\,(m, n)$	beta function
$B_y(m, n)$	incomplete beta function
bei	Bessel-imaginary function
ber	Bessel-real function
Ce, Se	modified Mathieu functions of first kind
Ci	cosine integral function
c	integration constant
ce	cosine-elliptic Mathieu function of first kind
cos	cosine function
cosh	hyperbolic cosine function
D^n	nth derivative
d	differential
d/dx	total derivative with respect to x
$\partial/\partial x$	partial derivative with respect to x
Ei	exponential integral function
e	base of natural logarithm = 2.71828
erf	error function
erfc	complementary error function
$\bar{f}$	mean value of function f, Laplace transform of f
fe, ge	Mathieu functions of second kind
H	heat of decomposition
$H^{(1)}$	Hankel function of first kind
$H^{(2)}$	Hankel function of second kind
h	unit surface conductance
I	modified Bessel function of first kind
I^n	nth integral
i	unit imaginary number
J	Bessel function of first kind
K	modified Bessel function of second kind
k	thermal conductivity
kei	Kelvin-imaginary function
ker	Kelvin-real function
L, l, r	intervals
l	ablation depth
ln	natural logarithm (base e)
log	common logarithm (base 10)
P	Legendre function of first kind
Q	Legendre function of second kind
q''	instantaneous heat flux
r	absolute value of complex number, radius
r_n, x_n, z_n, μ_n	roots
Si	sine integral function
se	sine-elliptic Mathieu function of first kind
sin	sine function
sinh	hyperbolic sine function
T	$t - t_0,\ (t - t_w)/(t_a - t_w)$
t	temperature
t_a	ambient temperature
t_d	decomposition temperature
t_w	wall (surface) temperature
t_0	initial temperature ($\theta = 0$)
tan	tangent function
tanh	hyperbolic tangent function

U normalized velocity = u/u_0

u, v velocities

x distance

Y Bessel function of second kind

y_c complementary function

y_p particular integral

Greek Symbols

α thermal diffusivity

β amplitude of complex number

β, γ acute angles of right triangle

Γ gamma function

Γ_y incomplete gamma function

γ Euler's constant = 0.57722

Δ, ϵ finite increments

η similarity variable

θ time

λ, τ dummy variables

λ mean value of a differential

μ separation constant

ν order

ξ moving coordinate = $x - v\theta = x - l(\theta)$

ρ density

Miscellaneous Symbols

$\mathcal{L}$ Laplace transform

Σ sum

$\int$ integral

lim   limit

!   factorial

n 0, 1, 2, . . .

$\sqrt{}$ square root

$>$ greater than

$<$ less than

∞ infinite

Section **2**

Thermophysical Properties

WARREN IBELE

Professor of Mechanical Engineering, University of Minnesota,
Minneapolis, Minnesota

ATOMIC WEIGHTS

TABLE 1. Relative Atomic Weights 1961
    (Based on the atomic mass of $^{12}C = 12$)*

Name	Symbol	Atomic number	Atomic weight	Name	Symbol	Atomic number	Atomic weight
Actinium	Ac	89		Hydrogen	H	1	1.00797†
Aluminium	Al	13	26.9815	Indium	In	49	114.82
Americium	Am	95		Iodine	I	53	126.9044
Antimony	Sb	51	121.75	Iridium	Ir	77	192.2
Argon	Ar	18	39.948	Iron	Fe	26	55.847‡
Arsenic	As	33	74.9216	Krypton	Kr	36	83.80
Astatine	At	85		Lanthanum	La	57	138.91
Barium	Ba	56	137.34	Lead	Pb	82	207.19
Berkelium	Bk	97		Lithium	Li	3	6.939
Beryllium	Be	4	9.0122	Lutetium	Lu	71	174.97
Bismuth	Bi	83	208.980	Magnesium	Mg	12	24.312
Boron	B	5	10.811†	Manganese	Mn	25	54.9380
Bromine	Br	35	79.909‡	Mendelevium	Md	101	
Cadmium	Cd	48	112.40	Mercury	Hg	80	200.59
Caesium	Cs	55	132.905	Molybdenum	Mo	42	95.94
Calcium	Ca	20	40.08	Neodymium	Nd	60	144.24
Californium	Cf	98		Neon	Ne	10	20.183
Carbon	C	6	12.01115†	Neptunium	Np	93	
Cerium	Ce	58	140.12	Nickel	Ni	28	58.71
Chlorine	Cl	17	35.453‡	Niobium	Nb	41	92.906
Chromium	Cr	24	51.996‡	Nitrogen	N	7	14.0067
Cobalt	Co	27	58.9332	Nobelium	No	102	
Copper	Cu	29	63.54	Osmium	Os	76	190.2
Curium	Cm	96		Oxygen	O	8	15.9994†
Dysprosium	Dy	66	162.50	Palladium	Pd	46	106.4
Einsteinium	Es	99		Phosphorus	P	15	30.9738
Erbium	Er	68	167.26	Platinum	Pt	78	195.09
Europium	Eu	63	151.96	Plutonium	Pu	94	
Fermium	Fm	100		Polonium	Po	84	
Fluorine	F	9	18.9984	Potassium	K	19	39.102
Francium	Fr	87		Praseodymium	Pr	59	140.907
Gadolinium	Gd	64	157.25	Promethium	Pm	61	
Gallium	Ga	31	69.72	Protactinium	Pa	91	
Germanium	Ge	32	72.59	Radium	Ra	88	
Gold	Au	79	196.967	Radon	Rn	86	
Hafnium	Hf	72	178.49	Rhenium	Re	75	186.2
Helium	He	2	4.0026	Rhodium	Rh	45	102.905
Holmium	Ho	67	164.930	Rubidium	Rb	37	85.47

TABLE 1. Relative Atomic Weights 1961
(Based on the atomic mass of $^{12}C = 12$)* (*Continued*)

Name	Symbol	Atomic number	Atomic weight	Name	Symbol	Atomic number	Atomic weight
Ruthenium	Ru	44	101.07	Thallium	Tl	81	204.37
Samarium	Sm	62	150.35	Thorium	Th	90	232.038
Scandium	Sc	21	44.956	Thulium	Tm	69	168.934
Selenium	Se	34	78.96	Tin	Sn	50	118.69
Silicon	Si	14	28.086†	Titanium	Ti	22	47.90
Silver	Ag	47	107.870‡	Tungsten	W	74	183.85
Sodium	Na	11	22.9898	Uranium	U	92	238.03
Strontium	Sr	38	87.62	Vanadium	V	23	50.942
Sulfur	S	16	32.064†	Xenon	Xe	54	131.30
Tantalum	Ta	73	180.948	Ytterbium	Yb	70	173.04
Technetium	Tc	43		Yttrium	Y	39	88.905
Tellurium	Te	52	127.60	Zinc	Zn	30	65.37
Terbium	Tb	65	158.924	Zirconium	Zr	40	91.22

*The values for atomic weights apply to elements as they exist in nature, without artificial alteration of their isotopic composition, and, further, to natural mixtures that do not include isotopes of radiogenic origin.

†Atomic weights so designated are known to be variable because of natural variations in isotopic composition. The observed ranges are:

Hydrogen	±0.00001	Oxygen	±0.0001
Boron	±0.003	Silicon	±0.001
Carbon	±0.00005	Sulfur	±0.003

‡Atomic weights so designated are believed to have the following experimental uncertainties:

Chlorine	±0.001	Bromine	±0.002
Chromium	±0.001	Silver	±0.003
Iron	±0.003		

SOURCE: "Pure and Applied Chemistry," vol. 5, pp. 260–261, 1962.

CONVERSION FACTORS

The following tables of conversion factors are convenient. In order to convert the numerical value of a property expressed in one of the units in the left-hand column of the table to the numerical value of the same property expressed in one of the units in the top row of the table, multiply the former value by the factor in the block common to both units.

In tables involving energy, *cal* denotes the thermochemical calorie; *IT cal* denotes the International Steam Table calorie. The thermochemical calorie (cal) equals 4.184 joule. The International Steam Table calorie (IT cal) equals 4.186 joule. The Btu is the International Steam Table Btu and it equals 1055.04 joule.

TABLE 2. Conversion Factors for Mass

	lbm	slugs	gm	kg	ton
1 lbm =	1	0.03108	453.59	0.45359	0.0005
1 slug =	32.174	1	1.4594×10^4	14.594	0.016087
1 gm =	2.2046×10^{-3}	6.8521×10^{-5}	1	10^{-3}	1.1023×10^{-6}
1 kg =	2.2046	6.8521×10^{-3}	10^3	1	1.1023×10^{-3}
1 ton =	2000	61.162	9.0718×10^5	907.18	1

SOURCE: Modified and extended from "Selected Values of Properties of Hydrocarbons," National Bureau of Standards.

TABLE 3. Conversion Factors for Density

	lbm/ft³	slug/ft³	lbm/in.³	lbm/gal	gm/cc
1 lbm/ft³ =	1	0.03108	5.787×10^{-4}	0.13368	0.01602
1 slug/ft³ =	32.174	1	0.1862	4.3010	0.51543
1 lbm/in.³ =	1728	53.706	1	231	27.680
1 lbm/gal =	7.4805	0.2325	4.329×10^{-3}	1	0.11983
1 gm/cc =	62.428	1.9403	0.03613	8.345	1

SOURCE: Modified and extended from "Selected Values of Properties of Hydrocarbons," National Bureau of Standards.

TABLE 4. Conversion Factors for Pressure

	lbf/ft²	lbf/in.²	atm	in. Hg	in. H₂O	mm Hg	bar
1 lbf/ft² =	1	0.006944	4.726×10^{-4}	0.014139	0.19243	0.3591	4.788×10^{-4}
1 lbf/in.² =	144	1	0.06805	2.036	27.71	51.715	0.06895
1 atm =	2116.2	14.696	1	29.921	407.18	760	1.01325
1 in. Hg =	70.726	0.49116	0.033421	1	13.608	25.40	0.03386
1 in. H₂O =	5.197	0.036092	0.002456	0.07348	1	1.8665	0.002488
1 mm Hg =	2.7845	0.019337	1.315×10^{-3}	0.03937	0.53577	1	1.333×10^{-3}
1 mm H₂O =	2.0886×10^{4}	14.504	0.98692	29.530	401.85	750.06	1

TABLE 5. Conversion Factors for Energy

	abs joule	cal	IT cal	Btu	int. kw-hr	hp-hr	ft-lb	liter-atm
1 abs joule =	1	0.239005	0.238848	0.947827×10^{-3}	2.77731×10^{-7}	3.72505×10^{-7}	0.737561	9.86896×10^{-3}
1 cal =	4.18401	1	0.999344	3.96572×10^{-3}	1.162028×10^{-6}	1.558566×10^{-6}	3.08596	4.12918×10^{-2}
1 IT cal =	4.18676	1.000657	1	3.96832×10^{-3}	1.162791×10^{-6}	1.559590×10^{-6}	3.08799	4.13189×10^{-2}
1 Btu =	1055.045	252.161	251.996	1	2.93018×10^{-4}	3.93010×10^{-4}	778.16	10.41220
1 int. kw-hr =	3,600,612	860,565	860,000	3412.76	1	1.341247	2,655,669	35534.3
1 hp-hr =	2,684,525	641,615	641.194	2544.46	0.745575	1	1,980,000	26493.5
1 ft-lb =	1.355821	0.324048	0.323836	1.285083×10^{-3}	3.76553×10^{-7}	5.05051×10^{-7}	1	1.338054×10^{-2}
1 liter-atm =	101.3278	24.2179	24.2020	0.0960412	2.81418×10^{-5}	3.77452×10^{-5}	74.7354	1

TABLE 6. Conversion Factors for Specific Energy

		abs joule/gm	cal/gm	IT cal/gm	Btu/lb	ft-lbf/lbm	int. kw-hr/gm	hp-hr/lb	ft^2/sec^2
abs joule/gm	=	1	0.2390	0.2388	0.4299	334.53	2.777×10^{-7}	1.690×10^{-4}	10763
cal/gm	=	4.184	1	0.9993	1.7988	1399.75	1.162×10^{-6}	7.069×10^{-4}	4.504×10^4
IT cal/gm	=	4.186	1.0007	1	1.8	1400.69	1.163×10^{-6}	7.074×10^{-4}	4.506×10^4
Btu/lb	=	2.326	0.5559	0.5556	1	778.16	6.460×10^{-7}	3.930×10^{-4}	25,037
ft-lbf/lbm	=	2.989×10^{-3}	7.144×10^{-4}	7.139×10^{-4}	1.285×10^{-3}	1	8.302×10^{-10}	5.051×10^{-7}	32.174
int. kw-hr/gm	=	3.610×10^6	860,565	860,000	1.548×10^6	1.2046×10^9	1	608.4	3.876×10^{10}
hp-hr/lb	=	5919	1414.5	1413.6	2545	1.980×10^6	0.001644	1	6.370×10^7
ft^2/sec^2	=	9.291×10^{-5}	2.220×10^{-5}	2.219×10^{-5}	3.994×10^{-5}	0.03108	2.580×10^{-11}	1.567×10^{-8}	1

TABLE 7. Conversion Factors for Specific Energy per Degree

		abs joule/gm-°K	cal/gm-°K	IT cal/gm-°K	Btu/lb-°R	w-sec/kg-°K
abs joule/gm-°K	=	1	0.2390	0.2388	0.2388	10^3
cal/gm-°K	=	4.184	1	0.9993	0.9993	4184
IT cal/gm-°K	=	4.186	1.0007	1	1	4186
Btu/lb-°R	=	4.186	1.0007	1	1	4186
w-sec/kg-°K	=	10^{-3}	2.390×10^{-4}	2.388×10^{-4}	2.388×10^{-4}	1

TABLE 8. Conversion Factors for Thermal Conductivity

		cal/sec-cm-°C	Btu/hr-ft-°F	Btu/hr-ft²-°F/in.	w/cm-°C
1 cal/sec-cm-°C	=	1	241.9	2903	4.183
1 Btu/hr-ft-°F	=	4.13×10^{-3}	1	12	0.0173
1 Btu/hr-ft²-°F/in.	=	3.45×10^{-4}	0.0833	1	1.44×10^{-3}
1 w/cm²-°C	=	0.239	57.8	694	1

TABLE 9. Conversion Factors for Dynamic Viscosity

		poise or g/cm-sec, or dyn-sec/cm²	lbm/ft-hr or pdl-hr/ft²	lbm/ft-sec or pdl-sec/ft²
1 poise	=	1	242	0.0672
1 lbm/ft-hr	=	4.13×10^{-3}	1	2.78×10^{-4}
1 lbm/ft-sec	=	14.87	3600	1

TABLE 10. Conversion Factors for Kinematic Viscosity

		ft²/hr	stokes	m²/hr	m²/sec
ft²/hr	=	1	0.25806	0.092903	2.58×10^{-5}
stokes	=	3.885	1	0.36	10^{-4}
m²/hr	=	10.764	2.778	1	2.778×10^{-4}
m²/sec	=	38,750	10^4	3600	1

CRITICAL CONSTANTS

TABLE 11. Critical Constants*

Compound	T_c, °K	p_c, atm	v_c, cm³/g-mole	z_c
NORMAL PARAFFIN				
HYDROCARBONS				
1. Methane	190.7	45.8	99	0.290
2. Ethane	305.4	48.2	148	0.285
3. Propane	369.9	42.0	200	0.277
4. Butane	425.17	37.47	255	0.274
5. Pentane	469.8	33.3	311	0.269
6. Hexane	507.9	29.9	368	0.264
7. Heptane	540.2	27.0	426	0.260
8. Octane	569.4	24.64	486	0.256
9. Decane	619.	20.8	602	0.247
ISOMERIC PARAFFIN				
HYDROCARBONS				
10. Isobutane	408.1	36.0	263	0.283
11. Isopentane	461.0	32.9	308	0.268
12. Neopentane	433.8	31.6	303	0.269
13. 2-Methyl pentane	497.9	29.95	367	0.269
14. 3-Methyl pentane	504.7	30.8	367	0.273
15. 2,2-Dimethyl butane	489.4	30.7	359	0.274
16. 2,3-Dimethyl butane	500.3	30.9	358	0.269
17. 2,2,4-Trimethyl pentane	544.1	25.4	482	0.274
OLEFIN HYDROCARBONS				
18. Ethylene	283.1	50.5	124	0.270
19. Propylene	365.1	45.4	181	0.274
20. 1,3-Butadiene	425.	42.7	221	0.271
21. 1-Butene	419.6	39.7	240	0.277
ETHERS				
22. Dimethyl ether	400.1	52.6	178	0.285
23. Methyl ethyl ether	437.9	43.4	221	0.267
24. Diethyl ether	467.	35.6	281	0.261
25. Ethyl propyl ether	500.6	32.1	339	0.265
ESTERS				
26. Methyl formate	487.2	59.2	172	0.255
27. Ethyl formate	508.5	46.3	229	0.257
28. Methyl acetate	506.9	46.3	228	0.254
29. Methyl propionate	530.6	39.3	282	0.255
30. Propyl formate	538.1	40.1	285	0.259
31. Ethyl acetate	523.3	37.8	286	0.252
32. Ethyl propionate	546.1	33.0	345	0.254
33. Methyl isobutyrate	540.8	33.9	339	0.259
34. Propyl acetate	549.4	32.9	345	0.252
ALCOHOLS				
35. Methyl alcohol	513.2	78.5	118	0.222
36. Ethyl alcohol	516.3	63.0	167	0.248
37. Propyl alcohol	537.3	50.2	220	0.251

*See footnote at end of table.

TABLE 11. Critical Constants* (*Continued*)

Compound	T_c, °K	p_c, atm	v_c, cm³/g-mole	z_c
ORGANIC SULFUR AND NITROGEN COMPOUNDS				
38. Hydrogen cyanide	456.7	53.2	139	0.197
39. Methyl mercaptan	470.0	71.4	149	0.276
40. Ethyl mercaptan	499.0	54.2	207	0.274
ORGANIC SULFUR AND NITROGEN COMPOUNDS				
41. Diethyl sulfide	557.0	39.1	323	0.276
42. Diethyl amine	496.7	36.6	297(calc.)	0.286
ORGANIC HALIDES				
43. Methyl chloride	416.3	65.9	143	0.276
44. Ethyl chloride	460.4	52.0	199(calc.)	0.274
45. Fluorobenzene	559.8	44.6	271	0.263
46. Dichloroethane 1,1	523.0	50.0	244(calc.)	0.279
47. Freon 13	302.0	38.2	180	0.278
48. Chlorobenzene	632.4	44.6	308	0.265
49. Chloroform	536.6	54.0	240	0.294
50. Freon 12	384.7	39.6	218	0.273
51. Carbontetrachloride	556.4	45.0	276	0.272
52. Bromobenzene	670.2	44.6	343	0.278
53. Iodobenzene	721.0	44.6	351	0.265
OTHER ORGANIC COMPOUNDS				
54. Acetylene	309.5	61.6	113	0.274
55. Ethylene oxide	468.0	71.0	138	0.255
56. Acetone	508.7	46.6	213	0.238
57. Acetic acid	594.8	57.1	171	0.200
58. Benzene	562.6	48.6	260	0.274
INORGANIC COMPOUNDS				
59. Hydrogen	33.3	12.8	65	0.304
60. Helium	5.26	2.26	58	0.304
61. Ammonia	405.5	111.3	72.5	0.243
62. Water	647.4	218.3	56	0.230
63. Neon	44.5	46.9	41.7	0.304
64. Carbon monoxide	133.0	34.5	93	0.294
65. Nitrogen	126.2	33.5	90	0.291
66. Nitric oxide	179.2	65.0	58	0.256
67. Oxygen	154.4	49.7	74	0.290
68. Phosphine	324.5	64.5	113	0.274
69. Hydrogen sulfide	373.6	88.9	98	0.284
70. Hydrogen chloride	324.6	81.5	87	0.266
71. Argon	151.2	48.0	75	0.290
72. Carbon dioxide	304.2	72.9	94	0.275
73. Nitrous dioxide	309.7	71.6	96.3	0.272
74. Nitrogen peroxide	431.0	100.0	82	0.232
75. Sulfur dioxide	430.7	77.8	122	0.269
76. Chlorine	417.0	76.1	124	0.276
77. Carbon disulfide	552.0	78.0	170	0.293

*See footnote at end of table.

TABLE 11. Critical Constants* (*Continued*)

Compound	T_c, °K	p_c, atm	v_c, cm³/g-mole	z_c
78. Sulfur trioxide	491.4	83.8	126	0.262
79. Krypton	209.4	54.3	92.2	0.291
80. Phosgene	455.0	56.0	190	0.291
81. Xenon	289.8	58.0	118.8	0.290
82. Stannic chloride	591.9	37.0	351	0.267

*Critical constant definitions: T_c = critical temperature; p_c = critical pressure; v_c = critical volume; z_c = critical compressibility factor.

SOURCE: "Generalized Thermodynamic Properties of Pure Fluids," A. L. Lydersen, R. A. Greenkorn, and O. A. Hougen, 1955. Courtesy of the University of Wisconsin Engineering Experiment Station.

THERMODYNAMIC PROPERTIES

Gases

TABLE 12. Saturated Ammonia

Temp	Pressure		Vol-ume	Den-sity	Enthalpy from −40°F			Entropy from −40°F	
	Abs psi	Gage psi	Va-por ft³/lb	Va-por lb/ft³	Liquid Btu/lb	Vapor Btu/lb	Latent Btu/lb	Liquid Btu/lb °F	Vapor Btu/lb °F
°F									
t	p	p_d	v_g	$1/v_g$	h_f	h_g	h_{fg}	s_f	s_g
− 60	5.55	18.6*	44.73	0.02235	−21.2	589.6	610.8	−0.0517	1.4769
− 55	6.54	16.6*	38.38	0.02605	−15.9	591.6	607.5	−0.0386	1.4631
− 50	7.67	14.3*	33.08	0.03023	−10.6	593.7	604.3	−0.0256	1.4497
− 45	8.95	11.7*	28.62	0.03494	− 5.3	595.6	600.9	−0.0127	1.4368
− 40	10.41	8.7*	24.86	0.04022	0.0	597.6	597.6	0.0000	1.4242
− 38	11.04	7.4*	23.53	0.04251	2.1	598.3	596.2	0.0051	1.4193
− 36	11.71	6.1*	22.27	0.04489	4.3	599.1	594.8	0.0101	1.4144
− 34	12.41	4.7*	21.10	0.04739	6.4	599.9	593.5	0.0151	1.4096
− 32	13.14	3.2*	20.00	0.04999	8.5	600.6	592.1	0.0201	1.4048
− 30	13.90	1.6*	18.97	0.05271	10.7	601.4	590.7	0.0250	1.4001
− 28	14.71	0.0	18.00	0.05555	12.8	602.1	589.3	0.0300	1.3955
− 26	15.55	0.8	17.09	0.05850	14.9	602.8	587.9	0.0350	1.3909
− 24	16.42	1.7	16.24	0.06158	17.1	603.6	586.5	0.0399	1.3863
− 22	17.34	2.6	15.43	0.06479	19.2	604.3	585.1	0.0448	1.3818
− 20	18.30	3.6	14.68	0.06813	21.4	605.0	583.6	0.0497	1.3774
− 18	19.30	4.6	13.97	0.07161	23.5	605.7	582.2	0.0545	1.3729
− 16	20.34	5.6	13.29	0.07522	25.6	606.4	580.8	0.0594	1.3686
− 14	21.43	6.7	12.66	0.07898	27.8	607.1	579.3	0.0642	1.3643
− 12	22.56	7.9	12.06	0.08289	30.0	607.8	577.8	0.0690	1.3600
− 10	23.74	9.0	11.50	0.08695	32.1	608.5	576.4	0.0738	1.3558
− 8	24.97	10.3	10.97	0.09117	34.3	609.2	574.9	0.0786	1.3516
− 6	26.26	11.6	10.47	0.09555	36.4	609.8	573.4	0.0833	1.3474
− 4	27.59	12.9	9.991	0.1001	38.6	610.5	571.9	0.0880	1.3433
− 2	28.98	14.3	9.541	0.1048	40.7	611.1	570.4	0.0928	1.3393
0	30.42	15.7	9.116	0.1097	42.9	611.8	568.9	0.0975	1.3352
2	31.92	17.2	8.714	0.1148	45.1	612.4	567.3	0.1022	1.3312
4	33.42	18.8	8.333	0.1200	47.2	613.0	565.8	0.1069	1.3273
6	35.09	20.4	7.971	0.1254	49.4	613.6	564.2	0.1115	1.3234
8	36.77	22.1	7.629	0.1311	51.6	614.3	562.7	0.1162	1.3195

*Inches of mercury below 1 atm.

TABLE 12. Saturated Ammonia (*Continued*)

Temp	Pressure		Vol-ume	Den-sity	Enthalpy from −40°F			Entropy from −40°F	
°F	Abs psi	Gage psi	Va-por ft³/lb	Va-por lb/ft³	Liquid Btu/lb	Vapor Btu/lb	Latent Btu/lb	Liquid Btu/lb °F	Vapor Btu/lb °F
t	p	p_d	v_g	$1/v_g$	h_f	h_g	h_{fg}	s_f	s_g
10	38.51	23.8	7.304	0.1369	53.8	614.9	561.1	0.1208	1.3157
12	40.31	25.6	6.996	0.1429	56.0	615.5	559.5	0.1254	1.3118
14	42.18	27.5	6.703	0.1492	58.2	616.1	557.9	0.1300	1.3081
16	44.12	29.4	6.425	0.1556	60.3	616.6	556.3	0.1346	1.3043
18	46.13	31.4	6.161	0.1623	62.5	617.2	554.7	0.1392	1.3006
20	48.21	33.5	5.910	0.1692	64.7	617.8	553.1	0.1437	1.2969
22	50.36	35.7	5.671	0.1763	66.9	618.3	551.4	0.1483	1.2933
24	52.59	37.9	5.443	0.1837	69.1	618.9	549.8	0.1528	1.2897
26	54.90	40.2	5.227	0.1913	71.3	619.4	548.1	0.1573	1.2861
28	57.28	42.6	5.021	0.1992	73.5	619.9	546.4	0.1618	1.2825
30	59.74	45.0	4.825	0.2073	75.7	620.5	544.8	0.1663	1.2790
32	62.29	47.6	4.637	0.2156	77.9	621.0	543.1	0.1708	1.2755
34	64.91	50.2	4.459	0.2243	80.1	621.5	541.4	0.1753	1.2721
36	67.63	52.9	4.289	0.2332	82.3	622.0	539.7	0.1797	1.2686
38	70.43	55.7	4.126	0.2423	84.6	622.5	537.9	0.1841	1.2652
40	73.32	58.6	3.971	0.2518	86.8	623.0	536.2	0.1885	1.2618
42	76.31	61.6	3.823	0.2616	89.0	623.4	534.4	0.1930	1.2585
44	79.38	64.7	3.682	0.2716	91.2	623.9	532.7	0.1974	1.2552
46	82.55	67.9	3.547	0.2819	93.5	624.4	530.9	0.2018	1.2519
48	85.82	71.1	3.418	0.2926	95.7	624.8	529.1	0.2062	1.2486
50	89.19	74.5	3.294	0.3036	97.9	625.2	527.3	0.2105	1.2453
52	92.66	78.0	3.176	0.3149	100.2	625.7	525.5	0.2149	1.2421
54	96.23	81.5	3.063	0.3265	102.4	626.1	523.7	0.2192	1.2389
56	99.91	85.2	2.954	0.3385	104.7	626.5	521.8	0.2236	1.2357
58	103.7	89.0	2.851	0.3508	106.9	626.9	520.0	0.2279	1.2325
60	107.6	92.9	2.751	0.3635	109.2	627.3	518.1	0.2322	1.2294
62	111.6	96.9	2.656	0.3765	111.5	627.7	516.2	0.2365	1.2262
64	115.7	101.0	2.565	0.3899	113.7	628.0	514.3	0.2408	1.2231
66	120.0	105.3	2.477	0.4037	116.0	628.4	512.4	0.2451	1.2201
68	124.3	109.6	2.393	0.4179	118.3	628.8	510.5	0.2494	1.2170
70	128.8	114.1	2.312	0.4325	120.5	629.1	508.6	0.2537	1.2140
72	133.4	118.7	2.235	0.4474	122.8	629.4	506.6	0.2579	1.2110
74	138.1	123.4	2.161	0.4628	125.1	629.8	504.7	0.2622	1.2080
76	143.0	128.3	2.089	0.4786	127.4	630.1	502.7	0.2664	1.2050
78	147.9	133.2	2.021	0.4949	129.7	630.4	500.7	0.2706	1.2020
80	153.0	138.3	1.955	0.5115	132.0	630.7	498.7	0.2749	1.1991
82	158.3	143.6	1.892	0.5287	134.3	631.0	496.7	0.2791	1.1962
84	163.7	149.0	1.831	0.5462	136.6	631.3	494.7	0.2833	1.1933
86	169.2	154.5	1.772	0.5643	138.9	631.5	492.6	0.2875	1.1904
88	174.8	160.1	1.716	0.5828	141.2	631.8	490.6	0.2917	1.1875
90	180.6	165.9	1.661	0.6019	143.5	632.0	488.5	0.2958	1.1846
92	186.6	171.9	1.609	0.6214	145.8	632.2	486.4	0.3000	1.1818
94	192.7	178.0	1.559	0.6415	148.2	632.5	484.3	0.3041	1.1789
96	198.9	184.2	1.510	0.6620	150.5	632.6	482.1	0.3083	1.1761
98	205.3	190.6	1.464	0.6832	152.9	632.9	480.0	0.3125	1.1733
100	211.9	197.2	1.419	0.7048	155.2	633.0	477.8	0.3166	1.1705
102	218.6	203.9	1.375	0.7270	157.6	633.2	475.6	0.3207	1.1677

TABLE 12. Saturated Ammonia *(Continued)*

Temp	Pressure		Vol-ume	Den-sity	Enthalpy from −40°F			Entropy from −40°F	
	Abs	Gage	Vapor	Vapor	Liquid	Vapor	Latent	Liquid Btu/lb	Vapor Btu/lb
°F	psi	psi	ft³/lb	lb/ft³	Btu/lb	Btu/lb	Btu/lb	°F	°F
t	p	p_d	v_g	$1/v_g$	h_f	h_g	h_{fg}	s_f	s_g
104	225.4	210.7	1.334	0.7498	159.9	633.4	473.5	0.3248	1.1649
106	232.5	217.8	1.293	0.7732	162.3	633.5	471.2	0.3289	1.1621
108	239.7	225.0	1.254	0.7972	164.6	633.6	469.0	0.3330	1.1593
110	247.0	232.3	1.217	0.8219	167.0	633.7	466.7	0.3372	1.1566
112	254.5	239.8	1.180	0.8471	169.4	633.8	464.4	0.3413	1.1538
114	262.2	247.5	1.145	0.8730	171.8	633.9	462.1	0.3453	1.1510
116	270.1	255.4	1.112	0.8996	174.2	634.0	459.8	0.3495	1.1483
118	278.2	263.5	1.079	0.9269	176.6	634.0	457.4	0.3535	1.1455
120	286.4	271.7	1.047	0.9549	179.0	634.0	455.0	0.3576	1.1427
122	294.8	280.1	1.017	0.9837	181.4	634.0	452.6	0.3618	1.1400
124	303.4	288.7	0.987	1.0132	183.9	634.0	450.1	0.3659	1.1372

SOURCE: Abridged, by permission, from "Tables of Thermodynamic Properties of Ammonia," U.S. Department of Commerce, National Bureau of Standards, Circular 142, 1945.

TABLE 13. Superheated Ammonia

[Absolute pressure, psi (saturation temperature in italics)]

Temp °F	50 *21.67*			60 *30.21*			70 *37.70*			80 *44.40*		
t	v	h	s	v	h	s	v	h	s	v	h	s
Sat.	5.710	618.2	1.2939	4.805	620.5	1.2787	4.151	622.4	1.2658	3.655	624.0	1.2545
30	5.838	623.4	1.3046									
40	5.988	629.5	1.3169	4.933	626.8	1.2913	4.177	623.9	1.2688			
50	6.135	635.4	1.3286	5.060	632.9	1.3035	4.290	630.4	1.2816	3.712	627.7	1.2619
60	6.280	641.2	1.3399	5.184	639.0	1.3152	4.401	636.6	1.2937	3.812	634.3	1.2745
70	6.423	646.9	1.3508	5.307	644.9	1.3265	4.509	642.7	1.3054	3.909	640.6	1.2866
80	6.564	652.6	1.3613	5.428	650.7	1.3373	4.615	648.7	1.3166	4.005	646.7	1.2981
90	6.704	658.2	1.3716	5.547	656.4	1.3479	4.719	654.6	1.3274	4.098	652.8	1.3092
100	6.843	663.7	1.3816	5.665	662.1	1.3581	4.822	660.4	1.3378	4.190	658.7	1.3199
110	6.980	669.2	1.3914	5.781	667.7	1.3681	4.924	666.1	1.3480	4.281	664.6	1.3303
120	7.117	674.7	1.4009	5.897	673.3	1.3778	5.025	671.8	1.3579	4.371	670.4	1.3404
130	7.252	680.2	1.4103	6.012	678.9	1.3873	5.125	677.5	1.3676	4.460	676.1	1.3502
140	7.387	685.7	1.4195	6.126	684.4	1.3966	5.224	683.1	1.3770	4.548	681.8	1.3598
150	7.521	691.1	1.4286	6.239	689.9	1.4058	5.323	688.7	1.3863	4.635	687.5	1.3692
160	7.655	696.6	1.4374	6.352	695.5	1.4148	5.420	694.3	1.3954	4.722	693.2	1.3784
170	7.788	702.1	1.4462	6.464	701.0	1.4236	5.518	699.9	1.4043	4.808	698.8	1.3874
180	7.921	707.5	1.4548	6.576	706.5	1.4323	5.615	705.5	1.4131	4.893	704.4	1.3963
190	8.053	713.0	1.4633	6.687	712.0	1.4409	5.711	711.0	1.4217	4.978	710.0	1.4050
200	8.185	718.5	1.4716	6.798	717.5	1.4493	5.807	716.6	1.4302	5.063	715.6	1.4136
210	8.317	724.0	1.4799	6.909	723.1	1.4576	5.902	722.2	1.4386	5.147	721.3	1.4220
220	8.448	729.4	1.4880	7.019	728.6	1.4658	5.998	727.7	1.4469	5.231	726.9	1.4304
240	8.710	740.5	1.5040	7.238	739.7	1.4819	6.187	738.9	1.4631	5.398	738.1	1.4467
260	8.970	751.6	1.5197	7.457	750.9	1.4976	6.376	750.1	1.4789	5.565	749.4	1.4626
280	9.230	762.7	1.5350	7.675	762.1	1.5130	6.563	761.4	1.4943	5.730	760.7	1.4781
300	9.489	774.0	1.5500	7.892	773.3	1.5281	6.750	772.7	1.5095	5.894	772.1	1.4933

TABLE 13. Superheated Ammonia *(Continued)*

[Absolute pressure, psi (saturation temperature in italics)]

Temp °F	90 *50.47*			100 *56.05*			120 *66.02*			140 *74.79*		
t	v	h	s	v	h	s	v	h	s	v	h	s
Sat.	3.266	625.3	1.2445	2.952	626.5	1.2356	2.476	628.4	1.2201	2.132	629.9	1.2068
50												
60	3.353	631.8	1.2571	2.985	629.3	1.2409						
70	3.443	638.3	1.2695	3.068	636.0	1.2539	2.505	631.3	1.2255			
80	3.529	644.7	1.2814	3.149	642.6	1.2661	2.576	638.3	1.2386	2.166	633.8	1.2140
90	3.614	650.9	1.2928	3.227	649.0	1.2778	2.645	645.0	1.2510	2.228	640.9	1.2272
100	3.698	657.0	1.3038	3.304	655.2	1.2891	2.712	651.6	1.2628	2.288	647.8	1.2396
110	3.780	663.0	1.3144	3.380	661.3	1.2999	2.778	658.0	1.2741	2.347	654.5	1.2515
120	3.862	668.9	1.3247	3.454	667.3	1.3104	2.842	664.2	1.2850	2.404	661.1	1.2628
130	3.942	674.7	1.3347	3.527	673.3	1.3206	2.905	670.4	1.2956	2.460	667.4	1.2738
140	4.021	680.5	1.3444	3.600	679.2	1.3305	2.967	676.5	1.3058	2.515	673.7	1.2843
150	4.100	686.3	1.3539	3.672	685.0	1.3401	3.029	682.5	1.3157	2.569	679.9	1.2945
160	4.178	692.0	1.3633	3.743	690.8	1.3495	3.089	688.4	1.3254	2.622	686.0	1.3045
170	4.255	697.7	1.3724	3.813	696.6	1.3588	3.149	694.3	1.3348	2.675	692.0	1.3141
180	4.332	703.4	1.3813	3.883	702.3	1.3678	3.209	700.2	1.3441	2.727	698.0	1.3236
190	4.408	709.0	1.3901	3.952	708.0	1.3767	3.268	706.0	1.3531	2.779	704.0	1.3328
200	4.484	714.7	1.3988	4.021	713.7	1.3854	3.326	711.8	1.3620	2.830	709.9	1.3418
210	4.560	720.4	1.4073	4.090	719.4	1.3940	3.385	717.6	1.3707	2.880	715.8	1.3507
220	4.635	726.0	1.4157	4.158	725.1	1.4024	3.442	723.4	1.3793	2.931	721.6	1.3594
230	4.710	731.7	1.4239	4.226	730.8	1.4108	3.500	729.2	1.3877	2.981	727.5	1.3679
240	4.785	737.3	1.4321	4.294	736.5	1.4190	3.557	734.9	1.3960	3.030	733.3	1.3763
250	4.859	743.0	1.4401	4.361	742.2	1.4271	3.614	740.7	1.4042	3.080	739.2	1.3846
260	4.933	748.7	1.4481	4.428	747.9	1.4350	3.671	746.5	1.4123	3.129	745.0	1.3928
280	5.081	760.0	1.4637	4.562	759.4	1.4507	3.783	758.0	1.4281	3.227	756.7	1.4088
300	5.228	771.5	1.4789	4.695	770.8	1.4660	3.895	769.6	1.4435	3.323	768.3	1.4243

[Absolute pressure, psi (saturation temperature in italics)]

Temp °F	160 *82.64*			180 *89.78*			200 *96.34*			220 *102.42*		
t	v	h	s	v	h	s	v	h	s	v	h	s
Sat.	1.872	631.1	1.1952	1.667	632.0	1.1850	1.502	632.7	1.1756	1.367	633.2	1.1671
90	1.914	636.6	1.2055	1.668	632.2	1.1853						
100	1.969	643.9	1.2186	1.720	639.9	1.1992	1.520	635.6	1.1809			
110	2.023	651.0	1.2311	1.770	647.3	1.2123	1.567	643.4	1.1947	1.400	639.4	1.1781
120	2.075	657.8	1.2429	1.818	654.4	1.2247	1.612	650.9	1.2077	1.443	647.3	1.1917
130	2.125	664.4	1.2542	1.865	661.3	1.2364	1.656	658.1	1.2200	1.485	654.8	1.2045
140	2.175	670.9	1.2652	1.910	668.0	1.2477	1.698	665.0	1.2317	1.525	662.0	1.2167
150	2.224	677.2	1.2757	1.955	674.6	1.2586	1.740	671.8	1.2429	1.564	669.0	1.2281
160	2.272	683.5	1.2859	1.999	681.0	1.2691	1.780	678.4	1.2537	1.601	675.8	1.2394
170	2.319	689.7	1.2958	2.042	687.3	1.2792	1.820	684.9	1.2641	1.638	682.5	1.2501
180	2.365	695.8	1.3054	2.084	693.6	1.2891	1.859	691.3	1.2742	1.675	689.1	1.2604
190	2.411	701.9	1.3148	2.126	699.8	1.2987	1.897	697.7	1.2840	1.710	695.5	1.2704
200	2.457	707.9	1.3240	2.167	705.9	1.3081	1.935	703.9	1.2935	1.745	701.9	1.2801
210	2.502	713.9	1.3331	2.208	712.0	1.3172	1.972	710.1	1.3029	1.780	708.2	1.2896
220	2.547	719.9	1.3419	2.248	718.1	1.3262	2.009	716.3	1.3120	1.814	714.4	1.2989
230	2.591	725.8	1.3506	2.288	724.1	1.3350	2.046	722.4	1.3209	1.848	720.6	1.3079
240	2.635	731.7	1.3591	2.328	730.1	1.3436	2.082	728.4	1.3296	1.881	726.8	1.3168

TABLE 13. Superheated Ammonia (*Continued*)

[Absolute pressure, psi (saturation temperature in italics)]

Temp °F	160 *82.64*			180 *89.78*			200 *96.34*			220 *102.42*		
t	v	h	s	v	h	s	v	h	s	v	h	s
Sat.	1.872	631.1	1.1952	1.667	632.0	1.1850	1.502	632.7	1.1756	1.367	633.2	1.1671
250	2.679	737.6	1.3675	2.367	736.1	1.3521	2.118	734.5	1.3382	1.914	732.9	1.3255
260	2.723	743.5	1.3757	2.407	742.0	1.3605	2.154	740.5	1.3467	1.947	739.0	1.3340
270	2.766	749.4	1.3838	2.446	748.0	1.3687	2.189	746.5	1.3550	1.980	745.1	1.3424
280	2.809	755.3	1.3919	2.484	753.9	1.3768	2.225	752.5	1.3631	2.012	751.1	1.3507
290	2.852	761.2	1.3998	2.523	759.9	1.3847	2.260	758.5	1.3712	2.044	757.2	1.3588
300	2.895	767.1	1.4076	1.561	765.8	1.3926	2.295	764.5	1.3791	2.076	763.2	1.3668
320	2.980	778.9	1.4229	2.637	777.7	1.4081	2.364	776.5	1.3947	2.140	775.3	1.3825
340	3.064	790.7	1.4379	2.713	789.6	1.4231	2.432	788.5	1.4099	2.203	787.4	1.3978
360							2.500	800.5	1.4247	2.265	799.5	1.4127
380							2.568	812.5	1.4392	2.327	811.6	1.4273

[Absolute pressure, psi (saturation temperature in italics)]

Temp °F	240 *108.09*			260 *113.42*			280 *118.45*			300 *123.21*		
t	v	h	s	v	h	s	v	h	s	v	h	s
Sat.	1.253	633.6	1.1592	1.155	633.9	1.1518	1.072	634.0	1.1449	0.999	634.0	1.1383
110	1.261	635.3	1.1621									
120	1.302	643.5	1.1764	1.182	639.5	1.1617	1.078	635.4	1.1473			
130	1.342	651.3	1.1898	1.220	647.8	1.1757	1.115	644.0	1.1621	1.023	640.1	1.1487
140	1.380	658.8	1.2025	1.257	655.6	1.1889	1.151	652.2	1.1759	1.058	648.7	1.1632
150	1.416	666.1	1.2145	1.292	663.1	1.2014	1.184	660.1	1.1888	1.091	656.9	1.1767
160	1.452	673.1	1.2259	1.326	670.4	1.2132	1.217	667.6	1.2011	1.123	664.7	1.1894
170	1.487	680.0	1.2369	1.359	677.5	1.2245	1.249	674.9	1.2127	1.153	672.2	1.2014
180	1.521	686.7	1.2475	1.391	684.4	1.2354	1.279	681.9	1.2239	1.183	679.5	1.2129
190	1.554	693.3	1.2577	1.422	691.1	1.2458	1.309	688.9	1.2346	1.211	686.5	1.2239
200	1.587	699.8	1.2677	1.453	697.7	1.2560	1.339	695.6	1.2449	1.239	693.5	1.2344
210	1.619	706.2	1.2773	1.484	704.3	1.2658	1.367	702.3	1.2550	1.267	700.3	1.2447
220	1.651	712.6	1.2867	1.514	710.7	1.2754	1.396	708.8	1.2647	1.294	706.9	1.2546
230	1.683	718.9	1.2959	1.543	717.1	1.2847	1.424	715.3	1.2742	1.320	713.5	1.2642
240	1.714	725.1	1.3049	1.572	723.4	1.2938	1.451	721.8	1.2834	1.346	720.0	1.2736
250	1.745	731.3	1.3137	1.601	729.7	1.3027	1.478	728.1	1.2924	1.372	726.5	1.2827
260	1.775	737.5	1.3224	1.630	736.0	1.3115	1.505	734.4	1.3013	1.397	732.9	1.2917
270	1.805	743.6	1.3308	1.658	742.2	1.3200	1.532	740.7	1.3099	1.422	739.2	1.3004
280	1.835	749.8	1.3392	1.686	748.4	1.3285	1.558	747.0	1.3184	1.447	745.5	1.3090
290	1.865	755.9	1.3474	1.714	754.5	1.3367	1.584	753.2	1.3268	1.472	751.8	1.3175
300	1.895	762.0	1.3554	1.741	760.7	1.3449	1.610	759.4	1.3350	1.496	758.1	1.3257
320	1.954	774.1	1.3712	1.796	772.9	1.3608	1.661	771.7	1.3511	1.544	770.5	1.3419
340	2.012	786.3	1.3866	1.850	785.2	1.3763	1.712	784.0	1.3667	1.592	782.9	1.3576
360	2.069	798.4	1.4016	1.904	797.4	1.3914	1.762	796.3	1.3819	1.639	795.3	1.3729
380	2.126	810.6	1.4163	1.957	809.6	1.4062	1.811	808.7	1.3967	1.686	807.7	1.3878
400				2.009	821.9	1.4206	1.861	821.0	1.4112	1.732	820.1	1.4024

TABLE 14. Air Tables* (for 1 pound)

T °F abs	t °F	h Btu/lb	p_r	u Btu/lb	v_r	ϕ Btu/lb °F
100	-360	23.7	0.00384	16.9	9643	0.1971
120	-340	28.5	0.00726	20.3	6123	0.2408
140	-320	33.3	0.01244	23.7	4170	0.2777
160	-300	38.1	0.01982	27.1	2990	0.3096
180	-280	42.9	0.0299	30.6	2230	0.3378
200	-260	47.7	0.0432	34.0	1715	0.3630
220	-240	52.5	0.0603	37.4	1352	0.3858
240	-220	57.2	0.0816	40.8	1089	0.4067
260	-200	62.0	0.1080	44.2	892	0.4258
280	-180	66.8	0.1399	47.6	742	0.4436
300	-160	71.6	0.1780	51.0	624	0.4601
320	-140	76.4	0.2229	54.5	532	0.4755
340	-120	81.2	0.2754	57.9	457	0.4900
360	-100	86.0	0.336	61.3	397	0.5037
380	- 80	90.8	0.406	64.7	347	0.5166
400	- 60	95.5	0.486	68.1	305	0.5289
420	- 40	100.3	0.576	71.5	270	0.5406
440	- 20	105.1	0.678	74.9	241	0.5517
460	0	109.9	0.791	78.4	215.3	0.5624
480	20	114.7	0.918	81.8	193.6	0.5726
500	40	119.5	1.059	85.2	174.9	0.5823
520	60	124.3	1.215	88.6	158.6	0.5917
540	80	129.1	1.386	92.0	144.3	0.6008
560	100	133.9	1.574	95.5	131.8	0.6095
580	120	138.7	1.780	98.9	120.7	0.6179
600	140	143.5	2.00	102.3	110.9	0.6261
620	160	148.3	2.25	105.8	102.1	0.6340
640	180	153.1	2.51	109.2	94.3	0.6416
660	200	157.9	2.80	112.7	87.3	0.6490
680	220	162.7	3.11	116.1	81.0	0.6562
700	240	167.6	3.45	119.6	75.2	0.6632
720	260	172.4	3.81	123.0	70.1	0.6700
740	280	177.2	4.19	126.5	65.4	0.6766
760	300	182.1	4.61	130.0	61.1	0.6831
780	320	186.9	5.05	133.5	57.2	0.6894
800	340	191.8	5.53	137.0	53.6	0.6956
820	360	196.7	6.03	140.5	50.4	0.7016
840	380	201.6	6.57	144.0	47.3	0.7075
860	400	206.5	7.15	147.5	44.6	0.7132
880	420	211.4	7.76	151.0	42.0	0.7189
900	440	216.3	8.41	154.6	39.6	0.7244
920	460	221.2	9.10	158.1	37.4	0.7298
940	480	226.1	9.83	161.7	35.4	0.7351
960	500	231.1	10.61	165.3	33.5	0.7403
980	520	236.0	11.43	168.8	31.8	0.7454

*See footnote at end of table.

TABLE 14. Air Tables* (for 1 pound) (*Continued*)

T °F abs	t °F	h Btu/lb	p_r	u Btu/lb	v_r	ϕ Btu/lb °F
1000	540	241.0	12.30	172.4	30.1	0.7504
1020	560	246.0	13.22	176.0	28.6	0.7554
1040	580	251.0	14.18	179.7	27.2	0.7602
1060	600	256.0	15.20	183.3	25.8	0.7650
1080	620	261.0	16.28	186.9	24.6	0.7696
1100	640	266.0	17.41	190.6	23.4	0.7743
1120	660	271.0	18.60	194.2	22.3	0.7788
1140	680	276.1	19.86	197.9	21.3	0.7833
1160	700	281.1	21.2	201.6	20.29	0.7877
1180	720	286.2	22.6	205.3	19.38	0.7920
1200	740	291.3	24.0	209.0	18.51	0.7963
1220	760	296.4	25.5	212.8	17.70	0.8005
1240	780	301.5	27.1	216.5	16.93	0.8047
1260	800	306.6	28.8	220.3	16.20	0.8088
1280	820	311.8	30.6	224.0	15.52	0.8128
1300	840	316.9	32.4	227.8	14.87	0.8168
1320	860	322.1	34.3	231.6	14.25	0.8208
1340	880	327.3	36.3	235.4	13.67	0.8246
1360	900	332.5	38.4	239.2	13.12	0.8285
1380	920	337.7	40.6	243.1	12.59	0.8323
1400	940	342.9	42.9	246.9	12.10	0.8360
1420	960	348.1	45.3	250.8	11.62	0.8398
1440	980	353.4	47.8	254.7	11.17	0.8434
1460	1000	358.6	50.3	258.5	10.74	0.8470
1480	1020	363.9	53.0	262.4	10.34	0.8506
1500	1040	369.2	55.9	266.3	9.95	0.8542
1520	1060	374.5	58.8	270.3	9.58	0.8577
1540	1080	379.8	61.8	274.2	9.23	0.8611
1560	1100	385.1	65.0	278.1	8.89	0.8646
1580	1120	390.4	68.3	282.1	8.57	0.8679
1600	1140	395.7	71.7	286.1	8.26	0.8713
1620	1160	401.1	75.3	290.0	7.97	0.8746
1640	1180	406.4	79.0	294.0	7.69	0.8779
1660	1200	411.8	82.8	298.0	7.42	0.8812
1680	1220	417.2	86.8	302.0	7.17	0.8844
1700	1240	422.6	91.0	306.1	6.92	0.8876
1720	1260	428.0	95.2	310.1	6.69	0.8907
1740	1280	433.4	99.7	314.1	6.46	0.8939
1760	1300	438.8	104.3	318.2	6.25	0.8970
1780	1320	444.3	109.1	322.2	6.04	0.9000
1800	1340	449.7	114.0	326.3	5.85	0.9031
1820	1360	455.2	119.2	330.4	5.66	0.9061
1840	1380	460.6	124.5	334.5	5.48	0.9091
1860	1400	466.1	130.0	338.6	5.30	0.9120
1880	1420	471.6	135.6	342.7	5.13	0.9150
1900	1440	477.1	141.5	346.8	4.97	0.9179
1920	1460	482.6	147.6	351.0	4.82	0.9208
1940	1480	488.1	153.9	355.1	4.67	0.9236
1960	1500	493.6	160.4	359.3	4.53	0.9264
1980	1520	499.2	167.1	363.4	4.39	0.9293

*See footnote at end of table.

TABLE 14. Air Tables* (for 1 pound) (*Continued*)

T °F abs	t °F	h Btu/lb	p_r	u Btu/lb	v_r	ϕ Btu/lb °F
2000	1540	504.7	174.0	367.6	4.26	0.9320
2020	1560	510.3	181.2	371.8	4.13	0.9348
2040	1580	515.8	188.5	376.0	4.01	0.9376
2060	1600	521.4	196.2	380.2	3.89	0.9403
2080	1620	527.0	204.0	384.4	3.78	0.9430
2100	1640	532.6	212	388.6	3.67	0.9456
2120	1660	538.2	220	392.8	3.56	0.9483
2140	1680	543.7	229	397.0	3.46	0.9509
2160	1700	549.4	238	401.3	3.36	0.9535
2180	1720	555.0	247	405.5	3.27	0.9561
2200	1740	560.6	257	409.8	3.18	0.9587
2220	1760	566.2	266	414.0	3.09	0.9612
2240	1780	571.9	276	418.3	3.00	0.9638
2260	1800	577.5	287	422.6	2.92	0.9663
2280	1820	583.2	297	426.9	2.84	0.9688
2300	1840	588.8	308	431.2	2.76	0.9712
2320	1860	594.5	319	435.5	2.69	0.9737
2340	1880	600.2	331	439.8	2.62	0.9761
2360	1900	605.8	343	444.1	2.55	0.9785
2380	1920	611.5	355	448.4	2.48	0.9809
2400	1940	617.2	368	452.7	2.42	0.9833
2420	1960	622.9	380	457.0	2.36	0.9857
2440	1980	628.6	394	461.4	2.30	0.9880
2460	2000	634.3	407	465.7	2.24	0.9904
2480	2020	640.0	421	470.0	2.18	0.9927
2500	2040	645.8	436	474.4	2.12	0.9950
2520	2060	651.5	450	478.8	2.07	0.9972
2540	2080	657.2	466	483.1	2.02	0.9995
2560	2100	663.0	481	487.5	1.971	1.0018
2580	2120	668.7	497	491.9	1.922	1.0040
2600	2140	674.5	514	496.3	1.876	1.0062
2620	2160	680.2	530	500.6	1.830	1.0084
2640	2180	686.0	548	505.0	1.786	1.0106
2660	2200	691.8	565	509.4	1.743	1.0128
2680	2220	697.6	583	513.8	1.702	1.0150
2700	2240	703.4	602	518.3	1.662	1.0171
2720	2260	709.1	621	522.7	1.623	1.0193
2740	2280	714.9	640	527.1	1.585	1.0214
2760	2300	720.7	660	531.5	1.548	1.0235
2780	2320	726.5	681	536.0	1.512	1.0256
2800	2340	723.3	702	540.4	1.478	1.0277
2820	2360	738.2	724	544.8	1.444	1.0297
2840	2380	744.0	746	549.3	1.411	1.0318
2860	2400	749.8	768	553.7	1.379	1.0338
2880	2420	755.6	791	558.2	1.348	1.0359
2900	2440	761.4	815	562.7	1.318	1.0379
2920	2460	767.3	839	567.1	1.289	1.0399
2940	2480	773.1	864	571.6	1.261	1.0419
2960	2500	779.0	889	576.1	1.233	1.0439
2980	2520	784.8	915	580.6	1.206	1.0458

*See footnote at end of table.

TABLE 14. Air Tables* (for 1 pound) (*Continued*)

T °F abs	t °F	h Btu/lb	p_r	u Btu/lb	v_r	ϕ Btu/lb °F
3000	2540	790.7	941	585.0	1.180	1.0478
3020	2560	796.5	969	589.5	1.155	1.0497
3040	2580	802.4	996	594.0	1.130	1.0517
3060	2600	808.3	1025	598.5	1.106	1.0536
3080	2620	814.2	1054	603.0	1.083	1.0555
3100	2640	820.0	1083	607.5	1.060	1.0574
3120	2660	825.9	1114	612.0	1.038	1.0593
3140	2680	831.8	1145	616.6	1.016	1.0612
3160	2700	837.7	1176	621.1	0.995	1.0630
3180	2720	843.6	1209	625.6	0.975	1.0649
3200	2740	849.5	1242	630.1	0.955	1.0668
3220	2760	855.4	1276	634.6	0.935	1.0686
3240	2780	861.3	1310	639.2	0.916	1.0704
3260	2800	867.2	1345	643.7	0.898	1.0722
3280	2820	873.1	1381	648.3	0.880	1.0740
3300	2840	879.0	1418	652.8	0.862	1.0758
3320	2860	884.9	1455	657.4	0.845	1.0776
3340	2880	890.9	1494	661.9	0.828	1.0794
3360	2900	896.8	1533	666.5	0.812	1.0812
3380	2920	902.7	1573	671.0	0.796	1.0830
3400	2940	908.7	1613	675.6	0.781	1.0847
3420	2960	914.6	1655	680.2	0.766	1.0864
3440	2980	920.6	1697	684.8	0.751	1.0882
3460	3000	926.5	1740	689.3	0.736	1.0899
3480	3020	932.4	1784	693.9	0.722	1.0916
3500	3040	938.4	1829	698.5	0.709	1.0933
3520	3060	944.4	1875	703.1	0.695	1.0950
3540	3080	950.3	1922	707.6	0.682	1.0967
3560	3100	956.3	1970	712.2	0.670	1.0984
3580	3120	962.2	2018	716.8	0.637	1.1000
3600	3140	968.2	2068	721.4	0.645	1.1017
3620	3160	974.2	2118	726.0	0.633	1.1034
3640	3180	980.2	2167	730.6	0.621	1.1050
3660	3200	986.1	2222	735.3	0.610	1.1066
3680	3220	992.12	2276	739.9	0.599	1.1083

SOURCE: Abridged from "Gas Tables" by J. H. Keenan and J. Kaye, by permission of John Wiley & Sons, Inc., New York, 1948.

*Air Table Definitions:

$$P_r = \exp\left[\frac{1}{R}\int_{T_0}^{T} c_p \frac{dT}{T}\right] \quad v_r = \frac{v}{v_0} = \frac{T}{T_0} \cdot \frac{1}{P_r}$$

$$\phi = \int_{T_0}^{T} c_p \frac{dT}{T}$$

TABLE 15. Thermodynamic Properties of Moist Air (Standard Atmospheric Pressure, 29.921 in. Hg)

Temp. t, °F	Humidity Ratio $W_s \times 10^5$	Volume ft³/lb dry air			Enthalpy Btu/lb dry air			Entropy Btu/°F-lb dry air			Condensed Water		
		v_a	v_{as}	v_s	h_a	h_{as}	h_s	s_a	s_{as}	s_s	Enthalpy Btu/lb h_w	Entropy Btu/°F-lb s_w	Vapor Pressure in. Hg $p_s \times 10^3$
-160	0.0002120	7.520	0.000	7.520	-38.504	0.000	-38.504	-0.10300	0.00000	-0.10300	-222.00	-0.4907	0.0001009
-155	0.0003869	7.647	0.000	7.647	-37.296	0.000	-37.296	-0.09901	0.00000	-0.09901	-220.40	-0.4853	0.0001842
-150	0.0006932	7.775	0.000	7.775	-36.088	0.000	-36.088	-0.09508	0.00000	-0.09508	-218.77	-0.4800	0.0003301
-145	0.001219	7.902	0.000	7.902	-34.881	0.000	-34.881	-0.09121	0.00000	-0.09121	-217.12	-0.4747	0.0005807
-140	0.002109	8.029	0.000	8.029	-33.674	0.000	-33.674	-0.08740	0.00000	-0.08740	-215.44	-0.4695	0.001004
-135	0.003586	8.156	0.000	8.156	-32.468	0.000	-32.468	-0.08365	0.00000	-0.08365	-213.75	-0.4642	0.001707
-130	0.006000	8.283	0.000	8.283	-31.262	0.000	-31.262	-0.07997	0.00000	-0.07997	-212.03	-0.4590	0.002858
-125	0.009887	8.411	0.000	8.411	-30.057	0.000	-30.057	-0.07634	0.00000	-0.07634	-210.28	-0.4538	0.004710
-120	0.01606	8.537	0.000	8.537	-28.852	0.000	-28.852	-0.07277	0.00000	-0.07277	-208.52	-0.4485	0.007653
-115	0.02571	8.664	0.000	8.664	-27.648	0.000	-27.648	-0.06924	0.00000	-0.06924	-206.73	-0.4433	0.01226
-110	0.04063	8.792	0.000	8.792	-26.444	0.000	-26.444	-0.06577	0.00000	-0.06577	-204.92	-0.4381	0.01939
-105	0.06340	8.919	0.000	8.919	-25.240	0.001	-25.239	-0.06234	0.00000	-0.06234	-203.09	-0.4329	0.03026
-100	0.09772	9.046	0.000	9.046	-24.037	0.001	-24.036	-0.05897	0.00000	-0.05897	-201.23	-0.4277	0.04666
-95	0.1489	9.173	0.000	9.173	-22.835	0.002	-22.833	-0.05565	0.00000	-0.05565	-199.35	-0.4225	0.07111
-90	0.2242	9.300	0.000	9.300	-21.631	0.002	-21.629	-0.05237	0.00001	-0.05236	-197.44	-0.4173	0.1071
-85	0.3342	9.426	0.000	9.426	-20.428	0.003	-20.425	-0.04913	0.00001	-0.04912	-195.51	-0.4121	0.1597
-80	0.4930	9.553	0.000	9.553	-19.225	0.005	-19.220	-0.04595	0.00001	-0.04594	-193.55	-0.4069	0.2356
-75	0.7196	9.680	0.000	9.680	-18.022	0.007	-18.015	-0.04280	0.00002	-0.04278	-191.57	-0.4017	0.3441
-70	1.040	9.806	0.000	9.806	-16.820	0.011	-16.809	-0.03969	0.00003	-0.03966	-189.56	-0.3965	0.4976
-65	1.491	9.932	0.000	9.932	-15.617	0.015	-15.602	-0.03663	0.00005	-0.03658	-187.53	-0.3913	0.7130
-60	2.118	10.059	0.000	10.059	-14.416	0.022	-14.394	-0.03360	0.00006	-0.03354	-185.47	-0.3861	1.0127
-55	2.982	10.186	0.000	10.186	-13.214	0.031	-13.183	-0.03061	0.00009	-0.03052	-183.39	-0.3810	1.4258
-50	4.163	10.313	0.001	10.314	-12.012	0.043	-11.969	-0.02766	0.00012	-0.02754	-181.29	-0.3758	1.9910
-48	4.747	10.364	0.001	10.365	-11.532	0.049	-11.483	-0.02649	0.00013	-0.02636	-180.44	-0.3738	2.2702
-46	5.406	10.414	0.001	10.415	-11.051	0.056	-10.995	-0.02532	0.00014	-0.02518	-179.59	-0.3717	2.5854
-44	6.149	10.465	0.001	10.466	-10.571	0.064	-10.507	-0.02416	0.00016	-0.02400	-178.73	-0.3696	2.9408

TABLE 15. Thermodynamic Properties of Moist Air (Standard Atmospheric Pressure, 29,921 in. Hg) (Continued)

Temp. t, °F	Humidity Ratio $W_s \times 10^5$	Volume ft³/lb dry air			Enthalpy Btu/lb dry air			Entropy Btu/°F-lb dry air			Condensed Water		
		v_a	v_{as}	v_s	h_a	h_{as}	h_s	s_a	s_{as}	s_s	Enthalpy Btu/lb h_w	Entropy Btu/°F-lb s_w	Vapor Pressure in. Hg $p_s \times 10^3$
−42	6.985	10.516	0.001	10.517	−10.090	0.073	−10.017	−0.02301	0.00019	−0.02282	−177.87	−0.3676	3.3408
−40	7.925	10.566	0.001	10.567	−9.609	0.083	−9.526	−0.02186	0.00021	−0.02165	−177.01	−0.3655	3.7906
−38	8.980	10.617	0.002	10.619	−9.129	0.094	−9.035	−0.02072	0.00024	−0.02048	−176.14	−0.3634	4.2958
−36	10.16	10.668	0.002	10.670	−8.648	0.106	−8.542	−0.01958	0.00026	−0.01932	−175.27	−0.3614	4.8626
−34	11.49	10.718	0.002	10.720	−8.168	0.121	−8.047	−0.01845	0.00030	−0.01815	−174.40	−0.3593	5.4980
−32	12.98	10.769	0.002	10.771	−7.687	0.136	−7.551	−0.01733	0.00034	−0.01699	−173.52	−0.3573	6.2093
−30	14.64	10.820	0.002	10.822	−7.207	0.154	−7.053	−0.01621	0.00038	−0.01583	−172.64	−0.3552	7.0046
−28	16.49	10.870	0.003	10.873	−6.726	0.173	−6.553	−0.01509	0.00043	−0.01466	−171.75	−0.3531	7.8928
−26	18.56	10.921	0.003	10.924	−6.246	0.196	−6.050	−0.01398	0.00048	−0.01350	−170.86	−0.3511	8.8838
−24	20.87	10.972	0.004	10.976	−5.765	0.219	−5.546	−0.01287	0.00054	−0.01233	−169.97	−0.3490	9.9885
−22	23.44	11.022	0.004	11.026	−5.285	0.246	−5.039	−0.01177	0.00061	−0.01116	−169.07	−0.3469	11.219
−20	26.30	11.073	0.005	11.078	−4.804	0.277	−4.527	−0.01067	0.00068	−0.00999	−168.17	−0.3449	12.587
−19	27.85	11.098	0.005	11.103	−4.564	0.293	−4.271	−0.01012	0.00072	−0.00940	−167.72	−0.3439	13.327
−18	29.48	11.124	0.005	11.129	−4.324	0.310	−4.014	−0.00958	0.00076	−0.00882	−167.26	−0.3428	14.107
−17	31.20	11.149	0.006	11.155	−4.083	0.328	−3.755	−0.00904	0.00080	−0.00824	−166.81	−0.3418	14.929
−16	33.01	11.174	0.006	11.180	−3.843	0.348	−3.495	−0.00850	0.00084	−0.00766	−166.35	−0.3408	15.795
−15	34.91	11.200	0.006	11.206	−3.603	0.368	−3.235	−0.00796	0.00089	−0.00707	−165.90	−0.3398	16.706
−14	36.92	11.225	0.007	11.232	−3.363	0.389	−2.974	−0.00743	0.00094	−0.00649	−165.44	−0.3387	17.666
−13	39.03	11.250	0.007	11.257	−3.123	0.412	−2.711	−0.00689	0.00099	−0.00590	−164.98	−0.3377	18.677
−12	41.25	11.275	0.008	11.283	−2.882	0.436	−2.446	−0.00636	0.00104	−0.00532	−164.52	−0.3367	19.740
−11	43.59	11.301	0.008	11.309	−2.642	0.461	−2.181	−0.00582	0.00109	−0.00473	−164.06	−0.3357	20.859
−10	46.06	11.326	0.008	11.334	−2.402	0.487	−1.915	−0.00529	0.00115	−0.00414	−163.60	−0.3346	22.035
−9	48.65	11.351	0.008	11.359	−2.162	0.514	−1.648	−0.00475	0.00121	−0.00354	−163.14	−0.3336	23.272
−8	51.37	11.376	0.009	11.385	−1.922	0.543	−1.379	−0.00422	0.00128	−0.00294	−162.67	−0.3326	24.573
−7	54.23	11.401	0.010	11.411	−1.681	0.574	−1.107	−0.00369	0.00135	−0.00234	−162.21	−0.3316	25.940
−6	57.24	11.427	0.010	11.437	−1.441	0.606	−0.835	−0.00316	0.00142	−0.00174	−161.74	−0.3305	27.377

Temp. t, °F	Humidity Ratio $W_s \times 10^4$	Volume ft³/lb dry air			Enthalpy Btu/lb dry air			Entropy Btu/°F-lb dry air			Condensed Water		
		v_a	v_{as}	v_s	h_a	h_{as}	h_s	s_a	s_{as}	s_s	Enthalpy Btu/lb h_w	Entropy Btu/°F-lb s_w	Vapor Pressure in. Hg $p_s \times 10^2$
-5	60.40	11.452	0.011	11.463	-1.201	0.639	-0.562	-0.00263	0.00149	-0.00114	-161.28	-0.3295	28.886
-4	63.71	11.477	0.012	11.489	-0.961	0.675	-0.286	-0.00210	0.00157	-0.00053	-160.81	-0.3285	30.472
-3	67.20	11.502	0.013	11.515	-0.721	0.712	-0.009	-0.00157	0.00165	0.00008	-160.34	-0.3275	32.137
-2	70.85	11.528	0.013	11.541	-0.480	0.751	0.271	-0.00105	0.00174	0.00069	-159.87	-0.3264	33.885
-1	74.69	11.553	0.014	11.567	-0.240	0.792	0.552	-0.00052	0.00183	0.00131	-159.40	-0.3254	35.720
0	78.72	11.578	0.015	11.593	0.000	0.835	0.835	0.00000	0.00192	0.00192	-158.93	-0.3244	37.645
1	8.295	11.604	0.015	11.619	0.240	0.880	1.120	0.00052	0.00202	0.00254	-158.46	-0.3234	3.9666
2	8.739	11.629	0.016	11.645	0.480	0.928	1.408	0.00104	0.00212	0.00316	-157.99	-0.3223	4.1785
3	9.204	11.654	0.017	11.671	0.721	0.977	1.698	0.00156	0.00223	0.00379	-157.52	-0.3213	4.4007
4	9.692	11.679	0.018	11.697	0.961	1.030	1.991	0.00208	0.00234	0.00442	-157.04	-0.3203	4.6337
5	10.20	11.705	0.019	11.724	1.201	1.085	2.286	0.00260	0.00246	0.00506	-156.57	-0.3193	4.8779
6	10.74	11.730	0.020	11.750	1.441	1.142	2.583	0.00312	0.00258	0.00570	-156.09	-0.3182	5.1339
7	11.30	11.756	0.021	11.777	1.681	1.202	2.883	0.00364	0.00271	0.00635	-155.61	-0.3172	5.4022
8	11.89	11.781	0.022	11.803	1.922	1.266	3.188	0.00415	0.00285	0.00700	-155.13	-0.3162	5.6832
9	12.51	11.806	0.024	11.830	2.162	1.332	3.494	0.00467	0.00299	0.00766	-154.65	-0.3152	5.9776
10	13.15	11.831	0.025	11.856	2.402	1.401	3.803	0.00518	0.00314	0.00832	-154.17	-0.3141	6.2858
11	13.83	11.857	0.026	11.883	2.642	1.474	4.116	0.00569	0.00330	0.00899	-153.69	-0.3131	6.6085
12	14.54	11.882	0.028	11.910	2.882	1.550	4.432	0.00620	0.00346	0.00966	-153.21	-0.3121	6.9462
13	15.28	11.907	0.029	11.936	3.123	1.630	4.753	0.00671	0.00363	0.01034	-152.73	-0.3111	7.2997
14	16.06	11.933	0.030	11.963	3.363	1.713	5.076	0.00721	0.00380	0.01101	-152.24	-0.3100	7.6696
15	16.87	11.958	0.032	11.990	3.603	1.800	5.403	0.00772	0.00399	0.01171	-151.76	-0.3090	8.0565
16	17.72	11.983	0.034	12.017	3.843	1.892	5.735	0.00822	0.00418	0.01240	-151.27	-0.3080	8.4612
17	18.61	12.009	0.035	12.044	4.083	1.988	6.071	0.00873	0.00438	0.01311	-150.78	-0.3070	8.8843
18	19.53	12.034	0.038	12.072	4.324	2.088	6.412	0.00923	0.00459	0.01382	-150.29	-0.3059	9.3267
19	20.51	12.059	0.040	12.099	4.564	2.192	6.756	0.00973	0.00481	0.01454	-149.80	-0.3049	9.7889
20	21.52	12.084	0.042	12.126	4.804	2.302	7.106	0.01023	0.00504	0.01527	-149.31	-0.3039	10.272
21	22.58	12.110	0.044	12.154	5.044	2.416	7.460	0.01073	0.00528	0.01601	-148.82	-0.3029	10.777

TABLE 15. Thermodynamic Properties of Moist Air (Standard Atmospheric Pressure, 29.921 in. Hg) (Continued)

Temp. t, °F	Humidity Ratio $W_s \times 10^4$	Volume ft³/lb dry air			Enthalpy Btu/lb dry air			Entropy Btu/°F-lb dry air			Condensed Water		
		v_a	v_{as}	v_s	h_a	h_{as}	h_s	s_a	s_{as}	s_s	Enthalpy Btu/lb h_w	Entropy Btu/°F-lb s_w	Vapor Pressure in. Hg $p_s \times 10^2$
22	23.69	12.135	0.046	12.181	5.284	2.536	7.820	0.01123	0.00553	0.01676	−148.33	−0.3018	11.305
23	24.85	12.160	0.049	12.209	5.525	2.661	8.186	0.01173	0.00579	0.01752	−147.84	−0.3008	11.856
24	26.06	12.186	0.051	12.237	5.765	2.792	8.557	0.01223	0.00607	0.01830	−147.34	−0.2998	12.431
25	27.33	12.211	0.054	12.265	6.005	2.929	8.934	0.01273	0.00635	0.01908	−146.85	−0.2988	13.032
26	28.65	12.236	0.057	12.293	6.245	3.072	9.317	0.01322	0.00665	0.01987	−146.35	−0.2977	13.659
27	30.03	12.262	0.059	12.321	6.485	3.221	9.706	0.01372	0.00696	0.02068	−145.85	−0.2967	14.313
28	31.47	12.287	0.062	12.349	6.726	3.377	10.103	0.01421	0.00728	0.02149	−145.36	−0.2957	14.966
29	32.97	12.312	0.065	12.377	6.966	3.540	10.506	0.01470	0.00761	0.02231	−144.86	−0.2947	15.709
30	34.54	12.338	0.068	12.406	7.206	3.709	10.915	0.01519	0.00796	0.02315	−144.36	−0.2936	16.452
31	36.17	12.363	0.071	12.434	7.446	3.887	11.333	0.01568	0.00832	0.02400	−143.86	−0.2926	17.227
32	37.88	12.388	0.075	12.463	7.686	4.072	11.758	0.01617	0.00870	0.02487	−143.36	−0.2916	18.035
32*	37.88	12.388	0.075	12.463	7.686	4.072	11.758	0.01617	0.00870	0.02487	0.04	0.0000	18.037
33	39.44	12.413	0.079	12.492	7.927	4.242	12.169	0.01666	0.00904	0.02570	1.05	0.0020	18.778
34	41.07	12.438	0.082	12.520	8.167	4.418	12.585	0.01715	0.00940	0.02655	2.06	0.0041	19.546
35	42.75	12.464	0.085	12.549	8.407	4.601	13.008	0.01764	0.00977	0.02741	3.06	0.0061	20.342
36	44.50	12.489	0.089	12.578	8.647	4.791	13.438	0.01812	0.01016	0.02828	4.07	0.0081	21.166
37	46.31	12.514	0.093	12.607	8.887	4.987	13.874	0.01861	0.01056	0.02917	5.07	0.0102	22.020
38	48.18	12.540	0.097	12.637	9.128	5.191	14.319	0.01909	0.01097	0.03006	6.08	0.0122	22.904
39	50.12	12.565	0.101	12.666	9.368	5.403	14.771	0.01957	0.01139	0.03096	7.08	0.0142	23.819
40	52.13	12.590	0.105	12.695	9.608	5.622	15.230	0.02005	0.01183	0.03188	8.09	0.0162	24.767
41	54.21	12.616	0.109	12.725	9.848	5.849	15.697	0.02053	0.01228	0.03281	9.09	0.0182	25.748
42	56.38	12.641	0.114	12.755	10.088	6.084	16.172	0.02101	0.01275	0.03376	10.09	0.0202	26.763
43	58.60	12.666	0.119	12.785	10.329	6.328	16.657	0.02149	0.01323	0.03472	11.10	0.0222	27.813
44	60.91	12.691	0.124	12.815	10.569	6.580	17.149	0.02197	0.01373	0.03570	12.10	0.0242	28.899
45	63.31	12.717	0.129	12.846	10.809	6.841	17.650	0.02245	0.01425	0.03670	13.10	0.0262	30.023

| Temp. t, °F | Humidity Ratio $W_s \times 10^3$ | Volume ft³/lb dry air | | | Enthalpy Btu/lb dry air | | | Entropy Btu/°F-lb dry air | | | Condensed Water | | Vapor Pressure in. Hg |
		v_a	v_{as}	v_s	h_a	h_{as}	h_s	s_a	s_{as}	s_s	Enthalpy Btu/lb h_w	Entropy Btu/°F-lb s_w	p_s
46	6.578	12.742	0.134	12.876	11.049	7.112	18.161	0.02293	0.01478	0.03771	14.10	0.0282	0.31185
47	6.835	12.767	0.140	12.907	11.289	7.391	18.680	0.02340	0.01534	0.03874	15.11	0.0302	0.32386
48	7.100	12.792	0.146	12.938	11.530	7.681	19.211	0.02387	0.01591	0.03978	16.11	0.0321	0.33629
49	7.374	12.818	0.151	12.969	11.770	7.981	19.751	0.02434	0.01650	0.04084	17.11	0.0341	0.34913
50	7.658	12.843	0.158	13.001	12.010	8.291	20.301	0.02481	0.01711	0.04192	18.11	0.0361	0.36240
51	7.952	12.868	0.164	13.032	12.250	8.612	20.862	0.02528	0.01774	0.04302	19.11	0.0381	0.37611
52	8.256	12.894	0.170	13.064	12.491	8.945	21.436	0.02575	0.01839	0.04414	20.11	0.0400	0.39028
53	8.569	12.919	0.178	13.097	12.731	9.289	22.020	0.02622	0.01906	0.04528	21.12	0.0420	0.40492
54	8.894	12.944	0.185	13.129	12.971	9.644	22.615	0.02669	0.01976	0.04645	22.12	0.0439	0.42004
55	9.229	12.970	0.192	13.162	13.211	10.01	23.22	0.02716	0.02047	0.04763	23.12	0.0459	0.43565
56	9.575	12.995	0.200	13.195	13.452	10.39	23.84	0.02762	0.02121	0.04883	24.12	0.0478	0.45176
57	9.934	13.020	0.208	13.228	13.692	10.79	24.48	0.02809	0.02197	0.05006	25.12	0.0497	0.46840
58	10.30	13.045	0.216	13.261	13.932	11.19	25.12	0.02855	0.02276	0.05131	26.12	0.0517	0.48558
59	10.69	13.071	0.224	13.295	14.172	11.61	25.78	0.02902	0.02357	0.05259	27.12	0.0536	0.50330
60	11.08	13.096	0.233	13.329	14.413	12.05	26.46	0.02948	0.02441	0.05389	28.12	0.0555	0.52159
61	11.49	13.121	0.242	13.363	14.653	12.50	27.15	0.02994	0.02527	0.05521	29.12	0.0574	0.54047
62	11.91	13.147	0.251	13.398	14.893	12.96	27.85	0.03040	0.02616	0.05656	30.12	0.0594	0.55994
63	12.35	13.172	0.261	13.433	15.134	13.44	28.57	0.03086	0.02708	0.05794	31.12	0.0613	0.58002
64	12.80	13.197	0.271	13.468	15.374	13.94	29.31	0.03132	0.02803	0.05935	32.12	0.0632	0.60073
65	13.26	13.222	0.282	13.504	15.614	14.45	30.06	0.03177	0.02901	0.06078	33.11	0.0651	0.62209
66	13.74	13.247	0.292	13.539	15.855	14.98	30.83	0.03223	0.03002	0.06225	34.11	0.0670	0.64411
67	14.24	13.273	0.303	13.576	16.095	15.53	31.62	0.03269	0.03106	0.06375	35.11	0.0689	0.66681
68	14.75	13.298	0.315	13.613	16.335	16.09	32.42	0.03314	0.03213	0.06527	36.11	0.0708	0.69019
69	15.28	13.323	0.327	13.650	16.576	16.67	33.25	0.03360	0.03323	0.06683	37.11	0.0727	0.71430
70	15.82	13.348	0.339	13.687	16.816	17.27	34.09	0.03405	0.03437	0.06842	38.11	0.0746	0.73915
71	16.39	13.373	0.351	13.724	17.056	17.89	34.95	0.03450	0.03554	0.07004	39.11	0.0765	0.76475
72	16.97	13.398	0.364	13.762	17.297	18.53	35.83	0.03495	0.03675	0.07170	40.11	0.0784	0.79112

*Extrapolated to represent metastable equilibrium with undercooled liquid.

TABLE 15. Thermodynamic Properties of Moist Air (Standard Atmospheric Pressure, 29.921 in. Hg) *(Continued)*

Temp. t, °F	Humidity Ratio $W_s \times 10^3$	Volume ft³/lb dry air			Enthalpy Btu/lb dry air			Entropy Btu/°F-lb dry air			Condensed Water		
		v_a	v_{as}	v_s	h_a	h_{as}	h_s	s_a	s_{as}	s_s	Enthalpy Btu/lb h_w	Entropy Btu/°F-lb s_w	Vapor Pressure in. Hg p_s
73	17.57	13.424	0.377	13.801	17.537	19.20	36.74	0.03540	0.03800	0.07340	41.11	0.0803	0.81828
74	18.19	13.449	0.392	13.841	17.778	19.88	37.66	0.03585	0.03928	0.07513	42.10	0.0821	0.84624
75	18.82	13.474	0.407	13.881	18.018	20.59	38.61	0.03630	0.04060	0.07690	43.10	0.0840	0.87504
76	19.48	13.499	0.422	13.921	18.259	21.31	39.57	0.03675	0.04197	0.07872	44.10	0.0859	0.90470
77	20.16	13.525	0.437	13.962	18.499	22.07	40.57	0.03720	0.04337	0.08057	45.10	0.0877	0.93523
78	20.86	13.550	0.453	14.003	18.740	22.84	41.58	0.03765	0.04482	0.08247	46.10	0.0896	0.96665
79	21.58	13.575	0.470	14.045	18.980	23.64	42.62	0.03810	0.04631	0.08441	47.10	0.0914	0.99899
80	22.33	13.601	0.486	14.087	19.221	24.47	43.69	0.03854	0.04784	0.08638	48.10	0.0933	1.0323
81	23.10	13.626	0.504	14.130	19.461	25.32	44.78	0.03899	0.04942	0.08841	49.09	0.0952	1.0665
82	23.89	13.651	0.523	14.174	19.702	26.20	45.90	0.03943	0.05105	0.09048	50.09	0.0970	1.1017
83	24.71	13.676	0.542	14.218	19.942	27.10	47.04	0.03987	0.05273	0.09260	51.09	0.0989	1.1379
84	25.55	13.702	0.560	14.262	20.183	28.04	48.22	0.04031	0.05446	0.09477	52.09	0.1007	1.1752
85	26.42	13.727	0.581	14.308	20.423	29.01	49.43	0.04075	0.05624	0.09699	53.09	0.1025	1.2135
86	27.31	13.752	0.602	14.354	20.663	30.00	50.66	0.04119	0.05807	0.09926	54.08	0.1043	1.2529
87	28.24	13.777	0.624	14.401	20.904	31.03	51.93	0.04163	0.05995	0.10158	55.08	0.1062	1.2934
88	29.19	13.803	0.645	14.448	21.144	32.09	53.23	0.04207	0.06189	0.10396	56.08	0.1080	1.3351
89	30.17	13.828	0.668	14.496	21.385	33.18	54.56	0.04251	0.06389	0.10640	57.08	0.1098	1.3779
90	31.18	13.853	0.692	14.545	21.625	34.31	55.93	0.04295	0.06596	0.10890	58.08	0.1116	1.4219
91	32.23	13.879	0.716	14.595	21.865	35.47	57.33	0.04339	0.06807	0.11146	59.07	0.1135	1.4671
92	33.30	13.904	0.741	14.645	22.106	36.67	58.78	0.04382	0.07025	0.11407	60.07	0.1153	1.5135
93	34.41	13.929	0.768	14.697	22.346	37.90	60.25	0.04426	0.07249	0.11675	61.07	0.1171	1.5612
94	35.56	13.954	0.795	14.749	22.587	39.18	61.77	0.04469	0.07480	0.11949	62.07	0.1188	1.6102
95	36.73	13.980	0.822	14.802	22.827	40.49	63.32	0.04513	0.07718	0.12231	63.07	0.1206	1.6606
96	37.95	14.005	0.851	14.856	23.068	41.85	64.92	0.04556	0.07963	0.12519	64.06	0.1224	1.7123
97	39.20	14.030	0.881	14.911	23.308	43.24	66.55	0.04600	0.08215	0.12815	65.06	0.1242	1.7654
98	40.49	14.056	0.911	14.967	23.548	44.68	68.23	0.04643	0.08474	0.13117	66.06	0.1260	1.8199

Temp. t, °F	Humidity Ratio $W_s \times 10$	Volume ft³/lb dry air v_a	v_{as}	v_s	Enthalpy Btu/lb dry air h_a	h_{as}	h_s	Entropy Btu/°F-lb dry air s_a	s_{as}	s_s	Condensed Water Enthalpy Btu/lb h_w	Entropy Btu/°F-lb s_w	Vapor Pressure in. Hg p_s
99	41.82	14.081	0.942	15.023	23.789	46.17	69.96	0.04686	0.08741	0.13427	67.06	0.1278	1.8759
100	43.19	14.106	0.975	15.081	24.029	47.70	71.73	0.04729	0.09016	0.13745	68.06	0.1296	1.9333
101	44.60	14.131	1.009	15.140	24.270	49.28	73.55	0.04772	0.09299	0.14071	69.05	0.1314	1.9923
102	46.06	14.157	1.043	15.200	24.510	50.91	75.42	0.04815	0.09591	0.14406	70.05	0.1332	2.0528
103	47.56	14.182	1.079	15.261	24.751	52.59	77.34	0.04858	0.09891	0.14749	71.05	0.1350	2.1149
104	49.11	14.207	1.117	15.324	24.991	54.32	79.31	0.04900	0.1020	0.1510	72.05	0.1367	2.1786
105	50.70	14.232	1.155	15.387	25.232	56.11	81.34	0.04943	0.1052	0.1546	73.04	0.1385	2.2439
106	52.34	14.258	1.194	15.452	25.472	57.95	83.42	0.04985	0.1085	0.1584	74.04	0.1403	2.3109
107	54.04	14.283	1.235	15.518	25.713	59.85	85.56	0.05028	0.1118	0.1621	75.04	0.1421	2.3797
108	55.78	14.308	1.278	15.586	25.953	61.80	87.76	0.05070	0.1153	0.1660	76.04	0.1438	2.4502
109	57.58	14.333	1.321	15.654	26.194	63.82	90.03	0.05113	0.1189	0.1700	77.04	0.1456	2.5225
110	59.44	14.359	1.365	15.724	26.434	65.91	92.34	0.05155	0.1226	0.1742	78.03	0.1472	2.5966
111	61.35	14.384	1.412	15.796	26.675	68.05	94.72	0.05197	0.1264	0.1784	79.03	0.1491	2.6726
112	63.33	14.409	1.460	15.869	26.915	70.27	97.18	0.05239	0.1302	0.1826	80.03	0.1508	2.7505
113	65.36	14.435	1.509	15.944	27.156	72.55	99.71	0.05281	0.1342	0.1870	81.03	0.1525	2.8304
114	67.46	14.460	1.560	16.020	27.397	74.91	102.31	0.05323	0.1384	0.1916	82.03	0.1543	2.9123
115	69.62	14.485	1.613	16.098	27.637	77.34	104.98	0.05365	0.1426	0.1963	83.02	0.1560	2.9962
116	71.85	14.510	1.668	16.178	27.878	79.85	107.73	0.05407	0.1470	0.2011	84.02	0.1577	3.0821
117	74.15	14.536	1.723	16.259	28.119	82.43	110.55	0.05449	0.1515	0.2060	85.02	0.1595	3.1701
118	76.52	14.561	1.782	16.343	28.359	85.10	113.46	0.05490	0.1562	0.2111	86.02	0.1612	3.2603
119	78.97	14.586	1.842	16.428	28.600	87.86	116.46	0.05532	0.1610	0.2163	87.02	0.1629	3.3527
120	0.8149	14.611	1.905	16.516	28.841	90.70	119.54	0.05573	0.1659	0.2216	88.01	0.1646	3.4474
121	0.8410	14.637	1.968	16.605	29.082	93.64	122.72	0.05615	0.1710	0.2272	89.01	0.1664	3.5443
122	0.8678	14.662	2.034	16.696	29.322	96.66	125.98	0.05656	0.1763	0.2329	90.01	0.1681	3.6436
123	0.8955	14.687	2.103	16.790	29.563	99.79	129.35	0.05698	0.1817	0.2387	91.01	0.1698	3.7452
124	0.9242	14.712	2.174	16.886	29.804	103.0	132.8	0.05739	0.1872	0.2446	92.01	0.1715	3.8493
125	0.9537	14.738	2.247	16.985	30.044	106.4	136.4	0.05780	0.1930	0.2508	93.01	0.1732	3.9558

TABLE 15. Thermodynamic Properties of Moist Air (Standard Atmospheric Pressure, 29.921 in. Hg) *(Continued)*

Temp. t, °F	Humidity Ratio $W_s \times 10$	Volume ft³/lb dry air			Enthalpy Btu/lb dry air			Entropy Btu/°F-lb dry air			Condensed Water		
		v_a	v_{as}	v_s	h_a	h_{as}	h_s	s_a	s_{as}	s_s	Enthalpy Btu/lb h_w	Entropy Btu/°F-lb s_w	Vapor Pressure in. Hg p_s
126	0.9841	14.763	2.323	17.086	30.285	109.8	140.1	0.05821	0.1989	0.2571	94.01	0.1749	4.0649
127	1.016	14.788	2.401	17.189	30.526	113.4	143.9	0.05862	0.2050	0.2636	95.00	0.1766	4.1765
128	1.048	14.813	2.482	17.295	30.766	117.0	147.8	0.05903	0.2113	0.2703	96.00	0.1783	4.2907
129	1.082	14.839	2.565	17.404	31.007	120.8	151.8	0.05944	0.2178	0.2772	97.00	0.1800	4.4076
130	1.116	14.864	2.652	17.516	31.248	124.7	155.9	0.05985	0.2245	0.2844	98.00	0.1817	4.5272
131	1.152	14.889	2.742	17.631	31.489	128.8	160.3	0.06026	0.2314	0.2917	99.00	0.1834	4.6495
132	1.189	14.915	2.834	17.749	31.729	133.0	164.7	0.06067	0.2386	0.2993	100.00	0.1851	4.7747
133	1.227	14.940	2.930	17.870	31.970	137.3	169.3	0.06108	0.2459	0.3070	101.00	0.1868	4.9028
134	1.267	14.965	3.029	17.994	32.211	141.8	174.0	0.06148	0.2536	0.3151	102.00	0.1885	5.0337
135	1.308	14.990	3.132	18.122	32.452	146.4	178.9	0.06189	0.2614	0.3233	103.00	0.1902	5.1676
136	1.350	15.016	3.237	18.253	32.692	151.2	183.9	0.06229	0.2695	0.3318	104.00	0.1918	5.3046
137	1.393	15.041	3.348	18.389	32.933	156.1	189.0	0.06270	0.2778	0.3405	105.00	0.1935	5.4446
138	1.439	15.066	3.462	18.528	33.174	161.2	194.4	0.06310	0.2865	0.3496	106.00	0.1952	5.5878
139	1.485	15.091	3.580	18.671	33.414	166.5	199.9	0.06350	0.2954	0.3589	107.00	0.1969	5.7342
140	1.534	15.117	3.702	18.819	33.655	172.0	205.7	0.06390	0.3047	0.3686	107.99	0.1985	5.8838
141	1.584	15.142	3.829	18.971	33.896	177.7	211.6	0.06430	0.3142	0.3785	108.99	0.2002	6.0367
142	1.636	15.167	3.961	19.128	34.136	183.6	217.7	0.06470	0.3241	0.3888	109.99	0.2018	6.1930
143	1.689	15.192	4.098	19.290	34.377	189.7	224.1	0.06510	0.3343	0.3994	110.99	0.2035	6.3527
144	1.745	15.218	4.239	19.457	34.618	196.0	230.6	0.06549	0.3449	0.4104	111.99	0.2051	6.5160
145	1.803	15.243	4.386	19.629	34.859	202.5	237.4	0.06589	0.3559	0.4218	112.99	0.2068	6.6828
146	1.862	15.268	4.539	19.807	35.099	209.3	244.4	0.06629	0.3672	0.4335	113.99	0.2084	6.8532
147	1.924	15.293	4.698	19.991	35.340	216.4	251.7	0.06669	0.3790	0.4457	114.99	0.2101	7.0273
148	1.989	15.319	4.862	20.181	35.581	223.7	259.3	0.06708	0.3912	0.4583	115.99	0.2117	7.2051
149	2.055	15.344	5.033	20.377	35.822	231.3	267.1	0.06748	0.4038	0.4713	116.99	0.2134	7.3867
150	2.125	15.369	5.211	20.580	36.063	239.2	275.3	0.06787	0.4169	0.4848	117.99	0.2150	7.5722
151	2.197	15.394	5.396	20.790	36.304	247.3	283.6	0.06827	0.4304	0.4987	118.99	0.2167	7.7616

152	2.271	15.420	5.587	21.007	36.545	255.9	292.4	0.06866	0.4445	0.5132	119.99	0.2183	7.9550
153	2.349	15.445	5.788	21.233	36.785	264.7	301.5	0.06906	0.4591	0.5282	120.99	0.2200	8.1525
154	2.430	15.470	5.996	21.466	37.026	273.9	310.9	0.06945	0.4743	0.5438	121.99	0.2216	8.3541
155	2.514	15.496	6.213	21.709	37.267	283.5	320.8	0.06984	0.4901	0.5599	122.99	0.2232	8.5599
156	2.602	15.521	6.439	21.960	37.508	293.5	331.0	0.07023	0.5066	0.5768	123.99	0.2248	8.7701
157	2.693	15.546	6.675	22.221	37.749	303.9	341.7	0.07062	0.5237	0.5943	124.99	0.2265	8.9846
158	2.788	15.571	6.922	22.493	37.990	314.7	352.7	0.07101	0.5415	0.6125	125.99	0.2281	9.2036
159	2.887	15.597	7.178	22.775	38.231	326.0	364.2	0.07140	0.5600	0.6314	127.00	0.2297	9.4271
160	2.990	15.622	7.446	23.068	38.472	337.8	376.3	0.07179	0.5793	0.6511	128.00	0.2313	9.6556
161	3.098	15.647	7.727	23.374	38.713	350.1	388.8	0.07218	0.5994	0.6716	129.00	0.2329	9.8876
162	3.211	15.672	8.020	23.692	38.954	363.0	402.0	0.07257	0.6204	0.6930	130.00	0.2345	10.125
163	3.329	15.698	8.326	24.024	39.195	376.5	415.7	0.07296	0.6423	0.7153	131.00	0.2361	10.367
164	3.452	15.723	8.648	24.371	39.436	390.5	429.9	0.07334	0.6652	0.7385	132.00	0.2377	10.614
165	3.581	15.748	8.985	24.733	39.677	405.3	445.0	0.07373	0.6892	0.7629	133.00	0.2393	10.866
166	3.716	15.773	9.339	25.112	39.918	420.8	460.7	0.07411	0.7142	0.7883	134.00	0.2409	11.123
167	3.858	15.799	9.708	25.507	40.159	437.0	477.2	0.07450	0.7405	0.8150	135.01	0.2426	11.385
168	4.007	15.824	10.098	25.922	40.400	454.0	494.4	0.07488	0.7680	0.8429	136.01	0.2441	11.652
169	4.163	15.849	10.508	26.357	40.641	471.8	512.4	0.07527	0.7969	0.8722	137.01	0.2457	11.925
170	4.327	15.874	10.938	26.812	40.882	490.6	531.5	0.07565	0.8273	0.9030	138.01	0.2473	12.203
171	4.500	15.900	11.391	27.291	41.123	510.4	551.5	0.07603	0.8592	0.9352	139.01	0.2489	12.486
172	4.682	15.925	11.870	27.795	41.364	531.3	572.7	0.07641	0.8927	0.9691	140.01	0.2505	12.775
173	4.875	15.950	12.376	28.326	41.605	553.3	594.9	0.07680	0.9281	1.0049	141.01	0.2521	13.069
174	5.078	15.975	12.911	28.886	41.846	576.5	618.3	0.07718	0.9654	1.0426	142.02	0.2537	13.369
175	5.292	16.001	13.475	29.476	42.087	601.1	643.2	0.07756	1.005	1.083	143.02	0.2553	13.675
176	5.519	16.026	14.074	30.100	42.328	627.1	669.4	0.07794	1.047	1.125	144.02	0.2568	13.987
177	5.760	16.051	14.710	30.761	42.569	654.7	697.3	0.07832	1.091	1.169	145.02	0.2584	14.304
178	6.016	16.076	15.386	31.462	42.810	684.1	726.9	0.07870	1.137	1.216	146.03	0.2600	14.628
179	6.288	16.102	16.104	32.206	43.051	715.2	758.3	0.07908	1.187	1.266	147.03	0.2616	14.958
180	6.578	16.127	16.870	32.997	43.292	748.5	791.8	0.07946	1.240	1.319	148.03	0.2631	15.294
181	6.887	16.152	17.689	33.841	43.534	783.9	827.4	0.07984	1.296	1.376	149.03	0.2647	15.636
182	7.218	16.177	18.565	34.742	43.775	821.9	865.1	0.08021	1.357	1.437	150.04	0.2662	15.985
183	7.572	16.203	19.504	35.707	44.016	862.5	906.5	0.08059	1.421	1.502	151.04	0.2678	16.340
184	7.953	16.228	20.513	36.741	44.257	906.2	950.5	0.08096	1.490	1.571	152.04	0.2693	16.702
185	8.363	16.253	21.601	37.854	44.498	953.2	997.7	0.08134	1.565	1.646	153.05	0.2709	17.071

TABLE 15. Thermodynamic Properties of Moist Air* (Standard Atmospheric Pressure, 29,921 in. Hg.) (Continued)

Temp. t, °F	Humidity Ratio $W_s \times 10$	Volume ft³/lb dry air			Enthalpy Btu/lb dry air			Entropy Btu/°F-lb dry air			Condensed Water		
		v_a	v_{as}	v_s	h_a	h_{as}	h_s	s_a	s_{as}	s_s	Enthalpy Btu/lb h_w	Entropy Btu/°F-lb s_w	Vapor Pressure in. Hg p_s
186	8.805	16.278	22.775	39.053	44.740	1004	1049	0.08171	1.645	1.727	154.05	0.2724	17.446
187	9.283	16.304	24.047	40.351	44.981	1059	1104	0.08208	1.731	1.813	155.05	0.2740	17.828
188	9.802	16.329	25.427	41.756	45.222	1119	1164	0.08245	1.825	1.907	156.06	0.2755	18.217
189	10.37	16.354	26.934	43.288	45.463	1184	1229	0.08283	1.928	2.011	157.06	0.2771	18.614
190	10.99	16.379	28.580	44.959	45.704	1255	1301	0.08320	2.039	2.122	158.07	0.2786	19.017
191	11.66	16.405	30.385	46.790	45.946	1332	1378	0.08357	2.161	2.245	159.07	0.2802	19.427
192	12.41	16.430	32.375	48.805	46.187	1418	1464	0.08394	2.296	2.380	160.07	0.2817	19.845
193	13.24	16.455	34.581	51.036	46.428	1513	1559	0.08431	2.444	2.528	161.08	0.2833	20.271
194	14.16	16.480	37.036	53.516	46.670	1619	1666	0.08468	2.609	2.694	162.08	0.2848	20.704
195	15.19	16.506	39.785	56.291	46.911	1737	1784	0.08505	2.794	2.879	163.09	0.2864	21.145
196	16.35	16.531	42.885	59.416	47.153	1871	1918	0.08542	3.002	3.087	164.09	0.2879	21.594
197	17.67	16.556	46.402	62.958	47.394	2022	2069	0.08579	3.238	3.324	165.10	0.2895	22.050
198	19.17	16.581	50.426	67.007	47.636	2195	2243	0.08616	3.507	3.593	166.10	0.2910	22.514
199	20.91	16.607	55.074	71.681	47.877	2395	2443	0.08653	3.817	3.904	167.11	0.2925	22.987
200	22.95	16.632	60.510	77.142	48.119	2629	2677	0.08689	4.179	4.266	168.11	0.2940	23.468

SOURCE: Abstracted by permission from 1967 ASHRAE Handbook of Fundamentals. Compiled by John A. Goff and S. Gratch.

*Moist Air Table Definitions:

W_s = humidity ratio of saturated air, lb of water vapor per lb dry air

v_a = specific volume of dry air under 14.696 psia pressure, ft³/lb

v_s = volume of saturated air, ft³/lb of dry air

v_{as} = $v_s - v_a$, ft³/lb dry air

h_a = specific enthalpy of dry air, Btu/lb. Zero enthalpy for dry air is taken at 0°F

h_s = enthalpy of saturated air, Btu/lb of dry air

h_{as} = $h_s - h_a$, Btu/lb of dry air

s_a = specific entropy of dry air, Btu/(lb)(°R). Zero entropy for dry air is taken at 0°F.

s_s = $s_s - s_a$, Btu/(lb dry air)(°R)

s_{as} = $s_s - s_a$, Btu/(lb dry air)(°R)

h_w = enthalpy of saturated liquid, Btu/lb of water

s_w = entropy of saturated liquid, Btu/(lb of water)(°R)

p_s = saturated vapor pressure, in Hg

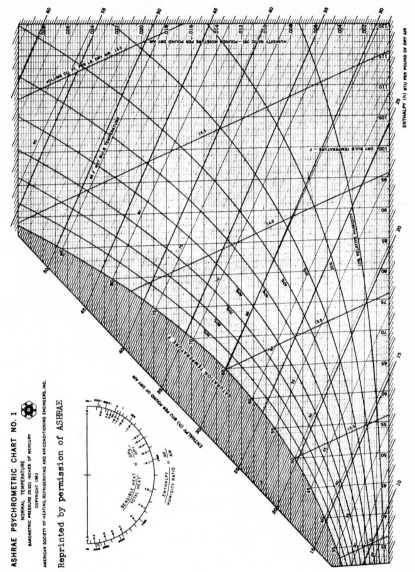

ASHRAE PSYCHROMETRIC CHART NO. 1

NORMAL TEMPERATURE
BAROMETRIC PRESSURE 29.921 INCHES OF MERCURY
COPYRIGHT 1963

AMERICAN SOCIETY OF HEATING, REFRIGERATING AND AIR-CONDITIONING ENGINEERS, INC.

Reprinted by permission of ASHRAE

Fig. 1. Psychrometric chart.

TABLE 16. Property Tables: Thermodynamic Properties of Dichlorodifluoromethane (Freon-12). Saturated Freon-12 Temperature Table

Temp. °F. t	Abs. Press. lb/in.² p	Specific volume		Enthalpy			Entropy		
		Sat. Liquid v_f	Sat. Vapor v_g	Sat. Liquid h_f	Evap. h_{fg}	Sat. Vapor h_v	Sat. Liquid s_f	Evap. s_{fg}	Sat. Vapor s_g
−60	5.358	0.01036	6.4774	−4.1919	74.885	70.693	−0.01021	0.18735	0.17714
−50	7.117	0.01046	4.9742	−2.1011	73.906	71.805	−0.00506	0.18039	0.17533
−40	9.308	0.01056	3.8750	0	72.913	72.913	0	0.17373	0.17373
−30	11.999	0.01067	3.0585	2.1120	71.903	74.015	0.00496	0.16733	0.17229
−20	15.267	0.01079	2.4429	4.2357	70.874	75.110	0.00983	0.16119	0.17102
−10	19.189	0.01091	1.9727	6.3716	69.824	76.196	0.01462	0.15527	0.16989
0	23.849	0.01103	1.6080	8.5207	68.750	77.271	0.01932	0.14956	0.16888
5*	26.483	0.01109	1.4580	9.6005	68.204	77.805	0.02165	0.14677	0.16842
10	29.335	0.01116	1.3241	10.684	67.651	78.335	0.02395	0.14403	0.16798
20	35.736	0.01130	1.0988	12.863	66.522	79.385	0.02852	0.13867	0.16719
30	43.148	0.01144	0.91880	15.058	65.361	80.419	0.03301	0.13347	0.16648
40	51.667	0.01159	0.77357	17.273	64.163	81.436	0.03745	0.12841	0.16586
50	61.394	0.01175	0.65537	19.507	62.926	82.433	0.04184	0.12346	0.16530
60	72.433	0.01191	0.55539	21.766	61.643	83.409	0.04618	0.11861	0.16479
70	84.888	0.01209	0.47818	24.050	60.309	84.359	0.05048	0.11386	0.16434
80	98.870	0.01228	0.41135	26.365	58.917	85.282	0.05475	0.10917	0.16392
86*	108.04	0.01240	0.37657	27.769	58.052	85.821	0.05730	0.10638	0.16368
90	114.49	0.01248	0.35529	28.713	57.461	86.174	0.05900	0.10453	0.16353
100	131.86	0.01269	0.30794	31.100	55.929	87.029	0.06323	0.09992	0.16315
110	151.11	0.01292	0.26769	33.531	54.313	87.844	0.06745	0.09534	0.16279
120	172.35	0.01317	0.23326	36.013	52.597	88.610	0.07168	0.09073	0.16241

*Standard ton temperature.
SOURCE: Freon Products Division, E. I. Du Pont de Nemours & Company. Copyrighted 1956 by E. I. Du Pont de Nemours & Co., Inc.

TABLE 17. Property Tables: Thermodynamic Properties of Dichlorodifluoromethane (Freon–12). Saturated Freon–12, Pressure Table

Abs. Press. lb/in.² p	Temp. °F t	Specific volume		Enthalpy			Entropy		
		Sat. Liquid v_f	Sat. Vapor v_g	Sat. Liquid h_f	Evap. h_{fg}	Sat. Vapor h_g	Sat. Liquid s_f	Evap. s_{fg}	Sat. Vapor s_g
5	-62.35	0.0103	6.9069	-4.682	75.114	70.432	-0.01144	0.18903	0.17759
10	-37.23	0.0106	3.6246	0.584	72.635	73.219	0.00138	0.17193	0.17331
15	-20.75	0.0108	2.4835	4.076	70.952	75.028	0.00947	0.16164	0.17111
20	-8.13	0.0109	1.8977	6.772	69.625	76.397	0.01550	0.15419	0.16969
30	11.11	0.0112	1.2964	10.925	67.527	78.452	0.02446	0.14343	0.16789
40	25.93	0.0114	0.9874	14.163	65.837	80.000	0.03119	0.13557	0.16676
50	38.15	0.0116	0.7982	16.861	64.388	81.249	0.03664	0.12933	0.16597
60	48.64	0.0117	0.6701	19.202	63.097	82.299	0.04125	0.12412	0.16537
70	57.90	0.0119	0.5772	21.289	61.917	83.206	0.04527	0.11962	0.16489
80	66.21	0.0120	0.5068	23.181	60.822	84.003	0.04886	0.11564	0.16450
90	73.79	0.0122	0.4514	24.924	59.789	84.713	0.05210	0.11207	0.16417
100	80.76	0.0123	0.4067	26.542	58.809	85.351	0.05507	0.10882	0.16389
120	93.29	0.0126	0.3389	29.494	56.965	86.459	0.06039	0.10301	0.16340
140	104.35	0.0128	0.2896	32.152	55.237	87.389	0.06506	0.09793	0.16299
160	114.30	0.0130	0.2522	34.591	53.589	88.180	0.06927	0.09336	0.16263
180	123.38	0.0133	0.2228	36.864	51.993	88.857	0.07311	0.08917	0.16228
200	131.74	0.0135	0.1989	39.002	50.437	89.439	0.07667	0.08528	0.16195
220	139.51	0.0137	0.1792	41.033	48.904	89.937	0.07999	0.08162	0.16161
240	146.77	0.0140	0.1625	42.973	47.388	90.361	0.08313	0.07813	0.16126
260	153.60	0.0142	0.1483	44.841	45.875	90.716	0.08611	0.07480	0.16091
280	160.06	0.0145	0.1359	46.650	44.358	91.008	0.08895	0.07158	0.16053
300	166.18	0.0147	0.1251	48.408	42.832	91.240	0.09169	0.06843	0.16012

SOURCE: Freon Products Division, E. I. Du Pont de Nemours & Company. Copyrighted 1956 by E. I. Du Pont de Nemours & Co., Inc.

2-31

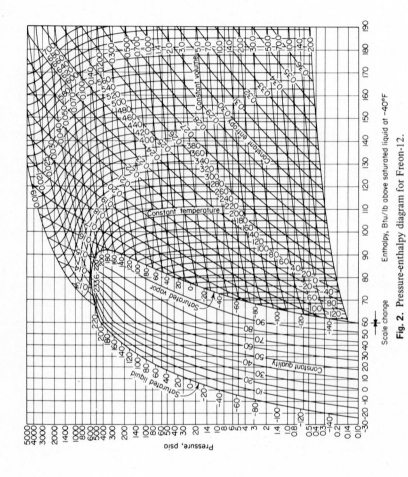

Fig. 2. Pressure-enthalpy diagram for Freon-12.

Enthalpy, Btu/lb above saturated liquid at −40°F

Pressure, psia

Scale change

Constant volume

Constant entropy

Constant temperature

Saturated vapor

Saturated liquid

Constant quality

TABLE 18. Property Tables: Thermodynamic Properties of Dichlorodifluoromethane (Freon-12). Properties of Superheated Freon-12 in °F

Abs. Press. lb/in.² (Sat. Temp.)		-40	-20	0	20	40	60	80	100	150	200	250	300
5 (-62.35)	v	7.3239	7.6938	8.0611	8.4265	8.7903	9.1528	9.5142	9.8747	10.773	11.668		
	h	73.291	75.909	78.582	81.309	84.090	86.922	89.806	92.738	100.272	108.079		
	s	0.18459	0.19069	0.19663	0.20244	0.20812	0.21367	0.21912	0.22445	0.23734	0.24964		
10 (-37.23)	v		3.7906	3.9809	4.1691	4.3556	4.5408	4.7248	4.9079	5.3627	5.8145	6.2643	
	h		75.526	78.246	81.014	83.828	86.689	89.596	92.548	100.123	107.957	116.031	
	s		0.17866	0.18471	0.19061	0.19635	0.20197	0.20746	0.21283	0.22579	0.23813	0.24993	
15 (-20.75)	v		2.4885	2.6201	2.7494	2.8770	3.0031	3.1281	3.2521	3.5592	3.8632	4.1651	
	h		75.131	77.902	80.712	83.561	86.451	89.383	92.357	99.972	107.835	115.929	
	s		0.17134	0.17751	0.18349	0.18931	0.19498	0.20051	0.20593	0.21895	0.23135	0.24317	
20 (- 8.13)	v			1.9390	2.0391	2.1373	2.2340	2.3295	2.4241	2.6573	2.8874	3.1155	
	h			77.550	80.403	83.289	86.210	89.168	92.164	99.820	107.712	115.826	
	s			0.17222	0.17829	0.18419	0.18992	0.19550	0.20095	0.21405	0.22649	0.23835	
25 (2.23)	v				1.6125	1.6932	1.7723	1.8502	1.9271	2.1161	2.3019	2.4857	2.6681
	h				80.088	83.012	85.965	88.950	91.968	99.667	107.588	115.723	124.061
	s				0.17414	0.18012	0.18591	0.19155	0.19704	0.21021	0.22269	0.23458	0.24593
30 (11.11)	v				1.3278	1.3969	1.4644	1.5306	1.5957	1.7553	1.9116	2.0658	2.2186
	h				79.765	82.730	85.716	88.729	91.770	99.513	107.464	115.620	123.973
	s				0.17065	0.17671	0.18257	0.18826	0.19379	0.20704	0.21957	0.23148	0.24286
35 (18.92)	v				1.1240	1.1850	1.2442	1.3021	1.3589	1.4975	1.6327	1.7659	1.8976
	h				79.434	82.442	85.463	88.504	91.570	99.357	107.338	115.516	123.885
	s				0.16761	0.17375	0.17968	0.18542	0.19100	0.20432	0.21690	0.22885	0.24024
40 (25.93)	v					1.0258	1.0789	1.1306	1.1812	1.3041	1.4236	1.5409	1.6568
	h					82.148	85.206	88.277	91.367	99.200	107.212	115.412	123.796
	s					0.17112	0.17712	0.18292	0.18854	0.20195	0.21457	0.22655	0.23797

TABLE 18. Property Tables: Thermodynamic Properties of Dichlorodifluormethane (Freon-12). Properties of Superheated Freon-12 in °F
(Continued)

Abs. Press. lb/in.² (Sat. Temp.)		-40	-20	0	20	40	60	80	100	150	200	250	300
50 (38.15)	v					0.80248	0.84713	0.89025	0.93216	1.0332	1.1307	1.2259	1.3197
	h					81.540	84.676	87.811	90.953	98.882	106.958	115.202	123.618
	s					0.16655	0.17271	0.17862	0.18434	0.19791	0.21064	0.22268	0.23414
60 (48.64)	v						0.69210	0.72964	0.76588	0.85247	0.93531	1.0159	1.0949
	h						84.126	87.330	90.528	98.558	106.700	114.989	123.438
	s						0.16892	0.17497	0.18079	0.19453	0.20736	0.21947	0.23098
70 (57.90)	v						0.58088	0.61458	0.64685	0.72325	0.79571	0.86578	0.93431
	h						83.552	86.832	90.091	98.228	106.439	114.775	123.257
	s						0.16556	0.17175	0.17768	0.19161	0.20455	0.21673	0.22828
80 (66.21)	v							0.52795	0.55734	0.62623	0.69095	0.75320	0.81386
	h							86.316	89.640	97.891	106.174	114.559	123.075
	s							0.16885	0.17489	0.18902	0.20207	0.21432	0.22592
90 (73.79)	v							0.46024	0.48749	0.55065	0.60941	0.66560	0.72016
	h							85.779	89.175	97.548	105.905	114.340	122.892
	s							0.16616	0.17234	0.18667	0.19984	0.21217	0.22381
100 (80.76)	v								0.43138	0.49009	0.54413	0.59549	0.64518
	h								88.694	97.197	105.633	114.119	122.707
	s								0.16996	0.18452	0.19782	0.21022	0.22191
120 (93.29)	v								0.34655	0.39896	0.44606	0.49025	0.53267
	h								87.675	96.471	105.076	113.670	122.333
	s								0.16559	0.18065	0.19421	0.20677	0.21856
140 (104.35)	v									0.33350	0.37584	0.41499	0.45226
	h									95.709	104.501	113.212	121.953
	s									0.17718	0.19104	0.20377	0.21567
160 (114.30)	v									0.28404	0.32301	0.35847	0.39190
	h									94.906	103.907	112.743	121.567
	s									0.17400	0.18819	0.20110	0.21312

180 (123.38)	v	0.24519	0.28176	0.31442	0.34492
	h	94.053	103.291	112.263	121.174
	s	0.17100	0.28556	0.19868	0.21081
200 (131.74)	v	0.21370	0.24860	0.27911	0.30730
	h	93.141	102.652	111.771	120.775
	s	0.16811	0.18311	0.19644	0.20870
220 (139.51)	v	0.18746	0.22130	0.25015	0.27648
	h	92.156	101.986	111.267	120.369
	s	0.16528	0.18079	0.19435	0.20674

SOURCE: Freon Products Division, E. I. Du Pont de Nemours & Company. Copyrighted 1956 by E. I. Du Pont de Nemours & Co., Inc.

TABLE 19. Saturated Steam: Temperature Table

Temp. t, °F	Abs. Press. lb/in.^{2} p	Specific volume			Enthalpy			Entropy		
		Sat. Liquid v_f	Evap. v_{fg}	Sat. Vapor v_g	Sat. Liquid h_f	Evap. h_{fg}	Sat. Vapor h_g	Sat. Liquid s_f	Evap. s_{fg}	Sat. Vapor s_g
32	0.0886	0.01602	3305.7	3305.7	0	1075.1	1075.1	0	2.1865	2.1865
36	0.1041	0.01602	2836.6	2836.6	4.03	1072.9	1076.9	0.0082	2.1645	2.1727
40	0.1217	0.01602	2445.1	2445.1	8.05	1070.5	1078.6	0.0162	2.1423	2.1585
50	0.1780	0.01602	1704.9	1704.9	18.07	1064.8	1082.9	0.0361	2.0891	2.1252
60	0.2561	0.01603	1208.1	1208.1	28.07	1059.1	1087.2	0.0555	2.0379	2.0934
70	0.3628	0.01605	868.9	868.9	38.05	1053.4	1091.5	0.0745	1.9887	2.0632
80	0.5067	0.01607	633.7	633.7	48.02	1047.8	1095.8	0.0932	1.9415	2.0347
90	0.6980	0.01610	468.4	468.4	58.00	1042.1	1100.1	0.1115	1.8958	2.0073
100	0.9487	0.01613	350.8	350.8	67.97	1036.4	1104.4	0.1295	1.8517	1.9812
110	1.274	0.01617	265.7	265.7	77.94	1030.9	1108.8	0.1471	1.8096	1.9567
120	1.692	0.01620	203.45	203.47	87.91	1025.3	1113.2	0.1645	1.7687	1.9332
130	2.221	0.01625	157.55	157.57	97.89	1019.5	1117.4	0.1816	1.7289	1.9105
140	2.887	0.01629	123.16	123.18	107.88	1013.7	1121.6	0.1984	1.6904	1.8888
150	3.716	0.01634	97.18	97.20	117.87	1007.8	1125.7	0.2149	1.6530	1.8679
160	4.739	0.01639	77.37	77.39	127.87	1002.0	1129.9	0.2311	1.6169	1.8480
170	5.990	0.01645	62.12	62.14	137.89	996.1	1134.0	0.2471	1.5819	1.8290
180	7.510	0.01651	50.26	50.28	147.91	990.2	1138.1	0.2629	1.5479	1.8108
190	9.336	0.01657	40.99	41.01	157.95	984.1	1142.1	0.2785	1.5147	1.7932
200	11.525	0.01663	33.65	33.67	167.99	977.8	1145.8	0.2938	1.4822	1.7760
210	14.123	0.01670	27.81	27.83	178.06	971.6	1149.6	0.3090	1.4507	1.7597
220	17.188	0.01677	23.14	23.16	188.14	965.2	1153.3	0.3239	1.4201	1.7440

230	20.78	0.01684	19.371	19.388	198.22	958.7	1156.9	0.3386	1.3900	1.7286
240	24.97	0.01692	16.307	16.324	208.34	952.1	1160.4	0.3531	1.3607	1.7138
250	29.82	0.01700	13.824	13.841	218.48	945.3	1163.8	0.3675	1.3320	1.6995
260	35.43	0.01708	11.754	11.771	228.65	938.6	1167.3	0.3817	1.3042	1.6859
270	41.85	0.01717	10.053	10.070	238.84	931.8	1170.6	0.3958	1.2770	1.6728
280	49.20	0.01726	8.634	8.651	249.06	924.6	1173.7	0.4096	1.2500	1.6596
290	57.55	0.01735	7.448	7.465	259.31	917.4	1176.7	0.4234	1.2237	1.6471
300	67.01	0.01745	6.454	6.471	269.60	910.1	1179.7	0.4370	1.1980	1.6350
310	77.68	0.01755	5.610	5.628	279.92	902.6	1182.5	0.4505	1.1727	1.6232
320	89.65	0.01765	4.897	4.915	290.29	895.0	1185.3	0.4637	1.1479	1.6116
330	103.03	0.01776	4.292	4.310	300.69	887.1	1187.8	0.4769	1.1234	1.6003
340	117.99	0.01788	3.771	3.789	311.14	879.2	1190.3	0.4900	1.0994	1.5894
350	134.62	0.01799	3.324	3.342	321.64	871.0	1192.6	0.5030	1.0757	1.5787
360	153.01	0.01811	2.940	2.958	332.19	862.5	1194.7	0.5159	1.0522	1.5681
370	173.33	0.01823	2.607	2.625	342.79	853.8	1196.6	0.5286	1.0291	1.5577
380	195.70	0.01836	2.318	2.336	353.45	844.9	1198.4	0.5413	1.0062	1.5475
390	220.29	0.01850	2.064	2.083	364.17	835.4	1199.9	0.5540	0.9835	1.5375
400	247.25	0.01864	1.8446	1.8632	374.97	826.2	1201.2	0.5664	0.9610	1.5274
410	276.72	0.01878	1.6508	1.6696	385.83	816.6	1202.4	0.5789	0.9390	1.5179
420	308.82	0.01894	1.4806	1.4995	396.78	806.7	1203.5	0.5912	0.9170	1.5082
430	343.71	0.01910	1.3303	1.3494	407.80	796.5	1204.3	0.6036	0.8953	1.4989
440	381.59	0.01926	1.1973	1.2166	418.91	785.9	1204.8	0.6159	0.8735	1.4894
450	422.61	0.01943	1.0796	1.0990	430.11	774.9	1205.0	0.6281	0.8518	1.4799

TABLE 19. Saturated Steam: Temperature Table (Continued)

Temp. t, °F	Abs. Press. lb/in.² p	Specific volume			Enthalpy			Entropy		
		Sat. Liquid v_f	Evap. v_{fg}	Sat. Vapor v_g	Sat. Liquid h_f	Evap. h_{fg}	Sat. Vapor h_g	Sat. Liquid s_f	Evap. s_{fg}	Sat. Vapor s_g
460	466.97	0.0196	0.9745	0.9941	441.42	763.6	1205.0	0.6403	0.8303	1.4706
470	514.70	0.0198	0.8808	0.9006	452.84	751.9	1204.7	0.6524	0.8088	1.4612
480	566.12	0.0200	0.7972	0.8172	464.37	739.8	1204.2	0.6646	0.7873	1.4519
490	621.44	0.0202	0.7219	0.7421	476.01	727.3	1203.3	0.6767	0.7658	1.4425
500	680.80	0.0204	0.6544	0.6748	487.80	714.2	1202.0	0.6888	0.7442	1.4330
510	744.55	0.0207	0.5932	0.6139	499.8	700.6	1200.4	0.7009	0.7225	1.4234
520	812.68	0.0209	0.5382	0.5591	511.9	686.5	1198.4	0.7132	0.7007	1.4139
530	885.20	0.0212	0.4885	0.5097	524.2	671.9	1196.1	0.7253	0.6789	1.4042
540	962.80	0.0214	0.4433	0.4647	536.6	656.7	1193.3	0.7375	0.6569	1.3944
550	1045.6	0.0218	0.4021	0.4239	549.3	640.9	1190.2	0.7498	0.6347	1.3845
560	1133.4	0.0221	0.3648	0.3869	562.2	624.1	1186.3	0.7622	0.6120	1.3742
570	1226.7	0.0224	0.3304	0.3528	575.4	606.5	1181.9	0.7737	0.5890	1.3627
580	1326.1	0.0228	0.2989	0.3217	588.9	588.1	1177.0	0.7872	0.5656	1.3528
590	1431.5	0.0232	0.2699	0.2931	602.6	568.8	1171.4	0.8000	0.5419	1.3419
600	1543.2	0.0236	0.2432	0.2668	616.8	548.4	1165.2	0.8130	0.5175	1.3305
610	1661.6	0.0241	0.2185	0.2426	631.5	526.6	1158.1	0.8263	0.4923	1.3186
620	1787.0	0.0247	0.1955	0.2202	646.5	503.7	1150.2	0.8398	0.4665	1.3063
630	1919.8	0.0253	0.1740	0.1993	662.2	478.8	1141.0	0.8535	0.4394	1.2931
640	2060.3	0.0260	0.1539	0.1799	678.7	452.0	1130.8	0.8681	0.4110	1.2791
650	2208.8	0.0268	0.1348	0.1616	696.0	422.7	1118.7	0.8832	0.3809	1.2641
660	2366.2	0.0278	0.1167	0.1445	714.4	390.5	1104.9	0.8991	0.3488	0.2479
670	2532.4	0.0290	0.0991	0.1281	734.6	353.3	1087.9	0.9161	0.3127	1.2288
680	2708.4	0.0305	0.0810	0.1115	757.2	310.0	1067.2	0.9352	0.2720	1.2072
690	2895.0	0.0328	0.0617	0.0945	784.2	254.9	1039.1	0.9579	0.2217	1.1796
700	3094.1	0.0369	0.0389	0.0758	823.9	171.7	995.6	0.9904	0.1481	1.1385
705.34*	3206.2	0.0541	0	0.0541	910.3	0	910.3	1.0645	0	1.0645

*Critical temperature.
SOURCE: Reprinted by permission from "Steam Tables," Combustion Engineering, Inc.

TABLE 20. Saturated Steam: Pressure Table

Abs. Press. lb/in.² p	Temp. °F t	Specific volume			Enthalpy			Entropy		
		Sat. Liquid v_f	Evap. v_{fg}	Sat. Vapor v_g	Sat. Liquid h_f	Evap. h_{fg}	Sat. Vapor h_g	Sat. Liquid s_f	Evap. s_{fg}	Sat. Vapor s_g
0.0886	32.00	0.01602	3305.7	3305.7	0	1075.1	1075.1	0	2.1865	2.1865
0.125	40.69	0.01602	2383.7	2383.7	8.74	1070.2	1078.9	0.0176	2.1388	2.1564
0.250	59.31	0.01603	1235.8	1235.8	27.38	1059.5	1086.9	0.0542	2.0414	2.0956
0.500	79.58	0.01607	641.71	641.73	47.60	1048.0	1095.6	0.0924	1.9434	2.0358
1	101.76	0.01614	333.77	333.79	69.72	1035.5	1105.2	0.1326	1.8443	1.9769
5	162.25	0.01641	73.584	73.600	130.13	1000.7	1130.8	0.2347	1.6090	1.8437
10	193.21	0.01659	38.445	38.462	161.17	982.1	1143.3	0.2834	1.5042	1.7876
14.696	212.00	0.01672	26.811	26.828	180.07	970.3	1150.4	0.3120	1.4446	1.7566
15	213.03	0.01672	20.303	20.320	181.11	969.6	1150.7	0.3135	1.4413	1.7548
20	227.96	0.01683	20.093	20.110	196.16	959.9	1156.1	0.3356	1.3959	1.7315
30	250.34	0.01700	13.746	13.763	218.83	945.2	1164.0	0.3680	1.3312	1.6992
40	267.24	0.01715	10.489	10.506	236.02	933.7	1169.7	0.3919	1.2844	1.6763
50	281.01	0.01727	8.505	8.522	250.09	923.9	1174.0	0.4110	1.2473	1.6583
60	292.71	0.01738	7.162	7.179	262.10	915.4	1177.5	0.4271	1.2166	1.6437
70	302.92	0.01748	6.193	6.210	272.61	907.9	1180.5	0.4409	1.1905	1.6314
80	312.03	0.01757	5.458	5.476	282.02	901.1	1183.1	0.4532	1.1677	1.6209
90	320.27	0.01766	4.880	4.898	290.57	894.8	1185.4	0.4641	1.1472	1.6113
100	327.83	0.01774	4.415	4.433	298.43	888.9	1187.3	0.4741	1.1287	1.6028
110	334.79	0.01782	4.032	4.050	305.69	883.3	1189.0	0.4832	1.1118	1.5950
120	341.26	0.01789	3.710	3.728	312.46	878.1	1190.6	0.4916	1.0963	1.5879
130	347.31	0.01796	3.437	3.455	318.81	873.2	1192.0	0.4995	1.0820	1.5815
140	353.03	0.01803	3.202	3.220	324.83	868.5	1193.3	0.5069	1.0686	1.5755
150	358.43	0.01809	2.998	3.016	330.53	863.9	1194.4	0.5138	1.0560	1.5698
160	363.55	0.01815	2.816	2.834	335.95	859.6	1195.5	0.5204	1.0442	1.5646
170	368.42	0.01821	2.656	2.674	341.11	855.2	1196.3	0.5266	1.0327	1.5593
180	373.08	0.01827	2.514	2.532	346.07	851.1	1197.2	0.5325	1.0220	1.5545
190	377.55	0.01833	2.386	2.404	350.83	847.2	1198.0	0.5382	1.0119	1.5501

Table 20. Saturated Steam: Pressure Table (*Continued*)

Abs. Press. lb/in.² p	Temp. °F. t	Specific volume			Enthalpy			Entropy		
		Sat. Liquid v_f	Evap. v_{fg}	Sat. Vapor v_g	Sat. Liquid h_f	Evap. h_{fg}	Sat. Vapor h_g	Sat. Liquid s_f	Evap. s_{fg}	Sat. Vapor s_g
200	381.82	0.01839	2.270	2.288	355.40	843.3	1198.7	0.5436	1.0021	1.5457
210	385.93	0.01844	2.165	2.183	359.80	839.6	1199.4	0.5488	0.9929	1.5417
220	389.89	0.01850	2.067	2.086	364.05	835.8	1199.9	0.5538	0.9838	1.5376
230	393.70	0.01855	1.9803	1.9989	368.16	832.2	1200.4	0.5585	0.9752	1.5337
240	397.40	0.01860	1.8990	1.9176	372.16	828.7	1200.9	0.5632	0.9669	1.5301
250	400.97	0.01866	1.8244	1.8431	376.04	825.4	1201.4	0.5677	0.9590	1.5267
260	404.43	0.01870	1.7555	1.7742	379.78	822.0	1201.8	0.5720	0.9513	1.5233
270	407.79	0.01875	1.6913	1.7101	383.43	818.8	1202.2	0.5761	0.9439	1.5200
280	411.06	0.01880	1.6316	1.6504	386.99	815.5	1202.5	0.5802	0.9365	1.5167
290	414.24	0.01885	1.5758	1.5947	390.47	812.4	1202.9	0.5841	0.9296	1.5137
300	417.33	0.01890	1.5237	1.5426	393.85	809.3	1203.2	0.5879	0.9228	1.5107
350	431.71	0.01912	1.3064	1.3255	409.70	794.7	1204.4	0.6057	0.8915	1.4972
400	444.58	0.0193	1.1416	1.1609	424.02	780.9	1204.9	0.6215	0.8635	1.4850
450	456.27	0.0195	1.0123	1.0318	437.18	767.8	1205.0	0.6357	0.8382	1.4739
500	467.00	0.0197	0.9077	0.9274	449.40	755.5	1204.9	0.6488	0.8153	1.4641
550	476.94	0.0199	0.8217	0.8416	460.83	743.6	1204.4	0.6609	0.7939	1.4548
600	486.21	0.0201	0.7494	0.7695	471.59	732.0	1203.6	0.6721	0.7739	1.4460
650	494.90	0.0203	0.6879	0.7082	481.73	721.0	1202.7	0.6826	0.7553	1.4379
700	503.09	0.0205	0.6347	0.6552	491.49	710.1	1201.6	0.6925	0.7376	1.4301
750	510.83	0.0207	0.5884	0.6091	500.8	699.4	1200.2	0.7019	0.7206	1.4225
800	518.20	0.0209	0.5476	0.5685	509.7	689.1	1198.8	0.7108	0.7047	1.4155
850	525.23	0.0210	0.5116	0.5326	518.3	678.9	1197.2	0.7194	0.6893	1.4087
900	531.94	0.0212	0.4794	0.5006	526.6	669.0	1195.6	0.7276	0.6746	1.4022
950	538.38	0.0214	0.4503	0.4717	534.6	659.2	1193.8	0.7355	0.6605	1.3960
1000	544.56	0.0216	0.4240	0.4456	542.4	649.5	1191.9	0.7431	0.6468	1.3899
1050	550.52	0.0218	0.4001	0.4219	550.0	640.0	1190.0	0.7504	0.6335	1.3839

Press.	Temp	v_f	v_{fg}	v_g	h_f	h_{fg}	h_g	s_f	s_{fg}	s_g
1100	556.26	0.0219	0.3783	0.4002	557.4	630.4	1187.8	0.7575	0.6205	1.3780
1150	561.81	0.0221	0.3583	0.3804	564.6	621.0	1185.6	0.7644	0.6079	1.3723
1200	567.19	0.0223	0.3397	0.3620	571.7	611.5	1183.2	0.7712	0.5955	1.3667
1250	572.39	0.0225	0.3228	0.3453	578.6	602.2	1180.8	0.7777	0.5835	1.3612
1300	577.43	0.0227	0.3067	0.3294	585.4	592.9	1178.3	0.7840	0.5717	1.3557
1350	582.32	0.0229	0.2918	0.3147	592.1	583.7	1175.8	0.7902	0.5602	1.3504
1400	587.07	0.0231	0.2780	0.3011	598.6	574.6	1173.2	0.7963	0.5489	1.3452
1450	591.70	0.0233	0.2652	0.2885	605.0	565.5	1170.5	0.8022	0.5379	1.3401
1500	596.20	0.0235	0.2530	0.2765	611.4	556.3	1167.7	0.8081	0.5269	1.3350
1550	600.59	0.0237	0.2416	0.2653	617.7	547.1	1164.8	0.8138	0.5160	1.3298
1600	604.87	0.0239	0.2309	0.2548	623.9	538.0	1161.9	0.8195	0.5054	1.3249
1650	609.05	0.0241	0.2207	0.2448	630.0	528.8	1158.8	0.8250	0.4948	1.3198
1700	613.12	0.0243	0.2111	0.2354	636.1	519.6	1155.7	0.8304	0.4843	1.3147
1750	617.11	0.0245	0.2020	0.2265	642.1	510.4	1152.5	0.8359	0.4740	1.3099
1800	621.00	0.0247	0.1933	0.2180	648.0	501.3	1149.3	0.8412	0.4639	1.3051
1850	624.82	0.0249	0.1850	0.2099	653.9	492.0	1145.9	0.8465	0.4537	1.3002
1900	628.55	0.0252	0.1770	0.2022	659.9	482.5	1142.4	0.8517	0.4434	1.2951
1950	632.20	0.0254	0.1695	0.1949	665.8	473.0	1138.8	0.8569	0.4332	1.2901
2000	635.78	0.0257	0.1622	0.1879	671.7	463.5	1135.2	0.8620	0.4231	1.2851
2100	642.73	0.0262	0.1486	0.1748	683.4	444.2	1127.6	0.8722	0.4029	1.2751
2200	649.42	0.0267	0.1359	0.1626	695.0	424.4	1119.4	0.8823	0.3826	1.2649
2300	655.87	0.0274	0.1240	0.1514	706.7	404.3	1111.0	0.8923	0.3624	1.2547
2400	662.09	0.0280	0.1130	0.1410	718.5	382.9	1101.4	0.9025	0.3413	1.2438
2500	668.10	0.0287	0.1026	0.1313	730.7	360.3	1091.0	0.9127	0.3195	1.2322
2600	673.91	0.0295	0.0924	0.1219	743.1	337.0	1080.1	0.9232	0.2973	1.2205
2700	679.54	0.0305	0.0818	0.1123	756.1	312.2	1068.3	0.9342	0.2740	1.2082
2800	684.98	0.0316	0.0716	0.1032	770.0	284.6	1054.6	0.9458	0.2486	1.1944
2900	690.26	0.0329	0.0612	0.0941	785.2	252.9	1038.1	0.9586	0.2199	1.1785
3000	695.37	0.0346	0.0503	0.0849	802.6	216.7	1019.3	0.9731	0.1876	1.1607
3100	700.29	0.0372	0.0380	0.0752	824.6	169.4	994.0	0.9916	0.1460	1.1376
3200	705.04	0.0443	0.0153	0.0596	871.3	75.3	946.6	1.0311	0.0647	1.0958
3206.2*	705.34	0.0541	0	0.0541	910.3	0	910.3	1.0645	0	1.0645

* Critical pressure

SOURCE: Reprinted by permission from "Steam Tables," Combustion Engineering, Inc.

TABLE 21. Superheated Steam (Read across to facing page)

Abs. Press. lb/in.² (Sat. Temp.)		Sat. Water	Sat. Steam	Temperature—					
				200°	250°	300°	350°	400°	450°
1 (101.76)	Sh			98.24	148.24	198.24	248.24	298.24	348.24
	v	0.0161	333.79	392.5	422.5	452.1	482.1	511.7	541.8
	h	69.72	1105.2	1149.2	1171.9	1194.4	1217.3	1240.2	1263.5
	s	0.1326	1.9769	2.0491	2.0822	2.1128	2.1420	2.1694	2.1957
5 (162.25)	Sh			37.75	87.75	137.75	187.75	237.75	287.75
	v	0.0164	73.600	78.17	84.24	90.21	96.26	102.19	108.23
	h	130.13	1130.8	1148.3	1171.1	1193.6	1216.6	1239.8	1263.0
	s	0.2347	1.8437	1.8710	1.9043	1.9349	1.9642	1.9920	2.0182
10 (193.21)	Sh			6.79	56.79	106.79	156.79	206.79	256.79
	v	0.0166	38.462	38.88	41.96	44.98	48.02	51.01	54.04
	h	161.17	1143.3	1146.7	1170.2	1192.8	1216.0	1239.3	1262.5
	s	0.2834	1.7876	1.7928	1.8271	1.8579	1.8875	1.9154	1.9416
14.696 (212.00)	Sh				38.00	88.00	138.00	188.00	238.00
	v	0.0167	26.828		28.44	30.52	32.61	34.65	36.73
	h	180.07	1150.4		1169.2	1192.0	1215.4	1238.9	1262.1
	s	0.3120	1.7566		1.7838	1.8148	1.8446	1.8727	1.8989
20 (227.96)	Sh				22.04	72.04	122.04	172.04	222.04
	v	0.0168	20.110		20.81	22.36	23.91	25.43	26.95
	h	196.16	1156.1		1168.0	1191.1	1214.8	1238.4	1261.6
	s	0.3356	1.7315		1.7485	1.7799	1.8101	1.8384	1.8646
40 (267.24)	Sh					32.76	82.76	132.76	182.76
	v	0.0172	10.506			11.04	11.84	12.62	13.40
	h	236.02	1169.7			1187.1	1211.9	1236.4	1259.6
	s	0.3919	1.6763			1.6997	1.7313	1.7606	1.7868
60 (292.71)	Sh					7.29	57.29	107.29	157.29
	v	0.0174	7.179			7.260	7.821	8.353	8.882
	h	262.10	1177.5			1181.8	1208.5	1234.0	1257.7
	s	0.4271	1.6437			1.6494	1.6834	1.7139	1.7407

Abs. Press. lb/in.² (Sat. Temp.)		Sat. Water	Sat. Steam	Temperature—					
				340°	360°	380°	400°	420°	450°
80 (312.03)	Sh			27.97	47.97	67.97	87.97	107.97	137.97
	v	0.0176	5.476	5.720	5.889	6.055	6.217	6.384	6.623
	h	282.02	1183.1	1200.0	1211.0	1221.2	1231.5	1240.3	1255.7
	s	0.4532	1.6209	1.6424	1.6560	1.6683	1.6804	1.6905	1.7077
100 (327.83)	Sh			12.17	32.17	52.17	72.17	92.17	122.17
	v	0.0177	4.433	4.520	4.663	4.801	4.936	5.070	5.266
	h	298.43	1187.3	1194.9	1207.0	1218.3	1228.4	1238.6	1253.7
	s	0.4741	1.6028	1.6124	1.6273	1.6409	1.6528	1.6645	1.6814
120 (341.26)	Sh				18.74	38.74	58.74	78.74	108.74
	v	0.0179	3.728		3.845	3.963	4.079	4.194	4.361
	h	312.46	1190.6		1202.7	1214.7	1225.4	1235.7	1251.4
	s	0.4916	1.5879		1.6028	1.6173	1.6299	1.6417	1.6592
140 (353.03)	Sh				6.97	26.97	46.97	66.97	96.97
	v	0.0180	3.220		3.258	3.364	3.467	3.567	3.715
	h	324.83	1193.3		1198.0	1210.6	1221.8	1232.9	1249.1
	s	0.5069	1.5755		1.5813	1.5965	1.6097	1.6225	1.6406
160 (363.55)	Sh					16.45	36.45	56.45	86.45
	v	0.0182	2.834			2.913	3.006	3.097	3.230
	h	335.95	1195.5			1206.0	1218.3	1230.0	1246.9
	s	0.5204	1.5646			1.5772	1.5917	1.6052	1.6241

Sh = superheat, °F   h = enthalpy, Btu/lb
v = specific volume, ft³/lb   s = entropy, Btu/°F-lb

(Read across from facing page)

Degrees Fahrenheit

500°	600°	700°	800°	900°	1000°	1100°	1200°
398.24	498.24	598.24	698.24	798.24	898.24	998.24	1098.24
571.3	630.9	690.6	750.2	809.8	869.4	929.1	988.7
1286.7	1333.9	1382.1	1431.0	1480.9	1531.4	1583.0	1635.4
2.2206	2.2673	2.3107	2.3512	2.3892	2.4251	2.4592	2.4918
337.75	437.75	537.75	637.75	737.75	837.75	937.75	1037.75
114.16	126.11	138.05	149.99	161.91	173.83	185.80	197.72
1286.1	1333.5	1381.8	1430.8	1480.6	1531.3	1582.9	1635.3
2.0429	2.0898	2.1333	2.1738	2.2118	2.2478	2.2820	2.3146
306.79	406.79	506.79	606.79	706.79	806.79	906.79	1006.79
57.02	63.01	68.99	74.96	80.92	86.89	92.88	98.85
1285.8	1333.3	1381.6	1430.6	1480.5	1531.2	˙1582.8	1635.2
1.9665	2.0135	2.0570	2.0975	2.1356	2.1716	2.2058	2.2384
288.00	388.00	488.00	588.00	688.00	788.00	888.00	988.00
38.75	42.83	46.91	50.97	55.03	59.09	63.19	67.25
1285.4	1333.0	1381.4	1430.5	1480.4	1531.1	1582.7	1635.1
1.9238	1.9709	2.0145	2.0551	2.0932	2.1292	2.1634	2.1960
272.04	372.04	472.04	572.04	672.04	772.04	872.04	972.04
28.45	31.46	34.46	37.44	40.43	43.42	46.43	49.41
1285.0	1332.7	1381.2	1430.3	1480.2	1531.0	1582.6	1635.1
1.8896	1.9368	1.9805	2.0211	2.0592	2.0952	2.1294	2.1620
232.76	332.76	432.76	532.76	632.76	732.76	832.76	932.76
14.16	15.68	17.19	18.69	20.18	21.68	23.20	24.69
1283.4	1331.6	1380.4	1429.6	1479.6	1530.6	1582.2	1634.8
1.8123	1.8600	1.9040	1.9447	1.9829	2.0191	2.0533	2.0860
207.29	307.29	407.29	507.29	607.29	707.29	807.29	907.29
9.398	10.42	11.44	12.44	13.44	14.44	15.45	16.45
1281.8	1330.4	1379.5	1428.9	1479.0	1530.2	1581.8	1634.4
1.7665	1.8146	1.8588	1.8996	1.9378	1.9741	2.0083	2.0410

Degrees Fahrenheit

500°	600°	700°	800°	900°	1000°	1100°	1200°
187.97	287.97	387.97	487.97	587.97	687.97	787.97	887.97
7.015	7.793	8.558	9.313	10.07	10.82	11.58	12.33
1280.2	1329.3	1378.5	1428.2	1478.4	1529.7	1581.4	1634.1
1.7339	1.7825	1.8268	1.8679	1.9062	1.9426	1.9768	2.0095
172.17	272.17	372.17	472.17	572.17	672.17	772.17	872.17
5.589	6.217	6.836	7.448	8.055	8.659	9.262	9.862
1278.6	1327.9	1377.5	1427.5	1478.0	1529.2	1581.0	1633.7
1.7080	1.7568	1.8015	1.8428	1.8814	1.9177	1.9520	1.9847
158.74	258.74	358.74	458.74	558.74	658.74	758.74	858.74
4.635	5.165	5.685	6.197	6.705	7.210	7.713	8.215
1276.7	1326.8	1376.7	1426.8	1477.2	1528.8	1580.6	1633.5
1.6863	1.7359	1.7809	1.8223	1.8608	1.8974	1.9317	1.9646
146.97	246.97	346.97	446.97	546.97	646.97	746.97	846.97
3.954	4.413	4.862	5.303	5.741	6.175	6.607	7.037
1275.0	1325.5	1375.7	1426.0	1476.6	1528.4	1580.2	1633.2
1.6683	1.7183	1.7635	1.8051	1.8437	1.8804	1.9147	1.9476
136.45	236.45	336.45	436.45	536.45	636.45	736.45	836.45
3.443	3.849	4.245	4.633	5.018	5.398	5.777	6.155
1273.2	1324.1	1374.7	1425.2	1476.0	1527.9	1579.8	1632.8
1.6522	1.7026	1.7482	1.7899	1.8287	1.8655	1.8999	1.9328

TABLE 21. Superheated Steam (*Continued*) (Read across to facing page)

Abs. Press. lb/in.² (Sat. Temp.)		Sat. Water	Sat. Steam	400°	420°	440°	460°	480°	500°
	Sh			26.92	46.92	66.92	86.92	106.92	126.92
180	v	0.0183	2.532	2.648	2.731	2.812	2.892	2.968	3.045
(373.08)	h	346.07	1197.2	1214.6	1226.8	1239.2	1250.2	1260.8	1271.5
	s	0.5325	1.5545	1.5751	1.5891	1.6030	1.6151	1.6265	1.6378
	Sh				18.18	38.18	58.18	78.18	98.18	118.18
200	v	0.0184	2.288	2.360	2.437	2.512	2.585	2.656	2.726
(381.82)	h	355.40	1198.7	1210.8	1223.7	1236.3	1247.9	1258.7	1269.4
	s	0.5436	1.5457	1.5599	1.5748	1.5889	1.6017	1.6133	1.6245
	Sh			10.11	30.11	50.11	70.11	90.11	110.11
220	v	0.0185	2.086	2.124	2.196	2.267	2.335	2.400	2.465
(389.89)	h	364.05	1199.9	1206.8	1220.1	1233.2	1245.2	1256.7	1267.5
	s	0.5538	1.5376	1.5457	1.5610	1.5757	1.5889	1.6013	1.6127
	Sh				22.60	42.60	62.60	82.60	102.60
240	v	0.0186	1.9176		1.995	2.062	2.126	2.187	2.247
(397.40)	h	372.16	1200.9		1216.6	1230.0	1242.5	1254.4	1265.7
	s	0.5632	1.5301		1.5482	1.5633	1.5770	1.5898	1.6017

Abs. Press. lb/in.² (Sat. Temp.)		Sat. Water	Sat. Steam	420°	440°	460°	480°	500°	520°
	Sh			15.57	35.57	55.57	75.57	95.57	115.57
260	v	0.0187	1.7742	1.8246	1.8876	1.9482	2.0063	2.0631	2.1185
(404.43)	h	379.78	1201.8	1212.8	1226.6	1239.5	1252.0	1263.6	1273.8
	s	0.5720	1.5233	1.5359	1.5514	1.5656	1.5790	1.5912	1.6017
	Sh			8.94	28.94	48.94	68.94	88.94	108.94
280	v	0.0188	1.6504	1.6780	1.7381	1.7957	1.8512	1.9048	1.9575
(411.06)	h	386.99	1202.5	1209.0	1223.2	1236.5	1249.4	1261.5	1272.2
	s	0.5802	1.5167	1.5241	1.5401	1.5547	1.5686	1.5813	1.5923
	Sh			2.67	22.67	42.67	62.67	82.67	102.67
300	v	0.0189	1.5426	1.5506	1.6082	1.6634	1.7164	1.7677	1.8172
(417.33)	h	393.85	1203.2	1205.2	1219.5	1233.4	1246.6	1259.2	1270.5
	s	0.5879	1.5107	1.5130	1.5291	1.5443	1.5585	1.5718	1.5834
	Sh				8.29	28.29	48.29	69.29	88.29
350	v	0.0191	1.3255		1.3472	1.3976	1.4460	1.4923	1.5377
(431.71)	h	409.70	1204.4		1210.3	1224.8	1239.5	1252.9	1265.5
	s	0.6057	1.4972		1.5038	1.5197	1.5355	1.5496	1.5626

Abs. Press. lb/in.² (Sat. Temp.)		Sat. Water	Sat. Steam	460°	480°	500°	520°	540°	560°
	Sh			15.42	35.42	55.42	75.42	95.42	115.42
400	v	0.0193	1.1609	1.1972	1.2422	1.2849	1.3269	1.3660	1.4042
(444.58)	h	424.02	1204.9	1216.5	1231.6	1245.9	1259.9	1272.4	1284.3
	s	0.6215	1.4850	1.4977	1.5140	1.5290	1.5434	1.5561	1.5678
	Sh			3.73	23.73	43.73	63.73	83.73	103.73
450	v	0.0195	1.0318	1.0401	1.0824	1.1230	1.1617	1.1982	1.2337
(450.27)	h	437.18	1205.0	1207.9	1223.7	1238.7	1253.8	1266.9	1280.0
	s	0.6357	1.4739	1.4771	1.4941	1.5099	1.5255	1.5387	1.5517
	Sh				13.00	33.00	53.00	73.00	93.00
500	v	0.0197	0.9274		0.9538	0.9926	1.0290	1.0636	1.0969
(467.00)	h	449.40	1204.9		1215.3	1231.4	1246.6	1261.1	1275.0
	s	0.6488	1.4641		1.4752	1.4922	1.5079	1.5225	1.5363

Sh = superheat, °F
v = specific volume, ft³/lb
h = enthalpy, Btu/lb
s = entropy, Btu/°F-lb

(Read across from facing page)

Degrees Fahrenheit

550°	600°	700°	800°	900°	1000°	1100°	1200°
176.92	226.92	326.92	426.92	526.92	626.92	726.92	826.92
3.229	3.410	3.765	4.112	4.455	4.794	5.132	5.468
1297.4	1322.8	1373.7	1424.5	1475.5	1527.4	1579.4	1632.5
1.6641	1.6886	1.7345	1.7765	1.8154	1.8522	1.8866	1.9196
168.18	218.18	318.18	418.18	518.18	618.18	718.18	818.18
2.895	3.059	3.381	3.697	4.005	4.311	4.616	4.919
1295.6	1321.4	1372.5	1423.9	1474.9	1526.6	1579.0	1632.1
1.6511	1.6761	1.7221	1.7646	1.8035	1.8402	1.8749	1.9079
160.11	210.11	310.11	410.11	510.11	610.11	710.11	810.11
2.621	2.772	3.067	3.354	3.637	3.916	4.193	4.469
1294.1	1320.0	1371.6	1422.9	1474.2	1526.4	1578.6	1631.8
1.6397	1.6647	1.7112	1.7536	1.7928	1.8298	1.8644	1.8974
152.60	202.60	302.60	402.60	502.60	602.60	702.60	802.60
2.392	2.532	2.805	3.069	3.330	3.586	3.841	4.095
1292.5	1318.6	1370.5	1422.1	1473.6	1525.9	1578.2	1631.5
1.6289	1.6541	1.7009	1.7435	1.7828	1.8199	1.8545	1.8876

Degrees Fahrenheit

550°	600°	700°	800°	900°	1000°	1100°	1200°
145.57	195.57	295.57	395.57	495.57	595.57	695.57	795.57
2.1991	2.3299	2.5833	2.8289	3.0701	3.3077	3.5437	3.7778
1290.8	1317.1	1369.5	1421.3	1473.0	1525.4	1577.8	1631.1
1.6188	1.6442	1.6914	1.7342	1.7737	1.8109	1.8456	1.8787
138.94	188.94	288.94	388.94	488.94	588.94	688.94	788.94
2.0334	2.1562	2.3932	2.6226	2.8475	3.0688	3.2883	3.5062
1289.1	1315.7	1368.5	1420.5	1472.4	1524.9	1577.4	1630.8
1.6093	1.6350	1.6826	1.7256	1.7652	1.8024	1.8372	1.8703
132.67	182.67	282.67	382.67	482.67	582.67	682.67	782.67
1.8896	2.0056	2.2286	2.4447	2.6547	2.8634	3.0670	3.2707
1287.4	1314.4	1367.4	1419.7	1471.8	1524.4	1577.0	1630.4
1.6004	1.6265	1.6742	1.7175	1.7572	1.7945	1.8294	1.8625
118.29	168.29	268.29	368.29	468.29	568.29	668.29	768.29
1.6016	1.7041	1.8991	2.0863	2.2687	2.4475	2.6246	2.8000
1282.9	1310.6	1364.7	1417.6	1470.2	1523.0	1576.0	1629.6
1.5801	1.6069	1.6556	1.6993	1.7395	1.7769	1.8120	1.8453

Degrees Fahrenheit

580°	600°	700°	800°	900°	1000°	1100°	1200°
135.42	155.42	255.42	355.42	455.42	555.42	655.42	755.42
1.4413	1.4777	1.6522	1.8179	1.9796	2.1367	2.2926	2.4475
1295.8	1307.0	1362.1	1415.5	1468.6	1521.5	1574.8	1628.8
1.5790	1.5897	1.6393	1.6835	1.7240	1.7615	1.7968	1.8304
123.73	143.73	243.73	343.73	443.73	543.73	643.73	743.73
1.2681	1.3013	1.4593	1.6092	1.7539	1.8951	2.0345	2.1720
1291.8	1303.1	1359.1	1413.4	1467.0	1520.3	1573.9	1627.8
1.5632	1.5740	1.6245	1.6694	1.7103	1.7481	1.7836	1.8171
113.00	133.00	233.00	333.00	433.00	533.00	633.00	733.00
1.1292	1.1600	1.3051	1.4417	1.5735	1.7016	1.8280	1.9532
1287.3	1299.3	1356.3	1411.2	1465.1	1518.8	1572.9	1627.3
1.5482	1.5596	1.6110	1.6564	1.6975	1.7356	1.7714	1.8052

TABLE 21. Superheated Steam (*Continued*) (Read across to facing page)

Abs. Press. lb/in.² (Sat. Temp.)		Sat. Water	Sat. Steam	Temperature—					
				500°	520°	540°	560°	580°	600°
550 (476.94)	Sh			23.06	43.06	63.06	83.06	103.06	123.06
	v	0.0199	0.8416	0.8851	0.9198	0.9530	0.9846	1.0151	1.0441
	h	460.83	1204.4	1223.4	1240.0	1254.8	1269.6	1282.9	1295.2
	s	0.6609	1.4548	1.4748	1.4919	1.5068	1.5215	1.5344	1.5461
600 (486.21)	Sh			13.79	33.79	53.79	73.79	93.79	113.79
	v	0.0201	0.7695	0.7945	0.8284	0.8605	0.8907	0.9194	0.9471
	h	471.59	1203.6	1215.6	1232.5	1248.3	1263.7	1278.1	1290.9
	s	0.6721	1.4460	1.4586	1.4760	1.4920	1.5072	1.5212	1.5334
700 (503.09)	Sh				16.91	36.91	56.91	76.91	96.91
	v	0.0205	0.6552		0.6830	0.7133	0.7419	0.7687	0.7941
	h	491.49	1201.6		1217.1	1234.7	1251.3	1266.8	1281.9
	s	0.6925	1.4301		1.4461	1.4638	1.4803	1.4953	1.5097

Abs. Press. lb/in.² (Sat. Temp.)		Sat. Water	Sat. Steam	Temperature—					
				520°	540°	560°	580°	600°	620°
800 (518.20)	Sh			1.80	21.80	41.80	61.80	81.80	101.80
	v	0.0209	0.5685	0.5714	0.6013	0.6288	0.6545	0.6785	0.7013
	h	509.7	1198.8	1200.3	1220.0	1238.2	1255.3	1271.4	1286.5
	s	0.7108	1.4155	1.4170	1.4369	1.4549	1.4715	1.4868	1.5009
900 (531.94)	Sh				8.06	28.06	48.06	68.06	88.06
	v	0.0212	0.5006		0.5123	0.5394	0.5644	0.5876	0.6094
	h	526.6	1195.6		1204.0	1224.2	1242.6	1260.0	1276.5
	s	0.7276	1.4022		1.4106	1.4306	1.4484	1.4649	1.4803
1000 (544.56)	Sh					15.44	35.44	55.44	75.44
	v	0.0216	0.4456			0.4665	0.4914	0.5141	0.5351
	h	542.4	1191.9			1208.8	1229.4	1248.2	1265.8
	s	0.7431	1.3899			1.4066	1.4266	1.4445	1.4610
1100 (556.26)	Sh					3.74	23.74	43.74	63.74
	v	0.0219	0.4002			0.4054	0.4304	0.4530	0.4736
	h	557.4	1187.8			1192.3	1214.9	1235.6	1254.5
	s	0.7575	1.3780			1.3824	1.4044	1.4241	1.4418

Abs. Press. lb/in.² (Sat. Temp.		Sat. Water	Sat. Steam	Temperature—					
				580°	600°	620°	640°	660°	680°
1200 (567.19)	Sh			12.81	32.81	52.81	72.81	92.81	112.81
	v	0.0223	0.3620	0.3793	0.4013	0.4219	0.4408	0.4585	0.4750
	h	571.7	1183.2	1200.2	1222.1	1242.6	1261.5	1279.2	1295.3
	s	0.7712	1.3667	1.3831	1.4040	1.4232	1.4405	1.4565	1.4707
1400 (587.07)	Sh				12.93	32.93	52.93	72.93	92.93
	v	0.0231	0.3011		0.3172	0.3388	0.3581	0.3760	0.3914
	h	598.6	1173.2		1191.8	1216.9	1239.2	1260.1	1278.2
	s	0.7963	1.3452		1.3629	1.3863	1.4068	1.4256	1.4416

Sh = superheat, °F   h = enthalpy, Btu/lb
v = specific volume, ft³/lb   s = entropy, Btu/°F-lb

(Read across from facing page)

Degrees Fahrenheit

650°	700°	750°	800°	900°	1000°	1100°	1200°
173.06	223.06	273.06	323.06	423.06	523.06	623.06	723.06
1.1132	1.1791	1.2428	1.3055	1.4262	1.5434	1.6590	1.7724
1324.9	1353.5	1381.6	1409.2	1463.6	1517.5	1571.7	1626.0
1.5735	1.5987	1.6224	1.6447	1.6862	1.7244	1.7603	1.7940
163.79	213.79	263.79	313.79	413.79	513.79	613.79	713.79
1.0123	1.0738	1.1332	1.1915	1.3032	1.4115	1.5179	1.6224
1321.4	1350.5	1379.0	1407.0	1461.8	1516.0	1570.5	1625.0
1.5615	1.5871	1.6112	1.6339	1.6757	1.7141	1.7502	1.7841
146.91	196.91	246.91	296.91	396.91	496.91	596.91	696.91
0.8534	0.9084	0.9608	1.0117	1.1096	1.2043	1.2965	1.3870
1314.3	1344.6	1373.7	1402.5	1458.2	1513.4	1568.2	1623.3
1.5396	1.5663	1.5908	1.6141	1.6567	1.6958	1.7321	1.7663

Degrees Fahrenheit

650°	700°	750°	800°	900°	1000°	1100°	1200°
131.80	181.80	231.80	281.80	381.80	481.80	581.80	681.80
0.7336	0.7838	0.8313	0.8770	0.9648	1.0486	1.1302	1.2105
1306.8	1338.4	1368.5	1397.8	1454.9	1510.5	1566.0	1621.4
1.5195	1.5473	1.5727	1.5964	1.6400	1.6794	1.7162	1.7506
118.06	168.06	218.06	268.06	368.06	468.06	568.06	668.06
0.6399	0.6866	0.7304	0.7720	0.8516	0.9277	1.0010	1.0727
1298.6	1331.8	1363.0	1392.9	1451.1	1507.8	1563.7	1619.3
1.5004	1.5296	1.5559	1.5801	1.6245	1.6647	1.7017	1.7362
105.44	155.44	205.44	255.44	355.44	455.44	555.44	655.44
0.5639	0.6085	0.6492	0.6879	0.7611	0.8306	0.8974	0.9626
1289.6	1324.9	1357.2	1388.0	1447.3	1504.7	1561.3	1617.5
1.4827	1.5138	1.5411	1.5660	1.6113	1.6520	1.6895	1.7244
93.74	143.74	193.74	243.74	343.74	443.74	543.74	643.74
0.5027	0.5445	0.5828	0.6190	0.6871	0.7511	0.8125	0.8724
1280.6	1317.8	1351.3	1383.0	1443.6	1501.7	1558.7	1615.4
1.4656	1.4984	1.5267	1.5523	1.5986	1.6398	1.6776	1.7128

Degrees Fahrenheit

700°	720°	750°	800°	900°	1000°	1100°	1200°
132.81	152.81	182.81	232.81	332.81	432.81	532.81	632.81
0.4907	0.5056	0.5271	0.5613	0.6251	0.6853	0.7423	0.7975
1310.3	1324.4	1345.0	1377.7	1439.5	1499.0	1556.6	1613.6
1.4838	1.4958	1.5131	1.5395	1.5867	1.6289	1.6671	1.7025
112.93	132.93	162.93	212.93	312.93	412.93	512.93	612.93
0.4063	0.4203	0.4401	0.4711	0.5283	0.5811	0.6313	0.6795
1294.9	1310.6	1332.8	1367.4	1431.9	1492.7	1551.7	1609.6
1.4562	1.4696	1.4882	1.5162	1.5654	1.6086	1.6476	1.6836

TABLE 21. Superheated Steam (*Continued*) (Read across to facing page)

Abs. Press. lb/in.² (Sat. Temp.)		Sat. Water	Sat. Steam	Temperature—					
				620°	640°	660°	680°	700°	720°
	Sh			15.13	35.13	55.13	75.13	95.13	115.13
1600	v	0.0239	0.2548	0.2730	0.2935	0.3114	0.3274	0.3421	0.3555
(604.87)	h	623.9	1161.9	1186.3	1213.7	1237.6	1258.9	1278.4	1295.7
	s	0.8195	1.3249	1.3477	1.3728	1.3943	1.4132	1.4302	1.4449
	Sh				19.00	39.00	59.00	79.00	99.00
1800	v	0.0247	0.2180		0.2405	0.2598	0.2764	0.2912	0.3045
(621.00)	h	648.0	1149.3		1183.7	1212.7	1237.6	1259.8	1279.6
	s	0.8412	1.3051		1.3367	1.3628	1.3848	1.4041	1.4211

Abs. Press. lb/in.² (Sat. Temp.)		Sat. Water	Sat. Steam	Temperature—					
				660°	680°	700°	720°	740°	760°
	Sh			24.22	44.22	64.22	84.22	104.22	124.22
2000	v	0.0257	0.1879	0.2162	0.2344	0.2498	0.2633	0.2752	0.2862
(635.78)	h	671.7	1135.2	1183.7	1214.3	1240.0	1262.4	1281.8	1299.6
	s	0.8620	1.2851	1.3289	1.3560	1.3783	1.3975	1.4138	1.4285
	Sh			10.58	30.58	50.58	70.58	90.58	110.58
2200	v	0.0267	0.1626	0.1773	0.1972	0.2138	0.2277	0.2399	0.2509
(649.42)	h	695.0	1119.4	1148.0	1185.8	1216.2	1241.8	1263.8	1283.6
	s	0.8823	1.2649	1.2906	1.3239	1.3504	1.3723	1.3907	1.4072
	Sh				17.91	37.91	57.91	77.91	97.91
2400	v	0.0280	0.1410		0.1646	0.1824	0.1974	0.2101	0.2214
(662.09)	h	718.5	1101.4		1152.2	1189.1	1219.4	1244.7	1267.1
	s	0.9025	1.2438		1.2887	1.3208	1.3467	1.3680	1.3865

Abs. Press. lb/in.² (Sat. Temp.)		Sat. Water	Sat. Steam	Temperature—					
				680°	700°	720°	740°	760°	780°
	Sh			6.09	26.09	46.09	66.09	86.09	106.09
2600	v	0.0295	0.1219	0.1323	0.1541	0.1706	0.1842	0.1961	0.2066
(673.91)	h	743.1	1080.1	1106.0	1157.0	1194.0	1223.9	1249.7	1272.2
	s	0.9232	1.2205	1.2433	1.2877	1.3193	1.3444	1.3657	1.3840
	Sh				15.02	35.02	55.02	75.02	95.02
2800	v	0.0316	0.1032		0.1267	0.1461	0.1615	0.1741	0.1851
(684.98)	h	770.0	1054.6		1115.9	1164.5	1201.5	1231.3	1257.0
	s	0.9458	1.1944		1.2476	1.2892	1.3203	1.3449	1.3658
	Sh				4.63	24.63	44.63	64.63	84.63
3000	v	0.0346	0.0849		0.0972	0.1236	0.1410	0.1546	0.1661
(695.37)	h	802.6	1019.3		1054.0	1130.3	1176.4	1211.4	1240.6
	s	0.9731	1.1607		1.1907	1.2559	1.2947	1.3236	1.3474
	Sh					14.96	34.96	54.96	74.96
3200	v	0.0443	0.0596			0.1024	0.1217	0.1368	0.1493
(705.04)	h	871.3	946.6			1089.8	1146.9	1189.3	1223.4
	s	1.0311	1.0958			1.2180	1.2660	1.3011	1.3288
	Sh					14.66	34.66	54.66	74.66
3206.2*	v	0.0541	0.0541			0.1018	0.1211	0.1363	0.1488
(705.34)	h	910.3	910.3			1088.5	1145.8	1188.5	1222.9
	s	1.0645	1.0645			1.2170	1.2652	1.3005	1.3284

*Critical pressure.   Sh = superheat, °F   h = enthalpy, Btu/lb
v = specific volume, ft³/lb   s = entropy, Btu/°F-lb

SOURCE: Reprinted by permission from "Steam Tables," Combustion Engineering, Inc.

(Read across from facing page)

Degrees Fahrenheit

740°	760°	780°	800°	900°	1000°	1100°	1200°
135.13	155.13	175.13	195.13	295.13	395.13	495.13	595.13
0.3682	0.3802	0.3919	0.4031	0.4554	0.5032	0.5482	0.5914
1311.8	1327.3	1342.2	1356.7	1424.1	1468.8	1547.0	1605.8
1.4585	1.4713	1.4834	1.4950	1.5465	1.5909	1.6308	1.6674
119.00	139.00	159.00	179.00	279.00	379.00	479.00	579.00
0.3168	0.3283	0.3393	0.3499	0.3986	0.4425	0.4834	0.5224
1297.4	1313.9	1329.9	1345.3	1416.0	1480.6	1542.0	1601.8
1.4360	1.4497	1.4627	1.4750	1.5290	1.5748	1.6155	1.6526

Degrees Fahrenheit

780°	800°	820°	850°	900°	1000°	1100°	1200°
144.22	164.22	184.22	214.22	264.22	364.22	464.22	564.22
0.2966	0.3067	0.3165	0.3305	0.3528	0.3940	0.4319	0.4678
1316.4	1332.7	1348.6	1371.1	1407.2	1473.5	1537.4	1598.6
1.4422	1.4552	1.4677	1.4851	1.5122	1.5592	1.6015	1.6395
130.58	150.58	170.58	200.58	250.58	350.58	450.58	550.58
0.2612	0.2710	0.2804	0.2939	0.3151	0.3540	0.3893	0.4226
1302.9	1319.5	1336.2	1360.3	1398.0	1467.9	1532.3	1594.3
1.4229	1.4362	1.4494	1.4679	1.4962	1.5458	1.5884	1.6269
117.91	137.91	157.91	187.91	237.91	337.91	437.91	537.91
0.2317	0.2415	0.2506	0.2636	0.2836	0.3205	0.3538	0.3848
1287.3	1306.5	1324.2	1349.6	1388.5	1460.8	1526.9	1589.8
1.4030	1.4183	1.4323	1.4519	1.4810	1.5323	1.5761	1.6152

Degrees Fahrenheit

800°	820°	850°	900°	950°	1000°	1100°	1200°
126.09	146.09	176.09	226.09	276.09	326.09	426.09	526.09
0.2164	0.2256	0.2380	0.2573	0.2754	0.2924	0.3237	0.3530
1293.1	1312.6	1338.9	1379.7	1418.1	1454.2	1521.6	1585.8
1.4008	1.4161	1.4364	1.4670	1.4947	1.5199	1.5645	1.6044
115.02	135.02	165.02	215.02	265.02	315.02	415.02	515.02
0.1950	0.2045	0.2166	0.2351	0.2521	0.2685	0.2983	0.3258
1279.8	1301.7	1329.3	1371.5	1410.4	1447.9	1517.0	1581.9
1.3840	1.4013	1.4226	1.4542	1.4823	1.5084	1.5542	1.5945
104.63	124.63	154.63	204.63	254.63	304.63	404.63	504.63
0.1763	0.1856	0.1978	0.2155	0.2322	0.2478	0.2763	0.3027
1266.0	1289.0	1319.0	1362.5	1403.4	1441.7	1512.4	1578.9
1.3677	1.3858	1.4090	1.4416	1.4711	1.4978	1.5447	1.5860
94.96	114.96	144.96	194.96	244.96	294.96	394.96	494.96
0.1596	0.1687	0.1810	0.1985	0.2145	0.2296	0.2570	0.2822
1251.1	1275.2	1307.7	1353.8	1395.7	1435.2	1507.6	1575.3
1.3510	1.3699	1.3950	1.4296	1.4598	1.4874	1.5353	1.5774
94.66	114.66	144.66	194.66	244.66	294.66	394.66	494.66
0.1591	0.1682	0.1805	0.1980	0.2140	0.2290	0.2564	0.2816
1250.6	1274.8	1307.3	1353.4	1395.3	1434.8	1507.3	1575.1
1.3506	1.3696	1.3947	1.4293	1.4596	1.4872	1.5351	1.5772

TABLE 22. Supercritical Pressure Steam

Psia		900°F	1000°F	1100°F	1200°F
3,500	v	0.1762	0.2058	0.2313	0.2546
	h	1340.7	1424.5	1496.6	1563.3
	s	1.4127	1.4723	1.5201	1.5615
4,000	v	0.1462	0.1743	0.1979	0.2192
	h	1314.4	1406.8	1482.9	1552.1
	s	1.3827	1.4482	1.4987	1.5417
5,000	v	0.1036	0.1303	0.1513	0.1696
	h	1256.5	1369.5	1455.0	1529.5
	s	1.3231	1.4034	1.4602	1.5066
6,000	v	0.0755	0.1013	0.1207	0.1370
	h	1192.5	1330.5	1427.2	1507.7
	s	1.2642	1.3622	1.4263	1.4765
8,000	v	0.0458	0.0673	0.0844	0.0985
	h	1074.4	1250.8	1369.8	1463.9
	s	1.1614	1.2867	1.3657	1.4243
10,000	v	0.0353	0.0495	0.0635	0.0760
	h	1012.9	1181.0	1313.3	1419.7
	s	1.1050	1.2243	1.3122	1.3784

SOURCE: Reprinted by permission from "Steam Tables," Combustion Engineering, Inc.

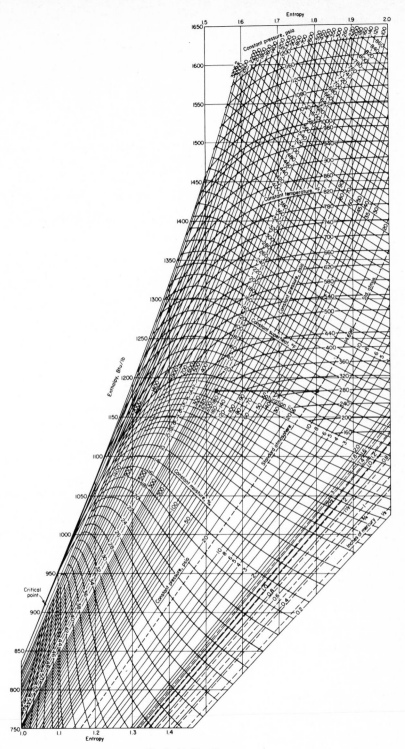

Fig. 3. Mollier diagram.

TABLE 23. Compressibility Factor: $Z = PV/RT$

$T, °K$	0.01 atm	0.1 atm	1 atm	10 atm	100 atm
Air					
100	0.99981	0.99813	0.98090		
300	1.00000	0.99997	0.99970	0.99717	0.9933
500	1.00000	1.00003	1.00034	1.00348	1.0393
1000	1.00000	1.00003	1.00033	1.00331	1.0333
1500	1.00001	1.00003	1.00024	1.00245	1.0244
2000	1.00167	1.00054	1.00035	1.00196	1.0188
3000		1.07007	1.0252	1.0095	1.0151
Argon					
100	0.99979	0.99782	0.9782		
300	0.99999	0.99994	0.99937	0.99382	0.9553
500	1.00000	1.00002	1.00018	1.00183	1.0224
1000	1.00000	1.00003	1.00026	1.00261	1.0265
1500	1.00000	1.00002	1.00020	1.00203	1.0203
2000	1.00000	1.00002	1.00016	1.00159	1.0159
3000	1.00000	1.00001	1.00011	1.00108	1.0108
4000	1.00000	1.00001	1.00008	1.00081	1.0081
5000	1.00000	1.00001	1.00006	1.00063	1.0063
Carbon dioxide					
200	0.99980	0.99805			
400	0.99998	0.99981	0.99817	0.9815	0.8155
600	0.99999	0.99997	0.99975	0.99763	0.9850
1000	1.00000	1.00002	1.00002	1.0022	1.0248
1500	1.00000	1.00002	1.00002	1.0025	1.0253
Carbon monoxide					
200	1.00000	0.99973	0.99730	0.97344	
500	1.00000	1.00004	1.00040	1.00411	1.0469
1000	1.00000	1.00004	1.00039	1.00388	1.0391
1500	1.00000	1.00003	1.00029	1.00287	1.0286
2000	1.00000	1.00002	1.00022	1.00221	1.0221
3000	1.00000	1.00002	1.00015	1.00148	1.0148

TABLE 23. Compressibility Factor:
$Z = PV/RT$ (Continued)

$T,\,^\circ K$	0.01 atm	0.1 atm	1 atm	10 atm	100 atm
Hydrogen					
50	0.9999	0.9992	0.9919	0.9186	
100	1.0000	1.0000	0.9998	0.9983	1.0560
200	1.0000	1.0001	1.0007	1.0068	1.0760
400	1.0000	1.0000	1.0005	1.0048	1.0486
600	1.0000	1.0000	1.0003	1.0034	1.0337
Nitrogen					
100	0.99982	0.9982	0.981		
300	1.00000	0.99998	0.99982	0.99838	1.0054
500	1.00000	1.00004	1.00041	1.00414	1.0461
1000	1.00000	1.00004	1.00036	1.00360	1.0365
1500	1.00000	1.00003	1.00026	1.00263	1.0264
2000	1.00000	1.00002	1.00020	1.00202	1.0202
3000	1.00000	1.00001	1.00014	1.00135	1.0135
Oxygen					
100	0.99978	0.99781	0.97724		
300	0.99999	0.99994	0.99939	0.99402	0.9541
500	1.00000	1.00002	1.00022	1.00222	1.0256
1000	1.00000	1.00003	1.00029	1.00288	1.0296
1500	1.00000	1.00002	1.00022	1.00221	1.0224
2000	1.00000	1.00002	1.00017	1.00173	1.0175
3000	1.00000	1.00001	1.00012	1.00117	1.0118

Water vapor					
	1 atm	*20 atm*	*100 atm*	*200 atm*	*300 atm*
400	0.98912				
500	0.99583	0.90274			
600	0.99790	0.95622	0.7180		
700	0.99880	0.97570	0.86977	0.70978	0.4944
800	0.99927	0.98527	0.92538	0.84804	0.7676
850	0.99942	0.98832	0.94181	0.88371	0.8249

SOURCE: Abridged by permission, from "Tables of Thermal Properties of Gases," U.S. Department of Commerce, National Bureau of Standards Circular 564, 1955.

Liquids

TABLE 24. Physical Properties of Some Liquid Metals
(Read across to facing page)

Element	Atomic weight	Melting point °C	Melting point °F	Latent heat of fusion, cal/g	Boiling point (760 mm Hg) °C	Boiling point (760 mm Hg) °F	Latent heat of vap., cal/g	Vapor pressure mm Hg	Vapor pressure °C	Vapor pressure °F	Density g/cm³	Density °C
Aluminum Al, 13	26.97	660.2	1220.4	96	2450	4442	3050	1	1537	2699	2.380	660
								10	1777	3231	2.369	700
								100	2081	3778	2.315	900
								200			2.261	1100
								400	2360	4280		
Antimony Sb, 51	121.76	630.5	1167	38.3	1440	2624	383	1	886	1627	6.49	640
								10	1033	1890	6.45	700
								100	1223	2233	6.38	800
								200	1288	2350	6.29	970
								400	1364	2487		
Bismuth Bi, 83	209.00	271.0	520	12.0	1477	2691	204.3	1	917	1683	10.03	300
								10	1067	1953	9.91	400
								100	1257	2295	9.66	600
								200	1325	2417	9.40	802
								400	1400	2552	9.20	962
Cadmiun Cd, 48	112.41	321	609	13.2	765	1409	286.4	1	394	741	8.01	330
								10	484	903	7.99	350
								100	611	1132	7.93	400
								200	658	1216	7.82	500
								400	711	1312	7.72	600
Cesium Cs, 55	132.91	28.5	83	3.766	705	1301	146.0	1	278	532		
								10	377	729		
								100	515	959	1.84	28
								200	570	1058		
								400	635	1175		
Gallium Ga, 31	69.72	29.92	85.86	19.16	1983	3601	1014	1	1315	2399	6.093	32.38
								10	1497	2727	5.905	301
								100	1726	3139	5.720	600
								200	1807	3285	5.604	806
								400	1895	3443	5.445	1100
Gold Au, 79	197.2	1063	1945	14.96	2966	5371	415	1	1869	3396	17.24	1100
								10	2154	3909	17.12	1200
								100	2521	4570	17.00	1300
								200	2657	4815		
								400	2808	5085		
Indium In, 49	114.76	156.4	313	6.807	2087	3789	468	1	1249	2280	7.026	164
								10	1466	2671	7.001	194
								100	1756	3193	6.974	228
								200	1863	3391	6.939	271
								400	1982	3600	6.916	300
Lead Pb, 82	207.21	327.4	621	5.89	1737	3159	204.6	1	987	1809	10.51	400
								10	1167	2133	10.39	500
								100	1417	2583	10.27	600
								200	1508	2746	10.04	800
								400	1611	2932	9.81	1000
Lithium Li, 3	6.94	179	354	158	1317	2403	4680	1	745	1373	0.507	200
								10	890	1634	0.490	400
								100	1084	1983	0.474	600
								200	1156	2113	0.457	800
								400	1236	2257	0.441	1000
Magnesium Mg. 12	24.32	651	1204	82.2	1103	2017	1337	1	621	1150	1.572	651
								10	743	1369	1.55	678
								100	909	1668	1.536	700
								200	967	1773	1.51	720
								400	1034	1893	1.47	750

* Na, 56 wt %, K, 44 wt %.
† Na, 22 wt %, K, 78 wt %.

TABLE 24. Physical Properties of Some Liquid Metals
(Read across from facing page)

Heat capacity		Viscosity		Thermal conductivity		Electrical resistivity		Surface tension		Vol.ch. on fusion, % of sol.vol.	Element
cal/g-°C	°C	centipoises (poises x 10²)	°C	$\dfrac{cal}{sec\text{-}cm\text{-}°C}$	°C	μΩ	°C	dyn/cm	°C		
	660	2.9	700	.247	700	19.6	657	520	20	6.6	
		1.4	800	.290	790	20.5	670				
0.259	to					21.3	735				Aluminum
						22.4	807				
	1000					23.2	870				
	650	1.296	702	0.052	630	117.00	627	383	635	-0.94	
		1.113	801			117.65	700	384	675		
0.0656	to	0.994	900	to		120.31	800	383	725		Antimony
		0.905	1002			123.54	850	380	800		
	950			0.05	730	131.00	900				
0.0340	271	1.662	304	0.041	300	128.9	300	376	300	-3.32	
0.0354	400	1.280	451	0.037	400	134.2	400	373	350		
0.0376	600	0.996	600	0.037	500	145.25	600	370	400		Bismuth
0.0397	800			0.037	600	153.53	750	367	450		
0.0419	1000			0.037	700			363	500		
	321	2.37	350	0.106	355	33.7	325	564	330	4.74	
		2.16	400	0.105	358	33.7	400	608	370		
0.0632	to	1.84	500	0.105	380	34.12	500	598	420		Cadmium
		1.54	600	0.119	435	34.82	600	611	450		
	700					35.78	700	600	500		
		0.6299	43.4			36.6	30			2.6	
		0.4753	99.6			37.0	37				
0.060	28.5	0.4065	140.5	0.044	mp						Cesium
		0.3750	168.0								
		0.3430	210.9								
	12.5	1.894	52.9	0.07		25.9	29.75	735	30	-3.1	
		1.029	301			27.2	30.3				
0.082	to	0.8783	402	to	mp	28.4	46.1		to		Gallium
		0.8113	500								
	200	0.6524	806	0.09					40		
	1063					31.34	1100	1128	1120	5.195	
						32.76	1200				
0.0355	to					34.17	1300				Gold
						35.58	1400				
	1300					37.00	1500				
				0.09		29.10	154.0		170	2.5	
						30.11	181.5				
0.0652	156.4			to	mp	31.87	222.0	340	to		Indium
						34.84	280.2				
				0.12					250		
0.039	327	2.116	441	0.039	330	94.6	327	442	350	3.6	
0.037	400	2.059	456	0.038	400	98.0	400	438	400		
0.037	500	1.700	551	0.037	500	107.2	600	438	450		Lead
		1.349	703	0.036	600	116.4	800	431	500		
		1.185	844	0.036	700	125.7	1000				
1.0	200	0.5918	183.4		218	45.25	230			1.5	
1.0	600	0.5749	193.2								
1.0	1000	0.5541	208.1	0.09	to						Lithium
		0.4917	250.8								
		0.4548	285.5		233						
								563	681	4.2	
0.317	651										
0.321	727							502	894		
0.332	927										Magnesium
0.337	1027										
0.342	1120										

‡Pb, 44.5 wt.%, Bi, 55.5 wt.%.

SOURCE: "Liquid Metals Handbook," 2d ed. (revised), Government Printing Office, Washington, D.C., 1954

TABLE 24. Physical Properties of Some Liquid Metals *(Continued)*

(Read across to facing page)

Element	Atomic weight	Melting point		Latent heat of fusion, cal/g	Boiling point (760 mm Hg)		Latent heat of vap., cal/g	Vapor pressure			Density	
		°C	°F		°C	°F		mm Hg	°C	°F	g/cm³	°C
Mercury Hg. 80	200.61	-38.87	-37.97	2.8	357	675	69.7	1	126.2	259	13.645	-20
								10	184.0	363	13.546	20
								100	261.7	503	13.352	100
								200	290.7	555	13.115	200
								400	323.0	613	12.881	300
Potassium K, 19	39.096	63.7	147	14.6	760	1400	496	1	342	648	0.819	100
								10	443	829	0.783	250
								100	581	1078	0.747	400
								200	635	1175	0.711	550
								400	696	1285	0.676	700
Rubidium Rb, 37	85.48	39.00	102	6.1	688	1270	212	1	294	561	1.475	39.0
								10	387	729		
								100	519	966		
								200	569	1056		
								400	628	1162		
Silver Ag, 47	107.88	960.5	1761	24.9	2212	4013	556	1	1357	2475	9.3	960.5
								10	1575	2857	9.26	1000
								100	1865	3389	9.20	1092
								200	1971	3580	9.10	1195
								400	2090	3794	9.00	1300
Sodium Na, 11	22.997	97.8	208	27.05	883	1621	1005	1	440	824	0.928	100
								10	548	1018	0.891	250
								100	696	1285	0.854	400
								200	752	1386	0.817	550
								400	815	1499	0.780	700
Thallium Tl, 81	204.39	303	577	5.04	1457	2655	189.9	1	825	1517	11.289	306.5
								10	983	1801	11.254	326.7
								100	1196	2185	11.250	330.0
								200	1274	2325	11.254	333.5
								400	1364	2487		
Tin Sn, 50	118.70	231.9	449	14.5	2270	4118	573	1	1492	2718	6.834	409
								10	1703	3097	6.761	523
								100	1968	3574	6.729	574
								200	2063	3745	6.671	648
								400	2169	3931	6.640	704
Zinc Zn, 30	65.38	419.5	787	24.4	906	1663	419.5	1	487	909	6.92	419.5
								10	593	1099	6.81	600
								100	736	1357	6.57	800
								200	788	1450		
								400	844	1551		
Sodium– potassium alloy*	28.1 (calc.)	19	66.2		825	1518		1	382	720	0.886	100
								10	490	914	0.850	250
								100	638	1180	0.814	400
								200	696	1285	0.778	550
								400	760	1400	0.742	700
Sodium– potassium alloy,† near eutectic	33.9 (calc.)	-11	12		784	1443		1	355	671	0.847	100
								10	458	856	0.811	250
								100	603	1117	0.775	400
								200	659	1218	0.739	550
								400	721	1330	0.703	700
Lead-bismuth alloy,‡ eutectic	208 (calc.)	125	257		1670	3038					10.46	200
											10.19	400
											9.91	600
											9.64	800
											9.36	1000

* Na, 56 wt %, K, 44 wt %.
† Na, 22 wt %, K, 78 wt %.

TABLE 24. Physical Properties of Some Liquid Metals
(Read across from facing page)

Heat capacity		Viscosity		Thermal conductivity		Electrical resistivity		Surface tension		Vol.ch. on fusion, % of sol.vol.	Element
cal/g-°C	°C	centipoises (poises × 10²)	°C	$\frac{cal}{sec\text{-}cm\text{-}°C}$	°C	$\mu\Omega$	°C	dyn/cm	°C		
0.03334	0	1.85	−20	0.0196	0	98.4	50	465	20	3.6	
0.03279	100	1.68	0	0.0231	60	103.2	100	454	112		
0.03245	200	1.55	20	0.0261	120	114.2	200	436	200		Mercury
0.03254	300	1.21	100	0.0279	160	127.5	300	405	300		
0.03256	450	1.01	200	0.0303	220	135.5	350	394	354		
0.1956	75	0.515	69.6	0.1073	200	13.16	64		100	2.41	
0.1887	200	0.331	167.4	0.1013	300	18.70	150				
0.1826	400	0.258	250	0.0956	400	25.00	250	86	to		Potassium
0.1825	600	0.191	400	0.0898	500	28.2	300				
0.1884	800	0.136	700	0.0846	600	31.4	350		150		
	39.0	0.6734	38	0.07	mp	23.15	50			2.5	
		0.6258	50			25.32	75				
0.0913	to	0.4884	99.7			27.47	100				Rubidium
		0.4133	140.5								
	126	0.3234	220.1	0.075	50						
	960.5	2.98	1200			17.0	1000	923	995	4.99	
						18.2	1100				
0.0692	to					19.4	1200				Silver
						20.5	1300				
	1300					21.0	1340				
0.3305	100	0.686	103.7	0.2055	100	9.65	100	206.4	100	2.5	
0.3200	200	0.504	167.6	0.1947	200	13.18	200	199.5	250		
0.3055	400	0.381	250	0.1809	300	14.90	250				Sodium
0.2998	600	0.269	400	0.1701	400	16.70	300				
0.3030	800	0.182	700	0.1596	500	18.44	350				
	303			0.059	350	74.0	303	401	327	3.2	
0.0367	to										Thallium
	500										
0.0580	250	1.91	240	0.08	240	47.6	231.9	526	300	2.6	
		1.67	300	0.081	292	51.4	400	522	350		
		1.38	400	0.079	417	56.8	600	518	400		Tin
		1.18	500	0.078	498	62.7	800	514	450		
0.0758	1100	1.05	600			68.6	1000	510	500		
0.1199	419.5	3.17	450	0.138	500	35.3	419.5	785	510	6.9	
0.1173	600	2.78	500	0.136	600	35.4	500	778	550		
0.1076	800	2.24	600	0.135	700	35.0	600	768	600		Zinc
0.1044	900	1.88	700			35.65	700	761	640		
0.1012	1000					35.70	800				
0.269	100	0.546	103.7	0.0617	100	33.0	50	110−	mp	2.5	
0.255	300	0.412	167.5	0.0633	200	35.5	100	100			Sodium-potassium alloy
0.249	500	0.316	250	0.0648	300	38.0	150		to		
0.248	600	0.230	400	0.0662	400	41.0	200				
0.253	800	0.161	700	0.0675	500				250		
0.238	0	0.468	103.7	0.0583	100	37.5	50	120−	mp	2.5	
0.217	200	0.359	167.5	0.0636	400	41.0	100	110			Sodium-potassium alloy
0.210	400	0.279	250			44.0	150		to		
0.209	600	0.205	400			47.0	200				
0.213	800	0.146	700						250		
	144	1.7	332	0.026	300	113	200	367	800	0.0	
		1.38	450	0.022	160	118	300	356	1000		Lead–bismuth alloy
0.035	to	1.29	500	0.023	200	123	400				
		1.23	550	0.024	240	128	500				
	358	1.17	600	0.027	320						

‡Pb, 44.5 wt.%, Bi, 55.5 wt.%.

SOURCE: "Liquid Metals Handbook," 2d ed. (revised), Government Printing Office, Washington, D.C., 1954.

TABLE 25. Property Values of Fluids in the Saturated State

t °F	p lb/ft³	c_p Btu/lb °F	ν ft²/sec	k Btu/hr ft °F	α ft²/hr	Pr	β $\frac{1}{R}$
			Water (H_2O)				
32	62.57	1.0074	1.925×10^{-5}	0.319	5.07×10^{-3}	13.6	
68	62.46	0.9988	1.083	0.345	5.54	7.02	0.10×10^{-3}
104	62.09	0.9980	0.708	0.363	5.86	4.34	
140	61.52	0.9994	0.514	0.376	6.02	3.02	
176	60.81	1.0023	0.392	0.386	6.34	2.22	
212	59.97	1.0070	0.316	0.393	6.51	1.74	
248	59.01	1.015	0.266	0.396	6.62	1.446	
284	57.95	1.023	0.230	0.395	6.68	1.241	
320	56.79	1.037	0.204	0.393	6.70	1.099	
356	55.50	1.055	0.186	0.390	6.68	1.004	
392	54.11	1.076	0.172	0.384	6.61	0.937	
428	52.59	1.101	0.161	0.377	6.51	0.891	
464	50.92	1.136	0.154	0.367	6.35	0.871	
500	49.06	1.182	0.148	0.353	6.11	0.874	
537	46.98	1.244	0.145	0.335	5.74	0.910	
572	44.59	1.368	0.145	0.312	5.13	1.019	
			Ammonia (NH_3)				
-58	43.93	1.066	0.468×10^{-5}	0.316	6.75×10^{-3}	2.60	
-40	43.18	1.067	0.437	0.316	6.88	2.28	
-22	42.41	1.069	0.417	0.317	6.98	2.15	
-4	41.62	1.077	0.410	0.316	7.05	2.09	
14	40.80	1.090	0.407	0.314	7.07	2.07	
32	39.96	1.107	0.402	0.312	7.05	2.05	
50	39.09	1.126	0.396	0.307	6.98	2.04	
68	38.19	1.146	0.386	0.301	6.88	2.02	1.36×10^{-3}
86	37.23	1.168	0.376	0.293	6.75	2.01	
104	36.27	1.194	0.366	0.285	6.59	2.00	
122	35.23	1.222	0.355	0.275	6.41	1.99	
			Carbon dioxide (CO_2)				
-58	72.19	0.44	0.128×10^{-5}	0.0494	1.558×10^{-3}	2.96	
-40	69.78	0.45	0.127	0.0584	1.864	2.46	
-22	67.22	0.47	0.126	0.0645	2.043	2.22	
-4	64.45	0.49	0.124	0.0665	2.110	2.12	
14	61.39	0.52	0.122	0.0635	1.989	2.20	
32	57.87	0.59	0.117	0.0604	1.774	2.38	
50	53.69	0.75	0.109	0.0561	1.398	2.80	
68	48.23	1.2	0.098	0.0504	0.860	4.10	7.78×10^{-3}
86	37.32	8.7	0.086	0 0406	0.108	28.7	
			Sulfur dioxide (SO_2)				
-58	97.44	0.3247	0.521×10^{-5}	0.140	4.42×10^{-3}	4.24	
-40	95.94	0.3250	0.456	0.136	4.38	3.74	
-22	94.43	0.3252	0.399	0.133	4.33	3.31	
-4	92.93	0.3254	0.349	0.130	4.29	2.93	
14	91.37	0.3255	0.310	0.126	4.25	2.62	
32	89.80	0.3257	0.277	0.122	4.19	2.38	
50	88.18	0.3259	0.250	0.118	4.13	2.18	
68	86.55	0.3261	0.226	0.115	4.07	2.00	1.08×10^{-3}
86	84.86	0.3263	0.204	0.111	4.01	1.83	
104	82.98	0.3266	0.186	0.107	3.95	1.70	
122	81.10	0.3268	0.174	0.102	3.87	1.61	

TABLE 25. Property Values of Fluids in the Saturated State
(Continued)

t °F	ρ lb/ft³	c_p $\frac{Btu}{lb\ °F}$	ν ft²/sec	k $\frac{Btu}{hr\ ft\ °F}$	α $\frac{ft^2}{hr}$	Pr	β $\frac{1}{R}$
			Methylchloride (CH₃Cl)				
-58	65.71	0.3525	0.344×10^{-5}	0.124	5.38×10^{-3}	2.31	
-40	64.51	0.3541	0.342	0.121	5.30	2.32	
-22	63.46	0.3564	0.338	0.117	5.18	2.35	
-4	62.39	0.3593	0.333	0.113	5.04	2.38	
14	61.27	0.3629	0.329	0.108	4.87	2.43	
32	60.08	0.3673	0.325	0.103	4.70	2.49	
50	58.83	0.3726	0.320	0.099	4.52	2.55	
68	57.64	0.3788	0.315	0.094	4.31	2.63	
86	56.38	0.3860	0.310	0.089	4.10	2.72	
104	55.13	0.3942	0.303	0.083	3.86	2.83	
122	53.76	0.4034	0.295	0.077	3.57	2.97	
		Dichlorodifluoromethane (Freon) (CCl₂F₂)					
-58	96.56	0.2090	0.334×10^{-5}	0.039	1.94×10^{-3}	6.2	1.46×10^{-3}
-40	94.81	0.2113	0.300	0.040	1.99	5.4	
-22	92.99	0.2139	0.272	0.040	2.04	4.8	
-4	91.18	0.2167	0.253	0.041	2.09	4.4	
14	89.24	0.2198	0.238	0.042	2.13	4.0	
32	87.24	0.2232	0.230	0.042	2.16	3.8	
50	85.17	0.2268	0.219	0.042	2.17	3.6	
68	83.04	0.2307	0.213	0.042	2.17	3.5	
86	80.85	0.2349	0.209	0.041	2.17	3.5	
104	78.48	0.2393	0.206	0.040	2.15	3.5	
122	75.91	0.2440	0.204	0.039	2.11	3.5	
		Eutectic calcium chloride solution (29.9% CaCl₂)					
-58	82.39	0.623	39.13×10^{-5}	0.232	4.52×10^{-3}	312	
-40	82.09	0.6295	26.88	0.240	4.65	208	
-22	81.79	0.6356	18.49	0.248	4.78	139	
-4	81.50	0.642	11.88	0.257	4.91	87.1	
14	81.20	0.648	7.49	0.265	5.04	53.6	
32	80.91	0.654	4.73	0.273	5.16	33.0	
50	80.62	0.660	3.61	0.280	5.28	24.6	
68	80.32	0.666	2.93	0.288	5.40	19.6	
86	80.03	0.672	2.44	0.295	5.50	16.0	
104	79.73	0.678	2.07	0.302	5.60	13.3	
122	79.44	0.685	1.78	0.309	5.69	11.3	
		Glycerin [C₃H₅(OH)₃]					
32	79.66	0.540	0.0895	0.163	3.81×10^{-3}	84.7×10^{-3}	
50	79.29	0.554	0.0323	0.164	3.74	31.0	
68	78.91	0.570	0.0127	0.165	3.67	12.5	0.28×10^{-3}
86	78.54	0.584	0.0054	0.165	3.60	5.38	
104	78.16	0.600	0.0024	0.165	3.54	2.45	
122	77.72	0.617	0.0016	0.166	3.46	1.63	
		Ethylene glycol [C₂H₄(OH₂)]					
32	70.59	0.548	61.92×10^{-5}	0.140	3.62×10^{-3}	615	
68	69.71	0.569	20.64	0.144	3.64	204	0.36×10^{-3}
104	68.76	0.591	9.35	0.148	3.64	93	
140	67.90	0.612	5.11	0.150	3.61	51	
176	67.27	0.633	3.21	0.151	3.57	32.4	
212	66.08	0.655	2.18	0.152	3.52	22.4	

TABLE 25. Property Values of Fluids in the Saturated State
           (Continued)

t °F	ρ lb/ft^3	c_p $\dfrac{\text{Btu}}{\text{lb °F}}$	ν ft^2/sec	k $\dfrac{\text{Btu}}{\text{hr ft °F}}$	α $\dfrac{\text{ft}^2}{\text{hr}}$	Pr	β $\dfrac{1}{R}$
				Engine oil (unused)			
32	56.13	0.429	0.0461	0.085	3.53×10^{-3}	47100	
68	55.45	0.449	0.0097	0.084	3.38	10400	0.39×10^{-3}
104	54.69	0.469	0.0026	0.083	3.23	2870	
140	53.94	0.489	0.903×10^{-3}	0.081	3.10	1050	
176	53.19	0.509	0.404	0.080	2.98	490	
212	52.44	0.530	0.219	0.079	2.86	276	
248	51.75	0.551	0.133	0.078	2.75	175	
284	51.00	0.572	0.086	0.077	2.66	116	
320	50.31	0.593	0.060	0.076	2.57	84	
				Mercury (Hg)			
32	850.78	0.0335	0.133×10^{-5}	4.74	166.6×10^{-3}	0.0288	
68	847.71	0.0333	0.123	5.02	178.5	0.0249	1.01×10^{-4}
122	843.14	0.0331	0.112	5.43	194.6	0.0207	
212	835.57	0.0328	0.0999	6.07	221.5	0.0162	
302	828.06	0.0326	0.0918	6.64	246.2	0.0134	
392	820.61	0.0375	0.0863	7.13	267.7	0.0116	
482	813.16	0.0324	0.0823	7.55	287.0	0.0103	
600	802	0.032	0.0724	8.10	316	0.0083	

SOURCE: E. R. G. Eckert and Robert M. Drake, Jr., "Heat and Mass Transfer," 2d ed., McGraw-Hill Book Company, Inc., New York, 1959.

TABLE 26. Surface Tension

(a) Liquids: Water and Alcohol in Contact with Moist Air, γ, dyn/cm

°C	H_2O	C_2H_5OH
-5	´76.4	
0	75.6	24.0
5	74.8	23.5
10	74.2	23.1
15	73.4	22.7
20	72.7	22.3
25	71.9	21.8
30	71.1	21.4
35	70.3	21.0
40	69.5	20.6
45	68.7	20.2
50	67.9	19.8
55	67.0	19.4
60	66.1	19.0
65	67.0	18.6
70	64.3	18.2
75	64.3	
80	62.5	
85	61.6	
90	60.7	
95	59.8	
100	58.8	

(b) Substances at Solidifying Point

Substance	Temperature solidification °C	γ dyn/cm
Antimony	432	249
Borax	1000	216
Carbonate of soda	1000	210
Chloride of sodium		116
Water	0	87.9
Selenium	217	71.8
Sulfur	111	42.1
Phosphorus	43	42.0
Wax	68	34.1
Platinum	2000	1691
Gold	1200	1003
Zinc	360	877
Tin	230	599
Mercury	-40	588
Lead	330	457
Silver	1000	427
Bismuth	265	1390
Potassium	58	371
Sodium	90	258

(c) Solutions of Salts in Water

Salt	% Salt	°C	γ dyn/cm
$BaCl_2$	0	30	71.1
	24.6	30	75.6
$CaCl_2$	0	30	71.1
	12.3	30	75.7
	31.9	30	86.4
HCl	0	20	73.0
	15	20	72.0
	25	20	70.7
KCl	0	30	71.1
	23.3	30	76.8
	21.1	18	77.7
NaCl	0	18	72.4
	7.6	18	74.8
	13.7	18	76.9
NH_4Cl	0	18	72.5
	11	18	74.9
K_2CO_3	0	30	71.1
	39.4	30	89.4
	53.6	30	107.2
Na_2CO_3	0	30	71.1
	10.5	30	73.9
	24.4	30	76.5
	63.1	30	80.6
KNO_3	0	18	72.6
	15.2	18	74.5
	21.5	18	75.4
$NaNO_3$	0	30	71.1
	35.6	30	78.4
	50.9	30	82.8
$CuSO_4$	0	30	71.1
	25.4	30	74.1
H_2SO_4	0	18	72.8
	12.7	18	73.5
	47.6	18	76.7
	80.3	18	71.2
	90	18	63.6
K_2SO_4	0	18	72.7
	9.1	18	74.6
HNO_3	7.2	20	73.1
	50	20	65.4
	70	20	59.4
NaOH	0	20	72.8
	10	20	77.3
	20	20	85.8
	30	20	95.1
KOH	0	18	72.8
	3.8	18	74.1
	7.8	18	75.5

TABLE 26. Surface Tension *(Continued)*

(d) Miscellaneous Liquids in Contact with Air

Liquid	Formula	°C	γ dyn/cm
Acetone	$(CH_3)_2CO$	20	23.7
Acetic acid	CH_3CO_2H	20	27.6
Amyl alcohol	$C_5H_{12}O$	20	24
Aniline	C_6H_7N	20	43
Benzene	C_6H_6	0	27
		20	28.9
Bromoform	$CHBr_3$	20	41.5
Butyric acid	$CH_3(CH_2)_2CO_2H$	15	26.7
Carbon disulfide	CS_2	20	32.3
Carbon tetrachloride	CCl_4	20	26.8
Chloroform	$CHCl_3$	20	27.2
Ether	$C_4H_{10}O$	20	17.01
Ethyl chloride	CH_3Cl	20	16.2
Glycerine	$C_3H_5(OH)_3$	18	63
Methyl alcohol	CH_3OH	20	22.6
Olive oil		18	33.1
Petroleum		25	26
Phenol	C_6H_6O	20	41.0
Propyl alcohol	$CH_3(CH_3)_1OH$	20	23
Silicon tetrachloride	$SiCl_4$	19	17.0
Toluene	C_7H_8	20	28.4
Turpentine		20	27

(e) Liquids with Air, Water, and Mercury

Liquid	Specific gravity	γ–Surface tension in dyn/cm of liquid in contact with–		
		Air	Water	Mercury
Water	1.9	75.0	.0	(392)
Mercury	13.595	513.0	392.0	0
Bisulfide of carbon	1.2687	30.5	41.7	(387)
Chloroform	1.498	(31.8)	26.8	(415)
Ethyl alcohol	.807	(24.1)		364
Olive oil	.918	34.6	18.6	317
Turpentine	.873	28.8	11.5	241
Petroleum	.870	29.7	(28.9)	271
Hydrochloric acid	1.10	(72.9)		(392)
Hyposulfite of soda solution	1.1248	69.9		429

SOURCE: "Smithsonian Physical Tables," 9th ed., prepared by William Elmer Forsythe, 1964. Reprinted by permission of the Smithsonian Institution.

Latent Heat of Vaporization

The results of Lydersen, Greenkorn, and Haugen* are used to obtain the latent heat of vaporization λ. In Fig. 4.1, values of λ/T_c are plotted at left as a function of critical state compressibility factor Z_c for different values of reduced temperature at saturation Tr_s and again (at right) as a function of Tr_s for various values of Z_c. Values of crititical temperature T_c and critical compressibility factor Z_c are found in Table 11.

As an example, estimate the molar heat of vaporization of acetone at (a) its normal boiling point and (b) at a pressure of 8.0 atm.

Boiling point at 1 atm = 329.7°K
at 8 atm = 406.3°K
From Table 11 T_c = 508.7°K
Z_c = 0.238
(a) Tr = 329.7/508.7 = 0.648

Using the right-hand graph of Fig. 4.1, a value between 14.1 and 14.2 is read for λ/T_c. This corresponds to a value of λ between 7160 cal/gm-mole and 7230 cal/gm-mole. The experimental value is 7231 cal/gm-mole.

(b) $Tr = \dfrac{406.3}{5087} = 0.798$

Either the right- or left-hand graph of Fig. 4.1 may be used. A value of between 11.4 and 11.5 is read for λ/T_c. This corresponds to a value of λ between 5800 cal/gm-mole and 5850 cal/gm-mole.

*A. L. Lydersen, R. A. Greenkorn, and O. A. Haugen, "Generalized Thermodynamic Properties of Pure Fluids," 1955. Courtesy of the University of Wisconsin Engineering Experiment Station.

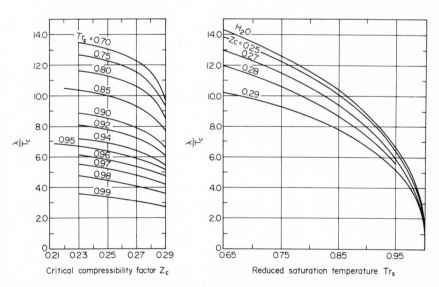

Fig. 4. Latent heats of vaporization (dimensionless).

Solids

TABLE 27. Physical Properties of Principal Alloy-forming Elements

	Atomic No.	Atomic weight	Density lb/in³	Melting point °F	Boiling point °F	Specific heat‡	Latent heat of fusion Btu/lb	Linear coef. of thermal exp. per °F $\times 10^{-8}$	Thermal conductivity Btu/sq ft hr in. °F	Electrical resistivity ohm-cm	Modulus of elasticity (tension) psi $\times 10^8$	Crystal structure‡	Transition temp °F	Symbol
Aluminum	13	26.97	0.09751	1220.4	3740	0.215	170	13.3	1740	2.655	10	FCC		Al
Antimony	51	121.76	0.239	1166.9	2620	0.049	68.9	4.7-6.0	131	39.0	11.3	Rhom		Sb
Arsenic	33	74.91	0.207	1497	1130	0.082	159	2.6		35	11	Rhom		As
Barium	56	137.36	0.13	1300	2980	0.068		(10)	1100	50	1.8	BCC		Ba
Beryllium	4	9.02	0.0658	2340	5020	0.52	470	6.9		5.9	37	HCP		Be
Bismuth	83	209.00	0.354	520.3	2590	0.029	22.5	7.4	58	106.8	4.6	Rhom		Bi
Boron	5	10.82	0.083	3812	4620	0.309		4.6		1.8×10^{12}		?		B
Cadmium	48	112.41	0.313	609.6	1409	0.055	23.8	16.6	639	6.83	8	HCP		Cd
Calcium	20	40.08	0.056	1560	2625	0.149	100	12	871	3.43	3	FCC/BCC	867	Ca
Carbon	6	12.010	0.0802	6700	8730	0.165		0.3-2.4	165	1375	0.7	Hex/D		C
Cerium	58	140.13	0.25	1460	4380	0.042	27.2			78		HCP/FCC/?	572/1328	Ce
Chromium	24	52.01	0.260	3350	4500	0.11	146	3.4	464	13	36	BCC/FCC	3344	Cr
Cobalt	27	58.94	0.32	2723	6420	0.099	112	6.8	479	6.24	30	HCP/FCC/HCP	783/2048	Co
Columbium	41	92.91	0.310	4380	5970	0.065		4.0		13.1	15	BCC		Cb
Copper	29	63.54	0.324	1981.4	4700	0.092	91.1	9.2	2730	1.673	16	FCC		Cu
Gadolinium	64	156.9	0.287						232			HCP		Gd
Gallium	31	69.72	0.216	85.5	3600	0.0977	34.5	10.1		56.8	1	O		Ga
Germanium	32	72.60	0.192	1756	4890	0.086	205.7	(3.3)		60×10^4	11.4	D		Ge
Gold	79	197.2	0.698	1945.4	5380	0.031	29.0	7.9	206	2.19	12	FCC		Au
Hafnium	72	178.6	0.473	3865	9700	0.0351		(3.3)		32.4	20	HCP/BCC	3540	Hf
Hydrogen	1	1.0080	3.026×10^{-6}	-434.6	-422.9	3.45	27.0		1.18			Hex		H
Indium	49	114.76	0.264	313.5	3630	0.057	12.2	18	175	8.37	1.57	FCT		In
Iridium	77	193.1	0.813	4449	9600	0.031	47	3.8	406	5.3	75	FCC		Ir
Iron	26	55.85	0.284	2802	4960	0.108	117	6.50	523	9.71	28.5	BCC/FCC/BCC	1663/2554	Fe
Lanthanum	57	138.92	0.223	1535	8000	0.0448				59	5	HCP/FCC/?	662/1427	La
Lead	82	207.21	0.4097	621.3	3160	0.031	11.3	16.3	241	20.65	2.6	FCC		Pb

Element	At. no.	At. wt.											Crystal structure		Symbol
Lithium	3	6.940	0.019	367	2500	0.79	286	31	494	11.7	1.7	1340/2010/2080	BCC		Li
Magnesium	12	24.32	0.0628	1202	2030	0.25	160	14	1100	4.46	6.5		HCP		Mg
Manganese	25	54.93	0.268	2273	3900	0.115	115	12		185	23		CCX/CCX/FCT?		Mn
Mercury	80	200.61	0.4896	-37.97	675	0.033	4.9	33.8	58	94.1	50		Rhom		Hg
Molybdenum	42	95.95	0.369	4750	8670	0.061	126	3.0	1020	5.17	30		BCC		Mo
Nickel	28	58.69	0.322	2651	4950	0.105	133	7.4	639	6.84			FCC		Ni
Nitrogen	7	14.008	0.042×10^{-3}	-346	-320.4	0.247	11.2		0.147				Hex		N
Osmium	76	190.2	0.813	4900	9900	0.031		2.6		9.5	80		HCP		Os
Oxygen	8	16.000	0.048×10^{-3}	-361.8	-297.4	0.218	5.9		0.171				C		O
Palladium	46	106.7	0.434	2829	7200	0.058	69.5	6.6	494	10.8	17		FCC		Pd
Phosphorus	15	30.98	0.0658	111.4	536	0.177	9.0	70		10^{17}			C		P
Platinum	78	195.23	0.7750	3224.3	7970	0.032	49	4.9	494	9.83	21	6 Forms	FCC		Pt
Plutonium	94	239	0.686	1229				50-65		150			?		Pu
Potassium	19	39.096	0.031	145	1420	0.177	26.1	46	697	6.15	0.5		BCC		K
Radium	88	226.05	0.18	1300									?		Ra
Rhenium	75	186.31	0.765	5733	10,700	0.0326	76	4.6	610	21	75		HCP		Re
Rhodium	45	102.91	0.4495	3571	8100	0.059		21	3	4.5	54		FCC		Rh
Selenium	34	78.96	0.174	428	1260	0.084	29.6	1.6-4.1	581	12	8.4	248	MC/Hex		Se
Silicon	14	28.06	0.084	2605	4200	0.162	607	10.9	2900	10^5	16		D		Si
Silver	47	107.88	0.379	1760.9	4010	0.056	45.0	39	929	1.59	11		FCC		Ag
Sodium	11	22.997	0.035	207.9	1638	0.295	49.5	36	1.83	4.2	1.3		BCC		Na
Sulphur	16	32.066	0.0748	246.2	832.3	0.175	16.7			2×10^{23}		204	Rhom/FCC		S
Tantalum	73	180.88	0.600	5420	9570	0.036	13.1	3.6	377	12.4	27		BCC		Ta
Tellurium	52	127.61	0.225	840	2530	0.047	9.1	9.3	41	2×10^5	6		Hex		Te
Thallium	81	204.39	0.428	577	2655	0.031	35.6	16.6	2700	18	1.2	446	HCP/BCC		Tl
Thorium	90	232.12	0.422	3348	8100	0.0355	26.1	6.2		18.6	11.4		FCC		Th
Tin	50	118.70	0.264	449.4	4120	0.054	187	13	464	11.5	6		D/BCT		Sn
Titanium	22	47.90	0.164	3074	6395	0.139	79	4.7	1190	47.8	16.8	55	HCP/BCC		Ti
Tungsten	74	182.92	0.697	6150	10,700	0.032	19.8	2.4	900	5.5	50	1650	BCC		W
Uranium	92	238.07	0.687	2065	7100	0.028		11.4	186	29	29.7	1229/1427	O/Tet/BCC		U
Vanadium	23	50.95	0.217	3452	5430	0.120	43.3	4.3	215	26	18.4	2822	BCC/?		V
Zinc	30	65.38	0.258	787	1663	0.092		9.4-22	784	5.92	12		HCP		Zn
Zirconium	40	91.22	0.23	3326	9030	0.066		3.1	116	41.0	11	1585	HCP/BCC		Zr

† Cal per g per deg C at room temperature equals Btu per lb per deg F at room temperature.

‡ FCC = face-centered cubic; BCC = body-centered cubic; C = cubic; HCP = hexagonal closest packing; Rhom = rhombohedral; Hex = hexagonal; FCT = face-centered tetragonal; O = orthorhombic; FCO = face-centered orthorhombic; CCX = cubic complex; D = diamond cubic; BCT = body-centered tetragonal. MC = monoclinic.

SOURCE: L.S. Marks, "Mechanical Engineers' Handbook," 6th ed., McGraw-Hill Book Company, Inc., New York, 1958.

TABLE 28. Property Values of Metals

Metals	Properties at 68°F				k, thermal conductivity, Btu/hr ft °F									
	ρ lb/ft³	c_p Btu/lb·°F	k Btu/hr·ft·°F	α ft³/hr	-148°F -100°C	32°F 0°C	212°F 100°C	392°F 200°C	572°F 300°C	752°F 400°C	1112°F 600°C	1472°F 800°C	1832°F 1000°C	2192°F 1200°C
Aluminum:														
Pure	169	0.214	118	3.262	124	117	119	124	132	144				
Al-Cu (Duralumin) 94-96 Al, 3-5 Cu, trace Mg	174	0.211	95	2.587	73	92	105	112						
Al-Mg (Hydronalium) 91-95 Al, 5-9 Mg	163	0.216	65	1.846	54	63	73	82						
Al-Si (Silumin) 87 Al, 13 Si	166	0.208	95	2.751	86	94	101	107						
Al-Si (Silumin, copper bearing) 86.5 Al, 12.5 St, 1 Cu	166	0.207	79	2.299	69	79	83	88	93					
Al-Si (Alusil) 78-80 Al, 20-22 Si	164	0.204	93	2.779	83	91	97	101	103					
Al-Mg-Si 97 Al, 1 Mg, 1 Si, 1 Mn	169	0.213	102	2.833		101	109	118						
Lead	710	0.031	20	0.908	21.3	20.3	19.3	18.2	17.2					
Iron:														
Pure	493	0.108	42	0.788	50	42	39	36	32	28	23	21	20	21
Wrought iron (C < 0.5%)	490	0.11	34	0.630		34	33	30	28	26	21	19	19	19
Cast iron (C ≈ 4%)	454	0.10	30	0.660										
Steel (C max ≈ 1.5%)														
Carbon steel C ≈ 0.5%	489	0.111	31	0.571		32	30	28	26	24	20	18	17	18
1.0%	487	0.113	25	0.454		25	25	24	23	21	19	17	16	17
1.5%	484	0.116	21	0.376		21	21	21	20	19	18	16	16	17
Nickel steel Ni ≈ 0%	493	0.108	42	0.785										
10%	496	0.11	15	0.279										
20%	499	0.11	11	0.204										
30%	504	0.11	7	0.126										
40%	510	0.11	6	0.108										
50%	516	0.11	8	0.140										
60%	523	0.11	11	0.191										

Iron (cont.):													
70%	531	0.11	15	0.258									21
80%	538	0.11	20	0.338									17
90%	547	0.11	27	0.448									
100%	556	0.106	52	0.882									
Invar Ni ≈ 36%	508	0.11	6.2	0.111	50								
Chrome steel Cr ≈ 0%	493	0.108	42	0.785									
1%	491	0.11	35	0.645	42	39	36	32	28	23	21	20	
2%	491	0.11	30	0.559	36	32	30	27	24	21	19	19	
5%	489	0.11	23	0.430	31	28	26	24	22	19	18	18	
10%	486	0.11	18	0.336	23	22	21	21	19	17	17	17	
20%	480	0.11	13	0.246	18	18	18	17	17	16	16	17	
30%	476	0.11	11	0.210	13	13	13	13	14	14	15	17	
Cr-Ni (chrome-nickel):													
15 Cr, 10 Ni	491	0.11	11	0.204									
18 Cr, 8 Ni (V2A)	488	0.11	9.4	0.172	9.4	10	10	11	11	13	15	18	
20 Cr, 15 Ni	489	0.11	8.7	0.161									
25 Cr, 20 Ni	491	0.11	7.4	0.140									
Ni-Cr (nickel-chrome):													
80 Ni, 15 Cr	532	0.11	10	0.172									
60 Ni, 15 Cr	516	0.11	7.4	0.129									
40 Ni, 15 Cr	504	0.11	6.7	0.118									
20 Ni, 15 Cr	491	0.11	8.1	0.151	8.1	8.7	8.7	9.4	10	11	13		
Cr-Ni-Al: 6 Cr, 1.5 Al, 0.5 Si (Sicromal 8)	482	0.117	13	0.230									
24 Cr, 2.5 Al, 0.5 Si (Sicromal 12)	479	0.118	11	0.194									
Manganese steel Ma = 0%	493	0.118	42	0.722									
1%	491	0.11	29	0.538	22	21	21	21	20	19			
2%	491	0.11	22	0.407									
5%	490	0.11	13	0.247									
10%	487	0.11	10	0.187									
Tungsten steel W = 0%	493	0.108	42	0.785									
1%	494	0.107	38	0.720	36	34	31	28	26	21			
2%	497	0.106	36	0.683									

TABLE 28. Property Values of Metals *(Continued)*

| Metals | Properties at 68°F | | | | k, thermal conductivity, Btu/hr ft °F | | | | | | | | | |
	ρ lb/ft³	c_p Btu/lb-°F	k Btu/hr-ft-°F	α ft²/hr	-148°F -100°C	32°F 0°C	212°F 100°C	392°F 200°C	572°F 300°C	752°F 400°C	1112°F 600°C	1472°F 800°C	1832°F 1000°C	2192°F 1200°C
Tungsten steel W = 0% *(cont.)*														
5%	504	0.104	31	0.591										
10%	519	0.100	28	0.539										
20%	551	0.093	25	0.484										
Silicon steel Si = 0%	493	0.108	42	0.785										
1%	485	0.11	24	0.451										
2%	479	0.11	18	0.344										
5%	463	0.11	11	0.215										
Copper:														
Pure	559	0.0915	223	4.353	235	223	219	216	213	210	204			
Aluminum bronze 95 Cu, 5 Al	541	0.098	48	0.903										
Bronze 75 Cu, 25 Sn	541	0.082	15	0.333										
Red Brass 85 Cu, 9 Sn, 6 Zn	544	0.092	35	0.699		34	41							
Brass 70 Cu, 30 Zn	532	0.092	64	1.322	51		74	83	85	85				
German silver 62 Cu, 15 Ni, 22 Zn	538	0.094	14.4	0.284	11.1		18	23	26	28				
Constantan 60 Cu, 40 Ni	557	0.098	13.1	0.237	12		12.8	15						

Material														
Magnesium:														
Pure	109	0.242	99	3.762	103	99	97	94	91					
Mg-Al (electrolytic) 6-8% Al, 1-2% Zn	113	0.24	38	1.397		30	36	43	48					
Mg-Mn 2% Mn	111	0.24	66	2.473	54	64	72	75						
Molybdenum	638	0.060	71	1.856	80	72	68	66	64	63	61	59	57	53
Nickel:														
Pure (99.9%)	556	0.1065	52	0.878	60	54	48	42	37	34	32	36	39	40
Impure (99.2%)	556	0.106	40	0.677		40	37	34	32	30				
Ni-Cr 90 Ni, 10 Cr	541	0.106	10	0.172		9.9	10.9	12.1	13.2	14.2	13.0			
80 Ni, 20 Cr	519	0.106	7.3	0.133		7.1	8.0	9.0	9.9	10.9				
Silver:														
Purest	657	0.0559	242	6.589	242	241	240	238						
Pure (99.9%)	657	0.0559	235	6.418	242	237	240	216	209	208				
Tungsten	1208	0.0321	94	2.430		96	87	82	77	73	65	44		
Zinc, pure	446	0.0918	64.8	1.591	66	65	63	61	58	54				
Tin, pure	456	0.0541	37	1.505	43	38.1	34	33						

SOURCE: E. R. G. Eckert and Robert M. Drake, Jr., "Heat and Mass Transfer," 2d ed., McGraw-Hill Book Company, Inc., New York, 1959.

TABLE 29. Property Values, Selected Nonmetals

Material	t °F	ρ lb/ft^3	c_p Btu/lb-°F	k hr-ft-°F	α ft^2/hr
Aerogel, silica	248	8.5		0.013	
Asbestos	-328	29.3		0.043	
Asbestos	32	29.3		0.090	
Asbestos	32	36.0	0.195	0.087	
Asbestos	212	36.0	0.195	0.111	
Asbestos	392	36.0		0.120	
Asbestos	752	36.0		0.129	
Asbestos	-328	43.5		0.090	
Asbestos	32	43.5		0.135	
Brick, dry	68	110–113	0.20	0.22–0.30	0.011–0.013
Bakelite	68	79.5	0.38	0.134	0.0044
Cardboard, corrugated				0.037	
Clay	68	91.0	0.21	0.739	0.039
Concrete	68	119–144	0.21	0.47–0.81	0.019–0.027
Coal, anthracite	68	75–94	0.30	0.15	0.005–0.006
Coal, powdered	86	46	0.31	0.067	0.005
Cotton	68	5	0.31	0.034	0.075
Cork, board	86	10		0.025	
Cork, expanded scrap	68	2.8–7.4	0.45	0.021	0.006–0.017
Cork, ground	86	9.4		0.025	
Diatomaceous earth	100	20.0		0.036	
Diatomaceous earth	1600	20.0		0.082	
Earth, coarse gravelly	68	128	0.44	0.30	0.0054
Felt, wool	86	20.6		0.03	
Fiber, insulating board	70	14.8		0.028	
Fiber, red	68	80.5		0.27	
Glass plate	68	169	0.2	0.44	0.013
Glass, borosilicate	86	139		0.63	
Glass, wool	68	12.5	0.16	0.023	0.011
Granite				1.0–2.3	
Ice	32	57	0.46	1.28	0.048
Marble	68	156–169	0.193	1.6	0.054
Rubber, hard	32	74.8		0.087	
Sandstone	68	135–144	0.17	0.94–1.2	0.041–0.049
Silk	68	3.6	0.33	0.021	0.017
Wood, oak radial	68	38–50	0.57	0.10–0.12	0.0043–0.0047
Wood, fir (20% moisture) radial	68	26.0–26.3	0.65	0.08	0.0048

SOURCE: L. S. Marks, "Mechanical Engineers' Handbook," 5th ed., McGraw-Hill Book Company, Inc., New York, 1951; W. H. McAdams, "Heat Transmission," 3d ed., McGraw-Hill Book Company, Inc., New York, 1954; E. R. G. Eckert and Robert M. Drake, Jr., "Heat and Mass Transfer," 2d ed., McGraw-Hill Book Company, Inc., New York, 1959.

TABLE 30. Specific Heats of Solids* (Expressed in Btu/lb-°F — gm-cal/gm-°C)

t, °F	t, °C	Pb	Zn	Al	Ag	Au	Cu	Ni	Fe	Co	Quartz
32	0	0.0306	0.0917	0.2106	0.0557	0.0305	0.0919	0.1025	0.1051	0.1023	0.1667
212	100	0.0315	0.0958	0.2225	0.0571	0.0312	0.0942	0.1132	0.1166	0.1079	0.2061
392	200	0.0325	0.0999	0.2344	0.0585	0.0320	0.0965	0.1241	0.1280	0.1138	0.2315
572	300	0.0335	0.1041	0.2463	0.0599	0.0327	0.0988	0.1352	0.1395	0.1192	0.2518
752	400	0.0328	0.1082	0.2582	0.0612	0.0334	0.1011	0.1295	0.1508	0.1249	0.2696
932	500	0.0328	0.1225	0.2702	0.0626	0.0341	0.1034	0.1310	0.1622	0.1305	0.2865
1112	600	0.0328	0.1233	0.2821	0.0640	0.0349	0.1057	0.1326	0.1737	0.1362	0.2624
1292	700	0.0328	0.1242	0.259	0.0654	0.0356	0.1080	0.1341	0.1853	0.1418	0.2714
1472	800	0.0328	0.1250	0.259	0.0668	0.0363	0.1103	0.1356	0.1741	0.1475	0.2806
1652	900	0.0328	0.1259	0.259	0.0082	0.0371	0.1126	0.1372	0.1805	0.1531	0.2897
1832	1000	0.0328	0.1267	0.259	0.076	0.0378	0.1149	0.1387	0.1505	0.1588	0.2989
2012	1100				0.076	0.0355	0.118	0.1403	0.1505	0.1644	0.3080
2192	1200				0.076	0.0355	0.118	0.1418	0.1505	0.1701	0.3172
2372	1300				0.076	0.0355	0.118	0.1434	0.1505	0.1757	0.3264
2552	1400							0.1449	0.1505	0.1814	0.3355
2732	1500							0.1455	0.1790	0.1425	0.3447
2912	1600							0.1455	0.1460	0.1425	0.3538
Melting points °F		621	786	1220	1760	1945	1981	2646	2795	2696	

*Calculated values from equations by K. K. Kelley, *U.S. Bur. Mines Bull.* 371 (1934). These values are the true specific heats for the particular physical state or allotropic modification existing at the indicated temperature.
SOURCE: William H. McAdams, "Heat Transmission," 3d ed., McGraw-Hill Book Company, Inc., New York, 1954.

TRANSPORT PROPERTIES

Gas Viscosity

TABLE 31. Experimental Gas Viscosity Data at One Atmosphere

T, °K	Micropoises	T, °K	Micropoises	T, °K	Micropoises	T, °K	Micropoises
Air [6]		*Ammonia [14]*		*Carbon dioxide [8]*		*Hydrogen [8]*	
100	71.26	298.1	100.2	*(Cont.)*		289.7	86.8
150	103.71	323.1	109.2	572	263.7	473	121.5
200	133.64	373.1	127.9	682	302.1	572	138.1
250	160.78	406.0	140.0	765	328.3	685	155.4
300	185.09	423.1	146.3	885	363.9	763	167.2
Air [8]		473.1	165.5	985	390.6	874	182.9
289.3	179.1	523.1	181.3	1097	419.1	986	198.2
572	292.0	573.1	198.6			1098	213.7
681	327.1	*Argon [7]*		*Carbon dioxide [9]*		*Helium [7]*	
763	351.8	100	83.90	273.1	136.3	100	94.70
873	381.1	150	122.20	553.1	256.0	150	123.65
986	412.4	200	159.40	607.1	275.6	200	149.60
1098	440.3	250	194.65	609.3	276.5	250	174.00
Air [9]		300	226.95	704.1	308.5	300	198.70
273	172.0	*Argon [1]*		705.4	309.0		
289.6	180.3	273	210.1	759.3	326.0	*Helium [2]*	
294.4	182.4	323	240.1	804.1	340.0	298.36	198.5
379.1	221.0	373	268.2	861.1	357.2	348.22	220.5
399.85	229.4	423	294.7	923.3	375.0	366.53	228.2
433.1	242.2	473	319.9	959.1	383.3	420.46	250.6
443.2	246.3	523	343.9	1002.5	394.0	447.96	261.8
541.1	282.0	573	366.7	1060.3	411.0	512.45	288.7
570.3	293.0	673	409.6	1071.1	414.7	*Helium [8]*	
609.1	305.5	773	449.4	1134.1	431.0	294.2	195.0
645.0	316.7	873	486.5	1201.1	445.5	554	299.8
783.2	357.5	973	521.4	1231.4	454.0	665	338.8
854.2	377.6	1073	554.5	1258.1	460.3	767	373.0
917.2	394.0	1173	586.0	1263.6	460.4	881	409.3
933.1	399.0	1273	616.0	1281.1	465.5	955	431.7
958.3	406.8	1373	644.8	1300.1	477.0	1088	470.3
982.3	410.5	1473	672.5	1351.1	481.5		
988.1	414.0	1573	699.1	1375.6	486.8	*Methane [6]*	
1046.1	428.3	1673	724.9	1553.1	525.4	100	40.28
1066.3	432.0	1773	749.8	1560.1	528.8	150	59.14
1082.1	436.5	1873	774.0	1678.1	552.3	200	77.78
1109.1	441.7	1973	797.5	1686.1	553.7	250	95.30
1147.3	452.4	2073	820.4			300	111.59
1206.1	464.3			*Freon 12–Calculated*		*Methane [8]*	
1207.2	465.3	*Carbon monoxide [7]*		*from equation*		296.7	110.3
1247.4	473.2	100	66.85	*in Table 34*		557	181.3
1252.1	476.5	150	97.90	*300*	*126*	653	202.6
1267.1	479.0	200	126.80	*350*	*143*	772	226.4
1341.1	494.0	250	153.80	400	158	*Methyl alcohol [15]*	
1363.4	499.5	300	178.45	450	172	384.4	125.9
1665.6	558.5	*Carbon dioxide [6]*				427.0	140.8
1685.6	564.5	200	101.50	*Hydrogen [6]*		461.9	152.7
1740.6	573.0	250	125.60	100	42.12	490.6	162.0
1747.6	573.2	300	149.51	150	56.03	523.1	172.5
1837.1	591.6	*Carbon dioxide [8]*		200	68.09	550.7	181.5
1845.1	590.5	288.7	144.3	250	79.25	584.6	192.1
				300	89.59		

TABLE 31. Experimental Gas Viscosity Data at One Atmosphere *(Continued)*

T,°K	Micropoises	T,°K	Micropoises	T,°K	Micropoises	T,°K	Micropoises
Nitrogen [6]		*Nitrogen [9] (Cont.)*		*Nitrous oxide [4]*		*Oxygen [4] (Cont.)*	
100	69.75	1122.1	428.0	*(Cont.)*		1130.05	523.0
150	100.83	1161.4	436.5	886.55	391.9	1166.15	538.2
200	129.54	1259.1	458.5	916.55	435.8	1203.15	548.0
250	155.46	1307.5	470.0	943.15	483.0	1278.35	567.7
300	178.57	1314.1	469.5	956.35	528.3	1292.15	571.5
Nitrogen [5]		1321.1	472.0	977.75	555.9		
273.2	166.2	1564.1	519.5	1048.25	631.4	*Sulfur dioxide [8]*	
293.7	177.1	1690.6	541.5	1174.75	736.0	290.4	124.2
357.2	205.8	1691.1	543.5	1296.35	764.0	566	244.7
400.3	223.1	1695.1	543.4			694	288.9
444.4	239.3	*Nitric oxide [6]*		*Oxygen [6]*		765	311.5
745.0	338.1	150	104.88	100	76.75	868	342.2
884.4	375.9	200	137.12	150	113.16	952	370.1
978.7	402.7	250	166.59	200	147.58	1096	410.0
1035.0	419.0	300	193.44	250	178.63		
Nitrogen [9]				300	207.06	*Water vapor [3]*	
273.1	166.3	*Nitrous oxide [6]*				373	121.1
550.6	274.4	200	99.87	*Oxygen [8]*		423	141.5
613.1	295.7	250	124.70	287.8	198.5	473	161.8
703.6	323.5	300	148.89	556	323.3	523	182.2
804.3	350.2			675	369.3	573	202.5
855.1	364.2	*Nitrous oxide [4]*		769	401.3	623	223
873.1	370.2	419.15	206.2	881	437.0	673	243
912.1	380.0	530.05	248.4	963	461.2	723	264
917.4	379.3	582.15	263.9	1102	501.2	773	284
977.3	395.0	635.95	287.5			823	304
996.9	400.5	684.75	312.9	*Oxygen [4]*		873	323
1085.1	421.6	739.85	326.3	960.25	460.8	923	325
		793.15	339.5	1041.95	488.0	973	365
				1080.25	512.0		

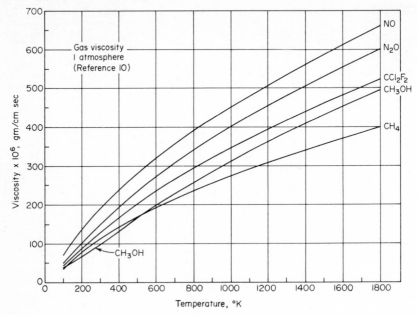

Fig. 5. Gas viscosity (NO, N_2O, CCl_2F_2, CH_3OH, CH_4).

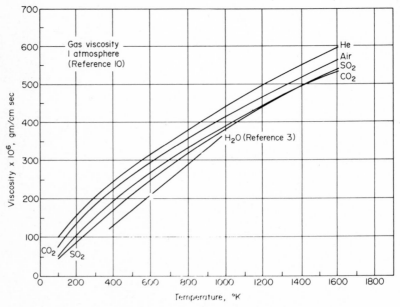

Fig. 6. Gas viscosity (He, Air, SO_2, CO_2, H_2O).

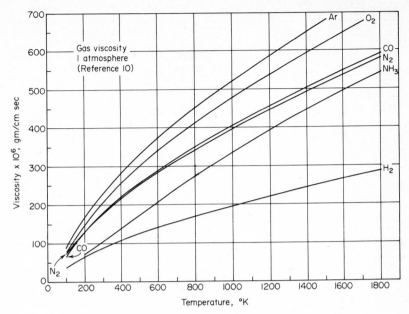

Fig. 7. Gas viscosity $(Ar, O_2, CO, N_2, NH_3, H_2)$.

TABLE 32. Summary of Data in Table 31

Gas	Reference	Temp. Range °K	Measurement Technique*	Reported Accuracy, %
Air	6	100–300	a	0.3-0.8
	8	289.3–1098	d	
	9	273–1845.1	b	1/1000-1/700
Ammonia	14	298.1–573.1	d	
Argon	1	273–2073	b	
	7	100–300	a	0.3-0.5
Carbon Monoxide	6	200–300	a	0.3-0.8
Carbon Dioxide	8	288.7–1097	d	
	9	273.1–1686.1	b	1/1000-1/700
Hydrogen	6	100–300	a	0.3-0.8
	8	289.7–1098	d	
Helium	2	298.36–512.45	a	0.2-0.5
	7	100–300	a	0.3-0.5
	8	294.4–1088	d	
Methane	6	100–300	a	0.3-0.8
	8	296.7–772	d	
Methyl Alcohol	15	384.4–584.6	d	
Nitrogen	5	273.2–1035.0	b	
	6	100–300	a	0.3-0.8
	9	273.1–1695.1	b	1/1000-1/700
Nitric Oxide	6	150–300	a	0.3-0.8
Nitrous Oxide	4	419.15–1296.35	d	
	6	200–300	a	0.3-0.8
Oxygen	4	960.15–1295.15	d	
	6	100–300	a	0.3-0.8
	8	287.8–1102	d	
Sulfur Dioxide	8	290.4–1096	d	
Water Vapor	3	373–973	c	1-3

*a–oscillating disk; b–capillary; c–data compilation; d–transpiration.

TABLE 33. Maximum Percent Deviation
of Graphical Values of Gas
Viscosity from Values
Reported in Table 31

Gas	Maximum Percent Deviation
Air	2
Ammonia	3
Argon	1
Carbon Monoxide	2
Carbon Dioxide	2
Freon-12	6
Hydrogen	2
Helium	4
Methane	4
Methyl Alcohol	2
Nitrogen	3
Nitric Oxide	1
Nitrous Oxide	2
Oxygen	2
Sulfur Dioxide	3

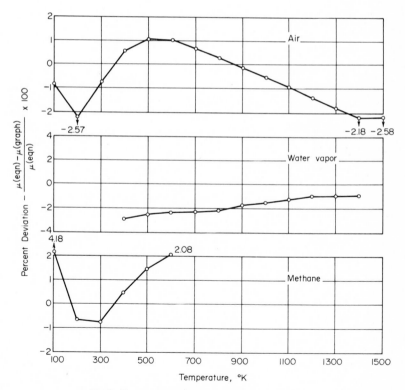

Fig. 8. Gas viscosity deviation plot (Air, H_2O, CH_4).

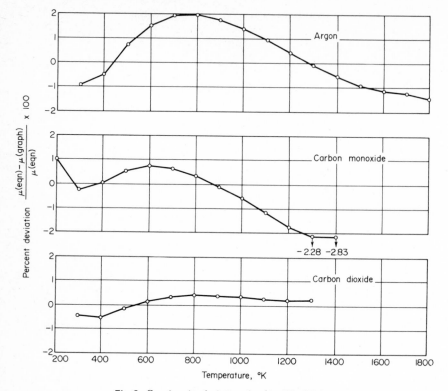

Fig. 9. Gas viscosity deviation plot (Ar, CO, CO_2).

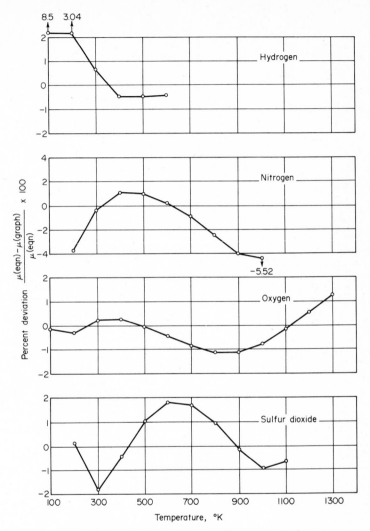

Fig. 10. Gas viscosity deviation plot (H_2, N_2, O_2, SO_2).

TABLE 34. Recommended Coefficients for Computing Dilute Gas Viscosities from the Power Series Equation $\mu(\text{micropoise}) = \alpha_0 + \alpha_1 T + \alpha_2 T^2 + \alpha_3 T^3 + \alpha_4 T^4 + \alpha_5 T^5$ (temperature in °K empirically fitted to the available experimental data)

Date of Pub.	Gas	Temp. Range of Exptl. Data °K	Precision of fit to Exptl. Data %	α_0	α_1 10^{-1}	α_2 10^{-4}	α_3 10^{-7}	α_3 10^{-11}	α_5 10^{-14}
12/62	Air	220–1850	± 2	4.0201	7.4582	-5.7171	2.9928	-6.2524	0
12/62	NH$_3$	200–700	± 3	1.7840	3.1447	.61487	0	0	0
12/61	Ar	220–2070	± 2	56.092	5.4646	2.0929	-5.3235	31.325	-6.0505
12/62	Cl$_2$	300–780	± 1	1.4329	4.6979	-0.37158	0.40541	0	0
12/64	CO	200–1400	± 1.5	23.64	5.87648	-2.73351	0.655858	0	0
12/64	CO$_2$	250–1380	± 1	-5.4384	5.9427	-2.5895	0.64557	0	0
12/64	CCl$_2$F$_2$	300–480	± 1	-5.87736	50.4220	-26.9735	0	0	0
6/63	H$_2$	100–680	± 3	2.105	4.774	-9.9784	16.1827	-128.024	38.164
9/60	CH$_4$	50–400	± 2	1.08617	4.08666	-1.381078	0	0	0
6/62	CH$_3$OH	200–580	± 1	4.0061	2.80911	0	0	0	0
12/64	N$_2$	220–1700	± 2	5.2239	7.2345	-5.8131	3.2167	-7.0913	0
6/62	NO	300–400	± ½	7.4049	8.0409	-4.6680	0	0	0
6/63	N$_2$O	200–1300	± 2%*	4.4787	4.9787	-0.81488	0	0	0
6/62	O$_2$	100–1300	± 2	-9.24353	9.41787	-9.02673	5.91612	-14.4572	0
6/62	SO$_2$	200–1100	± 2	33.4691	0.977464	10.8177	-13.4256	51.1165	0
12/62	H$_2$O	400–1480	± 5	9.65	2.9347	0.92109	-0.41666	0	0
1959	He†	4–1100	± 2	$\mu = 4.7744\ T^{0.6567}$					

SOURCE: "Thermophysical Properties Research Center Data Book," Vol. II.
* From 170-700 °K
† J. Kestin and W. Leidenfrost, "The Viscosity of Helium," *Physica*, **25**:537–555 (1959).

Deviation Plots for Gas Viscosity

The graphs given in Figs. 8, 9, and 10 represent deviations between the values given in Figs. 5, 6, and 7 and those calculated from the equation given in Table 34.

References for Gas Viscosities

1. C. F. Bonilla, S. J. Wang, and H. Weiner, The Viscosity of Steam, Heavy-Water Vapor, and Argon at Atmospheric Pressure up to High Temperatures, *ASME* 55-A-6.
2. J. Kestin and W. Leidenfrost, The Viscosity of Helium, *Physica,* 25:537 (1959).
3. *6th Int. Conf. on the Properties of Steam.*
4. C. J. G. Raw and C. P. Ellis, High Temperature Gas Viscosities. I. Nitrous Oxide and Oxygen, *J. Chem. Phys.,* 28:1198 (1958).
5. J. C. Westmoreland, Determination of Viscosity of Exhaust-Gas Mixtures at Elevated Temperatures, *NACA TN* 3180, 1954.
6. H. L. Johnston and K. E. McCloskey, Viscosities of Several Common Gases between 90K and Room Temperature, *J. Phys. Chem.,* 44:1038 (1940).
7. H. L. Johnston and E. R. Grilly, Viscosities of Carbon Monoxide, Helium, Neon, and Argon between 80° and 300°K Coefficients of Viscosity, *J. Phys. Chem.,* 46:948 (1942).
8. M. Trautz and R. Zink, Die Reibung, Wärmeleitung und Diffusion in Gasmischungen, *Ann. Physik,* 7:425 (1930).
9. V. Vasilesco, Recherches Expérimentales Sur la Viscosité des Gas aux Temperatures Élevées, *Ann. Physique,* 20:292 (1945).
10. R. A. Svehla, Estimated Viscosities and Thermal Conductivities of Gases at High Temperatures, *NASA Tech. Rept.* R-132, 1962.
11. M. Trautz and R. Heberling, Die Riebung, Wärmeleitung und Diffusion in Gasmischungen, *Ann. Physik,* 20:118 (1934).
12. "Thermophysical Properties Research Center Data Book," vol. 2, Purdue University, Lafayette, Indiana.
13. E. R. G. Eckert, W. E. Ibele, and T. F. Irvine, Jr., Prandtl Number, Thermal Conductivity, and Viscosity of Air-Helium Mixtures, *NASA TN* D-533, 1960.
14. M. Trautz and R. Heberling, Die Riebung, Wärmeleitung und Diffusion in Gasmischungen, *Ann. Physik,* 5:561 (1930).
15. T. Titani, The Viscosity of Vapours of Organic Compounds, Part III, *Bull. Chem. Soc. Japan,* 8:255 (1933).
16. R. S. Brokaw, Alignment Charts for Transport Properties, Viscosity, Thermal Conductivity, and Diffusion Coefficients for Non-Polar Gases and Gas Mixtures at Low Density, *NASA TR* R-81, 1961.
17. T. G. Cowling, The Theoretical Basis of Wassiljewa's Equation, *Proc. Roy. Soc. (London),* A263:186 (1961).
18. C. R. Wilke, A Viscosity Equation for Gas Mixtures, *J. Chem. Phys.,* 18:517, 1950.
19. J. O. Hirschfelder, C. F. Curtiss, and R. B. Bird, "Molecular Theory of Gases and Liquids," John Wiley & Sons, Inc., 1954.
20. R. S. Brokaw, Approximate Formulas for Viscosity and Thermal Conductivity of Gas Mixtures, *NASA TN* D-2502, 1964.
21. M. Trautz and H. E. Binkele, Die Reibung, Wärmeleitung und Diffusion von Gasmischungen, *Ann. Physik,* 5:561 (1930).
22. H. R. Heath, The Viscosity of Gas Mixtures, *Proc. Roy. Soc. (London),* B66:362 (1953).
23. A. O. Rietveld, A. Von Itterbeek, and G. T. Van Den Berg, Measurements of the Viscosity of Mixtures of Helium and Argon, *Physica,* 19:517 (1953).
24. M. Trautz and K. F. Kipphan, Die Reibung, Wärmeleitung und Diffusion von Gasmischungen, *Ann. Physik,* 2:763 (1929).

Gas Thermal Conductivity

TABLE 35. Experimental Gas Thermal Conductivity Data at One Atmosphere

t, °K	k 10⁵ cal/cm-sec-deg	t, °K	k 10⁵ cal/cm-sec-deg
Air [3]		*Argon [1] (Cont.)*	
273	5.75	873	9.33
373	7.30	973	10.02
473	8.83	1073	10.65
573	10.30	1173	11.24
673	11.67	1273	11.79
773	13.05	1373	12.30
873	14.40		
973	15.50	*Carbon monoxide [2]*	
1073	16.70	87.41	1.802
1173	17.73	97.87	2.039
1273	18.80	130.06	2.717
1373	19.75	145.31	3.054
		155.07	3.261
Air [9]		176.06	3.682
513	9.76	192.28	4.013
823	14.06	207.36	4.308
1033	16.7	222.62	4.604
1173	18.4	237.21	4.876
		253.12	5.162
Ammonia [3]		267.08	5.412
273	5.25	282.15	5.683
373	7.97	296.90	5.950
473	11.10	312.00	6.210
573	14.55	328.89	6.502
673	18.40	344.66	6.788
773	22.2	357.58	7.021
Argon [15]		376.96	7.372
311	4.28		
316	4.43	*Carbon monoxide [3]*	
327	4.54	273	5.46
336	4.60	373	7.05
577	6.97	473	8.65
586	7.03	573	10.14
661	7.68	673	11.53
669	7.83	773	12.80
777	8.64	873	13.95
787	8.70	973	15.15
916	9.58	1073	16.25
926	9.72	1173	17.40
1036	10.70		
1048	10.60	*Carbon dioxide [2]*	
1182	11.80	186.38	2.073
1189	11.65	200.86	2.285
1198	11.66	216.49	2.527
1201	11.75	230.68	2.752
		245.57	3.002
Argon [1]		259.60	3.245
273	3.96	274.22	3.505
373	5.08	287.75	3.754
473	6.09	302.38	4.031
573	6.99	311.24	4.199
673	7.83	333.87	4.649
773	8.59	347.32	4.932

TABLE 35. Experimental Gas Thermal Conductivity Data at One Atmosphere *(Continued)*

t, °K	$k \, 10^5$ cal/cm-sec-deg	t, °K	$k \, 10^5$ cal/cm-sec-deg
Carbon dioxide [2] Cont.		*Hydrogen* [2] (Cont.)*	
363.36	5.273	357.31	47.51
379.49	5.614	374.25	49.10
Carbon dioxide [4]		*Hydrogen [3]*	
300	3.9	273	40.3
400	5.8	373	49.9
500	7.8	473	58.6
600	9.7	573	67.0
700	11.7	673	75.6
800	13.5	773	83.6
900	15.2	873	92.0
1000	16.8	973	100.5
1100	17.9	1073	108.0
1200	18.6	1173	116.5
Carbon dioxide [3]		1273	124.6
273	3.56	1373	131.5
373	5.53	1473	138.5
473	7.53	*Helium† [2]*	
573	9.42	82.79	15.72
673	11.09	92.71	16.96
773	12.67	122.67	19.91
873	14.09	137.90	21.41
973	15.47	154.30	23.13
1073	16.80	169.89	24.72
1173	18.00	184.32	26.09
1273	19.20	200.24	27.50
1373	20.28	216.89	29.12
Freon-12 CCl_2F_2 [25]		231.74	30.42
423	2.68	248.42	31.87
473	3.17	264.36	33.19
523	3.83	280.53	34.50
Hydrogen* [2]		296.18	35.74
84.70	13.90	328.79	38.05
93.99	15.41	344.43	39.12
125.60	19.91	359.59	39.99
142.23	22.46	375.68	40.94
158.70	24.89	*Helium [5]*	
175.95	27.19	346.97	40.25
192.28	29.38	445.35	48.60
200.40	30.68	562.97	57.90
207.66	31.62	667.22	67.20
218.18	33.09	725.31	70.30
236.56	35.56	803.81	77.50
254.38	37.68	*Methane [3]*	
274.02	39.75	273	7.25
277.69	40.05	373	10.85
293.28	41.64	473	15.05
324.28	44.53	573	19.40
341.27	45.97	673	23.7

*For temperatures between 1200 and 2100°K, Ref. 8 gives $k = 10^{-6}[1431 + 1.257(T-1200)]$ cal/sec-cm-deg.

†For temperatures between 1200 and 2100°K, Ref. 8 gives $k = 10^{-6}[991 + 0.678(T-1200)]$ cal/sec-cm-deg.

TABLE 35. Experimental Gas Thermal Conductivity Data at One Atmosphere *(Continued)*

t, °K	$k\ 10^5$ cal/cm-sec-deg	t, °K	$k\ 10^5$ cal/cm-sec-deg
Methane [3] (Cont.)		*Nitric oxide [2] (Cont.)*	
773	28.3	283.03	5.860
873	33.2	298.80	6.168
973	39.0	313.38	6.444
Methyl alcohol [20]		330.56	6.752
339.0	4.34	346.25	7.056
352.2	4.70	361.97	7.361
358.1	4.85	376.77	7.630
382.9	5.44		
Nitrogen [16]		*Nitrous oxide [2]*	
303.7	6.38	190.25	2.715
323.2	6.66	206.88	2.438
589	10.62	225.80	3.760
677	11.95	232.06	2.870
690.7	11.95	249.05	3.172
765	12.80	262.40	3.421
775	12.90	275.60	3.669
973	15.20	290.59	3.958
981	15.10	303.43	4.217
1121	16.60	317.55	4.511
1134	16.90	332.42	4.829
Nitrogen [3]		348.08	5.150
273	5.72	364.11	5.505
373	7.25	378.45	5.836
473	8.72		
573	10.11	*Oxygen [2]*	
673	11.35	86.53	1.851
773	12.55	94.55	2.035
873	13.58	98.02	2.113
973	14.59	127.22	2.774
1073	15.51	130.99	2.861
1173	16.39	144.56	3.166
1273	17.20	151.24	3.316
1373	17.00	167.28	3.676
Nitrogen [4]		185.17	4.051
300	6.1	205.38	4.488
400	7.6	229.36	4.981
500	9.7	250.87	5.408
600	10.6	272.07	5.819
700	12.1	290.90	6.175
800	13.5	323.86	6.828
900	14.7	341.99	7.201
1000	15.7	357.6	7.547
1100	16.5	376.30	7.956
1200	17.1		
Nitric oxide [2]		*Oxygen [6]*	
129.86	2.786	300	6.3
147.82	3.165	400	8.2
174.57	3.719	500	9.8
191.51	4.077	600	11.3
208.40	4.420	700	12.7
227.76	4.793	800	14.1
245.51	5.139	900	15.5
266.20	5.544	1000	16.9
		1100	18.3
		1200	19.6

TABLE 35. Experimental Gas Thermal Conductivity Data at One Atmosphere *(Continued)*

$t, °K$	$k \, 10^5$ cal/cm-sec-deg	$t, °K$	$k \, 10^5$ cal/cm-sec-deg
Oxygen [3]		*Sulfur dioxide [7] (Cont.)*	
273	5.78	500	4.7
373	7.55	600	6.0
473	9.25	700	7.3
573	10.85		
673	12.50	*Water vapor [3]*	
773	14.05	373	5.89
873	15.39	473	8.00
973	16.69	573	10.63
1073	18.04	673	12.60
1173	19.29	773	15.48
1273	20.48	873	17.82
1373	21.71	973	20.61
Sulfur dioxide [7]		1073	22.98
300	2.3	1173	25.70
400	3.4	1273	28.40

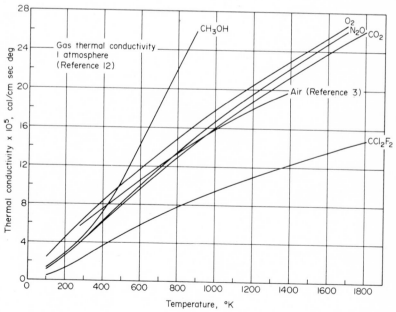

Fig. 11. Gas thermal conductivity (CH_3OH, O_2, N_2O, CO_2, Air, CCl_2F_2).

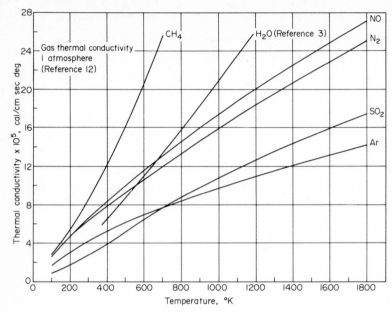

Fig. 12. Gas thermal conductivity (CH_4, H_2O, NO, N_2, SO_2, Ar).

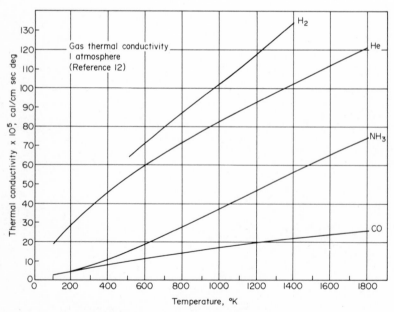

Fig. 13. Gas thermal conductivity (H_2, He, NH_3, CO).

TABLE 36. Summary of Data in Table 35

Gas	Reference	Temp. Range °K	Measurement Technique*	Reported Accuracy, %
Air	3	273–1373	a	2
	9	513–1173	c	
Ammonia	3	273–373	a	2
Argon	1	273–1373	a	2
	15	311–1201	a	
Carbon monoxide	2	87.41–376.96	a	0.5
		273–1173	a	2
Carbon dioxide	3	273–1373	a	2
	4	300–1200	b	
Freon-12	25	425–523	c	
Hydrogen	2	84.70–374.25	a	0.5
	3	273–1473	a	1–2
Helium	2	82.79–375.68	a	0.5
	5	346.97–803.81	a	
Methane	3	273.973	a	2
Methyl alcohol	20	339.0–382.9	a	1/300
Nitrogen	3	273–1373	a	2
	4	300–1200	b	
	16	303.7–1134	a	
Nitric oxide	2	129.86–376.77	a	0.5
Nitrous oxide	2	190.25–378.45	a	0.5
Oxygen	2	86.53–376.30	a	0.5
	3	273–1373	a	2
	6	300–1200	b	
Sulfur dioxide	7	300–700	b	
Water vapor	3	373–1273	a	2

*a—hot wire cell; b—line source technique; c—concentric cylinder.

TABLE 37. Maximum Percent Deviation of Graphical Values of Gas Thermal Conductivity from Values Reported in Table 35

Gas	Maximum Percent Deviation
Ammonia	20
Argon	6
Carbon Monoxide	5
Carbon Dioxide	12
Freon-12	35
Hydrogen	3
Helium	7
Methane	5
Methyl Alcohol	35
Nitrogen	8
Nitric Oxide	12
Nitrous Oxide	10
Oxygen	10
Sulfur Dioxide	15

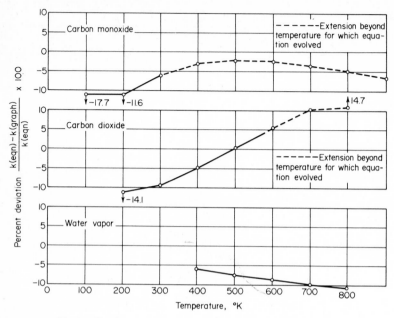

Fig. 14. Gas thermal conductivity deviation plot (CO, CO_2, H_2O).

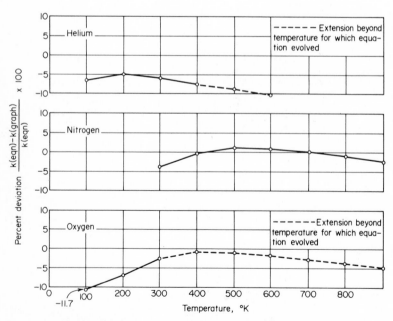

Fig. 15. Gas thermal conductivity deviation plot (He, N_2, O_2).

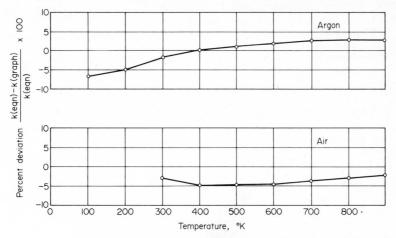

Fig. 16. Gas thermal conductivity deviation plot (Ar, Air).

TABLE 38. Constants for Equations to Calculate the Thermal Conductivity of Gases (cal/cm-sec-°K)

$$10^5 \, k = \frac{C_o T^{1/2}}{1 + C\tau \, 10^{C_1 \tau}} \quad , \quad \tau = \frac{1}{T}$$

Empirical Correlation Fits Data Previous to:	Gas	Ref	C_o	C	C_1	Temperature Range for Which Equation Evolved °K*
1964	N_2	10	0.5815	208.8	−12	273–1173
1964	Ar	10	0.3721	170.5	−8	90–1173
1960	Air	9	0.632	245	−12	90–1173
1951	He†	11	2.35	43.5	−10	15–375
1951	H_2 ‡	11	3.76	166.0	−10	90–575
1951	O_2	11	0.673	266.0	−10	90–375
1951	CO	11	0.586	218	−7.8	90–375
1951	CO_2	11	4.61	6212.0	−10	190–592
1951	H_2O	11	1.546	1737.3	−12	373–823

*The results of [9] suggest that these equations may be used for temperatures beyond the range indicated.

†For temperatures between 1200 and 2100°K, [8] gives $k = 10^{-6}$ [991 + 0.678 $(T − 1200)$] cal/cm-sec-°K.

‡For temperatures between 1200 and 2100°K, [8] gives $k = 10^{-6}$ [1434 + 1.257 $(T − 1200)$] cal/cm-sec-°K.

Deviation Plots for Gas Thermal Conductivity

The graphs given in Figs. 14, 15, and 16 represent deviations between the values given in Figs. 11, 12, and 13 and those calculated from the equation given in Table 38.

References for Gas Thermal Conductivities

1. K. L. Schäfer and F. W. Reiter, Eine Messmethode für die Ermittlung des Wärmeleitvermögens bei 1100°C, *Z. Electrochem.*, **61**:1230 (1957).

2. H. L. Johnston and E. R. Grilly, The Thermal Conductivity of Eight Common Gases between 80° and 380°K, *J. Chem. Phys.*, 14:233 (1946).
3. H. Geier and K. Schäfer, Wärmeleitfähigkeit von reinen Gasen und Gasgemischen zwischen 0° and 1200°C, *Allgem. Wärmetechnik,* 10:70 (1961).
4. A. A. Westenburg and N. de Hass, Gas Thermal-Conductivity Studies at High Temperatures. Line Source Technique and Results in N_2, CO_2, and N_2-CO_2 Mixtures, *Phys. Fluids,* 5:266 (1962).
5. L. S. Zaitseva, An Experimental Investigation of the Heat Conductivity of Monotonic Gases over Wide Temperature Intervals, *Sov. Phys.-Tech. Phys,* 4:444, April, 1959. [Translation of *J. Tech. Phys.*, Russian original vol. 29, no. 4:497 (1952).]
6. A. A. Westenburg and N. de Hass, Gas Thermal Conductivity Studies at High Temperatures. II. Results for O_2 and O_2-H_2 Mixtures, *Phys. Fluids,* 6:617 (1963).
7. B. Baker and N. de Hass, Gas Thermal Conductivity Studies at High Temperatures. III. Results for SO_2, *Phys. Fluids,* 7:1400 (1964).
8. N. C. Blais and J. B. Mann, Thermal Conductivity of Helium and Hydrogen at High Temperatures, *J. Chem. Phys.*, 32(5):1459 (1960).
9. R. G. Vines, Measurement of the Thermal Conductivities of Gases at High Temperatures, *J. Heat Transfer, Trans. ASME,* 82:48 (1960).
10. F. G. Keyes and R. G. Vines, The Thermal Conductivity of Nitrogen and Argon, *J. Heat Transfer, Trans. ASME,* C87:177 (1965).
11. F. G. Keyes, A Summary of Viscosity and Heat Conduction Data for He, A, H_2, O_2, N_2, CO, CO_2, H_2O, and Air, *Trans. ASME,* 73:589 (1951).
12. R. A. Svehla, Estimated Viscosities and Thermal Conductivities of Gases at High Temperatures, *NASA Tech. Rept.* R-132, 1962.
13. E. Thornton, Viscosity and Thermal Conductivity of Binary Gas Mixtures: Xenon-Krypton, Xenon-Argon, Xenon-Neon, and Xenon-Helium, *Proc. Phys. Soc. London,* 76:104 (1960).
14. E. Thornton and W. A. D. Baker, Viscosity and Thermal Conductivity of Binary Gas Mixtures: Argon-Neon, Argon-Helium, and Neon-Helium, *Proc. Phys. Soc. London,* 80:1171 (1962).
15. N. B. Vargaftik and N. Kh. Zimina, *Teplofiz. Vysokikh. Temp.,* 2:716 (1964); *High Temp.,* 2:645 (1964).
16. N. B. Vargaftik and N. Kh. Zimina, *Teplofiz. Vysokikh. Temp.,* 2:867 (1964); *High Temp.,* 2:782 (1964).
17. W. Ibele and D. Briggs, Prandtl Number Measurements and Transport Property Calculation for N_2-CO_2 Mixtures, *4th Symp. Thermophysical Properties* (to be published).
18. W. Ibele and G. Hingorani (to be submitted for publication).
19. R. G. Vines, Thermal Conductivity of Organic Vapours, *Aust. J. Chem.,* 6:1 (1953).
20. H. Ubisch, The Thermal Conductivity of Mixtures of Rare Gases at 29°C and 520°C, *Arkiv Fysik,* 16:93 (1959) (numerical values cited in [22]).
21. E. A. Mason and H. Ubisch, Thermal Conductivities of Rare Gas Mixtures, *Phys. Fluids,* 3:355 (1960).
22. R. S. Brokaw, Approximate Formulas for Viscosity and Thermal Conductivity of Gas Mixtures, *NASA TN* D-2502, 1964.
23. R. S. Brokaw, Alignment Charts for Transport Properties, Viscosity, Thermal Conductivity, and Diffusion Coefficients for Non-Polar Gases and Gas Mixtures at Low Density, *NASA TR* R-81, 1961.
24. F. G. Keyes, Thermal Conductivity of Gases, *Trans. ASME,* 70:809 (1954).

Gas Heat Capacity at Constant Pressure

TABLE 39. Heat Capacity at Constant Pressure of Gases at One Atmosphere C_p/R

t, °K	Ar	CCl₂F₂	CH₄	CH₃OH	CO	CO₂	H₂	H₂O	He	N₂	NH₃	NO	N₂O	O₂	SO₂	Air*	t, °K
100	2.500	4.780	4.000	4.323	3.501	3.512		4.006	2.500	3.500	4.003	3.886	3.530	3.501	4.032	3.5824	100
200	2.500	7.021	4.026	4.830	3.501	3.881		4.010	2.500	3.501	4.058	3.659	4.043	3.503	4.375	3.5062	200
300	2.500	8.721	4.295	5.531	3.505	4.460		4.040	2.500	3.503	4.281	3.590	4.655	3.534	4.803	3.5059	300
400	2.500	9.900	4.871	6.530	3.529	4.952	3.520	4.120	2.500	3.518	4.622	3.602	5.134	3.621	5.229	3.5333	400
500	2.500	10.706	5.574	7.563	3.583	5.346	3.527	4.236	2.500	3.558	5.000	3.667	5.515	3.739	5.600	3.5882	500
600	2.500	11.258	6.282	8.502	3.661	5.669	3.540	4.368	2.500	3.621	5.376	3.758	5.828	3.860	5.897	3.6626	600
700	2.500	11.644	6.951	9.327	3.749	5.938	3.562	4.508	2.500	3.699	5.738	3.853	6.088	3.967	6.127	3.7455	700
800	2.500	11.920	7.569	10.051	3.837	6.163	3.593	4.656	2.500	3.781	6.084	3.942	6.305	4.057	6.304	3.828	800
900	2.500	12.122	8.131	10.686	3.918	6.351	3.632	4.808	2.500	3.860	6.413	4.021	6.486	4.132	6.441	3.906	900
1000	2.500	12.274	8.635	11.245	3.991	6.509		4.962	2.500	3.932	6.722	4.089	6.638	4.194	6.550	3.979	1000
1100	2.500	12.391	9.084	11.735	4.054	6.643	3.677	5.114	2.500	3.998	7.010	4.147	6.765	4.246	6.636	4.046	1100
1200	2.500	12.482	9.482	12.165	4.110	6.756	3.726	5.262	2.500	4.056	7.275	4.197	6.872	4.290	6.707	4.109	1200
1300	2.500	12.555	9.832	12.543	4.158	6.852	3.777	5.404	2.500	4.107	7.517	4.239	6.962	4.328	6.765	4.171	1300
1400	2.500	12.613	10.140	12.875	4.199	6.934	3.829	5.538	2.500	4.151	7.737	4.275	7.040	4.363	6.814	4.230	1400
1500	2.500	12.661	10.410	13.167	4.235	7.004	3.880	5.663	2.500	4.190	7.935	4.306	7.107	4.395	6.855	4.289	1500
1600	2.500	12.700	10.649	13.424	4.266	7.065	3.931	5.780	2.500	4.224	8.113	4.333	7.164	4.426	6.891	4.352	1600
1700	2.500	12.734	10.859	13.650	4.294	7.118	3.979	5.887	2.500	4.254	8.274	4.356	7.215	4.455	6.922	4.418	1700
1800	2.500	12.762	11.044	13.851	4.318	7.164	4.026	5.987	2.500	4.281	8.419	4.377	7.260	4.483	6.950	4.487	1800
1900	2.500	12.785	11.208	14.029	4.339	7.205	4.070	6.079	2.500	4.304	8.549	4.395	7.299	4.511	6.975	4.566	1900
2000	2.500	12.806	11.354	14.187	4.358	7.242	4.112	6.164	2.500	4.325	8.667	4.411	7.335	4.539	6.997	4.662	2000
2100	2.500	12.823	11.483	14.328	4.375	7.274	4.152	6.242	2.500	4.344	8.773	4.425	7.367	4.567	7.017	4.781	2100
2200	2.500	12.839	11.599	14.454	4.390	7.303	4.189	6.314	2.500	4.360	8.869	4.438	7.395	4.594	7.036	4.947	2200
2300	2.500	12.852	11.703	14.567	4.404	7.329	4.224	6.381	2.500	4.375	8.956	4.450	7.422	4.621	7.053	5.179	2300
2400	2.500	12.864	11.796	14.668	4.416	7.353	4.257	6.443	2.500	4.389	9.035	4.461	7.446	4.647	7.069	5.484	2400
2500	2.500	12.875	11.880	14.760	4.427	7.375	4.288	6.500	2.500	4.401	9.107	4.471	7.468	4.673	7.084	5.882	2500
2600	2.500	12.884	11.955	14.843	4.437	7.395	4.318	6.553	2.500	4.413	9.172	4.480	7.488	4.699	7.099	6.40	2600
2700	2.500	12.892	12.024	14.918	4.447	7.413	4.346	6.603	2.500	4.423	9.232	4.489	7.508	4.724	7.112	7.06	2700
2800	2.500	12.900	12.086	14.987	4.456	7.430	4.372	6.649	2.500	4.433	9.287	4.497	7.526	4.748	7.125	7.87	2800
2900	2.500	12.906	12.143	15.049	4.464	7.445	4.397	6.692	2.500	4.442	9.338	4.504	7.542	4.771	7.137	8.86	2900
3000	2.500	12.913	12.194	15.106	4.471	7.460	4.421	6.733	2.500	4.450	9.384	4.511	7.558	4.794	7.149	9.96	3000

*See footnote at end of table.

TABLE 39. Heat Capacity at Constant Pressure of Gases at One Atmosphere C_p/R *(Continued)*

$t\,°K$	Ar	CCl_2F_2	CH_4	CH_3OH	CO	CO_2	H_2	H_2O	He	N_2	NH_3	NO	N_2O	O_2	SO_2	Air*	$t\,°K$
3100	2.500	12.918	12.242	15.158	4.478	7.474	4.444	6.771	2.500	4.457	9.427	4.518	7.573	4.816	7.160		3100
3200	2.500	12.923	12.285	15.206	4.485	7.486	4.465	6.807	2.500	4.464	9.467	4.524	7.588	4.837	7.171		3200
3300	2.500	12.928	12.325	15.250	4.491	7.499	4.486	6.841	2.500	4.471	9.504	4.530	7.601	4.858	7.182		3300
3400	2.500	12.932	12.361	15.290	4.497	7.510	4.505	6.873	2.500	4.477	9.538	4.535	7.614	4.877	7.192		3400
3500	2.500	12.936	12.395	15.327	4.502	7.521	4.524	6.903	2.500	4.483	9.570	4.541	7.627	4.896	7.202		3500
3600	2.500	12.939	12.427	15.362	4.508	7.531	4.542	6.932	2.500	4.489	9.600	4.546	7.639	4.913	7.212		3600
3700	2.500	12.942	12.455	15.394	4.513	7.541	4.559	6.960	2.500	4.494	9.628	4.551	7.651	4.930	7.222		3700
3800	2.500	12.945	12.482	15.424	4.517	7.550	4.576	6.986	2.500	4.499	9.654	4.556	7.662	4.946	7.231		3800
3900	2.500	12.948	12.507	15.451	4.522	7.559	4.592	7.011	2.500	4.504	9.678	4.560	7.673	4.961	7.240		3900
4000	2.500	12.951	12.530	15.477	4.526	7.568	4.608	7.035	2.500	4.508	9.701	4.565	7.683	4.976	7.250		4000
4100	2.500	12.953	12.552	15.501	4.531	7.576	4.623	7.058	2.500	4.513	9.723	4.569	7.694	4.989	7.259		4100
4200	2.500	12.955	12.572	15.523	4.535	7.584	4.637	7.080	2.500	4.517	9.743	4.573	7.704	5.002	7.267		4200
4300	2.500	12.957	12.591	15.544	4.538	7.592	4.651	7.102	2.500	4.521	9.763	4.577	7.714	5.015	7.276		4300
4400	2.500	12.959	12.609	15.564	4.542	7.599	4.665	7.122	2.500	4.525	9.781	4.581	7.723	5.026	7.285		4400
4500	2.500	12.961	12.625	15.582	4.546	7.606	4.678	7.142	2.500	4.528	9.798	4.585	7.733	5.037	7.293		4500
4600	2.500	12.963	12.641	15.599	4.549	7.614	4.691	7.161	2.500	4.532	9.815	4.589	7.742	5.048	7.302		4600
4700	2.500	12.964	12.655	15.616	4.553	7.620	4.704	7.180	2.500	4.535	9.831	4.593	7.751	5.058	7.310		4700
4800	2.500	12.966	12.669	15.631	4.556	7.627	4.717	7.198	2.500	4.539	9.845	4.596	7.760	5.068	7.319		4800
4900	2.500	12.967	12.682	15.645	4.559	7.634	4.729	7.216	2.500	4.542	9.860	4.600	7.769	5.078	7.327		4900
5000	2.500	12.968	12.694	15.659	4.563	7.640	4.740	7.233	2.500	4.545	9.873	4.604	7.778	5.087	7.335		5000

SOURCE: Roger A. Svehla, "Estimated Viscosities   and   Thermal Conductivities   of Gases   at High Temperatures," NASA Technical Report R-132, 1962.

*Data for air from "Tables of Thermal Properties of Gases," U.S. Department of Commerce, National Bureau of Standards, Circular 564, 1955.

Ordinary Diffusion Coefficients for Gases

TABLE 40. Experimental Gas Ordinary Diffusion Coefficients at One Atmosphere

$t, °K$	cm^2/sec	$t, °K$	cm^2/sec
Air-Carbon dioxide [20]		*Carbon dioxide-Argon [20]*	
276.2	0.1420	*276.2*	*0.1326*
317.2	0.1772	317.2	0.1652
Ammonia-Helium [23]		*Nitrogen-Nitrogen [7]*	
274.2	0.668	77.5	0.0168
308.2	0.783	194.5	0.104
331.1	0.881	273	0.185
Ammonia-Neon [23]		298	0.212
274.2	0.298	353	0.287
308.4	0.378	*Nitrogen-Xenon [17]*	
333.1	0.419	242.2	0.0854
Ammonia-Xenon [23]		274.6	0.1070
274.2	0.114	303.45	0.1301
308.4	0.145	334.2	0.1549
333.1	0.173	*Oxygen-Argon [24]*	
Argon-Argon [7]		293.2	0.200
77.5	0.0134	*Oxygen-Argon [16]*	
90	0.180	243.2	0.135
194.5	0.0830	274.7	0.168
273	0.156	304.5	0.202
295	0.178	334.0	0.239
353	0.249	*Oxygen-Helium [16]*	
Argon-Argon [12]		244.2	0.536
273	0.156	274.0	0.640
293	0.175	304.4	0.761
303	0.186	334.0	0.912
318	0.204	*Oxygen-Oxygen [7]*	
Argon-Helium [11]		77.5	0.0153
287.9	0.697	194.5	0.104
354.0	0.979	273	0.187
418.0	1.398	298	0.232
Argon-Helium [12]		353	0.301
273.0	0.640	*Oxygen-Water [4]*	
288.0	0.701	307.9	0.282
303.0	0.760	328.8	0.318
318.0	0.825	352.2	0.352
Argon-Xenon [12]		*Oxygen-Xenon [16]*	
273.0	0.0943	242.2	0.084
288.0	0.102	274.75	0.100
303.0	0.114	303.55	0.126
318.0	0.128	333.6	0.149
Argon-Xenon [13]		*Water-Air [3]*	
194.7	0.0508	289.9	0.244
273.2	0.0962	365.6	0.357
329.9	0.1366	372.5	0.377
378.0	0.1759	*Water-Carbon dioxide [3]*	
Carbon dioxide-Argon [25]		296.1	0.164
293	0.139	365.6	0.249

TABLE 40. Experimental Gas Ordinary Diffusion Coefficients at One Atmosphere (Continued)

t, °K	cm^2/sec	t, °K	cm^2/sec
Water-Carbon dioxide		*Neon-Neon [7]*	
[3] (Cont.)		77.5	0.0492
372.6	0.259	194.5	0.255
		273	0.452
Water-Carbon dioxide [4]		298	0.516
307.5	0.202	353	0.703
328.6	0.211		
352.4	0.245	*Neon-Xenon [14]*	
		273.0	0.186
Water-Helium [4]		288.0	0.202
307.2	0.902	303.0	0.221
328.5	1.011	318.0	0.244
352.5	1.121		
		Nitrogen-Argon [17]	
Water-Hydrogen [3]		244.2	0.1348
293.1	0.850	274.6	0.1689
322.7	1.012	303.55	0.1999
365.6	1.24	334.7	0.2433
365.6	1.26		
372.5	1.28	*Nitrogen-Helium [17]*	
		243.2	0.477
Hydrogen (trace)-Oxygen [2]		275.0	0.596
300	0.820	303.55	0.719
400	1.40	332.5	0.811
500	2.10		
600	2.89	*Helium-Nitrogen*	
700	3.81	*(20% N$_2$) [27]*	
800	4.74	190	0.305
900	5.74	298	0.712
		300	0.738
Hydrogen-Neon [10]		305	0.747
242.2	0.792	310	0.740
274.2	0.974	320	0.812
303.2	1.150	330	0.857
341.2	1.405	340	0.881
		350	0.946
Hydrogen-Xenon [10]		360	0.967
242.2	0.410	370	1.035
274.2	0.508	380	1.051
303.9	0.612	390	1.107
341.2	0.751	400	1.157
Methane-Methane [7]		*Helium-Nitrogen*	
90	0.0266	*(50% N$_2$) [27]*	
194.5	0.0992	190	0.310
273	0.206	298	0.725
298	0.240	300	0.751
353	0.318	305	0.758
		310	0.759
Methane-Methane [21]		320	0.827
298.2	0.235	330	0.879
353.6	0.315	340	0.899
382.6	0.360	350	0.966
		360	0.985
Methane-Water [4]		370	1.058
307.5	0.292	380	1.068
328.6	0.331	390	1.144
352.1	0.356	400	1.180
Neon-Argon [15]			
273.0	0.276		
288.0	0.300		
303.0	0.327		
318.0	0.357		

TABLE 40. Experimental Gas Ordinary Diffusion Coefficients at One Atmosphere *(Continued)*

$t,°K$	cm²/sec	$t,°K$	cm²/sec
Helium-Nitrogen		*Helium (trace)-Argon [8]*	
(100% N₂ extrapolated) [27]		300	0.76
190	0.317	400	1.26
298	0.740	500	1.86
300	0.766	600	2.56
305	0.774	700	3.35
310	0.775	800	4.23
320	0.845	900	5.20
330	0.902	1000	6.25
340	0.921	1100	7.38
350	0.989		
360	1.013	*Helium-Carbon dioxide [20]*	
370	1.086	276.2	0.5312
380	1.094	317.2	0.6607
390	1.168	346.2	0.7646
400	1.210		
		Helium-Carbon dioxide	
Helium-Oxygen (trace) [18]		*(trace) [18]*	
298	0.729	298	0.612
323	0.809	323	0.678
353	0.987	353	0.800
383	1.120	583	0.884
413	1.245	413	1.040
443	1.420	443	1.133
473	1.595	473	1.279
498	1.683	498	1.414
Helium-Xenon [12]		*Helium-Methyl alcohol*	
273.0	0.501	*(trace) [18]*	
288.0	0.550	423	1.032
303.0	0.604	443	1.135
318.0	0.655	463	1.218
		483	1.335
Hydrogen-Argon [10]		503	1.389
242.2	0.562	523	1.475
274.2	0.698		
303.9	0.830	*Helium-Neon [14]*	
341.2	1.010	273.0	0.906
		288.0	0.986
Hydrogen-Argon [11]		303.0	1.065
287.9	0.828	318.0	1.158
354.2	1.111		
418.0	1.714	*Helium-Nitrogen (trace) [18]*	
		298	0.687
Hydrogen (trace)-Argon [9]		323	0.766
295	0.83	353	0.893
448	1.76	383	1.077
628	3.21	413	1.200
806	4.86	443	1.289
958	6.81	473	1.569
1069	8.10	498	1.650
Helium-Argon (trace) [18]		*Helium (trace)-Nitrogen [1]*	
413	1.237	300	0.743
443	1.401	400	1.21
473	1.612	500	1.76
498	1.728	600	2.40

TABLE 40. Experimental Gas Ordinary Diffusion Coefficients at One Atmosphere *(Continued)*

$t, °K$	cm^2/sec	$t, °K$	cm^2/sec
Helium (trace)-Nitrogen [1]		*Carbon monoxide- Carbon*	
(Cont.)		*monoxide [22] (Cont.)*	
700	3.11	273.2	0.190
800	3.90	319.6	0.247
900	4.76	373.0	0.323
1000	5.69		
1200	7.74	*Carbon monoxide-*	
		Nitrogen [22]	
Carbon dioxide-Nitrogen		194.7	0.105
(trace) [1]		273.2	0.186
300	0.171	319.6	0.242
400	0.300	373.0	0.318
500	0.445		
600	0.610	*Carbon monoxide (trace)-*	
700	0.798	*Oxygen [2]*	
800	0.998	300	0.212
900	1.22	400	0.376
1000	1.47	500	0.552
1100	1.70	600	0.746
		700	0.961
Carbon dioxide-		800	1.22
Nitrogen [26]		*Helium-Air [20]*	
295	0.159	276.2	0.6242
1156	1.78	317.2	0.7652
1158	1.92	346.2	0.9019
1286	2.34		
1333	2.26	*Helium-Argon [20]*	
1426	2.55	276.2	0.6460
1430	2.72	317.2	0.7968
1469	2.85	346.2	0.9244
1490	2.92		
1653	3.32	*Helium-Argon (trace) [18]*	
		298	0.729
Carbon dioxide-		323	0.809
Nitrous oxide [19]		353	0.978
194.8	0.0531	383	1.122
273.2	0.0996		
312.8	0.1280	*Carbon dioxide-Argon [26]*	
362.6	0.1683	295	0.139
		1181	1.88
Carbon dioxide-Oxygen		1207	1.88
(trace) [2]		1315	2.38
300	0.160	1368	2.59
400	0.270	1383	2.13
500	0.400	1427	2.53
600	0.565	1445	2.66
700	0.740	1495	2.65
800	0.928	1503	2.84
900	1.14	1538	3.08
1000	1.39	1676	3.21
Carbon monoxide-		*Carbon dioxide-*	
Carbon monoxide [22]		*Carbon dioxide [7]*	
194.7	0.109	194.7	0.0500

TABLE 40. Experimental Gas Ordinary Diffusion
Coefficients at One Atmosphere *(Continued)*

$t,°K$	cm²/sec	$t,°K$	cm²/sec
Carbon dioxide-		*Carbon dioxide-*	
Carbon dioxide [7] (Cont.)		*Carbon dioxide [6] (Cont.)*	
273	0.9074	1330	2.38
298	0.113	1445	2.80
353	0.153		2.86
Carbon dioxide-		1450	2.56
Carbon dioxide [19]		1487	2.88
194.8	0.0516	1490	2.98
273.2	0.0970	1520	2.78
312.8	0.1248	1576	3.12
362.6	0.1644	1580	3.33
		1665	3.29
Carbon dioxide-		1680	3.50
Carbon dioxide [5]		*Carbon dioxide-Nitrogen [24]*	
233	0.0662	289	0.158
253	0.0794		
274	0.0925	*Water-Hydrogen [4]*	
293	0.1087	307.3	1.020
313	0.1239	328.6	1.121
333	0.1395	352.7	1.200
363	0.1613		
393	0.1876	*Water-Nitrogen [4]*	
423	0.2164	307.6	0.256
453	0.2477	328.6	0.303
483	0.2892	352.2	0.359
Carbon dioxide-			
Carbon dioxide [6]		*Xenon-Xenon [13]*	
296	0.109	194.7	0.0257
298	0.109	273.2	0.0480
1180	1.73	293.0	0.0443
	1.84	300.5	0.0576
1218	2.04	329.9	0.0684
		378.0	0.0900

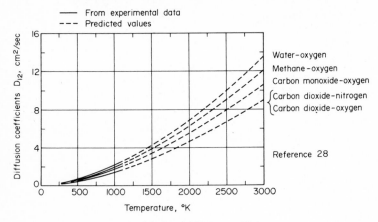

Diffusion coefficients calculated
by Lennard–Jones (6–12). Potential
trace of light gas.
I atmosphere

——— From experimental data
– – – Predicted values

Water–oxygen
Methane–oxygen
Carbon monoxide–oxygen
{Carbon dioxide–nitrogen
{Carbon dioxide–oxygen

Reference 28

Fig. 17. Calculated diffusion coefficients.

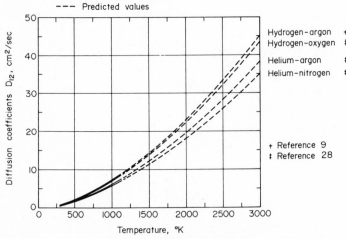

Diffusion coefficients calculated
by Lennard–Jones (6–12). Potential
trace of light gas.
I atmosphere

——— From experimental data
– – – Predicted values

Hydrogen–argon   +
Hydrogen–oxygen   ‡

Helium–argon   ‡
Helium–nitrogen   ‡

+ Reference 9
‡ Reference 28

Fig. 18. Calculated diffusion coefficients.

TABLE 41. Summary of Data in Table 40

Gases	Reference	Temp. Range °K	Measurement Techniques*	Reported Accuracy, %
Air-Carbon Dioxide	20	276.2–317.2	b	
Ammonia-Helium	23	274.2–333.1	c	
Ammonia-Neon	23	274.2–333.1	c	
Ammonia-Xenon	23	274.2–333.1	c	
Argon-Argon	7	77.5–353	c	2
	12	273–318	c	1
Argon-Helium	11	287.9–418.0	b	1.7
	12	273.0–318.0	c	1
Argon-Xenon	12	273.0–318.0	c	1
	13	194.7–378.0	b	1
Carbon Dioxide-Argon	25	293	c	
	20	276.2–317.2	b	
	26	295–1676	a	5
Carbon Dioxide-Carbon Dioxide	7	194.7–353	c	2
	19	194.8–362.6	b	1.5
	5	233–483	c	
	6	296–1680	a	
Carbon Dioxide-Nitrogen	24	289	b	
Carbon Dioxide-Nitrogen (trace)	1	300–1100	a	1.2†
Carbon Dioxide-Nitrogen	26	295–1653	a	5
Carbon Dioxide-Nitrous Oxide	19	194.8–362.6	b	1.5
Carbon Dioxide-Oxygen (trace)	2	300–1000	a	2
Carbon Monoxide-Carbon Monoxide	22	194.7–373.0	b	1
Carbon Monoxide-Nitrogen	22	194.7–373.0	b	1
Carbon Monoxide (trace)-Oxygen	2	300–800	a	2†
Helium-Air	20	276.2–346.2	b	
Helium-Argon	20	276.2–346.2	b	
Helium-Argon (trace)	18	298–498	e	
Helium (trace)-Argon	8	300–1100	a	1†
Helium-Carbon Dioxide	20	276.2–346.2	b	
Helium-Carbon-Dioxide (trace)	18	298–498	e	
Helium-Methyl-Alcohol (trace)	18	423–523	e	
Helium-Neon	14	273–318	c	1
Helium-Nitrogen (trace)	18	298–498	e	
Helium (trace)-Nitrogen	1	300–1200	a	1†
Helium-Nitrogen (20% N₂)	27	190–400	f	2
Helium-Nitrogen (50% N₂)	27	190–400	f	2
Helium-Nitrogen (100% N₂ extrapolated)	27	190–400	f	2
Helium-Oxygen (trace)	18	298–498	e	
Helium-Xenon	12	273.0–318.0	c	1
Hydrogen-Argon	10	242.2–341.2	c	1
	11	287.9–418.0	b	1.7
Hydrogen (trace)-Argon	9	295–1069	a	
Hydrogen (trace)-Oxygen	2	300–900	a	1†

See footnotes at end of table.

TABLE 41. Summary of Data in Table 40 *(Continued)*

Gases	Reference	Temp. Range °K	Measurement Techniques*	Reported Accuracy, %
Hydrogen-Neon	10	242.2–341.2	c	1
Hydrogen-Xenon	10	242.2–341.2	c	1
Methane-Methane	7	90–353	c	8
	21	298.2–382.6	c	
Methane-Water	4	307.5–352.1	d	
Neon-Argon	15	273.0–318.0	c	1
Neon-Neon	7	77.5–353	c	2
Neon-Xenon	14	273.0–318.0	c	1
Nitrogen-Argon	17	244.2–334.7	c	
Nitrogen-Helium	17	243.2–332.5	c	
Nitrogen-Nitrogen	7	77.5–353	c	3
Nitrogen-Xenon	17	242.2–334.2	c	
Oxygen-Argon	24	293.2	b	
	16	243.2–334.0	c	
Oxygen-Helium	16	244.2–334.0	c	
Oxygen-Oxygen	7	77.5–353	c	3
Oxygen-Water	4	307.9–352.2	d	
Oxygen-Xenon	16	242.2–333.6	c	
Water-Air	3	289.9–372.5	d	0.5
Water-Carbon Dioxide	3	296.1–372.6	d	1
	4	307.5–352.4	d	
Water-Helium	4	307.2–352.5	d	
Water-Hydrogen	3	293.1–372.5	d	2
	4	307.3–352.7	d	
Water-Nitrogen	4	307.6–352.2	d	
Xenon-Xenon	13	194.7–378.0	b	1

* a–point source; b–Loschmidt technique; c–two-bulb method; d–method of Stefan; e–chromatographic technique; f–interdiffusing streams.

† Maximum mean deviation of experimental points from fitted curve.

References for Ordinary Diffusion Coefficients

1. R. E. Walker and A. A. Westenberg, Molecular Diffusion Studies in Gases at High Temperatures. II. Interpretation of Results on the He-N$_2$ and CO$_2$-N$_2$ Systems, *J. Chem. Phys.*, **29**:1147 (1958).
2. R. E. Walker and A. A. Westenberg, Molecular Diffusion Studies in Gases at High Temperatures. IV. Results and Interpretation of the CO$_2$-O$_2$, CH$_4$-O$_2$, H$_2$-O$_2$, CO-O$_2$ and H$_2$O-O$_2$ Systems, *J. Chem. Phys.*, **32**:436 (1960).
3. M. Trautz and W. Müller, Die Reibung, Wärmeleitung and Diffusion in Gasmischungen. XXXIII. Die Korrektion der bisher mit der Verdampfungsmethode gemessenen Diffusionskonstanten, *Ann. Physik*, **22**:333 (1935).
4. F. A. Schwertz and J. E. Brow, Diffusivity of Water Vapor in Some Common Gases, *J. Chem. Phys.*, **19**:640 (1951).
5. K. Schäfer and P. Reinhard, Zwischenmolekulare Kräfte und die Temperaturabhängigkeit der Selbstdiffusion von CO$_2$, *Z. Naturforsch*, **18**:187 (1963).
6. G. Ember, J. R. Ferron, and K. Wohl, Self-Diffusion Coefficients of Carbon Dioxide at 1180°–1680°K, *J. Chem. Phys.*, **37**:891 (1962).
7. E. B. Winn, The Temperature Dependence of the Self-Diffusion Coefficients of Argon, Neon, Nitrogen, Oxygen, Carbon Dioxide, and Methane, *Phys. Rev.*, **80**:1024 (1950).
8. R. E. Walker and A. A. Westenberg, Molecular Diffusion Studies in Gases at High Temperature. III. Results and Interpretation of the He-A System, *J. Chem. Phys.*, **31**:319 (1959).
9. A. A. Westenberg and G. Frazier, Molecular Diffusion Studies in Gases at High Temperature. V. Results for the H$_2$-Ar System, *J. Chem. Phys.*, **36**:3499 (1962).
10. R. Paul and I. B. Srivastava, Mutual Diffusion of the Gas Pairs H$_2$Ne, H$_2$-Ar, and H$_2$-Xe at Different Temperatures, *J. Chem. Phys.*, **35**:1621 (1961).

11. R. A. Strehlow, The Temperature Dependence of the Mutual Diffusion Coefficient for Four Gaseous Systems, *J. Chem. Phys.,* **21**:2101 (1953).
12. K. P. Srivastava, Mutual Diffusion of Binary Mixtures of Helium, Argon, and Xenon at Different Temperatures, *Physica,* **25**:571 (1959).
13. I. Amdur and T. F. Schatzki, Diffusion Coefficients of the Systems Xe-Xe and A-Xe, *J. Chem. Phys.,* **27**:1049 (1957).
14. K. P. Srivastava and A. K. Barua, The Temperature Dependence of Interdiffusion Coefficient for Some Pairs of Rare Gases, *Indian J. Phys.,* **33**:229 (1959).
15. B. N. Srivastava and K. P. Srivastava, Mutual Diffusion of Pairs of Rare Gases at Different Temperatures, *J. Chem. Phys.,* **30**: 984 (1959).
16. R. Paul and I. B. Srivastava, Studies on Binary Diffusion of the Gas Pairs O_2-A, O_2-Xe, and O_2-He, *Indian J. Phys.,* **35**:465 (1961).
17. R. Paul and I. B. Srivastava, Studies on the Binary Diffusion of the Gas Pairs N_2-A, N_2-Xe, and N_2-He, *Indian J. Phys.,* **35**:523 (1961).
18. S. L. Seager, L. R. Geertson, and J. C. Giddings, Temperature Dependence of Gas and Vapor Diffusion Coefficients, *J. Chem. Eng. Data,* **8**:168 (1963).
19. I. Amdur, J. W. Irvine, Jr., E. A. Mason, and J. Ross, Diffusion Coefficients of the Systems CO_2-CO_2 and CO_2-N_2O, *J. Chem. Phys.,* **20**:436 (1952).
20. James N. Holsen and Mailand R. Strunk, Binary Diffusion Coefficients in Nonpolar Gases, *Ind. Eng. Chem. Fund.,* **3**:143 (1964).
21. C. R. Mueller and R. W. Cahill, Mass Spectrometric Measurement of Diffusion Coefficients, *J. Chem. Phys.,* **40**:651 (1964).
22. I. Amdur and L. M. Shuler, Diffusion Coefficients of the Systems CO-CO and CO-N_2, *J. Chem. Phys.,* **38**:188 (1963).
23. I. B. Srivastava, Mutual Diffusion of Binary Mixtures of Ammonia with He, Ne and Xe, *Indian J. Phys.,* **36**:193 (1962).
24. L. E. Boardman and N. E. Wild, The Diffusion of Pairs of Gases with Molecules of Equal Mass, *Proc. Roy. Soc.* **A162**:511 (1937).
25. L. Waldmann, Die Temperaturerscheinungen bei der Diffusion in ruhenden Gasen und ihre messtechnische Anwendung, *Z. Phys.,* **124**:2 (1947).
26. T. A. Pakurar and J. R. Ferron, Measurement and Prediction of Diffusivities to 1700°K in Binary Systems Containing Carbon Dioxide, *Univ. of Delaware Tech. Rept.* DEL-14-P, 1964.
27. J.-W. Yang, A New Method of Measuring the Mass Diffusion Coefficient and Thermal Diffusion Factor in a Binary Gas System, doctoral dissertation, Univ. of Minnesota, 1966.
28. R. E. Walker, L. Monchick, A. A. Westenberg, and S. Favin, High Temperature Gaseous Diffusion Experiments and Intermolecular Potential Energy Functions, *Planetary and Space Sci.,* **3**:221 (1961).

TABLE 42. Experimental Gas Thermal Diffusion Factors at One Atmosphere

$T, °K$	Percent	α_T	$T, °K$	Percent	α_T
Argon-Krypton [6]			*Helium-Carbon dioxide [4]*		
147		0.0286	200		0.801
185		0.0381	250		0.843
233		0.0571	300		0.866
293		0.0762	350		0.880
369		0.105	400		0.889
464		0.149	450		0.895
585		0.187	500		0.899
Argon-Krypton [7]			530		0.901
(A)			*Helium-Krypton [6]*		
	40	0.0475	117		0.37
	50	0.0548	147		0.41
	60	0.0642	185		0.43
	70	0.0681	233		0.45
	80	0.0725	293		0.45
Argon-Xenon [6]			369		0.45
185		0.0614	464		0.45
233		0.0707	585		0.45
293		0.085	*Helium-Krypton [7]*		
369		0.137	(He)		
464		0.175	30		
585		0.212		40	0.322
Argon-Xenon [7]				50	0.400
(A)				60	0.450
	30	0.0634		70	0.508
	40	0.0708	*Helium-Neon [6]*		
	50	0.0756	117		0.29
	60	0.0825	147		0.30
	70	0.0853	185		0.31
Helium-Argon [6]			233		0.32
117		0.33	293		0.32
147		0.34	369		0.32
185		0.37	464		0.32
233		0.38	585		0.32
293		0.39	*Helium-Neon [7]*		
369		0.39	(He)		
464		0.40		20	0.332
585		0.40		30	0.344
Helium-Argon [4]				40	0.432
200		0.651		50	0.388
250		0.664		60	0.433
300		0.671	*Helium-Nitrogen [8]*		
350		0.676	(He)		
400		0.678		45.4	0.210
450		0.680		50.1	0.237
500		0.681		54.6	0.263
550		0.682		58.8	0.286
590		0.683		62.9	0.307
Helium-Argon [7]				66.6	0.327
(He)				70.2	0.345
	10	0.278		73.4	0.362
	20	0.298		76.5	0.378
	30	0.314		79.3	0.394
	40	0.338		84.2	0.422
	50	0.372		88.1	0.447
				91.0	0.464

TABLE 42. Experimental Gas Thermal Diffusion Factors at One Atmosphere *(Continued)*

$T, °K$	Percent	α_T	$T, °K$	Percent	α_T
Helium-Xenon [6]			*Hydrogen (trace)-*		
185		0.43	*Krypton [2] (Cont.)*		
233		0.43	400		0.208
293		0.43	450		0.212
369		0.43	500		0.216
464		0.43	550		0.210
585		0.43			
			Hydrogen-Krypton (trace) [2]		
Helium-Xenon [7]			200		0.406
	(He)		250		0.456
	10	0.234	300		0.489
	20	0.264	350		0.513
	30	0.294	400		0.531
	40	0.337	450		0.545
	50	0.403	500		0.556
			550		0.565
Hydrogen-Argon [3]					
200		0.388	*Hydrogen-Nitrogen [5]*		
250		0.410		(N_2)	
300		0.425		2.5	0.475
350		0.435		5	0.470
400		0.443		7.5	0.429
				15	0.404
Hydrogen-Argon [5]				25	0.326
	(A)			35	0.329
	5	0.470		45	0.287
	15	0.418		55	0.274
	25	0.370		65	0.242
	35	0.342		75	0.235
	45	0.311		85	0.215
	55	0.289		95	0.206
	65	0.270			
	75	0.246	*Oxygen-Argon [5]*		
	85	0.227		(A)	
	95	0.197		12.5	0.050
				37.5	0.0485
				62.5	0.0486
Hydrogen-Carbon dioxide [4]				87.5	0.0482
200		0.351			
250		0.480	*Neon-Argon [6]*		
300		0.566	117		0.094
350		0.637	147		0.121
400		0.673	185		0.148
440		0.703	233		0.161
			293		0.175
Hydrogen-Carbon dioxide [5]			369		0.191
	(CO_2)		464		0.191
	3.13	0.410	585		0.191
	9.38	0.400			
	25	0.27	*Neon-Argon [7]*		
	50	0.24		(Ne)	
	75	0.21		20	0.146
	90.6	0.192		30	0.161
				40	0.170
Hydrogen (trace)-Krypton [2]				50	0.183
200		0.167		60	0.194
250		0.183			
300		0.194	*Neon-Krypton [6]*		
350		0.202	117		0.159

TABLE 42. Experimental Gas Thermal Diffusion
Factors at One Atmosphere *(Continued)*

$T, °K$	Percent	α_T	$T, °K$	Percent	α_T
Neon-Krypton [6]			*Nitrogen-Argon [5] (Cont.)*		
(Cont.)				58.4	0.0683
147		0.180		75	0.0695
185		0.212		91.7	0.0696
233		0.249			
293		0.292	*Xenon*[133] *(trace)-Argon [1]*		
369		0.312		300	0.15
464		0.324		400	0.22
585		0.339		500	0.26
				600	0.29
Neon-Krypton [7]				700	0.31
(Ne)					
	20	0.203	*Xenon*[133] *(trace)-Helium [1]*		
	30	0.212		300	0.97
	40	0.232		400	1.01
	50	0.267		500	1.04
	60	0.306		600	1.06
				700	1.08
Neon-Xenon [6]					
185		0.263	*Xenon*[133] *(trace)-Hydrogen [1]*		
233		0.281		300	0.56
293		0.300		400	0.69
369		0.330		500	0.77
464		0.367		600	0.82
585		0.403		700	0.86
Neon-Xenon [7]			*Xenon*[133] *(trace)-Neon [1]*		
(Ne)				300	0.57
	20	0.177		400	0.62
	30	0.200		500	0.65
	40	0.223		600	0.66
	50	0.253		700	0.68
	60	0.298			
			Xenon[133] *(trace)-Nitrogen [1]*		
Nitrogen-Argon [5]				300	0.22
(A)				400	0.30
	8.35	0.0719		500	0.35
	25	0.0708		600	0.39
	41.7	0.0721		700	0.41

TABLE 43. Summary of Data in Table 42

Gas	Reference	Temp. Range °K	Concentration	Reported Accuracy, %
Argon-Krypton	6	147–585	53.5% Argon	
	7	373–288*	40-80% Argon	
Argon-Xenon	6	185–583	56.4% Argon	
	7	373–288*	30-70% Argon	
Helium-Argon	6	117–585	51.2% Helium	
	4	200–590	Trace Argon	6†
	7	373–288*	10-50% Helium	
Helium-Carbon Dioxide	4	200–530	Trace Carbon Dioxide	6†

TABLE 43. Summary of Data in Table 42 (*Continued*)

Helium-Krypton	6	117–585	55% Helium	
	7	373–288*	30-70% Helium	
Helium-Neon	6	177–585	53.8% Helium	
	7	373–288*	20-60% Helium	
Helium-Nitrogen††	8	301	45.4-91% Helium	10
Helium-Xenon	6	185–585	53.6% Helium	
	7	373–288	10-50% Helium	
Hydrogen-Argon[3][7]	3	200–400	Trace Argon	10†
Hydrogen-Argon	5	293	5-95% Argon	
Hydrogen-Carbon Dioxide	4	200–440	Trace Carbon Dioxide	20†
	5	293	3.13-90.6% Carbon Dioxide	
Hydrogen-Krypton	2	200–550	Trace Hydrogen	
	2	200–550	Trace Krypton	
Hydrogen-Nitrogen	5	293	2.5-95% Nitrogen	
Oxygen-Argon	5	293	12.5-87.5% Argon	
Neon-Argon	6	117–585	51.2% Helium	
	7	373–288*	20-60% Neon	
Neon-Krypton	6	117–585	53% Neon	
	7	373–288	20-60% Neon	
Neon-Xenon	6	185–585	54.2% Neon	
	7	373–288	20-60% Neon	
Nitrogen-Argon	5	293	8.35-91.7% Argon	
Xenon[133] Argon	1	300–700	Trace Xenon[133]	5
Xenon[133]-Helium	1	300–700	Trace Xenon[133]	5
Xenon[133]-Hydrogen	1	300–700	Trace Xenon[133]	5
Xenon[133]-Neon	1	300–700	Trace Xenon[133]	5
Xenon[133]-Nitrogen	1	300–700	Trace Xenon[133]	5

Note: Two-bulb measurement technique except as indicated. *High and low temperature of two bulb apparatus. †Maximum average deviation from least square fit to data. †† Two interdiffusing streams.

References for Thermal Diffusion Factors

1. D. Heymann and J. Kistemaker, Thermal Diffusion of Xenon at Tracer Concentrations, *Physica*, **25**:556 (1959).
2. E. A. Mason, M. Islam, and Stanley Weissman, Thermal Diffusion and Diffusion in Hydrogen-Krypton Mixtures, *Phys. Fluids*, **7**:1011 (1964).
3. E. A. Mason, S. Weissman, and R. P. Wendt, Composition Dependence of Gaseous Thermal Diffusion Factors and Mutual Diffusion Coefficients, *Phys. Fluids*, **7**:174 (1964).
4. S. C. Saxena and E. A. Mason, Thermal Diffusion and the Approach to the Steady State in Gases: II, *Mol. Phys.*, **2**:379 (1959).
5. L. Waldmann, Die Temperaturerscheinungen bei der Diffusion in ruhenden Gasen und ihre messtechnische Anwendung, *Z. Phys.*, **124**:2 (1947).
6. K. E. Grew, Thermal Diffusion in Mixtures of the Inert Gases, *Proc. Roy. Soc. London*, **189A**:402 (1947).
7. B. E. Atkins, R. E. Bastick, and T. L. Iffs, Thermal Diffusion in Mixtures of the Inert Gases, *Proc. Roy. Soc. London*, **172A**:142 (1939).
8. J.-W. Yang, A New Method of Measuring the Mass Diffusion Coefficient and Thermal Diffusion Factor in a Binary Gas System, doctoral dissertation, Univ. of Minnesota, 1966.

Prandtl Number for Gases

TABLE 44. Prandtl Number of Common Gases* at One Atmosphere

T, °K	Air†	Ar	CH₄	CO	CO₂	H₂	H₂O‡	He	N₂	NH₃	NO	N₂O	O₂	SO₂
100		0.665	0.699	0.691	0.693			0.667	0.692		0.696	0.693	0.691	0.699
200	0.700	0.667	0.699	0.691	0.695			0.667	0.691	0.699	0.693	0.697	0.690	0.701
300	0.694	0.667	0.702	0.691	0.704			0.667	0.691	0.701	0.691	0.706	0.691	0.709
400	0.688	0.667	0.708	0.691	0.709		1.05	0.667	0.691	0.706	0.692	0.711	0.693	0.711
500	0.684	0.667	0.714	0.691	0.712	0.691	0.95	0.667	0.692	0.709	0.693	0.714	0.694	0.714
600	0.682	0.667	0.719	0.693	0.715	0.690	0.93	0.667	0.692	0.712	0.695	0.716	0.696	0.716
700	0.680	0.667	0.722	0.694	0.716	0.691	0.91	0.667	0.694	0.715	0.696	0.717	0.698	0.717
800	0.680	0.667	0.725	0.696	0.718	0.691	0.89	0.667	0.695	0.717	0.697	0.719	0.699	0.718
900	0.679	0.667	0.727	0.697	0.719	0.692	0.88	0.667	0.696	0.719	0.698	0.720	0.700	0.719
1000	0.679	0.667	0.729	0.697	0.720	0.693	0.88	0.667	0.697	0.721	0.699	0.721	0.700	0.720
1100	0.680	0.667	0.730	0.699	0.721	0.693		0.667	0.698	0.722	0.700	0.721	0.701	0.721
1200	0.680	0.667	0.731	0.700	0.721	0.694		0.667	0.699	0.724	0.701	0.722	0.702	0.721
1300		0.667	0.732	0.700	0.722	0.695		0.667	0.700	0.725	0.701	0.722	0.702	0.721

*See [10] for gas viscosity.
†K. M. Anderson, M.S. thesis, University of Minnesota, June, 1967.
‡See [3] for gas viscosity.

Thermal Conductivities of Metals and Alloys at Low Temperature

TABLE 45. Conductivities of Metals and Alloys at Low Temperatures Thermal Conductivity k, Btu/hr-ft-°F

Material	State or composition	−170°C −274°F	−160°C −256°F	−150°C −238°F	−125°C −193°F	−100°C −148°F	−75°C −103°F	−50°C −58°F	−25°C −13°F	0°C 32°F	18°C 64.4°F
Copper	Pure soft drawn	270	262	255	242	236	232	228	226	224	222
Silver	0.999 pure	241	242	242	256	244	243	242	242	238	236
Zinc	Pure, redistilled cast	67.8	67.4	66.9	65.4	65.6	65.1	65.1	65.1	65.1	65.1
Cadmium	Pure, redistilled cast	58.1	57.9	57.6	56.6	55.9	55.0	54.5	53.8	53.0	52.5
Aluminum	0.99 Al	127	124	123	119	119	119	120	121	121	122
Tin	Pure cast	47.2	46.5	45.8	44.1	42.6	41.6	40.6	39.7	38.8	38.0
Lead	Pure	22.5	22.3	22.0	21.6	21.1	20.6	20.6	20.3	20.3	20.1
Iron	0.994 Fe, 0.0015 Mn, 0.0013 Ni, 0.003 C	36.6	36.8	37.0	37.3	36.8	36.3	35.8	35.6	35.6	35.6
Nickel	0.99 Ni	31.0	31.2	31.4	31.9	32.4	32.9	33.2	33.6	33.9	33.9
Steel	Approx. 0.01 C	27.4	27.4	27.4	27.4	27.6	27.8	28.1	28.1	28.1	27.8
Brass	0.7 Cu, 0.3 Zn	42.4	43.8	45.0	48.4	51.6	54.4	56.9	59.1	61.5	62.9
German silver	0.62 Cu, 0.15 Ni, 0.22 Zn	10.2	10.4	10.7	10.9	11.4	11.9	12.4	13.1	13.6	14.3
Manganin	0.84 Cu, 0.04 Ni, 0.12 Mn	8.24	8.48	8.48	8.96	9.45	9.93	10.4	11.1	12.1	12.6
Lipowitz alloy	0.50 Bi, 0.25 Pb, 0.14 Sn, 0.11 Cd	10.3	10.3	10.3	10.3	10.3	10.4	10.4	10.7	10.7	10.7

SOURCE: C. H. Lees, *Phil. Trans. Roy. Soc. London,* **208A**:381 (1908); E. R. G. Eckert and Robert M. Drake, Jr., "Heat and Mass Transfer," 2nd ed., McGraw-Hill Book Company, Inc., New York, 1959.

Variation of Thermal Conductivities of Insulating Materials with Temperature

TABLE 46. Variation of Thermal Conductivities of Insulating Materials with Temperature

Material	Mean temperature, °F										Limiting use t, °F
	100	200	300	400	500	600	800	1000	1500	2000	
Asbestos (30 lb/ft³) laminated asbestos felt	0.097	0.110	0.117	0.121	0.123	0.125	0.130				
Approx. 40 laminations/in.	0.033	0.037	0.040	0.044	0.048						700
Approx. 20 laminations/in.	0.045	0.050	0.055	0.060	0.065						500
Corrugated asbestos (4 plies/in.)	0.050	0.058	0.069								300
85% magnesia (13 lb/ft³)	0.034	0.036	0.038	0.040							600
Diatomaceous earth, asbestos and bonder	0.045	0.047	0.049	0.050	0.053	0.055	0.060	0.065			1600
Diatomaceous earth brick	0.054	0.056	0.058	0.060	0.063	0.065	0.069	0.073			1600
Diatomaceous earth brick	0.127	0.130	0.133	0.137	0.140	0.143	0.150	0.158	0.176		2000
Diatomaceous earth brick	0.128	0.131	0.135	0.139	0.143	0.148	0.155	0.163	0.183	0.203	2500
Diatomaceous earth powder (density, 18 lb/ft³)	0.039	0.042	0.044	0.048	0.051	0.054	0.061	0.068			
Rock wool	0.030	0.034	0.039	0.044	0.050	0.057					

SOURCE: L. S. Marks, "Mechanical Engineers' Handbook," 5th ed., McGraw-Hill Book Company, Inc., New York, 1951; E. R. G. Eckert and Robert M. Drake, Jr., "Heat and Mass Transfer," 2nd ed., McGraw-Hill Book Company, Inc., New York, 1959.

ALIGNMENT CHARTS FOR TRANSPORT PROPERTIES: VISCOSITY, THERMAL CONDUCTIVITY, AND DIFFUSION COEFFICIENTS FOR DILUTE NONPOLAR GASES AND GAS MIXTURES*

Introduction

The kinetic theory of gases is developed to the point where it is now possible to calculate with good accuracy (precision of 2 percent or better) the transport properties of many nonpolar gases and gas mixtures at low pressures [1]. Though the early work of Maxwell and Boltzmann was soon followed by Chapman and Enskog, who obtained solutions of the kinetic theory equations for transport properties, only recently have the numerical values of the transport integrals been computed for intermolecular potentials representing the forces between real molecules [2, 3].

To avoid the rigorous but cumbersome formula for calculating the viscosity of gas mixtures, Wilke [4] developed an approximate formula which works well; it can be derived from the rigorous expression [5]. A similar approximation is derived for the thermal conductivity of a mixture of monatonic gases [5, 6], which combines with the expression [7] for internal energy transport in mixtures to give the conductivity of polyatomic gas mixtures.

The alignment charts described here permit rapid calculations according to these equations with a precision of 2 percent or better.

The charts for pure gas properties and the binary diffusion coefficients are based on collision integrals for the Lennard-Jones (12-6) potential [2], which faithfully describes the properties of many gases. This potential requires two parameters for a quantitative description of molecular behavior. More elaborate potentials involving three parameters improve agreement with experimental data only slightly [8].

Though the accuracy of viscosities and conductivities calculated at high temperatures may be poor, rigorous calculation of these properties is justified since in many engineering heat transfer correlations, systematic errors in viscosity and conductivity tend to cancel each other [6]. The present method deals only with the properties of nonpolar gases at low pressures. For gases at high density or liquids, the text of Reid and Sherwood [9] is recommended.

The following sections discuss the charts for the properties of pure gases and gas mixtures. The equations that form the basis of the charts are presented and referenced, and examples of chart use are presented. A clear, plastic straight edge, engraved with a straight line, will permit rapid and accurate reading of the charts.

Properties of Pure Gases

The alignment charts for pure gas properties are based on the Lennard-Jones (12-6) potential

$$\phi(r) = 4\epsilon \left[\left(\frac{\sigma}{r} \right)^{12} - \left(\frac{\sigma}{r} \right)^{6} \right] \tag{1}$$

which is a spherically symmetrical function describing the potential energy ϕ of a pair of interacting molecules as a function of intermolecular distance r. This potential is characterized by two parameters: the zero energy collision diameter σ, and the maximum energy of attraction ϵ. These parameters, together with the molecular weight, are required to calculate the viscosity and self-diffusion coefficients of pure gases. For estimating thermal conductivity, heat capacities are also required.

*This is a summary of R. S. Brokaw, *NASA Tech. Rept. R-81,* 1961.

Values of σ, ϵ/k_B (where k_B is the Boltzmann constant), and molecular weight for a number of species are shown in Table 49. Most of these values have been obtained with viscosity data; however, some estimates for unstable species found in dissociated atmospheric or combustion gases are also included.

In the event σ and ϵ/k_B are not available from transport-property measurements, they may be estimated from critical, boiling, or melting point properties as follows [1, p. 245]:

$$\left.\begin{array}{rl} \sigma \simeq & 0.841 \; V_c^{1/3} \; \overset{\circ}{\text{A}} \\[6pt] \simeq & 1.166 \; V_b^{1/3} \\[6pt] \simeq & 1.222 \; V_m^{1/3} \end{array}\right\} \tag{2}$$

$$\left.\begin{array}{rl} \epsilon/k_B \simeq & 0.77 \; T_c \\[6pt] \simeq & 1.15 \; T_b \\[6pt] \simeq & 1.92 \; T_m \end{array}\right\} \tag{3}$$

Here V_c, V_b, and V_m are the molar volumes of the gas at the critical point, the liquid at the boiling point, and the solid at the melting point; T_c, T_b, and T_m are the corresponding absolute temperatures. Estimates based on critical properties are preferred.

Next, the specific equations and charts for calculating the viscosity, thermal conductivity, and self-diffusion coefficients are considered.

Viscosity.   The viscosity of a pure gas may be written [1, Eq. (8.2–18)]:

$$\mu \times 10^6 = 26.693 \frac{\sqrt{MT}}{\sigma^2 \, \Omega^{(2,2)*}} \tag{4}$$

where μ is the viscosity in g/cm-sec, or poise. The quantity $\Omega^{(2,2)*}$ is a reduced collision integral that is a function of the reduced temperature kT/ϵ. Equation (4) may be rearranged as follows:

$$\mu \times 10^6 = 26.693 \left[\frac{1}{\sigma^2} \sqrt{\frac{M\epsilon}{k_B}}\right]\left[\frac{\sqrt{k_B T/\epsilon}}{\Omega^{(2,2)*}}\right] \tag{5}$$

The first bracketed quantity in Eq. (5) is a constant characteristic of the gas; the second quantity is a function of $k_B T/\epsilon$ alone. Bromley and Wilke [10] first used Eq. (5) as the basis of an alignment chart for calculating viscosity.

Figure 19 is an alignment chart based on Eq. (5). The scales for $(1/\sigma^2)\sqrt{M\epsilon/k_B}$ and viscosity are logarithmic. Hence, these scales can be used for values outside the tabulated range by merely multiplying or dividing by the appropriate power of 10. The reduced temperature scale is so arranged that, if it is read on the left, the viscosity is as read directly; if the reduced temperature scale is read on the right, the viscosity must be multiplied or divided by 10 as shown in the following three examples:

1. Calculate the viscosity of hydrogen at 27°C. From Table 49, $\epsilon/k_B = 38.0°\text{K}$; hence,

$$\frac{k_B T}{\epsilon} = \frac{27 + 273}{38} = 7.90$$

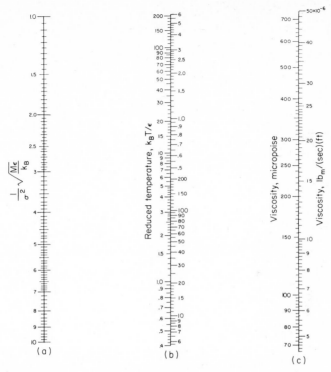

Fig. 19. Alignment chart for viscosity. If the reduced temperature is read on the right of scale *b*, divide the answer on scale *c* by 10. If the reduced temperature is read on the left of scale *b*, read the answer on scale *c* directly. If scale *a* is set 10 times its true value, then divide the answer on scale *c* by 10. If scale *a* is set 0.1 times its true value, then multiply the answer on scale *c* by 10.

Also from Table 49, $(1/\sigma^2)\sqrt{M\epsilon/k_B} = 1.0301$. This value is located on the $(1/\sigma^2)\sqrt{M\epsilon/k_B}$ scale. The point 7.9 is set on the left side of the $k_B T/\epsilon$ scale. Hence the viscosity is read directly from the chart as

$$\mu = 90.8 \text{ micropoises} = 6.08 \times 10^{-6} \text{ lbm/sec-ft}$$

A smoothed experimental value is 89.6 micropoises [11].

2. Estimate the viscosity of cadmium vapor at 893°K. The melting point of cadmium is 320.9°C. By Eq. (3),

$$\frac{\epsilon}{k_B} \simeq 1.92 (320.9 + 273.2) = 1141°K$$

The density of the solid at the melting point is 8.40 g/cm³. Since the atomic weight of cadmium is 112.4, the molar volume at the boiling point is 112.4/8.40 = 13.38 cm³/mole. From Eq. (2),

$$\sigma = 1.222 (13.38)^{1/3} = 2.90 \text{ Å}$$

and

$$\frac{1}{\sigma^2}\sqrt{\frac{M\epsilon}{k_B}} = \frac{1}{(2.90)^2}\sqrt{112.4 \times 1141} = 42.6$$

$$\frac{k_B T}{\epsilon} = \frac{893}{1141} = 0.783$$

To calculate the viscosity, the value 42.6 is located as 4.26 on the $(1/\sigma^2)\sqrt{M\epsilon/k_B}$ scale (thus, the answer must be multipled by 10). Because the point 0.783 is set on the right side of the $k_B T/\epsilon$ scale, the answer must be divided by 10. Therefore, $\mu = 564 \times (10/10) = 564$ micropoises. (This same viscosity is obtained by estimating from boiling-point properties.) An experimental value is 665 micropoises [12].

3. The viscosity of nitrogen at $0°C$ is 166.2 micropoises [11]. The boiling point of nitrogen is $-195.8°C$. Estimate the viscosity of nitrogen at (a) $100°K$ and (b) $1300°K$. From Eq. (3),

$$\frac{\epsilon}{k_B} \simeq 1.15(273.2 - 195.8) = 89°K$$

At $0°C$,

$$\frac{k_B T}{\epsilon} = \frac{273.2}{89} = 3.07$$

From this value and the experimental viscosity, $(1/\sigma^2)\sqrt{M\epsilon/k_B}$ is found to be 3.65. (This value differs somewhat from the value shown in Table 49 because in this example ϵ/k_B is estimated from the boiling point, rather than fitted to viscosity data over a range of temperature; values in Table 49 are to be preferred.) For $100°C$ and $1300°K$, $k_B T/\epsilon$ values are 1.123 and 14.61. From the chart, the corresponding viscosities are 69.0 and 473 micropoises; values from smoothed experimental data are 68.6 and 466 micropoises [11].

The errors in these three examples are typical. There is close agreement for hydrogen using force constants obtained from experimental viscosity over a range of temperature. Results are almost as good when the critical or boiling point is used to extrapolate from a single experimental measurement over a wide temperature range, as in the nitrogen example just given. Finally, the calculation of the viscosity of cadmium vapor based on melting-point properties is about 15 percent too low. This shows that in the absence of experimental data reasonable estimates can be made for substances as "unusual" as metal vapors.

Thermal Conductivity. The thermal conductivity of a pure monatomic gas is [1, Eq. (8.2−31)]:

$$k' \times 10^6 = 198.91 \frac{\sqrt{T/M}}{\sigma^2 \Omega^{(2,2)*}} \tag{6}$$

$$= \frac{15}{4}\frac{R}{M}\mu \times 10^6 \tag{7}$$

where k' is the monatomic thermal conductivity in cal/cm-sec-$°K$. In polyatomic gases, additional energy moves by the diffusional transport of internal energy. This contribution to the conductivity k'' is approximated by [6]:

$$k'' \simeq 0.88 \left(\frac{2}{5} \frac{C_p}{R} - 1 \right) k' \tag{8}$$

Here C_p/R is the ratio of the constant-pressure heat capacity to the gas constant. (Heat capacities are available in handbooks; C_p/R for some simple gases are tabulated over a wide range of temperature in [12].) Hence,

$$k = k' \left[1 + 0.88 \left(\frac{2}{5} \frac{C_p}{R} - 1 \right) \right] \tag{9}$$

Equation (6) may be rearranged in a fashion analogous to the viscosity expression (Eq. (5)) to form the basis of an alignment chart for calculating monatomic thermal conductivity. In this case, the parameter characterizing the gas is $(1/\sigma^2) \sqrt{\epsilon/Mk_B}$; values for a number of gases appear in Table 49.

Figure 20 is an alignment chart for calculating the thermal conductivity of gases based on Eqs. (6) and (9). The $(1/\sigma^2)\sqrt{\epsilon/Mk_B}$, $k_B T/\epsilon$ and monatomic conductivity scales are used in a fashion analogous to the viscosity chart. If, in calculating the polyatomic conductivity, the C_p/R scale is read on the right, the answer must be multiplied by 10. If read on the left, the answer is read from the thermal conductivity scale for C_p/R from 2.5 to 30; for C_p/R from 30 to 100, the answer must be multiplied by 100.

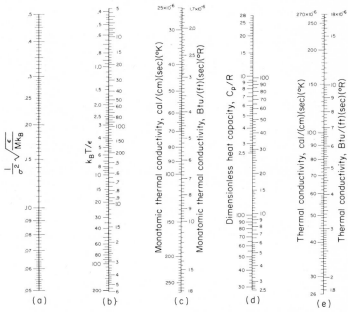

Fig. 20. Alignment chart for thermal conductivity. The restrictions on the a and b scales are the same as for Fig. 19. If the value set on scale c is 10 times its true magnitude, then divide the answer on scale e by 10. If the value set on scale c is 0.1 times its true magnitude, then multiply the answer on the e scale by 10. If, in determining the answer on the e scale, the d scale is read on the right, the answer must be multiplied by 10. If the d scale is read on the left, the answer is as read on the e scale for d values from 2.5 to 30; for d values from 30 to 100, the answer must be multiplied by 100.

The following three examples show some uses of Fig. 20. (As in the case of viscosity, molecular diameters can be estimated from molar volumes. This feature will not be reemphasized.)

1. Calculate the thermal conductivity of argon at $0°C$. From Table 49,

$$\frac{1}{\sigma^2} \sqrt{\frac{\epsilon}{Mk_B}} = 0.1478 \qquad \frac{\epsilon}{k_B} = 119.5$$

Hence

$$\frac{k_B T}{\epsilon} = \frac{273}{119.5} = 2.28$$

and, from Fig. 2,

$$k' = 39.2 \times 10^{-6} \text{ cal/cm-sec-}°K$$
$$= 2.65 \times 10^{-6} \text{ Btu/ft-sec-}°R$$

A smoothed experimental value is 39.05×10^{-6} cal/cm-sec-°K [11].

2. Calculate the thermal conductivity of carbon dioxide at $800°C$. From Table 49, $(1/\sigma^2)\sqrt{\epsilon/Mk_B} = 0.1365$, $\epsilon/k_B = 200°K$, so that

$$\frac{k_B T}{\epsilon} = \frac{1073}{200} = 5.37$$

From Fig. 20

$$k' = 68.8 \times 10^{-6} \text{ cal/cm-sec-}°K$$

At $800°C$, $C_p/R - 6.63$; therefore,

$$k = 169 \times 10^{-6} \text{ cal/cm-sec-}°K$$

A value from smoothed experimental data is 170×10^{-6} [13].

3. At $80°C$ the thermal conductivity of benzene vapor is 34.8×10^{-6} cal/cm-sec-°K [14]. Estimate the thermal conductivity of benzene at $160°C$.

At $80°C$ the heat capacity of benzene is 23.6 cal/deg-mole, so that $C_p/R = 11.88$. From Fig. 20,

$$k' = \frac{80.6}{10} \times 10^{-6} = 8.06 \times 10^{-6} \text{ cal/cm-sec-}°K$$

The boiling point of benzene is $80.1°C$, from which

$$\frac{\epsilon}{k_B} = 1.15\,(273.2 + 80.1) = 406°K$$

Thus, at $80°C$, $k_B T/\epsilon = 353/406 = 0.869$ and, from Fig. 20, $(1/\sigma^2)\sqrt{\epsilon/Mk_B} = 0.0760$. Now, for $160°C$ $k_B T/\epsilon = (160 + 273)/406 = 1.067$. From the chart,

$$k' = 9.90 \times 10^{-6}$$

At $160°C$, C_p = 28.8 cal/deg-mole and C_p/R − 14.5. Then

$$k = 52.0 \times 10^{-6} \text{ cal/cm-sec-°K}$$

An experimental value is 54.0×10^{-6} cal/cm-sec-°K [14].

At a given temperature, the viscosity and thermal conductivity are related through Eqs. (7) and (9); Fig. 21 is an alignment chart based on these equations. The scales with the exception of the dimensionless heat-capacity scale are logarithmic and can hence be used for any multiple of 10 of the scale values. The dimensionless heat-capacity scale is identical to that of Fig. 20. No units are shown since the chart may be used for several sets of units related in Table 47.

TABLE 47.

Viscosity units	Related thermal conductivity units
g/cm-sec or poise	cal/cm-sec-°K
lbm/sec-ft	Btu/ft-sec-°R
lbm/hr-ft	Btu/ft-hr-°R
lbm/sec-in.	Btu/in.-sec-°R
lbm/hr-in.	Btu/in.-hr-°R

The use of Fig. 21 is illustrated by the following three examples:

1. The viscosity of helium at $0°C$ is 186.3 micropoises [15]. What is the thermal conductivity of helium? The viscosity of helium is set at 18.6 on the viscosity scale; on this account, the answer must be multiplied by 10. The atomic weight of helium is 4.003; this is set as 40.0 on the molecular weight scale, so that the answer must be multiplied again by 10. From Fig. 21,

$$k' = 3.48 \times 10^2 \times 10^{-6} = 348 \times 10^{-6} \text{ cal/cm-sec-°K}$$

A smoothed experimental value is 338×10^{-6} [15].

2. The viscosity of carbon dioxide at $800°C$ is 408.0 micropoises [11]. Calculate the thermal conductivity.

The viscosity is set as 40.8, and therefore the answer must be multiplied by 10. Since the molecular weight of carbon dioxide is 44, the monatomic thermal conductivity of carbon dioxide is

$$k' = 6.92 \times 10 \times 10^{-6}$$
$$= 69.2 \times 10^{-6} \text{ cal/cm-sec-°K}$$

Since, at $800°C$, C_p/R = 6.63,

$$k = 170 \times 10^{-6} \text{ cal/cm-sec-°K}$$

As already noted, a smoothed experimental value is also 170×10^{-6}.

3. At $160°C$ the thermal conductivity of benzene is 54.0×10^{-6} cal/cm-sec-°K [14], and the dimensionless heat capacity is 14.5. Estimate the viscosity of benzene. From Fig. 21,

$$k' = 10.4 \times 10^{-6} \text{ cal/cm-sec-°K}$$

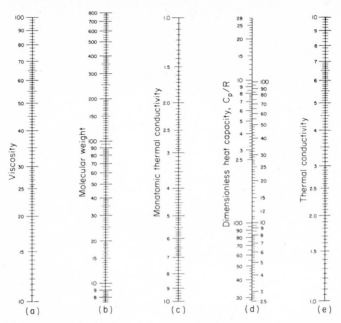

Fig. 21. Alignment chart relating viscosity and thermal conductivity. The restrictions on the d scale are the same as for Fig. 20. If viscosity on scale b is set 10 times its true magnitude, then divide the answer on scale c by 10. If molecular weight on scale b is set 10 times the true magnitude, then multiply the answer on scale c by 10, and so on with scale c. Note the special restrictions on scale d.

and since the molecular weight of benzene is 78.1,

$$\mu = 107 \text{ micropoises}$$

An experimental value is 109.5 micropoises [14].

The examples calculated with Figs. 20 and 21 show the results that can be obtained from Eqs. (6), (7), and (9). Equations (6) and (7) are quite reliable, as is reflected by the rather accurate calculations for the noble gases, argon and helium. The Eucken-type correction embodied in Eqs. (8) and (9) is an approximation that may lead to errors in conductivity on the order of 5 to 10 percent at low temperatures; as the temperature is raised, the errors decrease. Further, it is emphasized that these charts will greatly overestimate thermal conductivity for highly polar substances, such as water, ammonia, and methanol [14].

To calculate the thermal conductivity of polyatomic gas mixtures, it is necessary to know the internal thermal conductivity. (The properties of gas mixtures are treated in the next section.) If an experimental thermal conductivity is available, the internal thermal conductivity may be obtained as the difference between the experimental conductivity and the monatomic conductivity computed from Eqs. (6) and (7) (Fig. 20 or 21):

$$k'' = k - k' \qquad (10)$$

In the absence of experimental data, the internal conductivity k'' may be calculated from Eq. (8). An equivalent procedure is to use the heat capacity to compute the thermal conductivity with Fig. 20 or 21; the internal conductivity may then be obtained as the difference between the total and monatomic conductivities (Eq. (10)).

Self-diffusion Coefficient. The self-diffusion coefficient of a gas in cm^2/sec may be written [1, Eq. (8.2—46)]:

$$D = 0.002628 \frac{\sqrt{T^3/M}}{P\sigma^2 \Omega^{(1,1)*}} \tag{11}$$

A different reduced collision integral $\Omega^{(1,1)*}$ is used in calculating diffusion coefficients. The quotient of the diffusion coefficient and the molar volume DP/RT is independent of pressure and less markedly dependent on temperature. Thus the equation

$$\frac{DP}{RT} \times 10^6 = 32.03 \left[\frac{1}{\sigma^2}\sqrt{\frac{\epsilon}{Mk_B}}\right]\left[\frac{\sqrt{k_B T/\epsilon}}{\Omega^{(1,1)*}}\right] \tag{12}$$

forms a basis of the alignment chart for diffusion coefficients shown as Fig. 22. (In Eq. (12) the gas constant R is in cm^3/atm-deg-mole.) The parameter $(1/\sigma^2)\sqrt{\epsilon/Mk_B}$ characterizes the gas (also used for thermal conductivity). When Eq. (12) is multiplied by the molecular weight of a gas, the product of diffusion coefficient and density ρD is obtained. This quantity often appears in gas dynamic equations.

Actually, the self-diffusion coefficient is not usually an important transport property. The importance of Fig. 22 lies in the fact that it can be used to calculate binary diffusion coefficients by using properly averaged molecular weight, molecular diameter, and reduced temperature. This is discussed in the section on gas mixtures.

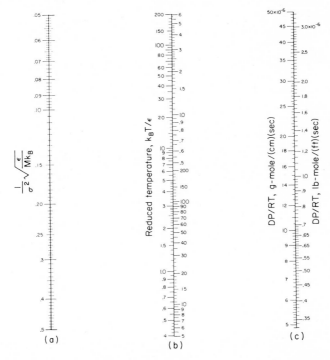

Fig. 22. Alignment chart for diffusion coefficient. Scales a and b are used in the same manner as outlined in Fig. 19.

Figure 22 may be used to calculate the self-diffusion coefficient as follows:
Calculate the self-diffusion coefficient of argon at 22°C and 1 atmosphere. From Table 49, $\epsilon/k_B = 119.5, (1/\sigma^2)\sqrt{\epsilon/Mk_B} = 0.1478$; hence $k_B T/\epsilon = (273.2 + 22)/119.5 = 2.47$. From Fig. 22, $DP/RT = 7.52 \times 10^{-6}$ g-mole/cm-sec. Hence,

$$D = (7.52 \times 10^{-6}) \frac{RT}{P} = 7.52 \times 10^{-6} \frac{82.06 \times 295.2}{1} = 0.182 \text{ cm}^2/\text{sec}$$

An experimental value is 0.180 cm²/sec [16].

Properties of Gas Mixtures

The charts of the previous section, devoted to calculation of the properties of pure gases, are in the main based on rigorous kinetic theory. The rigorous theory for the viscosity and thermal conductivity of gas mixtures is also well developed [1 (Eq. (8.2-25)), 7, 17], but the formulas are complex. Simpler approximations, derived from rigorous theory, are presented here. On the other hand, the binary diffusion coefficient equation is exact.

The viscosity of a gas mixture of ν components may be calculated by the approximate equation [7]:

$$\mu_{\text{mix}} = \sum_{i=1}^{\nu} \frac{\mu_i}{1 + \sum_{\substack{j=1 \\ j \neq 1}}^{\nu} \Phi_{ij} \frac{x_j}{x_i}} \tag{13}$$

where the μ_i are the viscosities of the component gases and x_i, x_j are mole fractions. The coefficients Φ_{ij} are given as a function of the viscosity and molecular-weight ratios of species i and j:

$$\Phi_{ij} = \frac{\left[1 + (\mu_i/\mu_j)^{1/2}(M_j/M_i)^{1/4}\right]^2}{2\sqrt{2}[1 + (M_i/M_j)]^{1/2}} \tag{14}$$

Equations (13) and (14) were first developed empirically by Wilke [4]; these equations have been related to the rigorous theory [5, 6].

Figures 23 and 24 are charts for calculating Φ_{ij} based on Eq. (14). It is believed that almost all cases can be obtained from Fig. 23; however, Fig. 24 is provided for large values of viscosity ratio.

Once Φ_{ij} has been obtained from Fig. 23 or 24, Φ_{ji} may be similarly obtained. Alternatively, it may be computed from the equation

$$\Phi_{ji} = \Phi_{ij} \left(\frac{\mu_j}{\mu_i}\right)\left(\frac{M_i}{M_j}\right) \tag{15}$$

The following example illustrates the use of Fig. 23 or 24 and Eqs. (13) and (15):
At 23°C the viscosities of hydrogen and carbon dioxide are 89.1 and 149.3 micropoises respectively (values cited in [18]). What is the viscosity of a mixture of 41.3 percent hydrogen in carbon dioxide?

Designating hydrogen as species 1 and carbon dioxide as species 2, $\mu_1/\mu_2 = 0.596$ and $M_1/M_2 = 0.0458$. From Fig. 23, $\Phi_{12} = 2.46$. Similarly, $\mu_2/\mu_1 = 1.676$ and $M_2/M_1 =$

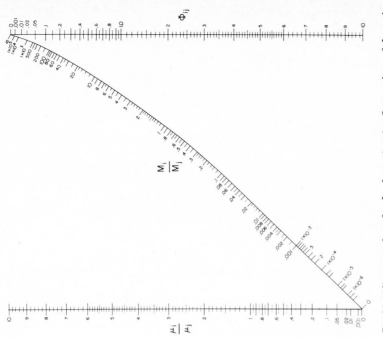

Fig. 24. Alignment chart for Φ_{ij} from viscosity (large values of Φ_{ij} and viscosity ratio).

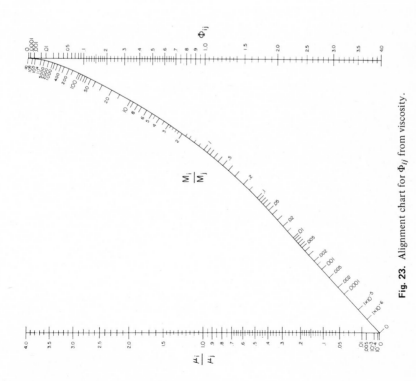

Fig. 23. Alignment chart for Φ_{ij} from viscosity.

21.82; from Fig. 23, Φ_{21} = 0.191. (Alternatively, from Eq. (15), Φ_{21} = 2.46 × 1.676 × 0.0458 = 0.1888.) Hence, from Eq. (13),

$$\mu_{mix} = \frac{89.1}{1 + 2.46\,[(1 - 0.413)/0.413]} + \frac{149.3}{1 + 0.191\,[0.413/(1 - 0.413)]}$$

$$= 151.6 \text{ micropoises}$$

This is in excellent agreement with both the experimental viscosity of 150.6 micropoises and the rigorously calculated viscosity of 150.8 micropoises [18].

Comparisons of viscosities from Eqs. (13) and (14) with rigorously calculated mixture viscosities suggest that the approximate formulas have a probable error of about 2 percent [6].

Thermal Conductivity. The thermal conductivity of a mixture of polyatomic gases may be divided into two portions:

$$k_{mix} = k'_{mix} + k''_{mix} \tag{16}$$

Here k'_{mix} represents the monatomic thermal conductivity of the mixture, whereas k''_{mix} accounts for the diffusional transport of internal energy.

The approximate formula for the monatomic mixture conductivity is [5]:

$$k'_{mix} = \sum_{i=1}^{\nu} \frac{k'_i}{1 + \displaystyle\sum_{\substack{j=1 \\ j\neq 1}}^{\nu} \Psi_{ij}\dfrac{x_j}{x_i}} \tag{17}$$

where k'_i are the monatomic thermal conductivities of the pure component gases. The coefficients Ψ_{ij} are found to be [6]:

$$\Psi_{ij} = \Phi_{ij}\left[1 + 2.41\,\frac{(M_i - M_j)(M_i - 0.142\,M_j)}{(M_i + M_j)^2}\right]$$

$$= \frac{\left[1 + (k'_i/k'_j)^{1/2}\,(M_i/M_j)^{1/4}\right]^2}{2\sqrt{2}\,[1 + (M_i/M_j)]^{1/2}} \times$$

$$\left[1 + 2.41\,\frac{(M_i - M_j)(M_i - 0.142\,M_j)}{(M_i + M_j)^2}\right] \tag{18}$$

Figures 25 and 26 are charts for calculating Ψ_{ij} as a function of the monatomic conductivity and molecular weight ratios; Fig. 25 is for small Ψ_{ij} and Fig. 26 is for large values. It should be noted that there is no simple form analogous to Eq. (15) relating to Ψ_{ij} and Ψ_{ji}.

The following example illustrates the use of Figs. 25 and 26 together with Eq. (17):

At 311°K the thermal conductivities of helium, argon, and xenon are 375.3, 43.8, and 13.5 × 10^{-6} cal/cm-sec-°K, respectively [19]. What is the thermal conductivity of a mixture of 39.01 percent helium, 36.75 percent argon, and 24.24 percent xenon?

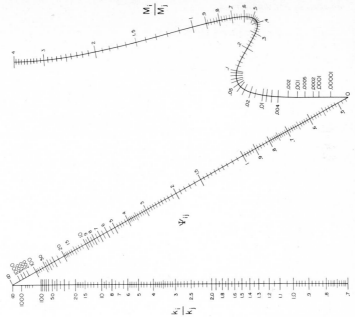

Fig. 26. Alignment chart for Ψ_{ij} from conductivity (large values of Ψ_{ij} and monatomic conductivity ratio).

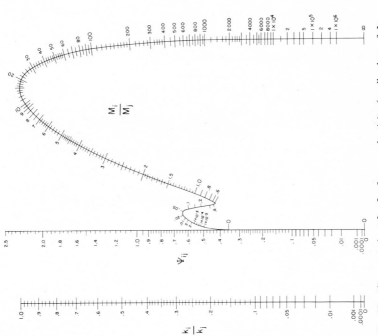

Fig. 25. Alignment chart for Ψ_{ij} from conductivity (small values of Ψ_{ij} and monatomic conductivity ratio).

Details of the calculation are presented in Table 48.

TABLE 48.

Component		k_i'/k_j'	M_i/M_j	Ψ_{ij} (Figs. 7 and 8)	x_j/x_i	$\Psi_{ij}\dfrac{x_i}{x_j}$
i	j					
He	Ar	8.57	0.1002	2.56	0.941	2.41
	Xe	27.8	0.03045	4.40	0.621	2.73
Ar	He	0.1167	9.98	0.757	1.062	0.805
	Xe	3.24	0.304	1.42	0.660	0.937
Xe	He	0.03596	32.80	0.408	1.609	0.656
	Ar	0.308	3.288	1.005	1.514	1.522

Equation (17) becomes

$$k'_{mix} = \frac{375.3}{1 + 2.41 + 2.73} + \frac{43.8}{1 + 0.805 + 0.937} + \frac{13.5}{1 + 0.656 + 1.522}$$

$$= 81.3 \times 10^{-6} \text{ cal/cm-sec-}^\circ K$$

A value of 84.5×10^{-6} cal/cm-sec-°K is found for this mixture both experimentally [19] and from rigorous theory [17].

A formula for k''_{mix}, the internal thermal conductivity, has been derived by Hirschfelder [7]. This expression, which embodies a treatment of the diffusional transport of internal energy, can be approximated by an expression analogous to Eq. (13) [6]:

$$k''_{mix} = \sum_{i=1}^{\nu} \frac{k_i''}{1 + \sum_{\substack{j=1 \\ j \neq 1}}^{\nu} \Phi_{ij}\dfrac{x_j}{x_i}} \tag{19}$$

where the k_i'' are the internal thermal conductivities of the pure component gases. If experimental conductivities for the pure gases are known, the k_i'' are best obtained as the difference between experimental and monatomic conductivities (Eq. (10)). The coefficient Φ_{ij} (the same coefficient required for mixture viscosity) can be expressed as a function of monatomic thermal conductivity and molecular weight ratios [6]

$$\Phi_{ij} = \frac{\left[1 + (k_i'/k_j')^{1/2}(M_i/M_j)^{1/4}\right]^2}{2\sqrt{2}[1 + (M_i/M_j)]^{1/2}} \tag{20}$$

Figures 27 and 28 are charts based on Eq. (20) (Fig. 27 for small Φ_{ij}, Fig. 28 for large Φ_{ij}). The coefficients Φ_{ij} and Φ_{ji} are related:

$$\Phi_{ji} = \Phi_{ij}\frac{k_j'}{k_i'} \tag{21}$$

The use of these relations is illustrated by the following two examples:

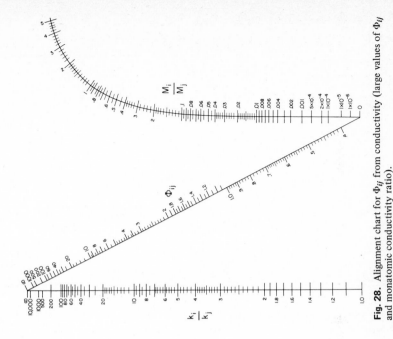

Fig. 28. Alignment chart for Φ_{ij} from conductivity (large values of Φ_{ij} and monatomic conductivity ratio).

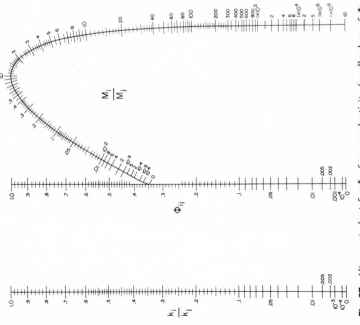

Fig. 27. Alignment chart for Φ_{ij} from conductivity (small values of Φ_{ij} and monatomic conductivity ratio).

TABLE 49. Constants for Transport Property Calculations

Gas	Reference No.	Molecular Weight	σ, Å	ϵ/k_B, °K†	$\dfrac{1}{\sigma^2}\sqrt{\dfrac{M\epsilon}{k_B}}$ (for viscosity)	$\dfrac{1}{\sigma^2}\sqrt{\dfrac{\epsilon}{Mk_B}}$ (for conductivity and diffusion)
Monatomic Gases						
He	1	4.003	2.576	10.22	0.964	0.2408
Ne	1	20.18	2.858	27.5	2.884	0.1429
Ar	11	39.94	3.421	119.5	5.90	0.1478
Kr	1	83.8	3.610	190	9.68	0.1155
Xe	1	131.3	4.055	229	10.55	0.0803
H	6	1.008	2.68	(38.0)	0.862	0.855
N		14.01	3.1	(91.5)	3.726	0.2659
O		16.00	2.90	(100)	4.757	0.2972
F		19.00	2.9	(112)	5.49	0.2887
Cl		35.46	3.6	(316)	8.17	0.2303
Br		79.9	3.9	(451)	12.48	0.1562
I		126.9	4.4	(466)	12.56	0.0990
Hg	1	200.6	2.898	851	49.19	0.2452
Metastable Excited States						
Ne*	6	20.18	7.36	(27.5)	0.4349	0.02155
Ar*	6	39.94	8.73	(119.5)	0.906	0.02270
Xe*	6	131.3	10.6	(229)	1.543	0.01175
Diatomic Gases						
Air	1	28.97	3.689	84	3.625	0.1251
H_2	1	2.016	2.915	38.0	1.030	0.511
N_2	11	28.02	3.681	91.5	3.737	0.1334
O_2	11	32.00	3.499	100	4.620	0.1444
F_2	1	38.00	3.41‡	112	5.61	0.1476
Cl_2	§	70.9	4.217	316	8.42	0.1187
Br_2	§	159.8	4.403	451	13.85	0.0867
I_2	§	253.8	5.179	466	12.82	0.0505
HCl¶	1	36.47	3.305	360	10.49	0.2876
HI	§	127.9	4.211	289	10.84	0.0848
CO	§	28.01	3.678	94.5	3.803	0.1358
NO	§	30.01	3.481	121	4.973	0.1657
Simple Polyatomic Gases						
CO_2	11	44.01	3.952	200	6.01	0.1365
N_2O	1	44.02	3.816	237	7.01	0.1593
NO_2	26	46.01	3.9	230	6.76	0.1470
CS_2	1	76.1	4.438	488	9.78	0.1286
SO_2 ¶	1	64.1	4.290	252	6.91	0.1077
COS	1	60.1	4.13	335	8.32	0.1384
BF_3	§	67.8	4.200	186	6.37	0.0939
BCl_3	§	117.2	5.668	201	4.778	0.04076
CH_4	1	16.04	3.796	144	3.335	0.2079
CF_4	§	88.0	4.662	134	4.996	0.0568
CCl_4	§	153.8	5.973	315	6.17	0.04011
SiF_4	§	104.1	4.880	172	5.62	0.0540
SF_6	§	146.1	5.128	222	6.85	0.04688

See footnotes at end of table.

TABLE 49. Constants for Transport Property Calculations *(Continued)*

Gas	Reference No.	Molecular Weight	σ, Å	ϵ/k_B, °K†	$\dfrac{1}{\sigma^2}\sqrt{\dfrac{M\epsilon}{k_B}}$ (for viscosity)	$\dfrac{1}{\sigma^2}\sqrt{\dfrac{\epsilon}{Mk_B}}$ (for conductivity and diffusion)
		Other Inorganic Gases				
$HgBr_2$	1	360.4	5.414	530	14.91	0.04138
HgI_2		454.4	5.625	698	17.80	0.03917
AsH_3		77.9	4.06	281	8.98	0.1152
$SnCl_4$		260.5	4.540	1550	30.83	0.1183
$SnBr_4$		438.4	6.666	465	10.16	0.02318
N_2O_4	26	92.0	4.74	383	8.36	0.0908
		Hydrocarbons				
C_2H_2	§	26.04	4.023	235	4.833	0.1856
C_2H_4	§	28.05	4.155	227	4.622	0.1648
C_2H_6	1	30.07	4.418	230	4.261	0.1417
C_3H_8		44.09	5.061	254	4.132	0.0937
$n\text{-}C_4H_{10}$		58.1	4.997	410	6.18	0.1064
$iso\text{-}C_4H_{10}$		58.1	5.341	313	4.727	0.0814
$n\text{-}C_5H_{12}$		72.2	5.769	345	4.742	0.0657
$n\text{-}C_6H_{14}$		86.2	5.909	413	5.40	0.0627
$n\text{-}C_8H_{18}$		114.2	7.451	320	3.443	0.03015
$n\text{-}C_9H_{20}$		128.2	8.448	240	2.458	0.01917
C_6H_{12}(cyclohexane) .		84.2	6.093	324	4.449	0.0528
C_6H_6		78.1	5.270	440	6.67	0.0855
		Other Organic Gases				
CH_3OH¶	1	32.04	3.585	507	9.92	0.3095
C_2H_5OH		46.07	4.455	391	6.76	0.1468
CH_3Cl¶		50.5	3.375	855	18.24	0.3612
CH_2Cl_2		84.9	4.759	406	8.20	0.0966
$CHCl_3$		119.4	5.430	327	6.70	0.0561
CCl_2F_2	27	120.9	5.110	288	7.15	0.0591
C_2N_2	1	52.0	4.38	339	6.92	0.1331

†Values in parentheses have been taken from the corresponding diatomic molecule or ground state atom.

‡σ is a mean value estimated by considering viscosity, thermal conductivity, critical temperature and pressure, and boiling-point density.

§Unpublished values obtained by R. A. Svehla, NASA Lewis Research Center, by a least-squares fit of viscosity data.

¶These gases are sufficiently polar that errors will be encountered in calculating properties of mixtures. (Predicted values will be too low.)

1. At $0°C$ the C_p/R for hydrogen and carbon dioxide are 3.44 and 4.32, respectively. What is the thermal conductivity of a 50 percent mixture of hydrogen in carbon dioxide?

From Fig. 20 together with the constants in Table 49, the monatomic thermal conductivities k' of hydrogen and carbon dioxide are 312×10^{-6} and 22.8×10^{-6} cal/cm-sec-°K, respectively. With hydrogen designated as species 1 and carbon dioxide as species 2, $k_1'/k_2' = 13.68$, $M_1/M_2 = 0.0458$; and $k_2'/k_1' = 0.0731$, $M_2/M_1 = 21.83$.

From Figs. 25 and 26, $\Psi_{12} = 3.05$ and $\Psi_{21} = 0.570$. From Eq. (17), the monatomic thermal conductivity is

$$k'_{mix} = \frac{312}{1 + 3.05\,(0.5/0.5)} + \frac{22.8}{1 + 0.570\,(0.5/0.5)} = 91.6 \times 10^{-6} \text{ cal/cm-sec-}^{\circ}\text{K}$$

From Fig. 20, together with the dimensionless heat capacities, the conductivities are found to be $k_1 = 416 \times 10^{-6}$, $k_2 = 37.3 \times 10^{-6}$. Then from Eq. (10), $k''_1 = 104 \times 10^{-6}$, $k''_2 = 14.5 \times 10^{-6}$ cal/cm-sec-$^{\circ}$K. From Fig. 28, $\Phi_{12} = 2.53$ and, from Fig. 27, $\Phi_{21} = 0.186$ (from Eq. (21), $\Phi_{21} = 2.53 \times 0.0731 = 0.185$). From Eq. (19), the internal thermal conductivity is

$$k''_{mix} = \frac{104}{1 + 2.53\,(0.5/0.5)} + \frac{14.5}{1 + 0.186\,(0.5/0.5)} = 41.7 \times 10^{-6} \text{ cal/cm-sec-}^{\circ}\text{K}$$

By Eq. (16), $k_{mix} = (91.6 + 41.7) \times 10^{-6} = 133.3 \times 10^{-6}$ cal/cm-sec-$^{\circ}$K. An experimental measurement for this mixture gives a conductivity 135×10^{-6} cal/cm-sec-$^{\circ}$K [20].

2. At 0°C the thermal conductivities of hydrogen and carbon dioxide are 404×10^{-6} and 36×10^{-6} cal/cm-sec-$^{\circ}$K, respectively. Calculate the thermal conductivity of a 50 percent mixture of hydrogen in carbon dioxide.

In this case, the monatomic thermal conductivities of the pure gases and the mixture are obtained exactly as in the preceding example; the internal thermal conductivities, however, are obtained from Eq. (10) with the experimental conductivities:

$$k''_1 = (404 - 312) \times 10^{-6} = 92 \times 10^{-6} \text{ cal/cm-sec-}^{\circ}\text{K}$$

$$k''_2 = (36 - 22.8) \times 10^{-6} = 13.2 \times 10^{-6}$$

The internal conductivity is

$$k''_{mix} = \frac{92}{1 + 2.53} + \frac{13.2}{1 + 0.186} = 37.2 \times 10^{-6} \text{ cal/cm-sec-}^{\circ}\text{K}$$

and from Eq. (16), $k_{mix} = (91.6 + 37.2) \times 10^{-6} = 128.8 \times 10^{-6}$ cal/cm-sec-$^{\circ}$K. As noted in the previous example, the experimental value is 135×10^{-6} cal/cm-sec-$^{\circ}$K. A value computed by using rigorous formulas is 135.0×10^{-6} cal/cm-sec-$^{\circ}$K [7].

In general, the procedure used in the second example is to be preferred, i.e., it is best to estimate the internal conductivities of the pure components from their thermal conductivities (when available) rather than their heat capacities. (The fact that this procedure gives poorer agreement with experiment in this example is fortuitous.)

Comparisons of calculations using these approximate mixture thermal-conductivity equations with rigorously computed mixture conductivities suggest that these approximations have a probable error of about 1.5 percent [6].

Finally, it should be noted that in chemically reacting gas mixtures there may be a very large increase in thermal conductivity due to the diffusional transport of chemical enthalpy. A general expression has been derived for the increase in thermal conductivity due to chemical reactions in gas mixtures at chemical equilibrium [21, 22]; this effect can account under some conditions for an eightfold increase in the thermal conductivity of nitrogen tetroxide vapor [6, 23] and a thirtyfold increase in the conductivity of hydrogen fluoride vapor [6, 24]. In order to compute the increase in thermal conductivity due to reaction, the gas composition, heats of reaction, and binary diffusion coefficients among the component gases are required. The next section deals with the calculation of binary diffusion coefficients.

Binary Diffusion Coefficient. The binary diffusion between two gases may be written [1, Eq. (8.2–44)]:

$$D_{12} = 0.002628 \frac{\sqrt{T^3(M_1 + M_2)/2 M_1 M_2}}{P\sigma_{12}^2 \, \Omega_{12}^{(1,1)*}} \qquad (22)$$

Here D_{12} is the binary diffusion coefficient between species 1 and 2, M_1 and M_2 are the molecular weights, and σ_{12} is a collision diameter characteristic of the unlike molecule interaction. In this case, the reduced collision integral $\Omega_{12}^{(1,1)*}$ is a function of a reduced temperature $k_B T/\epsilon_{12}$.

The parameters ϵ_{12}/k_B and σ_{12} may be approximated as the geometric and arithmetic means respectively

$$\frac{\epsilon_{12}}{k_B} = \left(\frac{\epsilon_1}{k_B} \frac{\epsilon_2}{k_B} \right)^{1/2} \qquad (23)$$

$$\sigma_{12} = \frac{1}{2}(\sigma_1 + \sigma_2) \qquad (24)$$

In addition, a reciprocal average molecular weight can be defined:

$$\frac{1}{M_{12}} = \frac{1}{2}\left(\frac{1}{M_1} + \frac{1}{M_2} \right) \qquad (25)$$

Then the quotient of the binary diffusion coefficient and the molar volume is given by Eq. (12) and Fig. 22, with ϵ, σ, and M average values calculated from Eqs. (23) to (25).

The following example illustrates the use of Fig. 22 and Eqs. (23) to (25) in calculating binary diffusion coefficients:

Calculate the hydrogen-nitrogen diffusion coefficient at 85°C and 1 atmosphere.

From the force constants from Table 49, together with Eqs. (23) to (25), $\sigma_{12} = 3.298$, $\epsilon_{12}/k_B = 59°K$, $M_{12} = 3.76$, and $(1/\sigma^2)\sqrt{\epsilon/Mk_B} = 0.364$. Therefore,

$$\frac{k_B T}{\epsilon} = \frac{85 + 273.2}{59} = 6.07$$

From Fig. 22, $DP/RT = 36.0 \times 10^{-6}$ g-mole/cm-sec. Hence

$$D_{12} = 36.0 \times 10^{-6} \frac{RT}{P} = 36.0 \times 10^{-6} \times \frac{82.06 \times 358.2}{1} = 1.058 \text{ cm}^2/\text{sec}$$

An experimental value is 1.052 cm²/sec [25].

Concluding Remarks

The alignment charts presented here are believed to represent the best general methods now available for computing the transport properties of pure nonpolar gases based on rigorous kinetic theory. On the other hand, although the binary diffusion coefficient calculations are rigorous, the formulas used for mixture viscosity and conductivity are approximations based on rigorous theory. They are believed to be accurate enough for all practical applications. However, they are not suitable for comparing experimental data with theory; for this purpose, rigorous calculations should be made.

Although these charts are not applicable to highly polar gases, they can probably be used for order-of-magnitude estimates. The mixture formulas should give approximately correct results for mixtures with traces of polar constituents.

The sole exception is the binary diffusion coefficient chart. This should be suitable for calculating diffusion coefficients between polar and nonpolar gases, provided the force constants σ_{12} and ϵ_{12}/k_B are properly chosen.

References

1. J. O. Hirschfelder, C. F. Curtiss, and R. B. Bird, Molecular Theory of Gases and Liquids, John Wiley & Sons, Inc., 1954.
2. J. O. Hirschfelder, R. B. Bird, and E. L. Spotz, The Transport Properties of Non-Polar Gases, *J. Chem. Phys.*, **16**:968 (1948).
3. E. A. Mason, Transport Properties of Gases Obeying a Modified Buckingham (Exp-Six) Potential, *J. Chem. Phys.*, **22**:169 (1954).
4. C. R. Wilke, A Viscosity Equation for Gas Mixtures, *J. Chem. Phys.*, **18**:517 (1950).
5. R. S. Brokaw, Approximate Formulas for the Viscosity and Thermal Conductivity of Gas Mixtures, *J. Chem. Phys.*, **29**: 391 (1958).
6. R. S. Brokaw, Energy Transport in High Temperature and Reacting Gases, *Air Force Office Sci. Res.–General Electric Co. Conf. Phys. Chem. in Aerodynamics and Space Flight, Philadelphia,* September 1–3, 1959.
7. J. O. Hirschfelder, *Proc. Joint Conf. Thermodynamic and Transport Properties of Fluids,* The Inst. Mech. Eng. (London), p. 133, 1958.
8. E. A. Mason and W. E. Rice, The Intermolecular Potentials for Some Simple Nonpolar Molecules, *J. Chem. Phys.*, **22**:843 (1954).
9. R. C. Reid and T. K. Sherwood, "The Properties of Gases and Liquids; Their Estimation and Correlation," McGraw-Hill Book Company, Inc., 1958.
10. L. A. Bromley and C. R. Wilke, Viscosity Behavior of Gases, *Ind. Eng. Chem.*, **43**:1641 (1951).
11. J. Hilsenrath, et al., Tables of Thermal Properties of Gases, Natl. Bur. Std. Circ. 564, November 1, 1955.
12. H. Braune, R. Basch, and W. Wentzel, The Viscosity of Some Gases and Vapors. II, Mercury, Cadmium, and Zinc, *Z. Phys. Chem.*, **A137**:447 (1928).
13. A. J. Rothman and L. A. Bromley, High Temperature Thermal Conductivity of Gases, *Ind. Eng. Chem.,* col. **47**:899 (1955).
14. R. G. Vines and L. A. Bennett, The Thermal Conductivity of Organic Vapors. The Relationship Between Thermal Conductivity and Viscosity and the Significance of the Eucken Factor, *J. Chem. Phys.*, **22**:360 (1954).
15. J. Hilsenrath and Y. S. Touloukian, The Viscosity, Thermal Conductivity, and Prandtl Number for Air, O_2, N_2, NO, H_2, CO, CO_2, H_2O, He and A, *Trans. ASME*, **76**:967 (1954).
16. F. Hutchinson, Self-Diffusion in Argon, *J. Chem. Phys.*, **17**: 1081 (1949).
17. C. Muckenfuss and C. F. Curtiss, Thermal Conductivity of Multicomponent Gas Mixtures, *J. Chem. Phys.*, **29**:1273 (1958).
18. J. O. Hirschfelder, R. Bird, and E. L. Spotz, The Transport Properties of Gases and Gaseous Mixtures, II, *Chem. Rev.*, **44**:205 (1949).
19. S. C. Saxena, Thermal Conductivity of Binary and Ternary Mixtures of Helium, Argon and Xenon, *Indian J. Phys.*, **31**:597 (1957).
20. T. L. Ibbs and A. A. Hirst, Thermal Conductivity of Gas Mixtures, *Proc. Roy. Soc. (London),* **123**:134 (1929).
21. J. N. Butler and R. S. Brokaw, Thermal Conductivity of Gas Mixtures in Chemical Equilibrium, *J. Chem. Phys.*, **26**:1636 (1957).
22. R. S. Brokaw, Thermal Conductivity of Gas Mixtures in Chemical Equilibrium, II, *J. Chem. Phys.*, **32**:1005 (1960).
23. K. P. Coffin and C. O'Neal, Jr., Experimental Thermal Conductivities of the $N_2O_4 \rightleftharpoons 2NO_2$ System, *NACA TN 4209,* 1958.
24. E. U. Franck and W. Spalthoff, Abnormal Heat Conductivity of Gaseous Hydrogen Fluoride, *Naturwiss.*, **40**:580 (1953).

25. R. E. Bunde, Studies of the Diffusion Coefficients of Binary Gas Systems, Wisconsin Univ., CM-850, August 8, 1955.
26. R. S. Brokaw, Correlation of Turbulent Heat Transfer in a Tube for the Dissociating System $N_2O_4 \rightleftharpoons 2NO_2$, *NACA RM* E57K19a, 1958.
27. J. W. Buddenberg and C. R. Wilke, Viscosities of Some Mixed Gases, *J. Phys. Colloid Chem.,* 55:1491 (1951).

Symbols

C_p/R dimensionless heat capacity
D diffusion coefficient, cm^2/sec
k_B Boltzmann constant, $erg/°K$
M molecular weight, g/g-mole
P pressure, atm
R universal gas constant, cal/mole-$°K$ (Eq. (7)), or cm^3-atm/mole-$°K$ (Eq. (12))
r intermolecular distance, Å (angstrom)
T absolute temperature, $°K$
V molar volume, cm^3/mole
x mole fraction
ϵ maximum energy of attraction, ergs (Eq. (1))
ϵ/k_B temperature characteristic of intermolecular attraction (Eq. (3), Table 49)
μ viscosity
k thermal conductivity
k' monatomic thermal conductivity
k'' internal thermal conductivity
σ zero energy collision diameter, A (Eqs. (1) and (2), Table 49)
Φ_{ij} coefficient for calculating the viscosity (Eq. (13)) and internal thermal conductivity (Eq. (19)) of mixtures
$\Phi(r)$ intermolecular potential energy (Eq. (1))
Ψ_{ij} coefficient for calculating the monatomic thermal conductivity of gas mixtures (Eq. (17))
$\Omega^{(1,1)^*}$ reduced collision integral for diffusion (Eqs. (12) and (22))
$\Omega^{(2,2)^*}$ reduced collision integral for viscosity (Eq. (4)) and thermal conductivity (Eq. (6))

Subscripts:

b boiling point
c critical point
i,j components i and j of a mixture
m melting point
mix entire mixture
1,2 components 1 and 2 of a mixture

TRANSPORT PROPERTIES FOR SELECTED BINARY GAS MIXTURES AT ONE ATMOSPHERE PRESSURE

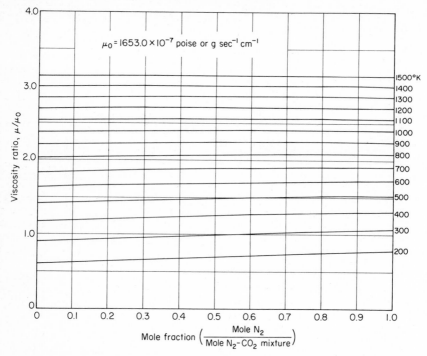

Fig. 29. Viscosity, N_2-CO_2 mixture (*taken from gas thermal conductivity Ref. 17*).

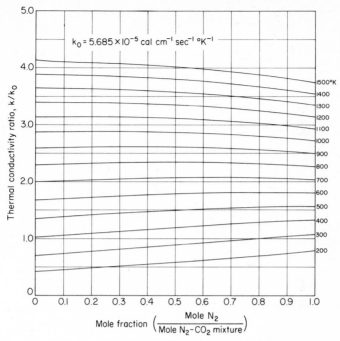

Fig. 30. Thermal conductivity, N_2-CO_2 mixture (*taken from gas thermal conductivity Ref. 17*).

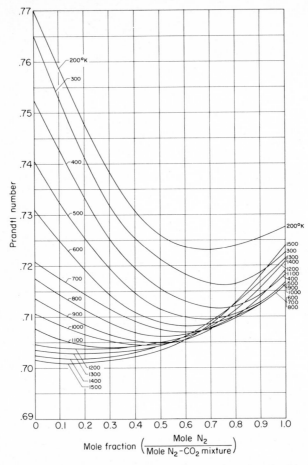

Fig. 31. Prandtl number, N_2-CO_2 mixture (*taken from gas thermal conductivity Ref. 17*).

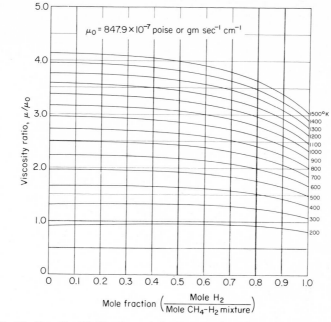

Fig. 32. Viscosity, CH_4-H_2 mixture (*taken from gas thermal conductivity Ref. 18*).

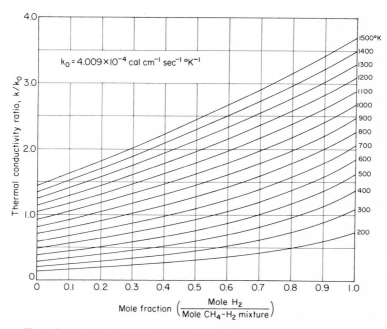

Fig. 33. Thermal conductivity, CH_4-H_2 mixture (*taken from gas thermal conductivity Ref. 18*).

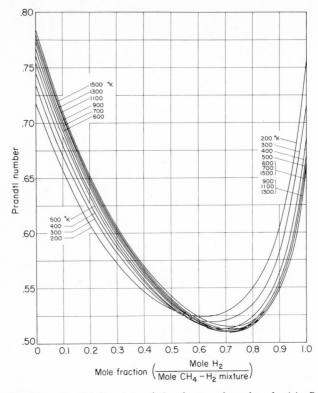

Fig. 34. Prandtl number, CH_4-H_2 mixture (*taken from gas thermal conductivity Ref. 18*).

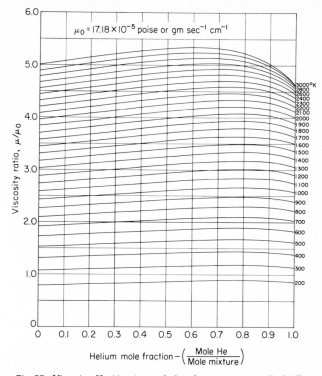

Fig. 35. Viscosity, He-Air mixture (*taken from gas viscosity Ref. 13*).

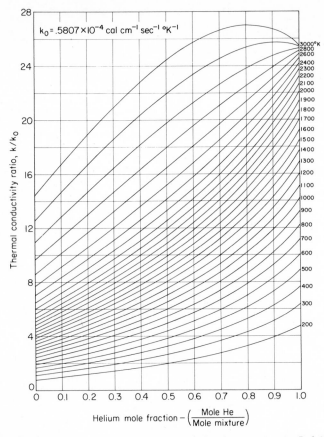

Fig. 36. Thermal conductivity, He-Air mixture (*taken from gas viscosity Ref. 13*).

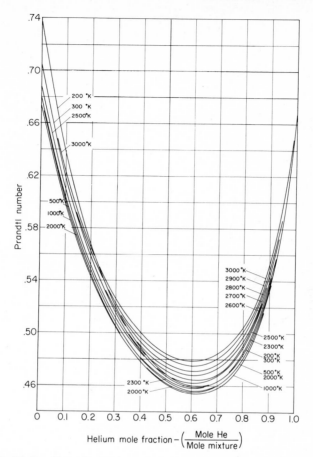

Fig. 37. Prandtl number, He-Air mixture (*taken from gas viscosity Ref. 13*).

HIGH-TEMPERATURE GAS PROPERTIES

Amdur and Mason [1] have used known intermolecular force laws to calculate virial coefficients of the equation of state and transport properties of the rare gases (He, Ne, Ar, Kr, Xe) and molecular nitrogen (N_2) at high temperatures. The force laws used for carrying out the calculations are derived from molecular beam-scattering measurements. The calculation techniques used are the same as those outlined by previous workers [2–4].

The results are tabulated in Tables 50 through 56. The effects of ionization are not considered. Only the translational degrees of freedom are considered over the temperature range from 1,000 to 15,000 °K. Hence, the results are for the condition of no dissociation or excitation.

It should be noted that the thermal conductivity values given are for the translational degrees of freedom only, i.e., frozen thermal conductivity (all internal degrees of freedom remain unchanged over the temperature range).

TABLE 50. Second Virial Coefficients at High Temperatures

T, °K	$B(T)$, cc/mole					
	He	Ne	Ar	Kr	Xe	N$_2$
1,000	9.4	14.2	19.2	16.5	11.4	19.5
1,500	8.4	13.5	21.6	24.0	28.0	30.9
2,000	7.6	12.9	23.0	26.9	35.1	31.9
2,500	7.1	12.4	23.6	28.3	38.7	32.5
3,000	6.6	12.0	23.5	29.2	39.6	32.3
3,500	6.2	11.7	23.2	29.8	40.4	32.0
4,000	5.9	11.4	22.8	29.8	41.2	31.7
4,500	5.6	11.1	22.4	29.7	42.1	31.3
5,000	5.4	10.8	22.1	29.6	42.9	30.9
5,500	5.2	10.6	21.7	29.3	42.8	30.4
6,000	5.0	10.3	21.4	29.0	42.8	29.9
6,500	4.8	10.1	21.0	28.7	42.6	29.4
7,000	4.7	9.9	20.7	28.4	42.5	28.8
7,500	4.5	9.7	20.3	28.1	42.3	28.2
8,000	4.4	9.5	20.1	27.7	42.1	27.6
8,500	4.3	9.4	19.8	27.4	41.9	27.1
9,000	4.2	9.2	19.6	27.0	41.7	26.6
9,500	4.1	9.1	19.4	26.7	41.4	26.1
10,000	4.0	8.9	19.2	26.4	41.2	25.7
11,000	3.8	8.7	18.8	25.8	40.9	24.9
12,000	3.7	8.4	18.4	25.3	40.6	24.1
13,000	3.6	8.2	18.0	24.8	40.3	23.4
14,000	3.4	8.0	17.7	24.3	40.0	22.8
15,000	3.3	7.9	17.4	23.8	39.8	22.1

TABLE 51. Third Virial Coefficients at High Temperatures

T, °K	$10^{-2}\,C(T)$, cc^2/mole^2					
	He	Ne	Ar	Kr	Xe	N$_2$
1,000	0.47	1.4	5.8	8.7	22	10
1,500	0.36	1.2	4.9	7.5	19	8.4
2,000	0.29	0.99	4.3	6.7	17	7.3
2,500	0.25	0.88	3.8	6.1	16	6.5
3,000	0.22	0.80	3.5	5.7	15	5.9
3,500	0.19	0.73	3.2	5.3	14	5.4
4,000	0.17	0.68	3.0	5.0	13	5.1
4,500	0.16	0.64	2.9	4.8	13	4.8
5,000	0.15	0.60	2.7	4.6	12	4.5
5,500	0.13	0.57	2.6	4.4	12	4.3
6,000	0.12	0.54	2.5	4.3	12	4.1
6,500	0.12	0.52	2.4	4.1	11	3.9
7,000	0.11	0.49	2.3	4.0	11	3.7
7,500	0.10	0.47	2.2	3.9	11	3.6
8,000	0.098	0.46	2.1	3.8	10	3.4
8,500	0.093	0.44	2.1	3.7	10	3.3
9,000	0.088	0.42	2.0	3.6	10	3.2
9,500	0.084	0.41	1.9	3.5	9.8	3.1
10,000	0.080	0.40	1.9	3.4	9.6	3.0
11,000	0.074	0.38	1.8	3.3	9.5	2.9
12,000	0.068	0.36	1.7	3.1	9.2	2.7
13,000	0.063	0.34	1.6	3.0	8.6	2.6
14,000	0.059	0.32	1.6	2.9	8.4	2.5
15,000	0.055	0.31	1.5	2.8	8.2	2.4

TABLE 52. Fourth Virial Coefficients at High Temperatures

T, °K	$10^{-3}\ D/T$, cc²/mole²					
	He	Ne	Ar	Kr	Xe	N₂
1,000	0.27	1.4	12	21	83	27
1,500	0.18	1.0	8.8	17	68	20
2,000	0.13	0.81	7.2	14	58	16
2,500	0.10	0.68	6.1	12	52	14
3,000	0.083	0.59	5.4	11	47	12
3,500	0.069	0.52	4.8	10	43	10
4,000	0.059	0.46	4.3	9.3	40	9.4
4,500	0.052	0.42	4.0	8.6	38	8.5
5,000	0.045	0.38	3.7	8.1	36	7.8
5,500	0.040	0.35	3.4	7.6	34	7.2
6,000	0.036	0.33	3.2	7.2	32	6.7
6,500	0.033	0.30	3.0	6.8	31	6.3
7,000	0.030	0.28	2.8	6.5	30	5.9
7,500	0.027	0.27	2.7	6.2	29	5.6
8,000	0.025	0.25	2.5	6.0	28	5.3
8,500	0.023	0.24	2.4	5.7	27	5.0
9,000	0.021	0.23	2.3	5.5	26	4.7
9,500	0.020	0.22	2.2	5.3	25	4.5
10,000	0.019	0.21	2.1	5.2	24	4.3
11,000	0.016	0.19	2.0	4.8	23	4.0
12,000	0.015	0.17	1.8	4.6	22	3.7
13,000	0.013	0.16	1.7	4.3	21	3.4
14,000	0.012	0.15	1.6	4.1	20	3.2
15,000	0.011	0.14	1.5	3.9	19	3.0

TABLE 53. Viscosities at High Temperatures

T, °K	$10^4\ \mu$, g/cm-sec					
	He	Ne	Ar	Kr	Xe	N₂
1,000	4.34	7.02	5.41	6.46	6.06	4.00
1,500	5.84	9.18	7.08	8.60	8.06	5.19
2,000	7.59	11.2	8.86	10.5	9.80	6.31
2,500	9.15	13.0	10.5	12.2	11.4	7.42
3,000	10.7	14.7	12.1	13.8	12.9	8.53
3,500	12.1	16.3	13.7	15.3	14.3	9.61
4,000	13.6	17.9	15.3	16.8	15.6	10.7
4,500	15.0	19.5	16.8	18.2	16.9	11.7
5,000	16.3	21.1	18.3	19.7	18.2	12.8
5,500	17.7	22.7	19.7	21.2	19.5	13.8
6,000	19.0	24.2	21.1	22.6	20.7	14.8
6,500	20.4	25.7	22.5	24.1	21.9	15.8
7,000	21.7	27.2	23.8	25.6	23.2	16.8
7,500	22.9	28.6	25.1	27.1	24.4	17.8
8,000	24.2	30.1	26.3	28.5	25.6	18.8
8,500	25.5	31.5	27.5	30.0	26.8	19.7
9,000	26.7	32.9	28.7	31.5	28.1	20.7
9,500	28.0	34.3	29.8	33.0	29.3	21.6
10,000	29.2	35.7	31.0	34.5	30.6	22.5
11,000	31.6	38.1	33.3	37.5	33.0	24.3
12,000	34.0	40.6	35.5	40.5	35.2	26.0
13,000	36.4	42.9	37.6	43.4	37.4	27.7
14,000	38.7	45.2	39.8	46.3	39.6	29.3
15,000	41.0	47.5	41.8	49.1	41.7	30.9

TABLE 54. Translational ("Frozen") Thermal Conductivities at High Temperatures

T, °K	$10^4 k°$, cal/cm-sec-deg					
	He	Ne	Ar	Kr	Xe	N_2
1,000	8.08	2.59	1.01	0.575	0.344	1.06
1,500	10.9	3.39	1.32	0.766	0.457	1.38
2,000	14.1	4.13	1.65	0.935	0.556	1.68
2,500	17.0	4.80	1.96	1.09	0.647	1.97
3,000	19.9	5.43	2.26	1.23	0.732	2.27
3,500	22.5	6.02	2.56	1.36	0.811	2.56
4,000	25.3	6.61	2.85	1.50	0.885	2.84
4,500	27.9	7.20	3.13	1.62	0.959	3.12
5,000	30.3	7.79	3.41	1.75	1.03	3.40
5,500	32.9	8.38	3.67	1.89	1.11	3.67
6,000	35.4	8.93	3.94	2.01	1.17	3.94
6,500	38.0	9.49	4.20	2.14	1.24	4.20
7,000	40.4	10.0	4.44	2.28	1.32	4.47
7,500	42.6	10.6	4.68	2.41	1.38	4.73
8,000	45.0	11.1	4.90	2.54	1.45	4.99
8,500	47.5	11.6	5.13	2.67	1.52	5.24
9,000	49.7	12.1	5.35	2.80	1.59	5.50
9,500	52.1	12.7	5.56	2.94	1.66	5.74
10,000	54.3	13.2	5.78	3.07	1.74	5.98
11,000	58.9	14.2	6.22	3.34	1.88	6.46
12,000	63.4	15.0	6.64	3.61	2.01	6.92
13,000	67.8	15.9	7.04	3.86	2.13	7.37
14,000	72.1	16.8	7.44	4.12	2.25	7.80
15,000	76.4	17.6	7.83	4.38	2.37	8.22

TABLE 55. Self-diffusion Coefficients at High Temperatures and 1 Atm Pressure

T, °K	D (1 atm), cm^2/sec					
	He	Ne	Ar	Kr	Xe	N_2
1,000	13.3	3.93	1.51	0.853	0.502	1.56
1,500	26.9	7.76	3.00	1.72	1.01	3.08
2,000	46.7	12.7	5.03	2.83	1.65	5.05
2,500	70.3	18.5	7.50	4.14	2.43	7.48
3,000	98.6	25.2	10.4	5.67	3.32	10.4
3,500	130	32.6	13.9	7.38	4.32	13.8
4,000	167	41.1	17.8	9.32	5.42	17.6
4,500	207	50.4	22.1	11.4	6.64	21.9
5,000	250	60.7	26.8	13.8	7.97	26.6
5,500	299	72.0	31.9	16.4	9.43	31.8
6,000	350	83.8	37.4	19.2	10.9	37.3
6,500	407	96.5	43.3	22.2	12.6	43.2
7,000	467	110	49.5	25.5	14.4	49.6
7,500	528	124	56.1	29.0	16.2	56.5
8,000	595	139	62.8	32.6	18.2	63.5
8,500	666	155	69.9	36.5	20.2	71.0
9,000	738	172	77.4	40.7	22.5	78.9
9,500	817	189	85.0	45.0	24.7	87.0
10,000	898	207	93.2	49.6	27.2	95.5
11,000	1070	244	111	59.6	32.4	113
12,000	1260	284	129	70.2	37.7	133
13,000	1460	325	148	81.5	43.4	153
14,000	1670	368	169	93.7	49.4	174
15,000	1890	414	190	107	55.7	197

TABLE 56. Reduced Isotopic Thermal Diffusion
Factors at High Temperatures

T, °K	α_0					
	He	Ne	Ar	Kr	Xe	N_2
1,000	0.33	0.51	0.48	0.45	0.43	0.58
1,500	0.33	0.51	0.46	0.44	0.48	0.54
2,000	0.33	0.51	0.45	0.43	0.49	0.51
2,500	0.33	0.50	0.43	0.42	0.49	0.49
3,000	0.33	0.50	0.43	0.41	0.49	0.47
3,500	0.33	0.50	0.42	0.40	0.49	0.45
4,000	0.33	0.50	0.41	0.39	0.48	0.44
4,500	0.33	0.50	0.41	0.39	0.48	0.42
5,000	0.33	0.50	0.41	0.38	0.47	0.41
5,500	0.33	0.50	0.41	0.37	0.47	0.40
6,000	0.33	0.50	0.41	0.37	0.46	0.39
6,500	0.33	0.50	0.41	0.36	0.46	0.38
7,000	0.33	0.50	0.40	0.36	0.46	0.37
7,500	0.33	0.50	0.40	0.35	0.45	0.37
8,000	0.33	0.50	0.40	0.34	0.45	0.36
8,500	0.33	0.50	0.40	0.34	0.44	0.36
9,000	0.33	0.50	0.40	0.33	0.44	0.35
9,500	0.33	0.50	0.40	0.33	0.44	0.35
10,000	0.33	0.50	0.40	0.32	0.44	0.35
11,000	0.33	0.50	0.40	0.31	0.43	0.35
12,000	0.33	0.50	0.40	0.30	0.43	0.35
13,000	0.33	0.49	0.40	0.30	0.43	0.35
14,000	0.33	0.49	0.40	0.30	0.43	0.35
15,000	0.33	0.49	0.40	0.30	0.43	0.35

References

1. I. Amdur and E. A. Mason, *Phys. Fluids,* **1**:370 (1958).
2. I. Amdur, *Science,* **118**:567 (1953).
3. J. O. Hirschfelder and M. A. Eliason, *Ann. N.Y. Acad. Sci.,* **67**:451 (1957).
4. E. A. Mason and J. T. Vanderslice, *Ind. Eng. Chem.,* **50**:1033 (1958).

TRANSPORT PROPERTIES OF POLAR GASES

Monchick and Mason [1] use the Stockmayer (12–6–3) potential* to calculate
transport properties of dilute polar gases. The potential parameters (E/k_B, σ, S) are
determined from experimental viscosities and dipole moments. A comparison of
calculated and experimental second virial coefficients are made in Table 60. The
authors also compare theoretical and experimental dispersion and induction energies
for added confirmation.

*

$$\phi(r) = 4E_0 \left[\left(\frac{\sigma_0}{r}\right)^{12} - \left(\frac{\sigma_0}{r}\right)^6 + \delta \left(\frac{\sigma_0}{r}\right)^3 \right]$$

where $\phi(r)$ = potential energy of interaction
r = distance of separation

All other parameters are defined on the following page.

To calculate polar gas transport properties, the following equations are used in conjunction with Tables 57, 58, and 59:

$$T^* = \frac{k_B T}{E_0} \quad \text{(dimensionless temperature)}$$

$$\mu = \frac{5}{16} \left(\frac{m k_B T}{\pi} \right)^{1/2} \frac{1}{\sigma_0^2 \langle \Omega^{(2,2)*} \rangle}$$

$$k^\circ = \frac{15}{4} k_B \mu$$

$$D_{11} = \frac{3}{8n} \left(\frac{k_B T}{\pi m} \right)^{1/2} \frac{1}{\sigma_0 \langle \Omega^{(1,1)*} \rangle}$$

where E_0 = potential well depth for Lennard Jones (12–6) potential
σ_0 = rigid sphere diameter
m = mass of molecule
k_B = Boltzmann constant
T = absolute temperature
μ = viscosity
k° = translational (frozen) thermal conductivity
D_{11} = self diffusion coefficient
$\bar{\mu}$ = dipole moment
δ_{max} = dimensionless quantity
n = molecular density
$\langle \Omega^{(1,1)*} \rangle$ = collision integral tabulated in Table 57
$\langle \Omega^{(2,2)*} \rangle$ = collision integral tabulated in Table 58

The necessary potential parameters are given in Table 59. The units given lead to viscosity units of gm/cm-sec, thermal conductivity units of erg/cm-sec-°K, and diffusion coefficient units of cm²/sec.

*The original paper presents slightly different equations for calculating viscosity k°, thermal conductivity, and the self-diffusion coefficient D_{11}, by including "f" quantities which are nearly 1.00. The results obtained by the equations given here are within a few percent of those obtained by the more rigorous equations.

TABLE 57. $\langle \Omega^{(1,1)*} \rangle$

T^* \ δ_{max}	0	0.25	0.50	0.75	1.0	1.5	2.0	2.5
0.1	4.0079	4.002	4.655	5.521	6.454	8.214	9.824	11.31
0.2	3.1300	3.164	3.355	3.721	4.198	5.230	6.225	7.160
0.3	2.6494	2.657	2.770	3.002	3.319	4.054	4.785	5.483
0.4	2.3144	2.320	2.402	2.572	2.812	3.386	3.972	4.539
0.5	2.0661	2.073	2.140	2.278	2.472	2.946	3.437	3.918
0.6	1.8767	1.885	1.944	2.060	2.225	2.628	3.054	3.474
0.7	1.7293	1.738	1.791	1.893	2.036	2.388	2.763	3.137
0.8	1.6122	1.622	1.670	1.760	1.886	2.198	2.535	2.872
0.9	1.5175	1.527	1.572	1.653	1.765	2.044	2.349	2.657

TABLE 57. $\langle \Omega^{(1,\,1)*} \rangle (Cont.)$

T^* \ δ_{max}	0	0.25	0.50	0.75	1.0	1.5	2.0	2.5
1.0	1.4398	1.450	1.490	1.564	1.665	1.917	2.196	2.478
1.2	1.3204	1.330	1.364	1.425	1.509	1.720	1.956	2.199
1.4	1.2336	1.242	1.272	1.324	1.394	1.573	1.777	1.990
1.6	1.1679	1.176	1.202	1.246	1.306	1.461	1.639	1.827
1.8	1.1166	1.124	1.146	1.185	1.237	1.372	1.530	1.698
2.0	1.0753	1.082	1.102	1.135	1.181	1.300	1.441	1.592
2.5	1.0006	1.005	1.020	1.046	1.080	1.170	1.278	1.397
3.0	0.95003	0.9538	0.9656	0.9852	1.012	1.082	1.168	1.265
3.5	0.91311	0.9162	0.9256	0.9413	0.9626	1.019	1.090	1.170
4.0	0.88453	0.8871	0.8948	0.9076	0.9252	0.9721	1.031	1.098
5.0	0.84277	0.8446	0.8501	0.8592	0.8716	0.9053	0.9483	0.9984
6.0	0.81287	0.8142	0.8183	0.8251	0.8344	0.8598	0.8927	0.9316
7.0	0.78976	0.7908	0.7940	0.7993	0.8066	0.8265	0.8526	0.8836
8.0	0.77111	0.7720	0.7745	0.7788	0.7846	0.8007	0.8219	0.8474
9.0	0.75553	0.7562	0.7584	0.7619	0.7667	0.7800	0.7976	0.8189
10.0	0.74220	0.7428	0.7446	0.7475	0.7515	0.7627	0.7776	0.7957
12.0	0.72022	0.7206	0.7220	0.7241	0.7271	0.7354	0.7464	0.7600
14.0	0.70254	0.7029	0.7039	0.7055	0.7078	0.7142	0.7228	0.7334
16.0	0.68776	0.6880	0.6888	0.6901	0.6919	0.6970	0.7040	0.7125
18.0	0.67510	0.6753	0.6760	0.6770	0.6785	0.6827	0.6884	0.6955
20.0	0.66405	0.6642	0.6648	0.6657	0.6669	0.6704	0.6752	0.6811
25.0	0.64136	0.6415	0.6418	0.6425	0.6433	0.6457	0.6490	0.6531
30.0	0.62350	0.6236	0.6239	0.6243	0.6249	0.6267	0.6291	0.6321
35.0	0.60882	0.6089	0.6091	0.6094	0.6099	0.6112	0.6131	0.6154
40.0	0.59640	0.5964	0.5966	0.5969	0.5972	0.5983	0.5998	0.6017
50.0	0.57626	0.5763	0.5764	0.5766	0.5768	0.5775	0.5785	0.5798
75.0	0.54146	0.5415	0.5416	0.5416	0.5418	0.5421	0.5424	0.5429
100.0	0.51803	0.5181	0.5182	0.5184	0.5184	0.5185	0.5186	0.5187

TABLE 58. $\langle \Omega^{(2,2)*} \rangle$

T^* \ δ_{max}	0	0.25	0.50	0.75	1.0	1.5	2.0	2.5
0.1	4.1005	4.266	4.833	5.742	6.729	8.624	10.34	11.89
0.2	3.2626	3.305	3.516	3.914	4.433	5.570	6.637	7.618
0.3	2.8399	2.836	2.936	3.168	3.511	4.329	5.126	5.874
0.4	2.5310	2.522	2.586	2.749	3.004	3.640	4.282	4.895
0.5	2.2837	2.277	2.329	2.460	2.665	3.187	3.727	4.249
0.6	2.0838	2.081	2.130	2.243	2.417	2.862	3.329	3.786
0.7	1.9220	1.924	1.970	2.072	2.225	2.614	3.028	3.435
0.8	1.7902	1.795	1.840	1.934	2.070	2.417	2.788	3.156
0.9	1.6823	1.689	1.733	1.820	1.944	2.258	2.596	2.933
1.0	1.5929	1.601	1.644	1.725	1.838	2.124	2.435	2.746
1.2	1.4551	1.465	1.504	1.574	1.670	1.913	2.181	2.451
1.4	1.3551	1.365	1.400	1.461	1.544	1.754	1.989	2.228
1.6	1.2800	1.289	1.321	1.374	1.447	1.630	1.838	2.053
1.8	1.2219	1.231	1.259	1.306	1.370	1.532	1.718	1.912
2.0	1.1757	1.184	1.209	1.251	1.307	1.451	1.618	1.795
2.5	1.0933	1.100	1.119	1.150	1.193	1.304	1.435	1.578
3.0	1.0388	1.044	1.059	1.083	1.117	1.204	1.310	1.428
3.5	0.99963	1.004	1.016	1.035	1.062	1.133	1.220	1.319

TABLE 58. $\langle \Omega^{(2,2)*} \rangle$*(Cont.)*

$T^* \diagdown \delta_{max}$	0	0.25	0.50	0.75	1.0	1.5	2.0	2.5
4.0	0.96988	0.9732	0.9830	0.9991	1.021	1.079	1.153	1.236
5.0	0.92676	0.9291	0.9360	0.9473	0.9628	1.005	1.058	1.121
6.0	0.89616	0.8979	0.9030	0.9114	0.9230	0.9545	0.9955	1.044
7.0	0.87272	0.8741	0.8780	0.8845	0.8935	0.9181	0.9505	0.9893
8.0	0.85379	0.8549	0.8580	0.8632	0.8703	0.8901	0.9164	0.9482
9.0	0.83795	0.8388	0.8414	0.8456	0.8515	0.8678	0.8895	0.9160
10.0	0.82435	0.8251	0.8273	0.8308	0.8356	0.8493	0.8676	0.8901
12.0	0.80184	0.8024	0.8039	0.8065	0.8101	0.8201	0.8337	0.8504
14.0	0.78363	0.7840	0.7852	0.7872	0.7899	0.7976	0.8081	0.8212
16.0	0.76834	0.7687	0.7696	0.7712	0.7733	0.7794	0.7878	0.7983
18.0	0.75518	0.7554	0.7562	0.7575	0.7592	0.7642	0.7711	0.7797
20.0	0.74364	0.7438	0.7445	0.7455	0.7470	0.7512	0.7569	0.7642
25.0	0.71982	0.7200	0.7204	0.7211	0.7221	0.7250	0.7289	0.7339
30.0	0.70097	0.7011	0.7014	0.7019	0.7026	0.7047	0.7076	0.7112
35.0	0.68545	0.6855	0.6858	0.6861	0.6867	0.6883	0.6905	0.6932
40.0	0.67232	0.6724	0.6726	0.6728	0.6733	0.6745	0.6762	0.6784
50.0	0.65099	0.6510	0.6512	0.6513	0.6516	0.6524	0.6534	0.6546
75.0	0.61397	0.6141	0.6143	0.6145	0.6147	0.6148	0.6148	0.6147
100.0	0.58870	0.5889	0.5894	0.5900	0.5903	0.5901	0.5895	0.5885

TABLE 59. Potential Parameters from Viscosity Data (Graphical Method)

Gas	$\bar{\mu}$ (Debye)	δ_{max}	12–6–3 potential $\sigma_0(\Lambda)$	ϵ_0/k_B (°K)	12–6 potential $\sigma_0(\Lambda)$	ϵ_0/k_B (°K)	Reference for Data
H_2O	1.85	1.0	2.52	775			e, i, j, p, y, z, bb, dd, ff
NH_3	1.47	0.7	3.15	358			e, k, o, p, r, aa, cc
HCl	1.08	0.34	3.36	328	3.305^a	360	g, m, o, u
HBr^o	0.80	0.14	3.41	417			g
HI	0.42	0.029	4.13	313	4.123^a	324	g, s, w
SO_2	1.63	0.42	4.04	347	4.026^b	363	c, l, n, o, q, s, t, cc
H_2S^c	0.92	0.21	3.49	343	3.591^b	309	f, o
$NOCl^c$	1.83	0.4	3.53	690	3.57^b	668	x
$CHCl_3$	1.013	0.07	5.31	355	5.430^a	327	d, e, h, p, v
CH_2Cl_2	1.57	0.2	4.52	483	4.748^b	398	p
CH_3Cl	1.87	0.5	3.94	414	4.151^b	355	e, n, p, ee
CH_3Br^c	1.80	0.4	4.25	382			n
$C_2H_5Cl^c$	2.03	0.4	4.45	423			ee
CH_3OH	1.70	0.5	3.69	417	3.666^b	452	v, ee
C_2H_5OH	1.69	0.3	4.31	431	4.370^b	415	d, v
$n\text{-}C_3H_7OH$	1.69	0.2	4.71	495			v
$i\text{-}C_3H_7OH$	1.69	0.2	4.64	518			v
$(CH_3)_2O$	1.30	0.19	4.21	432	4.264^b	412	n, aa
$(C_2H_5)_2O$	1.15	0.08	5.49	362	5.539^b	351	d, e, h, v, ee
$(CH_3)_2CO$	2.88	0.06	4.50	549	4.669^b	519	d, v, ee
CH_3COOCH_3	1.72	0.2	5.04	418	5.054^b	417	d, v
$CH_3COOC_2H_5$	1.78	0.16	5.24	499	6.163^b	531	d, e, v
Ne					2.858^a	27.5	
CCl_4					5.881^a	327	
CH_4					3.796^a	144	
C_2H_6					4.418^a	230	
C_3H_8					5.061^a	254	

TABLE 60. Comparison of Calculated and Experimental
Second Virial Coefficients

H$_2$O*			NH$_2$			CH$_3$Cl		
	$-B$, cc/mole			$-B$, cc/mole			$-B$, cc/mole	
T, °C	Calc	Expt	T, °C	Calc	Expt	T, °C	Calc	Expt
40	1040	976	−30	379	560	−34·	782	764
70	690	638	0	284	367	−23	702	668
100	497	450	30	224	261	−18	670	637
150	323	284	60	181	197	10	530	500
200	230	197	100	142	143	38	435	401
300	137	112	150	110	103	65	367	320
400	92	72	200	87	77	93	313	265
			250	71	59	121	271	214
						149	237	184
						177	209	155

*Least-square parameters used.

Reference

1. L. Monchick and E. A. Mason, *J. Chem. Phys.*, **35**:1676 (1961).

Footnotes to Table 59:
[a]MTGL, pp. 1110–1113.
[b]Redetermined.
[c]Scanty or uncertain data.
[d]K. Rappanecker, *Z. physik. Chem.*, **72**:695 (1910).
[e]H. Vogel, *Ann. Physik*, **43**:1235 (1914).
[f]A. O. Rankine and C. J. Smith, *Phil. Mag.*, **42**:615 (1921).
[g]H. Harle, *Proc. Roy. Soc. (London)*, **A100**:429 (1922).
[h]R. Suhrmann, *Z. Physik*, **14**:56 (1923).
[i]H. Speyerer, *Z. tech. Physik*, **4**:430 (1923).
[j]C. J. Smith, *Proc. Roy. Soc. (London)*, **A106**:83 (1924).
[k]R. S. Edwards and B. Worswick, *Proc. Phys. Soc. (London)*, **38**:16 (1925).
[l]M. Trautz and W. Weizel, *Ann. Physik*, **78**:305 (1925).
[m]M. Trautz and A. Narath, *Ann. Physik*, **79**:637 (1926).
[n]T. Titani, *Bull. Chem. Soc. Japan*, **5**:98 (1930).
[o]G. Jung and H. Schmick, *Z. physik. Chem.*, **B7**:130 (1930).
[p]H. Braune and R. Linke, *Z. physik. Chem.*, **A148**:195 (1930).
[q]M. Trautz and R. Zink, *Ann. Physik*, **7**:427 (1930).
[r]M. Trautz and R. Heberling, *Ann. Physik*, **10**:155 (1931).
[s]M. Trautz and H. Winterkorn, *Ann. Physik*, **10**:511 (1931).
[t]W. W. Stewart and O. Maass, *Can. J. Research*, **6**:453 (1932).
[u]J. R. Partington, *Physik. Z.*, **34**:289 (1933).
[v]T. Titani, *Bull. Chem. Soc. Japan*, **8**:255 (1933).
[w]M. Trautz and F. Ruf, *Ann. Physik*, **20**:127 (1934).
[x]M. Trautz and A. Freytag, *Ann. Physik*, **20**:135 (1934).
[y]W. Schugajew, *Physik. Z. Sowjetunion*, **5**:659 (1934).
[z]G. A. Hawkins, H. L. Solberg, and A. A. Potter, *Trans. Amer. Soc. Mech. Engrs.*, **57**:395 (1935).
[aa]A. B. van Cleave and O. Maass, *Can. J. Research*, **13B**:140 (1935).
[bb]D. L. Timroth, *J. Phys. (U.S.S.R.)*, **2**:419 (1940).
[cc]R. Wobser and F. Müller, *Kolloid-Beih.*, **52**:165 (1941).
[dd]F. G. Keyes, *J. Amer. Chem. Soc.*, **72**:433 (1950).
[ee]P. M. Craven and J. D. Lambert, *Proc. Roy. Soc. (London)*, **A205**:439 (1951).
[ff]C. F. Bonilla, S. J. Wang, and H. Weiner, *Trans. Amer. Soc. Mech. Engrs.*, **78**:1285 (1956). This paper gives smoothed data only; the original data are available as ADI 4545 from the Library of Congress.

Section **3**

Conduction

P. J. SCHNEIDER

Lockheed Missiles & Space Company, Sunnyvale, California

Equations and graphical data are given for the temperature state in selected heat-conduction systems composed of solid and decomposing materials exposed to various single and multiple thermal environments.

A. BODY TEMPERATURE RESPONSE

The end objective of heat-conduction analyses is to predict the history of *instantaneous temperature distribution* and/or the *fixed temperature pattern* within a body exposed to an environment whose temperature differs from that of the conducting solid. Once this body temperature response is calculated, other information such as surface and internal heat flow, thermal growth, and thermal stresses can be derived [1].

Calculation of material temperatures is a necessary first step in the selection of materials and in the design of structures that operate at suppressed or elevated temperature levels during their service life. For uninsulated structures, component temperature predictions are needed to determine if the structure will survive thermally, to calculate heat flow through the structure to adjacent elements or equipments that may be temperature-sensitive, to evaluate expansions and deflections, and to assess the effects on structural integrity of superimposing thermal stresses on mechanical stresses due to pressure, vibration, acoustic, and other applied loads. With composite, thermally protected structures, material temperature estimates are also needed to determine the degree of thermal protection achieved, to assess degradation of the thermal-protection materials used (oxidation, erosion, ablation, charring, etc.), to predict structural integrity of coatings, to evaluate bond-line survival and the strength response of load-bearing substructures whose mechanical properties are temperature-sensitive, and to calculate optimum material combinations, thicknesses, and weights.

A number of complications arise in predicting heat conduction in practical systems. These include problems of realistically specifying environmental boundary conditions, the requirement for handling complicated body geometry, and the need for flexibility in analyzing a variety of material systems that deviate widely from the concept of simple constant-property isotropic heat conductors. The nature of these problems and the character of input data required for their solution is discussed in the remainder of Sec. A.

1. Characteristics of the Problem

a. **Environmental Conditions.** The field of heat transfer deals largely with predicting the environment at a solid surface (thirteen of the nineteen sections in this handbook deal with this problem). These predictions constitute basic inputs to the body response problem.

Interactions. Generally the usual problem of heat conduction in solids cannot be separated from the associated problem of characterizing convective and radiative surface heating conditions. The two problems are cross-coupled due to *surface interactions* (mass injection, combustion, etc.) that show up as a dependence of surface heating on body response and vice versa. The strength of these environment/surface interactions depends on specific conditions: If the properties of a fluid flow are highly temperature-dependent, then surface conductance for convective heating of a submerged body is sensitive to surface temperature; when the surface of a body thermally decomposes and resulting products of decomposition enter the surrounding flow, the convective heat input is altered and this, in turn, alters decomposition rate of the body material. Such interactions may be sufficiently strong to require simultaneous solution of the heating and body temperature–response problems. Frequently, however, the environment and in-depth body problems can be rigorously or approximately decoupled and the two treated individually. In these cases, surface thermal boundary conditions can be *prescribed* once and for all; the problem remaining is to predict body temperature response subject to these independently specified environments.

Transients. The general problem of heat conduction has *time* as an independent variable. This is true because the prevailing conditions that cause heat to be transferred to the body surface must begin at some point in time. For this reason, several *transient* processes must be considered, namely transients in the heating or cooling environment and the resulting transient behavior of the body temperature field. For decoupled systems, the transient environment is specified independently as a time-dependent boundary condition. Even when environment is steady with respect to time, the body temperature is transient due to sudden exposure to the steady environment whose temperature differs from initial temperature state of the solid. After prolonged exposure, body temperature distribution also becomes steady; this represents a special case of *steady-state* heat conduction.

Condition of steady-state body temperature and heat conduction is always preceded by a *starting transient* of finite duration. In some applications, this starting transient damps out early in the heat-exchange process, and the body temperature field rapidly assumes a fixed unvarying pattern. In other applications, the steady-state condition may never be attained, and early and prolonged transients are the major periods of interest. Examples of both situations are illustrated in the following applications:

(1) Tube-wall temperature response during start-up of a fluid heat exchanger may be of negligible concern compared to steady value of wall temperature drop and its influence on the economics of a large industrial power or process plant; (2) conversely the start-up, sudden power surge, or shutdown of a cooled nuclear reactor may be critical to its structural integrity because during these periods of peak heating or cooling transients the temperature gradients that induce thermal stressing are maximum; (3) the skin temperature of a booster rocket fails to attain a steady-state condition during the engine's brief service life in ascent flight through the atmosphere; and (4) the heatshield of a ballistic entry body and the nozzle of a rocket engine both operate in transient state owing to time-dependent heat input and continuous thermal decomposition of the exposed material.

Initial and Final States. The exposure problem develops in one of several ways. Figure 1 gives the example of an infinite plate thermally insulated at the rear face and exposed at the front face to a convective heating environment. In Fig. 1a the plate is initially isolated and in *thermal equilibrium* with its surroundings at uniform temperature t_0; in this state the surface *heat flux* q'' (heat rate per unit surface area) is zero. In Fig. 1b the plate is suddenly exposed (time $\theta = 0$) to a second environment of elevated temperature t_a, the *unit surface conductance* for convective heating being h. This second environment may or may not be steady with respect to time.

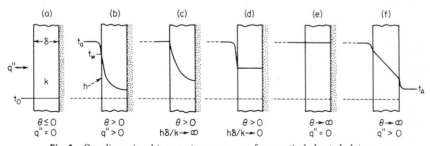

Fig. 1. One-dimensional temperature response of convectively heated plates.

The case of sudden exposure to a steady second environment is illustrated by the process of removing a hot billet from a steady-state furnace and immediately quenching the billet in a large oil bath. An analogous case occurs when the body remains in the equilibrium environment of Fig. 1a, but this environment suddenly begins to change with time and causes an accompanying surface heat input or rejection. This related situation occurs, for example, when an orbiting satellite or spacecraft enters a planetary atmosphere; the vehicle skin feels on entry a sudden change in thermal characteristics of its environment. A sudden change in environment is similarly felt by the nozzle liner of a rocket engine at ignition onset. A special case occurs when environmental temperature varies in a constant cyclic manner. After an initial starting transient, the body temperature also behaves cyclically. Once this cyclic body temperature is established, the problem is referred to as *quasi steady-state* heat conduction. Another special case develops where conditions of surface exposure remain fixed but sudden changes occur within the conducting material itself. If, for example, an electrically conducting body is initially in thermal equilibrium with a steady environment and an electric current is suddenly passed through it, then heat is generated *internally* and the body responds thermally even though environment temperature remains unchanged. Analogous situations occur in materials undergoing *phase changes* as during melting, or when sudden internal energy releases or absorptions occur as in liquids undergoing exothermic or endothermic *chemical reactions*. Finally, certain *viscoelastic* materials can be internally heated by conversion of mechanical energy into thermal energy during cyclic loading (e.g., vibration).

Instantaneous temperature distributions within a body depend on combined external and internal conditions. Figure 1c illustrates the fluid/solid temperature profile under conditions where convective surface conductance h is large compared to internal conductance of the body. In this case, exposed surface temperature t_w is identical to the bulk temperature t_a of the ambient fluid. The opposite extreme is indicated in Fig. 1d; here thermal conductivity k of plate material is large (and/or plate thickness δ is small), and the temperature gradient developed across plate is negligible compared to the fluid/surface temperature drop. The situation in Fig. 1c is referred to as a case of *infinite surface conductance*, while that in Fig. 1d is called *infinite internal conductance* (or, loosely, the *thin-skin* case). Figure 1e shows a limiting condition of thermal equilibrium and zero heat exchange after long exposure to a constant environment at t_a. This state is independent of material properties. For exposure to *multiple* environments such as convection plus radiation, equilibrium plate temperature is a combined function of the convective and radiative source (or sink) temperatures. When the plate is exposed to constant but different environments at each face surface, as in Fig. 1f, the long-time steady-state condition is represented by steady heat flow and constant temperature gradient across the plate.

b. Body Geometry. A number of elementary body shapes are illustrated in Fig. 2 (exposure to convective environment). These are typical of some geometric configurations that can represent reasonable approximations to actual hardware components, and that (owing to their relative simplicity) readily admit to *exact* solutions for temperature response.

The *infinite solid* $-\infty < x < \infty$ in Fig. 2a has "large" dimensions in the vertical, normal, and $\pm x$ directions. Internal plane $x = 0$ may represent a contact surface or contain a plane heat source or sink. The *semi-infinite solid* $x \geq 0$ in Fig. 2b contains an exposed face at $x = 0$. The insulated *plate* $0 \leq x \leq \delta$ in Fig. 2c has finite width. The two-layer solid $0 \leq x_1 \leq \delta_1$, $0 \leq x_2 \leq \delta_2$ in Fig. 2d may be a *composite* plate (finite δ_2) or a *composite semi-infinite solid* (large δ_2). The temperature drop at the interface of layers 1 and 2 suggests existence of finite *thermal contact resistance*.

The term "large" has important interpretations. Considering t_a and h uniform over $x = 0$ in Fig. 2b, the notion of large vertical and normal dimensions infers absence of y and z end effects that can influence local solid temperature. Without end effects, instantaneous body temperature depends only on x (*one-dimensional* conduction). When vertical dimension is finite (and/or ambient conditions vary vertically), body temperature depends on both x and y (*two-dimensional* conduction). If both vertical and normal dimensions are finite, then instantaneous solid temperature is a function of x, y, and z (*three-dimensional* conduction). One-dimensional conduction can occur, of course, with finite vertical and normal dimensions if all four sides are thermally insulated. The notion of largeness also depends on characteristics of the environment and on thermal properties of the solid. A plate, for instance, responds to a heating or cooling environment in early periods of exposure as if it were a semi-infinite solid.

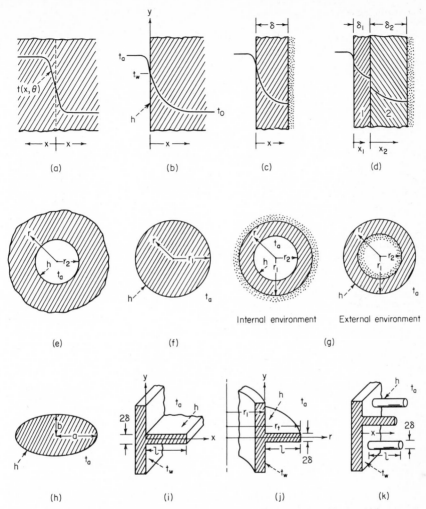

Fig. 2. Elementary body shapes: (a) infinite solid, (b) semi-infinite solid, (c) plate, (d) composite plate, (e) cylindrical and spherical cavity, (f) cylinder and sphere, (g) cylindrical and spherical shell, (h) ellipse and ellipsoid, (i) straight fin, (j) annular fin, (k) spine.

The *cylindrical* and *spherical cavities* $r \geq r_2$ in Fig. 2e are infinite solids with internal environment. The former may be an infinite plate with central hole and insulated faces or an infinitely long cylindrical hole in a large body, while the latter represents a large solid containing a spherical hole. These are one-dimensional systems only if t_a and h are uniform over $r = r_2$. The *cylinder* and *sphere* $0 \leq r \leq r_1$ in Fig. 2f are solid bodies externally heated or cooled. The cylinder may be infinitely long or represent a flat disk with insulated faces. The *cylindrical* and *spherical shells* $r_2 \leq r \leq r_1$ in Fig. 2g are further distinguished as to whether the outer or inner surfaces (or neither) are thermally insulated. The cylindrical shell may be an infinitely long tube or a flat ring with insulated faces. The *elliptical solids* in Fig. 2h can represent a long rod (or insulated disk) of elliptic cross-section or ellipsoids, a prolate spheroid if rotated about a and an oblate spheroid if rotated about b.

The solids in Fig. 2i, j, and k are called *extended surfaces*. This is a special class of finned and spined bodies whose purpose is to increase heat exchange between an otherwise plane surface and its environment.

c. Material Systems. Frequently the most difficult aspect of thermal/structural analysis is the fact that real materials and structures are more complicated than their idealized mathematical models. The nature and diversity of these modeling problems are indicated in Fig. 3 (see Secs. 16 and 19).

The simplest structure from a materials point of view is an infinite, single-layer wall constructed of a *monolithic* (single-component) material. It is imagined that the wall is its own load-bearing structure and that when exposed to a heating environment it acts as a simple *passive heat sink* (absorbing heat input solely by sensible heat capacity). Although this is the classical and most common problem, certain complicating effects often appear, such as heat loss at the unexposed face, reradiation from the heated face, and thermal properties of the wall that exhibit directional and temperature dependencies. Highly oriented and crystalline-like materials can be markedly *anisotropic* (e.g., thermally-deposited pyrolytic graphite can have a thermal-conductivity ratio as high as 200 to 1 in directions parallel and normal to its basal planes). In addition, some materials experience significant density changes due to expansion during heating or phase changes (ice, for example, is about 9 percent less dense than water), and all thermal properties may exhibit *hysteresis* effects when the material is thermally cycled in multiple heating exposures. If a material is semitransparent and surface heat input has a radiation component, then heat transfer within the *diathermanous* solid takes place by combined conductive/radiative mechanisms.

In many applications the simple monolithic wall is inadequate from strength, weight, or survival points of view. As a consequence, the single-layer wall may require coating to resist oxidation (or provide controlled surface optical properties), or certain material components may have to be combined into a nonhomogeneous material in order to optimize thermal/structural performance. Additionally, a second insulating layer may have to be provided to prevent overheating and strength loss of the monolithic load-bearing substructure. Such composited materials and structures introduce problems of physical/chemical compatibility and interface contact resistances.

Reinforced Composites. Modern materials technology has exploited the idea that combining single monolithic materials as *composites* can result in a stronger and lighter material when exposed to a given thermal environment [2]. It is known, for instance, that two layers of different homogeneous metals can combine with optimum thicknesses to absorb higher heat fluxes and a larger total heat load (without melting) than an equivalent weight of either one exposed singly.

Composite materials are made up of two or more macroscopic constituents (metallic, organic, or inorganic) bonded together with a distribution of constituents selected to optimize thermal and/or mechanical performance of the agglomerate. A few examples are illustrated in Fig. 3a. The *laminated* composite or sandwich (e.g., plywood) is made up of layer constituents. The *particulate* composite (e.g., concrete) is an aggregate of insoluble particles embedded in a basic *matrix* material. A *fiber* composite (e.g., bamboo) consists of filaments arranged in a selected pattern within a bonding matrix. Common fiber-matrix systems are glass, nylon, or carbon filaments embedded in synthetic thermoplastic or thermosetting resins (i.e., reinforced plastics). The reinforcing fibers may be oriented parallel, perpendicular or at an angle to exposed surface of the composite. Various orientations, including multidirectional patterns and short filaments randomly dispersed, give preferred thermal and strength characteristics to meet particular design requirements.

Composite materials are generally nonhomogeneous (large scale), porous, and anisotropic. Their directional character requires use of measured *effective* properties to

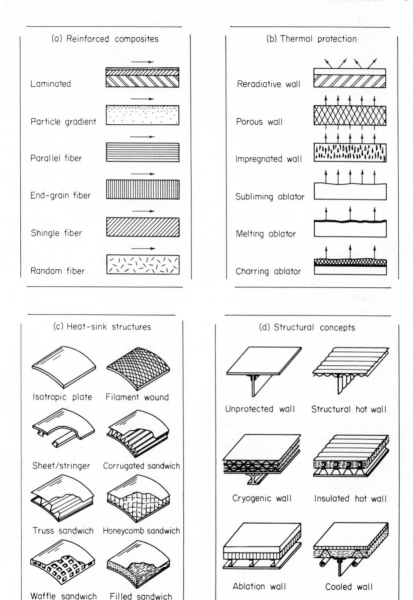

Fig. 3. Materials and structures.

account for nonisotropic and contact effects in predicting temperature response. Such materials often exhibit diathermancy when either one or both the reinforcement and matrix materials are transparent to radiative surface heating. Under these conditions, the analysis of wall temperature history must include conduction as well as absorption, scattering, and re-emission of radiation within the wall material.

Thermal Protection. For short exposure to severe thermal environments (or extended exposure to mild heating), a passive structure requires some form of thermal protection to assure material survival and to maintain structural components at temperature levels where adequate strength is retained. This protection is often accomplished by interposing thermal insulation between heating environment and the primary load-bearing substructure. However, even the insulation itself may not survive extreme heating without melting or experiencing serious physical deterioration. In this case some form of *active* thermal protection is required over and above passive sensible heat absorption which may be limited by melting or decomposition of the material.

A number of active heat-protection schemes are illustrated in Fig. 3b. The *reradiative wall* derives its thermal protection in a convective environment by operating at high surface temperatures. This produces two effects: (1) reduction in convective heating caused by reducing temperature potential for surface heat input, and (2) disposal of a portion of the absorbed heat load by reradiation. Under certain conditions, reradiation can substantially reduce net heat input (e.g., atmospheric entry vehicle reradiating to space) or be small due to configurational effects (e.g., nozzle throat). If the structure can tolerate operation near radiation equilibrium temperature (extended exposure), then it is only necessary to provide a high emissivity surface of low thermal-conductivity material to produce high surface temperatures and resulting high reradiation component. Cool structures require insulation between the hot reradiative surface and underlying structure.

A *porous wall* is maintained at structurally acceptable temperature levels by forcing a coolant through the porous matrix of wall material (e.g., sintered metal). Cooling is accomplished in two ways: (1) by flooding the wall internally with low-temperature fluid coolant, and (2) by mass-injection effects at the coolant exit surface (heated face). The coolant may be a liquid that accumulates and flows along the surface (*film cooling**) with subsequent phase-change heat absorption and gaseous vaporization into the main convective flow, or it may be a gas which is injected directly after passage through the porous wall. In either case a reduction in convective heat input is effected by mass injection and its accompanying velocity and mass-diffusion effects in the surrounding boundary layer (active *transpiration cooling**). Mass injection loses effectiveness if the external heating is predominately radiative.

The *impregnated wall* is a high-temperature open matrix or porous material infiltrated with a solid having a low melting point. Moderate wall temperatures are maintained by two effects: (1) progressive melting of the impregnant in the skeletal composite with heat absorption by phase change, and (2) extrusion and injection of melt at the heated surface with secondary phase change and accompanying gaseous transpiration effects. Whereas heat protection in an active porous-cooled wall is mechanically controlled by regulating coolant flow, the impregnated wall is *self-regulating* (passive) until the infiltrate is exhausted (higher heating loads stimulate higher melting rates, and melting ceases if heat flux falls below a certain level).

The *subliming ablator* is a monolithic material which progressively consumes itself (*ablates*) as surface material is heated to its sublimation temperature and gasifies into surrounding flow. Surface cooling results from phase change (latent heat of sublimation) and gaseous transpiration. Low-temperature sublimers (e.g., teflon) maintain low surface temperatures and cool interior, but reradiate only a small portion of the surface heat input. The cool unconsumed portion of the wall may act as a load-carrying structure, or the sublimer may be used to coat a primary metallic or plastic substructure. High-temperature sublimers (e.g., graphite) reradiate a large share of incident surface heating, but owing to relatively high conductivity the substructure must be protected against excessive conductive heat loads. Like the impregnated wall,

*See Sec. 17.

a subliming ablator is self-regulating. But unlike the former, which maintains its original thickness during consumption of infiltrate, the sublimer has a *receding surface* and continuously diminishing thickness. This may represent a design limitation such as complete loss of heat-protection material prior to end of sustained heating or detrimental configurational changes (e.g., trajectory disturbances with ablating entry body, or thrust changes in rocket engine with ablating nozzle throat).

The *melting ablator* absorbs energy over and above sensible heating due to phase change at its consumable surface (latent heat of fusion). Melt layer flows along the heated surface under action of viscous shear forces from the external flow, disposing of energy by convective transport and phase change at liquid surface (latent heat of vaporization) and gas injection. The higher surface temperatures of melting ablators (e.g., quartz) result in significant reradiation shielding.

The *charring ablator* represents a class of composites that combine many singular advantages of the aforementioned high-performance material systems [3]. Wood and cork are natural examples. Man-made charring ablators are generally composed of synthetic resins (phenolic, epoxy, etc.) reinforced with high-temperature fibers (glass, nylon, carbon, etc.) and cured under heat and pressure. The heat protection effects achieved under high heating loads are depicted in Fig. 4 where virgin (undecomposed) material is exposed to convective heat input which raises its surface to the decomposition temperature. This causes partial pyrolysis of the resin impregnant (binder) and subsequently the formation of a char layer of completely pyrolyzed resin held together in skeletal form by the high-temperature matrix (reinforcement) material; some reinforcements (e.g., nylon) completely volatilize and the remaining char consists only of a carbonaceous structure of pyrolyzed resin. Oxidation of the char (chemical consumption) and melting of reinforcement occurs at the char surface. After some period of heating, the pyrolysis zone separating char and virgin layers has receded well into

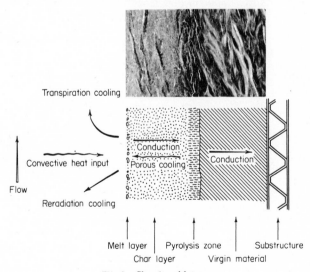

Fig. 4. Charring ablator.

the solid and char thickness has assumed a relatively constant value, with ablator surface being continuously oxidized, spalled, and melted away. Gaseous products of decomposition that originate in pyrolysis zone percolate through the char layer, absorbing energy by sensible heating and chemical cracking (convective transport partially offsetting heat conduction across char layer to pyrolysis zone), and ultimately leave the char surface to provide mass-transfer cooling of external flow. Since the char layer is highly carbonaceous, it sustains high surface temperatures and consequently provides significant reradiation cooling. Finally, the virgin material is of low thermal conductivity; this coupled with the fact that decomposition in the pyrolysis zone takes place at relatively low temperatures, serves to minimize heat conduction to the substructure. Performing as described, the charring ablator represents an effective means of self-regulated thermal protection under short, but transient and intense, convective heating loads.

Heat-sink Structures. A number of common heat-sink walls are sketched in Fig. 3c to indicate the variety of design approaches that exist for thermally passive structures. These structures are designed to absorb incident surface heat input while maintaining wall temperatures low enough for adequate strength retention and protection of adjacent or attached equipments from excessive heat gain (or loss).

Allowable wall temperatures vary widely as to application and whether strength or thermal protection is the dominant criterion. The titanium leading edge of a hypersonic aircraft, for example, being isolated and only moderately loaded, may be allowed to operate near 1500°F during accelerated flight. Aluminum panels on such a wing cannot operate at elevated temperatures because of heavy loads and serious strength degradation of aluminum above 500°F. Temperature control of unmanned space probes represents problems of narrow allowable temperature limits between about −30 and 110°F. Here the thermal sources are external solar radiation and internal heat generation from electronic components; since in a zero-gravity environment strength requirements are minimal, thermal degradation of internal equipments is the chief concern. The wide range of temperature/load limits described generate a variety of possible structures to satisfy requirements of thermal protection for structures and internal payloads.

The monolithic *isotropic plate* in Fig. 3c is the simplest structure to analyze thermally. Being unsupported by additional attachment structures, it represents the classical case of a simple heat sink exposed to a defined heating or cooling environment. *Filament-wound structures* represent composited materials of high-strength, continuous filaments netted in a prescribed pattern and embedded in thermosetting resin. Typical uses of these temperature/pressure-cured isotensoid (equally stressed) structures appear in light-weight pressure vessels, radomes, and solid-propellant rocket-motor cases. Such structures are designed to operate below decomposition temperature of the resin binder. The *sheet/stringer* combination where both sheet and stringer share the load is common. This requires consideration of the stringer heat-sink effect. *Corrugated, truss, honeycomb and waffle sandwich* constructions have higher temperature capabilities owing to separation of thermal and load-absorption functions. Their cellular design produces not only complicated heat-conduction paths but (if cells are unfilled) heat exchange between cover plates by coupled mechanisms of conduction (solid and gaseous), convection, and radiation (at elevated temperatures). The *filled sandwich* is a common design in which two cover plates enclose a blanket insulation material. Insulation is generally some form of low-density fiber, organic foam, or powder. Such loose fillers are complicated by entrapped interstitial gases, filler contact resistances, and combined heat-transfer mechanisms.

Structural Concepts. Typical wall structural concepts [4] are illustrated in Fig. 3d. The *unprotected wall* is a conventional isotropic plate stiffened by webs and used in low heat-flux environments that do not appreciably influence structural design. The *structural hot wall* uses corrugated sheet for added rigidity at elevated temperatures, but due to surface reradiation the structure does not require insulation to prevent excessive strength loss. The *cryogenic wall* is designed to insulate a vessel containing ultra low-temperature fluids (e.g., liquid hydrogen at −420°F) against heat gain from an exterior vessel environment. This is a *multiwall* concept in which multiple reflective radiation shields are interposed between low-density fibrous insulation. Although very low overall wall conductances are achieved in this way, a practical problem arises of minimizing *heat leaks* that occur in structural members that pass through the insulation for support, in edge joints, etc. These heat leaks tend to nullify the superior performance of isolated cryogenic insulations. The *insulated hot wall* is intended to operate in thermal environments severe enough to require passive thermal protection of a load-bearing substructure. Common designs depend on blanket insulation to achieve the required temperature drop. A typical *ablation wall* has consumable heat-protection material bonded to honeycomb (metal or plastic) which in turn is bonded to a primary structure that carries applied loads. One obvious feature of ablation material is its one-time use; since a portion of the material is consumed in accomplishing heat protection, it is essentially (unless refurbishable) a single-exposure material. The *cooled wall* absorbs a fraction of the incident surface heat flux and disposes of the remainder by active cooling of the substructure itself. Such designs are required for prolonged intense heating where application of ablative materials would either incur excessive weight penalties or be ruled out because of repetitive-exposure requirements.

2. Differential Equations of the Temperature Field

Fourier's law of heat conduction relates rate of heat flow $q = dQ/d\theta$ across conducting area A to the negative temperature gradient $-\partial t/\partial n$ in a direction n normal to A (Fig. 5) as

$$q_n = -kA \frac{\partial t}{\partial n} \tag{1}$$

where k is the *thermal conductivity* of the conducting material. The vector q is the *heat rate* (Btu/hr). Equation (1) is generally expressed in the alternate form

$$q_n'' = -k \frac{\partial t}{\partial n} \tag{2}$$

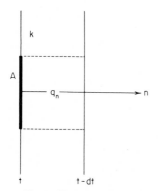

Fig. 5. Heat-flow vector.

where $q'' = q/A$ is the *heat flux* (Btu/hr-ft^2). If $t = t(x, \theta)$ the flux is transient, while if $t = t(x)$ only the flux is constant at a given station x.

The dependent variable temperature t is related to independent variables time θ and spatial location x, y, z in the conducting body by means of heat-balance *differential equations* derived from Fourier's law of heat conduction, Eq. (1). Although the form of these differential equations varies as to the nature of the conducting material (i.e., whether its conduction properties are constant or directional and temperature dependent), the basic character of their elementary forms is chiefly dependent on body shape and hence on the coordinate system chosen to represent body geometry.

a. Plane Boundaries. The spatial locations in a body with plane boundaries are located by points $p(x, y, z)$ expressed in *rectangular coordinates* (Fig. 6a). A partial differential equation expressing conservation of energy in this coordinate system is

$$\frac{\partial}{\partial x}\left(k\frac{\partial t}{\partial x}\right) + \frac{\partial}{\partial y}\left(k\frac{\partial t}{\partial y}\right) + \frac{\partial}{\partial z}\left(k\frac{\partial t}{\partial z}\right) + q''' = \rho C \frac{\partial t}{\partial \theta} \tag{3}$$

Here ρ is density of body material, C and k are thermal properties of specific heat and thermal conductivity for the conducting material, and q''' represents uniformly distributed *heat sources* or *sinks* (Btu/hr-ft^3). All four of these quantities may, in general, depend on temperature or location so $\rho = \rho(t, x, y, z)$, etc. Location dependence

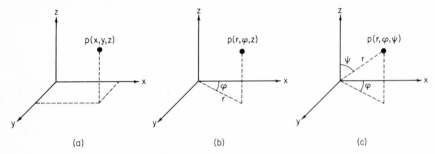

(a)   (b)   (c)

Fig. 6.   Coordinate systems: (a) rectangular, (b) cylindrical, (c) spherical.

arises because the conducting material may be both *nonisotropic* (directionally sensitive properties) and *nonhomogeneous* (location sensitive). If, for example, an infinite nongenerating plate has a regularly oriented grain structure with conductivities k_p parallel to and k_n normal to grain, then the partial-differential equation (y parallel and x perpendicular to plate face) is

$$(k_p \sin^2 \beta + k_n \cos^2 \beta) \frac{\partial^2 t}{\partial x^2} + (k_p \cos^2 \beta + k_n \sin^2 \beta) \frac{\partial^2 t}{\partial y^2}$$

$$+ (k_p - k_n) \sin 2\beta \frac{\partial^2 t}{\partial x \partial y} = \rho C \frac{\partial t}{\partial \theta}$$

where β is the angle between parallel laminations and the face surface (Fig. 65). If k is constant, Eq. (3) takes the simpler form

$$\frac{\partial^2 t}{\partial x^2} + \frac{\partial^2 t}{\partial y^2} + \frac{\partial^2 t}{\partial z^2} + \frac{q'''}{k} = \frac{1}{\alpha} \frac{\partial t}{\partial \theta} \tag{4}$$

where the derived property $\alpha = k/\rho C$ is called *thermal diffusivity*. For materials free of heat sources or sinks, Eq. (4) reduces to the *Fourier equation*

$$\frac{\partial^2 t}{\partial x^2} + \frac{\partial^2 t}{\partial y^2} + \frac{\partial^2 t}{\partial z^2} = \frac{1}{\alpha} \frac{\partial t}{\partial \theta} \tag{5}$$

In steady state $\partial t / \partial \theta = 0$ and Eq. (4) reduces to the *Poisson equation*

$$\frac{\partial^2 t}{\partial x^2} + \frac{\partial^2 t}{\partial y^2} + \frac{\partial^2 t}{\partial z^2} + \frac{q'''}{k} = 0 \tag{6}$$

Finally, if $q''' = 0$, Eq. (6) reduces to the *Laplace equation*

$$\frac{\partial^2 t}{\partial x^2} + \frac{\partial^2 t}{\partial y^2} + \frac{\partial^2 t}{\partial z^2} = 0 \tag{7}$$

For the special case of *porous-cooled* bodies (Fig. 3*b*), the above equations are modified for internal convective transport associated with forced coolant flow through the body. Considering one-dimensional conduction with constant coolant flow through the wall counter to *x* (see Fig. 10*a*), the governing partial-differential equation (assuming heat conduction only in solid matrix and equal coolant and solid temperatures within the wall itself) is

$$\frac{\partial^2 t}{\partial x^2} + \frac{G_c C_c}{k_e} \frac{\partial t}{\partial x} = \frac{1}{\alpha_e} \frac{\partial t}{\partial \theta} \tag{8}$$

where G_c is the coolant mass flow (rate per unit area). The second term in Eq. (8) is the added enthalpy transport. If *P* is *porosity* of the wall material (ratio of pore volume to total volume), then $\rho_e C_e = P \rho_c C_c + (1 - P) \rho_s C_s$ and $k_e = P k_c + (1 - P) k_s$ represent *effective* properties of the wall (subscript *e*) in terms of coolant properties (subscript *c*) and solid properties (subscript *s*).

The second-order partial-differential equations (4) through (8) are all *linear* in that each is of the first degree in dependent variable *t* and its derivatives. Exact solutions are therefore possible if, in general, the boundary conditions are also linear (Sec. 1.B).

b. Cylindrical Boundaries. The spatial locations in a body with cylindrical boundaries are located by points $p(r, \phi, z)$ expressed in *cylindrical coordinates* (Fig. 6*b*) through the relations $x = r \cos \phi, y = r \sin \phi$, and $z = z$, where ϕ is the *latitude* angle of *p*. The partial-differential equations (3) and (4) transformed to cylindrical coordinates are

$$\frac{1}{r} \frac{\partial}{\partial r} \left(r k \frac{\partial t}{\partial r} \right) + \frac{1}{r^2} \frac{\partial}{\partial \phi} \left(k \frac{\partial t}{\partial \phi} \right) + \frac{\partial}{\partial z} \left(k \frac{\partial t}{\partial z} \right) + q''' = \rho C \frac{\partial t}{\partial \theta} \tag{9}$$

$$\frac{\partial^2 t}{\partial r^2} + \frac{1}{r} \frac{\partial t}{\partial r} + \frac{1}{r^2} \frac{\partial^2 t}{\partial \phi^2} + \frac{\partial^2 t}{\partial z^2} + \frac{q'''}{k} = \frac{1}{\alpha} \frac{\partial t}{\partial \theta} \tag{10}$$

c. Spherical Boundaries. The spatial locations in a body with spherical boundaries are located by points $p(r, \phi, \psi)$ expressed in *spherical coordinates* (Fig. 6*c*) through the relations $x = r \cos \phi \sin \psi$, $y = r \sin \phi \sin \psi$, and $z = r \cos \psi$, where ψ is the *azimuth* angle of *p*. Equations (3) and (4) transformed to spherical coordinates are*

$$\frac{1}{r^2} \frac{\partial}{\partial r} \left(r^2 k \frac{\partial t}{\partial r} \right) + \frac{1}{r^2 \sin^2 \psi} \frac{\partial}{\partial \phi} \left(k \frac{\partial t}{\partial \phi} \right) + \frac{1}{r^2 \sin \psi} \left(k \sin \psi \frac{\partial t}{\partial \psi} \right) + q''' = \rho C \frac{\partial t}{\partial \theta} \tag{11}$$

$$\frac{\partial^2 t}{\partial r^2} + \frac{2}{r} \frac{\partial t}{\partial r} + \frac{1}{r^2 \sin^2 \psi} \frac{\partial^2 t}{\partial \phi^2} + \frac{1}{r^2} \frac{\partial^2 t}{\partial \psi^2} + \frac{1}{r^2 \tan \psi} \frac{\partial t}{\partial \psi} + \frac{q'''}{k} = \frac{1}{\alpha} \frac{\partial t}{\partial \theta} \tag{12}$$

Analysis of heat-conduction systems consists of obtaining exact or approximate solutions of these and other partial-differential equations subject to prescribed conditions on the initial temperature state of the conducting system and on thermal conditions at interfaces and surfaces of the body. Methods of exact solution are given in Sec. 1.B and approximate methods are outlined in Secs. 4 and 5.

*For transformation to *elliptic coordinates*, see Ref. 43 in Sec. 1.B.

3. Boundary Conditions

Complete solutions (integration) of governing differential equations for temperature response require specifying boundary conditions on the body temperature function regarding its initial state and its behavior at both internal and external boundaries of the body.

   a. Initial Condition. The initial boundary condition is on time θ; it describes *initial temperature state* of the body prior to heating or cooling ($\theta \leq 0$). In terms of a simple nongenerating plate, the initial condition is analytically described as some arbitrary temperature distribution $t(x,0) = t(x)$. The simplest case is a constant, $t(x,0) = t_0$ (Fig. 1a).

   b. Internal Conditions. Unless the body considered consists only of a single non-decomposing material, certain conditions internal to the body must be specified to characterize such effects as interface resistance and interface reactions.

   Interface Resistance. Two similar or dissimilar materials may be held together by an adhesive bond or by mechanical means (pressure, fasteners, etc.). If bonded, the *interface* adhesive represents a simple thermal resistance determined by its thickness and conductivity.

   At the interface between two materials held together in *unbonded contact*, there is a resistance to heat flow due to imperfect contact. Not only is the area of mutual contact between microscopic roughness asperities of the joint small in comparison to apparent surface area, but macroscopic waviness further diminishes the actual physical contact. Both random-point and large-scale contacts are enhanced by increasing pressure with which the two materials are held together. This produces increased contact by tightening the interference fit of hill-and-valley asperities, by overcoming waviness interference, and by elastically and plastically deforming the softer of the two materials. Such effects may be artificially produced by interposing an easily deformed, high-conductivity shim material at the otherwise bare interface.

   Heat transfer through an interface takes place by combined mechanisms of conduction across true contact points, conduction across entrapped interstitial fluid, and radiation across interstitial gaps. Resulting overall conductance of the joint is therefore a function of the materials in contact (conductivity, surface finish, flatness, and hardness), the contact pressure, the mean temperature and heat flux across joint, the nature of interstitial fluid (liquid, gas, vacuum), and the presence of oxide films or interface shim materials.

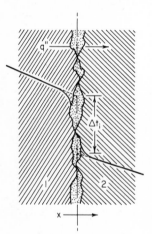

Fig. 7. Interface contact.

   The case of steady conduction across a rough contact between dissimilar materials 1 and 2 is illustrated in Fig. 7. There is an apparent temperature discontinuity in the immediate vicinity of the contact. A fictitious interface temperature drop Δt_i is defined by extrapolating to the centerline the actual linear temperature that exists in each material away from the contact disturbance. With steady heat flux $q'' = -k_1(dt/dx)_1 = -k_2(dt/dx)_2$, the unit *interface conductance* is defined as

$$h_i = \frac{q''}{\Delta t_i} \qquad (13)$$

For *perfect contact,* the temperature drop vanishes and $h_i \to \infty$; in this instance the internal boundary condition is that of temperature continuity, that is, $t_1 = t_2$. The value of h_i for imperfect contact may be positive or negative depending on whether the joint is less or more conducting than the materials that form the interface. Although in practice a positive h_i is generally the case, it is also possible for high-conductivity fluid to be trapped between low-conductivity solids and result in a lower temperature gradient in the fluid than the solid (here Δt_i is negative while q'' remains positive). A feeling for relative magnitude is given by the thermal conductance of a $\delta = 20$-mil air gap at $150°F$, computed as $k/\delta = 0.0167/(0.020/12) = 10$ Btu/hr-ft^2-°F.

Figure 8 shows examples of measured contact conductance data h_i as a function of contact pressure p for interface conditions identified in Table 1.

The data in Fig. 8 show the expected increase in h_i with increasing contact pressure. This effect is most pronounced for vacuum interface where contact conduction predominates. The relation $h_i \sim p^{2/3}$ (dashed line) is roughly what is expected from elastic deformation theory.

Aluminum pairs 1 and 2 in Fig. 8 show increased interface conductance with improved surface finish. Due to surface waviness, pair 3 exhibits lower conductance despite superior surface finish. Aluminum pairs 4, 5, and 6, although harder than pairs 1, 2, and 3, show higher h_i as a result of higher interface temperatures and because interstitial air improves overall conduction contribution. Aluminum pair 7 confirms the beneficial effect of inserting soft lead interface shim material to overcome waviness conductance loss in pair 3. Aluminum pair 8 illustrates, in comparison to pair 4, that use of shim material harder than aluminum degrades contact conductance. Stainless steel pairs 9, 10, and 11 indicate that sensitivity to high contact pressure increases as surface finishes improve.

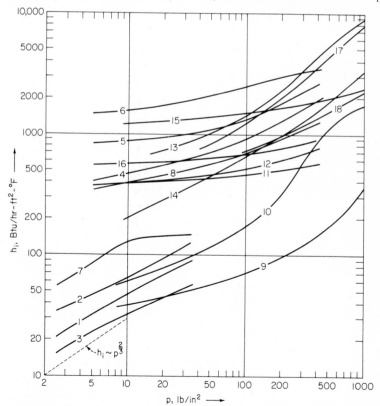

Fig. 8. Thermal interface conductance data.

TABLE 1. Interface Conditions for Conductance Data in Fig. 8

Curve	Material pair	RMS surface finish (μ in.)	Gap material	Mean contact temp. ($^\circ$F)	Ref.
1	aluminum (2024-T3)	48–65	vacuum (10^{-4} mm Hg)	110	5
2	aluminum (2024-T3)	8–18	vacuum (10^{-4} mm Hg)	110	5
3	aluminum (2024-T3)	6–8 (not flat)	vacuum (10^{-4} mm Hg)	110	5
4	aluminum (75S-T6)	120	air	200	6
5	aluminum (75S-T6)	65	air	200	6
6	aluminum (75S-T6)	10	air	200	6
7	aluminum (2024-T3)	6–8 (not flat)	lead foil (0.008 in.)	110	5
8	aluminum (75S-T6)	120	brass foil (0.001 in.)	200	6
9	stainless (304)	42–60	vacuum (10^{-4} mm Hg)	85	7
10	stainless (304)	10–15	vacuum (10^{-4} mm Hg)	85	7
11	stainless (416)	100	air	200	6
12	stainless (416)	100	brass foil (0.001 in.)	200	6
13	magnesium (AZ-31B)	50–60 (oxidized)	vacuum (10^{-4} mm Hg)	85	7
14	magnesium (AZ-31B)	8–16 (oxidized)	vacuum (10^{-4} mm Hg)	85	7
15	copper (OFHC)	7–9	vacuum (10^{-4} mm Hg)	115	7
16	stainless/aluminum	30/65	air	200	6
17	iron/aluminum	——	air	80	8
18	tungsten/graphite	——	air	270	8

In contrast to pair 8, the interface shim in pair 12 improves conductance (brass is softer than stainless steel). Magnesium pairs 13 and 14 suggest that oxidation of smooth surfaces can add thermal resistance sufficient to more than offset the usual effect of high surface finish; apparently oxide films have a greater effect on rough than on smooth surfaces. Copper pair 15 shows generally high conductance despite its vacuum interface. Joints of dissimilar materials, such as pairs 16, 17, and 18, are controlled in their contact-pressure sensitivity by the softer of the two materials. Another effect observed with dissimilar material interfaces is a dependence on direction of heat flow; this presumably arises from different temperature dependencies of both thermal and mechanical properties for the two materials and changes in interface warping when temperature pattern is inverted by reversing the direction of heat flow.

The three most important effects in practical joints are interface flatness, joint pressure, and mean interface temperature. Erratic contact conductances are observed in bolted, riveted, and spotwelded metallic joints where interface is generally distorted during assembly and subsequent heating [9]. Assembled joints are rarely well controlled in practice (relative to contact pressure and distortion), and thus caution is required in relating laboratory conductance data measured on carefully constructed interface assembles to prototype hardware. Owing to variable expansions and distortions, a high degree of variability in contact conductance can be expected between seemingly similar joints, and in a given fixed joint during transient heating.* In addition, significant hysteresis effects can occur in repeated heating and cooling of contact joints. In contrast to the expected result, adhesive bonding of joints generally reduces contact conductance due to large bond thickness (relative to contact gap) and the inherently low conductivity of most adhesives.

Data [11] on bonding of 20 μin. aluminum/aluminum surfaces with silicone-rubber, thermoplastic, polyester, and epoxy adhesives cured at 20 psi show contact conductances of 80, 150, 400, and 2000 Btu/hr-ft^2-$^\circ$F respectively. High-conductivity greases can increase conductance but may degrade at elevated temperatures or vaporize when exposed to vacuum environments.

Interface contact conductance h_i can be approximately calculated from properties of the joining surfaces and entrapped fluids [8]. Figure 9 shows a theory relating di-

*Contact conductance data on joints secured by bolts, screws, and rivets are reviewed in Ref. 10.

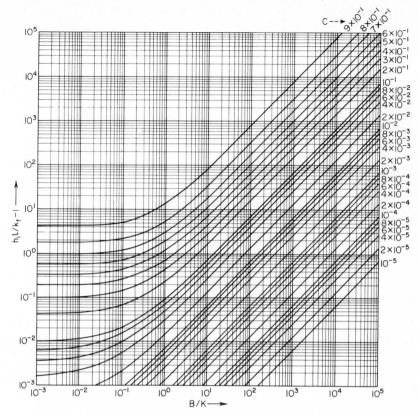

Fig. 9. Thermal interface conductance theory.

mensionless conductance $h_i l/k_f$ to *constriction number C, gap number B,* and *conductivity number K.* The procedure for using the theory and correlated measurements on the various input parameters [12] is as follows:

1. Calculate constriction number $C = \sqrt{p/M}$, where p is contact pressure and M is Meyer hardness of the softer contact material
2. Estimate effective gap thickness as $l = 3.56(l_1 + l_2)$ if $(l_1 + l_2) < 280\ \mu\text{in}$. (smooth contacts) or $l = 0.46(l_1 + l_2)$ if $(l_1 + l_2) > 280\ \mu$ in. (rough contacts), where l_1 and l_2 are mean (or rms) depths of surface roughness
3. Calculate gap number $B = 0.335 C^{0.315} (\sqrt{A}/l)^{0.137}$, where A is the interface area (one side)
4. Estimate equivalent conductivity of interstitial fluid as $k_f = k_0$ for liquids, evaluated at mean surface temperature $\bar{t}_i = (t_1 + t_2)/2$, or

$$k_f = \frac{k_0}{1 + 8\gamma(\nu/\bar{v})(a_1 + a_2 - a_1 a_2)/\text{Pr}(\gamma + 1)l a_1 a_2} + \frac{4\sigma k \epsilon_1 \epsilon_2 \bar{t}_i^3}{\epsilon_1 + \epsilon_2 - \epsilon_1 \epsilon_2}$$

for gases, where k_0 is gas conductivity at zero contact pressure; Pr is the Prandtl number, $\bar{v}$ is the mean molecular velocity, γ is the ratio of specific heats, and ν is the kinematic viscosity of the gas evaluated at $\bar{t}_i$; a and ϵ are the accommodation coefficient and emissivity of the contact surfaces evaluated at t_1 or t_2; and σ is the Stefan-Boltzmann radiation constant

5. Calculate conductivity number $K = k_f(k_1 + k_2)/2k_1 k_2$, where k_1 and k_2 are conductivities of the two solids evaluated at

$$\frac{t_1 + (k_1 t_1 + k_2 t_2)/(k_1 + k_2)}{2} \quad \text{or} \quad \frac{t_2 + (k_1 t_1 + k_2 t_2)/(k_1 + k_2)}{2}$$

6. Using B/K and C, enter Fig. 9 and determine interface contact conductance h_i.

Interface Reactions. In addition to sensible heat capacity, a material may absorb heat by endothermic *reaction* at its surface and/or internally. These effects can include *latent* heat capacities of melting (or sublimation, vaporization, etc.) or solidification (freezing) and *chemical* heat capacities such as dissociation. In melting and solidification of water or metals, for example, there is an internal moving plane separating the solid and melt where energy is absorbed or liberated as latent heat of fusion. In charring ablators (Fig. 4), internal energy is absorbed by pyrolysis and vaporization of solid constituents and by chemical cracking of evolved pyrolysis gases.

Considering an internal reaction interface $x = l$ separating phases 1 and 2 of a material (see Fig. 53b), the local heat-balance boundary condition for simple phase-change reaction is

$$-k_1 \left(\frac{\partial t_1}{\partial x}\right)_l + k_2 \left(\frac{\partial t_2}{\partial x}\right)_l = \rho_2 H \frac{dl}{d\theta} \tag{14}$$

Here H (Btu/lb) is the latent or chemical heat associated with the internal moving reaction plane. Equation (14) states that the difference in heat flux across reaction interface is equal to heat absorption $(+H)$ or heat liberation $(-H)$ at $x = l$. A short tabulation of approximate latent heats of melting (or sublimation) H_m and vaporization H_v for selected solids (1 atm) is given in Table 2; sensible heat absorption $H_s = C\Delta t$ and total energy absorbed from $32°F$ are also listed for comparison.*

TABLE 2. Latent Heats of Selected Solids

Solid	Heat absorbed (Btu/lb)				
	H_s (solid)†	H_m	H_s (liq)‡	H_v	Total
lead	18	10	85	371	484
bismuth	17	22	91	368	498
teflon	192	750	—	—	942
potassium	202	26	44	870	1,142
tin	23	26	198	1,030	1,277
ice	0	143	180	970	1,293
silver	97	451	125	1,010	1,683
sodium	52	49	415	1,810	2,326
tungsten	240	83	190	1,910	2,423
copper	179	88	202	2,060	2,529
magnesium	293	160	206	2,405	3,064
aluminum	308	170	559	4,070	5,107
silica	790	60	1,057	3,695	5,602
lithium	26	285	161	8,365	8,837
carbon	3,570	25,350	—	—	28,920

†From $32°F$ to melting (or sublimation) temperature.
‡From melting to vaporization temperatures.

*Accurate data on latent and chemical heats for various substances are tabulated in handbooks of chemistry (Sec. 1.B, Ref. 9)

The pyrolysis zone of a composited charring ablator (Fig. 4) represents an interface of finite thickness in which *depolymerization* of matrix reinforcement and resin impregnant gives rise to energy absorption and production of decomposition gases. Time-dependent pyrolysis (*p*) energy is $H_d dp_p/d\theta$, where H_d is *heat of decomposition* at pyrolysis temperature. Kinetics of the resin (*r*)/matrix (*m*) depolymerization process is described [13] as

$$\frac{dp_p}{d\theta} = \Gamma_r \frac{dp_r}{d\theta} + (1 - \Gamma_r) \frac{dp_m}{d\theta} \tag{15}$$

where Γ_r is volume fraction of resin in original composite. Each nth-order condensed-phase reaction i contributing to conversion of virgin-material density ρ_v to char density ρ_c is expressed by a specific rate equation

$$\frac{dp_i}{d\theta} = -B\rho_v \left(\frac{\rho_i - \rho_c}{\rho_v} \right)^n \tag{16}$$

The depolymerization rate coefficient B in Eq. (16) is temperature dependent; it is generally described by the simple *Arrhenius* equation

$$B = B_v e^{-E/Rt} \tag{17}$$

where E is activation energy and R is universal gas constant. Reaction order n, rate coefficient B, and activation energy E are obtained by kinetic analysis of thermogravimetric measurements on each constituent of the composited material.

c. Surface Conditions. Since the characteristic differential equations (Sec. 3.A.2) are of the second order, at least two independent surface boundary conditions must be prescribed. Figure 10 illustrates four classes of environmental conditions used to

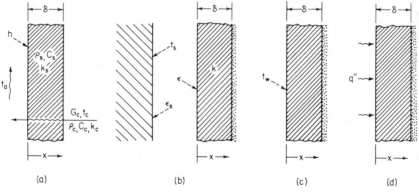

Fig. 10. Surface boundary conditions: (a) prescribed surface convection, (b) prescribed surface radiation, (c) prescribed surface temperature, (d) prescribed surface heat input.

characterize exposed-surface thermal conditions on a plate for $\theta > 0$. These four cases, along with designation of known and unknown quantities in each, are as follows:

Surface boundary condition	Known	Unknown
(a) Prescribed surface convection	t_a, h	t_w, q''
(b) Prescribed surface radiation	$t_s, \mathcal{F}$	t_w, q''
(c) Prescribed surface temperature	t_w	q''
(d) Prescribed surface heat flux	q''	t_w

Here t_a is *ambient fluid temperature*, t_w is *exposed-surface temperature*, t_s is absolute *radiation source (or sink) temperature* (°R), h is *convective unit surface conductance* (Btu/hr-ft²-°F), $\mathcal{F}$ is a combined *surface configuration-emissivity factor* for radiant interchange (function of geometry and relative positions of source and receiver surfaces and of their respective emissivities ϵ_s and ϵ), and q'' is *surface heat flux* (Btu/hr-ft²). Surface flux may be positive (input) or negative (rejection).

Prescribed Surface Convection. Figure 10a represents sudden exposure to an ambient condition of convective heat transfer prescribed in terms of h and t_a with $t_a \neq t_0$. The boundary condition is continuity of surface flux

$$h[t_a - t(0,\theta)] = -k \frac{\partial}{\partial x} t(0,\theta) \tag{18}$$

The back face of the plate may be thermally insulated. This *adiabatic* boundary condition is described as

$$\frac{\partial}{\partial x} t(\delta,\theta) = 0 \tag{19}$$

Table 3 lists typical order-of-magnitude values for unit surface conductance h obtained with flow of various fluids along nondecomposing surfaces (see Secs. 6, 7, 8, 12).

TABLE 3. Unit Surface Conductance for Selected Fluids

Fluid	Configuration	h (Btu/hr-ft²-°F)
air	free flow along vertical plate	1
air	forced flow along plate	10
air	flow in tube	20
steam	flow in tube	50
oil	flow in tube	100
exhaust gases	flow over turbine blades	200
water	flow in tube	500
exhaust gases	flow through rocket nozzle	1,000
water	flow along plate with boiling	1,000
steam	film condensation on plate	2,000
liquid metal	flow in tube	5,000
steam	dropwise condensation on plate	10,000
steam	high-speed condensing flow in tube	20,000

Once $t(x,\theta)$ is determined, the *instantaneous* surface heat flux is calculated from $t_w = t(0,\theta)$ using either side of the boundary condition in Eq. (18), i.e., *Newton's law* $q'' = h[t_a - t(0,\theta)]$ or *Fourier's law* [Eq. (2)] as

$$q'' = -k \frac{\partial}{\partial x} t(0,\theta) \tag{20}$$

Cumulative (total) surface heat input (Btu/ft²) is calculated from q'' as

$$Q'' = \int_0^\theta q''d\theta \tag{21}$$

If there is phase change and ablation at the surface, the boundary condition of Eq. (18) becomes

$$h[t_a - t(0,\theta)] = -k\frac{\partial}{\partial x}t(l,\theta) + \rho H\frac{dl}{d\theta} \tag{22}$$

where $x = l$ is instantaneous location of moving ablation surface. If the wall is porous cooled as indicated in Fig. 10a, k in Eqs. (18) and (20) is $k_e = Pk_c + (1 - P)k_s$, and the value of h is that associated with mass-transfer cooling (Sec. 16). The boundary condition at $x = \delta$ is determined not by adiabatic condition of Eq. (19) but by a condition of steady coolant flow through the wall $(\theta > 0)$ for which

$$G_cC_c[t(\delta,\theta) - t_c] = -k_e\frac{\partial}{\partial x}t(\delta,\theta) \tag{23}$$

where G_c is the coolant mass flow per unit area (lb/hr-ft^2). The upstream condition on coolant temperature $(x \gg \delta)$ is generally $t_c = t_0$.

Prescribed Surface Radiation. Figure 10b represents a sudden exposure to an ambient condition of radiant interchange between the surface of an *opaque* wall and an adjacent surface at constant source or sink temperature t_s. The boundary condition is prescribed in terms of t_s and $\mathcal{F}$ as

$$\sigma\mathcal{F}[t_s^4 - t^4(0,\theta)] = -k\frac{\partial}{\partial x}t(0,\theta) \tag{24}$$

where $\sigma = 17.3 \times 10^{-10}$ Btu/hr-ft^2-$^\circ$R^4 is the *Stefan-Boltzmann* radiation constant, and $\mathcal{F}$ is combined configuration-emissivity factor for multiple-surface radiation exchange. With $t(x,\theta)$ determined, net surface heat flux is calculated as $q'' = \sigma\mathcal{F}[t_s^4 - t^4(0,\theta)]$. Because the boundary condition in Eq. (24) is nonlinear, the process of radiative heating is different from that of radiation cooling. Some applications require combining case b with case a or d (see Sec. 15).

A thin insulated plate may receive heat q'' (constant or variable) by convection and/or radiation and reradiate (i.e., *emit*) energy from its exposed surface to a sink environment of effective temperature t_s. The condition of *radiation equilibrium* represents a special case where surface temperature increases to a level where all incident heat input is reradiated and the surface equilibrates to a unique temperature dependent only on $\mathcal{F}$ and t_s. In this situation, the radiation equilibrium temperature t_{re} is determined by a passive energy balance

$$q'' = \sigma\mathcal{F}(t_{re}^4 - t_s^4) \tag{25}$$

If the plate faces a space environment with $t_s = 0^\circ$R, then

$$t_{re} = \left(\frac{q''}{\sigma\epsilon}\right)^{1/4} \tag{26}$$

where ϵ is the *total hemispherical emissivity* of the surface (Fig. 11). When q'' is constant, t_{re} is a steady-state radiation-equilibrium temperature. Thermal inputs to the thin shell of a deep-space vehicle are external solar radiation q_S'' and internal power P_E from electronic equipments. A passive energy balance with $t_s = 0^\circ$R is $\alpha_S q_S'' + P_E = \sigma\epsilon t^4$, where α_S is an additional surface optical property of *solar absorptivity*. The average shell temperature

$$t = \left[\frac{1}{\sigma}\left(\frac{\alpha_S}{\epsilon}q_S'' + \frac{P_E}{\epsilon}\right)\right]^{1/4}$$

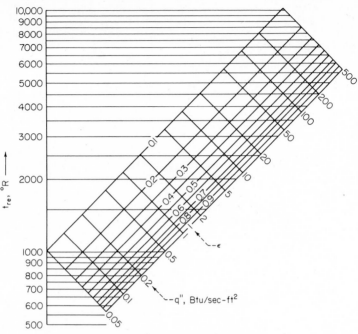

Fig. 11. Radiation equilibrium temperature ($t_s = 0°R$).

is thus seen to depend on not only ϵ alone but the ratio α_S/ϵ. Approximate values of these optical properties for typical temperature-control surfaces exposed to solar radiation [14] are listed in Table 4.

**TABLE 4. Emissivity and Solar Absorptivity
of Selected Surfaces**

Surface (500°R)	ϵ	α_S	α_S/ϵ
aluminum (buffed)	0.03	0.36	12
gold (plated)	0.03	0.30	10
gold (vacuum deposited)	0.03	0.24	8
stainless steel (polished)	0.05	0.40	8
aluminum foil (dull)	0.03	0.21	7
tantalum	0.08	0.48	6
beryllium (polished)	0.09	0.49	5.5
aluminum foil (shiny)	0.04	0.20	5.0
beryllium (milled)	0.11	0.49	4.5
titanium	0.20	0.80	4.0
René 41	0.11	0.39	3.5
chrome	0.08	0.24	3.0
nickel	0.18	0.45	2.5
silver	0.02	0.04	2.0
aluminum (polished)	0.05	0.10	2.0
aluminum (sandblasted)	0.40	0.60	1.5
black vinyl phenolic (dull)	0.84	0.93	1.1
black silicone paint (flat)	0.81	0.89	1.1
lamp black	0.95	0.95	1.0
grey silicone paint	0.96	0.53	0.55
white silicone paint (gloss)	0.75	0.26	0.35
magnesia	0.95	0.14	0.15

Prescribed Surface Temperature. In Fig. 10c the temperature of exposed surface $x = 0$ is prescribed as changing suddenly from its initial value at $\theta \leq 0$ to t_w which may be constant or a function of time. This is a special case of the convective or radiative boundary condition in which surface conductance is so large (e.g., boiling and condensing flows) that surface temperature instantaneously changes, on exposure, to $t_w = t_a$ or $t_w = t_s$.* The problem is to calculate $t(x, \theta)$ and therewith q'' from Eq. (20).

Prescribed Surface Heat Input. Figure 10d is the inverse of c; here q'' is a prescribed constant or function of time. The problem is to calculate $t(x, \theta)$ which contains unknown exposed surface temperature $t_w = t(0, \theta)$. The boundary condition itself is given by Eq. (20). If there is phase change and ablation at the surface, the boundary condition is modified as in Eq. (22).

● *Applications.* A large class of technically important systems have exposure conditions that can be *precalculated* in terms of surface heat-input functions. These include contact heaters and certain forms of intense convective and radiative heating that are largely independent of surface temperature. Figure 12 shows typical surface heat-flux

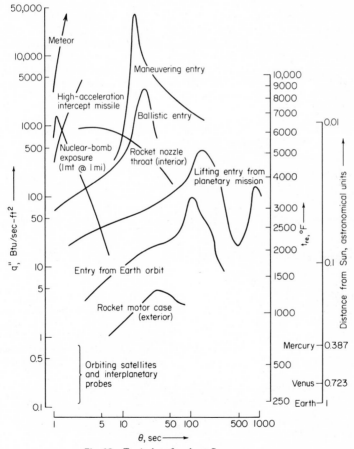

Fig. 12. Typical surface heat-flux exposures.

*Under certain conditions of intense heat input q'' to a thick, low-conductivity, reradiating material, heat conducted inward from the surface may be small and surface temperature may rapidly assume a radiation equilibrium value $t_{re} = (t_s^4 + q''/\sigma\epsilon)^{1/4}$.

histories for a number of intense heating applications. Also shown (for comparison with heat-flux scale) are steady-state values of solar radiation intensity at various distances from the sun,* and from Eq. (26) the steady radiation equilibrium temperature of a black body $(\epsilon = 1)$ exposed to the range of heat fluxes considered.

The heat-flux exposures illustrated in Fig. 12 are chosen from aerospace applications that exemplify a wide range of thermal environments and diverse problems of thermal design. For example, a space vehicle on a fly-by mission to Venus or Mercury is exposed to external radiation environments of less than 1 Btu/sec-ft^2; yet the problems of sensitive temperature control are critical to survival of delicate equipments aboard the spacecraft (e.g., solar cells, batteries, optics, detectors, gas and fuel storage, and stabilizing devices). Earth entry from an interplanetary mission represents opposite extreme; here the vehicle's initial kinetic energy may be of the order of 10^4 to 10^5 Btu/lb which, if all converted to thermal energy, would be more than sufficient to vaporize carbon. Although only a small portion of the converted thermal energy appears as net heat input to vehicle surface, the heat rates (10^3 to 10^4 Btu/sec-ft^2 for ballistic entry, 10^4 to 10^5 for maneuvering entry) cannot, in general, be accommodated without appreciable thermal degradation of all known materials (see Sec. 19).

A number of intense heating applications result in surface heat inputs of a *pulse* form. In aerodynamic heating, for example, a vehicle ascending the atmosphere experiences a peak in q'' owing to an increasing velocity coupled with decreasing atmospheric density; a similar (but higher) maximum occurs on entry where the velocity diminishes while density increases. Exposure to a nuclear detonation produces a pulse heat flux due to sudden buildup and decay of a thermally radiating source. Four useful forms of $q''(\theta)$ for heating pulses of duration D are:

$$\text{circular pulse;} \quad \frac{q''}{q''_{max}} = \sin^2 \pi \frac{\theta}{D} \quad , \quad Q'' = 0.5\, q''_{max} D$$

$$\text{power pulses;} \quad \frac{q''}{q''_{max}} = \left(3\frac{\theta}{D}\right)^{-3} e^{-(D/\theta - 3)} \quad , \quad Q'' = 0.5473\, q''_{max} D$$

$$\frac{q''}{q''_{max}} = \left(5\frac{\theta}{D}\right)^{-5} e^{-(D/\theta - 5)} \quad , \quad Q'' = 0.2795\, q''_{max} D$$

$$\text{exponential pulse;} \quad \frac{q''}{q''_{max}} = 10\left(\frac{\theta}{D}\right) e^{1 - 10\theta/D} \quad , \quad Q'' = 0.2717\, q''_{max} D$$

$$(27)$$

The symmetric circular pulse (Fig. 13) is useful for approximating ascent and ballistic-entry aerodynamic heating. The first asymmetric power pulse resembles heat input during lifting entry, characterized by strong initial heating followed by a long tail-off in heat flux; the second power pulse (see Planck's function, Sec. 1.B.1, Fig. 16) resembles hyperbolic-entry heating. The exponential pulse differs only in the ascending leg; here rise time in heat input is extremely short, characteristic of exposure to flash heating sources.

● *Nuclear Bomb Heating.* Total radiative heat input to a surface located distance d from a *nuclear fireball* is approximately [15]

$$Q'' = \frac{F_1 F_2 Y}{4\pi d^2}$$

---

*At mean Earth distance the unattenuated normal solar flux (above atmosphere) is about 445 Btu/hr-ft^2; this intensity varies between about 430 (July 1) to 460 (January 1).

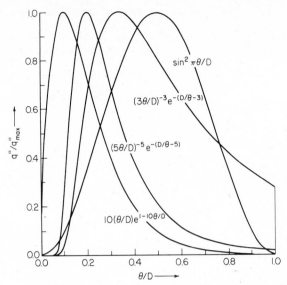

Fig. 13. Symmetric and asymmetric heating pulses.

where Y is the total energy yield of the spherical burst, F_1 is the fraction of this total appearing as thermal energy, and F_2 is the fraction of thermal energy transmitted through intervening atmosphere. From test explosions, $F_1 = 1/3$. The value of F_2 depends on the extent of atmospheric absorption and scattering [16] and on distance d; from test data [15], $F_2 = e^{-0.27\sqrt{d}}$ for distances less than 5 miles. Thus

$$Q'' = 3773 \, e^{-0.27\sqrt{d}} \, \frac{Y}{d^2} \; ; \qquad d \leq 5 \qquad (28)$$

where Y is the bomb yield in megatons,* d is the distance from the burst in miles, and Q'' is the total thermal radiation received in Btu/ft^2 .

Data on transient heat flux from a nuclear burst without *attenuation* $(F_2 = 1)$ have been curvefit [17] as

$$\frac{q''}{q''_{max}} = 0.574 \, e^{-0.35 \, \theta/\theta_1} + 0.090 \, e^{-1.40 \, \theta/\theta_1} + 39.181 \, e^{-2.45 \, \theta/\theta_1}$$
$$- 146.052 \, e^{-3.50 \, \theta/\theta_1} + 176.776 \, e^{-4.55 \, \theta/\theta_1} \qquad (29)$$
$$- 70.568 \, e^{-5.60 \, \theta/\theta_1}$$

In Eq. (29) the maximum flux q''_{max} and time θ_1 at which it occurs are given by

$$q''_{max} = 3991 \times 10^7 \, \frac{\sqrt{Y}}{d^2} \, , \qquad \theta_1 = 1.012\sqrt{Y} \qquad (30)$$

where q''_{max} = Btu/sec-ft^2 when Y =megatons and d = feet, and θ_1 = seconds when

*1 megaton = 1000 kiloton (TNT equivalent) = 10^5 calories = 3966 ∕ 10^9 Btu.

$|Y|$ = megatons. Thus, a 1-megaton detonation at a 1-mile distance can produce a maximum (unattenuated) heat flux of 1432 Btu/sec-ft² with a peak at approximately 1 second (Fig. 12); at five miles the peak intensity is 57, or about equivalent to the maximum radiant flux from a highly powered tungsten filament source.

• *Diathermanous Material.* A body may be *semitransparent* to radiation and, as such, absorb surface radiant heat at successive depths within the solid. If radiative surface flux is monochromatic and constant, absorption within such a *diathermanous* material may be approximately related to depth x as

$$q_x'' = q'' e^{-\mu x} \tag{31}$$

where q'' is the intensity of unreflected surface radiation and μ is the absorption coefficient. If none of the absorbed energy is lost from the surface, the boundary condition is $\partial t(0, \theta)/\partial x = 0$. For an opaque solid, $\mu \to \infty$ and all radiant energy is absorbed at surface $x = 0$ and then conducted inward.

B. SOLUTIONS OF HEAT CONDUCTION

Transient and steady-state solutions of heat conduction are summarized for cases that tend to reoccur in practice. Emphasis is placed on constant-property materials and plane geometry. Temperatures are given in equation form for those systems whose solutions are simple enough for repetitive calculations. Data for more complicated systems, obtained by computer or tedious evaluation of infinite-series solutions, are displayed in graphical form.

Various methods for deriving temperature-response solutions are given in Ref. 1.* An extensive collection of exact solutions is contained in Ref. 18. Graphical data alone are presented in Ref. 19.

1. Transient Heat Conduction

Transient heat conduction is characterized by *time-dependent* heat flow and temperature pattern within the conducting body.

In subdividing heat-conduction systems, a distinction is made between bodies that are thermally *thin* or *thick*. A thin body has insufficient internal thermal resistance to support temperature gradients, and as such instantaneous temperatures are uniform throughout the solid. Criteria for thermal thinness depend on material properties, wall thickness, boundary conditions, and exposure time (Sec. 3.B.1.b).

a. **Thin-wall Temperature Response.** A solid body of arbitrary shape (Fig. 14) is considered having density ρ, specific heat C, volume V, and surface area S. The solid may receive heat q'' per unit area at its surface and internally generate distributed heat q''' per unit volume of material. If the body has large internal conductance, then all heat received and generated diffuses instantaneously (without resistance) through the material, and body temperature rises uniformly throughout its volume. Under this *thin-wall* condition of zero body temperature gradients (denoted symbolically in terms of thermal conductivity as $k \to \infty$), the instantaneous heat balance is

$$q'' S + q''' V = \rho C V \frac{dt}{d\theta} \tag{32}$$

*Refer also to Secs. 1.B, 4, and 5.

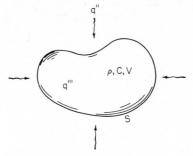

Fig. 14. Solid of arbitrary shape.

Quantities q'' and q''' may be either positive or negative, and each may be constant or depend on time θ. The nature of q'' is determined by the type of heating or cooling environment to which the body is exposed (surface boundary conditions).

Absence of internal body temperature gradients not only simplifies the mathematics of heat conduction but also affords treatment of otherwise more complicated (more realistic) systems. Since interior temperature is uniform at any given instant of time, the body may be sequentially exposed to different environments, with each exposure starting with a new uniform body temperature. In addition, an infinitely conducting plate cannot distinguish between heat addition at its front or rear face and between surface heat input and that generated internally. As such, exact solutions of temperature response for practical thin-wall systems can be obtained (such as combined-environment exposure) that are generally intractable for the corresponding thick-wall system.

Prescribed Surface Convection. For this condition, $q'' = h(t_a - t)$ in Eq. (32). If convective unit surface conductance h (Table 3) and ambient fluid temperature t_a are uniform on surface S, and if ρC for the material and q''' are constant, then body temperature $t(\theta)$ responds from its initial state $t(0) = t_0$ (the only admissible preexposure condition for $k \to \infty$) as

$$\frac{\vartheta}{\vartheta_a + p} = 1 - e^{-s\,\mathrm{BiFo}} \tag{33}$$

where

$$\vartheta = t - t_0 \quad , \quad \mathrm{Bi} = \frac{h\delta}{k}$$

$$p = \frac{q'''\delta}{hs} \quad , \quad \mathrm{Fo} = \frac{\alpha\theta}{\delta^2}$$

$$s = \frac{S\delta}{V} \quad , \quad \alpha = \frac{k}{\rho C}$$

Here Bi is the *Biot number* and Fo is the *Fourier number*, their product $\mathrm{BiFo} = h\theta/\rho C\delta$ being independent of k. Parameter s, containing some characteristic body dimension δ (e.g., half-thickness of a plate, radius of a cylinder or sphere), depends only on geometry, as illustrated for the following fundamental shapes (Fig. 2):

Geometry	s
infinite plate (thickness $= 2\delta$)	1
infinite elliptical rod ($b = \delta$)	$\dfrac{4}{\pi}E$
infinite square rod (side $= 2\delta$)	2
infinite cylinder (radius $= \delta$)	2
oblate spheroid ($b = \delta$)	$\dfrac{3}{2}F_1$
prolate spheroid ($b = \delta$)	$\dfrac{3}{2}\left(\dfrac{b}{a}\right)F_2$
cube (side $= 2\delta$)	3
short cylinder (radius $= \delta$, length $= 2\delta$)	3
sphere (radius $= \delta$)	3

Values of s for the three elliptical shapes depend on *eccentricity* $\epsilon = \sqrt{1 - (b/a)^2}$, where a and b are respectively semimajor and semiminor diameters of the ellipse. For the elliptical rod, $V = \pi ab$ and E is a complete elliptic integral

$$E = \int_0^{\pi/2} \sqrt{1 - \epsilon^2 \sin^2 x}\ dx.$$

If $a = b$, $\epsilon = 0$, and the elliptical rod becomes a cyclinder of radius b; when $a \to \infty$, $\epsilon = 1$, and the elliptical rod is a plate of semithickness b. For the *oblate* spheroid (ellipse rotated about b), $V = \frac{4}{3}\pi a^2 b$ and

$$F_1 = 1 + \left(\frac{1 - \epsilon^2}{2\epsilon}\right) \ln\left(\frac{1 + \epsilon}{1 - \epsilon}\right)$$

If $a = b$, the oblate spheroid becomes a sphere of radius b; when $a \to \infty$, the oblate spheroid is a plate of semithickness b. For the *prolate* spheroid (ellipse rotated about a), $V = \frac{4}{3}\pi ab^2$ and

$$F_2 = 1 + \left(\frac{1}{\epsilon\sqrt{1 - \epsilon^2}}\right) \sin^{-1} \epsilon$$

If $a = b$, the prolate spheroid is a sphere of radius b; when $a \to \infty$, the prolate spheroid is a circular rod of radius b. For all three elliptical shapes, $\delta = b$ in Eq. (33).

The temperature response of the above body shapes for the case of *symmetrical* heating or cooling of nongenerating materials ($p = 0$) is given in Fig. 15 (solid curves with specific heat C_0 evaluated at t_0). The plate may be of thickness 2δ and symmetrically heated at both faces, or considered to be of thickness δ and thermally insulated at its rear face.

In using Eq. (33), it is noted that at any instant of time the body may be removed from its environment and reexposed to a different heating or cooling medium without violating initial condition $t(0) = t_0$. For a completely insulated plate ($h = 0$), Eq. (33) is indeterminant; the solution in this case (independent of p) is $(k/q'''\delta^2)\vartheta = $ Fo. The response *time constant* θ_r for a given shape is the time required for the body to achieve a prescribed percentage of its steady-state value ($t_{ss} = t_a$). Selecting 86.5 percent as a criterion, the time constant of, say, an infinite cylinder of radius r_1 is determined from Fig. 15 to be $\theta_r = \rho C r_1/h$. *Instantaneous* heat gain $q = \rho C V dt/d\theta$ and *cumulative* heat gain $Q = \int_0^\theta q d\theta$ for an arbitrary thin body in a uniform environment are calculated from Eq. (33) as

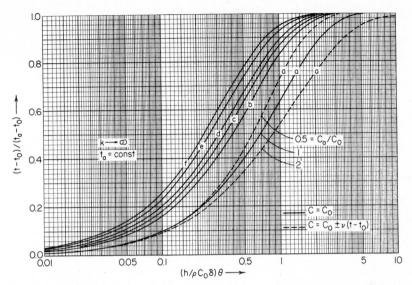

Fig. 15. Temperature response of infinitely conducting bodies suddenly exposed to uniform convective heating or cooling environments at t_a: (a) infinite plate (thickness = 2δ), (b) infinite elliptical rod ($b = \delta, \epsilon = 0.8$), (c) infinite square rod (side = 2δ), infinite cylinder (radius = δ), (d) oblate spheroid ($b = \delta, \epsilon = 0.8$), (e) prolate spheroid ($b = \delta, \epsilon = 0.8$), (f) cube (side = 2δ), short cylinder (radius = δ, length = 2δ), sphere (radius = δ)

$$\frac{q}{hS(\vartheta_a + p)} = e^{-s\mathrm{BiFo}} \quad , \quad \frac{Q}{\rho CV(\vartheta_a + p)} = 1 - e^{-s\mathrm{BiFo}}$$

By virtue of the thin-wall and uniform-environment assumptions, these expressions apply at all surface and internal body stations.

• *Variable Specific Heat.* If $C(t)$ varies *linearly* with temperature as $C = C_0 + \nu(t - t_0)$, body temperature response $\vartheta = t - t_0$ is given implicitly by

$$-s\mathrm{BiFo}_0 = \left[\frac{(C_a/C_0 - 1)p}{\vartheta_a} + \frac{C_a}{C_0}\right]\ln\left(1 - \frac{\vartheta}{\vartheta_a + p}\right) + \frac{(C_a/C_0 - 1)\vartheta}{\vartheta_a} \tag{34}$$

Here C_a and C_0 are specific heats evaluated at t_a and t_0, and $\mathrm{BiFo}_0 = (h/\rho C_0\delta)\theta$. If C is constant, $C_a = C_0$ and Eq. (34) reduces to Eq. (33). For a thin nongenerating plate, $p = 0$ and $s = 1$ in Eq. (34). Effects of linearly increasing $(+\nu)$ and linearly decreasing $(-\nu)$ specific heat of plate material are shown in Fig. 15 (dashed curves).

• *Transient Ambient Temperature.* Ambient fluid temperature may vary as a *power function* of time as $\vartheta_a = f\theta^n$. For a thin nongenerating plate:

$$n = \frac{1}{2}; \quad \frac{\sqrt{g}}{f}\vartheta = 2(\mathrm{BiFo})^{3/2}e^{-\mathrm{BiFo}}\left[\frac{1}{3} + \frac{\mathrm{BiFo}}{5} + \frac{(\mathrm{BiFo})^2}{7\cdot2!} + \frac{(\mathrm{BiFo})^3}{9\cdot3!} + \cdots\right]$$

$$n = 1; \quad \frac{g}{f}\vartheta = e^{-\mathrm{BiFo}} + \mathrm{BiFo} - 1 \tag{35}$$

$$n = 2; \quad \frac{g^2}{f}\vartheta = 2(1 - \mathrm{BiFo} - e^{-\mathrm{BiFo}}) + (\mathrm{BiFo})^2$$

In Eq. (35) $\vartheta = t - t_0$, $g = h/\rho C \delta$, and $\mathrm{BiFo} = g\theta$.

- *Harmonic Ambient Temperature.* Ambient fluid temperature may vary *harmonically* as $t_a = \bar{t}_a + t_a^+ \cos \omega\theta$ (Fig. 16), where t_a^+ is *amplitude* about the *mean* $\bar{t}_a$ and the *wave period* is $2\pi/\omega$. Consider a thin nongenerating plate of thickness δ asymmetrically

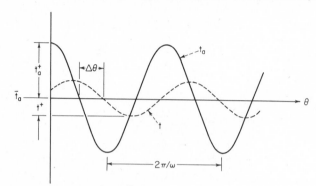

Fig. 16. Harmonic ambient temperature.

exposed at its front face to the above harmonic ambient temperature t_a with a uniform h_a and exposed at its rear face to an ambient of constant t_A and uniform h_A. With $t_1 = (h_a t_a + h_A t_A)/(h_a + h_A)$, $t_2 = h_a t_a^+/(h_a + h_A)$, $g_1 = (h_a + h_A)/\rho C \delta$, and $\mathrm{Bi_1 Fo} = g_1\theta$, the plate temperature is

$$\frac{t - t_1}{t_2} = \frac{1}{\sqrt{1 + (\omega/g_1)^2}} \cos\left(\frac{\omega}{g_1} \mathrm{Bi_1 Fo} - \psi_1\right) + \left[\frac{t_0 - t_1}{t_2} - \frac{1}{1 + (\omega/g_1)^2}\right] e^{-\mathrm{Bi_1 Fo}} \quad (36)$$

where $\psi_1 = \tan^{-1}(\omega/g_1)$ is the *phase angle* (time lag) between t_a and t. The *quasi steady-state* condition is represented by a $\mathrm{Bi_1 Fo}$ large enough so that the exponential term in Eq. (36) tends to zero; after this starting transient the plate temperature is also harmonic. If the rear face is insulated ($h_A = 0$),

$$\frac{t - \bar{t}_a}{t_a^+} = \frac{1}{\sqrt{1 + (\omega/g)^2}} \cos\left(\frac{\omega}{g} \mathrm{BiFo} - \psi\right) + \left[\frac{t_0 - \bar{t}_a}{t_a^+} - \frac{1}{1 + (\omega/g)^2}\right] e^{-\mathrm{BiFo}}$$

Figure 17a gives a starting transient for the case where $t_0 = \bar{t}_a$, and Fig. 17b gives the quasi-steady oscillation of plate temperature (independent of t_0) based on arbitrary time reference for $\omega\theta = 0$. The maxima and minima in plate temperature occur at $\omega\theta = \psi + n\pi$ and are equal to $(t^+ - \bar{t}_a)/t_a^+ = \pm 1/\sqrt{1 + (\omega/g)^2}$. Plate temperature lags ambient by an amount ψ. These quantities are shown in Fig. 17b for $\omega/g = 90°$. The instantaneous surface heat rate in the quasi steady state is

$$\frac{q''}{h t_a^+} = \mp \frac{\omega/g}{\sqrt{1 + (\omega/g)^2}} \sin\left[\left(\frac{\omega}{g}\right)\mathrm{BiFo} - \psi\right]$$

The heat is alternately rejected between $\omega\theta_1 = \psi$ and $\omega\theta_2 = \psi + \pi$ and added during the remaining half cycle. The half-cycle cumulative surface heating is

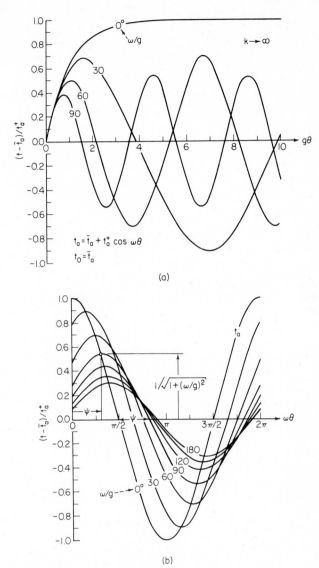

Fig. 17. Thin plate exposed to convective ambient whose temperature varies harmonically with time $(g = h/\rho C\delta)$: (a) plate temperature starting transient, (b) quasi-steady plate temperature oscillation.

$$\frac{Q''}{\rho C\delta t_a^+} = \frac{\mp 2}{\sqrt{1 + (\omega/g)^2}}$$

● *Aerodynamic Heating.* In aerodynamic heating of flight vehicles, both the unit surface conductance and ambient temperature may vary with time. Heat input for high-speed gas flow is $q'' = h(t_{aw} - t)$, where *adiabatic wall* (recovery) temperature t_{aw}

depends on *static* temperature t_a and *Mach number* Ma as $t_{aw} = t_a(1 + \beta Ma^2)$. For high-speed *turbulent* flow, $\beta = (Pr_a)^{1/3}(\gamma_a - 1)/2$, where Pr_a and γ_a are *Prandtl number* and *ratio of specific heats* for the gas flow. Local unit surface conductance for turbulent flow over a flat plate is roughly $h = \lambda Ma/\sqrt{1 + \beta Ma^2}$, where λ depends only on gas properties and location along surface. In level flight λ and β are constant and both h and t_{aw} depend only on flight velocity. For case of constant acceleration or deceleration at fixed altitude, Ma $= Ma_0 \pm j\theta$. Considering a thin insulated plate of thickness δ, its temperature response from the initial constant-altitude equilibrium condition $t_0 = t_a(1 + \beta Ma_0^2)$ is [20]

$$\frac{t}{t_a (A\beta)^2} = u^2 - 2(u - 1) + 2(u_0 - 1) e^{-(u - u_0)} \tag{37}$$

where $A = \rho C \delta j/\lambda$, $u = \sqrt{1 + \beta Ma^2}/A\beta$, and $u_0 = \sqrt{1 + \beta Ma_0^2}/A\beta$.

Prescribed Surface Radiation. For this condition, $q'' = \sigma \mathcal{F}(t_s^4 - t^4)$ in Eq. (32). The temperature response of a thin generating plate with insulated rear face and suddenly exposed at its front face to a constant *radiation* source or sink temperature t_s is given implicitly [Sec. 1.B, Eq. (68)] by

$$\frac{1}{2} \ln \left(\frac{T + N_1}{T - N_1} \right) \left(\frac{1 - N_1}{1 + N_1} \right) + \tan^{-1} \left(\frac{T}{N_1} \right) - \tan^{-1} \left(\frac{1}{N_1} \right) = 2M_0 N_1^3 Fo \tag{38}$$

where $T = t/t_0$, Fo $= \alpha\theta/\delta^2$, $N_1 = (q''\delta/\sigma\mathcal{F}t_0^4 + T_s^4)^{1/4}$, and $M_0 = \sigma\mathcal{F}t_0^3\delta/k$. Parameter M_0 is a *radiation number*, analogous to the Biot number for the convective boundary condition. The steady-state temperature is $T_{ss} = N_1$ or $t_{ss} = (q''\delta/\sigma\mathcal{F} + t_s^4)^{1/4}$.

For a nongenerating plate $(q''' = 0)$, $N_1 = T_s$ in Eq. (38) and $t_{ss} = t_s$. In the special case of $t_0 = 0°R$ with $q''' = 0$, the heating history is

$$\frac{1}{2} \ln \left(\frac{T - T_s}{T + T_s} \right) + \tan^{-1} \left(\frac{T}{T_s} \right) = 2M_0 T_s^3 Fo$$

If $t_0 \gg t_s$ with $q''' = 0$, the cooling history is

$$\frac{1}{2} \ln \left(\frac{T + T_s}{T - T_s} \right) + \tan^{-1} \left(\frac{T}{T_s} \right) - \frac{\pi}{2} = 2M_0 T_s^3 Fo$$

while if $t_s = 0°R$ the cooling history is given explicitly by

$$T = (1 + 3M_0 Fo)^{-1/3} \tag{39}$$

Combined Convection and Radiation. For simultaneous surface heat exchange by *convection and radiation*, $q'' = h(t_a - t) + \sigma\mathcal{F}(t_s^4 - t^4)$ in Eq. (32). Considering an insulated generating plate, its thin-wall temperature response on sudden exposure to these dual environments [Sec. 1.B, Eq. (70)] is

$$A_1 \ln\left(\frac{T - r_1}{1 - r_1}\right) + A_2 \ln\left(\frac{T - r_2}{1 - r_2}\right) + \frac{A_3}{2} \ln\left[\frac{(T - \eta)^2 + \phi^2}{(1 - \eta)^2 + \phi^2}\right]$$

$$+ \frac{A_4}{\phi}\left[\tan^{-1}\left(\frac{T - \eta}{\phi}\right) - \tan^{-1}\left(\frac{1 - \eta}{\phi}\right)\right] = M_0 \text{Fo} \tag{40}$$

where $T = t/t_0$, $\text{Fo} = \alpha\theta/\delta^2$, $M_0 = \sigma\mathcal{F}t_0^3\delta/k$, and

$$A_1 = \frac{1}{8\sqrt{\nu - \eta^2}\,[\eta\sqrt{\nu - \eta^2} + (\nu/2 + \eta^2)]}, \qquad A_3 = \frac{\eta}{\nu^2 + 8\eta^4}$$

$$A_2 = \frac{1}{8\sqrt{\nu - \eta^2}\,[\eta\sqrt{\nu - \eta^2} - (\nu/2 + \eta^2)]}, \qquad A_4 = \frac{\nu/2 - \eta^2}{\nu^2 + 8\eta^4}$$

$$r_1 = -\eta - \sqrt{\nu - \eta^2}, \qquad\qquad r_2 = -\eta + \sqrt{\nu - \eta^2}$$

$$\phi = \sqrt{\nu + \frac{\kappa}{4}}, \qquad \eta = \frac{\sqrt{\kappa}}{2}, \qquad \nu = \sqrt{N_2^4 + \frac{\kappa^2}{4}}$$

$$\kappa = \left(\frac{S^2}{2} + \sqrt{\frac{S^4}{4} + \frac{64}{27}N_2^{12}}\right)^{1/3} + \left(\frac{S^2}{2} - \sqrt{\frac{S^4}{4} + \frac{64}{27}N_2^{12}}\right)^{1/3}$$

$$S = \frac{h}{\sigma\mathcal{F}t_0^3}, \qquad N_2 = \left(\frac{q'''\delta}{\sigma\mathcal{F}t_0^4} + T_s^4 + ST_a\right)^{1/4} = (N_1^4 + ST_a)^{1/4}$$

The steady-state temperature is $T_{ss} = r_2 = \sqrt{(N_2^4 + \kappa^2/4)^{1/2} - \kappa/4} - \sqrt{\kappa}/2$. In the absence of surface convection ($h = 0$), Eq. (40) reduces to Eq. (38).

The maximum plate temperature attained in the mixed convective/radiative environment is attained for vanishing thermal capacity $\rho C = 0$ of the plate material (zero sensible heat absorption). This *radiation equilibrium temperature* is determined for a thin plate by the heat balance $h(t_a - t_{re}) + \sigma\mathcal{F}(t_s^4 - t_{re}^4) + q'''\delta = 0$, or

$$T_{re}^4 + ST_{re} = N_2^4 \tag{41}$$

Eq. (41) is readily solved by trial and error for the radiation equilibrium temperature T_{re} corresponding to given values of S and N_2 (see Fig. 63).

Prescribed Surface Heat Input. The heat input q'' in Eq. (32) may be prescribed directly in terms of time functions of zero or one parameter. Some useful zero-parameter *heat-input functions* (Fig. 13) are:

a. *step input*   ; $\dfrac{q''}{q''_{max}} = 1$, $\dfrac{Q''}{q''_{max} D} = 1$

b. *linear input*   ; $\dfrac{q''}{q''_{max}} = \tau$, $\dfrac{Q''}{q''_{max} D} = 0.5$

c. *linear input*   ; $\dfrac{q''}{q''_{max}} = (1 - \tau)$, $\dfrac{Q''}{q''_{max} D} = 0.5$

d. *circular pulse*   ; $\dfrac{q''}{q''_{max}} = \sin^2 \pi\tau$, $\dfrac{Q''}{q''_{max} D} = 0.5$

e. *power pulse*   ; $\dfrac{q''}{q''_{max}} = (3\tau)^{-3} e^{-(1/\tau - 3)}$, $\dfrac{Q''}{q''_{max} D} = 0.5473$

f. *power pulse*   ; $\dfrac{q''}{q''_{max}} = (5\tau)^{-5} e^{-(1/\tau - 5)}$, $\dfrac{Q''}{q''_{max} D} = 0.2795$

g. *exponential pulse*; $\dfrac{q''}{q''_{max}} = (10\tau) e^{1 - 10\tau}$, $\dfrac{Q''}{q''_{max} D} = 0.2717$

Here $\tau = \theta/D$ and Q'' is total input in heating duration D according to Eq. (21). The thin-wall temperature responses, $\vartheta = t - t_0$, of an insulated nongenerating plate exposed to these heat fluxes (Fig. 18) are:

a. $\dfrac{\rho C \delta}{Q''} \vartheta = \tau$ (42)

b. $\dfrac{\rho C \delta}{Q''} \vartheta = \tau^2$ (43)

c. $\dfrac{\rho C \delta}{Q''} \vartheta = \tau(2 - \tau)$ (44)

d. $\dfrac{\rho C \delta}{Q''} \vartheta = \tau - \left(\dfrac{1}{2\pi}\right) \sin 2\pi\tau$ (45)

e. $\dfrac{\rho C \delta}{Q''} \vartheta = 0.06767 \left(\dfrac{1}{\tau} + 1\right) e^{-(1/\tau - 3)}$ (46)

f. $\dfrac{\rho C \delta}{Q''} \vartheta = 0.006869 \left(\dfrac{1}{6\tau^3} + \dfrac{1}{2\tau^2} + \dfrac{1}{\tau} + 1\right) e^{-(1/\tau - 5)}$ (47)

g. $\dfrac{\rho C \delta}{Q''} \vartheta = 1.001 [1 - (10\tau + 1) e^{-10\tau}]$ (48)

Examples of one-parameter heat-input functions are as follows: When both the surface temperature potential and convective surface conductance vary *linearly* with time, as $(t_a - t) = m_1 \theta$ and $h = h_0 + n_1 \theta$, this is equivalent to prescribing heat flux $q'' = m\theta + n\theta^2$. For a thin nongenerating plate with this input

$$\dfrac{\rho C \delta}{Q''} \vartheta = J\tau^2 + (1 - J)\tau^3$$ (49)

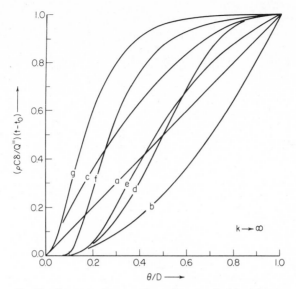

Fig. 18. Plate temperature response for prescribed heat inputs.

where $J = mD^2/2Q''$. If heat input varies *harmonically* as $q'' = \bar{q} + q^+ \cos \omega\theta$, the resulting temperature response of a thin nongenerating plate is

$$\frac{\rho C \delta \omega}{q^+} \vartheta = \left(\frac{\bar{q}}{q^+}\right) \omega\theta + \sin \omega\theta \tag{50}$$

By virtue of the thin-wall assumption, all the above solutions apply equally to cases of prescribed internal heat generation $q'''(\theta)$. This conversion is accomplished simply by replacing q'' with $q'''\delta$.

Combined Heat Input and Radiation. The temperature response given by Eq. (38) for a thin generating plate with the pure radiation boundary condition can be modified for calculating the response of a thin generating plate exposed to a *constant* heat input q'' while *reradiating* to a constant sink environment at t_s. In this case $N_1 = (q''/\sigma\mathcal{F}t_0{}^4 + q'''\delta/\sigma\mathcal{F}t_0{}^4 + T_s{}^4)^{1/4}$ and the steady-state plate temperature is $t_{ss} = (q''/\sigma\mathcal{F} + q'''\delta/\sigma\mathcal{F} + t_s{}^4)^{1/4}$ Figure 19 gives these data with $N_1{}^4$ as parameter. The heating curves (input exceeds loss) in Fig. 19*a* and cooling curves (loss exceeds input) in Fig. 19*b* apply equally to a thin plate receiving any combination of surface heat inputs. A satellite skin, for example, may continuously reradiate to space while receiving constant heat q''_E at its inner surface from internal heat-dissipating electrical equipments and constant solar heat q''_S at its outer surface (here $q'''\delta = 0$ and $q'' = q''_E + q''_S$).

Equation (40) may be used for a thin reradiating skin receiving constant q'' and backed by insulation (thickness δ_2, conductivity k_2) of negligible thermal capacity $(\rho_2 C_2 = 0)$ by replacing $q'''\delta$ with q'' and h with k_2/δ_2; then t_a is interpreted as a constant insulation rear-face temperature. Radiation equilibrium temperature of the skin in this instance is given exactly $(\rho_2 C_2 \neq 0)$ by Eq. (41).

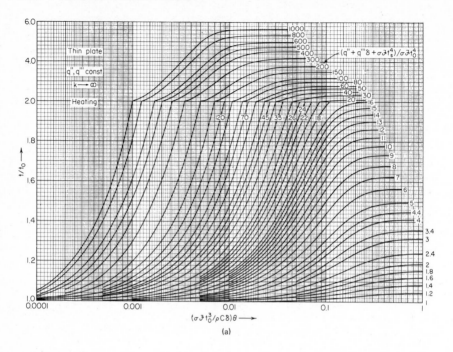

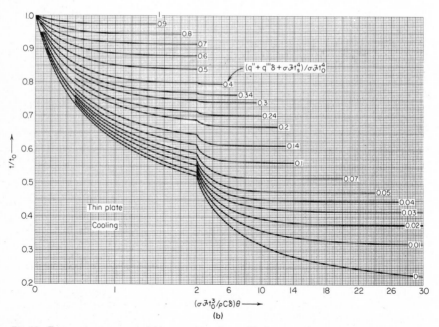

Fig. 19. Temperature response of thin generating plate suddenly exposed to constant heat input with surface reradiation: (a) heating, (b) cooling.

If a reradiating plate ($q''' = 0$) is heated by a *pulse* input (Fig. 13), its maximum temperature occurs prior to the end of heating. Figure 20a gives this maximum temperature [21] for a thin insulated plate suddenly exposed to a circular pulse $q'' = q''_{max} \sin^2 \pi\theta/D$ with surface reradiation to $t_s = 0°R$. For no reradiation ($\mathcal{F} = 0$), the maximum plate temperature is given by Eq. (45) as $t_{max}/t_0 = q''_{max} D/2\rho C\delta t_0 + 1$. For radiation equilibrium ($\rho C = 0$), the maximum plate temperature occurs at peak heating and is given by Eq. (26) as $t_{max}/t_0 = (q''_{max}/\sigma\mathcal{F}t_0^4)^{1/4}$. Figure 20b shows the maximum temperature data for special case of $t_0 = 0°F$.

b. Thick-wall Temperature Response. The conditions under which a plate becomes *thermally thick* relate not only to its material properties and thickness but also to imposed boundary conditions and exposure time. A plate exposed to convective or radiative environments always behaves as a semi-infinite solid in its early moments of heating; later it may transist to thick-wall behavior, and still later it may respond as a thin wall (Sec. B.1.a). The prescribed heat-input boundary condition leads to thick-wall criteria uniquely determined by material properties, plate thickness, and total heat load absorbed.

Prescribed Surface Convection. The Biot number $Bi = h\delta/k$ characterizes convective surface conductance and is a rough guide as to whether a given plate is to be treated as thermally thin or thick. This dimensionless group can be expressed as a ratio of surface to internal conductance K as $Bi = (hA)/(kA/\delta) = K_s/K_i$, where A is the surface area. If, say, $Bi < 0.1$, then internal conductance predominates and the plate responds as a simple capacitive heat sink. Conversely, for $Bi > 0.1$ surface conductance is controlling and internal temperature gradients develop. Thus, roughly, for plane geometry and the convection boundary condition:

$$\text{thin wall, } Bi \le 0.1 \quad ; \quad \text{thick wall, } Bi > 0.1 \tag{51}$$

The above limits do not constitute strict criteria because they ignore the aforementioned *time* effect. As a practical measure, a nongenerating plate initially at t_0 can be classed as thermally thin at a given instant of response if the temperature difference across its thickness is less than or equal to some prescribed value. Figure 21 shows thin-wall limits [22] for an insulated plate convectively heated at $x = 0$ and with successively more relaxed criteria on the temperature difference as $[\vartheta(0,\theta) - \vartheta(\delta,\theta)]/\vartheta(0,\theta) = 1, 5, 10\%$, where $\vartheta = t - t_0$. These data may be used to determine either maximum plate thickness δ that will give thin-wall response at a given instant θ (evaluate $FoBi^2$ and calculate δ from either Bi or Fo in Fig. 21) or the minimum time a plate of given thickness must be exposed before thin-wall response is attained (evaluate Bi and calculate θ from Fo in Fig. 21).

A thick plate initially responds to convective heating as a semi-infinite solid ($\delta \to \infty$). Minimum exposure time required for the insulated plate to cease responding as a semi-infinite solid (time at which rear-face temperature begins to rise above t_0) is approximately

$$Fo_{min} = 0.00756 \, Bi^{-0.3} + 0.02 \quad , \quad 0.001 \le Bi \le 1000$$

Following are graphical data for one-dimensional temperature response of thick plates and semi-infinite solids initially at uniform temperature t_0 and suddenly exposed to *convective heating* or *cooling* with uniform ambient temperature t_a and unit surface conductance h:*

● *Thick Plate.* Figure 22a - e gives heating or cooling history $(t - t_0)/(t_a - t_0)$ versus Fourier number $Fo = \alpha\theta/\delta^2$ with Biot number $Bi = h\delta/k$ as parameter for a thick insulated plate $0 \le x \le \delta$ at depths $x/\delta = 0, 0.02, 0.1, 0.4$ and 1 [23]. Dashed curves for $Bi = 0.1$ in Fig. 22a and e, given by Eq. (33) with $p = 0$ and $s = 1$, indicate

---

*Additional plots for plane bodies, as well as corresponding data for other geometric shapes and boundary conditions, are compiled in Ref. 19.

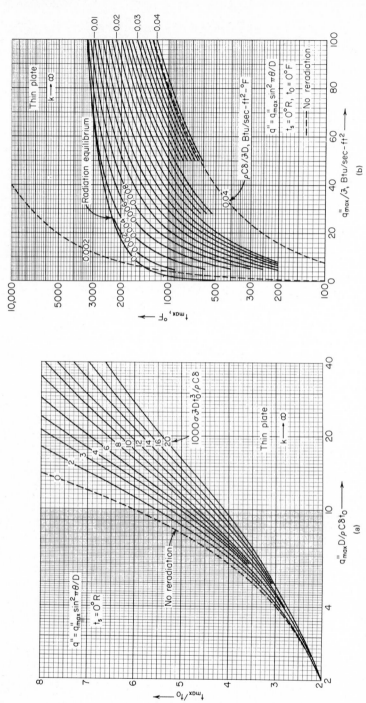

Fig. 20. Maximum temperature of thin insulated plate suddenly exposed to circular heat pulse with surface reradiation: (a) arbitrary initial plate temperature, (b) initial plate temperature $t_0 = 0°F$.

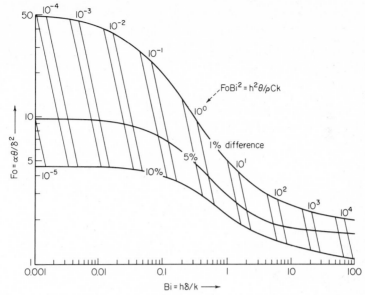

Fig. 21. Limits of thin-wall response for 1, 5, and 10% temperature difference across insulated plate.

that thin-wall response is approximately achieved for Bi ≤ 0.1 if Fo ≥ 10. Large Biot numbers (≥ 1000) approximate the case of instantaneous step change in exposed surface temperature from t_0 to t_a (see Fig. 32a).

• *Adiabatic Equilibration.* Heating or cooling of the plate may be abruptly terminated and the stored energy allowed to *equilibrate* while its surfaces are insulated. Temperature response of a plate undergoing adiabatic equilibration from an arbitrary initial temperature state $t(0, x) = f(x)$ is [18]

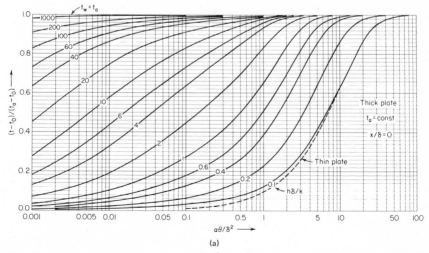

(a)

Fig. 22. Temperature response of thick plate ($0 \leq x \leq \delta$) with insulated rear face $x = \delta$ after sudden exposure to uniform convective environment t_a at $x = 0$: (a) $x/\delta = 0$.

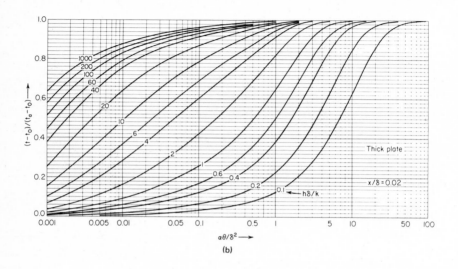

(b)

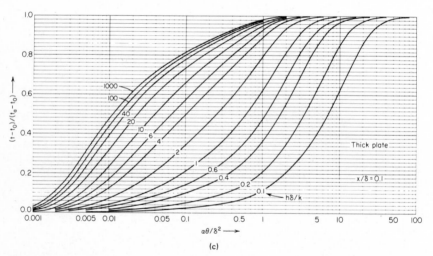

(c)

Fig. 22 (*Cont.*). Temperature response of thick plate $(0 \le x \le \delta)$ with insulated rear face $x = \delta$ after sudden exposure to uniform convective environment t_a at $x = 0$: (b) $x/\delta = 0.02$, (c) $x/\delta = 0.1$.

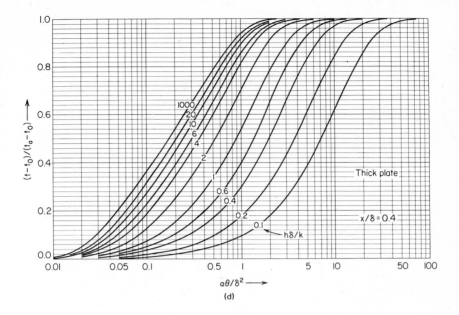

(d)

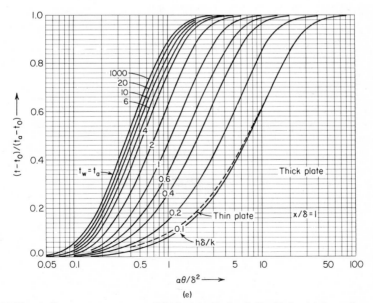

(e)

Fig. 22 (*Cont.*). Temperature response of thick plate ($0 \leq x \leq \delta$) with insulated rear face $x = \delta$ after sudden exposure to uniform convective environment t_a at $x = 0$: (d) $x/\delta = 0.4$, (e) $x/\delta = 1$.

$$t = \frac{2}{\delta} \sum_{n=1}^{\infty} e^{-(n\pi)^2 \mathrm{Fo}} \cos n\pi \frac{x}{\delta} \int_0^\delta f(x') \cos n\pi \frac{x'}{\delta} dx' + \frac{1}{\delta} \int_0^\delta f(x') dx'$$

Figure 23 gives equilibration history of both surfaces for the case where $t(x,0) = ae^{-bx} + c$. This exponential starting temperature is typical, for example, of a plate having absorbed X-rays (deposition time fast relative to conduction) at $x = 0$ [24].

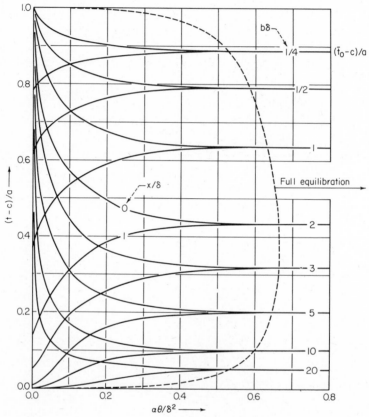

Fig. 23. Adiabatic equilibration of thick plate ($0 \le x \le \delta$) with starting temperature $t(x,0) = ae^{-bx} + c$.

Equilibration time (dashed curve) is approximately

$$\theta_{eqb} = \frac{\delta^2}{\pi^2 \alpha} \ln \left[4000 \, b\delta \left(\frac{1 - e^{-b\delta}}{b^2 \delta^2 + \pi^2} \right) \right]$$

• *Generating Plate.* Figure 24a–c gives the response $(k/q'''\delta^2)(t - t_0)$ versus $\mathrm{Fo} = \alpha\theta/\delta^2$ with $\mathrm{Bi} = h\delta/k$ as the parameter for an insulated thick plate $0 \le x \le \delta$ with uniform volume *heat sources* q''', the plate being convectively heated ($-q'''$) or cooled ($+q'''$) at $x = 0$ with $t_a = t_0$ [25]. Figure 24a gives the response at face surfaces $x/\delta = 0$ and 1 for $\mathrm{Fo} \le 0.2$. The case of $\mathrm{Bi} = 0$ represents no surface cooling ($h = 0$);

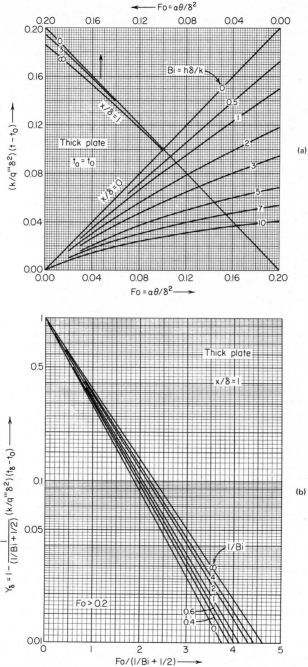

Fig. 24. Temperature response of thick generating plate $(0 \leq x \leq \delta)$ with insulated rear face $x = \delta$ after onset of internal heat generation and exposure to uniform convective environment $t_a = t_0$ at $x = 0$: (a) $x/\delta = 0, 1$ and $Fo \leq 0.2$, (b) $x/\delta = 1$ and $Fo > 0.2$.

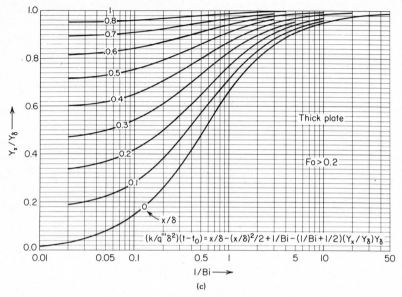

Fig. 24 (*Cont.*). Temperature response of thick generating plate ($0 \le x \le \delta$) with insulated rear face $x = \delta$ after onset of internal heat generation and exposure to uniform convective environment $t_a = t_0$ at $x = 0$: (c) $0 \le x/\delta \le 1$ and Fo > 0.2.

this is also a thin plate because q''' is uniform throughout the plate volume. Figure 24b gives rear-face response t_δ for Fo > 0.2 in terms of function Y_δ. Figure 24c gives response t_x for Fo > 0.2 at selected plate depths x/δ in terms of ratio Y_x/Y_δ.

• *Porous-cooled Plate.* Figure 25a, b gives response $(t - t_c)/(t_a - t_c)$ versus Fo $= \alpha_e \theta/\delta^2$ with *coolant flow number* $F = G_c C_c \delta/k_e$ as parameter and fixed Bi $= h\delta/k_e = 1$ for a thick *porous-cooled* plate $0 \le x \le \delta$. A steady flow of coolant with upstream (reservoir) temperature $t_c = t_0$ (Fig. 10a) enters the porous plate at rear face $x = \delta$ and exits at the convectively heated front face $x = 0$ [26]. Effective wall properties are $\rho_e C_e = P\rho_c C_c + (1 - P)\rho_s C_s$ and $k_e = Pk_c + (1 - P)k_s$, where P is porosity of wall material, and subscripts c and s refer to fluid coolant and wall solid. Figure 25a gives the response of heated face $x/\delta = 0$ and Fig. 25b shows similar data for rear face $x/\delta = 1$. Figure 25c gives the rear-face response if the front-face temperature is a constant $t_w(h \rightarrow \infty)$ [27]. Coolant flow parameter represents ratio of coolant to internal conductance as $F = (G_c C_c A)/(k_e A/\delta) = K_c/K_i$. Without coolant flow ($F = 0$), the response is given by Fig. 22a and e for Bi $= 1$. Steady-state plate temperatures (large Fourier number) are, by Eq. (124),

$$\frac{t - t_c}{t_a - t_c} = \frac{\text{Bi}}{\text{Bi} + F} e^{-F(x/\delta)}$$

• *Two-Layer Plate.* Figure 26a, b gives response functions for a two-layer plate, with the front layer $0 \le x_1 \le \delta_1$ being thick and exposed at $x_1 = 0$ to *convective* heating and in perfect contact with a rear layer $0 \le x_2 \le \delta_2$ that is thin and insulated at $x_2 = \delta_2$. Parameter $n = \rho_2 C_2 \delta_2/\rho_1 C_1 \delta_1$ represents the capacity ratio of the two layers. Figure 26a gives the temperature t_w of heated surface $x_1/\delta_1 = 0$ [28]. Figure

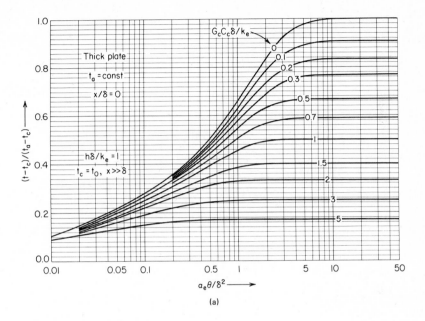

(a)

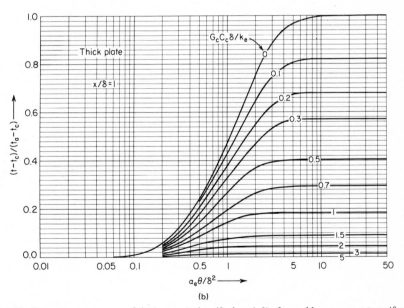

(b)

Fig. 25. Temperature response of thick porous plate ($0 \le x \le \delta$) after sudden exposure to uniform convective environment t_a with $\mathrm{Bi} = 1$ at $x = 0$ and cooled by steady flow of fluid through plate entering at $x = \delta$ from coolant reservoir at t_0: (a) $x/\delta = 0$, (b) $x/\delta = 1$.

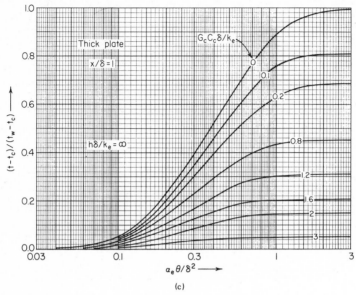

(c)

Fig. 25 (*Cont.*). Temperature response of thick porous plate ($0 \le x \le \delta$) after sudden exposure to uniform convective environment t_a with $\text{Bi} = \infty$ at $x = 0$ and cooled by steady flow of fluid through plate entering at $x = \delta$ from coolant reservoir at t_0: (c) $x/\delta = 1$.

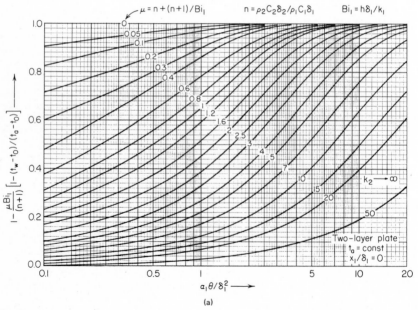

(a)

Fig. 26. Temperature response of thick plate ($0 \le x_1 \le \delta_1$) convectively heated at $x_1 = 0$ and in perfect contact at its rear face $x_1 = \delta_1$ with a thin plate ($0 \le x_2 \le \delta_2$) insulated at $x_2 = \delta_2$: (a) $x_1/\delta_1 = 0$.

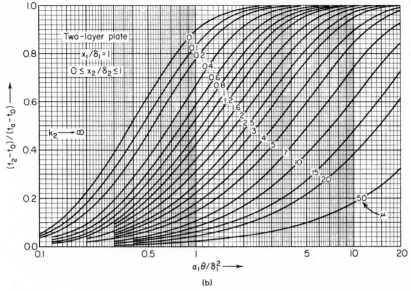

Fig. 26 (*Cont.*). Temperature response of thick plate ($0 \leq x_1 \leq \delta_1$) convectively heated at $x_1 = 0$ and in perfect contact at its rear face $x_1 = \delta_1$ with a thin plate ($0 \leq x_2 \leq \delta_2$) insulated at $x_2 = \delta_2$: (b) $x_1/\delta_1 = 1$, $0 \leq x_2/\delta_2 \leq 1$.

$26b$ gives temperature at interface $x_1/\delta_1 = 1[29]$; this is also the response of the complete rear layer $0 \leq x_2 \leq \delta_2$. If the rear layer is absent, $n = 0$ and $\mu = 1/Bi$ (Fig. 22); while if surface conductance is large, $Bi_1 \to \infty$ and $\mu = n$. The rear layer is thermally thin (relative to front layer) if, generally, $(k_2/\delta_2)/(k_1/\delta_1) \geq 100$. If $n \geq 1$ and $Bi_1 \geq 10$, the rear layer responds approximately as if dt_1/dx were constant, whereby

$$\frac{t_2 - t_0}{t_a - t_0} = 1 - \exp\left[\frac{-Fo_1}{(n + 1/2)(1/Bi_1 + 1)}\right]$$

for

$$\frac{Fo_1}{(n + 1/2)(1/Bi_1 + 1)} \leq 1.5$$

If $n \geq 10$ and $Bi_1 \geq 10$, the rear layer responds approximately as if $\rho_1 C_1 = 0$, and [30]

$$\frac{t_2 - t_0}{t_a - t_0} = 1 - \exp\left[\frac{-Fo_1}{n(1/Bi_1 + 1)}\right]$$

for

$$\frac{Fo_1}{(n + 1/2)(1/Bi_1 + 1)} \leq 1.5$$

Of interest is the insulation weight $W_1 = \rho_1 \delta_1$ required to protect a thin, load-bearing substructure from temperature rises exceeding $t_{2,\text{max}} - t_0$. Figure 27 shows these data derived from Fig. 26*b*. Insulation weight must be obtained by trial-and-error since W_1 appears explicitly in ordinate and implicitly in parameter.

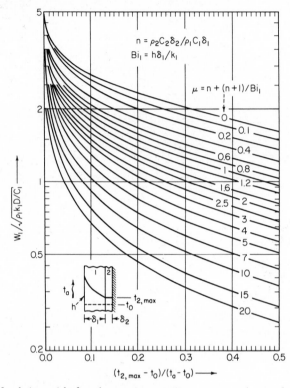

$$n = \rho_2 C_2 \delta_2 / \rho_1 C_1 \delta_1$$
$$Bi_1 = h\delta_1/k_1$$
$$\mu = n + (n+1)/Bi_1$$

Fig. 27. Insulation weight for substructure protection to $t_{2,\text{max}}$ in heating duration D.

- *Semi-infinite Solid.* Figure 28 gives the response $(t - t_0)/(t_a - t_0)$ versus $Bi\sqrt{Fo} = (h/k)\sqrt{\alpha\theta}$ with $(x/\delta)/2\sqrt{Fo} = x/2\sqrt{\alpha\theta}$ as the parameter for a plane semi-infinite solid $x \geq 0$ convectively heated or cooled at $x = 0$ [19].
- *Harmonic Ambient Temperature.* The quasi-steady temperature oscillation in a semi-infinite solid $x \geq 0$ exposed to *harmonic* ambient $t_a = \bar{t}_a + t_a^+ \cos\omega\theta$ (Fig. 16) is [31]

$$\frac{t - \bar{t}_a}{t_a^+} = \frac{e^{-L(x/\delta)}}{[1 + 2(L/Bi) + 2(L/Bi)^2]^{1/2}} \cos\left[2L^2 Fo - L\left(\frac{x}{\delta}\right) - \tan^{-1}\left(\frac{1}{1 + Bi/L}\right)\right] \quad (52)$$

where $L = \sqrt{\omega/2\alpha}\,\delta$, $Bi = h\delta/k$, and $Fo = \alpha\theta/\delta^2$.

Prescribed Surface Radiation. The *radiation number* $M_s = \sigma\mathscr{F}t_s^3\delta/k$ for heating or $M_0 = \sigma\mathscr{F}t_0^3\delta/k$ for cooling characterizes radiative surface conductance and roughly indicates whether a plate is thermally thin or thick during radiation exposure. This dimensionless group is the ratio of surface to internal conductance as $M_0 = (\sigma\mathscr{F}t_0^3 A)/$

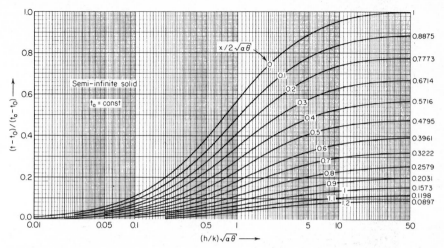

Fig. 28. Temperature response of semi-infinite solid $(x \geq 0)$ after sudden exposure to uniform convective environment t_a at $x = 0$.

$(kA/\delta) = K_s/K_i$. The thinness criterion for the radiative boundary condition is somewhat more restrictive than for convection [see Eq. (51)] and depends additionally on the ratio t_0/t_s. For a large range of this ratio:

$$thin\ wall,\ M \leq 0.02 \quad ; \quad thick\ wall,\ M > 0.02 \tag{53}$$

- *Thick Plate.* Figure 29a–f gives the heating response $(t - t_0)/(t_s - t_0)$ versus $\text{Fo} = \alpha\theta/\delta^2$ with the heating radiation number $M_s = \sigma\mathcal{F}t_s^3\delta/k$ as the parameter at the two face surfaces $x/\delta = 0$ and 1 for an insulated thick plate $0 \leq x \leq \delta$ being *radiatively heated* $(t_0 < t_s)$ by a uniform-temperature source t_s with heating ratios held constant at $t_0/t_s = 0$, 1/4, and 1/2. Figure 29g–l gives similar data for a thick plate *radiatively cooled* $(t_0 > t_s)$, the cooling radiation number being $M_0 = \sigma\mathcal{F}t_0^3\delta/k$ and the cooling ratios held constant at $t_0/t_s = \infty$, 4, and 2 [32].* As suggested in Eq. (53), the thin-wall response is nearly obtained for $M \leq 0.02$ if $\text{Fo} \geq 10$; this criterion becomes less restrictive as $t_0/t_s \to 1$. The large radiation numbers M_s or M_0 approximate (as with $\text{Bi} \to \infty$) the case of step change in exposed surface temperature from t_0 to t_s (see Fig. 32a).

The above charts can be extended to lower Fourier numbers $(\text{Fo} < 0.1)$ by noting that in the early response period a radiatively heated (or cooled) plate behaves as if it were exposed to constant surface heat input (or rejection) of magnitude $q'' = \sigma\mathcal{F}(t_s^4 - t_0^4)$. Using Fig. 40 in the $0 \leq \text{Fo} \leq 0.1$ range, determine $(k/q''\delta)(t - t_0)$ where $q''\delta/k = M_s t_s [1 - (t_0/t_s)^4]$ for heating or $q''\delta/k = M_0 t_0 [1 - (t_s/t_0)^4]$ for cooling. This approximation is accurate for surface and internal temperatures up to $t_w/t_s < 0.4$ during heating.

- *Sphere.* Figure 30a–c gives the cooling response t/t_0 versus $\text{Fo} = \alpha\theta/r_1^2$ with the cooling radiation number $M_0 = \sigma\mathcal{F}t_0^3r_1/k$ for the center $r/r_1 = 0$, midradius $r/r_1 = 1/2$, and surface $r/r_1 = 1$ of a solid sphere $0 \leq r \leq r_1$ radiatively cooled to a sink

---

*This reference also gives midplane response $(x/\delta = 1/2)$ and covers additional ratios $t_0/t_s = 3/4$ and 4/3.

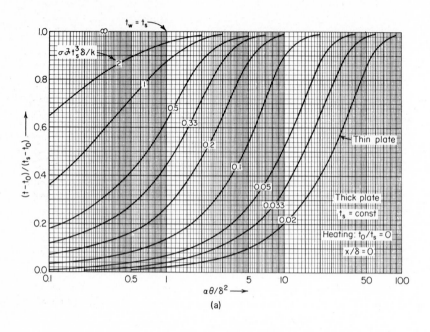

(a)

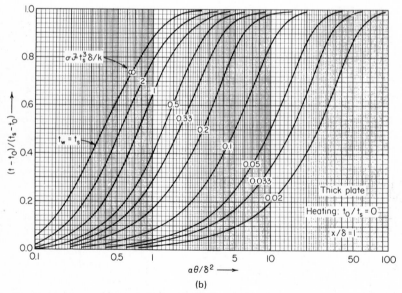

(b)

Fig. 29. Temperature response of thick plate $(0 \leq x \leq \delta)$ with insulated rear face $x = \delta$ after sudden exposure to uniform radiative environment t_s at $x = 0$: (a) heating, $t_0/t_s = 0$, $x/\delta = 0$, (b) heating, $t_0/t_s = 0$, $x/\delta = 1$.

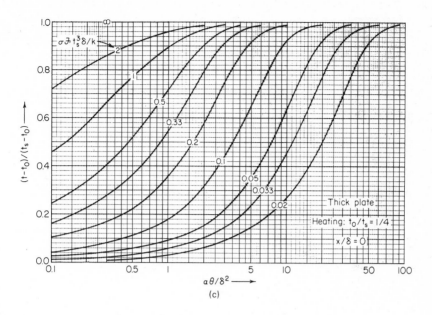

(c)

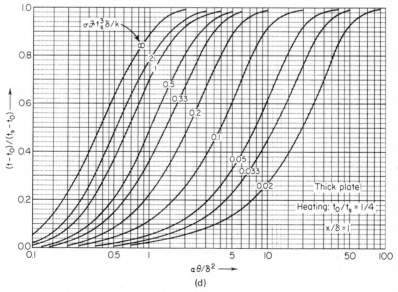

(d)

Fig. 29 (*Cont.*). Temperature response of thick plate ($0 \leq x \leq \delta$) with insulated rear face $x = \delta$ after sudden exposure to uniform radiative environment t_s at $x = 0$: (c) heating, $t_0/t_s = 1/4$, $x/\delta = 0$, (d) heating, $t_0/t_s = 1/4$, $x/\delta = 1$.

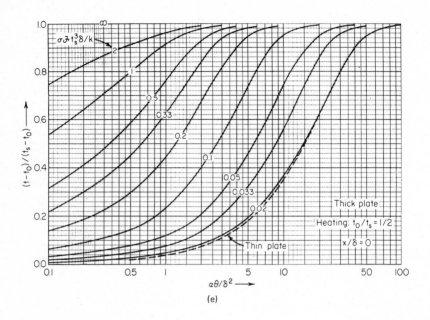

(e)

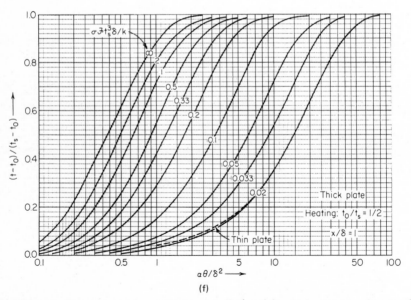

(f)

Fig. 29 (*Cont.*). Temperature response of thick plate ($0 \leq x \leq \delta$) with insulated rear face $x = \delta$ after sudden exposure to uniform radiative environment t_s at $x = 0$: (e) heating, $t_0/t_s = 1/2$, $x/\delta = 0$, (f) heating, $t_0/t_s = 1/2$, $x/\delta = 1$.

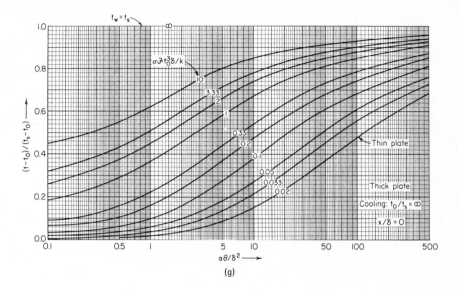

(g)

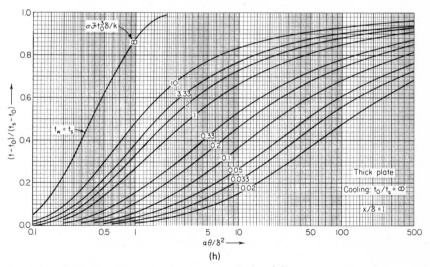

(h)

Fig. 29 (*Cont.*). Temperature response of thick plate ($0 \leq x \leq \delta$) with insulated rear face $x = \delta$ after sudden exposure to uniform radiative environment t_s at $x = 0$: (g) cooling, $t_0/t_s = \infty$, $x/\delta = 0$, (h) cooling, $t_0/t_s = \infty$, $x/\delta = 1$.

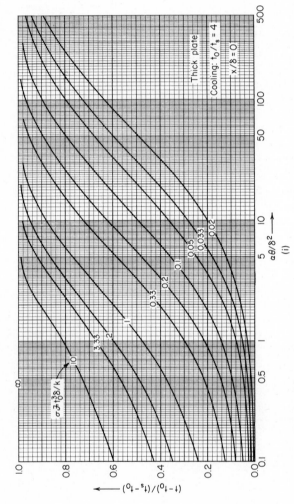

Fig. 29 (*Cont.*). Temperature response of thick plate ($0 \leq x \leq \delta$) with insulated rear face $x = \delta$ after sudden exposure to uniform radiative environment t_s at $x = 0$: (i) cooling, $t_0/t_s = 4$, $x/\delta = 0$.

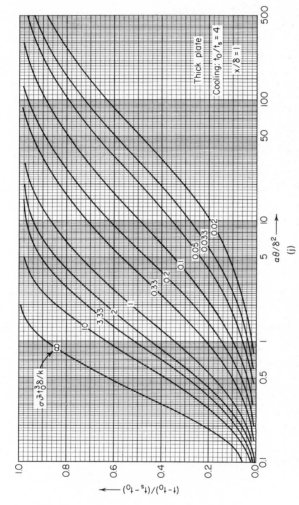

Fig. 29 (*Cont.*). Temperature response of thick plate ($0 \leq x \leq \delta$) with insulated rear face $x = \delta$ after sudden exposure to uniform radiative environment t_s at $x = 0$: (j) cooling, $t_0/t_s = 4$, $x/\delta = 1$.

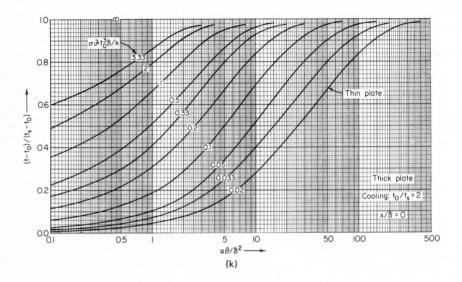

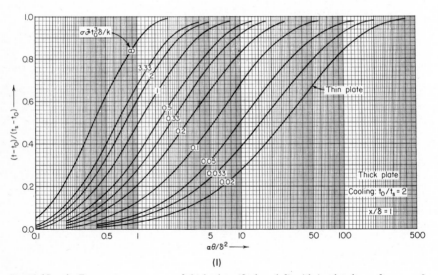

Fig. 29 (*Cont.*). Temperature response of thick plate ($0 \le x \le \delta$) with insulated rear face $x = \delta$ after sudden exposure to uniform radiative environment t_s at $x = 0$: (k) cooling, $t_0/t_s = 2$, $x/\delta = 0$, (l) cooling, $t_0/t_s = 2$, $x/\delta = 1$.

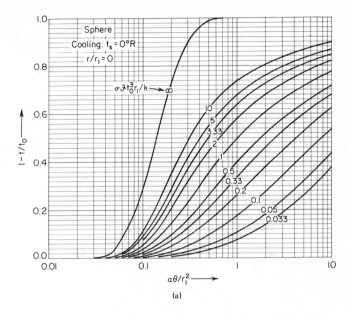

(a)

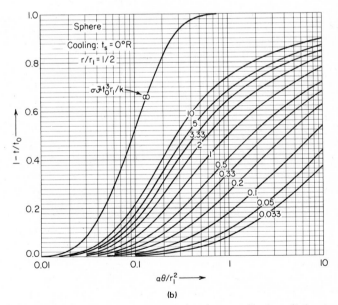

(b)

Fig. 30. Temperature response of solid sphere $(0 \leq r \leq r_1)$ cooling by radiation to a sink temperature of $0°R$: (a) $r/r_1 = 0$, (b) $r/r_1 = 1/2$.

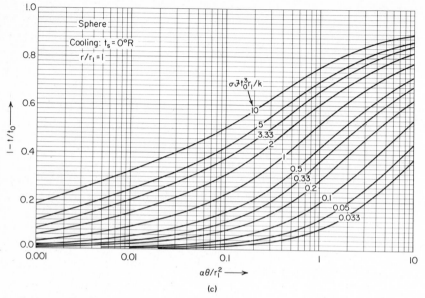

Fig. 30 (*Cont.*). Temperature response of solid sphere ($0 \leq r \leq r_1$) cooling by radiation to a sink temperature of $0°R$: (c) $r/r_1 = 1$.

environment at $t_s = 0°R$ [33]. Similar data for an infinitely long, solid cylinder are available for $t_s/t_0 = 0$, 1/4, 1/2, and 3/4 [34].

Prescribed Surface Temperature. If Biot number Bi or radiation numbers M_s or M_0 become sufficiently large, a suddenly exposed surface immediately assumes the convection or radiation ambient temperature t_a or t_s. This type of exposure represents a *step change* in surface temperature from initial value t_0 to t_w. The surface temperature t_w may also be any prescribed function of time.

● *Thick Bodies.* The temperature response $(t_{center} - t_0)/(t_w - t_0)$ versus Fo $= \alpha\theta/\delta^2$ at the center of various thick bodies of fundamental shape is given in Fig. 31 for the case of *step change* in surface temperature from t_0 for $\theta < 0$ to t_w for $\theta \geq 0$.*

● *Thick Plate.* Figure 32a gives the temperature response $(t - t_0)/(t_w - t_0)$ versus Fo $= \alpha\theta/\delta^2$ with x/δ as the parameter for an insulated thick plate $0 \leq x \leq \delta$ with a *step change* in surface temperature from t_0 to t_w at $x = 0$ [23]. Figure 32b gives the rear-face response at $x = \delta$ when this face is not insulated but exposed to a uniform-temperature convective environment $t_A = t_0$ [27].

● *Two-Layer Plate.* If a thick plate $0 \leq x_1 \leq \delta_1$ is backed (in perfect contact) by a thin plate $0 \leq x_2 \leq \delta_2$, Fig. 26 with $\mu = n$ (Bi$_1 \to \infty$) and $t_a = t_w$ should be used. The surface of a low-conductivity insulator may rapidly attain a constant radiation-equilibrium temperature $t_w = (q''/\sigma\epsilon)^{1/4}$ if it is suddenly exposed to input q'' when re-radiating to space. Figure 27 with $\mu = n$ gives the insulation weight required to limit the temperature rise of the thin underlying substructure to some permissible level $t_{2,max}$.

● *Linear Surface Temperature.* The temperature response $(t - t_0)/(t_w - t_0)$ versus Fo $= \alpha\theta/\delta^2$ with x/δ as the parameter is given in Fig. 33 for an insulated thick plate $0 \leq x \leq \delta$ ($h_A\delta/k = 0$) with surface temperature suddenly increasing *linearly*

*See Ref. 19 for data sources.

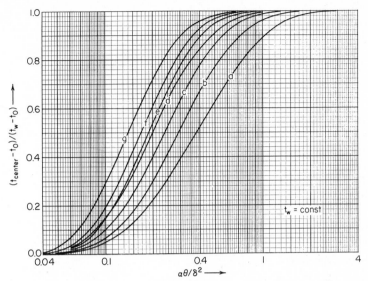

Fig. 31. Temperature response at center of thick bodies after sudden change in surface temperature from t_0 initially to t_w for $\theta \geq 0$: (a) infinite plate (thickness = 2δ), (b) infinite elliptical rod ($b = \delta, \epsilon = 0.8$), (c) infinite square rod (side = 2δ), (d) infinite cylinder (radius = δ), (e) cube (side = 2δ), (f) short cylinder (radius = δ, length = 2δ), (g) sphere (radius = δ).

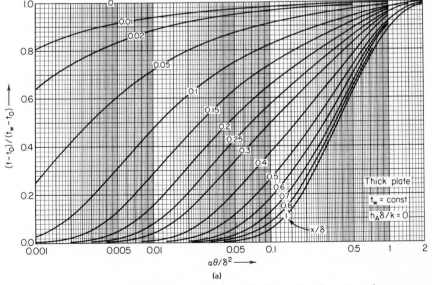

(a)

Fig. 32. Temperature response of thick plate ($0 \leq x \leq \delta$) after sudden change in surface temperature at $x = 0$ from t_0 to t_w: (a) insulated rear face.

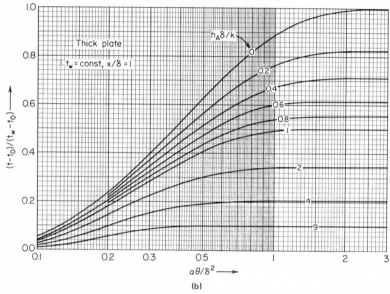

Fig. 32 (*Cont.*). Temperature response of thick plate ($0 \le x \le \delta$) after sudden change in surface temperature at $x = 0$ from t_0 to t_w : (b) uninsulated rear face ($t_A = t_0$).

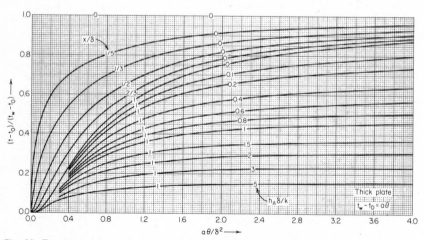

Fig. 33. Temperature response of thick plate ($0 \le x \le \delta$) with surface temperature t_w at $x = 0$ increasing linearly with time and rear face $x = \delta$ insulated or exposed to uniform convective environment t_0.

with time as $t_w - t_0 = a\theta$ [35]. The lower curves for finite $h_A\delta/k$ give the response of rear face $x = \delta$ when this surface is exposed to uniform-temperature convective environment $t_A = t_0$ [36].

• *Harmonic Surface Temperature.* The quasi-steady temperature oscillation in an insulated thick plate $0 \le x \le \delta$ with *harmonic* surface temperature $t_w = \bar{t}_w + t_w^+ \cos\omega\theta$ (Fig. 16) is [37]

$$\frac{t - \bar{t}_w}{t_w^+} = \left[\frac{f_s^2(X) + f_c^2(X)}{f_s^2(L) + f_c^2(L)}\right]^{1/2} \cos\left\{2L^2\,\text{Fo} - \tan^{-1}\left[\frac{f_s(L)f_c(X) - f_c(L)f_s(X)}{f_s(L)f_s(X) + f_c(L)f_c(X)}\right]\right\} \tag{54}$$

where $f_s(u) = \sinh u \sin u$, $f_c(u) = \cosh u \cos u$, $X = L(1 - x/\delta)$, and $L = \sqrt{\omega/2\alpha}\,\delta$. The maximum cumulative heat absorbed per half cycle occurs when $L = 1.2$.

• *Semi-infinite Solid.* The temperature history of a semi-infinite solid $x \ge 0$ with *step change* in surface temperature from t_0 to t_w is [1]

$$\frac{t - t_0}{t_w - t_0} = \text{erfc}\left(\frac{x}{2\sqrt{\alpha\theta}}\right) \tag{55}$$

where $\text{erfc}\,u = 1 - \text{erf}\,u$ is the complementary error function (Sec. 1.B.1). Both the temperature gradient $\partial t/\partial x$ and the heating rate $\partial t/\partial\theta$ at a fixed depth x in the solid have time maximums occurring at $x/2\sqrt{\alpha\theta} = 1/\sqrt{2} = 0.7071$:

$$\frac{\partial t}{\partial x} = -\frac{t_w - t_0}{\sqrt{\pi\alpha\theta}}\exp\left(-\frac{x^2}{4\alpha\theta}\right) \quad , \quad -\frac{x}{t_w - t_0}\left(\frac{\partial t}{\partial x}\right)_{\max} = \sqrt{\frac{2}{\pi e}} = 0.4839$$

$$\frac{\partial t}{\partial\theta} = \frac{t_w - t_0}{2\sqrt{\pi\alpha}}\frac{x}{\theta^{3/2}}\exp\left(-\frac{x^2}{4\alpha\theta}\right) \quad , \quad \frac{\theta}{t_w - t_0}\left(\frac{\partial t}{\partial\theta}\right)_{\max} = \sqrt{\frac{1}{2\pi e}} = 0.2420$$

These data are given in Fig. 34. The instantaneous and cumulative surface heat rates are

$$\frac{q''\delta}{k(t_w - t_0)} = \frac{1}{\sqrt{\pi\text{Fo}}} \quad , \quad \frac{Q''}{\rho C\delta(t_w - t_0)} = 2\sqrt{\frac{\text{Fo}}{\pi}}$$

When two semi-infinite solids, one at a uniform temperature t_1 and the other at t_2, are placed together in perfect *contact* (Fig. 35), their mutual interface temperature immediately assumes a steady value t_i. If solids are of identical material, t_i lies midway between t_1 and t_2; if not,

$$\frac{t_2 - t_i}{t_i - t_1} = \frac{k_1}{k_2}\sqrt{\frac{\alpha_2}{\alpha_1}}$$

or

$$\frac{t_i - t_1}{t_2 - t_1} = \frac{1}{1 + \sqrt{\rho_1 C_1 k_1 / \rho_2 C_2 k_2}} \tag{56}$$

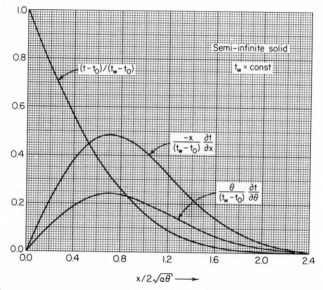

Fig. 34. Temperature response, temperature gradient, and heating rate in semi-infinite solid ($x \geq 0$) after sudden change in surface temperature from t_0 to t_w.

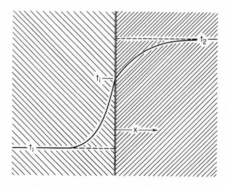

Fig. 35. Response of perfect contact.

The response of each solid is given by Eq. (55) as

$$\frac{t - t_i}{t_1 - t_i} = \mathrm{erf}\left(\frac{x}{2\sqrt{\alpha_1 \theta}}\right) \quad , \quad \frac{t - t_i}{t_2 - t_i} = \mathrm{erf}\left(\frac{x}{2\sqrt{\alpha_2 \theta}}\right)$$

● *Infinite Solids.* The temperature of a layer $-a \leq x \leq a$ in an infinite solid initially at t_0 is suddenly *pulsed* to a value t_w. This is equivalent to a heat source q of width $2a$ moving along an infinite plate with insulated faces and locally raising the plate temperature to t_w (e.g., butt welding). As the heat source moves on, the plate temperature equilibrates (Fig. 36) according to

$$\frac{t - t_0}{t_w - t_0} = \frac{1}{2}\left[\mathrm{erf}\left(\frac{a + x}{2\sqrt{\alpha\theta}}\right) + \mathrm{erf}\left(\frac{a - x}{2\sqrt{\alpha\theta}}\right)\right] \tag{57}$$

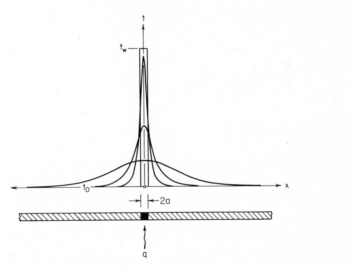

Fig. 36. Pulsed wall temperature.

The temperature response of a highly conductive sphere $0 \leq r \leq r_1$, initially at $t_{1,0}$ and embedded with perfect contact in an infinite solid $r > r_1$ of finite conductivity k_2 and initially at $t_{2,0}$, is given in Fig. 37 with $1/m = \rho_1 C_1/\rho_2 C_2$ as the parameter [38]. The sphere is isothermal if, approximately, $k_1/k_2 \geq 50$.

● *Product Solutions.* Internal and central temperature histories for two- or three-dimensional heat flow in simple bodies following step change in surface temperature can be obtained directly as products of one-dimensional solutions (Sec. 1.B.5). Figure 38 gives examples of such *product solutions*, where $T = (t - t_w)/(t_0 - t_w)$. The one-dimensional solutions for an infinite plate $T = P(\mathrm{Fo})$ and infinite cylinder $T = C(\mathrm{Fo_1})$ are given in Fig. 31 and for a semi-infinite solid $T = S(X)$ in Fig. 34.

● *Transient Surface Temperature.* The temperature response $(t - t_0)/(t_w - t_0)$ versus $(x/\delta)/2\sqrt{\mathrm{Fo}} = x/2\sqrt{\alpha\theta}$ for a semi-infinite solid $x \geq 0$ with surface temperature suddenly increasing as a *power function* of time $t_w - t_0 = a\theta^n$ is shown in Fig. 39 with n as parameter [18]. The value $n = 0$ corresponds to a step change (Fig. 34).

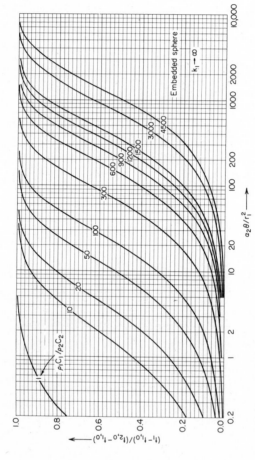

Fig. 37. Temperature response of solid sphere $0 \leq r \leq r_1$ with $k_1 \to \infty$ and initially at $t_{1,0}$ embedded in an infinite solid ($r > r_1$) initially at $t_{2,0}$.

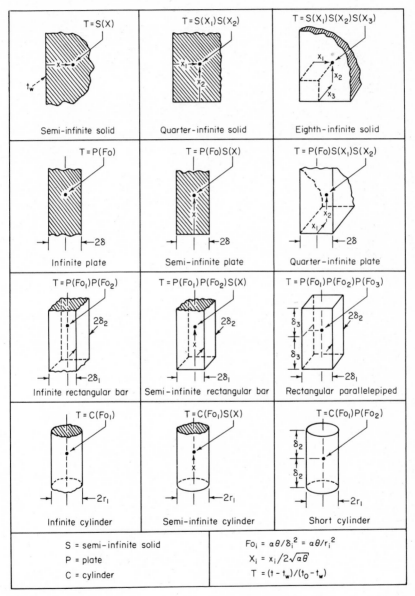

Fig. 38. Product solutions for internal and central temperatures in solids with step change in surface temperature.

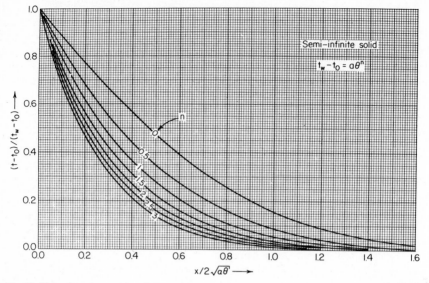

Fig. 39. Temperature response of semi-infinite solid ($x \geq 0$) with surface temperature t_w suddenly increasing as power function of time.

● *Harmonic Surface Temperature.* The quasi-steady temperature oscillation in a semi-infinite solid $x \geq 0$ with *harmonic* surface temperature $t_w = \bar{t}_w + t_w^+ \cos \omega\theta$ (Fig. 16) is given by Eq. (52) with Bi $\to \infty$ as

$$\frac{t - \bar{t}_w}{t_w^+} = e^{-L(x/\delta)} \cos L \left(2L\text{Fo} - \frac{x}{\delta} \right) \tag{58}$$

The *time lag* of internal temperature wave is $\sqrt{1/2\alpha\omega}\, x$. The wave propagates into the solid with a *velocity* of $\sqrt{2\alpha\omega}$ and a wavelength of $\pi\sqrt{8\alpha/\omega}$. The half-cycle instantaneous and cumulative heat rates at the surface are [1]

$$\frac{q''\delta}{kt_w^+} = \sqrt{\frac{\omega\delta^2}{\alpha}} \cos\left(\frac{\omega\delta^2}{\alpha}\text{Fo} + \frac{\pi}{4}\right) , \quad \frac{Q''}{\rho C \delta t_w^+} = 2\sqrt{\frac{\alpha}{\omega\delta^2}}$$

Prescribed Heat Input. Under certain conditions of convective or radiative heating the net heat input is insensitive to the body surface temperature. In this case q'' can be uniquely prescribed from known environment parameters.

● *Thick Plate.* Figure 40a gives the temperature response $(k/q''\delta)(t - t_0)$ versus Fo $= \alpha\theta/\delta^2$ with x/δ as the parameter for an insulated thick plate $0 \leq x \leq \delta$ which is suddenly exposed to a *constant* heat input q'' at $x = 0$ [39]. The dashed curve shows the corresponding response of a thin plate given by Eq. (42). This is also the instantaneous mean temperature $\bar{t}$ of a thick plate; it occurs at approximately $x/\delta = 0.43$. The curves in Fig. 40a can be extended to higher Fourier numbers by noting their linear character above Fo = 0.5, the exposed surface temperature being 1/3 above the mean and the insulated surface 1/6 below the mean. Thus, temperature drop across the plate following the starting transient is constant as $t_w - t_\delta = q''\delta/2k$, or

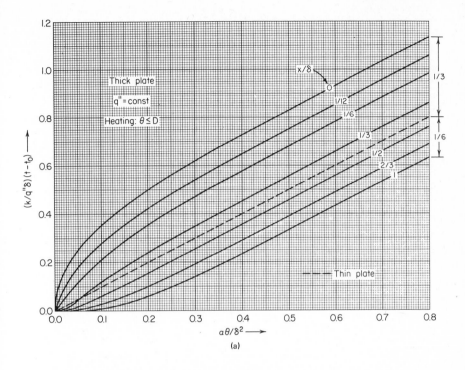

(a)

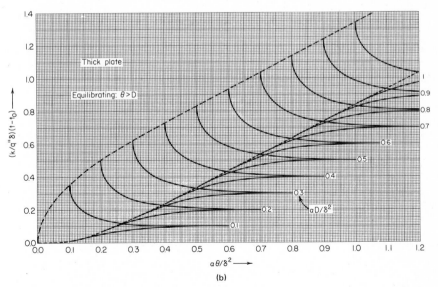

(b)

Fig. 40. Temperature response of thick plate ($0 \leq x \leq \delta$) with insulated rear face $x = \delta$ after sudden exposure to constant heat input q'' at $x = 0$: (a) heating, $\theta \leq D$, (b) equilibrating ($x = 0$ insulated), $\theta > D$.

$$\frac{k}{q''\delta}(t - t_0) = \text{Fo} + \frac{X^2}{2} - X + \frac{1}{3} \; ; \; \text{Fo} > \frac{1}{2}$$

where $\text{Fo} = \alpha\theta/\delta^2$ and $X = x/\delta$. The rear-face temperature becomes linear (+1%) when $\text{Fo} = 0.45$. The time difference between this and the occurrence of surface melting at front face is a maximum as $(\Delta\theta)_{\max} = 0.319(k^2/q''^2\alpha)(t_d - t_0)^2$ if plate thickness has the value $\delta_{\text{opt}} = 0.639(k/q'')(t_d - t_0)$.

Figure 40b shows how plate temperature equilibrates if heating is abruptly terminated and $x = 0$ is insulated at $\theta = D$ [40]. The equilibration time is roughly $\theta_{\text{eqb}} = \delta^2/2\alpha$.

Figure 41a, b gives surface responses when the rear face $x = \delta$ is not insulated but is exposed to a uniform-temperature convective environment $t_A = t_0$ [27].

● *Transient Heat Input.* If the heat input to the insulated plate increases *linearly* with time as $q'' = q''_{\max}\theta/D$, the temperature response after the starting transient is [41]

$$\frac{k}{q''\delta}(t - t_0) = \frac{\text{Fo}}{\text{Fo}'}\left(\frac{\text{Fo}}{2} - \frac{X^3}{12} + \frac{5X^2}{8} - X + \frac{5}{16}\right)$$

for $1/2 < \text{Fo} < \text{Fo}'$ and $\text{Fo}' \geq 1$, where $\text{Fo}' = \alpha D/\delta^2$ is the *modified* Fourier number (based on heating duration D). If the heat input decreases linearly as $q'' = q''_{\max}(1 - \theta/D)$, the temperature following the starting transient is (by superposition)

$$\frac{k}{q''\delta}(t - t_0) = \text{Fo} + \frac{X^2}{2} - X + \frac{1}{3} - \frac{\text{Fo}}{\text{Fo}'}\left(\frac{\text{Fo}}{2} - \frac{X^3}{12} + \frac{5X^2}{8} - X + \frac{5}{16}\right)$$

for $1/2 < \text{Fo} < \text{Fo}'$ and $\text{Fo}' \geq 1$.

Figure 42a, b gives the temperature response $(\pi\rho C\delta/Q'')(t - t_0)$ versus $2\pi\theta/D$ with the modified Fourier number $\text{Fo}'/2\pi = \alpha D/2\pi\delta^2$ as the parameter for surfaces $x/\delta = 0$ and 1 of an insulated thick plate $0 \leq x \leq \delta$ suddenly exposed to a *circular* heat pulse $q'' = q''_{\max}\sin^2\pi\theta/D$ at $x = 0$ [42].* After $\theta = D$ all plate temperatures equilibrate to $t_{ss} = t_0 + q''_{\max}D/2\rho C\delta$. The dashed curves in Fig. 42 represent the thin-wall response given by Eq. (45). The thin-wall criterion for circular pulse is approximately $\text{Fo}' \geq 2\pi$.

The parameters that determine heat-absorption capabilities of heat-sink materials can be uniquely correlated for a prescribed heat-input boundary condition by calculating $Q''/\sqrt{\rho CkD}\,(t_{w,\max} - t_0)$ as a function of $\delta/\sqrt{\alpha D}$, where $t_{w,\max}$ is the peak temperature of the heated surface (interpreted for a maximum total heat load Q'' as the decomposition or melting temperature t_d of the material). Figure 43 gives this general relationship for surface heating in form of step input or circular pulse of duration D. For a small $\delta/\sqrt{\alpha D}$, the plate behaves as a thick wall [Eqs. (42) and (45)], while for large $\delta/\sqrt{\alpha D}$ the plate absorbs heat as a semi-infinite solid. The plate is thermally thick if, approximately

$$\text{step input} \quad : \quad 0.2\sqrt{\alpha D} \leq \delta \leq 2.0\sqrt{\alpha D}$$

$$\text{circular pulse} : \quad 0.4\sqrt{\alpha D} \leq \delta \leq 1.4\sqrt{\alpha D}$$

In terms of heat rate q'' and heating duration D, the maximum energy that can be absorbed by an insulated plate of pure heat-sink material (Fig. 43) is

---

*Additional data for $x/\delta = 1/4$ and $1/2$ are given in Ref. 19.

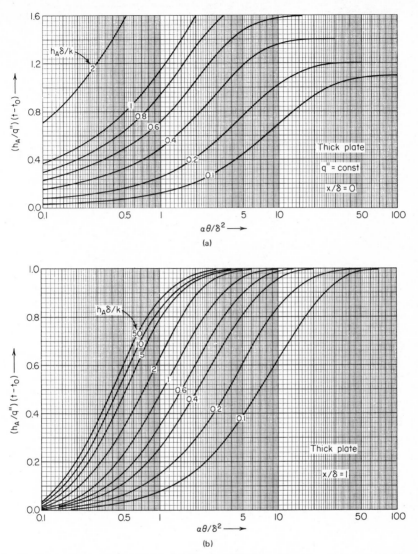

Fig. 41. Temperature response of thick plate ($0 \leq x \leq \delta$) after sudden exposure to a constant heat input q'' at $x = 0$ with rear face $x = \delta$ exposed to uniform convective environment $t_A = t_0$: (a) $x/\delta = 0$, (b) $x/\delta = 1$.

$$(q''\sqrt{D})_{max} = a\sqrt{\rho C k}(t_{w,max} - t_0) \begin{cases} step\ input & : \quad a = \dfrac{\sqrt{\pi}}{2} \\[2ex] circular\ pulse & : \quad a = \dfrac{3}{2} \end{cases} \tag{59}$$

Thus, for a given heat-input history the important parameter (ignoring surface reradiation) is $\sqrt{\rho C k}(t_d - t_0)$. Table 5 gives the approximate elevated-temperature properties and this derived quantity for selected heat-sink materials initially at $80°F$.

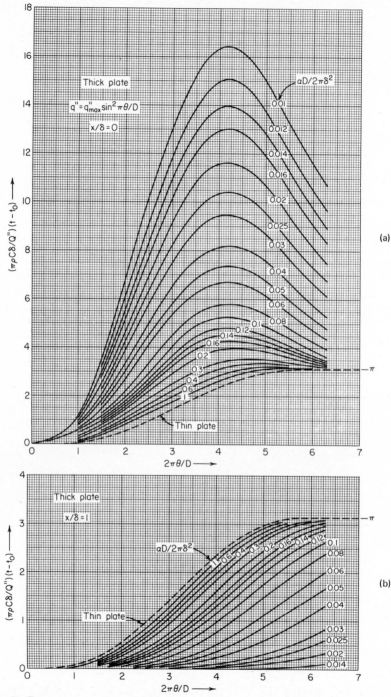

Fig. 42. Temperature response of thick plate ($0 \leq x \leq \delta$) with insulated rear face $x = \delta$ after sudden exposure to circular heat pulse $q''_{max} \sin^2 \pi\theta/D$ at $x = 0$: (a) $x/\delta = 0$, (b) $x/\delta = 1$.

TABLE 5. Thermal Properties of Selected Heat-sink Materials

Material	ρ	C	k	t_d	α	$\sqrt{\rho C k}(t_d - 80)$
	lb/ft^3	Btu/lb·°F	Btu/hr·ft·°F	°F	ft^2/hr	Btu/sec$^{1/2}$·ft^2
Aluminum (Al)	165	0.26	144	1220	3.36	1495
Alumina (Al$_2$O$_3$)	250	0.31	2.5	3722	0.0322	845
Beryllium (Be)	109	0.81	40	2343	0.454	2240
Beryllia (BeO)	172	0.50	10	4620	0.116	2210
Beryllium carbide (Be$_2$C)	152	0.31	1.3	3780	0.0276	483
Boron carbide (B$_4$C)	156	0.44	16	4410	0.233	2390
Boron nitride (BN)	135	0.45	15	5427	0.247	2690
Chromium (Cr)	445	0.12	40	2822	0.749	2115
Columbium (Cb)	535	0.072	40	4379	1.04	2815
Copper (Cu)	558	0.092	203	1981	3.96	3235
Graphite (ATJ)	108	0.47	16	6600	0.316	3100
Graphite (pyrolytic, a-direction)	137	0.47	42	6700	0.652	5740
Graphite (pyrolytic, c-direction)	137	0.47	0.68	6700	0.0106	730
Hastelloy (R-235)	511	0.11	15	2464	0.266	1155
Haynes alloy (25)	570	0.092	16	2425	0.305	1135
Inconel (X)	519	0.13	20	2540	0.297	1505
Iridium (Ir)	1400	0.031	34	4449	0.784	2800
Iron (Armco)	491	0.14	20	2794	0.291	1675
Magnesium (Mg)	112	0.25	60	1204	2.14	769
Magnesia (MgO)	224	0.33	4.9	5070	0.0663	1580
Molybdenum (Mo)	638	0.084	53	4752	0.990	4155
Nickel (Ni)	556	0.13	40	2651	0.553	2305
Platinum (Pt)	1340	0.032	40	3223	0.933	2175
Pyroceram®	162	0.31	1.7	2462	0.0338	367
Rhodium (Rh)	775	0.059	51	3571	1.12	2810
René (41)	515	0.16	10	2450	0.121	1135
Silicon (Si)	145	0.16	48	2570	2.07	1385
Silicon carbide (SiC)	198	0.12	10	4890	0.420	1240
Silicon nitride (Si$_3$N$_4$)	214	0.32	0.92	3452	0.0134	446
Steel (4130 alloy)	490	0.16	50	2700	0.637	2735
Steel (4340 alloy)	489	0.11	22	2740	0.410	1525
Steel (316 stainless)	492	0.12	8.3	2250	0.141	800

TABLE 5. Thermal Properties of Selected Heat-sink Materials (*Continued*)

Material	ρ	C	k	t_d	α	$\sqrt{\rho C k}(t_d - 80)$
	lb/ft³	Btu/lb·°F	Btu/hr·ft·°F	°F	ft²/hr	Btu/sec^½·ft²
Steel (410 stainless)	476	0.11	17	2723	0.324	1320
Tantalum (Ta)	1037	0.036	32	5425	0.858	3080
Tantalum carbide (TaC)	903	0.074	13	6840	0.195	3320
Tantalum nitride (Ta₂N)	854	0.052	8.3	5310	0.187	1670
Thorium (Th)	732	0.028	22	3348	1.07	1155
Thoria (ThO₂)	605	0.061	1.7	5486	0.0460	714
Titanium (Ti)	282	0.13	9.8	3074	0.268	945
Titania (TiO₂)	260	0.17	1.8	3294	0.0407	478
Titanium carbide (TiC)	307	0.20	9.9	5650	0.161	2290
Titanium nitride (TiN)	339	0.24	4.9	5342	0.0602	1755
Tungsten (W)	1208	0.032	40	6152	1.04	3980
Tungsten boride (WB)	954	0.071	25	5180	0.369	3500
Tungsten carbide (WC)	979	0.068	38	5000	0.571	4130
Vanadium (V)	381	0.13	17	3452	0.344	1630
Vanadium carbide (VC)	342	0.23	21	5130	0.267	3420
Zirconium (Zr)	415	0.069	9.6	3326	0.336	898
Zirconia (ZrO₂)	363	0.16	1.2	4928	0.0206	675
Zirconium boride (ZrB₂)	380	0.12	13	5430	0.285	3170
Zirconium carbide (ZrC)	418	0.12	12	6350	0.239	2565

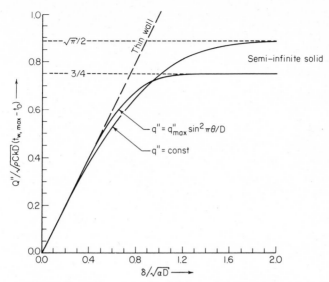

Fig. 43. Heat-sink performance with step and circular heat inputs.

● *Porous-Cooled Plate*. Figure 44a–c gives temperature response versus Fo $= \alpha_e \theta / \delta^2$ with the coolant flow number $F = G_c C_c \delta / k_e$ as the parameter for a thick porous-cooled plate $0 \leq x \leq \delta$ suddenly exposed to a *constant* heat input q'' at $x = 0$ [43]. Figure 44a gives the response $(k_e / q'' \delta) F(t - t_0)$ at the heated face $x/\delta = 0$ (coolant exit). Figure 44b gives $(k_e / q'' \delta) Fe^{F/2}(t - t_0)$ at the midplane $x/\delta = 1/2$, and Fig. 44c gives $(k_e / q'' \delta) Fe^{F}(t - t_0)$ at the rear face $x/\delta = 1$ (coolant entrance). Steady-state plate temperatures are, by Eq. (125), $(k_e / q'' \delta)(t - t_0) = (1/F) e^{-F(x/\delta)}$.

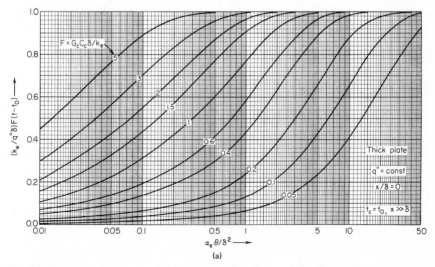

(a)

Fig. 44. Temperature response of thick porous plate $(0 \leq x \leq \delta)$ after sudden exposure to constant heat input q'' at $x = 0$ and cooled by steady flow of fluid through plate entering at $x = \delta$ from a coolant reservoir at t_0: (a) $x/\delta = 0$.

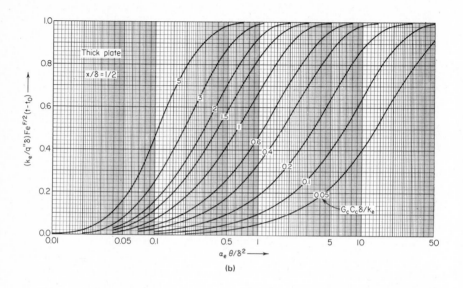

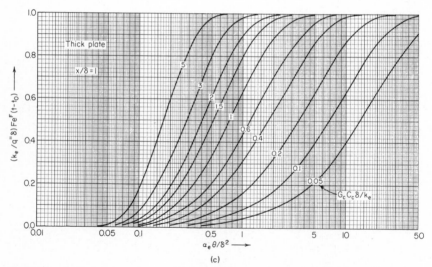

Fig. 44 (*Cont.*). Temperature response of thick porous plate ($0 \leq x \leq \delta$) after sudden exposure to constant heat input q'' at $x = 0$ and cooled by steady flow of fluid through plate entering at $x = \delta$ from a coolant reservoir at t_0: (b) $x/\delta = 1/2$, (c) $x/\delta = 1$.

● *Two-layer Plate.* Figure 45a–d gives the heated surface response $(k_1/q''\delta_1)(t_w - t_0)$ and the interface response $(t_i - t_0)/(t_w - t_0)$ versus $\mathrm{Fo}_1 = \alpha_1\theta/\delta_1^2$ for a thick two-layer plate, with the front layer $0 \leq x_1 \leq \delta_1$ being suddenly exposed at $x_1 = 0$ to a *constant* heat input q'' and in perfect contact with the rear layer $0 \leq x_2 \leq \delta_2$ insulated at $x_2 = \delta_2$ [44]. The parameters $(\delta_2/\delta_1)\sqrt{\alpha_1/\alpha_2}$ and $(k_2/k_1)\sqrt{\alpha_1/\alpha_2} = \sqrt{\rho_2 C_2 k_2/\rho_1 C_1 k_1}$ are the thickness and conductivity ratios for the two layers; values $(\delta_2/\delta_1)\sqrt{\alpha_1/\alpha_2} = 0$ and ∞ indicate, respectively, a single-layer plate $(\delta_2 = 0)$ and a two-layer semi-infinite solid $(\delta_2 \to \infty)$.*

Figure 46 gives the heated-surface and rear-layer responses for case of *constant* heat input and a thermally thin rear layer $(k_2 \to \infty)$ with $n = \rho_2 C_2\delta_2/\rho_1 C_1\delta_1$ as the parameter [27].

Figure 47a–c gives the heated-surface and rear-layer temperatures versus $2\pi\theta/D$ with the modified Fourier number $\mathrm{Fo}_1'/2\pi = \alpha_1 D/2\pi\delta_1^2$ as the parameter for the same insulated substructure system but exposed to a *circular* heat pulse $q'' = q''_{max}\sin^2 \pi\theta/D$ at $x_1 = 0$ [45] (the simpler case of no rear layer, $n = 0$, is given back in Fig. 42). The data for finite capacity ratios $n = \rho_2 C_2\delta_2/\rho_1 C_1\delta_1 = 0.1$, 1, and 10 are carried out to about two pulse durations, assuming adiabatic equilibration beyond $\theta = D$. Plate temperature rises $t - t_0$ are normalized by equilibrated values $t_{eqb} - t_0 = \delta_1 q''_{max}\mathrm{Fo}_1'/2$ $(n + 1)k_1$. As $k_1 \to \infty$, $\mathrm{Fo}_1' \to \infty$ and

$$\frac{t - t_0}{t_{eqb} - t_0} = \frac{1}{2\pi}\left(2\pi\frac{\theta}{D} - \sin 2\pi\frac{\theta}{D}\right)$$

Figure 47d shows rear-layer response $(x_1 = \delta_1)$ at the end of heating $(\theta = D)$ as a function of $\mathrm{Fo}_1'/2\pi$ with $n = 0$, 0.1, 0.5, 1, 5, 10, 50, and 100. Total equilibration times $2\pi\theta_{eqb}/D$ are given in Fig. 47e for the same values of n. Figure 48 shows the heat-sink capability of the two-layer system as compared to one layer when $n = 0$ (see Fig. 43).

● *Semi-infinite Solids.* Figure 49 gives the temperature response, temperature gradient, and the heating rate versus $(x/\delta)/2\sqrt{\mathrm{Fo}} = x/2\sqrt{\alpha\theta}$ for a semi-infinite solid $x \geq 0$ suddenly exposed to a *constant* surface heat input q''. The temperature history is [18]

$$\frac{k}{2q''\sqrt{\alpha\theta}}(t - t_0) = \frac{1}{\sqrt{\pi}}\exp\left(-\frac{x^2}{4\alpha\theta}\right) - \frac{x}{2\sqrt{\alpha\theta}}\,\mathrm{erfc}\left(\frac{x}{2\sqrt{\alpha\theta}}\right) = \mathrm{ierfc}\left(\frac{x}{2\sqrt{\alpha\theta}}\right) \qquad (60)$$

where $\mathrm{ierfc}\,u = I^1(\mathrm{erfc}\,u)$, the integral of complementary error function (Sec. 1.B.1, Fig. 12). The ratio of heat flux at any depth x to that at the surface is $q_x''/q'' = \mathrm{erfc}(x/2\sqrt{\alpha\theta})$.

A cylindrical hole of diameter d is buried in a semi-infinite solid (e.g., a thermocouple hole) with its axis normal to the surface and the closed end at a depth $x = l$. If d/l is small (line void), the surface temperature directly above the hole is

---

*A two-layer plate can be optimized with respect to choice of materials and thicknesses to accommodate higher values of heat input $q''\sqrt{D}$ than is possible by using separate equivalent weights of either material. This result is generally achieved if an outer layer of high melting temperature is backed by a material of higher conductivity and lower melting point. The quantitative aim is $\sqrt{\rho_2 C_2 k_2/\rho_1 C_1 k_1} > 1$ in combination with obtaining maximum allowable heated-surface and interface temperatures. Beryllium oxide over beryllium and molybdenum over copper are examples where, approximately, $\sqrt{\rho_2 C_2 k_2/\rho_1 C_1 k_1} = 2$ and $(t_d)_1/(t_d)_2 = 2$.

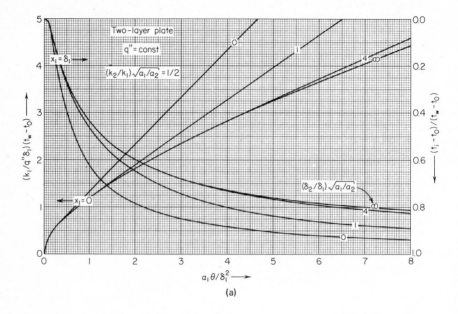

(a)

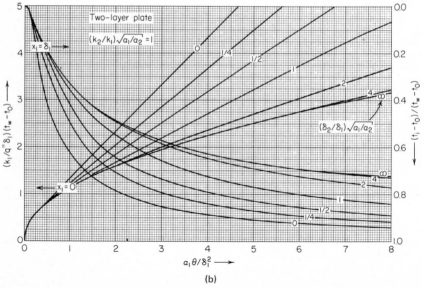

(b)

Fig. 45. Temperature response of thick plate $(0 \leq x_1 \leq \delta_1)$ exposed to constant heat input q'' at $x_1 = 0$ and in perfect contact at its rear face $x_1 = \delta_1$ with a thick plate $(0 \leq x_2 \leq \delta_2)$ insulated at $x_2 = \delta_2$: (a) $(k_2/k_1)\sqrt{\alpha_1/\alpha_2} = 1/2$, (b) $(k_2/k_1)\sqrt{\alpha_1/\alpha_2} = 1$.

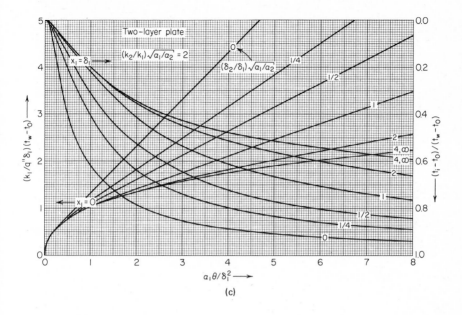

(c)

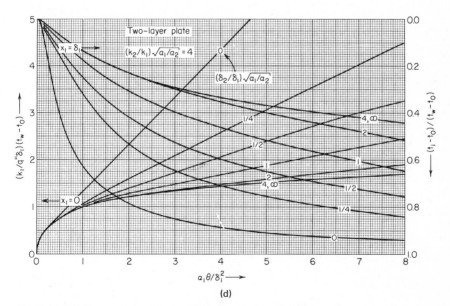

(d)

Fig. 45 (*Cont.*). Temperature response of thick plate $(0 \leq x_1 \leq \delta_1)$ exposed to constant heat input q'' at $x_1 = 0$ and in perfect contact at its rear face $x_1 = \delta_1$ with a thick plate $(0 \leq x_2 \leq \delta_2)$ insulated at $x_2 = \delta_2$: (c) $(k_2/k_1)\sqrt{\alpha_1/\alpha_2} = 2$, (d) $(k_2/k_1)\sqrt{\alpha_1/\alpha_2} = 4$.

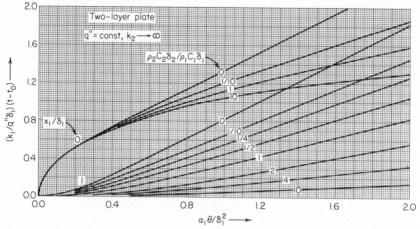

Fig. 46. Temperature response of thick plate ($0 \leq x_1 \leq \delta_1$) exposed to constant heat input q'' at $x_1 = 0$ and in perfect contact at its rear face $x_1 = \delta_1$ with a thin plate ($0 \leq x_2 \leq \delta_2$) insulated at $x_2 = \delta_2$.

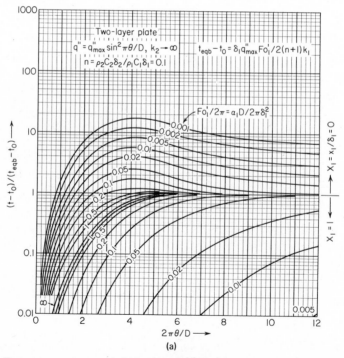

Fig. 47. Temperature response of thick plate ($0 \leq x_1 \leq \delta_1$) exposed to circular heat pulse $q''_{max} \sin^2 \pi\theta/D$ at $x_1 = 0$ and in perfect contact at its rear face $x_1 = \delta_1$ with a thin plate ($0 \leq x_2 \leq \delta_2$) insulated at $x_2 = \delta_2$: (a) $n = 0.1$.

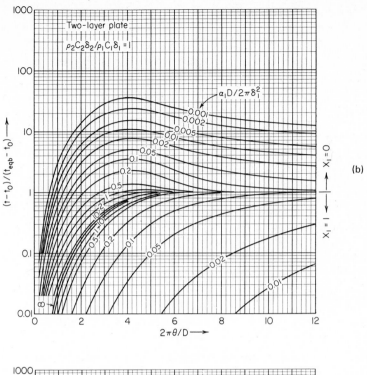

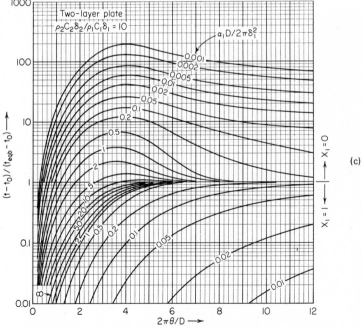

Fig. 47 (*Cont.*). Temperature response of thick plate ($0 \leq x_1 \leq \delta_1$) exposed to circular heat pulse $q''_{max} \sin^2 \pi\theta/D$ at $x_1 = 0$ and in perfect contact at its rear face $x_1 = \delta_1$ with a thin plate ($0 \leq x_2 \leq \delta_2$) insulated at $x_2 = \delta_2$: (b) $n = 1$, (c) $n = 10$.

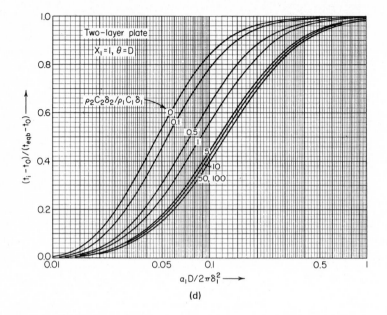

(d)

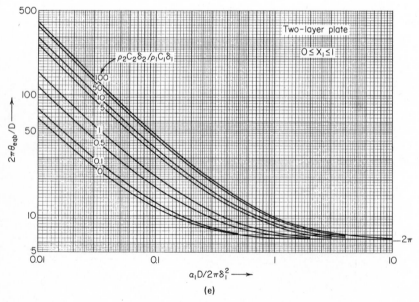

(e)

Fig. 47 (*Cont.*). Temperature response of thick plate ($0 \leq x_1 \leq \delta_1$) exposed to circular heat pulse $q''_{max} \sin^2 \pi\theta/D$ at $x_1 = 0$ and in perfect contact at its rear face $x_1 = \delta_1$ with a thin plate ($0 \leq x_2 \leq \delta_2$) insulated at $x_2 = \delta_2$: (d) $X_1 = 1$, $\theta = D$, (e) $0 \leq X_1 \leq 1$, $\theta = \theta_{eqb}$.

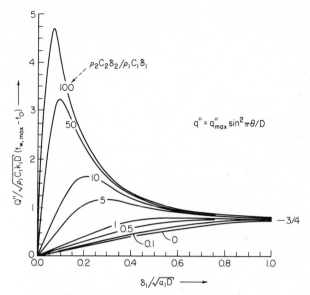

Fig. 48. Heat-sink performance of an insulated substructure.

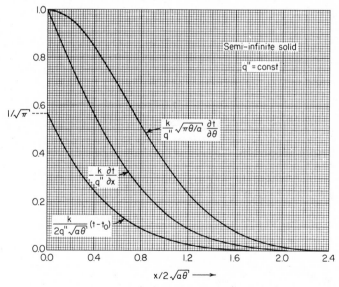

Fig. 49. Temperature response, temperature gradient, and heating rate in semi-infinite solid $(x \geq 0)$ after sudden exposure to constant surface heat input q''.

$$\frac{kl}{q''d^2}(t_{w,0} - t_{w,\infty}) = \frac{1}{8} \text{erfc}\left(\frac{1}{\sqrt{\text{Fo}_l}}\right) \tag{61}$$

where $\text{Fo}_l = \alpha\theta/l^2$ and $t_{w,\infty}$ is the surface temperature far removed from the hole given by Eq. (60) as $(k/q''l)(t_{w,\infty} - t_0) = 2\sqrt{\text{Fo}_l/\pi}$. The steady-state temperature *disturbance* is $(kl/q''d^2)(t_{w,0} - t_{w,\infty}) = 1/8$ with the surface distribution $(kl/q''d^2)(t_w - t_{w,\infty}) = 1/8\sqrt{(r/l)^2 + 1}$.

● *Transient Heat Input.* The temperature response $(kD/2Q''\sqrt{\alpha\theta})(t - t_0)$ versus $(x/\delta)/2\sqrt{\text{Fo}} = x/2\sqrt{\alpha\theta}$ with $\tau = \theta/D$ as the parameter is given in Fig. 50 for a semi-

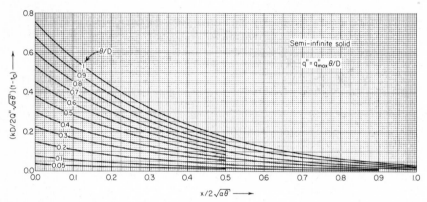

Fig. 50. Temperature response of semi-infinite solid ($x \geq 0$) after sudden exposure to linearly increasing surface heat input q''.

infinite solid $x \geq 0$ which is suddenly exposed to a surface heat input that increases *linearly* with time as $q'' = q''_{\max}\theta/D = 2(Q''/D)(\theta/D)$, where D is heating duration.

The surface temperature history of a semi-infinite solid $x \geq 0$ suddenly exposed to an arbitrary transient heat input is

$$\frac{\sqrt{\pi\rho Ck/D}}{q''_{\max}}(t_w - t_0) = \chi(\tau) \tag{62}$$

where $\tau = \theta/D$. The following tabulation gives the value of function $\chi(\tau)$ for various heat inputs:

a. *step input* ; $\dfrac{q''}{q''_{\max}} = 1$, $\chi = 2\sqrt{\tau}$ (63)

b. *step pulse* ; $\dfrac{q''}{q''_{\max}} = 1,\ 0 \leq \tau \leq \dfrac{1}{2}$, $\chi = 2\sqrt{\tau}$

$= 0,\qquad \tau > \dfrac{1}{2}$, $= 2\left(\sqrt{\tau} - \sqrt{\tau - \dfrac{1}{2}}\right)$ (64)

c. *linear input* ; $\dfrac{q''}{q''_{\max}} = \tau$, $\chi = \dfrac{4}{3}\tau^{3/2}$ (65)

d. *linear input*   ; $\dfrac{q''}{q''_{max}} = (1 - \tau)$, $\chi = 2\sqrt{\tau}\left(1 - \dfrac{2}{3}\tau\right)$ (66)

e. *power input*   ; $\dfrac{q''}{q''_{max}} = \tau^m$, $m > 0$, $\chi = \dfrac{2^{2m+1}(m!)^2}{(2m+1)!}\tau^{m+1/2}$ (67)

f. *power input*   ; $\dfrac{q''}{q''_{min}} = \tau^{-1/2}$, $\chi = \pi^*$ (68)

g. *exponential input*; $\dfrac{q''}{q''_{max}} = e^{-a(1-\tau)}$, $\chi = \sqrt{\dfrac{\pi}{a}}\, e^{-a(1-\tau)}\,\mathrm{erf}\,a\tau$ (69)

h. *triangular pulse*   ; $\dfrac{q''}{q''_{max}} = \dfrac{\tau}{a}$, $0 \le \tau \le a$, $\chi = \dfrac{4}{3a}\tau^{3/2}$

(70)

$= 1 - \dfrac{\tau - a}{1 - a}$, $= \dfrac{4}{3a}\left[\tau^{3/2} - \dfrac{(\tau - a)^{3/2}}{1 - a}\right]$

$a < \tau \le 1$

i. *circular pulse*   ; $\dfrac{q''}{q''_{max}} = \sin^2 \pi\tau$, $\chi = \dfrac{16}{15}\pi^2\tau^{5/2}\left[1 - \dfrac{(4\pi\tau)^2}{7\cdot 9}\right.$

(71)

$\left. + \dfrac{(4\pi\tau)^4}{7\cdot 9\cdot 11\cdot 13} \mp \cdots\right]$

The maximum surface temperature of a semi-infinite solid exposed to circular heat pulse is $t_{w,max} - t_0 = (2/3)\,q''_{max}\sqrt{D/\rho Ck}$; this occurs at $\theta/D = 2/3$. According to Fig. 42a, this is also the time at which the peak surface temperature occurs for an insulated plate $0 \le x \le \delta$ with circular pulse heating if $\alpha D/2\pi\delta^2 \le 0.1$.

• *Two-Layer Semi-Infinite Solid.* The temperature response $(k_2/2q''\sqrt{\alpha_2\theta})(t - t_0)$ versus $(x_2/\delta)/2\sqrt{Fo_2} = x_2/2\sqrt{\alpha_2\theta}$ with $m(x_2/\delta) = (\rho_2 C_2/\rho_1 C_1)x_2/\delta$ as the parameter is given in Fig. 51 for a thin plate $0 \le x_1 \le \delta$ backed (in perfect contact) by a semi-infinite solid $x_2 \ge 0$ after a sudden exposure to *constant* heat input q'' at $x_1 = 0$ [18]. The single curve (top scale) shows the response of the capacitive layer $0 \le x_1 \le \delta$ versus $m\sqrt{Fo_2} = (\rho_2 C_2/\rho_1 C_1)\sqrt{\alpha_2\theta}/\delta^2$ given by

$$\dfrac{k_2}{2q''\sqrt{\alpha_2\theta}}(t_1 - t_0) = \dfrac{1}{2m\sqrt{Fo_2}}\left[e^{m^2 Fo_2}\,\mathrm{erfc}(m\sqrt{Fo_2}) - 1\right] + \dfrac{1}{\sqrt{\pi}}$$ (72)

This is also the interface temperature t_i at $x_2 = 0$. The heat conducted across the interface into the back layer $x_2 \ge 0$ is

$$\dfrac{q''_i}{q''} = 1 - e^{m^2 Fo_2}\,\mathrm{erfc}(m\sqrt{Fo_2})$$ (73)

*Replace q''_{max} by q''_{min} in Eq. (62).

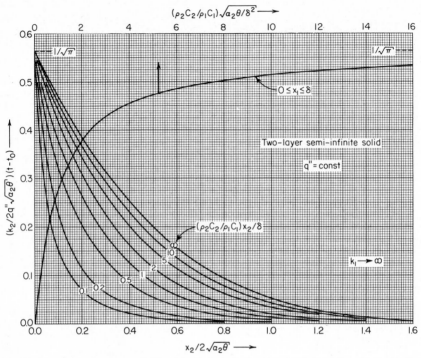

Fig. 51. Temperature response of thin plate ($0 \leq x_1 \leq \delta$) and semi-infinite backup ($x_2 \geq 0$) after sudden exposure to constant heat input q'' at $x_1 = 0$.

The function $e^{u^2} \text{erfc } u$ (reoccurring in heat conduction) is tabulated below.

u	$e^{u^2}\text{erfc } u$	u	$e^{u^2}\text{erfc } u$	u	$e^{u^2}\text{erfc } u$	u	$e^{u^2}\text{erfc } u$
0	1	0.4	0.6708	1.5	0.3216	5	0.1107
0.05	0.9460	0.5	0.6157	2	0.2554	6	0.09278
0.1	0.8965	0.6	0.5678	2.5	0.2108	7	0.07980
0.2	0.8090	0.8	0.4891	3	0.1790	8	0.06998
0.3	0.7346	1	0.4276	4	0.1370	∞	0

• *Diathermanous Solid.* A semi-infinite solid $x \geq 0$ is suddenly exposed to a *constant* radiant heat flux q''. If the wall material is partially transparent to radiation, surface flux is absorbed exponentially within the solid according to Eq. (31) as $q''_x = q'' e^{-\mu x}$. The temperature response of the *diathermanous* solid in the absence of surface heat loss is [46] *

$$\frac{k}{2q''\sqrt{\alpha\theta}}(t - t_0) = \text{ierfc}\left(\frac{x}{2\sqrt{\alpha\theta}}\right) - \frac{\exp[-2(\mu\sqrt{\alpha\theta})(x/2\sqrt{\alpha\theta})]}{2\mu\sqrt{\alpha\theta}}[1 - \qquad (74)$$

*This solution also holds for an insulated semi-infinite solid with internal heat generation distributed exponentially with depth x.

$$\exp{(\mu \sqrt{\alpha\theta})^2}] - \frac{\exp{(\mu\sqrt{\alpha\theta})^2}}{4\mu\sqrt{\alpha\theta}} \left\{ \exp\left[-2(\mu\sqrt{\alpha\theta})\frac{x}{2\sqrt{\alpha\theta}}\right] \operatorname{erfc}\left(\frac{x}{2\sqrt{\alpha\theta}} - \mu\sqrt{\alpha\theta}\right) \right.$$

$$\left. - \exp\left[2(\mu\sqrt{\alpha\theta})\frac{x}{2\sqrt{\alpha\theta}}\right] \operatorname{erfc}\left(\frac{x}{2\sqrt{\alpha\theta}} + \mu\sqrt{\alpha\theta}\right) \right\}$$

$$(74)$$
(Cont.)

The surface temperature is

$$\frac{k}{2q''\sqrt{\alpha\theta}}(t_w - t_0) = 0.5642 - \frac{1 - \exp[(\mu\sqrt{\alpha\theta})^2]\operatorname{erfc}(\mu\sqrt{\alpha\theta})}{2\mu\sqrt{\alpha\theta}}$$

The effect of diathermancy is indicated in Fig. 52 as a suppression of surface temperatures and an increase in internal temperatures as absorption coefficient μ decreases. For perfectly opaque solids, $\mu = \infty$ and Eq. (74) reverts to Eq. (60).

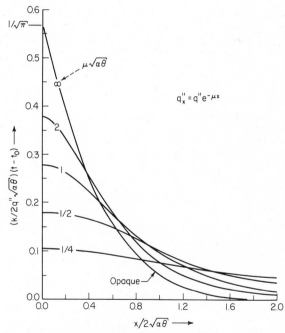

Fig. 52. Instantaneous temperature distributions in semi-infinite diathermanous solid.

- *Stationary Heat Sources.* An infinite solid $-\infty < (x, y, z) < \infty$ may contain permanently located heat sources suddenly activated at $\theta = 0$. Temperature histories for *point source* Q located at (x_0, y_0, z_0), *line source* Q' located parallel to z (Fig. 6a) at (x_0, y_0), and *plane source* Q'' located parallel to y-z at (x_0) are:

$$\frac{\rho C(\pi\alpha\theta)^{3/2}}{Q}(t - t_0) = \frac{1}{8}\exp\left[-\frac{(x - x_0)^2 + (y - y_0)^2 + (z - z_0)^2}{4\alpha\theta}\right] \qquad (75)$$

$$\frac{\rho C\,(\pi\alpha\theta)}{Q'}\,(t - t_0) = \frac{1}{4}\exp\left[-\frac{(x - x_0)^2 + (y - y_0)^2}{4\alpha\theta}\right] \tag{76}$$

$$\frac{\rho C\,(\pi\alpha\theta)^{1/2}}{Q''}\,(t - t_0) = \frac{1}{2}\exp\left[-\frac{(x - x_0)^2}{4\alpha\theta}\right] \tag{77}$$

The form of the point-source function in Eq. (75) is shown in Sec. 1.B.1, Fig. 15.
  ● *Moving Heat Sources.* The temperature response of an infinite body containing a *point source* q moving at a constant velocity v in x direction (Sec. 1.B.5, Fig. 30) is

$$\frac{kr}{q}\,(t - t_0) = \frac{1}{4\pi}\exp\left[-\frac{v}{2\alpha}(\xi + r)\right] \tag{78}$$

where $\xi = x - v\theta$ and $r = \sqrt{\xi^2 + y^2 + z^2}$. A thin plate $0 \le z \le \delta$ contains a *line source* q' oriented parallel to z (through the plate) and moving in x direction at constant velocity v. If the plate is insulated at one face and cooled at the opposite face by convection to an ambient at t_0, its temperature response is

$$\frac{k}{q'}\,(t - t_0) = \frac{1}{2\pi}\left[\exp\left(-\frac{v}{2\alpha}\right)\right]K_0\left[\sqrt{\left(\frac{v}{2\alpha}\right)^2 + \frac{h}{k\delta}}\,\sqrt{\xi^2 + y^2}\right] \tag{79}$$

where K_0 is the zero-order modified Bessel function of the second kind (Sec. 1.B.3). The case $h = 0$ corresponds to a line source moving in the plate with both faces insulated (or in an infinite solid). If the plate contains a *plane source* q'' moving in x direction, then

$$\frac{k}{q''}\sqrt{\left(\frac{v}{2\alpha}\right)^2 + \frac{h}{k\delta}}\,(t - t_0) = \frac{1}{2}\exp\left\{\left[\sqrt{\left(\frac{v}{2\alpha}\right)^2 + \frac{h}{k\delta}} - \frac{v}{2\alpha}\right]\xi\right\} \quad;\quad \xi < 0$$

$$= \frac{1}{2}\exp\left\{-\left[\sqrt{\left(\frac{v}{2\alpha}\right)^2 + \frac{h}{k\delta}} + \frac{v}{2\alpha}\right]\xi\right\} \quad;\quad \xi > 0 \tag{80}$$

The maximum temperature occurs at $\xi = 0$. As before, $h = 0$ represents a completely insulated plate (or plane source in an infinite solid); in this case $(\rho Cv/q'')(t - t_0) = 1$ behind the source and $(\rho Cv/q'')(t - t_0) = \exp[-(v/\alpha)\xi]$ ahead of the source.
  c. Phase Changes. Figure 53 shows some cases of a semi-infinite body $x \ge 0$ undergoing internal or surface phase changes owing to cooling below solidification temperature or heating above decomposition (or melting) temperature of the body material. Case (a) is an example of liquid *solidification* (freezing) under action of surface heat removal (e.g., ice production or solidification of a casting). Case (b) represents thermal/chemical *decomposition* of a solid without removal of degraded material 1 from remaining virgin material 2 (e.g., charring of wood or reinforced plastics). Case (c) represents *ablation* of solids where products of decomposition are removed on formation (e.g., melting glass or subliming teflon). In each case shown, the phase change is

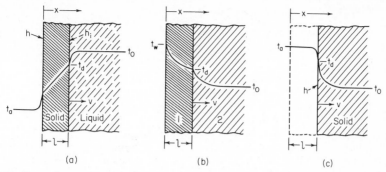

Fig. 53. Phase changes in semi-infinite solid: (a) solidification, (b) decomposition, (c) ablation.

assumed to occur at fixed decomposition temperature t_d; in addition, the phase-change reaction is considered to go to completion at one and only one t_d. The problem is to determine the amount of material l solidified, decomposed, melted, sublimed, or otherwise consumed.

● *Solidification.* Case (a) in Fig. 53 considers uniform convective cooling and accompanying *solidification* of a *plane* semi-infinite liquid body $x \geq 0$ initially at t_0. The thermal capacity of the solidified layer is ignored. With the added assumption that heat transfer between liquid and solid occurs by convection with a constant interface conductance h_i, the solidification depth l is given implicitly by [47]

$$\Theta = \frac{1}{N^2} \ln\left[\frac{N-1}{N(L+1)-1}\right] - \frac{L}{N} \tag{81}$$

where $L = (h/k)l$, $\Theta = (h^2/\rho Hk)(t_d - t_a)\theta$, and $N = (h_i/h)(t_0 - t_d)/(t_d - t_a)$. Here ρ and k are properties of the solid and H is latent heat of the phase change (Table 2). The mass solidified is ρl and the heat liberated at the liquid/solid interface is $Q'' = \rho l H$. The depth/time relationship of Eq. (81) is shown in Fig. 54 as L versus Θ with N as the parameter. The free-surface wall temperature is calculated from L as $(t_w - t_a)/(t_d - t_a) = 1/(L+1)$. If the liquid is initially at solidification temperature ($t_0 = t_d$), then $N = 0$ and the solution is explicit as $L = \sqrt{2\Theta + 1} - 1$.

Solidification can also occur in or on cylindrical and spherical bodies. If $t_0 = t_d$, the solidification thickness $l = r_2 - r$ in a thin *tube* of inner radius r_2 initially filled with liquid at t_d and cooled by an external ambient at t_a (ignoring tube resistance) is

$$\Theta = \frac{Bi^2}{2}\left[\left(\frac{1}{Bi} + \frac{1}{2}\right)(1 - R^2) - R^2 \ln\frac{1}{R}\right] \quad ; \quad R \leq 1 \tag{82}$$

where $Bi = hr_2/k$ and $R = r/r_2$. The tube contents are completely frozen ($R = 0$) when $\Theta = (Bi^2/2)(1/Bi + 1/2)$. Solidification *on* a tube surrounded by liquid at t_d and cooled internally by ambient at t_a is

$$\Theta = \frac{Bi^2}{2}\left[\left(\frac{1}{Bi} - \frac{1}{2}\right)(R^2 - 1) + R^2 \ln R\right] \quad ; \quad R \geq 1 \tag{83}$$

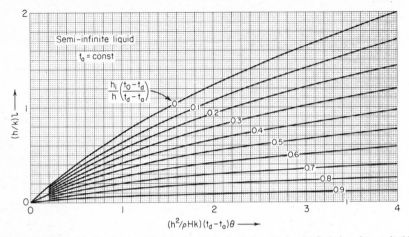

Fig. 54. Solidification depth in semi-infinite liquid ($x \geq 0$) for which solidified phase has negligible thermal capacity and is exposed to convective environments t_a at free surface $x = 0$ and t_0 at liquid/solid interface $x = l$.

Similar results for solidification *in* or *on* a thin *spherical shell* are

$$\Theta = \frac{Bi^2}{3}\left[\left(\frac{1}{Bi} - 1\right)(1 - R^3) + \frac{3}{2}(1 - R^2)\right] \quad ; \quad R \leq 1 \qquad (84)$$

$$\Theta = \frac{Bi^2}{3}\left[\left(\frac{1}{Bi} + 1\right)(R^3 - 1) - \frac{3}{2}(R^2 - 1)\right] \quad ; \quad R \geq 1 \qquad (85)$$

Case (a) is extended by assuming that $t_0 = t_d$ as well as taking into account the finite heat capacity of the solidified phase. Figure 55a gives the free-surface temperature response t_w/t_d versus $(a^2/\rho Ck)\theta$ with H/Ct_d as the parameter for a solidifying semi-infinite liquid initially at t_d and with surface *convective* heat removal $q'' = at_w$ with a constant [48]. Figure 55b gives the history of the solidification depth $(a/k)l$. Figure 55c, d gives similar results with surface *radiative* heat removal $q'' = at_w^4$.

• *Decomposition.* Case (b) in Fig. 53 considers *decomposition* (or melting) of a semi-infinite solid $x \geq 0$ after a sudden change in surface temperature from t_0 to $t_w(>t_d)$. The properties on each side of the moving decomposition plane $x = l$ are constant but different, being $\rho_1 C_1$ and k_1 for the char layer and $\rho_2 C_2$ and k_2 for the virgin material. If char and virgin materials remain in perfect contact, their temperature response is*

$$\frac{t - t_d}{t_w - t_d} = 1 - \frac{erf(x/2\sqrt{\alpha_1\theta})}{erf(a/2\sqrt{\alpha_1})} \quad ; \quad 0 \leq x \leq l$$

$$\frac{t - t_0}{t_d - t_0} = \frac{erfc(x/2\sqrt{\alpha_2\theta})}{erfc(a/2\sqrt{\alpha_2})} \quad ; \quad x \geq l \qquad (86)$$

Neumann's solution [18]. Equation (86) can be modified to account for swelling of solid during heating (cork and rubber-base materials) or cooling (freezing water); see Refs. 18 and 31.

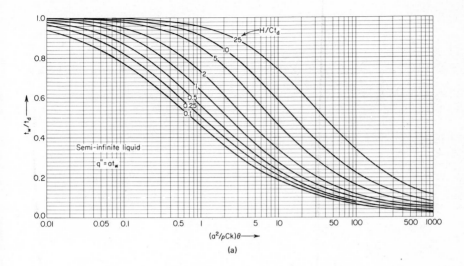

(a)

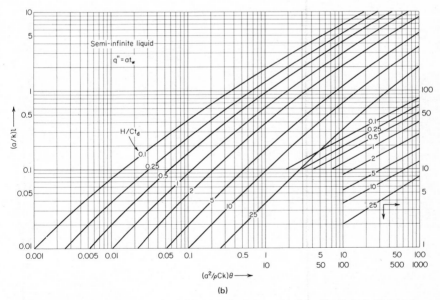

(b)

Fig. 55. Temperature response and solidification depth in semi-infinite liquid $(x \geq 0)$ initially at t_d: (a) surface temperature for convective cooling $q'' = at_w$ at $x = 0$, (b) solidification depth for convective cooling $q'' = at_w$ at $x = 0$.

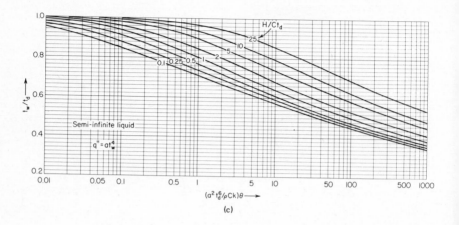

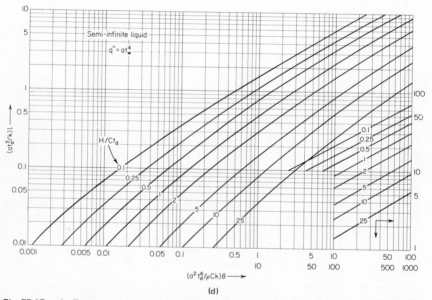

Fig. 55 (*Cont.*). Temperature response and solidification depth in semi-infinite liquid ($x \geq 0$) initially at t_d : (c) surface temperature for radiative cooling $q'' = at_w^4$ at $x = 0$, (d) solidification depth for radiative cooling $q'' = at_w^4$ at $x = 0$.

where constant a is evaluated (e.g., graphically) from the implicit relation

$$\frac{\exp(-a^2/4\alpha_1)}{\text{erf}(a/2\sqrt{\alpha_1})} - \sqrt{\frac{\rho_2 C_2 k_2}{\rho_1 C_1 k_1}} \left(\frac{t_d - t_0}{t_w - t_d}\right) \frac{\exp(-a^2/4\alpha_2)}{\text{erfc}(a/2\sqrt{\alpha_2})} = \frac{\sqrt{\pi \rho_1/C_1 k_1}\ H}{2(t_w - t_d)}\ a \qquad (87)$$

Char depth l and velocity v of char front are calculated from a as*

$$l = a\sqrt{\theta} \quad , \quad v = \frac{a}{2\sqrt{\theta}} \qquad (88)$$

Figure 56a gives the temperature response $(t - t_d)/(t_w - t_d)$ versus $(x/\delta)/2\sqrt{\text{Fo}_1} = x/2\sqrt{\alpha_1}\theta$ with $a/2\sqrt{\alpha_1}$ as the parameter for char layer $0 \leq x \leq l$. Similar data for virgin

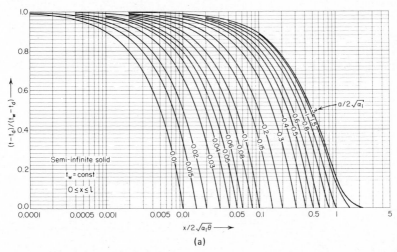

Fig. 56. Temperature response of decomposing semi-infinite solid $(x \geq 0)$ after sudden change in surface temperature from t_0 to t_w $(> t_d)$: (a) $0 \leq x \leq l$.

layer $x \geq l$ are given in Fig. 56b. Two special cases occur if initial or surface temperatures equal t_d:

$$t_0 = t_d \quad ; \quad \frac{a}{2\sqrt{\alpha_1}} \exp\left(\frac{a^2}{4\alpha_1}\right) \text{erf}\left(\frac{a}{2\sqrt{\alpha_1}}\right) = \frac{C_1(t_d - t_w)}{\sqrt{\pi}\,H}$$

$$t_w = t_d \quad ; \quad \frac{a}{2\sqrt{\alpha_2}} \exp\left(\frac{a^2}{4\alpha_2}\right) \text{erfc}\left(\frac{a}{2\sqrt{\alpha_2}}\right) = \frac{C_2(t_d - t_0)}{\sqrt{\pi}\,H} \qquad (89)$$

Values of a for these simpler cases are given in Fig. 56c.

Case (b) can be extended [18] to include *solidification* of semi-infinite liquid against a containing semi-infinite solid (e.g., freezing of a casting in a mold with thick, heat-

---

*Equations (86), (87), and (88) also apply to solidification of a liquid $x \geq 0$ $(t_w < t_d < t_0)$ providing subscripts 1 and 2 on thermal properties and temperature differences $t_d - t_0$ and $t_w - t_d$ are interchanged.

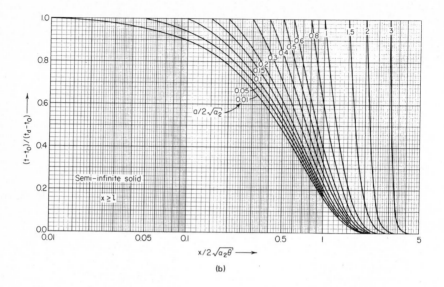

(b)

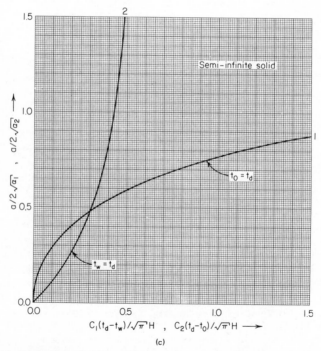

(c)

Fig. 56 (*Cont.*). Temperature response of decomposing semi-infinite solid ($x \geq 0$) after sudden change in surface temperature from t_0 to $t_w (> t_d)$: (b) $x \geq l$, (c) values of a for decomposition depth $l = a\sqrt{\theta}$ when $t_0 = t_d$ or $t_w = t_d$.

conducting walls). If liquid, initially at t_0, is poured against the container (melting temperature $> t_0$) initially at t_a (Fig. 57)

$$\frac{t - t_a}{t_d - t_a} = \frac{1 + \mathrm{erf}(x/2\sqrt{\alpha_w \theta})}{1 + \sqrt{\rho_w C_w k_w / \rho_1 C_1 k_1} \ \mathrm{erf}(a/2\sqrt{\alpha_1})} \quad ; \quad -x \geq 0$$

$$\frac{t - t_a}{t_d - t_a} = \frac{1 + \sqrt{\rho_w C_w k_w / \rho_1 C_1 k_1} \ \mathrm{erf}(x/2\sqrt{\alpha_1 \theta})}{1 + \sqrt{\rho_w C_w k_w / \rho_1 C_1 k_1} \ \mathrm{erf}(a/2\sqrt{\alpha_1})} \quad ; \quad 0 \leq x \leq l \qquad (90)$$

$$\frac{t - t_d}{t_0 - t_d} = 1 - \frac{\mathrm{erfc}\,[x/2\sqrt{\alpha_2}\,\theta + a(\rho_1/\rho_2 - 1)/2\sqrt{\alpha_2}]}{\mathrm{erfc}\,[a(\rho_1/\rho_2)/2\sqrt{\alpha_2}]} \quad ; \quad x \geq l$$

where a is calculated from

$$\frac{\exp(-a^2/4\alpha_1)}{\sqrt{\rho_1 C_1 k_1 / \rho_w C_w k_w} + \mathrm{erf}(a/2\sqrt{\alpha_1})} - \sqrt{\frac{\rho_2 C_2 k_2}{\rho_1 C_1 k_1}} \left(\frac{t_0 - t_d}{t_d - t_a}\right) \frac{\exp[-a^2(\rho_1/\rho_2)^2/4\alpha_2]}{\mathrm{erfc}\,[a(\rho_1/\rho_2)/2\sqrt{\alpha_2}]}$$

$$= \frac{\sqrt{\pi \rho_1 / k_1 C_1} \ H}{2(t_d - t_a)} a$$

$$(91)$$

and $l = a\sqrt{\theta}$. The initial liquid/solid interface w-2 will not chill sufficiently to solidify

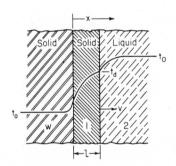

Fig. 57. Solidification in a mold.

liquid phase [compare Eq. (91) for $a = 0$ with Eq. (56) for the interace response] unless the initial liquid temperature is $t_0 < t_d + (t_d - t_a)\sqrt{\rho_w C_w k_w / \rho_2 C_2 k_2}$.

- *Ablation.* Case (c) in Fig. 53 considers heating of a semi-infinite solid $x \geq 0$ to surface temperatures in excess of t_d and continuous *ablation* (e.g., removal by aerodynamic shear forces) of the thermally degraded or melted material as it forms. Figure 58*a*–*c* gives the ablation depth $l/v_{ss}\theta_d$ versus time $\theta/\theta_d - 1$ with $(t_a - t_d)/(t_d - t_0)$ as the parameter and $(h/\rho H)(t_a - t_d)\sqrt{\theta_d/\alpha} = 1/2, 1$, and 2 for *convective* heating $(t_a > t_d)$ at surface $x = 0$ [49]. Quantity θ_d is the time required for the surface to attain decomposition temperature t_d; this preablation time can be obtained from Fig. 28 or calculated from the implicit equation

$$\frac{t_d - t_0}{t_a - t_0} = 1 - \exp\left[\left(\frac{h}{k}\right)^2 \alpha \theta_d\right] \mathrm{erfc}\left(\frac{h}{k} \ \sqrt{\alpha \theta_d}\right) \qquad (92)$$

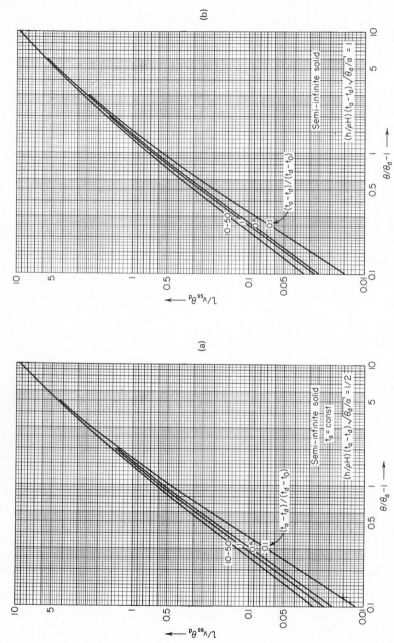

Fig. 58. Ablation depth of semi-infinite solid ($x \geq 0$) after sudden exposure to convective environment t_a ($> t_d$): (a) $(h/\rho H)(t_a - t_d)\sqrt{\theta_d/\alpha} = 1/2$, (b) $(h/\rho H)(t_a - t_d)\sqrt{\theta_d/\alpha} = 1$.

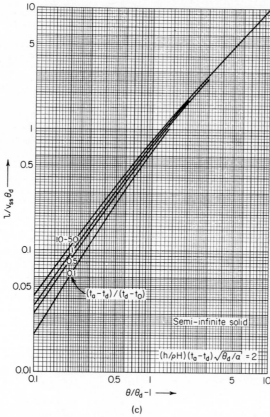

Fig. 58 (*Cont.*). Ablation depth of semi-infinite solid $(x \geq 0)$ after sudden exposure to convective environment $t_a (> t_d)$: (c) $(h/\rho H)(t_a - t_d)\sqrt{\theta_d/\alpha} = 2$.

At some time in excess of θ_d, ablation velocity v is constant and the surface recedes at a constant rate. This quasi-steady ablation velocity is

$$v_{ss} = \frac{h(t_a - t_d)}{\rho [C(t_d - t_0) + H]} \tag{93}$$

and the associated quasi-steady temperature distribution in the surviving solid is

$$\frac{t - t_0}{t_d - t_0} = \exp\left[-\frac{v_{ss}}{\alpha}(x - l)\right] \tag{94}$$

The quasi-steady ablation heat absorption is $q''_{ss} = \rho H v_{ss}$ and the total sensible heat in the remaining solid is $Q''_{ss} = k(t_d - t_0)/v_{ss}$.

Case (c) can be solved for boundary condition of *constant* surface heat input q''. Figure 59 gives ablation depth $l/v_{ss}\theta_d$ versus time $\theta/\theta_d - 1$ with $\sqrt{\pi} C(t_d - t_0)/2H$ as the parameter [50]. The preablation time is, from Eq. (63),

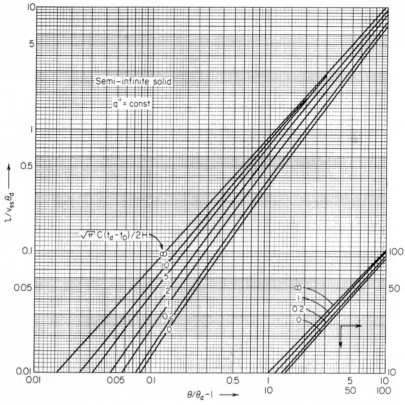

Fig. 59. Ablation depth of semi-infinite solid $(x \geq 0)$ after sudden exposure to constant surface heat input q''.

$$\theta_d = \frac{\pi}{4} \rho C k \left(\frac{t_d - t_0}{q''}\right)^2 \tag{95}$$

The quasi-steady ablation velocity is

$$v_{ss} = \frac{q''}{\rho [C(t_d - t_0) + H]} \tag{96}$$

and the quasi-steady temperature distribution in the surviving solid is given by Eq. (94). If the specific heat and conductivity of the solid vary linearly as $C = C_0(1 + b_1 t)$ and $k = k_0(1 + b_2 t)$, then quasi-steady ablation velocity, temperature distribution, and total sensible heat in the remaining solid are

$$v_{ss} = \frac{q''}{\rho \left\{ C_0 (t_d - t_0) [(b_1/2)(t_d - t_0) + 1] + H \right\}}$$

$$\ln\left(\frac{t - t_0}{t_d - t_0}\right) + \left(2\frac{b_2}{b_1} - 1\right)\ln\left[\frac{(b_1/2)(t - t_0) + 1}{(b_1/2)(t_d - t_0) + 1}\right] = -\frac{v_{ss}}{\alpha_0}(x - l)$$

$$Q'' = \left[\frac{k_0}{v_{ss}}(t_d - t_0)\right]\left[\frac{b_2}{2}(t_d - t_0) + 1\right]$$

The quantity $H^* = q''/\rho v_{ss}$ is termed *heat of ablation*.

The general problem of predicting material ablation in a hyperthermal environment requires defining a number of complex material-degradation processes and material/environment interactions (Fig. 4). These include thermal decomposition in depth (rather than only at the surface as in Fig. 53c), production of a variety of decomposition gas species, percolation of these gases through a solid-phase surface char layer, and chemical interaction between decomposition gases and the porous char near and at the char surface. Other surface effects include nonequilibrium multicomponent mass transfer, combustion, char reradiation, and char consumption. The latter may occur by char sublimation (due to surface temperature), gaseous char erosion (due to surface shear forces), char oxidation (due to oxygen diffusion in boundary layer), and intermittent char spallation (due to a combination of vibration associated with external and boundary-layer noise, internal thermal stresses associated with char temperature gradient, and internal mechanical forces associated with decomposition-gas pressure gradient).

The above problem requires a variety of detailed thermal-, chemical-, and mechanical-property data, and elaborate computer programs for generating coupled finite-difference solutions of the temperature and ablation histories [13]. Various approximate analyses are available, however, for predicting general behavior during quasi-steady ablation of materials that degrade in depth. A simplified quasi-steady solution considering in-depth reactions [e.g., Eq. (15)] but no internal transpiration or property changes in the decomposition layer is given in Ref. [51]. The case of simple decomposition with char-layer formation is treated in Ref. [52]. The in-depth reaction coupled with char-layer formation and intermittent char spallation is considered in Ref. [53].

Figure 60 shows quasi-steady erosion data on a variety of thermal-protection materials for exposure in air-arc test facilities generating plasmas of enthalpy 500–10,000 Btu/lb. The independent variable $q''\sqrt{p}$ characterizes the environment in terms of cold-wall heat rate and model stagnation pressure, and the dependent variable ρv_{ss} represents the mass ablation rate in terms of virgin-material density and steady-state rate of surface recession. Heat-sink materials *tantalum* (a) and *tungsten* (b) melt at fixed temperatures and thus suffer accelerated erosion ρv_{ss} over narrow ranges of the environment parameter $q''\sqrt{p}$. All of the remaining materials show a continuous dependence on $q''\sqrt{p}$, characteristic of decomposing materials that are consumed by some combination of melting, charring, sublimation, oxidation, and spallation.

Teflon (c), a low-pressure sublimer, derives its heat protection by solid/gas phase change and accompanying surface blowing. Since it does not decompose in depth to form an insulating decomposition layer, its mass ablation is large. *Asbestos/phenolic* (d) is a reinforced plastic that decomposes in depth but produces a weak char. *Cork* (e) and the *silicone elastomers* (f) and (g) tend to swell at low $q''\sqrt{p}$ and generate weak char layers that spall in elevated environments. The reinforcement in *nylon/phenolic* (h) completely volatilizes and erosion increases as a uniform power of $q''\sqrt{p}$. *Quartz/phenolic* (i), by contrast, forms a liquid layer due to reinforcement melting. A pronounced increase in erosion occurs from runoff of this melt layer under the action of high surface shear at elevated $q''\sqrt{p}$. The remaining four materials are carbon-based. *Graphite/phenolic* (j) experiences low erosion due to high char conductivity that promotes lower surface temperature (reduced surface oxidation) and deeper pyrolysis (increased blowing). *Carbon/phenolic* (k) produces a structurally stable char of somewhat lower conductivity. Although its chemical consumption is correspondingly higher than graphite/phenolic, it can be used at an elevated $q''\sqrt{p}$ where the latter suffers excessive mechanical erosion. *Carbon/carbon* (l) represents a composite of carbon fiber reinforcement with internal carbon deposits as binder. This material approximates the ablative performance of homogenous graphite while retaining the thermal-shock resistance of composited carbon/phenolic. Pure crystalline *graphite* (m) erodes least (except at a low $q''\sqrt{p}$ where convective blocking from blowing is minimal) because its high density minimizes both chemical consumption and surface spallation. However, in contrast to reinforced materials, graphites are brittle and therefore susceptible to large-scale thermal-strain failure over an extended range of $q''\sqrt{p}$.

The shaded areas in Fig. 60 form a corridor between which all carbonaceous ablators should perform if mechanical erosion effects are absent. As a minimum, these materials should suffer more ablation than that caused by rate-controlled carbon sublimation alone. As a maximum, they should

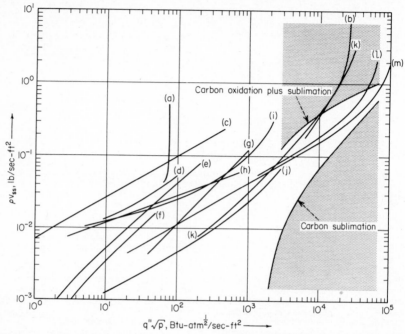

$q'' \sqrt{p}$, Btu-atm$^{\frac{1}{2}}$/sec-ft^2 ⟶

Fig. 60. Steady-state ablation measurements in arc-heated air environments: (a) tantalum (1040 lb/ft^3), (b) tungsten (1210 lb/ft^3), (c) teflon (135 lb/ft^3), (d) asbestos/phenolic (85-115 lb/ft^3), (e) cork (30 lb/ft^3), (f) silicone elastomer (30 lb/ft^3), (g) silicone elastomer (75-105 lb/ft^3), (h) nylon/phenolic (75 lb/ft^3), (i) quartz/phenolic (100 lb/ft^3), (j) graphite/phenolic (80 lb/ft^3), (k) carbon/phenolic (90 lb/ft^3), (l) carbon/carbon (85-100 lb/ft^3), (m) graphite (100 lb/ft^3).

erode less than that caused by combined carbon oxidation and thermochemical-equilibrium sublimation. This upper bound can be exceeded by superimposing other surface-recession mechanisms such as mechanical erosion. Another mechanism, evident with homogeneous carbon at high pressures, occurs when the triple point is exceeded and accelerated erosion results from combined carbon melting and carbon liquid vaporization.

2. Steady Heat Conduction

Steady-state heat conduction is characterized by constant heat flow and a stationary temperature pattern within the conducting body.

a. One-dimensional Systems. As in transient heat flow (Sec. 3.B.1), a steady system is treated as thermally thin or thick depending on whether surface or internal resistance dominates.

Thin Wall. The temperature gradients both through and along a wall are negligible providing the wall is of high relative internal conductance and is exposed to uniform surface environments.

• *Multiple Radiation Environments.* The steady temperature of a thin generating plate simultaneously exposed at either or both faces to n constant-temperature radiative source and/or sink environments at $t_{s,n}$ is

$$t = \left[\frac{\sum_{1}^{n} \left(\mathcal{F} t_s^4 \right)_n + q''' \delta/\sigma}{\sum_{1}^{n} (\mathcal{F})_n} \right]^{1/4} \tag{97}$$

For a single radiative environment at one face (other face insulated), $n = 1$ and $t = (t_s^4 + q''\delta/\sigma\, \mathcal{F})^{1/4}$. If the plate is exposed to *constant* heat input q'' at one face and reradiates to multiple sink environments at either or both faces, replace $q'''\delta$ in Eq. (97) by $q'''\delta + q''$.

- *Channel Radiation.* Two thin parallel plates with surface emissivities ϵ_1 and ϵ_2 (Table 4) receive *constant* but different heat inputs q_1'' and q_2'' at their outer surfaces (Fig. 61a).

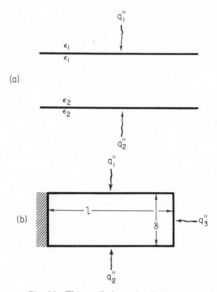

Fig. 61. Thin-wall channel radiation.

Considering inner-surface radiation exchange between plates (across nonabsorbing medium) and outer-surface reradiation to space ($t_s = 0°R$), the steady plate temperatures are [54]

$$\frac{t_1}{(q_2''/\sigma\epsilon_2)^{1/4}} = \left[\frac{(\epsilon^*/\epsilon_1)(q_1''/q_2'' + 1) + (\epsilon_2/\epsilon_1)(q_1''/q_2'')}{(\epsilon^*/\epsilon_1 + 1)(\epsilon^*/\epsilon_2 + 1) - (\epsilon^*)^2/\epsilon_1\epsilon_2}\right]^{1/4}$$

$$\frac{t_2}{(q_2''/\sigma\epsilon_2)^{1/4}} = \left[\frac{(\epsilon^*/\epsilon_1)(q_1''/q_2'' + 1) + 1}{(\epsilon^*/\epsilon_1 + 1)(\epsilon^*/\epsilon_2 + 1) - (\epsilon^*)^2/\epsilon_1\epsilon_2}\right]^{1/4} \tag{98}$$

where $\epsilon^* = 1/(1/\epsilon_1 + 1/\epsilon_2 - 1)$. If the space between the plates is filled with a solid of perfect conductance ($k \to \infty$) and $\epsilon_1 = \epsilon_2 = \epsilon$, then $t_{1,2}/(q_2''/\sigma\epsilon)^{1/4} = [(q_1''/q_2'' + 1)/2]^{1/4}$. If the filler is a perfect insulator ($k = 0$), $t_1/(q_2''/\sigma\epsilon)^{1/4} = (q_1''/q_2'')^{1/4}$ and $t_2/(q_2''/\sigma\epsilon)^{1/4} = 1$.

The two parallel plates may be closed at both ends by thin webs, with the outer surface of one web being insulated and the other web receiving *constant* heat input q_3'' (Fig. 61b). If all surface emissivities are assumed equal, and if internal radiative reflections and heat conduction along surfaces are neglected, then the steady temperature of the heated web is

$$\frac{t_3}{(q_3''/\sigma\epsilon)^{1/4}} = \left[(1 - c)\left(\frac{q_1'' + q_2''}{2q_3''}\right) + c\right]^{1/4} \tag{99}$$

where

$$c = \frac{0.5[a^2b^3 - (a + 1.5)ab^2 + ab + 0.75]}{(0.5a + 1)ab^3 - (0.5a^2 + 1.75a + 1.5)b^2 + (0.5a + 1.5)b + 0.375}$$

Here $a = \delta/l$ and $b = (a - \sqrt{a^2 + 1} + 1)/2a$, δ being plate spacing and l web separation. If the enclosed volume is filled with a perfect conductor, all four surfaces have a steady temperature $t/(q_3''/\sigma\epsilon)^{1/4} = \{[a + (q_1'' + q_2'')/q_3'']/(a + 2)\}^{1/4}$.

● *Radiation Shields.* Net radiative heat exchange between two surfaces is reduced by interposing thin, high-conductivity shields with low surface emissivities. The steady temperature of a *cylindrical shield* 2 placed between cylindrical surfaces 1 and 3 (Fig. 62) that exchange heat only by radiation (across nonabsorbing medium) is [55]

$$\frac{t_2}{t_3} = \left[\frac{A(t_1/t_3)^4 + B(r_3/r_2)}{A + B(r_3/r_2)}\right]^{1/4} \tag{100}$$

where $A = 1/\epsilon_3 + (1/\epsilon_{23} - 1)(r_3/r_2)$ and $B = 1/\epsilon_{21} + (1/\epsilon_1 - 1)(r_2/r_1)$. If $t_1 > t_3$, the

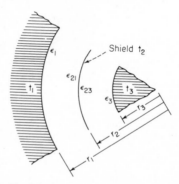

Fig. 62. Cylindrical or spherical radiation shield.

shield reduces heat transfer to inner cylinder (e.g., solar-heating protection of spacecraft), while if $t_1 < t_3$ the shield reduces heat loss from the inner cylinder (e.g., guarding of furnace heating elements). Heat loss for the latter is

$$q_3' = \frac{2\pi r_3 \sigma}{A + B(r_3/r_2)} (t_3^4 - t_1^4) \tag{101}$$

The ratio of heat loss ($t_1 < t_3$) or heat gain ($t_1 > t_3$) by the inner cylinder without and with a radiation shield is

$$\frac{(q_3')_{\text{without}}}{(q_3')_{\text{with}}} = 1 + \frac{(1/\epsilon_{21} + 1/\epsilon_{23} - 1)(r_3/r_2)}{1/\epsilon_3 + (1/\epsilon_1 - 1)(r_3/r_1)}$$

These expressions hold also for a *spherical shield* providing $2\pi r_3$ is replaced by $4\pi r_3^2$ and radius ratios r_3/r_1 and r_3/r_2 are squared.

If $\epsilon_1 = \epsilon_{21} = \epsilon_{23} = \epsilon_3 = \epsilon$, the steady temperature of an infinite *plane shield* interposed between plane surfaces ($r_1 = r_2 = r_3$) is

$$\frac{t_2}{t_3} = \left[\frac{(t_1/t_3)^4 + 1}{2} \right]^{1/4} \qquad (102)$$

If $t_1/t_3 \to 0, t_2 = 0.84 t_3$; if $t_1/t_3 \to \infty, t_2 = 0.84 t_1$. The heat loss for $t_1 < t_3$ is

$$q_3'' = \frac{\sigma \epsilon}{2(2 - \epsilon)} (t_3^4 - t_1^4) \qquad (103)$$

For n radiation shields between plane walls at t_a and t_b $(< t_a)$, all surfaces having equal emissivity ϵ and all spaces between shields being evacuated,*

$$q_a'' = \frac{\sigma \epsilon}{(n + 1)(2 - \epsilon)} (t_a^4 - t_b^4) \qquad (104)$$

Equation (104) shows that heat loss is proportional to $1/(n + 1)$. Thus, the first shield reduces unshielded heat loss q'' to $q''/2$, the second shield reduces $q''/2$ to $q''/3$, etc.

• *Combined Radiation and Convection.* The steady temperature of a thin generating plate simultaneously exposed to a radiation source or sink environment t_s at one face and a convective heating or cooling environment t_a at the same or opposite face is [see Eq. (40)]

$$t = \sqrt{\left(a_1 + \frac{a_2^2}{4} \right)^{1/2} - \frac{a_2}{4} - \frac{\sqrt{a_2}}{2}} \qquad (105)$$

where

$$a_1 = t_s^4 + a_3 t_a + \frac{q''' \delta}{\sigma \mathcal{F}} \quad , \quad a_3 = \frac{h}{\sigma \mathcal{F}}$$

$$a_2 = \left[\frac{a_3^2}{2} + \sqrt{\frac{a_3^4}{4} + \frac{64}{27} a_1^3} \right]^{1/3} + \left[\frac{a_3^2}{2} - \sqrt{\frac{a_3^4}{4} + \frac{64}{27} a_1^3} \right]^{1/3}$$

In the absence of convection, $a_3 = 0 = a_2$ and $t = (t_s^4 + q''' \delta/\sigma \mathcal{F})^{1/4}$ as in Eq. (97) with $n = 1$. In the absence of radiation ($\mathcal{F} = 0$), the steady plate temperature is $(k/q''' \delta^2)(t - t_a) = 1/\text{Bi}$, where $\text{Bi} = h\delta/k$.

Since the plate is thermally thin, q''' in Eq. (105) may be interpreted to include any combination of *constant* surface heat inputs (+) or rejections (−). Figure 63 gives data for the case of a thin generating plate convectively heated at one face by environment

_____

*If rarified gas remains between the shields, a gas conduction component is added according to

$$(n + 1)q_a'' = \frac{\sigma \epsilon}{2 - \epsilon} (t_a^4 - t_b^4) + \frac{3}{4} \rho_b \bar{v}_b \kappa (\sqrt{t_a t_b} - t_b)$$

where ρ_b and $\bar{v}_b$ are the density and mean velocity of residual molecules (evaluated at t_b) and κ is Boltzmann's constant.

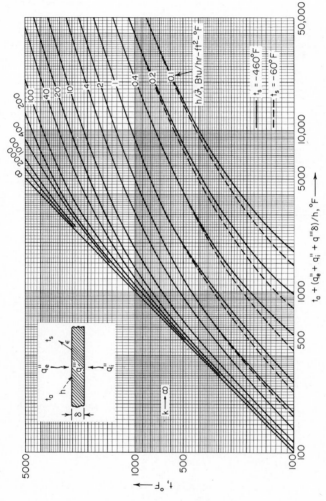

Fig. 63. Steady temperature of thin generating plate exposed to constant convective and prescribed heat-input loads and reradiating to constant sink environment.

at t_a, reradiating at the same face to sink environment t_s (=0 or 400°R), and receiving constant external and internal surface heat inputs q''_e and q''_i.

Thick Wall. Figure 64 illustrates a steady heat rate q conducted through a multi-layer series-composite wall separating convective environments at uniform temperatures t_a and t_A ($< t_a$). The unit surface conductances at exposed faces are uniform as h_a and h_A. The left face shows a thin *coating* (e.g., scale deposit) having unit conductance h_s*, and interface between materials 1 and 2 shows finite *contact resistance* with unit interface conductance h_i (Fig. 7).

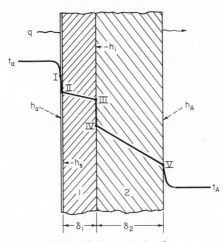

Fig. 64. Series composite plate.

● *Plate.* Steady one-dimensional heat rate conducted through a plane wall is, by Eq. (1),

$$q = -kA \frac{dt}{dx}$$

where A is the conducting area *normal* to vector q. For a single-layer plate of constant thermal conductivity k, uniform thickness δ, and with steady face temperatures t_1 at $x = 0$ and t_2 ($< t_1$) at $x = \delta$,

$$q = \frac{kA}{\delta}(t_1 - t_2) = K(t_1 - t_2) \tag{106}$$

Here $K = kA/\delta$ is the *wall conductance*. The corresponding steady temperature distribution in the plate is

$$\frac{t - t_2}{t_1 - t_2} = 1 - \frac{x}{\delta} \tag{107}$$

● *Directional Conductivity.* When the wall material is crystalline, laminated, or has a distinct grain structure, its conductivity will have different values parallel and

---

*The thickness of solids precipitated on clean surfaces emersed in a flow depends on the type of fluid and its temperature and flow rate. Typical values of h_s for scale deposite range from 100 (oils, hard water) to 2000 Btu/hr-ft^2-°F (organic vapors, distilled water). Design data are given in Ref. [56].

normal to grain. If the grain is *oblique* to the face surface (Fig. 65), net q in Eq. (1) is not normal to this face plane. In terms of parallel and normal conductivities k_p and k_n,

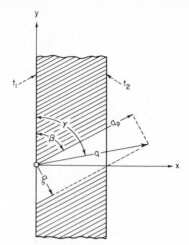

Fig. 65. Plate with oblique grain orientation.

apparent conductivities through and along the anisotropic plate are

$$k_x = k_p \sin^2 \beta + k_n \cos^2 \beta$$
$$k_y = k_n \sin^2 \beta + k_p \cos^2 \beta \tag{108}$$

where β is the grain angle relative to plate surface. The ratio of heat-rate components along and across the grain is $q_p/q_n = (k_p/k_n) \tan \beta$; as such the net q vector lies at an angle

$$\gamma = \tan^{-1} \left(\frac{k_n}{k_p} \cot \beta \right) \tag{109}$$

from face surface $x = 0$ [31].

● *Variable Conductivity.* If the thermal conductivity is *linearly* dependent on local temperature as

$$k = k_0(1 + a_0 t) \quad , \quad \bar{k} = k_0 \left[1 + \frac{a_0}{2} (t_1 + t_2) \right] \tag{110}$$

then the heat flow across plate $0 \le x \le \delta$ is

$$q = k_0 \left[1 + \frac{a_0}{2} (t_1 + t_2) \right] \frac{A}{\delta} (t_1 - t_2) = \frac{\bar{k} A}{\delta} (t_1 - t_2) \tag{111}$$

Thus, Eq. (106) applies if k is evaluated at the arithmetic mean plate temperature $(t_1 + t_2)/2$. The temperature distribution is

$$a_0 t = \left\{ (1 + a_0 t_1)^2 - [(1 + a_0 t_1)^2 - (1 + a_0 t_2)^2] \frac{x}{\delta} \right\}^{1/2} - 1 \qquad (112)$$

• *Series Composite Plate.* Heat flow through the two-layer plate of Fig. 64 with *series* resistances is

$$q = UA(t_a - t_A) \qquad (113)$$

where U is thermal *transmittance* (the reciprocal sum of the separate series *resistivities*) calculated as*

$$\frac{1}{U} = \frac{1}{h_a} + \frac{1}{h_s} + \frac{\delta_1}{k_1} + \frac{1}{h_i} + \frac{\delta_2}{k_2} + \frac{1}{h_A} \qquad (114)$$

Internal temperatures are determined by the resistance concept. For example,

$$t_{III} = t_a - \frac{q}{A}\left(\frac{1}{h_a} + \frac{1}{h_s} + \frac{\delta_1}{k_1} \right) = t_A + \frac{q}{A}\left(\frac{1}{h_A} + \frac{\delta_2}{k_2} + \frac{1}{h_i} \right)$$

• *Generating Plate.* If *uniform* heat generation q''' is distributed in a thick plate $0 \le x \le \delta$ whose surfaces are exposed to convective environments t_a and $x = 0$ and t_A at $x = \delta$,

$$\frac{t - t_a}{t_A - t_a} = \left[\frac{(2/Bi_A + 1)\zeta + 1}{1/Bi_a + 1/Bi_A + 1} \right]\left(\frac{x}{\delta} + \frac{1}{Bi_a} \right) - \zeta\left(\frac{x}{\delta} \right)^2 \qquad (115)$$

where $Bi_a = h_a\delta/k$, $Bi_A = h_A\delta/k$, and $\zeta = q'''\delta^2/2k(t_A - t_a)$. If $Bi_a \to \infty$ and $Bi_A \to \infty$ (constant surface temperatures t_1 at $x = 0$ and t_2 at $x = \delta$),

$$\frac{t - t_1}{t_2 - t_1} = (\zeta + 1)\left(\frac{x}{\delta} \right) - \zeta\left(\frac{x}{\delta} \right)^2 \qquad (116)$$

where $\zeta = q'''\delta^2/2k(t_2 - t_1)$. If $Bi_A = 0$ (back face insulated),

$$\frac{k}{q'''\delta^2}(t - t_a) = \frac{x}{\delta} - \frac{1}{2}\left(\frac{x}{\delta} \right)^2 + \frac{1}{Bi_a} \qquad (117)$$

Finally, if $q''' = 0$,

$$\frac{t - t_a}{t_A - t_a} = \frac{x/\delta + 1/Bi_a}{1/Bi_a + 1/Bi_A + 1} \qquad (118)$$

*Although a composite wall with *parallel* resistances is at least two dimensional, it is often idealized by summing assumed parallel heat flows.

If a plate with surface temperature t_1 at $x = 0$ and insulated at $x = \delta$ generates distributed internal heat that increases *linearly* with local temperature as

$$q''' = q_0'''(1 + b_0 t) \tag{119}$$

then plate temperature distribution is

$$\frac{b_0 t + 1}{b_0 t_1 + 1} = \frac{\cos\sqrt{\lambda}\,(1 - x/\delta)}{\cos\sqrt{\lambda}} \tag{120}$$

where $\lambda = q_0''' \delta^2 b_0 / k$. If heat generation *decreases* linearly as $q''' = q_0'''(1 - b_0 t)$,

$$\frac{b_0 t - 1}{b_0 t_1 - 1} = \frac{\cosh\sqrt{\lambda}\,(1 - x/\delta)}{\cosh\sqrt{\lambda}} \tag{121}$$

- *Diathermanous Plate.* A partially transparent plate $0 \leq x \leq \delta$ is irradiated at $x = 0$ with a *constant* flux q'' which is absorbed exponentially within the plate [see Eq. (31)] as $q_x'' = q'' e^{-\mu x}$. If the back face $x = \delta$ is convectively cooled by an ambient at t_A and no front-face heat leak occurs,

$$\frac{k}{q''\delta}(t - t_A) = \frac{1}{\mu\delta}[e^{-\mu\delta} - e^{-\mu\delta(x/\delta)}] + \left(1 - \frac{x}{\delta}\right) + \frac{1}{\text{Bi}}(1 - e^{-\mu\delta}) \tag{122}$$

where $\text{Bi} = h_A \delta / k$.

- *Porous-cooled Plate.* A generating plate $0 \leq x \leq \delta$ is cooled by a steady mass flow G_c of coolant through its *porous* structure, with the fluid coolant entering the plate at $x = \delta$ (Fig. 10a). If the upstream reservoir temperature of the coolant is t_c $(x \gg \delta)$, the steady plate temperature is

$$\frac{k_e}{q'''\delta^2}(t - t_c) = \frac{1}{F}\left(\frac{1 - e^{-Fx/\delta}}{F} - \frac{x}{\delta} + 1\right) \tag{123}$$

Here $F = G_c C_c \delta / k_e$ and $k_e = P k_c + (1 - P)k_s$, with P representing the porosity of the wall material and k_c and k_s the conductivities of the coolant and solid. For a nongenerating plate with coolant exit face $x = 0$ convectively heated by an ambient at t_a

$$\frac{t - t_c}{t_a - t_c} = \frac{\text{Bi}}{\text{Bi} + F}e^{-F(x/\delta)} \tag{124}$$

where $\text{Bi} = h\delta/k_e$. As $\text{Bi} \to \infty$, this reduces to $(t - t_c)/(t_1 - t_c) = e^{-F(x/\delta)}$. If the nongenerating plate is exposed to constant heat input q'' at $x = 0$,

$$\frac{k_e}{q''\delta}(t - t_c) = \frac{1}{F}e^{-F(x/\delta)} \tag{125}$$

- *Cylindrical Shell.* Steady one-dimensional heat rate conducted through a *cylindrical* system of length L (disregarding end effects) is

$$q = -kA_r \frac{dt}{dr} \qquad (126)$$

where $A_r = 2\pi r L$. For a single-layer *tube* of constant conductivity k and with steady surface temperatures t_1 at $r = r_1$ and $t_2 (< t_1)$ at $r = r_2$,

$$q = \frac{2\pi k L}{\ln(r_2/r_1)} (t_1 - t_2) \qquad (127)$$

The corresponding steady temperature distribution in the tube wall is

$$\frac{t - t_2}{t_1 - t_2} = 1 - \frac{\ln(r/r_1)}{\ln(r_2/r_1)} \qquad (128)$$

- *Variable Conductivity.* If k varies *linearly* with t as in Eq. (110),

$$q = k_0 \left[1 + \frac{a_0}{2}(t_1 + t_2) \right] \left[\frac{2\pi L}{\ln(r_2/r_1)} \right] (t_1 - t_2) = \frac{2\pi \bar{k} L}{\ln(r_2/r_1)} (t_1 - t_2) \qquad (129)$$

and

$$a_0 t = \left\{ (1 + a_0 t_1)^2 - [(1 + a_0 t_1)^2 - (1 + a_0 t_2)^2] \frac{\ln(r/r_1)}{\ln(r_2/r_1)} \right\}^{1/2} - 1 \qquad (130)$$

- *Series Composite Cylindrical Shell.* The heat flow through the two-layer tube of Fig. 66 with series resistances is

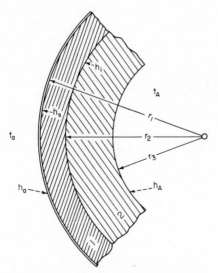

Fig. 66. Series composite tube or spherical shell.

$$q = UA_1(t_a - t_A) \tag{131}$$

where transmittance U, based (arbitrarily) on outer surface area $A_1 = 2\pi r_1 L$, is

$$\frac{1}{U} = \frac{1}{h_a} + \frac{1}{h_s} + \frac{r_1 \ln(r_1/r_2)}{k_1} + \frac{r_1}{h_i r_2} + \frac{r_1 \ln(r_2/r_3)}{k_2} + \frac{r_1}{h_A r_3} \tag{132}$$

● *Maximum Heat Rate Through Tube.* Since increasing the outer radius r_1 of a single-layer tube (fixed r_2) diminishes radial conductance while simultaneously increasing surface conductance, a *critical* outer radius exists for which heat exchange between inner and outer ambients is at a maximum. This critical outer radius is

$$\left(r_1\right)_c = \frac{k}{h_a} \tag{133}$$

or $\left(\mathrm{Bi}_a\right)_c = 1$. Figure 67 shows this effect for the special case of infinite internal surface conductance ($t = t_2$ at $r = r_2$).

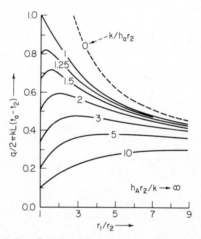

Fig. 67. Steady heat flow through single-layer tube with inner surface at t_2.

A pipe carrying elevated-temperature fluid ($t_A > t_a$) and covered with insulation may lose more (rather than less) heat if type (conductivity) and thickness of outer insulation are improperly chosen. For a two-layer pipe as in Fig. 66, $\left(r_1\right)_c = k_1/h_a$. To be effective in reducing heat loss, the outer radius of insulation must be chosen to satisfy the condition

$$\frac{1}{k_1} \ln\left(\frac{r_1}{r_2}\right) + \frac{1}{r_1 h_a} > \frac{1}{r_2 h_a}$$

i.e., combined insulation and surface resistance must exceed original surface resistance of uninsulated pipe.

● *Generating Tube.* If there is uniform heat generation q''' distributed in a tube $r_2 \le r \le r_1$ with r_2 insulated and r_1 exposed to convective environment at t_a,

$$\frac{k}{q'''r_1^2}(t - t_a) = \frac{1}{2}\left\{\frac{1}{2}\left[1 - \left(\frac{r}{r_1}\right)^2\right] - \left(\frac{r_2}{r_1}\right)^2\left[\ln\left(\frac{r_1}{r}\right) + \frac{1}{\text{Bi}_a}\right] + \frac{1}{\text{Bi}_a}\right\} \quad (134)$$

where $\text{Bi}_a = h_a r_1/k$. If, instead, r_1 is insulated and r_2 is exposed to t_A,

$$\frac{k}{q'''r_2^2}(t - t_A) = \frac{1}{2}\left\{\frac{1}{2}\left[1 - \left(\frac{r}{r_2}\right)^2\right] + \left(\frac{r_1}{r_2}\right)^2\left[\ln\left(\frac{r}{r_2}\right) + \frac{1}{\text{Bi}_A}\right] - \frac{1}{\text{Bi}_A}\right\} \quad (135)$$

where $\text{Bi}_A = h_A r_2/k$.

When a tube with surface temperatures t_1 at $r = r_1$ and t_2 at $r = r_2$ generates distributed heat q''' that increases *linearly* with local temperature as in Eq. (119),

$$b_0 t = \frac{1}{\Psi}\left[(b_0 t_2 + 1)Y_0(\sqrt{\lambda}) - (b_0 t_1 + 1)Y_0\left(\sqrt{\lambda}\frac{r_2}{r_1}\right)\right]J_0\left(\sqrt{\lambda}\frac{r}{r_1}\right)$$

$$+ \frac{1}{\Psi}\left[(b_0 t_1 + 1)J_0\left(\sqrt{\lambda}\frac{r_2}{r_1}\right) - (b_0 t_2 + 1)J_0(\sqrt{\lambda})\right]Y_0\left(\sqrt{\lambda}\frac{r}{r_1}\right) - 1 \quad (136)$$

where $\Psi = J_0(\sqrt{\lambda}r_2/r_1)Y_0(\sqrt{\lambda}) - J_0(\sqrt{\lambda})Y_0(\sqrt{\lambda}r_2/r_1)$ and $\lambda = q_0'''r_1^2 b_0/k$, the J_0 and Y_0 being zero-order Bessel functions of the first and second kind (Sec. 1.B.3).

● *Generating Cylinder.* The steady temperature of a convectively cooled solid cylinder $0 \le r \le r_1$ generating *uniform* distributed heat q''' is, from Eq. (134),

$$\frac{k}{q'''r_1^2}(t - t_a) = \frac{1}{4}\left[1 - \left(\frac{r}{r_1}\right)^2 + \frac{2}{\text{Bi}}\right]$$

where $\text{Bi} = hr_1/k$. For q''' *linearly* dependent on t as in Eq. (119),

$$\frac{b_0 t + 1}{b_0 t_a + 1} = \frac{J_0(\sqrt{\lambda}r/r_1)}{J_0(\sqrt{\lambda}) - (\sqrt{\lambda}/\text{Bi})J_1(\sqrt{\lambda})} \quad (137)$$

where the J_0 and J_1 are zero- and first-order Bessel functions of the first kind. For a constant surface temperature t_1 at $r = r_1$, $\text{Bi} \to \infty$ and, by Eq. (137),

$$\frac{b_0 t + 1}{b_0 t_1 + 1} = \frac{J_0(\sqrt{\lambda}r/r_1)}{J_0(\sqrt{\lambda})}$$

If the heat generation *decreases* linearly with local temperature as $q''' = q_0'''(1 - b_0 t)$,

$$\frac{b_0 t - 1}{b_0 t_1 - 1} = \frac{I_0(\sqrt{\lambda}r/r_1)}{I_0(\sqrt{\lambda})} \quad (138)$$

where I_0 is a zero-order modified Bessel function of the first kind.

● *Generating Composite Cylinder.* A solid cylinder $0 \le r \le r_2$ of conductivity k_2 generates uniform distributed heat q_2''' and is covered in perfect contact by a tube

$r_2 \leq r \leq r_1$ of conductivity k_1 that generates a uniform distributed heat q_1'''. If the composite is convectively cooled at $r = r_1$,

$$\frac{4k_1}{q_1''' r_1^2}(t - t_a) = 2\left(\frac{r_2}{r_1}\right)^2 \left(\frac{q_2'''}{q_1'''} - 1\right)\left[\ln\left(\frac{r_1}{r_2}\right) + \frac{1}{Bi_1}\right]$$

$$- \left(\frac{q_2'''}{q_1'''}\right)\left(\frac{k_1}{k_2}\right)\left[\left(\frac{r}{r_1}\right)^2 - \left(\frac{r_2}{r_1}\right)^2\right] - \left(\frac{r_2}{r_1}\right)^2 + \frac{2}{Bi_1} + 1 \ ; \ \ 0 \leq r \leq r_2 \tag{139}$$

for the cylinder, and

$$\frac{4k_1}{q_1''' r_1^2}(t - t_a) = 2\left(\frac{r_2}{r_1}\right)^2 \left(\frac{q_2'''}{q_1'''} - 1\right)\left[\ln\left(\frac{r_1}{r}\right) + \frac{1}{Bi_1}\right] - \left(\frac{r}{r_1}\right)^2 + \frac{2}{Bi_1} + 1 \ ; \ \ r_2 \leq r \leq r_1 \tag{140}$$

for the tube, where $Bi_1 = hr_1/k_1$.

• *Radial Flow in Plate.* A *thin* plate of thickness δ contains a circular *hole* of radius r_1. If the surface of the hole is maintained at a steady temperature t_1 and the plate surfaces are exposed to uniform convective environments t_a at one face and t_A at the other,

$$\frac{t - t_\infty}{t_1 - t_\infty} = \frac{K_0(\eta r/\delta)}{K_0(\eta r_1/\delta)} \tag{141}$$

where $t_\infty = (h_a t_a + h_A t_A)/(h_a + h_A)$ is the plate temperature at $r \to \infty$ and $\eta = \sqrt{(h_a + h_A)\delta/k}$. If the hole contains a heat source of strength q,

$$\frac{k\delta}{q}(t - t_\infty) = \frac{K_0(\eta r/\delta)}{2\pi(\eta r_1/\delta) K_1(\eta r_1/\delta)} \tag{142}$$

where K_0 and K_1 are zero- and first-order modified Bessel functions of the second kind.

If, instead of a hole, the plate has a thin cylindrical *rod* of radius r_1 and conductivity k_1 attached to its 'a' side, the rod being exposed to t_a with uniform surface conductance h_1, the plate temperature at the junction point is

$$\frac{t_1 - t_\infty}{t_a - t_\infty} = \frac{1}{1 + \nu K_1(\eta r_1/\delta)/K_0(\eta r_1/\delta)} \tag{143}$$

where $\nu = \sqrt{2[(h_a + h_A)/h_1](k/k_1)\delta/r_1}$.*

---

*If the rod is considered an uninsulated thermocouple of equivalent diameter $2r_1$ fused to the plate, $t_1 - t_\infty$ represents the absolute thermocouple error; i.e., the difference between what the thermocouple registers and what the true plate temperature would be in absence of the thermocouple [1].

Radial temperature distribution in a thin *disk* of thickness δ and diameter $2r_1$, with one face insulated and the other exposed to uniform convective environment t_a, and with its periphery $r = r_1$ held at steady temperature t_1, is

$$\frac{t - t_a}{t_1 - t_a} = \frac{I_0(\sqrt{\mathrm{Bi}}\, r/\delta)}{I_0(\sqrt{\mathrm{Bi}}\, r_1/\delta)} \tag{144}$$

where $\mathrm{Bi} = h\delta/k$.

- *Spherical Shell.* Steady one-dimensional heat rate conducted through a *spherical* system is given by Eq. (126) with $A_r = 4\pi r^2$. For a single-layer spherical shell of constant conductivity k and with steady surface temperatures t_1 at $r = r_1$ and t_2 ($< t_1$) at $r = r_2$,

$$q = \frac{4\pi k r_1 r_2}{(r_2 - r_1)}(t_1 - t_2) \tag{145}$$

Steady temperature distribution in the shell is

$$\frac{t - t_2}{t_1 - t_2} = \frac{1 - r_2/r}{1 - r_2/r_1} \tag{146}$$

Critical outer radius of a single-layer spherical shell (fixed r_2) is

$$(r_1)_c = \frac{2k}{h_a} \tag{147}$$

- *Series Composite Spherical Shell.* Heat flow through a two-layer spherical shell (Fig. 66) with *series* resistances is given by Eq. (131) with outer surface area $A_1 = 4\pi r_1^2$ and transmittance

$$\frac{1}{U} = \frac{1}{h_a} + \frac{1}{h_s} + \frac{r_1^2(r_1 - r_2)}{k_1 r_1 r_2} + \frac{r_1^2}{h_i r_2^2} + \frac{r_1^2(r_2 - r_3)}{k_2 r_2 r_3} + \frac{r_1^2}{h_A r_3^2} \tag{148}$$

- *Generating Sphere.* Steady temperature of a convectively cooled solid sphere $0 \le r \le r_1$ generating *uniform* distributed heat q''' is

$$\frac{k}{q''' r_1^2}(t - t_a) = \frac{1}{6}\left[1 - \left(\frac{r}{r_1}\right)^2 + \frac{2}{\mathrm{Bi}}\right] \tag{149}$$

where $\mathrm{Bi} = hr_1/k$.

- *Irradiated Spherical Shell.* A *thin* shell may be uniformly irradiated from one side (e.g., exposure to solar heating) and develop temperature gradients along its surface due to skin heat conduction to cooler (shaded) zones. Figure 68 gives steady temperature at three locations $\phi = 0°$, $90°$, and $180°$ on a thin nonrotating spherical shell [57]; effects included are external exposure to uniform solar flux q_S'', internal surface radiation, and external reradiation to space ($t_s = 0°R$). The normalizing temperature in Fig. 68 is

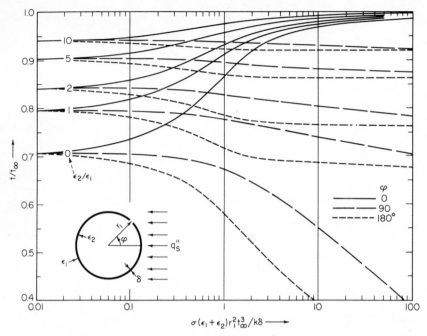

Fig. 68. Steady temperature of thin, nonrotating spherical shell in uniform radiation field.

$$t_\infty = \left[\frac{q_S'' \alpha_S}{\sigma \epsilon_1} \left(\frac{\epsilon_1/\epsilon_2 + 1/4}{\epsilon_1/\epsilon_2 + 1} \right) \right]^{1/4}$$

where α_S is solar absorptivity of outer shell surface (Table 4).*

b. **Extended Surfaces.** When a wall is convectively or radiatively cooled, attaching *extended surfaces* such as fins and spines (Fig. 2) serves to artificially increase the wall area and, as such, to increase heat loss over that of the original plane surface [1].

Convective Fins and Spines. Ambient fluid temperature t_a and unit surface conductance h are assumed uniform over wall and fin surfaces, and conductivity k of fin material is considered sufficiently large that temperature gradients develop only in the x direction.

• *Straight Rectangular Fin.* Steady temperature $t(x)$ along the width l of a *straight* fin with rectangular profile (Fig. 2i) and uninsulated at $x = l$ is

$$\frac{t - t_a}{t_w - t_a} = \frac{Ne^{-ml(2 - x/l)} + e^{-ml(x/l)}}{Ne^{-2ml} + 1} \tag{150}$$

where t_w is the fixed wall (and fin base) temperature, $N = (mk - h)/(mk + h)$, and $m = \sqrt{h/k\delta}$. Heat conducted through base of fin (semi-thickness δ) is

*Reference 57 also considers stationary cylindrical shells. Rotating cylindrical and spherical shells are treated in Refs. 17 and 58.

$$\frac{q'_w}{\sqrt{hk\delta}\,(t_w - t_a)} = \frac{2\,(\tanh ml + \sqrt{\text{Bi}})}{\sqrt{\text{Bi}}\,\tanh ml + 1} \tag{151}$$

where $\text{Bi} = h\delta/k$. If $\text{Bi} < 1$ the fin will cool the wall surface $(t_a < t_w)$, while if $\text{Bi} > 1$ the fin will have an unfavorable insulating effect and so defeat its purpose. If fin tip $x = l$ is insulated,

$$\frac{t - t_a}{t_w - t_a} = \frac{\cosh ml(1 - x/l)}{\cosh ml} \tag{152}$$

and

$$\frac{q'_w}{\sqrt{hk\delta}\,(t_w - t_a)} = 2\,\tanh ml \tag{153}$$

Fin-tip heat loss can be approximately corrected for when using Eqs. (152) and (153) by taking the fin width as $l + \delta$ if $\text{Bi} \leq 1/4$.

If δ is fixed, maximum heat rejection occurs, according to Eq. (153), for an infinitely wide fin $l \to \infty$. However, if *profile area* $A = 2\,\delta l$ is fixed, the maximum heat loss corresponds to a unique set of dimensions δ and l. These *optimum dimensions* are $\delta_{\text{opt}} = 0.4988\,(A^2h/k)^{1/3} = 1.264\,(q'_w)^2/hk(t_w - t_a)^2$ and $l_{\text{opt}} = A/2\delta_{\text{opt}} = 1.596\,q'_w/h(t_w - t_a)$.

• *Straight Triangular Fin.* If the thickness of a straight fin varies *linearly* as $y = \delta_1(1 - x/l)$, where δ_1 is base semi-thickness,

$$\frac{t - t_a}{t_w - t_a} = \frac{I_0(2m_1 l\sqrt{1 - x/l})}{I_0(2m_1 l)} \tag{154}$$

and

$$\frac{q'_w}{\sqrt{hk\delta_1}\,(t_w - t_a)} = \frac{2I_1(2m_1 l)}{I_0(2m_1 l)} \tag{155}$$

The optimum dimensions are $(\delta_1)_{\text{opt}} = 0.8355\,(A^2h/k)^{1/3} = 1.655\,(q'_w)^2/hk(t_w - t_a)^2$ and $l_{\text{opt}} = A/(\delta_1)_{\text{opt}} = 1.684\,q'_w/h(t_w - t_a)$.

• *Annular Rectangular Fin.* For an *annular* fin of rectangular profile (Fig. 2j)

$$\frac{t - t_a}{t_w - t_a} = \frac{I_1(mr_t)\,K_0(mr) + K_1(mr_t)\,I_0(mr)}{I_0(mr_1)\,K_1(mr_t) + I_1(mr_t)\,K_0(mr_1)} \tag{156}$$

and

$$\frac{q_w}{r_1\sqrt{hk\delta}\,(t_w - t_a)} = \frac{4}{\pi}\left[\frac{I_1(mr_t)\,K_1(mr_1) - I_1(mr_1)\,K_1(mr_t)}{I_0(mr_1)\,K_1(mr_t) + I_1(mr_t)\,K_0(mr_1)}\right] \tag{157}$$

Optimum dimensions [59] are given graphically in Fig. 69. For required wall heat rejection q_w and given the inner fin radius r_1, enter Fig. 69 to determine the optimum fin semithickness δ_{opt} and optimum fin *volume* $V_{opt} = 2\pi\delta(r_t^2 - r_1^2)$, and from the latter calculate optimum fin width $l_{opt} = r_t - r_1$.

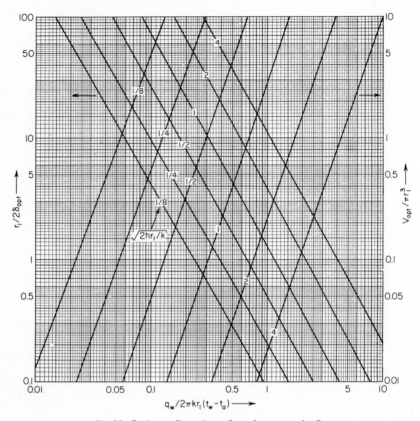

Fig. 69. Optimum dimensions of annular rectangular fin.

- *Rectangular Spine.* The steady temperature along an infinitely long spine (*pin fin*) of circular cross section and rectangular profile is

$$\frac{t - t_a}{t_w - t_a} = \exp\left(-\sqrt{2}\,ml\,\frac{x}{l}\right) \tag{158}$$

where δ is cross-section radius. Heat conducted across spine base is

$$\frac{q_w}{\sqrt{hk\delta^3}\,(t_w - t_a)} = \sqrt{2}\,\pi \tag{159}$$

compared to $q_w = \pi h\delta^2(t_w - t_a)$ for the same base area but with the spine removed. Thus, for the rectangular spine to cool, $Bi < 2$. If the spine is of finite length l (Fig. 2k),

$$\frac{t - t_a}{t_w - t_a} = \frac{\cosh\sqrt{2}\,ml(1 - x/l)}{\cosh\sqrt{2}\,ml} \tag{160}$$

and

$$\frac{q_w}{\sqrt{hk\delta^3}\,(t_w - t_a)} = \sqrt{2}\,\pi\,\tanh\sqrt{2}\,ml \tag{161}$$

Fin and Spine Efficiency. The ratio of heat dissipated over the total fin or spine *surface area* S to that which would be rejected if the fin or spine were isothermal $(k \to \infty)$ at t_w is referred to as *efficiency e* of the extended surface. If h and t_a are constant,

$$e = \frac{1}{S(t_w - t_a)}\int^S (t - t_a)\,dS = \frac{\bar{t} - t_a}{t_w - t_a} \tag{162}$$

where $\bar{t}$ is the mean temperature of the extended surface. The actual heat rejection (numerator) can be alternatively calculated as the heat conducted through fin or spine base. Using either definition, heat dissipated from the wall area occupied by the extended surface is $q_w = ehS(t_w - t_a)$.

Table 6 gives efficiencies in terms of parameter ml for ten fins and spines of various profile shapes, each with insulated tips [60]. Efficiencies of fins ①, ③, and ④ with optimum dimensions (maximum heat rejection q_w for given profile area A) are e_{opt} =

TABLE 6. Equations for Efficiency of Convective Fins and Spines

Straight fins	① Rectangular $y = \delta$		$e = \dfrac{1}{ml}\tanh ml, \quad m = \sqrt{h/k\delta}$	$A = 2\delta l$
	② Parabolic $y = \delta_1(1 - x/l)^{\frac{1}{2}}$		$e = \dfrac{1}{m_1 l}\dfrac{I_{\frac{2}{3}}(\frac{4}{3}m_1 l)}{I_{-\frac{1}{3}}(\frac{4}{3}m_1 l)}, \quad m_1 = \sqrt{h/k\delta_1}$	$A = \frac{4}{3}\delta_1 l$
	③ Triangular $y = \delta_1(1 - x/l)$		$e = \dfrac{1}{m_1 l}\dfrac{I_1(2m_1 l)}{I_0(2m_1 l)}$	$A = \delta_1 l$
	④ Parabolic $y = \delta_1(1 - x/l)^2$		$e = \dfrac{2}{\sqrt{4(m_1 l)^2 + 1} + 1}$	$A = \frac{2}{3}\delta_1 l$
Circular fins	⑤ Rectangular $y = \delta$		$e = \dfrac{2}{ml(r_t/r_1 + 1)}\left[\dfrac{I_1(mr_1)K_1(mr_t) - I_1(mr_t)K_1(mr_1)}{I_0(mr_1)K_1(mr_t) + I_1(mr_t)K_0(mr_1)}\right]$	$A = 2\delta(r_t - r_1)$
	⑥ Hyperbolic $y = \delta_1(r_1/r)$		$e = \dfrac{2}{m_1 l(r_t/r_1 + 1)}\left[\dfrac{I_{\frac{2}{3}}(\frac{2}{3}m_1 r_1)I_{\frac{2}{3}}(\frac{2}{3}m_1 r_t\sqrt{r_t/r_1}) - I_{\frac{2}{3}}(\frac{2}{3}m_1 r_t\sqrt{r_t/r_1})I_{\frac{2}{3}}(\frac{2}{3}m_1 r_1)}{I_{-\frac{1}{3}}(\frac{2}{3}m_1 r_1)I_{\frac{2}{3}}(\frac{2}{3}m_1 r_t\sqrt{r_t/r_1}) - I_{\frac{2}{3}}(\frac{2}{3}m_1 r_t\sqrt{r_t/r_1})I_{-\frac{1}{3}}(\frac{2}{3}m_1 r_1)}\right]$	$A = 2\delta_1 r_1 \ln\frac{r_t}{r_1}$

TABLE 6. Equations for Efficiency of Convective Fins and Spines (*Continued*)

Spines (circular cross-section)				
⑦ Rectangular $y = \delta$	2δ		$e = \dfrac{1}{\sqrt{2}\,ml}\,\tanh\sqrt{2}\,ml$	$V = \pi\delta^2 l$
⑧ Parabolic $y = \delta_1(1 - x/l)^{\frac{1}{2}}$	$2\delta_1$		$e = \dfrac{2}{(\frac{4}{3}\sqrt{2}\,m_1 l)}\,\dfrac{I_1(\frac{4}{3}\sqrt{2}\,m_1 l)}{I_0(\frac{4}{3}\sqrt{2}\,m_1 l)}$	$V = \dfrac{\pi}{2}\delta_1^2 l$
⑨ Triangular $y = \delta_1(1 - x/l)$			$e = \dfrac{4}{(2\sqrt{2}\,m_1 l)}\,\dfrac{I_2(2\sqrt{2}\,m_1 l)}{I_1(2\sqrt{2}\,m_1 l)}$	$V = \dfrac{\pi}{3}\delta_1^2 l$
⑩ Parabolic $y = \delta_1(1 - x/l)^2$			$e = \dfrac{2}{\sqrt{\frac{8}{9}(m_1 l)^2 + 1} + 1}$	$V = \dfrac{\pi}{5}\delta_1^2 l$

62.7, 59.4, and 49.6 percent.* These values are independent of what material is used if fins have optimum dimensions. A fin has least volume (minimum weight) when its profile shape is such that $t(x)$ is linear. This occurs for parabolic fin ④ when $ml = \sqrt{2}$, or $e = 50$ percent; this particular fin has dimensions $\delta_1 = 2(q_w')^2/hk(t_w - t_a)^2$ and $l = 2q_w'/h(t_w - t_a)$.

Efficiencies of the ten extended surfaces in Table 6 are shown in Fig. 70 with

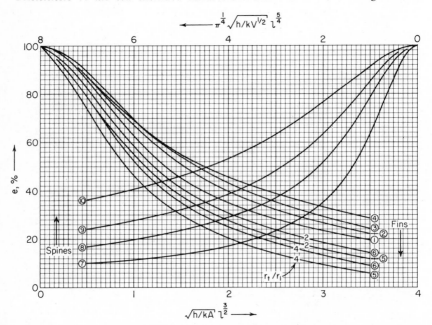

Fig. 70. Efficiency of convective fins and spines.

*A good approximation for efficiency of a *square* fin with side length b on a tube of outer radius r_1 (or *sheet* fin pierced by a rectangular array of tubes) is obtained by using e for the annular rectangular fin ⑤ with S based on equivalent surface area ($r_t = b/\sqrt{\pi}$).

fins compared on basis of equal width l and profile area A (or equal ratios l^3/A) and spines on basis of equal width l and volume V (or equal ratios l^5/V). In each case, tapering the extended surface (placing more material where temperature potential $t - t_a$ is highest) favors efficiency, but at the expense of increasing root attachment area.

Radiative Fins. The fin is considered as a grey surface radiating in vacuum (no convection) to a uniform effective sink temperature t_s, and without radiative interactions between adjacent fins and between a fin and its wall-attachment surface. The net heat radiated from an elemental width dx of both fin surfaces (each of emissivity ϵ) is then $dq = 2\sigma\epsilon l(t^4 - t_s^4)dx$. Both ϵ and t_s can be adjusted to accommodate such effects as radiant input (e.g., solar heating) at one or both fin surfaces.

• *Straight Rectangular Fin and Spine.* Steady heat dissipation from a straight rectangular fin of *infinite width* is [61]

$$\frac{q'_w}{k\delta t_w n} = 2\sqrt{\frac{2}{5}\left[4\left(\frac{t_s}{t_w}\right)^5 - 5\left(\frac{t_s}{t_w}\right)^4 + 1\right]} \tag{163}$$

where $n = \sqrt{\sigma\epsilon t_w^3/k\delta}$ (analogous to m for convective fins). As $x \to \infty$, the fin temperature approaches t_s. If each surface is irradiated and absorbs normal flux q'', then $t_s = (\alpha q''/2\sigma\epsilon)^{1/4}$, where α is surface absorptivity. When $t_s = 0°R$, the temperature distribution and heat rejection are

$$\frac{t}{t_w} = \left(\frac{3}{\sqrt{10}}nx + 1\right)^{-2/3} \quad , \quad \frac{q'_w}{k\delta t_w n} = 2\sqrt{\frac{2}{5}} \tag{164}$$

If the radiative fin is to cool, $\sigma\epsilon t_w^3\delta/k < 0.4$. Corresponding results for an *infinitely long* circular spine with $t_s = 0°R$ are

$$\frac{t}{t_w} = \left(\frac{3}{\sqrt{5}}nx + 1\right)^{-2/3} \quad , \quad \frac{q_w}{k\delta^2 t_w n} = \frac{2\pi}{\sqrt{5}} \tag{165}$$

with the condition that $\sigma\epsilon t_w^3\delta/k < 0.8$ if the spine is to cool.

• *Finite Rectangular Fin.* The steady temperature along a straight rectangular fin of *finite width* l, insulated at $x = l$ and radiating to $t_s = 0°R$, is [62]

$$B_{1-(t_l/t)^5}(0.3, 0.5) - \sqrt{10}\left(\frac{t_l}{t_w}\right)^{3/2} nl\left(1 - \frac{x}{l}\right) = 0 \tag{166}$$

Here t_l is tip temperature ($x = l$) and $B_y(a, b)$ is an incomplete beta function (Sec. 1.B.1). Substituting $t(0) = t_w$ gives t_l (Fig. 71 with $w = q''/\sigma\epsilon t_w^4 = 0$) as

$$\left(\frac{t_l}{t_w}\right)^{-3/2} B_{1-(t_l/t_w)^5}(0.3, 0.5) = \sqrt{10}\,nl \tag{167}$$

The rate of heat rejection is thus

$$\frac{q'_w}{k\delta t_w n} = 2\sqrt{\frac{2}{5}\left[1 - \left(\frac{t_l}{t_w}\right)^5\right]} \tag{168}$$

• *Fin Efficiency.* For a radiating fin with constant ϵ and t_s,

$$e = \frac{1}{S(t_w^{\ 4} - t_s^{\ 4})} \int^S (t^4 - t_s^{\ 4})dS \qquad (169)$$

The heat dissipated by the fin is then $q_w = e\sigma\epsilon S(t_w^{\ 4} - t_s^{\ 4})$. The efficiency of a straight rectangular fin with $t_s = 0°\mathrm{R}$ is

$$e = \frac{1}{nl}\sqrt{\frac{2}{5}\left[1 - \left(\frac{t_l}{t_w}\right)^5\right]} \qquad (170)$$

where t_l is calculated from Eq. (167). As $nl \to \infty$, $t_l \to 0°\mathrm{R}$ and $e = \sqrt{2/5}\,/\,nl$.

If each side of the rectangular fin is irradiated and absorbs normal flux q'', then Eqs. (168) and (170) become

$$\frac{q_w'}{k\delta t_w n} = 2\sqrt{\frac{2}{5}\left[1 - \left(\frac{t_l}{t_w}\right)^5 - 5w\left(1 - \frac{t_l}{t_w}\right)\right]} \qquad (171)$$

and

$$e = \frac{1}{nl}\sqrt{\frac{2}{5}\left[1 - \left(\frac{t_l}{t_w}\right)^5 - 5w\left(1 - \frac{t_l}{t_w}\right)\right]} \qquad (172)$$

with t_l given in Fig. 71 for finite $w = q''/\sigma\epsilon t_w^{\ 4}$ [63].

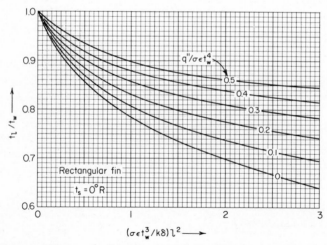

Fig. 71. Tip temperature ($x = l$) of straight rectangular fin absorbing surface flux q'' and radiating to space.

Efficiencies of radiating straight and annular rectangular fins ($l = r_t - r_1$) are compared in Fig. 72 [61, 64]. Optimum dimensions of a straight rectangular fin with $t_s = 0°\mathrm{R}$ [65] are $\delta_{\mathrm{opt}} = 0.924\,(q_w')^2/\sigma\epsilon k t_w^{\ 5}$ and $l_{\mathrm{opt}} = 0.884\,q_w'/\sigma\epsilon t_w^{\ 4}$; the correspond-

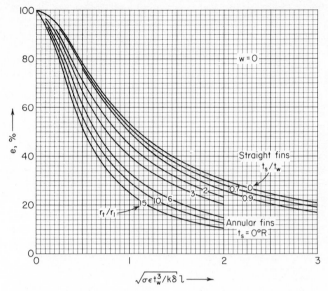

Fig. 72. Efficiency of rectangular radiative fins.

ing efficiency is e_{opt} = 56.9 percent. The optimum dimensions of a straight triangular fin with t_s = 0°R [66] are $(\delta_1)_{opt}$ = 1.149 $(q'_w)^2/\sigma\epsilon k t_w{}^5$ and l_{opt} = 0.960 $q'_w/\sigma\epsilon t_w{}^4$; these occur at e_{opt} = 52.1 percent.*

c. Two-dimensional Systems. The majority of these steady, two-dimensional systems require solution of the Laplace equation $\partial^2 t/\partial x^2 + \partial^2 t/\partial y^2 = 0$. Where exact solutions are unknown, approximate analytical and empirical methods exist for calculation of specific cases.†

A steady heat flow q per unit area A through an infinite single-layer plate of thickness δ and with constant properties and uniform surface temperatures t_1 and t_w ($< t_1$) is expressed in the following variety of ways:

$$q = K(t_1 - t_w) = \frac{1}{R}(t_1 - t_w) = ks(t_1 - t_w)$$

Here $K = kA/\delta$ is *thermal conductance*, $R = \delta/kA$ is *thermal resistance*, and $s = A/\delta$ is *geometric shape factor*. The conduction data for specified two-dimensional shapes can be expressed explicitly as q or implicitly in terms of K, R, or s.

Buried Objects and Cavities. Table 7 gives known solutions for steady conduction between the isothermal surface of a buried object or cavity and a remote isothermal surface of the surrounding solid [18, 68].

In all cases the enveloping mass is infinite in the direction normal to diagram. The heat flow for bodies extending infinitely in this direction (or of finite length with adiabatic ends) is q', the rate per unit normal length. All equations for q and q' hold equally for heating ($t_w > t_1$) or cooling ($t_w < t_1$). If the surface condition at $x = 0$ is one of convection to a uniform ambient at t_a (rather than isothermal wall t_w), this can be roughly accommodated by increasing submergence depth x to $x + k/h$, where h is the unit surface conductance.

*Minimum-weight radiating fins and spines are treated in Refs. 63 and 67.
†See Secs. 4 and 5.

TABLE 7. Equations for Heat Loss from Buried Objects and Cavities

Thin vertical strip in semi-infinite solid		
	$q'/k(t_1 - t_w) \doteq 2.38/(x/l)^{0.24}$	$\frac{1}{2} < \frac{x}{l} < 12$
Thin horizontal strip in semi-infinite solid		
	$q'/k(t_1 - t_w) \doteq 2.94/(x/l)^{0.32}$	$\frac{1}{2} < \frac{x}{l} < 12$
Thin rectangular plate in semi-infinite solid		
	$q/ka(t_1 - t_w) = \pi/\ln(4a/b)$	$x = 0$
	$= 2\pi/\ln(2\pi x/b)$	$a \gg b, x > 2b$
	$= 2\pi/\ln(4a/b)$	$x \gg a$
Thin circular disk in semi-infinite solid		
	$q/kD(t_1 - t_w) = 2$	$x = 0$
	$= 4$	$x \gg D$
Rectangular hole in semi-infinite solid		
	$q'/k(t_1 - t_w) = (a/2b + 5.7)/\ln(3.5x/a^{\frac{1}{4}}b^{\frac{3}{4}})$	$a > b$
Hemispherical hole in surface of semi-infinite solid		
	$q/kD(t_1 - t_w) = \pi$	

Spherical hole in semi-infinite solid		
	$q/kD(t_1 - t_w) = 2\pi/(1 - D/4x)$	$x > D/2$
	$= 2\pi$	$x \gg D$
Vertical circular hole in surface of semi-infinite solid		
	$q/kL(t_1 - t_w) = 2\pi/\ln(4L/D)$	$L > D$
Horizontal circular hole in semi-infinite solid		
	$q/kL(t_1 - t_w) = 2\pi/\ln(4x/D)$	$D < x < L$
	$= 2\pi/\ln(2L/D)$	$x \gg D$

TABLE 7. Equations for Heat Loss from Buried Objects and Cavities (*Continued*)

Infinite circular hole in semi-infinite solid		
	$q'/k(t_1 - t_w) = 2\pi/\cosh^{-1}(2x/D)$	$x \approx D$
	$= 2\pi/\ln\left[2x/D + \sqrt{(2x/D)^2 - 1}\right]$	$x > 2D$
	$= 2\pi/\ln(4x/D)$	$x \gg D$
Two infinite circular holes in semi-infinite solid		
	$q'_1/k(t_1 - t_w) = 2\pi\left\{ \dfrac{\ln\left(\frac{4x_1}{D}\right)\ln\left(\frac{4x_2}{d}\right) + \left[\ln\sqrt{\dfrac{(x_1+x_2)^2 + l^2}{(x_1-x_2)^2 + l^2}}\right]^2\right]^{-1}}{\ln\left(\frac{4x_2}{d}\right) - \left(\frac{t_2 - t_w}{t_1 - t_w}\right)\ln\sqrt{\dfrac{(x_1+x_2)^2 + l^2}{(x_1-x_2)^2 + l^2}}} \right\}$ (Interchange subscripts for q'_2)	$D > d$ $x_1 > 2D$ $x_2 > 2d$
Row of infinite circular holes at equal depth in semi-infinite solid		
	$q'/k(t_1 - t_w) = 2\pi/\ln\left[(2l/\pi D)\sinh(2\pi x/l)\right]$ (Any one hole)	$x > D$

Row of infinite circular holes in midplane of infinite plate		
	$q'/k(t_1 - t_w) = 2\pi/\ln\left[(2l/\pi D)\sinh(\pi\delta/l)\right]$ (Any one hole)	$\delta > D$
Two infinite circular holes in infinite solid		
	$q'/k(t_1 - t_2) = 2\pi/\cosh^{-1}\left[(4l^2 - D^2 - d^2)/2Dd\right]$	$D > d$
Horizontal toroidal hole in semi-infinite solid		
	$q/kD(t_1 - t_w) = 2\pi^2/\ln\left\{8\left(\frac{D}{d}\right)\left[\dfrac{\ln(2D/x)}{\ln(8D/d)} + 1\right]\right\}$ $= 2\pi^2/\ln(8D/d)$	$\dfrac{d}{2} \ll x < \dfrac{D}{2}$ $x \gg D$
Circular hole in square solid		
	$q'/k(t_1 - t_w) \doteq 2\pi/\ln(1.08 l/D)$	$l > D$
Eccentric circular hole in cylindrical solid		
	$q'/k(t_1 - t_w) = 2\pi/\cosh^{-1}\left[(D^2 + d^2 - 4l^2)/2Dd\right]$	$D > d$

Two-phase Materials. An otherwise continuous and homogeneous material may have discontinuous *inclusions* of a second homogeneous material that are dispersed in some regular array (Fig. 3*a*). This second phase may consist, for example, of dispersed particles, insulating microballoons, reinforcing fibers, or voids (porosity). Neglecting inclusion geometry and dispersion pattern of the second phase, effective conductivity can be derived by simply imagining the dispersed phase to be represented by lumped parallel or series resistances;

$$k_e = (1 - \Gamma)k_1 + \Gamma k_2 \quad (parallel)$$

$$k_e = \frac{k_1 k_2}{\Gamma k_1 + (1 - \Gamma)k_2} \quad (series)$$

Here k_1 and k_2 are thermal conductivity of parent (first phase) and dispersed (second phase) materials respectively, and Γ is volume fraction of the latter.

A better approximation assumes the composite is subdivided into groups of elemental parallel resistances that are summed in series (*parallel/series*) or, alternatively, subdivided into groups of series resistances summed in parallel (*series/parallel*) [69]:

- *Square Inclusions in Square Array.* For *square* inclusions of side D arranged in a two-dimensional square pattern of spacing d;

$$\frac{k_e}{k_1} = \frac{\kappa + \beta(1 - \kappa)}{\kappa + \beta(1 - \beta)(1 - \kappa)} \quad (parallel/series)$$

$$\frac{k_e}{k_1} = \frac{1 - \beta(1 - \beta)(1 - \kappa)}{1 - \beta(1 - \kappa)} \quad (series/parallel)$$

where $\kappa = k_1/k_2$ and $\beta = D/d$. If the square inclusions are pores of conductivity k_2, then porosity is $P = \beta^2$.

- *Cubic Inclusions in Square Array.* For *cubic* inclusions of side D arranged in a square pattern of spacing d;

$$\frac{k_e}{k_1} = \frac{\kappa + \beta^2(1 - \kappa)}{\kappa + \beta^2(1 - \beta)(1 - \kappa)} \quad (parallel/series)$$

$$\frac{k_e}{k_1} = \frac{1 - \beta(1 - \beta^2)(1 - \kappa)}{1 - \beta(1 - \kappa)} \quad (series/parallel)$$

Here $P = \beta^3$ for cubic pores.

- *Circular Inclusions in Square Array.* For *circular* inclusions of diameter D arranged in a square pattern of spacing d (parallel/series);

$$\frac{k_e}{k_1} = \left\{1 - \beta\left[1 + \frac{\gamma}{\sqrt{1 - \gamma^2}} \ln\left(\frac{1 - \sqrt{1 - \gamma^2}}{\gamma}\right)\right]\right\}^{-1} \quad (\gamma^2 < 1)$$

$$\frac{k_e}{k_1} = \left\{1 - \beta\left[1 + \frac{\gamma}{\sqrt{\gamma^2 - 1}}\left(\tan^{-1}\frac{1}{\sqrt{\gamma^2 - 1}} + \frac{\pi}{2}\right)\right]\right\}^{-1} \quad (\gamma^2 > 1)$$

where $\gamma = (1/\beta)\kappa/(1 - \kappa)$. Here $P = (\pi/4)\beta^2$ for circular pores.

- *Spherical Inclusions in Cubic Array.* For *spherical* inclusions of diameter D arranged in a three-dimensional cubic pattern of spacing d (parallel/series);

$$\frac{k_e}{k_1} = \left\{ 1 - \beta \left[1 - \frac{\gamma'}{2\sqrt{\gamma'+1}} \ln\left(\frac{\sqrt{\gamma'+1}+1}{\sqrt{\gamma'+1}-1} \right) \right] \right\}^{-1} \qquad (\kappa < 1)$$

$$\frac{k_e}{k_1} = \left\{ 1 - \beta \left[1 + \frac{\gamma'}{\sqrt{-\gamma'-1}} \tan^{-1}\left(\frac{1}{\sqrt{-\gamma'-1}} \right) \right] \right\}^{-1} \qquad (\kappa > 1)$$

where $\gamma' = (4/\pi\beta^2)\kappa/(1-\kappa)$. Here $P = (\pi/6)\beta^3$ for spherical pores.

Constrictions. If the steady heat flux q'_x flowing in an infinite solid encounters a thin insulated *crack* $-l \le y \le l$, $-\infty \le z \le \infty$, its distribution about the crack is

$$\frac{q'_x}{q'_{x=\infty}} = \left[\frac{x^2 + y^2}{\sqrt{x^4 + 2(y^2 + l^2)x^2 + (y^2 - l^2)^2}} \right]^{1/2} \qquad (173)$$

where $q'_{x=\infty}$ is undisturbed flux at large $\pm x$ ($q'_{y=\infty}$ assumed zero). At crack tips $x = 0$, $y = \pm l$ the flux is infinite. Along the central flow line ($y = 0$) Eq. (173) gives a flux ratio $q'_x/q'_{x=\infty} = (x/l)\sqrt{(x/l)^2 + 1}$. Thus, at a point l ahead of (or behind) the crack the flux is 70.7 percent of $q'_{x=\infty}$; at a point five crack widths away the flux recovers to 99.5 percent.

The foregoing heat flow, though locally altered by a finite material discontinuity (flow constriction), is otherwise unaffected if the conducting medium is infinite. If the solid is not infinite, such geometric discontinuities reduce the overall body conductance. Figure 73 shows an infinite plate (or strip) with four different *cut patterns* normal to heat-flux vector q'. Exact and measured conductance data K [70]

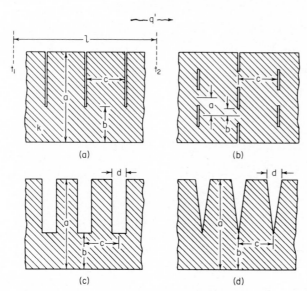

Fig. 73. Wall cuts normal to heat flow: (a) regular cuts, (b) staggered cuts, (c) rectangular grooves, (d) vee grooves.

for these cut patterns (plate and cut boundaries adiabatic) are tabulated below, normalized by conductance of the uncut wall $K_{uncut} = ka/l$:

3-124 **Handbook of Heat Transfer**

(a)	K/K_{uncut}		
c/a	$b/a=1/2$	1/4	1/8
4	0.902	0.760	0.646
2	0.818	0.618	0.480
1	0.704	0.465	0.339
1/2	0.610	0.363	0.235
1/4	0.564	0.303	0.182

(b)	K/K_{uncut}		
c/a	$b/a=1/2$	1/4	1/8
3/2	0.747	0.520	0.381
3/4	0.575	0.315	0.217
1/2	0.430	0.209	0.146
1/3	0.296	0.113	0.080

(c)		K/K_{uncut}		
c/a	d/a	$b/a=1/2$	1/4	1/8
3	1/8	0.822	0.634	0.478
	3/8	0.767	0.546	0.374
	5/8	0.721	0.480	0.311
	12/8	0.599	0.339	0.190
3/2	1/16	0.740	0.486	0.342
	3/16	0.698	0.430	0.284
	5/16	0.661	0.390	0.242
	12/16	0.570	0.299	0.164
1/2	0	0.627	0.403	0.270
	1/16	0.561	0.312	0.186
	3/16	0.527	0.276	0.152

(d)		K/K_{uncut}		
c/a	d/a	$b/a=1/4$	1/8	1/16
3/2		0.402	0.507	
		0.469		0.362
		0.502		0.310
		0.866	0.479	
		1.010		0.338
		1.083		0.254
1/2	1/2	0.329	0.213	0.141

An approximation for a wall with a *single* rectangular groove is [71]

$$q' = \frac{k}{(1 - d)/a + (d/b) + (4/\pi)\ln[\sec(\pi/2)(1 - b/a)]}(t_1 - t_2) \qquad (174)$$

Corner. The distribution of heat flux $q'(x, y)$ through an *infinite corner* (Fig. 74) with isothermal surfaces t_1 and t_2 $(< t_1)$ is [18]

$$q' = k\left\{\frac{x}{l} + \frac{y}{\delta} + \frac{2}{\pi}\left[\ln\left(\frac{l^2 + \delta^2}{4l\delta}\right) + \frac{\delta}{l}\tan^{-1}\left(\frac{l}{\delta}\right) + \frac{l}{\delta}\tan^{-1}\left(\frac{\delta}{l}\right)\right]\right\}(t_1 - t_2) \quad (175)$$

If the corner were ignored, $q' = k(x/l + y/\delta)(t_1 - t_2)$; thus, the corner increases the heat loss by

$$\frac{2}{\pi}k\left[\ln\left(\frac{l^2 + \delta^2}{4l\delta}\right) + \frac{\delta}{l}\tan^{-1}\left(\frac{l}{\delta}\right) + \frac{l}{\delta}\tan^{-1}\left(\frac{\delta}{l}\right)\right](t_1 - t_2)$$

For a corner formed by walls of equal thickness $(l = \delta)$, this increase amounts to $0.559k(t_1 - t_2)$.

Parallelepiped Shell. Empirical conductance data for closed *rectangular shells* (Fig. 75) having six isothermal inner surfaces at t_1 and six isothermal outer surfaces at t_2 are [72]:

1. Equal side, floor, and ceiling thicknesses, $\delta_s = \delta_f = \delta_c = \delta$:

$$For \quad a, b, c > \frac{\delta}{5}, \quad K = k\left[\frac{S_1}{\delta} + 2.16(a + b + c) + 1.20\delta\right] \qquad (176)$$

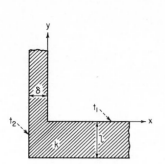

Fig. 74. Infinite corner.

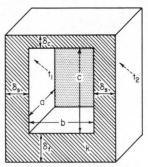

Fig. 75. Closed rectangular shell.

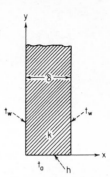

Fig. 76. Semi-infinite plate.

For $a, b > \dfrac{\delta}{5}$ and $c < \dfrac{\delta}{5}$, $K = k\left[\dfrac{S_1}{\delta} + 1.86\,(a + b) + 0.35\delta\right]$ $\qquad$ (177)

For $a > \dfrac{\delta}{5}$ and $b, c < \dfrac{\delta}{5}$, $K = k\left[\dfrac{6.41a}{\ln(S_2/S_1)}\right]$ $\qquad$ (178)

For $a, b, c < \dfrac{\delta}{5}$, $K = k\left(\dfrac{0.79\sqrt{S_1 S_2}}{\delta}\right)$ $\qquad$ (179)

Here S_1 and S_2 are the total inner and outer surface areas of the shell. The second and last terms in Eqs. (176) and (177) represent shape-factor corrections for the edges and corners respectively.

2. Unequal side, floor, and ceiling thicknesses, $\delta_s \neq \delta_f \neq \delta_c$:

For $a, b, c > \dfrac{\delta_{max}}{5}$, $K = k\left[\dfrac{4ac}{\delta_s} + ab\left(\dfrac{1}{\delta_f} + \dfrac{1}{\delta_c}\right) + 1.08\,(a + b)\right.$

$$\left. + 2.16\,(a + c) + 0.4\,(4\delta_s + \delta_f + \delta_c)\right] \qquad (180)$$

Plate. A *semi-infinite* plate $0 \le x \le \delta$, $y \ge 0$ has steady face temperatures t_w at $x = 0, \delta$ and its edge face $y = 0$ exposed to uniform convective heating or cooling (Fig. 76). The following tabulation gives selected data [73] on the integrated heat flux q' along edge face $y = 0$ and temperatures $t(x, 0)$ along this face as a function of Biot number Bi $= h\delta/k$:

				$q'/k(t_w - t_a)$ at $y = 0$					
$h\delta/k$	0.001	0.01	0.1	1	10	20	100	200	1000
	0.001	0.01	0.097	0.792	3.107	3.971	5.984	6.827	8.603
x/δ	$(t - t_a)/(t_w - t_a)$ at $y = 0$								
0.05		0.999	0.989	0.906	0.552		0.111		
0.1		0.998	0.982	0.854	0.406		0.069		
0.2		0.997	0.973	0.788	0.275		0.036		
0.3		0.997	0.968	0.750	0.219		0.027		
0.5		0.996	0.964	0.723	0.186		0.022		

These data also refer to a thin, semi-infinite rectangular strip with adiabatic surfaces and longitudinal edges $x = 0, \delta$ at t_w.

Rib. An infinite *rib* of thickness δ and with surface temperature t_w projects from an infinite plate (same material) of thickness l and with inner surface at t_w. Figure 77 shows distributions of steady temperature along rib centerline y for uniform convective heating or cooling at outer plate surface $y = 0$ [74].

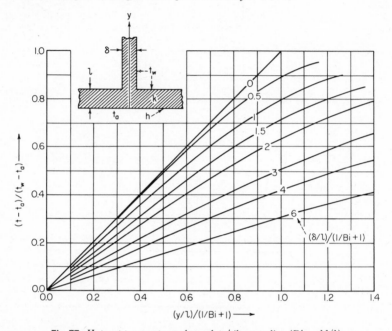

Fig. 77. Hotspot temperatures along plate/rib centerline (Bi = hl/k).

Sandwich Structures. Many heat-sink walls consist of a *sandwich* structure where face cover sheets enclose some form of rigid (structural) or loose insulation (Fig. 3c). Heat transfer through these walls is by combined two-dimensional conduction, convection, and radiation. Generally, the cover sheets represent negligible thermal resistance compared to filler, and as such an *effective* conductivity is sufficient to characterize wall insulation performance.

● *Honeycomb Sandwich.* Figure 78 gives effective thermal conductivity k_e as a function of mean face temperature $\bar{t} = (t_1 + t_2)/2$ for various square-cell honeycomb core materials brazed to thin, high-conductivity cover sheets at t_1 and t_2 [75]. The following tabulation gives common core sizes and honeycomb gages in order of decreasing effective conductivity for a given core material:

core type	square cell size (in.)	honeycomb gage (mil)	decrease in k_e (%)
3–20	3/16	2	0
3–15	3/16	1.5	32
4–20	1/4	2	35
4–15	1/4	1.5	59
3–10	3/16	1	68
6–20	3/8	2	76

core type	square cell size (in.)	honeycomb gage (mil)	decrease in k_e (%)
4–10	1/4	1	82
6–15	3/8	1.5	90
6–10	3/8	1	100

The last column gives the approximate percent decrease in k_e between core types 3–20 and 6–10; these data are generally applicable for all mean face temperatures and core materials shown in Fig. 78.

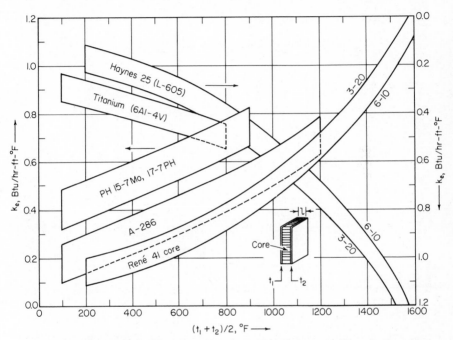

Fig. 78. Effective thermal conductivity of brazed honeycomb (square cell) sandwich structure.

A simplified expression based on an approximate one-dimensional, decoupled heat-transfer analysis of honeycomb-core sandwich panels (experimentally verified by measurements on brazed hexagonal honeycomb to 1500°F) is given by [76]

$$q'' = \frac{k_c}{l}\frac{A_c}{A}(t_1 - t_2) + \frac{k_a}{l}(t_1 - t_2) + 0.664\sigma\epsilon^{1.63(l/d+1)^{-0.89}}\left(\frac{l}{d} + 0.3\right)^{-0.69}(t_1^4 - t_2^4)$$

(181)

Here k_c is the conductivity of core material, k_a is the conductivity of entrapped air, l is the core height (Fig. 78), d is the equivalent honeycomb diameter (core circumference/π), A_c/A is the core solidity (ratio of conduction area to total area), and σ is the Stefan-Boltzmann constant. First, second, and third terms in Eq. (181) account, respectively, for separate contributions of steady-state core conduction, air conduction (core convection neglected), and core radiation.

• *Insulation Sandwich.* Figure 79 gives effective thermal conductivity k_e of evacuated foams, fibers, and multilayer (shielded) insulations used in cryogenic and elevated-temperature structures [77]. Conductivity of air versus temperature is shown

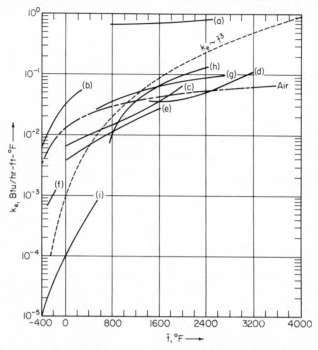

Fig. 79. Effective thermal conductivity of evacuated foams, powders, fibers, and multilayer insulations: (a) silicon carbide foam (17 lb/ft³), (b) Styrofoam (2 lb/ft³), (c) zirconia powder (150 lb/ft³), (d) thoria powder (16 lb/ft³), (e) ADL-17 powder (16 lb/ft³), (f) perlite powder (10 lb/ft³), (g) Dynaquartz fiber (5 lb/ft³), (h) Fiberfrax multilayer felt, tantalum shields (5 lb/ft³), (i) Linde S14 multilayer felt, aluminum shields (5 lb/ft³).

for reference. Effective conductivity of foams (a,b) is dictated by solid conduction. This contribution diminishes in powders and fiberous insulators (c-g) because the solid is discontinuous (contact resistance). Insulation is further improved by interposing foil radiation shields (h,i) to attenuate radiation. At high temperatures, all insulators are dominated by radiative transfer; in this instance, heat flux goes as $q \sim \bar{t}^4$, so effective conductivity behaves roughly as $k_e \sim \bar{t}^3$.

Effective conductivity of loose or rigid porous insulations is reasonably predictable. Analysis of low-temperature, fiberous multilayer insulation is treated in Ref.[78]. The physics of multiple radiation shields in cryogenic insulations is given in Ref. [79]. Various analytical models for predicting high-temperature insulation performance of foams, powders, and refractory flakes are reviewed in Ref.[80]. Effective conductivity for many of these porous insulations are adequately correlated by a simple decoupled model such as

$$k_e = (1 - P)k_s + Pk_g + 4d\sigma\bar{t}^3 \qquad (182)$$

In Eq. (182), $\bar{t}$ is the mean material temperature, k_s is the conductivity of solid at $\bar{t}$, k_g is the conductivity of entrained gases at $\bar{t}$, P is the material porosity, and d is the pore diameter.

REFERENCES

1. P. J. Schneider, "Conduction Heat Transfer," Addison-Wesley Publishing Co., Inc., Reading, Mass., 1957.
2. Editors, The Promise of Composites, *Materials in Design Engineering,* **58**:79–126 (1963).
3. Anon., Materials for Space Operations, *NASA SP-27,* Washington, D.C., Dec. 1962.
4. Anon., Structures for Space Operations, *NASA SP-28,* Washington, D.C., Dec. 1962.
5. E. Fried and F. A. Costello, Interface Thermal Contact Resistance Problem in Space Vehicles, *ARS Journal,* **32**:237–243 (1962).
6. M. E. Barzelay, et al., Effect of Pressure on Thermal Conductance of Contact Joints, *NACA TN 3295* (Syracuse Univ.), Washington, D.C., May 1955.
7. E. Fried, Study of Interface Thermal Contact Conductance, *General Electric Co. 64SD652,* May 1964 (also *ARS* Paper No. 1690-61, April 1961).
8. H. Fenech and W. H. Rohsenow, Prediction of Thermal Conductance of Metallic Surfaces in Contact, *J. Heat Transfer, Trans. ASME,* **85**:15–24 (1963).
9. M. E. Barzelay, Range of Interface Thermal Conductance for Aircraft Joints, *NASA TN D-426* (Syracuse Univ.), Washington, D.C., May 1960.
10. D. C. Andrew, An Investigation of the Thermal Conductance of Bolted Joints, *RAE TN WE46,* Ministry of Aviation, London, Jan. 1964.
11. D. M. Lewis and H. J. Sauer, Jr., Thermal Resistance of Adhesive Bonds, *J. Heat Transfer, Trans. ASME,* **87**:310–311 (1965).
12. T. N. Veziroglu, Correlation of Thermal Contact Conductance Experimental Results, *Prog. Astron. Aero.,* **20**, Academic Press, Inc., New York, 1967.
13. K. M. Kratsch, L. F. Hearne, and H. R. McChesney, Theory for the Thermophysical Performance of Charring Organic Heat-Shield Composites, *Lockheed Missiles & Space Co. 803099,* Oct. 1963.
14. R. E. Gaumer, Problems of Thermal Control Materials in Space Environment, *ASM Golden Gate Metals Conf.,* Feb. 1962.
15. S. Glasstone (ed.), The Effects of Nuclear Weapons, *U.S. Dept. of Defense,* U.S. Gov't. Printing Office, Washington, D.C., June 1957.
16. J. C. Rogers and T. Miller, Survey of the Thermal Threat of Nuclear Weapons (unclassified), *U.S. Dept. of Defense,* U.S. Dept. of Commerce, Washington, D.C., Nov. 1963.
17. W. E. Olmstead, L. A. Peralta, and S. Raynor, Transient Radiation Heating of a Rotating Cylindrical Shell, *AIAA Journal,* **1**:2166–2168 (1963).
18. H. S. Carslaw and J. C. Jaeger, "Conduction of Heat in Solids," 2d ed., Oxford University Press, London, 1959.
19. P. J. Schneider, "Temperature Response Charts," John Wiley & Sons, Inc., New York, 1963.
20. L. A. Schmit, The Influence of Aerodynamic Heating on the Structural Design of High Speed Aircraft, part II, Massachusetts Institute of Technology, 1955.
21. D. M. Tellep, Temperature Response of Thin Skins Under Low Heat Flux Conditions, *Lockheed Missiles & Space Co. IAD 349,* Sept. 1958.
22. J. R. Katz, When Is a Skin Thermally Thin?, *U.S. Naval Ordnance Test Station TN 4061-10,* May 1958.
23. C. J. Thorne, Temperature Tables. Part I. One-Layer Plate, One Space Variable, Linear, *U.S. Naval Ordnance Test Station NAVORD 5562 (NOTS 1756),* July 1957.
24. P. J. Schneider, Thermal Equilibration, *J. Spacecraft and Rockets,* **4**:402–404 (1967).
25. M. P. Heisler, Temperature Charts for Internal Heat Generation, *Trans. ASME,* **78**:1187–1192 (1956).
26. P. J. Schneider and J. J. Brogan, Temperature Response of a Transpiration-Cooled Plate, *ARS Journal,* **32**: 233–236 (1962).
27. M. D. Mikhaylov, Transient Temperature Fields in Shells, *NASA TT F-552,* Washington, D.C., June 1969.
28. W. H. Holter and J. H. Grover, Insulation Temperature for the Transient Heating of an Insulated Infinite Metal Slab, *ARS Journal,* **30**:907–908 (1960).
29. J. H. Grover and W. H. Holter, Solution of the Transient Heat-Conduction Equation for an Insulated Infinite Metal Slab, *Jet Propulsion,* **27**:1249–1252 (1957).
30. R. S. Harris, Jr. and J. R. Davidson, An Analysis of Exact and Approximate Equations for the Temperature Distribution in an Insulated Thick Skin Subject to Aerodynamic Heating, *NASA TN D-519,* Washington, D.C., Jan. 1961.

31. E. R. G. Eckert and R. M. Drake, Jr., "Heat and Mass Transfer," 2d ed., McGraw-Hill Book Company, Inc., New York, 1959.
32. R. D. Zerkle and J. E. Sunderland, The Transient Temperature Distribution in a Slab Subject to Thermal Radiation, *J. Heat Transfer, Trans. ASME,* **87**:117–133 (1965).
33. D. L. Ayers, Transient Cooling of a Sphere in Space, *J. Heat Transfer, Trans. ASME,* **92**:180–182 (1970).
34. D. L. Ayers, The Transient Temperature Distribution in a Radiating Cylinder, *ASME* Paper No. 67-HT-71, Aug. 1967.
35. M. Jakob, "Heat Transfer," vol. I, John Wiley & Sons, Inc., New York, 1949.
36. M. L. Anthony, Temperature Distribution in Slabs with a Linear Temperature Rise at One Surface, part I, *Proceedings of the General Discussion on Heat Transfer, IME and ASME,* Sept. 1951.
37. H. Gröber, S. Erk, and U. Grigull, "Fundamentals of Heat Transfer," McGraw-Hill Book Company, Inc., New York, 1961.
38. N. Konopliv and E. M. Sparrow, Transient Heat Conduction in Nonhomogeneous Spherical Systems, *Wärme-und Stoffübertragung,* **3**:197–210 (1970).
39. W. A. Brooks, Jr., Temperature and Thermal-Stress Distributions in Some Structural Elements Heated at a Constant Rate, *NACA TN-4306,* Washington, D.C., 1958.
40. A. B. Newman, The Temperature-Time Relations Resulting From the Electrical Heating of the Face of the Slab, *Trans. AIChE,* **30**:598–613 (1934).
41. C. D. Coulbert et al., Temperature Response of Infinite Flat Plates and Slabs to Heat Inputs of Short Duration at One Surface, Dept. of Engineering, University of California, Los Angeles, April 1951.
42. J. A. Kuhn, Temperature Distribution in a Finite Slab Caused by a Heat Rate of the Form $q(\theta) = (q_{max}/2)(1 - \cos \omega\theta)$, *Lockheed Missiles & Space Co. TXN/548,* Jan. 1961.
43. A. R. Mendelsohn, Transient Temperature of a Porous-Cooled Wall, *AIAA Journal,* **1**:1449–1451 (1963).
44. H. Halle and D. E. Taylor, Transient Heating of Two-Element Slabs Exposed to a Plane Heat Source, *Chicago Midway Laboratories CML-59-M-1,* Chicago, May 1959.
45. P. J. Schneider and C. F. Wilson, Jr., Pulse Heating and Equilibration of an Insulated Substructure, *J. Spacecraft and Rockets,* **7**:1457–1460 (1970).
46. H. C. Hottel and C. C. Williams III, Transient Heat Flow in Organic Materials Exposed to High-Intensity Thermal Radiation, *ASME Heat Transfer & Fluid Mechanics Institute,* 1954.
47. A. L. London and R. A. Seban, Rate of Ice Formation, *Trans. ASME,* **65**:771–778 (1943).
47. A. L. London and R. A. Seban, Rate of Ice Formation, *Trans. ASME,* **65**:000 (1943).
48. R. R. Cullom and W. H. Robbins, One-Dimensional Heat-Transfer Analysis of Thermal-Energy Storage for Solar Direct-Energy-Conversion Systems, *NASA TN D-2119,* Washington, D.C., March 1964.
49. J. E. Sunderland and R. J. Grosh, Transient Temperature in a Melting Solid, *J. Heat Transfer, Trans. ASME,* **83**:409–414 (1961).
50. H. G. Landau, Heat Conduction in a Melting Solid, *Quart. Appl. Math.,* **8**:81–94 (1950).
51. J. J. Kauzlarich, Ablation of Reinforced Plastic for Heat Protection, *J. Appl. Mech.,* **32**:177–182 (1965).
52. R. J. Barriault and J. Yos, Analysis of the Ablation of Plastic Heat Shields That Form a Charred Surface Layer, *ARS Journal,* **30**:823–829 (1960).
53. P. J. Schneider, T. A. Dolton, and G. W. Reed, Mechanical Erosion of Charring Ablators in Ground-Test and Re-Entry Environments, *AIAA Journal,* **6**:64–72 (1968).
54. R. A. Anderson and W. A. Brooks, Jr., Effectiveness of Radiation as a Structural Cooling Technique for Hypersonic Vehicles, *J. Aero/Space Sci.,* **27**:41–48 (1960).
55. W. P. Manos, D. E. Taylor, and A. J. Tuzzolino, Thermal Protection of Structural, Propulsion, and Temperature-Sensitive Materials for Hypersonic and Space Flight, *Chicago Midway Laboratories WADC TR 59-366,* part II, Chicago, July 1960.
56. W. H. McAdams, "Heat Transmission," 3d ed., McGraw-Hill Book Company, Inc., New York, 1954.
57. J. W. Tatom, Shell Radiation, *ASME* Paper No. 60-WA-234, Nov. 1960.
58. W. E. Olmstead and S. Raynor, Solar Heating of a Rotating Spherical Space Vehicle, *Int. J. Heat Mass Transfer,* **5**:1165–1177 (1962).
59. A. Brown, Optimum Dimensions of Uniform Annular Fins, *Int. J. Heat Mass Transfer,* **8**:655 (1965).
60. K. A. Gardner, Efficiency of Extended Surface, *Trans. ASME,* **67**:621–631 (1945).

61. S. Lieblein, Analysis of Temperature Distribution and Radiant Heat Transfer Along a Rectangular Fin of Constant Thickness, *NASA TN D-196*, Washington, D.C., Nov. 1959.
62. C.-Y. Liu, On Minimum-Weight Rectangular Radiating Fins, *J. Aerospace Sci.*, **27**:871–872 (1960).
63. O. N. Favorskii and I. S. Kadaner, Methods of Analyzing Fundamental Elements of Cooler-Radiators, in "Questions of Heat Exchange in Space" (Russian), Vysshaia Shkola, Moscow, 1967.
64. R. L. Chambers and E. V. Somers, Radiation Fin Efficiency for One-Dimensional Heat Flow in a Circular Fin, *J. Heat Transfer, Trans. ASME*, **81**:327–329 (1959).
65. J. G. Bartas and W. H. Sellers, Radiation Fin Effectiveness, *J. Heat Transfer, Trans. ASME*, **82**:73–75 (1960).
66. J. E. Wilkins, Jr., Minimizing the Mass of Thin Radiating Fins, *J. Aero/Space Sci.*, **27**:145–146 (1960).
67. J. E. Wilkins, Jr., Minimum Mass Thin Fins for Space Radiators, *Proc. Heat Transfer and Fluid Mechanics Institute, 1960*, Stanford University, June 1960.
68. S. S. Kutateladze, "Fundamentals of Heat Transfer" (English ed.), Academic Press, Inc., New York, 1963.
69. J. J. Duga, Electrical and Thermal Transport Models for Analysis of Reinforced Composites, *Battelle Memorial Institute ONR TR No. 1*, July 1966.
70. S. Katzoff, Similitude in Thermal Models of Spacecraft, *NASA TN D-1631*, Washington, D.C., April 1963.
71. E. L. Knuth, An Engineering Approximation for Resistances of Certain Two-Dimensional Conductors, *Int. J. Heat Mass Transfer*, **7**:270–272 (1964).
72. I. Langmuir, E. Q. Adams, and F. S. Meikle, Flow of Heat Through Furnace Walls, *Trans. Amer. Electrochem. Soc.*, **24**:53–84 (1913).
73. J. H. Vansant and M. B. Larson, Heat Transfer From a Semi-Infinite Rectangular Strip, *J. Heat Transfer, Trans. ASME*, **85**:191–192 (1963).
74. J. G. Bartas, Gas-Side Wall Temperatures in Rib-Backed Liquid-Cooled Combustion Chambers, *Jet Propulsion*, **27**:784–786 (1957).
75. Anon., Designing Brazed Sandwich Structure, *Rohr Corporation*, 1962.
76. R. T. Swann and C. M. Pittman, Analysis of Effective Thermal Conductivities of Honeycomb-Core and Corrugated-Core Sandwich Panels, *NASA TN D-714*, Washington, D.C., April 1961.
77. M. L. Minges, Passive Insulations for Reentry Vehicles, *Space Aeronautics*, **42**:78–85 (1964).
78. D. I-J. Wang, Multiple Layer Insulations, in "Aerodynamically Heated Structures" (P. E. Glaser, ed.), Prentice-Hall, Inc., Englewood Cliffs, N.J., 1962.
79. A. G. Emslie, Fundamental Limitations of Multilayered Radiation Shields, in "Aerodynamically Heated Structures" (P. E. Glaser, ed.), Prentice-Hall, Inc., Englewood Cliffs, N.J., 1962.
80. A. E. Wechsler and M. A. Kritz, Investigation and Development of High Temperature Insulation Systems, *Air Force Materials Laboratory TR-65-138*, Ohio, June 1965.

NOMENCLATURE

English

A	conducting area, profile area, ft^2
a	semi-major diameter of ellipse, ft
a	accommodation coefficient
B	rate coefficient, 1/hr
B	gap number
Bi	Biot number = $h\delta/k$ (typical)
B_y	incomplete beta function
b	semi-minor diameter of ellipse, ft
C	specific heat, Btu/lb-°F
C	infinite cylinder, constriction number
D	diameter, ft
D	heating duration, hr
d	distance, diameter, ft
d/dx	total derivative with respect to x
$\partial/\partial x$	partial derivative with respect to x
E	activation energy, Btu/mole

E complete elliptic integral

e base of natural logarithm = 2.71828

e fin efficiency

erf   error function

erfc   complementary error function = 1 − erf

F coolant flow number = $G_c C_c \delta / k_e$

Fo   Fourier number = $\alpha\theta/\delta^2$ (typical)

Fo′   modified Fourier number = $\alpha D/\delta^2$

G mass flow rate, lb/hr-ft^2

g $h/\rho C\delta$, 1/hr

H latent heat, sensible heat, Btu/lb

H^* heat of ablation, Btu/lb

h unit surface conductance, Btu/hr-ft^2-$^\circ$F

I_n nth-order modified Bessel function of first kind

ierfc   integral of complementary error function

J $mD^2/2Q''$

J_n nth-order Bessel function of first kind

K thermal conductance = hA, kA/δ, Btu/hr-$^\circ$F

K conductivity number

K_n nth-order modified Bessel function of second kind

k thermal conductivity, Btu/hr-ft-$^\circ$F

L length, ft

L $\sqrt{\omega/2\alpha}\,\delta$, hl/k

l length, width, depth, ft

ln   natural logarithm (base e)

M Meyer hardness

M_0 radiation number = $\sigma \mathscr{F} t_0^3 \delta/k$ (typical)

Ma   Mach number

m $\sqrt{h/k\delta}$, 1/ft

m $\rho_2 C_2 / \rho_1 C_1$

N $(h_i/h)(t_0 - t_d)/(t_d - t_a)$, $(mk - h)/(mk + h)$

N_1 $(q''\delta/\sigma \mathscr{F} t_0^4 + T_s^4)^{1/4}$

N_2 $(N_1^4 + ST_a)^{1/4}$

n $\sqrt{\sigma \epsilon t_w^3/k\delta}$, 1/ft

n $\rho_2 C_2 \delta_2 / \rho_1 C_1 \delta_1$, positive integers, reaction order

P power, Btu/hr-ft^2

P infinite plate, porosity

Pr   Prandtl number

p pressure, lb/ft^2

p $q'''\delta/hs$, $^\circ$F

p point

Q cumulative heat, Btu

Q'' cumulative heat flux, Btu/ft^2

q instantaneous heat rate, Btu/hr

q'' instantaneous heat flux, Btu/hr-ft^2

q''' internal volume heat source, Btu/hr-ft^3

R gas constant, Btu/mole-$^\circ$R

R thermal resistance = $1/hA$, δ/kA, hr-$^\circ$F/Btu

R normalized radius = r/r_2

r radial coordinate, $\sqrt{\xi^2 + y^2 + z^2}$, ft

S surface area, ft^2

S semi-infinite solid, $h/\sigma \mathscr{F} t_0^3$

s	shape factor = A/δ, ft
s	geometry factor = $S\delta/V$
T	normalized temperature = t/t_0
t	temperature, $^\circ$F or $^\circ$R
U	thermal transmittance, Btu/hr-ft^2-$^\circ$F
V	volume, ft^3
v	velocity, ft/hr
W	unit weight = $\rho\delta$, lb/ft^2
w	$q''/\sigma\epsilon t_w^4$
X	$x/2\sqrt{\alpha\theta}$, $L(1 - x/\delta)$, normalized distance = x/δ
x, y, z	rectangular coordinates, ft
Y	energy yield of explosion, megatons
Y_n	nth-order Bessel function of second kind

Greek

α	thermal diffusivity = $k/\rho C$, ft^2/hr
α	radiation absorptivity
β	grain angle, $(\mathrm{Pr}_a)^{1/3}(\gamma_a - 1)/2$, D/d
Γ	volume fraction
γ	angle of heat-flux vector in nonisotropic solid, ratio of specified heats, $(1/\beta)\kappa/(1 - \kappa)$
γ'	$(4/\pi\beta^2)\kappa/(1 - \kappa)$
Δ	finite difference
δ	thickness, ft
ϵ	radiation emissivity, eccentricity of ellipse = $\sqrt{1 - (b/a)^2}$
ζ	$q''\delta^2/2k(t_A - t_a)$
η	$\sqrt{(h_a + h_A)\delta/k}$
Θ	$(h^2/\rho Hk)(t_d - t_a)\theta$
θ	time, hr
κ	k_1/k_2, Boltzmann constant
λ	$q_0''\delta^2 b_0/k$, $q_0'' r_1^2 b_0/k$
μ	radiation absorption coefficient, 1/ft
μ	micro, $n + (n + 1)/\mathrm{Bi}_1$
ν	kinematic viscosity, ft^2/hr
ν	$\sqrt{2[(h_a + h_A)/h_1](k/k_1)\delta/r_1}$
ξ	$x - v\theta$, ft
ρ	density, lb/ft^3
σ	Stefan-Boltzmann radiation constant = 17.3×10^{-10}, Btu/hr-ft^2-$^\circ$R^4
τ	normalized time = θ/D
ϕ	latitude angle in cylindrical coordinates, degrees
ψ	azimuth angle in spherical coordinates, phase angle, degrees
$\omega/2\pi$	harmonic frequency, cycles/hr
ϑ	temperature difference = $t - t_0$, $^\circ$F
$\mathcal{F}$	radiation configuration–emissivity factor

Subscripts

A, a	convective ambients
a	air
aw	adiabatic wall
c	ceiling, coolant, core, critical
d	decomposition
E	electrical

e	effective, external
eqb	equilibrium
f	floor, fluid
g	gas
i	interface, positive integers
m	matrix, melting
max	maximum
min	minimum
n	normal direction
0	initial ($\theta \leq 0$), zero contact pressure
∞	infinite
opt	optimum
p	parallel direction, pyrolysis
r	radial, resin, response
re	radiation equilibrium
S	solar
s	radiation sink, scale deposit, sensible, side, solid, surface
ss	steady state
t	tip
v	vaporization, virgin
w	wall container, wall surface

Superscripts

+	amplitude
--	mean value
$'$	per unit length
$''$	per unit area
$'''$	per unit volume

Section 4

Methods of Obtaining Approximate Solutions

P. RAZELOS

Associate Professor of Engineering Science
Richmond College of the CUNY

Various methods of obtaining approximate solutions of the steady-state and transient heat transfer equations are given.

A. STEADY STATE

Problems involving systems with stationary temperature fields are treated.

1. Introduction

a. Finite-difference Approximations.   Approximate solutions of the heat transfer equations are obtained by replacing the differential equation with finite differences, and solving the resulting set of algebraic equations. The temperature then will be determined at only a finite number of discrete points which are called *nodes*. The substitution of the derivatives by their finite differences can be achieved as follows:

Consider first a function of one variable $T(x)$, which is assumed analytic. The Taylor's theorem relating the value of the function at the points x and $x + h$ states that

$$T(x + h) = T(x) + hT'(x) + \left(\frac{h^2}{2!}\right)T''(x) + \cdots \tag{1}$$

Suppose we know the value of the function at a set of equally spaced points (Fig. 1)

$$\ldots x - 2h, x - h, x, x + h, x + 2h, \ldots$$

and we want to approximate the derivative at x by values of the function at these points. Then from Eq. (1) one may obtain

$$T'(x) = \frac{T(x + h) - T(x)}{h} \tag{2}$$

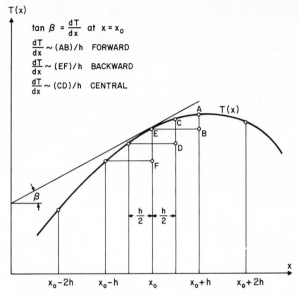

Fig. 1. Approximation of the first derivative.

If T_n is the value of the function at the point $x_0 + nh$, then

$$T'_n = \frac{T_{n+1} - T_n}{h} \qquad (3)$$

If $-h$ is substituted for h in Eq. (1), one obtains an approximation of the derivative with values of the function at the points x and x-h

$$T'_n = \frac{T_n - T_{n-1}}{h} \qquad (4)$$

A more accurate approximation of the derivative is obtained if the values of the function at the points x and $h/2$ and $x - h/2$ are considered

$$T'_n = \frac{T_{n+\frac{1}{2}} - T_{n-\frac{1}{2}}}{h} \qquad (5)$$

The numerators in Eqs. (3), (4), and (5) are the forward, backward, and central first-order differences, respectively, of the function $T(x)$. The usual notation is

$$\Delta T_n = T_{n+1} - T_n \qquad \text{(6) Forward}$$

$$\nabla T_n = T_n - T_{n-1} \qquad \text{(7) Backward}$$

$$\delta T_n = T_{n+\frac{1}{2}} - T_{n-\frac{1}{2}} \qquad \text{(8) Central}$$

Higher order differences are readily obtained in the same manner; thus, the second-order finite differences are

$$\Delta(\Delta T_n) = \Delta^2 T = \Delta T_{n+1} - \Delta T_n = T_{n+2} - T_{n+1} + T_n \tag{9}$$

$$\nabla(\nabla T_n) = \nabla^2 T_n = \nabla T_n - \nabla T_{n-1} = T_n - 2T_{n-1} + T_{n-2} \tag{10}$$

$$\delta(\delta T_n) = \delta^2 T_n = \delta T_{n+\frac{1}{2}} - \delta T_{n-\frac{1}{2}} = T_{n+1} - 2T_n + T_{n-1} \tag{11}$$

As an example, consider the function $T(x) = 3x^2$; the successive values, with their successive first-, second-, and third-order forward differences, are presented in Table 1.

TABLE 1. Forward Differences

x	1	2	3	4	5	6	. . .
$T(x)$	3	12	27	48	75	102	. . .
ΔT	9	15	21	27	33	. . .	
$\Delta^2 T$	6	6	6	6	. . .		
$\Delta^3 T$	0	0	0	. . .			

An approximation of the second derivative is now obtained as the ratio of one of the previous second differences divided by h^2, e.g.,

$$T_n'' = \frac{\delta^2 T_n}{h^2} = \frac{T_{n+1} - 2T_n + T_{n-1}}{h^2} \tag{12}$$

If T is a function of two variables, $T(x, y)$, its value at the points $x = x_0 + mh$, $y = y_0 + nk$, can be conveniently written using a double script notation

$$T(x, y) = T_{m,n} = T(x_0 + mh, y_0 + nk) \tag{13}$$

Here h and k are constant intervals in the x and y direction, respectively, forming a rectangular array of mesh points. Approximation of the partial derivatives can be obtained using the Taylor's expansion which relates the value of $T(x, y)$ to the value $T(x + h, y + k)$ [1].

In steady-state two-dimensional heat transfer, the temperature satisfies either Laplace's equation

$$\frac{\partial^2 T}{\partial x^2} + \frac{\partial^2 T}{\partial y^2} = 0 \tag{14}$$

or Poisson's equation

$$\frac{\partial^2 T}{\partial x^2} + \frac{\partial^2 T}{\partial y^2} = f(x, y) \tag{15}$$

In either case, if an approximate solution is desired, one should replace the partial derivatives by the finite differences. Let us assume a rectangular mesh of points with $h = k$. Consider the point (x, y) and its surrounding four points $(x + h, y)$, $(x - h, y)$,

$(x, y + h)$, and $(x, y - h)$ shown in Fig. 2. The left-hand sides of Eqs. (14) and (15) can be approximated by [1]

$$\frac{T_{m+1,n} + T_{m-1,n} + T_{m,n+1} + T_{m,n-1} - 4T_{m,n}}{h^2} \tag{16}$$

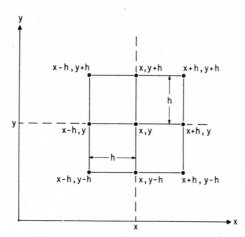

Fig. 2. Two-dimensional Cartesian system.

It is convenient to consider the numerator in Eq. (16) as an operator acting on $T_{m,n}$ which is usually referred to as the Laplacian 5-point difference operator [2]. It is denoted by H and it can be schematically represented by the stencil

$$H = \begin{array}{|c|c|c|} \hline 0 & 1 & 0 \\ \hline 1 & -4 & 1 \\ \hline 0 & 1 & 0 \\ \hline \end{array}$$

One may think of this stencil as centered at the position (m, n); then the number of each position is the coefficient multiplying the value of $T(x, y)$ at that position.

Suppose one has to solve the Laplace or Poisson's equation in a rectangular region. Then using Eq. (16), an approximate solution for the temperature at the points (mh, nh) can be obtained by solving the set of algebraic equations $H(T_{m,n}) = 0$, or $H(T_{m,n}) = h^2 f_{m,n}$.

Consider further the points $(m +1, n + 1)$, $(m + 1, n - 1)$, $(m - 1, n + 1)$, and $(m - 1, n - 1)$. It is clear that these form a configuration similar to that of the points in the H operator,* so that the operator $2X$ defined as

$$2X = \begin{array}{|c|c|c|} \hline 1 & 0 & 1 \\ \hline 0 & -4 & 0 \\ \hline 1 & 0 & 1 \\ \hline \end{array}$$

should also be an approximation to the Laplace's differential operator. This is easily shown by virtue of the Taylor's theorem [2].

*There are several other ways to approximate $\delta^2(T_{m,n})$. For a complete description see L. Collatz, "The Numerical Treatment of Differential Equations," 3d ed., p. 542, translated by P. G. Williams, Springer-Verlag, New York Inc., 1966 [35].

A more accurate finite difference approximation to the Laplacian differential operator can be obtained by properly combining the H and $2X$ operators. The best combination is given by the nine-point operator K defined as follows [2]

$$
K = 4H + 2X =
\begin{array}{|c|c|c|}
\hline
1 & 4 & 1 \\
\hline
4 & -20 & 4 \\
\hline
1 & 4 & 1 \\
\hline
\end{array}
$$

The extension to three-dimensional problems is obvious; thus, the Laplace's equation will be replaced by the algebraic equation referring to a three-dimensional mesh, with three subscripts m, n, r defining the position of the node. Thus

$$
\frac{\partial^2 T}{\partial x^2} + \frac{\partial^2 T}{\partial y^2} + \frac{\partial^2 T}{\partial z^2}
$$

$$
= \frac{T_{m+1,n,r} + T_{m-1,n,r} + T_{m,n+1,r} + T_{m,n,r+1} + T_{m,n,r-1} - 6T_{m,n,r}}{h^2} = 0 \quad (17)
$$

Any one of the previous methods can be used to obtain the set of algebraic equations satisfied by the nodal temperatures. The choice depends on the nature of the problem, the accuracy required, and the means available for the solutions of these equations. These equations are satisfied by the interior points of an orthogonal mesh. The treatment of the boundaries and other than square nets is explained in the following sections.

b. Formulation of the Nodal Equations. A numerical solution, in contrast to an analytical one, determines the temperatures only in preselected discrete points in the system in which heat flow occurs. Each of these points represents a region, i.e., a volume and its bounding surface, in the system, and its temperature is taken as the representative temperature of that region. Heat flow at the boundary of a region is taken as the summation of the flows from the adjacent regions.

The first step in a numerical analysis is to subdivide the system into suitable regions, locate the reference points (nodes) at the center of each region, and assign a reference number to each node. If the problem demands the temperature at certain points, these points should be included among the reference points. The number of points would depend on the required accuracy and it is advantageous for hand calculations to start with the *crudest* possible subdivision. Occasionally it would be necessary to go to a finer subdivision but the previous results are useful as a starting point of the future calculations.

In certain cases one may subdivide the system physically into regions. Otherwise, an arbitrary geometrical subdivision can be used according to some scheme which best will simplify the work.

Example 1. Physical subdivision. Consider a building wall consisting of an outer layer of facing brick, followed by a layer of common brick and an inner layer of gypsum plaster. The thermal conductivities and the thickness of the various layers are given and the temperatures and the corresponding heat transfer coefficients of the outside and the inside air are given. It is required to find the surface and interface temperatures.

Using a natural physical subdivision, one may consider the following five regions (Fig. 3): Outside air, facing brick, common brick, gypsum plaster, and inside air. (Note that the nodes are located at the points where the temperature is needed.)

Example 2. Geometric Subdivisions. Consider a circular pin fin of diameter D, length L, and thermal conductivity k, exposed to a gas of temperature T_a, with a base temperature T_1. The average heat transfer coefficient over its surface is h. It is required to find the temperature distribution in the fin.

In this case, one may subdivide the fin geometrically in arbitrary regions (in this case, four). It is convenient to use regions of equal length, with the possible exception of the last one (Fig. 4).

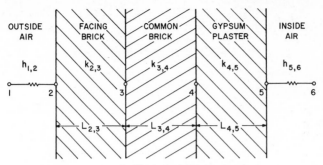

Fig. 3. Physical subdivision.

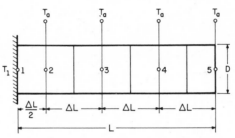

Fig. 4. Geometric subdivision.

After completing the subdivision of the system, the thermal conductance K_{ij} between the node i and its adjacent j nodes must be calculated. The rate of heat flow q_{ij} (Btu/h) from any node j to the adjacent node i is

$$q_{ij} = K_{ij}(T_j - T_i) \tag{18}$$

The heat balance for the node i gives the following equation

$$\sum_j q_{ij} = \sum_j K_{ij}(T_j - T_i) = 0 \tag{19}$$

(The summation is extended to the j adjacent nodes.) If heat q''' (Btu/hr-unit volume) is generated in the system, the term $q'''V_i$ (V_i being the volume assigned to node i) should be added in the left-hand side of Eq. (19). If n points have been selected, one then obtains n algebraic equations as in Eq. (19); the object of a numerical analysis is to provide methods of solving these equations.

The thermal conductances are calculated using the formulas

Interior points conduction: $$K_{ij} = k\left(\frac{A_{ij}}{L_{ij}}\right) \tag{20}$$

Boundary points: $$K_{ij} = h_j A_{ij} \tag{21}$$

where k is the thermal conductivity, A_{ij} the average surface perpendicular to the direction of heat flow, h the average heat transfer coefficient on the surface A_{ij}, and L_{ij} the distance between the nodes.

Example 3.   Heat balance equation for points 2 and 3 of Example 1.   Considering one square foot of surface, one may obtain the following

$$K_{1,2} = h_{1,2} \text{ Btu/hr-}^\circ\text{F}$$

$$K_{2,3} = \frac{k}{L_{2,3}} \text{ Btu/hr-}^\circ\text{F}$$

$$K_{3,4} = \frac{k}{L_{3,4}} \text{ Btu/hr-}^\circ\text{F}$$

Point 2

$$h_{1,2}(T_1 - T_2) + \left(\frac{k}{L_{2,3}}\right)(T_3 - T_2) = 0 \tag{22}$$

Point 3

$$\left(\frac{k}{L_{2,3}}\right)(T_2 - T_3) - \left(\frac{k}{L_{3,4}}\right)(T_4 - T_3) = 0 \tag{23}$$

Example 4.   Heat balance equation for points 2 and 3 of Example 2

$$K_{1,2} = 2K_{2,3} = 2K_{3,4} = k\left(\frac{\pi D^2}{4}\right)\left(\frac{2}{\Delta L}\right)$$

$$K_{2,a} = K_{3,a} = h\pi D\,(\Delta L)$$

Point 2

$$K_{1,2}(T_1 - T_2) + K_{2,a}(T_a - T_2) + K_{2,3}(T_3 - T_2) = 0 \tag{24}$$

Point 3

$$K_{2,3}(T_2 - T_3) + K_{3,a}(T_a - T_3) + K_{3,4}(T_4 - T_3) = 0 \tag{25}$$

2. Relaxation Process

a. **Fundamentals.**   The relaxation process is a numerical method for solving a set of algebraic equations such as the approximate heat-conduction Eq. (19). The general method was first used by R. Southwell in 1935 for the solution of framework problems and was applied to heat transfer problems by H. W. Emmons [3]. The fundamental ideas of the method are given below.

Assume a value for each of the unknown temperatures T_i. Substitute these values into Eq. (19); the right-hand side is not zero any more but has a value, let us say R_i, which is called the *residual*. Any change ΔT_i in T_i will change R_i by $-(\sum_j K_{ij})\Delta T_i$ and the R_j (j's are the adjacent nodes to i) by $(K_{ij})\Delta T_i$. Thus a change in a temperature causes a redistribution of the residuals which follows a fixed pattern. The desirable results will be the reduction of all residuals to zero. The relaxation technique accomplishes this by successive changes in the T_i's. The method should be carried out according to the following general rules.

1. Compute and record all residuals which result from the assumed values of the temperatures.

2. Choose the node i at which the $R_i/\sum_j K_{ij}$ is the largest, and relax it (i.e., reduce R_i to zero by changing T_i by an amount equal to minus $R_i/\sum_j K_{ij}$).

3. Compute the new residual R_j (due to the change of T_i) of the adjacent nodes.

4. Again select the largest $R_i/\Sigma K_{ij}$ and repeat steps 2 and 3; continue in this fashion until all residuals are reduced as nearly to zero as accuracy requires. The recording of the new temperatures and residuals should be done systematically, and many forms of tables have been suggested. One such table, which might not be the shortest one but has the merit of logic and clarity, provides for each node two columns, one marked "temperature" and the other "residual" (Table 4). In the first line all assumed temperatures and the corresponding residuals are recorded. Any new line will comprise only the new (changed) temperature and the new residuals. To assist the computations, one may prepare an operation table which indicates the changes ΔR_j in R_j for a unit change in the T_i's.

TABLE 2. Operation Table

	ΔR_1	ΔR_2	ΔR_n
$\Delta T_1 = 1$	$-\sum K_{1,j}$	$K_{2,1}\;\cdots$	K_{n1}
$\Delta T_2 = 1$	$K_{1,2}$	$-\sum K_{2,j}\;\cdots$	K_{n2}
$\vdots$			
$\Delta T_n = 1$	K_{1n}	$K_{2n}\;\cdots$	$-\sum K_{nj}$

Occasionally it is convenient to divide each equation by $\sum_j K_{ij}$ (especially when these sums are different); thus, the coefficient of T_i is -1 and the largest residual can be detected immediately. The new operation table will be as follows:

TABLE 3. New Operation Table

	ΔR_1	ΔR_2		ΔR_n
$\Delta T_1 = 1$	-1	$D_{1,2}$	$\cdots$	D_{1n}
$\Delta T_2 = 1$	$D_{2,1}$	-1	$\cdots$	D_{2n}
$\vdots$				
$\Delta T_n = 1$	D_{n1}	D_{n2}	$\cdots$	-1

In this case

$$D_{ij} = \frac{K_{ij}}{\sum_j K_{ij}} \neq D_{ji} \tag{26}$$

The choice of the assumed arbitrary values of the nodal temperatures will influence the length of computation; the closer our original assumption is to the correct value, the fewer steps we shall have to make. One should always first examine carefully the problem at hand and try to make a good guess; of course, this guess depends on the experience and intuition of the calculator. Some helpful suggestions are given in the various examples. As a rule, one should never assume an interior temperature higher or lower than a boundary temperature in a steady-state problem with no heat sources. In general, an infinite number of steps will be required to reduce all residuals to zero. Practically, one stops when the relaxation of any residual (step 3 above) will produce a change in its corresponding temperatures less than a prescribed one, and when all residuals are rather *evenly* distributed as to sign. The *even* distribution of the residual sign is not always possible if the above-described procedure is followed. However, it can be made possible (and it is *recommended* to do so) by using the modifications of the method explained in Sec. A.3. When all steps are completed, a final check over the results should be made, referring to the original equations. It is recommended *not to wait* until the last step, but to check the result in a few intermediate steps.

b. One-dimensional Conduction. Thin Fins. As a first example consider the pin fin of Example 2.

Example 4. The pin is 3.5 in. long and 0.5 in. in diameter. Its base is at a uniform temperature of $100°F$ and is exposed to an ambient temperature of $0°F$, through an average heat transfer coefficient of 3 Btu/hr-ft^2-$°F$. The fin is made from a material with a thermal conductivity of 24 Btu/hr-ft-$°F$. The tip of the fin is assumed to be insulated. With these data and the subdivision shown in Fig. 4 the various conductances are

$$K_{1,2} = k\pi\left(\frac{D^2}{4}\right)\left(\frac{1}{L_{1,2}}\right) = \frac{\pi}{4} \text{ Btu/hr-}°F$$

$$K_{2,3} = K_{3,4} = K_{4,5} = \frac{K_{1,2}}{2} = \frac{\pi}{8} \text{ Btu/hr-}°F$$

$$K_{2,a} = h\pi D(\Delta L) = \frac{\pi}{96} \text{ Btu/hr-}°F$$

$$K_{a,3} = K_{a,4} = K_{a,2} = 2K_{a,5} = \frac{\pi}{96} \text{ Btu/hr-}°F$$

The resulting equations are

Point 2
$$\frac{1}{4}(T_1 - T_2) + \frac{1}{96}(T_a - T_2) + \frac{1}{8}(T_3 - T_2) = 0$$

Point 3
$$\frac{1}{8}(T_2 - T_3) + \frac{1}{96}(T_a - T_3) + \frac{1}{8}(T_4 - T_3) = 0$$

Point 4
$$\frac{1}{8}(T_3 - T_4) + \frac{1}{96}(T_a - T_4) + \frac{1}{8}(T_5 - T_4) = 0 \qquad (27)$$

Point 5
$$\frac{1}{8}(T_4 - T_5) + \frac{1}{2 \times 96}(T_a - T_5) = 0$$

Substituting for $T_1 = 100$ and $T_a = 0$, one then obtains the following equations

$$2400 + 12T_3 - 37T_2 = 0$$

$$12T_2 + 12T_4 - 25T_3 = 0$$

$$12T_3 + 12T_5 - 25T_4 = 0 \qquad (28)$$

$$24T_4 - 25T_5 = 0$$

TABLE 4. Temperature Distribution in a Pin

Line		ΔR_2		ΔR_3		ΔR_4		ΔR_5
1	$\Delta T_2 = 1$	-1		0.48		0		0
2	$\Delta T_3 = 1$	0.324		-1		0.48		0
3	$\Delta T_4 = 1$	0		0.48		-1		0.96
4	$\Delta T_5 = 1$	0		0		0.48		-1
5	T_2	R_2	T_3	R_3	T_4	R_4	T_5	R_5
6	95	-4.22	80	-0.80	70	-0.4	65	2.2
7	90.78	0		-2.83				
8		-0.92	77.17	0		-1.76		
9						.7	67.2	0
10				-0.34	69.3	0		-0.67
11	89.86	0		-0.78				
12		-0.25	76.39	0		-0.37		
13						-0.69	66.53	0
14				-0.33	68.61	0		-0.66
15						-0.32	65.87	0
16		-0.36	76.06	0		-0.48		
17				-0.23	68.13	0		-0.46
18						-0.22	65.41	0
19	89.50	0		-0.40				
20		-0.13	75.66	0		-0.41		
21				-0.17	67.72	0		-0.39
22						-0.19	65.02	0
23	89.5	-0.13	75.7	-0.16	67.7	-0.19	65.0	0
24	90.0		75.3		67.0		64.3	

In this case it is better to divide each equation by the corresponding coefficient of the temperature appearing with negative sign ($\sum_j K_{ij}$), resulting in the following system

$$64.86 + 0.324\,T_3 - T_2 = 0$$
$$0.48\,T_2 + 0.48\,T_4 - T_3 = 0$$
$$0.48\,T_3 + 0.48\,T_5 - T_4 = 0$$
$$0.96\,T_4 - T_5 = 0$$

$$(29)$$

Let us assume the temperatures of 95, 80, 70, and 65°F for the nodes 2, 3, 4, 5, respectively, and let us say that the computations should end when the relaxation of any residual produces a change in its corresponding temperature of less than a quarter of a degree (i.e., when all residuals have reduced to a value of less than 0.25). As it can be seen from Table 4, where all the steps of the calculations are shown, this has been accomplished after 17 steps (line 22). On top of the relaxation table there is the operation table. All calculated temperatures have been recorded (the numbers are rounded off) on line 23. For comparison the exact values of the temperatures are shown on line 24, calculated according to Eq. (160) in Section 3.

The pin's efficiency, which equals its average temperature, calculated from the above approximate solution is

$$e = \frac{89.5 \times 1 + 75.7 \times 1 + 67.7 \times 1 + 65.5 \times 0.5}{100 \times 3.5} = 75.8\%$$

$$(30)$$

The exact solution according to the equation (given in Table 6, Sec. 3) gives an efficiency of 75.8%. The agreement between the exact and the approximate solution is excellent in spite of the crude subdivision.

Example 5.   An annular fin has an inner radius of 1 in. and an outer radius of 5 in. The fin has a uniform thickness $2w = 1/12$ in. and is composed of a material with thermal conductivity $k = 36$ Btu/hr-ft-°F. The base of the fin is maintained at 100°F and the surrounding fluid is at 0°F. If the average heat transfer coefficient is 2 Btu/hr-ft²-°F, determine the temperature distribution and the efficiency of the fin. Assume the tip of the fin to be insulated.

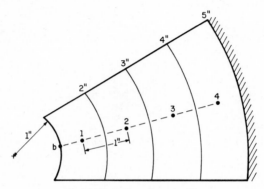

Fig. 5. Sectioning of annular fin.

Let us subdivide the fin in four equal sections as shown in Fig. 5 and locate the nodes 1, 2, 3, 4 at the center of each section. The thermal conductances are calculated taking as an average area of heat flow between nodes the thickness times the arithmetic mean of their corresponding radii. Thus

$$K_{ij} = 2\pi k \left(\frac{r_i + r_j}{2}\right) \cdot \left(\frac{1}{\Delta r_{ij}}\right)(2w) \tag{31}$$

The conductance between a node and the surrounding fluid is twice the product of the film coefficient times the average area corresponding to the node (twice because heat is transferred from both sides of the fin). For example, typical conductances are

$$K_{b,1} = 2\pi k \left(\frac{r_b + r_1}{2}\right)\left(\frac{1}{\Delta r_{b,1}}\right)2w$$

$$= 2\pi 36 \left(\frac{1 + 1.5}{2}\right)\left(\frac{1}{0.5}\right)\left(\frac{1}{144}\right) = 1.25\pi \text{ Btu/hr-°F} \tag{32}$$

$$K_{1,a} = 2\pi \frac{2^2 - 1^2}{144} \cdot h = \frac{\pi}{12} \text{ Btu/hr-°F} \tag{33}$$

The calculated values of the conductances are:

$$K_{1,b} = 1.25 \text{ Btu/hr-°F} \qquad K_{1,a} = \tfrac{1}{12} \text{ Btu/hr-°F}$$

$$K_{1,2} = 1 \quad\ \text{ Btu/hr-°F} \qquad K_{2,a} = \tfrac{5}{36} \text{ Btu/hr-°F}$$

$$K_{2,3} = 1.5 \ \text{ Btu/hr-°F} \qquad K_{3,a} = \tfrac{7}{36} \text{ Btu/hr-°F}$$

$$K_{3,4} = 2 \quad \text{Btu/hr-}°\text{F} \quad K_{4,a} = \tfrac{9}{36} \text{ Btu/hr-}°\text{F}$$

(All conductances have been divided by π.) The corresponding equations for the nodes are

$$1.25(T_b - T_1) + (T_2 - T_1) + \tfrac{1}{12}(T_a - T_1) = 0$$

$$(T_1 - T_2) + 1.5(T_3 - T_2) + \tfrac{5}{36}(T_a - T_1) = 0 \tag{34}$$

$$1.5(T_2 - T_3) + 2(T_4 - T_3) + \tfrac{7}{36}(T_a - T_3) = 0$$

$$2(T_3 - T_4) + \tfrac{9}{36}(T_a - T_4) = 0$$

Substituting the values of T_b and T_a into Eq. (34), one then obtains the following system

$$4500 + 36T_2 - 84T_1 = 0$$

$$36T_1 + 54T_3 - 95T_2 = 0 \tag{35}$$

$$54T_2 + 72T_4 - 133T_3 = 0$$

$$72T_3 - 81T_4 = 0$$

In this case it is also better to divide each equation with the coefficient of the temperature appearing with negative sign $(\sum_j K_{ij})$; thus one obtains

TABLE 5. Temperature Distribution in an Annular Fin

Line		ΔR_1		ΔR_2		ΔR_3		ΔR_4
1	$\Delta T_1 = 1$	-1		0.379		0		0
2	$\Delta T_2 = 1$	.429		-1		.406		0
3	$\Delta T_3 = 1$	0		.568		-1		.889
4	$\Delta T_4 = 1$	0		0		.541		-1
5	T_1	R_1	T_2	R_2	T_3	R_3	T_4	R_4
6	70	7.7	55	-2.91	45	-3.74	35	5
7	77.7	0		-0.19				
8						-1.03	40	0
9				-0.78	43.97	0		-0.92
10						-0.50	38.08	0
11		-0.33	54.22	0		-0.82		
12			-0.47		43.15	0		-0.73
13						-0.39	38.35	0
14		-0.53	53.75	0		-0.58		
15				-0.33	42.57	0		-0.52
16	76.64	0		-0.53				
17		-0.23	53.22	0		-0.22		
18						-0.40	37.83	0
19				-0.23	42.17	0		-0.36
20						-0.19	37.47	0
21		-0.33	52.99	0		-0.28		
22	76.31	0		-0.13				
23				-0.29	41.85	0		-0.25
24		-0.12	52.7	0		-0.1		
25						-0.24	37.22	0
26				-0.14	41.61	0		-0.21
27						-0.11	37.01	0
28	76.3		52.7		41.6		37	
29	74.1		50.2		38.9		34.6	

$$53.571 + 0.429T_2 - T_1 = 0$$

$$0.379T_1 + 0.568T_3 - T_2 = 0 \tag{36}$$

$$0.406T_2 + 0.541T_4 - T_3 = 0$$

$$0.889T_3 - T_4 = 0$$

Let us assume the values of 70, 55, 45, and 35°F for the nodes 1, 2, 3, and 4, respectively, and let us say that we shall stop the calculations when the relaxation of any residual produces a change in its corresponding temperature of less than 0.2°F. All the steps of the calculations are shown in Table 5. On line 28 the calculated values are shown and on line 29, for comparison, the exact values calculated according to Eq. (156) (Sec. 3.B.) are given. The fin's efficiency can be calculated from the above approximate solution, and is equal to

$$e = \frac{3 \times 76.3 + 5 \times 52.7 + 7 \times 41.6 + 9 \times 37}{(3 + 5 + 7 + 9)10} = 46.5\% \tag{37}$$

(The numbers 3, 5, 7, and 9 are proportional to the volume assigned to each node.) The exact value, calculated according to the equation (given in Table 6, Sec. 3), is equal to 45.0%. The agreement between the exact and the approximate solution is satisfactory in spite of the crude subdivision.

c. Two-dimensional Conduction. Cartesian Coordinates. Two-dimensional steady-state heat-flow problems can be approximately solved using the relaxation process. Usually square nets are used and the nodal equations are obtained with the methods explained in Secs. 1.a and 1.b. The most frequent boundary conditions encountered in practical applications are those of prescribed surface temperatures, Newtonian cooling, or prescribed heat flux. The application of the method is explained in the following examples where all types of boundary conditions are considered. Some useful suggestions for facilitating the computations and selecting the starting-up values of the nodal temperatures are also made. However, the calculator *should always use the modifications* of the method explained in Sec. A.3 which greatly shorten the length of computations.

Example 6. Prescribed Surface Temperature. A steel slab with thermal conductivity of 18 Btu/hr-ft-°F has its two faces perfectly insulated and the other four kept at constant temperature. The dimensions and temperatures are shown in Fig. 6. Determine the temperature distribution within the slab. Let us take $\Delta x = \Delta y = 1$ ft such that one has to solve for eight interior, equally spaced nodes indicated by the numbers 1 to 8. In this case, where the subdivision of the system in exact squares is possible, and the boundary temperatures are prescribed, the nodal equations can be written down immediately as has been shown in Sec. 1.a, using, for example, the operator H and the equation

$$HT_i = 0 \tag{38}$$

Therefore the following equations are obtained

$$800 + 100 + T_2 + T_3 - 4T_1 = 0$$

$$T_1 + 100 + 700 + T_4 - 4T_2 = 0$$

$$800 + T_1 + T_4 + T_5 - 4T_3 = 0$$

$$T_3 + T_2 + 700 + T_6 - 4T_4 = 0 \tag{39}$$

$$800 + T_3 + T_6 + T_7 - 4T_5 = 0$$

$$T_5 + T_4 + 700 + T_8 - 4T_6 = 0$$

$$800 + 800 + T_5 + T_8 - 4T_7 = 0$$

$$T_7 + T_6 + 700 + 800 - 4T_8 = 0$$

The same equations would have been obtained using the heat-balance method explained in Sec. 1.b. For example, taking the heat balance for node 1 (for a unit length in the third direction) one may obtain the following equation

$$\frac{k(\Delta y)(800 - T_1)}{\Delta x} + \frac{k(\Delta x)(100 - T_1)}{\Delta y} + \frac{k(\Delta y)(T_2 - T_1)}{\Delta x} + \frac{k(\Delta x)(T_3 - T_1)}{\Delta y} = 0 \quad (40)$$

Equation 40, after having been divided by $k(\Delta y)/\Delta x$, becomes identical with the first equation in Eq. (39) for $\Delta x = \Delta y$

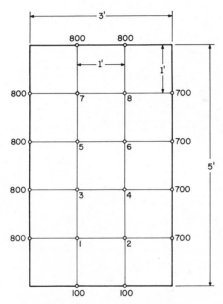

Fig. 6. Sectioning of steel slab.

In this case the lengthy operations table can be omitted; clearly any change ΔT_i in a temperature T_i changes its own residual R_i by -4 and the residual of the adjacent nodes by 1. During the course of computation then, instead of the operation table one may use Fig. 6 and see immediately which are the influenced adjacent nodes. For example, if T_4 should be changed by 5 in order to reduce its own residual to zero, then looking at Fig. 6 we can see that R_2, R_3, and R_6 should be changed by 5.

To start the calculations, let us assume the values of 450, 400, 550, 500, 650, 600, 750, and 700°F for the temperatures at the nodes 1 through 8, respectively. Let us say that the calculations should end when each residual is reduced to a value such that it causes a change to its own temperature of less than 1°F, i.e., when all residuals are reduced to a value of less than 4. The sequence of the computations is shown in Table 6. Notice that in each step the residual is not always reduced to zero but to a value which is less than four, so that the change ΔT_i in temperature T_i is always the integer value of $R_i/4$. This way we avoid the decimals which in this case are beyond the required accuracy. For example, in line 2 the temperature T_4 has been changed by 62°F which is the integer of $R_4/4 = 250/4 = 62.5$. Similarly in line 3, T_3 has been changed by 66°F which is the integer of $R_3/4 = 65.5$. It is suggested that, even if the required accuracy is

TABLE 6. Temperature Distribution in a Two-dimensional Slab

Line	T_1	R_1	T_2	R_2	T_3	R_3	T_4	R_4	T_5	R_5	T_6	R_6	T_7	R_7	T_8	R_8
1	450	50	400	150	550	200	500	250	650	100	600	150	750	-50	700	50
2		50		212		262	562	2		100		212		-50		50
3		116		212	616	-2		68		166		212		-50		50
4		169	453	0		-2		121		166		212		-50		50
5		169		0		-2		174		219	653	0		-50		103
6		169		44		53		174	705	-1		55		5		103
7		169		86	606	97		-2		-1		99		5		103
8	492	1		86		139		-2		-1		99		5		103
9		36		86	651	-1		33		34		99		5		103
10		36		86		-1		33		34		125		31	726	-1
11		36		86		-1		64		65	684	1		31		30
12		58	475	-2		-1		86		65		1		31		30
13		58		20		21	628	-2		65		23		31		30
14		58		35		37		-2	721	1		39		47		30
15	507	-2		35		52		-2		1		39		47		30
16		11		35	664	0		11		14		39		47		30
17		11		35		0		11		26		39	762	-1		42
18		11		35		0		11		26		50		10	737	-2
19		11		35		0		24	697	39		-2		10		11
20		11		35		10		24	731	-1		8		20		11
21		20	484	-1		10		33		-1		8		20		11

Pt	n	Elev	n	Elev	n	Elev	n	Elev	n	Elev	n	Elev	n	Elev	n	Elev
22	20		7		18		1	636	-1		16		20		11	
23	20	512	7		18		1		4		16		0	767	16	
24	0		12		23	670	7		4		16		0		16	
25	5		12		-1		11		10		16		0		16	
26	6		12		-1		11		14		0		5		20	
27	6		12		-1		11		14		5		9		0	742
28	6		0		3		14		-2		9	701	9		0	
29	9		4	487	3		-2		-2		9		9		0	
30	9		3		7		1		-2	735	13		1		3	
31	9		3		7		1	640	1		1		1	769	5	
32	9		5		7	672	1		3		1	704	1		5	
33	1	514	5		9		3		3		1		2		5	
34	3		1	488	1		4		5		2		2		5	
35	4		1		1		4		5		2		3		5	
36	4		2		2		4		1	736	3		3		1	
37	0	515	2		3		4		1		4		3		1	
38	0		3		3		0		1		4		3		1	
39	0		4		4		1		2		0	705			2	
40	1		4		0	673	2	641	3							743
41	1				0											
42	2	515	0	489	0	673	3	641	3	736	0	705	3	769	2	743
43		515.6		488.4		680.0		647.8		742.2		709.9		776.3		749.0

less than a degree, decimals not to be used, at least at the beginning, where large residuals exist. It is not necessary to keep the entire table which can be very lengthy and difficult to handle; instead, it can be erased after a check has been made. On line 42 the final values are shown and on line 43 the exact values calculated using the series solution [4]. Despite the small number of nodes, the error is approximately 1%.

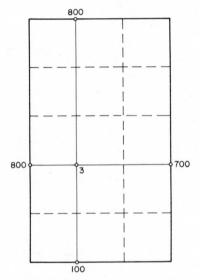

Fig. 7. Slab with one interior node.

The assigned initial values of the nodal temperatures are somewhat arbitrary, with their magnitude following a logic sequence and limits. For example, T_1 being larger than T_2 but smaller than T_3, etc., and, of course, all values lie between the limits of 800 and 100. One way of obtaining initial values is to consider one node at a time. Assume first that there is only one node, obtain its temperature; then, using this temperature, assume two nodes, etc. Use an interior node as the first node. For example, suppose there is only the node 3 (Fig. 7). The heat balance for this node yields the following equation

$$\frac{k(2.5\,\Delta y)(800 - T_3)}{\Delta x} + \frac{k(2.5\,\Delta y)(700 - T_3)}{\Delta x}$$

$$+ \frac{k(1.5\,\Delta x)(800 - T_3)}{3\Delta y} + \frac{k(1.5\,\Delta x)(100 - T_3)}{2\Delta y} = 0 \quad (41)$$

From Eq. (41) the value $T_3 = 670$ is obtained.

Next consider the system with two nodes 3 and 4 (Fig. 8) and obtain the temperature T_4 by the following equation

$$\frac{k(2.5\,\Delta y)(670 - T_4)}{\Delta x} + \frac{k(2.5\,\Delta y)(700 - T_4)}{\Delta x} + \frac{k(\Delta x)(800 - T_4)}{3\Delta y} + \frac{k(\Delta x)(100 - T_4)}{2\Delta x} = 0 \quad (42)$$

or

$$T_4 = 641$$

Continuing in this fashion the temperatures at the nodes 5, 6, 1, 2, 8, and 7 have been calculated

TABLE 7. Temperature Distribution in a Two-dimensional Slab

T_1	R_1	T_2	R_2	T_3	R_3	T_4	R_4	T_5	R_5	T_6	R_6	T_7	R_7	T_8	R_8
494	78	484	−1	670	0	641	8	745	−19	718	−38	773	1	748	−1
515		489		673		641		736		705		769		743	

and their values, together with their corresponding residuals, are shown in Table 7. For comparison the last line 42 of Table 6 has been added. It is evident that with this method very good starting-up values have been obtained.

The heat flowing out of the system can be readily calculated by considering the nodes 1, 2, 6, and 8 which have temperatures higher than their adjacent boundaries. Thus

$$q' = k[(515 - 100) + (489 - 100) + (705 - 700) + (743 - 100)] \qquad (43)$$

or

$$q' = 18 \times 852 = 15,336 \text{ Btu/hr-ft}$$

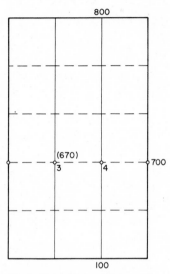

Fig. 8. Slab with two interior nodes.

In a similar way the heat flow into the system can be obtained by considering the nodes 1, 2, 3, 4, 5, 7, and 8. Thus

$$q' = k[(800 - 515) + (700 - 489) + (800 - 673) + (700 - 641) + (800 - 736)$$
$$+ 2(800 - 769) + (800 - 743)] \qquad (44)$$

or

$$q' = 18 \times 865 = 15,570 \text{ Btu/hr-ft}$$

The flow of heat in and out agrees fairly well.

Example 7. (Newtonian Cooling) Determine the temperature distribution in a rectangular 3-ft by 3-ft slab of thermal conductivity $k = 20$ Btu/ft-hr-°F, where its three sides are at constant temperatures 600, 300, and 200°F, while the fourth faces a temperature of 100°F through a heat transfer coefficient $h = 10$ Btu/hr-ft²·°F.

The subdivision of the system is shown in Fig. 9. The equations for the nodes 3, 4, 5, and 6 can be written down immediately as in the previous example, as follows

$$600 + T_1 + T_4 + T_5 - 4T_3 = 0$$

$$T_3 + T_2 + 200 + T_6 - 4T_4 = 0$$

$$600 + T_3 + T_6 + 300 - 4T_5 = 0 \tag{45}$$

$$T_5 + T_4 + 200 + 300 - 4T_6 = 0$$

The heat balance for node 1 yields the following equation

$$\frac{k(\Delta y)(600 - T_1)}{2\Delta x} + h(\Delta x)(100 - T_1) + \frac{k(\Delta y)(T_2 - T_1)}{2\Delta x} + \frac{k(\Delta x)(T_3 - T_1)}{\Delta y} = 0 \tag{46}$$

Substituting the values of k, h, Δx, Δy, one then obtains

$$600 + 100 + T_2 + 2T_3 - 5T_1 = 0 \tag{46a}$$

Similarly, the equation for node 2 is

$$T_1 + 100 + 200 + 2T_4 - 5T_2 = 0 \tag{46b}$$

Equations (45), (46a), and (46b) can be solved using the previous method.

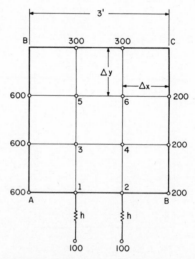

Fig. 9. Rectangular slab with Newtonian cooling.

Example 8. (General boundary conditions and heat generation) Determine the temperature distribution in the rectangular 4-ft by 4-ft slab ABCD of thermal conductivity $k = 20$ Btu/hr-ft-°F, where the faces AB, BC are insulated, on DC there is a heat input q'' of 400 Btu/hr-ft^2, AD faces an ambient temperature of 100°F through a heat transfer coefficient $h = 10$ Btu/hr-ft^2-°F, and its 2-ft by 2-ft center portion generates heat $q''' = 800$ Btu/hr-ft^3. The subdivision of the system is shown in Fig. 10 ($\Delta x = \Delta y = 1.0$ ft). Considering the third dimension to be equal to unity, some typical nodal equations are derived as follows
Node 1

$$h\left(\frac{\Delta y}{2}\right)(100 - T_1) + q''\left(\frac{\Delta x}{2}\right) + k\left(\frac{\Delta y}{2\Delta x}\right)(T_2 - T_1) + k\left(\frac{\Delta x}{2\Delta y}\right)(T_6 - T_1) = 0 \tag{47}$$

Substituting the values of h, k, q'', Δx, Δy into Eq. (47), one obtains

$$140 + 2T_2 + 2T_6 - 5T_1 = 0 \tag{47a}$$

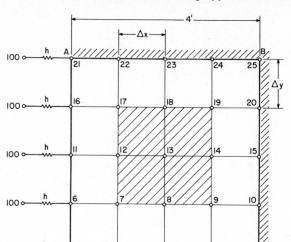

Fig. 10. Rectangular slab with three types of boundary conditions and heat generation.

Node 3

$$k\left(\frac{\Delta y}{2\Delta x}\right)(T_2 - T_3) + q''(\Delta x) + k\left(\frac{\Delta y}{2\Delta x}\right)(T_4 - T_3) + k\left(\frac{\Delta x}{\Delta y}\right)(T_8 - T_3) = 0 \quad (48)$$

or

$$40 + T_2 + T_4 + 2T_8 - 4T_3 = 0 \quad (48a)$$

Node 5

$$k\left(\frac{\Delta y}{2\Delta x}\right)(T_4 - T_5) + q''\left(\frac{\Delta x}{2}\right) + k\left(\frac{\Delta x}{2\Delta y}\right)(T_{10} - T_5) = 0 \quad (49)$$

or

$$20 + T_4 + T_{10} - 2T_5 = 0 \quad (49a)$$

Node 7

$$T_6 + T_2 + T_8 + T_{12} - 4T_7 + q'''\frac{\Delta x}{2}\frac{\Delta y}{2}\frac{1}{k} = 0 \quad (50)$$

or

$$T_6 + T_2 + T_8 + T_{12} - 4T_7 + 10 = 0 \quad (50a)$$

Node 8

$$T_7 + T_3 + T_9 + T_{13} - 4T_8 + 20 = 0 \quad (51)$$

Node 13

$$T_{12} + T_8 + T_{14} + T_{18} - 4T_{13} + 40 = 0 \quad (52)$$

Node 21

$$h\left(\frac{\Delta y}{2}\right)(100 - T_{21}) + k\left(\frac{\Delta x}{2}\right)\left(\frac{1}{\Delta y}\right)(T_{16} - T_{21}) + k\left(\frac{\Delta y}{2}\right)\left(\frac{1}{\Delta x}\right)(T_{22} - T_{21}) = 0$$

(53)

or

$$100 + 2T_{16} + 2T_{22} - 5T_{21} = 0 \tag{53a}$$

The equations for the rest of the nodes can be readily obtained in a similar manner.

d. Cylindrical Coordinates. Problems involving systems of circular shapes should be solved using cylindrical coordinates. Laplace's equation in a cylinder when its variation in the third direction is neglected takes the following form

$$\frac{\partial^2 T}{\partial r^2} + \frac{1}{r}\left(\frac{\partial T}{\partial r}\right) + \frac{1}{r^2}\left(\frac{\partial^2 T}{\partial \phi^2}\right) = 0 \tag{54}$$

Consider the circular domain $ABCD$, shown in Fig. 11, which is subdivided in sections with equal increments Δr and $\Delta \phi$. The nodal equation for the typical node 1 is obtained if we consider the finite differences of Eq. (54). The first term will be approximated by $(T_5 + T_3 - 2T_1)/\Delta r^2$, the second term, if the average of the first forward and backward differences is considered, will become $(T_5 - T_3)/2r_1\Delta r$, and the third term will give $(T_1 + T_4 - 2T_1)/r_1^2\Delta\phi^2$.

Therefore the nodal equation for node 1 becomes

$$\frac{T_5 + T_3 - 2T_1}{\Delta r^2} + \frac{T_5 - T_3}{2r_1\Delta r} + \frac{T_2 + T_4 - 2T_1}{r_1^2\Delta\phi^2} = 0 \tag{55}$$

This equation is also obtained using the energy balance, i.e., the sum of thermal currents directed toward the node 1, being equal to zero. Thus

$$k(T_2 - T_1)\left(\frac{\Delta r}{r_1\Delta\phi}\right) + \frac{k(T_3 - T_1)(r_1 - \Delta r/2)\Delta\phi}{\Delta r} + k(T_4 - T_1)\left(\frac{\Delta r}{r_1\Delta\phi}\right)$$

$$+ \frac{k(T_5 - T_1)(r_1 + \Delta r/2)\Delta\phi}{\Delta r} = 0 \tag{56}$$

Rearranging terms in Eq. (56), one readily obtains Eq. (55). For a boundary node, for example: 7, which faces an ambient temperature T_a through a heat transfer coefficient h, the nodal equation is readily obtained as follows

$$\frac{k(T_5 + T_8 - 2T_7)(\Delta r/2)}{r_7\Delta\phi} + \frac{k(T_2 - T_7)(r_7 - \Delta r/2)\Delta\phi}{\Delta r} + hr_7\Delta\phi(T_a - T_7) = 0 \tag{57}$$

In many instances it is advantageous to transform Eq. (54) using the new variable $\xi = \ln r$ instead of r. Then Eq. (54) takes the form

$$\frac{\partial^2 T}{\partial \xi^2} + \frac{\partial^2 T}{\partial \phi^2} = 0 \tag{58}$$

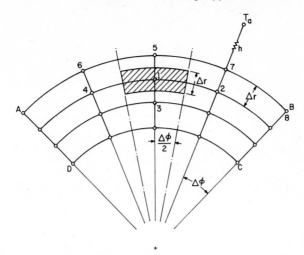

Fig. 11. Two-dimensional cylindrical coordinates.

Equation (58) is identical with the one in Cartesian coordinates and can be treated in exactly the same manner. However, it should be pointed out that equal increments of $\Delta\xi$ do not correspond to equal increments in Δr and the temperature is determined in points whose intervening distances are not constant. The following example illustrates the method.

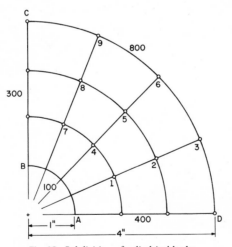

Fig. 12. Subdivision of cylindrical body.

Example 9. The dimensions and boundary temperatures of a cylindrical body are given in Fig. 12. Its temperature distribution is desired ($1'' \leq r \leq 4''$, $0 \leq \phi \leq \pi/2$).

Let us consider four sections in the ϕ direction; then $\Delta\phi = \pi/2 = 0.3927$. The total $\Delta \ln r = (\ln 4 - \ln 1) = 1.4863$. Therefore in the r direction there are three sections with $\Delta\xi = 0.3927$ and a fourth one equal to $(1.4863 - 3 \times 0.3927) = 0.3082$. The system has now been transformed as shown in Fig. 13. The nodal equations are now simpler and can be readily obtained. For example, the heat balance for nodes 5 and 3 gives

$$T_2 + T_6 + T_8 + T_4 - 4T_5 = 0 \tag{59}$$

$$(T_2 - T_3) + \frac{0.3927}{0.3082}(800 - T_3) + \left[1 + \frac{0.3082}{2(0.3927)}\right](400 + T_6 - 2T_3) = 0 \tag{59a}$$

The location or the corresponding points in the original system is determined by calculating the radii r_1, r_2, and r_3 from the corresponding ξ_1, ξ_2, ξ_3. Thus

$$\xi_1 = 0.3927 \qquad r_1 = 1.481$$

$$\xi_2 = 0.7854 \qquad r_2 = 2.193$$

$$\xi_3 = 1.178 \qquad r_3 = 3.248$$

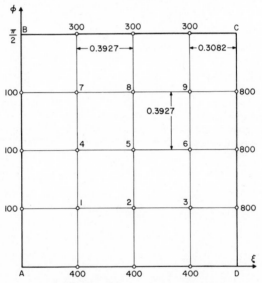

Fig. 13. Transformed cylindrical body.

3. Modifications of the Basic Method

Three modifications of the basic method of relaxation are now introduced, aiming at the speeding-up of the calculations. The application of the appropriate modification cannot be prescribed by definite rules, but it is left entirely to the computer's judgment. These speeding-up devices can be applied to a problem simultaneously; nevertheless, for the better understanding of the reader, they are treated separately.

a. Overrelaxation. This modification is useful when, at some stage of the calculations, there is an uneven distribution of the residuals (i.e., they are of the same sign). If the basic method is to be followed, the liquidation of the largest residual will increase the absolute value of the residuals of the adjacent nodes, while further reduction of these residuals will cause an increase in the value of the residual which was liquidated in the previous step. Thus the values of the residuals oscillate and the process converges slowly. This can be seen clearly from Table 6. In the first step (line 2) the residual R_4 is reduced from 250 to 2, while in the following three steps its value is increased again to 174, then is reduced to - 2 and, again during the next five steps, is increased to 86. Inspection of Table 6 reveals a similar behavior of the values of the other residuals. These observations suggest that in the first step the temperature

TABLE 8. Temperature Distribution in a Two-dimensional Slab, Determined by Overrelaxation

	T_1	R_1	T_2	R_2	T_3	R_3	T_4	R_4	T_5	R_5	T_6	R_6	T_7	R_7	T_8	R_8
1	450	50	400	150	550	200	500	250	650	100	600	150	750	−50	700	50
2		50		290	670	340	640	−310		100		290		−50		50
3		170		−70		−140		−190		220		290		−50		50
4		260	490	−70		−140		−100		220		290		−50		50
5		260		−70		−140		0		320		−110		−50		150
6		260		−70		−60		0	730	0	700	−30		30		150
7	515	0		−5		5		0		0		−30		30	740	150
8		0		−5		5		0		0		10	770	70		−10
9		0		−5		5		0		20		10		−10		10
10		0		−5		10		5		0		15		−5		10
11		0		−5		10		5	735	5	705	5		−5		15
12		0		−5		−2		8		5		−1		−1	744	−1
13		3		−5	673	0		8		8		−1		−1		−1
14		3		−5		2		0	737	0		−1		1		−1
15		3		−3		−2	642	0		0		3		1		−1
16	515	3	490	−3	673	−2	642	0	737	0	705	3	770	1	744	−1

T_4 should be *overrelaxed* thus making the value of R_4 negative such that subsequent changes of adjacent nodes will have a favorable effect on the value of R_4. In Table 8 the overrelaxation of the temperatures of the different nodes is shown step by step. Comparison of Tables 6 and 8 shows that using this device we have obtained the same results in 15 steps instead of the 41 steps required previously. It should be understood that underrelaxation can also be applied during the computations of the residuals having negative values.

   b. Block Relaxation. Block relaxation is a very effective speeding device and it should always be used for the solutions of Laplace's and Poisson's equations. It modifies the basic method to the extent that instead of reducing one residual to zero in each step, the sum of residuals comprised in a selected block is reduced to zero, thus obtaining an even distribution of the residuals. In a square network, a unit change in the temperature of the nodes in the block brings about a change of 1 to the residuals of the nodes surrounding the block and of -1 or -2 or -3 to the residuals of the nodes contained in the block. These changes are shown schematically in Figs. 14 and 15 where the dotted line indicates the bounding contour of the block.

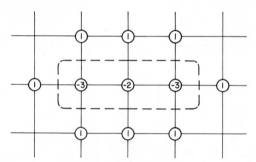

Fig. 14. Block relaxation.

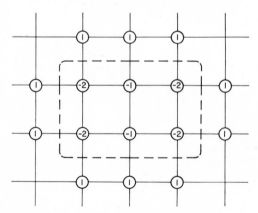

Fig. 15. Block relaxation.

   The block relaxation can be performed as follows: First, select the block (usually including nodes that have residuals of the same sign). Then add up all the residuals in the block and change the temperature of each node in it by an amount which is equal

TABLE 9. Temperature Distribution in a Two-dimensional Slab, Determined by Block Relaxation

	T_1	R_1	T_2	R_2	T_3	R_3	T_4	R_4	T_5	R_5	T_6	R_6	T_7	R_7	T_8	R_8	Block
1	450	50	400	150	550	200	500	250	650	100	600	150	750	-50	700	50	3, 4, 5, 6
2		138		238	638	24	588	74	738	-76	688	-26		38		138	1, 2
3	513	-51	463	49		87		137		-76		-26		38		130	3, 4
4		-14		86	675	-24	625	26		-39		11		38		138	
5		-14	485	86		-24		26		-39		46		73	735	-2	
6		8		-2		-24		48		-39		46		73		-2	
7		8		-2		-24		48		-21		46		1		16	
8		8		14		-8	641	0		-5	704	-2	768	1		32	4, 6
9		8		14		-8		0		-5		6		9	743	0	
10	517	-4	489	2		-4		4		-5		6		9		0	1, 2
11		-4		2		-4		4		-3		8	770	1		2	
12		-4		4		-2	643	-2		-1	706	0		1		4	4, 6
13		-3	490	0		-2		-1		-1		1		2	744	0	2, 8
14	517	-3	490	0	675	-2	643	-1	738	-3	706	1	770	2	744	0	

to the sum of the residuals divided by the number of nodes surrounding the block. The individual changes of the residuals of all nodes influenced by the block operation are readily obtained from a schematic diagram of the block similar to those of Figs. 14 and 15.

In Table 9 the solution of Example 6 is shown. The numbers assigned to the nodes included in the block, which is relaxed at each step, are shown in the last column.

c. **Group Relaxation.** This speeding device consists of the simultaneous change of the temperatures of a selected group of nodes. It differs from the block relaxation by the fact that these changes are not the same but, rather, they depend on the particular situation at hand. With this technique one may achieve an improved, but similar to the previous, distribution of the residuals in one step. As an example, consider line 1 of Table 6. Suppose that one selects the group comprising the nodes 1 through 6 and change each temperature by $R_i/4$. An improved distribution of the residuals, yet similar to the one of line 1, is obtained in line 2 of Table 10. The application of this method follows no set rules and greatly depends on the computer's experience.

TABLE 10. Group Relaxation

	T_1	R_1	T_2	R_2	T_3	R_3	T_4	R_4
1	450	50	400	150	550	200	500	250
2	462	90	438	98	600	99	562	128

	T_5	R_5	T_6	R_6	T_7	R_7	T_8	R_8
1	650	100	600	150	750	-50	700	50
2	675	88	638	85		-25		88

4. Nets Other than Squares

The subdivision of a system in squares is not always possible and in certain cases it is advantageous to use other types of relaxation nets, such as triangular, hexagonal, or even irregular nets. Moreover, in many practical problems, while a square net is used for the interior of a body, the irregular boundaries require special treatment. In all these cases, the thermal conductances between nodes can be readily calculated by a unified method [5].

Let us consider a portion of a system, as shown in Fig. 16, which has been subdivided into irregular triangular sections with the restriction that all interior angles are less than or equal to 90°. To calculate the thermal conductances between nodes, one must bring the perpendicular bisectors of the sides of the triangles, which subdivide the region in polygons surrounding each node. The thermal conductance between nodes, say, 1 and 2 (see Fig. 16) is equal to [5].

$$K_{1,2} = \frac{k L_{a,b}}{L_{1,2}} \tag{60}$$

Using simple geometric relationships, Eq. (60) becomes

$$K_{1,2} = \frac{(\cot\beta_1 + \cot\beta_2)k}{2} \tag{61}$$

The volume assigned to each node is equal to the area of the surrounding polygon. It is preferable to subdivide bodies of triangular shape into similar triangular sections. An example is shown in Fig. 17. The thermal conductances between node 1 and

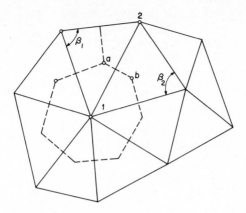

Fig. 16. System subdivided in irregular triangular sections.

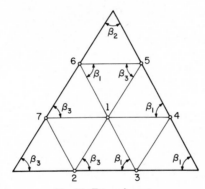

Fig. 17. Triangular net.

the surrounding nodes 2, 3, 4, 5, 6, and 7 are, according to Eq. (61),

$$K_{1,2} = K_{1,5} = k \cot\beta_1$$
$$K_{1,3} = K_{1,6} = k \cot\beta_2 \qquad (62)$$
$$K_{1,4} = K_{1,7} = k \cot\beta_3$$

Thus the heat balance yields the following equation

$$\cot\beta_1(T_2 + T_5) + \cot\beta_2(T_4 + T_7) + \cot\beta_3(T_3 + T_6)$$
$$- 2(\cot\beta_1 + \cot\beta_2 + \cot\beta_3)T_1 = 0 \quad (63)$$

A regular hexagonal net is formed if $\beta_1 = \beta_2 = \beta_3 = 60°$ where Eq. (63) is simplified to

$$T_2 + T_3 + T_4 + T_5 + T_6 + T_7 - 6T_1 = 0 \qquad (64)$$

If one of the angles, say β_2, is equal to $90°$, an orthogonal net is formed and the nodal equation becomes

$$\cot\beta_1(T_2 + T_5) + \cot\beta_3(T_3 + T_6) - 2(\cot\beta_1 + \cot\beta_3)T_1 = 0 \qquad (65)$$

In an anisotropic material with thermal conductivities k_x, k_y along the x and y directions, the use of an orthogonal net with

$$\frac{\Delta x}{\Delta y} = \left(\frac{k_x}{k_y}\right)^{1/2} \qquad (66)$$

gives simple nodal equations.

For example, the equation for node 1 in Fig. 18 is

$$k_y \frac{\Delta x}{\Delta y}(T_2 + T_4 - 2T_1) + k_x \frac{\Delta y}{\Delta x}(T_3 + T_5 - 2T_1) = 0 \qquad (67)$$

If k_x, k_y, Δx, Δy are related according to Eq. (66), then Eq. (67) becomes

$$T_2 + T_3 + T_4 + T_5 - 4T_1 = 0 \qquad (68)$$

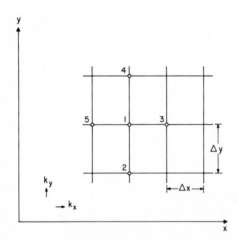

Fig. 18. Anisotropic material.

Boundary nodes are treated in a similar way, with the previously described method. A portion of a boundary is shown in Fig. 19. The thermal conductance between the nodes 1 and 2 is equal to [5]

$$K_{1,2} = \frac{k \cot\beta}{2} \qquad (69)$$

To satisfy boundary conditions which involve specified heat fluxes or Newtonian cooling, the required surface element associated with a boundary node (say node 1, Fig. 19) is taken equal to $L_{a,b}$ [5]. Square nets terminating in irregular boundaries, as the one shown in Fig. 20, are easily handled with this method. Thus,

$$K_{1,3} = k \frac{\cot\beta_1 + \cot\beta_2}{2} \qquad (70)$$

or

$$K_{1,3} = k \frac{S_{1,2}/S_{1,3} + 1/S_{1,3}}{2} = k \frac{1 + S_{1,2}}{2S_{1,3}} \tag{70a}$$

Similarly

$$K_{1,2} = k \frac{1 + S_{1,3}}{2S_{1,2}}$$

$$K_{1,5} = k \frac{1 + S_{1,2}}{2} \tag{71}$$

$$K_{1,4} = k \frac{1 + S_{1,3}}{2}$$

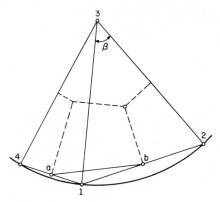

Fig. 19. Boundary nodes of irregular net.

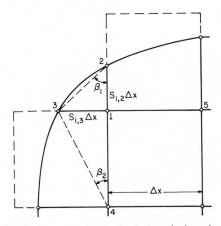

Fig. 20. Square net terminating in irregular boundary.

5. Composite Media: Nonuniform Properties

a. Composite Media. In many practical applications, heat is transferred through a composite structure made of a number of different materials. The nodal equations for such cases can be obtained by the heat balance method. As an example, consider the structure shown in Fig. 21 consisting of two materials having thermal conductivities k_1, k_2. In the interface there is a thermal contact resistance which is indicated by an equivalent heat transfer coefficient h. If there is good thermal contact (for example, sweated or welded materials) between the two materials ($h = \infty$) the temperatures of the nodes 1, 2, and 4 are equal to the temperatures of the nodes 3, 6, and 8, respectively (Fig. 21). In this case the nodal equation for node 1 located in the interface is as follows

$$(k_1 + k_2)\left(\frac{\Delta y}{2\Delta x}\right)(T_2 - T_1) + k_2\left(\frac{\Delta x}{\Delta y}\right)(T_7 - T_1)$$

$$+ (k_1 + k_2)\left(\frac{\Delta y}{2\Delta x}\right)(T_4 - T_1) + k_1\left(\frac{\Delta x}{\Delta y}\right)(T_5 - T_1) = 0 \quad (72)$$

If there is a thermal contact resistance between the materials the nodal equation for the same node becomes

$$k_1\left(\frac{\Delta y}{2\Delta x}\right)(T_2 - T_1) + h(\Delta x)(T_3 - T_1) + k_1\left(\frac{\Delta y}{2\Delta x}\right)(T_4 - T_1)$$

$$+ k_1\left(\frac{\Delta x}{\Delta y}\right)(T_5 - T_1) = 0 \quad (73)$$

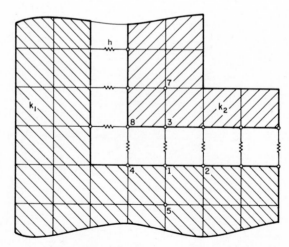

Fig. 21. Two-dimensional composite system.

For one-dimensional problems, the total conductances between isothermal faces can be obtained by adding the individual parallel conductances. If the conductances are in series, the reciprocal of the total conductance is equal to the sum of the reciprocal of the individual conductances. As an illustration consider the structure shown in Fig. 22. The faces AB and CD are at constant temperatures; therefore, they can be represented by two nodes, say 1 and 2. Consider the auxiliary nodes a, b, c, d, e, and f. The total conductance between the node 1 and 2 is equal to

$$K_{1,2} = K_{a,b} + K_{c,d} + K_{e,f} \tag{74}$$

where

$$\frac{1}{K_{a,b}} = \frac{L_1}{l_1 k_1} + \frac{L_2}{l_2 k_2} \tag{75a}$$

$$K_{c,d} = \frac{k_1 l_2}{L_1 + L_2} \tag{75b}$$

$$K_{e,f} = \frac{k_3 l_3}{L_1 + L_2} \tag{75c}$$

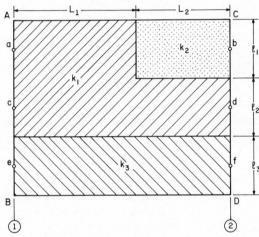

Fig. 22. One-dimensional composite system.

b. Nonuniform Properties. The thermal properties of the materials are temperature dependent (see Sec. 2.D). In steady-state problems where large temperature differences might exist, the variation of thermal conductivity with temperature should be considered. In this case, the thermal conductances between nodes can be calculated as follows

$$K_{ij} = \frac{[k(T_i) + k(T_j)]}{2} \cdot \frac{A_{ij}}{L_{ij}} \tag{76}$$

If the thermal conductivity varies linearly with the temperature, then

$$K_{ij} = k\left(\frac{T_i + T_j}{2}\right) \frac{A_{ij}}{L_{ij}} \tag{77}$$

To obtain numerical solutions to such problems one may proceed as follows:

1. Assume an average k and solve the problem.
2. Using the solution obtained in step 1, modify the conductances according to Eq. (76) and solve the problem again.
3. Repeat step 2 (always using the latest solution) until a satisfactory small change in the K_{ij}'s is found.

Problems which involve only prescribed surface temperatures and insulated surfaces can be best solved by the following procedure [4, 6]:

1. Calculate the function F, defined as follows:

$$F = \int_0^T k(T)dT \tag{78}$$

This function satisfies Laplace's equation; i.e., in an orthogonal Cartesian system

$$\frac{\partial^2 F}{\partial x^2} + \frac{\partial^2 F}{\partial y^2} = 0$$

2. Using Eq. (78), calculate the values of F at the boundaries from the prescribed surface temperatures.
3. Solve the problem in terms of F in an identical manner to that which would have been used to solve for T if k was constant.
4. Using Eq. (78), convert the nodal values F_i to the corresponding T_i.

If $k(T)$ is a linear function of temperature, then F becomes proportional to k^2. In this case, one may proceed through steps 2, 3, and 4 by considering the function k^2. Equation (78) could be integrated if $k(T)$ is given by a formula. If $k(T)$ is given by a table one has to perform a numerical integration. (For numerical integration see Refs. 1 and 7.)

6. Effect of Mesh Size; Advance to a Finer Net

a. Effect of Mesh Size. The approximate solutions of Laplace- and Poisson-type equations obtained by the relaxation technique (or any other numerical method) are affected by the mesh size used to subdivide the region where heat transfer takes place. There are "two types" of errors committed in an approximate solution, and these errors are affected differently by the choice of the mesh size, and the finite difference scheme used [7].

The first type of error is due to the fact that the relaxation method gives an approximate solution of the algebraic equations. An estimate of this error for a square net of size h and for the five- and nine-point formula is given in Ref. 7; it is shown there that these errors do not exceed the quantities $RL^2/4h^2$ and $RL^2/24h^2$ respectively, where R is the maximum residual (in absolute value) and L is the radius of a circle enclosing the region.

The second type of error is committed by using the solution of the difference equation instead of the differential equation. The inherent error can be found by expressing the difference quotients in terms of derivatives. For square nets of size h, the errors for the five- and nine-point formula are bounded by quantities which are proportional to h^2 and h^6, respectively [7]. The choice of the mesh size and the finite difference scheme depend on the problem at hand. As it was suggested earlier, one may start with a coarser net and then advance to a finer net. If T_1 and T_2 are the values of a temperature at a point calculated using two different size nets, h_1 and h_2, respectively, then one may obtain a good approximation for the temperature at this point using the following equations [7, 8, 9]

Five-point formula
$$T = \frac{T_2 - (h_2/h_1)^2 T_1}{1 - (h_2/h_1)^2} \tag{79}$$

Nine-point formula
$$T = \frac{T_2 - (h_2/h_1)^6 T_1}{1 - (h_2/h_1)^6} \tag{80}$$

If $h_2 = \frac{1}{2} h_1$, then Eqs. (79) and (80) become as follows:

$$T = \frac{4}{3} T_2 - \frac{1}{3} T_1 \tag{79a}$$

$$T = \frac{64}{63} T_2 - \frac{1}{63} T_1 \tag{80a}$$

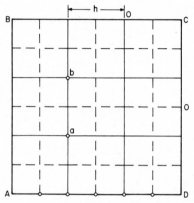

Fig. 23. Rectangular plate with AD at 1000°F and all other faces at 0°F.

Example 10. The square plate shown in Fig. 23 has its face AD at temperature 1000°F and its three other faces at temperature 0°F. The temperatures at the points a and b have been obtained using different methods and two mesh sizes. The results are summarized in Table 11.

TABLE 11. Comparison of Temperatures Determined by Different Methods

Method	T_a	T_b	% Error	
			a	b
Five-point, mesh size h	375.00	125.00	1.78	4.8
Five-point, mesh size $h/2$	378.79	121.21	0.79	1.64
Equation (79a)	380.05	119.95	0.46	0.58
Nine-point, mesh size h	381.00	119.00	0.21	0.22
Exact solution	381.79	119.26		
(Fourier series solution)				

In this example the algebraic equations have been solved exactly so that there is no residual error involved. In order to use the nine-point formula, the temperatures at the points A and D are required; these temperatures were taken as the average value of the temperatures of the faces to which they belong (that is, 500°F). From the last column where the absolute percent error is given with respect to the exact solution, the superiority of the nine-point formula and the improvement obtained by halving the mesh size are obvious. However, in halving the mesh size, the improvement is not as drastic as in the previous example, whereas the labor required for the solution is increased considerably.

b. Advance to a Finer Net. In certain cases it is advantageous to use different mesh size for different parts of the system. For example, if the temperature distribution at some portion of the system is required more accurately than the rest of the system, then in this portion one should use a finer net. The principles of the asymmetrical nets explained in Sec. A.4 can be applied for the interconnection of two different mesh size nets [5]. As an example, consider the two square nets A and B with mesh size h and $2h$, respectively. There are two ways of interconnecting the two nets shown in Figs. 24 and 25. In Fig. 24 the interconnection has been obtained through nodes a and b. Following the principles of Sec. A.4 the equations for the nodes a and 3 are as follows

Node a
$$\frac{T_1 - T_a}{2} + (T_6 - T_a) + (T_7 - T_a) + \frac{T_3 - T_a}{2} + (T_2 - T_a) = 0 \qquad (81)$$

Node 3
$$\frac{T_a - T_3}{2} + \frac{T_b - T_3}{2} + (T_4 - T_3) + (T_2 - T_3) = 0 \qquad (82)$$

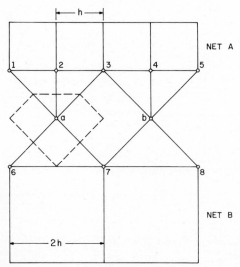

Fig. 24. Interconnection of two different size square nets.

Similar equations are readily obtained for the other nodes. A different way of interconnecting the two nets is shown in Fig. 25.

7. Conformal Mapping and Graphical Solutions

The theory of complex variables provides the means of obtaining solutions to two-dimensional steady-state heat transfer problems in many situations where the usual analytical procedures (see Sec. I.C.5) are not applicable. The most useful aspect of this theory is its ability of transforming a region in which heat transfer takes place (where, for example, complicated boundary conditions prohibit the use of analysis) into another region, so that the problem can be easily solved. This transformation, which preserves the magnitude and sense of the angles between isothermal boundaries, is called conformal mapping [10]. This theory also provides the fundamentals for graphical solutions.

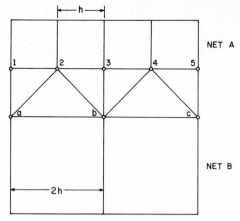

Fig. 25. Alternate way of interconnecting two different size square nets.

a. Analytical Functions and Orthogonality. The following definition of analytic function is given in standard books of complex variables. A function $f(z)$ of the complex variable $z = x + iy$ is analytic at a point z_0 if and only if its derivative exists at every point in some neighborhood of z_0 [10]. For example, every polynomial $P(z)$ in z is analytic at every point in the complex plane.

The functions $\phi(x, y)$, $\psi(x, y)$ which are related to z by the expression

$$w = \phi(x, y) + i\psi(x, y) = f(x + iy) = f(z) \tag{83}$$

are called conjugate functions.

If $f(z)$ is analytic, the following relations hold

$$\frac{\partial \phi}{\partial x} + i\frac{\partial \psi}{\partial x} = f'(z) \tag{84a}$$

$$\frac{\partial \phi}{\partial y} + i\frac{\partial \psi}{\partial y} = if'(z) \tag{84b}$$

Conjugate functions satisfy the Cauchy-Riemann conditions (Eqs. (85a) and (85b)) which are readily obtained by separating the real and imaginary parts from the above equations:

$$\frac{\partial \phi}{\partial x} = \frac{\partial \psi}{\partial y} \tag{85a}$$

$$\frac{\partial \phi}{\partial y} = -\frac{\partial \psi}{\partial x} \tag{85b}$$

Conditions (85a) and (85b) imply that the contours ϕ = constant, ψ = constant are orthogonal, i.e., perpendicular to each other at the points of intersection. (See Fig. 26.) Conjugate functions also satisfy Laplace's equation. This can be demonstrated by partially differentiating Eq. (85a) with respect to x and Eq. (85b) with respect to y and then adding the resulting equations. Thus

$$\frac{\partial^2 \phi}{\partial x^2} + \frac{\partial^2 \phi}{\partial y^2} = 0 \tag{86a}$$

similarly one obtains
$$\frac{\partial^2 \psi}{\partial x^2} + \frac{\partial^2 \psi}{\partial y^2} = 0 \qquad (86b)$$

The analyticity of $f(z)$ guarantees the inverse relationship [10]
$$z = F(w) \qquad (87)$$

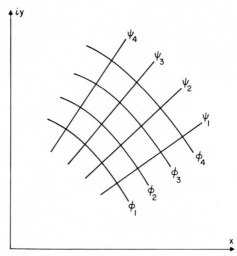

Fig. 26. Conjugate functions in the complex plane.

b. Mapping of Laplace's Equation. The following method can be used to obtain solutions of heat transfer problems which are governed by Laplace's equation.

Using Eq. (83) which relates the variable $w = \phi + i\psi$ to $z = x + iy$, one may uniquely transform (map) a curve C on the z plane to a curve S in the w plane. This transformation is called conformal mapping; it possesses the following properties [10]:

1. The angle γ between two curves C_1, C_2 in the z plane, which are transformed into the curves S_1, S_2 in the w plane, is preserved in magnitude and sense (see Fig. 27).
2. A function T which satisfies Laplace's equation in the z plane satisfies Laplace's equation in the w plane. Thus

$$\frac{\partial^2 T}{\partial x^2} + \frac{\partial^2 T}{\partial y^2} = 0 \qquad\qquad \frac{\partial^2 T}{\partial \phi^2} + \frac{\partial^2 T}{\partial \psi^2} = 0$$

3. If T has a constant value T_0 on a boundary C in the z plane, then T attains the same constant value on the transformed boundary S, in the w plane.
4. If the normal derivative of T vanishes on a boundary C in the z plane, it also vanishes on the transformed boundary S, in the w plane.

According to the above properties the differential equation, isothermal and adiabatic boundary conditions remain unchanged under this transformation. Other boundary conditions may transform into conditions which are substantially different from the original ones. Therefore this method is most useful for problems involving isothermal and adiabatic boundaries.

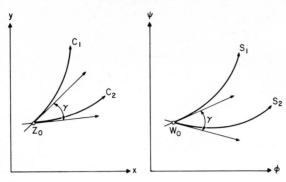

Fig. 27. Conformal mapping.

The following example is cited to illustrate the method.

Example 11. Determine the temperature distribution in the quadrant shown in Fig. 28. This problem cannot be solved by analytical methods because of the mixed boundary conditions on the x axis. Applying the transformation [10]

$$w = \sin^{-1}(z) \tag{88}$$

the quadrant is transformed into a strip as shown in Fig. 29. In the w plane one then has to solve a simple one-dimensional problem. Thus

$$T = T_1 + \frac{(T_2 - T_1)}{\sin^{-1} a} \phi \tag{89}$$

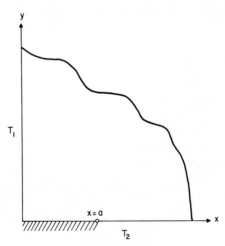

Fig. 28. z plane.

Using Eq. (88) to transform back to the original coordinates, one obtains the following solution [10]

$$T = T_1 + \frac{(T_2 - T_1)}{\sin^{-1} a} \sin^{-1} \frac{1}{2} \left[\sqrt{(x + a)^2 + y^2} - \sqrt{(x - a)^2 + y^2} \right] \tag{90}$$

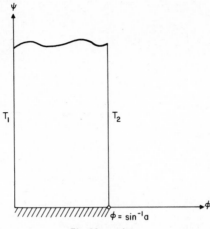

Fig. 29. w plane.

Although this technique is very powerful, the determination of the appropriate transformation is not always easy to obtain. Moreover, the inverse problem, i.e., transforming to the original coordinates, is in general a difficult problem. A large number of conformal transformations are presented in Ref. 11.

c. Graphical Solutions. In two dimensional heat conduction problems where the required accuracy does not justify lengthy numerical calculations, the following graphical solution may be used. This method, which is most useful when the region is bounded by isothermal and insulated boundaries (for nonisothermal boundaries, see Refs. 12–15), is based on the properties of the conjugate functions briefly discussed earlier; thus, if we identify in Eq. (83) the function $\phi = T$ the temperature, then ψ must be a path of constant heat flow and therefore related to the heat-flow rate q. This relationship can be demonstrated with the aid of Fig. 30; there n and s are the normal and tangential directions at a point of the isotherm T.

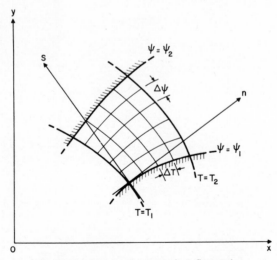

Fig. 30. Isotherms and constant heat flow paths.

The heat flux q'' is equal to

$$q'' = -k\frac{\partial T}{\partial n} = -k\cos\beta\frac{\partial T}{\partial x} - k\sin\beta\frac{\partial T}{\partial y} \tag{91}$$

Using the conditions of Eq. (85), Eq. (91) becomes

$$q'' = -k\frac{\partial T}{\partial n} = -k\cos\beta\frac{\partial \psi}{\partial y} + k\sin\beta\frac{\partial \psi}{\partial x} = -k\frac{\partial \psi}{\partial s} \tag{92}$$

Therefore

$$\frac{q''}{k} = -\frac{\partial T}{\partial n} = -\frac{\partial \psi}{\partial s} \tag{93}$$

Integrating Eq. (93) along s from ψ_1 to ψ_2, one obtains the relationship

$$\psi_1 - \psi_2 = \frac{q''}{kL} \tag{94}$$

where L is the dimension in the third direction. Now, taking small increments of T and ψ, Eq. (93) becomes

$$\frac{\Delta T}{\Delta n} = \frac{\Delta \psi}{\Delta s} \tag{95}$$

Using Eq. (95) and the orthogonality between T and ψ, one may construct a network consisting of isotherms and constant flow paths. If $\Delta n = \Delta s$ the region is subdivided in curvilinear squares. A freehand construction (which is usually called flux plotting) of the foregoing mentioned network satisfies the following requirements:

1. Isothermal and flow lines form a network of curvilinear squares. (Curvilinear rectangles satisfying Eq. (95) can also be used.)
2. Flow lines are perpendicular to isothermal boundaries.
3. Isotherms are perpendicular to insulated boundaries.
4. Diagonals of curvilinear squares bisect each other at 90° and bisect the corners.

Graphical solutions have been used to obtain the shape factor for regions bounded between two isothermal and two adiabatic surfaces. If the isothermal surfaces are at temperatures T_1, T_2, the shape factor is defined by the equation (see Sec. 3)

$$s = \frac{q}{k(T_1 - T_2)} \tag{96}$$

Suppose now that the region has been divided in N_q flow tubes and N_t temperature increments. Thus

$$\frac{q}{k} = N_q \Delta\psi = s(T_1 - T_2) = sN_t \Delta t \tag{97}$$

Therefore

$$s = \frac{N_q}{N_t} \cdot \frac{\Delta\psi}{\Delta T} \tag{98}$$

If the system has been subdivided in curvilinear squares with $\Delta\psi = \Delta T$, Eq. (98) becomes

$$s = \frac{N_q}{N_t} \qquad (98a)$$

It can be readily shown that if the isothermal and adiabatic surfaces are interchanged, the new shape factor s' is equal to $1/s$.

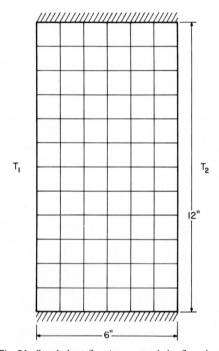

Fig. 31. Steady heat flow in a rectangle by flux plot.

For example, consider the system shown in Fig. 31 divided by $N_t = 6$ temperature increments; thus $\Delta T = (T_1 - T_2)/6$, $\Delta n = 6\text{in.}/6$. In order to have $\Delta s = \Delta n = 6/6$ one must take $N_q = 12$. Therefore, the shape factor according to Eq. (98a) is equal to

$$s = \frac{N_q}{N_t} = \frac{12}{6} = 2$$

The same result is obtained by taking $Nq = 6$; now $\Delta n = 12/6$ and s, by virtue of Eq. (98), becomes

$$s = \frac{6}{6} \cdot \frac{(12/6)}{(6/6)} = 2$$

A more complicated system is shown in Fig. 32. The faces AFE and BCD are temperatures T_1 and T_2, respectively, while the faces AB and DE are insulated. There are $N_t = 8$ temperature increments and $N_q = 24$ flow tubes; thus $s = 3$. The shape factor for this system can be calculated according to Eq. (174) in Sec. 3, resulting in the value $s = 3.04$.

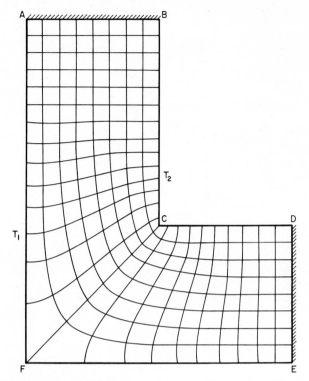

Fig. 32. Temperature distribution and constant heat flow paths by flux plot.

8. Three-dimensional Conduction

Numerical solutions of three-dimensional problems can be obtained using the techniques discussed earlier. (See Secs. A.2 and A.3.) However, because of the large number of nodes usually involved in such problems, hand calculations become tedious; their solution most often requires the aid of a high-speed digital computer. The nodal equations are formulated by the methods introduced in Sec. A.1; the following three examples show the derivation of the nodal equations for the Cartesian, cylindrical, and spherical coordinate systems.

a. Cartesian Coordinate System.   Consider the system shown in Fig. 33. Using the energy-balance method, one may derive the following equation for the temperature T_0 of an interior node

$$k \frac{\Delta x \Delta z}{\Delta y} (T_1 + T_3 - 2T_0) + k \frac{\Delta y \Delta z}{\Delta x} (T_2 + T_4 - 2T_0) + k \frac{\Delta x \Delta y}{\Delta z} (T_5 + T_6 - 2T_0) = 0$$

$$(99)$$

By selecting $\Delta x = \Delta y = \Delta z$, Eq. (99) becomes

$$T_1 + T_2 + T_3 + T_5 + T_6 - 6T_0 = 0 \qquad (100)$$

If there is a heat generation q''' in the system, the term $q''' \Delta x \Delta y \Delta z$ should be added in the left-hand side of Eq. (99). The thermal conductance of the boundary node (for example, node 6) is, according to Eq. (21), equal to

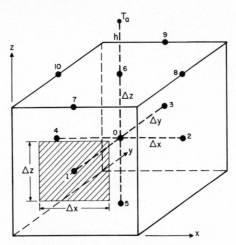

Fig. 33. Three-dimensional Cartesian system.

$$K_{a6} = h\Delta y\Delta x \tag{101}$$

h being the heat transfer coefficient. The nodal equation for this node is readily shown to be

$$h\Delta y\Delta x(T_a - T_6) + k\frac{\Delta y\Delta x}{\Delta z}(T_0 - T_6) + k\frac{\Delta x\Delta z}{2\Delta y}(T_7 + T_9 - 2T_6)$$

$$+ k\frac{\Delta y\Delta z}{2\Delta x}(T_8 + T_{10} - 2T_6) = 0 \tag{102}$$

where T_a is the ambient temperature.

b. Cylindrical Coordinates. The system to be considered is shown in Fig. 34. The energy-balance gives the following equation for the temperature T_0

$$k\frac{(r_0 + \Delta r/2)\Delta\phi\Delta z}{\Delta r}(T_2 - T_0) + k\frac{(r_0 - \Delta r/2)\Delta\phi\Delta z}{\Delta r}(T_4 - T_0)$$

$$+ k\frac{\Delta r\Delta z}{\Delta\phi r_0}(T_1 + T_3 - 2T_0) + k\frac{r_0\Delta\phi\Delta r}{\Delta z}(T_5 + T_6 - 2T_0) = 0 \tag{103}$$

If there is heat generation q''', the term $q'''r_0\Delta\phi\Delta r\Delta z$ should be added in the left-hand side of the above equation. Equation (103), after some rearrangement of terms, takes the following simpler form

$$\frac{T_2 + T_4 - 2T_0}{\Delta r^2} + \frac{T_2 - T_4}{2r_0\Delta r} + \frac{T_1 + T_3 - 2T_0}{r_0^2\Delta\phi^2} + \frac{T_5 + T_6 - 2T_0}{\Delta z^2} = 0 \tag{104}$$

Thermal conductances for boundary nodes can be obtained using Eq. (21). For example, if the nodes 2 and 6 are located at the boundary where the system exchanges heat with the surroundings through a heat transfer coefficient h, their thermal conductances are

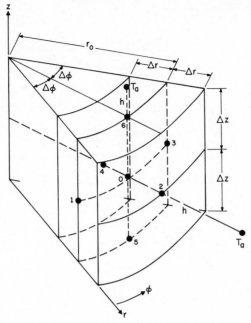

Fig. 34. Three-dimensional cylindrical system.

$$K_{a2} = h(r_0 + \Delta r)\Delta\phi\Delta z \tag{105a}$$

$$K_{a6} = hr_0\,\Delta\phi\Delta r \tag{105b}$$

c. Spherical Coordinates.   The nodes with their respective coordinates are shown in Fig. 35. In order to derive the difference equation for the temperature T_0, the conductances between the various nodes are calculated according to Eq. (20) as follows:

$$K_{0,1} = k\frac{(r - \Delta r/2)^2\Delta\phi\Delta\psi\,\sin\psi}{\Delta r} \tag{106a}$$

$$K_{0,2} = k\frac{r\Delta r\Delta\phi\,\sin(\psi + \Delta\psi/2)}{r\Delta\psi} \tag{106b}$$

$$K_{0,3} = k\frac{(r + \Delta r/2)^2\Delta\phi\Delta\psi\,\sin\psi}{\Delta r} \tag{106c}$$

$$K_{0,4} = \frac{kr\Delta r\Delta\phi\,\sin(\psi - \Delta\psi/2)}{r\Delta\psi} \tag{106d}$$

$$K_{0,5} = K_{0,6} = k\frac{r\Delta r\Delta\psi}{r\Delta\phi\,\sin\psi} \tag{106e}$$

Thus the equation for the temperature T_0 becomes

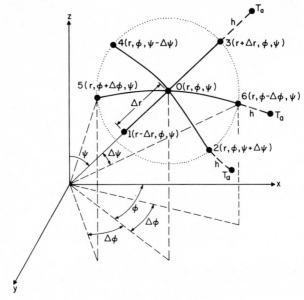

Fig. 35. Three-dimensional spherical system.

$$k\Delta r\Delta\phi\Delta\psi \, \sin\psi \left[\frac{(r - \Delta r/2)^2}{\Delta r^2} (T_1 - T_0) + \frac{(r + \Delta r/2)^2}{\Delta r^2} (T_3 - T_0) \right.$$

$$+ \frac{\sin(\psi + \Delta\psi/2)}{\Delta\psi^2 \, \sin\psi} (T_2 - T_0) + \frac{\sin(\psi - \Delta\psi/2)}{\Delta\psi^2 \, \sin\psi} (T_4 - T_0)$$

$$\left. + \frac{1}{\Delta\phi^2 \, \sin^2\psi} (T_5 + T_6 - 2T_0) \right] = 0 \quad (107)$$

If there is heat generation q''' in the system, the term $q'''r^2\Delta r\Delta\phi\Delta\psi \, \sin\psi$ should be added in the left-hand side of Eq. (107). Boundary conductances are calculated according to Eq. (21). For example, if, in the system shown in Fig. 35, the nodes 3, 2, and 6 are located at the boundaries where the system exchanges heat with the surroundings through a heat transfer coefficient h, their respective conductances are

$$K_{a,3} = h(r + \Delta r)^2 \, \Delta\phi\Delta\psi \, \sin\psi \quad (107a)$$

$$K_{a,2} = hr\Delta r\Delta\phi \, \sin(\psi + \Delta\psi) \quad (107b)$$

$$K_{a,6} = hr\Delta r\Delta\psi \quad (107c)$$

B. UNSTEADY STATE

Unsteady-state (or transient) heat transfer problems are characterized by time-dependent heat fluxes and temperature fields. Exact solutions to most practical transient heat transfer problems are rather difficult if not impossible to obtain. Thus, the necessity of obtaining an approximate solution to such problems has motivated a considerable effort for developing methods for achieving approximate solutions. The most commonly used of these methods can, in general, be classified in the following four categories:

1. Method of finite differences
2. Graphical methods
3. Approximate analytical methods
4. Analog methods

The choice of the method to be used for obtaining an approximate solution depends on the experience of the individual calculator, the means available, the complexity and size of the problem, and the accuracy required. The first three methods are discussed in the following sections while the fourth one is treated elsewhere. (See Sec. 5.)

1. Method of Finite Differences

This method has attracted most of the attention in recent years,* primarily because it is easily adapted to digital computations; consequently, this is the only method that can be used to obtain an approximate solution if a high degree of accuracy is required or if the problem at hand is of considerable size and complexity. In principle, by discretizing the space and time, one using this method has to solve algebraic equations (nodal equations) instead of the partial differential equation. The nodal equations can be derived by purely mathematical methods or by energy considerations, as it is demonstrated in the following:

The first step one has to take in order to obtain the nodal equations is to subdivide the region under consideration into sections, locate a node in the center of each section, and then calculate the conductances between adjacent nodes. (Details of this procedure are shown in Secs. A.1 and A.4.) Consider the energy balance for the node i, which is surrounded by j adjacent nodes, during a small time interval Δt. During this time interval we assume that the temperatures of the j nodes remain constant. Therefore the heat flux into the node will cause a change ΔT_i to its temperature. Using Eq. (19) one may then obtain the following expression

$$\sum_j K_{ij}[T_j(t) - T_i(t)] = \rho c V_i \frac{\Delta T_i}{\Delta t} \tag{108}$$

where ρ and c are the density and specific heat of the system and V_i the volume assigned to the node i. If there is heat generation in the system, the term $q'''V_i$ should be added in the left-hand side of Eq. (108). The above expression can be used in general to obtain the nodal equation for any interior node. Equations for boundary nodes can be obtained using Eq. (21) as it will be shown shortly. (See Sec. B.1.e.)

The difference ΔT_i may be taken in many ways which will yield different forms of Eq. (108). The most widely used finite difference schemes are discussed in the following. (Detailed discussion of this subject is given in Refs. 16 and 17).

a. Explicit Equation. The simplest form of Eq. (108) is obtained using the forward difference for $\Delta T_i = T_i(t + \Delta t) - T_i(t)$. In this case one may obtain explicitly the temperature $T_i(t + \Delta t)$ from the knowledge of the temperatures T_i, T_j at time t, from the following equation

$$T_i(t + \Delta t) = \frac{\Delta t}{\rho c V_i}\left[\sum_j K_{ij}T_j(t) + \left(\frac{\rho c V_i}{\Delta t} - \sum_j K_{ij}\right)T_i(t)\right] \tag{109}$$

Equation (109) for some of the systems discussed earlier (Sec. A.1, A.2, and A.4) takes the following forms:

*There is a vast literature on the subject; see Ref. 16.

Cartesian One-dimensional.

$$MT_m(t + \Delta t) = T_{m+1}(t) + T_{m-1}(t) + (M - 2)T_m(t) \tag{110}$$

where α is the thermal diffusivity, $x = m\Delta x$ and $M = \Delta x^2/\alpha\Delta t$, the Fourier modulus.
Cartesian Two-dimensional. (square net $\Delta x = \Delta y$)

$$MT_{m,n}(t + \Delta t) = T_{m+1,n}(t) + T_{m-1,n}(t) + T_{m,n+1}(t)$$
$$+ T_{m,n-1}(t) + (M - 4)T_{m,n}(t) \tag{111}$$

where $x = m\Delta x$, $y = n\Delta y$, $M = \Delta x^2/\alpha\Delta t = \Delta y^2/\alpha\Delta t$.
Cartesian Three-dimensional. ($\Delta x = \Delta y = \Delta z$)

$$MT_{m,n,r}(t + \Delta t) = T_{m+1,n,r}(t) + T_{m-1,n,r}(t) + T_{m,n+1,r}(t) + T_{m,n-1,r}(t)$$
$$+ T_{m,n,r+1}(t) + T_{m,n,r-1}(t) + (M - 6)T_{m,n,r} \tag{112}$$

Hexagonal System. (See Fig. 17.) In this case where $\beta_1 = \beta_2 = \beta_3 = 60°$, the conductances are

$$K_{1,2} = K_{1,3} = K_{1,4} = K_{1,5} = K_{1,6} = K_{1,7} = k \cot 60 = k/\sqrt{3}$$

The volume element $V_1 = 3\sqrt{3}\,\Delta x^2/2$, where Δx is the side of the hexagon. Therefore Eq. (109) takes the form

$$M'T_1(t + \Delta t) = T_2(t) + T_3(t) + T_4(t) + T_5(t) + T_6(t) + T_7(t) + (M' - 6)T_1(t) \tag{113}$$

where $M' = 9\Delta x^2/2\alpha\Delta t$.

Equation for cylindrical and spherical coordinates can be readily obtained using Eqs. (103) and (107), respectively.

b. Implicit Equation.* A different form of Eq. (108) is obtained, assuming that the temperatures of the nodes T_j, T_i remain constant during Δt and equal to its value at $t + \Delta t$. Using the energy balance for the node i, one obtains the following equation

$$\sum_j K_{ij}[T_j(t + \Delta t) - T_i(t + \Delta t)] = \rho c V_i \frac{T_i(t + \Delta t) - T_i(t)}{\Delta t} \tag{114}$$

In this case the backward difference for ΔT_i has been used instead of the forward. Rearranging terms in Eq. (114), the following expression is obtained

$$\frac{\Delta t}{\rho c V_i}\left[\frac{\rho c V_i}{\Delta t} + \sum_j K_{ij}\right]T_i(t + \Delta t) - \sum_j K_{ij}T_j(t + \Delta t) = T_i(t) \tag{115}$$

The essential difference between this method (Eq. (115)) and the previous method (Eq. (109)) is the fact that in this case, in order to obtain the temperatures at time $t + \Delta t$, one has to solve at each time step a set of algebraic equations (equal to the total number of nodes), instead of one.

Applying Eq. (115) to the Cartesian and hexagonal system treated previously, Eqs. (110)–(113) become

*See Refs. 17–20.

$$(M + 2) T_m(t + \Delta t) - T_{m-1}(t + \Delta t) - T_{m+1}(t + \Delta t) = MT_m(t) \qquad (110a)$$

$$(M + 4) T_{m,n}(t + \Delta t) - T_{m-1,n}(t + \Delta t) - T_{m+1,n}(t + \Delta t)$$
$$- T_{m,n+1}(t + \Delta t) - T_{m,n-1}(t + \Delta t) = MT_{m,n}(t) \quad (111a)$$

$$(M + 6) T_{m,n,r}(t + \Delta t) - T_{m-1,n,r}(t + \Delta t) - T_{m+1,n,r}(t + \Delta t)$$
$$- T_{m,n+1,r}(t + \Delta t) - T_{m,n-1,r}(t + \Delta t) - T_{m,n,r+1}(t + \Delta t)$$
$$- T_{m,n,r-1}(t + \Delta t) = MT_{m,n,r}(t) \quad (112a)$$

$$(M + 6) T_1(t + \Delta t) - T_2(t + \Delta t) - T_3(t + \Delta t) - T_4(t + \Delta t)$$
$$- T_5(t + \Delta t) - T_6(t + \Delta t) - T_7(t + \Delta t) = MT_1(t) \quad (113a)$$

c. Mid-difference.* Crank and Nicholson [21] expressed the difference ΔT_i by taking the arithmetic average of the forward and backward difference. In this case Eq. (108) becomes

$$\rho c V_i \frac{[T_i(t + \Delta t) - T_i(t)]}{\Delta t} = \frac{1}{2} \Bigg\{ \sum_j K_{ij}[T_j(t) - T_i(t)]$$

$$+ \sum_j K_{ij}[T_j(t + \Delta t) - T_i(t + \Delta t)] \Bigg\} (116)$$

Rearranging terms in Eq. (116) we obtain the following expression

$$\left(\frac{\rho c V_i}{\Delta t} + \frac{1}{2} \sum_j K_{ij} \right) T_i(t + \Delta t) - \frac{1}{2} \sum_j K_{ij} T_j(t + \Delta t)$$

$$= \left(\frac{\rho c V_i}{\Delta t} - \frac{1}{2} \sum_j K_{ij} \right) T_i(t) + \frac{1}{2} \sum_j K_{ij} T_j(t) \qquad (117)$$

This is an implicit equation relating the temperatures T_i and T_j at $t + \Delta t$ to the temperatures T_i and T_j at time t; therefore a set of algebraic equations has to be solved at each time step. Applying Eq. (117) to the one-dimensional Cartesian system, Eq. (110) becomes

$$(2M + 2) T_m(t + \Delta t) - T_{m+1}(t + \Delta t) - T_{m-1}(t + \Delta t) = T_{m-1}(t)$$
$$+ T_{m+1}(t) + (2M - 2) T_m(t) \quad (110b)$$

d. Generalized Six-point Formula.† A more general formula is obtained by expressing ΔT_i as a weighted average of the forward and backward differences. In this case, Eq. (108) takes the form

* See Refs. 17, 21, and 22.
† See Ref. 17.

$$\rho c V_i \frac{[T_i(t + \Delta t) - T_i(t)]}{\Delta t} = \lambda \sum_j K_{ij}[T_j(t + \Delta t) - T_i(t + \Delta t)]$$

$$+ (1 - \lambda) \sum_j K_{ij}[T_j(t) - T_i(t)] \quad (118)$$

where λ is a positive constant (usually $0 \le \lambda \le 1$). For example, the forward, backward, and mid-difference schemes can be obtained from Eq. (118) by selecting λ equal to 0, 1, or ½, respectively. Applying Eq. (118) to the one-dimensional Cartesian system, Eq. (110) becomes

$$(M + 2\lambda) T_m(t + \Delta t) - \lambda T_{m+1}(t + \Delta t) - \lambda T_{m-1}(t + \Delta t) = (1 - \lambda) T_{m+1}(t)$$

$$+ (1 - \lambda) T_{m-1}(t) + [M - 2(1 - \lambda)] T_m(t) \quad (110c)$$

e. Finite Difference Equations for Boundary Nodes. The most frequent boundary conditions encountered in practice are those of Newtonian cooling and solid radiation.* (In certain problems, boundary temperatures or heat fluxes may be specified.) The derivation of the nodal equation for boundary nodes is shown in the following section. Although examples of only some one- and two-dimensional Cartesian systems are treated, the method can be applied to any other suitable subdivision. (Some examples are given in Ref. 23.)

Convective Boundaries. Consider the system shown in Fig. 36 as having constant properties, ρ, c, k. The energy balance for a small time interval Δt gives the following equation for the temperature rise ΔT_0 of the boundary node

$$\rho c V_0 \frac{\Delta T_0}{\Delta t} = \frac{k}{\Delta x} [T_1(t) - T_0(t)] + h[T_a(t) - T_0(t)] \quad (119)$$

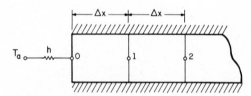

Fig. 36. One-dimensional convective boundary surface.

where h is the heat transfer coefficient and T_a the specified ambient temperature. Equation (119) takes the following forms if the forward or backward difference is used

Forward
$$\frac{M}{2} T_0(t + \Delta t) = T_1(t) + N T_a(t) + \left(\frac{M}{2} - 1 - N\right) T_0(t) \quad (119a)$$

Backward
$$\frac{M}{2} T_0(t) = \left(\frac{M}{2} + 1 + N\right) T_0(t + \Delta t) - T_1(t + \Delta t) - N T_a(t + \Delta t)$$

$$(119b)$$

where $N = h\Delta x/k$.

* See Sec. 3.A.3c.

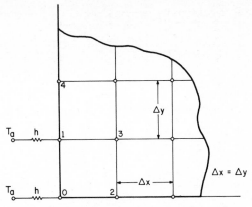

Fig. 37. Two-dimensional convective boundary surface.

For the two-dimensional system shown in Fig. 37 having constant properties ρ, c, k, there are two types of boundary nodes, one located at the corner (node 0) and another node located at the surface (node 1). The nodal equations for these nodes are readily shown to be

Node 0

Forward $$\frac{M}{2} T_0(t + \Delta t) = T_1(t) + T_2 + 2N T_a(t) + \left(\frac{M}{2} - 2 - 2N\right) T_0(t) \qquad (120a)$$

Backward $$\frac{M}{2} T_0(t) = \left(\frac{M}{2} + 2 + 2N\right) T_0(t + \Delta t) - T_1(t + \Delta t) - T_2(t + \Delta t)$$

$$- 2N T_a(t + \Delta t) \quad (120b)$$

Node 1

Forward $$M T_1(t + \Delta t) = T_4(t) + T_0(t) + 2T_3(t) + 2N T_a(t) + (M - 4 - 2N) T_1(t) \qquad (121a)$$

Backward $$M T_1(t) = (M + 4 + 2N) T_1(t + \Delta t) - T_4(t + \Delta t) - T_0(t + \Delta t)$$

$$- 2T_3(t + \Delta t) - 2N T_a(t + \Delta t) \quad (121b)$$

Radiating Boundaries. The heat flux at the surface in the system shown in Fig. 36 is equal to $\sigma F (T_a^{\,4} - T_0^{\,4})$, where σ is the Stefan-Boltzmann radiation constant, F the combined configuration-emissivity factor, and T_a the radiation source temperature. The energy balance for a small time increment Δt gives the following equation for the temperature rise ΔT_0

$$\rho c \frac{\Delta x}{2} \frac{\Delta T_0}{\Delta t} = \frac{k}{\Delta x} [T_1(t) - T_0(t)] + \sigma F [T_a^{\,4}(t) - T_0^{\,4}(t)] \qquad (122)^*$$

*In Eqs. (122)–(125) the temperatures are expressed in Rankine or Kelvin scale.

Using the *forward* difference for ΔT_0 we obtain the following expression

$$\frac{M}{2} T_0(t + \Delta t) = T_1(t) + \left[\frac{M}{2} - R T_0^3(t) - 1\right] T_0(t) + R T_a^4(t) \qquad (123)$$

where $R = \sigma F \Delta x / k$.

If the backward difference is used, the resulting equation is nonlinear, comprising terms of the fourth power of the unknown temperature. In Ref. 24 the following approximation has been suggested for the linearization of the equation

$$T_0^4(t + \Delta t) = 4T_0^3(t) T_0(t + \Delta t) - 3T_0^3(t) \qquad (124)$$

Using Eq. (124) and a similar expression for the radiation source temperature T_a^4, the *backward* difference becomes

$$\frac{M}{2} T_0(t) + 3R [T_0^3(t) - T_a^3(t)] = \left[\frac{M}{2} + 1 + 4RT_0^3(t)\right] T_0(t + \Delta t)$$

$$- T_1(t + \Delta t) - 4R T_a^3(t) T_a(t + \Delta t) \qquad (125)$$

f. Composite Media. Finite difference equations for composite media are readily obtained using the energy method. As an example consider the composite system shown in Fig. 38, consisting of two one-dimensional media, with perfect thermal contact in the interface AA. The energy balance for the node m, located in the interface, yields the following equation

$$\left(\frac{\rho_1 c_1 \Delta x_1}{2} + \frac{\rho_2 c_2 \Delta x_2}{2}\right) \frac{\Delta T_m}{\Delta t} = \frac{k_1}{\Delta x_1} (T_{m-1} - T_m) + \frac{k_2}{\Delta x_2} (T_{m+1} - T_m) \qquad (126)$$

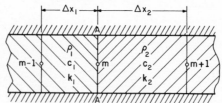

Fig. 38. One-dimensional composite system.

Using the same Fourier modulus $M_1 = M_2$ for both sides, Eq. (126) becomes

$$\frac{M}{2} \left(\frac{k_1}{\Delta x_1} + \frac{k_2}{\Delta x_2}\right) \frac{\Delta T_m}{\Delta t} = \frac{k_1}{\Delta x_1} T_{m-1} + \frac{k_2}{\Delta x_2} T_{m+1} - \left(\frac{k_1}{\Delta x_1} + \frac{k_2}{\Delta x_2}\right) T_m \qquad (127)$$

with

$$\frac{\Delta x_1}{\Delta x_2} = \left(\frac{\alpha_1}{\alpha_2}\right)^{1/2} \qquad (128)$$

If the forward time difference is used in Eq. (127), the temperature $T_m(t + \Delta t)$ can be determined from the following explicit equation

$$\frac{M}{2}\left(\frac{k_1}{\Delta x_1} + \frac{k_2}{\Delta x_2}\right) T_m(t + \Delta t) = \frac{k_1}{\Delta x_1} T_{m-1}(t) + \frac{k_2}{\Delta x_2} T_{m+1}(t)$$

$$+ \left(\frac{k_1}{\Delta x_1} + \frac{k_2}{\Delta x_2}\right)\left(\frac{M}{2} - 1\right) T_m(t) \qquad (129)$$

2. Convergence, Stability, and Accuracy

In numerical computations, generally the term "error" refers to the difference between an approximate solution, obtained by the methods described in Sec. B.1, and the exact solution of the original partial differential equation. There are two types of errors, which have a combined effect on the above-mentioned difference of the two solutions. The first type is due to the replacement of the derivatives by finite differences, and it is called the "truncation error." Analytical investigation has shown that the truncation error depends on the given initial temperature distribution, the boundary conditions, the choice of the finite difference scheme, and the magnitude of the Fourier modulus M, used for the computation [17]. The degree to which the approximate solution approaches the exact solution, as the space and time increments Δx, Δt, become smaller and smaller, is called the "convergence" of the finite difference representation. The second type are numerical errors, the so-called "round-off errors," which are caused from the inability in practical computations to carry out an infinite number of decimal places. Apart from the truncation and round-off errors, there is a more serious problem associated with certain finite difference schemes, namely that of "stability." As an illustration, consider the following example.

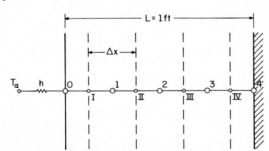

Fig. 39. Subdivision of plate for transient-numerical computations.

Example 12. An infinite plate of length $L = 1$ ft, having thermal diffusivity $\alpha = 0.20$ ft^2/hr, is initially at temperature 0°F. At time $t = 0$, one face of the plate is brought to a temperature of 100°F, while the other face is insulated. The temperature distribution as function of time is to be determined numerically. We begin by subdividing the plate into five equal sections as shown in Fig. 39. (Here h is very large so that $T_a = T_1 = 100$°F). We perform the calculations using the explicit finite difference (Eq. (110)) with two different values of $M, M = 1$ and $M = 4$. The corresponding time steps calculated from the expression $\Delta t = \Delta x^2/\alpha M$, are 12 and 3 minutes, respectively. The results of the computations are shown in Tables 12 and 13.

Inspection of Table 12 reveals that the computed values of the temperatures oscillate with an increasing magnitude as the computation progresses. This phenomenon, which is called "instability," has nothing to do with the round-off errors

TABLE 12. Plate Temperature Response with M = 1.

t	Nodal point temperature (°F)					
min	0	1	2	3	4	5
0	100	0	0	0	0	0
12	100	100	0	0	0	0
24	100	0	100	0	0	0
36	100	200	−100	100	0	0
48	100	−200	400	−200	100	0
60	100	700	−800	700	−300	
72	100	−1400	2200	−1800	1200	

TABLE 13. Slab Temperature Response with M = 4.

t(min)	Nodal point temperature (°F)					
	0	1	2	3	4	5
0	100	0	0	0	0	0
3	100	25.0000	0	0	0	0
6	100	37.5000	6.2500	0	0	0
9	100	45.3125	12.5000	1.5625	0	0
12	100	50.7812	17.9687	3.9063	0.3906	0
15	100	54.8828	22.6562	6.5430	1.1719	0.1953
18	100	58.1054	26.6845	9.2285	2.2705	0.6836
21	100	60.7238	30.1757	11.8530	3.6133	1.4771
24	100	62.9058	33.2321	14.3737	5.1392	2.5452
27	100	64.7609	35.9359	16.7797	6.7993	3.8422
30	100	66.3644	38.3531	19.0736	8.5551	5.3207
Exact solution [4] 30 min.	100	65.1878	37.1830	18.0124	8.2897	5.4963

(there are no round-off errors in the cited example); it is a property of the finite difference Eq. (110) and the given initial and boundary conditions. It should be noted, though, that round-off errors tend to grow with time in unstable schemes, while they decay if the conditions of stability are satisfied. For example, the last two lines of Table 13 indicate an excellent agreement between the numerically computed values and those obtained from the exact solution, despite the presence of round-off errors. It becomes evident from the foregoing example that we cannot select arbitrarily the space and time increments Δx, Δt, but that they should be so, chosen as to satisfy certain stability criteria, which have been established in extensive analytical investigations [17, 24–28]. Simple stability criteria, adequate for practical applications, may be developed by the following elementary argument. Let us consider the explicit difference Eq. (109):

$$T_i(t + \Delta t) = \frac{\Delta t}{\rho c V_i}\left[\sum_j K_{ij}T_j(t) + \left(\frac{\rho c V_i}{\Delta t} - \sum_j K_{ij}\right)T_i(t)\right] \qquad (109)$$

The coefficients of $T_j(t)$ are positive. If the coefficient of $T_i(t)$ was negative, then the higher the temperature $T_i(t)$ at the present time t, the lower the temperature $T_i(t + \Delta t)$ at the future time $t + \Delta t$. This would be in violation of thermodynamic principles. Therefore Δt (for a given space subdivision) should be chosen so that the coefficient of $T_i(t)$ is positive or zero. Applying this simple rule, one may establish the following restrictions on M, for the various forward difference equations developed earlier.

Interior Nodes.

One-dimensional Cartesian	$M \geq 2$	(130a)
Two-dimensional Cartesian	$M \geq 4$	(130b)
Three-dimensional Cartesian	$M \geq 6$	(130c)
Hexagonal System	$M \geq 4/3$	(130d)

Boundary Nodes (Convective).

One-dimensional Cartesian	$M \geq 2(1 + N)$	(131a)
Two-dimensional Cartesian	$M \geq 2(2 + N)$	(131b)
Two-dimensional Cartesian (Corner Node)	$M \geq 4(1 + N)$	(131c)

The stability for radiating boundaries is discussed in Ref. 24. Conditions of Eq. (131) are unfavorable for systems with a large ratio of surface conductance to the internal conductance, i.e., large values of N. For example, according to Eq. (131a), the time step Δt must be chosen such that

$$\Delta t \leq \frac{L^2}{2\alpha} \frac{1}{n(n + B_i)} \tag{132}$$

where L is the characteristic length, $n = L/\Delta x$, the total number of space subdivisions, and $B_i = hL/k$, the Biot number. Thus, for large B_i, Δt must be chosen unnecessarily small, resulting in a considerable increase in labor without always increasing the accuracy. A simple way to overcome this difficulty (though less accurate) is either to neglect the capacitance associated with the surface node or to locate the nodes in a different position as shown in Fig. 39 (nodes I, II, III, IV). The resulting stability criteria are

$$M \geq \frac{2N + 1}{N + 1} \tag{133a}$$

and

$$M \geq \frac{2 + 3N}{N + 2} \tag{133b}$$

respectively. The surface temperature can be obtained by using the steady-state energy balance in the previous two arrangements. Thus

$$T_0(t) = T_a(t) + \frac{1}{1 + N} [T_1(t) - T_a(t)] \tag{134a}$$

$$T_0(t) = T_a(t) + \frac{2}{2 + N} [T_I(t) - T_a(t)] \tag{134b}$$

More sophisticated techniques for the numerical treatment of the boundary conditions have been developed [25, 29]. In the discussion of Ref. 29, it is shown that a suitable substitution of Eq. (119a), which is stable for all M, is the following expression

$$T_0(t + \Delta t) = (1 - F) T_a(t) + \frac{[(N + 1) F - 1]}{N} T_0(t) + \left(\frac{1 - F}{N}\right) T_1(t) \qquad (135)$$

where

$$F = \exp\left(\frac{N^2}{M}\right) \text{erfc}\left(\frac{N}{\sqrt{M}}\right) \qquad (136)*$$

Stability criteria have been developed for the generalized six-point formula. It can be shown [17] that Eq. (110c) is stable for all

$$M \geq 2(1 - 2\lambda) \qquad (137)$$

Since $M > 0$, the restriction of Eq. (137) indicates that for $\frac{1}{2} \leq \lambda \leq 1$ the resulting finite difference scheme is always stable.

With regard to the accuracy of a numerical solution, the following remarks can be useful for practical applications, though they by no means cover the entire subject (see Refs. 16–29). Round-off errors have little or no effect in the solution if the restrictions of stability are observed; they accumulate roughly proportionally to the square root of the number of computing steps [17]. For example, in hand calculations, carrying three decimals is quite adequate for one hundred steps; if one uses an automatic computing machine which is capable of carrying a large number of decimals (7 to 14) the number of steps can be increased substantially without appreciable effect of the round-off errors. The predominant effect on the accuracy is due to the truncation error. It can be shown [17] that for interior nodes, if one uses the generalized six-point formula (Eq. (110c)), this error can be expressed as follows

$$|e_t| \leq \frac{\alpha^2 \Delta t}{2} \frac{\partial^4 T}{\partial x^4} \left[1 - 2\lambda - \frac{M}{6}\right] + A\Delta t^2 + B\Delta x^4 \qquad (138)$$

where e_t is the truncation error, T represents the exact solution, and A and B are constants. According to Eq. (138), the truncation errors for the three most commonly used finite-difference schemes are

Forward $|e_t| \leq A_1 \Delta t + B_1 \Delta x^2$ (139a)

Backward $|e_t| \leq A_2 \Delta t + B_2 \Delta x^2$ (139b)

Mid-difference $|e_t| \leq A_3 \Delta t^2 + B_2 \Delta x^2$ (139c)

If $0 \leq \lambda < \frac{1}{2}$ higher accuracy can be obtained by selecting

$$M = 6(1 - 2\lambda) \qquad (140)$$

in this case the truncation error of the forward-difference $(\lambda = 0, M = 6)$ becomes

$$|e_t| \leq A_4 \Delta t^2 + B_4 \Delta x^4 \qquad (141)$$

It should be mentioned that, in general, the boundary conditions introduce error of higher order of magnitude. For the practical case where the system is at uniform

---

*The function F is tabulated in Ref. 4.

temperature and suddenly a change occurs at the boundaries, one may begin the calculations with a smaller time step (for example, $M = 6$) and after some time, shift to a larger time step (for example, $M = 4$).

3. Graphical Methods

Graphical methods for obtaining approximate solutions of transient heat conduction problems were first introduced by Binder [30] and Schmidt [31] (they are known as the "Binder-Schmidt method"), with further improvements by Nessi and Nisolle [32]. In recent years, with the development of automatic computing machines, the interest in such methods has diminished, though these methods can provide quick answers and also have the advantages of simplicity and visualization. One-dimensional problems in Cartesian cylindrical and spherical coordinates, composite structures, and time-dependent boundary conditions can be easily handled by this method. The accuracy of a graphical solution (apart from the truncation error) depends on the choice of scale and the precision of the graphic construction. The basic principles of the method are explained in the following sections.

a. Plates.

Interior Nodes (Prescribed Surface Temperature). The explicit difference equation for plates is written as follows (see Eq. (110))

$$T_m(t + \Delta t) = \frac{1}{M}[T_{m+1}(t) + T_{m-1}(t)] + \left(\frac{M - 2}{M}\right)T_m(t) \tag{142}$$

Binder and Schmidt employed the value of $M = 2$, which reduces Eq. (142) to the simple form

$$T_m(t + \Delta t) = \frac{1}{2}[T_{m+1}(t) + T_{m-1}(t)] \tag{143}$$

The temperature T_m at $t + \Delta t$ is then, according to Eq. (143), equal to the arithmetic mean of the temperatures T_{m+1} and T_{m-1} at time t. The graphical construction of $T_m(t + \Delta t)$ for given $T_{m+1}(t)$ and $T_{m-1}(t)$ is shown in Fig. 40. The intersection C of the vertical line through the node m, with the straight line connecting A and B,

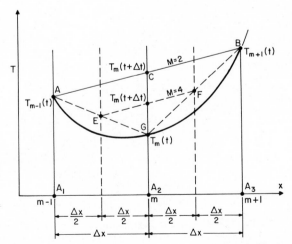

Fig. 40. Graphical solution of unsteady heat flow in plane wall using $M = 2$ and $M = 4$.

determines the value of $T_m(t + \Delta t)$. Indeed, from simple geometric considerations one can readily show that

$$\overline{A_2 C} \;=\; T_m(t + \Delta t) \;=\; \tfrac{1}{2}(\overline{A_1 A} + \overline{A_3 B}) \;=\; \tfrac{1}{2}[T_{m-1}(t) + T_{m+1}(t)] \qquad (143a)$$

where $\overline{A_2 C}$, $\overline{A_1 A}$, and $\overline{A_3 B}$ are the lengths of line segments which in an appropriate scale represent the values of the corresponding temperatures. Nessi and Nisolle extended the method using values of M different from two. (See also Ref. 33.) For example, if $M = 4$, Eq. (142) becomes

$$T_m(t + \Delta t) \;=\; \tfrac{1}{2}\left[\tfrac{1}{2}[T_{m-1}(t) + T_m(t)] + \tfrac{1}{2}[T_{m+1}(t) + T_m(t)]\right] \qquad (144)$$

Therefore $T_m(t + \Delta t)$ in this case is equal to the arithmetic mean of two arithmetic means. The graphical construction of $T_m(t + \Delta t)$ is self-explanatory in Fig. 40. As an illustration of the Binder-Schmidt method, the graphical solution of the problem presented in Example 12 (Sec. B.2) is shown in Fig. 41.

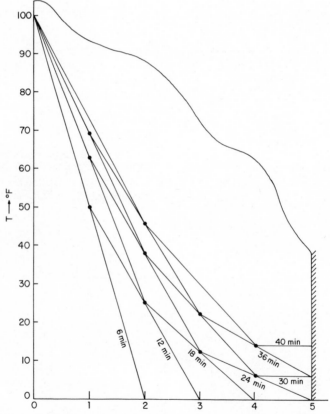

Fig. 41. Graphical solution of unsteady heat flow by the Binder-Schmidt method for the plate shown in Fig. 39.

Boundary Nodes (Prescribed Ambient Temperature). If there is convection at the surface with a heat transfer coefficient h, the previous method will be applied to

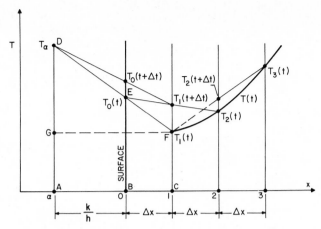

Fig. 42. Graphical solution of unsteady heat flow in a plate with convection on the surface.

obtain the temperatures of the interior nodes, while the surface temperature is determined according to Eq. (134a). Graphically this can be achieved by extending the medium by a distance equal to k/h (Fig. 42). Suppose $T_1(t)$ and $T_a(t)$ are given, then the intersection E, of the vertical line through the surface node 0, and the straight line connecting the points D and F determines the temperature $T_0(t)$. This statement can be verified by considering the following simple geometric relationships in the triangle DFG

$$\frac{(\overline{BE} - \overline{FC})}{(\overline{DA} - \overline{FC})} = \frac{\Delta x}{\Delta x + (k/h)} \tag{145}$$

If $\overline{FC}$ and $\overline{DA}$ are proportional to the values of the temperatures $T_1(t)$ and $T_a(t)$ respectively, then

$$\overline{BE} = T_1(t) + \left[\frac{\Delta x}{\Delta x + (k/h)}\right][T_a(t) - T_1(t)] \tag{146}$$

Equation (146), after some algebraic manipulations and the substitution of $h\Delta x/k$ by N, becomes

$$\overline{BE} = T_a(t) + \frac{1}{N + 1}[T_1(t) - T_a(t)] \tag{147}$$

A comparison between Eqs. (147) and (134) reveals that indeed $\overline{BE} = T_1(t)$. Knowing $T_0(t)$, one proceeds to find $T_1(t + \Delta t)$ as shown in Fig. 42. Some refinements of this method can be found in Ref. 34.

   Composite Media. The previous methods can be applied also to composite media, with slight modifications for the treatment of the interface and one of the boundaries. Consider the composite system consisting of two plates with properties α_1, k_1 and α_2, k_2 respectively, as shown in Fig. 43. In the absence of a thermal contact resistance, the interface temperature is given by Eq. (129) which for $M = 2$ takes the form

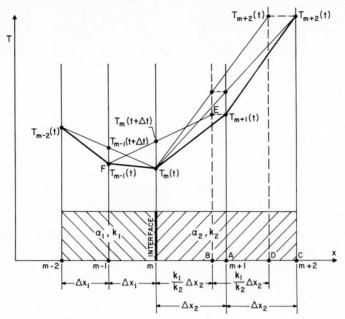

Fig. 43. Graphical solution of unsteady heat flow in composite plate.

$$\frac{T_m(t + \Delta t) - T_{m-1}(t)}{\Delta x_1} = \frac{1}{[(k_1/k_2)\,\Delta x_2]}\,[T_{m+1}(t) - T_m(t + \Delta t)] \quad (148)$$

with the space subdivisions Δx_1 and Δx_2 satisfying Eq. (128). (The time step Δt is the same for both sides.) According to Eq. (148), $T_m(t + \Delta t)$ can be determined graphically using the straight line procedure, as shown in Fig. 43, by simply relocating the node $m + 1$ from the point A which is at distance Δx_2 from the interface to the point B located at a distance $(k_1/k_2)\Delta x_2$. Moreover, since graphically determined temperatures are independent of the choice of scale used, it is suggested that the nodes of the second medium be located at distances $(k_1/k_2)\Delta x_2$ instead of Δx_2. Suppose now that "one" medium has been subdivided in an integer number of equal space increments Δx_2; then, since Δx_1 is not independent of Δx_2, see Eq. (128); one cannot, in general, subdivide the other medium in an integer number of equal subdivisions. In this case, one should take the last increment near the surface larger as shown in Fig. 44; that is, the node 1 is located at distance $\mu \Delta x_1$ from the surface, where $1 \le \mu < 2$. The equation which relates $T_1(t + \Delta t)$ to $T_0(t)$, $T_1(t)$ and $T_2(t)$ is obtained by the energy method and it is readily shown to be

$$T_1(t + \Delta t) = \frac{2}{M(1 + \mu)}\left[T_2(t) + \left(\frac{1}{\mu}\right)T_0(t) + (1 + \mu)\left(\frac{M}{2} - \frac{1}{\mu}\right)T_1(t)\right] \quad (149)*$$

Equation (149) for $M = 2$ simplifies to

$$T_1(t + \Delta t) = \left(\frac{1}{1 + \mu}\right)\left[T_2(t) + \left(\frac{1}{\mu}\right)T_0(t)\right] + \left(\frac{\mu - 1}{\mu}\right)T_1(t) \quad (150)$$

*Equation (149) should be used in connection with numerical computations in composite media.

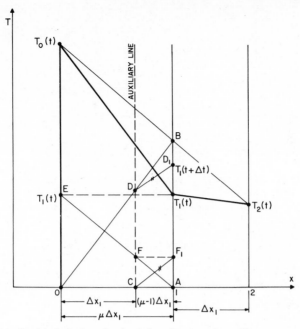

Fig. 44. Graphical solution of unsteady heat flow in a plate with uneven spacing.

The graphical solution of Eq. (150) is shown in Fig. 44. The point B, being the intersection of the vertical line through the node 1 and the straight line connecting $T_0(t)$ and $T_2(t)$, determines the line segment $\overline{AB}$ which is equal to

$$\overline{AB} = \left(\frac{1}{1+\mu}\right)[T_0(t) + \mu T_2(t)] \qquad (151)$$

At the point C located at distance Δx_1 from the surface, a vertical line has been drawn which intersects the line $\overline{OB}$ at the point D. From the similar triangles OCD and OAB we obtain that

$$\overline{CD} = \frac{1}{\mu} AB = \left(\frac{1}{1+\mu}\right)\left[T_2(t) + \frac{1}{\mu}T_0(t)\right] \qquad (152)$$

Similarly, the line segment $\overline{CF}$ is equal to

$$\overline{CF} = \left(\frac{\mu - 1}{\mu}\right)\overline{OE} = \left(\frac{\mu - 1}{\mu}\right)T_1(t) \qquad (153)$$

Therefore

$$T_1(t + \Delta t) = \overline{CF} + \overline{CD} \qquad (154)$$

Taking now $\overline{AF}_1 = \overline{AF}$ and $\overline{DD}_1$ parallel to $\overline{CF}_1$, the point D_1 is determined such that $\overline{AD}_1 = T_1(t + \Delta t)$. If there is a heat transfer coefficient at the surface the method remains unchanged since $T_0(t)$ is determined by the steady-state solution from $T_a(t)$ and $T_1(t)$.

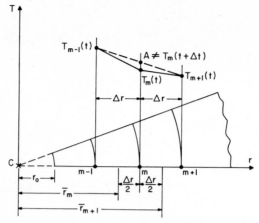

Fig. 45. Subdivision of cylindrical wall.

b. Cylindrical Shells. Consider a portion of a hollow cylinder subdivided into equal space increments Δr as shown in Fig. 45. The explicit difference equation for the node m is (Eq. (103))

$$\frac{k}{\Delta r}\left[\left(T_{m-1}(t) - T_m(t)\right)\left(r_m - \frac{\Delta r}{2}\right) + \left(T_{m+1}(t) - T_m(t)\right)\left(r_m + \frac{\Delta r}{2}\right)\right]$$

$$= \frac{\Delta r \rho c r_m}{\Delta t}[T_m(t + \Delta t) - T_m(t)] \quad (155)$$

With $M = \Delta r^2/\alpha \Delta t = 2$, Eq. (155) becomes

$$T_m(t + \Delta t) = \frac{1}{2}\left[T_{m-1}(t)\left(1 - \frac{\Delta r}{2r_m}\right) + T_{m+1}(t)\left(1 + \frac{\Delta r}{2r_m}\right)\right] \quad (156)$$

or

$$T_m(t + \Delta t) = \frac{1}{2}\left[T_{m-1}(t)\frac{\bar{r}_m}{r_m} + T_{m+1}(t)\frac{\bar{r}_{m+1}}{r_m}\right] \quad (157)$$

where $\bar{r}_m = \frac{1}{2}(r_m + r_{m-1})$ and $\bar{r}_{m+1} = \frac{1}{2}(r_m + r_{m+1})$.

Equations (156) and (157) indicate that the straight line procedure is not applicable in this case, because $T_m(t + \Delta t)$ depends more on $T_{m+1}(t)$ and less on $T_{m-1}(t)$. Therefore, in order to apply the Binder-Schmidt method in cylindrical coordinates, a new distorted length scale ξ must be used, such that the node $m + 1$ will appear closer to m than the node $m - 1$, with their distances $\Delta \xi_{m+1}$ and $\Delta \xi_m$ (see Fig. 46) being proportional to $\bar{r}_{m+1}/r_m$ and $\bar{r}_m/r_m$, respectively. It can be readily shown that this scale is $\xi = \ln(r)$. In Fig. 46 the graphical construction of $T_m(t + \Delta t)$ is shown. Note that the scale is normalized to the inner radius; that is, $\xi = \ln(r/r_0)$. For the case where

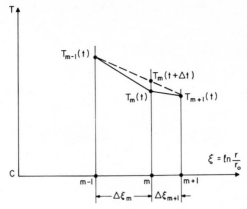

Fig. 46. Graphical solution of unsteady heat flow in a hollow cylinder (r_o = inner radius).

there is a heat transfer coefficient at the surface, the same straight line procedure can be used; it suffices only to locate the ambient temperature $T_a(t)$ at a distance from the surface equal to k/hr_w where r_w is the radius of the surface (inner or outer radius). Composite hollow cylinders are treated in a manner identical to that of composite plates. (See Sec. B.3.a.)

c. Spherical Shells. Consider a spherical shell subdivided into equal space increment Δr. The explicit finite difference equation for the node m is as follows (Eq. (107))

$$k \frac{(r_m - \Delta r/2)^2}{\Delta r^2} [T_{m-1}(t) - T_m(t)] + \frac{(r_m + \Delta r/2)^2}{\Delta r^2} [T_{m+1}(t) - T_m(t)]$$

$$= \frac{\rho c r_m^2}{\Delta t} [T_m(t + \Delta t) - T_m(t)] \quad (158)$$

Rearranging terms in Eq. (158), we obtain the following expression

$$T_m(t + \Delta t) = \frac{1}{M} \left[\frac{\bar{r}_m^2}{r_m^2} T_{m-1}(t) + \frac{\bar{r}_{m+1}^2}{r_m^2} T_{m+1}(t) + \left(M - \frac{\bar{r}_m^2 \ \bar{r}_{m+1}^2}{r_m^2} \right) T_m(t) \right] \quad (159)$$

where $M = \Delta r^2/\alpha\Delta t$, $\bar{r}_m = \frac{1}{2}(r_m + r_{m-1})$, $\bar{r}_{m+1} = \frac{1}{2}(r_m + r_{m+1})$. With the above definitions of $\bar{r}_m$, $\bar{r}_{m+1}$, the coefficient of $T_m(t)$ can be written as

$$M - 2 - \frac{1}{2} \left(\frac{\Delta r}{r_m} \right)^2 \simeq M - 2 \quad (160)$$

Therefore, using the value at $M = 2$, Eq. (159) simplifies to

$$T_m(t + \Delta t) = \frac{1}{2} \left[\left(\frac{\bar{r}_m}{r_m} \right)^2 T_{m-1}(t) + \left(\frac{\bar{r}_{m+1}}{r_m} \right)^2 T_{m+1}(t) \right] \quad (161)$$

According to Eq. (161), the value of $T_m(t + \Delta t)$ is equal to the arithmetic mean of $T_{m-1}(t)$ and $T_{m+1}(t)$ if a new scale $\xi = 1/r$ is used. Therefore, in order to apply the Binder-Schmidt method to unsteady heat flow in spherical shells, the vertical lines which represent the temperatures must be taken at distances $1/r$ instead of r. As in the previous case of cylindrical shells, the scale should be normalized to the inner radius; i.e., one should take as $\xi = r_0/r$.

4. Approximate Analytical Methods

Several methods have been developed for obtaining approximate solutions of heat transfer problems in an analytical form (see Refs. 36 and 37 and references cited there). They can be applied to both linear and nonlinear problems yielding simple solutions with an accuracy adequate for engineering purposes. The most extensively used of these methods to be discussed here are the Integral method first introduced by Goodman [38] and the Biot's variational principle [39]. An example using the method of collocation is also given. The fundamental techniques of the methods are illustrated in connection with some simple one-dimensional problems.

   a. **Integral Method.** Integral methods were first introduced by von Karman and Pohlhausen [40] in order to obtain solutions of boundary value problems in fluid mechanics. The application of the method to heat transfer problems is explained here by determining the temperature distribution in a semi-infinite medium which is initially at uniform temperature T_0,* while at time $t = 0$ a variety of boundary conditions are imposed at the surface $x = 0$.

   Let $T(x, t)$ be the temperature above the constant temperature T_0; then the differential equation and the initial conditions satisfied by the temperature $T(x, t)$ are

$$\frac{\partial}{\partial x}\left(k\frac{\partial T}{\partial x}\right) = \rho c\,\frac{\partial T}{\partial t} \qquad \begin{matrix} x > 0 \\[4pt] t > 0 \end{matrix} \qquad (162)$$

$$T(x, 0) = 0 \qquad x > 0 \qquad (163)$$

Introduced now is the penetration distance $\delta(t)$ such that for $x > \delta(t)$ the system is unaffected by the applied boundary conditions; i.e., the system is at equilibrium temperature and, hence, there is no heat transfer beyond the point $x = \delta(t)$.

   Multiplying Eq. (162) by dx and integrating from $x = 0$ to $x = \delta$ we obtain the following expression

$$-k\left.\frac{\partial T}{\partial x}\right|_{x=0} = \frac{d}{dt}(\rho c\delta\bar{T}) \qquad (164)$$

$\left(k\left.\dfrac{\partial T}{\partial x}\right|_{x=\delta} \text{ has been assumed equal to zero.} \right)$

Where $\bar{T}$ is the space average temperature defined as

$$\bar{T} = \frac{1}{\delta}\int_0^\delta T(x, t)\,dx \qquad (165)$$

Equation (164) is called the heat-balance integral relating the heat transfer at the boundary to the time-rate change of the average energy of the system. To complete the solution one may proceed as follows:

   1. Assume a suitable profile for the temperature distribution for $0 \le x \le \delta$.
   2. Substituting the assumed $T(x, t)$ into Eqs. (165) and (164), one obtains a

---

*This method is applied almost exclusively to one-dimensional problems involving systems with uniform initial temperature [37].

differential equation with respect to δ which is further solved subject to initial conditions $\delta(0) = 0$.

3. Substituting the value $\delta(t)$ into the assumed temperature profile, one obtains $T[x, \delta(t)]$.

The temperature distribution is usually approximated by a polynomial in x, although trigonometric or exponential approximations may also be used. To illustrate the method, assume that the temperature T can be approximated by an mth degree polynomial

$$T = \sum_{r=0}^{m} b_r(t) x^r \qquad 0 \le x \tag{166}$$

There are $m + 1$ constants $b_0, b_1, \ldots, b_m$ which must be determined. Three relationships can be obtained: one from the boundary conditions at $x = 0$ and two from the boundary conditions at $x = \delta$; that is, the temperature and heat transfer are zero at this point. The remaining relationships can be obtained by imposing additional boundary conditions. In this case we assume that the first $m - 1$ derivatives are zero at $x = \delta$

$$\left. \frac{d^m T}{dx^m} \right|_{x=\delta} = 0 \qquad m = 0, 1 \ldots m - 1 \tag{167}$$

Introducing Eq. (166) into the m Eqs. (167) and solving the resulting algebraic equations, one obtains a general expression for b_r

$$b_r = T_s \left(\frac{-1}{\delta} \right)^r \tag{168}$$

where T_s is the surface temperature: $T_s = T(0, t)$. Therefore

$$T = T_s \left(1 - \frac{x}{\delta} \right)^m \tag{169}$$

The temperature T_s can be determined from the boundary conditions at $x = 0$. Consider now the following examples.

Heat Input at $x = 0$, Prescribed Function of Time, $F(t)$. The boundary conditions at $x = 0$ are

$$-k \left. \frac{\partial T}{\partial x} \right|_{x=0} = F(t) \qquad t > 0 \tag{170}$$

By virtue of Eq. (169) the following is obtained

$$T_s = \frac{\delta F(t)}{mk} \tag{171}$$

With the assumed temperature profile, the average temperature $\overline{T}$ is readily obtained

$$\overline{T} = \frac{T_s}{m + 1} \tag{172}$$

Substituting Eqs. (172) and (170) into Eq. (164) and also using the relationship of Eq. (171) the following differential equation is obtained

$$F(t) = \frac{1}{\alpha} \frac{d}{dt} \left[\frac{\delta^2 F(t)}{m(m+1)} \right] \tag{173}$$

where α is the thermal diffusivity. Integrating Eq. (173), taking into account that $\delta(0) = 0$, one obtains

$$\delta = \left\{ \frac{\alpha m(m+1)}{F(t)} \int_0^t F(\xi)\,d\xi \right\}^{1/2} \tag{174}$$

If $F(t)$ is constant equal to F_0, then

$$\delta = \sqrt{m(m+1)\alpha t} \tag{175}$$

and the surface temperature

$$T_s = \frac{F_0}{k} \sqrt{\frac{m+1}{m}\alpha t} \tag{176}$$

The exact solution gives the following expression for the surface temperature [4]

$$T_{se} = \frac{F_0}{k} \sqrt{\frac{4}{\pi}\alpha t} \tag{177}$$

Selecting $m = 4$, the resulting error, if Eq. (176) is used instead of Eq. (177), is less than 1%.

Convection at the Surface $x = 0$. In this case the semi-infinite medium at time $t = 0$ starts to receive heat from a source which is at temperature $T_a(t)$ (above the uniform temperature T_0) through a heat transfer coefficient h. The boundary conditions now are

$$k \frac{\partial T}{\partial x} = h[T_s - T_a(t)] \qquad x > 0 \tag{178}$$
$$t > 0$$

Using the previously assumed temperature profile Eq. (169), the following is obtained

$$T_s = \left(\frac{N}{m+N} \right) T_a(t) \tag{179}$$

where

$$N = \frac{h\delta}{k} \tag{180}$$

Therefore

$$\overline{T} = \frac{T_s}{m + 1} = \frac{NT_a(t)}{(m + N)(m + 1)} \tag{181}$$

Substituting the above expressions into Eq. (164), the following differential equation is obtained

$$\left(\frac{m}{N + m}\right) hT_a(t) = \frac{\rho c}{m + 1} \cdot \frac{d}{dt}\left(\frac{\delta N}{N + m}\right) \tag{182}$$

Substituting further δ from Eq. (180), one may obtain the following equation

$$(N + m) \frac{d}{d\beta}\left(\frac{N^2 T_a(t)}{N + m}\right) = m(m + 1) T_a(t) \tag{183}$$

where

$$\beta = \frac{h^2}{k^2}\alpha t \tag{184}$$

If $T_a(t)$ is constant, Eq. (183) can be integrated subject to initial conditions $N(0) = 0$, yielding the following implicit relationship between N and β

$$\frac{1}{m(m + 1)}\left[\frac{N(N + 2m)}{2} - m^2 \ln\left(\frac{N + m}{m}\right)\right] = \beta \tag{185}$$

Therefore the temperature can be obtained approximately by the equation

$$T = \left(\frac{N}{N + m}\right)\left(1 - \frac{B_i}{N}\right)^m \tag{186}$$

where

$$B_i = \frac{hx}{k} \tag{187}$$

The exact solution of this problem is [4]

$$\frac{T_e}{T_a} = \text{erfc}\left(\frac{B_i}{2\beta}\right) - \exp(B_i + \beta) \, \text{erfc}\left(\frac{B_i}{2\beta} + \beta\right) \tag{188}$$

The surface temperature $(x = 0, B_i = 0)$ is equal to

$$\frac{T_{se}}{T_a} = 1 - \exp(\beta) \, \text{erfc}(\sqrt{\beta}) \tag{189}$$

For comparison, in Table 14 some values of T_s/T_a (Eq. (179)) and T_{se}/T_a (Eq. (189)) versus N for $m = 2$ and $m = 3$ are presented together with the percent error involved. It can be seen that the approximate results are fairly accurate.

TABLE 14. Comparison of Surface Temperature from Exact and Approximate Integral Solutions for a Semi-infinite Medium

N	m = 2			m = 3		
	T_{se}/T_a	T_s/T_a	% error	T_{se}/T_a	T_s/T_a	% error
0.5	0.188	0.200	1	0.139	0.143	2.8
1.0	0.319	0.333	4.2	0.248	0.250	0.8
2.0	0.486	0.500	0.8	0.390	0.400	2.5
3.0	0.591	0.600	1.5	0.503	0.500	-0.6
4.0	0.659	0.667	1.2	0.577	0.571	-1.05
5.0	0.709	0.714	0.7	0.633	0.625	-1.28
6.0	0.748	0.750	0.27	0.673	0.667	-0.59
7.0	0.777	0.778	0.13	0.709	0.700	-1.29
8.0	0.796	0.800	0.5	0.736	0.727	-1.24

Prescribed Surface Temperature $T_s(t)$. In this case using the temperature profile given by Eq. (169), the surface condition becomes

$$k \frac{dT}{dx} = -\left(\frac{mk}{\delta}\right) T_s(t) \tag{190}$$

Substituting Eqs. (172) and (190) into Eq. (164), the following differential equation is obtained

$$\delta \frac{d[\delta T_s(t)]}{d(\alpha t)} = m(m + 1) T_s(t) \tag{191}$$

If $T_s(t)$ is constant, δ can be readily found by integrating Eq. (191). Thus

$$\delta = [2m(m + 1)\alpha t]^{1/2} \tag{192}$$

Therefore

$$\frac{T}{T_s} = \left\{1 - \frac{x}{[2m(m + 1)\alpha t]^{1/2}}\right\}^m \tag{193}$$

The exact solution of this problem is [4]

$$\frac{T_e}{T_s} = \mathrm{erfc}\left(\frac{x}{2\sqrt{\alpha t}}\right) \tag{194}$$

For comparison, some values of T/T_s calculated according to Eq. (193) with $m = 2$ and T_e/T_s versus $x/2\sqrt{\alpha t}$ are plotted in Fig. 47. One may observe that the approximate results are fairly accurate.

Several other examples involving finite, linear, and nonlinear systems are presented in Ref. 37.

   b. Biot's Variational Principle. Variational principles have been used extensively in recent years in order to obtain approximate analytical solutions of heat-conduction problems ([41] and references cited there). The thermodynamic analogy to Hamilton's principle in mechanics is due to Biot [39]. The variational method is applicable to both linear and nonlinear multidimensional systems. The fundamental aspects of the method are explained here in connection with the one-dimensional Cartesian systems. Consider a system of length L (finite or infinite) initially at some equilibrium temperature T_0, while at time $t = 0$ certain boundary conditions are imposed. Let H represent a heat-flow vector (here H has only one component) whose time rate of change $\partial H/\partial t$ is the heat flux q''. From energy considerations the temperature T is related to H by the following expression

$$-\rho c T = \frac{\partial H}{\partial x} \qquad x > 0 \qquad (195)$$

where T is the temperature above the equilibrium temperature T_0. In order to apply the variational method one may proceed as follows:

1. Assume a suitable function of n parameters $q_1, q_2, \ldots, q_n, x, t$ to represent H, that is,

$$H = H(q_1, q_2, \ldots, q_n, x, t) \qquad (196)$$

   These parameters are called the generalized coordinates.
2. From Eq. (195) determine T^* as

$$T = T(q_1, q_2, \ldots, q_n, x, t) \qquad (197)$$

3. Evaluate the thermal potential V defined by the following definite integral

$$V = \frac{\rho c}{2} \int_0^L T^2 \, dx = V(q_1, q_2, \ldots, q_n, t) \qquad (198)$$

4. Evaluate the dissipation function D defined by the following definite integral

$$D = \frac{1}{2k} \int_0^L \left(\frac{\partial H}{\partial t}\right)^2 dx + \frac{1}{2h}\left(\frac{\partial H}{\partial t}\right)^2_{x=L} - \frac{1}{2h}\left(\frac{\partial H}{\partial t}\right)^2_{x=0} \qquad (199)$$

therefore

$$D = D(q_1, q_2, \ldots, q_n, \dot{q}_1, \dot{q}_2, \ldots, \dot{q}_n, t) \qquad (199a)$$

where h is the heat transfer coefficient and $\dot{q}$ means differentiation with respect to time.
5. Construct the generalized thermal force Q_i defined as

$$Q_i = T\frac{\partial H}{\partial q_i}\bigg|_{x=L} - T\frac{\partial H}{\partial q_i}\bigg|_{x=0} \qquad (200)$$

or

$$Q_i = Q_i(q_1, q_2, \ldots, q_n, t) \qquad (200a)$$

It is shown [39] that the quantities V, D, and Q_i satisfy a differential equation

*The steps 1 and 2 can be interchanged.

similar to the well-known Lagrange's equation in mechanics. Therefore

$$\frac{\partial V}{\partial q_i} + \frac{\partial D}{\partial \dot{q}_i} = Q_i \tag{201}$$

Equation (201) constitutes a set of n simultaneous differential equations from which the parameters q_i can be found.

As an example, consider the semi-infinite medium initially at temperature T_0 while at time $t = 0$ the surface $x = 0$ is raised to a temperature T_s above T_0. In this case the length L is taken equal to the penetration distance $\delta(t)$ (see Sec. B.4.a) which also serves as the parameter q_1. Assume H to be represented by the following expression

$$H = a\left(1 - \frac{x}{\delta}\right)^3 \tag{202}$$

where a is a constant that can be determined from the boundary conditions at $x = 0$. By virtue of Eq. (195) we obtain

$$T = \frac{3a}{\rho c \delta}\left(1 - \frac{x}{\delta}\right)^2 \tag{203}$$

since for $x = 0$, $T = T_s$; the constant a is equal to

$$a = \frac{\rho c T_s \delta}{3} \tag{204}$$

The thermal potential V is equal to

$$V = \frac{\rho c T_s^2}{2}\int_0^\delta \left(1 - \frac{x}{\delta}\right)^4 dx = \frac{\rho c T_s^2 \delta}{10} \tag{205}$$

The heat flux $\partial H/\partial t$ is equal to

$$\frac{\partial H}{\partial t} = \frac{\rho c T_s}{3}\left[\dot{\delta}\left(1 - \frac{x}{\delta}\right)^3 + 3\dot{\delta}\frac{x}{\delta}\left(1 - \frac{x}{\delta}\right)^2\right] \tag{206}$$

thus

$$\left(\frac{\partial H}{\partial t}\right)^2 = \left[\frac{\rho c T_s \dot{\delta}}{3}\right]^2 \left(1 - \frac{x}{\delta}\right)^6 + 6\left(\frac{x}{\delta}\right)\left(1 - \frac{x}{\delta}\right)^5 + 9\left(\frac{x}{\delta}\right)^2\left(1 - \frac{x}{\delta}\right)^4 \tag{207}$$

Substituting Eq. (207) into Eq. (199) and performing the integration, the dissipation function D is determined. Thus

$$D = \frac{13}{630k}(\rho c T_s)^2 \delta \dot{\delta}^2 \tag{208}$$

The thermal force Q_1 is equal to

$$Q_1 = T_s \left[\left(\frac{\partial H}{\partial \delta} \right)_{x=\delta} - \left(\frac{\partial H}{\partial \delta} \right)_{x=0} \right] = \frac{\rho c T_s}{3} \tag{209}$$

The partial derivatives of V and D with respect to δ and $\dot{\delta}$ are

$$\frac{\partial V}{\partial \delta} = \frac{\rho c T_s^2}{10} \tag{210}$$

$$\frac{\partial D}{\partial \dot{\delta}} = \frac{13}{315} \delta \dot{\delta} \frac{(\rho c T_s)^2}{k} \tag{211}$$

Therefore, according to Eq. (201), δ should satisfy the following differential equation

$$\frac{13}{315} \delta \frac{d\delta}{dt} = \frac{7\alpha}{30} \tag{212}$$

where α is the thermal diffusivity. The solution of Eq. (212), subject to the initial condition $\delta(0) = 0$, yields the following expression for δ

$$\delta = \left(\frac{147}{13} \alpha t \right)^{1/2} = 3.363 (\alpha t)^{1/2} \tag{213}$$

therefore the temperature distribution is

$$\frac{T}{T_s} = \left(1 - \frac{x}{3.363 \sqrt{\alpha t}} \right)^2 \tag{214}$$

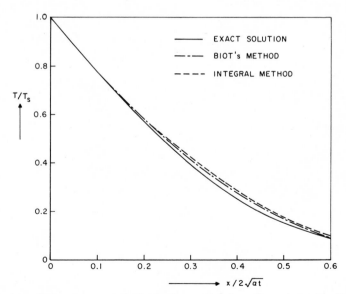

Fig. 47. Approximate temperature distribution in a semi-infinite medium calculated by the integral and Biot's method.

For comparison some of the values of T/T_s versus $x/2\sqrt{\alpha t}$, determined using Eq. (214), have been plotted in Fig. 47 together with the exact solution; the approximate solution agrees fairly well with the exact solution.

For additional examples involving finite, linear, and nonlinear systems with a variety of boundary conditions, one may consult Refs. 38 and 41.

c. The Method of Collocation. The method of collocation is probably the simplest method of solving differential equations. In many cases the method, despite its simplicity, yields quite accurate results. Heat transfer problems having nonzero initial conditions can be solved by this method. The principles of the method can be summarized in the following step-by-step procedure.

1. Assume a temperature distribution

$$T = T(x, y, z, a_1, a_2, \ldots, a_n) \tag{215}$$

which depends on n parameters $a_1(t), a_2(t), \ldots, a_n(t)$. The number n is chosen to be greater than the number of boundary conditions given.

2. Determine the relationships between the parameters $a_1(t), a_2(t), \ldots, a_n(t)$ from the given boundary conditions, i.e., the temperature T satisfies exactly the boundary conditions.*

3. The parameters are further determined by satisfying the differential equation at preselected points (collocation points). The number of collocation points is equal to the number of parameters minus the number of the given boundary conditions.

The method will be illustrated by the following example. Consider a slab of thermal conductivity k, diffusivity α, length L, having an initial temperature distribution $T(x, 0) = f(x)$ and insulated at $x = 0$. At time $t = 0$ the slab is brought into contact, through a heat transfer coefficient h, with a fluid of temperature $T_a = 0$. It is desired to determine the temperature distribution as a function of time. The temperature satisfies the following differential equation and boundary conditions

$$\frac{\partial^2 T}{\partial \xi^2} = \frac{\partial T}{\delta \tau} \tag{216}$$

$$\frac{\partial T}{\partial \xi} = 0 \qquad \xi = 0 \tag{217}$$

$$\frac{\partial T}{\partial \xi} = -NT \qquad \xi = 1 \tag{218}$$

where

$$\xi = \frac{x}{L} \tag{219a}$$

$$\tau = \frac{\alpha t}{L^2} \tag{219b}$$

$$N = \frac{hL}{k} \tag{219c}$$

---

*The collocation method based on this condition is called the "interior" method in contrast to the "boundary" method where the assumed temperature T satisfies the differential equation instead (see Ref. 35, p. 28).

Assume a biquadratic profile in order to satisfy the symmetry condition (Eq. (217))

$$T = a_1 + a_2 \xi^2 + a_3 \xi^4 \tag{220}$$

Here there is one more boundary condition and three parameters. Therefore we must select two collocation points, for example, the point $\xi = 0$, $\xi = 1$. Substituting the above expression in Eq. (218), the following relationship between a_1, a_2, a_3 is obtained

$$2a_2 + 4a_3 = -N(a_1 + a_2 + a_3) \tag{221}$$

Satisfying Eq. (216) at the collocation points $\xi = 0, \xi = 1$, one obtains the following two differential equations, respectively,

$$2a_2 = \frac{da_1}{d\tau} \tag{222}$$

$$2a_2 + 12a_3 = \frac{d}{d\tau}(a_1 + a_2 + a_3) \tag{223}$$

From Eqs. (222) and (223) one may obtain the auxiliary equation

$$12a_3 = \frac{d}{d\tau}(a_2 + a_3) \tag{223a}$$

Eliminating further the parameters a_1, a_2 from Eqs. (221)–(223), one obtains the following differential equation for the parameter a_3

$$\frac{d^2 a_3}{d\tau^2} + (12 + 5N)\frac{da_3}{d\tau} + 12 N a_3 = 0 \tag{224}$$

Equation (224) is a simple linear ordinary differential equation with constant coefficients. Therefore

$$a_3 = k_1 \exp(\rho_1 \tau) + k_2 \exp(\rho_2 \tau) \tag{225}$$

where k_1 and k_2 are constants which will be determined from the initial conditions and ρ_1 and ρ_2, the two real distinct and negative roots of the following quadratic equation

$$\rho^2 + (12 + 5N)\rho + 12 N\rho = 0 \tag{226}$$

The roots ρ_1, ρ_2 are

$$\rho_1 = -\left[\frac{12 + 5N + (144 + 25N^2 + 72N)^{1/2}}{2} \right] \tag{226a}$$

$$\rho_2 = -\left[\frac{12 + 5N - (144 + 25N^2 + 72N)^{1/2}}{2} \right] \tag{226b}$$

To determine the constants k_1 and k_2, one applies the initial conditions at the points $\xi = 0, \xi = 1$. Thus the following expressions are obtained

$$a_1(0) = f(0) \tag{227}$$

$$a_1(0) + a_2(0) + a_3(0) = f(1) \tag{228}$$

As an example, consider the case where $f(x) = T_i$ a constant. After some algebraic manipulation one may obtain the following expressions for the constants k_1 and k_2

$$k_1 = \frac{T_i N \rho_1}{2(\rho_2 - \rho_1)} \tag{229a}$$

$$k_2 = \frac{T_i N \rho_2}{2(\rho_2 - \rho_1)} \tag{229b}$$

The parameters a_2 and a_1 can be further determined from Eq. (223a) and (222) respectively, thus having the temperature T as a function of time and space.

For comparison let us evaluate the surface temperature $T_s = T(\xi = 1)$ which is equal to

$$T_s = a_1 + a_2 + a_3 \tag{230}$$

By virtue of Eqs. (221) and (223a), the following expression for T_s is obtained

$$\frac{T_s}{T_i} = \frac{\rho_1(N + \rho_2) \exp(\rho_1 \tau) + \rho_2(N + \rho_1) \exp(\rho_2 \tau)}{N(\rho_2 - \rho_1)} \tag{231}$$

One may observe that the absolute value of ρ_1 is always greater than or equal to 12. Therefore the predominant term in Eq. (231) is the second term appearing in the numerator. Thus T_s can be expressed as follows

$$\frac{T_s}{T_i} = \frac{\rho_2(N + \rho_1)}{N(\rho_1 - \rho_2)} \exp(\rho_2 \tau) \tag{232}$$

The exact solution is given in a form of series [4]. Taking only the first term of the series, the exact surface temperature T_{se} is given by the following equation

$$\frac{T_{se}}{T_i} = \frac{2N}{N(N + 1) + \alpha_1^2} \exp(-\alpha_1^2 \tau) \tag{233}$$

where α_1 is the first root of the equation [4]

$$\alpha_1 \tan \alpha_1 = N \tag{234}$$

Equation (233) is an adequate approximation of the exact surface temperature since the square of the second root α_2 varies from 9.87 to 22.2 [4, page 491]. A comparison between the exact and approximate solution for the surface temperature, using different values of N, is given in Table 15. In Table 16, the exact and approximate solutions of the surface temperatures are given for $N = 1$ and $0.2 \leq \tau \leq 3$.

TABLE 15. Comparison between Exact and Approximate Surface Temperature for a Slab Determined by the Collocation Method, Using Different Values of N

N	T_s/T_i	T_{se}/T_i
0.1	$0.959 \exp(-0.096\,\tau)$	$0.967 \exp(-0.097\,\tau)$
0.5	$0.848 \exp(-0.426\,\tau)$	$0.850 \exp(-0.427\,\tau)$
1	$0.725 \exp(-0.738\,\tau)$	$0.730 \exp(-0.740\,\tau)$
2	$0.551 \exp(-1.151\,\tau)$	$0.559 \exp(-1.160\,\tau)$
3	$0.438 \exp(-1.412\,\tau)$	$0.447 \exp(-1.422\,\tau)$
4	$0.361 \exp(-1.578\,\tau)$	$0.370 \exp(-1.599\,\tau)$
5	$0.307 \exp(-1.700\,\tau)$	$0.315 \exp(-1.726\,\tau)$
10	$0.172 \exp(-2.000\,\tau)$	$0.179 \exp(-2.042\,\tau)$

TABLE 16. Comparison between Exact and Approximate Surface Temperature for a Slab Determined by the Collocation Method Using Different Values of τ and $N = 1$

τ	T_s/T_i	T_{se}/T_i	% error
0.2	0.6258	0.6296	0.60
0.3	0.5813	0.5846	0.56
0.4	0.5400	0.5430	0.55
0.6	0.4660	0.4683	0.49
0.8	0.4019	0.4038	0.47
1	0.3468	0.3482	0.40
1.5	0.2398	0.2406	0.33
2	0.1658	0.1661	0.18
3	0.0793	0.0793	0.0

The results shown in Table 16 indicate that despite the fact that only two collocation points have been selected, the error is less than 1.0%. Additional examples are given in Ref. 38. Some refinements of the method appear in Ref. 42.

REFERENCES

1. M. G. Salvadori and M. L. Baron, "Numerical Methods in Engineering," Prentice-Hall, Inc., New York, 1962.
2. W. E. Milne, "Numerical Solution of Differential Equations," John Wiley & Sons, Inc., New York, 1953.
3. H. W. Emmons, The Numerical Solution of Heat-Conduction Problems, *Trans. ASME*, **65**:607 (1943).
4. H. S. Carslaw and J. C. Jaeger, "Conduction of Heat in Solids," Oxford University Press, New York, 1959.
5. R. H. Macneal, An Asymmetrical Finite Difference Network, *Quart. Appl. Math.*, **2**:295 (1953).
6. T. S. Nickerson and G. M. Dusinberre, Heat Transfer through Thick Insulation on Cylindrical Enclosures (see discussion by H. G. Elrod), *Trans. ASME*, **70**:903 (1948).
7. J. B. Scarborough, "Numerical Mathematical Analysis," 4th ed., The Johns Hopkins University Press, Baltimore, 1958.
8. P. C. Richardson, The Difference Equation Method for Solving the Dirichlet Problem, *Phil. Trans. Roy. Soc. London*, **210A**:307 (1910).
9. D. M. Young, ORDVAC Solutions of the Dirichlet Problem, *J. Assoc. Comput. Math.*, **2**:137 (1955).

10. R. V. Churchill, "Introduction to Complex Variables and Applications," McGraw-Hill Book Company, New York, 1948.
11. H. Kober, "Dictionary of Conformal Representations," Dover Publications, Inc., 1957.
12. E. Pollman, Temperature and Stresses on Hollow Blades for Gas Turbines (translated from German), *NACA TM No. 1183,* Washington, 1947.
13. O. Lutz, Graphical Determination of Wall Temperatures for Heat Transfer through Walls of Arbitrary Shape (translated from German), *NACA TM No. 1820,* Washington, 1950.
14. O. Lutz, Graphical Determination of Wall Temperatures during Heat Transfer, *R.T.P. Translation No. 2500,* Durand Reprinting Committee, California Institute of Technology.
15. P. J. Schneider, "Conduction Heat Transfer," Addison-Wesley Publishing Co., Inc., Reading, Mass, 1955.
16. J. Douglas, Jr., A Survey of Numerical Methods for Parabolic Differential Equations, *Advances in Computers,* 2:1, Academic Press (1961).
17. R. D. Richtmyer and K. W. Morton, "Difference Methods for Initial Value Problems," Interscience Publishers, Inc., New York, 1967.
18. P. Laasonen, Über eine Methode zur Lösung der Wärmeleitungsleichung, *Acta Math.,* 81:309 (1949).
19. G. Liebmann, A New Electrical Analog Method for the Solution of Transient Heat-Conduction Problems, *Trans. ASME,* 78:655 (1956).
20. G. Liebmann, Solution of Transient Heat Transfer Problems by the Resistance Network Method, *ASME Paper No. 55-A-61,* November 1955.
21. J. Crank and P. Nicholson, A Practical Method for Numerical Integration of Solutions of Partial Differential Equations of Heat-Conduction Type, *Proc. Cambr. Philos. Soc.,* 43:50 (1947).
22. G. Leppert, A Stable Numerical Solution for Transient Heat Flow, *ASME Paper No. 53–F-4,* October 1953.
23. G. M. Dusinberre, "Heat Transfer Calculations by Finite Differences," International Textbook Company, Scranton, Pa., 1961.
24. G. R. Gaumer, The Stability of Three Finite Difference Methods of Solving for Transient Temperatures, *ARS Journal,* 32:1595 (1962).
25. C. M. Fowler, Analysis of Numerical Solutions of Transient-Flow Problems, *Quart. Appl. Math.,* 3:361 (1946).
26. G. G. O'Brien, M. A. Hyman, and S. Kaplan, A Study of the Numerical Solution of Partial Differential Equations, *J. Math. Phys.,* 29:223 (1950).
27. F. B. Hildebrand, On the Convergence of Numerical Solutions of the Heat-Flow Equations, *J. Math. Phys.,* 31:35 (1952).
28. G. W. Evans, R. Brousseau, and R. Kierstead, Stability Considerations for Various Difference Equations Derived for the Linear Heat Conduction Equation, *J. Math. Phys.,* 34:267 (1955).
29. H. G. Elrod, Jr., New Finite Difference Technique for the Solution of Heat Conduction Equation, Especially Near Surfaces with Convective Heat Transfer, *Trans. ASME,* 79:1519 (1957).
30. L. Binder, "Aussere Wärmeleitung und Erwärmung electrischer Machinen," Dissertation, Techn. Hochschule München, Wilhelm Knapp, Halle, pp. 20–26, 1911.
31. E. Schmidt, Über die Auswendung der Differenzenrechnung auf technische Anheiz-und Abkühlungsprobleme, *Beitr. tech. Mechanik und tech. Physik,* August Föppl Festschrift, Berlin, 1924.
32. A. Nessi and A. Nisolle, "Méthodes graphiques pour l'étude des installation de chauffage, etc.," Paris, Dumond, 1929.
33. B. Paul, "Generalization of the Schmidt Graphical Method for Transient Heat Conduction," *ARS Journal,* 32:1098 (1962).
34. M. Jacob, "Heat Transfer," vol. 1, John Wiley & Sons, Inc., New York, 1956.
35. L. Collatz, "The Numerical Treatment of Differential Equations," 3d ed., translated by P. G. Williams, Springer-Verlag, New York Inc., New York, 1966.
36. B. A. Boley and J. H. Weiner, "Theory of Thermal Stresses," John Wiley & Sons, Inc., New York, 1960.
37. T. R. Goodman, Integral Methods for Nonlinear Heat Transfer, "Advances in Heat Transfer" (T. F. Irvine, Jr. and J. P. Hartnett, eds.), vol. 1, Academic Press, Inc., New York, 1964.
38. T. R. Goodman, The Heat Balance Integral and Its Application to Problems Involving a Change of Phase, *Trans. ASME,* 80:335 (1958).

39. M. A. Biot, New Methods in Heat Flow Analysis with Applications to Flight Structures, *J. Aeronautical Sci.*, **24**:857 (1957).
40. H. Schlichting, "Boundary Layer Theory," 4th ed., McGraw-Hill Book Company, New York, 1960.
41. T. J. Lardner, Biot's Variational Principle in Heat Conduction, *J. Amer. Inst. Aero. Astro.*, **1**:196 (1963).
42. J. U. Ojalvo and F. D. Linzer, Improved Point Matching Techniques, *Quart. J. Mech. Appl. Math.*, **18**:41 (1965).

NOMENCLATURE

English

A area, ft^2

a_i parameters in Eq. (215)

B_i Biot number

b_r coefficients in Eq. (166)

c specific heat at constant pressure, Btu/lb-°F

D diameter, ft

D normalized thermal conductance

D dissipation function, Btu-°F/ft^2

d/dx total derivative

$\partial/\partial x$ partial derivative

e base of natural logarithm = 2.71828, efficiency

erf   error function

erfc   complementary error function = 1 – erf

F combined configuration factor for radiation

$F(t)$ surface heat flux, Btu/hr-ft^2

H finite difference operator

H heat flow vector, Btu/ft^2

h heat transfer coefficient, Btu/hr-ft^2-°F

h length increment, ft

k, k_x, k_y thermal conductivity, Btu/hr-ft-°F

    length increment, ft

K thermal conductance, Btu/hr-°F

K finite difference operators

L length, ft

l width, ft

l_n natural logarithm

M Fourier modulus for Cartesian cylindrical and spherical coordinates

M' Fourier modulus for hexagonal system

m dimensionless rectangular coordinate

N Nusselt number

N_g number of heat flow tuber

N_t number of temperature increments

n dimensionless rectangular coordinate

Q_i generalized thermal force, Btu-°F/ft^3

q rate of heat flow, Btu/hr

q_i generalized coordinates, ft

q' rate of heat flow per unit length, Btu/hr-ft

q'' heat flux, Btu/hr-ft^2

q''' internal volume heat source, Btu/hr-ft^3

R residual, °F

r radial coordinate, ft

r dimensionless rectangular coordinate

S dimensionless length
T temperature, °F, °R
t time, hr
V volume, ft^3
V thermal potential, Btu-°F/ft^2
w fin thickness, ft
w complex variable transformation
x rectangular coordinate, ft
$2X$ finite difference operator
y rectangular coordinate, ft
z rectangular coordinate, ft
z complex variable

Greek

α thermal diffusivity = $k/\rho c$, ft^2/hr
β angle
Δ forward finite difference
∇ backward finite difference
δ central finite difference
δ thickness, ft
λ constant in Eq. (118)
μ normalized distance
ξ transformed cylindrical coordinate = ln r
π 3.1416
ρ density, lb/ft^3
ρ roots of the quadratic equation Eq. (226)
σ Stefan-Boltzmann radiation constant, Btu/hr-ft^2-°R^4
ϕ latitude angle in cylindrical coordinates
ϕ function of complex variable
ψ azimuth angle in spherical coordinates
ψ function of complex variable

Subscripts

a ambient
i, j, k, m, n, r indicate location of the node
s surface

Superscripts

$-$ mean value
$'$ first derivative
$'$ per unit length
$''$ per unit area
$'''$ per unit volume

Section **5**

Analog Methods

VICTOR PASCHKIS

Consultant, Professor Emeritus of Mechanical Engineering and
Formerly Director of the Heat and Mass Flow Analyzer Laboratory,
Columbia University, New York City, N.Y.

A. INTRODUCTION

1. Definition of Analog Methods

If problems in two or more fields of physics can be described by the same
equations, these fields may be said to be "analogous." In order to solve a problem in

one field (that of the *prime system*), one may carry out experiments in an analogous field (that of the *analogous system*) and interpret the results in terms of the prime system. In carrying out a computation by the analogy method, there is usually a one-to-one relationship between components of the analogous system and those of the prime system.

Electric analogy is the most widely used for heat problems, although other analogies are useful and will be described briefly in Secs. C and D. However, since electric analogy methods are predominant, it is well to stress the difference between the *direct analogy*, which will be discussed here, and the *indirect* analog computers, which are widely used, though rarely for heat conduction problems. Technical literature is confused by not distinguishing between these two fields of analogs. The indirect analogs (the word "indirect" is usually omitted from papers dealing with them) are *differential analyzers*, solving ordinary differential equations and having as main components operational amplifiers which perform certain mathematical operations; there is no direct relationship between each component and physical parts of the prime system. Here the term "analog" will henceforth refer to direct analogs.*

2. Two Approaches to Analog Procedure

The two approaches are explained at the hand of electrical analogy. Let Fig. 1 represent an insulated rod in which heat flow can occur only in the x direction. Then the temperature-space-time relationship is given by Eq. (1)

$$k \frac{\partial^2 T}{\partial x^2} = c_p \rho \frac{\partial T}{\partial t} \tag{1}$$

Now let Fig. 1 represent a noninductive electric cable; then the voltage-space-time relationship is given by Eq. (2)

$$\frac{1}{\rho_e} \frac{\partial^2 e}{\partial x^2} = c_e \frac{\partial e}{\partial t} \tag{2}$$

A comparison of the two equations shows that they are identical. Thus one can use an electrical cable as an *analogous system* for the rod (*prime system*). Since it is inconvenient to seek appropriate cables for each case, one can resort to a method, common in electrical engineering, namely lumping. As indicated by dotted lines in Fig. 1, the rod is divided into n ($n = 6$ in the illustration) sections; the electrical properties of the cable in each of the n sections are thought of as concentrated either in the center or along the axis of the section. A circuit shown in Fig. 2 results.

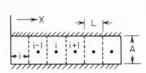

Fig. 1. Insulated rod, heat flow in x direction.

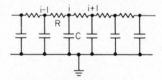

Fig. 2. Circuit representation of cable.

The sequence of logical steps is:

Prime system

---

*Confusion in nomenclature is being combatted in part by the proposed IRE Standards [1] which define the "direct analogy computer as a simulator in which the individual elements have a scalar, usually one to one, relationship to represent another physical system."

Analogous system (established on the basis of identity of equations)
Replacing continuous system by lumped system
Measurements on lumped system
Interpretation of results in terms of continuous prime system

However, a completely different conceptual approach is possible that leads to practically the same method. Replacing Eq. (1) by a finite difference approximation while keeping time continuous results in Eq. (3), written for node i:

$$\frac{kA}{L^2}[(T_{i-1} - T_i) - (T_i - T_{i+1})] = c_p \rho \frac{dT}{dt} \tag{3}$$

This equation represents the heat balance for section i in Fig. 1. Now, writing Kirchhoff's equation for node i in Fig. 2 leads to

$$\frac{1}{R}[(e_{i-1} - e_i) - (e_i - e_{i+1})] = C\frac{de_i}{dt} \tag{4}$$

Again, comparison of Eq. (3) with Eq. (4) shows identity. In this approach no use is made of the analogy between the equations for prime and analogous systems. Instead, the logical sequence is as follows:

Prime system
Replacing partial difference equation by finite difference approximation (sectioning or lumping)
Design of an electric circuit which may be described by the same equation as the finite difference equation for the prime system
Measurements on this circuit
Interpretation of results in terms of prime system

The latter concept also allows analogy methods to be applied to cases without a directly analogous continuous physical system. In Sec. B.5 this concept is used to present an analog for heat exchangers.

B. Electric Analogies

1. Classification

Three types of analogs are used for *pure conduction* problems: geometric, network, and combined. Geometric analogs (Sec. B.2) are capable of solving only steady-state problems (Laplace equation); networks (Sec. B.3) can be used either for steady-state or transient problems; and combined analogs (Sec. B.4) serve only for the solution of transient or steady-state problems with distributed sources or sinks (Fourier and Poisson equations). While in most transient analogs time is treated continuously (Beuken), there is one analog technique (Liebmann) using a pure resistance network (Sec. B.3.b) where time is also discretized; thus one can also establish a sequence of analog methods on the basis of degree of discretization:

time, resistance, and capacitance discretized: R-network, Liebmann
resistance and capacitance discretized: R-C network, Beuken
capacitance discretized: 'Cognac', Paschkis

2. Geometric Methods

A large variety of methods can be used, but only two have found wide application: tanks and conducting sheets.
   a. **Tanks.** A tank is made having the shape of the prime system. The walls of the tank must be electrically nonconducting, and they must be chemically neutral and

impermeable with respect to the liquid resistance material that fills the tank. After the tank is filled, appropriate electrodes are placed on the walls of the tank permitting the representation of boundary conditions. The voltage distribution in the tank is directly proportional to the temperature distribution in the prime system, and the currents are proportional to the rate of heat flow.

● *Resistance Material.* Water is frequently used; if necessary, it is made conducting by adding salt. Because of electrolytic effects, only alternating current can be used. In connection with the 'Cognac' (Sec. B.4), a resistance material, a solution of tetra-butyl-ammonium picrate in Dowtherm, was developed. This material can be used with fair accuracy with direct current.

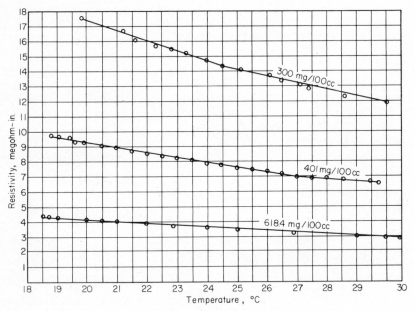

Fig. 3.   Resistivity as a function of temperature.

This solution is quite stable and does not disassociate to any noticeable degree, even under prolonged exposure to direct current of up to 100 volts. Figure 3 shows, for three different concentrations, the resistivity plotted against temperature; the marked temperature dependence makes it necessary to operate in an air-conditioned room. The stability of the solution is illustrated in Fig. 4 in which the voltage-current characteristic is shown. The test cell was held at 100 volts dc continuously; the characteristic was taken at frequent intervals and did not change over a 20-day period.

Tanks may be made of wood. In order to make them impermeable to the liquid they can be coated by paraffin for water or salt-water solutions; a coating of Bondmaster M 648T (2 parts) and Hardener CM-16 (1 part), both products of Pittsburgh Plate Glass Co., is used with Dowtherm [2]. The tank must be made geometrically similar, that is, with a given geometric factor to the shape of the prime system. The geometric factor is the ratio of a tank dimension to the equivalent dimension in the prime system. The tank size can be chosen arbitrarily. To reduce cost (of the tank, of electrodes, and with Dowtherm also of the liquid), small tanks are suggested. On the other hand, the larger the tank, the less critical is the positioning of the probe by which voltages are determined, and the less stringent are the accuracy requirements in the actual shaping of the tank.

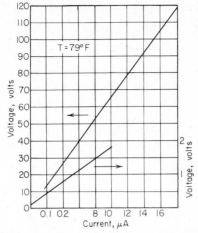

Fig. 4. Voltage-current characteristic, 400 mg/100 cc.

● *Boundary Conditions.* Isothermal surfaces are represented by covering the appropriate face of the tank by a metal electrode which is connected to a voltage corresponding to the temperature. If a face is completely insulated, no electrode is applied. A nonisothermal surface can result from one of the following boundary conditions: surface temperature prescribed as function of space; heat transfer coefficient between surface and a uniform ambient temperature prescribed; rate of heat flow through the surface prescribed, either uniform (Btu/ft²-hr are constant) or as function of position. In all these cases the surface should be covered by a number of smaller electrodes.

The electrodes should be separated by insulating fins (Fig. 5). The system has been analyzed in detail only for the two-dimensional case, with a heat transfer resistance prescribed [3].

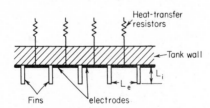

Fig. 5. Arrangement of electrodes and fins in tank wall.

The number of electrodes required is smaller if only the heat-flow rate should be established, and larger if the temperature distribution over the surface must also be known. In the latter case, the number of electrodes required depends on the curvature of the heat-flow lines at the surface [2]. Electrodes can be screwed to the tank wall. Fins, in two-dimensional arrangements, can be fastened as shown in Fig. 6. The fins F extend above the bath level L. They are separated by blocks B which are pressed together by a bolt S. For three-dimensional problems, the electrodes become small squares surrounded on all four sides by insulating fins. The basic design for the insulating fins can be the same as for two-dimensional arrangements. The vertical fins are then notched at regular intervals, corresponding to the size of the square electrodes; horizontal fin strips notched in egg-crate style are used to interlock with the vertical fins, leaving between them square openings for the electrodes. Heat transfer coefficients are represented by adding external resistors, as indicated in Fig. 5. Better accuracy is achieved if tank and fins are so designed that the shape of the prime system is obtained by the plane which connects the tips of all fins, rather than connecting all the bottom ends of the fins or the electrodes. The resistance of the liquid contained within L_e and L_i should be considered part of the heat transfer resistance.

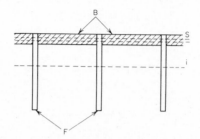

Fig. 6.   Fastening of fins.

● *Electrode Material.* Copper electrodes can be used with tap water and alternating current. They should be sandpapered or sandblasted repeatedly to reduce contact resistance. However, with Dowtherm and direct current, the accuracy is influenced by electrode material. Noble metals give a lower back-emf than common metals. (Back-emf is the voltage below which no measurable current flows through the liquid.) For paladinized palladium the back-emf is below 0.05 volt; for nickel it is about 0.2 volt. Except for readings near ground potential, the error is tolerable [2].

● *Scale Factors.* No question of scale factor is involved if the only task is to determine temperature distribution. Percentages of the maximum applied voltage correspond to the percentages of the maximum applied temperature difference. However, if rates of heat flow in addition to temperature are of interest, four scale factors (voltage, current, resistance, geometric factors) apply as follows

$$V_F = \frac{e}{T} \quad I_F = \frac{i}{q} \quad R_F = \rho_e k \quad G_F = \frac{L_e}{L} \tag{5}$$

The four factors are connected by the relationship

$$V_F = R_F I_F G_F \tag{6}$$

Three of the four factors can be chosen arbitrarily; the fourth follows from Eq. (6). The resistors representing heat transfer coefficients can be computed from Eq. (7)

$$R_f = \frac{R_F}{h} \frac{1}{A_S G_F} \tag{7}$$

Here R_f is the resistance representing the heat transfer coefficient, A_S the area in the prime system assigned to the one resistor. If a surface with an area of 1 ft^2 (in the prime system) is covered by nine electrodes of equal size, each electrode being connected to one boundary resistor, then $A_S = 1/9$ ft^2.

The value R_f thus computed comprises both the resistance of the liquid between the isolating fins and the external resistor. If the electrode has an actual area of 1.5 in.^{2}, the fin a depth of 1.2 in., and the liquid a resistivity of 10 megohm-inches, then the resistance of the liquid within the isolating fins has a resistance of $(10 \times 1.2)/1.5 = 8$ megohm; this value would have to be subtracted from the value of R_f, computed in Eq. (7), in order to find the external resistance.

● *Current Measurements.* These are performed at the surface by inserting an ammeter into the appropriate electrode circuit. Within the tank one cannot carry out current measurements because the resistor is multidimensional. However, if one determines the voltage gradient, as described below, one can determine the heat flux in a given direction

$$q'' = \frac{de}{dx} \frac{1}{\rho_e} \frac{G_F^2}{I_F}$$

(8)

● *Voltage Measurements.* One must be careful if the liquid resistor has high resistivity. Voltmeters of sufficient impedance must be used (possibly by inserting an isolating amplifier) to avoid shunting of the tank liquid by the instrument. Frequently, one must apply a potentiometer circuit, having in equilibrium infinite resistance. Positioning of the probe at specified locations in the tank can be done manually. Since this is usually too cumbersome, a carriage with x-y movement of the probe by means of a precision drive is desired. Frequently it is necessary to determine isotherms instead of measuring voltages (temperatures) at specified locations. A voltage is present on a bridge circuit and then the carriage, to which the needle probe is attached, is moved so that the potentiometer is always kept at zero. The movement of the carriage is reproduced on a paper through a pantograph, this providing the isotherms directly.

The Laplace equation applies to so many different fields that literature regarding the tanks is not limited to heat-flow papers. An exhaustive bibliography is given by T. J. Higgins [4].

Two-dimensional problems involving bodies with conductivity varying with position have been handled in tanks with depths varying according to the local conductivity.

b. Conducting Sheets. For two-dimensional problems the use of conducting sheets is attractive from many aspects. Providing odd shapes is easy; one simply cuts the sheet to an appropriate scale according to the shape of the prime system. Sheets are characterized by ohm-square, indicating resistance between two opposite sides of squares. One can easily see that the value of ohm-square is independent of the size of the square: if the resistivity of the conducting layer of the sheet is ρ_e and the thickness of the layer ν, then

$$R_{\text{square}} = \frac{\rho_e}{\nu}$$

(9)

Metal foil is sometimes used as material for the conducting sheet. But as Eq. (9) shows, one must use extremely thin foil in order to achieve high values of R_{square}. Also useful is paper covered with a conducting layer of graphite. Teledeltos paper (manufactured by Western Union) and conducting paper (manufactured by the Riegel Paper Corporation) are two examples, the former having resistances of about 1,000 and 10,000 ohm-square respectively. Common paper has also been used [5]. Early investigators used blotting paper made conductive by moistening with water or salt water. With the exception of the latter, conducting sheets can be used with either direct or alternating current.

The applied voltage is limited by the permissible wattage, which in turn is limited by the permissible temperature rise of the sheet. For a rectangle where electrodes are separated by a distance L the maximum voltage which may be applied is

$$e_{\text{max}} = \sqrt{W_A R_{\text{square}}} L$$

(10)

where W_A is the permissible watt density of the sheet. There is no general expression for the maximum voltage for irregular shapes; one can estimate the value by approximating the shape by a rectangle of comparable size.

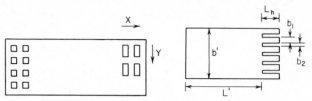

Fig. 7. Thermal-conductivity representation of anisotropic material.

Fig. 8. Heat transfer coefficient representation.

Electrodes are usually painted on with silver paint; another method is to place the sheet on a smooth base and to press electrodes in the form of metal strips onto it.

Local changes in thermal conductivity can be represented by cutting holes into conducting sheets [6]. Figure 7 is indicative of the arrangement which can also be used to represent anisotropic material. In the left half of the figure, the square holes indicate isotropical increase of the resistance compared to the middle portion of the figure. Toward the right end, the rectangular holes represent an anisotropic material, with thermal conductivity in the x direction higher than in the y direction. The smaller the holes, the more accurate is the representation; however, no studies have been made regarding either the effect of the size of the holes or the accuracy of punching on the accuracy of the results.

In order to represent heat transfer coefficients either one can paint electrodes on the edge which represents the surface exposed to a heat transfer coefficient, or one can cut in the paper, extending the length in one direction as shown in Fig. 8.

The length L_h of the strips follows from Eq. (11)

$$L_h = \frac{k}{h} G_F \tag{11}$$

It should be noted that the larger b_1 is, the less accurate the solution; but the smaller b_1 is made, the less acceptable it is to disregard the width b_2 of the cut in comparison with b_1; if one introduces b_1, Eq. (11) becomes

$$L_h = \frac{k}{h} G_F \frac{b_1}{b_1 + b_2} \tag{12}$$

From this equation it would appear that one can shorten L_h by making the cuts wider (increasing b_2). But this leads to errors at the foot of the extension. One can, of course, follow the same procedure as for the tank by representing the heat transfer resistance by two parts: (1) extend the body proper by cutting slots in the sheet (in the tank it was the liquid between the fins), and (2) apply an electrode connecting to an exterior resistor.

When currents have to be measured, as in the case of conducting sheets, the following relationship exists between the scale factors

$$V_F = R_F I_F \tag{13}$$

Note that for two-dimensional problems

$$R_F = R_{\text{square}} \cdot k \tag{14}$$

and that in Eq. (13) G_F does not appear (compare with Eq. (6)).

3. Network Analogies

a. R-C Networks. These have been briefly introduced in Sec. A.2. However, the circuit in Fig. 2 is not the only possible way to represent the rod illustrated in Fig. 1. Figure 9 illustrates three methods to represent a single section by using lumped circuit elements. Circuit (a), used in T lumping, places the capacitance in the center of the section; in Π lumping [circuit (b)] capacitances are on either end of the section, while in L lumping [two alternatives as shown in circuit (c)] the entire capacitance is placed at one end of the resistor. As indicated by the letters in any of the three methods, both the total resistance and the total capacitance are the same. As Fig. 10 shows, any one of the three lumping methods leads on the inside of the circuit to the same sizes of capacitors and resistors; dotted lines indicate connections from section to section. Differences occur only at the surface; therefore it has also been suggested to designate T and Π lumping as *resistor terminated* and *capacitor terminated* respectively.

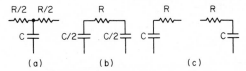

Fig. 9. Lumped circuit elements.

Heating or cooling at the surfaces, according to Newton's law, can be represented by adding a resistor at the surfaces. In case of T lumping, this "boundary resistor" R_f can be combined with the last half-resistor of the body network (Fig. 11). If, however, conditions at the surface are important, Π lumping should be used to give more accurate results at the surface.

If the insulation is removed from the sides of the rod, in Fig. 1, heat will flow not only in the x direction but also in directions perpendicular to x. If we assume the cross section A in that figure to be rectangular, and two sides of the rectangle to be insulated, two-dimensional heat flow is encountered. Using T lumping, a circuit as shown in Fig. 12 will result. The straight lines are resistors. The negative poles of all capacitors are connected to ground (not shown in the figure). While the resistors in the x and y directions are shown here to be equal, this is not necessarily the case, as will be explained below.

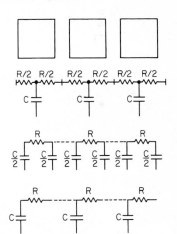

Fig. 10. Lumping methods.

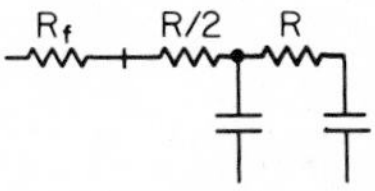

Fig. 11. T lumping.

From this circuit it is only one step to a circuit for three-dimensional problems. Figure 13 shows one node for such a three-dimensional network, again for T lumping. A capacitor is placed at the intersection of six resistors, which represent the heat flow along the three axes of a cartesian coordinate system. Assuming an equal number m of sections in each of the three directions (x, y, and z), the number of components for one-, two-, and three-dimensional networks are shown in Table 1.

It is not possible to give general instructions regarding the desirable value of m. Choice depends on the desired accuracy, the emphasis put on computation for early times, complexity of shape, composition of prime system (one material, or composite body), etc. But even with quite limited values, of say $m = 6$ or $m = 10$, the required

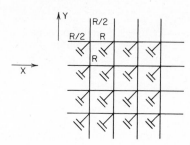

Fig. 12. T-lumping circuit—two-dimensional.

TABLE 1. Number of Components (for m Sections in Each Direction)

Number of Dimensions	Number of Resistors		Number of Capacitors	
	T Lumping	II Lumping	T Lumping	II Lumping
1	$m + 1$	m	m	$m + 1$
2	$2m(m + 1)$	$2m(m + 1)$	m^2	$(m + 1)^2$
3	$3m^2(m + 1)$	$3m(m + 1)^2$	m^3	$(m + 1)^3$

Fig. 13. Three-dimensional network node with T lumping.

number of resistors for three-dimensional networks very quickly becomes so high that consideration of the Cognac is worthwhile.

• *Determination of Scale Factors.* Five scale factors have to be found: one for the time scale T_F; two for the components R_F and C_F, and two for interpreting the readings V_F and I_F. The first three are related as shown by Eq. (15)

$$T_F = R_F C_F \qquad (15)$$

Still another relationship exists between R_F, V_F, and I_F

$$V_F = R_F I_F \qquad (16)$$

The resistance and capacitance scale factors R_F and C_F can be defined using the first concept (Sec. A.2) by reference to a noninductive electric cable having the same dimensions as the prime system and having the electric properties of ρ_e (electric resistivity) and c_e (electric capacity). Using this concept, the scale factors are defined by

$$R_F = \rho_e k \qquad (17)$$

$$C_F = \frac{c_e}{c_p \rho} \qquad (18)$$

It is very important to understand that ρ_e and c_e are not the electric properties of the prime system, but are those of a fictitious cable of dimensions equal to those of the prime system.

Avoiding the introduction of the "fictitious cable," one can define R_F and C_F by referring to the resistance and capacitance representing a one-dimensional section of cross-sectional area A and of length L. If Fig. 14 shows the section and the representation by circuit elements, one can relate R_F and C_F with R_1 and C_1, by Eqs. (19) and (20) respectively

$$R_1 = \frac{R_F}{k} \frac{L}{A} \tag{19}$$

$$C_1 = C_F (c_p \rho) LA = C_F (c_p) V \tag{20}$$

where V is the volume of the section.

● *Computation of a Circuit.*   Find the value T_F, which is defined by

$$T_F = \frac{t_e}{t_h} \tag{21}$$

where subscripts e and h stand for "electric" and "heat" respectively.

The lengths of time to be considered in the (thermal) prime systems are known. The desired length of the computing run depends on the design of the analog and depends on the availability of components (resistors and capacitors). Also, if no assembled computer is available and a circuit is put together as need arises, the choice of t_e must of necessity be influenced by the available measuring equipment. Actual values of computing time t_e vary over a wide range. Values of 5 to 10 minutes are sometimes used; on the other hand, very short times of a few milliseconds are also employed.

Longer times require more electric capacitance and resistance and are thus more expensive. On the other hand, they allow simpler measuring techniques. And above all, the longer computing times are essential if problems involving temperature-dependent properties are to be handled and the property-temperature functions differ from case to case; short-time computers with millisecond values for electric time do not conveniently allow one to vary a change in properties from one computation to the next.

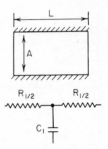

Fig. 14.   Circuit elements.

Once the electric time has been chosen (and thus T_F determined), one proceeds to find R_F and C_F. Either may be chosen arbitrarily; the other follows from Eq. (15). Since a number of factors favor working with as large a capacitance as is available, one convenient method is as follows: Determine the total volume V_{tot} of the prime system (it is here assumed that it is a uniform, i.e., noncomposite body; the reader can readily extend the technique to the case of a composite body). By applying the total volume instead of LA in Eq. (20) and using the total available capacitance instead of C_1, one can find the value of C_F. Before using it, it is helpful to decrease it slightly so that the value of C_1 for a frequently occurring section be of convenient magnitude (not fractional values, which may be time-consuming to set up). Once C_F has been thus established, one finds R_F from Eq. (15). Then it is helpful to see if the resulting values of R are practical, based on Eq. (19). If not, the value of C_F must be changed: if resistances turn out too large, a shorter computing time (lower value of T_F) must be employed. If the resistances turn out too small, one can reduce the value of C_F.

Once C_F and R_F have been established, one can select either V_F or I_F. The two scale factors are defined by Eqs. (22) and (23) respectively

$$V_F = \frac{e}{\Delta T} \tag{22}$$

where ΔT is the greatest temperature difference occurring at any time in the system and

$$I_F = \frac{i}{q} \tag{23}$$

where q is the rate of heat flow (Btu/hr).

In most cases the former will be chosen: the maximum temperature differences in the prime system and the maximum voltage with which the analog can work are known, and thus V_F follows readily. Then from Eq. (16) one finds I_F. If the resulting currents are too small for measurement or too big (overloading of the resistors), it may become necessary to return to R_F; a change in the latter may require new choices of C_F and possibly even of T_F.

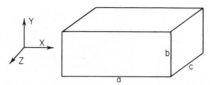

Fig. 15. Section in an orthogonal coordinate system.

Once the scale factors are established, all the resistance and capacitance values have to be computed; Eqs. (19) and (20) can be used. Consider one section in a cartesian coordinate system as shown in Fig. 15; the resistors and capacitor in a three-dimensional network with a resistance factor R_F, a capacitance factor C_F representing a prime system with a thermal conductivity k, and a volumetric specific heat $c_p\rho$ would be as follows

$$R_x = \frac{R_F}{k} \frac{a}{bc}$$

$$R_y = \frac{R_F}{k} \frac{b}{ac} \tag{24}$$

$$R_z = \frac{R_F}{k} \frac{c}{ab}$$

$$C = (C_F c_p \rho)\, abc$$

• *Cylinder and Sphere.* Both a cylinder of great length (no axial flow) and a sphere may be considered one-dimensional problems, provided there is symmetry around the entire circumference. If a cylindrical cross section is divided into an odd number of equal spaces, one obtains an arrangement as shown in Fig. 16.

The resistors $R_1, R_2, R_3, \ldots, R_n$ have values in the ratio of the respective ratios of L/A (according to Eq. (19)). Now the value of L is always equal to Δr, if the capacitances are placed in the middle of each section. And the value of A is proportional to r_n. Thus one finds

$$R_1 : R_2 : R_3 \ldots = 1 : 1/3 : 1/5 \ldots$$

The first resistor (R_1) is found as follows

$$R_1 = \frac{R_F}{k} \frac{\Delta r}{0.5\,\Delta r} \tag{25}$$

The capacitances are proportional to the area of each section. The capacitance C_0 is

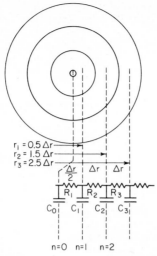

Fig. 16. Cylindrical cross section with odd number of spaces.

to be found from Eq. (26)

$$C_0 = (C_F c_p \rho) \left(\frac{\Delta r}{2}\right)^2 \qquad (26)$$

Subsequent capacitors become larger as the cross-sectional areas increase; the ratios of areas (and therefore capacitors) are

$$C_0 : C_1 : C_2 : C_3 : \ldots = 0.25 : 2 : 4 : 6 \ldots \qquad (27)$$

Two-dimensional problems involving cylinders can be of two different types: cylinders with uniform conditions at the periphery, but of finite length; or long cylinders (no axial flow), but with circumferentially uneven conditions (e.g., different ambient temperatures, or heat transfer coefficient not uniform over the entire periphery; or local heat sources). For the first type of radial symmetry, but with axial heat flow, the same two-dimensional network as in Fig. 12 is used; however, if the sections are equal, resistors and capacitors are different in the different sections. The radial resistors decrease from the axis outward as shown above in the ratio $1 : 1/3 : 1/5$.

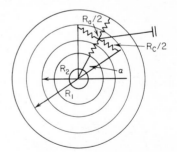

Fig. 17. Sectioning of cylinders without circular symmetry.

The axial resistors, being inversely proportional to the cross-sectional area, vary in the reverse ratio as the capacitors

$$R_0 : R_1 : R_2 \ldots = 1 : 1/8 : 1/16 : 1/24 \ldots \qquad (28)$$

If circular symmetry does not exist, sectioning as shown in Fig. 17 applies. Resistors and capacitors are to be computed from Eqs. (19) and (20). The values for area, length, and volume for radial and circumferential resistors are shown in Table 2.

TABLE 2. Equations for Area, Length, and Volume for Circumferentially Nonuniform Sectioning

	Area	Length
Resistor R_a (Fig. 17)	$0.5(R_1 + R_2)^2 \, \pi\alpha/360$	$R_1 - R_2$
Resistor R_c (Fig. 17)	$(R_1^2 - R_2^2)\pi$	$0.5(R_1 + R_2)\,\pi\alpha/360$
Volume	$(R_1^2 - R_2^2)\,\pi\alpha/360$	

In case of circumferential sectioning, as in Fig. 17, small section size is more important than in the case of rectangular sections. If a circumferentially uneven distribution occurs in a problem together with finite length, the problem becomes three-dimensional.

In computations dealing with long processes, it is sometimes important to study both early as well as longer times. One can save appreciable time by not running the entire computation with a value of T_F, which would allow accurate study at the early times; instead, one can change the time scale in the middle of the run. This is simply accomplished by changing all resistance values at the given moment. Although the time factor depends not only on the resistances but also on the capacitance factor, one cannot change the time factor during a run by changing the capacitances because one would gain or lose the charge stored in the units at the moment of switching.

● *Boundary Conditions.* These can be represented as follows: if surface temperatures are prescribed, voltages are impressed at the surface nodes. Such voltages need not be uniform over all the surface nodes, and with appropriate instruments the surface voltage can be made a function of time. If the boundary condition provides that a given heat flux should enter or leave the surface, this is to be represented by current being fed into or drained off the surface nodes. The amount of current to be fed into or drained off each node follows from the current factor; in order to use it one has to establish the heat flux for each node, which presumably is proportional to the area represented by this node. If Newtonian cooling (or heating) exists, i.e., if a constant heat transfer coefficient exists in the prime system, so that at all times the heat flux through the surface is proportional to the temperature difference between surface and ambient, a resistor is added at the surface, the value of which is given by

$$R_f = R_F(hA) \tag{29}$$

with A, the surface area represented by the section, terminating in the surface node.

If the body comprises heat sources or sinks, current should be fed into or drained out of all sections lying within the range of the sink. If the productivity of a source is q''' Btu/ft^3-hr and a node covers a volume of V in.^{3}, then the current fed into that node should correspond to a heat flux of $q'''V$, and can be computed directly from Eq. (23).

● *Composite Bodies.* For such cases it is well to remember that R_F and C_F are given for an entire computation; in handling different materials within a problem, the respective values of k, c_p, and ρ must be introduced. Two sections of equal geometry but comprising different material will have resistances and capacitances described by Eq. (30). (Subscripts 1 and 2 refer to the two materials.)

$$R_1 k_1 = R_2 k_2$$

$$\frac{C_1}{C_2} = \frac{(c_p\rho)_1}{(c_p\rho)_2} \tag{30}$$

All problems mentioned so far in the text were assumed to be linear; and all properties were assumed to be independent of temperature (and of time); only properties that vary with location (composite bodies) have been mentioned. One of the strong points of analog computers is that they can readily handle nonlinear problems in which properties change with temperature (or with time).

- *Step Curves.* In order to represent such bodies, the curve relating the property (conductivity, volumetric specific heat, heat transfer coefficient) to temperature (Fig. 18) is replaced by a step curve. If the properties are given as a function of time, a stepping switch can provide for a change of steps at the appropriate times. If, however, the properties are given as a function of temperature, a property changer is provided to change the value of property at each node when the nodal voltage reaches one of the values shown in Fig. 18. If in a given problem several

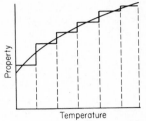

Fig. 18. Property as a function of temperature.

properties are given as functions of temperature, one saves equipment if the steps (according to Fig. 18) of all the properties are made at the same increments of voltage (temperature). Here, as in the case of composite bodies, one should remember that the scale factors R_F and C_F remain constant: if two values of conductivity are k_a and k_b and the two values of volumetric specific heat are $(c_p\rho)_a$ and $(c_p\rho)_b$, then the electric resistors and capacitors for these steps are defined by

$$\frac{R_a}{R_b} = \frac{k_b}{k_a} \tag{31}$$

$$\frac{C_a}{C_b} = \frac{(c_p\rho)_a}{(c_p\rho)_b}$$

The direct analogy established earlier obviates the need for writing the equations for each case; this may result in appreciable time saving if conditions are complex.

- *Moving Boundary Problems.* A group of problems involving considerable mathematical complexity are those dealing with moving boundaries; such moving-boundary problems occur when liquids freeze. Such problems can be solved very readily on the analogs and may serve as examples of how to translate such problems into analog language. Two types of conditions exist, exemplified by pure metals freezing at a given and known temperature, and noneutectic alloys which freeze over a range of temperatures. The conditions to be represented are: (1) at a given temperature (pure metals), or over a given temperature range (alloys), a certain amount of heat is liberated, and (2) thermal properties of the liquid and of the solid are different.

The case of noneutectic alloys is discussed first. The heat which is liberated is called *heat of fusion;* it is represented in the analog as a temporary increase of specific heat. A convenient circuit is shown in Fig. 19. This figure represents the last node in the metal circuit and the first node in a circuit, representing a container (mold) which is

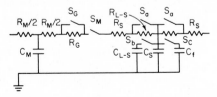

Fig. 19. Circuit representation of the heat of fusion (noneutectic alloy).

at a lower temperature than the metal and extracts heat from the liquid metal during freezing. For the sake of simplicity, the "mold circuit" (R_M and C_M) is shown for constant properties. The metal circuit is designed for the normally prevailing condition, with the conductivity of the liquid metal lower than that of the solid and the specific heat of the liquid higher than that of the solid. The resistors R_S correspond to the solid conductivity, while the resistors R_{L-S} correspond to the decrement of liquid versus solid conductivity. C_S corresponds to the solid volumetric specific heat and C_{L-S} to the difference between liquid and solid specific heats. Values for these components are to be computed according to the procedure of *varying properties* given above. Heat of fusion is liberated between two temperatures, the liquidus and the solidus temperature, and is computed as follows: let λ be the heat of fusion (Btu/lb), ΔT the temperature range of solidification (difference between liquidus and solidus temperature), and V the volume of the section; then the value of C_f, the capacitor representing heat of fusion, is found from

$$C_f = \frac{C_F \rho \lambda}{\Delta T} V \qquad (32)$$

Operation of the circuit is as follows: Capacitors C_{L-S} and C_S are precharged to a voltage corresponding to the initial temperature (pouring temperature) while C_f is precharged to a value corresponding to the liquidus. All switches except S_b and S_g are open. In order to start the computation, S_M is closed. When the nodal voltage has dropped to the liquidus value, S_c is closed, thus connecting the "heat of fusion capacitor" C_f into the circuit. If the change of properties is not defined more closely, one can make the change when the nodal temperature equals the arithmetic mean of liquidus and solidus temperatures. The change is carried out by closing switches S_a and opening S_b. When the solidus is reached at any node, the heat of fusion capacitor is taken out of the circuit by opening S_c. No mention was made of switch S_G and resistor R_G. As the metal cools it shrinks in the mold and thus a gap is formed between the metal and the mold. This gap represents a thermal resistance; the switch S_G is to be opened at the time when the gap is formed. If enough information is available to represent the change of gap resistance, R_G can of course be made a variable resistor. This resistance is to be considered as composed of two parallel branches, one representing conduction across the gap, the resistance increasing as the gap widens; and the other representing radiation across the gap, with the resistance changing in accordance with the fourth-power law.

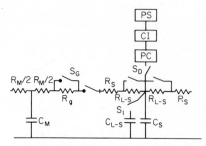

Fig. 20. Heat of fusion circuit (pure metal).

From Eq. (32) one sees that the value of C_f increases as the solidification range becomes smaller. If the range becomes zero, as in pure metals, the value of C_f becomes infinity. Then a different circuit (Fig. 20) should be used. If ΔT in Eq. (32) is so small that it makes C_f uncomfortably large, one may use, as an approximation, the circuit in Fig. 20, introducing the heat of fusion at the middle of the solidification range. It will be noted in this figure that there is no capacitor C_f and hence no switch S_c. The other components remain unchanged (see Fig. 19) and are not explained again. However, new equipment is added as follows: A power supply PS operates at a constant voltage corresponding to the solidification (constant) temperature. As the node reaches this temperature, property changer PC closes the switch S_D, which was originally open, and current flow from PS and is integrated in the current integrator CI. When the current-time integral Q_f as measured by CI reaches a value which corresponds to λ in Eq. (33), switch S_D is opened again

$$Q_f = \lambda \rho V T_F I_F \tag{33}$$

• *Output.* Results of analog computation usually comprise voltage-time or current-time curves, representing temperature-time and heat-flux–time curves. Instruments cannot be connected directly into the circuit because their inherent resistance would falsify the results. Particularly it is not possible to insert a voltmeter V directly into the circuit (Fig. 21a) because its resistance would represent a shunt to the capacitance C and thus provide a serious leakage path. Instead, a dc amplifier A should be inserted (Fig. 21b) which, in turn, feeds the voltmeter V. Depending on size and quality of the capacitors and on the length of computing run, the input impedance of A must be sufficiently high to avoid unacceptable drain. Moreover, zero drift must be kept low. If, in a large network a number of temperature curves are taken, one can gain a dual advantage by time-sharing the isolating amplifiers; less equipment is needed and the drain on the node is reduced to the time actually required to carry out the measurements.

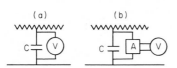

Fig. 21.   Output instrumentation.

Fig. 22. Recording micro-ammeter.

Measurements of currents are more difficult to make if, as normally occurs, recordings are to be taken. Recording microammeters which operate off ground are not readily available. (One is shown in Fig. 22.) Instead one can use a difference amplifier, shown in the same figure, to measure the potential difference between two adjacent nodes. This difference divided by the resistance between the nodes equals the current. Difference amplifiers, where neither of the input terminals is grounded and which possess sufficient stability and leakage resistance of the input to ground, are not easily obtained. They frequently depend on the average voltage level of operation, and since this voltage level usually changes during the course of a computation, these instruments are sensitive. Computers may be driven by power supplies having a constant voltage; if heat flux is prescribed, a power supply is required yielding a given current independent of the momentary impedance of the network at hand.

Occasionally, the required output does not comprise the entire time-temperature curve but only a record of the time when a specific temperature was attained. Again, the solidification problems mentioned above are a case in point. In order to study the progress of the solidification front it is necessary to record for each node in the network the time at which it reaches the solidification temperature. This can be accomplished by connecting to each node a pair of counters. All counters are activated at regular intervals (e.g., every one-half second, one second, etc.) by a timer. One of the counter circuits is then broken when solidification starts, the other when it stops. For alloys, start and stop coincide with reaching liquidus and solidus temperatures, respectively; for pure metals, the times can be those when the node first reaches the solidification (constant) temperature and when the heat of fusion has been transmitted. The opening of the counter circuits can be accomplished by means of the property changers which in the examples and elsewhere are frequently in the circuit for other purposes.

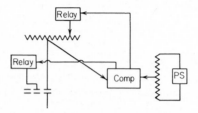

Fig. 23. Comparison of nodal and reference voltages.

Several designs of property changers have been developed, yet the basic perform-ance is always the same: means are provided to compare continuously the nodal voltage with a reference voltage (Comp≡ Comparator in Fig. 23). As the difference approaches zero, the output of the comparator activates relays which perform two tasks: (1) they set the components representing the properties to the values according to the step-curve (Fig. 18), and (2) they provide that the comparator compares the nodal voltage with the next reference voltage, corresponding to the next step on the step curve.

The network itself is composed of resistors and capacitors. Probable errors due to value tolerances of these components are smaller, percentagewise, than the tolerances themselves because of the averaging effect. If a given percentage tolerance (±) is guaran-teed, components may be expected to straddle this value according to a probability curve, and one may expect that neighboring sections will randomly have components either above or below the nominal value. Components with 1 percent accuracy are easily obtainable and the price advantage in using less accurate components is small. Generally, the use of more accurate components is not warranted because thermal properties are not known with sufficient accuracy to make more precise computations worthwhile. More emphasis should be put on stability (aging) and, for capacitors, particularly on very high resistance and minimal soaking. The leakage resistance, commonly expressed in terms of a time constant (seconds= capacitance of unit in microfarads times the leakage resistance in megohms) is, next to the value of the capacitance itself, the determining characteristic of capacitors to be used in analogs. Capacitors using plastics (polystyrene, mylar, etc.) as dielectrics have time constants of 10^5 to 10^6 sec. Such capacitors usually are also sufficiently free of soaking effects to be acceptable for analogs.

Magnitude of capacitors and resistors for analog work depends on the desired length of computing runs. For a nominal length of 3 to 5 minutes, a resistance adjustable between 100 ohms and 10 megohms and a total capacitance of perhaps 5 to 20 microfarads per node have proved satisfactory. The desired capacitance per node increases as the total number of nodes decreases: in large networks frequently not all the nodes need maximum capacitance, and thus a pool of capacitances can be provided from which assignments are made to the several nodes.

A quite extensive bibliography on this type of computer has been published by S. Zandstra and A. J. Bovy [7]. The first information was published by Beuken [8] and the first in the United States by Paschkis [9]. The Heat and Mass Flow Analyzer Laboratory at Columbia University published a bibliography of its work in this field. References 10 to 13 deserve special mention.

b. R-R Networks. Good capacitors are expensive and were in earlier times almost unobtainable. These difficulties led Liebmann [14, 15] to develop a pure resistance network for the solution of transient problems. Referring to Eq. (4), the dimensions show that the right side also must have the dimension of amperes (because the left side has dimensions of V/R). A current of a given magnitude can be obtained by imposing a voltage on a resistor. Liebmann thus developed the circuit element shown in Fig. 24. The resistors R have the same function as those in Fig. 10, with the resistor R_c replacing the capacitance. After discretizing time, one can, for any given value of R_c, adjust the voltage e in such a manner that the voltage difference $e_i - e$ will produce a current through R_c equal to the mean current which would flow during the same time interval into a capacitance. This is the principle underlying the Liebmann technique. One could, of course, keep e constant instead and adjust the value of R_c. The important feature is the discretization of time and the avoidance of using a capacitance. Variable parameters can be handled readily since one can use as much time as is required to reset resistance values when the need arises as a result of different thermal properties. Essentially this technique carries out by network calculation the same approximation to the heat transfer equation as is used on digital computers. The speed of operation can be improved by automatically scanning all points in the network and automatically adjusting voltages. Yet the slowness of operation and the awkward repeated adjustments have limited the application of this technique, particularly in the light of rapid advances in the field of digital machines.

Fig. 24. Liebmann circuit element.

4. Combined Methods (Cognac)

The large number of resistors required for three-dimensional problems and the unavailability of purely geometric methods for the solution of transient problems have led to the development of a combined method. In this method, thermal conduction is represented in a continuum, while heat storage is represented in lumped form, as in the network technique.

An array of metallic capacitance connectors is immersed in a tank described in connection with geometric steady-state methods. For two-dimensional studies the connectors are cylinders

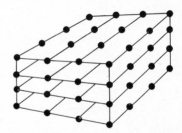

Fig. 25. Array of spheres for Cognac.

and for three-dimensional studies spheres are used. Figure 25 indicates such an array; for clarity, not all spheres are shown. To each capacitance connector, a capacitor located outside the tank is connected by an insulated wire. The negative poles of all capacitors are connected to the negative pole of the power supply (PS) (Fig. 26); again, for clarity, only nine connectors are shown. The boundary conditions are as discussed for the geometric method. For two-dimensional tanks it is easy to provide notched plates at top and bottom to hold the cylindrical connectors in place (Fig. 27). Spheres when used (Fig. 28) can be strung on rods R that keep the spheres apart by spacers S. Or, since spheres are heavier than the liquid resistors, one can use nylon or Teflon thread to suspend them from the top. Special designs have to be developed (e.g., Fig. 29) for odd shapes when there is no continuous row of vertical spheres, one on top of the other. While best results would be obtained with noble metal capacitance connectors, very satisfactory work has been carried out using simple steel or nickel connectors.

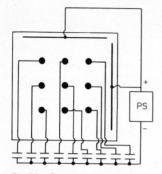

Fig. 26. Capacitor and power supply circuit.

Both theoretical and experimental work [2, 16, 17] has shown that the accuracy is worst at nodes at the corners; as one proceeds to interior points, or to points along the surface but away from the corner, the accuracy improves very rapidly. If the time-temperature curve at a given point is needed, one can then provide that the measured point is an interior node. But this method may lead to expensive setups; since connector spacing should be uniform, this arrangement greatly increases the required number of capacitor units. Let the voltage at point A in Fig. 30 be required. The solid circles show the connectors originally planned. Now, to improve the accuracy, A should become an interior node. If one adds connectors, making the previous first node into the

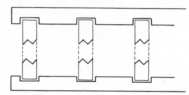

Fig. 27. Connector arrangement (two-dimensional tank).

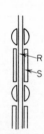

Fig. 28. Sphere arrangement.

Fig. 29. Design of Cognac for odd shapes.

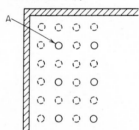

Fig. 30. Connector layout.

second position, as indicated by the broken line circles, the number of required capacitors is increased $2^3 = 8$ fold. This difficulty can be overcome by a local change of the scale factor. At the point of special interest (be it the surface, as shown, or anywhere on the inside of the tank) an insulated insert I is used as shown in Fig. 31. This insert precludes the presence of resistor fluid in its space. A number of electrodes is mounted on the surface; six are shown in the example, numbered from 1 to 6. Then a second tank E is manufactured, enlarging the part of the tank blocked out in the original tank. Electrodes are fastened on the corresponding faces of the tank E. They will, of course, be larger than on I, but the same number and the same relative sizes must be used; if for some reason the lengths of the electrodes on E are not equal, similar (i.e., proportional) lengths must be used in E. Corresponding electrodes on I and E are connected by copper wires. Since E is made to a larger scale, one can place more connectors inside. The desired point A now becomes an inside point, yielding higher accuracy. The saving in capacitor units is obvious.

The same technique can be applied if a composite body has to be studied. Make two or more tanks, depending on the number of materials used in the prime system; the several tanks are then connected by using surface electrodes and connecting wires, as shown in Fig. 31. The scale factors are similar to those in lumped networks (Eqs. (15) to (21)) but are sufficiently different to warrant separate treatment. There are six factors, R_F, C_F, G_F, T_F, V_F, and I_F. Of these, C_F, V_F, and I_F have been defined by

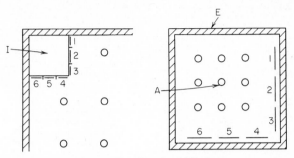

Fig. 31. Use of insulated inserts.

Eqs. (18), (22), and (23) respectively. But R_F is now not only related to the electric resistivity of the liquid resistors but also to the geometry of spacing of capacitance connectors. If λ is the ratio of sphere diameter to the sphere spacing (center-to-center), Fig. 32 shows the relationship of F_λ / λ. The resistance factor is now given by

$$R_F = k\rho_e F_\lambda \tag{34}$$

Derivation of these relationships is presented in Ref. 16.

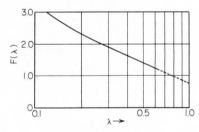

Fig. 32. $F(\lambda)$ as a function of λ.

The geometric factor G_F is the ratio of length in the tank to length in the prime system for any pair of comparable sides. The time factor T_F also relates to the size of the tank

$$T_F = \frac{t_e}{t_h} = R_F C_F G_F^2 \tag{35}$$

This equation yields important information. For a given size of the prime system and a desired time factor, one can reduce the size of necessary capacitance by using more resistance fluid: by making G_F larger. Also, the number of nodes does not appear in the scale factors; thus the change in spacing, as indicated in Figs. 30 and 31, increases the number of capacitor units but not the total capacitance used.

5. Electric Analogs for Combined Heat and Mass Flow

The equations describing heat transfer in heat exchangers with or without heat storage can be solved conveniently on analog computers, although analogous physical continuous systems which follow the same equations are not always known.

   a. Tubular Heat Exchangers: Steady-state Condition [18]. Let Fig. 33 be a section from a tubular heat exchanger, with subscript h denoting the hot and subscript

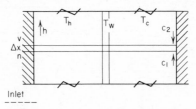

Fig. 33.   Tubular heat-exchanger section.

c the cold fluid; there are two arrows for the cold fluid so as to use the one figure for parallel flow (c_1) and counterflow (c_2). Taking somewhere along the length of the exchanger an element Δx and letting its length approach zero, one obtains from heat balances for the hot (Eq. (36)) and cold (Eq. (37)) fluids

$$-W_h c_{ph} \frac{dT_h}{dx} = UP(T_h - T_c) \tag{36}$$

$$\pm W_c c_{pc} \frac{dT_c}{dx} = UP(T_h - T_c) \tag{37}$$

U is the heat transfer coefficient per unit length, and P is the heated perimeter.

   In Eq. (37) for the cold side, the positive sign holds for the parallel and the negative sign for the counterflow exchanger.

   Speaking first of a parallel flow exchanger and a circuit, as shown in Fig. 34, the following equations hold

$$-C_h \frac{de_h}{dt} = \frac{1}{R}(e_h - e_c) \tag{38}$$

$$\pm C_c \frac{de_c}{dt} = \frac{1}{R}(e_h - e_c) \tag{39}$$

Comparison of these equations with Eqs. (36) and (37) shows full identity. The corresponding items are:

$W c_p$ (sometimes referred to as "water value")	capacity (C)
UP	resistance (R)
x (length of tube)	time

   b. Counterflow Exchangers. For counterflow exchangers, the circuit of Fig. 35 can be used, provided that by adjustment of resistor R_2 the current i_2 is at all times held at twice the value of i_1. The proof for this statement can be found by equating the left sides of Eqs. (38) and (39); since both sides are negative, current can flow out of both capacitors simultaneously only by providing an additional drain, and since the two currents i_1 and i_3 are equal, current i_2 must be twice as large as i_1.

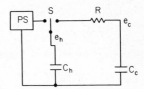

Fig. 34. Parallel flow exchanger circuit.

Operation of the circuits is based on the concept that time corresponds to length; hence computation has to start with the initial conditions of both fluids at the same end of the exchanger. It is convenient to use the entrance temperature of the cold fluid as a reference so that the difference between hot and cold fluid entrance temperatures is represented by "100%" of the voltage. In case of the parallel flow exchanger, this simply requires precharge of C_h (Fig. 34) to this voltage; computation is started by throwing switch S from PS to R; the cold capacitor may be connected to R all the time since it is initially uncharged. Charge

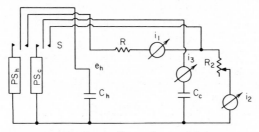

Fig. 35. Counterflow exchanger circuit.

flows from C_h to C_c and the voltages e_h and e_c are observed. In case of the counterflow exchanger, the cold fluid is exiting at the entrance end of the hot fluid. Therefore both C_h and C_c must be disconnected from R up to the start of the computation; both capacitors must be precharged to voltages corresponding to the gas temperatures at that end: i.e., the entrance temperature of the hot fluid and the exit fluid temperature of the cold one. If this value is not known (for the case when the length of the exchanger is prescribed and the exit temperatures are to be found), a trial and error procedure must be followed, assuming first an exit temperature. If this does not yield a zero entrance temperature at the specified length, the computation must be repeated with a different assumption of the exit conditions. The condition that at all times $i_2 = 2i_1$ is achieved by adjusting R_2.

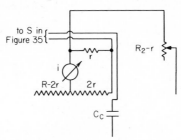

Fig. 36. Simplified circuit for Fig. 35.

In order to obviate the need to read (in Fig. 35) two ammeters (i_1 and i_2), the connections between R, C_c, and R_2 can be changed as illustrated in Fig. 36, omitting both ammeters shown in Fig. 35, and introducing a single instrument i which is kept constantly at zero reading by adjusting $R_2 - r$.

Variable properties can be handled as in the case of pure conduction problems: capacitances and resistances are changed in steps according to the voltage at the corresponding point. In case of changing heat transfer coefficients, R may be divided into three parts representing the heat transfer coefficients on the hot and cold sides and in the tube wall. This kind of circuitry can be extended to heat exchangers dealing with more than two fluids [18].

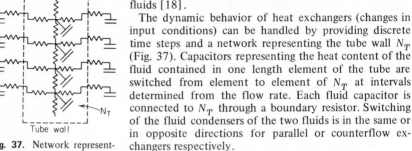

Fig. 37. Network representing the tube wall.

The dynamic behavior of heat exchangers (changes in input conditions) can be handled by providing discrete time steps and a network representing the tube wall N_T (Fig. 37). Capacitors representing the heat content of the fluid contained in one length element of the tube are switched from element to element of N_T at intervals determined from the flow rate. Each fluid capacitor is connected to N_T through a boundary resistor. Switching of the fluid condensers of the two fluids is in the same or in opposite directions for parallel or counterflow exchangers respectively.

Discretizing of time is a disadvantage and, moreover, requires elaborate switching procedure. It can be obviated by introducing certain active elements: cathode followers with a gain of 1.000. (Since commercial cathode followers have a gain somewhat below unity, the devices with gain of 1.000 are here called voltage-dependent voltage sources (VDVS).) Circuits as shown in Figs. 38 and 39 represent sections of counterflow and parallel flow heat exchangers, respectively. Contrary to the previous treatment, the length of the exchanger is now discretized;

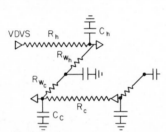

Fig. 38. Counterflow heat exchanger circuit.

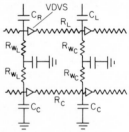

Fig. 39. Parallel flow heat exchanger circuit.

electric resistance represents both the mass rate of flow × specific heat and the heat transfer coefficient × area, while capacitance represents the product of the weight × specific heat. The following scale factors apply

$$C_F = \frac{C}{c_p V \rho} \tag{40}$$

$$R_F = RUA = RWc_p \tag{41}$$

$$T_F = C_F R_F \tag{42}$$

$$V_F = \frac{e}{\Delta T} \tag{43}$$

$$I_F = \frac{i}{q} \tag{44}$$

Note that specific heat appears in two scale factors (Eqs. (40) and (41)) and that (in Figs. 38 and 39) values of R_h and R_c use the second definition in Eq. (41), while the first definition in this equation applies to resistors R_{wh} and R_{wc}. Each of these two resistors is considered composed of two parts: one is the heat transfer coefficient on the respective side (hot or cold), the other is half of the wall resistance. Capacitors C_h and C_c are computed on the basis of the volume of the fluid in the respective section of the exchanger, while C_w represents the heat storage capacity of the wall section. Dynamic behavior is studied by changing input voltage at the appropriate time (for a change of input temperature) or flow resistors (for a change in flow rate). If this change (as frequently will be the case) also influences the heat transfer coefficients, the values of R_{wh} or R_{wc} have to be changed simultaneously. For a detailed discussion see Ref. 19.

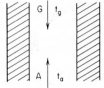

Fig. 40. Regenerator circuit.

Heat exchangers with heat storage (regenerators) can again be studied by two techniques: discretizing time or with continuous time. A typical case is presented by open hearth regenerators. During the "gas period" hot gas enters at a constant temperature at the top and leaves at a lower temperature which gradually increases during the gas period. Upon completion of the latter, air starts flowing through the flue in the opposite direction, absorbing heat stored in the brickwork (Fig. 40). The walls of the flue are represented by a normal conduction network (N_B in Fig. 41). Each horizontal row of nodes ends in a resistor representing the heat transfer coefficient between wall and gas or wall and air: these resistors, marked R_f, can be changed between the gas and the air period and differ at each level according to the prevailing temperature of gas and brick. Both gas and air periods (t_g and t_a respectively) are divided into n time steps; the length of one time step is then $t_g/n = t_a/n = \Delta t$.

The value of a gas capacitor is determined by using the C_F from the flue wall determination

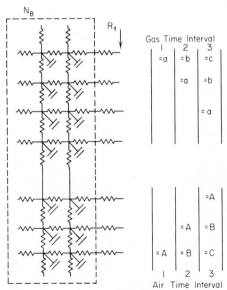

Fig. 41. Normal conduction circuit for regenerators.

$$C_g = C_F c_{pg} w_g \Delta t_g \tag{45}$$

All gas capacitors are precharged to a voltage corresponding to the gas entrance temperature; all air capacitors are discharged before entering into the circuit. During the several time steps of the gas period, gas capacitors a, b, c are connected successively with different levels of the N_B across the appropriate level of R_f. Then, after the last gas capacitor has been disconnected from the lowest level of N_B, the first (uncharged) air capacitor A is connected to this lowest level and, as it is moved upwards in successive air-time steps, air capacitors B and C follow, as shown in Fig. 41. The procedure continues until *cyclic equilibrium* has been reached, i.e., until two consecutive time cycles yield the same temperatures. The circuit for nondiscretized time is shown in Fig. 42.

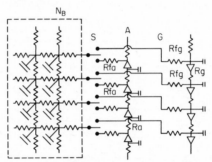

Fig. 42. Nondiscretized time circuit for regenerators.

The network to represent the bricks is, as before, N_B. The capacitors in the air and gas networks which correspond essentially to the heat exchanger circuits (Fig. 39) are computed from Eq. (40), using the value of R_F in determining N_B, and the resistors are computed from Eq. (41). By means of the row of switches S, which are operated simultaneously, either the *gas row* or the *air row* are connected to N_B each time through the appropriate heat transfer resistance R_f. Refer to a detailed study [20] showing both circuitry and result.

6. Analogies for Radiation View Factors

Heat transfer by radiation between solid black surfaces depends, among other factors, on the geometries involved. The determination of "view factors" is complicated for all but the simplest shapes. However, view factors can be determined by electric analogy and can be represented together with simultaneous conduction in a single circuit. Let black body I radiate to body II (Fig. 43). Consider both bodies divided into sections, as shown in the figure. If the sections are small enough to be considered elemental, the radiative heat exchange of each with all others can be computed from Eq. (46) (see Fig. 44 for notations)

$$q = \frac{\sigma(T_I^4 - T_{II}^4) \cos\phi_1 \cos\phi_2 \, dA_I \, dA_{II}}{\pi L^2} \tag{46}$$

The resistance is found by dividing by $(T_I - T_{II})$, and is a function of the temperatures of I and II. Now provide resistors between each of the sections of I and all the other sections of both I and II, and between each section of II and all the sections of I (since II is straight, none of the sections of II "sees" any of the other sections of II). All resistors (a total of 35) are computed according to the angles of Eq. (46). They must be adjusted either manually or automatically, according to the temperature of each section. (Temperatures here are, of course, in absolute units of either Rankin or Kelvin, depending on whether the British or metric system is used.) Heat sources are introduced appropriately; if one of the bodies is an electric resistor, current is fed in. For work along this line see Paschkis [21] and Oppenheim [22]; Kreith [23] also refers to networks to simulate radiation problems.

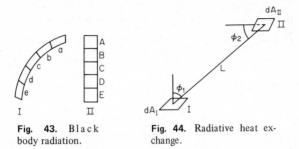

Fig. 43. Black body radiation.

Fig. 44. Radiative heat exchange.

C. HYDRAULIC ANALOGS

The Laplace equation applies, among others, to the flow of highly viscous fluids in a narrow channel. Thus so-called "fluid mappers" can be used for the solution of the steady-state problem with or without heat sources and sinks. Experimental requirements are so high that this technique can hardly qualify as a "computer technique"; it is more a tool for qualitative illustration [24, 25].

More important, although much less used than electric analogs, are hydraulic analogs used for the study of transient problems, as first described by Moore [26]. Following the concept of electric analogs, "resistors" are replaced by capillary tubes and "capacitors" by reservoirs. If the latter is made of transparent material the fluid level in the reservoir is an excellent visual indication of the potential at each "capacitor."

Moore's original paper is written as a progress report, with no discussion of scale factors and only tentative information regarding design. A history of the development and a bibliography of 25 items are included in a paper by Scott [27]. The system, as represented in Fig. 45, consists of standpipes acting as capacitors (C) and conduits (R) acting as resistors.

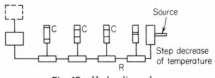

Fig. 45. Hydraulic analog.

Conduits must be such that only laminar flow occurs in them. Capillary tubes have been used where resistance is given by length, but Scott describes other devices which can be more easily adjusted. However, all resistors must be calibrated because their resistance cannot be computed.

Heat sources and sinks may be represented by lateral extensions to the standpipe. These extensions have to be filled or emptied before the level can rise above or below the point of connection. Such addition is indicated at the last capacitance in Fig. 45 and presents several difficulties. In principle, it is possible to extend the method to two dimensions; but the connection of conduits presents appreciable experimental difficulties. The same system has been applied to the study of the steady-state behavior of heat exchangers [28].

D. OTHER ANALOGS

Some analogs of other design are available occasionally for the solution of steady-state problems; their practical significance for engineering purposes in heat flow is so limited that only two examples are given.

One method is a soap film, stretched over a boundary of the shape of the prime system. Boundary potentials are reproduced by providing a rim over a flat plate. The contour of the rim corresponds to the contour of the prime system; the height of the rim to the boundary potential. A film deposited over this rim assumes the shape of the temperature field represented by the rim [29, 30].

Another method of solving the Laplace equation is by stretching a sheet of rubber over an appropriate rim. Karplus [29, p. 273] and Mindlin and Salvadori [30, p. 732] deal with this method; both give additional references. These methods have limited use for heat transfer because they are applicable only if the surface temperatures are prescribed, not for either of the two other boundary conditions.

E. USE OF INDIRECT ANALOG COMPUTERS
TO SOLVE HEAT CONDUCTION PROBLEMS

Heat conduction computations can also be carried out with active element (indirect) analogs or differential analyzers [31, 32]. Before discussing this method, the use of active element analogs for addition and multiplication will be briefly explained. Such analogs are based on operational amplifiers. These amplifiers operate with direct current and have a high input impedance (but not as high as those used for voltage measurements in direct analogs) and a gain of about 15,000.

An operational amplifier usually not only provides for the mathematical operation, but also for a sign change.

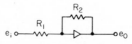

Fig. 46. Input and output voltages; indirect analog.

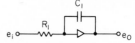

Fig. 47. Output voltage at time zero; indirect analog.

In Fig. 46, the voltage relationship (output e_o/input e_i) with both measured to ground is

$$e_o = -\frac{R_2}{R_1} e_i \tag{47}$$

In Fig. 47

$$e_o = -\frac{1}{R_1 C_1} \int_0^{t_1} e_i \, dt + e_{o,t=0} \tag{48}$$

Here $e_{o,t=0}$ is the output voltage at time zero. In the arrangement of Fig. 48, the output voltage finally becomes

$$e_o = -R_2 \left(\frac{e_1}{R'_1} + \frac{e_2}{R''_1} + \frac{e_3}{R'''_1} \right) \tag{49}$$

If $R'_1 = R''_1 = R'''_1 = R_2$, then

$$e_o = -(e_1 + e_2 + e_3) \tag{49a}$$

This device directly adds the several voltages. One can divide voltages by appropriately choosing the R_2/R_1 ratio in Eq. (47), but one cannot, in any simple way, subtract voltages.

Turning to the solution of the heat transfer equations, consider first the steady-state case. For a symmetrical grid (where the dimensions of each section are identical along each of the three coordinates), the voltage at each node equals 1/2, 1/4, or 1/6 of the sum of the voltages at contiguous nodes for one-, two-, and three-dimensional problems respectively.

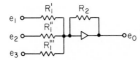

Fig. 48. Output voltage; indirect analog.

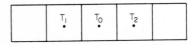

Fig. 49. Rod subdivided into a number of sections.

Consider (Fig. 49) a rod subdivided into a number of sections (five are shown in the figure). In steady state, the temperature T_0 of the third section can be approximated by

$$T_0 = 0.5(T_1 + T_2) \tag{50}$$

Similarly, in two- or three-dimensional problems, the section with a temperature T_0 would be surrounded by four or six sections respectively; the temperature T_0 would relate to the surrounding temperatures by Eq. (51) (two-dimensional) or Eq. (52) (three-dimensional)

$$T_0 = \frac{1}{4}(T_1 + T_2 + T_3 + T_4) \tag{51}$$

$$T_0 = \frac{1}{6}(T_1 + T_2 + T_3 + T_4 + T_5 + T_6) \tag{52}$$

By using summing circuits as shown in Fig. 48, one circuit for each node, such a problem can be simulated on an indirect analog computer. Since there is no advantage against a pure resistance network, the circuits are not shown in detail.

For transient problems, however, the use of indirect analogs avoids the use of relatively large and high quality capacitors; an operational amplifier must be used instead of each capacitor. Thus if an indirect analog is available or finds other uses an indirect analog may be helpful. Figure 50 shows part of a one-dimensional network; Fig. 51 part of a two-dimensional one. Dotted lines indicate five adjacent sections numbered 0 to 4; contiguous sections are indicated by numbers 5 to 12 in the corners. The circuit is complete only for section 0; in sections 1 to 4 part of the connections are indicated only by letters and numbers (e.g., h5 or h6), indicating the point in the next section which is not shown in the figure to which connections would be made.

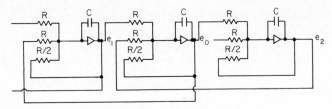

Fig. 50. One-dimensional network.

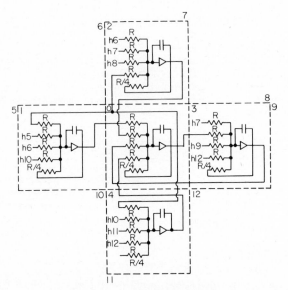

Fig. 51. Two-dimensional network.

The four resistors R of each section are the sum of the voltages of the four adjacent sections, while the resistor $R/4$ subtracts $1/4$ of the voltage of the proper node. Thus

$$e_0 = RC \int (e_1 + e_2 + e_3 + e_4 - 4e_0)dt \qquad (53)$$

The scale factors for R and C are the same as those explained for the direct analogs. Since indirect analogs usually operate with a large time factor, values for R and C are relatively small.

REFERENCES

1. *IRE Trans.*, Feb. 1962, special issue on electronic computers. (see footnote on page 5-2).
2. S. Dicker, "Representation of Surface Conductance with Two Dimensional Continuous Analogs," doctoral dissertation, Columbia University, New York, 1961.
3. V. Paschkis, Combined Geometric and Network Analog Computer for Transient Heat Flow, *ASME J. Heat Trans.*, **81**:144 (1959).
4. T. J. Higgins, *Appl. Mech. Rev.*, **9**:1–4, 49, 55 (1956); **10**:49–54, 331–335, 443–448 (1957); **11**:203–206 (1958).
5. E. B. Dahlin, *Rev. Sci. Insts.*, **24**:951–953 (1954).
6. C. F. Kayan, *Refrig. Engr.*, **54**:103–151 (1947).

7. S. Zandstra and A. J. Bovy, Bibliography. Literature with respect to the theory, design and applications of the Beuken Model to the analysis of transient heat-flow problems. Library, Techn. University Delft, Sec. Ed., 1963.

8. C. L. Beuken, Wärmeverluste Bei Periodisch Betriebenen Elektroöfen; Comptes Rendus du Congrès International des Applications Electrocalorifiques et Electrochimiques, Scheveningen, Union Internationale d'Electrothermie, Paris, 1936, p. 204.

9. V. Paschkis and H. D. Baker, *Trans. ASME,* **64**:105 (1942).

10. V. Paschkis, Theoretical Thermal Studies of Steel Ingot Solidification, *Trans. ASME,* **38**:117 (1947).

11. V. Paschkis, Thermal Considerations in Foundry Work, exchange paper to the British Institute of Foundrymen from AFS, 1951.

12. V. Paschkis, Heat and Mass Flow Analyzer, *Scientific Monthly,* **73**(2):81, Aug. 1951

13. V. Paschkis, The Heat and Mass Flow Analyzer—A Simulator for the Study of Heat Conduction, *Annales de l'Association Internationale pour le Calcul Analogique,* **1**:4–18 (1963).

14. G. Liebmann, *Trans. ASME,* **178**:655 (1956).

15. G. Liebmann, *Brit. J. Appl. Phys.,* **4**:143 (1953).

16. P. Razelos and H. G. Elrod, ASME Preprint 67-WA-HT-14.

17. P. Razelos and V. Paschkis, ASME Preprint 67-WA-HT-15.

18. V. Paschkis and M. P. Heisler, Chemical Engineering Progress Symposium, AIChE, No. 5, **40**:65 (1953).

19. V. Paschkis and J. W. Hlinka, *Trans. N.Y. Acad. Sci.,* (II)**19**(8):714 (1957).

20. J. W. Hlinka, F. S. Puhr, and V. Paschkis, A Comprehensive Method for the Thermal Design of Regenerators, *Iron and Steel Engineer,* **38**:59–76 (1961).

21. V. Paschkis, *Elektrotechnik u. Maschinenbau,* **52**:54 (1936).

22. A. K. Oppenheim, *Trans. ASME,* **78**:725–735 (1956).

23. F. Kreith, "Heat Transfer," International Textbook Company, Scranton, Pa., 1958.

24. A. D. Moore, *J. Appl. Phys.,* **20**:790 (1949).

25. A. D. Moore, *Trans. AIEE,* **1**:69:1615 (1950).

26. A. D. Moore, *Ind. Eng. Chem.,* **28**:704–708 (1936).

27. R. F. Scott, *Geotechnique (London),* **7**:55–72 (1956).

28. S. I. Juhasz and F. C. Hooper, *Proc. Sec. Natl. Cong. Appl. Mech., 1954,* pp. 805–810.

29. W. J. Karplus, "Analog Simulation," p. 273, McGraw-Hill Book Company, New York, 1958.

30. R. D. Mindlin and M. G. Salvadori, in "Handbook of Experimental Stress Analysis" (M. Metenyi, ed.), p. 732, John Wiley & Sons, Inc., New York, 1950.

31. R. M. Howe and V. S. Haneman, *Proc. IRE,* **41**:1497 (1953).

32. W. J. Karplus, "Analog Simulation," McGraw-Hill Book Company, New York, 1958; a hybrid computer comprising active elements and digital computers described by W. J. Karplus, *IEEE Trans. Electronic Computers,* p. 597 (1964).

NOMENCLATURE

a, b, c spatial dimensions
A cross-sectional area
c_p specific heat
C capacitance
e voltage
G_F geometric factor
h heat transfer coefficient
i current
I_F current factor
k thermal conductivity
L length (thickness) of body subjected to one-dimensional heat flow
L_e length in tank
q'' heat flow
Q_f total heat of solidification (Btu)
r radius

R resistance
R_f boundary resistance
R_F resistance factor
t time
t_e electric time
t_h heat time
T temperature
U heat transfer coefficient per unit length
V volume of section
V_F voltage factor
W mass rate of flow
w weight of one section
x spatial coordinate
λ heat of fusion (Btu/lb)
r thickness of conducting layer
ρ density
ρ_e electic resistivity
σ coefficient of black body radiation (Boltzmann constant)

Section **6**

Free-convection Boundary-layer Heat Transfer

ROBERT D. CESS

State University of New York, Stony Brook, New York

INTRODUCTION

Free convective motions are the consequence of an external body force (such as gravitational and centrifugal forces) acting upon a fluid within which density variations occur. For single component fluids such density variations are due to temperature gradients within the fluid, and these are in turn induced by heat addition to or heat subtraction from the fluid. This heat transfer often occurs at a bounding surface, and usually the prediction of such surface phenomenon is of primary interest.

The present chapter is concerned with free convection heat transfer in boundary layer flows. Since fluid velocities in free convection are generally quite low, then, unlike forced convection, a large percentage of physically practical problems are laminar flow problems. Major emphasis will, in this chapter, be directed towards laminar boundary layers, with only limited information being presented for turbulent free convection.

Two excellent survey articles on free convection have been written by Ostrach [1] and Ede [2]. The intent of the present chapter is to cover the principal features of free convection boundary layers, and for a more complete survey of the literature, the reader is referred to [1] and [2].

BASIC EQUATIONS

Attention will, for the time being, be restricted to two-dimensional geometries, and the coordinate system is illustrated in Fig. 1. At any location along the surface, the tangent to the surface is at the angle ϕ with reference to the body force g. Although the notation here is that conventionally used for the gravitational body force, g can denote any externally applied body force. In the following development, it will be assumed that $T_w > T_\infty$, and that the x-component of the body force acts in the negative x-direction as illustrated in Fig. 1. The resulting equations will, of course, apply for $T_w < T_\infty$ by simply reversing the sign of the body force. Thus, when $T_w < T_\infty$ the body force is $-g$. In either case the flow is in the positive x-direction.

The boundary layer equations expressing conservation of mass, momentum, and energy may be written respectively, as

$$\frac{\partial(\rho v_x)}{\partial x} + \frac{\partial(\rho v_y)}{\partial y} = 0 \tag{1}$$

$$\rho\left(v_x \frac{\partial v_x}{\partial x} + v_y \frac{\partial v_x}{\partial y}\right) = \frac{\partial}{\partial y}\left(\mu \frac{\partial v_x}{\partial y}\right) - \frac{dp}{dx} - \rho g \cos\phi \tag{2}$$

$$\rho c_p\left(v_x \frac{\partial T}{\partial x} + v_y \frac{\partial T}{\partial y}\right) = \frac{\partial}{\partial y}\left(k \frac{\partial T}{\partial y}\right) \tag{3}$$

where the assumption of steady laminar flow has been invoked. The above equations are not, of course, the complete conservation equations, but constitute the boundary-layer form of the conservation equations. A formal development of this is given by Ostrach [1] and the above are asymptotic equations for large Grashof number.

In addition to the need for describing the temperature dependence of the physical properties appearing in Eqs. (1), (2), and (3), it is necessary to evaluate the presssure gradient dp/dx. Since dp/dx is independent of y, this evaluation may be accomplished in the conventional manner by evaluating Eq. (2) at the outer region of the boundary layer. Since, for free convection, this region corresponds to $v_x = 0$, then from Eq. (2)

$$\frac{dp}{dx} = -\rho_\infty g \cos\phi$$

such that Eq. (2) becomes

$$\rho\left(v_x \frac{\partial v_x}{\partial x} + v_y \frac{\partial v_x}{\partial y}\right) = \frac{\partial}{\partial y}\left(\mu \frac{\partial v_x}{\partial y}\right) + (\rho_\infty - \rho)g \cos\phi \tag{4}$$

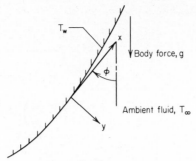

Fig. 1. Coordinate system.

Evaluation of the quantity $(\rho_\infty - \rho)$ is most readily accomplished by considering two special, but widely useful, cases. The first involves a perfect gas, for which it follows that, since p is constant across the boundary layer

$$\rho_\infty - \rho = \frac{p}{RT_\infty} - \frac{p}{RT} = \frac{\rho}{T_\infty}(T - T_\infty)$$

Thus, since the coefficient of thermal expansion for the ambient gas is $\beta = 1/T_\infty$, then

$$\rho_\infty - \rho = \beta\rho(T - T_\infty) \tag{5}$$

where, again, it is emphasized that β is a constant corresponding to the ambient temperature T_∞.

The second special case is that of a liquid. Recalling that

$$\beta = -\frac{1}{\rho}\left(\frac{\partial \rho}{\partial T}\right)_p$$

then, since pressure is constant across the boundary layer, $d\rho = -\beta\rho dT$. For most liquids the temperature dependence of both β and ρ is very slight, such that, to a good approximation, the above equation may be integrated directly to yield

$$\rho_\infty - \rho = \beta\rho(T - T_\infty) \tag{6}$$

This is identical to Eq. (5), with the exception of the interpretation of the product $\beta\rho$. For a perfect gas, Eq. (5) is exact, and β corresponds to the ambient temperature T_∞ while ρ refers to the local temperature within the boundary layer. Regarding Eq. (6), the product $\beta\rho$ is assumed to be constant for liquids, and thus Eq. (6) is an approximation. For either case the momentum equation has the same form, and from Eq. (4) this is

$$\rho\left(v_x \frac{\partial v_x}{\partial x} + v_y \frac{\partial v_x}{\partial y}\right) = \frac{\partial}{\partial y}\left(\mu \frac{\partial v_x}{\partial y}\right) + g\beta\rho(T - T_\infty)\cos\phi \tag{7}$$

The system of conservation equations consisting of Eqs. (1), (7), and (3) may now be replaced by a system of two equations through the introduction of the stream function ψ, which is defined by

$$v_x = \frac{\rho_\infty}{\rho} \frac{\partial \psi}{\partial y} \qquad v_y = -\frac{\rho_\infty}{\rho} \frac{\partial \psi}{\partial x} \tag{8}$$

This automatically satisfies Eq. (1), while Eqs. (7) and (3) become

$$\frac{\partial \psi}{\partial y} \frac{\partial}{\partial x}\left(\frac{\rho_\infty}{\rho} \frac{\partial \psi}{\partial y}\right) - \frac{\partial \psi}{\partial x} \frac{\partial}{\partial y}\left(\frac{\rho_\infty}{\rho} \frac{\partial \psi}{\partial y}\right) = \frac{\partial}{\partial y}\left[\frac{\mu}{\rho_\infty} \frac{\partial}{\partial y}\left(\frac{\rho_\infty}{\rho} \frac{\partial \psi}{\partial y}\right)\right] + g\beta \frac{\rho}{\rho_\infty}(T - T_\infty)\cos\phi \tag{9}$$

$$\rho c_p \left(\frac{\partial \psi}{\partial y} \frac{\partial T}{\partial x} - \frac{\partial \psi}{\partial x} \frac{\partial T}{\partial y}\right) = \frac{\partial}{\partial y}\left(k \frac{\partial T}{\partial y}\right) \tag{10}$$

The boundary conditions for the above system of equations must now be specified. Since continuum momentum and energy transport have been inferred, then, assuming an impermeable surface, it follows that at the surface

$$\frac{\partial \psi}{\partial y} = \frac{\partial \psi}{\partial x} = 0 \ , \quad T = T_w \ ; \quad y = 0 \tag{11}$$

With respect to conditions at the outer edge of the boundary layer, it should be noted that boundary layer theory consists of asymptotically matching an inner region (boundary layer) with an outer region (inviscid flow) [3]. Furthermore, the thickness of the boundary layer for free convection is of the order $x(\mathrm{Gr})^{-1/4}$, where the Grashof number is defined as*

$$\mathrm{Gr} = \frac{g\beta(T_w - T_\infty)x^3}{\nu^2}$$

while the outer region consists of a stationary isothermal fluid. Hence the statement of asymptotic matching is that

$$\frac{\partial \psi}{\partial y} = 0 \ , \quad T = T_\infty \ ; \quad \frac{y}{x}(\mathrm{Gr})^{1/4} \to \infty \tag{12}$$

Equations (9)–(12), together with appropriate expressions for the temperature variation of physical properties, completely describe the boundary layer formulation for laminar free convection.

LAMINAR FLOW: VERTICAL FLAT PLATE

Laminar free convection along a vertical flat plate will be employed as an illustrative model for the purpose of examining the principal features of free convection heat transfer. The physical geometry and coordinate system are illustrated in Fig. 2, and restriction will presently be made to constant fluid properties. For this case, Eqs. (9) and (10) become

*When $T_w < T_\infty$, then as previously discussed the sign of the body force is reversed and Gr remains a positive quantity.

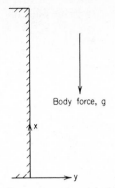

Fig. 2. Coordinate system for vertical flat plate.

$$\frac{\partial \psi}{\partial y} \frac{\partial^2 \psi}{\partial x \partial y} - \frac{\partial \psi}{\partial x} \frac{\partial^2 \psi}{\partial y^2} = \nu \frac{\partial^3 \psi}{\partial y^3} + g\beta (T - T_\infty) \tag{13}$$

$$\frac{\partial \psi}{\partial y} \frac{\partial T}{\partial x} - \frac{\partial \psi}{\partial x} \frac{\partial T}{\partial y} = \alpha \frac{\partial^2 T}{\partial y^2} \tag{14}$$

Temperature-dependent fluid properties, turbulent flow, and other geometries will be considered later.

Isothermal Plate. Attention will first be directed toward the boundary condition of a uniform temperature on the plate surface. This results in the well-known similarity transformation described by

$$\eta = c \frac{y}{x^{1/4}} \qquad c = \left[\frac{g\beta (T_w - T_\infty)}{4\nu^2} \right]^{1/4}$$

$$f(\eta) = \frac{\psi(x, y)}{4\nu c x^{3/4}} \qquad \Theta(\eta) = \frac{T - T_\infty}{T_w - T_\infty}$$

such that Eqs. (13) and (14) transform to

$$f''' + 3ff'' - 2(f')^2 + \Theta = 0 \tag{15}$$

$$\Theta'' + 3\mathrm{Pr}\, f\Theta' = 0 \tag{16}$$

while the boundary conditions follow from Eqs. (11) and (12) to be

$$f = f' = 0, \quad \Theta = 1; \quad \eta = 0 \tag{17a}$$

$$f' = 0, \quad \Theta = 0; \quad \eta \to \infty \tag{17b}$$

Note that the velocity component v_x is given by

$$v_x = 4\nu c^2 x^{1/2} f'(\eta)$$

If a reference velocity v_r is defined at some value of η, then

$$v_r \sim 4\nu c^2 x^{1/2} \sim [g\beta(T_w - T_\infty)x]^{1/2}$$

and the corresponding Reynolds number is

$$\text{Re} = \frac{v_r x}{\nu} \sim (\text{Gr})^{1/2}$$

Thus, for free convection $(\text{Gr})^{1/2}$ plays the role of a Reynolds number. The local heat flux at a location x on the plate is described by

$$q''(x) = -k\left(\frac{\partial T}{\partial y}\right)_{y=0}$$

Expressing this in terms of the local Nusselt number

$$\text{Nu} = \frac{q''x}{k(T_w - T_\infty)}$$

then

$$\frac{\text{Nu}}{(\text{Gr})^{1/4}} = -\frac{\Theta'(0)}{\sqrt{2}} \tag{18}$$

A large number of numerical solutions to the system of equations (15)–(17) have been performed, and the corresponding values of $\Theta'(0)$ as a function of Prandtl number are tabulated by Ede [2]. In addition to these results, LeFevre [4] has obtained limiting solutions for $\text{Pr} \to 0$ and $\text{Pr} \to \infty$. The low Prandtl number asymptote corresponds to neglecting the viscous term (that is, f''') in Eq. (15), while the high Prandtl number limit is obtained by deleting the nonlinear inertia terms in Eq. (15). The results are

$$\frac{\text{Nu}}{(\text{Gr})^{1/4}} = 0.600\,(\text{Pr})^{1/2} \quad ; \quad \text{Pr} \to 0 \tag{19}$$

$$\frac{\text{Nu}}{(\text{Gr})^{1/4}} = 0.503\,(\text{Pr})^{1/4} \quad ; \quad \text{Pr} \to \infty \tag{20}$$

The numerical results for the local Nusselt number, taken from Ede [2] and Ostrach [5], as well as Sparrow and Gregg [6], are listed in Table 1 for low and moderate Prandtl number and in Table 2 for moderate and large Prandtl number. Note that the dimensionless grouping in Table 1 follows Eq. (19), while that in Table 2 corresponds to Eq. (20). In this way, the effect of Prandtl number is suppressed for each of the two Prandtl number ranges.

Variable Surface Temperature. In many applications of practical importance, the surface temperature will be nonuniform. The effect of a nonuniform surface temperature upon free convection for the geometry of a vertical flat plate will now be considered.

For the general case of an arbitrary variation of surface temperature with position $T_w(x)$, a similarity solution does not exist, i.e., Eqs. (13) and (14) do not reduce to ordinary differential equations. There do, however, exist certain classes of surface temperature variations which do yield similarity solutions, and the specific case for

TABLE 1. Nusselt Number Results
for Low and Moderate
Pr

Pr	$Nu/(Pr^2 Gr)^{1/4}$
0	0.600
0.003	0.583
0.01	0.567
0.03	0.555
0.09	0.517
0.5	0.443
0.72	0.421
1.0	0.401

TABLE 2. Nusselt Number Results
for Moderate and High
Pr

Pr	$Nu/(Pr Gr)^{1/4}$
1	0.401
2	0.426
5	0.451
10	0.464
100	0.490
1,000	0.500
∞	0.503

which

$$T_w(x) - T_\infty = N x^n \tag{21}$$

will be treated here. As illustrated by Sparrow and Gregg [7], this corresponds to the similarity transformation

$$\eta = c \frac{y}{x^{1/4}} \qquad\qquad c = \left(\frac{g\beta N}{4\nu^2} \right)^{1/4}$$

$$f(\eta) = \frac{\psi(x, y)}{x^{(n+3)/4}} \frac{1}{4\nu c} \qquad \Theta(\eta) = \frac{T - T_\infty}{T_w - T_\infty}$$

and Eqs. (13) and (14) transform to

$$f''' + (n + 3)f''f - (2n + 2)(f')^2 + \Theta = 0 \tag{22}$$

$$\Theta'' + (n + 3)\,Pr f\Theta' - 4n\,Pr f'\Theta = 0 \tag{23}$$

subject to the boundary conditions

$$f = f' = 0 , \quad \Theta = 1 ; \quad \eta = 0 \tag{24a}$$

$$f' = 0 , \quad \Theta = 0 ; \quad \eta \to \infty \tag{24b}$$

It readily follows that the local Nusselt number is again expressed by

$$\frac{Nu}{(Gr)^{1/4}} = -\frac{\Theta'(0)}{2}$$

Before discussing numerical solutions to Eqs. (22)–(24), it is of interest to investigate the special case of $n = -0.6$, for which Eq. (22) reduces to

$$\frac{d^2\Theta}{d\eta^2} + \frac{12}{5} \Pr \frac{d}{d\eta} (f\Theta) = 0 \tag{25}$$

Upon integrating this equation once, and noting that $\Theta'(\infty) = 0$, then

$$\Theta'(0) = 0$$

Thus, even though the temperature difference is finite, free convection heat transfer along the plate surface is zero. For $n < -0.6$ one finds that the surface heat transfer is negative.

A similar result occurs for forced convection across a flat plate with a surface temperature variation described by Eq. (21). In this case negative heat transfer occurs for $n < -0.5$ [8]. In both situations an S-shaped temperature profile results when the heat transfer is negative. There appears to be a parallel between these two cases and the velocity boundary layer for flow about an arbitrary body. In this latter case, the boundary layer equations predict that the surface shear stress can, in the presence of an adverse pressure gradient, pass through zero and become negative, with the subsequent achievement of an S-shaped velocity profile. The standard and logical explanation here is that as the zero shear stress position is approached, the boundary layer equations become invalid, and beyond that location boundary layer flow no longer exists; that is, separated flow occurs.

It appears that a similar argument applies to the thermal boundary layer. When $n < -0.6$ for free convection, it may easily be shown that the condition $\partial T/\partial x \ll \partial T/\partial y$ cannot be fulfilled across the entire boundary layer. Since the satisfaction of this condition is a requirement for the validity of the thermal boundary layer equation, then evidently boundary layer heat transfer does not exist under the above stated condition.

TABLE 3. Nusselt Number Results for
$$T_w(x) - T_\infty = Nx^n$$

n	$\mathrm{Nu}/(\mathrm{PrGr})^{1/4}$	
	$\Pr = 0.7$	$\Pr = 1.0$
3	0.749	0.770
2	0.676	0.698
1	0.575	0.597
0.5	0.502	0.521
0.2	0.440	0.457
0	0.386	0.401
-0.2	0.313	0.325
-0.5	0.120	0.128

Local Nusselt number results, as obtained from the numerical solution of Eqs. (22) and (23) by Sparrow and Gregg [7], are listed in Table 3 for Prandtl number of 0.7 and 1.0. These results may additionally be interpreted as a prescribed variation of surface heat flux $q''(x)$. For example, from Eqs. (18) and (21) it follows that

$$q''(x) = -kNc\,\Theta'(0)x^{(5n-1)/4}$$

Thus, if the heat flux is prescribed as being proportional to x^r, then the results of Table 3 apply with

$$n = \frac{4r + 1}{5}$$

The important case of a uniform surface heat flux ($r = 0$, $n = 1/5$) has been treated in more detail by Sparrow and Gregg [9]. Letting Nu_q and Nu_T denote the local Nusselt numbers for a uniform surface heat flux and a uniform surface temperature, respectively, then Table 4 illustrates the effect of Prandtl number upon these two boundary conditions. It is seen that the influence of Prandtl number upon the ratio $\mathrm{Nu}_q/\mathrm{Nu}_T$ is quite small.

TABLE 4. Nusselt Number Results for a Uniform Surface Heat Flux and a Uniform Surface Temperature

Pr	$\mathrm{Nu}_q/(\mathrm{Gr})^{1/4}$	$\mathrm{Nu}_T/(\mathrm{Gr})^{1/4}$	$\mathrm{Nu}_q/\mathrm{Nu}_T$
0.1	0.189	0.164	1.15
1	0.457	0.401	1.14
10	0.933	0.826	1.13
100	1.71	1.55	1.12

Blowing and Suction. As in forced convection [8], there are several physical situations in free convection that can be formulated in terms of a blowing or suction boundary condition at the plate surface. Such problems, for example, involve blowing or suction through a porous surface, and free convection along a subliming surface. In the present section, attention will be primarily directed to the effect of an imposed blowing or suction velocity at the plate surface. Again referring to forced convection [8], other types of surface mass transfer problems may often be formulated in terms of this basic solution. Constant fluid properties and an isothermal plate surface will be assumed.

With blowing or suction at the plate surface, the impermeable surface boundary condition $v_y = 0$ for $y = 0$ is replaced by $v_y = v_w$, where v_w is the blowing or suction velocity (positive for blowing and negative for suction). Consideration will first be given to the case for which v_w is uniform over the surface. This results in a nonsimilarity boundary layer problem, and a second objective of this section is to illustrate methods of solving for such nonsimilarity problems.

For surface blowing or suction, the boundary layer equations described by Eqs. (13) and (14), as well as the matching conditions of Eq. (12), remain unchanged, while at the surface the boundary conditions are

$$\frac{\partial \psi}{\partial y} = 0 , \quad \frac{\partial \psi}{\partial x} = -v_w , \quad T = T_w ; \quad y = 0 \tag{26}$$

Recalling that v_w is constant, and defining the dimensionless quantities

$$\eta = c \frac{y}{x^{1/4}} \qquad \qquad \xi = \frac{v_w x^{1/4}}{4\nu c}$$

$$f(\xi, \eta) = \frac{\psi(x, y)}{4\nu c x^{3/4}} \qquad \Theta(\xi, \eta) = \frac{T - T_\infty}{T_w - T_\infty}$$

where c is defined in the expression preceding Eq. (15), then Eqs. (13) and (14) transform to

$$\frac{\partial^3 f}{\partial \eta^3} + 3f \frac{\partial^2 f}{\partial \eta^2} - 2\left(\frac{\partial f}{\partial \eta}\right)^2 + \xi \frac{\partial f}{\partial \xi} \frac{\partial^2 f}{\partial \eta^2} - \xi \frac{\partial f}{\partial \eta} \frac{\partial^2 f}{\partial \xi \partial \eta} = -\Theta \tag{27}$$

$$\frac{1}{\text{Pr}} \frac{\partial^2 \Theta}{\partial \eta^2} + 3f \frac{\partial \Theta}{\partial \eta} + \xi \frac{\partial f}{\partial \xi} \frac{\partial \Theta}{\partial \eta} - \xi \frac{\partial f}{\partial \eta} \frac{\partial \Theta}{\partial \xi} = 0 \tag{28}$$

while the boundary conditions, Eqs. (12) and (26), yield

$$\frac{\partial f}{\partial \eta} = 0 , \quad \frac{3}{4} f + \frac{\xi}{4} \frac{\partial f}{\partial \xi} = 0 , \quad \Theta = 1 ; \quad \eta = 0 \tag{29a}$$

$$\frac{\partial f}{\partial \eta} = 0 , \quad \Theta = 0 ; \quad \eta \to \infty \tag{29b}$$

Then, for $v_w = 0$, $\xi = 0$, and it is clear that Eqs. (27)–(29) reduce to the similarity problem described by Eqs. (15)–(17). For blowing or suction, however, the above system of equations cannot be transformed to ordinary differential equations, such that a nonsimilarity analysis must be employed. In the present section three methods of solving Eqs. (27)–(29), which tend to complement one another, will be illustrated.

The first method is the rather conventional expansion about a known similarity solution, having the form

$$f(\xi, \eta) = f_0(\eta) + f_1(\eta)\xi + \cdots \tag{30}$$

$$\Theta(\xi, \eta) = \Theta_0(\eta) + \Theta_1(\eta)\xi + \cdots \tag{31}$$

Upon substituting these expressions into Eqs. (27) and (28) and collecting like powers of ξ, ordinary differential equations are obtained describing the functions f_0, f_1, Θ_0, and Θ_1. Since $\xi = 0$ corresponds to an impermeable plate, it is obvious that Eqs. (30) and (31) constitute a first-order expansion about the case of $v_w = 0$, and that the equations describing f_0 and Θ_0 are simply Eqs. (15) and (16). Numerical solutions of the equations describing f_1 and Θ_1 have been obtained by Sparrow and Cess [10] for Pr = 0.72, and the resulting expression for the Nusselt number is

$$\frac{\text{Nu}}{(\text{Gr})^{1/4}} = 0.357 - 0.990\xi + \cdots \tag{32}$$

Since blowing (positive ξ) tends to thicken the boundary layer, this results in the decrease in heat transfer relative to that for no blowing ($\xi = 0$) illustrated above. The reverse explanation, of course, applies for suction (negative ξ).

A second solution to Eqs. (27)–(29) consists of an asymptotic solution for large negative ξ (large suction). This may be obtained by transforming Eqs. (27) and (28) from the independent variables ξ and η to ξ and τ, where

$$\tau = -\frac{v_w}{\nu} y = -4\xi\eta$$

From this transformed system of equations, it may readily be shown that asymptotic solutions exist of the form

$$f(\xi, \tau) = -\xi + 0\left(\frac{1}{\xi^3}\right) \tag{33}$$

$$\Theta(\xi, \tau) = e^{-Pr\tau} + 0\left(\frac{1}{\xi^8}\right) \tag{34}$$

It in turn follows that

$$\frac{Nu}{(Gr)^{1/4}} = -2\sqrt{2}\,Pr\,\xi + 0\left(\frac{1}{\xi^7}\right) \tag{35}$$

is the asymptotic solution for large negative values of ξ.

An additional solution to Eqs. (27) and (28) involves the approximation of local similarity. Characteristically, boundary layers have "poor memories", such that conditions at a given location along the plate surface are not severely influenced by conditions upstream. Specifically, the assumption of local similarity implies that the temperature and velocity profiles within the boundary layer vary slowly with ξ, such that ξ takes on the role of a parameter rather than a variable. The assumption of local similarity is thus expressed as

$$f(\xi, \eta) \simeq f(\eta)$$
$$\Theta(\xi, \eta) \simeq \Theta(\eta)$$

and Eqs. (27) and (28) reduce to

$$f''' + 3ff'' - 2(f')^2 = -\Theta \tag{36}$$

$$\frac{1}{Pr}\Theta'' + 3f\Theta' = 0 \tag{37}$$

with the boundary conditions

$$f' = 0, \quad f = -\frac{4\xi}{3}, \quad \Theta = 1; \quad \eta = 0 \tag{38a}$$

$$f' = 0, \quad \Theta = 0; \quad \eta \to \infty \tag{38b}$$

The Nusselt number is again given by

$$\frac{Nu}{(Gr)^{1/4}} = -\frac{\Theta'(0)}{2}$$

Equations (37)–(38) have been solved numerically, although in a somewhat different context, by Eichhorn [11] for a large range of values of ξ and $Pr = 0.73$, which differs only slightly from $Pr = 0.72$ for Eq. (32). These results, as well as the small ξ solution of Eq. (32) and the large negative ξ result given by Eq. (35), are illustrated in Fig. 3. It may be seen that for suction (negative ξ) the local similarity solution constitutes a logical joining of the limiting small ξ and large ξ solutions.

It is worth mentioning that both the asymptotic solution for large negative ξ and the local similarity solution can be derived without requiring that v_w be

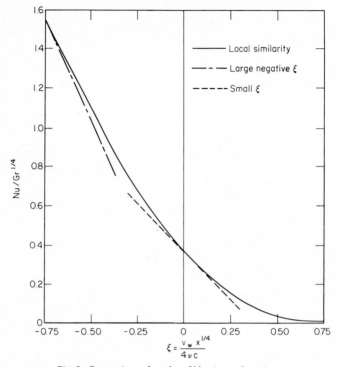

Fig. 3. Comparison of results of blowing and suction.

uniform.* For the specific case in which v_w varies as $x^{-1/4}$, ξ = constant, and Eqs. (36)–(38) constitute an exact representation to the blowing-suction problem. This is, in fact, the case considered by Eichhorn [11]. Since the local similarity solution is exact for this particular dependence of v_w with x, while it appears to be a reasonably good approximation for v_w = constant, then it would seem that the local similarity solution should be a useful approximation for an arbitrary variation of v_w with x.

Variable Fluid Properties. The isothermal vertical plate will again be utilized in order to illustrate the effect of variable fluid properties upon free convection heat transfer. As previously discussed, for perfect gases $\beta = 1/T_\infty$, so that the temperature dependency of the coefficient of thermal expansion does not influence free convection heat transfer for a perfect gas. In addition, together with the perfect gas equation of state, a first approximation to the temperature dependence of μ, k, and c_p may, for gases, be written as

$$\mu \sim T \ , \ k \sim T \ , \ c_p = \text{constant} \tag{39}$$

It follows that the Prandtl number

$$\text{Pr} = \frac{\nu}{\alpha} = \frac{\mu c_p}{k}$$

is correspondingly independent of temperature. As illustrated by Sparrow and Gregg [12], free convection heat transfer for a perfect gas which obeys Eq. (39) coincides

---

*For example, Eichhorn's derivation [11] of the asymptotic solution for large negative ξ applies for any variation of v_w with x.

with that for a constant property fluid. Thus, results of the previous sections, with the exception of blowing and suction, apply directly to the present variable property problem. A similar situation holds for forced convection over a flat plate.

A second type of gas considered by Sparrow and Gregg [12] consisted of a perfect gas which obeyed

$$\mu \sim T^{3/4} \quad , \quad k \sim T^{3/4} \quad , \quad c_p = \text{constant} \tag{40}$$

Again, the Prandtl number is constant. The variation of μ and k for monatomic and diatomic gases generally lies somewhere between Eqs. (39) and (40). For the property variations described by Eq. (40), Sparrow and Gregg have obtained heat transfer results for an isothermal and impermeable plate, and these results are listed in the first column in Table 5, where the subscript w denotes that the physical properties in Nu and Gr are evaluated at the wall temperature. For a gas represented by Eq. (39), the corresponding value would be independent of T_w/T_∞ and equal to 0.353.

TABLE 5. Heat Transfer Results for Variable Physical Properties and $\Pr = 0.7$

T_w/T_∞	$\mathrm{Nu}_w/(\mathrm{Gr}_w)^{1/4}$, exact	$\mathrm{Nu}_w/(\mathrm{Gr}_w)^{1/4}$, Eq. (42)
4	0.371	0.368
2	0.363	0.362
½	0.339	0.339
¼	0.323	0.321

Sparrow and Gregg [12] have further suggested a reference temperature approach, whereby a variable property problem may be treated as a constant property one with physical properties (except β, since $\beta = 1/T_\infty$ for a perfect gas) evaluated at the reference temperature

$$T_r = T_w - 0.38(T_w - T_\infty) \tag{41}$$

As an example, for the property variations described by Eq. (40) and $\Pr = 0.7$, it readily follows that

$$\frac{\mathrm{Nu}_w}{(\mathrm{Gr}_w)^{1/4}} = 0.353 \left[1 - 0.38 \left(1 - \frac{T_w}{T_\infty} \right) \right]^{-1/8} \tag{42}$$

Heat transfer results predicted by this equation are also listed in Table 5, and it is seen that the reference temperature approach is quite accurate for this case.

Additional property variations have been considered by Sparrow and Gregg [12] representative of both gases and liquid metals. Again the reference temperature, as defined by Eq. (41), proved to be a useful means of treating variable property problems. In the case of liquid metals, β was evaluated at the reference temperature.

TURBULENT FLOW: VERTICAL FLAT PLATE

There are two restrictions which must be imposed upon the preceding laminar boundary layer analyses. The first pertains to Grashof numbers which are smaller than that for which the boundary layer equations are valid. Recall that the free convection boundary layer equations are asymptotic equations for large Grashof number. A very crude rule of thumb is that boundary layer theory becomes invalid for $PrGr < 10^4$. The second restriction involves the transition from laminar to turbulent flow. As in forced convection, there is no neat and simple means of predicting this transition. Again resorting to a very crude rule of thumb, a turbulent boundary layer is often assumed to occur for $PrGr > 10^9$. In the present section the turbulent boundary layer on a vertical flat plate will be discussed.

An analysis of turbulent free convection on an isothermal vertical plate has been performed by Eckert and Jackson [13]. This involved use of the integral method, by which the equations expressing conservation of momentum and energy are integrated across the boundary layer, and assumed velocity and temperature profiles are employed in order to evaluate the boundary layer thickness. For turbulent flow it is necessary to employ additional information in the integral method formulation, and Eckert and Jackson [13] chose surface shear stress and heat flux expressions which are commonly utilized in forced convection. Their result for turbulent free convection from an isothermal plate is

$$Nu = 0.0295 (Pr)^{7/15}(Gr)^{2/5}(1 + 0.494 Pr^{2/3})^{-2/5} \tag{43}$$

Employing the same type of analysis, Siegal [14] has treated the case of a uniform surface heat flux, for which

$$Nu = 0.0293 (Pr)^{7/15}(Gr)^{2/5}(1 + 0.444 Pr^{2/3})^{-2/5} \tag{44}$$

The assumptions which have been employed in arriving at Eqs. (43) and (44) preclude application to either large or small Prandtl numbers, and their main applicability is to gases. For air ($Pr = 0.72$), Eqs. (43) and (44) both yield the result

$$Nu = 0.025 (PrGr)^{2/5} \tag{45}$$

This insensitivity of the Nusselt number upon surface boundary conditions for turbulent flow is also characteristic of forced convection.

Several empirical correlations have been proposed for turbulent free convection from a vertical plate. Fishenden and Saunders [15] suggest

$$Nu = 0.12 (PrGr)^{1/3} \tag{46}$$

for gases. It may easily be verified that the 1/3 exponent results in the coexistence of a uniform surface temperature and a uniform surface heat flux. For $PrGr = 10^{10}$, Eq. (46) yields a Nusselt number which is 24% above that predicted by Eq. (45), while for $PrGr = 10^{12}$ this difference is -9%. Existing experimental data are not sufficiently accurate to indicate which of the two expressions is preferable.

OTHER GEOMETRIES

A limited number of empirical correlations are available for geometries other than the vertical flat plate. These correlations pertain to isothermal surfaces, and they are formulated in terms of the total heat transfer from the surface. Denoting the total heat transferred by q, an overall Nusselt number $\overline{Nu}$ is defined as

$$\overline{Nu} = \frac{qL}{kA(T_w - T_\infty)}$$

where L is a characteristic dimension of the geometry, and A is the surface area. The overall Nusselt number $\overline{Nu}$ is, in turn, correlated in terms of a Grashof number based upon this characteristic dimension; that is

$$Gr = \frac{g\beta(T_w - T_\infty)L^3}{\nu^2}$$

For example, with reference to the isothermal plate, since

$$\frac{q}{A} = \frac{1}{L}\int_0^L q''(x)dx$$

for a unit width, it follows that for laminar flow $\overline{Nu} = (4/3)Nu$, where Nu is the local Nusselt number at $x = L$. For turbulent flow, use of Eq. (45) yields $\overline{Nu} = (5/6)Nu$, while from Eq. (46) $\overline{Nu} = Nu$.

Existing empirical correlations for $\overline{Nu}$, applicable to geometries other than the vertical plate, are expressed in forms similar to those previously discussed for the plate. For laminar boundary layer flow ($10^4 < PrGr < 10^9$), the overall Nusselt number is expressed by

$$\overline{Nu} = c_1(PrGr)^{1/4} \tag{47}$$

where c_1 is a function of geometry and Prandtl number. For the limited Prandtl number range of gases, c_1 may be regarded as independent of Pr, and values of c_1 for several geometries are listed in Table 6.

TABLE 6. Values of c_1 for Laminar Free Convection of Gases

Geometry	c_1	Reference
Horizontal cylinder, L = diameter	0.47	[15]
Sphere, L = diameter	0.49	[16]
Horizontal plate facing upward (downward if cooled), L = width	0.54	[15]
Horizontal plate facing downward (upward if cooled), L = width	0.27	[17]

For turbulent free convection ($PrGr > 10^9$), the empirical correlation is of the form

$$\overline{Nu} = c_2(PrGr)^{1/3} \tag{48}$$

Again with application to the Prandtl number range of gases, values of c_2 are listed in Table 7.

TABLE 7. Values of c_2 for Turbulent Free
Convection of Gases

Geometry	c_2	Reference
Horizontal cylinder, L = diameter	0.10	[15]
Horizontal plate facing upward (downward if cooled), L = width	0.14	[15]

REFERENCES

1. S. Ostrach, Laminar Flows with Body Forces, in "High Speed Aerodynamics and Jet Propulsion," vol. 4, "Theory of Laminar Flows," sec. F, Princeton University Press, Princeton, N.J., 1964.
2. A. J. Ede, Advances in Free Convection, in "Advances in Heat Transfer," vol. 4, p. 1, Academic Press, New York, 1967.
3. M. VanDyke, "Perturbation Methods in Fluid Mechanics," Academic Press, New York, 1964.
4. E. J. LeFevre, Laminar Free Convection from a Vertical Plane Surface, *Ninth International Congress of Applied Mechanics, Brussels, Belgium,* Paper 1-168, 1956.
5. S. Ostrach, An Analysis of Free-Convection Flow and Heat Transfer about a Flat Plate Parallel to the Direction of the Generating Body Force, *NACA Report* 1111, 1953.
6. E. M. Sparrow and J. L. Gregg, Low Prandtl Number Free Convection, *Z. agnew. Math. u. Phys.,* 9:383 (1958).
7. E. M. Sparrow and J. L. Gregg, Similar Solutions for Free Convection from a Nonisothermal Vertical Plate, *Trans. Am. Soc. Mech. Engrs.,* 80:379 (1958).
8. E. R. G. Eckert and R. M. Drake, Jr., "Heat and Mass Transfer," McGraw-Hill Book Company, New York, 1959.
9. E. M. Sparrow and J. L. Gregg, Laminar Free Convection from a Vertical Plate with a Uniform Surface Heat Flux, *Trans. Am. Soc. Mech. Engrs.,* 78:435 (1956).
10. E. M. Sparrow and R. D. Cess, Free Convection with Blowing or Suction, *Jour. of Heat Transfer,* 83:387 (1961).
11. R. Eichhorn, The Effect of Mass Transfer on Free Convection, *Jour. of Heat Transfer,* 82:260 (1960).
12. E. M. Sparrow and J. L. Gregg, The Variable Fluid-Property Problem in Free Convection, *Trans. Am. Soc. Mech. Engrs.,* 80:879 (1958).
13. E. R. G. Eckert and T. W. Jackson, Analysis of Turbulent Free-Convection Boundary Layer on Flat Plate, *NACA Report* 1015, 1951.
14. R. Siegal, Analysis of Laminar and Turbulent Free Convection from a Smooth Vertical Plate with Uniform Heat Dissipation per Unit Surface Area, *G.E. Report* R54GL89, 1954.
15. M. Fishenden and O. A. Saunders, "An Introduction to Heat Transfer," Oxford University Press, London, 1950.
16. R. J. Bromham and Y. R. Mayhew, "Free Convection from a Sphere in Air," *International Jour. of Heat and Mass Transfer,* 5:83 (1962).
17. W. H. McAdams, "Heat Transmission," McGraw-Hill Book Company, New York, 1954.

Section **7**

Forced Convection, Internal Flow in Ducts

WILLIAM M. KAYS

Mechanical Engineering Department,
Stanford University, Stanford, California

H. C. PERKINS

Aerospace and Mechanical Engineering Department,
The University of Arizona, Tucson, Arizona

INTRODUCTION

This chapter, devoted to the fluid mechanics and flow inside of pipes and ducts, is organized according to duct geometry, i.e., circular tubes, annuli and parallel plates, and other noncircular geometries. The first part of each section deals with the hydrodynamics of the particular geometry, and the second part with the heat transfer.

The reader will first find information, for laminar and turbulent flow, on the friction factor, velocity profile, and pressure drop for both fully developed flow and flow in the entrance region. Heat transfer information, under laminar and turbulent flow conditions, as a function of the pertinent nondimensional numbers and thermal boundary conditions, then follows.

A variety of specialized topics is given in later sections, including topics on the effects of variable properties, rotating flows, unsteady state, non-Newtonian fluids, internal heat sources, frictional dissipation, and others. The important effects of superimposed natural convection, circumferential heat flux variations, and roughness and artificial turbulence promoters are treated as special problems under the circular tube section, since most of the available results deal with the circular tube case.

The authors have attempted to give complete information on topics which are well in hand so that reference to other literature is not required. On topics which are still under active investigation, information is provided to indicate when an effect is important and, when possible, to indicate the magnitude of the effect. Extensive references on these incomplete topics are given so that the reader may apprise himself of the work available.

An attempt has been made to use information from readily available journals insofar as possible; however, it was necessary to refer to research reports and other limited circulation publications to complete some sections.

The authors found some difficulty in using a few of the cited references particularly with regard to nomenclature, clarity of presentation, and meaning. Any difficulties in this chapter, are, however, the sole responsibility of the authors and they would appreciate learning of misprints, misinterpretations, and misinformation.

A. CIRCULAR CROSS SECTION

1. Hydrodynamics

a. Friction Coefficients. The Fanning friction factor f is defined in terms of the wall shear stress as

$$f = \frac{\tau_w}{\rho \bar{V}^2 / 2g_c} \tag{1}$$

where $\bar{V}$ is the *mean* velocity. For *laminar,* fully developed, constant property flow in a circular tube the friction factor is given by

$$f = \frac{16}{\mathrm{Re}} \tag{2}$$

where Re is the Reynolds number

$$\mathrm{Re} = \frac{\bar{V} D \rho}{\mu} = \frac{DG}{\mu}$$

Langhaar [1] has solved for the velocity development in the laminar hydrodynamic entry length. Friction coefficients from his solution are plotted in Fig. 1 in the form $(f\mathrm{Re}) = \phi\{\mathrm{Re}/(x/D)\}$.

Three friction coefficients are indicated, f, f_x, and $\bar{f}_{\mathrm{app}}$. f_x is termed the *local* friction coefficient, and is based on the actual local wall shear stress at x, through Eq. (1). In computing pressure drop in a tube, the integrated mean wall shear stress from $x = 0$ to the point of interest is of more utility then the local shear stress, and the f plotted is the mean friction coefficient from $x = 0$ to L. Part of the pressure drop in

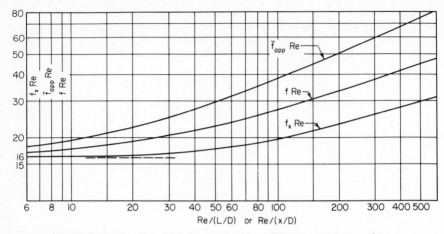

Fig. 1. Friction coefficients in the entrance of a circular tube with laminar flow.

the entrance region of a tube is attributable to an increase in the total fluid momentum flux, associated with the development of the velocity profile, and pressure drop calculations must take into consideration any variation in momentum flux as well as the effects of surface shear forces. The combined effects of surface shear and momentum flux have been incorporated in a single *apparent* mean friction coefficient $\bar{f}_{app}$. The pressure drop from 0 to x can then be evaluated from

$$\Delta P = \frac{4\bar{f}_{app}\, \rho \bar{V}^2}{2g_c}\, \frac{L}{D} \tag{3}$$

The approach of f_x towards $16/Re$ on Fig. 1 is a measure of the development of the velocity profile towards a fully developed profile.

For *turbulent* fully developed flow in a smooth pipe, the Fanning friction factor is given by the Karman-Nikuradse relation

$$\frac{1}{\sqrt{4f}} = -0.8 + 0.87\,\ln Re\sqrt{4f} \tag{4}$$

In the range $30{,}000 < Re < 1{,}000{,}000$, this equation is closely approximated by

$$f = 0.046\,Re^{-0.2} \tag{5}$$

The Blasius equation may be used in the range $5{,}000 < Re < 30{,}000$. This is given by

$$f = 0.079\,Re^{-0.25} \tag{6}$$

In a rough pipe, Knudsen and Katz [2] recommend

$$\frac{1}{\sqrt{f}} = 1.74\,\ln\frac{D}{e} + 2.28 \;;\quad \frac{r/e}{Re\sqrt{f}} > 0.005 \tag{7}$$

The complete friction factor plot is shown in Fig. 2. Moody [3] has presented the relative roughness for pipes of different materials. Figure 3 shows this.

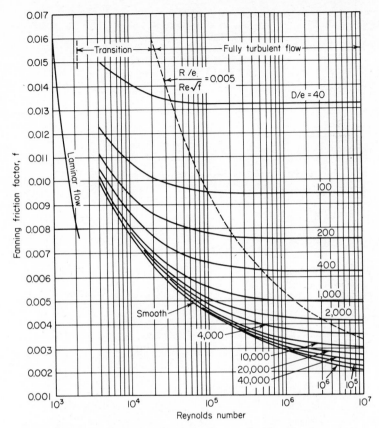

Fig. 2. Fanning friction factor plot for smooth and rough tubes.

Deissler [4] has presented results for the friction coefficient in the *turbulent entry length*. The following has been abstracted from his work. The results are based on solution of the momentum integral equation for flow in a circular tube, with uniform velocity at the entrance, and constant properties. Three friction factors are defined. The first is the *local* f_x based on local shear stress. The second is the *local* f_p based on static pressure gradient, i.e., it includes the change in momentum flux. The third is the integration of the second, and can be used to calculate the total pressure drop. Figure 4 shows these where

$$f_x = \frac{\tau_w}{\rho \bar{V}^2 / 2g_c} \tag{8}$$

where τ_w is the local shear stress

$$f_p = -\frac{D(dp/dx)}{2\rho \bar{V}^2} \tag{9}$$

$$\bar{f}_{app.} = \frac{\Delta P}{(x/D)2\rho \bar{V}^2} \tag{10}$$

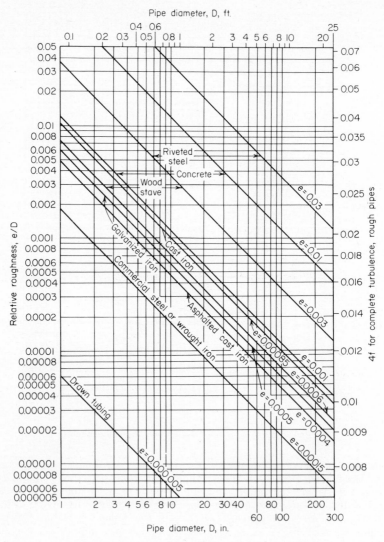

Fig. 3. Relative roughness of pipes, from Moody [3].

Latzko [5] has predicted the length required for the friction factor to become constant as

$$\frac{L}{D} = 0.623\,(\mathrm{Re})^{0.25} \tag{11}$$

which appears consistent with Deissler's results.

b. Velocity Distributions. For *fully developed laminar flow* in a tube the velocity profile is given by the parabolic relation

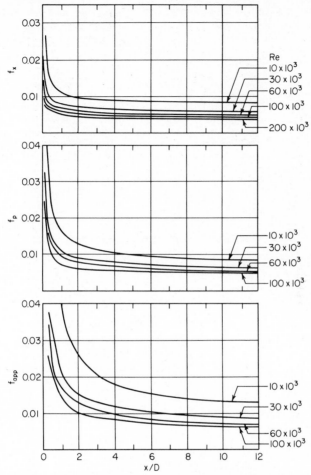

Fig. 4. Friction coefficients in the entrance of a circular tube with turbulent flow; f_x local friction factor based on wall shear; f_p local friction factor based on static pressure gradient; f_{app} average friction coefficient based on static pressure drop.

$$v_x = 2\overline{V} \left[1 - \left(\frac{r}{r_w} \right)^2 \right] \tag{12}$$

The *hydrodynamic entry length* for laminar flow has been solved by Langhaar [1]. Table 1 shows the velocity profile development as a function of r/r_w and $(x/D)/Re$ based on his results. This velocity profile development is plotted in Fig. 5. Langhaar found that for the centerline velocity to reach 0.99 of the downstream centerline velocity $L/D = 0.0575\ Re$.

Campbell and Slattery [6] have redone this work with some changes and include a comparison of various authors' efforts on this subject.

Fully Developed Turbulent Flow. At Reynolds numbers above about 2000, the laminar motion tends to become unstable and a transition to a *turbulent flow* will generally occur. The local time-mean velocity may then be expressed as

TABLE 1. Hydrodynamic Development for Laminar Flow in a Circular Tube, $v_x/\overline{V}$ versus $x/D/Re$. (Calculated from Langhaar [1])

r/r_w	$v_x/\overline{V}$									
	0.9	0.8	0.7	0.6	0.5	0.4	0.3	0.2	0.1	0.0
$(x/D)/Re$										
0.000	1.000	1.000	1.000	1.000	1.000	1.000	1.000	1.000	1.000	1.000
0.000175	0.960	1.087	1.098	1.103	1.104	1.104	1.104	1.104	1.104	1.104
0.000350	0.903	1.090	1.130	1.137	1.140	1.140	1.140	1.140	1.140	1.140
0.000700	0.814	1.068	1.151	1.179	1.189	1.191	1.193	1.193	1.193	1.193
0.00140	0.718	1.028	1.162	1.222	1.248	1.260	1.264	1.266	1.267	1.267
0.00350	0.609	0.952	1.152	1.265	1.330	1.368	1.390	1.403	1.409	1.412
0.00700	0.530	0.883	1.120	1.282	1.389	1.460	1.510	1.539	1.555	1.560
0.0140	0.462	0.813	1.088	1.290	1.442	1.556	1.637	1.690	1.721	1.735
0.0210	0.426	0.778	1.062	1.289	1.462	1.601	1.704	1.777	1.818	1.832
0.0280	0.410	0.760	1.042	1.283	1.477	1.631	1.748	1.828	1.876	1.892
0.0350	0.401	0.740	1.037	1.280	1.484	1.650	1.774	1.860	1.913	1.931
∞	0.380	0.720	1.020	1.280	1.500	1.680	1.820	1.920	1.990	2.000

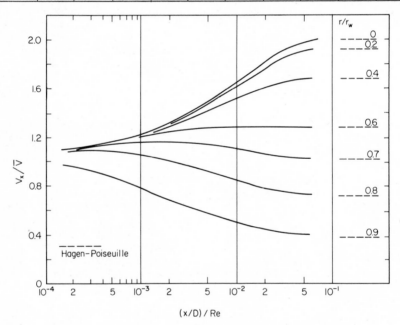

Fig. 5. Laminar velocity profiles in the entrance region of a circular tube.

$$v_x = f(y, \tau_w, \nu, \rho)$$

By application of dimensional analysis, this relation can be reduced to a function of two-dimensionless groups, as follows:

$$\frac{v_x}{\sqrt{g_c\tau_w/\rho}} = f\left[\frac{y\sqrt{g_c\tau_w/\rho}}{\nu}\right] \tag{13}$$

The term $\sqrt{g_c\tau_w/\rho}$ is sometimes spoken of as the *shear velocity*, since it has the dimensions of velocity. For brevity, the symbols u^+ and y^+ will be used for the two-dimensionless groups. Note that u^+ is simply a dimensionless velocity, and y^+ is a dimensionless distance from the wall. Thus

$$u^+ = \frac{v_x}{\sqrt{g_c \tau_w / \rho}} \quad ; \quad y^+ = \frac{y \sqrt{g_c \tau_w / \rho}}{\nu}$$

and then

$$u^+ = f(y^+)$$

Various investigators have described the turbulent velocity profile near the wall (the "law of the wall") in terms of one, two, and three separate regions. Martinelli [7] interpreted the extensive measurements of Nikuradse [8] in terms of three distinct regions. For $y^+ < 5$ the data appear to approach the relation $u^+ = y^+$ closely, and this was then termed the "laminar sublayer." For $y^+ > 30$ a logarithmic curve fits the data well and this region was termed the "turbulent core." An intermediate region was called the "buffer layer." The complete universal profile is then given by

$$y^+ < 5 \qquad u^+ = y^+ \tag{14}$$

$$5 < y^+ < 30 \qquad u^+ = -3.05 + 5.00 \ln y^+ \tag{15}$$

$$y^+ > 30 \qquad u^+ = 5.5 + 2.5 \ln y^+ \tag{16}$$

Other equations have also been used to describe the velocity profiles. Among these are those of Deissler [9] for the two regions

$$0 < y^+ < 26$$

$$y^+ = \sqrt{\frac{\tau_w}{2m}} \, \exp\left(\frac{1}{2} m u^{+2}\right) \mathrm{erf}\left(\sqrt{\frac{m}{2}} \, u^+\right) \tag{17}$$

where $m = 0.0119$, and

$$u^+ = 3.8 + 2.78 \ln y^+ \tag{18}$$

for $y^+ > 26$.

A single equation for the velocity profile has been proposed by Spalding [10] in the form

$$y^+ = u^+ + \frac{1}{E}\left\{ e^{ku^+} - 1 - ku^+ - \frac{(ku^+)^2}{2!} - \frac{(ku^+)^3}{3!} - \frac{(ku^+)^4}{4!} \right\} \tag{19}$$

where $k = 0.407$ and $E = 10.1$.

A plot of these velocity profiles is shown in Fig. 6. Data are from Deissler and from Laufer [11].

Turbulent Velocity Profiles in the Entrance Region. Knudsen and Katz [2] comment that no general relationship is available to predict the distance required for a fully developed turbulent velocity profile. In general, a greater distance is required for a rounded entrance than for a sharp-edged entrance because of the laminar

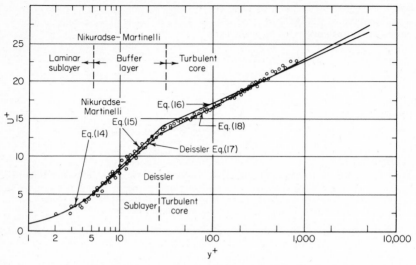

Fig. 6. Generalized velocity distribution for adiabatic turbulent flow.

boundary layer formation in the former. Barbin and Jones [12] have concluded for turbulent flow that the wall shear stress is developed by 15 diameters, in agreement with the results of Deissler in Sec. A.1.a; however, for their data taken with air, the velocity profile was still changing at an x/D of 40. Figure 7 presents this data for which transition was artificially promoted by gluing sand grains to the pipe wall some two inches ($x/D = 0.25$) downstream of the leading edge of the pipe. The pipe was located in the throat of a smooth 4:1 area contraction such that an annular space was located between the pipe and the contraction.

Transition Flow.   Below a Reynolds number of 2,200, flow in a circular tube is laminar; above Reynolds numbers of 10,000, the flow will normally be turbulent.

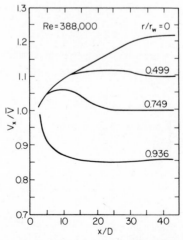

Fig. 7. Turbulent velocity profiles in the entrance region of a circular tube.

Between 2,200 and 10,000 the flow regime is termed transitional and the velocity profile is neither parabolic nor the usual turbulent profile.

Rotta [13] and Lindgren [14] have performed extensive experiments on flow through circular tubes at Reynolds numbers in and around the transition region. They found the transition from a laminar flow pattern to a turbulent flow at a given Reynolds number occurred continually as the axial distance increased. Other work has shown that transition is affected by the entrance configuration (Boelter et al. [15] and Mills [16]), the roughness of the tube wall (Schlichting [17]), the axial location in the tube (Rotta [13], Lindgren [14], Senecal and Rothfus [18]), natural convection effects when heating or cooling is taking place (Scheele et al. [19]), and the change of viscosity when large heating rates occur (Perkins and Worsoe-Schmidt [20]). Nondimensional u^+ versus y^+ profiles based on Senecal's work are given in McEligot et al. [21].

c. Pressure Drop. Pressure-drop information for flow through both single-tube and multiple-tube systems, such as flow through heat exchangers, is presented in this section. Information on the effects of valves, fittings, bends, etc., also of importance to the designer, is given. Noncircular geometries as well as circular tubes are treated here in order to present the information in a convenient and concise form. Special cases are also covered later under the appropriate section on the particular geometry.

The model for the pressure-drop analysis is presented in Fig. 8 in which an abrupt entrance and exit are shown. The equations to be presented apply to this model and are not restricted to a particular geometry. They are applicable to single-tube flow by reducing various terms.

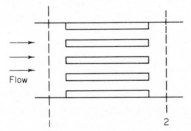

Fig. 8. Heat exchanger core model for pressure drop analysis. G is based on the minimum free flow area in the core.

The general relation for flow-stream pressure-drop calculations is

$$\frac{\Delta P}{P_{in}} = \frac{G^2}{2g_c} \frac{1}{\rho_{in} P_{in}} \left[(K_c + 1 - \sigma^2) + 2\left(\frac{\rho_1}{\rho_2} - 1\right) + 4f\frac{L}{D}\frac{\rho_1}{\bar{\rho}} - (1 - \sigma^2 + K_e)\frac{\rho_1}{\rho_2} \right] \quad (20)$$

where K_c and K_e represent entrance and exit loss coefficients, σ is the free flow/frontal area ratio ($\sigma = 1$ for a single tube), and $\bar{\rho}$ is given by

$$\frac{1}{\bar{\rho}} = \frac{1}{L} \int_0^L \frac{1}{\rho} dx$$

The terms represent in order: the entrance effect, the flow acceleration effect for a compressible fluid, the usual friction pressure drop, and the exit effect. The entrance effect is taken here to include both pressure losses due to the expansion of the fluid following a vena-contracta, and *changes in pressure due to the change in momentum as the velocity profile is established* in the smaller tubes. Consequently, the friction factor f is defined on the basis of the *mean wall shear;* i.e, see Fig. 1 for laminar flow in a circular tube.

Alternative methods involve the use of $\bar{f}_{app}$ in which the pressure drop due to momentum changes in the tube entrance is lumped together with both the increased wall shear stress in the entrance region and the usual downstream wall shear to establish the total pressure drop (exclusive of *exit* effects), *for an incompressible flow,* as

$$\Delta P = 4\bar{f}_{app} \frac{G^2}{2g_c\rho} \frac{x}{D} \tag{21}$$

Another commonly used method to account for entrance pressure drop, where the initial velocity profile is uniform, and there is no abrupt contraction, is to define a K which includes both the effect of momentum change and increased wall shear in the entrance region, but separates out the downstream fully developed flow pressure drop due to wall shear to give

$$\frac{\Delta P}{\frac{1}{2}\rho\bar{V}^2} = 4f \frac{L}{D} + K \tag{22}$$

where f is the friction factor for fully developed flow. This method is used in later sections of this chapter dealing with specific noncircular geometries.

Note that in the absence of an abrupt entrance and exit and for an incompressible flow, Eq. (20) reduces to the relation

$$\frac{\Delta P}{\frac{1}{2}\rho\bar{V}^2} = 4f \frac{L}{D} \tag{23}$$

For *laminar flow in a circular tube* the mean friction factor for use in Eq. (20) is available from Fig. 1 and the K_c and K_e values are given in Fig. 9. Alternatively, Eq. (22) may be used with $K = 1.33$. For *turbulent flow* the results of Deissler, treated in Sec. A.1.a, may be used with Eq. (21), or the small effects of increased wall shear in the entrance region may be neglected and Eq. (20) used with the usual fully developed downstream friction factor and the K_c values from Fig. 9 which include the effect of change in momentum flux.

Abrupt Contraction and Expansion Pressure Loss Coefficients. The typical heat transfer application involves a flow contraction at the entrance and a flow expansion at the exit. Often an abrupt (right angle) contraction and expansion are used, and these introduce flow-stream pressure drop. This problem has been experimentally and analytically treated by Kays [22]. The present section presents a brief summary of the pertinent conclusions, and includes information on the entrance and exit loss coefficients K_c and K_e.

Equation (20) may be reduced (by restricting considerations to the first term in the bracket) to investigate only the entrance pressure drop, which is made up of two parts. The first is the pressure drop which would occur due to flow area change alone, without friction. The second is the loss due to irreversible free expansion that follows the abrupt contraction. The entrance pressure drop may be expressed as

$$\frac{\Delta P_{ent}}{\rho} = \frac{\bar{V}^2}{2g_c}(1 - \sigma^2) + K_c \frac{\bar{V}^2}{2g_c} \tag{24}$$

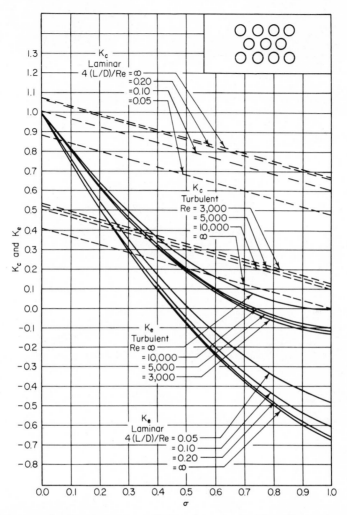

Fig. 9. Entrance and exit pressure loss coefficients for a circular tube with abrupt contraction entrance and abrupt expansion exit.

The irreversible component of the pressure drop is contained in the abrupt contraction, or entrance, coefficient K_c.

The exit *pressure rise* is similarly broken into two parts. The first is the pressure rise which occurs due to the area change alone, without friction, and is identical to the corresponding term in the entrance pressure drop. The second is the pressure loss associated with the irreversible free expansion following an abrupt expansion, and this term in the present case subtracts from the other. Thus

$$\frac{\Delta P_{\text{exit}}}{\rho} = \frac{\overline{V}^2}{2g_c}(1 - \sigma^2) - K_e \frac{\overline{V}^2}{2g_c} \tag{25}$$

where K_e is the abrupt expansion, or exit, coefficient.

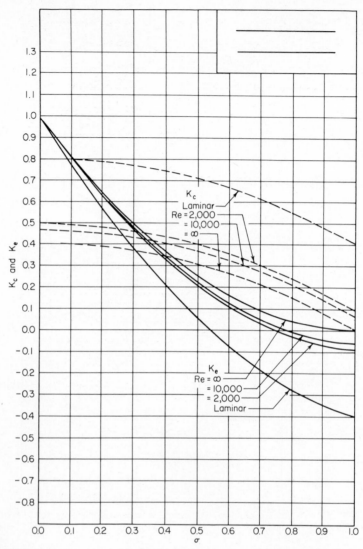

Fig. 10. Entrance and exit pressure coefficients for a flat duct with abrupt contraction entrance and abrupt expansion exit.

These coefficients have been established analytically for a number of simple entrance and exit geometries, and are presented graphically in Figs. 9 to 12.

The coefficients have been evaluated on the assumption of essentially uniform velocity in the large duct leading to the tube, or tubes, of uniform velocity in the leaving duct, but of a fully established velocity profile in the small tubes, with the exception that in the case of circular tubes partially developed flow is also included. The assumption that the velocity profile in the larger duct can be treated as uniform is generally justified, because in a multiple tube system the Reynolds number in the smaller tubes is invariably many times smaller than that in the entering and leaving

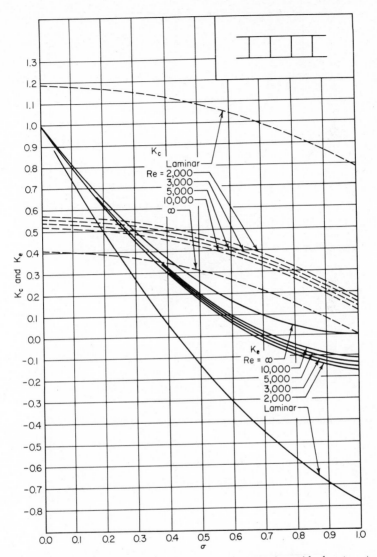

Fig. 11. Entrance and exit pressure loss coefficients for a square duct with abrupt contraction entrance and abrupt expansion exit.

ducts. It is only at low Reynolds numbers that velocity profile has any appreciable influence. In applying the data, however, it should be remembered that the coefficients already include pressure change associated with change in velocity profile. Thus as noted previously the friction factor used in Eq. (20) to evaluate the overall pressure drop should be defined on the basis of mean wall shear and not on the basis of overall pressure drop.

Since data cover only a few simple geometries, the designer must exercise judgment in the application of these results. In particular, the only case that is rigorously applicable to very short tubes is that of circular tubes. The others have been

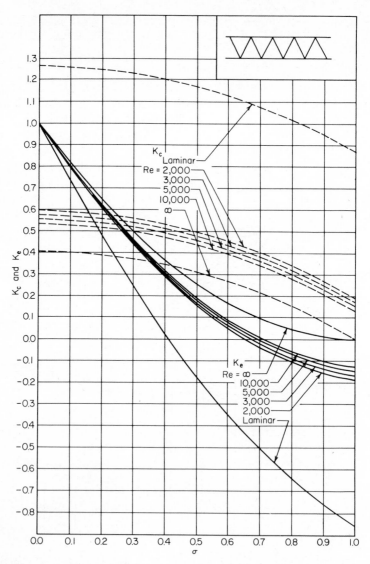

Fig. 12. Entrance and exit pressure loss coefficients for a triangular tube with abrupt contraction entrance and abrupt expansion exit.

worked out only for the case of tubes of sufficient length so that there is an essentially fully established velocity profile at the tube exit. A less than fully established velocity profile will result in entrance coefficients K_c that are lower, and exit coefficients K_e that are higher, than those obtained for the fully established condition.

Further information is available in Kays and London [23], the SAE Manual [24], and the articles by Benedict and co-workers [25].

Radius-bend Losses. The frictional mechanical energy loss due to 90-degree bends may be calculated according to the equation

$$\frac{\Delta P}{\rho} = K_b \frac{\overline{V}^2}{2g_c} \tag{26}$$

Figure 13 shows values of K_b as a function of bend geometry. Values are shown plotted against the ratio of bend radius (from center of bend to centerline of pipe) to inside pipe diameter.

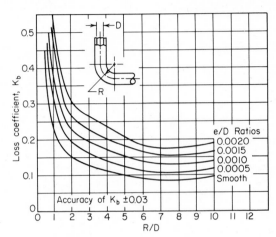

Fig. 13. Loss coefficients for 90° bends. (*Reprinted from "Standards of the Hydraulic Institute," 8th ed., Hydraulic Institute, New York, 1947.*)

Figure 14 shows the effect of bend angle on K_b at Re $= 2.25 \times 10^5$.

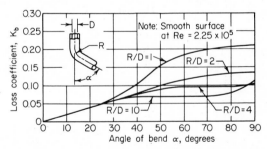

Fig. 14. Loss coefficients for bends versus angle of bend in smooth tubes. (*Reprinted from "Standards of the Hydraulic Institute," 8th ed., Hydraulic Institute, New York, 1947.*)

Miter-bend Losses. The frictional losses due to miter bends may be calculated by means of the following equation

$$\frac{\Delta P}{\rho} = K_s \frac{\overline{V}^2}{2g_c} \qquad \text{or} \qquad \frac{\Delta P}{\rho} = K_r \frac{\overline{V}^2}{2g_c} \tag{27}$$

Figure 15 gives the values of K_s and K_r as functions of the angle of the miter bend and for various combinations of bends connected by a straight pipe. The values are given for a Reynolds number of 2.25×10^6.

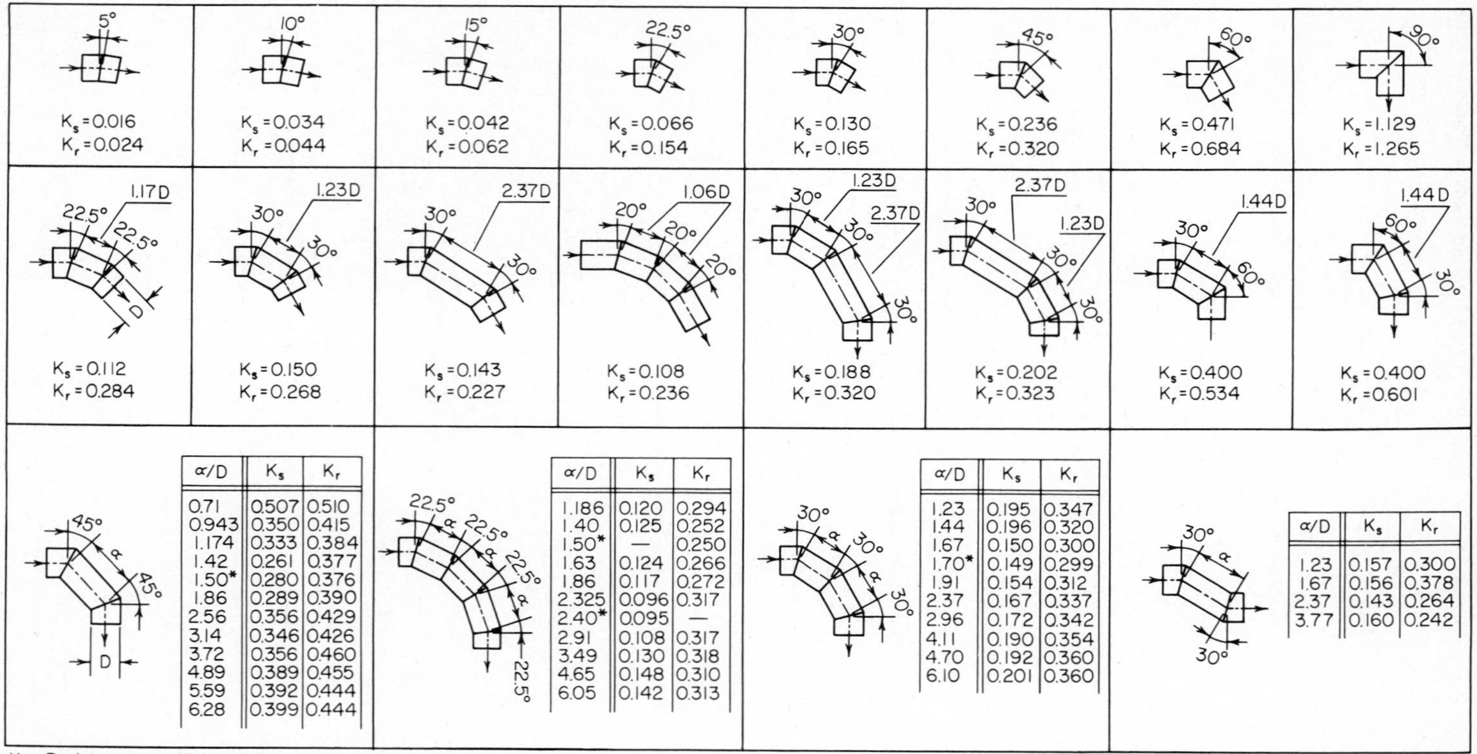

5°
K_s = 0.016
K_r = 0.024
10°
K_s = 0.034
K_r = 0.044
15°
K_s = 0.042
K_r = 0.062
22.5°
K_s = 0.066
K_r = 0.154
30°
K_s = 0.130
K_r = 0.165
45°
K_s = 0.236
K_r = 0.320
60°
K_s = 0.471
K_r = 0.684
90°
K_s = 1.129
K_r = 1.265
1.17D
22.5°
22.5°
D
K_s = 0.112
K_r = 0.284
1.23D
30°
30°
K_s = 0.150
K_r = 0.268
2.37D
30°
30°
K_s = 0.143
K_r = 0.227
1.06D
20°
20°
20°
K_s = 0.108
K_r = 0.236
1.23D
2.37D
30°
30°
30°
K_s = 0.188
K_r = 0.320
2.37D
1.23D
30°
30°
30°
K_s = 0.202
K_r = 0.323
1.44D
30°
60°
K_s = 0.400
K_r = 0.534
1.44D
60°
30°
K_s = 0.400
K_r = 0.601
45°
α
45°
D
22.5°
α
22.5°
α
22.5°
α
22.5°
30°
α
30°
α
30°
30°
α
30°

α/D	K_s	K_r
0.71	0.507	0.510
0.943	0.350	0.415
1.174	0.333	0.384
1.42	0.261	0.377
1.50*	0.280	0.376
1.86	0.289	0.390
2.56	0.356	0.429
3.14	0.346	0.426
3.72	0.356	0.460
4.89	0.389	0.455
5.59	0.392	0.444
6.28	0.399	0.444

α/D	K_s	K_r
1.186	0.120	0.294
1.40	0.125	0.252
1.50*	—	0.250
1.63	0.124	0.266
1.86	0.117	0.272
2.325	0.096	0.317
2.40*	0.095	—
2.91	0.108	0.317
3.49	0.130	0.318
4.65	0.148	0.310
6.05	0.142	0.313

α/D	K_s	K_r
1.23	0.195	0.347
1.44	0.196	0.320
1.67	0.150	0.300
1.70*	0.149	0.299
1.91	0.154	0.312
2.37	0.167	0.337
2.96	0.172	0.342
4.11	0.190	0.354
4.70	0.192	0.360
6.10	0.201	0.360

α/D	K_s	K_r
1.23	0.157	0.300
1.67	0.156	0.378
2.37	0.143	0.264
3.77	0.160	0.242

K_s = Resistance coefficient for smooth surface
K_r = Resistance coefficient for rough surface, e/D = 0.0022

*Optimum value of α interpolated

Fig. 15. Loss coefficients for Miter Bends of various angles and straight sections. (*Reprinted from "Standards of the Hydraulic Institute," 8th ed., Hydraulic Institute, New York, 1947.*)

Losses Due to Valves and Fittings.   The frictional pressure loss through valves and fittings can be calculated according to the equation

$$\frac{\Delta P}{\rho} = K_v \frac{\overline{V}^2}{2g_c} \qquad (28)$$

Values of K_v are given in Table 2.

TABLE 2. Pressure Drop for Turbulent Flow through Screwed Valves and Fittings

Type of valve or fitting	Pressure drop*	
	Equivalent L/D	Equivalent velocity head (K_v)
Couplings	2	0.04
Unions	2	0.04
Gate valves†		
Open	7	0.1
¾ open	40	0.8
½ open	200	4
¼ open	800	15
Globe valves,† bevel seat		
Open	350	6
½ open	550	10
Composition disk		
Open	330	6
½ open	500	9
Plug disk		
Open	500	9
¾ open	700	13
½ open	2000	35
¼ open	6000	110
Angle valves†		
Open	170	3
Y or blow-off valves†		
Open	170	3
Check valves‡		
Swing	110	2
Disk	500	10
Ball	3500	65
Water meters		
Disk	400	8

*Pressure drop is expressed as a straight pipe of same nominal size as the fitting. Where the pressure drop is expressed as L/D, L is the equivalent length (feet) of straight pipe and D is the inside diameter (feet). Where the pressure drop is expressed as velocity heads, the velocity in the pipe is based on the nominal diameter of the fitting.

†Flow direction through valves has negligible effect on pressure drop.

‡Values apply only when check valve is fully open, which generally is attained at pipe velocities over 3 ft/sec for water.

SOURCE: Reprinted from "Velocity Head Simplified Flow Computation," by C. E. Lapple, by permission from *Chem. Engr.*, 56, May 1949.

2. Heat Transfer

a. Laminar Flow.

Fully Developed Velocity and Temperature Profiles.   Heat transfer to (or from) a steady laminar flow of a constant property fluid in a circular tube has been handled analytically for a considerable number of boundary conditions. The simplest possible case is that of a fully developed velocity and temperature profile, which can occur at points far removed from the tube entrance. The fully developed temperature profile can exist for only certain particular temperature boundary conditions. Two of these boundary conditions of considerable technical importance are those of *constant surface temperature* everywhere, and *constant heat rate* per unit of tube length.

In either case, the heat transfer coefficient h can be expressed conveniently in terms of a nondimensional *Nusselt number*

$$\text{Nu} = \frac{hD}{k} \tag{29}$$

where $\bar{h}$ is the heat transfer coefficient, related to the heat flux at the wall by $q_w'' = h(T_w - T_b)$. T_w is the wall temperature; T_b is the bulk, or mixed-mean temperature; k is the fluid thermal conductivity; and D is the tube diameter.

A unique feature of the laminar flow, fully developed profile problem is that Nusselt number is a function only of the type of heating boundary condition. Thus:

Constant Surface Temperature

$$\text{Nu} = 3.656 \tag{30}$$

Constant Heat Rate

$$\text{Nu} = 4.364 \tag{31}$$

An alternative representation of these results makes use of a nondimensional *Stanton number* St rather than a *Nusselt number*. By simple dimensional reasoning, it may be shown that

$$\text{Nu} = \text{St} \, \text{Pr} \, \text{Re} \tag{32}$$

where $\text{St} = h/(\bar{V}\rho c_p)$, or $h/(Gc_p)$; $\text{Pr} = \mu c_p/k$, the Prandtl number; $\text{Re} = D\bar{V}\rho/\mu$, DG/μ, the Reynolds number. Then, for *constant surface temperature*

$$\text{St} \, \text{Pr} = \frac{3.656}{\text{Re}} \tag{33}$$

and for *constant heat rate*

$$\text{St} \, \text{Pr} = \frac{4.364}{\text{Re}} \tag{34}$$

The above results have been obtained neglecting *axial heat conduction, frictional heating effects, and buoyancy effects* which will be discussed in later sections.

Thermal Entry Length, Fully Developed Velocity Profile.* Solutions of more general applicability than given in the preceding section can be obtained for the case of a fully developed (parabolic) velocity profile throughout the tube, but with a uniform temperature profile at the tube entrance $(x = 0)$. Such solutions then include the region where the temperature profile is developing, the *thermal entry length.* For the particular cases of *constant surface temperature* (Graetz problem), and *constant*

---

*The corresponding solution for the case of a uniform velocity throughout the length of the tube ("slug-flow," a purely hypothetical case) will be found as a special case of a later discussion of the effects of axial conduction.

heat rate, the Nusselt number starts indefinitely high at $x = 0$, but approaches the values given in the preceding section as $x \rightarrow \infty$.

For *constant surface temperature,* the results of Sellars, Tribus, and Klein [26] can be expressed in an eigenvalue solution as follows

$$
\text{Nu}_x = \frac{\displaystyle\sum_{n=0}^{\infty} G_n e^{-\lambda_n^2 x^+}}{2 \displaystyle\sum_{n=0}^{\infty} \frac{G_n}{\lambda_n^2} e^{-\lambda_n^2 x^+}}
\tag{35}
$$

where $x^+ = (x/r_w)/\text{RePr}$ (nondimensional tube length) and λ_n^2 and G_n are eigenvalues and corresponding constants, given in Table 3.

TABLE 3. **Eigenvalues and Constants for Circular Tube, Constant Surface Temperature, Thermal Entry Length**

n	λ_n^2	G_n
0	7.312	0.749
1	44.62	0.544
2	113.8	0.463
3	215.2	0.414
4	348.5	0.382

for $n > 2$, $\lambda_n = 4n + 8/3$,
$G_n = 1.01276\, \lambda_n^{-1/3}$

The subscript x on the Nusselt number indicates that it, and thus the heat transfer coefficient, is a *local* value, as opposed to a *mean* Nusselt number and *mean* heat transfer coefficient. The latter, which is defined as

$$
\text{Nu}_m = \frac{1}{x} \int_0^x \text{Nu}_\xi \, d\xi
\tag{36}
$$

(ξ is a dummy length variable; thus $\text{Nu}_\xi = \text{Nu}_x$) is frequently useful in heat exchanger design. It is given by

$$
\text{Nu}_m = \frac{1}{2x^+} \ln \left[\frac{1}{8 \displaystyle\sum_{n=0}^{\infty} \frac{G_n}{\lambda_n^2} e^{-\lambda_n^2 x^+}} \right]
\tag{37}
$$

As the fully developed temperature profile condition is approached at large values of x^+, the series become increasingly more rapidly convergent, until ultimately only the first term, $n = 0$, is significant. Note then that, from Eq. (35)

$$
\text{Nu}_x = \text{Nu}_\infty = \frac{\lambda_0^2}{2}
\tag{38}
$$

In this case, since $\lambda_0^2 = 7.312$ from Table 3, $Nu_\infty = 3.656$, as previously given in Eq. (30). However, this relation will also be used later for different shaped tubes, and for turbulent flow.

The above discussed convergence behavior may also be used to determine the necessary length of pipe for a close approach to fully developed conditions. For $x^+ > 0.1$, the first term only may be used as an approximation, and thus $x^+ = (x/r_w)/(\mathrm{Re\,Pr}) = 0.1$ may be used as a criterion for fully developed conditions.

The eigenvalues and constants given in Table 3 can also be used to calculate directly the bulk mean fluid temperature (cup mixing, or mixed-mean temperature), and the local surface heat flux, as functions of distance from the tube entrance

$$\theta_b = 8 \sum_{n=0}^{\infty} \frac{G_n}{\lambda_n^2} e^{-\lambda_n^2 x^+} \tag{39}$$

where

$$\theta_b = \frac{T_w - T_b}{T_w - T_e}$$

T_e is the uniform entering fluid temperature

$$q'' = \frac{2k}{r_w} \sum_{n=0}^{\infty} G_n e^{-\lambda_n^2 x^+} (T_w - T_e) \tag{40}$$

Table 4 gives values of Nu_x and Nu_m evaluated from Eqs. (35) and (37). Figure 16 shows how Nu_x varies with x^+, while Fig. 17, as presented by Grigull and Tratz [27], shows graphically how the temperature field behaves.

TABLE 4. Nusselt Numbers, Constant Surface Temperature, Thermal Entry Length

x^+	Nu_x	Nu_m
0	∞	∞
0.001	12.86	22.96
0.004	7.91	12.59
0.01	5.99	8.99
0.04	4.18	5.87
0.08	3.79	4.89
0.10	3.71	4.66
0.20	3.66	4.16
∞	3.66	3.66

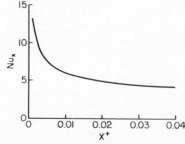

Fig. 16. Variation of local Nusselt number in the thermal entry region of a tube with constant surface temperature.

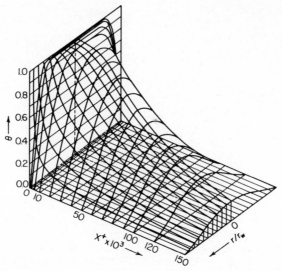

Fig. 17. Graphical presentation of the temperature field in the thermal entry region of a tube with constant surface temperature.

For *constant heat rate* per unit of tube length, the results of Siegel, Sparrow, and Hallman [28] can be expressed in an eigenvalue solution as follows

$$\mathrm{Nu}_x = \frac{1}{\dfrac{1}{\mathrm{Nu}_\infty} - \dfrac{1}{2} \displaystyle\sum_1^\infty \dfrac{e^{-\gamma_m^2 x^+}}{A_m \gamma_m^4}} \tag{41}$$

Nu_∞ is the Nusselt number for fully developed conditions, in this case 4.364. The eigenvalues and constants are given in Table 5. Figure 18 shows how the bulk mean fluid temperature T_b and the wall temperature T_w vary under these conditions. Table 6 gives values of Nu_x evaluated from Eq. (41).

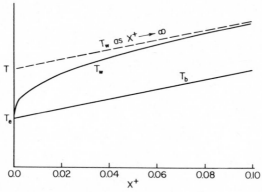

Fig. 18. Temperature variations in the thermal entry region of a tube with constant heat rate per unit of tube length.

TABLE 5. Eigenvalues and Constants for Circular Tube, Constant Heat Rate, Thermal Entry Length

m	γ_m^2	A_m
1	25.68	7.630×10^{-3}
2	83.86	2.058×10^{-3}
3	174.2	0.901×10^{-3}
4	296.5	0.487×10^{-3}
5	450.9	0.297×10^{-3}

for $m > 5$, $\gamma_m = 4m + 4/3$

$\quad A_m = 0.358 \, \gamma_m^{-2.32}$

TABLE 6. Nusselt Numbers, Constant Heat Rate, Thermal Entry Length

x^+	Nu_x
0	∞
0.002	12.00
0.004	9.93
0.010	7.49
0.020	6.14
0.040	5.19
0.100	4.51
∞	4.36

Combined Thermal and Hydrodynamic Entry Length. If *both* the velocity and temperature are uniform at the tube entrance, the velocity and temperature profiles develop simultaneously, resulting in Nusselt numbers that are always higher than in the preceding case. This is the case that is more frequently met in technical applications, but the differences are generally only significant for a fluid with Prandtl number less than about 10.

For *constant surface temperature,* a very approximate numerical solution for $Pr = 0.7$ was presented by Kays [29], and was extended by Goldberg [30] to a range of Prandtl numbers. Their results for the *mean* Nusselt number, Nu_m, are shown on Fig. 19.

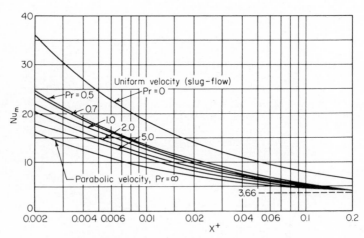

Fig. 19. Variation of the mean Nusselt number in the combined thermal and hydrodynamic entry region of a tube with constant surface temperature.

A more accurate solution has been obtained by Ulrichson and Schmitz [31], but only for $Pr = 0.7$, and the *local,* rather than the mean, Nusselt number is given. These results are shown on Fig. 20.

For *constant heat rate* per unit of tube length, the *local* results of Heaton, Reynolds, and Kays [32] are believed to be quite accurate, and are presented in Table 7 and Fig. 21.

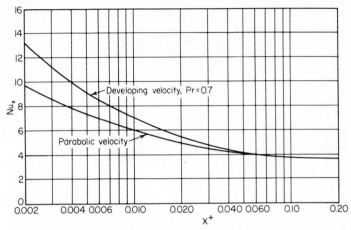

Fig. 20. Variation of the local Nusselt number in the combined thermal and hydrodynamic region of a tube with constant surface temperature for Pr = 0.7.

TABLE 7. Nusselt Numbers, Constant Heat Rate, Combined Thermal and Hydrodynamic Entry Length

Pr	x^+	Nu_x
0.01	0.002	24.2
	0.010	12.0
	0.020	9.10
	0.10	6.08
	0.20	5.73
	∞	4.36
0.70	0.0002	51.9
	0.002	17.8
	0.010	9.12
	0.020	7.14
	0.10	4.72
	0.20	4.41
	∞	4.36
10.0	0.0002	39.1
	0.002	14.3
	0.010	7.87
	0.020	6.32
	0.10	4.51
	0.20	4.38
	∞	4.36

Effects of Axial Conduction.   In all of the preceding solutions, the *conduction* of heat in the axial direction has been neglected, i.e., it has been assumed that conduction in the axial direction is negligible relative to the axial energy transport by the bulk movement of the fluid. The significant nondimensional parameter in considering the influence of axial conduction is the Peclet Number Pe which is the product of the Reynolds and the Prandtl numbers.

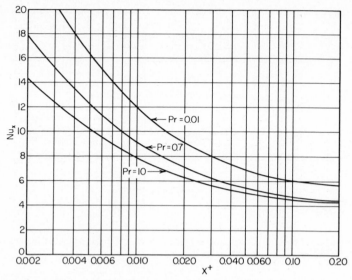

Fig. 21. Variation of the local Nusselt number in the combined thermal and hydrodynamic entry region of a tube with constant heat rate per unit of length.

Both Schneider [33] and Singh [34] have suggested, on the basis of analytical studies, that for Pe > 100 axial conduction effects can be neglected; experimental results with liquid metals appear to confirm this conclusion, although a slightly higher Peclet number of 160 seems indicated.

The results of Singh are based on a thermal entry length solution with constant surface temperature and a uniform fluid temperature at the tube entrance, but with a uniform velocity ("slug-flow") *throughout* the tube. At very low Prandtl numbers (liquid metals), the temperature profile develops very much more rapidly than the velocity profile, so the "slug-flow" solution is frequently a useful approximation. Actually turbulent flow of a very low Prandtl number fluid also approaches this condition.

Singh's solution contains in addition some results for *uniform internal heat generation.* The local Nusselt number, and the bulk mean temperature may be evaluated from the following equations

$$
\text{Nu}_x = \frac{\displaystyle\sum_{n=0}^{\infty}\left(1 - \frac{Q}{\alpha_n^2}\right) e^{-\lambda_n^2 x^+} + Q}{\displaystyle\sum_{n=0}^{\infty}\left(1 - \frac{Q}{\alpha_n^2}\right)\frac{e^{-\lambda_n^2 x^+}}{\alpha_n^2} + \frac{Q}{8}}
\tag{42}
$$

$$
\theta_b = \sum_{n=0}^{\infty} \frac{4\left(1 - \dfrac{Q}{\alpha_n^2}\right)}{\alpha_n^2} e^{-\lambda_n^2 x^+} + \frac{Q}{8}
\tag{43}
$$

where

$$
Q = \frac{q'''r_w^2}{k(T_e - T_w)}
$$

and λ_n^2 and α_n^2 are tabulated in Table 8.

**TABLE 8. Eigenvalues and Constants for Circular Tube, Constant Surface Tempera-
ture, Slug Flow, Including Axial Conduction and Uniform Internal Heat
Generation**

Pe = RePr	λ_0^2	α_0^2	λ_1^2	α_1^2	λ_2^2	α_2^2
1	1.956	5.783	5.043	30.47	8.168	74.89
5	4.844	5.783	17.80	30.47	32.54	74.89
10	5.483	5.783	24.48	30.47	49.94	74.89
50	5.770	5.783	30.11	30.47	72.77	74.89
100	5.780	5.783	30.38	30.47	74.33	74.89
1000	5.783	5.783	30.47	30.47	74.88	74.89
∞	5.783	5.783	30.47	30.47	74.89	74.89
	λ_3^2	α_3^2	λ_4^2	α_4^2	λ_5^2	α_5^2
1	11.30	139.0	14.44	222.9	17.58	326.6
5	47.77	139.0	63.19	222.9	78.72	326.6
10	78.08	139.0	107.5	222.9	137.5	326.6
50	132.1	139.0	206.0	222.9	292.4	326.6
100	137.2	139.0	218.2	222.9	316.5	326.6
1000	139.0	139.0	222.9	222.9	326.5	326.6
∞	139.0	139.0	222.9	222.9	326.6	326.6

b. Transition Flow. The reader is referred to Sec. A.1.b, on the hydrodynamics of transition flow, for information on the variables affecting transition. Heat transfer results are uncertain in the transition region because of the large number of parameters which determine when and how transition occurs.

Below Reynolds numbers of 2,200 the heat transfer results will be laminar (Sec. A.2.a); and above 10,000 the results will normally be turbulent (Sec. A.2.c). In between the results are difficult to predict, and it is necessary to consult the literature for specific information. Pertinent references are Dalle-Donne and Bowditch [35], McEligot et al. [21], and Kays and London [23].

c. Turbulent Flow.

Fully Developed Velocity and Temperature Profiles. Under conditions of constant-property moderate-velocity turbulent flow in a tube, convective heat transfer behavior may be expressed in terms of a Nusselt number, which under certain temperature boundary conditions will approach a value independent of tube length at points well downstream from the entrance. As was the case for laminar flow, the two boundary conditions of most technical interest for which such an asymptotic Nusselt number is obtained are for a *constant heat rate* per unit of tube length, and a *constant surface temperature.*

The *constant heat rate* Nusselt number is always greater than the *constant surface temperature* Nusselt number, but with the exception of the *liquid metals* (very low Prandtl number) the difference is generally much smaller than for laminar flow, and becomes quite negligible for Pr > 1.00. The ratio of these Nusselt numbers, designated $\mathrm{Nu}_H/\mathrm{Nu}_T$, has been computed by Sleicher and Tribus [36], and is shown on Fig. 22 plotted as a function of Reynolds number and Prandtl number.

The remainder of this section will be primarily concerned with the *constant heat rate* asymptotic Nusselt number, and Fig. 22 can be used to obtain Nu_T from Nu_H.

It should also be pointed out at this time that for turbulent flow, the *thermal entry length* (and the combined entry length) is characteristically much shorter than for a laminar flow (again with the exception of the low Prandtl number liquid metals), and thus the asymptotic values of the Nusselt number are frequently used directly in heat transfer design without reference to the entry length behavior.

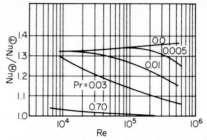

Fig. 22. Ratio of the Nusselt number for constant heat rate to Nusselt number for constant surface temperature for turbulent fully developed conditions in a circular tube.

The Nusselt number for turbulent flow in a smooth tube may be expressed as a function of the Reynolds number based on tube diameter, and the Prandtl number of the fluid. There is a vast amount of experimental data reported in the literature, and there are numerous theories based on heat and momentum transfer analogies. The procedure to be used here will be to present, for illustrative purposes only, some selected experimental data which are believed to have been obtained under reasonably ideal conditions, and to show how these data compare with some selected theories. Then some complete tables, graphs, and equations, based on these theories, will be presented as recommended for design.

On Fig. 23 there are presented experimental data obtained for air (Pr = 0.7) in a smooth tube by Kays and Leung [37].

The results of the analytic solution of Sparrow, Hallman, and Siegel [38] are shown for comparison, together with an analytic solution developed by Kays and Leung [37], based on a theory very similar to that used by Sparrow et al.

On Fig. 24 and Fig. 25 there is shown the analytic solution of Deissler [39], which provides a theory for fluids of high Prandtl number. In this case, the Stanton number, $St = h/(Gc_p)$, is used in place of the Nusselt number, and is plotted as a function of Prandtl number with Reynolds number as parameter. The experimental data points shown are from various sources, and include air, water, ethylene glycol and water, and sodium hydroxide. The data points for Pr > 100 are from various mass transfer experiments.

Liquid metal heat transfer data tend to scatter widely evidently because of the substantial influence of impurities. In general, the earlier experiments yielded somewhat lower Nusselt numbers than the more recent data that have been obtained under conditions where continuous purification is employed. In commercial applications, it appears likely that performance similar to the earlier data would be obtained. Figure 26 shows the recent data of Skupinski, Tortel, and Vautrey [40] for NaK (Pr = 0.015), which appear to have resulted from very carefully conducted experiments, with continuous purification. Shown also is an empirical equation proposed by Skupinski et al.

$$Nu = 4.82 + 0.0185\,Pe^{0.827} \tag{44}$$

where Pe (Peclet number) = RePr. (In liquid metal heat transfer, the Nusselt number is very nearly a function of the Peclet number alone, rather than Reynolds and Prandtl numbers separately.)

There is also shown on Fig. 26 the results of the analysis of Kays and Leung [37] and also Dwyer [41], reference to which will be made below. Table 9 gives a complete set of *constant heat rate* Nusselt numbers for a fully developed flow in a

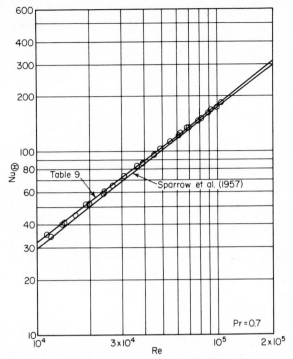

Fig. 23. Comparison of data and analysis for air, Pr = 0.7, for the constant heat rate Nusselt number for fully developed turbulent flow in a circular tube.

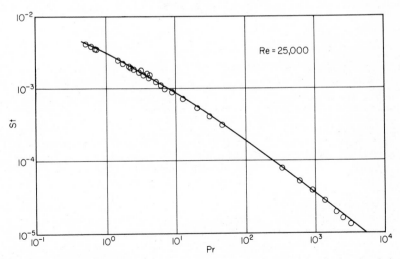

Fig. 24. Comparison of data and analysis for the Stanton number for fully developed turbulent flow in a circular tube at Re = 25,000.

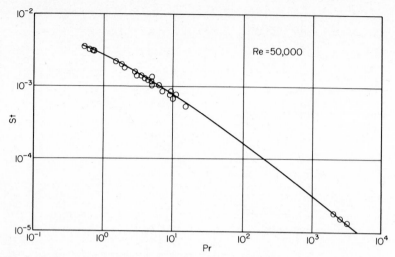

Fig. 25. Comparison of data and analysis for the Stanton number for fully developed turbulent flow in a circular tube at Re = 50,000.

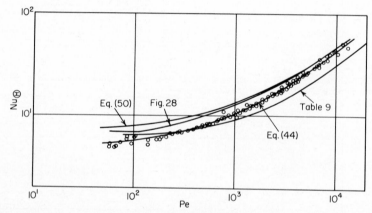

Fig. 26. Comparison of data and analysis for the Nusselt number for turbulent flow of liquid metals in a circular tube. Equations and tables are described in the text.

smooth circular tube, covering a wide range of Reynolds numbers, and the complete spectrum of Prandtl number. This tabulation was developed by Kays and Leung, using a single consistent theory and computing procedure over the entire range of variables. The equation of Deissler [39] was used to compute the thermal diffusivity in the sublayers, and this tabulation is consistent with the solution of Deissler shown on Figs. 24 and 25. Figure 23 demonstrates the correspondence between the theory and experiment in the gas Prandtl number range; the liquid metal results are lower than those reported by Skupinski et al. (see Fig. 26), because certain critical constants in this analysis were based on the earlier experimental data. However, the results in Table 9 may actually be more characteristic of design applications than the experimental data obtained under ideal laboratory conditions.

TABLE 9. Nusselt Numbers, Fully Established Turbulent Flow in a Circular Tube, Fully Developed Constant Heat Rate

Pr \ Re	Laminar	10^4	3×10^4	10^5	3×10^5	10^6
0.0	4.364	6.30	6.64	6.84	6.95	7.06
0.001		6.30	6.64	6.84	7.08	8.12
0.003		6.30	6.64	7.10	8.14	12.8
0.01		6.43	7.00	8.90	14.2	30.5
0.03		6.90	9.10	15.9	32.4	80.5
0.5		26.3	57.3	142	340	895
0.7		31.7	70.7	178	430	1150
1.0		37.8	86.0	222	543	1470
3.0		61.5	149	404	1030	2900
10		99.8	248	690	1810	5220
30		141	362	1030	2750	8060
100		205	522	1510	4030	12000
1000		380	975	2830	7600	22600

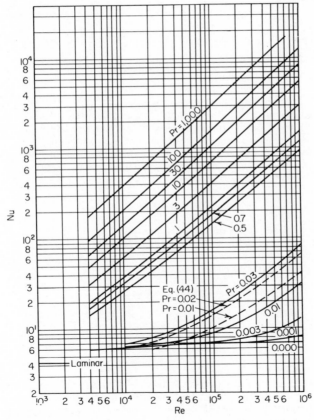

Fig. 27. Constant heat rate Nusselt number for fully developed turbulent velocity and temperature profiles in a circular tube.

The results in Table 9 are also shown graphically on Fig. 27, but in the liquid metal region the Skupinski equation has been added.

Dwyer [41] has studied the available liquid metal experimental data extensively, and suggests that under conditions of extreme purity, Nusselt number will be considerably higher than given in any of the preceding graphs. Dwyer's recommended results for very low Prandtl number fluid are shown on Fig. 28.

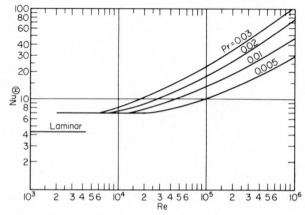

Fig. 28. Constant heat rate Nusselt number for liquid metals in turbulent flow in a circular tube, according to Dwyer.

Algebraic equations are used extensively to represent turbulent convective heat transfer data, rather than tables or graphs. Some of the best known of these equations will be presented here; in some cases they fit the experimental data discussed above very well, while in others they are merely a good approximation.

One of the earliest equations used for fully developed turbulent flow heat transfer in a smooth tube is the so-called *Dittus-Boelter equation,* Dittus and Boelter [42], given by McAdams [43] as

$$Nu = 0.023 \, Re^{0.8} \, Pr^{0.4} \tag{45}$$

This equation was based on experimental data covering the Prandtl number range 0.7 to 120, Reynolds numbers from 10,000 to 120,000 and $L/D > 60$. It is still a very good approximation to the available experimental data for this range of variables.

Another popular equation is the *Colburn equation* [44]

$$St Pr^{2/3} = 0.023 \, Re^{-0.2} \tag{46}$$

or

$$Nu = 0.023 \, Re^{0.8} \, Pr^{1/3} \tag{47}$$

(Note that Nu = StPrRe.)

The so-called *Seider-Tate* equation is essentially the same as the Colburn equation, but a different scheme is used for correcting for the influence of temperature-dependent fluid properties. The latter subject will be treated separately later.

The adequacy of these equations may be checked by reference to Table 9, which for Pr $\geq$ 0.5 is believed to be accurate within a very few percent. The difficulty with equations of the type of Eqs. (45) to (47) is that they are based on the assumption that the influence of Reynolds number and Prandtl number can be expressed by simple powers. This is a good approximation for Reynolds number, but only a fair approximation for Prandtl number. In the *gas* Prandtl number range, 0.5 to 1.0, where the range of Prandtl number is small, a simple power on Prandtl number is a good approximation, but slight modification of these equations gives a better fit to the data. In this range there begins to be a measurable difference between *constant heat rate and constant surface temperature data,* so two equations are suggested, as follows,

$$\text{Nu}_{\textcircled{H}} = 0.022 \, \text{Re}^{0.8} \, \text{Pr}^{0.6} \tag{48}$$

$$\text{Nu}_{\textcircled{T}} = 0.021 \, \text{Re}^{0.8} \, \text{Pr}^{0.6} \tag{49}$$

For *liquid metals* (very low Pr), two popular equations are the *Lyon equation* [45] for *constant heat rate,* and the *Seban-Shimazaki equation* [46] for constant surface temperature

$$\text{Nu}_{\textcircled{H}} = 7.0 + 0.025 \, \text{Pe}^{0.8} \tag{50}$$

$$\text{Nu}_{\textcircled{T}} = 4.8 + 0.025 \, \text{Pe}^{0.8} \tag{51}$$

It will be noted that the Lyon equation yields results that are close to the recommendation of Dwyer [41], as shown on Fig. 28. The Seban-Shimazaki equation is based on the same theory. Thus, these equations will probably predict optimistic Nusselt numbers, except perhaps under conditions of extreme purity.

Thermal Entry Length, Fully Developed Velocity Profile. If the velocity profile is fully developed at the point where heat transfer begins, and this assumes an adiabatic starting length in which the velocity can develop, thermal entry length solutions can be developed in much the same manner as for the corresponding laminar flow problem.

Sleicher and Tribus [36] solved the case of *constant surface temperature,* with fluid temperature uniform at the tube entrance. This is a series solution and can be presented in precisely the same form as the corresponding laminar flow solution. Thus Eqs. (35) and (37) to (39) are all directly applicable. The eigenvalues and constants for these series λ_n^2 and G_n are given in Table 10 for two values of Prandtl number, 0.01 and 0.70, and for Reynolds numbers from 50,000 to 200,000. Unlike the corresponding laminar flow problem, the eigenvalues and constants are functions of Reynolds number and Prandtl number.

The basic turbulent flow data and turbulent heat transfer theory used in the Sleicher and Tribus analysis is essentially the same as used in the fully developed temperature profile analysis, the results of which were presented in Table 9 and Fig. 27, with the exception of some numerical differences at Pr = 0.70. Thus, at Pr = 0.01, the same question arises about the effects of impurities as discussed before, and the results in Table 10 should be regarded as perhaps typical of what might be encountered in commercial applications, rather than in very clean laboratory experiments.

Note that the Nusselt number for a fully developed temperature profile can be obtained from Eq. (38), that is, $\text{Nu}_\infty = \lambda_0^2/2$. Note further that the data in Table 4 may be used to determine the actual length of pipe necessary to obtain essentially fully developed conditions (see discussion in connection with the corresponding laminar flow problem).

Sparrow, Hallman, and Siegel [38] present a solution to the thermal entry length problem for *constant heat rate* per unit of tube length. This is again a series solution,

TABLE 10. Eigenvalues and Constants for Circular Tube, Turbulent Flow, Thermal Entry Length

n, m	Pr	Re	λ_n^2	G_n	γ_m^2	A_m	
0	0.01	50,000	11.7	1.11			
1			65.0	0.950	34.5	6.45	$\times 10^{-3}$
2			163	0.880	113	1.81	$\times 10^{-3}$
3			305	0.835	237	0.828	$\times 10^{-3}$
4			491	0.802	406	0.468	$\times 10^{-3}$
5			722	0.777	621	0.299	$\times 10^{-3}$
0	0.01	100,000	13.2	1.30			
1			74.0	1.04	40.9	5.39	$\times 10^{-3}$
2			190	0.935	135	1.40	$\times 10^{-3}$
3			360	0.869	284	0.625	$\times 10^{-3}$
4			583	0.823	488	0.326	$\times 10^{-3}$
5			860	0.788	746	0.204	$\times 10^{-3}$
0	0.01	200,000	16.9	1.70			
1			96.0	1.24	55.5	3.86	$\times 10^{-3}$
2			247	1.07	181	0.989	$\times 10^{-3}$
3			467	0.978	376	0.449	$\times 10^{-3}$
4			757	0.901	642	0.252	$\times 10^{-3}$
5			1116	0.847	980	0.160	$\times 10^{-3}$
0	0.70	50,000	235	28.6			
1			2640	5.51	1947	7.51	$\times 10^{-5}$
2			7400	3.62	5230	1.97	$\times 10^{-5}$
3					9875	0.800	$\times 10^{-5}$
4					15900	0.402	$\times 10^{-5}$
5					23270	0.228	$\times 10^{-5}$
0	0.70	100,000	400	49.0			
1			4430	9.12	3557	4.12	$\times 10^{-5}$
2			12800	5.66	9530	1.08	$\times 10^{-5}$
3					17970	0.443	$\times 10^{-5}$
4					28840	0.226	$\times 10^{-5}$
5					42230	0.130	$\times 10^{-5}$
0	10.0	50,000					
1					2.736×10^4	5.021	$\times 10^{-6}$
2					7.316×10^4	1.21	$\times 10^{-6}$
3					13.73×10^4	0.436	$\times 10^{-6}$
4					21.96×10^4	0.187	$\times 10^{-6}$
5					31.98×10^4	0.0879	$\times 10^{-6}$
0	10.0	100,000					
1					5.04×10^4	2.78	$\times 10^{-6}$
2					13.46×10^4	0.697	$\times 10^{-6}$
3					25.28×10^4	0.268	$\times 10^{-6}$
4					40.46×10^4	0.126	$\times 10^{-6}$
5					59.04×10^4	0.0656	$\times 10^{-6}$

and can be presented in precisely the same form as for the corresponding laminar flow problem, Eq. (41). The eigenvalues and constants γ_m^2 and A_m are given in Table 10 for a variety of Prandtl numbers and Reynolds numbers. (Actually the results for Pr = 0.01 in Table 10 have been computed from the Sleicher and Tribus data, rather than from Sparrow et al.) Nu_∞ in Eq. (41) is the value for fully developed constant heat rate, as given in Table 9 and Fig. 27. However, alternatively Nu_∞ for constant heat rate may be computed from constant surface temperature theory, using superposition methods

$$\mathrm{Nu}_{H,\infty} = \frac{1}{16 \sum\limits_{n=0}^{\infty} \dfrac{G_n}{\lambda_n^4}} \tag{52}$$

The constant heat rate results may also be used to determine the length of pipe necessary to obtain essentially fully developed conditions, although the results will be approximately the same as determined from the constant surface temperature thermal entry solution.

Figures 29, 30, and 31 show the influence of Reynolds number and Prandtl number on the thermal entry length behavior for constant heat rate, as computed from the above results.

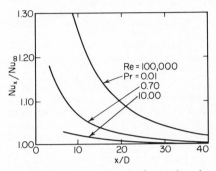

Fig. 29. Nusselt numbers in a thermal entry length of a circular tube, constant heat rate.

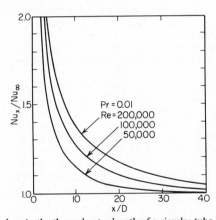

Fig. 30. Nusselt numbers in the thermal entry length of a circular tube, constant heat rate.

Combined Thermal and Hydrodynamic Entry Length. The entry-length problem where both velocity *and* temperature are uniform at the tube entrance is somewhat different in turbulent flow than in laminar flow. In either case it is necessary to have a *nozzle or bellmouth entrance* in order to approximate this boundary condition. If the nozzle and tube walls are smooth, and if the free-stream turbulence level is low, a laminar boundary layer will initially form even though the Reynolds number (based on tube diameter) is very much greater than the value of 2300 that is generally used as a criterion for a turbulent flow. In this case a transition to a turbulent flow will

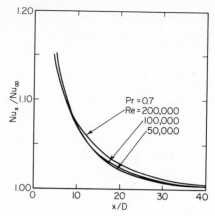

Fig. 31. Nusselt numbers in the thermal entry length of a circular tube, constant heat rate.

take place further down the tube. Because of the initial laminar boundary layer, the mean value of the Nusselt number actually may not be significantly higher than the fully developed turbulent flow value. On the other hand, if a boundary-layer trip is provided at an appropriate point in the tube near the entrance, it is possible to have a turbulent boundary layer virtually from the beginning of the tube, and then the Nusselt number will not only become high in the entry region, but will be higher than for the corresponding purely thermal entry region.

Deissler [4] has computed the combined thermal and hydrodynamic entry length behavior for turbulent flow of a fluid with Pr = 0.73 for both constant heat rate and constant surface temperature conditions. The results are shown on Figs. 32 and 33. Note again that a nozzle entrance with boundary-layer trip would be necessary to realize these results.

Boelter, Young, and Iversen [15] report the results of experiments with *air* flowing in a steam-jacketed tube with various types of entrance, including abrupt contractions and pipe bends. (See also Mills [16].) Both of the latter result in boundary-layer separation (stall) and very high heat transfer coefficients after the boundary layer reattaches. (See also Krall and Sparrow [47].) The boundary layer is apparently always turbulent if the Reynolds number is sufficiently high to result in a turbulent flow far downstream. Some of the results of Boelter et al. are shown on Fig. 34. The appropriate value of Nu_∞ can be taken from Figs. 22 and 27.

In heat exchange design, the *mean* Nusselt number is often of more utility than the local Nusselt number shown on this figure. The mean Nusselt number for air (or any gas) can be evaluated by integration of these curves, and for $L/D > 20$ it may be expressed by the following equation

$$\frac{Nu_m}{Nu_\infty} = 1 + \frac{C}{L/D} \tag{53}$$

where C has been computed to be: fully developed velocity profile—1.4; abrupt contraction entrance—6; 180° round bend—6.

The value for the fully developed velocity profile can also be computed from Eq. (37) and the data in Table 10. For Pr = 0.7, a value of $C = 1.0$ results. At other Prandtl numbers, C is a function of Reynolds number, but in the gas region it is virtually independent of Reynolds number for both the simple thermal entry length, and the combined entry length.

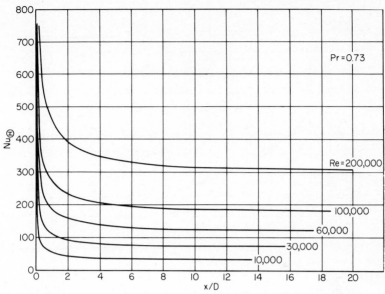

Fig. 32. Nusselt numbers in the combined thermal and hydrodynamic entry length of a circular tube, constant heat rate.

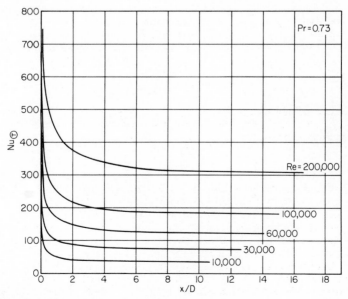

Fig. 33. Nusselt numbers in the combined thermal and hydrodynamic entry length of a circular tube, constant surface temperature.

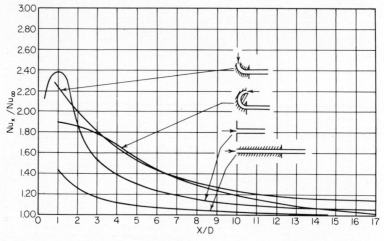

Fig. 34. Measured local Nusselt numbers in the entry region of a circular tube for various entry configurations, air with constant surface temperature.

The constant C can also be computed for the nozzle-entrance-with-trip from the results of Deissler [4] (Figs. 32 or 33). A value of approximately 2.1 results.

d. Effect of Circumferential Heat Flux Variation. In all of the preceding sections, the heating is assumed to be axially symmetric. It is quite possible to have a variation of heat flux around the tube periphery, and this will be accompanied by a variation in local Nusselt number.

In this discussion, it is assumed that the heat transfer coefficient h in the Nusselt number is defined on the basis of the difference between the *local* wall surface temperature, and the bulk mean fluid temperature.

For *constant heat rate* per unit of tube length, but varying circumferentially according to a *cosine law*, the local Nusselt number may be evaluated from

$$\text{Nu}(\phi) = \frac{\text{Nu}(1 + b\cos\phi)}{1 + (S_1 b\,\text{Nu}/2)\cos\phi} \qquad (54)$$

where $\text{Nu}(\phi)$ is the local Nusselt number, a function of angle ϕ; Nu is the corresponding value of the Nusselt number for axially symmetric heating; the prescribed flux is described by

$$q_w''(\phi) = q_w''(1 + b\cos\phi) \qquad (55)$$

where q_w'' is the *average* heat flux, and S_1 is a function of Reynolds number and Prandtl number. (It is simply equal to 1.00 for a laminar flow.)

The function S_1 has been evaluated by Reynolds [48] for the case of fully developed velocity and temperature profiles. These functions are tabulated in Table 11 for both laminar and turbulent flow and for a wide variety of Reynolds and Prandtl numbers. These results are exact for laminar flow, and for turbulent flow they are based on the same data as used by Kays and Leung [37]. Thus, they are consistent with the results in Table 9, but are subject to the same possible inaccuracies at very low Prandtl number as discussed in connection with Table 9.

The original reference also provides a method for handling any arbitrarily specified circumferential heat flux variation that can be approximated by a Fourier expansion.

TABLE 11. Circumferential Heat Flux Functions S_1 for Fully Developed Turbulent Flow in a Circular Tube with Constant Heat Rate

Re / Pr	Laminar	10^4	3×10^4	10^5	3×10^5	10^6
0	1.000	1.000	1.000	1.000	1.000	1.000
0.001		1.000	1.000	0.999	0.974	0.901
0.003		0.999	0.994	0.957	0.831	0.473
0.01		0.991	0.952	0.733	0.409	0.161
0.03		0.923	0.699	0.348	0.145	0.0535
0.7		0.121	0.0490	0.0180	0.00721	0.00275
3		0.0448	0.0178	0.00629	0.00246	0.000902
10		0.0239	0.00931	0.00322	0.00123	0.000438
30		0.0151	0.00582	0.00199	0.000751	0.000166
100		0.00994	0.00383	0.00130	0.000486	0.000166
1000		0.00513	0.00198	0.000667	0.000248	0.0000841

The problem has also been extended by Reynolds [49] to include *conduction* in the peripheral direction in the wall.

The problem of a prescribed circumferentially varying surface temperature has been treated by Sparrow and Lin [50].

e. Axial Variation of Surface Temperature and/or Heat Flux. The thermal entry length solutions for both laminar and turbulent flow can be used to build up solutions for any arbitrarily specified axial variation of surface temperature or heat flux, using superposition. This is possible because of the linearity of the applicable energy differential equation. Although in principle the method can be used in the velocity entry length, results are only at present available for a *fully developed velocity* throughout the tube.

The following equations are given by Kays [51].

If the fluid temperature is a uniform T_e at some point $x^+ = 0$, and then the *surface temperature* is specified as some arbitrary

$$T_w = T_w(\xi)$$

where ξ is a dummy variable, dimensionally the same as x^+ and running from 0 to x^+, then the local wall surface heat flux at any desired value of x^+ may be calculated from

$$q_w''(x^+) = -\frac{k}{r_0}\left[\int_0^{x^+} \Theta_{r^+}(x^+ - \xi, 1)\left(\frac{dT_w}{d\xi}\right)d\xi + \sum_{i=1}^{k} \Theta_{r^+}(x^+ - \xi_i, 1)\,\Delta T_{w_i}\right] \quad (56)$$

where

$$\Theta_{r^+}(x^+ - \xi, 1) = -2\sum_{n=0}^{\infty} G_n e^{-\lambda_n^2(x^+ - \xi)} \quad (57)$$

The integral in Eq. (56) is employed wherever there is a continuous variation of surface temperature, and the summation is employed whenever there is a discontinuity (step) in surface temperature. Positive ΔT_{w_i} is a step *increase* in surface temperature.

The eigenvalues and constants in Eq. (57) λ_n^2 and G_n are obtained from the series solutions to the constant surface temperature thermal entry-length problem which are given in previous sections for both laminar and turbulent flow.

If the fluid temperature is a uniform T_e at some point $x^+ = 0$, and the *local heat flux* is specified as some arbitrary

$$q_w'' = q_w''(\xi)$$

then the local wall surface temperature may be calculated from

$$T_w(x^+) - T_e = \frac{r_o}{k} \int_0^{x^+} g(x^+ - \xi) \, q_w''(\xi) \, d\xi \tag{58}$$

where

$$g(x^+ - \xi) = 4 + \sum_{m=1}^{\infty} \frac{\exp[-\gamma_m^2 (x^+ - \xi)]}{\gamma_m^2 A_m} \tag{59}$$

The *bulk mean* fluid temperature may be evaluated from

$$T_b(x^+) - T_e = \frac{4r_o}{k} \int_0^{x^+} q_w''(\xi) \, d\xi \tag{60}$$

The eigenvalues and constants in Eq. (59) are obtained from the series solutions to the thermal entry-length constant heat-rate problem, which are given in previous sections for both laminar and turbulent flow.

The computing procedures described above assume that the eigenvalues and constants for the appropriate thermal entry length solutions are known. It is sometimes more convenient to work directly with the Nusselt numbers derived from the thermal entry length solutions rather than with the infinite series. If Nu $=$ Nu(x^+) is known for the *constant heat rate thermal entry length problem*, i.e., a step in heat flux at $x^+ = 0$ where the fluid temperature is a uniform T_e, then the wall surface temperature for any arbitrarily specified $q_w''(\xi)$ may be computed from

$$T_w(x^+) - T_b(x^+) = \frac{2r_o}{k} \left[\int_0^{x^+} \frac{1}{\text{Nu}(x^+ - \xi)} \left(\frac{dq_w''}{d\xi} \right) d\xi + \sum_{i=1}^{k} \frac{1}{\text{Nu}(x^+ - \xi_i)} (\Delta q_w'')_i \right] \tag{61}$$

The bulk mean temperature T_b is evaluated from Eq. (60) as before.

f. Turbulence Promoters. Channels with turbulence promoters are frequently found in practice. The system design may be such that turbulence promotion is inherent, as with a rough surface or an inlet geometry that creates a swirling flow. More frequently, the intent is to augment heat transfer by providing artificial protrusions on the tube wall or inserting a swirl generator. This section is concerned with basic heat transfer data and correlation for turbulence promoters. A complete survey and evaluation of augmentative techniques are given in Sec. 10.

Literature surveys on the subject of roughness effects (Bhattacharyya [88]) and on swirl flow (Gambill and Bundy [79]) have been prepared. Table 12 shows a summary of many of the more pertinent experiments that have been performed. The table includes information on the geometry, coolant, element type, etc., and is given according to the type of technique used, such as roughness, swirl, etc. The wide variety of experiments is evident.

From the Moody plot (Fig. 2), it can be seen that roughness increases the friction factor, with an asymptotic value of f as a function of the parameter e/D being reached at large Reynolds numbers. Some effect on the heat transfer would also be expected due to the break up of the sublayers by the rough protrusions. It would thus seem that roughness should have a larger effect on heat transfer at high Prandtl numbers where the heat transfer resistance is more concentrated in the sublayers than at low Prandtl numbers. In general it has been found that artificial and natural roughness do increase the heat transfer; however, they increase the friction factor even more. As a consequence, the pumping power requirements and the allowable pressure drop must also be considered when these techniques are used.

One variable of interest with roughness is obviously the ratio of roughness height e to tube diameter. Edwards and Sheriff [63] have noted that, to be effective, a roughness element must penetrate the laminar sublayer ($y^+ = 5$), and for full effectiveness the roughness height should be larger than the combined laminar and buffer layer thickness ($y^+ = 30$). There is no increase in heat transfer for greater heights, but the friction factor continues to increase.

A second variable is the pitch-to-height ratio of the roughness elements. This variable appears to be particularly important in annuli with only the inner surface roughened (Kaul and Von Kiss [69], Burgoyne et al. [72], and Gomelauri [74]).

The reader is cautioned to check the nomenclature used in each reference consulted since a variety of definitions of diameter, bulk velocity, Reynolds number, etc., are used.

Natural Roughness. Experimental data for naturally rough pipe (cast iron or galvanized) have been obtained by Nunner [52] and Smith and Epstein [54]. The rough pipes have higher friction factors and heat transfer coefficients than smooth tubes. However, this type of roughness does not appear to be well enough defined or reproducible to permit application of these data in design.

Artificial Roughness. Artificial roughness may be categorized as a "natural type" (Dipprey and Sabersky [53]), and the various artificial types as listed in Table 12.

At least two authors (Sams [76], Kolar [59]) have achieved a correlation of roughness effects on heat transfer using the shear velocity u^+ in the Reynolds number. The recent results of Kolar, using 60° triangular roughness elements, are correlated in this manner with Re_T given by $u^+ D/\nu$. The diameter is the volumetric diameter defined by $D = \sqrt{4 \, \mathrm{Vol.}/\pi L}$ and $u^+ = \overline{V}\sqrt{f/2}$ where f is the Fanning friction factor. The data are correlated by

$$\frac{\mathrm{Nu}}{\mathrm{Pr}^{0.4}} = 0.102 \, \mathrm{Re}_T^{0.914} \tag{62}$$

The results of Cope [56], Dipprey and Sabersky [53], and Nunner [52] are also correlated by this equation. A fall-off from the correlation at high Reynolds numbers is attributed by Kolar to the fact that the roughness elements act in part as coolant fins in the stream, with the temperature decreasing along their length, particularly at high Reynolds numbers.

A second method of correlation for natural and artificial roughness has been suggested by Dipprey and Sabersky [53]. This correlation is based on a "heat transfer similarity law." The experiments were run with water in a tube roughened by a granular material giving three-dimensional roughness. Dipprey and Sabersky correlate their data in the fully rough region by

$$\left[8.48 + \frac{f/(2\mathrm{St}) - 1}{(f/2)^{1/2}} \right] \mathrm{Pr}^{-0.44} = 5.19 \, (e^*)^{0.2} \tag{63}$$

where $e^* = \mathrm{Re}\sqrt{f/2} \, e_s/D$ and the diameter is defined as $D = \sqrt{4 \, \mathrm{Vol.}/\pi L}$. The e_s/D represents equivalent sand grain roughness based on friction measurements. The

TABLE 12. Summary of Experiments on Augmenting Convective Heat Transfer

Reference	Coolant	Element Type	Pitch/Height	Height/Diameter	Reynolds Number	Comments
A. Roughness, Circular Tubes						
Nunner, W. [52]	air	square rings, rounded rings	2–81.7	0.33–0.008	$500-8 \times 10^4$	constant wall temperature, volumetric diameter definition
Dipprey, D.F. Sabersky, R.H. [53]	water	natural		0.0024–0.049	$6 \times 10^4 - 5 \times 10^5$	$1.2 < \text{Pr} < 5.9$ constant heat flux, volumetric diameter definition
Smith, J.W. Epstein, N. [54]	air	natural		0.0053–0.0158	$10^4 - 8 \times 10^4$	
Grass, G. [55]	water	longitudinal and/or transverse grooves, thread roughness	12.5 31	.005 .0012		
Cope, W.F. [56]	water	knurled elements	0.73–1.15	.063, .034, .011	2000–65,000	volumetric diameter definition
Sams, E.W. [57]	air	thread roughness		.008, 0.012, 0.018	$10^3 - 3.5 \times 10^5$	constant heat flux, large temperature difference, volumetric diameter definition
Savage, D.W. Myers, J.E. [58]	water	single ring	.66–infinity	.041–0.165	$10^4 - 1.3 \times 10^5$	$2.8 < \text{Pr} < 8$ temperature profile only partially developed
Kolar, V. [59]	air, water	$60°$ triangular thread	1.6	0.55, 037, .019	$45 \times 10^3 - 1.45 \times 10^5$	$0.71 < \text{Pr} < 5.52$ constant wall temperature, volumetric diameter definition
Gargand, J. Paumard, G. [60]	CO_2	square rings, triangular rings	5–15 10	.004–.016	$2 \times 10^5 - 3.5 \times 10^6$	

B. Roughness, Rectangular Tubes and Annuli

Reference	Coolant	Element Type	Pitch/Height	Height/Diameter	Reynolds Number	Comments
Cowin, M. [61]	air			0.00594-0.0174	2×10^9-2×10^5	friction only
Lancet, R.T. [62]	air	round protusions	10		5,000-30,000	5:1 aspect ratio, uniform heat flux, 0.010" height in a 0.145" by 0.045" duct
Edwards, F.J. Sheriff, N. [63]	air	single wire element	16-192	0.0169	2.5×10^5	friction only for multiple elements
Brauer, H. [64]	water		1-5.9	0.375-0.0697	$200-10^5$	$D_i/D_o = 0.777, 0.578$
Malherbe, J.M. [65]	air	various	1.25-70.0	0.02-0.06	5×10^3-5×10^4	$D_i/D_o = 0.60$
Draycott, A. Lawther, K.R. [66]	air	wire wound and machined elements	2-55.5	0.0103-0.0442	2×10^5	$D_i/D_o = 0.484$ reasonable agreement with Nunner's equation
Walker, V. Rapier, A.C. [67]	air	ribs on inner, heated tube	2.5-30	0.004-0.016	2×10^5-5×10^5	$D_i/D_o = 1.571$ within 13% of Nunner's equation
Kemeny, G. A. Cyphers, J.A. [68]	water	ribs, grooves with wires on inner, heated, surface	10, 20	0.0244-0.0281	2×10^5	$D_i/D_o = 0.726$
Kaul, V. von Kiss, M. [69]	air	ribs on inner, heated surface	10	0.0333-0.0533	4×10^4-4×10^5	$D_i/D_o = 0.500, 0.333$ agreement with Nunner's equation
Bennett, A.W. Kearsey, H.A. [70]	super heated steam	Helical grooves on inner, heated, surface	15	0.014	10^5-4×10^5	$D_i/D_o = 0.680$

TABLE 12. Summary of Experiments on Augmenting Convective Heat Transfer *(Continued)*

Reference	Coolant	Element Type	Pitch/Height	Height/Diameter	Reynolds Number	Comments
B. Roughness, Rectangular Tubes and Annuli (Cont.)						
Holcomb, R.S. Lynch, F.D. [71]	air	wires	3.23–24.0	0.0127–0.0216	2×10^5	$D_i/D_o = 0.602, 0.590$ friction only
Burgoyne, T.B. Burnett, P. Wilkie, D. [72]	air	inner wall ribs	2.5–50	0.0014–.028	$7 \times 10^4 - 3.5 \times 10^5$	$D_i/D_o = 0.52$ inner surface heated, design charts in reference
Kattchee, N. Mackewicz, W.N. [73]	water and nitrogen	wires wrapped on each element	10–61.6	0.0235–0.0751	5×10^4	heat transfer - nitrogen friction - water
Gomelauri, W. [74]	oil, water	square and rounded rings	8.2–59.0	0.012–.030	$5 \times 10^3 - 9 \times 10^4$	constant heat flux, peak in heat transfer at P/e about 13
Gargand, J. Paumard, G.	air, CO_2	rings, transverse fins, etc. on inner tube			$1.5 \times 10^4 - 3 \times 10^5$	constant heat flux
C. Wire Inserts						
Nagaoka, Z. Watanabe, A. [75]	water	coiled wires		0.03–0.11	$1.5 \times 10^4 - 3 \times 10^4$	
Sams, E.W. [76]	air	coiled wires		0.030–0.063	$10^4 - 10^5$	constant heat flux, 360° pitch to diameter from 1–168
Kreith, F. Margolis, D. [77]	air, water	coiled wires		0.12	$10^4 - 10^5$	360° pitch to diameter from 3.4–14.6

D. Vortex (Swirl) Flow

Reference	Coolant	Element Type	360° Pitch/Diameter	Reynolds Number	Comments
Colburn, A.P. King, W.J. [78]	air	twisted tape inserts	1.14–6.10	$5 \times 10^3 – 10^5$	cooling data
Gambill, W.R. Bundy, R.D. [79, 80]	air, water, ethylene, glycol	twisted tape inserts	4.5–∞ (st. fin)	$10^3 – 5 \times 10^5$	constant heat flux
Smithberg, E. Landis, F. [81]	air, water	twisted tape inserts	3.6–∞ (st. fin)	5000–50,000	constant wall temperature
Evans, S.I. Sarjant, R.J. [82]	air	twisted tape inserts	5.8–11.8	$6 \times 10^3 – 2 \times 10^4$	tape did not touch walls
Kreith, F. Margolis, D. [83]	air, water	twisted tape inserts	5.0–14.6	$10^4 – 10^5$	
Ibragimov, M.H. Nomofelov, E.V. Subbotin, V.I. [84]	air	twisted tape inserts	4.2–9.1	$2 \times 10^4 – 7 \times 10^4$	
Seymour, E.V. [85]	air	twisted tape inserts	various	various	optimum twist at 5.0 for air in a 0.87" tube
Koch, R. [86]	air	twisted tape inserts	5.0–22.0	3500–85,000	

E. Blunt Body Inserts

Reference	Coolant	Element Type	Spacing	Reynolds Number	Comments
Evans, L.B. Churchill, S.W. [87]	water	discs and streamline bodies normal to flow direction	disc spacing from 2–12 diameters and disc to D ratio from 0.625–0.875, streamline spacing 4–12 D's, diameter ratio 0.625–0.875	$5 \times 10^3 – 6 \times 10^4$	constant heat flux
Koch, R. [86]	air	disc normal to stream		$10^4 – 4 \times 10^4$	see description in Reference

results of Nunner [52] for two-dimensional ring-type roughness are also correlated in this manner using a coefficient of 6.37 in place of 5.19.

Figures 35 and 36 show design charts developed by Dipprey and Sabersky for a Prandtl number of 6. Information at other Prandtl numbers is not yet available.

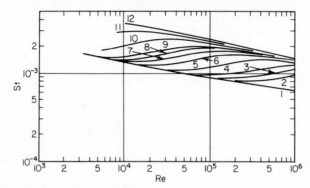

Fig. 35. Design chart for the Stanton number as a function of roughness at a Prandtl number of 6. e_s/D : (1) smooth; (2) 0.0004; (3) 0.0006; (4) 0.001; (5) 0.002; (6) 0.004; (7) 0.006; (8) 0.008; (9) 0.01; (10) 0.02; (11) 0.04; (12) 0.08.

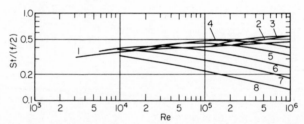

Fig. 36. Ratio of Stanton number to friction coefficient as a function of roughness for Pr = 6. e_s/D : (1) smooth; (2) 0.0004; (3) 0.0008; (4) 0.004; (5) 0.008; (6) 0.02; (7) 0.04; (8) 0.08.

Dipprey and Sabersky were particularly concerned with the effects of Prandtl number. Part of their conclusions were: "... there is a limit for any combination of Reynolds and Prandtl number beyond which increases in roughness, while increasing the friction, will no longer increase the heat transfer coefficient.

"At high Prandtl numbers large improvements in Stanton number due to roughening can be achieved with little or no loss in the ratio of heat transfer to friction coefficient."

Still a third correlation has been suggested by Nunner [52]. Nunner's results are well correlated by

$$\text{Nu} = \frac{(f/2)\,\text{RePr}}{1 + 1.5\,\text{Re}^{-1/8}\,\text{Pr}^{-1/6}\,[\text{Pr}(f/f_s) - 1]} \qquad (64)$$

where f_s is the smooth-tube friction factor. This equation has successfully correlated air results of other authors, but appears to fail for Prandtl numbers away from gas values. It is not recommended for Prandtl numbers larger than 1.

Artificial Roughness—Wire Coil Inserts. Wire coils inserted into tubes have also been used as roughness and swirl generation devices. A listing of the various studies is given in Table 12.

Turbulence Promoter Techniques in Annuli. The annulus with the inner surface roughened is a geometry of particular interest to reactor designers. Several studies (Table 12) have been made with short ribs on the *inner* tube of the annulus to promote turbulence and increase heat transfer. In this case, the variables of interest are e/D_e and p/e, the relative roughness height and the pitch-to-roughness ratio, respectively. Hall [89] has suggested a correlation technique for this geometry in which the equivalent diameter is calculated separately for the inner and outer surfaces, and is defined as $D_e = 4A/P$, where A is the flow area between the surface in question and the *radius of maximum velocity,* and P is the perimeter of the *wall surface,* i.e., the perimeter of the surface of maximum velocity is not included.

In a like manner the velocity in the inner surface Reynolds number is the mean velocity between the inner surface and the maximum velocity surface, while for the outer surface Reynolds number the mean velocity between the outer and maximum velocity surfaces is used. Burgoyne et al. [72] and Wilkie [90] have presented design charts for this geometry with rectangular fin roughness elements on the inner annular surface. Some of these are presented in Figs. 37 to 40. The equivalent diameter D_e is defined using the Hall correlation, as is the velocity used in the inner Reynolds number.

The results for the *smooth* inner tube are

$$f_{\text{smooth}} = 0.09 \, \text{Re}_{\text{inner}}^{-0.245}$$

$$\text{St}_{\text{smooth}} = 0.034 \, \text{Re}_{\text{inner}}^{-0.215}$$

Since the design charts are presented in terms of ratios of rough-to-smooth values, the smooth value correlation must be known. These figures are included to *illustrate* the effects of the variables on both the friction and heat transfer results. The recent results of Wilkie [90], to be discussed, indicate that other variables are also important to these results.

Other results from this study are shown in Fig. 41. The maximum Stanton number is found to be almost independent of Reynolds number; however, it is a strong

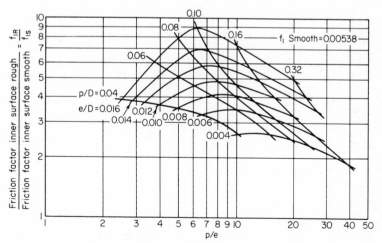

Fig. 37. Friction factor relative to smooth surface value at a Reynolds number of 10^5 for flow through an annulus with ribs on the inner surface. p = rib pitch, e = rib height.

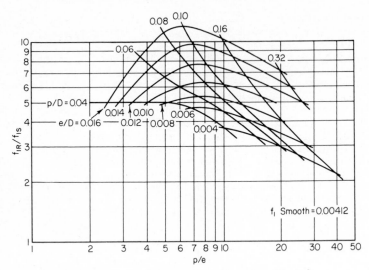

Fig. 38. Same as Fig. 37, but at a Reynolds number of 3×10^5.

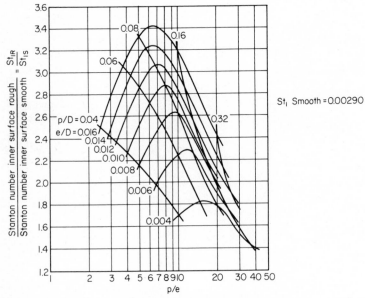

Fig. 39. Stanton number relative to smooth surface value at a Reynolds number of 10^5 for flow through an annulus with ribs on the inner surface. p = rib pitch, e = rib height.

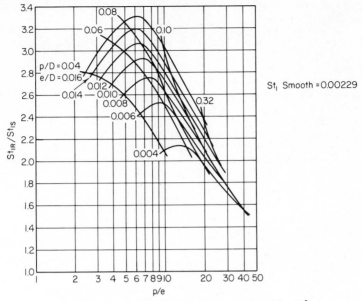

Fig. 40. Same as Fig. 39, but at a Reynolds number of 3×10^5.

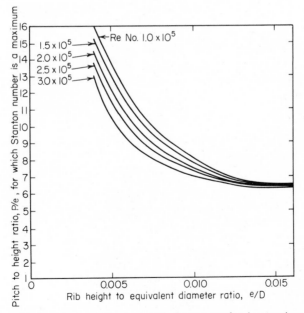

Fig. 41. Values of independent variables for which Stanton number is a maximum for flow in annuli with ribs on the inner surface.

function of the p/e ratio and the e/D_e ratio. The Reynolds number in this figure is the inner Reynolds number.

Wilkie has shown that several other variables are important to this work. He found that the ratio of roughness element width to roughness height w/e is an important parameter. The information on Figs. 37 to 40 is for $0.5 < w/e < 2.5$. Wilkie has data in greater detail for square ribs. The effect of w/e may cause as much as a 30 percent change in the information on the figures, particularly at small p/e ratios.

Additionally it has been found that rounding the corners of the roughness elements reduces the friction factors up to 8 percent. Further information on rib shape, rib orientation (i.e., helix or transverse), and experimental correlation is in the references.

A second type of correlation for the rough inner tube annulus has been suggested by Gomelauri [74]. His results are less detailed than those of Burgoyne et al., but may be used to give an indication of the increased heat transfer. The correlation is

$$\text{Nu}_f = 0.0218 \, \text{Re}_f^{0.8} \, \text{Pr}_f^{0.4} \left(\frac{\text{Pr}_f}{\text{Pr}_w}\right)^{0.25} \exp\left[f_n\left(\frac{p}{e}\right)\right] \tag{65}$$

where

$$f_n \frac{p}{e} = 0.5 \frac{(p/e)_{\text{opt}}}{p/e} \; ; \quad \frac{p}{e} \geq \left(\frac{p}{e}\right)_{\text{opt}}$$

$$= 0.85 \frac{p/e}{(p/e)_{\text{opt}}} \; ; \quad \frac{p}{e} \leq \left(\frac{p}{e}\right)_{\text{opt}}$$

He finds

$$\left(\frac{p}{e}\right)_{\text{opt}} = 13$$

The data cover the range of variables

$$6 \times 10^3 < \text{Re} < 9 \times 10^4$$

$$0.72 < \text{Pr} < 80$$

$$6.7 < \frac{p}{e} < 67.7$$

$$0.013 < \frac{e}{D_e} < 0.08$$

The D_e used here is the usual equivalent diameter for the annulus $D_o - D_i$. Data correlated by this equation are those of Gomelauri [74], Nunner [52], Brauer [64], and Fedynsky [91]. All data are for two-dimensional ring-type roughness on the inner wall with square, triangular, and rounded rings. The subscript f refers to properties evaluated at a film reference temperature given by one-half the sum of the wall and bulk temperatures (see Sec. H).

Vortex and Swirl Effects. Several experiments have been run on the effects of putting twisted tape, swirl generators, etc. into tubes. A review article on this topic has been published by Gambill and Bundy [79]. Results from this article are shown

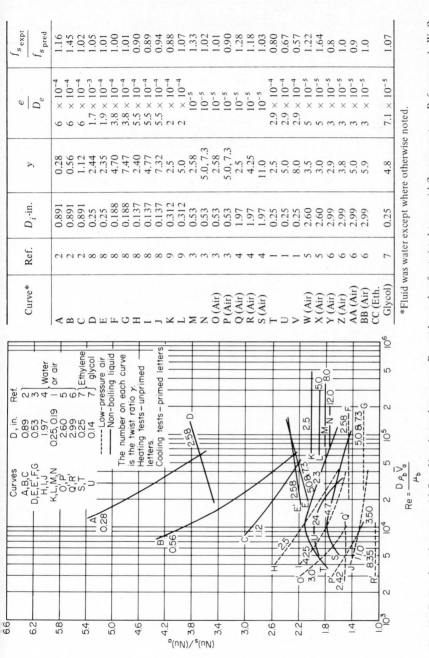

Curve*	Ref.	D_i-in.	y	$\dfrac{e}{D_e}$	$\dfrac{f_{s\,\text{expt}}}{f_{s\,\text{pred}}}$
A	2	0.891	0.28	6×10^{-4}	1.16
B	2	0.891	0.56	6×10^{-4}	1.45
C	2	0.891	1.12	6×10^{-4}	1.02
D	8	0.25	2.44	1.7×10^{-3}	1.05
E	8	0.25	2.35	1.9×10^{-4}	1.01
F	8	0.188	4.70	3.8×10^{-4}	1.00
G	8	0.188	7.47	3.8×10^{-4}	1.01
H	8	0.137	2.40	5.5×10^{-4}	0.90
I	8	0.137	4.77	5.5×10^{-4}	0.89
J	8	0.137	7.32	5.5×10^{-4}	0.94
K	9	0.312	2.5	2×10^{-4}	0.88
L	9	0.312	5.0	2×10^{-4}	1.07
M	3	0.53	2.58	10^{-5}	1.33
N	3	0.53	5.0, 7.3	10^{-5}	1.02
O (Air)	3	0.53	2.58	10^{-5}	1.01
P (Air)	3	0.53	5.0, 7.3	10^{-5}	0.90
Q (Air)	4	1.97	2.5	10^{-5}	1.28
R (Air)	4	1.97	4.25	10^{-5}	1.18
S (Air)	4	1.97	11.0	10^{-5}	1.03
T	1	0.25	2.5	2.9×10^{-4}	0.80
U	1	0.25	5.0	2.9×10^{-4}	0.67
V	1	0.25	8.0	2.9×10^{-4}	0.57
W (Air)	5	2.60	3.5	5×10^{-5}	1.22
X (Air)	5	2.60	3.0	5×10^{-5}	1.64
Y (Air)	6	2.99	2.9	3×10^{-5}	0.8
Z (Air)	6	2.99	3.8	3×10^{-5}	1.0
AA (Air)	6	2.99	5.0	3×10^{-5}	0.9
BB (Air)	6	2.99	5.9	3×10^{-5}	1.0
CC (Eth. Glycol)	7	0.25	4.8	7.1×10^{-5}	1.07

*Fluid was water except where otherwise noted.

Fig. 42. Ratio of swirl to nonswirl flow Nusselt numbers versus Reynolds number for twisted tape swirl flow generators. References: 1—W. R. Gambill et al. [94]; 2—N. D. Greene [95]; 3—F. Kreith and D. Margolis [83]; 4—R. Koch [86]; 5—A. P. Colburn and W. J. King [78]; 6—S. I. Evans and R. J. Sarjant [82]; 7—W. R. Gambill and R. D. Bundy [80]; 8—Unpublished ORNL data [96]; 9—R. Viskanta [97].

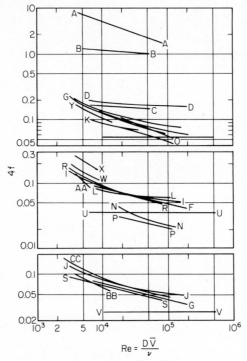

Fig. 43. Friction factors versus Reynolds number for twisted tape swirl flow generators. All data and references are the same as in Fig. 42.

on Fig. 42, where the ratio of the swirl-to-axial Nusselt number, both based on the axial flow diameter D_i, is plotted against the axial Reynolds number. The twist ratio is the number of diameters per $180°$ twist, and is the parameter shown on the curves. The curves A, B, C represent results obtained for a twisted-tape about a rod coincident with the tube flow axis. Figure 43 includes friction factor information taken from Gambill and Bundy. The friction factors would be about 0.6 as large if plotted on D_e instead of D_i. These results are correlated roughly, as noted on the legend for the figure, by the relation

$$4(f_S - f_{axial})D_e = \frac{0.21}{y^{1.31}}\left[\frac{Re_{D_e}}{2000}\right]^{-n} \tag{66}$$

where y is the number of inside diameters per $180°$ twist, and $n = 0.81 \exp[-1700 \; e/D_e]$, where e/D_e is the relative roughness. The Re_{D_e} has the hydraulic diameter in it including the perimeter of the twisted tape.

Other work has been done by Smithberg and Landis [81]. Their friction results, based on a *smooth tube*, are correlated by

$$f = \left[0.046 + 2.1\left(\frac{H}{D} - 0.5\right)^{-1.2}\right]Re_{D_e}^{-n} \tag{67}$$

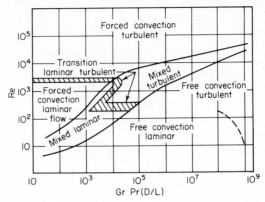

Fig. 44. Regimes for free, forced, and mixed convection for flow through vertical tubes $(10^2 < \mathrm{Pr}(D/L) < 1)$.

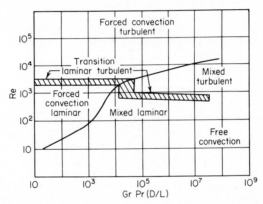

Fig. 45. Regimes for free, forced, and mixed convection for flow through horizontal tubes $(10^2 < \mathrm{Pr}(D/L) < 1)$.

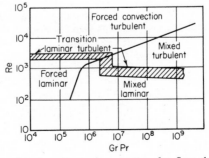

Fig. 46. Regimes for free, forced, and mixed convection for flow through horizontal tubes $(10^2 < \mathrm{Pr}(D/L) < 1)$.

where $n = 0.2[1 + 1.7(H/D)^{-0.5}]$ and H/D is the number of inside diameters *per 360° twist*. These results cover the range of parameters

$$3.6 < \frac{H}{D} < 22$$

$$5000 < \mathrm{Re}_{D_e} < 50,000$$

Gambill and Bundy [79, 80] have correlated their ethylene glycol and water data on a buoyant parameter with the results

$$\left\{ \frac{\mathrm{Nu}_{\mathrm{swirl}}}{\mathrm{Re}^{0.89}\,\mathrm{Pr}^{0.6}} \right\}_{\mathrm{bulk}} = 0.00675 \left\{ \frac{\rho_f \beta_f \Delta T u_a^2}{y^2 D_i} \right\}^{0.0344} \tag{68}$$

where the units of the right side are in lb/ft^3, ft/sec, and ft and y is the 180° twist ratio. Both Nu and Re are based on the tube diameter and the axial velocity is used in the Reynolds number.

The remarkable thing to note here is that over the parametric range of 10^3 to 10^6 the right-hand side of the equation varies from only 0.009 to 0.011, so that the Nusselt number is almost independent of the twist ratio.

This correlation includes the fin effect of the twisted tape in the heat transfer coefficient. Smithberg and Landis [81] give information for separating out the fin effect. Their results, when correlated in this form, vary from about 0.014 to 0.007 for $0.7 < \mathrm{Pr} < 15$.

Blunt-body Promoters.   Blunt-body and ring-type inserts have also been tried as a means of improving heat transfer performance. The ring-type elements have been successful in increasing the ratio of augmented to unaugmented heat transfer coefficient at the lower Reynolds numbers. However, blunt inserts have not been particularly successful. With blunt inserts a large increase in pressure drop is noted.

Economics of Augmentation Techniques.   The economics of using heat transfer promoters has been discussed by Evans and Churchill [92]. Other information is included in Kolar [59], Gambill and Bundy [79], and Spalding and Lieberam [93]. Section 20 presents an evaluation of some of the data mentioned here in terms of a plot of the ratio of the augmented to unaugmented heat transfer coefficient at the same pumping power.

g. Superimposed Natural Convection Effects.   Natural convection may play an important role in forced convection problems when the velocity and temperature profiles become sufficiently altered by natural convection effects to change significantly the friction and heat transfer characteristics. This is most likely to occur with laminar flow. Metais and Eckert [98] have classified the flow regimes and their results are shown in Figs. 44 to 46. These charts may be used to determine whether superimposed natural convection is an important effect.

Figure 44 shows the flow regimes for *vertical* tube flow. The Grashof number $\mathrm{Gr} = g\rho^2 D^3 \beta \Delta T/\mu^2$ is based on tube diameter and the difference between the wall and fluid bulk temperatures. The limits of the free and forced convection regimes are defined in such a way that the actual heat flux under the combined influence of the forces does not deviate by more than 10 percent from the heat flux that would be caused by the external forces alone, or by the body forces alone. The results are applicable for *both constant heat flux and constant wall temperature boundary conditions.*

Figures 45 and 46 contain results of the survey for flow through a *horizontal* tube in which the gravitational body forces are, therefore, normal to the pressure gradients. These figures are for *uniform wall temperature*. Metais and Eckert comment that the available experiments do not permit the establishment of a limit between the mixed convection and free convection regimes. The correlations are equally good regardless of whether the factor D/L is included into the parameter on the abscissa or not.

A design chart from Worsoe-Schmidt and Leppert [99] to facilitate determining the pertinent parameters for superimposed natural convection is shown on Fig. 47. The Gr^* is a modified Grashof number defined as $Gr^* = 8g\rho_0^2 r_w^3/\mu_0^2$ where the subscript 0 refers to conditions at start of heating. This graph is particularly helpful in using Worsoe-Schmidt's results which are included in this section.

Specific information and correlations for special geometries follow:

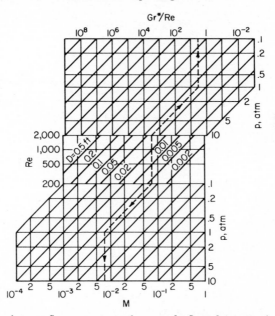

Fig. 47. Relation between flow parameters and pressure for flow of air in circular tubes at 80°F.

Laminar Flow—Constant Heat Flux. Iqbal and Stachiewicz [100] have studied the case of combined free and forced convection inside an *inclined circular tube* with fully developed laminar flow and a constant wall heat flux. Their results for Nusselt number as a function of Rayleigh number $Ra = \beta g r_w^4 \rho^2 c_p (dT/dx)/\mu k$ with the inclined angle as a parameter are shown in Fig. 48 and 49. 0° represents a horizontal tube. The results are for relatively low values of the RaRe parameter.

Velocity and temperature profiles resulting from this study are given in the reference. The effects of superimposed natural convection are particularly noticeable on the temperature profile for the horizontal tube.

Mori et al. [101] have experimentally studied the *horizontal constant heat flux tube* for laminar flow of air at large value of the RaRe product. Figure 50 presents their Nusselt results, given as the ratio of the Nusselt number with finite RaRe to that of zero RaRe, for fully developed laminar flow. Velocity and temperature profiles are given in the paper for a RaRe of about 10^5. They note that the *critical Reynolds number for transition from laminar to turbulent flow* is dependent on both

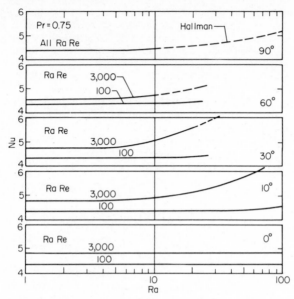

Fig. 48. Nusselt number versus Rayleigh number for combined free and forced convection in an inclined circular tube for Pr = 0.75, 0° is horizontal.

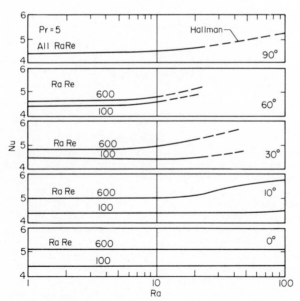

Fig. 49. Nusselt number versus Rayleigh number for combined free and forced convection in an inclined circular tube for Pr = 5, 0° is horizontal.

the secondary flow and the initial turbulence level. Specific critical Reynolds number results are given in the paper. Their Rayleigh number is based on the tube radius.*

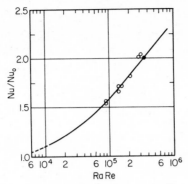

Fig. 50. Nusselt number versus the RaRe product for laminar flow with superimposed natural convection effects in a horizontal tube. Data for air.

The above references include work only for fully developed laminar flow conditions. McComas and Eckert [102] have noted that in a *horizontal tube* the free convection effects do not start at the tube inlet but rather require their own starting length. There may thus be significant entrance effects before information given above is applicable.

Both Hallman [103] and Worsoe-Schmidt and Leppert [99] have studied the *vertical-upflow–constant-heat-flux* tube, the former for water, the latter for air. Hallman's results, for fully developed laminar flow, are shown in Fig. 51. These are low temperature, essentially constant property results. Hallman's earlier paper [104] includes results for uniform internal heat sources, and presents fully developed velocity and temperature profiles. From the results of Fig. 51, it is seen that a

*Note that the nomenclature in the original reference is in error.

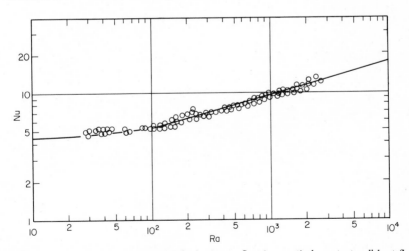

Fig. 51. Fully developed Nusselt numbers for laminar upflow in a vertical constant wall heat flux tube, as a function of the Rayleigh number.

Rayleigh number of about 50 causes a 10 percent increase in the fully developed Nusselt number.

Worsoe-Schmidt and Leppert numerically solved the case of a *uniform wall heat flux with fully developed velocity* but including the thermal entrance region. These results are for vertical upflow of air at a q^+ of 5, where $q^+ = r_w q''/k_0 T_0$, subscript 0 refers to the thermal inlet condition and the temperature is on the absolute scale. Figures 52 and 53 present their results for the local Nusselt number, local friction factor, and local wall-to-bulk temperature difference. M_o is the inlet Mach number, and bulk properties are used.

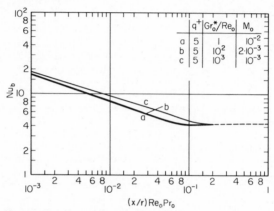

Fig. 52. Local Nusselt numbers for superimposed forced and natural convection with uniform heat flux and upflow in a vertical circular tube.

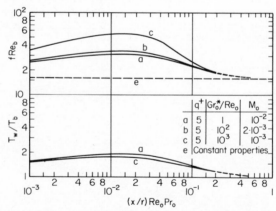

Fig. 53. Local friction factors and wall to bulk temperature ratios for superimposed forced and natural convection with uniform heat flux and upflow in a vertical circular tube.

They conclude that the Nusselt number and temperature profile are not grossly affected by the superimposed natural convection effects, but that there are substantial effects on the friction factor. Further, the effects appear to be largely limited to the usual thermal entrance region, at least for these values of q^+.

These results are compatible with those of Hallman when it is noted that, for gases, heating increases μ and k but decreases ρ and β. Consequently, with the large heating rates investigated here the "fully developed" Rayleigh number has reached a value negligible in its effect on the forced convection heat transfer. It appears that for gases only for low heating rates will the Rayleigh number approach large values in the thermally fully developed laminar region of a tube.

Effects of Natural Convection on Transition to Turbulent Flow. Hallman [103] has noted the apparent onset of flow instabilities due to natural convection effects in vertical upflow and has proposed a tentative correlation for transition to turbulent flow based on a modified Rayleigh number. The correlation is shown in Fig. 54. The triangles are from the results of Scheele et al. [19]. These results are for water in a constant heat-flux tube, with the transition criterion for Hallman's data based on instabilities in wall temperature, and that of Scheele et al. based on dye-trace instabilities. The wall temperatures would presumably show a transition before the stream fluctuations are large enough to break up a dye-trace.

A similar effect has been noted, also with water, by Brown [105] who found that for values of Gr/Re between 300 and 500 the critical Reynolds numbers dropped from the usual value of about 2000 to something between 30 and 50.

It is important to note that the properties of gases and liquids are affected by temperature in opposite ways. Thus, heating of a gas in a tube tends to decrease the Reynolds number and stabilize the flow, whereas the opposite is true for a liquid. The Rayleigh number is affected in the same manner. The results of Hallman, Worsoe-Schmidt and Leppert, and of Brown are consistent when the property variations are considered.

Laminar Flow—Constant Wall Temperature. Martinelli and Boelter [106] studied laminar flow heat transfer in *vertical tubes* with constant wall temperature, and

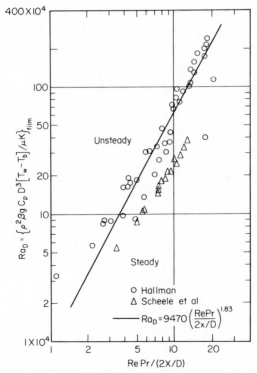

Fig. 54. Effects on transition to turbulent flow due to superimposed natural convection with heating in upflow in a vertical tube with a uniform wall heat flux. Ra_D based on film properties.

proposed the relation

$$\text{Nu}_{am} = 1.75\,F_1 \left[\text{Gz}_{am} + 0.0722\,F_2 \left(\frac{\text{GrPr}}{L/D}\right)_w^{0.84}\right]^{1/3} \tag{69}$$

where the factors F_1 and F_2 are given in Fig. 55.

The equation applies for (1) Gz_{am} determined using fluid properties evaluated at the *arithmetic mean* bulk temperature; (2) the Gr and Pr evaluated at the *tube wall temperature;* (3) the temperature difference in the Gr as $(T_w - T_\infty)$, *the initial* temperature difference; (4) heating with upward flow and cooling with downward flow; (5) cooling with upward flow and heating with downward flow with the +0.0722 changed to -0.0722. The Gr is based on tube diameter. The equation was developed from the data on oils and water.

A prediction of natural convection effects for vertical flow for liquids and gases including the effects of variable viscosity is available in Pigford [107]. Data which are not in agreement with the Martinelli-Boelter correlation when the equation was applied *locally with air* have been determined by Jackson et al. [108] for a tube of 17 diameters length.

Combined free and forced convection for laminar flow in *horizontal tubes* with constant wall temperature has been discussed by Eubank and Proctor [108]. They determined a tentative empirical correlation

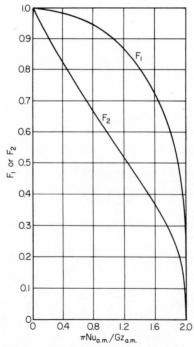

Fig. 55. Required functions for use in the Martinelli-Boelter correlation for superimposed natural convection effects in a vertical constant wall temperature tube.

$$\left(\frac{\bar{h}D}{k}\right)_{a.m.}\left(\frac{\mu_w}{\mu_b}\right)^{0.14} = 1.75\left[\frac{\pi}{4}(RePr)_b\frac{D}{L} + 0.04\left(\frac{D}{L}GrPr\right)_b^{0.75}\right]^{1/3} \qquad (70)$$

where Gr is based on the tube diameter and the *arithmetic mean* temperature. The equation was developed from data for oils and holds in the ranges

$$3 \times 10^5 < GrPr < 8.6 \times 10^8 \quad ; \quad 140 < Pr < 15,200 \quad ; \quad 49 < \frac{L}{D} < 235$$

Information on *local coefficients* for a constant temperature horizontal pipe with laminar flow of water is available in Martin and Carmichael [110].

Turbulent Flow. Figures 44 and 45 may be used to determine whether natural convection effects are significant with turbulent flow for a particular case of interest. In general for turbulent flow natural convection effects are negligible. The results of Ojalvo and Grosh [111] for uniform wall heat flux with upflow in a vertical tube, with fully developed velocity and temperature conditions, indicate a decreasing effect with increasing Prandtl number.

Parallel Plates. The case of laminar vertical flow between parallel plates with a linear temperature variation along the wall has been treated by Ostrach [112]. The horizontal fully developed laminar flow case, also with a linear axial temperature distribution, has been discussed by Gill and Casal [113].

B. CIRCULAR TUBE ANNULI AND FLOW BETWEEN PARALLEL PLATES

1. Hydrodynamics

a. Friction Coefficients.

Laminar Flow. Lundgren et al. [114] have solved for the fully developed Fanning friction factor in annuli. Table 13 presents their results. The friction coefficient in the hydrodynamic entrance region has been treated by Heaton et al. [115, 32]. Figure 56 shows the Fanning friction factor based on the local wall shear stress. The plotted friction factor is for the inner wall, that is, it is based on the inner wall shear stress. The outer wall friction factors lie between the curves for r^* equal to 1.0 and 0.0, so that an interpolation is possible for outer wall friction coefficients. Here $r^* = r_i/r_0$ and $z = x/(D_e Re)$. The "equivalent diameter" D_e is also known as the "hydraulic diameter," defined as 4(flow area)/(perimeter). In this case $D_e = D_0 - D_i$.

This information is also given in tabular form in Tables 14 and 15. The latter includes the parallel plate results of Han [116]. Subscript i refers to results based on the inner wall shear stress, subscript 0 on outer wall conditions.

Turbulent Flow. Several authors have suggested that turbulent friction factors in annuli are dependent on the radius ratio (Meter and Bird [117], Walker, Whan, and Rothfus [118]). The available data, however, show that this dependence is a small one. Brighton and Jones [119] suggest that the friction factor

TABLE 13. Fully Developed Friction Factors for Laminar Flow in Annuli

r_i/r_0	fRe
0.0001	17.94
0.001	18.67
0.01	20.03
0.05	21.56
0.10	22.34
0.15	22.79
0.20	23.09
0.30	23.46
0.40	23.68
0.60	23.89
0.80	23.98
0.90	23.99
1.00	24.00

TABLE 14. Friction Factors Results for the Laminar Entry Region of an Annulus

γ_0	X/D_e Re	$r^* = 0.50$		X/D_e Re	$r^* = 0.25$	
		$(f \cdot Re)_i$	$(f \cdot Re)_o$		$(f \cdot Re)_i$	$(f \cdot Re)_o$
50.0	$0.0_3 1267$	110.8	107.6	$0.0_4 5234$	164.7	156.9
40.0		91.08	87.77		134.9	127.0
30.0		71.50	68.06		105.2	97.12
25.0		61.86	58.32		90.43	82.25
20.0	0.001020	52.43	48.72	$0.0_3 4171$	75.85	67.47
18.0		48.76	44.96		70.09	61.59
16.0		45.19	41.27		64.41	55.75
14.0		41.76	37.68		58.83	49.97
13.0		40.11	35.93		56.11	47.10
12.0		38.52	34.23		53.43	44.27
11.0		36.98	32.58		50.82	41.47
10.0	0.003371	35.53	30.99	0.001798	48.30	38.71
9.0		34.16	29.48		45.89	36.02
8.0		32.89	28.05		43.62	33.40
7.0		31.74	26.72		41.51	30.90
6.5		31.21	26.11		40.54	29.70
6.0	0.006088	30.71	25.52	0.003997	39.62	28.53
5.5		30.24	24.97		38.77	27.42
5.0		29.82	24.46		37.98	26.36
4.5		29.43	23.99		37.27	25.36
4.0	0.008590	29.07	23.55	0.006405	36.63	24.43
3.4		28.70	23.10		35.96	23.42
3.0	0.01046	28.49	22.83	0.008362	35.58	22.82
2.4		28.21	22.48		35.09	22.03
2.0	0.01318	28.06	22.29	0.01134	34.83	21.59
1.8		28.00	22.21		34.72	21.40
1.6		27.94	22.14		34.62	21.23
1.4		27.89	22.07		34.54	21.07
1.2		27.84	22.02		34.46	20.94
1.0	0.01968	27.80	21.97	0.01679	34.40	20.83
0	∞	27.72	21.86	∞	34.25	20.56

γ_0	X/D_e Re	$r^* = 0.10$		X/D_e Re	$r^* = 0.05$	
		$(f \cdot Re)_i$	$(f \cdot Re)_o$		$(f \cdot Re)_i$	$(f \cdot Re)_o$
50.0	$0.0_4 3539$	206.4	186.5	$0.0_4 3151$	235.1	196.4
40.0		170.5	150.5		197.0	158.4
30.0		134.8	114.7		159.0	120.5
25.0		117.1	96.77		140.1	101.6
20.0	$0.0_3 2729$	99.44	78.93	$0.0_3 2400$	121.4	82.75
18.0		92.48	71.81		114.0	75.23
16.0		85.59	64.73		106.6	67.73
14.0		78.82	57.68		99.45	60.27
13.0		75.50	54.18		95.93	56.55
12.0		72.24	50.69		92.49	52.85
11.0		69.06	47.23		89.15	49.17
10.0	0.001277	65.98	48.80	0.001144	85.95	45.52
9.0		63.05	40.42		82.94	41.91
8.0		60.30	37.10		80.17	38.35
7.0		57.79	33.87		77.85	34.87
6.5		56.66	32.29		76.70	33.17
6.0	0.003209	55.61	30.76	0.003015	75.78	31.50
5.5		54.66	29.27		74.99	29.87
5.0		53.81	27.83		74.37	28.30
4.5		53.08	26.46		73.90	26.79

TABLE 14. Friction Factors Results for the Laminar Entry Region of an Annulus *(Continued)*

γ_0	X/D_e Re	$r^* = 0.10$		X/D_e Re	$r^* = 0.05$	
		$(f \cdot \mathrm{Re})_i$	$(f \cdot \mathrm{Re})_o$		$(f \cdot \mathrm{Re})_i$	$(f \cdot \mathrm{Re})_o$
4.0	0.005652	52.47	25.16	0.005551	73.60	25.35
3.4		51.89	23.72		73.45	23.75
3.0	0.007813	51.59	22.85	0.007903	73.47	22.77
2.4		51.26	21.70		73.63	21.47
2.0	0.01129	51.12	21.05	0.001181	73.80	20.72
1.8		51.06	20.76		73.90	20.39
1.6		51.02	20.50		73.99	20.09
1.4		50.98	20.27		74.08	19.82
1.2		50.95	20.06		74.17	19.59
1.0	0.01792	50.93	19.89	0.01944	74.25	19.39
0	∞	50.89	19.49	∞	74.46	18.92

γ_0	X/D_e Re	$r^* = 0.02$		X/D_e Re	$r^* = 0.001$	
		$(f \cdot \mathrm{Re})_i$	$(f \cdot \mathrm{Re})_o$		$(f \cdot \mathrm{Re})_i$	$(f \cdot \mathrm{Re})_o$
50.0	$0.0_4 2945$	292.1	202.3	$0.0_4 2810$	1330	206.0
40.0		251.8	163.1		1256	166.1
30.0		211.4	124.0		1179	126.2
25.0		191.1	104.5		1139	106.3
20.0	$0.0_3 2221$	171.0	85.04	$0.0_3 2092$	1100	86.43
18.0		163.0	77.27		1085	78.49
16.0		155.2	69.52		1071	70.58
14.0		147.5	61.80		1059	62.68
13.0		143.8	57.95		1054	58.75
12.0		140.2	54.12		1051	54.82
11.0		136.8	50.31		1049	50.92
10.0	0.001070	133.6	46.52	0.001021	1049	47.03
9.0		130.7	42.77		1053	43.17
8.0		128.2	39.06		1061	39.34
7.0		126.3	35.42		1075	35.57
6.5		125.7	33.64		1085	33.72
6.0	0.002932	125.2	31.88	0.002978	1097	31.89
5.5		125.0	30.17		1112	30.09
5.0		125.1	28.50		1130	28.34
4.5		125.5	26.89		1151	26.64
4.0	0.005624	126.1	25.36	0.006135	1175	25.00
3.4		127.3	23.64		1208	23.15
3.0	0.008232	128.2	22.58	0.009424	1231	22.00
2.4		129.8	21.16		1266	20.45
2.0	0.01271	130.9	20.35	0.01535	1289	19.56
1.8		131.4	19.98		1299	19.16
1.6		131.9	19.65		1310	18.79
1.4		132.3	19.36		1319	18.46
1.2		132.8	19.10		1327	18.17
1.0	0.02163	133.1	18.88	0.02762	1335	17.92
0	∞	134.0	18.36	∞	1353	17.34

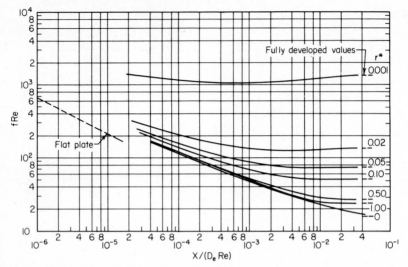

Fig. 56. Inner wall laminar friction factor in the entry region of an annulus.

is independent of r_i/r_0 for their data for $0.0625 < r_i/r_0 < 0.562$. For their results, the friction factors, as determined with water, were about 6 to 8 percent below the results of Rothfus [120], and 6 to 8 percent higher than the generally accepted values for smooth pipe flow for Reynolds number between 4000 to 17,000. The friction factors for air flow were about 1 to 10 percent higher than pipe flow values, having little dependence on the radius ratio. Their water and air data are presented in Fig. 57.

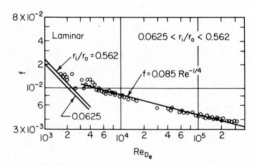

Fig. 57. Friction factors for flow in annuli.

The recommendation is to use the circular tube correlation plus 10 percent. This is approximated by a Blasius type formula

$$f = 0.085 \left(\mathrm{Re}_{D_e} \right)^{-0.25} \tag{71}$$

for $6000 < \mathrm{Re} < 300{,}000$ where $D_e = D_0 - D_i$.

TABLE 15. Friction Factor Results
for the Laminar Entry
Region of Parallel Plates

x/D_e Re	$f \cdot$ Re
$0.0_4 431$	168.4
$0.0_3 209$	88.89
$0.0_3 354$	73.14
$0.0_3 686$	57.60
0.00159	42.63
0.00260	35.73
0.00338	32.60
0.00448	29.78
0.00529	28.76
0.00567	27.83
0.00644	26.97
0.00733	26.21
0.00845	25.56
0.00910	25.27
0.00983	25.01
0.01067	24.77
0.01114	24.67
0.01165	24.57
0.01221	24.47
0.01283	24.40
0.01352	24.32
0.01427	24.25
0.01518	24.20
0.01611	24.14
0.01695	24.10
0.02059	24.06
∞	24.00

This recommendation is consistent with the data of Jonsson and Sparrow [121].

b. Velocity Distribution

Laminar Flow. Lundgren et al. [114] have presented the fully developed velocity for the annulus as

$$\frac{v_x}{\overline{V}} = 2 \left[\frac{r_0^2 - r^2 - 2r_m^2 \ln(r_0/r)}{r_0^2 + r_i^2 - 2r_m^2} \right], \quad r_m^2 = \frac{r_0^2 - r_i^2}{2 \ln(r_0/r_i)} \tag{72}$$

in which r_i and r_0 are, respectively, the inner and outer radii of the annulus, and r is the radial coordinate. The position of maximum velocity (zero shear) is denoted by r_m.

The hydrodynamic entrance velocity profiles for annuli may be found in Sparrow and Lin [122] and Heaton et al. [32], and for parallel plates in Han [116].

Figure 58 presents the hydrodynamic entry length results for the axial position where $v_{x,max} = 0.99 v_{x,max,\infty}$.

The velocity profile development has been obtained by Sparrow and Lin [122] and is shown in Fig. 59 for several radius ratios.

Transitional Flow. Walker et al. [118] include some information on the hydrodynamic transition region between laminar and turbulent flow. Walker and Rothfus [123] present velocity profiles for water in an annulus for Reynolds numbers in the transition region.

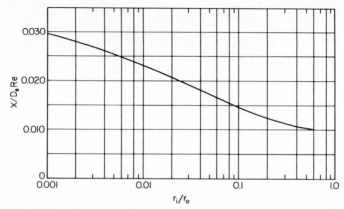

Fig. 58. Laminar hydrodynamic entry lengths for an annulus based on $v_{x,\max} = 0.99 v_{x,\max,\infty}$.

Turbulent Flow. Turbulent velocity profiles, from the air data of Brighton and Jones [119], are shown in Fig. 60. Here Y_0 is the distance from the surface of the *outer* pipe, $r_0 - r$, so that the value $Y_0 = 0$ represents data at the outer wall. Other profiles are given by Knudsen and Katz [2], and Okiishi and Serovy [124].

Brighton and Jones include data on the eddy diffusivity of momentum and the law-of-the-wall for annuli. Other law-of-the-wall data are available in Knudsen and Katz, and Rothfus et al. [125].

Leung and Kays [126] have made a recommendation for the location of maximum velocity in *turbulent* flow in annuli. They recommend the equation

$$\frac{r_m - r_i}{r_0 - r_i} = \left(\frac{r_i}{r_0}\right)^{0.343} \tag{73}$$

[where r_m is the radius of maximum velocity (zero shear)] which gives a value below the analytical *laminar* result. The data of Brighton and Jones [119], and Jonsson and Sparrow [127], agree well with this equation.

Turbulent Hydrodynamic Entry. There is some information available on the turbulent velocity entry length in annuli. Hart and Lawther [128] suggest that annular flow develops more quickly than pipe flow. This is in contrast to the data of Rothfus et al. [129] who note that the outer wall shear stress was not yet fully developed for their results at a distance of 250 equivalent diameters. Olson and Sparrow's data [130] appear to show entry lengths of less than 50 diameters, with 30 diameters sufficient for a close approach to fully developed conditions. The latter results were for both rounded and square entrance configurations.

c. Overall Pressure Drop. Overall pressure drop results, including the laminar entrance pressure drop, have been abstracted from the articles of Heaton et al. [32] and Sparrow and Lin [122]. Sparrow and Lin suggest the pressure drop in an annulus may be calculated from Eq. (20), repeated here as

$$\frac{\Delta P}{\frac{1}{2}\rho \overline{V}^2} = 4f\frac{x}{D_e} + K \tag{74}$$

where f is the downstream fully developed friction factor, K includes both the effects of momentum change between the entrance and downstream section as well as the accumulated increment in wall shear between a developing and a fully developed flow.

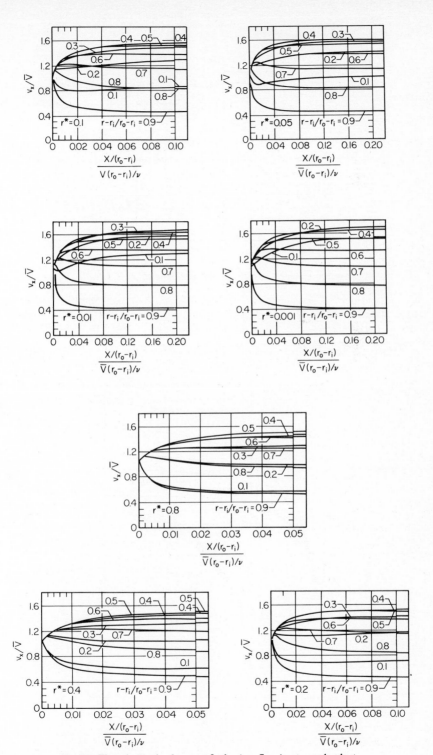

Fig. 59. The velocity development for laminar flow in an annular duct.

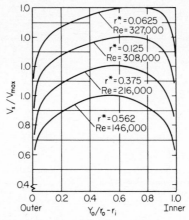

Fig. 60. Turbulent velocity profiles in an annulus, $r^* = r_i/r_0$.

The inlet velocity is assumed to be uniform. Figure 61 presents the $K(x)$ values for various annuli. Similar information is available in Fig. 62 where the overall pressure drop is directly plotted against $x/(D_e \text{ Re})$. Note that $4z = 4x/(D_e \text{ Re}) = [x/(r_0 - r_i)]/[\bar{V}(r_0 - r_i)/\nu]$.

d. Effects of Eccentricity. A certain amount of eccentricity must always be accepted in the circular tube annulus system, and occasionally the annulus may be

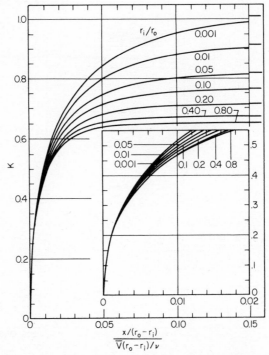

Fig. 61. Pressure drop factor K due to laminar flow development in an annulus.

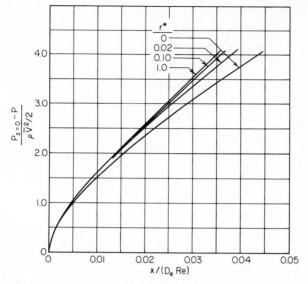

Fig. 62. Overall pressure drop with laminar flow in the entrance region of an annulus.

designed deliberately eccentric. Eccentricity causes the local shear stress, and thus the local friction coefficient, to vary circumferentially around each of the two surfaces.

Snyder and Goldstein [131] have calculated a few cases for *fully developed laminar flow,* and Jonsson and Sparrow [121] present experimental data for *turbulent flow.*

The geometry under consideration is shown on Fig. 63. Let the shear stress on the inner surface be designated at τ_i, and on the outer surface τ_0. Both of these stresses will, under eccentric conditions, be functions of angular position θ. On Figs. 64 and 65, local shear stresses calculated by Snyder and Goldstein are shown as functions of θ for an annulus radius ratio of 0.5, $r^* = r_i/r_0$, and three values of eccentricity, $e^* = e/(r_0 - r_i)$, all

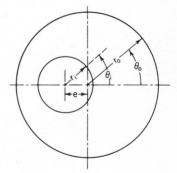

Fig. 63. Geometry for the eccentric annulus problem.

for fully developed laminar flow. In each case, the local shear stress is normalized by dividing by an *average* shear stress with respect to θ.

The average shear stresses can be used to define friction coefficients on each of the inner and outer surfaces

$$f_i = \frac{\tau_{i,\text{ave}}}{\rho \overline{V}^2/2g_c} \tag{75}$$

$$f_0 = \frac{\tau_{0,\text{ave}}}{\rho \overline{V}^2/2g_c} \tag{76}$$

An *overall average* friction coefficient can also be defined as

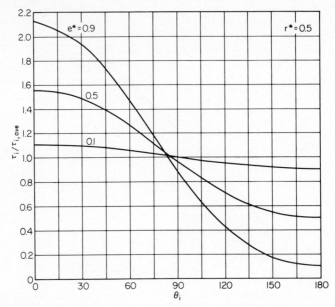

Fig. 64. Local shear stress distributions on the inner wall for fully developed laminar flow in an eccentric annulus.

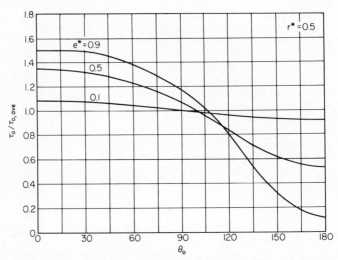

Fig. 65. Local shear stress distributions on the outer wall for fully developed laminar flow in an eccentric annulus.

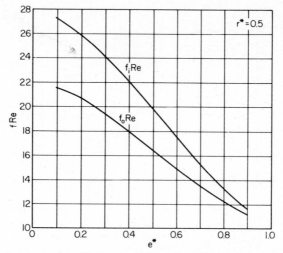

Fig. 66. Inner and outer surface friction coefficients for fully developed laminar flow in an eccentric annulus.

$$f = \frac{r^* f_i + f_0}{1 + r^*} \tag{77}$$

The overall average friction coefficient is of particular significance because it can be used to evaluate pressure drop. For an incompressible fluid the pressure gradient then becomes

$$\frac{dP}{dx} = -\frac{2f}{(r_0 - r_i)}\left(\frac{\rho \bar{V}^2}{2g_c}\right) = -\frac{4f}{D_e}\left(\frac{\rho \bar{V}^2}{2g_c}\right) \tag{78}$$

On Fig. 66 the inner and outer surface friction coefficients are plotted as functions of eccentricity for a radius ratio of 0.5. Actually it is the product fRe that is plotted, where $\mathrm{Re} = D_e \bar{V}\rho/\mu$, and $D_e = 2(r_0 - r_i)$. On Fig. 67, the *overall average* friction coefficients are plotted as a function of eccentricity for *two* values of r^*, 0.50 and 0.83. Of particular significance is the fact that the effect of r^* in this range is small.

Jonsson and Sparrow [121] present rather extensive data on *turbulent flow*, all determined experimentally. In Table 16 there is tabulated the necessary entry length x/D_e for a 2 percent approach to a fully developed velocity gradient. Significant is the fact that eccentricity increases the hydrodynamic entry length.

On Fig. 68 the *overall average* friction factor for fully developed flow is given as a function of e^* for three values of r^*. It was found that in all cases f varies as close to the -0.18 power of Reynolds number; hence $f\mathrm{Re}^{0.18}$ is plotted.

2. Heat Transfer

a. Laminar Flow

Fully Developed Velocity and Temperature Profiles. Under conditions of *constant heat rate* per unit of tube length, or *constant surface temperature* with respect to length, it is possible to have a fully developed temperature profile and a Nusselt number that is invariant with tube length. In a laminar flow, the fully developed Nusselt number will be independent of Reynolds number and Prandtl number. However, for a

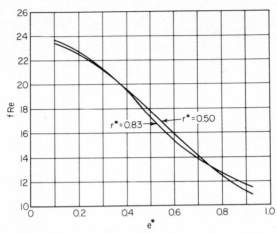

Fig. 67. Overall average friction coefficient for fully developed laminar flow in an eccentric annulus.

TABLE 16. Entry Length x/D_e Necessary for 2 Percent Approach to Fully Developed Pressure Gradient

r^*	e^*			
	0.0	0.5	0.9	1.0
0.281	29	32	38	38
0.561	26	38	59	78
0.750	28	50	69	91

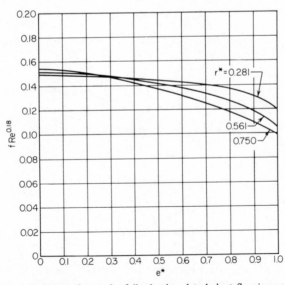

Fig. 68. Overall average friction factors for fully developed turbulent flow in an eccentric annulus.

concentric circular tube annulus there are two Nusselt numbers of interest, one for the inner surface and one for the outer surface.

Another degree of freedom in the annulus system results from the fact that the two surfaces can be heated (or cooled) independently, i.e., the heat fluxes on the two surfaces can be varied independently. This in turn affects the values of the two Nusselt numbers. We will employ the subscript *ii* to designate conditions on the inner surface when the inner surface *alone* is heated, and *oo* to designate conditions on the outer surface when the outer surface *alone* is heated (opposite surface in either case insulated). The single subscript, *i* or *o*, refers to conditions on the inner and outer surface, respectively, under *any* condition of simultaneous heating at both surfaces.

For the case of *constant heat rate* per unit of tube length,* it is possible to express Nu_i and Nu_o, the Nusselt numbers on the inner and outer surfaces respectively, for any heat flux ratio on the two surfaces, in terms of Nu_{ii} and Nu_{oo}, and a pair of *influence coefficients* θ_i^* and θ_o^*. The following equations are derived by Lundberg, McCuen, and Reynolds [132]

$$Nu_i = \frac{Nu_{ii}}{1 - (q_o''/q_i'')\theta_i^*} \tag{79}$$

$$Nu_o = \frac{Nu_{oo}}{1 - (q_i''/q_o'')\theta_o^*} \tag{80}$$

where

$$Nu_i = \frac{h_i D_e}{k}$$

$$Nu_o = \frac{h_o D_e}{k}$$

$$D_e = 4r_h = 2(r_0 - r_i)$$

$$q_i'' = h_i(T_i - T_b)$$

$$q_o'' = h_o(T_o - T_b)$$

The following equation for the temperature difference between the inner and outer surfaces can also be readily derived from the above equations

$$T_i - T_o = \frac{D_e}{k}\left[q_i''\left(\frac{1}{Nu_{ii}} + \frac{\theta_o^*}{Nu_{oo}}\right) - q_o''\left(\frac{1}{Nu_{oo}} + \frac{\theta_i^*}{Nu_{ii}}\right)\right] \tag{81}$$

Values of Nu_{ii}, Nu_{oo}, θ_o^*, and θ_i^* for *constant heat rate* are given in Table 17 as functions of tube radius ratio $r^* = r_i/r_o$. These results were calculated by Lundberg et al.

Positive q'' is taken as heat transfer *to* the fluid. The heat flux ratio can be either positive or negative.

TABLE 17. Circular Tube Annulus, Constant Heat
Rate, Fully Developed Velocity and
Temperature Profiles

r^*	Nu_{ii}	Nu_{oo}	θ_i^*	θ_o^*
0	∞	4.364	∞	0
0.05	17.81	4.792	2.18	0.0294
0.10	11.91	4.834	1.383	0.0562
0.20	8.499	4.883	0.905	0.1041
0.40	6.583	4.979	0.603	0.1823
0.60	5.912	5.099	0.473	0.2455
0.80	5.58	5.24	0.401	0.299
1.00	5.385	5.385	0.346	0.346

The case of *constant surface temperature* on one surface, with the other surface *insulated,* has also been computed by Lundberg et al. The Nusselt numbers on the active (noninsulated) surfaces are given in Table 18.

TABLE 18. Circular Tube Annulus,
Constant Surface Tem-
perature on One Surface,
the Other Insulated,
Fully Developed Veloc-
ity and Temperature
Profiles

r^*	Nu_i	Nu_o
0	∞	3.66
0.05	17.46	4.06
0.10	11.56	4.11
0.25	7.37	4.23
0.50	5.74	4.43
1.00	4.86	4.86

There are still other cases of interest that result in a fully developed temperature profile. For flow between parallel planes, $r^* = 1.00$, $Nu = 7.5407$ for the case of *both* surfaces at the same constant temperature. For constant and identical heat rates on the two surfaces of a parallel planes duct, $Nu = 8.235$.

Thermal Entry Length, Fully Developed Velocity Profile, Fundamental Solutions. Because of the two separate surfaces in the circular tube annulus system, the thermal entry length problem involves many more possible boundary conditions than does the simple circular tube. Lundberg, McCuen, and Reynolds [132] have shown that it is possible to reduce the problem to just *four fundamental solutions,* which can then be combined using superposition techniques to yield a solution for any desired boundary conditions. In this section, we will present the fundamental solutions and their use for certain families of thermal entry length problem. In Sec. B.2.d it will be shown how the fundamental solutions can be used to obtain solutions for any arbitrarily specified variations of boundary conditions in the axial direction.

In all of this discussion, it should be noted that heating is assumed to be axisymmetric, and the tubes concentric. Asymmetric heating and eccentricity are treated in separate sections.

The fundamental solutions are described as follows:

1. **First Kind.** One surface at a constant temperature, different from the uniform entering fluid temperature; the opposite surface temperature maintained constant at the entering fluid temperature.

   There are two solutions of the first kind, one for each of the two surfaces.
2. **Second Kind.** One surface with a constant heat flux; the other surface insulated; two solutions.
3. **Third Kind.** One surface at a constant temperature, different from the uniform entering temperature; the other surface insulated; two solutions.
4. **Fourth Kind.** One surface with a constant heat flux; the other surface maintained constant at the entering fluid temperature; two solutions.

The nomenclature used in describing these solutions is as follows:

$\theta_{lj}^{(k)}$ nondimensional wall surface temperature. $l = i$ or o, referring to the particular wall at which the temperature is evaluated. $j = i$ or o, referring to the wall at which $T \neq T_e$, or $q'' \neq 0$ (that is, the wall with a nonzero boundary condition). $k = 1, 2, 3,$ or 4, *kind* of solution.

$\theta_{bj}^{(k)}$ nondimensional bulk mean fluid temperature.* $j = i$ or o, referring to the wall at which $T \neq T_e$, or $q'' \neq 0$.

$\phi_{lj}^{(k)}$ nondimensional wall surface heat flux, where l, j, and k have the same significance as above.

Example. $\theta_{io}^{(2)}$ is the temperature on the *inner surface* when there is a constant heat flux on the *outer surface,* and this is a fundamental solution of the second kind.

$$\bar{x} = \frac{x/D_e}{\text{Re}\,\text{Pr}}$$

(Note that $\bar{x}$ is not the same as x^+ used for the simple circular tube.)

$$D_e = 2(r_o - r_i)$$

$$\text{Re} = \frac{D_e G}{\mu}$$

$$r^* = \frac{r_i}{r_o}$$

Values of $\theta_{lj}^{(k)}$, $\theta_{bj}^{(k)}$, and $\phi_{lj}^{(k)}$ have been calculated by Lundberg et al. [132] and are given in Table 19 as functions of r^* and $\bar{x}$. The following thermal entry length problems can be solved employing the given equations, which have been derived from superposition theory, and the results in Table 19. (These results can *also* be used to solve for arbitrarily specified axial variations in surface temperature and heat flux; the methods are described in Sec. B.2.d.)

Constant but unequal wall temperatures

$$
\left.
\begin{aligned}
T &= T_o \quad \text{on the outer wall} \\
T &= T_i \quad \text{on the inner wall}
\end{aligned}
\right\} \quad \text{for } \bar{x} \geq 0
$$

$$T = T_e \quad \text{for all } r \text{ for } \bar{x} < 0$$

Then

---

*Note that in the original reference the subscript b is replaced by m.

TABLE 19. Laminar Flow in a Circular Tube Annulus, Fundamental Solutions

r^*	$\bar{x}$	$\phi_{ii}^{(1)}$	$\phi_{oi}^{(1)}$	$\theta_{bi}^{(1)}$	$\phi_{oo}^{(1)}$	$\phi_{io}^{(1)}$	$\theta_{bo}^{(1)}$
0.05	0.0005	35.4		0.0045	13.1		0.039
	0.001	30.4		0.0076	10.1		0.060
	0.005	22.0		0.0265	5.36		0.167
	0.01	19.4	-0.0025	0.0461	3.95	-0.042	0.253
	0.05	14.7	-0.329	0.147	1.59	-6.58	0.593
	0.10	13.3	-0.544	0.191	0.915	-10.9	0.732
	0.50	12.7	-0.634	0.210	0.634	-12.7	0.790
	∞	12.7	-0.634	0.210	0.634	-12.7	0.790
0.10	0.0005	26.1		0.0066	13.3		0.037
	0.001	22.0		0.0110	10.3		0.058
	0.005	15.0		0.0362	5.46		0.161
	0.01	12.9	-0.0030	0.0613	4.04	-0.0379	0.245
	0.05	9.23	-0.431	0.183	1.67	-4.31	0.515
	0.10	8.20	-0.686	0.234	1.02	-6.86	0.701
	0.50	7.82	-0.782	0.253	0.782	-7.82	0.747
	∞	7.82	-0.782	0.253	0.782	-7.82	0.747
0.25	0.0005	19.5		0.0113	13.7		0.034
	0.001	15.8		0.0183	10.6		0.053
	0.005	9.98		0.0564	5.72		0.144
	0.01	8.24	-0.00763	0.0923	4.28	-0.0292	0.225
	0.05	5.31	-0.640	0.253	1.88	-2.56	0.528
	0.10	4.57	-0.974	0.312	1.28	-3.90	0.635
	0.50	4.33	-1.08	0.331	1.08	-4.33	0.669
	∞	4.33	-1.08	0.331	1.08	-4.33	0.669
0.50	0.0005	16.7		0.0168	14.2		0.029
	0.001	13.3		0.0266	11.1		0.045
	0.005	7.93		0.0784	6.09		0.128
	0.01	6.34	-0.0116	0.125	4.62	-0.0232	0.198
	0.05	3.71	-0.884	0.323	2.20	-1.77	0.471
	0.10	3.07	-0.132	0.390	1.62	-2.63	0.563
	0.50	2.89	-0.144	0.410	1.44	-2.89	0.590
	∞	2.89	-0.144	0.410	1.44	-2.89	0.590
1.00	0.00025	19.1		0.0145			
	0.0025	8.64		0.0660			
	0.01	5.24	-0.0170	0.162			
	0.05	2.76	-1.24	0.399			
	0.125	2.08	-1.92	0.490			
	0.50	2.00	-2.00	0.500			
	∞	2.00	-2.00	0.500			

r^*	$\bar{x}$	$\theta_{ii}^{(2)} - \theta_{bi}^{(2)}$	$\theta_{oi}^{(2)} - \theta_{bi}^{(2)}$	$\theta_{oo}^{(2)} - \theta_{bo}^{(2)}$	$\theta_{io}^{(2)} - \theta_{bo}$	$\theta_{bi}^{(2)}$	$\theta_{bo}^{(2)}$
0.05	0.0005	0.0259	-0.0_4952	0.0591	-0.00190	0.0_4952	0.00190
	0.001	0.0301	-0.0_3190	0.0748	-0.00381	0.0_3190	0.00381
	0.005	0.0413	-0.0_3952	0.125	-0.0190	0.0_3952	0.0190
	0.01	0.0465	-0.00190	0.152	-0.0380	0.00190	0.0381
	0.05	0.0553	-0.00565	0.203	-0.113	0.00952	0.190
	0.10	0.0561	-0.00610	0.208	-0.122	0.0190	0.381
	0.50	0.0561	-0.00613	0.209	-0.123	0.0952	1.90
	∞	0.0561	-0.00613	0.209	-0.123	∞	∞
0.10	0.0005	0.0331	-0.0_3181	0.0586	-0.00182	0.0_3181	0.00182
	0.001	0.0399	-0.0_3364	0.0741	-0.00364	0.0_3364	0.00364
	0.005	0.0584	-0.00182	0.124	-0.0182	0.00182	0.0182
	0.01	0.0671	-0.00362	0.150	-0.0363	0.00364	0.0364
	0.05	0.0825	-0.0107	0.202	-0.107	0.0182	0.182
	0.10	0.0839	-0.0116	0.207	-0.116	0.0364	0.364
	0.50	0.0840	-0.0116	0.207	-0.116	0.182	1.82
	∞	0.0840	-0.0116	0.207	-0.116	∞	∞

TABLE 19. Laminar Flow in a Circular Tube Annulus, Fundamental Solutions
(Continued)

r^*	$\bar{x}$	$\theta_{ii}^{(2)} - \theta_{bi}^{(2)}$	$\theta_{oi}^{(2)} - \theta_{bi}^{(2)}$	$\theta_{oo}^{(2)} - \theta_{bo}^{(2)}$	$\theta_{io}^{(2)} - \theta_{bo}$	$\theta_{bi}^{(2)}$	$\theta_{bo}^{(2)}$
0.25	0.0005	0.0429	-0.0_3400	0.0575	-0.00160	0.0_3400	0.00160
	0.001	0.0530	-0.0_3800	0.0725	-0.00320	0.0_3800	0.00320
	0.005	0.0828	-0.00397	0.121	-0.0160	0.00400	0.0160
	0.01	0.0978	-0.00794	0.147	-0.0319	0.00800	0.0320
	0.05	0.126	-0.0235	0.198	-0.0940	0.0400	0.160
	0.10	0.129	-0.0254	0.203	-0.102	0.0800	0.320
	0.50	0.129	-0.0255	0.204	-0.102	0.400	1.60
	∞	0.129	-0.0255	0.204	-0.102	∞	∞
0.50	0.0005	0.0490	-0.0_3667	0.0560	-0.00133	0.0_3667	0.00133
	0.001	0.0611	-0.0_2133	0.0704	-0.00267	0.0_2133	0.00267
	0.005	0.0987	-0.0_2667	0.117	-0.0133	0.0_2667	0.0133
	0.01	0.119	-0.0133	0.142	-0.0266	0.0133	0.0267
	0.05	0.157	-0.0392	0.193	-0.0784	0.0667	0.133
	0.10	0.162	-0.0425	0.198	-0.0851	0.133	0.267
	0.50	0.162	-0.0428	0.199	-0.0855	0.667	1.33
	∞	0.162	-0.0428	0.199	-0.0855	∞	∞
1.00	0.00025	0.0425	-0.00050			0.00050	
	0.0025	0.0893	-0.00500			0.00500	
	0.01	0.134	-0.0199			0.0200	
	0.05	0.180	-0.0589			0.100	
	0.125	0.186	-0.0642			0.250	
	0.50	0.186	-0.0643			1.00	
	∞	0.186	-0.0643			∞	

r^*	$\bar{x}$	$\phi_{ii}^{(3)}$	$\theta_{oi}^{(3)}$	$\theta_{bi}^{(3)}$	$\theta_{io}^{(3)}$	$\phi_{oo}^{(3)}$	$\theta_{bo}^{(3)}$
0.05	0.005	35.2		0.0041		13.2	0.038
	0.001	30.3		0.0072		10.2	0.060
	0.005	22.0		0.0261		5.36	0.166
	0.01	19.4	0.0_319	0.0456	0.0014	3.95	0.253
	0.05	14.6	0.078	0.170	0.343	1.57	0.616
	0.10	12.3	0.217	0.297	0.694	0.718	0.823
	0.50	3.24	0.793	0.814	0.999	0.0015	0.999
	∞	0	1.000	1.000	1.000	0	1.000
0.10	0.0005	26.1		0.0065		13.3	0.037
	0.001	21.9		0.0109		10.3	0.058
	0.005	15.0		0.0361		5.47	0.161
	0.01	12.9	0.0_332	0.0612	0.00153	4.04	0.245
	0.05	9.16	0.101	0.214	0.336	1.64	0.604
	0.10	7.37	0.269	0.363	0.684	0.770	0.813
	0.50	1.37	0.864	0.881	0.999	0.0019	1.000
	∞	0.00	1.000	1.000	1.000	0	1.000
0.25	0.0005	19.4		0.0113		13.7	0.034
	0.001	15.8		0.0183		10.6	0.053
	0.005	9.98		0.0564		5.72	0.147
	0.01	8.24	0.0_3575	0.0923	0.00159	4.28	0.225
	0.05	5.25	0.145	0.293	0.313	1.84	0.568
	0.10	3.88	0.361	0.474	0.649	0.928	0.781
	0.50	0.366	0.940	0.950	0.998	0.00412	0.999
	∞	0.00	1.000	1.000	1.000	0	1.000
0.50	0.0005	16.7		0.0166		14.2	0.029
	0.001	13.3		0.0265		11.1	0.045
	0.005	7.93		0.0782		6.09	0.128
	0.01	6.34	0.0_3871	0.125	0.00147	4.26	0.198
	0.05	3.65	0.188	0.369	0.280	2.15	0.518
	0.10	2.47	0.444	0.570	0.599	1.18	0.733
	0.50	0.116	0.974	0.980	0.996	0.0105	0.998
	∞	0.00	1.00	1.00	1.00	0.00	1.00

TABLE 19. Laminar Flow in a Circular Tube Annulus, Fundamental Solutions
(Continued)

r^*	$\bar{x}$	$\phi_{ii}^{(3)}$	$\theta_{oi}^{(3)}$	$\theta_{bi}^{(3)}$	$\theta_{io}^{(3)}$	$\phi_{oo}^{(3)}$	$\theta_{bo}^{(3)}$
1.00	0.00025						
	0.0025	8.64		0.066			
	0.01	5.24	0.00119	0.163			
	0.05	2.70	0.236	0.449			
	0.125	1.29	0.630	0.734			
	0.50	0.0337	0.990	0.993			
	∞	0.00	1.00	1.00			

r^*	$\bar{x}$	$\theta_{ii}^{(4)}$	$\phi_{oi}^{(4)}$	$\theta_{bi}^{(4)}$	$\theta_{oo}^{(4)}$	$\phi_{io}^{(4)}$	$\theta_{bo}^{(4)}$
0.05	0.0005	0.0251		0.0_4952	0.0622		0.00190
	0.001	0.0298		0.0_3190	0.0794		0.00380
	0.005	0.0425		0.0_3952	0.144		0.0190
	0.01	0.0484	-0.0_470	0.00190	0.190		0.0375
	0.05	0.0647	-0.0172	0.00842	0.393	-1.55	0.185
	0.10	0.0723	-0.0347	0.0128	0.576	-4.33	0.348
	0.50	0.0788	-0.0500	0.0166	1.31	-15.8	1.01
	∞	0.0788	-0.0500	0.0166	1.58	-20.0	1.25
0.10	0.0005	0.0328		0.0_3182	0.0611		0.00182
	0.001	0.0399		0.0_3364	0.0782		0.00364
	0.005	0.0602		0.00182	0.142		0.0182
	0.01	0.0707	-0.0_3152	0.00364	0.187		0.0363
	0.05	0.101	-0.0337	0.0161	0.383	-1.01	0.176
	0.10	0.115	-0.0684	0.0246	0.554	-2.69	0.324
	0.50	0.128	-0.0999	0.0323	1.14	-8.64	0.838
	∞	0.128	-0.100	0.0323	1.28	-10.0	0.956
0.25	0.0005	0.0428		0.0_3400	0.0592		0.00160
	0.001	0.0535		0.0_3800	0.0758		0.00320
	0.005	0.0868		0.00400	0.137		0.0160
	0.01	0.106	-0.0_3397	0.00800	0.179		0.0320
	0.05	0.165	-0.0782	0.0357	0.357	-0.580	0.153
	0.10	0.198	-0.162	0.0578	0.503	-1.44	0.271
	0.50	0.231	-0.250	0.0764	0.884	-3.76	0.585
	∞	0.231	-0.250	0.0765	0.924	-4.00	0.618
0.50	0.0005	0.0496		0.0_3667	0.0570		0.00133
	0.001	0.0623		0.00133	0.0729		0.00267
	0.005	0.105		0.00667	0.130		0.0133
	0.01	0.132	-0.0_3729	0.0133	0.169		0.0265
	0.05	0.223	-0.140	0.0604	0.325	-0.376	0.125
	0.10	0.278	-0.299	0.968	0.443	-0.888	0.215
	0.50	0.346	-0.498	0.142	0.681	-1.95	0.400
	∞	0.347	-0.500	0.142	0.693	-2.00	0.409
1.00	0.00025	0.0430		0.00050			
	0.0025	0.0943		0.0050			
	0.01	0.154	-0.00120	0.0200			
	0.05	0.279	-0.236	0.0921			
	0.125	0.394	-0.630	0.174			
	0.50	0.497	-0.990	0.248			
	∞	0.500	-1.00	0.250			

$$q_i'' = \frac{k}{D_e}\left[(T_i - T_e)\phi_{ii}^{(1)} + (T_o - T_e)\phi_{io}^{(1)}\right] \tag{82}$$

$$q_o'' = \frac{k}{D_e}\left[(T_i - T_e)\phi_{oi}^{(1)} + (T_o - T_e)\phi_{oo}^{(1)}\right] \tag{83}$$

$$T_b - T_e = (T_i - T_e)\theta_{bi}^{(1)} + (T_o - T_e)\theta_{bo}^{(1)} \tag{84}$$

Constant but unequal wall heat fluxes

$$\left.\begin{array}{l} q'' = q_i'' \text{ on the inner wall} \\ q'' = q_o'' \text{ on the outer wall} \end{array}\right\} \text{ for } \bar{x} \geq 0$$

$$T = T_e \text{ for all } r \text{ for } \bar{x} < 0$$

(Note that q_i'' and q_o'' can be either *positive into the fluid or negative out of the fluid.*)
Then

$$T_i - T_e = \frac{D_e}{k}\left[q_i''\theta_{ii}^{(2)} + q_o''\theta_{io}^{(2)}\right] \tag{85}$$

$$T_o - T_e = \frac{D_e}{k}\left[q_i''\theta_{oi}^{(2)} + q_o''\theta_{oo}^{(2)}\right] \tag{86}$$

$$T_b - T_e = \frac{D_e}{k}\left[q_i''\theta_{bi}^{(2)} + q_o''\theta_{bo}^{(2)}\right] \tag{87}*$$

Constant heat flux on one wall and constant temperature on the other

$$\left.\begin{array}{l} T = T_a \text{ on wall } a \\ q'' = q_c'' \text{ on wall } c \end{array}\right\} \text{ for } \bar{x} \geq 0$$

$$T = T_e \text{ for all } r \text{ for } \bar{x} < 0$$

Then

$$T_c - T_e = (T_a - T_e)\theta_{ca}^{(3)} + \frac{D_e}{k}q_c''\theta_{cc}^{(4)} \tag{88}$$

$$q_a'' = q_c''\phi_{ac}^{(4)} + \frac{k}{D_e}\phi_{aa}^{(3)} \tag{89}$$

$$T_b - T_e = (T_a - T_e)\theta_{ba}^{(3)} + \frac{D_e}{k}q_c''\theta_{bc}^{(4)} \tag{90}$$

In any of these cases a heat transfer coefficient and a Nusselt number can be readily defined and evaluated in terms of the fundamental solutions. For example, consider the case of a constant heat flux on the inner surface, and zero heat flux on the outer. We can develop a heat transfer coefficient and Nusselt number for the inner tube (which we would call Nu_{ii}, that is, the Nusselt number on the inner tube when the inner tube alone is heated) as follows:

From the "constant but unequal wall heat flux" formulation

*Note that $\theta_{bi}^{(2)}$ and $\theta_{bo}^{(2)}$ are given by Eqs. (94) and (95) as well as in the tabulations.

$$T_i - T_e = \frac{D_e}{k} q_i'' \theta_{ii}^{(2)} , \quad \text{since } q_o'' = 0$$

$$T_b - T_e = \frac{D_e}{k} q_i'' \theta_{bi}^{(2)}$$

Subtracting

$$T_i - T_b = \frac{D_e}{k} q_i''(\theta_{ii}^{(2)} - \theta_{bi}^{(2)})$$

But by definition

$$q_i'' = h(T_i - T_b)$$

Thus

$$1 = \frac{D_e h}{k}(\theta_{ii}^{(2)} - \theta_{bi}^{(2)})$$

or

$$Nu_{ii} = \frac{1}{\theta_{ii}^{(2)} - \theta_{bi}^{(2)}}$$

For the case of *flow between parallel planes*, $r^* = 1.00$, the thermal entry length Nusselt numbers for *constant and equal wall temperatures*, and *constant and equal wall heat fluxes*, are of particular interest. These have been computed from the fundamental solutions and are given in Tables 20 and 21. Note that these are *local* Nusselt numbers.

TABLE 20. Local Nusselt Numbers for Laminar Flow between Parallel Planes, Thermal Entry Length

Equal Wall Temperatures	
$\bar{x} = (x/D_e)/RePr$	Nu
2.5×10^{-4}	19.72
2.5×10^{-3}	9.951
1.0×10^{-2}	7.741
1.5×10^{-2}	7.582
2.5×10^{-2}	7.543
5.0×10^{-2}	7.541
7.5×10^{-2}	7.541
1.0×10^{-1}	7.541
1.25×10^{-1}	7.541
1.5×10^{-1}	7.541
2.5×10^{-1}	7.541
5.0×10^{-1}	7.541
∞	7.54072

TABLE 21. Local Nusselt Numbers for Laminar Flow between Parallel Planes, Thermal Entry Length

Equal Wall Heat Fluxes	
$\bar{x} = (x/D_e)/RePr$	Nu
2.5×10^{-4}	23.79
2.5×10^{-3}	11.86
1.0×10^{-2}	8.803
1.5×10^{-2}	8.439
2.5×10^{-2}	8.263
5.0×10^{-2}	8.236
7.5×10^{-2}	8.235
1.25×10^{-1}	8.235
2.5×10^{-1}	8.235
5.0×10^{-1}	8.235
∞	8.23529

Thermal Entry Length, Fully Developed Velocity Profile, Eigenfunction Solutions for Flow between Parallel Planes. All of the thermal entry length solutions described in the preceding section are available in the original references in infinite series form. Of particular interest is the result for *flow between parallel planes* with both surfaces at constant and equal temperatures. The results of Cess and Shaffer [133] are given in Table 22. These results are applicable to the series equations given for the circular tube, i.e., to Eqs. (35) and (37) to (40), provided that in each case x^+ is replaced by $2(x/D_e)/(\text{Re Pr})$, where D_e is equal to twice the plate spacing.

TABLE 22. Eigenvalues and Constants,
Laminar Flow between
Parallel Planes with Both
Surfaces at Equal
Constant Temperatures

n	λ_n^2	G_n
0	15.09	1.717
1	171.3	1.139
2	498	0.952

for higher n, $\lambda_n = \dfrac{16n}{\sqrt{3}} + \dfrac{20}{3\sqrt{3}}$

$$G_n = 2.68 \lambda_n^{-1/3}$$

Combined Thermal and Hydrodynamic Entry Length. Heaton, Reynolds, and Kays [32] present *Fundamental Solutions of the Second Kind* for the complete family of concentric circular tube annuli for the case of uniform temperature and velocity at the tube entrance, for Pr = 0.01, 0.70, 10.0. $\bar{x}$ is again defined as

$$\bar{x} = \frac{x/D_e}{\text{Re Pr}}$$

where

$$D_e = 2(r_o - r_i)$$

$$\text{Re} = \frac{D_e G}{\mu}$$

The results are given in Table 23, where the nomenclature is as described in connection with the discussion of the fundamental solutions.

Equations (85) to (87) are again applicable so that any case of constant but unequal wall heat flux can be readily handled. Nusselt numbers can be defined if desired, as described following Eq. (90). In fact, equations of the form of Eqs. (79) and (80) can be easily generated from the tabulated results, starting with Eqs. (85) to (87). In particular, it is worth noting that

$$\text{Nu}_{ii} = \frac{1}{\theta_{ii}^{(2)} - \theta_{bi}^{(2)}} \tag{91}$$

$$\text{Nu}_{oo} = \frac{1}{\theta_{oo}^{(2)} - \theta_{bo}^{(2)}} \tag{92}$$

TABLE 23. Laminar Flow in a Circular Tube Annulus, Fundamental Solutions of the Second Kind for Combined Thermal and Hydrodynamic Entry Length

$r^* = 0.02$
$\mathrm{Pr} = 10.00$

$\bar{x}$	$\theta_{ii} - \theta_{bi}$	$\theta_{oi} - \theta_{bi}$	$\theta_{oo} - \theta_{bo}$	$\theta_{io} - \theta_{bo}$	θ_{bi}	θ_{bo}
0.00050	0.0163	$-0.0_4 392$	0.0510	-0.00196	$0.0_4 3922$	0.001961
0.0010	0.0185	$-0.0_4 784$	0.0676	-0.00392	$0.0_4 7843$	0.003922
0.0050	0.0239	$-0.0_3 392$	0.1197	-0.0191	$0.0_3 3922$	0.01961
0.010	0.0263	$-0.0_3 784$	0.1483	-0.0391	$0.0_3 7843$	0.03922
0.050	0.0302	-0.00236	0.205	-0.118	0.003922	0.1961
0.10	0.0306	-0.00255	0.211	-0.127	0.007843	0.3922
0.50	0.0306	-0.00256	0.211	-0.128	0.03922	1.9608
∞	0.0306	-0.00256	0.211	-0.128	∞	∞

$r^* = 0.02$
$\mathrm{Pr} = 0.70$

$\bar{x}$	$\theta_{ii} - \theta_{bi}$	$\theta_{oi} - \theta_{bi}$	$\theta_{oo} - \theta_{bo}$	$\theta_{io} - \theta_{bo}$	θ_{bi}	θ_{bo}
0.00050	0.0161	$-0.0_4 392$	0.0407	-0.00196	$0.0_4 3922$	0.001961
0.0010	0.0187	$-0.0_4 784$	0.0553	-0.00392	$0.0_4 7843$	0.003922
0.0050	0.0244	$-0.0_3 392$	0.1075	-0.0196	$0.0_3 3922$	0.01961
0.010	0.0267	$-0.0_3 784$	0.1363	-0.0392	$0.0_3 7843$	0.03922
0.050	0.0302	-0.00236	0.1992	-0.0118	0.003922	0.1961
0.10	0.0306	-0.00255	0.210	-0.127	0.007843	0.3922
0.50	0.0306	-0.00256	0.211	-0.128	0.03922	1.9608
∞	0.0306	-0.00256	0.211	-0.128	∞	∞

$r^* = 0.02$
$\mathrm{Pr} = 0.01$

$\bar{x}$	$\theta_{ii} - \theta_{bi}$	$\theta_{oi} - \theta_{bi}$	$\theta_{oo} - \theta_{bo}$	$\theta_{io} - \theta_{bo}$	θ_{bi}	θ_{bo}
0.00050		$-0.0_4 392$	0.0296	-0.00196	$0.0_4 3922$	0.001961
0.0010		$-0.0_4 784$	0.0413	-0.00392	$0.0_4 7843$	0.003922
0.0050	0.0247	$-0.0_3 392$	0.0841	-0.0196	$0.0_3 3922$	0.01961
0.010	0.0275	$-0.0_3 784$	0.1112	-0.0392	$0.0_3 7843$	0.03922
0.050	0.0312	-0.00201	0.1676	-0.125	0.003922	0.1961
0.10	0.0311	-0.00205	0.1768	-0.128	0.007843	0.3922
0.50	0.0309	-0.00232	0.1913	-0.128	0.03922	1.9608
∞	0.0306	-0.00256	0.211	-0.128	∞	∞

$r^* = 0.10$, $P_r = 10.00$

0.00050	0.0327	-0.0_3182	0.0507	-0.00182	0.0_31818	0.001818
0.0010	0.0395	-0.0_3364	0.0671	-0.00364	0.0_33636	0.003636
0.0050	0.0585	-0.00182	0.1179	-0.0182	0.001818	0.01818
0.010	0.0671	-0.00363	0.1458	-0.0363	0.003636	0.03636
0.050	0.08245	-0.0107	0.200	-0.107	0.01818	0.1818
0.10	0.08390	-0.0116	0.206	-0.116	0.03636	0.3636
0.50	0.08399	-0.0116	0.207	-0.116	0.1818	1.818
∞	0.08399	-0.0116	0.207	-0.116	∞	∞

$r^* = 0.10$, $P_r = 0.70$

0.00050	0.0301	-0.0_3182	0.0405	-0.00182	0.0_31818	0.001818
0.0010	0.0380	-0.0_3364	0.0551	-0.00364	0.0_33636	0.003636
0.0050	0.0591	-0.00182	0.1069	-0.0182	0.001818	0.01818
0.010	0.0697	-0.00364	0.1353	-0.0364	0.003636	0.03636
0.050	0.0824	-0.0107	0.1960	-0.107	0.01818	0.1818
0.10	0.0839	-0.0116	0.206	-0.116	0.03636	0.3636
0.50	0.0840	-0.0116	0.207	-0.116	0.1818	1.818
∞	0.0840	-0.0116	0.207	-0.116	∞	∞

$r^* = 0.10$, $P_r = 0.01$

0.00050		-0.0_3182	0.0296	-0.00182	0.0_31818	0.001818
0.0010	0.0562	-0.0_3364	0.0413	-0.00364	0.0_33636	0.003636
0.0050	0.0676	-0.00182	0.0843	-0.0182	0.001818	0.01818
0.010	0.0847	-0.00364	0.1116	-0.0364	0.003636	0.03636
0.050	0.0854	-0.00955	0.1684	-0.113	0.01818	0.1818
0.10	0.0845	-0.0102	0.1744	-0.116	0.03636	0.3636
0.50	0.0840	-0.0110	0.1902	-0.116	0.1818	1.818
∞		-0.0116	0.2069	-0.116	∞	∞

TABLE 23. Laminar Flow in a Circular Tube Annulus, Fundamental Solutions of the Second Kind for Combined Thermal and Hydrodynamic Entry Length (Continued)

$r^* = 0.50$
$Pr = 10.00$

$\bar{x}$	$\theta_{ii} - \theta_{bi}$	$\theta_{oi} - \theta_{bi}$	$\theta_{oo} - \theta_{bo}$	$\theta_{io} - \theta_{bo}$	θ_{bi}	θ_{bo}
0.00050	0.0465	-0.0_3667	0.0503	-0.00133	0.0_36667	0.001333
0.0010	0.0592	-0.00133	0.0660	-0.00267	0.001333	0.002667
0.0050	0.0980	-0.00667	0.1143	-0.0133	0.006667	0.01333
0.010	0.1186	-0.0133	0.1410	-0.0266	0.01333	0.02667
0.050	0.1574	-0.0392	0.1925	-0.0784	0.06667	0.1333
0.10	0.1615	-0.0425	0.1979	-0.0851	0.1333	0.2667
0.50	0.1618	-0.0428	0.1985	-0.0855	0.6667	1.333
∞	0.1618	-0.0428	0.1985	-0.0855	∞	∞

$r^* = 0.50$
$Pr = 0.70$

$\bar{x}$	$\theta_{ii} - \theta_{bi}$	$\theta_{oi} - \theta_{bi}$	$\theta_{oo} - \theta_{bo}$	$\theta_{io} - \theta_{bo}$	θ_{bi}	θ_{bo}
0.00050	0.0388	-0.0_3667	0.0403	-0.00133	0.0_36667	0.001333
0.0010	0.05203	-0.00133	0.0547	-0.00267	0.001333	0.002667
0.0050	0.09551	-0.00667	0.1058	-0.0133	0.006667	0.01333
0.010	0.1174	-0.0133	0.1334	-0.0267	0.01333	0.02667
0.051	0.1575	-0.0392	0.1899	-0.0784	0.06667	0.1333
0.10	0.1616	-0.0425	0.1975	-0.0851	0.1333	0.2667
0.50	0.1618	-0.0428	0.1985	-0.0855	0.6667	1.333
∞	0.1618	-0.0428	0.1985	-0.0855	∞	∞

$r^* = 0.50$
$Pr = 0.01$

$\bar{x}$	$\theta_{ii} - \theta_{bi}$	$\theta_{oi} - \theta_{bi}$	$\theta_{oo} - \theta_{bo}$	$\theta_{io} - \theta_{bo}$	θ_{bi}	θ_{bo}
0.00050		-0.0_3667	0.0296	-0.00133	0.0_36667	0.001333
0.0010		-0.00133	0.0413	-0.00267	0.001333	0.002667
0.0050		-0.00667	0.0847	-0.0133	0.006667	0.01333
0.010	0.1060	-0.0133	0.1123	-0.0267	0.01333	0.02667
0.050	0.1563	-0.0392	0.1702	-0.0820	0.06667	0.1333
0.10	0.1608	-0.0425	0.1786	-0.0855	0.1333	0.2667
0.50	0.1618	-0.0428	0.1879	-0.0855	0.6667	1.333
∞	0.1618	-0.0428	0.1986	-0.0855	∞	∞

Parallel Plates ($r* = 1.0$)

$\bar{x}$	Pr = 10.00		Pr = 0.70		Pr = 0.01		
	$\theta_{oo} - \theta_{bo}$	$\theta_{io} - \theta_{bo}$	$\theta_{oo} - \theta_{bo}$	$\theta_{io} - \theta_{bo}$	$\theta_{oo} - \theta_{bo}$	$\theta_{io} - \theta_{bo}$	θ_{bo}
0.00050	0.0495	-0.00100	0.0398	-0.00100	0.0296	-0.00100	0.001000
0.0010	0.0643	-0.00200	0.0541	-0.00200	0.0413	-0.00200	0.0020000
0.0050	0.1087	-0.0100	0.1040	-0.0100	0.0855	-0.0100	0.01000
0.010	0.1335	-0.0199	0.1302	-0.0200	0.1136	-0.0200	0.02000
0.050	0.1803	-0.0589	0.1802	-0.0589	0.1733	-0.0655	0.1000
0.10	0.1854	-0.0639	0.1854	-0.0639	0.1808	-0.0680	0.2000
0.50	0.1857	-0.0643	0.1857	-0.0643	0.1841	-0.0646	1.000
∞	0.1857	-0.0643	0.1857	-0.0643	0.1857	-0.0643	∞

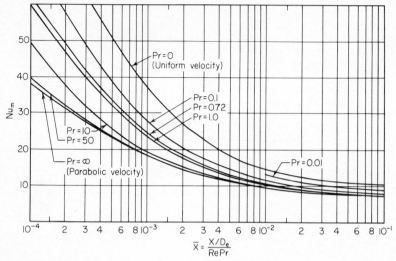

Fig. 69. Mean Nusselt number for laminar flow between parallel plates with constant and equal temperatures on both surfaces.

Hwang and Fan [134] present finite difference solutions for flow between parallel planes ($r^* = 1.00$) for constant and equal heat flux on both surfaces, and constant and equal temperatures on both surfaces. The former can be readily evaluated from the fundamental solutions of the second kind given above, but the *constant and equal surface temperature* case is, of course, different. The latter is given on Fig. 69. The *mean* Nusselt number, defined by Eq. (36), is plotted as a function of $\bar{x}$.

This problem has also been treated by Siegel and Sparrow [135], Han [136], and Stephan [137].

b. Turbulent Flow

Fully Developed Velocity and Temperature Profiles. Numerous experimental studies have been made of heat transfer in the annulus system; analytic studies have been fewer, primarily because of lack of knowledge of the apparent shear stress distribution in turbulent flow. In this section, we will present first the results of Kays and Leung [37], who carried out perhaps the most comprehensive study, and then will present some recommendations for application at very low Prandtl number (liquid metals) that differ somewhat from the Kays and Leung results.

Kays and Leung carried out experiments in an annulus, using *air* as a working medium, for various values of radius ratio, with *constant heat rate* per unit of tube length, but with various heat flux ratios at the two surfaces, including the two limiting cases of only one side heated. They then developed an analytic solution for *constant heat rate*, under fully developed conditions, based upon empirical data for the shear stress ratio, and the best available data for the eddy diffusivity at various Prandtl numbers. In addition, certain critical constants were evaluated from the experimental data for air. This analysis is internally consistent with the analysis for flow in a circular tube discussed in Sec. A.2.c, and incorporates the equation of Deissler [39] for the eddy diffusivity in the sublayers. The eddy diffusivity ratio at low Prandtl numbers is based on some of the earlier liquid metal experiments, and for this reason additional more recent results for low Prandtl number application will be presented.

The results are presented in the form of Nu_{ii} and Nu_{oo} and the two influence co-efficients θ_i^* and θ_o^*. These are given in Table 24 for a wide range of Reynolds and Prandtl numbers, and for radius ratios $r*$ of 0.10, 0.20, 0.50, 0.80, and 1.00 (parallel planes). These results are then directly applicable to Eqs. (79) to (81), that is, heat transfer rates and/or temperature differences can be easily computed for any heat-flux-ratio on the two surfaces, including heat transfer into one surface and out of the other.

Nusselt numbers for one side alone heated for $r* = 0.20$ are shown graphically on Figs. 70 and 71.

Figures 72 and 73 show some examples of a comparison of this analysis with the experimental data of Kays and Leung for air with very small temperature differences. Note particularly that the analysis tends to overpredict Nu_{ii} at Reynolds numbers below about 30,000.

Judd and Wade [138] present experimental data for Nu_{ii} for the flow of *water* in a constant heat-rate system for six different values of $r*$. The data in Table 24 agree quite satisfactorily with these results.

Dwyer [41] concludes that the more recent experimental data for liquid metals, obtained under conditions of continuous purification, yield Nusselt numbers considerably higher than the earlier data (see corresponding discussion in Sec. A.2.c for flow of liquid metals in circular tubes). Figures 74 and 75 have been computed from Dwyer's equations, and are applicable for Prandtl numbers of the order of magnitude

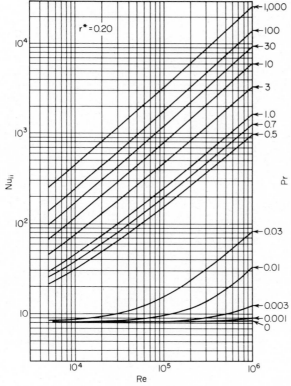

Fig. 70. Nu_{ii} for fully developed turbulent flow through an annulus (inner surface heated, outer surface insulated).

TABLE 24. Nusselt Numbers and Influence Coefficients, Fully Developed Constant Heat Rate, Turbulent Flow in a Circular Tube Annulus

$r^* = 0.10$, Heating from outer tube

Re	10^4		3×10^4		10^5		3×10^5		10^6	
Pr	Nu_{oo}	θ^*_o	Nu_{oo}	θ^*_o	Nu_{oo}	θ^*_o	Nu_{oo}	θ^*_o	Nu_{oo}	θ^*_o
0	6.00	0.077	6.12	0.079	6.32	0.081	6.50	0.084	6.68	0.085
0.001	6.00	0.077	6.12	0.079	6.40	0.082	6.60	0.082	7.20	0.082
0.003	6.00	0.077	6.24	0.081	6.55	0.083	7.34	0.082	10.8	0.071
0.01	6.13	0.076	6.50	0.081	7.80	0.077	12.1	0.067	26.4	0.052
0.03	6.45	0.076	7.95	0.075	13.7	0.065	28.2	0.051	71.8	0.036
0.5	24.8	0.039	53.4	0.032	134	0.028	320	0.025	860	0.022
0.7	29.8	0.032	66.0	0.028	167	0.024	409	0.022	1,100	0.020
1.0	36.5	0.026	81.8	0.023	212	0.021	520	0.019	1,430	0.017
3	61.5	0.013	147	0.013	395	0.012	1,000	0.012	2,830	0.011
10	99.2	0.006	246	0.006	685	0.006	1,780	0.006	5,200	0.006
30	143	0.003	360	0.003	1,030	0.003	2,720	0.003	8,030	0.003
100	205	0.002	525	0.002	1,500	0.002	4,030	0.002	12,100	0.002
1,000	378		980		2,850		7,600		23,000	

$r^* = 0.10$, Heating from core tube

Re	10^4		3×10^4		10^5		3×10^5		10^6	
Pr	Nu_{ii}	θ^*_i	Nu_{ii}	θ^*_i	Nu_{ii}	θ^*_i	Nu_{ii}	θ^*_i	Nu_{ii}	θ^*_i
0	11.5	1.475	11.5	1.502	11.5	1.500	11.5	1.460	11.6	1.477
0.001	11.5	1.475	11.5	1.502	11.5	1.480	11.7	1.462	12.3	1.410
0.003	11.5	1.475	11.5	1.475	11.7	1.473	12.6	1.391	17.0	1.124
0.01	11.8	1.482	11.8	1.442	13.5	1.323	19.4	1.090	39.0	0.760
0.03	12.5	1.472	14.1	1.330	21.8	1.027	42.0	0.760	103	0.526
0.5	40.8	0.632	81.0	0.486	191	0.394	443	0.339	1,160	0.294
0.7	48.5	0.512	98.0	0.407	235	0.338	550	0.292	1,510	0.269
1.0	58.5	0.412	120	0.338	292	0.286	700	0.256	1,910	0.232
3	93.5	0.202	206	0.175	535	0.162	1,300	0.152	3,720	0.148
10	140	0.089	328	0.081	890	0.078	2,300	0.078	6,700	0.077
30	195	0.041	478	0.039	1,320	0.038	3,470	0.038	10,300	0.040
100	272	0.017	673	0.015	1,910	0.015	5,030	0.016	15,200	0.018
1,000	486	0.004	1,240	0.003	3,600	0.003	9,600	0.004	28,700	0.004

$r^* = 0.20$, Heating from outer tube

Pr \ Re	10^4		3×10^4		10^5		3×10^5		10^6	
	Nu_{oo}	θ^*_o	Nu_{oo}	θ^*_o	Nu_{oo}	θ^*_o	Nu_{oo}	θ^*_o	Nu_{oo}	θ^*_o
0	5.83	0.140	5.92	0.145	6.10	0.151	6.16	0.152	6.35	0.157
0.001	5.83	0.140	5.92	0.144	6.10	0.151	6.30	0.154	6.92	0.153
0.003	5.83	0.140	6.00	0.146	6.22	0.150	6.90	0.150	10.2	0.136
0.01	5.95	0.140	6.20	0.146	7.40	0.144	11.4	0.131	24.6	0.102
0.03	6.22	0.140	7.55	0.140	12.7	0.125	26.3	0.098	80.0	0.074
0.5	22.5	0.071	51.5	0.064	130	0.055	310	0.049	823	0.044
0.7	29.4	0.063	64.3	0.055	165	0.049	397	0.044	1,070	0.040
1.0	35.5	0.051	80.0	0.046	206	0.042	504	0.039	1,390	0.035
3	60.0	0.026	145	0.026	390	0.024	980	0.024	2,760	0.023
10	98.0	0.013	243	0.013	680	0.012	1,750	0.012	4,980	0.012
30	142	0.004	360	0.006	1,030	0.006	2,700	0.006	7,850	0.006
100	205	0.003	520	0.003	1,500	0.003	4,000	0.003	12,000	0.003
1,000	380	0.001	980	0.001	2,830	0.001	7,500	0.001	22,500	0.001

$r^* = 0.20$, Heating from core tube

Pr \ Re	10^4		3×10^4		10^5		3×10^5		10^6	
	Nu_{ii}	θ^*_i	Nu_{ii}	θ^*_i	Nu_{ii}	θ^*_i	Nu_{ii}	θ^*_i	Nu_{ii}	θ^*_i
0	8.40	1.009	8.30	1.028	8.30	1.020	8.30	1.038	8.30	1.020
0.001	8.40	1.009	8.40	1.040	8.30	1.020	8.40	1.014	8.90	0.976
0.003	8.40	1.009	8.40	1.027	8.50	1.025	9.05	0.980	12.5	0.834
0.01	8.50	1.000	8.60	1.018	9.70	0.944	14.0	0.796	33.6	0.748
0.03	9.00	1.012	10.1	0.943	15.8	0.771	31.7	0.600	81.0	0.374
0.5	31.2	0.520	64.0	0.398	157	0.333	370	0.295	980	0.262
0.7	38.6	0.412	79.8	0.338	196	0.286	473	0.260	1,270	0.235
1.0	46.8	0.339	99.0	0.284	247	0.248	600	0.229	1,640	0.209
3	77.4	0.172	175	0.151	465	0.143	1,150	0.137	3,250	0.135
10	120	0.120	290	0.074	800	0.072	2,050	0.073	6,000	0.077
30	172	0.036	428	0.034	1,210	0.035	3,150	0.036	9,300	0.038
100	243	0.014	617	0.014	1,760	0.015	4,630	0.016	13,800	0.016
1,000	448	0.004	1,400	0.002	3,280	0.002	8,800	0.004	26,000	0.003

TABLE 24. Nusselt Numbers and Influence Coefficients, Fully Developed Constant Heat Rate, Turbulent Flow in a Circular Tube Annulus *(Continued)*

$r^* = 0.50$, Heating from outer tube

Re \ Pr	10^4		3×10^4		10^5		3×10^5		10^6	
	Nu_{oo}	θ^*_o	Nu_{oo}	θ^*_o	Nu_{oo}	θ^*_o	Nu_{oo}	θ^*_o	Nu_{oo}	θ^*_o
0	5.66	0.281	5.78	0.294	5.80	0.296	5.83	0.302	5.95	0.310
0.001	5.66	0.281	5.78	0.294	5.80	0.296	5.92	0.302	6.40	0.304
0.003	5.66	0.281	5.78	0.294	5.85	0.294	6.45	0.301	9.00	0.278
0.01	5.73	0.281	5.88	0.289	6.80	0.289	10.3	0.264	22.6	0.217
0.03	6.03	0.279	7.05	0.284	11.6	0.258	24.4	0.214	64.0	0.163
0.5	22.6	0.162	49.8	0.142	125	0.123	298	0.111	795	0.098
0.7	28.3	0.137	62.0	0.119	158	0.107	380	0.097	1,040	0.090
1.0	34.8	0.111	78.0	0.101	200	0.092	490	0.085	1,340	0.078
3	60.5	0.059	144	0.058	384	0.055	960	0.054	2,730	0.052
10	100	0.028	246	0.028	680	0.028	1,750	0.028	5,030	0.028
30	143	0.013	365	0.013	1,030	0.014	2,700	0.014	8,000	0.015
100	207	0.006	530	0.006	1,500	0.006	4,000	0.006	12,000	0.006
1,000	387	0.001	990	0.001	2,830	0.001	7,600	0.001	23,000	0.001

$r^* = 0.50$, Heating from core tube

Re \ Pr	10^4		3×10^4		10^5		3×10^5		10^6	
	Nu_{ii}	θ^*_o	Nu_{ii}	θ^*_o	Nu_{ii}	θ^*_o	Nu_{ii}	θ^*_o	Nu_{ii}	θ^*_o
0	6.28	0.620	6.30	0.632	6.30	0.651	6.30	0.659	6.30	0.654
0.001	6.28	0.620	6.30	0.632	6.30	0.651	6.40	0.659	6.75	0.644
0.003	6.28	0.620	6.30	0.632	6.40	0.656	6.85	0.637	9.40	0.585
0.01	6.37	0.622	6.45	0.636	7.30	0.623	10.8	0.540	23.2	0.427
0.03	6.75	0.627	7.53	0.598	12.0	0.533	24.8	0.430	65.5	0.333
0.5	24.6	0.343	52.0	0.292	130	0.253	310	0.229	835	0.208
0.7	30.9	0.300	66.0	0.258	166	0.225	400	0.206	1,080	0.185
1.0	38.2	0.247	83.5	0.218	212	0.208	520	0.183	1,420	0.170
3	66.8	0.129	152	0.121	402	0.115	1,010	0.114	2,870	0.111
10	106	0.059	260	0.059	715	0.059	1,850	0.059	5,400	0.061
30	153	0.028	386	0.027	1,080	0.028	2,850	0.031	8,400	0.032
100	220	0.006	558	0.006	1,600	0.006	4,250	0.007	12,600	0.007
1,000	408	0.002	1,040	0.002	3,000	0.002	8,000	0.002	24,000	0.002

r* = 0.80, Heating from outer tube

Pr \ Re	Nu_{oo} (10^4)	θ_o^*	Nu_{oo} (3×10^4)	θ_o^*	Nu_{oo} (10^5)	θ_o^*	Nu_{oo} (3×10^5)	θ_o^*	Nu_{oo} (10^6)	θ_o^*
0	5.65	0.379	5.70	0.386	5.75	0.398	5.80	0.407	5.85	0.409
0.001	5.65	0.379	5.70	0.386	5.75	0.398	5.88	0.406	6.25	0.407
0.003	5.65	0.379	5.70	0.386	5.84	0.397	6.35	0.407	8.80	0.374
0.01	5.75	0.381	5.85	0.386	6.72	0.390	9.95	0.361	21.0	0.286
0.03	6.10	0.388	6.90	0.380	11.1	0.339	23.2	0.290	62.0	0.216
0.5	22.4	0.225	48.0	0.191	121	0.169	292	0.153	790	0.136
0.7	28.0	0.192	61.0	0.166	156	0.150	378	0.136	1,020	0.122
1.0	34.8	0.159	76.5	0.141	197	0.129	483	0.120	1,330	0.111
3	61.3	0.083	142	0.079	382	0.078	960	0.076	2,730	0.073
10	100	0.039	243	0.039	670	0.039	1,740	0.040	5,050	0.040
30	146	0.019	365	0.019	1,040	0.020	2,720	0.021	8,000	0.022
100	209	0.008	533	0.008	1,500	0.009	4,000	0.009	12,000	0.010
1,000	385	0.002	1,000	0.002	2,870	0.002	7,720	0.002	23,000	0.002

r* = 0.80, Heating from core tube

Pr \ Re	Nu_{ii} (10^4)	θ_i^*	Nu_{ii} (3×10^4)	θ_i^*	Nu_{ii} (10^5)	θ_o^*	Nu_{ii} (3×10^5)	θ_i^*	Nu_{ii} (10^6)	θ_o^*
0	5.87	0.489	5.90	0.505	5.92	0.515	5.95	0.525	5.97	0.528
0.001	5.87	0.489	5.90	0.505	5.92	0.515	6.00	0.518	6.33	0.516
0.003	5.87	0.489	5.90	0.505	6.03	0.485	6.40	0.504	8.80	0.468
0.01	5.95	0.485	6.07	0.506	6.80	0.493	10.0	0.452	21.7	0.382
0.03	6.20	0.478	7.05	0.485	11.4	0.445	23.0	0.357	61.0	0.276
0.5	22.9	0.268	49.5	0.250	123	0.214	296	0.193	800	0.174
0.7	28.5	0.244	62.3	0.212	157	0.186	384	0.172	1,050	0.160
1.0	35.5	0.200	78.3	0.181	202	0.166	492	0.154	1,350	0.140
3	63.0	0.108	145	0.102	386	0.097	973	0.096	2,750	0.093
10	102	0.051	248	0.051	693	0.052	1,790	0.051	5,150	0.051
30	147	0.027	370	0.027	1,050	0.028	2,750	0.029	8,100	0.030
100	215	0.010	540	0.010	1,540	0.010	4,050	0.011	12,100	0.012
1,000	393	0.002	1,000	0.002	2,890	0.002	7,700	0.002	23,000	0.002

TABLE 24. Nusselt Numbers and Influence Coefficients, Fully Developed Constant Heat Rate, Turbulent Flow in a Circular Tube Annulus *(Continued)*

$r^* = 1.00$, Parallel plates

Re	10^4		3×10^4		10^5		3×10^5		10^6	
Pr	Nu	θ^*	Nu	θ^*	Nu	θ^*	Nu	θ^*	Nu	θ^*
0	5.70	0.428	5.78	0.445	5.80	0.456	5.80	0.460	5.80	0.468
0.001	5.70	0.428	5.78	0.445	5.80	0.456	5.88	0.460	6.23	0.460
0.003	5.70	0.428	5.80	0.445	5.90	0.450	6.32	0.450	8.62	0.422
0.01	5.80	0.428	5.92	0.445	6.70	0.440	9.80	0.407	21.5	0.333
0.03	6.10	0.428	6.90	0.428	11.0	0.390	23.0	0.330	61.2	0.255
0.5	22.5	0.256	47.8	0.222	120	0.193	290	0.174	780	0.157
0.7	27.8	0.220	61.2	0.192	155	0.170	378	0.156	1,030	0.142
1.0	35.0	0.182	76.8	0.162	197	0.148	486	0.138	1,340	0.128
3	60.8	0.095	142	0.092	380	0.089	966	0.087	2,700	0.084
10	101	0.045	214	0.045	680	0.045	1,760	0.045	5,080	0.046
30	147	0.021	367	0.022	1,030	0.022	2,720	0.023	8,000	0.024
100	210	0.009	514	0.009	1,520	0.010	4,030	0.010	12,000	0.011
1,000	390	0.002	997	0.002	2,880	0.002	7,650	0.002	23,000	0.002

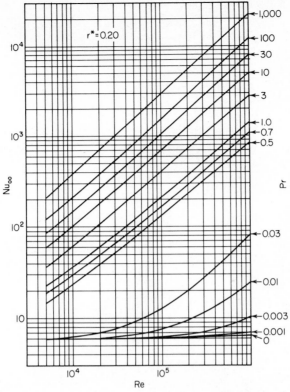

Fig. 71. Nu_{oo} for fully developed turbulent flow through an annulus (outer surface heated, inner surface insulated).

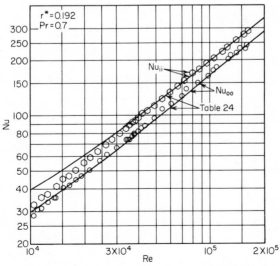

Fig. 72. Comparison of constant heat rate analysis for fully developed turbulent flow through an annulus with the data of Kays and Leung [37].

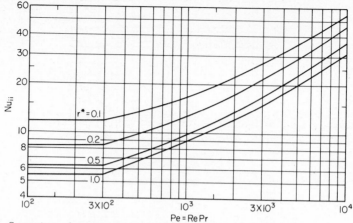

Fig. 73. Comparison of constant heat rate analysis for fully developed turbulent flow through an annulus with the data of Kays and Leung [37].

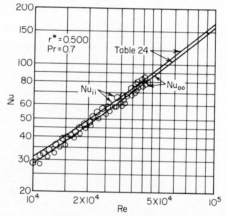

Fig. 74. Nu_{ii} for turbulent flow of a liquid metal through an annulus.

of 0.01. In commercial applications, it is not certain whether the higher Nusselt numbers given in these figures will actually be obtained. The experimental data of Duchatelle and Vautrey [139] for flow of NaK($0.022 < Pr < 0.01$) between parallel planes ($r* = 1.00$), with constant heat rate on one side and the other side insulated, were obtained in very carefully conducted experiments with continuous purification. On Fig. 76, these data are compared with the results of Dwyer from Fig. 74, and also the corresponding results from Table 24. It is apparent that the results from Table 24 are conservative at the higher values of Peclet number.

Additional analytic results for $r* = 1.00$ will be found in Tables 26 and 27.

An approximation that has been used for turbulent flow heat transfer to a liquid metal is the so-called "slug-flow" approximation. This is an analytic solution based on a *uniform velocity* distribution, and a pure molecular conduction heat transfer mechanism. Presumably this is what is approached as $Pr \to 0$ and a uniform velocity profile is substituted for the only slightly rounded turbulent flow velocity profile. Hartnett and Irvine [140] present computed "slug-flow" Nusselt

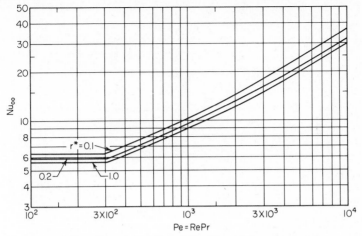

Fig. 75. Nu_{oo} for turbulent flow of a liquid metal through an annulus.

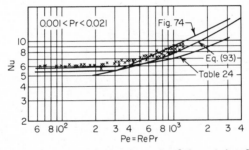

Fig. 76. Comparison of data of Duchatelle and Vautrey [139] for turbulent flow of a liquid metal between parallel plates ($r^* = 1.00$) with various analyses. Experiments were for constant heat rate on one wall and the other wall insulated.

numbers for fully developed temperature profiles for the annulus system for *axially constant and equal surface temperature* on both surfaces; *axially constant heat flux*, but *equal temperature* on both surfaces; and *constant heat flux on one surface with the other surface insulated*. These are given on Figs. 77 to 79, respectively.

The "slug" Nusselt number for the first two cases is the *average* for the two surfaces, i.e., the *total* heat transfer rate from both surfaces is the product of the heat transfer coefficient from the Nusselt number, the temperature difference between the surfaces and the bulk fluid, and the *total* surface area on *both* surfaces. Hartnett and Irvine suggest that these results can be used in conjunction with the following equation as a good approximation for liquid metals

$$Nu = \frac{2}{3} Nu_s + 0.015 \, Pe^{0.8} \tag{93}$$

For the case of $r^* = 1.00$, parallel planes, and constant heat-rate heating from one side, Eq. (93) is plotted on Fig. 76 where it can be compared with experimental data, and with the results in Table 24 and Fig. 74.

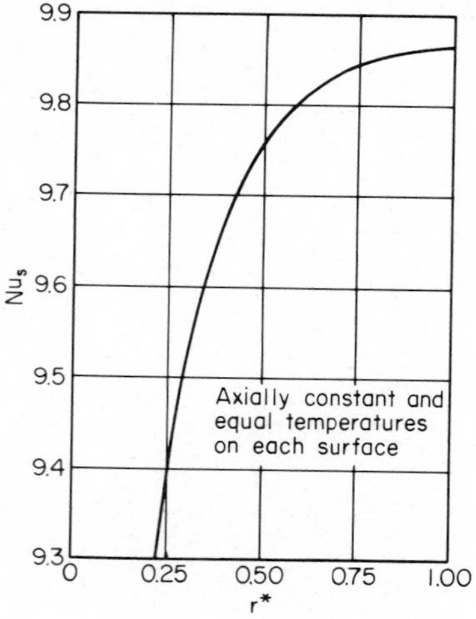

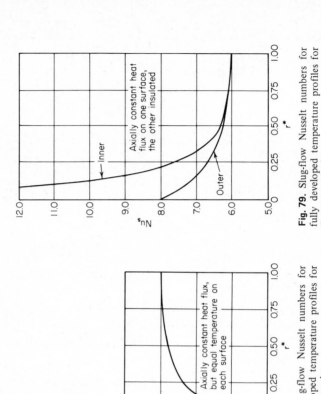

Fig. 79. Slug-flow Nusselt numbers for fully developed temperature profiles for flow in an annulus.

Fig. 78. Slug-flow Nusselt numbers for fully developed temperature profiles for flow in an annulus.

Fig. 77. Slug-flow Nusselt numbers for fully developed temperature profiles for flow in an annulus.

Thermal Entry Length, Fully Developed Velocity Profile. Fundamental solutions of the second kind (constant heat rate per unit of tube length) have been obtained *experimentally* by Kays and Leung for air (Pr = 0.7) for four values of r^* and a number of different values of Reynolds number. These are presented in Figs. 80 to 83.

These results are directly applicable to Eqs. (85) to (87) which were presented in connection with the corresponding laminar flow results. Thus, any combination of heat flux on the two surfaces can be calculated, and local heat transfer coefficients and local Nusselt numbers can also be readily evaluated.

In using these results, the nondimensional bulk-mean temperature is needed in addition to what is given on the figures. $\theta_{bi}^{(2)}$ and $\theta_{bo}^{(2)}$ may be evaluated from the following equations

$$\theta_{bi}^{(2)} = \frac{4\bar{x}r^*}{(1+r^*)} = \frac{4(x/D_e)r^*}{\mathrm{Re}\,\mathrm{Pr}(1+r^*)} \tag{94}$$

$$\theta_{bo}^{(2)} = \frac{4\bar{x}}{(1+r^*)} = \frac{4(x/D_e)}{\mathrm{Re}\,\mathrm{Pr}(1+r^*)} \tag{95}$$

For *flow between parallel planes* ($r^* = 1.00$), eigenfunction solutions for the turbulent flow thermal entry-length problem have been developed by Hatton and Quarmby [141, 142]. Two cases are considered: (1) one side at constant temperature and the other side insulated (fundamental solution of the third kind), and (2) one side at constant heat rate and the other side insulated (fundamental solutions of the second kind).

Since fundamental solutions of the first and fourth kinds are not available, it is not possible to solve for all the boundary conditions indicated for the corresponding laminar flow problem, Eqs. (82) to (94). However, for *one side at constant temperature and the other side insulated,* the simple thermal entry-length problem can be completely handled, and the necessary eigenvalues and constants are given in Table 25 for various values of Reynolds number and Prandtl number. These results can be used directly in Eqs. (35) to (40), provided that x^+ in the series expressions is defined as $(x/D_e)/(\mathrm{Re}\,\mathrm{Pr})$, and D_e, the hydraulic diameter, is twice the plate separation.

The *constant heat-rate* results can be used to solve for the thermal entry-length behavior for any desired combination of heat flux on the two surfaces. Although available in eigenfunction form in the original references, the fundamental solutions, for convenience, have been calculated for presentation here. Nu_{11} and θ_1^* are given in Tables 26 and 27 as functions of Reynolds number and Prandtl number.* These results can be used directly in Eqs. (79) to (81) for any specified heat flux ratio, or the nondimensional fundamental solution temperatures can be evaluated from the following equations

$$\theta_{11}^{(2)} - \theta_{b1}^{(2)} = \frac{1}{\mathrm{Nu}_{11}} \tag{96}$$

$$\theta_{21}^{(2)} - \theta_{b1}^{(2)} = -\frac{\theta_1^*}{\mathrm{Nu}_{11}} \tag{97}$$

$$\theta_{b1}^{(2)} = 2\bar{x} = 2\frac{x/D_e}{\mathrm{Re}\,\mathrm{Pr}} \tag{98}$$

*For flow between parallel planes, the two surfaces are designated as 1 and 2 rather than i and o.

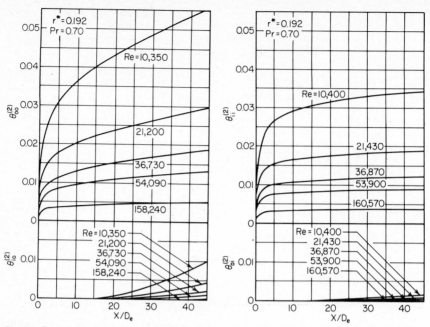

Fig. 80. Experimental results of Kays and Leung for the thermal entry length, for a fully developed turbulent velocity profile, in an annulus. Results are presented as the non-dimensional fundamental solutions of the second kind.

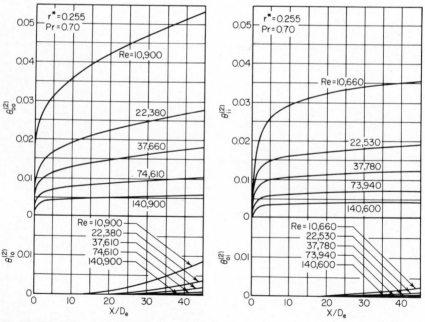

Fig. 81. Experimental results of Kays and Leung for the thermal entry length, for a fully developed turbulent velocity profile, in an annulus. Results are presented as the non-dimensional fundamental solutions of the second kind.

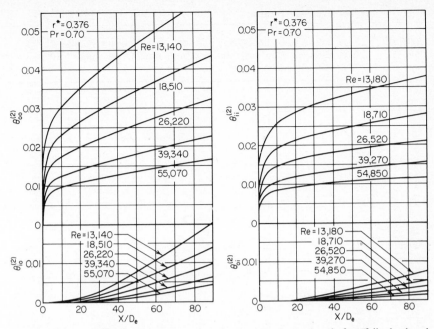

Fig. 82. Experimental results of Kays and Leung for the thermal entry length, for a fully developed turbulent velocity profile, in an annulus. Results are presented as the non-dimensional fundamental solutions of the second kind.

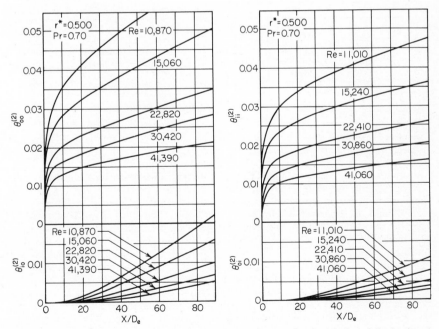

Fig. 83. Experimental results of Kays and Leung for the thermal entry length, for a fully developed turbulent velocity profile, in an annulus. Results are presented as the non-dimensional fundamental solutions of the second kind.

TABLE 25. Eigenvalues and Constants, Turbulent Flow between Parallel Planes, One Side at Constant Temperature, the Other Side Insulated, Thermal Entry Length*

Pr	0.1		0.1		0.1	
Re	7096		73612		494576	
n	λ_n^2	G_n	λ_n^2	G_n	λ_n^2	G_n
0	16.7	1.86	59.4	6.87	228	27.1
1	158	1.34	611	3.78	2685	10.9
2	450	1.03	1788	2.54	8200	6.47
3	893	0.859	3535	2.14	16210	5.37
4	1473	0.794	5830	2.04	26700	5.10
5	2201	0.762	8660	1.95	39820	4.85
Pr	1.0		1.0		1.0	
Re	7096		73612		494576	
n	λ_n^2	G_n	λ_n^2	G_n	λ_n^2	G_n
0	52.1	6.28	296	36.1	1303	160
1	656	2.17	4400	9.44	2244	33.0
2	2067	1.16	14100	4.85	7360	16.3
3	4175	0.939	22150	3.73	14890	12.6
4	6775	0.936	46850	3.27	24530	10.9
5	9945	0.981	70100	2.87	36660	9.55
Pr	10.0		10.0		10.0	
Re	7096		73612		494576	
n	λ_n^2	G_n	λ_n^2	G_n	λ_n^2	G_n
0	153	19.1	1038	129.4	5180	648
1	4880	1.81	37200	10.2	199600	46.9
2	17000	0.914	132000	4.45	696000	20.2
3	34250	0.865	265600	3.26	1427000	14.5
4	53800	1.01	442000	2.73	2368000	11.8
5	76700	1.18	663000	2.35	3556000	9.65

*Let $x^+ = (x/D_e)/(RePr)$ to use these results with Eqs. (35) to (40).

Note that for Pr = 0.01, and Pr = 0.10, the two tables overlap. Table 26 was calculated assuming that the eddy diffusivity for heat and momentum are equal regardless of Prandtl number, Hatton and Quarmby [141]. Table 27, on the other hand, was calculated by Hatton and Quarmby [142] using the eddy diffusivity ratios of Azer and Chao [143]. The difference represents the presently existing uncertainty in the eddy diffusivity ratio for very low Prandtl number fluids, and is related directly to the problem of purity in liquid metal experiments as discussed in connection with fully developed flow. Tentatively, it can be concluded that the behavior at low Prandtl numbers indicated in Table 26 might be obtainable under conditions of extreme purity, while the data indicated in Table 27 may be more representative of commercial applications. The results in Table 27 are reasonably consistent with the data of Table 24 for $r^* = 1.00$, while Table 26 corresponds more closely with Fig. 74.

The results of Table 26 are compared with NaK experimental data of Duchatelle and Vautrey [139] on Fig. 84. The experimental Prandtl number varied, but averaged about 0.02, with Reynolds number varying up to about 60,000. Heating was from one side only, and heat was added at a constant rate per unit of duct length.

Combined Thermal and Hydrodynamic Entry Length.   There appear to be little data available for the annulus system under these conditions. Duchatelle and Vautrey did

TABLE 26. Nusselt Numbers and Influence Coefficients, Turbulent Flow between Parallel Planes, One Side at Constant Heat Rate, the Other Insulated, Thermal Entry Length Solution (Equal Eddy Diffusivities Assumed Throughout)

			Pr = 0.01			
Re	7104		73712		495164	
x/D_e	Nu_{11}	θ_1^*	Nu_{11}	θ_1^*	Nu_{11}	θ_1^*
1	8.33	0.233	23.5	0.076	60.2	0.058
3	6.52	0.378	16.1	0.133	45.1	0.063
10	6.11	0.417	11.3	0.284	32.0	0.131
30	6.10	0.417	9.36	0.399	24.8	0.265
100	6.10	0.417	9.13	0.414	21.9	0.349
300	6.10	0.417	9.13	0.414	21.8	0.353
			Pr = 0.1			
Re	7096		73612		494576	
x/D_e	Nu_{11}	θ_1^*	Nu_{11}	θ_1^*	Nu_{11}	θ_1^*
1	19.7	0.056	75.2	0.018	241	0.005
3	14.3	0.122	56.2	0.016	194	0.023
10	10.7	0.267	42.4	0.115	155	0.062
30	9.44	0.352	34.8	0.233	132	0.147
100	9.34	0.359	32.1	0.290	120	0.219
300	9.34	0.359	32.1	0.291	120	0.219
			Pr = 1.0			
Re	7096		73612		494576	
x/D_e	Nu_{11}	θ_1^*	Nu_{11}	θ_1^*	Nu_{11}	θ_1^*
1	47.3	0.013	234	0.005	940	0.000
3	37.9	0.033	203	0.018	851	0.009
10	31.5	0.089	177	0.049	761	0.030
30	28.0	0.173	160	0.114	697	0.077
100	27.1	0.200	152	0.155	661	0.123
			Pr = 10.0			
Re	7096		73612		494576	
x/D_e	Nu_{11}	θ_1^*	Nu_{11}	θ_1^*	Nu_{11}	θ_1^*
1	102	0.004	602	0.004	2925	0.000
3	88.6	0.012	575	0.008	2829	0.003
10	81.9	0.027	550	0.018	2724	0.010
30	78.6	0.057	532	0.041	2640	0.027
100	77.5	0.070	522	0.057	2590	0.045

obtain data for flow between parallel planes ($r^* = 1.00$) for NaK with a combined entry length, and these are given on Fig. 85. The operating conditions are approximately the same as discussed in connection with the preceding figure, and it will be noted by comparison that there is a rather substantial effect of the hydrodynamic entry length.

The lack of entry length data, especially at low Prandtl numbers, for the annulus system for all radius ratios other than $r^* = 1.00$ is not so restricting as might appear.

TABLE 27. Nusselt Numbers and Influence Coefficients, Turbulent Flow between Parallel Planes, One Side at Constant Heat Rate, the Other Insulated, Thermal Entry Length Solution (Eddy Diffusivity Results of Azer and Chao Used)

	Pr = 0.01					
Re	7104		73712		495164	
x/D_e	Nu_{11}	θ_1^*	Nu_{11}	θ_1^*	Nu_{11}	θ_1^*
3	6.23	0.377	14.3	0.117	37.4	0.060
10	5.76	0.424	9.47	0.251	25.5	0.105
30	5.76	0.424	7.32	0.400	18.7	0.220
100	5.76	0.424	6.86	0.438	15.1	0.354
300	5.76	0.424	6.86	0.438	14.7	0.374

	Pr = 0.10					
Re	7104		73712		495164	
x/D_e	Nu_{11}	θ_1^*	Nu_{11}	θ_1^*	Nu_{11}	θ_1^*
3	13.0	0.110	49.3	0.048	183	0.043
10	9.28	0.248	36.2	0.099	146	0.069
30	7.65	0.371	28.6	0.215	122	0.150
100	7.40	0.391	25.0	0.310	108	0.238
300	7.40	0.391	24.7	0.317	106	0.248

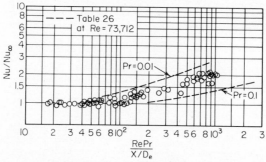

Fig. 84. Comparison of the experimental thermal entry results of Duchatelle and Vautrey for flow of a liquid metal between parallel plates, with the analysis shown in Table 26. Dashed line, Table 26, at Re = 73,712.

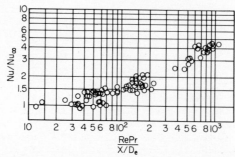

Fig. 85. Data of Duchatelle and Vautrey for the combined thermal and hydrodynamic entry length for flow of NaK between parallel planes with one wall heated and the other wall insulated.

In turbulent flow, as may be verified by examination of the data for fully developed conditions, the parallel planes performance differs little from that of an annulus with $r*$ down to at least 0.50.

c. Effect of Circumferential Heat Flux Variation. Sutherland and Kays [144] have considered the problem of both laminar and turbulent flow in the concentric annulus system where the heat flux varies circumferentially on either or both the inner and outer surfaces. The calculations were made for *fully developed constant axial heat rate* conditions, but for any specified circumferential distribution of heat flux that can be expressed in a Fourier expansion. In the case of turbulent flow, the velocity and eddy diffusivity distributions were taken from Kays and Leung [37], so that the results are completely consistent with the axisymmetric heating data given in Table 24. The laminar flow calculations are exact.

For brevity only, the first terms of the expansion will be presented here, which means that only *cosine* variations of heat flux can be calculated. However, this covers quite adequately the great majority of cases that might be encountered in practice.

The results of this analysis are based on the following scheme. Let θ be an angular coordinate. Suppose the heat flux on each of the inner and outer surfaces can be expressed in the form

$$q_i''(\theta) = q_i''(1 + b_i \cos \theta) \tag{99}$$

$$q_o''(\theta) = q_o''(1 + b_o \cos \theta) \tag{100}$$

where b_i and b_o are constants, and q_i'' and q_o'' are *averages* around the circumference.

Then a *local* heat transfer coefficient can be defined at each position θ and on each surface as

$$h_i(\theta) = \frac{q_i''(\theta)}{T_i(\theta) - T_b} \tag{101}$$

$$h_o(\theta) = \frac{q_o''(\theta)}{T_o(\theta) - T_b} \tag{102}$$

where T_b is the *bulk-mean* fluid temperature.

From these, local Nusselt numbers can be defined

$$\mathrm{Nu}_i(\theta) = \frac{D_e h_i(\theta)}{k} \quad , \quad \mathrm{Nu}_o(\theta) = \frac{D_e h_o(\theta)}{k}$$

The local Nusselt numbers on each surface at each position θ can then be evaluated from the following

$$\mathrm{Nu}_i(\theta) = \frac{\mathrm{Nu}_{ii}(1 + b_i \cos \theta)}{(1 + \mathrm{Nu}_{ii} b_i R_{ii} \cos \theta) - [(q_o''/q_i'')(\theta_i^* - \mathrm{Nu}_{ii} b_o R_{io} \cos \theta)]} \tag{103}$$

$$\mathrm{Nu}_o(\theta) = \frac{\mathrm{Nu}_{oo}(1 + b_o \cos \theta)}{(1 + \mathrm{Nu}_{oo} b_o R_{oo} \cos \theta) - [(q_i''/q_o'')(\theta_o^* - \mathrm{Nu}_{oo} b_i R_{oi} \cos \theta)]} \tag{104}$$

where Nu_{ii} and Nu_{oo} and the influence coefficients θ_i^*, θ_o^* are obtained from the corresponding axially symmetric solutions, Table 24 for turbulent flow, and Table 17 for laminar flow.

The functions R_{ii}, R_{oo}, R_{io}, R_{oi} are given in Table 28 as functions of $r*$, Re, and Pr.†

Equations (103) and (104) are applicable to both laminar and turbulent flow, constant heat rate per unit of tube length, and for both the entry length situations as well as for fully developed velocity and temperature profiles. However, the functions R given in Table 28 are available only for the fully developed condition.

d. Axial Variation of Surface Temperature and Heat Flux.

Use of Fundamental Solutions. The following procedures have been developed by Lundberg, McCuen, and Reynolds [132] to use the *fundamental solutions* described in Secs. B.2.a and B.2.b for various types of problem in which heat flux and/or surface temperature is arbitrarily specified as a function of axial distance along the tube. (It should be noted that this procedure can also be used for the simple *circular tube*. The procedure described for the circular tube in Secs. A.2.a and A.2.c involves direct use of the eigenfunction results. To use the method described here, on the other hand, it is necessary to tabulate $\phi_{oo}^{(1)}$ and $\theta_{oo}^{(2)}$, using the appropriate thermal entry length solutions and Eqs. (83) and (85). In these cases, $\phi_{oi}^{(1)}$ and $\theta_{oi}^{(2)}$ are both simply zero. Note that the θ_{r+} used with the eigenfunction results is not the same as $\theta_{oo}^{(2)}$ used here, although it is closely related to $\phi_{oo}^{(1)}$.)

The necessary fundamental solutions are available only for *hydrodynamically fully developed flow* (velocity profile fully developed at the beginning of heat transfer). Complete solutions are available for *laminar flow* and are given in Table 19. For *turbulent flow* fundamental solutions only of the second kind are given for a fluid with Pr = 0.7 on Figs. 80 to 83. For *turbulent flow* between parallel planes ($r* = 1.00$), fundamental solutions only of the second kind are given in Tables 19 and 20 for Pr = 0.01, 0.1, 1.0, and 10.0. These are given in the form of Nusselt numbers and influence coefficients, but the necessary functions can be readily evaluated using Eqs. (96) to (98).

Three types of problem with arbitrarily specified axial boundary conditions can be handled, as follows:

1. *Wall temperatures prescribed arbitrarily* on each surface.‡
Given

$$T_i(\bar{x}) \quad \text{for} \quad \bar{x} \geq 0 \quad (\text{inner wall})$$

$$T_o(\bar{x}) \quad \text{for} \quad \bar{x} \geq 0 \quad (\text{outer wall})$$

$$T = T_e \quad \text{for all } r \text{ for } \bar{x} < 0$$

Let ξ be a dummy variable going from 0 to $\bar{x}$. Then

$$q_i''(\bar{x}) = \frac{k}{D_e} \left\{ \int_{\xi=0}^{\xi=\bar{x}} \phi_{ii}^{(1)}(\bar{x} - \xi) dT_i(\xi) + \int_{\xi=0}^{\xi=\bar{x}} \phi_{io}^{(1)}(\bar{x} - \xi) dT_o(\xi) \right\} \quad (105)$$

$$q_o''(\bar{x}) = \frac{k}{D_e} \left\{ \int_{\xi=0}^{\xi=\bar{x}} \phi_{oi}^{(1)}(\bar{x} - \xi) dT_i(\xi) + \int_{\xi=0}^{\xi=\bar{x}} \phi_{oo}^{(1)}(\bar{x} - \xi) dT_o(\xi) \right\} \quad (106)$$

$$T_b(\bar{x}) - T_e = \int_{\xi=0}^{\xi=\bar{x}} \theta_{bi}^{(1)}(\bar{x} - \xi) dT_i(\xi) + \int_{\xi=0}^{\xi=\bar{x}} \theta_{bo}^{(1)}(\bar{x} - \xi) dT_o(\xi) \quad (107)$$

†The following correspondence exists between the functions given in Table 28 and in the original reference

$$R_{ii} = R_{1_i}(r*) \; ; \; R_{io} = R_{1_o}(r*) \; ; \; R_{oo} = R_{1_o}(1) \; ; \; R_{oi} = R_{1_i}(1)$$

‡Note that all boundary conditions are assumed to be axially symmetric.

TABLE 28. Circumferential Heat Flux Functions, Fully Developed Constant Heat Rate in a Concentric Circular Tube Annulus

$r^* = \dfrac{r_i}{r_o}$	Pr	Re	R_{ii}	R_{io}	R_{oo}	R_{oi}
0.20	LAMINAR		0.135	0.260	0.677	0.0521
	0	10^4	0.135	0.260	0.677	0.0521
		10^5	0.135	0.260	0.677	0.0521
		10^6	0.135	0.260	0.677	0.0521
	0.01	10^4	0.133	0.256	0.688	0.0513
		10^5	0.119	0.220	0.553	0.0440
		10^6	0.0404	0.0544	0.134	0.0109
	1.00	10^4	0.0244	0.0207	0.0616	0.00413
		10^5	0.00436	0.00267	0.00928	0.000535
		10^6	0.000662	0.000338	0.00127	0.000675
	10.00	10^4	0.00839	0.00172	0.0129	0.000341
		10^5	0.00127	0.000235	0.00185	0.0000471
		10^6	0.000173	0.0000320	0.000244	0.0000064
0.50	LAMINAR		0.833	1.33	1.67	0.667
	0	10^4	0.833	1.33	1.67	0.667
		10^5	0.833	1.33	1.67	0.667
		10^6	0.833	1.33	1.67	0.667
	0.01	10^4	0.821	1.32	1.64	0.657
		10^5	0.710	1.13	1.41	0.565
		10^6	0.183	0.272	0.344	0.136
	1.00	10^4	0.0764	0.0995	0.137	0.0484
		10^5	0.0114	0.0131	0.0193	0.00655
		10^6	0.00162	0.00166	0.00252	0.00119
	10.00	10^4	0.135	0.00774	0.0186	0.00425
		10^5	0.00197	0.00115	0.00272	0.000575
		10^6	0.000270	0.000158	0.000360	0.0000794
0.80	LAMINAR		9.11	11.1	11.4	8.89
	0	10^4	9.11	11.1	11.4	8.89
		10^5	9.11	11.1	11.4	8.89
		10^6	9.11	11.1	11.4	8.89
	0.01	10^4	8.99	11.0	11.2	8.78
		10^5	7.75	10.0	10.3	7.56
		10^6	1.89	2.29	2.36	1.84
	1.00	10^4	0.681	0.830	0.865	0.648
		10^5	0.0934	0.110	0.116	0.0880
		10^6	0.0120	0.0140	0.0148	0.0112
	10.00	10^4	0.0656	0.0643	0.0749	0.0559
		10^5	0.00917	0.00965	0.0111	0.00770
		10^6	0.00127	0.00132	0.00152	0.00105
0.90	LAMINAR		42.9	47.4	47.6	42.6
	0	10^4	42.9	47.4	47.6	42.6
		10^5	42.9	47.4	47.6	42.6
		10^6	42.9	47.4	47.6	42.6
	0.01	10^4	42.3	46.8	47.0	42.1
		10^5	36.5	40.3	40.5	36.3
		10^6	8.90	9.82	9.89	8.84
	1.00	10^4	3.16	3.54	3.57	3.12
		10^5	0.429	0.471	0.477	0.424
		10^6	0.0546	0.0597	0.0606	0.0538
	10.00	10^4	0.281	0.280	0.290	0.271
		10^5	0.0386	0.0413	0.0428	0.0371
		10^6	0.00531	0.00564	0.00584	0.00509

These integrals are to be interpreted in the Stieltjes sense, i.e., for evaluation they may be replaced by an integral plus a summation, where the integral is used for continuous changes in surface temperature, and the summation is used for discontinuities. In the case of continuous changes in surface temperature $dT(\xi)$ may be replaced by $[dT(\xi)/d\xi]d\xi$. If the integrals are to be evaluated numerically, as would generally be the case, the integral is replaced by a summation with $dT(\xi)$ replaced by $\Delta T(\xi)$, which is simply a step in surface temperature of height ΔT at location ξ, that is, the problem is treated as a series of small discontinuities in surface temperature.

2. *Wall heat fluxes prescribed arbitrarily* on each surface.

Given

$$q_i''(\bar{x}) \text{ for } \bar{x} \geq 0 \text{ (inner wall)}$$
$$q_o''(\bar{x}) \text{ for } \bar{x} \geq 0 \text{ (outer wall)}$$
$$T = T_e \text{ for all } r \text{ for } \bar{x} < 0$$

Let ξ be a dummy variable going from 0 to $\bar{x}$

$$T_i(\bar{x}) - T_e = \frac{D_e}{k} \left\{ \int_{\xi=0}^{\xi=\bar{x}} \theta_{ii}^{(2)}(\bar{x} - \xi) dq_i''(\xi) + \int_{\xi=0}^{\xi=\bar{x}} \theta_{io}^{(2)}(\bar{x} - \xi) dq_o''(\xi) \right\} \quad (108)$$

$$T_o(\bar{x}) - T_e = \frac{D_e}{k} \left\{ \int_{\xi=0}^{\xi=\bar{x}} \theta_{oi}^{(2)}(\bar{x} - \xi) dq_i''(\xi) + \int_{\xi=0}^{\xi=\bar{x}} \theta_{oo}^{(2)}(\bar{x} - \xi) dq_o''(\xi) \right\} \quad (109)$$

$$T_b(\bar{x}) - T_e = \frac{D_e}{k} \left\{ \int_{\xi=0}^{\xi=\bar{x}} \theta_{bi}^{(2)}(\bar{x} - \xi) dq_i''(\xi) + \int_{\xi=0}^{\xi=\bar{x}} \theta_{bo}^{(2)}(\bar{x} - \xi) dq_o''(\xi) \right\} \quad (110)$$

Again the integrals are to be interpreted in the Stieltjes sense.

3. *Heat flux prescribed* on wall (c) and *temperature prescribed* on wall (a).

Given

$$T_a(\bar{x}) \text{ for } \bar{x} \geq 0$$
$$q_c''(\bar{x}) \text{ for } \bar{x} \geq 0$$
$$T = T_e \text{ for all } r \text{ for } \bar{x} < 0$$

Let ξ be a dummy variable going from 0 to $\bar{x}$. Then

$$T_c(\bar{x}) - T_e = \int_{\xi=0}^{\xi=\bar{x}} \theta_{ca}^{(3)}(\bar{x} - \xi) dT_a(\xi) + \frac{D_e}{k} \int_{\xi=0}^{\xi=\bar{x}} \theta_{cc}^{(4)}(\bar{x} - \xi) dq_c''(\xi) \quad (111)$$

$$q_a''(\bar{x}) = \frac{k}{D_e} \int_{\xi=0}^{\xi=\bar{x}} \phi_{aa}^{(3)}(\bar{x} - \xi) dT_a(\xi) + \int_{\xi=0}^{\xi=\bar{x}} \phi_{ac}^{(4)}(\bar{x} - \xi) dq_c''(\xi) \quad (112)$$

$$T_b(\bar{x}) - T_e = \int_{\xi=0}^{\xi=\bar{x}} \theta_{ba}^{(3)}(\bar{x} - \xi) dT_a(\xi) + \frac{D_e}{k} \int_{\xi=0}^{\xi=\bar{x}} \theta_{bc}^{(4)}(\bar{x} - \xi) dq_b''(\xi) \quad (113)$$

Note that problem 3 includes, as a special case, the rather common case of a specified temperature distribution on one wall with the other wall *insulated*, $q_c'' = 0$.

Use of Eigenfunction Solutions. For two cases: (1) *laminar* flow between parallel planes with each surface at the same constant temperature; and (2) *turbulent* flow between parallel planes with one surface at a constant temperature and the other surface insulated, the functions λ_n^2 and G_n have been tabulated (see Tables 22 and 25 respectively). These thermal entry-length solutions can also be used to build up solutions for any arbitrarily specified surface temperature distribution. However, for (1) both surfaces must be at the same temperature, and for (2) one surface must be insulated. In *either* case

$$q_w''(x^+) = \frac{4k}{D_e} \int_{\xi=0}^{\xi=x^+} \left\{ \sum_{n=0}^{\infty} G_n \exp[-\lambda_n^2(x^+ - \xi)] \right\} dT_w(\xi) \quad (114)$$

$$T_b(x^+) - T_e = \frac{2D_e}{k} \int_{\xi=0}^{\xi=x^+} q_w''(\xi) d\xi \quad (115)$$

where T_w is the specified surface temperature, q_w'' is the wall heat flux, and x^+ is as defined in the tabulations of λ_n^2 and G_n.

Virtually all of the thermal entry-length solutions, except those obtained by experiment, are available in eigenfunction form in the original references. However, the fundamental solutions will usually be found to be more useful for handling variable boundary conditions.

e. Effect of Eccentricity. Although annular systems may occasionally be deliberately designed with eccentricity, the major importance of some knowledge of the effects of eccentricity is to determine the influence of the small amount of eccentricity that is inevitable in any annulus system. This problem has not been extensively studied, either analytically or experimentally. A few results, both analytic and experimental, will be given here, and the reader will be referred to still other results in the literature.

Snyder [145] considered *slug-flow,* i.e., flow with uniform velocity throughout, with fully developed *constant heat rate* per unit of tube length, with the inner surface maintained at a constant temperature circumferentially, and the outer surface insulated. The slug-flow solutions are useful as an approximation for very low Prandtl number fluids in *turbulent* flow (see discussion of liquid metal heat transfer in Sec. B.2.b), and in the present case the influence of eccentricity on laminar flow heat transfer is probably quite similar.

On Fig. 86, Snyder's calculated results for the local heat flux on the inner tube are given as a function of angular position for two values of eccentricity. (See Fig. 61 and Sec. B.1.c for a description of the eccentric annulus nomenclature.) On Fig. 87 an *average* slug-flow Nusselt number for the inner surface is given as a function of eccentricity for one value of radius ratio, 0.515. The average Nusselt number is based on the average of the local heat flux with respect to circumferential position, as given in Fig. 86. The tendency of the average Nusselt number to *decrease* with increasing eccentricity is the characteristic behavior of the eccentric annulus.

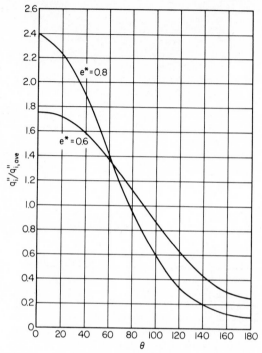

Fig. 86. Local heat flux on the inner tube for slug flow in an eccentric annulus with a fully developed constant heat rate on the inner tube and a constant circumferential temperature, and the outer surface insulated.

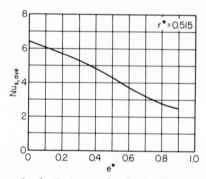

Fig. 87. Average Nusselt number for the inner surface for slug flow in an eccentric annulus. Heating conditions as on Fig. 86.

Experimental data for the flow of water in an eccentric annulus is presented by Judd and Wade [138]. Heating is from the inner surface, with the outer surface insulated, and heat was supplied at a constant rate per unit of surface area. Fully developed conditions were attained. Nusselt numbers were measured only at the two extremes of spacing, i.e., at the point of minimum separation, and at the point of maximum separation. The results are given on Fig. 88, where the local Nusselt numbers are divided by the corresponding

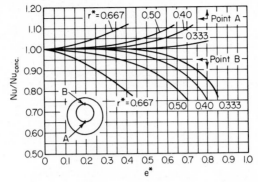

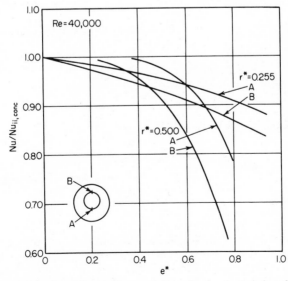

Fig. 88. Experimental results of Judd and Wade for the fully developed turbulent flow of water in an eccentric annulus as normalized on the concentric annulus results. Heating from the inner surface, outer surface insulated.

results (at same Reynolds number and Prandtl number) for the completely concentric annulus. Although average Nusselt numbers were not measured, it is apparent that there is a *small* tendency for the average Nusselt number to *decrease* with eccentricity.

Figures 89 and 90 show some experimental data obtained by Leung, Kays, and Reynolds [146] for the flow of air in an eccentric annulus with fully developed constant heat rate. Measurements were made for heating from both the inner and outer surfaces independently with the opposite surface insulated. The results are presented in the same form as the Judd and Wade results. In this case, the average Nusselt number (not shown) decreases with eccentricity considerably more than do the results of Judd and Wade, but not as much as for the

Fig. 89. Experimental results of Leung, Kays, and Reynolds for the turbulent flow of air in an eccentric annulus with fully developed constant heat rate. Heating from the inner surface, with the outer surface insulated.

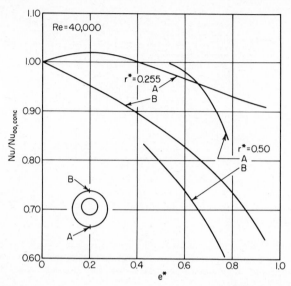

Fig. 90. Experimental results of Leung, Kays, and Reynolds for the turbulent flow of air in an eccentric annulus with fully developed constant heat rate. Heating from the outer surface with the inner surface insulated.

slug-flow results. This suggests a substantial Prandtl number influence on this phenomenon. The Leung et al. data also suggest a considerable effect of Reynolds number, with large eccentricity effects at high Reynolds number.

Deissler and Taylor [147] present an analytic solution for the influence of eccentricity with turbulent flow in an annulus, but their results predict a decrease in average Nusselt number which is very much greater than is observed experimentally. Yu and Dwyer [148, 149] present analytic solutions for turbulent flow of a *liquid metal*. Both are for fully developed constant heat rate per unit of length on the inner surface only, but the first paper considers uniform heat flux circumferentially as well, while the second considers uniform temperature circumferentially. The second boundary conditions correspond with those of the slug-flow solutions discussed above, and the results for the influence of eccentricity are very similar.

C. RECTANGULAR AND TRIANGULAR DUCTS

1. Hydrodynamics

a. Friction Coefficients

Laminar–Triangular. Experimental results for fully developed laminar flow friction factors have been determined by Carlson and Irvine [150] for various apex angles in an isosceles triangle. An analysis for this case has been completed by Sparrow [151] which shows good agreement with the data. The results of both references are shown in Fig. 91.

Turbulent–Triangular. The experimental results of Carlson and Irvine are presented in Fig. 92 for turbulent friction factors in an isosceles triangle. The correlation achieved is

$$4f = \frac{C}{Re^{0.25}} \tag{116}$$

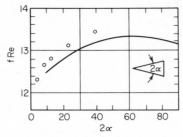

Fig. 91. Laminar friction factor--Reynolds number product for an isosceles triangle versus the apex angle.

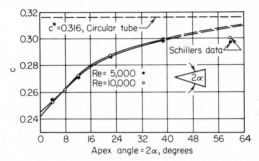

Fig. 92. Correlation parameter for turbulent friction factors for isosceles triangular ducts.

where C is taken from the figure. The data reach a Reynolds number of 30,000 and were not well correlated by the use of the hydraulic diameter. Carlson and Irvine note that the circular tube correlation, with the hydraulic diameter, predicts values some 20 percent high for a 4 degree apex angle, and 5 percent high at a 38 degree angle. Schiller's data [152] for an equilateral triangle are also shown on the figure.

Laminar–Rectangular. Kays [51] presents results for the fully developed laminar friction coefficients in rectangular ducts as shown in Fig. 93.

Turbulent–Rectangular. A very complete survey by Hartnett et al. [153] is available for turbulent friction coefficients in rectangular ducts. Figure 94 shows these results compared to Eq. (6) in the turbulent region. The agreement with Kays in the laminar region, and with Eq. (6) using a hydraulic diameter in the turbulent region, is good. Hartnett et al. conclude that for Reynolds numbers from 6×10^3 to 5×10^5, the circular tube correlation accurately represents the friction coefficient for flow through rectangular ducts for any and all aspect ratios. The more recent results of Maurer and LeTourneau [154] for water flow in a rectangular channel of 11:1 aspect ratio are also correlated with the circular tube results by the use of an equivalent hydraulic diameter.

For preliminary design purposes, the circular tube correlation with an equivalent diameter and the appropriate equivalent roughness parameter is recommended for calculating turbulent friction factors in rectangular ducts.

b. Velocity Distributions

Laminar–Triangular. The velocity in the cross section of an isosceles triangle has been determined by Sparrow [151]. The velocity at any point r, θ, (see Fig. 95) may

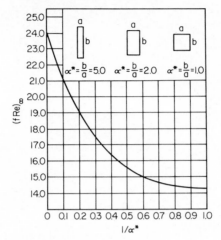

Fig. 93. Laminar friction factor-Reynolds number product for a rectangular duct versus the aspect ratio.

be calculated from Eq. (117) as

$$\frac{v_x}{(s^2/4\mu)(-dp/dx)} = \left(\frac{r}{s}\right)^2\left(\frac{\cos 2\theta}{\cos 2\alpha} - 1\right) + \sum_{n=1}^{\infty} C_n\left(\frac{r}{s}\right)^{\frac{(2n-1)\pi}{2\alpha}}\cos\left[\frac{(2n-1)\pi\theta}{2\alpha}\right] \tag{117}$$

The constants C_n are listed in Table 29 as functions of the half-opening angle α. Velocity profiles may be plotted from Sparrow's results. Figure 95 shows the variation of velocity along various radial lines (that is, lines of constant θ) corresponding to a duct-opening angle of 40 degrees. Experimental velocity plots across the width of an isosceles triangle are in Eckert and Irvine [155].

Formulas for the point laminar velocity for the right isosceles triangle and the equilateral triangle, taken from Knudsen and Katz [2], are given below. The origin of the coordinate system is shown in Fig. 96.

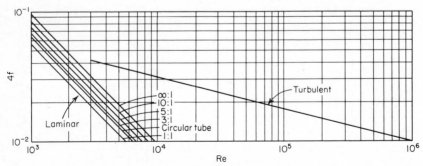

Fig. 94. Correlation of available data for rectangular duct friction coefficients using D_e, solid line for Re $> 10^4$ is the circular tube correlation.

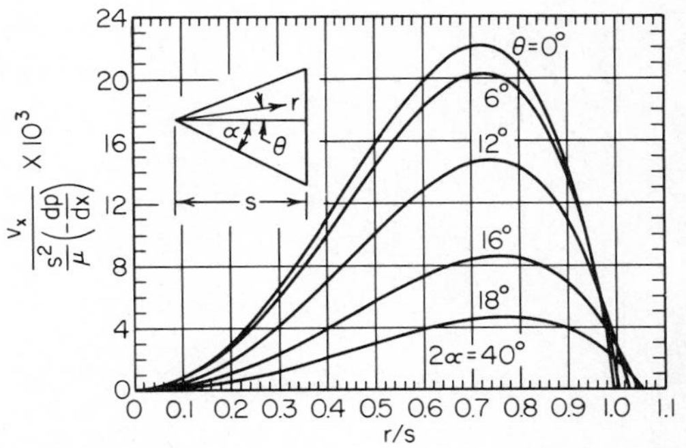

Fig. 95. Velocity profiles for an isosceles triangular duct with a 40° opening angle. Coordinate origin as shown.

TABLE 29. Constants Required to Calculate the Laminar Velocity Distribution in an Isosceles Triangle

α	C_1	C_2	C_3	C_4	C_5
5°	−0.01577	0.4027×10^{-3}	-0.6979×10^{-4}	0.2108×10^{-4}	-0.834×10^{-5}
10°	−0.06514	0.1086×10^{-2}	-0.1486×10^{-3}	0.3692×10^{-4}	-0.122×10^{-4}
15°	−0.1561	0.1563×10^{-2}	-0.1650×10^{-3}	0.3331×10^{-4}	-0.913×10^{-5}
20°	−0.3069	0.1569×10^{-2}	-0.1257×10^{-3}	0.2025×10^{-4}	-0.4554×10^{-5}
22.5°	−0.4155	0.1411×10^{-2}	-0.9415×10^{-4}	0.1347×10^{-4}	-0.272×10^{-5}
25°	−0.5568	0.1083×10^{-2}	-0.6024×10^{-4}	0.759×10^{-5}	-0.14×10^{-5}
30°	−1	0	0	0	0
35°	−1.922	-0.1630×10^{-2}	0.306×10^{-4}	-0.234×10^{-5}	0.3×10^{-6}
40°	−4.755	-0.3748×10^{-2}	0.157×10^{-4}	-0.17×10^{-5}	0.1×10^{-6}

α	C_6	C_7	C_8	C_9	C_{10}
5°	0.386×10^{-5}	-0.199×10^{-5}	0.109×10^{-5}	-0.63×10^{-6}	0.38×10^{-6}
10°	0.478×10^{-5}	-0.209×10^{-5}	0.99×10^{-6}	-0.49×10^{-6}	0.25×10^{-6}
15°	0.300×10^{-5}	-0.111×10^{-5}	0.44×10^{-6}	-0.19×10^{-6}	0.75×10^{-7}
20°	0.124×10^{-5}	-0.39×10^{-6}	0.12×10^{-6}	-0.5×10^{-7}	0.1×10^{-7}
22.5°	0.67×10^{-6}	-0.19×10^{-6}	0.5×10^{-7}	-0.3×10^{-7}	
25°	0.3×10^{-6}	-0.9×10^{-7}	0.2×10^{-7}	-0.2×10^{-7}	
30°	−1	0	0	0	
35°	-0.6×10^{-7}				
40°	-0.6×10^{-7}				

7-113

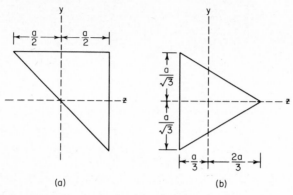

Fig. 96. Coordinate systems for noncircular ducts for use with Eqs. (118) and (119). (1) Right isosceles triangle, (2) equilateral triangle.

Right isosceles triangle

$$u = -\frac{dP}{dx}\frac{g_c}{2\mu}\left\{\frac{1}{2}(z^2 + y^2) + zy - \frac{a}{2}(z + y)\right.$$

$$+\frac{4a^2}{\pi^3}\sum_{n=0}^{\infty}\frac{(-1)^n}{(2n+1)^3\,\sinh[(2n+1)\pi/2]}\left[\sinh\frac{(2n+1)\pi y}{a}\cos\frac{(2n+1)\pi z}{a}\right.$$

$$\left.\left. + \sinh\frac{(2n+1)\pi z}{a}\cos\frac{(2n+1)\pi y}{a}\right]\right\} \quad (118)$$

Equilaterial triangle

$$u = -\frac{dP}{dx}\frac{g_c}{2\mu}\left[\frac{1}{2}(z^2 + y^2) - \frac{1}{2a}(z^3 - 3zy^2) - \frac{2a^2}{27}\right] \quad (119)$$

Laminar–Rectangular. The fully developed velocity distribution for a rectangular duct of cross-sectional dimensions 2a and 2b is given in Lundgren et al. [114] as

$$u_{\infty} = -\frac{1}{2\mu}\frac{dp}{dx}\left[b^2 - y^2 + \frac{4}{b}\sum_{n=0}^{\infty}\frac{(-1)^{n+1}}{N_n^3}\frac{\cosh N_n z}{\cosh N_n a}\cos N_n y\right] \quad (120)$$

in which $N_n = (2n + 1)\pi/2b$ and

$$-\frac{1}{\mu}\frac{dp}{dx} = \frac{4ab\bar{V}}{(4/3)\,ab^3 - (8/b)\,\Sigma N_n^{-5}\,\tanh N_n a}$$

The origin of the coordinates is at the center of the duct.

Combined Laminar-Turbulent Flow. An interesting effect was noted in the Eckert and Irvine [155] studies. These revealed the fact that, over the whole tube length, laminar and turbulent flow coexisted side by side over a wide range of

Reynolds numbers. The flow near the base of the tube was found to be turbulent, that near the apex laminar. Figure 97 shows the results of this experiment. The velocity profile measurement location was at 71 hydraulic diameters from the entrance for Duct 1, and 66 for Duct 2, while the smoke probe for flow visualization measurements was located at 24 and 35 diameters for Ducts 1 and 2, respectively.

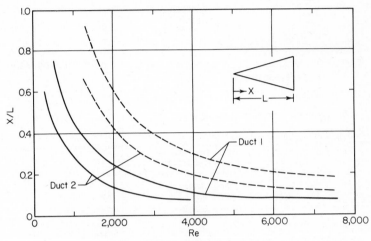

Fig. 97. Results of coexisting laminar and turbulent flow in triangular ducts. Solid line from flow visualization measurements, dotted line from velocity profile measurements.

Hydrodynamic Entrance Region. Hartnett et al. [153] present information on the hydrodynamic entrance region for laminar flow in rectangular tubes. It was found that the entrance configuration (e.g., abrupt or smooth) had little effect on the hydrodynamic entry length for Reynolds numbers less than 2000, but had a large effect above this value because of the transition to turbulent flow. Hartnett et al. correlate entry lengths using the relationship

$$\frac{X}{D_e} = C \cdot \text{Re}$$

where C is approximately 0.033, 0.046, 0.057 for the aspect ratios 10:1, 5:1, and 1:1 respectively. The experimental results thus demonstrate that there is an effect of the aspect ratio on the magnitude of the entrance length, with the low aspect ratios requiring a greater length for the attainment of a fully established velocity field.

Information of a similar nature for an isosceles triangular duct with a 23 degree apex angle is available in Eckert and Irvine [156].

Transition to Turbulence. Both Hartnett et al. and Eckert and Irvine give information on the critical Reynolds number in noncircular ducts. Tables 30 and 31 present their results. It is important to note that the critical Reynolds number is greatly affected by the entrance configuration, noted as abrupt or smooth, and that with acute angular ducts there can be a coexistence of laminar and turbulent flow as previously noted.

TABLE 30. Critical Reynolds Numbers for
Rectangular Ducts, Aspect Ratio

b/a		Re_{crit}
1:1	(smooth)	4300
1:1	(abrupt)	2200
3:1	(smooth)	6000
5:1	(smooth)	7000
5:1	(abrupt)	2500
10:1	(smooth)	4400
10:1	(abrupt)	2500

TABLE 31. Values of Re_{crit} for Noncircular Ducts as Reported by Various
Investigators

Investigator	Geometry	Re_{crit}
Davies and White [157]	Rectangles, $b/a = 40$	2800
R. J. Cornish [158]	Rectangles, $b/a = 2.92$	2800
J. Nikuradse [159]	Rectangle $b/a = 3.50$	2800
	Equilateral triangle	2800
	Right isosceles triangle	2800
	Acute triangle	2000–2360
	Trapezoid	2000–2360
	One notch circle	2000–2360
L. Schiller [160]	Rectangle $b/a = 3.52$	1600
	Square	2100
Lea and Tadros [161]	Rectangle $b/a = 2.36$	2000
	$b/a = 2.09$, $b/a = 1$	2000
	Annulus $d_2/d_1 = 2$	1765
Allen and Grundberg [162]	Rectangle $b/a = 3.92$	2400
Washington and Marks [163]	Rectangles, $b/a = 40$ and 20	2400
Eckert and Irvine [156]	Rectangle $b/a = 3$	5000
		6000—Smooth Entrance
	Isosceles triangle, $23°$	1800—Abrupt Entrance

Entrance Pressure Drop Effects. Lundgren et al. [114] have presented results for
the increased pressure drop due to entrance effects with laminar flow. Their results
are in the form of Eq. (22). Figures 98 and 99 give values of K, based on a
uniform inlet velocity, for isosceles triangular ducts and rectangular ducts. This is
the fully developed value of K, and represents the total effect of the entire
entrance region on the pressure drop, including both the momentum change
accompanying the development of the velocity profile, and the increased wall shear
stress in the entrance region.

Velocity Distributions for Turbulent Flow. The classical point velocity profiles
of Nikuradse [159] are shown in Fig. 100.

Turbulent Entry Lengths. Hartnett et al. [153] discuss the problem of determining turbulent hydrodynamic entry lengths in a rectangular channel. For a
smooth entry, the flow in the inlet region can remain laminar for tube Reynolds
number considerably greater than 2200, thus causing a longer hydrodynamic inlet
length than with an abrupt entrance. For an abrupt entrance, they note that
approximately 40 hydraulic diameters were required to arrive at fully established

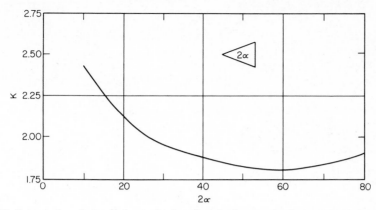

Fig. 98. Pressure drop factor K due to laminar flow development in the entrance region of an isosceles triangular duct.

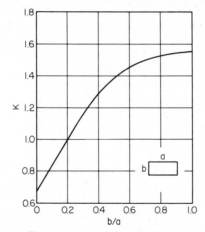

Fig. 99. Pressure drop factor K due to laminar flow development in the entrance region of a rectangular duct.

flow for a Reynolds number of approximately 3000. Above a Reynolds number of 4000, the abrupt entrance gave entry lengths of less than 20 diameters for the ducts studied, that is, 1:1, 5:1, 10:1.

2. Heat Transfer

a. Laminar Flow. There are a larger number of possible thermal boundary conditions in rectangular and triangular ducts, in contrast to circular tubes, because of the influence of conduction around the tube periphery.

Following Hartnett and Irvine [140] the boundary conditions will be classified according to the cases shown in Fig. 101. These are:

case A: Wall temperature constant in flow direction and around periphery.
case B: Constant heat input per unit length, and constant peripheral wall temperature at a given axial position.

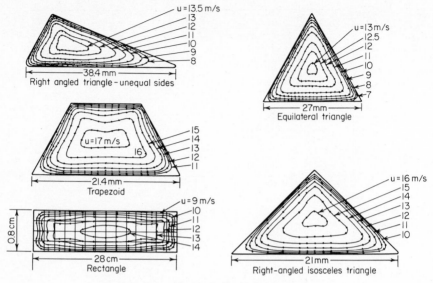

Fig. 100. Point velocity profiles in noncircular ducts.

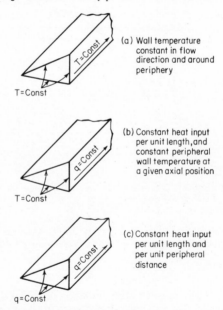

(a) Wall temperature constant in flow direction and around periphery

(b) Constant heat input per unit length, and constant peripheral wall temperature at a given axial position

(c) Constant heat input per unit length and per unit peripheral distance

Fig. 101. Thermal boundary conditions of importance in noncircular ducts.

case C: Constant heat input per unit length and per unit peripheral distance.

Many cases of interest fall between conditions B and C. To classify these, a conduction parameter is defined as: $\phi = k_w b/k_{\text{fluid}} D_e$, where b is the wall thickness. As ϕ gets large, conditions tend to case B; as ϕ approaches 0, case C is approached.

Fully Developed Velocity and Temperature Profiles. Heat transfer results for *fully developed* laminar velocity and temperature profiles for noncircular ducts are given in Table 32.

TABLE 32. Nusselt Numbers for Fully Developed Laminar Velocity and Temperature Profiles in Tubes of Various Cross Sections

	Case	Boundary condition		
		A	B	C
	Circle	3.66	4.364	
a ☐ b	$b/a = 1.0$	2.98	3.61	3.09
	$b/a = 1.4$	3.08	3.73	
	$b/a = 2.0$	3.39	4.12	3.02
	$b/a = 3.0$	3.96	4.79	
	$b/a = 4.0$	4.44	5.33	2.93
	$b/a = 8.0$	5.60	6.49	2.90
	$b/a = 10.0$		6.78	2.90
	$b/a = \infty$	7.54	8.235	
〰		4.86	5.385	
△		2.47	3.11	

Results for boundary conditions *A* and *B* are from Kays [51], for *C* from Sparrow and Siegel [164]. One additional case from Sparrow and Siegel gives a Nusselt number of 4.96 for an aspect ratio of 10, case *C*, but with the short sides of the duct completely insulated.

Thermal Entrance Length. Results for condition *A* are partially available from Dennis et al. [165] in Table 33. These results are used with Eqs. (35) to (38) to determine the Nusselt number, etc. Here α^* is the ratio of the long side to short side of the rectangle.

TABLE 33. Infinite Series Solution Functions for the Rectangular Tube Family, Constant Surface Temperature, Case A Thermal Entry Length

$1/\alpha^*$	1.000	0.667	0.500	0.250	0.125	0.000
λ_0^2	5.96	6.25	6.78	8.88	11.19	15.09
λ_1^2	35.54					171.3
λ_2^2	78.9					498
G_0	0.598	0.627	0.669	0.839	1.030	1.717
G_1	0.462					1.139
G_2	0.138					0.952

Some information on the thermal entry region for a square duct with boundary condition *C* is available in Sparrow and Siegel [166].

b. Turbulent Flow. Deissler and Taylor [167] have predicted the heat transfer and friction parameters for an equilateral triangle and a square duct for the case of constant axial heat flux. Their analytical method has been compared by other authors (Carlson and Irvine [150], Hartnett et al. [153], Eckert and Irvine [168]) to experimental data for ducts with more acute angles without conclusive results. In the absence of definitive experimental results, the Deissler and Taylor results are suggested as applicable to the equilateral triangle and the square duct. These results are shown in Figs. 102 and 103. The hydraulic diameter is used in Nu and Re. Comparison with the data of Lowdermilk et al. [169] is good. Deissler and Taylor use the conduction parameter $k_{\text{fluid}} r / k_w b$ where r is the distance between the corner and the midpoint of the wall. Further specific heat transfer results, based on experiment follow.

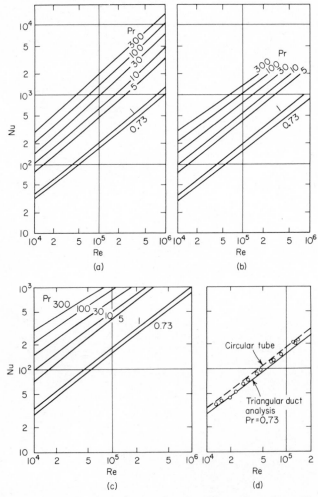

Fig. 102. Predicted average Nusselt number as a function of Reynolds number and Prandtl number for flow in a triangular passage. Uniform axial heat flux. (a) $k r / k_w b = 0$ (uniform peripheral wall temperature); (b) $k r / k_w b = 0.005$; (c) $k r / k_w b = 0.010$; (d) $k r / k_w b = 0.025$ (data of Lowdermilk et al. [169]).

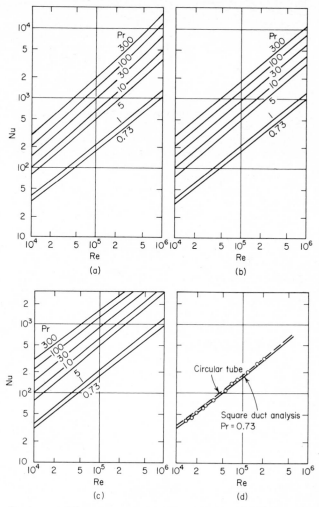

Fig. 103. Predicted average Nusselt number as a function of Reynolds number and Prandtl number for flow in a square passage. Uniform axial heat flux. (a) $kr/k_w b = 0$ (uniform peripheral wall temperature); (b) $kr/k_w b = 0.005$; (c) $kr/k_w b = 0.010$; (d) $kr/k_w b = 0.025$ (data of Lowdermilk et al. [169]).

Triangle. As noted in the section on fluid dynamics in noncircular ducts, one may observe laminar effects in narrow corners even at turbulent Reynolds numbers. Consequently, with an acute-angle duct one must be careful in applying analytical methods at low but turbulent Reynolds numbers.

The experimental results of Eckert and Irvine [168] for air, are shown in Fig. 104, for an isosceles triangular duct with an apex angle of 11.5 degrees, five inches on the long side and one inch on the short side. Here the average Nusselt number is based on the average heat transfer coefficient which is based, in turn, on the average wall temperature around the periphery. Bulk properties are used.

Most of these results lie in the transition region for Reynolds number, the highest Reynolds number being 24,000.

Rectangle. Table 34 summarizes the rectangular duct heat transfer results presently available. For parallel plate results refer to the section on annuli (Sec. B.2.b).

The problem of heat transfer with turbulent flow in rectangular ducts is not yet completely resolved. The *local* heat transfer results of Levy et al. [170] for water, Fig. 105, were taken for large property variations and with some bowing in the thin rectangular channels; the thermal boundary condition was a constant axial heat flux with a ϕ of approximately 8.

The average heat transfer results of Gambill and Bundy for water, obtained in a rounded corner, rectangular channel, are well correlated by the original Sieder-Tate equation with a constant of 0.027 and the $(\mu_b/\mu_w)^{0.14}$ correction, that is,

$$Nu_b = 0.027 \, Re_b^{0.8} Pr_b^{1/3} \left(\frac{\mu_b}{\mu_w}\right)^{0.14} \tag{121}$$

These results were determined for a thermal boundary condition of constant axial heat flux and with a ϕ of approximately 60.

Results with air, see Table 34, using film-temperature properties (see Sec. H) correlate with the 0.023 coefficient in the Dittus-Boelter equation, as in the circular case, that is,

$$Nu = 0.023 \, Re^{0.8} Pr^{0.4} \tag{122}$$

The recommendation for preliminary design, using rectangular ducts, is to use a circular tube correlation employing the hydraulic diameter in the Reynolds number and Nusselt number.

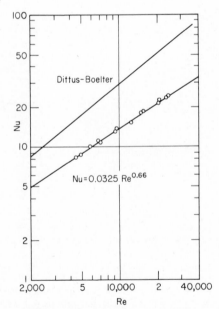

Fig. 104. Average cross sectional Nusselt numbers for an $11.5°$ isosceles triangular duct; $k_w \, b/k D_e$ = 24, x/D_e = 114.

TABLE 34. Summary of Rectangular Channel Heat Transfer Experiments

Reference	Fluid	Aspect Ratio	Thermal Condition	Correlation based on hydraulic diameter
Lowdermilk et al. [169]	air	1:1, 5:1	Constant axial heat flux, $\phi \simeq 35$	Same as circular tube with a constant of 0.023 at large L/D
Washington and Marks [163]	air	20:1, 40:1		0.023 constant with circular tube correlation all right
Levy et al. [170]	water	25:1	Constant axial heat flux, $\phi \simeq 8$	Results some 30% below circular tube correlation of Sieder-Tate
Lancet [62]	air	5:1	Close to case B	Same as circular tube with a constant of 0.023
Gambill and Bundy [171]	water	10:1	Constant axial heat flux, $\phi \simeq 60$	Sieder-Tate correlation all right

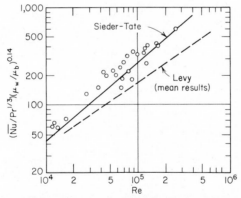

Fig. 105. Average heat transfer results of Gambill and Bundy for thin rectangular channels compared to results of Levy et al. and the Sieder-Tate equation.

Liquid Metal Results. Hartnett and Irvine [140] recommend an approximate correlation for flow in noncircular ducts which should be applicable to liquid metal flow. The correlation is based on correcting slug-flow Nusselt numbers, and is given as

$$\text{Nu} = 2/3 \, \text{Nu}_s + 0.015 \, (\text{Pe})^{0.8} \tag{123}$$

Table 35 presents the slug-flow Nusselt numbers for several geometries. Results for other geometries are shown in Figs. 106 to 109.

TABLE 35. Slug Nusselt Numbers for Simple Geometries

Geometry	Boundary Condition	Slug Nusselt Number
Circle	A	5.80
	B	8.0
Square	A	$\pi^2/2 = 4.93$
	B	7.03
Equilateral triangle	A	—
	B	6.67
Infinite slot	A	$\pi^2 = 9.87$
	B	12
Infinite slot with one wall insulated	A	$\pi^2/2 = 4.93$
	B	6
90-degree isosceles triangle	A	—
	B	6.55

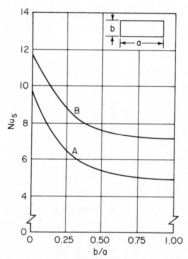

Fig. 106. Slug-flow Nusselt numbers for rectangular channels; thermal condition A, isothermal axially and peripherally; B isothermal peripherally and uniform heat flux axially.

Hartnett and Irvine note that the results of Fig. 109 for the isosceles triangular duct were obtained from a velocity distribution solution and the results are therefore somewhat questionable at small angles. They recommend that the approximate isosceles triangular duct results be used only above 30 degrees, and the circular sector be utilized at smaller angles.

Thermal Entrance Effects. Levy et al. [170] found for their 25:1 aspect ratio rectangular duct that the combined thermal and hydraulic entry-length-to-diameter ratio was 55 to 65 diameters, in contrast to the much smaller circular tube values, on the order of 15 diameters, for constant property flows. Levy recommends using entry length values of 40 to 60 diameters for Reynolds numbers from 10,000 to 100,000 for rectangular ducts. Similar results have been determined for the isosceles triangle with an 11.5 degree apex angle by Eckert

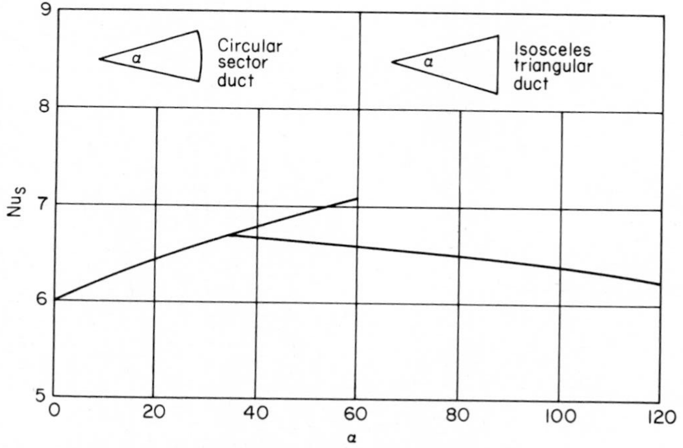

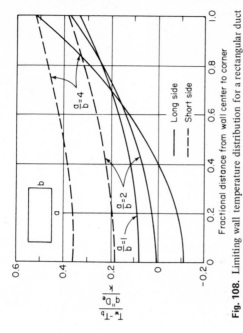

Fig. 107. Slug-flow Nusselt numbers for circular sector duct and isosceles triangular duct, thermal conditions B.

Fig. 108. Limiting wall temperature distribution for a rectangular duct with constant heat flux everywhere, case C, with slug flow.

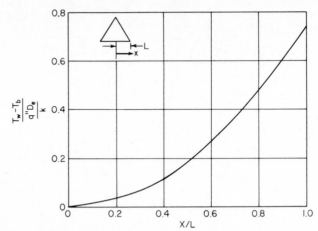

Fig. 109. Limiting wall temperature distributions for equilateral triangular duct with constant heat flux everywhere, case C, with slug flow.

and Irvine [168]. They note that at their high Reynolds number runs, approximately 24,000, the thermal results do not become fully developed even at an L/D_e of 114.

Consequently, one may expect that with flow in noncircular ducts, thermal entry regions longer than those in circular tubes will occur. It should be noted, however, that the two experiments above represent what might be termed highly noncircular ducts; that is, 25:1 aspect ratio and 11.5 degree apex angle. For less extreme noncircular ducts a shorter thermal entry length should result.

c. Peripheral Effects. In noncircular ducts, peripheral effects will, in general, be more important than in a circular duct because of the variation of the heat transfer coefficient around the perimeter. As previously noted, most cases of interest fall between the thermal boundary conditions *B* (peripheral wall temperature constant) and *C* (peripheral wall heat flux constant) so that a conduction parameter is required to describe the thermal conditions.

Laminar flow results are available for the peripheral variations of temperature and heat flux for conditions *B* and *C*, for wedge-shaped passages, in the work of Eckert et al. [172]. These results may be used to approximate those of an isosceles triangle for small apex angles. Results for a geometry of interest to nuclear engineers, rectangular ducts with heating on the broad sides only, are available in Siegel and Savino [173].

Figure 110 shows the wall temperature distributions for *turbulent flow* in an 11.5 degree apex angle isosceles triangle, as determined by Eckert and Irvine [168]. The duct wall was electrically heated so that distribution of the heat flux around the periphery of the duct was determined by uniform heat generation and by conduction within the duct walls. The conduction parameter ϕ is equal to 24 for these results.

Measurements indicate a temperature variation from two to three times the difference between the average wall temperature and the fluid bulk temperature. The laminar solution on the figure is from Eckert et al. [172]. The local heat transfer coefficient around the tube wall varied from less than 0.1 of the average value, to over 2 times the average value.

The case of uniform heating on the broad sides only of rectangular ducts, with turbulent flow, has been treated by Novotny et al. [174]. Their results for air flow in ducts of aspect ratios 1:1, 1:5, and 1:10 are shown in Fig. 111. The solid line represents a parallel-plate analysis and the dashed line a circular-tube analysis. The results agree closely with the analyses, and are essentially independent of aspect ratio.

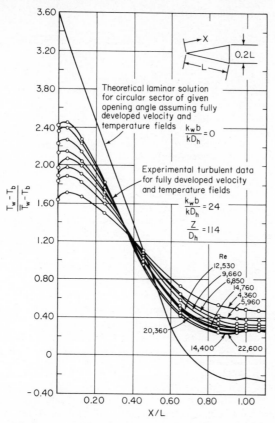

Fig. 110. Dimensionless peripheral wall temperature distributions in an isosceles triangle.

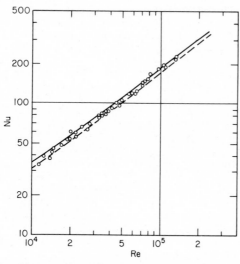

Fig. 111. Fully developed heat transfer results in rectangular ducts with heating on the broad wall only: dashed line—circular tube analysis; solid line—parallel plate analysis.

The hydraulic diameter here includes the entire wetted perimeter, and the heat transfer coefficient is based on the spanwise average wall-minus-bulk-temperature difference.

Sparrow et al. [175] have presented data for the case of asymmetric heating on the two broad sides of a 5:1 rectangular duct. The short sides were insulated. They found that for the case of one insulated broad wall, with heating on the other broad wall, the Nusselt numbers were 10 to 20 percent below the symmetrically heated case. With unequal heating rates on both walls, the more strongly heated wall, with heat flux q_1'', exhibited a Nusselt number slightly below the symmetric case, while the Nusselt number of the lesser heated wall, with $q_2'' = (1/2) q_1''$, was 15 percent higher than the symmetric case. The Nusselt number is defined here on the basis of the hydraulic diameter, with the entire wetted perimeter, and an h based on the heat flux of the wall of interest and the spanwise average temperature difference.

These results were correlated to the symmetric case (Novotny et al. [174]) by using an average heat transfer coefficient defined as

$$\bar{h} = \frac{q_1'' + q_2''}{\left(T_w - T_b\right)_1 + \left(T_w - T_b\right)_2} \tag{124}$$

d. Axial Variation of Surface Temperature and Heat Flux. Refer to general equations under laminar flow circular tube (Sec. A.2.e).

e. Roughness Effects. Refer to the circular tube section (Sec. A.2.f) for information on roughness and turbulence promoters. The results available specifically for roughness effects in a rectangular duct are those of Lancet [62].

f. Superimposed Free and Forced Convection. Han [176] has solved for the case of superimposed natural convection effects on forced convection in a vertical square duct under the case B thermal boundary condition. His results, for flow upward, and a linear density variation with temperature, are shown in Table 36 where the Rayleigh number is based on D_e.

TABLE 36. Influence of Free Convection on
Forced Convection Nusselt
Number for Square Tubes

Ra_D	Nu
0	3.61
π^4	3.69
$10\pi^4$	4.27
$100\pi^4$	9.46

D. MISCELLANEOUS SHAPES

1. Longitudinal Flow between Cylinders

Laminar–Hydrodynamics. An analytical solution has been obtained by Sparrow and Loeffler [177] for the friction factor, pressure drop, and shear stress for longitudinal fully developed laminar flow between cylinders in triangular and square arrays. Figure 112 shows the geometry of interest. The hydraulic diameter, used here in presenting the results, is given by

$$D_e = \frac{4\,(\text{flow area})}{(\text{wall perimeter})} \tag{125}$$

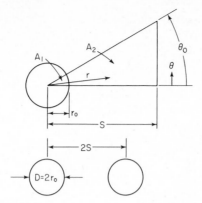

Fig. 112. Geometry for square and triangular arrays.

For the triangular array, θ_0 = 30 degrees and

$$D_e = r_0 \left[2.20 \left(\frac{s}{r_0} \right)^2 - 2.0 \right] \tag{126}$$

For the square array, θ_0 = 45 degrees and

$$D_e = r_0 \left[\frac{8}{\pi} \left(\frac{s}{r_0} \right)^2 - 2.0 \right] \tag{127}$$

Figure 113 presents the results for the Fanning friction factor (see Sec. A.1.a for definition) versus the ratio s/r_0. The Reynolds number is defined as $\bar{V} D_e / \nu$,

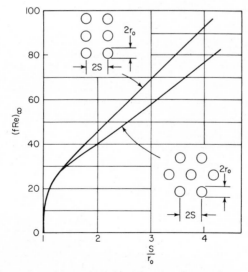

Fig. 113. Friction factor–Reynolds number relation based on D_e, as a function of spacing ratio for fully developed laminar flow parallel to tube bundles.

where D_e is the equivalent diameter, and $\bar{V}$ is the average velocity through the element shown in Fig. 113. The friction factor is defined in terms of the average shear stress; local shear stress distributions are given in the reference. The local velocity for the triangular array is given by

$$v_x = \frac{\sqrt{3}}{\pi} s^2 \left(-\frac{1}{\mu}\frac{dp}{dx}\right) \ln\frac{r}{r_0} - \frac{1}{4}\left(-\frac{1}{\mu}\frac{dp}{dx}\right)(r^2 - r_0^2) + \sum_{i=1}^{l} G_i\left(r^{6i} - \frac{r_0^{12i}}{r^{6i}}\right)\cos 6i\theta \quad (128)$$

where

$$\Delta_i = G_i\frac{6is^{6i}}{[(-1/\mu)(dp/dx)]s^2}$$

For the square array the velocity is

$$v_x = \frac{2}{\pi} s^2 \left(-\frac{1}{\mu}\frac{dp}{dx}\right) \ln\frac{r}{r_0} - \frac{1}{4}\left(-\frac{1}{\mu}\frac{dp}{dx}\right)(r^2 - r_0^2) + \sum_{i=1}^{l} G_i\left(r^{4i} - \frac{r_0^{8i}}{r^{4i}}\right)\cos 4i\theta \quad (129)$$

where

$$\delta_i = G_i\frac{4is^{4i}}{[(-1/\mu)(dp/dx)]s^2}$$

Table 37 presents the results for Δ_i and δ_i. Note that G_i can be evaluated from the tabulated values of δ_i.

TABLE 37. Constants for Calculating Velocity Distributions for Laminar Flow through Square and Triangular Tube Arrays

			(a) Values of Δ_j				
s/r_0	Δ_1	Δ_2	Δ_3	Δ_4	Δ_5	Δ_6	Δ_7
4.0	-0.0505	-0.0008	0.0000				
2.0	-0.0505	-0.0008	0.0000				
1.5	-0.0502	-0.0007	0.0000				
1.2	-0.0469	0.0007	0.0002	0.0000			
1.1	-0.0416	0.0028	0.0004	0.0000			
1.05	-0.0368	0.0043	0.0003	-0.0001	0.0000		
1.04	-0.0357	0.0046	0.0002	-0.0001	0.0000		
1.03	-0.0345	0.0049	0.0002	-0.0001	0.0000		
1.02	-0.0332	0.0051	0.0000	-0.0001	0.0000		
1.01	-0.0319	0.0052	-0.0001	-0.0002	0.0000		
1.00	-0.0305	0.0053	-0.0003	-0.0002	0.0000		
			(b) Values of δ_j				
s/r_0	δ_1	δ_2	δ_3	δ_4	δ_5	δ_6	δ_7
4.0	-0.1253	-0.0106	-0.0006	0.0000			
2.0	-0.1250	-0.0105	-0.0006	0.0000			
1.5	-0.1225	-0.0091	-0.0002	0.0000			
1.2	-0.1104	-0.0024	-0.0015	0.0003	0.0001	0.0000	
1.1	-0.0987	0.0036	0.0029	0.0005	0.0000		
1.05	-0.0904	0.0073	0.0032	0.0002	-0.0001	0.0000	

Turbulent—Hydrodynamics. Deissler and Taylor's [167] results for the Fanning friction factor for fully developed turbulent flow in a triangular and square array are shown in Fig. 114. The Reynolds number is defined using the bulk velocity and the hydraulic diameter.

Laminar-Heat Transfer. Sparrow et al. [178] have investigated the problem of heat transfer with fully developed laminar flow between cylinders in an equilateral triangular array. The Nusselt number based on D_e for the case of uniform heat transfer per unit length and constant temperature circumferentially (case B) is shown in Fig. 115. The nomenclature is shown on Fig. 112.

Circumferential effects are shown in Fig. 116, where the ratio of the local heat transfer rate to the circumferential average rate is plotted versus angle for the case of a uniform circumferential surface temperature. It is seen that for spacing ratios of 1.5 to 2.0 or greater there is little difference between the case of circumferentially uniform temperature and circumferentially uniform heat flux.

Turbulent Heat Transfer. Table 38 from Sutherland and Kays [179, 180] gives the results of various investigators for air and water for turbulent flow parallel to tube bundles. Figure 117 shows the results of Deissler and Taylor [185], based on D_e, for both triangular (also called delta) and square arrays. Figure 118 presents these results in a different form illustrating that for small pitch-to-diameter ratios the heat transfer

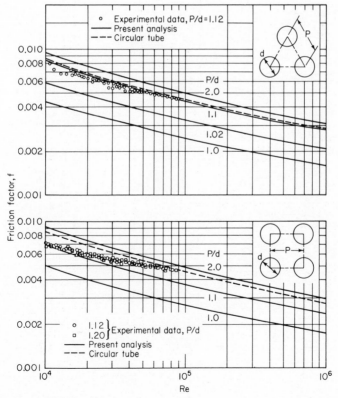

Fig. 114. Friction factor versus Reynolds number for turbulent flow through triangular and rectangular array. Reynolds number based on D_e.

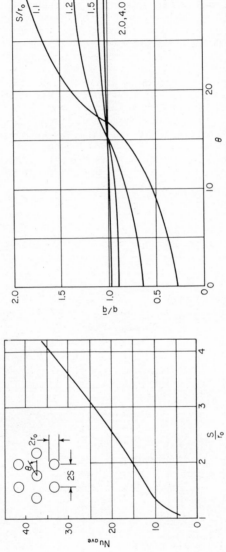

Fig. 115. Nusselt numbers based on D_e as a function of spacing ratio for fully developed laminar flow. Flow parallel to equilateral triangular arrays with constant heat rate per unit of tube length.

Fig. 116. Peripheral variation of wall heat transfer for a uniform peripheral surface temperature for fully developed laminar flow parallel to triangular arrays.

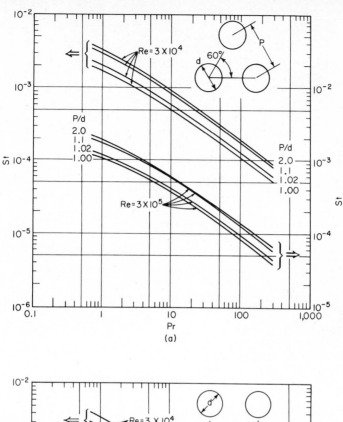

(a)

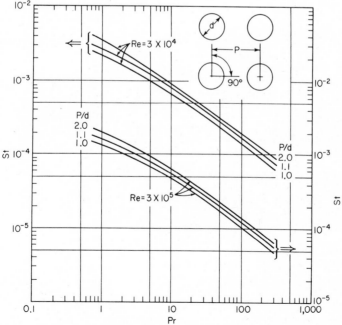

Fig. 117. Heat transfer in uniformly heated parallel rod arrays with hydrodynamically and thermally developed turbulent flows. Nu and Re based on D_e.

TABLE 38. Summary of Experiments on Heat Transfer in Turbulent Flow Through Parallel Rod Delta Arrays

Investigator	Test Section	P/d	Pr	Reynolds Number	Best Estimate of Data *	$Nu/Pr^{0.6}$†	
						$Re = 10^4$‡	$Re = 10^5$
Wantland [181]	Simulates infinite array with uniform heating, square and delta array	1.107	3 to 6	2,000 to 13,000	$Nu/Pr^{0.4}$ $= 0.0066\ Re^{0.91}$	28.8‡	145‡
Dingee and Chastain [182]	Nine-rod array with uniform heating, square and delta array	1.12 1.20 1.27	1.18 1.75 1.75 1.18 1.75	30,000 to 400,000	$Nu = 0.022\ Re^{0.8}$ $Nu = 0.026\ Re^{0.8}$ $Nu = 0.0275\ Re^{0.8}$ $Nu = 0.030\ Re^{0.8}$ $Nu = 0.033\ Re^{0.8}$	31.6 29.4 31.2 43.1 37.4	199 186 197 272 236
Miller et al. [183]	Simulates infinite array with one rod heated	1.46	1.1 to 2.75	90,000 to 700,000	$Nu/Pr^{1/3}$ $= 0.036\ Re^{0.8}$	43.6 to 55.6	275 to 351§
Hoffman et al. [184]	Seven-rod array with uniform heating, delta array	1.71	0.7	75,000	$St\ Pr^{2/3} = 0.0031$ to 0.0034	50.9 to 55.8	321 to 352
Deissler and Taylor [185]	Analysis of infinite array with uniform heating	1.0 1.02 1.10 2.0	0.7	30,000 and 300,000	43 and 252 52 and 284 75 and 420 84 and 462	53 64 93 104	312 351 520 572
Subbotin et al. [186]	Simulates infinite array with uniform heating	1.0 1.1 1.2 1.4 1.5	Water	400 to 50,000 not given¶	$Nu = 0.01\ Re^{0.8}$ $Nu = 0.025\ Re^{0.8}$ $Pr^{0.43}$ $Nu = 0.025\ Re^{0.8}$ $Pr^{0.43}$ $Nu = 0.032\ Re^{0.8}$ $Pr^{0.43}$ $Nu = 0.032\ Re^{0.8}$ $Pr^{0.43}$	15.8	100 240 250 320 320

| Palmer, Swanson [187] | Seven rod cluster, delta array | 1.015 | 0.7 | 10,000 to 60,000 | | | |
| Sutherland, Kays [180] | Simulate infinite array with one rod heated, delta array | 1.15 1.25 | 0.7 | 7,000 to 200,000 | $Nu = 0.023\ Re^{0.8}\ Pr^{0.4}$ | 49.2 | 288 |

* The correlation reported by the investigator or the best evaluation of the data as reported in graphical form.

† Examining the results from analytical treatments of turbulent flow heat transfer, cf. Kays and Leung, would suggest the best choice for the Prandtl number is 0.6 for the range of values $0.5 \leq Pr \leq 2.0$. This has been confirmed experimentally with steam where the Prandtl number varied from 1.0 to 1.7. For Prandtl numbers greater than 2, the exponent decreases with decreasing Reynolds number.

‡ This result is computed with the 0.4 exponent on Prandtl number because of the high Prandtl number and low Reynolds number range of the data reported. Beyond the reported Reynolds number range extrapolation is made parallel to the 1.15 data reported herein. The point extrapolated from is, at $Re = 10^4$, based on examination of the reported data.

§ The spread of some values is due to uncertainty of the Prandtl number for the data reported.

¶ This data was reported as the coefficient for the Reynolds number, i.e., $Nu/Pr^{1/3} = C\ Re^{0.8}$.

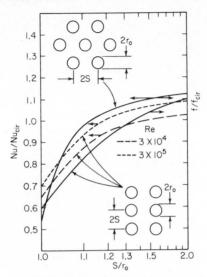

Fig. 118. Normalized heat transfer and friction results for fully developed turbulent flow parallel to a bank of rods or tubes, for a uniform axial heat flux and isothermal circumference. For triangular arrays the Reynolds number influence is small and the Nusselt number behavior is virtually the same as the friction behavior. The Prandtl number influence is small for these results ($0.7 \leq Pr \leq 3.0$).

coefficient is lower than for the circular tube, while for larger spacings it is greater (based on the equivalent diameter). The ratio of Nu/Nu_{cir} is virtually the same as f/f_{cir} for the triangular array, and is essentially independent of both Prandtl and Reynolds numbers. For the square array there is some difference between the Nusselt and friction ratios and also there is a weak dependence on Reynolds number.

Generalizations which can be made for turbulent flow through tube bundles may be summarized as:

1. The Nusselt number, based on D_e, increases with increasing rod spacing.
2. Peripheral temperature effects may be important at small P/d ratios and low Reynolds number, but for $P/d > 1.1$ there should be little significant variation. The same is true of the circumferential heat transfer coefficient variation.
3. Without spacers designed for flow mixing, there is little mixing transverse to the flow.
4. Bowing of the tubes may have a large effect on the heat transfer and on peripheral variations of T and q''.

The air data of Palmer and Swanson [187] taken at an x/D_e of 38, and greater, showed no entry effects. The air data of Hoffman et al. [184] showed a combined thermal and hydrodynamic entry length of some 15 to 30 diameters for a constant axial heat flux boundary condition.

Liquid Metals. Maresca and Dwyer [188] have presented an analysis for heat transfer from parallel rod bundles for an equilateral triangular array and a uniform longitudinal heat flux. Their results apply for $70 < Pe < 10^4$ and $1.3 < P/d < 3.0$.

Design curves have been developed from their results for the equilateral triangular array. Figure 119 presents these curves which are based on D_e. The straight-line portion represents the region in which Dwyer suggests that the

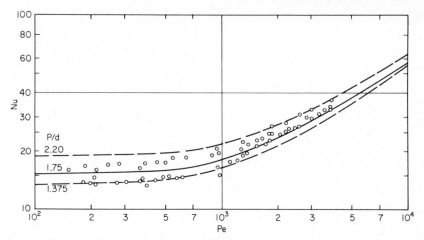

Fig. 119. Nusselt number versus Peclet number for fully developed longitudinal flow through an equilateral triangular array. Nu and Re based on D_e.

turbulent eddies lose their thermal energy while in transit owing to the large thermal conductivities of the liquid metals. Recent data of Nimmo and Dwyer [189] for mercury in parallel flow in a 13-rod bundle with $P/d = 1.750$ are shown on the figure. The data show good agreement with the analysis. Some information on entry effects with liquid metal flow in bundles is available in Maresca and Dwyer [188].

Discussion pertinent to those low Prandtl number heat transfer results is given in Sec. A.2.c.

2. Ellipses

Lundgren et al. [114] have presented hydrodynamic information for laminar flow in an elliptical duct. The fully developed velocity profile for such a duct is given by

$$\frac{v_x}{\overline{V}} = 2\left[1 - \left(\frac{x}{a}\right)^2 - \left(\frac{y}{b}\right)^2\right] \tag{130}$$

in which a and b are the semiaxes of the ellipse. With this one finds K, the total pressure drop entrance parameter, to have a value of 1.33 (see Eq. (71)). Additionally

$$f = \frac{8}{\mathrm{Re}} D_e^2 \left[\frac{1}{a^2} + \frac{1}{b^2}\right] \tag{131}$$

where

$$D_e = \frac{\pi b}{E(\lambda)} \quad \text{and} \quad \lambda = \frac{(a^2 - b^2)^{1/2}}{a}$$

where E is the elliptic integral of the second kind.

Figure 120 shows Tao's results for the Nusselt number versus aspect ratio b/a for the case of uniform peripheral wall temperature and a constant wall temperature

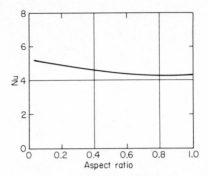

Fig. 120. Nusselt number versus aspect ratio for fully developed laminar flow through an elliptical tube.

gradient in the axial direction. Note that $b/a = o$ does not correspond to a physical situation and does not represent the parallel plate case.

3. Finned Tubing and Annuli

Hilding and Coogan [190] have obtained data on the effects of inner longitudinal fins in *circular tubing*. They conclude that multiple interruptions of fin surface, to break up the flow, improve the relative heat transfer--flow friction performance in the laminar flow and transition region, but the same advantage does not appear for turbulent flow. They also note that large increases in heat transfer may be accomplished by introducing internal fins, but the increase in heat transfer is accompanied by a greater percentage increase in pressure drop.

Finned Annuli. Results for fully developed laminar flow in finned annuli are given in Sparrow et al. [191]. The velocity distribution is given by the following

$$\frac{v_x}{(r_i^2/4\mu)(-dp/dx)} = \frac{2}{\pi} \sum_{n=1}^{\infty} \frac{1/n}{1 + (2\alpha/n\pi)^2} \left[1 - (-1)^n \left(\frac{r_0}{r_i}\right)^2 \right] \frac{\cosh(n\pi/\alpha)(\theta_0/2 - \theta)}{\cosh(n\pi/\alpha)(\theta_0/2)} \sin\frac{n\pi X}{a}$$

$$+ \frac{2}{\pi} \sum_{n=1}^{\infty} \frac{1}{n} [1 - (-1)^n] \left[\frac{\sinh(n\pi/\theta_0)(\alpha - X) + (r_0/r_i)^2 \sinh(n\pi X/\theta_0)}{\sinh(n\pi\alpha/\theta_0)} \right] \sin\frac{n\pi\theta}{\theta_0} - \left(\frac{r}{r_i}\right)^2 \quad (132)$$

in which

$$X = \ln\left(\frac{r}{r_i}\right) , \quad \alpha = \ln\left(\frac{r_0}{r_i}\right)$$

The friction results are given in Fig. 121. For small angles, Table 39 presents the friction results. The Reynolds number is based upon the average velocity and the hydraulic diameter.

For turbulent flow Fig. 122 shows the friction results of Lorenzo and Anderson [192]. These results are based on an opening angle of 8--13 degrees, and a radius ratio of about 0.65. The equivalent diameter of annuli containing longitudinal fins is four times the free cross-sectional area divided by the wetted perimeter.

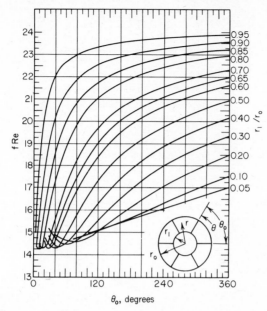

Fig. 121. Friction factor results for fully developed laminar flow in internally finned annular ducts.

TABLE 39. Friction Factor Reynolds Number Product for Laminar Flow through Internally Finned Annular Ducts

$f \cdot Re$	$\theta_0 \rightarrow$	$5°$	$10°$	$15°$	$20°$	$30°$	$40°$	$50°$	$60°$
$r_i/r_o =$	0.05	13.42	13.55	13.68	13.81	14.03	14.22	14.38	14.52
	0.1	14.47	14.49	14.51	14.54	14.59	14.62	14.66	14.69
	0.2	16.34	16.04	15.79	15.57	15.21	14.94	14.75	14.65
	0.3	17.80	17.07	16.47	15.98	15.24	14.78	14.54	14.47
	0.4	18.75	17.51	16.56	15.84	14.90	14.48	14.38	14.49
	0.5	19.14	17.37	16.13	15.30	14.47	14.33	14.52	14.88
	0.6	18.94	16.68	15.35	14.64	14.29	14.60	15.14	15.72
	0.65	18.60	16.16	14.91	14.38	14.40	14.97	15.65	16.32
	0.7	18.08	15.56	14.52	14.26	14.71	15.52	16.33	17.04
	0.8	16.42	14.42	14.33	14.84	16.15	17.28	18.16	18.86
	0.85	15.32	14.26	14.91	15.82	17.39	18.52	19.34	19.95
	0.9	14.35	14.98	16.36	17.51	19.07	20.05	20.71	21.18
	0.95	15.06	17.62	19.17	20.14	21.26	21.88	22.28	22.56

Heat transfer results for finned annuli are available in Clark and Winston [193]. Figure 123 presents these. The Reynolds number is based on D_e, L is the heated length of finned tube, and p is the wetted perimeter of the channel between two longitudinal fins. Beyond a Reynolds number of 60,000, the circular tube equations are recommended, using an equivalent diameter D_e.

Transverse Finned Annuli. Friction coefficients for transverse finned annuli have been presented by Braun and Knudsen [194].

Heat transfer results for water have been obtained by Knudsen and Katz [195]. Detailed information is also available in Knudsen and Katz [2].

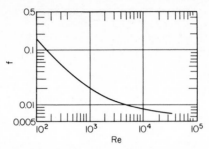

Fig. 122. Friction factors for annuli with longitudinal fins.

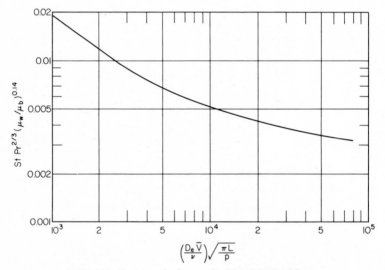

Fig. 123. Heat transfer for annuli with longitudinal fins.

4. Circular Sectors

Sparrow and Haji-Sheikh [196, 197] have presented a summary of results for the hydrodynamics and heat transfer in circular sectors. Figure 124 illustrates the two types of circular sectors possible, denoted here as chord sectors and radius sectors.

Table 40, from Ref. 197, presents results based on D_e for the *chord sectors*. K is the factor to account for the incremental pressure drop due to flow development in the hydrodynamic region (Sec. A.1.c), and α is the angular coordinate, chosen so that $\alpha = 180$ degrees is a circle, and 90 degrees is a semicircle. $\overline{Nu}$ corresponds to a uniform peripheral temperature boundary condition and $\overline{\overline{Nu}}$ to a uniform peripheral heat flux. Both cases are for uniform axial heat flux. Velocity and temperature profiles are given in Ref. 197 as well as in Eckert et al. [172].

Friction factor, pressure drop, and heat transfer results from Ref. 196 are presented in Fig. 125 for the *radial circular sector*, as well as for the two triangular geometries which are similar to the radial sector. D_e is used in the Reynolds and Nusselt numbers.

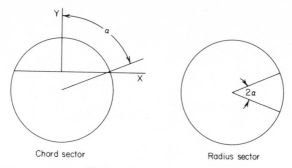

Chord sector   Radius sector

Fig. 124. Types of circular sectors.

TABLE 40. Friction and Heat Transfer Results for Laminar Flow through Chord Circular Sectors. $\overline{Nu}$, **Uniform Peripheral Temperature, and** $\overline{\overline{Nu}}$, **Uniform Peripheral Heat Flux**

α	$4f \cdot Re$	K	$\overline{Nu}$	$\overline{\overline{Nu}}$
$0°$	62.22	1.740	3.580	0.00
$5°$	62.23	1.739	3.608	0.01316
$10°$	62.24	1.734	3.616	0.05247
$20°$	62.30	1.715	3.648	0.2071
$30°$	62.39	1.686	3.696	0.4558
$40°$	62.51	1.650	3.756	0.7849
$60°$	62.76	1.571	3.894	1.608
$90°$	63.06	1.463	4.089	2.923
$120°$	63.26	1.385	4.228	3.882
$150°$	63.66	1.341	4.328	4.296
$180°$	64.00	1.333	4.364	4.364

5. Other Shapes

The reader is referred to "Compact Heat Exchangers" by Kays and London [23] for information on a wide variety of noncircular geometries. Some specific cases follow.

Tapered Passages. Heat transfer results for fully developed laminar flow in plane walled tapered passages are available in Sparrow and Starr [198].

Right-angle Bend. The effect of a right-angle bend on heat transfer in a pipe for laminar and turbulent flow has been discussed by Ede [199, 200].

Odd Corners. Gunn and Darling [201] have investigated the geometries shown in Fig. 126. Table 41 shows the Reynolds number, based on D_e, for the start of transition, and also shows the laminar friction factor results. Turbulent results were correlated by Eq. (133) where k_c/k_n represents the ratio of constants in the *laminar* friction results, that is, k_c equals 16 for a circular tube. The subscript c is circular, subscript n noncircular

$$\frac{f_c}{f_n} = \left[\frac{k_c}{k_n}\right]^{0.45 \, \exp(-[Re - 3000]/10^6)}$$

$$(133)$$

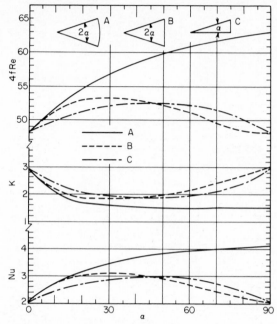

Fig. 125. Fully developed friction, pressure drop, and heat transfer results for laminar flow through radial circular sectors, isosceles triangles, and right triangles. Equivalent diameter used throughout.

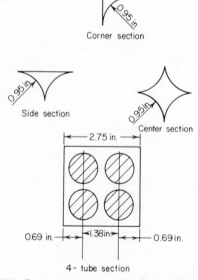

Fig. 126. Geometries investigated by Gunn and Darling.

TABLE 41. Transition Reynolds Numbers
and Laminar Friction Factors
for Odd Corner Geometries

Section	Reynolds number of transition	$f \cdot Re$ (laminar)
Circular	2200	16
Corner	1100	7.06
Side	1000	6.50
Center	900	6.50
4-tube	2000	14.5

E. HEAT SOURCE SOLUTIONS AND DISSIPATION EFFECTS

Laminar Flow–Circular Tube. The temperature distribution with a uniform heat source, with a constant temperature wall, for both a parabolic and slug-flow velocity profile, has been determined by Topper [202]. The results for a parabolic profile are given in Fig. 127.

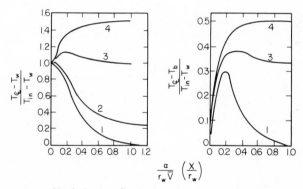

Fig. 127. Temperature profiles for laminar flow with uniform heat generation and an isothermal wall.
(1) $q''' r_w^2/k(T_{in} - T_w) = 0$; (2) $q''' r_w^2/k(T_{in} - T_w) = 1$; (3) $q''' r_w^2/k(T_{in} - T_w) = 4$;
(4) $q''' r_w^2/k(T_{in} - T_w) = 6$.

Sparrow and Siegel [203] have solved the *circular tube laminar flow* problem for a *constant wall heat flux* including heat sources in the fluid. They separate the problem into one with (1) a uniform heat source with the tube wall insulated, plus (2) a uniform heat flux at the wall with no sources. Additionally they have solved for a parabolic radial source variation.

The combined problem solution may be found by superposition as

$$T_w - T_b = \left\{ \frac{T_{q''',w} - T_{q''',b}}{\left(T_{q''',w} - T_{q''',b}\right)_{fd}} \right\} \left[T_{q''',w} - T_{q''',b}\right]_{fd}$$

$$+ \left\{ \frac{T_{q'',w} - T_{q'',b}}{\left(T_{q'',w} - T_{q'',b}\right)_{fd}} \right\} \left[T_{q'',w} - T_{q'',b}\right]_{fd} \tag{134}$$

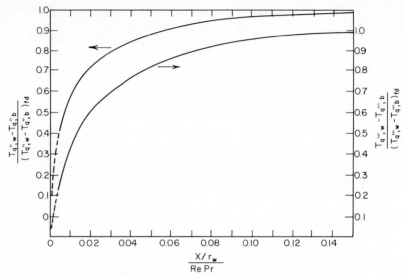

Fig. 128. Dimensionless wall temperature variation for a uniform internal heat source in an insulated tube and for a uniform heat flux at the tube wall with no sources. Laminar flow, circular tube.

Figure 128 presents the bracketed ratios. The following fully developed results are for the heat generation and uniform wall heat flux terms respectively

$$\left[T_{q''',w} - T_{q''',b}\right]_{fd} = \frac{1}{16} \frac{q''' r_w^2}{k} \tag{135}$$

$$\left[T_{q'',w} - T_{q'',b}\right]_{fd} = \frac{11}{24} \frac{q'' r_w}{k} \tag{136}$$

where q''' refers to the heat source strength. The bulk temperature variation may then be found from an energy balance as

$$T_b = T_i + \frac{4(x/D)}{Re\,Pr}\left(\frac{q''' r_w^2}{k} + \frac{2q'' r_w}{k}\right) \tag{137}$$

The local wall temperature is then calculated from Eqs. (134) to (137). For the fully developed case the results reduce to a Nusselt number, based on wall heat flux, as

$$Nu_{fd} = \frac{2}{11/24 + q''' r_w/16q''} \tag{138}$$

This analysis has been compared to experimental results by Inman [204] with good agreement noted.

  Turbulent Flow—Circular Tube. Siegel and Sparrow [205] have solved the case of turbulent flow in a circular tube with variable heat flux and a source varying with x. They used equal eddy diffusivities and Deissler's [206] turbulent velocity profile. For the case of a uniform heat flux and a uniform heat source, superposition gives Eq. (134). Numerical values of the first term may be read from Fig. 129 and the second term from Fig. 130.

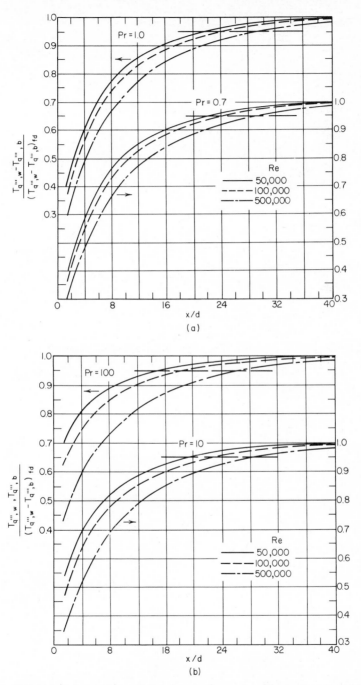

Fig. 129. Wall to bulk temperature variation for a uniform internal heat source in an insulated circular tube with turbulent flow.

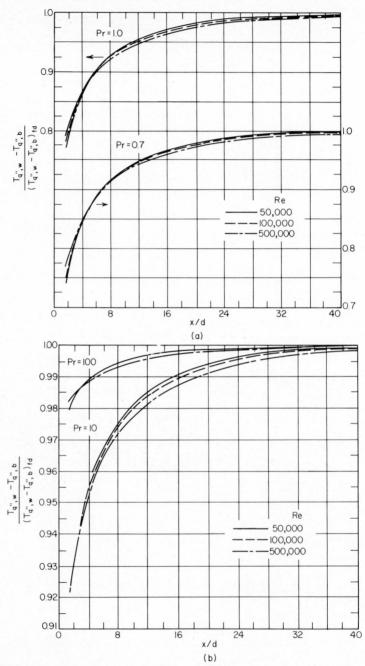

Fig. 130. Wall to bulk temperature variation for a uniform wall heat flux and no internal source in a circular tube with turbulent flow.

The fully developed results required for Eq. (134) are given in Figs. 131 and 132. From an energy balance Eq. (137) again results. From the energy balance and Eq. (134) the wall temperature as a function of x can be calculated.

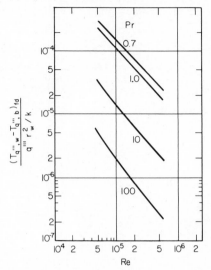

Fig. 131. Fully developed wall to bulk temperature difference for a uniform internal heat source in an insulated circular tube with turbulent flow.

Data verifying the uniform source solution with the adiabatic wall are available in Kinney and Sparrow [207].

Laminar Flow-Parallel Plates. The case of a heat-generating fluid in laminar flow between parallel plates has been analyzed for both a prescribed wall temperature and a prescribed wall heat flux. The analysis of Sparrow et al. [208] also includes the thermal entrance region.

For fully developed laminar flow between parallel plates with a uniform heat source and a uniform wall heat flux, the Nusselt number is given by

$$Nu = \frac{4}{17/35 + (3/35) \, q''' a/q''} \tag{139}$$

where a is the half-width between the plates.

Laminar Flow–Rectangular Duct. Experimental results for laminar forced convection of a heat-generating fluid in rectangular passages (1:1, 4:1, 8:1) with adiabatic walls have been presented by Strite and Inman [209]. They conclude that the longitudinal variation in wall temperature can be predicted from the analytical parallel plate channel results when the half-length of the short side is used as the characteristic dimension.

Other Geometries. Information on the equilateral triangle and the elliptical duct with heat generation are given in Tao [210] and Tyagi [211]. Some of these follow under the subsection below.

Dissipation Effects. A variety of geometries have been investigated by Tyagi for the case of a uniform heat source plus dissipation in a uniformly heated tube. His results have been amended by Perkins [212] to give the following.

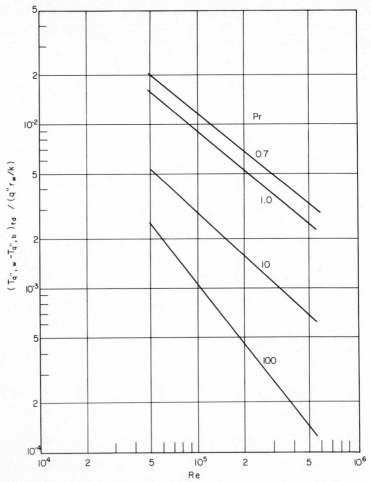

Fig. 132. Fully developed wall to bulk temperature difference for a uniform wall heat flux and no internal source in a circular tube with turbulent flow.

Circular Tube

$$\text{Nu} = \frac{48}{11} \frac{1 + 8a + \beta}{1 + (64/11)a + (5/11)\beta} \qquad (140)$$

$$a = \frac{-1}{8(1 + 4q''/q'''D)}$$

$$\beta = \frac{-32\mu\overline{V}^2}{D^2(q''' + 4q''/D) \cdot J \cdot g_c}$$

where a is the heat source parameter and β is the dissipation parameter. For the zero dissipation case, $\beta = 0$, these values agree with those already presented.

Parallel Plates

$$\text{Nu} = \frac{140}{17} \frac{1 + 3a + \beta}{1 + (42/17)a + (8/17)\beta}$$

$$a = \frac{-1}{3(1 + 4q''/q'''D_e)} \tag{141}$$

$$\beta = \frac{-48\mu\overline{V}^2}{D_e^2(q''' + 4q''/D_e) \cdot J \cdot g_c}$$

where D_e equals twice the plate spacing.

Equilateral Triangle

$$\text{Nu} = \frac{28}{9} \frac{1 + (20/3)a + \beta}{1 + (40/9)a + \beta}$$

$$a = \frac{-1}{6.667(1 + 4q''/q'''D_e)} \tag{142}$$

$$\beta = \frac{-26.66\mu\overline{V}^2}{D_e^2(q''' + 4q''/D_e)}$$

where $D_e = 2a$; side length $= 2\sqrt{3}\,a$.

Ellipse. For this case see Tyagi [211].

F. CURVED DUCTS AND COILS

1. Hydrodynamics—Coils

Mori and Nakayama [213] have predicted laminar friction results for flow in curved ducts, while Ito [214] has presented experimental results for laminar and turbulent friction coefficients. Ito's results are correlated in the *laminar* region by

$$\frac{f_c}{f_s} = \frac{21.5K}{(1.56 + \log_{10} K)^{5.73}} \tag{143}$$

for $2000 > K > 13.5$ and where subscript c refers to "coiled" or "curved," and subscript s refers to "straight." $K = \text{Re}\sqrt{r_w/R}$ and is called the Dean number. R is the radius of curvature of the pipe, and r_w is the pipe radius. The following empirical formula was deduced from the data in the *turbulent* regime

$$4f_c\left(\frac{R}{r_w}\right)^{1/2} = 0.029 + 0.034\left[\text{Re}\left(\frac{r_w}{R}\right)^2\right]^{-0.25} \tag{144}$$

This equation gives good agreement with experimental results in the range

$$300 > \text{Re}\left(\frac{r_w}{R}\right)^2 > 0.034$$

Below 0.034 the friction factor of a curved pipe practically coincides with that of a straight pipe. For values of $\text{Re}(r_w/R)^2$ greater than 6

$$4f_c \left(\frac{R}{r_w}\right)^{1/2} = \frac{0.316}{\left[\text{Re}\,(r_w/R)^2\right]^{1/5}} \tag{145}$$

Using the Blasius relation (Sec. A.1.a), one may express this as

$$\frac{f_c}{f_s} = 1.0 \left[\text{Re}\left(\frac{r_w}{R}\right)^2\right]^{1/20} \tag{146}$$

Ito gives the relation

$$\text{Re}_{\text{crit}} = 2\left(\frac{r_w}{R}\right)^{0.32} \times 10^4 \tag{147}$$

for the transition Reynolds number. This equation gives good agreement with experimental results in the range

$$15 < \frac{R}{r_w} < 860$$

For $R/r_w > 8.6 \times 10^2$, the critical Reynolds number for a curved pipe is equal to that for a straight pipe.

Ito's isothermal turbulent friction factors have been verified by Rogers and Mayhew [215], using water in coiled pipes, as has his criterion for transition. The laminar results were analytically verified by Mori and Nakayama [213]. Representative velocity and temperature profiles for laminar flow, as determined with a hot-wire anemometer, are in Mori and Nakayama.

2. Heat Transfer—Coils

Mori and Nakayama [213] have given analytical results for the average peripheral Nusselt number with laminar flow in a coil. As a first approximation, the Nusselt number ratio is given by

$$\left(\frac{\text{Nu}_c}{\text{Nu}_s}\right)_I = 0.1979\frac{K^{1/2}}{\zeta} \tag{148}$$

where $K = \text{Re}\sqrt{r_w/R}$, and ζ is given by Eq. (149) for $\text{Pr} \geq 1$, and Eq. (150) for $\text{Pr} \leq 1$

$$\zeta = \frac{2}{11}\left[1 + \sqrt{\left(1 + \frac{77}{4}\cdot\frac{1}{\text{Pr}^2}\right)}\right] \tag{149}$$

$$\zeta = \frac{1}{5}\left[2 + \sqrt{\left(\frac{10}{\text{Pr}^2} - 1\right)}\right] \tag{150}$$

The second approximation is obtained by correcting the first approximation as follows. For the case of $\text{Pr} \geq 1$

$$\left(\frac{Nu_c}{Nu_s}\right)_{II} = \left(\frac{Nu_c}{Nu_s}\right)_I \cdot \cfrac{1}{1 + \cfrac{37.05}{\zeta}\left[\cfrac{1}{40} - \cfrac{17}{120}\zeta + \left(\cfrac{1}{10\zeta} + \cfrac{13}{30}\right)\cfrac{1}{10Pr}\right]K^{-1/2}} \qquad (151)$$

For the case of $Pr \le 1$

$$\left(\frac{Nu_c}{Nu_s}\right)_{II} = \left(\frac{Nu_c}{Nu_s}\right)_I \cdot \cfrac{1}{1 - \cfrac{37.05}{\zeta}\left[\cfrac{\zeta^2}{12} + \cfrac{1}{24} - \cfrac{1}{120\zeta} - \left(\cfrac{4}{3}\zeta - \cfrac{1}{3\zeta} + \cfrac{1}{15\zeta^2}\right)\cfrac{1}{20Pr}\right]K^{-1/2}}$$

$$(152)$$

For large Prandtl numbers, the Nusselt number tends to depend only on the intensity of secondary flow and shows little change with Prandtl number. The experimental data for air obtained by Mori and Nakayama are in good agreement with the analysis. The oil data of Seban and McLaughlin [216], using the mean value of the Nusselt numbers at the inner and outer wall, are also in good agreement.

Turbulent heat transfer in coils has been investigated by Rogers and Mayhew and Seban and McLaughlin, the former with water in steam-heated pipes, the latter with air in electrically heated pipes. Seban and McLaughlin's data are correlated by

$$Nu\,Pr^{-0.4} = \frac{f}{2}\,Re \qquad (153)$$

which from Ito's friction data can be reduced to

$$Nu\,Pr^{-0.4} = 0.023\,Re^{0.85}\left(\frac{r_w}{R}\right)^{0.1} \qquad (154)$$

Rogers and Mayhew's data are correlated by the same equation with a constant of 0.021. The Nusselt number is the average of the peripheral values.

Curved Channels. The effect of curvature on the difference of the heat transfer coefficients on the two walls of a parallel-plate channel has been discussed by Kreith [217]. Data are also given in Seban and McLaughlin [216] for circular tube-flow in which it was noted that the outside h was about twice the inside h for their larger coil ($R_{coil} = 15$ in., $r_{tube} = 0.145$ in.) and about four times the inside value for a coil of 2.46-in. radius and 0.145-in. tube radius. The Reynolds number range was $10^4 < Re < 10^5$.

G. CONVERGING AND DIVERGING DUCTS (NOZZLES AND DIFFUSERS)

Flow in nozzles and other variable area ducts presents a problem that might be classified as either "duct-flow," or "external-boundary-layer-flow," depending upon whether the boundary layers are thin and therefore unaffected by the proximity of opposite walls, or are sufficiently thick so that they actually interact along the centerline of the duct. Flow in nozzles, which is perhaps the most important technical application in this category of problem, is usually essentially an external boundary layer problem. Although external boundary layers are extensively treated elsewhere in this book, the nozzle and diffuser problems are so closely related to the other duct problems treated in this chapter that their inclusion here is not felt to be redundant. (It should be noted that the hydrodynamic and thermal entry length problems considered in this chapter are also at least partially external boundary layer problems.)

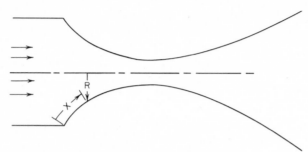

Fig. 133. Geometry for flow through a nozzle.

The duct geometry under consideration is shown in Fig. 133. It will be assumed that the boundary layers are sufficiently thin so that they do not interact, and that there is a definable point $x = 0$ where the boundary layer originates. The axial velocity $\bar{V}$ is then assumed to be uniform over the duct cross section at this point. Actually, these are not particularly critical idealizations, and the heat transfer behavior farther downstream is not strongly affected by uncertainty in the location of the origin, or lack of free-stream velocity uniformity.

It is also assumed that the stagnation temperature of the fluid T_{stag} is uniform over the central core of the flow, and along the duct. This is consistent with the idealization of a thin boundary layer.

The duct shown in Fig. 133 is assumed to be an internal body of revolution, completely described by some $R(x)$ which is given. Note that the family of possible ducts includes the simple two-dimensional case, for which R is indefinitely large. R then merely cancels out of the equation to be presented.

The primary flow variable will be the free-stream *mass velocity G*. In a *one-dimensional idealization,* G is simply the mass flow rate divided by local duct cross-sectional area. It is also the product of the free-stream velocity (assumed uniform in the one-dimensional idealization) and the fluid density. If information is available on the *two-dimensional* flow over the central core of the duct (i.e., outside the boundary layer), then G is the *local* mass velocity just outside of the boundary layer, the product of the local free-stream velocity and density.

Laminar Boundary Layer. The following approximate equation for the local Stanton number is developed by Smith and Spalding [218], and is based on exact wedge-flow laminar boundary layer solutions, constant fluid properties, and a *constant surface temperature.* It is applicable to accelerating flows, and mildly decelerating flows

$$\mathrm{St} \ = \ \frac{C_1 \mu^{1/2} R G^{C_2}}{\left[\displaystyle\int_0^x G^{C_3} R^2 d\xi \right]^{1/2}} \tag{155}$$

where $G = G(\xi) = G(x)$

$\quad\quad R = R(\xi) = R(x)$

$\quad\quad \text{St} = h/(Gc_p)$

$\quad\quad G = \bar{V}\rho = \dot{m}/A_c$

where A_c is the local duct cross-section area.

The constants C_1, C_2, C_3 are given as functions of Prandtl number in Table 42. The local Stanton number, and thus the local heat transfer coefficient, at any point along the duct wall is evaluated by integrating the term in the denominator from the assumed boundary layer origin to the point x in question, and inserting the local values of R and G, at x, in the numerator.

TABLE 42. Constants for Use in Eq. (155)

Pr	C_1	C_2	C_3
0.7	0.418	0.435	1.87
0.8	0.384	0.450	1.90
1.0	0.332	0.475	1.95
5.0	0.117	0.595	2.19
10.0	0.073	0.685	2.37

For *liquids*, it is recommended that a correction for the temperature dependence of the viscosity be made *locally* by including the following factor on the right-hand side

$$\left(\frac{\mu_w}{\mu_b}\right)^{-0.14}$$

The viscosity term in Eq. (155) should be evaluated at bulk, or free-stream temperature, that is, μ_b.

For *gases*, at high velocities and/or moderately large temperature differences, it is recommended that Eq. (155) be used as it stands, evaluating the viscosity coefficient at local *free-stream static temperature* (which will be here termed T_b). However, the local *heat flux* at the wall should be evaluated from

$$q_w'' = h(T_{aw} - T_w) \tag{156}$$

where T_{aw}, the *adiabatic wall temperature*, is evaluated from

$$T_{aw} = T_b + \text{Pr}^{1/2}(T_{stag} - T_b)$$

and

$$T_{stag} = T_b + \frac{\bar{V}^2}{2g_c Jc_p}$$

(where T_{stag} is assumed constant through the duct). With these modifications, the equation is applicable at supersonic as well as subsonic velocities.

Transition to a Turbulent Boundary Layer. A transition to a *turbulent* boundary layer will generally occur if the Reynolds number based on the *momentum thickness* of the boundary layer Re_m becomes greater than 200–400. The following is a fair approximation for Re_m for a laminar boundary layer. It is based on *constant properties*, but is usually adequate for a transition criterion for varying properties. It is applicable to accelerating flows, and mildly decelerating flows, and again assumes a boundary layer origin at some $x = 0$

$$\text{Re}_m = \frac{0.67\rho^{5/2}}{RG^2\mu^{1/2}} \left[\int_0^x \frac{G^5 R^2}{\rho^5} d\xi \right]^{1/2} \tag{157}$$

where ρ is the free-stream fluid density which should be evaluated *locally* for a gas, i.e., it is a function of local pressure and T_b.

Turbulent Boundary Layer. The following equation is based on a method of solving the energy integral equation of the boundary layer developed by (among others) Ambrok [219]. In the form presented here, it is particularized to a *gas* which may be at either subsonic or supersonic velocities. Modification for use with *liquids* is indicated below.

The wall surface temperature T_w may vary along the surface in any arbitrary manner, although the equation is not highly accurate for *abrupt* changes in surface temperature. It is an excellent approximation for accelerating flow so long as the momentum thickness Reynolds number is well above the critical value, and it is applicable to mildly decelerating flows. Very strong accelerations can, however, cause a retransition back to a laminar boundary layer. The following is a tentative criterion for retransition. Let

$$K = \frac{\mu\rho}{G^2} \frac{d}{dx}\left(\frac{G}{\rho}\right)$$

Then if $K > 3 \times 10^{-6}$ a retransition to a laminar boundary layer may occur.

For $1 \times 10^{-6} < K < 3 \times 10^{-6}$ the acceleration may tend to suppress the Stanton number somewhat below that calculated from the equation given below. For $K < 1 \times 10^{-6}$ the equation gives excellent results even for supersonic nozzles.

(It is again assumed that the thermal boundary layer originates at $x = 0$.)

$$\text{St} = \frac{0.0295}{\text{Pr}^{0.45}} \frac{R^{1/4}(T_w - T_{aw})^{1/4}\mu^{1/5}(T_w/T_b)^n}{\left[\int_0^x R^{5/4}(T_w - T_{aw})^{5/4} G\, d\xi \right]^{1/5}} \tag{157}$$

where

$n = -0.5$ for $(T_w/T_b) > 1$

$n = -0.25$ for $(T_w/T_b) < 1$

$T_{aw} = T_b + \text{Pr}^{1/3}(T_{stag} - T_b)$

$T_w - T_{aw} = fn(x) = fn(\xi)$

$q''_w = h(T_{aw} - T_w)$

$\text{St} = h/(Gc_p)$

All properties are to be evaluated at the *local* value of T_b.

Equation (157) should also be a good approximation for a *liquid*, but the exponent on Pr should be changed to 0.6, and the factor $(T_w/T_b)^n$ should be replaced by $(\mu_w/\mu_b)^{-0.14}$.

Additional information on heat transfer in supersonic nozzles may be found in Bartz [220], Back, Massier, and Gier [221], Benser and Graham [222], Witte and Harper [223], and Fortini and Ehlers [224].

H. EFFECTS OF TEMPERATURE VARYING PROPERTIES

1. Introduction

The heat transfer and flow friction solutions considered in the previous sections have been based on the assumption that the fluid properties remain constant throughout the flow field. When applied to real heat transfer problems this is obviously an idealization, since the transport properties of most fluids vary with temperature, and thus will vary over the flow cross section of a tube. In this section the results of a number of analytic solutions and experiments in which this influence has been investigated will be examined.

The temperature-dependent property problem is further complicated by the fact that the properties of different fluids behave differently with temperature. For *gases,* the specific heat varies only slightly with temperature, but the viscosity and thermal conductivity increase as about the 0.8 power of the absolute temperature (in the moderate temperature range). Furthermore, the density varies inversely with the first power of the absolute temperature. On the other hand, the Prandtl number $\mu c_p/k$ does not vary significantly with temperature.

For most *liquids* the specific heat and thermal conductivity are relatively independent of temperature, but the viscosity decreases very markedly with temperature. This is especially true of oils, but even for water the viscosity is very temperature dependent. The density of liquids, on the other hand, varies little with temperature. The Prandtl number of liquids varies with temperature in much the same manner as viscosity.

There are a number of variable property analyses in the literature, as well as a considerable body of experimental data. The general effect of transport properties varying with temperature is to yield distorted velocity and temperature profiles, and thus different friction and heat transfer coefficients than would be obtained if properties were constant.

Figure 134 shows the distortion of the laminar velocity profiles due to heating or cooling. Figure 135 shows turbulent velocity and temperature profiles for a gas with a Prandtl number of 1, with significant property variation, as calculated by Deissler [4]. Beta is a heat-flux parameter and is given by

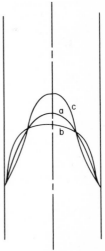

$$\frac{q_w''\sqrt{\tau_w/\rho_w}}{c_{p_w}\tau_w T_w}$$

Fig. 134. Distortion of the laminar velocity profile in a heated or cooled tube when the viscosity of the fluid depends on the temperature. (*a*) parabolic profile, (*b*) heating of a liquid or cooling of a gas, (*c*) cooling of a liquid or heating of a gas.

For engineering applications it has been found convenient to employ the constant property analytic solutions, or experimental data obtained with small temperature differences, and then to apply some kind of correction to account for properties variation. Fortunately most of the variable properties results indicate that fairly simple corrections will generally suffice.

Two schemes for correction of the constant property results are in common use. In the *reference temperature* method, some temperature is found at which the properties appearing in the nondimensional groups (Re, Pr, Nu, etc.) may be evaluated such that the constant property results may be used to evaluate variable property behavior. Typically this may be the surface temperature, or a temperature part way between the surface temperature and the bulk temperature, although there is no general rule. In the

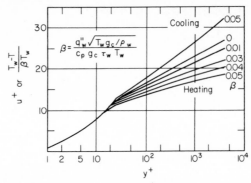

Fig. 135. Predicted generalized velocity and temperature distributions including the effects of heating and cooling, for a Prandtl number of 1.

property ratio method, all properties are evaluated at the bulk temperature, and then all of the variable properties effects are lumped into a function of a ratio of some pertinent property evaluated at the surface temperature to that property evaluated at the bulk temperature.

For *liquids,* where the viscosity variation is responsible for most of the effect, it is found that equations of the following type are often excellent approximations

$$\frac{\text{Nu}}{\text{Nu}_{\text{c.p.}}} = \frac{\text{St}}{\text{St}_{\text{c.p.}}} = \left(\frac{\mu_w}{\mu_b}\right)^n \qquad (158)$$

$$\frac{f}{f_{\text{c.p.}}} = \left(\frac{\mu_w}{\mu_b}\right)^m \qquad (159)$$

All properties in the nondimensional parameters are evaluated at local bulk temperature and the subscript c.p. refers to the appropriate constant property solution or small temperature difference experimental result. The viscosity μ_w is the viscosity evaluated at surface temperature, while μ_b is evaluated at bulk temperature. The exponents m and n are functions of geometry and the type of flow.

For *gases,* the viscosity, thermal conductivity, and density are all functions of *absolute* temperature. Fortunately the absolute temperature dependence tends to be similar for different gases, although the similarity does break down at the temperature extremes. Thus it is found that the temperature-dependent properties effects can usually be adequately correlated by the equations

$$\frac{\text{Nu}}{\text{Nu}_{\text{c.p.}}} = \frac{\text{St}}{\text{St}_{\text{c.p.}}} = \left[\frac{T_w}{T_b}\right]^n \qquad (160)$$

$$\frac{f}{f_{\text{c.p.}}} = \left[\frac{T_w}{T_b}\right]^m \qquad (161)$$

2. Laminar Flow in Tubes—Liquids

Deissler [225] carried out the calculations for a circular tube for a fluid for which the viscosity varies with temperature as

$$\mu \propto T^{-1.6}$$

which corresponds approximately with the behavior of liquid metals. When the results are put in the form of Eqs. (158) and (159), we obtain

$$n = -0.14$$

which is a figure that has also been used extensively to correlate experimental data on laminar flow of liquids.

For the friction coefficient, Deissler's results indicate a slightly different exponent depending upon whether the liquid is being heated or cooled

$$\frac{\mu_w}{\mu_b} > 1.0 \quad ; \quad m = +0.50 \text{ (cooling)}$$

$$\frac{\mu_w}{\mu_b} < 1.0 \quad ; \quad m = +0.58 \text{ (heating)}$$

K. T. Yang [226] has considered the laminar thermal entry problem in a circular tube for both a constant surface temperature and constant heat rate. Yang employed an integral technique which he demonstrates yields results very close to the exact eigenvalue solution for constant viscosity, and he assumed that the viscosity dependence upon temperature could be expressed as

$$\frac{\mu}{\mu_w} = \frac{1}{1 + \gamma\theta}$$

where $\theta = (T - T_w)/(T_{in} - T_w)$ and γ is a parameter.

Yang considered only the heat transfer behavior, and his results for both constant surface temperature and constant heat rate can be very well approximated by

$$n = -0.11$$

This result (which refers to the local conductance) applies to both the thermal entry length and the subsequent fully developed region. The difference in exponent, -0.11 as compared to -0.14, is evidently related to the different form of viscosity-temperature function employed, and the choice of exponent then should depend upon the viscosity-temperature relation of the particular fluid to be employed. In any event, the difference is small.

3. Turbulent Flow in Tubes—Liquids

Fully developed turbulent flow of a liquid in a circular tube, with constant heat rate and viscosity a function of temperature, has been considered by Deissler [39]. An iterative procedure for solution of the momentum and energy differential equations was employed as in Deissler's laminar solution.

It was assumed that $\epsilon_H = \epsilon_M$, which limits the results to Pr ≥ 1.00. The assumed viscosity-temperature relation was

$$\frac{\mu}{\mu_w} = \left(\frac{t}{t_w}\right)^d$$

where t is in °F.

Values of $d = -1$ and -4 were used, and the results were approximately the same.

When the results are put in the form of Eqs. (158) and (159) Table 43 is obtained. The trends of these results suggest that for very low Prandtl number m and n take on approximately their laminar magnitudes.

TABLE 43. Viscosity Ratio Exponents for Fully Developed Turbulent Flow in a Circular Tube, Liquids

Pr	$(\mu_w/\mu_b) > 1.0$ (cooling)		$(\mu_w/\mu_b) < 1.0$ (heating)	
	m	n	m	n
1	0.12	-0.19	0.092	-0.20
3	0.087	-0.21	0.063	-0.27
10	0.052	-0.22	0.028	-0.36
30	0.029	-0.21	0.000	-0.39
100	0.007	-0.20	-0.04	-0.42
1000	-0.018	-0.20	-0.12	-0.46

Experimentally $n = -0.14$ has been used often (Sieder-Tate [227]) to correlate heat transfer data, and since such experiments have generally been confined to the moderate Prandtl number range, and the experimental scatter is substantial, the data do not really contradict these results. Allen and Eckert [228] find the 0.14 power adequate for water at high Reynolds numbers, 10^5, with a smaller power for correlation at lower Reynolds numbers. Their results also indicate a greater exponent m than given by the Deissler results. For water at a Prandtl number of 8 their data are correlated by $m = 0.25$ which is essentially equivalent to evaluating properties at the wall temperature. Similar results have been found by Maurer and LeTourneau [154].

4. Laminar Flow in Tubes—Gases

In this case there are reasonably complete analytic solutions available, as well as a considerable body of experimental data. Deissler [225] investigated this problem analytically assuming a fully developed velocity profile and temperature profile. However, experimental data are only in moderate agreement with these results (Kays and Nicoll [229], Davenport and Leppert [230]). Worsoe-Schmidt and Leppert [99] have solved the applicable differential equations using a finite-difference technique and including the radial velocity component. Their results are for a gas, with the properties of air, in the temperature range from 500 to 3000°R. The analysis is based on a circular tube with *uniform temperature at the tube entrance* and *a fully developed velocity profile*. Heating and cooling with a constant surface temperature, and heating with a constant heat flux, are considered. The results show a small effect of temperature ratio on Nusselt number, but a substantial effect on friction coefficient. Both near the entrance and downstream the results can be correlated for heating by

$$n = 0.0 \quad ; \quad m = 1.00 \quad 1 < \frac{T_w}{T_b} < 3$$

and for cooling by

$$n = 0.0 \quad ; \quad m = 0.81 \quad 0.5 < \frac{T_w}{T_b} < 1$$

The experiments of Kays and Nicoll have verified that n is approximately 0.0 over the temperature-ratio range from 0.5 to 2.0. The Davenport and Leppert results for temperature ratios from 0.0 to 2.2 yield $m = 1.35$, in fair agreement with the analytical value, especially considering the difficulty of the experiment.

The Worsoe-Schmidt solution shows that the effects of appreciable gas property variation are limited to the region usually designated as the thermal entrance region. These results are shown in Figs. 136 to 139. These are for air in laminar flow with no natural convection effects. Details on the velocity and temperature profiles are given in the reference.

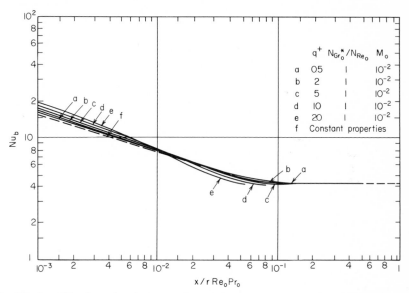

Fig. 136. Local Nusselt numbers for laminar flow of air in pure forced convection with uniform heat flux. $\mathrm{Nu}_b = 2q'' r_w/k_b[T_w - T_b]; N^*_{\mathrm{Gr},0} = 8g\rho_0^2 r_w^3/\mu_0^2; q^+ = q'' r_w/k_0 T_0; M_0 = \mathrm{Mach\ \#}; 0 = $ inlet conditions.

5. Turbulent Flow in Tubes—Gases
(Temperature-ratio Method)

The most significant analytic treatment is due to Deissler and Eian [231], who considered fully developed turbulent flow in a circular tube with constant heat rate. The calculation procedure is essentially the same as described for turbulent flow of a liquid, except of course, the temperature dependence of thermal conductivity and density are included as well as viscosity. Deissler did not put his results in the temperature-ratio form. However, when put in the form of Eqs. (160) and (161) the results may be approximated by

$$\frac{T_w}{T_b} > 1.0 \quad ; \quad n = -0.34$$

$$\frac{T_w}{T_b} < 1.0 \quad ; \quad n = -0.19$$

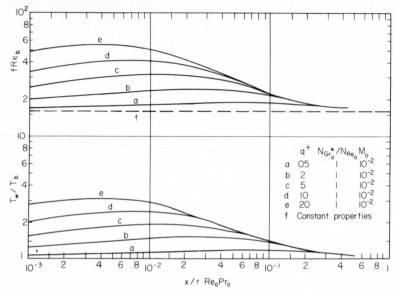

Fig. 137. Local friction coefficients and wall-to-bulk temperature ratios for laminar flow of air in pure forced convection with uniform wall heat flux.

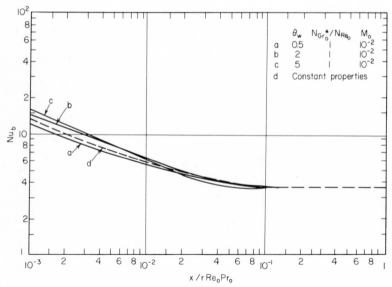

Fig. 138. Local Nusselt numbers for laminar flow of air in pure forced convection with uniform wall temperature.

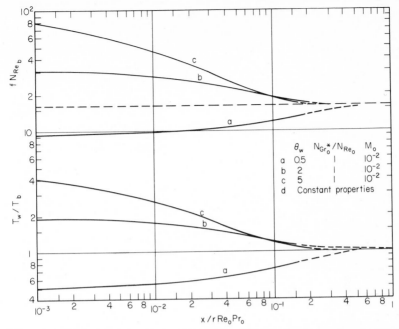

Fig. 139. Local friction coefficients and wall to bulk temperature ratios for pure forced convection with uniform wall temperature.

There is a considerable body of experimental data, see Table 44, for $(T_w/T_b) > 1.0$ which suggests a somewhat stronger effect of temperature ratio. The data of Humble et al. [232] for air yield $n = -0.55$. The extensive data of McCarthy and Wolf [233] for helium and hydrogen also yield $n = -0.55$. On the other hand, the experiments of Barnes [234] for air, helium, and carbon dioxide yield values of n that differ for the different gases and are somewhat lower. In view of this evidence, and the experiments of Table 44, it is believed that $n = -0.55$ for $(T_w/T_b) > 1.0$ represents the best conclusion that can be drawn.

The cooling experiment is much more difficult to carry out and little experimental data are available. Nicoll and Kays [239] carried out experiments with air down to $(T_w/T_b) = 0.38$ and could detect no effect of temperature ratio, that is, $n = 0.0$, although experimental uncertainty was high. Zellnik and Churchill [240] have also determined $n = 0.0$ for ratios to 0.25.

Experimental data for the friction coefficients is not extensive, and is difficult to interpret. The available data are not completely consistent, but it is apparent that for $(T_w/T_b) > 1.0$ the value of m from the analysis is considerably too large. McEligot et al. [21] conclude that $m = -0.1$ on the basis of their own experiments to a temperature ratio of 2.5, and on an examination of the data of Lel'chuk and Dyadyakin [238]. The local friction results of Perkins and Worsoe-Schmidt [20] for temperature ratios to 7.5 were not well correlated on bulk properties. A correlation was achieved, following a suggestion in Kutateladze and Leont'ev [241], as

$$\frac{f_b}{f_{iso, \text{ wall Re}}} = \left[\frac{2}{(T_w/T_b)^{1/2} + 1}\right]^2$$

TABLE 44. Selected Variable Properties, Turbulent Heat-Transfer Experiments

Investigators	Gas	$T_w/T_{b_{max}}$	x/D	Reference Temperature	Correlation Constant	Exponent on T_w/T_b	Entry Parameter	Remarks
Overall Average Coefficients								
Humble et al. [232]	Air	3.5	30–120	wall	0.034	0	$(L/D)^{-0.1}$ to 60	Modified Re
McCarthy and Wolf [233]	Hydrogen, Helium	9.9, 9.9, 9.9	21–67, 21–67, 21–67	film, wall, bulk	0.045, 0.045, 0.045	-0.7, 0, -0.55	$(L/D)^{-0.15}$, $(L/D)^{-0.15}$, $(L/D)^{-0.15}$	$Re_f = 4\dot{m}/\pi D\mu_f$, Modified Re
Taylor and Kirchgessner [235]	Helium	2.7	60, 92	film	0.021	0	$1 + (L/D)^{-0.7}$	
Incremental Average Coefficients								
Taylor and Kirchgessner [235]	Helium	3.9	60, 92	film, wall	0.021, 0.0265	0, 0		Modified Re, Modified Re
Taylor [236]	Hydrogen, Helium	5.6	77	film	0.021	0		Modified Re
McCarthy and Wolf [233]	Hydrogen, Helium	11.1	21–67	wall, bulk	0.025, 0.025	0, -0.55	0	Modified Re
Local Coefficients								
Weiland [237]	Helium, Hydrogen	2.83	250	film, wall	0.021, 0.021	0	Downstream only, Downstream only	Modified Re, Modified Re
McEligot et al. [21]	Air, Helium, Nitrogen	2.5	160	bulk	0.021	-0.5	$1 + (x/D)^{-0.7}$	Modified Re
Lel'chuk and Dyadyakin [238]	Air	2.4	147	wall, bulk	0.025	graphic	x/D dependence	
Dalle-Donne and Bowditch [35]	Air, Helium	2.4	180, 351	wall, bulk, entry	0.0208, 0.022, 0.022	0, x/D in exponent on T_w/T_b $(T_w/T_{ent})^{-0.30}$		Modified Re
Perkins and Worsoe-Schmidt [20]	Nitrogen	7.5	160	wall, bulk	0.023, 0.024	0, -0.7	$1 + (x/D)^{-0.7}(T_w/T_b)^{0.7}$, $1 + (x/D)^{-0.7}(T_w/T_b)^{0.7}$	Modified Re

with the friction factors normalized on a *constant property value evaluated at the wall Reynolds number for the nonisothermal case.*

These results may be closely approximated by

$$\frac{f_b}{f_{\text{iso, wall Re}}} = \left(\frac{T_w}{T_b}\right)^{-0.6} \tag{162}$$

The wall Reynolds number used here is given by

$$\frac{4\dot{m}}{\pi D \mu_w} \frac{T_b}{T_w}$$

and is the so-called "modified wall Reynolds number." For a perfect gas this reduces to $\bar{V} D \mu_w / \rho_w$ using the bulk velocity but evaluating density and viscosity at wall temperature.

There have apparently been no experiments for friction factors for temperature ratio less than 1.0.

Summary. On the basis of these results the following is recommended for turbulent flow of a gas in a circular tube

$$\frac{T_w}{T_b} > 1.0$$ $n = -0.55$

m use 0.1 (all bulk properties);
or Eq. (162) (wall property normalization)

$$\frac{T_w}{T_b} < 1$$

$n = 0.0$

$m = -0.1$ (tentative)

6. Noncircular Tubes

Information on flow through noncircular tubes with variable properties is incomplete, but it is relatively certain that, for turbulent flow, tube geometry has little effect. For laminar flow, tube shape probably has some effect, but in lieu of better data, it is recommended that the circular tube results be used.

7. Turbulent Flow in Tubes—Gases
(Reference Temperature Method)

Deissler and Presler [242] have presented a summary of the reference temperatures for several gases, along with the necessary constant property solution. The reference temperature T_x is defined as

$$T_x = T_b + x(T_w - T_b) \tag{163}$$

The analysis is a two-region analysis with the assumption of equal eddy diffusivities, and constant shear stress and heat flux across the tube equal to the wall values.

Figure 140 gives the reference temperature for helium, argon, hydrogen, carbon dioxide, and air as a function of Reynolds number. Their results may be approximated by $Nu_{0.4} = Re_{0.4}^{3/4}/31$ where the 0.4 refers to the reference temperature evaluated with $x = 0.4$.

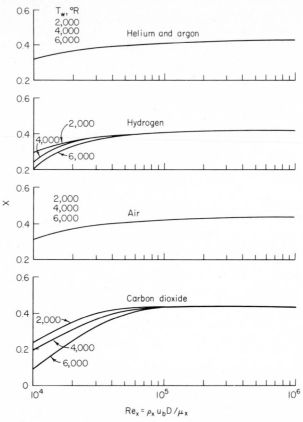

Fig. 140. Values of x for use in calculating reference temperatures for evaluating properties for turbulent flow of gases in tubes.

Note that this method requires an iterative solution to achieve an answer since the proper value of x depends on several quantities that are dependent on the final calculated value of the wall temperature.

8. Correlations

Specific correlations are sometimes required for various reference temperatures. Four such follow.

1. *Wall Temperature Correlation.* See subsection 5 for definition of Re_w (Perkins and Worsoe-Schmidt [20])

$$Nu_w = 0.023\,Re_w^{0.8}\,Pr_w^{0.4}\left[1 + \left(\frac{x}{D}\right)^{-0.7}\left(\frac{T_w}{T_b}\right)^{0.7}\right] \qquad (164)$$

for

$$5000 < \text{Re}_w < 10^5 \qquad \text{to } \frac{x}{D} = 24 \text{ then drop}$$

$$1.0 < \frac{T_w}{T_b} < 7.5 \qquad \frac{x}{D} \text{ term}$$

2. *Film Correlation* (Miller and Taylor [243]), that is, $x = 0.5$ in the reference temperature method

$$\text{Nu}_f = 0.021 \left(\frac{GD}{\mu_f}\right)^{0.8} \left(\frac{T_b}{T_f}\right)^{0.8} \text{Pr}_f^{0.4} \tag{165}$$

3. *Bulk Correlation* (Perkins and Worsoe-Schmidt [20])

$$\text{Nu}_b = 0.024 \,\text{Re}_b^{0.8} \text{Pr}_b^{0.4} \left(\frac{T_w}{T_b}\right)^{-0.7} \left[1 + \left(\frac{x}{D}\right)^{-0.7} \left(\frac{T_w}{T_b}\right)^{0.7}\right]$$

$$1.2 < \frac{x}{D} < 40 \text{ then drop } \frac{x}{D} \text{ term} \tag{166}$$

$$1 < \frac{T_w}{T_b} < 7.5 \,; \ 10^4 < \text{Re}_b < 2.5 \times 10^5$$

Dalle-Donne and Bowditch [35], as amended by Taylor [244]

$$\text{Nu}_b = 0.022 \,\text{Re}_b^{0.8} \text{Pr}_b^{0.4} \left(\frac{T_w}{T_b}\right)^{-[0.29 + 0.0019 \, x/D]}$$

$$10^4 < \text{Re} < 10^5 \tag{167}$$

$$18 < \frac{x}{D} < 316$$

Note that Eqs. (164), (166), and (167) included entry length corrections.

9. Supercritical Fluids

Laminar. Koppel and Smith [245] have analytically solved the specific example of CO_2 in a 0.194-in. tube, 20 in. long at inlet Reynolds number of 100 and 1000 for the case of a uniform wall heat flux.

Turbulent—Velocity and Temperature Profiles. Velocity and temperature profiles have been measured for turbulent flow of CO_2 near the critical point by Wood and Smith [246]. They conclude:

1. The temperature profiles flatten significantly when the bulk-mean temperature passes through the transposed critical temperature (temperature for the given pressure, at which c_p is a maximum).
2. Velocity profiles show a possible maximum at a radial position not on the center of the tube.
3. A maximum in the local heat transfer coefficient occurs when the bulk fluid temperature passes through the transposed critical temperature.

Heat Transfer. Swenson et al. [247] have proposed a correlation in the supercritical region for fully established turbulent flow which is applicable to their water data and to the carbon dioxide data of Bringer and Smith [248], Koppel and Smith [249], and Wood and Smith [246]. It also correlates the water data of Dickinson and Welch [250].

The correlation is for local heat transfer downstream results with fully established turbulent flow. Entrance effects are not included and in their experiment there was also a hydrodynamic entry length before the start of heating. The correlation, based on wall properties, is

$$\frac{hD}{k_w} = 0.00459 \left(\frac{DG}{\mu_w}\right)^{0.923} \left[\left(\frac{i_w - i_b}{T_w - T_b}\right) \frac{\mu_w}{k_w}\right]^{0.613} \left(\frac{v_b}{v_w}\right)^{0.231} \tag{168}$$

where i is the unit enthalpy of the fluid and v is the specific volume. The range of variables covered for water is

$$27,000 < \text{Re}_w < 680,000$$

$$0.10 < \frac{v_b}{v_w} < 0.98$$

$$3300 < p < 6000 \text{ psia}$$
$$65,000 < q'' < 578,000$$
$$11°F < \Delta T < 513°F$$

For carbon dioxide

$$75,000 < \text{Re}_w < 3.1 \times 10^6$$

$$10° < \Delta T < 410°F$$

The correlation shows good agreement with the analysis of Goldmann [251] up to wall temperatures of 950°F, after which an increasing divergence occurs. However, Goldmann's analysis was based on property information available at that time, and subsequent results show that his thermal conductivity values are too low at the high temperatures.

Two possible generalizations for supercritical heat transfer are:

1. Koppel and Smith [249] found that h was strongly affected by property variation near the critical point. h was found to increase as the bulk temperature passed through the transposed critical temperature.
2. Swenson et al. [247] note that the thermal entry length may be extremely long because of the large property variation. They also note that unusual effects may occur in the thermal entrance region because of these variations and in fact it is difficult to separate out the usual thermal entrance effect from variable property effects.

Heat Transfer–Hydrogen. Hess and Kunz [252] have presented a correlation for hydrogen in supercritical conditions in turbulent flow as

$$\text{Nu}_f = 0.0208 \left(\frac{\rho_f \bar{V} D}{\mu_f}\right)^{0.8} \text{Pr}_f^{0.4} \left(1 + 0.01457 \frac{v_w}{v_b}\right) \tag{169}$$

where subscript f refers to a reference temperature evaluation with properties evaluated at the arithmetic average of the wall and bulk temperatures. This correlation is essentially a pipe-flow correlation using a "modified film Reynolds number" (film properties, but bulk velocity) and corrected by a kinematic viscosity ratio.

As pointed out by Hendricks and Hsu in discussion of the paper, one must be very much aware of the domain of applicability of the correlation. It applies only in the downstream region, here $x/D > 30$, and there is some question as to the applicability when T_{bulk} is less than the temperature where the specific heat attains a maximum.

Other information on supercritical hydrogen in turbulent flow is available in Hendricks et al. [253] where a correlation is achieved using the two-phase flow parameters of Martinelli and Nelson.

I. POROUS WALL DUCTS

While transpiration cooling is a problem involving mass transfer as well as heat transfer, it is appropriate to include the available information on the basic porous-duct laminar and turbulent flow solutions in this chapter on heat transfer in pipes and ducts. The reader is referred to references on transpiration cooling to pursue this topic to greater depth.

The information available is not yet complete; however, the laminar and turbulent flow cases are covered in sufficient depth to indicate the effects of blowing and suction.

1. Laminar Flow

Parallel Plates. Because the case of porous parallel plates has been treated in greater generality than the circular tube, the former will be treated first. A very complete analytical paper by Carter and Gill [254] is available for laminar flow, with constant blowing or suction, in vertical and horizontal parallel plates, with fully developed thermal and hydrodynamic flow.* The analysis includes combined free and forced convection effects, and presents velocity and temperature profiles with heat transfer and friction results. The thermal boundary condition is a linearly increasing wall-temperature distribution. Figure 141 presents their results for Nusselt number in vertical flow, including natural convection effects as a parameter. The Nusselt number is based on d, the *full width between the plates*, not D_e; the wall Reynolds number is dv_w/ν and the Rayleigh number is $d^3 g\beta(dT/dx)/\nu\alpha$ where β is the expansion coefficient. The results for Ra and $Re_{wall} = 0$ agree with the usual fully developed downstream result of Nu = 4.11. Negative Re_w represents *injection*. As expected, injection reduces the Nusselt number.

Figure 142 shows the effect of Prandtl number on the heat transfer with blowing for the case of no natural convection effects. Injection causes only minor effects at liquid metal Prandtl numbers, but causes large effects at large Prandtl numbers. Note that since Ra = 0 for these results the flow may be either vertical or horizontal.

Terrill [255] has solved for the case of parallel porous plates both at equal and uniform wall temperatures. His results for fully developed conditions are presented in Table 45. Here the wall Reynolds number is again based on the channel width, not

*Note that with injection (or suction) the bulk velocity increases (or decreases) along the channel. A fully developed condition may still exist, however, if the *nondimensional* velocity and temperature profiles become invariant in the flow direction. For constant properties and constant v_w this can occur for the injection and suction cases. This is similar to the case discussed in Sec. A.2.a where it was pointed out that for flow in a tube a fully developed thermal condition can exist for the two cases of uniform wall heat flux and uniform wall temperature. In either case, an entrance length of duct is required before fully developed conditions are reached.

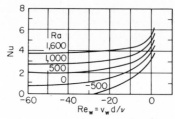

Fig. 141. Nusselt number versus wall injection Reynolds number for vertical flow through parallel porous plates, Pr = 1, including the effects of natural convection.

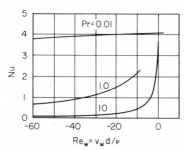

Fig. 142. Prandtl number effect on heat transfer as a function of wall injection Reynolds number with no natural convection effects, Ra = 0.

TABLE 45. Fully Developed Nusselt Numbers for Uniform Wall Temperature Parallel Porous Plates

	Injection				Suction				
$\mathrm{Re}_w\mathrm{Pr}$	-6	-4	-2	-1	0	1	2	4	6
Nu	1.5	2.1	2.9	3.3	3.77	4.3	4.9	6.1	7.6

D_e, and is dv_w/ν. Positive values of $\mathrm{Re}_w\mathrm{Pr}$ are for suction, negative for injection. The Nusselt number is hd/k.

Horton and Yuan [256] have discussed the hydrodynamic development in a parallel porous plate channel with injection, and they find that injection increases the rate of growth of the boundary layer and thus reduces the entrance length required for fully developed flow. It would thus seem that Terrill's and Carter and Gill's results would be applicable to a duct with a shorter x/D than for flow without injection, since presumably the thermal development is also enhanced by injection. For flow without injection the thermal results may be considered developed when $2(x/D_e)/(\mathrm{Re}\,\mathrm{Pr}) = 0.1$.

Note that the linearly increasing wall temperature results do not show nearly as rapid a drop-off of Nusselt numbers as do the constant wall temperature results. Whether this is a real effect as the analysis shows remains to be verified by experiment.

The parallel plate case for uniform injection at one wall and suction at the other wall has been treated by Lee and Gill [257] for both laminar and turbulent flow with a linear axial wall temperature distribution.

Summary. The parallel plate results show that injection decreases the Nusselt number and suction increases it. Similar conclusions can be drawn for the friction factor (Carter and Gill [254]), although the effect is less pronounced.

Circular Tubes. Yuan and co-workers have dealt with the laminar and turbulent porous circular tube problem in a series of papers. The *laminar constant property* hydrodynamic problem for uniform wall velocity is treated by Yuan and Finkelstein [258], the corresponding heat transfer problem is treated by the same authors [259], the variable properties laminar problem for air injected into air is given by Yuan and Peng [260], and the variable properties *laminar problem for hydrogen into air* is treated by Peng and Yuan [261].

Some information on the hydrodynamic entry problem for a circular tube is given by Weissberg [262] and Hornbeck et al. [263].

The information in Figs. 143 to 146 from Peng and Yuan [261] is for hydrogen injected in air, and air into air. The flow is *fully developed* before uniform injection begins. The tube is at a *constant wall temperature* of 1000°R, the inlet bulk-to-wall temperature ratio is 1.3, and the coolant is assumed to enter the tube at the wall temperature. The particular axial location of these figures is at $x/D = 50$. The ratio $\rho_w v_w / \rho_{in} V_1$ is based on V_1 the Poiseuille *centerline* velocity at the inlet, and the *friction* factor is the Fanning value *divided* by 4. It is important to notice that Yuan defines the friction factor $\tau_w / \frac{1}{2} \rho V_1^2$, that is, with the *inlet* centerline velocity. Consequently his results show an increasing friction factor with injection. Since the local bulk velocity is increasing due to the injection, the local friction coefficients based on the local bulk mean velocity would be lower than those shown here.

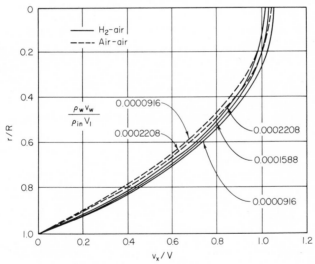

Fig. 143. Velocity distributions for injection in an isothermal circular pipe with laminar flow, $x/D = 50$, Re $= 1000$.

The nondimensional axial velocity distributions shown in Fig. 143 exhibit the expected increase in local velocity due to injection. Note that while the velocity gradients for flow with hydrogen are larger than those with air, the friction factors are less for hydrogen than air. Peng and Yuan attribute this to the reduced viscosity with hydrogen coolant. The zero injection Nusselt number on Fig. 146 does not reach the usual constant temperature Nusselt number value, since the flow is not fully developed thermally.

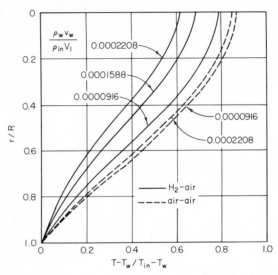

Fig. 144. Temperature distribution for injection in an isothermal circular pipe with laminar flow, $x/D = 50$, Re $= 1000$.

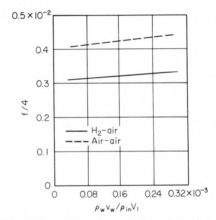

Fig. 145. Friction coefficients for injection in an isothermal circular pipe with laminar flow, $x/D = 50$, Re $= 1000$.

2. Turbulent Flow

For the case of uniform blowing Yuan and Barazotti [264] have presented experimental velocity and temperature profiles for data obtained in a short test section. Burnage [265] gives velocity data for suction, and Olsen and Eckert [266] present a complete hydrodynamic study for injection. Figure 147 shows Olsen and Eckert's results for the local-to-average velocity ratio at a given x/D location. Their tube had an 18 diameter long test section, and the flow was fully developed upon entering this injection section.

The normalized velocity profiles indicate that most of the adjustment due to injection occurred within the first four diameters. Velocities near the tube wall are

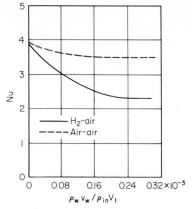

Fig. 146. Nusselt numbers for injection in an isothermal circular pipe with laminar flow, x/D = 50, Re = 1000.

reduced with an increase in the velocities at the centerline. They note that discernible changes in the core velocities were still taking place after 18 diameters. Similar results for the case of zero entrance velocity, i.e., injection into an empty tube, are given in the paper.

The ratio of *local* friction coefficients with injection to those for fully developed flow at the same local Reynolds number (experimentally determined in the same tube to account for roughness effects) is shown in Fig. 148. It is seen that injection markedly decreases the friction coefficient.

The only available *turbulent flow heat transfer* results for uniform injection in a porous circular tube are those of Churchill and Stubbs [267]. They show a decrease in Nusselt number with injection, the rate of decrease lessening with larger injection rates. The Nusselt number was decreased approximately 20

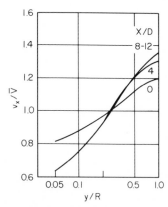

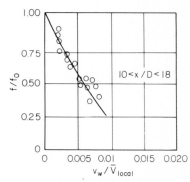

Fig. 147. Normalized velocity profiles for turbulent flow in a circular tube with an injection rate of $v_w \rho / \overline{V}_{in} \rho_{in}$ = 0.0121.

Fig. 148. Normalized local friction factors for turbulent flow in a circular porous tube with fully developed flow at the entrance to the porous tube.

percent at a value of $v_w/\overline{V}$ of 0.004. These results do not encourage the use of Eckert and Olsen's friction data to predict heat transfer results from Reynolds analogy.

J. SUPERSONIC FLOW IN A DUCT

Bialokoz and Saunders [268] have shown that for *established turbulent* flow conditions, for air at Mach numbers to 1.6, the average Stanton number can be predicted by

$$St_b = 0.022\, Re_b^{-0.2}\, Pr_b^{-0.6}\left(\frac{T_w}{T_b}\right)^{-0.5} \tag{170}$$

where subscript b refers to properties evaluated at bulk mean temperature. This is essentially the same result as the subsonic turbulent pipe-flow correlation. Caution is suggested in using this equation unless one has established conditions, since the supersonic flow problem in a duct may result in an external boundary layer problem rather than in an internal duct problem (Saunders and Calder [269]).

The above correlation was established for *heating* of the gas, and $1.2 \times 10^5 < Re < 4.3 \times 10^5$, with the heat transfer coefficient based on the difference between the wall and *adiabatic wall* temperature, that is,

$$q_w'' = h(T_w - T_{aw})$$

where

$$T_{aw} = T_b + Pr^{1/3}(T_{stagnation} - T_b) \tag{171}$$

and

$$(T_{stagnation} - T_b) = \frac{G^2}{2g_c\rho^2 c_p}$$

K. UNSTEADY-STATE EFFECTS

This section is concerned with the influence on the heat transfer characteristics of (1) a transient change in wall temperature or heat flux, (2) a transient change in the flow rate, or (3) a transient change in both flow rate and the thermal conditions. (See also Sec. L.)

Laminar Flow. For hydrodynamically fully developed laminar flow in a tube, the response to a step change in wall temperature with time is shown in Fig. 149. These results from Siegel [270] are for the case where $T_w = T_{in}$ (inlet temperature) initially, and then at time zero a step change is made in T_w.

Further information for this type of laminar problem is available only for flow between parallel plates. The result for a step change in wall temperature, with steady fully developed flow, from the condition $T_w = T_{in}$ at time zero, is shown for parallel plates, of half width a, in Fig. 150. (See Ref. 270.)

For the case when the channel walls in the heated section are initially at a constant temperature, *different from* that in the isothermal hydrodynamic entrance, so that there is steady-state heat transfer initially, the response to a step change in pumping pressure is shown in Fig. 151. This solution represents a transient in flow velocity as the velocity approaches a new steady-state value. The wall temperature remains constant throughout the whole process.

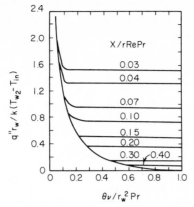

Fig. 149. Transient variation in wall heat flux following a step change in wall temperature for laminar flow in a circular tube, $x^+ = x/r\,\mathrm{Re}\,\mathrm{Pr}$. Initially $T_{w,1} = T_{in}$; finally $T_w = T_{w,2} > T_{in}$.

The response to a combined step change in both pressure gradient, i.e., velocity, and wall temperature from an *unheated initial condition* is shown in Fig. 152. These results are from Perlmutter and Siegel [271].

The laminar slug-flow case has been solved for parallel plates by Perlmutter and Siegel [272] as shown in Figs. 153 and 154. These results include the only available information on the wall temperature response to a step change in heat flux. For the parallel plate results, a is the duct half-width.

Superposition techniques may be used to handle the problem of response to an arbitrary change in wall conditions, or to solve for the response to a change from

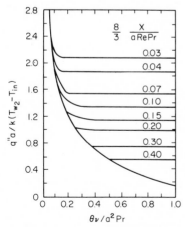

Fig. 150. Transient wall heat flux after a step change in wall temperature for laminar flow between parallel flat plates. Initially $T_{w,1} = T_{in}$; finally $T_w = T_{w,2} > T_{in}$.

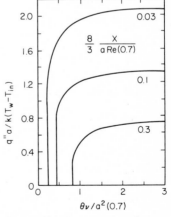

Fig. 151. Transient variation, from a steady-state heat transfer condition, after a step change in pressure gradient with wall temperature held constant ($T_{w,2} = T_{w,1}$) for laminar flow between parallel plates, $\mathrm{Pr} = 0.7$, initially $u = u_1 = 0$; finally $u = u_2$.

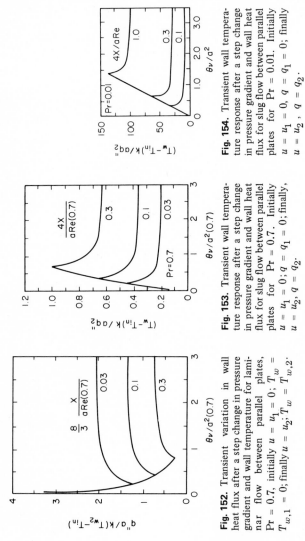

Fig. 152. Transient variation in wall heat flux after a step change in pressure gradient and wall temperature for laminar flow between parallel plates, $Pr = 0.7$, initially $u = u_1 = 0$; $T_w = T_{w,1} = 0$; finally $u = u_2$; $T_w = T_{w,2}$.

Fig. 153. Transient wall temperature response after a step change in pressure gradient and wall heat flux for slug flow between parallel plates for $Pr = 0.7$. Initially $u = u_1 = 0$; $q = q_1 = 0$; finally, $u = u_2$, $q = q_2$.

Fig. 154. Transient wall temperature response after a step change in pressure gradient and wall heat flux for slug flow between parallel plates for $Pr = 0.01$. Initially $u = u_1 = 0$, $q = q_1 = 0$; finally $u = u_2$, $q = q_2$.

initial heating conditions rather than from the nonheating initial conditions treated in these references.

For the case of a pulsating laminar slug-flow between parallel plates, for either constant wall temperature or a uniform heat flux, solutions have been obtained by Siegel and Perlmutter [273].

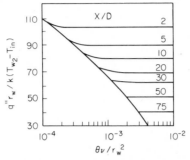

Fig. 155. Transient heat transfer response to a step change in wall temperature for turbulent flow through a circular tube, $Pr = 0.7$, $Re = 100,000$. Both fluid and wall at the inlet temperature initially, $T_{w,1} = T_{in}$; finally $T_w = T_{w,2} > T_{in}$.

The same authors [274] have solved the slug-flow, parallel-plate problem for a wall heat-flux variable with position and time. A specific exponential case is worked out. This case, with the effect of finite wall heat capacity, has been treated by Siegel [275]. The example included there is for a sinusoidal variation, in time, of wall heat flux.

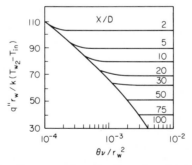

Fig. 156. Transient heat transfer response to a step change in wall temperature for turbulent flow through a circular tube, $Pr = 10$, $Re = 100,000$. Both fluid and wall at the inlet temperature initially, $T_{w,1} = T_{in}$; finally $T_w = T_{w,2} > T_{in}$.

Turbulent Flow. The response to a step change in wall temperature, according to Sparrow and Siegel [276], for hydrodynamically fully established turbulent flow in a circular tube, is shown in Figs. 155 and 156. These results are for the case of both fluid and wall at the fluid inlet temperature before time zero; the case of the wall at a temperature different from the fluid inlet temperature before time zero can be obtained from this solution by superposition. The time to reach steady state, defined as the time required for 95 percent of the total change, is shown in Fig. 157.

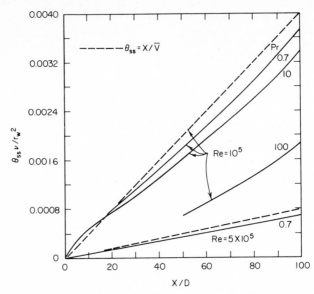

Fig. 157. Time required to reach steady state after a step change in wall temperature for turbulent flow through a circular tube.

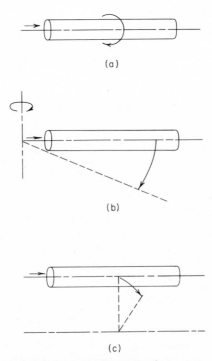

(a)

(b)

(c)

Fig. 158. Geometries for rotating duct systems.

L. ROTATING DUCTS

The duct or pipe in which a fluid is flowing may rotate about various axes, and applications involving rotating ducts commonly occur in rotating machinery. Figure 158 shows three examples of rotating ducts, each distinguished by a different axis or rotation.

An *axially* rotating duct is illustrated by Fig. 158a, that is, the duct rotates about its centerline while fluid flows through the duct. A *radially* rotating duct is shown on Fig. 158b. This is the classic "lawn sprinkler" problem. Figure 158c shows a duct rotating about an axis *parallel* to the tube axis. Internal cooling tubes in generator rotors sometimes correspond to this type.

Convective heat transfer data for rotating ducts are very sparse. The little that are available provide some idea of the effects that can be expected due to rotation.

Axially Rotating Ducts (Fig. 158a).   Some experimental results of Cannon [277] are shown in Fig. 159. The data were taken with air flowing in a 60-inch long pipe, 1 inch in diameter. The surface temperature was maintained constant, and the overall mean (log-mean) heat transfer coefficient was determined. Velocity and temperature were uniform at the entrance to the heated rotating tube.

The most significant effect of rotation is seen to be in the transition region between laminar and turbulent flow. Rotation tends to stabilize the laminar flow so that transition occurs at higher Reynolds numbers. In fact, it is found that a turbulent flow can be made to return to a laminar flow upon entering a rotating section of pipe.

Radially Rotating Ducts (Fig. 158b).   Trefethen [278] has suggested that the secondary flows which occur in *radially* rotating ducts are similar in nature to those caused by any transverse body force, in turbulent as well as laminar flow. Thus he suggests applying bent, or curved, tube correlations to the case of radially rotating ducts. For curved tubes, the usual correlation parameter is $\text{Re}(r_w/R)^{1/2}$ where r_w is the tube radius, and R the bend radius. For *rotating* tubes, he suggests $\text{Re}(2r_w\omega/\overline{V})^{1/2}$. He finds for laminar flow that the friction factor is about 13 percent below Ito's [214] curved-tube laminar results when presented in this manner. In turbulent flow,

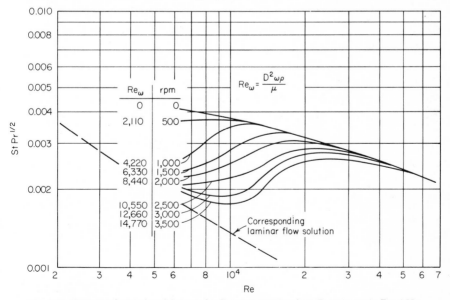

Fig. 159. Heat transfer results of Cannon for flow in a rotating duct. Rotation as in Fig. 158a.

Trefethen's results are some 2 to 5 percent below Ito's results. Correlations from this reference are shown in Figs. 160 and 161.

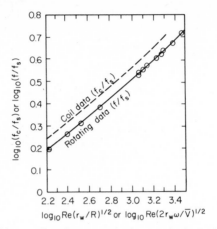

Fig. 160. Comparison of laminar flow friction results for curved tubes with data for radial rotating tubes.

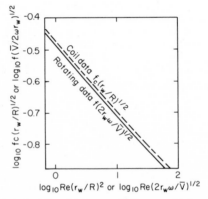

Fig. 161. Comparison of turbulent flow friction results for curved tubes with data of radial radial rotating tubes.

Parallel Axis Rotating Ducts (Fig. 158c).   An analytic solution for fully developed *laminar* flow in a vertical circular tube with constant heat rate per unit of length, for the rotation case shown in Fig. 158c, is given by Morris [279]. In this solution the effects noted are entirely due to the action of centrifugal, coriolis, and gravity body forces on density gradients within the fluid arising from temperature gradients. In the absence of such gradients, the flow is again simple fully developed laminar flow. Thus this case differs from the previous two cases where the phenomena discussed occur independently of density gradients.

M. NON-NEWTONIAN FLUIDS

The previous sections of this chapter have dealt with the hydrodynamics and heat transfer characteristics of Newtonian fluids. Newtonian fluids are those that have a linear relationship between the shear stress and the velocity gradient (or rate of strain). The shear stress τ is equal to zero when dv_x/dy equal zero. The viscosity, given by the ratio of shear stress to velocity gradient, is independent of the velocity gradient (or rate of strain), but may be a function of temperature and pressure.

Gases, and liquids such as water, usually exhibit Newtonian behavior. However, many fluids, such as colloidal suspension and polymeric solutions, do not follow the linear shear stress-velocity gradient relation, and these are called non-Newtonian fluids.

Metzner [280, 281] presents surveys of the hydrodynamics and heat transfer characteristics of non-Newtonian fluids. The reader seeking information in depth is referred to these, and also to Wilkinson [282], and Rohsenow [283]. Only the fundamental laminar and turbulent tube-flow cases will be covered in this section.

The classification of fluid behavior by Metzner [281] is shown in Table 46. According to this scheme, Newtonian fluids are a subclass of purely viscous fluids. This section will deal only with non-Newtonian fluids in Category 1, purely viscous fluids. Metzner points out that often the fluids in Categories 2 and 3 may be treated as purely viscous fluids as an approximation.

TABLE 46. Types of Fluid Behavior

Category 1	Purely Viscous Fluids	
(a)	μ	= constant, independent of rate of strain (Newtonian fluid)
(b)	μ	= decreasing function of the invariants of the strain-rate tensor; includes the "Pseudoplastic," "Bingham plastic," and "Shear-thinning" materials
(c)	μ	= increasing function of the invariants of the strain-rate tensor; includes the "dilatants" and "shear-thickening" materials
Category 2	Time-dependent fluids	
Category 3	Viscoelastic fluids	
Category 4	More complex materials	

Various equations have been used to describe the purely viscous non-Newtonian flow behavior. The more common of these are:
1. The power law (sometimes called the Ostwald equation)

$$\tau = K(u')^n \tag{172}$$

where u' is the shear rate. For Newtonian fluids $n = 1$ and $K = \mu/g_c$
2. The three parameter Eyring-Powell equation.

Other equations are discussed in the references cited. The behavior of many non-Newtonian fluids may be treated using the power-law model. Those that have exponents n less than one are called "pseudoplastic." Those characterized by an n greater than one are termed "dilatant." Values for the constant n in the power-law model are available in Metzner [284] and Bird et al. [285].

The power-law model will be used as the basic model in this discussion.

1. Hydrodynamics—Circular Tube

Laminar Flow. Following Metzner [281], one may express the *laminar friction factor* as

$$f = \frac{16}{\text{Re}'} \tag{173}$$

where

$$\text{Re}' = D^{n'}\overline{V}^{2-n'}\frac{\rho}{\gamma}$$

and

$$\gamma = g_c K' 8^{n'-1}$$

The coefficient

$$n' = \frac{d(\log \tau_w)}{d(\log 8\overline{V}/D)} \tag{174}$$

or

$$\tau_w = K'\left(\frac{8\overline{V}}{D}\right)^{n'}$$

Relating these expressions to the power-law results, one finds

$$K' = K\left[\frac{3n' + 1}{4n'}\right]^n \left[\frac{8\overline{V}}{D}\right]^{n-n'} \tag{175}$$

If the fluid obeys the power-law formulation, then $n = n'$ and

$$K' = K\left[\frac{3n + 1}{4n}\right]^n \tag{176}$$

and

$$\text{Re}' = \frac{D^n V^{2-n}\rho}{g_c(K/8)[(6n + 2)/n]^n}$$

Note that the case of $n = 1$ gives the Newtonian fluid.

Information on the effects of heating on the laminar friction factor is available in Hanks and Christiansen [286].

For power-law fluids Metzner [280] shows that the laminar velocity distribution is

$$\frac{v_x}{\overline{V}} = \frac{1 + 3n}{1 + n}\left[1 - \left(\frac{r}{r_w}\right)^{\frac{n+1}{n}}\right] \tag{177}$$

where $\overline{V}$ is the bulk velocity. For $n = 1$ this reduces to the Newtonian laminar parabola.

Turbulent Flow. Results for the friction factor for the power-law fluid in *turbulent flow* are given in Fig. 162 where n' is given in Eq. (174), and for power-law fluids $n' = n$. More detailed information is available in Metzner [280] and includes approximate algebraic expressions for f.

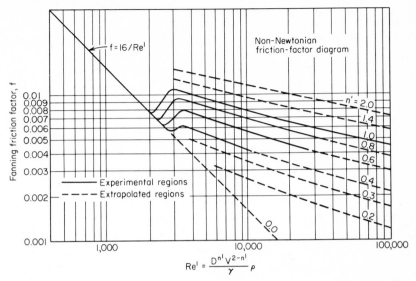

$$Re^l = \frac{D^{n^l} V^{2-n^l}}{\gamma} \rho$$

Fig. 162. Friction factor–Reynolds number design chart for purely viscous non-Newtonian fluids, after Metzner [280].

Metzner [281] suggests that the turbulent non-Newtonian velocity profiles are substantially the same as those of Newtonian fluids, although it should be noted that the measurements are mostly in the turbulent core region and have not been made in the region close to the wall (Bogue and Metzner [287], Eissenburg and Bogue [288]).

2. Heat Transfer

Laminar Flow. The asymptotic, downstream, Nusselt number for both the isothermal wall condition and the uniform wall heat flux are shown in Fig. 163 from Rohsenow [283]. The results are based on work by Beek and Eggink [289].

The approach of the *local* Nusselt number to the fully developed value for the *thermal entry* problem, with constant properties, is shown in Fig. 164 for both the uniform temperature and uniform wall heat flux cases (Rohsenow [283]).

The *combined velocity and thermal entry problem* has been treated by McKillop [290] for the case of a power-law fluid with $n = 0.5$. These results are shown in Table 47. The nomenclature here is slightly different from that used by Metzner.

$$x_0 = \frac{x}{r_w Re'Pr'} \qquad Pr' = \frac{g_c K c_p}{k}\left(\frac{r_w}{\overline{V}}\right)^{1-n}$$

$$Re' = \overline{V}^{2-n} r_w^n \rho / g_c K \tag{178}$$

$$\tau = K\left[\frac{du}{dr}\right]^n$$

The thermal entrance problem for the case of variable viscosity has been treated by Christiansen and Craig [291], who solved the non-Newtonian Graetz problem.

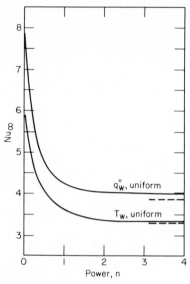

Fig. 163. Downstream fully developed Nusselt number for laminar flow of a power law, non-Newtonian fluid in round tubes.

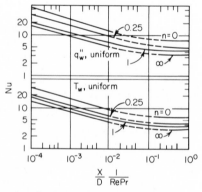

Fig. 164. Local Nusselt numbers for laminar flow of a power law, non-Newtonian, fluid in the thermal entrance region of a circular tube.

When *natural convection* effects are important, Oliver and Jensen [292] recommend for the constant wall temperature, *horizontal* tube

$$\mathrm{Nu}_{amt}\left(\frac{\gamma_w}{\gamma_b}\right)^{0.14} = 1.75\left[Gz + 0.0083\,(\mathrm{Gr}_w^{\cdot}\,\mathrm{Pr}_w)^{0.75}\right]^{1/3} \tag{179}$$

where $\gamma = g_c K' 8^{n'-1}$. (See also Sec. A.2.g for other information on superimposed free and forced convection.) The Grashof and Prandtl numbers here are evaluated at the *wall* conditions for both temperature and shear rate. Equation (179) is recommended

TABLE 47. Local and Mean Nusselt Numbers for a Power-law Fluid with $n = 0.5$ for the Combined Hydrodynamic and Thermal Entry Length

Constant wall temperature			Constant heat flux		Constant wall temperature			Constant heat flux	
x_0	Nu_x	Nu_m	Nu_x	Nu_m	x_0	Nu_x	Nu_m	Nu_x	Nu_m
		$Pr' = 1.0$					$Pr' = 10$ (cont.)		
0.008	10.05	18.27	15.43	27.98	0.10	4.30	6.08	5.40	7.93
0.014	8.71	14.60	12.07	22.04	0.12	4.19	5.78	5.23	7.50
0.019	7.95	12.98	10.77	19.29	0.15	4.08	5.45	5.06	7.03
0.039	6.31	9.95	8.29	14.28	0.20	4.00	5.10	4.90	6.51
0.059	5.48	8.57	7.15	12.02	0.30	3.95	4.73	4.79	5.97
0.079	4.95	7.72	6.45	10.70			$Pr' = 100$		
0.099	4.61	7.13	5.97	9.79					
0.12	4.39	6.69	5.64	9.12	0.001	8.01	11.08	10.26	15.59
0.15	4.18	6.20	5.31	8.39	0.0145	7.02	9.83	8.95	13.50
0.20	4.04	5.67	5.02	7.57	0.0195	6.41	9.03	8.16	12.23
0.30	3.96	5.11	4.82	6.68	0.0395	5.26	7.36	6.66	9.72
		$Pr' = 10$			0.0595	4.75	6.56	6.00	8.57
					0.0795	4.47	6.07	5.62	7.87
0.005	11.32	16.55	14.55	24.52	0.0995	4.29	5.73	5.38	7.39
0.010	8.68	13.31	11.22	18.83	0.1195	4.18	5.12	5.21	7.04
0.020	6.64	10.45	8.52	14.32	0.1495	4.08	5.20	5.05	6.66
0.040	5.32	8.17	6.77	10.94	0.1995	3.99	4.91	4.90	6.23
0.060	4.78	7.13	6.06	9.43	0.2995	3.95	4.59	4.79	5.76
0.080	4.49	6.51	5.66	8.53					

only when natural convection is important. (The subscript amt refers to the arithmetic mean temperature.)

The effects of viscous dissipation and internal heat generation are discussed in Metzner [281].

Turbulent Flow. The transition region may be estimated from the friction factor data on Fig. 162. No heat transfer experiments have been run in this region. Metzner [281] recommends the use of the Newtonian analysis of Friend and Metzner [293] to provide a turbulent non-Newtonian heat transfer prediction. This reduces to

$$St = \frac{f/2}{1.20 + 11.8\sqrt{f/2}\,(Pr_w - 1)(Pr_w)^{-0.33}} \tag{180}$$

with $(Pr)(Re)^2(f) > 5 \times 10^5$. The friction factor is taken from Fig. 162. For non-Newtonian fluids, the generalized Reynolds number Re' and the *wall* Prandtl number are used. The wall Prandtl number uses the non-Newtonian viscosity determined at the shear stress conditions at the tube wall. The available data show a 20–25 percent spread about this equation.

The problem of circumferential temperature and heat flux variations for laminar flow of a non-Newtonian power-law fluid in a circular tube, with uniform axial heat flux, has been discussed by Inman [294].

Friction factors for annuli from the work of Fredrickson and Bird [295] are reported in Metzner [280] in detail. Friction results in a duct are given by Wheeler and Wissler [296].

The results discussed in this section have been for purely viscous fluids following the power-law model for non-Newtonian behavior. Results by Matsuhisa and Bird [297] cover many of the above results with the Ellis fluid model. Other results for nonpower-law models are given by Rohsenow [283].

TABLE 48. Fractional Heat Losses by Convection and Radiation for a Short Tube ($x/D = 5$)

		ϵ	$t_{r,e}$	Convection out/heat in	Radiation to inlet reservoir/heat in	Radiation to outlet reservoir/heat in
Effect of ϵ ($H = 0.8$, St $= 0.1$, $t_{r,i} = t_{g,i} = 1.5$)		0.01	$1.546 = t_{g,e}$	0.734	0.136	0.130
		0.1	$1.528 = t_{g,e}$	0.440	0.282	0.278
		1	$1.522 = t_{g,e}$	0.345	0.329	0.326
Effect of St ($H = 0.8$, $t_{r,i} = t_{g,i} = 1.5$)	St $= 0.005$	0.01	$1.523 = t_{g,e}$	0.736	0.133	0.131
	0.005	1	$1.511 = t_{g,e}$	0.346	0.328	0.326
	0.02	0.01	$1.591 = t_{g,e}$	0.728	0.144	0.128
	0.02	1	$1.543 = t_{g,e}$	0.341	0.333	0.326
Effect of H (St $= 0.01$, $t_{r,i} = t_{g,i} = 1.5$)	$H = 0.2$	0.01	$1.575 = t_{g,e}$	0.300	0.357	0.343
	0.2	1	$1.527 = t_{g,e}$	0.108	0.449	0.443
	1.5	0.01	$1.529 = t_{g,e}$	0.873	0.065	0.062
	1.5	1	$1.517 = t_{g,e}$	0.513	0.245	0.242
Effect of $t_{g,i}$ ($H = 0.8$, St $= 0.01$, $t_{r,i} = t_{g,i}$)	$t_{g,i} = 0.5$	0.01	$0.558 = t_{g,e}$	0.928	0.035	0.037
	0.5	1	$0.544 = t_{g,e}$	0.708	0.144	0.148
	3.0	0.01	$3.025 = t_{g,e}$	0.394	0.323	0.283
	3.0	1	$3.006 = t_{g,e}$	0.090	0.459	0.451
Effect of $t_{r,e}$ ($H = 0.8$, St $= 0.01$, $t_{r,i} = t_{g,i} = 1.5$)		0.01	0	0.719	0.095	0.186
		1	0	0.287	0.306	0.407
		0.01	5	1.782	4.861	-5.643
		1	5	1.993	4.767	-5.760

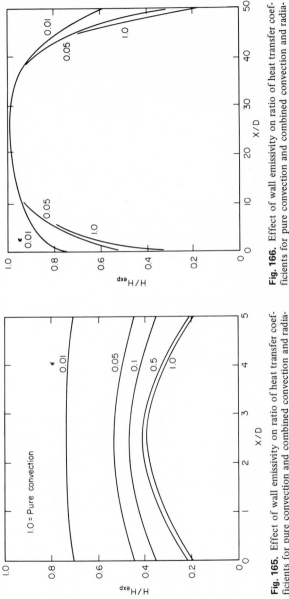

Fig. 166. Effect of wall emissivity on ratio of heat transfer coefficients for pure convection and combined convection and radiation for a long tube, $x/D = 50$, $H = 0.8$, St $= 0.01$, $t_{r,i} = t_{g,i} = 1.5$, $t_{r,e} = t_{g,e}$.

Fig. 165. Effect of wall emissivity on ratio of heat transfer coefficients for pure convection and combined convection and radiation for a short tube, $x/D = 5$, $H = 0.8$, St $= 0.01$, $t_{r,i} = t_{g,i} = 1.5$, $t_{r,e} = t_{g,e}$.

TABLE 49. Fractional Heat Losses by Convection and Radiation for a Long Tube ($x/D = 50$)

	ϵ	$t_{r,e}$	Convection out/heat in	Radiation to inlet reservoir/heat in	Radiation to outlet reservoir/heat in
Effect of ϵ ($H = 0.8$, St $= 0.01$, $t_{r,i} = t_{g,i} = 1.5$)	0.01	$2.066 = t_{g,e}$	0.906	0.039	0.055
	0.1	$2.047 = t_{g,e}$	0.875	0.053	0.072
	1	$2.043 = t_{g,e}$	0.869	0.056	0.075
Effect of St ($H = 0.8$, $t_{r,i} = t_{g,i} = 1.5$) St $= 0.01$	0.01	$2.066 = t_{g,e}$	0.906	0.039	0.055
St $= 0.01$	1	$2.043 = t_{g,e}$	0.869	0.056	0.075
St $= 0.02$	1	$2.556 = t_{g,e}$	0.845	0.059	0.096
Effect of H (St $= 0.01$, $t_{r,i} = t_{g,i} = 1.5$) $H = 0.2$	0.01	$2.743 = t_{g,e}$	0.497	0.240	0.263
$H = 0.2$	1	$2.691 = t_{g,e}$	0.476	0.251	0.273
$H = 0.8$	0.01	$2.066 = t_{g,e}$	0.906	0.039	0.055
$H = 0.8$	1	$2.043 = t_{g,e}$	0.869	0.056	0.075
Effect of $t_{g,i}$ ($H = 0.8$, St $= 0.01$, $t_{r,i} = t_{g,i}$) $t_{g,i} = 1.5$	0.01	$2.066 = t_{g,e}$	0.906	0.039	0.055
$t_{g,i} = 1.5$	1	$2.043 = t_{g,e}$	0.869	0.056	0.075
3	0.01	$3.467 = t_{g,e}$	0.748	0.122	0.130
3	1	$3.444 = t_{g,e}$	0.705	0.141	0.154
Effect of $t_{r,e}$ ($H = 0.8$ St $= 0.01$ $t_{r,i} = t_{g,i} = 1.5$)	0.01	0	0.895	0.039	0.066
	1	0	0.851	0.056	0.093
	0.01	4.5	1.108	0.040	−0.148
	1	4.5	1.116	0.056	−0.172

N. COMBINED RADIATION AND CONVECTION

Siegel and Perlmutter [298] have investigated the problem of a transparent gas flowing through a circular tube with a gray diffusely radiating wall. This paper is concerned with the heat loss out of the ends of a tube by radiation superimposed on the convection mechanism. They assume that the convective heat transfer coefficient and the wall heat flux are constant. Table 48 and Fig. 165 show their heat transfer results for a short tube of $x/D = 5$; similar results for an x/D of 50 are shown in Table 49 and Figs. 166 and 167.

Here H is $(h/q'')(q''/\sigma)^{1/4}$ where σ is the Stefan-Boltzmann constant; St is the Stanton number; subscript r represents reservoir conditions, i the inlet, e the exit, and g the gas condition. The ratio H/H_{exp} is the correction factor to reduce the experimental h to one of pure forced convection.

Other information on combined convection and radiation may be found in Chen [299], Keshock and Siegel [300], and Viskanta [301] for parallel plates, Nichols [302] for annuli in the entrance region, and deSoto [303] for laminar flow of a nontransparent gas in the entrance region of a tube. The text by Sparrow and Cess [304] also includes information on these problems.

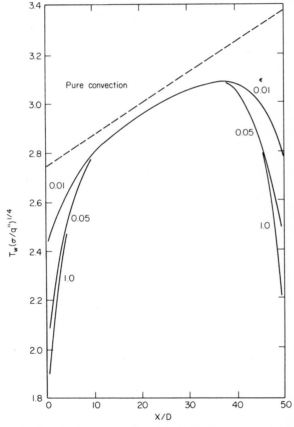

Fig. 167. Effect of wall emissivity on the temperature distribution in a long tube, conditions same as Fig. 166.

REFERENCES

1. H. L. Langhaar, *Trans. ASME,* **64**:A55 (*J. Appl. Mech.,* **9**) (1942).
2. J. G. Knudsen and D. L. Katz, "Fluid Dynamics and Heat Transfer," McGraw-Hill Book Company, New York, 1958.
3. L. F. Moody, *Trans. ASME,* **66**:671 (1944).
4. R. G. Deissler, *NACA TN* 3016, 1953.
5. H. Latzko, *Z. Angew. Math. u. Mech.,* **2**:26 (1921).
6. W. D. Campbell and J. E. Slattery, *J. Bas. Engr., Trans. ASME,* **85**:41 (1963).
7. R. C. Martinelli, *Trans. ASME,* **69**:947 (1947).
8. J. Nikuradse, Forsch. Arb. Ing.–Wes. No. 356 (p. 593, Schlichting, 6th ed.), 1932.
9. R. G. Deissler, *NACA TN* 2138, 1950.
10. D. B. Spalding, "International Developments in Heat Transfer," *Conf. Int. Developments in Heat Transfer,* part II, pp. 439–446, ASME, 1961.
11. J. Laufer, *NACA TN* 2954, 1953.
12. A. R. Barbin and J. B. Jones, *J. Bas. Engr., Trans. ASME,* **85**:28 (1963).
13. J. Rotta, *Ing. Arch.,* **24**:258 (1956).
14. E. R. Lindgren, *Arkiv. Fysik,* **7**: 293 (see also vols. 12, 15, 16, 18) (1953).
15. L. M. K. Boelter, G. Young, and H. W. Iversen, *NACA TN* 1451, 1948.
16. A. F. Mills, *J. Mech. Engr. Sci.,* **4**:63 (1962).
17. H. Schlichting, "Boundary Layer Theory," McGraw-Hill Book Company, New York, 1968.
18. V. E. Senecal and R. R. Rothfus, *Chem. Engr. Prog.,* **49**:533 (1953).
19. G. F. Scheele, E. M. Rosen, and T. J. Hanratty, *Can. J. Chem. Eng.,* **38**:67 (1960).
20. H. C. Perkins and P. Worsoe-Schmidt, *Int. J. Heat Mass Transfer,* **8**:1011 (1965).
21. D. M. McEligot, P. M. Magee, and G. Leppert, *J. Heat Transfer, Trans. ASME,* **87**:67 (1965).
22. W. M. Kays, *Trans. ASME,* **72**:1067 (1950).
23. W. M. Kays and A. L. London, "Compact Heat Exchangers," McGraw-Hill Book Company, New York, 1964.
24. "Aerospace Applied Thermo-Dynamics Manual," part A, Incompressible Flow, SAE, 1962.
25. R. P. Benedict, N. A. Carlucci, and S. D. Swetz, *J. Engr. Power, Trans. ASME,* **88**:73 (1966).
26. J. R. Sellars, M. Tribus, and J. S. Klein, *Trans. ASME,* **78**:441 (1956).
27. U. Grigull and H. Tratz, *Int. J. Heat Mass Transfer,* **8**:669 (1965).
28. R. Siegel, E. M. Sparrow, and T. M. Hallman, *Appl. Sci. Res.,* **A7**:386 (1958).
29. W. M. Kays, *Trans. ASME,* **77**:1265 (1955).
30. P. Goldberg, M.S. thesis, Mech. Engr. Dept., Massachusetts Institute of Technology, 1958.
31. D. L. Ulrichson and R. A. Schmitz, *Int. J. Heat Mass Transfer,* **8**:253 (1965).
32. H. S. Heaton, W. C. Reynolds, and W. M. Kays, *Int. J. Heat Mass Transfer,* **7**:763 (1964).
33. P. J. Schneider, *Trans. ASME,* **79**:765 (1957).
34. S. N. Singh, *Appl. Sci. Res.,* **A7**:325 (1958).
35. M. Dalle-Donne and F. H. Bowditch, *Nuc. Engr.,* **8**:20 (1963).
36. C. A. Sleicher and M. Tribus, *Trans. ASME,* **79**:789 (1957).
37. W. M. Kays and E. Y. Leung, *Int. J. Heat Mass Transfer,* **6**:537 (1963).
38. E. M. Sparrow, T. M. Hallman, and R. Siegel, *Appl. Sci. Res.,* **A7**:37 (1957).
39. R. G. Deissler, *NACA TN* 3145, 1954.
40. E. Skupinski, J. Tortel, and L. Vautrey, *Int. J. Heat Mass Transfer,* **8**:937 (1965).
41. D. E. Dwyer, *AIChE Journal,* **9**:261 (1963).
42. F. W. Dittus and L. M. K. Boelter, *Univ. of Calif. Pubs. Engr.,* **2**:443 (1930).
43. L. H. McAdams, "Heat Transmission," p. 219, McGraw-Hill Book Company, New York, 1954.
44. A. P. Colburn, *Trans. AIChE,* **29**:174 (1933).
45. R. N. Lyon, "Liquid Metals Handbook," NAVEXOS P-733 (Rev.), 1952.
46. R. A. Seban and T. T. Shimazaki, *Trans. ASME,* **73**:803 (1951).
47. K. M. Krall and E. M. Sparrow, *J. Heat Transfer, Trans. ASME,* **88**:131 (1966).
48. W. C. Reynolds, *Int. J. Heat Mass Transfer,* **6**:445 (1963).
49. W. C. Reynolds, *Int. J. Heat Mass Transfer,* **6**:925 (1963).
50. E. M. Sparrow and S. H. Lin, *Int. J. Heat Mass Transfer,* **6**:800 (1963).
51. W. M. Kays, "Convective Heat and Mass Transfer," McGraw-Hill Book Company, New York, 1966.
52. W. Nunner, *VDI Forschungsheft* 455, **B22**:5 (1956) (AERE Lib./Trans. 786).

53. D. F. Dipprey and R. H. Sabersky, *Int. J. Heat Mass Transfer*, **6**:329 (1963).
54. J. W. Smith and N. Epstein, *AIChE Journal*, **3**:241 (1957).
55. G. Grass, *Atom Kernenergie*, **3**:328 (1958) (AEC-tr-3641).
56. W. F. Cope, *Proc. Inst. Mech. Engrs.*, **145**:99 (1941).
57. E. W. Sams, *NACA RM-E-52D17*, 1952.
58. D. W. Savage and J. E. Myers, *AIChE Journal*, **9**:694 (1963).
59. V. Kolar, *Int. J. Heat Mass Transfer*, **8**:639 (1965).
60. J. Gargand and G. Paumard, Rapport CEA (France) R2464, 1964.
61. M. Cowin et al., *TRG Report* 539 (W) UKAEA, 1963.
62. R. T. Lancet, *J. Heat Transfer, Trans. ASME*, **81**:168 (1959).
63. F. J. Edwards and N. Sheriff, "International Developments in Heat Transfer," *Conf. Int. Developments in Heat Transfer*, vol. 2, pp. 415–425, ASME, 1961.
64. H. Brauer, *Mannesman Forschungsberichte* 109, 1961.
65. J. M. Malherbe, F. CEA 2283, Centre d'Etudes Nucleaires de Saclay, 1963.
66. A. Draycott and K. R. Lawther, "International Developments in Heat Transfer," *Conf. Int. Developments in Heat Transfer*, pp. 543–552, ASME, 1961.
67. V. Walker and A. C. Rapier, Brit. Nucl. Energy Soc., *Symp. Advanced Gas-Cooled Reactor*, p. 181, 1963.
68. G. A. Kemeny and J. A. Cyphers, *J. Heat Transfer, Trans. ASME*, **83**:189 (1961).
69. V. Kaul and M. Von Kiss, *Neue Technik*, **6**:297 (1964).
70. A. W. Bennett and H. A. Kearsey, *AERE-R4350*, 1964.
71. R. S. Holcomb and F. D. Lynch, *ORNL* 3445, 1963.
72. T. B. Burgoyne, P. Burnett, and D. Wilkie, *UKAEA TRG Report* 781 (W), 1964.
73. N. Kattchee and W. N. Mackewicz, *ASME Paper* 63-Ht-1, 1963.
74. W. Gomelauri, *Int. J. Heat Mass Transfer*, **7**:653 (1964).
75. Z. Nagaoka and A. Watanabe, *Proc. 7th Int. Cong. Refrigeration*, vol. 3, p. 221, 1936.
76. E. W. Sams, *TID* 7529, pt. 1, book 2, p. 390, 1957.
77. F. Kreith and D. Margolis, *Appl. Sci. Res.*, **A8**: 457 (1959).
78. A. P. Colburn and W. J. King, *Ind. Eng. Chem.*, **23**:910 (1931).
79. W. R. Gambill and R. D. Bundy, *ASME Paper* 62-HT-42, 1962.
80. W. R. Gambill and R. D. Bundy, *AIChE Journal*, **9**:55 (1963).
81. E. Smithberg and F. Landis, *J. Heat Transfer, Trans. ASME*, **86**:39 (1964).
82. S. I. Evans and R. J. Sarjant, *J. Inst. Fuel*, **24**:216 (1951).
83. See Ref. 77.
84. M. H. Ibragimov, E. V. Nomofelov, and V. I. Subbotin, *Teploenergetika*, **8**:57 (1961).
85. E. V. Seymour, *Trans. Inst. Chem. Engrs.*, **41**:159 (1963).
86. R. Koch, *VDI–Forschungsheft* 469, **B29**: 1 (1958).
87. L. B. Evans and S. W. Churchill, *Chem. Eng. Prog. Symp. Series*, vol. 59, no. 41, p. 36, 1963.
88. A. Bhattacharyya, *Aktiebolaget Atomenergi Arbets Rapport* (Sweden) RPL-711, 1964.
89. W. B. Hall, *J. Mech. Engr. Sci.*, 4:287 (1962).
90. D. Wilkie, *Int. Heat Transfer Conf.*, Chicago, Ill., papers 1 and 2, 1966.
91. O. S. Fedynsky, Problems of Heat Transfer, "Problemy Energetiki," p. 53, Energetich, Inst. Akad. Nauk SSSR, Moscow, 1959.
92. L. B. Evans and S. W. Churchill, *Chem. Engr. Prog.*, **58**:55 (1962).
93. D. B. Spalding and A. Lieberam, *Chemie-Ing.-Techn.*, **37**:335 (1965).
94. W. R. Gambill, R. D. Bundy, R. W. Wansbrough, *Chem. Engr. Prog., Symp. Series*, **57**(32):127 (1961).
95. N. D. Greene, Convair Aircraft, personal communication to W. R. Gambill, 1960.
96. Unpublished ORNL data.
97. R. Viskanta, *Nuc. Sci. Engr.*, **10**:202 (1961).
98. B. Metais and E. R. G. Eckert, *J. Heat Transfer, Trans. ASME*, **86**:295 (1964).
99. P. Worsoe-Schmidt and G. Leppert, *Int. J. Heat Mass Transfer*, **8**:1281 (1965).
100. M. Iqbal and J. W. Stachiewicz, *J. Heat Transfer, Trans. ASME*, **88**:109 (1966).
101. Y. Mori, K. Futagami, S. Tokuda, and M. Nakamura, *Int. J. Heat Mass Transfer*, **9**:453 (1966).
102. S. T. McComas and E. R. G. Eckert, *J. Heat Transfer, Trans. ASME*, **88**:147 (1966).
103. T. M. Hallman, *NASA TN D* 1104, 1961.
104. T. M. Hallman, *Trans. ASME*, **78**:1831 (1956).
105. W. G. Brown, *VDI Forschungsheft* 480, Ausgabe B, **26**:1 (1960).
106. R. C. Martinelli and L. M. K. Boelter, Univ. Calif. (Berkeley), *Publ. Engr.*, **5**:23 (1942).

107. R. L. Pigford, *Chem. Engr. Prog., Symp. Series,* **51**:79 (1955).
108. T. W. Jackson, W. B. Harrison, and W. C. Boteler, *Trans. ASME,* **80**:739 (1958).
109. O. C. Eubank and W. S. Proctor, MS thesis, Chem. Engr. Dept., Massachusetts Institute of Technology, 1951.
110. J. J. Martin and M. B. Carmichael, *ASME Paper* 55-A-30, 1955.
111. M. S. Ojalvo and R. J. Grosh, *ANL* 6528, 1962.
112. S. Ostrach, *NACA TN* 3141, 1954.
113. W. N. Gill and E. D. Casal, *AIChE Journal,* **8**:513 (1962).
114. T. S. Lundgren, E. M. Sparrow, and J. B. Starr, *J. Bas. Engr., Trans. ASME,* **86**:620 (1964).
115. H. S. Heaton, W. C. Reynolds, and W. M. Kays, *TR-AHT-5,* Mech. Engr. Dept., Stanford University, Calif., 1962.
116. L. S. Han, *J. Appl. Mech., Trans. ASME,* **27**:403 (1960).
117. D. M. Meter and R. B. Bird, *AIChE Journal,* **7**:41 (1961).
118. J. E. Walker, G. A. Whan, and R. R. Rothfus, *AIChE Journal,* **3**:485 (1957).
119. J. A. Brighton and J. B. Jones, *J. Bas. Engr., Trans. ASME,* **86**:835 (1964).
120. R. R. Rothfus, Doctoral dissertation, Carnegie Institute of Technology, Pittsburgh, Pa., 1948.
121. V. K. Jonsson and E. M. Sparrow, *J. Fl. Mech.,* **25**:1:65 (1966).
122. E. M. Sparrow and S. H. Lin, *J. Bas. Engr., Trans. ASME,* **86**:827 (1964).
123. J. E. Walker and R. R. Rothfus, *AIChE Journal,* **5**:51 (1959).
124. T. H. Okiishi and G. K. Serovy, *ASME Paper* 64–WA/Fe-32, 1964.
125. R. R. Rothfus, J. E. Walker, and G. A. Whan, *AIChE Journal,* **4**:240 (1958).
126. E. Y. Leung and W. M. Kays, *Int. J. Heat Mass Transfer,* **6**:537 (1963).
127. V. K. Jonsson and E. M. Sparrow, *J. Bas. Engr., Trans. ASME,* **88**:550 (1966).
128. J. A. Hart and K. R. Lawther, Discussion to J. A. Brighton and J. B. Jones, 1964.
129. See Ref. 125.
130. R. M. Olson and E. M. Sparrow, *AIChE Journal,* **9**:766 (1963).
131. W. T. Snyder and G. A. Goldstein, *AIChE Journal,* **11**:462 (1965).
132. R. E. Lundberg, P. A. McCuen, and W. C. Reynolds, *Int. J. Heat Mass Transfer,* **6**:495 (1963).
133. R. D. Cess and E. C. Shaffer, *J. Aero-Space Sci.,* **26**:538 (1959).
134. C. L. Hwang and L. T. Fan, *Appl. Sci. Res.,* **A13**:401 (1964).
135. R. Siegel and E. M. Sparrow, *AIChE Journal,* **5**:73 (1959).
136. L. S. Han, "International Developments in Heat Transfer," pp. 591–597, ASME, 1961.
137. K. Stephan, *Chem. Ing. Tech.,* **13**:773 (1959).
138. R. L. Judd and J. H. T. Wade, "Heat Transfer and Fluid Mechanics Institute," pp. 272–288, Stanford University Press, Stanford, Calif., 1963.
139. L. Duchatelle and L. Vautrey, *Int. J. Heat Mass Transfer,* **7**:1017 (1964).
140. J. P. Hartnett and T. F. Irvine, *AIChE Journal,* **3**:313 (1957).
141. A. P. Hatton and A. Quarmby, *Int. J. Heat Mass Transfer,* **6**:903 (1963).
142. A. P. Hatton and A. Quarmby, *Int. J. Heat Mass Transfer,* **7**:817 (1964).
143. N. Z. Azer and B. T. Chao, *Int. J. Heat Mass Transfer,* **3**:121 (1960).
144. W. A. Sutherland and W. M. Kays, *Int. J. Heat Mass Transfer,* **7**:1187 (1964).
145. W. T. Snyder, *AIChE Journal,* **9**:503 (1963).
146. E. Y. Leung, W. M. Kays, and W. C. Reynolds, *Report* AHT-4, Thermosciences Division, Dept. Mech. Engr., Stanford University, Stanford, Calif., 1962.
147. R. G. Deissler and M. F. Taylor, *NACA TN* 3451, 1955.
148. W. Yu and O. E. Dwyer, *Nuc. Sci. Engr.,* **24**:105 (1966).
149. W. Yu and O. E. Dwyer, *Nuc. Sci. Engr.,* **27**:1 (1967).
150. L. W. Carlson and T. F. Irvine, *J. Heat Transfer, Trans. ASME,* **83**:441 (1961).
151. E. M. Sparrow, *AIChE Journal,* **8**:599 (1962).
152. L. Schiller, *Zeit. Ang. Math. und Mech.,* **3**:2 (1923).
153. J. P. Hartnett, J. C. Y. Koh, and S. T. McComas, *J. Heat Transfer, Trans. ASME,* **84**:82 (1962).
154. G. W. Maurer and B. W. LeTourneau, *J. Heat Transfer, Trans. ASME,* **86**:627 (1964).
155. E. R. G. Eckert and T. F. Irvine, *Trans. ASME,* **78**:709 (1956).
156. E. R. G. Eckert and T. F. Irvine, Fifth Midwestern Conference on Fluid Mech., 1957.
157. S. C. Davies and C. M. White, *Proc. Roy. Soc. (London),* **A119**:92 (1928).
158. R. J. Cornish, *Proc. Roy. Soc. (London),* **A120**:691 (1928).
159. J. Nikuradse, *Ingenieur Archiv.,* **1**:306 (1930).
160. See Ref. 152.

161. F. C. Lea and A. G. Tadros, *Phil. Mag.*, **7**:1235 (1931).
162. J. Allen and N. D. Grundberg, *Phil. Mag.*, **7**:490 (1937).
163. L. Washington and W. M. Marks, *Ind. Engr. Chem.*, **29**:337 (1937).
164. E. M. Sparrow and R. Siegel, *J. Heat Transfer, Trans. ASME*, **81**:157 (1959).
165. S. C. R. Dennis, A. McD. Mercer, and G. Poots, *Quart. Appl. Math.*, **17**:285 (1959).
166. E. M. Sparrow and R. Siegel, *Int. J. Heat Mass Transfer*, **1**:161 (1960).
167. R. G. Deissler and M. F. Taylor, *NASA* TR-R-31, 1959.
168. E. R. G. Eckert and T. F. Irvine, *J. Heat Transfer, Trans. ASME*, **82**:125 (1960).
169. W. H. Lowdermilk, W. F. Weiland, and J. N. B. Livingood, *NACA RM* E53J07, 1954.
170. S. Levy, R. A. Fuller, and R. O. Niemi, *J. Heat Transfer, Trans. ASME*, **81**:129 (1959).
171. W. R. Gambill and R. D. Bundy, *Nuc. Sci. Engr.*, **18**:69 (1964).
172. E. R. G. Eckert, T. F. Irvine, and J. J. Yen, *Trans. ASME*, **80**:1433 (1958).
173. R. Siegel and J. M. Savino, *J. Heat Transfer, Trans. ASME*, **87**:59 (1965).
174. J. L. Novotny, S. T. McComas, E. M. Sparrow, and E. R. G. Eckert, *AIChE Journal*, **10**:466 (1964).
175. E. M. Sparrow, J. R. Lloyd, and C. W. Hixon, *J. Heat Transfer, Trans. ASME*, **88**:170 (1966).
176. L. S. Han, *J. Heat Transfer, Trans. ASME*, **81**:121 (1959).
177. E. M. Sparrow and A. L. Loeffler, *AIChE Journal*, **5**:325 (1959).
178. E. M. Sparrow, A. L. Loeffler, and H. A. Hubbard, *J. Heat Transfer, Trans. ASME*, **83**:415 (1961).
179. W. A. Sutherland and W. M. Kays, *General Electric Report* AD-4637, San Jose, Calif., 1965.
180. W. A. Sutherland and W. M. Kays, *J. Heat Transfer, Trans. ASME*, **88**:117 (1966).
181. J. L. Wantland, *Reactor Heat Transfer Conf.*, New York, TID-7529, pt. 1, p. 525, 1956.
182. D. A. Dingee and J. W. Chastain, *Reactor Heat Transfer Conf.*, New York, TID-7529, pt. 1, p. 462, 1956.
183. P. Miller, J. J. Byrnes, and D. M. Benforado, *AIChE Journal*, **2**:226 (1956).
184. H. W. Hoffman, J. L. Wantland, and W. S. Stelzman, "International Developments in Heat Transfer," pp. 553–560, ASME, 1961.
185. R. G. Deissler and M. F. Taylor, *Reactor Heat Transfer Conf.*, New York, TID-7529, pt. 1, p. 416, 1956.
186. V. I. Subbotin, P. A. Vshakov, P. L. Kirillov, M. K. Ibraginov, M. N. Ivanovsky, E. V. Nomophilov, D. M. Ovechkin, and V. P. Sorokin, *Third U.N. Int. Conf. Peaceful Uses of Atomic Energy*, A (Conf.), 28/P/328, 1964.
187. L. D. Palmer and L. L. Swanson, "International Developments in Heat Transfer," pp. 535–543, ASME, 1961.
188. M. W. Maresca and O. E. Dwyer, *J. Heat Transfer, Trans. ASME*, **86**:180 (1964).
189. B. Nimmo and O. E. Dwyer, *J. Heat Transfer, Trans. ASME*, **87**:312 (1965).
190. W. E. Hilding and C. H. Coogan, *Symp. Air-Cooled Heat Exchangers, ASME*, pp. 57–85, 1964.
191. E. M. Sparrow, T. S. Chen, and V. K. Jonsson, *Int. J. Heat Mass Transfer*, **7**:583 (1964).
192. B. de Lorenzo and E. D. Anderson, *Trans. ASME*, **67**:697 (1945).
193. L. Clark and R. E. Winston, *Chem. Engr. Prog.*, **51**:147 (1955).
194. F. W. Braun and J. G. Knudsen, *Chem. Engr. Prog.*, **48**:517 (1952).
195. J. G. Knudsen and D. L. Katz, *Chem. Engr. Prog.*, **46**:490 (1950).
196. E. M. Sparrow and A. Haji-Sheikh, *J. Heat Transfer, Trans. ASME*, **87**:426 (1965).
197. E. M. Sparrow and A. Haji-Sheikh, *J. Heat Transfer, Trans. ASME*, **88**:351 (1966).
198. E. M. Sparrow and J. B. Starr, *J. Appl. Mech., Trans. ASME*, **32**:684 (1965).
199. A. J. Ede, "International Developments in Heat Transfer," pp. 634–642, ASME, 1961.
200. A. J. Ede, *Int. Heat Transfer Conf., Chicago, Ill.*, Paper 9, 1966.
201. D. J. Gunn and C. W. Darling, *Trans. Inst. Chem. Engrs.*, **41**:163 (1963).
202. L. Topper, *Chem. Engr. Sci.*, **5**:13 (1956).
203. E. M. Sparrow and R. Siegel, *Nuc. Sci. Engr.*, **4**:239 (1958).
204. R. M. Inman, *Int. J. Heat Mass Transfer*, **5**:1053 (1962).
205. R. Siegel and E. M. Sparrow, *J. Heat Transfer, Trans. ASME*, **81**:280 (1959).
206. R. G. Deissler, *NACA Rept.* 1210, 1955.
207. R. B. Kinney and E. M. Sparrow, *J. Heat Transfer, Trans. ASME*, **88**:314 (1966).
208. E. M. Sparrow, J. L. Novotny, and S. H. Lin, *AIChE Journal*, **9**:797 (1963).
209. M. Strite and R. M. Inman, *NASA TN* D-3039, 1965.
210. L. N. Tao, *J. Heat Transfer, Trans. ASME*, **83**:466 (1961).
211. V. P. Tyagi, *J. Heat Transfer, Trans. ASME*, **88**:161 (1966).

212. Perkins, H. C., see p. 169, Ref. 211.
213. Y. Mori and W. Nakayama, *Int. J. Heat Mass Transfer,* 8:67 (1965).
214. H. Ito, *J. Bas. Engr., Trans. ASME,* 81:123 (1959).
215. G. F. C. Rogers and Y. R. Mayhew, *Int. J. Heat Mass Transfer,* 7:1207 (1964).
216. R. A. Seban and E. F. McLaughlin, *Int. J. Heat Mass Transfer,* 6:387 (1963).
217. F. Kreith, *Heat Transfer Fluid Mechanics Institute,* pp. 111-122, ASME, 1953.
218. A. G. Smith and D. B. Spalding, *J. Roy. Aero. Soc.,* 62:60 (1958).
219. F. S. Ambrok, *Sov. Phys. Tech. Phys.,* 2:1979 (1957).
220. D. Bartz, "Advances in Heat Transfer," vol. 2, pp. 2–109, Academic Press, New York, 1965.
221. L. H. Back, P. F. Massier, and H. L. Gier, *Int. J. Heat Mass Transfer,* 7:549 (1964).
222. W. A. Benser and R. W. Graham, *ASME Paper* 62-AV-22, 1962.
223. A. B. Witte and E. Y. Harper, *AIAA Journal,* 1:443 (1963).
224. A. Fortini and R. C. Ehlers, *NASA TN* D-1743, 1963.
225. R. G. Deissler, *NACA TN* 2410, 1951.
226. K. T. Yang, *J. Heat Transfer, Trans. ASME,* 84:353 (1962).
227. E. N. Sieder and G. E. Tate, *Ind. Engr. Chem.,* 28:1429 (1936).
228. R. W. Allen and E. R. G. Eckert, *J. Heat Transfer, Trans. ASME,* 86:301 (1964).
229. W. M. Kays and W. B. Nicoll, *J. Heat Transfer, Trans. ASME,* 85:191 (1963).
230. M. E. Davenport and G. Leppert, *J. Heat Transfer, Trans. ASME,* 87:191 (1965).
231. R. G. Deissler and C. S. Eian, *NACA TN* 2629, 1952.
232. L. V. Humble, W. H. Lowdermilk, and L. G. Desmon, *NACA Report* 1020, 1951.
233. J. R. McCarthy and H. Wolf, Rocketdyne Corp., *Report* RR-60-12, Canoga Park, California, 1960.
234. J. F. Barnes, *Natl. Gas Turbine Establishment Rept.* R-241, Pyestock, Hants, England, 1960.
235. M. F. Taylor, and T. A. Kirchgessner, *NASA TN* D-133, 1959.
236. M. F. Taylor, *Heat Transfer and Fluid Mechanics Institute,* p. 251, Stanford University Press, Stanford, Calif., 1963.
237. W. F. Weiland, *AIChE Chem. Engr. Prog. Symp. Series, 61,* no. 60, pp. 97–105, Chicago, Ill., 1965.
238. V. D. Lel'chuk, and B. V. Dyadyakin, *USAEC Translation* AEC-Tr-4511, 1962.
239. W. B. Nicoll and W. M. Kays, *Dept. of Mech. Engr. TR* 43, Stanford University, Stanford, Calif., 1959.
240. H. E. Zellnik and S. Churchill, *AIChE Journal,* 4:37 (1958).
241. S. S. Kutateladze and A. I. Leont'ev, "Turbulent Boundary Layers in Compressible Gases," Trans. D. B. Spalding, Arnold, London, 1964.
242. R. G. Deissler and A. F. Presler, "International Developments in Heat Transfer," pp. 579–589, ASME, 1961.
243. J. V. Miller and M. F. Taylor, *NASA TN* D-2594, 1965.
244. M. F. Taylor, *NASA TN* D-2595, 1965.
245. L. B. Koppel and J. M. Smith, *J. Heat Transfer, Trans. ASME,* 84:151 (1962).
246. R. D. Wood and J. M. Smith, *AIChE Journal,* 10:180 (1964).
247. H. S. Swenson, J. R. Carver, and C. R. Kakarala, *J. Heat Transfer, Trans. ASME,* 87:477 (1965).
248. R. P. Bringer and J. M. Smith, *AIChE Journal,* 3:49 (1957).
249. L. B. Koppel and J. M. Smith, "International Developments in Heat Transfer," *Conf. Int. Developments in Heat Transfer,* pp. 585–590, ASME, 1961.
250. N. L. Dickinson and C. P. Welch, *Trans. ASME,* 80:747 (1958).
251. K. Goldmann, "International Developments in Heat Transfer," *Conf. Int. Developments in Heat Transfer,* pp. 561–568, ASME, 1961.
252. H. L. Hess and H. R. Kunz, *J. Heat Transfer, Trans. ASME,* 87:41 (1965).
253. R. C. Hendricks, R. W. Graham, Y. Y. Hsu, and A. A. Medeiros, *ARS Journal,* 32:244 (1962).
254. L. F. Carter and W. N. Gill, *AIChE Journal,* 10:330 (1964).
255. R. M. Terrill, *Int. J. Heat Mass Transfer,* 8:1491 (1965).
256. T. E. Horton and S. W. Yuan, *Appl. Sci. Res.,* A14:223 (1965).
257. S. M. Lee and W. N. Gill, *AIChE Journal,* 10:896 (1964).
258. S. W. Yuan and A. B. Finkelstein, *Trans. ASME,* 78:719 (1956).
259. S. W. Yuan and A. B. Finkelstein, *Jet Propulsion,* 28:171 (1958).
260. S. W. Yuan and Y. Peng, "International Developments in Heat Transfer," *Conf. Int. Developments in Heat Transfer,* pp. 717-724, ASME, 1961.

261. Y. Peng and S. W. Yuan, *J. Heat Transfer, Trans. ASME*, **87**:252 (1965).
262. H. L. Weissberg, *Phy. Fluids*, **2**:510 (1959).
263. R. N. Hornbeck, W. T. Rouleau, and F. Osterle, *Phy. Fluids*, **6**:1649 (1963).
264. S. W. Yuan and A. Barazotti, *Heat Transfer and Fluid Mechanics Institute*, pp. 25–39, Stanford Univ. Press, Stanford, Calif., 1958.
265. H. Burnage, *Publ. Scientifiques et Techniques du Ministre de l'Air, Tn* 114, Travaux de l'Institut de Mechanique Statistique de la Turbulence, 1962.
266. R. M. Olsen and E. R. G. Eckert, *J. Appl. Mech., Trans. ASME*, **33**:7 (1966).
267. S. W. Churchill and H. E. Stubbs, *AFOSR-TR*-56-58, *ASTIA* AD115-001, 1957.
268. J. A. Bialokoz and O. A. Saunders, *Proc. Inst. Mech. Engrs.*, **170**:389 (1956).
269. O. A. Saunders and P. H. Calder, *Proc. Heat Transfer Fluid Mechanics Inst.*, p. 91, 1951.
270. R. Siegel, *J. Appl. Mech., Trans. ASME*, **27**:241 (1960).
271. M. Perlmutter and R. Siegel, *Int. J. Heat Mass Transfer*, **3**:94 (1961).
272. M. Perlmutter and R. Siegel, *J. Heat Transfer, Trans. ASME*, **83**:432 (1961).
273. R. Siegel and M. Perlmutter, *J. Heat Transfer, Trans. ASME*, **84**:111 (1962).
274. R. Siegel and M. Perlmutter, *J. Heat Transfer, Trans. ASME*, **85**:358 (1963).
275. R. Siegel, *Int. J. Heat Mass Transfer*, **6**:607 (1963).
276. E. M. Sparrow and R. Siegel, *J. Heat Transfer, Trans. ASME*, **82**:170 (1960).
277. J. N. Cannon, doctoral dissertation, Stanford University, Stanford, Calif., 1965.
278. L. M. Trefethen, *J. Bas. Engr., Trans. ASME*, **81**:132 (1959).
279. W. D. Morris, *J. Fluid Mech.*, **21**:453 (1965).
280. A. B. Metzner, "Handbook of Fluid Mechanics," McGraw-Hill Book Company, New York, 1961.
281. A. B. Metzner, "Advances in Heat Transfer," vol. 2, Academic Press, New York, 1965.
282. W. L. Wilkinson, "Non-Newtonian Fluids," Pergamon Press, New York, 1960.
283. W. M. Rohsenow, "Developments in Heat Transfer," Massachusetts Institute of Technology Press, Cambridge, Mass., 1964.
284. A. B. Metzner, "Advances in Chemical Engineering," vol. 1, Academic Press, New York, 1956.
285. R. B. Bird, W. E. Stewart, and E. N. Lightfoot, "Transport Phenomena," John Wiley & Sons, Inc., New York, 1960.
286. R. W. Hanks and E. B. Christiansen, *AIChE Journal*, **7**:519 (1961).
287. D. C. Bogue and A. B. Metzner, *Ind. Eng. Chem. Fundamentals*, **2**:143 (1963).
288. D. M. Eissenburg and D. C. Bogue, *AIChE Journal*, **10**:723 (1964).
289. N. J. Beek and R. Eggink, *De Ingenieur*, **74**:81 (1962).
290. A. A. McKillop, *Int. J. Heat Mass Transfer*, **7**:853 (1964).
291. E. B. Christiansen and S. E. Craig, *AIChE Journal*, **8**:154 (1962).
292. D. R. Oliver and V. G. Jensen, *Chem. Eng. Sci.*, **19**:115 (1964).
293. W. L. Friend and A. B. Metzner, *AIChE Journal*, **4**:393 (1958).
294. R. M. Inman, *NASA TN* D-2674, 1965.
295. A. G. Fredrickson and R. B. Bird, *Ind. Engr. Chem.*, **50**:347 (1958).
296. J. A. Wheeler and E. H. Wissler, *AIChE Journal*, **11**:207 (1965).
297. S. Matsuhisa and R. B. Bird, *AIChE Journal*, **11**:588 (1965).
298. R. Siegel and M. Perlmutter, *Int. J. Heat Mass Transfer*, **5**:639 (1962).
299. J. C. Chen, *AIChE Journal*, **10**:253 (1964).
300. E. G. Keshock and R. Siegel, *J. Heat Transfer, Trans. ASME*, **86**:341 (1964).
301. R. Viskanta, *Appl. Sci. Res.*, **A13**:291 (1964).
302. L. D. Nichols, *Int. J. Heat Mass Transfer*, **8**:589 (1965).
303. S. deSoto, *AIAA Paper* 66-136, 1966.
304. E. M. Sparrow and R. D. Cess, "Radiation Heat Transfer," Brooks/Cole Publ. Company, Belmont, Calif., 1966.

Section **8**

Forced Convection, External Flows

MORRIS W. RUBESIN

Palo Alto, California

AND

MAMORU INOUYE

Los Gatos, California

INTRODUCTION

In the current era of large electronic computers, many complex problems in convection are being solved precisely by numerical solution of equations expressing basic principles. Keen insight into the fine points of such problems, however, requires extensive parametric studies that are computer time consuming and, therefore, the numerical approach is usually applied only to very few examples. Further, these numerical programs occupy so much computer storage space that their use as subroutines within generalized systems studies is often impractical. Thus, there still exists a need for general formulas and data correlations that can be used in preliminary design, in systems studies where convection is only one of many inputs, and in creative design where inventiveness is based on understanding the influences of the variables of a problem. This section provides many of these tools for the case of forced convection over bodies.

Specifically, theoretical equations and correlations of data are presented for evaluating the local rate of heat transfer between the surface of a body and an encompassing fluid at different temperatures and in relative motion. *Forced convection* requires either the fluid to be pumped past the body, as for a model in a wind tunnel, or the body to be propelled through the fluid, as an aircraft in the atmosphere. The methods presented apply equally to either situation when velocities are expressed relative to the body. Gravity forces are usually negligible under these conditions. Further, the contents of this chapter are confined to those conditions where the fluid behaves as a continuum.

The evaluation of forced convection to bodies has become a major problem in many aspects of modern technology. A few examples of applications include the following: thermally de-icing aircraft surfaces; turbine blade cooling; and protecting high-performance aircraft, missile nose cones, and reentry bodies from intense aerodynamic heating. Formulas for evaluating convective heat transfer rates are generally established through a combination of theoretical analysis and experimentation. Analysis is almost universally based on "boundary-layer theory"—the mathematical solution of conservation equations of individual species, overall mass, momentum, and energy that are applicable to the region of fluid adjacent to the surfaces of bodies where the effects of shear, heat conduction, and species diffusion are controlling. The experimentation involves the measurement of solid and fluid temperatures and of heat flux in a multitude of ways. In this section, emphasis will be placed on the theoretical foundations of the science of convection because this approach provides a systematic basis for understanding the interrelationships and correlations of data for a variety of problems. Comparisons of theoretical results will be made in many cases with corresponding data to assess the validity, or degree of accuracy, of the prescribed theoretical or experimental correlation techniques. Little emphasis will be given, however, to the details of specific experimental setups or the techniques of measuring heat fluxes or temperatures.

1. Boundary-layer Concept. When observations are made of the behavior of fluid convecting heat to a body, it is noticed that the fluid immediately adjacent to the body surface is brought to rest and attains the surface temperature, except in those cases where densities are exceedingly low. Further, the shearing forces within the fluid retard additional layers of fluid above the surface. As the distance normal to the surface is increased, these forces diminish and the velocity parallel to the surface approaches an asymptotic value. In the limit, the shear forces vanish and the fluid behaves as if it were inviscid. Similarly, the temperature varies with distance from the surface and approaches an asymptotic value. Thus, the heat flux normal to the surface, associated with the temperature gradient in that direction, is relatively large in the vicinity of the surface and diminishes to a negligibly small value away from the surface, approaching the behavior of a nonconducting fluid. The observation of this behavior early in the 20th Century led Ludwig Prandtl [1] to define the "boundary-layer concept." Here the flow field about a body is considered to be divided into two distinct regimes: the inner region adjacent to the surface where shear and heat transmission are controlling phenomena—the boundary layer; and the outer region where the gradients in flow properties are so small as to render the effects of shear and heat transfer negligible. Thus, forced convection processes take place within a boundary layer that is located between the surface and the inviscid flow field.

One predominant feature of the boundary layer is that the variation of flow properties as affected by shear and heat conduction occurs over a very narrow region; i.e., the boundary layer is very thin compared with a characteristic dimension of the body such as its length. Since the mass flux parallel to the surface is usually smaller in the boundary layer than in the inviscid flow, the presence of the boundary layer diverts or displaces the inviscid flow so that it behaves as if it were flowing over a body usually slightly larger than the actual body. This outward displacement of the effective body surface, defined as the "displacement thickness," is negligibly small in most cases. The inviscid region, therefore, generally can be evaluated independently of the boundary-layer growth and, in fact, is the first step in the evaluation of convective heat transfer rates (§11). An interdependence of these regions does occur, however, on slender bodies in high-speed flight at low Reynolds numbers (treated in Sec. 9) and on the rear portion of bodies or near protruding aircraft control surfaces where rising surface pressures cause the boundary layer to separate, i.e., the surface streamline moves out away from the body. Another equally important consequence of the "thinness" of the boundary layer is the considerable mathematical simplification of the general equations of the transport of mass, momentum, and energy within this region. These simplifications made otherwise intractable equations amenable to analysis and permitted the development of a considerable amount of boundary-layer theory even prior to the advent of large electronic computers.

When the density and transport properties within this thin region are constant, the extent of the temperature and velocity profiles differ depending on the particular Prandtl number of the fluid. This has led to a terminology that distinguishes between the thickness of the "flow boundary layer," the "thermal boundary layer," and, additionally, the "concentration or diffusional boundary layer" when the latter occurs from transpiration, film, or ablation cooling systems that transfer foreign species into the boundary layer. For Prandtl and Lewis numbers near unity, these boundary-layer thicknesses are of comparable magnitude.

For more realistic fluid properties (ρ, k, μ, c_p), which vary because of their temperature or concentration dependence, there is an interrelationship between the velocity, temperature, and concentration profiles where a variation in one will affect all the others. Thus, no meaningful distinction can be made between the thicknesses of the individual layers far from the surface. Near the surface, however, the relative slopes of the individual profiles are related, in part, to the fluid properties in a qualitative fashion similar to that which occurs with a constant property fluid.

2. Laminar, Turbulent, and Transitional Boundary Layers. Sometimes the fluid in the boundary layer flows such that adjacent layers, or laminas, moving at different velocities act as if they "glide" over one another. Under these conditions, the shear, heat transfer, and diffusion are produced by molecular motions across the interface between adjacent layers. This type of flow is called "laminar" flow and the transport properties governing the convection in such boundary layers are the coefficients of viscosity, thermal conductivity, and diffusion.

If for a given fluid in laminar flow, the velocity or the distance along the body is increased, resulting in a larger Reynolds number, the laminas of the fluid begin to oscillate. These oscillations grow into bursts of eddying motion that flow downstream at velocities slightly less than the velocity at the edge of the boundary layer and grow with time or the distance along the surface. Further downstream, new bursts appear that coalesce with the growing upstream bursts. At some point, the boundary layer is filled with the eddying bursts and becomes a fully turbulent boundary layer. This type of boundary layer possesses a three-dimensional eddying motion that is random and is characterized by a scale (or size) and intensity of fluctuations of the eddies. The transport of momentum (shear), energy, and the individual species is achieved predominantly by these eddying motions. The presence of the stationary surface damps the eddying motion and consequently, at the surface, the shear, heat transfer, and diffusion are again controlled by the molecular processes. The turbulent boundary layer is characterized, therefore, by inner regions where molecular motions predominate, outer regions where eddying motions are controlling, and a blending region to which both types of motions contribute.

The preceding description of the onset of a fully turbulent boundary layer indicated a phase that is neither laminar nor fully turbulent. This transitional boundary layer, because of its complexity, has not received the attention given the others. Knowledge of the behavior of the transitional boundary layer is quite important, however, because often it extends over a major portion of the body surface. Statistical methods of superposing the characteristics of laminar and turbulent boundary layers to account for the behavior of the transitional region are given in later portions of this section.

3. Boundary Conditions. The boundary-layer characteristics at a station are determined, generally, by upstream conditions in both the inviscid flow and on the surface. The inviscid flow forms the outer boundary condition of the boundary layer, and the important characteristics there are the local velocity parallel to the surface, the local thermodynamic state conditions, and, behind curved shock waves, the level of vorticity. From the viewpoint of boundary layer transition, the sound intensity and turbulence level are also important.

The surface conditions that affect the local boundary layer characteristics are the upstream distributions of temperature or heat flux and of any coolant introduced either tangent or normal to the surface through backward facing slots or a porous surface. An ablating surface provides a continuous mass transfer with zero momentum tangential to the surface. Additional surface boundary conditions of importance, especially with regard to transition, are the surface roughness and flutter or compliance to external fluctuating pressure fields.

4. Three-dimensional Effects. When a body has either a three-dimensional shape or is axisymmetric but oriented at an angle to the free-stream direction, the inviscid flow velocity vector along the surface may not align itself with the natural coordinates of the body surface. Under these conditions, the local boundary layer characteristics depend on the two components of the inviscid velocity vector at the surface along the natural coordinates of the body. Generally, the inviscid flow streamlines near the surface are curved, and the velocity vectors within the boundary layer tend to rotate away from the inviscid streamline as the surface is approached. These effects produce a complex behavior of the boundary layer that markedly affects the convective heat transfer. Attention is given to the evaluation of some of these effects in subsequent sections.

DEFINITION OF TERMS

The theoretical results and correlation equations presented in this section are expressed in terms defined and explained in this subsection.

5. Types of Fluids. The variety of flows and surface mass transfer conditions offered requires consideration of different fluids, e.g., liquids or gases, with properties that may vary depending on the local thermodynamic conditions. The properties in the fundamental equations are as general as possible; however, particular theoretical solutions and experiments have restrictions as stated. The pertinent thermodynamic and molecular transport properties include ρ, e, i, c_v, c_p, μ, k, and $\mathfrak{D}$. For turbulent flows, it is also necessary to include eddy coefficients that depend on the structure of the flows as well as the aforementioned properties.

An incompressible fluid with constant thermal and transport properties represents liquids and gases in low speed flow provided the temperature differences are small.

The internal energy is given by $e = c_v T$, where c_v for a liquid is the same for both constant volume and constant pressure processes.

When the temperature differences are large, the variation in thermal and transport properties cannot be neglected. The internal energy is then defined as

$$e = \int_0^T c_v(T)dT \tag{1}$$

A *thermally perfect gas* is the model employed for the flow of gases with moderately large temperature differences or Mach numbers greater than 0.2. Its properties are variable and depend, in general, upon temperature and pressure.

The density is given by the perfect gas law

$$\rho = \frac{p}{RT}$$

The specific heats c_v and c_p are solely temperature dependent. The internal energy, which includes translation, rotation, vibration, and electronic excitation, is defined by Eq. (1). The enthalpy is defined by

$$i = e + \frac{p}{\rho} = \int_0^T c_p(T)dT$$

The speed of sound, which is the speed at which small disturbances are propagated through the gas, is given by

$$a = \sqrt{\left(\frac{\partial p}{\partial \rho}\right)_S} \tag{2}$$

and is a function of temperature only.

The transport properties μ and k are temperature dependent only.

A *calorically perfect gas* is a special case of a thermally perfect gas with constant specific heats such that the internal energy and enthalpy are given by $e = c_v T$ and $i = c_p T$, respectively. The ratio of specific heats $\gamma = c_p/c_v$ is also constant and is sufficient to determine the thermodynamic properties for inviscid flow.

The speed of sound is given by

$$a = \sqrt{\gamma RT}$$

A *mixture of thermally perfect gases* may be used to describe air and other gases at high temperatures. All the species, which may include molecules, atoms, ions, and electrons, are assumed to be at the same temperature and to contribute partial pressures in proportion to their molar concentrations.

The density is given by

$$\rho = \frac{p}{T \sum_i K_i R_i}$$

where K_i is the mass concentration and R_i is the gas constant for the ith species.

The enthalpy is given by

$$i = \sum_i K_i i_i$$

where

$$i_i = \int_0^T c_{p_i} dT + i_i^0$$

with i_i^0 representing the heat of formation of the ith species at zero absolute temperature.

The speed of sound is again defined by Eq. (2), but is not just a function of temperature alone.

When the mixture is in thermodynamic equilibrium, the species concentrations are obtained using statistical and quantum mechanics [2]. The thermodynamic properties are functions of two variables as for a perfect gas, but are not expressible in equation form. Thermodynamic equilibrium occurs when the elemental concentrations remain constant throughout the flow field, and the chemical reaction times are much shorter than the flow times.

6. Dimensionless Groups of Flow and Transport Parameters. The number of variables in the expressions for convective heating may be reduced by forming dimensionless groups as correlation parameters. When the fluid properties vary, the state conditions used are important and must be clearly defined. In this section, both reference state and flow conditions will be indicated by subscripts only when they are invariant.

Mach number is a measure of the compressibility effects and is the ratio of the local velocity to the local speed of sound $M = V/a$.

Pressure coefficient (sometimes called Euler number) is the difference between surface and free-stream pressure normalized by the dynamic pressure

$$C_p = \frac{p_w - p_\infty}{\frac{1}{2} \rho_\infty V_\infty^2} \tag{3}$$

Reynolds number is the ratio of the inertial force to the viscous force $Re_L = \rho V L/\mu$ where various reference conditions may be used to evaluate ρ, V, and μ. Reynolds numbers based on the concentration, momentum, energy, and displacement thicknesses are obtained by setting $L = \Delta$, θ, Γ, and δ^*, respectively.

Eckert number is a measure of the energy dissipated by friction and is the ratio of the kinetic energy to a reference enthalpy $Ec = V^2/2i$.

Prandtl number is the ratio of the momentum diffusivity to the thermal diffusivity, $Pr = \mu c_p/k$. For a single species gas there is no ambiguity in defining Pr. For a multispecies gas in chemical equilibrium, it is convenient to use a total Prandtl number Pr_T based on total thermal conductivity and specific heat (see §32). For a multispecies chemically frozen gas, the appropriate frozen Prandtl number Pr_F is defined in §34.

Lewis number is the ratio of the mass diffusivity to the thermal diffusivity in a binary mixture, $Le = \rho \mathcal{D} c_p/k$. It is sometimes more convenient to utilize the Schmidt number $Sc = \nu/\mathcal{D} = Pr/Le$.

7. Surface Shear Parameters. For a Newtonian fluid, the shear stress is proportional to the velocity gradient. At the wall

$$\tau_{w1} = \mu_w \left(\frac{\partial u_1}{\partial x_2}\right)_w$$

where x_2 is the distance normal to the surface.
The local skin friction coefficient is defined as

$$\frac{c_f}{2} = \frac{\tau_w}{\rho_e u_e^2}$$

The average skin friction coefficient over a flat plate of area A is given by

$$\frac{C_f}{2} = \frac{\displaystyle\int^A \tau_w \, dA}{\rho_e u_e^2 A}$$

8. Surface Heat Transfer Parameters. Since the velocity vanishes at the surface, the heat transfer from the surface to the fluid occurs by the mechanism of conduction. Fourier's law of conduction yields for the fluid

$$q_w = -k_w \left(\frac{\partial T}{\partial x_2}\right)_w \tag{4}$$

For low speed flows and a constant surface temperature, Newton's Law of Cooling states that the surface heat flux is proportional to the temperature difference $T_w - T_e$. The heat flux may be normalized by a convection term to yield a dimensionless local heat transfer coefficient called the Stanton number

$$c_h = \frac{q_w}{\rho_e u_e c_p (T_w - T_e)}$$

The average Stanton number for a flat plate is given by

$$C_h = \frac{\displaystyle\int^A q_w \, dA}{\rho_e u_e c_p (T_w - T_e) A}$$

Alternatively, the heat flux may be normalized by a conduction term to yield the Nusselt number

$$\text{Nu} = \frac{q_w L}{k(T_w - T_e)} = c_h \left(\frac{\rho u_e L}{\mu}\right)\left(\frac{\mu c_p}{k}\right)$$

which is a product of the Stanton, Reynolds, and Prandtl numbers.

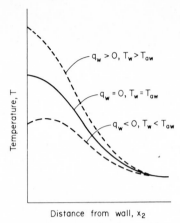

Distance from wall, x_2

Fig. 1. Temperature profiles in a high-speed boundary layer with viscous dissipation of flow energy into heat (T_{aw} = temperature of an insulated wall).

For high speed flows, these heat transfer relations must be modified to account for the temperature rise in the boundary layer caused by the viscous dissipation of energy as depicted in Fig. 1. The surface heat flux is still given by Eq. (4), but the condition of zero heat flux now occurs when the wall temperature is equal to the recovery or adiabatic wall temperature T_{aw}.

The recover factor is defined as the ratio of the temperature at an insulated wall to the value for an adiabatic compression

$$r(0) \; = \; \frac{T_{aw} - T_e}{u_e^2/2c_p}$$

and is largely a function of the Prandtl number.

Newton's Law of Cooling is modified to make the heat flux proportional to the temperature difference $T_w - T_{aw}$ such that $q_w > 0$ for $T_w > T_{aw}$ and $q_w < 0$ for $T_w < T_{aw}$ (see Fig. 1).

For the case of constant c_p, the Stanton number is defined as

$$c_h \; = \; \frac{q_w}{\rho u_e c_p (T_w - T_{aw})}$$

If the specific heat varies appreciably, the enthalpy is more appropriate than the temperature in defining the recovery factor

$$r(0) \; = \; \frac{i_{aw} - i_e}{u_e^2/2}$$

and the Stanton number

$$c_h \; = \; \frac{q_w}{\rho u_e (i_w - i_{aw})}$$

With mass transfer or variable surface temperature, the above heat transfer parameters must be modified as described in the pertinent subsequent sections.

9. Surface Mass Transfer Parameters. If thermal diffusion is negligible, the diffusional flux of species i in a binary mixture depends only on its concentration gradient. Fick's Law yields at the surface

$$j_{iw} = -\rho_w \mathcal{D}_w \left(\frac{\partial K_i}{\partial x_2}\right)_w$$

A dimensionless species mass transfer coefficient is defined as

$$c_{m_i} = \frac{j_{iw}}{\rho u_e (K_{iw} - K_{ie})}$$

10. Reynolds' Analogy. On a flat plate with zero pressure gradient, constant surface temperature, and $Pr = 1$, the local Stanton number and skin friction coefficient are related by

$$c_h = \frac{c_f}{2}$$

This relationship is known as Reynolds' analogy and is valid for both laminar and turbulent boundary layers. For $Le = 1$ also, the analogy may be extended to the species mass transfer coefficient to yield

$$c_{m_i} = c_h = \frac{c_f}{2}$$

For a fluid with $Pr \neq 1$ and $Le \neq 1$, the analogy must be modified to include functions of Pr and Le. For example, Colburn [3] found empirically that

$$c_h \simeq \frac{c_f}{2} Pr^{-2/3}$$

The convenience of these relationships is utilized in many of the correlations which follow.

EVALUATION OF THE INVISCID FLOW

General Considerations

The first step in evaluating the convective heating of a body is to establish the character of the inviscid flow field surrounding it. This involves more disciplines as the speed increases. For low speeds, the fluid may be considered incompressible, and the problem is treated by classical hydrodynamics [4] as strictly a mathematical one. For high subsonic and supersonic speeds, the compressibility of the fluid must be considered as well as possible discontinuities in flow field properties (i.e., shock waves and slip lines), and the problem is treated by gas dynamics [5], involving knowledge of the physical properties of the fluid. For hypersonic speeds, dissociation and ionization of the fluid must be considered, and the problem is treated by physical gas dynamics [6], requiring knowledge of the local chemical composition and the physical properties of the component species.

The general equations that govern the motion of an inviscid, nonconducting, and nonradiating fluid will be presented initially. Solutions of these equations will be discussed for incompressible flow and for high-subsonic and supersonic flow of a perfect gas. In the latter case the thermodynamic properties of the fluid are characterized by γ, the ratio of specific heats or isentropic exponent, which will be assumed constant. In the presence of only inertial and elastic forces, dynamic similarity of two flow fields requires equality of Mach number. Solutions to the inviscid flow problem will be dependent, therefore, on the body shape and orientation, γ, and free-stream Mach number.

Exact solutions, either analytical or numerical, of the inviscid flow equations exist only for a few simple configurations. These solutions yield flow properties throughout the field including the location of any shock waves. For the purpose of calculating convective heat transfer, however, it is necessary to know only the flow properties at the body surface, which corresponds to the outer edge of the boundary layer. Consequently, the results from exact solutions are presented as distributions of surface pressure coefficient C_p as functions of the body shape, γ, and M_∞. Approximate methods and correlations for surface pressure coefficient that are applicable to more general body shapes are also presented. Earlier handbooks [7, 8] and the Durand volumes [9] may be consulted for further study.

11. Equations of Motion. The equations governing the steady motion of an inviscid fluid in thermodynamic equilibrium are expressed in generalized three-dimensional orthogonal coordinates. The metric coefficients h_1, h_2, h_3 are related to particular coordinate systems as discussed in § 12.

The equation for continuity of mass flow is

$$\frac{\partial}{\partial x_1}(h_2 h_3 \rho u_1) + \frac{\partial}{\partial x_2}(h_1 h_3 \rho u_2) + \frac{\partial}{\partial x_3}(h_1 h_2 \rho u_3) = 0 \tag{5}$$

Euler's equations for the conservation of momentum in the three coordinate directions are

x_1 direction

$$\rho\left[\frac{u_1}{h_1}\frac{\partial u_1}{\partial x_1} + \frac{u_2}{h_2}\frac{\partial u_1}{\partial x_2} + \frac{u_3}{h_3}\frac{\partial u_1}{\partial x_3} + \frac{u_2}{h_1 h_2}\left(u_1\frac{\partial h_1}{\partial x_2} - u_2\frac{\partial h_2}{\partial x_1}\right)\right.$$

$$\left. + \frac{u_3}{h_1 h_3}\left(u_1\frac{\partial h_1}{\partial x_3} - u_3\frac{\partial h_3}{\partial x_1}\right)\right] = -\frac{1}{h_1}\frac{\partial p}{\partial x_1} \tag{6}$$

x_2 direction

$$\rho\left[\frac{u_1}{h_1}\frac{\partial u_2}{\partial x_1} + \frac{u_2}{h_2}\frac{\partial u_2}{\partial x_2} + \frac{u_3}{h_3}\frac{\partial u_2}{\partial x_3} + \frac{u_3}{h_2 h_3}\left(u_2\frac{\partial h_2}{\partial x_3} - u_3\frac{\partial h_3}{\partial x_2}\right)\right.$$

$$\left. + \frac{u_1}{h_1 h_2}\left(u_2\frac{\partial h_2}{\partial x_1} - u_1\frac{\partial h_1}{\partial x_2}\right)\right] = -\frac{1}{h_2}\frac{\partial p}{\partial x_2} \tag{7}$$

x_3 direction

$$\rho\left[\frac{u_1}{h_1}\frac{\partial u_3}{\partial x_1} + \frac{u_2}{h_2}\frac{\partial u_3}{\partial x_2} + \frac{u_3}{h_3}\frac{\partial u_3}{\partial x_3} + \frac{u_1}{h_1 h_3}\left(u_3\frac{\partial h_3}{\partial x_1} - u_1\frac{\partial h_1}{\partial x_3}\right)\right.$$

$$\left. + \frac{u_2}{h_2 h_3}\left(u_3\frac{\partial h_3}{\partial x_2} - u_2\frac{\partial h_2}{\partial x_3}\right)\right] = -\frac{1}{h_3}\frac{\partial p}{\partial x_3} \tag{8}$$

Bernoulli's equation is obtained by integrating Euler's equation along a streamline

$$\frac{V^2}{2} + \int_{p_\infty}^{p}\frac{dp}{\rho} = \frac{u_1^2 + u_2^2 + u_3^2}{2} + \int_{p_\infty}^{p}\frac{dp}{\rho} = \frac{V_\infty^2}{2} \tag{9}$$

The equation for the conservation of energy is

$$\rho\left(\frac{u_1}{h_1}\frac{\partial e}{\partial x_1} + \frac{u_2}{h_2}\frac{\partial e}{\partial x_2} + \frac{u_3}{h_3}\frac{\partial e}{\partial x_3}\right) = \frac{p}{\rho}\left(\frac{u_1}{h_1}\frac{\partial \rho}{\partial x_1} + \frac{u_2}{h_2}\frac{\partial \rho}{\partial x_2} + \frac{u_3}{h_3}\frac{\partial \rho}{\partial x_3}\right) \tag{10}$$

For a compressible fluid, Eq. (10) can be combined with the momentum equations (6) through (8) to obtain an equation for the total enthalpy

$$\frac{u_1}{h_1}\frac{\partial I}{\partial x_1} + \frac{u_2}{h_2}\frac{\partial I}{\partial x_2} + \frac{u_3}{h_3}\frac{\partial I}{\partial x_3} = 0 \tag{11}$$

A solution of Eq. (11) for inviscid, nonconducting, nonradiating flow is that the total enthalpy be constant, that is,

$$I = e + \frac{u_1^2 + u_2^2 + u_3^2}{2} + \frac{p}{\rho} = \text{constant} \tag{12}$$

Hence, Eq. (10) may be replaced by Eq. (12).

The equation of state defines the thermodynamic properties which are functions of two independent thermodynamic variables, e.g. (see §5),

$$e = e(p, \rho) \tag{13}$$

The solution of the inviscid flow field entails solving the above equations for the six unknowns u_1, u_2, u_3, p, ρ, and e as functions of the spatial coordinates x_1, x_2, and x_3, with appropriate boundary conditions. The normal component of velocity must vanish at the body surface, and the flow properties must approach uniform flow values far from the body.

12. Coordinate Systems. The choice of coordinate systems depends on the body shape and the method selected to solve the above equations. The metric coefficients h_1, h_2, and h_3 are defined in terms of a length element as

$$(dl)^2 = h_1^2(dx_1)^2 + h_2^2(dx_2)^2 + h_3^2(dx_3)^2 \tag{14}$$

TABLE 1. Coordinate Systems and Equations

	Rectangular x, y, z
Coordinates	$x_1 = x$ $x_2 = y$ $x_3 = z$
Metric coefficients	$h_1 = 1$ $h_2 = 1$ $h_3 = 1$
Velocity components	$u_1 = u$ $u_2 = v$ $u_3 = w$
Continuity Eq.	$\dfrac{\partial}{\partial x}(\rho u) + \dfrac{\partial}{\partial y}(\rho v) + \dfrac{\partial}{\partial z}(\rho w) = 0$
Momentum Eqs. x_1 direction	$\rho\left(u\dfrac{\partial u}{\partial x} + v\dfrac{\partial u}{\partial y} + w\dfrac{\partial u}{\partial z}\right) + \dfrac{\partial p}{\partial x} = 0$
x_2 direction	$\rho\left(u\dfrac{\partial v}{\partial x} + v\dfrac{\partial v}{\partial y} + w\dfrac{\partial v}{\partial z}\right) + \dfrac{\partial p}{\partial y} = 0$
x_3 direction	$\rho\left(u\dfrac{\partial w}{\partial x} + v\dfrac{\partial w}{\partial y} + w\dfrac{\partial w}{\partial z}\right) + \dfrac{\partial p}{\partial z} = 0$
Energy Eq.	$\rho\left(u\dfrac{\partial e}{\partial x} + v\dfrac{\partial e}{\partial y} + w\dfrac{\partial e}{\partial z}\right) - \dfrac{p}{\rho}\left(u\dfrac{\partial \rho}{\partial x} + v\dfrac{\partial \rho}{\partial y} + w\dfrac{\partial \rho}{\partial z}\right) = 0$
Stream function $\psi(x_1, x_2)$	$\rho u = \rho_\infty \dfrac{\partial \psi}{\partial y}$ $\rho v = -\rho_\infty \dfrac{\partial \psi}{\partial x}$
Potential function	$u = \dfrac{\partial \Phi}{\partial x}$ $v = \dfrac{\partial \Phi}{\partial y}$ $w = \dfrac{\partial \Phi}{\partial z}$
Potential Eq.	$\left[1 - \dfrac{1}{a^2}\left(\dfrac{\partial \Phi}{\partial x}\right)^2\right]\dfrac{\partial^2 \Phi}{\partial x^2} + \left[1 - \dfrac{1}{a^2}\left(\dfrac{\partial \Phi}{\partial y}\right)^2\right]\dfrac{\partial^2 \Phi}{\partial y^2}$ $+ \left[1 - \dfrac{1}{a^2}\left(\dfrac{\partial \Phi}{\partial z}\right)^2\right]\dfrac{\partial^2 \Phi}{\partial z^2} - \dfrac{2}{a^2}\left(\dfrac{\partial \Phi}{\partial x}\dfrac{\partial \Phi}{\partial y}\dfrac{\partial^2 \Phi}{\partial x \partial y}\right.$ $+ \dfrac{\partial \Phi}{\partial y}\dfrac{\partial \Phi}{\partial z}\dfrac{\partial^2 \Phi}{\partial y \partial z} + \left.\dfrac{\partial \Phi}{\partial x}\dfrac{\partial \Phi}{\partial z}\dfrac{\partial^2 \Phi}{\partial x \partial z}\right) = 0$

and, in general, may be functions of all the spatial coordinates [10]. However, for commonly used coordinate systems, the metric coefficients are relatively simple. In Table 1 the metric coefficients and equations of motion are presented for rectangular, cylindrical, and spherical polar coordinates. The cylindrical coordinates x, r, and ϕ are applicable to axisymmetric flow with flow properties independent of the meridian angle ϕ and with $u_3 = v_\phi = 0$. In this case with the exception of the continuity equation, the equations of motion are identical to those for a two-dimensional flow in rectangular coordinates with y and v replaced by r and v_r, respectively.

13. Stream Function. For steady flow, a streamline is the path traced by a fluid particle. Since there is no mass flow across a streamline, a function can be introduced that represents the mass flow between streamlines or stream surfaces. Mathematically, the stream function is obtained by integrating the continuity equation (Eq. (5)), which results in a pair of stream functions for a general three-dimensional flow.

TABLE 1. Coordinate Systems and Equations (*Continued*)

Cylindrical polar $r,\ \theta,\ z$

$x_1 = r$	$x_2 = \theta$	$x_3 = z$
$h_1 = 1$	$h_2 = r$	$h_3 = 1$
$u_1 = v_r$	$u_2 = v_\theta$	$u_3 = w$

$$\frac{\partial}{\partial r}\left(\rho v_r\, r\right) + \frac{\partial}{\partial \theta}\left(\rho v_\theta\right) + \frac{\partial}{\partial z}\left(\rho w r\right) = 0$$

$$\rho\left(v_r\frac{\partial v_r}{\partial r} + \frac{v_\theta}{r}\frac{\partial v_r}{\partial \theta} + w\frac{\partial v_r}{\partial z} - \frac{v_\theta^2}{r}\right) + \frac{\partial p}{\partial r} = 0$$

$$\rho\left(v_r\frac{\partial v_\theta}{\partial r} + \frac{v_\theta}{r}\frac{\partial v_\theta}{\partial \theta} + w\frac{\partial v_\theta}{\partial z} + \frac{v_r v_\theta}{r}\right) + \frac{1}{r}\frac{\partial p}{\partial \theta} = 0$$

$$\rho\left(v_r\frac{\partial w}{\partial r} + \frac{v_\theta}{r}\frac{\partial w}{\partial \theta} + w\frac{\partial w}{\partial z}\right) + \frac{\partial p}{\partial z} = 0$$

$$\rho\left(v_r\frac{\partial e}{\partial r} + \frac{v_\theta}{r}\frac{\partial e}{\partial \theta} + w\frac{\partial e}{\partial z}\right) - \frac{p}{\rho}\left(v_r\frac{\partial \rho}{\partial r} + \frac{v_\theta}{r}\frac{\partial \rho}{\partial \theta} + w\frac{\partial \rho}{\partial z}\right) = 0$$

$$\rho v_r\, r = \rho_\infty \frac{\partial \psi}{\partial \theta} \qquad \rho v_\theta = -\rho_\infty \frac{\partial \psi}{\partial r}$$

$$v_r = \frac{\partial \Phi}{\partial r} \qquad v_\theta = \frac{1}{r}\frac{\partial \Phi}{\partial \theta} \qquad w = \frac{\partial \Phi}{\partial z}$$

$$\left[1 - \frac{1}{a^2}\left(\frac{\partial \Phi}{\partial r}\right)^2\right]\frac{\partial^2 \Phi}{\partial r^2} + \left[1 - \frac{1}{a^2 r^2}\left(\frac{\partial \Phi}{\partial \theta}\right)^2\right]\frac{1}{r^2}\frac{\partial^2 \Phi}{\partial \theta^2} + \left[1 - \frac{1}{a^2}\left(\frac{\partial \Phi}{\partial z}\right)^2\right]\frac{\partial^2 \Phi}{\partial z^2}$$
$$+ \frac{1}{r}\frac{\partial \Phi}{\partial r} - \frac{2}{a^2 r^2}\left(\frac{\partial \Phi}{\partial r}\frac{\partial \Phi}{\partial \theta}\frac{\partial^2 \Phi}{\partial r \partial \theta} + \frac{\partial \Phi}{\partial \theta}\frac{\partial \Phi}{\partial z}\frac{\partial^2 \Phi}{\partial \theta \partial z} + r^2\frac{\partial \Phi}{\partial r}\frac{\partial \Phi}{\partial z}\frac{\partial^2 \Phi}{\partial r \partial z}\right.$$
$$\left. - \frac{1}{r}\frac{\partial \Phi}{\partial r}\frac{\partial^2 \Phi}{\partial \theta^2}\right) = 0$$

For two-dimensional flow in generalized coordinates x_1, x_2 the stream function is defined by

$$h_2 h_3\, \rho u_1 = \rho_\infty \frac{\partial \psi}{\partial x_2} \qquad h_1 h_3\, \rho u_2 = -\rho_\infty \frac{\partial \psi}{\partial x_1} \tag{15}$$

and satisfies Eq. (5) identically. The stream functions for various coordinates are shown in Table 1. The introduction of the stream function reduces by one the number of unknowns at the expense of increasing the order of the differential equations. This stratagem may be helpful in analytical solutions but not necessarily so in numerical solutions.

TABLE 1. Coordinate Systems and Equations (*Continued*)

	Cylindrical polar x, r, φ
Coordinates	$x_1 = x$ $x_2 = r$ $x_3 = \phi$
Metric coefficients	$h_1 = 1$ $h_2 = 1$ $h_3 = r$
Velocity components	$u_1 = u$ $u_2 = v_r$ $u_3 = v_\phi$
Continuity Eq.	$\dfrac{\partial}{\partial x}(\rho\, ur) + \dfrac{\partial}{\partial r}(\rho v_r r) + \dfrac{\partial}{\partial \phi}(\rho v_\phi) = 0$
Momentum Eqs. x_1 direction	$\rho\left(u\dfrac{\partial u}{\partial x} + v_r\dfrac{\partial u}{\partial r} + \dfrac{v_\phi}{r}\dfrac{\partial u}{\partial \phi}\right) + \dfrac{\partial p}{\partial x} = 0$
x_2 direction	$\rho\left(u\dfrac{\partial v_r}{\partial x} + v_r\dfrac{\partial v_r}{\partial r} + \dfrac{v_\phi}{r}\dfrac{\partial v_r}{\partial \phi} - \dfrac{v_\phi^2}{r}\right) + \dfrac{\partial p}{\partial r} = 0$
x_3 direction	$\rho\left(u\dfrac{\partial v_\phi}{\partial x} + v_r\dfrac{\partial v_\phi}{\partial r} + \dfrac{v_\phi}{r}\dfrac{\partial v_\phi}{\partial \phi} + \dfrac{v_r v_\phi}{r}\right) + \dfrac{1}{r}\dfrac{\partial p}{\partial \phi} = 0$
Energy Eq.	$\rho\left(u\dfrac{\partial e}{\partial x} + v_r\dfrac{\partial e}{\partial r} + \dfrac{v_\phi}{r}\dfrac{\partial e}{\partial \phi}\right) - \dfrac{p}{\rho}\left(u\dfrac{\partial \rho}{\partial x} + v_r\dfrac{\partial \rho}{\partial r} + \dfrac{v_\phi}{r}\dfrac{\partial \rho}{\partial \phi}\right) = 0$
Stream function $\psi(x_1,\,x_2)$	$\rho\, ur = \rho_\infty \dfrac{\partial \psi}{\partial r}$ \qquad $\rho v_r r = -\rho_\infty \dfrac{\partial \psi}{\partial x}$
Potential function	$u = \dfrac{\partial \Phi}{\partial x}$ \qquad $v_r = \dfrac{\partial \Phi}{\partial r}$ \qquad $v_\phi = \dfrac{1}{r}\dfrac{\partial \Phi}{\partial \phi}$
Potential Eq.	$\left[1 - \dfrac{1}{a^2}\left(\dfrac{\partial \Phi}{\partial x}\right)^2\right]r\dfrac{\partial^2 \Phi}{\partial x^2} + \left[1 - \dfrac{1}{a^2}\left(\dfrac{\partial \Phi}{\partial r}\right)^2\right]r\dfrac{\partial^2 \Phi}{\partial r^2} + \left[1 - \dfrac{1}{a^2 r^2}\left(\dfrac{\partial \Phi}{\partial \phi}\right)^2\right]\dfrac{1}{r}\dfrac{\partial^2 \Phi}{\partial \phi^2}$ $+ \dfrac{\partial \Phi}{\partial r} - \dfrac{2}{a^2}\left[r\dfrac{\partial \Phi}{\partial x}\dfrac{\partial \Phi}{\partial r}\dfrac{\partial^2 \Phi}{\partial x \partial r} + \dfrac{1}{r}\dfrac{\partial \Phi}{\partial r}\dfrac{\partial \Phi}{\partial \phi}\dfrac{\partial^2 \Phi}{\partial r \partial \phi} + \dfrac{1}{r}\dfrac{\partial \Phi}{\partial x}\dfrac{\partial \Phi}{\partial \phi}\dfrac{\partial^2 \Phi}{\partial x \partial \phi} + \dfrac{1}{r^2}\dfrac{\partial \Phi}{\partial r}\left(\dfrac{\partial \Phi}{\partial \phi}\right)^2\right] = 0$

14. Irrotational Flow and Potential Function. The vorticity is related to the angular momentum of a fluid particle and has components in the three coordinate directions

$$\omega_1 = \frac{1}{h_2 h_3}\left[\frac{\partial}{\partial x_2}(h_3 u_3) - \frac{\partial}{\partial x_3}(h_2 u_2)\right]$$

$$\omega_2 = \frac{1}{h_1 h_3}\left[\frac{\partial}{\partial x_3}(h_1 u_1) - \frac{\partial}{\partial x_1}(h_3 u_3)\right] \tag{16}$$

$$\omega_3 = \frac{1}{h_1 h_2}\left[\frac{\partial}{\partial x_1}(h_2 u_2) - \frac{\partial}{\partial x_2}(h_1 u_1)\right]$$

TABLE 1. Coordinate Systems and Equations (*Continued*)

Spherical polar $\quad r, \theta, \phi$

$x_1 = r$	$x_2 = \theta$	$x_3 = \phi$
$h_1 = 1$	$h_2 = r$	$h_3 = r \sin \theta$
$u_1 = v_r$	$u_2 = v_\theta$	$u_3 = v_\phi$

$$\frac{\partial}{\partial r}\left(\rho v_r\, r^2 \sin\theta\right) + \frac{\partial}{\partial \theta}\left(\rho v_\theta r \sin\theta\right) + \frac{\partial}{\partial \phi}\left(\rho v_\phi\, r\right) = 0$$

$$\rho\left(v_r \frac{\partial v_r}{\partial r} + \frac{v_\theta}{r}\frac{\partial v_r}{\partial \theta} + \frac{v_\phi}{r\sin\theta}\frac{\partial v_r}{\partial \phi} - \frac{v_\theta^2 + v_\phi^2}{r}\right) + \frac{\partial p}{\partial r} = 0$$

$$\rho\left(v_r \frac{\partial v_\theta}{\partial r} + \frac{v_\theta}{r}\frac{\partial v_\theta}{\partial \theta} + \frac{v_\phi}{r\sin\theta}\frac{\partial v_\theta}{\partial \phi} - \frac{v_\phi^2 \cot\theta}{r} + \frac{v_r\, v_\theta}{r}\right) + \frac{1}{r}\frac{\partial p}{\partial \theta} = 0$$

$$\rho\left(v_r \frac{\partial v_\phi}{\partial r} + \frac{v_\theta}{r}\frac{\partial v_\phi}{\partial \theta} + \frac{v_\phi}{r\sin\theta}\frac{\partial v_\phi}{\partial \phi} + \frac{v_r\, v_\phi}{r} + \frac{v_\theta\, v_\phi \cot\theta}{r}\right) + \frac{1}{r\sin\theta}\frac{\partial p}{\partial \phi} = 0$$

$$\rho\left(v_r \frac{\partial e}{\partial r} + \frac{v_\theta}{r}\frac{\partial e}{\partial \theta} + \frac{v_\phi}{r\sin\theta}\frac{\partial e}{\partial \phi}\right) - \frac{p}{\rho}\left(v_r \frac{\partial p}{\partial r} + \frac{v_\theta}{r}\frac{\partial p}{\partial \theta} + \frac{v_\phi}{r\sin\theta}\frac{\partial p}{\partial \phi}\right) = 0$$

$$\rho v_r\, r^2 \sin\theta = \rho_\infty \frac{\partial \psi}{\partial \theta} \qquad \rho v_\theta\, r\sin\theta = -\rho_\infty \frac{\partial \psi}{\partial r}$$

$$v_r = \frac{\partial \Phi}{\partial r} \qquad v_\theta = \frac{1}{r}\frac{\partial \Phi}{\partial \theta} \qquad v_\phi = \frac{1}{r\sin\theta}\frac{\partial \Phi}{\partial \phi}$$

$$\left[1 - \frac{1}{a^2}\left(\frac{\partial \Phi}{\partial r}\right)^2\right] r^2 \sin\theta \frac{\partial^2 \Phi}{\partial r^2} + \left[1 - \frac{1}{a^2 r^2}\left(\frac{\partial \Phi}{\partial \theta}\right)^2\right]\sin\theta \frac{\partial^2 \Phi}{\partial \theta^2}$$

$$+ \left[1 - \frac{1}{a^2 r^2 \sin^2\theta}\left(\frac{\partial \Phi}{\partial \phi}\right)^2\right]\frac{1}{\sin\theta}\frac{\partial^2 \Phi}{\partial \phi^2} + 2r\sin\theta \frac{\partial \Phi}{\partial r} + \cos\theta \frac{\partial \Phi}{\partial \theta}$$

$$- \frac{1}{a^2}\left[2\left(\sin\theta \frac{\partial \Phi}{\partial r}\frac{\partial \Phi}{\partial \theta}\frac{\partial^2 \Phi}{\partial r \partial\theta} + \frac{1}{r^2\sin\theta}\frac{\partial \Phi}{\partial \theta}\frac{\partial \Phi}{\partial \phi}\frac{\partial^2 \Phi}{\partial \theta \partial\phi} + \frac{1}{\sin\theta}\frac{\partial \Phi}{\partial r}\frac{\partial \Phi}{\partial \phi}\frac{\partial^2 \Phi}{\partial r \partial\phi}\right)\right.$$

$$- \frac{\sin\theta}{r}\frac{\partial \Phi}{\partial r}\left(\frac{\partial \Phi}{\partial \theta}\right)^2 - \frac{\cot\theta}{r^2\sin\theta}\frac{\partial \Phi}{\partial \theta}\left(\frac{\partial \Phi}{\partial \phi}\right)^2 - \left.\frac{1}{r\sin\theta}\frac{\partial \Phi}{\partial r}\left(\frac{\partial \Phi}{\partial \phi}\right)^2\right] = 0$$

Irrotational flow is defined as one in which the vorticity is everywhere zero, that is, $\omega_1 = \omega_2 = \omega_3 = 0$. These conditions of irrotational flow are satisfied identically by a potential function defined as

$$u_1 = \frac{1}{h_1}\frac{\partial \Phi}{\partial x_1} \qquad u_2 = \frac{1}{h_2}\frac{\partial \Phi}{\partial x_2} \qquad u_3 = \frac{1}{h_3}\frac{\partial \Phi}{\partial x_3} \tag{17}$$

Substitution of Eqs. (17) into Eqs. (6) to (8) and the elimination of p yields for the potential function

$$\left[1 - \frac{1}{a^2 h_1^2}\left(\frac{\partial\Phi}{\partial x_1}\right)^2\right]\frac{h_2 h_3}{h_1}\frac{\partial^2\Phi}{\partial x_1^2} + \left[1 - \frac{1}{a^2 h_2^2}\left(\frac{\partial\Phi}{\partial x_2}\right)^2\right]\frac{h_3 h_1}{h_2}\frac{\partial^2\Phi}{\partial x_2^2}$$

$$+ \left[1 - \frac{1}{a^2 h_3^2}\left(\frac{\partial\Phi}{\partial x_3}\right)^2\right]\frac{h_1 h_2}{h_3}\frac{\partial^2\Phi}{\partial x_3^2} + \frac{\partial\Phi}{\partial x_1}\frac{\partial}{\partial x_1}\left(\frac{h_2 h_3}{h_1}\right) + \frac{\partial\Phi}{\partial x_2}\frac{\partial}{\partial x_2}\left(\frac{h_3 h_1}{h_2}\right)$$

$$+ \frac{\partial\Phi}{\partial x_3}\frac{\partial}{\partial x_3}\left(\frac{h_1 h_2}{h_3}\right) + \frac{1}{a^2}\left\{\frac{h_2 h_3}{h_1^4}\left(\frac{\partial\Phi}{\partial x_1}\right)^3\frac{\partial h_1}{\partial x_1} + \frac{h_3 h_1}{h_2^4}\left(\frac{\partial\Phi}{\partial x_2}\right)^3\frac{\partial h_2}{\partial x_2} + \frac{h_1 h_2}{h_3^4}\left(\frac{\partial\Phi}{\partial x_3}\right)^3\frac{\partial h_3}{\partial x_3}\right.$$

$$- 2\left[\frac{h_3}{h_1 h_2}\frac{\partial\Phi}{\partial x_1}\frac{\partial\Phi}{\partial x_2}\frac{\partial^2\Phi}{\partial x_1\partial x_2} + \frac{h_1}{h_2 h_3}\frac{\partial\Phi}{\partial x_2}\frac{\partial\Phi}{\partial x_3}\frac{\partial^2\Phi}{\partial x_2\partial x_3} + \frac{h_2}{h_3 h_1}\frac{\partial\Phi}{\partial x_1}\frac{\partial\Phi}{\partial x_3}\frac{\partial^2\Phi}{\partial x_1\partial x_3}\right]$$

$$+ \frac{h_3}{h_1 h_2}\frac{\partial\Phi}{\partial x_1}\frac{\partial\Phi}{\partial x_2}\left[\frac{\partial\Phi}{\partial x_2}\frac{1}{h_2}\frac{\partial h_2}{\partial x_1} + \frac{\partial\Phi}{\partial x_1}\frac{1}{h_1}\frac{\partial h_1}{\partial x_2}\right] + \frac{h_1}{h_2 h_3}\frac{\partial\Phi}{\partial x_2}\frac{\partial\Phi}{\partial x_3}\left[\frac{\partial\Phi}{\partial x_3}\frac{1}{h_3}\frac{\partial h_3}{\partial x_3}\right.$$

$$\left. + \frac{\partial\Phi}{\partial x_2}\frac{1}{h_2}\frac{\partial h_2}{\partial x_3}\right] + \frac{h_2}{h_1 h_3}\frac{\partial\Phi}{\partial x_1}\frac{\partial\Phi}{\partial x_3}\left[\frac{\partial\Phi}{\partial x_1}\frac{1}{h_1}\frac{\partial h_1}{\partial x_3} + \frac{\partial\Phi}{\partial x_3}\frac{1}{h_3}\frac{\partial h_3}{\partial x_1}\right]\right\} = 0 \quad (18)$$

The potential functions and Eq. (18) for various coordinates are shown in Table 1.

For a compressible fluid the condition of irrotational flow is tantamount to a flow with constant entropy everywhere.

Incompressible Flow

Incompressible flow represents the limiting case when the speed of sound is much larger than the flow velocity; i.e., the Mach number is small. Under these conditions, the change in density between the free stream and the stagnation point can be expressed from Bernoulli's equation (Eq. (9)) as $\Delta\rho/\rho_\infty \sim \frac{1}{2}M_\infty^2$. If changes in the density of 2 percent can be neglected, the flow is essentially incompressible for $M_\infty <$ 0.2. Hence, C_p is a function of the body shape alone and not of the free stream properties.

15. Equations. Since the density is a known constant, the continuity and momentum equations suffice to yield the velocity and pressure distributions. The assumption of constant density also implies constant internal energy or temperature as a possible solution to the energy equation (Eq. (10)).

With a uniform free stream, the flow is irrotational so that a potential function can be defined. Equation (18) simplifies with $a^2 \gg u_i^2$ ($i = 1, 2,$ or 3) to

$$\frac{\partial}{\partial x_1}\left(\frac{h_2 h_3}{h_1}\frac{\partial\Phi}{\partial x_1}\right) + \frac{\partial}{\partial x_2}\left(\frac{h_3 h_1}{h_2}\frac{\partial\Phi}{\partial x_2}\right) + \frac{\partial}{\partial x_3}\left(\frac{h_1 h_2}{h_3}\frac{\partial\Phi}{\partial x_3}\right) = 0 \quad (19)$$

which is Laplace's equation in generalized coordinates. The form of this equation for various coordinates can be obtained from Table 1. Laplace's equation is linear so that simple solutions may be superimposed to satisfy the boundary conditions. For example, solutions for sources and sinks may be added to obtain the flow around various shapes.

Bernoulli's equation (Eq. (9)) can be written as

$$\frac{u_1^2 + u_2^2 + u_3^2}{2} + \frac{p}{\rho_\infty} = \frac{V_\infty^2}{2} + \frac{p_\infty}{\rho_\infty} = \frac{p_{st}}{\rho_\infty} \tag{20}$$

If the flow is planar, that is, $h_3 = 1$, and derivatives with respect to x_3 vanish, Eq. (19) reduces to

$$\frac{\partial}{\partial x_1}\left(\frac{h_2}{h_1}\frac{\partial \Phi}{\partial x_1}\right) + \frac{\partial}{\partial x_2}\left(\frac{h_1}{h_2}\frac{\partial \Phi}{\partial x_2}\right) = 0$$

which is identical to the equation obtained by substituting the stream function (Eq. (15)) into Eq. (16) for $\omega_3 = 0$

$$\frac{\partial}{\partial x_1}\left(\frac{h_2}{h_1}\frac{\partial \psi}{\partial x_1}\right) + \frac{\partial}{\partial x_2}\left(\frac{h_1}{h_2}\frac{\partial \psi}{\partial x_2}\right) = 0$$

Hence, for incompressible planar flows, both the stream and potential functions satisfy Laplace's equation. The theory of complex variables and conformal transformation can then be used to solve the flow around various shapes.

Given a potential function that satisfies Laplace's equation (Eq. (19)), the velocity components can be found from Eq. (17), and the pressure coefficient can be obtained from Eqs. (3) and (20)

$$C_p = 1 - \frac{u_1^2 + u_2^2 + u_3^2}{V_\infty^2}$$

16. Exact Solutions for Planar Flow. The flow field around two-dimensional shapes $(h_3 = 1, u_3 = 0, \partial/\partial x_3 = 0)$ may be obtained by either superposition of potential functions for simple flows or conformal transformation [11]. Solutions for several simple shapes will be considered now.

Circular Cylinder. The potential function in cylindrical polar coordinates (Table 1 and Fig. 2) is

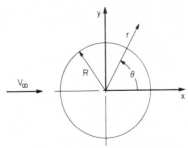

Fig. 2. Coordinate system for flow normal to a circular cylinder or for flow over a sphere.

$$\Phi = V_\infty \left(\frac{r}{R} + \frac{R}{r} \right) R \cos\theta$$

The corresponding velocity and surface pressure distributions are

$$\left(v_\theta\right)_{r=R} = -2V_\infty \sin\theta$$

$$C_p = 1 - 4\sin^2\theta$$

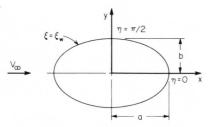

Fig. 3. Coordinate system for flow normal to an elliptic cylinder.

Elliptic Cylinder. The potential function in elliptic coordinates ξ, η ([11, pp. 131--132] and Fig. 3) is

$$\Phi = V_\infty \left[b\sqrt{\frac{a+b}{a-b}} \, e^{-\xi} + \sqrt{a^2 - b^2} \, \cosh\xi \right] \cos\eta$$

where

$$x = \sqrt{a^2 - b^2} \, \cosh\xi \, \cos\eta$$
$$y = \sqrt{a^2 - b^2} \, \sinh\xi \, \sin\eta$$

On the surface of the ellipse, $\xi = \xi_w$ and

$$e^{-\xi_w} = \sqrt{\frac{a-b}{a+b}}$$

$$x = a\cos\eta \qquad y = b\sin\eta$$

The velocity and surface pressure distributions are

$$\left(v_\eta\right)_{\xi=\xi_w} = -\frac{V_\infty(a+b)\sin\eta}{\sqrt{a^2 - (a^2 - b^2)\cos^2\eta}}$$

$$C_p = 1 - \frac{(a+b)^2 \sin^2\eta}{a^2 - (a^2 - b^2)\cos^2\eta}$$

Wedge. The flow over a wedge of half angle θ_w is obtained by conformal transformation [11, pp. 104--105]. The velocity and pressure distributions along the surface expressed in cylindrical polar coordinates (Fig. 4) are

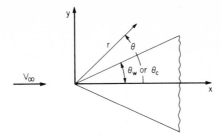

Fig. 4. Coordinate system for flow over a wedge or a cone at zero angle of attack.

$$\left(v_r\right)_{\theta = \theta_w} = cr^{\theta_w/(\pi - \theta_w)}$$

$$C_p = 1 - \frac{c^2 r^{2\theta_w/(\pi - \theta_w)}}{V_\infty^2}$$

The solution is restricted to the tip region because v_r increases without limit for large values of r. As a result, the constant c cannot be expressed in terms of the free-stream conditions and wedge angle but must be determined empirically. These flows are of interest because they provide boundary conditions that permit exact solutions to the boundary-layer equations as will be shown later. Note that for $\theta_w = \pi/2$, or stagnation-point flow, the velocity varies linearly from the stagnation point.

Airfoils. The flows around various wing sections may be obtained by conformal transformation as described in Ref. 12, where solutions are tabulated in Appendix I for several NACA families of airfoils. For example, the velocity and pressure distributions for a typical airfoil are shown in Fig. 5.

17. Exact Solutions for Axisymmetric Flow. The flow field around axisymmetric shapes ($v_\phi = 0$, $\partial/\partial\phi = 0$) may be obtained by superposition of potential functions for simple flows, but conformal transformation is not applicable. Some solutions for simple shapes follow.

Sphere. The potential function for a sphere in spherical polar coordinates (Table 1 and Fig. 2) is

$$\Phi = \frac{V_\infty}{2}\left(\frac{R^2}{r^2} + \frac{r}{R}\right) R \cos\theta$$

The corresponding velocity and pressure distributions are

$$\left(v_\theta\right)_{r = R} = -\frac{3}{2}V_\infty \sin\theta$$

$$C_p = 1 - \frac{9}{4}\sin^2\theta$$

Rankine Body. Superposition of a source and sink of equal strength on a uniform stream results in the flow around a family of closed shapes known as Rankine bodies [11, pp. 57–60]. The potential function in cylindrical coordinates is

$$\Phi = V_\infty x + \frac{m}{4\pi}\left[\frac{1}{\sqrt{(x - x_0)^2 + r^2}} - \frac{1}{\sqrt{(x + x_0)^2 + r^2}}\right]$$

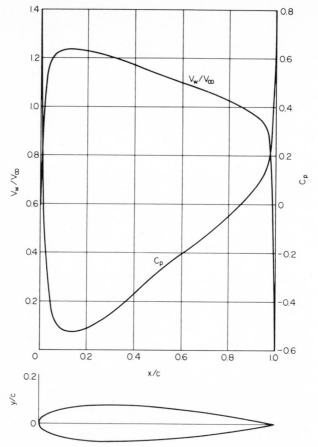

Fig. 5. Velocity and pressure distribution on NACA 0015 wing section, incompressible flow [12].
(*Reprinted with permission from Dover Publications, Inc.*)

where for a body with semiaxes a and b, x_0 is obtained from the solution of

$$\frac{(a^2 - x_0^2)^2}{a} = b^2 \sqrt{b^2 + x_0^2}$$

and the strength of the source or sink m is obtained from

$$\frac{m}{V_\infty} = \frac{\pi b^2}{x_0} \sqrt{b^2 + x_0^2}$$

The procedure for finding the body coordinates and surface velocity distribution may be found in Ref. 11.

18. Approach for General Configuration. Although the equations governing incompressible flow are linear (§15), solutions are not readily obtainable because of the difficulty in satisfying the boundary conditions.

Exact numerical solutions are now possible through the use of high-speed computers. A computer program developed recently for calculating exactly the flow around arbitrary body shapes is described in Ref. 221. Excellent agreement is obtained with analytic solutions and with experimental results.

Approximate solutions of the equations governing incompressible flow may be obtained by simplifying the boundary conditions as, for example, by applying them on the chord line of an airfoil rather than on the surface [13]. This approximation is valid for thin airfoils where the velocity perturbations relative to the free stream are, in general, small. Development of this technique to predict pressure distributions on airfoils including angle of attack effects is discussed in Ref. 12.

Subsonic Flow

Subsonic flow, as defined here, covers the regime where the fluid can no longer be considered incompressible, and the free-stream Mach number is less than a critical value which depends on the body shape and for which the velocity at some point on the surface becomes sonic. The compressibility of the fluid is characterized by the ratio of specific heats and Mach number. Hence, C_p is a function of the body shape, γ, and M_∞.

19. Equations. With a uniform free stream, the flow is irrotational and isentropic in the absence of shock waves. It is possible to define a potential function, but the resulting differential equation (Eq. (18)) is nonlinear and not readily tractable. The basic continuity and momentum equations (Eq. (5) to (8)) apply for subsonic flow without any simplifications. The energy equation is not required in solving for the velocity and pressure distributions since the condition of isentropic flow automatically satisfies Eq. (10). The temperature distribution may be obtained from the adiabatic flow relation (Eq. (12)). Given the pressure coefficient distribution along the surface, the perfect gas and isentropic flow relations required to calculate the other flow properties are summarized in Table 2. See Ref. 14 for a complete set of equations and tables.

20. Approximate Compressibility Corrections. The lack of exact solutions necessitates approximate methods which yield corrections for compressibility effects that may be applied to either incompressible flow theory or experimental data [8, N. Rott: Chap. 75, Subsonic flow], [12, Chap. 9], [15, W. R. Sears: Sec. C, Small perturbation theory]. The simplest type of rule relates flow at different Mach numbers over the same configuration. Other similarity rules relate flow over geometrically affine shapes.

The Prandtl-Glauert rule for planar flow is obtained from linearized theory which assumes small flow disturbances and yields a linear equation for the perturbation velocity potential. The pressure coefficient for subsonic flow around a planar body is related to the value for incompressible flow as follows: At any location on a given profile, C_p is increased by the factor $1/\sqrt{1 - M_\infty^2}$, that is,

$$\frac{C_p}{\left(C_p\right)_{M_\infty = 0}} = \frac{1}{\sqrt{1 - M_\infty^2}} \tag{21}$$

In view of the assumptions, this rule is applicable to thin profiles at low Mach numbers.

The Kármán-Tsien rule for planar flow [16] is obtained by transforming the equations of motion to the hodograph plane with the velocity components as the independent variables and by making the tangent-gas approximation which assumes that $dp/d(1/\rho)$ is a constant evaluated at free-stream conditions. The equations for

TABLE 2. Compressible Flow Relations for Surface Streamline

Static pressure

$$\frac{p_w}{p_\infty} = 1 + \frac{1}{2}\, \gamma M_\infty^2 C_p$$

Stagnation pressure

Subsonic flow

$M_\infty < 1$

$$\frac{p_{st}}{p} = \left(1 + \frac{\gamma-1}{2} M^2\right)^{\gamma/(\gamma-1)}$$

Supersonic flow with attached shock wave

$M_\infty > 1$

Use subsonic flow relation with p, M obtained from wedge or conical flow at tip.

Supersonic flow with detached shock wave

$M_\infty > 1$

Use $C_{p_{st}}$ from Fig. 7 or

$$\frac{p_{st}}{p_\infty} = \left[\frac{(\gamma+1)M_\infty^2}{2}\right]^{\gamma/(\gamma-1)} \left[\frac{\gamma+1}{2\gamma M_\infty^2 - (\gamma-1)}\right]^{1/(\gamma-1)}$$

Flow properties required for convective heating calculations

	Assumptions
$M = \sqrt{\dfrac{2}{\gamma-1}\left[\left(\dfrac{p_{st}}{p}\right)^{(\gamma-1)/\gamma} - 1\right]}$	Isentropic flow
$T = T_\infty \left(\dfrac{1 + \dfrac{\gamma-1}{2} M_\infty^2}{1 + \dfrac{\gamma-1}{2} M^2}\right)$	Adiabatic flow
$\rho = p / RT$	Thermally perfect gas
$V = M\sqrt{\gamma RT}$	Calorically perfect gas

compressible flow are then reduced to those for an equivalent incompressible flow over a slightly different profile. The pressure coefficients are related as follows

$$\frac{C_p}{\left(C_p\right)_{M_\infty=0}} = \frac{1}{\sqrt{1 - M_\infty^2} + \dfrac{M_\infty^2}{1 + \sqrt{1 - M_\infty^2}}\dfrac{\left(C_p\right)_{M_\infty=0}}{2}} \tag{22}$$

The effects of the tangent-gas approximation and the change in profile tend to cancel each other so that Eq. (22) can be used directly to calculate the compressibility correction for a given profile. For a given value of $\left(C_p\right)_{M_\infty=0}$, the value of C_p for $M_\infty > 0$ may be calculated from Eq. (22) or obtained from Fig. 6 by following a constant $\left(C_p\right)_{M_\infty=0}$ line. (The limiting line indicates where the velocity becomes sonic on the surface.) For small values of $\left(C_p\right)_{M_\infty=0}$, Eq. (22) reduces to the Prandtl-Glauert rule (Eq. (21)). In general, the Kármán-Tsien rule is valid for thin profiles and is widely used.

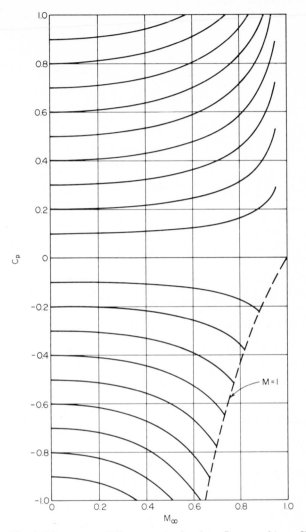

Fig. 6. Kármán-Tsien compressibility correction for planar flow over thin profiles.

Supersonic Flow

Supersonic flow is defined as the regime where the free-stream Mach number is greater than unity. The flow field is characterized by a shock wave at the leading edge or nose of the body, and in general, this bow wave is curved and decays to a Mach wave at a large distance from the body. (The Mach wave is the result of a vanishingly weak disturbance and its inclination to the free stream is given by $\sin^{-1} 1/M_\infty$.) For blunted bodies the bow wave is detached, and for pointed bodies it may be either attached or detached. In reality all pointed bodies are blunted because of machining limitations and viscous effects, but the effects of bluntness may be neglected in many cases. The strength of the shock is measured by the increase in entropy or decrease in total pressure across the shock. The total pressure ratio p_{st}/p_{t_∞} is a minimum for a normal shock and increases to unity as the shock decays to a Mach wave. For a supersonic flow the surface pressure coefficient is a function of the body shape, γ, and M_∞. One other input, e.g., the stagnation pressure for the surface stream-line, is required to completely define the flow properties at the edge of the boundary layer.

21. Equations. For the general case of a curved bow wave, the vorticity is not zero between the body and shock wave. Thus, the flow is not irrotational, and a potential function cannot be used in an exact solution. The basic continuity, momentum, and energy equations (Eqs. (5) to (8) and (10)) are applicable in regions where the flow properties are continuous. The flow is adiabatic across shock waves and isentropic on streamlines on either side for the inviscid fluid being considered in this section. These conditions may be used in lieu of the energy equation (Eq. (10)). The Rankine-Hugoniot relations [14] must be used across the bow shock and any imbedded shocks caused by a compression corner or boundary-layer separation.

Given the pressure-coefficient distribution, the perfect gas and isentropic flow relations required to find the other flow properties on a body streamline are summarized in Table 2 (see Ref. 14 for tabulated values). In general, the stagnation pressure for a particular streamline is determined by the shock angle where the streamline crosses the shock wave. For plane or axisymmetric flow over a pointed nose, the surface stagnation pressure is equal to the value for wedge or conical flow at the tip. For a rounded-nose body with a detached bow wave, the stagnation pressure for a body streamline is equal to its value behind a normal shock. The stagnation-point pressure coefficient for $\gamma = 1.4$ is shown in Fig. 7.

22. Simple Flows. Exact solutions exist for a few simple supersonic flows. In addition to providing answers to actual flows, these solutions can either be used in approximate methods or to verify them.

Wedge. Superposition of a transverse velocity onto a normal shock wave yields conditions behind an oblique shock wave where the free stream is deflected by an

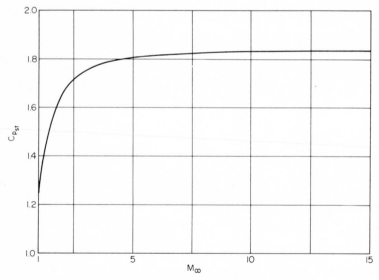

Fig. 7. Stagnation point pressure coefficient, $\gamma = 1.4$.

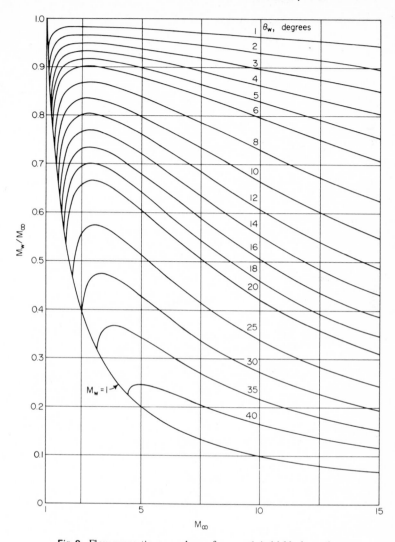

Fig. 8. Flow properties on wedge surface, $\gamma = 1.4$. (*a*) Mach number.

angle θ_w (Fig. 4). The equations together with solutions presented as tables and charts may be found in Refs. 14 and 17. These results are applicable to the flow over a wedge provided the Mach number behind the shock is greater than unity. The Mach number and pressure coefficient on the wedge for $\gamma = 1.4$ are shown in Fig. 8 as functions of θ_w and M_∞. For a given value of M_∞, the flow becomes subsonic for a limiting deflection angle, and the shock wave detaches for slightly larger deflection angles. The wedge flow results are also applicable to wings with wedge cross sections that may be asymmetric or at angle of attack provided supersonic flow is maintained on both surfaces. The C_p and M_w are used in the equations of Table 2 to define local boundary-layer edge properties.

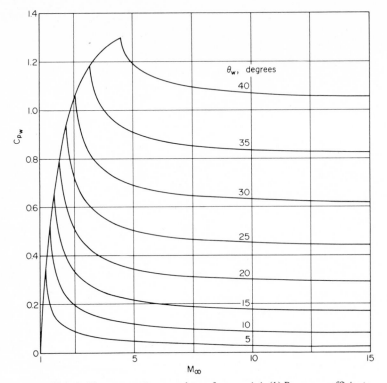

Fig. 8 (*Cont.*).   Flow properties on wedge surface, $\gamma = 1.4$. (*b*) Pressure coefficient.

Cone. With the equations of motion written in terms of spherical polar coordinates (Table 1 and Fig. 4), conical flow occurs when the flow properties are independent of the radial distance r and are just functions of θ. The shock wave and the cone surface are formed by rays with $\theta = \theta_s$ and θ_c, respectively. Conical flow solutions were first obtained by Taylor and MacColl and later by Kopal [18]. See Refs. 14 and 19 for tables and charts. The surface pressure coefficient and Mach number for $\gamma = 1.4$ are shown in Fig. 9 as functions of θ_c and M_∞. Similarly to the flow over a wedge, the flow becomes subsonic for a limiting cone angle, and the shock wave detaches for slightly larger cone angles. The equations of Table 2 are used with the information from these charts to define boundary-layer edge properties.

Prandtl-Meyer Flow. A Prandtl-Meyer expansion occurs when a uniform planar flow ($M \geq 1$) expands around a sharp corner. With the equations of motion written in cylindrical polar coordinates (see Table 1), the flow properties are independent of the radial distance r and just functions of θ. The Prandtl-Meyer angle is defined as [14]

$$\nu = \sqrt{\frac{\gamma + 1}{\gamma - 1}} \ \tan^{-1}\sqrt{\frac{\gamma - 1}{\gamma + 1}(M^2 - 1)} - \cos^{-1}\left(\frac{1}{M}\right)$$

and is shown in Fig. 10 as a function of the Mach number for $\gamma = 1.4$.

Given an expansion of angle $\Delta\theta$ (Fig. 11), the final and initial Mach numbers M_f and M_i are related to their respective Prandtl-Meyer angles by

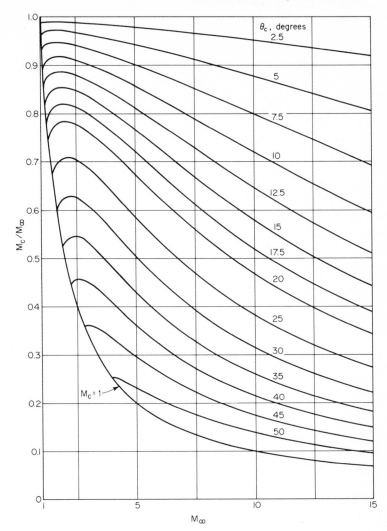

Fig. 9. Flow properties on the surface of a cone at zero angle of attack, $\gamma = 1.4$. (a) Mach number.

$$\nu(M_f) \;=\; \nu(M_i) \;+\; \Delta\theta \tag{23}$$

The other flow properties can be calculated from the isentropic flow relationships in Table 2 (or see Ref. 14).

Yawed Cylinder. The concept advanced by Jones [20], which is not limited to just supersonic flow, may be used to solve the flow over a yawed cylinder of infinite length as shown in Fig. 12. Superposition of a transverse velocity w_∞ onto a planar flow in the x, y plane results in the flow over a cylinder yawed at an angle Λ. The surface pressure distribution is undisturbed by this Galilean transformation of the

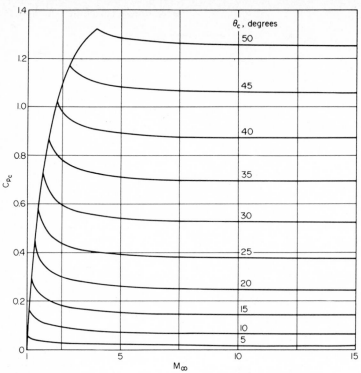

Fig. 9 (*Cont.*). Flow properties on the surface of a cone at zero angle of attack, $\gamma = 1.4$. (*b*) Pressure coefficient.

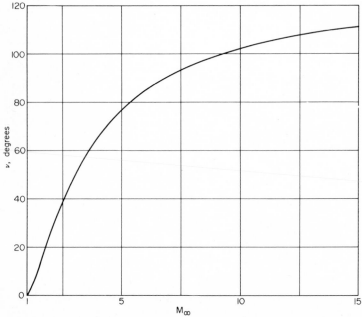

Fig. 10. Relationship between the Prandtl-Meyer angle and the local Mach number, $\gamma = 1.4$.

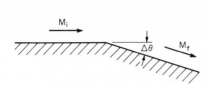

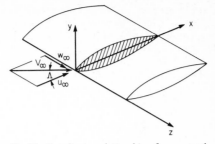

Fig. 11. Expansion angle in Prandtl-Meyer flow.

Fig. 12. Coordinate relationships for a yawed cylinder of infinite length.

axes. In order to utilize this concept, the free-stream velocity V_∞ is resolved into x and z components, normal and tangent to the leading edge, respectively, that is,

$$u_\infty = V_\infty \cos \Lambda \qquad w_\infty = V_\infty \sin \Lambda$$

The flow in the x, y plane is then solved as a two-dimensional problem. The resultant velocity is obtained by adding vectorially $u(x, y)$, $v(x, y)$, w_∞ which results in a three-dimensional streamline pattern with

$$V = \sqrt{u^2 + v^2 + w_\infty^2}$$

23. Approximate Methods. The impossibility of exact solutions for flow fields in general necessitates the use of approximate methods. In many problems these methods provide rapid answers with sufficient engineering accuracy.

Newtonian flow is the flow model originally proposed by Newton for blunt bodies in incompressible flow and which has been found applicable in supersonic flow [21, p. 129]. The pressure on a surface inclined to the stream by an angle θ_w is assumed merely to be the result of the destruction of the normal component of the free-stream momentum, that is,

$$C_p = 2 \sin^2 \theta_w$$

This concept is correct for the limiting case of $\gamma = 1$ and $M_\infty = \infty$ where the bow wave lies on the surface.

Modified Newtonian theory includes a factor to insure the correct pressure at the stagnation point and is expressed as

$$C_p = \frac{p_{st} - p_\infty}{\frac{1}{2} \rho_\infty V_\infty^2} \sin^2 \theta_w = C_{p_{st}} \sin^2 \theta_w \qquad (24)$$

where $C_{p_{st}}$ is shown in Fig. 7 for $\gamma = 1.4$.

Modified Newtonian-Busemann theory includes a correction for the difference in pressure between the shock wave and surface due to the centrifugal force and is expressed as

$$C_p = C_{p_{st}} \left(1 - \frac{k + 3}{k + 2} \cos^2 \theta_w \right) \qquad (25)$$

where $k = 1$ for two-dimensional flow and $k = 2$ for axisymmetric flow.

The tangent wedge and cone method assumes that the local surface pressure on a two- or three-dimensional body is equal to the pressure on a wedge or cone, respectively, with the same surface inclination [21, pp. 522–526]. The surface pressure distribution is then just a function of the local surface angle and the free-stream conditions and is obtained from either Fig. 8 or Fig. 9. The surface stagnation pressure depends on whether the nose is pointed or rounded as shown in Table 2. This method, valid for a thin shock layer with nearly constant pressure and flow angles across the layer, is exact for the limiting case of $\gamma = 1$ and $M_\infty = \infty$.

The shock expansion method is applicable to pointed bodies with attached bow waves. It neglects the interaction of waves within the shock layer (see Ref. 22 and Ref. 8, A. J. Eggers, Jr., C. A. Syvertson: Chap. 78, Hypersonic Flow). The flow properties at the vertex are obtained from a wedge or conical flow solution (Fig. 8 or 9) depending on whether the flow is two-dimensional or axisymmetric. The flow properties along the surface are then found by assuming the surface to be composed of a series of expansion corners which in the limit form a continuous surface. The results for a Prandtl-Meyer expansion (Fig. 10 and Eq. (23)) can then be used to obtain the surface pressure distribution by substituting for $\Delta\theta$, the difference between the vertex and local surface angle.

The blast-wave analogy assumes that the flow behind the bow wave of a blunt-nosed plane or axisymmetric body is analogous to the flow behind a plane or cylindrical blast wave with the x and y coordinates corresponding to time and distance from the blast center, respectively [21, pp. 74–92]. The decay in pressure with distance x downstream of a blunted nose is thus similar to the decay of pressure with time behind the blast wave. The method is not applicable either in the nose region where the details of the flow are important or far downstream where the shock wave decays to a Mach wave. The blast wave analogy provides the form of correlation equations to be fitted with empirical constants. For example, the pressure distribution over cylindrical bodies with various nose shapes has been correlated in Ref. 23 and generalized in Ref. 24 to include the effect of different values of γ as follows

$$\frac{p_w}{p_\infty} = \frac{0.85\, f_1(\gamma)\, M_\infty^{\,2} C_D^{\,1/2}}{x/d} + 0.55$$

where

$$f_1(\gamma) = 0.098\,(\gamma - 1)^{0.395}$$

and d is the cylinder diameter. The effect of the nose shape appears only in its drag coefficient C_D. This equation may be used for real gas flows with dissociated species by substituting the value of γ for stagnation point conditions [25].

Small perturbation theory is applicable as in subsonic flow to slender bodies where the velocity disturbances are small compared with the free-stream velocity. Application of this theory to two- and three-dimensional and axisymmetric flows is discussed in Ref. 8, (M. D. Van Dyke, Chap. 77, Supersonic Flow) and Ref. 15 (M. A. Heaslet and H. Lomax, Sec. D, Supersonic and Transonic Small Perturbation Theory). A useful result is that C_p for a sharp two-dimensional shape is dependent only on the local surface inclination θ_w relative to the free stream [26], that is,

$$C_p = \frac{2\theta_w}{\sqrt{M_\infty^2 - 1}}$$

24. Exact Numerical Solutions. Numerical methods for solving the flow field around arbitrary body shapes are discussed in Refs 21 and 15 (A. Ferri: Sec. G,

The Method of Characteristics). Computer programs developed at the NASA Ames Research Center for this purpose are described in Ref. 27. A brief discussion of the calculation methods follows.

The programs solve numerically the exact equations of motion (Eqs. (5) to (7) and (10)) for the flow of a perfect gas or a real gas in thermodynamic equilibrium about plane or axisymmetric bodies at zero angle of attack. For blunt-nosed bodies an inverse method is used for the subsonic-transonic region wherein the calculations proceed from an assumed shock-wave shape to the corresponding body shape. A one-parameter family of shock shapes [28] in effect permits a direct solution of the problem for spherical and ellipsoidal nose shapes for $y = 1.1$ to 1.6667 and for $M_\infty > 4$. Numerical difficulties associated with the method preclude solutions for lower Mach numbers. The inverse method for the blunted nose is used to provide data along a line joining the body and the shock wave and on which the flow is supersonic. Corresponding data for sharp-nosed bodies may be obtained from either wedge or conical flow solutions.

The method of characteristics is used to continue the calculations into the supersonic region. This method utilizes the property that along characteristic or Mach lines, the equations of motion reduce to ordinary differential equations. These equations, in conjunction with the energy equation (Eq. (12)) and the condition that the flow be isentropic on streamlines, can be integrated numerically commencing with flow conditions along a noncharacteristic line and with a prescribed body shape. The presence of expansion or compression corners and imbedded shocks requires special treatment as described in Ref. 27.

Flow fields around various configurations have been calculated by the above and other similar programs for a wide range of free-stream conditions. Comparisons of computed and experimental pressure distributions show good agreement when boundary-layer displacement effects are negligible. Pressure distributions for a number of simple shapes are presented in this section together with correlation equations and approximate methods wherever applicable.

Hemicylinder. The pressure distributions for $y = 1.4$ and $M_\infty = 4$--10 can be correlated by a trigonometric function similar to Eq. (25)

$$\frac{C_p}{C_{p_{st}}} = 1 - 1.02 \sin^2 \frac{s}{R_n} + 0.120 \sin^4 \frac{s}{R_n} \qquad (26)$$

This equation is not valid near the shoulder as shown in Fig. 13a because of the effect of M_∞ in numerical solutions (dashed lines). Modified Newtonian theory is very good over the forward portion but predicts free-system pressure at the shoulder. The addition of the centrifugal correction makes the comparison worse. (See Refs. 29 and 30 for additional pressure distributions.)

The stagnation-point velocity gradient corresponding to Eq. (26) is given by

$$\frac{R_n}{V_\infty} \frac{dV}{ds} = \frac{1.43}{V_\infty} \sqrt{\frac{p_{st} - p_\infty}{\rho_{st}}} \qquad (27)$$

This result is essentially the same as for modified Newtonian theory.

Hemisphere. The pressure distribution for $y = 1.4$ and $M_\infty = 3$--10 can be correlated by an equation similar to Eq. (26)

$$\frac{C_p}{C_{p_{st}}} = 1 - 1.19 \sin^2 \frac{s}{R_n} + 0.235 \sin^4 \frac{s}{R_n} \qquad (28)$$

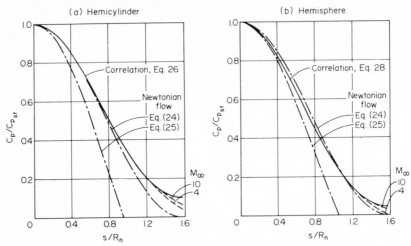

Fig. 13. Pressure distribution on hemicylinder and hemisphere, $\gamma = 1.4$.

Again, the equation is not valid near the shoulder as shown in Fig. 13b because of the effect of M_∞ in the numerical solutions (dashed lines). Modified Newtonian theory predicts pressures on the forward portion either higher or lower than Eq. (28) depending on whether or not the centrifugal correction is included. See Ref. 31 for additional pressure distributions. Real gas solutions [32] may be correlated by an equation similar to Eq. (28) [25].

The stagnation-point velocity gradient is correlated by

$$\frac{R_n}{V_\infty} \frac{dV}{ds} = \frac{1.54}{V_\infty} \sqrt{\frac{p_{st} - p_\infty}{\rho_{st}}} \tag{29}$$

Circularly Blunted Wedge. The pressure distributions for $\gamma = 1.4$ and $M_\infty = 4, 6, 10$ are shown in Fig. 14 for a blunted flat plate and for wedges with half-angles from 10 to 30°. For the cylindrical portion, the pressure distribution is identical to that shown in Fig. 13a. On the flat portion there is a slight recompression, and then the pressure decays slowly to the sharp wedge value indicated at the right edge of the figures. These results may be applied to blunted wedges at angle of attack or to asymmetric wedges provided the subsonic region is confined to the circular nose and not affected by the afterbody [30].

Spherically Blunted Cone. The pressure distributions for $\gamma = 1.4$ and $M_\infty = 4, 6, 10$ are shown in Fig. 15 for a hemisphere cylinder and for cones with half-angles from 10 to 40°. For the spherical portion, the pressure distribution is idential to that shown in Fig. 13b. For small cone angles, the pressure on the conical portion decays slowly to the sharp cone value indicated at the right edge of the figures. For large cone angles, the pressure rises quickly to the sharp cone value. Additional solutions may be found in Refs. 33 and 34.

Cone Cylinder. The conical portion is not affected by the cylindrical afterbody, and the surface Mach number and pressure coefficient may be obtained from Fig. 9. The pressure on the cylinder immediately behind the corner may be calculated from a Prandtl-Meyer expansion. Downstream of this location the surface pressure approaches asymptotically the free-stream pressure in a way determinable only by

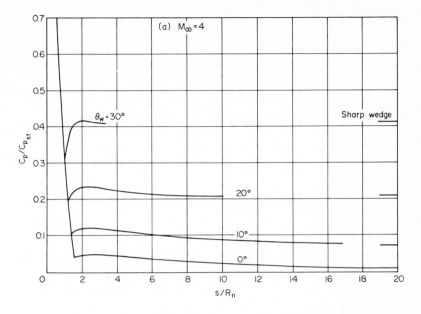

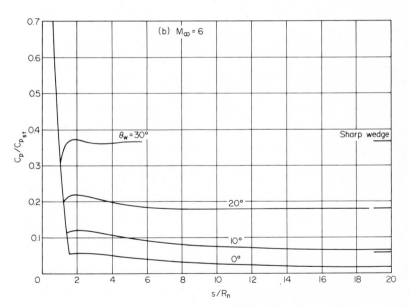

Fig. 14. Pressure distribution on circularly blunted wedge, $\gamma = 1.4$.

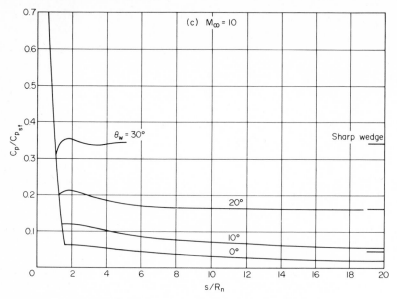

Fig. 14 (*Cont.*). Pressure distribution on circularly blunted wedge, $\gamma = 1.4$.

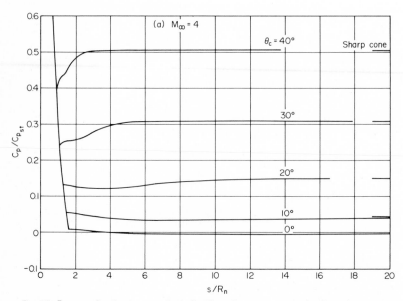

Fig. 15. Pressure distribution on spherically blunted cone at zero angle of attack, $\gamma = 1.4$.

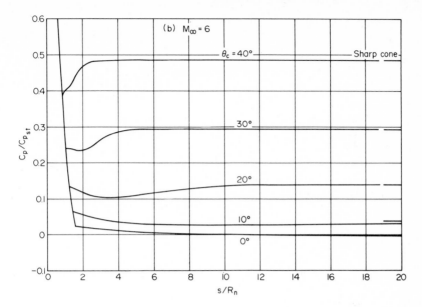

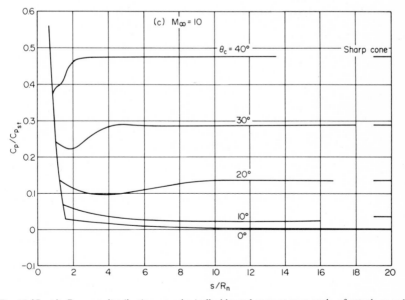

Fig. 15 (*Cont.*). Pressure distribution on spherically blunted cone at zero angle of attack, $\gamma = 1.4$.

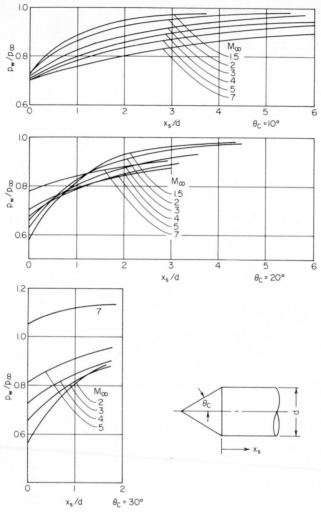

Fig. 16. Pressure distribution on the cylindrical portion of a cone-cylinder at zero angle of attack, $\gamma = 1.4$ [35].

numerical calculations [27]. Results from Ref. 35 for $\theta_c = 10°$, $20°$, $30°$ and $M_\infty = 1.5$ to 7 are shown in Fig. 16.

Ogive Cylinder. Numerical solutions may be obtained by the method of characteristics beginning with a cone solution at the vertex [27]. Surface pressures for $3 \le M_\infty \le 12$ and fineness ratios $3 \le l/d \le 12$ have been correlated [36] using the hypersonic similarity rule [37] where $K = M_\infty d/l$ is the similarity parameter. Note that ogive cylinders of different length are not truly affinely related bodies. The surface pressure is shown in Fig. 17 as a function of K for various x/l locations on the ogive and cylinder. The shock expansion method may also be used to predict the surface pressures for $K > 1$ [38].

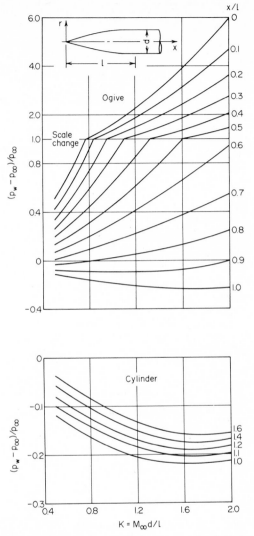

Fig. 17. Pressure distribution on an ogive-cylinder at zero angle of attack, $\gamma = 1.4$ [36].

25. Approximate Numerical Methods. The preceding numerical solutions have been limited to body shapes that are either pointed or smooth at the nose and to flow fields that can be expressed in terms of two spatial coordinates—plane or axisymmetric flow. Therefore, various approximate methods have been proposed for more general body shapes [21]. A few will be mentioned here.

The method of integral relations for inviscid flow is analogous to the boundary-layer integral method discussed in §35. The partial differential equations (Eqs. (5) to (7)) for inviscid flow may be reduced to ordinary differential equations by assuming the functional variation of the flow properties in one of the coordinate directions. This method was proposed by Dorodnitsyn and used by Belotserkovskii [39] to solve

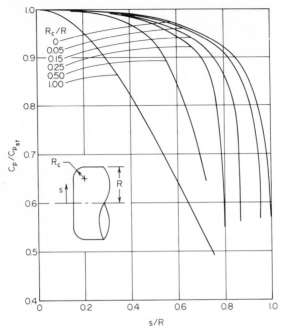

Fig. 18. Pressure distribution on a flat-faced cylinder with a rounded shoulder, air, $M_\infty = 10.5$, $p_\infty = 4.09$ lb/ft^2, $T_\infty = 90.7°$F, $\alpha = 0$ [42].

the flow in the subsonic region of blunt bodies. Generally, the shock layer is divided into strips, and the flow properties between the body and shock wave are represented by polynomials. The number of strips determines the degree of the polynomial. For one strip the polynomial is a straight line; for two strips it is a quadratic, etc. With this approximation the problem reduces to the integration of ordinary differential equations along the surface with initial conditions prescribed along the stagnation line. The number of equations and the number of initial conditions are both proportional to the number of strips. Some of the initial conditions must be iterated to obtain a solution for a given body shape. As a result, the method becomes quite complex for more than a few strips.

Solutions for various shapes may be found in Refs. 39 to 41. Calculations were made in Ref. 42 for flat-faced cylinders with various shoulder radii for $M_\infty = 10.5$. Surface pressures from two-strip solutions are shown in Fig. 18. A computer program and results applicable to blunted cones with sonic corners are described in Ref. 43.

The method of small perturbations may be used to calculate the flow around axisymmetric bodies at small angle of attack. The flow properties are assumed to differ by only small linear perturbations from their values at zero angle of attack. Solutions were obtained numerically for sharp cones [44, 45]. The linearized method of characteristics incorporates the same assumptions to calculate the supersonic flow over axisymmetric bodies at small angles of attack. A computer program developed at the NASA Ames Research Center [46] has been shown to be useful for sharp-nosed and spherically blunted bodies at angles of attack less than 5°. Since flow properties in the supersonic region of the shock layer are required as input, the method is at present limited to sharp noses where these properties may be obtained by the method of Ref. 44 and to spherical noses where they may be obtained from the axisymmetric solution [27]. The calculations are performed in the plane of symmetry; the variation

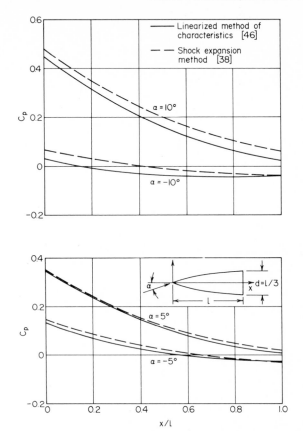

Fig. 19. Surface pressures in the symmetry plane of an ogive, fineness ratio l/d = 3 and γ = 1.4, M_∞ = 5.05 [46].

of properties between the leeward and windward planes is assumed to be either a cosine or sine function of the meridian angle ϕ.

Surface pressures in the symmetry plane for an ogive of fineness ratio l/d = 3 are shown in Fig. 19 for M_∞ = 5.05. For $\alpha = \pm 5°$, the linearized method of characteristics [46] yields pressures in agreement with the shock expansion method [38] and experiment. For $\alpha = \pm 10°$, the results of Ref. 46 are slightly lower than the predictions of Ref. 38, which agree with experiment.

Surface pressures in the symmetry plane for a spherically blunted $30°$ cone are shown in Fig. 20 for M_∞ = 10 and $\alpha = \pm 5°$. Fair agreement is indicated with the results of a three-dimensional method of characteristics program ($\S$ 26).

26. Approach for General Configuration. The problem of solving the flow field around an arbitrary three-dimensional configuration is a formidable one, even with the advanced computing machines available today. The added dimension requires considerably more computer time and storage. Restriction of the problem to an axisymmetric body at angle of attack permits little simplification. Consequently, approximate methods in $\S$ 23 and $\S$ 25 are favored for rapid engineering answers.

For the nose region, modified Newtonian theory ($\S$ 23) has proven useful for predicting pressures on surfaces inclined at least $20°$ to the free stream. An empirical method utilizing a circular shock profile and correlations for the shock standoff

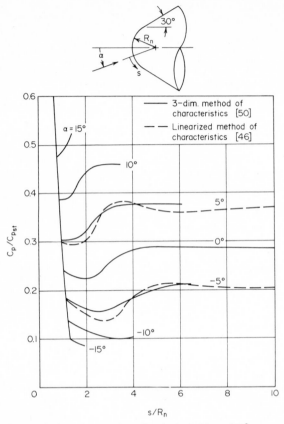

Fig. 20. Surface pressures in the symmetry plane of a spherically blunted $30°$ cone, $\gamma = 1.4$, $M_\infty = 10$.

distance and surface pressure has been quite successful in predicting pressure and shock shapes for a variety of configurations [47]. Recently, several numerical methods have been reported that solve the unsteady flow equations and obtain in the limit, the steady flow about axisymmetric bodies at angle of attack [48, 49].

For the afterbody region, the shock expansion and tangent wedge and cone methods (§ 23) may be used to predict surface pressures. The method of characteristics is applicable to three-dimensional flow but only recently has been incorporated in a program with reasonable computing times and storage requirements [50]. For example, the surface pressure distribution in the symmetry plane of a spherically blunted $30°$ cone is shown in Fig. 20 for $M_\infty = 10$ and angles of attack up to $\pm 15°$. The circumferential variations in cross flow angle and surface pressure at three axial stations for $\alpha = 10°$ are shown in Fig. 21.

STEADY-STATE BOUNDARY-LAYER EQUATIONS

Partial Differential Equations

The behavior of a multicomponent fluid within a boundary layer is governed by a system of partial differential equations representing the conservation of overall mass,

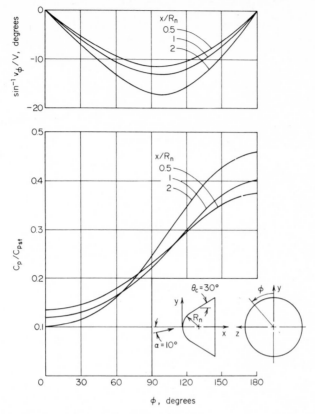

Fig. 21. Circumferential variations in cross flow angle and surface pressure on a spherically blunted cone at angle of attack, $\gamma = 1.4, M_\infty = 10$.

individual species, momentum in three directions, and energy. These equations are expressed here in terms of generalized three-dimensional orthogonal coordinates where a line element in space is given by Eq. (14) [51, 52]. In the boundary-layer expressions that follow, x_2 is always the coordinate normal to the body surface.

These equations are based on the standard order of magnitude arguments of boundary-layer theory [53] involving the following physical assumptions:

1. The boundary-layer thickness is very small compared with a characteristic length of the body.
2. The velocity component normal to the surface is much smaller than the velocity components parallel to the surface.
3. Transport coefficients (diffusion coefficients, viscosity, and thermal conductivity) are very small quantities, i.e., the Reynolds number is large.

Consistent with assumption 1, the metric coefficients (h_1, h_2, h_3) can be treated as being independent of x_2.

In addition to the usual molecular transport terms, the equations also contain terms that are introduced by turbulence, where dependent variables are the sum of a time average value and a randomly fluctuating value, for example,

$$u_1 = \bar{u}_1 + u_1'$$

The resulting turbulence terms are correlations in time of two or more of the randomly varying quantities. Details of the derivation of these equations are given in Ref. 54. The statistical correlations of the fluctuating quantities that are neglected in the turbulent transport terms are triple correlations, derivatives of correlations parallel to the surface, and correlations involving fluctuating components of molecular transport mechanisms, for example, μ' or k'.

27. Overall Continuity of Mass. The conservation of total mass flowing into and out of a unit volume at steady state is

$$\frac{\partial}{\partial x_1}(h_2 h_3 \bar{\rho}\bar{u}_1) + \frac{\partial}{\partial x_2}(h_1 h_3 \overline{\rho u_2}) + \frac{\partial}{\partial x_3}(h_1 h_2 \bar{\rho}\bar{u}_3) = 0 \tag{30}$$

where

$$\overline{\rho u_2} = \bar{\rho}\bar{u}_2 + \overline{\rho' u_2'} \tag{31}$$

Since $\bar{u}_2$, the velocity normal to the surface, is small, the statistical correlation represented by the second term on the right side of Eq. (31), although small itself, cannot be neglected as are comparable quantities in the other terms of Eq. (30).

28. Conservation of Species or Diffusion Equation. The conservation of species i in a multicomponent system is represented by

$$\frac{\bar{\rho}\bar{u}_1}{h_1}\frac{\partial \bar{K}_i}{\partial x_1} + \frac{\bar{\rho}\bar{u}_2}{h_2}\frac{\partial \bar{K}_i}{\partial x_2} + \frac{\bar{\rho}\bar{u}_3}{h_3}\frac{\partial \bar{K}_i}{\partial x_3} = -\frac{1}{h_2}\frac{\partial}{\partial x_2}(j_{i2l} + j_{i2t}) + \bar{\psi}_i \tag{32}$$

where $\bar{K}_i = \bar{\rho}_i/\bar{\rho}$ is the local mass concentration. The terms j_{i2l} and j_{i2t} represent the diffusional flux in the direction normal to the surface (subscript 2) by laminar and turbulent flow mechanisms, respectively. The term $\bar{\psi}_i$ represents the mass rate of chemical formation of species i per unit volume. For a real gas in chemical equilibrium and with uniform elemental composition, Eq. (32) is automatically satisfied. The chemical composition in this case can be related directly by statistical mechanics to the local pressure and temperature. For a gas that is frozen chemically ($\bar{\psi}_i = 0$), the local chemical composition results from a balance between the convection and diffusion of the species.

29. Diffusion Flux Vector. *For a chemically frozen ternary mixture* of species the laminar diffusion flux vector for species 1 is expressed in terms of binary diffusion coefficients as [55]

$$
\begin{aligned}
j_{12l} = &\left[\frac{\bar{K}_1}{M_1}\mathcal{D}_{23} + \frac{\bar{K}_2}{M_2}\mathcal{D}_{13} + \frac{\bar{K}_3}{M_3}\mathcal{D}_{12}\right]^{-1}\left\{\frac{\bar{K}_1}{M_1}\mathcal{D}_{23}\left[\mathcal{D}_{12} + \bar{K}_3(\mathcal{D}_{13} - \mathcal{D}_{12})\right]\right. \\
&+ \frac{\bar{K}_2}{M_2}\mathcal{D}_{13}\left[\mathcal{D}_{12} + \frac{\bar{K}_1\bar{K}_3}{\bar{K}_2}(\mathcal{D}_{12} - \mathcal{D}_{23})\right] + \frac{\bar{K}_3}{M_3}\mathcal{D}_{12}\left[\mathcal{D}_{13} + \bar{K}_1(\mathcal{D}_{23} - \mathcal{D}_{13})\right]\bigg]\frac{\bar{\rho}}{h_2}\frac{\partial \bar{K}_2}{\partial x_2} \\
&+ \left[\frac{\bar{K}_1}{M_1}\mathcal{D}_{23}\left[\mathcal{D}_{13} + \bar{K}_2(\mathcal{D}_{12} - \mathcal{D}_{13})\right] + \frac{\bar{K}_2}{M_2}\mathcal{D}_{13}\left[\mathcal{D}_{12} + \bar{K}_1(\mathcal{D}_{23} - \mathcal{D}_{12})\right]\right. \\
&\left.+ \frac{\bar{K}_3}{M_3}\mathcal{D}_{12}\left[\mathcal{D}_{13} + \frac{\bar{K}_1\bar{K}_2}{\bar{K}_3}(\mathcal{D}_{13} - \mathcal{D}_{23})\right]\bigg]\frac{\bar{\rho}}{h_2}\frac{\partial \bar{K}_3}{\partial x_2}\right\} - \frac{\mathcal{D}_1^T}{h_2}\frac{\partial \ln \bar{T}}{\partial x_2}
\end{aligned} \tag{33}
$$

Similar expressions for j_{22l} and j_{32l} are obtainable from Eq. (33) by interchanging the indices representing the individual species. The first terms on the right side of Eq. (33) represent the diffusion caused by concentration gradients, while the last term represents thermal diffusion. Note that diffusion introduced by pressure gradients and external forces has been neglected as in most forced convection studies.

For a binary mixture $(K_3 \equiv 0)$, the most common case considered in boundary-layer theory, Eq. (33) simplifies to

$$j_{12l} = -\frac{\bar{\rho}\,\mathcal{D}_{12}}{h_2}\frac{\partial \bar{K}_1}{\partial x_2} - \frac{\mathcal{D}_1^T}{h_2}\frac{\partial \ln \bar{T}}{\partial x_2} \tag{34}$$

because $\bar{K}_1 + \bar{K}_2 = 1$. In the absence of thermal diffusion, Eq. (34) reduces to Fick's Law.

For a ternary mixture with all the binary diffusion coefficients assumed equal, that is, $\mathcal{D}_{12} = \mathcal{D}_{13} = \mathcal{D}_{23} = \mathcal{D}$, Eq. (33) reduces to

$$j_{12l} = -\frac{\bar{\rho}\,\mathcal{D}}{h_2}\frac{\partial \bar{K}_1}{\partial x_2} - \frac{\mathcal{D}_1^T}{h_2}\frac{\partial \ln \bar{T}}{\partial x_2} \tag{35}$$

because $\bar{K}_1 + \bar{K}_2 + \bar{K}_3 = 1$. Thus, the assumption of equal binary diffusion coefficients for all combinations of pairs of species results in the diffusion of a particular species being dependent only on the gradient of that species and not on the gradients of all other species as in the case of the general ternary mixture (Eq. (33)). This major simplification holds for mixtures containing any number of species and is the reason for the popularity of the equal binary diffusion coefficient assumption when a multispecies system is considered.

For a ternary mixture with one species at a much lower concentration than the other two, the diffusion flux for the dilute species 1 simplifies to

$$j_{12l} = -\frac{\bar{\rho}\,\mathcal{D}_{12}\,\mathcal{D}_{13}\,(\bar{K}_2/M_2 + \bar{K}_3/M_3)}{(\bar{K}_2/M_2)\,\mathcal{D}_{13} + (\bar{K}_3/M_3)\,\mathcal{D}_{12}}\frac{1}{h_2}\frac{\partial \bar{K}_1}{\partial x_2} - \frac{\mathcal{D}_1^T}{h_2}\frac{\partial \ln \bar{T}}{\partial x_2} \tag{36}$$

and for one of the dense species (for example, 3) to

$$j_{32l} = -\frac{\bar{\rho}\,\mathcal{D}_{23}}{h_2}\frac{\partial \bar{K}_3}{\partial x_2} - \frac{\mathcal{D}_3^T}{h_2}\frac{\partial \ln \bar{T}}{\partial x_2} \tag{37}$$

Thus, Eq. (36) indicates that the diffusion flux of the dilute species depends on its own concentration gradient but with an effective binary diffusion coefficient composed of a combination of the binary diffusion coefficients with each of the two dense species. The dense species diffuse as in a binary mixture composed of these species alone.

For mixtures of more than three species, the Stefan-Maxwell equation is most convenient for relating diffusion vectors and concentration and temperature gradients. The diffusion coefficients that appear in this equation are the binary diffusion coefficients for the individual species grouped in pairs. In the absence of pressure gradients and external body forces, the Stefan-Maxwell equation for each species i of a mixture of ν species is [55]

$$\frac{1}{\overline{\rho}} \sum_{\substack{j=1 \\ j \neq i}}^{\nu} \frac{\overline{K}_j}{\mathcal{D}_{ij} M_j} \left[\frac{j_{j2l} + (\mathcal{D}_j^{T}/h_2)(\partial \ln \overline{T}/\partial x_2)}{\overline{K}_j} - \frac{j_{i2l} + (\mathcal{D}_i^{T}/h_2)(\partial \ln \overline{T}/\partial x_2)}{\overline{K}_i} \right]$$

$$= \sum_{j=1}^{\nu} \frac{\overline{K}_j}{M_j} \left(\frac{1}{\overline{K}_i h_2} \frac{\partial \overline{K}_i}{\partial x_2} - \frac{1}{\overline{K}_j h_2} \frac{\partial \overline{K}_j}{\partial x_2} \right) \quad (38)$$

Again if $\mathcal{D}_{ij} = \mathcal{D}$ for all species pairs, Eq. (38) is satisfied by Eq. (35).

The turbulent diffusion flux vector is

$$j_{i2t} = \overline{(\rho u_2)' K_i'} = - \frac{\overline{\rho} \epsilon_{Di}}{h_2} \frac{\partial \overline{K}_i}{\partial x_2} \quad (39)$$

The term on the far right is in the Boussinesq form where a turbulent eddy diffusivity is employed in the manner analogous to a laminar diffusivity. Note in turbulent diffusion there is an absence of the analog to thermal diffusion and the flux of a species depends only on the fluctuation in concentration of that species.

30. Conservation of Momentum. The conservation of momentum equations in the x_1, x_2, x_3, directions are

$$\frac{\overline{\rho} \overline{u}_1}{h_1} \frac{\partial \overline{u}_1}{\partial x_1} + \frac{\overline{\rho} \overline{u}_2}{h_2} \frac{\partial \overline{u}_1}{\partial x_2} + \frac{\overline{\rho} \overline{u}_3}{h_3} \frac{\partial \overline{u}_1}{\partial x_3} + \frac{\overline{\rho} \overline{u}_1 \overline{u}_3}{h_1 h_3} \frac{\partial h_1}{\partial x_3} - \frac{\overline{\rho} \overline{u}_3^2}{h_1 h_3} \frac{\partial h_3}{\partial x_1}$$

$$= -\frac{1}{h_1} \frac{\partial \overline{p}}{\partial x_1} + \frac{1}{h_2} \frac{\partial}{\partial x_2} (\tau_{12l} + \tau_{12t}) \quad (40)$$

$$\frac{\partial \overline{p}}{\partial x_2} = 0 \quad (41)$$

$$\frac{\overline{\rho} \overline{u}_1}{h_1} \frac{\partial \overline{u}_3}{\partial x_1} + \frac{\overline{\rho} \overline{u}_2}{h_2} \frac{\partial \overline{u}_3}{\partial x_2} + \frac{\overline{\rho} \overline{u}_3}{h_3} \frac{\partial \overline{u}_3}{\partial x_3} + \frac{\overline{\rho} \overline{u}_1 \overline{u}_3}{h_1 h_3} \frac{\partial h_3}{\partial x_1} - \frac{\overline{\rho} \overline{u}_1^2}{h_1 h_3} \frac{\partial h_1}{\partial x_3}$$

$$= -\frac{1}{h_3} \frac{\partial \overline{p}}{\partial x_3} + \frac{1}{h_2} \frac{\partial}{\partial x_2} (\tau_{32l} + \tau_{32t}) \quad (42)$$

The shear terms $\tau_{12l}, \tau_{32l}, \tau_{12t}, \tau_{32t}$ are expressed as

$$\tau_{12l} = \frac{\mu}{h_2} \frac{\partial \overline{u}_1}{\partial x_2} \qquad \tau_{32l} = \frac{\mu}{h_2} \frac{\partial \overline{u}_3}{\partial x_2} \quad (43)$$

and

$$\tau_{12t} = -\overline{(\rho u_2)' u_1'} = \frac{\overline{\rho} \epsilon_{M1}}{h_2} \frac{\partial \overline{u}_1}{\partial x_2} \quad (44)$$

$$\tau_{32t} = -\overline{(\rho u_2)' u_3'} = \frac{\overline{\rho} \epsilon_{M3}}{h_2} \frac{\partial \overline{u}_3}{\partial x_2} \tag{45}$$

where the center members in Eqs. (44) and (45) represent the Reynolds shear stresses and the right members are expressed in terms of eddy viscosities. Note that the eddy viscosities are not necessarily the same in the x_1 and x_3 directions.

31. Conservation of Energy.   The conservation of energy equation is

$$\frac{\overline{\rho}\overline{u}_1}{h_1} \frac{\partial \overline{I}}{\partial x_1} + \frac{\overline{\rho}\overline{u}_2}{h_2} \frac{\partial \overline{I}}{\partial x_2} + \frac{\overline{\rho}\overline{u}_3}{h_3} \frac{\partial \overline{I}}{\partial x_3} = -\frac{1}{h_2} \frac{\partial}{\partial x_2} (q_{2l} + q_{2t})$$

$$+ \frac{1}{h_2} \frac{\partial}{\partial x_2} [\overline{u}_1(\tau_{12l} + \tau_{12t}) + \overline{u}_3(\tau_{32l} + \tau_{32t})] \tag{46}$$

where $\overline{I} = \overline{i} + (\overline{u}_1^2 + \overline{u}_3^2)/2$. The first term on the right represents the conduction of heat by molecular and turbulent processes and the second, the rate of work done by the combined shearing forces.

32. Energy Flux Vector.   The flux term for the laminar transport of energy is given by

$$q_{2l} = -\frac{k}{h_2} \frac{\partial \overline{T}}{\partial x_2} + \sum_i j_{i2l} \overline{i}_i + \frac{R\overline{T}}{\overline{\rho}} \sum_{i,j} \frac{M}{M_i M_j} \frac{\mathcal{D}_i^T}{\mathcal{D}_{ij}} \overline{K}_j \left(\frac{j_{i2l}}{\overline{K}_i} - \frac{j_{j2l}}{\overline{K}_j} \right) \tag{47a}$$

and is the sum of the transport of energy by conduction (temperature gradients), by diffusion of species with different specific enthalpies, and by the Dufour or diffusion-thermo effect, which is the reciprocal process of thermal diffusion. The second term is of major importance in the evaluation of energy transfer to a body in a real gas at elevated temperatures or in a chemically frozen mixture which includes an inert species introduced through the body surface. The last term is usually small except when some very light species are present such as in cooling systems where light gases have been introduced into the boundary layer. The value of the mean molecular weight M is given by

$$M = \left(\sum \frac{\overline{K}_i}{M_i} \right)^{-1} \tag{47b}$$

The total thermal conductivity concept is convenient for a real gas in chemical equilibrium and where uniform elemental concentrations exist throughout the flow field. The heat flux is expressed as

$$q_{2l} = -\frac{k_T}{h_2} \frac{\partial \overline{T}}{\partial x_2} \tag{48}$$

The basis for this equation is shown in the following where for simplicity the effects of thermal diffusion and diffusion-thermo are omitted. The diffusion vector is

$$j_{i2l} = \frac{\overline{n}^2}{\overline{\rho}} \sum_j \frac{m_i m_j D_{ij}}{h_2} \frac{\partial}{\partial x_2} \left(\frac{\overline{n}_j}{\overline{n}} \right) \tag{49}$$

Here, D_{ij} is a multicomponent diffusion coefficient and is not to be confused with the binary diffusion coefficient [55]. For constant elemental composition and chemical equilibrium, the number or molal concentration $\bar{n}_j/\bar{n}$ is a function only of temperature and pressure so that Eq. (49) becomes

$$j_{i2l} = \left\{ \frac{\bar{n}^2}{\bar{\rho}} \sum_j m_i m_j D_{ij} \left[\frac{\partial}{\partial \bar{T}} \left(\frac{\bar{n}_j}{n} \right) \right]_p \right\} \frac{1}{h_2} \frac{\partial \bar{T}}{\partial x_2} \tag{50}$$

when the pressure is constant across the boundary layer. Thus

$$j_{i2l} = \frac{f(\bar{p},\bar{T})}{h_2} \frac{\partial \bar{T}}{\partial x_2}$$

so that Eq. (47a) becomes

$$q_{2l} = -\left[k - \sum_i f(\bar{p},\bar{T})\, \bar{i}_i \right] \frac{1}{h_2} \frac{\partial \bar{T}}{\partial x_2}$$

which reduces to Eq. (48) when the total thermal conductivity k_T is expressed as the sum of the usual thermal conductivity and the effects of species diffusion.

The corresponding total specific heat is defined as

$$c_{p_T} = \frac{\partial \bar{i}}{\partial \bar{T}} .$$

The transport of heat by turbulent processes for a multicomponent mixture is expressed as

$$q_{2t} = \sum_i \bar{K}_i \overline{(\rho u_2)' i_i'} + \sum_i \bar{i}_i \overline{(\rho u_2)' K_i'}$$

$$= c_{p_F} \overline{(\rho u_2)' T'} + \sum_i \bar{i}_i \overline{(\rho u_2)' K_i'}$$

$$= -\frac{\bar{\rho} c_{p_F} \epsilon_H}{h_2} \frac{\partial \bar{T}}{\partial x_2} - \sum_i \frac{\bar{i}_i \bar{\rho} \epsilon_{Di}}{h_2} \frac{\partial \bar{K}_i}{\partial x_2} \tag{51}$$

The first term on the right represents the "conduction" of heat by turbulent processes (c_{pF}, the frozen specific heat is defined in Eq. (66)) and the second the transport of heat by turbulent diffusion, the summation of the two effects being similar to the corresponding terms for laminar flow.

For the gas in chemical equilibrium and with uniform elemental composition, the turbulent heat transfer term becomes

$$q_{2t} = -\frac{\bar{\rho} c_{p_T} \epsilon_H}{h_2} \frac{\partial \bar{T}}{\partial x_2} \tag{52}$$

33. Summary of Equations for Gas in Chemical Equilibrium. For a gas with uniform elemental composition in chemical equilibrium and where eddy viscosities ϵ_{M1} and ϵ_{M3} are assumed equal, the boundary-layer differential equations are

$$\frac{\partial}{\partial x_1}(h_2 h_3 \rho u_1) + \frac{\partial}{\partial x_2}(h_1 h_3 \rho u_2) + \frac{\partial}{\partial x_3}(h_1 h_2 \rho u_3) = 0 \tag{53}$$

$$K_i = K_i(p, i) \tag{54}$$

$$\frac{\rho u_1}{h_1}\frac{\partial u_1}{\partial x_2} + \frac{\rho u_2}{h_2}\frac{\partial u_1}{\partial x_2} + \frac{\rho u_3}{h_3}\frac{\partial u_1}{\partial x_3} + \frac{\rho u_1 u_3}{h_1 h_3}\frac{\partial h_1}{\partial x_3} - \frac{\rho u_3^2}{h_1 h_3}\frac{\partial h_3}{\partial x_1}$$

$$= -\frac{1}{h_1}\frac{\partial p}{\partial x_1} + \frac{1}{h_2}\frac{\partial}{\partial x_2}\left(\frac{\mu + \rho \epsilon_M}{h_2}\frac{\partial u_1}{\partial x_2}\right) \tag{55}$$

$$\frac{\partial p}{\partial x_2} = 0 \tag{56}$$

$$\frac{\rho u_1}{h_1}\frac{\partial u_3}{\partial x_1} + \frac{\rho u_2}{h_2}\frac{\partial u_3}{\partial x_2} + \frac{\rho u_3}{h_3}\frac{\partial u_3}{\partial x_3} + \frac{\rho u_1 u_3}{h_1 h_3}\frac{\partial h_3}{\partial x_1} - \frac{\rho u_1^2}{h_1 h_3}\frac{\partial h_1}{\partial x_3}$$

$$= -\frac{1}{h_3}\frac{\partial p}{\partial x_3} + \frac{1}{h_2}\frac{\partial}{\partial x_2}\left(\frac{\mu + \rho \epsilon_M}{h_2}\frac{\partial u_3}{\partial x_2}\right) \tag{57}$$

$$\frac{\rho u_1}{h_1}\frac{\partial I}{\partial x_1} + \frac{\rho u_2}{h_2}\frac{\partial I}{\partial x_2} + \frac{\rho u_3}{h_3}\frac{\partial I}{\partial x_3} = \frac{1}{h_2}\frac{\partial}{\partial x_2}\left\{\left(\frac{\mu}{\mathrm{Pr}_T} + \frac{\rho \epsilon_M}{\mathrm{Pr}_t}\right)\frac{1}{h_2}\frac{\partial I}{\partial x_2}\right.$$

$$+ \left[\mu\left(1 - \frac{1}{\mathrm{Pr}_T}\right) + \rho \epsilon_M\left(1 - \frac{1}{\mathrm{Pr}_t}\right)\right]\frac{1}{h_2}\frac{\partial}{\partial x_2}\left(\frac{u_1^2 + u_3^2}{2}\right)\right\} \tag{58}$$

Because the turbulent flow terms are expressed in the Boussinesq form, all the symbols represent time-averaged quantities, and there is no need to retain the superscript bar symbol. (Note that $\rho u_2 = \overline{\rho u_2}$ as in Eq. (31).) The Prandtl numbers are defined as

$$\mathrm{Pr}_T = \frac{\mu c_{p_T}}{k_T} \tag{59}$$

$$\mathrm{Pr}_t = \frac{\epsilon_M}{\epsilon_H} \tag{60}$$

Equation (54) represents the evaluation of the local species concentrations by statistical thermodynamics from the local pressure and enthalpy. Equations (53) to (58) together with the relations for the thermodynamic and molecular transport properties in terms of the local pressure and enthalpy constitute the complete set of equations for an entirely laminar flow. To include the effects of turbulence it is necessary, in addition, to define the functional forms of ϵ_M and ϵ_H.

34. Summary of Equations for Multicomponent, Chemically Frozen Gas. For the case of a chemically frozen multicomponent mixture of species, Eqs. (53) and (55) to (57) apply; however, new forms of Eqs. (54) and (58) are required. In general, the species diffusion equation is

$$\frac{\rho u_1}{h_1}\frac{\partial K_i}{\partial x_1} + \frac{\rho u_2}{h_2}\frac{\partial K_i}{\partial x_2} + \frac{\rho u_3}{h_3}\frac{\partial K_i}{\partial x_3} = -\frac{1}{h_2}\frac{\partial}{\partial x_2}\left(j_{i2l} - \rho\epsilon_{Di}\frac{1}{h_2}\frac{\partial K_i}{\partial x_2}\right) \tag{61}$$

which for a binary mixture simplifies to

$$\frac{\rho u_1}{h_1}\frac{\partial K_1}{\partial x_1} + \frac{\rho u_2}{h_2}\frac{\partial K_1}{\partial x_2} + \frac{\rho u_3}{h_3}\frac{\partial K_1}{\partial x_3} = \frac{1}{h_2}\frac{\partial}{\partial x_2}\left[\left(\frac{\mu\,\mathrm{Le}}{\mathrm{Pr}_F} + \frac{\rho\epsilon_M\mathrm{Le}_t}{\mathrm{Pr}_t}\right)\frac{1}{h_2}\frac{\partial K_1}{\partial x_2} + \frac{\mathcal{D}_1^{\,T}}{h_2}\frac{\partial\ln T}{\partial x_2}\right] \tag{62}$$

when expressed in terms of species 1. The corresponding general energy equation is

$$\frac{\rho u_1}{h_1}\frac{\partial I}{\partial x_1} + \frac{\rho u_2}{h_2}\frac{\partial I}{\partial x_2} + \frac{\rho u_3}{h_3}\frac{\partial I}{\partial x_3} = \frac{1}{h_2}\frac{\partial}{\partial x_2}\Bigg\{\left(\frac{\mu}{\mathrm{Pr}_F} + \frac{\rho\epsilon_M}{\mathrm{Pr}_t}\right)\frac{1}{h_2}\frac{\partial I}{\partial x_2} + \sum_{i=1}^{\nu}\left[-j_{i2l}\right.$$

$$-\left[\frac{\mu}{\mathrm{Pr}_F} + \frac{\rho\epsilon_M}{\mathrm{Pr}_t}(1-\mathrm{Le}_t)\right]\frac{1}{h_2}\frac{\partial K_i}{\partial x_2}\Bigg]i_i$$

$$+\left[\mu\left(1-\frac{1}{\mathrm{Pr}_F}\right) + \rho\epsilon_M\left(1-\frac{1}{\mathrm{Pr}_t}\right)\right]\frac{1}{h_2}\frac{\partial}{\partial x_2}\left(\frac{u_1^{\,2}+u_3^{\,2}}{2}\right)\Bigg\} \tag{63}$$

which for a binary mixture simplifies to

$$\frac{\rho u_1}{h_1}\frac{\partial I}{\partial x_1} + \frac{\rho u_2}{h_2}\frac{\partial I}{\partial x_2} + \frac{\rho u_3}{h_3}\frac{\partial I}{\partial x_3} = \frac{1}{h_2}\frac{\partial}{\partial x_2}\Bigg\{\left(\frac{\mu}{\mathrm{Pr}_F} + \frac{\rho\epsilon_M}{\mathrm{Pr}_t}\right)\frac{1}{h_2}\frac{\partial I}{\partial x_2}$$

$$-(i_1-i_2)\left[\frac{\mu}{\mathrm{Pr}_F}(1-\mathrm{Le}) + \frac{\rho\epsilon_M}{\mathrm{Pr}_t}(1-\mathrm{Le}_t)\right]\frac{1}{h_2}\frac{\partial K_1}{\partial x_2} + (i_1-i_2)\frac{\mathcal{D}_1^{\,T}}{h_2}\frac{\partial\ln T}{\partial x_2}$$

$$+\left[\mu\left(1-\frac{1}{\mathrm{Pr}_F}\right) + \rho\epsilon_M\left(1-\frac{1}{\mathrm{Pr}_t}\right)\right]\frac{1}{h_2}\frac{\partial}{\partial x_2}\left(\frac{u_1^{\,2}+u_3^{\,2}}{2}\right)\Bigg\} \tag{64}$$

In the above equations the superscript F means "frozen" and

$$\mathrm{Pr}_F = \frac{\mu c_{p_F}}{k} \tag{65}$$

where

$$c_{p_F} = \sum_i K_i c_{p_i} \tag{66}$$

The Lewis numbers are defined as

$$\text{Le} = \frac{\rho c_{p_F} \mathcal{D}}{k} \tag{67}$$

$$\text{Le}_t = \frac{\epsilon_D}{\epsilon_H} \tag{68}$$

It is the system of Eqs. (53) to (64) that forms the basis for most of the boundary-layer theory used in the evaluation of the convective heat transfer rates to the surfaces of submerged bodies. Note that diffusion-thermo effects have been neglected in Eqs. (63) and (64).

Solutions of the aforementioned equations are subject to boundary conditions at the surface and at the outer edge of the boundary layer. The requirement of uniform elemental composition for a fluid in chemical equilibrium implies zero mass transfer of foreign species at the surface. For a frozen multicomponent mixture, the concentration of each species at the boundary-layer edge is usually uniform.

Integral Equations

Solutions of the equations of §33 and §34 have only been obtained for a limited number of cases restricted largely to laminar flow, simple body shapes, and simple surface distributions of temperature and mass transfer rates. For more general problems, von Kármán devised a method where the boundary-layer equations are satisfied, not locally, but in the mean over the entire boundary-layer thickness. Here, the boundary-layer equations are first integrated with respect to the coordinate normal to the surface which transforms the dependent variables from single quantities to specific integrals. The boundary-layer equations now define the rate of change of these integrals over the surface of the body. The process reduces the number of dimensions of the problem by one; thus two-dimensional partial differential equations transform into ordinary differential equations with integral forms of dependent variables. Further, the number of equations involved is reduced by one because the overall continuity of mass equation is combined with each of the other conservation equations. These integral equations can be written as follows where j_{iw}, τ_{w1}, q_w are given by the previously defined laminar-flow expressions (turbulence vanishes at the surface). The first of these equations applies for a chemically frozen mixture of species. The remainder apply either to chemically frozen or equilibrium mixtures, the main distinction being in the use of the appropriate form of q_w, either Eq. (47a) or (48).

35. Summary of Integral Equations
Diffusion of Species i

$$\frac{j_{iw}}{\rho_e u_e (K_{iw} - K_{ie})} + \frac{\rho_w u_{2w}}{\rho_e u_e} = \frac{u_{1e}}{u_e} \left\{ \frac{1}{h_1} \frac{\partial \Delta_{i1}}{\partial x_1} + \frac{\Delta_{i1}}{h_1} \frac{\partial}{\partial x_1} [\ln(h_3 \rho_e u_{1e}(K_{iw} - K_{ie}))] \right\}$$

$$+ \frac{u_{3e}}{u_e} \left\{ \frac{1}{h_3} \frac{\partial \Delta_{i3}}{\partial x_3} + \frac{\Delta_{i3}}{h_3} \frac{\partial}{\partial x_3} [\ln(h_1 \rho_e u_{3e}(K_{iw} - K_{ie}))] \right\} \tag{69}$$

Momentum in x_1 direction

$$\frac{\tau_{w1}}{\rho_e u_{1e}^2} + \frac{\rho_w u_{2w}}{\rho_e u_{1e}} = \frac{1}{h_1}\frac{\partial \theta_{11}}{\partial x_1} + \left[\frac{1}{h_3 h_1}\frac{\partial h_3}{\partial x_1} + \frac{1}{\rho_e h_1}\frac{\partial \rho_e}{\partial x_1} + \frac{1}{u_{1e}h_1}\frac{\partial u_{1e}}{\partial x_1}\left(2 + \frac{\delta_1^*}{\theta_{11}}\right)\right]\theta_{11}$$

$$+ \frac{u_{3e}}{u_{1e}}\left\{\frac{1}{h_3}\frac{\partial \theta_{31}}{\partial x_3} + \left[\frac{1}{h_1 h_3}\frac{\partial h_1}{\partial x_3}\left(2 + \frac{\delta_3^*}{\theta_{31}}\right) + \frac{1}{\rho_e h_3}\frac{\partial \rho_e}{\partial x_3} + \frac{1}{u_{3e}h_3}\frac{\partial u_{3e}}{\partial x_3}\right.\right.$$

$$\left.\left. + \frac{1}{u_{1e}h_3}\frac{\partial u_{1e}}{\partial x_3}\left(1 + \frac{\delta_3^*}{\theta_{31}}\right)\right]\theta_{31}\right\} - \left(\frac{u_{3e}}{u_{1e}}\right)^2 \frac{1}{h_3 h_1}\frac{\partial h_3}{\partial x_1}\left(1 + \frac{\delta_3^*}{\theta_{33}}\right)\theta_{33} \quad (70)$$

Momentum in x_3 direction

$$\frac{\tau_{w3}}{\rho_e u_{3e}^2} + \frac{\rho_w u_{2w}}{\rho_e u_{3e}} = \frac{1}{h_3}\frac{\partial \theta_{33}}{\partial x_3} + \left[\frac{1}{h_1 h_3}\frac{\partial h_1}{\partial x_3} + \frac{1}{\rho_e h_3}\frac{\partial \rho_e}{\partial x_3} + \frac{1}{u_{3e}h_3}\frac{\partial u_{3e}}{\partial x_3}\left(2 + \frac{\delta_3^*}{\theta_{33}}\right)\right]\theta_{33}$$

$$+ \frac{u_{1e}}{u_{3e}}\left\{\frac{1}{h_1}\frac{\partial \theta_{13}}{\partial x_1} + \left[\frac{1}{h_3 h_1}\frac{\partial h_3}{\partial x_1}\left(2 + \frac{\delta_1^*}{\theta_{13}}\right) + \frac{1}{\rho_e h_1}\frac{\partial \rho_e}{\partial x_1} + \frac{1}{u_{1e}h_1}\frac{\partial u_{1e}}{\partial x_1}\right.\right.$$

$$\left.\left. + \frac{1}{u_{3e}h_1}\frac{\partial u_{3e}}{\partial x_1}\left(1 + \frac{\delta_1^*}{\theta_{13}}\right)\right]\theta_{13}\right\} - \left(\frac{u_{1e}}{u_{3e}}\right)^2 \frac{1}{h_1 h_3}\frac{\partial h_1}{\partial x_3}\left(1 + \frac{\delta_1^*}{\theta_{11}}\right)\theta_{11} \quad (71)$$

Energy

$$\frac{q_w}{\rho_e u_e (i_w - I_e)} + \frac{\rho_w u_{2w}}{\rho_e u_e} = \frac{u_{1e}}{u_e}\left\{\frac{1}{h_1}\frac{\partial \Gamma_1}{\partial x_1} + \frac{\Gamma_1}{h_1}\frac{\partial}{\partial x_1}\left[\ln(h_3 \rho_e u_{1e}(i_w - I_e))\right]\right\}$$

$$+ \frac{u_{3e}}{u_e}\left\{\frac{1}{h_3}\frac{\partial \Gamma_3}{\partial x_3} + \frac{\Gamma_3}{h_3}\frac{\partial}{\partial x_3}\left[\ln(h_1 \rho_e u_{3e}(i_w - I_e))\right]\right\} \quad (72)$$

The new dependent variables are:
Concentration Thickness

$$\Delta_{i1} = \int_0^{x_{2e}} \frac{\rho u_1}{\rho_e u_{1e}}\left(\frac{K_i - K_{ie}}{K_{iw} - K_{ie}}\right)h_2 dx_2 \qquad \Delta_{i3} = \int_0^{x_{2e}} \frac{\rho u_3}{\rho_e u_{3e}}\left(\frac{K_i - K_{ie}}{K_{iw} - K_{ie}}\right)h_2 dx_2$$

Momentum Thicknesses

$$\theta_{11} = \int_0^{x_{2e}} \frac{\rho u_1}{\rho_e u_{1e}} \left(1 - \frac{u_1}{u_{1e}}\right) h_2 dx_2 \qquad \theta_{31} = \int_0^{x_{2e}} \frac{\rho u_3}{\rho_e u_{3e}} \left(1 - \frac{u_1}{u_{1e}}\right) h_2 dx_2$$

$$\theta_{33} = \int_0^{x_{2e}} \frac{\rho u_3}{\rho_e u_{3e}} \left(1 - \frac{u_3}{u_{3e}}\right) h_2 dx_2 \qquad \theta_{13} = \int_0^{x_{2e}} \frac{\rho u_1}{\rho_e u_{1e}} \left(1 - \frac{u_3}{u_{3e}}\right) h_2 dx_2$$

Energy Thicknesses

$$\Gamma_1 = \int_0^{x_{2e}} \frac{\rho u_1}{\rho_e u_{1e}} \left(\frac{I - I_e}{i_w - I_e}\right) h_2 dx_2 \qquad \Gamma_3 = \int_0^{x_{2e}} \frac{\rho u_3}{\rho_e u_{3e}} \left(\frac{I - I_e}{i_w - I_e}\right) h_2 dx_2$$

Coefficients appearing in three of the above equations are:

Displacement Thicknesses

$$\delta_1^* = \int_0^{x_{2e}} \left(1 - \frac{\rho u_1}{\rho_e u_{1e}}\right) h_2 dx_2 \qquad \delta_3^* = \int_0^{x_{2e}} \left(1 - \frac{\rho u_3}{\rho_e u_{3e}}\right) h_2 dx_2$$

The term u_e in Eqs. (69) and (72) is the component parallel to the surface of the inviscid flow velocity V_e. (For a thin boundary layer $u_e \approx V_e$.) The value of x_{2e} at the outer edge of the boundary layer appears as the upper limit of integration in the new dependent variables. Because the boundary-layer functions u, ρ, etc., approach their free-stream values asymptotically, the boundary-layer thickness x_{2e} chosen is arbitrary. This is not a problem, however, because the integral quantities converge when x_{2e} exceeds certain values.

Equation (69) indicates that the growth in the concentration thicknesses of species i in the two directions along the surface is dependent on the sum of the mass transfer coefficient (the first term on the left) and the normalized surface injection rate (second term on the left). Similarly for Eqs. (70) and (71), the momentum thickness in a given direction along the surface is dependent on the skin friction coefficient in that direction and normalized injection rates. Note that θ_{13} and θ_{31} represent momentum loss in a given direction by fluid motion normal to that direction.

In the energy equation (Eq. (72)), a quantity appears that can be related to the local heat transfer coefficient, or Stanton number

$$\frac{q_w}{\rho_e u_e (i_w - I_e)} = c_h \left[\frac{i_w - i_{aw}}{i_w - I_e}\right] \qquad (73)$$

For the case where Pr = 1, the term in the brackets is equal to unity. For blunt bodies, it remains nearly unity even when the Prandtl number differs from unity.

When a coordinate system is chosen such that $u_{3e} = 0$ everywhere (streamwise coordinate system), the terms in the above equations that contain u_{3e} as a factor do not necessarily vanish since the products $u_{3e}\theta_{31}$, $u_{3e}\theta_{13}$, etc., retain values other than zero. For example,

$$\lim_{u_{3e} \to 0} [\rho_e u_{1e} u_{3e} \theta_{13}] = \int_0^{x_{2e}} \rho u_1 (u_{3e} - u_3) h_2 \, dx_2 = - \int_0^{x_{2e}} \rho u_1 u_3 h_2 \, dx_2$$

where u_3 is only zero at the surface and outer edge of the boundary layer.

LAMINAR BOUNDARY LAYER ON A FLAT PLATE

The most studied configuration for forced convection has been the "flat plate," a surface at constant pressure with a sharp leading edge. The simplicity of this configuration so facilitates the solution of the boundary-layer equations, even for a variety of surface boundary conditions, that the bulk of heuristic theoretical boundary-layer research is identified with the flat plate. These results are useful because much that is learned can be extended to more realistic body shapes and applied directly to plate-like surfaces (e.g., supersonic aircraft wings or fins having wedge cross sections and attached shock waves) or to generally shaped bodies where pressure gradients are small.

Impervious Surface with Uniform Temperature

36. Governing Differential Equations. The boundary condition of zero mass transfer results in a uniform elemental composition throughout a boundary layer. For a fluid as general as a gas in chemical equilibrium, Eqs. (53) to (58) are then appropriate. The flat plate configuration reduces the metric coefficients to $h_1 = h_2 = h_3 = 1$, and the two-dimensional character of the flow requires $u_3 \equiv 0$, $\partial/\partial x_3 = 0$. The simplified differential equations for all-laminar flow are

$$\frac{\partial}{\partial x_1} (\rho u_1) + \frac{\partial}{\partial x_2} (\rho u_2) = 0 \tag{74}$$

$$\rho u_1 \frac{\partial u_1}{\partial x_1} + \rho u_2 \frac{\partial u_1}{\partial x_2} = \frac{\partial}{\partial x_2} \left(\mu \frac{\partial u_1}{\partial x_2} \right) \tag{75}$$

$$\rho u_1 \frac{\partial I}{\partial x_1} + \rho u_2 \frac{\partial I}{\partial x_2} = \frac{\partial}{\partial x_2} \left[\frac{\mu}{\mathrm{Pr}_T} \frac{\partial I}{\partial x_2} + \mu \left(1 - \frac{1}{\mathrm{Pr}_T} \right) \frac{\partial (u_1^2/2)}{\partial x_2} \right] \tag{76}$$

with the boundary conditions

$$
\left.
\begin{aligned}
& x_1 = 0, \quad x_2 > 0; && u_1 = u_{1e}, \quad I = I_e \\
& x_1 > 0, \quad x_2 \to \infty; && u_1 \to u_{1e}, \quad I \to I_e \\
& \quad\quad\quad\;\; x_2 = 0; && u_1 = 0, \quad u_2 = 0 \\
& && I = i_w = \text{constant or } \partial i/\partial x_2 = 0 \text{ for } I = i_{aw}
\end{aligned}
\right\} \tag{77}
$$

The leading edge of the plate is located at $x_1 = 0$. The surface boundary conditions at $x_2 = 0$ reflect the assumed conditions of zero mass transfer, a prescribed uniform temperature including the case of zero heat flux, and an implied condition of a smooth surface.

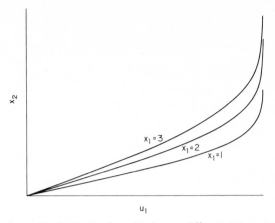

Fig. 22. Velocity profiles in a laminar boundary layer at different stations on a flat plate.

37. Boundary-layer Similarity Concept. A stream function ψ defined as

$$\rho u_1 = \rho_e \frac{\partial \psi}{\partial x_2} \qquad \rho u_2 = -\rho_e \frac{\partial \psi}{\partial x_1} \tag{78}$$

immediately satisfies Eq. (74), and Eq. (75) acquires a form amenable to separation of variables with $\psi = A(x_1)B(x_2)$. This direct approach, however, leads to contradictions with the boundary conditions (77). The concept of "boundary-layer similarity," perhaps the most essential element of boundary-layer theory, was discovered by Prandtl [1] in an effort to overcome this difficulty. Two factors contributed toward this concept. First was the observation in early experiments that the boundary-layer velocity profiles possessed a systematic thickening with increasing distance along the plate as indicated in Fig. 22, where x_1 is expressed in multiples of length. Second was the similarity of the boundary conditions (77) at the leading edge ($x_1 = 0$, $x_2 > 0$) with those at the outer edge of the boundary layer ($x_1 > 0$, $x_2 \to \infty$). Both of these factors are consistent with the idea that it is possible to redefine the coordinates of the boundary-layer profile to achieve a single profile for all stations on the plate as is indicated in Fig. 23. This is accomplished mathematically by transforming the independent variables x_1, x_2 to new independent variables $\zeta(x_1)$ and $\eta(x_1, x_2)$, choosing a properly normalized dependent variable, and showing that the boundary conditions can be satisfied when the normalized dependent variable is a function of η alone. These operations reduce the original partial differential equations to an ordinary differential equation in η with consistent boundary conditions. The resulting mathematical expression for the velocity profile is

$$\frac{u_1(x_1, x_2)}{u_{1e}} = f'(\eta) \quad \text{where} \quad \eta = \frac{x_2}{b(x_1)} \tag{79}$$

with u_{1e}, the velocity at the edge of the boundary layer, as the scaling factor for u_1 and $b(x_1)$, a quantity proportional to the boundary-layer thickness, as the scaling factor for x_2. Depending on surface boundary conditions, corresponding relationships

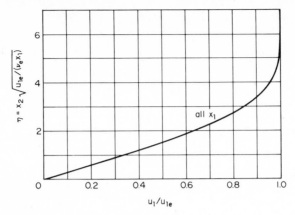

Fig. 23. Similar velocity profile in the laminar boundary layer on a flat plate, constant fluid properties.

can result for the other dependent variables, the total enthalpy and concentration. This concept of similarity and the use of the transformation of variables to seek "similarity variables" is the crux of almost every existing boundary-layer analysis.

38. Fluid with Constant Properties. When the density and viscosity in Eqs. (74) and (75) are constant, the velocity field is independent of the temperature field represented by Eq. (76). Blasius [56] transformed Eqs. (74) and (75) to similarity form by utilizing the new independent variables

$$\zeta = x_1 \qquad \eta = x_2 \sqrt{\frac{u_{1e}}{\nu_e x_1}} \tag{80}$$

The stream function

$$\psi = m(\zeta) f(\eta)$$

yields velocity components

$$\frac{u_1}{u_{1e}} = f'(\eta) \qquad \frac{u_2}{u_{1e}} = \frac{1}{2} \frac{[\eta f'(\eta) - f(\eta)]}{\sqrt{Re_{x_1 e}}} \tag{81}$$

where $f(\eta)$ is the solution of the ordinary differential equation

$$f''' + \frac{1}{2} f f'' = 0 \tag{82}$$

with the boundary conditions

$$\left. \begin{array}{l} \eta = 0 \; ; \quad f = 0 \, , \; f' = 0 \\ \eta \to \infty \; ; \quad f' \to 1 \end{array} \right\} \tag{83}$$

The u_1 velocity profile is shown in Fig. 23. The velocity ratio reaches a value of 0.99 at a boundary-layer thickness of

$$\frac{\delta}{x_1} = \frac{5}{\sqrt{Re_{x_1 e}}} \tag{84}$$

The local skin friction coefficient is

$$\frac{c_f}{2} = \frac{f''(0)}{\sqrt{\rho_e u_{1e} x_1/\mu_e}} = \frac{0.332}{\sqrt{Re_{x_1 e}}} \tag{85}$$

For a flat plate, the momentum integral equation (Eq. (70)) reduces to

$$\frac{\tau_{w1}}{\rho_e u_{1e}^2} = \frac{d\theta_{11}}{dx_1} = \frac{c_f}{2} \tag{86}$$

which integrates to yield

$$\frac{\theta_{11}}{x_1} = \frac{1}{x_1} \int_0^{x_1} \frac{c_f}{2} \, dx_1 = \frac{0.664}{\sqrt{Re_{x_1 e}}} \tag{87}$$

Equation (87) indicates that on a flat plate the average skin friction coefficient is merely the momentum thickness at the end of the plate divided by the length or equal to twice the local skin friction coefficient at the trailing edge

$$\frac{C_f}{2} = \frac{\theta_{11}}{x_1} = 2\left(\frac{c_f}{2}\right) \tag{88}$$

It is often useful to employ the momentum thickness as the independent variable in a problem. Here

$$\frac{c_f}{2} = \frac{0.220}{Re_{\theta_{11} e}} \tag{89}$$

Also, the boundary-layer form factor which appears in Eq. (70) is

$$\frac{\delta_1^*}{\theta_{11}} = 2.59 \tag{90}$$

Experimental verification of the Blasius theory has been hindered by the difficulty in reproducing the ideal flat-plate boundary conditions in the laboratory. Whenever uniform pressure was attained and the effects of the real leading edge accounted for, however, it was found that the preceding calculated results were always verified to within the accuracy of the experiment.

Pohlhausen [57] utilized the Blasius velocity distribution to evaluate the convective heating processes within the constant property boundary layer on a flat plate. He solved two problems:

1. The convective heat transfer rate to a plate with uniform surface temperature for fluid speeds sufficiently low to make viscous dissipation negligible.

2. The temperature attained by an insulated plate (zero surface heat transfer) when exposed to a high-speed stream where viscous dissipation is important.

The latter is the "plate thermometer" or "adiabatic wall problem." Eckert and Drewitz [58] showed that the general problem of heat transfer to a uniform surface temperature plate in constant-property high-speed flow is merely the superposition of the two Pohlhausen solutions.

For a uniform surface temperature plate in a low speed flow ($u_1^2 \ll 2c_pT$), Eq. (76) simplifies to

$$\rho u_1 \frac{\partial T}{\partial x_1} + \rho u_2 \frac{\partial T}{\partial x_2} = \frac{\mu}{\text{Pr}} \frac{\partial^2 T}{\partial x_2^2} \tag{91}$$

with the boundary conditions

$$x_1 = 0 \,, \; x_2 > 0 \,; \quad T = T_e$$
$$x_1 > 0 \,, \; x_2 = 0 \,; \quad T = T_w$$
$$x_2 \to \infty \,; \quad T = T_e$$

When Eq. (91) is transformed to the independent variables (80) and the new normalized dependent variable

$$Y_0(\eta) = \frac{T - T_e}{T_w - T_e} \tag{92}$$

is introduced, the ordinary homogeneous differential equation that results is

$$Y_0'' + \tfrac{1}{2} \text{Pr} f Y_0' = 0 \tag{93}$$

where f is the Blasius stream function. The transformed boundary conditions are

$$\left. \begin{array}{l} \eta = 0 \,, \quad Y_0 = 1 \\[4pt] \eta \to \infty \,, \quad Y_0 \to 0 \end{array} \right\} \tag{94}$$

Solutions for Pr = 0.5 and 1.0 are shown in Fig. 24 as solid curves. The abscissa of this figure is the thermal boundary-layer thickness parameter η_{th} consisting of the Blasius boundary-layer similarity parameter multiplied by $\text{Pr}^{1/3}$. The close agreement of the two solid curves suggests for Pr near unity that the thermal boundary-layer thickness where $Y_0 = 0.01$ is inversely proportional to approximately $\text{Pr}^{1/3}$ or

$$\frac{\delta_{th}}{x_1} = \frac{5 \, \text{Pr}^{-1/3}}{\sqrt{\text{Re}_{x_1 e}}} = \frac{\delta}{x_1} \, \text{Pr}^{-1/3} \tag{95}$$

Thus, fluids with Pr less than unity have thermal boundary layers that are thick relative to their flow boundary layers. Conversely, fluids with Pr greater than unity have relatively thin thermal boundary layers.

This latter condition suggests a particularly simple solution of Eq. (93) for very large Pr [59] because the temperature variations occur where the velocity distribution is still linear in η (see Fig. 23 for $\eta < 2.5$). The linear velocity condition in Eq. (93) permits expressing Y_0 explicitly in terms of η_{th}. The solution for this case of large Pr is shown as the dashed line in Fig. 24 and agrees quite well with the calculations based

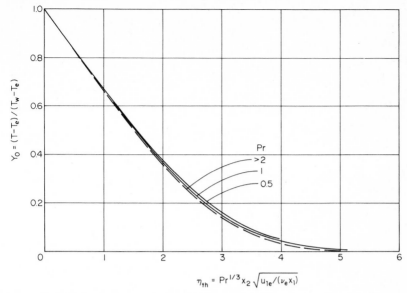

$$\eta_{th} = Pr^{1/3}x_2\sqrt{u_{1e}/(\nu_e x_1)}$$

Fig. 24. Temperature distributions in the laminar boundary layer on a flat plate at uniform temperature–constant property, low-speed flow.

on the more exact velocity distributions for Pr near unity. This agreement indicates that Eq. (95) is applicable over a large range of Pr from values characteristic of gases to those for heavy oils.

The local Stanton number found by Pohlhausen is represented very well by

$$c_h = \frac{0.332\,Pr^{-2/3}}{\sqrt{Re_{x_1e}}} \tag{96}$$

a form consistent with the parameter η_{th}.

The modified Reynolds analogy from Eqs. (85) and (96) is

$$c_h = \frac{c_f}{2}Pr^{-2/3} \tag{97}$$

The excellent agreement of this formula with the precise numerical results of Pohlhausen [57] over a large range of Pr is indicated in Table 3 and shown graphically

**TABLE 3. Error in Modified
Reynolds Analogy
of Eq. (97)**

Pr	% error
0.5	1.6
0.725	0.7
1.00	0
2.00	−1.0
≫ 1	−2.1

in Fig. 26 (solid curves are the numerical results). The dashed line, labeled $1.02\,\mathrm{Pr}^{-2/3}$, results from the analysis employing a linear velocity distribution throughout the boundary layer [60]. Equation (96) has been shown to be consistent with experimental results through a successive series of data correlations dating back to Colburn [3]. (This reference has also shown that Eq. (97) applies, at least to normal engineering accuracy, to turbulent flow on flat plates and within pipes, as well, for Pr corresponding to gases and water.)

For the flat plate in low-speed flow, the energy integral Eq. (72) simplifies to

$$c_h = \frac{d\Gamma_1}{dx_1} \tag{98}$$

On integration, this equation yields the energy thickness normalized by x_1 as

$$\frac{\Gamma_1}{x_1} = \frac{1}{x_1} \int_0^{x_1} c_h \, dx_1 = C_h = 2c_h \tag{99}$$

which is equal to the average Stanton number or twice the local Stanton number.

The local Stanton number expressed in terms of the Reynolds number based on energy thickness is

$$c_h = \frac{0.220\,\mathrm{Pr}^{-4/3}}{\mathrm{Re}_{r_{1_e}}} \tag{100}$$

For an insulated plate in high-speed flow with constant properties, Eq. (76) combined with Eq. (75) reduces to

$$\rho u_1 \frac{\partial T}{\partial x_1} + \rho u_2 \frac{\partial T}{\partial x_2} = \frac{\mu}{\mathrm{Pr}} \frac{\partial^2 T}{\partial x_2^{\,2}} + \frac{\mu}{c_p} \left(\frac{\partial u_1}{\partial x_2} \right)^2 \tag{101}$$

with boundary conditions

$$\left. \begin{array}{lll} x_1 = 0, & x_2 > 0; & T = T_e \\[2mm] x_1 > 0, & x_2 \to \infty, & T = T_e \\[2mm] & x_2 = 0, & \partial T/\partial x_2 = 0 \end{array} \right\} \tag{102}$$

When the independent variables are transformed to the Blasius variables and a new dependent variable

$$r(\eta) = \frac{T - T_e}{u_{1e}^{\,2}/2c_p} \tag{103}$$

is introduced into Eq. (101), the ordinary inhomogeneous equation that results is

$$r'' + \frac{1}{2}\,\mathrm{Pr}\,fr' = -2\,\mathrm{Pr}\,f''^2 \tag{104}$$

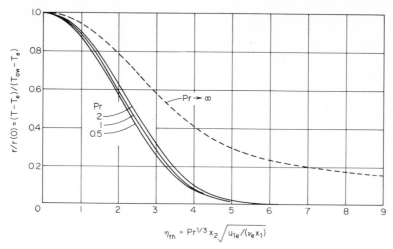

Fig. 25. Temperature profiles in the laminar boundary layer on an insulated plate—constant property, high-speed flow.

with boundary conditions

$$\eta = 0 \ , \ r' = 0$$
$$\eta \to \infty \ , \ r \to 0 \tag{105}$$

The solutions of this problem are indicated in Fig. 25. The temperature distributions shown are based on calculations employing exact velocity distributions for Pr near unity and a linear velocity distribution for very large Pr. These temperatures have been normalized by the temperature rise at the surface and the abscissa is the η_{th} utilized in Fig. 24.

Figure 25 shows less correlation for different Pr than was exhibited for the uniform surface temperature case in Fig. 24. The implication here is that the thermal boundary layer produced by viscous dissipation grows at a rate different from $Pr^{-1/3}$ for all but the very large values of Pr. For Pr near unity the growth factor is closer to $Pr^{-.28}$.

The adiabatic wall temperature (recovery temperature) is given by

$$T_{aw} = T_e + r(0) \frac{u_{1e}^2}{2c_p} \tag{106}$$

Figure 26 shows the dependence of the recovery factor $r(0)$ on Pr as given by Refs. [57] and [60] (solid line). In the region $0.5 < Pr < 2$ the formula

$$r(0) = Pr^{1/2} \tag{107}$$

represents the calculated values to within 1 percent. For Pr = 7, Eq. (107) yields results high by 5.4 percent. The dotted line in Fig. 26 represents an extrapolation of the exact numerical results for Pr < 15 to approach asymptotically the limiting value

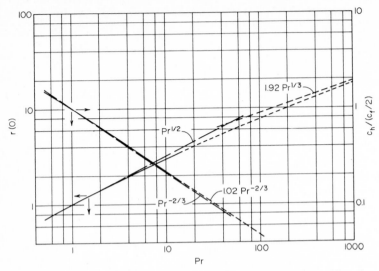

Fig. 26. Influence of Prandtl number on the recovery factor and modified Reynolds analogy for a laminar boundary layer on a flat plate.

$$r(0) \;=\; 1.92\,\mathrm{Pr}^{1/3} \tag{108}$$

resulting explicitly when a linear velocity distribution exists throughout the thermal boundary layer. The very high value of Pr at the region where the two solutions approach is consistent with the spread in temperature distributions indicated in Fig. 25.

For a uniform surface temperature plate in high-speed flow, the temperature distribution within the boundary layer is expressed by a superposition of the two Pohlhausen solutions [58]. This is permissible because the energy equation (Eq. (76)) with constant properties is linear in temperature. The general solution of the energy equation is the sum of the general solution of the homogeneous equation (Eq. (91)) and a particular solution of the inhomogeneous equation (Eq. (101))

$$T - T_e \;=\; (T_w - T_{aw})\,Y_0(\eta) \;+\; \frac{u_{1e}^2}{2c_p}\,r(\eta) \tag{109}$$

The heat transfer rate is therefore expressed in terms of the driving potential $T_w - T_{aw}$. The appropriate Stanton number is again represented by Eq. (96), T_{aw} is given by Eq. (106), and $r(0)$ is given by Eq. (107) or Fig. 26.

39. Liquids with Variable Viscosity. When the temperature difference between a liquid and a surface becomes significant, it is necessary in the evaluation of convective heating to consider the temperature dependence of the viscosity across the boundary layer. Calculations of convective heating are made [61] for a liquid whose viscosity varies as

$$\frac{\mu}{\mu_w} \;=\; \left(\frac{T_w + T_c}{T + T_c}\right)^b$$

where b and T_c are constants. The boundary-layer equations are solved through a transformation of independent variables identical in form with Eq. (80), but with the kinematic viscosity ν evaluated at the surface temperature. The resulting transformed momentum equation is

$$\left(\frac{\mu}{\mu_w} f''\right)' + \frac{1}{2} ff'' = 0$$

and the energy equation, where viscous dissipation has been neglected, is identical with Eq. (93) but with Pr evaluated at the wall temperature. In Ref. 61 the form of the solution requires a choice of the constant b ($b = 3$ in most of the examples) but avoids the necessity of predetermining an explicit value of the constant T_c. The skin friction and heat transfer are expressed directly in terms of the viscosity ratio across the boundary layer μ_w/μ_e and the Prandtl number at the surface.

Figure 27 shows the velocity distributions in a boundary layer of a liquid with $Pr_w = 100$ (e.g., sulfuric acid at room temperature). For this Prandtl number, the thermal boundary layer penetration into the liquid is much less than the flow boundary layer, and the regions where viscosity variations occur are confined close to the surface. The curve corresponding to $\mu_w/\mu_e = 1$ is the Blasius solution (see Fig. 23). The curve labeled $\mu_w/\mu_e = 0.23$ corresponds to a heated surface where the low viscosity near the surface requires steeper velocity gradients to maintain a continuity of shear with the outer portion of the boundary layer. The heated free stream cases reveal the opposite effects. In general, the outer portions of the flow boundary layers act similarly to the velocity distribution of the Blasius case except for their being displaced in or out by the effects that have taken place in the thermal boundary layer.

The temperature profiles for different Pr_w and μ_w/μ_e are indicated in Fig. 28. Note that the curves for $Pr_w = 10$ apply equally well for greater Prandtl numbers because of the use of the thermal boundary layer thickness parameter as the abscissa (see Fig. 24).

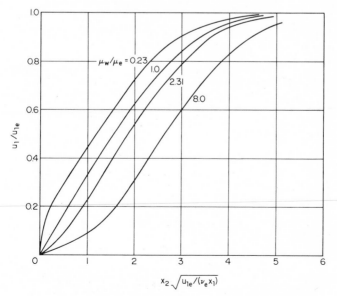

Fig. 27. Velocity profiles in the laminar boundary layer of a liquid, $Pr_w = 100$ [61].

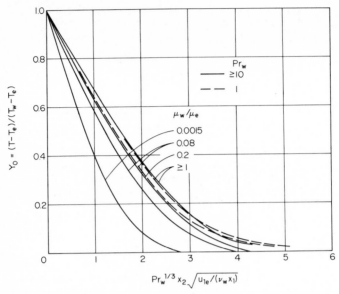

Fig. 28. Temperature profiles in the laminar boundary layer of a liquid [61].

The effects of the viscosity variation across the boundary layer on the surface shear stress and heat flux are shown in Fig. 29. The shear stress is normalized by the value obtained from the Blasius solution with free stream properties. The heat flux is normalized by the Pohlhausen value with the viscosity and Prandtl number evaluated

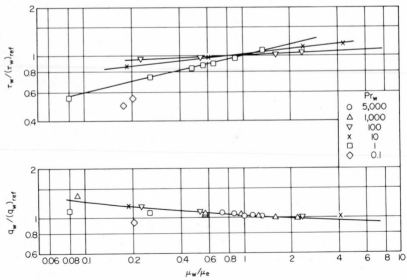

Fig. 29. Effects of viscosity changes across a laminar liquid boundary layer on surface shear and heat flux—reference shear from Blasius solution with free-stream properties, reference heat flux from Pohlhausen solution with wall properties [61].

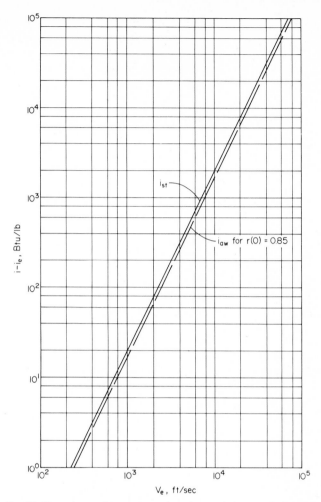

Fig. 30. Dependence of stagnation and recovery enthalpies on flight speed.

at the wall temperature. Note at the higher Prandtl numbers the wall shear becomes less dependent on the fluid properties.

40. Ideal Gases at High Temperatures. In convective processes encountered currently in aerodynamics and astronautics, such large temperature variations occur that the gas properties can no longer be treated as constant. The rise of stagnation enthalpy in excess of the free stream enthalpy for the range of speeds from sonic to meteor entry is shown in Fig. 30 as a solid line. Enthalpy is used here because it is a more meaningful measure of energy potential than temperature when specific heat variations occur. The rise in the recovery enthalpy of a flat plate is also indicated (dashed line) although the plate is an unrealistic aerodynamic shape for the higher velocities. Historically, the variations in air properties were considered for increasingly complex gas behavior from an ideal gas to a real gas, including the effects of dissociation and ionization. The real gas behavior is so complex that numerical analysis is the only means of introducing it into boundary layer problems. Because the cases of simpler gas behavior are directly applicable to many aircraft problems and are guides for correlating the numerical results of the real-gas computations, examples of ideal gas solutions will be presented first.

Equations (74) through (76) are converted to a convenient form through the Howarth-Dorodnitsyn transformation of the independent variables [62] from x_1, x_2 to ζ, η as follows

$$\zeta = u_{1e} \mu_r \rho_r x_1$$

$$\eta = \sqrt{\frac{u_{1e}}{\nu_r x_1}} \int_0^{x_2} \frac{\rho}{\rho_r} dx_2 \qquad (110)$$

The subscript r represents a reference condition for the properties, usually the conditions at the edge of the boundary layer or at the surface.

The transformed momentum and energy equations are

$$(C_r f'')' + \frac{1}{2} ff'' = 0 \qquad (111)$$

$$\left(\frac{C_r g'}{\mathrm{Pr}}\right)' + \frac{1}{2} fg' + \frac{u_{1e}^2}{I_e} \left[C_r\left(1 - \frac{1}{\mathrm{Pr}}\right) f'f''\right]' = \zeta f' \frac{\partial g}{\partial \zeta} \qquad (112)$$

where

$$f' = \frac{u_1}{u_{1e}} \qquad g = \frac{I}{I_e} \qquad C_r = \frac{\mu\rho}{\mu_r \rho_r} \qquad \frac{\partial}{\partial \eta} = (\ )' \qquad (113)$$

with the boundary conditions

$$\eta = 0 \ ; \quad f = f' = 0 \ , \ g = i_w/I_e$$

$$\eta \to \infty \ ; \quad f' \to 1 \ , \quad g \to 1$$

The term on the right side of Eq. (112) is required when the surface temperature is nonuniform, a problem treated in § §43 to 45. In the present subsection, for the case of constant surface temperature, it can be neglected. The fluid properties are introduced into these equations only through C_r and Pr, which involve combinations of individual physical properties. Certain values of C_r and Pr permit simplifications that lead to useful general relationships.

Prandtl Number Equal to Unity. If Pr = 1, considerable simplification results. Equation (112) acquires a form identical with Eq. (111) with g being analogous to f'. A solution of the energy equation, therefore, is directly expressible in terms of the velocity distribution as

$$g = \left(1 - \frac{i_w}{I_e}\right) f' + \frac{i_w}{I_e} \qquad (114)$$

after the boundary conditions at the surface and boundary-layer edge have been introduced. When the wall enthalpy equals the total enthalpy, Eq. (114) indicates that $g = 1$; i.e., the total enthalpy is constant throughout the boundary layer. For other wall enthalpies, the local total enthalpy is linearly dependent on the local velocity. The corresponding static enthalpy distribution is

$$\frac{i}{i_e} = 1 + \frac{i_w - i_e}{i_e}(1 - f') + \frac{u_{1e}^2}{2i_e}(1 - f'^2) \tag{115}$$

Reynolds analogy is given by

$$c_h = \frac{c_f}{2} \tag{116}$$

with the requirement that the recovery enthalpy be

$$i_{aw} = i_e \quad \text{or} \quad r(0) = 1$$

Note that $c_f/2$ in Eq. (116) differs in magnitude from that given by Eq. (85) because of the departure from the Blasius solution by the existence of $C_r(\eta)$ in Eq. (111).

The use of Pr = 1 results in the following:

1. The energy equation is replaced by Eq. (115) for evaluating local properties in the momentum equation.
2. The skin friction coefficient and Stanton number are directly related.

Viscosity-Density Product Equal to a Constant. If $C_r = \overline{C}$ and Pr are constant, a natural transformation suggests itself [63] where η is replaced by

$$\eta_c = \frac{\eta}{\sqrt{\overline{C}}} \tag{117}$$

and

$$F'(\eta_c) = f'(\eta) \tag{118}$$

$$G(\eta_c) = g(\eta) \tag{119}$$

Equations (111) and (112) become

$$F''' + \frac{1}{2}FF'' = 0 \tag{120}$$

$$G'' + \frac{1}{2}\mathrm{Pr}\,FG' + \frac{u_{1e}^2}{i_e}(\mathrm{Pr} - 1)(F'F'')' = \mathrm{Pr}\,\zeta F'\frac{\partial G}{\partial \zeta} \tag{121}$$

with boundary conditions

$$\eta_c = 0 \; ; \quad F = F' = 0 \,, \; G = G_w(\zeta)$$
$$\eta_c \to \infty \; ; \quad F' \to 1 \qquad , \; G \to 1$$

The assumption of constant C_r, therefore, permits separation of the momentum equation from its dependence on the energy equation and results in an energy equation that is linear in G so that general solutions can be obtained from a superposition of individual solutions.

Equation (120) with its boundary conditions is the Blasius problem again. For the case of uniform surface temperature, the right side of Eq. (121) vanishes and the energy equation is satisfied by

$$\frac{i}{i_e} = \frac{i_w - i_{aw}}{i_e} Y_0(\eta_c) + \frac{u_{1e}^2}{2 i_e} r(\eta_c) + 1 \tag{122}$$

where Y_0 and r are obtainable from Figs. 24 and 25 when η_{th} is adjusted according to Eq. (117). The recovery factor $r(0)$ is independent of $\overline{C}$ and, therefore, is identical with the constant property value given in Fig. 26 and Eq. (107).

The local skin friction coefficient can be written as

$$\frac{c_f}{2} = \frac{0.332}{\sqrt{Re_{x_1 e}}} \frac{\mu_w \rho_w / \mu_e \rho_e}{\sqrt{\overline{C}}} = \left(\frac{c_f}{2}\right)_i \frac{\mu_w \rho_w / \mu_e \rho_e}{\sqrt{\overline{C}}} \tag{123}$$

The term $(c_f/2)_i$ represents the skin friction coefficient corresponding to a constant property boundary layer at the same local length Reynolds number. Similarly, the Stanton number is given by

$$c_h = \frac{0.332}{\sqrt{Re_{x_1 e}}} \frac{\mu_w \rho_w / \mu_e \rho_e}{\sqrt{\overline{C}}} Pr^{-2/3} = c_{h_i} \frac{\mu_w \rho_w / \mu_e \rho_e}{\sqrt{\overline{C}}} \tag{124}$$

where c_{h_i} is the corresponding constant property Stanton number.

Another quantity that follows directly from the constant $\overline{C}$ solution is the boundary-layer form factor [63] (evaluated for $Pr = 0.72$)

$$\frac{\delta_1^*}{\theta_{11}} = 2.59 + 3.19 \frac{u_{1e}^2}{2 i_e} + 2.95 \frac{i_w - i_{aw}}{i_e} \tag{125}$$

On slender bodies the quantity $u_{1e}^2/2i_e$ in Eqs. (122) and (125) can be replaced by $[(\gamma - 1)/2]M_e^2$. Also, for conditions where the specific heat remains constant, the enthalpy terms are replaceable by $c_p T$.

For a gas that satisfies the perfect gas equation of state and whose viscosity obeys the equation

$$\frac{\mu}{\mu_e} = \frac{T}{T_e} \tag{126}$$

where T_e is the reference condition, the constant $\overline{C}$ becomes

$$\overline{C} = C_e = \frac{\mu \rho}{\mu_e \rho_e} = \frac{\mu_w \rho_w}{\mu_e \rho_e} \equiv 1 \tag{127}$$

Thus, Eqs. (123) and (124) indicate that the skin friction coefficient and Stanton number remain equal to their constant property values. In terms of these dimensionless transfer coefficients the effects of the linear dependence of viscosity on temperature just cancel those of the perfect gas variation of the density. It should be noted, however, that the density variation itself still affects the boundary-layer thickness through the second of Eqs. (110); this is also evident in Eq. (125).

For a constant temperature plate, Chapman and Rubesin [63] modified Eq. (126) to

$$\frac{\mu}{\mu_e} = C_{ew} \frac{T}{T_e} \tag{128}$$

in order to approximate better the actual viscosity distribution near the surface. Equation (128) with the perfect gas equation of state yields

$$\overline{C} = C_{ew} = \frac{\mu_w \rho_w}{\mu_e \rho_e}$$

and Eqs. (123) and (124) become

$$\frac{c_f}{2} = \left(\frac{c_f}{2}\right)_i \sqrt{\frac{\mu_w \rho_w}{\mu_e \rho_e}} \tag{129}$$

$$c_h = \frac{c_f}{2} \Pr^{-2/3} \tag{130}$$

$$c_h = c_{h_i} \sqrt{\frac{\mu_w \rho_w}{\mu_e \rho_e}} \tag{131}$$

The above skin friction relationship was deduced intuitively many years earlier by von Kármán [64], who assumed local wall properties would control the skin friction law when property variations occur. Thus, Eq. (129) is equivalent to

$$\left(\frac{c_f}{2}\right)_w = \frac{0.332}{\sqrt{\rho_w u_{1e} x_1/\mu_w}}$$

where

$$\tau_w = \rho_w u_{1e}^2 \left(\frac{c_f}{2}\right)_w$$

Similarly, since Pr has been assumed constant in Eq. (121), the modified Reynolds analogy also applies under these conditions with

$$c_{h_w} = \left(\frac{c_f}{2}\right)_w \Pr^{-2/3}$$

where

$$q_w = c_{h_w} \rho_w u_{1e} (i_w - i_{aw})$$

Power Law Viscosity. The early aerodynamic investigations [65--69] that solved equations equivalent to Eqs. (111) and (112) expressed the viscosity with the approximate relationship

$$\frac{\mu}{\mu_e} = \left(\frac{T}{T_e}\right)^n \tag{132}$$

The density was assumed to follow the ideal equation of state, and Pr was assumed constant. The skin friction coefficients for an insulated flat plate predicted by these investigators are shown in Fig. 31. The results given by Eq. (129) with the wall temperature evaluated from Eqs. (106) and (107) are also included. A comparison of these results reveals the following:

1. For a given Prandtl number the value of the viscosity exponent has a major effect on the skin friction behavior with Mach number.
2. A linear distribution of viscosity ($n = 1$) yields the same skin friction coefficient as with constant properties (see Eq. (85)).
3. For a fixed viscosity exponent, changes in Pr between 0.72 and 1.0 have negligible effect on the skin friction coefficient.
4. The use of wall properties in the Blasius formula tends to underestimate the skin friction coefficient on an insulated plate by a few percent.
5. The reduction of laminar boundary-layer skin friction at flight Mach numbers of aircraft is moderate; it is less than 17 percent for $M_e < 5$.

Sutherland Law Viscosity. At moderate enthalpies, in the absence of any dissociation, the gas has the following properties:

1. The gas is a single species that satisfies the ideal equation of state.
2. The viscosity is described by the Sutherland formula.
3. The Prandtl number is slowly varying.

Crocco [60] solved equations equivalent to Eqs. (111) and (112) for the above stated properties, but with Pr set equal to a constant. The combination of properties 1 and 2 yields the quantity C_r in Eqs. (111) and (112) as

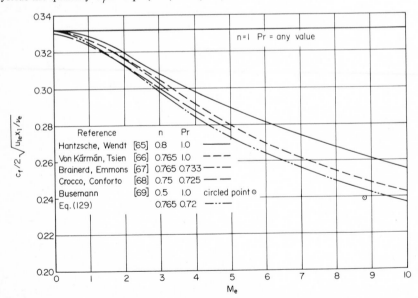

Fig. 31. Local laminar boundary layer skin friction coefficient on an insulated flat plate from early aerodynamic investigations, n = viscosity exponent in Eq. (132).

$$C_e = \frac{\mu\rho}{\mu_e\rho_e} = \sqrt{\frac{T}{T_e}}\left(\frac{1+\theta}{T/T_e + \theta}\right) \qquad (133)$$

where $\theta = T_c/T_e$ and T_c is the Sutherland constant corresponding to the specific gas. Values of θ are indicated in Table 4 for a variety of gases and boundary-layer edge temperatures characteristic of those occurring in the stratosphere, under room conditions, and in products of combustion. The range of θ from 0.03 to 3.09 is covered by Crocco's numerical results for $\theta = 0$, $1/3$, 1, and 3. Because enthalpy is employed as the thermodynamic dependent variable in Eq. (112) to account for specific heat variations, it is necessary to express C_e in terms of enthalpy rather than temperature as in Eq. (133). This is no problem when attention is confined to a specific gas where the enthalpy and temperature are uniquely related at specified pressures. Crocco, however, chose to avoid this approach because it would confine his results to specific gases and thermodynamic conditions. To retain generality, he made the assumption that the temperature ratio in Eq. (133) can be replaced by the enthalpy ratio i/i_e without introducing serious errors.

TABLE 4. Prandtl Number and Sutherland Constant for Gases [60]

Gas	Pr $T = 230°K$	Sutherland constant T_c °K	$\theta = T_c/T_e$		
			T_e 218°K	T_e 300°K	T_e 3000°K
H_2	0.717	90	0.413	0.300	0.030
CO	0.765	104	0.477	0.347	0.035
N_2	0.739	112	0.514	0.373	0.037
Air	0.725	116	0.532	0.387	0.039
O_2	0.731	131	0.601	0.437	0.044
CO_2	0.805	266	1.220	0.887	0.089
H_2O	1.08	673	3.09	2.24	0.224

A major result of Crocco's numerical solutions was the discovery that the functional dependence of the local enthalpy on the local velocity is independent of the particular law of viscosity employed. Thus, $i(f')$ found for the simplified case of $C_e = 1$ applies for all values of θ. The conclusions deduced from this discovery are that the modified Reynolds analogy of Eq. (97) or Fig. 26 and the recovery factor expression of Eq. (107) or Fig. 26 apply to all gases, regardless of their viscosity laws as long as Prandtl number is constant. Another significant consequence of this discovery is that it simplifies the solution of Eqs. (111) and (112) by avoiding either a simultaneous solution of two differential equations or a sequential iteration process. The simpler process uses Eq. (122) and the Blasius solution to relate i and f'. Then C_e is evaluated in terms of f' through Eq. (133) with T/T_e replaced by i/i_e and with the proper θ, and Eq. (111) is solved to yield the final velocity distribution. The local enthalpy distribution in terms of the local velocity is given by

$$\frac{i}{i_e} = 1 + \frac{i_w - i_{aw}}{i_e}Y_0(f') + \frac{u_{1e}^2}{2i_e}r(0)\left[\frac{r(f')}{r(0)}\right] \qquad (134)$$

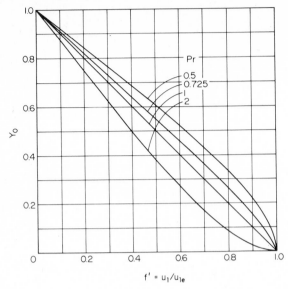

Fig. 32. Laminar boundary layer enthalpy profile function on a uniform temperature flat plate, $C_e = 1$ [60].

where the enthalpy profile functions Y_0 and $r/r(0)$ are plotted in Figs. 32 and 33 for several Prandtl numbers. Many authors have argued that Eq. (115) can be modified to account for Prandtl number deviations from unity by replacing I_e by i_{aw} and multiplying the last term by $r(0)$. A comparison with Eq. (134) reveals that this suggestion is equivalent to retaining Y_0 and $r/r(0)$ characteristic of $Pr = 1$ for all

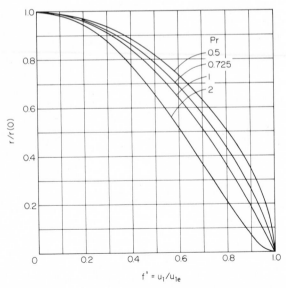

Fig. 33. Laminar boundary layer enthalpy profile function on an insulated flat plate, $C_e = 1$ [60].

Prandtl numbers. Reference to Figs. 32 and 33 indicates the errors introduced by this procedure. For example, on a plate at constant temperature, the local temperature at $f' = u_1/u_{1e} = 0.5$ is 10 percent higher at Pr = 0.725 from Fig. 32 than would be given by the aforementioned rule.

In the use of the boundary-layer integral equations, the form of the velocity profile in terms of x_2 or η is often assumed. The velocity ratio u_1/u_{1e} then becomes the variable of integration, and the enthalpy distributions needed to evaluate the local properties are obtainable from Figs. 32 and 33. Correlation equations that fit these results within 0.015 of the ordinates are

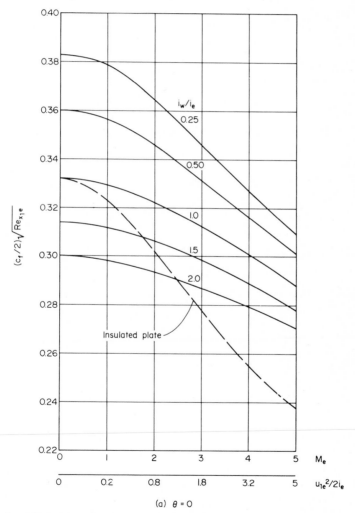

(a) $\theta = 0$

Fig. 34. Local laminar boundary layer skin friction coefficient on a flat plate at uniform temperature, Pr = 0.725 [60].

$$Y_0 = 1 - Pr^{1/3}\frac{u_1}{u_{1e}} - (1 - Pr^{1/3})\left(\frac{u_1}{u_{1e}}\right)^{6.3\,Pr^{-1/2}} \tag{135}$$

$$\frac{r}{r(0)} = 1 - Pr^{1/2}\left(\frac{u_1}{u_{1e}}\right)^2 - (1 - Pr^{1/2})\left(\frac{u_1}{u_{1e}}\right)^{7.3\,Pr^{-0.38}} \tag{136}$$

Crocco's results for the local skin-friction coefficient are shown in Fig. 34. The abscissa is expressed in terms of the Eckert number $u_{1e}^2/2i_e$, or in terms of the local Mach number M_e if the free stream specific heat is constant and $\gamma = 1.4$. Figure

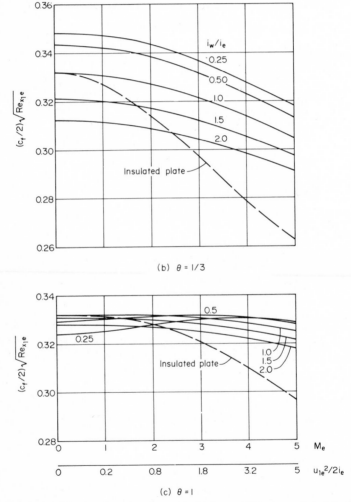

(b) $\theta = 1/3$

(c) $\theta = 1$

Fig. 34 (*Cont.*). Local laminar boundary layer skin friction coefficient on a flat plate at uniform temperature, Pr = 0.725 [60].

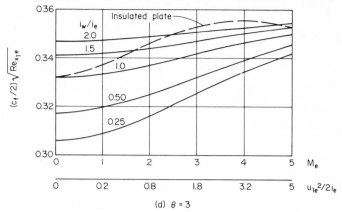

(d) $\theta = 3$

Fig. 34 (*Cont.*). Local laminar boundary layer skin friction coefficient on a flat plate at uniform temperature, Pr = 0.725 [60].

34*a, b, c,* and *d* corresponds to values of $\theta = 0, 1/3, 1,$ and 3, respectively. The individual curves represent values for constant surface temperatures where $i_w/i_e = 0.25, 0.50, 1.0, 1.5, 2.0,$ and for an insulated plate. Although these results are based on calculations with Pr = 0.725, reference to Fig. 31 indicates the insensitivity of the local skin friction coefficient to variations of Pr about unity, thereby indicating the generality of these results in terms of Prandtl numbers corresponding to gases.

The local convective heating to a uniform temperature plate is evaluated as follows:

1. The value of θ is calculated for the particular gas and from the temperature at the edge of the boundary layer.

2. The Reynolds number is evaluated.

3. The local skin friction coefficient is found from Fig. 34.

4. The local Stanton number is obtained from the modified Reynolds analogy of Eq. (97) or Fig. 26.

5. The adiabatic wall temperature or enthalpy is calculated from Eq. (106) or

$$i_{aw} = i_e + r(0) \frac{u_{1e}^2}{2}$$

(137)

with the recovery factor obtained from Eq. (107) or Fig. 26.

6. The local convective heat flux is calculated from

$$q_w = \rho_e u_{1e} c_h \left[i_w - i_e - r(0) \frac{u_{1e}^2}{2} \right]$$

(138)

For slender aircraft flying in the stratosphere, the temperature at the edge of the boundary layer is $-67.6°F$ and the value of θ for air based on the Sutherland constant is 0.505. Van Driest [70] repeated Crocco's analysis for these conditions and Pr = 0.75. Graphs of the local velocity, temperature, and Mach number profiles for an extensive range of conditions are presented in Ref. 70. The local skin friction coefficient is indicated in Fig. 35 for M_e up to 20, and for wall temperatures $0.25 \leq i_w/i_e \leq 6$ and for an insulated plate. Two examples of the solutions based on a constant value of $C_e = C_{ew}$ are indicated for comparison with the insulated plate curve (circled points). Note the use of wall properties underestimates the skin friction here by about 5 percent. The difference between 0.332 and the ordinate of the $i_w/i_e = 1$ curve represents the error in using the $C_e = C_{ew}$ solution for this surface boundary condition. Errors exceeding 10 percent are indicated at Mach numbers greater than 6 for this cooled surface condition.

Velocity Distributions. A comparison of the velocity distributions obtained from the van Driest and the Chapman-Rubesin theories is indicated in Fig. 36a and b for an insulated and cooled plate, respectively. The Chapman-Rubesin theory is based on the assumption that $C_e = C_{ew}$ is constant throughout the boundary layer, c_p equals a constant, and Pr = 0.72. The difference in Pr used in the two theories should not produce a noticeable deviation in these results. In spite of the rather drastic assumption of $C_e = C_{ew}$, the good agreement between the results from the two theories, except for the cooled plate at $M_e = 16$, warrants the use of the convenient Chapman-Rubesin theory for establishing reasonably accurate boundary-layer velocity distributions, especially in the range of Mach numbers where the effects of dissociation are negligible ($M_e \leq 8$). The velocity distributions are obtained from

$$x_2 \sqrt{\frac{u_{1e}}{\nu_e x_1}} = \sqrt{C_{ew}} \left[\eta_c + \frac{u_{1e}^2}{2i_e} \bar{F}(\eta_c) + \frac{i_w - i_{aw}}{i_e} \bar{Y}_0(\eta_c) \right] \tag{139}$$

and the functional dependence (see Fig. 23)

$$\frac{u_1}{u_{1e}} = f'(\eta_c)$$

The following functions are listed in Table 5

$$\bar{F}(\eta_c) = \int_0^{\eta_c} r(\eta)\, d\eta \tag{140}$$

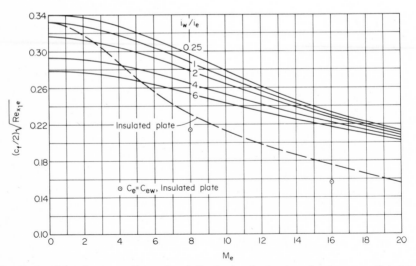

Fig. 35. Local skin friction coefficient for air flowing in a laminar boundary layer on a flat plate, Pr = 0.75, $\theta = 0.505$ [70].

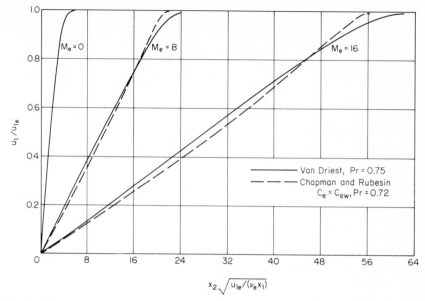

Fig. 36. Velocity profiles in a laminar boundary layer on a flat plate, $\theta = 0.505$. (a) insulated surface.

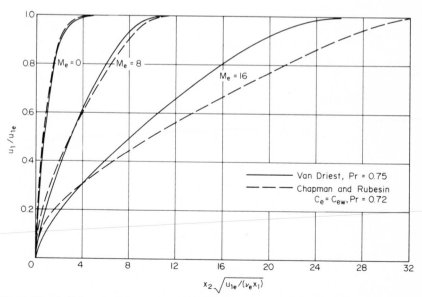

Fig. 36 (*Cont.*). Velocity profiles in a laminar boundary layer on a flat plate, $\theta = 0.505$. (b) $i_w/i_e = 1/4$.

TABLE 5. Laminar Boundary-layer Characteristic Functions for $Pr_T = 0.72$, $C_e = C_{ew}$

η_c	r	$\bar{F}$	Y_0	Y_1	Y_2	Y_3	Y_4	Y_5	Y_{10}	$\bar{Y}_0$	$\bar{Y}_1$	$\bar{Y}_2$	$\bar{Y}_3$	$\bar{Y}_4$	$\bar{Y}_5$	$\bar{Y}_{10}$
.0	0.845	0	1.000	1.000	1.000	1.000	1.000	1.000	1.000	0	0	0	0	0	0	0
0.4	0.832	0.336	0.882	0.807	0.765	0.735	0.711	0.690	0.617	0.376	0.362	0.352	0.346	0.342	0.338	0.322
0.8	0.798	0.662	0.764	0.627	0.556	0.507	0.468	0.438	0.338	0.706	0.648	0.616	0.594	0.576	0.560	0.510
1.2	0.736	0.970	0.647	0.469	0.384	0.329	0.289	0.258	0.167	0.988	0.866	0.802	0.760	0.726	0.698	0.608
1.6	0.653	1.250	0.535	0.338	0.253	0.202	0.167	0.142	0.075	1.224	1.026	0.928	0.866	0.816	0.778	0.656
2.0	0.557	1.492	0.430	0.234	0.159	0.118	0.091	0.074	0.032	1.418	1.140	1.010	0.928	0.866	0.820	0.676
2.4	0.454	1.694	0.333	0.156	0.096	0.066	0.047	0.036	0.012	1.570	1.218	1.062	0.964	0.894	0.840	0.684
2.8	0.352	1.854	0.249	0.100	0.056	0.034	0.024	0.017	0.005	1.686	1.268	1.092	0.984	0.908	0.852	0.688
3.2	0.259	1.976	0.179	0.062	0.031	0.018	0.011	0.008	0.002	1.770	1.300	1.108	0.994	0.914	0.856	0.688
3.6	0.182	2.040	0.123	0.037	0.017	0.009	0.005	0.003	0.001	1.830	1.318	1.116	1.000	0.918	0.858	
4.0	0.121	2.124	0.081	0.021	0.009	0.004	0.002	0.001	0.000	1.872	1.330	1.122	1.002	0.920	0.858	
4.4	0.076	2.162	0.051	0.011	0.004	0.002	0.001	0.001		1.898	1.336	1.124	1.004	0.920		
4.8	0.045	2.186	0.031	0.006	0.002	0.001	0.000	0.000		1.914	1.340	1.126	1.004			
5.2	0.025	2.202	0.018	0.003	0.001	0.000				1.924	1.342	1.126				
5.6	0.014	2.210	0.009	0.001	0.000					1.928	1.342					
6.0	0.007	2.214	0.005	0.000						1.930						
6.4	0.003	2.216	0.003							1.932						
7.4	0.000	2.216	0.000							1.932						

$$\bar{Y}_0(\eta_c) = \int_0^{\eta_c} Y_0(\eta)\,d\eta \tag{141}$$

Corresponding temperature profiles in the boundary layer are obtained from the velocity profiles through Eq. (134) together with either Figs. 32 and 33 or the correlation equations (135 and 136).

41. Air as a Real Gas in Chemical Equilibrium. The upper end of the enthalpy range indicated in Fig. 30 introduces Prandtl number variations and the nonideal effects of dissociation and ionization in the behavior of equilibrium air. Several studies [71–76] have determined the effects of these property variations on the behavior of the laminar boundary layer for successively increasing speeds. A characteristic common to these theories because of the complexity of the behavior of air at elevated enthalpies is the reliance on completely numerical computation of a relatively limited number of examples. The results, in general, are not markedly different from the ideal gas cases. The variations in Pr_T cause the recovery factor in the heat flux equation (Eq. (138)) to be dependent on the surface temperature and Mach number at the edge of the boundary layer. When this relatively small effect is taken into account, the skin friction and heat transfer coefficients exceed the van Driest results of the previous section by less than 15 percent for enthalpies characteristic of flight speeds less than 25,000 feet per second and wall temperatures below the sublimation temperature of carbon ($6000°R$). When ionization takes place, however, a marked increase occurs in both the skin friction and heat transfer.

Errors are inherent in the above solutions because of the uncertainties in the transport properties of air at very high temperatures. Theories of Refs. 74 and 76 employ total properties from different sources, while Ref. 75 accounts for equilibrium air by using frozen properties and assuming Le = 1 in the diffusional heat flux contribution. A comparison of skin friction and heat transfer coefficients reveals differences of less than 10 percent between the results of Refs. 74 and 75 and only a few percent between the results of Refs. 75 and 76. Thus, prior to ionization the errors in convective heating predictions caused by property uncertainties are rather small. With the onset of ionization, large errors may be introduced because of the current large uncertainty of the thermal conductivity of ionized air as influenced primarily by the charge transfer cross section of atomic nitrogen. Hence, the marked increase in heat transfer rate with the presence of ionization [76] can only be considered qualitatively correct at present.

A technique for correlating the results for the convective heating behavior of any ideal gas and real air is given in the following section.

42. Reference Enthalpy Method. The behavior of the skin friction coefficient indicated in Figs. 34 and 35 can be correlated to a very good approximation by the modified incompressible formula

$$\frac{c_f}{2} = \frac{0.332}{\sqrt{Re_{x_1 e}}}\sqrt{\frac{\rho'\mu'}{\rho_e \mu_e}} \tag{142}$$

where the properties designated with the prime are evaluated at a reference enthalpy i' and the boundary-layer edge pressure. This correlation technique was expressed originally in terms of a reference temperature T' [71, 77–79] and later as a reference enthalpy i' to account for variations in specific heat [73, 80].

A convenient form of the reference enthalpy is

$$\frac{i'}{i_e} = a + b\frac{i_w}{i_e} + c\frac{i_{aw}}{i_e}$$

Upon evaluation of the coefficients a, b, and c based on the van Driest values of skin friction in Fig. 35, it is found that they differ from those given in the earlier references and that they contain a small Mach number dependence. In addition, Wilson [74] related the reference enthalpy to the enthalpy averaged with respect to the velocity rather than the normal distance in the boundary layer. Because $i(u_1)$ is Prandtl number dependent, Wilson suggested a particular Prandtl number dependence for the coefficients a, b, c. For the range of Pr_T between 0.6 and 1.0, this dependence turns out to be even smaller than the effects of the varying Mach number. The approximate nature of the technique does not warrant the refinements of the small M_e and Pr_T corrections to the coefficients. Therefore, the following formula is adopted

$$\frac{i'}{i_e} = 0.32 + 0.50 \frac{i_w}{i_e} + 0.18 \frac{i_{aw}}{i_e} \tag{143}$$

A convenient form of the skin friction coefficient compatible with the Crocco and van Driest formulations is

$$\frac{c_f}{2} = \frac{0.332}{\sqrt{Re_{x_1e}}} \sqrt{\left(\frac{i'}{i_e}\right)^{1/2} \frac{1 + \theta}{i'/i_e + \theta}} \tag{144}$$

Comparisons of results based on Eqs. (143) and (144) and the earlier more common forms of the reference enthalpy with the results of Figs. 34 and 35 are indicated in Tables 6 and 7. The figures of merit indicated in these tables represent the rms errors of the formulas at the matrix of points $M_e = 0$ and 5, $i_w/i_e = 0.25, 2.0$, and i_{aw}/i_e in Fig. 34 and at $M_e = 0, 5$, and 10, $i_w/i_e = 0.25, 6$, and i_{aw}/i_e in Fig. 35.

TABLE 6. Comparison of Skin Friction Coefficients from Reference Enthalpy Methods with van Driest's Results for Air (Fig. 35)

Source	Reference enthalpy Eq.	% rms error
Rubesin, Johnson [77]	$\dfrac{i'}{i_e} = 1 + 0.58\left(\dfrac{i_w}{i_e} - 1\right) + 0.032\,M_e^2$	1.8
Young, Janssen [71]	$M_e \leq 5.6:\ \dfrac{i'}{i_e} = 1 + 0.58\left(\dfrac{i_w}{i_e} - 1\right) + 0.032\,M_e^2$ $M_e \geq 5.6:\ \dfrac{i'}{i_e} = 0.7 + 0.58\left(\dfrac{i_w}{i_e} - 1\right) + 0.023\,M_e^2$	2.7
Eckert [79]	$\dfrac{i'}{i_e} = 1 + 0.50\left(\dfrac{i_w}{i_e} - 1\right) + 0.038\,M_e^2$	2.0
Eq. (143)	$\dfrac{i'}{i_e} = 1 + 0.50\left(\dfrac{i_w}{i_e} - 1\right) + 0.031\,M_e^2$	0.64

TABLE 7. Comparison of Skin Friction Coefficients from Reference Enthalpy Methods with Crocco's Results (Fig. 34)

Source	Reference enthalpy Eq.	% rms error			
		$\theta = 0$	$\theta = 1/3$	$\theta = 1$	$\theta = 3$
Eckert [79]	$\dfrac{i'}{i_e} = 0.28 + 0.50 \dfrac{i_w}{i_e} + 0.22 \dfrac{i_{aw}}{i_e}$	2.3	1.1	0.65	0.83
Eq. (143)	$\dfrac{i'}{i_e} = 0.32 + 0.50 \dfrac{i_w}{i_e} + 0.18 \dfrac{i_{aw}}{i_e}$	1.2	0.30	0.50	0.64

From Table 6 for air, it can be seen that Eq. (143) gives some improvement in comparison with the van Driest results over the older formulas modified with temperature replaced by enthalpy. The formulas in this table are in their original form which employed the Mach number rather than the recovery enthalpy. Although the coefficients in the formulas differ, the ultimate correlation between the van Driest results and Eq. (144) is quite good with any of the reference enthalpy formulas. This behavior is further substantiated in Table 7 where the widely used Eckert formula is compared with Eq. (143) for the entire range of θ evaluated by Crocco.

At the high speeds where air behaves as a real gas, Wilson [74] shows that an equation equivalent to Eqs. (142) and (143) yields skin friction coefficients that agree with those found from numerical integrations of the boundary-layer equations to within 5 percent for total enthalpies corresponding to free stream speeds up to 25,000 feet per second. Similar close agreement is achieved between the use of the Eckert reference enthalpy and the results of Cohen [75].

For real air, the total Prandtl number varies in a fluctuating manner with enthalpy distribution across the boundary layer. In view of this behavior, it would not be expected a priori that evaluation of Pr_T at the reference enthalpy would be appropriate for evaluating the recovery factor from Eq. (107) or the modified Reynolds analogy from Eq. (97). Comparison with the numerical results of Refs. 71, 72, and 74, however, reveals that this interpretation of the reference enthalpy technique yields results of recovery factor correct to within 2.5 percent and the Reynolds' analogy to within 5 percent. Note that Wilson [74] suggests the use of Pr_{T_w} evaluated at wall enthalpy in the Reynolds' analogy. Comparison of this method with the use of Pr_T' evaluated at the reference enthalpy for surface temperatures below the sublimation temperature of carbon reveals little difference. Because Pr_T' rather than Pr_w or Pr_e yields better agreement with the modified Reynolds' analogy of van Driest [72], the consistent use of Pr_T' in both the recovery factor and Reynolds' analogy as suggested by Eckert is still appropriate. Further, when the assumption Le $= 1$ in Ref. 75 is interpreted as equivalent to the assumption that $Pr_F = Pr_T$, the use of Pr_T' based on the reference enthalpy for the recovery factor and Reynolds' analogy is again borne out by these independent calculations to the accuracies quoted previously.

With the onset of ionization, the reference enthalpy technique yields results that begin to depart from the exact calculations, and recourse to the latter [75, 76] is recommended for accurate predictions.

Impervious Surface with Nonuniform Temperature

The previous subsection was devoted to uniform temperature plates. In practice, however, this ideal condition seldom occurs, and it is necessary to account for the effects of surface temperature variations along the plate on the local and average convective heat transfer rates. This is required especially in the regions directly downstream of surface temperature discontinuities; e.g., at seams between dissimilar structural elements in poor thermal contact. These effects cannot be accounted for by merely utilizing heat transfer coefficients corresponding to a uniform surface plate coupled with the local enthalpy or temperature potentials. Such an approach not only leads to serious errors in magnitude of the local heat flux, but can yield the wrong direction, i.e., whether the heat flow is into or out of the surface.

For the boundary-layer equations to apply precisely to this problem, it is necessary that the temperature gradients along the surface be smaller than those across the boundary layer. This condition is usually satisfied for continuous nonuniform surface temperatures. For discontinuous surface temperatures, the boundary-layer equations will given erroneous results in the immediate vicinity of the discontinuity; however, the affected region along the plate only corresponds to a Reynolds number of about 100 based on the distance from the discontinuity, which is a negligible portion of the normal laminar boundary-layer length of run. Thus, the appropriate boundary-layer equations are identical to Eqs. (74) through (76), but with the more general boundary condition at $x_2 = 0$ of $i_w = i_w(x_1)$. Use of the transformation variables (Eq. (110)) leads to Eqs. (111) and (112), where the right side of the latter must now be retained to satisfy the variable surface temperature boundary condition.

When the property parameters C_r and Pr_T are temperature dependent, Eqs. (111) and (112) comprise a system of coupled, nonlinear, inhomogeneous partial differential equations that are not amenable to analysis. When both C_r and Pr_T are constant, however, several critical simplifications occur. It is fortunate that the importance of accounting for surface temperature variations is greatest when the overall driving potential $T_{aw} - T_w$ is of modest magnitude, i.e., when use of constant C_r and Pr_T is quite reasonable. The momentum equation (Eq. (120)) becomes independent of the energy equation and the transformed velocity function F' remains a function of η_c alone. The energy equation is linear in total energy G, and a series of solutions of Eq. (121) can be superimposed to satisfy the boundary condition $G_w(x_1)$. A similarity solution of Eq. (121) is presented in the following subsection.

43. Polynomial Surface Temperature Distribution. In Ref. 63, Eqs. (120) and (121) are solved with $Pr_T = 0.72$ for a surface enthalpy distribution given by

$$\frac{i_w(x_1)}{i_e} = \frac{i_{aw}}{i_e} + \sum_n a_n \left(\frac{x_1}{L}\right)^n \tag{145}$$

The method of solution expresses G in Eq. (121) as the sum of:

1. The solution of the entire inhomogeneous equation for the particular boundary condition of zero heat transfer to the surface.
2. The series of solutions of the homogeneous portion of Eq. (121) where the surface temperature is expressed as

$$\frac{i_w}{i_e} = a_n \left(\frac{x_1}{L}\right)^n$$

with a_n a constant and n a positive number.

The latter is the only functional form of surface temperature distribution satisfying finite boundary conditions at $x_1 = 0$ that permits expressing the enthalpy as the product of a function of ζ (or x_1) and a function of η. This classical separation of variables in the solution of the partial differential equation (Eq. (121)) was first demonstrated by Schuh. In Eq. (145), n is one of a set of positive numbers, either integer or fractional, whichever leads to a good fit of the prescribed surface temperature distribution with a reasonable number of terms in the summation.

The results of this analysis are expressed by the following formulas:
Enthalpy distribution through the boundary layer

$$\frac{i}{i_e} = 1 + \frac{u_{1e}^2}{2i_e} r(\eta_c) + \sum_n a_n \left(\frac{x_1}{L}\right)^n Y_n(\eta_c) \tag{146}$$

Boundary-layer profile and thickness parameter

$$x_2 \sqrt{\frac{u_{1e}}{\nu_e x_1}} = \sqrt{\bar{C}_{ew}} \left[\eta_c + \frac{u_{1e}^2}{2i_e} \bar{r}(\eta_c) + \sum_n a_n \left(\frac{x_1}{L}\right)^n \bar{Y}_n(\eta_c) \right] \tag{147}$$

Heat flux from the wall

$$q_w(x_1) = \rho_e u_{1e} c_h \frac{C_{ew}(x_1)}{\bar{C}_{ew}} \left[i_e \sum_n a_n \left(\frac{x_1}{L}\right)^n H_n \right] \tag{148}$$

The quantities $r(\eta_c)$, $Y_n(\eta_c)$, $\bar{r}(\eta_c)$, $\bar{Y}_n(\eta_c)$ are tabulated in Table 5 for $n = 0, 1, 2, 3,$ 4, 5, and 10. The value of H_n, the effective potential function, is presented in Table 8 and plotted in Fig. 37 to facilitate interpolation to fractional powers of n. Equations (146) and (147) are direct extensions of their counterparts for a uniform surface temperature plate. The parameter $\bar{C}_{ew}$ in Eqs. (147) and (148), although rigorously considered a constant throughout the boundary layer, is usually chosen at the average surface temperature up to the point x_1. Equation (148) for the local heat flux from the wall has the same form as Eq. (138) where c_h corresponds to the Stanton number on a uniform temperature plate, and the bracketed quantity represents the driving potential appropriate to a nonuniform surface temperature. Thus, the effect of a nonuniform surface temperature is primarily on the effective enthalpy potential for driving the local heat flux. An intuitive approach for including the property variation effects in the constant C_e solution is to use the reference enthalpy technique for evaluating c_h locally (local surface temperature in Eq. (143)) and setting $C_{ew}(x_1) = \bar{C}_{ew}$ in Eq. (148). Thus, the constant C_e solutions are used primarily in establishing H_n and the functional form of the driving potential.

TABLE 8. Effective Potential Function for Exponentially Varying Surface Temperature Distribution, $Pr_T = 0.72$, Eq. (148)

n	H_n	$\dfrac{\Gamma\left(\frac{4}{3}n + 1\right) \Gamma\left(\frac{2}{3}\right)}{\Gamma\left(\frac{4}{3}n + \frac{2}{3}\right)}$
0	1.00	1.00
1	1.65	1.61
2	2.02	1.95
3	2.29	2.21
4	2.52	2.41
5	2.70	2.59
10	3.40	3.25

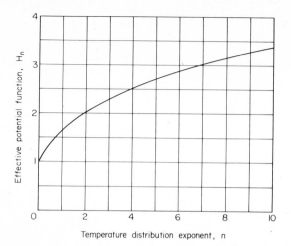

Fig. 37. Effective potential function for exponentially varying surface temperature distributions, see Eq. (148).

Either a prescribed $i_w(x_1)$ or $q_w(x_1)$ may be used to establish the required values of n and a_n. For example, if the surface temperature required to establish a constant heat flux along the surface is sought, Eq. (148) indicates that $n = 1/2$ alone is appropriate when it is recalled that $c_h \sim x_1^{-1/2}$ when property variations along the surface are neglected. Thus, from Eq. (145), $(i_w - i_{aw})/i_e \sim (x_1/L)^{1/2}$ on a constant heat flux plate.

44. Stepwise and Arbitrary Surface Temperature Distribution. It is often inconvenient to express the surface temperature distribution as a polynomial series, especially if temperature discontinuities occur. Thus, the technique of the previous subsection has to be modified to permit a more general variation of surface temperature. Again, the property parameters C_e and Pr_T must be constant to permit a superposition of individual solutions of the energy equation so that the sum satisfies the surface temperature distribution. In this case, the solutions that are superimposed occur on a plate that is at the recovery temperature from the leading edge to a point $x_1 = s$ and at a different constant temperature for $x_1 > s$. This stepwise discontinuous surface temperature accounts for the growth of a thermal boundary layer that starts downstream of the momentum boundary layer which always begins at the leading edge (see Ref. 81 for a summary of the early work).

It has been shown that for property variations for which superposition of solutions is permitted, a series of solutions corresponding to a step in surface temperature can be utilized to represent an arbitrary surface temperature [82]. This approach is identical with the Duhamel method used in heat conduction problems to satisfy time-dependent boundary conditions with a series of solutions involving sudden changes in surface boundary conditions [83]. The resulting convective heat flux distribution expressed in terms of the surface enthalpy distribution is

$$q_w(x_1) = \rho_e u_{1e} c_h(x_1, 0) \int_0^{x_1} \frac{c_h(x_1, s)}{c_h(x_1, 0)} d[i_w(s) - i_{aw}] \qquad (149)$$

Here, $c_h(x_1, s)$ represents the local Stanton number on a plate at a uniform temperature for $x_1 > s$ and preceded by an unheated portion of length s, and $c_h(x_1, 0)$ is the Stanton number on a plate with a uniform temperature over its entire length. The Stieltjes integral represents the enthalpy potential corrected for nonuniform surface effects. In terms of a Riemann integral, Eq. (149) expands to

$$q_w(x_1) = \rho_e u_{1e} c_h(x_1, 0) \left\{ [i_w(0) - i_{aw}] + \int_0^{x_1} \frac{c_h(x_1, s)}{c_h(x_1, 0)} \frac{d[i_w(s) - i_{aw}]}{ds} ds \right.$$

$$\left. + \sum_{j=1}^{k} \frac{c_h(x_1, s_j)}{c_h(x_1, 0)} [i_w(s_j^+) - i_w(s_j^-)] \right\} \quad (150)$$

The first term in parentheses in the enthalpy potential arises from the difference between the leading edge enthalpy of the plate and the recovery enthalpy. The integral term accounts for the portions where continuous surface enthalpy variations occur. The last term sums over a k number of discontinuous jumps in surface enthalpy that may occur downstream of the leading edge. The terms $i_w(s_j^-)$ and $i_w(s_j^+)$ represent the surface enthalpy just upstream and downstream of s_j where the jth jump in enthalpy occurs.

The effect of a stepwise discontinuity in surface temperature on a flat plate can be expressed as

$$\frac{c_h(x_1, s)}{c_h(x_1, 0)} = \left[1 - \left(\frac{s}{x_1} \right)^{3/4} \right]^{-1/3} \quad (151)$$

This closed form equation was obtained through similarity solutions of the energy equation by investigators who assumed that the velocity profile in the boundary layer is linear in η_c for the case of constant C_e and Pr_T or in x_2 for the case of constant fluid properties [84, 85]. Other approximate methods involving solutions of the energy integral equation (Eq. (72)) with a variety of assumed velocity and enthalpy distributions also resulted in Eq. (151). Note that the right side does not contain terms involving the fluid properties, a direct consequence of C_e and Pr_T being assumed constant throughout the boundary layer. Again, an intuitive approach to include property variations is to use the local surface enthalpy in the reference enthalpy technique for evaluating $c_h(x_1, 0)$. The stepwise discontinuous surface temperature solutions are used solely to define the functional form of the enthalpy potential appropriate to an arbitrary surface temperature. A plot of Eq. (151) is given in Fig. 38 ($\beta = 0$ for a flat plate).

The accuracy of the approximate formula (Eq. (151) is assessed by applying it in Eq. (150) to the problem of a polynomial surface temperature distribution and comparing the results with the exact solution of the previous section. By this process it is found that H_n in Eq. (148) may be approximated by gamma functions or a beta function as

$$H_n = \frac{\Gamma\left(\frac{4}{3}n + 1\right)\Gamma\left(\frac{2}{3}\right)}{\Gamma\left(\frac{4}{3}n + \frac{2}{3}\right)} = \frac{4}{3} n B\left(\frac{4}{3}n, \frac{2}{3}\right)$$

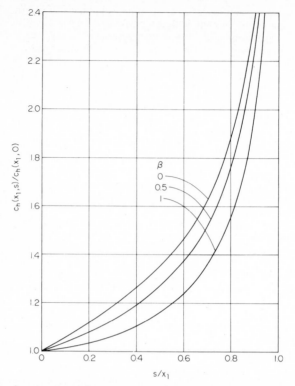

Fig. 38. Effect of a stepwise surface temperature discontinuity on the local Stanton number, Eq. (151) for a flat plate ($\beta = 0$) and Eq. (240) for a surface with a pressure gradient ($\beta > 0$).

This approximation is quite good for a flat plate, resulting in errors of less than 4.5 percent for $n < 10$ (see Table 8).

When $c_h(x_1, 0)$ is based on some average temperature along the plate so that its change with distance is proportional to $x_1^{-1/2}$ and the only temperature discontinuity occurs at the leading edge, Eq. (150) can be integrated directly to yield the average heat flux defined as

$$\bar{q}_w(x_1) = \frac{1}{x_1} \int_0^{x_1} q_w(x_1') \, dx_1' \tag{152}$$

which becomes

$$\bar{q}_w(x_1) = 2\rho_e u_{1e} c_h(x_1, 0) \left\{ i_w(0) - i_{aw} + \int_0^{x_1} \left[1 - \left(\frac{x_1'}{x_1} \right)^{3/4} \right]^{2/3} \frac{d[i_w(x_1') - i_{aw}]}{dx_1'} \, dx_1' \right\} \tag{153}$$

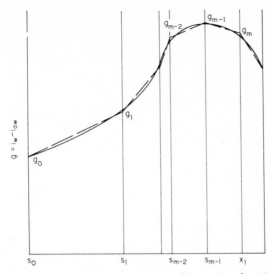

Fig. 39. Surface temperature distribution approximated by a series of continuous ramps.

A particularly convenient technique for evaluating Eqs. (150) and (153) is presented in Ref. 86. The continuous portions of the surface temperature are approximated by a series of m connected straight lines as indicated in Fig. 39. The advantage of the straight line approximation is that the derivatives of the surface enthalpy in the integrands become constants over the axial distance covered by a single line segment. The other portions of the integrands are universal functions, independent of any particular temperature distribution. Thus, for the case where the only temperature discontinuity occurs at the leading edge, Eq. (150) becomes

$$q_w(x_1) = \rho_e u_{1e} c_h(x_1,0) \left\{ g_0 + \frac{g_1 - g_0}{s_1/x_1} G\left(\frac{s_1}{x_1}\right) + \frac{g_2 - g_1}{s_2/x_1 - s_1/x_1} \left[G\left(\frac{s_2}{x_1}\right) - G\left(\frac{s_1}{x_1}\right) \right] \right.$$

$$\left. + \cdots + \frac{g_m - g_{m-1}}{1 - s_{m-1}/x_1} \left[G(1) - G\left(\frac{s_{m-1}}{x_1}\right) \right] \right\} \quad (154)$$

where

$$G\left(\frac{s_n}{x_1}\right) = \int_0^{s_n/x_1} (1 - \alpha^{3/4})^{-1/3} d\alpha$$

and g_n represents the ordinate values of $i_w(s_n) - i_{aw}$ at the end points s_n of the line segments that best fit the continuous temperature distribution (see Fig. 39). Note $s_m = x_1$ and $n \le m$. The function $G(s_n/x_1)$ is tabulated in Ref. 86 and is shown in Fig. 40 ($\beta = 0$ for a flat plate). Similarly, Eq. (153) becomes

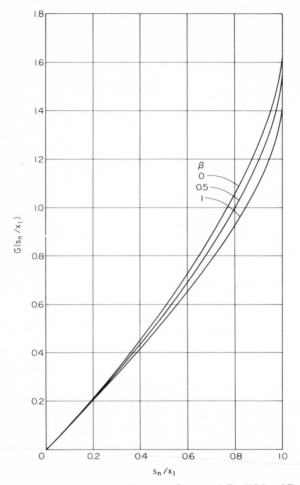

Fig. 40. Laminar boundary layer local heat transfer integral, Eq. (154) and Eq. (243).

$$\bar{q}_w(x_1) = 2\rho_e u_{1e} c_h(x_1,0) \left\{ g_0 + \frac{g_1 - g_0}{s_1/x_1} H\left(\frac{s_1}{x_1}\right) + \frac{g_2 - g_1}{s_2/x_1 - s_1/x_1}\left[H\left(\frac{s_2}{x_1}\right) - H\left(\frac{s_1}{x_1}\right)\right]\right.$$

$$\left. + \cdots + \frac{g_m - g_{m-1}}{1 - s_{m-1}/x_1}\left[H(1) - H\left(\frac{s_{m-1}}{x_1}\right)\right]\right\} \quad (155)$$

where

$$H\left(\frac{s_n}{x_1}\right) = \int_0^{s_n/x_1} (1 - \alpha^{3/4})^{2/3} \, d\alpha$$

The function $H(s_n/x_1)$, tabulated in Ref. 86, is shown in Fig. 41.

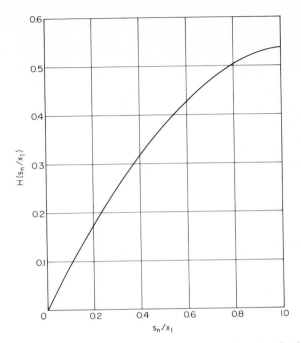

Fig. 41. Laminar boundary layer average heat transfer integral for a flat plate, Eq. (155) [86].

The preceding theory has been verified by several experiments. For example, in Ref. 87 local heat transfer rates were measured in the presence of ramp-like temperature distributions that began downstream of the leading edge (see insert in Fig. 42). The data shown in Fig. 42 agree with the theory (solid line) to within ±10 percent, the estimated accuracy of the data. The dot-dashed line in the figure represents the use of a local temperature potential in estimating the heat flux and yields large errors for this particular form of the surface temperature distribution. Had the leading edge been raised above the recovery temperature, the error in neglecting the variable surface temperature effects would have diminished. In general, if continuous variations in the surface temperature or enthalpy are large compared to the overall driving potential, the variable surface temperature methods must be utilized. For discontinuous surface temperatures much smaller variations are important.

45. Stepwise and Arbitrary Heat Flux Distribution. It is often necessary to evaluate the surface temperature distribution resulting from a prescribed heat flux distribution. If the heat flux distribution can be represented by a polynomial, Eq. (148) can be used to establish a_n, and Eq. (145) directly yields the resulting surface enthalpy or temperature distribution. Note, for surface heat flux distributions varying as x_1^n, the corresponding surface enthalpy varies as $x_1^{n+1/2}$.

For heat flux distributions that cannot be represented by a polynomial, the superposition of solutions yields the surface enthalpy distribution as

$$i_w(x_1) - i_{aw} = \frac{0.207}{\rho_e u_{1e} c_h(x_1, 0) x_1} \int_0^{x_1} \frac{q_w(x_1')}{\left[1 - (x_1'/x_1)^{3/4}\right]^{2/3}} \, dx_1' \qquad (156)$$

Equation (156) results from the solution of a Volterra integral equation composed of

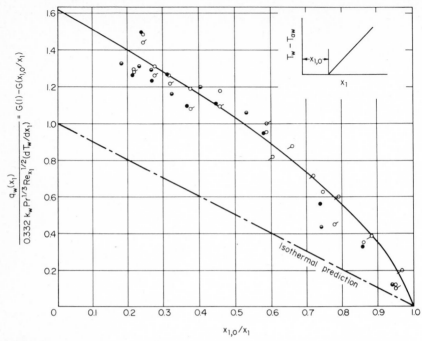

Fig. 42. Comparison of data and theory for a flat plate with a delayed ramp surface temperature distribution [87].

Eqs. (149) and (151) [88]. For the case of a stepwise heat flux distribution where $q_w = 0$ for $x_1 < s$ and $q_w = q_{w0}$ for $x_1 \geq s$, Eq. (156) reduces to

$$i_w(x_1) - i_{aw} = \frac{0.730\, q_{w0}\, I_z\left(\frac{1}{3}, \frac{4}{3}\right)}{\rho_e u_{1e} c_h(x_1, 0)} \tag{157}$$

where

$$I_z(a, b) = \frac{B_z(a, b)}{B_1(a, b)}$$

and $B_z(a, b)$ is the incomplete beta function [89, 133]. The argument of the beta function is

$$z = 1 - \left(\frac{s}{x_1}\right)^{3/4}$$

Hanna and Meyers [90] have argued that the approximate nature of Eq. (151) does not warrant the mathematical complication of the use of Eqs. (156) and (157). They solve the problem of the step in heat flux directly by utilizing the energy integral equation (Eq. (72)) and assuming the velocity and enthalpy profiles in terms of the local momentum and thermal boundary-layer thicknesses are the same as on a uniform surface temperature plate. They found the analog to Eq. (157) to be

$$i_w(x_1) - i_{aw} = \frac{0.722\, q_{w0}}{\rho_e u_{1e} c_h(x_1, 0)} \left(1 - \frac{s}{x_1}\right)^{1/3} \qquad (158)$$

where the constant, evaluated in Ref. 91, varies by less than ±1 percent for the range of Prandtl numbers from 0.7 to 100. The difference between Eqs. (157) and (158) is 6 percent at x_1/s of 1.05, 3 percent at x_1/s of 2, and less than 2 percent for x_1/s greater than 3.5. The experimental results in Ref. 92 could not distinguish these small differences and agree generally quite well with the theories.

The analog to Eq. (156) is

$$i_w(x_1) - i_{aw} = \frac{0.722}{3\rho_e u_{1e} c_h(x_1, 0) x_1} \int_0^{x_1} \frac{q_w(x_1')}{(1 - x_1'/x_1)^{2/3}}\, dx_1' \qquad (159)$$

The approximate technique of Ref. 86 can be applied rather simply to Eq. (159). If the $q_w(x_1)$ distribution up to the point x_1 is represented by a series of m straight line segments, as is the temperature distribution in Fig. 39, and s_n, q_{wn} represent the coordinates of the end points of the line segments, the expression for the local enthalpy at x_1 is

$$i_w(x_1) - i_{aw} = \frac{0.722}{\rho_e u_{1e} c_h(x_1, 0)} \left\{ q_{w0} + \frac{3}{4} \sum_{n=0}^{m} \frac{\left[\left(1 - \dfrac{s_{n-1}}{x_1}\right)^{4/3} - \left(1 - \dfrac{s_n}{x_1}\right)^{4/3}\right]}{\dfrac{s_n}{x_1} - \dfrac{s_{n-1}}{x_1}} \times (q_{wn} - q_{w(n-1)}) \right\} \qquad (160)$$

Surface with Mass Transfer

An effective method of alleviating the intense convective heating of surfaces exposed to very high enthalpy streams is by means of mass transfer cooling systems. The coolant is introduced, usually in gaseous form, into the hot boundary layer from the surface being protected. Such cooling systems have several advantages:

1. The coolant can absorb heat through phase change or endothermic reactions within or at the protected surface.
2. The introduction of the coolant with zero tangential momentum relative to the surface reduces boundary-layer velocities and the attendant convective heat exchange rates.
3. The coolant continues to absorb heat well within the boundary layer and thus reaches temperatures much higher than the surface temperature, the upper limit of internal convective cooling systems.
4. The energy absorbed by the coolant does not have to be removed but is dumped overboard.

Mass transfer cooling is particularly applicable to leading edges of wings, fins, and nose tips of aircraft; turbine blades; and to reentry capsules and missile nose cones. The types of systems include transpiration, ablation, and film cooling.

A transpiration cooling system is characterized by a porous surface material that remains intact with the coolant being pumped through the pores toward the hot boundary layer. The coolant may be a gas or a liquid that changes phase within the porous surface or after it emerges from the surface. This system operates best when pore sizes are so small that the coolant leaves at the surface temperature of the porous solid. These systems are complicated in that they require coolant storage vessels, pumps, controls, distribution ducts, and filters to avoid pore clogging. It is also difficult to fabricate strong, aerodynamically smooth porous surfaces. Another drawback of these systems is that they are unstable because a clogged pore resulting from local overheating seals off the flow of coolant and causes local failure. The advantages of transpiration cooling systems are their versatility in the choice of coolant and coolant distribution, the reusability of the system, and the retention of intact aerodynamically contoured surfaces.

An ablation cooling system is one where the gas entering the boundary layer has been generated by the thermal destruction of a sacrificial solid thermally protecting an underlying structure. The simplest ablation mechanism is the sublimation of a homogeneous material. More complex ablation involves the thermal degradation of composite materials such as reinforced plastics where a heat absorbing pyrolysis occurs below the surface. The gaseous products of pyrolysis pass through a carbonaceous char gaining additional sensible heat or chemical heat through endothermic reactions and then pass into the boundary layer to absorb additional heat. Ablation cooling has the enormous advantage of being self-controlled and requiring no active elements. The disadvantages are that surface dimensions are usually altered, part of the char is eroded mechanically by shear forces rather than through heat absorbing phase change, and the heavy gaseous products are often not as effective in blocking the incoming convective heat as light gases. Furthermore, ablation systems are generally not reusable. For short time applications, dimensional stability has been achieved in ablation systems by employing porous refractory metal surfaces that have been impregnated with lower melting temperature metals that absorb heat by melting, vaporizing, and transpiring through the porous refractory matrix.

A film cooling system is one where a surface is protected by a film of coolant introduced into the boundary layer from either a finite length of porous surface or a slot upstream. A liquid is often used as the coolant to absorb heat by vaporization as it is drawn along the surface by the main stream gas. A severe limitation on such systems is the requirement that gravity or inertial forces be in a direction to keep the liquid film stable and against the surface. In addition, a film cooling system requires all the active elements of a transpiration cooling system. Its main advantage over the latter is in the simpler construction of limited areas of porous surfaces or slots and in localized ducting.

The boundary-layer behavior over a continuously transpiration-cooled surface and an ablation-cooled surface is generally the same, differing only as a result of the specific chemical identity of the coolant. The effects of a porous surface, when the pore size is below some threshold dimension that is a small fraction of the local boundary-layer thickness, and of the flow of liquid char over ablating surfaces do not appreciably alter the behavior of the boundary layer and can be neglected in design considerations. Thus, boundary-layer theory with continuous mass injection is applicable to both types of cooling systems. Further, results of experiments involving transpiration systems can be utilized in the prediction of the behavior of ablation systems. In film cooling systems, because of the discontinuities formed by slots or limited porous regions, the boundary-layer profiles at various stations along the surface are dissimilar so that prediction methods are quite complex and rely on experimental data or rather complicated numerical analyses.

The techniques treated in this section deal with the injection of air into air, resulting in a gas in chemical equilibrium with uniform elemental composition, or of a foreign gas into air, where chemical reactions are frozen.

46. Uniform Surface Temperature

Air as Coolant. When the boundary-layer and coolant gases are the same, the equations controlling boundary-layer behavior are Eqs. (74) through (76). The mass injection at the surface simply alters the boundary conditions (Eq. (77)) at the wall to be

$$x_1 > 0, \ x_2 = 0 \ ; \quad u_1 = 0, \ u_2 = u_{2w}(x_1)$$

$$I = \text{constant} = i_w \text{ or } i_{aw} \left(\frac{\partial i}{\partial x_2} = 0 \right)$$

If the requirement of boundary-layer similarity is imposed, i.e., $u_1/u_{1e} = f'(\eta)$, it is found that the boundary condition for u_2 transforms according to Eq. (110) to

$$f(0) = -\frac{2 \rho_w u_{2w}}{\rho_e u_{1e}} \sqrt{\frac{\rho_e u_{1e} x_1}{\mu_e}} \tag{161}$$

For $f(0)$ to be independent of x_1 (the similarity requirement), $\rho_w u_{2w}$ must be proportional to $x_1^{-1/2}$. A simple heat balance on an element of a porous surface with a transpired gas, or of a subliming surface, reveals that this distribution of gaseous injection is uniquely compatible with the requirements of a constant surface temperature and similar enthalpy profiles. Thus, $\rho_w = \text{constant}$, and $u_{2w} \sim x_1^{-1/2}$. This mass injection distribution has practical importance because the porous surface can operate at its maximum allowable temperature everywhere, thereby minimizing the coolant required.

The boundary-layer equations, together with these boundary conditions, were solved in a series of similarity theories such as those described in § § 38 and 40, beginning with solutions for constant properties and progressing to ideal gases with variable properties. A rather complete bibliography of these theories and corroborating experiments is given in Ref. 93. The assumption of constant $C_e = \mu\rho/\mu_e\rho_e$ proved equally useful in this problem as with the impervious plate in extending constant property solutions to high Mach number cases where air still behaves as an ideal gas. The results shown here are predominantly from Refs. 94 through 96.

The velocity profiles in a boundary layer with surface mass transfer are shown in Fig. 43 for different values of $f(0)$, defined by Eq. (161). In addition to thickening the boundary layer, blowing promotes S-shaped velocity profiles characteristic of a boundary layer approaching separation. Separation, $\left(\partial u_1/\partial x_2\right)_w = 0$, with blowing occurs for $f(0) = -1.238$ [94]. An S-shaped profile is also less stable in terms of the growth of turbulence producing velocity waves; therefore, surface blowing decreases transition Reynolds numbers [97].

The enthalpy or temperature distribution functions Y_0 and $r/r(0)$ (see Eq. (134)) are shown in Figs. 44 and 45 as functions of $f(0)$. Note that increased blowing also thickens the thermal boundary layer. Whereas for no surface mass transfer a Prandtl number less than unity causes each part of the thermal boundary-layer profile to extend farther from the surface than the corresponding points of the velocity profile, under conditions of blowing this trend reverses in the inner portions of the thermal boundary layer. This means that the heat transfer rate ($\sim Y_0'(0)$) drops off less rapidly with the blowing rate than does the shear ($\sim f''(0)$). The function $r/r(0)$ in Fig. 45 applies for Pr = 0.72.

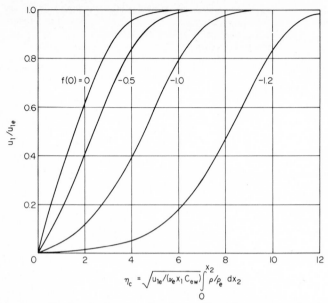

Fig. 43. Velocity profiles in a laminar boundary layer with surface mass transfer, $C_e = C_{ew}$, $f(0)$ defined by Eq. (161) [96].

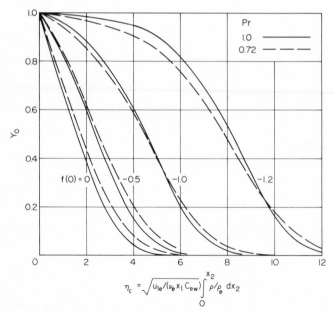

Fig. 44. Enthalpy profile function in the laminar boundary layer on a uniform temperature flat plate with surface mass transfer, $C_e = C_{ew}$.

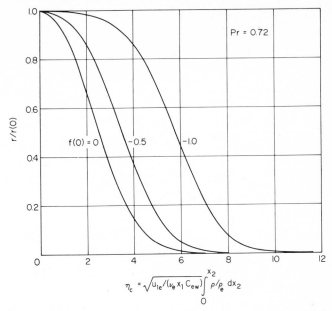

$$\eta_c = \sqrt{u_{1e}/(\nu_e x_1 C_{ew})} \int_0^{x_2} \rho/\rho_e \, dx_2$$

Fig. 45. Enthalpy profile function in the laminar boundary layer on an insulated flat plate with surface mass transfer, $C_e = C_{ew}$.

The effect of surface mass transfer on the recovery factor is shown in Fig. 46 for three different Prandtl numbers. For $Pr < 1$, increased blowing rates result in the reduction of the recovery factor.

The velocity and temperature distribution functions can be related to the physical coordinates x_1, x_2 instead of η_c through the use of Eq. (139). The effect of blowing on $\bar{r}(\eta_c)$ and $\bar{Y}_0(\eta_c)$, defined by Eqs. (140) and (141), is indicated in Fig. 47.

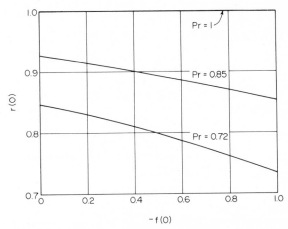

$-f(0)$

Fig. 46. Effect of surface mass transfer on the laminar boundary layer recovery factor on a flat plate.

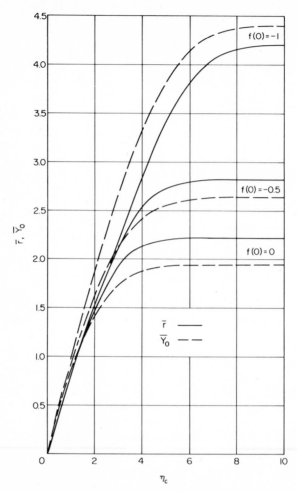

Fig. 47. Effect of surface mass transfer on $\bar{r}$ and $\bar{Y}_0$, see Eqs. (139) through (141).

The effectiveness of mass injection in reducing the local skin friction and Stanton number is indicated in Fig. 48. The solid line represents c_f and also c_h when Pr = 1 and Reynolds analogy holds. The dashed line represents c_h for Pr = 0.72. The most significant result indicated by this figure is the large reduction in both skin friction and heat transfer with surface mass transfer. It is this feature that has made mass transfer cooling systems so attractive. Another important point is the breakdown of the modified Reynolds analogy for Pr = 0.72 where

$$\frac{c_h \mathrm{Pr}^{2/3}}{c_f/2} \neq 1$$

for $f(0) < 0$. The Stanton number diminishes less with blowing than does the skin friction coefficient, making the modified Reynolds analogy factor a function of $f(0)$

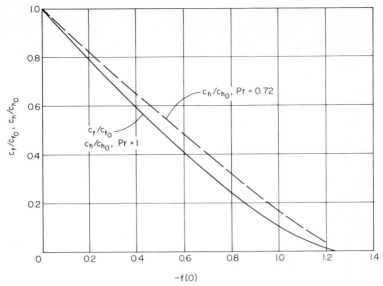

Fig. 48. Effectiveness of mass transfer in reducing local skin friction coefficient and Stanton number on a flat plate with a laminar boundary layer.

that is greater than unity. This result is consistent with the temperature distributions shown in Fig. 44. Use of the Reynolds' analogy for $-0.4 < f(0) < 0$, however, results in errors of less than 10 percent.

Single Foreign Gas as the Coolant. The effectiveness of air injection in reducing convective heat flux stimulated investigations into the use of other coolants. With the introduction of a foreign species into the boundary layer, the boundary-layer equations of §34 reduce to the continuity equation (Eq. (74)), the momentum equation (Eq. (75)), the energy equation

$$\rho u_1 \frac{\partial I}{\partial x_1} + \rho u_2 \frac{\partial I}{\partial x_2} = \frac{\partial}{\partial x_2} \left\{ \frac{\mu}{\Pr_F} \left[\frac{\partial I}{\partial x_2} - (i_1 - i_2)(1 - \mathrm{Le}) \frac{\partial K_1}{\partial x_2} \right] \right.$$

$$\left. + \left[\mu \left(1 - \frac{1}{\Pr_F} \right) \frac{\partial (u_1^2/2)}{\partial x_2} \right] \right\} \quad (162)$$

and the diffusion equation

$$\rho u_1 \frac{\partial K_1}{\partial x_1} + \rho u_2 \frac{\partial K_1}{\partial x_2} = \frac{\partial}{\partial x_2} \left(\frac{\mu \, \mathrm{Le}}{\Pr_F} \frac{\partial K_1}{\partial x_2} \right) \quad (163)$$

Here, $\Pr_F/\mathrm{Le} = \mathrm{Sc}$ is the Schmidt number, K_1 is the coolant mass concentration, and i_1 and i_2 are the coolant and air enthalpies, respectively. These equations are the bases for the bulk of the analyses involving foreign gas transpiration. The effects of thermal diffusion and diffusion-thermo (Eq. (47)) are mentioned later in this section.

The boundary conditions for Eqs. (162) and (163) are identical with those for air as the coolant except for the additional ones required for the species diffusion equation. The main stream does not contain any of the coolant species, and at the surface the coolant concentration is uniform. These additional boundary conditions are

$$x_1 = 0 , \quad x_2 \geq 0 ; \quad K_1 = 0$$

$$x_1 > 0 , \quad \begin{array}{l} x_2 \to \infty; \quad K_1 = 0 \\ x_2 = 0 , \quad K_1 = K_{1w} \end{array}$$

The value of K_{1w} is dependent on the blowing rate, and some hypothesis is required to establish this dependency. Most authors have assumed a zero net mass transfer of air into the surface. This requires that the air carried away from the surface by the mass motion normal to the surface be balanced by the diffusion of air towards the surface. Since $K_1 = 1 - K_2$ and $j_1 = -j_2$, this balance of air motion can be expressed directly in terms of the coolant properties as

$$\rho_w u_{2w} = - \frac{\rho_w \mathcal{D}_{12}}{1 - K_{1w}} \left(\frac{\partial K_1}{\partial x_2} \right)_w = \frac{j_{1w}}{1 - K_{1w}} \tag{164}$$

Equation (164) is the required relationship between the blowing rate of the coolant and its concentration at the wall. When the diffusion flux is expressed in terms of the mass transfer coefficient (see § 9), the concentration of the coolant at the wall is given by

$$K_{1w} = \frac{1}{1 + (\rho_e u_{1e}/\rho_w u_{2w}) c_{m_1}} \tag{165}$$

Note that the mass transfer coefficient is defined differently in Ref. 98.

Although the total heat flux at the surface in a binary gas is composed of the sum of a conduction term and a diffusion term, the results of analyses are expressed solely in terms of the heat conduction term. The reason is that this term is equal to the heat gained by the coolant in passing from its reservoir to the surface in contact with the boundary layer. This simple heat balance is

$$k_w \left(\frac{\partial T}{\partial x_2} \right)_w = \rho_w u_{2w} (i_{1w} - i_{1c}) \tag{166}$$

where the subscript c represents the initial coolant condition in the reservoir. Further, the recovery factor is defined in terms of the surface temperature that results in zero heat conduction at the surface for a given mass transfer rate. This adiabatic condition is achieved in experiments by setting $T_{1c} = T_w$ for a given mass flow rate, i.e., the coolant gains no heat. In terms of the Stanton number, the heat balance indicated by Eq. (166) can be expressed as either

$$\frac{c_h}{c_{h_0}} = \left(\frac{\rho_w u_{2w}}{\rho_e u_{1e} c_{h_0}} \right) \left(\frac{i_{1w} - i_{1c}}{i_{2aw} - i_{2w}} \right) \tag{167}$$

or

$$\frac{q_w}{q_{w_0}} = \left(\frac{\rho_w u_{2w}}{\rho_e u_1 e^{c_{h_0}}}\right)\left(\frac{i_{1w} - i_{1c}}{i_{2aw_0} - i_{2w}}\right) \tag{168}$$

The subscript 0 in these equations signifies zero mass transfer at the surface ($f(0) = 0$). The subscript 2 indicates that the enthalpy potential contributing to the heat flux by conduction alone is dependent only on the specific heat of air and not of the coolant.

The historical pattern of the analyses for a binary gas followed that described earlier for the impervious plate and for the case of air injection, beginning with constant property solutions and progressing to those where the individual gases are ideal but where the transport and thermodynamic properties of the mixture are temperature and concentration dependent. The boundary-layer equations are again transformed according to the variables of Eq. (110), or their equivalents, to ordinary differential equations with η as the independent variable.

For constant properties, the ordinary differential equations governing the boundary layer are Eqs. (82), (93), and the diffusion equation

$$\phi'' + \frac{1}{2}\, \mathrm{Sc}\, f\phi' = 0 \tag{169}$$

where

$$\phi = \frac{K_1}{K_{1w}} \tag{170}$$

The similarity of Eqs. (169) and (93) and of the boundary conditions indicates $\phi = Y_0$ when Sc = Pr. Thus the curves of Fig. 44 also represent ϕ when Sc = 1 and 0.72. Other consequences of the similarity of these equations are

$$\frac{c_{m_1}}{c_{m_{1_0}}} = \frac{c_h}{c_{h_0}} \tag{171}$$

at a given mass transfer rate when Sc = Pr, and the modified Reynolds analogy

$$c_{m_{1_0}} = \frac{c_{f_0}}{2}\, \mathrm{Sc}^{-2/3} = c_{h_0}\, \mathrm{Le}^{2/3} \tag{172}$$

In these equations, Sc and Le correspond to a very dilute mixture of the coolant in air. Because Sc and Le are rapidly varying functions of the coolant concentration, especially for coolants that are light or heavy relative to air, these constant property solutions are limited to cases of low-speed air flow and small surface mass transfer rates. Psychrometers and cooling towers represent applications where these conditions prevail.

For variable fluid properties, the partial differential equations (74), (75), (162), and (163) were solved, consistent with the mass transfer boundary conditions, in a series of analyses summarized in Ref. 98. Because these analyses involve complex numerical integrations of the system of differential equations for each set of boundary-layer edge and surface conditions, only a limited number of such solutions exist. A careful comparison of these solutions in Ref. 98 resulted in useful engineering correlations.

The correlation parameter employed was suggested by the work of Baron [99], who expressed the product of viscosity and density of a binary gas mixture as

$$\frac{\mu\rho}{\mu_e\rho_e} = \left(\frac{\mu_2\rho_2}{\mu_e\rho_e}\right)\left(\frac{\mu\rho}{\mu_2\rho_2}\right) = C_e\lambda \tag{173}$$

where C_e is the Chapman-Rubesin constant for air only and λ is the ratio of the $\mu\rho$ product of the mixture to the local air value; λ is relatively insensitive to temperature and a function only of the concentration. For these assumptions η_c, Eq. (117), again is appropriate as the similarity variable; hence, the boundary-layer characteristics are expressible in terms of a parameter $(\rho_w u_{2w}/\rho_e u_{1e})\sqrt{u_{1e}x_1/\nu_e C_e}$. For convective heating problems, it is convenient to replace this mass transfer parameter by another proportional to it and related to the local Stanton number in the absence of mass transfer, viz., $\rho_w u_{2w}/\rho_e u_{1e} c_{h_0}$, which appears in Eqs. (167) and (168). The Stanton number c_{h_0} is evaluated by the techniques described in § § 40 and 42. If the reference enthalpy method is employed, the recovery enthalpy used for evaluating the reference conditions corresponds to the impervious plate value. The preceding mass transfer parameter is directly analogous to the one utilized in the generalized correlations of Ref. 98.

The correlations of the effects of mass transfer of a foreign gas into an air boundary layer are shown in Figs. 49 through 53. The lines in these figures represent the mean of a band of discrete points from individual numerical solutions for $T_e = 392°R$. Figures 49 and 50 show the effects of mass transfer of a foreign gas on the local Stanton number and recovery factor. For those cases where the difference between the surface and free stream temperature is small compared to the temperature rise by frictional heating, the heat flux is proportional to the product of the Stanton number and the recovery factors. The effect of mass transfer on this heat flux is shown in Fig. 51, obtained from Figs. 49 and 50. Figures 52 and 53 show the effect of mass transfer

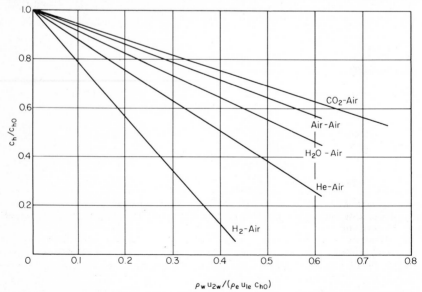

Fig. 49. Reduction of the local Stanton number by surface mass transfer of a foreign gas [98]. *(Reprinted with permission from Pergamon Press.)*

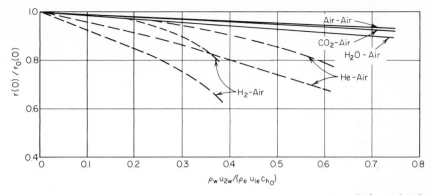

Fig. 50. Effect of foreign gas surface mass transfer on the local recovery factor [98] and [100]. *(Reprinted with permission from Pergamon Press.)*

on the local skin friction and mass transfer coefficients, respectively. Although the exact solutions shown in Fig. 48 indicate some nonlinearity in the drop-off of c_h and c_f, this effect is neglected in the correlations, and c_h, q_w, and c_f are assumed to decrease linearly with surface mass transfer rate of a particular coolant. An important result from Figs. 49 through 52 is that lighter gases are more effective in reducing the transport coefficients and surface heat flux. For a range of calculations including Mach numbers as high as 12 and surface temperatures from free stream (392° R) to recovery temperatures, the maximum departure from these correlation lines of individual solutions is ±15 percent for q_w and ±25 percent for c_h and c_f. The bulk of the discrete numerical results lie within about one-third of these band widths. The maximum spread of results is obtained with the mass transfer of hydrogen, the spread is smaller with helium and much smaller with the heavier gases. The differences

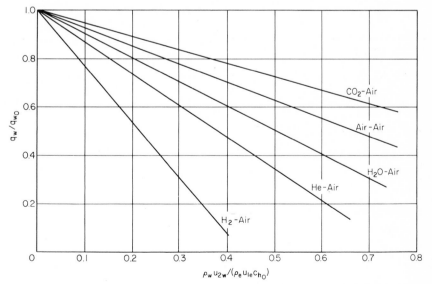

Fig. 51. Reduction of the local heat flux by surface mass transfer of a foreign gas, $T_w - T_e < r(0)u_{1e}^2/2c_{p_1}$ [98].

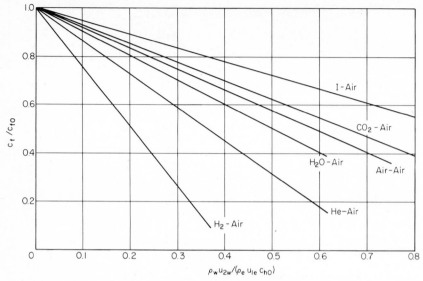

Fig. 52. Reduction of the local skin friction coefficient by surface mass transfer of a foreign gas [98]. (*Reprinted with permission from Pergamon Press.*)

between two solutions for the recovery factor with light gas mass transfer shown in Fig. 50 result from the use of different gas properties. It has been shown that either a very cold or very hot free stream widens the range of variations in recovery factor shown in Fig. 50 [100]. The effect of this is to widen the spread of heat flux about the correlation line to ±20 percent.

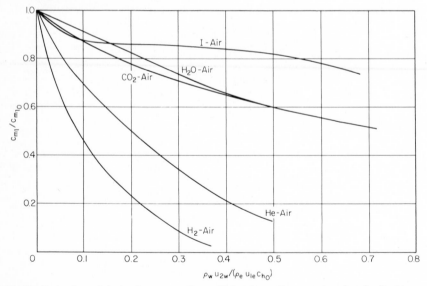

Fig. 53. Reduction of the local mass transfer coefficient by surface mass transfer of a foreign gas [98]. (*Reprinted with permission from Pergamon Press.*)

Experiments conducted to test these theoretical results [101, 102] showed that both c_h and $r(0)$ for the mass transfer of air agreed with the predictions. For light gases, in particular helium, it was found that $r(0)$ rose with mass transfer rate rather than falling as in Fig. 50, and c_h first rose with mass transfer to about $c_h/c_{h_0} = 1.07$ at $\rho_w u_{2w}/\rho_e u_{1e} c_{h_0} = 0.2$ and then fell off in a band of data lying roughly an increment of $c_h/c_{h_0} = 0.2$ higher than the correlation curve of Fig. 49. Similar results in earlier experiments led to theories that include the effects of thermal diffusion and diffusion-thermo in the mass transfer and heat flux vectors. (See §32.) These calculations are extremely complex so that only a few cases were considered [103, 104]. For helium injection, it was found that the general trend of the data was predicted; however, the magnitude of the effects could not be explained. For example, Ref. 104 shows helium injection to be less effective than the correlation curve of Fig. 49 for $\rho_w u_{2w}/\rho_e u_{1e} c_{h_0} < 0.1$ but then c_h/c_{h_0} falls off rapidly to be only 5 percent higher than the correlation curve for $\rho_w u_{2w}/\rho_e u_{1e} c_{h_0} > 0.11$. This behavior suggests possible inaccuracies in the magnitude of the thermal diffusion coefficients used in the theories. In Ref. 104, it is shown that diffusion-thermo effects introduce an adiabatic wall temperature in excess of the free-stream temperature even in the absence of frictional heating. Although this zero frictional heating recovery temperature is an unimportant phenomenon under intense aerodynamic heating conditions where mass transfer cooling is required, it may provide an accurate means for measuring the thermal diffusion properties of gases.

In view of the current state of knowledge, the following procedures are recommended for predicting the effects of surface mass transfer. For mass transfer of air, Fig. 48 or the correlation curves of Figs. 49 through 53 are appropriate. The latter curves are also useful for gases having molecular weights greater than or equal to air. Water vapor can be expected to be more effective than indicated by these correlations; however, in the absence of extensive calculations or experimental data the correlation curves yield conservative design information. These correlations can be represented by

$$\frac{c_h}{c_{h_0}} = 1 - 0.72 \left(\frac{M_2}{M_1}\right)^{1/3} \frac{\rho_w u_{2w}}{\rho_e u_{1e} c_{h_0}} \tag{174}$$

$$\frac{q_w}{q_{w_0}} = 1 - 0.75 \left(\frac{M_2}{M_1}\right)^{1/3} \frac{\rho_w u_{2w}}{\rho_e u_{1e} c_{h_0}} \tag{175}$$

$$\frac{c_f}{c_{f_0}} = 1 - 0.95 \left(\frac{M_2}{M_1}\right)^{1/3} \frac{\rho_w u_{2w}}{\rho_e u_{1e} c_{h_0}} \tag{176}$$

For helium, reliance on experimental data is recommended [102]. The mean of the recovery factor data obtained under wind tunnel conditions can be represented by

$$\frac{r(0)}{r_0(0)} = 1 + 0.2 \frac{\rho_w u_{2w}}{\rho_e u_{1e} c_{h_0}} \tag{177}$$

and the blockage of the incoming convective heating by

$$\frac{c_h}{c_{h_0}} = 1.28 - 1.4 \left(\frac{\rho_w u_{2w}}{\rho_e u_{1e} c_{h_0}}\right) \tag{178}$$

for $\rho_w u_{2w}/\rho_e u_{1e} c_{h_0} > 0.15$.

The effect of the mass injection of hydrogen is currently an enigma. Hydrogen possesses thermal diffusion and diffusion-thermo properties similar to those of helium, and in view of the differences between theory and experiment for helium, the hydrogen theory is also suspect. Unfortunately, no flat plate experiments have been conducted with hydrogen injection to confirm the theory. In addition, hydrogen cannot be expected to remain inert when injected into high temperature air. The following equation which corresponds to the correlation line in Fig. 51 yields gross estimates of the heat blockage by hydrogen

$$\frac{q_w}{q_{w_0}} = 1 - 2.3 \frac{\rho_w u_{2w}}{\rho_e \, u_{1e} c_{h_0}} \tag{179}$$

Figures 49 through 51 are particularly useful for obtaining the mass transfer rate required in a transpiration cooling system. Usually the following quantities are specified: the coolant and its reservoir conditions, the porous surface material and its maximum allowable surface temperature, and the inviscid flow conditions outside the boundary layer. For cases where the difference between the surface temperature and the boundary-layer edge temperature is small compared to the temperature rise by frictional heating, Fig. 51 can be used directly. For the specified conditions, the factor $(i_{1w} - i_{1c})/(i_{2aw_0} - i_{2w})$ can be readily established from the thermodynamic properties of the coolant and air and Eq. (137) for i_{2aw_0}. This factor is used as the slope of a straight line which represents Eq. (168) and passes through the origin of Fig. 51. The intercept of this line with the appropriate heat blockage curve on the figure is the operating condition required to yield the specified surface temperature. The abscissa of this point yields the required local mass transfer rate. For cases with large differences between the surface and the boundary-layer edge temperature, the effect of recovery factor variations on heat transfer will be attenuated, and an alternate approach is required to establish the required mass transfer rate. The factor $(i_{1w} - i_{1c})/(i_{2aw} - i_{2w})$ based on $r_0(0)$ is used to establish the operating point as described above but with Fig. 49, where the heat balance line represents Eq. (167). The operating mass transfer rate can be used in Fig. 50 to establish $r(0)$, and the slope of the line representing Eq. (167) can be adjusted for the new recovery factor. Normally a few iterations will yield a converged solution. Once the operating mass transfer rate is established, c_f is obtained from Fig. 52 and c_{m_1} from Fig. 53. The mass transfer coefficient yields the diffusion rate (see §9) and the K_{1w} from Eq. (165).

Multiple Foreign Species as Coolant. The behavior of a coolant composed of mixtures of He, Ar, Xe, and N_2 injected into a nitrogen boundary layer has been analyzed in Ref. 105. The four species boundary layer is solved with unequal binary diffusion coefficients and employs Eq. (38), in addition to the boundary-layer equations cited in the previous subsection. Although the problem lends itself to similarity transformation and involves ordinary differential equations, it is sufficiently complex mathematically that only two flow conditions were treated in Ref. 105. Examination of the results reveals that at a given mass transfer rate the skin friction coefficient is correlated for different coolants within ±5 percent using the mean molecular weight of the coolant. Thus, for a coolant composed of a mixture of ν gases, the correlation equation (Eq. (176)) can be utilized with the average molecular weight of the coolant at reservoir conditions defined as

$$M_1 = M_{avg} = \frac{1}{\displaystyle\sum_i^\nu \frac{K_{ic}}{M_i}} \tag{180}$$

Further, the use of the average molecular weight of the coolant in the correlation equations of the previous section predicts the operating surface temperatures found in the examples of Ref. 105 at $f(0) = -0.4$ to ±15 percent, with the largest deviations surprisingly associated with the examples of the pure gas coolants.

47. Uniform Surface Injection. Although a mass transfer distribution yielding a uniform surface temperature is most efficient, it is much easier to construct a porous surface with a uniform mass transfer distribution. Libby and Chen [106] have considered the effects of uniform foreign gas injection on the temperature distribution of a porous flat plate. For these conditions, however, boundary-layer similarity does not hold. Libby and Chen extended the work of Iglisch [107] and Lew and Fanucci [108] where direct numerical solutions of the partial differential equations were employed. In Ref. 106 the transformations of Eq. (110) are used to obtain partial differential equations containing derivatives with respect to ζ as well as η. It is assumed that $C_e = \mu\rho/\mu_e\rho_e$, Le, and Pr_F are all equal to unity, that thermal diffusion and diffusion-thermo effects are absent, and that all binary diffusion coefficients in a multicomponent mixture are equal.

As an example of the equations employed, the transformed momentum equation in terms of the modified stream function f where $f_\eta = u_1/u_{1e}$ is

$$f_{\eta\eta\eta} + ff_{\eta\eta} = 2\zeta(f_\eta f_{\zeta\eta} - f_\zeta f_{\eta\eta}) \tag{181}$$

The left side of the equation is the Blasius equation. The boundary conditions are

$$f_\eta(0,\eta) = f_B'(\eta), \quad \text{Blasius function} \qquad f_\eta(\zeta,\infty) = 1$$

$$f(\zeta,0) = -\frac{\rho_w u_{2w}}{\sqrt{2}\,\rho_e u_{1e}}\sqrt{\frac{\rho_e u_{1e} x_1}{\mu_e}} = \epsilon(\zeta) \qquad f_\eta(\zeta,0) = 0 \tag{182}$$

Note that under conditions of uniform mass transfer, the boundary condition for f at the wall differs basically from that of Eq. (161) in that $f(\zeta,0)$ varies as $x_1^{1/2}$ since $\rho_w u_{2w}$ is constant.

An analytical solution of Eq. (181) and its boundary condition was obtained in Ref. 106 by expanding f as

$$f(\zeta,\eta) = f_B(\eta) + \sum_{n=1}^{\infty}\epsilon^n N_n(\eta) \tag{183}$$

which leads to a series of ordinary differential equations in N_n. Numerical solutions for $N_n(\eta)$, $N_n'(\eta)$, and $N_n''(\eta)$ are given in Ref. 106 for $n = 1$ through 5. The resulting skin friction coefficient is shown in Fig. 54 as a solid line and is compared with that corresponding to the similarity distribution for mass transfer obtained from Fig. 48. The use of the similarity skin friction coefficient based on the local mass transfer rate and Reynolds number results in values that are low for uniform mass transfer.

For uniform surface mass transfer, a heat balance on the porous surface shows a nonuniform surface temperature. Thus, the resultant Stanton number is affected by both upstream surface mass transfer and nonuniform surface temperature distributions and is no longer a useful parameter. Both of these effects can be combined in a heat balance to yield the surface enthalpy distribution, conveniently expressed in terms of the total enthalpy and the coolant reservoir enthalpy

$$\frac{I_e - i_w(x_1)}{I_e - i_c} = \sum_{n=1}^{\infty}\epsilon^n G_n(0) \tag{184}$$

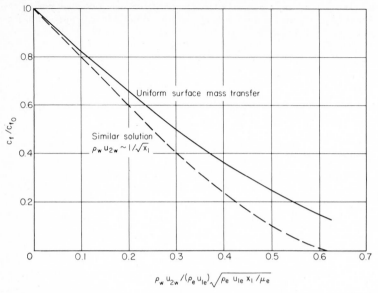

Fig. 54. Effect of surface mass transfer distribution on the reduction of the local skin friction coefficient.

where the G_n are terms in the analytical expansion for the solution of the transformed energy equation which are given in Ref. 106 for $n = 1$ through 4. With Le = 1, the diffusion equation is analogous to the energy equation, and, therefore,

$$\frac{K_{iw}(x_1) - K_{ie}}{K_{ic} - K_{ie}} = \frac{I_e - i_w(x_1)}{I_e - i_c} = \sum_{n=1}^{\infty} \epsilon^n G_n(0) \qquad (185)$$

The function representing Eq. (184) is shown in Fig. 55. For uniform mass injection, then, the wall enthalpy is near the recovery enthalpy when x_1 is small and approaches the reservoir coolant enthalpy when x_1 is large. Similarly, the coolant species concentrations rise continuously from their outer edge values and approach the reservoir coolant concentrations. To convert the surface enthalpy to temperature it is necessary to solve

$$i_w = \sum_i K_{iw} i_i(T_w) \qquad (186)$$

which reduces to

$$T_w = \frac{i_w}{\sum_i K_{iw} c_{p_{iw}}} = \frac{i_w}{c_{p_{Fw}}} \qquad (187)$$

for species with constant specific heat.

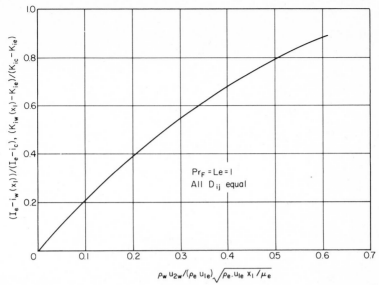

Fig. 55. Surface temperature and coolant concentration distribution along a plate with uniform foreign gas injection, Le = Pr = C_e = 1, all binary diffusion coefficients equal [106]. (*Reprinted with permission from The Physics of Fluids.*)

48. Arbitrary Distribution of Surface Injection. Often in the design of a transpiration cooling system, it is convenient to use discrete areas of porous surface. Unfortunately, the problem of arbitrary injection distributions is most complex because of the non-similar character of the boundary layer. The methods employed in the solution of these problems include numerical solution of finite-difference forms of the boundary-layer partial differential equations [109, 110], numerical solutions of higher order forms of the boundary-layer integral equations [111], and perturbation solutions of differential equations such as Eq. (181) [112]. The latter analytical approach is confined to cases with C_e = Le = Pr_F = 1 and with equal binary diffusion coefficients. Second-order solutions, however, require the use of an electronic computer.

Film Cooling

Film cooling systems provide protection to a surface exposed to high enthalpy streams by injecting a coolant into the hot boundary layer upstream of the surface. Injection can take place through local porous sections or slots at various angles to the surface. The coolant may be the same gas as in the boundary layer, a foreign gas, or a liquid.

49. Film Cooling with Upstream Transpiration of Gas. In the upstream porous section, more coolant is provided than is required to maintain safe surface temperatures. The excess coolant is entrained in the boundary layer close to the surface and is carried downstream, providing an insulating layer between the hot free stream gas and the surface. This layer is dissipated by laminar mixing while flowing with the boundary layer such that protection is afforded over a limited distance. Because of the discontinuous nature of the surface injection, the boundary layer downstream of the discontinuity is far from similar. Either approximate techniques employing the boundary-layer integral equations [113, 114], or numerical solutions of higher order integral equations [115] or of finite difference forms of the partial

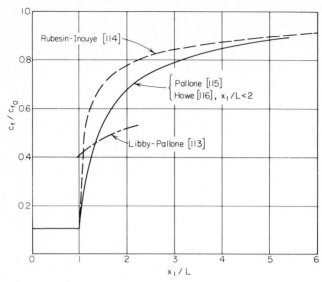

Fig. 56. Comparison of various methods for evaluating the local skin friction coefficient on a flat plate with a porous upstream section, $x_1 < L$.

differential equations [116] are required for evaluating the local convection to the surface.

Solutions have been obtained for injection of air into air with similarity conditions prevailing over the porous section, $0 \leq x_1 \leq L$. (See §46.) The local skin friction coefficient over the solid portion of the plate $x_1 > L$ with $f(0) = -1$ in the porous section is shown in Fig. 56 for the various calculation techniques. The local skin friction coefficient is normalized by the value in the absence of upstream injection but with the same surface temperature distribution. The agreement between the numerical solutions of Refs. 115 and 116 indicates that either higher order integral relations or finite difference solutions yield accurate results. The earlier integral equation calculations either missed the continuous transition at $x_1 = L$ [113] or indicated too quick a return to the impervious plate condition [114]. The skin friction distribution from Ref. 115 for a range of upstream coolant rates is shown in Fig. 57.

The temperature distribution on an insulated surface downstream of the porous section at a uniform temperature T_{wL} is shown in dimensionless form in Fig. 58 for three coolant rates. The curves labeled $f(0) = -0.5$ and -1 are from Ref. 115 and correlate within a few percent the two examples evaluated there for $M_e = 3$ and 6. In these calculations, $C_e = \mu\rho/\mu_e\rho_e$ was assumed constant and $Pr_F = 0.72$. The curve labeled $f(0) = 0$ was calculated from Eq. (159) and shows the effect of a cold upstream surface without coolant injection. The coolant injection distorts the boundary-layer velocity and temperature profiles so as to afford greater downsteam protection.

The surface temperature distribution resulting from injection of a mixture of foreign gases can be estimated by use of Fig. 58. If it is assumed that $C_e = \mu\rho/\mu_e\rho_e =$ constant, $Pr_F = Le = 1$, all binary diffusion coefficients are equal, and thermal diffusion is negligible, the diffusion equation for each species and the energy equation in terms of the total enthalpy of the mixture are directly analogous for $x_1 > L$. Also,

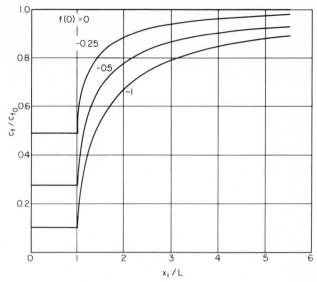

Fig. 57. Effect of the mass transfer rate in the upstream porous section ($x_1 < L$) on the skin friction coefficient over the impervious portion of a flat plate.

the boundary condition for the diffusion equation on the downstream impervious plate requires $\left(\partial K_i/\partial x_2\right)_w = 0$ which is compatible with $\left(\partial I/\partial x_2\right)_w = 0$ for the energy equation. Thus, when the rather small effect of Pr on the curves of Fig. 58 is disregarded and $f(0)$ is replaced by $f(0)(M_2/\overline{M}_c)^{1/3}$ the ordinate can be replaced by

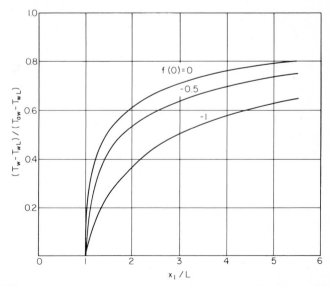

Fig. 58. Effect of upstream transpiration cooling on the temperature distribution of the impervious portion of a flat plate.

$$\frac{K_{iw} - K_{iw_L}}{K_{ie} - K_{iw_L}} = \frac{i_w - i_{wL}}{I_e - i_{wL}}$$

downstream of the porous section. The results of §46 yield K_{iw_L} and i_{w_L}. Thus, both i_w and K_{iw} are known and the surface temperature distribution may be calculated from Eq. (187).

50. Tangential Slot Injection of a Gas. A typical slot for coolant injection on a flat surface (see Fig. 59) is characterized by two dimensions, the step height h and the slot height s. The velocity distributions at the slot exit plane in the primary boundary-layer gas u_1 and in the slot u_c are shown schematically. The complexity of these velocity distributions restricts boundary-layer analysis to numerical integration techniques or to rather severe approximations [117, 118]. Because of the physical difficulty of attaining completely laminar slot cooling [119], no data are available to assess the accuracy of these approximate analyses. Curves for the enthalpy distribution on an insulated plate downstream of a slot when u_1 and u_c and their corresponding total enthalpies are uniform are given in Ref. 117. Possible effects of a nonuniform u_1 in the boundary layer of the mainstream are indicated in Ref. 118.

Cone in Supersonic Flow

The preceding solutions for a flat plate may be applied to a cone in supersonic flow through the Mangler transformation [120], which in its most general form relates the boundary-layer flow over an arbitrary axisymmetric body to an equivalent flow over a two-dimensional body. This transformation is contained in Eq. (201), which results in transformed axisymmetric momentum and energy equations equivalent to the two-dimensional equations (Eqs. (207) and (209)). Hence, solutions of these equations are applicable to either a two-dimensional or an axisymmetric flow, the differences being contained solely in the coordinate transformations.

For the case of an arbitrary pressure distribution, it is just as convenient to solve the axisymmetric problem directly. When the solution for the equivalent two-dimensional flow already exists, however, as for a flat plate, then the results for the corresponding axisymmetric problem can be obtained by direct comparison. This correspondence exists for a cone in supersonic flow when the surface pressure is constant. (See §22.) Solutions of Eqs. (207) and (209) for a flat plate expressed in terms of ζ and η may then be applied to a cone. Illustrative examples are presented in the following subsections.

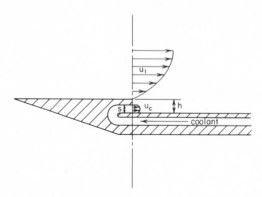

Fig. 59. Sketch of the flow characteristics in the vicinity of a tangential slot.

51. Impervious Surface with Uniform Temperature. Transformations (201) for a flat plate ($k = 0$) or a cone ($k = 1$) become

$$\zeta = \frac{\rho_w \mu_w u_{1e}}{\mu_r^2} \left(\frac{\sin\theta_c}{L}\right)^{2k} \frac{x_1^{2k+1}}{2k+1}$$

$$\eta = \sqrt{\frac{(2k+1)u_{1e}}{2\rho_w \mu_w x_1}} \int_0^{x_2} \rho \, dx_2 \tag{188}$$

with ρ_w, μ_w, u_{1e}, equal to constants, $\bar{\zeta} = 0$, and $r = x_1 \sin\theta_c$. The transformed momentum and energy equations (Eqs. (207) and (209)) are essentially the same as Eqs. (111) and (112) for a flat plate. For the same wall and boundary-layer edge conditions then, the solutions for $f(\eta)$ and $g(\eta)$ are the same for a flat plate or a cone. These results may be expressed in terms of physical variables as

$$x_2 = \sqrt{\frac{2\rho_w \mu_w x_1}{(2k+1)u_{1e}}} \int_0^\eta \frac{1}{\rho} \, d\eta \tag{189}$$

$$\tau_w = \mu_w \left(\frac{\partial u_1}{\partial x_2}\right)_w = \rho_e u_{1e}^2 \sqrt{\frac{(2k+1)\mu_e}{2\rho_e u_{1e} x_1}\left(\frac{\mu_w \rho_w}{\mu_e \rho_e}\right)} f''(0) \tag{190}$$

$$q_w = -k_w \left(\frac{\partial T}{\partial x_2}\right)_w = -\frac{k_w l_e}{\mu_w c_{p_w}} \sqrt{\frac{(2k+1)\mu_e}{2\rho_e u_{1e} x_1}\left(\frac{\mu_w \rho_w}{\mu_e \rho_e}\right)} g'(0) \tag{191}$$

For a given value of x_1 and the same flow properties, the boundary layer on a cone ($k = 1$) is thinner by a factor of $1/\sqrt{3}$, and the surface shear stress and heat transfer are larger by a factor of $\sqrt{3}$. The local skin friction and heat transfer coefficients are related similarly

$$\frac{(c_f)_{\text{cone}}}{(c_f)_{\text{flat plate}}} = \frac{(c_h)_{\text{cone}}}{(c_h)_{\text{flat plate}}} = \sqrt{3} \tag{192}$$

The local and average coefficients for a cone are related as follows

$$\left(\frac{C_f}{c_f}\right)_{\text{cone}} = \left(\frac{C_h}{c_h}\right)_{\text{cone}} = \frac{4}{3} \tag{193}$$

These relationships may be used to obtain cone flow results from the flat plate results of §§ 38, 40--42. Real gas solutions for air obtained in this manner are given in Ref. 76.

52. Impervious Surface with Nonuniform Temperature. Transformations (Eq. (188)) are applicable to flows with nonuniform surface temperature provided a linear viscosity law is assumed ($\mu\rho$ = constant). The flat plate results of §§43--45 for constant Pr_T may be applied to a cone with the requirement that the surface boundary conditions be the same in terms of ζ. For a flat plate $\zeta \sim x_1$ and for a cone $\zeta \sim x_1^3$. Therefore, the flat plate results must be modified such that lengths in the x_1

direction must be replaced by x_1^3 to obtain the cone results. For example, Eq. (151), which expresses the effect of a stepwise surface temperature for a flat plate, becomes for a cone

$$\left[\frac{c_h(x_1, s)}{c_h(x_1, 0)}\right]_{\text{cone}} = \left[1 - \left(\frac{s}{x_1}\right)^{9/4}\right]^{-1/3} \tag{194}$$

Similar expressions may be derived for the effect of an arbitrary heat flux distribution.

53. Mass Transfer with Uniform Surface Temperature. The results of §51 may be extended to include mass transfer from a uniform temperature surface. Similarity requires that $f(0)$ be constant, which determines the blowing distribution along the surface. The normal velocity component from Eq. (211) is

$$\frac{u_{2w}}{u_{1e}} = -\sqrt{\frac{(2k + 1)\mu_w}{2\rho_w u_{1e} x_1}} \, f(0) \tag{195}$$

For a cone $u_{2w} \sim x_1^{-1/2}$ as for a flat plate but is larger by the factor $\sqrt{3}$ for given values of x_1 and $f(0)$ (thus, the blowing parameter $\rho_w u_{2w}/\rho_e u_{1e} c_{h_0}$ remains unchanged). With this difference, the flat plate results of §46 may be applied to a cone according to the equations given in §51.

For a nonsimilar blowing distribution, for example, u_{2w} = constant, Eq. (195) is not applicable. Solutions to this problem may be found in Ref. 121.

LAMINAR BOUNDARY LAYER WITH PRESSURE GRADIENT

Gas with Uniform Elemental Composition in Chemical Equilibrium

Except for a few configurations as a flat plate and wedges or cones in supersonic flow, the pressure varies over the body surface as determined by inviscid flow theory. (See §§16–26.) The influence of pressure gradients on forced convection in laminar boundary layers is presented in this subsection. Axisymmetric bodies at zero angle of attack and yawed cylinders of infinite length will be treated together to illustrate their close relationship (see Fig. 60). Because boundary-layer theory requires negligible pressure gradients across the boundary layer, the techniques presented here apply only to those bodies whose local surface radius of curvature (R in Fig. 60) is large compared with the boundary-layer thickness, thereby minimizing centrifugal forces.

54. Governing Differential Equations. In the absence of foreign gas injection, the boundary layer can be considered to have uniform elemental composition and be in chemical equilibrium. Hence, the equations of §33 are applicable. For an axisymmetric body, the metric coefficients are $h_1 = 1$, $h_2 = 1$, $h_3 = r$ (Fig. 60a), and the partial derivative of any quantity with respect to x_3 ($\partial/\partial x_3$) and u_3 are equal to zero. For a yawed cylinder of infinite length, the metric coefficients are $h_1 = h_2 = h_3 = 1$, and again $\partial/\partial x_3 = 0$ because the flow is identical in all planes normal to the cylinder axis.

For these configurations, Eqs. (53), (55), (57), and (58) reduce to

$$\frac{\partial}{\partial x_1}(\rho u_1 r^k) + \frac{\partial}{\partial x_2}(\rho u_2 r^k) = 0 \tag{196}$$

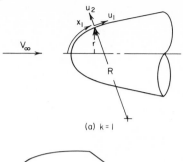

(a) k = 1

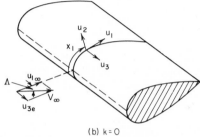

(b) k = 0

Fig. 60. Sketch of coordinates employed for related two dimensional flows. (a) Axisymmetric body; (b) yawed cylinder of infinite length.

$$\rho u_1 \frac{\partial u_1}{\partial x_1} + \rho u_2 \frac{\partial u_1}{\partial x_2} = -\frac{dp}{dx_1} + \frac{\partial}{\partial x_2}\left(\mu \frac{\partial u_1}{\partial x_2}\right) \tag{197}$$

$$\rho u_1 \frac{\partial u_3}{\partial x_1} + \rho u_2 \frac{\partial u_3}{\partial x_2} = \frac{\partial}{\partial x_2}\left(\mu \frac{\partial u_3}{\partial x_2}\right) \qquad \text{(for } k = 0 \text{ only)} \tag{198}$$

$$\rho u_1 \frac{\partial I}{\partial x_1} + \rho u_2 \frac{\partial I}{\partial x_2} = \frac{\partial}{\partial x_2}\left[\frac{\mu}{\Pr_T} \frac{\partial I}{\partial x_2} + \mu\left(1 - \frac{1}{\Pr_T}\right)\frac{\partial}{\partial x_2}\left(\frac{u_1^2 + u_3^2}{2}\right)\right] \tag{199}$$

where $k = 1$ for the axisymmetric body and $k = 0$ for the yawed cylinder.
The boundary conditions analogous to Eq. (77) are

$$x_1 = 0, \; x_2 > 0; \qquad u_1 = u_{1e}(0),$$
$$I = I_e$$
$$u_3 = u_{3e} = V_\infty \sin\Lambda \; \text{ for } k = 0$$

$$x_1 > 0, \; x_2 \to \infty; \qquad u_1 \to u_{1e}(x_1)$$
$$I \to I_e$$
$$u_3 \to u_{3e} = V_\infty \sin\Lambda \; \text{ for } k = 0 \tag{200}$$

$$x_2 = 0; \qquad u_1 = u_3 = 0$$

$$u_2 = 0 \text{ for an impervious surface or } u_{2w}(x_1)$$
$$\text{with transpiration.}$$

$$I = i_w(x_1) \text{ for } q_w \neq 0 \text{ or } \frac{\partial I}{\partial x_2} = 0 \text{ for } I = i_{aw}(x_1)$$

The yaw angle Λ is the complement of the angle between the free-stream direction and the cylinder axis (see Fig. 60).

55. Transformations of Variables and Equations. The extensions of transformations (110) to include the effects of pressure gradients are

$$\zeta = \frac{1}{\mu_r^2} \int_0^{x_1} \rho_w \mu_w u_{1e} \left(\frac{r}{L}\right)^{2k} dx_1 = \zeta(x_1)$$

$$\eta = \frac{u_{1e}(r/L)^k}{\mu_r \sqrt{2(\zeta - \bar{\zeta})}} \int_0^{x_2} \rho \, dx_2$$

(201)

where μ_r is a reference viscosity, and L is a reference length introduced to make ζ and η dimensionless [75, 122]. The function $\bar{\zeta}(\zeta)$ is yet to be determined and is a key element in the extension of similarity solutions to flows where the inviscid boundary conditions do not permit boundary-layer similarity. (Note the change in symbols employed here from those of Refs. 75 and 122.)

The dependent variables are defined as

$$f_\eta = \frac{u_1}{u_{1e}(x_1)} \qquad g = \frac{I}{I_e} \qquad w = \frac{u_3}{u_{3e}} \quad \text{(for } k = 0 \text{ only)}$$

(202)

Additional parameters that enter the equations are

$$t = \frac{i}{I_e} = g - (1 - t_s)w^2 - (t_s - t_e)f_\eta^2$$

(203)

where

$$t_s = 1 - \frac{u_{3e}^2}{2I_e}$$

(204)

and

$$t_e = 1 - \frac{u_{1e}^2 + u_{3e}^2}{2I_e}$$

(205)

The pressure gradient parameter is defined as

$$\beta = \frac{2(\zeta - \bar{\zeta})}{u_{1e}} \frac{t_s}{t_e} \frac{du_{1e}}{d\zeta}$$

(206)

Both t_e and β are functions of x_1 through their dependence on $u_{1e}(x_1)$

The transformed partial differential equations of momentum and energy are

$$\left(C_w f_{\eta\eta}\right)_\eta + \left(1 - \frac{d\bar{\zeta}}{d\zeta}\right) f f_{\eta\eta} = \beta\left[f_\eta^2 - \frac{g}{t_s} + \left(\frac{1 - t_s}{t_s}\right)w^2 - \frac{t_e}{t_s}\left(\frac{\rho_e}{\rho} - \frac{t}{t_e}\right)\right]$$

$$+ 2(\zeta - \bar{\zeta})(f_\eta f_{\eta\zeta} - f_\zeta f_{\eta\eta})$$

(207)

$$\left(C_w w_\eta\right)_\eta + \left(1 - \frac{d\bar{\zeta}}{d\zeta}\right) f w_\eta = 2(\zeta - \bar{\zeta})(f_\eta w_\zeta - f_\zeta w_\eta) \tag{208}$$

$$\left(\frac{C_w}{\Pr_T} g_\eta\right)_\eta + \left(1 - \frac{d\bar{\zeta}}{d\zeta}\right) f g_\eta = \left\{ C_w \left(\frac{1}{\Pr_T} - 1\right) \left[(t_s - t_e)\left(f_\eta^{\,2}\right)_\eta + (1 - t_s)(w^2)_\eta\right] \right\}_\eta$$
$$+ 2(\zeta - \bar{\zeta})(f_\eta g_\zeta - f_\zeta g_\eta) \tag{209}$$

The boundary conditions for zero mass transfer are

$$\eta = 0 ; \quad f(\zeta, 0) = f_\eta(\zeta, 0) = w(\zeta, 0) = 0$$
$$g(\zeta, 0) = g_w(\zeta) = t_w(\zeta) \text{ for } q_w \neq 0$$
$$\text{and } g_\eta(\zeta, 0) = 0 \text{ for } q_w = 0 \tag{210}$$
$$\eta \to \infty ; \quad f_\eta(\zeta, \infty) = w(\zeta, \infty) = g(\zeta, \infty) \to 1$$

With mass transfer, $f(\zeta, 0)$ depends on $u_{2w}(x_1)$ as follows

$$\frac{u_{2w}(x_1)}{u_{1e}(x_1)} = -\frac{\mu_w}{\mu_r} \left(\frac{r}{L}\right)^k \left[\frac{f(\zeta, 0)}{\sqrt{2\zeta}} + \sqrt{2\zeta}\, f_\zeta(\zeta, 0)\right] \tag{211}$$

Similar Solutions

56. Similarity Criteria and Reduced Equations. The partial differential equations (Eqs. (207) through (209)) are not amenable to solution except by numerical methods utilizing high-speed computers. Considerable simplifications can result, as in the case of the flat plate, if these equations are reduced to ordinary differential equations through the similarity concept where the dependent variables f, w, and g are assumed functions of η alone. Equations (207) through (209) become for $\bar{\zeta} = $ constant

$$(C_w f'')' + ff'' = \beta \left[f'^2 - \frac{g}{t_s} + \left(\frac{1 - t_s}{t_s}\right) w^2 - \frac{t_e}{t_s}\left(\frac{\rho_e}{\rho} - \frac{t}{t_e}\right)\right] \tag{212}$$

$$(C_w w')' + fw' = 0 \tag{213}$$

$$\left(\frac{C_w}{\Pr_T} g'\right)' + fg' = \left\{ C_w \left(\frac{1}{\Pr_T} - 1\right) [(t_s - t_e)(f'^2)' + (1 - t_s)(w^2)']\right\}' \tag{214}$$

Consistent with the similarity assumption, none of the terms that appear in these equations or in the related boundary conditions can be dependent on x_1 or ζ, that is, β, t_e, g_w, as well as $\bar{\zeta}$ must be constant. The parameter t_e, defined by Eq. (205), however, violates this requirement when u_{1e} varies with x_1. The similarity assumption is also violated by the terms that contain t_e explicitly in Eqs. (212) and (214) and by the gas properties C_w, $\Pr_T$, ρ_e/ρ, that can be expressed in terms of t, which in turn depends on t_e through Eq. (203). Consequently, exact similar solutions are not possible under general flow conditions.

Exact similar solutions are possible for stagnation regions where $u_{1e} \approx 0$ and t_e is a constant and equal to unity for an axisymmetric body and to t_s for a yawed cylinder. The terms involving t_e drop out of Eqs. (212) through (214), and similarity occurs for constant i_w and β.

For similar flows, the pressure gradient parameter expressed as follows must be constant

$$\beta = \frac{2}{u_{1e}} \frac{t_s}{t_e} \frac{du_{1e}}{dx_1} \frac{\displaystyle\int_0^{x_1} \rho_w \mu_w u_{1e} \left(\frac{r}{L}\right)^{2k} dx_1}{\rho_w \mu_w u_{1e} \left(\frac{r}{L}\right)^{2k}} \tag{215}$$

In a stagnation region, the fluid properties are nearly constant, and $u_{1e} \sim x_1$; also $r \approx x_1$. Thus, $\beta = 1/2$ for an axisymmetric body, and $\beta = 1$ for a yawed cylinder.

The skin friction coefficient and Stanton number are defined under the conditions of similarity on axisymmetric and two-dimensional bodies as follows:

The component of the shear stress in the x_i direction is given by

$$\tau_{wi} = \sqrt{\frac{\mu_w \rho_w t_s u_{ie}^2}{\beta t_e} \frac{du_{1e}}{dx_1} \frac{\partial}{\partial \eta}\left(\frac{u_i}{u_{ie}}\right)\Big|_w} \tag{216}$$

where the subscript $i = 1$ or 3. The skin friction coefficient is defined as

$$\frac{c_{fwi}}{2} = \frac{\tau_{wi}}{\rho_w u_{1e} u_{ie}} \tag{217}$$

with

$$\frac{c_{fw1}}{2} \sqrt{\frac{\rho_w u_{1e} x_1}{\mu_w}} = \sqrt{\frac{t_s}{\beta t_e} \frac{x_1}{u_{1e}} \frac{du_{1e}}{dx_1}} \, f_w'' \tag{218}$$

$$\frac{c_{fw3}}{2} \sqrt{\frac{\rho_w u_{1e} x_1}{\mu_w}} = \sqrt{\frac{t_s}{\beta t_e} \frac{x_1}{u_{1e}} \frac{du_{1e}}{dx_1}} \, w_w' \tag{219}$$

Alternative forms of these equations that are sometimes more convenient are

$$\frac{c_{fw1}}{2} \sqrt{\frac{\rho_w u_{1e} x_{eff}}{\mu_w}} = \frac{1}{\sqrt{2}} f_w'' \tag{220}$$

$$\frac{c_{fw3}}{2} \sqrt{\frac{\rho_w u_{1e} x_{eff}}{\mu_w}} = \frac{1}{\sqrt{2}} w_w' \tag{221}$$

where

$$x_{eff} = \frac{\int_0^{x_1} \rho_w \mu_w u_{1e} \left(\frac{r}{L}\right)^{2k} dx_1}{\rho_w \mu_w u_{1e} \left(\frac{r}{L}\right)^{2k}}$$ (222)

The corresponding Stanton number expressions for a surface at constant temperature are

$$c_{hw} \sqrt{\frac{\rho_w u_{1e} x_1}{\mu_w}} = \sqrt{\frac{t_s}{\beta t_e} \frac{x_1}{u_{1e}} \frac{du_{1e}}{dx_1}} \frac{1}{Pr_{Tw}} \frac{g'_w}{g_w - g_{aw}}$$ (223)

$$c_{hw} \sqrt{\frac{\rho_w u_{1e} x_{eff}}{\mu_w}} = \frac{1}{\sqrt{2}} \frac{1}{Pr_{Tw}} \frac{g'_w}{g_w - g_{aw}}$$ (224)

These equations can be expressed in terms of a Nusselt number (§8) by noting that

$$Nu_w = c_{hw} Re_w Pr_{Tw}$$ (225)

57. Ideal Gas with Viscosity-Density Product and Prandtl Number Equal to Unity and Uniform Surface temperature. For the case of an ideal gas with $C_w = Pr = 1$, similarity is possible away from the stagnation region of a body. Equations (212) and (214) for an axisymmetric body or a cylinder normal to the free stream reduce to

$$f''' + ff'' = \beta(f'^2 - g)$$ (226)

$$g'' + fg' = 0$$ (227)

when u_{1e} satisfies

$$\frac{u_{1e}}{\sqrt{t_e}} = A\zeta^{\beta/2}$$ (228)

These equations are equivalent to those solved in Ref. 123 for a uniform surface temperature. Examples of the extensive solutions in Refs. 75 and 123 are presented in this section.

Figure 61 shows the velocity distribution in the boundary layer of an axisymmetric body or an unswept cylinder for a cold wall ($g_w = 0$). (Note that the value of η used here is a factor of $\sqrt{2}$ smaller than the one employed in sections on the flat plate.) An accelerating free stream ($\beta > 0$, u_{1e} increasing) reduces the flow boundary-layer thickness and increases the velocity gradient rather uniformly through the boundary layer. A decelerating free stream ($\beta < 0$) thickens the flow boundary layer rather severely and causes the velocity distribution to acquire an "S" shape. Eventually, the boundary layer will separate; that is, $f''_w = 0$. For the boundary conditions of Fig. 61, two solutions are possible for negative values of β near separation. It is argued in Ref. 123 that true similarity with negative β cannot occur physically because Eq. (228)

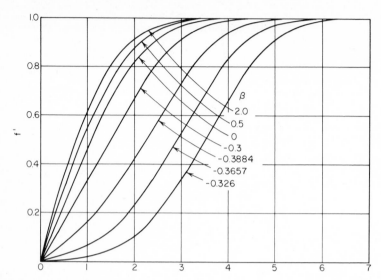

Fig. 61. Similar velocity distributions for body with surface pressure gradient, β defined by Eq. (228), $g_w = 0$, $t_s = 1$ [123].

would require $u_{1e} \to \infty$ as $\zeta \to 0$. Thus, similar solutions with negative β can only be approached after a period of nonsimilar flow, and depending on the conditions, one or the other of the similar solutions for a given β can be attained. In Ref. 123 experimental evidence from Ref. 124 for turbulent flow is cited for the possible existence of double valued flow-field behavior. The velocity profiles shown here are characteristic of those for either a cooled surface or one at the total enthalpy of the fluid. For a heated surface and $\beta > 0$, it is possible for the velocity ratio f' in the outer portion of the boundary layer to attain a value greater than unity before approaching unity at the outer edge. The physical reason for this is that the acceleration of lower density fluid by the favorable pressure gradient exceeds the retardation by the viscous forces.

The total enthalpy distribution in terms of the velocity is shown in Fig. 62 for $g_w = 0$ and $t_s = 1$. A pressure gradient can cause significant departures from the Crocco relationship, Eq. (114), represented by the straight line labeled $\beta = 0$. The latter is often used for approximate calculations even when pressure gradients are present.

Figure 63 shows the wall shear parameter f_w'' required to evaluate the local skin friction coefficients by Eqs. (218) or (220). These curves apply for the case where $t_s = 1$. The effect of sweep for the two-dimensional case will be discussed in §62 where nonsimilar conditions are considered. The double-valued nature of f_w'' for a cooled surface ($g_w = 0$) for β near separation ($f_w'' = 0$) is evident. Generally, f_w'' is more sensitive to variations in β for a hot surface. In fact, for cold wall conditions ($g_w = 0$), the variation of f_w'' with β for $\beta > 0$ is quite modest. Also, a cooled surface also tends to retard separation, that is, $f_w'' = 0$ at a smaller value of β.

The heat transfer parameter required in Eqs. (223) or (224) to calculate Stanton number is shown in Fig. 64 for $t_s = 1$. It should be noted that because of the similarity of Eqs. (213) and (214) for $\text{Pr} = 1$

$$w_w' = \frac{g_w'}{g_w - g_{aw}} \tag{229}$$

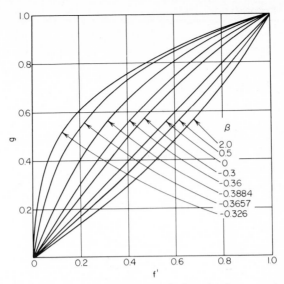

Fig. 62. Enthalpy and velocity relationship within similar boundary layers on a body with pressure gradient, β defined by Eq. (228), $g_w = 0$, $t_s = 1$ [123].

Hence, the ordinate in Fig. 64 can also be used in conjunction with Eqs. (219) or (221) to calculate the cross flow skin friction coefficient for cases of very small yaw angles ($t_s \approx 1$). Note that g_{aw} is equal to unity because the solution of Eq. (214) with Pr = 1 and an insulated surface is $g \equiv 1$. Although the trends exhibited in Figs. 63 and 64 are generally similar, it must be cautioned that such large variations in the Reynolds analogy factor occur that the latter is no longer a useful concept. The heat transfer parameter for a cooled surface shows a rather small variation with β for $\beta \geq 1/2$, a fact first utilized in Ref. 125 to obtain relatively simple expressions for the local heat flux to blunt bodies in hypersonic flow.

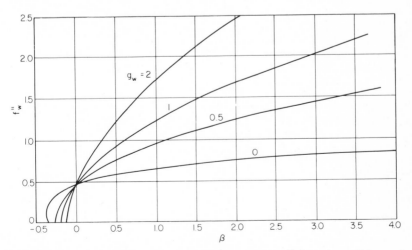

Fig. 63. Effect of pressure gradient on the skin friction parameter f_w'' for various wall temperature levels, $t_s = 1$ [123].

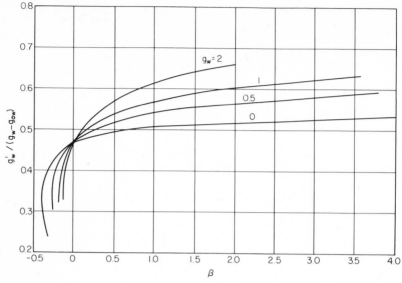

Fig. 64. Effect of pressure gradient on the heat transfer parameter g'_w for various wall temperature levels, $t_s = 1, Pr = 1$ [123].

58. Cylinder Normal to the Free Stream, Fluid with Constant Properties. For constant fluid properties, Eqs. (212) and (214) reduce to

$$f''' + ff'' = \beta(f'^2 - 1) \tag{230}$$

$$g'' + Pr fg' = 2 Pr \zeta f'g_\zeta \tag{231}$$

For a cylinder normal to the free stream, the inviscid velocity distribution is given by

$$u_{1e} = Ax_1^m \tag{232}$$

with

$$\beta = \frac{2m}{1 + m} \tag{233}$$

The term on the right-hand side of Eq. (231) has been added to account for a nonuniform surface temperature.

Similar solutions for Prandtl numbers other than unity may be obtained from Eqs. (230) and (231) or their equivalent. A major simplification is the independence of the momentum equation (230) from the energy equation (231), which makes f independent of g. Also, the linear form of the energy equation in g permits handling arbitrary surface temperature distributions as in the case of the flat plate. (See §§43–45.)

Solutions of the momentum equation (Eq. (230) [126] yield velocity distributions generally similar to those of Fig. 61 and the skin friction parameter f''_w shown by the line labeled $g_w = 1$ in Fig. 63. The skin friction coefficient is given by

$$\frac{c_{f_1}}{2} \sqrt{\frac{\rho_e u_{1e} x_1}{\mu_e}} = \frac{f''_w}{\sqrt{2-\beta}} \tag{234}$$

For a uniform surface temperature, solutions of the energy equation (Eq. (231)) [127, 128] can be expressed as

$$c_h(x_1,0) \sqrt{\frac{\rho_e u_{1e} x_1}{\mu_e}} = \frac{\mathrm{Pr}^{-a}}{\sqrt{2-\beta}} \left(\frac{g'_w}{g_w - g_{aw}} \right)_{\mathrm{Pr}=1} \tag{235}$$

where the heat transfer parameter in parentheses on the right is given by the $g_w = 1$ line in Fig. 64. The exponent a for the Prandtl number is given in Table 9 as a function of β.

TABLE 9. Exponent of Pr in Eq. (235)

β	-0.199	0	1	1.6
a	0.746	0.673	0.645	0.633

In Ref. 127, an ingenious set of transformations is employed to evaluate the recovery factor away from the stagnation line. The results for $\beta = 1$ show a significant departure (≈ -10 percent for $\mathrm{Pr} = 0.7$) from $r(0) = \mathrm{Pr}^{1/2}$. These values, however, do not agree with calculations performed in Ref. 129. Perhaps the discrepancy is due to the evaluation of $r(0)$ in Ref. 127 by taking the derivative of a function. Slight errors in the function itself could easily account for a 10 percent error in the derivative. For accuracies of $r(0)$ within a few percent [129], it is recommended that

$$r(0) = \mathrm{Pr}^{1/2} \tag{236}$$

be employed for all β.

For a polynomial surface temperature distribution (§43) represented as

$$\frac{T_w - T_{aw}}{T_{st}} = \sum_n a_n \left(\frac{x_1}{L} \right)^n \tag{237}$$

where L is a reference length and a_n an arbitrary coefficient, the heat flux distribution is given by

$$q_w(x_1) = \rho_e u_{1e} c_p c_h(x_1,0) T_{st} \sum_n a_n \left(\frac{x_1}{L} \right)^n H_{\beta n} \tag{238}$$

The values of the effective potential function $H_{\beta n}$ are given to a few percent for $\mathrm{Pr} > 0.7$ by [84]

$$H_{\beta n} = \frac{\Gamma\left[\frac{4}{3} n \left(1 - \frac{\beta}{2}\right) + 1\right] \Gamma\left(\frac{2}{3}\right)}{\Gamma\left[\frac{4}{3} n \left(1 - \frac{\beta}{2}\right) + \frac{2}{3}\right]} \tag{239}$$

or by using $n(1 - \beta/2)$ as the abscissa in Fig. 37. Values of $H_{\beta n}$ given by this approach are quite consistent with those tabulated in Ref. 128. As m in Eq. (232) becomes large, $\beta \to 2$ and $H_{\beta n} \to 1$ for all n. Under these steep velocity gradient conditions, the local temperature potential acts as the driving potential for the heat flux, and the effect of the upstream surface temperature variation becomes negligible.

For a stepwise and arbitrary surface temperature distribution, the local heat flux distribution is given by Eq. (150). The term $c_h(x_1, 0)$ represents the Stanton number for the same flow conditions but with a uniform surface temperature. Equation (150) was derived formally with the assumption that i_{aw} is uniform along the surface; however, small continuous variations in i_{aw} or T_{aw} are permissible. The value of i_{aw} at $x_1 = 0$ is used in the first term on the right side of Eq. (150), and the local value of i_{aw} is employed within the integral. Although Eq. (150) appears to be reciprocal in variations of i_w and i_{aw}, this is not the case. Changes in the same direction of both i_w and i_{aw} do not necessarily cancel because a change in i_{aw} takes effect gradually downstream.

The kernel function $c_h(x_1, s)/c_h(x_1, 0)$ in Eq. (150) represents the behavior of the heat transfer coefficient after a jump in wall temperature at $x_1 = s$. This function was obtained in Refs. 84 and 85 by solving the energy equation with the assumption of a linear velocity profile and is given by

$$\frac{c_h(x_1, s)}{c_h(x_1, 0)} = \left[1 - \left(\frac{s}{x_1}\right)^a\right]^{-b} \tag{240}$$

where

$$a = \frac{3}{2(2 - \beta)} \qquad b = \frac{1}{3} \tag{241}$$

The analogous function for a flat plate is given by Eq. (151). These results were improved upon in Ref. 130 where a power law velocity profile was assumed, $u_1/u_{1e} = (x_2/x_{2e})^d$ with d found to best fit Hartree's calculations [126] as listed in Table 10. The form of the kernel function (Eq. (240)) is the same but the exponents are changed to

$$a = \frac{2 + d}{(1 + d)(2 - \beta)} \qquad b = \frac{1}{2 + d} \tag{242}$$

Use of these values to calculate the heat flux distribution from Eqs. (150) and (240) yields excellent agreement with the exact solutions of Refs. 63, 128, and 131. Hence, for $m \geq 0$, values of a and b given by Eq. (242) are preferred to those given by Eq. (241). For $m = 0$, there is little difference between the kernel functions (240) based on the two different sets of exponents. Values of the kernel function Eq. (240) are shown in Fig. 38 for $\beta = 0.5$ and 1. The effect of the upstream temperature jump decays more rapidly with increasing β. As $\beta \to 2$, the kernel function becomes unity for all $s/x_1 < 1$ as is seen directly from the functional form of Eqs. (240) and (242).

When the only surface temperature discontinuity occurs at the leading edge, the technique described in §44 of fitting the temperature distribution with a series of

TABLE 10. Exponent d for Velocity Function, $u_1/u_{1e} = (x_2/x_{2e})^d$

m	0	1/9	1/3	1	4
β	0	0.2	0.5	1	1.6
d	0.88	0.86	0.80	0.76	0.66

straight line segments can also be applied to a body with a pressure gradient. The function $G(s_n/x_1)$ for this case is given by

$$G\left(\frac{s_n}{x_1}\right) = \int_0^{s_n/x_1} (1 - \alpha^a)^{-b} d\alpha \tag{243}$$

with a and b given by Eq. (242) and is plotted in Fig. 40 for $\beta = 0.5$ and 1 [132]. Functions required for evaluating the total heat flux up to a point x_1 using Eq. (155) are also given in Ref. 132.

For a prescribed arbitrary heat flux distribution, the techniques described in §45 can be applied to a body with a pressure gradient. First, consider a polynomial heat flux distribution. The coefficients of the polynomial can be evaluated from Eq. (238) where it is noted that the terms ahead of the series are proportional to $x_1^{(\beta-1)/(2-\beta)}$. For example, with a uniform heat flux, the only value of n in the series would be

$$n = \frac{1 - \beta}{2 - \beta}$$

If $q_w \sim x_1^p$, the unique value of n is

$$n = p + \frac{1 - \beta}{2 - \beta}$$

These values of n and the a_n appropriate to the heat flux values in Eq. (238) when substituted together in Eq. (237) yield the surface temperature distribution.

For an arbitrary heat flux distribution, the resulting surface temperature distribution is given by

$$T_w(x_1) - T_{aw}(x_1) = \frac{a}{\Gamma(1-b)\,\Gamma(b)\rho_e u_{1e} c_p c_h(x_1,0)x_1} \int_0^1 \frac{q_w(x_1')}{[1 - (x_1'/x_1)^a]^{1-b}} dx_1' \tag{244}$$

where a and b are given by Eq. (242).

An approximate technique for integrating Eq. (244) is suggested in Ref. 132 where $q_w(x_1)$ is represented by a series of steps at constant values. Thus, for $0 \le x_1 \le s_1$, $q_w = q_1$; for $s_1 \le x_1 \le s_2$, $q_w = q_2$; .... The values of $s_1, s_2, \ldots$ can be spaced arbitrarily to achieve the best fit to the prescribed $q_w(x_1)$. The last value s_n is equal to x_1, the position where the surface temperature is being evaluated. Equation (244) becomes

$$T_w(x_1) - T_{aw}(x_1) = \frac{a}{\Gamma(1-b)\,\Gamma(b)\rho_e u_{1e} c_p c_h(x_1,0)} \left[(q_1 - q_2) P\left(\frac{s_1}{x_1}\right) \right.$$

$$\left. + (q_2 - q_3) P\left(\frac{s_2}{x_1}\right) + \cdots + (q_{n-1} - q_n) P\left(\frac{s_{n-1}}{x_1}\right) + q_n P(1) \right] \tag{245}$$

Values of the function

$$P\left(\frac{s}{x_1}\right) = \int_0^{s/x} (1 - \alpha^a)^{-(1-b)} d\alpha \tag{246}$$

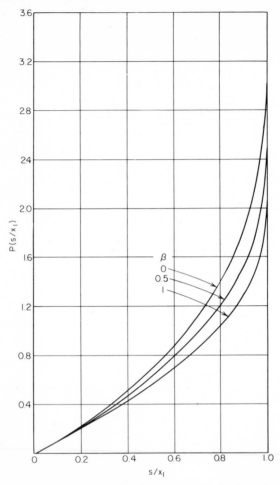

Fig. 65. Values of function $P(s/x_1)$ defined by Eq. (246) [132].

are given in Ref. 132 and plotted in Fig. 65 for $\beta = 0$, 0.5, and 1. For cases of rapid but continuous changes of $q_w(x_1)$ on a flat plate ($\beta = 0$), Eq. (160) is preferred to Eq. (245). For values of β other than those shown in Fig. 65, interpolation may be employed or use may be made of the results of Ref. 133 where

$$P\left(\frac{s}{x_1}\right) = B\left(\frac{1}{a}, b\right)\left[1 - I_z\left(\frac{1}{a}, b\right)\right]$$ (247)

The term B is the complete beta function, and I_z is the incomplete beta function ratio with argument

$$z = 1 - \left(\frac{s}{x_1}\right)^a$$ (248)

where a and b are obtained from Eq. (242).

59. Stagnation Point of Axisymmetric Body in High-Speed Flow. The axisymmetric stagnation point has received attention by many investigators because of its importance in the assessment of the convective heat load of missile nose cones and atmospheric entry vehicles. The speeds involved in these applications produce stagnation enthalpies (see Fig. 30) where real gas behavior must be considered in the evaluation of the forced convection. Because of the very complex behavior of the physical properties of real gases, a characteristic common to all the studies is the reliance on numerical solutions followed by correlations of the results in terms of parameters involving the fluid properties. In addition to air, other gases have been treated because of the current interest in the exploration of the planetary neighbors of earth. The contents of this section are confined to gases in chemical equilibrium with uniform elemental composition and to flows where boundary-layer similarity occurs, viz., the immediate vicinity of the stagnation point and with a uniform surface temperature. The effects of uniformly distributed surface mass transfer of the same gas as exists in the free stream are also included.

Equations (212) and (214) with $t_s = t_e = 1$, $t = g$ have been solved for real air in Ref. 75 and in Refs. 134 through 138 with the latter references utilizing the concept of total properties k_T, c_{p_T}, Pr_T. The air properties of Ref. 2 were employed in all the studies except that of Ref. 138, which employed properties evaluated recently in Refs. 139 and 140, where careful consideration has been given to the effect of dominant resonant charge exchange cross sections in establishing the thermal conductivity of ionized nitrogen.

For speeds under 30,000 ft/sec, which represent relatively moderate entry conditions into the earth's atmosphere without appreciable ionization, the numerical results of Ref. 75 for 10^{-4} atm $< p_{st} < 10^2$ atm and $540°R < T_w < 3100°R$ are correlated by

$$\frac{Nu_w}{Pr_{T_w}^{0.4} Re_w^{0.5}} = 0.767 \left(\frac{\mu_e \rho_e}{\mu_w \rho_w}\right)^{0.43} \tag{249}$$

where

$$Nu_w = \frac{q_w c_{pw} L}{(i_w - I_e) k_w}$$

and

$$Re_w = \frac{\left(\frac{du_{1e}}{dx_1}\right) L^2 \rho_w}{\mu_w}$$

This correlation in terms of the property ratio $\mu_e \rho_e / \mu_w \rho_w$ is a direct extension of the results of Ref. 141, where calculations were made with "frozen" physical properties and constant Lewis number through the boundary layer for air treated as an equivalent binary gas. Modifications to the correlation (Eq. (249)) were suggested in Ref. 141 to account for Le $\neq$ 1. It is argued in Ref. 75, however, that the actual variation of Le through the boundary layer for a binary mixture equivalent to air is from 1.4 to about 0.6 under typical entry conditions so that a Lewis number correction based on a single value is unwarranted and, as the results based on total properties indicate, not required. From §57, it can be seen that the constant on the right side of Eq. (249) agrees with the ideal-gas calculations for $\beta = 0.5$ and

$g_w = 1$, and the exponent of the Prandtl number is in fair agreement with the value found in the constant property case. (See §58.)

At the stagnation point (where $u_{1e} = 0$) the recovery enthalpy is identical to the stagnation enthalpy even for $Pr_T \neq 1$. Thus, from the definitions of Nu_w and Re_w, the local heat flux at the stagnation point in air flow for the speed range up to 30,000 ft/sec can be expressed as

$$q_w = \left(\frac{Nu_w}{\sqrt{Re_w}}\right) \frac{I_e(g_w - 1)}{Pr_{Tw}} \sqrt{\mu_w \rho_w} \sqrt{\frac{du_{1e}}{dx_1}} \tag{250}$$

or from Eq. (249) as

$$q_w = 0.767 \, Pr_{Tw}^{-0.6} (\mu_e \rho_e)^{0.43} (\mu_w \rho_w)^{0.07} \sqrt{\frac{du_{1e}}{dx_1}} \, I_e(g_w - 1) \tag{251}$$

The units used in these equations are for q_w, Btu/sec-ft^2 ; ρ, lb/ft^3 ; μ, lb/sec-ft; u_{1e}, ft/sec; x_1, ft; I_e, Btu/lb. Here, a negative value of q_w represents heat flux toward the body.

At speeds greater than 10,000 ft/sec, where $I_e \approx V_\infty^2/2$, Eq. (251) can be represented in much simpler form when a relatively cold surface temperature (below dissociation temperature) is assumed

$$q_w \sqrt{\frac{R_n}{p_{st}}} = 121 \left(\sqrt{\frac{T_w}{492}} \frac{1.38}{T_w/492 + 0.38}\right)^{0.07} \sqrt{\frac{R_n}{V_\infty} \frac{du_{1e}}{dx_1}} \, \bar{U}^{2.21} (g_w - 1) \tag{252}$$

The dimensionless velocity is expressed as $\bar{U} = V_\infty/10,000$ ft/sec and T_w in °R, p_{st} in atm. Equation (29) is recommended for evaluating the velocity gradient.

For stagnation point heating in gases other than air, correlations similar to Eq. (250) were obtained in Ref. 136 for speeds up to 30,000 ft/sec. The correlation is of the form

$$\frac{Nu_w}{\sqrt{Re_w}} = A \left(\frac{\mu_e \rho_e}{\mu_w \rho_w}\right)^B \tag{253}$$

with coefficients A and B given in Table 11 for the various gases considered. There is rather close agreement between Eq. (249) and Eq. (253) for air. The heat flux is obtained from Eq. (250).

For speeds above 30,000 ft/sec, where the total enthalpy reaches values where ionization significantly lowers the viscosity μ_e, the correlation for air (Eq. (249))

TABLE 11. Coefficients for Eq. (253)
[136]

Gas	A	B
Air	0.718	0.475
N_2	0.645	0.398
H_2	0.675	0.358
CO_2	0.649	0.332
A	0.515	0.110

from Ref. 75 begins to break down. Similar behavior was observed on pointed cones in Ref. 76. Similarly, it was found for argon in Ref. 136 that only the solutions for the lowest wall temperatures were correlated well by the property parameter $\mu_e \rho_e / \mu_w \rho_w$. Thus, extrapolation of the Nusselt number relations beyond the range of $\mu_e \rho_e / \mu_w \rho_w$ actually used in the correlations could yield significant errors. Two alternate approaches avoid this problem. In Ref. 135 the intermediate Nusselt number correlation was bypassed, and a correlation for air was achieved directly in terms of the heat flux for speeds up to 50,000 ft/sec as follows

$$q_w \sqrt{\frac{R_n}{p_{st}}} = 119 \sqrt{\frac{R_n}{V_\infty} \frac{du_{1e}}{dx_1}} \, \bar{U}^{2.19} (g_w - 1) \qquad (254)$$

It will be noted that this equation is quite similar to Eq. (252). The implication of this similarity is that the large variations in μ_e associated with ionization, which is not present in the range of velocities for which Eq. (252) is valid, can be ignored in the evaluation of heat flux. In fact, it is systematically demonstrated in Ref. 136 that the surface heat flux is quite insensitive to variations of the gas properties at the boundary-layer edge and is controlled instead by the gradients of these properties near the surface. Apparently, correlations such as Eqs. (249) or (253) result because in the speed range where they are applicable the physical properties vary monotonically through the boundary layer, and their derivatives in the inner portion of the boundary layer are related to the ratio of properties across the boundary layer. The correlation based on an average gradient of properties near the wall in terms of properties existing at the onset of dissociation (or ionization for monatomic gases) is shown in Fig. 66 for all the gases outlined in Ref. 136. The resulting heat flux expression, using strong shock relationships, is given by

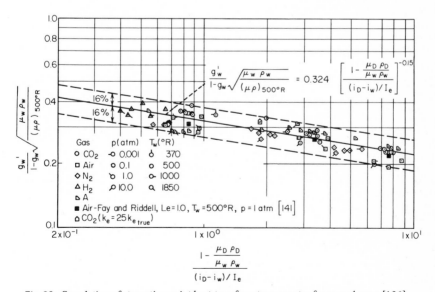

Fig. 66. Correlation of stagnation-point heat transfer rate parameter for several gases [136].

$$q_w \sqrt{\frac{R_n}{P_{st}}} = \mathcal{F} \sqrt{\frac{R_n}{V_\infty} \frac{du_{1e}}{dx_1}} \bar{U}^{2.2} (g_w - 1) \tag{255}$$

where the units are the same as in Eqs. (251) and (252).

The term $\mathcal{F}$ represents the product of quantities

$$\mathcal{F} = 2.07 \times 10^5 \left[\frac{1}{Pr_{Tw}} \right] \sqrt{\frac{(\mu\rho)_{500°R}}{P_{st}}} \left[\frac{1 - \mu_D \rho_D / \mu_w \rho_w}{i_D - i_w} \right]^{-0.15} \tag{256}$$

which is independent of the stagnation pressure level and the wall temperature and can be evaluated just once for a given gas composition. Values of $\mathcal{F}$ for a variety of gases are given in Table 12. In general, $\mathcal{F}$ for a gas mixture lies between the values for the individual constituent species, but not necessarily in direct proportion to their respective mole fractions.

TABLE 12. Coefficient $\mathcal{F}$ Defined in Eq. (256)

Gas (% Vol.)	$\mathcal{F}$
Air	121
N_2	121
CO_2	141
A	165
91% N_2–10% CO_2	120
50% N_2–50% CO_2	134
40% N_2–10% CO_2–50% A	144
65% CO_2–35% A	142

From the value of $\mathcal{F}$ for air in Table 12 and the form of Eq. (255), it is apparent that the correlation of Ref. 136 yields results essentially identical to Eq. (254) taken from Ref. 135 and to Eq. (252) derived by directly extrapolating the lower speed range equation of Ref. 75. Thus, it is recommended that Eq. (255) with the coefficients of Table 12 be utilized to predict stagnation point heat flux at speeds greater than 10,000 ft/sec. At lower speeds, Eqs. (250) and (253) with the coefficients of Table 11 are appropriate. A comparison of these techniques with existing stagnation point measurements is shown in Fig. 67a–d. It should be noted that convection predominated over radiation in these measurements despite speeds up to 50,000 ft/sec because of the small nose radii for the models. For larger body dimensions ($\sim$ feet), shock layer radiation begins to compete with convection at speeds of about 25,000 ft/sec and becomes the predominant heating mechanism at higher speeds.

60. Stagnation Line on a Cylinder in High-speed Flow. The stagnation line on a uniform temperature cylinder of infinite length with its axis either normal to the free stream or swept back at angle Λ is characterized by boundary-layer similarity solutions with $\beta = 1$. The solution in Ref. 75 of Eqs. (212) through (214) for $\beta = 1$ yields the correlation

$$\frac{Nu_w}{Pr_{Tw}^{0.4} Re_w^{0.5}} = 0.57 \left(\frac{\mu_e \rho_e}{\mu_w \rho_w} \right)^{0.45} \tag{257}$$

for $i_w / I_e < t_s$ (see Eq. (204) for definition of t_s) and $V_\infty \cos \Lambda < 29,000$ ft/sec.

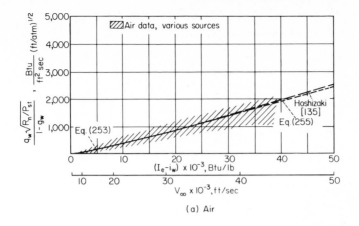

(a) Air

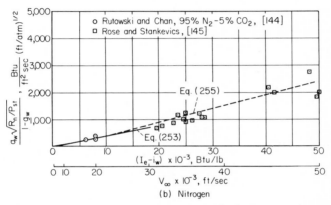

(b) Nitrogen

Fig. 67. Comparison of heat transfer rates predicted by Eq. (255) and the coefficients of Table 12 with data [136].

Although Eq. (257) was established from calculations that employed real air properties, the resulting coefficient and exponents are consistent with those for low-speed calculations for either constant properties or ideal gases. (See §§57 and 58.) In terms of heat flux, Eq. (257) becomes

$$q_w = 0.57 \, \mathrm{Pr}_{Tw}^{-0.6} (\mu_e \rho_e)^{0.45} (\mu_w \rho_w)^{0.05} \sqrt{\frac{du_{1e}}{dx_1}} (i_w - I_{aw}) \qquad (258)$$

where the recovery enthalpy I_{aw} is given by

$$I_{aw} = I_e - [1 - \mathrm{Pr}_{Tw}^{0.5}] \frac{u_{3e}^2}{2} = I_e - [1 - \mathrm{Pr}_{Tw}^{0.5}] \frac{V_\infty^2 \sin^2 \Lambda}{2} \qquad (259)$$

and a negative q_w represents heat flow into the body. An alternate form of Eq. (258), valid for $V_\infty > 10,000$ ft/sec and having the same form as Eq. (252) for an axisymmetric stagnation point in air flow, is

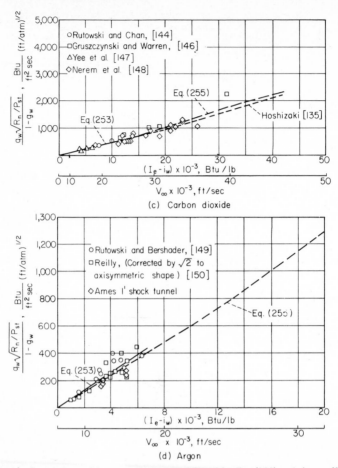

Fig. 67 (*Cont.*). Comparison of heat transfer rates predicted by Eq. (255) and the coefficients of Table 12 with data [136].

$$q_w \sqrt{\frac{R_n}{p_{st}}} = 87.3 \left[\left(\frac{T_w}{492} \right)^{1/2} \frac{1.38}{T_w/492 + 0.38} \right]^{0.05} \sqrt{\frac{R_n}{V_\infty} \frac{du_{1e}}{dx_1}} \; \bar{U}^{2.2} (g_w + 0.15 \sin^2 \Lambda - 1)$$

(260)

with the following units: q_w, Btu/sec-ft^2; R_n, ft; p_{st}, atm; T_w, °R; V_∞, ft/sec; and $\bar{U} = V_\infty/10,000$ ft/sec.

The velocity gradient in Eq. (260) is obtained from Eq. (27). In these equations, p_{st} and ρ_{st} are the inviscid flow conditions on the stagnation line of the swept cylinder. For an ideal gas in hypersonic flow, the inviscid flow relationships are particularly simple, and Eq. (260) shows that the heat flux is reduced with sweep by approximately $\cos^{3/2} \Lambda$. Equation (260) may be extended to gases other than air by setting the quantity in braces equal to unity and replacing the coefficient 87.3 by 0.72 $\mathcal{F}$, where $\mathcal{F}$ is given in Table 12.

61. Mass Transfer in Stagnation Region. As on a flat plate, surface mass transfer is an effective means for alleviating convective heating in the stagnation regions of axisymmetric bodies and cylinders. The effect of surface mass transfer of a gas with the same elemental composition as the free stream will be treated here. Foreign gas injection will be treated in §64.

Consistent with the similarity requirements following Eqs. (212) through (214), the surface mass transfer rate is given by

$$\rho_w u_{2w} = -f(0)\sqrt{\frac{\mu_w \rho_w t_s}{\beta t_e}\frac{du_{1e}}{dx_1}} \tag{261}$$

In the vicinity of the stagnation region and with a uniform surface temperature, the terms under the radical sign are constants. Thus, boundary-layer similarity, that is, $f(0)$ being independent of x_1, requires a uniform mass injection rate along the surface rather than one varying as $x_1^{-1/2}$ as on a flat plate. A convenient correlation parameter, as in the case of the flat plate, is

$$B = \frac{\rho_w u_{2w}}{\rho_e u_{1e} c_{h_0}} = \frac{\rho_w u_{2w}(i_w - I_{aw})}{q_{w_0}} = -\frac{f(0)\,Pr_{Tw}}{\sqrt{\beta}\left(Nu_w/\sqrt{Re_w}\right)_0} \tag{262}$$

where the subscript 0 denotes zero surface mass transfer.

The effect of surface mass transfer in reducing the Stanton number is indicated in Fig. 68. Note that for the stagnation point on a body of revolution or an unswept cylinder, the recovery enthalpy is equal to the total enthalpy. Hence, $q_w \sim c_h$ so that Fig. 68 indicates the reduction of the heat flux as well as the Stanton number. The line representing the axisymmetric stagnation point correlates all the gases listed in Table 11 to within a few percent [137]. Although the curves for the cylinder are

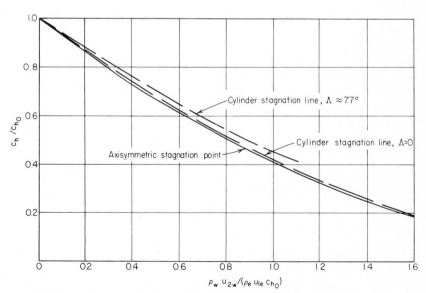

Fig. 68. The reduction of Stanton number in stagnation regions by surface mass transfer.

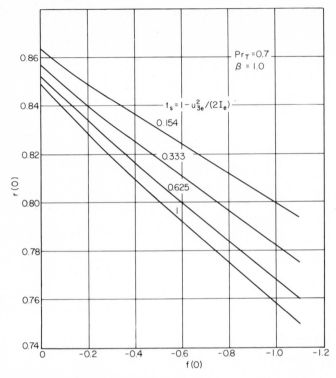

Fig. 69. The effect of surface mass transfer on the recovery factor on the stagnation line of a yawed cylinder of infinite length [129].

based on air calculations, the correlation for the various gases for the axisymmetric stagnation point implies that the cylinder results can be applied to other gases when the injected and free stream gas are the same. Note that in the coordinate system of Fig. 68, the effect of sweep is quite small.

Along the stagnation line of a swept cylinder, the recovery enthalpy is less than the stagnation enthalpy. Thus, the effect of surface mass transfer on the recovery factor, as shown in Fig. 69, should be considered in establishing the proper driving potential for the heat flux [129]. The heat flux for a swept cylinder normalized by its value with zero mass transfer is

$$\frac{q_w}{q_{w_0}} = \frac{c_h}{c_{h_0}} \left[\frac{1 - (1 - r(0)) \sin^2 \Lambda - g_w}{1 - (1 - r_0(0)) \sin^2 \Lambda - g_w} \right] \tag{263}$$

Because the term containing the recovery factor depends on $\sin^2 \Lambda$, modification of the recovery factor to account for surface mass transfer need be made only for large sweep angles. The graphical procedure for using these figures in establishing the required mass flow rate to yield a prescribed surface temperature is given in §46.

Nonsimilar Solutions

The criteria for boundary-layer similarity specified in §56 are not satisfied away from stagnation regions for usual aerodynamic configurations. Extensive computer programs have been developed to solve the partial differential boundary-layer equations in finite difference form under these conditions [109, 110]. For many engineering applications, however, approximate methods yield sufficiently accurate convective heating predictions. The method recommended here for use with a favorable pressure gradient ($\beta > 0$) is based on the use of the energy integral equation together with existing similar boundary-layer solutions [75].

62. Formulation of Nonsimilar Boundary-layer Equations. In the absence of surface mass transfer, the energy integral equation (Eq. (72)) for an axisymmetric body at zero angle of attack or an infinite cylinder normal to the stream reduces to

$$\frac{q_w}{\rho_e u_{1e}(i_w - I_e)} = \frac{d\Gamma_1}{dx_1} + \Gamma_1 \frac{d}{dx_1}\left\{\ln[r^k \rho_e u_{1e}(i_w - I_e)]\right\} \tag{264}$$

In terms of the transformed variables of §55, Eq. (264) simplifies to

$$\frac{(dg/d\eta)_w}{\text{Pr}_{Tw}\sqrt{2(\zeta - \bar{\zeta})}} = \frac{d}{d\zeta}\left[(1 - t_w)\sqrt{2(\zeta - \bar{\zeta})}\,\Gamma_{tr}\right] \tag{265}$$

where

$$\Gamma_{tr} = \int_0^\infty f_\eta\left(\frac{1 - g}{1 - g_w}\right) d\eta \tag{266}$$

Under boundary-layer similarity conditions, $\bar{\zeta}$, t_w, and Γ_{tr} are independent of ζ, and Eq. (265) yields

$$(1 - t_w)\Gamma_{tr} = \frac{(dg/d\eta)_w}{\text{Pr}_{Tw}} \tag{267}$$

for a given value of β. The key hypothesis in the method of Ref. 75 is that the relationship between the local energy thickness and heat flux under strict similarity conditions, as represented by Eq. (267), can be applied *locally* in Eq. (265) even when similarity no longer occurs. This leads to the definition of $\bar{\zeta}$ in Eq. (206) and, in turn, yields an expression for an effective local value of β for nonsimilar boundary layers as

$$\beta = \frac{2}{u_{1e}} \frac{t_s}{t_e} \frac{du_{1e}}{dx_1} \frac{\int_0^{x_1} \rho_w \mu_w u_{1e}\left(\frac{r}{L}\right)^{2k} \Gamma^{*2} dx_1}{\rho_w \mu_w u_{1e}\left(\frac{r}{L}\right)^{2k} \Gamma^{*2}} \tag{268}$$

where

$$\Gamma^* = \frac{g'_w/\text{Pr}_{Tw}}{\left(g'_w/\text{Pr}_{Tw}\right)_{st}} \tag{269}$$

and the local heat flux in terms of the stagnation point heat flux on the body is given by

$$\frac{q_w}{(q_w)_{st}} = \sqrt{\frac{\mu_w \rho_w /\beta t_e}{\left(\mu_w \rho_w /\beta t_e\right)_{st}\left(du_{1e}/dx_1\right)_{st}}}\ \Gamma^* \tag{270}$$

Because Γ^* appears in the integrand of Eq. (268) and is dependent on β, iteration is required in the use of Eqs. (268) through (271) for evaluating the local heat flux distribution $q_w(x_1)$. This iteration is avoided in the simple local similarity method of Ref. 142 by setting Γ^* equal to unity in Eq. (268). If Γ^* is set equal to unity in both Eqs. (268) and (270), the method of Ref. 125 results.

In evaluating the local heat flux by means of Eqs. (268) through (271), it is first necessary to evaluate the stagnation point heat flux by the methods of §59 or 60. Then it is necessary to establish g'_w in terms of t_s and the local values of β, t_e, and g_w. Cohen [75] gives extensive tables of g'_w (called ζ'_w in Ref. 75) and the shear parameters required in Eqs. (218) through (221) for a large range of values for β, t_e, g_w, and t_s. Even more convenient is the equation in Ref. 75 that correlates the numerical results for g'_w to within 15 percent for $V_\infty \cos \Lambda \le 29{,}000$ ft/sec, $g_w/t_s \le 1$, and $0 \le \beta \le 4$. This equation is

$$\Gamma^* = \frac{g'_w/\mathrm{Pr}_{Tw}}{\left(g'_w/\mathrm{Pr}_{Tw}\right)_{st}} = (1.033)^k \left(\frac{1 + P\beta^N}{Q + R\beta^N}\right)\left[1 + 0.050\,(1 - t_s)\left(1 - \frac{t_e}{t_s}\right)\left(\frac{\beta - 1}{0.2\beta + 1}\right)\right]$$

$$\times \left[1.1 - 0.1625\,\frac{t_e}{t_s} + 0.0625\left(\frac{t_e}{t_s}\right)^2\right]\left[\frac{0.85 + 0.15t_e - g_w}{0.85 + 0.15t_s - g_{w_{st}}}\right]\left[\frac{\mathrm{Pr}_{Tw}}{\mathrm{Pr}_{Tw,\,st}}\right]^{-0.6}$$

$$\times \left[\frac{1.0213\,(g_w/t_E)^{0.3329} - 0.0213}{1.0213\left(g_w/t_E\right)_{st}^{0.3329} - 0.0213}\right] \tag{271}$$

where $k = 0$ or 1 on a cylinder or axisymmetric body, respectively, and $t_E = i_E/I_e$ where $i_E = 2.119 \times 10^8$ ft^2/sec^2 or 8465 Btu/lb. Values of P, Q, and N are plotted in Fig. 70 as functions of the yaw parameter t_s defined by Eq. (204), and $R = 1 + P - Q$. Note that when $\beta = 1$, the main yaw parameter $(1 + P\beta^N)/(Q + R\beta^N) = 1$ for all values of t_s. The groups of terms containing g_w and Pr_{Tw} account primarily for the effect of the surface temperature level on the properties of the fluid, but do not account for variable surface temperature effects as discussed in §§43 and 44. Thus, only the most gradual surface temperature variations are treated by Eq. (271).

63. Comparison of Nonsimilar Boundary-layer Theory with Data. Predictions by the methods described in §62 are compared with data in Figs. 71 and 72 [122]. The ordinate in these figures $[q_w/(g_w - g_{aw})]/[q_w/(g_w - 1)]_{st}$ accounts for the departure of g_{aw} from unity at stations away from the stagnation point for air data. Figure 71 represents the heat flux distribution on a circular cylinder normal to the stream. The solid line is the iterated nonsimilar boundary-layer solution of Eqs. (268) through (271); the dashed line is the method of Ref. 142 with $\Gamma^* = 1$ in Eq. (268); and the dot-dashed line is the method of Ref. 125 with $\Gamma^* = 1$ in both Eqs. (268) and (270). Although the data exhibit considerable scatter, especially beyond 70° away from the stagnation line, they favor the complete theory. The method of Ref. 142, however, yields results quite close to the more complete theory and is more convenient to use. Similar conclusions can be drawn from Fig. 72 for a flat-faced cylinder. The shape of

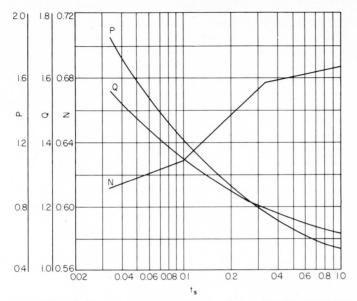

Fig. 70. Effect of the yaw parameter t_s on functions P, Q, and N of Eq. (271) [75].

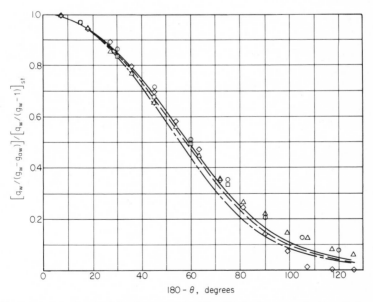

Fig. 71. Comparison of predicted and measured heat flux distributions on a circular cylinder normal to a stream. ———— Eqs. (268) through (271), iterated; — — — Kemp, Rose, and Detra [142]; ——— – ——— – Lees [125], [122].

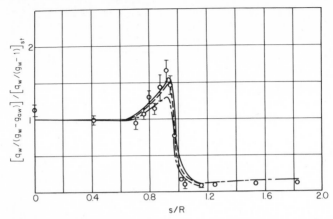

Fig. 72. Comparison of predicted and measured heat flux distributions on a flat-faced cylinder. ———— Eqs. (268) through (271), iterated; —— —— —— Kemp, Rose, and Detra [142]; —— – —— – Lees [125], [122].

this body introduces a very rapid change of β in the vicinity of the shoulder and is a rather severe test of the theory. The theoretical stagnation point values of heat flux are used as the reference in this figure because the model was known to give slightly erroneous values there [122]. In general, it can be concluded that the nonsimilar boundary-layer theories are quite accurate for predicting the heat flux to bodies with a favorable pressure gradient $(\beta > 1)$.

Foreign Gas Mass Transfer

64. Stagnation Region. The evaluation of the effectiveness of a transpiration cooling system utilizing foreign gas in the vicinity of the stagnation point requires consideration of nonequilibrium, multicomponent effects in the boundary layer. For a binary mixture composed of an inert coolant gas and air and in the immediate vicinity of the stagnation point where viscous dissipation can be ignored, the energy equation, based on Eqs. (46) and (47), is

$$\rho u_1 \frac{\partial I}{\partial x_1} + \rho u_2 \frac{\partial I}{\partial x_2} = \frac{\partial}{\partial x_2}\left[k \frac{\partial T}{\partial x_2} - j_1(i_1 - i_2) - j_1 \alpha \frac{M}{M_1 M_2} RT \right] \tag{272}$$

where subscripts 1 and 2 refer to the coolant and air, respectively. The diffusion vector is

$$j_1 = -\rho \mathcal{D}_{12}\left[\frac{\partial K_1}{\partial x_2} + \alpha K_1(1 - K_1) \frac{1}{T} \frac{\partial T}{\partial x_2} \right] \tag{273}$$

where

$$\alpha = \frac{\mathcal{D}_1^T}{\rho \mathcal{D}_{12}(1 - K_1)K_1} \tag{274}$$

is the thermal diffusion factor, a quantity fairly insensitive to local concentrations in the mixture of species. Both the effects of thermal diffusion (the last term on the right of Eq. (273)) and diffusion-thermo (the last term on the right of Eq. (272)) are included in these equations. Because the local coolant concentration affects the fluid properties, it is necessary to consider also the diffusion equation resulting from Eq. (32)

$$\rho u_1 \frac{\partial K_1}{\partial x_1} + \rho u_2 \frac{\partial K_1}{\partial x_2} = - \frac{\partial j_1}{\partial x_2} \tag{275}$$

The overall continuity and momentum equation remain as Eqs. (196) and (197).

The techniques for solving Eqs. (272) through (275), simultaneously with the momentum and continuity equations, employ the same similarity arguments as were used for the equilibrium boundary layer to transform the partial differential equations to ordinary differential equations. A major difference resulting from the introduction of diffusion-thermo effects is in the boundary condition involving the heat gained by the coolant. Instead of Eq. (166), this boundary condition is now

$$k_w \left(\frac{\partial T}{\partial x_2}\right)_w - \left(j_1 \alpha \frac{M}{M_1 M_2} RT\right)_w = \rho_w u_{2w} (i_{1w} - i_{1c}) \tag{276}$$

For the adiabatic condition, when the coolant does not gain any heat, both sides of Eq. (276) vanish, and a negative value of α (usually associated with coolant gas of low molecular weight) results in a negative value of $\left(\partial T/\partial x_2\right)_w$. Thus, the temperature drops with distance from the surface so that an adiabatic wall temperature in excess of the boundary-layer edge temperature is realized even in the absence of frictional heating in the low-speed stagnation region.

The solution of this problem is quite formidable, even in the transformed coordinate system; consequently, only a limited number of examples have been evaluated, e.g., Refs. 151 and 152. Corresponding experiments are reported in Refs. 153 and 154.

Figure 73 shows the adiabatic wall temperature resulting from diffusion-thermo effects introduced by surface mass transfer of either helium or Freon-13 at the stagnation point of a hemisphere in a $M_\infty = 8$ air stream. The surface mass transfer rate has been divided by the Stanton number that occurs at the stagnation point under the same flow conditions and surface temperature but in the absence of surface mass transfer. The ordinate is the ratio of the adiabatic wall temperature to the local boundary-layer edge temperature, which in this case is the stagnation temperature. In the absence of diffusion-thermo effects, T_{aw}/T_{st} would remain identically equal to unity for all mass transfer rates. For helium, the theoretical curves [152] and the experimental data [153] both indicate that the adiabatic wall temperature increases with the surface mass transfer rate as was suggested by Eq. (276) for the negative value of α characteristic of helium. The heavy gas, Freon-13, yields opposite results, as expected. The agreement between the theory and experiment is well within the error bands of the data. The adiabatic wall temperature data of Ref. 154 for tests conducted on a porous cylinder normal to the stream, further substantiates the theories of Refs. 151 and 152. The coordinates of Fig. 73 minimize the differences introduced by various body shapes and free stream flow conditions; thus, Fig. 73 is representative of the effects of diffusion-thermo processes on the adiabatic wall temperature of blunt bodies in general. In fact, when the local values of the zero mass transfer heat flux are employed in the abscissa, it has been found that the results of Fig. 73 yield reasonable predictions of the data of Ref. 154 at stations on the cylinder away from the stagnation line.

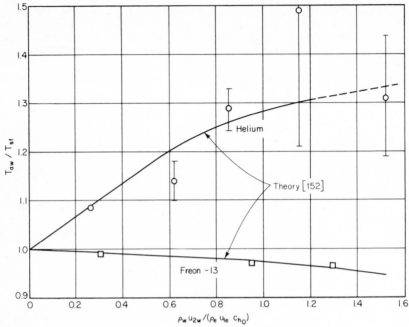

Fig. 73. Adiabatic wall temperature resulting from diffusion-thermo effects caused by mass transfer of foreign gases at a stagnation point.

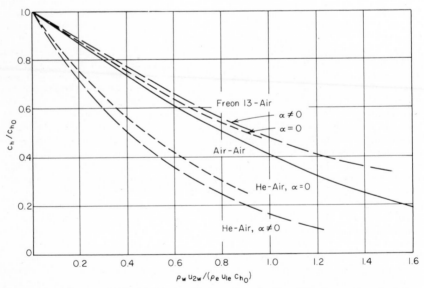

Fig. 74. The influence of diffusion-thermo effects on the reduction of Stanton number at a stagnation point by foreign gas surface mass transfer [152]. (*Reprinted with permission from Pergamon Press.*)

Figure 74 [152] shows the influence of the diffusion-thermo effect on the reduction of the Stanton number by surface mass transfer for the same flow conditions as for Fig. 73. The Stanton number with mass transfer is based on Eq. (138) with the adiabatic wall temperature obtained from Fig. 73 for those cases containing the diffusion-thermo effect. A curve representing the injection of air into air is included as a basis of comparison. The coordinates employed in this figure tend to remove the effects of surface temperature level for the range $0.4 < T_w/T_{st} < 1.0$. The main parameter of the curves is the composition of the inert coolant gas. It is quite evident that diffusion-thermo effects for both helium and Freon have a minor influence on the behavior of the Stanton number. The data of Ref. 153, although they exhibit considerable scatter, generally favor the curves of Fig. 74 for $\alpha = 0$, as do the data of Ref. 154. The latter reference suggests the calculations of Ref. 151 for helium yield too high values of Stanton number due to the values of thermal conductivity used. For gases other than helium, comparison with the curves of Fig. 74 further suggests the results of Ref. 151 are high. The values of Fig. 74 and the data of Ref. 153 indicate that the effect of foreign gas injection on the Stanton number can be determined in general from the air-air case if the abscissa is modified to $(\rho_w u_{2w}/\rho_e u_{1e} c_{h_0})(M_2/M_1)^{1/4}$. The reduction of Stanton number is given by Ref. 137 as

$$\frac{c_h}{c_{h_0}} = 1 - 0.72\left[\frac{\rho_w u_{2w}}{\rho_e u_{1e} c_{h_0}}\left(\frac{M_2}{M_1}\right)^{1/4}\right] + 0.13\left[\frac{\rho_w u_{2w}}{\rho_e u_{1e} c_{h_0}}\left(\frac{M_2}{M_1}\right)^{1/4}\right]^2 \quad (277)$$

where subscripts 1 = coolant and 2 = air.

In §46 it was indicated that the diffusion-thermo effect has no practical impact on the aerodynamic heating of aircraft surfaces that behave as flat plates. The reason is that frictional heating and the diffusion-thermo effect are additive effects on the adiabatic wall temperature, and frictional heating is by far the dominant process. In the stagnation region, however, the boundary-layer velocities are low, and frictional heating is negligible. Here, the effects of diffusion-thermo processes on the adiabatic wall temperature have the magnitudes indicated in Fig. 73. For wall temperatures much less than the stagnation temperature, the effect of mass transfer on the surface heat flux is proportional to the product of the ordinates of Figs. 73 and 74. Under wind tunnel conditions, where surface temperatures are not usually small compared with the stagnation temperatures, the diffusion-thermo effect will be magnified, and for wall temperatures near the stagnation temperature, reversals from the expected direction of heat flow can result with light coolants.

A convenient technique for using the coordinate system of Figs. 73 and 74 in establishing coolant mass flow requirements for prescribed surface and coolant temperatures is described in §46.

TURBULENT BOUNDARY LAYER ON A FLAT PLATE

In the preceding sections on the laminar boundary layer, emphasis is placed on theoretical results and the means by which they are derived. Experimental data are used primarily to corroborate the theoretical results. For the turbulent boundary layer, however, the converse procedure must be followed. Here, experimental data are the foundations upon which all theories are based. The role of theory is not to develop results from first principles, but to provide systematic techniques for critically interpreting existing data so as to permit generalized correlations and logical extrapolations to conditions where data do not exist. The underlying reason for this approach is the lack of understanding of the physical bases of the turbulent mechanisms for transport of mass, momentum, and energy. Although the flux vectors are represented formally by Eqs. (39), (44), (45), and (51), the eddy diffusivities which act as coefficients of property gradients in these equations are not unique point values of fluid properties as in the laminar case but are also dependent on the dynamical behavior of the boundary layer.

Impervious Surface with Uniform Temperature

65. Differential Equations.   For the turbulent boundary layer, Eqs. (75) and (76) resulting from the general Eqs. (55) and (58) are modified to

$$\rho u_1 \frac{\partial u_1}{\partial x_1} + \rho u_2 \frac{\partial u_1}{\partial x_2} = \frac{\partial}{\partial x_2}\left[(\mu + \rho\epsilon_M)\frac{\partial u_1}{\partial x_2}\right] \tag{278}$$

and

$$\rho u_1 \frac{\partial I}{\partial x_1} + \rho u_2 \frac{\partial I}{\partial x_2} = \frac{\partial}{\partial x_2}\left\{ \left(\frac{\mu}{\mathrm{Pr}_T} + \frac{\rho\epsilon_M}{\mathrm{Pr}_t}\right)\frac{\partial I}{\partial x_2} \right.$$

$$\left. + \left[\mu\left(1 - \frac{1}{\mathrm{Pr}_T}\right) + \rho\epsilon_M\left(1 - \frac{1}{\mathrm{Pr}_t}\right)\right]\frac{\partial(u_1^2/2)}{\partial x_2} \right\} \tag{279}$$

The continuity equation is represented by Eq. (74), but with ρu_2 representing the sum, $\overline{\rho}\overline{u}_2 + \overline{\rho'u_2'}$. These equations have not been solved in this general form and are presented here merely to illustrate the approximations that are inherent in the methods to be described subsequently.

66. Fluid with Constant Properties.   *Mean velocity profile data* obtained in turbulent boundary layers when the fluid properties are uniform (low speeds and small temperature differences between the free stream and the surface) are described in the comprehensive reviews of Refs. 155 and 156. These data reveal that the turbulent boundary layer does not possess a similar form $(u_1/u_{1e} = f(x_2/\delta))$ at different stations along the plate as in the case of the laminar boundary layer. At a given station, the turbulent boundary layer is characterized by two regions [157], with velocity profiles described by the "law of the wall" and the "law of the wake" after Coles [158, 159].

The region near the wall possesses a universal velocity profile when the data are correlated in terms of the coordinates $u^+ = u_1/u_\tau$ and $y^+ = u_\tau' x_2/\nu$ as indicated in Fig. 75 for smooth surfaces [156]. The quantity $u_\tau = \sqrt{\tau_w/\rho}$ is called the friction velocity. The layer immediately adjacent to the surface $(y^+ \gtrsim 5)$ is called the laminar sublayer where, because of the presence of the surface, the turbulence has been damped into a fluctuating laminar flow. Here, viscosity predominates over the eddy viscosity, and the velocity distribution may be approximated by

$$u^+ = y^+ \tag{280}$$

For $y^+ > 30$, the turbulent processes completely dominate the local shear, and the resulting correlation can be represented by

$$u^+ = 2.43 \ln y^+ + 5.0 \tag{281}$$

Intermediate between these limits is the buffer layer, where both shear mechanisms are important. The essential feature of this data correlation is that the wall shear

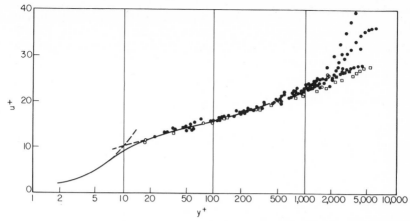

Fig. 75. Universal velocity profile for an incompressible turbulent boundary layer near the surface ("law of the wall") [156]. *(Reprinted with permission from Academic Press, Inc.)*

completely controls the turbulent boundary-layer velocity distribution in the vicinity of the wall. So dominant is the effect of the wall shear that even when pressure gradients along the surface are present, the velocity distributions near the surface are essentially coincident with the data obtained on plates with uniform surface pressure [160]. Within this region, the local shear stress remains within about 5 to 10 percent of the surface shear stress for a flat plate. It will be noted that this shear variation is often ignored in turbulent boundary-layer theory.

Correlation of the velocity distribution for the outer portion of the boundary layer—the wake region—is shown in Fig. 76 [157] for a plate with uniform pressure. The decrement of velocity relative to that of the free stream correlates with the distance from the surface normalized by the boundary-layer thickness δ. Although δ

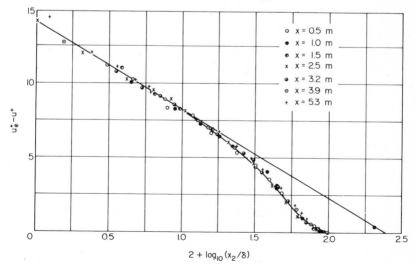

$$2 + \log_{10}(x_2/\delta)$$

Fig. 76. Velocity decrement for an incompressible turbulent boundary layer away from the surface ("law of the wake") [157].

is rather arbitrary because of the asymptotic approach of the velocity to its free stream value, no serious error results in these correlations if a consistent definition such as $u_1(\delta) = 0.995\, u_{1e}$ is adopted. For $0.01 < x_2/\delta < 0.1$, the correlation is linear on the logarithmic plot. The velocity distribution in the entire wake region of a fully developed turbulent boundary layer on a flat plate can be represented by extending Eq. (281) to

$$u^+ = 2.43 \ln y^+ + 5.0 + 1.35 w \,(x_2/\delta) \tag{282}$$

where $w(x_2/\delta)$ is the Coles function [159] indicated in Fig. 77. This function can be approximated quite accurately by $1 - \cos(\pi x_2/\delta)$ [54].

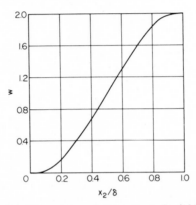

Fig. 77. Coles' "law of the wake" function in Eq. (282) [159].

Clauser [156] made two observations regarding turbulent boundary-layer velocity profiles that have direct influence on convective heating analyses. First, he noted that the region characterized by the "law of the wall" equilibrates very quickly after it is disturbed by some external force. Thus, this region is nearly in equilibrium with its local surface boundary conditions. In contrast, the region characterized by the "law of the wake" possesses a long memory of the upstream events. Clauser also found that the eddy diffusivity in the wake region is essentially independent of the distance from the surface x_2 and is related to the local boundary-layer thickness which reflects the growth of the boundary layer from the leading edge. The resulting expression for the eddy diffusivity in the wake region on a flat plate or a surface with an equilibrium turbulent boundary layer $((\delta_1^*/\tau_w)(dp/dx_1) = \text{constant})$ is

$$\frac{\bar{\epsilon}}{\nu} = 0.018 \left(\frac{u_{1e}\, \delta_1^*}{\nu}\right) \tag{283}$$

where $\bar{\epsilon} = \nu + \epsilon_M$.

Because the linear regions in the logarithmic plots of Figs. 75 and 76 overlap, it can be deduced that a unique relationship must exist between the skin friction on a flat plate and the Reynolds number based on the boundary-layer thickness [156]

$$\sqrt{\frac{2}{c_f}} = 5.6 \log_{10} \mathrm{Re}_\delta \sqrt{\frac{c_f}{2}} + 7.4 \tag{284}$$

Other relationships obtained from the wake correlation for the flat plate may be expressed in terms of a boundary-layer thickness

$$\bar{\Delta} = \int_0^\infty \frac{u_1 - u_{1e}}{u_\tau} \, dx_2 \tag{285}$$

that is independent of an arbitrary selection of δ. The following result

$$\frac{\delta_1^*}{\bar{\Delta}} = \sqrt{\frac{c_f}{2}} \tag{286}$$

$$\frac{\theta_{11}}{\bar{\Delta}} = \sqrt{\frac{c_f}{2}} \left(1 - 6.8 \sqrt{\frac{c_f}{2}} \right) \tag{287}$$

The dependence on the skin friction coefficient in the above equations indicates that the turbulent boundary layer on a flat plate is nonsimilar, in contrast to the corresponding laminar boundary layer. Equations (286) and (287) lead, for δ defined at $u_1 = 0.995 \, u_{1e}$, to

$$\frac{\delta}{\bar{\Delta}} = 0.28 \tag{288}$$

and to a boundary-layer form factor

$$\frac{\delta_1^*}{\theta_{11}} = \left(1 - 6.8 \sqrt{\frac{c_f}{2}} \right)^{-1} \tag{289}$$

Local skin friction coefficient laws can be derived directly from these velocity profile parameters. In terms of Eqs. (284) through (288)

$$\sqrt{\frac{2}{c_f}} = 2.43 \ln \left(\frac{5.87 \, \mathrm{Re}_{\theta_{11e}}}{1 - 6.8\sqrt{c_f/2}} \right) \tag{290}$$

If a mean value of δ_1^*/θ_{11} is employed for $0.001 < c_f/2 < 0.004$, a simple equation for $c_f/2$ on a flat plate in terms of the momentum thickness Reynolds number results from Eq. (290)

$$\frac{c_f}{2} = \frac{1}{(2.43 \ln \mathrm{Re}_{\theta_{11e}} + 5.0)^2} \tag{291}$$

This form of equation originated with von Kármán; however, the coefficients appearing here are slightly different because they are based on more recent data [156]. The resulting skin friction coefficients are 7 percent and 0.2 percent less than the von Kármán values at $\mathrm{Re}_{\theta_{11e}} = 10^3$ and 10^5, respectively. It is often convenient to evaluate the local skin friction coefficient with the following simple equations

$$\frac{c_f}{2} = \frac{0.0128}{\mathrm{Re}_{\theta_{11e}}^{1/4}} \qquad \mathrm{Re}_{\theta_{11e}} < 4000 \tag{292}$$

$$\frac{c_f}{2} = \frac{0.0065}{\mathrm{Re}_{\theta_{11e}}^{1/6}} \qquad \mathrm{Re}_{\theta_{11e}} > 4000 \qquad (293)$$

after Blasius and Falkner [162], respectively. Equations (291) through (293) are compared with recent data [163] in Fig. 78.

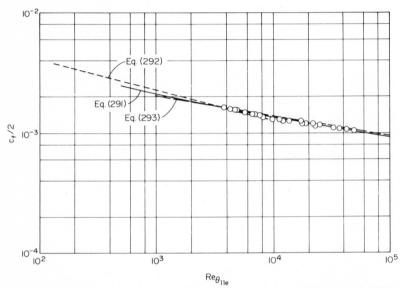

Fig. 78. Comparison of local skin friction coefficients predicted by Eqs. (291) through (293) with the experimental data of [163].

On an impervious flat plate with uniform pressure, the momentum integral equation (70) reduces to

$$\frac{\tau_{w1}}{\rho_e u_{1e}^2} = \frac{d\theta_{11}}{dx_1} \qquad (294)$$

This equation leads directly to an expression for the local length Reynolds number

$$\mathrm{Re}_{x_{1e}} = \int_0^{\mathrm{Re}_{\theta_{11e}}} \frac{d\,\mathrm{Re}_{\theta_{11e}}}{c_f/2} \qquad (295)$$

where the boundary layer has zero thickness at $x_1 = 0$. For a laminar boundary layer, the evaluation of Eq. (295) is quite direct through the use of a relationship such as Eq. (89). For the turbulent boundary layer, however, integration of Eq. (295) from the leading edge must account for the extent that the fully turbulent boundary layer has been preceded by a laminar boundary layer and, often more important, the extent of the transitional zone (see §2). This conceptually proper approach is quite complex, and the techniques for its application, e.g., described in §77, are somewhat uncertain. The usual approach is to assume that the laminar and transitional zones are so short that the turbulent boundary layer acts as if it originated at the leading edge of the plate. With this assumption, for length Reynolds numbers of a few million, Eqs. (292) and (295) combine to yield

$$\frac{c_f}{2} = \frac{0.0296}{\text{Re}_{x_{1e}}^{1/5}} \tag{296}$$

For length Reynolds numbers in the tens of millions, the upstream portion of the boundary layer is not significant, and Eq. (293) may be used in Eq. (295) all the way from $\text{Re}_{\theta_{11e}} = 0$ to yield

$$\frac{c_f}{2} = \frac{0.0131}{\text{Re}_{x_{1e}}^{1/7}} \tag{297}$$

A more complicated equation proposed by Schultz-Grunow [157] and which covers the full Reynolds number range is given by

$$\frac{c_f}{2} = \frac{0.185}{\left(\log_{10} \text{Re}_{x_{1e}}\right)^{2.584}} \tag{298}$$

Heat transfer to a uniform temperature plate in low speed flow is best found from the Reynolds analogy, the relationship between heat transfer and skin friction evaluated through analyses utilizing the empirical velocity distributions cited earlier. Knowledge of the flow field, which is independent of the temperature field when the fluid properties are constant, can be used directly to define the temperature field for a variety of thermal conditions and to evaluate the resulting convective heat transfer rates. For low speed, constant property flow, Eq. (279) reduces to

$$\rho c_p u_1 \frac{\partial T}{\partial x_1} + \rho c_p u_2 \frac{\partial T}{\partial x_2} = \frac{\partial}{\partial x_2} \left[\left(\frac{\mu}{\text{Pr}_T} + \frac{\rho \epsilon_M}{\text{Pr}_t} \right) c_p \frac{\partial T}{\partial x_2} \right] \tag{299}$$

Early solutions of this equation were based on the premise that the rate of change of temperature or velocity in the x_1 direction is small relative to the change of the same properties in the x_2 direction (i.e., Couette flow). This assumption and the continuity equation result in the left sides of Eqs. (278) and (299) being set equal to zero so that the momentum and energy equations reduce to

$$\tau_w = (\mu + \rho \epsilon_M) \frac{du_1}{dx_2} \tag{300}$$

and

$$\frac{q_w}{c_p} = - \left(\frac{\mu}{\text{Pr}_T} + \frac{\rho \epsilon_M}{\text{Pr}_t} \right) \frac{dT}{dx_2} \tag{301}$$

Here, x_2 may be eliminated from Eqs. (300) and (301), and direct integration yields

$$c_h = \frac{c_f/2}{\displaystyle\int_0^1 \left(\frac{\mu + \rho \epsilon_M}{\mu/\text{Pr}_T + \rho \epsilon_M/\text{Pr}_t} \right) d \left(\frac{u_1}{u_{1e}} \right)} \tag{302}$$

Reference 155 gives an excellent review of the hypotheses employed by various analysts to empirically define the integrand and the intermediate limits of integration in Eq. (302) for the different regions within the turbulent boundary layer. The most widely used formula resulting from integrating Eq. (302) is due to von Kármán [164] and includes the laminar sublayer, the buffer layer, and the fully turbulent portion of the law of the wall extended to the edge of the boundary layer.

Deissler [165] replaced the empirical velocity distribution employed by von Kármán in the laminar sublayer and buffer layer with one that was calculated for an eddy viscosity gradually increasing with distance from the wall, viz., $\epsilon_M = n^2 u_1 x_2$ where n is an empirical constant. The resulting velocity distribution and gradients normal to the surface are continuous until the fully turbulent portion is reached (Eq. (281)), where a sudden change in gradient occurs. Deissler assumed that the local heat flux was constant through the boundary layer and that $Pr_t = 1$. Use of his eddy viscosity expression near the surface and that corresponding to Eq. (281) away from the surface permitted solving Eq. (302) numerically for a large range of Prandtl numbers.

Because the wake region encompasses more of the boundary layer as the Reynolds number is increased, later analyses attempted to account for the wake region in different ways. Van Driest [166] retained the Couette flow approximation but permitted the shear in the boundary layer to approach zero at the outer edge. He also introduced nonunity turbulent Prandtl numbers to permit his recovery factor expression for air to agree with data, consistent with the findings of an earlier analysis [167]. The van Driest expression is

$$
\frac{c_f/2}{c_h} = Pr_t \left\{ 1 + 5\sqrt{\frac{c_f}{2}} \left\{ \frac{1 - Pr_t}{5K} \left[\frac{\pi^2}{6} + \frac{3}{2}(1 - Pr_t) \right] + \left(\frac{Pr_T}{Pr_t} - 1 \right) \right. \right.
$$

$$
\left. \left. + \ln \left[1 + \frac{5}{6}\left(\frac{Pr_T}{Pr_t} - 1 \right) \right] \right\} \right\} \tag{303}
$$

where $0.7 \leq Pr_t \leq 1$ and $K = 0.4$. Equation (303) reduces to the von Kármán analogy for $Pr_t = 1$.

Spalding [168] abandoned the Couette flow hypothesis and solved Eq. (299) through the use of the von Mises transformation obtaining

$$
\frac{\partial T}{\partial x^+} = \frac{1}{\epsilon_u^+ u^+} \frac{\partial}{\partial u^+} \left(\frac{\epsilon_h^+}{\epsilon_u^+} \frac{\partial T}{\partial u^+} \right) \tag{304}
$$

where

$$
x^+ = \int_0^{Re_{x_{1e}}} \sqrt{\frac{c_f}{2}} \, dRe_{x_{1e}}
$$

$$
\epsilon_u^+ = 1 + \frac{\epsilon_M}{\nu} \tag{305}
$$

$$
\epsilon_h^+ = \frac{1}{Pr_T} + \frac{1}{Pr_t} \frac{\epsilon_M}{\nu}
$$

By assuming the shear is constant through the boundary layer, Spalding achieved a particularly simple expression for the eddy diffusivity appropriate to the entire region of the law of the wall

$$\epsilon_u^+ = 1 + \frac{K}{E}\left[e^{Ku^+} - 1 - Ku^+ - \frac{(Ku^+)^2}{2!} - \frac{(Ku^+)^3}{3!}\right] \tag{306}$$

The corresponding eddy thermal conductivity chosen by Spalding is

$$\epsilon_h^+ = \frac{1}{Pr_T} + \frac{1}{Pr_t}\frac{K}{E}\left[e^{Ku^+} - 1 - Ku^+ - \frac{(Ku^+)^2}{2!} - \frac{(Ku^+)^3}{3!}\right] \tag{307}$$

where $K = 0.4$ and $9 < E < 10$ are constants determined empirically from the law of the wall. Later, Spalding [182] set $E = 12$ in a skin friction law. A quantity Pr_{tot} represents a Prandtl number based on the sum of the laminar and turbulent transport mechanisms, viz.,

$$Pr_{tot} = \frac{\epsilon_u^+}{\epsilon_h^+} = \frac{1 + \epsilon_M/\nu}{1/Pr_T + (1/Pr_t)(\epsilon_M/\nu)} \tag{308}$$

Although this approach is a complete solution of the energy equation, it contains inconsistencies in that the eddy diffusivities employed throughout the boundary layer are based entirely on the law of the wall. Thus, the solutions provided by this approach are best when the thermal boundary layer is much thinner than the flow boundary layer ($Pr_T \gg 1$) or when the thermal boundary layer is initiated later than the flow boundary layer as a step in the wall temperature downstream from the leading edge of the plate. (See §44.)

Ferrari [169] also solved the von Mises form of the energy equation but with the velocity distributions in the inner and outer portions of the boundary layer represented by polynomials of a velocity potential function. This analysis is applicable to gases in that it accounts for compressibility, but the solutions may be in error for high Prandtl numbers because of the rather approximate fit to the law of the wall.

Spalding [170] examined the results of other investigators who followed his approach and deduced that Pr_t must differ from unity to yield agreement with experimental data which show a Reynolds number dependence of the Reynolds analogy factor. The requirement of a nonunity turbulent Prandtl number for gases seems to be independent of the details of the analyses used to define the Reynolds analogy factor. From velocity distributions in gases, Spalding suggested $Pr_t = 0.9$. He showed, further, how previous solutions can be adjusted to accommodate non-unity Pr_t and found that the numerical results can be fitted with the interpolation formula

$$\frac{c_f/2}{c_h} = Pr_{tot}\sqrt{\frac{c_f}{2}}\left\{6.76\left(\frac{x^+}{Pr_t}\right)^{1/9} + 11.57\left[\left(\frac{Pr_T}{Pr_t}\right)^{3/4} - 1\right]\right\} \tag{309}$$

In a recent publication, Spalding [161] extended his analysis to include the wake region of the boundary layer. The numerical results for heat transfer rates to a plate agree with those given by Eq. (309) to a few percent. This would be expected because

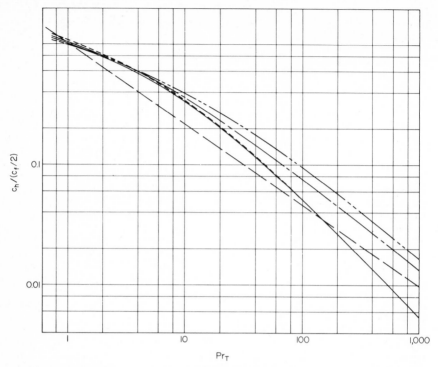

Fig. 79. Reynolds analogy factor for incompressible turbulent boundary layer. —— von Kármán analogy, Eq. (303) with $Pr_t = 1.0$; —— —— Colburn analogy, $Pr_T^{-2/3}$; — — — van Driest analogy, Eq. (303) with $Pr_t = 0.86$; —— — Spalding analogy, Eq. (309) with $Pr_t = 0.9$; —— — — Deissler analogy, $Pr_t = 1.0$.

at high Prandtl numbers the thermal boundary layer is confined to the inner portion of the flow boundary layer and is insensitive to the flow characteristics in the outer portion of the flow boundary layer.

The numerical results of the various Reynolds analogy factors are compared in Fig. 79 for laminar Prandtl numbers ranging from those characteristic of gases to those of oils and for $Re_{x_{1e}} = 10^7$. Results for very low laminar Prandtl numbers, characteristic of liquid metals, are not shown because the assumptions for the velocity distributions in the various analyses are not valid when the thermal boundary layer is much thicker than the flow boundary layer. A comparison of the von Kármán analogy with the van Driest results indicates that use of $Pr_t = 0.89$ rather than $Pr_t = 1$ affects the results only for laminar Prandtl numbers less than about 10 and causes the Reynolds analogy factor to exceed unity when $Pr_T = 1$. These observations are also evident from the form of Eq. (309). The Deissler and Spalding results, although crossing in the vicinity of $Pr_T = 1$, both approach an asymptotic limit proportional to $Pr_T^{-3/4}$ for very large Prandtl numbers. Most significant is the departure of all the analyses from the Colburn analogy $Pr_T^{-2/3}$. An explanation of this can be deduced from the Deissler analysis [165] where the Reynolds analogy is also applied to the case of pipe flow through the calculation of bulk-mean properties. Here, the analytical results agree reasonably well with the Colburn analogy. Thus, a deductive extension of the behavior of turbulent pipe flow data and laminar boundary-layer theory on a flat plate, both of which yield a Reynolds analogy factor of $Pr_T^{-2/3}$, to the turbulent boundary layer on a flat plate yields erroneous results, particularly for $Pr_T \leq 10$.

Lack of heat transfer data on plates at high Prandtl numbers precludes an accurate assessment of the methods shown in Fig. 79. Pipe flow data favor the analyses of Deissler and Spalding; however, distinguishing between these analyses depends merely on a judgment regarding the accuracy of specific experiments leading to the empirical constants in each. The differences are indicative of the basic uncertainty inherent in these results. Either of these methods may be used at high Prandtl numbers, with the choice being governed by conservatism in a specific design.

For Prandtl numbers slightly less than unity (gases), it is quite obvious that the Colburn analogy is at variance with the analytical results. For $Pr_T = 0.7$, excellent low-speed data [164] indicate values of $2c_h/c_f \simeq Pr_T^{-0.4} \simeq 1.15$. Other data in air at higher speeds [171], though less accurate and sometimes containing disturbances such as small pressure gradients, favor the value $2c_h/c_f = 1.2$. Since these data were obtained over a small Prandtl number range, it is difficult to ascertain empirically the influence of Pr_t and the functional dependence on Pr_T. Accordingly, it is recommended that the Spalding and van Driest results, represented by Eqs. (309) and (303) be utilized for $Pr_T \sim 1$ in view of their general agreement with the data cited.

For high speed flow, frictional dissipation occurs in the turbulent boundary layer because of combined viscous and turbulent shear mechanisms. When the fluid properties are constant, Eq. (279) can be transformed by the von Mises transformation to an equation that is similar to Eq. (304) except that a term proportional to u_{1e}^2 is added to the right side. Although the resulting equation is inhomogeneous, it is separable into a homogeneous part accounting for heat transfer (Eq. (304)) and an inhomogeneous part relating to the insulated case or the recovery temperature [169]. As with the laminar boundary layer, the resulting heat flux expression is

$$q_w = \rho_e u_{1e} c_h (i_w - i_{aw}) \tag{310}$$

where

$$i_{aw} = i_e + r(0) \frac{u_{1e}^2}{2} \tag{311}$$

and $r(0)$ is the recovery factor. From Couette flow analyses

$$r(0) = \int_0^1 \frac{\mu + \rho \epsilon_M}{\mu/Pr_T + \rho \epsilon_M/Pr_t} \, d\left(\frac{u_1}{u_{1e}}\right)^2 \tag{312}$$

When Eq. (312) is solved under the same assumptions employed in Eq. (302) [166, 167, 169], it is found that the recovery factor for air ($Pr_T = 0.7$, $Pr_t = 1$) experiences a marked Reynolds number dependence, contrary to a rather profuse accumulation of data that yield values of $r(0)$ between 0.87 and 0.89. Thus, solutions of Eq. (312), rather than yielding values of $r(0)$, have been used to estimate the turbulent Prandtl number necessary to eliminate the dependence of $r(0)$ on $Re_{x_{1e}}$. It is in this manner that van Driest found $Pr_t = 0.89$ for use in Eq. (303). Values of $r(0)$ within a percent of most existing data can be obtained by using

$$r(0) = Pr_T^{1/3} \tag{313}$$

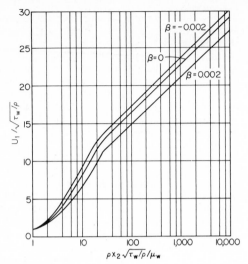

Fig. 80. Universal velocity profiles for a liquid with $\mu/\mu_w = (T/T_w)^{-4}$; $\beta = \dfrac{c_h}{\sqrt{c_f/2}} \dfrac{T_w - T_e}{T_w}$ [165].

67. Liquids with Variable Viscosity. Deissler [165] considered a fluid where the viscosity is temperature dependent but all the other fluid properties remain constant. Solutions for $Pr_T = 10$ and

$$\frac{\mu}{\mu_w} = \left(\frac{T}{T_w}\right)^{-4} \tag{314}$$

are shown in Figs. 80 and 81. The ordinate in Fig. 81 is defined as

$$t^+ = \frac{\sqrt{c_f/2}}{c_h}\left(\frac{T_w - T}{T_w - T_e}\right) \tag{315}$$

and the parameter

$$\beta = \frac{c_h}{\sqrt{c_f/2}}\left(\frac{T_w - T_e}{T_w}\right) = \frac{1}{t_e^+}\left(\frac{T_w - T_e}{T_w}\right) \tag{316}$$

It is noted in Fig. 80 that the effects of variable viscosity cause the velocity distributions outside the buffer layer to displace while remaining essentially parallel, so that

$$\frac{u_1}{\sqrt{\tau_w/\rho}} = A \ln\left(\frac{\sqrt{\tau_w/\rho}\ \rho x_2}{\mu_w}\right) + B(\beta) \tag{317}$$

where $B(\beta)$ is shown in Fig. 82. The parallel curves lead to the conclusion that $c_f/2$

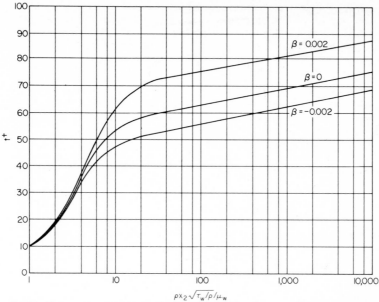

$$\rho x_2 \sqrt{\tau_w/\rho}/\mu_w$$

Fig. 81. Universal temperature profiles for a liquid with velocity profiles as in Fig. 80.

can be computed from $c_f\left(\mathrm{Re}_{x_{1e}}\right)$ relationships, e.g., Eqs. (296) or (297), if the Reynolds number employed is replaced by

$$\mathrm{Re}_{\mathrm{eff}} \simeq \frac{\rho_e u_{1e} x_1}{\mu_w} e^{B(\beta) - B(0)} \tag{318}$$

A simple iterative procedure for evaluating the local heat transfer rate is begun by assuming a value of β. The effective Reynolds number is evaluated from Fig. 82 and Eq. (318). Equation (296) or (297) yields $c_f/2$. The ordinate in Fig. 80 set equal to $\sqrt{2/c_f}$ yields the limiting value at the edge of the boundary layer of $\sqrt{\tau_w/\rho}\,\rho\delta/\mu_w$ that is then used in Fig. 81 to obtain a value of t_e^+. This value of t_e^+ and the wall and

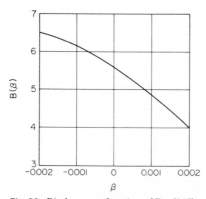

Fig. 82. Displacement function of Eq. (317).

boundary-layer edge temperatures are substituted in Eq. (316) to yield the next approximation for β. The procedure is repeated until convergence on a value of β is achieved. The local heat flux is then found from

$$q_w = \rho c_p u_{1e} T_w \beta \sqrt{\frac{c_f}{2}} \tag{319}$$

68. Ideal Gases at High Temperature. Three fundamentally different approaches have been applied to the treatment of the turbulent boundary layer with variable fluid properties; all are restricted to air behaving as an ideal, calorically perfect gas. First, the Couette flow solutions have been extended to permit variations in viscosity and density. Second, mathematical transformations, analogous to Eq. (110) for a laminar boundary layer, have been used to transform the variable property turbulent boundary-layer differential equations into constant property equations in order to provide a direct link between the low-speed boundary layer and its high-speed counterpart. Third, empirical correlations have been found that directly relate the variable property results to incompressible skin friction and Stanton number relationships. Examples of the latter are reference temperature or enthalpy methods analogous to those used for the laminar boundary layer, and the method of Spalding and Chi [182]. At present, contradictions exist between predictions based on these methods, and a paucity of definitive experiments makes the choice of the design method somewhat arbitrary. To avoid a dependence of subsequent formulations of the effects of pressure gradients and transition on any particular method, all the methods considered are expressed in a common generalized form that is then used in later sections. This form can also serve to provide insight into the sources of the differences between the methods.

The Couette flow analyses [172–175] involve the simultaneous solution of the momentum equation (Eq. (300)) and the energy equation

$$q_w + u_1(\mu + \rho \epsilon_M) \frac{du_1}{dx_2} = -\left(\frac{\mu}{Pr_T} + \frac{\rho \epsilon_M}{Pr_t} \right) \frac{dT}{dx_2} c_p \tag{320}$$

Equation (320) differs from Eq. (301) in that it contains the frictional dissipation term necessary at high speeds. In Ref. 172, it is assumed that both the laminar and turbulent Prandtl numbers are unity; thus, Eq. (115) can be directly extended to the turbulent boundary layer. As a result of this assumption, the energy equation is directly related to the momentum equation ($c_h = c_f/2$), and the properties ρ, μ are expressible in terms of u_1. Equation (300) is then directly integrable once ϵ_M is formulated and the shear is assumed constant through the boundary layer and equal to its value at the wall. Van Driest [172] neglected the sublayer and the contribution of viscosity in the shear stress and used the Prandtl mixing length relationship $l = Kx_2$ in place of the eddy viscosity. Later van Driest [174] modified his analysis to include the von Kármán form of the mixing length

$$l = K \frac{\partial u_1/\partial x_2}{\partial^2 u_1/\partial x_2^2}$$

where $K = 0.4$ in both forms. The solution of Eq. (300) results in a velocity profile with the local wall shear as a parameter. This profile is introduced into the reduced form of the momentum integral equation (Eq. (294)) to yield $c_f(Re_{x_{1e}})$. Deissler and Loeffler [175] performed an analysis equivalent to the second of van Driest's theories

[174] but included the effects of viscosity and an eddy viscosity model that extended to the surface (see p. 8-146). Further, they allowed the laminar and turbulent Prandtl numbers to deviate from unity. These changes required that Eqs. (300) and (320), as well as the momentum integral equation, be solved numerically. The resulting Stanton numbers contained modifications to the Reynolds analogy, cited earlier for $M_e = 0$. Values of the recovery factor were also evaluated and although they showed a Reynolds number dependence not evident in data (a lower turbulent Prandtl number than the 0.93 used might have eliminated most of this), the work revealed that $r(0) = 0.9$ is valid within a few percent over a large Mach number range.

The boundary-layer transformations [176--178] imply that a formal mathematical transformation of the variable property turbulent boundary-layer partial differential equations into equivalent constant property equations is physically valid even though our understanding of the local turbulent processes in either case is quite poor. A similar transformation [179] results from an examination of the boundary layer as a whole through the integral equations where the local shear is not a factor except at the wall where the functional form is known. The most general transformation, by Coles [178], introduces an additional degree of freedom not contained in Eq. (110) for the laminar boundary layer. This may be necessary because of the essential nonsimilar character of the turbulent boundary layer. The Coles transformation is represented by

$$\tilde{\psi}(\tilde{x}_1, \tilde{x}_2) = \sigma(x_1)\psi(x_1, x_2)$$

$$d\tilde{x}_1 = \xi(x_1)\, dx_1 \tag{321}$$

$$\tilde{\rho}\, d\tilde{x}_2 = \eta(x_1)\rho\, dx_2$$

where the superscript tilde represents a constant property case. (Crocco [179] is less restrictive by utilizing a low speed but variable property case as the reference.) In terms of the transformations (321), the velocities in the two systems are

$$\tilde{\rho}\,\tilde{u}_1 = \frac{\partial\tilde{\psi}}{\partial\tilde{x}_2} \qquad \rho u_1 = \frac{\partial\psi}{\partial x_2} \tag{322}$$

so that

$$\tilde{u}_1 = \frac{\sigma}{\eta}\, u_1 \tag{323}$$

The skin friction coefficients are related as

$$\tilde{c}_f = \left(\frac{\rho_e \mu_e}{\rho_w \mu_w}\right)\left(\frac{\tilde{\mu}}{\sigma\mu_e}\right)c_f \tag{324}$$

Further, from the definition of the momentum thickness,

$$\frac{\tilde{\rho}\,\tilde{u}_{1e}\,\tilde{\theta}_{11}}{\tilde{\mu}} = \left(\frac{\sigma\mu_e}{\tilde{\mu}}\right)\frac{\rho u_{1e}\,\theta_{11}}{\mu_e} \tag{325}$$

The transformation variable $\tilde{\mu}/\sigma$ is eliminated between Eqs. (324) and (325) to yield Coles' "law of corresponding stations"

$$\tilde{c}_f\, \mathrm{Re}_{\tilde{\theta}_{11}} = \frac{\rho_e \mu_e}{\rho_w \mu_w}\, c_f\, \mathrm{Re}_{\theta_{11}} \tag{326}$$

To complete the transformation, it is necessary to evaluate $\tilde{\mu}/\sigma$ in Eq. (324). Coles identified $\tilde{\mu}/\sigma$ with the viscosity based on the mean temperature of a layer near the surface that is contained within the "law of the wall" region but extends beyond the laminar sublayer. This region is called the "substructure" and its mean temperature T_s is evaluated by Coles as

$$\frac{T_s}{T_e} = \frac{T_w}{T_e} + 17.2\sqrt{\frac{\tilde{c}_f}{2}}\left(1 + r(0)\frac{\gamma - 1}{2}M_e^2 - \frac{T_w}{T_e}\right) - 305\left(\frac{\tilde{c}_f}{2}\right)r(0)\frac{\gamma - 1}{2}M_e^2 \qquad (327)$$

The recovery factor $r(0)$ has been added in Eq. (327) as a first-order modification to account for the departure of Prandtl number from unity [179]. Finally, from Eq. (324)

$$c_f = \tilde{c}_f\left(\frac{\mu_w \rho_w}{\mu_e \rho_e}\right)\left(\frac{\mu_e}{\mu_s}\right) \qquad (328)$$

where the viscosity μ_s is evaluated at the temperature T_s from Eq. (327). The term $\tilde{c}_f$ acts as a parameter in establishing $c_f\left(\text{Re}_{\theta_{11}}\right)$ for given values of M_e and T_w/T_e.

The empirical methods using a reference temperature are represented by Refs. 79, 180, and 181. Eckert [79] employs the same coefficients as indicated in Table 6 for both laminar and turbulent boundary layers. Sommer and Short [181] utilize the form of T'/T_e for the laminar boundary layer but modify the coefficients to agree with skin friction data on insulated plates in wind tunnels and their own free-flight data for cold wall conditions. Their expression is

$$\frac{T'}{T_e} = 1 + 0.45\left(\frac{T_w}{T_e} - 1\right) + 0.035\,M_e^2 \qquad (329)$$

for use in an incompressible skin friction law of the form $c_f'\left(\text{Re}_{x_1}\right)$, where the superscript prime means that all the fluid properties are evaluated at the temperature T'.

The Spalding and Chi method [182] relates the variable and constant property skin friction coefficients by means of van Driest's theory and relates the corresponding Reynolds number through empirical relationships involving the surface, free stream, and recovery temperatures. Although a large body of data was used in the correlation, the method suffers because of a lack of selectivity in the choice of data and an equal weighting of all data even though some contain systematic errors.

Generalized Coordinate Transformation. Each of the previously mentioned methods can be thought to provide coordinate transformation rules for extending constant property skin friction and Stanton number formulas to account for property variations as performed by Spalding and Chi. Thus, if the low-speed constant property formulas can be expressed functionally as

$$\tilde{c}_f = f_1\left(\tilde{\text{Re}}_{\tilde{\theta}_{11}}\right) \qquad (330)$$

and

$$\tilde{c}_f = f_2\left(\tilde{\text{Re}}_{x_1}\right) \qquad (331)$$

the generalized formulas are

$$c_f F_c = f_1\left(F_{R_\theta} \mathrm{Re}_{\theta_{11}}\right) \tag{332}$$

$$c_f F_c = f_2\left(F_{R_x} \mathrm{Re}_{x_{1e}}\right) \tag{333}$$

In general, the coordinate transformation factors F_c, F_{R_θ}, F_{R_x} are functions of M_e, T_w/T_e, $\mathrm{Re}_{x_{1e}}$, and T_e. (If the viscosity is expressed as a power law function of temperature, the dependence on T_e can be eliminated.) From the von Kármán momentum integral equation (70) for a flat plate

$$\frac{c_f}{2} = \frac{d\,\mathrm{Re}_{\theta_{11}}}{d\,\mathrm{Re}_{x_{1e}}} \tag{334}$$

it follows that

$$F_{R_x} = \frac{F_{R_\theta}}{F_c} \tag{335}$$

The expressions for F_c and F_{R_θ} given by the various methods are presented in Table 13. Values of these functions obtained from the van Driest theory [174] are shown in Figs. 83 and 84 for $T_e = 400°R$, $0 < M_e < 10$, and $T_w/T_e = 1, 2, 3, 5$, T_{aw}/T_e. It is noted that only the Coles method allows F_c and F_{R_θ} to depend on the Reynolds number (through the parameter $\tilde{c}_f$).

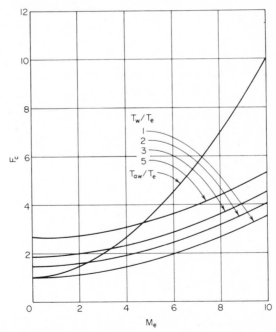

Fig. 83. Compressible turbulent boundary layer transformation parameter for skin friction coefficient, $r(0) = 0.9$, $T_e = 400°R$.

TABLE 13. Expressions for F_c and F_{R_θ} for Air as an Ideal Gas

Method	F_c	F_{R_θ}	Supplemental formulas and notes
Eckert [79]	ρ_e/ρ'	μ_e/μ'	$\dfrac{T'}{T_e} = 0.28 + 0.50\,\dfrac{T_w}{T_e} + 0.22\,\dfrac{T_{aw}}{T_e}$
Sommer Short [181]	ρ_e/ρ'	μ_e/μ'	$\dfrac{T'}{T_e} = 0.36 + 0.45\,\dfrac{T_w}{T_e} + 0.19\,\dfrac{T_{aw}}{T_e}$
van Driest (von Kármán mixing length) [174]	$\dfrac{T_{aw}/T_e - 1}{(\sin^{-1}\alpha + \sin^{-1}\beta)^2}$	μ_e/μ_w	$\alpha = \dfrac{T_{aw}/T_e + T_w/T_e - 2}{\sqrt{(T_{a\omega}/T_e - T_w/T_e)^2 + 4(T_w/T_e)(T_{aw}/T_e - 1)}}$ $\beta = \dfrac{T_{aw}/T_e - T_w/T_e}{\sqrt{(T_{aw}/T_e - T_w/T_e)^2 + 4(T_w/T_e)(T_{aw}/T_e - 1)}}$
Coles [178]	$\left(\dfrac{\mu_e \rho_e}{\mu_w \rho_w}\right)\left(\dfrac{\mu_s}{\mu_e}\right)$	μ_e/μ_s	$\dfrac{T_s}{T_e} = \dfrac{T_w}{T_e} + 17.2\sqrt{\dfrac{\bar{c}_f}{2}}\sqrt{\dfrac{\bar{c}_f}{2}\left(\dfrac{T_{aw}}{T_e} - \dfrac{T_w}{T_e}\right) - 305\left(\dfrac{\bar{c}_f}{2}\right)\left(\dfrac{T_{aw}}{T_e} - 1\right)}$ α, β, same as van Driest
Spalding Chi [182]	$\dfrac{T_{aw}/T_e - 1}{(\sin^{-1}\alpha + \sin^{-1}\beta)^2}$	$\dfrac{(T_{aw}/T_e)^{0.772}}{(T_w/T_e)^{1.474}}$	$\mu \sim T^{0.76}$ assumed in empirical exponent of F_{R_θ}

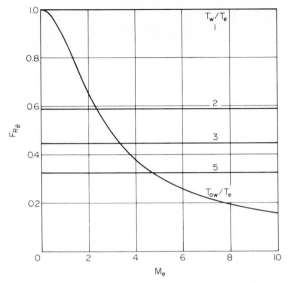

Fig. 84. Compressible turbulent boundary layer transformation parameter for momentum thickness Reynolds number, $r(0) = 0.9$, $T_e = 400°R$.

The recommended design method for Mach numbers up to supersonic ($M_e < 4$) and adiabatic wall conditions ($T_w = T_{aw}$) is arbitrary since all the methods yield essentially the same results. The design methods for hypersonic Mach numbers ($M_e > 4$) and cold wall conditions ($T_w < T_{aw}$) should be based on conservatism. Comparison of the available skin friction data reveal differences of ±20 percent on surfaces near adiabatic wall temperature and as much as a factor of 2 for highly cooled walls ($T_w \sim 0.2 T_{aw}$). If only the most recent skin friction data for $0.3 < T_w/T_{aw} < 1.0$ are considered, particularly those where $Re_{\theta_{11}}$ was obtained from boundary-layer surveys, a recent evaluation [222] favors the Coles and van Driest methods. The Sommer and Short, and Spalding and Chi methods may underpredict the skin friction as much as 30 percent. For very cold walls ($T_w < 0.3 T_{aw}$), none of the methods predict the effect of wall temperature ratio in the available skin friction data. The van Driest skin friction predictions are shown in Fig. 85 for $T_e = 400°R$, $0 < M_e < 10$, and $T_w/T_e = 1, 2, 3, 5, T_{aw}/T_e$.

The heat transfer is obtained by using a Reynolds analogy factor in combination with a skin friction prediction. For Mach numbers up to supersonic and near adiabatic wall conditions, the recommended Reynolds analogy factor is the same as for low speed flow, viz., 1.16. Any of the previously discussed skin friction theories may be used. For hypersonic Mach numbers and cold wall conditions, there are indications that the Reynolds analogy factor is more nearly equal to unity [222]. Use of this value and either the van Driest or the Coles theory for skin friction result in heat transfer predictions within ±10 percent of the data for $5 < M_e < 7.5$ and $0.1 < T_w/T_{aw} < 0.6$ [222]. Alternatively, to similar accuracy one can use the usually accepted Reynolds analogy factor of 1.16 and the Spalding and Chi skin friction theory.

Impervious Surface with Nonuniform Temperature

Nonuniform surface temperatures affect the convective heat transfer in a turbulent boundary layer similarly as in a laminar case except that the turbulent boundary layer responds in shorter downstream distances.

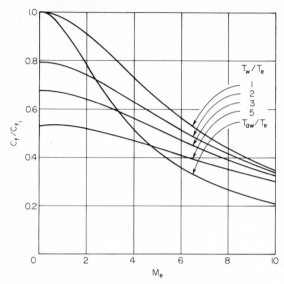

Fig. 85. Local skin friction coefficient for a compressible turbulent boundary layer on a flat plate, $r(0) = 0.9$, $T_e = 400°\,\text{R}$.

69. Stepwise and Arbitrary Surface Temperature Distribution. As in §44, the heat transfer to surfaces with arbitrary temperature variations is obtained by superposition of solutions for convective heating to a uniform temperature surface preceded by a surface at the recovery temperature of the fluid (Eq. (150)). For the superposition to be valid, it is necessary that the energy equation be linear in T or i, which imposes restrictions on the types of fluid property variations that are permitted. In the turbulent boundary layer, it is generally required that the fluid properties remain constant; however, under the assumption that boundary-layer velocity distributions are expressible in terms of the local stream function rather than x_2 for ideal gases, the energy equation is also linear in T [169].

The effect of a stepwise discontinuity in surface temperature on a flat plate is expressible [82] as

$$\frac{c_h(x_1, s)}{c_h(x_1, 0)} = \left[1 - \left(\frac{s}{x_1} \right)^{\frac{28m + 39}{40}} \right]^{-\frac{7 + 28m}{39 + 28m}} \qquad (336)$$

where m is an empirically evaluated constant related to the effect on the local heat transfer rate of the growth of the flow boundary layer. The basis of Eq. (336) is the solution of the energy integral equation (72) for assumed velocity and temperature distributions of the form $(x_2/\delta)^{1/7}$ and $(x_2/\delta_{th})^{1/7}$, respectively, where δ_{th} is the thermal boundary-layer thickness. Scesa's data [183] for air suggest that m is nearly zero, the value which was adopted in Ref. 82. Closer examination of Scesa's data and an improved analysis by Seban indicates $m = -3/28$. This value has been substantiated independently by analysis and experiments in air [184] and by analogous mass transfer experiments [185]. Therefore, for gases

$$\frac{c_h(x_1,s)}{c_h(x_1,0)} = \left[1 - \left(\frac{s}{x_1} \right)^{9/10} \right]^{-1/9} \tag{337}$$

For fluids other than gases, use of the Spalding relationship (309) results in

$$\frac{c_h(x_1,s)}{c_h(x_1,0)} = \frac{\dfrac{5.62\,\mathrm{Re}_{x_{1e}}^{1/10}}{\mathrm{Pr}_t^{1/9}} + 11.57 \left[\left(\dfrac{\mathrm{Pr}}{\mathrm{Pr}_t} \right)^{3/4} - 1 \right]}{\dfrac{5.62\,\mathrm{Re}_{x_{1e}}^{1/10}}{\mathrm{Pr}_t^{1/9}} \left[1 - \left(\dfrac{s}{x_1} \right)^{9/10} \right]^{1/9} + 11.57 \left[\left(\dfrac{\mathrm{Pr}}{\mathrm{Pr}_t} \right)^{3/4} - 1 \right]} \tag{338}$$

A short distance downstream from the step in surface temperature, for $\mathrm{Pr} = 0.7$ and $\mathrm{Pr}_t = 0.9$, the second term in both the numerator and denominator of Eq. (338) becomes comparatively small, and Eq. (338) reduces to Eq. (337). For very high Prandtl numbers, the Stanton number behind a step in surface temperature is essentially the same as on a uniform temperature surface.

The intuitive approach for including variable fluid properties is to evaluate the local $c_h(x_1,0)$ from the variable property techniques of §68, employing the local surface temperature or enthalpy as a parameter. Equations (337) and (338) are used primarily to define the form of the enthalpy potential appropriate to an arbitrary surface temperature. (See Eq. (150).)

The approximate numerical methods expressed by Eqs. (154) and (155), where the temperature distribution is approximated by a series of ramps, can be applied to the turbulent boundary layer by replacing the G and H functions by

$$S\left(\frac{s_n}{x_1} \right) = \int_0^{s_n/x_1} (1 - \alpha^{9/10})^{-1/9}\, d\alpha \tag{339}$$

and

$$R\left(\frac{s_n}{x_1} \right) = \int_0^{s_n/x_1} (1 - \alpha^{9/10})^{8/9}\, d\alpha \tag{340}$$

respectively. These functions are tabulated in Ref. 86 and are shown in Figs. 86 and 87.

A numerical method permitting sudden jumps as well as ramps in the approximated surface temperature distribution is given in Ref. 186. However, one of the functions involved becomes infinite at small distances downstream of the temperature jumps, and must be evaluated carefully. It is more convenient to use the method represented by Eqs. (339) and (340), with the sudden jump in temperature approximated by a ramp with large, but finite, slope. The method of Ref. 186 is particularly suited, however, to the problem described in the following section.

70. Stepwise and Arbitrary Heat Flux Distribution. A corollary of the problem treated in the previous subsection is the problem of finding the surface temperature

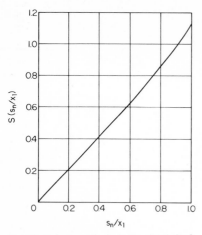

Fig. 86. Function $S(s_n/x_1)$, Eq. (339) [86].

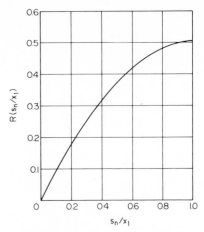

Fig. 87. Function $R(s_n/x_1)$, Eq. (340) [87].

distribution (or enthalpy distribution) resulting from a prescribed heat flux distribution. The superposition of solutions based on Eq. (337) yields

$$i_w(x_1) - i_{aw} = \frac{0.0979}{\rho_e u_{1e} c_h(x_1,0)} \int_0^1 \frac{q_w(s/x_1)}{\left[1 - (s/x_1)^{9/10}\right]^{8/9}} \, d\left(\frac{s}{x_1}\right) \tag{341}$$

Through a coordinated analytical and experimental program, Reynolds et al. [186] demonstrated the validity of Eq. (341). In addition, they provide an excellent numerical procedure for evaluating Eq. (341) for heat flux distributions that can be represented by a series of steps and ramps. The crux of this method is the manner of expressing the heat flux distribution as

$$q_w(x_1) = \sum_{n=1}^{N-1} m_n x_1 \left(\frac{s_{n+1}}{x_1} - \frac{s_n}{x_1}\right) + \sum_{j=1}^{J} b_j \quad , s_N = x_1 \tag{342}$$

where m_n is related to the slope of the nth ramp, and b_j is the height of the jth step in heat flux ahead of the point at which the temperature or enthalpy is to be evaluated. These terms are illustrated by an example heat flux distribution in Fig. 88, where $J = 2$ and $N = 4$. Note that the slope of the third ramp, for example, is $m_1 + m_2 + m_3$.

For the heat flux distribution (Eq. (342)), the surface enthalpy at x_1 is

$$i_w(x_1) - i_{aw} = \frac{q_w(x_1) - x_1 \sum_{n=1}^{N} m_n D\left(\frac{s_n}{x_1}\right) - \sum_{j=1}^{J} b_j E\left(\frac{s_j}{x_1}\right)}{\rho_e u_{1e} c_h(x_1,0)} \tag{343}$$

where the functions D and E are given in Fig. 89, and m_n and b_j are evaluated from the approximated heat flux distribution.

Surface with Mass Transfer

The most effective technique for thermally protecting missile nose cones, solid rocket nozzles, or turbine blades, for example, has been mass transfer cooling involving transpiration or ablation (see page 8-91). Because of the high Reynolds numbers encountered in many of these applications, the surface roughness or discrete gas jetting introduced by ablating or porous surfaces often hastens transition to turbulent boundary layers. Although the effect of mass transfer on the turbulent boundary layer is not completely understood, especially at high Mach numbers, sufficient information exists to permit rational engineering design.

71. Transpiration, Constant Fluid Properties. The largest body of information available concerns the transpiration of air into a low speed air stream [187, 188]. Under these conditions, the fluid properties in the boundary layer are essentially constant, and the turbulent boundary layer can be described mathematically with Eqs. (278) and (299). In addition, when small quantities of a foreign species are introduced into the boundary layer for diagnostic purposes or by evaporation, the local foreign species concentration in the absence of thermal diffusion is given by

$$\rho u_1 \frac{\partial K_1}{\partial x_1} + \rho u_2 \frac{\partial K_1}{\partial x_2} = \frac{\partial}{\partial x_2} \left[\left(\frac{\mu}{\mathrm{Sc}} + \frac{\rho \epsilon_M}{\mathrm{Sc}_t} \right) \frac{\partial K_1}{\partial x_2} \right] \tag{344}$$

where the laminar and turbulent Schmidt numbers are $\mathrm{Sc} = \nu/\mathfrak{D}$ and $\mathrm{Sc}_t = \epsilon_M/\epsilon_D$, respectively. The boundary conditions at the surface are $u_2 = u_{2_w}(x_1)$ and $T = T_w(x_1)$, where one of these is specified and the other is determined from the heat balance represented by Eq. (167) or (168).

Reference 187 presents a rather complete solution of the above equations that is guided strongly by empirical knowledge. For example, the solution is based on a boundary layer that is divided into inner and outer regions characterized by the laws of the wall and wake, respectively. The inner region is treated as a Couette flow where changes with respect to x_1 are neglected and $x_2/\delta \ll 1$. The outer region involves the complete two-dimensional equations. This procedure is a direct extension of Clauser's model for an impervious surface [156].

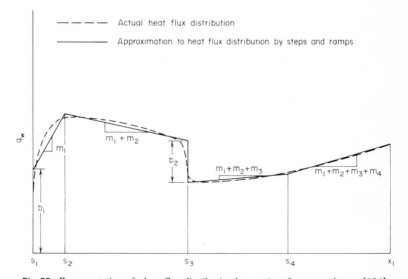

Fig. 88. Representation of a heat flux distribution by a series of ramps and steps [186].

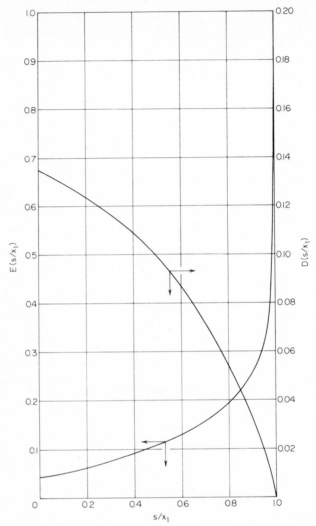

Fig. 89. Functions $D(s/x_1)$ and $E(s/x_1)$ for use in Eq. (343) [186].

In the inner region, the momentum equation reduces to

$$1 + v_0^+ u^+ = \left(1 + \frac{\epsilon_M}{\nu}\right) \frac{du^+}{dy^+} \tag{345}$$

where the term involving $v_0^+ \equiv u_{2w}/(u_1 \sqrt{c_f/2})$ is added to account for the local increase in shear over its wall value caused by the surface mass transfer. The essential element in the solution of Eq. (345) is the choice of the form of ϵ_M/ν, which in Ref. 187 was related to a mixing length defined by

$$\frac{dl^+}{dy^+} = \frac{Ky^+ - l^+}{y_a^+} (1 + v_0^+ u^+)^{1/2} \tag{346}$$

and where $\epsilon_M/\nu = l^{+2}(du^+/dy^+)$. The quantities $K = 0.4$ and $y_a^+ = 11.8$ are constants independent of the transpiration rate. For the case of zero blowing, Eq. (346) yields values consistent with more sophisticated analyses [189, 223] for an impervious plate. In terms of the mixing length, Eq. (345) is

$$\frac{du^+}{dy^+} = \frac{-1 + \sqrt{1 + 4l^{+2}(1 + v_0^+ u^+)}}{2l^{+2}} \tag{347}$$

which is solved numerically, simultaneously with Eq. (346). The temperature and concentration distributions in the inner region are then found from the solution of

$$\frac{dT^+}{dy^+} = \frac{Pr(1 + v_0^+ T^+)}{1 + (Pr/Pr_t)\, l^{+2}\, (du^+/dy^+)} \tag{348}$$

and

$$\frac{dK_1^+}{dy^+} = \frac{Sc\,(1 + v_0^+ K_1^+)}{1 + (Sc/Sc_t)\, l^{+2}\, (du^+/dy^+)} \tag{349}$$

where

$$T^+ = \frac{T - T_w}{T_e - T_w}\, \frac{\sqrt{c_f/2}}{c_h} \tag{350}$$

and

$$K_1^+ = \frac{K_1 - K_{1w}}{K_{1e} - K_{1w}}\, \frac{\sqrt{c_f/2}}{c_m} \tag{351}$$

Figure 90 shows a comparison of the computed and experimental concentrations of a trace of helium transpired along with air. A value of $Sc_t = 0.75$ was found to give the best agreement with the data in the law of the wall region ($y^+ < 40$). Note that $Pr_t = 0.75$ was also adopted in this analysis, although a corresponding comparison with temperature distribution data is not as consistent as the concentration data shown in Fig. 90.

In the outer region corresponding to the wake, the complete Eqs. (278) and (299) are solved along with the continuity equation (Eq. (74)). When the eddy viscosity is expressed according to Eq. (283), with no alteration for the effects of surface mass transfer, the boundary-layer equations transform into similarity form in the same manner as for a laminar boundary layer. The similarity variable is $\tilde{y} = x_2/\delta_{sim}^*$, where δ_{sim}^* is a displacement thickness associated with the "outer" solution extended to the wall. For the equations in the wake region to be similar, δ_{sim}^* must be linearly dependent on x_1, the surface mass transfer rate must be uniform, and the surface temperature or foreign species concentration must vary exponentially with x_1. For the "outer" solution to yield velocity profiles in the wake region similar in form to data, it is necessary that a slip velocity be assumed at the surface as a boundary condition. The resulting velocity distribution is expressible as

$$\frac{u_1}{u_{1e}} = G\left(\tilde{y}, \frac{u_{2w}}{u_{1e}}, \frac{c_f}{2}, \frac{u_{1w}}{u_{1e}}, \frac{d\delta_{sim}^*}{dx_1}\right) \tag{352}$$

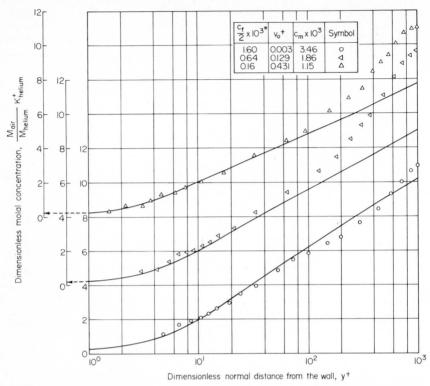

$\frac{c_f}{2} \times 10^{3*}$	v_0^+	$c_m \times 10^3$	Symbol
1.60	0.003	3.46	$\circ$
0.64	0.129	1.86	$\triangleleft$
0.16	0.431	1.15	$\triangle$

Dimensionless molal concentration, $\dfrac{M_{air}}{M_{helium}}\dfrac{K_{helium}^+}{}$

Dimensionless normal distance from the wall, y^+

Fig. 90. Comparison of the predicted dimensionless concentration profiles with data near the surface [187], ($Sc = 0.21$, $Sc_t = 0.75$).

A wide variety of solutions for $u_1(\tilde{y})$ is found by arbitrarily assigning values to two of the parameters and finding the values of the others from explicit solutions of the momentum equation.

The inner and outer solutions are matched by selecting values for $c_f/2$ and u_{2w}/u_{1e} from a large set of solutions. This immediately yields $\tilde{u}(\tilde{y})_{outer}$ and permits evaluating $u^+(y^+)_{inner}$ from Eqs. (346) and (347). For the inner solution, the eddy viscosity ($\bar{\epsilon} = \epsilon_M + \nu$) is expressed locally as

$$\frac{\bar{\epsilon}}{\nu} = 1 + l^{+2}\frac{du^+}{dy^+}$$

so that once a match point is chosen, $\bar{\epsilon}/\nu$ is known for the outer solution and Eq. (283) permits evaluating Re_{δ^*}. The distinction between δ^*_{sim} and δ^*_1 for the complete boundary layer is dropped at this point, causing less than a 5 percent error in Re_{δ^*}. Given Re_{δ^*}, the inner solution can then be expressed as $\tilde{u}(\tilde{y})_{inner}$, where

$$\tilde{u} = u^+\sqrt{\frac{c_f}{2}}$$

and

$$\tilde{y} = \frac{y^+}{\sqrt{Re_{\delta_1^*}} \sqrt{c_f/2}}$$

when $\delta_{sim}^* \simeq \delta_1^*$. The match point is adjusted until it lies on the intersection of the inner and outer solutions in the same coordinate system. The resulting inner and outer velocity distribution is then employed in Eq. (348) and the similarity energy equation in the outer region, respectively, to yield the temperature distribution and heat flux at the wall. An identical procedure is used with the diffusion equation. Good agreement between a matched inner and outer solution with data for the concentration of a trace amount of helium injected along with air into a boundary layer is shown in Fig. 91.

The main theoretical results appropriate to the direct evaluation of skin friction drag, convective heat transfer rates, and mass transfer rates are shown in Figs. 92 through 98. The transport coefficients c_f, c_h, c_{m_i}, are expressed in terms of Reynolds numbers based on the thickness of a corresponding property (see §35) and on the length of run along the plate, x_1, where the boundary layer is turbulent from the leading edge. The parameter of each curve is u_{2w}/u_{1e}, which is constant in the figures based on $Re_{x_{1e}}$ but can be considered a local value in the others.

Figure 92 shows $\frac{c_f}{2}\left(Re_{\theta_{11}}\right)$ for both surface blowing and suction. The zero mass transfer case agrees extremely well with the experimental results indicated in Fig. 78. For the curves representing suction ($u_{2w}/u_{1e} < 0$), a limiting line is shown where the local suction rate just balances the local skin friction coefficient, which results in a constant boundary-layer thickness beyond that point according to the momentum integral equation (Eq. (70)) reduced for this case to

$$\frac{c_f}{2} + \frac{u_{2w}}{u_{1e}} = \frac{d\theta_{11}}{dx_1} \tag{353}$$

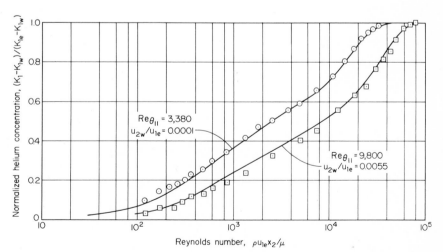

Fig. 91. Comparison of the prediction of the entire turbulent boundary layer concentration profiles with data [187].

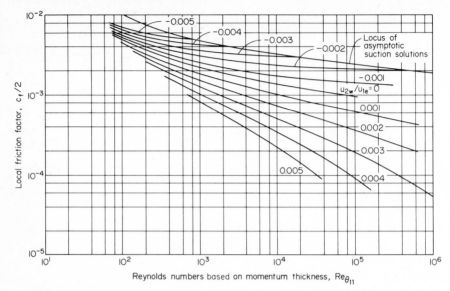

Fig. 92. Predicted skin friction coefficient with and without transpiration in terms of momentum thickness Reynolds number [187].

These calculated results are within 10 percent of the mean and well within the uncertainty envelope of the data cited in Ref. 190 for $10^3 < Re_{\theta_{11}} < 10^4$.

Figure 93 represents $\dfrac{c_f}{2}\left(Re_{x_{1e}}\right)$ on a flat plate with uniform u_{2w}/u_{1e} and where the turbulent boundary layer begins at the leading edge. It should be noted that the

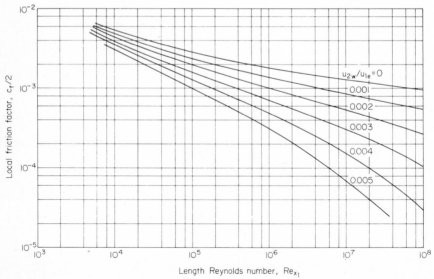

Fig. 93. Predicted skin friction coefficient with and without transpiration in terms of length Reynolds number [187].

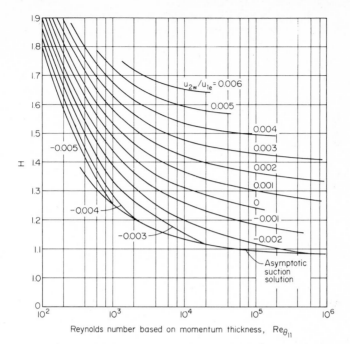

Fig. 94. Effect of mass transfer on the boundary layer shape factor, $H = \delta_1^*/\theta_{11}$ [187].

effectiveness of uniform mass transfer in reducing the local skin friction, increases with increasing Reynolds number. Note also that the skin friction results for zero mass transfer are in excellent agreement with the previously cited correlation equations.

The theoretically predicted form factor $H = \delta_1^*/\theta_{11}$ is shown in Fig. 94. The form factor is quite sensitive to the mass transfer rate as well as the Reynolds number. Consideration of this variation is of considerable importance in the evaluation of the turbulent boundary-layer growth on a surface with a pressure gradient (Eq. (70)).

Curves similar to those described above are given for the local heat and mass transfer coefficients in Figs. 95 through 98. The heat transfer coefficients apply for laminar and turbulent Prandtl numbers of 0.71 and 0.75, respectively. The values shown are uniformly 10 percent higher than recent experimental results [188]. The reason for this discrepancy is the use of $Pr_t = 0.75$ instead of 0.9, which reflects the uncertainty of determining Pr_t from temperature profiles. It is recommended that these curves be employed because of their convenience, with either a 10 percent reduction applied or with the realization of the conservatism contained therein. In the absence of extensive data on mass transfer rates, the curves of c_{m_i} should be used as they appear.

Reference 190 shows that the relationships represented in Figs. 92 and 95 (also Fig. 97 by inference) can be utilized locally for nonuniform mass transfer rate distributions that yield constant surface temperatures. For large variations in injection rate or surface temperature distribution, the methods of Ref. 190 are suggested.

72. Transpiration, Variable Fluid Properties. The fluid properties within a boundary layer may vary because of large temperature differences caused by frictional dissipation at high Mach numbers or because the coolant introduced through the surface possesses properties quite different from those of the main

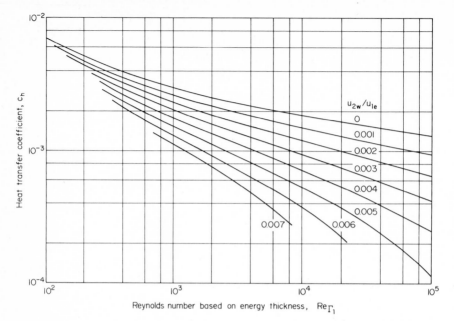

Reynolds number based on energy thickness, Re_{Γ_1}

Fig. 95. Predicted Stanton number with and without transpiration in terms of energy thickness Reynolds number, $Pr = 0.71$, $Pr_t = 0.75$ [187].

stream. Our current understanding of these effects for a turbulent boundary layer is rather incomplete primarily due to the limited available data. Further, the uncertainties cited earlier for the impervious surface in high Mach number flow carry over to the case with surface mass transfer.

Early theories for transpiration of air into air [173, 191] were based on the solution of Eqs. (345) and (348), wherein the resulting boundary-layer profiles were

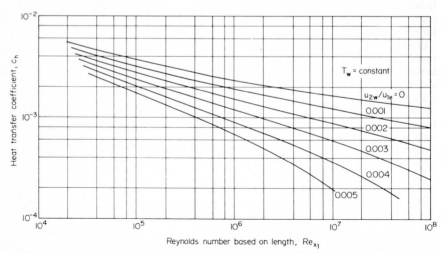

Reynolds number based on length, Re_{x_1}

Fig. 96. Predicted Stanton number with and without transpiration in terms of length Reynolds number, $Pr = 0.71$, $Pr_t = 0.75$ [187].

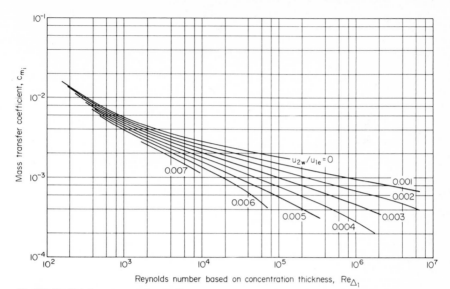

Fig. 97. Predicted mass transfer coefficient with and without uniform transpiration in terms of concentration thickness Reynolds number, $Sc = 0.2$, $Sc_t = 0.75$ [187].

utilized in the von Kármán momentum equation (Eq. (70)) to express the results in terms of position on the plate. Reference 173 extended the Reynolds analogy to include mass transfer by defining a two-part boundary layer consisting of a laminar sublayer and a fully turbulent core. Here, $l^+ = 0$ in the sublayer ($y^+ < y_a^+$), and $l^+ = 0.4\,y^+$ and $\mu = 0$ in the fully turbulent region. The density was permitted to vary with temperature. The effect of foreign gas injection in a low speed boundary layer was studied in Ref. 192, and all these theories were improved upon in Ref. 193.

The aforementioned theories fail to predict the Mach number effects exhibited by skin friction and heat transfer data [194, 195]. If the skin friction coefficient with

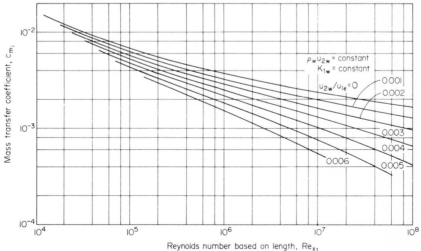

Fig. 98. Predicted mass transfer coefficient with and without uniform transpiration in terms of length Reynolds number, $Sc = 0.2$, $Sc_t = 0.75$ [187].

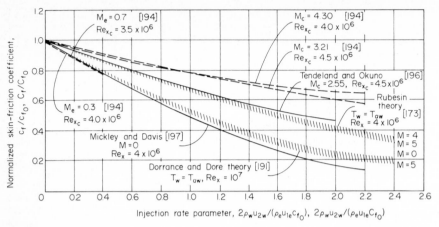

Injection rate parameter, $2\rho_w u_{2w}/(\rho_e u_{1e} c_{f_0})$, $2\rho_w u_{2w}/(\rho_e u_{1e} c_{f_0})$

Fig. 99. Comparison of early skin friction theories and experiments with air injection at supersonic speeds.

mass transfer is normalized by the value for an impervious plate under the same flow conditions, surface temperature, and length Reynolds number, it is found that a plot such as Fig. 99 for uniform blowing is insensitive to the particular value of the Reynolds number. This insensitivity applies to both the local and average skin friction coefficients and to a cone and a plate. Since the theories obviously fail to predict the Mach number effects of data from Refs. 194, 196, and 197 shown in Fig. 99, it is recommended that the experimental curves be utilized directly in estimating the effect of air injection on skin friction at Mach numbers greater than zero. The effects of foreign gas injection (He and Freon-12) can be obtained from the plots in Ref. 194.

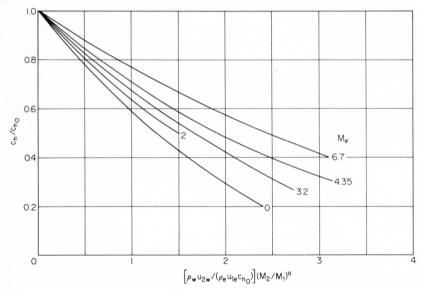

$$\left[\rho_w u_{2w}/(\rho_e u_{1e} c_{h_O})\right](M_2/M_1)^a$$

Fig. 100. Compressibility effects on the reduction of the Stanton number by surface mass transfer on bodies with zero axial pressure gradient and including effects of foreign gas injection.

Surface mass transfer affects the Reynolds analogy factor so that the curves of c_h/c_{h_0} differ from those of Fig. 99. Again, the purely empirical approach is adopted for establishing c_h. Figure 100 represents a composite of the results of several experiments with air injection [188, 198–201]. In the construction of this figure, the data was smoothed and inconsistencies between experiments were resolved in an arbitrary but systematic way. Although the data also indicate that the recovery factor is reduced by surface mass transfer, the uncertainty of much of the data suggests adoption of the conservative approach of retaining the impervious surface value, $r(0) = Pr^{1/3}$ or about 0.9. The technique for using this figure to establish required mass transfer rates for specified flight conditions, allowable surface temperatures, and specific coolants is described on p. 8-104. The value of c_{h_0} results from the methods of §68.

For foreign gas injection, the definition of the abscissa in Fig. 100 is modified to $\left(\rho_w u_{2w}/\rho_e u_{1e} c_{h_0}\right)(M_2/M_1)^a$ where subscripts 1 and 2 refer to the coolant and air, respectively, and $a = 0.35$ when $M_1 > M_2$, $a = 0.6$ when $M_1 < M_2$. The graphical technique for using this figure to establish $\rho_w u_{2w}$ is to determine the intersection of the curves in Fig. 100 with a straight line drawn from the origin with slope

$$\frac{i_{1w} - i_{1c}}{i_{2aw} - i_{2w}} \left(\frac{M_1}{M_2}\right)^a$$

The situation regarding the recovery factor for foreign gas injection is somewhat confused. Data [198] show a reduction in the recovery factor with injection rate until a point is reached where the trend reverses. With helium injection, the subsequent rise in recovery factor at high injection rates causes it to exceed the impervious plate value, $r(0) = 0.9$. Calculations of the recovery factor at supersonic speeds including diffusion-thermo effects [202] show a continual reduction in recovery factor with increased helium injection. Until these discrepancies are resolved experimentally, it is recommended that $r(0) = 0.9$ be used under the conditions of foreign gas injection. At low speeds, in the absence of frictional heating, it is necessary to modify the adiabatic wall temperature to account for diffusion-thermo effects in the sublayer [203].

73. Film Cooling. The techniques of film cooling described in §§49 and 50 have been applied to the turbulent boundary layer in a series of studies summarized in Ref. 204. The results are strictly applicable to the region relatively far downstream from the end of injection where the flow near the surface behaves as a boundary layer. Just downstream of the region of coolant injection, the main stream gas and the coolant are either flowing independently of each other or are mixing. The flow in this region depends on the configuration of the coolant slot or porous zone and is quite complex; therefore, generalization is most difficult. The boundary-layer region, however, behaves relatively independently of the particular manner in which the coolant is introduced into the boundary layer. If a surface is insulated and does not radiate a significant amount of energy, the temperature distribution of this surface T_{wi} can be expressed in terms of T_{aw}, the adiabatic wall temperature of the free stream, T_c, the coolant temperature at the point of injection, and a function z given by

$$z = \left(\frac{x_1}{s}\right)\left(\frac{\rho_e u_{1e}}{\rho_c u_c}\right)\left(\frac{u_c s}{\nu_c}\right)^{-1/4}$$

where x_1 is the distance downstream from the end of injection, s is the slot dimension (Fig. 59), and ρ_c, u_c, and ν_c are the density, velocity, and kinematic viscosity of the coolant at the point of injection. For a flush slot or porous section, s represents the stream-wise dimension of the slot. The bulk of the simpler theories show for air injection into air that

$$\frac{T_{wi} - T_{aw}}{T_c - T_{aw}} = c_1 z^{-0.8} \left(\frac{\mu_c}{\mu_e}\right)^{0.2} \tag{354}$$

where $3.09 < c_1 < 5.44$. For $z < c_1^{1.25} (\mu_c/\mu_e)^{0.25}$, $T_{wi} = T_c$.

Low speed data correlate quite well with the form of Eq. (354) at large z, except that c_1 shows a dependence on $\rho_c u_c/\rho_e u_{1e}$ as indicated in Fig. 101. At $z \simeq 10$ or lower, better results are achieved with

$$\frac{T_{wi} - T_{aw}}{T_c - T_{aw}} = \frac{1}{1 + (1/c_1) z^{0.8} (\mu_e/\mu_c)^{0.2}} \tag{355}$$

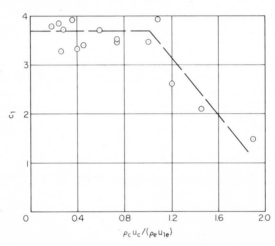

Fig. 101. Dependence of c_1 in Eq. (354) on mass flow rates [204]. (*Reprinted with permission from Pergamon Press.*)

With foreign gas injection, Eq. (355) also represents the coolant concentration along the wall

$$K_{1w} = \frac{T_{wi} - T_{aw}}{T_c - T_{aw}}$$

For small amounts of helium injection (low values of $\rho_c u_c/\rho_e u_{1e}$) into low speed air, Eq. (355) is modified to [205]

$$\frac{T_{wi} - T_{aw}}{T_c - T_{aw}} = \frac{3.73}{1.66 + (c_{p2}/c_{p1}) z^{0.8} (\mu_e/\mu_c)^{0.2}} \tag{356}$$

where c_{p_1} and c_{p_2} are the specific heats of helium and air, respectively. At high subsonic free-stream velocities, the step height is found to affect the constants in these equations, and direct use of the data of Refs. 206 and 207 is recommended.

If the temperature of the surface behind the injection slot differs from the insulated wall value given by Eqs. (354) through (356), heat flow to or from the surface will occur. The local magnitude of this heat flux can be approximated by utilizing in Eq. (138) a Stanton number in the absence of fluid injection and the enthalpy difference between the surface and the insulated wall, that is,

$$i_w - i_{wi} = K_{1w} c_{p_1} (T_w - T_{wi}) + (1 - K_{1w}) c_{p_2} (T_w - T_{wi}) \tag{357}$$

where T_{wi} and K_{1w} are found from Eqs. (355) and (356).

TURBULENT BOUNDARY LAYER
WITH PRESSURE GRADIENT

The effect of a favorable pressure gradient ($dp/dx_1 < 0$) on turbulent boundary-layer convection can be estimated rather simply by considering the boundary layer at a station on the body to grow locally as if it were on a flat plate with the same boundary-layer edge conditions and energy thickness Γ_1. In this approach, the upstream history of the boundary layer reflects itself solely in the local value of the energy thickness. Both constant property and variable property solutions will be considered in the following sections.

74. Stanton Number–Energy Thickness Relation. The fluid properties in the boundary layer will first be considered constant, and the local skin friction coefficient will be expressed as a power law on a flat plate. (See §66.)

$$\frac{\tilde{c}_f}{2} = \frac{A}{\left(\tilde{Re}_{x_{1e}}\right)^n} = \frac{d\tilde{\theta}_{11}}{dx_1} = \frac{d\tilde{Re}_{\theta_{11}}}{d\tilde{Re}_{x_{1e}}} \tag{358}$$

which integrates to yield

$$\frac{\tilde{c}_f}{2} = \frac{C_1}{\left(\tilde{Re}_{\theta_{11}}\right)^m} \tag{359}$$

where A, n, C_1, m, are the constants of the Blasius or Falkner equations (Eqs. (292) and (269) or (293) and (297)). These constants are related by

$$C_1 = \frac{A^{1/(1-n)}}{(1-n)^{n/(1-n)}} \tag{360}$$

and

$$m = \frac{n}{1-n} \tag{361}$$

The analogous relationship between Stanton number and the energy thickness Reynolds number follows from the energy integral equation (Eq. (72)), specialized to an impervious plate at constant temperature as

$$\frac{\tilde{q}_w}{\tilde{\rho} \tilde{u}_{1e} (\tilde{i}_w - \tilde{i}_e)} = \frac{d\tilde{\Gamma}_1}{dx_1} = \frac{d\tilde{Re}_{\Gamma_1}}{d\tilde{Re}_{x_{1e}}} \tag{362}$$

where the superscript tilde again refers to constant fluid properties (see Eqs. (322) through (326)). In addition, $\tilde{i}$ is defined as

$$\frac{\tilde{i}}{\tilde{u}_{1e}^2/2} = \frac{i}{u_{1e}^2/2} \tag{363}$$

The left side of Eq. (362) is equal to the product of Stanton number and the quantity $\tilde{z}$ defined as

$$\tilde{z} = \frac{\tilde{i}_{aw} - \tilde{i}_w}{\tilde{i}_e - \tilde{i}_w} \tag{364}$$

or

$$z = \frac{i_{aw} - i_w}{i_e - i_w}$$

when Eq. (363) is invoked. If the recovery factor $r(0)$ and the Reynolds analogy factor $2c_h/c_f = 1/s$ are constants, combination of Eqs. (358), (359), and (362) yields

$$\tilde{c}_h = \frac{C_1 \tilde{z}^m}{\left(\tilde{Re}_{r_1}\right)^m s^{1+m}} \tag{365}$$

In terms of the boundary-layer coordinate transformation parameters (§68), Eq. (365) may be extended to include variable fluid properties when written as

$$F_c c_h = \frac{C_1 z^m}{\left(Re_{r_1} F_{R\theta}\right)^m s^{1+m}} \tag{366}$$

The requirement (Eq. (363)) permits setting $F_{R\theta} = F_{R_r}$ so that the formulas of Table 13 and Figs. 83 and 84 are directly applicable in Eq. (366). This equation relates the local Stanton number to the local Reynolds number based on the energy thickness and is assumed to apply locally on bodies with favorable pressure gradients as well as on flat plates.

75. Solution of Energy Integral Equation on Body with Pressure Gradient. The energy integral equation (Eq. (72)) for an impervious axisymmetric body at zero angle of attack and at a uniform temperature is

$$\frac{q_w}{\rho_e u_{1e}(i_w - i_e)} = \frac{d\Gamma_1}{dx_1} + \Gamma_1 \frac{d}{dx_1}[\ln(r\rho_e u_{1e})] \tag{367}$$

When Eqs. (366) and (367) are combined, an ordinary differential equation results with Γ_1 and x_1 as the dependent and independent variables, respectively. Direct integration yields $\Gamma_1(x_1)$, which is substituted into Eq. (366) to yield the final expression for the local Stanton number on the body

$$c_h(x_1) = \frac{A}{s(\rho_e u_{1e} x_{\text{eff}}/\mu_e)^n F_c^{1-n} F_{R\theta}^n} \tag{368}$$

where

$$x_{\text{eff}} = \frac{F_c F_{R_\theta}^{\,n/(1-n)}}{\rho_e u_{1e} (zr\mu_e^{\,n})^{1/(1-n)}} \int_0^{x_1} \frac{\rho_e u_{1e} (zr\mu_e^{\,n})^{1/(1-n)}}{F_c F_{R_\theta}^{\,n/(1-n)}} \, dx_1 \tag{369}$$

If the Coles method is used to evaluate F_c and F_{R_θ} (see Table 13), it is necessary to relate these quantities to a local Reynolds number. In keeping with the approximate nature of Eqs. (368) and (369), it is adequate to use $\rho_e u_{1e} x_1/\mu_e$ to define the local F_c and F_{R_θ} in the integrand of Eq. (369). For the other methods, the local values of F_c and F_{R_θ} in Eq. (369) can be found from the equations of Table 13. For the method of van Driest, Figs. 83 and 84 are convenient for local values of T_e near 400°R.

76. Pointed Cone. Equations (368) and (369) can be used to obtain the local Stanton number on a cone quite easily. If F_c and F_{R_θ} do not vary with Reynolds number the only variable in the integrand of Eq. (369) is the radius $r = x_1 \sin\theta_c$. There results

$$x_{\text{eff}} = \frac{1}{(x_1 \sin\theta_c)^{1/(1-n)}} \int_0^{x_1} (x_1 \sin\theta_c)^{1/(1-n)} \, dx_1 = \left(\frac{1-n}{2-n}\right) x_1$$

and from Eq. (368)

$$\frac{\left(c_h\right)_{\text{cone}}}{\left(c_h\right)_{\text{flat plate}}} = \left(\frac{2-n}{1-n}\right)^n$$

For Reynolds numbers of about 10^6, $n = 0.2$ and $\dfrac{\left(c_h\right)_{\text{cone}}}{\left(c_h\right)_{\text{flat plate}}} = 1.12$, and for Reynolds numbers greater than 10^7, $n = 1/7$ and $\dfrac{\left(c_h\right)_{\text{cone}}}{\left(c_h\right)_{\text{flat plate}}} = 1.18$. These results are quite consistent with van Driest's rule [208] of using half the Reynolds number in the flat plate equation.

TRANSITIONAL BOUNDARY LAYER ON A FLAT PLATE

Because the transition zone from a laminar to a turbulent boundary layer often covers a major portion of the exposed surface of a body, it is necessary to be able to predict the rapidly changing convection rate in this zone. The position of the onset of turbulence and the extent of the transition zone for a specific configuration depends on many factors such as the scale and spectral content of the free stream turbulence and sound field, the free stream Mach number, and the surface characteristics of smoothness, waviness, temperature, compliance, and mass transfer. To date, there is no universal correlation of these factors that will permit the prediction of the position and extent of the transition zone. What is presented here is a technique for predicting transitional boundary-layer convection on a plate, given the position of the transition region. If the transition zone is not well known, one design approach is to arbitrarily assign a series of positions of the onset of turbulence and to set the length of the transition zone equal to the length of the fully laminar boundary layer. The sensitivity of the final design to changes in the position of transition must be determined; a high degree of sensitivity suggests the need for careful experimentation with prototype models.

77. Statistical Distribution of Heat Flux in the Transition Zone. The contents of this section are an extension of the work of Ref. 209 to include the effects of variable fluid properties. The ideas employed are based on the observations of Ref. 210 that on a flat plate the distribution of β, the fraction of time a surface point is covered by a fully turbulent boundary layer, is closely approximated by a Gaussian integral curve throughout the transition zone, that is,

$$\beta(x_1) = \int_0^{x_1} P(s) \frac{ds}{\sigma} \tag{370}$$

where σ is the standard deviation of the transition location about its mean, $\bar{s}$, and $P(s)$ is the probability that transition initiates between s and $s + ds$

$$P(s) = \frac{1}{\sqrt{2\pi}} \exp\left[-\frac{1}{2}\left(\frac{s - \bar{s}}{\sigma}\right)^2\right] \tag{371}$$

Use of Eq. (371) requires the transition to take place sufficiently downstream so that the boundary layer is fully laminar at all times near the leading edge ($\bar{s} > 2\sigma$)

If the term $q_{w_t}(x_1, s)$ is defined as the heat flux at point x_1 in a turbulent boundary layer with instantaneous transition from laminar to turbulent flow at s, and $q_{wl}(x_1)$ is defined as the heat flux at point x_1 with a laminar boundary layer beginning at the leading edge, the heat flux caused by the intermittency of turbulence in the transition zone is then

$$q_w(x_1) = [1 - \beta(x_1)] q_{wl}(x_1) + \int_0^{x_1} P(s) q_{wt}(x_1, s) \frac{ds}{\sigma} \tag{372}$$

The first term is the product of the laminar heat flux and the fraction of time the boundary layer is laminar at x_1. The second term accounts for both the fraction of time the boundary layer is turbulent and the effect of the moving transition location. The term $q_{wt}(x_1, s)$ is sufficiently complex mathematically that Eq. (372) is normally solved by numerical integration. If it is assumed that the energy thickness remains unchanged as the laminar boundary layer changes instantaneously into a turbulent boundary layer, then

$$\frac{q_{wt}(x_1, s)}{q_{wt}(x_1, 0)} = \left\{1 - \frac{s}{x_1}\left[1 - \frac{36.9}{Pr^{1/3}}\left(\frac{z_l}{z_t}\right)^{5/4} \frac{F_c F_{R_\theta}^{1/4}}{\left(\frac{\mu' \rho'}{\mu_e \rho_e}\right)^{5/8} \left(\frac{\rho_e u_{le} s}{\mu_e}\right)^{3/8}}\right]\right\}^{-1/5} \tag{373}$$

where

$$\frac{z_l}{z_t} = \frac{i_e + \mathrm{Pr}^{1/2}(u_{1e}^2/2) - i_w}{i_e + \mathrm{Pr}^{1/3}(u_{1e}^2/2) - i_w} \tag{374}$$

The form of Eq. (373) applies for $\mathrm{Re}_{x_{1e}} < 4 \times 10^6$, where the Blasius skin friction equation (Eq. (85)) is reasonably accurate and $2c_h/c_f = \mathrm{Pr}^{-2/3}$ and $\mathrm{Pr}^{-0.4}$ for laminar and turbulent flow, respectively. It also uses the laminar reference enthalpy approach to define $\mu'\rho'/\mu_e\rho_e$ (§42) and uses the turbulent boundary-layer transformations F_c and F_{R_θ} which are assumed insensitive to Reynolds number variations (§68). Thus, given M_e, T_e, T_w, $\bar{s}$, σ, Eq. (372) provides the distribution of heat flux in the prescribed transition zone by techniques consistent with those for the fully laminar and fully turbulent boundary layers given in §§42 and 68.

LAMINAR AND TURBULENT BOUNDARY LAYERS ON THREE-DIMENSIONAL BODIES

The general three-dimensional effects described briefly in §4 and represented by the equations in §§33 and 34 constitute a most formidable problem in convective heating studies. Comprehensive reviews are presented in Refs. 52 and 211 through 214. In its most general form, this problem is much too complex to yield to relatively simple analysis and must be solved either by extensive computer calculations or by experimentation, preferably both. Existing analyses deal either with special configurations that contain inherent simplifications or with general shapes but where simplifying physical hypotheses are imposed. Examples of the first class include solutions of the laminar boundary layer on the stagnation line of an infinitely long swept cylinder (§60), the windward generator of a sharp cone at angle of attack [215], and the three-dimensional stagnation point where the orthogonal generators have different radii of curvature [216]. The second class of solutions employs a system of coordinates where the x_1 direction is the projection onto the surface of the streamlines in the inviscid flow and the x_3 direction is orthogonal to the first. In this coordinate system, $u_{3e} = 0$ and because $u_{3w} = 0$ also, a most practical approach is to neglect u_3 everywhere in the boundary layer; this results in a major simplification of the problem for both laminar and turbulent boundary layers. Details of solutions using the streamwise coordinate system with zero cross flow are given in the following sections.

78. Laminar Boundary Layers on Pointed Bodies at Small Angle of Attack.
Reference 215 presents solutions for convective heating from a laminar boundary layer on the windward ray of a pointed cone with $\theta_c = 5°$, $7.5°$, and $10°$ at small angles of attack in ideal air. Boundary-layer similarity arguments identical to those employed for the two-dimensional boundary layer are carried over directly to this case. As an example, the results for a $5°$ half-angle cone at $M_\infty = 3.1$ are shown as the solid curve in Fig. 102 where the ordinate is the local heat flux normalized by its value at the same point for the cone at zero angle of attack. Sizeable increases in heat transfer result from small changes in the angle of attack for this slender body.

A quick, approximate evaluation of the heat flux on the windward side of a pointed body at angle of attack can be achieved by resolving the free stream velocity into an axial component, $V_\infty \cos\alpha$, and a cross-flow component, $V_\infty \sin\alpha$. The axial flow component is used to calculate Stanton numbers along the body by the techniques described earlier for an axisymmetric body. The cross-flow component is used to calculate Stanton numbers around the body by the methods described in §62 for flow over a circular cylinder normal to the stream, where the radius is taken equal to

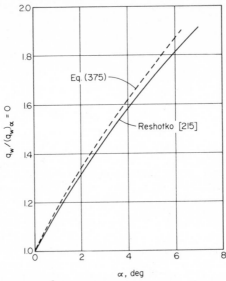

Fig. 102. Effect of angle of attack on the laminar boundary layer heat flux on the windward ray of a pointed cone, $M_\infty = 3.1$, $\theta_c = 5°$.

the local radius of the body normal to the axis. The resulting heat flux on the windward side of the body is well represented by

$$q_w = \sqrt{\left(c_{h_{\text{axisymmetric}}} \rho_\infty V_\infty \cos\alpha\right)^2 + \left(c_{h_{\text{cyl}}} \rho_\infty V_\infty \sin\alpha\right)^2} \left(T_\infty + \frac{V_\infty^{\,2}}{2} - T_w\right) \qquad (375)$$

where a recovery factor of unity is assumed. The dashed line in Fig. 102 represents the results of applying Eq. (375) to the windward ray of the cone and shows good agreement with the boundary-layer analysis.

79. Laminar Boundary Layers at a Three-dimensional Stagnation Point. The heat transfer rates for a laminar boundary layer at a three-dimensional stagnation point, where the radii of curvature are different on the principal orthogonal generators, have been evaluated through an exact solution of the three-dimensional boundary-layer equations in Refs. 216 and 217. The more complete of these solutions [216] includes surface mass transfer of the same gas in the free stream and also permits the stagnation region to have the shape of either a torus or a saddle. For the latter configuration, there exists an inflow toward the stagnation point on one generator and an outflow away from the stagnation point on the orthogonal generator (e.g., the fairing on a nacelle support). The techniques for solving the three-dimensional boundary-layer equations are quite similar to those employed for the two-dimensional and axisymmetric cases. The distance normal to the surface is transformed to a variable similar to the one defined by Eq. (201) but with the fluid properties referred to the boundary-layer edge. When the viscosity-density ratio $C_e = \mu\rho/\mu_e\rho_e$ and the Prandtl number are constants, the resulting transformed boundary-layer equations are similar to Eqs. (212) through (214) with $t_e = t_s = \beta = 1$. An additional equation must be included for flow in the x_3 direction. The mass flow is expressed by Eq. (261) with the fluid properties again referred to the boundary-layer edge. A key parameter of the solutions is the ratio of the velocity gradients along the principal orthogonal generators, evaluated at the stagnation point

$$c = \frac{\left(du_{3e}/dx_3\right)_{st}}{\left(du_{1e}/dx_1\right)_{st}} \qquad (376)$$

For an axisymmetric body c, of course, is unity. The effects of the value of c and of the injection parameter $f(0)$ on the stagnation point heat transfer rate q_w, normalized by q_{w_1} for $c = 1$ and $f(0) = 0$, is shown in Fig. 103 for a cold wall ($g_w = T_w/T_{st} = 0$). The line represents earlier impervious wall solutions from Ref. 217. With fixed rates of mass transfer, the heat flux decreases with increased magnitude of the lateral radius of curvature from the axisymmetric case ($c = 1$) to the infinitely long cylinder ($c = 0$). The heat flux continues to decrease when the shape of the stagnation region becomes a saddle ($c < 0$) rather than a torus until $c = -0.5$. Further decreases in the magnitude of the lateral radius of curvature causes the heat transfer rate to rise again.

80. Axisymmetric Approximation for Laminar and Turbulent Boundary Layers on Three-dimensional Bodies. When the external inviscid flow streamlines over a three-dimensional body are projected onto the surface, a coordinate system is defined where $x_2 = 0$ represents the surface, $x_3 =$ constant represents the projections of the streamlines on the surface, and $x_1 =$ constant represents the family of surfaces orthogonal to the other coordinates. The origin is usually the stagnation point, and for application of the techniques recommended here, it is necessary that the surface pressure decrease or remain constant in the x_1 direction.

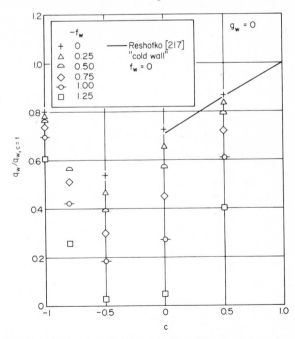

Fig. 103. Heat flux in the laminar boundary layer at a three-dimensional stagnation point, including the effects of surface mass transfer [216]. (*Reprinted with permission from AIAA.*)

From a careful examination of the form of the pressure gradient term when Eq. (57) for the laminar case is written in terms of the coordinate system defined above, Vaglio-Laurin [218] arrived at the conclusion that the contribution of u_3 can be neglected in evaluating the convective heat flux, provided the surface is highly cooled and $u_{1e}^2/(I_e - i_w) \ll 1$. These requirements are satisfied on blunt bodies in hypersonic flight. Fannelop [219] later solved the three-dimensional laminar boundary-layer equations on a spherically blunted cone through a perturbation analysis that introduces a first-order correction for the cross flow u_3, where the zero-order calculations apply for $u_3 = 0$, the "axisymmetric analogy." Although the numerical results are confined to a single case of a cooled $10°$ cone at $M_\infty = 10$ for angles of attack below $10°$, the first-order corrections to the convective heating are shown to be small, and zero-order calculations agree quite well with existing data [220]. Thus, from these analyses, one can justify use of the axisymmetric analogy to estimate convective heat flux for high-speed laminar boundary layers. Since the turbulent boundary layer is less sensitive to disturbances than the laminar boundary layer, the axisymmetric analogy should hold also for this case. Unfortunately, there are no comprehensive turbulent boundary layer data to test this conclusion, and the current recommendation must be regarded as tentative.

When $u_3 = 0$ everywhere, all the equations governing boundary-layer behavior in the streamline coordinate system reduce to the axisymmetric form in terms of x_1 and x_2. Note that $u_{3e}\theta_{13}$, $u_{3e}\theta_{31}$, etc., all equal zero when $u_3 = 0$ so that Eqs. (69) through (72) take on their two-dimensional form. For small cross flow, $u_3 \neq 0$ everywhere within the boundary layer although $u_{3e} = u_{3w} = 0$. Therefore, terms such as $u_{3e}\theta_{13}$ must be retained for first-order corrections. In the axisymmetric case, $h_3 = r$, the local radius normal to the axis of the body; whereas, in the three-dimensional case with zero cross flow, it is related to the spacing of the external streamlines projected onto the surface. Thus, the methods of §§62 and 75 can be applied directly for a zero-order approximation once the streamline spacing is established as a function of x_1 from calculations of the inviscid flow field where the metric coefficient h_3 replaces r. For example, with the cross-flow angles shown in Fig. 21, the method of isoclines can be used to determine the streamline shapes for a blunted cone at angle of attack.

REFERENCES

1. L. Prandtl, Motion of Fluids with Very Little Viscosity, *NACA Tech. Mem.* 452, 1928, transl. of Über Flussigkeitsbewegung bei sehr kleiner Reibung, *Verh. III int. Math. Kongr.* Heidelberg, pp. 484–491, 1904.
2. C. F. Hansen, Approximations for the Thermodynamic and Transport Properties of High-Temperature Air, *NASA Tech. Rept.* R-50, 1959.
3. A. P. Colburn, A Method for Correlating Forced Convection Heat Transfer Data and a Comparison with Fluid Friction, *Trans. Amer. Inst. Chem. Eng.*, **29**:174 (1933).
4. H. Lamb, "Hydrodynamics," 6th ed., Dover Publications, Inc., New York, 1945.
5. A. H. Shapiro, "The Dynamics and Thermodynamics of Compressible Fluid Flow," vols. 1 and 2, The Ronald Press Co., New York, 1958.
6. W. G. Vincenti and C. H. Kruger, Jr., "Introduction to Physical Gas Dynamics," John Wiley & Sons, Inc., New York, 1965.
7. V. L. Streeter (ed.), "Handbook of Fluid Dynamics," McGraw-Hill Book Company, New York, 1961.
8. W. Flügge (ed.), "Handbook of Engineering Mechanics," McGraw-Hill Book Company, New York, 1962.
9. W. F. Durand (ed.), "Aerodynamic Theory, a General Review of Progress," vols. 1–6, Dover Publications, Inc., New York, 1963.
10. P. Moon and D. E. Spencer, "Field Theory Handbook Including Coordinate Systems Differential Equations and Their Solutions," Springer, Berlin, 1961.
11. V. L. Streeter, "Fluid Dynamics," McGraw-Hill Book Company, New York, 1948.

12. I. H. Abbott and A. E. Von Doenhoff, "Theory of Wing Sections," Dover Publications, Inc., New York, 1959.
13. M. M. Munk, Elements of the Wing Section Theory and of the Wing Theory, *NACA Rept.* 191, 1924.
14. Ames Research Staff, Equations, Tables, and Charts for Compressible Flow, *NACA Rept.* 1135, 1953.
15. W. R. Sears (ed.), "General Theory of High Speed Aerodynamics," vol. 6 of "High Speed Aerodynamics and Jet Propulsion," Princeton University Press, Princeton, N.J., 1954.
16. T. von Kármán, Compressibility Effects in Aerodynamics, *J. Aeronaut. Sci.,* 8:337 (1941).
17. J. S. Dennard and P. B. Spencer, Ideal-gas Tables for Oblique-Shock Flow Parameters in Air at Mach Numbers from 1.05 to 12.0, *NASA Tech. Note* D-2221, 1964.
18. Staff of the Computing Section, Center of Analysis (under direction of Z. Kopal), (a) Tables of Supersonic Flow around Cones, (b) Tables of Supersonic Flow around Yawing Cones, *M.I.T. Tech. Repts.* 1 and 3, 1947.
19. J. L. Sims, Tables for Supersonic Flow around Right Circular Cones at Zero Angle of Attack, *NASA Spec. Pub.* SP-3004, 1964.
20. R. T. Jones, Wing Plan Forms for High-speed Flight, *NACA Rept.* 863, 1945.
21. W. D. Hayes and R. F. Probstein, "Hypersonic Flow Theory," vol. 1, "Inviscid Fluids," Academic Press, Inc., New York, 1966.
22. P. S. Epstein, On the Air Resistance of Projectiles, *Proc. Nat. Acad. Sci.,* 17:532 (1931).
23. V. Van Hise, Analytic Study of Induced Pressure on Long Bodies of Revolution with Varying Nose Bluntness at Hypersonic Speeds, *NASA Tech. Rept.* R-78, 1960.
24. D. M. Kuehn, Experimental and Theoretical Pressures on Blunt Cylinders for Equilibrium and Nonequilibrium Air at Hypersonic Speeds, *NASA Tech. Note* D-1979, 1963.
25. L. H. Jorgensen, Aerodynamics of Planetary Entry Configurations in Air and Assumed Martian Atmospheres, *J. Spacecraft Rockets,* 3:1024 (1966).
26. J. Ackeret, Air Forces on Airfoils Moving Faster than Sound, *NACA Tech. Mem.* 317, 1925, translation of *Z. Flugtech. Motorluftschiffahrt,* 16:72 (1925).
27. M. Inouye, J. V. Rakich, and H. Lomax, A Description of Numerical Methods and Computer Programs for Two-dimensional and Axisymmetric Supersonic Flow over Blunt-nosed and Flared Bodies, *NASA Tech. Note* D-2970, 1965.
28. H. Lomax and M. Inouye, Numerical Analysis of Flow Properties about Blunt Bodies Moving at Supersonic Speeds in an Equilibrium Gas, *NASA Tech. Rept.* R-204, 1964.
29. F. B. Fuller, Numerical Solutions for Supersonic Flow of an Ideal Gas around Blunt Two-dimensional Bodies, *NASA Tech. Note* D-791, 1961.
30. J. W. Cleary and J. A. Axelson, Theoretical Aerodynamic Characteristics of Sharp and Circularly Blunt Wedge Airfoils, *NASA Tech. Rept.* R-202, 1964.
31. M. D. Van Dyke and H. D. Gordon, Supersonic Flow Past a Family of Blunt Axisymmetric Bodies, *NASA Tech. Rept.* R-1, 1959.
32. M. Inouye, Blunt Body Solutions for Spheres and Ellipsoids in Equilibrium Gas Mixtures, *NASA Tech. Note* D-2780, 1965.
33. J. W. Cleary, An Experimental and Theoretical Investigation of the Pressure Distribution and Flow Fields of Blunted Cones at Hypersonic Mach Numbers, *NASA Tech. Note* D-2969, 1965.
34. J. F. Roberts, C. H. Lewis, and M. Reed, Ideal Gas Spherically Blunted Cone Flow Field Solutions at Hypersonic Conditions, *AEDC Tech. Rept.* 66-121, 1966.
35. R. F. Clippinger, J. H. Giese, and W. C. Carter, Tables of Supersonic Flow about Cone Cylinders, part I; Surface Data, *Ballistic Research Lab. Rept.* 729, 1950.
36. V. J. Rossow, Applicability of the Hypersonic Similarity Rule to the Pressure Distributions which Include the Effects of Rotation for Bodies of Revolution at Zero Angle of Attack, *NACA Tech. Note* 2399, 1951.
37. H. S. Tsien, Similarity Laws of Hypersonic Flows, *J. Math. Phys.,* 25:247 (1946).
38. A. J. Eggers, Jr. and R. C. Savin, A Unified Two-dimensional Approach to the Calculation of Three-dimensional Hypersonic Flows with Application to Bodies of Revolution, *NACA Rept.* 1249, 1955.
39. O. M. Belotserkovskii, The Calculation of Flow over Axisymmetric Bodies with a Detached Shock Wave, Computation Center, Acad. Sci., Moscow, USSR, 1961; transl. and ed. J. F. Springfield, *AVCO Corp.* RAD-TM-62-64, 1962.

40. O. M. Belotserkovskii and P. I. Chushkin, The Numerical Solution of Problems in Gas Dynamics, in M. Holt (ed.), "Basic Developments in Fluid Dynamics," vol. 1, pp. 1--126, Academic Press, Inc., New York, 1965.

41. O. M. Belotserkovskii (ed.), Supersonic Gas Flow around Blunt Bodies, *NASA Tech. Transl.* F-453, 1967; trans. of Obtekaniye Zatuplennykh Tel Sverkhzvukovym Potokom Gaza. Teoreticheskoye i Eksperimental 'noye Issledovaniya," Computer Center of the Acad. of Sci. USSR, Moscow, 1966.

42. M. Inouye, J. G. Marvin, and A. R. Sinclair, Comparison of Experimental and Theoretical Shock Shapes and Pressure Distributions on Flat-faced Cylinders at Mach 10.5, *NASA Tech. Note* D-4397, 1968.

43. J. C. South, Jr., Calculation of Axisymmetric Supersonic Flow Past Blunt Bodies with Sonic Corners Including a Program Description and Listing, *NASA Tech. Note* D-4563, 1968.

44. A. H. Stone, On Supersonic Flow Past a Slightly Yawing Cone, *J. Math. Phys.*, **27**:67 (1948).

45. J. L. Sims, Tables for Supersonic Flow Around Right Circular Cones at Small Angle of Attack, *NASA Spec. Pub.* 3007, 1964.

46. J. V. Rakich, Numerical Calculation of Supersonic Flows of a Perfect Gas over Bodies of Revolution at Small Angles of Yaw, *NASA Tech. Note* D-2390, 1964.

47. G. E. Kaattari, Predicted Gas Properties in the Shock Layer Ahead of Capsule-type Vehicles at Angles of Attack, *NASA Tech. Note* D-1423, 1962.

48. I. O. Bohachevsky and R. E. Mates, A Direct Method for Calculation of the Flow about an Axisymmetric Blunt Body at Angle of Attack, *AIAA J.*, **4**:776 (1966).

49. G. Moretti and G. Bleich, Three-dimensional Flow around Blunt Bodies, *AIAA J.*, **5**:1557 (1967).

50. J. V. Rakich, Three-dimensional Flow Calculations by the Method of Characteristics, *AIAA J.*, **5**:1906 (1967).

51. W. D. Hayes, The Three-dimensional Boundary Layer, *U.S. Naval Ordnance Test Station Tech. Mem.* RRB-105, 1950.

52. J. C. Cooke and M. G. Hall, Boundary Layers in Three-dimensions, in "Progress in Aeronautical Sciences, Boundary Layer Problems" (D. Küchemann and A. Ferri, eds.), vol. 2, pp. 221--285, Pergamon Press, New York, 1962.

53. H. Schlichting, "Boundary Layer Theory," 4th ed., transl. by J. Kestin, McGraw-Hill Book Company, New York, 1960.

54. J. O. Hinze, "Turbulence: An Introduction to Its Mechanism and Theory," McGraw-Hill Book Company, New York, 1959.

55. J. O. Hirschfelder, C. F. Curtiss, and R. B. Bird, "Molecular Theory of Gases and Liquids," John Wiley & Sons, Inc., New York, 1954.

56. H. Blasius, The Boundary Layer in Fluids with Little Friction, *NACA Tech. Mem.* 1256, 1950; transl. of Grenzschichten in Flüssigkeiten mit kleiner Reibung, *Z. Math. Physik*, **56**:1 (1908).

57. E. Pohlhausen, Der Wärmeaustausch zwischen festen Körpern und Flüssigkeiten mit kleiner Reibung und kleiner Wärmeleitung, *Z. angew. Math. Mech.*, **1**:115 (1921).

58. E. R. G. Eckert and O. Drewitz, The Heat Transfer to a Plate in Flow at High Speed, *NACA Tech. Mem.* 1045, 1943; transl. of Der Wärmeübergang an eine mit grosser Geschwindigkeit längs angeströmte Platte, *Forschung*, **11**:116 (1940).

59. G. W. Morgan and W. H. Warner, On Heat Transfer in Laminar Boundary Layers at High Prandtl Numbers, *J. Aeronaut. Sci.*, **23**:937 (1956).

60. L. Crocco, Lo Strato Limite Laminare nei Gas, *Monografie Sci. di Aeronaut. 3*, Rome (1946); transl. as *North American Aviation Aerophys. Lab. Rept.* APL/NAA/CF-1038, 1948.

61. R. A. Seban: Laminar Boundary Layer of a Liquid with Variable Viscosity, in "Heat Transfer Thermodynamics and Education, Boelter Anniversary Volume" (H. A. Johnson, ed.), pp. 319--329, McGraw-Hill Book Company, New York, 1964.

62. L. Howarth, Concerning the Effect of Compressibility in Laminary Boundary Layers and Their Separation, *Proc. Roy Soc. (London)*, **A194**:16 (1948).

63. D. R. Chapman and M. W. Rubesin, Temperature and Velocity Profiles in the Compressible Laminar Boundary Layer with Arbitrary Distribution of Surface Temperature, *J. Aeronaut. Sci.*, **16**:547 (1949).

64. T. von Kármán, The Problem of Resistance in Compressible Fluids, *Atti del Convegno della Fondazione Alessandro Volta 1935*, pp. 223--326, 1936.

65. W. Hantzsche and H. Wendt, Die Laminare Grenzschicht an der ebenen Platte mit und ohne Wärmeübergang unter Berücksichtigung der Kompressibilität, *Jahrb. deut. Luftfahrt-Forsch.*, 1:40 (1942).

66. T. von Kármán and H. S. Tsien, Boundary Layer in Compressible Fluids, *J. Aeronaut. Sci.*, 5:227 (1938).

67. J. G. Brainerd and H. W. Emmons, Effect of Variable Viscosity on Boundary Layers with a Discussion of Drag Measurements, *J. Appl. Mech.*, 9:1 (1942).

68. L. Crocco, Sullo Strato Limite Laminare nei Gas Lungo una Lamina Plana, *Rend. Mat. Univ. Roma*, 2:138 (1941).

69. A. Busemann, Gasströmung mit Laminarer Grenzschicht entlang einer Platte, *Z. angew Math. Mech.*, 15:23 (1935).

70. E. R. van Driest, Investigation of Laminar Boundary Layer in Compressible Fluids Using the Crocco Method, *NACA Tech. Note* 2597, 1952.

71. G. B. W. Young and E. Janssen, The Compressible Boundary Layer, *J. Aeronaut. Sci.*, 19:229 (1952).

72. E. R. van Driest, The Laminar Boundary Layer with Variable Fluid Properties, *North American Aviation Rept.* AL-1866, 1954.

73. M. F. Romig and F. J. Dore, Solutions of the Compressible Laminar Boundary Layer Including the Case of a Dissociated Free Stream, *Convair Rept.* ZA-7-012, 1954.

74. R. E. Wilson, Real-gas Laminar-Boundary-layer Skin Friction and Heat Transfer, *J. Aerospace Sci.*, 29:640 (1962).

75. N. B. Cohen, Boundary-layer Similar Solutions and Correlation Equations for Laminar Heat Transfer Distribution in Equilibrium Air at Velocities up to 41,000 Feet per Second, *NASA Tech. Rept.* R-118, 1961.

76. G. T. Chapman, Theoretical Laminar Convective Heat Transfer and Boundary Layer Characteristics on Cones at Speeds of 24 km/sec, *NASA Tech. Note* D-2463, 1964.

77. M. W. Rubesin and H. A. Johnson, A Summary of Skin Friction and Heat Transfer Solutions of the Laminar Boundary Layer on a Flat Plate, *Proc. 1948 Heat Transfer and Fluid Mechanics Institute;* also *Trans. ASME*, 71:383 (1949).

78. A. D. Young, Boundary Layers, in "Modern Developments in Fluid Dynamics; High Speed Flow" (L. Howarth, ed.), vols. 1 and 2, chap. 10, p. 422, Oxford University Press, New York, 1953.

79. E. R. G. Eckert, Survey on Heat Transfer at High Speeds, *WADC Tech. Rept.* 54-70, 1954.

80. E. R. G. Eckert, Engineering Relations for Heat Transfer and Friction in High Velocity Laminar and Turbulent Boundary Layer Flow over Surfaces with Constant Pressure and Temperature, *Trans. ASME*, 78:1273 (1956).

81. M. Tribus and J. Klein, Forced Convection from Non-isothermal Surfaces, in "Heat Transfer," p. 211, University of Michigan Press, Ann Arbor, 1953.

82. M. W. Rubesin, The Effect of an Arbitrary Surface Temperature Variation along a Flat Plate on the Convective Heat Transfer in an Incompressible Turbulent Boundary Layer, *NACA Tech. Note* 2345, 1951.

83. H. S. Carslaw and J. C. Jaeger, "Conduction of Heat in Solids," 2d ed., Oxford University Press, New York, 1959.

84. R. Bond, Heat Transfer to a Laminar Boundary Layer with Nonuniform Free Stream Velocity and Nonuniform Wall Temperature, University of California, 1950.

85. M. J. Lighthill, Contributions to the Theory of Heat Transfer through a Laminar Boundary Layer, *Proc. Roy. Sci. (London)*, A202:359 (1950).

86. J. P. Hartnett, E. R. G. Eckert, R. Birkebak, and R. L. Sampson, Simplified Procedures for the Calculation of Heat Transfer to Surfaces with Nonuniform Temperatures, *WADC Tech. Rept.* 56-373, 1956.

87. R. Eichhorn, E. R. G. Eckert, and A. D. Anderson, An Experimental Study of the Effects of Nonuniform Wall Temperature on Heat Transfer in Laminar and Turbulent Axisymmetric Flow along a Cylinder, *WADC Tech. Rept.* 58-33, 1958.

88. M. Tribus and J. Klein, Forced Convection through a Laminar Boundary Layer over an Arbitrary Surface with an Arbitrary Surface Temperature Variation, *J. Aeronaut. Sci.*, 22:62 (1955).

89. D. C. Baxter and W. C. Reynolds, Fundamental Solutions for Heat Transfer from Nonisothermal Flat Plates, *J. Aerospace Sci.*, 25:403 (1958).

90. O. T. Hanna and J. E. Meyers, Heat Transfer in Boundary Layer Flows Past a Flat Plate with a Step-in Wall Heat Flux, *Chem. Eng. Sci.,* **17**:1053 (1962).
91. E. M. Sparrow and S. H. Lin, Boundary Layers with Prescribed Heat Flux–Application to Simultaneous Convection and Radiation, *Int. J. Heat Mass Transfer,* **8**:437 (1965).
92. W. J. McCroskey, The Effect of a Stepwise Distribution of Heat Transfer on the Compressible Flow over a Flat Plate, *Int. J. Heat Mass Transfer,* **9**:593 (1966).
93. B. M. Leadon, The Status of Heat Transfer Control by Mass Transfer for Permanent Structures, in "Aerodynamically Heated Structures" (P. E. Glaser, ed.), p. 171, Prentice-Hall, Inc., New York, 1962.
94. H. W. Emmons and D. Leigh, Tabulation of the Blasius Function with Blowing and Suction, *Harvard Univ. Combustion Aerodynamics Lab. Tech. Rept.* 9, 1953.
95. J. P. Hartnett and E. R. G. Eckert, Mass Transfer Cooling in the Laminar Boundary Layer with Constant Fluid Properties, *Trans. ASME,* **79**:247 (1957).
96. G. M. Low, The Compressible Laminar Boundary Layer with Fluid Injection, *NACA Tech. Note* 3404, 1955.
97. C. C. Pappas and A. F. Okuno, Measurements of Skin Friction of the Compressible Turbulent Boundary Layer on a Cone with Foreign Gas Injection, *J. Aerospace Sci.,* **27**:321 (1960).
98. J. F. Gross, J. P. Hartnett, D. J. Masson, and C. Gazley, Jr., A Review of Binary Laminar Boundary Layer Characteristics, *Int. J. Heat Mass Transfer,* **3**:198 (1961).
99. J. R. Baron, The Binary Boundary Layer Associated with Mass Transfer Cooling at High Speed, *M.I.T. Naval Supersonic Lab. Rept.* 160, 1956.
100. E. R. G. Eckert, A. A. Hayday, and W. J. Minkowycz, Heat Transfer, Temperature Recovery and Skin Friction on a Flat Plate with Hydrogen Release into a Laminar Boundary Layer, *Int. J. Heat Mass Transfer,* **4**:17 (1961).
101. C. J. Scott, Experimental Investigations of Laminar Heat Transfer and Transition with Foreign Gas Injection, *Univ. Minn. Rosemount Lab. Res. Rept.* 174, 1960.
102. C. C. Pappas and A. F. Okuno, Heat-transfer Measurements for Binary Gas Laminar Boundary Layers with High Rates of Injection, *NASA Tech. Note* D-2473, 1964.
103. J. R. Baron, Thermodynamic Coupling in Boundary Layers, *ARS J.,* **32**:1053 (1962).
104. E. M. Sparrow, W. J. Minkowycz, and E. R. G. Eckert, Mass-transfer Cooling of a Flat Plate with Various Transpiring Gases, *AIAA J.,* **3**:1341 (1965).
105. P. A. Libby and P. Sepri, Laminar Boundary Layer with Complex Composition, *Phys. Fluids,* **10**:2138 (1967).
106. P. A. Libby and K. Chen, Laminar Boundary Layer with Uniform Injection, *Phys. Fluids,* **8**:568 (1965).
107. R. Iglisch, Exact Calculations of Laminar Boundary Layers in Longitudinal Flow over a Flat Plate with Homogeneous Suction, *NACA Tech. Mem.* 1205, 1949, transl. of Exakte Berechnung der laminaren Grenzschicht an der längsangeströmten ebenen Platte mit homogener Absaugung, Schriften der Deutschen Akademie der Luftfahrtforschung, Band 8 B, Heft 1, 1944.
108. H. G. Lew and J. B. Fanucci, On the Laminar Compressible Boundary Layer over a Flat Plate with Suction or Injection, *J. Aeronaut. Sci.,* **22**:589 (1955).
109. A. M. O. Smith and D. W. Clutter, Machine Calculation of Compressible Laminar Boundary Layers, *AIAA J.,* **3**:639 (1965).
110. E. P. Bartlett and R. M. Kendall, Nonsimilar Solution of the Multicomponent Laminar Boundary Layer by an Integral Matrix Method, *Aerotherm Corp. Rept.* 66-7, pt. 3, 1967.
111. H. E. Bethel, Approximate Solution of the Laminar Boundary-layer Equations with Mass Transfer, *AIAA J.,* **6**:220 (1968).
112. P. A. Libby, The Laminar Boundary Layer with Arbitrarily Distributed Mass Transfer, *Int. J. Heat Mass Transfer,* **9**:1109 (1966).
113. P. A. Libby and A. J. Pallone, A Method for Analyzing the Heat Insulating Properties of the Laminar Compressible Boundary Layer, *J. Aeronaut. Sci.,* **21**:825 (1954).
114. M. W. Rubesin and M. Inouye, Theoretical Study of the Effect of Upstream Transpiration Cooling on the Heat Transfer and Skin-friction Characteristics of a Compressible, Laminar Boundary Layer, *NACA Tech. Note* 3969, 1957.
115. A. J. Pallone, Nonsimilar Solutions of the Compressible Laminar Boundary Layer Equations with Applications to the Upstream Transpiration Cooling Problem, *J. Aerospace Sci.,* **28**:449, 492 (1961).
116. J. T. Howe, Some Finite Difference Solutions of the Laminar Compressible Boundary Layer Showing Effects of Upstream Transpiration Cooling, *NASA Mem.* 2-26-59A, 1959.

117. P. A. Libby and J. A. Schetz, Approximate Analysis of the Slot Injection of a Gas in Laminar Flow, *AIAA J.*, **1**:1056 (1963).
118. J. A. Schetz and J. Jannone, Initial Boundary Layer Effects on Laminar Flows with Wall Slot Injection, *J. Heat Transfer*, **87**:157 (1965).
119. J. A. Schetz and H. E. Gilreath, Tangential Slot Injection in Supersonic Flow, *AIAA J.*, **5**:2149 (1967).
120. W. Mangler, Zusammenhang zwischen ebenen und rotationssymmetrischen Grenzschichten in kompressiblen Flüssigkeiten, *Z. angew. Math. Mech.*, **28**:97 (1948).
121. P. A. Libby, Laminar Boundary Layer on a Cone with Uniform Injection, *Phys. Fluids*, **8**:2216 (1965).
122. I. E. Beckwith and N. B. Cohen, Application of Similar Solutions to Calculations of Laminar Heat Transfer on Bodies with Yaw and Large Pressure Gradient in High Speed Flow, *NASA Tech. Note* **D-625**, 1961.
123. C. B. Cohen and E. Reshotko, Similar Solutions for the Compressible Laminar Boundary Layer with Heat Transfer and Pressure Gradient, *NACA Rept.* 1293, 1956.
124. F. H. Clauser, Turbulent Boundary Layers in Adverse Pressure Gradients, *J. Aeronaut. Sci.*, **21**:91 (1954).
125. L. Lees, Laminar Heat Transfer over Blunt-nosed Bodies at Hypersonic Flight Speeds, *Jet. Propul.*, **26**:259 (1956).
126. D. R. Hartree, On an Equation Occurring in Falkner and Skan's Approximate Treatment of the Equation of the Boundary Layer, *Proc. Cambridge Phil. Soc.*, **33**:223 (1937).
127. A. N. Tifford, The Thermodynamics of the Laminar Boundary Layer of a Heated Body in a High-speed Gas Flow Field, *J. Aeronaut. Sci.*, **12**:241 (1945).
128. S. Levy, Heat Transfer to Constant-property Laminar Boundary Layer Flows with Power-function Free-stream Velocity and Wall-temperature Variation, *J. Aeronaut. Sci.*, **19**:341 (1952).
129. I. E. Beckwith, Similar Solutions for the Compressible Boundary Layer on a Yawed Cylinder with Transpiration Cooling, *NASA Tech. Rept.* R-42, 1959.
130. D. R. Davies and D. E. Bourne, On the Calculation of Heat and Mass Transfer in Laminar and Turbulent Boundary Layers, I. The Laminar Case, *Quart. J. Mech. Appl. Math.*, **9**:457 (1956).
131. A. N. Tifford and S. T. Chu, Heat Transfer in Laminar Boundary Layers Subject to Surface Pressure and Temperature Distributions, *Proc. 2d Midwestern Conf. Fluid Mechanics*, p. 363, 1952.
132. J. P. Hartnett, E. R. G. Eckert, and R. Birkebak, Calculation of Convection Heat Transfer to Non-isothermal Surfaces Exposed to a Fluid Stream with Wedge Type Surface Pressure Gradient, *WADC Tech. Rept.* 57-753, 1958.
133. K, Pearson, "Tables of the Incomplete Gamma Function," Cambridge University Press, Cambridge, 1948.
134. A. Pallone and W. Van Tassell, Stagnation Point Heat Transfer for Air in the Ionization Regime, *ARS J.*, **32**:436 (1962).
135. H. Hoshizaki, Heat Transfer in Planetary Atmospheres at Super-satellite Speeds, *ARS J.*, **32**:1544 (1962).
136. J. G. Marvin and G. S. Deiwert, Convective Heat Transfer in Planetary Gases, *NASA Tech. Rept.* R-224, 1965.
137. J. G. Marvin and R. B. Pope, Laminar Convective Heating and Ablation in the Mars Atmosphere, *AIAA J.*, **5**:240 (1967).
138. P. DeRienzo and A. J. Pallone, Convective Stagnation-point Heating for Re-entry Speeds up to 70,000 fps Including Effects of Large Blowing Rates, *AIAA J.*, **5**:193 (1967).
139. S. Bennett, J. M. Yos, C. F. Knopp, J. Morris, and W. L. Bade, Theoretical and Experimental Studies of High-temperature Gas Transport Properties, *AVCO Corp.* RAD-TR-65-7, 1965.
140. W. F. Ahtye, A Critical Evaluation of Methods for Calculating Transport Coefficients of Partially Ionized Gas, *NASA Tech. Mem.* X-54, 1964.
141. J. A. Fay and F. R. Riddell, Theory of Stagnation Point Heat Transfer in Dissociated Air, *J. Aerospace Sci.*, **25**:73 (1958).
142. N. H. Kemp, P. H. Rose, and R. W. Detra, Laminar Heat Transfer around Blunt Bodies in Dissociated Air, *J. Aerospace Sci.*, **26**:421 (1959).
143. J. T. Howe and Y. S. Sheaffer, Effects of Uncertainties in the Thermal Conductivity of Air on Convective Heat Transfer for Stagnation Temperature up to 30,000°K, *NASA Tech. Note* D-2678, 1964.
144. R. W. Rutowski and K. K. Chan, Shock Tube Experiments Simulating Entry into Planetary Atmospheres, *Lockheed Missiles and Space Co.* LMSD 288139, vol. 1, pt. 2, 1960.

145. P. H. Rose and J. O. Stankevics, Stagnation-point Heat-transfer Measurements in Partially Ionized Air, *AIAA J.,* **1**:2752 (1963).
146. J. S. Gruszczynski and W. R. Warren, Measurements of Hypervelocity Stagnation Point Heat Transfer in Simulated Planetary Atmospheres, *General Electric Space Sci. Lab.* R63SD29, 1963.
147. L. Yee, H. E. Bailey, and H. T. Woodward, Ballistic Range Measurements of Stagnation-point Heat Transfer in Air and Carbon Dioxide at Velocities up to 18,000 feet per second, *NASA Tech. Note* D-777, 1961.
148. R. M. Nerem, C. J. Morgan, and B. C. Graber, Hypervelocity Stagnation Point Heat Transfer in a Carbon Dioxide Atmosphere, *AIAA J.,* **1**:2173 (1963).
149. R. W. Rutowski and D. Bershader, Shock Tube Studies of Radiative Transport in an Argon Plasma, *Phys. Fluids,* **7**:568 (1964).
150. J. P. Reilly, Stagnation Point Heating in Ionized Monatomic Gases, *M.I.T. Fluid Mechanics Lab. Pub.* 64-1 (AFOSR 5442), 1964.
151. E. M. Sparrow, W. J. Minkowycz, and E. R. G. Eckert, Diffusion Thermo Effects in Stagnation Point Flow of Air with Injection of Gases of Various Molecular Weights into the Boundary Layer, *AIAA J.,* **2**:652 (1964).
152. J. R. Baron, Thermal Diffusion Effects in Mass Transfer, *Int. J. Heat Mass Transfer,* **6**:1025 (1963).
153. A. F. Gollnick, Jr., An Experimental Study of Thermal Diffusion Effects on a Partially Porous Mass Transfer Cooled Hemisphere, *Int. J. Heat Mass Transfer,* **7**:699 (1964).
154. O. E. Tewfik, E. R. G. Eckert, and L. S. Jurewicz, Diffusion Thermo Effects on Heat Transfer from a Cylinder in Cross Flow, *AIAA J.,* **1**:1537 (1963).
155. J. Kestin and P. D. Richardson, Heat Transfer Across Turbulent Incompressible Boundary Layers, *Int. J. Heat Mass Transfer,* **6**:147 (1963).
156. F. H. Clauser, The Turbulent Boundary Layer, in "Advances in Applied Mechanics," vol. 4 (H. L. Dryden et al., eds.), Academic Press, Inc., New York, 1956.
157. F. Schultz-Grunow, New Frictional Resistance Law for Smooth Plates, *NACA Tech. Mem.* 986, 1941), translation of Neues Widerstandsgesetz fur glatte Platten, *Luftfahrforschung,* **17**: 239 (1940).
158. D. Coles, The Law of the Wall in Turbulent Shear Flow, in Sonderdruck aus "50 Jahre Grenzschichtforschung" (H. Goertler and W. Tollmien, eds.), Verlag Friedr. Vieweg and Sohn, Braunschweig, Germany, 1955.
159. D. Coles, The Law of the Wake in Turbulent Boundary Layer, *J. Fluid Mech.,* **1**:191 (1956).
160. H. Ludwieg and W. Tillman, Investigation of the Wall Shearing Stress in Turbulent Boundary Layers, *NACA Tech. Mem.* 1285, 1950, transl. of Untersuchungen über die Wandschubspannung in turbulenten Reibungsschichten, *Ing.-Arch.,* **17**:288 (1949).
161. D. B. Spalding, A Unified Theory of Friction, Heat Transfer, and Mass Transfer in the Turbulent Boundary Layer and Wall Jet, Aeronautical Research Council (England), ARC-CP-829 (1965).
162. C. C. Lin (ed.), "Turbulent Flows and Heat Transfer," vol. 5 of "High Speed Aerodynamics and Jet Propulsion," Princeton University Press, Princeton, N.J., 1959.
163. D. W. Smith and J. H. Walker, Skin-friction Measurements in Incompressible Flow, *NASA Tech. Rept.* R-26, 1959.
164. W. C. Reynolds, W. M. Kays, and S. J. Kline, Heat Transfer in the Turbulent Incompressible Boundary Layer, I–Constant Wall Temperature, *NASA Mem.* 12-1-58W, 1958.
165. R. G. Deissler, Analysis of Turbulent Heat Transfer, Mass Transfer, and Friction in Smooth Tubes at High Prandtl and Schmidt Numbers, *NACA Rept.* 1210, 1954.
166. E. R. van Driest, The Turbulent Boundary Layer with Variable Prandtl Number, *North American Aviation Rept.* Al-1914, 1954.
167. M. W. Rubesin, A Modified Reynolds Analogy for the Compressible Turbulent Boundary Layer on a Flat Plate, *NACA Tech. Note* 2917, 1953.
168. D. B. Spalding, Heat Transfer to a Turbulent Stream from a Surface with a Stepwise Discontinuity in Wall Temperature, "International Developments in Heat Transfer," *Conf. Int. Developments in Heat Transfer,* pt. 2, pp. 439–446, ASME, 1961.
169. C. Ferrari, Effect of Prandtl Number on the Heat Transfer Properties of a Turbulent Boundary Layer When the Temperature Distribution along the Wall Is Arbitrarily Assigned, *Z. angew. Math. Mech.,* **36**:116 (1956).
170. D. B. Spalding, Contribution to the Theory of Heat Transfer Across a Turbulent Boundary Layer, *Int. J. Heat Mass Transfer,* **7**:743 (1964).

171. A. Seiff, Examination of the Existing Data on the Heat Transfer of Turbulent Boundary Layers at Supersonic Speeds from the Point of View of Reynolds Analogy, *NACA Tech. Note* 3284, 1954.

172. E. R. van Driest, Turbulent Boundary Layer in Compressible Fluids, *J. Aeronaut. Sci.,* **18**:145 (1951).

173. M. W. Rubesin, An Analytical Estimation of the Effect of Transpiration Cooling on the Heat Transfer and Skin Friction Characteristic of a Compressible Turbulent Boundary Layer, *NACA Tech. Note* 3341, 1954.

174. E. R. van Driest, The Problem of Aerodynamic Heating, *Aeronaut. Eng. Rev.,* **15**:26 (1956).

175. R. G. Deissler and A. L. Loeffler, Analysis of Turbulent Flow and Heat Transfer on a Flat Plate at High Mach Numbers with Variable Fluid Properties, *NASA Tech. Rept.* R-17, 1959.

176. A. Mager, Transformation of the Compressible Turbulent Boundary Layer, *J. Aeronaut. Sci.,* **25**:305 (1958).

177. O. R. Burggraf, The Compressibility Transformation and Turbulent Boundary Layer Equations, *J. Aerospace Sci.,* **4**:434 (1962).

178. D. Coles, The Turbulent Boundary Layer in a Compressible Fluid, *Rand Corp. Rept.* R-403-PR, 1962.

179. L. Crocco, Transformation of the Compressible Turbulent Boundary Layer with Heat Exchange, *AIAA J.,* **1**:2723 (1963).

180. R. J. Monaghan, On the Behavior of Boundary Layers at Supersonic Speeds, *5th Int. Aeronaut. Conf., Los Angeles, Calif.,* June 20–23, 1955.

181. S. C. Sommer and B. J. Short, Free-flight Measurements of Turbulent Boundary Layer Skin Friction in the Presence of Severe Aerodynamic Heating at Mach Numbers from 2.8 to 7.0, *NACA Tech. Note* 3391, 1955.

182. D. B. Spalding and S. W. Chi, The Drag of a Compressible Turbulent Boundary Layer on a Smooth Flat Plate with and without Heat Transfer, *J. Fluid Mech.,* **18**:117 (1964).

183. S. Scesa, Experimental Investigation of Convective Heat Transfer to Air from a Flat Plate with a Stepwise Discontinuous Surface Temperature, thesis, University of California, 1951.

184. W. C. Reynolds, W. M. Kays, and S. J. Kline, Heat Transfer in the Turbulent Incompressible Boundary Layer, II–Step Wall Temperature Distribution, *NASA Mem.* 12-2-58W, 1958.

185. H. H. Sogin and R. J. Goldstein, Turbulent Transfer from Isothermal Spanwise Strips on a Flat Plate, "International Developments in Heat Transfer," *Conf. Int. Developments in Heat Transfer,* pt. 2, pp. 447–453, ASME, 1961.

186. W. C. Reynolds, W. M. Kays, and S. J. Kline, Heat Transfer in the Turbulent Incompressible Boundary Layer, III–Arbitrary Wall Temperature and Heat Flux, *NASA Mem.* 12-3-58W, 1958.

187. R. M. Kendall, M. W. Rubesin, T. J. Dahm, and M. R. Mendenhall, Mass, Momentum, and Heat Transfer within a Turbulent Boundary Layer with Foreign Gas Mass Transfer at the Surface, Part 1, Constant Fluid Properties, *Itek Corp. Vidya Div. Rept.* 111, 1964.

188. R. J. Moffat and W. M. Kays, The Turbulent Boundary Layer on a Porous Plate; Experimental Heat Transfer with Uniform Blowing and Suction, *Stanford Univ., Dept. Mech. Eng. Rept.* HMT-1, 1967.

189. J. Rotta, Velocity Law of Turbulent Flow Valid in the Neighborhood of a Wall, *Ing.-Arch.,* **18**:277 (1956).

190. D. G. Whitten, W. M. Kays, and R. J. Moffat, The Turbulent Boundary Layer on a Porous Plate; Experimental Heat Transfer with Variable Suction, Blowing, and Surface Temperature, *Stanford Univ., Dept. Mech. Eng. Rept.* HMT-3, 1967.

191. W. H. Dorrance and F. J. Dore, The Effect of Mass Transfer on the Compressible Turbulent Boundary Layer Skin Friction and Heat Transfer, *J. Aeronaut. Sci.,* **21**:404 (1954).

192. M. W. Rubesin and C. C. Pappas, An Analysis of the Turbulent Boundary Layer Characteristics on a Flat Plate with Distributed Light Gas Injection, *NACA Tech. Note* 4149, 1958.

193. E. L. Knuth and H. Dershin, Use of Reference States in Predicting Transport Rates in High-speed Turbulent Flows with Mass Transfer, *Int. J. Heat Mass Transfer,* **6**:999 (1963).

194. C. C. Pappas and A. F. Okuno, Measurements of Skin Friction of the Compressible Turbulent Boundary Layer on a Cone with Foreign Gas Injection, *J. Aerospace Sci.,* **27**:321 (1960).

195. P. Baronti, H. Fox, and D. Soll, A Survey of the Compressible Turbulent Boundary Layer with Mass Transfer, *Astronautica Acta,* **13**:239 (1967).

196. T. Tendeland and A. F. Okuno, The Effect of Fluid Injection on the Compressible Turbulent

Boundary Layer–The Effect on Skin Friction of Air Injected into the Boundary Layer of a Cone at $M = 2.7$, *NACA Res. Mem.* A56DO5, 1956.

197. H. S. Mickley and R. S. Davis, Momentum Transfer for Flow over a Flat Plate with Blowing, *NACA Tech. Note* 4017, 1957.

198. C. C. Pappas and A. F. Okuno, Measurements of Heat Transfer and Recovery Factor of a Compressible Turbulent Boundary Layer on a Sharp Cone with Foreign Gas Injection, *NASA Tech. Note* D-2230, 1964.

199. E. R. Bartle and B. M. Leadon, The Effectiveness as a Universal Measure of Mass Transfer Cooling for a Turbulent Boundary Layer, *Proc. 1962 Heat Transfer and Fluid Mechanics Inst.*, pp. 27-41, Stanford University Press, Stanford, 1962.

200. J. E. Danberg, Characteristics of the Turbulent Boundary Layer with Heat and Mass Transfer at Mach Number 6.7, *Proc. 5th U.S. Navy Symp. Aeroballistics*, U.S. Naval Ordnance Lab., 1961.

201. C. J. Scott, G. E. Anderson, and D. R. Elgin, Laminar, Transitional and Turbulent Mass Transfer Cooling Experiments at Mach Numbers from 3 to 5, *Univ. Minnesota Inst. Tech. Res. Rept.* 162, 1959.

202. C. J. Scott and V. K. Jonsson, Diffusion Thermo Effects in Turbulent Mass Transfer, *Int. J. Heat Mass Transfer,* **10**:1618 (1967).

203. O. E. Tewfik, E. R. G. Eckert, and C. J. Shirtliffe, Thermal Diffusion Effects on Energy Transfer in a Turbulent Boundary Layer with Helium Injection, *Proc. 1962 Heat Transfer and Fluid Mechanics Inst.*, pp. 42-61, Stanford University Press, Stanford, 1962.

204. J. L. Stollery and A. A. M. El-Ehwany, A Note on the Use of a Boundary Layer Model for Correlating Film-cooling Data, *Int. J. Heat Mass Transfer,* **8**:55 (1965).

205. R. J. Goldstein, R. B. Rask, and E. R. G. Eckert, Film Cooling with Helium Injection into an Incompressible Air Flow, *Int. J. Heat Mass Transfer,* **9**:1341 (1966).

206. S. S. Papell and A. M. Trout, Experimental Investigation of Air Film Cooling Applied to an Adiabatic Wall by Means of an Axially Discharging Slot, *NASA Tech. Note* D-9, 1959.

207. J. E. Hatch and S. S. Papell, Use of a Theoretical Flow Model to Correlate Data for Film Cooling or Heating an Adiabatic Wall by Tangential Injection of Gases of Different Fluid Properties, *NASA Tech. Note* D-130, 1959.

208. E. R. van Driest, Turbulent Boundary Layer on a Cone in a Supersonic Flow at Zero Angle of Attack, *J. Aeronaut. Sci.,* **19**:55 (1952).

209. W. C. Reynolds, W. M. Kays, and S. J. Kline, Heat Transfer in the Turbulent Incompressible Boundary Layer, IV–Effect of Location of Transition and Prediction of Heat Transfer in a Known Transition Region, *NASA Mem.* 12-4-58W, 1958.

210. G. B. Schubauer and P. S. Klebanoff, Contribution on the Mechanics of Boundary Layer Transition, *NACA Rept.* 1289, 1956.

211. F. K. Moore, Three-dimensional Boundary Layer Theory, in "Advances in Applied Mechanics," vol. 4 (H. L. Dryden and T. von Kármán, eds.), pp. 159--228, Academic Press, Inc., New York, 1956.

212. L. G. Crabtree, D. Küchemann, and L. Sowerby, Three-dimensional Boundary Layers, in "Laminar Boundary Layers" (L. Rosenhead, ed.), pp. 409--492, Oxford University Press, New York, 1963.

213. K. Stewartson, "The Theory of Laminar Boundary Layers in Compressible Fluids," pp. 99--121, Oxford University Press, London, 1964.

214. A. Mager, Three-dimensional Laminar Boundary Layers, in "Theory of Laminar Flows" (F. K. Moore, ed.), vol. 4 of "High Speed Aerodynamics and Jet Propulsion," Princeton University Press, Princeton, N.J., 1964.

215. E. Reshotko, Laminar Boundary Layer with Heat Transfer on a Cone at Angle of Attack in a Supersonic Stream, *NACA Tech. Note* 4152, 1957.

216. P. A. Libby, Heat and Mass Transfer at a General Three-dimensional Stagnation Point, *AIAA J.,* **5**:507 (1967).

217. E. Reshotko, Heat Transfer to a General Three-dimensional Stagnation Point, *Jet Propul.,* **26**:58 (1958).

218. R. Vaglio-Laurin, Laminar Heat Transfer on Three-dimensional Blunt Nosed Bodies in Hypersonic Flows, *ARS J.,* **29**:123 (1959).

219. T. K. Fannelop, A Method of Solving the Three-dimensional Laminar Boundary Layer Equations with Application to a Lifting Re-entry Body, *AIAA J.,* **6**:1075 (1968).

220. J. W. Cleary, Effects of Angle of Attack and Nose Bluntness on the Hypersonic Flow over Cones, AIAA Paper 66-414, Los Angeles, June 1966.

221. J. L. Hess and A. M. O. Smith, Calculation of Potential Flow about Arbitrary Bodies, in "Progress in Aeronautical Sciences" (D. Küchemann, ed.), vol. 8, pp. 1--138, Pergamon Press, Inc., New York, 1967.
222. E. J. Hopkins, M. W. Rubesin, M. Inouye, E. R. Keener, G. G. Mateer, and T. E. Polek, Summary and Correlation of Skin-friction and Heat-transfer Data for a Hypersonic Turbulent Boundary Layer on Simple Shapes, *NASA Tech. Note* D-5089, 1969.
223. E. R. van Driest, On Turbulent Flow near a Wall, *J. Aeronaut. Sci.,* **23**:1007 (1956).

NOMENCLATURE

a	speed of sound
a	exponent in Eq. (235), see also Table 9
a	exponent, p. 8-171, Fig. 100
a_n	coefficients in Eq. (145)
a_n	coefficients in Eq. (237)
a, b	semimajor and minor axes of ellipse or Rankine body
a, b	see Eq. (241)
a, b, c	constants in reference enthalpy equation p. 8-79
A	surface area
A	constant in Eq. (228)
A	constant in Eq. (232)
A	constant in Eq. (317)
A	constant in Eq. (358)
A, B	coefficients in Eq. (253), Table 11
b	quantity proportional to boundary-layer thickness, Eq. (79)
b	exponent in viscosity law for liquids
b_j	height of jth step in heat flux, Eq. (342)
B	beta function
B	correlation parameter with mass transfer, Eq. (262)
$B(\beta)$	function used in Eq. (317), Fig. 82
$B_z(a, b)$	incomplete beta function
c	constant in wedge solution, p. 8-21
c	airfoil chord
c	ratio of velocity gradients, Eq. (376)
c_f	local skin friction coefficient
c_{f_w}	local skin friction coefficient in terms of wall properties, Eq. (217)
c_h	local Stanton number
c_{h_w}	local Stanton number in terms of wall properties
$c_h(x_1, s)$	local Stanton number on a plate with a temperature jump at s
c_m	mass transfer coefficient
c_p	specific heat per unit mass at constant pressure
c_v	specific heat per unit mass at constant volume
c_1	constant in Eq. (354)
$\overline{C}$	constant value of C_r
C_D	nose drag coefficient
C_f	average skin friction coefficient
C_h	average Stanton number
C_p	pressure coefficient
C_r	normalized viscosity-density product, Eq. (113)
C_1	constant defined in Eq. (359)
d	body diameter
d	exponent in power law velocity profile, p. 8-122
D	function given in Fig. 89
D_{ij}	multicomponent diffusion coefficient
$\mathcal{D}$	binary diffusion coefficient

$\mathfrak{D}^T$	thermal diffusion coefficient
$\mathfrak{D}_{ij}$	binary diffusion coefficient of species i in species j
e	internal energy per unit mass
E	turbulent boundary-layer constant, p. 8-147
E	function given in Fig. 89
Ec	Eckert number
f	dimensionless boundary-layer stream function, Blasius stream function, Eq. (81)
f_1	function in blast wave analogy, p. 8-32
F	dimensionless boundary-layer stream function, Eq. (118)
$F_c, F_{R_\theta}, F_{Rx}$	turbulent boundary-layer transformation functions, $\widetilde{c}_f/c_f$, $\widetilde{Re}_{\theta_{11}}/Re_{\theta_{11}}, \widetilde{Re}_{x_1}/Re_{x_{1e}}$ respectively
$\mathfrak{F}$	function defined by Eq. (256)
g	normalized total enthalpy, Eq. (113)
g_n	surface enthalpy distribution functions, p. 8-87
G	normalized total enthalpy, Eq. (119)
G	function used in Eq. (154), Fig. 40
G_n	function used in Eq. (184)
h	step height
h_1, h_2, h_3	metric coefficients for x_1, x_2, x_3 coordinate system
H	function, p. 8-88, Fig. 41
H	form factor, δ_1^*/θ_{11}
H_n	effective potential function, Eq. (148)
$H_{\beta n}$	effective potential function, Eq. (239)
i	enthalpy per unit mass
i_E	reference value of enthalpy, 2.119×10^8 ft^2/sec^2 or 8465 Btu/lb
i_i^0	heat of formation of ith species at $T = 0°$R
I	total enthalpy per unit mass
I_z	incomplete beta function ratio, p. 8-90
j	diffusion flux
J	number of steps in heat flux distribution, Eq. (342)
k	thermal conductivity
k	number of temperature jumps
k	index, 0 for two-dimensional flow and 1 for axisymmetric flow
K	hypersonic similarity parameter
K	mass concentration
K	mixing length constant, Eq. (303)
K^+	normalized mass concentration, Eq. (351)
l	length
l	length of ogive
l	mixing length
l^+	normalized mixing length, Eq. (346)
L	reference length
L	length of porous section in film cooling
Le	Lewis number
m	strength of source or sink
m	mass of molecule
m	exponent in Eq. (232)
m	constant in Eq. (336)
m	constant defined by Eq. (361)
m_n	slope of nth ramp, Eq. (342)
M	Mach number
M	molecular weight, mean molecular weight, Eq. (47b)

$\overline{M}_c$	mean molecular weight of coolant
M_f, M_i	final and initial Mach numbers in Prandtl-Meyer expansion
n	number of moles
n	exponent in power law viscosity, Eq. (132)
n	power in Eq. (145)
n	gamma function parameter
n	Deissler empirical constant, p. 8-146
n	constant in Eq. (358)
N	function in Eq. (271)
N	number of ramps in heat flux distribution, Eq. (342)
N_n	function in Eq. (183)
Nu	Nusselt number, p. 8-9 and p. 8-125
p	pressure
p	exponent, p. 8-123
P	function defined by Eq. (246)
$P(s)$	probability of transition at s, Eq. (371)
P, Q, R	functions in Eq. (271)
Pr	Prandtl number
Pr_t	turbulent Prandtl number
$\mathrm{Pr}_{\mathrm{tot}}$	Prandtl number based on sum of laminar and turbulent transport mechanisms, Eq. (308)
q	heat flux
$\overline{q}$	average heat flux (over space)
q_{wn}	value of q_w at end of nth step
q_{w0}	constant value of q_w for $x_1 \geq s$, Eq. (157)
$r(\eta)$	dimensionless temperature variable, Eq. (103)
$\overline{r}(\eta_c)$	dimensionless temperature variable, Eq. (140)
$r(0)$	recovery factor
r, θ, z	cylindrical polar coordinates
r, θ, ϕ	spherical polar coordinates
R	gas constant
R	body radius
R	surface radius of curvature
R	function defined by Eq. (340)
R_c	corner radius on flat-faced cylinder
R_n	nose radius
$\mathrm{Re}_{\mathrm{eff}}$	effective Reynolds number, Eq. (318)
Re_L	Reynolds number based on reference length L
s	distance along surface measured from stagnation point
s	distance to location of surface temperature discontinuity
s	slot height, Fig. 59
s	reciprocal of the Reynolds analogy factor
s	distance from leading edge to transition from laminar to turbulent flow
s_j^-, s_j^+	upstream and downstream of a surface temperature jump
S	entropy
S	function defined by Eq. (339)
Sc	Schmidt number
Sc_t	turbulent Schmidt number
t	normalized enthalpy, Eq. (203)
t^+	normalized temperature parameter, Eq. (315)
t_e	normalized enthalpy, Eq. (205)
t_E	normalized enthalpy, i_E/I_e
t_s	normalized enthalpy, Eq. (204)

T	temperature
T^+	normalized temperature parameter, Eq. (350)
T_c	constant in viscosity law for liquids
T_c	Sutherland constant
u^+	normalized velocity, u_1/u_τ
u_τ	friction velocity, $\sqrt{\tau_w/\rho}$
u_1, u_2, u_3	velocity components in x_1, x_2, x_3 directions
u, v, w	velocity components in x, y, z directions
u, v_r, v_ϕ	velocity components in x, r, ϕ directions
$\bar{U}$	normalized free stream velocity, $V_\infty/10,000$ ft/sec
v_r, v_θ, w	velocity components in r, θ, z directions
v_r, v_θ, v_ϕ	velocity components in r, θ, ϕ directions
v_η	velocity component in η direction
v_0^+	velocity normal to surface normalized by friction velocity
V	velocity
w	normalized cross-flow velocity, Eq.(202)
w	Coles wake function, Fig. 77
x^+	normalized x_1 distance, Eq. (305)
x_{eff}	length defined by Eq. (222)
x_{eff}	length defined by Eq. (369)
x_s	distance measured from cone-cylinder junction
x_0	constant in solution for Rankine body, p. 8-22
x_1, x_2, x_3	generalized orthogonal coordinates
x, y, z	rectangular coordinates
x, r, ϕ	cylindrical polar coordinates
y^+	normalized distance from wall, $u_\tau x_2/\nu$
y_a^+	constant, Eq. (346)
Y_n	function in Eq. (146)
$Y_0(\eta)$	dimensionless temperature function, Eq. (92)
$\bar{Y}_0(\eta_c)$	dimensionless temperature function, Eq. (141)
z	argument of beta function
z	function used in film cooling, p. 8-171
$z, \bar{z}$	dimensionless enthalpy, Eq. (364)
z_1/z_t	enthalpy ratio, Eq. (374)
α	angle of attack
α	thermal diffusion factor, Eq. (274)
β	pressure gradient parameter, Eq. (206)
β	normalized temperature parameter, Eq. (316)
β	fraction of time a surface point is covered by a fully turbulent boundary layer, Eq. (370)
γ	ratio of specific heats
Γ	gamma function
Γ	energy thickness
Γ^*	ratio of transformed energy thicknesses, Eq. (269)
Γ_{tr}	transformed energy thickness, Eq. (266)
δ^*	displacement thickness
δ^*_{sim}	displacement thickness for similarity solution, p. 8-163
δ_{th}	thermal boundary-layer thickness
Δ	concentration thickness
$\bar{\Delta}$	turbulent boundary-layer thickness, Eq. (285)
$\Delta\theta$	expansion angle in Prandtl-Meyer flow
ϵ	blowing function, Eq. (182)
$\bar{\epsilon}$	sum of molecular and eddy viscosity, $\nu + \epsilon_M$

ϵ_D	turbulent eddy diffusivity
ϵ_h^+	normalized eddy thermal conductivity, Eq. (305)
ϵ_H	eddy thermal conductivity
ϵ_M	eddy viscosity
$\epsilon_{M1}, \epsilon_{M3}$	eddy viscosities in x_1, x_3 directions
ϵ_u^+	normalized eddy viscosity, Eq. (305)
$\bar{\zeta}$	function in Eq. (201)
ζ, η	transformed boundary-layer variables
η	function in Coles transformation, Eq. (321)
η_c	modified transformed boundary-layer variable, Eq. (117)
η_{th}	thermal boundary-layer thickness parameter
θ	momentum thickness
θ	ratio of Sutherland constant to edge temperature, p. 8-71
λ	normalized viscosity-density product, Eq. (173)
Λ	sweepback angle
μ	dynamic viscosity
ν	Prandtl-Meyer angle
ν	number of species in Eq. (38)
ν	kinematic viscosity
ξ, η	elliptic coordinates
ρ	density
σ	function in Coles transformation, Eq. (321)
σ	standard deviation of transition location about its mean
τ	shear stress
ϕ	normalized mass concentration, Eq. (170)
Φ	potential function
ψ	stream function
$\bar{\psi}_i$	mass rate of chemical formation of species i per unit volume
$\omega_1, \omega_2, \omega_3$	vorticity components in x_1, x_2, x_3 directions

Subscripts

aw	adiabatic wall
B	Blasius function
c	cone
c	initial coolant condition in reservoir
D	condition before dissociation, Eq. (256)
e	boundary-layer edge
F	based on frozen properties
i	species i
i	x_i direction
i	constant property case
l	laminar
r	reference condition
s	shock
s	mean substructure temperature, Eq. (327)
st	stagnation point of body
t	turbulent
t_∞	free stream stagnation conditions
T	based on total properties
w	wall
w	wedge
0	zero mass transfer
1,2	coolant and air, respectively
∞	free stream

Superscripts

–	time average value
′	randomly fluctuating value
′	reference enthalpy or temperature condition
′	dummy variable
∼	constant property case

Section **9**

Rarefied Gases

WARREN H. GIEDT
Professor of Mechanical Engineering,
University of California, Davis

D. ROGER WILLIS
Professor of Aeronautical Sciences,
University of California, Berkeley

INTRODUCTION

At very low pressure the distance traveled by the molecules of a gas between collisions becomes relatively large. As a result, the gas no longer behaves like a continuum. Instead, its basic molecular character becomes predominant. The following properties then become important in describing and evaluating energy exchanges.

Mean Free Path. This is the average distance traveled by an atom or molecule of a gas between collisions. It is dependent on the gas state and the frame of reference from which the system is observed. For a gas at rest in local thermodynamic equilibrium it is defined by

$$\lambda = \frac{\mu}{\rho}\left(\frac{\pi}{2RT}\right)^{1/2} \tag{1a}$$

where μ is the viscosity, ρ the density, T the temperature, and R the gas constant per unit mass. A specific expression of this formula applicable to air in the range of 180° to 3400° Rankine is obtained by substituting the Sutherland law for the viscosity. This yields

$$\lambda = \frac{(2.08 \times 10^{-7})T^2}{(T + 198.6)p} \tag{1b}$$

in which λ is in inches, T in °R and p in inches of mercury.

For a gas whose molecules are represented by rigid spheres of diameter σ, the mean free path in an equilibrium gas is

$$\lambda = \frac{m}{\sqrt{2}\,\pi\rho\sigma^2} \tag{2}$$

where m is the mass of the molecule. The variation of λ with altitude [1] shown in Fig. 1 is based on this relationship using the best presently available information on density variation and an effective collision diameter for air molecules evaluated from sea level viscosity measurements.

Knudsen Number. Departure from continuum flow occurs when λ becomes significant compared to a pertinent dimension L of the flow field. The ratio of these two quantities is an important parameter called the Knudsen Number

$$\text{Kn} = \frac{\lambda}{L} \tag{3}$$

L may be a body dimension or a length characteristic of the flow such as a boundary layer thickness.

Distribution Function and Moments. In a rarefied monatomic gas the state of the gas is described by a distribution function $f(x, y, z, \xi_x, \xi_y, \xi_z, t)$ such that the number of molecules at time t having velocities in the infinitesimal ranges ξ_x to $\xi_x + d\xi_x$, ξ_y to $\xi_y + d\xi_y$ and ξ_z to $\xi_z + d\xi_z$ in the elemental volume between x and $x + dx$, y and $y + dy$ and z and $z + dz$ is given by $f(x, y, z, \xi_x, \xi_y, \xi_z, t)dxdydz\,d\xi_x\,d\xi_y\,d\xi_z$. It is often convenient to use vectors $\mathbf{r}$ and $\boldsymbol{\xi}$ to represent the position and molecular velocity and to write f as $f(\mathbf{r}, \boldsymbol{\xi}, t)$.

Various weighted integrals over all molecular velocities yield moments of the distribution function. The simpler moments correspond to conventional macroscopic variables, e.g.,

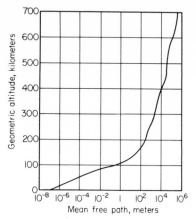

Fig. 1. Variation of molecular mean free path with altitude. The curve [1] represents average conditions in the upper atmosphere.

Number density	$n(\mathbf{r}, t) = \displaystyle\int f d^3\xi$	(4)
Density	$\rho(\mathbf{r}, t) = m n(\mathbf{r}, t)$	(5)
Velocity	$\mathbf{u}(\mathbf{r}, t) = \dfrac{1}{\rho}\displaystyle\int m\xi f d^3\xi$	(6)
Temperature	$T(\mathbf{r}, t) = \dfrac{1}{3\rho R}\displaystyle\int m(\xi - \mathbf{u})^2 f d^3\xi$	(7)
Energy flux component	$q_i(\mathbf{r}, t) = \displaystyle\int \dfrac{1}{2} m(\xi - \mathbf{u})^2 (\xi_i - u_i) f d^3\xi$	(8)

in which the shorthand notation

$$\int d^3\xi = \int_{-\infty}^{\infty} d\xi_x \int_{-\infty}^{\infty} d\xi_y \int_{-\infty}^{\infty} d\xi_z$$

has been employed and R is the gas constant per unit mass.

Maxwell Equilibrium Distribution Function. For a spatially homogeneous monatomic gas in equilibrium and moving with a bulk translational velocity $\mathbf{u}$, the distribution function is given exactly by the Maxwellian distribution

$$f(\mathbf{r}, \xi, t) = f^0(\xi) = n(2\pi RT)^{-3/2} \exp\left[-\frac{(\xi - \mathbf{u})^2}{2RT} \right] \qquad (9)$$

FLOW CHARACTERIZATION CRITERIA AND PARAMETERS

Rarefied gas flows are characterized by the degree of rarefaction present. Although no universally acceptable and applicable criteria have been devised, a problem is usually considered to be rarefied if the appropriate Knudsen number is not negligibly small everywhere in the flow field. Both the mean free path λ, which depends on the frame of reference and varies inversely with density, and the reference length L must be selected with care.

The Knudsen number is related to more conventional dimensionless parameters by the relation

$$\text{Kn} = \frac{\lambda}{L} = \frac{(\pi \gamma/2)^{1/2} M}{\text{Re}_L} \tag{10}$$

where M is the Mach number given by $M = u/(\gamma RT)^{1/2} = u/a$ and Re_L is the Reynolds number $\text{Re} = \rho u L/\mu$. When there is no net flow a more convenient definition is

$$\text{Kn} = \frac{(\pi \gamma/2)^{1/2}}{\text{Re}_L^*} \tag{11}$$

where $\text{Re}_L^* = \rho a L/\mu$.

Classical continuum flow described by the Navier Stokes equations and the Fourier heat conduction law represents the limiting case when the Knudsen number is everywhere negligibly small. In estimating the Knudsen number(s) in such a case the smallest appropriate values of L should be used. These are often a boundary layer or viscous layer thickness.

When the Knudsen number is everywhere large the flow is called free molecular. The average distance traveled by molecules between collisions is at least several times greater than all other dimensions in the problem. In estimating the Knudsen number in these cases the largest appropriate value of L should be used, often a body dimension.

Between these two limits the flow is called transitional. Neither theory nor experiment has been well established for any but very simple geometries.

For supersonic or hypersonic flow past a blunt body a suitable rarefaction parameter for low Knudsen number is

$$\text{Kn}_2 = \frac{\lambda_2}{L} = \frac{\mu(T_2)}{\rho_2 (2RT_2/\pi)^{1/2} L} \tag{12a}$$

or assuming no significant change in molecular weight before and after the shock

$$\text{Kn}_2 = \frac{\lambda_\infty}{L} \frac{\rho_\infty}{\rho_2} \left(\frac{T_\infty}{T_2}\right)^{1/2} \frac{\mu(T_2)}{\mu(T_\infty)} \tag{12b}$$

where subscript 2 denotes conditions behind the normal shock and subscript ∞ free-stream conditions. $\mu(T_2)$ and $\mu(T_\infty)$ are the viscosities at these two temperatures. This parameter can be significantly different from the free-stream Knudsen number (λ_∞/L), and depends crucially on the nature of the viscosity temperature relationship. The proportional change, from the continuum results, of heat transfer and drag coefficients will be of the same magnitude as Kn_2.

For high Knudsen number flow past a blunt body it is not appropriate to consider a shock as existing in front of the body. A useful parameter, suggested by theory and found reasonable for correlating wind tunnel data, is

$$Kn_r = \left(\frac{T_r}{T_s}\right)^{1/2} \frac{\mu(T_s)}{\rho_\infty (2RT_s/\pi)^{1/2}L} \tag{13a}$$

$$= \frac{\lambda_\infty}{L}\left(\frac{T_r}{T_s}\right)^{1/2}\left(\frac{T_\infty}{T_s}\right)^{1/2}\frac{\mu(T_s)}{\mu(T_\infty)} = \frac{\lambda_r}{L} \tag{13b}$$

where T_s is the stagnation temperature and T_r is a temperature characteristic of the molecules leaving the body. For a thermal accommodation coefficient α (defined in the following section) close to unity T_r will be close to T_w, the actual wall temperature. For values of α less than unity T_r can be estimated from $T_r = (1 - \alpha)T_s + \alpha T_w$. The quantity λ_r can be regarded as the mean free path for molecules leaving the body and colliding with incoming free-stream molecules. Except when $T_r \ll T_s$, i.e., a highly cooled body, the two parameters defined in Eqs. (12) and (13) will be of the same order. The proportional change from the free molecular results of heat transfer and drag coefficients will be of the same magnitude as $1/Kn_r$. For $\gamma = 7/5$ and a linear variation of viscosity with temperature (reasonable approximations for unheated low-density wind tunnel conditions) $1/Kn_r \simeq 2.2/Kn_\infty M$. The mean free path in the atmosphere which serves as a basis for calculating appropriate rarefaction parameters is presented in Fig. 1.

The actual estimation of Kn_2 and Kn_r for high-altitude flight at high speeds corresponding to reentry conditions is complicated by nonequilibrium effects in the shock layer and the lack of reliable data for viscosity (or collision cross sections) at high temperatures and low pressures. The problem is further complicated by possible molecular dissociation, excitation, and ionization which may occur when the molecules strike the surface.

Example of Estimating Mean Free Paths. Consider flow in an unheated low-density wind tunnel using air as the working medium. The stagnation pressure and temperature are respectively 4×10^{-3} inch of mercury and $530°R$. A body cooled by liquid nitrogen is placed in the flow where the free stream Mach number is 2. The body temperature is $250°R$, and complete thermal accommodation is assumed. To determine whether the flow around the body is in the continuum, transition, or free molecular region it is necessary to calculate λ_2 and λ_r.

From perfect gas relations for isentropic flow (14) with $\gamma = 7/5$ the free-stream conditions are found to be

$$p_\infty = (0.1278)(4 \times 10^{-3}) = 5.112 \times 10^{-4} \text{ inch Hg}$$
$$T_\infty = (0.5556)(530) = 294.5°R$$
$$p_2 = (4.500)(5.112 \times 10^{-4}) = 2.300 \times 10^{-3} \text{ inch Hg}$$
$$T_2 = (1.688)(294.5) = 497.1°R$$

Substituting in Eq. (1b) yields

$$\lambda_\infty = \frac{(2.08 \times 10^{-7})(294.5)^2}{(294.5 + 198.6)(5.112 \times 10^{-4})} = 7.15 \times 10^{-2} \text{ inch}$$

Similarly for conditions just downstream of the shock wave at the stagnation region

$$\lambda_2 = \frac{(2.08 \times 10^{-7})(497.1)^2}{(497.1 + 198.6)(2.300 \times 10^{-3})} = 3.22 \times 10^{-2} \text{ inch}$$

From Eq. (13b) using $T_r = T_w = 250°R$ and $\mu = CT^{3/2}/(T + 198.6)$

$$\lambda_r = \lambda_\infty \left(\frac{294.5}{530}\right)^{1/2} \left(\frac{250}{530}\right)^{1/2} \frac{(530)^{3/2}}{530 + 198.6} \frac{294.5 + 198.6}{(294.5)^{3/2}} = 5.98 \times 10^{-2} \text{ inch}$$

For this modest Mach number it is seen that the mean free path in the free stream is approximately twice as large as the mean free path behind the shock wave at the stagnation point. λ_r is intermediate between these two. More dramatic differences would be obtained for higher Mach numbers, but the situation is then frequently complicated by nonequilibrium effects.

Estimation of Knudsen Number.   Consider the body in the previous example to be a cylinder 1.0 inch in diameter normal to the flow. The first evaluation of a Knudsen number would be made using the body diameter and the mean free path of molecules in the free stream λ_∞. This yields a value of 7.15×10^{-2} which indicates that the flow is definitely not free molecular. However, it is not small enough to conclude that continuum conditions exist everywhere in the flow field.

We therefore must investigate local values. The crucial region is near the stagnation point. Here a pertinent dimension is the stand-off distance Δ of the shock wave formed in front of the cylinder based on continuum flow. This can be approximated by

$$\frac{\Delta}{r} = \frac{\rho_\infty}{\rho_2}$$

where r is the cylinder radius and ρ_∞/ρ_2 is the density ratio across the shock wave along the stagnation streamline. This leads to $\Delta = 0.5(1/2.667) = 0.1875$ in. In calculating the Knudsen number in this region λ_2 is the appropriate mean free path. This gives $\text{Kn}_2 = (3.22 \times 10^{-2})/0.1875 = 0.172$, which suggests that deviations from continuum flow are to be expected.

If the cylinder diameter is 0.1 inch, the value of Δ is 0.0187 inch. In this case $\text{Kn}_2 = 1.72$, and $1/\text{Kn}_r = d/\lambda_r = 0.1/(5.98 \times 10^{-2}) = 1.67$. As neither of these parameters is small, the flow is not close to either continuum or free molecular conditions. The flow is, therefore, in the transition region.

Energy Transfer between Gas and Surface.   The local energy transfer at a surface is calculated from the net flux of energy due to the molecules incident on and the molecules leaving the surface per unit time and area. In terms of the distribution function this is given by

$$\int \left(\frac{1}{2} m \xi^2\right) \xi_n f \, d^3\xi \quad \text{evaluated at the surface}$$

where ξ_n is the component of the molecular velocity normal to the surface. This requires a knowledge of f for all molecular velocities. Such complete knowledge is rarely available. Even when f is completely known for incoming molecules, as in the case of free molecular flow, the interaction between the molecules and the surface is sufficiently complex so that the distribution function for outgoing molecules cannot be obtained in detail at the present time. For many engineering purposes, however, it is sufficiently accurate to consider the molecules leaving the surface to have a Maxwellian distribution characterized by the mean surface velocity (usually zero) and a temperature T_r. This temperature is not in general the surface temperature T_w.

Thermal Accommodation Coefficient.   When the main concern is with the energy transfer at the surface a detailed knowledge of f is not required for the outgoing molecules. Instead, a thermal accommodation coefficient α is employed. It is defined by

$$\alpha = \frac{E_i - E_r}{E_i - E_w} \tag{14}$$

where E_i is the incident flux, E_r the flux reemitted from the surface, and E_w the flux that would leave the surface if the reemitted molecules were in complete Maxwellian equilibrium at T_w. All these fluxes (E_i, E_r, E_w) are positive quantities. In terms of the net energy transfer

$$\left(E_{net}\right)_{actual} = \alpha \left(E_{net}\right)_{Maxwellian,\ T_w} \tag{15}$$

Implicit in the definition of α is the assumption that the details of the distribution function of the incoming molecules are unimportant. This is certainly not strictly true and recent experimental techniques and theoretical models are beginning to account in detail for the energy transfer.

The parameter α is dependent on the gas and surface in question and is very sensitive to any adsorbed molecules, oxide layers, or other surface contamination. However, molecularly clean surfaces rarely occur outside the laboratory, and the values of α given in Table 1 are applicable for engineering calculations for surfaces that have not been specially prepared with a view to obtaining a molecularly clean or controlled surface. In Table 2 data are given for clean surfaces. The variation of α with temperature is discussed by Thomas [2].

TABLE 1. Thermal Accommodation Coefficients for Engineering Surfaces, $T_w \simeq T_{gas} \simeq 300°K$

Gas	Surface	α
Air	flat lacquer on bronze	0.88 to 0.89
Air	polished bronze	0.91 to 0.94
Air	machined bronze	0.89 to 0.93
Air	etched bronze	0.93 to 0.95
Air	polished cast iron	0.87 to 0.93
Air	machined cast iron	0.87 to 0.88
Air	etched cast iron	0.87 to 0.96
Air	polished aluminum	0.87 to 0.95
Air	machined aluminum	0.95 to 0.97
Air	etched aluminum	0.89 to 0.97

SOURCE: M. L. Wiedmann and P. R. Trumpler, Thermal Accommodation Coefficient, *Trans. ASME*, 68:57, 1946.

The surface interaction is also dependent on the energy of the incident molecules and the temperature of the surface. Unfortunately most of the data has been obtained in circumstances where the incoming molecules are essentially in a Maxwellian equilibrium with a temperature only slightly different from the surface temperature.

Temperature Jump and Velocity Slip. When the Knudsen number is small but not negligible it has been proposed that problems can be analyzed by using the Navier Stokes equations with the following boundary conditions at the surface

$$T_{gas} - T_{wall} = \frac{2 - \alpha}{\alpha} \left[\frac{2(\gamma - 1)}{\gamma + 1} \right] \frac{1}{R\rho (2RT_w/\pi)^{1/2}} (-q_n) \tag{16}$$

$$u_{t\ gas} - u_{t\ wall} = \frac{2 - \sigma}{\sigma} \frac{1}{\rho (2RT_w/\pi)^{1/2}} (\tau) + \left(\frac{3}{4}\right) \frac{Pr(\gamma - 1)}{\gamma \rho R T_w} (-q_t) \tag{17}$$

TABLE 2. Thermal Accommodation Coefficients for
Clean Surfaces, $T_w \simeq T_{gas}$

Gas	Surface	Temperature, °K	α
He	Al	305	0.073
He	Be	305	0.145
He	K	298	0.083
He	Li	80	0.024
He	Mo	298	0.0261
He	Na	298	0.090
He	Pt	305	0.038
Ne	Al	305	0.159
Ne	Be	305	0.315
Ne	Fe	300	0.053
Ne	K	298	0.199
Ne	Li	80	0.049
Ne	Mo	298	0.055
Ne	Na	298	0.198
A	K	298	0.444
A	Li	80	0.290
A	Mo	298	0.3155
A	Na	298	0.459
Kr	Li	80	0.400
Kr	Mo	298	0.510
Xe	Mo	298	0.956

SOURCE: F.O. Goodman and H. Y. Wachman, "Formula for
Thermal Accommodation Coefficient," AFOSR 66-0295, *MIT
Fluid Dynamics Res. Lab. Rept.* 66-1, 1966.

where q_n, q_t are the normal and tangential heat transfer components in the gas and τ is the (viscous) stress component corresponding to the skin friction. The parameter σ is the coefficient of tangential momentum accommodation defined analogously to α by the relation

$$\sigma = \frac{\tau_i - \tau_r}{\tau_i} \tag{18}$$

where τ_i and τ_r are the incident and reflected tangential momentum fluxes. The net skin friction τ is given by $\sigma \tau_i$. The parameter σ is close to unity for engineering surfaces providing most of the molecules do not strike the surface with grazing incidence.

When the boundaries are planar and there are no temperature gradients parallel to the surface, these relations reduce to the more conventional relations

$$T_{gas} - T_{wall} = \frac{2 - \alpha}{\alpha} \left(\frac{2\gamma}{\gamma + 1} \right) \frac{\lambda}{Pr} \frac{dT}{dy} \tag{19}$$

$$u_{t\,gas} - u_{t\,wall} = \frac{2 - \sigma}{\sigma} \lambda \left(\frac{du_t}{dy} \right) \tag{20}$$

where y is the direction normal to the boundary.

In the case of a monatomic gas the quantity $2\gamma/[(\gamma + 1)Pr] = 15/8$.

The term in Eq. (17) proportional to $(-q_t)$ is associated with the phenomena of thermal creep, which can be of importance in causing variation in pressure along tubes, e.g., pitot tubes, when there is a temperature gradient along the wall.

Equations (16) to (19) are not exact and detailed calculations suggest that the factors $(2 - \sigma)/\sigma$ and $(2 - \alpha)/\alpha$ need small modification. However, the use of the

Navier Stokes equations with these boundary conditions has no strong theoretical foundation so these relations should be sufficiently accurate.

It should be noted that only for very simple situations (e.g., planar geometry, flow through cylindrical pipes) have solutions to the full Navier Stokes equations been obtained. For slip and temperature jump to be of concern the Reynolds number cannot be very high and the classical boundary layer plus inviscid flow solutions will not be valid. Obtaining appropriate solutions to the Navier Stokes equations is then often more challenging and at least as important as including the slip boundary conditions.

FREE MOLECULE FLOW

Free molecule flow obtains when the minimum appropriate Knudsen number for the flow around a body is large. The effect of molecular collisions can then be neglected. Consequently mass, momentum, and energy fluxes of the molecules incident on and reemitted from a surface, can be calculated independently.

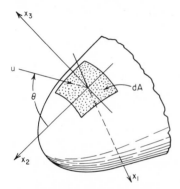

Fig. 2. Free molecule flow incident on an elemental area.

A value of the Knudsen number for which free molecule flow can definitely be assumed has not been established. It will depend on the body geometry as well as the stream characteristics. The proportional change from free molecule flow results was indicated previously (see page 5) to be of the order of $1/Kn_r$. It is recommended that this parameter be used in determining the applicability of the free molecule flow results given below.

Analysis for Convex Surfaces. For surface geometries from which reemitted molecules cannot strike another part of the surface, and when incoming and outgoing flows are in equilibrium, the fluxes can be evaluated using the Maxwellian distribution function. For the element of the outer surface of the body shown in Fig. 2, it is convenient to define a Cartesian coordinate system by taking x_1 normal to the surface and x_2 in the plane defined by the incident flow u. The distribution function (Eq. (9)) then becomes

$$ f = \frac{\rho}{m(2\pi RT_\infty)^{3/2}} \exp\left[-\frac{(\xi_1 - u\sin\theta)^2 + (\xi_2 + u\cos\theta)^2 + \xi_3^2}{2RT_\infty} \right] \tag{21} $$

The number flux dN_i incident on dA is determined from

$$ dN_i = \int_{-\infty}^{\infty} d\xi_3 \int_{-\infty}^{\infty} d\xi_2 \int_{0}^{\infty} \xi_1 f\, d\xi_1\, dA \tag{22} $$

Introducing the "molecular speed ratio" $S = u/\sqrt{2RT_\infty}$ (a Mach number based on the most probable molecular speed*) and integrating yields

$$ dN_i = \frac{\rho}{m}\sqrt{\frac{RT}{2\pi}} \left\{ e^{-(S\sin\theta)^2} + \sqrt{\pi}\,(S\sin\theta)[1 + \mathrm{erf}(S\sin\theta)] \right\} dA \tag{23} $$

*By using the molecular speed ratio, which can also be expressed as $S = \sqrt{\gamma/2}\,M$, and a modified recovery factor r' or Stanton number St' as defined on page 12, only a single curve is needed to show the variation of r' or St' with free-stream velocity for gases with different values of γ (Figs. 3 and 4).

The pressure p_i and shear stress τ_i produced by the incident fluxes normal and tangential to the surface are

$$p_i = \int_{-\infty}^{\infty} d\xi_3 \int_{-\infty}^{\infty} d\xi_2 \int_0^{\infty} m\xi_1^2 f d\xi_1$$

$$= \frac{\rho u^2}{2\sqrt{\pi}\, S^2} \left\{ (S\sin\theta)\, e^{-(S\sin\theta)^2} + \sqrt{\pi}\left[\frac{1}{2} + (S\sin\theta)^2 \right][1 + \text{erf}(S\sin\theta)] \right\} \qquad (24)$$

$$\tau_i = \int_{-\infty}^{\infty} d\xi_3 \int_{-\infty}^{\infty} d\xi_2 \int_0^{\infty} m\xi_1 \xi_2 f d\xi_1$$

$$= \frac{-\rho u^2 \cos\theta}{2\sqrt{\pi}\, S} \left\{ e^{-(S\sin\theta)^2} + \sqrt{\pi}\,(S\sin\theta)[1 + \text{erf}(S\sin\theta)] \right\} \qquad (25)$$

To account for the transfer of rotational and vibrational energy, the incident energy flux is considered in two parts $E_{i,\text{tr}}$ due to translational energy and $E_{i,\text{int}}$ which is associated with internal rotation and vibration.

$$E_{i,\text{tr}} = \int_{-\infty}^{\infty} d\xi_3 \int_{-\infty}^{\infty} d\xi_2 \int_0^{\infty} \frac{1}{2} m\xi^2 \xi_1 f d\xi_1$$

$$= \rho RT \sqrt{\frac{RT}{2\pi}} \left\{ (S^2 + 2)\, e^{-(S\sin\theta)^2} + \sqrt{\pi}\left(S^2 + \frac{5}{2} \right)(S\sin\theta)[1 + \text{erf}(S\sin\theta)] \right\} \qquad (26)$$

To evaluate $E_{i,\text{int}}$ it is assumed that each internal mode possesses $\frac{1}{2}kT$ units of energy per molecule. Then

$$E_{i,\text{int}} = \frac{j}{2} mRT \frac{dN_i}{dA}$$

$$= \frac{\rho(5 - 3\gamma)}{\sqrt{\pi}\,(\gamma - 1)} \left(\frac{RT}{2} \right)^{3/2} \left\{ e^{-(S\sin\theta)^2} + \sqrt{\pi}\,(S\sin\theta)[1 + \text{erf}(S\sin\theta)] \right\} \qquad (27)$$

where j denotes the number of internal degrees of freedom and is related to γ by $j = (5 - 3\gamma)/(\gamma - 1)$.

The effect of reemitted molecules on the pressure, shear stress, and heat flux q is accounted for by considering

$$p = p_i + p_r \qquad (28)$$

$$\tau = \tau_i - \tau_r \qquad (29)$$

$$\frac{q}{A} = E_i - E_r \qquad (30)$$

Expressing the reemitted quantities in terms of the accommodation coefficients defined by Eqs. (14) and (18) and $\sigma' = (p_i - p_r)/(p_i - p_w)$ gives

$$p_r = (1 - \sigma')p_i + \sigma' p_w \tag{31}$$

$$\tau_r = (1 - \sigma)\tau_i \tag{32}$$

$$E_r = (1 - \alpha)E_i + \alpha E_w \tag{33}$$

p_w and E_w denote the fluxes of normal momentum and energy which would be carried by a reemitted stream of molecules in complete equilibrium at the surface temperature T_w.

The relations for p, τ, and q are

$$
p = \frac{\rho u^2}{2S^2}\left\{\left(\frac{2 - \sigma'}{\sqrt{\pi}}\, S \sin\theta + \frac{\sigma'}{2}\sqrt{\frac{T_w}{T_\infty}}\right)e^{-(S\sin\theta)^2}\right.
$$

$$
\left. + \left[(2 - \sigma')\left(S^2 \sin^2\theta + \frac{1}{2}\right) + \frac{\sigma'}{2}\sqrt{\frac{\pi T_w}{T_\infty}}\, S \sin\theta\right]\left[1 + \mathrm{erf}(S \sin\theta)\right]\right\} \tag{34}
$$

$$
\tau = -\frac{\sigma \rho u^2 \cos\theta}{2S\sqrt{\pi}}\left\{e^{-(S\sin\theta)^2} + \sqrt{\pi}\, S \sin\theta\,[1 + \mathrm{erf}(S\sin\theta)]\right\} \tag{35}
$$

$$
\frac{q}{A} = \alpha\rho RT\sqrt{\frac{RT_\infty}{2\pi}}\left\{\left[S^2 + \frac{\gamma}{\gamma + 1} - \frac{\gamma + 1}{2(\gamma - 1)}\frac{T_w}{T_\infty}\right]\left[e^{-(S\sin\theta)^2}\right.\right.
$$

$$
\left.\left. + \sqrt{\pi}\, S \sin\theta\,[1 + \mathrm{erf}(S\sin\theta)]\right] - \frac{1}{2}e^{-(S\sin\theta)^2}\right\} \tag{36}
$$

Heat Transfer to Typical Bodies. In integrating Eq. (36) over the total surface of a body to obtain the overall heat transfer, it is usually necessary to assume T_w and α to be constant to obtain a solution in closed form. For a surface which is completely diffusely reflecting and for which internal heat conduction is rapid, T_w = constant and $\alpha = \sigma = \sigma' = 1$ may be quite accurate. Although no actual surfaces are known to be completely specularly reflecting, characteristics for this extreme case would be predicted by using $\alpha = \sigma = \sigma' = 0$.

The quantities needed to characterize the overall heat transfer are the net heat flow q between the stream and body and the equilibrium surface temperature T_{aw} when $q = 0$. Typical results are (see Refs. 3 to 5):

1. Flat plate at angle of attack β with A the total area of both sides of the plate

$$
T_{aw} = T_\infty\left\{1 + \frac{\gamma - 1}{\gamma + 1}\left[2S^2 + 1 - \frac{1}{1 + \sqrt{\pi}[S \sin\beta\,\mathrm{erf}(S \sin\beta)]\,e^{(S\sin\beta)^2}}\right]\right\} \tag{37}
$$

$$
q = \frac{\alpha A \rho (\gamma + 1)}{\gamma - 1}\sqrt{\frac{RT_\infty}{8\pi}}\, R\,(T_{aw} - T_w)\left\{e^{-(S\sin\beta)^2} + \sqrt{\pi}\,[S \sin\beta\,\mathrm{erf}(S \sin\beta)]\right\} \tag{38}
$$

2. Sphere of diameter D

$$T_{aw} = T_\infty \left[1 + \frac{(\gamma - 1)S^2}{\gamma + 1} - \frac{2S(2S^2 + 1)(S + \text{ierfc } S) + (2S^2 - 1)\text{ erf } S}{2S^2(S + \text{ierfc } S) + \text{erf } S} \right] \quad (39)$$

$$q = \frac{\alpha\rho(\gamma + 1)\pi D^2}{4(\gamma - 1)} \left[\sqrt{\frac{RT_\infty}{2}} \, R \, \frac{T_{aw} - T_w}{S} \left(S^2 + S \text{ ierfc } S + \frac{1}{2}\text{erf } S \right) \right] \quad (40)$$

where

$$\text{ierfc}(x) = \frac{2}{\sqrt{\pi}} \int_0^x dz \int_0^z e^{-t^2} \, dt$$

3. Right circular cylinder of diameter D normal to stream (q per unit length)

$$T_{aw} = T_\infty \left[1 + \frac{\gamma - 1}{\gamma + 1} \frac{(3S^2 + 2S^4) I_0(S^2/2) + (S^2 + 2S^4) I_1(S^2/2)}{(1 + S^2) I_0(S^2/2) + S^2 I_1(S^2/2)} \right] \quad (41)$$

$$q = \frac{\alpha(\gamma + 1)}{\gamma - 1} D\rho \sqrt{\frac{\pi RT_\infty}{8}} \, R(T_{aw} - T_w) e^{-(S^2/2)} \left[(1 + S^2) I_0\left(\frac{S^2}{2}\right) + S^2 I_1\left(\frac{S^2}{2}\right) \right] \quad (42)$$

where I_0 and I_1 are modified Bessel functions.

The above results are presented graphically in Figs. 3 and 4 (from Ref. 3) in terms of a modified recovery factor r' and modified Stanton modulus St'. These are defined by

$$r' = \frac{T_{aw} - T_\infty}{T_s - T_\infty} \frac{\gamma + 1}{\gamma} \quad (43)$$

$$\text{St'} = \frac{q}{A\rho u c_p (T_{aw} - T_w)} \frac{\gamma}{\alpha(\gamma + 1)} \quad (44)$$

Also included are curves for forward and rearward facing flat plates, hemi-cylinders, and hemispheres.

Note that the results for a flat plate at an angle of attack β apply to any surface with the same constant angle of incidence to the free stream. This means that the heat transfer characteristics of a wedge or cone of semiapex angle β will be the same as for a flat plate at this angle of incidence, and the heat transfer rate to a cylinder of any shape axially aligned with the flow will be the same as to a flat plate parallel to the flow.

From Fig. 3 it is apparent that the modified recovery factors approach the same constant value of 2.0 for all geometries. Consequently, for $S \geq 3.0$ it is possible to write $r' = 2.0$ or $r = 2\gamma/(\gamma + 1)$ (5/4 for a monatomic gas and 7/6 for a diatomic gas).

Similarly, from Fig. 4 it can be seen that for higher velocities Stanton numbers also become independent of S or Mach number for all bodies except the flat plate parallel to the flow. Equations for the asymptotic values of the Stanton numbers and the limiting lower Mach numbers are listed in Table 3.

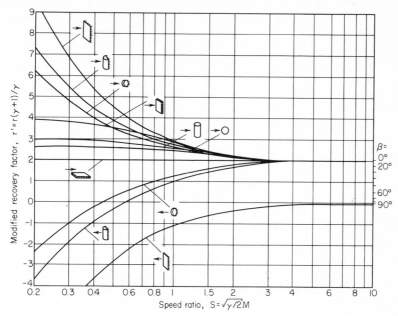

Fig. 3. Modified recovery factors for a flat plate, sphere, and transverse cylinder in free molecule flow [3].

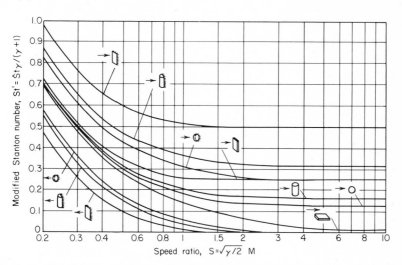

Fig. 4. Modified Stanton numbers for a flat plate, sphere, and transverse cylinder in free molecule flow [3].

Example of Free Molecular Heat Transfer. A polished aluminum sphere 3 feet in diameter is moving at a velocity of 20,000 ft/sec and an altitude of 475,000 feet (145 kilometers). Determine the rate of convective heat transfer to the sphere if its surface temperature is uniform and equal to 100°F. From Fig. 1, $\lambda_\infty = 40$ meters. Therefore $Kn_\infty = 40/(3/3.28) = 43.7$ based on the sphere diameter. This definitely establishes the flow as free molecular. The heat rate is specified as Eq. (44)

TABLE 3. Equations for Stanton Number in Free Molecule Flow

Body geometry	Lower limit of Mach number	Stanton number $(h/\rho u c_p)$
Flat plate parallel to flow	none	$0.169\alpha\,(\gamma + 1)/\gamma M$
Cone	$M = 5$	$0.50\alpha\,\sin\beta\,(\gamma + 1)/\gamma^*$
Cylinder	$M = 3$	$0.16\alpha\,(\gamma + 1)/\gamma$
Sphere	$M = 3$	$0.125\alpha\,(\gamma + 1)/\gamma$
Stagnation Point	$M = 1.8$	$0.50\alpha\,(\gamma + 1)/\gamma$

*Applicable only for $\beta > 10°$

$$q = \frac{St'\rho u_\infty c_p A\alpha\,(T_{aw} - T_w)(\gamma + 1)}{\gamma}$$

or

$$q = St\rho u_\infty c_p A\alpha\,(T_{aw} - T_w)$$

From Eq. (43)

$$T_{aw} - T_\infty = r'\frac{\gamma}{\gamma + 1}(T_s - T_\infty)$$

from which

$$T_{aw} = r'(T_s - T_\infty)\frac{\gamma}{\gamma + 1} + T_\infty = r'\frac{u_\infty^2}{2g_c c_p J}\frac{\gamma}{\gamma + 1} + T_\infty$$

For the very high velocity involved it is permissible to neglect T_w in comparison with T_{aw}. On the same basis T_∞ can be neglected in calculating T_{aw} from Eq. (43). Therefore

$$T_{aw} \simeq r'\left(\frac{u_\infty^2}{2g_c c_p J}\right)\frac{\gamma}{\gamma + 1}$$

and

$$q \simeq St'r'\alpha A\rho_\infty\frac{u_\infty^3}{2g_c J}$$

or

$$q \simeq \frac{St r\alpha A\rho_\infty u_\infty^3}{2g_c J}$$

$$S = \frac{u_\infty}{\sqrt{2RT_\infty}}$$

$$= \frac{20,000}{\sqrt{2\,(32.2)(53.3)1440}} \qquad (T_\infty \text{ and } \rho_\infty \text{ from ref. 1})$$

$$= \frac{20,000}{2220} = 9.0$$

From Fig. 3

$$r' = 2.0$$

From Fig. 4

$$St' = 0.125$$

From Table I, α is taken as 0.9. Then

$$\frac{q}{A} \simeq \frac{(0.125)(2.0)(0.9)(1.55 \times 10^{-10})(20 \times 10^3)^3}{2\,(32.2)(778)} = 0.0056\ \text{Btu/ft}^2\text{sec}$$

The radiant energy from the sun incident on a surface normal to the incoming radiation is about 0.124 Btu/sec-ft^2. A polished aluminum surface might absorb about 5 percent of this, or 0.0062 Btu/sec-ft^2. The free molecular aerodynamic heating rate is seen to be of comparable magnitude.

TRANSITIONAL FLOW

Subsonic. A procedure recommended for calculating heat transfer in subsonic transitional flow has been proposed by Sherman [6]. This method has been very successful in correlating available data. It is based on first nondimensionalizing the quantity of interest in such a manner that the nondimensional coefficient, say $F(M, \text{Re})$, becomes independent of Re in the free molecular limit. For heat transfer this coefficient is usually the Stanton number. The free molecular limit defined by

$$F_{fm}(M) = \lim_{\text{Re}\to 0} F(M, \text{Re}) \tag{45}$$

and the continuum limit

$$F_c(\text{Re}) = \lim_{M\to 0} F(M, \text{Re}) \tag{46}$$

must both be obtained. The former is often available from theory, the latter from experiment or theory. Sherman then plots results in the form of $F(M, \text{Re})/F_{fm}(M)$ as a function of F_c/F_{fm} and finds excellent correlation of data for various Reynolds numbers and subsonic Mach numbers. (F_c/F_{fm} is essentially a modified Knudsen number.) When there are insufficient data available to construct such a curve, the interpolation formula

$$\frac{F(M, \text{Re})}{F_{fm}} = \frac{1}{1 + (F_{fm}/F_c)} \tag{47}$$

is used.

For heat conduction between stationary *concentric cylinders* of radii R_1 and R_2 ($R_2 > R_1$) this interpolation formula gives

$$\frac{q}{q_{fm}} = \frac{1}{1 + \alpha(4/15)(R_1/\lambda)\,\ln(R_2/R_1)} \tag{48}$$

when the relative temperature difference between the cylinders is small, and q is the heat transfer per unit area. The corresponding problem for stationary *concentric spheres* gives

$$\frac{q}{q_{fm}} = \frac{1}{1 + \alpha(4/15)(R_1/\lambda)(1 - R_1/R_2)} \tag{49}$$

For both cases

$$q_{fm} = \alpha\left(\frac{2}{\pi}\right)^{1/2} \bar{\rho}(R\bar{T})^{1/2} R(T_1 - T_2) \tag{50}$$

where $\bar{\rho}$ and $\bar{T}$ are the mean density and temperature respectively. Both of these results have been verified by approximate kinetic theory methods.

The appropriate expression for the case of *parallel walls* is obtained from either of the above equations if $R_2 - R_1$ remains finite while $R_1 \to \infty$. This yields

$$\frac{q}{q_{fm}} = \frac{1}{1 + \alpha(4/15)(R_2 - R_1)/\lambda} \tag{51}$$

Results for subsonic convective heat transfer to a sphere are shown in Fig. 5a. A curve for the continuum Stanton number is plotted in Fig. 5b.

Example of Calculating Subsonic Heat Transfer. Helium flows at a Mach number of 0.5 and a Reynolds number of 10 about a sphere. The Stanton number is required for these conditions assuming a thermal accommodation coefficient of 0.9.

Since helium is monatomic the speed ratio is given by

$$S = \sqrt{\frac{\gamma}{2}}\, M = \sqrt{\frac{5}{6}}\,(0.5) = 0.456$$

From Fig. 4 for this speed ratio

$$St'_{fm} = 0.315$$

and hence

$$St_{fm} = \frac{\alpha(\gamma + 1)}{\gamma}\, St'_{fm} = (0.9)\left(\frac{8}{5}\right)(0.315) = 0.454$$

This would be the value for Re $\to$ 0 and M = 0.5.

From Fig. 5b for continuum conditions, i.e., Re = 10 and M $\to$ 0,

$$St_c = 0.466$$

Hence, $St_{fm}/St_c = 0.974$ and from Eq. (47)

$$St = \frac{0.454}{1 + 0.974} = 0.230$$

Supersonic Flow Around Cylinders. Hot wire measurements in low-density streams have provided recovery factor and heat transfer data which span the entire region between free molecule and continuum flow. A summary of results for both subsonic

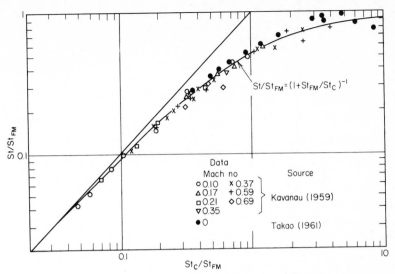

Fig. 5a. Heat transfer to spheres in subsonic flow (according to Sherman [6]).

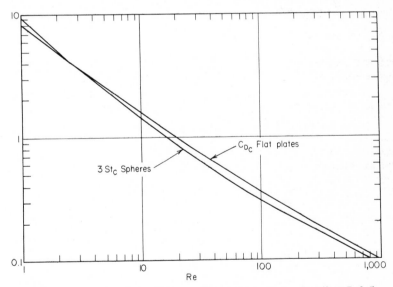

Fig. 5b. Heat transfer and skin friction coefficients for continuum flow (from Ref. 6).

and supersonic flow is shown in Fig. 6 [7] in terms of the average Nusselt number versus the stagnation Reynolds number. Here $Nu_s = hD/k_s$ and $Re_s = \rho_\infty u_\infty D/\mu_s$ where the subscript s denotes stagnation conditions. Note that for near continuum flow when a shock wave will form in front of the cylinder with supersonic flow, $\rho_\infty u_\infty = \rho_2 u_2$, $k_s \simeq k_2$, and $\mu_s = \mu_2$, where the subscript 2 denotes conditions just downstream of the shock at the stagnation streamline.

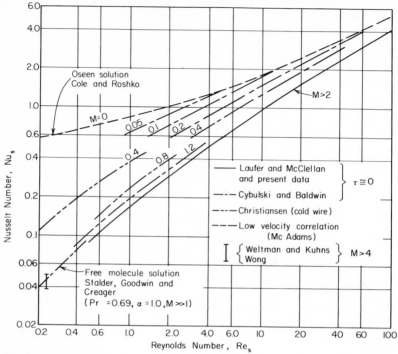

Fig. 6. Average Nusselt number variation for a cylinder in the transition region (according to Dewey [7]).

The Nusselt number and the recovery temperature ratio for a cylinder normal to a free molecule flow can be expressed as

$$\text{Nu}_s = \frac{\gamma - 1}{2\pi^{3/2}} \alpha \text{Re}_s \text{Pr}_s \frac{G(S)}{S} \tag{52}$$

$$\eta_f = \frac{T_{aw}}{T_s} = \left(1 + \frac{\gamma + 1}{2} M^2\right) \frac{F(S)}{G(S)} \tag{53}$$

where the functions $F(S) G(S)$ depend on the speed ratio S and the number of excited degrees of freedom of the gas. As $M \to \infty$, Eqs. (52) and (53) approach the limiting forms (Figs. 3 and 4)

$$\text{Nu}_s = \left(\frac{\gamma + 1}{2\gamma\pi}\right) \alpha \text{Re}_s \text{Pr}_s \tag{52a}$$

and

$$\eta_f = \frac{T_{aw}}{T_s} = \frac{2\gamma}{\gamma + 1} \tag{53a}$$

Thus as for continuum flow the Nusselt number and recovery temperature ratio become independent of Mach number as M becomes large. This suggests that for hypersonic flow Nu_s should be a function of Re_s only. This prediction is verified by the data. Note that for $\text{Re}_s \sim 100$ the data tend to fair into the correlation curve for low speed results.

The independence of the recovery temperature ratio in the free molecule and continuum regions suggests that a normalized recovery ratio

$$\bar{\eta} = \left.\frac{\eta - \eta_c}{\eta_f - \eta_c}\right|_{M \gg 1} \tag{54}$$

is a unique function of the Knudsen number for M sufficiently large. The available data presented in this manner in Fig. 7 [7] bear this out.

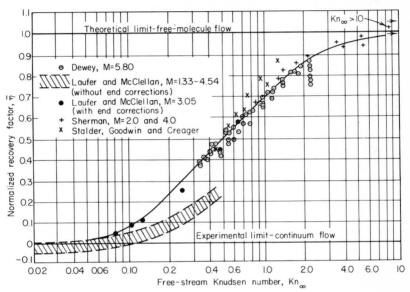

Fig. 7. Variation of normalized recovery temperature with Knudsen number (according to Dewey [7]).

Supersonic Flow Around Spheres. Data for heat transfer to spheres in a supersonic rarefied flow are limited and do not extend to the free molecule end of the transition region. As indicated in Fig. 8 (based on [8]) the available data (in terms of the shock Nusselt numbers) tend to identify a limiting curve which fairs into the free molecule flow predictions at low Reynolds numbers and into the continuum low-speed correlation curve at high Reynolds numbers (curves representing subsonic data are included for comparison).

The upper curves in Fig. 8 which approach the limiting Nusselt number value of 2.0 were obtained by solution of the continuum energy equation assuming constant properties, slip, and temperature jump boundary conditions. Reasonable agreement with the data results at Reynolds numbers above 100. For lower values, however, it appears that additional rarefaction effects will have to be considered in theoretical analyses.

Measurements indicate that the overall recovery factor for spheres depends primarily on Reynolds number and is insensitive to Mach number. For Reynolds numbers less than 600, recovery factor measurements in air increase from a value of 0.92 to above 1.0 and indicate a trend toward the theoretical value of 1.17 for free molecular flow.

The experimental results presented in Figs. 5 to 8 are for conditions in which the temperatures and temperature differences between the free stream and model were moderate. Information for ultra-high speeds and large temperature differences is not

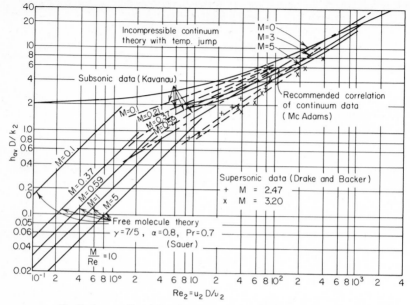

Fig. 8. Average Nusselt number variation for a sphere in the transition region.

available. It should be noted, however, that such phenomena as dissociation, ionization, and surface accommodation variation may have substantial effects.

SUPERSONIC NEAR CONTINUUM FLOW

Local Heat Transfer Distribution Around a Cylinder. The difficulty in uniquely defining the flow of a rarefied gas stream (i.e., whether it is in the free molecule, transition, or continuum region) is illustrated by considering the conditions which exist locally around a body. Examples of the pressure, recovery factor, and heat transfer distributions around a cylinder normal to a rarefied air stream are shown in Figs. 9 to 11 (from Ref. 9). These measurements were made in a low-density wind tunnel for which the stagnation temperature and pressure were around 70°F and 2 psi, respectively. At the outer edge of the boundary layer at the cylinder stagnation point, the mean free path is of the order of 0.5×10^{-3} inch. The Knudsen number based on the model diameter of 0.5 inch is thus 10^{-3} or 10^{-2} based on the stagnation point shock standoff distance $\Delta \simeq 0.05$ inch. In contrast on the rear half of the cylinder the pressure is equal to or less than the static pressure in the approaching stream. The mean free path is about 50 times longer and the Knudsen number, λ/Δ, is ~ 0.05.

The above values indicate that, at the stagnation point, and over a substantial part of the front half of the cylinder, departure from continuum flow will be negligible. Near the 90° location, however, slip and temperature jump effects would be expected. In this region the pressure and heat transfer coefficient have decreased to around 10 percent of their stagnation point values. The contribution of the portion of the cylinder from the 90° to the 180° region to the overall drag and heat transfer will consequently be small. From a practical point of view, therefore, high accuracy in this region is not critical.

An empirical formula for the pressure coefficient distribution is given in Fig. 9. This expression provides a better representation of measured distributions up to Mach numbers of 6.0 than does the modified Newtonian relation

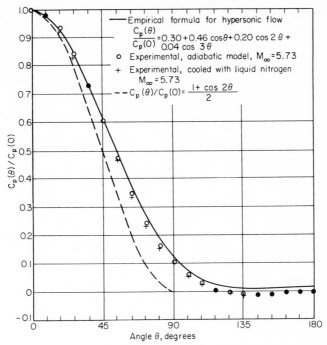

Fig. 9. Pressure variation around a cylinder normal to a supersonic rarefied air stream [9].

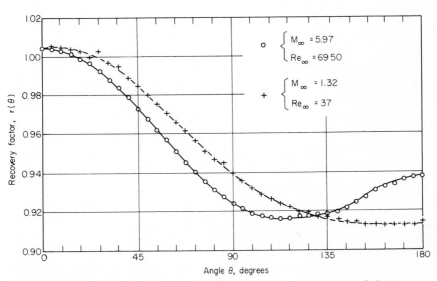

Fig. 10. Recovery factor distributions around a cylinder normal to a supersonic rarefied air stream [9].

$$\frac{C_p(\theta)}{C_p(0)} = \frac{1 + \cos 2\theta}{2} \tag{55}$$

where $C_p(\theta)$ is defined as $(p_w - p_\infty)/(1/2)\rho_\infty u_\infty^2$. The comparison is shown in Fig. 9.

The recovery factors shown in Fig. 10 are defined in terms of the local adiabatic wall temperatures and the free-stream temperature, i.e., $r(\theta) = [T_{aw}(\theta) - T_\infty]/(T_s - T_\infty)$. Measured values decrease from around 1.0 at the stagnation point to a value around 0.92 over most of the rear half of the cylinder. Distributions are relatively insensitive to Reynolds and Mach numbers. Although the data points in Fig. 10 indicate that the recovery factor is slightly greater than 1.0 near the stagnation point, this is considered to be due to small variations in test conditions rather than the incipience of rarefaction effects.

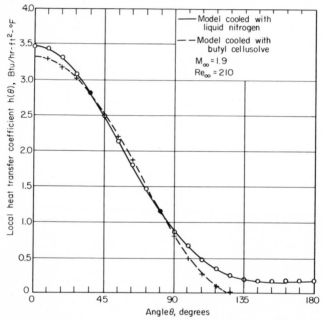

Fig. 11. Local heat transfer coefficient distributions around a cylinder normal to a supersonic rarefied air stream [9].

The heat transfer coefficient distribution in terms of the stagnation point value has been shown to be independent of the free stream Mach and Reynolds numbers throughout the Mach number range of 1 to 6 and the Reynolds number range of 35 to 4100. An empirical equation describing the ratio of the local heat transfer coefficient to the stagnation point value is given in Fig. 12. This expression represents the data to within ± 6 percent.

The recommended relation for predicting the stagnation point values is presented in Fig. 13. Here the fluid properties are evaluated at conditions just behind the normal shock. Dividing the Nusselt number by the square root of the nondimensional velocity gradient at the edge of the boundary layer, $(\tilde{u}_e')^{1/2} = [d(u_e/u_2)/d(x/D)]^{1/2}$, accounts for the Mach number effect. $\tilde{u}_e'$ can be evaluated from Eq. (55) or the empirical equation given in Fig. 9.

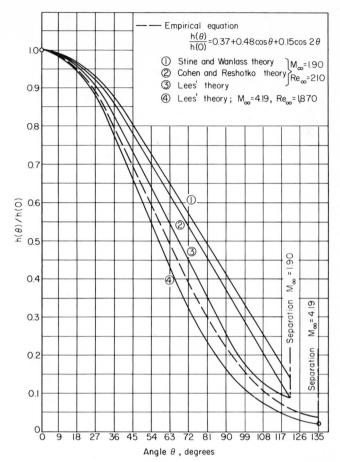

Fig. 12. Heat transfer coefficient ratio around a cylinder normal to a supersonic rarefied air stream [9].

As can be seen in Fig. 12, the variation of the local heat transfer coefficient around a cylinder in these Reynolds and Mach number ranges is satisfactorily predicted (particularly on the forward half) by a method developed by Lees [10]. This result is probably fortuitous, however. His analysis is based on continuum flow and includes simplifying assumptions which apparently have appropriate corrective or over-corrective effects. For example, the theoretical curve (labeled 4) for $M = 4.19$ begins to drop below the experimental results at $\theta \simeq 27°$. Rarefaction effects, which are not included in the theory, would be expected to cause the measured heat transfer to be lower than predicted for continuum flow. More appropriate analytical results are not available.

The Hemisphere Cylinder. The flow between the shock wave and an axially symmetric body in a supersonic stream can be divided into inviscid and viscous regions. Under continuum conditions the viscous (boundary layer) region is very thin compared to the inviscid. As a result the viscous shear forces near the surface are dominant. As the density of the flow decreases, however, the boundary layer thickens. The shear stresses parallel to the surface decrease, and the velocity gradients near the outer edge of the boundary layer can be of the same order as the vorticity introduced into

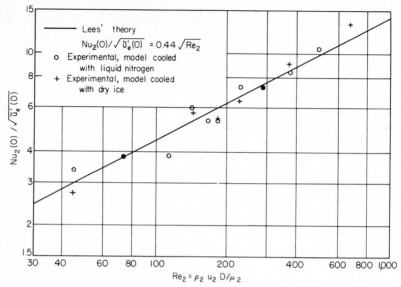

Fig. 13. Heat transfer coefficient at stagnation point of a cylinder normal to a supersonic rarefied air stream [9].

the inviscid flow due to the shock wave. The vorticity will then influence the heat transfer. The effect can be accounted for by using the vorticity in the inviscid flow region as an outer boundary condition for the boundary layer. An alternate procedure is to treat the flow between the body surface and the shock wave as a single region in which viscosity plays an important role. The influence of other second order effects such as slip and temperature jump, boundary layer displacement of the flow, and body curvature must also be considered [11 and 12]. For air in equilibrium at temperatures below the dissociation level the heat transfer can be expressed by [11]

$$\frac{Nu_2}{\sqrt{\widetilde{u}'_e}} = 0.67 \sqrt{Re_2} \tag{56}$$

where properties are evaluated at conditions downstream of the normal shock. This relation is plotted in Fig. 14. Experiments have shown that the modified Newtonian pressure distribution $(p_w - p_\infty)/(p_s - p_\infty) = \cos^2 \theta$ is satisfied. In this case the non-dimensional velocity gradient at outer edge of the boundary layer $\widetilde{u}'_e$ is equal to

$$\widetilde{u}'_e = \frac{2}{u_2} \sqrt{\frac{2(p_s - p_\infty)}{\rho_s}} \tag{57a}$$

Expressing this in terms of M_∞ yields

$$\widetilde{u}'_e = 2 M_\infty (\gamma + 1) \left\{ \left[\frac{M_\infty^2 - 1}{1 + (\gamma - 1)/2 M_\infty^2} \right] \left[\frac{1}{2\gamma M_\infty^2 - (\gamma - 1)} \right] \right\}^{1/2} \tag{57b}$$

For $M_\infty \to \infty$, $(\widetilde{u}'_e)^{1/2} \to 2.52$, when $\gamma = 1.4$.

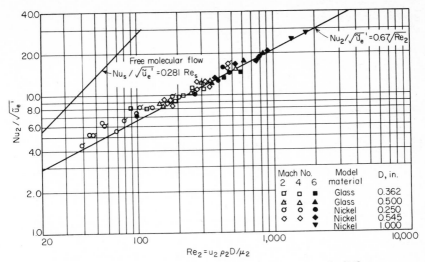

Fig. 14. Near continuum supersonic stagnation point heat transfer [11].

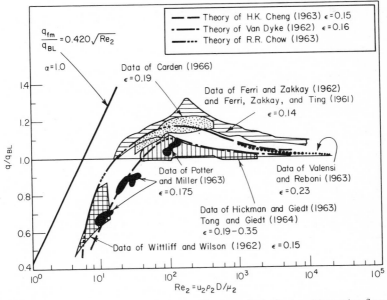

Fig. 15. Summary of hypersonic stagnation point heat transfer in the transition region (based on Refs. 11 and 12.)

At high temperatures the effects of viscosity variation, dissociation, diffusion, surface catalytic efficiency, and nonequilibrium must be considered. Present information is not adequate to account accurately for these factors. However significant progress has been made. For example, assuming equilibrium conditions, numerical solution of the continuum boundary layer equations [13] indicated that heat transfer depended only on the total variation of $\rho\mu$ across the boundary layer, and the Lewis number. The recommended empirical result for the stagnation point was

$$\frac{Nu_2}{\sqrt{\widetilde{u_e'}}} = 0.67\sqrt{Re_2}\left(\frac{\rho_s\mu_s}{\rho_w\mu_w}\right)^{0.4}\left[1 + (L^{0.52} - 1)\left(\frac{h_D}{h_s}\right)\right] \tag{58}$$

where h_D, the dissociation enthalpy per unit mass in the external flow, is given by

$$h_D = \sum_i c_{is}(-h_i^0) \tag{59}$$

For a Prandtl number of 0.71 the expression for the heat rate is then

$$\frac{q}{A} = 0.94(\rho_w\mu_w)^{0.1}(\rho_s\mu_s)^{0.4}\left[1 + (L^{0.52} - 1)\left(\frac{h_D}{h_s}\right)\right](h_s - h_w)\sqrt{\widetilde{u_e'}\left(\frac{D}{u_2}\right)} \tag{60}$$

Although available shock tube measurements tend to confirm this result, assessment of its accuracy must await precise determination of viscosity variation at high temperatures.

The curve for free molecule flow in Fig. 14 is based on assuming $\alpha = 1$ and that incoming molecules are reemitted at the free-stream temperature. q_{fm}/A then equals $(1/2)\rho_\infty u_\infty^3$. Letting $T_s - T_\infty = (1/2)(u_\infty^2/c_p)$, the average heat transfer coefficient $h = u_\infty\rho_\infty c_p$. Rearranging, introducing $\sqrt{\widetilde{u_e'}} = 2.52$, and substituting $Pr = 0.71$ yields

$$\frac{Nu}{\sqrt{\widetilde{u_e'}}} = 0.281\,Re_s \tag{61}$$

in which the subscript s indicates properties are evaluated at stagnation conditions.

Data for Mach numbers of 2, 4, and 6 included on Fig. 14 begin to deviate from the continuum curve around $Re_2 = 400$. Whereas this is in agreement with some theoretical studies, other experimental and analytical results indicate deviation from the continuum curve at higher Reynolds numbers. Figure 15 summarizes the present status of hypersonic stagnation point heat transfer in the transition region compared with continuum boundary layer curve extrapolated to low Reynolds numbers. The increase due to second order effects appears to rise to a maximum and then decrease as free molecule flow is approached. The disagreement about the magnitude of these effects is at present unresolved. The curve for free molecule flow represents the ratio of Eqs. (61) to (56), and is thus based on $\gamma = 1.4$ and $Pr = 0.71$.

Measurements of the local recovery factor around a hemisphere-cylinder indicate a variation similar to that for a cylinder normal to the flow (see Fig. 16). For Reynolds numbers (Re_2) down to approximately 50 and Mach numbers up to 6, $r(\theta)$ defined as $[T_{aw}(\theta) - T_\infty]/(T_s - T_\infty)$ decreases from 1.0 at the stagnation point to around 0.93 at the 90° location. The curve in Fig. 16 represents available measurements to within 2 percent.

Local heat-rate distributions compared with the stagnation point value are shown in Fig. 17. They can be satisfactorily predicted by Lees' theory [10] except possibly near the shoulder. Heating rates at the 90° location are between 10 and 20 percent of the stagnation point value.

Example of Heat Transfer to a Hemisphere in the Transition Region. A hemisphere-capped cylinder is traveling through the atmosphere at an altitude of 300,000 feet with a velocity of 5,000 ft/sec. Specify the aerodynamic heat transfer distribution to the hemisphere cap if the cylinder is 3 feet in diameter and the direction of motion is parallel to its axis. The surface temperature is uniform and equal to 100°F.

The free-stream Mach number is determined from

$$M = \frac{u}{a} = \frac{u}{\sqrt{\gamma g_c R T_\infty}} = \frac{5{,}000}{\sqrt{1.4\,(32.2)53.3(333)}} = \frac{5{,}000}{895} = 5.59$$

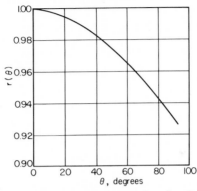

Fig. 16. Recovery factor variation on a hemisphere (based on Ref. 11).

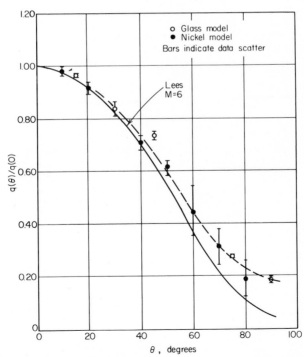

Fig. 17. Heat transfer distribution around a hemisphere for $M_\infty = 6$ [11].

From Ref. 14 for $\gamma = 1.4$

$$\frac{T_s}{T_\infty} = 7.25$$

Since T_∞ at this altitude (1) is $333°R$, $T_s = 7.25(333) = 2415$. This is not high enough to cause dissociation of the air molecules; hence Eq. (56) is applicable at the stagnation point. However, there is significant variation of the specific heat capacities of air in the range between 333 and $2415°R$. Correction of the stagnation or total temperature for this (according to Ref. 14) leads to $T_s = 2265°R$. The temperature ratio across the shock at the stagnation point is found to be $T_2/T_\infty = 7.016 (0.948) = 6.65$, and $T_2 = 2215°R$. From the Sutherland relation $\mu_2 = 9.97 \times 10^{-7}$ lb-sec/ft^2.

Since $\rho_\infty u_\infty = \rho_2 u_2$

$$\mathrm{Re}_2 = \rho_\infty u_\infty \frac{D}{\mu_2} = \frac{1.488 \times 10^{-7}(5,000)3}{(9.97 \times 10^{-7})(32.2)} = 69.5$$

Equation (57a) gives $\tilde{u}'_e = 5.65$. Equation (56) then yields

$$\mathrm{Nu}_2 = 0.67\sqrt{(69.5)(5.65)} = 13.25$$

and

$$h_s = \mathrm{Nu}_2 k_2/D = 13.25\,(0.051)/3 = 0.225 \ \mathrm{Btu/ft^2\text{-}hr\text{-}°R}$$

The stagnation point heating rate is

$$\frac{q_{oBL}}{A} = h_s(T_{aw} - T_w)$$

The recovery factor $r(\theta)$ is 1.0 and $T_{aw} = 333 + 1.0(2265 - 333) = 2265°R$. Thus

$$\frac{q_{oBL}}{A} = 0.225\,(2265 - 560) = 384 \ \mathrm{Btu/ft^2 hr}$$

From Fig. 15 at $\mathrm{Re}_2 = 69.5$ results of the measured and predicted effect of vorticity on the stagnation point heating rate vary from 0 to 20 percent. Taking the average of 10 percent, $q_o = 1.1\, q_{oBL} = 422 \ \mathrm{Btu/ft^2}$-hr.

The heating rate at any θ is then calculated by multiplying values of q_θ/q_o obtained from the dashed curve in Fig. 17 by q_o. At $\theta = 60°$, for example, $q = 0.44\,(422) = 186$ Btu/ft^2-hr.

REFERENCES

1. *U.S. Standard Atmosphere,* Government Printing Office, Washington, 1962.
2. L. B. Thomas, in "Rarefied Gas Dynamics" (C. L. Brundin, ed.), vol. 1, p. 155, Academic Press, Inc., 1967.
3. A. K. Oppenheim, *J. Aeronaut. Sci.,* 20:49(1953).
4. S. A. Schaaf and P. L. Chambré, Flow of Rarefied Gases, in "Fundamentals of Gas Dynamics," p. 687, Princeton University Press, Princeton, N. J., 1961.
5. S. A. Schaaf, Mechanics of Rarefied Gases, in "Handbuch der Physik," vol. 8, no. 2, p. 591, Springer, Berlin, 1962.
6. F. S. Sherman, in "Rarefied Gas Dynamics" (J. A. Laurmann, ed.), vol. 2, p. 228, Academic Press, Inc., New York, 1963.
7. C. F. Dewey, Jr., *J. ARS,* 31:1709(1961).

8. R. M. Drake and G. H. Backer, *Trans. ASME,* **74**:1241(1952).
9. O. K. Tewfik and W. H. Giedt, *J. Aerospace Sci.,* **27**:722(1960).
10. L. Lees, *Jet Prop.,* **26**:259(1956).
11. R. S. Hickman and W. H. Giedt, *AIAA Journal,* **1**:665(1963).
12. J. L. Potter, in "Rarefied Gas Dynamics" (C. L. Brundin, ed.), vol. 2, p. 881, Academic Press, Inc., New York, 1967.
13. J. A. Fay and F. A. Riddell, *J. Aeronaut. Sci.,* **25**:73(1958).
14. "Equations, Tables and Charts for Compressible Flow," *NACA Rept.* 1135, 1953.

NOMENCLATURE

a acoustic velocity
A area
c mass concentration
E energy flux
f distribution function
g_c dimensional constant
h enthalpy
h^o enthalpy of formation
j number of internal degrees of freedom
k Boltzmann's constant
Kn Knudsen number
L characteristic length
m mass of molecule
M Mach number
n number density
N number flux
p pressure
Pr Prandtl number
q energy flux
R gas constant per unit mass
Re Reynolds number
r vector position
r recovery factor
r' modified recovery factor (Eq. 43)
S speed ratio
St Stanton number
St′ Modified Stanton number (Eq. 44)
T temperature
t time
u mean (macroscopic) vector velocity
x, y, z position coordinates
α thermal accommodation coefficient
β angle of attack
γ ratio of specific heats
η ratio of T_{aw} to T_s (Eq. 53)
λ mean free path
μ viscosity
ξ molecular vector velocity
ξ_x, ξ_y, ξ_z molecular velocity components
ρ mass density
σ molecular diameter
σ tangential momentum accommodation coefficient
σ' normal momentum accommodation coefficient
τ viscous stress

Subscripts

aw adiabatic wall
BL boundary layer
 i incident
 i ith species
 n normal
 r reflected
 s stagnation
 t tangential
 w wall
 2   behind normal shock
 ∞ free stream

Section **10**

Techniques to Augment Heat Transfer

ARTHUR E. BERGLES
Professor of Mechanical Engineering,
Georgia Institute of Technology

A. INTRODUCTION

Most of the ever-increasing research effort in heat transfer is devoted to analyzing what might be called the standard situation. However, the development of high-performance thermal systems has also stimulated interest in methods to augment or intensify heat transfer. The performance of conventional heat exchangers can be substantially improved by a number of augmentative techniques. On the other hand, certain systems, particularly those in space vehicles, may require an augmentative device for successful operation. In recent years, a great deal of research effort has been devoted to developing apparatus and performing experiments to define the conditions under which an augmentative technique will improve heat (and mass) transfer. The more effective and feasible techniques have graduated from the laboratory to full-scale industrial equipment.

The first goal of this work is to briefly survey the literature pertinent to each augmentative technique so as to provide preliminary guidance to potential users. This survey is abstracted from a much more extensive work recently prepared [1].

It does not appear to be possible to establish a generally applicable selection criterion for the use of augmentative techniques due to the large number of factors which enter into the ultimate decision. Most of the considerations revolve around economics: development cost, initial cost, operating cost, maintenance cost, etc.; but there are other factors such as reliability and safety. In any event, due to the exploratory nature of most of the studies reported herein, all of the necessary information is not available. For example, pumping power and transducer power data are generally not reported for vibration studies.

As a preliminary guide to the selection of a technique, much of the data for turbulence promoters will be presented in terms of a criterion relevant to the pumping power or operating cost; this is a major consideration since no external power is required. The first three augmentative techniques—surface promoters, displaced promoters, and vortex flow—include those geometrical configurations generally referred to as turbulence promoters. Since both heat transfer and pressure-drop data are usually available for single-phase flow in channels, it is possible to compare the performance of the augmented channel and the standard channel at equal pumping power. Correlations for friction and heat transfer data with turbulence promoters are summarized in Sec. 7.A.2.f. Consider the typical data for turbulence promoters noted in Fig. 1. The promoters produce a sizeable elevation in the heat

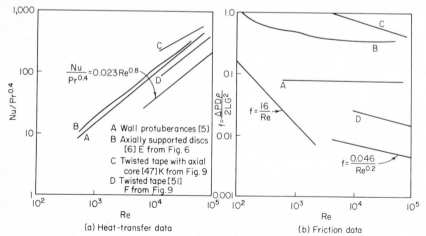

(a) Heat-transfer data   (b) Friction data

Fig. 1. Typical data for turbulence promoters inserted inside tubes.

transfer coefficient at constant velocity; however, there is generally a greater percentage increase in the friction factor. A partial answer to the question of whether it pays to use the promoter is provided by the following ratio

$$\left(\frac{h_a}{h_o}\right)_{P,D,L} = f \text{ (Re, Pr, promoter geometry)} \tag{1}$$

This ratio, which can be easily computed and plotted as a function of Re and geometry (assuming negligible change in Pr), provides an indication of the benefit of modifying the existing equipment. Assume, for example, that $\left(h_a/h_o\right)_P = 2$ for a particular promoter geometry. For a given exchanger, with the surface under consideration controlling the heat transfer, the heat transfer rate could be doubled for a constant temperature difference, or for the same q, the ΔT could be halved.

B. SURFACE PROMOTERS

1. Surface Roughness with Single-phase Flow

Surface roughness was one of the first techniques to be considered seriously as a means of augmenting forced-convection heat transfer. Initially it was speculated that elevated heat transfer coefficients might accompany the relatively high friction factors characteristic of rough conduits. However, since the commercial roughness is not well defined, artificial surface roughness has been employed. Although the enhanced heat transfer with surface promoters is often due in part to the fin effect, it is difficult to separate the fin contribution. For the data discussed here, the promoted heat transfer coefficient is referenced to the base or nominal surface area. Discussions of fins, which would be classed as large-scale protuberances, are contained in standard reference works as well as in Sec. 7.D.3.

Several studies have considered machined roughness on the inside of tubes. As shown in Fig. 2, the data of Cope [2] indicate that the roughest knurl is the most

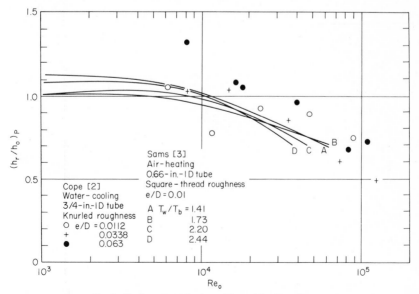

Fig. 2. Performance of tubes with machined roughness.

favorable; but in general the performance is improved over that of a smooth tube only at low flow rates. The data of Sams [3] shown in this figure further suggest that machined roughness is not particularly effective.

A unique sand-grain-type roughness was produced in tubes, and extensively investigated by Dipprey and Sabersky [4]. The tubes were fabricated by electrolytic deposition of nickel on a mandrel coated with sand grains. The mandrel was subsequently dissolved, leaving a tube with an inner surface resembling a close-packed sand grain roughness. Figure 3 indicates that this type of roughness has excellent characteristics, with $\left(h_r/h_o\right)_P$ approaching 2 for the higher Prandtl numbers.

Nunner [5] and Koch [6] considered rings of a rectangular or half-round shape (Fig. 1) inserted at various spacings along a tubular test section. It was found [1] that the optimum performance $\left(\left(h_r/h_o\right)_P \sim 1.5\right)$ was obtained with a spacing-to-protuberance height of about 10. Nagaoka and Watanabe [7] tested coiled wire

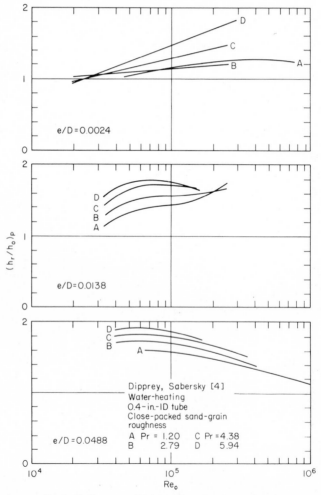

Fig. 3. Performance of tubes with sand-grain roughness.

inserts, which are relatively easy to fabricate, and obtained data which exhibit similar performance characteristics. Edwards and Sheriff [8] investigated increases in h and f in the vicinity of single wires placed in a rectangular channel. It was concluded that the wire height must exceed the laminar sublayer thickness before it becomes effective. It appears that more studies of this type will be needed to establish optimum promoter geometries. Mantle [9] has suggested that flow visualization techniques may be sufficient to determine optimum promoter configurations in this type of geometry.

The frequently used annular geometry is generally more suitable for the application of surface roughness. Machined surfaces are relatively easy to produce, and increased friction affects only a portion of the wetted surface. The results of Kemeny and Cyphers [10] for a helical groove and a helical protuberance are given in Fig. 4. The

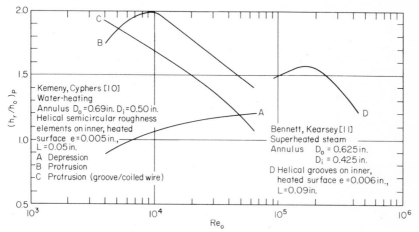

Fig. 4. Performance of annuli with roughness.

grooved surface is not effective in general, although there is a tendency to improve with increasing Re. The inferior performance of the coiled wire assembly compared to the integral protrusion is probably due to poor thermal contact between the wire and the groove. The results of Bennett and Kearsey [11] for superheated steam flowing in an annulus are included in Fig. 4. The data of Brauer [12] suggest that the optimum L/e for an annular geometry is about 3.

Durant et al. [13] summarized an extensive investigation of heat transfer in annuli with the inner, heated tube roughened by means of diamond knurls. It was found that heat transfer coefficients for the knurled annuli were up to 75 percent higher than those of the smooth annuli when compared at equal pumping power.

A comprehensive investigation of rough surfaces in complex geometries is summarized by Draycott and Lawther [14]. Annuli were used to survey the friction and heat transfer characteristics of 21 machined and wire-wound heater elements. Certain of these surfaces were apparently quite favorable from a performance standpoint. Some of the surfaces were apparently quite favorable from a performance standpoint. Kjellström and Larsson [15] reported a study of annuli having 24 different roughness geometries. The annular data were transformed to the rod cluster geometry, and correlated by analogy expressions.

Due to the wide variety of geometrical arrangements, it has not been possible to develop a unified theory or general correlation for the effects of surface roughness on single-phase heat transfer. Analogy solutions have been found useful for relating h to f

for particular types of roughness [4, 5, 15]. Analogy correlations for a variety of roughness configurations have not been particularly successful [16].

The preceding discussion indicates that certain types of roughness can be of considerable benefit in improving heat transfer performance. Under nonuniform flow or thermal conditions, however, it may be advantageous to roughen only that portion of the heating surface which has a higher heat flux or lower heat transfer coefficient. In many cases the overall pressure drop will not be greatly affected by roughening the hot spot. Any of the foregoing roughness types are then of interest since they produce large increases in heat transfer coefficient over the smooth tube value at equal flow rates. The partial roughening technique has been considered for gas-cooled reactors where the thermal limit is reached only in the downstream portion of the core due to the axial heat flux variation [17]. One scheme for achieving selective roughening in large diameter pipes involves sandblasting through a smaller tube which is transversed inside the pipe [18].

2. Surface Promoters in Boiling and Condensation

As discussed in the chapter on boiling heat transfer, surface material and finish have a strong effect on nucleate and transition pool boiling; thus optimum surface conditions might be selected for a system operating in these boiling regimes. Certain types of fouling and oxidation, apparently those which improve the wettability, produce significant increases in pool boiling critical heat flux (CHF). Surface treatment has also been used to remedy the notorious nucleation instability of wetting liquid metals.

A novel technique for promoting nucleate boiling has been proposed by Young and Hummel [19]. Spots of Teflon or other nonwetting material, either on the heated surface or in pits, were found to promote nucleation as shown in Fig. 5. Relatively

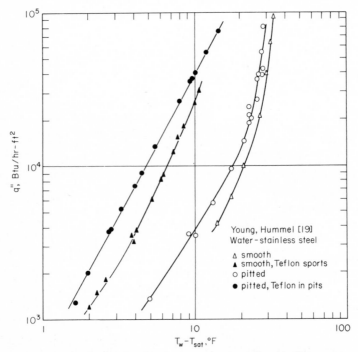

Fig. 5. Influence of surface treatment on saturated pool boiling.

low superheat is required to activate the nonwetting cavities represented by the spots. The spots are placed in such a manner that the bubble area of influence includes the whole surface, so the net result is a low average superheat. Extensive studies of this surface promoter are reported in Ref. 20.

Surface conditions do not appear to significantly alter the boiling curve for reasonably high flow velocities; however, certain surface finishes improve flow boiling CHF. Durant et al. [13, 21] have demonstrated that there is a substantial increase in subcooled CHF with relatively coarse mechanical roughness. It was also suggested that the critical fluxes for the rough tubes were higher than those for smooth tubes at comparable pumping power. Several investigators have demonstrated that bulk boiling CHF can be improved by 50 to 100 percent with various surface processing: Bernstein et al. [22] (irregular diameter tubing and slotted helical inserts), Swenson et al. [23] (helical ribs), Janssen et al. [24] (sandblasting), Rouvillois [25] (sandblasting), and Quinn [26] (machined protuberances). Studies of post CHF, or dispersed flow film boiling, indicate that roughness elements increase the heat transfer coefficient [26, 27].

Large-scale roughness elements, or fins, have been considered for numerous boiling applications. Katz et al. [28], for example, found that it is possible to get at least twice the heat transfer with fins for the same overall temperature difference and length of tube. Massive fins have been used to facilitate the cooling of high-power-density electronic tubes [29]. Fins have also been found effective in evaporators [30] and steam boiler tubes [22]; however, there are practical difficulties involved in fabrication and cleaning of these internally finned tubes.

As noted in Sec. 12, surface treatment has been extensively investigated for the promotion of dropwise condensation. Numerous promoters have been found effective; however, there are practical problems relating to permanence and compatibility with the rest of the system. Plating with noble metals and coating with thin nonwetting films have been considered in recent investigations.

C. DISPLACED PROMOTERS

1. Single-phase Heat Transfer

In certain cases it may be desirable to leave the heat transfer surface intact and achieve the augmentation by disturbing the flow near the heated surface. Koch [6] placed thin rings or discs in a tube. The typical basic data shown in Fig. 1 translate to performance data in Fig. 6, where it is seen that rings substantially improve heat transfer in the lower Reynolds number range. These data, as well as the data of Evans [31] and Thomas [32] (rings in annuli), indicate that discs are not particularly effective.

2. Flow Boiling

Janssen et al. [24] reported on bulk boiling CHF with displaced turbulence promoters. Flow-disturbing rings were located on the outer tube of an annular test section. Critical heat fluxes for quality boiling with the rough liner are seen in Fig. 7 to be as much as 60 percent greater than those for the smooth liner. This is to be expected since the rings strip the liquid from the inactive surface, thereby increasing the film flow rate on the heated surface. Moeck et al. [33] performed an extensive investigation of CHF for annuli with rough outer tubes. Steam-water mixtures were introduced at the test-section inlet so as to obtain high outlet steam qualities. It was found that the critical heat flux increased as the roughness height (0.050 in. max.) increased and spacing (1.5 to 4.5 in.) decreased, with a maximum increase of over 600 percent based on similar inlet conditions. The pressure drop with the most optimum promoter was about six times the smooth annulus value for similar inlet conditions at

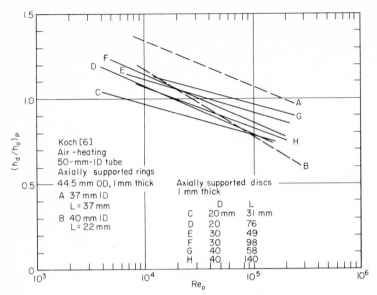

Fig. 6. Performance of tubes with ring and disc inserts.

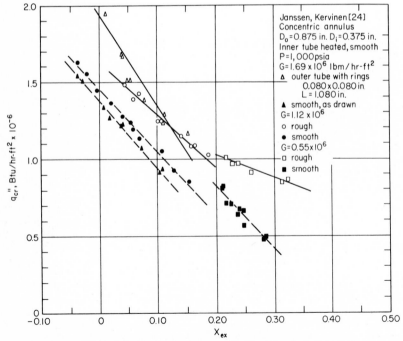

Fig. 7. Effect of displaced promoters on critical heat flux for bulk boiling.

the critical condition. Rough liners were also found to produce significant increases in CHF for rods located in a square channel [34].

Similar promoters have been applied successfully to small rod bundles. As reported by Quinn [35], rings of stainless-steel wire, e = 0.044 in. and L = 1 in., were spot-welded to the channel wall of a two-rod assembly. Both CHF and film-boiling heat transfer coefficients were improved.

D. VORTEX FLOW

1. Various Methods of Producing Vortex Flow

It has been established for nearly 50 years that swirling the flow will improve heat transfer in duct flow; however, it is only in the past decade that comprehensive investigations of swirl flow have been reported. Generation of swirl flow has been accomplished by coiled wires, spiral fins, propellers, coiled tubes, inlet vortex generators, and twisted tapes. Virtually all of these arrangements have been shown to improve single-phase and boiling heat transfer at the expense of increased pumping power. Heat transfer coefficients are relatively high for vortex flow due to increased velocity, secondary flow produced by the radial body force when favorable density gradients are present, and a fin effect with certain continuous swirl generators.

Coiled wires and short spiral fins produce a certain amount of rotation in the flow; however, their primary effect would appear to be an increase of turbulence at the heated surface. Accordingly, the discussion of these augmentative devices has been included in the section on surface roughness.

Propellers spaced along the flow channel have also been considered by several investigators [6, 36]. The manufacture and installation of these inserts would appear to be too elaborate to justify the modest improvement in performance.

Heat transfer is definitely improved when the flow channel is formed into a helix. Seban and McLaughlin [37] presented correlations for laminar and turbulent flow. Detailed analytical and experimental results for both laminar and turbulent flow of gases and liquids have been reported by Mori and Nakayama [38]. Improvements of up to 50 percent in turbulent heat transfer coefficients can be realized with helix-to-tube diameter ratios of about 20. Further single-phase data for coiled tubes are given in Sec. 7.F.2.

Coiled tubes were recently suggested as a means of improving boiler performance. Carver et al. [39] reported substantial improvement in bulk-boiling CHF with coils of 16 in. and 65 in. radii. A variation on the coiled-tube technique was investigated with considerable success at Pratt and Whitney [22]. Standard tubing was formed in a wave-shaped or serpentine pattern. In the high-quality bulk boiling region, heat transfer coefficients were found to be much higher than those for straight tubes at comparable conditions, whereas pressure drop was not greatly increased. Due to these excellent characteristics, this scheme was chosen for a zero-gravity boiler and tested successfully with bulk boiling of potassium [40].

In a summary report of their studies with inlet vortex generators, Gambill and Greene [41] demonstrated that tubes with spiral-ramp and tangential-slot vortex generators could handle extremely large heat fluxes under subcooled boiling conditions. The now-classic $q_{cr}'' = 55 \times 10^6$ Btu/hr-ft^2 for subcooled nucleate boiling was obtained using a tangential-slot generator together with a short test section. It was suggested that this arrangement was superior to a straight flow on the basis of equal pumping power. Goldmann [42] also reported data for a spiral-ramp vortex generator which indicated that CHF was improved by about 20 percent at equal pumping power for relatively low values of pumping power. Mayinger et al. [43] have reported significant increases in CHF for both subcooled and bulk boiling when twisted tapes were used to generate vortex motion at the inlet of short tubes.

With inlet vortex generators there is a pronounced effect of heated length on CHF or heat transfer coefficient due to the vortex decay. Blum and Oliver [44] reported data relating to the axial distribution of heat transfer coefficient for air and carbon dioxide in a tube with a tangential-slot vortex generator at the inlet. Friction and heat transfer data for various length tubes with tangential-slot generators were reported by Alimov [45].

Twisted-tape inserts have been used as continuous vortex generators in many industrial applications. Fabrication is relatively simple, and the inserts may be used to improve the performance of existing equipment. Most of the studies of vortex augmentation have considered twisted tapes.

2. Twisted-tape Inserts in Single-phase Flow

Many investigations have been made to determine the heat transfer and pressure-drop characteristics for various single-phase fluids in tape-generated swirl flow. Studies with air and other gases include Refs. 6, 36, 46, 50, and 55, water is considered in Refs. 47–51; liquid metals in Ref. 49, and glycol in Ref. 52. The survey papers by Gambill et al. [53, 54] and Bergles [1] demonstrate clearly the considerable disagreement among investigators regarding heat transfer and friction data for full-length twisted-tape assemblies.

Several correlations have been suggested to account for the various factors influencing the heat transfer improvement in swirl flow. The most general expressions account for the increased turbulence due to spiralling flow, increased circulation created with heating due to the large centrifugal force, and the tape fin effect. The work of Thorsen and Landis [55] is a comprehensive extension of an earlier development by Smithberg and Landis [50], and their semianalytical prediction method appears to account for all of their observed variations in h_s with the swirl flow of air subjected to large radial temperature gradients. Constant property heat transfer and friction prediction equations were developed as complicated functions of Re, Pr, and curvature of the flow path at the tube wall. These equations were then corrected for the significant radial property variation and buoyancy effects, for heating and cooling cases. The effect of the extended surface was considered in the data reduction.

The correlation recently suggested by Lopina and Bergles [51] is relatively simple, yet also accounts satisfactorily for the several mechanisms. This correlation postulates

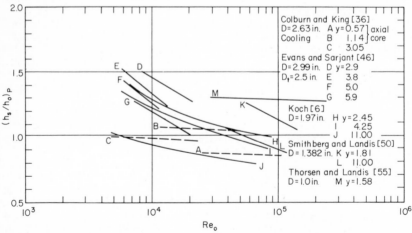

Fig. 8. Performance of twisted-tape vortex generators with air.

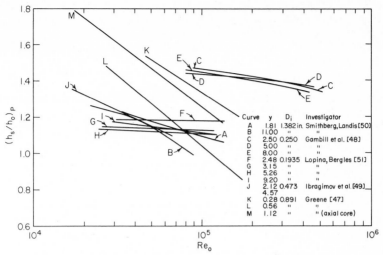

Fig. 9. Performance of twisted-tape vortex generators with single-phase water.

that the average heat transfer coefficient can be represented essentially as a superposition of heat transfer coefficients for spiral convection and centrifugal convection. For water this correlation is

$$\frac{h_s D_e}{k} = F \left\{ 0.023 \left[1 + \left(\frac{\pi}{2y}\right)^2\right]^{0.4} \mathrm{Re}_e^{0.8} + 0.193 \left[\left(\frac{\mathrm{Re}_e}{y}\right)^2 \frac{D_e}{D_i} \beta (T_w - T_b) \mathrm{Pr}\right]^{1/3} \right\} \quad (2)$$

where the terms can be recognized as modified forms of the conventional relations for turbulent flow in tubes and turbulent free convection. The fin factor F, representing the ratio of total heat transfer to the heat transferred by the walls alone, can be estimated from conduction calculations. The value of F is close to 1.0 for a loose tape fit and may be as high as 1.25 for a tight tape fit. With the use of appropriate convective constants and fin factors, Eq. (2) predicted generally to within 10 percent extensive data for heating of water [50, 51]. Water-cooling data of Ref. 51 were predicted to within 10 percent by Eq. (2) with the centrifugal convection term deleted.

It is appropriate to conclude this discussion of single-phase data by presenting a constant pumping power comparison. Actual friction and heat transfer data from the various investigations are utilized in the computations leading to Fig. 8, for air, and Fig. 9, for water. Due to the diversity already noted in heat transfer and friction data, the performance curves exhibit rather wide scatter; however, the general consensus is that the tapes provide a substantial improvement in performance. Tape twist is, of course, an important parameter; however, even for similar geometries, differences are to be expected due to variations in the fin effect and centrifugal convection effect. The data for axial core assemblies for both air (A, B) and water (K, L, and M) suggest that the tightest twist ratio is not necessarily the best. On the basis of a similar pumping power comparison for regular tape inserts, Seymour [56] found that a tape twist of 2.5 was optimum for air.

Several studies [36, 56] have considered tapes which do not extend the length of the heated section. For uniformly heated tubes, the performance is less than that for full-length tapes, and would not be used. However, intermittent tapes are particularly useful for cases where nonuniform heat fluxes are involved. The tapes can be placed at the hot-spot location, producing the desired improvement in heat transfer with little effect on the overall pressure drop. This has recently been used to eliminate the burnouts caused by degeneration in heat transfer in certain supercritical boiler systems [57].

3. Twisted-tape Inserts in Boiling

A recent study has determined that the boiling curve for swirl flow with subcooled boiling is essentially the same as in straight flow [51]. However, it was found that the swirl flow pressure drop was as much as a factor of three lower than straight flow pressure drop for comparable flow conditions. This suggests that swirl flow is very effective when compared with straight flow on a pumping power basis. Vortex flow is particularly effective in elevating the critical heat flux in subcooled boiling. The data of Gambill et al. [48] shown in Fig. 10 indicate that CHF for swirl flow is about twice that for straight flow at the same test-section pumping power.

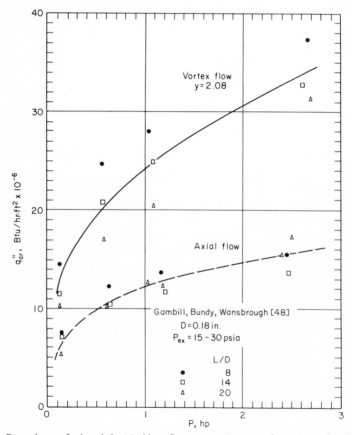

Fig. 10. Dependence of subcooled critical heat flux on pumping power for vortex and straight flow.

Twisted tapes have been found to be particularly effective in improving the performance of evaporators and elevating CHF for high power boiler tubes. Both water and liquid metals have been extensively investigated under conditions approximating those found in nuclear reactors, space power plants, and environmental control systems.

Annular flow is the predominant flow regime with swirl flow, and the stability of the annular liquid film is enhanced by the centrifuging of all liquid to the wall. Tests by Bernstein et al. [22] with water indicated that twisted tapes of both plain and and perforated types were effective in delaying dryout at high vapor qualities. Similar results were reported by Pai and Pasint [58], who installed loose fitting tapes to avoid collection of impurities which might cause corrosion. Figure 11 presents data of

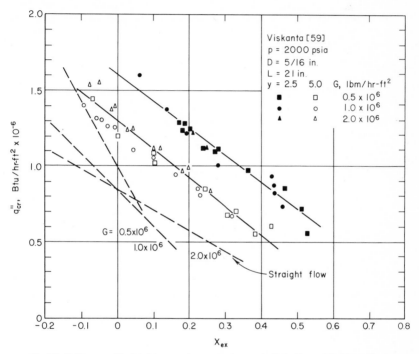

Fig. 11. Influence of twisted-tape vortex generators on bulk boiling critical heat flux.

Viskanta [59] which illustrate that twisted tapes are effective in increasing critical heat flux over a wide quality range. Workers at SNECMA [60, 61] studied various geometries: round tubes, annuli (six tapes), and rod clusters (four heated rods with nine tapes). Typical increases in CHF of 30 percent were reported, although improvements of over 60 percent were obtained with the annular geometry. Recent in-pile tests of a vortex fuel element indicated excellent mechanical behavior and absence of corrosion [62]. It was suggested that the volumetric power generation of boiling water reactors might be doubled by this technique.

Two-phase heat transfer beyond the critical condition (two-phase film boiling or post burnout condition) can be improved by use of twisted-tape inserts. Liquid droplets centrifuged to the heated surface improve the heat transfer in the classic Leidenfrost fashion. Nitrogen data exhibiting this behavior were reported by Burke and Rawdon [63]. A similar flow regime is observed with mercury, and vortex generators of the axial core type have been shown to be effective in improving performance of the first stage of once-through mercury boilers [27].

E. HEATED SURFACE VIBRATION

1. Single-phase Heat Transfer

It has long been recognized that transport processes can be significantly affected by inherent or induced oscillations. In general, sufficiently intense oscillations improve heat transfer; however, decreases in heat transfer have been recorded on both a local and an average basis. A wide range of effects is to be expected as a result of the large number of variables necessary to describe the vibrations and the convective conditions.

In discussing the interactions between vibration and heat transfer, it is convenient to distinguish between vibrations that are applied to the heat transfer surface and those that are imparted to the fluid. The most direct approach is to vibrate the surface mechanically, usually by an electrodynamic vibrator or a motor-driven eccentric. In order to achieve an adequate amplitude of vibration, frequencies are generally well below 1000 Hz.

The predominant geometry employed in vibrational studies has been the horizontal, heated cylinder vibrating either horizontally or vertically. When the ratio of the amplitude to the diameter is large, it is reasonable to assume that the convection process which occurs in the vicinity of the cylinder is quasi-steady. The heat transfer may then be described by conventional correlations for steady convection. In order to achieve the quasi-steady convection, characterized by $a/D \gg 1$, it is necessary to use small diameter cylinders. The data presented in Fig. 12 illustrate this situation. The data fall into three rather distinct regions depending on the intensity of vibration: the region of low Re_v where free convection dominates, a transition region where free convection and the "forced" convection due to vibration interact, and finally the region of dominant forced convection. The data in this last region are in good agreement with a standard correlation for forced flow normal to a cylinder.

When cylinders of large diameter, typical of those found in heat exchange equipment, are used, a different type of behavior is expected. When $a/D \leq 1$, there is no longer a significant displacement of the cylinder through the fluid to provide enthalpy transport. The natural convection should then dominate. However, where

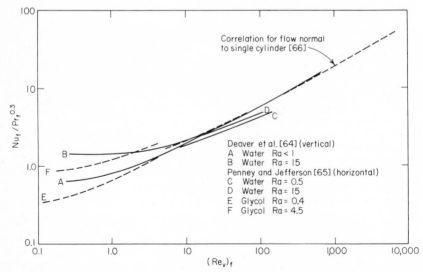

Fig. 12. Influence of mechanical vibration on heat transfer from horizontal cylinders — $a/D \gg 1$.

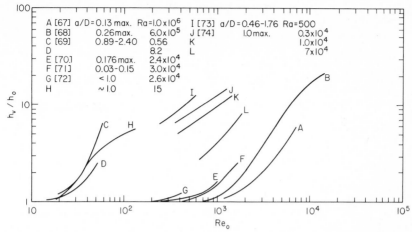

Fig. 13. Influence of mechanical vibration on heat transfer from horizontal cylinders – $a/D < 1$.

the vibrational intensity reaches a critical value, a secondary flow, commonly called acoustic or thermoacoustic streaming, develops, which is able to effect a net enthalpy flux from the boundary layer. Since the coordinates of Fig. 12 are inappropriate to describe streaming data, a simple heat transfer coefficient ratio is used in Fig. 13 to indicate typical improvements in heat transfer observed under these conditions. The heat transfer coefficient remains at the natural convection value until a critical intensity is reached, then increases with increasing intensity. The rate of improvement in heat transfer appears to decrease as Re_v is increased. If these data were plotted on the coordinates of Fig. 12, they would lie below the quasi-steady prediction, except at very high Re_v where they are generally higher.

Several studies have been concerned with the effects of transverse or longitudinal vibrations on heat transfer from vertical plates. Analyses indicate that laminar flow is virtually unaffected; however, experimental observations indicate that turbulent flow is induced by sufficiently intense vibrations. The improvement in heat transfer appears to be rather small, with the largest values of $h_v/h_o < 1.6$ [75].

From an efficiency standpoint, it is important to note that the improvement in heat transfer coefficient with vibration may be quite dramatic (Figs. 12 and 13) but it is only relative to natural convection. The average velocities are actually quite low; for example $4af \simeq 2$ ft/sec for the highest intensity data of Mason and Boelter [68] in Fig. 12. For most systems, it would appear to be more convenient and economical to provide steady forced flow to achieve the desired increase in heat transfer coefficient.

Substantial improvements in heat transfer have been recorded when vibration of the heated surface is used in forced-flow systems. No general correlation has been obtained; however, this is not surprising in view of the diverse geometrical arrangements. Figure 14 presents representative data for heat transfer to liquids. The effect on heat transfer varies from slight degradation to over 300 percent improvement, depending on the system and the vibrational intensity. One of the problems is cavitation when the intensity becomes too large. As indicated by A and B, the vapor blanketing causes a sharp degradation of heat transfer.

These experiments indicate that it is feasible to apply vibrations to practical heat-exchanger geometries; however, economic evaluation is difficult since sufficient data are not available. No comparative pressure drop data appear to have been reported for forced flow. In any case, it appears that the overriding consideration would be the cost of the vibrational equipment and the power to run it. Ogle and

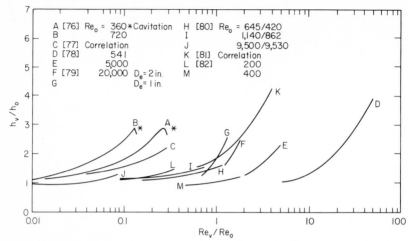

Fig. 14. Effect of surface vibration on heat transfer to liquids with forced flow.

Engel [80] found for one of their runs that about 20 times as much energy was supplied to the vibrator as was gained in improved heat transfer. Even though the vibrator mechanism was not optimized in this particular investigation, this result suggests that heat-surface vibration will not be practical.

2. Boiling and Condensation

Experiments by Bergles [74] have established that vibrations have little effect on subcooled or saturated pool boiling. It was found that the coefficients characteristic of single-phase vibrational data govern the entry into boiling conditions. Once boiling is fully established, however, there is no discernible effect of vibration. The maximum increase in critical heat flux was about 10 percent at an average velocity of 10 in./sec. Recent experiments by Parker et al. [83, 84], run over the frequency range 50–2000 Hz, have further confirmed that fully developed nucleate boiling is essentially unaffected by vibration.

In an interesting variation of their ultrasonic fluid vibration experiments, Markels et al. [85] applied 20–1000 Hz electrical signals to a steam-heated tube used for pool boiling isopropanol. The voltage was adjusted to maintain an SPL of 174 dB near the tube surface. Negligible change in the critical heat flux was observed; however, heat transfer in the transition boiling region was improved by as much as a factor of five.

Raben et al. [79] have reported what appears to be the only study of flow surface boiling with heated-surface vibration. It was noted that there was a large improvement at low heat flux, but that the improvement decreased with increasing heat flux. This is consistent with the pool boiling results which indicate no improvement in the region of fully established boiling.

F. FLUID VIBRATION

1. Single-phase Heat Transfer

In many applications, it is difficult to apply surface vibration due to the large mass of the heat transfer apparatus. The alternative technique is then utilized, whereby vibrations are applied to the fluid and focused toward the heated surface. The generators which have been employed range from the flow interrupter to the piezoelectric transducer, thus covering the range from pulsations of 1 Hz to ultrasound of 10^6 Hz. The description of the interaction between fluid vibrations and heat transfer is even more complex than in the case of surface vibration. In particular, the vibrational variables are more difficult to define due to the remote placement of the generator. Under certain conditions, the flow fields may be similar for both fluid and surface vibration, and analytical results can be applied to both types of data.

A great deal of research effort has been devoted to studying the effects of sound fields on heat transfer from horizontal cylinders to air. Intense plane sound fields of the progressive or stationary type have been generated by loudspeakers or sirens. The sound fields have been oriented axially and transversely, in either the horizontal or vertical plane. With plane transverse fields directed transversely, improvements of 100–200 percent over natural convection heat transfer coefficients were obtained by Holman et al. [86, 87], Fand and Kaye [88], and Lee and Richardson [89].

It is commonly observed that increases in average heat transfer occur at a sound pressure level of about 134–140 dB (well above the normal human tolerance of 120 dB), and are associated with the formation of an acoustically induced flow (acoustic or thermoacoustic streaming) near the heated surface. Large circumferential variations in heat transfer coefficient are present [90], and it has been observed that local improvements in heat transfer occur at intensities well below those which affect the average heat transfer [91]. Correlations have been proposed for individual experiments; however, an accurate correlation covering the limits of free convection and fully developed vortex motion has not been developed.

In general, it appears that acoustic vibrations yield relatively small improvements in heat transfer to gases in free convection. From a practical standpoint, a relatively simple forced-flow arrangement could be substituted to obtain equivalent improvements.

When acoustic vibrations are applied to liquids, heat transfer may be improved by acoustic streaming as in the case of gases. With liquids, though, it is possible to operate with ultrasonic frequencies due to favorable coupling between a solid and a liquid. At frequencies of the order of a megacycle, another type of streaming called crystal wind may be developed. These effects are frequently encountered; however, intensities are usually high enough to cause cavitation which may become the dominant mechanism.

Seely [92], Zhukauskas [93], Larson and London [94], Robinson et al. [95], Fand [96], Gibbons and Houghton [97], and Li and Parker [98] have demonstrated that natural convection heat transfer to liquids can be improved from 30 to 450 percent by use of sonic and ultrasonic vibrations. In general, cavitation must occur before significant improvements in heat transfer are noted. In spite of these improvements there appears to be some question regarding the practical aspects of acoustic augmentation. When the difficulty of designing a system to transmit acoustic energy to a large heat transfer surface is considered, it appears that forced flow or simple mechanical agitation will be a more attractive means of improving natural convection heat transfer.

Low-frequency pulsations have been produced in forced convection systems by partially damped reciprocating pumps and interrupter valves. Quasi-steady analyses suggest that heat transfer will be improved in transitional or turbulent flow with sufficiently intense vibrations [1]. However, heat transfer coefficients are usually higher than predicted, apparently due to cavitation. Figure 15 indicates the improvements that have been reported for pulsating flow in channels. The improvement is most significant in the transitional range of Reynolds numbers, as might be expected since the pulsations force the transition to turbulent flow. Interrupter valves are a particularly simple means of generating the pulsations. The valves must be located directly upstream of the heated section to produce cavitation, which appears to be largely responsible for the improvement in heat transfer [102, 103].

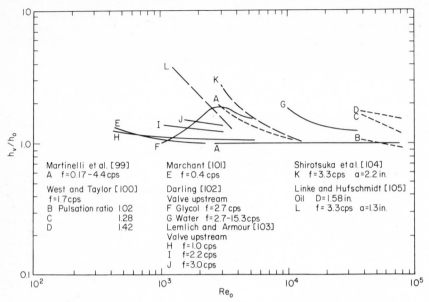

Fig. 15. Effect of upstream pulsations on heat transfer to liquids flowing in pipes.

A wide variety of geometrical arrangements and more complex flow fields are encountered when sound fields are superimposed upon forced flow of gases. In general, the improvement is dependent upon the relative strengths of the acoustic streaming and the forced flows. As indicated in Ref. 1, the reported improvements in average heat transfer are limited to about 100 percent. Typical data obtained with sirens or loudspeakers placed at the end of a channel are indicated in Fig. 16. The improvements in average heat transfer coefficient are generally significant only in the transition range where the vibrating motion acts as a turbulence trigger.

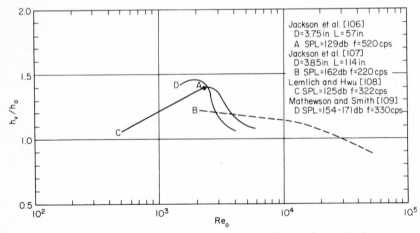

Fig. 16. Influence of acoustic vibrations on heat transfer to air flowing in tubes.

The data of Zhukauskas et al. [93] and Larson and London [94] suggest that ultrasonic vibration has no effect on forced-convection heat transfer once the flow velocity is raised to about 1 ft/sec. However, Bergles [110] demonstrated that lower frequency vibrations (80 Hz) can produce improvements of up to 50 percent. This experiment was carried out at higher surface temperatures where it was possible to achieve cavitation.

2. Two-phase Heat Transfer

The available evidence indicates that fully established nucleate pool boiling is unaffected by ultrasonic vibration, apparently due to the dominance of bubble agitation and attenuation of acoustic energy by the vapor [111, 112]. However, due to enhanced vapor removal, CHF can be improved by about 50 percent [85, 111, 116]. Transition and film boiling can be substantially improved, since the vibration has a strong tendency to destabilize film boiling [113, 85].

In channel flow it is usually necessary to locate the transducer upstream or downstream of the test channel, with the result that the sound field is greatly attenuated. Tests with 80 Hz vibrations [110] indicate no improvement of subcooled boiling heat transfer or critical heat flux. Romie and Aronson [114], using ultrasonic vibrations, found that subcooled critical heat flux was unaffected. Even where intense ultrasonic vibrations were applied to the fluid in the immediate vicinity of the heated surface, boiling heat transfer was unaffected [115]. The severe attenuation of the acoustic energy by the two-phase coolant appears to rule out this technique for flow boiling systems.

Mathewson and Smith [109] investigated the effects of acoustic vibrations on condensation of isopropanol vapor flowing downward in a vertical tube. A siren was used to generate a sound field of up to 176 dB at frequencies ranging from 50–330 Hz. The maximum improvement in condensing coefficient was found to be about 60 percent at low vapor flow rates. The condensate film under these conditions was normally laminar; thus, an intense sound field produced sufficient agitation in the vapor to cause turbulent conditions in the film. The effect of the sound field was considerably diminished as the vapor flow rate increased.

G. ELECTROSTATIC FIELDS

1. Single-phase Heat Transfer

Electrohydrodynamic (EHD) augmentation involves the use of electrical fields to influence heat transfer to dielectric* fluids. Generally speaking, electrostatic fields can be directed to cause greater bulk mixing of fluid in the vicinity of the heat transfer surface. The electrostatic interactions which produce this fluid motion can be of several kinds depending on the field geometry and type of fluid.

Senftleben and Braun [117] investigated the influence of a radial electric field on free convection from a heated horizontal wire. The 0.03-mm wire was located in a concentric tube which was maintained at a high voltage relative to the wire, so as to produce a highly nonuniform electrical field. Up to 50 percent improvement in heat transfer was recorded with gases, including air, oxygen, and C_2H_5Cl. Arajs and Legvold [118] reported data for gaseous N_2, CO_2, NH_3, SO_2, and CCl_3F taken with a similar radial field geometry. Heat transfer coefficients were apparently increased over 40 times for the SO_2 system with maximum field strengths in the vicinity of 10^5 volts/cm.

---

*The term dielectric is used in a practical sense to denote a substance which allows such a small amount of current to exist for large voltages that magnetic effects can be neglected and power expenditures are small.

The data of Choi [119] as shown in Fig. 17 indicate a 200 percent improvement in natural convection heat transfer with Freon. Bonjour and co-workers [120] utilized a parallel-wire geometry which also gave a nonuniform electrostatic field. Their data indicate, for example, that heat transfer coefficients can be increased by 400 percent

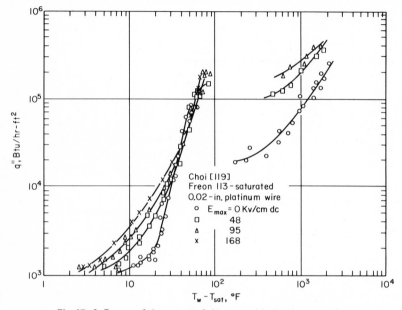

Fig. 17. Influence of electrostatic fields on pool boiling heat transfer.

for ethyl ether in natural convection with a field strength of 160 kv/cm. The improvements obtained in these experiments appear to be due to a combination of dielectrophoreses (due to polarization of a dielectric material) and electrophoresis or corona wind (due to presence of a free charge).

Marco and Velkoff [121] studied the interactions between an electrostatic field and the natural-convection boundary layer where a fine-wire electrode was used with horizontal (heated side down) and vertical flat plates. Due to corona wind, the average heat transfer coefficient for the horizontal 6-in. × 10-in. plate was increased by 500 percent when a 0.004-in. wire was placed 0.78 in. below the centerline of the plate.

An early study by Wahlert and Croft [122] indicated that the heat transfer coefficient to air flowing in an annulus could be improved by about 10 percent when about 40,000 vdc was maintained across the annular gap. It was suggested by Berger and co-workers [123–125] that EHD might be particularly effective in gas-cooled reactors where the coolant is partially ionized by nuclear radiation.

Velkoff [126, 127] produced a corona discharge in the gas flowing inside a tube by locating a fine wire electrode concentrically in the tube. A maximum increase in the heat transfer coefficient of 100 percent was noted for laminar flow conditions. Moss and Grey [128] obtained comparable improvements in laminar heat transfer to nitrogen flowing in a similar tubular apparatus.

Schmidt and Leidenfrost [129] applied a radial electric field to the fully developed laminar flow of transformer oil in a horizontal annulus with the inner surface heated. Improvements in heat transfer of over 400 percent were recorded.

2. Boiling and Condensation

Most of the recent work with EHD has been concerned with systems involving change of phase. The data of Choi [119] presented in Fig. 17 illustrate the effects which have been obtained in saturated pool boiling. The apparatus consisted of a horizontal electrically heated wire located concentrically within a charged cylinder. As noted previously, the heat transfer coefficient is elevated at low wall superheat with the result that boiling is not observed until relatively high heat fluxes. Once boiling is initiated there is little effect of the electrical field. However, CHF is elevated substantially, and large increases in the film boiling heat transfer coefficient are obtained. Similar results were obtained by Bonjour et al. [120] when ac fields were applied to a wide variety of fluids, including water, hexane, and liquid nitrogen.

Durfee and co-workers [130] have conducted an extensive series of tests to evaluate the feasibility of applying EHD augmentation to boiling water reactors. Tests with water at 300 psia in electrically heated annuli indicated that wall temperatures for flow bulk boiling were slightly reduced through application of the field. Increases in CHF were observed for all pressures, flow rates, and inlet subcoolings, with the improvement being generally in the 15 to 40 percent range for applied voltages up to 3000 V. On the basis of limited pressure-drop data, it was suggested that greater steam energy flow was obtained with the EHD system than with the conventional system at the same pumping power.

Velkoff and Miller [131] investigated the effect of uniform and nonuniform electric fields on laminar film condensation of Freon-113 on a vertical plate. With screen grid electrodes providing a uniform electrical field over the entire plate surface, a 150 percent increase in the heat transfer coefficient was obtained with a power expenditure of a fraction of a watt. Choi [132, 133] recently reported data for condensation of Freon-113 on the outside wall of an annulus in the presence of a radial electric field. With the maximum applied voltage of 30 kV, the average heat transfer coefficients for a 1-in. o.d. by 0.5-in. i.d. annulus were increased by 100 percent.

H. ADDITIVES

1. Boiling Systems with Trace Additives

The working fluid for a heat transfer system is usually specified by the process or chosen on the basis of its desirable properties. An additive is then desired which will essentially preserve the desirable properties of the working fluid while still improving the heat transfer.

For liquid systems, additives appear to be beneficial only when boiling is involved. Dissolved gas might be considered as an additive since it improves heat transfer in certain systems. Surface degassing, which is initiated when wall temperatures are below the saturation temperature, produces an agitation comparable to that of boiling. In a recent study, Behar et al. [134] found that the wall superheat for saturated pool boiling of nitrogen-pressurized metaterphenyl was reduced by as much as 50°F; while in subcooled flow boiling, the reduction was as much as 30°F. In general, the surface degassing is only effective at lower heat fluxes; once the nucleate boiling becomes well established, there is negligible reduction in the wall superheat.

Trace additives have been extensively investigated in pool boiling, and to a lesser extent in subcooled flow boiling. As noted in Ref. 1, a great many additives have been investigated, and some have been found to produce substantial improvements in heat transfer. With the proper concentration of certain additives (wetting agents, alcohols), increases of about 20 to 40 percent in the heat transfer coefficient for saturated nucleate pool boiling can be realized [135–139]. Most additives increase CHF, but the concentration of the additive and heater geometry are extremely important.

Typical results of van Stralen et al. [140, 141] as shown in Fig. 18 indicate a sharp increase in CHF at some low concentration of 1-pentanol, and rather rapid decrease as the concentration is increased. The optimum concentration varies with the mixture, and to some extent with the pressure. For a similar water-pentanol system, Carne [142] obtained an increase of only 25 percent in CHF with a 1/8-in. heater, which is small compared to the 240 percent increase that van Stralen obtained with a 0.008-in. heater.

In subcooled flow boiling, the improvement (if any) in heat transfer is modest. Leppert et al. [143] found that the main advantage of alcohol-water mixtures was an improvement in smoothness of boiling. The influence of a volatile additive on subcooled flow boiling CHF in tubes has been investigated by Bergles and Scarola [144]. As shown in Fig. 19, there is a distinct reduction in CHF at low subcooling with the addition of 1-pentanol.

In general, the improvements in heat transfer and CHF offered by additives are not sufficient to make them useful for practical systems. There are difficulties involved in maintaining the desired concentration, particularly when the additive is volatile. In many cases, the additives, even in small concentrations, are somewhat corrosive and require special piping or seals.

2. Gas-Solid Suspensions

Dilute gas-solid suspensions have been considered as working fluids for gas turbine and nuclear reactor systems.

Solid particles in the micron-to-millimeter size range are dispersed in the gas stream at loading ratios w_s/w_g ranging from 1 to 15. The solid particles, in addition to giving the mixture a higher heat capacity, are highly effective in promoting enthalpy transport near the heat-exchange surface. Heat transfer is further enhanced at high temperatures by means of the particle-surface radiation. Extensive experimental work was undertaken at Babcock and Wilcox to obtain detailed heat transfer and pressure-drop information as well as operating experience. Summary articles by Rhode et al. [145] and Schluderberg et al. [146] elaborate on the conclusions of this work. Heat transfer coefficients for heating were improved by as much as a factor of ten through the addition of graphite. The suspensions were also shown to be far superior to gas coolants on the basis of pumping power requirements, especially when twisted-tape inserts were used. There was relatively little settling, plugging, or erosion in the system. With helium suspensions, however, there was serious fouling of the loop coolers, which was attributed to Brownian particle motion induced by temperature gradient.

Abel and co-workers [147] demonstrated that the cold-surface deposition is a very serious problem with micronized graphite. This occurred with both helium and nitrogen suspensions and could be alleviated only with very high gas velocities. An economic comparison was presented in terms of a system pumping power to heat transfer rate ratio as a function of gas flow rate. This comparison indicated that the pure gas was generally more effective than the suspension at both low and high gas flow rates. In all probability, the loop heater is very effective; however, this gain is offset by the low performance of the cooler.

A comprehensive analysis of much of the data for dilute gas-solid suspension was reported by Pfeffer et al. [148]. Correlations for both heat transfer coefficient and friction factor were developed. These investigators recently presented a feasibility study of using suspensions as the working fluid in a Brayton space power generation cycle [149].

I. ADDITIONAL AUGMENTATIVE TECHNIQUES

The techniques discussed in the preceding sections would appear to represent the major methods for augmenting convective heat transfer. It is appropriate to mention several other techniques which have been explored.

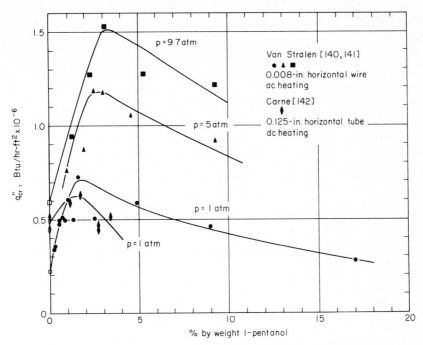

Fig. 18. Dependence of critical heat flux for saturated pool boiling on additive concentration.

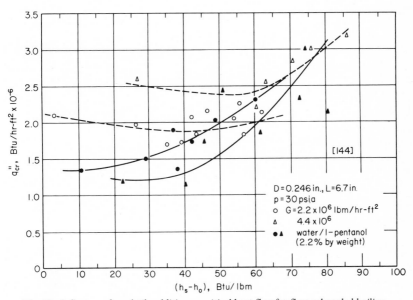

Fig. 19. Influence of a volatile additive on critical heat flux for flow subcooled boiling.

a. Injection and Suction. Injection and suction have been considered primarily in connection with retarding of heat transfer to bodies subject to aerodynamic heating. On the augmentative side, there has been some thought given to intensifying heat transfer by injecting gas through a porous heat-transfer surface. The bubbling produces an agitation similar to that of nucleate boiling. Gose et al. [150, 151] bubbled gas through sintered or drilled heated surfaces so as to simulate nucleate pool and flow boiling. Sims et al. [152] analyzed the pool data and found that Kutateladze's pool boiling relationship correlated the porous-plate data quite well. For their limited forced-circulation tests with a sintered pipe, Gose et al. found that heat transfer coefficients were increased by as much as 500 percent in laminar flow and about 50 percent in turbulent flow. Kudirka [153] found that heat transfer coefficients for flow of ethylene glycol in porous tubes were increased by as much as 130 percent by injection of air. The practical application of injection would appear to be rather limited due to the difficulty of supplying and removing the gas.

Bankoff [154] suggested that heat transfer coefficients in film boiling could be substantially improved by continuously removing vapor through a porous heated surface. Subsequent experimental work [155, 156] demonstrated that coefficients could be increased by as much as 150 percent, provided that a porous block was placed on the surface to stabilize the flow of liquid toward the surface. Wayner and Kestin [157] extended this concept to nucleate boiling and found that wall superheats could be maintained at about 5°F for heat fluxes over 100,000 Btu/hr-ft^2. The requirements of a porous heated surface and a flow control element would appear to limit the application of suction boiling.

b. Capillary Devices. The best known example of an augmentative technique involving capillary forces is the heat pipe; however, the heat pipe is usually thought of as an augmenter of conduction heat transfer. Another technique which relies on capillary forces involves the application of wicking materials to heated surfaces. Wicking is usually considered for situations where coolant would be unable to reach the heater surface without the wicking material. Such situations might arise in aircraft undergoing violent maneuvers or spacecraft operating in a near-zero-gravity environment. Under certain circumstances it might be possible to augment heat transfer under standard conditions by means of wicking.

Tests by Costello et al. [158, 159] indicated that CHF could be raised by as much as 200 percent when the wicking was not too dense and a narrow channel was maintained at the top for easy escape of vapor.

In a recent investigation, Gill [160] spiralled wicking around a cylindrical heater. The boiling curve was generally displaced to lower superheat than the normal curve, to a degree dependent on the diameter and pitch of the wicking. The stable film boiling region was investigated by quenching a copper calorimeter in liquid nitrogen. It appeared that the capillary action was effective in transporting liquid through the vapor film to the heated surface, since the heat transfer coefficient was increased by about 100 percent.

c. Electromagnetic Fields. When magnetic fields are applied to fluids with a sufficiently large electrical conductivity, the fluid motion is retarded by the magnetic force. Heat transfer coefficients thus tend to decrease [161]. Singer [162] has suggested that magnetic fields can be used with electric fields to augment condensation of liquid metals. An analysis indicated that laminar film condensation rates could be improved by a factor of ten through the electromagnetic pumping.

d. Suspensions of Particles Which Undergo Phase Change. Under certain conditions there may be advantages of using liquids which contain small cells of a low melting point material. The mixture is non-Newtonian in character, and depends upon the liquid, or plasma, for convection, while the enthalpy transport is primarily via the cells. The concept of this "Cyto Dynamic" fluid is due to Katz, who recently conducted an experimental and analytical investigation of free-convection heat transfer [163]. It was found that the heat transfer coefficient for free convection between parallel horizontal plates was as much as two times the normal value when the cells were added to water. However, it does not appear to be feasible to use these fluids for natural convection due to the narrow range of stable operation. There may be some forced-circulation applications where the advantages of increased heat capacity and constant temperature outweigh the disadvantages of fluid cost and special pumping equipment.

J. CONCLUSIONS AND PROSPECTS FOR THE FUTURE

The foregoing summary suggests that a number of the augmentative techniques have practical utility at the present time. In general, those methods which require no external power, surface promoters, displaced promoters, vortex flow, and additives, are the most attractive, and should be more thoroughly investigated. However, it would be unwise to discourage further research into the other techniques since new equipment developments may permit practical application of vibrations, electrostatic fields, or even suction boiling. Furthermore, it is by no means clear that the reported investigations have considered a wide enough range of variables to fully demonstrate what constitutes favorable circumstances for augmentation. In any event, future investigations should be along more fundamental lines. For the most part, adequate data exist to demonstrate the effects; what is now needed is analytical work to explain the effects so that existing data can be correlated and guidelines established for the pertinent range of variables in experimental work. In certain cases, particularly in boiling, the standard situation must be better understood before progress in augmentation can be made.

One of the more interesting areas that has not really been explored might be termed compound augmentation. Spiralling fins and vortex flow of gas-solid suspensions appear to be the only reported attempts to combine augmentative techniques. The other possibilities are numerous, and it is interesting to speculate on the results. For example, if rough surfaces are combined with displaced promoters or vortex flow, the final result may be very favorable. However, considerable care must be exercised in mounting an investigation of this type, since the number of variables can be very large.

REFERENCES

1. A. E. Bergles, Survey and Evaluation of Techniques to Augment Convective Heat and Mass Transfer, "Progress in Heat and Mass Transfer," vol. 1, pp. 331–424, Pergamon Press, Oxford, 1969.
2. W. F. Cope, The Friction and Heat Transmission Coefficients of Rough Pipes, *Proc. Instn. Mech. Engrs.*, **145**:99–105 (1941).
3. E. W. Sams, Experimental Investigation of Average Heat Transfer and Friction Coefficients for Air Flowing in Circular Tubes Having Square-thread-type Roughness, *NACA* RME 52D17, 1952.
4. D. F. Dipprey and R. H. Sabersky, Heat and Momentum Transfer in Smooth and Rough Tubes at Various Prandtl Numbers, *JPL Tech. Rept.* 32-269, 1962. Also, *Int. J. Heat Mass Transfer,* **6**:329 (1963).
5. W. Nunner, Wärmeübergang und Druckabfall in Rauhen Rohren, *ForschHft. Ver. dt. Ing.* 455, **B22**:5 (1956). Also, *AERE Lib./Trans.* 786 (1958).
6. R. Koch, Druckverlust und Wärmeübergang bei Verwirbelter Strömung, *ForschHft. Ver. dt. Ing.* 469, **B24**:1 (1958).
7. Z. Nagaoka and A. Watanabe, Maximum Rate of Heat Transfer with Minimum Loss of Energy, *Proc. 7th Int. Cong. Refrigeration,* **3**:221 (1936).
8. F. J. Edwards and N. Sheriff, The Heat Transfer and Friction Characteristics for Forced Convection Air Flow over a Particular Type of Rough Surface, "International Developments in Heat Transfer," pp. 415–425, ASME, New York, 1961.

9. P. L. Mantle, A New Type of Roughened Heat Transfer Surface Selected by Flow Visualization Techniques, *Proc. 3d Int. Heat Transfer Conf.*, AIChE, New York, 1:45 (1966).
10. G. A. Kemeny and J. A. Cyphers, Heat Transfer and Pressure Drop in an Annular Gap with Surface Spoilers, *J. Heat Transfer,* 83:189 (1961).
11. A. W. Bennett and H. A. Kearsey, Heat Transfer and Pressure Drop for Superheated Steam Flowing through an Annulus with One Roughened Surface, *AERE-R* 4350 (1964).
12. H. Brauer, Flow Resistance and Heat Transfer in Annuli with Roughened Inner Tubes, Mannesman Forschungsberichte 109/1961 (1961). Also, *AAEC/Trans.* 13.
13. W. S. Durant, R. H. Towell, and S. Mirshak, Improvement of Heat Transfer to Water Flowing in an Annulus by Roughening the Heated Wall, *Chem. Eng. Prog. Symp., Ser. 61,* 60:106 (1965).
14. A. Draycott and K. R. Lawther, Improvement of Fuel Element Heat Transfer by Use of Roughened Surfaces and the Application to a 7-Rod Cluster, "International Developments in Heat Transfer," pp. 543–552, ASME, 1961.
15. B. Kjellström and A. E. Larsson, Improvement of Reactor Fuel Element Heat Transfer by Surface Roughness, *AE* 271 (1967).
16. V. Kolář, Heat Transfer in Turbulent Flow of Fluids through Smooth and Rough Tubes, *Int. J. Heat Mass Transfer,* 8:639 (1965).
17. G. B. Melese, Comparison of Partial Roughening of the Surface of Fuel Elements with Other Ways of Improving Performance of Gas-cooled Nuclear Reactors, *GA* 4624 (1963).
18. Anon., Heat Transfer Capability, *Mech. Engng.,* 89:55 (1967).
19. R. K. Young and R. L. Hummel, Improved Nucleate Boiling Heat Transfer, *Chem. Engng. Prog.,* 60:53 (1964).
20. R. K. Young, Pool Boiling with Precise Arrays of Active Sites, doctoral dissertation in chemical engineering and applied chemistry, University of Toronto, 1968.
21. W. S. Durant and S. Mirshak, Roughening of Heat Transfer Surfaces as a Method of Increasing Heat Flux at Burnout, *DP* 380 (1959).
22. E. Bernstein, J. P. Petrek, and J. Meregian, Evaluation and Performance of Once-through, Zero-gravity Boiler Tubes with Two-phase Water, *PWAC* 428 (1964).
23. H. S. Swenson, J. R. Carver, and G. Szoeke, The Effects of Nucleate Boiling Versus Film Boiling on Heat Transfer in Power Boiler Tubes, *J. Engng. Power,* 84:365 (1962).
24. E. Janssen and J. A. Kervinen, Burnout Conditions for Single Rod in Annular Geometry, Water at 600 to 1400 psia, *GEAP* 3899 (1963).
25. A. Rouvillois, Heat Transfer Improvement in Boiling Water Reactors, *SNECMA Quart. Rept.* 16, *EURAEC*-1059 (1964).
26. E. P. Quinn, Transition Boiling Heat Transfer Program, *5th Quart. Prog. Rept., GEAP* 4608 (1964).
27. A. J. Sellers, G. M. Thur, and M. K. Wong, Recent Developments in Heat Transfer and Development of the Mercury Boiler for the SNAP-8 System, *Proc. Conf. Application of High Temperature Instrumentation to Liquid-Metal Experiments,* ANL-7100, pp. 573–632, 1965.
28. D. L. Katz, J. E. Meyers, E. H. Young, and G. Balekjian, Boiling Outside Finned Tubes, *Petrol. Refiner.,* 34:113 (1955).
29. C. Berutheut, The Vapotron Technique, *Compagnie Francoise Thomson-Houston Revue Technique,* 24:53 (1956).
30. J. G. Lavin and E. H. Young, Heat Transfer to Evaporating Refrigerants in Two-phase Flow, AIChE, Preprint 21e for National Meeting, 1964.
31. L. B. Evans and S. W. Churchill, The Effect of Axial Promoters on Heat Transfer and Pressure Drop inside a Tube, *Chem. Engng. Prog. Symp Ser.* 59, 41:36 (1963).
32. D. G. Thomas, Enhancement of Forced Convection Heat Transfer Coefficient Using Detached Turbulence Promoters, *Industrial and Engineering Chemistry Process Design and Development,* 6:385 (1967).
33. E. O. Moeck, G. A. Wikhammer, I. P. L. Macdonald, and J. G. Collier, Two Methods of Improving the Dryout Heat Flux for High Pressure Steam/Water Flow, AECL-2109, 1964.
34. L. S. Tong, R. W. Steer, A. H. Wenzel, M. Bogaardt, and C. L. Spigt, Critical Heat Flux of a Heater Rod in the Center of Smooth and Rough Square Sleeves, and in Line-contact with an Unheated Wall, *ASME Paper* 67-WA/HT-29 (1967).
35. E. P. Quinn, Transition Boiling Heat Transfer Program, *6th Quart. Prog. Rept.,* GEAP 4646 (1964).
36. A. P. Colburn and W. J. King, Heat Transfer and Pressure Drop in Empty, Baffled, and Packed Tubes, *Ind. Engng. Chem.,* 23:910 (1931).

37. R. A. Seban and E. F. McLaughlin, Heat Transfer in Tube Coils with Laminar and Turbulent Flow, *Int. J. Heat Mass Transfer*, **6**:387 (1963).
38. Y. Mori and W. Nakayama, Study on Forced Convective Heat Transfer in Curved Pipes, *Int. J. Heat Mass Transfer*, **10**:681 (1967).
39. J. R. Carver, C. R. Kakarala, and J. S. Slotnik, Heat Transfer in Coiled Tubes in Two-phase Flow, *TID* 20983 (1964).
40. E. Bernstein, J. P. Petrek, G. J. Rose, and J. J. Horan, Experimental Results of Forced Convection Boiling Potassium Heat Transfer and Pressure Drop Tests, *PWAC* 429 (1964).
41. W. R. Gambill and N. D. Greene, Boiling Burnout with Water in Vortex Flow, *Chem. Engng. Prog.*, **54**:68 (1958).
42. K. Goldmann, Improved Heat Transfer by Application of Centrifugal Forces, *NDA* 2-79 (1958).
43. F. Mayinger, O. Schad, and E. Weiss, Investigations into the Critical Heat Flux in Boiling, *Maschinenfabrik Augsburg-Nürnberg Final Rept.* 09.03.01, 1966.
44. H. A. Blum and L. R. Oliver, Heat Transfer in a Decaying Vortex System, *ASME* 66-WA/HT-62 (1966).
45. P. S. Alimov, Hydraulic Resistance and Heat and Mass Transfer in Vortex Flow, *Inz. Fis. Zh.*, **10**:437 (1966).
46. S. I. Evans and R. J. Sarjant, Heat Transfer and Turbulence in Gases Flowing inside Tubes, *J. Inst. Fuel*, **24**:216 (1951).
47. N. D. Greene, Convair Aircraft, Private communication to W. R. Gambill, cited in Ref. 7, May, 1960.
48. W. R. Gambill, R. D. Bundy, and R. W. Wansbrough, Heat Transfer, Burnout, and Pressure Drop for Water in Swirl Flow through Tubes with Internal Twisted Tapes, *ORNL* 2911 (1960). Also, *Chem. Engng. Prog. Symp. Ser. 57*, **32**:127 (1961).
49. M. H. Ibragimov, E. V. Nomofelov, and V. I. Subbotin, Heat Transfer and Hydraulic Resistance with the Swirl-type Motion of Liquid in Pipes, *Teploenergetika*, **8**:57 (1961).
50. E. Smithberg and F. Landis, Friction and Forced Convection Heat Transfer Characteristics in Tubes with Twisted Tape Swirl Generators, *J. Heat Transfer*, **86**:39 (1964).
51. R. F. Lopina and A. E. Bergles, Heat Transfer and Pressure Drop in Tape Generated Swirl Flow, MIT Engineering Projects Lab. Rept. 70281-47, 1967. Also, *J. Heat Transfer*, **91**:434 (1969).
52. W. R. Gambill and R. D. Bundy, High-Flux Heat Transfer Characteristics of Pure Ethylene Glycol in Axial and Swirl Flow, *AIChE Journal*, **9**:55 (1963).
53. W. R. Gambill and R. D. Bundy, An Evaluation of the Present Status of Swirl Flow Heat Transfer, ORNL 61-4-61 (1961). Also, *ASME* 61-HT-42 (1962).
54. H. F. Poppendiek and W. R. Gambill, Helical, Forced-flow Heat Transfer and Fluid Dynamics in Single and Two-phase Systems, *Proc. 3rd Int. Conf. Peaceful Uses of Atomic Energy*, vol. 8, pp. 274–282, New York, 1965.
55. R. Thorsen and F. Landis, Friction and Heat Transfer Characteristics in Turbulent Swirl-flow Subjected to Large Transverse Temperature Gradients, *J. Heat Transfer*, **90**:87 (1968).
56. E. V. Seymour, A Note on the Improvement in the Performance Obtainable from Fitting Twisted-tape Turbulence-promoters to Tubular Heat Exchangers, *Trans. Instn. Chem. Engrs.*, **41**:159 (1963).
57. B. S. Shiralker and P. Griffith, The Deterioration in Heat Transfer to Fluids at Supercritical Pressure and High Heat Fluxes, MIT Engineering Projects Lab. Rept. 70332-55 (1968).
58. R. H. Pai and D. Pasint, Research at Foster Wheeler Advances Once-Through Boiler Design, *Electric Light and Power*, **43**(1):66 (1965).
59. R. Viskanta, Critical Heat Flux for Water in Swirling Flow, *Nuclear Sci. Eng.*, **10**:202 (1961).
60. C. Moussez, Adaptation of Vortex Flow to a Biphase Liquid Gas Mixture, Two-phase Flow Problems, *Proc. Meeting of the Working Group Heat Transfer*, *TID* 7994, pp. 89–107, 1963.
61. A. Rosuel, G. Soorioux, J. Dolle, G. Tournier, and A. Beghin, Ecoulements Giratoires dans l'Eau Bouillante, *SNECMA* Final Rept. 26 (1966).
62. Anon., High-Performance Nuclear Reactor, *Mech. Engng.*, **89**(10):64 (1967).
63. J. C. Burke and A. H. Rawdon, An Experimental Study of Heat Transfer to Two-phase Film-boiling Nitrogen, *ASME* 65-HT-37 (1965).
64. F. K. Deaver, W. R. Penney, and T. B. Jefferson, Heat Transfer from an Oscillating Horizontal Wire to Water, *J. Heat Transfer*, **84**:251 (1962).
65. W. R. Penney and T. B. Jefferson, Heat Transfer from an Oscillating Horizontal Wire to Water and Ethylene Glycol, *J. Heat Transfer*, **88**:359 (1966).

66. W. H. McAdams, "Heat Transmission," 3d ed., McGraw-Hill Book Company, New York, 1954.

67. R. C. Martinelli and L. M. K. Boelter, The Effect of Vibration on Heat Transfer by Free Convection from a Horizontal Cylinder, *Proc. 5th Int. Cong. Appl. Mech.*, pp. 578–584, John Wiley & Sons, Inc., New York, 1938. Also, ASH&VE Journal Section, *Heating, Piping, and Air Conditioning*, 11:525 (1939).

68. W. E. Mason and L. M. K. Boelter, Vibration—Its Effect on Heat Transfer, *Pwr. Pl. Engng.*, 44:43 (1940).

69. R. Lemlich, Effect of Vibration on Natural Convective Heat Transfer, *Ind. and Engng. Chem.*, 47:1175 (1955); Errata 53:314 (1961).

70. C. Teleki, R. M. Fand, and J. Kaye, The Influence of Vertical Vibration on the Rate of Heat Transfer from a Horizontal Cylinder in Air, *WADC TN* 59-357 (1960). Also, R. M. Fand and J. Kaye, The Influence of Vertical Vibrations on Heat Transfer by Free Convection from a Horizontal Cylinder, "International Developments in Heat Transfer," pp. 490–498, ASME, 1961.

71. R. M. Fand and E. M. Peebles, A Comparison of the Influence of Mechanical and Acoustical Vibrations on Free Convection from a Horizontal Cylinder, *J. Heat Transfer*, 84:268 (1962). Also, *ARL TR*-148, pt. II (1961).

72. A. J. Shine, Comments on paper by Deaver et al., *J. Heat Transfer*, 84:225 (1962).

73. R. Lemlich and M. A. Rao, The Effect of Transverse Vibration on Free Convection from a Horizontal Cylinder, *Int. J. Heat Mass Transfer*, 8:27 (1965).

74. A. E. Bergles, The Influence of Heated-Surface Vibration on Pool Boiling, *J. Heat Transfer*, 91:152 (1969).

75. V. D. Blankenship and J. A. Clark, Experimental Effects of Transverse Oscillations on Free Convection of a Vertical, Finite Plate, *J. Heat Transfer*, 86:159 (1964).

76. J. A. Scanlan, Effects of Normal Surface Vibration on Laminar Forced Convection Heat Transfer, *Ind. Engng. Chem.*, 50:1565 (1958).

77. R. Anantanarayanan and A. Ramachandran, Effect of Vibration on Heat Transfer from a Wire to Air in Parallel Flow, *Trans. ASME*, 80:1426 (1958).

78. I. A. Raben, The Use of Acoustic Vibrations to Improve Heat Transfer, *Proc. 1961 Heat Transfer and Fluid Mech. Inst.*, pp. 90–97, Stanford University Press, 1961.

79. I. A. Raben, G. E. Cummerford, and G. E. Neville, An Investigation of the Use of Acoustic Vibrations to Improve Heat Transfer Rates and Reduce Scaling in Distillation Units Used for Saline Water Conversion, Final Rept., Contract No. 14-01-188 Project 2-919-1, Southwest Research Institute, 1962, Office of Saline Water Research and Development Progress Rept. No. 65 (1962).

80. J. W. Ogle and A. J. Engel, The Effect of Vibration on a Double-Pipe Heat Exchanger, AIChE Preprint No. 59 for 6th Natl. Heat Transfer Conf., 1963.

81. I. I. Palyeyev, B. D. Kachnelson, and A. A. Tarakanovskii, Study of Process of Heat and Mass Exchange in a Pulsating Stream, *Teploenergetika*, 10:71 (1963).

82. E. D. Jordan and J. Steffans, An Investigation of the Effect of Mechanically Induced Vibrations on Heat Transfer Rates in a Pressurized Water System, The Catholic University of America, *NYO*-2655-1 (1965).

83. F. C. McQuiston and J. D. Parker, Effect of Vibration on Pool Boiling, *ASME* 67-HT-49 (1967).

84. D. C. Price and J. D. Parker, Nucleate Boiling on a Vibrating Surface, *ASME* 67-HT-58 (1967).

85. M. Markels, R. L. Durfee, and R. Richardson, Annual Progress Report of Methods to Increase Burnout Heat Transfer, *USAEC NYO*-9500 (1960).

86. J. P. Holman and T. P. Mott-Smith, The Effects of Constant-Pressure Sound Fields on Free Convection Heat Transfer from a Horizontal Cylinder, *J. Aero/Space Sci.*, 26:188 (1959).

87. A. L. Sprott, J. P. Holman, and F. L. Durand, An Experimental Study of the Effects of Strong Progressive Sound Fields on Free-Convection Heat Transfer from a Horizontal Cylinder, *ASME* 60-HT-19 (1960). Also, *WADC, TR* 59-717 (1959).

88. R. M. Fand and J. Kaye, The Influence of Sound on Free Convection from a Horizontal Cylinder, *J. Heat Transfer*, 83:133 (1961). Also, *WADC TN* 59-18 (1959).

89. B. H. Lee and P. D. Richardson, Effect of Sound on Heat Transfer from a Horizontal Circular Cylinder at Large Wavelength, *J. Mech. Engng. Sci.*, 7:127 (1965).

90. R. M. Fand, J. Roos, P. Cheng, and J. Kaye, The Local Heat-Transfer Coefficient around a Heated Horizontal Cylinder in an Intense Sound Field, *J. Heat Transfer*, 84:245 (1962).

91. P. D. Richardson, Local Details of the Influence of a Vertical Sound Field on Heat Transfer from a Circular Cylinder, *Proc. 3d Int. Heat Transfer Conf.*, AIChE, vol. 3, pp. 71–77, 1966.

92. J. H. Seely, Effect of Ultrasonics on Several Natural Convection Cooling Systems, Masters thesis, Syracuse University, 1960. Also, *Int. Bus. Mach. Tech. Note TN* 00.456 (1960).

93. A. A. Zhukauskas, A. A. Shlanchyauskas, and Z. P. Yaronees, Investigation of the Influence of Ultrasonics on Heat Exchange between Bodies in Liquids, *J. Engng. Physics*, **4**:58 (1961).

94. M. B. Larson and A. L. London, A Study of the Effects of Ultrasonic Vibrations on Convection Heat Transfer to Liquids, *ASME* 62-HT-44 (1962).

95. G. C. Robinson, C. M. McClude III, and R. Hendricks, Jr., The Effects of Ultrasonics on Heat Transfer by Convection, *Amer. Ceram. Soc. Bull.*, **37**:399 (1958).

96. R. M. Fand, The Influence of Acoustic Vibrations on Heat Transfer by Natural Convection from a Horizontal Cylinder to Water, *J. Heat Transfer*, **87**:309 (1965).

97. J. H. Gibbons and G. Houghton, Effects of Sonic Vibrations on Boiling, *Chem. Engng. Sci.*, **15**:146 (1961).

98. K. W. Li and J. D. Parker, Acoustical Effects on Free Convective Heat Transfer from a Horizontal Wire, *J. Heat Transfer*, **89**:277 (1967).

99. R. C. Martinelli, L. M. Boelter, E. B. Weinberg, and S. Takahi, Heat Transfer to a Fluid Flowing Periodically at Low Frequencies in a Vertical Tube, *Trans. ASME*, **65**:789 (1943).

100. F. B. West and A. T. Taylor, The Effect of Pulsations on Heat Transfer, *Chem. Engng. Prog.*, **48**:34 (1952).

101. J. H. Marchant, Discussion of Paper by R. C. Martinelli et al., *Trans. ASME*, **65**:789 (1943).

102. G. B. Darling, Heat Transfer to Liquids in Intermittent Flow, *Petroleum*, **180**:177 (1959).

103. R. Lemlich and J. C. Armour, Forced Convection Heat Transfer to a Pulsed Liquid, AIChE Preprint No. 2 for Sixth Natl. Heat Transfer Conf., 1963.

104. T. Shirotsuka, N. Honda, and Y. Shima, Analogy of Mass, Heat and Momentum Transfer to Pulsation Flow from Inside Tube Wall, *Kagaku-Kikai,* **21**:638 (1957).

105. W. Linke and W. Hufschmidt, Wärmeübergang bei Pulsierender Strömung, *Chem. Ing. Tech.*, **30**:159 (1958).

106. T. W. Jackson, W. B. Harrison, and W. C. Boteler, Free Convection, Forced Convection, and Acoustic Vibrations in a Constant Temperature Vertical Tube, *J. Heat Transfer,* **81**:68 (1959).

107. T. W. Jackson, K. R. Purdy, and C. C. Oliver, The Effects of Resonant Acoustic Vibrations on the Nusselt Number for a Constant Temperature Horizontal Tube, "International Developments in Heat Transfer," pp. 483–489, ASME, New York, 1961.

108. R. Lemlich and C. K. Hwu, The Effect of Acoustic Vibration on Forced Convective Heat Transfer, *AIChE Journal*, **7**:102 (1961).

109. W. F. Mathewson and J. C. Smith, Effect of Sonic Pulsation on Forced Convective Heat Transfer to Air and on Film Condensation of Isopropanol, *Chem. Engng. Prog. Symp. Ser. 59*, **41**:173 (1963).

110. A. E. Bergles, The Influence of Flow Vibrations on Forced-Convection Heat Transfer, *J. Heat Transfer*, **86**:559 (1964).

111. S. E. Isakoff, Effect of an Ultrasonic Field on Boiling Heat Transfer—Exploratory Investigation, *Heat Transfer and Fluid Mechanics Institute Preprints*, p. 16, Stanford University, 1956.

112. S. W. Wong and W. Y. Chon, Effects of Ultrasonic Vibrations on Heat Transfer to Liquids by Natural Convection and by Boiling, *AIChE Journal*, **15**:281 (1969).

113. D. A. DiCicco and R. J. Schoenhals, Heat Transfer in Film Boiling with Pulsating Pressures, *J. Heat Transfer*, **86**:457 (1964).

114. F. E. Romie and C. A. Aronson, Experimental Investigation of the Effects of Ultrasonic Vibrations on Burnout Heat Flux to Boiling Water, *ATL* A-123, July (1961).

115. A. E. Bergles and P. H. Newell, Jr., The Influence of Ultrasonic Vibrations on Heat Transfer to Water Flowing in Annuli, *Int. J. Heat Mass Transfer*, **8**:1273 (1965).

116. A. P. Ornatskii and V. K. Shcherbakov, Intensification of Heat Transfer in the Critical Region with the Aid of Ultrasonics, *Teploenergetika*, **6**:84 (1959). Translation in Ref. 85.

117. H. Senftleben and W. Braun, Der Einfluss Elektrischer Felder auf den Wärmeström in Gasen, *Zeitschrift für Physik*, **102**:480 (1936).

118. S. Arajs and S. Legvold, Electroconvectional Heat Transfer in Gases, *J. Chem. Phys.*, **29**:531 (1958).

119. H. Choi, Electrohydrodynamic Boiling Heat Transfer, Tufts Univ. Mech. Engng. Rept. 61-12-1, December, 1961. Also, doctoral dissertation in mechanical engineering, MIT, 1962.

120. E. Bonjour, J. Verdier, and L. Weil, Electroconvection Effects on Heat Transfer, *Chem. Engng. Prog.,* **58**:63 (1962).
121. S. M. Marco and H. R. Velkoff, Effect of Electrostatic Fields on Free-Convection Heat Transfer from Flat Plates, *ASME* 63-HT-9 (1963).
122. M. R. Wahlert and H. O. Croft, The Electrostatic Effect and the Heat Transmission of a Tube, University of Iowa Studies in Engineering, Bulletin 25, 1941.
123. F. Berger and V. Stach, Increase of Heat Transfer in a Gas-cooled Reactor, *Proc. 2d Int. Conf. Peaceful Uses of Atomic Energy, Geneva,* vol. 7, pp. 751–757, 1958.
124. V. Stach, Influence of Electric Field on the Cooling Gas Flow in a Nuclear Reactor, *Int. J. Heat Mass Transfer,* **5**:445 (1962).
125. F. Berger and L. Derian, The Influence of a Direct Electric Field on the Heat Transfer to Cooling CO_2 at Higher than Atmospheric Pressures in a Nuclear Reactor, *Proc. 3d Int. Conf. Peaceful Uses of Atomic Energy, New York,* vol. 8, pp. 355–361, 1965.
126. H. R. Velkoff, An Exploratory Investigation of the Effects of Ionization on the Flow and Heat Transfer with a Dense Gas, ASD-TDR-63-842 (1963).
127. H. R. Velkoff, An Analysis of the Effect of Ionization on the Laminar Flow of a Dense Gas in a Channel, RTD-TDR-63-4009 (1963).
128. R. A. Moss and J. Grey, Heat Transfer Augmentation by Steady and Alternating Electric Fields, *Proc. 1966 Heat Transfer and Fluid Mechanics Institute,* Stanford University Press, pp. 210–235, 1966.
129. E. Schmidt and W. Leidenfrost, Der Einfluss Elektrischer Felder auf den Wärmetransport in Flüssigen Elektrischen Nichtleitern, *ForschHft. Ver dt. Ing.,* **19**:65 (1953).
130. R. L. Durfee, C. R. Nichols, J. M. Spurlock, and M. Markels, Jr., Boiling Heat Transfer with Electrical Field (EHD), *NYO* 2404-76 (1966).
131. H. R. Velkoff and J. H. Miller, Condensation of Vapor on a Vertical Plate with a Transverse Electrostatic Field, *J. Heat Transfer,* **87**:197 (1965).
132. H. Y. Choi and J. M. Reynolds, Study of Electrostatic Effects on Condensing Heat Transfer, *AFFDL* TR-65-51 (1966).
133. H. Y. Choi, Electrohydrodynamic Condensation Heat Transfer, *ASME* 67-HT-39 (1967).
134. M. Behar, M. Courtaud, R. Ricque, and R. Semeria, Fundamental Aspects of Subcooled Boiling with and without Dissolved Gases, *Proc. 3d Int. Heat Transfer Conf., AIChE, New York,* vol. 4, pp. 1–11, 1966.
135. M. Jakob and W. Linke, Der Wärmeübergang beim Verdampfen von Flüssigkeiten an Senkrechten und Waagerechten Flächen, *Phys. Zeitschrift,* **36**:267 (1935).
136. T. H. Insinger, Jr. and H. Bliss, Transmission of Heat to Boiling Liquids, *Trans. AIChE,* **36**:491 (1940).
137. A. I. Morgan, L. A. Bromley, and C. R. Wilke, Effect of Surface Tension on Heat Transfer in Boiling, *Ind. Engng. Chem.,* **41**:2767 (1949).
138. E. K. Averin and G. N. Kruzhilin, The Influence of Surface Tension and Viscosity on the Conditions of Heat Exchange in the Boiling of Water, *Isvest. Akad. Nauk SSSR. Otdel. Tekh. Nauk,* **10**:131 (1955).
139. A. J. Lowery, Jr. and J. W. Westwater, Heat Transfer to Boiling Methanol—Effect of Added Agents, *Ind. Engng. Chem.,* **49**:1445 (1957).
140. W. R. van Wijk, A. S. Vos, and S. J. D. van Stralen, Heat Transfer to Boiling Binary Liquid Mixtures, *Chem. Engng. Sci.,* **5**:68 (1956).
141. S. J. D. van Stralen, Heat Transfer to Boiling Binary Liquid Mixtures, *Brit. Chem. Engng.,* part I, 4:8–17; part II, 4:78–82 (1959).
142. M. Carne, Some Effects of Test Section Geometry, in Saturated Pool Boiling, on the Critical Heat Flux for Some Organic Liquids and Liquid Mixtures, AIChE Preprint 6 for Seventh Natl. Heat Transfer Conf., August, 1964.
143. G. Leppert, C. P. Costello, and B. M. Hoglund, Boiling Heat Transfer to Water Containing a Volatile Additive, *Trans. ASME,* **80**:1395 (1958).
144. A. E. Bergles and L. S. Scarola, Effect of a Volatile Additive on the Critical Heat Flux for Surface Boiling of Water in Tubes, *Chem. Engng. Sci.,* **21**:721 (1966).
145. G. K. Rhode, D. M. Roberts, D. C. Schluderberg, and E. E. Walsh, Gas-suspension Coolants for Power Reactors, *Proc. Amer. Power Conf.,* vol. 22, pp. 130–137 (1960).
146. D. C. Schluderberg, R. L. Whitelaw, and R. W. Carlson, Gaseous Suspensions—A New Reactor Coolant, *Nucleonics,* **19**:67 (1961).
147. W. T. Abel, D. E. Bluman, and J. P. O'Leary, Gas-Solids Suspensions as Heat-carrying Mediums, *ASME* 63-WA-210 (1963).

148. R. Pfeffer, S. Rossetti, and S. Lieblein, Analysis and Correlation of Heat Transfer Coefficient and Friction Factor Data for Dilute Gas-Solid Suspensions, *NASA TN* D-3603 (1966).

149. R. Pfeffer, S. Rossetti, and S. Lieblein, The Use of a Dilute Gas-Solid Suspension as the Working Fluid in a Single Loop Brayton Space Power Generation Cycle, *AIChE* Paper 49c, presented at 1967 national meeting.

150. E. E. Gose, E. E. Peterson, and A. Acrivos, On the Rate of Heat Transfer in Liquids with Gas Injection through the Boundary Layer, *J. Appl. Phys.,* **28**:1509 (1957).

151. E. E. Gose, A. Acrivos, and E. E. Peterson, Heat Transfer to Liquids with Gas Evolution at the Interface, paper presented at AIChE annual meeting, 1960.

152. G. E. Sims, U. Aktürk, and K. O. Evans-Lutterodt, Simulation of Pool Boiling Heat Transfer by Gas Injection at the Interface, *Int. J. Heat Mass Transfer,* **6**:531 (1963).

153. A. A. Kudirka, Two-phase Heat Transfer with Gas Injection through a Porous Boundary Surface, ANL-6862 (1964). Also, *ASME* 65-HT-47 (1965).

154. S. G. Bankoff, Taylor Instability of an Evaporating Plane Interface, *AIChE Journal,* **7**:485 (1961).

155. P. C. Wayner, Jr. and S. G. Bankoff, Film Boiling of Nitrogen with Suction on an Electrically Heated Porous Plate, *AIChE Journal,* **11**:59 (1965).

156. V. K. Pai and S. G. Bankoff, Film Boiling of Nitrogen with Suction on an Electrically Heated Horizontal Porous Plate: Effect of Flow Control Element Porosity and Thickness, *AIChE Journal,* **11**:65 (1965).

157. P. C. Wayner, Jr. and A. S. Kestin, Suction Nucleate Boiling of Water, *AIChE Journal,* **11**:858 (1965).

158. C. P. Costello and E. R. Redeker, Boiling Heat Transfer and Maximum Heat Flux for a Surface with Coolant Supplied by Capillary Wicking, *Chem. Eng. Prog. Symp. Ser.* **59**:104 (1963).

159. C. P. Costello and W. J. Frea, The Role of Capillary Wicking and Surface Deposits in the Attainment of High Pool Boiling Burnout Heat Fluxes, *AIChE Journal,* **10**:393 (1964).

160. R. S. Gill, Pool Boiling in the Presence of Capillary Wicking Materials, master's thesis in mechanical engineering, M.I.T., 1967.

161. E. M. Sparrow and R. D. Cess, The Effect of a Magnetic Field on Free Convection Heat Transfer, *Int. J. Heat Mass Transfer,* **3**:267 (1961).

162. R. M. Singer, Laminar Film Condensation in the Presence of an Electromagnetic Field, *ASME* 64-WA/HT-47 (1964).

163. L. Katz, Natural Convection Heat Transfer with Fluids Using Suspended Particles which Undergo Phase Change, doctoral dissertation in mechanical engineering, M.I.T., 1967.

NOMENCLATURE

a vibrational amplitude

c_p specific heat

D channel diameter

D_e channel hydraulic diameter

D_i inside diameter of tube or annulus

D_o outside diameter of annulus

E_{max} maximum electrostatic field strength

e protrusion height

F fin factor used in Eq. (2)

f friction factor,
vibrational frequency

G mass velocity

h heat transfer coefficient, enthalpy

k thermal conductivity

L channel heat length
protrusion spacing

P pumping power

p pressure

ΔP pressure drop

q rate of heat transfer
q'' heat flux
q_{cr}'' critical heat flux
SPL sound pressure level
T temperature
ΔT $= T_w - T_b$ or $T_w - T_\infty$
V average velocity
w_g gas mass rate of flow
w_s solid mass rate of flow
y tube diameters per $180°$ tape twist
X vapor quality
β volumetric coefficient of expansion
μ dynamic viscosity
ν kinematic viscosity
ρ density

Dimensionless Groups

Gr   Grashof number $= g\beta\Delta T L^3/\nu^2$
Nu   Nusselt number $= hD/k$
Pr   Prandtl number $= c_p\mu/k$
Ra   Rayleigh number $= $ Gr $\cdot$ Pr
Re   Reynolds number $= VD/\nu,\ VL/\nu$

Subscripts

a augmentative data
b bulk fluid condition
∞ free stream condition
d displaced promoter data
e extended surface data, evaluated with hydraulic diameter
ex   condition at outlet of channel
f film fluid condition $(T_w + T_b)/2$
in   condition at inlet of channel
o nonaugmentative data
P evaluated at constant pumping power
r rough surface
s swirl flow data
sat   saturation condition
v vibration data
w wall condition

Section **11**

Electric and Magnetic Fields

MARY F. ROMIG

Engineering Consultant, Topanga, California

INTRODUCTION

When an electromagnetic field is applied to an electrically conducting fluid, body forces are induced and dissipative mechanisms are created which generate unique problems in the field of heat transfer. The subject designated as magnetohydrodynamic heat transfer is usually abbreviated as MHD heat transfer. It can be roughly divided into two sections: one which contains problems in which the heating is an incidental byproduct of the electromagnetic fields, and one embracing studies in which the primary use of electromagnetic fields is to control the heat transfer. The first group includes such devices as magnetohydrodynamic generators and accelerators, and, to a lesser degree, pumps and flowmeters. These are broadly classified as channel flows, although most operating designs consist of variable-area ducts. The second group includes natural convection flows and aerodynamic heating; here the geometric configurations are varied. Both of these areas share a lack of experimental verification of the existing theory, and also suffer from the complete absence of a reliable theory of turbulent heat transfer.

Because of the restrictions on length for a handbook article it is impossible to discuss all the aspects of MHD heat transfer, in particular the special problems encountered in engineering applications. Rather, this article will present a condensed theoretical treatment of three general areas of interest in MHD heat transfer: channel flows, free convection, and boundary-layer flows. A comprehensive review and discussion of these basic areas are given in Ref. 1, along with an extensive bibliography of research published up to 1962. The review article by Kantrowitz [2] briefly covers the problems associated with working designs for generators and propulsion devices, while the proceedings of the several annual conferences on engineering magnetohydrodynamics [3] and power generation [4] contain more specific examples of techniques used in the solution of various heating problems encountered in the production of more efficient designs. Reference 4, in particular, describes research being carried on in Europe and Japan. Journals such as the *Physics of Fluids, AIAA Journal* (and the predecessor *ARS Journal*), and the *Journal of Heat Transfer* (ASME, Series C) also occasionally contain relevant articles. More current results in engineering MHD are reported in corporation reports, such as those released by the AVCO-Everett Research Laboratory (Everett, Mass.); the Missile and Space Division, Space Sciences Laboratory, of the General Electric Co., Valley Forge, Pa.; and Westinghouse Research Laboratories, Pittsburgh, Pa.

Basic Equations and Assumptions. The equations and examples discussed herein are based on the concept of a continuum fluid. This implies that the electronic mean free path is small compared to the characteristic length in the problem; furthermore, the fluid is assumed electrically neutral in a local sense, and there are no large relative motions of the charged particles which could induce electric fields or cause the transport coefficients to be anisotropic. Unfortunately, these conditions are not ordinarily satisfied in many practical applications.

In most operating devices the magnetic field must be so large (in order to produce appreciable forces), or the pressure so low (because of high velocities) that the electrons are free to spiral many times about the field lines between collisions with heavier particles. These electrons will thus have a component of motion perpendicular to the applied fields. For very large collision times, a current will be produced perpendicular to the applied electric field. The field produced by the motion is called the Hall field. The angle which the current makes with the electric field is $\tan^{-1}\Omega$, where Ω is the ratio of the gyration frequency of the electrons, eB/m_e, to the collision frequency. Therefore, for high magnetic fields and low pressures, Hall effects will be predominant. This motion of charges in a preferential direction can also cause the transport coefficients to become dependent on the direction of the magnetic field.

However, when anisotropy and Hall currents are generated in the working medium the equations become so complex that in many cases a general solution is impossible. For this reason, much more research has been done in the area of generator and accelerator design on the development of refractory wall materials, or on various electrode configurations, etc., than on extensive theoretical calculations. Many examples of the alternative techniques used for construction of more efficient devices are given in Refs. 3 and 4.

Since the governing equations of MHD heat transfer result from a combination of two disciplines (electromagnetic theory and fluid mechanics) the ensuing equations are necessarily complex and difficult to treat except for several simplified geometries and flows. For the purpose of generalization, they are given here in vector form; particular examples will be given in later sections.

The application of an electromagnetic field to an electrically conducting fluid cannot create mass, so the continuity equation retains its usual form of*

$$\nabla \cdot (\rho \mathbf{V}) = 0 \tag{1}$$

The symbols are defined in the nomenclature; MKS (meter-kilogram-sec) units are used.

The momentum equation must contain all the forces which are induced by the electromagnetic fields, i.e., the ponderomotive, magnetostrictive, electrostrictive, and electrostatic forces. The ponderomotive force is the same force which commonly acts on a rigid current-carrying wire in a magnetic field. It is present here because the electrically conducting fluid elements are crossing magnetic field lines. Magneto-striction and electrostriction are both defined (for a nonferromagnetic medium) as an elastic deformation of the fluid. When the magnetic permeability and electric susceptibility are functions of the fluid density they must be included; however, in most cases of practical interest these parameters are constants. This does not hold true for polar liquids and gases, where the dielectric coefficient depends on both temperature and density; however, the forces generated by electrostriction are generally insignificant except in free-convection flows [1]. Hence the momentum equation can be written

$$\rho \frac{d\mathbf{V}}{dt} = -\nabla p + \tau_{ij} + \mathbf{f} + \mathbf{j} \times \mathbf{B} \tag{2}$$

where τ_{ij} is the stress tensor and the last term is the ponderomotive force.

The energy equation must satisfy the first law of thermodynamics and Maxwell's laws for a moving medium. This equation is formidible when the magnetoelectro-strictive terms are important. Chu [5] has given a complete exposition of the derivation, as has Lykoudis, discussed in Ref. 1. The electromagnetic field contributes only one term explicitly; this is the Joule heating. However, because the field alters the fluid-dynamic parameters, terms such as the viscous dissipation can contribute substantially to the heating. The state variables are not affected by electromagnetic fields as long as magnetoelectrostriction is not a factor, and the energy equation becomes

$$\rho \left[\frac{d\mathcal{U}}{dt} + p \frac{d}{dt}\left(\frac{1}{\rho}\right) \right] = -\nabla \cdot \mathbf{q} + \Phi + W + \frac{j^2}{\sigma} \tag{3}$$

where Φ is the viscous dissipation and the last term is the Joule heating. Other forms of Eq. (3) are

$$\rho C_p \frac{dT}{dt} = -\nabla \cdot \mathbf{q} + \Phi + W + \frac{j^2}{\sigma} \tag{3a}$$

for an incompressible fluid, and

$$\rho \frac{dh}{dt} = \frac{dp}{dt} - \nabla \cdot \mathbf{q} + \Phi + W + \frac{j^2}{\sigma} \tag{3b}$$

---

*Only steady-state solutions will be discussed here.

for a compressible fluid in terms of enthalpy. If the stagnation enthalpy is defined in the usual manner as $h_s = h + V^2/2$, and the pressure term is eliminated by using the momentum equation, then

$$\rho \frac{dh_s}{dt} = -\nabla \cdot \mathbf{q} + W + (\Phi + \boldsymbol{\tau} \cdot \mathbf{V}) + \mathbf{j} \cdot \mathbf{E} \qquad (3c)$$

which shows that the stagnation enthalpy is increased by the power addition $\mathbf{j} \cdot \mathbf{E}$. When there is no applied electric field, the stagnation enthalpy is unaltered by the presence of the magnetic field, except by its influence on the viscous term.

The phenomenological law which determines the current $\mathbf{j}$ is given by Ohm's law for a moving medium [5]

$$\mathbf{j} = \sigma(\mathbf{E} + \mathbf{V} \times \mathbf{B}) \qquad (4)$$

where the term $\mathbf{V} \times \mathbf{B}$ is the electric field induced by the motion of the fluid across field lines.

One final equation is necessary to complete the system hydrodynamically; it involves the magnetic field. Since an applied field alters the fluid flow by the ponderomotive force, the fluid in turn will react on the field to relieve the force. Hence the magnetic field appearing in Eqs. (2) to (4) is the *resultant,* or total, field present in the fluid. Its behavior is determined from the first of the Maxwell equations given below for time-dependent fields and constant permeability and susceptibility (this particular equation is also called Faraday's law)

$$\nabla \times \mathbf{B} = \mu_e \mathbf{j}$$
$$\nabla \times \mathbf{E} = 0$$
$$\nabla \cdot \mathbf{B} = \nabla \cdot \mathbf{E} = 0 \qquad (5)$$

This, along with Ohm's law, gives

$$\nabla \times \mathbf{B} = \mu_e \mathbf{j}_0 + \mu_e \sigma(\mathbf{E} + \mathbf{V} \times \mathbf{B}) \qquad (6)$$

where $\mathbf{j}_0$ is the current present in any external generating solenoid. If σ and ρ are constant, then Eq. (6) reduces to

$$(\mathbf{V} \cdot \nabla)\mathbf{B} = (\mathbf{B} \cdot \nabla)\mathbf{V} + \frac{1}{\mu_e \sigma} \nabla^2 \mathbf{B} \qquad (7)$$

Dimensionless Parameters. The system of equations given above can be simplified considerably under certain geometric or flow conditions. This is not easily seen unless the system is written in dimensionless form. The arbitrary reference conditions will be given the subscript 0 and either refer to values in the undisturbed flow or are taken external to the environment. The momentum equation, for example, becomes (with f taken as the buoyant force, and with the current eliminated using Ohm's law)

$$\rho^* \frac{d\mathbf{V}^*}{dt} = -\nabla p^* + \frac{Gr}{Re^2} \theta + \frac{1}{Re} \tau_{ij}^* + SK(\sigma \mathbf{E}^* \times \mathbf{B}^*) + S(\sigma \mathbf{V}^* \times \mathbf{B}^* \times \mathbf{B}^*) \qquad (2a)$$

where starred terms are dimensionless; the pressure is referred to the dynamic pressure $\rho_0 V_0^2$ and the temperature is

$$\theta = \frac{T - T_1}{T_2 - T_1} \tag{8}$$

where T_1 and T_2 are arbitrarily constant temperatures.

Two of the parameters in Eq. (2a) are already familiar: Reynolds and Grashof numbers. The new parameters are

$$S = \frac{\sigma_0 B_0^2 L}{\rho_0 V_0} = \text{magnetic interaction parameter}$$

$$= \frac{\text{ponderomotive force}}{\text{inertia force}} \tag{9}$$

and

$$K = \frac{E_0}{V_0 B_0} = \text{generator (pump) coefficient}$$

$$= \frac{\text{applied electric field}}{\text{induced electric field}} \tag{10}$$

The sign of K is negative when the applied field is opposite to the induced field. S determines the induced ponderomotive force due to the interaction of the flow with the field, and SK determines the applied force, which would act even if the system were at rest. Thus if $SK > 1$ in an inviscid flow the velocity is affected by the field.

When viscous forces predominate, either $SRe > 1$ or $SKRe > 1$ in order to change the flow. The product

$$SRe = \frac{\sigma_0 B_0^2 L}{\mu_0} = M^2 = \frac{\text{ponderomotive force}}{\text{viscous force}} \tag{11}$$

defines the square of the Hartmann number M. Although the above equation indicates that M^2 is the more natural parameter, and that the symbol H would better denote the connection with Hartmann, the above usage has been established in MHD and is followed here. (It is also likely that H, if used for the Hartmann number, would become confused with the magnetic field $\mathbf{H}$.) As M increases, the field more greatly affects the boundary-layer structure.

The dimensionless energy equation, written here for a perfect nonreacting gas and isothermal wall, is

$$\rho^* \frac{d\theta}{dt} = -\frac{1}{PrRe} \nabla \cdot \mathbf{q}^* + \frac{\epsilon}{Re} \Phi^* + \epsilon W^* + \mathbf{V}^* \cdot \nabla p^* + S[\sigma(K E^* + \mathbf{V}^* \times \mathbf{B}^*)^2] \tag{3d}$$

where $W^* = W_0 L/\rho_0 V_0^3$ is a dimensionless heat source. The usual parameters in the above equation are the Prandtl and Eckert numbers, based on reference conditions. Note that the Joule heating will vanish only if

$$KE^* = -\mathbf{V}^* \times \mathbf{B}^* \tag{12}$$

that is, *only if the electric field as experienced by an observer moving with velocity* $\mathbf{V}$ *vanishes.* It is obvious that increases in ϵ, S, and K will produce larger internal heating from the electromagnetic fields.

Finally, the magnetic induction equation (which is remarkably similar to the vorticity equation) becomes

$$(\mathbf{V}^* \cdot \nabla)\mathbf{B}^* = (\mathbf{B}^* \cdot \nabla)\mathbf{V}^* + \frac{1}{Re_m} \nabla^2 \mathbf{B}^* \tag{7a}$$

where

$$Re_m = V_0 L \sigma \mu_e = \text{magnetic Reynolds number} \tag{13}$$

in which $1/\sigma\mu_e$ can be thought of as a magnetic "kinematic viscosity." Re_m acts like a Reynolds number; when $Re_m \to 0$, the field lines are unaffected by the flow and the induced magnetic field vanishes. Many of the heat transfer problems discussed here utilize this assumption, which is valid for most working fluids. When $Re_m \to \infty$ the lines are frozen into the fluid; problems of this type are usually encountered in fusion devices and in astrophysics.

Transport Properties. In order to determine the heat transfer, suitable assumptions must be made about the transport properties. Under the assumption of this article, that of a continuum fluid concept, the electromagnetic field will not affect the scalar behavior of the transport properties. If, however, individual motions were allowed, the drift due to electron and ion motion under strong electromagnetic fields causes the viscosity and conductivities to become dependent on the direction of the field. Nonisotropic transport properties are discussed in most books on plasma physics [6].

For liquid metals, the usual thermodynamic transport properties are found in most handbooks. For high-temperature air, an excellent survey is given in Ref. 7.

The electrical conductivity is a direct factor in determining the magnitude of the interaction parameters S, M, and Re_m. In most gases the temperature must be prohibitively high before strong interaction is obtained. Rather than encounter the attendant heating problems, many workers use air or other gases seeded with an easily ionized noncorrosive substance such as cesium, or they utilize the high conductivities

TABLE 1. DC Conductivity in mho/m, for Air in Thermodynamic Equilibrium* ($\rho_0 = 1.293$ g/cm^3)

ρ/ρ_0 T, °K	1	10^{-1}	10^{-2}	10^{-3}
3000	0.00654	0.0236	0.0682	0.172
3500	0.114	0.341	0.766	1.59
4000	0.766	1.99	3.98	7.61
4500	3.41	7.95	14.2	19.9
5000	10.8	18.8	35.5	50.5
5500	26.2	47.7	71	91
6000	50.9	88.4	128	151
6500	85.2	145	202	264
7000	128	213	301	463
7500	176	298	455	752
8000	228	398	740	1180
8500	290	536	1140	1710
9000	392	739	1590	2270
9500	534	1020	2060	2760
10,000	725	1392	2530	
11,000	1340	2530		
12,000	2440			

*Based on calculations made at The RAND Corporation.

and low atom-ion temperatures associated with nonequilibrium ionization [2]. Space prohibits a comprehensive listing of conductivities, but common values of σ vary from 20 mho/m for saturated salt water and 80 mho/m for 31 percent HNO_3 to 10^6 mho/m for liquid mercury. Seeded gas conductivities vary from about 10 to 10^3 for $T < 3500°K$. Seed mixtures usually consist of 0.1 to 1 percent mole percent Cs or K in some combustible fuel. The conductivity of unseeded air, which is given in Table 1, roughly follows an exponential law for degrees of ionization less than 0.1 percent and $\sigma \sim T^{3/2}$ thereafter [8]. It is suggested that Refs. 2 to 4, 8, and 9 be consulted for the particular values encountered in experimental applications.

CHANNEL FLOWS

The analysis of MHD flow through ducts has received considerable theoretical attention because of the simple geometry and practical applications in the design of generators, accelerators, shock tubes, pumps, and flowmeters. Heat is generated by internal dissipation as well as through the electric currents in the walls.

In a simple generator a dc magnetic field is applied normal to a moving, conducting gas. When electrodes are connected to walls perpendicular to B and V the induced electric field produces currents in the fluid and in an external load. The efficiency depends on how much of the generated power, which varies as σV^2 per unit volume, is delivered to the load. Losses occur through Joule and viscous dissipation, wall heating, voltage drops at the electrodes, and end effects. The heat generated varies as σV.

Hall effects are always present in generators. They can be circumvented by segmenting the electrodes and connecting each pair to a separate load (removing the continuous path for circulation of the Hall current). Some generators, however, work on the Hall principle; however the ones in which the electric field is perpendicular to the plasma motion are called Faraday generators [2]. Another practical limitation which exists in gases with low degrees of ionization occurs when the transfer of momentum between ions and neutrals becomes ineffective; in this case the ponderomotive force $j \times B$, which is initially carried by the electrons and transferred electrically to the ions, will not be given to the bulk of the gas. Ion slip (as this is called) occurs when the product of the degree of ionization and the ratio of mean free path to characteristic length becomes of order 1.

The length of the generator determines the magnitude of wall losses. Because of the finite term taken for the forces to interact, the channel can become very long if the transfer of momentum is inefficient. The lower limits on temperature and pressure are set by the cost of the magnet (i.e., a larger magnetic field is necessary if the degree of ionization is low) and by the degree of ion slip and Hall currents permissible. The upper limits on both are determined by the amount of heat loss which can be tolerated. Estimates of heat transfer are generally made for particular designs by performing heat balances and by using nonmagnetic values for the convective heat transfer coefficient. (See, e.g., the discussion in Ref. 3.)

In an accelerator the electromagnetic field is reversed and energy is transferred to the gas by Joule heating and the ponderomotive force. A simple example is the well-known plasma jet. The velocity achieved by the usual expansion through a nozzle can be increased by acceleration with the ponderomotive force of an applied magnetic field. However, high heat transfer rates at the boundaries limit the maximum velocity, although recently magnetic containment has been considered [2]. Other applications for this type of device have been found in wind tunnels, shock tubes, space vehicle propulsion, and welding torches.

A liquid metal flowmeter utilizes the voltage induced by the applied field; the potential drop across the electrodes indicates the flow rate. Heat transfer is not likely to be a problem unless the device is used in a heat exchanger. Closed-cycle MHD systems have also found practical application in the cooling circuits of high-temperature nuclear reactors. When placed before the heat exchangers they improve the efficiency of the reactor for power generation [4].

Theoretical treatment of actual design devices is difficult, for, in general, the flows are not strictly one dimensional. However, the following discussion of one-dimensional flows illustrates the behavior in most simple channels with the exception of effects near the electrodes and at the ends. It is assumed here that the fluid density is high enough so that ion slip or Hall currents do not enter. (These problems are discussed in more detail in Refs. 3 and 4.)

One-dimensional Hartmann Flow. The Hartmann problem (named after its discoverer) approximates one-dimensional flow between parallel plates of spacing d (see Fig. 1) which is considered a channel if the sidewalls are placed far enough apart. A constant magnetic field B_0 is applied normal to the plates and an electric field E_z to the side walls.*

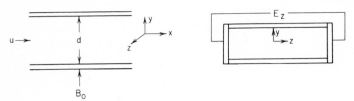

Fig. 1. Channel flow.

From Maxwell's equations it is obvious that $E_z = \text{const} = E_0$ everywhere; since the applied magnetic field is in the y direction, the induced field is in the $(-x)$ direction, opposite to the flow. Written in dimensionless form, the MHD equations are

$$0 = -\frac{\partial p^*}{\partial X} - SK - SU - \frac{1}{\text{Re}}\frac{d^2U}{dY^2} \tag{14}$$

$$0 = -\frac{\partial p^*}{\partial Y} + J_z SB_x \tag{15}$$

$$J_z = K + U = -\frac{1}{\text{Re}_m}\frac{dB_x^*}{dY} \tag{16}$$

where $J_z = j_z/\sigma \bar{u} B_0$, and where $\bar{u}$ is the mean velocity; all other quantities are made dimensionless with respect to B_0, E_0, and the channel radius. The energy equation will be discussed later. Equation (14) is immediately integrable

$$U = \left(-K - \frac{\partial p^*/\partial X}{S}\right)\left(1 - \frac{\cosh MY}{\cosh M}\right) = \frac{M}{1 - \tanh M}\left(1 - \frac{\cosh MY}{\cosh M}\right) \tag{17}$$

The mean current is

$$\bar{J} = \frac{1}{2}\int_{-1}^{1} J_z \, dY = K + 1 \tag{18}$$

and the total current can be written

*The problem when the plates are electrically conducting is discussed in Ref. 1. For the present we will not consider boundary effects at the electrodes.

$$J_z = K + U = \bar{J} - (1 - U) = \bar{J} + J_c \tag{19}$$

or as a sum of mean and circulating currents. When $K = -1$, the walls are electrically insulating and the only currents flowing are circulatory. When $K = 0$ there is no applied electrical field and the channel is short circuited. The theoretical maximum operating condition for a generator is when $K = 1$.

The induced magnetic field can generally be neglected. Integration of Eq. (16) gives

$$B_x^* = -\mathrm{Re}_m \left[KY + \frac{M}{M - \tanh M} \left(Y - \frac{\sinh MY}{M \cosh M} \right) \right] \tag{20}$$

so that $B_x^*(\mathrm{max}) = B_x^*(\pm 1)$. For liquid mercury this equals 0.815 $d\bar{u}$ (MKS), and is an order of magnitude less for ionized air. Since B_x^* is symmetric with respect to the origin, the net ponderomotive force in the Y direction vanishes and there is no net pressure in the Y direction.

The sketch in Fig. 2 shows that as M increases U becomes more flat, so that convection and viscous dissipation near the wall are increased. In order to discuss the effect of M on the heat transfer we assume that either (a) $\bar{u} = \mathrm{const}$ as M and K vary, so that $\partial p^*/\partial X$ must be adjusted to keep the mass flow constant; or (b), $K = \mathrm{constant}$ and M varies, hence E_0 is varied so that the parameter $E_0 d\sqrt{\sigma}/2\bar{u}M\sqrt{\mu}$ remains constant.

The energy equation for constant-density flow is

$$U \frac{\partial \theta}{\partial X} = \frac{1}{\mathrm{PrRe}} \frac{\partial^2 \theta}{\partial Y^2} + \frac{M^2}{\mathrm{Re}} \epsilon J_z^2 + \frac{\epsilon}{\mathrm{Re}} \left(\frac{\partial U}{\partial Y} \right)^2 \tag{21}$$

where $W = 0$. A heat balance gives

$$\frac{\partial \theta}{\partial X} = \mathrm{const} = A = \frac{(\epsilon/\mathrm{Re}) \int_0^1 [M^2 J_z + (\partial U/\partial Y)^2] dY}{\rho C_p \bar{u} \Delta T} + \frac{W_0}{\rho C_p \bar{u} \Delta T} \tag{22}$$

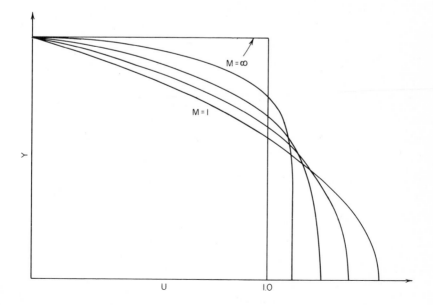

where W_0 is any heat added externally. The three mechanisms which affect the heat transfer are: convection $AU(Y)$, internal heat distribution J_z^2, and the total heat input $\rho C_p \bar{u} \Delta T A - W_0$.

Figure 2 indicates that convection near the wall increases as M increases. If there were no internal heating, the temperature would become uniform as M increases. (If the wall temperature is constant, this term vanishes.) The internal heat due to viscous dissipation depends only on M. It is maximum as the wall and increases with M, that is, its ratio to the mean viscous dissipation is

$$\left(\frac{dU}{dY}\right)^2 = \frac{M^4}{(M - \tanh M)^2} \frac{\sinh^2 MY}{\cosh^2 M} \tag{23}$$

The ohmic heating J_z^2 depends on both M and K. When $K = -1$, circulating currents produce internal heat very close to the walls, and as $M \to \infty$ the viscous dissipation and ohmic heating are of the same order. (However, the *maximum* ohmic dissipation is larger than the *mean* viscous dissipation by a factor of M^2, which implies that the latter can be neglected in analyses of the heating.) The current at the wall will vanish when $K = 0$. Then, most of the ohmic heating is at the center of the channel, and as $M \to \infty$ the variation becomes uniform across the channel. The net result of both viscous and ohmic heating in this case is to also produce a uniform heat source throughout the channel. For other values of K the ohmic heating will be dominant. For K less than zero the bulk of ohmic heating will take place at the walls, and for positive K, the heating will be larger in the center of the channel.

The heat transfer for a constant wall temperature is

$$q = \frac{2\mu \bar{u}^2}{d} M^2 \left[(K + 1)^2 + \frac{M \sinh 2M - \cosh 2M + 1}{2(M \cosh M - \sinh M)^2} \right] \tag{24}$$

When $M = K = 0$, $q = 6\mu \bar{u}^2/d$. The ratio $q/q_{M = K = 0}$ is plotted in Fig. 3. The dashed lines show the effect of neglecting viscous dissipation, since this term is often omitted in order to simplify the problem of finding the mean temperature when there are heat sources or when the walls are conducting. When the device is electrically insulated ($K = -1$), the neglect of viscous dissipation incurs errors of a factor of 2 in the heat transfer.

Integration of Eq. (22) to obtain the mean temperature cannot be done in general except when $T_w(X)$ and W_0 are specified. These cases are discussed in Ref. 1. Caution should be used when solving this equation, for it is easy to incur algebraic errors because of the number of terms in the resulting equations.

Two-dimensional Flow. In most large devices the flow in the central core will be inviscid and a boundary layer will form at the walls (Fig. 4). The treatment of the boundary layer is difficult because the individual behavior of the charged particles becomes more important near the electrode surface. Furthermore, channel boundary layers, unlike those treated later, grow in an electrically-conducting "free-stream," along with a velocity gradient, pressure gradient, and applied electric field. Because of this complexity, few attempts have been made to solve the equations analytically; rather, the attention of research workers has turned to other techniques [3, 4].

An electrode boundary layer is sketched in Fig. 4. Here the magnetic field is transverse to the flow and the electric field normal to it. The problem of heat transfer is accentuated in this device because the currents will be highest near the wall for an accelerator. Furthermore, in a gas, the decrease in the temperature near a cooled electrode will decrease the electric conductivity and increase the Joule dissipation, all of which leads to a higher temperature gradient and an increase in heat transfer. There

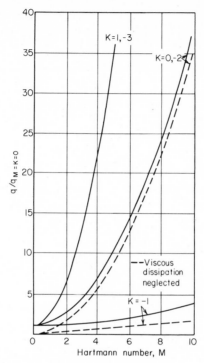

Fig. 3. Convective heat transfer in channel flow [1].

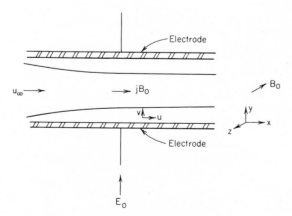

Fig. 4. Accelerator.

will also be an additional source of energy carried to the wall by the electrons; near the electrode, the assumption of quasi-neutrality breaks down and the heat transfer will be affected by the energy gain (or loss) of particles in the potential sheath at the electrodes.

Studies discussed [1] show that the largest effects occur at low Mach numbers and highly accelerated flow, and that the Joule heating tends to thin out the thermal boundary layer and prevent internal heat from reaching the wall. The temperature excess developed in the boundary layer can increase the heat transfer by an order of magnitude.

Axisymmetric Flow. Shock tubes and plasma jets are also used in studying MHD flows, although not explicitly in the region of heat transfer. A suggested configuration is sketched in Fig. 5, where the ponderomotive force due to radial and axial magnetic

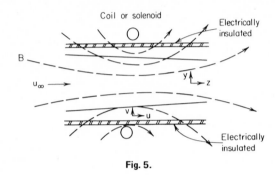

Fig. 5.

fields interacts with the axial and radial velocity components to inhibit the flow toward the wall, thus reducing the heat transfer. No electric field is applied. Some experiments discussed [1] indicate that for $M \sim 10^4$ and when B is in the same direction as the flow, the magnetic field evidently provides a shield against the transfer of heat, since the convective heating is reduced 15 percent below its value when $B = 0$. In the antiparallel or opposite direction, however, the field is ineffective in reducing the heat transfer. Theoretically, there should be no difference in the mode of operation as far as the forces on the fluid are concerned, unless one were to consider Hall currents, which were small in this case. So far, this effect has not been explained.

FREE CONVECTION

Among the classical hydrodynamic flows which already contain a body force is free convection. It is of interest to see what effect the additional (ponderomotive) force will have. Practical application of these flows may be found in heat exchangers utilizing liquid metal coolants; in the area of thermal instability, there may be application to the study of boiling heat transfer.

Before these topics are discussed, it should be mentioned that magneto-electrostrictive forces can become important in free convection flows, because the inertial and shear forces are small by comparison. In liquids or gases with polar molecules the dielectric susceptibility depends on density and temperature so that the resulting force due to any electric field gradient behaves like a buoyant force. Electrostriction can increase the heat transfer due to natural convection by as much as 50 percent if the molecules of the working fluid carry a permanent dipole moment. Other electrostatic effects have been noticed in natural convection flows of certain nonconducting oils, e.g., paraffin, beeswax, and castor oil, under strong electric fields such as those in an oil-cooled transformer. These special topics are discussed at much more length in Ref. 1.

Thermal Instability. A magnetic field will inhibit the onset of convection in a liquid heated from below, because when the convection tends to bend the magnetic force lines, the tension $\mu_e H^2$ leads to a force which opposes any further bending. This is illustrated by placing a container of heated liquid mercury on the edge of a magnetic pole piece. The portion of the liquid in the normal field will remain stable, while the part in the oblique or weaker field will exhibit Benard cells.

The theory of thermal stability has been developed by Chandrasehkar and many supportive experiments have been carried out by Nakagawa, as discussed in Ref. 1. The critical Rayleigh number at which convection first occurs is a function of the boundary conditions at the surface of the liquid, and of the Hartmann number: $M = B \cos\theta d\sqrt{\sigma/\mu}$, where d is the depth of the liquid and θ the angle of incidence of the magnetic field. As $M \to \infty$, $\mathrm{Ra}_c \to \pi^2 M^4$, where the Rayleigh number is defined as $\mathrm{Ra} = \mathrm{Pr} \cdot \mathrm{Gr}$.

Figure 6 illustrates the variation of critical Rayleigh number with Hartmann number as predicted by theory and as verified by the experiments of Nakagawa. The region to the left of the curve is unstable and that to the right is stable. Although the upper surface of the liquid in these experiments was "free," a contaminant film formed which prevented any motion of the surface; hence the data points are compared with the theory for rigid boundaries. This film may have prevented any velocity fluctuations, but it is not clear whether the temperature fluctuations are inhibited as well. Further work remains to be done in this area.

Natural Convection Flows. It would be expected from an investigation of the basic equations that the application of a magnetic field normal to a heated wall would reduce the heat transferred from the plate to the fluid. This action will also reduce the fluid velocity, which may not be desirable in all cases; furthermore, the magnetic field can produce inflection points in the velocity profile if the geometry is not

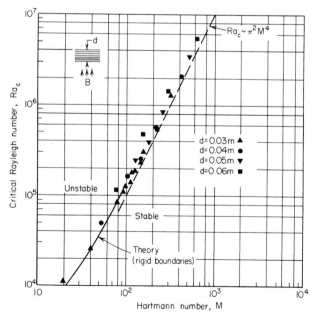

Fig. 6. Thermal stability for liquid mercury heated from below [1].

symmetric. This destabilizing effect may lower the point of transition to turbulent convection.

Heated Vertical Plate. The basic equations indicate that this problem is formidable unless the momentum, energy, and magnetic equations can be uncoupled. In practice, the following assumptions facilitate solution:

1. The magnetic Reynolds number is small.
2. Viscous and Joule dissipation are neglected.
3. Semi-incompressible fluid; the density is constant except in its contribution to the buoyant forces.
4. The applied electric field is zero.

These assumptions are physically realistic for a single plate, and the neglect of the magnetic Reynolds number is probably more justifiable here than in most cases; in fact, for most practical values of B_0 (less than 500 gauss or 0.05 weber/m^2), the induced field is less than the earth's magnetic field.

In order to neglect viscous and Joule dissipation, the Eckert number must be small, and (from Eq. (3a)), $M^2(K + 1)^2 < 1/\epsilon$, or, if the system is short-circuited to satisfy the electromagnetic boundary conditions, then $M^2 < 1/\epsilon$, which is true for relatively high magnetic fields.

Under these assumptions, the basic equations can be written in terms of the plate length and conditions at the edge of the layer as

$$\frac{\partial U}{\partial X} + \frac{\partial V}{\partial Y} = 0$$

$$U\frac{\partial U}{\partial X} + V\frac{\partial U}{\partial Y} = \frac{Gr}{Re^2}\theta + \frac{1}{Re}\frac{\partial^2 U}{\partial Y^2} - U\left(\frac{B}{B_0}\right)^2 S \tag{25}$$

$$U\frac{\partial \theta}{\partial X} + V\frac{\partial \theta}{\partial Y} = \frac{1}{PrRe}\frac{\partial^2 \theta}{\partial Y^2}$$

where $X = x/L$ is along the plate and $Y = y/L$ is normal to it. The temperatures are referred to the adiabatic temperature and the wall temperature. These equations may be solved by a series expansion when B is constant, or by similarity techniques when $B(x)$ assumes the appropriate form.

Solutions for a constant magnetic field can be obtained by a series expansion in powers of the parameter ΛX^2, where

$$\Lambda = \frac{2M^2}{\sqrt{Gr}} = \frac{\text{ponderomotive force}}{\text{buoyant force} \times \text{inertia force}} \tag{26}$$

is called the Lykoudis number; it will be on the order of one. The results of Sparrow and Cess (as discussed in Ref. 1) can be written in terms of the mean Nusselt number over the plate

$$\frac{\overline{Nu}}{(Gr)^{1/4}} = \frac{\overline{Nu_0}}{(Gr)^{1/4}} - 0.3145\,\theta_1'\Lambda \tag{27}$$

where $\overline{Nu_0} = -0.404\,\theta_0'(Gr)^{1/4}$ is the mean Nusselt number without a magnetic field, and θ' is the temperature gradient at the wall.

When similar solutions are desired, the magnetic field must vary as $X^{-1/4}$ for the equations to be independent of X under a similarity transformation. The solutions obtained by Lykoudis, as discussed in Ref. 1, are plotted along with the series solutions in Fig. 7. It appears that the mean heat transfer is not reduced as effectively when the magnetic field is variable. This is especially true at low Prandtl numbers and for $\Lambda \leq 1$. Values of local heat transfer are in better agreement for ΛX^2 up to and

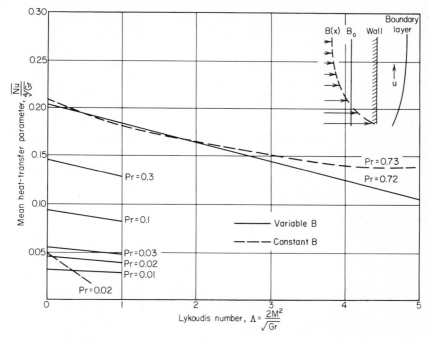

Fig. 7. Mean heat transfer for vertical heated plate in magnetic field [1].

including unity for $Pr = 0.73$. For higher values of ΛX^2, the results of the constant field solution are lower; this is probably due to upstream influence, which becomes more important as the Prandtl number decreases, or as the boundary layer becomes more transparent to heat. Physically, the large variable magnetic field at the leading edge of the plate has little effect on the heat transfer because the velocities are low; further up the plate, the magnetic field has decreased below the values used in the series solution and the ponderomotive force is again less than it would be for the constant-field case. The upstream heat transfer is thus greater than that for a constant field, and the net heat transfer to the fluid is larger when the magnetic field varies.

Parallel Vertical Plates. The solution for parallel plates is more complex because of the addition of the additional parameter of the wall-temperature difference. The geometry, with electrically insulated walls, is identical to the Hartmann problem discussed before. The problem can be solved by series expansion in terms of the parameter $\kappa = Pr Gr \beta g d / C_p$. Although the product $Pr \cdot Gr$ is large, the coefficient $\beta g d / C_p$ is small for liquid mercury or sodium. Solutions for this problem were made by Poots and are discussed at length in Ref. 1. He showed that the dissipative portion of the total heat transfer decreased as the Hartmann number increased, and that the heat transfer was primarily due to conduction, even at zero Hartmann number. These results are sketched in Fig. 8. Here, Nu_d represents the dissipative portion of the total heat transfer $Nu = Nu_c + \kappa Nu_d + O(\kappa^2)$, where Nu_c is the conductive portion of the heating and where $K \leq 1$.

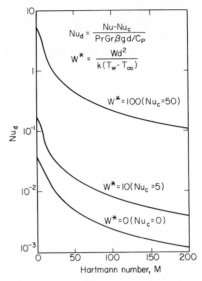

Fig. 8. Reduction in dissipative heat transfer for free convection between parallel plates [1].

BOUNDARY-LAYER FLOWS

In high-velocity flow past an aerodynamic body the air in the boundary layer will become heated either because of viscous dissipation or shocks. If the velocity is large enough, ionization will be produced and the air will become electrically conducting. Application of magnetic field normal to the wall will permit an interaction to occur between the boundary-layer flow and the field, and will cause a reduction in the velocity and a decrease in skin friction drag and heat transfer.

However, the reduction of aerodynamic heat transfer and the control of flow by hydromagnetic means have proved to be less promising than when they were first considered in 1957. The principal drawback has been the increment in vehicle weight caused by the large magnetic field strengths necessary to control naturally ionized air. Now, however, since superconducting coils have been developed, the strength-to-weight ratio of magnets has decreased somewhat.* Also, the velocities associated with space-vehicle reentry from superorbital missions is large enough so that significant ionization can be obtained without seeding the air, at least at the stagnation point of a blunt body.

All boundary-layer flows will be discussed in this section, including the flat plate and the stagnation point. In order to gain insight into what is often an intractible problem, the simpler problem of Couette flow will be discussed first at some length.

Couette Flow. Couette flow, or one-dimensional shear flow, is produced when a viscous fluid between two infinite parallel plates is set into motion by the relative velocity of one of the walls; it is often experimentally studied as the flow between two concentric cylinders when the spacing between the cylinders is small in comparison to the radii (see Fig. 9). Analyses of MHD Couette flow have been made by Bleviss and Leadon, and are discussed at length in Ref. 1.

Incompressible Flow. Mathematically, this flow is the same as Hartmann flow, with the important exception that the boundary conditions on the fluid are different.

---

*Still, cryogenic coolant supplies must be carried, and furthermore, the field area of these magnets is still small although the line density is very large compared to conventional magnets of the same size.

In order to simulate a boundary-layer flow, it is necessary to assume that the pressure gradient is zero and there is no applied electric field; then some shorting arrangement must be provided for the currents, such as that in Fig. 9c. The Hartmann equations become

$$\frac{d^2U}{dY^2} - M^2U = 0 \tag{28}$$

$$J_z SB_x^* = \frac{\partial p^*}{\partial Y} \tag{29}$$

$$J_z = U = -\frac{1}{\mathrm{Re}_m} \frac{dB_x^*}{dY} \tag{30}$$

where all quantities are referred to the upper wall at displacement δ from the lower, stationary wall, and where $u(\delta) = u_\infty$. The velocity is immediately obtained from Eq. (28)

$$U = \frac{\sinh MY}{\sinh M} \tag{31}$$

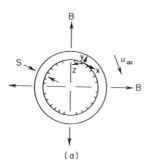

(a)

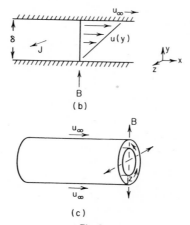

(b)

(c)

Fig. 9.

From Eq. (30), it is seen that J_z and U have the same variation with Y, as sketched in Fig. 10.

Visualizing the induced magnetic field B_z^* as the field induced by a current sheet of mean density $\bar{J}$, then B_x^* will be antisymmetric at the walls, obtaining equal and opposite values. Thus, integration of Eq. (30) gives

$$B_x^* = \frac{\text{Re}_m}{2} \left(\frac{1 + \cosh M - 2 \cosh MY}{M \sinh M} \right) \tag{32}$$

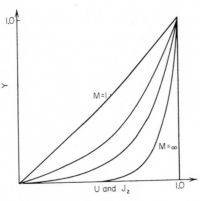

Fig. 10.

The field is zero at the point η determined by $\int_0^\eta J_z dY = \int_\eta^1 J_z dY$. This integral condition gives η in terms of hyperbolic functions

$$\cosh M\eta = \frac{\cosh M + 1}{2} \tag{33}$$

For the channel flow discussed earlier, the null-point was always at the center because J_z was symmetric. Here, as the Hartmann number increases, the current density becomes confined to a thin layer near the wall. Because $J_z \cdot B_x^*$ is not symmetric as it was in the channel-flow problem, a hydrostatic pressure is induced by the ponderomotive force. Defining Δp^* as $p^*(1) - p^*(0)$, integration across the flow yields

$$\Delta p^* = \frac{\text{Re}_m}{\text{Re}} \left(\frac{\cosh M - 1}{2 \sinh^2 M} \right) \tag{34}$$

which is negligible as long as $\text{Re}_m \ll \text{Re}$. For liquid mercury this ratio is on the order of 10^{-6}; for air, where the local dynamic pressures are low, this term could become significant.

The heat transfer to the lower wall was not changed by the addition of a magnetic field. Writing the energy equation in terms of the temperature difference between the walls, $\theta = (T - T_w)/(T_\infty - T_w)$

$$\frac{\partial^2 \theta}{\partial Y^2} = -\text{Pr} \cdot \epsilon \left[M^2 U^2 + \left(\frac{dU}{dY} \right)^2 \right] \tag{35}$$

and integrating twice

$$\theta = Y + \frac{Pr \cdot \epsilon}{2}\left(Y - \frac{\sinh^2 MY}{\sinh^2 M}\right) \qquad (36)$$

This reduces to

$$\theta_0 = Y + \frac{Pr \cdot \epsilon}{2}(Y - Y^2) \qquad (37)$$

when $M \to 0$. The heat transfer is

$$q_w = \frac{\Delta Tk}{\delta} + \frac{\mu u_\infty^2}{2\delta} \qquad (38)$$

for both equations, hence the heat transfer to the lower wall is unaffected by the magnetic field.

At the upper wall

$$q_\infty = \frac{\mu u_\infty^2}{\delta}\left(\frac{2M}{\tanh M} - 1\right) - \frac{k\Delta T}{\delta} \qquad (39)$$

so that heat will flow from the fluid to the upper wall when

$$Pr \cdot \epsilon > \frac{2}{(2M/\tanh M) - 1} \geq 2 \qquad (40)$$

where $Pr \cdot \epsilon > 2$ for the same to occur when $M = 0$. The increase in heat transfer is linear with M as M increases.

Probably the most important contribution from this analysis of incompressible Couette flow is the fact that Reynolds' analogy does not hold in MHD shear flows. This is not unexpected, because of the additional energy source in the energy equation. Defining a recovery temperature as $T_r = T_\infty + Pr u_\infty^2/2C_p T_\infty$, and letting $C_H = 1/PrRe$, then

$$\frac{C_f}{C_H} = \frac{2 Pr M}{\sinh M} \qquad (41)$$

The drag is increased because of the action of the ponderomotive force on the fluid. This is also evident from the channel flow analysis discussed earlier, where the pressure drop had to be increased as M grew larger in order to maintain the same mass flow.

Hypersonic Couette Flow. Bleviss also studied the case of Couette flow for an ionized compressible gas. When reasonable assumptions were made for the air properties ($Pr = $ constant and μ varied as the Sutherland law), it was found that the electrical conductivity has a peculiar effect on the heat transfer and skin friction: for certain temperature levels in the shear flow, these parameters became multi-valued functions of the applied field strength. This behavior was traced to the variation of σ with enthalpy [8] in the shear layer, and will be discussed later. The heat transfer in compressible Couette flow is slightly increased by the magnetic field. This is due primarily to the fact that Couette flow is unable to grow in the normal direction as does an ordinary boundary layer.

Flat-plate Boundary Layers. This section covers flow over flat plates, in which the free stream may or may not be conducting. "Wedge" MHD flows, for which $u_\infty \sim x^m$, are discussed in Ref. 1 and in the subsequent section on stagnation flow.

Incompressible Flow. The pioneer work of Rossow, discussed in Ref. 1, provided the basis for many later studies in this field. Here, a small magnetic Reynolds number was assumed so that the induced magnetic field did not enter into the equations of motion. Both the Rayleigh problem and the incompressible boundary layer were analyzed and the equations were solved by a series expansion in powers of $\sqrt{SX}$, where $X = x/\delta$. Terms higher than $(SX)^2$ were neglected. When the free stream was conducting, the velocity varied as $\partial u / \partial x = -\sigma B_0^2/\rho$. The skin friction decreased as S increased; the heat transfer for the case $T_w = T_\infty$ varied as

$$\frac{Nu}{Pr\sqrt{Re_x}} = 0.332 - 0.342SX - \cdots \tag{42}$$

where both the Reynolds number and S are based on the free stream velocity.

An investigation was also made for the case when σ varied throughout the boundary layer; a linear variation was chosen

$$\sigma = \sigma_0 \left(\frac{u_\infty - u}{u_\infty} \right) \tag{43}$$

so that now u_∞ is not affected by the magnetic field. The heat transfer is

$$\frac{Nu}{Pr\sqrt{Re_x}} = 0.332 - 0.103\,SX - \cdots \tag{44}$$

which shows a reduction of significantly less magnitude than previously.

If it is assumed that $\sigma \sim \exp(T - \text{const})^{1/2}$ and the Blasius temperature distribution is used to relate σ with velocity, then the heat transfer is

$$\frac{Nu}{Pr\sqrt{Re_x}} = 0.332 - 1.216SX \tag{45}$$

Comparing the three equations on the basis of the same free-stream kinetic energy indicates that the latter equation shows considerably more reduction in heat transfer than the other two. Neither of the variable-conductivity solutions show any effect of the hysteresis discovered by Bleviss in the Couette flow problem. This is due to the arbitrary relationships assumed between conductivity, temperature, and velocity, and introduces the possibility that a properly seeded gas at relatively low velocities might produce the heat transfer reductions predicted above.

Compressible Flow. When the electric conductivity varies as some power of enthalpy, as it would in the compressible boundary layer, hysteresis effects are produced just as in the hypersonic Couette flow problem discussed earlier. This was discovered by Bush, and is discussed in Ref. 1 at more length.

The boundary-layer equations were solved using a similarity transformation, hence $B_0 \sim 1/\sqrt{x}$ (just as in the mass-transfer problem). Using enthalpy and pressure as state variables, Bush expressed the Chapman-Rubesin variable as

$$\frac{\rho\mu}{\rho_0\mu_0} = \left(\frac{h}{h_0} \right)^{-0.35} \tag{46}$$

and the product of conductivity and viscosity

$$\frac{\sigma \mu}{\sigma_T \mu_0} \sim \text{const} \left(\frac{h}{h_0}\right)^{4.79}$$

(47)

$$\sim 0 \text{ for } T < 1000°K$$

where the constant was varied to account for the variation of σ with pressure, and where the subscript 0 refers to conditions at 222°K; $\sigma_T = 100$ mho/m.

The solution obtained by Bush showed the behavior illustrated in Fig. 11, where both skin friction drag and heat transfer are plotted as functions of the interaction parameter $S = 2\sigma_T B_0^2 L / \rho_0 \sqrt{h_0}$, where $\sqrt{h_0} = u_0 = 472$ m/sec under the Hantzsche-Wendt boundary-layer transformation used by Bush. As u_∞ increases for a fixed value of S, τ or q at first increase. This occurs because the initial low-temperature boundary layer is nonconducting. As u_∞ increases further, magnetic effects occur and both of these drop below their nonmagnetic values until a point is reached where any further increase in velocity (above u_1) would cause a discontinuous drop in τ or q to the

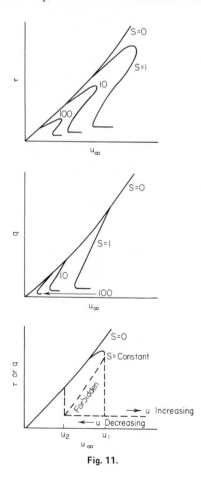

Fig. 11.

lower branch of the curve. If the velocity were decreased at this point, the functions would follow the lower curve until the point u_2 where nonmagnetic behavior again predominates, and would then move up to the upper curve. The region between is mathematically unstable.

This behavior is traced back to the rapid initial variation of σ with h; as the air begins to ionize, electron production is primarily from the reaction $N + O \rightleftharpoons NO^+ + e$. For higher temperatures, above $h = 2600$ cal/gram, NO^+ production decreases until enthalpies above 9400 cal/gram where both N and O will ionize and σ will again increase.

Although the results of calculations show that convective heat transfer could be decreased by as much as 80 percent, the analysis did not include any low-pressure effects; furthermore, the velocities necessary for obtaining these reductions *and* minimizing low-pressure effects would be on the order of 6000 m/sec or larger, a value which is unlikely to be reached with conventional flat-winged vehicles. However, some applications may exist in hypersonic glider entry. (See the discussion under "wedge" flows in Ref. 1.)

Before turning to stagnation point flows, some remarks must be made about these boundary-layer analyses. First, when the solution is pressure-limited by Hall and ion slip effects (to about 10^3 atm or 40,000 m altitude), the neglect of induced magnetic fields, that is, $Re_m \sim 0$, is justified. However, the assumption of local thermodynamic equilibrium over the plate is not easily verified. This assumption implies that the chemical reaction times are short compared to the time it takes an air particle to enter the boundary layer and traverse the body. If the ionization time is much larger, then there will be no magnetic effects. If it is equal to or greater than the dissociation time, then the assumption of equilibrium will produce maximum magnetic effects unless the ionization time is less than the traverse time. Here, the equilibrium solution would *underestimate* the effects. Very little research has been done on ionization times in viscous-heated flows. In actual hypersonic flight, diffusion of heat from the reacting gas will also contribute to the heat transfer. However, recombination heating at the wall will probably not be significant for low degrees of ionization.

Work done to date [1] indicates that reduction of flat plate aerodynamic heating is not promising for naturally ionized air. However, much remains to be done with seeding and nonequilibrium techniques which would produce more efficient cooling at lower flight velocities. However, the addition of magnetic forces to the boundary layer will increase the *total* drag [1], while reducing the skin friction drag, and would probably contribute to heat in the wake regions.

Stagnation Flows. A critical region of heat transfer in hypersonic flight is at the stagnation point. For a typical earth satellite reentry at velocities of 8000 m/sec, the maximum heating rate to a 1-m body is 7850 w/m^2 (80 Btu/ft^2-sec); if $u_\infty = 11,000$ m/sec, the heating rate increases by a factor of 3.

Because a large percentage of the kinetic energy of the airstream is converted into heat by the normal bow shock, a source of ionized air is readily available for $u_\infty > 7000$ m/sec. The sketch in Fig. 12 shows that if there is no applied electric field and a magnetic field is located so that it is perpendicular to the flow, the ponderomotive force acts opposite to the tangential velocity, since j flows circumferentially.

It is immediately obvious from Fig. 12 that magnetic interaction will cause the shock standoff distance to increase, since a larger volume is needed for passage for the air; the heat transfer and skin friction will decrease because the velocity gradients are lower; the pressure will be hardly affected because the normal momentum is virtually unchanged; and the total drag will increase because another force has been imposed on the fluid.

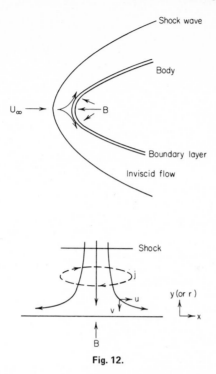

Fig. 12.

To solve the problem of stagnation point heating, the entire flow must be determined, since both the inviscid and viscous regions are conducting. It can be assumed that the conductivity and density are constant in these regions; this is a good approximation for altitudes below 60,000 m. The inviscid solution provides the velocity gradient and the viscous solution the enthalpy gradient at the wall. These, along with a function of density and viscosity (which is unaffected by B_0), determine the stagnation heating

$$\dot{q} = k\left(\frac{\partial T}{\partial y}\right)_w = \left(2\rho_e\mu_e\right)_s^{1/2}\left(\frac{du_e}{dx}\right)^{1/2} h_w' \qquad (48)$$

The velocity gradient can be obtained from either a Newtonian or similarity solution of the inviscid flow. (Both of these are fully discussed in Ref. 1.) The former permits closed form solutions for the flow characteristics, while the latter permits more refinement in the shock shape. In both techniques it is necessary to specify the size of the magnetic Reynolds number. The comparison made in Ref. 1 of the various solutions indicates that when the magnetic interaction parameter is defined at the body

$$S_b = \frac{\sigma B_b^2 R_b}{\rho_\infty u_\infty} \qquad (49)$$

then the particular form of the magnetic field along the stagnation streamline does not greatly affect the velocity gradient. The Newtonian values obtained by Lykoudis [10] are listed in Table 2. These solutions were made for $Re_m \ll 1$, and agree with the more exact solutions within about 5 percent for $S_b < 10$.

TABLE 2. Dimensionless Velocity Gradient for Spheres and Cylinders $(du_e/dx)/(du_e/dx)_0$ **(Newtonian Flow)***

Interaction parameter	Sphere	Cylinder	
$S_b = \dfrac{\sigma B_b^2 R_b}{\rho_\infty u_\infty}$	independent of $\dfrac{\rho_\infty}{\rho_e}$	$\dfrac{\rho_\infty}{\rho_e} = 0.2$	$\dfrac{\rho_\infty}{\rho_e} = 0.5$
0	1	1	1
2	0.810	0.71	0.88
4	0.656	0.473	0.764
6	0.534		0.657
8	0.450		0.565
10	0.388		0.490

*Adapted from data given in Ref. 1.

The enthalpy gradient is determined from a solution of the stagnation-point boundary-layer equations. These fall into the class of "wedge" flows for which u varies as x^m; for axisymmetric flows in $m = -1/3$. This dictates the form of the magnetic field (for similar solutions) to be $B_x \sim x^{m-1/2}$. All solutions disussed in Ref. 1 show that the effects of compressibility and type of flow have negligible effect on the enthalpy gradients as long as the electric conductivity is constant in the boundary layer. The enthalpy gradient increases as the field increases because the radial nature of the magnetic field leads to fuller profiles. However this increase is more than offset by the reduction in the velocity gradient. Table 3 lists the results

TABLE 3. Enthalpy Gradient at the Cooled Wall of a Sphere for Variable Conductivity* $(\sigma/\sigma_e) = (h/h_e)^n$

Interaction parameter	h'_w/h'_{w_0}			
$S_e = \dfrac{\sigma_e B_b^2}{\rho_e (du_e/dx)}$	$n = 0$	$n = 3$	$n = 5$	$n = 10$
0.1	1.02			
0.3	1.024			
0.7	1.033			
1.0	1.040	1.077		
2	1.063	1.142		
5	1.110	1.260		
8	1.143	1.338		
10	1.159	1.378	1.432	1.495
20	1.210	1.530	1.607	1.703
40	1.268	1.709	1.817	1.958
60	1.304	1.825	1.960	2.128
100	1.350	1.975	2.157	2.355

*Adapted from the data of Ref. 1.

obtained by Bush [11] for variable conductivity, and results for constant conductivity, all taken from Ref. 1. Note that here, the proper parameter for boundary-layer data is that based on conditions at the edge of the boundary layer. The two interaction parameters are related by

$$S_e = S_b \left(\frac{\rho_\infty}{\rho_e}\right) \frac{du_e/dx}{(du_e/dx)_0} \frac{R_b \left(du_e/dx\right)_0}{u_\infty} \tag{50}$$

where $\left(du_e/dx\right)_0$ is given in Eq. (52).

The heat transfer for a sphere is given in Fig. 13. The magnetic fields were different

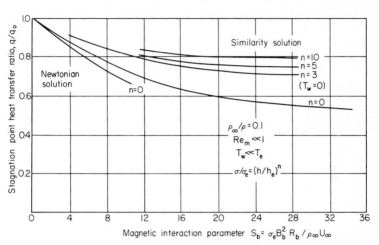

Fig. 13. Reduction in stagnation point heat transfer [1].

for these curves; the Newtonian solution used a constant field, and the similarity solution a dipole field and a variable conductivity across the boundary layer. Note that in this case the conductivity in the parameter S_b is evaluated at the edge of the boundary layer rather than at the body. The Newtonian solution can be approximated by the simple equation [10]

$$\frac{q}{q_0} = \left[\frac{du_e/dx}{(du_e/dx)_0}\right]^{0.4} \tag{51}$$

where, for a sphere and cylinder, respectively

$$\left(\frac{du_e}{dx}\right)_0 = \frac{u_\infty}{R_b}\left(1 + \frac{\rho_\infty/\rho_e}{1 + \sqrt{8\rho_\infty/3\rho_e}}\right)\sqrt{\frac{8\rho_\infty}{3\rho_e}}$$

$$\left(\frac{du_e}{dx}\right)_0 = \frac{u_\infty}{R_b}\left(1 + \frac{\rho_\infty/\rho_e}{\sqrt{1 - 3\rho_\infty/\rho_e}} \cosh^{-1}\frac{1}{\sqrt{3\rho_\infty/\rho_e}}\right)\sqrt{\frac{3\rho_\infty}{\rho_e}} \tag{52}$$

This approximation overestimates the reduction in heat transfer only slightly for $S_b < 5$, and does not exceed 10 percent until $S_b \gg 10$. The difference would be even less if an average magnetic field were used (rather than the value at the wall).

Notice in Fig. 13 that when the exponent n exceeds 5 an upper limit exists for the reduction in heat transfer, so that a further increase in S_b will not permit any further reduction in q/q_0. The experimental data for variation of σ with h in air indicates that $\sigma \sim h^{5.44}$ is a good fit; thus this limit will probably be encountered under actual flight conditions. Reductions of 20 percent in the heat transfer can be obtained, however, if $S_b \geq 15$.

The data in Fig. 13 pertain to convective heating only. It is well known that radiative heating can be more significant than the convective for hypersonic stagnation flow around very large bodies. Using the concept that the energy radiated is proportional to the volume of gas at the stagnation point, we can write on the basis of continuity

$$\frac{q_R}{q_{R_0}} = \left[\frac{(du_e/dx)_0}{du_e/dx} \right]^m \tag{53}$$

where the subscript 0 refers as usual to the case when no magnetic field is present.* Writing the total heat as

$$q_{tot} = q + q_R = q_0 \left[\frac{du_e/dx}{(du_e/dx)_0} \right]^{0.4} + q_{R_0} \left[\frac{du_e/dx}{(du_e/dx)_0} \right]^{-m} \tag{54}$$

and optimizing with respect to the ratio of velocity gradients, we find that when

$$\frac{du_e/dx}{du_e/dx_0} = \left(\frac{m q_{R_0}}{0.4 q_0} \right)^{1/(m+0.4)} \tag{55}$$

the *total* heat transfer is reduced. Since $du_e/dx < (du_e/dx)_0$ then $q_{R_0}/q_0 < 0.4/m$ for a reduction to occur. Now

$$q_0 \sim const \left(\frac{\rho_\infty}{R_b} \right)^{1/2} u_\infty^3 \tag{56}$$

and

$$q_{R_0} \sim const (\rho_\infty)^{3/2} R_b u_\infty^{10} \tag{57}$$

approximately, hence

$$\frac{q_{R_0}}{q_0} = const \, \rho_\infty R_b^{3/2} u_\infty^7 < 0.4/m \tag{58}$$

For a cylinder $q_{R_0}/q_0 < 0.533$, and for a sphere $q_{R_0}/q_0 < 0.8$ in order for the magnetic field to reduce total heating. Equation (56) implies that the radius for a body with minimum convective heating is such that

---

*The shock detachment distance is proportional to $(du_e/dx)^{-m}$, where $m = \frac{3}{4}$ for two- and $\frac{1}{2}$ for three-dimensional flow.

$$\frac{q_{R_0}}{q_0} < 0.5 \tag{59}$$

Hence if the body shape is designed so that *convective* heating is a minimum, then the magnetic field will not increase the total heat transfer.

However, Eq. (57) is only true for certain regions of radiative heating. A similar analysis should be performed for the particular flight conditions expected and an accurate empirical formula should be used for q_{R_0} (see, for example, Ref. 12) before magnetic effects on the radiative heat transfer can be safely neglected. This factor is certainly more important at the higher altitudes and velocities associated with reentry from space missions. However, to the author's knowledge, a refined analysis has not yet been published.

Further effects which should be investigated in connection with reentry of this type are nonequilibrium ionization and catalytic wall effects. Lykoudis, as discussed in Ref. 1, showed that the term governing recombination at the wall was increased by the action of a magnetic field, thus increasing the heat transfer to the wall.

In general, the remarks made in this section imply that a Newtonian approximation can be used for the analysis of MHD heat transfer when σ = constant. In fact, if an average magnetic field were used in the parameter S (perhaps the mean value between the wall and shock), and the boundary layer were seeded so that the critical layers near the wall were conducting, then the Newtonian approach would suffice for engineering estimates. For problems where wall effects are predominant (such as for the Langmuir probe), a more detailed analysis is necessary [13].

CONCLUDING REMARKS

In every area of MHD heat transfer discussed herein it is evident that there is a serious lack of experimental data to verify the theoretical work. In generator and accelerator design, experiments are largely restricted to obtaining working machines of better efficiency. Therefore, the gross heat transfer is consciously minimized, but since a detailed theory for any given design is rarely available, no comparison can be made. The simpler designs for which the theory exists are not efficient or practical and hence few experiments have been made in these. Experiments in magnetoaerodynamic heating have also been scanty [1], largely because early research indicated that only slight reductions in heat transfer could be obtained. However, these investigations were made for ballistic reentry and not for the larger velocities associated with reentry from planetary missions, for example, $V \sim 20{,}000$ m/sec for a Mars mission. In this area, both radiative and convective MHD heating should be investigated, including low-pressure and nonequilibrium effects.

On the theoretical side, there is a serious lack of work in the area of turbulent flow. As any active participant in heat transfer engineering knows, turbulence is present in almost every design. It is not likely that any progress will be made in the complete utilization of magnetic techniques unless extensive experimental and theoretical work is also done in this area.

REFERENCES

1. M. F. Romig, The Influence of Electric and Magnetic Fields on Heat Transfer to Electrically Conducting Fluids, in "Advances in Heat Transfer," (J. P. Hartnett and T. F. Irvine, eds.), vol. 1, Academic Press, Inc., New York, 1964.
2. A. Kantrowitz, *Astronautics and Aeronautics*, 3:52 (1965).
3. C. Mannal and N. W. Mather (eds.), "Engineering Aspects of Magnetohydrodynamics," Columbia University Press, New York, 1962, and subsequent volumes.

4. OECD/ENEA, "Magnetohydrodynamic Electrical Power Generation," McGraw-Hill Book Company, New York, 1964.
5. B. T. Chu, *Phys. Fluids,* **2**:473 (1959).
6. L. Spitzer, "Physics of Fully Ionized Gases," 2d ed., John Wiley & Sons, Inc., Interscience Publishers, Inc., New York, 1962.
7. B. Ragent and C. E. Noble, Jr., High Temperature Transport Coefficients of Selected Gases, VIDYA Rept. No. 52, Sept. 1961.
8. A. Sherman, *ARS Journal,* **30**:559 (1960); see also L. Lamb and S. C. Lin, *J. Appl. Phys.,* **28**:754 (1957).
9. W. D. Weatherford, Jr., J. C. Tyla, and P. M. Ku, Properties of Inorganic Energy Conversion and Heat-Transfer Fluids for Space Application, *USAF-WADD* TR-61-96, Nov. 1961.
10. P. S. Lykoudis, *J. Aerospace Sci.,* **28**:541 (1961).
11. W. B. Bush, *J. Aerospace Sci.,* **28**:610 (1961).
12. R. M. Nerem, Shock Layer Radiation During Hypervelocity Re-entry, *Proc. AIAA Entry Technology Conf.,* pp. 158–170, Oct. 12–14, 1964.
13. L. Talbot, *Phys. Fluids,* **3**:289 (1960).

NOMENCLATURE

B magnetic induction

C_f skin friction coefficient, $2\tau/\rho V^2$

C_H heat transfer coefficient, $q/\rho C_p V\Delta T$

C_p specific heat at constant pressure

d unit of length

e unit of electric charge

E electric field intensity

f arbitrary force

g acceleration due to gravity

Gr   Grashof number

h enthalpy

h_s stagnation enthalpy

H magnetic field intensity

j current (J, dimensionless)

k coefficient of thermal conductivity or Boltzmann's constant

K generator coefficient, Eq. (10)

L unit of length

m mass

M Hartmann number, Eq. (11)

Nu   Nusselt number, $qL/k\Delta T$

p pressure

Pr   Prandtl number

q heat flux

r, R radial coordinate

Ra   Rayleigh number, Ra = (Pr)(Gr)

Re   Reynolds number

Re_m magnetic Reynolds number, Eq. (13)

S magnetic interaction parameter, Eq. (9)

T temperature, °K

u, v velocity components (U, V, dimensionless)

$\mathfrak{U}$ internal energy

V velocity

W heat source

x, y, z coordinates (X, Y, Z, dimensionless)

β coefficient of thermal expansion

δ boundary-layer thickness

Δ difference
ϵ Eckert number
θ dimensionless temperature
Λ Lykoudis number, Eq. (26)
μ viscosity
μ_e magnetic permeability
ρ density
σ electric conductivity
τ shear stress
τ_{ij} shear-stress tensor
Φ viscous dissipation

Subscripts

0 reference conditions and conditions without electromagnetic field present
∞ free-stream or adiabatic conditions
b body
e edge of boundary layer
s shock
w wall

Superscripts

* dimensionless
— mean value

Section **12**

Condensation

Part **A**

Film Condensation

WARREN M. ROHSENOW

Professor of Mechanical Engineering
Massachusetts Institute of Technology

INTRODUCTION

The development of the present-day theory and the prediction of heat transfer rates associated with film condensation had its origin with the analyses presented by Nusselt in 1916 [1].

Subsequently, many modifications of this analysis have appeared, each one relaxing further the restrictive assumptions of the Nusselt analysis. For most fluids, the simple analysis predicts rather well the heat transfer associated with film condensation; however, for liquid metals, it has been found necessary to include a resistance or temperature drop at the liquid vapor interface. This resistance becomes more significant at low pressures.

Until recently, condensation in forced convection inside of tubes has been essentially an empirical art. During the past year some success has been obtained in an analysis for this process.

When a cold surface temperature T_w below saturation temperature T_s is exposed to a vapor, either saturated or superheated, liquid condensate may form on the surface. The exact nature of the condensation mechanism on a clean surface is not well established. One suggestion visualizes the surface to embody small cavities containing liquid; this small liquid surface may be the site at which condensation begins. A finite amount of subcooling would be required, the magnitude depending on the curvature of the small liquid surface.

If the liquid does not wet the surface macroscopically, the condensate forms liquid droplets and runs off the surface. Then the surface is covered with alternate patches of dry and wet spots. The dry patches have been shown to be dry.

If, on the other hand, the liquid wets the surface macroscopically, a continuous liquid layer covers the condensing surface. Commercially, this type of condensation predominates, since dropwise condensation on almost all surfaces ultimately changes to film-type condensation. Since heat transfer coefficients with film condensation are the smaller, commercial condensers are sized assuming film condensation will prevail.

DESCRIPTION OF FILM CONDENSATION

The sketch in Fig. 1 represents a liquid condensate film flowing down a vertical cold plate. The velocity distribution is shown, an approximate parabolic shape with the velocity gradient at the liquid-vapor interface being determined by the motion of the vapor (vapor shear stress at the liquid surface). A freely falling film with all of the liquid introduced at the top falls with a uniform thickness. Here liquid forms at the interface where condensation occurs all along the film, resulting in a film thickness δ that increases down the plate.

The resulting temperature distribution at a position z below the top of the plate is shown. At the wall heat is removed and $(q/A)_w = -k_l(\partial T/\partial y)_w$. If the flow in the liquid is laminar, the temperature distribution is very nearly linear; so as a first approximation, the temperature gradient at the wall may be estimated as $(T_v - T_w)/\delta$.

At the liquid-vapor interface, there is a continual interchange of molecules being condensed from the vapor and molecules evaporating from the surface. With a net flow of molecules toward the surface, there exists, even with a pure saturated vapor, a temperature drop $(T_s - T_v)$—a kind of heat transfer resistance at the interface—whose magnitude depends on the fluid being condensed, on the saturation pressure, and on the rate of condensation. This interface resistance seems to be negligible for common fluids such as water and Freon, condensing around atmospheric pressure. However, at low pressures, it can assume great significance for any fluid, particularly liquid metals. Only recently has this interface resistance been measured experimentally for a liquid metal.

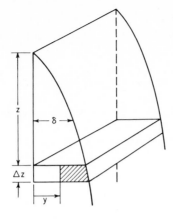

Fig. 1. Sketch of condensate film.

CONDENSATION AT THE LIQUID-VAPOR INTERFACE

For many years, simple kinetic theory analyses have suggested the existence of the temperature drop at the liquid-vapor interface when condensation occurs. Schrage [3] modified the simple theory; we use his suggested result here.

In a stationary container of molecules, the rate of flow of mass (molecules) passing in either direction (to the right or to the left) through an imagined plane is given by [3]

$$m\frac{N}{A} = \rho \left(\frac{RT}{2\pi}\right)^{1/2} = \left(\frac{1}{2\pi R}\right)^{1/2} \frac{p}{T^{1/2}} \tag{1}$$

where m is the mass of a molecule and N/A is the flux of molecules, number/hr-ft^2.

If there is a progress velocity V_y toward the plane $w/A = \rho V_y$, then

$$m\frac{N}{A} = \left(\frac{1}{2\pi R}\right)^{1/2} \frac{p}{T^{1/2}} \Gamma \tag{2}$$

where

$$\Gamma = e^{-\eta^2} + \pi^{1/2}\eta(1 + \operatorname{erf}\eta) \simeq 1 + \sqrt{\pi}\,\eta \;, \quad \text{for small } \eta \;. \tag{3a}$$

$$\eta = \frac{V_y}{\sqrt{(2RT)}} = \frac{w/A}{\rho\sqrt{2RT}} \tag{3b}$$

At a liquid-vapor interface not all of the molecules striking the surface will actually condense. We define σ as that fraction of the molecules striking the surface that actually do condense. The quantity σ has many alternative names, two of which are "condensation coefficient" and "accommodation coefficient." In a similar way, we may define an evaporation coefficient σ_e which is the ratio of the flux of molecules actually leaving a surface to the flux given by Eq. (1).

At a condensing surface such as in Fig. 1, Schrage [3] visualized the saturated vapor stream at T_s moving toward the surface at a progress flow rate of w/A and the counterflow of molecules of T_v from the surface as being the flow equivalent to molecules in a stationary container. Then, from Eqs. (1) and (2) with σ and σ_e, the net mass flow rate toward the liquid-vapor interface is expressed as

$$\frac{w}{A} = \left(\frac{1}{2\pi R}\right)^{1/2} \left(\sigma\Gamma \frac{P_s}{T_s^{1/2}} - \sigma_e \frac{P_v}{T_v^{1/2}}\right)$$

At equilibrium when there is no net condensation ($w/A = 0$), then $T_v = T_s$ and $P_v = P_s$; also $\eta = 0$ and $\Gamma = 1.0$. Clearly, then, at equilibrium $\sigma_e = \sigma$. Under nonequilibrium condition with net condensation, if it is further assumed that $\sigma_e = \sigma$, then the preceding equation becomes

$$\frac{w}{A} = \left(\frac{1}{2\pi R}\right)^{1/2} \sigma\left(\Gamma \frac{P_s}{T_s^{1/2}} - \frac{P_v}{T_v^{1/2}}\right) = \sqrt{\frac{1}{2\pi R}} \frac{2\sigma}{2 - \sigma} \frac{P_v}{\sqrt{T_s}} \left(\frac{P_s - P_v}{P_v} - \frac{T_s - T_v}{2T_v}\right) \quad (4)$$

The quantity σ in this theoretical development is interpreted as an accommodation coefficient. Magnitudes of σ obtained from condensation heat transfer measurements may in fact include effects of departure from this idealized theory. An alternative approach to this problem is presented by Bornhorst [4] on the basis of irreversible thermodynamics for coupled heat and mass flux at the liquid-vapor interface. The magnitude of σ determined in the range where precision of measurement is adequate and in the absence of noncondensable gas appears to be unity (11).

APPROXIMATE ANALYSIS OF FILM CONDENSATION

A simple analysis neglecting momentum changes in the liquid film and neglecting the interface resistance ($T_v = T_s$) leads to results that agree well with experimental data for liquid nonmetals at moderate to high pressures. This type of analysis may be used for a variety of surface geometries.

In Fig. 1, assume there is no shear stress at the liquid interface. A force balance on the element between z, δ, and dz neglecting momentum changes is

$$\mu \frac{dv_z}{dy} = g(\rho - \rho_v)(\delta - y) \quad (5)$$

With $v_z = 0$ at $y = 0$, this integrates to

$$v_z = \frac{g(\rho - \rho_v)}{\mu} \left(\delta y - \frac{y^2}{2}\right) \quad (6)$$

or the liquid flow rate in the film per unit width is

$$\Gamma = \int_0^\delta \rho v_z \, dz = \frac{g\rho (\rho - \rho_v) \delta^3}{3\mu} \tag{7}$$

Then

$$\frac{d\Gamma}{d\delta} = \frac{g\rho (\rho - \rho_v) \delta^2}{\mu} \tag{8}$$

A good approximation for laminar flow in the liquid film is a linear temperature distribution. Then an energy balance on the control volume $\delta \cdot \Delta s$ is

$$\frac{q}{A} = \frac{k}{\delta}(T_s - T_w) = \frac{d\Gamma}{dz}\left[h_{fg} + \frac{1}{\Gamma} \int_0^\delta \rho v_z c (T_s - T) dy \right] \tag{9}$$

For an assumed linear temperature distribution

$$\frac{T_s - T}{T_s - T_w} = 1 - \frac{y}{\delta} \tag{10}$$

Then with Eqs. (6) and (10), Eq. (9) becomes

$$\frac{k}{\delta}(T_s - T_w) = \frac{d\Gamma}{dz}\left[h_{fg} + \frac{3}{8} c (T_s - T_w) \right] \tag{11}$$

Solve for $d\Gamma$ from Eqs. (8) and (11) and equate to obtain

$$\delta^3 d\delta = \frac{k\mu (T_s - T_w)}{g\rho (\rho - \rho_v) h'_{fg}} dz \tag{12}$$

where

$$h'_{fg} \equiv h_{fg} + \frac{3}{8} c (T_s - T_w)$$

Integrate Eq. (12) from $\delta = 0$ at $z = 0$ with uniform $(T_s - T_w)$ to obtain

$$\delta = \sqrt[4]{\frac{4k\mu z (T_s - T_w)}{g\rho (\rho - \rho_v) h'_{fg}}} \tag{13}$$

Since

$$h_z = \frac{q/A}{T_s - T_w} = \frac{k}{\delta}$$

$$h_z = \sqrt[4]{\frac{g\rho(\rho - \rho_v)k^3 h'_{fg}}{4z\mu(T_s - T_w)}} \tag{14}$$

The average h between $z = 0$ and L is

$$h = \frac{1}{L}\int_0^L h_z\, dz \tag{15}$$

or

$$h = 0.943\sqrt[4]{\frac{g\rho(\rho - \rho_v)k^3 h'_{fg}}{L\mu(T_s - T_w)}} \tag{16}$$

This is the approximate result for vertical flat plate and vertical cylindrical tubes of diameter larger than around 1/8 inch. Film thickness δ of laminar films are of the order of magnitude of hundredths of inches, therefore the effect of curvature is negligible except for very small diameters. For condensation inside of small diameter tubes the vapor velocity entering the tube is necessarily very high, hence this simple analysis must be modified to include this effect.

For a flat plate *inclined* at an angle ϕ to the horizontal plane, the preceding analysis and Eqs. (13) through (16) apply with g replaced by $g \sin\phi$.

The simplified analysis above may be applied to many other geometries. The following are results for a few such cases.

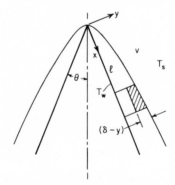

Fig. 2. Vertical cone.

VERTICAL CONE

A force balance on the element of the thin liquid layer shown in Fig. 2 for condensation on a cone is

$$\mu\frac{dv_x}{dy}2\pi x \sin\theta\, dx = (\rho_l - \rho_v)(\delta - y)g\cos\theta\, 2\pi x \sin\theta\, dx$$

Following the procedure of the previous approximate analysis leads to

$$\frac{3}{4}\frac{d(\delta^4)}{dx} + \frac{\delta^4}{x} = \frac{3k\mu(T_s - T_w)}{h'_{fg}g\cos\theta\rho_l(\rho_l - \rho_v)} \equiv K \tag{17}$$

This is a homogeneous differential equation with the solution

$$\delta = \sqrt[4]{\frac{4}{7}Kx} \tag{18}$$

Since the liquid films will be thin $h \approx k/\delta$, or

$$h_x = 0.875\sqrt[4]{\frac{g\cos\theta\rho(\rho - \rho_v)k^3 h'_{fg}}{x\mu(T_s - T_w)}} \tag{19}$$

and the average coefficient

$$h = \frac{1}{\pi\sin\theta L^2}\int_0^L h_x 2\pi x \sin\theta\, dx$$

or

$$h = \sqrt[4]{\frac{g\cos\theta\,\rho(\rho - \rho_v)k^3 h'_{fg}}{L\mu(T_s - T_w)}} \tag{20}$$

where

$$h'_{fg} = h_{fg} + \frac{3}{8}c(T_s - T_w)$$

ROTATING DISK

For a disk rotating about a vertical axis (Fig. 3), the force balance (shear and centrifugal) on the element of fluid at radius r is

$$2\pi r dr\,\mu\frac{dv}{dy} = r\omega^2\rho(\delta - y)2\pi r dr$$

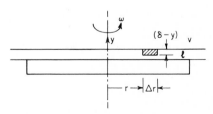

Fig. 3. Rotating disk.

Following the preceding method of analysis leads to a predicted uniform liquid film thickness and heat transfer coefficient on a flat disk

$$\delta = \sqrt[4]{\frac{3k\Delta T\mu}{2h'_{fg}\omega^2\rho^2}} \tag{21}$$

$$h = \sqrt[4]{\frac{2k^3 h_{fg}\omega^2\rho^2}{3\Delta T\mu}} \tag{22}$$

where ω is angular rotation in radians per hour. Note that ω^2 here replaces g/D in the preceding results (Eqs. (16) and (20)).

A more complete analysis of this problem was made by Sparrow and Gregg [6] including momentum and convection terms. Experimental results of Beatty and Nandapurkar [7] for methanol, ethanol, and R-113 at atmospheric pressure confirm the form of the equation, but fell approximately 25 percent below the prediction of Eq. (20).

BOTTOM OF A CONTAINER

If the bottom of a container (Fig. 4) with insulating walls is held at $T_w < T_{sat}$, and if the container holds vapor at T_{sat} and there is a vapor supply to maintain a constant pressure in the container, then the transient rate of condensation at the bottom may be approximated by assuming that a quasi-steady-state conduction heat transfer takes place and that an approximately linear temperature distribution exists in the condensed liquid. Then an energy balance is

$$\frac{q}{A} = \frac{k}{\delta}(T_s - T_w) = \rho h_{fg}\frac{d\delta}{dt}$$

Integrating from $\delta = 0$ at $t = 0$ yields

$$\delta = \sqrt{\frac{2k(T_s - T_w)}{\rho h_{fg}}t} \tag{23}$$

where t is in hours and δ in feet. The condensate layer thickness increases with $\sqrt{t}$.

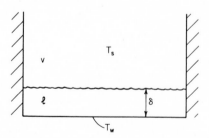

Fig. 4. Condensation on bottom of container.

ZERO GRAVITY

In a zero gravity condition, condensate may form at a cold surface and not run off, just as in the preceding case. Consider a cold circular tube in zero gravity (Fig. 5).

Making the quasi-steady-state assumption used in the preceding problem, an energy balance is

$$\frac{q}{L} = \frac{2\pi k(T_s - T_w)}{\ln(R_\delta/R)} = 2\pi R_\delta \rho h_{fg} \frac{dR_\delta}{dt}$$

Integrating from $R_\delta = R$ at $t = 0$

$$\left[\left(\frac{R_\delta}{R}\right)^2 - 1\right]\left(\ln\frac{R_\delta}{R} - \frac{1}{2}\right) = \frac{2k\Delta T}{\rho h_{fg} R^2} t$$

and

$$h = \frac{q}{\Delta T\, 2\pi\, RL} = \frac{k}{R\,\ln(R_\delta/R)} \tag{24}$$

From the first equation determine R_δ vs. t and from the second h vs. t.

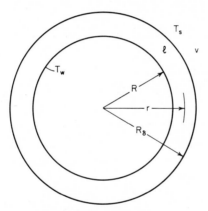

Fig. 5. Tube in zero gravity.

The preceding are examples of how to perform the approximate analysis for various geometries. Assumptions included uniform wall temperatures, saturated vapor, neglect of the resistance at the liquid-vapor interface, neglect of effect of momentum and convection effects, and neglect of variation of properties with temperature. With these assumptions the results are quite good for liquid nonmetals when $c\Delta T/h_{fg}$ is small, perhaps less than 0.2 to 0.5.

VERTICAL FLAT PLATE

This problem was reevaluated by Rohsenow [5] to account for the nonlinear temperature distribution. The modified result is

$$h = 0.943 \sqrt[4]{\frac{g\rho(\rho - \rho_v)k^3(h_{fg} + 0.68c\Delta T)}{L\mu(T_s - T_w)}} \tag{25}$$

which is the equation recommended for predicting heat transfer performance of vertical flat plates and larger size vertical tubes for $Pr > 0.5$ and $cT/h_{fg} \leq 1.0$.

The effect of variation of viscosity with temperature may be approximated by using Eq. (25) with properties k, ρ, and c evaluated at the following temperature [8]

$$T_{ref} = T_w + 0.31(T_s - T_w) \tag{26}$$

This was shown [9] to provide results within 0.2 percent of a more exact solution for atmospheric steam when $T_s - T_w$ is as much as 50F.

The problem was solved as a boundary-layer problem including momentum and convection terms, both of stationary vapor and a zero shear stress at the interface (Figs. 6 and 12). Here the results for $Pr > 0.5$ may be used directly but for $Pr < 0.03$ must be combined with Eq. (4) to include the interfacial resistance.

An interface heat transfer coefficient may be defined as

$$h_i = \frac{q/A}{T_s - T_i} = \frac{(w/A)h_{fg}}{T_s - T_i}$$

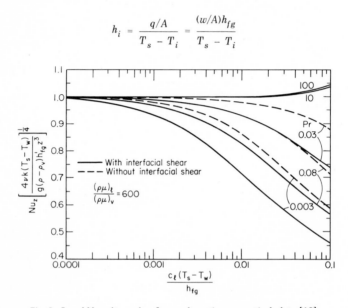

Fig. 6. Local Nusselt number for condensation on vertical plate [10].

Then from Eq. (4) for w/A

$$h_i = \left(\frac{1}{2\pi R}\right)^{1/2} \frac{\sigma h_{fg}}{T_s - T_i}\left(\Gamma \frac{P_s}{\sqrt{T_s}} - \frac{P_i}{\sqrt{T_i}}\right) \tag{27}$$

This may be approximated, when $\eta < 0.1$ of Eq. (3) as follows

$$h_i \simeq \frac{2\sigma}{2 - \sigma}\sqrt{\frac{1}{2\pi R}} \frac{h_{fg}^2}{v_{fg}T_v^{3/2}} \tag{28}$$

The overall h for condensing liquid metal is obtained by determining h for the liquid film from Fig. 6 with $(T_s - T_w)$ replaced by $(T_i - T_w)$ and h_i from either Eq. (27) or (28). Then

$$\frac{1}{h} = \frac{1}{h(\text{Fig. 6})} + \frac{1}{h_i} \tag{29}$$

For liquid nonmetals the $(1/h_i)$ term is negligible except at very low pressures where the vapor saturation temperature is near the freezing temperature or triple point and the condensation rates are very low because of small ΔT.

Analysis of all available data for condensation coefficient σ suggests that it should have a magnitude of unity in Eqs. (27) and (28). This data is shown collected on a single plot of σ vs. pressure in Fig. 7. At pressures above 0.1 atm, the magnitude of $(T_s - T_i)$ is small compared with $(T_i - T_w)$; therefore it is difficult to measure with precision. The apparent decrease in σ at higher pressure is probably due to lack of precision of measurement and/or traces of noncondensable gas released from vessel walls at the higher temperatures. The open data points in Fig. 7 were obtained with nickel or stainless steel condensing blocks. When copper condensing blocks were used (closed points) the precision in measuring the quantity $(T_s - T_i)$ is greatly improved and σ remained unity out to higher pressures. The data of Sakhuja was obtained by collecting traces of noncondensable gas at a secondary condenser and bleeding it from the system. When the bleed line was closed, data similar to the open points could be obtained. Subsequent bleeding caused σ to return to unity. The conclusion from these various experiments is that the true magnitude of σ is unity and that traces, indeed small, of noncondensable gas can cause the apparent σ to fall below unity at the high temperatures.

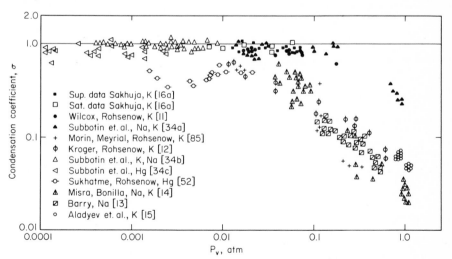

Fig. 7. Survey of condensation coefficient data [16a].

Experimental data generally agree with the equations suggested here or are slightly higher. If waves or ripples on the liquid vapor are present, the net effect is to increase the measured h above that predicted by the theory by as much as 20 percent. Kapitza [17] has shown for water that waves begin to form for $4\Gamma/\mu \geq 33$.

The preceding equations have been developed for the case of uniform T_w. If ΔT varies linearly from ΔT_t at the top to ΔT_b at the bottom of a vertical surface, McAdams [18] suggests the following correction factor F_t

$\Delta T_b / \Delta T_t$	0.5	1.0	2.0	5.0
F_t	0.96	1.0	1.06	1.15

where the h from Eq. (16) using ΔT_{avg} is multiplied by F_t to get the corrected h_{avg}.

EFFECT OF INTERFACE SHEAR STRESS AND TRANSITION TO TURBULENT FILM

In a confined space such as inside of tubes, outside of tube bundles, or in arrays of parallel cold plates, the vapor must enter at either end. Because of the great difference between liquid and vapor density the vapor velocity may be significant and may result in a large shear stress τ_v along the liquid-vapor interface. Equation (5) then is modified to include a τ_v term. The analysis proceeds in the same manner to the following results [19]

$$z^* = (\delta^*)^4 \pm \frac{4}{3} \tau_v^* (\delta^*)^3$$

$$\frac{4\Gamma}{\mu} \frac{1}{1 - \rho_v/\rho} = \frac{4}{3} (\delta_L^*)^3 \pm 2\tau_v^* (\delta_L^*)^2$$

$$h^* = \frac{h}{k} \left(\frac{\nu^2}{g} \right)^{1/3} = \frac{4}{3} \frac{(\delta_L^*)^3}{z_L^*} \pm 2\tau_v^* \frac{(\delta_L^*)^2}{z_L^*} \tag{30}$$

where

$$\delta^* = \delta \left(\frac{g}{\nu^2} \right)^{1/3}$$

$$z^* = \frac{4z\Delta T}{Pr} \frac{c}{h_{fg}'} \left(\frac{g}{\nu^2} \right)^{1/3} \frac{1}{1 - \rho_v/\rho} \tag{31}$$

$$\tau_v^* = \frac{\tau_v}{g(\rho - \rho_v)(\nu^2/g)^{1/3}}$$

Here the plus sign (+) is for downward flow and (−) for upward flow of vapor.

Any two of the dimensionless quantities may be eliminated since there are three equations in Eq. (30). The results for downward vapor flow are plotted in Fig. 8 to the left of the dotted curve and are the same on each graph, e.g., independent of Pr. In these curves, $\rho_v \ll \rho$ hence, $(1 - \rho_v/\rho) \approx 1.0$.

When the vertical plate or tube is long enough or the vapor flow high enough, the laminar liquid film changes to turbulent flow. Figure 9 represents the estimated film Reynolds number $4\Gamma/\mu$ at transition. A freely falling film ($\tau_v = 0$) is observed [21] to change to turbulent flow at $4\Gamma/\mu = 1800$. Then for this condition the shear stress at

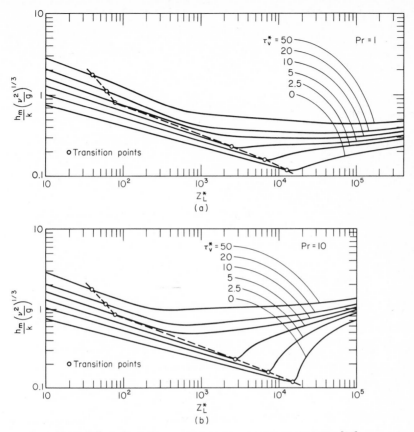

Fig. 8. Effect of turbulence and vapor shear stress on condensation [20].

the wall is calculated from Eq. (30) with $\tau_v = 0$. The steep curve up to $\tau_v^* \simeq 11$ (Fig. 9) results from assuming that the wall shear stress remains the same at transition, independent of the magnitude of τ_v.

For $\tau_v^* > 11$ the film thickness is very small. Here we assume, after Carpenter [22], that $\delta^+ = (\delta/\nu)\sqrt{\tau_0/\rho} = 6$. Experimental data are limited, but agreement with Carpenter's ethanol data was within 25 percent for $4\Gamma/\mu$ at $\tau_v^* = 110$.

The heat transfer associated with turbulent film condensation with significant vapor shear was analyzed (Rohsenow, Webber, and Ling [20]) by assuming that the universal velocity distribution applies and then using the methods based on the analogy between momentum and heat transfer. The details of the analysis are omitted here, but the results for downward vapor flow are shown in Fig. 8 for Pr = 1 and 10.

In these graphs (Fig. 8) a uniform τ_v was assumed. Actually τ_v varies along the length because of the mass removed by condensation. Lehtinen [23] performed a stepwise analysis and determined that τ_v should be calculated for the following average mass velocity of the vapor inside of vertical tubes

$$G_{vm} = 0.4(G_{\text{top}} + 1.5 G_{\text{bottom}}) \tag{32}$$

If all of the vapor condenses ($G_{\text{bottom}} = 0$), this reduces to

$$G_{vm} = 0.4\,G_{\text{top}} \tag{33}$$

Then the power average τ_v is calculated as

$$\tau_v = \frac{fG_{vm}^2}{2\rho_v} \tag{34}$$

where f is taken from the data of Bergelin et al. [24] shown plotted in Fig. 10 where σ_w/σ is the ratio of surface tension of water and the particular fluid being condensed, Γ_L corresponds to mean liquid flow, and ρ is the liquid density.

For condensation of a particular fluid on a vertical surface of a given geometry and magnitude of T_v and T_w, the value of Pr and z_L^* are readily calculated. Interpolation of Fig. 8 will give a value of $(h_m/k)(\nu^2/g)^{1/3}$ from which h_m is obtained and $(q/A)_{\text{av}}$ is calculated. This value of h_m includes the effect of a laminar film on the upper part of the surface and, if present, the effect of a turbulent film on the remainder of the surface.

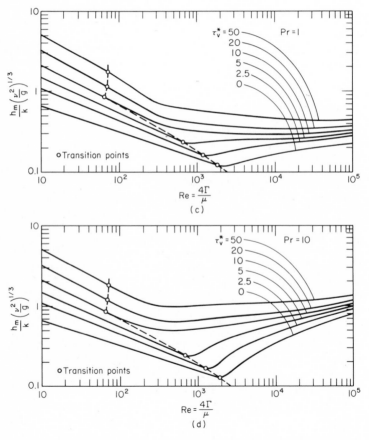

Fig. 8 (*Cont.*). Effect of turbulence and vapor shear stress on condensation [20].

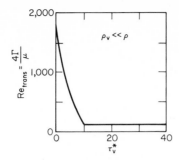

Fig. 9. Transition Reynolds number for a liquid film [20] .

In determining the condensation rate for a given size surface, the calculation requires a trial-and-error type calculation with these curves, calculating the performance for several magnitudes of τ_v^* and interpolating.

In analyzing test data, the condensation rate is often measured. Then $(h_m/k)(\nu^2/g)^{1/3}$ is read directly from the curve at the particular $4\Gamma/\mu$, corresponding to the condensation rate.

The results of the analysis leading to the graphs of Fig. 8 were shown [20] to agree well with the experimental data of Carpenter and Colburn [25] which was taken for fluids whose Pr was in a limited range of 2 to 5.

An alternative analysis of these effects of τ_v and turbulence was presented by Dukler [26] with essentially similar results.

CONDENSATION ON HORIZONTAL TUBES

The condensate film on the outside of horizontal tubes flows around the tube and off the bottom in a sheet, shown exaggerated in Fig. 11. The liquid film actually is very thin, so that the simple analysis used on the vertical plate applies here except that g is replaced by $g \sin\phi$ and the average value of h follows from integration over the range of ϕ values from 0 to 180°.

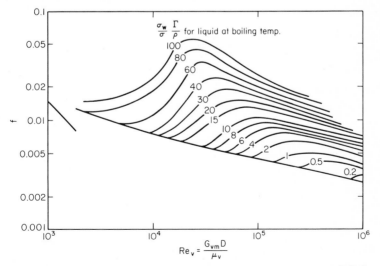

Fig. 10. Friction for gas (air) flowing in a tube with a liquid layer on the wall [24] .

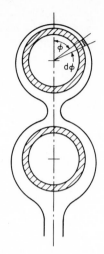

Fig. 11. Condensate film on horizontal tubes.

The details of the analysis are omitted here but the result equivalent to Eq. (16) is as follows for the average h on the top tube

$$h = 0.728 \left[\frac{g\rho (\rho - \rho_v) k^3 h'_{fg}}{D\mu \, \Delta T} \right]^{1/4} \tag{35}$$

The result of the boundary-layer analysis, assuming zero shear stress at the liquid-vapor interface (Sparrow and Gregg [6]), and of the integral equation analysis, assuming stagnant vapor (Chen [27]), are shown in Fig. 12. The same graph also

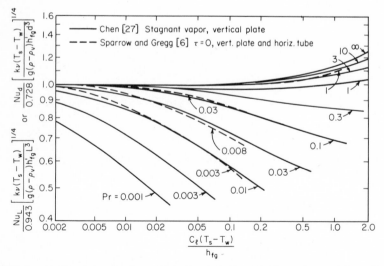

Fig. 12. Average Nusselt number for condensation on horizontal tubes and vertical plates [6, 27].

applies to the average h on a vertical flat plate with the ordinate interpreted as indicated. Inspection of Fig. 12 shows Eq. (35) is adequate for $\text{Pr} \geq 1$ for $c\Delta T/h_{fg} < 1$.

The heat transfer coefficient on a horizontal tube decreases from a maximum value at $\phi = 0$ to essentially zero for $\phi = 180°$. The condensing rate on the upper half of the tube is 46 percent greater than on the lower half.

If we compare Eq. (16) with Eq. (35), the same average heat transfer coefficient exists on vertical plates and horizontal tubes for identical fluid conditions and ΔT when $0.728/D^{1/4} = 0.943/L^{1/4}$, or

$$L = 2.78 D \tag{36}$$

TUBE BANK

Nusselt [28] analyzed condensation on vertical banks of horizontal tubes (Fig. 11). Here all of the condensate dropping from any tube is assumed to fall on the next lower tube. The result for the heat transfer coefficient averaged over all n tubes in a vertical bank is

$$h = 0.728 \left[\frac{g\rho(\rho - \rho_v)k^3 h'_{fg}}{nD\mu\,\Delta T} \right]^{1/4} \tag{37}$$

This result indicates that the average h for n tubes equal the h for the tube divided by $n^{1/4}$. Actually, experimental data for banks of tubes fall well above the results of Eq. (37).

Chen [27] suggested that since the liquid film was subcooled, an average of $3/8(T_s - T_w)$, it is possible that additional condensation occurs on the liquid layer between tubes. Assuming all of this subcooling is removed, he provided the result

$$h = 0.728 \left[1 + 0.2\,\frac{c\Delta T}{h_{fg}}(n - 1) \right]\left[\frac{g\rho(\rho - \rho_v)k^3 h'_{fg}}{nD\mu\,\Delta T} \right]^{1/4} \tag{38}$$

which is a good approximation provided $(c\Delta T/h_{fg}) < 2$.

Equation (38) agrees well with most of the available data for condensation on banks of tubes. Most experimental data for pure vapors agree with or are slightly higher than the equations presented in this chapter. Departures are usually attributed to such phenomena as external vibrations causing ripples in the liquid surface, splashing off tubes in a bank, or uneven runoff due to bowing or inclination of tubes.

Condensation on horizontal finned tubing [29–31] may be predicted with

$$h = 0.689 \left(\frac{k^3\rho^2 g}{\mu} \right)^{1/4} \left(\frac{h_{fg}}{\Delta T D_{eq}} \right)^{1/4} \tag{39}$$

where

$$\frac{1}{D_{eq}^{1/4}} = 1.3\,\frac{A_s\phi}{A_{ef}(L_{mf})^{1/4}} + \frac{A_p}{A_{ef}D^{1/4}} \tag{40}$$

and A_s = actual fin area
   A_p = horizontal tube area
   ϕ = fin efficiency
A_{eff} = $A_s \phi + A_p$
L_{mf} = (area of one side of fin)/(outside diameter of fin)
   Obviously Eq. (39) applies only to cases in which the condensate liquid will drain off the tube. For very close fin spacing the surface tension may be sufficient to hold liquid between the fins resulting in very much lower rates of heat transfer.

FORCED CONVECTION ACROSS A CYLINDER

Condensation on the outside of a cylinder with a cross-flow velocity of the vapor has not been analyzed. A series of test results were obtained by Fishman [32] for saturated steam (2 < p < 13 psia) flowing in a horizontal duct 1 inch high and

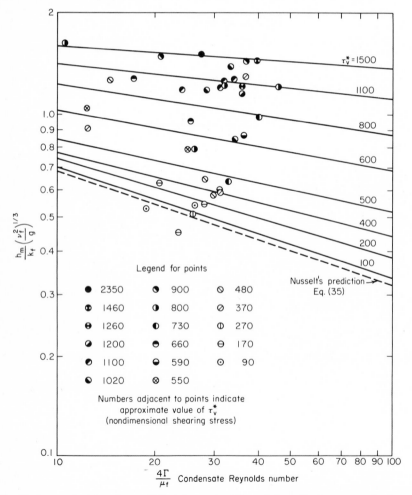

Fig. 13. Average heat transfer coefficient for condensation of steam flowing across a horizontal tube [32].

5 inches wide with the horizontal cold tube placed centrally across the duct. The test data points are plotted in Fig. 13 and "best" straight lines drawn for constant magnitudes of τ_v^*. Here

$$\tau_v^* = \frac{g_0 \tau_v}{g(\rho - \rho_v)(\nu^2/g)^{1/3}}$$

$$\tau_v = C_D \rho_v \frac{V^2}{2g_0}$$

$$\Gamma = \frac{(q/A) h_{fg}}{L} \tag{41}$$

Here C_D is the drag coefficient for flow of a single-phase fluid across cylinders [33].

The lower dotted line of Fig. 13 represents Eq. (35) for zero vapor velocity. The results presented in Fig. 13 should be used with caution and should not be extrapolated. Although dimensionless groups have been used, only data for steam in the range of pressure between 2 and 13 psia are available.

CONDENSATION INSIDE HORIZONTAL TUBES, LOW CONDENSATION RATES OR SHORT TUBES

At low condensation rates or with short tubes, the vapor velocities are low, permitting the liquid condensate to run off in a stratified condition. Specifically, the condensate formed in the upper portion of the tube (Fig. 14) flows around the periphery to the bottom layer which flows off in a kind of open-channel type flow. Chato [35] investigated this problem and suggested that this flow regime exists when the inlet Reynolds number of the vapor is below 35,000. Chato presents a detailed analysis of the hydrodynamic and the heat transfer processes. His test data for R-113 agree well with his suggested analytical predictions. At the lower vapor velocities (inlet $Re_v < 35,000$) the analytical and experimental results agree well with

$$h_m = 0.555 \left[\frac{g\rho (\rho - \rho_v) k^3 h'_{fg}}{\mu \Delta T D} \right]^{1/4} \tag{42}$$

Chato also investigated the effect of inclining the tube to the horizontal and found the heat transfer coefficient to increase to a maximum of about 10 to 20 percent at inclination angles to horizontal of about 10 to 20°. His suggested analytical results agreed well with the experimental data.

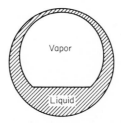

Fig. 14. Stratified condensate flow inside a horizontal tube.

Comparison of Eqs. (16) and (42) suggests that the heat transfer coefficient for condensation on the inside of a horizontal tube is greater than that for a vertical tube when $L > (0.943/0.555)^4 D$ or

$$L > 8.4 D \qquad (43)$$

CONDENSATION INSIDE LONG TUBES

When the condensation rate or the length is large, the flow regime becomes annular with a vapor core; at even higher vapor velocities, the vapor core may contain a mist of liquid droplets.

Extending the analysis leading to Fig. 8, Altman, Staub, and Norris [36] present an analytical procedure for predicting the condensing heat transfer coefficient in the higher velocity range. The method leads to implicit equations that must be solved simultaneously in each case. Experimental data for R-22 verified the analytical prediction procedure to within +20 percent and –10 percent.

Condensing with this same flow regime was studied by Akers and Rosson [37]. When $\text{Re}_v > 20,000$ and $\text{Re}_L > 5000$ the data over a wide range of variables is correlated to around ±50 percent by Akers et al. [38] as

$$\frac{hD}{k_L} = 0.026 \, \text{Pr}_L^{1/3} \left[\frac{D}{\mu_L} \left(G_v \left(\frac{\rho_L}{\rho_v} \right)^{1/2} + G_L \right) \right]^{0.8} \qquad (44)$$

In this turbulent flow regime the heat transfer coefficient increases with liquid flow rate but is independent of temperature differences. Data does not agree well with this prediction. The method presented next is recommended.

Most refrigeration condensers operate by condensing inside of long vertical or horizontal tubes. Generally for most of the tube the flow regime is annular flow. At the higher end of the design flow range there may be liquid droplets in the core—mist annular flow—and at the end of the tube where condensation is nearly complete there may be a stratified flow. Condenser performance over most of the practical range of operation can be well predicted by an annular flow model (Fig. 15). Attempts have been made to correlate average heat transfer coefficients for condensing inside of tubes without much success. Data and prediction can disagree by over 100 percent. The following procedure calculates local heat transfer coefficients along the tube and from them determines an average coefficient. This procedure has been found to agree with data within ±10 percent.

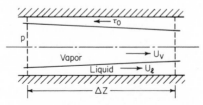

Fig. 15. Control volume for entire tube.

The momentum equation for the entire tube is

$$\frac{dP}{dz} = \underbrace{\tau_0 \frac{S}{A_z}} - \underbrace{\frac{g \sin\theta}{g_0} [\alpha\rho_v + (1-\alpha)\rho_l]} + \underbrace{\frac{1}{g_0 A_z} \frac{d}{dz} (V_v w_v + V_l w_l)} \qquad (45)$$

$$\left(\frac{dP}{dz}\right)_f \qquad + \qquad \left(\frac{dP}{dz}\right)_g \qquad + \qquad \left(\frac{dP}{dz}\right)_m$$

which shows the pressure gradient composed of three components associated with friction, gravity, and momentum. In horizontal tubes $\sin\theta = 0$.

The friction pressure gradient or wall shear stress can be predicted by the Lockhart-Martinelli method [60, 61]

$$\left(\frac{dP}{dz}\right)_f \frac{g_0 D}{G^2/\rho_v} = -\tau_0 \frac{S}{A_z} \frac{g_0 D}{G^2/\rho_v} = -0.09 \left(\frac{GD}{\mu_v}\right)^{-0.2} \left[x^{1.8} \right.$$

$$+ 5.7 \left(\frac{\mu_l}{\mu_v}\right)^{0.0523} (1-x)^{0.470} x^{1.33} \left(\frac{\rho_v}{\rho_l}\right)^{0.261}$$

$$+ 8.11 \left(\frac{\mu_l}{\mu_v}\right)^{0.105} (1-x)^{0.94} x^{0.86} \left.\left(\frac{\rho_v}{\rho_l}\right)^{0.522} \right] \qquad (46)$$

The gravity term when the tube is not horizontal is given by

$$\left(\frac{dP}{dz}\right)_g \frac{g_0 D}{G^2/\rho_v} = \frac{g \sin\theta D}{G^2/\rho_v} [\alpha\rho_v + (1-\alpha)\rho_l] \qquad (47)$$

where Zivi [62] suggests void fraction α is related to quality by the following approximation

$$\frac{1}{\alpha} = 1 + \frac{1-x}{x} \left(\frac{\rho_v}{\rho_l}\right)^{2/3} \qquad (48)$$

For momentum

$$\left(\frac{dP}{dz}\right)_m \frac{g_0 D}{G^2/\rho_v} = -D \frac{dx}{dz} \left[2x + (1-2x)\left(\frac{\rho_v}{\rho_l}\right)^{1/3} + (1-2x)\left(\frac{\rho_v}{\rho_l}\right)^{2/3} - 2(1-x)\left(\frac{\rho_v}{\rho_l}\right) \right] \qquad (49)$$

where in Eq. (45)

$$W_v = GA_z x = A_z \alpha V_v \rho_v$$
$$W_l = GA_z(1-x) = A_z(1-\alpha)V_l \rho_l \qquad (50)$$

Here A_z is tube cross-sectional area and S is the perimeter.

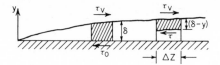

Fig. 16. Control volume for liquid film.

The following is the momentum equation applied only to the liquid layer moving along the wall (Fig. 16) where τ_v shear stress and V_i velocity exist at the interface

$$-\left(\frac{dP}{dz}\right)A_{z_l} + \tau_v S_v - \tau_0 S + \frac{g \sin\theta}{g_0}\rho_l A_{z_l} = \frac{1}{g_0}\left[\frac{d(V_l W_l)}{dz} - V_i \frac{dW_l}{dz}\right] \qquad (51a)$$

or

$$\tau_0 = F_0 \frac{A_{z_l}}{S} + \tau_v \frac{S_v}{S} \qquad (51b)$$

where

$$F_0 \equiv -\frac{dP}{dz} + \frac{g \sin\theta}{g_0}\rho_l - \frac{1}{g_0 A_{z_l}}\left[\frac{d(V_l W_l)}{dz} - V_i \frac{dW_l}{dz}\right] \qquad (51c)$$

Since for most of the tube length the liquid film is thin, $A_{z_l}/S \approx \delta$ and $S_v/S \approx 1$; then Eq. (51) becomes

$$\tau_0 = F_0 \delta + \tau_v \qquad (52)$$

The quantities F_0 and τ_v may be written in terms of x and α by substituting Eqs. (48) and (50) into Eq. (51c)

$$F_0 = -\left(\frac{dP}{dz}\right) + \frac{g \sin\theta}{g_0}\rho_l - \frac{G^2}{g_0 \rho_v}\frac{dx}{dz}\left\{\frac{1}{1-\alpha}\left(\frac{\rho_v}{\rho_l}\right)^{1/3} - \left[\frac{(1-x)(2-\beta)}{(1-\alpha)^2}\right]\left(\frac{\rho_v}{\rho_l}\right)\right\} \qquad (53)$$

where $\beta \equiv V_i/V_l$ and is calculated from the universal velocity distribution with the result shown in Fig. 17 with β as a function of δ^+.

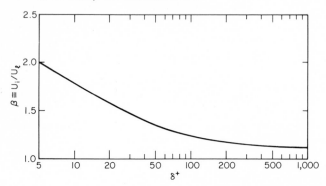

Fig. 17. β vs. δ^+.

Bae et al. [63] determined the temperature distribution and hence heat flux by the momentum-heat transfer analogy using

$$\tau = \frac{\rho_l}{g_0}(\nu_l + \epsilon_m)\frac{dv_z}{dy}$$

$$\frac{q}{A} = \rho_l c_l(\alpha_l + \epsilon_h)\frac{dT}{dy} \tag{54}$$

the universal velocity distribution and $\epsilon_m = \epsilon_h$. Further the following approximation was made

$$\tau = F_0(\delta - y) + \tau_v \tag{55}$$

The results for the local heat transfer coefficient h_z and film Reynolds number Re_l can be simplified to the following form (84):

$$h_z \equiv \frac{q/A}{T_v - T_w}$$

$$0.1 < F(X_{tt}) < 1 \qquad \frac{Nu\,F_2}{Pr_l\,Re_l} = F(X_{tt}) \tag{56}$$

$$1 < F(X_{tt}) < 20 \qquad \frac{Nu\,F_2}{Pr_l Re_l^{0.9}} = [F(X_{tt})]^{1.15} \tag{57}$$

where

$$F(X_{tt}) \equiv 0.15[X_{tt}^{-1} + 2.85 X_{tt}^{-0.476}] \tag{58}$$

$$X_{tt} \equiv \left(\frac{\mu_l}{\mu_v}\right)^{0.1}\left(\frac{1-x}{x}\right)^{0.9}\left(\frac{\rho_v}{\rho_l}\right)^{0.5} \tag{59}$$

$$Re_l < 50 \qquad F_2 = 0.707\,Pr_l\,Re_l^{0.5} \tag{60a}$$

$$50 < Re_l < 1125 \qquad F_2 = 5\,Pr_l + 5\ln[1 + Pr_l(0.09636\,Re_l^{0.585} - 1)] \tag{60b}$$

$$Re_l > 1125 \qquad F_2 = 5\,Pr_l + 5\ln(1 + 5\,Pr_l) + 2.5\ln(0.0031\,Re_l^{0.812}) \tag{60c}$$

Further for

$$Re_l \equiv \frac{(1 - x)GD}{\mu_l} = \frac{4\Gamma}{\mu_l} \qquad (61)$$

The length of tube Δz required to change the quality Δx from an energy balance is

$$\Delta z = \frac{Gh_{fg}D\Delta x}{4h_z \Delta T} \qquad (62)$$

After calculating h_z local in the tube the average coefficient of heat transfer h_m may be calculated by either of the following equivalent relations

$$h_m = \frac{1}{L} \int_0^L h_z \, dz \qquad (63a)$$

$$\frac{1}{h_m} = \frac{1}{(1 - x_e)} \int_{x_e}^1 \frac{dx}{h_z} \qquad (63b)$$

for an inlet quality of 1.0 and exit quality x_e.

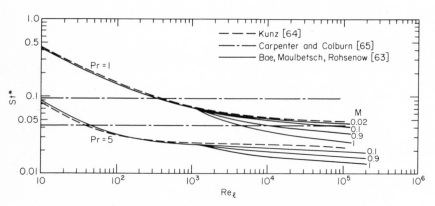

Fig. 18. St^* vs. Re_l and comparison of correlations.

It is possible to construct a graph of St^* vs. Re_l for various values of M and Pr as shown in Fig. 18 for Pr of 1 and 5. For assumed magnitudes of Pr, δ^+, and M, calculate Re_l from Eq. (62), F_2 from Eq. (58), and St^* from Eq. (56). Then the curves can be drawn as shown in Fig. 18.

SAMPLE CALCULATION

Given conditions

$$G = 250{,}000 \text{ lbm/ft}^2\text{-hr}$$
$$T_{\text{sat}} = 86°F$$
$$T_0 = 76°F$$

physical properties (from Du Pont Table of F-22)

Viscosity	μ_l	= 0.557 lbm/hr-ft
	μ_v	= 0.0322 lbm/hr-ft
Conductivity	k_l	= 0.0495 Btu/hr-ft-°F
Specific heat	C_p	= 0.305 Btu/lbm-°F
Latent heat	h_{fg}	= 76.470 Btu/lbm
Density	ρ_l	= 73.278 lbm/ft³
	ρ_v	= 3.1622 lbm/ft³

Pr = 3.43
D = 0.493 in.

Assuming that complete condensation occurs in the tube, the quality change is divided into 20 steps. A sample calculation will be done for the quality change from 72.5 to 67.5 percent. The local heat transfer coefficient at $x = 0.7$ will be considered as the average value in this quality change.

From Eq. (46)

$$\left(\frac{dP}{dz}\right)_f = -16.96 \text{ lbf/ft}^2/\text{ft}$$

Also from Eq. (46) with $S/A_z = D/4 = 0.0103$ ft,

$$\tau_0 = -\frac{A_z}{S}\left(\frac{dP}{dz}\right)_f = 0.174 \text{ lbf/ft}^2$$

From Eq. (57)

$$u_\tau = 992 \text{ ft/hr}$$

Take for a first trial $D(dx/dz) = -0.001$. From Eq. (49)

$$\left(\frac{dP}{dz}\right)_m = 1.36 \text{ lbf/ft}^2/\text{ft}$$

For a horizontal tube $(dP/dz)_g = 0$. From Eq. (45)

$$\frac{dP}{dz} = -16.96 + 1.36 = -15.60 \text{ lbf/ft}^2/\text{ft}$$

From Eq. (61)

$$\text{Re}_l = \frac{(1 - 0.7)(250,000)(0.493)}{(0.557)(12)} = 5532$$

From Eq. (60c)

$$F_2 = 34.71$$

From Eq. (59)

$$X_{tt} = 0.1287$$

From Eq. (58)

$$F(X_{tt}) = 2.299$$

From Eq. (57)

$$\text{Nu} = 602$$

or

$$h_z = \frac{\text{Nu}\, k_l}{D} = 725$$

Since

$$\frac{q}{A} = h_z \Delta T = \frac{\pi}{4} \frac{D^2 G h_{fg}}{\pi D} \frac{\Delta x}{\Delta z}$$

we have

$$D \frac{\Delta x}{\Delta z} = \frac{4 h_z \Delta T}{G h_{fg}} = \frac{4(725)(10)}{(250,000)(76.470)} = 0.00152$$

Recalculate using this magnitude instead of 0.001. The final result is $(dp/dz)_m = 2.07$. From Eq. (62)

$$\Delta z = \frac{(\Delta x)(D)}{0.00152} = \frac{(0.05)(0.493)}{(0.00152)(12)} = 1.34 \text{ ft}$$

the increment of length required to change the quality from 72.5 percent to 67.5 percent.

A similar calculation should be made for each Δx of 5 percent to determine the corresponding h_z and Δz. A plot of h_z and x vs. z may be constructed. Also P vs. x or z may be plotted.

FILM CONDENSATION ON THE UNDERSIDE OF A HORIZONTAL AND INCLINED PLATE

When vapor condenses on the underside of a horizontal plate, a liquid film collects and in the quasi-steady state is continually unstable, falling off in somewhat random arrangement of drops that have approximate cosine shapes before detaching from the liquid. Visually the film appears to be the reverse of the film boiling process on the upper side of a horizontal plate. Following an analysis similar to the one Berenson [39] performed for film boiling, Gerstmann and Griffith [40] analyzed this condensation process and made measurements of condensation of R-113 and water at 1 atm which resulted in and determined the magnitude of the coefficient in

$$\tau < 10^{-8} \qquad\qquad Nu = 0.81\,\tau^{-0.193} \qquad\qquad (64a)$$

$$10^{-6} > \tau > 10^{-8} \qquad Nu = 0.69\,\tau^{-0.20} \qquad\qquad (64b)$$

where

$$Nu \equiv \frac{h}{k}\sqrt{\frac{g_0\sigma}{(\rho - \rho_v)g\,\cos\theta}}$$

$$\tau = \frac{k\mu\,\Delta T}{\rho(\rho - \rho_v)g\,\cos\theta\,h'_{fg}[g_0\sigma/g(\rho - \rho_v)\,\cos\theta]^{3/2}}$$

where θ is the angle with the horizontal plane. For the horizontal plate $\theta = 0$ and $\cos\theta = 1$.

For condensation on the underside of plates inclined slightly to the horizontal, $\theta < 15°$ and $\Delta T < 100F$, data for water and R-113 is correlated by

$$Nu = \frac{0.90\,\tau^{-1/6}}{1 + 1.1\,\tau^{1/6}} \qquad\qquad (65)$$

In both Eqs. (64) and (65) $k\,\Delta T \ll \mu\,h'_{fg}$.

EFFECT OF ELECTROSTATIC FIELD
ON FILM CONDENSATION

The sketch inset in Fig. 19 shows a cooled 1-inch diameter tube with R-113 condensing on the inside wall. Along the center line there is an 0.5-inch diameter electrode in the vapor space. A high-voltage difference of various magnitudes is placed across the electrode and the tube. Since R-113 vapor is a poor electrical conductor, only a very minute amount of electric current flows. The radial change in the electric field strength produces an inward radial force on the dipoles induced in the molecules. This acts as an effective radial force field such as a gee field.

The development of electrohydrodynamic instability of the condensate film visually resembled that of a liquid film flowing down the outside surface of a vertical, rotating cylinder. The wavelengths of the electrohydrodynamic waves suggested a polarized rather than a free-charge interface. A typical set of heat transfer data for a constant condensing rate of 90.4 lb/hr is shown in Fig. 19 (Choi [41]), where the ordinate is the ratio of the average condensation heat transfer coefficient with and without the electric field and the abscissa is the electric field intensity.

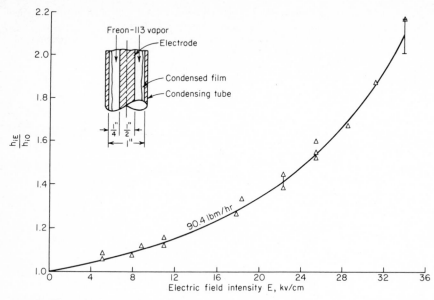

Fig. 19. Effect of radial electrostatic field on condensation heat transfer coefficient inside a vertical tube [41].

EFFECT OF SUPERHEATED VAPOR

When the vapor is superheated and the temperature of the cold surface is above the saturation temperature, no condensation occurs; the vapor is cooled, becoming less superheated. Heat transfer follows the pattern of any single phase fluid in either natural or forced convection. When the surface temperature is below saturation and film condensation takes place, the condensation rate is only slightly increased by the superheat in the vapor.

Consider the analysis of the condensation process on the vertical plate, Fig. 1. If the vapor is superheated the enthalpy change of Eqs. (11) through (20) must include the superheat enthalpy of the vapor; so h_{fg}' is replaced by

$$h_{fg}'' = c_v (T_v - T_{sat}) + h_{fg} + \frac{3}{8} c (T_{sat} - T_w) \tag{66}$$

The heat transfer is still governed by conduction $(T_{sat} - T_w)$ through the liquid film and calculated by

$$\frac{q}{A} = h(T_{sat} - T_w) \tag{67}$$

The condensate rate is calculated from

$$\frac{w}{A} = \frac{1}{h_{fg}''} \frac{q}{A} \tag{68}$$

The ratio of heat transfer coefficients for superheated vapor to that with saturated vapor is

$$\frac{h}{h_{sat}} = \left[\frac{h'_{fg} + c_v(T_v - T_{sat})}{h'_{fg}}\right]^{1/4} \tag{69}$$

Experiments for condensation on a vertical surface (Spencer and Ibele [42] and Still [66]) indicate Eq. (69) represents a minimum improvement in h since natural convection effects in the vapor result in downward vapor motion at the surface and thinner liquid films. For water and R-113 at 25F superheat Eq. (69) was exceeded by 4 percent and at 90F superheat by 12 percent.

For condensing potassium at up to 100F superheat, Eq. (69) predicts essentially no effect of superheat. Experiments [16a] showed Eq. (27) or (28) and Eq. (29) predicted the results when $\sigma = 1.0$ and T_s and P_s were taken as the saturation temperature and system pressure regardless of the amount of superheat.

EFFECT OF NONCONDENSABLE GAS

When a condensable vapor is condensing in the presence of a noncondensable gas, the vapor must diffuse through the gas, requiring a decrease in vapor partial pressure toward the liquid-vapor interface. Thus, interface saturation temperature is significantly below the temperature of the main vapor-gas mixture (Fig. 20a). This combined mass and heat transfer process requires a lengthy trial-and-error stepwise calculation to predict performance accounting for this effect.

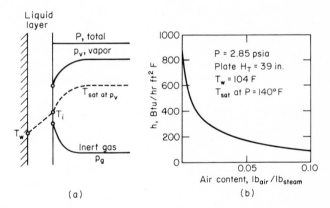

Fig. 20. Effect of noncondensable gas on heat transfer coefficient for condensing vapor [47].

A significant decrease in h results from the presence of very small amounts of noncondensable gas. Figure 20b shows this effect.

In most condensing equipment, such as power plant steam condensers, provision is made to bleed off noncondensable gases that leak into the system. There are, however, cases in which noncondensable gases must be tolerated, such as in separating ammonia from air by condensation. Design of such condensers must allow for the presence of noncondensables.

When a small amount of gas is added to the system the condensation process becomes diffusion limited and the gas accumulates at the cold surface with a partial pressure variation as in Fig. 20. In this case the important quantity is the total mass of gas present W instead of mass ratio of gas-to-vapor since the latter quantity is a function of system volume. An effective thickness of gas t may also be used

$$t = \frac{WR_g T_{avg}}{Ap} \tag{70}$$

where A is the area of condensing surface, p is total pressure, $T_{avg} = \frac{1}{2}(T_{sat} + T_i)$, R_g is gas constant.

The equation representing diffusion of the vapor through the gas is

$$\frac{1}{p - p_v} \frac{dp_v}{dy} = \frac{-w_v R_v T}{D_{12} p} \tag{71}$$

where w_v is the rate of vapor flow and

$$D_{12} = C \frac{T^{3/2}}{p}$$

$$C \equiv \frac{0.0069}{(V_1^{1/3} + V_2^{1/3})^2} \sqrt{\frac{1}{M_1} + \frac{1}{M_2}} \tag{72}$$

is the diffusivity given by Gilliland in Sherwood [44]. Here V_1 and V_2 are atomic volumes, some of which are listed in Table 1.

TABLE 1. Atomic Volume V

Air	29.9	N_2 in secondary amines	12.0
Antimony	24.2	Oxygen, molecule (O_2)	7.4
Arsenic	30.5	Oxygen coupled to two other elements:	
Bismuth	48.4	In aldehydes and ketones	7.4
Bromine	27.0	In methyl esters	9.1
Carbon	14.8	In ethyl esters	9.9
Chlorine, terminal as in R—Cl,	21.6	In higher esters and ethers	11.0
medial as in R—CHCl—R'	24.6	In acids	12.0
Chromium	27.4	In union with S, P, N.	8.3
Fluorine	8.7	Phosphorus	27.0
Germanium	34.5	Silicon	32.0
Hydrogen, molecule (H_2)	14.3	Sulfur	25.6
In compounds	3.7	Tin	42.3
Iodine	37.0	Titanium	35.7
Nitrogen, molecule (N_2)	15.6	Vanadium	32.0
N_2 in primary amines	10.5	Zinc	20.4

An alternative method of calculating D_{12} is given by Hirschfelder and Curtiss [45] and is not reproduced here.

Equations (71) and (72) may be integrated with $T = T_{avg} = \frac{1}{2}(T_{sat} - T_i)$ to obtain the partial pressure distribution for the vapor

$$\frac{p - p_v}{p - p_{vi}} = \exp\left(\frac{w_v R_v y}{C T_{avg}^{1/2}}\right) \tag{73}$$

As a first approximation this leads to

$$\frac{w_v}{A} = \frac{p - p_{vi}}{(W/A)} \frac{Dp_v}{R_v R_g T_{sat}^2}$$

and

$$h = \frac{q''}{T_{sat} - T_i} = \frac{p - p_{vi}}{(W/A)} \frac{Dp_v[h_{fg} + c_{\rho v}(T_{sat} - T_i)]}{T_{sat}^2 R_g R_v (T_{sat} - T_i)} \tag{74}$$

This is a heat transfer coefficient from the vapor to the interface.

The prediction of Eq. (74) was verified by experiments of Kroger [43] for condensing potassium in the presence of argon and helium on the underside of a horizontal surface to minimize convection effects and also the resistance of the liquid film. For condensation on a vertical surface the analysis, Eqs. (5) to (16), is readily modified to include the effect represented by Eq. (74). As an approximation, use Eq. (16) to determine an average h for the liquid film and Eq. (74) to represent an average h for the vapor region. Then

$$\frac{1}{h} = \frac{T_{sat} - T_w}{q''} = \frac{1}{h(\text{Eq. 29})} + \frac{1}{h(\text{Eq. 74})} \tag{75}$$

is an expression to calculate an average T_i, and the q'' and h may then be calculated from Eq. (75). A more detailed analytical study of this problem is given by Sparrow and Lin [46].

In general the procedure outlined above will give conservative results. Because of the temperature (and hence) density gradient at the cold surface, natural convection effects become significant at moderate ΔT; then the actual heat transfer rates will be greater than those predicted as above.

For condensation in heat exchangers, such as shell and tube, Colburn and Hougen [56] presented a stepwise calculation procedure most commonly used by designers. Kern [55] described the procedure in detail and presents numerical calculations for a number of designs. Bras [57] extended the Colburn-Hougen procedure by developing a direct graphical calculation for determining the interface temperature T_i throughout the exchanger. The method presented here was developed by Votta [58] and is really the Colburn-Hougen stepwise method modified to simplify the calculation for determining interface temperature.

The analysis neglects the sensible heat transferred in the mass transfer process. Colburn and Drew [59] found this is negligible for most condenser designs. At an element of area along the exchanger

$$\frac{dq}{dA} = h_g(T_v - T_i) + k_g h_{fg}(p_v - p_i) = U'(T_i - T_L) = U(T_v - T_L) \tag{76}$$

where T_v, T_i, and T_L are temperatures of vapor, liquid-vapor interface, and cooling liquid on the other side of the wall; p_v and p_i are taken as saturation pressure corresponding to T_v and T_i; h_g and k_g are the heat and mass transfer coefficients for the particular gas mixture flow; U' is a partial overall heat transfer coefficient including cooling water side, dirt deposit, tube wall, and condensate film; and U is the overall heat transfer coefficient vapor to cooling water. The Colburn-Hougen method consists of evaluating T_i from Eq. (76) at many places along the exchanger; the U is also known and the total area is calculated by

$$A = \int_0^{q_{total}} \frac{dq}{U\Delta T}$$

(77)

when dq over the element of area equals the enthalpy change of the mixture including amount of vapor condensed.

Votta [58] suggests defining a new potential F defined by

$$\frac{dq}{dA} = h_g(F_v - F_i)$$

(78)

Note F has the dimensions of temperature. Then from Eq. (76)

$$F_v - F_i = (T_v - T_i) + \frac{k_g}{h_g} h_{fg}(p_v - p_i)$$

(79)

From the Chilton-Colburn analog

$$\frac{k_g}{h_g} = \frac{1}{M_m c_{p_m} p_{lm}} \left(\frac{Pr}{Sc}\right)^{2/3}$$

(80)

where M_m is molecular weight of the mixture
c_{p_m} is specific heat of the mixture
p_{lm} is log mean of the gas pressure

$$p_{lm} \equiv \frac{(p - p_i) - (p - p_v)}{\ln[(p - p_i)/(p - p_v)]}$$

(81)

where p is the total pressure. Then

$$\frac{k_g}{h_g} = \frac{\ln[(p - p_i)/(p - p_v)]}{M_m c_{p_m}(p_v - p_i)} (Le)^{2/3}$$

(82)

where Le = Lewis number = Pr/Sc = $D_{12}\rho c_p/k$. Let

$$R \equiv \frac{h_{fg}}{M_m c_{p_m}} Le^{2/3}$$

(83)

Then combining Eqs. (79) and (82)

$$F_v - F_i = T_v - T_i + R \ln \frac{p - p_i}{p - p_v}$$

(84)

The function F may be calculated for saturated mixtures of any two gases above an arbitrary reference state 0, since R is a function only of the state of the mixture. Then

$$F_n - F_0 = T_n - T_0 + R \ln \frac{p - p_0}{p - p_n} \qquad (85)$$

Table 2 presents magnitudes of F as a function of T for saturation states of various mixtures above a reference state at $32°F$. Total pressure = 1 atm.

TABLE 2

Temp. °F	Air-Water		Air-Toluene		Air-Benzene		Helium-Water		Carbon dioxide-Water	
	R	F	R	F	R	F	R	F	R	F
32		0		0		0		0		0
40	3,110	15	1,220	11	1,273	25	2,560	14	3,390	16
50	3,078	37	1,225	27	1,270	55	2,560	34	3,320	38
60	2,990	62	1,210	45	1,259	91	2,560	56	3,230	65
80	2,895	132	1,195	86	1,226	186	2,590	124	3,150	139
100	2,830	239	1,160	143	1,194	322	2,640	229	3,030	252
120	2,690	400	1,150	217	1,193	561	2,800	414	2,940	430
140	2,680	679	1,130	330	1,164	940	2,950	738	2,820	710
160	2,670	1,138	1,110	490	1,132	1,722	3,330	1,420	2,705	1,153
170			1,100	591	1,124	2,782				
180	2,640	2,020					3,760	2,810	2,640	2,018
200	2,860	4,538					5,380	8,300	2,585	4,118

Then the calculation is made by selecting temperature intervals, T_v, along the heat exchanger and calculating T_i from

$$h_g(F_v - F_i) = U'(T_i - T_L) \qquad (86)$$

or

$$T_i = T_L + \frac{h_g}{U'}(F_v - F_i) \qquad (87)$$

For each interval in T_v the heat load Δq is calculated from the enthalpy change between the saturated mixture states. Then since $\Delta q = h_g(F_v - F_i)\Delta A$

$$A = \sum \frac{\Delta q}{h_g(F_v - F_i)} \qquad (88)$$

Actually the process of constructing the Votta table of F vs. T magnitudes replaces the trial-and-error calculation of the Colburn-Hougen procedure.

Part **B**

Dropwise Condensation

PETER GRIFFITH
Professor of Mechanical Engineering
Massachusetts Institute of Technology

INTRODUCTION

Dropwise condensation occurs when the receding contact angle, as measured through the liquid, is greater than zero. (The receding contact angle is the one at the trailing edge of a drop which is running down the surface.) Compared to film condensation, it is a much better way to transfer heat. For film condensation, one expects heat transfer coefficients of the order of 1000 Btu/hr-ft^2-$^\circ$F. For dropwise condensation heat transfer coefficients of 10,000 to 80,000 Btu/hr-ft^2-$^\circ$F are common. In order to maintain dropwise condensation on a condensing surface it is necessary to treat the surface with a substance which is strongly attracted to the condensing surface and repels the water. This substance is called a promoter.

Usually when dropwise condensation occurs, other heat transfer resistances in the apparatus govern the heat transfer rate. A quick estimate of the dropwise condensation heat transfer coefficient for water can be made from Fig. 21. The conditions under which these data were taken are indicated in Table 3. All these curves refer to water condensing at pressures close to 1 atm.

In the remainder of this chapter we shall look at the mechanism of dropwise condensation, the correlation and prediction of dropwise condensation heat transfer coefficients, and list the substances which can be used to promote dropwise condensation.

12-34

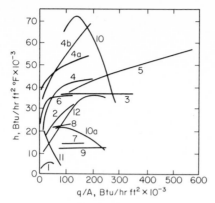

Fig. 21. Dropwise condensation heat transfer data from a variety of sources [67]. See Table 3 for the key to the curves.

MECHANISM OF DROPWISE CONDENSATION

All the condensation which occurs takes place on drops on the surface. These drops generally cover more than 90 percent of the surface area. The heat (or enthalpy) which is transferred must pass through a series of resistances. They are as follows:

First, the vapor must diffuse through a layer of noncondensable gas concentrated near the surface of the drop. Condensation then occurs on the drop which is slightly below saturation temperature. The heat released by the condensation is transferred through the drop to the metal surface and then through the metal surface to the cooling medium. A schematic diagram of the heat transfer through a drop and the resulting temperature distribution is shown in Fig. 22. The drops, upon which the condensation occurs, originate in cracks and pits on the surface where pockets of liquid (nuclei) remain after the surface has been cleared by the passage of a drop.

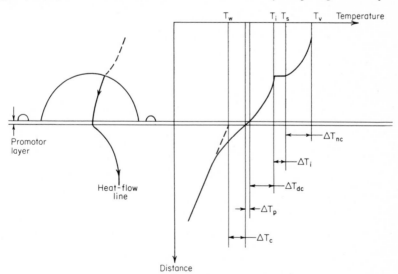

Fig. 22. Schematic diagram of heat flow lines through a drop showing the various temperature drops occurring in dropwise condensation (nucleation effects ignored).

TABLE 3. Variation of Heat Transfer Coefficient with Heat Flux (Fig. 21)

Curve	Authors	Orientation	Surface material	Surface finish	Promoter	Venting arrangement	Gas concentration
1	Fatica and Katz [73]	Vertical	Copper	Mirror smooth	Stearic acid	Unknown	Unknown
2	Hampson and Ozisik [74]	Vertical	Copper	Mirror smooth	Benzyl mercaptan and oleic acid mixed	Continuous bleed	Unknown
3	Wenzel [75]	Vertical	Chromium plated on brass	Highly polished	Oleic acid	Unknown	Unknown
4 b, c	Tanner et al. [76]	Vertical	Copper	Mirror smooth	Dioctadecyl Disulphide	Blow past surface, various velocities	2 ppm
5	LeFevre and Rose [77]	Vertical	Copper	Mirror smooth	Dioctadecyl Disulphide	Close by venting	"Very small"
6	Graham [67]	Vertical	Copper	Mirror smooth	Dioctadecyl Disulphide	Blow past surface	"Very small"
7	Nagle et al. [78]	Vertical tubes	Chromium plated on copper	Unknown	Oleic acid	Continuous bleed	Unknown
8	Gnam [79]	Vertical	Copper	Unknown	Oleic acid	Blow past surface	Unknown
9	Fitzpatrick et al. [80]	Vertical tubes	Copper	Unknown	Benzyl mercaptan	Continuous bleed	Unknown
10, 10a	Shea and Krase [81]	Vertical	Copper	Highly polished	Benzyl mercaptan	Blow past surface	Unknown
11	Welch and Westwater [82]	Vertical	Copper	Polished	Cupric oleate	Unknown	Unknown
12	Kast [83]	Vertical	Copper	Unknown	Benzyl mercaptan	Blow past surface	Unknown

The drop nucleation site density is immense. Reference 67 indicates a nucleation site density of $200 \times 10^6 / \text{cm}^2$ at a pressure of 34 mm of Hg and a wall subcooling of $0.5°F$. The minimum nucleation site size is about 1 micron for these conditions.

The major resistance to heat transfer, for gas free steam, is the drop itself. For an experimentally determined drop population it has been found that most of the heat is transferred by drops less than 100 microns in diameter. The large, visible drops cover the surface, impeding the heat transfer, and grow largely by agglomeration with the small drops growing at their edge. The resistance to heat transfer offered by these large drops is so great that only negligible heat is transferred through them. In addition to the resistance to heat transfer due to the drops themselves, there is a resistance due to noncondensables, to interfacial mass transfer effects, and to constriction of the heat flow lines in the surface. There is also a temperature drop penalty due to the fact that a finite temperature difference is needed to nucleate a drop.

THE CORRELATION AND PREDICTION OF DROPWISE CONDENSATION

If we confine our attention to temperature differences greater than $0.5°F$, the nucleation effects become negligible and we can write that

$$\frac{1}{h} = \frac{1}{h_{nc}} + \frac{1}{h_i} + \frac{1}{h_{dc}} + \frac{1}{h_p} + \frac{1}{h_c} \tag{89}$$

Equation (89) gives the overall heat transfer coefficient for dropwise condensation. Each term in Eq. (89) includes a heat transfer coefficient referred to the metal surface area. The meaning and recommendations for evaluating each of these terms are as follows.

Noncondensable Gas Heat Transfer Coefficient (h_{nc}). This is referred to the metal surface area. It is often the most important factor in Eq. (89) and, at this time, cannot be evaluated for an untested system. The reason is that the specification of the amount of air in the incoming steam does not fix the amount of air in the condenser. The venting arrangement and the large-scale circulation pattern in the condenser are also very important. The lowest curves on Fig. 21 certainly reflect substantial amounts of air being present even though precautions usually were taken to prevent air from being a problem.

In order to insure that effects due to noncondensables are negligible, it is necessary to blow some of the steam past the condensing surface and maintain an appreciable velocity while doing so. It is suggested that if degassed water is used, a vapor velocity past the surface in excess of 3 ft/sec should be maintained and the flow passages designed so that gas pockets cannot form. The effect of noncondensable gas on the heat transfer can then be neglected. For this case

$$h_{nc} = \infty \tag{90}$$

Interfacial Mass Transfer Heat Transfer Coefficient (h_i). This is referred to the metal surface area. When mass is transferred at a finite rate at an interface, there must be a temperature drop at that interface. For a plane surface and for water at 1 atm the overall heat transfer coefficient must be lower. In addition, this heat transfer coefficient is a function of the pressure. Therefore, for water from the calculations of Ref. 67

$$h_i = 3.7 \times 10^5 \left(\frac{p}{p_1}\right)^{0.65} \tag{91}$$

where h_i is in Btu/hr-ft^2-$^\circ$F.

Drop Conduction Heat Transfer Coefficient (h_{dc}). This is referred to the metal surface area. It is the governing factor for heat transfer with degassed water on copper. Because the drop population varies with both temperature and surface orientation both these factors enter in the determination of h_{dc}. For a vertical surface the following equation is recommended for h_{dc}

$$h_{dc} = 2,600 + 200 \, T_s \tag{92}$$

where 77°F $< T_s < 212^\circ$F and

$$h_{dc} = 45,000 \quad \text{when}$$
$$T_s > 212^\circ\text{F} \tag{93}$$

Temperatures are in degrees F and heat fluxes in Btu/hr-ft^2-$^\circ$F. If steam at a high velocity sweeps by the surface the drops will be smaller and h_{dc} larger. When the surface is inclined the drop conduction heat transfer coefficient changes. Figure 23 is a composite plot of data from several sources showing the effect of inclination on the drop conduction heat transfer coefficient. For other than vertical surfaces, multiply the result from Eqs. (92) or (93) by the factor from Fig. 23.

Surprisingly, it does not appear that surface finish is a very important variable. Smooth tubes appear to give slightly higher dropwise condensation heat transfer coefficients than rough ones largely because the maximum size of the drops on the surface is somewhat smaller. Given the uncertainties about the other factors in dropwise condensation it is suggested that this variable be ignored.

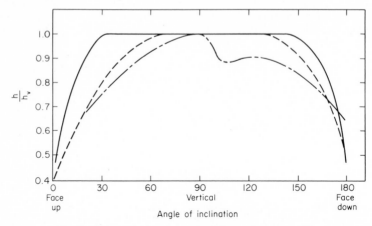

Fig. 23. Effect of inclination on dropwise condensation [67].

Promoter Heat Transfer Coefficient (h_p). In general it takes only a monolayer of promoter to obtain dropwise condensation. The resistance offered by such a thin promoter layer is entirely negligible. If more than a monolayer of promoter is applied to the surface, the excess promoter is soon washed off and the resistance to heat transfer again becomes negligible. The only exception to this is a "permanent" promoter such as Teflon [72]. For this promoter the thermal conductivity is low and the minimum layer thickness is well over a monolayer so the heat transfer resistance is appreciable. For the permanent promoters, then

$$h_p = \frac{k_p}{t_p} \tag{94}$$

For Teflon the thermal conductivity is 0.1 Btu/hr-ft²-°F. The minimum Teflon layer thickness that gives good dropwise condensation is about 6×10^{-5} inch so the minimum Teflon promoter heat transfer coefficient is approximately 20,000 Btu/hr-ft²-°F. (See the subsection on promotion.)

Constriction Heat Transfer Coefficient (h_c). It arises from the fact that the heat transfer rate is exceedingly nonuniform on the surface. Because of this, the heat flow lines are crowded at certain points on the surface. The temperature drop penalty associated with this depends on the drop population and the heat transfer surface thermal conductivity. For copper, the heat transfer resistance is not large but for poor conductors such as stainless steel it can be the most important single resistance. Figure 24 displays data from two sources and shows clearly the decrease in the overall heat transfer coefficient is associated with decreasing the thermal conductivity of the underlying metal.

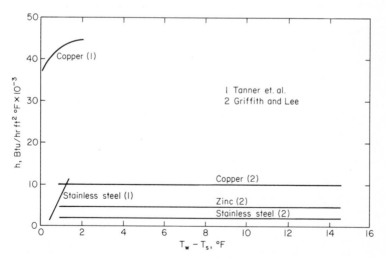

Fig. 24. Effect of condensing surface thermal conductivity on the dropwise condensation heat transfer coefficient. The data of Griffith and Lee was taken with appreciable air present [67].

Precise prediction of the constriction heat transfer coefficient is difficult because its value, again, depends on the drop population. This, in turn, is a function of pressure. An equation developed in Ref. 69 is given in simplified form below as

$$h_c = 3 \frac{k_m}{r_d} \tag{95}$$

The symbol k_m is the thermal conductivity of the metal surface. The r_d in the denominator of Eq. (95) is, physically, the average size of the inactive drops. On a surface where the drops are small when they are removed, r_d will be small. By back calculation with the data of Fig. 23 and from the calculations of the data in Ref. 67, Table 4 has been constructed. It is recommended that h_c be calculated with r_d lying between the limits of the values tabulated in Table 4. The last value of Table 4 is calculated from data taken on a system in which an appreciable amount of air was present. This has the effect of increasing the apparent average size of the inactive drops and giving an inordinately large r_d. For this reason values of r_d closer to the values of Ref. 67 are likely to be found in practice for gas-free steam.

TABLE 4. Heat Transfer Average Drop Radii for Substituting in Eq. (95)

Orientation	Temperature °F	r_d ft	Source
Vertical	212	0.10×10^{-2}	Ref. 67
Vertical	88	0.13×10^{-2}	Ref. 67
Horizontal down	212	1.26×10^{-2}	Fig. 23

PROMOTION OF DROPWISE CONDENSATION

A considerable amount of work has been done on dropwise condensation promotion and a number of satisfactory promoters have been found. A satisfactory promoter must repel water, stick tenaciously to the substrate, prevent oxidation, yet be sufficiently mobile so that flaws in the promoter coating will heal themselves.

Table 3 includes the promoter used and all these are certainly possibilities. However, most of the data reported on Fig. 21 was not taken with the object of testing promoters; therefore some of the promoters are less than satisfactory. A recent study [70] reviewed the literature and tested the most promising promoters for copper. This list includes all the promoters tested in this program. These promoters are to be added to the system as indicated in Table 5. They are appropriate for copper and copper-rich alloy condenser tubes. The condenser was a single, vertical tube condensing evaporated sea water. The tube was 0.625 inch OD, 56 inches long, and vented to the atmosphere. For comparison, the condenser on which these data were taken had an overall heat transfer coefficient of 700 Btu/hr-ft²-°F when film condensation was used.

Two permanent promoters have shown some promise. Teflon [67, 72] and gold [71]. The Teflon must be deposited in a layer 1.5 microns thick and heat treated. The manufacturer provides instructions. The gold promoter is adaptable to any surface that can be gold plated. An exceedingly thin layer is put on the surface and, as long as the surface remains clean, dropwise condensation will persist. Tests with sea water and pure water on a gold plated surface have shown dropwise condensation continues after thousands of hours of condensing [71].

TABLE 5. Summary of Performance of Drop-by-drop (Dropwise Condensation) Promoters in Chronological Order [70]

Promoter	Solution	How injected	Typical U*	Comments
Silicone Dow Corning 1107	1% in isopropyl alcohol	20 ml/hr for 5 hr		Small amount of very poor quality drop-by-drop extending about one-fourth way down condenser tube.
$(C_{12}H_{25}S)_4Si$	1% in kerosene	20 ml/hr for several days		About 10 hr required to produce 100% drop-by-drop. Excellent quality.
$(C_{12}H_{25}S)_4Si$	0.1% in kerosene	20 ml/hr for 8 hr		Maintained excellent drop-by-drop obtained above. Condensate smells like kerosene.
Octanoic acid	1% in isopropyl alcohol	45 ml/hr for 3½ hr		Very poor drop-by-drop covered tube. Cleaned residual oxide off tube.
Stearic acid	1% in isopropyl alcohol	45 ml/hr for 2 hr	1,100	Slightly better than octanoic acid.
Oleic acid	1% in isopropyl alcohol	65, 45, 25 ml/hr ea. for 1 hr	1,300	Slightly better than stearic acid but still of poor quality.
Cupric oleate	Sat'd. sol. in isopropyl alcohol	25 ml/hr for 2½ hr	1,400	Slightly better than oleic. Still poor quality. One-third of tube reverted to film type 10 min after ceasing injection.
Montan wax	Seawater, contacted sol'n. wax in octanoic acid	45 ml/hr for 5½ hr		No drop-by-drop condensation. Montan wax extremely insoluble in seawater.
Montan wax	Sat'd. sol'n. (~0.1%) in octanoic acid	Various rates inter-mittently	1,500	Excellent drop-by-drop. Would last ~2 hr after stopping injection.
Montan wax	Pure wax	Passed feed seawater through packed bed of wax		No drop-by-drop condensation, wax too insoluble.
Montan wax	Sat'd. sol'n. in iso-propyl alcohol	65 ml/hr for 1½ hr		Very poor drop-by-drop. Did not seem characteristic of montan wax.

TABLE 5. Summary of Performance of Drop-by-drop (Dropwise Condensation) Promoters in Chronological Order (*Cont.*)

Promoter	Solution	How injected	Typical U*	Comments
Montan wax	Sat'd. sol'n. in conc. H_2SO_4	65 ml/hr for 1½ hr		Very poor drop-by-drop. Wax may have been decomposed.
Montan wax	Sat'd. sol'n. in oleic acid	45 ml/hr for 3 hr		Good drop-by-drop 30 min after start. Like wax in octanoic except cleaning not as good.
$(C_{12}H_{25}S)_4Si$	0.1% in octanoic acid	27 ml in 35 min	1,500	Excellent to good drop-by-drop condensation continued for 86 hr. No new promoter added. About one-fourth film type at end. Tube rather clean after standing over weekend.
Montan wax + $(C_{25}H_{25}S)_4Si$	0.1% in coconut oil containing 10% octanoic acid	To give 1 ppm coconut oil in seawater	900	Excellent foam suppression. Promotion started at top and slowly spread to three-eighths of tube after 43 hr. Eighteen more hours no change.
$(C_{12}H_{25}S)_4Si$	1% in octanoic acid	6 ml intermittently injected over 1 hr	1,500	Very good drop-by-drop. Some deposited on oxide is inferior to that on clean metal.
Montan wax	Liquid wax at ~100°C	0.033 cc into boiler in increments		Almost no drop-by-drop.
Montan wax	0.1% in octanoic acid	6.9 cc at rate corresponding to 4 ppm octanoic acid		After 6 hr about 40% of tube drop-by-drop.

Removed tube and cleaned by scrubbing with NH_4OH—$(NH_4)_2CO_3$ solution. Replaced and flooded with acidified seawater. Gave 100% film type

$(C_{18}H_{37}S)_4Si$	1% in octanoic acid	Add 20 ml intermittently over 1¼ hr	1,500 (1,700)±	After ~3 ml 95% good drop-by-drop. After 20 ml 99% excellent drop-by-drop. Ran during days only for 40 hr condensing. 95% excellent at end. Almost no tarnishing after 12 days including two weekends.
Octanoic acid	Pure acid	0.1 ml directly into boiler		Poor short-lived drop-by-drop on entire tube.

$(C_{18}H_{37}S)_4Si$	Pure liquid	0.03 ml directly into boiler		Only two very small spots of drop-by-drop condensation appeared.
Lauric acid	Pure acid heated to liquefy	0.1 ml directly into boiler		Almost exactly like octanoic, perhaps a little longer lived.
$(C_{18}H_{37}S)_4Si$	1% in octanoic acid	4 ml in 1½ min	$(1,700)\pm$	100% excellent drop-by-drop in 3 min. Tube had been carefully cleaned in place in ~1 hr after oxidizing in air using a little chlorine to remove residual promoter followed by sulfur dioxide (all in steam). 100% clean and film type before promoter add'n. After 100% drop-by-drop, promoter was then 100% removed in chlorine to film type in 1 min. Tube darkened. Sulfur dioxide add'n. produced pink 100% film type tube in 2 min which oxidized rapidly on standing in air.
$(C_{18}H_{37}S)_4Si$	1% in octanoic acid	0.1 ml added directly to boiler		Chlorine-sulfur dioxide cleaning before add'n. Excellent drop-by-drop almost instantly (~10 sec) on lower half of tube. Good on top half.
$(C_{18}H_{37}S)_4Si$	1% in octanoic acid	1.0 ml directly into boiler		100% excellent drop-by-drop condensation; good oxidation resistance.
$(C_{18}H_{37}S)H$	1% in octanoic acid	1.0 ml directly into boiler		Chlorine-sulfur dioxide cleaning before add'n.‡ 100% excellent drop-by-drop condensation. On standing in air, oxidation resistance inferior to thio silanes.

*For comparison, $U_{film\ type}$ = 700 Btu/hr·ft²·°F. These represent typical values corrected to a flux of ~42,000 Btu/hr·ft² cooling water velocity 7 ft/sec, and are for outside surface of tube.

†This value is at a reduced but more nearly typical flux of ~14,000 Btu/hr·ft². The film type coefficient is ~900.

‡The procedure involving cleaning with sulfur dioxide with short bursts of chlorine and direct addition of about 1 ml of 1% thio silane in octanoic acid was repeated many times with the same result. Oxygen must be kept away from the tube after cleaning and before promotion or the promoter will deposit on the oxide, resulting mainly in poor promoter life and decreased oxidation resistance. Traces of chlorine must not be allowed to contact the promoted tube or the promoter will be partially removed and oxidation will proceed.

REFERENCES

1. W. Nusselt, *VDI Zeitschrift*, **60**:541, 569 (1916).
2. S. Sukhatme and W. M. Rohsenow, *TR* 9167, Department of Mechanical Engineering, Massachusetts Institute of Technology, May 1964; also S. Sukhatme, doctoral dissertation, Department of Mechanical Engineering, Massachusetts J Institute of Technology, June 1964.
3. R. W. Schrage, "A Theoretical Study of Interphase Mass Transfer," Columbia University Press, New York, 1953.
4. W. Bornhorst, doctoral dissertation, Department of Mechanical Engineering, Massachusetts Institute of Technology, June 1966.
5. W. M. Rohsenow, *Trans. ASME*, **78**:1645 (1956).
6. E. M. Sparrow and J. L. Gregg, Condensation on a Rotating Disk, *J. Heat Transfer, Trans. ASME*, **C81**:249 (1959); **C82**:71 (1960); **C83**:101 (1961).
7. K. O. Beatty, Jr. and S. S. Nandapurkar, Condensation on a Horizontal Rotating Disk, *3d Nat. Heat Transfer Conf. ASME/AIChE, Storrs, Conn.*, Preprint 112, Aug. 1959.
8. W. J. Minkowycz and E. M. Sparrow, Condensation Heat Transfer in the Presence of Non-condensables, Interface Resistance, Superheating, Variable Properties and Diffusion, *Int. J. Heat Mass Transfer*, **9**:1125 (1966).
9. G. Poots and R. G. Miles, Effect of Variable Properties on Laminar Film Condensation of Steam, *Int. J. Heat Mass Transfer*, **10**:1677 (1967).
10. J. C. Y. Koh, E. M. Sparrow, and J. P. Hartnett, Two Phase Boundary Layer in Laminar Film Condensation, *Int. J. Heat Mass Transfer*, **2**:69 (1961).
11. S. J. Wilcox and W. M. Rohsenow, Film Condensation Using Copper Condensing Block for Precise Wall Temperature Measurement, *J. Heat Transfer, Trans. ASME*, **C92**:359 (1970).
12. D. G. Kroger and W. M. Rohsenow, Heat Transfer During Film Condensation of Potassium Vapor, M.I.T. Rept. 75239-42, Dec. 1966.
13. R. E. Barry, doctoral dissertation, University of Michigan, 1965.
14. B. Misra and C. F. Bonilla, *Chem. Eng. Prog. Sym.*, **(18)**52:7 (1956).
15. I. T. Aladyev et al., Thermal Resistance of Phase Transition with Condensation of Potassium Vapor, *Proc. 3d Int. Heat Transfer Conf.*, 1966.
16. D. A. Labuntsov and S. I. Smirnov, Heat Transfer in Condensation of Liquid Metal Vapors, *Proc. 3d Heat Transfer Conf.*, 1966.
16a. R. Sakhuja, "Effect of Superheat on Film Condensation of Potassium," doctoral dissertation, Department of Mechanical Engineering, Massachusetts Institute of Technology, Sept. 1970.
17. P. L. Kapitza, *Zh. Ekper. Teoret. Fiz.*, **18**:1 (1948).
18. W. H. McAdams, "Heat Transmission," 2d ed., p. 261, McGraw-Hill Book Company, New York, 1942.
19. W. M. Rohsenow and H. Y. H. Choi, "Heat, Mass and Momentum Transfer," secs. 10.1 and 10.3, Prentice-Hall, Inc., Englewood Cliffs, N.J., 1961.
20. W. M. Rohsenow, J. H. Webber, and A. T. Ling, *Trans. ASME*, **78**:1637 (1956).
21. W. H. McAdams, "Heat Transmission," 3d ed., p. 334, McGraw-Hill Book Company, New York, 1954.
22. F. G. Carpenter, doctoral dissertation, Department of Chemical Engineering, University of Delaware, 1948.
23. J. A. Lehtinen, doctoral dissertation, Department of Mechanical Engineering, Massachusetts Institute of Technology, June 1957.
24. O. P. Bergelin, P. K. Kegel, F. G. Carpenter, and C. Gasley, Heat Transfer Fluid Mech. Inst., Berkeley, Calif., 1949.
25. F. G. Carpenter and A. P. Colburn, London discussion of heat transfer, ASME, July 1951.
26. A. E. Dukler, Fluid Mechanics and Heat Transfer in Vertical Falling Film Systems, *3d Nat. Heat Transfer Conf. ASME/AIChE, Storrs, Conn.*, Preprint 101, Aug. 1959.
27. M. M. Chen, *J. Heat Transfer, Trans. ASME*, **C83**:55 (1961).
28. See Ref. 1.
29. K. O. Beatty and D. L. Katz, Condensation of Vapors on Outside of Finned Tubes, *Chem. Engr. Prog.*, **44**:55 (1948).
30. D. L. Katz, R. B. Williams, and E. L. Tyner, Jr., A Study of Possible Uses for Finned Tubes in Utility Power Plants, Project M 592, Engineering Research Institute, University of Michigan, Jan. 1949.
31. D. L. Katz, P. E. Hope, S. C. Datsko, and D. B. Robinson, Condensation of Freon-12 with

Finned Tubes, Part I, Single Horizontal Tubes; Part II, Multiple Condensers, *Refrig. Engr.*, pp. 211, 315, March/April 1947.

32. H. M. Fishman, An Experimental Deterimination of the Effect of Vapor Velocity on Condensation on the Outside of Horizontal Tubes, thesis, Department of Mechanical Engineering, Massachusetts Institute of Technology, June 1960.

33. See Ref. 19, Fig. 4.17.

34a. V. I. Subbotin, M. N. Ivanovskii, V. P. Sorokin, and V. A. Chulkov, *Teplofizika Vysokih Temperatur*, **4**:616 (1964).

34b. V. I. Subbotin, N. V. Bakulin, M. N. Ivanovskii, and V. P. Sorokin, *Teplofizika Vysokih Temperatur*, **5** (1967).

34c. V. I. Subbotin, M. N. Ivanovskii, and A. I. Milovanov, Condensation Coefficient for Mercury, *Atomnaya Energia*, **24** (1968).

35. J. C. Chato, *J. Am. Soc. Heating Refrig. Aircond. Engrs.*, p. 52, Feb. 1962.

36. M. Altman, F. W. Staub, and R. H. Norris, *3d Nat. Heat Transfer Conf., ASME/AIChE, Storrs, Conn.*, Preprint 115, Aug. 1959.

37. W. W. Akers and H. F. Rosson, *3d Nat. Heat Transfer Conf., ASME/AIChE, Storrs, Conn.*, Preprint 114, Aug. 1959.

38. W. W. Akers, O. K. Crosser, and H. A. Deans, *Proc. 2d Nat. Heat Transfer Conf., ASME/AIChE*, Aug. 1958.

39. P. J. Berenson, *J. Heat Transfer, Trans. ASME*, **C83**:351 (1961).

40. J. Gerstmann and P. Griffith, Effect of Surface Instability on Laminar Film Condensation, *M.I.T. Heat Transfer Lab. Rept.* 5050-36; *Int. J. Heat Mass Transfer*, **10**:567 (1967).

41. H. Y. Choi, Electrostatic Effects of Condensing in a Vertical Tube, *Tufts Univ. Dept. of Mech. Engr. Rept.* 64-1, Feb. 1964.

42. D. L. Spencer and W. E. Ibele, Laminar Film Condensation of a Saturated and Superheated Vapor, *3d. Int. Heat Transfer Conf., Chicago*, vol. II, p. 337, Aug. 1966.

43. D. G. Kroger, Heat Transfer During Film Condensation of Potassium Vapor, *M.I.T. Heat Transfer Lab. Rept.* 75239-42, Sept. 1966.

44. T. K. Sherwood, "Absorption and Extraction," p. 18, McGraw-Hill Book Company, New York, 1937.

45. J. O. Hirschfelder and C. F. Curtiss, "Molecular Theory of Gases and Liquids," John Wiley & Sons, Inc., New York, 1954.

46. E. M. Sparrow and S. H. Lin, Condensation in the Presence of a Non-condensable Gas, *J. Heat Transfer, Trans. ASME*, **C86**:430 (1963).

47. E. Langen, *Forschung a.d. Geb. d. Ingenieurwes*, **2**:359 (1931).

48. See Ref. 18, pp. 355--356.

49. See Ref. 34a.

50. R. E. Barry, doctoral dissertation, University of Michigan, 1965.

51. See Ref. 14.

52. S. P. Sukhatme and W. M. Rohsenow, Heat Transfer During Film Condensation of a Liquid Metal Vapor, *Trans. ASME*, **88**:19 (1966).

53. See Ref. 15.

54. D. A. Labuntsov and S. I. Smirnov, Heat Transfer in Condensation of Liquid Metal Vapors, *Proc. 3d Int. Heat Transfer Conf.*, 1966.

55. D. Q. Kern, "Process Heat Transfer," pp. 339--351, McGraw-Hill Book Company, New York, 1950.

56. A. P. Colburn and O. A. Hougen, *Ind. Eng. Chem.*, **26**:1178 (1934).

57. G. H. P. Bras, *Chem. Engr. Sci.*, **6**:277 (1957).

58. F. Votta, Jr., Condensing from Vapor Gas Mixtures, *Chem. Engrg.*, **71**:223 (1964).

59. A. P. Colburn and T. B. Drew, *AIChE Trans.*, **33**:197 (1937).

60. R. W. Lockhart and R. C. Martinelli, Proposed Correlation of Data for Isothermal Two-Phase, Two-Component Flow in Pipes, *Chem. Engr. Prog.*, **45**:39 (1949).

61. M. Soliman, J. R. Schuster, and P. J. Berenson, A General Heat Transfer Correlation for Annual Flow Condensation, *J. Heat Transfer, Trans. ASME*, **C90** (1968).

62. S. M. Zivi, Estimation of Steady-State Steam Void-Fraction by Means of the Principle of Minimum Entropy Production, *J. Heat Transfer, Trans. ASME*, **C86**:247 (1964).

63. S. Bae, J. D. Maulbetsch, and W. M. Rohsenow, Refrigerant Forced Convection Condensation in Horizontal Tubes, *Heat Transfer Lab., M.I.T., Repts.* 59 and 64 (to be published in *ASHRAE Journal*).

64. H. R. Kunz and S. Yerazunis, An Analysis of Film Condensation, Film Evaporation, and Single-Phase Heat Transfer for Liquid Prandtl Numbers From 10^{-3} to 10^4, *Heat Transfer Conference, Seattle, Washington,* Paper 67-HT-1, 1967.
65. E. F. Carpenter and A. P. Colburn, The Effect of Vapor Velocity on Condensation inside Tubes, *Proc. General Discussion of Heat Transfer, I. Mech. E. and ASME,* 1951.
66. D. E. Still, Film Condensation of Superheated Freon-113 Vapor, thesis, Mechanical Engineering Department, Massachusetts Institute of Technology, Nov. 1970.
67. C. Graham, The Limiting Heat Transfer Mechanisms of Dropwise Condensation, doctoral dissertation, Mechanical Engineering Department, Massachusetts Institute of Technology, Cambridge, Mass., March 1969.
68. K. Nabavian and L. A. Bromley, Condensation Coefficient of Water, *Chem. Eng. Sci.,* **18**:651 (1963).
69. B. B. Mikic, On the Mechanism of Dropwise Condensation, *Int. J. Heat Mass Transfer,* **12**:1311 (1969).
70. L. A. Bromley, J. W. Ponter, and S. M. Read, Promotion of Drop-by-Drop Condensation of Steam from Seawater on a Vertical Copper Tube, *AIChE J.,* **14**:245 (1968).
71. R. A. Erb and E. Thelan, Dropwise Condensation on Hydrophobic Metal and Metal Sulfide Surfaces, *First Int. Symp. Water Desalination, Washington, D.C.,* Oct. 3–9, 1965.
72. Teflon 30, TFE: Fluorocarbon resin manufactured by duPont.
73. N. Fatica and D. L. Katz, Dropwise Condensation, *Chem. Engr. Prog.,* **45**:661 (1949).
74. H. Hampson and N. Ozizik, An Investigation into Condensation of Steam, *Proc. Inst. Mech. Eng.,* **7**:282 (1952).
75. H. Wenzel, Versuch über Tropfenkondensation, *Allg. Wärmetech,* **8**:53 (1957).
76. D. W. Tanner, D. Pope, C. J. Potter, and D. West, Heat Transfer in Dropwise Condensation of Low Steam Pressures in the Absence and Presence of Non-condensable Gas, *Int. J. Heat Mass Transfer,* **11**:181 (1968).
77. E. J. Le Fevre and J. W. Rose, An Experimental Study of Heat Transfer by Dropwise Condensation, *Int. J. Heat Mass Transfer,* **8**:1117 (1965).
78. W. M. Nagle, G. S. Bays, L. M. Blenderman, and T. B. Drew, Heat Transfer Coefficients during Dropwise Condensation of Steam, *Trans. Amer. Inst. Chem. Eng.,* **31**:593 (1935).
79. E. Gnam, *Tropfenkondensation von Wasserdampf,* Forschungshefte, Verein Deutsch. Ingenieure, **17**:382 (1937).
80. J. P. Fitzpatrick, S. Baum, and W. H. McAdams, Dropwise Condensation of Steam on Vertical Tubes, *Trans. Amer. Inst. Chem. Eng.,* **35**:97 (1939).
81. F. L. Shea and N. W. Krase, Dropwise and Filmwise Condensation of Steam, *Trans. Amer. Inst. Chem. Eng.,* **35**:463 (1940).
82. J. F. Welch and J. W. Westwater, Microscopic Study of Dropwise Condensation, "International Developments in Heat Transfer," *Conf. Int. Developments in Heat Transfer,* p. 302, ASME, 1961.
83. W. Kast, Heat Transfer with Drop Condensation, *Chemie-Ingenieur Technik,* **31**:169 (1963).
84. D. P. Traviss, A. B. Baron, and W. M. Rohsenow, Forced Convection Condensation inside Tubes, *M.I.T. Heat Transfer Lab. Rept.* 74, July 1971 (to be published in *ASHRAE Journal*).
85. P. M. Meyrial, M. M. Morin, S. J. Wilcox, and W. M. Rohsenow, *4th Int. Heat Transfer Conf.,* Versailles, 1970, Vol. VI, paper Cs 1.1.

NOMENCLATURE

C_D drag coefficient, Eq. (41)

c, c_p specific heat at constant pressure, liquid

D diameter

F defined in Eq. (78)

G mass velocity

g acceleration of gravity, 32.17 ft/sec^2

h heat transfer coefficient

h_c constriction heat transfer coefficient in Btu/hr-ft^2-°F

h_{dc} drop conduction heat transfer coefficient in Btu/hr-ft^2-°F

h_i interfacial heat transfer coefficient in Btu/hr-ft^2-°F

h_p promoter heat transfer coefficient in Btu/hr-ft^2-°F
h_{nc} noncondensable gas heat transfer coefficient in Btu/hr-ft^2-°F
h_v heat transfer coefficient for a vertical surface in Btu/hr-ft^2-°F
h_{fg} latent heat
h'_{fg} $h_{fg} + (3/8) c (T_s - T_w)$
k thermal conductivity of liquid
k_m thermal conductivity of the metal in Btu/hr-ft-°F
k_p thermal conductivity of the promoter in Btu/hr-ft-°F
k_v thermal conductivity of vapor
L length
m mass of a molecule
N/A molecular flux
Nu   Nusselt number, hL/k
n number of horizontal tubes in a vertical bank
Pr   Prandtl number, $c\mu/k$
P pressure in lb/in.^{2}
P_1 pressure at one atmosphere in lb/in.^{2}
q/A heat flux in Btu/hr-ft^2-°F
r_d drop radius from Table 4 for Eq. (95) in ft
R gas constant for particular gas
Re   Reynolds number, $4\Gamma/\mu$
t_p promoter thickness
T temperature
ΔT_c temperature drop due to constriction resistance
ΔT_{dc} temperature drop due to conduction
ΔT_i temperature drop at interface
ΔT_p temperature drop across the promoter
ΔT_{nc} temperature drop due to noncondensable gas
U overall coefficient of heat transfer
V velocity
V_y velocity perpendicular to a surface
Γ defined by Eq. (3a); also by Eq. (7)
δ thickness of liquid film
η defined by Eq. (3b)
θ $(T - T_s)/(T_w - T_s)$
μ viscosity of liquid
ν kinematic viscosity, μ/ρ
ρ density of liquid
ρ_v density of vapor
σ accommodation coefficient for condensation
σ_e evaporation coefficient
τ_v shear stress of liquid-vapor interface
θ angle with horizontal plane

Subscripts

m average value or metal
s properties at saturation pressure
v or i properties at liquid-vapor interface temperature
w properties at wall temperature

Superscript

$*$ defined in Eqs. (30) and (31)

Section 13

Boiling

WARREN M. ROHSENOW

Mechanical Engineering Department,
Massachusetts Institute of Technology,
Cambridge, Massachusetts

The process of evaporation associated with vapor bubbles in a liquid is called boiling. Here attention will be focused on boiling at solid heated surfaces of interest in engineering applications.

When a pool of liquid at saturation temperature is heated by an electrically heated wire or flat plate, data for q/A vs. $(T_w - T_{sat})$ usually appear as shown by the lower solid curve in Fig. 1. As ΔT increases, the initial natural circulation regime AB changes to the nucleate boiling regime. The appearance of the first bubble at B requires a significant finite superheat.

As ΔT or heat flux is increased, more and more nucleation sites become active producers of bubbles. A peak nucleate boiling heat flux (called critical heat flux, CHF) is reached when bubbles stream forth from so many nucleation sites that the liquid is unable to flow to the heat surface. With electric heating (q/A the independent variable) the operating condition at point C of Fig. 1 changes rapidly to point E' which is at a very much higher temperature. A "burnout" occurs if point E' is above the melting point of the solid surface. If it is below, then operation can be maintained along the curve ED which is called the "film boiling" regime because a layer of vapor continually blankets the surface and liquid does not contact the surface.

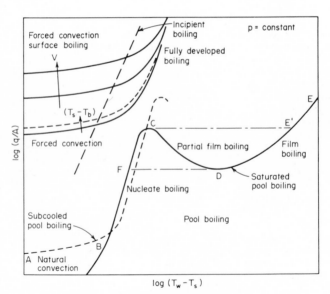

Fig. 1. Regimes in boiling heat transfer.

If the heat flux is reduced below point D at the minimum point on the curve, the operating condition reverts to nucleate boiling at point F. Then power again must be increased to point C before the process can revert to film boiling.

To operate in the transition region C-D between nucleate and film boiling, the surface temperature should be the independent variable accomplished by a condensing vapor, etc. In this region the process may be alternating between nucleate and film boiling, with the liquid touching the surface intermittently.

The dashed line shown for pool boiling of a subcooled liquid lies near the curve for saturated liquids; it may lie either to the left or right depending on different natural convection effects resulting from different surface geometries.

Experimental data for forced convection—inside of ducts or across wires and plates—when plotted on the same coordinates, appear as shown in Fig. 1. If the liquid temperature T_b is subcooled, the heat flux is represented by

$$\left(\frac{q}{A}\right)_{\text{before boiling}} = h[(T_w - T_s) + (T_s - T_b)] \tag{1}$$

At high-heat flux, the forced convection curves for a particular pressure and at various subcoolings and velocities appear to merge on log-log plots into a single curve called the "fully developed boiling curve." For some systems this fully developed region lies approximately on an extension of the pool boiling line for the same surface.

The literature on boiling heat transfer is now quite voluminous. For references in addition to those discussed here the reader is referred to the quite complete collection of annotated bibliography assembled by Gouse [1].

NUCLEATION

In carefully cleaned test systems, heating of pure liquids can produce high superheat at temperatures up to a limit of stability where homogeneous nucleation (vapor formation) takes place. For the case of heat transfer associated with boiling at solid heating surfaces, observed wall temperatures are very much lower than these. Further observations of nucleate boiling show bubble streams emerging at single spots on the surface. Microscopic observation of these spots [1] revealed a scratch or a cavity where a bubble formed. Earlier, Corty and Foust [3a], Bankoff [3b], and others developed the postulate that bubbles at a heating surface emerge from cavities in which a gas or vapor phase preexists (Fig. 2). As heat is added, more vapor forms in the cavity (Fig. 2a), and a bubble emerges and departs (Fig. 2b) after which liquid closes in over the cavity, trapping vapor which is the source of the next bubble. Surveys of nucleation phenomena are found in Refs. 4 and 5.

To put these notions on a quantitative basis, consider the equilibrium and growth of a spherical bubble in a large body of liquid. The following equilibrium conditions

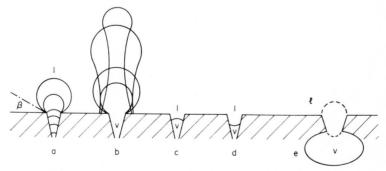

Fig. 2. The formation of bubbles of vapor over cavities in a heated surface.

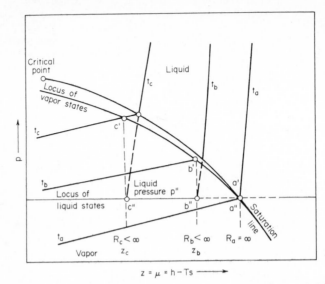

Fig. 3. Approximation of vapor-liquid equilibrium at a curved interface of a vapor bubble in a one-component system.

apply: equal and uniform temperatures in both phases equal chemical potential ($\mu = h - TS$) in both phases, and

$$p_v - p_l = \frac{2\sigma}{r} \qquad (2)$$

With these the equilibrium conditions may be shown graphically in Fig. 3 where states b', b'', and c' and c'' are corresponding vapor and liquid states for two different radii. The production of a Fig. 3 is rather tedious. A simplified procedure evolves from the observation that the locus of vapor states is very close to the saturation line. If it is assumed that the vapor temperature is the plane-surface saturation temperature corresponding to the pressure of the vapors p_v, a much simpler analysis follows.

Calculate the ΔT corresponding to $p_v - p_l$ given by Eq. (1), integrating the Clausius-Clapeyron relation

$$\frac{dT}{dp} = \frac{T v_{fg}}{h_{fg}} \qquad (3)$$

along the p-T saturation curve.

The following are results of such an integration for various assumptions:

If $h_{fg}/v_{fg}T$ is constant and $T \approx T_{sat}$

$$T_v = T_{sat}\left(1 + \frac{2\sigma}{r}\frac{v_{fg}}{h_{fg}}\right) \qquad (4a)$$

If h_{fg}/v_{fg} is constant

$$T_v = T_{sat} \exp\left(\frac{2\sigma}{r}\frac{v_{fg}}{h_{fg}}\right) \tag{4b}$$

If $v_{fg} \approx v_g = R_v T/p$

$$T_v = \frac{T_{sat}}{1 - (T_{sat}R_v/h_{fg})\ln(1 + 2\sigma/r p_{sat})} \tag{4c}$$

Same as Eq. (4c) with $2\sigma/p_l r \ll 1$

$$T_v = \frac{T_{sat}}{1 - (T_{sat}R_v/h_{fg})(2\sigma/r p_{sat})} \tag{4d}$$

Other alternatives are as follows:

With Eq. (1) use an empirical fit of data for the saturation curve as $\log_{10} p_{(psia)} = A - B/T_{(°R)}$. With Eq. (1) this may be solved by trial and error, or with Eq. (4c) above

$$\Delta T = \frac{T_{sat}^2}{B}\log_{10}\left(1 + \frac{2\sigma}{r p_{sat}}\right)\left[1 + \frac{T_{sat}}{B}\log_{10}\left(1 + \frac{2\sigma}{r p_{sat}}\right)\right] \tag{5}$$

for steam $A = 6.945, B = 3880.81$ at 14.7 psia; $A = 6.5928, B = 3608.17$ at 1000 psia.

With Eq. (1) use tabular values from properties tables for the saturation curve.

A comparison of results of the above calculation methods is given in Table 1 for $r = 0.0005$ inches and water at 14.7 and 1000 psia.

The above equations represent an approximate expression for the superheat required for equilibrium of a bubble of radius r. Nuclei of radius greater than r from Eq. (5) should become bubbles and grow; those of smaller radius should collapse.

Some experiments have shown that at a heated surface in water at atmospheric pressure, boiling begins at around 30°F above saturation temperature. For this condition Eq. (5) predicts an equilibrium bubble radius of 10^{-4} in. This is about 10,000 times larger than the maximum cavity size expected from molecular fluctuations [14]. Volmer estimated a cavity formation rate of size 10^{-4} in. to

TABLE 1. Comparison of Calculation Methods

Method	$p = 14.7$ psia $T_{sat} = 212°F$	$p = 1000$ psia $T_{sat} = 544.61°F$
Eq. (4a)	$T_v = 216.60°F$	T_v 544.66°F
Eq. (4b)	216.63	644.66
Eq. (4c)	216.50	544.68
Eq. (4d)	216.71	544.68
Eq. (5)	216.39	544.66
Properties table	216.46	544.66

be approximately one per cubic inch per hour. From this, we readily conclude that free vapor nuclei arising from molecular fluctuations are not important as nucleation cavities. On the other hand, gas nuclei will be very significant as nucleation cavities.

If inert gas molecules are present in the liquid and, hence, in the bubble, Eqs. (1) and (5) are modified

$$p_v - p = \frac{2\sigma}{r} - p_g \tag{6}$$

$$T - T_{sat} = \frac{R_v T^2}{p \, h_{fg}} \left(\frac{2\sigma}{r} - p_g \right) \tag{7}$$

which shows that the superheat required for a given bubble to grow is decreased by the presence of the gas partial pressure p_g.

The preceding calculation procedure relates to nucleation at a cavity in the heating surface in the following way. Griffith [5, 7] showed that the radius of curvature of a vapor interface for positions inside and outside of the cavity (Fig. 2a) may be plotted as in Fig. 4 as $1/r$ vs. vapor volume. In Eqs. (4), (6), and (7) the superheat required to cause a bubble to grow is determined by $1/r$. The maximum point in the $1/r$ curve (Fig. 4) then determines the minimum superheat required for a bubble to grow at this cavity. The significant feature of Fig. 4 is that while the shape of the curve depends on cavity geometry and contact angle β, the magnitude of r at the maximum point of the curve is equal to the radius of the cavity for any $\beta \leq 90°$. Hence the cavity radius should determine the required superheat for nucleation for liquids which wet the surface.

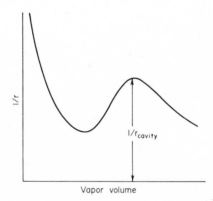

Fig. 4. Interface curvature vs. bubble volume for cavity shown in Fig. 2a.

Equation (6) has been verified experimentally by Griffith and Wallis [7] for single cavities in a uniform temperature system.

It is possible for a cavity to continue to contain vapor even when the surface is highly subcooled. If the contact angle is such that the curvature is as shown in Fig. 2d, the vapor pressure may be very much less than the liquid pressure as the radius decreases down into the cavity. Here Eq. (2) has a minus sign, and vapor will be present at temperatures very much less than the normal saturation temperature of the liquid. Hence this cavity is immediately ready to produce bubbles on its next heating. In a similar way, reentrant cavities, such as the one shown in Fig. 2e, may also withstand high subcooling without collapsing the cavity. The cavity in Fig. 2c would immediately fill with liquid upon cooling.

INCIPIENT BOILING IN FORCED COVECTION

Consider a liquid flowing in a tube. As the heat flux or wall temperature is raised, boiling or nucleation begins at a particular value of the wall temperature. We wish to investigate this condition.

The following procedure was developed by Bergles [8] based on a suggestion of Hsu and Graham [9]. In this flow inside the tube the expression for heat flux is

$$\frac{q}{A} = h(T_w - T_{\text{liq}}) = -k_1 \left(\frac{\partial T}{\partial y}\right)_{y=0} \tag{8}$$

Here h is the heat transfer coefficient which is a function of the geometry, the fluid properties, and the flow rate; k_1 is the liquid thermal conductivity. For a given liquid temperature, both the temperature gradient $(\partial T/\partial y)_{y=0}$ and the wall temperature T_w increase as the heat flux increases. A series of curves representing the temperature distribution very near the heated wall is shown for increasing heat flux in Fig. 5. Also shown in Fig. 5 is a curve labeled T_g^*, which is a plot of Eq. (4) with the radius of the cavity plotted as the distance from the heated surface. A possible theory: Nucleation takes place when the temperature curve in the liquid is tangent to the curve representing Eq. (4). The implication is that the surface contains cavities of various sizes (Fig. 10) and when the temperature at the outer surface of the bubble reaches the critical value given by Eq. (4), the bubble grows at the cavity whose radius is represented by the distance between the wall and the point of intersection.

At this point of tangency, the radius of the first cavity to nucleate (solving Eqs. (4a) and (8) simultaneously) is

$$r_{\text{nucl}} = \sqrt{\frac{2\sigma T v_{fg} k}{h_{fg}(q/A)}}$$

and the heat flux at incipient boiling is predicted by this postulate to be

$$\left(\frac{q}{A}\right)_{\text{incip}} = \frac{h_{fg} k}{8\sigma T v_{fg}} (T_w - T_{\text{sat}})^2$$

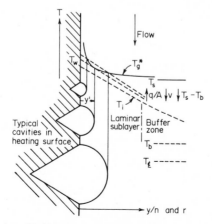

Fig. 5. Initiation of bubble growth in forced-convection surface boiling [8].

For heat fluxes greater than this incipient condition, the radii which are active are on either side of the magnitude given by the above equation, intersections of the upper dashed line and the T_g^* curve of Fig. 5.

In commercially prepared surfaces a wide range of cavity sizes can be expected. Incipient boiling, then, should be independent of surface condition for most commercially finished surfaces. The present analysis should therefore be valid in most practical applications. It is noted, however, that the position and slope of the remainder of the boiling curve should be dependent on the size range of active nuclei.

Equations (4) and (8) could be differentiated to determine the conditions at the point of tangency. However, this procedure does not lead to an explicit formulation for the incipient heat flux $(q/A)_i$. Consequently, a graphical solution was found to be much more convenient. The results for $(q/A)_i$ were calculated for water over a pressure range of 15 to 2,000 lbf/in.^{2} abs and can be expressed quite accurately as follows

$$\left(\frac{q}{A}\right)_i = 15.60\, p^{1.156}(T_w - T_s)^{2.30/p^{0.0234}} \tag{9}$$

where q/A is in Btu/hr-ft^2, p is in lbf/in.^{2} abs, and the temperatures are in °F. Equation (9) is shown plotted for water in Fig. 6. Similar expressions may be calculated for other fluids.

It should be noted that although velocity does not appear in Eq. (9) the equation is valid for all fluid velocities.

As heat flux is increased above the condition for incipience (the upper dotted line) a range of cavity sizes, as shown, become nucleated. A word of caution is in order here. The preceding calculation procedure presumes that cavities exist at sizes out to the point of tangency. At low heat fluxes, as in natural convection, the slope of the temperature distribution (Eq. (8)) may be so small that the point of tangency occurs at cavity sizes larger than those present. Then the heat flux and wall superheat must be raised until the two curves intersect at the largest available active cavity. Hence for low heat flux cases as in natural convection, the procedure usually predicts superheats and (q/A) lower than the observed magnitude.

The intersection of the two curves of Fig. 5 (Eqs. (4a) and (8)) at the maximum available cavity r_{max} gives

$$\left(\frac{q}{A}\right)_{incip} = \frac{k_l}{r_{max}}(T_v - T_{sat}) - \frac{2\sigma T v_{fg} k}{h_{fg} r_{max}^2}$$

A number of other suggestions have been made for predicting the incipient boiling condition. Most of them reduce to the scheme shown in Fig. 5—except the T_l is plotted vs. y/n with T_g^* plotted vs. r. In the above n is taken as unity. In these other proposals the suggested magnitudes of n range from 0.67 to 2. Brown and Bergles [26] tried these suggestions with experimental data for various fluids and found n to range between 1 and 3.

The above procedure then appears to provide an estimate of the lower ΔT and $(q/A)_i$ at incipience. Actual ΔT and $(q/A)_i$ may be greater than these predicted magnitudes. These higher wall superheats may be required to provide sufficient bubble growth acceleration to remove it from the wall. Also thermocapillarity as described later may provide convection currents not accounted for in the above model.

Another study of nucleation [10a] attempted to account for flow patterns around the bubble surface protruding from the cavity, in forced convection. From

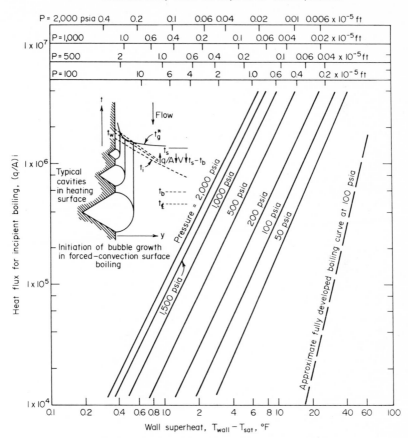

Fig. 6. Bergles-Rohsenow criteria [39] for incipient boiling

$$\frac{q}{A}\,i = 15.60\,p^{1.156}\left(\frac{(T_{\text{wall}} - T_{\text{sat}})^{2.30}}{p^{0.0234}}\right)$$

comparison with only a small amount of data this study led to the following prediction for wall superheat in forced convection

$$\left(T_w - T_{\text{sat}}\right)_{\text{incip}} = 1.33\left(\frac{q/A}{k}\right)^{0.4}\left(\frac{\tau_w}{\rho_l \nu^2}\right)^{0.1}\left(\frac{2\sigma T_s v_{fg}}{h_{fg}}\right)^{0.6}$$

More study of this with comparison with a wider variety of data is needed.

With liquid metals, which have very high thermal conductivities, this same situation exists even in forced convection. For comparison, curves for both water and sodium are shown in Fig. 7 for incipient boiling conditions. Very much greater superheats are required in the liquid metals [10].

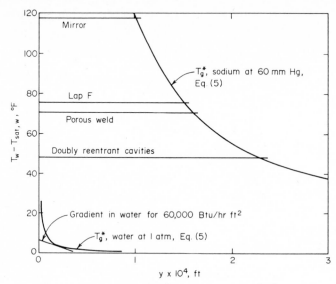

Fig. 7. Incipient boiling for sodium and water [10].

The thermal conductivity of liquid metals is so large that the temperature gradient in the liquid in the vicinity of the bubble at a cavity is essentially zero. Then the first cavity to nucleate should be at r_{max} if it is not inactive (filled with liquid). Then incipient boiling superheat should be predicted by Eq. (4a) with $r = r_{max}$.

Chen [10b] investigated the postulate that cavities could be deactivated to a predictable size by preconditioning by raising the pressure and temperature to specified levels of subcooling. This would establish a "negative" radius of the cavity from Eq. (4a) on the assumption the oxidation in the cavity would prevent further collapse to smaller radii. On subsequent reduction of pressure and increasing the degree of superheat, the radius, now taken as positive in Eq. (4a), should predict the superheat required for nucleation at the surface. The same formalism was reexplained [10c] in terms of reentrant cavities (Fig. 2e) which eliminates the necessity of requiring oxidation to aid the process. Most test results fall below the predictions of this theory; however, Singer and Holtz [10d] showed that this discrepancy was probably due to the presence of noncondensible gas. Their data showed that after extended periods of boiling to drive gas off the surface the measured superheats agreed well with the theory. Also surface finish had no effect on the results.

POSSIBLE MECHANISMS OF HEAT TRANSFER IN BOILING

As heat flux to a saturated liquid is increased beyond the incipient boiling point, the number of spots on the surface from which bubble columns rise increases. Photographs of Hsu and Graham suggest that between bubbles in pool boiling a thermal layer builds up and each bubble carries away with it the liquid in a region $2D_b$, thus pumping away the liquid superheat. In the region between the bubbles at the lower heat fluxes the heat transfer may be of the order of that associated with natural convection. Calculations based on this hypothesis were verified at low heat fluxes by Han and Griffith [11].

As heat flux is increased the nucleation site density increases and the columns with discrete bubbles change to columns of continuous vapor (Moissis and Berenson

[12]). The zones between these vapor columns become "return passages" bringing liquid to the heating surface where it is converted to vapor in the columns. As heat flux is increased further this situation becomes unstable and the heating surface becomes starved for liquid causing the transition to film boiling, point C of Fig. 1.

A typical history of the temperature at a cavity under a bubble is shown in Fig. 8a. At point A the bubble begins to grow on the surface. The major portion of the bubble grows in an essentially nonviscous way, except near the wall where viscosity predominates. Measurements of Cooper and Lloyd [13] on toluene suggest that a thin liquid layer is left on the surface under the bubble (Fig. 8b). It is presumed that viscous force retards the motion of the liquid next to the wall as the bubble grows, thus leaving behind a liquid layer. At point B this layer has evaporated and the dry surface rises in temperature. At point C the bubble has departed and colder liquid covers the surface reducing its temperature to D. Then the surface continues to rise in temperature to point A when a new bubble forms. Calculations of heat flux from the surface suggest that major contributions to the heat flux occur while the thin liquid film evaporates under the bubble AB and when the colder liquid rushes into contact with the surface CDA.

Cooper [13a] showed that this microlayer is significant only at low pressure. At pressures above 1 atm for fluids such as water, hydrocarbons, and cryogens the bubble growth is thermally controlled (not dynamically controlled); then the microlayer does not greatly influence the growth and the heat transfer.

Measurements [14] of temperature at a cavity in pool boiling of sodium suggest that for liquid metals the B-C portion of the curve is essentially nonexistent and CD nearly coincides with AB. Here the primary mechanism of heat transfer is to the liquid which flows in behind a departing bubble. Because of the high thermal conductivity of the liquid metal the rate of rise of surface temperature is much slower and bubble frequencies are an order of magnitude less than in nonmetals.

Boiling from a heated surface in highly subcooled liquids may be accompanied by a different effect which predominates. Cumo [15] has observed photographically long "jets" of hot liquid streaming forth in front of bubbles forming at surfaces in subcooled liquid. From heat transfer measurements and photographs in highly subcooled liquids Brown [10] suggests that because of evaporation at the hot side and condensation at the cold side of the bubble, a small temperature difference exists around the bubble. This results in a slightly lower surface tension at the outer edge of the bubble producing significant flow of the interface which induces large flows of the surrounding liquid to form the observed jets. Here the bubble appears to induce large convection currents which are heated by the hot surface. This process has been called thermocapillarity [10]. It should have its most influence in boiling of subcooled liquids in forced convection, but is no doubt present in some degree at all times.

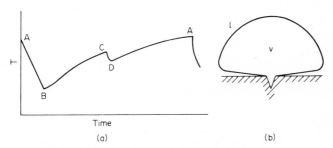

Fig. 8. Temperature of wall under a bubble.

Thermocapillarity may tend to remove rapidly the thin layer of liquid under a bubble in Fig. 8b, reducing the time between points A and B in Fig. 8a. Also the convection induced by thermocapillarity may require wall superheats higher than those predicted by Fig. 6, as observed in some test data.

Probably all of the preceding mechanisms occur simultaneously and participate to varying degrees of importance in the heat transfer process under the various conditions. The boundaries where one or the other of the mechanisms predominate are not now well established.

BUBBLE PARAMETERS

For liquids which wet the heating surface, the size of bubbles at departure from the heating surface has been studied by a number of people. Fritz [16] and Wark [17] equated buoyancy and surface tension to determine the following expression for departure diameter

$$D_b = C_d \beta \left[\frac{2 g_0 \sigma}{g(\rho_l - \rho_v)} \right]^{1/2} \tag{10}$$

where C_d was found experimentally to be 0.0148 for H_2O and H_2 bubbles.

Mikic et al. [19c] and Lien and Griffith [19d] studied bubble growth analytically and experimentally. Their results show clearly that all bubbles start out their growth dynamically controlled and in later stages become thermally controlled, vaporization being governed by heat conduction. Mikic [19c] shows that for a bubble growing at a wall at $T_w > T_{sat}$ into a liquid $T_b < T_{sat}$

$$\frac{dR^+}{dt^+} = \left[t^+ + 1 + \theta \left(\frac{t^+}{t^+ + t_w^+} \right)^{1/2} \right]^{1/2} - (t^+)^{1/2} \tag{11}$$

where

$$R^+ \equiv R \frac{A}{B^2} \quad ; \quad t^+ = t \frac{A^2}{B^2}$$

$$A \equiv \left[b \frac{(T_w - T_{sat}) h_{fg} \rho_v}{T_{sat} \rho_l} \right]^{1/2} \quad ; \quad B \equiv \left[\frac{12}{\pi} \frac{Ja^2 k_l}{\rho_l c_l} \right]^{1/2}$$

$$Ja \equiv \frac{(T_w - T_{sat}) \rho_l c_l}{h_{fg} \rho_v} \quad ; \quad \theta \equiv \frac{T_w - T_b}{T_w - T_{sat}}$$

Here for the bubble growing at the wall $b = \pi/7$. Mikic [19e] shows that the waiting time between bubbles may be approximated by

$$t_w = \frac{\rho_l c_l}{\pi k_l} \left\{ \frac{(T_w - T_b) R_c}{T_w - T_{sat} \left[1 + (2 \sigma v_{fg}/R_c h_{fg}) \right]} \right\}^2$$

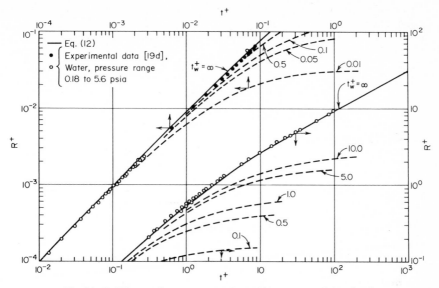

Fig. 8A. Bubble growth curves, comparison with experimental data [19c].

where R_c is the radius of the cavity. Mikic [19c] integrated Eq. (11) for various magnitudes of the parameters neglecting R_c, which is the bubble radius at $t = 0$, compared with R throughout the growth; the results are shown graphically in Fig. 8A.

For the dynamically controlled growth $t^+ \ll 1$ and $t_w^+ \gg t^+$, Eq. (11) integrates to

$$R^+ = t^+ \tag{11a}$$

For the thermally controlled region $t^+ \gg 1$, Eq. (11) and the curves of Fig. 8A are approximated by

$$R^+ = t^{+1/2} - \theta\left[\left(t^+ + t_w^+\right)^{1/2} - t_w^{+1/2}\right]$$

For a bubble growing in an initially uniformly superheated liquid ($t_w^+ \to \infty$) from an initial radius greater than the critical radius given by Eq. (4a), Eq. (11) integrates to

$$R^+ = \frac{2}{3}\left[(t^+ + 1)^{3/2} - t^{+3/2} - 1\right] \tag{12}$$

which is valid for both regimes of growth. Here $b = 2/3$ in A.

For the dynamically controlled region $t^+ \ll 1$, Eq. (12) becomes

$$R^+ = t^+ \tag{12a}$$

which is the Rayleigh solution, and is identical to Eq. (11a).

For the thermally controlled region $t^+ \gg 1$, Eq. (12) becomes

$$R^+ = \sqrt{t^+} \tag{12b}$$

which is the result obtained by Plesset and Zwick [19a] and Scriven [18].

Lien and Griffith [19d] performed definitive experiments for bubble growth in superheated water over a pressure range of 0.18 to 5.6 psia, superheats in the range of 15 to 28°F, and $58 < Ja < 2690$. For pressures less than around 0.4 psia the duration of the bubble growth period remained in the dynamically controlled region $R^+ = t^+$, and for pressures above around 5 psia the dynamically controlled region existed for only a short time and practically all of the growth was governed by heat conduction. Between these pressures the early stage of growth was dynamically controlled and the later stage thermally controlled. The midrange of this region exists when $t^+ \approx 1$. The results agreed with the prediction of Eq. (12).

Applying Eq. (12) to bubble growth in sodium [19d] suggests that for superheats around 30°F the thermally controlled region governs for pressures above around 1.7 atm and the dynamically controlled region governs the pressures less than around 0.3 atm.

Experimental data for growth in nonuniform temperature fields for which waiting time t_w is recorded is limited but does agree [19e] with the prediction of Eq. (11a). Data for bubble growth on heated surfaces at reduced pressures do exist [20] but without observed magnitudes of t_w. These data fall below the prediction of Eq. (12), suggesting that Eq. (11) and Fig. 8A should apply and that the effect of waiting time on the data is significant.

The preceding solutions and results are for lower pressures where $\rho_v \ll \rho_l$. At higher pressures where bubble growth is thermally controlled, Scriven [18] showed the effect of going to higher pressures (significant magnitudes of ρ_v/ρ_l) is given by the curves of Fig. 8B where

$$R^+ = C\sqrt{t^+} \tag{12c}$$

Bubble departure diameters as given by Eq. (10) do not agree well with data. Cole and Rohsenow [20a] correlated the departure diameters D_b for various fluids as

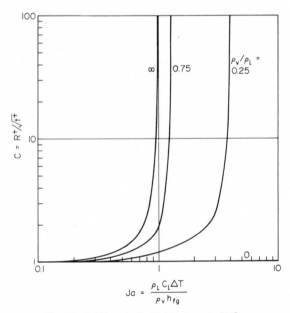

Fig. 8B. Bubble growth at high pressures [18].

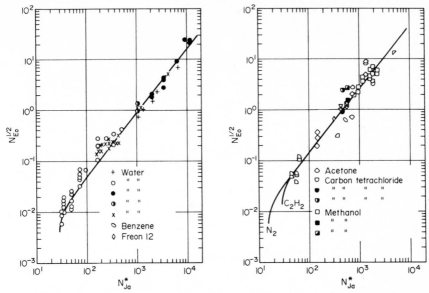

Fig. 8C. *Left*: departure diameter correlation for water. *Right*: departure diameter correlation for other liquids.

shown in Fig. 8C. The straight line portion of the two curves where $p/p_{crit} < 0.2$ is given by

$$\text{Water:} \quad Eo^{1/2} = (1.5 \times 10^{-4})(Ja^*)^{5/4}$$

$$\text{Other fluids:} \quad Eo^{1/2} = (4.65 \times 10^{-4})(Ja^*)^{5/4}$$

(13)

where

$$Eo \equiv \frac{g(\rho_l - \rho_v)D_b^2}{g_0\sigma}$$

$$Ja^* \equiv \frac{\rho_l c_l T_{sat}}{\rho_v h_{fg}}$$

The prediction of bubble frequency for growth at a heating surface is not well established. Early Jakob and Linke suggest $fD_b = 920$ ft/hr for hydrogen and water bubbles. Later Zuber and Peebles and Garber [21] observed

$$fD_b = 1.18\left(\frac{t_c}{t_c + t_w}\right)\left[\frac{\sigma g_c g(\rho_l - \rho_v)}{\rho_l^2}\right]^{1/4}$$

(14)

Cole [20b] showed from data that $1.18\,(t_c/(t_c + t_w))$ ranged between 0.15–1.4, raising serious doubt regarding the validity of the above equation.

Later Ivey [20c] showed that the frequency-diameter relation depended on the regime of bubble growth

$$\text{Dynamically controlled:} \qquad D_b f^2 = \text{constant}$$

$$\text{Thermally controlled:} \qquad D_b f^{1/2} = \text{constant}$$

In the intermediate region the exponent on f changes from 2 to ½. Mikic's analysis [19c] for the thermally controlled regime leads to

$$\frac{D_b f^{1/2}}{2Ja\sqrt{\pi\alpha}} = \left(\frac{t_c}{t_c + t_w}\right)^{1/2} + \left(\frac{t_w}{t_c + t_w}\right)^{1/2} - 1$$

where t_c and t_w are bubble contact time and waiting time between bubbles, and $(t_c + t_w) = 1/f$. Over the range $0.2 < [t_c/(t_c + t_w)] < 0.8$. This becomes

$$D_b f^{1/2} = \frac{3}{4}\sqrt{\pi\alpha_l}\, Ja \pm 10\% \tag{14a}$$

In the dynamically controlled regime, Cole [20d] suggested

$$D_b f^2 = \frac{4}{3}\frac{g(\rho_l - \rho_v)}{C_{\text{drag}}\rho_l} \tag{14b}$$

which for steam at 1 atm and $C_{\text{drag}} \approx 1$ and $\rho_v \ll \rho_l$ reduces to

$$D_b f^2 = 1.32\, g \tag{14c}$$

Information on f vs. D_b is not well documented. The above relations should be considered to be approximate.

Effects of forced convection on D_b and f are not well established. Koumoutsos et al. [20e] showed D_b to decrease linearly with increase in velocity. No general prediction method is available.

As q/A increases the number of active nucleating cavities increases. Staniszewski [22] observed

$$\frac{q}{A} \sim n^m \tag{15}$$

where $m = 1$ at low q/A and decreases to around ½ at high q/A.

FACTORS AFFECTING POOL BOILING DATA

In the following there is presented the results of experimental observations showing the effect on pool boiling data (Fig. 1) of changing the magnitude of various properties and conditions. Except as noted, all data discussed are for liquids which wet the heating surface.

Pressure and temperature difference have a marked effect on all regimes of boiling; Fig. 9 shows representative data. Imagine a cavity size distribution, number of cavities of a particular size range vs. diameter of cavity, represented by the curve sketched in Fig. 10. The first cavity to be activated has an approximate diameter as calculated for the point of tangency in Fig. 5. Then, in accordance with Eq. (15), as q/A is increased, n must increase and more cavities must become activated at diameters spread on either side of D_{crit} (Fig. 10). As shown in Fig. 5, this requires greater wall superheat.

The slope of the q/A vs. ΔT curve expresses the change in ΔT necessary to increase the number of activated points sufficiently to accommodate the new heat flux.

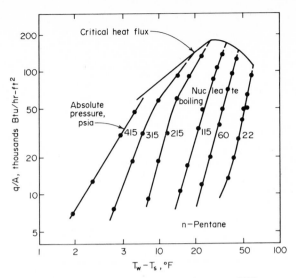

Fig. 9. Effect of pressure on pool boiling curve [23].

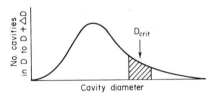

Fig. 10. Cavity-size distribution.

Although the slope is predominantly in the neighborhood of 3, observations are available with resulting slopes as low as unity for contaminated surfaces and as high as approximately 25 for clean, polished surfaces. The actual slope depends upon the uniformity or distribution of the size and shape of the nucleating cavities, Fig. 10.

Griffith and Wallis [7] obtained data for a number of active nucleation sites vs. wall superheat and calculated the radius corresponding to this wall superheat from Eq. (5). For boiling water, methanol, and ethanol on the same surface, a single curve resulted for number of active nucleation sites vs. this radius, suggesting the existence of a characteristic cavity size distribution (Fig. 10) for that surface.

Effect of Pressure

Pool boiling heat flux data plotted vs. wall superheat result in curves as shown in Fig. 9. At higher pressures the data lie at lower superheat magnitudes. Of significance in explaining this shift to lower wall superheat is that the fact that for all fluids Eq. (4) predicts lower ΔT_{sh} required to activate given size cavities as pressure is increased. Figure 11 presents results for pool boiling of sodium showing the shift to lower ΔT_s with pressure.

Figure 9 also shows typical results for the effect of pressure on the critical (or peak) heat flux in pool boiling. This $(q/A)_{crit}$ typically goes through a maximum at a pressure well below the critical pressure.

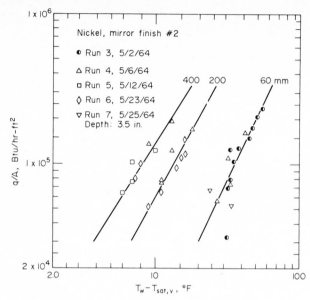

Fig. 11. Effect of pressure on nucleate pool boiling of sodium [24].

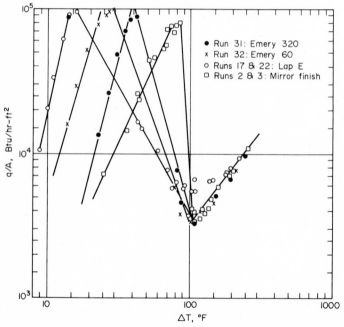

Fig. 12. Copper pentane test results; effect of roughness [25].

Effect of Surface Finish

As expected, surface finish can shift the position of the boiling curve markedly, probably because changing surface finish changes the cavity-size distribution (Fig. 10). The initial controlled experiments studying this effect were performed by Corty [2]. Figures 12 and 13 are typical of results for nonmetals and metals. The data for the rougher surfaces lie to the left at lower wall superheat, presumably because active cavity sizes are smaller on the smoother surfaces.

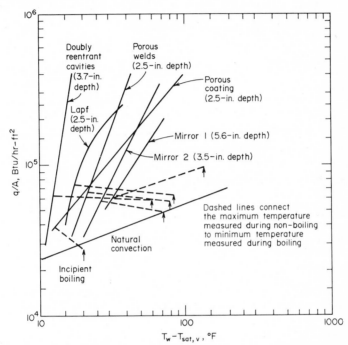

Fig. 13. Comprehensive plot of sodium heat transfer data at an average pressure of 65 mm Hg [24].

The data in Fig. 14 show an interesting effect of direction of surface finishing (or scratch) orientation on boiling. Here axial and circumferential finishing of a horizontal tube produce quite different results at low heat fluxes. At low heat flux, the liquid-vapor interface of bubbles moving circumferentially on the surface tend to fill the scratch or cavities with liquid, deactivating them. With axial scratches vapor tends to be trapped in the cavities, thus keeping them active. For this reason the horizontal scratches require lower superheat to produce the same heat flux. At higher heat fluxes this effect disappears because the bubbles tend to move perpendicularly away from the surface. Then the roughness effect of Figs. 12 and 13 predominates.

Of significance is the observation from Fig. 12 that surface finish does not influence the position of the data in film boiling, nor does it affect the position of the minimum heat flux in film boiling.

The influence of surface finish on the critical heat flux is shown in Fig. 12 to be rather small—about a 20 percent reduction in $(q/A)_{crit}$ for very smooth surfaces.

Aging

Experience teaches that aged surfaces have higher required ΔT for a given q/A. On metallic surfaces a scale or deposit may form from the boiling liquid or a film may

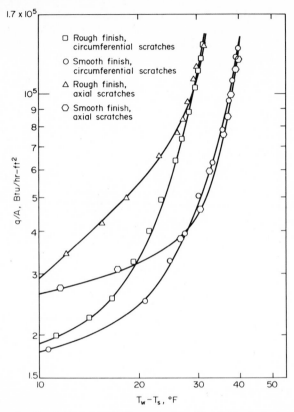

Fig. 14. Influence of surface finish on pool boiling water from the outside of a horizontal tube [26].

form from oxidation or other chemical reaction. In either case the vapor-trapping cavity may shrink, necessitating higher superheat for activation.

Surface Coatings or Deposits

In addition to the effect described above in connection with aging there obviously will exist a temperature drop associated with conduction across the coating or deposit. Since coatings may be very thin and their properties not known, this additional temperature difference is usually included in the ΔT when plotting boiling data. Including this constant, conduction resistance causes q/A vs. ΔT curves to be farther to the right and at lower slopes than those for clean surfaces. Curves for surfaces with various coatings are spread apart even farther than those of Fig. 12.

Various investigators [25, 27–29] have reported data for the effect of surface deposits on critical heat flux. Though there is some disagreement, the following conclusions are suggested. Coated surfaces—mechanically or by oxidation—may have higher $(q/A)_{crit}$ for small diameter cylinders, by as much as a factor of 2 to 3; this increase seems to decrease as diameter increases, disappearing at diameter larger than ½ inch. No appreciable effect is noticed for flat plates.

For clean surfaces not prone to severe oxidation, such as chromel, silver, stainless steel, nickel, etc., data for $(q/A)_{crit}$ scatter with a band of about ±20% for a given wire as well as for different materials [28].

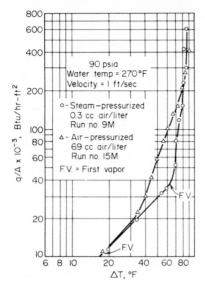

Fig. 15. Effect of dissolved gas on boiling curve [30].

Effect of Gases

Noncondensable gases dissolved in the water tend to come out of solution at the hot surface. In general they tend to move the q/A vs. ΔT curve to the left and to reduce the magnitude of $(q/A)_{crit}$.

Figure 15 shows the results when fresh liquid containing the gas is continually brought to the heating surface by forced convection. Gas bubbles may appear at temperatures well below the normal boiling point and additional convection, probably induced by thermocapillarity, may produce the higher heat fluxes as shown.

Figure 16 shows results for pool boiling. It is assumed that sandblasting the surface caused air to be present in the nucleation cavities. As time goes on after heating is

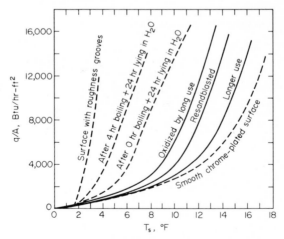

Fig. 16. Effect of surface sandblasting on pool boiling [31].

started, the boiling curve moves to the right to higher ΔT as the gas is pumped out of the cavities by the bubbles.

Another phenomenon which appears to be associated with the presence of gas in the nucleating cavities is a kind of unstable boiling or bumping, thus far observed only in boiling of liquid metals [32, 33]. At moderate heat fluxes in nucleate boiling of sodium a thermocouple in the heating surface near the liquid-solid interface is observed to behave as shown in Fig. 17. The process seems to change back and forth between nucleate boiling and natural convection as active cavities become deactivated and reactivated. After a time during which gas was being removed from the cavity and the liquid, this oscillation ceases and steady natural convection takes place. In the absence of gas the performance follows the natural convection curve until the cavity is nucleated (Eq. (4)) and nucleate boiling occurs above this heat flux and the average ΔT is much lower, Fig. 18, though the surface temperature varies considerably under each bubble in liquid metals.

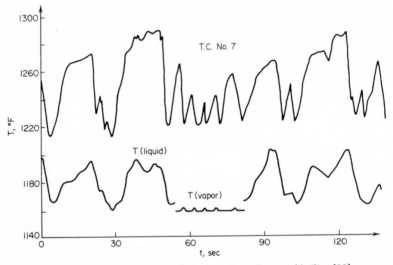

Fig. 17. Temperature variations during bumping sodium pool boiling [33].

This minimum heat flux for nucleate boiling in the absence of gas is also the maximum heat flux for which the unstable boiling will occur when gas is present. This limit is predicted by Shai [33].

The effect of noncondensable gases on the critical heat flux in pool boiling appears to be most significant at high subcooling (Fig. 21) and not very significant at temperatures near the normal boiling point.

Hysteresis

The overshooting of the wall temperature on increase of q/A in natural convection and significant drop in ΔT when boiling begins is known as a hysteresis effect. This overshooting is more pronounced with liquid metals, particularly alkali metals which wet practically all solid surfaces extremely well. This tends to deactivate cavities as discussed in connection with Fig. 2c.

With liquid nonmetals this overshooting occurs less frequently since cavities such as those of Fig. 2d are usually present and remain active cavities even after a system is cooled down. In nonmetals, if some of the cavities become deactivated, the hysteresis

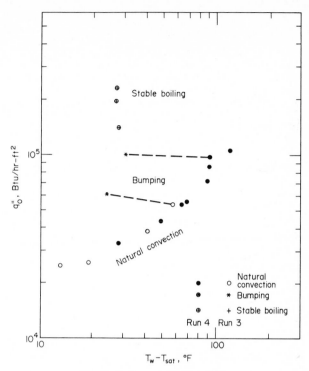

Fig. 18. Sodium pool boiling [33].

overshoot of Fig. 18 can occur [2]; however, on reducing the heat flux in the nucleate boiling region the data follows smoothly down the boiling curve to natural convection.

Size and Orientation of Heating Surface

There appears to be very little effect of size of wire or of orientation of the (q/A) vs. ΔT curve in the nucleate boiling region. This geometry effect, however, is pronounced in the nonboiling natural convection region which is the left asymptote of the curve and is predicted by the normal single-phase natural convection data. Also, in the film boiling regime geometry influences the data (curves for smaller wire diameters are to the left).

The critical heat flux (CHF) appears to be influenced more pronouncedly by geometry. There may be as much as a 25 percent variation in CHF in flat-plate data while the CHF for vertical heaters is less than for horizontal [34, 35]. The difference may be more for larger heater sizes. Critical heat flux appears to increase with wire size [28, 35, 36, 36a] from 0.01 inch up to 0.05 to 0.1 inch and level off beyond that size. For sizes below 0.01 inch down to 0.003 inch the magnitude of CHF levels off again [36a]. In the range where CHF changes with D, the bubble sizes are of the order of the wire diameter. Houchin and Lienhard [36b] measured reduced critical heat flux magnitudes for thin-ribbon heaters with thickness down to as low as 0.00035 inch. Data for various thicknesses of aluminum, nickel, silver, tantalum, and stainless steel were correlated with $\rho c \delta$ product where δ is the ribbon thickness with one side insulated. Data for water with dc heating decreased from $(q/A)_{crit} = 330,000$ down to 200,000 Btu/hr-ft^2, when $\rho c \delta$ decreased from 0.06 to 0.0015 Btu/ft^2-$^\circ$F.

They also found a more pronounced decrease with ac heating. For $\rho c\delta > 0.06$, both ac and dc gave the same $(q/A)_{crit} = 330,000$ independent of $\rho c\delta$.

Agitation

The effect of agitation such as with a propeller is quite similar to the velocity effect shown in Fig. 1. Curves for greater rpm are higher up on the graph. This effect was studied early [37] and more recently [38]. Also the critical heat flux increases with increasing rpm. No attempt to correlate this information is made, since the effect is greatly influenced by shape, size, and position of the agitation.

Subcooling

The effect of subcooling on the position of the boiling curve appears to depend on the geometry for convection. Figure 19 shows data for water [39] boiling on an 0.0645-inch-diameter stainless steel tube. Here the boiling curves lie farther to the right as subcooling is increased. The left asymptotes are for natural convection and should have higher q/A for greater subcool at equal $T_w - T_s$.

In contradiction with this are results [40] for a horizontal flat nichrome heater, 4.25 inches square and 0.0165 inch thick (Fig. 20). Here the boiling curves lie farther to the left as subcooling increases.

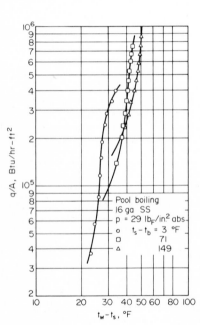

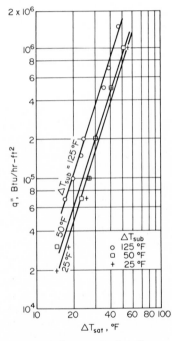

Fig. 19. Influence of subcooling on pool boiling, horizontal tube [39].

Fig. 20. Influence of subcooling on pool boiling, horizontal plate [40].

The differences between the results of Figs. 19 and 20 may be due to different convection geometries or they may be due to different nucleating conditions of the surface.

Critical heat flux increases markedly as subcooling increases as shown in Fig. 21. Additional data are also available [42–44].

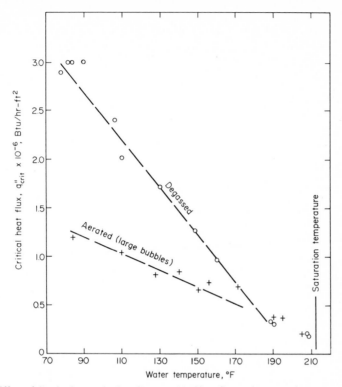

Fig. 21. Effect of dissolved gas and subcooling on critical heat flux under atmospheric pressure [41].

Nonwet Surfaces

When the liquid does not wet the heating surface, very large vapor bubbles form and cover much larger portions of the heating surface [45, 46]. The net result is very greatly increased magnitudes of $T_w - T_s$ required to transfer a given q/A. The critical heat flux is also very much reduced by factors of 10 to 20.

Gravitational Field

At the NASA laboratories, a ribbon heater immersed in a beaker of water was photographed during free-fall conditions, and the results were reported by Siegel and Usiskin [48]. In each case the heat fluxes were in the nucleate boiling range under a normal gravity field. In the free-fall condition with the lower heat fluxes bubbles grew while remaining attached to the ribbon. At the higher heat fluxes a very large vapor volume formed around the ribbon. This suggests that nucleate boiling is essentially nonexistent under zero-g conditions.

Subsequently, Siegel and Usiskin added a small amount of friction to the free-fall system raising g to approximately 0.09. They reported verbally that under these conditions nucleate boiling appeared to continue throughout the fall. This indicates that only a small g field is needed to maintain nucleate boiling.

Merte and Clark [49] report the results of tests on a heated surface at the bottom of a pool. The system was placed in a centrifuge and rotated so that the resultant acceleration field was normal to the surface. Their boiling tests covered a range of 1

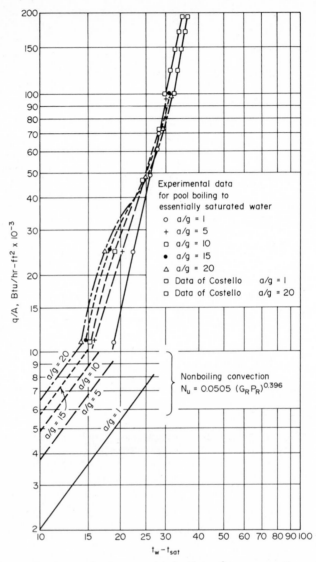

Fig. 22. Influence of system acceleration normal to heating surface on convection and pool boiling [49].

to 21 g with the results replotted as shown in Fig. 22. There seems to be very little effect on the position of the q/A vs. $T_w - T_{sat}$ curves at the higher heat fluxes. The displacement of the curves at the lower heat fluxes is probably due to the effect of superimposed natural convection.

COMPOSITE OF BOILING DATA

Temperature differences ($T_w - T_{sat}$) associated with nucleate boiling are small (1 to $100°F$). In many applications this resistance is not of primary importance; hence an

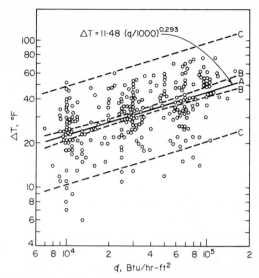

Fig. 23. Nucleate pool boiling of organic liquids at 1 atm: A, regression line; B, 95 percent confidence limits on regression line; C, 95 percent probability limits on ΔT predicted by regression [50].

order of magnitude for this temperature difference is of interest. Figure 23 shows a composite of nucleate boiling data for various organic fluids and surfaces [50]. Required superheat for boiling liquid metals is somewhat higher (Figs. 17 and 18). The conditions for the 276 data points of Fig. 23 are combinations of:

1. *Liquids:* Acetone, benzene, *i*-butanol, *n*-butanol, carbon disulfide, carbon tetrachloride, diethyl ether, diphenyl, ethanol, ethyl acetate, Freon-12$_R$, Freon-113$_R$, heptane, *n*-hexane, methanol, methyl chloroform, *n*-pentane, *i*-propanol, styrene, *m*-terphenyl, *o*-terphenyl, toluene.
2. *Surfaces:* Brass, chromium plate, copper, gold, Inconel, nickel, nickel plate, platinum, stainless steel, vitreous enamel, zinc (crystal).
3. *Surface Treatment:* None; No. 36, 60, 120 (lapped), 150, 200, and 320 grit polishes; mirror finish; acid and steel-wool cleaned; annealed and unannealed; fresh and aged.
4. *Geometry:* 0.005--1.5-inch diameter cylinders (horizontal and vertical); ¾-3¾-inch diameter disks (vertical, horizontal, facing up); ¾ × ¾-inch and × 4-inch plates (vertical, horizontal, facing up, and facing down).

CORRELATION OF POOL BOILING HEAT TRANSFER DATA

Attempts have been made by various people to correlate pool boiling heat transfer data. The logic leading to the various forms of the correlations will be omitted here and suggested equations presented along with comments on their applicability.

Referring to Figs. 12 and 13, one must inescapably conclude that any correlation equation which embodies only properties of the fluid (liquid or vapor) cannot be a "universal" correlation for all fluids or, for that matter, for any particular fluid. As a minimum the coefficient (and possibly even the exponents) must change in magnitude as the character of the solid surface changes, since the data for, say, Fig. 12 are all for the same fluid conditions; only the solid surface condition has changed.

This point cannot be overemphasized. This has led to a great deal of confusion among boiling heat transfer researchers and users. At the present moment there is no satisfactory way to include quantitatively the effect of the solid surface in any correlation equation; in spite of this, some researchers present correlation equations with a fixed magnitude for the coefficient. The best that can be accomplished is to correlate the effect of pressure for a given fluid and solid surface.

Many of the proposed correlations were developed by analyzing a simplified model of boiling leading to some dimensionless groups of quantities. Various forms of bubble Nusselt numbers, bubble Reynolds numbers, and Prandtl number appear in these equations. One of the early correlations [51] employed such groups with the characteristic dimension D given by Eq. (10). The equation proposed is

$$\frac{c_l(T_w - T_{sat})}{h_{fg}} = C_{sf} \left\{ \frac{q/A}{\mu_l h_{fg}} \left[\frac{g_0 \sigma}{g(\rho_1 - \rho_v)} \right]^{1/2} \right\}^r \left(\frac{c_l \mu_l}{k_l} \right)^s \tag{16}$$

where C_{sf} should be a function of the particular fluid heating surface combination. From Fig. 24a or 24b the exponent $r = 0.33$. A cross plot of $c_l \Delta T/h_{fg}$ vs. N_{Pr} for constant values of the ordinate shows $s = 1.0$ for water, but $s = 1.7$ for all other fluids. The final correlation is shown in Fig. 24c which results in $C_{sf} = 0.013$ with a spread of approximately ±20%. This process was repeated for other data with the results shown in Table 2.

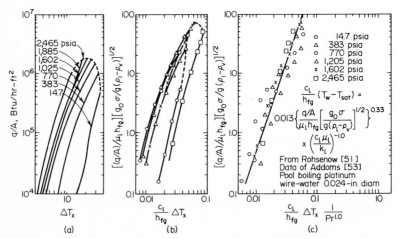

Fig. 24. Correlation of nucleate boiling data for water [51].

It should be emphasized that accurate values of fluid properties are essential to obtain a correlation or in using Eq. (16). Also the heating surface should be clean. The presence of a contamination or deposit of the heating surface can shift the relative position and slope of the curve thus changing r and s in addition to C_{sf}. It should also be noted that Eq. (16) includes a $g^{1/6}$ term which came from the expression for D in Eq. (10). As noted in Fig. 22, there appears to be no effect of g on the position of the boiling curve; therefore Eq. (16) should be used with $|g| = |g_0|$.

Later Forster and Zuber [57] used similar dimensionless groups, but with $\dot{R}R$ of Eq. (13) as the velocity times characteristic dimension in the Reynolds number and $2\sigma/\Delta p$ as the characteristic dimension in the Nusselt number.

TABLE 2. Correlation of Eq. (16) with $r = 0.33$

Surface combination	C_{sf}	s
Water-nickel [52]	0.006	1.0
Water-platinum [53]	0.013	1.0
Water-copper [53]	0.013	1.0
Water-brass [54]	0.006	1.0
CCl_4-copper [55]	0.013	1.7
Benzene-chromium [56]	0.010	1.7
n-pentane-chromium [56]	0.015	1.7
Ethyl alcohol-chromium [56]	0.0027	1.7
Isopropyl alcohol-copper [55]	0.0025	1.7
35% K_2CO_3-copper [55]	0.0054	1.7
50% K_2CO_3-copper [55]	0.0027	1.7
n-butyl alcohol-copper [55]	0.0030	1.7

*Numbers in brackets refer to source of data.

Forster and Greif [58] modified the Forster and Zuber equation by not linearizing the Δp vs ΔT relationship and suggested

$$\frac{q/A}{h_{fg}\rho_v}\left(\frac{2\sigma}{a_t\,\Delta p}\right)^{1/2}\left(\frac{\rho_L}{\Delta p}\right)^{1/4} = C_2\left\{\frac{\rho_L}{\mu}\left[\frac{c\rho_L\,(\pi a_t)^{1/2}\,T_{sat}}{(h_{fg}\rho_v)^2\,788}\Delta p\right]^2\right\}^{5/8}\mathrm{Pr}^{1/3} \quad (17)$$

where C_2 was suggested to be 0.0012 from data for water at 1 and 50 atm, n-butyl alcohol at 50 psia, analine at 35 psia, and mercury for 1 and 3 atm. Equation (17) in the following form approximates the data

$$\frac{q}{A} = K_{sf}\left(\frac{a_t c\rho_L\,T_S}{h_{fg}\rho_v\,\sigma^{1/2}}\right)\left[\frac{cT_S(a_t)^{1/2}}{(h_{fg}L\rho_v)^2}\right]^{1/4}\left(\frac{\rho_L}{\mu}\right)^{5/8}\mathrm{Pr}_l^{1/3}\,(\Delta P)^2 \quad (18)$$

For this limited amount of data K_{sf} was suggested to be 43×10^{-6}. Recently this equation was compared with other data. For ethanol-chromium data of Cichelli-Bonilla [56] it was found that K_{sf} should be 82×10^{-6} and for water-platinum data of Addoms [53], $K_{sf} = 142 \times 10^{-6}$. Obviously the coefficient does indeed change in magnitude for various surface-fluid combinations. Equation 18 needs more comparison with data.

Equation (18) is identical to the earlier Forster and Zuber equation, except that Δp and ΔT are related by Eq. (6). Also Eq. (18) yields a varying exponent of q/A vs. ΔT, increasing with ΔT in the range from 2 to 4. In many cases the curve cuts across natural boiling data at a lower slope.

Other correlations have been suggested by Gilmour [59], McNeilly [60], and Levy [61]. The Gilmour correlation contains a size effect not verified by experiment. The Levy procedure is dimensional and employs an empirical curve around which data scatter by a factor of five or more, which is about the same variation observed in C_{sf} of Table 2. (Readers of the Levy paper as originally published should note the ordinate of its Fig. 4 should be read multiplied by 10^{-5} instead of 10^{-6}.)

Various Russian workers have suggested correlation equations in the following form [61a]

$$\mathrm{Nu}_* = A\,\mathrm{Pr}^{n_1}\mathrm{Pe}_*^{n_2}K_p^{\,n_3}K_t^{\,n_4}\,\mathrm{Ar}_*^{n_5} \quad (19)$$

TABLE 3. Correlations by Russian Authors

Authors [63a]	A	n_1	n_2	n_3	n_4	n_5
M. A. Kichigan and N. Yu. Tobilevich	1.04×10^{-4}	0	0.7	0.7	0	0.125
S. S. Kutateladze	7.0×10^{-4}	-0.35	0.7	0.7	0	0
V. M. Borishanskiy and F. P. Minchenko	8.7×10^{-4}	0	0.7	0.7	0	0
G. N. Kruzhilin and Ye. K. Averin	0.082	-0.5	0.7	0	0.377	0
D. A. Labuntsov	0.125	-0.32	0.65	0	0.35	0

$$\text{Nu}_* \equiv \frac{h}{k} \left[\frac{g_0 \sigma}{g(\rho_l - \rho_v)} \right]^{1/2}$$

$$\text{Pe}_* \equiv \frac{q/A}{\alpha_t \rho_v h_{fg}}$$

$$K_p \equiv \frac{p}{[g\sigma(\rho_l - \rho_v)/g_0]^{1/2}}$$

(20)

$$K_t \equiv \frac{(\rho_v h_{fg})^2}{Jc_l T_s \rho_l [g\sigma(\rho_l - \rho_v)/g_0]^{1/2}}$$

$$\text{Ar}_* \equiv \frac{g}{\nu^2} \left(\frac{g_0 \sigma}{g(\rho_l - \rho_v)} \right)^{3/2} \left(1 - \frac{\rho_v}{\rho_l} \right)$$

It should be emphasized again that the coefficient A in Eq. (19) is not a constant, but varies with the surface fluid combination.

Equation (16) has correlated the pressure effect for a variety of pool boiling data. The state of our knowledge is such that the coefficient C_{sf} must be determined from limited data for each fluid-surface combination. This, of course, is true for any of the other pool-boiling correlations.

It should also be emphasized that the actual metal of the surface is perhaps less important than the surface character as represented by the cavity-size distribution of Fig. 10, which is unknown for practically all of the surface tested.

Mikic [63b] has attempted to show how the cavity-size distribution for any surface influences the position of the q/A vs. ΔT boiling curve in pool boiling. Starting with the description of the boiling process as outlined on pages 13-10 and 13-11 and assuming that the number of cavities of radius greater than r is expressible by

$$n = \left(\frac{r_s}{r} \right)^m$$

(20a)

where r_s is the radius of the largest cavity present and r_s and m are determined from cavity-size distribution measurements (Brown and Bergles [63c]). The following expressions are obtained

$$\frac{q}{A} = \frac{A_{nc}}{A} \left(\frac{q}{A}\right)_{nc} + \left(\frac{q}{A}\right)_b \tag{20b}$$

where A_{nc}/A is the fraction of the area where bubbles are not being formed, $(q/A)_{nc}$ is the natural convection heat transfer at this area, and $(q/A)_b$ is given by

$$\frac{(q/A)_b}{\mu_l h_{fg}} \sqrt{\frac{g_0 \sigma}{g(\rho_l - \rho_v)}} = B(\phi\Delta T)^{m+1} \tag{20c}$$

where

$$B \equiv \left(\frac{r_s J}{2}\right)^m 2\sqrt{\pi} \frac{g_0^{11/8}}{g^{9/8}} C_2^{5/3} C_3^{1/2} \tag{20d}$$

$$\phi^{m+1} \equiv \frac{k_l^{1/2} \rho_l^{17/8} c_l^{19/8} h_{fg}^{(m-23/8)} \rho_v^{(m-15/8)}}{\mu_l [(\rho_l - \rho_v)]^{9/8} \sigma^{(m-11/8)} T_s^{(m-15/8)}} \tag{20e}$$

Here $C_2 = 0.00015$ for water and 0.000465 for other fluids (Eq. (13)) and $C_3 = 0.6$ (Eq. (14)); r_s and m are to be determined from the cavity-size distribution (Eq. (20a)). Note that B is solely a function of cavity-size distribution and ϕ is a function of fluid properties except for the exponent m. Further if the cavity-size distribution has a slope of m, then the q/A vs. ΔT curve should have a slope of $m + 1$.

For most fluids $(q/A)_{nc} \ll (q/A)_b$ in Eq. (20b) and may be neglected.

Cavity-size distribution for most surfaces for which data is available has not been measured. In these cases the boiling data may be used to determine m and B. Mikic [63b] shows that Eq. (20c) used in this way does correlate existing boiling data.

CORRELATION OF CRITICAL HEAT FLUX, POOL BOILING

As heat flux is increased in nucleate boiling, more nucleation sites become active. Equations for correlating the critical flux data emerge from models which suggest (1) as nucleation sites increase, a critical "bubble packing" is reached to produce vapor blanketing [65], (b) as nucleation sites increase, the liquid flow to the heating surface is restricted sufficiently to produce vapor blanketing [62], and (3) as heat flux and number of nucleation sites increase, bubbles in a column overtake each other producing a vapor column in which there are liquid droplets which fall back to the surface if horizontal. When the vapor velocity is sufficient to carry these liquid droplets away from the surface, vapor blanketing occurs [61a]. Regardless of the model used, all correlations thus far proposed contain the relation between $(q/A)_{crit}/\rho_v h_{fg}$, an average vapor velocity normal to the surface and some function of, or measure of, pressure level. The following are some correlations which have been proposed for liquid nonmetals.

Zuber [62] (modified)

$$\frac{(q/A)_{crit}}{\rho_v h_{fg}} = 0.13 \left[\frac{\sigma(\rho_l - \rho_v)g g_0}{\rho_v^2}\right]^{1/4} \left[\frac{\rho_l}{\rho_l + \rho_v}\right]^{1/2} \tag{21}$$

Here 0.18 agrees better with the data than 0.13 as originally suggested by Zuber. Kutateladze [63]

$$\frac{(q/A)_{crit}}{\rho_v h_{fg}} = 0.16 \left[\frac{\sigma(\rho_l - \rho_v)g\,g_0}{\rho_v^2} \right]^{1/4} \tag{22}$$

Chang and Snyder [64]

$$\frac{(q/A)_{crit}}{\rho_v h_{fg}} = 0.145 \left[\frac{\sigma(\rho_l - \rho_v)g\,g_0}{\rho_v^2} \right]^{1/4} \left[\frac{\rho_l + \rho_v}{\rho_l} \right]^{1/2} \tag{23}$$

Rohsenow and Griffith [65]

$$\frac{(q/A)_{crit}}{\rho_v h_{fg}} = 143 \left(\frac{g}{g_0} \right)^{1/4} \left[\frac{\rho_l - \rho_v}{\rho_v} \right]^{0.6} \quad \text{ft/hr} \tag{24}$$

which is dimensional but agrees well with most data.

The quantities in the right-hand side of Eqs. (21)–(23) are essentially the same except when they are very close to the critical pressure. The Zuber equation with 0.18 instead of 0.13 is recommended.

Lienhard and Shrock [66] and Lienhard and Watanabe [67] extended the Zuber-type equation by applying the Law of Corresponding States, relating various parameters to conditions at the critical pressure and temperature. The suggested correlation is of the following form

$$\frac{(q/A)_{crit}}{(g_0)^{1/4} \left[p_c/(\rho_l - \rho_v) \right] (8p_c/3RT_c)^{3/4}} = f_1(R') \cdot f_2\left(\frac{p}{p_c} \right) \tag{25}$$

where f_1 is a function of geometry alone and f_2 is a function of reduced pressure alone. Here

$$R' = L \left[\frac{g(\rho_l - \rho_v)}{\sigma} \right]^{1/2} \tag{26}$$

where L is a characteristic length-like diameter for wires.

For liquid metals Noyes and Lurie [68] suggest

$$\left(\frac{q}{A} \right)_{crit} = 0.16 \rho_v h_{fg} \left| \frac{\sigma(\rho_l - \rho_v)g\,g_0}{\rho_v^2} \right|^{1/4} + K_{NL} \tag{27}$$

where $K_{NL} = 400,000$ for sodium (0.2 to 22 psia) data. Here the first group of quantities is the Kutateladze (Eq. (22)) and the K_{NL} is a constant which perhaps has different magnitudes for each fluid. Potassium (0.15 to 22 psia) data of Balzheiser and Colver [69] is correlated by Eq. (27) with $K_{NL} = 300,000$. In terms of pressure in psia, these equations become

Sodium: $\left(\dfrac{q}{A}\right)_{crit} = (4.0 + 1.75\,p^{0.457}) \times 10^5$

$$\text{(28)}$$

Potassium: $\left(\dfrac{q}{A}\right)_{crit} = (3.0 + 1.00\,p^{0.444}) \times 10^5$

The above equations apply to saturated liquid. For subcooled liquids Zuber [70] modified Eq. (21) as follows

$$\frac{(q/A)_{crit\ sub}}{(q/A)_{crit\ sat}} = 1 + \frac{5.3}{\rho_v h_{fg}}\sqrt{k_l \rho_l c_l}\left[\frac{\sigma(\rho_l - \rho_v)g\,g_0}{\rho_v^2}\right]^{-1/8}\left[\frac{g(\rho_l - \rho_v)}{g_0\sigma}\right]^{1/4}(T_{sat} - T_{liq}) \quad \text{(29)}$$

which was tested and agreed with data for water and ethyl alcohol at pressures below 142 psia. Ivey and Morris [42] suggest a simpler relation

$$\frac{(q/A)_{crit\ sub}}{(q/A)_{crit\ sat}} = 1 + \frac{0.1}{\rho_v h_{fg}}\left(\frac{\rho_v}{\rho_l}\right)^{1/4} c_l \rho_l (T_{sat} - T_{liq}) \quad \text{(30)}$$

which was tested against and agreed with data for water, ethyl alcohol, ammonia, CCl_4, and iso-octane over a pressure range of 4.5 to 500 psia.

In Eqs. (21) through (27) the $(q/A)_{crit}$ is predicted to vary with $g^{1/4}$. Test data for wires in water in a drop tower [48] fall on or above a curve $(q/A)_{crit} \sim g^{1/4}$ in the range $0.02 < (g/g_0) < 1$. The data points above the curve may be due to a transient condition resulting from the short drop time. Adams [71] using cylindrical graphite heaters in a centrifugal field found $(q/A)_{crit} \sim g^{0.15}$ for $1 < g < 10$ and $(q/A)_{crit} \sim g^{0.25}$ for $10 < g < 100$. Bessant and Jones [72] found the exponent to drop from 0.27 to 0.14 as pressure increased from 14.7 to 300 psia.

In the equations $(q/A)_{crit} \sim \sigma^{1/4}$ limited data of Adams [71] show for vertical heating surfaces (2 to 4 inches high) $(q/A)_{crit} \sim \sigma$.

None of the correlation equations include an effect of the condition of the solid surface. Berenson [25] and Ivey and Morris [28] obtain less than a 20 percent variation in $(q/A)_{crit}$ for a wide variety of clean surface materials and surface finishes. Oxidized or coated surfaces which tend to increase wettability, tend to have higher magnitudes of $(q/A)_{crit}$ [28, 73].

POOL BOILING OF CRYOGENIC FLUIDS

Cryogenic fluids—liquid oxygen, nitrogen, hydrogen, helium, etc.—in pool boiling behave like noncryogenic fluids. The position of the q'' vs. $T_w - T_s$ curve depends strongly on the nature of the solid surface. A survey of such data is presented by Zuber and Fried [74], Richards et al. [75], and Clark [76]. Figure 25 shows a composite plot of a wide variety of pool boiling and also film boiling data for oxygen. Brentani and Smith [77] attempted to correlate all of this data with a modified equation of Kutateladze

$$q'' = 4.87 \times 10^{-11}\left[\frac{c_{pl}}{h_{fg}\rho_v}\right]^{1.5}\left[\frac{k_l \rho_l^{1.282}\,p^{1.75}}{\sigma^{0.906}\mu_l^{0.626}}\right](T_w - T_s)^{2.5} \quad \text{(31)}$$

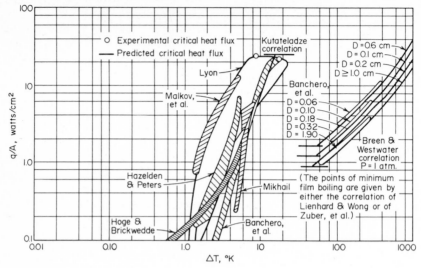

Fig. 25. Experimental nucleate and film pool boiling of oxygen at 1 atm compared with the predictive correlations of Kutateladze, and Breen and Westwater [77].

which represents the solid line through the central portion of the data. Quite obviously the coefficient of the equation must vary by a factor of 10 or so to accommodate all of the data. Similar composite data for nitrogen, hydrogen, and helium [77] exist.

Data for critical heat flux of the various cryogens are shown to be correlated [77–79a] rather well by Eqs. (21), (22), and (24) except at pressures approaching the critical pressure where the data fall significantly below the predictions.

Clark [76] shows that a modified form of Eq. (16) also correlates the cryogenics boiling data

$$\frac{q/A}{h_{fg}\mu_1}\left[\frac{g_0\sigma}{g(\rho_1-\rho_v)}\right]^{1/2} = 3.25\,(10^5)\left[\frac{C_{p1}\Delta T}{h_{fg}}\left(\frac{T/T_c}{\text{Pr}_1}\right)^{1.8}\right]^{2.89} \tag{31a}$$

Here Eq. (16) was modified by including an additional pressure effect empirically in the form of T/T_c, the exponent on the Prandtl number was changed, and single magnitude of C_{sf} was selected.

FORCED CONVECTION

Forced convection here refers to flow inside of tubes, annuli, parallel plates, outside of rod bundles, etc. The flow is a two-phase flow of liquid and its vapor. Adiabatic flow occurs in a variety of flow regimes, but two-phase flow with heat addition usually occurs as shown in Fig. 26, which represents liquid entering subcooled where boiling takes place at the wall (*local boiling*). At some place along the tube the liquid reaches the saturation point. Beyond this point the process is called *bulk boiling*. At low quality (below 5 to 8 percent) and in the subcooled region the flow is bubbly flow. At higher qualities the flow becomes annular with a thin liquid layer on the wall and a vapor core, which may or may not contain significant amounts of liquid

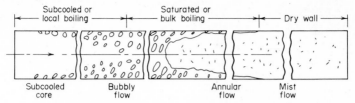

Fig. 26. Flow regimes in a vertical heated channel.

droplets in the center. In the annular flow regime there may or may not be nucleate boiling. If the liquid film on the wall is thin enough, heat is transferred by conduction through the liquid and vaporization occurs at the liquid-vapor interface. Under these conditions the wall temperature is below that required for nucleation or activation of the cavities. At higher heat fluxes nucleation will take place under these thin liquid layers at the wall. A more detailed study of flow patterns is given by Tippets [80]. The simplified description above is adequate for classifying heat transfer data.

At some point down the channel the heated wall may become dry. Beyond this point wall temperature usually rises significantly for a particular heat flux. This is the critical condition. The cause of the dryout may be intensive nucleation if in the nucleation regime, or tearing off of the liquid film by high velocity of the vapor which increases proportionately down the channel.

We separate our discussion of forced convection boiling into the low quality and subcooled regions (essentially bubbly flow) and the high quality regions (essentially annular flow). The region beyond the critical condition is a mist-flow, dry-wall regime and is handled separately. This may be called *film boiling* in forced convection.

Low-quality and Subcooled Forced Convection

Data in this regime typically follow a pattern as shown in the upper left corner of Fig. 1. No boiling occurs to the left of the input of the incipient boiling line, determined as discussed in convection with Figs. 5 and 6. This curve is applicable for any velocity and subcooling. Beyond this line nucleate boiling occurs and the curves, regardless of velocity or subcooling, appear to merge into a fully developed boiling curve, the location of which is quite insensitive to the magnitude of either velocity or subcooling. Evidence of this is seen in Fig. 27 where data for various velocities and subcooling converge at high heat flux to a single fully developed boiling curve.

Figure 27 also shows data for pool boiling taken on the same surface which was obtained by cutting open one of the tubes axially and boiling on the inner surface (half of the tube), the outer suface being insulated. Note that the lower end of the pool boiling data appears to lie on an extension downward of the fully developed boiling curve drawn with a 3 to 1 slope. In this case, for pool boiling on the inner curved half tube it is suspected that induced convection effects caused the pool boiling data at the higher heat flux to have a slope greater than 3 to 1 and lie above the dashed curve. There is insufficient evidence to conclude that the pool boiling data will always lie on an extension of the fully developed boiling curve.

The above observations led to the following two suggestions for drawing the forced convection boiling curve for a particular velocity and subcooling. On a graph of q'' vs. $T_w - T_s$ (Fig. 28) draw the forced convection nonboiling curve q''_{FC} for a particular velocity and subcooling (Eq. (1)). Then, from the procedure of Fig. 5 or, for water (Fig. 6 or Eq. (9)), locate the point of incipient boiling. Then draw the fully developed boiling curve q''_B, if known, at a slope of 3. This may be approximated by the pool boiling curve, if known, for the same surface. Locate the magnitude q_{Bi} immediately below the incipient boiling point. Revising a suggestion of Kutateladze [81], Bergles and Rohsenow [39] suggest the following interpolation equation

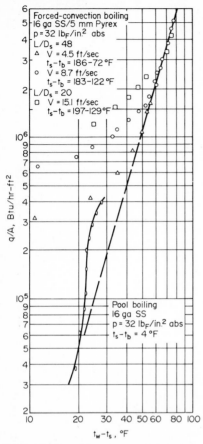

Fig. 27. Forced-convection surface-boiling data and pool-boiling data for stainless-steel tube [39].

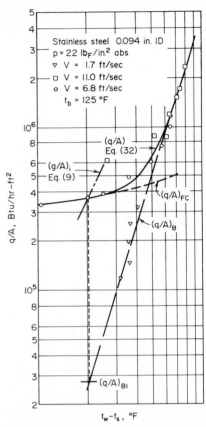

Fig. 28. Recommended procedure for construction of curve for forced-convection surface boiling.

$$\frac{q''}{q''_{FC}} = \left[1 + \frac{q''_B}{q''_{FC}} \left(1 - \frac{q''_{Bi}}{q''_B} \right)^2 \right]^{1/2} \qquad (32)$$

which is a curve asymptotic to the q''_{FC} curve at the incipient boiling point and asymptotic to the q''_B at high heat flux. An alternative suggestion [39] is a simple additive or superposition equation having the same asymptotes

$$q'' = q''_{FC} + q''_B - q''_{Bi} \qquad (33)$$

Either equation follows the data reasonably well.

The data in Fig. 14 show the strong effect of roughness and scratch orientation, when filling the system, on the position of the boiling curve in pool boiling. Tests were performed on these same surfaces for forced convection in an annulus around the

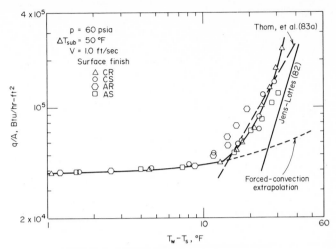

Fig. 29. Flow surface boiling data—low heat flux [26].

tube. In this case (Fig. 29) (Brown [26]) the data show very little effect of surface finish in forced convection. This may result from the fact that the position of the incipient boiling point as predicted by the procedure of Fig. 5 is independent of surface finish as long as cavities exist of the size represented by the point of tangency in Fig. 5. This fact may be the explanation for why surface finish has less influence in forced convection boiling. Nevertheless, the actual shape of the cavity-size distri-bution curve (Fig. 10) for a particular surface probably does influence the position of the fully developed boiling curve in forced convection.

Based on the assumption that surface characteristics do not influence significantly the position of the fully developed boiling curve in forced convection, the following two equations have been proposed for subcooled and low-quality forced convection of water

$$\text{Jens and Lottes [82].} \quad T_w - T_s = \frac{60}{e^{p/900}} \left(\frac{q''}{10^6} \right)^{1/4} \tag{34}$$

$$\text{Weatherhead [83].} \quad T_w - T_s = (90 - 0.127\,T_s) \left(\frac{q''}{10^6} \right)^{1/4} \tag{34a}$$

where T_s is in °F and p in psia. Both equations are shown in Fig. 30 along with some data [84]. More recently Thom et al. [83a] suggested

$$T_w - T_s = \frac{72}{e^{p/1260}} \left(\frac{q''}{10^6} \right)^{1/2} \tag{35}$$

which appears to be in better agreement with the data of Fig. 29, even at low pressures.

High-quality Forced Convection

Farther along a flow passage (Fig. 26) at qualities above 5 to 10 percent the flow regime becomes annular. As more vapor is generated along the tube the vapor velocity progressively increases down the tube, thus increasing the heat transfer coefficient.

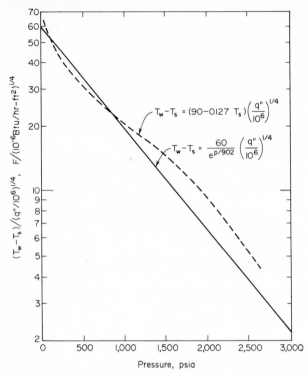

y-axis: $(T_w - T_s)/(q''/10^6)^{1/4}$, F/($10^{-6}$Btu/hr-ft^2)$^{1/4}$

x-axis: Pressure, psia

$$T_w - T_s = (90 - 0.127\, T_s)\left(\frac{q''}{10^6}\right)^{1/4}$$

$$T_w - T_s = \frac{60}{e^{P/902}}\left(\frac{q''}{10^6}\right)^{1/4}$$

Fig. 30. Jens-Lottes and Weatherhead equations for fully developed boiling curve for low-quality or subcooled forced convection.

This effect is shown in Fig. 31 for R-22 refrigerant. The effect is more pronounced at the higher mass flow rates. At high flow rates, high qualities and lower heat flux, the liquid layer on the wall may be thin enough to keep the wall temperature below the nucleation temperature. Then no actual boiling takes place. Heat is transferred by conduction across the liquid layer and evaporation takes place at the liquid vapor interface [86]. Evidence of suppressed nucleation is shown in Fig. 32 by Bennett et al.

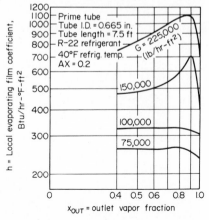

y-axis: h = Local evaporating film coefficient, Btu/hr-hr-°F-ft^2

Prime tube
Tube I.D. = 0.665 in.
Tube length = 7.5 ft
R-22 refrigerant
40°F refrig. temp.
$\Delta X = 0.2$

$G = 225,000$ (lb/hr-ft^2)

150,000

100,000

75,000

x-axis: x_{OUT} = outlet vapor fraction

Fig. 31. Variation of h with vapor fraction for forced-convection boiling [85].

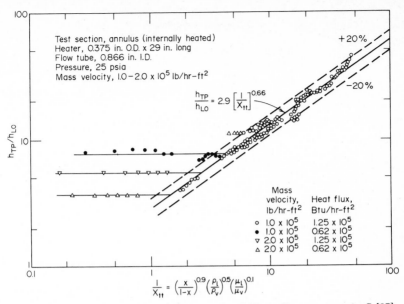

Fig. 32. Variation of heat transfer coefficient ratio with Martinelli parameter, series G [87].

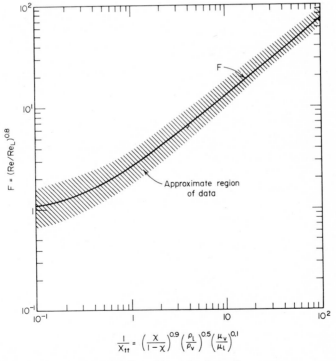

Fig. 33. Reynolds number factor F.

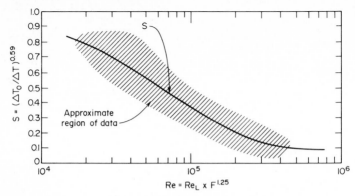

Fig. 34. Suppression factor S.

[87]. The data to the left are at low quality and are in the nucleation regime. The right-hand portion of the data for higher quality are in the no-nucleation regime and appear to be correlatable by the Lockhart-Martinelli two-phase flow parameter x_{tt}.

Various investigators have suggested a variety of correlation equations which are gathered together in Table 4. Some equations refer the actual h in the two-phase region to h_{LO} or h_{LP} with correction factors F to account for nucleation. The Chen equation is recommended for the lower quality region and Dengler and Addoms for the higher quality regions.

As in the case of pool boiling a kind of hysteresis has been observed in forced convection boiling. Figure 35 shows some data obtained at CISE [10] for the mist-annular flow regime. On increasing the heat flux, the wall temperature follows the curve a-b interpreted as a nonboiling curve. At point b the wall temperature dropped to point c. Further heating went to point d and subsequent successive

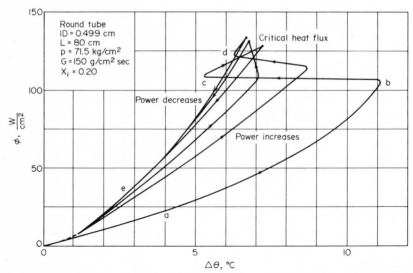

Fig. 35. Example of the wall to coolant temperature difference below CHF, as a multivalued function of the heat flux [101].

TABLE 4. High-quality Forced-convection Evaporation Correlations (Yadigaroglu and Bergles)

Author	Guerrieri and Talty [92]	Collier, Lacey, and Pulling [90]	Wright [89]
Fluid	Methane, cyclohexane, benzene, pentane, heptane	Water	Water
Direction of flow	Up (nat. circ.)	Up	Down
Channel geometry	Circular	Annular	Circular
Diameter, D (in.)	0.75, 1.0	0.625/0.866, 0.375/0.551	0.719, 0.472
Length, L (ft)	6.0, 6.5	4.0, 2.42	5.67, 4.69
Pressure level, p (psia)	Atmospheric	15.5–80	15.8–68.2
Mass flux, $G/10^5$ (lb/hr-ft²)	Around 3.5	0.987–7.94	3.96–25.2
Heat flux, q'' (Btu/hr-ft²)	2170–17,400	31,700–252,000	13,800–88,000
Quality range, x	0–0.116	0–0.659	0–0.19
Recommended correlation	$F = 0.187 \left(\dfrac{r_b}{\delta}\right)^{-5/9}$ r_b from $\Delta T = \dfrac{R}{M}\dfrac{T_w T_{sat}}{h_{fg}} \ln\left(1 + \dfrac{2\sigma}{r_b p}\right)$, $\delta = 10\,\dfrac{\mu_l}{\rho_l}\sqrt{\dfrac{4\rho_l}{(dp/dl)_{TPF}\,Dg_c}}$	$\dfrac{h}{h_{LP}} = A\left(\dfrac{1}{X_{tt}}\right)^n F$ $A = 2.065 - 3.772$ $n = 0.491 - 0.749$ Recommend: $A = 2.167$, $n = 0.699$ No recommendation for F The correlations apply to "convective" regime only, in general. The correction coefficient F is intended to take into account the nucleation effects. $(dp/dl)_{TPF}$ calculated from Lockhart and Martinelli (friction only) [lb/ft³] $\delta =$ thickness of lam. film adj. the wall	$A = 2.721$ $n = 0.581$
Properties evaluated at	T_{sat}	$T_f = T_{sat} + 0.33\,\Delta T_{sat}$	T_{sat}
No. of data points/runs	192/96	1542/153	1272/106
Error in correlating data	88% within ±20%	26.9% (std. deviation)	15.7% (average)
Equation	36a	36b	36c

13-41

TABLE 4. High-quality Forced-convection Evaporation Correlations (Yadigaroglu and Bergles) (*Continued*)

Author	Dengler and Addoms [88]	Chen [97]
Fluid	Water	Water, methanol, cyclohexane, pentane, heptane, benzene
Direction of flow	Up	Up and down
Channel geometry	Circular	Circular and annular
Diameter, D (in.)	1.0	Large range
Length, L (ft)	20	Large range
Pressure level, p (psia)	7.5–40	8–505
Mass flux, $G/10^5$ (lb/hr·ft^2)	0.44–10	Large range
Heat flux, q'' (Btu/hr·ft^2)	2100–21,800	2000–760,000
Quality range, x	0–0.70	0–0.71
Recommended correlation	$$\frac{h}{h_{LO}} = 3.5 \left(\frac{1}{X_{tt}}\right)^{0.5}$$ for convective region only: $0.25 < X_{tt} < 70$ To take into account nucleation, multiply by F $$F = 0.67 \left\{ \Delta T - \Delta T_i \left[\left(\frac{\delta p}{\delta T}\right)_{sat} \frac{D}{\sigma}\, T_w\right]^{0.1} \right\}$$ if $F > 1$, $= 1$ otherwise $$\Delta T_i = 10(\bar{V})^{0.3} \qquad [\bar{V}] = \text{ft/sec}$$ $$\bar{V} = G\bar{v}/3600 \simeq G v_l/3600\,(1 - \alpha)$$ α from measurements of Lockhart and Martinelli	$h = h_{\text{mic}} + h_{\text{mac}}$ $$= 0.00122\, \frac{k_l^{0.79}\, c_l^{0.45}\, \rho_l^{0.49}\, g_c^{0.25}}{\sigma^{0.5}\, \mu_l^{0.29}\, h_{fg}^{0.24}\, \rho_v^{0.24}}$$ $$\cdot (\Delta T)^{0.24} (\Delta p)^{0.75}\, S + h_{LP} F$$ with: Δp = diff. in sat. vap. press. corr. to $\Delta T = T_w - T_{sat}$ $$\frac{\Delta T\, h_{fg}}{T_{sat}\, v_{lg}}$$ $g_c = 32.2(3600)^2$, S and F given in Fig. 33, 34
Properties evaluated at	T_{sat}	T_{sat}
No. of data points/runs	150	Data of many investigators
Error in correlating data	85% within ±20% for convective region	85% within ±15.1%
Equation	37	38

	Schrock and Grossman [91]		Wright [89]
Author	Schrock and Grossman [91]		Wright [89]
Fluid	Water		Water
Direction of flow	Up		Down
Channel geometry	Circular		Circular
Diameter, D (in.)	0.1162-0.4317		0.719, 0.472
Length, L (ft)	1.3-3.3		5.67, 4.69
Pressure level, p (psia)	42-505		15.8-68.2
Mass flux, $G/10^5$ (lb/hr-ft^2)	1.76-32.8		3.96-25.2
Heat flux, q'' (Btu/hr-ft^2)	60,000-1,450,000		13,800-88,000
Quality range, x	0-0.57		0-0.19
Recommended correlation	$$\frac{h}{h_{LO}} = C_1 \text{Bo} + C_2 \left(\frac{1}{X_{tt}}\right)^{2/3}$$ $C_1 = 7400$ $C_2 = 1.11$ for $0 < x < 0.50$ Correlation presented also in graphical form	$C_1 = 6690$ $C_2 = 2.36$ Use h_{LP} instead of h_{LO}	$h = 4.192\,(\text{Re}_{LP})^{0.455}\, q^{-0.289}\, x^{0.379}\, \text{Pr}_l^{0.4}$ Wright presents a total of 12 different correlations which all fit his data.
Properties evaluated at	T_{sat}		T_{sat}
No. of data points/runs	~900/~180		1272/106
Error in correlating data	35%	14.9% (average)	10.8% (average)
Equation	39	40	41

TABLE 4. High-quality Forced-convection Evaporation Correlations (Yadigaroglu and Bergles) (*Continued*)

Author	Pierre [99]	Altman, Norris, and Staub [100]	Sachs and Long [100a]
Fluid	F-12, F-22 (F-11, CH$_3$Cl)*	F-22	F-11
Direction of flow	Horizontal	Horizontal	Up
Channel geometry	Circular	Circular	Annular
Diameter, D (in.)	0.471–0.708	0.343	0.5/0.97
Length, L (ft)	11.7–27	8	1.0
Pressure level, p (psia)	Atmospheric	Atmospheric	Atmospheric
Mass flux, $G/10^5$ (lb/hr-ft^2)		0.585–4.63	0.16–0.8
Heat flux, q'' (Btu/hr-ft^2)			5000–23,000
Quality range, x	0.9 or 11°F sup	0.198–25°F sup.	
Recommended correlation	$\mathrm{Nu} = C\left[(\mathrm{Re}_{LO})^2\left(\dfrac{J\Delta x^h fg}{L}\right)^{0.5}\right]^{\nu}$ for $x_{in} = 0 - 0.56$ $x_{out} = 0.08 - 1.0$ $\mathrm{Re}_{LO}^2 K_f = 10^9 - 0.7 \times 10^{12}$ $C = 0.0009$, $\nu = 0.5$ for $x_{in} < 0.5$ $x_{out} = 1.0 - 11°F$ sup $\mathrm{Re}_{LO}^2 K_f = 10^9 - 0.7 \times 10^{12}$ $C = 0.0082$, $\nu = 0.4$ Valid for average h along the tube	for $x_{out} < 0.8$ $\mathrm{Re}_{LO}^2 K_f = 2.5 \times 10^{10} - 1.5 \times 10^{12}$ $C = 0.0225$, $\nu = 3/8$ For $x_{in} > 0.9$: $\mathrm{Nu}_v = 0.021\left(\dfrac{GD}{\mu_v}\right)^{0.8}\mathrm{Pr}_v^{0.4}$ Valid locally (replace $\Delta x/L$ by dx/dL)	$\mathrm{Nu} = 0.020\,\mathrm{Re}_v^{0.8}\,\mathrm{Pr}^{1/3}\left(\dfrac{D_2}{D_1}\right)^{0.53}$ where the Nu is calculated using the hydr. diam. $D_2 > D_1$ $\mathrm{Re}_v = \dfrac{V_v d_v \rho_l}{\mu_l}$ The actual velocity of the vapor phase V_v and the hydr. diam. of the vapor core d_v are obtained from void fraction measurements.
Properties evaluated at	T_{sat}	T_{sat}	T_{sat}
No. of data points/runs			27
Error in correlating data	-10% to +20% (-20%+35%)*		
Equation	42	43	44

*Using data of other investigators

General definitions and symbols

Subscripts

l liquid

v vapor

LO all liquid flowing in tube

LP only liquid phase flowing in tube

b bulk

sat saturation

i at point of suppression of nucleation
     (or at incipience of nucleation)

$\Delta T = T_w - T_b$, where $T_b = T_{sat}$

$\Delta T_i = (T_w - T_b)_i$

α Void fraction

$Bo = \dfrac{q''}{h_{fg} G}$ Boiling number

$Pr_l = \dfrac{c_l \mu_l}{k_l}$ Liquid Prandtl number

$Re_{LO} = \dfrac{GD}{\mu_l}$ $Re_{LP} = \dfrac{G(1 - x)D}{\mu_l}$ Reynolds numbers

$Nu = \dfrac{hD}{k_l}$ Nusselt number

$K_f = \dfrac{J \Delta x h_{fg}}{\Delta L}$

$h_{LO} = 0.023 \dfrac{k_l}{D} \left(\dfrac{GD}{\mu_l} \right)^{0.8} Pr_l^{0.4}$ Heat transfer coefficient

$h_{LP} = 0.023 \dfrac{k_l}{D} \left[\dfrac{G(1 - x)D}{\mu_l} \right]^{0.8} Pr_l^{0.4}$ Heat transfer coefficient

$\dfrac{1}{X_{tt}} = \left(\dfrac{x}{1 - x} \right)^{0.9} \left(\dfrac{v_v}{v_l} \right)^{0.5} \left(\dfrac{\mu_v}{\mu_l} \right)^{0.1}$ Martinelli parameter

reduction in heat flux caused the data to follow down curve d-c-e, interpreted as a curve representing the presence of nucleation or boiling. A possible interpretation is that nucleation is delayed along the path a-b because cavities have become deactivated.

Critical Heat Flux in Forced Convection

Our ability to predict and describe the conditions at the critical heat flux is far inferior for boiling with forced convection than for pool boiling. Most test data are taken with uniformly heated test sections for flow inside round tubes, rectangular channels and annuli; and some data are taken for flow along rod bundles. Some data are available for nonuniform heating—cosine distribution and discontinuous distributions.

Typical of test results in the vicinity of burnout is the curve shown in Fig. 36 with four regions identified [102]. Beyond a critical heat flux, the surface temperature begins to rise sharply and to oscillate significantly (Region III) until the wall is completely dry and film boiling is established (Region IV). The detailed description of what occurs along this curve is intimately associated with the two-phase flow regime that exists, the vapor quality and the total flow rate in a particular geometry. Regardless of the detailed description, the critical condition is associated with a sharp reduction in ability to transfer heat from the surface, e.g., a reduced heat flux at a given temperature difference or an increased temperature difference at the same heat flux.

Imagine the curve of Fig. 36 to represent the test results taken at constant inlet enthalpy, constant flow rate, and uniform heat flux in a channel of uniform cross section. Region I will usually be a region of no boiling, that is, ordinary single-phase forced convection. We focus attention on the exit condition. The description of Regions II and III depends largely on the flow rate.

At *low flow rate* it is possible for vapor to form at stray nucleation sites and, because the heat flux would be low, to establish annular flow where evaporation can

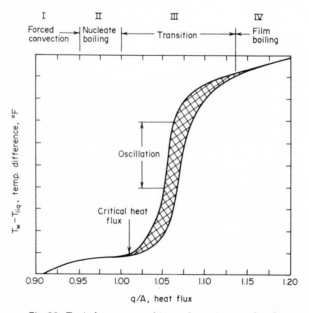

Fig. 36. Typical temperature history during burnout [102].

continue at the liquid-vapor interface by pure conduction through the annular liquid film without nucleate boiling. In any case, Region II is characterized by an increased heat transfer coefficient (decrease in the slope of the curve) because of the increased velocity caused by the presence of the vapor. As the quality increases down the channel, the vapor velocity may become large enough to "tear" the liquid off the wall, resulting in the temperature oscillation associated with Region III. At sufficiently high heat flux, the wall would be dry, and a fog flow would exist in the core, resulting in Region IV that might be called film boiling.

At *higher flow rates*, the condition in Region II is identified as nucleate boiling. If the flow rate is sufficiently high, the critical heat flux can occur at very low exit-vapor quality, in which case the mechanism in Region III is similar to that associated with pool-boiling burnout. Here, in Region IV, the core may be primarily liquid in bubbly flow with a vapor film separating the liquid from the wall.

The two cases perhaps represent extremes of all possible flowing conditions at the critical heat flux that might occur in nuclear reactors. In the low-flow case, it is possible that no nucleate boiling ever really existed, and the critical heat flux was in fact not a departure from nucleate boiling. Also, the heat fluxes under these flow conditions would be low enough to result in a wall temperature in Region IV that would be well below the melting point of the wall; hence, no physical burnout would result. At the other extreme (high flow rate) the critical heat flux would represent a condition just below a departure from nucleate boiling. Also, the heat flux would be rather high, causing the wall temperature in Region IV to be well above the melting point of the wall; hence, physical burnout would result.

In the light of the above discussion, terms such as "burnout," "departure from nucleate boiling," "maximum heat flux," and so forth should be discarded in favor of the more noncommittal "critical heat flux" defined as the heat flux just below the point, on a curve such as that of Fig. 36, where the wall temperature begins to rise sharply.

In the preceding discussion we identify the point that we would like to call the "critical heat flux point." Detecting and measuring the corresponding critical heat flux $(q/A)_{crit}$ are quite another matter. Some detectors respond to amplitude of temperature oscillation of a temperature-measuring device usually placed on the outside surface of the test section and close to the exit; others respond to rate of rise of such a temperature indicator; still others respond to rate of rise of the ratio of the voltage drop over the last quarter of the electrically heated test section to the voltage drop over the entire test section. Clearly, each of these dectectors, set to be tripped at various magnitudes of temperature oscillation or various magnitudes of rate of rise of temperature or voltage ratio, can indicate wide differing magnitudes of $(q/A)_{crit}$ for a particular test condition. The possible variation is not known, but a $\pm 10\%$ variation is not unreasonable to expect.

The magnitude of $(q/A)_{crit}$ seems to be seriously affected by the system dynamics of the entire test loop. It was demonstrated early in the testing programs [103] that even with subcooled liquids introduced at the inlet it is necessary to place a well-throttled valve just ahead of the test section in order to obtain reproducible results for $(q/A)_{crit}$. More recently [107], the quantitative reduction of $(q/A)_{crit}$ due to insufficient inlet throttling was measured in a particular system. In the late 1950s, data appeared for test sections with inlet conditions in the quality regime [104--107] by mixing liquid and vapor streams. Also, compressible volumes [108] were intentionally introduced at the test-section inlet. In both these latter cases, system dynamics or instabilities were found to influence significantly the magnitude of measured $(q/A)_{crit}$.

The heat transfer mechanism associated with the $(q/A)_{crit}$ is intimately related to the kind of two-phase flow regime existing at its occurrence.

All of these complicating effects have made it virtually impossible to describe adequately the conditions existing when $(q/A)_{crit}$ occurs and to obtain adequate correlations of the data for $(q/A)_{crit}$. Some correlation equation have been suggested for particular fluids (primarily water), specific geometries, and for limited ranges of operating conditions.

INFLUENCE OF SYSTEM INSTABILITIES
ON $(q/A)_{crit}$

The data in Fig. 37 were taken [108] on a once-through flow system with an expander vessel teed to the flow path just ahead of the test section. The upper curve is obtained when the expander vessel is filled with well-subcooled water, or is not in the system. In this case, the flow is quite stable. The lower curve is obtained when the expander vessel is either with a noncondensible gas, saturated vapor, or the test liquid at saturation temperature. In these cases, flow oscillates rather violently, presumably due to the compressible volume in the expander.

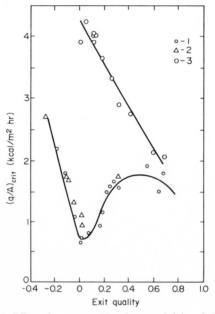

Fig. 37. Effect of a compressible volume on $(q/A)_{crit}$ [108].

Another operating condition that produces similar results and instabilities is with the inlet to the test section in the vapor quality regime. The research groups at Harwell (England) and CISE (Italy) have operated test systems with inlet quality by mixing liquid and vapor streams ahead of the test section.

In this test section, the data for $(q/A)_{crit}$ are influenced greatly by the system oscillations and are not representative of the stable forced-convection-boiling processes. To show this, the CISE group ran tests at similar operating conditions but attempted to reduce or eliminate the system oscillations by placing orifices just ahead of the test section. Figure 38 shows these results. Curve A is obtained when there were no orifices between the test-section inlet and the point of mixing the liquid and vapor streams. Curve B is obtained when the two orifices are placed in series upstream of the heated section. The $(q/A)_{crit}$ in the dip region was increased probably because

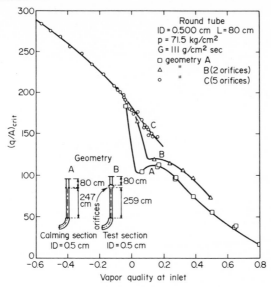

Fig. 38. Effect of throttling at inlet [107].

of the damping of the system oscillations. When five orifices were placed just ahead of the test section, the damping of the system oscillations, particularly as they influenced the *flow* in the test section, must have been nearly complete, because the resulting curve *A* is characteristic of data obtained when a system is in stable operation.

We conclude here that data for $(q/A)_{crit}$ taken in a system that is experiencing unstable flow is significantly lower than that which would be observed if a stable flow condition existed. The amount of the reduction in $(q/A)_{crit}$ will vary from system to system. The only data representative of the test-section boiling process that can be correlated or interpreted are data taken in stable operation.

VARIABLES AFFECTING CRITICAL HEAT FLUX

An energy balance for a particular fluid (properties) at a pressure (p) and mass velocity (G) in a duct (L, D_e) with an inlet enthalpy (h_i) is

$$h_e - h_i = \frac{4}{GD_e} \int_0^L q'' \, dz \tag{45}$$

For the case of uniform heat flux this becomes

$$h_e - h_i = \frac{4L}{GD_e} q'' \tag{46}$$

Experiments to determine q''_{crit} for a particular fluid may be performed by setting the magnitudes of the following independent variables: L, D_e, G, h_i, p. Then the power to test section is raised until the critical condition is reached. At this condition the exit enthalpy may be considered to be $h_{e,crit}$. Then, in functional form

$$h_{e,\text{crit}} = f_1(G, h_i, L, D_e, p) \tag{47}$$

or for uniform heat flux

$$q''_{\text{crit}} = f_2(G, h_i, L, D_e, p) \tag{48}$$

These two formulations assume the existence of the energy balance, Eq. (46), which may be used to replace any one of the variables in Eqs. (47) or (48). For example, h_i may be replaced by h_e using Eq. (46); then

$$q''_{\text{crit}} = f_3(G, h_e, L, D_e, p) \tag{49}$$

The tests for q''_{crit} could have been run taking h_e instead of h_i as an independent variable; then there would have been an $h_{i,\text{crit}}$ corresponding to q''_{crit}.

Other quantities, related to those above, may be useful in correlating data:

$\Delta h_i = h_{\text{sat}} - h_i$, inlet subcooling enthalpy
L_s, saturation length measured downstream from the position where the heat added
 would bring the fluid to the saturated liquid state at equilibrium
X_e, exit quality (related to h_e)
X_i, inlet quality (related to h_i); note X_i and X_e may both be negative if subcooled
W total power added to heated section
W_s, power added over length L_s
W_{crit}, power added to location where critical condition occurs
$W_{s,\text{crit}}$, power added from point where liquid is saturated to point where critical
 condition occurs

Critical Heat Flux Data

Data obtained when flow oscillations exist are greatly influenced by the nature of the entire flow system. In order to exclude that data (Figs. 37 and 38) only the data taken with subcooled inlet or very high quality inlet should be considered as providing stable, reproducible data independent of a particular flow circuit. This puts severe limitations on any test system in attempting to cover the entire range of variables desired.

A common way to present critical heat flux data for uniform q''_{crit} is shown in Fig. 39 as q''_{crit} vs. h_e (or X_e) with lines of constant G for L, D_e, and p constant.

Tong [109] pointed out that because two relations—the critical condition relations and the energy balance—must be satisfied, the magnitude of q''_{crit} cannot vary with one variable alone. In Fig. 39 h_i varies along each of the curves. Alternatively, then it is possible to plot the same data against h_i as in Fig. 40. Note the maximum X_e in Fig. 39 corresponding to saturated liquid at inlet. In order to obtain data beyond these points longer test sections must be used.

Tong [109a] further shows the difficulty of showing trends in q''_{crit} because at least two variables always change simultaneously. Figures 41 and 42 show lines of constant tube diameter as parameters. One plot shows larger q''_{crit} for larger D, the other shows smaller q''_{crit} for larger D. The dashed line for $L = 60$ inches in Fig. 42 is the same data on Fig. 41 at $G = 10^6$ lb/hr-ft². Similarly Tong shows the different way q''_{crit} varies with pressure when h_e is held constant, as in Fig. 43, and when h_i is held constant, as in Fig. 44.

The preceding data was for water. Figure 45 shows data for R-12 refrigerant plotted as in Fig. 39 for water.

In the subcooled boiling [115] region, Figs. 46 and 47 show the effect of diameter and length on q''_{crit} in small diameter tubes. In Fig. 46 the data fall asymptotically to a constant value for $D > 0.25$ in.

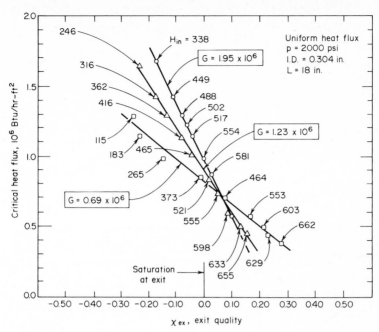

Fig. 39. Mass velocity effect on critical heat flux (local condition concept) [83].

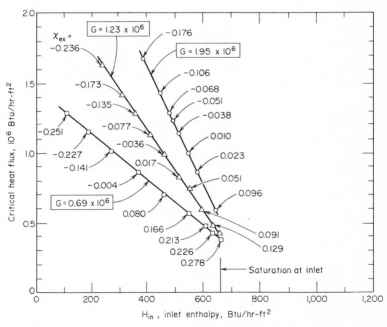

Fig. 40. Mass velocity effect on critical heat flux (system parameter concept) (same data as Fig. 39) [83].

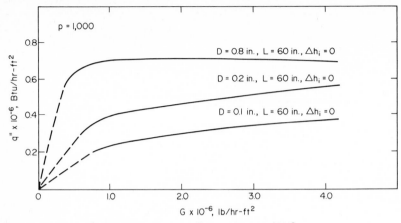

Fig. 41. Parametric effects on burnout [111].

Critical Heat Flux Correlations for Uniform q''

No forced-convection critical heat flux correlation has been evolved that applies universally to data for all fluids and all geometries. This discussion will first direct its attention toward correlations for a particular fluid and a particular flow cross-section shape and uniform heat flux distribution.

The correlations suggested prior to 1958 are summarized by De Bertoli et al. [116] and Tong [109b]. Some are purely empirical, others attempt to include some fluid

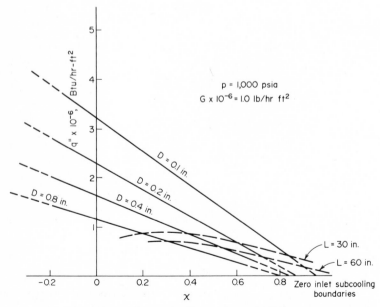

Fig. 42. Parametric effects on burnout [111].

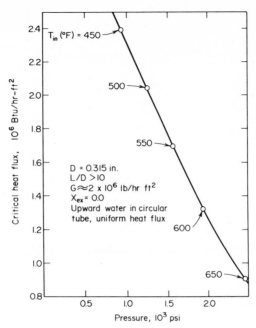

Fig. 43. Effect of pressure and inlet subcooling on critical heat flux [113].

properties by resorting to a description of a postulated mechanism and a limited dimensional analysis. Most of these earlier correlations are limited to rather small ranges of variables, and some even show the wrong influence of flow rate at higher qualities. Here the discussion is limited to the more recently proposed correlations.

It should again be emphasized that much of the available data may have been taken with flow oscillations in the system, particularly that taken with inlet quality greater

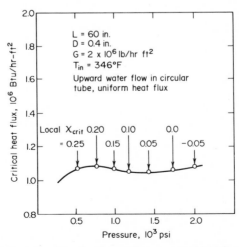

Fig. 44. Pressure and local enthalpy effects on critical heat flux [113].

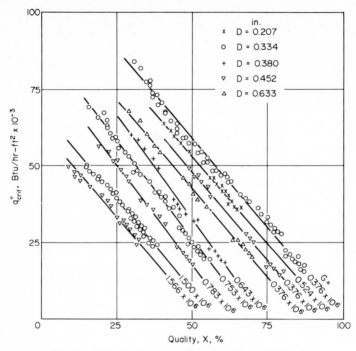

Fig. 45. Critical heat flux for Freon-12 at 155 psia [114].

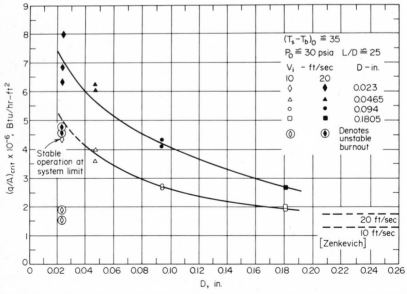

Fig. 46. Dependence of burnout heat flux on tube diameter and velocity for surface boiling [115].

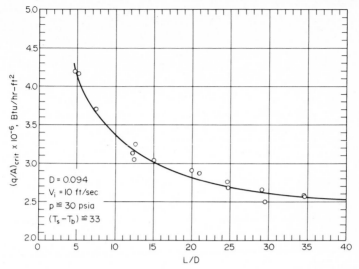

Fig. 47. Dependence of burnout heat flux on length-diameter ratio [115].

than zero, as shown in Figs. 37 and 38. These data should not be included in an attempt to correlate successfully the critical heat flux data.

Tong [109c] reviews the various idealized physical models employed by various people in attempting to arrive at a logical correlation equation. In the subcooled region these models focus on the forces acting on a bubble at the wall and on a superheat layer near the wall. In the mist-annular flow regime (quality greater than 5 to 10 percent) the critical condition is visualized to occur when the liquid film dries out or disappears from the wall. This occurs when the rate of evaporation plus the entrainment exceeds the rate of deposition of liquid droplets from the core. Obviously these processes are greatly influenced by the vapor velocity and take place either with or without actual nucleation at the wall. If nucleation (nucleate boiling) is taking place in the liquid layer, the resulting disruption of the surface of the layer will cause the wall to become dry of liquid at a vapor velocity lower than that required when nucleation is suppressed.

Tong [109d] presents and reviews many of the proposed correlations, which have been developed primarily for water as the fluid. Only a few will be presented here.

Correlation of q''_{crit},
Subcooled Forced Convection

For subcooled liquid, Gambill [117] developed an additive correlation

$$q''_{crit} = q''_{b,crit} + q''_{c,crit} \qquad (50)$$

where

$$q''_{b,crit} = 0.145\, h_{fg} \rho_v \left[\frac{\sigma g_0 g (\rho_l - \rho_v)}{\rho_v^2} \right]^{1/4} \left[1 + \left(\frac{\rho_l}{\rho_v} \right)^{0.923} \left(\frac{c_l \Delta T_{sub}}{25\, h_{fg}} \right) \right] \qquad (51)$$

$$q''_{c,\text{crit}} = h_c (T_{w,\text{crit}} - T_{\text{liq}}) \tag{52}$$

and $T_{w,\text{crit}}$ is evaluated from Bernath's [118] relation

$$T_{w,\text{crit}} = 102 \ln p - 97 \left(\frac{p}{p + 15} \right) - \frac{V}{2.22} + 32 \tag{53}$$

where p is absolute pressure in psia, and V is flow velocity in ft/sec assuming all liquids flow. The magnitude of h_c in Eq. (52) is the heat transfer coefficient, assuming all liquids flow in the channel.

When compared with low pressure, small tube data for water (Bergles [115]) Eq. (50) predicted results too low.

Another empirical correlation for the subcooled region for water only was developed at Westinghouse APD (Tong [121])

$$q''_{\text{crit}} = (0.23 \times 10^6 + 0.094\, G)(3 + 0.01\, \Delta T_{\text{sub}})$$

$$\times (0.435 + 1.23\, e^{-0.0093 L/D_e})(1.7 - 1.4\, e^{-a}) \tag{54}$$

where

$$a \equiv 0.532 \left(\frac{\rho_l}{\rho_v} \right)^{1/3} \left(\frac{h_{\text{sat}} - h_i}{h_{fg}} \right)^{3/4}$$

Equation (54) correlates 95 percent of the water data to ±20 percent for the following ranges of variables: tube, rectangular channels, rod bundles (not annulus), uniform heat flux:

$$G = 0.2 \times 10^6 \text{ to } 8 \times 10^6 \text{ lb/hr-ft}^2$$
$$p = 800 \text{ to } 2700 \text{ psia}$$
$$L/D_e = 21 \text{ to } 365$$
$$h_i > 300 \text{ Btu/lb}$$
$$\Delta T_{\text{sub}} = 0 \text{ to } 228°F$$
$$D_e = 0.1 \text{ to } 0.54 \text{ in.}$$
$$q''_{\text{crit}} = 0.4 \times 10^6 \text{ to } 4 \times 10^6 \text{ Btu/lb-ft}^2$$

Correlation of q''_{crit} Quality Regions

Macbeth Correlation. An attempt has been made to correlate the data for one particular fluid, water, using only those independent and dependent variables of Eqs. (46) to (49) for specific geometries (Macbeth [119]).

In order to correlate conveniently the data with Eqs. (46) to (49), it is necessary to seek simplification of the functional relation. By careful examination of the world's data, excluding all data points taken with inlet conditions in the quality range that are suspected of being associated with flow instabilities, Macbeth found the $(q/A)_{\text{crit}}$ to be related linearly with Δh_i. The equation was further simplified by employing the local-condition hypothesis that $(q/A)_{\text{crit}}$ depends only on the conditions at the exit in a uniformly heated test section as it approaches the critical condition from the nucleate boiling or the wetted-wall condition. Then Eq. (49) should not include L; therefore

$$\left(\frac{q}{A}\right)_{crit} = f_2(G, h_e, D, p) \tag{55}$$

Using data from carefully planned experiments at Winfrith, Barnett [120] shows these equations may be written in the following form

$$\left(\frac{q}{A}\right)_{crit} = \frac{A + 1/4\, CD(G \times 10^{-6})\Delta h_i}{1 + CL}$$

or

$$\left(\frac{q}{A}\right)_{crit} = A - 1/4\, CD(G \times 10^{-6})h_{fg}x_e \tag{56}$$

where A and C are functions of D, G, and p.

Macbeth [119] applied these equations to practically all available data for uniformly heated round tubes and rectangular channels, using only data for which the inlet condition was subcooled, and no flow oscillations were reported. The coefficients A and C were determined empirically. Macbeth found he could present the results more simply by dividing the representative equations into two zones—high and low flow regions, as defined by Fig. 48.

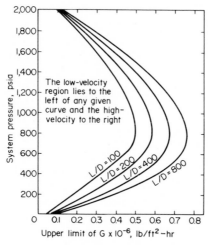

Fig. 48. Approximate boundary limits of the low-velocity and high-velocity burnout regimes for round tubes [119].

In the *low-flow* regime for *uniformly heated round tubes,* the critical heat flux is given by

$$\left(\frac{q}{A}\right)_{crit} \times 10^{-6} = 0.00633\, h_{fg}D^{-0.1}(G \times 10^{-6})^{0.51}(1 - x_e) \tag{57}$$

TABLE 5. Optimized y Values and rms Errors for High-velocity Regime Burnout Data for Round Tubes (Eq. (59))

Pressure* (psia)	y_0	y_1	y_2	y_3	y_4	y_5	Rms error (%)	No. of test data points
15	1.12	-0.211	-0.324	0.0010	-1.4	-1.05	13.8	88
250 (nom.)	1.77	-0.553	-0.260	0.0166	-1.4	-0.937	4.7	237
530 (nom.)	1.57	-0.566	-0.329	0.0127	-1.4	-0.737	5.7	170
1000	1.06	-0.478	-0.179	0.0085	-1.4	-0.555	7.4	405
1570 (nom.)	0.720	-0.527	0.024	0.0121	-1.4	-0.096	3.4	133
2000	0.627	-0.268	0.192	0.0093	-1.4	-0.343	9.0	362
2700 (nom.)	0.0124	-1.45	0.489	0.0097	-1.4	-0.529	4.7	37

Conversion factors: 1 psia = 0.0731 kg/cm^2 = 0.06805 atm.

TABLE 6. Optimized y Values and rms Errors for High-velocity Regime Burnout Data for Rectangular Channels (Eq. (59))

Pressure* (psia)	y_0	y_1	y_2	y_3	y_4	y_5	Rms error (%)	No. of test data points
600	23.4	-0.472	-3.29	0.123	-1.4	-3.93	6.1	22
800	0.445	-1.01	-0.384	0.0096	-1.4	-0.0067	12.9	28
1200	1.88	-0.081	-0.526	0.0036	-1.4	-1.29	4.9	42
2000	0.546	-0.315	-0.056	0.0027	-1.4	-0.725	9.4	359

*1 psia = 0.07031 kg/cm^2 = 0.06805 atm.

In the *low-flow* regime for *uniformly heated narrow channels of one-inch width*, the critical heat flux is given by

$$\left(\frac{q}{A}\right)_{crit} \times 10^{-6} = 0.264\, h_{fg} S^{1.73} (G \times 10^{-6})^{-0.1} (1 - x_e) \qquad (58)$$

where S is the internal spacing between the flat heating surfaces in a rectangular channel.

In the high-flow regime, $(q/A)_{crit}$ is given by Eq. (56) with A and C given by

$$A = y_0 D^{y_1} (G - 10^{-6})^{y_2}$$

$$C = y_3 D^{y_4} (G \times 10^{-6})^{y_5} \qquad (59)$$

where optimized y values are given in Table 5 for various pressures for water.

For rectangular channels, replace ¼ by 0.555 and D by S in Eq. (55). The optimized values are given in Table 6.

In the range of reactor design, these equations appear to predict the $(q/A)_{crit}$ to well within 10 percent rms error; however, they should not be used beyond the range of the magnitudes of D, G, L, and p for which data exist. Also Δh_i must be greater than zero. These equations are not applicable to tube bundles and annuli.

For uniform heat flux the group at CISE [107] suggest the following form of equation

$$q''_{crit} = (a - x_e) \frac{G h_{fg}}{4b} \qquad (60)$$

where

$$a = \frac{1 - p/p_{cr}}{1.11 (G/10^6)^{1/3}} \quad ; \quad a \le 1$$

$$b = 62 \left(\frac{p_{cr}}{p} - 1\right)^{0.4} D^{0.4} \left(\frac{G}{10^6}\right) \qquad (61)$$

Ranges of variables for which 84 percent of the data is predicted to ±15 percent are

$$D > 0.28''$$
$$p = 640 \text{ to } 2130 \text{ psia}$$
$$G/10^6 = 0.74 (1 - p/p_a)^3 \text{ to } 2.95$$
$$x_i = -(a/b)(L/D) \text{ to } 0.5a.$$

Note that Eqs. (60) and (56) have the same form but the coefficients are slightly different.

Tong [121] also presents an empirical correlation for the region when exit quality is greater than zero

$$h_{crit} - h_i = 0.529\, \Delta h_i + [0.825 + 2.3 \exp(-17 D_e)]\, h_{fg} \exp\left(-\frac{1.5\, G}{10^6}\right)$$

$$-0.41\, h_{fg} \exp\left(-\frac{0.0048\, L}{D_e}\right) - 1.12\, h_{fg} \frac{\rho_v}{\rho_l} + 0.548\, h_{fg} \qquad (62)$$

where D_e is in inches, h in Btu/lb, and G in lb/hr-ft^2.

For 95 percent of the data a ±25 percent, the following are the limits of variables considered: tubes, rectangular channels, rod bundles (not annuli):

$$G = 0.4 \times 10^6 \text{ to } 2.5 \times 10^6 \text{ lb/hr-ft}^2$$
$$p = 800 \text{ to } 2250 \text{ psia}$$
$$L = 9 \text{ to } 76 \text{ inches}$$
$$D_e = 0.1 \text{ to } 0.54 \text{ inches}$$
$$h_i = 400 \text{ Btu/lb to sat.}$$
$$q''_{\text{crit}} = 0.1 \times 10^6 \text{ to } 1.8 \times 10^6 \text{ Btu/hr-ft}^2$$
$$x_e = 0. \text{ to } 90\%$$

Quite a number of other correlation equations have been proposed. A survey of many of these is presented by Tong [109d]. The ones selected for presentation here appear to agree with data over a wider range of variables. Nevertheless all of these correlations have shortcomings. Where possible it is best to use actual data for a system operating close to the desired conditions. Collections of such data are found in Refs. 109d and 121 through 124.

A Limiting Vapor Velocity. In the higher quality regions the simultaneous effects of evaporation, droplet deposition, and entrainment suggest a critical magnitude of vapor flow. Some of the British data [125] suggest a critical $G_{v,\text{crit}} \approx 0.45 \times 10^6$ lb/hr-ft² to ±30 percent for $G = 1$ to 3×10^6 and $p = 200$ to 1400 psia. Mozharov [126] suggests the following dimensionless relation for this critical flow velocity

$$\left(\frac{G_{v,\text{crit}}^2 D}{g_0 \sigma \rho_v} \right)^{1/2} = 115 \left(\frac{x}{1-x} \right)^{1/4} \tag{63}$$

The actual process of liquid film dryout is undoubtedly much more complicated than these two simple relations suggest and is surely significantly influenced by whether or not nucleation is taking place at the wall.

Effect of Nonuniform Heat Flux

The effect of nonuniform axial heat flux distribution on the critical condition is of interest in nuclear reactor and combustion chamber cooling jacket design. Tong [109d] surveys the various test data. More recently, Todreas [127] obtained data for a variety of flux distributions including discontinuous spikes for water at lower pressures in aluminum tubes. Figure 49 shows the location of the critical condition (actual tube rupture) for various flux shapes. Here M is the ratio of maximum-to-minimum heat flux along the tube.

With uniform heat flux the critical condition occurs at the outlet end of the tube, but with nonuniform flux it sometimes occurs upstream from the outlet.

For uniform heat flux, since the critical condition occurs at the tube exit and since the energy balance Eq. (46) is a simple relation, an adequate correlation can be obtained in any of the forms of Eqs. (47), (48), or (49), or for that matter in terms of total power W added instead of q''. These formulations are essentially all equivalent for uniform flux distribution. This is not so for nonuniform flux distribution.

Earlier, Tong [121, 109d] suggested that nonuniform flux critical condition in the quality regime could also be predicted by Eq. (62) where $h_{\text{crit}} - h_i$ is a measure of the integrated power added.

The group at CISE [107] showed that both uniform and nonuniform heat flux critical data may be correlated by use of boiling length L_s from where the liquid is calculated to be saturated to the dryout location and the total or integrated power W_s added over the boiling length L_s. The correlation equation is written as

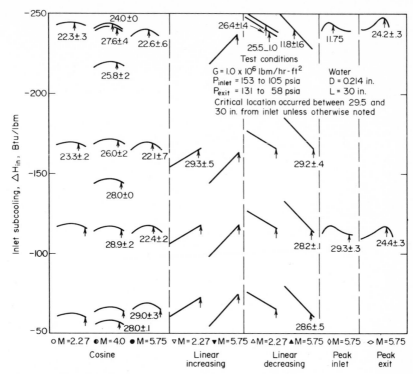

Fig. 49. Critical locations (inches from inlet) for the nonuniform axial flux distributions investigated at $G = 1.0 \times 10^6$ lbm/hr-ft^2 [127].

$$\frac{W_s}{wh_{fg}} = \frac{a}{1 + b\,D/L_s} \tag{64}$$

where a and b are given by Eq. (61) for water, w is flow rate (lb/hr), and $W_s = wx_{crit}h_{fg}$. For Refrigerant-12, at a reduced pressure of 0.27, Eq. (64) correlated the data with the 62 in Eq. (61) for b replaced by 118.

Following the completely empirical approach of Macbeth, Kirby [128] correlated available data for nonuniform heat flux as

$$q''_{crit} = A - B\,X_{crit} \tag{65}$$

where q'' varies with length and q''_{crit} and X_{crit} are the appropriate magnitudes at the critical location. The data determined magnitudes of the constants expressed as

$$A = Y_1 G^{Y_2} D^{Y_3}$$
$$B = Y_4 G^{Y_5} D^{Y_6} \tag{66}$$

where the magnitudes of Y_n are given in Table 7. The agreement with data is rather good. The following ranges of variables were covered

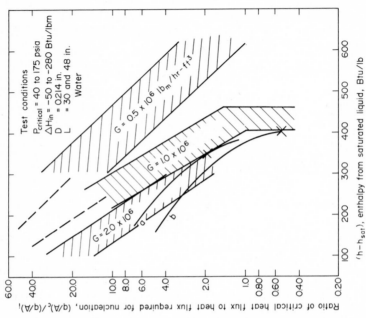

Fig. 51. Comparison of critical flux results for the range of mass velocities investigated [127].

Test conditions
$P_{critical}$ = 40 to 175 psia
ΔH_{in} = −50 to −280 Btu/lbm
D = 0.214 in.
L = 30 and 48 in.
Water

$G = 0.5 \times 10^6 \; lb_m/hr\text{-}ft^3$

$G = 1.0 \times 10^6$

$G = 2.0 \times 10^6$

a

b

$(h-h_{sat})$, enthalpy from saturated liquid, Btu/lb

Ratio of critical heat flux to heat flux required for nucleation, $(q/A)_c/(q/A)_i$

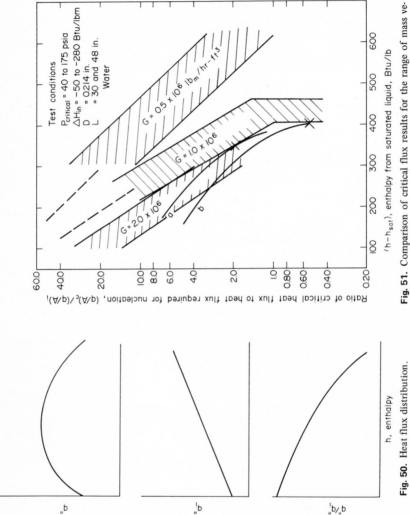

q''

q''_i

q''/q''_i

h, enthalpy

Fig. 50. Heat flux distribution.

TABLE 7. Optimized Constants

	560	1000	1250	1500	1600	1800	2000
Y_1	2.411	0.7468	0.988	0.7423	0.776	0.616	6.18
Y_2	-0.271	-0.1285	-0.191	-0.1057	-0.118	-0.155	0.231
Y_3	0.0*	-0.544	0.0*	-0.2335	-0.452	-0.293	2.907
Y_4	3.511	1.344	2.918	2.384	3.130	2.086	50.8
Y_5	0.345	0.526	0.287	0.6910	0.739	0.743	0.827
Y_6	0.0*	-0.2278	0.0*	0.0903	-0.296	-0.695	4.228

*No diameter variation.

	560	1000	1250	1500	1600	1800	2000
No. of expts.	34	269	22	79	98	80	69
Rms % error	3.66	5.73	3.37	4.82	3.42	5.95	6.60

Overall rms deviation = 5.30%.

$$D = 0.37 \text{ to } 1.11 \text{ inches}$$
$$L = 39 \text{ to } 192 \text{ inches}$$
$$G = 0.25 \text{ to } 7.01 \times 10^6 \text{ lb/hr-ft}^2$$

Figure 50 shows a plot of heat flux variation along a tube and also a plot of incipient heat flux q_i'' as calculated from, say, Eq. (8) and Fig. 5. Of course, this incipient heat flux would occur at a wall temperature different from that actually existing at any position, except where $q'' = q_i''$ at any point. The lower curve is a plot of the ratio of q''/q_i'' vs. position along the tube. This can be placed anywhere on the $q'' = q_i''$ vs. h plot merely by changing the level of heat flux and the magnitude of the inlet enthalpy. This lower curve of Fig. 50 is an operating curve whose shape and position depend upon the actual flux distribution.

Todreas [127] found that the data at the actual burnout or tube rupture location for various flux shapes, including those with sharp spikes would all fall within a narrow band on a q_{crit}''/q_i'' vs. h_{crit} at burnout as shown in Fig. 51. For q_{crit}''/q_i'' below a magnitude around 1, no nucleation occurs and the burnout data appeared to fall on a nearly vertical line which for a given G is a particular value of G_v or vapor quality, lending support to the suggestion that there may be a critical G_v at which the liquid layer on the wall dries out. At higher magnitudes of q_{crit}''/q_i'', nucleation occurs and the burnout takes place at a lower h or x or G_v, suggesting that the presence of nucleation disrupts the liquid film and the wall dries out at a lower magnitude of G_v.

This same type of plot tends to show how burnout sometimes occurs at the tube exit and sometimes upstream. Since the operating curve can be placed anywhere on this plot, there is shown in Fig. 51 two operating curves, a and b, for $G = 1 \times 10^6$. For curve a, the critical local G_v (or h) is reached at an upstream position, nucleation being present. For curve b the critical G_v (or h) is reached at the end of the tube; here nucleation was suppressed. This same reasoning shows why tubes with heat flux spikes sometimes burn out upstream and sometimes at the exit.

The same phenomenon was shown for flux distributions with spikes [129] for water data at high pressures.

Effect of Unheated Walls of h_{crit}

For flow in an annulus with one wall unheated, or flow along clusters of rod bundles in an unheated container wall, the curve of q_{crit}'' vs. h_e as in Fig. 39 is shifted to lower h_e or quality. This is understandable, since the unheated wall contributes no vapor generation, hence, for a particular local critical condition at the heated wall the averaged vapor generation (or h_e) is less by an amount which depends on the ratio of the heated perimeter P_H to the total flow perimeter P. In fact, Becker et al. [130] found that critical heat flux data for various numbers of rod bundles and annulus sizes followed a definite linear pattern (Fig. 52) when the exit critical enthalpy is plotted vs. P_H/P. In Fig. 52 the ordinate is the ratio of the quality at the critical condition in the particular geometry to that in a round tube with the entire perimeter heated.

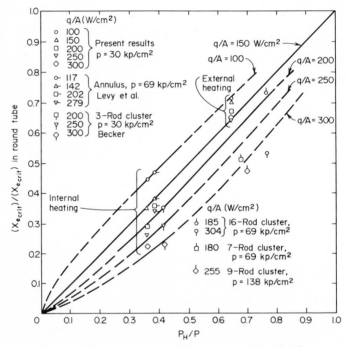

Fig. 52. Effect of perimeter ratio on critical conditions [130].

LIQUID METALS

Critical heat flux data for liquid metals is very limited in quantity. Noyes and Lurie [68] present data for sodium in an annulus covering the range from around 70°F subcooled to around 15 percent exit quality. The resulting curves look very much like the lower curve of Fig. 37 suggesting the presence of severe instability. The lower curve in Fig. 37 resulted from a compressible volume at the inlet of the test section. The Noyes and Lurie system had a short heated length with a compressible volume in the condenser located very close to the exit. Noncondensible gases were allowed to collect in this condenser. It is quite likely that this compressible volume caused premature burnout as in Fig. 37.

MISCELLANEOUS EFFECTS

For flow of subcooled water in vertical annuli, Durant and Mirshak [131] report increases in q''_{crit} by as much as 100 percent. For water in annuli in the high-quality, annular flow regime, Janssen [132] reports decreases in q''_{crit} of as much as 50 percent for sand-blasted (300 microinch) surfaces. On the other hand, Moussez [109e] reports a 20 percent increase in q''_{crit} when roughness was increased from 20 to 200 microinches.

In contrast to results for pool boiling, the effect of dissolved gas in flow boiling q''_{crit} has been found to be insignificant [109f].

As a means of delaying the film dryout, twisted ribbons in tubes and vortex flow at the inlet have produced increases in q''_{crit} [133, 134].

FILM BOILING

Film boiling as a cooling process has not had wide commercial applications because of the accompanying high surface temperatures: it may find use in the future as better materials become available. It is, however, often encountered in chemical process equipment and in cryogenic systems.

In film boiling, vapor is generated at the liquid-vapor interface by conduction and radiation from the heating surface through the vapor film. Using the hydrodynamic instability of the liquid-vapor boundary, Zuber [135] arrived at the following equation representing the minimum heat flux for film boiling

$$\left(\frac{q}{A}\right)_{min} = 0.09 \rho_{vf} h_{fg} \left[\frac{g(\rho_l - \rho_v)}{\rho_l + \rho_v}\right]^{1/2} \left[\frac{g_0\sigma}{g(\rho_l - \rho_v)}\right]^{1/4} \qquad (67)$$

where ρ_{vf} is the vapor density film temperature and the other properties are evaluated at saturation temperature. The coefficient 0.09 was determined empirically by Berenson [136] from data such as those shown in Fig. 12. It is significant that the data for any surface finish converge on this point and in the film boiling region are independent of surface finish. In the transition boiling region the data are influenced by the same factors that influence nucleate boiling data, suggesting that the film does occasionally touch the heating surface.

Extending Zuber's analysis, Berenson arrives at the following expression for ΔT at the minimum heat flux point

$$\Delta T_{min} = 0.127 \frac{\rho_{vf} h_{fg}}{k_{vf}} \left[\frac{g(\rho_l - \rho_v)}{\rho_l + \rho_v}\right]^{2/3} \left[\frac{g_0\sigma}{g(\rho_l - \rho_v)}\right]^{1/2} \left[\frac{\mu_f}{g_0(\rho_l - \rho_v)}\right]^{1/3} \qquad (68)$$

The stable film regime was studied experimentally and analytically by Bromley [137, 138] for horizontal tubes and vertical plates. By balancing the buoyant and frictional forces on the vapor flowing in the film on the outside of a horizontal tube, Bromley arrived at the following equation representing the h_c associated with conduction alone

$$h_c = 0.62 \left[\frac{k_v^3 \rho_v (\rho_l - \rho_v) g(h_{fg} + 0.4c_{pv} \cdot \Delta T)}{D_0 \mu_v (T_w - T_{sat})}\right]^{1/4} \qquad (69)$$

Radiation contributes to the heat transfer and increases the vapor film thickness, reducing the effective contribution of the conduction. The total heat transfer coefficient is given by

$$h = h_c \left(\frac{h_c}{h}\right)^{1/3} + h_r \tag{70}$$

where h_r is calculated for radiation between two parallel planes.

For forced convection flow of the liquid across the tube, Bromley [138] suggests the following equation when $V_\infty \geq 2\sqrt{gD_0}$

$$h_c = 2.7 \sqrt{\frac{V_\infty k_v \rho_v (h_{fg} + 0.4c_{pv}\,\Delta T)}{D_0(\Delta T)}} \tag{71}$$

and

$$h = h_c + \frac{7}{8} h_r \tag{72}$$

Film boiling on a horizontal surface was analyzed by Berenson [136] and compared with data for pentane (Fig. 12), CCl_4, benzene, and ethyl alcohol. The following equation results

$$h = 0.425 \left[\frac{k_{vf}^3 \rho_{vf}(\rho_l - \rho_v)g(h_{fg} + 0.4c_{pv}\,\Delta T)}{\mu_f(\Delta T)\sqrt{g_0\sigma/g(\rho_l - \rho_v)}}\right]^{1/4} \tag{73}$$

Note this is identical in form with Eq. (69) with D_0 replaced by $\sqrt{g_0\sigma/g(\rho_l - \rho_v)}$ which is proportional to bubble diameter (Eq. (10)).

Data for film boiling of cryogenic horizontal cylinders was correlated [77] by the Breen and Westwater [139] equation

$$\left(\frac{q}{A}\right)_{cond} = \left\{4.94\,\frac{\rho_l - \rho_v}{\sigma^{1/8}} + 0.115\left[\frac{\sigma^{3/8}}{D(\rho_l - \rho_v)^{1/8}}\right]\right\}\left(\frac{k_v^3 h'_{fg}\rho_v}{\mu_v}\right)^{1/4}(T_w - T_s)^{3/4}$$

$$h'_{fg} = \frac{[h_{fg} + 0.340c_{pv}(T_w - T_s)]^2}{h_{fg}} \tag{74}$$

The minimum heat flux in film boiling on horizontal cylinders is correlated [77] by the Lienhard and Wong [140] equation

$$\left(\frac{q}{A}\right)_{min_{cond}} = 0.114\left(\frac{h_{fg}\rho_v}{D}\right)\left[\frac{2g(\rho_l - \rho_v)}{\rho_l - \rho_v} + \frac{4\sigma}{D^2(\rho_l - \rho_v)}\right]^{1/2}\left[\frac{g(\rho_l - \rho_v)}{g_0\sigma} + \frac{2.0}{D^2}\right]^{-3/4} \tag{75}$$

Radiation for $T < 425°R$ is usually $< 5\%$ of q''_{total}.

For quenching of spheres Frederking and Clark [141]

$$\frac{hD}{k} = 0.15\,Ra^{1/3} \tag{76}$$

For Ra > 5×10^7 where

$$Ra \equiv \left\{ \frac{D^3 g(\rho_l - \rho_v)}{\nu_v^2 \rho_v} \, Pr_v \left[\frac{h_{fg}}{c_{p,v}(T_w - T_{sat})} + 0.4 \right] \frac{a}{g} \right\}$$

where a is local acceleration, equal to g for a stationary sphere on earth.

In flowing forced convection systems, heat transfer beyond the point where the liquid film dries out is essentially a film boiling process. Here the vapor flows down the tube carrying along liquid droplets. The process is described by Quinn [142] in Fig. 53. In the region of the dry-out the liquid film oscillates back and forth with accompanying temperature fluctuations (the transition region). In the next region the heat flux and wall temperatures are steady and the ratio of $q''/(T_w - T_{sat})$ may go through a minimum depending on the flow rate.

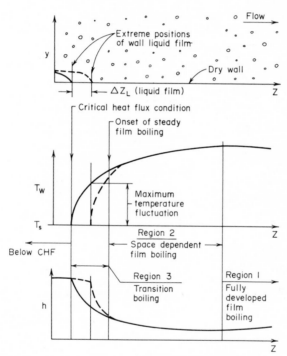

Fig. 53. Heat transfer behavior beyond the CHF (uniform heat flux) [142].

The process of film boiling was studied in more detail by Laverty [143] and Forslund [144] for nitrogen. Typical measurements of wall temperature are shown in Fig. 54 for various heat fluxes. The curves merge asymptotically to the wall temperature predicted for 100 percent vapor flow far downstream well beyond the point where enough heat has been added to evaporate all of the vapor. Clearly the vapor must have been superheated with liquid droplets entrained. The solid line asymptotes are the predicted wall temperatures for 100 percent vapor flow and the arrows show the location where the quality (X_E') would have been 100 percent had

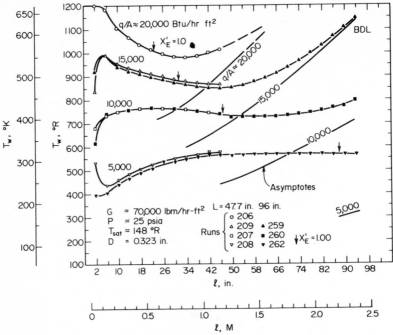

Fig. 54. Wall temperatures: film boiling, forced convection. Liquid nitrogen. 0.323-inch ID tube [144].

there been no superheating of the vapor. These curves plus visual observations show that liquid droplets exist far downstream of the $X'_E = 1.00$ point; hence significant vapor superheat was present.

Since vapor is generated along the tube the vapor flow accelerates and drags the liquid droplets along. Forslund [144] predicts wall temperatures and heat fluxes along the tube by calculating the heat transfer from the wall to the vapor and then from the vapor to the droplets. In the accelerating vapor stream, the droplets are taken to break up at a critical Weber number, $(V_v - V_l)^2 \rho_v D/\sigma = 7.5$, which determines local droplet diameters with drag coefficients on spheres. The prediction process involves a stepwise calculation of all conditions down the tube. Calculated wall temperatures, droplet diameters, and vapor superheat agree well with observations. Since the calculation procedure is somewhat complicated it is not reproduced here.

Knowledge of heat transfer in this beyond-dryout region is limited. Data over limited ranges of variables have been correlated by simple dimensionless equations, but no universally applicable prediction scheme has evolved other than the calculation procedure by Forslund [144].

Additional data and empirical correlations applicable over limited ranges of data are summarized by Tong [109g] and Silvestri [145].

OTHER SURVEYS OF BOILING
HEAT TRANSFER

The reader is referred to the following other surveys of boiling heat transfer by Gouse [1], Rohsenow [147], Tong [109], Silvestri [145], Leppert and Pitts [146], Westwater [148], and Dwyer [149].

REFERENCES

1. S. W. Gouse, An Index to Two-Phase Gas-Liquid Flow Literature, *M.I.T. Rept.* 9, 1966, Cambridge, Mass.
2. J. W. Westwater and P. H. Strenge, *Chem. Eng. Prog. Symp. Series,* vol. 29, p. 95, 1959.
3a. C. Corty and A. S. Foust, Surface Variables in Nucleate Boiling, *Chem. Eng. Prog. Symp. Series,* vol. 17, p. 51, 1955.
3b. S. G. Bankoff, *Proc. Heat Transfer Fluid Mechanics Inst.,* Stanford University Press, 1956.
4. W. M. Rohsenow, Nucleation with Boiling Heat Transfer, *Ind. Eng. Chem.,* **58**:1 (1966).
5. P. Griffith, Nucleation and Bubble Formation in Boiling, *Symp. Boiling Heat Transfer, Proc. Inst. Mech. Engrs.,* vol. 180, part 1 and 3C, 1965-66.
6. I. Shai, Doctoral dissertation, Massachusetts Institute of Technology, Cambridge, Mass., 1967.
7. P. Griffith and J. D. Wallis, *Chem. Eng. Prog. Symp. Series,* vol. 30, p. 49, 1960.
8. A. E. Bergles and W. M. Rohsenow, *J. Heat Transfer, Trans. ASME,* **86**:365 (1964).
9. Y. Y. Hsu and R. W. Graham, *NASA Tech. Note* TNP-594, May 1961.
10. P. J. Marto and W. M. Rohsenow, *J. Heat Transfer, Trans. ASME,* **C88**:2, 183 (1966).
10a. D. B. R. Kenning and M. G. Cooper, Flow Patterns Near Nuclei and the Initiation of Boiling During Forced Convection Heat Transfer, *Boiling Symp., Inst. Mech. Engrs.,* vol. C180, London, Sept. 1965.
10b. J. C. Chen, Incipient Boiling Superheats in Liquid Metals, *J. Heat Transfer, Trans. ASME,* **C90**:303 (1968).
10c. C. W. Deane IV and W. M. Rohsenow, Mechanism and Behavior of Nucleate Boiling in Alkali Liquid Metals, *M.I.T. Heat Transfer Lab. Rept.* DSR 76303-65, Oct. 1969.
10d. R. M. Singer and R. H. Holtz, A Study of Incipient Nucleation of Liquid Sodium, *4th Int. Heat Transfer Conf.,* Versailles, France, Sept. 1970.
11. C. Y. Han and P. Griffith, Mechanism of Heat Transfer in Nucleate Pool Boiling, *M.I.T. Heat Transfer Lab. Rept.* 19, 1962.
12. R. Moissis and P. J. Berenson, Hydrodynamic Transitions In Nucleate Boiling, *J. Heat Transfer, Trans. ASME,* **C85**:221 (1963).
13. M. G. Cooper and J. P. Lloyd, Transient Local Heat Flux in Nucleate Boiling, *3d Int. Heat Transfer Conf.,* vol. 3, pp. 193–203, Chicago, Aug. 1966.
13a. M. G. Cooper, Microlayer and Bubble Growth In Nucleate Pool Boiling, *Int. J. Heat Mass Transfer,* **12**:895 (1969).
14. I. Shai, Mechanism of Nucleate Pool Boiling to Sodium, *M.I.T. Heat Transfer Lab. Rept.* 76303-45, Jan. 1967.
15. M. Cumo, Personal communication, C.N.E.N., C.S.N., Rome, Italy.
16. W. Fritz, Maximum Volume of Vapor Bubbles, *Phys. Z.,* **36**:379 (1935).
17. J. W. Wark, The Physical Chemistry of Flotation, I, *J. Phys. Chem.,* **37**:623 (1933).
18. L. E. Scriven, On the Dynamics of Phase Growth, *Chem. Eng. Sci.,* **10**:1 (1959).
19a. M. S. Plesset and J. A. Zwick, The Growth of Vapor Bubbles in Superheated Liquids, *J. Appl. Phys.,* **25**:493 (1954).
19b. H. K. Forster and N. Zuber, Growth of a Vapor Bubble in a Superheated Fluid, *J. Appl. Phys.,* **25**:474 (1954).
19c. B. B. Mikic, W. M. Rohsenow, and P. Griffith, On Bubble Growth Rates, *Int. J. Heat Mass Transfer,* **13**:657 (1970).
19d. Y. Lien and P. Griffith, "Bubble Growth at Reduced Pressure," doctoral dissertation, Massachusetts Institute of Technology, 1969.
19e. B. B. Mikic and W. M. Rohsenow, Bubble Growth Rates in Non-Uniform Temperature Fields, in "Progress in Heat and Mass Transfer," vol. 2, p. 283, Pergamon Press, 1969.
20. R. Cole and H. L. Shulman, Bubble Growth Rates at High Jakob Numbers, *Int. J. Heat Mass Transfer,* **9**:1377 (1966).
20a. R. Cole and W. M. Rohsenow, Correlation of Bubble Departure Diameters for Boiling of Saturated Liquids, *Chem. Eng. Prog. Symp. Series, AIChE,* vol. 65, pp. 92, 211–213, 1969.
20b. R. Cole, *AIChE Journal,* **13**:779 (1967).
20c. H. J. Ivey, Relationships Between Bubble Frequency, Departure Diameter and Rise Velocity In Nucleate Boiling, *Int. J. Heat Mass Transfer,* **10**:1023 (1967).
20d. R. Cole, Photographic Study of Boiling in Region of Critical Heat Flux, *AIChE Journal,* **6**:533 (1960).
20e. N. Koumoutsos, R. Moissis, and A. Spyridonos, Study of Bubble Departure In Forced Convection Boiling, *J. Heat Transfer, Trans. ASME,* **C90**:223 (1968).

21. N. Zuber, "Hydrodynamic Aspects of Boiling Heat Transfer," doctoral dissertation, University of California at Los Angeles, 1959.
22. B. E. Staniszewski, Nucleate Boiling Bubble Growth and Departure, *Tech. Rept.* 16, DSR 7673, Office of Naval Research Contract NONR-1841 (39), M.I.T. Heat Transfer Lab., Aug. 1959.
23. M. T. Cichelli and C. F. Bonilla, *Trans. AIChE,* **41**:755 (1945).
24. P. J. Marto and W. M. Rohsenow, Effects of Surface Conditions on Nucleate Pool Boiling of Sodium, *Trans. ASME,* **88**:196 (1966).
25. P. Berenson, "Transition Boiling Heat Transfer from a Horizontal Surface," doctoral dissertation, Massachusetts Institute of Technology, 1960; also *M.I.T. Heat Transfer Lab. Tech. Rept.* 17, March 1960.
26. W. Brown, "Study of Flow Surface Boiling," doctoral dissertation, Massachusetts Institute of Technology, 1967.
27. C. P. Costello and W. J. Frea, A Salient Non-hydrodynamic Effect on Pool Boiling Burnout of Small Semi-cylindrical Heaters, AIChE Preprints 15, *6th Nat. Heat Transfer Conf.,* Boston, 1963.
28. H. J. Ivey and D. J. Morris, The Effect of Test Section Parameters on Saturation Pool Boiling Burnout at Atmospheric Pressure, AIChE Preprint 160, Chicago, 1962.
29. E. A. Farber and R. L. Scorah, Heat Transfer to Water Boiling Under Pressure, *Trans. ASME,* **70**:369 (1948).
30. W. H. McAdams, W. E. Kennel, C. S. Minden, C. Rudolf, and J. E. Dow, Heat Transfer at High Rates to Water with Surface Boiling, *Ind. Eng. Chem.,* **41**:1945 (1959).
31. M. Jakob, *Mech. Eng.,* **58**:643 (1936).
32. P. J. Marto and W. M. Rohsenow, Nucleate Boiling Instability of Alkali Metals, *Trans. ASME,* **88**:183 (1966).
33. I. Shai, Mechanism of Nucleate Pool Boiling to Sodium, *M.I.T. Heat Transfer Lab. Rept.* 76303-45, Jan. 1967.
34. L. Bernath, Prediction of Heat Transfer Burnout, *Heat Transfer Symp.,* AIChE Nat. Meet., Louisville, Ky., 1955.
35. L. Bernath, A Theory of Local-Boiling Burnout and its Application to Existing Data, *Chem. Eng. Prog. Symp. Series,* vol. 56, no. 30, pp. 95–116, 1960.
36. C. Costello, C. O. Bock, and C. C. Nichols, A Study of Induced Convective Effects on Saturated Pool Boiling Burnout, *7th Nat. Heat Transfer Conf.,* AIChE Preprint No. 7, Cleveland, 1964.
36a. C. C. Pitts and G. Leppert, *Int. J. Heat Mass Transfer,* **9**:365 (1966).
36b. W. R. Houchin and J. H. Lienhard, Boiling Burnout in Low Thermal Capacity Heaters, *ASME Paper* 66-WA/HT-40, Nov. 1966.
37. L. Austin, *Mitt. Forsch,* **7**:75 (1903).
38. F. S. Pramuk and J. W. Westwater, *Chem. Eng. Prog. Symp. Series,* vol. 52, no. 18, p. 79, 1956.
39. A. E. Bergles and W. M. Rohsenow, *J. Heat Transfer, Trans. ASME,* **C86**:365 (1964).
40. E. E. Duke and V. E. Shrock, *Fluid Mech. Heat Transfer Inst.,* pp. 130–145, June 1961.
41. M. E. Ellion, A Study of the Mechanism of Boiling Heat Transfer, *Jet Propulsion Lab. Memo* 20-88, California Institute of Technology, 1954.
42. H. J. Ivey and D. J. Morris, On the Relevance of the Vapor-Liquid Exchange Mechanism for Subcooled Boiling Heat Transfer at High Pressure, *UK Rept.* AEEW-R-137, Winfrith, 1962.
43. S. S. Kutateladze, *Otdelenie Tekhnicheskikh Nauk,* **4**:529 (1951).
44. S. S. Kutateladze and L. L. Schneiderman, Experimental Study of Influence of Temperature of Liquid on Change in Rate, in "Problems of Heat Transfer during a Change of State," pp. 95–100, *USAEC Rept.* AEC-tr-3405, 1953.
45. M. Jakob and W. Fritz, *Forsch. Giebete. Ing.,* **2**:434 (1931).
46. E. K. Averin, *AERE Lib/Trans.* 562, from *Izvest. Akad. Nauk SSSR (OTN),* **3**:116 (1954).
47. M. Jakob and W. Linke, *Forsch. Giebete. Ing.,* **4**:75 (1933).
48. R. Siegel and C. Usiskin, Photographic Study of Boiling in Absence of Gravity, *J. Heat Transfer, Trans. ASME,* **81**:3 (1959).
49. J. Merte, Jr., and J. A. Clark, Study of Pool Boiling in an Accelerating System, *University of Michigan Rept.* 2646-21-T, Tech. Rept. 3, Nov. 1959.
50. R. J. Armstrong, Temperature Difference in Nucleate Boiling, *Int. J. Heat Mass Transfer,* **9**:1148 (1966).

51. W. M. Rohsenow, A Method of Correlating Heat Transfer Data for Surface Boiling of Liquids, *Trans. ASME,* **74**:969 (1952).

52. W. M. Rohsenow and J. A. Clark, "Heat Transfer and Pressure Drop Data for High Heat Flux Densities to Water at High Sub-Critical Pressures," Heat Transfer Fluid Mech. Inst., Stanford, Calif., 1951.

53. J. N. Addoms, "Heat Transfer at High Rates to Water Boiling Outside Cylinders," doctoral dissertation, Massachusetts Institute of Technology, 1948.

54. D. S. Cryder and A. C. Finalborgo, Heat Transmission from Metal Surfaces to Boiling Liquids: Effect of Temperature of the Liquid on Film Coefficient, *Am. Inst. Chem. Eng.,* 33:346 (1937).

55. E. L. Piret and H. S. Isbin, Two-phase Heat Transfer in Natural Circulation Evaporators, *AIChE Heat Transfer Symp.,* St. Louis, Dec. 1953.

56. M. T. Cichelli and C. F. Bonilla, *Trans. AIChE,* **41**:755 (1945).

57. K. Forster and N. Zuber, *AIChE Journal,* **1**:531 (1955).

58. K. Forster and R. Greif, Heat Transfer to a Boiling Liquid: Mechanism and Correlations, *UCLA Dept. of Eng. Prog. Rept.* 7, 1958, *J. Heat Transfer, Trans. ASME,* **C81**:43 (1959).

59. C. H. Gilmour, Nucleate Boiling—A Correlation, *Chem. Eng. Prog.,* **54**:77 (1958).

60. M. J. McNeilly, A Correlation of the Rates of Heat Transfer to Nucleate Boiling Liquids, *J. Imp. Coll. Chem. Eng. Soc.,* 7:18 (1953).

61. S. Levy, Generalized Correlation of Boiling Heat Transfer, *J. Heat Transfer, Trans. ASME,* **C81**:37 (1959).

61a. G. W. Wallis, *Int. Heat Transfer Conf.,* vol. 2, Boulder, Colo., Aug. 1961.

62. N. Zuber, Hydrodynamic Aspects of Boiling Heat Transfer, *USAEC Rept.* AECU-4439, 1959; doctoral dissertation, University of California at Los Angeles, 1959.

63. S. S. Kutateladze, Heat Transfer in Condensation and Boiling, *USAEC Rept.* AEC=tr-3770, 1952.

63a. S. S. Kutateladze, "Fundamentals of Heat Transfer," p. 362, Academic Press, Inc., New York, 1963.

63b. B. B. Mikic and W. M. Rohsenow, New Correlation of Pool Boiling Data Including the Effect of Heating Surface Characteristics, *J. Heat Transfer, Trans. ASME,* **91**:245 (1969).

63c. W. T. Brown, Jr., and A. E. Bergles, "A Study of Flow Surface Boiling," doctoral dissertation, Massachusetts Institute of Technology, 1967.

64. Y. P. Chang and N. W. Snyder, Heat Transfer in Saturated Boiling, *Chem. Eng. Prog. Symp. Series,* vol. 56, no. 30, pp. 25–38, 1960.

65. W. M. Rohsenow and P. Griffith, Correlation of Maximum Heat Transfer Data for Boiling of Saturated Liquids, *Chem. Eng. Prog. Symp. Series,* vol. 52, no. 18, p. 47, 1956.

66. J. H. Lienhard and V. E. Schrock, *J. Heat Transfer, Trans. ASME,* **C85**:265 (1963).

67. J. H. Lienhard and K. Watanabe, *J. Heat Transfer, Trans. ASME,* **C88**:94 (1966).

68. R. C. Noyes and H. Lurie, Boiling Sodium Heat Transfer, *3d Int. Heat Trans. Conf.,* vol. 5, pp. 92–100, Chicago, Aug. 1966.

69. R. E. Balzheiser et al., Investigation of Liquid Metal Boiling Heat Transfer, *Wright-Patterson AFB Rept.* RTD-TDR-63-4130, 1963.

69a. C. T. Sciance, C. P. Colver, and C. M. Sliepcevich, *Cryogenics Eng. Conf.,* Paper C-4, Boulder, Colo., 1966.

70. N. Zuber, M. Tribus, and J. W. Westwater, The Hydrodynamic Crisis in Pool Boiling of Saturated and Subcooled Liquids, in "International Developments in Heat Transfer," pt. II, pp. 230–236, ASME, 1961.

71. J. M. Adams, A Study of the Critical Flux in an Accelerating Pool Boiling System, NSF G-19697, University of Washington, 1962.

72. W. R. Beasant and H. W. Jones, The Critical Heat Flux in Pool Boiling under Combined Effect of High Acceleration and Pressure, *UK Rept.* AEEW-R 275, Winfrith, 1963.

73. C. P. Costello and W. J. Frea, A Salient Non-hydrodynamic Effect on Pool Boiling Burnout of Small semi-cylindrical Heaters, AIChE Preprint 15, *6th Nat. Heat Transfer Conf.,* Boston, 1963.

74. N. Zuber and E. Fried, Two Phase Flow and Boiling Heat Transfer to Cryogenic Liquids, Amer. Rocket Soc. Propellants, *Combustion & Liquid Rocket Conf.,* Arpil 1961.

75. R. J. Richards, W. G. Steward, and R. B. Jacobs, Survey of Literature on Heat Transfer to Cryogenic Fluics, *NBS TN* 122, Boulder Labs., Oct. 1961.

76. J. A. Clark, in "Cryogenic Technology" (R. W. Vance, ed.), chap. 5, John Wiley & Sons, Inc.,

New York, 1963; also in "Advances in Heat Transfer" (Hartnett and Irvine, eds.), vol. 5, pp. 325–505, Academic Press, Inc., New York, 1968.

77. E. G. Brentani and R. V. Smith, Nucleate and Film Pool Boiling Design Correlations for O_2, N_2, H_2 and He, in "Advances in Cryogenic Engineering: Proceedings" (K. D. Timmerhaus, ed.), vol. 10, pp. 325–341, Plenum Publishing Corp., New York, 1965.

78. D. N. Lyon, in "Advances in Cryogenic Engineering: Proceedings" (K. D. Timmerhaus, ed.), vol. 10, pp. 371–329, Plenum Publishing Corp., New York, 1965.

79. D. N. Lyon, P. G. Kosky, and B. N. Harman, in "Advances in Cryogenic Engineering: Proceedings" (K. D. Timmerhaus, ed.), vol. 9, pp. 77–87, Plenum Publishing Corp., New York, 1964.

79a. J. A. Clark, Cryogenic Heat Transfer, in "Advances in Heat Transfer" (Hartnett and Irvine, eds.), vol. 5, pp. 325–517, Academic Press, Inc., New York, 1968.

80. F. E. Tippets, Critical Heat Fluxes and Flow Patterns in High Pressure Boiling Water Flows, *ASME Paper* 62-WA-162, 1962.

81. S. S. Kutateladze, Boiling Heat Transfer, *Int. J. Heat Mass Transfer,* **4**:31 (1961).

82. W. H. Jens and P. A. Lottes, Analysis of Heat Transfer, Burnout, Pressure Drop and Density Data for High Pressure Water, *USAEC Rept.* ANL-4627, 1951.

83. R. J. Weatherhead, Nucleate Boiling Characteristics and the Critical Heat Flux Occurrence in Subcooled Axial-Flow Water System, *USAEC Rept.* ANL-6675, 1962.

83a. J. R. S. Thom, W. M. Walker, T. A. Fallon, and G. F. S. Reising, Boiling in Subcooled Water During Flow In Tubes and Annuli, *Proc. Inst. Mech. Engr.,* **3C180**:226 (1965–66).

84. H. Buchberg, F. E. Romie, R. Lipkis, and M. Greenfield, Heat Transfer, Pressure Drop, and Burnout Studies with and without Surface Boiling for Deaerated and Gassed Water at Elevated Temperatures in a Forced Flow System, *Proc. Heat Transfer Fluid Mech. Inst., 1951,* pp. 171–191, Stanford University Press, 1951.

85. S. W. Anderson, *J. Heat Transfer, Trans. ASME,* **82**:196 (1960).

86. G. F. Hewitt, H. A. Kearsey, P. M. C. Lacey, and D. J. Pulling, Burnout and Nucleation in Climbing Film Flow, *UK Rept.* AERE-R 4374, 1963.

87. J. A. R. Bennett, J. G. Collier, H. R. C. Pratt, and J. D. Thornton, Heat Transfer to Two-Phase Gas-Liquid Systems, Part I: Steam-Water Mixtures in the Liquid-Dispersed Region in an Annulus, *Trans. Inst. Chem. Engrs. (London),* **39**:113 (1961).

88. C. E. Dengler and J. N. Addoms, *Chem. Eng. Prog. Symp. Series,* vol. 52, no. 18, pp. 95–103, 1956.

89. R. M. Wright, Downflow Forced Convection Boiling of Water in Uniformly Heated Tubes, *USAEC Rept.* UCRL-9744, 1961.

90. J. G. Collier, P. M. C. Lacey, and D. J. Pulling, *Trans. Inst. Chem. Engrs.,* **42** (1964).

91. V. E. Shrock and L. M. Grossman, Forced Convection Boiling Studies, Forced Convection Vaporization Project, *USAEC Rept.* TID-14632, 1959.

92. S. A. Guerrieri and R. D. Talty, *Chem. Eng. Prog. Symp. Series,* vol. 52, no. 18, pp. 69–77, 1956.

93. V. E. Shrock and L. M. Grossman, Local Pressure Gradients in Forced Convection Vaporization, *Nucl. Sci. Eng.,* **3**:245 (1959).

94. Boiling Symposium, ASHRAE Semiannual Meeting, Houston, 1966.

95. J. B. Chaddock, Heat Transfer, in "Handbook of Fundamentals," chap. 3, pp. 52–53, ASHRAE, 1967.

96. E. R. Michaud and J. B. Chaddock, Forced Convection of Refrigerants, phase 2, *Purdue Univ. Herrick Lab. Rept.* HL-106, Aug. 1966.

97. J. C. Chen, A Correlation for Boiling Heat Transfer to Saturated Fluids in Convective Flow, *ASME Paper* 63-HT-34, 1963.

98. J. C. Chen, Existence of Non-Equilibrium, Inverse Temperature Profiles in Boiling Two-Phase Flow, *BNL* 9296, 1965.

99. B. Pierre, *Kylteknisk Tidskrift,* **3**:129 (1957); also *Svenska Flaktfabriken Review,* vol. 2, no. 1 (1955).

100. M. Altman, R. H. Norris, and F. W. Staub, *J. Heat Transfer, Trans. ASME,* **82**:189 (1960).

100a. P. Sachs and R. A. K. Long, Correlation for Heat Transfer in Stratified Two-Phase Flow with Vaporization, *Int. J. Heat Mass Transfer,* **2**:222 (1960).

101. M. Silvestri, in "Advances in Heat Transfer" (T. F. Irvine and J. P. Hartnett, eds.), vol. 1, Academic Press, Inc., New York, 1968.

102. S. Levy and A. P. Bray, Reliability of Burnout Calculations in Nuclear Reactors, *Nuclear News,* p. 3, Feb. 1963.

103. W. M. Rohsenow and J. A. Clark, Heat Transfer and Pressure Drop Data for High Heat Flux Densities to Water at High Sub-Critical Pressures, *Heat Transfer Fluid Mech. Inst.*, Stanford, Calif., 1951.

104. A. W. Bennett, J. G. Collier and P. M. C. Lacey, *AERE*-R3804, Aug. 1961.

105. M. Silvestri, *Int. Heat Transfer Conf.*, Paper 39, Boulder, Colo., 1961.

106. A. W. Bennett, J. G. Collier, and P. M. C. Lacey, *AERE*-R3934, part III, 1963.

107. S. Bertoletti, G. P. Gaspari, C. Lombardi, C. Peterlongo, M. Silvestri, and F. A. Tacconi, Heat Transfer Crisis in Steam-Water Mixtures, *Energia Nucleare*, 12 (1965).

108. I. T. Aladyev, Z. L. Miropolsky, V. E. Doroshchuk, and M. A. Styrikovich, in "International Developments in Heat Transfer," vol. 2, ASME, Boulder, Colo., 1961.

109a. L. S. Tong, "Boiling and Two-Phase Flow," p. 183, John Wiley & Sons, Inc., New York, 1965.

109b. p. 172.

109c. pp. 137–155.

109d. pp. 155–178.

109e. p. 188.

109f. pp. 194–195.

109g. pp. 130–134.

110. R. J. Weatherhead, Nucleate Boiling Characteristics and the Critical Heat Flux Occurrence in Subcooled Axial-Flow Water System, *USAEC Rept.* ANL 6675, 1962.

111. R. V. Macbeth, Forced Convection Burnout in Simple, Uniformly Heated Channels: A Detailed Analysis of World Data, *European Atomic Energy Society Symposium on Two-Phase Flow, Steady State Burnout and Hydrodynamic Instability,* Sweden, 1963.

112. I. T. Aladyev, Z. L. Miropolsky, V. E. Doroschuk, and M. A. Styrikovich, Boiling Crisis in Tubes, in "International Developments in Heat Transfer," pt. 2, pp. 237–243, ASME, 1961.

113. R. V. Macbeth, Burnout Analysis, Pt. 2, The Basic Burnout Curve, *UK Rept.* AEEW-R 167, Winfrith, 1963; also Pt. 3, The Low Velocity Burnout Regime, AEEW-R 222, 1963; Pt. 4, Application of Local Conditions Hypothesis to World Data for Uniformly Heated Round Tubes and Rectangular Channels, AEEW-R 267, 1963.

114. G. J. Kirby, Burnout in Climbing Film Two-Phase Flow—Theories, *UK Rept.* AEEW-R 470, 1960.

115. A. E. Bergles, "Forced Convection Surface Boiling Heat Transfer and Burnout in Tubes of Small Diameter," doctoral dissertation, Massachusetts Institute of Technology, 1962.

116. De Bertoli, S. J. Green, B. W. LeTourneau, M. Troy, and A. Weiss, Forced Convection Heat Transfer Burnout Studies for Water in Rectangular Channels and Round Tubes at Pressures above 500 psi, *WAPD*-188, Bettis Atomic Power Lab., 1958.

117. W. R. Gambill, Generalized Prediction of Burnout Heat Flux for Flowing Subcooled, Wetting Liquids, AIChE Preprint 17, *5th Nat. Heat Transfer Conf.*, Houston, 1962.

118. L. Bernath, A Theory of Local Boiling Burnout and its Application to Existing Data, *Chem. Eng. Prog. Symp. Series,* vol. 56, no. 30, pp. 95–116, 1960.

119. R. V. Macbeth, Burnout Analysis. Part 4: World Data for Uniformly Heated Round Tubes and Rectangular Channels, *AEEW-R* 267, Winfrith, 1963; also Part 3, *AEEW-R 222, 1963, and Part 2, AEEW-R* 167, 1963.

120. P. G. Barnett, An Investigation into the Validity of Certain Hypotheses Implied by Various Burnout Correlations, *AEEW-R* 214, 1963.

121a. L. S. Tong, H. B. Currin, and F. C. Engel, DNB (Burnout) Studies in an Open Lattice Core, *USAEC Rept.* WCAP-3736, 1964.

121b. L. S. Tong, H. B. Currin, and A. G. Thorp, New Correlations Predict DNB Conditions, *Nucleonics,* 21:43 (1963).

122. R. P. Stein and P. A. Lottes, Status of Boiling Burnout for Reactor Design, in "Selected Topics in Reactor Technique" (L. Link, ed.), *USAEC Rept.* TID-8540, 1964.

123. See Ref. 119.

124. H. Firstenberg et al., Compilation of Experimental Forced Convection Quality Burnout Data with Calculated Reynolds Numbers, *NDA*-2131-16, 1960.

125. A. W. Bennett, J. G. Collier, and P. M. C. Lacey, Part 3, *AERE*-3934, 1963.

126. N. A. Mozharov, An Investigation into the Critical Velocity at which a Moisture Film Breaks away from the Wall of a Steam Pipe, *Teploenerg,* 6:50 (1959), DSIR-trans—RTS-1581.

127. N. E. Todreas and W. M. Rohsenow, Effect of Non-Uniform Axial Heat Flux Distribution, *M.I.T. Rept.* 9843-37, Sept. 1965; also *3d Int. Heat Transfer Conf.*, ASME-AIChE, Chicago, vol. III, pp. 78–85, 1966.

128. G. J. Kirby, New Correlation of Non-Uniformly Heated Tubes Burnout Data, *AEEW-R*500, 1966.
129. N. E. Todreas, Effect of Local Flux Peaks on Critical Heat Flux in Annular Two-phase Flow, presented at American Nuclear Society Meeting, Pittsburgh, Nov. 1966.
130. K. M. Becker and P. Persson, Analysis of Burnout Conditions for Flow of Boiling Water in Vertical Round Ducts, AE-113, *Aktiebolaget Atomenergi*, Stockholm, Sweden (1963); see also AE-114 (1963).
131. W. S. Durant and S. Mirshak, Roughening of Surfaces to Increase Burnout Heat Flux, *USAEC Rept. DP*-380, 1959; and *DPST-60-284, 1960*.
132. E. Janssen, S. Levy, and J. A. Kervinen, Investigations of Burnout—Internally Heated Annulus Cooled by Water at 600 to 1450 psia, *ASME Paper* 63–WA-149, 1963.
133. W. R. Gambill and N. D. Greene, Boiling Burnout with Water in Vortex Flow, *Chem. Eng. Prog.,* **54**:68 (1958).
134. R. Viskanta, Critical Heat Flux for Water in Swirling Flow, *Nucl. Sci. Eng.,* **10**:202 (1961).
135. N. Zuber and M. Tribus, Further Remarks on the Stability of Boiling Heat Transfer, *UCLA Rept.* 58-5, 1958.
136. P. Berenson, Transition Boiling Heat Transfer from a Horizontal Surface, AIChE Paper 18, *ASME-AIChE Heat Transfer Conf.,* Buffalo, Aug. 1960.
137. L. A. Bromely, *Chem. Eng. Prog.,* **46**:221 (1950).
138. L. A. Bromley et al., *Ind. Eng. Chem.,* **45**:2639 (1953).
139. B. P. Breen and J. W. Westwater, *Chem. Eng. Prog.,* **58**:67 (1962); also AIChE Preprint 19, *5th Nat. Heat Transfer Conf.,* Houston, 1962.
140. J. H. Lienhard and P. T. Y. Wong, *ASME Paper* 63–HT-3, 1963.
141. T. H. K. Frederking and J. A. Clark, Natural Convection Film Boiling on a Sphere, in "Advances in Cryogenic Engineering: Proceedings" (K. D. Timmerhaus, ed.), vol. 8, Plenum Publishing Corp., New York, 1962.
142. E. P. Quinn, Forced Flow Heat Transfer Beyond the Critical Heat Flux, *ASME Paper* 66-Wa/Ht-36, Nov. 1966.
143. W. F. Laverty and W. M. Rohsenow, Film Boiling of Saturated Nitrogen Flowing in a Vertical Tube, *ASME Paper* 65-WA/HT-26.
144. R. Forslund and W. M. Rohsenow, Dispersed Flow Film Boiling, *M.I.T. Rept.* 75312-44, Jan. 1967.
145. M. Silvestri, Two Phase Annular Dispersed Flow, in "Advances in Heat Transfer" (T. F. Irvine, Jr. and J. P. Hartnett, eds.), vol. 1, pp. 355–446, Academic Press, Inc., New York, 1964.
146. G. Leppert and C. C. Pitts, Boiling, in "Advances in Heat Transfer" (T. F. Irvine, Jr. and J. P. Hartnett, eds.), vol. I, pp. 185–265, Academic Press, Inc., New York, 1964.
147. W. M. Rohsenow, Boiling Heat Transfer, in "Developments in Heat Transfer" (W. M. Rohsenow, ed.), pp. 169–260, The M.I.T. Press, Cambridge, Mass., 1964.
148. J. W. Westwater, Boiling of Liquids, in "Advances in Chemical Engineering" (T. B. Drew and J. W. Hoopes, Jr., eds.), vols. 1 and 2, Academic Press, Inc., New York, 1956, 1958.
149. O. E. Dwyer, Recent Developments in Liquid Metal Heat Transfer, *Atomic Energy Review,* vol. 4, no. 1, International Atomic Energy Agency, Vienna, 1966.

NOMENCLATURE

A	area of heating surface
A_b	surface area of a bubble
a_t	thermal diffusivity, $k/\rho c_p$
C_{sf}	coefficient of Eq. (16) which depends on the nature of the heating-surface-fluid combination
c_1	specific heat of saturated liquid, Btu/lbm-°F
D	tube diameter
D_b	diameter of the bubble as it leaves the heating surface, ft
D_e	equivalent diameter = 4(flow area)/perimeter
f	frequency of bubble formation, 1/hr
G	mass velocity in a channel or tube, lbm/hr-ft^2
G_b	mass velocity of bubbles at their departure from the heating surface, lbm/hr-ft^2

g	acceleration of gravity
g_0	conversion factor, 4.17×10^8 lbm-ft/hr²-lbf
$H, \bar{h}$	enthalpy, Btu/lb
h_{fg}	latent heat of evaporation, Btu/lbm
h	$(q/A)/\Delta T$, film coefficient of heat transfer, Btu/hr-ft²-°F
J	778 ft-lb/Btu
J_a	Jakob number, $\rho_l c_l (T_h - T_{sat})/\rho_v h_{fg}$
k_1	thermal conductivity of saturated liquid, Btu/hr-ft-°F
L	length
M	molecular weight
Nu	Nusselt number $= hd/k$
Nu_b	bubble Nusselt number
n	number of points of origin of bubble columns per ft² of heating surface
Pr_1	Prandtl number $= C_1 \mu_1/k_1$
p_c	critical pressure
p_L	pressure on liquid side of interface of radius r
p_v	pressure on vapor side of interface of radius r
$q/A, q''$	heat transfer rate per unit heating surface area, Btu/hr-ft²
$(q/A)_b$	heat transfer rate to bubble per unit heating surface while bubble remains attached to the surface, Btu/hr-ft²
R	gas constant; also bubble radius
Re_b	bubble Reynolds number
r	radial distance
T_b, T_{liq}, T_l	temperature of liquid, °F
T_w	temperature of heating surface, °F
T_{sat}, T_s	saturation temperature, °F
t	time
v	volume, ft³/lb
W, w	flow rate, lb/hr
X_{tt}	two-phase flow parameter, defined in Eq. (36)
z	depth
β	bubble contact angle, defined in Fig. 2
$\Delta T_{subcool}$	$T_{sat} - T_{liq}$
α	thermal diffusivity, ft²/hr
μ_l	viscosity of saturated liquid, lbm/ft-hr
ρ	density of saturated liquid, lbm/ft³
ρ_v	density of saturated vapor, lbm/ft³
σ, σ_{lv}	surface tension of liquid-vapor interface, lbf/ft
σ_{sl}	surface tension of solid-liquid interface, lbf/ft
σ_{vs}	surface tension of vapor-solid interface, lbf/ft
ν	kinematic viscosity, μ/ρ

Two-phase Flow

PETER GRIFFITH
Professor of Mechanical Engineering
Massachusetts Institute of Technology
Cambridge, Massachusetts

VOID FRACTION, QUALITY, AND SLIP RELATIONS

In general, when two phases flow in a pipe they do not move at the same velocity. The continuity equation for each phase fixes a relation between the phase velocities V_f and V_g, the void fraction α, and the flowing quality x. This relation is as follows. First, the definitions of the phase velocities

$$V_f = \frac{Q_f}{A(1 - \alpha)} \tag{1}$$

$$V_g = \frac{Q_g}{A(\alpha)} \tag{2}$$

Using these definitions it is possible to develop the following relationship between the velocity ratio, the void, and the quality

$$\left(\frac{1 - \alpha}{\alpha}\right) = \left(\frac{V_g}{V_f}\right)\left(\frac{\rho_g}{\rho_f}\right)\left(\frac{1 - x}{x}\right) \tag{3}$$

OVERALL PRESSURE DROP AND VOID CALCULATION METHODS

The pressure drop in a straight pipe is the result of the action of three factors: the wall friction force, the gravity force, and momentum changes. The general equation for pressure drop in a straight pipe is

$$\Delta P = \Delta P_f + \Delta P_g + \Delta P_m \tag{4}$$

Three overall methods will be presented for calculating the terms in the above equation: the homogeneous method, the Thom Method, and the Martinelli method. The recommendations when to use each of these calculation methods are as follows:

Homogeneous Method. Use this method where simplicity and an analytical expression are needed. Precision is low for low velocities or low pressure where the gravity contribution to the pressure drop is large. Precision improves for high quality and high pressure. The homogeneous method is used as the basis of the calculation of pressure drop for fittings.

Thom Method. The Thom method is most accurate and convenient for adiabatic and nonadiabatic systems involving water, or other fluids less viscous than water. Mass velocities should be greater than 0.5×10^6 lb/hr-ft^2 as all the data used to establish the correlation were taken at mass velocities above this. For systems involving water above a pressure of 250 psia, this method gives the highest precision.

Martinelli Method. For adiabatic, low-pressure systems involving fluids other than water, the Martinelli method is recommended. The bulk of the data behind it were taken in unheated horizontal pipes with a very wide variety of liquids at about one atmosphere pressure with air as the gas phase.

Homogeneous Method, Adiabatic Pipes

The density of a homogeneous two-phase mixture is calculated assuming the velocity of each phase is the same. That is

$$v = \frac{1}{\rho} = v_f + x v_{fg} \tag{5}$$

The friction pressure drop is determined from

$$\Delta P_f = 4f\left(\frac{L}{D}\right)\frac{G^2 v}{2g_0} \tag{6}$$

The specific volume is calculated from Eq. (5). The mass velocity is calculated from the total flow rate and pipe cross-sectional area. Thus

$$G = \frac{w}{A} \tag{7}$$

The friction factor is determined from the Moody curve in Sec. 7 assuming pure liquid is flowing in the pipe at the mixture mass velocity. The friction factor so determined will generally be low at qualities below 70 percent and will be too high at qualities above 70 percent.

The gravity pressure drop is calculated from

$$\Delta P_g = \rho \left(\frac{g}{g_0}\right) L \sin\theta \tag{8}$$

The density ρ is evaluated from Eq. (5). The angle θ is measured from the horizontal.

The momentum pressure drop is a result of either a pressure drop or heat transfer giving rise to a change in the volume flow rate. It is

$$\Delta P_m = \frac{G^2}{g_0} (v_2 - v_1) \tag{9}$$

The first term in the bracket is the specific volume at the discharge, the second term is the specific volume at the entrance. Both are evaluated using Eq. (5).

Heated Pipes. In principle, it is possible to integrate Eqs. (6) and (8) and evaluate (9) for any distribution of heat flux in a heated pipe. In practice, however, two factors make this difficult. First, the pressure drop is often large enough so that the drop static pressure causes an appreciable amount of flashing. Second, the functional relations arising from the integration are so inconvenient that it is inconvenient to work them out numerically. Both these difficulties can be circumvented by dividing the pipe up into a series of short sections and using average values of the quality for each. Sum the three pressure drop contributions for each section of the pipe and keep a running trace of the local pressure. With this calculation scheme Eqs. (6), (8), and (9) are very easy to use.

The Thom scheme, which is described in the next section, is easier to use than the homogeneous method on heated pipes. However, at low pressures where Thom does not apply, the homogeneous method is often the most convenient. A check is necessary, however, to see if two-phase choking occurs. If the drop in pressure is more than 30% of the initial pressure, it is recommended that the methods for checking critical flows described later in this section be used to see that no choking does actually occur. If the check indicates that choking does occur, the assumed flow rate is too high.

Method of Thom

Thom's calculation method [1] represents an improvement over the homogeneous method, in two respects. First, the velocities of the two phases are allowed to differ, with the ratio of the velocities being a function of pressure alone. Second, the friction factor is allowed to become a function of quality. Both the velocity ratio variation and the friction factor variation are determined from data taken on heated and adiabatic steam water systems in pipes between 0.2 and 2 inches in diameter, at all pressures, and orientated both horizontally or vertically. The mass velocity is above 0.5×10^6 lbm/ft^2. In this section the method of computing the pressure drop for unheated pipes will be given first. Then the method for computing uniformly heated pipes will be given. The basic equation for evaluation of the pressure drop is still Eq. (4).

Adiabatic Pipes. The friction pressure drop is evaluated by computing that which would occur if saturated liquid were flowing alone at the *mixture* mass velocity and multiplying by a dimensionless multiplier. This is shown in Eq. (10). The multiplier r_f is given as a function of pressure and quality in Fig. 1.

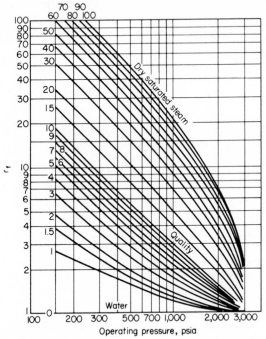

Fig. 1. Friction multiplier for substitution in Eq. (10). This curve is for unheated tubes with steam and water flowing at the indicated pressure [1].

$$\Delta P_f = 4f\left(\frac{L}{D}\right)\left(\frac{G^2 v_f}{2g_0}\right) r_f \tag{10}$$

The friction factor in Eq. (10) is evaluated from the Moody graph in Sec. 7.

In order to calculate the gravity pressure drop it is necessary to know the void fraction. The velocity ratio is given in Fig. 2. Equation (3) relates the velocity ratio to the void. The void, in turn, is used in Eq. (11) to determine the gravity pressure drop. The equation is

$$\Delta P_g = [\rho_f(1 - \alpha) + \rho_g(\alpha)] L \frac{g}{g_0} \sin\theta \tag{11}$$

Figure 3 shows the void fraction to be used in Eq. (11) while the curve of Fig. 2 shows the corresponding velocity for several different pressures (to be used in Eq. (3)).

The acceleration pressure drop is evaluated using Eq. (9). For unheated pipes it is normally small.

Heated Pipes. For uniformly heated pipes, starting with saturated liquid at the entrance, the friction contribution to the pressure drop has been integrated and is presented in terms of a multiplier. The multiplier is a function of quality and pressure for water. It is obtained from Fig. 4 and is used in Eq. (12)

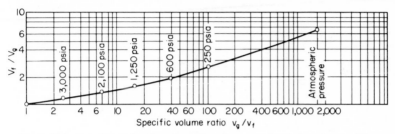

Fig. 2. Velocity ratio as a function of specific volume ratio for steam and water [1].

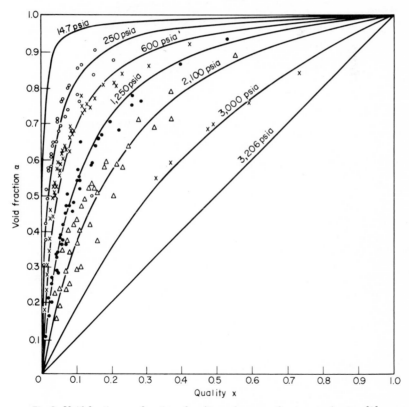

Fig. 3. Void fraction as a function of quality and pressure for steam and water. [1].

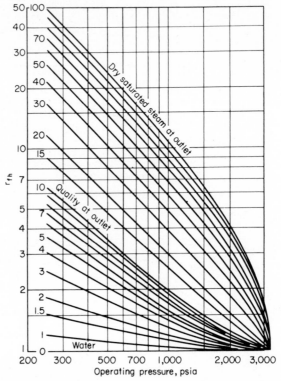

Fig. 4. Friction multiplier for heated tubes for substituting in Eq. (12). The heat flux is uniform and saturated liquid enters the test section [1].

$$\Delta P_{fh} = 4f\left(\frac{L}{D}\right)\frac{G^2 v_f}{2g_0} r_{fh} \tag{12}$$

The friction factor is evaluated from the Moody graph in Sec. 7 assuming saturated water is flowing at the mixture mass velocity.

The gravity contribution is evaluated assuming saturated water is flowing at the mixture mass velocity and using the multiplier obtained from Fig. 5 in Eq. (13)

$$\Delta P_{gh} = \rho_f \frac{g}{g_0} L(\sin\theta) r_{gh} \tag{13}$$

The angle θ is measured from the horizontal. The gravity pressure drop is positive for up-flow and negative for down-flow.

The momentum pressure drop is evaluated from Eq. (9) [3].

The Thom calculation method, for heated pipes in particular, is so convenient to use that it is desirable to be able to extend it to other fluids. For this reason, it is suggested that instead of pressure as the parameter on Figs. 1 through 5, the density ratio be used as the independent variable. For liquids much more viscous than cold water, this cannot be correct, so that the Martinelli method, given next, is recommended for these.

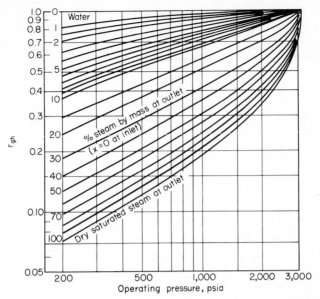

Fig. 5. Gravitational multiplier for heated tubes for substitution into Eq. (13). The heat flux is uniform and saturated liquid enters the test section [1].

The Thom calculation method for water above 200 psia and above a mass velocity of 0.5×10^6 lb/hr-ft² predicts the overall pressure drop to an accuracy of $\pm 20\%$.

The Martinelli Calculation Technique for Adiabatic Pipes

The Martinelli [2] two-phase pressure drop calculation procedure has been supplanted at high pressure by that of Thom. At pressures close to one atmosphere pressure for a very wide variety of liquids, the Martinelli procedure is still the best. The bulk of the data behind it was taken in horizontal pipes between 0.06 and 1.0 inches in diameter at about one-atmosphere pressure. The basic pressure drop equation is still Eq. (4).

The friction pressure drop is determined by first calculating the pressure drop for either the liquid flowing alone, or the gas flowing alone in the pipe at its mass velocity. That is

$$\Delta P_{fa} = 4f\left(\frac{L}{D}\right)\frac{G_f^2 v_f}{2g_0} \tag{14}$$

$$\Delta P_{ga} = 4f\left(\frac{L}{D}\right)\frac{G_g^2 v_g}{2g_0} \tag{15}$$

The two-phase friction pressure drop is then determined by multiplying either the liquid-only pressure drop, or the gas-only pressure drop, by the appropriate multiplier ϕ_f^2 or ϕ_g^2. This multiplier is given in Fig. 6 as a function of the variable X. The following procedure is used to determine the variable X.

Determine the Reynolds numbers for the liquid flowing alone and for the gas flowing alone in the pipe, each at its own mass velocity. These Reynolds numbers are defined as

$$N_{Ref} = \frac{G_f D}{\mu_f} \tag{16}$$

$$N_{Reg} = \frac{G_g D}{\mu_g} \tag{17}$$

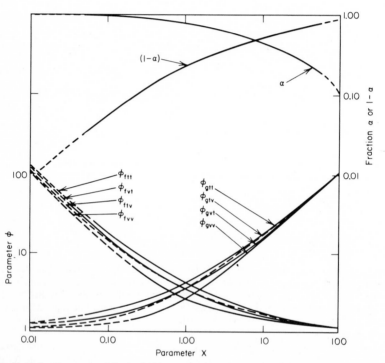

Fig. 6. Void fraction and adiabatic friction multipliers for all fluids at about one atmosphere pressure [2].

Go to Table 1 and determine which of the four combinations of viscous or turbulent flow for each phase exists for your problem. Go then to Table 2 and determine the coefficients for substituting into Eq. (18)

$$X^2 = \frac{N_{Reg}^m}{N_{Ref}^n} \; \frac{C_f}{C_g} \left(\frac{w_f}{w_g}\right)^2 \frac{\rho_g}{\rho_f} \tag{18}$$

Finally, go to Fig. 6 and determine the multiplier. For example, if one uses the liquid as the basis for the pressure drop calculation, and both N_{Ref} and N_{Reg} are turbulent, the friction pressure drop will be

$$\Delta P_f = \Delta P_{fa} \Phi_{ftt}^2 \tag{19}$$

TABLE 1

Flow Mechanism	Symbol	N_{Ref} is	N_{Reg} is
Liquid Gas			
Turbulent – Turbulent	t - t	>2000	>2000
Viscous – Turbulent	v - t	<1000	>2000
Turbulent – Viscous	t - v	>2000	<1000
Viscous – Viscous	v - v	<1000	<1000

For the other three possibilities of vt, tv, or vv, one simply uses the appropriate multiplier in an equation similar to Eq. (19). In general it is best to base the pressure drop calculation on phase which has the largest pressure drop when flowing alone in the pipe.

TABLE 2

	t - t	v - t	t - v	v - v
n	0.2	1	0.2	1.0
m	0.2	0.2	1.0	1.0
C_f	0.046	16	0.046	16
C_g	0.046	0.046	16	16

The gravity contribution is evaluated from Eq. (20) and Fig. 6

$$\Delta P_g = [(\alpha)\rho_g + (1 - \alpha)\rho_f] \frac{g}{g_0} L \sin\theta \tag{20}$$

For determining the momentum pressure drop it is recommended again that Eq. (9) be used.

Heated Pipes. The Thom calculation method is most easily used for calculating heated pipes. For low pressures and fluids appreciably different from water, the Martinelli curves should be used. It is suggested that the pipe be broken up into a series of short sections and the separate contributions to the pressure drop of Eqs. (9), (19), and (20) be evaluated. An average quality for each section can be used in the calculation and the result summed up. If there is an appreciable drop in pressure, the true specific volumes and qualities should be used in Eqs. (9), (19), and (20). If more than 30% of the initial pressure is lost, choking might have occurred and the assumed initial pressures and flow rates are impossible. For water, the methods given at the end of this section should be used to see if it has actually occurred.

FITTINGS

It is necessary to have calculation methods for handling pressure drop in fittings as well as straight pipe. Recommendations are based on data taken from a variety of sources, but here primarily Ref. 4 is used.

Pressure losses in fittings are determined experimentally by setting the fitting in question up with long approach and discharge pipes. These pipes are fitted with a large number of manometers. Upstream and downstream of the fitting, where the flow is fully developed, the pressure distribution is measured. This pressure distribution is extrapolated back to the fitting, and the resulting step in pressure is the loss ascribed to the fitting. A schematic of the experiment is shown in Fig. 7 and some actual pressure-length traces for an expansion [5] are shown in Fig. 8.

The measured pressure losses in fittings have been correlated by expressing losses in velocity heads and using the homogeneous model as the basis of the calculation

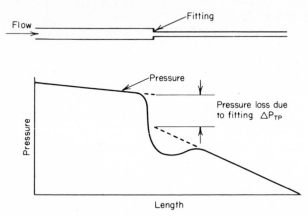

Fig. 7. Pressure as a function of length for a fitting, in this case a contraction. The difference between the extrapolated, fully developed, pressure gradients defines the pressure loss due to the fitting.

scheme. It is desirable to use velocity heads rather than equivalent lengths, as roughness should not enter into pressure drop for fittings. The homogeneous model is used as the basis of the calculation scheme for simplicity.

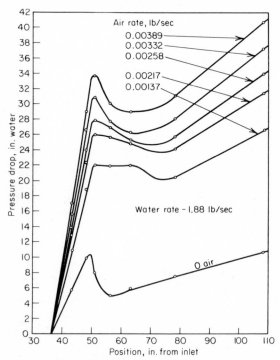

Fig. 8. Pressure-length curves for a sudden expansion, showing the extent of the developing region with two phases flowing. This expansion takes place in a horizontal channel $0.25'' \times 2''$ to $0.5'' \times 2''$ with air and water at one atmosphere [5].

In a typical experimental test set-up, a preheater is used to give the desired quality. The liquid flow rate is fixed and the extrapolated pressure drop across the fitting determined. The preheater is turned on to give the desired quality and the pressure drop again determined. In this way the ratio between the two-phase pressure drop and the saturated-liquid-only pressure drop can easily be determined.

The calculation procedure to determine the pressure loss in a fitting is as follows. Calculate the pressure drop for saturated liquid flowing alone in the fitting at the mixture mass velocity. (See Sec. 7.) Determine the two-phase pressure drop from

$$\frac{\Delta P_{tp}}{\Delta P_{fo}} = \left(1 + C \; x \; \frac{v_{fg}}{v_f}\right) \tag{21}$$

In this equation, x is the quality and C is a dimensionless coefficient to be obtained from Table 3. If the homogeneous theory were perfect for determining the pressure drop, C would be one. It is recommended that outside the region covered by Table 3 C be taken as one. In treating the data this way, the orientation of the fitting has not been found to affect pressure drop.

TABLE 3.

Fitting	Pressure	Quality range	C
Bend—short radius	$P/P_{cr} < 0.015$	0-10%	1.5
short radius	$P/P_{cr} > 0.1$	0-50%	4
long radius	$P/P_{cr} > 0.1$	0-50%	2.2
Tee (serving as bend)	$P/P_{cr} > 0.1$	0-50%	1.6
Gate valves	$P/P_{cr} > 0.1$	0-50%	1.5
Globe valves	Not to be used in two-phase piping		
Contractions	$0.015 < P/P_{cr} < 1$	0-50%	1.0
Expansions	$0.015 < P/P_{cr} < 1$	0-50%	1.1
Orifices—short			
($L/D < 0.5$)	$P/P_{cr} < 0.1$	0-50%	0.8

It should be pointed out that entrance lengths for two-phase flow are much larger than for single-phase flow in the same fitting. This can be seen on Fig. 8 by comparing the lower, single-phase curve with the other curves. When a series of fittings are used in a piping tree with insufficient length between them to insure that the flow will be fully developed, the total pressure loss is less than it would be if the flow had a chance to become fully developed.

EFFECTS DEPENDENT ON FLOW REGIMES

The pressure drop calculation procedures given in the first section all ignore the flow regime. In most cases this is satisfactory as only an overall answer is needed. However, at low velocity in vertical and inclined pipes, the assumptions embodied in the homogeneous, Thom, and Martinelli procedure are not adequate. In particular, the assumption that the density is a function of quality and pressure alone becomes a poor one. In the limit, it is easy to see this is the case, if one considers the problem of bubbling gas through a stagnant liquid. In this case the quality "x" is 100 percent while the pressure drop is virtually equal to the head of liquid through which bubbling is taking place. The two-phase pressure drop evaluated, using any of the above three procedures, would be seriously in error. In this section some flow regime maps will be presented. After that some recommendations will be made for evaluating the void in the low velocity, slug, and bubbly flow regimes. After that, brief mention will be made of stratified flow in inclined pipes, and roughness effects.

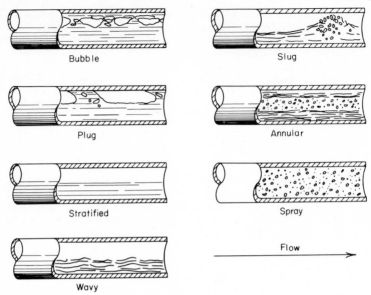

Fig. 9. Flow regimes in horizontal pipes [6].

Flow Regime Maps

The boundaries between flow regimes are not sharp and the pictures used to describe them represent idealized descriptions of a very complex distribution of phases.

Figure 9 shows the flow regimes which have been identified by Baker [6] in a horizontal pipe, and Fig. 10 is the flow regime map. The slug and plug regimes are

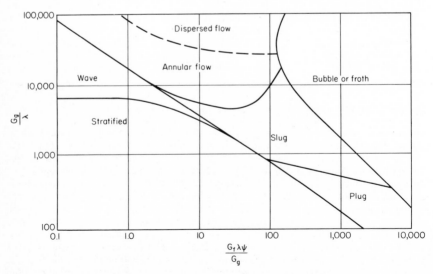

Fig. 10. Flow regime map for a horizontal pipe [6].

very unsteady with large fluctuations in pressure and void while the other flow regimes are steady.

The flow regimes that have been identified in a vertical pipe with up-flow are illustrated in Fig. 11. Figure 12 is a flow regime map due to Kosterin [7], for air and water. Figure 13 is a flow regime map due to Hewitt [8] for steam and water at 1,000 psia. It is suggested that for other fluids in vertical pipes these maps be redrawn using the Baker coordinates.

Bubbly and Slug-flow Void at Low Velocity

At mixture velocities which are comparable to the bubble rise velocity, the expressions for void given by the homogeneous, Thom, and Martinelli methods can be

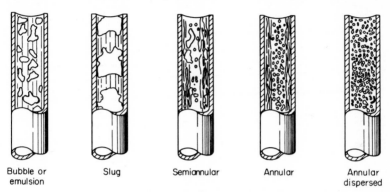

| Bubble or emulsion | Slug | Semiannular | Annular | Annular dispersed |

Fig. 11. Flow regimes for vertical up-flow.

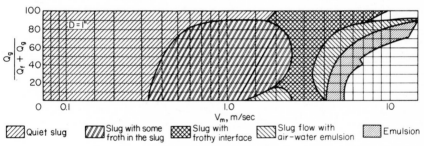

Fig. 12. Flow regime map for vertical up-flow of air and water in a 1-inch pipe at one atmosphere [7].

seriously in error. Figure 14 shows the rise velocities of single bubbles in a large volume of liquid [9]. In tubes, the bubble rise velocity depends on the tube diameter and properties in a complex way. Figure 15 [10] is a composite of bubble rise velocity for all wetting fluids in circular tubes of any diameter. For liquids no more viscous than cold water, in tubes greater than one-half inch in diameter, the upper asymptote is appropriate so that

$$V_b = 0.35\sqrt{gD} \qquad (22)$$

Both Eq. (22) and the data of Fig. 14 give bubble rise velocities of approximately 1 ft/sec.

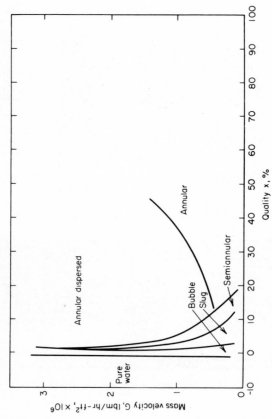

Fig. 13. Flow regimes for steam and water at 1,000 psia in vertical up-flow in a pipe 0.5″ diameter [8].

On the basis of void data taken in pipes ranging from 0.5 to 18 inches in diameter, and at pressures up to 2,000 psia, in water and several other fluids, it is suggested that bubbly and slug-flow void be calculated using [11]

$$\alpha = \frac{Q_g}{1.2(Q_f + Q_g) + (0.35\sqrt{gD})A} \tag{23}$$

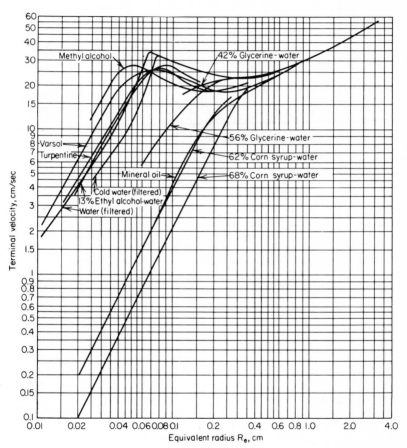

Fig. 14. Bubble rise velocities taken for air bubbles rising in a large tank of stagnant liquid for a variety of liquids [9].

Inclined Pipes

Inclined pipes with up-flow at low velocities and qualities experience either stratified or slug flow. In heated tubes this can give rise to overheating on the top of the tube and failure due to either melting or thermal fatigue. To avoid this, it is recommended [12] that the mass velocity in inclined heated tubes be kept above 0.5 x 10^6 lbm/hr-ft^2 and the angle of inclination be kept above 10 degrees from the horizontal.

Roughness Effects on Two-phase Flow Pressure Drop

The effects of pipe roughness on two-phase pressure drop depend on flow regimes. For bubbly and slug flow, where the liquid is continuous, the effects are the same as for single-phase flow. For annular flow the pressure drop is largely unaffected by the roughness. For very rough pipes at fixed gas flow rate, the pressure drop is sometimes

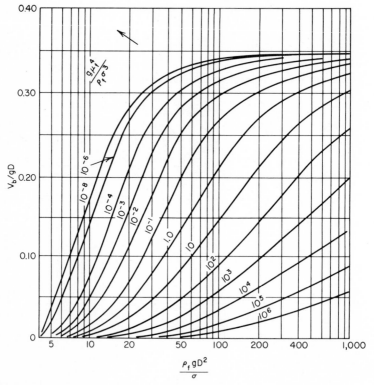

Fig. 15. Bubble rise velocities in vertical tubes as a function of the liquid properties [10].

actually reduced by the addition of liquid. Under the circumstances, it is suggested that the smooth pipe friction factor always be used for two-phase pressure-drop calculation whether the pipe is rough or not.

THE CRITICAL FLOW OF A TWO-PHASE MIXTURE

The critical flow of a two-phase mixture is defined as the flow which takes place when a drop in the discharge plenum pressure for the pipe no longer results in an increase in the flow rate through the pipe. It does not bear a simple relation to the velocity of sound in the mixture. Figure 16 is an illustration of the kind of apparatus in which critical flow measurements are made along with a typical pressure profile. In general, the pressure in the receiving plenum is below the pressure of the end of the tube.

Unlike the critical flow of a single phase, the critical flow rate for two phases depends on more than the thermodynamic state of the system. It depends on the flow regime also. In addition, for short tubes, one finds substantial departures from thermal equilibrium. Because of these complications, the critical flow of a saturated,

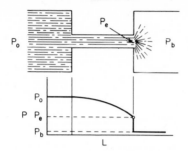

Fig. 16. Schematic of the critical flow experiment showing the location of the various pressures.

single component, adiabatic flashing mixture is divided into three regions. Following Fauske [13], the following procedure is recommended.

$L/D = 0$ **(Orifices).** There is no critical flow and the flow rate is computed from Eq. (24). This equation is simply the equation for the flow through a sharp-edged orifice

$$G = 0.61 \sqrt{2g_0(P_0 - P_b)\rho_f} \tag{24}$$

$0 < L/D < 12$ **(Nozzles and Short Pipes).** Use Fig. 17 and Eq. (24) with P_b replaced with P_c from Fig. 17. There is some flashing in this region with substantial departures from thermal equilibrium.

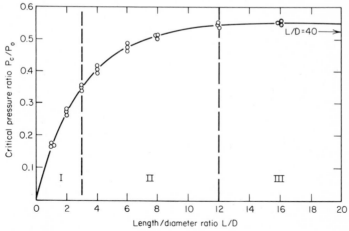

Fig. 17. Critical pressure ratio as a function of L/D for water at all pressures. Saturated water enters the tube [13].

$L/D > 12$ **(Pipes).** Use Fig. 18 for water along with Fig. 17 to relate the critical pressure to the initial pressure. As the pipes get longer the ratio P_c/P_0 drops slowly due to wall friction, but no reliable data has been taken for water at the longer L/D's. In this region there is a close approach to thermodynamic equilibrium with an annular

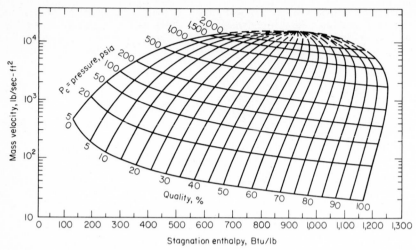

Stagnation enthalpy, Btu/lb

Fig. 18. Critical mass velocity as a function of stagnation enthalpy for water in pipe of all sizes [13].

or annular-dispersed flow regime and appreciable slip between the phases. The recommended techniques have been proven for water on pressures from a little below atmospheric up to 600 psia in pipes of diameter from 0.25 to 9 inches.

Bolstad [14] reports some critical flow measurements from Refrigerant 12 flashing in a long capillary tube. Figure 19 shows one typical pressure and temperature profile. Figure 20 shows the variation of critical flow rate with discharge pressure P_b. The critical flow rate is a weak function of L/D from the point at which the Refrigerant

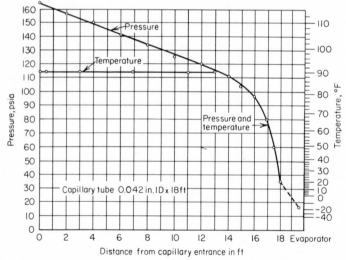

Distance from capillary entrance in ft

Fig. 19. Temperature and pressure as a function of length for a tube with Refrigerant 12 flashing. Tube diameter is 0.042 inches. Vaporization begins at 13 feet where the pressure drops below the saturation temperature [14].

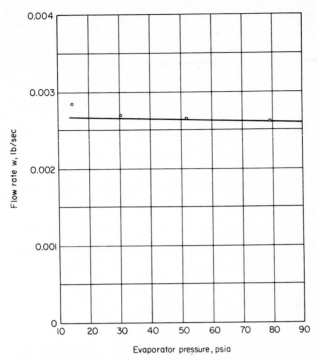

Fig. 20. Critical flow rate as a function of the discharge plenum pressure showing the flow rate is almost independent of the discharge plenum pressure. The fluid is Refrigerant 12 in a 12-foot tube, 0.042 inches in diameter. At the inlet the temperature is 90°F and the pressure is 140 psia [14].

12 drops to the saturation pressure to the discharge and a much stronger function of the inlet temperature. For fixed vaporizing L/D and inlet temperature, the critical mass velocity is almost independent of the tube diameter.

REFERENCES

1. J. R. S. Thom, Prediction of Pressure Drop During Forced Circulation Boiling of Water, *Int. J. Heat Mass Transfer,* **7**:709 (1964).
2. R. W. Lockhart and R. C. Martinelli, Proposed Correlation of Data for Isothermal Two-Phase Two-Component Flow in Pipes, *Chem. Eng. Prog.,* **45**:39 (1947).
3. G. B. Andeen and P. Griffith, Momentum Flux in Two-Phase Flow, *ASME Paper* 67-HT-32.
4. D. E. Fitzsimmons, Two-Phase Pressure Drop in Piping Components, HW-08970 Rev. 1, March 20, 1964. (A report by GE for the AEC available from the Office of Technical Services, Department of Commerce.)
5. B. L. Richardson, Some Problems in Horizontal Two-Phase Two-Component Flow, ANL-5949, December, 1952.
6. O. Baker, Designing Pipelines for Simultaneous Flow of Oil and Gas, Handbook Section, "Pipeline Engineer," February 1960, PH69.
7. S. I. Kosterin, Study of Influence of Tube Diameter and Position upon Hydraulic Resistance and Flow Structure of Gas Liquid Mixtures, *Izvestiya Akad. Nauk SSSR Otdelenie Tekhnicheskikh. Nauk,* **12**:1824 (1949).
8. A. W. Bennett, C. F. Hewitt, H. A. Kearsey, R. K. F. Kays, and P. M. C. Lacey, Flow Visualisation Studies of Boiling at High Pressure, *AERE* R-4874, Harwell, England, 1965.
9. W. F. Haberman and R. K. Morton, An Experimental Investigation of the Drag and Shape of

Air Bubbles Rising in Various Liquids, David W. Taylor Model Basin Report No. 802, Sept. 1953.

10. E. T. White and R. H. Beardmore, The Rise Velocity of Single Cylindrical Bubbles through Liquids Contained in Vertical Tubes, *Chem. Eng. Sci.,* **17**:351 (1962).

11. N. Zuber and J. A. Findlay, Average Volumetric Concentration in Two-Phase Flow Systems, *J. Heat Transfer, Trans. ASME,* **C87**:453 (1965).

12. M. A. Styrikovich and Z. L. Miropol'skii, The Temperature Regime for Operational Horizontal and Inclined Steam Generating Tubes at High Pressure, in "Hydrodynamics and Heat Transfer During Boiling in High Pressure Boilers" (M. A. Styrikovich, ed.). Also translated in AEC-tr-4490.

13. H. K. Fauske, "The Discharge of Saturated Water Through Tubes," Chem. Eng. Prog. Symp. Series, **59**:210 (1965).

14. M. M. Bolstad and R. C. Jordan, Theory and Use of the Capillary Tube Expansion Device, *J. ASRE,* **6**:518 (1948).

NOMENCLATURE

A	area, ft^2
C	dimensionless constant defined in Eq. (21) and tabulated in Table 3
$\left.\begin{array}{l}C_f \\ C_g\end{array}\right\}$	dimensionless constants from Table 2 to be substituted into Eq. (18)
D	diameter, ft
f	friction factor, dimensionless
g	acceleration of gravity, ft/hr^2
G	mass velocity defined in Eq. (7), lbm/hr-ft^2
G_f	liquid mass velocity, lbm/hr-ft^2
G_g	gas phase mass velocity, lbm/hr-ft^2
g_0	gravitational constant, ft-lbm/hr^2-lbf
L	tube length, ft
$\left.\begin{array}{l}m \\ n\end{array}\right.$	dimensionless constants from Table 2 to be substituted into Eq. (18)
N_{Ref}	dimensionless Reynolds number defined in Eq. (16)
N_{Reg}	dimensionless Reynolds number defined in Eq. (17)
P	pressure, lbf/ft^2
P_b	pressure in the receiving plenum in the critical flow experiment, Fig. 16, lbf/ft^2
P_c	critical pressure in the critical flow experiment, lbf/ft^2
P_{cr}	thermodynamic critical pressure, lbf/ft^2
P_e	pressure just inside the tube in the critical flow experiment of Fig. 16, lbf/ft^2
P_0	stagnation pressure in the critical flow experiment, lbf/ft^2
Q_f	liquid volume flow rate, ft^3/hr
Q_g	gas phase volume flow rate, ft^3/hr
R_e	equivalent radius of a bubble, equal to the radius of the spherical bubble of the same volume, cm
r_f	dimensionless friction multiplier for adiabatic pipes with water flowing to be evaluated from Fig. 1 and substituted in Eq. (10)
r_{fh}	dimensionless friction multiplier for uniformly heated pipes entering as saturated liquid for substitution into Eq. (12)
r_{gh}	dimensionless gravity multiplier for uniformly heated pipes entering as saturated liquid for substitution into Eq. (13)
v	specific volume, ft^3/lbm
V_b	bubble rise velocity, ft/hr
V_f	true velocity of the liquid phase, Eq. (1), ft/hr
V_g	true velocity of the gas phase, Eq. (2), ft/hr

v_f specific volume of the liquid, ft³/lbm

v_{fg} $v_g - v_f$ at saturation conditions, ft³/lbm

v_g specific volume of the gas phase, ft³/lbm

v_1 specific volume entering the pipe evaluated from Eq. (5), ft³/lbm

v_2 specific volume leaving the pipe, evaluated from Eq. (5), ft³/lbm

V_m $(Q_f + Q_g)/A$ mixture velocity for Fig. 12, m/sec

w weight flow, lbm/hr

w_t weight flow of liquid, lbm/hr

w_g weight flow of gas phase, lbm/hr

X Martinelli parameter, Eq. (18), dimensionless

x quality equal to weight fraction of gas flowing, dimensionless

Delta Quantities

ΔP pressure drop, equal to final pressure less the initial pressure, lbf/ft²

ΔP_f friction pressure drop, lbf/ft²

ΔP_{fa} adiabatic friction pressure drop with liquid only flowing at its mass velocity, Eq. (14), lbf/ft²

ΔP_{fh} friction pressure drop two phases flowing in a uniformly heated pipe with saturated liquid at the inlet, see Fig. 4 and Eq. (12), lbf/ft²

ΔP_{fo} pressure drop with saturated liquid only flowing at the mixture mass velocity, lbf/ft²

ΔP_g gravity pressure drop, lbf/ft²

ΔP_{gh} gravity pressure drop for a uniformly heated pipe with saturated inlet conditions, lbf/ft²

ΔP_m momentum pressure drop evaluated from Eq. (9), lbf/ft²

ΔP_{TP} two-phase pressure drop occurring in a fitting, lbf/ft²

Greek Symbols

α void fraction, dimensionless

ρ density, lbm/ft³

ρ_f liquid density, lbm/ft³

ρ_g gas density, lbm/ft³

θ angle of inclination measured from the horizontal

σ surface tension measured, dyn/cm

μ_f liquid viscosity measured, lbm/hr-ft except for the definition of ψ where it is in centipoises

μ_g gas phase viscosity measured, lbm/hr-ft

ψ Baker flow regime map parameter equal to

$$\left[\left(\frac{73}{\sigma}\right)\mu_f\left(\frac{62.3}{\rho_f}\right)^2\right]^{1/3} \text{ in (centipoises)}^{1/3}$$

λ Baker flow regime map parameter equal to

$$\left[\left(\frac{\rho_g}{0.075}\right)\cdot\left(\frac{\rho_f}{62.3}\right)\right]^{1/2}$$

ϕ Martinelli pressure drop multipliers to be obtained from Fig. 6 and substituted into equations of the type of Eq. (19). f in the subscript refers to the liquid phase, g refers to the gas phase, t refers to turbulent, v refers to viscous. Dimensionless

Section 15

Radiation

Part A

Relations and Properties

E. R. G. ECKERT
University of Minnesota, Minneapolis, Minnesota

BASIC RELATIONS

Heat transfer by radiation can be thought of as transport of energy by photons (packages of energy) being released from excited atoms and traveling on straight paths until they are absorbed by some other atoms. Alternatively it can also be visualized as energy transport in the form of electromagnetic waves. Parameters describing radiation are the photon or wave velocity c, the wavelength λ, and frequency ν. These are connected by the relation $c = \lambda\nu$. In the concept of radiation as traveling photons, the energy E of a photon is connected with the frequency ν by Planck's relation $E = h\nu$, with h indicating Planck's constant. Frequency and wavelength are used interchangeably. Experimentalists prefer the wavelength because it can be measured more readily. In analytical work, the frequency is preferred because it does not change when a radiant beam travels from one medium to another while wavelength and velocity undergo, in general, a sudden change at the interface.

The velocity of radiation in vacuum is independent of frequency and has the value $c_0 = 2.9977 \times 10^8$ m/s. All media occurring in nature have wave velocities which are smaller and depend to some degree on frequency. These velocities c are connected with the velocity c_0 in vacuum by the equation $n_0 = c_0/c$ with n_0 indicating the refractive index. Air and gases in general have a refractive index which differs only in the fourth decimal from one. Solids like glass or quartz, and liquids like water have considerably larger refractive indices (of an order of 1.5). For these materials, the variation of n_0 with wavelength is also larger (for quartz, as an example, from 1.45 to 1.6). The refractive index is connected through Snell's law with the change in direction of a radiant beam traveling from vacuum through a smooth interface into the medium under consideration. It is $n_0 = \sin\beta_0/\sin\beta$ when β_0 indicates the angle between the approaching beam and the surface normal and β the angle between the refracted beam and the surface normal. Electromagnetic waves are transverse waves. The state of polarization has to be stated to describe a radiant beam completely.

Radiation Field

A characteristic feature of the radiation process is the fact that the energy transport occurs along straight lines as long as refraction can be neglected. Consequently the following way to describe a radiation field is convenient. We select a point in the space under consideration and an area element dA arbitrarily located at this point (Fig. 1). The energy flux per unit time through dA and within a pencil with infinitely

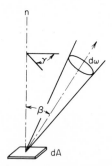

Fig. 1. Definition of intensity of radiant beam.

small solid angle $d\omega$ in a direction characterized by the angle β towards the surface normal and by the azimuth angle γ may be denoted by $d\Phi$. The following equation then defines a monochromatic intensity K_ν of the radiant flux

$$d\Phi = K_\nu \cos\beta \, dA \, d\omega \, d\nu \qquad (1)$$

Using alternatively the wavelength to characterize the radiation, one can also write

$$d\Phi = K_\lambda \cos\beta \, dA \, d\omega \, d\lambda \qquad (2)$$

The radiation field is described if the monochromatic flux intensities K_ν or K_λ are known for all points in the field in all directions and for all frequencies or wavelengths.

In a nonemitting and nonabsorbing medium, the intensity K_ν will be constant along each ray. However, it increases or decreases along the radiant beam when the medium emits or absorbs radiation. In this case the radiant flux increases by the emission process in the specified direction according to the equation

$$d^2\Phi_e = j_\nu \rho \, dV d\omega d\nu \tag{3}$$

in which j_ν is the coefficient of emission or the energy which is emitted per unit mass into a unity solid angle and within a unity frequency range. The notation $d^2\Phi_e$ may indicate that the flux in Eq. (3) is by an order of magnitude smaller than the flux in Eq. (2). The absorption process is described by the equation

$$d^2\Phi_a = -\kappa_\nu \rho \, ds d\Phi \tag{4}$$

with κ_ν indicating the coefficient of absorption. Some media scatter radiation. The intensity of the radiant beam is decreased by this process according to the equation

$$d^2\Phi_s = -\sigma_\nu \rho ds d\Phi \tag{5}$$

with σ_ν denoting the coefficient of scatter. Radiation is called thermal when the radiation properties j_ν, κ_ν, σ_ν are functions of the nature of the medium and its local thermodynamic state, in other words, when local thermodynamic equilibrium exists or is at least closely approached.

Kirchhoff's Law

G. R. Kirchhoff has derived important relations between the radiation parameters discussed above for thermal radiation by applying the laws of thermodynamics to media in thermodynamic equilibrium [1]. He showed that in an isotropic, homogeneous medium in thermodynamic equilibrium the radiant flux intensity $K_{\nu,e}$ is independent of direction and location and is connected with the radiation properties of that medium by the equation

$$K_{\nu,e} = \frac{j_\nu}{\kappa_\nu} \tag{6}$$

By consideration of a system of several media in thermodynamic equilibrium, he found that the parameters on both sides of the equation

$$c^2 K_{\nu,e} = c^2 \frac{j_\nu}{\kappa_\nu} \tag{7}$$

are now independent of the nature of the medium and are functions of temperature and frequency only. This relation can also be expressed by a set of different parameters which describe the behavior of radiation at the interface between two media. Consider for this purpose two media, 1 and 2, and a monochromatic beam of radiation approaching the interface in medium 1 in a specified direction. The radiation is partly reflected at the interface and the ratio of the reflected to the incident energy may be denoted by ρ_ν. The rest enters medium 2 in which a fraction α_ν is absorbed and the fraction τ_ν travels on and is finally absorbed in other media or in medium 1. In equilibrium, the relation

$$\rho_\nu + \alpha_\nu + \tau_\nu = 1 \qquad (8)$$

must hold. If medium 2 absorbs radiant energy, then it will also emit radiation and the intensity of the radiant beam emitted in medium 2 and traveling into medium 1 in a direction opposite to the incident beam considered before, may be denoted by i_ν. Kirchhoff's law then states

$$\frac{i_\nu}{\alpha_\nu} = K_{\nu e} \qquad (9)$$

$K_{\nu,e}$ is the equilibrium radiant flux intensity in medium 1. A medium which does not reflect any energy on its surface and does not transmit any energy is called a black body. Such a body then has an absorptance $\alpha_\nu = 1$. Equation (9) shows that the intensity $i_{\nu,b}$ which a black body 2 emits into a medium 1 is equal to $K_{\nu,e}$. Kirchhoff's law is then often written in the form

$$\alpha_\nu = \frac{i_\nu}{i_{\nu,b}} = \epsilon_\nu \qquad (10)$$

ϵ_ν is called emittance or emissivity of medium 2 into medium 1. It has to be kept in mind that absorptance and emittance depend in principle on both media 1 and 2, whereas $i_{\nu,b}$ is a property of medium 1. Equation (7) shows that the expression $c^2 \times i_{\nu,b}$ is independent of the nature of the medium and a function of temperature and frequency only. The above expression states that the intensity of black body radiation at a prescribed temperature and frequency depends, through the wave velocity c, on the medium into which the black radiation is emitted. Kirchhoff's law is especially important through the fact that it connects radiation properties in various media.

Black Body Radiation

The laws describing black body radiation will at first be presented for radiation emitted from such a body into vacuum because most of the laws have been derived analytically for such a condition. M. Planck obtained the following relation for the monochromatic intensity from his quantum theory

$$i_{\nu,b} = \frac{2h\nu^3}{c_0^2(e^{h\nu/kT} - 1)} \qquad (11)$$

Planck's constant h and Boltzmann's constant k in this equation are universal constants. In order to convert this equation to the monochromatic intensity expressed as a function of wavelength, the following relations have to be considered

$$i_{\lambda,b}\, d\lambda = i_{\nu,b}\, d\nu \qquad d\lambda = -\frac{c_0}{\nu^2}\, d\nu \qquad (12)$$

The second relation follows from $\lambda = c/\nu$ considering c to be a constant. Planck's equation as a function of wavelength is

$$i_{\lambda,b} = \frac{2hc_0^2}{\lambda^5\left(e^{hc_0/k\lambda T} - 1\right)} = \frac{2C_1}{\lambda^5\left(e^{C_2/\lambda T} - 1\right)} \qquad (13)$$

Numerical values of the constants C_1 and C_2 are listed together with some constants which will be introduced later in Table 1. Figure 2 presents curves of the intensity of monochromatic black body radiation into vacuum as a function of wavelength for a number of temperatures. The wavelength range in which radiation is visible to the eye is indicated in the figure by the dashed area and it can be recognized that for temperatures with which one is generally concerned in engineering, the bulk of the radiation occurs at wavelengths larger than the visible ones (in the infrared range). The wavelength distribution of the radiation coming from the sun corresponds approximately to black body radiation at a temperature of $6500°$K. At this temperature, approximately one-third of the total energy radiated is in the visible range.

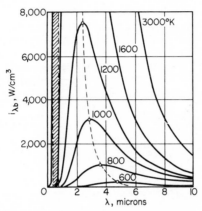

Fig. 2. Monochromatic intensity of black radiation in vacuum as function of wavelength.

The ratio of $i_{\lambda, b}/i_{\lambda, b, \max}$ on any isotherm is according to Eq. (13) a function of $\lambda T/C_2$ only. The relation between both parameters is listed in Table 2. The maximum $i_{\lambda, b, \max}$ of the intensity occurs at a wavelength which is described by Wien's law

$$\lambda T = C_3 \tag{14}$$

This relation can be derived from Eq. (13). The value of the maximum itself can be calculated from the equation

$$i_{\lambda, b, \max} = C_4 T^5 \tag{15}$$

The constants C_3 and C_4 are listed in Table 1. The intensity $i_{\lambda, b}$ can, therefore, be easily obtained from Tables 1 and 2.

In calculations of radiant heat transfer one frequently needs the energy emitted by a black body in a certain wavelength range. For this reason, Table 2 contains a third column listing the value $i_{b, O - \lambda}/C_5 T^4$. The term $i_{b, O - \lambda}$ is defined by the equation

$$i_{b, O - \lambda} = \int_0^\lambda i_{\lambda, b} d\lambda \tag{16}$$

and expresses, therefore, the intensity of radiation emitted by a black body at temperature T in the wavelength range from zero to λ. The intensity emitted in the range between two wavelengths is obtained by subtraction of the corresponding values of $i_{b, O - \lambda}$ in Table 2. The constant C_5 is again found in Table 1. The total energy radiated by a black body into unity solid angle in the whole wavelength range from 0 to ∞ is obtained by integration of Eq. (13) over all wavelengths. The result is

$$i_b = \int_0^\infty i_{\lambda, b} d\lambda = \frac{\sigma}{\pi} T^4 \tag{17}$$

The radiation e_b into all directions in space called emissive power is for black body radiation π times the radiation intensity i_b as can be found by a simple integration. In this way Stefan-Boltzmann's law

$$e_b = \sigma T^4 \tag{18}$$

is obtained. The Stefan-Boltzmann constant σ is listed in Table 1.

TABLE 1. Radiation Constants in Planck's, Wien's, Stefan-Boltzmann's Equations,* and Table 2

C_1	0.18892×10^8 Btu-μ^4/hr-ft^2	0.59525×10^{-8} Wm^2
C_2	25,896 μR	1.4387 cm-K
C_3	5215.6 μR	0.28976 cm-K
σ	0.17141×10^{-8} Btu/hr-ft^2-R^4	5.6688×10^{-8} $W/m^2 K^4$
C_4	0.68788×10^{-13} Btu/hr-ft^2-μR^5	4.0949×10^{-6} $W/m^3 K^5$
C_5	0.84019×10^{-10} Btu/hr-ft^2-R^4	0.27787×10^{-8} $W/m^2 K^4$

*According to N. W. Snyder, *Trans. ASME,* **76**:537 (1954).

For gases, the velocity c is practically equal to the velocity c_0 in vacuum; correspondingly, the above relations describe to a very good approximation also the behavior of radiation emitted from a black body into a gas. Solids and liquids, however, have refractive indices which are considerably larger than 1 and which depend also on wavelength. For such media, the monochromatic intensity of black body radiation is given by

$$i_{\nu, b} = \frac{2h\nu^3 n_0{}^2}{c_0{}^2(e^{h\nu/kT} - 1)} \tag{19}$$

which follows readily from the discussion at the end of the preceding section. Wien's law cannot be derived for black radiation in such a medium and the integration of Eq. (19) does not lead to Stefan-Boltzmann's law, when n_0 is now a function of ν.

Radiation Properties

The radiation properties, which have been introduced before together with the laws of black body radiation, are the basis for calculations of radiative heat exchange. In the following discussions of such properties, λ will be used as parameter because almost all of the experimental information is available in this form.

Most solids absorb radiation entering through their surfaces very strongly so that practically all of the radiation is absorbed in a very thin layer below the surface. In metals, the thickness of this layer is only a fraction of a micron. In electrically nonconducting materials, fractions of a millimeter are usually sufficient. This can be recognized from Tables 3 and 4. Exceptions are a few solid substances like glass, quartz, rock salt, and most liquids in the visible and near infrared range. However, these substances also absorb very strongly in the bulk of the infrared as indicated by

TABLE 2. Monochromatic Intensities $i_{\lambda,b}$ and Intensities $i_{b,O-\lambda}$ in the Wavelength Range between O and λ for Black Radiation in Vacuum ($i_{\lambda,b,max} = C_4 T^5$) Constants C Are Listed in Table 1)

$\lambda T/C_2$	$i_{\lambda,b}/i_{\lambda,b,max}$	$i_{b,O-\lambda}/C_5 T^4$	$\lambda T/C_2$	$i_{\lambda,b}/i_{\lambda,b,max}$	$i_{b,O-\lambda}/C_5 T^4$
0.0600	0.00350	0.32176E-03	0.0910	0.12763	0.29741E-01
0.0610	0.00424	0.40367E-03	0.0920	0.13618	0.32537E-01
0.0620	0.00509	0.50239E-03	0.0930	0.14501	0.35517E-01
0.0630	0.00607	0.62051E-03	0.0940	0.15412	0.38688E-01
0.0640	0.00719	0.76087E-03	0.0950	0.16350	0.42054E-01
0.0650	0.00846	0.92657E-03	0.0960	0.17314	0.45622E-01
0.0660	0.00990	0.11210E-02	0.0970	0.18303	0.49397E-01
0.0670	0.01151	0.13477E-02	0.0980	0.19317	0.53385E-01
0.0680	0.01332	0.16106E-02	0.0990	0.20355	0.57590E-01
0.0690	0.01532	0.19138E-02	0.1000	0.21415	0.62018E-01
0.0700	0.01754	0.22617E-02	0.1010	0.22496	0.66672E-01
0.0710	0.01998	0.26589E-02	0.1020	0.23598	0.71558E-01
0.0720	0.02265	0.31104E-02	0.1030	0.24719	0.76679E-01
0.0730	0.02557	0.36212E-02	0.1040	0.25858	0.82041E-01
0.0740	0.02875	0.41966E-02	0.1050	0.27014	0.87645E-01
0.0750	0.03219	0.48421E-02	0.1060	0.28185	0.93496E-01
0.0760	0.03591	0.55635E-02	0.1070	0.29372	0.99597E-01
0.0770	0.03990	0.63666E-02	0.1080	0.30571	0.10595E+00
0.0780	0.04419	0.72575E-02	0.1090	0.31783	0.11256E+00
0.0790	0.04876	0.82423E-02	0.1100	0.33006	0.11943E+00
0.0800	0.05364	0.93274E-02	0.1110	0.34238	0.12656E+00
0.0810	0.05882	0.10519E-01	0.1120	0.35479	0.13395E+00
0.0820	0.06431	0.11824E-01	0.1130	0.36727	0.14160E+00
0.0830	0.07011	0.13248E-01	0.1140	0.37982	0.14952E+00
0.0840	0.07622	0.14799E-01	0.1150	0.39241	0.15771E+00
0.0850	0.08264	0.16482E-01	0.1160	0.40505	0.16616E+00
0.0860	0.08937	0.18305E-01	0.1170	0.41771	0.17488E+00
0.0870	0.09642	0.20274E-01	0.1180	0.43038	0.18387E+00
0.0880	0.10377	0.22396E-01	0.1190	0.44306	0.19313E+00
0.0890	0.11142	0.24677E-01	0.1200	0.45573	0.20266E+00
0.0900	0.11938	0.27123E-01	0.1210	0.46839	0.21246E+00

TABLE 2. Monochromatic Intensities $i_{\lambda,b}$ and Intensities $i_{b,O-\lambda}$ in the Wavelength Range between O and λ for Black Radiation in Vacuum ($i_{\lambda,b,max} = C_4 T^5$, Constants C Are Listed in Table 1) (Continued)

$\lambda T/C_2$	$i_{\lambda,b}/i_{\lambda,b,max}$	$i_{b,O-\lambda}/C_5 T^4$	$\lambda T/C_2$	$i_{\lambda,b}/i_{\lambda,b,max}$	$i_{b,O-\lambda}/C_5 T^4$
0.1220	0.48102	0.22252E.00	0.1890	0.99007	0.13617E+01
0.1230	0.49361	0.23285E+00	0.1900	0.99166	0.13827E+01
0.1240	0.50616	0.24345E+00	0.1910	0.99310	0.14037E+01
0.1250	0.51865	0.25432E+00	0.1920	0.99440	0.14248E+01
0.1260	0.53107	0.26544E+00	0.1930	0.99556	0.14459E+01
0.1270	0.54343	0.27683E+00	0.1940	0.99657	0.14670E+01
0.1280	0.55569	0.28849E+00	0.1950	0.99745	0.14882E+01
0.1290	0.56787	0.30040E+00	0.1960	0.99820	0.15093E+01
0.1300	0.57995	0.31256E+00	0.1970	0.99881	0.15305E+01
0.1310	0.59192	0.32499E+00	0.1980	0.99929	0.15517E+01
0.1320	0.60378	0.33766E+00	0.1990	0.99965	0.15729E+01
0.1330	0.61552	0.35059E+00	0.2000	0.99988	0.15940E+01
0.1340	0.62713	0.36376E+00	0.2010	0.99999	0.16152E+01
0.1350	0.63861	0.37718E+00	0.2020	0.99998	0.16364E+01
0.1360	0.64995	0.39084E+00	0.2030	0.99985	0.16576E+01
0.1370	0.66114	0.40474E+00	0.2040	0.99961	0.16788E+01
0.1380	0.67218	0.41887E+00	0.2050	0.99925	0.17000E+01
0.1390	0.68307	0.43324E+00	0.2060	0.99878	0.17212E+01
0.1400	0.69380	0.44784E+00	0.2070	0.99821	0.17424E+01
0.1410	0.70436	0.46266E+00	0.2080	0.99753	0.17635E+01
0.1420	0.71476	0.47770E+00	0.2090	0.99674	0.17847E+01
0.1430	0.72499	0.49296E+00	0.2100	0.99586	0.18058E+01
0.1440	0.73503	0.50844E+00	0.2110	0.99488	0.18269E+01
0.1450	0.74490	0.52413E+00	0.2120	0.99380	0.18480E+01
0.1460	0.75459	0.54003E+00	0.2130	0.99262	0.18690E+01
0.1470	0.76409	0.55613E+00	0.2140	0.99136	0.18901E+01
0.1480	0.77341	0.57242E+00	0.2150	0.99000	0.19111E+01
0.1490	0.78254	0.58892E+00	0.2160	0.98856	0.19321E+01
0.1500	0.79147	0.60561E+00	0.2170	0.98703	0.19530E+01
0.1510	0.80021	0.62248E+00	0.2180	0.98542	0.19739E+01
0.1520	0.80876	0.63954E+00	0.2190	0.98372	0.19948E+01

0.1530	0.81711	0.65677E+00	0.2200	0.98195	0.20156E+01
0.1540	0.82527	0.67418E+00	0.2210	0.98010	0.20364E+01
0.1550	0.83322	0.69176E+00	0.2220	0.97818	0.20572E+01
0.1560	0.84098	0.70951E+00	0.2230	0.97618	0.20779E+01
0.1570	0.84854	0.72742E+00	0.2240	0.97411	0.20986E+01
0.1580	0.85590	0.74549E+00	0.2250	0.97197	0.21192E+01
0.1590	0.86307	0.76371E+00	0.2260	0.96977	0.21398E+01
0.1600	0.87003	0.78209E+00	0.2270	0.96750	0.21603E+01
0.1610	0.87679	0.80060E+00	0.2280	0.96516	0.21808E+01
0.1620	0.88336	0.81926E+00	0.2290	0.96277	0.22012E+01
0.1630	0.88972	0.83806E+00	0.2300	0.96031	0.22216E+01
0.1640	0.89589	0.85699E+00	0.2350	0.94720	0.23228E+01
0.1650	0.90187	0.87605E+00	0.2400	0.93283	0.24224E+01
0.1660	0.90764	0.89523E+00	0.2450	0.91742	0.25205E+01
0.1670	0.91322	0.91453E+00	0.2500	0.90112	0.26169E+01
0.1680	0.91861	0.93395E+00	0.2550	0.88411	0.27115E+01
0.1690	0.92381	0.95348E+00	0.2600	0.86653	0.28043E+01
0.1700	0.92882	0.97312E+00	0.2650	0.84850	0.28952E+01
0.1710	0.93363	0.99286E+00	0.2700	0.83014	0.29842E+01
0.1720	0.93826	0.10127E+01	0.2750	0.81155	0.30712E+01
0.1730	0.94270	0.10326E+01	0.2800	0.79283	0.31562E+01
0.1740	0.94696	0.10527E+01	0.2850	0.77405	0.32393E+01
0.1750	0.95104	0.10728E+01	0.2900	0.75529	0.33204E+01
0.1760	0.95493	0.10930E+01	0.2950	0.73660	0.33994E+01
0.1770	0.95865	0.11133E+01	0.3000	0.71805	0.34765E+01
0.1780	0.96218	0.11337E+01	0.3050	0.69968	0.35517E+01
0.1790	0.96555	0.11541E+01	0.3100	0.68152	0.36249E+01
0.1800	0.96874	0.11746E+01	0.3150	0.66362	0.36962E+01
0.1810	0.97176	0.11952E+01	0.3200	0.64599	0.37656E+01
0.1820	0.97461	0.12158E+01	0.3250	0.62867	0.38331E+01
0.1830	0.97982	0.12365E+01	0.3300	0.61168	0.38989E+01
0.1840	0.98219	0.12572E+01	0.3350	0.59502	0.39628E+01
0.1850	0.98439	0.12780E+01	0.3400	0.57871	0.40251E+01
0.1860	0.98643	0.12989E+01	0.3450	0.56277	0.40856E+01
0.1870	0.98833	0.13198E+01	0.3500	0.54719	0.41444E+01
0.1880		0.13407E+01	0.3550	0.53199	0.42016E+01

TABLE 2. Monochromatic Intensities $i_{\lambda,b}$ and Intensities $i_{b,0-\lambda}$ in the Wavelength Range between 0 and λ for Black Radiation in Vacuum $\left(i_{\lambda,b,max} = C_4 T^5 \right)$, Constants C Are Listed in Table 1) *(Continued)*

$\lambda T/C_2$	$i_{\lambda,b}/i_{\lambda,b,max}$	$i_{b,0-\lambda}/C_5 T^4$	$\lambda T/C_2$	$i_{\lambda,b}/i_{\lambda,b,max}$	$i_{b,0-\lambda}/C_5 T^4$
0.3600	0.51716	0.42572E+01	0.6950	0.09045	0.59364E+01
0.3650	0.50271	0.43112E+01	0.7000	0.08845	0.59459E+01
0.3700	0.48864	0.43638E+01	0.7050	0.08651	0.59552E+01
0.3750	0.47494	0.44148E+01	0.7100	0.08461	0.59642E+01
0.3800	0.46161	0.44645E+01	0.7150	0.08277	0.59731E+01
0.3850	0.44864	0.45127E+01	0.7200	0.08097	0.59818E+01
0.3900	0.43604	0.45596E+01	0.7250	0.07923	0.59903E+01
0.3950	0.42380	0.46052E+01	0.7300	0.07752	0.59986E+01
0.4000	0.41190	0.46495E+01	0.7350	0.07587	0.60067E+01
0.4050	0.40035	0.46925E+01	0.7400	0.07425	0.60147E+01
0.4100	0.38914	0.47344E+01	0.7450	0.07268	0.60225E+01
0.4150	0.37826	0.47750E+01	0.7500	0.07115	0.60301E+01
0.4200	0.36770	0.48146E+01	0.7550	0.06965	0.60375E+01
0.4250	0.35745	0.48530E+01	0.7600	0.06820	0.60449E+01
0.4300	0.34751	0.48904E+01	0.7650	0.06678	0.60520E+01
0.4350	0.33787	0.49267E+01	0.7700	0.06540	0.60590E+01
0.4400	0.32852	0.49620E+01	0.7750	0.06405	0.60659E+01
0.4450	0.31945	0.49964E+01	0.7800	0.06274	0.60726E+01
0.4500	0.31066	0.50298E+01	0.7850	0.06146	0.60792E+01
0.4550	0.30214	0.50622E+01	0.7900	0.06021	0.60856E+01
0.4600	0.29387	0.50938E+01	0.7950	0.05899	0.60919E+01
0.4650	0.28586	0.51246E+01	0.8000	0.05780	0.60981E+01
0.4700	0.27810	0.51544E+01	0.8050	0.05664	0.61042E+01
0.4750	0.27057	0.51835E+01	0.8100	0.05551	0.61101E+01
0.4800	0.26327	0.52118E+01	0.8150	0.05441	0.61160E+01
0.4850	0.25619	0.52393E+01	0.8200	0.05333	0.61217E+01
0.4900	0.24933	0.52661E+01	0.8250	0.05228	0.61273E+01
0.4950	0.24268	0.52922E+01	0.8300	0.05126	0.61328E+01
0.5000	0.23624	0.53176E+01	0.8350	0.05026	0.61381E+01
0.5050	0.22999	0.53423E+01	0.8400	0.04928	0.61434E+01
0.5100	0.22393	0.53664E+01	0.8450	0.04833	0.61486E+01

0.5150	0.21805	0.53898E+01	0.8500	0.04739	0.61537E+01
0.5200	0.21235	0.54126E+01	0.8550	0.04649	0.61586E+01
0.5250	0.20683	0.54348E+01	0.8600	0.04560	0.61635E+01
0.5300	0.20147	0.54565E+01	0.8650	0.04473	0.61683E+01
0.5350	0.19627	0.54775E+01	0.8700	0.04388	0.61730E+01
0.5400	0.19123	0.54981E+01	0.8750	0.04306	0.61776E+01
0.5450	0.18634	0.55181E+01	0.8800	0.04225	0.61821E+01
0.5500	0.18160	0.55376E+01	0.8850	0.04146	0.61866E+01
0.5550	0.17700	0.55566E+01	0.8900	0.04069	0.61909E+01
0.5600	0.17253	0.55751E+01	0.8950	0.03994	0.61952E+01
0.5650	0.16820	0.55932E+01	0.9000	0.03920	0.61994E+01
0.5700	0.16400	0.56108E+01	0.9050	0.03848	0.62035E+01
0.5750	0.15992	0.56280E+01	0.9100	0.03778	0.62075E+01
0.5800	0.15596	0.56447E+01	0.9150	0.03709	0.62115E+01
0.5850	0.15212	0.56610E+01	0.9200	0.03642	0.62154E+01
0.5900	0.14838	0.56769E+01	0.9250	0.03576	0.62192E+01
0.5950	0.14476	0.56925E+01	0.9300	0.03511	0.62230E+01
0.6000	0.14124	0.57076E+01	0.9350	0.03449	0.62267E+01
0.6050	0.13783	0.57224E+01	0.9400	0.03387	0.62303E+01
0.6100	0.13451	0.57369E+01	0.9450	0.03327	0.62339E+01
0.6150	0.13129	0.57510E+01	0.9500	0.03268	0.62374E+01
0.6200	0.12816	0.57647E+01	0.9550	0.03211	0.62408E+01
0.6250	0.12511	0.57781E+01	0.9600	0.03154	0.62442E+01
0.6300	0.12216	0.57912E+01	0.9650	0.03099	0.62475E+01
0.6350	0.11929	0.58040E+01	0.9700	0.03045	0.62507E+01
0.6400	0.11650	0.58165E+01	0.9750	0.02992	0.62539E+01
0.6450	0.11378	0.58287E+01	0.9800	0.02941	0.62571E+01
0.6500	0.11115	0.58407E+01	0.9850	0.02890	0.62602E+01
0.6550	0.10858	0.58523E+01	0.9900	0.02841	0.62632E+01
0.6600	0.10609	0.58637E+01	0.9950	0.02792	0.62662E+01
0.6650	0.10367	0.58748E+01	1.0000	0.02745	0.62691E+01
0.6700	0.10131	0.58857E+01	1.0100	0.02653	0.62749E+01
0.6750	0.09902	0.58963E+01	1.0200	0.02565	0.62804E+01
0.6800	0.09679	0.59067E+01	1.0300	0.02480	0.62857E+01
0.6850	0.09462	0.59168E+01	1.0400	0.02399	0.62909E+01
0.6900	0.09251	0.59267E+01	1.0500	0.02322	0.62959E+01

TABLE 2. Monochromatic Intensities $i_{\lambda,b}$ and Intensities $i_{b,O-\lambda}$ in the Wavelength Range between O and λ for Black Radiation in Vacuum ($i_{\lambda,b,\max} = C_4 T^5$, Constants C Are Listed in Table 1) *(Continued)*

$\lambda T/C_2$	$i_{\lambda,b}/i_{\lambda,b,\max}$	$i_{b,O-\lambda}/C_5 T^4$	$\lambda T/C_2$	$i_{\lambda,b}/i_{\lambda,b,\max}$	$i_{b,O-\lambda}/C_5 T^4$
1.0600	0.02247	0.63008E+01	1.5400	0.00596	0.64230E+01
1.0700	0.02175	0.63054E+01	1.5500	0.00582	0.64242E+01
1.0800	0.02106	0.63100E+01	1.5600	0.00568	0.64254E+01
1.0900	0.02040	0.63144E+01	1.5700	0.00555	0.64266E+01
1.1000	0.01976	0.63186E+01	1.5800	0.00542	0.64278E+01
1.1100	0.01915	0.63228E+01	1.5900	0.00530	0.64289E+01
1.1200	0.01856	0.63267E+01	1.6000	0.00518	0.64300E+01
1.1300	0.01799	0.63306E+01	1.6100	0.00506	0.64311E+01
1.1400	0.01745	0.63344E+01	1.6200	0.00495	0.64322E+01
1.1500	0.01692	0.63380E+01	1.6300	0.00484	0.64332E+01
1.1600	0.01642	0.63416E+01	1.6400	0.00473	0.64342E+01
1.1700	0.01593	0.63450E+01	1.6500	0.00463	0.64352E+01
1.1800	0.01546	0.63483E+01	1.6600	0.00453	0.64362E+01
1.1900	0.01501	0.63515E+01	1.6700	0.00443	0.64372E+01
1.2000	0.01457	0.63547E+01	1.6800	0.00433	0.64381E+01
1.2100	0.01415	0.63577E+01	1.6900	0.00424	0.64390E+01
1.2200	0.01374	0.63607E+01	1.7000	0.00415	0.64399E+01
1.2300	0.01335	0.63635E+01	1.7100	0.00406	0.64408E+01
1.2400	0.01298	0.63663E+01	1.7200	0.00397	0.64416E+01
1.2500	0.01261	0.63690E+01	1.7300	0.00389	0.64424E+01
1.2600	0.01226	0.63717E+01	1.7400	0.00381	0.64433E+01
1.2700	0.01192	0.63742E+01	1.7500	0.00373	0.64441E+01
1.2800	0.01159	0.63767E+01	1.7600	0.00365	0.64448E+01
1.2900	0.01128	0.63792E+01	1.7700	0.00358	0.64456E+01
1.3000	0.01097	0.63815E+01	1.7800	0.00350	0.64464E+01
1.3100	0.01067	0.63838E+01	1.7900	0.00343	0.64471E+01
1.3200	0.01039	0.63860E+01	1.8000	0.00336	0.64478E+01
1.3300	0.01011	0.63882E+01	1.8100	0.00329	0.64485E+01
1.3400	0.00984	0.63903E+01	1.8200	0.00323	0.64492E+01
1.3500	0.00958	0.63924E+01	1.8300	0.00316	0.64499E+01
1.3600	0.00933	0.63944E+01	1.8400	0.00310	0.64505E+01

1.3700	0.00909	0.63964E+01
1.3800	0.00886	0.63983E+01
1.3900	0.00863	0.64001E+01
1.4000	0.00841	0.64019E+01
1.4100	0.00820	0.64037E+01
1.4200	0.00799	0.64054E+01
1.4300	0.00779	0.64071E+01
1.4400	0.00760	0.64087E+01
1.4500	0.00741	0.64103E+01
1.4600	0.00723	0.64118E+01
1.4700	0.00705	0.64134E+01
1.4800	0.00688	0.64148E+01
1.4900	0.00671	0.64163E+01
1.5000	0.00655	0.64177E+01
1.5100	0.00640	0.64190E+01
1.5200	0.00625	0.64204E+01
1.5300	0.00610	0.64217E+01

1.8500	0.00304	0.64512E+01
1.8600	0.00298	0.64518E+01
1.8700	0.00292	0.64525E+01
1.8800	0.00286	0.64531E+01
1.8900	0.00280	0.64537E+01
1.9000	0.00275	0.64543E+01
1.9100	0.00270	0.64548E+01
1.9200	0.00265	0.64554E+01
1.9300	0.00259	0.64560E+01
1.9400	0.00255	0.64565E+01
1.9500	0.00250	0.64570E+01
1.9600	0.00245	0.64576E+01
1.9700	0.00240	0.64581E+01
1.9800	0.00236	0.64586E+01
1.9900	0.00231	0.64591E+01
∞	0.00000	0.64939E+01

Example: $0.32176E-03 = 0.32176 \times 10^{-3}$.

**TABLE 3. Transmittance of Metal Foils for
Monochromatic Radiation with
Wavelength λ***

λ	Silver	Gold s (μ)†	Platinum
2.5		0.0272	0.0352
2.0		0.0238	0.0331
1.5	0.0222	0.0244	0.0309
1.2	0.0214	0.0249	0.0300
1.0	0.0230	0.0266	0.0284
0.8	0.0236	0.0281	0.0274
0.7	0.0233	0.0311	0.0268

*After E. Hagen and H. Rubens, *Ann. Phys.*,
8:432 (1902).
†s denotes the thickness in microns within which
the intensity of entering radiation has decreased to
1/10 of the original value.

**TABLE 4. Transmittance of Electric Nonconductors
for Incident Black Radiation at 95°C***

	s† (mm)
Glass	0.3
Typewriter paper	0.06
Writing paper	0.07
Tracing paper	0.09
Mica	0.25

*After E. Eckert, *Forschung Geb. Ingenieurw.*, 9:251
(1938).
†s denotes the thickness in mm within which the intensity
of incident radiation is decreased to 1/10 of the original
value.

the curve for optical glass in Fig. 3. The arsensulfid glass has been developed
especially for good transmittance in the infrared. According to Kirchhoff's law, the
radiation emitted by such substances through their surface originates also in this thin
layer. For such media, with a thickness greater than the values mentioned above, the
transmittance can be set equal to zero and the following equations connect the
monochromatic absorptance, reflectance, and emittance

$$\alpha_\lambda + \rho_\lambda = 1 \ , \quad \alpha_\lambda = \epsilon_\lambda \tag{20}$$

The second expression presents Kirchhoff's law, which was derived assuming a
medium of constant temperature. It can, however, be applied to situations where the
temperature of the medium varies normal to the surface provided the variation within
the thin layer which participates in the radiative exchange is small. The radiation
properties then can be uniquely related to the surface temperature and one talks in
this sense often of absorbing and emitting surfaces. Radiation properties of solids and
liquids are usually listed in the various handbooks for radiation into air. This is also
the case for the values presented in this and the following sections. These values can
be used for radiation into any gas or into vacuum, whereas they are different if one is
concerned with radiation into a liquid or solid.

Gases are much weaker emitters and absorbers than solids and liquids. As a
consequence, much thicker layers are involved in the radiative exchange process and

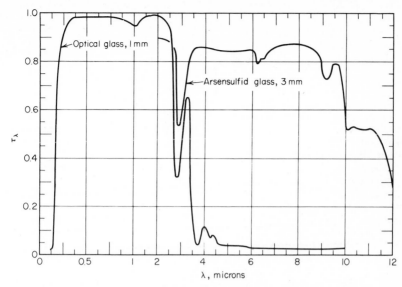

Fig. 3. Monochromatic transmittance of glass as function of wavelength [2, 4].

absorptances, emittances, and transmittances are useful radiation properties only as long as one is concerned with gas bodies of uniform temperature; otherwise one has to use the absorption and scattering coefficients as defined by Eqs. (4) and (5). The reflection of the radiant beam at the interface between two gases is very small and usually neglected. At the interface to a solid or liquid, the reflection process is usually attributed to the solid or liquid so that the reflectance of gases can be set equal to zero.

RADIATION PROPERTIES OF SOLIDS AND LIQUIDS

Emittance

The emittance of surfaces has been defined by the equation

$$\epsilon_\lambda = \frac{i_\lambda}{i_{\lambda, b}} \tag{21}$$

More exactly, this is the *monochromatic directional emittance* concerned with radiation leaving the surface in a specified direction. In addition to this emittance, a *hemispherical emittance* is often listed in reference books which is concerned with radiation emitted in all directions of the hemispherical space. It is defined as

$$\epsilon_\lambda = \frac{e_\lambda}{e_{\lambda, b}} \tag{22}$$

A special case of the directional emittance is the *normal emittance* $\epsilon_\perp$ which is concerned with the radiation in the direction of the surface normal. Another distinction has to be made between the *monochromatic emittance* and the *total emittance* referring to the total radiation regardless of wavelength. It is connected with the monochromatic emittance by the relation

$$\epsilon = \frac{\displaystyle\int_0^\infty \epsilon_\lambda\, i_{\lambda,b}\, d\lambda}{\displaystyle\int_0^\infty i_{\lambda,b}\, d\lambda} \tag{23}$$

Electromagnetic theory shows that solids and liquids have in the infrared a characteristic different behavior, depending on whether they are electric conductors or nonconductors.

Metals (Electric Conductors). Metals as electric conductors have small emittance values, as can be seen in Table 5. The radiation properties of electric conductors are mainly determined by the free electrons. As a consequence, there exists a connection between the emittance and the specific electric resistance of a metal. The emittance increases with increasing electric resistance. Because of the very thin layer beyond the surface which participates in the radiation, the emittance values depend very strongly on surface condition. A thin oxide layer, not visible to the eye, or a film of impurities or surface roughness changes the emittance considerably. Thicker oxide layers can increase the emittance by an order of magnitude, as various values in Table 5 indicate.

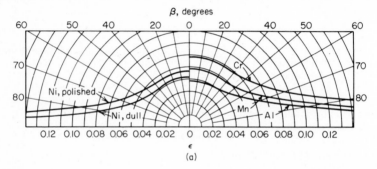

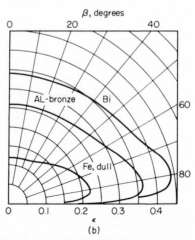

Fig. 4. Directional distribution of emittance of smooth metal surfaces (after E. Schmidt and E. Eckert, *Forsch. Geb. Ingenieurw.*, **6**:175 (1935)).

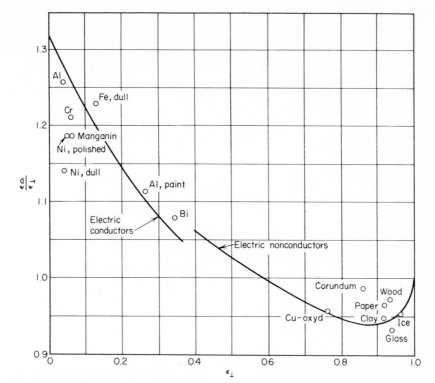

Fig. 5. Ratio of hemispherical to normal emittance (after E. Schmidt and E. Eckert, *Forsch. Geb. Ingenieurw.*, **6**:175 (1935), with a correction of the curve for electric nonconductors after M. Jakob, "Heat Transfer," vol. I, p. 52, John Wiley & Sons, Inc., New York, 1949.

The impossibility to describe the surface condition exactly accounts for the fact that emittance values listed in various reference books differ sometimes strongly. The directional distribution of the emittance for a few metals is shown in Fig. 4a and b. It will be observed that the directional emittance for metals increases with increasing angle β. Electromagnetic theory which is in agreement with the measured results presented in Fig. 4 indicates that for angles β close to 90°, the emittance decreases again and reaches the value zero at $\beta = 90°$. The directional distribution curves permit calculation of the ratio of the hemispherical to the normal emittance.

Figure 5 presents this ratio as obtained by electromagnetic theory for electric conductors and nonconductors and values measured for a number of metal surfaces. This figure is useful to convert normal emittances listed in Table 5 to hemispherical values. For larger values of the angle of the radiant beam measured toward the surface normal, the radiation is polarized in the sense that the intensity of the component whose electric field fluctuates parallel to the plane of incidence (the plane containing the surface normal and the radiant beam) is considerably larger than the intensity of the component with the electric field fluctuating normal to this plane. Figure 6 indicates the size of the emittance ϵ_p and ϵ_n characterizing the two components according to a simplified calculation based on electromagnetic theory.

Figure 7 presents values of monochromatic emittances as function of wavelength. The figure indicates that the monochromatic emittance for metals decreases in the infrared range in a regular way with increasing wavelength. This behavior is also predicted by electromagnetic theory. The titanium surface which has been heated for

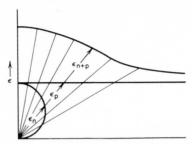

Fig. 6. Emittance of metal surface for radiation polarized normal (ϵ_n) and parallel (ϵ_p) to plane of incidence (after E. Schmidt, *Gesundheits-Ing.*, Beihefte, Ser. 1, Nr. 20, 1927).

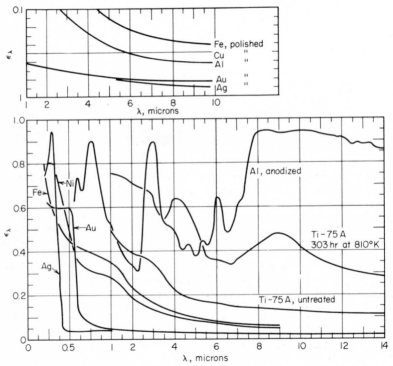

Fig. 7. Monochromatic normal emittance of metal surfaces as function of wavelength (from Ref. 4 and after J. T. Bevans, J. T. Gier, and R. V. Dunkle, *Trans. ASME*, **80**:80 (1958)).

303 hours at 810°K and the anodized aluminum surface behave differently as a consequence of the fact that these metal surfaces are covered by an oxide layer. Their behavior is therefore approaching one which is characteristic for electric nonconductors. In the visible and near infrared range, the behavior of the emittance for metal surfaces is quite irregular. Electromagnetic theory does not apply in this range anymore and an estimate of the emittance can be better made from the appearance of the surface to the eye of the observer when it reflects light. Partly as a consequence of this wavelength dependence and partly as a consequence of the fact that the monochromatic emittance depends on temperature, the total emittance increases with increasing surface temperature. This can be seen in Fig. 8. The increase in the normal

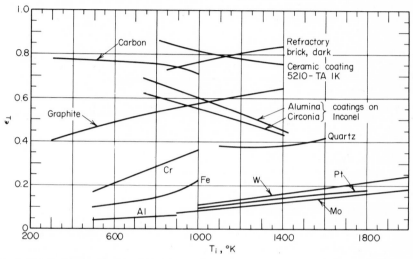

Fig. 8. Total normal emittance of electric conductors and nonconductors as function of surface temperature [3, 4, 5].

emittance for the metals aluminum, tungsten, platinum, and molybdenum is quite regular and indicative of the general behavior of metals. According to electromagnetic theory, the normal emittance of a metal is connected with the electric resistivity ρ by the equation [2]

$$\epsilon_n = 0.576\sqrt{\rho T} - 0.124\,\rho T \tag{24}$$

(ρ in Ωcm, T in °K). For pure metals, the electric resistivity changes with temperature like

$$\rho = \rho_0 \frac{T}{273} \tag{25}$$

(ρ_0 resistivity at $T = 273$°K). This results in a temperature dependence of the emittance

$$\epsilon_n = 0.0348\sqrt{\rho_0}\,T^{3/2} - 0.00045\,\rho_0 T^2 \tag{26}$$

For good electric conductors, the first term in these equations is dominant. The fact that the emittance of iron starts to increase at a faster rate beyond a temperature of 800°K has to be attributed to the gradual formation of a thin oxide layer.

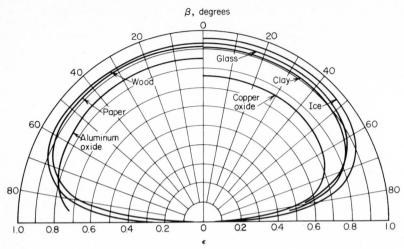

Fig. 9. Directional distribution of emittance of electric nonconductors (after E. Schmidt and E. Eckert, *Forsch. Geb. Ingenieurw.*, **6**:175 (1935)).

Nonconductors. Electrically nonconducting materials have in the infrared generally large values of the emittance. Such values are listed in Table 5 for materials of engineering importance. The distribution in the various directions in space is also different from that of metals and can be seen in Fig. 9. Monochromatic emittances for a few nonconductors with engineering importance are presented in Fig. 10. It is found that in the infrared range almost all electric nonconductors have large values of monochromatic emittance. Magnesium oxide is an exception. In the visible range and in the near infrared, the emittance values can again be estimated from their

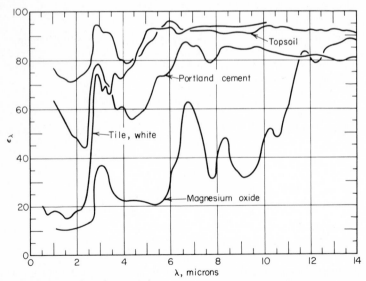

Fig. 10. Monochromatic normal emittance of electric nonconductors as function of wavelength (after W. Sieber, *Z. Tech. Physik,* **22**:130 (1941), and R. V. Dunkle and J. T. Gier, *Progr. Rep.* DA-11-190-ENG-3, Institute of Engineering Research, University of California, Berkeley, 1953).

TABLE 5. Emittance $\epsilon_\perp$ in the Direction of the Surface Normal for Various Materials at Temperature t *

Surface	$t, °F$	$\epsilon_\perp$
Metals		
Aluminum		
Bright rolled	338	0.039
Bright rolled	932	0.050
Polished, 98% pure	200	0.05
Aluminum-Paint	212	0.20–0.40
Bismuth, bright	176	0.340
Chrome, polished	302	0.058
Copper, polished	68	0.030
Lightly oxidized	68	0.037
Scraped	68	0.070
Black oxidized	68	0.78
Oxidized	268	0.76
Gold, polished	266	0.018
Polished	752	0.022
Iron		
Pure polished	200	0.06
Bright etched	302	0.128
Bright abrased	68	0.24
Red rusted	68	0.61
Hot rolled	68	0.77
Hot rolled	266	0.60
Heavily crusted	68	0.85
Heat-resistant oxidized	176	0.613
Heat-resistant oxidized	392	0.639
Cast iron bright	200	0.21
Cast iron oxidized	200	0.61
Black iron oxide	200	0.56
Lead		
Pure polished	100	0.05
Gray oxidized	68	0.28
Mercury		
Pure clean	200	0.12
Molybdenum		
Polished	200	0.06
Nickel		
Bright matte	212	0.041
Polished	212	0.045
Platinum		
Pure polished	200	0.05
Black	200	0.93
Silumin		
Cast polished	302	0.186
Silver	68	0.020
Tin		
Bright tinned iron sheet	100	0.08
Tungsten		
Polished	200	0.04
Zinc		
Matte	200	0.21
Gray oxidized	68	0.23–0.28

TABLE 5. Emittance $\epsilon_{\perp}$ in the Direction of the Surface Normal for Various Materials at Temperature t (*Continued*)

Surface	$t, {}^{\circ}F$	$\epsilon_{\perp}$
Alloys		
Brass		
Polished	100	0.05
Freshly rubbed with emery	100	0.21
Dull	100	0.22
Oxidized	100	0.46
Manganin, bright rolled	245	0.048
Paints		
Varnish, dark glossy	100	0.89
Lacquer, clear on bright copper	100	0.07
White on clear copper, thin coat	100	0.85
White	212	0.925
Black matte	176	0.970
Enamel	68	0.85–0.95
Bakelite	176	0.935
Gold enamel	200	0.37
Red lead paint	212	0.93
Oil on polished iron		
0.0008-in. thick	100	0.22
0.0080-in. thick	100	0.81
Pigments		
Acetylene soot	125	0.99
Camphor soot	125	0.98
Lampblack	125	0.94
Candle soot	125	0.95
Platinum black	125	0.91
Red (Fe_2O_3)	125	0.96
Green (Cr_2O_3)	125	0.95
White (Al_2O_3)	125	0.98
White (MgO)	125	0.97
Miscellaneous		
Asbestos paper	500	0.94
Brick, mortar, plaster	68	0.93
Clay		
Fired	158	0.91
Corundum		
Emery rough	176	0.855
Glass	194	0.940
Hoar frost		
(0.1–0.2-mm thick)	100	0.98
Ice		
Smooth, water	32	0.966
Rough crystals	32	0.985
Limestone	100	0.95
Marble		
White	100	0.95
Mica	100	0.75
Paper	203	0.92
Plaster of Paris		
(½ mm)	100	0.91
Porcelain	68	0.92–0.94
Quartz		
Fused, rough	100	0.93

TABLE 5. Emittance $\epsilon_\perp$ in the Direction of the Surface Normal for Various Materials at Temperature t (*Continued*)

Surface	$t, °F$	$\epsilon_\perp$
Miscellaneous (continued)		
Refractory brick, ordinary	2000	0.59
White	2000	0.29
Dark chrome	2000	0.98
Rubber		
Grey, soft, rough	100	0.86
Sandstone	100	0.83
Tar paper	68	0.93
Velvet		
Black	100	0.97
Water		
(0.1-mm or more thick)	100	0.96
Waterglass	68	0.96
Wood		
Beech	158	0.935
Oak, planed	100	0.90
Spruce, sanded	100	0.82

*From E. R. G. Eckert and R. M. Drake, "Heat and Mass Transfer," McGraw-Hill Book Company, Inc., New York, 1959, and J. R. Singham, *Int. J. Heat Mass Transfer*, 5:67 (1962).

appearance to the eye. Bright surfaces have low emittance values like the white tile in Fig. 10, whereas surfaces appearing dark to the eye have high emittance values in the visible and near infrared range. This behavior determines the way in which the total emittance changes with temperature. For surfaces which appear bright to the eye, one can expect a decrease of ϵ with increasing temperature, whereas surfaces with a dark appearance will exhibit an increase or a small drop. Figure 8 verifies this rule.

Absorptance

The monochromatic absorptance is connected with the monochromatic emittance by Kirchhoff's relation

$$\alpha_\lambda = \epsilon_\lambda \tag{27}$$

This relation holds strictly for a monochromatic, directional, polarized ray. In practice, the state of polarization has to be considered only when metal surfaces and large angles of incidence are involved. Hemispherical absorptances are equal to the hemispherical emittances only when the intensity of the incident flux is independent of the angle of incidence.

Much more restricted is Kirchhoff's equation

$$\alpha = \epsilon \tag{28}$$

for total emittance and absorptance. It holds only under the following conditions:

1. For $T_i = T_s$ (T_i temperature of the body from which the incident radiation comes, T_s surface temperature of the absorbing or emitting medium).
2. When ϵ and α are independent of wavelength (for gray body radiation).
3. The absorptance of a surface at temperature T_s is equal to the emittance of the same surface at the temperature T_i provided the incident radiation has the same distribution over the wavelength range as a black body and the emittance is independent of temperature.

4. For metals, the absorptance at the temperature T_s is equal to the emittance at the temperature $T = \sqrt{T_i T_s}$ according to electromagnetic theory [3].

Figure 11 presents total absorptance values of various materials with engineering importance at room temperature irradiated by black radiation at a temperature T_i listed on the abscissa. The figure shows that the absorptance of aluminum increases with temperature, a fact which is in agreement with the increase of emittance values of

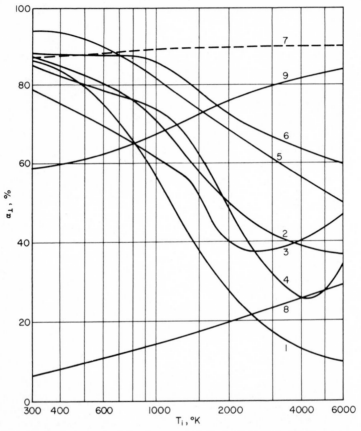

Fig. 11. Total normal absorptance of various materials for incident black radiation at temperature T_i: 1—white fire clay, 2—asbestos, 3—cork, 4—wood, 5—porcelain, 6—concrete, 7—roof shingles, 8—aluminum, 9—graphite. (*After W. Sieber, Z. Tech. Physik., 22:130 (1941)*).

metals in Fig. 8 and is characteristic for metal surfaces. Electric nonconductors exhibit, in general, a decrease of absorptance with temperature. This decrease is especially pronounced for surfaces which appear white to the eye and causes the absorptance of such surfaces to be smaller than that of metal surfaces at very high temperatures of the incident radiation. This means, for example, that a white painted surface absorbs less solar radiation than a metal surface.

Reflectance

The reflectance in Eqs. (8) and (20) is concerned with the energy reflected on a surface regardless of how it is distributed in space. This follows from the derivation of the equations. In radiative exchange calculations, one is often concerned with the angular distribution of the reflected radiation; correspondingly, various definitions of reflectance are required. The reflectance in Eqs. (8) and (20) may be called angular-hemispherical reflectance because it considers radiation incident under a specified angle being reflected into all directions in space. It is defined by the equation

$$\rho_{ah} = \frac{d\phi_{r,\circ}}{d\phi_i} = \frac{d\phi_{r,\circ}}{K_i \cos\beta_i \, d\omega} \tag{29}$$

($d\phi = d\Phi/dA$, see Eq. (1)). In describing the directional distribution of reflected radiation, it is of advantage to consider two extreme situations. Reflection is mirror-like or specular if a radiant pencil approaching the surface under an angle β_i towards the surface normal is reflected under the same angle and within an opening angle of the same magnitude. A perfectly diffuse surface is one which reflects an incident beam into all directions of space with an intensity independent of the angle of the reflected beam. It is convenient to use two different definitions for the reflectance for both kinds of surfaces. The specular reflectance is the ratio of the intensities of the reflected to the incident radiation according to equation

$$\rho_s = \frac{K_r}{K_i} \tag{30}$$

The biangular reflectance used for a diffuse surface is defined as the ratio of the intensity of the reflected radiation to the specific energy flux $d\phi$ of the incident radiation

$$\rho_{ba} = \frac{dK_r}{d\phi_i} = \frac{dK_r}{K_i \cos\beta_i \, d\omega_i} \tag{31}$$

The parameters in Eqs. (30) and (31) are defined such that their values are independent of the solid angle of the incident beam as long as this angle is small.

The angular-hemispherical reflectance is difficult to measure and it is often preferred to determine a reflectance value which may be called hemispherical-angular reflectance. This value is concerned with the energy reflected in a specific direction when the surface is irradiated from all directions in space. Its proper definition is

$$\rho_{ha} = \frac{\pi K_r}{\phi_{i,\circ}} \tag{32}$$

It can be shown with the use of Helmholtz' theorem that this reflectance is equal to the angular-hemispherical reflectance provided the irradiation has an intensity independent of direction. The directional distribution of reflected radiation is presented for a few materials with engineering importance in Fig. 12a and b. The incident black radiation at approximately 280°C temperature arrived at the surface within a cone with a 6° opening angle and at near normal incidence. The optics of the experimental setup was such that radiation reflected from a specular surface is measured within an opening angle of 11°. The reflection of the anodized aluminum surface can therefore be considered as perfectly specular; other surfaces indicate a gradual transition to diffuse radiation. The wood surface in Fig. 12b exhibits nearly perfect diffuse reflection.

Reflected radiation is mainly determined by the surface roughness and by the wavelength of the incident radiation. It is to be expected that a surface acts specularly

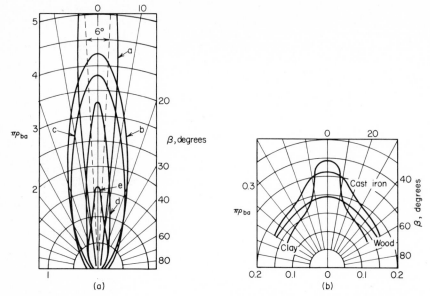

Fig. 12. Biangular reflectance of various materials for black radiation at approximately 280°C with near normal incidence; a–aluminum, anodized, b–aluminum paint, c–iron, scraped, d–black iron, e–copper oxide (after E. Eckert, *Forsch. Geb. Ingenieurw.*, 7:265 (1936)).

when the ratio of the mean roughness dimension k to the wavelength λ is small. It acts diffusely when this ratio is large. Experiments studying the directional distribution of monochromatic reflected radiation are, therefore, more revealing than measurements of total reflectance as shown in Fig. 12. Such information is, however, very scarce [4, 5]. On the surfaces which have been investigated, it was found that transition from the specular to the diffuse character occurs on a surface when the parameter k/λ changes by a factor of an order of ten. The surface roughness is expected to be the dominant factor on metals, whereas the spatial distribution of reflected radiation on a nonconducting surface may also be partially determined by internal reflections in the radiating layer.

REFERENCES

1. M. Planck, "The Theory of Radiation," Dover Publications, Inc., New York, 1959. (English translation of German text, "Vorlesungen über die Theorie der Wärmestrahlung," Leipzig, 1923.)
2. E. Schmidt and E. Eckert, *Forsch. Geb. Ingenieurw.*, 6:175 (1935).
3. E. Eckert, *Forsch. Geb. Ingenieurw.*, 7:265 (1936).
4. R. C. Birkebak and E. R. G. Eckert, "Effects of Roughness on Angular Distribution of Monochromatic Reflected Radiation," *J. Heat Transfer*, 87:85 (1965).
5. K. E. Torrance and E. M. Sparrow, "Biangular Reflectance of an Electric Nonconductor as a Function of Wavelength and Surface Roughness," *J. Heat Transfer*, 87:283 (1965).

LITERATURE ON RADIATION PROPERTIES

1. "International Critical Tables," McGraw-Hill Book Company, Inc., New York, 1927.
2. Landolt-Börnstein, "Zahlenwerte und Funktionen aus Physik, Chemie, Astronomie, Geophysik, Technik," Springer-Verlag, Berlin, 1950.
3. G. G. Gubareff, J. E. Janssen, and R. H. Torborg, "Thermal Radiation Properties Survey," Honeywell Research Center, Minneapolis-Honeywell Regulator Company, Minneapolis, Minn., 1960.

4. J. D'Ans and E. Lax, "Taschenbuch für Chemiker und Physiker," Springer-Verlag, Berlin, 1949.
5. J. R. Singham, Tables of Emissivities of Surfaces, *Int. J. Heat Mass Transfer,* **5**:67 (1962).
6. E. R. G. Eckert and R. M. Drake, "Heat and Mass Transfer," McGraw-Hill Book Company, Inc., New York, 1959.
7. Y. S. Touloukian, "Retrieval Guide to Thermophysical Properties Research Literature," McGraw-Hill Book Company, Inc., New York, 1960.

SUPPLEMENTARY REFERENCES

The following books on radiative heat transfer were published after completion of this section:

E. M. Sparrow and R. D. Cess, "Radiation Heat Transfer," Brooks/Cole Publishing Company, Belmont, Calif., 1966.

J. A. Wiebelt, "Engineering Radiation Heat Transfer," Holt, Rinehart and Winston, Inc., New York, 1966.

H. C. Hottel and A. F. Sarofim, "Radiative Transfer," McGraw-Hill Book Company, New York, 1967.

T. J. Love, "Radiative Heat Transfer," Charles E. Merrill Books, Inc., Columbus, Ohio, 1968.

R. Siegel and J. R. Howell, "Thermal Radiation Heat Transfer," NASA SP-164, vols. I and II, National Aeronautic and Space Administration, Washington, D.C., 1968 and 1969.

NOMENCLATURE

General

c	wave velocity
F	angle factor
F	view factor
F	shape factor
h	Planck's constant
k	Boltzmann's constant
n	refractive index
λ	wavelength
ν	frequency
ρ	density
σ	Stefan-Boltzmann constant
ω	solid angle
ω	wave number ($\omega = 1/\lambda$)

Within a Medium

U	radiant energy
u	radiation density, $u = dU/dV$ (V = volume)
Φ	radiant flux, $\Phi = dU/dt$ (t = time)
ϕ	specific radiant flux, $\phi = d\Phi/dA$ (A = area)
K	intensity of radiant flux, $K = d\phi/d\omega \cos\beta$ (β = angle between surface normal and radiant beam)
j	coefficient of emission, $d\Phi_e = j\rho \, dV \, d\omega$
κ	mass absorption coefficient, $d\Phi_a/\Phi = -\rho\kappa ds$ (s = radiant beam)
σ	mass scattering coefficient, $d\Phi_s/\Phi = -\rho\sigma ds$

(monochromatic values indicated by subscript ν or λ)

At Interface

B	radiosity, radiant flux leaving interface per unit area and time, $B = \oint K \cos\beta \, d\omega$
H	irradiation, radiant flux arriving at interface per unit area and time, $H = \oint K \cos\beta \, d\omega$
i	intensity of emitted radiant flux per unit area A_n and unit ω (same definition as K, but referring to emitted radiation)

e emissive power, $e = \oint i \cos\beta d\omega$

ϵ emittance (emissivity), $\epsilon = i/i_b$

α absorptance (absorptivity), $\alpha = d\phi_{a,\,\Omega}/d\phi_i = d\phi_{a,\,\Omega}/K_i \cos\beta_i d\omega_i$

τ transmittance (transmissivity), $\tau = d\phi_{\tau,\,\Omega}/d\phi_i = d\phi_{\tau,\,\Omega}/K_i \cos\beta_i d\omega_i$

ρ reflectance (reflectivity) with following subdivisions:

ρ_s specular reflectance (reflectivity), $\rho_s = K_r/K_i$ (for specular surfaces)

ρ_{ba} biangular reflectance, $\rho_{ba} = dK_r/d\phi_i = dK_r/K_i \cos\beta_i d\omega_i$ (for diffuse surfaces)

ρ_{ah} angular-hemispherical reflectance, $\rho_{ah} = d\phi_{r,\,\Omega}/d\phi_i = d\phi_{r,\,\Omega}/K_i \cos\beta_i d\omega_i$

ρ_{ha} hemispherical-angular reflectance, $\rho_{ha} = \pi K_r/\phi_{i,\,\Omega}$

ρ_{bh} bihemispherical reflectance, $\rho_{bh} = \phi_{r,\,\Omega}/\phi_{i,\,\Omega}$

Subscripts

a absorbed

b black body

e emitted

i incident

n normal to radiant beam

r reflected

s specular

κ scattered

τ transmitted

ν, λ monochromatic

Ω hemispherical

Part B

Radiant Interchange between Surfaces Separated by Nonabsorbing and Nonemitting Media

E. M. SPARROW

Heat Transfer Laboratory, Department of Mechanical Engineering
University of Minnesota, Minneapolis, Minnesota

In this part, methods will be described for computing the heat transfer resulting from the exchange of thermal radiation between surfaces that are separated by a radiatively nonparticipating medium. The selection of a particular method from among the various available computation procedures depends on the actual radiation properties of the participating surfaces (i.e., black or non-black, gray or non-gray, diffuse or specular, etc.), on the accuracy to which the results are desired, and on the computational effort that can be invested. The following cases will be discussed: first, black surfaces; second, gray, diffusely emitting and reflecting surfaces; third, quasi-gray,* diffusely emitting and reflecting surfaces; and fourth, partially or completely specular (mirror-like) surfaces. For engineering applications, radiant

*Gray over a finite wavelength range.

interchange calculations are most frequently performed within the framework of the second case. However, more exact determinations may require the application of methods outlined under the third and fourth cases.

An essential first step in all such calculations is to envision a complete enclosure made up of all surfaces that can radiatively interact with the surface (or surfaces) of interest. The purpose of the enclosure concept is to ensure that radiant energy arriving from all directions in space is fully accounted. One or more of the surfaces of the enclosure may not be material surfaces, for example, an open window. Each of such surfaces may be assigned equivalent radiation properties and an equivalent black body temperature that corresponds to the rate at which radiant energy passes through the surface into the enclosure.

Before proceeding with a general account of available computation procedures, results will be quoted for a basic situation which frequently recurs in practice; namely, a body having surface temperature T_1 and surface area A_1 completely surrounded by an isothermal enclosure (temperature T_2) of appreciably larger dimensions. The rate of heat loss at surface 1 is

$$\frac{Q_1}{A_1} = \epsilon_1 \sigma T_1^{\,4} - \alpha_1 \sigma T_2^{\,4} \tag{1}$$

in which ϵ and α respectively denote the emittance and absorptance. When the surface 1 is a gray body, then $\alpha_1 = \epsilon_1$.

BLACK SURFACES

Consider an enclosure containing N black surfaces respectively having absolute temperatures $T_1, T_2, \ldots, T_N$ (°F abs or °K) and surface areas $A_1, A_2, \ldots, A_N$. The net rate of heat loss by radiation Q_j from any typical surface j is

$$\frac{Q_j}{A_j} = \sigma T_j^{\,4} - \sum_{i=1}^{N} \sigma T_i^{\,4} F_{j-i} \tag{2}$$

wherein σ is the Stefan-Boltzmann constant. This expression applies for any $j = 1, 2, \ldots, N$. Each of the quantities F_{j-i} is a dimensionless number that is variously termed an angle factor, a shape factor, a geometrical factor, a view factor, or a configuration factor. In physical terms, F_{j-i} represents the fraction of the radiant energy leaving area A_j that arrives at area A_i. In general, for any surface j in the enclosure, energy conservation requires that $\sum_{i=1}^{N} F_{j-i} = 1$, from which it follows that

$$\frac{Q_j}{A_j} = \sum_{i=1}^{N} \sigma (T_j^{\,4} - T_i^{\,4}) F_{j-i} \tag{3}$$

A detailed discussion of methods for determining angle factors is presented in a subsequent section of this article.

In some applications, the heat flux may be prescribed at one or more of the surfaces; for instance, at a no-flux or adiabatic surface $Q = 0$. Correspondingly, the temperatures of these surfaces are unknown. Consider first the case where the heat transfer rate Q_k is prescribed at one of the surfaces k, while the temperatures are prescribed at all other surfaces. Then, the unknown temperature at surface k is calculable from

$$\sigma T_k^4 = \frac{\displaystyle\sum_{i=1}^{N} \sigma T_i^4 F_{k-i} + (Q_k/A_k)}{1 - F_{k-k}} \tag{4}$$

in which $i = k$ is excluded from the summation. Once T_k^4 is determined, the heat transfer rates at the other surfaces follow by direct evaluation of Eqs. (2) or (3).

Suppose next that the heat fluxes Q_k and Q_n are prescribed at surfaces k and n along with temperatures $T_1, \ldots, T_N$ at the other surfaces of the enclosure. The unknown values of σT_k^4 and σT_n^4 are found by simultaneous solution of the following pair of *linear* algebraic equations

$$\sigma T_k^4 (1 - F_{k-k}) - \sigma T_n^4 F_{k-n} = \sum_{i=1}^{N} \sigma T_i^4 F_{k-i} + \frac{Q_k}{A_k} \tag{5a}$$

$$-\sigma T_k^4 F_{n-k} + \sigma T_n^4 (1 - F_{n-n}) = \sum_{i=1}^{N} \sigma T_i^4 F_{n-i} + \frac{Q_n}{A_n} \tag{5b}$$

in which $i = k$ and $i = n$ are excluded from the summations. Once these temperatures have been found, the heat transfer rates of all other surfaces are immediately calculable from Eqs. (2) or (3).

The generalization to cases wherein the heat flux is prescribed at three or more of the surfaces follows in a straightforward manner.

GRAY, DIFFUSELY EMITTING AND REFLECTING SURFACES

Radiation heat transfer computations for engineering systems are commonly carried out under the following assumptions: First, the enclosure can be subdivided into a finite number of isothermal surfaces. Second, the surfaces are gray body emitters, absorbers, and reflectors. Third, the emitted radiation and the reflected radiation leaving any surface in the enclosure both have directional distributions that are diffuse, that is, Lambert's Cosine Law is obeyed. Fourth, the magnitude of the radiant energy leaving any surface is uniform over that surface. The gray body condition leads to the relationships

$$\alpha = \epsilon \qquad \rho = 1 - \epsilon \tag{6}$$

where α, ϵ, and ρ respectively represent the absorptance, emittance, and reflectance. Moreover, the third and fourth assumptions permit the use of the same angle factors for gray surfaces as are used for black surfaces.

As a consequence of the foregoing assumptions, the surface heat transfer rates can be computed with a minimal knowledge of the radiation properties of the participating surfaces; only the hemispherical emittance ϵ of each surface is required. However, real surfaces seldom achieve the idealized behavior that is postulated. As more information about directional distributions and wavelength dependence becomes available, more accurate heat transfer computations will be possible.

The computation procedure for gray-diffuse enclosures can be phrased in various forms depending on the scope of the problem and on the available computational aids (for example, a digital computer). In the following paragraphs, consideration will be given to: (a) enclosures with few surfaces, (b) enclosures with many surfaces, and (c) the electric-circuit analogy. In paragraph (d) a generalization is described which takes account of nonisothermal surfaces and at the same time removes the aforementioned fourth assumption.

In dealing with gray-diffuse enclosures, it is convenient to introduce the radiosity B, which is the rate at which radiant energy leaves a surface per unit area. The radiosity of any surface j is related to the absolute temperature T_j and the net heat loss Q_j as follows

$$\frac{Q_j}{A_j} = \frac{\epsilon_j \sigma T_j^4 - (1 - \rho_j)B_j}{\rho_j} \tag{7}$$

For a gray surface, this becomes

$$\frac{Q_j}{A_j} = \frac{\epsilon_j}{1 - \epsilon_j}(\sigma T_j^4 - B_j) \tag{7a}$$

It is evident that if T_j is prescribed, Q_j is readily calculable if B_j is known. Thus, it is only necessary to provide computational equations for determining B_j.

a. **Few Surfaces.** At each surface of the enclosure, an equation can be written expressing the fact that the flux of radiant energy leaving the surface is the sum of the emitted radiation plus the reflected radiation. In mathematical terms, these equations take the form

$$B_1 = \epsilon_1 \sigma T_1^4 + \rho_1[B_1 F_{1-1} + B_2 F_{1-2} + \cdots + B_N F_{1-N}] \tag{8a}$$

$$B_2 = \epsilon_2 \sigma T_2^4 + \rho_2[B_1 F_{2-1} + B_2 F_{2-2} + \cdots + B_N F_{2-N}] \tag{8b}$$

. . . .

and for a typical surface j

$$B_j = \epsilon_j \sigma T_j^4 + \rho_j\left[\sum_{i=1}^{N} B_i F_{j-i}\right] \tag{8c}$$

where the ρ_j is equal to $1 - \epsilon_j$ in view of the gray body assumption. If there are N surfaces, then N equations of the form Eq. (8c) are written. Moreover, when the temperatures are prescribed there are N unknowns $B_1, B_2, \ldots, B_N$. Thus, the problem resolves itself into the task of solving N linear, inhomogeneous algebraic equations. The angle factors F_{j-i} are the very same that apply to radiant interchange between black surfaces. Methods for determining angle factors are described in detail in a later section of this article.

If there are relatively few surfaces, say two, three, four, or even five, then one can write down the appropriate equations and obtain solutions for the B_j by elementary numerical techniques, perhaps aided by a desk calculator. The corresponding Q_j then follow directly by evaluation of Eq. (7).

Further, if there are relatively few unknowns, then Eqs. (8) can be solved algebraically, giving

$$B_j = \sum_{i=1}^{N} C_{ji} \sigma T_i^4 , \quad j = 1, 2, \ldots, N \tag{9}$$

wherein the C_{ji} are constants that depend on the radiation properties of the surfaces and on the angle factors. Substituting this into Eq. (7), there follows

$$\frac{Q_j}{A_j} = \sum_{i=1}^{N} D_{ji} \sigma T_i^4 \quad , \quad j = 1, 2, \ldots, N \tag{10}$$

in which the D_{ji} are another set of constants that also depend on the radiation properties and angle factors. Equations (10) are in an especially convenient form for carrying out repetitive heat transfer computations in a given enclosure in which a succession of temperature values is assigned at the various surfaces. Computations of this type are encountered in transient heat transfer analysis. The essential simplification afforded by Eqs. (10) is that the basic set of linear algebraic equations need not be solved again and again as the temperatures of the problem are assigned different values.

If the heat flux rather than the temperature is prescribed at one of the surfaces, then the aforementioned computation method continues to apply with only slight modifications. Suppose that Q_k is prescribed at surface k. Then, from Eq. (7)

$$\epsilon_k \sigma T_k^4 = \rho_k \left(\frac{Q_k}{A_k}\right) + (1 - \rho_k) B_k \tag{11}$$

which becomes, for a gray surface

$$\epsilon_k \sigma T_k^4 = (1 - \epsilon_k) \left(\frac{Q_k}{A_k}\right) + \epsilon_k B_k \tag{11a}$$

With this, the quantity $\epsilon_k \sigma T_k^4$ can be eliminated from the radiant flux balance, Eq. (8c) with $j = k$, giving

$$B_k = \frac{Q_k}{A_k} + \sum_{i=1}^{N} B_i F_{k-i} \tag{12}$$

Equation (12) properly expresses the radiant flux balance for a surface at which the heat flux is prescribed.

In general, then, equations of the Eq. (8c) form are written for all surfaces j at which the temperature is prescribed; while equations of the form (12) are written for all surfaces at which the heat flux is prescribed. For an enclosure made up of N surfaces, this formulation yields N linear algebraic equations from which the N radiosities $B_1, B_2, \ldots, B_N$ can be determined numerically. With these, the heat flux or the surface temperature (whichever is unknown) can be respectively computed from Eqs. (7) and (11).

b. Many Surfaces. When a large number of surfaces is involved, the computation procedure outlined in the foregoing paragraph can be employed provided that a digital computer is available for solving the algebraic equations. This approach is fully satisfactory when numerical results for a given enclosure (that is, fixed radiation properties and angle factors) are to be obtained for only a single set of prescribed surface temperatures or heat fluxes. However, when repetitive computations involving a succession of surface temperatures or heat fluxes are to be carried out, it is uneconomic to re-solve the basic set of linear algebraic equations for each set of thermal conditions. Such repetitive computations may arise, for instance, in connection with a study of thermal transients. As shown below, the basic set of linear equations need be solved only once for a given enclosure; provided that use is made of the capabilities of the digital computer.

To illustrate the method, consider the case in which the surface temperatures are prescribed and the surface heat fluxes are to be computed. It is convenient to rewrite the radiant flux balances Eq. (8c) in the form

$$\sigma T_j^4 = \sum_{i=1}^N A_{ji} B_i \quad , \quad j = 1, 2, \ldots, N \tag{13}$$

in which

$$A_{ji} = \frac{\delta_{ji} - \rho_j F_{j-i}}{\epsilon_j} \tag{13a}$$

$$\delta_{ji} = 1 \text{ when } j = i \text{ and } \delta_{ji} = 0 \text{ when } j \neq i \tag{13b}$$

From the coefficients A_{ji}, one can form an N by N array of numbers, denoted by A

$$A = \begin{Vmatrix} A_{11} & A_{12} & \cdots & A_{1N} \\ A_{21} & A_{22} & \cdots & A_{2N} \\ \ldots & \ldots & \ldots & \ldots \\ \ldots & \ldots & \ldots & \ldots \\ A_{N1} & A_{N2} & \cdots & A_{NN} \end{Vmatrix} \tag{14}$$

A is called a matrix.

Modern digital computers, as part of their library of standard programs, have a routine called matrix inversion. Thus, given a matrix of numbers such as A, the machine will provide another matrix of numbers C which is called the inverse of A

$$C = A^{-1} = \begin{Vmatrix} C_{11} & C_{12} & \cdots & C_{1N} \\ C_{21} & C_{22} & \cdots & C_{2N} \\ \ldots & \ldots & \ldots & \ldots \\ \ldots & \ldots & \ldots & \ldots \\ C_{N1} & C_{N2} & \cdots & C_{NN} \end{Vmatrix} \tag{15}$$

In terms of the coefficients C_{ji} of the inverse matrix, one can write

$$B_j = \sum_{i=1}^N C_{ji} \sigma T_i^4 \quad , \quad j = 1, 2, \ldots, N \tag{16}$$

Thus, if the C_{ji} have been determined, the B_j are calculable for any prescribed set of temperatures T_j merely by evaluating Eq. (16). Moreover, if the B_j from Eq. (16) are introduced into Eq. (7), there follows

$$\frac{Q_j}{A_j} = \sum_{i=1}^{N} D_{ji}\sigma T_i^4 \quad , \quad j = 1, 2, \ldots, N \qquad (17)$$

in which the D_{ji} are a set of constants that are simply related to the C_{ji}. It is evident that Eq. (17) is ideally suited for repetitive-type computations in which the temperatures T_j are assigned a sequence of different values.

Now consider the situation in which temperatures are prescribed at some of the surfaces and heat fluxes are prescribed at the others. Suppose that the surfaces $1 \le j \le N_1$ have prescribed temperature, and surfaces $(N_1 + 1) \le j \le N$ have prescribed heat flux. For the former, Eqs. (13), (13a), and (13b) continue to apply, while for the latter

$$\frac{Q_j}{A_j} = \sum_{i=1}^{N} A_{ji} B_i \quad , \quad j = (N_1 + 1), (N_1 + 2), \ldots, N \qquad (18)$$

in which

$$A_{ji} = \delta_{ji} - F_{j-i} \qquad (18a)$$

and δ_{ji} has the same meaning as in Eq. (13b).

Next, one forms a matrix A as shown in Eq. (14); the first N_1 lines of the matrix are evaluated from Eq. (13a), while the last $(N - N_1)$ lines are evaluated from Eq. (18a). The inversion of the A matrix yields the C matrix as shown in Eq. (15). Once the C_{ji} are known, the radiosities B_j are calculable as follows

$$B_j = \sum_{i=1}^{N_1} C_{ji}\sigma T_i^4 + \sum_{N_1+1}^{N} C_{ji}\left(\frac{Q_i}{A_i}\right) \qquad (19)$$

Finally, for the surfaces $1 \le j \le N_1$, the heat transfer is computed by substituting the B_j from Eq. (19) into Eq. (7); on the other hand, for the surfaces $(N_1 + 1) \le j \le N$, the surface temperature is determined by introducing the B_j from Eq. (19) into Eq. (11).

c. The Electric-circuit Analogy. The radiant flux balance equations can be rephrased in a form that suggests an analogy with an electric circuit composed of resistances, voltage sources, and current sources [1]. For gray surfaces with prescribed temperatures, the flux balance equations can be written as

$$\sum_{i=1}^{N} \frac{B_i - B_j}{(A_j F_{j-i})^{-1}} + \frac{\sigma T_j^4 - B_j}{(1 - \epsilon_j)/\epsilon_j A_j} = 0 \qquad (20a)$$

while for gray surfaces with prescribed heat flux, the flux balance equations can be stated as

$$\sum_{i=1}^{N} \frac{B_i - B_j}{(A_j F_{j-i})^{-1}} + Q_j = 0 \qquad (20b)$$

One may associate the radiosities B and the emissive powers σT^4 with potentials, and the quantities $1/AF$ and $(1 - \epsilon)/\epsilon A$ with resistances. Then, terms such as

$$\frac{B_i - B_j}{(A_j F_{j-i})^{-1}} \quad \text{and} \quad \frac{\sigma T_j^4 - B_j}{(1 - \epsilon_j)/\epsilon_j A_j}$$

are clearly identifiable as currents. Moreover, Q_j is also a current.

Thus, Eqs. (20a) and (20b) express conservation of current at a node. Each term in the summation represents a current flowing from node i into node j. The term $(\sigma T_j^4 - B_j)/[(1 - \epsilon_j)/\epsilon_j A_j]$ is a current flowing into node j through an external connection that links an externally applied potential σT_j^4 with node j; the resistance of the connection is $(1 - \epsilon_j)/\epsilon_j A_j$. Q_j represents a current flowing into node j from an external current source.

In light of the foregoing interpretation, an electric-circuit analogy can be constructed from which the radiosities B_j are determinable by measurement of the voltages at the various nodes. Once the B_j are known, the surface heat fluxes or surface temperatures (whichever are unknown) are computed by evaluating Eqs. (7) or (11).

d. Nonuniform Surface Temperature and Nonuniform Radiosity. The occurrence of large temperature variations along the walls of the enclosure may necessitate the subdivision of the enclosure into a large number of surfaces. The limit of such a process of subdivision results in infinitesimal surfaces. Correspondingly, the surface temperature, the radiosity, and the heat flux become functions of the *local* position on the enclosure wall. Moreover, the summation that formerly appeared in the radiant flux balance now becomes an integral. Consequently, the radiant flux balance at any typical point x_0 on the wall of the enclosure takes the form

$$B(x_0) = \epsilon(x_0)\sigma T^4(x_0) + \rho(x_0) \int_{\text{all } x} B(x) \, dF_{x_0 - x} \tag{21}$$

wherein x is the running coordinate on the enclosure wall.

Equation (21) is called an integral equation because the unknown function $B(x)$ appears under the integral sign. To achieve a solution, this equation must be satisfied at all points x_0 on the wall of the enclosure. Once $B(x)$ has been solved for, the local heat flux is found by evaluation of Eq. (7).

Regardless of the nature of the surface temperature variation, the integral equation formulation takes account of the surface variations of radiosity B. In the formulations that were discussed earlier, wherein the enclosure was subdivided into a finite number of finite surfaces, the radiosity was assumed to be uniform along each surface. This assumption is rarely satisfied in practice.

Further details relating to the use of the integral equation formulation are given by Sparrow [2]. A typical application of this approach to engineering-type problems may be found in Ref. 3.

QUASI-GRAY SURFACES

The essential characteristic of a gray surface is that the spectral (or monochromatic) emittance is the same for all wavelengths. Actual materials rarely, if ever, approach this ideal behavior. Notwithstanding this, accurate results for many engineering systems can be obtained by applying the gray body assumption. This may happen when the radiant energy participating in the interchange is essentially confined to a finite wavelength range in which the spectral emittance is nearly uniform.

The gray body model becomes prone to significant error when the radiant energy spans a wide wavelength range that encompasses the visible as well as the infrared. Such a situation may occur, for example, when surfaces emit infrared radiation and are simultaneously irradiated by a high-temperature source such as the sun. A precise accounting of the wavelength dependence of the radiation properties can, in principle, be achieved by performing the radiant interchange computations monochromatically and then integrating over the entire wavelength range. However, the detailed spectral information needed to carry out such computations is not now available for the vast majority of engineering materials. Moreover, such computations constitute a major numerical undertaking.

In many cases, it may be sufficient to subdivide the active wavelength range into several finite bands and to assume that within each band the participating surfaces are gray. To illustrate the method, a two-band model will be discussed here. It is evident that as the number of bands is increased, the band approximation approaches a monochromatic description of the process.

Consider a situation in which essentially all the radiant energy is contained within a finite wavelength range $\Delta\lambda$. It will be assumed that $\Delta\lambda$ can be subdivided into two bands $\Delta\lambda'$ and $\Delta\lambda''$ such that the radiation properties are essentially constant within each band. The emittances ϵ' and ϵ'' respectively corresponding to the two bands are defined as

$$\epsilon' = \frac{e'}{e'_b} \quad , \quad \epsilon'' = \frac{e''}{e''_b} \tag{22}$$

wherein e' is the emissive power of a surface in the range $\Delta\lambda'$, and e'_b is the emissive power of a black surface in the same range. Corresponding definitions apply to e'' and e''_b. It should be emphasized that neither e'_b nor e''_b is equal to σT^4. Within each band, the gray body condition gives

$$\alpha' = \epsilon' \quad , \quad \rho' = 1 - \epsilon' \quad \text{and} \quad \alpha'' = \epsilon'' \quad , \quad \rho'' = 1 - \epsilon'' \tag{23}$$

At each surface of the enclosure, a separate radiant flux balance can be written for each band, thus

$$B'_j = \epsilon'_j e'_{bj} + (1 - \epsilon'_j) \sum_{i=1}^{N} B'_i F_{j-i} \qquad j = 1, 2, \ldots, N \tag{24a}$$

$$B''_j = \epsilon''_j e''_{bj} + (1 - \epsilon''_j) \sum_{i=1}^{N} B''_i F_{j-i} \qquad j = 1, 2, \ldots, N \tag{24b}$$

These equations are based on diffuse distributions of the emitted and reflected radiation. When the surface temperatures are prescribed, then e'_{bj} and e''_{bj} are known (e.g., Dunkle [4]). Correspondingly, the $B'_1, B'_2, \ldots, B'_N$ are completely independent of the $B''_1, B''_2, \ldots, B''_N$. Thus, the set of N equations represented by Eq. (24a) can be solved independently of Eq. (24b), and vice versa. Any of the various methods of attack outlined in prior sections of this article can be used without modification. Once the B' and B'' have been found, the net heat transfer rate is evaluated from

$$\frac{Q_j}{A_j} = \frac{\epsilon'_j}{1 - \epsilon'_j} (e'_{bj} - B'_j) + \frac{\epsilon''_j}{1 - \epsilon''_j} (e''_{bj} - B''_j) \tag{25}$$

It is evident from the foregoing that when the surface temperatures are prescribed, the use of the band approximation causes no essential change in the computation procedure compared to that for the gray enclosure. The situation is somewhat different when the heat flux is prescribed rather than the temperature. By considering

Eq. (25), it is evident that when Q_j is prescribed, the quantities B'_j and B''_j are no longer independent of each other.

The case where the heat flux is prescribed at only one of the surfaces, say $j = k$, is analyzed as follows: The flux balance equations (24a) and (24b) are solved in the following form

$$B'_j = \sum_{i=1}^{N} C'_{ji} e'_{bi} \quad , \quad j = 1, 2, \ldots, N \tag{26a}$$

$$B''_j = \sum_{i=1}^{N} C''_{ji} e''_{bi} \quad , \quad j = 1, 2, \ldots, N \tag{26b}$$

This form has already been discussed in connection with Eqs. (9) and (16). In particular, for surface k

$$B'_k = C'_{kk} e'_{bk} + \sum_{i=1}^{N} C'_{ki} e'_{bi} \tag{27a}$$

$$B''_k = C''_{kk} e''_{bk} + \sum_{i=1}^{N} C''_{ki} e''_{bi} \tag{27b}$$

where $i = k$ is excluded from the summation. These equations contain three unknowns: B'_k, B''_k, and the temperature T_k (e'_{bk} and e''_{bk} are functions of T_k). In addition, Eq. (25) with $j = k$ provides a third relationship between the unknowns when Q_k is prescribed. These three equations may be solved simultaneously with the aid of Ref. 4 which gives information on e'_b and e''_b as functions of T. Once e'_{bk} and e''_{bk} have been found, then the B' and B'' can be computed from Eqs. (26a) and (26b) for each surface in the enclosure; the corresponding heat fluxes are then evaluated from Eq. (25). The case in which the heat flux is prescribed at two or more surfaces of the enclosure is analyzed by a straightforward generalization of the foregoing procedure.

The present discussion has been limited to a two-band model in order to facilitate a concise presentation. It is evident that the use of multiband models offers nothing that is conceptually different from that described here. However, it should be noted that the required computational effort becomes greater as the number of bands is increased.

SPECULARLY REFLECTING SURFACES

Although radiant interchange computations are usually carried out under the assumption that the participating surfaces are diffuse reflectors, there is experimental evidence that the directional component in the specular direction may also contribute. The general problem of radiant interchange wherein the surfaces have arbitrary directional distributions of emitted and reflected radiation is very formidable indeed. Moreover, detailed directional distributions are not now available for most engineering materials. Furthermore, the directional distributions are highly sensitive to surface conditions such as roughness and the presence of contaminants.

A model which permits some assessment of directional effects employs the postulate of diffusely emitting surfaces having reflectances that can be subdivided into specular and diffuse components, thus

$$\rho = \rho_d + \rho_s \tag{28}$$

In principle, for a gray enclosure composed of such surfaces, the equations of radiant interchange can be written as

$$B_j = \epsilon_j \sigma T_j^4 + \rho_{dj} \sum_{i=1}^{N} B_i E_{j-i} \quad , \quad j = 1, 2, \ldots, N \tag{29}$$

in which E_{j-i} is an exchange factor, and ρ_{dj} is the diffuse component of the reflectance of surface j. In general, each E_{j-i} is composed of a sum of terms; each one of such terms is a diffuse angle factor times a specular reflectance raised to a power. The heat flux is related to the radiosity and emissive power as follows

$$\frac{Q_j}{A_j} = \frac{\epsilon_j (1 - \rho_{sj}) \sigma T_j^4 - \epsilon_j B_j}{1 - \epsilon_j - \rho_{sj}} \tag{30}$$

where ρ_{sj} is the specular component of the reflectance.

For an enclosure composed of surfaces of arbitrary geometrical configuration, the determination of the exchange factors is a difficult undertaking. However, if the surfaces having specular components (that is, $\rho_s \neq 0$) are plane, then the task of finding the exchange factors is significantly simplified. In this case, use is made of a basic property of plane mirrors; namely, that radiant energy (or light) reflected from a plane mirror appears to come from an image surface located behind the mirror. The distance between the image surface and the mirror is identical to the distance at which the object surface is located in front of the mirror. Moreover, if the radiant energy incident on the mirror comes from a diffuse source, the specularly reflected radiation will appear to emanate from a diffuse surface situated at the image of the source. As a consequence of the aforementioned property, one can construct an exchange factor such as E_{j-i} by making use of diffuse angle factors between surface j and the images of surface i that are formed in the plane mirrors. The procedure is illustrated in detail by Sparrow [2] and will not be dwelt upon here.

When the surfaces having specular components of reflection are curved, the aforementioned image method does not appear to afford significant simplification. The case of curved surfaces has been treated by Perlmutter and Siegel [5] and Lin and Sparrow [6]. The first of these demonstrates a method for finding exchange factors appropriate to radiant transport through an open-ended circular tube that connects isothermal environments. The second seeks to formulate a general method of analysis which is illustrated by application to radiant interchange in conical and circular cylindrical cavities. The formulation is somewhat lengthy and is omitted here; details may be found in the references.

ANGLE FACTORS

A knowledge of the numerical values of angle factors is a requisite for the computation of radiant interchange between surfaces. An angle factor is a dimensionless quantity representing the fraction of the radiant energy leaving one surface that arrives at a second surface. When the radiant energy leaving a surface is diffusely distributed (that is, Lambert's Cosine Law is obeyed), the following mathematical definitions apply.

For two infinitesimal surfaces dA_1 and dA_2

$$dF_{dA_1 - dA_2} = \frac{\cos\beta_1 \cos\beta_2 \, dA_2}{\pi r^2} \tag{31a}$$

$$dA_1 dF_{dA_1 - dA_2} = dA_2 dF_{dA_2 - dA_1} \tag{31b}$$

For an infinitesimal surface dA_1 and a finite surface A_2

$$F_{dA_1 - A_2} = \int_{A_2} \frac{\cos\beta_1 \cos\beta_2}{\pi r^2} \, dA_2 \tag{32a}$$

$$dA_1 F_{dA_1 - A_2} = A_2 dF_{A_2 - dA_1} \tag{32b}$$

For two finite surfaces A_1 and A_2

$$F_{A_1 - A_2} = \frac{1}{A_1} \int_{A_1} \int_{A_2} \frac{\cos\beta_1 \cos\beta_2}{\pi r^2} \, dA_1 dA_2 \tag{33a}$$

$$A_1 F_{A_1 - A_2} = A_2 F_{A_2 - A_1} \tag{33b}$$

In the foregoing equations, the first subscript affixed to F or dF represents the surface from which the radiant energy leaves; the second subscript represents the surface upon which the radiant energy is incident.

The angles β_1 and β_2 and the distance r are illustrated in Fig. 1. In terms of a rectangular coordinate system, these quantities can be evaluated as

$$\cos\beta_1 = \frac{l_1(x_2 - x_1) + m_1(y_2 - y_1) + n_1(z_2 - z_1)}{r} \tag{34a}$$

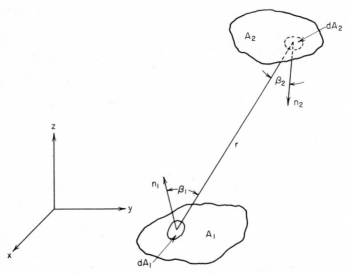

Fig. 1. Nomenclature and coordinates for angle factor computation.

$$\cos\beta_2 = \frac{l_2(x_1 - x_2) + m_2(y_1 - y_2) + n_2(z_1 - z_2)}{r} \qquad (34b)$$

$$r^2 = (x_2 - x_1)^2 + (y_2 - y_1)^2 + (z_2 - z_1)^2 \qquad (34c)$$

where (x_1, y_1, z_1) are the coordinates of a point on surface 1 and (x_2, y_2, z_2) are the coordinates of a point on surface 2. The l, m, and n are direction cosines. Specifically, l_1 is the cosine of the angle between the x axis and the surface normal at position (x_1, y_1, z_1); m_1 is the cosine of the angle between the y axis and the surface normal at (x_1, y_1, z_1); and so forth.

When the expressions for $\cos\beta_1$, $\cos\beta_2$, and r^2 are substituted into the defining equations (31a), (32a), and (33a), angle factors can, in principle, be computed provided that the direction cosines can be expressed as a function of the coordinates. The integrals involved in Eqs. (32a) and (33a) are frequently difficult to carry out analytically and numerical integration may be necessary.

In many cases, it is not necessary to deal with Eqs. (31) to (33) in order to obtain numerical values of angle factors. There exists in the literature a large body of numerical and algebraic information for surface arrangements that often recur in engineering practice. Information of this type will be presented later in this article. There are also analytical methods alternate to Eqs. (31) to (33) as well as certain shorthand procedures that are generally easier to apply; these will be discussed below. Finally, graphical and analogy methods are available [7–10], but these will not be treated here.

CONTOUR INTEGRAL REPRESENTATION

The area integral appearing in Eq. (32a) usually implies a double integration, while the double area integral of Eq. (33a) usually implies a quadruple integration. A reduction in the multiplicity of the integration can be achieved by employing contour integrals instead of area integrals [11]. The reduction in multiplicity is especially advantageous when numerical evaluation is required. The contour integral representation is as follows

$$F_{dA_1 - A_2} = l_1 \oint_{C_2} \frac{(z_2 - z_1)dy_2 - (y_2 - y_1)dz_2}{2\pi r^2} + m_1 \oint_{C_2} \frac{(x_2 - x_1)dz_2 - (z_2 - z_1)dx_2}{2\pi r^2}$$

$$+ n_1 \oint_{C_2} \frac{(y_2 - y_1)dx_2 - (x_2 - x_1)dy_2}{2\pi r^2} \qquad (35)$$

$$F_{A_1 - A_2} = \frac{1}{2\pi A_1} \left(\oint_{C_1} \oint_{C_2} \ln r\, dx_1 dx_2 + \ln r\, dy_1 dy_2 + \ln r\, dz_1 dz_2 \right) \qquad (36)$$

It is evident that Eq. (35) involves only single integrals and Eq. (36) only double integrals.

In Eq. (35), the integration is carried out around the boundary C_2 of the area A_2. (x_2, y_2, z_2) are coordinates on the contour C_2; these vary during the integration. On the other hand, the coordinates (x_1, y_1, z_1) of dA_1 are constant during the integration, as are the direction cosines l_1, m_1, and n_1. The distance r between dA_1 and the boundary points of C_2 is expressed by Eq. (34c).

If the element dA_1 is aligned so that its normal lies along one of the coordinate axes, then two of the direction cosines are zero and correspondingly, two of the three terms that appear in Eq. (35) may be deleted (e.g., if the normal is aligned along the y direction, $l_1 = n_1 = 0$). Further, if any coordinate is constant along any part of the contour C_2, then the corresponding differentials are zero and portions of the integrands are thus deleted. Therefore, by proper choice of the coordinate system, Eq. (35) yields a relatively simple expression for the angle factor.

Each one of the three integrals appearing in Eq. (35) is, in itself, an angle factor. For instance, the first integral is the angle factor between an element dA_1 whose direction cosines are $l_1 = 1$, $m_1 = n_1 = 0$ and the area A_2. The other integrals are similarly interpreted. When viewed in this way, Eq. (35) states that the angle factor for an element with arbitrary direction cosines l_1, m_1, and n_1 can be built up by superposing the angle factors for three basic elements whose direction cosines are, respectively, (1, 0, 0), (0, 1, 0), and (0, 0, 1).

In Eq. (36), the integrations are performed around the contours C_1 and C_2 that respectively bound the areas A_1 and A_2. The coordinates (x_1, y_1, z_1) and (x_2, y_2, z_2) correspond to points on the respective boundaries, and r is the distance between such boundary points.

Further details relating to the application of the contour integration method are available [11].

ANGLE FACTOR ALGEBRA

An unknown angle factor for some surface arrangement of interest can often be computed by summing and/or differencing known angle factors that pertain to different surface arrangements. This procedure, known as angle factor algebra, is illustrated in Fig. 2. Consider the computation of the angle factor between a disk,

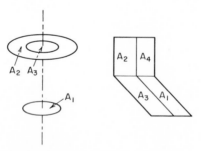

Fig. 2. Illustrations of angle factor algebra.

area A_1, and a coaxial annulus, area A_2. The radiant energy arriving at the annulus A_2 from the disk A_1 is the difference between the radiant energy passing from A_1 to disk ($A_2 + A_3$) and that passing from A_1 to disk A_3. Furthermore, angle factors for interchange between coaxial disks of arbitrarily different radii are available in the literature. Thus

$$F_{A_1 - A_2} = F_{A_1 - (A_2 + A_3)} - F_{A_1 - A_3} \qquad (37)$$

where both terms on the right-hand side are known.

An illustration involving plane rectangular surfaces is shown in the right-hand sketch of Fig. 2. For this case, it can be shown that [12]

$$F_{A_1 - A_2} = \frac{1}{2A_1}\left[(A_1 + A_3) F_{(A_1 + A_3) - (A_2 + A_4)} - A_1 F_{A_1 - A_4} - A_3 F_{A_3 - A_2}\right] \qquad (38)$$

All the angle factors appearing within the brackets pertain to the well-known case of perpendicular rectangles sharing a common edge.

An expanded discussion of angle factor algebra may be found in Refs. 8 and 12.

ELONGATED SURFACES

Particularly simple relationships exist for surfaces that are greatly elongated in one of their dimensions. For example, the surfaces pictured in Fig. 3 are very long in the

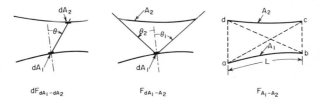

Fig. 3. Surface elongated in the direction normal to the plane of the page.

direction into and out of the plane of the page. For interchange between two such surfaces having infinitesimal arc lengths (left-hand sketch), the angle factor is

$$dF_{dA_1 - dA_2} = \frac{1}{2} d(\sin \theta) \tag{39a}$$

wherein θ is formed by the normal to dA_1 and the connecting line between dA_1 and dA_2. By direct integration of Eq. (39a), the angle factor for interchange between an infinitesimal element and one having finite arc length is

$$F_{dA_1 - A_2} = \frac{1}{2} (\sin \theta_1 + \sin \theta_2) \tag{39b}$$

The angles θ_1 and θ_2 are formed by the normal at dA_1 and the connecting lines drawn to the extremities of A_2 (center sketch).

The angle factor for interchange between elements A_1 and A_2 having finite arc lengths (right-hand sketch) is conveniently derived by Hottel's string method [13]. In this method, strings are imagined to be tightly stretched between the end points of A_1 and A_2. The angle factor $F_{A_1 - A_2}$ is computed by adding the lengths of the crossed strings and subtracting away the lengths of the uncrossed strings, and then dividing by twice the arc length of A_1. In terms of the notation of the figure, this is

$$F_{A_1 - A_2} = \frac{1}{2L} [(\overline{ac} + \overline{bd}) - (\overline{ad} + \overline{bc})] \tag{39c}$$

ANGLE FACTOR CATALOGUE

Graphical and algebraic angle factor information will now be presented for the surface configurations that are shown schematically in Fig. 4a and b. The first of these figures pertains to interchange between pairs of finite surfaces, while the second figure pertains to interchange between an infinitesimal element and a finite surface. Numerical values of $F_{A_1 - A_2}$ and $dF_{dA_1 - A_2}$ are presented in Figs. 5 through 12 for several of the aforementioned configurations. Algebraic expressions from which the angle factors can be computed are stated in the equations that follow. Information for other configurations may be found in Refs. 8 and 14.

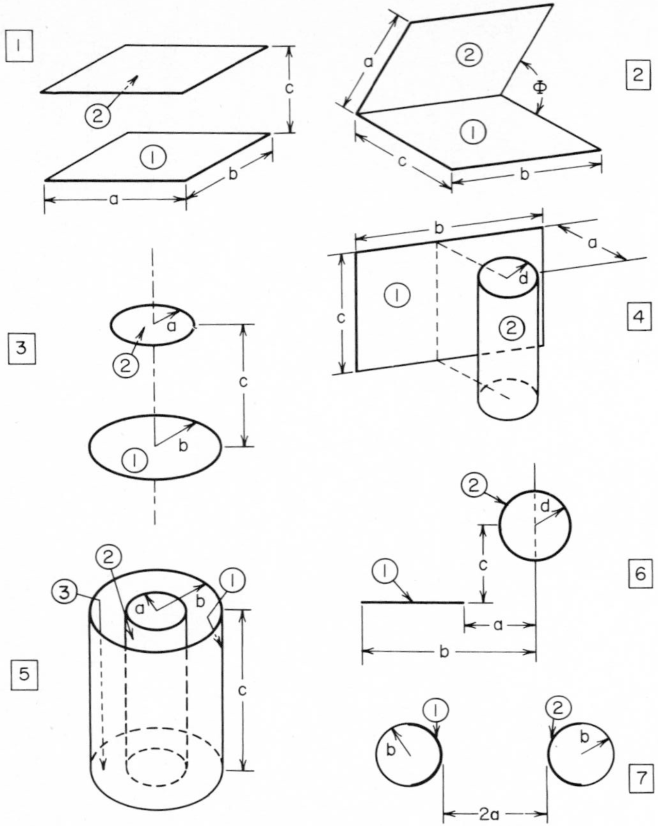

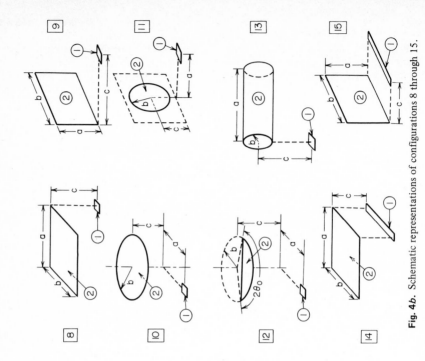

Fig. 4b. Schematic representations of configurations 8 through 15.

Fig. 4a. Schematic representations of configurations 1 through 7.

Configuration 1 $X = a/c,\ Y = b/c$

$$F_{A_1 - A_2}\left(\frac{\pi XY}{2}\right) = \ln\left[\frac{(1 + X^2)(1 + Y^2)}{1 + X^2 + Y^2}\right]^{1/2} + Y\sqrt{1 + X^2}\ \tan^{-1}\left(\frac{Y}{\sqrt{1 + X^2}}\right)$$

$$+ X\sqrt{1 + Y^2}\ \tan^{-1}\left(\frac{X}{\sqrt{1 + Y^2}}\right) - Y\tan^{-1}Y - X\tan^{-1}X$$

Configuration 2 $X = a/b,\ Y = c/b,\ Z = X^2 + Y^2 - 2XY\cos\Phi$

$$F_{A_1 - A_2}(\pi Y) = -\frac{\sin 2\Phi}{4}\left[XY\sin\Phi + \left(\frac{\pi}{2} - \Phi\right)(X^2 + Y^2) + Y^2\tan^{-1}\left(\frac{X - Y\cos\Phi}{Y\sin\Phi}\right)\right.$$

$$+ X^2\tan^{-1}\left(\frac{Y - X\cos\Phi}{X\sin\Phi}\right)\bigg] + \frac{\sin^2\Phi}{4}\left\{\left(\frac{2}{\sin^2\Phi} - 1\right)\ln\left[\frac{(1 + X^2)(1 + Y^2)}{1 + Z}\right]\right.$$

$$+ Y^2\ln\left[\frac{Y^2(1 + Z)}{(1 + Y^2)Z}\right] + X^2\ln\left[\frac{X^2(1 + X^2)^{\cos 2\Phi}}{Z(1 + Z)^{\cos 2\Phi}}\right]\bigg\}$$

$$+ Y\tan^{-1}\left(\frac{1}{Y}\right) + X\tan^{-1}\left(\frac{1}{X}\right) - \sqrt{Z}\tan^{-1}\left(\frac{1}{\sqrt{Z}}\right)$$

$$+ \frac{\sin\Phi\sin 2\Phi}{2} X\sqrt{1 + X^2\sin^2\Phi}\left[\tan^{-1}\left(\frac{X\cos\Phi}{\sqrt{1 + X^2\sin^2\Phi}}\right)\right.$$

$$+ \tan^{-1}\left(\frac{Y - X\cos\Phi}{\sqrt{1 + X^2\sin^2\Phi}}\right)\bigg] + \cos\Phi\int_0^Y \sqrt{1 + \xi^2\sin^2\Phi}$$

$$\times\left[\tan^{-1}\left(\frac{X - \xi\cos\Phi}{\sqrt{1 + \xi^2\sin^2\Phi}}\right) + \tan^{-1}\left(\frac{\xi\cos\Phi}{\sqrt{1 + \xi^2\sin^2\Phi}}\right)\right]d\xi$$

Configuration 3 $X = a/c,\ Y = c/b,\ Z = 1 + (1 + X^2)Y^2$

$$F_{A_1 - A_2} = \tfrac{1}{2}(Z - \sqrt{Z^2 - 4X^2Y^2})$$

Configuration 4 $X = a/d,\ Y = b/d,\ Z = c/d$

$$A = Z^2 + X^2 + \xi^2 - 1\ ,\quad B = Z^2 - X^2 - \xi^2 + 1$$

$$F_{A_1 - A_2} = \frac{2}{Y}\int_0^{Y/2} f(\xi)d\xi$$

$$f(\xi) = \frac{X}{X^2 + \xi^2} - \frac{X}{\pi(X^2 + \xi^2)} \left\{ \cos^{-1}\frac{B}{A} - \frac{1}{2Z} \left[\sqrt{A^2 + 4Z^2} \cos^{-1}\left(\frac{B}{A\sqrt{X^2 + \xi^2}} \right) \right.\right.$$

$$\left.\left. + B \sin^{-1}\left(\frac{1}{\sqrt{X^2 + \xi^2}} \right) - \frac{\pi A}{2} \right] \right\}$$

Configuration 5 $X = b/a$, $Y = c/a$, $A = Y^2 + X^2 - 1$, $B = Y^2 - X^2 + 1$

$$F_{A_1 - A_2} = \frac{1}{X} - \frac{1}{\pi X} \left\{ \cos^{-1}\frac{B}{A} - \frac{1}{2Y} \left[\sqrt{(A + 2)^2 - (2X)^2} \cos^{-1}\frac{B}{XA} + B \sin^{-1}\frac{1}{X} - \frac{\pi A}{2} \right] \right\}$$

$$F_{A_1 - A_1} = 1 - \frac{1}{X} + \frac{2}{\pi X} \tan^{-1}\left(\frac{2\sqrt{X^2 - 1}}{Y} \right)$$

$$- \frac{Y}{2\pi X} \left\{ \frac{\sqrt{4X^2 + Y^2}}{Y} \sin^{-1}\left[\frac{4(X^2 - 1) + (Y^2/X^2)(X^2 - 2)}{Y^2 + 4(X^2 - 1)} \right] \right.$$

$$\left. - \sin^{-1}\left(\frac{X^2 - 2}{X^2} \right) + \frac{\pi}{2}\left(\frac{\sqrt{4X^2 + Y^2}}{Y} - 1 \right) \right\}$$

$$F_{A_1 - A_3} = \frac{1}{2}\left(1 - F_{A_1 - A_2} - F_{A_1 - A_1} \right)$$

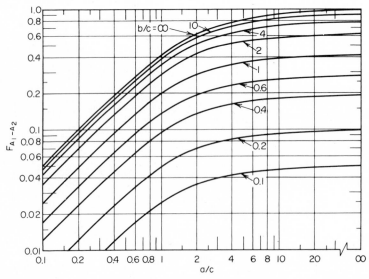

Fig. 5. Angle factors for configuration 1.

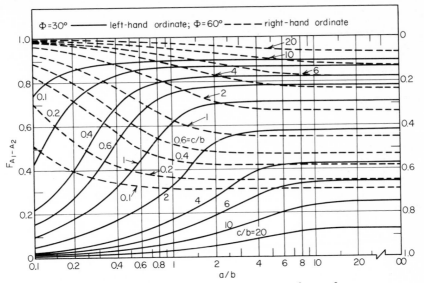

Fig. 6a. Angle factors for configuration 2, $\Phi = 30°$ and $60°$.

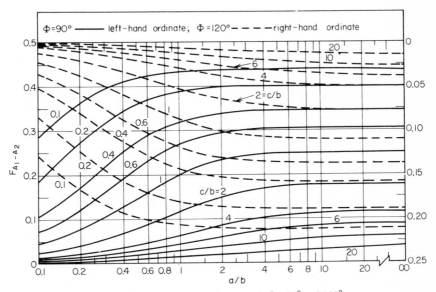

Fig. 6b. Angle factors for configuration 2, $\Phi = 90°$ and $120°$.

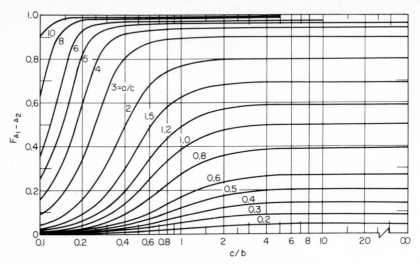

Fig. 7. Angle factors for configuration 3.

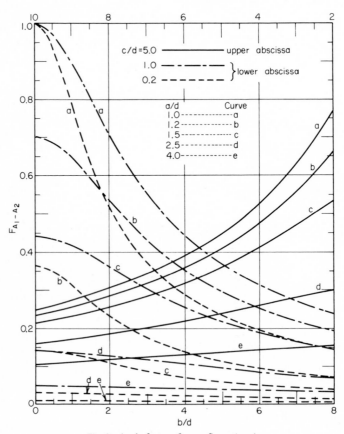

Fig. 8. Angle factors for configuration 4.

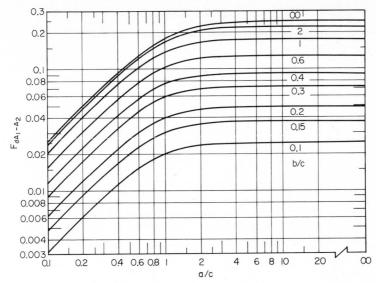

Fig. 9. Angle factors for configuration 8.

Configuration 6 $X = c/d,\ Y = a/d,\ Z = b/d$

$$F_{A_1 - A_2} = \frac{1}{Z - Y}\left(\tan^{-1}\frac{Z}{X} - \tan^{-1}\frac{Y}{X}\right)$$

Configuration 7 $X = 1 + (a/b)$

$$F_{A_1 - A_2} = \frac{2}{\pi}\left(\sqrt{X^2 - 1} - X + \frac{\pi}{2} - \cos^{-1}\frac{1}{X}\right)$$

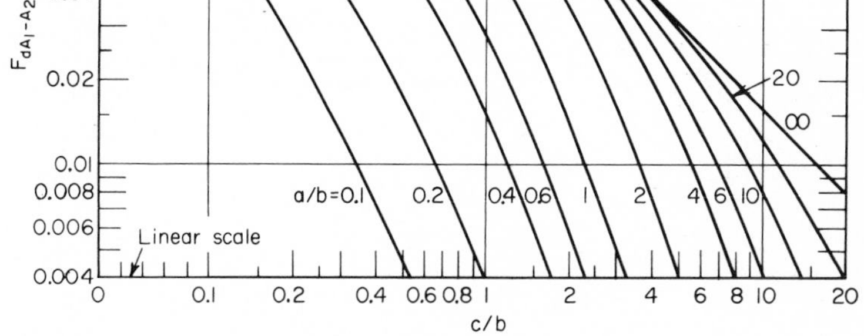

Fig. 10. Angle factors for configuration 9.

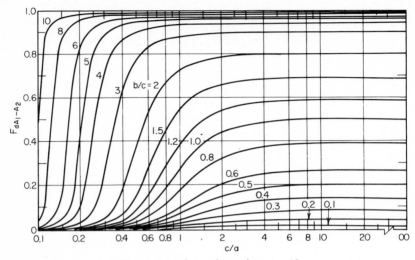

Fig. 11. Angle factors for configuration 10.

Configuration 8 $X = a/c$, $Y = b/c$

$$F_{dA_1 - A_2} = \frac{1}{2\pi}\left[\frac{X}{\sqrt{1 + X^2}}\ \tan^{-1}\left(\frac{Y}{\sqrt{1 + X^2}}\right) + \frac{Y}{\sqrt{1 + Y^2}}\ \tan^{-1}\left(\frac{X}{\sqrt{1 + Y^2}}\right)\right]$$

Configuration 9 $X = a/b$, $Y = c/b$, $A = 1/\sqrt{X^2 + Y^2}$,

$$F_{dA_1 - A_2} = \frac{1}{2\pi}\left(\tan^{-1}\frac{1}{Y} - AY\ \tan^{-1} A\right)$$

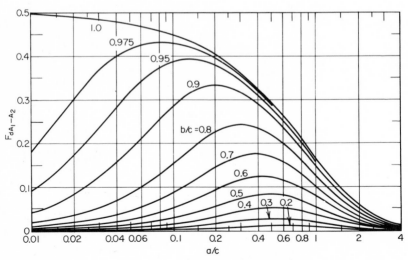

Fig. 12. Angle factors for configuration 11.

Configuration 10 $X = c/a$, $Y = b/c$, $Z = 1 + (1 + Y^2)X^2$

$$F_{dA_1 - A_2} = \frac{1}{2}\left(1 - \frac{Z - 2Y^2 X^2}{\sqrt{Z^2 - 4Y^2 X^2}}\right)$$

Configuration 11 $X = a/c$, $Y = b/c$

$$F_{dA_1 - A_2} = \frac{X}{2}\left[\frac{1 + Y^2 + X^2}{\sqrt{(1 + Y^2 + X^2)^2 - 4Y^2}} - 1\right]$$

Configuration 12 $X = a/b$, $Y = c/b$, $Z = \left[(1 + X^2 + Y^2)^2 - 4X^2\right]^{1/2}$

$$F_{dA_1 - A_2} = \frac{\theta_0}{2\pi} + \frac{1 - X^2 - Y^2}{\pi Z}\tan^{-1}\left[\frac{Z \tan(\theta_0/2)}{1 + X^2 + Y^2 - 2X}\right]$$

$$+ \frac{X - \cos\theta_0}{\pi\sqrt{(X - \cos\theta_0)^2 + Y^2}}\tan^{-1}\left[\frac{\sin\theta_0}{\sqrt{(X - \cos\theta_0)^2 + Y^2}}\right]$$

Configuration 13 $X = a/b$, $Y = c/b$, $A = (1 + Y)^2 + X^2$, $B = (1 - Y)^2 + X^2$

$$F_{dA_1 - A_2} = \frac{1}{\pi Y}\tan^{-1}\left(\frac{X}{\sqrt{Y^2 - 1}}\right) + \frac{X}{\pi}\left\{\frac{A - 2Y}{Y\sqrt{AB}}\tan^{-1}\left[\sqrt{\frac{A(Y - 1)}{B(Y + 1)}}\right]\right.$$

$$\left. - \frac{1}{Y}\tan^{-1}\left(\sqrt{\frac{Y - 1}{Y + 1}}\right)\right\}$$

Configuration 14 $X = b/c$, $Y = a/c$

$$F_{dA_1 - A_2} = \frac{1}{\pi X}\left[\sqrt{1 + X^2}\ \tan^{-1}\left(\frac{Y}{\sqrt{1 + X^2}}\right) - \tan^{-1}Y + \frac{XY}{\sqrt{1 + Y^2}}\tan^{-1}\left(\frac{X}{\sqrt{1 + Y^2}}\right)\right]$$

Configuration 15 $X = a/b$, $Y = c/b$, $Z = X^2 + Y^2$

$$F_{dA_1 - A_2} = \frac{1}{\pi}\left\{\tan^{-1}\frac{1}{Y} + \frac{Y}{2}\ln\left[\frac{Y^2(Z + 1)}{(Y^2 + 1)Z}\right] - \frac{Y}{\sqrt{Z}}\tan^{-1}\left(\frac{1}{\sqrt{Z}}\right)\right\}$$

REFERENCES

1. A. K. Oppenheim, Radiation Analysis by the Network Method, *Trans. ASME*, **78**:725 (1956).
2. E. M. Sparrow, Radiation Heat Transfer between Surfaces, in "Advances in Heat Transfer" (J. P. Hartnett and T. F. Irvine, Jr., eds.), Academic Press, Inc., New York, 1965.
3. E. M. Sparrow, E. R. G. Eckert, and T. F. Irvine, Jr., The Effectiveness of Radiating Fins with Mutual Irradiation, *J. Aerospace Sci.*, **28**:763 (1961).
4. R. V. Dunkle, Thermal Radiation Tables and Applications, *Trans. ASME*, **65**:537 (1954).
5. M. Perlmutter and R. Siegel, Effect of Specularly Reflecting Gray Surface on Thermal Radiation through a Tube and from Its Heated Wall, *J. Heat Transfer*, **C85**:55 (1963).

6. S. H. Lin and E. M. Sparrow, Radiant Interchange among Curved Specularly Reflecting Surfaces; Application of Cylindrical and Conical Cavities, *J. Heat Transfer,* **C87**:299 (1965).

7. E. R. G. Eckert and R. M. Drake, "Heat and Mass Transfer," McGraw-Hill Book Company, Inc., New York, 1959.

8. D. C. Hamilton and W. R. Morgan, Radiant Interchange Configuration Factors, *NACA TN* 2836 (1952).

9. M. Jakob, "Heat Transfer," vol. 2, John Wiley & Sons, Inc., New York, 1957.

10. F. Kreith, "Radiation Heat Transfer," International Textbook Company, Scranton, Pa., 1962.

11. E. M. Sparrow, A New and Simpler Formulation for Radiative Angle Factors, *J. Heat Transfer,* **C85**:81 (1962).

12. A. J. Chapman, "Heat Transfer," The Macmillan Company, New York, 1960.

13. H. C. Hottel, Radiant Heat Transmission, in "Heat Transmission" (by W. H. McAdams), 3d ed., chap. 3, McGraw-Hill Book Company, Inc., New York, 1954.

14. H. Leuenberger and R. A. Pearson, Compilation of Radiation Shape Factors for Cylindrical Assemblies, *ASME Paper* 56-A-144, ASME Annual Meeting, New York, 1956.

Part C

Radiation Exchange in an Enclosure with a Participating Gas

R. V. DUNKLE

Division of Mechanical Engineering, C.S.I.R.O.
Highett, Victoria, Australia

INTRODUCTION

The computation of the radiant heat exchange in an enclosure containing an absorbing and radiating gas is inherently an extremely complex problem. It is necessary to make certain idealizations and approximations to arrive at a workable solution. The method of analysis is essentially the same as that for the enclosure containing a nonparticipating medium, as discussed in Part B of this section by E. M. Sparrow [1], but is considerably complicated by the presence of the participating gas.

In this section two methods of analysis will be presented, the first method based on a grey system approximation and the second on a band energy approximation. In both systems, the gas will be assumed to be isothermal and homogeneous. Each wall in the enclosure is assumed to be isothermal and of uniform radiosity. All surfaces are assumed to be perfectly diffuse. The technique for dealing with specular surfaces is covered by Sparrow [1] for the empty enclosure, while a paper by Dunkle [2] treats a gas-filled enclosure with some specular surfaces as well as a diathermanous wall. In the analysis presented herein, all surfaces are assumed to be opaque, so that all radiation leaving a surface (radiosity) is either emitted by or reflected from the surface.

The radiation leaving the surface is distributed continuously over the spectrum, although the intensity varies due to the interaction of several factors. The emitted radiation has a spectral and directional distribution fixed by the temperature, composition, and surface structure of the wall. As the surface is assumed diffuse, the intensity of emitted radiation is independent of direction; this assumption is also usually made for emission from specular surfaces in complex systems. The spectral and directional distributions for the reflected radiation are in general much more complex, although for diffuse surfaces the intensity is independent of direction.

The radiation leaving the surface passes through the gas to other surfaces. As discussed in the section on gas radiation by Penner, emission and absorption by the gas are discontinuous and occur only in certain spectral regions. If the spectral intensity of a black body at the gas temperature is greater than that leaving the wall, then the spectral intensity will be augmented in passing through the gas; if less then the intensity will decrease. The rate of change with distance depends upon the spectral absorption coefficient of the gas.

The spectral absorption coefficient of gases varies extremely rapidly with wavelength and is very small outside the active absorption bands so that the spectral distribution after passing through a mass of gas is very irregular. However, these peaks and dips tend to be smoothed out somewhat by absorption and emission at wall surfaces. For example, if the wall is black, all incident radiation is absorbed and the spectral distribution of radiation leaving the wall is that of an ideal radiator at the wall temperature. On the other hand, for a perfect reflector, there is no alteration in the wavelength distribution, only in the directional distribution. Due to the band character of gaseous radiation, radiant heat exchange calculations based on total properties are only exact for enclosures consisting of black surfaces. However, the method presented here results in a good approximation for grey-walled enclosures.

THE GREY ENCLOSURE, TOTAL RADIATION METHOD

The total emittance and absorptance correlations of Hottel [3] as presented in the chapter on gaseous radiation are used in this method. The correlation equations of Bevans [4] can also be used. Each wall surface is assumed to be grey, in the case of non-grey surfaces the total emittance of the wall temperature is used in the calculations and equated to the absorptance.

The method given here is based upon a calculation of the radiation exchange between each surface element and the gas. Next a heat balance is written for each surface element or node. The term node comes from the thermal radiation circuit concept [7], and refers to the junction or connection between two or more circuit elements. Thus each surface element is joined by a conductance to a surface node which in turn is linked by conductances to all other surface and gas nodes in the system. Two conditions may apply to each node, either the temperature is fixed or the net heat flux is specified. For each node the heat balance yields an equation in terms of the radiosities of all the nodes in the system. The resulting set of equations, similar to those for the enclosure with a nonparticipating gas, can be solved by standard techniques as outlined in the section by Sparrow.

RADIATION FROM A SURFACE TO THE GAS

Radiation leaving the jth surface and passing through the gas to the ith surface is attenuated by absorption, the absorptance α_{gji} depending upon the mean beam length r_{ji} between the two surfaces. A short section on mean beam lengths is included at the end of this section for convenient reference [3, 5]. The portion of the

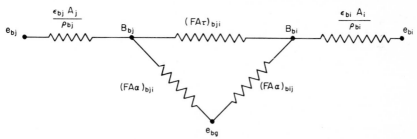

Fig. 1. Circuit illustrating conductances for radiant interchange between surfaces j, i, and the gas g.

radiation which is reflected from surface i is further attenuated in passing back in all directions through the gas. The path length for absorption on this return is equal to the mean beam length r_{mb} of the enclosure as seen from surface i. The total path length over which this radiation has suffered depletion is now $(r_{ij} + r_{mb})$, the gas absorptance for this path length being α_{gjie} or the absorptance over the path from j to i and back through the enclosure e. The radiation is again reflected from the walls of the enclosure and undergoes further absorption, the total path length over which absorption has occurred now being $(r_{ji} + r_{mb} + r_{mb})$, the absorptance over this path being α_{gjiee}. The fraction of energy leaving surface j in the direction of surface i which is actually absorbed by the gas is called the effective absorptance α'_{gji} of the gas between nodes j and i for radiation from j and is given

$$\alpha'_{ji} = \alpha_{gji} + \rho_i(\alpha_{gjie} - \alpha_{gji}) + \rho_i\rho_e(\alpha_{gjiee} - \alpha_{gjie}) + \cdots \tag{1}$$

where ρ_e is the mean reflectance of the enclosure as seen from i, but is approximated by the area-averaged reflectance of the enclosure. These absorptances are calculated for the given gas temperature and composition for a body at the temperature of surface j.

This equation indicates that if surface i is black the effective absorptance depends only on the mean beam length between surfaces j and i. In most cases only the first two or three terms in Eq. (1) need to be considered. The limiting case is when all surfaces in the enclosure are perfect reflectors, in which case the effective mean beam length is infinite. This approach to the problem of radiation exchange with the gas is necessary because of the discontinuous nature of the gas absorption spectra.

An effective emittance for the gas between nodes j and i for radiation to surface j is arrived at similarly

$$\epsilon'_{gji} = \epsilon_{gji} + \rho_i(\epsilon_{gjie} - \epsilon_{gji}) + \rho_i\rho_e(\epsilon_{gjiee} - \epsilon_{gjje}) + \cdots \tag{2}$$

The net radiation exchange between surface j and the gas is obtained by summing over all nodes

$$q_{jg} = \sum_{i=1}^{N} \epsilon_j(FA)_{ji}(\alpha'_{gji}e_j - \epsilon'_{gji}e_g) \tag{3}$$

This equation is zero for all cases where the enclosure contains only a nonabsorbing and nonemitting gas. It will be noted that the configuration factor in Eq. (3) is written in the abbreviated form $(FA)_{ij}$, the following reciprocity relationships holding: $(FA)_{ij} = (FA)_{ji} = F_{ji}A_j = F_{ij}A_i$.

FIXED TEMPERATURE SURFACE

The heat balance equation simply states that the net heat loss by radiation is equal to the net heat exchange with the gas plus the sum of the heat exchanges with the other nodes in the system. The main approximation here is that the transmittance for radiation through the gas from surface j to i is the same as that for black body radiation from a source at temperature T_j. If j is a black body this is rigorously true, but if ϵ_j is small some error will be introduced. The heat balance equation is

$$q_j = \frac{\epsilon_j A_j}{\rho_j}(e_j - B_j) \tag{4}$$

and

$$q_j = q_{jg} + \sum_{i=1}^{N}(FA)_{ji}(\tau_{ji}B_j - \tau_{ij}B_i) \tag{5}$$

From Eqs. (4) and (5)

$$B_j\left[\frac{\epsilon_j A_j}{\rho_j} - \sum_{\substack{i=1\\i\neq j}}^{N}(FA\tau)_{ji}\right] = \frac{\epsilon_j A_j e_j}{\rho_j} - q_{jg} + \sum_{\substack{i=1\\i\neq j}}^{N}(FA\tau)_{ij}B_i \tag{6}$$

As all the radiation which leaves surface j either is absorbed by the gas or arrives at the other surfaces, it can be shown that

$$\sum_{i=1}^{N}(FA\tau)_{ji} + \sum_{i=1}^{N}(FA\alpha)_{ji} = A_j \tag{6a}$$

or

$$\sum_{\substack{i=1\\i\neq j}}^{N}(FA\tau)_{ji} = A_j - \sum_{i=1}^{N}(FA\alpha)_{ji} - \rho_j(FA\tau)_{jj} \tag{6b}$$

Substitution of Eq. (6b) into Eq. (6) and multiplying through by ρ_j results in the following equation for the radiosity of the jth surface

$$B_j = \frac{\epsilon_j A_j e_j - \rho_j q_{jg} + \rho_j \sum_{\substack{i=1\\i\neq j}}^{N}(FA\tau)_{ij}B_j}{A_j - \rho_j \sum_{i=1}^{N}(FA\alpha)_{ji} - \rho_j(FA\tau)_{jj}} \tag{7}$$

SURFACES OF FIXED HEAT FLUX

In this case the energy balance equations for the jth node are

$$q_j = \frac{\epsilon_j A_j}{\rho_j}(e_j - B_j) \tag{8}$$

and

$$q_j = \sum_{i=1}^{N}[\epsilon_j(FA)_{ji}(\alpha'_{gji}e_j - \epsilon'_{gji}e_g) + (FA\tau)_{ji}B_j - (FA\tau)_{ij}B_i] \tag{9}$$

From Eqs. (8) and (9)

$$\left[\sum_{\substack{i=1 \\ i\neq j}}^{N}(FA\tau)_{ji} + \sum_{i=1}^{N}\epsilon_j(FA)_{ji}\alpha'_{gji}\right]B_j = \left[1 - \sum_{i=1}^{N}(FA)_{ji}\alpha'_{gji}\frac{\rho_j}{A_j}\right]q_j$$

$$+ \sum_{i=1}^{N}\epsilon_j(FA)_{ji}\epsilon'_{gji}e_g + \sum_{\substack{i=1 \\ i\neq j}}^{N}(FA\tau)_{ij}B_i \tag{10}$$

or, upon solving for B_j

$$B_j = \frac{\left[A_j - \rho_j\sum_{i=1}^{N}(FA)_{ji}\alpha'_{gji}\right]\dfrac{q_j}{A_j} + \epsilon_j e_g\sum_{i=1}^{N}(FA)_{ji}\epsilon'_{gji} + \sum_{\substack{i=1 \\ i\neq j}}^{N}(FA\tau)_{ij}B_i}{A_j - \sum_{i=1}^{N}(FA\alpha)_{ji} - (FA\tau)_{jj} + \epsilon_j\sum_{i=1}^{N}(FA)_{ji}\alpha'_{gji}} \tag{11}$$

For an enclosure with N surfaces there will result N equations either of the form of Eq. (7) for surfaces of known temperature or of the form of Eq. (11) for surfaces of specified heat flux.

These equations are solved simultaneously for the radiosities of each node. For the nodes of specified temperature the net heat flux can then be found from Eq. (4) and the net heat exchange with the gas from Eq. (3). For a node of fixed heat flux, the emissive power, and consequently the temperature, is found by rewriting Eq. (4) into

$$e_j = \frac{\rho_j q_j}{\epsilon_j A_j} + B_j \tag{12}$$

Once e_j is known the net heat flux from this node to the gas is also found from Eq. (3). It should be remembered that the radiation leaving a surface is taken as positive in this analysis and a negative sign implies radiation towards the surface.

NON-GREY ENCLOSURES

The method to be presented here is known as the "band energy approximation" [6]. Essentially, the spectrum is broken up into a number of bands, the band limits being fixed in most cases by the properties of the radiating gases, although in some instances further spectral divisions may be introduced to cope with the varying spectral characteristics of the walls. The band energy approximation is preferred to other methods because it is relatively simple and straightforward and the Oppenheim network method [7] can be used as a visual aid in setting up the problem.

Unfortunately, the fundamental band absorption correlations necessary for the band energy approximation method are, in many cases, not available. Adequate information is available only for carbon dioxide [8] and some diatomic gases. However, Edwards and Nelson [9] have presented a method for rapid calculation of radiant energy transfer in enclosures with non-grey walls and isothermal water and/or carbon dioxide gas. They introduce the concept of a fractional band absorption which, in combination with the absorptances and emittances of Hottel [3], yields an approximate method for non-grey enclosures. Edwards [10] has recently produced narrow band correlations for water, carbon dioxide, and methane. These correlations can be used for more accurate interchange calculations, but as about 250 individual bands are involved for each gas the correlations will not be presented herein.

It should be noted that the use of the band energy absorption method can result in significant errors for strong bands with walls of high reflectance if band limits are chosen incorrectly.

BAND ABSORPTION

The band absorption A_b is defined by the integral with respect to wave number (frequency) of the spectral absorptance of the gas for the specified state and path length, or

$$A_b = \int_{\Delta \nu_b} \alpha_\nu \, d\nu \tag{13}$$

The band limits in this integration have to be sufficiently broad to include the wings of the band as far out as a significant contribution to absorption exists. This depends again on the state and mass path lengths, and use of these wide band widths in calculations can lead to errors. As radiation travels from one surface to another absorption will be nearly complete in the center of the band while little depletion will occur near the outer limits. Thus, upon reflection, the absorptance will be significantly different for the reflected radiation than for the initial beam traversing the gas.

RADIATION ABSORPTION IN BAND

The energy absorbed by an absorption band of the gas from radiation leaving surface j in the direction of surface i per unit area of j is given by

$$\int_{\Delta \nu_b} B_{\nu j} \alpha_{\nu g j i} \, d\nu \simeq \bar{B}_{\nu j} \int_{\Delta \nu_b} \alpha_{\nu g j i} \, d\nu = \bar{B}_{\nu j} A_{b j i} \tag{14}$$

$$\bar{B}_\nu = \frac{1}{\Delta \nu_b} \int_{\Delta \nu_b} B_\nu \, d\nu = \frac{B_b}{\Delta \nu_b} \tag{15}$$

The removal of $B_{\nu j}$ from beneath the integral can introduce significant error if either the reflectance of the surface varies rapidly over the band width or if the reflectance of the wall is high so that the radiosity varies rapidly over the band due to the nonuniform spectral distribution of the incident radiation after traversing the gas from the other surfaces in the enclosure. The first effect is usually small, particularly if the band widths are not too great. The second effect is minimized by proper selection of band limits as pointed out by Edwards [8].

The band emittance of the gas ϵ_{bgji} between two surfaces is defined as the fraction of the black body radiation in a given band which is actually emitted by the gas volume bounded by the given pair of surfaces and further bounded by the limit of the bundle of rays which intersect both surfaces. The spectral emittance of the gas for a given path length is, furthermore, equal to the spectral absorptance for the same configuration.

The irradiation of either surface i or j by radiation from the gas volume specified above and in a given band is

$$G_{bgji} = \int_{\Delta\nu_b} e_{\nu g}\epsilon_{\nu gji}d\nu \tag{16}$$

$$G_{bgji} \simeq \bar{e}_{\nu g}\int_{\Delta\nu_b} \alpha_{\nu gji}d\nu = \bar{e}_{\nu g}A_{bji} = \frac{e_{bg}}{\Delta\nu_b}A_{bji} = \epsilon_{bji}e_{bg} \tag{17}$$

Furthermore, the relationship between the band absorption, band absorptance, and band emittance is seen to be

$$\epsilon_{bji} = \alpha_{bji} = \frac{A_{bji}}{\Delta\nu_b} \tag{18}$$

The band absorptance in this expression is computed for the geometric mean beam length between the two surfaces. Although a small error is consequently introduced [3, 5, 8, 11] it is thought that the simplification resulting from the use of the geometric beam length far outweighs the slight loss in accuracy.

FRACTIONAL BAND ABSORPTION, BAND ABSORPTANCES, BAND TRANSMITTANCE, AND BAND LIMITS

The fractional band absorption is defined by Edwards [9] as the ratio of the energy absorbed in a given band from a source at gas temperature to the total radiation absorbed by the gas from this source over the same path length. Of course the definition is equally applicable to emittances.

Thus

$$b_b = \frac{A_{bji}\bar{e}_{bg}}{\epsilon_g e_g} = \frac{\epsilon_{bji}e_{bg}}{\epsilon_g e_g} = \frac{\alpha_{bji}e_{bg}}{\epsilon_g e_g} \tag{19}$$

So that

$$\alpha_{bji} = 1 - \tau_{bji} = \frac{b_b\epsilon_b e_g}{e_{bg}} \tag{20}$$

The advantage of the fractional band absorption is that it allows the approximation of the band absorption of water and carbon dioxide from the total emissivity data of

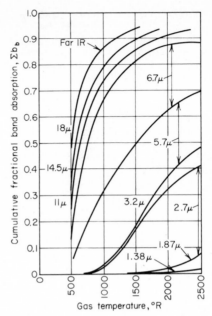

Fig. 2. Fractional band absorption b_b for water at $x\mathrm{Pr} = 1.27$ atm-ft.

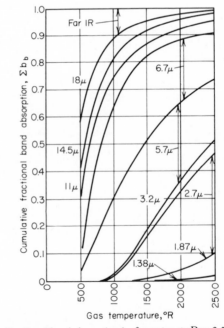

Fig. 3. Fractional band absorption b_b for water at $x\mathrm{Pr} = 2.54$ atm-ft.

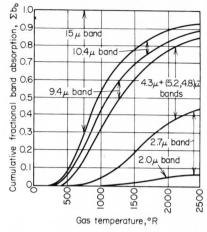

Fig. 4. Fractional band absorption b_b for carbon dioxide at $x\mathrm{Pr} = 10$ atm-ft.

Hottel [3] in the absence of reliable band absorption correlations. The band absorption correlations for carbon dioxide [8] can be used for most problems, but broad band correlations are not available for water vapor. While the fractional band absorption varies rapidly with temperature, it is relatively insensitive to variations in path length, concentration, and pressure. Thus a few plots of b_b versus temperature at different values of mass path length are sufficient to approximate the radiant transfer for most conditions. Values of the fractional band absorption taken from the paper by Edwards and Nelson [9] are plotted in Figs. 2 and 3 for water and Figs. 4 to 6 for carbon dioxide. In these figures the cumulative fractional band absorption has been plotted starting from the short wave end of the spectrum. Thus at $1000°R$ from Fig. 4 for carbon dioxide there would be negligible absorption in the 2.0-μ band, about 0.04 of the total absorption would occur in the 2.7-μ band, and $(0.32–0.04)$ or 0.28 of the total absorption will take place in the combined 4.2-, 4.8-, and 5.2-μ bands.

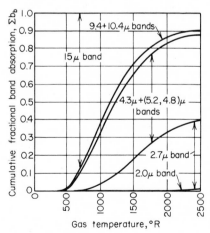

Fig. 5. Fractional band absorption b_b for carbon dioxide at $x\mathrm{Pr} = 1.0$ atm-ft.

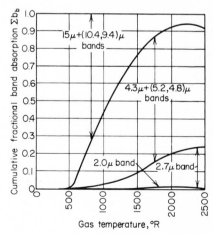

Fig. 6. Fractional band absorption b_b for carbon dioxide at $x\mathrm{Pr} = 0.10$ atm-ft.

Edwards [8] has shown that improper selection of band limits can result in serious errors. In particular, very wide band limits are to be avoided when there is appreciable absorption in the band. The band limits assigned by experimenters such as Howard, Burch, and Williams [12], Edwards [13], or Nelson [14] are generally such that they include the widest band encountered at high temperatures, pressures, and mass path lengths. Thus such band limits are usually inappropriate for interchange calculations using the band energy approximation.

Since the band widths change with temperature and mass path length, it is not possible to set fixed band limits to be used for all temperatures and enclosure sizes. However, according to Edwards [13], large errors can be avoided by selecting band limits such that the band transmittance τ_{bij} is approximately equal to the band mass path length exponent m_b for power law correlations or to

$$\frac{D_b}{A_{bji}} \text{ for logarithmic correlations}$$

General guides [9] for water and carbon dioxide are summarized in Tables 1 and 2. If fixed band limits are recommended in Table 1 or 2 the band absorptance can be obtained as follows.

FIXED BAND LIMITS

1. The geometric mean beam length is found between the given pair of surfaces.
2. The total emittance of the gas is obtained from Hottel's curves for this mean beam length.
3. The fractional band absorption for the band is obtained from Figs. 2 to 6 as the case may be.
4. The band energy fraction e_{bg}/e_g corresponding to the gas temperature is found from the radiation tables.
5. The band absorption and transmittance are found from Eq. (20).

BAND LIMITS NOT FIXED

If only one band limit or a band center is given in Table 1 or 2 the procedure is modified as follows:

TABLE 1. Approximate Criteria for Band Limit Selection (H_2O)

Band b	Lower limit cm^{-1}	Center cm^{-1}	Upper limit cm^{-1}	τ_{gb}
Far infrared rotational structure	Say 0		500	0
18μ	500			0.2
14.5μ	625		770	
11.0μ	770		1100	
6.7μ			1610	0.2
5.7μ	1610			0.3
3.2μ	2650		3300	
2.7μ		3800		0.3
1.87μ	4620		6200	
1.38μ	6200		8100	

1. The geometric mean beam length is found for all pairs of surfaces.
2. The total emittance of the gas is found for all paths.

3. The fractional band absorption is found from Figs. 2 to 6.
4. The band transmittance listed in Table 1 or 2 is used in Eq. (20) together with items 2 and 3 to fix the fraction of energy in the band and hence the band limits.

For a mixture containing both water and carbon dioxide an approximate procedure is to determine the band limits as explained above. Overlapping of strong bands such as in the 2.7-μ region will be seen to occur. The overlapped region is only counted once as otherwise the absorptance would exceed unity.

TABLE 2. Approximate Criteria for Band Limit Selection (CO_2)

Band b	Lower limit cm^{-1}	Center cm^{-1}	Upper limit cm^{-1}	τ_{gb}
15μ		667		0.11
10.4μ	849		1013	
9.4μ	1013		1141	
7.5μ	1141		1485	
5.2μ	1830		1995	
4.8μ	1995		2169	
4.3μ			2460	0.11
2.7μ			3830	0.12
2.0μ	4400		6000	

BAND ENERGY METHOD

Figure 1 illustrates the conductances for radiant exchange between surfaces j, i, and the gas. The problem here is similar in character to the band method for the non-grey enclosure without active gas discussed in the section by Sparrow. For surfaces of specified temperature, the black body emissive power of each band is known. The band radiosity of this surface node in terms of the band radiosities of all other surfaces in the system and the band emissive power of the gas is then given by

$$B_{bj} = \frac{1}{A_j - \rho_{bj}(FA\tau)_{bjj}} \left[\epsilon_{bj} A_j e_{bj} + \rho_{bj} e_{bg} \sum_{i=1}^{N} (FA\alpha)_{bji} + \rho_{bj} \sum_{\substack{i=1 \\ i \neq j}}^{N} (FA\tau)_{bji} B_{bi} \right] \quad (21)$$

Thus there is an equation of this form for each surface of known temperature. The net band heat flux from surface j is given by

$$q_{bj} = (e_{bj} - B_{bj}) \frac{\epsilon_{bj} A_j}{\rho_{bj}} = q_{bjg} + \sum q_{bji} \quad (22)$$

Where the net band heat flux to the gas from node j is given by

$$q_{bjg} = \sum_{i=1}^{N} (FA\alpha)_{bji} (B_{bj} - e_{bg}) \quad (23)$$

And the net band heat flux to the other surface nodes in the enclosure is

$$\sum q_{bji} = \sum_{i=1}^{N} (FA\tau)_{bji} (B_{bj} - B_{bi}) \tag{24}$$

The total heat flux from the surface is simply obtained by summing the heat fluxes in the individual bands, or

$$q_j = \sum_{b=1}^{n} q_{bj} \tag{25}$$

Unlike the total radiation case, a simple equation does not follow for a surface with a specified heat flux. Instead, it is necessary to use Eq. (21) for the radiosity of these surfaces. This means that a temperature must be initially assigned to each surface. The band heat fluxes for this assumed temperature are obtained from Eq. (22) and the total heat flux from Eq. (25). This heat flux is then compared to the specified heat flux and the surface temperature corrected to yield the correct heat flux. Reiteration may be necessary for better accuracy.

It is thus seen that the complete solution of the radiant energy exchange on the band basis involves the solution of n sets of N simultaneous equations with the band radiosities of each surface as the unknowns. The band emissive powers, emittances, and reflectances of each surface and the band geometrical absorptances and transmittances between each pair of surfaces are required as input information. As pointed out above, if some surfaces have fixed heat fluxes rather than fixed temperatures, the solution is further complicated by the need to first assume temperatures for these surfaces followed by a temperature correction to yield the proper heat flux and repetition of the solution until all restrictions are satisfied. At this point the net heat transfer rate to the gas can be obtained by summing the contributions from each surface as indicated in the following expression

$$q_g = \sum_{j=1}^{N} \sum_{b=1}^{n} q_{bjg} \tag{26}$$

The total heat fluxes for each surface are obtained from Eq. (22). Hence it is seen that all radiant heat fluxes and temperatures in the system are now established.

GEOMETRIC ABSORPTION FACTOR

Strictly speaking, the geometrical absorption factor should be used for computing the radiation absorbed by the gas [5, 6]. This is defined as follows

$$(FA\alpha)_{ij} = \int_{A_i} \int_{A_j} \frac{\alpha(r) \cos \theta_i \cos \theta_j \, dA_j dA_i}{\pi r^2} \tag{27}$$

However, if a linear absorption law is substituted in Eq. (27) it is found that the geometrical absorption factor can be replaced by the product of the configuration factor $(FA)_{ij}$ with an absorptance α_{gij} where this absorptance is that corresponding to the geometric mean beam length between the two surfaces. Use of the geometric mean beam length has the great advantage that it is purely a geometric function which is readily calculated and tabulated or plotted for various geometries. On the other hand, if other absorption laws are substituted into Eq. (27), the resulting expressions are extremely difficult to treat analytically and introduce additional parameters into any correlations.

GEOMETRIC MEAN BEAM LENGTHS

The geometric mean beam length between two surfaces is expressed by [5]

$$r_{ij} = \frac{1}{(FA)_{ij}} \int_{A_i} \int_{A_j} \frac{\cos \theta_j \cos \theta_i \, dA_j \, dA_i}{\pi r} \tag{28}$$

The geometric mean beam length of an enclosure r_{mb} is given by four times the ratio of the volume to the total surface area, or

$$r_{mb} = \frac{4V}{A} \tag{29}$$

Hottel [3] reports mean beam lengths for several geometries. These are tabulated in Table 3. The precise value of the mean beam length depends upon the form of the absorption law, but it has been demonstrated [5, 8, 11] that little error occurs if the geometric mean beam length is used to calculate the absorption.

TABLE 3. Geometric Mean Beam Lengths for Gas Radiation

Shape	Characterizing dimension X	r/X
Sphere	Diameter	2/3
Infinite cylinder	Diameter	1
Semi-infinite cylinder, radiating to center of base	Diameter	
Right-circular cylinder, height = diameter, radiating to center of base	Diameter	
Same, radiating to whole surface	Diameter	2/3
Infinite cylinder of half-circular cross section. Radiating to spot on middle of flat side	Radius	
Rectangular parallelepipeds:		
1:1:1 (cube)	Edge	2/3
1:1:4, radiating to 1 x 4 face		0.90
radiating to 1 x 1 face	Shortest edge	0.86
radiating to all faces		0.89
1:2:6, radiating to 2 x 6 face		1.18
radiating to 1 x 6 face	Shortest edge	1.24
radiating to all faces		1.20
1:∞:∞ (infinite parallel planes)	Distance between planes	2
Space outside infinite bank of tubes with centers on equilateral triangles; tube diameter = clearance	Clearance	3.4
Same as preceding, except tube diameter = one-half clearance	Clearance	4.45
Same, except tube centers on squares: diameter = clearance	Clearance	4.1

Dunkle [5] has tabulated geometric mean beam lengths for pairs of rectangles that are either parallel or at right angles to each other.

For parallel equal rectangles of sides a and b separated by a distance c, the geometric mean beam lengths are expressed in terms of the ratios $\eta = a/c$ and $\beta = b/c$, and are tabulated in Table 4 and plotted in Fig. 7.

For perpendicular rectangles with a common edge b and sides a and c, the geometric mean beam length parameter Z_r is determined by the ratios $\alpha = a/b$ and $\gamma = c/b$. These configuration factors and mean beam length parameters for this geometry are given in Table 5.

TABLE 4. Geometric Mean Beam Length Ratios and Configuration Factors, Parallel Equal Rectangles

β:		0	0.1	0.2	0.4	0.6	1.0	2.0	4.0	6.0	10.0	20.0
η												
0	r/c	1.000	1.001	1.003	1.012	1.025	1.055	1.116	1.178	1.205	1.230	1.251
	F											
0.1	r/c	1.001	1.002	1.004	1.013	1.026	1.056	1.117	1.179	1.207	1.234	1.255
	F		0.00316	0.00626	0.01207	0.01715	0.02492	0.03514	0.04210	0.04463	0.04671	0.04829
0.2	r/c	1.003	1.004	1.006	1.015	1.028	1.058	1.120	1.182	1.210	1.235	1.256
	F		0.00626	0.01240	0.02391	0.03398	0.04941	0.06971	0.08353	0.08859	0.09272	0.09586
0.4	r/c	1.012	1.013	1.015	1.024	1.037	1.067	1.129	1.192	1.220	1.245	1.267
	F		0.01207	0.02392	0.04614	0.06560	0.09554	0.13513	0.16219	0.17209	0.18021	0.18638
0.6	r/c	1.025	1.026	1.028	1.037	1.050	1.080	1.143	1.206	1.235	1.261	1.282
	F		0.01715	0.03398	0.06560	0.09336	0.13627	0.19341	0.23271	0.24712	0.25896	0.26795
1.0	r/c	1.055	1.056	1.058	1.067	1.080	1.110	1.175	1.242	1.272	1.300	1.324
	F		0.02492	0.04941	0.09554	0.13627	0.19982	0.28588	0.34596	0.36813	0.38638	0.40026
2.0	r/c	1.116	1.117	1.120	1.129	1.143	1.175	1.246	1.323	1.359	1.393	1.421
	F		0.3514	0.06971	0.13513	0.19342	0.28588	0.41525	0.50899	0.54421	0.57338	0.59563
4.0	r/c	1.178	1.179	1.182	1.192	1.206	1.242	1.323	1.416	1.461	1.505	1.543
	F		0.04210	0.08353	0.16219	0.23271	0.34596	0.50899	0.63204	0.67954	0.71933	0.74990
6.0	r/c	1.205	1.207	1.210	1.220	1.235	1.272	1.359	1.461	1.513	1.564	1.609
	F		0.4463	0.08859	0.17209	0.24712	0.36813	0.54421	0.67954	0.07324	0.77741	0.84713
10.0	r/c	1.230	1.232	1.235	1.245	1.261	1.300	1.393	1.505	1.564	1.624	1.680
	F		0.04671	0.09270	0.18021	0.25896	0.38638	0.57361	0.71933	0.77741	0.82699	0.86563
20.0	r/c	1.251	1.253	1.256	1.267	1.282	1.324	1.421	1.543	1.609	1.680	1.748
	F		0.04829	0.09586	0.18638	0.26795	0.40026	0.59563	0.74990	0.86563	0.95125	0.9079
∞	r/c	1.272	1.274	1.277	1.289	1.306	1.349	1.452	1.584	1.660	1.745	1.832
	F		0.04988	0.09902	0.19258	0.27698	0.41421	0.61803	0.78078	0.84713	0.09499	0.95125

TABLE 5. Configuration Factor and Mean Beam Length Functions for Rectangles at Right Angles

$$\phi = \frac{FA}{b^2} \qquad Z_r = \frac{FAR}{abc} \qquad \alpha = \frac{a}{b} \qquad \gamma = \frac{a}{b}$$

α		γ:— 0.05	0.10	0.20	0.4	0.6	1.0	2.0	4.0	6.0	10.0	20.0	∞
0.02	ϕ	0.007982	0.008875	0.009323	0.009545	0.009589	0.009628	0.009648	0.009653	0.009655	0.009655	0.009655	0.009655
	Z_r	0.17840	0.12903	0.08298	0.04995	0.03587	0.02291	0.01263	0.006364	0.004288	0.002594	0.001305	
0.05	ϕ	0.014269	0.018601	0.02117	0.02243	0.02279	0.02304	0.02316	0.02320	0.02321	0.02321	0.02321	0.02321
	Z_r	0.21146	0.18756	0.13834	0.08953	0.06627	0.04372	0.2364	0.01234	0.008342	0.005059	0.002549	
0.10	ϕ		0.02819	0.03622	0.04086	0.04229	0.04325	0.04376	0.04390	0.04393	0.04394	0.04394	0.04395
	Z_r		0.20379	0.17742	0.12737	0.09795	0.06659	0.03676	0.01944	0.013184	0.008018	0.004049	
0.20	ϕ			0.05421	0.06859	0.07377	0.07744	0.07942	0.07999	0.08010	0.08015	0.08018	0.08018
	Z_r			0.18854	0.15900	0.13028	0.09337	0.05356	0.02890	0.01972	0.012047	0.006103	
0.40	ϕ				0.10013	0.11524	0.12770	0.13514	0.13736	0.13779	0.13801	0.13811	0.13814
	Z_r				0.16255	0.14686	0.11517	0.07088	0.03903	0.02666	0.01697	0.008642	
0.60	ϕ					0.13888	0.16138	0.17657	0.18143	0.18239	0.18289	0.18311	0.18318
	Z_r					0.14164	0.11940	0.07830	0.04467	0.03109	0.02025	0.010366	
1.0	ϕ						0.20004	0.23285	0.24522	0.24783	0.24921	0.24980	0.25000
	Z_r						0.11121	0.08137	0.04935	0.03502	0.02196	0.01175	
2.0	ϕ							0.29860	0.33462	0.34386	0.34916	0.35142	0.35222
	Z_r							0.07086	0.04924	0.03670	0.02401	0.01325	
4.0	ϕ								0.40544	0.43104	0.44840	0.45708	0.46020
	Z_r								0.04051	0.03284	0.02320	0.01300	
6.0	ϕ									0.46932	0.49986	0.51744	0.52368
	Z_r									0.02832	0.02132	0.01272	
10.0	ϕ										0.5502	0.5876	0.6053
	Z_r										0.01759	0.01146	
20.0	ϕ											0.6608	0.7156
	Z_r											0.008975	

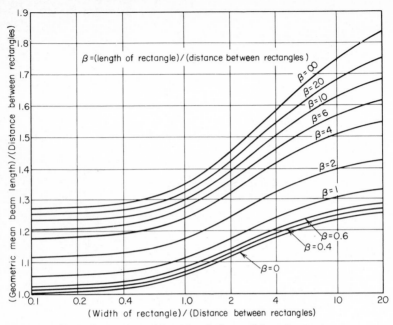

Fig. 7. Geometric mean beam lengths for parallel equal rectangles.

FLUX ALGEBRA

The rules of flux algebra for computing configuration factors for other shapes are outlined in the report of Morgan and Hamilton [15] as well as in most texts on illumination and heat transfer. The same rules apply to manipulation of the FAr group as apply to FA. As an example, the mean beam length will be computed from a horizontal square one foot on each edge to an equal vertical square which is located above the horizontal square.

The mean beam length between surface 1 and 3 (Fig. 8) is required, and is found as follows:

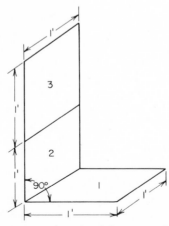

Fig. 8. Rectangles at right angles, mean beam length between surfaces 1 and 3 to be determined.

Configuration factors

$$(FA)_{1-3} = (FA)_{1-(2+3)} - (FA)_{1-2}$$

$$(FA)_{1-2} : a = 1, b = 1, c = 1, \alpha = 1, \gamma = 1$$

$$(FA)_{1-2} = \phi_{1-2}\, b^2 = 0.2004 \times 1 = 0.2004 \text{ (Table 5)}$$

$$(FA)_{1-(2+3)} : a = 1, b = 1, c = 2, \alpha = 1, \gamma = 2$$

$$(FA)_{1-(2+3)} = 0.23285$$

hence

$$(FA)_{1-3} = 0.23285 - 0.2004 = 0.03245$$

Mean beam length factor

$$(FAr)_{1-3} = (FAr)_{1-(2+3)} - (FAr)_{1-2}$$

$$(FAr)_{1-2} = (abc)Z_{1-2} = 0.11121 \text{ (table 5)}$$

$$(FAr)_{1-(2+3)} = (abc)Z_{1-(2+3)} = 0.08137 \times 2 = 0.16274$$

$$(FAr)_{1-3} = 0.16274 - 0.11121 = 0.05153$$

and

$$r_{1-3} = \frac{(FAr)_{1-3}}{(FA)_{1-3}}$$

or

$$r_{1-3} = \frac{0.05153}{0.03245} = 1.57 \text{ ft}$$

SUMMARY

The information and methods presented in this section together with the material in the sections by Sparrow and Penner enable one to compute the radiant heat exchange in complex enclosures containing radiating gases. It should be realized that the methods are approximate, but the accuracy is adequate for most engineering work. For many cases, the grey enclosure approach is satisfactory and much simpler than the band energy approach.

REFERENCES

1. E. M. Sparrow, Radiant Interchange between Surfaces Separated by Nonabsorbing and Nonemitting Media, page 15-29 of this book.
2. R. V. Dunkle, Radiant Interchange in an Enclosure with Specular Surfaces and Enclosures with Windows or Diathermanous Walls, in "Heat Transfer, Thermodynamics and Education," Boelter Anniversary Volume (H. A. Johnson, ed.), p. 133, McGraw-Hill Book Company, Inc., New York 1964.
3. H. C. Hottel, Radiation Heat Transmission, in "Heat Transmission," by W. H. McAdams, chap. 4, 3d ed., McGraw-Hill Book Company, Inc., New York, 1954.

4. J. T. Bevans, Correlation of the Total Emissivity of Carbon Dioxide and Water Vapor, *ASME Paper* 60-WA-175.
5. R. V. Dunkle, Geometric Mean Beam Lengths for Radiant Heat-Transfer Calculations, *J. Heat Transfer, Trans. ASME*, **C86**:75 (1964).
6. J. T. Bevans and R. V. Dunkle, Radiant Interchange Within an Enclosure, *J. Heat Transfer, Trans. ASME*, **C82**:1 (1960).
7. A. K. Oppenheim, Radiation Analysis by the Network Method, *Trans. ASME*, **78**:725 (1956).
8. D. K. Edwards, Radiation Interchange in a Nongray Enclosure Containing an Isothermal Carbon-Dioxide--Nitrogen Gas Mixture, *J. Heat Transfer, Trans. ASME*, **C84**:1 (1962).
9. D. K. Edwards and K. E. Nelson, Rapid Calculation of Radiant Energy Transfer between Nongray Walls and Isothermal H_2O or CO_2 Gas, *J. Heat Transfer, Trans. ASME*, **C84**:273 (1962).
10. D. K. Edwards, Studies of Infrared Radiation in Gases, University of California at Los Angeles, *Department of Engineering Report* 62-65, 1963.
11. J. A. Wiebelt, Comparison of Geometric Absorption Factors with Geometric Mean Beam Lengths, *J. Heat Transfer, Trans. ASME*, **C85**:287 (1963).
12. J. N. Howard, D. L. Burch, and D. Williams, Near Infrared Transmission Through Synthetic Atmospheres, *Geophysical Research Paper* 40, Air Force Cambridge Research Center, 1953; also *J. Opt. Soc. Am.*, **46**:186 (1956).
13. D. K. Edwards, Absorption by Infrared Bands of Carbon Dioxide Gas at Elevated Pressures and Temperatures, *J. Opt. Soc. Am.*, **50**:617 (1960).
14. K. E. Nelson, Experimental Determination of the Band Absorptivity of Water Vapor at Elevated Pressures and Temperatures, MS thesis, University of California, Berkeley, Calif., 1959.
15. D. C. Hamilton and W. R. Morgan, Radiant Interchange Configuration Factors, National Advisory Committee for Aeronautics, *Tech. Note* 2836, 1952.

NOMENCLATURE

a side of rectangle
A area, ft^2
A band absorption, cm^{-1}
b side of rectangle
b fractional band absorption
B radiosity, Btu/hr-ft^2
c side of rectangle or distance between parallel rectangles
D coefficient in logarithmic absorption law
e black body emissive power, Btu/hr-ft^2
F shape or configuration factor
G irradiation, Btu/hr-ft^2
n number of individual bands in spectrum
N number of surfaces in enclosure (subdivisions)
P total pressure, atmospheres
q heat transfer rate, Btu/hr
r beam length
T temperature, $^\circ R$
V volume of enclosure
x mol fraction, active gas
Z_r correlation factor for mean beam lengths
α absorptance
α ratio of sides of rectangles for mean beam length correlation
β ratio of sides of rectangles for mean beam length correlation
γ ratio of sides of rectangles for mean beam length correlation
ϵ emittance
θ angle from normal to surface
λ wavelength, microns

ν wave number (frequency), cm^{-1}
ρ reflectance
τ transmittance

Subscripts

b band
e enclosure
g gas
i surface or node i
j surface or node j
mb mean beam length of enclosure
ν spectral value

Part **D**

Equilibrium Radiation Properties of Gases*

S. S. PENNER

Department of the Aerospace and Mechanical Engineering Sciences
University of California, San Diego, La Jolla, California

The spectral (monochromatic) black body radiancy is the energy in erg/cm^2-sec, in the frequency range between ν (sec^{-1}) and $\nu + d\nu$ (sec^{-1}), emitted into a solid angle of 2π steradians, and is given by the Planck black body distribution law

$$B_\nu^\circ d\nu = 3.742 \times 10^{-5} \left(\frac{\nu}{c}\right)^3 \frac{d(\nu/c)}{[\exp 1.439\,(\nu/c)/T] - 1}$$

where $c = 2.998 \times 10^{10}$ cm/sec is the velocity of light and the temperature T is expressed in °K. For an *equilibrium emitter* at the temperature T, $B_\nu^\circ d\nu$ represents an

*Supported by the Physics Branch of the Office of Naval Research under Contract No. Nonr-2216(00), NR 015-401. Reproduction in whole or in part is permitted for any purpose of the United States Government. The author is indebted to Mr. P. Varanasi for assistance with the numerical compilations contained in this section. This chapter was written in July 1964, except for minor revisions and the discussion given under "Total Emissivities of H_2O and CO_2," which was written in April 1968. For a much more extensive account of opacity calculations, we refer to "Radiation and Reentry," by S. S. Penner and D. B. Olfe, Academic Press, Inc., New York, N.Y., 1968.

upper bound for any substance. The spectral radiancy of a non-black emitter B_ν is obtained on multiplying $B_\nu^{\,\circ}$ by the spectral emissivity $\epsilon_\nu \leq 1$, i.e.,

$$B_\nu = \epsilon_\nu B_\nu^{\,\circ}$$

The total radiancy of a black body is

$$\int_0^\infty B_\nu^{\,\circ} d\nu \equiv B^\circ = 5.670 \times 10^{-5}\, T^4 \text{ erg/cm}^2\text{-sec}$$

and the total radiancy B of a non-black emitter is obtained on multiplying B° by the total emissivity ϵ, i.e.,

$$B = \epsilon B^\circ$$

The determination of ϵ_ν and of ϵ from fundamental spectroscopic data or from experimental measurements forms the subject of the following discussion. The related spectral and total absorptivities (α_ν and α, respectively) are also considered briefly.

For diatomic and polyatomic gases at equilibrium, ϵ_ν is nonzero in the infrared region of the spectrum because of the presence of infrared vibration-rotation bands (in the wavelength range from about 1 to 20μ) and of pure rotation lines (at longer wavelengths). The vibration-rotation bands are produced by the simultaneous occurrence of motion of atomic nuclei in molecules with respect to each other and rotation of the entire molecular complex. It is convenient to classify the nuclear motion as corresponding to fundamental, overtone, harmonic, or combination bands, depending on the change in molecular configuration involved. Since the black body radiance decreases very rapidly at wavelengths that are appreciably shorter than the maximum value, which is determined by the Wien displacement law, it follows that the contributions of the vibration-rotation bands to ϵ dominate the radiant-energy emission from heated gases at temperatures less than a few thousand degrees Kelvin.

As the temperature is raised, the maximum value of $B_\nu^{\,\circ}$ shifts to higher frequencies, i.e., to shorter wavelengths and the nonzero values of ϵ_ν in the visible and ultraviolet regions of the spectrum make the principal contributions to ϵ. In the visible and ultraviolet, ϵ_ν is nonzero because of the occurrence of electronic band systems in molecules and electronic transitions in atoms corresponding to changes in the states of excitation of electrons in molecules and in atoms. At still higher temperatures, the gas composition is dominated by atoms and ions and continuum radiation is emitted from heated plasmas. This radiation is produced by the motion of electrons in the fields of atoms and ions.

A full discussion of the origin of the equations and correlations summarized on the following pages requires the complete apparatus of atomic and molecular spectroscopy, as well as consideration of selected topics in theoretical physics. Fortunately, the essential results of these studies can often be stated quite simply and in useful form.

RELATION BETWEEN TOTAL EMISSIVITIES AND ABSORPTIVITIES

For *equilibrium* radiation, spectral absorptivities α_ν and spectral emissivities ϵ_ν are identically equal. For *weakly absorbing* (i.e., transparent) gases, irrespective of the band structure, the total absorptivity α is related to the total emissivity ϵ through the expression

$$\alpha(T_s \to T_g, X) \simeq \frac{T_s}{T_g} \epsilon(T_s, X) \tag{1}$$

where T_s is the temperature of a black or grey light source and the emitting or absorbing gas is at the temperature T_g and optical depth X ($\equiv$ gas pressure $\times$ geometric path length). Equation (1) is expected to hold, within a few percent, at temperatures up to several thousand degrees Kelvin. Equation (1), as well as several additional expressions given in this section, were first found empirically [1–3] and were subsequently derived theoretically [4, 5]. The following scaling laws apply for transparent gases:

$$\lim_{X \to 0} \alpha(T, \delta X) = \delta \alpha(T, X)$$

$$\lim_{X \to 0} \epsilon(T, \delta X) = \delta \epsilon(T, X)$$

The general relation [5] between α and ϵ is of the form

$$\alpha(T_s \to T_g, X) = \left(\frac{T_g}{T_s}\right)^{\beta - 1} \epsilon\left[T_s, X\left(\frac{T_s}{T_g}\right)^{\beta} ; a_{T_g}\right] \tag{2}$$

where the parameter β depends on the assumed band model and a_{T_g} identifies a spectral line shape parameter at the temperature T_g.

At elevated total pressures ($p \gtrsim 1$ atm for CO_2, $p \gtrsim 4$ atm for H_2O, $p \gtrsim 15$ atm for CO, etc.), the spectral line structure of absorbing vibration-rotation bands is removed by line broadening. In this case, $\beta = 3/2$ and the dependence of ϵ on the line shape parameter a_{T_g} disappears.

At low pressures (i.e., pressures smaller by about a factor of ten than those specified in the preceding paragraph) and moderate temperatures (that is, $T_g \lesssim 3000°K$ unless the pressures are exceedingly small), the spectral lines of molecules do not overlap and may be approximated by a dispersion contour. In this case, $\beta = 1$ and the explicit dependence on the line-shape parameter a_{T_g} again disappears. Additional spectral lines will normally contribute as the gas is heated in such a way that the spacing between lines varies as $T^{-\eta}$. Equation (2) with $\beta = 1$ should then be replaced by the expression

$$\alpha(T_s \to T_g, X) \simeq \left(\frac{T_g}{T_s}\right)^{\eta} \epsilon\left[T_s, X\left(\frac{T_s}{T_g}\right)^{\eta + 1}\right] \tag{3}$$

Numerical values for η are not normally available; however, η is expected to be a small positive number.

For randomly distributed dispersion lines *at moderate or large optical depths,*

$$\alpha(T_s \to T_g, X) \simeq \left(\frac{T_g}{T_s}\right)^{1/2} \epsilon\left[T_s, X\left(\frac{T_s}{T_g}\right)^{2 - \eta}\right] \tag{4}$$

Equation (4) is particularly useful for molecules with complex spectral structure (for example, H_2O, CO_2) and is the preferred expression for gas mixtures at moderate temperatures.

Several additional cases have been treated theoretically but cannot be discussed adequately without detailed reference to topics in molecular spectroscopy. We refer, therefore, to the literature [5] for further elaboration on the relation between absorptivity and emissivity.

Hottel's preferred empirical expression for CO_2 may be obtained from Eq. (4) by setting $\eta = 4/3$ and noting that

$$\epsilon\left[T_s, X\left(\frac{T_s}{T_g}\right)^{2-\eta}\right] \equiv f(T_s)\sqrt{X\left(\frac{T_s}{T_g}\right)^{2-\eta}}$$

for the relatively large optical depths for which experimental data are available. Hence

$$\alpha(T_s \rightarrow T_g, X) \simeq \left(\frac{T_g}{T_s}\right)^{1/2}\left(\frac{T_s}{T_g}\right)^{(1-\eta)/2} f(T_s)\sqrt{X\left(\frac{T_s}{T_g}\right)}$$

$$= \left(\frac{T_g}{T_s}\right)^{2/3}\epsilon\left[T_s, X\left(\frac{T_s}{T_g}\right)\right] \tag{5}$$

which is essentially the functional form used originally to correlate the experimental data [1]. For $X \gtrsim 0.1$ ft-atm, the preferred relation for H_2O is Eq. (4) with $\eta = 1$; for $X \lesssim 0.1$ ft-atm, a better fit to the empirical results is obtained by using Eq. (2) with $\beta = 2$ [5]. The best choices of the correlation formulas for given molecules are expected to be (weak) functions of T_s and T_g.

TOTAL EMISSIVITIES OF H_2O AND CO_2

Useful charts for estimating the total emissivities of H_2O and CO_2 were prepared, some years ago, by Eckert, Hottel, and others [6–9]. The validity of these charts remains largely unchanged to this day and no important additions are required to the data. As the result of intensive recent theoretical studies [10–17], considerable understanding has been gained of the theoretical calculation of gas emissivities from first principles using fundamental spectroscopic data. The theoretical calculations, as well as the spectroscopic constants on which they are based, are now sufficiently precise to verify all of the empirically determined results which are reproduced in Figs. 1 to 4.

Successful procedures for *ab initio* calculations of total emissivities of CO_2 and H_2O from spectroscopic data have been developed [13, 14]. These procedures involve judicious use of an overlapping line model for symmetrical bands and calculations of sums of band intensities in the pseudoharmonic oscillator approximation [13–15]. Line-structure corrections are introduced by the use of statistical procedures, which lead to very accurate prediction of the curves shown in Fig. 3 [14]. Even corrections for overlapping between vibration-rotation bands belonging to CO_2 and H_2O (see Fig. 4) can be calculated by the use of simplified procedures [16], although some uncertainties remain at low temperatures because of difficulties in the theoretical calculation of absorption coefficients for H_2O in the pure rotation region [17]. Some of the discrepancies noted may well be the result of errors in the empirical correlations shown in Fig. 4 at low temperature [17].

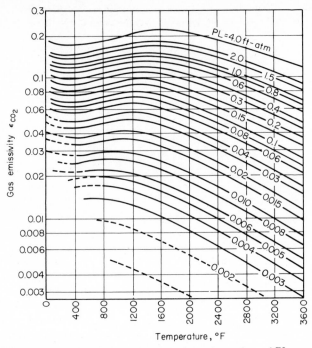

Fig. 1. Emissivity of carbon dioxide vs. temperature for various values of PL; reproduced from H. C. Hottel [6]. These data should be approximately valid if the total pressure exceeds about $0.5(T/300)$ atm at the temperature T in $°K$; $P =$ partial pressure of CO_2 in atm, $L =$ geometric path length in ft.

SPECTRAL AND TOTAL EMISSIVITIES FOR DIATOMIC MOLECULES AT MODERATE TEMPERATURES

In recent years, a number of papers have been published on experimental measurements and on theoretical calculations for diatomic molecules relating to spectral absorption coefficients P_ω, to spectral emissivities $\epsilon_\omega = 1 - \exp(-P_\omega X)$, and total emissivities ϵ. (A selected list [18–23] appear in the references.) The empirical data are too incomplete to justify inclusion in a handbook and the theoretical studies too complex for exhaustive summary. For these reasons, we content ourselves here with the listing of representative examples which should serve to indicate the nature of the available information. The data given in this section constitute the dominant contributions at relatively low temperatures ($T \lesssim 3000°K$) where the important emission arises from the infrared vibration-rotation bands [18–22], with much smaller contributions from pure rotation bands. The theoretical calculations are customarily performed in the "weak-line" or "strong-line" approximations using a variety of molecular band models, including those mentioned in connection with the discussion on the relation between absorptivities and emissivities. In the "weak-line" approximation, the spectral emissivities and absorption coefficients are determined by the optical depth $X = pl$; in the "strong-line" approximation, the important parameter is p^2l. These two cases arise because the radiant-energy contribution from

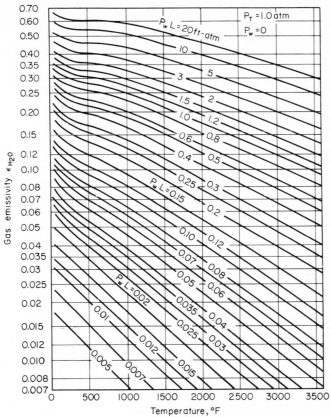

Fig. 2. Emissivity of water vapor; reproduced from R. B. Egbert [7] ; P_w = partial pressure of H_2O in atm, L = geometric path length in ft. The data listed in Fig. 2 apply only at atmospheric pressure in the limit as the partial pressure P_w of H_2O goes to zero; an approximate correction for the actual partial pressure of water may be made by using the data of Fig. 3.

an isolated spectral line with dispersion contour is roughly proportional to $X = pl$ (for $SX/2\pi b \stackrel{<}{\sim} 2/\pi$) or to $pl^{1/2}$ (for $SX/2\pi b > 2/\pi$).*

Representative theoretical results, which were derived from numerical computations for CO, NO, HCl and HF, are summarized in Figs. 5 to 9 for the spectral regions including the fundamental vibration-rotation bands.

A simple analytical expression may be obtained for reproducing the essential features of the band contours at elevated pressures where the line width is not small compared with the line spacing and the rotational fine structure has been removed

*Here S (in cm^{-2}-atm^{-1}) is a spectroscopic constant and stands for the integrated intensity of a spectral line $= \int_{line} P_\omega d\omega$, b (in cm^{-1}) = half-width of a dispersion line such that the spectral absorption coefficient (in cm^{-1}-atm^{-1}) is

$$P_{|\omega - \omega_0|} = \frac{S}{\pi} \frac{b}{(\omega - \omega_0)^2 + b^2}$$

at the wave number displacement $\omega - \omega_0$ from the line center at ω_0 (in cm^{-1}). Also p = partial pressure of gas in atm and ℓ = path length in cm.

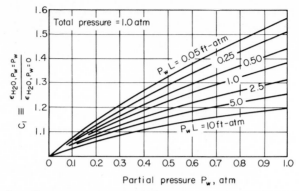

Fig. 3. Factor C_1 vs. P_w for various values of $P_w L$; reproduced from R. B. Egbert [7]. The factor C_1 is used to correct the data of Fig. 2 for the effect of partial pressure P_w of water vapor on spectral line shape and hence on emissivity; P_w = partial pressure of H_2O in atm, L = geometric path length in ft. As $P_w L$ is decreased from 0.05 ft-atm to about 0.008 ft-atm, C_1 must decrease toward unity.

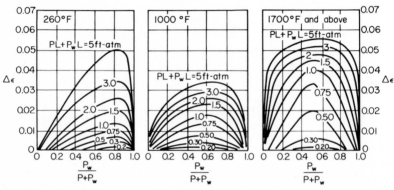

Fig. 4. The decrease $\Delta\epsilon$ in total emissivity in mixtures of H_2O and CO_2 (reproduced from R. B. Egbert [7]), that is, $\epsilon = \epsilon_{CO_2} + \epsilon_{H_2O} - \Delta\epsilon$.

[19]. Thus, to the rigid-rotator, harmonic-oscillator approximation, the following expression is obtained [19] for the spectral absorption coefficient (in cm^{-1}-atm^{-1}) at the wave number ω (in cm^{-1}) in a fundamental (that is, $0 \to 1$) band:

$$P_{\omega,0\to1} \simeq 1.19 \times 10^7 f_{0\to1} \frac{273}{T} \frac{1 - \exp(-1.44\omega/T)}{1 - \exp(-1.44\omega_{0\to1}^*/T)} \frac{0.72\,|\omega - \omega_{0\to1}^*|}{B_e T}$$

$$\times \exp\left[-\frac{0.36(\omega - \omega_{0\to1}^*)^2}{B_e T}\right] \quad (6)$$

Similarly, for the first overtone (i.e., the $0 \to 2$ band),

$$P_{\omega,0\to2} \simeq P_{\omega,0\to1}^r \frac{|\omega - \omega_{0\to2}^*|}{|\omega - \omega_{0\to1}^*|} \frac{\exp\left\{-(0.36/B_e T)\left[(\omega - \omega_{0\to2}^*)^2 - (\omega - \omega_{0\to1}^*)^2\right]\right\}}{1 - \exp(-1.44\,\omega_{0\to1}^*/T)} \quad (7)$$

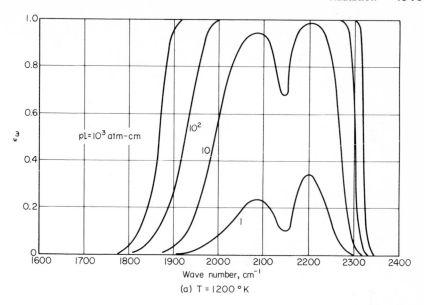

(a) T = 1200 °K

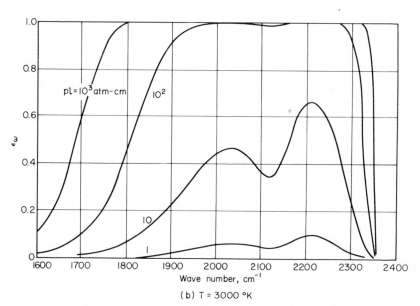

(b) T = 3000 °K

Fig. 5. Spectral emissivities of CO (weak-line approximation) [21].

Here $f_{0\to1}$ is an empirically determined spectroscopic constant (see Table 1); the temperature T is expressed in °K; $\omega^*_{0\to1}$ and $\omega^*_{0\to2}$ are wave numbers for rotationless transitions in the fundamental band and first overtone, respectively; B_e (in cm^{-1}) is the rotational constant for a molecule (see Table 1); r is a dimensionless, temperature-dependent ratio of the f numbers (see Table 1), that is,

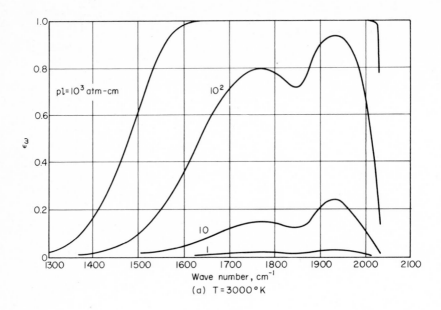

(a) T = 3000 °K

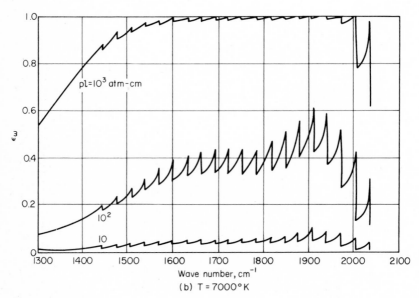

(b) T = 7000 °K

Fig. 6. Spectral emissivities of NO (weak-line approximation) [21]. In first approximation, the spectral emissivities ϵ_w should be replaced by the term $\{1 - \exp[(128/70) \ln(1 - \epsilon_w)]\}$, where ϵ_w represents the values plotted in the figures; this change allows for recent revisions in the integrated intensities. For sufficiently small values of ϵ_w, the spectral emissivities are thus increased by the factor (128/70).

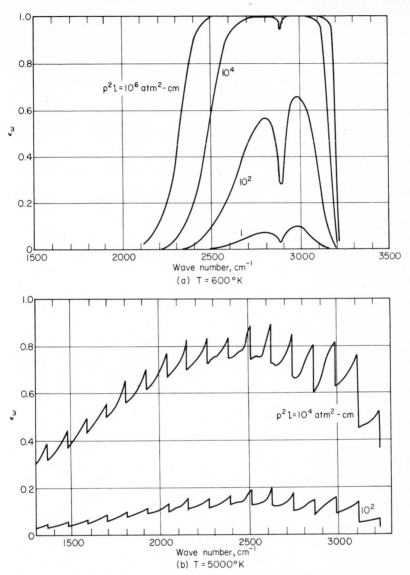

Fig. 7. Spectral emissivities of HCl (strong-line approximation) [21].

$$r = \frac{f_{0 \to 2}}{f_{0 \to 1}} \left[1 + \exp\left(-\frac{1.44 \, \omega^*_{0 \to 1}}{T} \right) \right] \tag{8}$$

The band contours computed from Eqs. (6) and (7) are symmetrical but otherwise reproduce the more exact curves fairly well. The total emissivities (including the fundamentals and first overtone bands) per unit optical depth X (in cm-atm) for diatomic gases for which $1 - \exp(-P_\omega X)$ may be approximated by $P_\omega X$ (that is, for

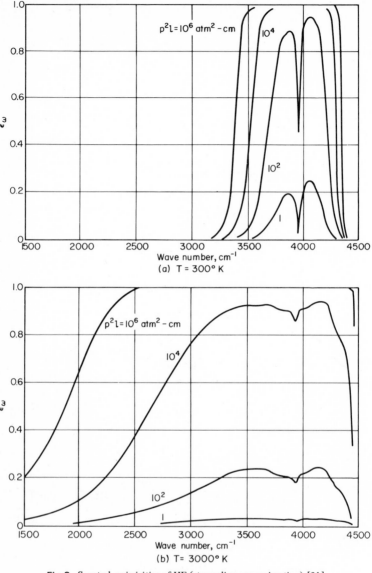

Fig. 8. Spectral emissivities of HF (strong-line approximation) [21].

"transparent gases") are given (within a few percent) by the relation

$$\frac{\epsilon}{X} \simeq \frac{3.66 \times 10^6 f_{0 \to 1}}{\omega_{0 \to 1}^*} \left(\frac{1.44\,\omega_{0 \to 1}^*}{T}\right)^4 \frac{273}{T} \frac{\exp(-1.44\,\omega_{0 \to 1}^*/T)}{1 - \exp(-1.44\,\omega_{0 \to 1}^*/T)}$$

$$\times \left[1 + 8r\,\frac{\exp(-1.44\,\omega_{0 \to 1}^*/T)}{1 - \exp(-1.44\,\omega_{0 \to 1}^*/T)}\right] \quad (9)$$

TABLE 1. Useful Spectroscopic Constants for Diatomic Molecules

Molecule	Change in vibrational quantum number	ω^* (cm^{-1})†	B_e (cm^{-1})†	f
CO	$0 \rightarrow 1$	2143	1.931	1.1×10^{-5} ‡
	$0 \rightarrow 2$	4259		7.5×10^{-8} ‡
NO	$0 \rightarrow 1$	1876	1.705	5.2×10^{-6} ¶
	$0 \rightarrow 2$	3724		9.6×10^{-8} §
HCl	$0 \rightarrow 1$	2886	10.59	7.3×10^{-6} §
	$0 \rightarrow 2$	5668		1.7×10^{-7} §
HBr	$0 \rightarrow 1$	2559	8.473	2.5×10^{-6} §
	$0 \rightarrow 2$	5028		3.2×10^{-8} §
OH	$0 \rightarrow 1$	3568	18.87	4.0×10^{-6} ‡
	$0 \rightarrow 2$	6970		1.7×10^{-7} ‡

†Reproduced from G. Herzberg, "Spectra of Diatomic Molecules," D. Van Nostrand Co., New York, 1963.
‡Reproduced from Ref. 10; the result for the $0 \rightarrow 1$ transition of CO is practically identical with the recent measurements of D. E. Burch and D. Williams, *Appl. Optics*, 1:587 (1962).
§S. S. Penner and D. Weber, *J. Chem. Phys.*, 21:649 (1953).
¶P. Varanasi and S. S. Penner, *JQSRT*, 7:279 (1967).

SPECTRAL AND BAND EMISSIVITIES FOR H_2O AND CO_2 AT MODERATE TEMPERATURES

Selected measurements and calculations of spectral absorption coefficients are available on H_2O [24–32] and CO_2 [33–36]. Favorite research tools for high-temperature studies (that is, $T > 1000°K$) are the shock tube and the nozzle exhausts of small rocket motors [26]. At temperatures up to about 1500°K, equilibrium measurements on heated cells have been used extensively [24, 25, 28, 30, 31].

The experimental results reproduced in Fig. 9 should serve as an illustration of the complexity of the spectral structure of polyatomic molecules.

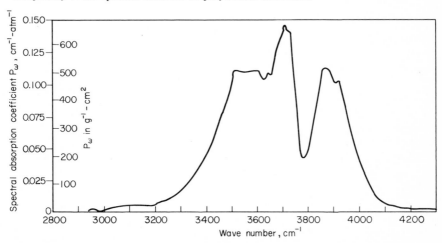

Fig. 9. The spectral absorption coefficient of water vapor in the 2.7-μ region at 1000°K as a function of wave number. The experimental points were determined at 5-cm^{-1} intervals at pressures sufficiently high to remove the rotational fine structure. Reproduced from Goldstein [25].

The theoretical calculation of spectral, band, and total emission or absorption for polyatomic molecules is a formidable theoretical and computational problem if an attempt [33, 34] is made to evaluate accurately detailed spectral structure and line-by-line contributions. On the other hand, approximate calculations can be performed with not much greater difficulty than for diatomic molecules [24, 37, 41, 42].

The *ab initio* calculations should be contrasted with the approach followed by Edwards [38] who has derived empirical expressions for the band absorption of CO_2 and has shown that this procedure leads to useful predictions of total emissivities by a direct comparison with the measurements of Hottel and Mangelsdorf [39] and of Eckert [40].

At the present time, there is sufficient information available on integrated intensities [24–32] for polyatomic molecules, and the theoretical studies [24, 33–37, 41, 42] are sufficiently well developed to permit good (within 20 percent) *a priori* calculations on band and total absorption or emission, as well as useful predictions of the detailed spectral band structure. Unfortunately, considerable skill in analysis and detailed knowledge of topics in quantitative spectroscopy are required for these studies, especially for the successful prediction of spectral properties.

The following simplified procedure has been found [41] to be useful for CO_2 and should serve as an illustration of a method for employing spectroscopic data for the calculation of spectral and band absorption of a polyatomic molecule. Implicit in the calculations are the following assumptions: the band contours may be approximated by a symmetric profile; harmonic oscillator approximations may be used for the sums of band intensities even for combination bands; the pressures are sufficiently high to justify the assumption that the rotational fine structure has been removed. A full examination of the implications of these assumptions cannot be made without detailed consideration of topics in molecular spectroscopy.

The band absorption [37] of the ith band is

$$A_i \equiv \int_{i\text{th band}} [1 - \exp(-P_\omega X)]d\omega \simeq \left(\frac{4B_{e,i}}{\sqrt{\gamma_i}}\right)I(K_i) \tag{10}$$

where

$$I(K_i) = \int_0^\infty \left[1 - \exp\left(-K_i \xi e^{-\xi^2}\right)\right]d\xi \tag{11}$$

(see Fig. 10 for a plot of $I(K)$ as a function of K),

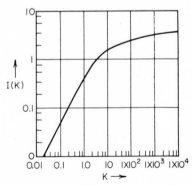

Fig. 10. Plot of $I(K)$ as a function of K.

$$I(K_i) \simeq \frac{K_i}{2} \text{ for small } K_i \text{ (this approximation is good to 2\% for } K \leq 0.05) \quad (11a)$$

$$I(K_i) \simeq 1.11\sqrt{2.303 \log(1.21 K_i)} \text{ for large and intermediate values of } K_i \quad (11b)$$

In Eq. (10), P_ω denotes the spectral absorption coefficient (in $cm^{-1}\text{-atm}^{-1}$) at the wave number ω, X is the optical depth in cm-atm, $y_i = 1.44 B_{e,i}/T$ if $B_{e,i}$ is the rotational constant in cm^{-1} of the ith band, and T represents the temperature in $^\circ$K, $K_i = \alpha X \sqrt{\gamma_i}/2 B_{e,i}$ with α identifying the integrated intensity for the band in $cm^{-2}\text{-atm}^{-1}$ at the temperature T. Equation (10) yields an upper bound for A_i if the ith band is an isolated band because the rotational fine structure has been assumed to be completely smeared out. For two or more partially overlapping bands in the jth spectral region, we may obtain both upper and lower bounds for

$$A_j = \int\limits_{j\text{th spectral region}} [1 - \exp(-P_\omega X)] d\omega \quad (12)$$

by noting that

$$\left(\frac{4\bar{B}_e}{\sqrt{\bar{\gamma}}}\right) I\left(\sum_i K_i\right) \leq A_j \leq \sum_i \left(\frac{4 B_{e,i}}{\sqrt{\gamma_i}}\right) I(K_i) \quad (13)$$

where the summation extends over all of the bands contributing to the jth region and $\bar{B}_e$ and $\bar{\gamma}$ are (arithmetic) average values for the contributing bands in the jth region. The lower bound in Eq. (13) corresponds to the case in which the various contributing bands are practically coincident; the upper bound corresponds to the case where strong contributions from different bands fall in different spectral regions.

TABLE 2. Spectroscopic Data Required for the Theoretical Calculation of Band Absorption of CO_2 in the 2.7-μ Region

Band center (cm^{-1})	B_e (cm^{-1})	T $(^\circ K)$	α $(cm^{-2}\text{-atm}^{-1})$
3716	0.3906	300	42.3
3609			28.5
3716		600	21.2
3609			19.5
3716		900	14.1
3609			21.8
3716		1200	13.8
3609			24.2
3716		1500	12.5
3609			28.3
3716		2000	11.65
3609			33.5
3716		2500	10.1
3609			38.8

TABLE 3. Comparison of Measured and Calculated Values of
A for the 2.7-μ Region*†

T	X	A observed*	$\sum_i (4B_{e,i}/\sqrt{\gamma_i}) I(K_i)$
(°K)	(cm-atm)	(cm^{-1})	(cm^{-1})
1200	15.5	197	216
	7.75	121	151
1273	11.7	169	178
1500	15.5	208	219
	7.75	128	153
2000	15.5		280
	7.75		189
2500	15.5		303
	7.75		209

*Literature citations to the experimental studies, as well as a comparison of experimental data with results derived from extensive numerical calculations, may be found in Malkmus [33].
†Reproduced from Penner and Varanasi [13].

The upper bound in Eq. (13) is generally a reasonable approximation for molecular bands, provided harmonic bands are treated as coincident (see below).

As an illustration of the utility of Eq. (13), we present a detailed comparison between theoretical and experimental results obtained for the 2.7μ bands of CO_2. The required spectroscopic data are listed in Table 2 and the results of the calculations are summarized in Table 3.

Extensive data on $\mathcal{Q}$ for CO_2 may be found in Table 11-9 of Penner [24]; probably the most reliable measurements on $\mathcal{Q}$ for H_2O are those reported by Goldstein [25]. The numerical value of $B_e = 0.3906$ cm^{-1} should be used for CO_2; proper choice of an appropriate value of $B_{e,i}$ for H_2O is difficult (see, e.g., the discussion of pages 156–157 of Ref. 10) although a numerical value of about 11.6 cm^{-1} does permit useful predictions of band intensities for water [42].

SPECTRAL EMISSIVITIES FOR DIATOMIC
MOLECULES AT ELEVATED TEMPERATURES

Theoretical calculations of spectral emissivities for diatomic molecules at elevated temperatures may be performed quite readily if suitable simplifications are introduced [43, 44]. However, most of the analytical studies on gases at temperatures above about 3000°K, where electronic transitions dominate radiant-energy emission, have been performed by the use of machine computer programs [45–47]. The results given in Fig. 11 are indicative of the types of data that are obtained by using approximate, explicit theoretical relations [43] (solid curve) or numerical calculations including an averaging procedure over a wide wave-number interval [45] (dotted curve).

The spectral absorption coefficient at the wave number ω is given by

$$P_\omega = \sum_{v''} \sum_{v'} P_{\omega; v'v''}$$

where the subscript $v'v''$ identifies the contributions made by the $v' \to v''$ vibrational transition from the upper electronic state (identified by the superscript prime) to the lower electronic state (identified by the superscript double prime). For each band $v'v''$ in the electronic band system, the contribution is of the form

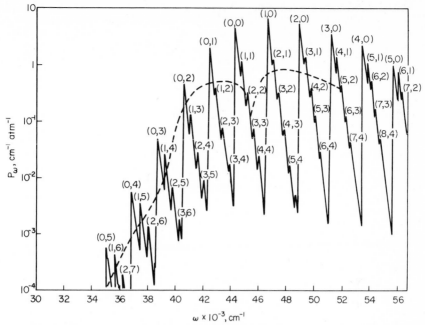

Fig. 11. Approximate values for the spectral absorption coefficients of the NO γ-band system at 2000°K. The solid curves were obtained from a simple analytical representation for the spectral absorption coefficient; the dotted curves are taken from Ref. 45. Band heads for transitions from the upper electronic state with vibrational quantum number v' to the lower electronic state with vibrational quantum number v'' are identified by the symbol (v', v''); reproduced from Patch, Shackleford, and Penner [43].

$$\ln P_{\omega;\, v'v''} \;=\; \ln P_{\omega;\, v'v'';\, h} \,-\, D\omega$$

where

$$D \;\equiv\; \frac{1.44\, B''_e}{(B'_e - B''_e)\, T}$$

and, to the order of approximation used in the analysis [43]

$$P_{\omega;\, v'v'';\, h} \;=\; 9.35 \times 10^{+5}\, \frac{f q_{v'v''} B_{e''}}{Q_{v''} |B''_e - B'_e|}\, \exp\!\left(-\frac{1.44\, \omega_{v''}}{T}\right)\ \text{cm}^{-1}\,\text{atm}^{-1}$$

represents the value of the spectral absorption coefficient at the band head (identified by the subscript h). Values of the dimensionless electronic f numbers, of the dimensionless vibrational overlap integrals $q_{v'v''}$, of the vibrational wave numbers $\omega_{v''}$, and of the rotational constants B''_e or B'_e may be found in standard texts on spectroscopy. The vibrational partition function is given by the approximate expression

$$Q_{v''} \;\simeq\; \left[\, 1 - \exp\!\left(-\frac{1.44\, \omega_{v''}}{T}\right)\right]^{-1}$$

TABLE 4. Required Spectroscopic Constants for the Theoretical Calculation of Spectral Absorption Coefficients of the NO γ-bands (Transition $A^2 \Sigma^+ \to X^2 \Pi$)

$$\omega_{v''} = 1876 \text{ cm}^{-1}*$$

$$B'_e = 1.9952*$$

$$B''_e = 1.7046*$$

$$f = 0.0025 \dagger \ddagger$$

Partial Table of Vibrational Overlap Integrals $q_{v'v''}$ §

v' \ v''	0	1	2	3	4	5	6	7
0	0.277	0.305	0.218	0.133	0.065	0.028	0.010	0.003
1	0.367	0.056	0.024	0.133	0.160	0.126	0.070	0.034
2	0.275	0.087	0.169	0.015	0.020	0.092	0.125	0.106
3	0.099	0.274	0.001	0.112	0.094	0.000	0.026	0.085
4	0.026	0.193	0.174	0.053	0.020	0.109	0.056	0.000
5	0.005	0.077	0.270	0.048	0.127	0.000	0.062	0.082
6	0.002	0.010	0.105	0.234	0.006	0.127	0.022	0.026
7	0.000	0.001	0.030	0.192	0.149	0.018	0.049	0.077

*From G. Herzberg, "Spectra of Diatomic Molecules," p. 558, D. Van Nostrand Co., New York, 1950.

†D. Weber and S. S. Penner, *J. Chem. Phys.*, **26**:860 (1957).

‡G. W. Bethke, *J. Chem. Phys.*, **31**:662 (1959).

§D. J. Flinn, R. J. Spindler, S. Fifer, and M. Kelley, *JQSRT*, **4**:271 (1964). The data used in the theoretical calculations of Fig. 11 are based on earlier, less precise, estimates by B. Kivel, H. Mayer, and H. Bethe, *Ann. Phys.*, **2**:57 (1957).

The numerical data required for the theoretical calculation of $P_{\omega; v'v''; h}$ and of $P_{\omega; v'v''}$ for the NO γ bands are summarized in Table 4.

SPECTRAL EMISSIVITIES FOR HEATED ATOMS

The theoretical calculation of spectral absorption coefficients for heated atoms is easily accomplished if a hydrogenic approximation is employed [48–51]. More exact computations generally require the use of an elaborate physical and analytical apparatus [52–54]. The utility of employing a semianalytical first approximation may be judged from the illustrative data shown in Fig. 12 for nitrogen at a temperature of 5 eV $\simeq$ 62,000°K.

Spectral absorption coefficients per centimeter for atoms at very high temperatures may be computed from the following approximate expressions for photon energies less than the first ionization potential [51]:

$$\kappa(x) = \kappa'(x)[1 - \exp(-x)]$$

$$x = 1.24 \times 10^{-4} \frac{\omega}{\theta}$$

ω is in cm^{-1}, $\theta = T (°K)/11,600$ is the temperature in eV, and

$$\kappa'(x) = \frac{2.43 \times 10^{-37} N^2}{\theta^{7/2} x^3} \bar{m} \left(\bar{m}^2 + \frac{1}{4} \right) \exp(x)$$

where N is the total number density of neutral atoms in cm^{-3} from which the plasma

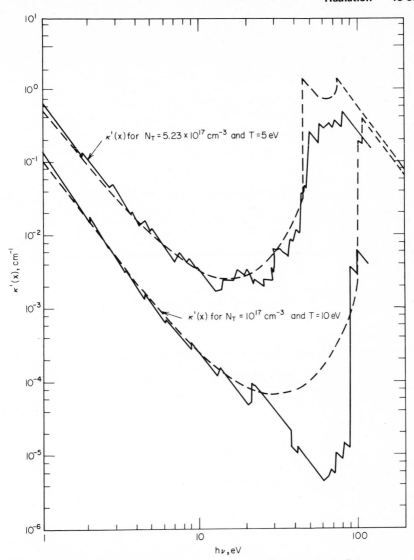

Fig. 12. The spectral absorption coefficient per unit length $\kappa'(x)$ where $x = h\nu/kT$, without the induced emission term, for nitrogen calculated by employing a simplified procedure based on a hydrogenic approximation [51] (dotted curves) and using the best available methods (solid curves) [52], for $T = 5$ eV, total number densities $N_T = 5.23 \times 10^{17}$ cm^{-3} and $T = 10$ eV, $N_T = 10^{17}$ cm^{-3}.

is formed and $\bar{m}$ denotes the number of electrons per atom present at equilibrium. In first approximation, $\bar{m}$ is obtained by plotting the ionization potential I (in eV) as a function of the number of electrons stripped from the atom core ($m = 0$ for the neutral atom, $m = 1$ for the singly ionized atom, etc.) and requiring that

$$I_{\bar{m}-(1/2)} = \theta \ln\left(3.0 \times \frac{10^{21}\theta^{3/2}}{N\bar{m}}\right)$$

a condition which is easily satisfied by graphical means.

The specified procedure for calculating $\kappa(x)$ includes only the continuum radiation (bound-free and free-free radiation) in a modified hydrogenic approximation [49–51]. The line contributions are generally about as important as the continuum contributions and can only be estimated approximately with great difficulty [52–54].

SPECTRAL AND TOTAL EMISSIVITIES
FOR HEATED AIR

As the result of interest in radiant-heat transfer to reentry vehicles, the theoretical calculation of spectral emissivities for heated air has become one of the most active research areas in high-temperature spectroscopy [55–59]. These data are supported, to some extent, by recently performed measurements using especially the shock tube [60, 61] and an electric arc [62].

In view of the high current level of activity in this field, it is clearly premature to attempt an authoritative summary of results. For making rough estimates, the data of Kivel and Bailey [58] are in most convenient form although they are less accurate than the results compiled in Refs. 56 and 57. Representative results taken from Gilmore's recent compilation [59] are reproduced in Figs. 13 to 15. The data plotted

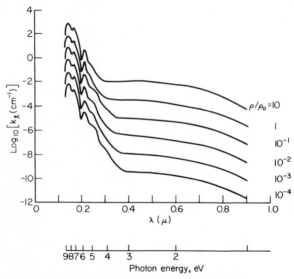

Fig. 13. The spectral absorption coefficient for air at $2000°$K per unit length, k_l (cm^{-1}), as a function of wavelength in microns ($1\mu = 10^{-4}$ cm) or photon energy (in eV) for various densities ($\rho_0 = 1.29 \times 10^{-3}$ g/cm^3); data taken from Gilmore [59].

in Fig. 13 show the spectral absorption coefficient per unit length in equilibrated air at $2000°$K as a function of wavelength for various densities. The graphs in Fig. 14 provide information on the Planck mean absorption coefficient $\bar{k}_{l,Pl}$ in cm^{-1} as a function of temperature and density, where $\bar{k}_{l,Pl}$ may be identified with the total emissivity per unit length when the total emissivity is smaller than about 0.2. The curves in Fig. 15 show the total emission rate of radiant energy from unit volume of heated air, as a function of temperature and density, neglecting self-absorption, induced emission, and infrared emission beyond 2.5μ.

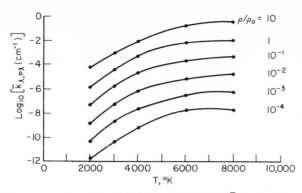

Fig. 14. The Planck mean absorption coefficient per unit length, $\bar{k}_{l,Pl}$ (cm^{-1}), as a function of temperature and density ($\rho_0 = 1.29 \times 10^{-3}$ g/cm^3) for equilibrated air; data taken from Gilmore [59].

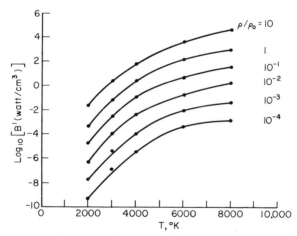

Fig. 15. The total emission rate of radiant energy per unit volume, B' (watt/cm^3), as a function of temperature and density ($\rho_0 = 1.29 \times 10^{-3}$ g/cm^3) for equilibrated air, neglecting self-absorption, induced emission, and infrared emission beyond 2.5μ; data taken from Gilmore [59].

MISCELLANEOUS RELATED TOPICS

There is a significant number of important additional applications requiring spectroscopic data on heated gases or solids. Examples are the calculation of radiant-heat transfer in nuclear reactors using hydrogen as a driving fluid [63], transmission of radiation through the atmosphere [64], Mie scattering and absorption of spherical carbon particles [65] and of spherical particles of aluminum oxide or magnesium oxide [66], semiempirical fits of band absorption to CH$_4$ [67], direct comparisons with experimental data [68] of theoretical studies [69] performed in the weak-line and strong-line approximation, etc.

REFERENCES

1. H. C. Hottel and H. G. Mangelsdorf, *Trans. Am. Inst. Chem. Eng.*, **31**:517 (1935).
2. H. C. Hottel and R. B. Egbert, *Trans. Am. Inst. Chem. Eng.*, **38**:531 (1942).

3. H. C. Hottel and A. T. Sarofim, "Fundamental Research in Heat Transfer," pp. 139–159, Pergamon Press, Ltd., London, 1963.

4. S. S. Penner and A. Thomson, *J. Appl. Phys.,* **28**:614 (1957).

5. S. S. Penner, D. B. Olfe, and A. Thomson, "Thermodynamic and Transport Properties of Gases, Liquids and Solids," pp. 2–13, McGraw-Hill Book Company, Inc., New York, 1959.

6. H. C. Hottel, Radiant Heat Transmission, in "Heat Transmission" (W. H. McAdams, ed.), chap. 3, McGraw-Hill Book Company, Inc., New York, 1942.

7. R. B. Egbert, Doctoral dissertation, Massachusetts Institute of Technology, Cambridge, Mass., 1941.

8. See Ref. 1.

9. E. Eckert, *VDI Forschungsheft,* **387**:1 (1937).

10. S. S. Penner, "Quantitative Molecular Spectroscopy and Gas Emissivities," Addison-Wesley Publishing Co., Reading, Mass., 1959.

11. G. N. Plass, *J. Opt. Soc. Am.,* **48**:690 (1958); **50**:868 (1960).

12. M. Lapp, L. D. Gray, and S. S. Penner, Equilibrium Emissivity Calculations for CO_2, in "International Developments in Heat Transfer," part 4, pp. 812–819, ASME, New York, 1961.

13. S. S. Penner and P. Varanasi, *JQSRT,* **4**:799 (1964).

14. S. S. Penner and P. Varanasi, *JQSRT,* **5**:391 (1965).

15. L. D. Gray and S. S. Penner, *JQSRT,* **5**:611 (1965).

16. S. S. Penner and P. Varanasi, *JQSRT,* **6**:181 (1966).

17. S. S. Penner and P. Varanasi, *JQSRT,* **7**:687 (1967).

18. See Ref. 10.

19. S. S. Penner, K. G. P. Sulzmann, and C. B. Ludwig, *JQSRT,* **1**:96 (1961).

20. S. S. Penner and L. D. Gray, *J. Opt. Soc. Am.,* **51**:460 (1961).

21. W. Malkmus and A. Thomson, *JQSRT,* **2**:17 (1962).

22. V. R. Stull and G. N. Plass, *J. Opt. Soc. Am.,* **50**:1279 (1960).

23. S. A. Golden, *JQSRT,* **2**:201 (1962).

24. See Ref. 10.

25. R. Goldstein, *JQSRT,* **4**:343 (1964).

26. C. C. Ferriso and C. B. Ludwig, *JQSRT,* **4**:215 (1964).

27. R. W. Patch, *JQSRT,* **5**:137 (1965).

28. U. P. Oppenheim and Y. Ben-Aryeh, *J. Opt. Soc. Am.,* **53**:344 (1963).

29. P. J. Wyatt, V. R. Stull, and G. N. Plass, *Appl. Optics,* **3**:229 (1964).

30. R. H. Tourin, *J. Opt. Soc. Am.,* **51**:1225 (1961).

31. D. E. Burch, D. Gryvnak, and D. Williams, *Appl. Optics,* **1**:759 (1962).

32. D. M. Gates, R. F. Calfee, and D. W. Hansen, *Appl. Optics,* **2**:1117 (1963).

33. W. Malkmus, *J. Opt. Soc. Am.,* **54**:751 (1964).

34. W. Malkmus, *J. Opt. Soc. Am.,* **53**:951 (1963).

35. V. R. Stull, P. J. Wyatt, and G. N. Plass, *Appl. Optics,* **3**:243 (1964).

36. V. R. Stull, P. J. Wyatt, and G. N. Plass, The Infrared Absorption of CO_2, Final Report, vol. 3, *SSD-TDR-62-127,* Aeronutronic Division of Ford Motor Company, Newport Beach, Calif., January 1963.

37. See Ref. 12.

38. D. K. Edwards and W. A. Menard, *Appl. Optics,* **3**:621, 847 (1964).

39. See Ref. 1.

40. See Ref. 9.

41. See Ref. 13.

42. See Ref. 14.

43. R. W. Patch, W. L. Shackleford, and S. S. Penner, *JQSRT,* **2**:263 (1962).

44. J. C. Keck, J. C. Kamm, B. Kivel, and T. Wentink, *Ann. Phys.,* **7**:1 (1959).

45. B. H. Armstrong, J. Sokoloff, R. W. Nicholls, D. H. Holland, and R. E. Meyerott, *JQSRT,* **1**:143 (1961).

46. R. G. Breene, Jr. and M. C. Nardone, *JQSRT,* **2**:273 (1962).

47. B. Kivel and K. Bailey, *AVCO Research Report* 21, Everett, Mass., 1957.

48. D. H. Menzel and C. L. Pekeris, *Mon. Not. Roy. Astr. Soc.,* **96**:77 (1935).

49. Yu. P. Raizer, *Sov. Phys. JETP,* **9**:1124 (1959).

50. Yu. P. Raizer, *Sov. Phys. JETP,* **10**:769 (1960).

51. S. S. Penner and M. A. Thomas, *AIAA Journal,* **2**:1572 (1964).

52. B. H. Armstrong, Mean Absorption Coefficients of Air, Nitrogen, and Oxygen from 22,000° to 220,000°, *Report LMSD-4979,* Lockheed Missiles and Space Division, Palo Alto, Calif., July, 1959; see also B. H. Armstrong and R. E. Meyerott, *Phys. Fluids,* **3**:138 (1960).

53. M. J. Seaton, *Rev. Mod. Phys.*, **30**:979 (1958).
54. V. N. Zhigulev, Ye. A. Romishevskii, and V. K. Vertushkin, *Inzhenernii Zhurnal*, 1:60 (1961); translation in *AIAA Journal*, 1:1473 (1963).
55. D. R. Churchill, S. A. Hagstrom, and R. K. M. Landshoff, *JQSRT*, **4**:291 (1964).
56. See Ref. 45.
57. See Ref. 46.
58. See Ref. 47.
59. F. R. Gilmore, Approximate Radiation Properties of Air Between 2000 and 8000°K, *Memorandum* RM-3997-ARPA, The RAND Corporation, Santa Monica, Calif., March 1964.
60. J. C. Keck, R. A. Allen, and R. L. Taylor, *JQSRT*, **3**:335 (1963).
61. W. H. Wurster, *JQSRT*, **3**:355 (1963).
62. H. N. Olsen, *JQSRT*, **3**:305 (1963).
63. D. B. Olfe, *JQSRT*, 1:105 (1961).
64. A. R. Boileau, *Appl. Optics*, **3**:570 (1964).
65. V. R. Stull and G. N. Plass, *J. Opt. Soc. Am.*, **50**:121 (1960).
66. E. Bauer and D. J. Carlson, *JQSRT*, **4**:363 (1964); G. N. Plass, *Appl. Optics*, **3**:867 (1964).
67. See Ref. 38, p. 847.
68. D. E. Burch and D. Williams, *Appl. Optics*, **3**:55 (1964).
69. G. N. Plass, *Appl. Optics*, **3**:859 (1964).

Section 16

Ablation

H. HURWICZ

Chief Advance Technology Engineer

J. E. ROGAN

Branch Chief

Aero/Thermodynamics and Nuclear Effects
Research and Development, Advance Systems and Technology
McDonnell Douglas Astronautics Company-West

ACKNOWLEDGMENT

The authors benefited by the many fruitful discussions on the topics of interest with Messrs. C. L. Arne, D. J. Chow, K. M. Kratsch, A. W. Maddox, M. Thomas, all of McDonnell Douglas Corp., Huntington Beach, California, and Mr. M. M. Sherman of the Space and Re-entry Systems Division of Philco-Ford Corp. Their contribution to this work is gratefully acknowledged.

INTRODUCTION

Engineering applications involving high-speed atmospheric flight or high-temperature propulsive devices such as reentry bodies or rocket nozzles created a need for high-temperature thermal protection systems. Various types of such systems are discussed in Sec. 10. The ablative systems have been found to offer a practical, often indispensable, solution to many of the reentry problems in the manned or unmanned applications, e.g., Refs. 1 and 2. It is felt thus that a modern Handbook of Heat Transfer would not be complete without a separate discussion of the heat and mass transfer phenomena occurring in the ablation process with major emphasis on hypersonic flight. However, it should be noted that much detailed and particularly pertinent information for this problem is contained also in Secs. 3, 8, 10, 15, and 16 and will not be repeated here. Ablation has been defined in many historical and functional ways [3, 4]. For the purpose of this text, ablation is the removal of mass (material) caused by thermochemical and mechanical processes. One should not be, however, misled into sole consideration of mass removal out of the context of the overall problem of the response of the material, the mechanisms by which it absorbs or rejects heat, the boundary-layer phenomena, and mass and energy balance at the surface. It is also important to distinguish between dimensional and total ablation, the former being characterized by surface recession and the latter by total mass loss.

TABLE 1. Dissociating, Compressible, Viscous, Multicomponent, Environmental Flow (with Equal Binary Diffusion Coefficients)

State of local boundary layer flow → / Local heat flux solutions (degree of complexity) ↓	Stagnation region	Laminar region	Turbulent region
Quasi-steady local flow region—aerodynamic heat flux (perfect gas)		Solutions with Prandtl number, $P_r \neq 1$ and Lewis number, $L_r \neq 1$ only	
I. Semi-empirical solutions for simple shapes—flat plate		$-\dot{q}_w = 0.332 (R_e^*)^{-1/2} (P^*)^{-2/3} \rho^* u_e^* (I_r - h_w)$ Eckert's ref. enth. meth.	$-\dot{q}_w = 0.0296 (R_e^*)^{-0.2} (P^*)^{-2/3} \rho^* u_e^* (I_r - h_w)$ Eckert's ref. enth. meth.
Wedge		$-\dot{q}_w = 0.332 \dfrac{u_e \rho^*}{(P^*)^{2/3}(R_e^*)^{1/2}} [I_r - h_w + (L^{1/2} - 1)h_c]$ Eckert's ref. enth. meth.	$-\dot{q}_w = 0.144 \dfrac{u_e \rho^* I^*}{(\log_{10} R_e^*)^{2.45}} [I_r - h_w + (L^{1/3} - 1)h_c]$ Eckert's ref. enth. meth.
Cone		$-\dot{q}_w = 0.575 \dfrac{u_e \rho^*}{(P^*)^{2/3}(R_e^*)^{1/2}} [I_r - h_w + (L^{1/2} - 1)h_c]$ Eckert's ref. enth. meth.	$-\dot{q}_w = 0.166 \dfrac{u_e \rho^* I^*}{(\log_{10} R_e^*)^{2.45}} [I_r - h_w + (L^{1/3} - 1)h_c]$ Eckert's ref. enth. meth.
Swept cyl.	$-\dot{q}_{w1}(0) = 0.576 \dfrac{(\rho_e \mu_e)^{0.44}}{P^{0.6}(\rho_w \mu_w)^{-0.06}} \left(\dfrac{du_e}{ds}\right)^{0.5} \left[1 + (L^{0.52} - 1)\dfrac{h_c}{I_e}\right](I_r - h_w)$	$-\dot{q}_w(\theta) = -\dot{q}_{w1}(0)[0.37 + 0.48\cos\theta + 0.15\cos2\theta]$	$-\dot{q}_w(\theta) = 0.0823 \left(\dfrac{du_e}{d_s}\right)^{0.2} \left[\dfrac{P_\theta}{P_0}\dfrac{(\rho_e\mu_e)_\theta^{0.88}(\rho_w\mu_w)_\theta^{0.12}}{(\rho_e\mu_e)_0^{0.88}(\rho_w\mu_w)_0^{0.12}}\right]^{0.9} \dfrac{(\rho_e\mu_e)^{0.704}}{(\rho_w\mu_w)^{-0.096}} \left(\dfrac{u_\infty \sin\Delta}{\mu_e}\right)(I_r - h_w)$
Hemisphere	$-\dot{q}_w(0)$ Defers to solutions III-stagnation	$-\dot{q}_w(\theta) = -\dot{q}_w(0)\, 2\theta\sin\theta \dfrac{\left\{\left(1-\dfrac{p_e}{p_0}\right)\left(\theta^2 - \dfrac{\theta}{2}\sin4\theta + \dfrac{1-\cos4\theta}{8}\right) + \dfrac{4p_e}{p_0}\left(\theta^2 - \theta\sin2\theta + \dfrac{1-\cos2\theta}{2}\right)\right\}^{1/2}}{\left(1 - \dfrac{p_e}{p_0}\right)\cos^2\theta + \dfrac{p_e}{p_0}}$	$-\dot{q}_w(\theta) = 0.0405 P(\theta)^{-2/3}\left(\dfrac{\mu_e}{s}\right)^{0.2}(\rho u_e^*)^{0.8}\left(\dfrac{p_\theta}{p_0}\right)^{0.075}\left(1 + 0.58\dfrac{h_c}{I_r}\right)(I_r - h_w)$

TABLE 1. Dissociating, Compressible, Viscous, Multicomponent, Environmental Flow (with Equal Binary Diffusion Coefficients) (Continued)

State of local boundary layer flow → Local heat flux solutions (degree of complexity) ↓	Stagnation region	Laminar region	Turbulent region
II. Converging-diverging rocket nozzle (3 dim. axisymmetric) (Bartz, Sibulkin or ref. Enthalpy methods)	▨	Nozzle flow primarily turbulent ———→	Typically Bartz' equation (decoupled from gas-phase or surface chemistry) $$C_{H_0} \sim \frac{0.026}{P_e u_e}\frac{\mu_e^{0.2}}{P^{0.6}}\left(\frac{P_c g}{C^*}\right)^{0.8}(D^*_{r_c})^{-0.1}\left(\frac{A_*}{A}\right)^{0.9}\sigma$$
III. 2 dim. and 3 dim. axisymmetric similarity and "locally" similar solutions	$$\beta = \frac{1}{k+1} \quad (\text{i.e., } \beta = 1, 1/2)$$ Dissociating only, $P \neq 1$, $L \neq 1$ (Fay and Riddell Equations) "Frozen" gas phase $$-\dot{q}_w(0) = 0.763\,P^{-0.6}(\rho_e\mu_e)^{1/2}\left(\frac{du_e}{ds}\right)_0^{1/2}$$ $$\left(\frac{\rho_w\mu_w}{\rho_e\mu_e}\right)^{0.1}(I_e - I_w)$$ $$\left[1 + (L^{0.63}-1)\frac{h_c}{I_e}\right]$$	Lee's – Dorodnitsyn similarity transformation performed on conservation equations, namely: $$\bar{s} = \int_0^s \rho_e\mu_e u_e r_0^{2k}\,ds; \quad \eta = \frac{\rho_e u_e r_0^k}{(2\bar{s})^{1/2}}\int_0^y \frac{\rho}{\rho_e}\,dy$$ $$k = \begin{cases} 0\text{-}2 \text{ dimensional} \\ 1\text{-}3 \text{ dimensional axisymmetric} \end{cases}$$ $$\beta\left(\text{pressure gradient parameter} \sim \frac{dp_e}{ds} = \frac{2\bar{s}}{u_e}\frac{du_e}{d\bar{s}}\right)$$ *Frozen gas phase and catalytic surface solutions only* $$-\dot{q}_w = \frac{\mu_e^k r_0}{(2\bar{s})^{1/2}}[G(\infty;P)]^{-1}\rho_e u_e\big[(I_t)_r - (I_t)_w\big]\left\{1 + \frac{G(\infty;P)}{G(\infty;P/L)}\frac{\sum_i h_i^0[(c_i)_e - (c_i)_w]}{(I_t)_e - (I_t)_w}\right\}$$ $(G\infty;P)$ and $G(\infty;P/L)$ are known numerically integrated solutions and explicit functions of β, P, L and $f(0)$ (total ablation mass flux). $f(0)$ is a diffusion controlled or reaction rate function of $\dot{m}_w$ *Equilibrium gas phase with ablation requires numerical solution of the ordinary differential equations of conservation* *Nonequilibrium gas phase solutions not implemented*	*Frozen gas phase and catalytic surface solutions only* $$-\dot{q}_w = \frac{C_f}{2}\rho_e u_e \frac{B'}{(1+B'\lambda)(1+B'\lambda B'\lambda+1)^{P-1}-1}$$ $$+\frac{(1+B'\lambda(1+B'\lambda)^{P-1}-1}{(1+B'\lambda)(1+B'\lambda)^{(P/L)-1}-1}\sum_i h_i^0\big[(c_i)_e - (c_i)_w\big]\Bigg\}\big\{(I_t)_r - (I_t)_w\big\}$$ $$C_f = 2\left[\left(\frac{1+(c_0)_w}{1+(c_0)_e}\right)^{1/\tau} - 1\right]\left(\frac{1}{1+B'\lambda}\right)C_{l_0}(0)$$ correction for dissociation $$\frac{C_{l_0}(0)}{2}\frac{d\theta}{ds} + \theta\left(\frac{2+H_f}{u_e}\frac{du_e}{ds} + \frac{1}{P_e}\frac{dp_e}{ds} + \frac{1}{r_0}\frac{dr_0}{ds}\right)$$ θ = von Kármán momentum integral $C_{l_0}(0)$ evaluated by Von Driest or Spalding and Chi methods H_f = "form factor" B' is a known function of $\dot{m}_w$ λ = ratio of laminar sublayer and fully turbulent layer, edge velocities *Equilibrium and nonequilibrium gas-phase solutions not implemented*

Equilibrium gas phase

$$-\dot{q}_w(0) = 0.763\, P^{-0.6}(\rho_e\mu_e)^{1/2}\left(\frac{du_e}{ds}\right)_0^{1/2}\left(\frac{\rho_w\mu_w}{\rho_e\mu_e}\right)^{0.1}(I_e - h_w)\left[1 + (L^{0.52} - 1)\frac{h_c}{I_e}\right]$$

Ablation solutions – (Defers to Solutions III–Laminar)

Otherwise simultaneous solution of ordinary differential equations of conservation

IV. 2 dim. and 3 dim. axisymmetric *nonsimilar solutions* $\left\{\begin{array}{l}\beta = 0 \text{ or } \beta \neq 0\\ P = 1 \text{ or } P \neq 1\\ L = 1 \text{ or } L \neq 1\end{array}\right\}$ Including unequal binary diffusion coefficients	(Defers to Solutions III – stagnation)	Frozen and equilibrium gas phase requires simultaneous numerical solution of the ordinary differential equations resulting from Lee's–Dorodnitsyn "similarity" transformation on the conservation equations (nonsimilar terms are retained)	Not implemented
V. 3 dimensional solutions completely *uncoupled* from surface ablation and subsurface reactions	(Defers to Solutions III – stagnation)	Only the *nonablating* surface case is implemented by simultaneous numerical solution of the partial differential conservation equations	Only the *nonreacting* gas phase, *nonablating* surface case is implemented by iterative numerical integral methods using empirical flow variable distributions, otherwise *not implemented*

TABLE 1. Dissociating, Compressible, Viscous, Multicomponent, Environmental Flow (with Equal Binary Diffusion Coefficients) *(Continued)*

State of local boundary layer flow → / Local heat flux solutions (degree of complexity) ↓	Stagnation region	Laminar region	Turbulent region
Surface Region— (one-dimensional)			
1. Energy balance			

$$\dot{q}_s = \underbrace{C_H \rho_e u_e \left[(I_r) - (h_I)_w \right]}_{\text{(standard convective contribution)}} + \underbrace{C_H \rho_e u_e \left[\frac{C_{H_d}}{C_H} \sum_i h_i^0 \left[(c_i)_e - (c_i)_w \right] \right]}_{\text{(diffusion contribution)}} - C_H \rho_e u_e \left[\frac{C_{H_d}}{C_H} \sum_{k'} \beta^* \frac{(\dot{m}_{k'})_w}{\dot{m}_w} L_{k'} \right]$$

(conduction into surface) k' = solid species of surface material   (vaporization or sublimation of char material)

$$- \underbrace{C_H \rho_e u_e \left[\frac{C_{H_d}}{C_H} \beta^* h_w \right]}_{\text{(gases leaving the wall)}} + \underbrace{C_H \rho_e u_e \left[\frac{C_{H_d}}{C_H} \sum_k \beta^* \frac{(\dot{m}_k)_w}{\dot{m}_w} (h_k)_w \right]}_{\text{(pyrolysis off-gasing and char reactions)}} - \underbrace{\epsilon_w \sigma T_w^4}_{\text{reradiation}} + \underbrace{\alpha_w \dot{q}_{rad}}_{\substack{\text{shock layer} \\ \text{radiation}}} - \underbrace{\dot{q}_w}_{\substack{\text{(From boundary} \\ \text{layer solutions)}}}$$

k = ablator species

II. Coupling considerations	(Defers to laminar)	Laminar	Turbulent
A. Convective, C_H		$C_H = \dfrac{\mu_e r_0^k}{(2\bar{s})^{1/2}} [G(\infty; P)]^{-1}$	$C_{iH} = \dfrac{C_{f_0}}{2} \dfrac{1}{1 + B'} \left\{ \dfrac{B'}{(1 + B'\lambda)(1 + B'\lambda\beta'\lambda + 1)^{P-1} - 1} \right\}$
B. Diffusive, C_{H_d}		$C_{H_d} = \dfrac{\mu_e r_0^k}{(2\bar{s})^{1/2}} G\left(\infty; \dfrac{P}{L} \right)^{-1}$	$C_{H_d} = \dfrac{C_{f_0}}{2} \dfrac{1}{1 + B'\lambda} \left\{ \dfrac{B'}{(1 + B'\lambda\beta'\lambda + 1)^{(P/L)-1} - 1} \right\}$
C. Blowing factor, β^*		$\beta^* = -f(0) G\left(\infty; \dfrac{P}{L} \right)$	$\beta^* = [(1 + B'\lambda\beta'\lambda + 1)^{(P/L)-1} - 1]\left(\dfrac{\bar{M}_{BL}}{\bar{M}_{inj}} \right)^{0.46}$

$G(\infty; P)$ and $G(\infty; P/L)$ are functions of $f(0)$

III. Surface reactions

A. Diffusion controlled

$$a_n A_n + b_n B_n \overset{K_n(T_w)}{\rightleftharpoons} c_n C_n + d_n D_n \quad (n = 1, 2, \ldots, N) \text{ equations}$$

$K_n(T_w)$ is equilibrium constant

a_n, b_n, c_n, and d_n are the stoichiometric coefficients for the reaction

A_n = one of the kth species originating in the ablator

B_n = a boundary layer originating species

B. Reaction rate controlled

$$a_m A_m + b_m B_m \longrightarrow c_m C_m + d_m D_m \quad (m = 1, 2, \ldots, M) \text{ equations}$$

IV. Mass balance

$$\dot{m}_w = \sum_{k \neq k'} (\dot{m}_k)_w + \sum_k (\dot{m}_k)_w = \sum_{j'} (\dot{m}_{j'})_w$$

(total injected mass flux)   (gaseous pyrolysis products)   (char erosion flux)   (elemental mass flux)

$(\dot{m}_{k \neq k'})_w$ is determined from in depth rate controlled reactions

where

$$(\dot{m}_{j'})_w = \sum_k \epsilon_{j'k} (\dot{m}_k)_w$$

j' = elements originating in ablator

k = species originating in ablator

$\epsilon_{j'k}$ = (fraction of mass of species k contributed by elements j') = $M_{j'}/M_k$

V. Blowing factor β^* determination

① $\beta^* = \dfrac{\dot{m}_w}{\rho_e u_e C_{H_d}}$

A. Diffusion-controlled regime

Equations ① to ⑦ are solved simultaneously along with mass Balance expressions IV to arrive at β^* with unknowns $\dot{m}_w$ and $(\dot{m}_k)_w$ eliminated among the expressions

② $(\overline{C}_j)_w = \dfrac{(\overline{C}_j)_e}{1 + \beta^*}$ (J = elements originating in boundary layer)

③ $\overline{C}_j = \sum_i \epsilon_{ji} C_i$ (for all J and J' elements)   here ϵ_{ji} = fraction of mass of species i contributed by element j

④ $(\overline{C}_{J'})_w = \dfrac{(C_{J'})_e + (\dot{m}_{J'})_w \, \beta^*/\dot{m}_w}{1 + \beta^*}$ (J' = elements originating in ablator)

TABLE 1. Dissociating, Compressible, Viscous, Multicomponent, Environmental Flow (with Equal Binary Diffusion Coefficients) (*Continued*)

State of local boundary layer flow → Local heat flux solutions (degree of complexity) ↓	Stagnation region	Laminar region	Turbulent region
A. Diffusion-controlled regime (*Continued*)	(5) $p_e = \sum_i (p_i)_w$ Dalton's partial pressure law (6) $C_i = \dfrac{p_i M_i}{\sum_i p_i M_i}$ (M_i is mol. wt. of ith species) (7) $K_n(Tw) = \dfrac{(p_{c_n})_w^{c_n}(p_{D_n})_w^{d_n}}{(p_{A_n})_w^{a_n}(p_{B_n})_w^{b_n}}$ ($N = 1, 2, \ldots, N$ equations)		
B. Reaction rate-controlled regime	Equations (1) and (9) are solved simultaneously with Mass balance expressions IV to arrive at β^* directly (9) $(\dot{m}_{k'})_w =$ constant $p_e^{\nu} e^{E_{k'}/A_1 Tw}(p_{B_1})_w^{\nu_1}(p_{B_2})_w^{\nu_2} \cdots$ ($k' = 1, 2, \ldots, Q$ equations)		
In-depth response— (one-dimensional-transient) I. Reacting parabolic heat conduction equation	$\rho_s C_s \dfrac{\partial T}{\partial t} = \dfrac{\partial}{\partial y}\left(\kappa \dfrac{\partial T}{\partial y}\right) + \dot{m}_g \bar{C}_{pg}\dfrac{\partial T}{\partial y} + \Delta H_{dp}\dfrac{\partial p_{dp}}{\partial t} + \sum_i \Delta H_{c_i}\dfrac{\partial p_{c_i}}{\partial t} + \dfrac{\partial F}{\partial y}$ where $\bar{C}_{pg} = \sum_i C_{gi} C_{pgi}$		
II. Continuity equation	$\dfrac{\partial \rho_s}{\partial t} = \dfrac{\partial \dot{m}_g}{\partial y}$ with $\rho_{dp} + \rho_c = \rho_s$ and $(\dot{m}_g)_w = (\dot{m}_{k'\,k'})_w$		
III. Surface coupling considerations	σ (surface erosion rate) $= \dfrac{\sum_{k'}(\dot{m}_{k'})_w}{\bar{\rho}_k'}$ surface boundary condition $\dot{q}_s = -\left(\kappa \dfrac{\partial T}{\partial y}\right)_w$		
IV. Typical charring ablator decomposition reaction	$\Gamma\underbrace{\left(E_{a_1} R_{b_1} P_{c_1}\right)_{d_1} + (1-\Gamma)\left(E_{a_2} R_{b_2} P_{c_2}\right)_{d_2}}_{\text{virgin composite}} \xrightarrow{\Delta H_{dp}} \underbrace{e(T) R_2 + l(T)(E) R_4 + g(T) E_2 R_2 + l(T) P}_{\text{pyrolysis gaseous products}} + \underbrace{E_n}_{\substack{\text{solid} \\ \text{char}}}$		

V. Decomposition rate expressions for charring ablator (Arrhenius rate expressions)

$$\frac{\partial \rho_{dp}}{\partial t} = -\left\{ \Gamma k_{11} \rho_{11}(0) \left[\frac{\rho_{11}(t) - (f_c/\Gamma)\rho_{virgin}}{\rho_{11}(0)}\right]^{n_{11}} e^{-E_{11}/RT} + \Gamma k_{12} \rho_{12}(0) \left[\frac{\rho_{12}(t)}{\rho_{12}(0)}\right]^{n_{12}} e^{-E_{12}/RT} + (1 - \Gamma)k_2 \rho_2(0) \left[\frac{\rho_2(t)}{\rho_2(0)}\right]^{n_2} e^{-E_2/RT} \right\} \text{ where } f_c = \frac{\rho_{char}}{\rho_{virgin}}$$

VI. Typical deposition reactions

$$2E_2 R_2 \xrightarrow{\Delta H_{c_1}} ER_4 + 3E \text{ (solid)}$$

$$ER_4 \xrightarrow{\Delta H_{c_2}} 2R_2 + E \text{ (solid)}$$

VII. Deposition rate or "coking" expressions (typically Arrhenius)

$$\frac{\partial \rho_c}{\partial t} = \frac{\partial \rho_{c_1}}{\partial t} + \frac{\partial \rho_{c_2}}{\partial t}$$

with

$$\frac{\partial \rho_{c_i}}{\partial t} = -k_{c_i} \rho_{c_i}(0) \exp\left(-\frac{E_{c_i}}{RT}\right)(\rho_e)^{n_{c_i}} \left[\frac{\rho_{c_i}(t)}{\rho_{c_i}(0)}\right]^{n_{c_i}} \quad (i = 1, 2)$$

VIII. In-depth mass flux

$$\dot{m}_{k \neq k'} \int_{backface}^{\sigma} \frac{\partial \rho_s}{\partial t}\, dy$$

IX. "Glowing"

$$\frac{\partial F}{\partial y} = \alpha(J - J_e) + a(1 - R_0)F_0 \left\{ \frac{e^{-a(y-\sigma)} + R_y e^{-a(2\lambda - y - \sigma)}}{1 - R_0 R_y e^{-2a(\lambda-\sigma)}} \right\},$$

where σ = char surface coordinate, λ = backface coordinate

with $\dfrac{\partial^2 J}{\partial y^2} = a^2(J - J_e)$; $\dfrac{\partial J}{\partial y} = -\dfrac{a^2 F_1}{a}$; $J_e = 4\eta_e^2 \sigma T^4$ Boundary conditions: $J(y = \sigma) = \alpha_w \dot{q}_{rad}$;

$$F_1(\sigma) = -\frac{\alpha}{a}\left(\frac{1 - R_0}{1 + R_0}\right)\alpha_w \dot{q}_{rad}; \text{ and}$$

$$J(y = \lambda) = F_1(\lambda) = 0$$

X. Moving boundary coordinate transformation

$$\xi = \frac{y - \sigma(t)}{\lambda - \sigma(t)}$$

XI. Recommended numerical solution method   Implicit finite difference (e.g., Crank-Nicholson Method)

XII. Assumptions

1. One continuous heat equation across entire ablator
2. Net pyrolysis gases after coking appear instantaneously at surface
3. Pyrolysis gases always in thermal equilibrium with char
4. Virgin material either decomposes completely to gases or partially leaving a solid residue capable of melting, vaporization, sublimation, or heterogeneous reactions

5. Internal radiation implicit in conductivity
6. Pyrolysis gas velocities negligible compared to u_∞
7. One-dimensional heat and mass flow, normal to surface
8. Thermal conductivity and density of pyrolysis gases negligible compared to those of solid phase

9. Thermal conductivity and specific heat represented as linear functions of condensed phase density or empirical functions of temperature
10. Pressure gradient throughout ablator small enough to be neglected
11. Decomposition and coking are assumed to follow simple Arrhenius' rate expressions.

Unlike more established disciplines, it is not possible to make definitive recommendation of formulas, equations, and definitive solutions to the various transfer regimes encountered in ablation. This is due to the complexity of the ablation process which involves heat and mass transfer interactions between the environmental medium and a solid but decomposing object over which it flows with the attendant chemical reactions and changes in phase. The complexities of the process are best illustrated in Table 1 which indicates the various regimes involved in ablation as well as the state of the art.

Thus it is clear that no simple formula can be provided and that room for differences in approach must be made. One is tempted [4, 5] to provide such simple solutions, but in the long run more damage than benefit is reaped. The approach here (as well as in practical analysis) will be then to state the physical problems in mathematical terms and essentially formulate the boundary-value problem, for the solution and simulation on high-speed computers. The solutions to the problems (if available) will be given by reference and approximate methods or solutions will be given where applicable. It should be noted that by and large the coupling of the environment with the transient material response and structural interaction has been accomplished in one-dimensional solutions. Beyond this, the field is fertile for progress. This is not to say that other coupled solutions have not been achieved as will be seen in the text, but rather to emphasize that the ablation process is not yet completely understood in the context of the various modes of heat and mass transfer within and without an ablating body. Practical applications of ablation are further discussed in Sec. 19.

Several review papers [5–10] may be consulted for extensive bibliographies on the subject matter. The general state of the art is reviewed in the following sections.

A. STATE OF THE ART

The current state-of-the-art treatments of thermochemical response of ablators which are surface coupled, with the possibility of artificial mass injection, to a hypersonic, compressible, reacting, viscous flow field, follow three approaches. Common to most are the basic assumptions of a quasi-steady environmental flow field confined to either the two-dimensional or axisymmetric geometry and coupled to a transient (usually) one-dimensional analysis of the ablator's interior [11].

Latest analyses show concern for coupling of various boundary-layer phenomena among themselves, as well as with the ablator response [11, 12]. The three approaches commonly found are:

1. Heat-of-ablation Techniques

These highly empirical, highly test-condition-data-dependent methods are reasonably economical and time saving. They are successful when applied to conditions which do not appreciably depart from the test conditions during which the empirical relationships were derived. Extrapolation to other conditions in view of the highly nonlinear coupling that exists among the many simultaneous phenomena normally occurring between the boundary layer and ablator surface is not reliable and is not recommended [5, 10].

2. Coupled Convective Heat and Mass Transfer Coefficients

Frozen-gas phase chemistry assumption for boundary-layer species dissociation is combined with the artifice of surface-confined infinite-rate equilibrium chemistry for their recombination and/or interaction with the ablator surface products. When no boundary-layer species interaction is expected with the ablator products then the frozen-gas phase constraint on the dissociation and surface confinement to recombination can be lifted. Equilibrium dissociation and recombination throughout the boundary layer are then assumed. When the assumption of unity Lewis number is allowed, the energy and specie-concentration equations become uncoupled resulting in a convective heat transfer coefficient completely independent of mass transfer and chemical reactions. Also, somewhat unrealistic solutions for heat transfer are produced whose only dependence on chemical reaction are through the implicit appearance of the chemical reaction enthalpies in the total enthalpy terms [13]. The speci-concentration equations are then left unaltered to assume "frozen," equilibrium, or nonequilibrium chemistry solutions in the gas phase. When the additional assumption of unity Prandtl number is made, the viscous interaction is decoupled from the heat transfer expressions allowing complete local similarity between thermal and velocity profiles.

The treatments of this approach display the effects of mass injection alone through direct coupling of the heat and mass transfer coefficients or alternately rearranged into blowing parameters with or without material chemical reaction with the boundary-layer species.

In-depth ablator analyses utilize a numerical transient solution of the parabolic heat transfer differential equation [14–19]. Additional terms account for the internal mass transport and cracking, energy transfer by depolymerization and pyrolysis, energy transfer by deposition in the char, and finally transfer due to transparency to external radiation. The instantaneous mass and/or density quantities are obtained from the Arrhenius rate expressions. The deposition term [20] accounts for char densification and is sometimes referred to as coking. The actual process is highly pressure and pressure-gradient dependent and suggests strongly that chemical equilibrium as well as thermal equilibrium could be maintained between the porous char and the passing pyrolysis gases. It should be noted, however, that the questionable assumption of negligible pressure gradient is retained.

Hybridization between the in-depth analytical expressions and the wide spectrum of numerical solution is noted. Once in a finite difference form, many numerical schemes are available for actual solutions (Sec. 1.B.) and some are explicit, as Crank-Nicholson, Runge-Kutta, etc. These overall approaches permit retention of a reasonable detailed account of the significant reaction chemistry [21] and represent a manageable compromise between the theoretical analysis and rapid numerical solution. Practical engineering solutions are thus obtained; shortcomings of the many assumptions should be noted.

3. General Theoretical Models

Attempts are made here to account for finite rate nonequilibrium chemistry throughout the boundary layer and at the surface, simple transpiration to massive blowing effects [23, 24], semigrey optically thick boundary layers, surface melting and spallation [25], and in-depth chemical equilibrium between pyrolysis gases and the porous char. Primarily emphasized in the majority of these analyses is the fact that "only in the nonequilibrium consideration where chemical reactions proceed at a finite rate can the true boundary layer coupling manifest its most interesting characteristics" [22]. Here laminar boundary layer nonsimilarity may be considered [20, 26] as well as multicomponent variational transport coefficients [20, 27]. Turbulent boundary-layer treatment appears, however, to be more conservative [11, 23, 24]. Conversion to rapid numerical techniques is economically prohibitive for the sophisticated solutions, while occasionally engineering solutions with moderate computation times are feasible. Claims have been made of adaptation to numerical methods [20]; however, experimental data are sparse and the results remain unchallenged. As in less sophisticated analyses the various interacting phenomena are presented in option form. A variety of sophisticated treatments on each option are available [26, 28]. In contrast to the preceding approaches where generalization of the problem allowed engineering solutions, the general theoretical model options are generally assembled so as to maintain a reasonable compromise between complexity and reality. However, the analyses do remain theoretical, venturing not far beyond the differential equation stage. The advantages are profound, however, in providing the framework for studying new mechanisms influencing engineering decisions and in setting the limits of physical reality. At the same time, the current liabilities are apparent, as engineering solution of today's rocket nozzle or reentry problems is not possible with this approach.

The second approach will be presented here in more detail as representing the more accepted (although by no means unique) approach to the ablation-through-aerodynamic-heating problem [12].

B. ENVIRONMENT

1. General Considerations and Applications

Ablation as generally defined occurs in many natural conditions. Its mechanisms are, however, taken advantage of principally in thermal protection systems operating in very high temperature environments. These are created in and by hypersonic flight, reentry into the atmosphere of the earth or another planet, launch vehicles exiting from the atmosphere, and in large rocket engines. It will be necessary to review here some of the laws of fluid mechanics, and heat and mass transfer processes of relevance to ablation because of the often specialized uses of environmental relations as adapted to this process. The properties required to establish the environment will also be reviewed. It will be, however, assumed that the reader has either acquainted himself with other chapters of this handbook treating the same topics or will refer to them for more thorough treatment.

Transportation of hardware and personnel from one point to another at hypersonic speeds demands protection of both the capsule and of the propulsive device from the hyperthermal environments. The specific transport requirements dictate a mission and subsequently a vehicle and trajectory(s) which in turn determine the environment. Factors considered in the selection and analysis of the ablation system include: (1) overall system constraints and limitations, (2) trajectory and related parameters (initial reentry velocity and angle), (3) vehicle-oriented parameters (optimum aerodynamic shape, aerodynamic coefficients), (4) aerodynamic heating and pressures on all vehicle surfaces—considering free molecular flow, laminar flow, boundary-layer transition, turbulent flow, radiative heating, and radiation-convective coupling, and (5) characteristics and response (mass loss and temperature history) of the ablators under review. The accomplishment of the above tasks is greatly facilitated through the use of current computers and often involves iterations of the effects. However, on occasion, manual calculations are acceptable and may be accomplished utilizing Refs. 29 to 35.

Figure 1 [36] depicts some typical flight trajectories and delineates some of the important heating regimes for various applications requiring analysis of ablation.

The problems presenting themselves to a rocket nozzle designer are similar. The primary difference between rocket nozzle and reentry environments can be seen by referring to the conventional convective heat flux equation

$$q_0 = \bar{\bar{h}}(T_e - T_w) \tag{1}$$

For a high-performance reentry body, the maximum value of the heat transfer coefficient ($\bar{h}$) is of the order of 0.1 Btu/ft^2-sec-$^\circ$F while T_e is approximately 20,000°F. In rocket nozzles, however, $\bar{h}$ is normally about 0.5 while T_e is about

$5500°\,F.$ Thus, it can be seen that little can be done to decrease q_0 for reentry conditions while for rocket nozzles a sufficiently high wall temperature may reduce the convective heat input to essentially zero. This is, in fact, the reason why high temperature materials are quite effective in rocket nozzle applications, in spite of their relatively high thermal conductivity. It is also the reason why accurate knowledge of the ablation temperature is essential, and high temperatures are desirable.

Another reason why rocket nozzle environments may be unfriendly to conventional ablation materials is because of the small benefit that is obtained via transpiration. This can again be attributed to the relatively small $(T_e - T_w)$ or $(I_e - h_w)$.

2. Properties

Determination of the local environment of a reentry body or rocket nozzle liners requires definition of various physical, thermodynamic, transport, and radiative properties. These properties are needed in the determination of flow fields and various heat and mass transfer phenomena.

a. Atmospheric Properties. A clear understanding of atmospheric deviations is required as the initial step. In principle, one should examine the actual atmospheric properties. For an entry body, this would be a function of altitude on the day and at the time of the flight. Consideration should be given to such factors as moisture content, foreign particles (dust, smog), specific concentration of various chemical species, as well as the gross properties of temperature, density, and pressure. This approach is not practical in design analysis since these properties cannot be determined accurately in advance. Instead, a series of "standard atmospheres" which represent means, have been postulated over the years. The recent, most commonly used standard atmosphere is reported in Ref. 37. Its shortened version is contained in Table 2. Similar information for atmospheres other than earth has been postulated [38], and data for man-made environments in propulsive devices may be found in, e.g., Ref. 39.

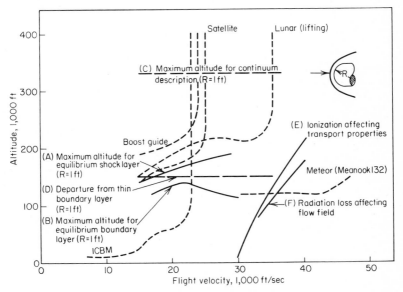

Fig. 1. Typical flight trajectory domains.

TABLE 2. 1962 ICAO Atmosphere

Altitude, ft	Pressure, lb/ft^2	Temperature, °R	Density, slug/ft^3
0	2.1162E 03	518.67	2.3769E-03
10,000	1.4556E 03	483.03	1.7556E-03
20,000	9.7327E 02	447.42	1.2673E-03
30,000	6.2966E 02	411.84	8.9069E-04
40,000	3.9313E 02	389.97	5.8728E-04
50,000	2.4361E 02	389.97	3.6892E-04
60,000	1.5103E 02	389.97	2.2561E-04
70,000	9.3727E 01	392.25	1.3920E-04
80,000	5.8511E 01	397.69	8.5710E-05
90,000	3.6778E 01	403.14	5.3147E-05
100,000	2.3272E 01	408.57	3.3182E-05
110,000	1.4837E 01	418.38	2.0659E-05
120,000	9.6013E 00	433.58	1.2900E-05
130,000	6.3094E 00	448.76	8.1907E-06
140,000	4.2061E 00	463.92	5.2818E-06
150,000	2.8418E 00	479.07	3.4557E-06
160,000	1.9419E 00	487.17	2.3221E-06
170,000	1.3297E 00	487.17	1.5901E-06
180,000	9.0836E-01	478.55	1.1058E-06
190,000	6.1549E-01	467.77	7.6654E-07
200,000	4.1344E-01	457.00	5.2704E-07
210,000	2.7435E-01	437.71	3.6514E-07
220,000	1.7843E-01	416.21	2.4975E-07
230,000	1.1348E-01	394.73	1.6749E-07
240,000	7.0403E-02	373.27	1.0988E-07
250,000	4.2480E-02	351.83	7.0339E-08
260,000	2.4843E-02	330.41	4.3802E-08
270,000	1.4177E-02	325.17	2.5400E-08
280,000	8.0860E-03	325.17	1.4487E-08
290,000	4.6143E-03	325.17	8.2668E-09
300,000	2.6430E-03	332.95	4.6245E-09
310,000	1.5504E-03	349.41	2.5850E-09
320,000	9.3293E-04	365.86	1.4855E-09
330,000	5.7338E-04	384.43	8.6891E-10
340,000	3.6337E-04	411.86	5.1398E-10
350,000	2.3733E-04	439.29	3.1473E-10

Example: $2.1162E03 = 2.1162 \times 10^3 = 2116.2$.

b. Thermodynamic Properties. A second and very important set of properties needed in the determination of local environment results from the selection of the equation of state model. For example, below velocities of 5,000 to 7,000 feet per second in the air, the perfect gas relationship may be used with considerable success. However, above these velocities a more sophisticated state model is required.

References 40 to 46 show in tabular and graphical form the equilibrium thermodynamic properties ranging to 24,000°K. Reference 45 is recommended for use. A typical chart [45] is shown in Table 3. For other media Ref. 47 is available.

c. Transport Properties. These properties, although as important as the others shown, are not as well defined numerically. The available data are at best only approximations. Various models which differ in the number and degree of species interactions are shown in Refs. 46, 48, and 50. Reference 45 is recommended below temperatures of 10,000°K; above 10,000°K only care and good judgment can, at the present time, be recommended. The summary tables of Hansen [45] for thermal

TABLE 3. Thermodynamic Properties for Air

$T,°K$	(a) Compressibility, Z				(b) Dimensionless enthalpy, ZH/RT			
	Pressure, atm				Pressure, atm			
	100	1.0	0.1	0.0001	100	1.0	0.1	0.0001
1000	1.000	1.000	1.000	1.000	3.65	3.65	3.65	3.65
2000	1.000	1.000	1.001	1.016	3.92	3.92	3.93	4.41
3000	1.003	1.026	1.072	1.200	4.13	4.61	5.55	8.19
4000	1.633	1.163	1.198	1.287	4.70	6.74	7.28	0.82
5000	1.118	1.214	1.252	1.910	5.73	7.10	7.96	23.46
6000	1.189	1.316	1.529	2.008	6.38	8.70	12.93	22.54
7000	1.243	1.607	1.904	2.088	6.95	13.20	18.34	22.29
8000	1.311	1.896	2.091	2.446	8.16	16.73	18.43	28.65
9000	1.512	1.903	2.050	3.282	10.20	17.01	18.09	43.74
10000	1.718	2.012	2.149	3.843	12.42	16.84	18.85	50.96
11000	1.876	2.111	2.351	3.960	13.77	17.13	21.31	49.48
12000	1.965	2.232	2.691	3.903	14.28	19.16	23.36	45.26
13000	2.017	2.426	3.135	3.908	14.30	20.32	30.90	43.97
14000	2.062	2.700	3.527	3.900	14.34	23.29	34.97	41.61
15000	2.113	3.028	3.769	4.000	14.49	26.66	36.53	39.60

	(c) Dimensionless entropy, ZS/R				(d) Dimensionless specific heat at presure, ZC_p/R			
1000	23.7	28.3	30.6	37.5	3.96	3.96	3.96	3.96
2000	26.6	31.2	33.5	40.9	4.34	4.41	3.57	11.57
3000	28.4	33.5	36.9	47.5	5.03	9.62	16.61	5.41
4000	30.2	37.3	40.6	51.6	8.24	11.05	6.91	10.78
5000	32.4	39.2	42.9	68.9	10.53	9.58	18.98	47.49
6000	31.2	42.2	49.7	72.2	9.05	26.95	57.60	11.43
7000	35.8	48.3	57.6	75.4	12.59	49.61	31.18	36.82
8000	38.0	53.9	60.1	85.0	21.42	28.27	13.68	121.60
9000	41.1	56.2	61.9	104.3	31.05	11.83	18.90	173.32
10000	44.5	57.8	64.6	116.6	31.43	16.53	34.11	60.29
11000	47.1	59.7	69.0	119.9	22.82	24.46	59.25	19.92
12000	48.8	62.4	75.4	121.3	16.39	37.23	87.57	12.82
13000	50.0	66.0	82.9	122.3	14.57	53.70	96.15	11.47
14000	51.1	70.5	89.4	123.1	15.52	69.41	73.42	11.16
15000	52.2	75.6	93.4	123.9	18.02	75.95	44.65	11.08

conductivity, viscosity, and Prandtl numbers are shown in Table 4. Other media are treated in Refs. 48 and 51.

d. Radiative Properties. The knowledge of radiative properties of atmospheres is particularly important at very high-speed flight in each atmosphere ($v > 35,000$ fps) and especially so for blunt bodies. The importance of radiative properties of other atmospheres depends of course on the gas composition present or assumed. The discussion of the air properties below will be typical.

Since most of the mechanisms of radiation in high-temperature air have been identified, a pretty good agreement between experiment and theory is possible. A complete description of the spectral absorption coefficient of air must include band radiation, photodissociation, photoionization, photodetachment, atomic lines, and brehmstrahlung. Often, atomic-lines belonging to multiplets may be grouped together, since for many conditions of interest the spacing between lines of a particular multiplet is of the same order as the line half-widths. The f-numbers for UV lines may be taken from Ref. 52 and the visible lines from Ref. 53. In addition, the hydrogenic model [54] can be used to define lines lying close to the photoelectric edges; this model may be used down to the first NBS tabulated line, after which, for

TABLE 4. Transport Properties of Air

	(a) Coefficient of viscosity			
$T, °K$	Ratio η/η_0			
	Pressure, atm			
	100	1.0	0.1	0.0001
1000	1.000	1.000	1.000	1.000
2000	1.000	1.000	1.000	1.000
3000	1.000	1.000	1.000	1.000
4000	1.003	1.016	1.020	1.032
5000	1.022	1.013	1.051	1.181
6000	1.050	1.090	1.148	1.256
7000	1.089	1.208	1.294	1.264
8000	1.143	1.312	1.371	1.072
9000	1.238	1.425	1.396	0.517
10000	1.361	1.415	1.375	0.118
11000	1.467	1.412	1.267	0.029
12000	1.549	1.394	1.040	0.012
13000	1.581	1.274	0.711	0.008
14000	1.599	1.082	0.408	0.007

	(b) Coefficient of thermal conductivity			
$T, °K$	Ratio κ/κ_0			
	Pressure, atm			
	100	1.0	0.1	0.001
1000	1.100	1.100	1.100	1.100
2000	1.177	1.177	1.251	3.99
3000	1.421	3.20	5.48	1.465
4000	2.69	2.99	1.000	15.03
5000	3.07	3.29	7.18	11.84
6000	1.930	10.19	18.74	6.14
7000	4.69	15.69	8.32	26.9
8000	8.21	7.80	3.42	81.8
9000	10.90	3.26	14.03	93.8
10000	9.87	10.95	28.2	25.6
11000	6.81	19.17	48.3	5.81
12000	4.79	30.9	65.4	2.29
13000	5.78	43.3	63.4	1.622
14000	11.41	52.3	43.5	1.586
15000	14.95	52.4	24.3	1.754

	(c) Prandtl number			
$T, °K$	Pressure, atm			
	100	1.0	0.1	0.0001
1000	0.756	0.756	0.756	0.756
2000	0.773	0.773	0.766	0.614
3000	0.740	0.627	0.636	0.714
4000	0.640	0.762	0.759	0.387
5000	0.702	0.611	0.381	0.003
6000	0.763	0.602	0.736	0.455
7000	0.593	0.796	0.986	0.361
8000	0.620	0.983	0.648	0.322
9000	0.730	0.807	0.382	0.200
10000	0.886	0.429	0.348	0.0576
11000	0.955	0.382	0.327	0.0213
12000	0.908	0.355	0.292	0.0143
13000	0.525	0.333	0.227	0.0121
14000	0.421	0.302	0.144	0.0108
15000	0.394	0.341	0.253	0.0819

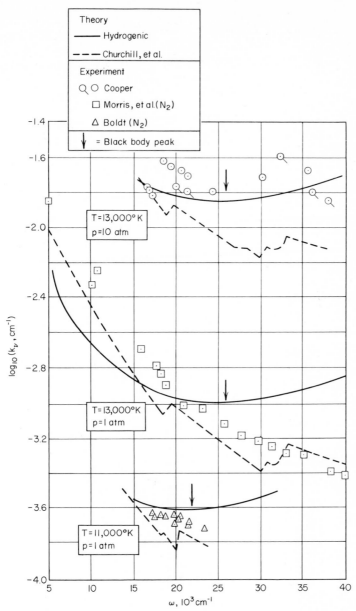

Fig. 2. A comparison of theoretical and experimental values for the continuum spectral absorption coefficient of air.

lower wave numbers, only the lines tabulated for that species should be considered. Under typical entry conditions where radiation is important, lines of N, N+, O, and O+ contribute.

The atomic continuum remains a somewhat poorly described element in radiative transfer, with the contribution of negative ions particularly unresolved. Figure 2 shows some of the experimental data for air and N_2 continuum. The quantum-mechanical computations [55] give an accurate description of measured N_2 radiation [56, 57], but are less convincing for air. For complex gas mixtures, the modified-hydrogenic model [58] is probably sufficiently accurate to calculate continuum radiation in air. This method involves the replacement of the complex mixture of ionized air species of all varieties with two simple hydrogen-like ions of charge $\bar{m} + 1/2$ and $\bar{m} - 1/2$ where $\bar{m}$ is the mean charge in the plasma. More details concerning the technique can be found in Dirling, Rigdon, and Thomas [59].

Some investigators have included atomic-line radiation using "equivalent width" representations [60]; this technique is of general use where the flow and the radiation are uncoupled so that the temperature profile is a predetermined function. Average band representations [61, 67] are of little use for computing combined radiation-flow field containing a multitude of lines of grossly different half-widths, locations and strengths such as occur in air shocklayer radiation. Perhaps an adequate line transfer model will emerge someday, but for the next several years the engineer must rely heavily upon detailed spectral calculations of radiative properties. Such a method is recommended for application to cases of high-velocity atmospheric entry.

The spectral absorption coefficient for air is shown in Fig. 3 for $T = 12,500°F$ and

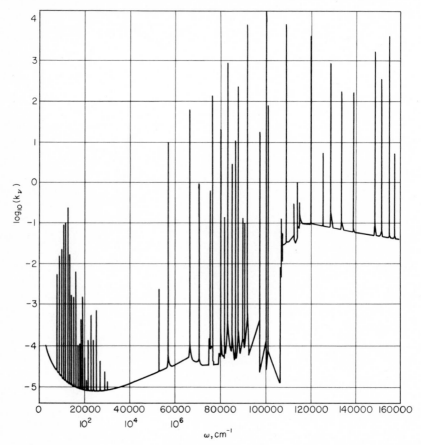

Fig. 3. Spectral absorption coefficient for high temperature air ($T = 12,500°F$, $P = 0.032$ ATM).

$P = 0.032$ atm. It is apparent that the radiation under those conditions will be dominated by atomic lines. Even though very narrow, these lines are strong enough to contribute over 90 percent to the Planck mean absorption coefficient. Also apparent in Fig. 3 are the photoelectric edges, as extended by close-by lines. At these conditions, molecular band radiation is not important.

3. Boundary Layer

The general conservation equations in three-dimensional, time-dependent vector form for a transparent compressible, viscous, reacting, boundary layer flow field with simple equal binary diffusion coefficients among the components (using Prandtl approximations and Fick law assumed to apply) were shown in Ref. 12 and previous sections of this handbook.

The reduced equations, for the purposes of this section, are first presented in their most generalized form, thus encompassing all options. The purpose is to show how these expressions merely reduce to a simpler form of the same solution when documenting cases containing fewer or less complex options. The only exception will be in the case of stagnation point heating where the simplified geometry allows much greater liberties with the other options. The text is further divided into laminar boundary layer and the turbulent boundary layer sections. This is by nature of the quite divergent derivation paths pursued from the original equations to the final solutions, although there exists not only close similarity in the original equation forms but also remarkable duality in the I_2 final expressions for heat transfer to the surface. Because of the strong effect of mass addition on the velocity and enthalpy profiles, the dependence of the surface heat transfer rate on the Lewis number is quite sensitive for different surface mass fluxes [14]. It is now largely acknowledged that the effect of the "frozen" gas phase assumption on the heat and mass transfer coefficients and the overall heat transfer rate solution comes closer to reality than that of assuming unity Lewis and Prandtl numbers in spite of the accompanying wider latitude for finite rate gase phase chemistry and the resulting sophisticated species profiles [13]. For this reason, the "frozen" gas phase criteria and associated solutions will be projected here.

a. Laminar Boundary Layer and Stagnation Point Solutions. Employing the accepted Prandtl concept of the boundary-layer hypothesis along with the quasi-steady-state assumption restricted to the two-dimensional or axisymmetric flow, the above vector equations for laminar flow reduce to [12, 14]

$$\frac{\partial(\rho u r_0^k)}{\partial s} + \frac{\partial(\rho v r_0^k)}{\partial y} = 0 \qquad \text{Continuity} \qquad (2)$$

$$\rho u \frac{\partial C_i}{\partial s} + \rho v \frac{\partial C_i}{\partial y} = \frac{\partial}{\partial y}\left(\rho D_{12} \frac{\partial C_i}{\partial y}\right) + \dot{w}_i \qquad \text{Specie} \qquad (3)$$

$$\rho u \frac{\partial u}{\partial s} + \rho v \frac{\partial u}{\partial y} = -\frac{\partial p}{\partial s} + \frac{\partial}{\partial y}\left(\mu \frac{\partial u}{\partial y}\right) \qquad \text{Momentum} \qquad (4)$$

$$\rho u \frac{\partial I}{\partial s} + \rho v \frac{\partial I}{\partial y} = \frac{\partial}{\partial y}\left[\frac{\mu}{P}\frac{\partial I}{\partial y} + \mu\left(1 - \frac{1}{P}\right)\frac{1}{2}\frac{\partial u^2}{\partial y}\right] - \frac{\partial}{\partial y}\left[\left(\frac{1}{L} - 1\right)\rho D_{12}\sum_i h_i \frac{\partial C_i}{\partial y}\right] \qquad \begin{array}{c}\text{Energy}\\ (5)\end{array}$$

Thus, surface curvature is incorporated, without interruption, into the originating laminar partial differential equations.

For the laminar case, all efforts expended in the search for a more general means of reducing the partial differential equations of the boundary layer to total-differential equations involved the finding of a new coordinate system. This has led to the similarity transformation known by a variety of names including Lee's-Dorodnitsyn and others [14]. Even in the cases where similarity is neither expected nor desired, this transformation is nonetheless performed and the resulting nonsimilar terms are retained as perturbations. The analysis then proceeds on as one of nonsimilarity. The nonsimilar terms are quite often small and can certainly be examined individually in this form and further eliminated in most classes of problems. They are significant only in the case of massive blowing (or suction), severe pressure gradients (or curvatures), and surface discontinuities. It can safely be rationalized otherwise that they are negligible. The Lee's-Dorodnitsyn transformation is expressed as [12, 14]

$$\bar{s} = \int_0^s \rho_e u_e \mu_e r_0^{2k} ds \qquad (6)$$

where the subscript e denotes the boundary layer edge value

$$\eta = \frac{\rho_e u_e r_0^k}{\sqrt{2\bar{s}}} \int_0^y \frac{\rho}{\rho_e} dy \qquad (7)$$

Utilizing this transformation and the conservation equations along with the further assumptions of negligible streamwise variations of the boundary-layer edge values of the total enthalpy and specie mass concentrations, one obtains the three familiar reduced equations in terms of dimensionless variables, namely [14]

$$(Cf'') + f'' + \beta[g_f - (f')^2] = 0 \qquad \text{Momentum} \qquad (8)$$

$$\left(\frac{C}{S} z_i'\right) + fz_i' = -\frac{2\bar{s}\dot{w}_i}{\rho\rho_e u_e^2 \mu_e r_0^{2k}(C_i)_e} \qquad \text{Specie-continuity} \qquad (9)$$

$$\left(\frac{C}{P} g'\right)' + fg' = \left[\frac{C}{S}\left(\frac{1}{L} - 1\right) \sum_i \frac{h_i(C_i)_e}{I_e} z_i'\right]' + \frac{u_e^2}{I_e}\left[\left(\frac{1}{P} - 1\right) Cf'f''\right]' \qquad \text{Energy} \qquad (10)$$

or

$$\left(\frac{C}{P} g_g'\right) + fg_f' = \frac{u_e^2}{(I_f)_e}\left[\left(\frac{1}{P} - 1\right) Cf'f''\right]' + \sum_i \frac{h_i^0}{(I_f)_e}\left[\frac{2\bar{s}\dot{w}_i}{\rho\rho_e u_e^2 \mu_e r_0^{2k}}\right] \qquad \text{Energy} \qquad (11)$$

Here $C = \rho\mu/\rho_e\mu_e$, namely, the Chapman-Rubesin parameter.

The Eckert reference enthalpy expression [63] can usually be put to good use here rendering C effectively constant across the boundary layer.

Stagnation Point. The stagnation point provides such simplifying geometry that when considering the nonablating body option the above equations can be extended to accept the "frozen," equilibrium, or nonequilibrium gas phase options. This in turn leads to their direct solution in the form of the Fay and Riddell expressions [64]

$$\left(-\dot{q}_w\right)_f = 0.763 P^{-0.6} (\rho_e \mu_e)^{1/2} \left[\left(\frac{du_e}{ds}\right)_0\right]^{1/2} \left(\frac{\rho_w \mu_w}{\rho_e \mu_e}\right)^{0.1} (I_e - h_w) \left[1 + (L^{0.63} - 1) \frac{h_c}{I_e}\right]$$

(12)

in the frozen limit, or

$$\left(-\dot{q}_w\right)_{eq} = 0.763 P^{-0.6} (\rho_e \mu_e)^{1/2} \left[\left(\frac{du_e}{ds}\right)_0\right]^{1/2} \left(\frac{\rho_w \mu_w}{\rho_e \mu_e}\right)^{0.1} (I_e - h_w) \left[1 + (L^{0.52} - 1) \frac{h_c}{I_e}\right]$$

(13)

in the equilibrium limit.

Thus, it is noted that only the laminar stagnation point will allow consideration of nonequilibrium chemistry throughout the gas phase provided there is no surface ablation region. It should be noted that at points other than the stagnation region, while considering no surface ablation reactions, the only other option possible in addition to the "frozen" gas phase chemistry is the one of equilibrium–infinite-rate chemistry. For equilibrium gas-phase reaction solutions the specie's conservation equations must be multiplied by ϵ_{ij}, defined as the mass fraction of element j in species i and the resulting terms summed over all species i, otherwise known as the Shvab-Zeldovich transformation [14]. This effectively removes the specie annihilation and creation term $\dot{w}_i$. Of course, as stated, one must assume no surface interaction or ablation otherwise

$$\left(\sum_i \epsilon_{ij} \dot{w}_i\right) \neq 0$$

at the surface. Also, in the energy equation C_i must be replaced with

$$\tilde{C}_j = \sum_i \epsilon_{ij} C_i$$

When the "frozen" chemistry gas-phase option is considered the $\dot{w}_i$ in the above equations only appears at the surface. That is, $\dot{w}_i = 0$ throughout the gas phase. This affords not only ease in numerical solution but other important options such as ablation and general mass injection may be considered. When the frozen gas-phase assumption is made then it may also be assumed that diffusion predominantly determines the atom concentration profiles in the boundary layer. Comparison of the extremes of the two Fay and Riddell expressions for the stagnation point shows this to be a good assumption. In fact, it is this precedent that supports popular use of the "frozen" gas phase approximation when extended to the turbulent boundary-layer regions. Typically then, the "frozen" gas-phase reaction and catalytic surface assumption is used to remove the $\dot{w}_i$ (that is, Damkohler number $\to \infty$) and enable simultaneous numerical solution of the above reduced laminar equations for various β [65, 66].

Solutions then involve the kernel

$$G(\eta; Z) = \int_0^\eta \frac{Z}{C} \exp\left(-\int_0^{\eta'} \frac{Z}{C} f d\eta''\right) d\eta'$$

(14)

(where $Z = P$ or S) such that

$$g_f = \left(g_f\right)_w + \left[1 + \left(g_f\right)_w\right] \frac{G(\eta; P)}{G(\infty; P)} \tag{15}$$

and

$$z_i = \left(z_i\right)_w + \left[1 + \left(z_i\right)_w\right] \frac{G(\eta; S)}{G(\infty; S)} \tag{16}$$

Yielding for a heat transfer coefficient (or Stanton number)

$$C_H = \mu_e \frac{r_0^k}{(2\bar{s})^{1/2}} \left[G(\infty; P)\right]^{-1} \tag{17}$$

and

$$-\dot{q}_w = C_H \rho_e u_e \left(I_f\right)_r - \left(I_f\right)_w \left\{ 1 + \frac{G(\infty; P)}{G(\infty; S)} \frac{\sum_i h_i^0 \left[\left(c_i\right)_e - \left(c_i\right)_w\right]}{\left[\left(I_f\right)_r - \left(I_f\right)_w\right]} \right\} \tag{18}$$

where

$$\left(I_f\right)_r = \left(I_f\right)_e + (r - 1) \frac{u_e^2}{2} \tag{19}$$

with

$$r \approx P^{1/2} \tag{20}$$

The numerical solutions then give $G(\infty; Z)$ as a function of β and $f(o)$ with

$$f(o) = -\frac{\dot{m}_w (2\bar{s})^{1/2}}{\rho_e u_e \mu_e r_0^k} \tag{21}$$

β is commonly called the "pressure gradient parameter." Typically the entire equation is rewritten as

$$-\dot{q}_w = C_H \rho_e u_e \left\{ \left(I_f\right)_r - \left(h_f\right)_w + \frac{C_{H_d}}{C_H} \sum_i h_i^0 \left[\left(c_i\right)_e - \left(c_i\right)_w\right] \right\} \tag{22}$$

where by definition

$$C_{H_d} = \frac{G(\infty; P)}{G(\infty; S)} C_H \tag{23}$$

b. Turbulent Boundary Layer. Again employing the accepted Prandtl concept of the boundary-layer hypothesis along with the temporal mean approximation plus a small linear fluctuation about the mean as an expression of the turbulent variables, and further as a first-order approximation, restricted to the flat-plate two-dimensional flow case with zero-pressure gradient, the original vector equations when applied to turbulent flow reduce to [12, 67]

$$\frac{\partial \overline{\rho u}}{\partial s} + \frac{\partial \overline{\rho v}}{\partial y} = 0 \qquad \text{Continuity} \tag{24}$$

$$\overline{\rho u}\frac{\partial \overline{u}}{\partial s} + \overline{\rho v}\frac{\partial \overline{u}}{\partial y} = \frac{\partial}{\partial y}\left[(\mu + \mu_T)\frac{\partial \overline{u}}{\partial y}\right] \qquad \text{Momentum} \tag{25}$$

$$\overline{\rho u}\frac{\partial \overline{C}_i}{\partial s} + \overline{\rho v}\frac{\partial \overline{C}_i}{\partial y} = \frac{\partial}{\partial y}\left[(\rho D_{12} + \rho D_r)\frac{\partial \overline{C}_i}{\partial y}\right] + \overline{\dot{w}}_i \qquad \text{Specie} \tag{26}$$

$$\overline{\rho u}\frac{\partial \overline{I}}{\partial s} + \overline{\rho v}\frac{\partial \overline{I}}{\partial y} = \frac{\partial}{\partial y}\left[\left(\frac{\mu}{P} + \frac{\mu_T}{P_T}\right)\frac{\partial \overline{I}}{\partial y}\right] + \frac{\partial}{\partial y}\left\{\left[\frac{\kappa}{C_{p_f}}(L-1) + \frac{\kappa_T}{C_{P_f}}(L_T-1)\right]\sum_i \overline{h}_i\frac{\partial \overline{C}_i}{\partial y}\right\}$$

$$+ \frac{\partial}{\partial y}\left\{\left[\mu\left(1 - \frac{1}{P}\right) + \mu_T\left(1 - \frac{1}{P_T}\right)\right]\frac{\partial \overline{u}^2/2}{\partial y}\right\} \qquad \text{Energy} \tag{27}$$

Where

$$\mu_T = -\frac{\overline{(\rho v)'u'}}{\partial \overline{u}/\partial y} \qquad \text{Turbulent viscosity coefficient} \tag{28}$$

$$\kappa_T = -\frac{\overline{(\rho v)'h'_{f_i}}}{\left(C_{p_i}/C_{p_f}\right)(\partial \overline{T}/\partial y)} \qquad \text{Turbulent thermal conductivity} \tag{29}$$

$$D_T = \frac{1}{\overline{\rho}}\frac{\overline{(\rho v)'C'_i}}{\partial \overline{C}_i/\partial y} \qquad \text{Turbulent diffusion coefficient} \tag{30}$$

with L_T defined as

$$C_{p_f}\frac{\overline{\rho}D_T}{\kappa_T}$$

and P_T defined as

$$C_{p_f}\frac{\mu_T}{\kappa_T}$$

here the prime ($'$) denotes fluctuation contributions and the barred quantity ($^-$)

represents the temporal mean. Further, requiring use of von Kármán's law of similitude and mixing length theory along with the well-known laminar sublayer–fully turbulent layer hypothesis, one may extract from these equations an expression for the heat transfer from a compressible reacting, viscous turbulent boundary-layer accommodating mass injection [12]. The heat transfer coefficient in this form is left in terms of the local skin friction coefficient approached via a suitable modified Reynolds analogy. This is largely the work of Dorrance [67]. Although the turbulent expression derivations implicitly suggest certain restrictions on the value of the Lewis and Prandtl numbers (the simplest case suggesting $L = 1$, $P = 1$) by virtue of the appearance of the Crocco-Integral like expressions or the integrating-factor-derived expressions; this, to the contrary, is not the case. The *a priori* existence of these expressions is assumed to be valid without the Lewis and Prandtl numbers necessarily being unity.

Typically this is represented as [12]

$$-\dot{q}_w = C_H \rho_e u_e \left\{ \left(I_f\right)_r - \left(h_f\right)_w + \frac{C_{H_d}}{C_H} \sum_i h_i^0 \left[\left(C_i\right)_e - \left(C_i\right)_w\right] \right\} \tag{31}$$

where
$$\frac{C_{H_d}}{C_H} = \frac{(1 + B')(1 + B'\lambda)^{P-1} - 1}{(1 + B')(1 + B'\lambda)^{S-1} - 1} \tag{32}$$

and
$$C_H = \frac{C_f}{2} \left\{ \frac{B'}{(1 + B')(B'\lambda + 1)^{P-1} - 1} \right\} \qquad \text{Modified Reynolds analogy} \tag{33}$$

with
$$C_f = \frac{C_{f_0}}{1 + B'\lambda} \tag{34}$$

$$B' = \frac{B^o}{1 - B^o\lambda} \tag{35}$$

$$\lambda = 11.5 \sqrt{\frac{T'}{T_e} \frac{C_{f_0}}{2}} \approx \frac{u_\lambda}{u_e} \tag{36}$$

$$\frac{T'}{T_e} = 1 + 0.038 M_e^2 + 0.5\left(\frac{T_w}{T_e} - 1\right) \tag{37}$$

T' being Eckert's reference temperature

$$B^o = \tilde{B}^o \left(\frac{\overline{M}_{inj}}{\overline{M}_{BL}}\right)^{0.46} \tag{38}$$

$\overline{M}_{inj}$ being the mean mol. wt. of the injected species whereas, $\overline{M}_{BL}$ being the mean mol. wt. of the boundary layer specimen.

$$\tilde{B}^o = \frac{\dot{m}_w}{\rho_e u_e} \frac{2}{C_{f_0}} \tag{39}$$

$$C_{f_0} = \left[2 \left(\frac{1 + (C_o)_w}{1 + (C_o)_e} \right)^{1/7} - 1 \right] C_{f_0}(o) \tag{40}$$

where [ ] represents Dorrance's correction for dissociation of O_2 in the boundary layer [67].

$C_{f_0}(o)$ may then be extracted from the well-known von Kármán momentum integral, more notably as

$$C_{f_0}(o) = \frac{d\theta}{ds} \quad \text{(for the flat plate case)} \tag{41}$$

with

$$\theta = \frac{\rho_w}{\rho_e} \int_0^1 \frac{\rho}{\rho_w} z(1 - z) \left(\frac{dz}{dy} \right)^{-1} dz \tag{42}$$

and

$$z = \frac{u}{u_e} \tag{43}$$

Accounting for compressibility one may employ either the expansion kernel of Van Driest's Method I or Method II [68] or from the method of Spalding and Chi [69]. Typically, the Method I of Van Driest yields an iterative solution for $C_{f_0}(o)$ from an expression such as [68]

$$\frac{1}{[C_f(o)(T_w/T_e)]^{1/2}} \left(\frac{\arcsin \alpha + \arcsin \beta}{A} \right) = 4.15 \left\{ \log_{10} R_e C_{f_0}(o) \right.$$
$$\left. - 1.26 \log_{10} \left(\frac{T_w}{T_e} \right) \right\} + 1.70 \tag{44}$$

where

$$\alpha = \frac{2A^2 - B}{(B^2 + 4A^2)^{1/2}} \qquad \beta = \frac{B}{(B^2 + 4A^2)^{1/2}} \tag{45}$$

$$A^2 = \left(\frac{T_w}{T_e} \right)^{-1} \frac{\gamma - 1}{2} M_e^2 \qquad B = \left(\frac{T_w}{T_e} \right)^{-1} \left[\frac{\gamma - 1}{2} M_e^2 + 1 \right] - 1 \tag{46}$$

and

$$R_e = \frac{\rho_e u_e}{\mu_e} s \qquad \text{with } s \text{ measured from the virtual leading edge} \atop \text{extrapolated from the point of transition.} \tag{47}$$

Here also

$$\left(I_f\right)_r = \left(I_f\right)_e + (r - 1)\,\frac{u_e^2}{2} \tag{48}$$

with

$$r \approx P^{1/3} \tag{49}$$

It is generally viewed that at this stage it is better to incorporate the effects of gradual curvature or pressure gradient as a perturbation to the first-order approximations of the turbulent flat plate solutions. It may be argued from a practical engineering point of view, that by the time one passes the transition point and enters well into the turbulent region on most ablating shapes, the curvature is either gradual or practically flat plate.

However, $C_{f_0}(o)$ or C_f when referred to other than the flat-plate--zero-pressure gradient case, must be extracted from the generalized von Kármán momentum integral as

$$\frac{C_f}{2} = \frac{d\theta}{ds} + \theta \left(\frac{2 + H_f}{u_e}\frac{du_e}{ds} + \frac{1}{\rho_e}\frac{d\rho_e}{ds} + \frac{1}{r_0}\frac{dr_0}{ds}\right) \tag{50}$$

where $H_f = \delta^*/\theta$ is recognized as the "form factor" [67] with

$$\delta^* \approx \int_0^\delta \left(1 - \frac{\rho u}{\rho_e u_e}\right) dy \tag{51}$$

Here, analytical approximations of semiempirical determinations usually lead to acceptable "form factors" depending on the body shape or known pressure gradient.

Thus, one sees that the turbulent boundary layer expression may be manipulated into a parallel or analogous form to the laminar boundary-layer, thereby facilitating a convenience in the surface analyses mechanisms [12]. This form is recognized as a general expression with surface coupling contained within the C_H and C_{H_d}, namely

$$-\dot{q}_w = C_H \rho_e u_e \left\{\left(I_f\right)_r - \left(h_f\right)_w + \frac{C_{H_d}}{C_H}\sum_i h_i^{\,o}\left[\left(C_i\right)_e - \left(C_i\right)_w\right]\right\} \tag{52}$$

thereby capable of accounting for the options of gas-phase dissociation, pure mass injection, and ablation with surface chemical reactions or simply pure aerodynamic heat conduction for both turbulent and laminar environments. The various options in terms of contractions of the above solution correspond to the conditions encountered for ablating and nonablating bodies in hyperthermal environments.

4. Nonablating Body

a. No Surface Chemical Reactions, Boundary-layer Dissociation, Mass Injection. In this option the above expression reduces for both turbulent and laminar flow to

$$-\dot{q}_w = C_H \rho_e u_e \left[\left(I_f\right)_r - \left(h_f\right)_w\right] \tag{53}$$

with

$$C_H = \frac{C_{f_0}(o)}{2} P^{-2/3} \qquad (54)$$

b. Dissociated Boundary Layer.
1. For the stagnation point region Fay-Riddell expressions are recommended.
2. For laminar-flow regions the general expression reduces to

$$-\dot{q}_w = C_H \rho_e u_e \left\{ \left(I_f\right)_r - \left(h_f\right)_w + L^{2/3} \sum_i h_i^{\,o} \left[\left(C_i\right)_e - \left(C_i\right)_w\right] \right\} \qquad (55)$$

3. The turbulent expression becomes

$$-\dot{q}_w = \rho_e u_e P^{-2/3} \frac{C_{f_0}(o)}{2} \left[2 \left(\frac{1 + \left(C_0\right)_w}{1 + \left(C_0\right)_e} \right)^{1/7} - 1 \right] \left\{ \left(I_f\right)_r - \left(h_f\right)_w \right.$$

$$\left. + L^{2/3} \sum_i h_i^{\,o} \left[\left(C_i\right)_e - \left(C_i\right)_w\right] \right\} \qquad (56)$$

c. Radiation-Convection Coupling.
For entry velocities above 35,000 fps, or for very large entry vehicles, the radiation field in the thermal layer can be strong enough to alter the flow field. Figure 4 shows typical effects of the strong radiation upon the stagnation region. The lower temperatures in the thermal layer, caused by radiative energy loss, cause a contraction of the shock layer. The steep tails near the shock are caused by strong air line radiation in the vacuum ultraviolet spectrum, and become more pronounced as entry conditions become more severe. Precursor, or forward-traveling, radiation in the vacuum ultraviolet spectrum is reabsorbed in the thin ambience, resulting in preheating the air. At this time, preheating effects have not been accurately treated. The display in Fig. 4 is based upon a crude energy balance; however, the nature of the effect is well illustrated.

Estimates of radiation from high-temperature shock layers must consider:

1. Strong line radiation in the visible and ultraviolet spectrum
2. Radiative cooling of the shock layer
3. Radiative contraction of the shock layer
4. Preheating of the ambient air

Early estimates of entry heating caused by radiation from the high-temperature shock layer were based upon emissivity tables [70--72]. More recent work [73] has included atomic-line radiation based upon coulombic oscillator strengths; this data has been verified for wavelengths greater than 0.2 micron for the case of thin shock layers by comparison with shock-tube data [74, 75]. However, such data has not by itself been adequate for correlating radiation from thick high-temperature shock layers. At present there seems little doubt that an accurate analysis must include the effects of radiative cooling and line self-absorption. The problem of radiation-convection coupling becomes more complicated when mass transfer is included.

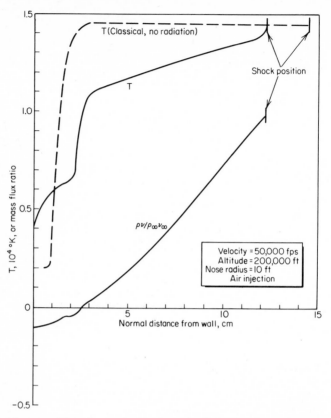

Fig. 4. Shock-layer conditions for massive blowing and radiation coupling in the stagnation region.

5. Ablating Body

a. Surface Chemical Reactions with Mass Injection but No Dissociating Boundary Layer.

1. For the laminar case

$$-\dot{q}_w = C_H \rho_e u_e \left\{ \left(l_f \right)_r - \left(h_f \right)_w - \frac{C_{H_d}}{C_H} \sum_i h_i^{,o} \left(C_i \right)_w \right\} \tag{57}$$

2. For the turbulent case the same expression as for the laminar case except that the term

$$2 \left(\frac{1 + \left(C_0 \right)_w}{1 + \left(C_0 \right)_e} \right)^{1/7} - 1 \tag{58}$$

is missing from the C_{f_0}. That is C_{f_0} is replaced with $C_{f_0}(o)$ [67].

b. Surface Chemical Reactions with Mass Injection and Boundary-layer Dissociation. For both the turbulent and laminar cases the expression

$$-\dot{q}_w = C_H \rho_e u_e \left\{ \left(I_f \right)_r - \left(h_f \right)_w + \frac{C_{H_d}}{C_H} \sum_i h_i^{\,o} \left[\left(C_i \right)_e - \left(C_i \right)_w \right] \right\} \qquad (59)$$

and its respective supporting expressions remain as presented earlier. In the specialized case of an ablating rocket nozzle liner, the equation of Bartz [77] has been successfully used in determination of the heat transfer.

c. Boundary-layer Species Distribution with Ablation. The boundary-layer model assumed here used the artifice of "frozen" gas-phase chemistry thereby leaving diffusion to predominately determine the species profiles or distributions. This does not affect the first-order expression for the heat transfer coefficient appreciably; however, proper transport parameters such as Prandtl number, Lewis number, etc., must still be calculated. To arrive at perturbation terms accounting for nonequilibrium chemistry effects, a more realistic exhibition of boundary-layer species distribution is required.

The influence of ablation on boundary-layer species distributions is of consequence in numerous disciplines involving the physics of reentry. Species concentrations are needed not only in transport property calculations, but also in convective heating and radiative transport numerical solutions, observables calculations, etc., as well. In the absence of definitive reaction kinetic data for polyatomic hydrocarbons involved in most heat shield materials, chemical equilibrium constraints may be employed to gain insight into the significant physical processes. Figure 5 illustrates the complex chemistry embodied in ablation of composite charring materials. It shows the large effect of injection of gaseous products (including alkali metals) from an organic charring ablator at the stagnation point of a blunt body. Close to ablator surface, polyatomic hydrocarbons are predominant; all of the boundary-layer oxygen has reacted to form carbon monoxide (see discussion on specialized ablation models). Away from the ablator surface, temperature is sufficiently elevated to cause dissociation of these products and formation of oxides. In the outer part of the boundary-layer dissociation and ionization processes produce diatomic molecules, ions, and electrons. It is significant that the easily ionized sodium contaminate is the chief source of electrons through most of the boundary layer. Thus a detailed accounting of accurate specie profiles is ultimately needed—whatever the degree of boundary-layer heat transfer analysis.

d. Effect of High Ablation Rates of Strongly Radiating Shock Layers. For entry velocities above 45,000 ft/sec, stagnation-point heating can be primarily caused by strong atomic line and continuum radiation. Under these conditions, ablation rates result in mass fluxes from the heat shield on the order of 20 percent of the air flux through the shock. Classically, such conditions result in "blown-off" shock layers, with the convective heating reduced many orders of magnitude below the value that would occur without blowing.

When such massive blowing is caused by extreme radiative heating, however, the classical blown-off shock layer no longer exists. Instead, strong absorption of the shock-layer radiation occurs in the volume of blown-off gases, and an appreciable rise in temperature results. A typical solution for the stagnation-point flow in such a case is given in Fig. 6 temperatures [78] in the layer of gases flowing from the wall $(\rho v / \rho_\infty v_\infty < 0)$ rise toward a value one-half of the stagnation temperature. Convective heating rises correspondingly to a measurable fraction of the total heat input, although it is still small. On the other hand, 20 percent reductions in radiative heating over the case of no blowing are quite possible.

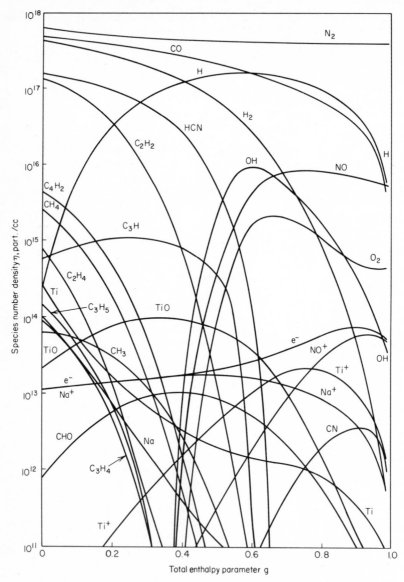

Fig. 5. Stagnation point boundary layer equilibrium specie concentration distribution.

The data of Fig. 6 is for air injection from the wall. Since most of the coupling radiation occurs in the ultraviolet spectrum, ablation-products injection will not offer any significant alteration of these phenomena, although they will be somewhat accentuated. These conclusions are borne out by Coleman, Lefferdo, Hearne, and Vojvodich [79] where radiative coupling to ablating heat shields is studied; reductions in radiative heating from ablation-products injection for the cases considered by Rigdon, Dirling, and Thomas [78] reach 30 percent.

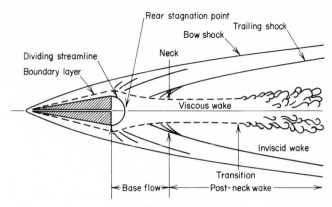

Fig. 6. Flow field regions.

6. Downstream and Other Effects of Ablation Products

As previously shown, ablative products injected into a hot boundary layer undergo thermochemical reactions which alter the chemical composition, transport properties, temperature profile, and observable characteristics of the local flow field. The complications encountered in properly accounting for these effects, as well as the previous boundary-layer and ablation histories, render the problem unmanageable. The ablation process itself may affect the vehicle characteristics by changing its aerodynamic shape and stability which in turn would affect the environment and subsequent heat transfer. In addition, the ablating body can enhance the optical radiation from the gas-cap and trailing wake and increase the number of charged particles in the flow field. (Figure 6 schematically illustrates the wake flow-field regions.) Detailed discussion of these effects is, however, beyond the scope of this handbook.

It must be said in closing the subject of environment that the problem of boundary-layer transition from laminar to turbulent flow is almost always tacitly avoided or circumvented with some form of extrapolation from a "turbulent" virtual leading edge a span of distance $s*$ whose Reynolds number $\rho_e u_e s*/\mu_e \sim 10^6$.

C. MECHANISMS

1. General Considerations

The environmental parameters associated with hypervelocity flight in planetary atmospheres affecting ablation mechanisms are: gas temperature, enthalpy, mode of energy transport, total heat load, peak heating rate, shape of the heat pulse, duration of the pulse, external and internal pressure forces, aerodynamic shear, particle impingement, mechanical and acoustical vibration, inertial and dynamic forces, and the chemical reactivity [80]. Of primary concern here is the means of accommodating the energy converted to heat, i.e., the thermal protective system. This is accomplished in three ways: (1) absorption, (2) radiation, and (3) reduction of the aerodynamic heat prior to its reaching the vehicle [81]. The environment was discussed in the preceding section.

The requirement to minimize body weight has often led to the choice of materials purposely designed to undergo internal endothermic reactions leading to the formation of a carbonaceous char when heated and to vaporize at the surface exposed to the hypersonic environment. The energy absorption associated with such processes, together with the reduction of the heat transfer coefficient (previously defined in Sec. B) as a result of injection of the products of degradation and the vaporized species into the boundary layer have proved to be extremely effective in providing reliable thermal protection systems [19].

It has been found necessary [5] for efficient design to have an accurate mathematical description of the complex heat and mass transfer phenomena associated with such processes inside of the material and an efficient numerical procedure for implementing these calculations based on the mathematical model. In practice, the ablation phenomena are most often encountered in situations where the environmental parameters are nonlinear functions of time. This fact, coupled with the heat shield physically being of finite thickness, makes it necessary to perform transient rather than steady-state calculations [19]. The fully transient solution includes the decomposition in depth with the attendant density and temperature gradients, internal transpiration and radiation generation, material redeposition, source terms for convective and/or radiative input, and provides for a variety of boundary conditions or surface phenomena [5]. A detailed description of this type of solution is given in Sec. B. With the aid of present day computers, this is practical even for quite complex systems.

It has been shown [12, 13, 19] that an intricate relation between the environment and material response indeed exists and that the solution of the problem must account for this.

Although the transient problem can be solved with certain degree of reliability, for certain combinations of environment and materials, or for steady-state experiments, a steady-state solution is deemed adequate (and is considerably less expensive in terms of computing time). Often, in preliminary designs phases where much of the information generated is not used in the final design, the costs associated with the transient analysis cannot be justified. One such condition where this type of analysis (i.e., neglecting in-depth phenomena) can be effectively utilized is in the investigation of the ablation requirements for lifting vehicles which enter from orbit and sustain long constant altitude trajectories and minimal decelerations. Here the ablative process remains in the diffusion-limited regime and involves the least intricate of mechanisms. In this regime, char formation due to pyrolysis and decomposition are limited and the ablative process may be described by the simplest of models. In low heating environments (i.e., heat fluxes less than 75 Btu/ft^2-sec), the surface recession has been shown experimentally to be insignificant (e.g., for DC-325 with 15 percent microballoons, the recession rate is less than 0.001 inch per second) and thus may be neglected. The formation of the char and the influence of pyrolysis on material properties may also be considered inconsequential with respect to a material insulating capability. An assumption of this nature inherently postulates an energy balance of the contradicting effects of pyrolysis whereby blockage of heat transfer by gaseous mass flux is compensated for by neglecting the increase of thermal conductivity due to char formation (for the low density class of ablators). Hence, based on these premises in low heating environments, a simplified steady-state model may be utilized for estimating ablation requirements. Other conditions where similar approaches can be taken to reduce the problem to a steady-state solution are discussed in Refs. 82 to 84. A general discussion of aerodynamic heating of ablating bodies was given in previous sections.

2. Surface Equations and Mechanisms— Transient Ablation

The several energy transfer mechanisms between the boundary-layer gas and the ablating surface are shown in the sketch in Fig. 6a. The energy balance at the surface establishes the boundary condition for solution of the boundary-layer equations and the boundary condition of the transient subsurface equations. The energy balance yields

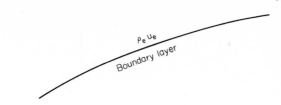

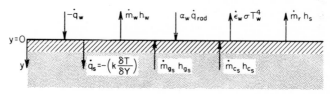

Fig. 6a.

$$\dot{q}_s = -\dot{q}_w - \dot{m}_w h_w + \sum_{k'}\left(\dot{m}_{k'}\right)_w\left(h_{k'}\right)_w + \sum_{k''}\left(\dot{m}_{k''}\right)_w\left(h_{k''}\right)_w - \epsilon_w \sigma T_w^{\,4} + \alpha_w \dot{q}_{rad} - \dot{m}_r h_s$$
(60)

where the mass balance across the surface is given by

$$\dot{m}_w + \dot{m}_r = \dot{m}_{c_s} + \dot{m}_{g_s}$$
(61)

with
$$\dot{m}_{c_s} = \sum_{k'}\left(m_{k'}\right)_w \qquad h_{c_s} = \frac{1}{\dot{m}_{c_s}}\sum_{k'}\left(\dot{m}_{k'}\right)_w\left(h_{k'}\right)_w$$

and
$$\dot{m}_{g_s} = \sum_{k}\left(\dot{m}_{k''}\right)_w \qquad h_{g_s} = \frac{1}{\dot{m}_{g_s}}\sum_{k''}\left(\dot{m}_{k''}\right)_w\left(h_{k''}\right)_w$$

The dependency of both C_H and C_{H_d} on $\dot{m}_w$ and implicitly on h_w is seen in the boundary-layer expressions as presented earlier. For the surface energy balance it is expedient to define

$$\beta^* = \frac{\dot{m}_w}{\rho_e u_e C_{H_d}}$$
(62)

Substituting for $-\dot{q}_w$ from the boundary-layer expressions presented earlier, one observes

$$\dot{q}_s = \underbrace{\rho_e u_e C_H\left[\left(I_f\right)_r - \left(h_f\right)_w\right]}_{} + \underbrace{\rho_e u_e C_H \left\{ \frac{C_{H_d}}{C_H}\left[\sum_i h_i^{\,\circ}\left[\left(C_i\right)_e - \left(C_i\right)_w\right]\right]\right.}_{\displaystyle -\dot{q}_w}$$

Surface	Convection sensible	Chemical energy transfer
conduction	energy transfer with	associated with surface
term	C_H dependent on $\dot{m}_w$	reactions from mass transfer

(63)

$$+ \sum_{k'} \beta^* \frac{\left(\dot{m}_{k'}\right)_w}{\dot{m}_w} \left(h_{k'}\right)_w + \sum_{k''} \beta^* \frac{\left(\dot{m}_{k''}\right)_w}{\dot{m}_w} \left(h_{k''}\right)_w$$

$$\underbrace{\qquad}\qquad\underbrace{\qquad}$$

Energy transfer to   Energy transfer to the
the surface by   surface by internal
condensed char   gas products
material

$$\left. - \beta^* h_w \right]\Bigg\} \qquad - \epsilon_w \sigma T_w^{\,4} \qquad + \alpha_w \dot{q}_{\text{rad}} \qquad - \dot{m}_r h_s \qquad \begin{array}{c} (63) \\ (\text{Cont.}) \end{array}$$

$$\underbrace{\qquad}\qquad\underbrace{\quad}\quad\underbrace{\quad}\quad\underbrace{\quad}$$

Energy transfer away   Re-radiation   Absorption   Energy transfer
from surface a nor-   of incident   by mechanical
mal gas mass flow   radiation   erosion

The above equation includes the effect of a nonunity Lewis number and is applicable for both laminar and turbulent flow. The energy balance couples the boundary-layer solution through species concentrations and the transfer coefficients to the in-depth solutions of the charring ablator analysis.

Equation (63) is general although, as presented, thermochemical equilibrium is facilitated in the boundary layer and at the ablating surface. However, surface mass ablation rates $\dot{m}_{c_s} \left[\left(m_k \right)_w \right]$ can be reaction-rate controlled by reactions between the condensed ablator char reactants and the boundary-layer species. At low surface temperatures $(\dot{m}_{c_s} - \dot{m}_r)$ is often determined from an Arrhenius rate expression and the reactions are rate controlled. At higher temperatures, the rate of injection of $\dot{m}_w$ into the boundary layer can be controlled solely by the diffusion rate of the boundary-layer reactants to the reacting ablator surface. In that case, first β^* and then $\dot{m}_w$ must be solved for independently in terms of thermochemical equilibrium reactions involving the surface mass balance species [12]. Here β^* and $\dot{m}_w$ are found through the simultaneous solution of all surface thermochemical equilibrium reactions expressed in terms of partial pressures and equilibrium constants, equations of conservation of elemental species at the surface, and Dalton's partial pressure law.

The equations for conservation of elemental species come directly from the conservation of specie equations and the catalytic surface reaction boundary condition [14, 67]. Fick's law must be assumed in order to demonstrate the same solution form in the turbulent case [12]. However, at all times the boundary-layer reactants are assumed to be undergoing simultaneous equilibrium dissociation-recombination reactions among themselves at the surface. In fact, a continuum between the two extremes of diffusion-limited and rate-limited reactions may be effected as long as the processes are confined to the surface. Therefore, reactions between some ablator species and the boundary reactants may be diffusion-limited while others are simultaneously rate-limited [11, 12, 18].

The ablation mechanisms, variously termed sublimation, vaporization, and/or melting with spallation, are characteristically represented as change of phase reactions with energy absorption. The sublimation or vaporization is generally associated with the chemical reactions between ambient boundary-layer constituents, pyrolysis gases, and the subliming surface material. Consequently the heats of phase change are most often included with chemical enthalpies of formation of the surface char condensed phase species. These phase change reactions effectively decrease the chemical reaction enthalpies.

The ablation mechanism termed mechanical erosion includes liquid layer flow, mechanical spallation, and mechanical shear of surface material. Liquid layer ablation is discussed in the next subsection. Mechanical spallation of charring ablators can be caused by the internal pressure of the gases percolating through the char [96]. The stresses can exceed the tensile strength of the char for many low-char strength ablators. As a consequence, a part or most of the char can separate from the virgin material. Mechanical shear ablation is characterized by small solid char particles leaving the surface due to surface shear forces. Both of these processes can be expedited by the impingement of small solid particles in the boundary layer as frequently occurs in rocket nozzle erosion. Mechanical shear ablation is normally associated with more severe environmental conditions and thinner char layers. Equation (63) may not be familiar in the form readily presented but can be rearranged to a familiar form. An alternate choice for the surface energy balance is the following expression [18, 19]

$$\dot{q}_s = \underbrace{C_{H_0}\rho_e u_e\left[\left(I_f\right)_r - \left(h_f\right)_w\right]}_{\dot{q}_c} - \epsilon_w \sigma T_w^4 + \alpha_\omega \dot{q}_{rad} + \underbrace{\dot{q}_{chemical} + \dot{m}_s H - \dot{m}_r h_c}_{} + \underbrace{\dot{q}_{blocking}}_{}$$

Pure convection term
with C_{H_0} bearing no
dependence on $\dot{m}_n$ or h_w

Mass transfer, vaporization,   blocking
and surface chemistry term   term

(64)

where

$$\dot{q}_{chemical} = C_{H_d}\rho_e u_e\left\{\sum_{i \neq k} h_i^o\left[\left(C_i\right)_e - (1 + \beta^*)\left(C_i\right)_w\right]\right.$$

$$\left. + \sum_k h_k^o\left[\frac{\left(\dot{m}_k\right)_w}{\dot{m}_w}\beta^* - (1 + \beta^*)\left(C_k\right)_w\right]\right\}$$

and

$$\dot{m}_s H = \sum_{k'}\left(\dot{m}_{k'}\right)_w\left(h_{k'f}\right)_w + \sum_{k''}\left(\dot{m}_{k''}\right)_w\left(h_{k''f}\right)_w$$

$$\dot{q}_{blocking} = \gamma_2\left(\dot{m}_{c_s} + \dot{m}_{g_s}\right)\left[\left(I_f\right)_r - \left(h_f\right)_w\right]$$ (65)

where

$$\dot{m}_{c_s} = \sum_{k'}\left(\dot{m}_{k'}\right)_w \qquad \dot{m}_{g_s} = \sum_{k''}\left(\dot{m}_{k''}\right)_w \qquad \text{and} \qquad \dot{m}_w = \dot{m}_{c_s} + \dot{m}_{g_s} - \dot{m}_r$$

with

$$\gamma_2 = f\left[\left(\frac{\overline{C}_{p_g}}{\overline{C}_{p_{atmos}}} \quad \text{or} \quad \frac{\overline{M}_g}{\overline{M}_{atmos}}\right)\frac{\dot{m}_{c_s} + \dot{m}_{g_s}}{\rho_e u_e C_{H_0}}\right]$$ (66)

The "blocking" term arises from expressing $\dot{q}_c$ in terms of C_{H_0} instead of C_H where C_{H_0} is the Stanton number C_H with zero mass injection (that is, $\dot{m}_w = 0$). This relieves the responsibility of determining the pure convective Stanton number from solutions to species-coupled conservation equations. Empirical relations frequently facilitate the "blocking" term.

In most cases ϵ_w, σ, and α_w are constants while $\dot{q}_{rad}(t)$ is a tabulated function resulting from the shock-layer solutions. As stated earlier, quite frequently $\dot{q}_c$ is evaluated by Eckert's reference enthalpy method for a few simple shapes, both for turbulent and laminar environments [18, 19].

3. Ablation with Liquid Layer Formation

This mechanism may be considered as an intermediate process to the ablation with only surface mass transfer (discussed above) and the even more complex process involving decomposition in depth discussed below. The case of simple melting with the liquid layer instantaneously swept from the surface or total vaporization at the liquid-gas interface for very high viscosity materials may be treated by the methods of the previous section. However, cases of combined melting and evaporation of glassy materials where the liquid layer partially flows around the body and partially evaporates require a more complex model of this mechanism. Exact solutions for the transient state arbitrary heat flux are given [85–87] at stagnation as well as around the body for arbitrary velocity and temperature gradient in the liquid layer. This solution is coupled with internal radiation transport [88] and a review of the problem is given in Ref. 5.

Motion

$$\frac{\partial p^*}{\partial x} = \frac{\partial T}{\partial y} \tag{67}$$

where the generalized pressure p^* is given by

$$\frac{\partial p^*}{\partial x} = \frac{\partial p}{\partial x} - \rho g_1 \tag{68}$$

Continuity

$$\frac{\partial(\rho u r^k)}{\partial x} + \frac{\partial(\rho v r^k)}{\partial y} = \frac{-\partial(r^k p)}{\partial t} \tag{69}$$

where $k = 0$ for a two-dimensional body and $k = 1$ for a three-dimensional axisymmetric body.

Energy

$$\rho c \frac{\partial T}{\partial t} + \rho v c \frac{\partial T}{\partial y} = \frac{\partial}{\partial y}\left(\kappa \frac{\partial T}{\partial y}\right) \tag{70}$$

reduced to such simple forms due to the slowness of the u velocity, the thinness of the liquid layer, and the neglect of $u(\partial T/\partial x)$ with respect to $v(\partial T/\partial x)$, are used in conjunction with the further reduced form of the energy equation

$$\rho C_p \frac{\partial T}{\partial t} = \frac{\partial}{\partial x}\kappa \frac{\partial T}{\partial x}\bigg|_{s(t) < x < L} \tag{71}$$

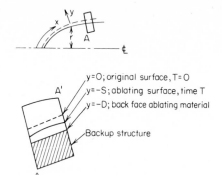

Fig. 7. Coordinates for liquid layer model.

and with boundary conditions (as modified to reflect two-phase flow)

$$-\kappa \frac{\partial T}{\partial x}\bigg|_{x=s} = q_0(t) - \sigma_\epsilon T^4 - f_1 \rho [h_v + \eta(H_e - h_w)]$$

$$= q_{or} - f_1 m[h_v + \eta(H_e - h_w)] \quad (72)$$

$$\dot{s}(t) = 0 \quad \text{for} \quad T(s,t) < T_w \quad (73)$$

$$\dot{s}(t) \geq 0 \quad \text{for} \quad T(s,t) = T_w \quad (74)$$

$$\frac{\partial T}{\partial x}\bigg|_{x=L} = 0 \quad (75)$$

and initial condition

$$T - T_0 \big|_{t=0} \quad (76)$$

For specific results of the liquid layer analysis and bibliography the reader is referred to Refs. 85 to 89. It should be noted that the methods recommended are applicable to any materials displaying liquid layer flow around the body and not just glassy materials. The importance of the internal temperature distribution is particularly noted because of the strong effect of temperature on viscosity of the liquid layer and thus on ablation rates.

4. Ablator Subsurface Equations and In-depth Mechanisms

In this area technology has exerted considerable effort to arrive at physically meaningful and reasonably accurate solutions [11, 12, 17–19]. Actually, the surface equations and the coupled boundary-layer solutions merely represent a set of boundary conditions setting the stage for the complex transient parabolic heat conduction equation describing the "in-depth" phenomena, that is,

$$\dot{q}_s = -\left(\kappa \frac{\partial T}{\partial y}\right)_w \quad (77)$$

The usual assumptions constraining the in-depth solutions are as follows:

1. One continuous equation applies across the entire ablator to include the deposition, porous char, decomposition (or reaction), and virgin material regions.
2. The pyrolyzing material either decomposes completely into gaseous species or leaves partially a solid residue capable of melting or entering into heterogeneous reactions at the surface.
3. The mass flux of in-depth pyrolysis gaseous products is sensed instantaneously at the surface after deposition and/or decomposition reactions.
4. The pyrolysis gases are always in thermal equilibrium with the porous char.
5. The density of the pyrolysis gases is assumed small in comparison to the density of the condensed solid phase.
6. Pyrolysis gas enthalpies are considered known functions of local temperature and external pressure.
7. The thermal conductivity and specific heat of the solids can be represented as linear functions of the condensed phase density and functions of temperature.
8. The ablation model assumes the one-dimensional transient heat and mass flow to be normal to the surface.
9. The velocity-dependent terms appearing in the parabolic transient heat transfer equation such as those due to viscous stress, friction, mixing, etc., are considered small enough in comparison to the remaining terms so as to be neglected.
10. The ablating surface will serve as the reference frame for the moving coordinate system.
11. Thermal decomposition and deposition are assumed to follow simple Arrhenius' rate expressions.
12. Internal radiation of the char is frequently accounted for in implicit form through the thermal conductivity.
13. External radiation or "glowing" from a transparent char is usually represented as a gradient expression added to the heat transfer equation.

The preceding are generally accepted as the standard assumptions accompanying an in-depth ablation solution. However, other options exist presenting internal pressure distribution effects, transpiration cooling, and models for separating the internal radiation contributions.

Upon exercising the above assumptions one may obtain directly the following transient in-depth heat conduction equation in one-dimensional partial differential equation form [12, 90]

$$\underbrace{\rho_s C_{p_s} \frac{\partial T}{\partial t}}_{\substack{\text{Sensible}\\\text{energy}\\\text{exchange}}} = \underbrace{\frac{\partial}{\partial y}\left(\kappa \frac{\partial T}{\partial y}\right)}_{\substack{\text{pure}\\\text{conduction}}} \quad \underbrace{+ \dot{m}_g \overline{C}_{p_g} \frac{\partial T}{\partial y}}_{\substack{\text{Energy exchange due}\\\text{to mass transfer and}\\\text{cracking}}}$$

$$\underbrace{+ \Delta H_{dp} \frac{\partial \rho_{dp}}{\partial t}}_{\substack{\text{Energy exchange due to}\\\text{decomposition (i.e.,}\\\text{pyrolysis and}\\\text{depolymerization)}}} \quad \underbrace{+ \Delta H_c \frac{\partial \rho_c}{\partial t}}_{\substack{\text{Energy exchange}\\\text{due to deposition}\\\text{(i.e., coking)}}} \quad \underbrace{+ \frac{\partial F}{\partial y}}_{\substack{\text{External radiation}\\\text{exchange due to char}\\\text{transparency}}} \qquad (78)$$

with the auxiliary equations

$$\frac{\partial \rho_s}{\partial t} = \frac{\partial \dot{m}_g}{\partial y} \quad \text{and} \quad \rho_{dp} + \rho_c = \rho_s \tag{79}$$

Enthalpy of decomposition ΔH_{dp} and enthalpy of deposition ΔH_c are $f(P_e, T)$. Typically, an example decomposition reaction might be [91, 92]

$$\underbrace{\Gamma \left(E_{a_1} R_{b_1} P_{c_1}\right)_{d_1} + (1 - \Gamma)\left(E_{a_2} R_{b_2} P_{c_2}\right)_{d_2}}_{\text{Virgin composite}} \xrightarrow{\Delta H_{dp}}$$

$$\underbrace{e(T)R_2 + f(T)ER_4 + g(T)E_2R_2 + 1(T)P}_{\text{Pyrolysis gaseous products}} + \underbrace{E_n}_{\substack{\text{solid} \\ \text{char}}} \tag{80}$$

such that $\rho_{dp} = \Gamma(\rho_{11} + \rho_{12}) + (1 - \Gamma)\rho_2$ and Γ being the volume fraction where ρ_{12} and ρ_2 are completely converted into gaseous products.

Then the governing Arrhenius rate expression might be, for instance,

$$\frac{\partial \rho_{dp}}{\partial t} = - \left\{ \Gamma k_{11} \rho_{11}(0) \left[\frac{\rho_{11}(t) - (fc/\Gamma)\rho_{\text{virgin}}}{\rho_{11}(0)} \right]^{n_{11}} \exp\left(-\frac{E_{11}}{RT}\right) \right.$$

$$\left. + \Gamma k_{12} \rho_{12}(0) \left[\frac{\rho_{12}(t)}{\rho_{12}(0)} \right]^{n_{12}} \exp\left(-\frac{E_{12}}{RT}\right) + (1 - \Gamma)k_2 \rho_2(0) \left[\frac{\rho_2(t)}{\rho_2(0)} \right]^{n_2} \exp\left(-\frac{E_2}{RT}\right) \right\} \tag{81}$$

where $f_c = \rho_{\text{char}}/\rho_{\text{virgin}}$.

Similarly for a typical deposition reaction which might be

$$2E_2R_2 \text{ (pyrolysis gas)} \xrightarrow{\Delta H_c} ER_4 \text{ (further pyrolysis gas)} + 3E \text{ (solid char)}$$

the governing Arrhenius rate expression would probably be

$$\frac{\partial \rho_c}{\rho t} = -k_c \rho_c(0) \exp\left(-\frac{E_c}{RT}\right)(\rho_c)^{n_c} \left[\frac{\rho_c(t)}{\rho_c(0)} \right]^{n_c} \tag{82}$$

Note the arrow in the above deposition reaction shows going to completion as stated in the approximations. A reverse reaction would involve equilibrium constants and partial pressure dependence.

Considering this as the complete picture then conservation requires, as seen by the auxiliary equations, that

$$\frac{\partial \dot{m}g}{\partial y} = \frac{\partial \rho_s}{\partial t} = \frac{\partial \rho_{dp}}{\partial t} + \frac{\partial \rho_c}{\partial t} \tag{83}$$

Finally the char transparency radiation term as proposed [93] would be quite adequate for this typical analysis, namely,

$$\frac{\partial F}{\partial y} = \alpha(J - J_e) + a(i - R_0)F \frac{e^{-a(y - \acute{\sigma})} + R_0 e^{-a(2 - y - \acute{\sigma})}}{1 - R_0 R_y e^{-2a(\lambda - \acute{\sigma})}} \tag{84}$$

where J is solved for the auxiliary set of equations

$$\frac{\partial^2 J}{\partial y^2} = a^{2(J - J_e)} \tag{85}$$

$$\frac{\partial J}{\partial y} = -\frac{a^2 F_1}{\alpha} \tag{86}$$

$$J_e = 4\mu_e^2 \sigma T^4 \tag{87}$$

with boundary conditions

$$J(y = \acute{\sigma}) = \alpha_W \dot{q}_{rad} \tag{88}$$

and

$$F_1(s) = -\frac{\alpha}{a} \frac{1 - R_0}{1 + R_0} \alpha_W \dot{q}_{rad} \tag{89}$$

and

$$J(y = \lambda) = F_1(\lambda) = 0 \tag{90}$$

All of these equations are then assembled for transformation of their coordinates to the moving boundary-reference frame in finite difference form and solved numerically on the computer [12, 94]. The solutions provide accurate histories and profiles throughout the ablator as well as a realistic prediction of the decomposition and deposition process.

5. Mechanical Removal

It is known that some ablation materials under certain external or internal high-pressure conditions may either experience mechanical spallation of the surface or other failures resulting in mechanical removal of material. The mechanisms or failure criteria for such ablation processes have been studied [95, 96], criteria assumed and postulated, and their results incorporated in the thermal ablation models. It is felt, however, that the mechanisms and phenomena are not sufficiently understood to be discussed in any detail in this text.

6. Specialized Ablation Models

The sample model of ablation material exhibited earlier to represent the typical in-depth decomposition and deposition mechanisms was symbolically that of phenolic nylon in absence of mechanical removal. The symbolic pyrolysis gases released by decomposition of this model were a four component mixture, namely ER_4, E_2R_2, R_2, and P representing respectively methane (CH_4), acetylene (C_2H_2), hydrogen (H_2), and a combination of assumed nonparticipating gases (i.e., nonreacting with B.L. species) such as nitrogen, hydrogen cyanide, and ammonia [92]. In this model the completely volatile nylon reinforcement was represented by ρ_2 whereas the ρ_{12} (completely volatile) and ρ_{11} combination represented the density of the phenolic resin with Γ being the volume fraction of phenolic resin in the entire virgin plastic composite [12, 20, 91]. With the equilibrium surface reactions between these pyrolysis gaseous products and the B.L. species in mind, the choice of which components are representative or most significant becomes largely an art depending on the range of surface temperatures [97]. The problem of handling *consecutive* equilibrium surface reactions involves no more than writing one combined stoichiometric expression and the associated effective equilibrium constant. On the other hand, *simultaneous* equilibrium surface reactions momentarily appearing as primary, secondary, and tertiary in some order, only to have their roles reversed or permuted as the temperature changes, present the only real challenge for choice of original pyrolysis components. This certainly would seem to preclude the four component model from even representing the ablation of phenolic nylon, much less the general ablator.

However, knowledge of the thermochemical properties of materials such as teflon or carbon and the graphites and their respective ablation behavior in such veritable detail permits specialized handling. This greatly simplifies the ablation equations, further facilitating inclusion of more detailed chemistry [98]. For instance, for carbon (graphite C_3) one may have the following homogeneous and/or heterogeneous surface reactions while diffusion limited [99, 100]

$\frac{1}{3} C_3(s) \overset{K_1}{\rightleftharpoons} C(g)$ (Sublimation or vaporization)   Consecutive Phase Reaction

$C(g) + O \overset{K_2}{\rightleftharpoons} CO$ (Primary)

$C(g) + \frac{1}{2} O_2 \overset{K_3}{\rightleftharpoons} CO$ (Secondary)   Simultaneous Homogeneous

$C(g) + O_2 \overset{K_4}{\rightleftharpoons} CO_2$ (Tertiary)

$CO + O \overset{K_5}{\rightleftharpoons} CO_2$ (Primary)

$CO + \frac{1}{2} O_2 \overset{K_6}{\rightleftharpoons} CO_2$ (Secondary)   Consecutive Homogeneous

$C + CO_2 \overset{K_7}{\rightleftharpoons} 2CO$ (Tertiary)

$\frac{1}{3} C_3(s) + O \overset{K_8}{\rightleftharpoons} CO$ (Primary)

$\frac{1}{3} C_3(s) + \frac{1}{2} O_2 \overset{K_9}{\rightleftharpoons} CO$ (Secondary)   Simultaneous Heterogeneous

$\vdots$

etc.

(s) denotes solid phase; (g) denotes gaseous phase. The partial pressure for graphite C_3 is generally given as a function of T which can be entered directly into the equilibrium expressions. In this case the surface reactions with carbon double for the surface erosion and the "in-depth" ablation of the graphite.

In the case for graphite where the surface reactions and hence ablation reaction is reaction-rate limited the following heterogeneous completion reactions are representative, namely

$$\frac{1}{3} C_3(s) + O \longrightarrow CO$$

Simultaneous completion
(reaction-rate limited)

$$\frac{1}{3} C_3(s) + \frac{1}{2} O_2 \longrightarrow CO$$

The mass flux of $C_3(s)$ leaving the surface by interaction with O and O_2 being represented by the combined rate law and Arrhenius expression

$$\dot{m}_{C_3} = \text{const } p_e^{n_1} \left[\frac{\bar{M}_{C_3}}{\bar{M}_O} \left(C_O\right)_w \right]^{n_2} \left[\frac{\bar{M}_{C_3}}{\bar{M}_{O_2}} \left(C_{O_2}\right)_w \right]^{n_3} \exp\left(\frac{E_{C_3}}{RT}\right) \tag{91}$$

Similarly for teflon whose entire "in-depth" ablation and surface erosion occur all in one reaction on the surface, the following vaporization of the teflon monomer and subsequent consecutive oxidation reaction, can be expressed as [101]

$$\left(C_2F_4\right)_N(s) \longrightarrow N\left(CF_2\right)_2(s) \rightleftharpoons N\left(CF_2\right)_2(g) \rightleftharpoons 2N(CF_2)(g)$$

$$(CF_2)(g) + O \overset{K_{10}}{\rightleftharpoons} COF_2$$

(with arrows in both directions for diffusion-limited and forward direction only for rate-limited reactions).

The basic advantage of considering the teflon and/or carbon models separately from the generalized model lies in the feature of the in-depth transient heat conduction equation being essentially free of all terms with the exception of the heat conduction and "sensible" absorption. All reactions are confined to the surface area.

The silicone elastomers conversely pursue exactly the opposite route. Namely, ablation occurs primarily through the in-depth decomposition mechanisms with relatively sparse surface erosion reactions. The porous silica char produced provides a highly insulative layer that primarily reradiates energy back to the environmental flow field [102]. The basic in-depth reactions consist of unzipping of the highly cross-linked bonds of the elastomer with the release of a few basic pyrolysis gases such as water vapor, carbon monoxide and methane [103]. A highly porous silica char, SiO_2, is left being further capable of melting and vaporization. The SiO_2 may reduce to the carbide and monoxide later capable of entering into reaction with the B.L. constituents forming the ioxide again [103–105].

The silica-resin composites present the combined complexities of the silicone elastomeric in-depth reactions with the phenolic-nylon's phenolic resin decomposition and surface reaction characteristics [16, 105].

The carbon-graphite resin composites essentially proceed with the same or comparable in-depth reaction kinetics as the phenolic nylon model with the exception of having char-erosion surface reactions similar to the pure graphite model [16, 98].

D. MATERIALS PERFORMANCE AND TESTING

The importance of materials characterization in the study of ablation phenomena cannot be overemphasized. Since most applications require specialized materials and since each environment interacts differently with the ablator, it may be said that each ablation theory is fathered by development of an improved ablator. Whereas a more detailed discussion of various ablative materials, their properties, and optimum performance is given in Sec. 20, it is desirable here to review factors which affect the ablation performance of materials, sensitivity of ablation to various material and environment parameters, and means of the ablation testing.

The materials classes of interest are:

1. refractories and ceramics
2. plastics
3. elastomers
4. composites

Energy of decomposition, mass transfer blocking of convective heating, thermal diffusivity, and surface ablation temperature are some of the ablation parameters influencing the ablative material selection. Table 5 summarizes application of ablative materials and the general importance of selection criteria. The surface ablation temperatures are indicated to be high, moderate, or low as characteristic of efficient ablative materials for the application. The heating exposure time strongly influences insulative requirements.

The following paragraphs identify the sensitivity of ablation parameters and for determining selection the interrelationship to material properties, environmental effects and thermostructural consideration.

1. Material Properties Effects

Material properties have a broad influence on mass ablation (and surface recession) and on thermal insulative efficiency. The thermal insulation requirements are defined by the temperature which the structure and heat shield bond are capable of withstanding. Total weight requirements result from the anticipated mass loss and insulation demand.

An energy balance across the ablating surface provides insight into the effects of material properties on mass ablation rates

$$q_0 - L(\rho v)_w \Delta H - (\rho v)_w \Delta Q - q_r = \kappa \frac{\partial T}{\partial y}\bigg|_w - (\rho v)_c \int_{T_0}^{T_w} c_{p_s} dt - (\rho v)_{rp} H_{rp} \quad (92)$$

The above relationship disregards liquid-flow, mechanical-shear, and spallation ablation. For quasi-steady ablation the terms on the right-hand side of the equation disappear.

Rewriting Eq. (92)

$$(\rho v)_w = \frac{q_0 - q_r}{L\Delta H + \Delta Q} \quad (93)$$

the terms influencing the mass ablation rate become apparent. An increase in the boundary-layer enthalpy potential increases both the convective heating rate q_0 and the mass transfer blocking term $L\Delta H$. In addition, ΔH has a significant effect on combustion reactions which will be discussed in Sec. D.2. Mass ablation rates and surface recession rates are reduced by high-surface reradiation rates. Consequently, high emissivity coefficients and surface ablation temperatures are important material properties. The mass-transfer blocking parameter is dependent on molecular weights or specific heat of ablation products as previously discussed. As a result ablative

TABLE 5. Ablative Material Applications and Selection Criteria

Application	Heating exposure time (seconds)	Weight per unit surface area	Mass ablation rate	Thermal insulative properties	Surface ablation temperature	Surface recession
1. High-performance conical reentry vehicles ($w/c_D A > 1500$ psf):	15–40					
a. Nosecap		Unimportant	Less important	Unimportant	High	Important
b. Conical section		Important	Important	Important	High	Less important
c. Aft cover		Important	Unimportant	Important	Low to moderate	Unimportant
2. Low-ballistic coefficient reentry vehicles ($w/c_D A < 1500$ psf):	20–100					
a. Nosecap		Less important	Important	Less important	High	Important
b. Other regions		Important	Important	Important	Moderate to high	Unimportant
c. Aft cover		Important	Unimportant	Important	Low	Unimportant
3. Low to intermediate range ballistic missiles	10–50					
a. Nosecap		Less important	Important	Less important	Moderate to high	Less important
b. Other regions		Important	Important	Important	Low to moderate	Unimportant
c. Aft cover		Important	Unimportant	Important	Low	Unimportant

4. Lifting, satellite, and planetary entry	300–2000	Important	Unimportant	Important	Moderate	Important
5. Launch and booster insulation	5–200	Important	Unimportant	Important	Low	Unimportant
6. Rocket nozzles						
a. Throat	1–100	Less important	Unimportant	Less important	High	Important
b. Other regions		Important	Less important	Important	High	Less important
7. Hypersonic interceptor missiles	5–100					
a. Nosecap and control fin leading edges		Unimportant	Less important	Less important	High	Important
b. Control fins and forward surface regions		Important	Important	Important	Moderate to high	Unimportant
c. Other regions		Important	Less important	Important	Low to moderate	Unimportant

materials which vaporize or decompose into low-molecular-weight or high-specific-heat ablation products are more efficient than other ablative materials.

The ablation rates are most strongly influenced by the thermochemical response of the ablative material as represented by the term ΔQ. The heat of vaporization and/or energy of decomposition, combustion reactions, and the sensible enthalpy at the surface ablation temperature are included in ΔQ

$$\Delta Q = \frac{(\rho v)_c}{(\rho v)_w} H_v + \frac{(\rho v)_c}{(\rho v)_w} \int_{T_0}^{T_w} c_{p_s} dt + \frac{(\rho v)_{rp}}{(\rho v)_c} H_{rp} - \Delta E_c \tag{94}$$

where ΔE_c represents the energy released by combustion reactions between the gaseous surface material, pyrolysis products, and environment.

To evaluate ΔQ the thermochemical properties of the ablative material must be known. These include the chemical constituents, both elemental and molecular. For charring ablators the constituents must be determined for the char, original material, and gaseous pyrolysis products. The surface ablation temperature must be defined. For most ablative materials the evaluation of ΔQ or the equivalent terms is a complex problem which must be coupled to boundary-layer solutions and results of a themochemical, digital computer program. Above surface temperatures of about $2000°R$ thermochemical equilibrium between gaseous surface material, pyrolysis products, and environment is the most widely accepted assumption for evaluating the thermochemistry.

The contribution of combustion reactions to ΔQ is shown by comparisons of ablative performance of a combustible and noncombustible ablative material. The ablation rate comparisons for a high-performance reentry are shown in Fig. 9. The two ablative materials are carbon phenolic and silica phenolic. Carbon phenolic forms a carbon char which is susceptible to oxidation; and silica phenolic forms a char in which the principal constituent is silica (SiO_2) which vaporizes and decomposes into

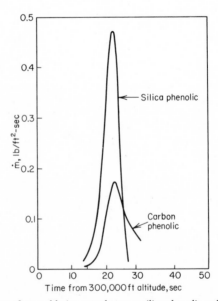

Fig. 8. Comparison of mass ablation rates between silica phenolic and carbon phenolic.

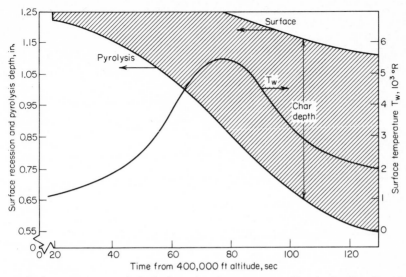

Fig. 9. Surface recession, pyrolysis depth, and surface temperature histories for phenolic nylon ablation.

$SiO + \frac{1}{2}O_2$ absorbing about 5000 Btu/lb in the process. The released oxygen combines with pyrolysis products to form CO, CO_2, and H_2O. Assuming quasi-steady ablation $\Delta Q \simeq 4000$ Btu/lb. For carbon phenolic the char is very susceptible to oxidation. In the oxidation-diffusion controlled regime for low to moderate heating rates $\Delta Q \simeq 4000$ Btu/lb, i.e., combustion reactions give a net heat input at the surface with recession and Eq. (93) is invalid and the methods presented previously are used to predict the mass ablation. However, for higher heating rates ΔQ increases at peak heating to values near 7,000 Btu/lb and Eq. (93) is valid. The susceptibility of the carbon phenolic to oxidation is most pronounced at later flight times.

The mass ablation comparisons in Fig. 8 show that although carbon phenolic is more susceptible to oxidation total mass loss is greater for the silica phenolic. The reason is that surface ablation temperatures are greater for carbon phenolic, thus increasing surface reradiation and reducing the convective heating (also, $\epsilon_s = 0.9$ for carbon phenolic and $\epsilon_s = 0.5$ for silica phenolic). The mass transfer blocking of convective heating is less for carbon phenolic because of lower ablation rate and enthalpy potential.

Mass ablation rates are enhanced by liquid flow ablation, spallation of the char layer, and mechanical removal of the char layer. Silica base materials such as silica phenolic, phenolic glass laminates, quartz, etc., are susceptible to liquid flow surface recession. However, it is only important for nosecap applications and other regions subjected to significant pressure and shear gradients. A detrimental characteristic of many charring ablators is low-char strength. The application of ablative materials, such as the epoxies and elastomerics, is consequently restricted to low-shear and low-pressure regions or to vehicles subjected to less severe heating environments. Shear or spallation of the char is also an important consideration to high-strength charring ablators, such as carbon phenolic, for nosetip application.

Mass ablation may not be critical for some applications. Ablative material selection is strongly sensitive to material properties which influence thermal insulative performance. A numerical description of the transient heat conduction relationship indicates the sensitivity to material properties. An energy balance on lamina i of thickness Δb gives the following relationship for the rate of temperature rise

$$\Delta b_i \rho_i c_{p_i} \frac{\Delta T_i}{\Delta b_i} = \frac{2k_i k_{i-1}}{k_{i-1}\Delta b_i + k_i \Delta b_{i-1}} (T_{i-1} - T_i) + \frac{2k_i k_{i+1}}{k_{i+1}\Delta b_i + k_i \Delta b_{i+1}} (T_{i+1} - T_i)$$

$$- (\rho v)_{rp} \Delta b_i \left(\frac{H_{rp_{i-1}} - H_{r_i}}{\Delta b_{i-1} + \Delta b_i} - \frac{H_{rr_{i+1}} - H_{rp_i}}{\Delta b_{i+1} + \Delta b_i} \right) + \dot{\rho}_i \Delta b_i \Delta H_{Di}$$

$$- \dot{s} \rho_i c_{p_i} \Delta b_i \left(\frac{T_{i-1} - T_i}{\Delta b_{i-1} + \Delta b_i} + \frac{T_{i+1} - T_i}{\Delta b_{i+1} + \Delta b_i} \right) \qquad (95)$$

where

$$\dot{\rho}_i = f_B \rho_0 \left[\frac{\rho_i - \rho_0(1 - f_B + \gamma f_B)}{f_B \rho_0 (1 - \gamma)} \right]^n ke^{-E*/T_i} \qquad (96)$$

In the above relationship the last term is included because the coordinate system is fixed to the receding surface. As a general rule the specific heat and thermal conductivity are temperature and density dependent. Consequently, specific-heat and thermal-conductivity data are required for both char and undecomposed ablative material of the charring ablators.

The dominant material property in Eq. (95) is the thermal conductivity. The specific heats for most ablative materials have nearly identical values.

Heat conducted into the ablative material is absorbed by pyrolysis decomposition and by the mass transfer toward the surface. The decreased density due to pyrolysis has an effect on the thermal insulation performance which is more than counterbalanced by internal mass-transfer energy absorption. However, internal pyrolysis results in the thermal conductivity shifting from low-undecomposed material values to higher values for the char. A rather misleading example of the influence of high-char thermal conductivity on phenolic-nylon ablation is shown in Fig. 9. The thermal conductivity data are from Wilson [106]. Although surface recession is low, the buildup of the char thickness near the end of flight is about 0.5 inch. Pyrolysis occurs in the temperature range of 1000 to 1500°R. The char density is about 0.30 times the original density of 72 lb/ft³; consequently, internal mass transfer removed a significant fraction of heat conducted in at the surface.

The insulative performance of ablation materials is a strong function of thermal conductivity. For the charring ablators the char thermal conductivity dominates as in the above example and in Swann, Dow, and Tomkins [107]. The thermal conductivities of several ablative materials are presented in Fig. 10 as functions of temperature. The experimental correlation for phenolic nylon [106] is compared with an unpublished correlation derived with a charring ablator computer program which was used to correlate experimental ablation data. The data shown in Fig. 10 are representative of different classes of ablative materials. Tungsten represents the refractory metals. Graphites represent a class of materials by themselves. Boron nitride is a representative refractory ceramic. Carbon phenolic is a high-temperature and char strength composite. Phenolic nylon is representative of the plastic composites, and purple blend MOD-5 is one of a class of materials specifically

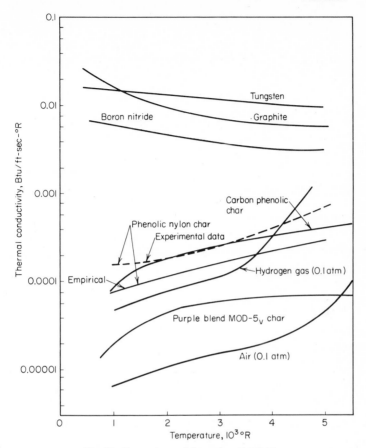

Fig. 10. Comparisons of thermal conductivities.

tailored to provide low-thermal conductivity. Hydrogen gas and air thermal conductivities are shown for comparison purposes.

For applications requiring good insulative properties coupled to long heating exposure time, namely, lifting reentry vehicles, ablative plastic or elastomeric composites have been developed with the following properties to give improved insulative and ablative performance: (1) surface oxidation resistance; (2) low density, (3) low-char density, (4) low-molecular weight pyrolysis products, (5) fiber reinforcement for char strength, and (6) spherical voids. Examples of these materials are purple-blend MOD-5, and phenolic cork.

Pyrolysis decomposition using the Arrhenius rate equation has limited effects on ablative material performance provided the decomposition temperature is adequately duplicated, i.e., from about 800 to 1500°R.

For materials which form thick chars evidence of "cracking" of gaseous hydrocarbons has been found through observed increases in densities of chars near the ablating surface. The effect on the ablative material char is to increase char strength and at the same time to increase char thermal conductivity. Significant alteration of material performance through decreases in char permeability and increase in internal pressure can result if the char layer spalls.

Strength properties of the ablative materials are important for many applications. Low-char strength precludes many ablative materials from application to high-performance reentry vehicles. Low-char strength leads to high-surface recession rates caused by mechanical removal. The mechanical removal phenomena are not yet well understood. Thus the failure criteria, or parameters of significance, cannot be well defined.

2. Environmental Effects

As indicated above, ablation performance is sensitive to environmental effects. An examination of Eq. (93) shows that an increase in convective (q_0) or radiative (q^R) heating rate increases ablation rates. However, if there is proportionate increase in reradiation or in the terms $L\Delta H + \Delta Q$ there may not be an increase. Examples of this effect are oxidation-diffusion controlled ablation for carbon base materials and increases in heating caused by higher enthalpy. These changes in environmental heating increase the ablation efficiency of the materials [107].

One effect of boundary-layer enthalpy potential on ablation efficiency is shown by the mass-transfer blocking term $L\Delta H$. However, for ablative combustion reaction the enthalpy potential has a greater importance. In Eq. (94) combustion reactions are included with the term ΔE_c. Equations (93) and (94) can be combined and combustion reactions can be accounted for [108]

$$\bar{q}_c = \frac{K_{oe} h_c}{\Delta H}$$ (97)

in the following relationship

$$(\rho v)_w = \frac{q_0(1 + \bar{q}_c) - q_r}{L\Delta H(1 + \bar{q}_c) + \Delta E}$$ (98)

where

$$\Delta E = \frac{(\rho v)_c}{(\rho v)_w} H_v + \frac{(\rho v)_c}{(\rho v)_w} \int_{T_v}^{T_w} C_{ps}\, dT + \frac{(\rho v)_{vP}}{(\rho v)_w} H_{vP}$$ (99)

h_c = heat of combustion per unit mass of oxygen
K_{oe} = mass fraction of oxygen at the outer edge of the boundary layer

The term $\bar{q}_c$ gives an upper bound on the effect of combustion by assuming that all oxygen diffusing toward the surface reacts with the material vapor. For silica base materials this is not true and in a rocket nozzle environmental oxygen may be fixed as carbon monoxide.

In Eq. (98) the combustion contribution to heating and to the mass transfer blocking is shown to be independent of the boundary-layer enthalpy potential (note that $q_0 = N_{co}\Delta H$). Consequently, combustion heating reactions are less important for high-enthalpy potential heating environments that reentry vehicles experience over major portions of their flight. The side variation in material efficiency that results with enthalpy variations is shown by Swann, Dow, and Tomkins [107].

Environmental pressure and shear forces strongly influence ablation performance reentry vehicles and to a lesser extent for throats of rocket nozzles. The charring ablators are recognized to have char strengths considerably lower than the strength of the undecomposed materials. Consequently, those materials which form weak

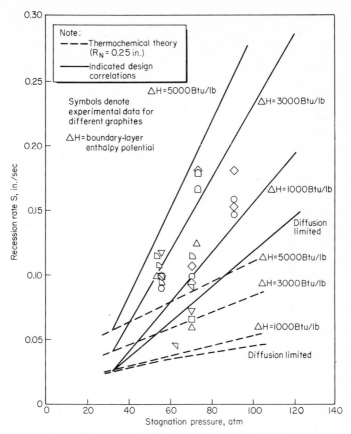

Fig. 11. Environmental effects on recession rates of graphite.

chars are restricted to vehicles that experience low pressure and shear environment, the high-char strength materials, i.e., carbon phenolic and silica phenolic have been widely used on high-performance reentry vehicles. Experimental graphite ablation data are presented in Fig. 11. Both environmental enthalpy and stagnation pressure effects are included in the data. Comparisons with thermochemical theory are also presented.

3. Testing

a. Simulation. Ablation testing is the method by which the analytical models are verified, many material properties are determined, and candidate heat shield materials are screened and tested. Because of the importance of testing in providing preflight design verification and confidence, it is obviously desirable to perform these tests in an accurately simulated if not duplicated flight environment. For many reasons (primarily facility power requirements and model size limitations) this is usually impossible, and tests must be run under several types of partial flight simulation. These results can then be evaluated by means of previously proven mathematical ablation models, or by overlapping the tests in such a way that a coherent composite picture of the important phenomena can be constructed [109].

Ballistic and lifting entry vehicles, designed for various missions, experience a wide range of entry environments. Two convenient parameters for illustrating these environments, which depend on flight performance rather than on geometry, are total enthalpy and stagnation point pressure. Figure 12 is a plot of pressure enthalpy with overlays of typical entry-vehicle trajectories and maps of approximate simulation facility performance. Two important conclusions may be immediately reached:

1. No one facility can duplicate a complete flight environment, although large portions of some lifting-body trajectories can be closely approximated.

2. High pressure and high-enthalpy effects cannot be duplicated simultaneously.

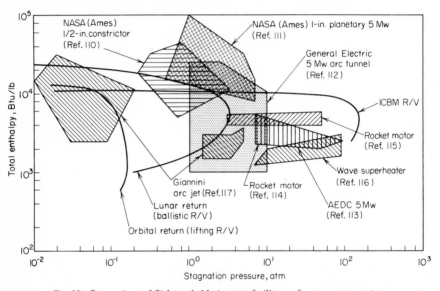

Fig. 12. Comparison of flight and ablation test facility performance parameters.

For simulation, steady-state ablation is usually desired, since this condition provides the necessary material parameters and evaluation for the analytical models. For this reason, the facilities most applicable to entry ablation studies are the plasma-arc tunnel and the rocket-motor exhaust, both of which have operating times ranging from several seconds to several minutes. Facilities such as the ballistic range and shock tube are of only slight interest because of their very short test times—approximately 10 milliseconds and 100 microseconds, respectively.

Because of simulation difficulties, test conditions must be carefully selected to obtain meaningful results. Plasma-arc facilities can frequently provide adequate simulation up to entry velocities of 30,000 ft/sec. The important simulation parameters for most ablation performance and screening tests are the heat flux and the enthalpy [118]. Except as noted below, ablation performance is only slightly affected by the pressure.

Efficient performance of the glassy ablators depends on the retention of the liquid melt layer, so the aerodynamic shear levels of flight should also be duplicated, if possible. The ablation performance of charring ablators is known to be reduced under very high pressures or pressure-gradients which tend to mechanically remove the char layer. For flight vehicles which will experience unusually high local pressures, such as low-drag ballistic entry vehicles, it may be necessary to simulate the pressure and shear levels as well as the heat flux and enthalpy. However, since this amounts to full-scale flight duplication, these conditions must be simulated in separate tests and applied with much caution to the flight environments.

At the very high entry velocities which will be experienced in interplanetary return missions, the radiant heating from the shock-heated air can be equal to or significantly greater than the convective heating rates. Since the efficiency of most ablative materials depends on the heat blockage effect of the transpiring gases which will have little or no effect on the radiant heating, ablation performance will be reduced. Some plasma-arc facilities are available [117] which can superimpose radiant heating on the convective heating, and these must be employed for performance evaluation of materials to be used in this type of environment.

   b. Interpretation of Test Data. Several types of experimental data must be examined in order to obtain an appropriate set of material-dependent characteristics for use in any given analytical model of the ablation and heat transfer processes.

Generally, ablation data are obtained from experiments in which both energy and mass transfer effects play a significant role. Useful quantitative information for mathematical models will be obtained only from near steady-state ablation experiments [119]. For an experiment to be interpreted as steady-state, for a charring ablator, the post-test char thickness must be small relative to the sum of the post-test char thickness and the total recession, or

$$\dot{y} = \frac{\Delta y}{\Delta t} = \frac{\Delta (S + \delta c)}{\Delta t} \qquad (100)$$

where y = surface coordinate
   t = time
   δ_c = char depth measured from the outer surface

As a further consequence of the definition of steady state, one arrives at a relationship between the weight loss, length loss, and the initial and final weights and lengths which must hold for any experiment conducted under steady-state assumptions

$$L_i \Delta W = W_i \Delta L \qquad (101)$$

This results in the criterion (F) for steady-state ablation reported in Ref. 119

$$F = \frac{L_i \Delta W}{W_i \Delta L} \qquad (102)$$

or in terms of the post-test char thickness (δc), char density (ρc), and virgin material density (ρo)

$$F^* = \frac{\Delta L + (1 - \rho c/\rho o)\rho c}{\Delta L} \qquad (103)$$

An excellent discussion on other parameters which can be obtained from arc-jet and turbulent pipe ablation tests is available [119, 120].

   In addition to the surface, recession, mass loss, etc., an accurate description of the kinetics of the char forming and surface reactions is required. The most common type of experiment for this purpose is the linear thermogravimetric analysis experiment where measurements are obtained for sample weight and temperature under conditions such that the sample temperature is a linear function of temperature.

Other experimental data necessary to completely verify the physical-mathematical model is the density, thermal conductivity, and specific heat of both the char and the virgin material. With the exception of the char thermal conductivity, these measurements can usually be adequately taken in any well equipped thermodynamics laboratory. In the case of the char, the conductivity is normally obtained by the conventional steady-state laboratory techniques and is therefore not necessarily representative of the transient characteristics which would be experienced during high-speed flight. Since the decomposition kinetics of an ablator are heating rate dependent, the material or char composition and, hence, the thermophysical properties, will also be heating rate dependent. Some methods of determining the thermal properties of a char under transient heating conditions [121] consist of systematically determining the properties which will cause the computed temperature histories to match the temperatures measured at several locations in an ablating arc-jet specimen.

It is well to reemphasize that the ablation experiments provide the means for verification of physical-mathematical models, rather than providing direct-design data. The models are then used in simulation of the flight response on the computer. There are no "handy" shortcuts to this method.

REFERENCES

1. R. B. Erb and S. Jacobs, Entry Performance of the Mercury Spacecraft Heat Shield, AIAA Entry Tech. Conf. Pub. CP-9, Oct. 1964.
2. Anon., *Aircraft and Missiles,* **3**:40 (1960).
3. S. I. Freedman, in "Developments in Heat Transfer" (W. M. Rohsenow, ed.), p. 87, The M.I.T. Press, Cambridge, Mass., 1964.
4. C. J. Katsikas, G. K. Castle, and J. S. Higgins, Ablation Handbook—Entry Materials Data and Design, *AFML* 66-262, Nov. 1966.
5. H. Hurwicz, W. C. Broding, and A. J. Hanawalt, An Approach to Re-entry Thermal Protection/System Development, Bumble Bee Conf., St. Louis, Mo., Nov. 1963.
6. D. L. Schmidt, Ablative Plastics and Elastomers in Chemical Propulsion Environments, *AFML-TR*-65-4, April 1965.
7. D. L. Schmidt, Ablative Polymers for Hypersonic Atmospheric Flight, *AFML-TR*-66-78, May 1966.
8. L. Steg and H. Lew, Hypersonic Ablation, in "The High Temperature Aspects of Hypersonic Flow" (W. Nelson, ed.), pp. 629-680, The Macmillan Co., New York, 1964.
9. M. C. Adams, Recent Advances in Ablation, *J. Am. Rocket Soc.,* **29**:625 (1959).
10. Thermal Protection Systems, *ML-TDR*-64-82, Jan. 1965.
11. H. Hurwicz, Aerothermochemistry Studies in Ablation, "Combustion and Propulsion Fifth AGARD Colloquium," Pergamon Press, Inc., New York, 1963.
12. A. O. Taylor, A Transient Reaction Kinetics Analysis on Ablation Through Aerodynamic Heating, *General Dynamics Convair Report GDC-ERR-AN*-1029, Dec. 1966.
13. H. R. Enkenhus, The Effect of Variable Lewis Number on Heat Transfer in a Binary Gas, *AIAA Journal,* **2**:747 (1964).
14. L. Lees, Convective Heat Transfer with Mass Addition and Chemical Reactions, "Combustion and Propulsion Third AGARD Colloquium," Pergamon Press, Inc., New York, 1958.
15. S. M. Scala and L. M. Gilbert, The Sublimation of Graphite at Hypersonic Speeds, AIAA Entry Tech. Conf., Pub. CP-9, Oct. 1964.
16. R. T. Swann, C. M. Pitman, and J. C. Smith, One-Dimensional Numerical Analysis of the Transient Response at Thermal Protection Systems, *NASA TN-D*-2976, Sept. 1965.
17. C. L. Arne and D. N. Kopec, Charring Ablator Multiphase Processes Computer Program, *Douglas Rept. SM*-49310, June 17, 1960.
18. L. H. Hillberg, The Convective Heating and Ablation Program (CHAP), *Boeing Rept. D*2-36402-1, May 31, 1966.

19. T. R. Munson, R. E. Mascola, J. D. Brown, R. J. Spindler, and J. Mugerman, An Advanced Analytical Program for Charring Ablators—Final Report Vol. I, *ACO Space Systems Div. Rept. AVSSD*-0172-67-RR.

20. R. M. Kendall, E. P. Bartlett, R. A. Rindel, and C. B. Mayer, An Analysis of the Coupled Chemically Reacting Boundary Layer and Charring Ablator, *Aerotherm Final Rept.* 66-7.

21. R. T. Achard, Fundamental Relationships for Ablation and Hyperthermal Heat Transfer, *Air Force Res. and Tech. Div. Flight Dynamics Lab. Rept. AFFDL-TR*-66-25, April 1966.

22. Paul M. Chung, "Advances in Heat Transfer," vol. 2, Academic Press, Inc., New York, 1965.

23. L. W. Woodruff and G. C. Lorenz, Hypersonic Turbulent Transpiration Cooling Including Downstream Effects, *AIAA Journal,* 4:969 (1966).

24. T. N. Stevenson, Turbulent Boundary Layer with Transpiration, *AIAA Journal,* 2:1500 (1964).

25. P. J. Schneider, T. A. Dolton, and G. W. Reed, Mechanical Erosion of Charring Ablators in Ground Test and Re-Entry Environments, *AIAA Journal,* 6:64 (1968).

26. A. M. O. Smith and N. A. Jaffe, General Method for Solving the Laminar Non-equilibrium Boundary-Layer Equations of a Dissociation Gas, *AIAA Journal,* 4:611 (1966).

27. R. M. Kendall, R. A. Rindal, and E. P. Bartlett, A Multicomponent Boundary Layer Chemically Coupled to an Ablating Surface, *AIAA Journal,* 5:1063 (1967).

28. W. H. Carden, Heat Transfer in Non-equilibrium Dissociated Hypersonic Flow with Analysis and Second-Order Effects, *AIAA Journal,* 4:1704 (1966).

29. C. J. Katsikas, Ablation Handbook—Entry Materials Data and Design, *AFML* 66-262, Nov. 1966.

30. M. Rauscher, "Introduction to Aeronautical Dynamics," John Wiley & Sons, Inc., New York, 1953.

31. H. J. Allen and A. J. Eggers, Jr., A Study of the Motion of Aerodynamic Heating of Ballistic Missiles Entering the Earth's Atmosphere at High Supersonic Speeds, *NACA Rept.* 1381, 1958.

32. W. H. T. Loh, "Dynamics and Thermodynamics of Planetary Entry," Prentice-Hall, Inc., Englewood Cliffs, N.J., 1963.

33. R. W. Truitt, "Fundamentals of Aerodynamic Heating," The Ronald Press Co., New York, 1960.

34. R. H. Weatherford and S. Sayano, Techniques for Rapid Aerodynamic Heat Transfer Calculations, *Douglas Rept. SM*-45932, May 1964.

35. J. N. Nielsen, "Missile Aerodynamics," McGraw-Hill Book Company, New York, 1960.

36. K. K. Cheng, Recent Advances in Hypersonic Flow Research, *AIAA Journal,* 1:295 (1963).

37. Anon., "U.S. Standard Atmosphere," U.S. Government Printing Office, Washington, D.C., Dec. 1962.

38. G. Fjeldbo, C. Fjeldbo, and V. R. Eshelman, Models for the Atmosphere of Mars Based on the Mariner IV Occulation Experiment, *J. Geophys. Res.,* 71:2307 (1966).

39. J. R. Khegel and G. R. Nickerson, Axisymmetric Two-Phase Perfect Gas Performance Program, *TRW/Systems Report* 02874-6006-R000, April 1967. Proposal under NASA Contract NAS 9-4358.

40. J. Hilsenrath, M. Klein, and H. W. Woolley, Tables of Thermodynamic Properties of Air Including Dissociation and Ionization from $1500°K$ to $15,000°K$, Arnold Engineering and Development Center, *AEDC TR*-59-20, Dec. 1959.

41. W. J. Little, Mollier Diagram for Air, Von Kármán Gas Dynamics Facility, ARO, Inc., *AEDC-TDR*-63-190, Sept. 1963.

42. F. W. Fenter and H. B. Gibbons, The Thermodynamic Properties of High Temperature Air, *Chance Vought Research Center Rept.* RE-1R-14, June 1961.

43. W. E. Moeckel and K. C. Weston, Composition and Thermodynamic Properties of Air in Chemical Equilibrium, *NACA TN* 4265, April 1958.

44. F. R. Gilmore, Equilibrium Composition and Thermodynamic Properties of Air to $24,000°K$, *Rand Corporation Rept.* RM 154B, 1955.

45. C. F. Hansen, Approximations for the Thermodynamic and Transport Properties of High Temperature Air, *NASA TR-R*-50, 1959.

46. T. E. Chappel, Approximations for the Thermodynamic and Transport Properties of High Temperature Air, *Martin Marietta Corp. Rept. OR* 2109, *EC*, 1961.

47. F. Bosworth, C. Cook, L. Gilbert, and S. Scala, Normal Shock Parameters for the Martian Atmosphere, *GE* R63SD12.

48. J. M. Yos, Transport Properties of Nitrogen, Hydrogen, Oxygen and Air to 30,000K, *AVCO/RAD-TM*-63-7, March 1963.

49. S. Bennet, J. M. Yos, D. F. Knopp, J. Morris, and W. L. Bade, Theoretical and Experimental Studies of High Temperature Gas Transport Properties, Final Report, *AVCO/RAD-TR*-65-7, May 1965; also *AVCO/RAD-SR*-65-35, Feb. 1965.

50. T. Peng and A. L. Pindroh, An Improved Calculation of Gas Properties of High Temperature Air, *Boeing Doc. D2* 11722, Feb. 1962.

51. M. Thomas, Jr., The High Temperature Transport Properties of Carbon Dioxide, *Douglas Rept. SM*-37790, July 1960.

52. W. L. Wiese, M. W. Smith, and B. M. Glennon, Atomic Transition Probabilities, vol. I. Hydrogen Through Neon, National Bureau of Standards, *NSRDS-NBS* 4, May 1966.

53. H. R. Griem, "Plasma Spectroscopy," McGraw-Hill Book Company, New York, 1964.

54. J. C. Stewart and K. D. Pyatt, Jr., Theoretical Study of Optical Properties—Photon Absorption Coefficients, Opacities, and Equations of State of Light Elements, Including the Effect of Lines, *Report GA*-2528, vol. 1, *AFSWC-TR*-61-71, vol. 1, General Atomic Div., General Dynamics Corp., Sept. 1961.

55. D. R. Churchill, B. H. Armstrong, and K. G. Mueller, Absorption Coefficients of Heated Air: A Compilation to 24,000°K, *Technical Report AFWL-TR*-65-132, vols. 1 and 2, Kirtland Air Force Base, Air Force Weapons Lab., New Mexico, Oct. 1965.

56. J. C. Morris, C. F. Knopp, R. U. Krey, R. W. Liebermann, and R. L. Garrison, Radiation of Oxygen and Nitrogen and Thermal Conductivity of Nitrogen up to 15,000°K, AIAA Paper 66-182, AIAA Plasmadynamics Conf., Monterey, Calif., March 1966.

57. G. Bolt, The Recombination and "Minus" Continua of Nitrogen Atoms, *A. Physik,* **154**:330 (1959).

58. S. S. Penner and M. Thomas, Approximate Theoretical Calculations of Continuum Opacities, *AIAA Journal,* **2**:1672 (1964).

59. R. B. Dirling, Jr., W. S. Rigdon, and M. Thomas, Stagnation-Point Heating Including Spectral Radiative Transfer, in "Proc. 1967 HTFMI," Stanford Univ. Press, 1967.

60. F. S. Simmons, Radiances and Equivalent Widths of Lorentz Lines for Nonisothermal Paths, *JQSRT,* **7**:11 (1967).

61. A. R. Curtis, Discussion of a Statistical Model for Water-Vapor Absorption by R. M. Goody, *J. Roy. Met. Soc.,* **78**:638 (1952).

62. W. L. Goodson, The Evaluation of Infra-red Radiative Fluxes Due to Atmospheric Water Vapor, *J. Roy. Met. Soc.,* **79**:367 (1953).

63. E. R. G. Eckert, Survey of Boundary Layer Heat Transfer at High Velocities and High Temperatures, *WADC Tr* 59-624, April 1960.

64. J. A. Fay and F. R. Riddell, Theory of Stagnation Point Heat Transfer in Dissociated Air, *J. Aeronaut. Sci.,* **25**:73 (1958).

65. D. F. Deney, Jr., Use of Local Similarity Concepts in Hypersonic Viscous Interaction Problems, *AIAA Journal,* **1**:20 (1963).

66. C. B. Cohen and E. Reshotho, Similar Solutions for the Compressible Laminar Boundary Layer with Heat Transfer and Pressure Gradient, *NACA Rept.* 1293, 1956.

67. W. H. Dorrance, "Viscous Hypersonic Flow," McGraw-Hill Book Company, New York, 1962.

68. E. R. Van Driest, Turbulent Boundary Layer in Compressible Fluids, *J. Aeronaut. Sci.,* **8**:145 (1951).

69. D. B. Spalding and S. W. Chi, The Drag of a Compressible Turbulent Boundary Layer on a Smooth Flat Plate With and Without Heat Transfer, *J. Fluid Mech.,* **18**:117 (1964).

70. B. Kivel and K. Bailey, Tables of Radiation from High-Temperature Air, *AVCO Res. Rept.* 21, 1957.

71. M. Nardone, R. G. Breene, S. Zeldin, and T. R. Riethof, Radiance of Species in High Temperature Air, *General Electric Rept.* R63SD3, June 1963.

72. R. E. Meyerott, J. Sokoloff, and R. W. Micholls, Absorption Coefficients of Air, *Geophysical Research Paper 58*, GRD-TN-277, July 1960.

73. R. A. Allen, Air Radiation Graphs: Spectrally Integrated Fluxes Including Line Contributions and Self Absorption, AVCO—Everett Res. Lab. Report 230, Sept. 1965.

74. R. M. Nerem and G. H. Stickford, Shock Tube Studies of Equilibrium Air Radiation, *AIAA Journal,* **3**:1011 (1965).

75. J. C. Morris, C. F. Knopp, R. U. Krey, R. W. Liebermann, and R. L. Garrison, Radiation of

Oxygen and Nitrogen and Thermal Conductivity of Nitrogen up to 15,000°K, AIAA Paper 66-182, AIAA Plasmadynamics Conf., Monterey, Calif., March 1966.

76. L. M. Biberman, B. S. Norobyev, G. E. Norman, and I. T. Yalrubou, Radiation in Hypersonic Flow, *Kosmicheskie Issledovania,* 2:441 (1964).

77. D. G. Elliott, D. R. Bartz, and S. Silver, Calculation of a Turbulent Boundary Layer Growth and Heat Transfer in Axisymmetric Nozzles, *Jet Propulsion Laboratory Report* 32-387, California Institute of Technology, Feb. 15, 1963.

78. W. S. Rigdon, R. B. Dirling, Jr., and M. Thomas, Radiative and Convective Heating During Atmospheric Entry, March 1967. NASA contractor's report to be published.

79. W. D. Coleman, J. M. Lefferdo, L. F. Hearne, and N. S. Vojvodich, A Study of the Effects of Environmental and Ablator, and Performance Uncertainties on Heat Shielding Requirements for Blunt and Slender Hyperbolic-Entry Vehicles, Paper 68-154, AIAA 6th Aerospace Sciences Meeting, Jan. 1968.

80. D. L. Schmidt, Ablative Polymers for Hypersonic Atmospheric Flight, *AFML-TR*-66-78, May 1966.

81. R. T. Archard, Fundamental Relationships for Ablation and Hyperthermal Heat Transfer, *AFFDL-TR*-66-25, April 1966.

82. X. X. Katsikas, G. K. Castle, and J. S. Higgins, Ablation Handbook Entry Materials Data and Design, *AFML*-66-262, Nov. 1966.

83. J. D. Brown and F. A. Shukis, An Approximate Method for Design of Thermal Protection Systems, IAS 30th Ann. Meet., New York, Jan. 1962.

84. H. Hidalgo, Ablation of Glassy Material Around Blunt Bodies of Revolution, *ARS Journal,* 30:806 (1960).

85. R. G. Fleddermann and H. Hurwicz, Analysis of Transient Ablation and Heat Conduction Phenomena at a Vaporizing Surface, *Chemical Engineering Symposium Series,* 57:2 (1960); also *AVCO RAD-TR*-9(7)-60-9, April 1960.

86. M. Zlothnick and B. Nordquist, Calculation of Transient Ablation, Int. Heat Transfer Conference, Boulder, Colo., Sept. 1961; also *AVCO RAD*-9-TM-60-83, Jan. 1961.

87. H. Hurwicz and R. G. Fleddermann, Computer Simulation of Transient Ablation and Heat Conduction Phenomena at a Vaporizing Surface, ARS Space Flight Report to the Nation, New York, Paper No. 2209-61, Oct. 1961; also *AVCO-RAD-TR*-9(7)-60-9 April 1960.

88. R. J. Spindler and H. Hurwicz, Transient Heat Flow in Translucent Non-Gray Ablating Materials, Proc. Ballistic Missiles and Space Technology Conf., vol. 2, Pergamon Press, Inc., New York, 1961; also *AVCO-RAD-TR*-9-(7)-59-17, Oct. 1959.

89. H. Hidalgo, Ablation of Glassy Material Around Blunt Bodies of Revolution, *J. Am. Rocket Soc.,* 30:806 (1960).

90. C. W. Stroud, A Study of the Chemical Reaction Zone in Charring Ablators During Thermal Degradation, M.S. thesis, Virginia Polytechnic Institute, Blacksburg, Va., Sept. 1965.

91. K. M. Kratsch, L. F. Hearne, and M. R. McChesney, Theory for the Thermophysical Performance of Charring Organic Heat-Shield Composites, *Lockheed Missiles and Space Co. Rept. LMSC*-80309-2-60-63-7, Oct. 18, 1963.

92. S. M. Scala and L. M. Gilbert, Thermal Degradation of a Char-Forming Plastic During Hypersonic Flight, *ARS Journal,* 32:917 (1962).

93. T. R. Munson and R. J. Spindler, Transient Thermal Behavior of Decomposing Materials, part I—General Theory and Application to Convective Heating, IAS Paper 62-30, IAS 30th Annual Meeting, New York, Jan. 22-24, 1962.

94. G. M. Dusinberre, Numerical Methods for Transient Heat Flow, *Trans. ASME,* 67:703 (1945).

95. P. J. Schneider, T. A. Dolton, and G. W. Reed, Charlayer Structural Response in High-performance Ballistic Re-Entry, AIAA Paper 66-424, June 1966.

96. R. D. Mathieu, Mechanical Spallation of Charring Ablators in Hyperthermal Environments, *AIAA Journal,* 2:1621 (1964).

97. C. L. Arne, Ablative Materials Subject to Combustion and Thermal Radiation Phenomena, *Douglas Paper* 1851, Jan. 1964.

98. R. J. Barriault and J. Yos, Analysis of the Ablation of Plastic Heat Shields that Form a Charred Surface Layer, *J. Am. Rocket Soc.,* 30:823 (1960).

99. J. W. Metzger, W. J. Engel, and N. S. Diaconis, Oxidation and Sublimation of Graphite in Simulated Re-Entry Environments, *AIAA Journal,* 5:451 (1967).

100. J. A. Moore and M. Zlotnick, Combustion of Carbon in an Air Stream, *J. Am. Rocket Soc.,* 31:1388 (1961).

101. S. M. Scala and N. S. Diacomis, The Stagnation Point Ablation of Teflon During Hypersonic Flight, *J. Aeronaut. Sci.*, **27**:140 (1960).
102. W. E. Welsch and K. E. Starner, Low Density Ablation Materials Survey, Air Force Ballistic Systems and Air Force Space Systems Division, *Aerospace Corp. Rept. SSD-TR-66-35 TDR-669* (6240-10)-5, Jan. 1966.
103. L. Steg, Materials for Re-Entry Heat Protection of Satellites, *ARS Journal*, **30**:815 (1960).
104. N. Beecher and R. E. Rosenweig, Ablation Mechanisms in Plastics with Inorganic Reinforcement, *J. Am. Rocket Soc.*, **31**:532 (1961).
105. M. Ladacki, J. V. Hamilton, and S. N. Cobz, Heat of Pyrolysis of Resin in Silica Phenolic Ablator, *AIAA Journal*, **4**:1798 (1966).
106. R. G. Wilson, Thermophysical Properties of Six Charring Ablators From 140° to 700°K, and Two Chars From 800° to 3000°K, *NASA TN D-2991*, Oct. 1965.
107. R. T. Swann, M. B. Dow, and S. S. Tomkins, Analysis of the Effects of Environmental Conditions on the Performance of Charring Ablators, AIAA Publication CP-9, AIAA Entry Technology Conference, Williamsburg & Hampton, Va., Oct. 12–14, 1964.
108. J. P. Hartnett and E. R. G. Eckert, Mass Transfer Cooling with Combustion in a Laminar Boundary Layer, *Proc. 1958 Heat Transfer and Fluid Mech. Inst.*, University of California, Berkeley, June 1958.
109. J. Cordero, F. W. Diederich, and H. Hurwicz, Aerothermodynamic Test Techniques for Re-Entry Structures and Materials, *Aerospace Eng.*, **22**:166 (1963).
110. C. E. Shepard, D. N. Ketner, and J. W. Vorreiter, High Enthalpy Plasma Generator for Entry Heating Simulation, *Proc. 13th Ann. Meeting*, Institute of Environmental Sciences, April 1967.
111. Anon., Arc Heaters and MHD Accelerators for Aerodynamic Purposes, part I, *AGARDograph* 84, Sept. 1964.
112. Anon., Hyperthermal Arc Facility, General Electric Co., Phila. (no date).
113. St. George A. Brown and J. B. Patton, Calibration and Operation of the Linde Model N4000 Arc Heater with 1/2- and 3/8-Inch-Diameter Constrictors, *AEDC-TR-65-102*, ARO, Inc., June 1965.
114. J. R. Stetson, Philco-Ford $H_2 F_2$ Test Facility, Part I, Description and Capabilities, Philco-Ford Corp. Space and Re-Entry Systems Division, 4462-67-84, March 1967.
115. Anon., Gas Thermodynamic and Transport Properties of Re-Entry Test Facilities, General Electric Co., Malta Test Station, *Rept. MD* 63-1, March 1963.
116. Anon., Wave Superheater Hypersonic Tunnel—Description and Capabilities, Cornell Aeronaut. Lab., May 1965.
117. Anon., Hyperthermal Test Facility, Giannini Scientific Corp., Santa Ana Division, *Bulletin* 15-009, June 1965.
118. S. Georgiev, The Relative Merits of Various Test Facilities with Regard to Simulation of Hypersonic Ablation Phenomena, *Proc. Nat. Symp. Hypervelocity Techniques*, Denver, Colo., IAS, pp. 162–174, Oct. 1960.
119. An Advanced Analytical Program for Charring Ablators—Final Report—Vol. 1, *AVSSD-0172-67-RR*.
120. G. P. Johnson and J. E. Wuerer, Re-Entry Vehicle Nose-Tip Environmental Simulation for Ablation Testing, *Douglas Paper* 4075, Aerospace Nose-Tip Tech. Symp., June 1966.
121. J. P. Brazel, R. A. Tanzilli, and R. A. Begany, Determination of the Thermal Performance of Char Under Heating Conditions Simulating Atmospheric Entry, AIAA Paper No. 65-640, Sept. 1965.

NOMENCLATURE

Letter symbol

a Attenuation coefficient for the radiant flux through transparent char used in Eqs. (84) to (89)

A, B Terms in Van Driest's solution for the local turbulent skin-friction coefficient defined in Eq. (46)

B' A rearranged form of the turbulent blowing factor seen in Eq. (35)

$B°$ The turbulent blowing factor accounting for injection of unlike gases seen in Eq. (38)

$\tilde{B}^{\circ}$ Pure turbulent mass-injection blowing factor as seen in Eq. (39)

C Chapman-Rubesin density viscosity parameter $\rho\mu/\rho_e\mu_e$

C_f Local skin-friction coefficients for turbulent boundary layer with dissociation, mass injection, and surface reactions included

C_{f_0} Local skin-friction coefficient for turbulent boundary layer accounting for dissociation only

$C_{f_0}(o)$ Local skin-friction coefficient for turbulent boundary layer without dissociation, mass injection, or surface reactions

C_H Local (dimensionless) pure convective heat transfer coefficient or Stanton numbers including dissociation, mass injection and surface reactions

C_{H_d} Local (dimensionless) diffusion (mass transfer) Stanton number

C_{H_0} Local (dimensionless) cold-wall heat transfer coefficient or Stanton number uncoupled from species solution

C_i Specie mass fraction or concentration of ith specie ρ_i/ρ

C_{p_f} Specific heat for composite mixture of boundary-layer species at constant pressure

C_{p_i} Specific heat for ith species at constant pressure

C_{p_s} Specific heat of condensed phase in the decomposing ablator

$C_{p_{\mathrm{atmos}}}$ Specific heat of mixture of boundary-layer species C_{p_f}

$\overline{C}_{pg}$ Specific heat for composite mixture of gaseous products of depolymerization at constant pressure

D_{12} Molecular binary-diffusion coefficient (laminar)

D_T Eddy binary-diffusion coefficient (turbulent) defined by Eq. (30)

E_{11} Activation energy used in Arrhenius' rate expression (Eq. (81)) for the representative solid-phase resin component ρ_{11}

E_{12} Activation energy used in Arrhenius' rate expression (Eq. (81)) for the representative completely volatile resin component ρ_{12}

E_2 Activation energy used in Arrhenius' rate expression (Eq. (81)) for the representative completely volatile nylon reinforcement ρ_2

E_{C_3} Activation energy for Arrhenius' rate expression describing rate limited surface erosion or removal by oxidation of graphite C_3

E_c Activation energy used in Arrhenius' rate expression describing a form of deposition interaction or coking for the ablation model (Eq. (82))

E^* Activation energy used in Arrhenius' rate expression (divided by gas const.) of Eq. (96)

$f_{(\eta)}$ Reduced (dimensionless) stream function used with Lee's Dorodnitsyn laminar similarity transformation such that $df/d\eta = u/u_e$

$f(o)$ Value of $f(\eta)$ evaluated at surface yielding mass-injection expression for the laminar-boundary layer (Eq. (21))

f_B Mass fraction of pyrolyzing material used in Eq. (96)

f_c Ratio of char density to virgin plastic density used in Eq. (81) of model ablator

f_1 Melt or evaporation fraction $\dot{m}_y/\dot{m}_w$ giving the ratio of the rate of evaporation to the total ablation rate in the liquid layer equation (Eq. (72))

F External radiation potential from "glowing" internal ablation reactions as a result of char transparency (Eq. (84)); also term defined by Eq. (102)

F^* Term defined by Eq. (103)

F_1 Term defined by Eq. (89)

$g(\eta)$ Reduced (dimensionless) total enthalpy function; refers to I/I_e in the laminar-boundary layer solutions

$g_f(\eta)$ Reduced (dimensionless) total "sensible" enthalpy function $I_f/\left(I_f\right)_e$ in the laminar-boundary layer solutions

$G(\eta; Z)$ Integral relation (or kernel) used in the solution of the similarity transformed laminar-boundary layer equations (Eq. (14))

h Enthalpy function defined as $\displaystyle\sum_i C_i\left(\int_0^T C_i\, dt + h_i^{\,o}\right)$

h_f Sensible enthalpy function defined as $\displaystyle\sum_i C_i \int_0^T C_{p_i}\, dt$

h_c Combined chemical enthalpy of all surface-coupled boundary-layer chemically reaction species; also heat of combustion per unit mass of oxygen as used in Eq. (97)

h_i Specie enthalpy function $\displaystyle\int_0^T C_{p_i}\, dt + h_i^{\,o}$

$h_i^{\,o}$ Chemical enthalpy of formation of the ith species

h_v Combined heat of melting and vaporization (also includes combustion and internal reaction effects if not otherwise accounted for) in the liquid layer expression (Eq. (72))

h_w Wall enthalpy (at ablator–boundary-layer interface)

$\bar{h}$ Heat transfer coefficient (Eq. (1))

H Lumped enthalpy of mass transfer and vaporization or sublimation term (Eq. (64))

H_e Total enthalpy at boundary layer edge as used in Eq. (72); same as I_e

H_{rp} Pyrolysis gas enthalpy as used in Eq. (92)

I Total enthalpy function defined as $\displaystyle\sum_i C_i\left(\int_0^t C_{p_i}\, dt + h_i^{\,o}\right) + \frac{u^2}{2}$

I_f Total "sensible" enthalpy function defined as $\displaystyle\sum_i C_i \int_0^t C_{p_i}\, dt + \frac{u^2}{2}$

J Variable in the "glowing" expression of Eqs. (85) to (90)

Je Defined by Eq. (87)

k Frequency factor used in Arrhenius' rate expression of Eq. (96); also exponent of γ_o to signify two dimensional ($K = o$) or axisymmetrical ($k = 1$)

k_{11} Frequency coefficient used in Arrhenius' rate expression (Eq. (81)) for the representative solid-phase resin component ρ_{11}

k_{12} Frequency coefficient used in Arrhenius' rate expression (Eq. (81)) for the representative completely volatile resin component ρ_{12}

k_2 Frequency coefficient used in Arrhenius' rate expression (Eq. (81)) for the representative completely volatile nylon reinforcement ρ_2

k_c Frequency coefficient used in Arrhenius' rate expression (Eq. (82)) describing a form of deposition interaction or coking for the ablation model

K_{o_e} Mass fraction of oxygen at the outer edge of the boundary layer as used by Eq. (97), same as $\left(C_o\right)_e$

L Molecular Lewis number (laminar); mass-transfer blocking parameter used in Eq. (92)

L_i Initial length of an ablating test model; see Eq. (101)

L_T Eddy (turbulent) Lewis number defined by $C_{pf}\overline{\zeta D_T}/\kappa_T$

$\dot{m}_{C_3}$ Mass flux due to surface erosion or removal by oxidation of graphite C_3 when in the rate-limited temperature regime

$\dot{m}_{c_s}, \dot{m}_s$ Mass flux due to solid-surface erosion as used by Eqs. (61) and (65)

$\dot{m}_g, \dot{m}_{g_s}$ Composite mass flux of gaseous products of decomposition

$\dot{m}_k$ Mass flux of ablation gas products, includes both $\dot{m}_{k'}$ and $\dot{m}_{k''}$

$\dot{m}_{k'}$ Mass flux of solid species of surface material

$\dot{m}_{k''}$ Mass flux of in-depth pyrolysis gas species

$\dot{m}_w$ Total injected mass flux from the ablating surface

$\dot{m}_r$ Mass flux due to mechanical erosion (liquid or solid)

M_e Local Mach number at boundary-layer edge

$\overline{M}_{BL}$ Mean molecular weight of boundary-layer species

$\overline{M}_{C_3}$ Molecular weight of graphite C_3 (Eq. (91))

$\overline{M}_{inj}$ Mean molecular weight of injected species

$\overline{M}_O$ Molecular weight of atomic oxygen O (Eq. (91))

$\overline{M}_{O_2}$ Molecular weight of molecular oxygen O_2 (Eq. (91))

n Order of reaction in the rate law expression (Eq. (96))

n_1 Pressure exponent appearing in rate law expression (Eq. (91))

n_{11} Reaction order exponent used in rate law expression for the representative solid-phase resin component ρ_{11} (Eq. (81))

n_{12} Reaction order exponent used in rate law expression for the representative completely volatile resin component ρ_{12} (Eq. (81))

n_2 Reaction order exponent used in rate law expression for the representative completely volatile nylon reinforcement ρ_2 (Eq. (81)); also reaction order exponent for oxidation of graphite by O in rate law expression (Eq. (91))

n_3 Reaction order exponent for oxidation of graphite by O_2 in rate law expression (Eq. (91))

n_c Local partial pressure of char deposition exponent in coking rate law expression (Eq. (82))

n_c' Reaction order exponent for a rate law expression describing a form of deposition for the ablation model (Eq. (82))

p Local pressure

p_c Local partial pressure of coking char material (Eq. (82))

p_e Local boundary-layer edge pressure

p^* Generalized pressure defined by Eq. (68)

P Molecular Prandtl number

P_T Eddy (turbulent) Prandtl number

$\dot{q}_{\text{blocking}}$ Heat transfer rate contribution due to "heat blockage" by mass transfer (Eq. (64))

$\dot{q}_{\text{chemical}}$ Heat transfer rate contribution due to chemical reactions in surface energy balance equation (Eq. (64))

$\dot{q}_o$ Cold wall pure convective heat transfer rate evaluated by Eckert's reference enthalpy method for expression Eq. (72) and Eq. (98); also stagnation point heat transfer rate

$\dot{q}_{or}$ Radiative plus convective heat transfer rate as used in Eq. (72)

$\dot{q}_r$ Net radiative heat transfer rate into surface $\dot{q}_{rr} - q^R$ or $-\epsilon_w \sigma T_w^4 + \alpha_w \dot{q}_{rad}$

$\dot{q}_{rad}$ Radiative heat flux to surface from environment

$\dot{q}_{rr}$ Heat flux reradiated from surface back to environment given as $-\epsilon_w \sigma T_w^4$

$\dot{q}^R$ Radiative heat flux absorbed by surface from environment given also as $\alpha_w \dot{q}_{rad}$

$\dot{q}_s$ Net heat transfer rate or flux through surface to the interior of ablator $= -(\kappa\, dt/dy)_w$ defined by Eqs. (63), (64), and (77)

$\dot{q}_w$ Aerodynamic heat transfer rate to the surface of the ablator from the boundary layer

r Recovery factor $\approx P^{1/2}$ (for lam. B.L.) and $\approx P^{1/3}$ (for turb. B.L.)

r_0^k Axial radius of body curvature in meridianal plane for the laminar boundary-layer solutions

R Universal gas constant

R_e Local Reynolds number used in turbulent boundary-layer solution (Eq. (44))

R_0 Reflectivity of the char at the surface of the ablator appearing in "glowing" term (Eq. (84))

R_y Reflectivity of the char material in depth at location of evaluation of dF/dy appearing in "glowing" term (Eq. (84))

s Special curvilinear coordinate parallel to the body curvature and boundary-layer flow streamwise; also used as surface erosion coordinate (Eq. (100))

$\bar{s}$ Similarity transformed (Lee's Dorodnitsyn) or "stretched" body coordinate used in laminar boundary-layer solutions

s Molecular Schmidt number

t Time

T Local temperature

T' Eckert's reference temperature

T_e Local temperature at boundary-layer edge

T_w Local "wall" or surface temperature

u Streamwise component of flow-field velocity vector v parallel to s direction

v Transverse component of flow-field velocity vector v normal to s direction and parallel to y direction

$\mathbf{V}$ Boundary-layer flow-field velocity vector

$\dot{w}_i$ Annihilation or creation of species (i.e., mass rate of change of specie) through chemical reactions within the boundary layer

W_i Initial weight of ablating test model, see Eq. (101)

x Same as s coordinate, used in liquid layer equations (Eqs. (67) to Eq. (75))

y Spacial curvilinear coordinate normal to s coordinate and transverse to ablator surface

z Dimensionless variable expressing ratio of local turbulent stream velocity u to its boundary layer edge value and used in the Von Kármán momentum integral, namely u/u_e

z_i Reduced (dimensionless) specie function; refers to $C_i / (C_i)_e$

Z Dimensionless parameter representing either S (Schmidt) or P (Prandtl) numbers in Eq. (14)

Greek symbol

α Absorptance in depth used in "glowing" expression (Eq. (84))

α_w Absorptance of surface of the ablator

α, β Terms in Van Driest's solution for the local turbulent skin friction coefficient defined in Eq. (45)

β Pressure gradient parameter, $2\bar{s}/Me \, dMe/d\bar{s}$ used in laminar similarity solutions, see Eq. (8)

β^* Laminar and turbulent boundary-layer mass injection parameter defined by Eq. (62)

γ Specific heat ratio C_p/C_i for ideal gas; also given as the char mass fraction of pyrolyzing material in Eq. (96)

Γ Volume fraction of a degradating polymer in a virgin ablator composite (represented as phenolic nylon for ablator model)

δ "Velocity" profile boundary layer thickness used in Eq. (51)

δ_c Char depth measured from the outer surface of an ablation test model; see Eq. (100)

δ^* Displacement thickness defined in Eq. (51)

Δb Ablator nodal thickness used in transient heat conduction solution described by Eq. (95)

ΔE_c Energy released by combustion reactions between the gaseous surface material, pyrolysis products, and environment combined into one term as used in Eq. (94)

ΔH Boundary layer enthalpy potential for the free stream gas as used in Eq. (92)

ΔH_c Enthalpy of deposition as used in Eq. (78)

ΔH_{dp} Enthalpy of depolymerization and pyrolysis as used in Eq. (78)

ΔH_{D_i} Energy absorbed by pyrolysis as used in Eq. (95)

ΔL Length loss in an ablation test model defined in Eq. (101)

ΔQ Net energy of decomposition plus sensible enthalpy with original temperature and state as reference (Eq. (94))

Δt Change in time for an ablation model defined in Eq. (100)

ΔW Weight loss for an ablation test model defined in Eq. (101)

ϵ_{ij} Defined as the mass fraction of element j in species i used in the Schvah-Zeldovich transformation

ϵ_w Ablator surface emissivity

η Similarity transformed (Lee's Dorodnitsyn) or reduced body coordinate y used in laminar boundary-layer solutions; also transpiration coefficient used in liquid-layer equation (Eq. (72))

θ Momentum thickness defined by Von Kármán's momentum integral (Eq. (42))

k Molecular (for gas species) thermal conductivity

k_T Eddy (turbulent) thermal conductivity as defined by Eq. (29)

λ Refers to the value of the velocity ratio U_λ (turbulent laminar-sublayer value) to U_e (fully turbulent edge value) used in Eq. (36)

μ Molecular viscosity coefficient

μ_T Eddy (turbulent) viscosity coefficient defined by Eq. (28)

ρ Local composite density of all species (gaseous: both boundary-layer and pyrolysis products)

ρ_{11} Instantaneous condensed phase density of the representative solid-phase resin component of the model ablator (Eq. (81))

ρ_{12} Instantaneous condensed phase density of the representative depolymerized and completely volatile resin component of the model ablator (Eq. (81))

ρ_2 Instantaneous condensed phase density of the representative depolymerized and completely volatile nylon reinforcement component of the model ablator (Eq. (81))

ρ_c Instantaneous condensed phase density due to a form of deposition represented by Eq. (82)

ρ_{dp} Instantaneous weighted sum of local polymer condensed phase densities $(\rho_{11}, \rho_{12}, \rho_2)$

ρ_i Density of the ith species in the boundary layer of the same unit volume as ρ

ρ_o Density of original ablative material before decomposition used in Eq. (96)

ρ_s Total instantaneous condensed phase densities of local polymer and deposition contributions (Eq. (83)

$\dot{\rho}_i$ Rate of change of density in the ith decomposing nodal lamina of an ablating material as described by Eq. (96)

$(\rho v)_c$ Mass flux of decomposing char material used in Eq. (92); same as $\dot{m}_c$, $\dot{m}_s$, $\dot{m}_{k'}$

$(\rho v)_{rp}$ Mass flux of pyrolysis gases used in Eq. (92); same as $\dot{m}_g$, $\dot{m}_{gs}$, $\dot{m}_k$

$(\rho v)_w$ Total ablation mass flux used in Eq. (92); same as $\dot{m}_w$

σ Stefan-Boltzmann constant used in Eq. (60)

σ' Eroded surface depth used in Eq. (84)

Subscripts

e Denotes boundary-layer edge value

f Denotes sensible enthalpy term

i Denotes ith species, initial value; or ith nodal laminar

k' Denotes ablator char solid-phase species

k'' Denotes ablator pyrolysis gas species

j, k, l Refers to orthagonal rectangular coordinates (three dimensional)

o Refers to stagnation point value or virgin material

r Denotes recover value

r_p Pyrolysis product

s Denotes condensed phase or net surface value

w Denotes "wall" or ablator surface value

∞ Denotes free stream value

$(')$ prime   Denotes d/dn

$(-)$ barred   Denotes temporal mean (turbulent B.L.)

Section **17**

Mass Transfer Cooling

J. P. HARTNETT
University of Illinois at Chicago Circle, Chicago, Ill.

INTRODUCTION

The term mass transfer cooling includes transpiration cooling, film cooling with a liquid, and film cooling with a gas as well as various ablation schemes. A pictorial representation of various forms of mass transfer cooling is given in Fig. 1. With the exception of film cooling with a gas all of the systems are physically similar. The major difference in methods 1*b* through 1*d* is that the mass transfer distribution may be independently controlled for the systems 1(*b*) and 1(*c*) while for the other systems the mass transfer rate is set by the thermodynamics of the system.

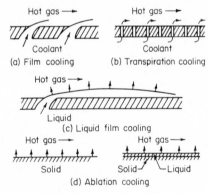

Fig. 1. Various mass transfer cooling schemes.

In light of these facts, the first section of this chapter will deal with transpiration cooling and then follow with a brief note on the applicability of these results to liquid film cooling and ablation. The chapter will conclude with a section of gaseous film cooling. Throughout the first section of the chapter the effects of suction will also be discussed.

A. TRANSPIRATION COOLING

1. Forced Convection Laminar Flow

a. The Flat Plate with Constant Properties. The system of differential equations describing the physical situation where a constant property gas flows in laminar motion over a porous flat plate with mass addition into the boundary layer (Fig. 1b) is

Continuity
$$\frac{\partial u}{\partial x} + \frac{\partial v}{\partial y} = 0 \tag{1}$$

Momentum
$$\rho u \frac{\partial u}{\partial x} + \rho v \frac{\partial u}{\partial y} = \mu \frac{\partial^2 u}{\partial y^2} \tag{2}$$

Energy
$$\rho c_p u \frac{\partial T}{\partial x} + \rho c_p v \frac{\partial T}{\partial y} = k \frac{\partial^2 T}{\partial y^2} + \mu \left(\frac{\partial u}{\partial y}\right)^2 \tag{3}$$

Species
$$\rho u \frac{\partial Y}{\partial x} + \rho v \frac{\partial Y}{\partial y} = \rho D_{12} \frac{\partial^2 Y}{\partial y^2} \tag{4}$$

Strictly speaking, in a constant property flow there is no need for the species equation. However, it can be assumed that the injected gas, although having the same transport and thermodynamic properties as the free stream gas, is given special identification (e.g., it may be an isotope of the free stream gas or it may be "tagged" with a radioactive tracer). The resulting mass fraction profile and the value of the mass fraction at the wall may be of value in actual binary flows when the free stream and the secondary gas are not markedly different in physical properties (e.g., nitrogen injected into air).

This system of equations requires seven boundary conditions. The following boundary conditions lead to a set of ordinary differential equations

$$\left.\begin{array}{c} u = 0 \\ T = T_w \\ v \sim u_e/\sqrt{Re_x} \end{array}\right\} \text{ at } y = 0 \tag{5}$$

$$\left.\begin{array}{c} u = u_e \\ T = T_e \\ Y = 0 \end{array}\right\} \text{ at } y \to \infty \tag{6}$$

The next and last boundary condition states that there is no net flow of boundary-layer fluid into the plate surface. This fixes the value of the mass fraction at the surface $Y_{y=0}$

$$v = -\frac{D_{12}}{1 - Y}\frac{\partial Y}{\partial y} \text{ at } y = 0 \tag{7}$$

Note that the resulting solution is restricted to the case where the distribution of the injected mass varies as $x^{-1/2}$. This blowing distribution is selected since it gives rise to a system of ordinary differential equations. For this physical system the resulting velocity distribution, local skin friction coefficient and local dimensionless heat transfer coefficients are shown in Figs. 2 to 4 [1].

For mass transfer to a laminar boundary on a flat plate the following conclusions may be drawn from Fig. 2:

1. Mass addition to a zero-pressure gradient boundary layer results in an S-shaped velocity profile. Since this is known to be an unstable type of profile. mass transfer is

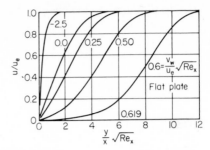

Fig. 2. Dimensionless velocity distribution u/u_e for constant property laminar flow over a flat plate for various values of the blowing parameter $(v_w/u_e)\sqrt{Re_x}$ [1].

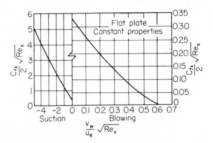

Fig. 3. Local skin-friction coefficient for constant property laminar flow over a flat plate for various values of the blowing parameter $(v_w/u_e)\sqrt{Re_x}$ [1].

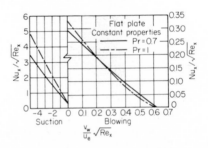

Fig. 4. Local dimensionless heat transfer coefficient $Nu_x/\sqrt{Re_x}$ for constant property laminar flow over a flat plate for various values of the blowing parameter $(v_w/u_e)\sqrt{Re_x}$ [1].

destabilizing (i.e., mass addition to a laminar boundary layer flow on a flat plate may cause the boundary layer to become turbulent). This conclusion does not apply to flows with favorable pressure gradient.

Conversely, removal of mass from a zero-pressure gradient boundary layer (i.e., suction) is stabilizing.

2. The maximum value of the dimensionless blowing parameter $(\rho_w v_w/\rho_e u_e)\sqrt{Re_x}$ is 0.619. Beyond this value the boundary-layer equations do not describe the flow.

The local shearing stress may be determined from Fig. 3 and the equation

$$\tau_w = \left(\frac{c_f}{2}\right)\rho u_e^2 \tag{8}$$

The local heat transfer rate is calculated as

$$q = h(T_w - T_{aw}) \tag{9}$$

The local heat transfer coefficient is given on Fig. 4 as a function of the blowing or suction rate. The adiabatic wall or recovery temperature T_{aw} is determined from the relation

$$T_{aw} = T_e + r\frac{u_e^2}{2c_p} = T_e\left(1 + r\frac{\gamma - 1}{2}Ma_e^2\right) \tag{10}$$

In Eq. (10) the temperatures are absolute values in °R or °K. The recovery factor r is given in Fig. 5. Note that the recovery temperature approaches the free-stream temperature for low-velocity flow and is equal to it in the limiting case of zero Mach number.

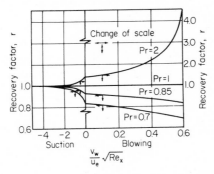

Fig. 5. Recovery factor for constant property laminar flow over a flat plate for various values of the blowing parameter $(v_w/u_e)\sqrt{Re_x}$ [1].

b. The Flat Plate with Variable Properties including Foreign Gas Injection into an Air-boundary Layer. The effect of variable physical properties is taken into account by the use of the Eckert reference method [2]. First the heat transfer coefficient, skin-friction coefficient and the recovery factor are determined for a solid surface exposed to the same free-stream conditions and held at the same surface temperature as the mass transfer cooled plate. The reference temperature T^* is first calculated from

$$T^* = T_w + 0.5\left(T_w - T_e\right) + 0.22\left(T_{aw_0} - T_e\right) \tag{11}$$

where

$$T_{aw_0} = T_e + r_0^*\frac{u_e^2}{2c_p^*} \tag{12}$$

and

$$r_0^* = \sqrt{\text{Pr*}} \tag{13}$$

The physical properties of the free-stream gas are known as functions of temperature and pressure and it is assumed that the wall temperature and the free-stream velocity and temperature are prescribed. The Prandtl number Pr* and the specific heat c_p^* are to be evaluated at the reference temperature T^*. An initial estimate of these two properties is made, leading to a value for T_{aw_0} and for T^*. New values of c_p^* and Pr* and T^* may now be determined since T^* is known. The calculation is repeated until a consistent set of values of c_p^*, Pr*, and T^* is achieved. The local skin-friction coefficient c_{f_0} and local Stanton number C_{H_0} are then calculated from

$$\frac{c_{f_0}}{2} = \frac{0.332}{\sqrt{u_e x/\nu^*}} \tag{14}$$

and

$$C_{H_0} = \frac{c_{f_0}}{2} (\text{Pr*})^{-2/3} \tag{15}$$

The reference temperature method has been shown to be valid for air, nitrogen, carbon dioxide, and hydrogen [3, 4].

The local skin-friction coefficient c_f and the local Stanton number C_H in the presence of mass transfer with air as the free-stream gas for various injectant gases are shown in Figs. 6 and 7 in a normalized form c_f/c_{f_0} and C_H/C_{H_0} as a function of the dimensionless mass transfer rate

$$\frac{\rho_w v_w}{\rho^* u_e} \sqrt{\frac{u_e x}{\nu^*}}$$

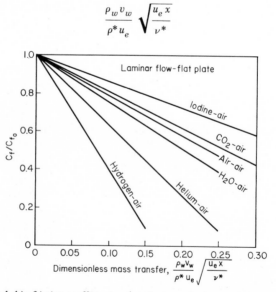

Fig. 6. Normalized skin friction coefficient c_f/c_{f_0} for transpiration cooling in a laminar boundary layer on a flat plate as a function of the dimensionless mass transfer for foreign gas injection into an air boundary layer [5].

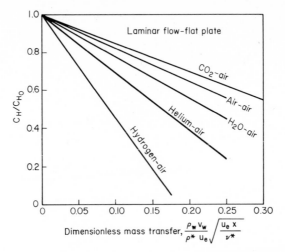

Fig. 7. Normalized Stanton number C_H/C_{H_0} for transpiration cooling in a laminar boundary layer on a flat plate as a function of the dimensionless mass transfer for foreign gas into an air boundary layer [5].

where ρ^* and ν^* are density and kinematic viscosity of the free stream gas evaluated at T^* [5]. The normalized recovery factor r/r_0 is given on Fig. 8 [5]. The local shearing stress and the local heat transfer rate are then calculated.

$$\tau_w = \frac{c_{f_0}}{2} \left(\frac{c_f}{c_{f_0}} \right) \rho^* u_e^2 \qquad (16)$$

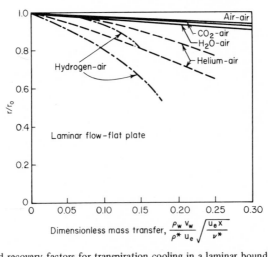

Fig. 8. Normalized recovery factors for transpiration cooling in a laminar boundary layer on a flat plate as a function of the dimensionless mass transfer [5].

$$q = C_{H_0}\left(\frac{C_H}{C_{H_0}}\right)\rho^* c_p^* u_e (T_w - T_{aw}) \tag{17}$$

The recovery temperature T_{aw} is calculated from

$$T_{aw} = T_e + \left(\frac{r}{r_0}\right) r_0^* \frac{u_e^2}{2 c_p^*} \tag{18}$$

For other coolant gases injected into an air boundary layer it is recommended that the following approximate formulas be used if the flow is laminar

$$\frac{c_f}{c_{f_0}} = 1 - 2.08 \left(\frac{m_2}{m_1}\right)^{1/3} \left(\frac{m^*}{\rho^* u_e} \sqrt{\frac{u_e x}{\nu^*}}\right) \tag{19}$$

$$\frac{C_H}{C_{H_0}} = 1 - 1.90 \left(\frac{m_2}{m_1}\right)^{1/3} \left(\frac{\dot{m}}{\rho^* u_e} \sqrt{\frac{u_e x}{\nu^*}}\right) \tag{20}$$

$$\frac{r}{r_0} = 1 \ \text{(or alternatively Fig. 11)} \tag{21}$$

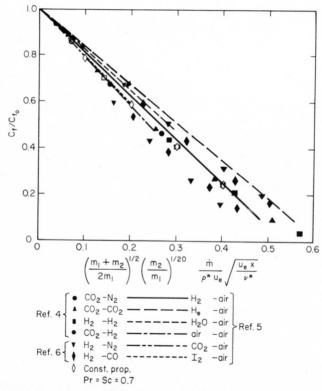

Fig. 9. Normalized skin-friction coefficient c_f/c_{f_0} for mass transfer into foreign gas-free streams on a flat plate as a function of the dimensionless mass transfer rate [4].

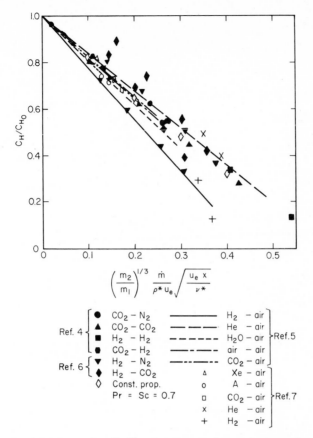

Fig. 10. Normalized Stanton number C_H/C_{H_0} for mass transfer into foreign gas-free streams on a flat plate as a function of the dimensionless mass transfer rate [4].

c. The Flat Plate with Variable Properties for Free-stream Gases Other than Air.

For free-stream gases other than air there are analytical predictions of normalized skin-friction coefficients, Stanton numbers, and recovery factors for carbon dioxide, hydrogen, and nitrogen free-streams. These results together with those reported above for an air free-stream are shown on Figs. 9 to 11. Table 1 summarizes the range of variables covered in these studies [4]. Again, all properties in the function plotted along the abscissa are those of the free-stream gas.

The wall mass fraction values which may be of value in the case of ablation are shown in Figs. 12 and 13 [4, 5].

The calculation procedure outlined in this section will yield reasonable estimates of the heat transfer provided that the wall temperature is considerably lower than the recovery temperature, a condition encountered in most flight applications. However, if the wall is close to the recovery temperature, and if the coolant gas is much lighter than the free-stream gas, the influence of diffusion thermo becomes important.

Influence of Diffusion Thermo. In those cases where the secondary gas differs from the main-stream gas there exists a mass flow due to the temperature gradient, the so-called thermal diffusion or Soret effect, and a heat flow due to the

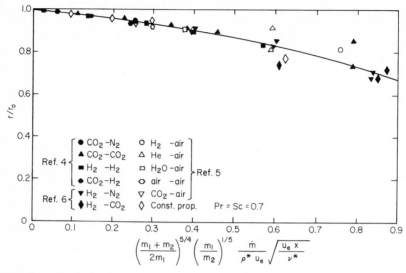

Fig. 11. Normalized recovery factor r/r_0 for mass transfer into a foreign gas-free stream on a flat plate as a function of the mass transfer rate [4].

TABLE 1. Gas Combinations and Flow Parameters Applicable to Figs. 9–11

	Gas pair	Ma_e	$T_e, °K$	T_w/T_e
Ref. 4	$CO_2 - N_2$	0, 4, 8, 12	218	2, 6
	$CO_2 - H_2$	0, 4, 8, 12	218	2, 6
	$CO_2 - CO_2$	0, 4, 8, 12	218	2, 6
	$H_2 - H_2$	0, 4, 8, 12	218	2, 6
Ref. 6	$H_2 - N_2$	0, 4, 8, 12	218	2, 6
	$H_2 - N_2$	4	555	1, 2, 3
	$H_2 - N_2$	0, 4	1110	0.5
	$H_2 - CO_2$	0, 4, 8, 12	218	2, 6
	$H_2 - CO_2$	4, 8	555	1, 2, 3
	$H_2 - CO_2$	0, 4	1110	0.5
Ref. 5	$H_2 - Air$	0	218	0.5, 2, 6
	$H_2 - Air$	12	218	2, 6
	$H_2 - Air$	0, 3, 6	218	1
	$He - Air$	0, 3, 6	218	1
	$H_2O - Air$	0, 3, 6	218	1
	$Air - Air$	0, 3, 6	218	1
	$CO_2 - Air$	0	218	1
	$I_2 - Air$	0	218	1
Ref. 7	$Xe - Air$	0	297, 1666	0.25, 1.1
	$A - Air$	0	297, 1666	0.25, 1.1
	$CO_2 - Air$	0	297, 1666	0.25, 1.1
	$He - Air$	0	297, 1666	0.25, 1.1
	$H_2 - Air$	0	297, 1666	0.25, 1.1

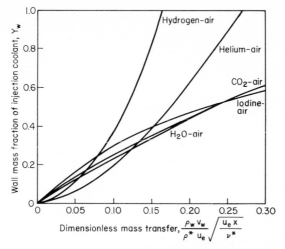

Fig. 12. Wall mass fraction of injected coolant for foreign gas injection into an air boundary layer on a flat plate as a function of the mass transfer rate [5].

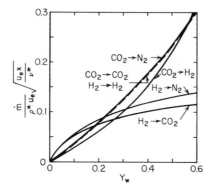

Fig. 13. Wall mass fraction of injected coolant for mass transfer into a foreign gas-free stream on a flat plate as a function of the mass transfer rate [4].

concentration gradient, the diffusion thermo or Dufour effect. The former is not of critical importance, but the latter is very important when the wall temperature is close to the recovery value (or equivalently if the wall temperature is close to the free-stream value under low-velocity conditions). The adiabatic wall temperature is defined as the temperature where the conduction heat flow is equal and opposite to the diffusion thermo

$$k_w \left(\frac{\partial T}{\partial y} \right)_w = \rho_w v_w \left(RT \frac{k_T}{Y} \right)_w \tag{22}$$

Calculations of this adiabatic wall temperature have been carried out for low-velocity conditions for an air free-stream for a number of injectant gases [7]. The results are shown on Fig. 14. In general, if a lightweight gas is injected, the recovery or adiabatic

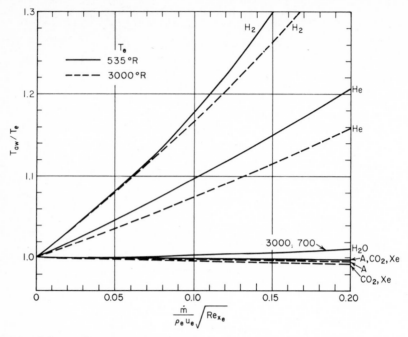

Fig. 14. Adiabatic wall temperatures resulting from diffusion thermo for various gases injected into a laminar air boundary layer on a flat plate [7].

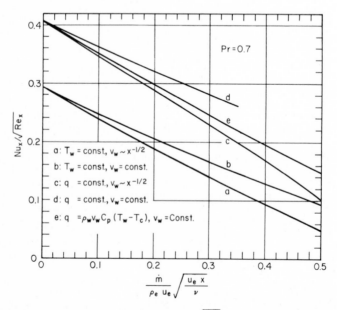

Fig. 15. Dimensionless heat transfer coefficient $Nu_x/\sqrt{Re_x}$ for constant property transpiration cooling on a flat plate as a function of the blowing rate for various boundary conditions [8].

wall temperature is greater than the free-stream temperature due to the diffusion thermo effect and, conversely, if a heavy coolant is used, the recovery temperature is lower than the free-stream value. For gases shown the heat transfer may still be calculated from Eq. (9) where the adiabatic wall temperature is determined from Fig. 14. For high velocities no results are available and caution is in order if the wall temperature is close to recovery conditions especially if lightweight coolants are to be used.

d. The Flat Plate—Nonsimilar Solutions. All of the aforementioned solutions assumed a special lengthwise distribution of the injected mass, namely $v_w \sim x^{-1/2}$, and further assumed a constant wall temperature since these assumptions lead to a system of ordinary differential equations. Other distributions of interest, such as constant injection along the length of the plate or constant heat input along the plate, have been studied by Sparrow for a constant property fluid with a Prandtl number 0.7. These results apply to such systems as air into air, N_2 into air, N_2 into N_2. Reference 8 gives the dimensionless local heat transfer coefficient $(hx/k)/\sqrt{(u_e x)/\nu}$ as a function of the dimensionless injection rate $(\dot{m}/\rho_e u_e)\sqrt{u_e x/\nu_e}$ on Fig. 15 for the following cases

a) T_w = constant $\qquad$ $v_w \sim x^{-1/2}$

b) T_w = constant $\qquad$ v_w = constant

c) q = constant $\qquad$ $v_w \sim x^{-1/2}$

d) q = constant $\qquad$ v_w = constant

e) $q = \dot{m}c_p(T_w - T_c)$ $\qquad$ v_w = constant

The last boundary condition is of special interest for it considers the realistic case when there is uniform mass injection along the length of the plate. The coolant leaves the reservoir at a constant temperature T_c as in Fig. 16, and the local wall temperature T_w and the local heat rate q are determined by the external flow conditions.

Figure 15 may also be used for variable properties by taking advantage of the Eckert reference procedure described earlier. If foreign gas coolants are used with boundary conditions of the type given in Fig. 15, it is suggested that the heat transfer be first calculated by the procedure developed earlier for the constant wall temperature case with $v_w \sim x^{-1/2}$. This result may then be corrected using Fig. 15 by assuming that the ratio of the heat transfer coefficients $h/(h)_{T_w = \text{constant}, v_w \sim x^{-1/2}}$ is the same for foreign gas injection as for constant property injection provided it is evaluated at the same value of $(\rho_w v_w/\rho^* u_e)\sqrt{u_e x/\nu^*}$.

Another case of engineering interest as described on Fig. 17 involves transpiration into a laminar boundary layer on a flat plate under the condition that there is a solid

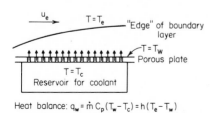

Heat balance: $q_w = \dot{m} C_p(T_w - T_c) = h(T_e - T_w)$

Fig. 16. Heat balance on a transpiration-cooled flat plate.

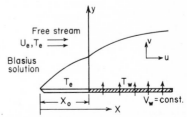

Fig. 17. Transpiration cooling with an unheated solid starting length.

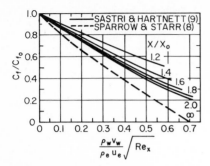

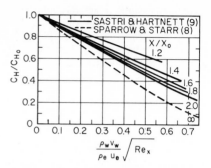

Fig. 18. Normalized skin-friction coefficients c_f/c_{f_0} for transpiration into a laminar boundary layer on a flat plate with an unheated starting length [9].

Fig. 19. Normalized Stanton numbers C_H/C_{H_0} for transpiration into a laminar boundary layer on a flat plate with an unheated starting length [9].

starting length. The solutions for the reduced skin-friction coefficients and Stanton numbers are presented on Figs. 18 and 19 [9]. Although the solutions were obtained for a Prandtl number of unity they may be applied to a Prandtl number of 0.7 with engineering accuracy.

e. Plane Stagnation Flow with Constant Properties. The velocity distributions, skin-friction coefficients, and dimensionless heat transfer coefficients for constant property laminar flow in the immediate vicinity of a stagnation line on a two-dimensional body are shown on Figs. 20 to 22 [1]. The blowing (or suction) is constant (i.e., v_w = constant) and the wall is at a constant temperature. It is important to note that in contrast to the flat-plate case the velocity distributions do not reveal inflection points with mass injection. It would appear that blowing is not destabilizing in the presence of a favorable pressure gradient. Further the velocity boundary layer remains attached even at large blowing rates and the skin-friction coefficient remains finite. On the other hand the heat transfer goes to zero which means that the thermal boundary lifts off the surface. The shearing stress may be calculated from Eq. (8) and Fig. 21. The heat transfer is calculated from Eq. (9) using Fig. 22 for the determination of the heat transfer coefficient. Here it should be noted that T_{aw} is equal to the free-stream temperature T_e.

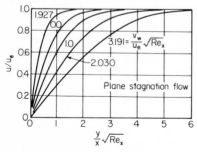

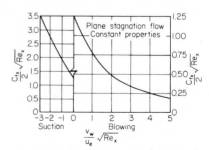

Fig. 20. Velocity distributions for constant property transpiration cooling in the neighborhood of a plane stagnation point [1].

Fig. 21. Dimensionless skin-friction coefficient as a function of the mass transfer rate for constant property transpiration cooling in the neighborhood of a plane stagnation point [1].

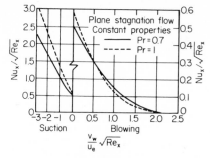

Fig. 22. Dimensionless heat transfer coefficient Nu_x/Re_x as a function of the mass transfer rate for constant property transpiration cooling in the neighborhood of a plane stagnation point [1].

f. Plane Stagnation Flow with Variable Properties including Diffusion Thermo Effects. Studies have been carried out for the injection of several foreign gases in the region of the plane stagnation point. It has been found for light gases such as hydrogen or helium that the influence of diffusion thermo becomes very important in the case where the wall temperature is close to the free-stream temperature, a situation which may frequently be encountered, particularly in wind-tunnel studies. The heat transfer is again calculated from Eq. (9)

$$q = h(T_w - T_{aw}) \tag{9}$$

In the presence of diffusion thermo the adiabatic wall temperature (corresponding to the case where there is zero net heat transfer to the wall) is not equal to the free-stream temperature even at low velocities. For light gases the adiabatic wall temperature may be considerably higher than the free-stream value. The predicted adiabatic wall results for helium injected into an air boundary layer are given on Fig. 23 for free-stream temperatures ranging from 570 to 5000°K. Good agreement with

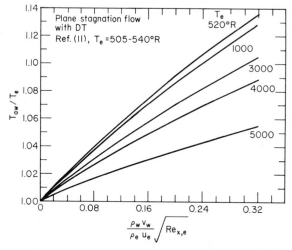

Fig. 23. Adiabatic wall temperatures resulting from diffusion thermo when helium is injected into a plane stagnation flow of air.

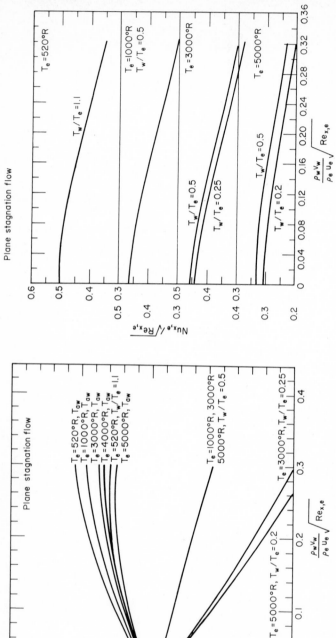

Fig. 25. Dimensionless heat transfer coefficients as a function of the mass transfer rate for helium injected into a plane stagnation flow of air.

Fig. 24. Normalized skin-friction coefficients as a function of the mass transfer rate for helium injected into a plane stagnation flow of air.

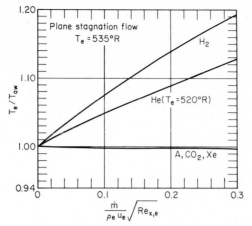

Fig. 26. Adiabatic wall temperatures resulting from diffusion thermo for various gases injected into a plane stagnation flow of air [12].

experimental results [11] is demonstrated. The corresponding normalized skin-friction coefficient and Nusselt numbers are given on Figs. 24 and 25 for a wide range of free-stream and wall-temperature levels. Although helium transpiration in general reduces the Nusselt number, it may result in an increase in the skin friction. An increase in skin friction occurs when the wall temperature is greater than or very nearly equal to the adiabatic wall temperature. On the other hand, when the wall is highly cooled, helium transpiration reduces the skin-friction coefficient over its solid wall value.

Similar results are available for hydrogen, argon, carbon dioxide, and xenon injected into an air boundary layer, although the range of wall temperatures and free-stream temperatures is limited [12]. These are shown in Figs. 26 to 28. In general it is only the lighter gases which demonstrate an appreciable influence on the adiabatic wall temperatures. Figures 29 and 30 give representative values of the wall concentration of the injected foreign gas as a function of the blowing rate.

g. Axisymmetric Stagnation Flow. For a constant property flow in the region of a stagnation point on a three-dimensional body, the solid wall Nusselt value is given by the relation

$$\left(\frac{hx}{k}\right) = 0.767\sqrt{\frac{u_e x}{\nu}}\ (\text{Pr})^{0.4} \tag{23}$$

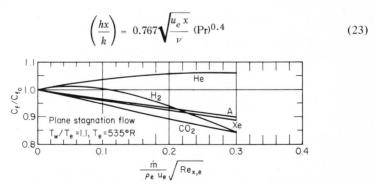

Fig. 27. Normalized skin-friction coefficient as a function of mass transfer for foreign gas injection into a plane stagnation flow of air [12].

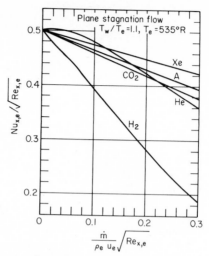

Fig. 28. Dimensionless heat transfer coefficient as a function of the mass transfer rate for foreign gas injected into a plane stagnation flow of air [12].

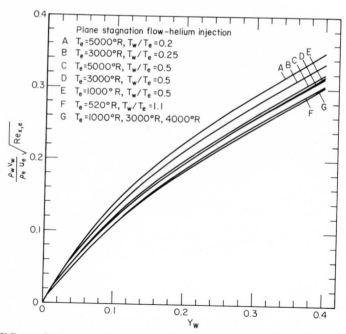

Fig. 29. Wall mass fraction of helium as a function of mass transfer rate for transpiration cooling in a plane stagnation flow of air [10].

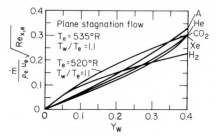

Fig. 30. Wall mass fraction of foreign gas as a function of mass transfer rate for transpiration cooling in a plane stagnation flow of air [10].

The influence of mass transfer on this value is given on Fig. 31 for a Prandtl number of 0.7 [13]. The heat transfer is calculated from Eq. (9) with the adiabatic wall temperature T_{aw} being equal to the free-stream temperature T_e for low-speed flow and to the total temperature for high-speed flow.

In a binary system the same difficulties arise as in the two-dimensional stagnation flow, namely the influence of diffusion thermo. The adiabatic wall temperature, skin-friction coefficients, and Nusselt numbers for helium injected into an air-boundary layer are presented in Figs. 32 through 34 [10]. Other injectant gases are shown in Figs. 35 through 38 [12]. The heat transfer is now calculated from Eq. (9) using the appropriate value of T_{aw} determined from Fig. 32. Wall mass fraction values for the injected coolant are presented on Figs. 39 and 40.

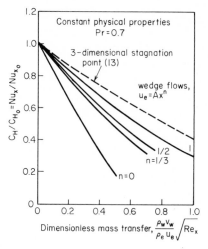

Fig. 31. Dimensionless Stanton or Nusselt numbers $(C_H/C_{H_0}) = Nu_x/Nu_{x_0}$ as a function of the mass transfer rate for constant property laminar flow in the region of a three-dimensional stagnation point and for laminar wedge flows [5].

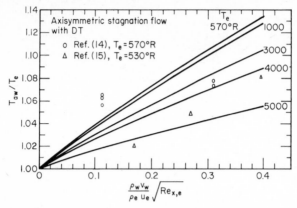

Fig. 32. Adiabatic wall temperatures for helium transpiration into an air boundary layer in the region of an axisymmetric stagnation flow of air [10].

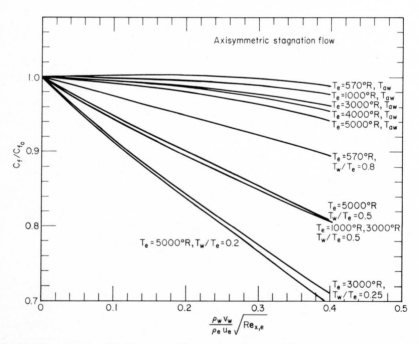

Fig. 33. Dimensionless skin-friction coefficients c_f/c_{f_0} for helium transpiration into an air boundary layer in the region of an axisymmetric stagnation point as a function of the blowing rate [10].

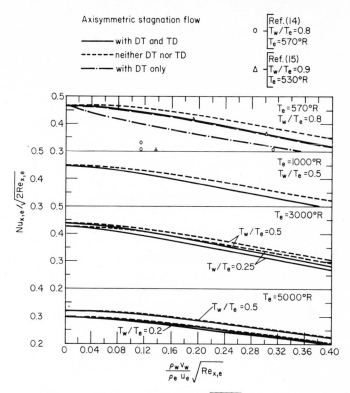

Fig. 34. Dimensionless heat transfer coefficients $\mathrm{Nu}_{x,e}/\sqrt{2\mathrm{Re}_{x,e}}$ for helium transpiration into an air boundary layer in the region of an axisymmetric stagnation point as a function of the blowing rate [10].

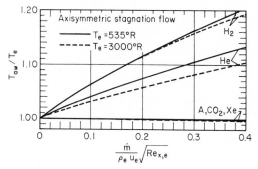

Fig. 35. Adiabatic wall temperature for transpiration of foreign gases into an axisymmetric stagnation flow of air as a function of blowing rate [12].

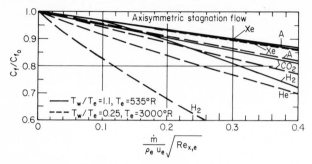

Fig. 36. Reduced skin-friction coefficients c_f/c_{f_0} for transpiration of foreign gases into an axisymmetric stagnation flow of air as a function of blowing rate [12].

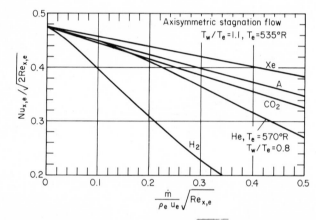

Fig. 37. Dimensionless heat transfer coefficient $Nu_{x,e}/\sqrt{2Re_{x,e}}$ for transpiration of foreign gases into an axisymmetric stagnation flow of air as a function of blowing rate $(T_W/T_e) = 1.1$, $T_e = 535°R$ [12].

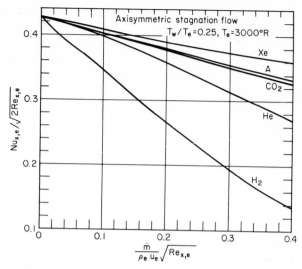

Fig. 38. Dimensionless heat transfer coefficient $\mathrm{Nu}_{x,e}/\sqrt{2\mathrm{Re}_{x,e}}$ for transpiration of foreign gases into an axisymmetric stagnation flow of air as a function of blowing rate $(T_w/T_e) = 0.25$, $T_e = 3000°\mathrm{R}$ [12].

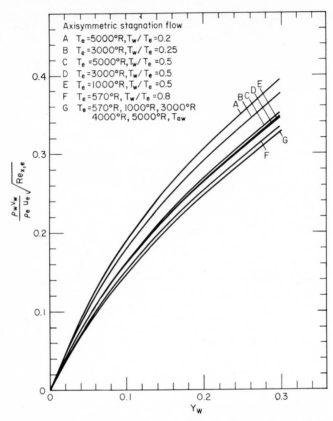

Fig. 39. Wall mass fraction of helium as a function of the blowing rate for transpiration cooling in an axisymmetric stagnation flow of air [10].

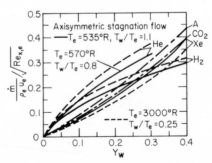

Fig. 40. Wall mass fraction of injected foreign gas as a function of the blowing rate for transpiration cooling in an axisymmetric flow of air [12].

h. Wedge Flows, $u_e = cx^n$, T_w = constant. The general class of flows described as wedge flows or Falkner-Skan flows [16] which treat low-velocity laminar flow over a two-dimensional wedge, have been analyzed for constant physical properties including the influence of mass transfer [17]. The Nusselt number for the solid-wall case for constant wall temperature is shown as a function of the exponent n (or equivalently the wedge angle $\beta\pi$) on Fig. 41. The normalized Stanton or Nusselt value is given on Fig. 31 as a function of the blowing rate [17].

i. Generalized Flows, Two-dimensional and Axially Symmetric Three-dimensional Flows with Mass Transfer. Heat transfer and skin-friction coefficients are available [18] for the case where the free-stream velocity is given by

$$u_e \sim \sqrt{\frac{T_e}{T_0}}\, \xi^m \tag{24}$$

and

$$\beta = \frac{2\xi}{u_e}\frac{du_e}{d\xi}\left(\frac{T_0}{T_e}\right) = 2m \tag{25}$$

Here

$$\xi = \int_0^x \rho_w \mu_w u_e r_w^{2j}\, dx \tag{26}$$

$j = 0$ for two-dimensional flow

$j = 1$ for axisymmetric three-dimensional flow

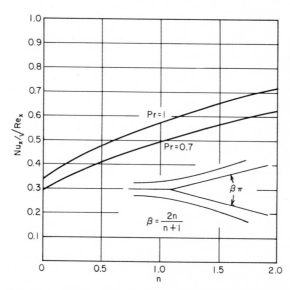

Fig. 41. Dimensionless heat transfer coefficients $Nu_x/\sqrt{Re_x}$ for constant property wedge flows $u_e = cx^n$ as a function of n.

Although rigorously developed for the system of boundary conditions which lead to similar solutions (i.e., the normal velocity $v_w \sim (u_e/x)^{1/2}$ and T_w is constant), it is proposed [19] that the following simplified graphical representation be used as an engineering approximation for other cases. It is assumed that the free-stream velocity and temperatures are given and that the wall temperature is a prescribed constant value and further that the blowing distribution is specified. The results may be used for variable property conditions for air-air, nitrogen-nitrogen, CO_2 into CO_2 etc.

First determine the dimensionless local parameters $\bar{\xi}$ and β

$$\bar{\xi} = \frac{\int_0^x \rho_w \mu_w u_e r_w^{2j} dx}{\rho_w \mu_w u_e r_w^{2j} x} \tag{27}$$

and

$$\beta = 2\bar{\xi}\left(\frac{x}{u_e}\frac{du_e}{dx}\right)\frac{T_0}{T_e} \tag{28}$$

where $j = 0$ for two-dimensional flows
$j = 1$ for axisymmetric three-dimensional flow
and x is the distance from the leading edge or from the stagnation point. The constant-property solid-wall Nusselt values, skin-friction coefficients, and recovery factors may then be determined from Figs. 42 to 45. The starred quantities indicate that all physical properties are to be evaluated at the reference temperature T^* as mentioned above

$$T^* = T_w + 0.5(T_w - T_e) + 0.22(T_{aw_0} - T_e) \tag{11}$$

$$T_{aw_0} = T_e + r_0^* \frac{u_e^2}{2c_p^*} \tag{12}$$

where r_0^* is obtained from Fig. 45. The influence of variable physical properties on the solid-wall skin friction and heat transfer is determined from Figs. 46 and 47. Here σ is the Eckert or dissipation parameter

$$\sigma = \frac{u_e^2}{2H_e} = \frac{u_e^2}{2c_{pe}T_0} = \frac{[(\gamma_e - 1)/2]Ma_e^2}{1 + [(\gamma_e - 1)/2]Ma_e^2} \tag{29}$$

and $t_w = T_w/T_0$, the ratio of the local absolute wall temperature to the absolute local free-stream total temperature, while ω is the viscosity temperature exponent $\mu \sim T^\omega$. For air the value of ω is approximately 0.7 for typical wind-tunnel conditions and 0.5 for conditions encountered in hypersonic flight.

Finally, the influence of the blowing on the heat transfer, skin-friction coefficient, and recovery factors are given in Figs. 48 to 52. The local skin-friction coefficient and

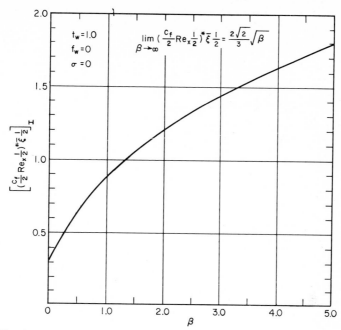

Fig. 42. Skin-friction coefficient as a function of the dimensionless pressure gradient β for constant property flows [19].

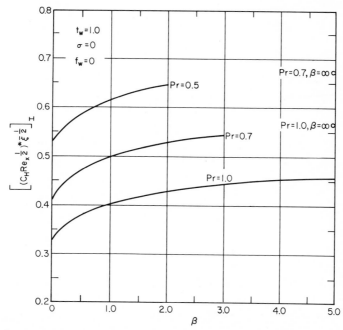

Fig. 43. Stanton number as a function of the dimensionless pressure gradient β for constant property flows [19].

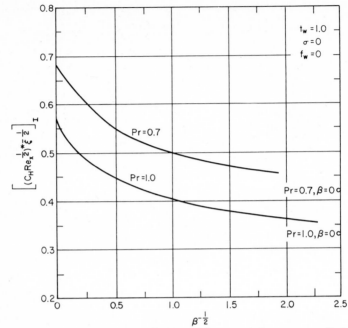

Fig. 44. Solid-wall Stanton number as a function of the asymptotic pressure gradient parameter $\beta^{-1/2}$ [19].

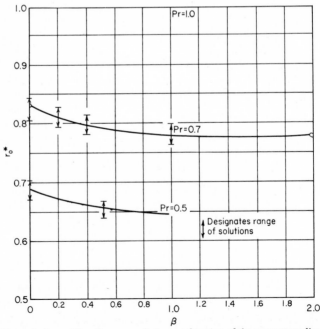

Fig. 45. Dimensionless solid-wall recovery factor as a function of the pressure gradient β [19].

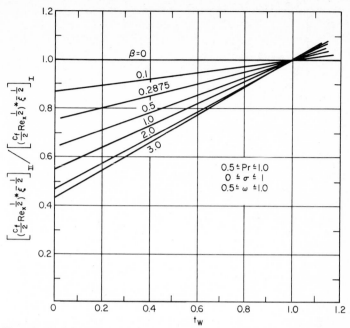

Fig. 46. Normalized solid-wall skin-friction coefficient as a function of wall temperature ratio t_w and pressure gradient β [19].

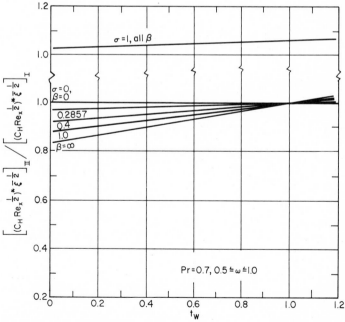

Fig. 47. Normalized solid-wall Stanton number as a function of the wall temperature ratio t_w, pressure gradient β, and the Eckert number σ [19].

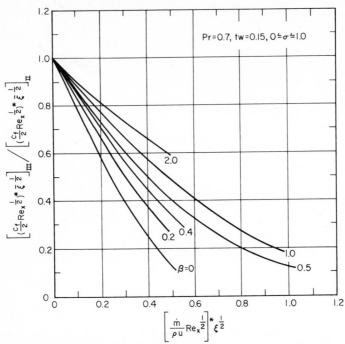

Fig. 48. Normalized skin-friction coefficient as a function of the mass transfer rate and pressure gradient for a temperature ratio of 0.15 [19].

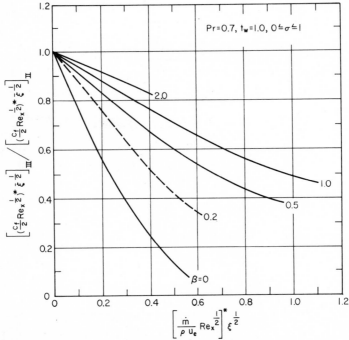

Fig. 49. Normalized skin-friction coefficient as a function of the mass transfer rate and pressure gradient for a temperature ratio of 1.0 [19].

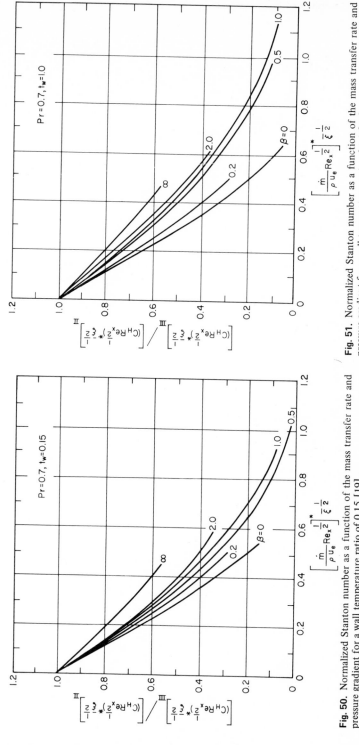

Fig. 51. Normalized Stanton number as a function of the mass transfer rate and pressure gradient for a wall temperature ratio of 1.0 [19].

Fig. 50. Normalized Stanton number as a function of the mass transfer rate and pressure gradient for a wall temperature ratio of 0.15 [19].

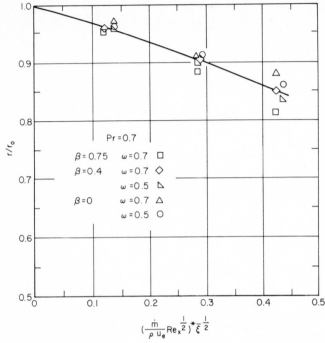

Fig. 52. Normalized recovery factor as a function of the mass transfer rate for a Prandtl number of 0.7 [19].

heat transfer coefficient are then determined. Here the following special subscript notation is used:

I — Corresponds to constant property solid wall conditions, $t_w = 1$, $\sigma = 0$, $f_w = 0$

II — Corresponds to variable property solid wall conditions, t_w, σ, $f_w = 0$ where t_w and σ are specified

III — Corresponds to variable property conditions with mass transfer, but with the wall held at the same temperature and with the same free-stream conditions as II, t_w, σ, f_w

$$
\frac{c_f^*}{2} = \underbrace{\left(\frac{c_f^*}{2} \sqrt{Re_x^*} \sqrt{\overline{\xi}} \right)_I}_{\text{Fig. 42}} \cdot \underbrace{\frac{\left[(c_f^*/2) \sqrt{Re_x^*} \sqrt{\overline{\xi}} \right]_{II}}{\left[(c_f^*/2) \sqrt{Re_x^*} \sqrt{\overline{\xi}} \right]_I}}_{\text{Fig. 46}} \cdot \underbrace{\frac{\left[(c_f^*/2) \sqrt{Re_x^*} \sqrt{\overline{\xi}} \right]_{III}}{\left[(c_f^*/2) \sqrt{Re_x^*} \sqrt{\overline{\xi}} \right]_{II}}}_{\text{Figs. 48 \& 49}} \cdot \frac{1}{\sqrt{Re_x^*} \sqrt{\overline{\xi}}} \tag{30}
$$

and

$$
C_H^* = \underbrace{\left(C_H^* \sqrt{Re_x^*} \sqrt{\overline{\xi}} \right)_I}_{\text{Fig. 43 or 44}} \cdot \underbrace{\frac{\left(C_H^* \sqrt{Re_x^*} \sqrt{\overline{\xi}} \right)_{II}}{\left(C_H^* \sqrt{Re_x^*} \sqrt{\overline{\xi}} \right)_I}}_{\text{Fig. 47}} \cdot \underbrace{\frac{\left(C_H^* \sqrt{Re_x^*} \sqrt{\overline{\xi}} \right)_{III}}{\left(C_H^* \sqrt{Re_x^*} \sqrt{\overline{\xi}} \right)_{II}}}_{\text{Figs. 50 \& 51}} \cdot \frac{1}{\sqrt{Re_x^*} \sqrt{\overline{\xi}}} \tag{31}
$$

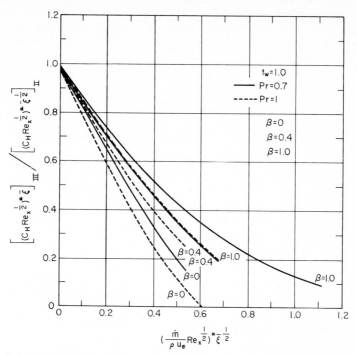

Fig. 53. Influence of Prandtl number on reduced Stanton number as a function of mass transfer rate and pressure gradient [19].

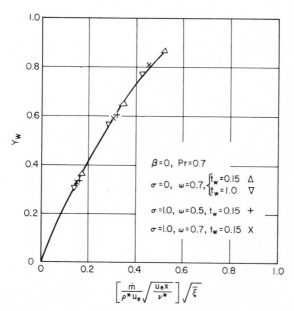

Fig. 54. Wall mass fraction value as a function of the dimensionless mass transfer rate for $\beta = 0$ and Prandtl number $= 0.7$ [19].

The local shearing stress and heat transfer are then calculated

$$\tau_w = \frac{c_f^*}{2} \rho^* u_e^2 \tag{8a}$$

$$q = C_H^* \rho^* c_p^* u_e (T_w - T_{aw}) \tag{17}$$

and

$$T_{aw} = T_e + \left(\frac{r}{r_0}\right) \cdot (r_0) \cdot \frac{u_e^2}{2c_p^*} \tag{18}$$

$$\underbrace{\hspace{1.5cm}}_{\text{Fig. 52}} \underbrace{\hspace{1.5cm}}_{\text{Fig. 45}}$$

It may be noted that these solutions, given on Figs. 42 to 52, are applicable to low-velocity two-dimensional flow $u_e = cx^n$ where the following relationships apply

$$\bar{\xi} = \frac{1}{n+1} \tag{32}$$

$$n = \frac{\beta}{2-\beta} \quad \text{or} \quad \beta = \frac{2n}{1+n} \tag{33}$$

Thus the flat-plate case $(n = 0)$ corresponds to $\beta = 0$ and $\bar{\xi} = 1$ and the plane stagnation case $(n = 1)$ to $\beta = 1$ and $\bar{\xi} = 1/2$. This particular class of wedge low-velocity flows is confined to values of β less than 2 which corresponds to an infinite pressure gradient condition. The same restriction does not apply to the more general class of flows given by $u_e \sim \sqrt{(T_e/T_0)}\,\xi^m$ and any value of β is possible.

Further it may be noted that the axisymmetric stagnation point is also given on these figures. In this case $u_e \sim x$ and $r_w \sim x$ and thus $\bar{\xi} = 1/4$ and $\beta = 1/2$.

There is little influence of Prandtl number over the range of Pr = 0.5, to 1.0 on the reduced skin-friction values shown in Figs. 46, 48, and 29. However, the heat transfer is sensitive to the Prandtl number as may be seen on Fig. 53.

The knowledge of the wall mass fraction as a function of blowing rate is useful for ablation and evaporation studies. This information is given on Figs. 54 to 56 for pressure gradients of $\beta = 0, 0.5$, and 1.0.

j. Supersonic Laminar Flow over a Cone. Supersonic laminar flow over a cone gives rise to constant pressure along the surface of the cone and the heat transfer results are obtainable from Figs. 38 to 46. In this case u_e = constant and $r_w \sim x$. Thus $\bar{\xi} = 1/3$ and $\beta = 0$.

k. Laminar Flow over a Circular Cylinder. Analytical and experimental results are available for the transpiration of air into the laminar boundary layer on the forward region of a circular cylinder [20, 21]. The analysis and experiment both correspond to a constant blowing distribution and a constant coolant temperature within the reservoir. The surface temperature and heat flux are then determined by free-stream conditions. Figure 57 shows the dimensionless temperature distribution along the surface of the cylinder for various blowing rates. There are two different curves for each condition, one corresponding to the velocity distribution proposed by Hiemenz and the other found by Schmidt and Wenner. Comparable results for the Nusselt number are given on Fig. 56. Other methods including the one described in Section 8 are available for determining the heat transfer for laminar flow over cylinders and other two-dimensional geometries [23].

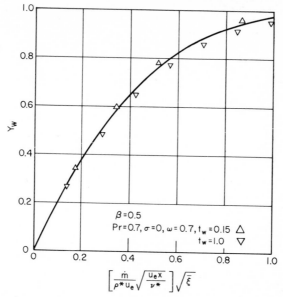

Fig. 55. Wall mass fraction value as a function of the dimensionless mass transfer rate for $\beta = 0.5$ and Prandtl number = 0.7 [19].

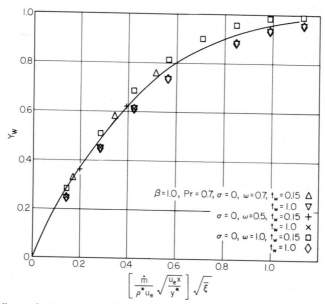

Fig. 56. Wall mass fraction value as a function of the dimensionless mass transfer rate for $\beta = 1.0$ and Prandtl number = 0.7 [19].

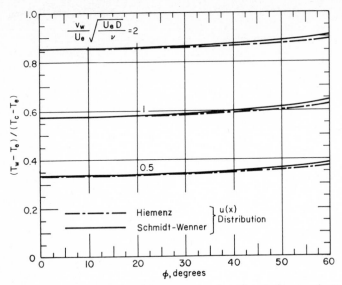

Fig. 57. Dimensionless temperature distribution as a function of angle for several rates of mass transfer through a porous cylinder in cross flow [20].

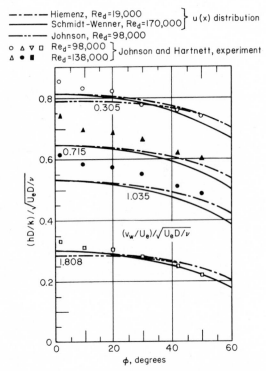

Fig. 58. Analytical and experimental heat transfer results for mass transfer from a porous cylinder in cross flow [21].

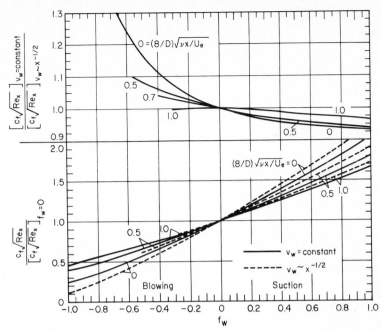

Fig. 59. Influence of curvature and mass transfer on the skin-friction coefficient for laminar flow along a porous cylinder [24].

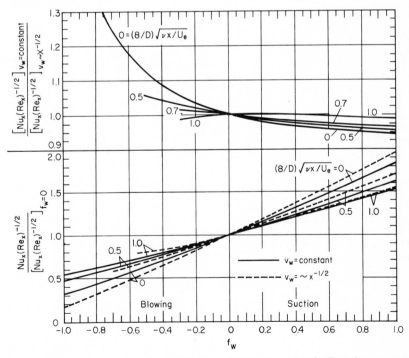

Fig. 60. Influence of curvature and mass transfer on the Nusselt number for flow along a porous cylinder [25].

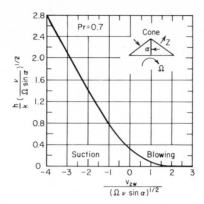

Fig. 61. Nusselt number as a function of mass transfer for a rotating cone.

l. Longitudinal Flow along a Circular Cylinder. The influence of curvature on the skin-friction and heat transfer coefficients for longitudinal flow along a circular cylinder is given in Figs. 59 [24] and 60 [25] for the case of mass transfer. Two blowing distributions are considered, one corresponding to the similar solution where $v_w \sim x^{-1/2}$ and the other to the realistic but nonsimilar case where v_w is a constant. The solid-wall values of the heat transfer and skin friction for $v \sim x^{-1/2}$ may be obtained from Figs. 41 to 44. The heat transfer results although strictly applicable to a Prandtl number of 0.7 should yield reasonable estimates of curvature effects over the Prandtl number range of 0.5 to 1.0.

m. Rotating Cone. The heat transfer coefficient for a porous cone rotating in an otherwise quiescent fluid with mass transfer to or from the surface is given on Fig. 61 for the constant property laminar case with a Prandtl number of 0.7 [26, 27]. The normal velocity at the surface $v_{z,w}$ is assumed proportional to $(\Omega \nu)^{1/2}$ and the surface temperature is constant. The special case of a rotating disk is obtained by setting $\sin \alpha = 1$. The local heat transfer rate is determined from the relation $q = h(T_w - T_e)$. These same references [26, 27] contain information on the local shearing stress and torque.

2. Free-convection Laminar Flow

a. The Horizontal Circular Cylinder. Cess has treated the influence of mass transfer on free convection for a single component system in Sec. 6. If a foreign gas is introduced into an air boundary layer the influence of thermo diffusion discussed earlier in this chapter becomes of critical importance. Experiments and analytical studies have been carried out for helium, hydrogen, carbon dioxide, and Freon-12 injected into a free-convection air boundary layer in the vicinity of the plane stagnation point of a horizontal circular cylinder [28]. The stagnation point is at the lowest point on the cylinder circumference for helium and hydrogen injection while for carbon dioxide and Freon-12 it occurred at the highest point on the circumference. This difference occurred since the temperature differences and blowing rates were such that the buoyancy forces created an upflow for the first two gases and conversely a downflow was created for carbon dioxide and Freon-12 injection. The adiabatic wall temperatures are given on Fig. 62 while the Nusselt numbers are shown on Figs. 63 and 64. Some data are also presented at positions around the porous cylinder. The local heat transfer rate is calculated from the relation $q = h(T_w - T_{aw})$. Results are also available for water vapor and argon injected into the plane stagnation region of an air free convection boundary layer [12, 31].

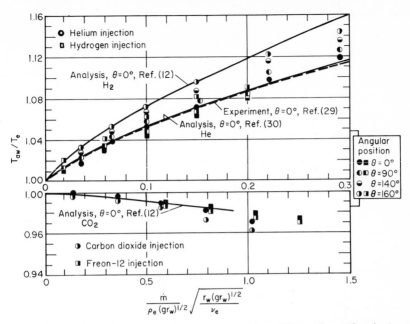

Fig. 62. Dimensionless adiabatic wall temperatures resulting from diffusion thermo for a horizontal cylinder in free convection [28].

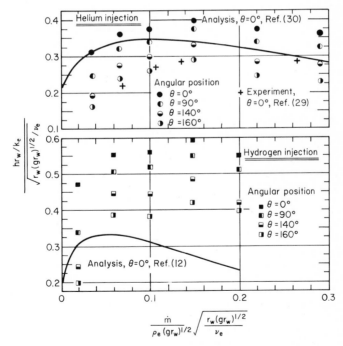

Fig. 63. Dimensionless Nusselt number as a function of mass transfer for helium and hydrogen injected into a free-convection air boundary layer on a horizontal cylinder [28].

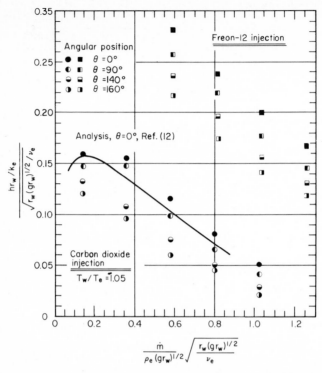

Fig. 64. Dimensionless Nusselt number as a function of mass transfer for Freon-12 and carbon dioxide injected into a free-convection air boundary layer on a horizontal cylinder [28].

3. Forced-convection Turbulent Flow

a. The Flat Plate with Constant Properties, Air to Air. The experimental studies of W. M. Kays and his colleagues [32, 33] and of Torii et al. [34] provide a basis for predicting heat transfer and skin friction in turbulent low-speed air-boundary layers on a flat plate geometry with blowing or suction. The local Stanton numbers are shown on Figs. 65 and 66 as a function of the length Reynolds number Re_x for a wide range of blowing and suction values F. The values shown are applicable for uniform blowing F = constant and $F \simeq x^{-0.2}$. For the constant blowing or suction distribution with a constant temperature of the coolant in the reservoir the wall temperature varies slightly along the length of the plate. The $F \simeq x^{-0.2}$ distribution leads to an isothermal condition along the plate surface. Similar results for the skin-friction coefficient are given in Fig. 67 and they may be used for F = constant or for $F \simeq x^{-0.2}$.

The local Stanton number is also given in Fig. 68 as a function of the enthalpy thickness Reynolds number Re_Δ and the blowing parameter F. Here Re_Δ is defined as $(u_e \Delta/\nu)$ where

$$\Delta = \int_0^\infty \frac{u}{u_e} \frac{T - T_e}{T_w - T_e} \, dy \tag{34}$$

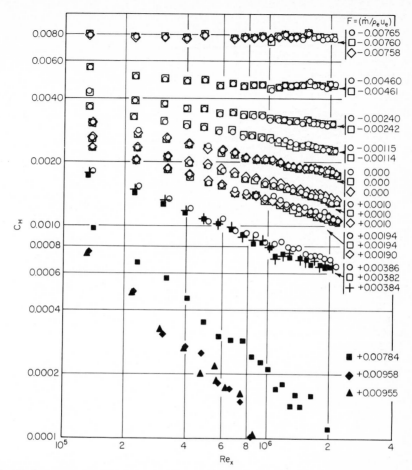

Fig. 65. Stanton number as a function of the length Reynolds number for various rates of mass transfer into a turbulent boundary layer on a flat plate [32].

This representation is more general than that shown in Figs. 65 and 66 and may be used for modest pressure gradients and for varying blowing or suction. Finally for the flat plate the normalized Stanton number $\left(C_H/C_{H_0}\right)_{Re_\Delta}$ is shown on Fig. 69 as a function of the blowing parameter F/C_{H_0}. Here the subscript Re_Δ means that the Stanton number with blowing $(C_H)_{Re_\Delta}$ is evaluated at the same enthalpy thickness Reynolds numbers as the solid plate $\left(C_{H_0}\right)_{Re_\Delta}$. A similar presentation is given for the skin-friction coefficient except that the momentum thickness $\theta = \int_0^\infty (u/u_e)(1 - u/u_e)\,dy$ replaces the enthalpy thickness and the abscissa now reads $F/(c_{f_0}/2)$.

b. The Flat Plate with Variable Blowing, Air to Air. Consider a turbulent boundary layer on a flat plate with the plate temperature approximately constant along the length. The initial section of the plate is solid while at some distance downsteam the plate is porous and uniform blowing occurs beyond this point. The

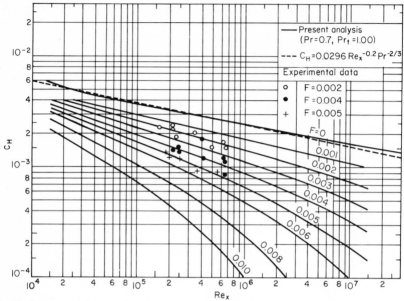

Fig. 66. Stanton number as a function of the length Reynolds number for various rates of mass transfer into a turbulent boundary layer on a flat plate [34].

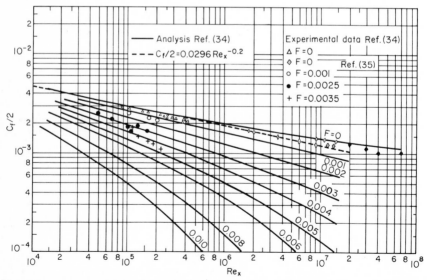

Fig. 67. Dimensionless skin-friction coefficient as a function of the length Reynolds number for various rates of mass transfer into a turbulent boundary layer on a flat plate [34].

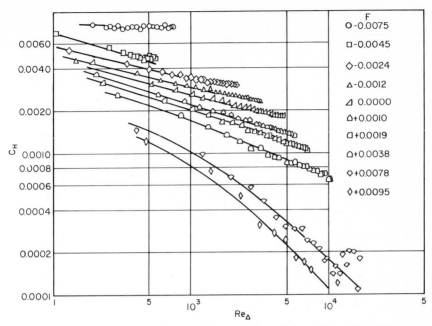

Fig. 68. Dimensionless Stanton number as a function of the enthalpy thickness Reynolds number for various rates of mass transfer into a turbulent boundary layer on a flat plate.

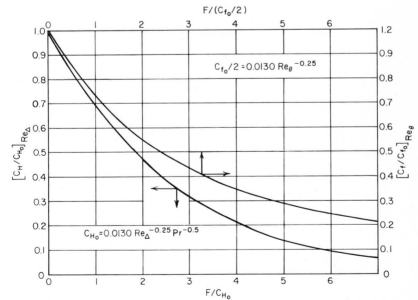

Fig. 69. Normalized Stanton number and normalized skin-friction coefficient as a function of the dimensionless mass transfer rate.

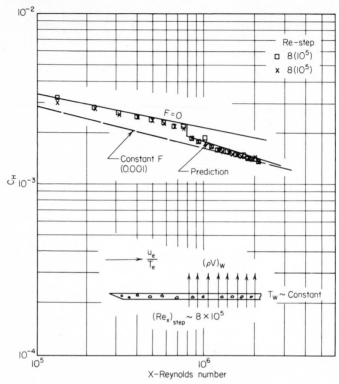

Fig. 70. Stanton number as a function of the length Reynolds number for a step change in mass injection rate for a turbulent boundary layer on a flat plate [33].

resulting Stanton numbers C_H, as a function of the length Reynolds number, are shown on Figs. 70 to to 72 for steps in the blowing rate F of 0.001, 0.002, and 0.004. It is seen that there is very rapid adjustment to local conditions.

A variation is shown in Fig. 73 wherein the solid portion of the plate is held at the same temperature as the free stream while at some distance downstream the plate temperature is suddenly changed to a new value and constant blowing simultaneously begins. Again within a short distance downstream the Stanton number adjusts to the same value as for constant blowing from the leading edge. In all these figures (70 to 73) the local Stanton number may be predicted from the integral relationship

$$\frac{d\mathrm{Re}_\Delta}{d\mathrm{Re}_x} = C_H + F \tag{35}$$

Here the Stanton number is taken to be a unique function of Re_Δ and F as given on Fig. 68. If the turbulent boundary layer begins at the leading edge, the equation reads

$$\mathrm{Re}_\Delta = \int_0^{\mathrm{Re}_x} \{C_H + F\}\, d\mathrm{Re}_x \tag{36}$$

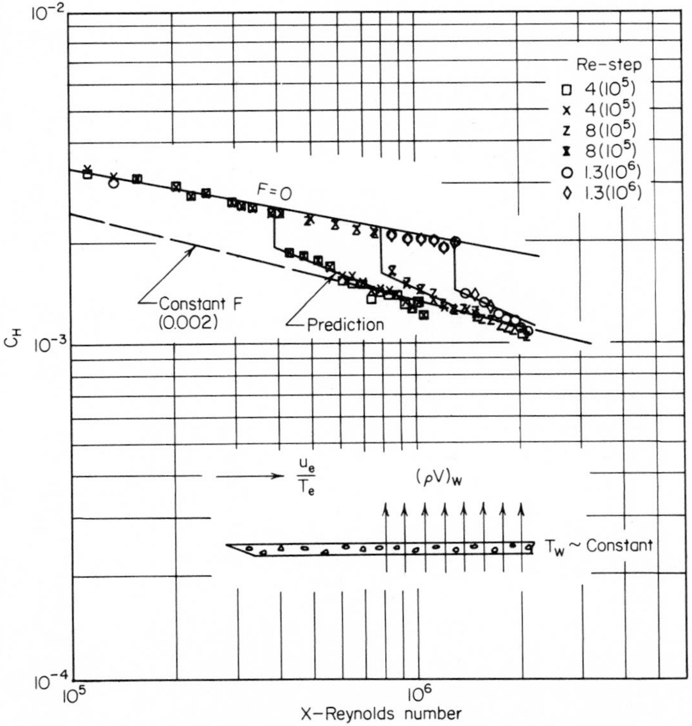

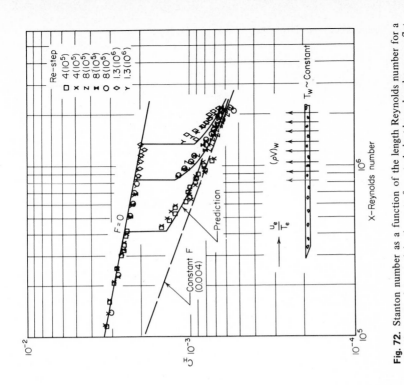

Fig. 72. Stanton number as a function of the length Reynolds number for a step change in the mass injection rate for a turbulent boundary layer on a flat plate [33].

Fig. 71. Stanton number as a function of the length Reynolds number for a step change in mass injection rate for a turbulent boundary layer on a flat plate [33].

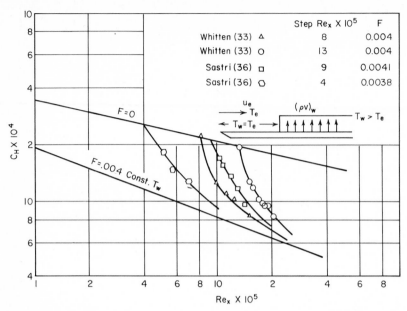

Fig. 73. Stanton number as a function of the length Reynolds number for a simultaneous step in the blowing distribution and in the wall temperature; turbulent boundary layer on a flat plate [36].

and now the Stanton number may be estimated at some incremental distance downstream, the value of Re_Δ then determined, and the Stanton number reevaluated. This process is repeated at this station until the initial guess and final value of the local Stanton number are in agreement. This procedure is continued step by step along the plate. More details may be found in Ref. 33.

c. The Turbulent Boundary Layer on a Flat Plate at High Velocity, Air to Air. The Stanton numbers given above may be used for high-velocity flow, or alternatively the Stanton number results of Rubesin and Inouye (Sec. 8) may be adopted. The recovery factor is somewhat reduced by blowing as shown on Fig. 74. The recovery or adiabatic wall temperature is determined by the relation

$$T_{aw} = T_e + r_0^* \frac{r}{r_0} \frac{u_e^2}{2c_p^*}$$ (18)

The solid-wall recovery factor is presented in Sec. 8 and c_p^* is the specific heat evaluated at the reference temperature. The heat transfer is then determined from

$$q = C_H \cdot (\rho^* u_e c_p^*)(T_w - T_{aw})$$ (17)

d. The Flat Plate with Helium Injected into Air. The reduced Stanton number C_H/C_{H_0} as a function of the blowing parameter F/C_{H_0} for helium injected into air is given in Fig. 75. The solid-wall Stanton number C_{H_0} may be determined by methods outlined in Sec. 8. At low velocities the influence of diffusion thermo is important for light gas injection, resulting in an adiabatic wall temperature substantially higher than the free-stream temperature. Experimental values of this adiabatic wall temperature as

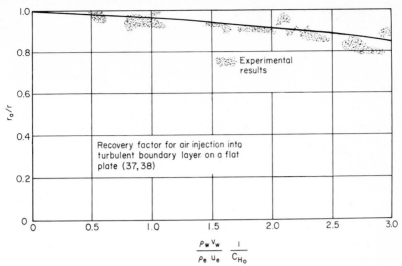

Fig. 74. Normalized recovery factor as a function of the mass transfer rate from air injected into a turbulent air boundary layer on a flat plate.

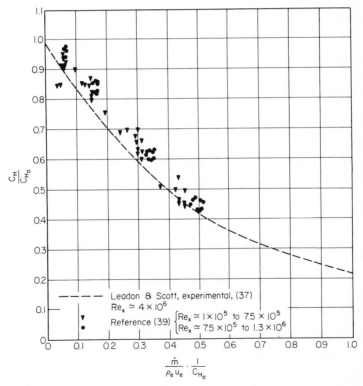

Fig. 75. Reduced Stanton number as a function of the mass transfer rate for helium injected into a turbulent boundary layer on a flat plate [37].

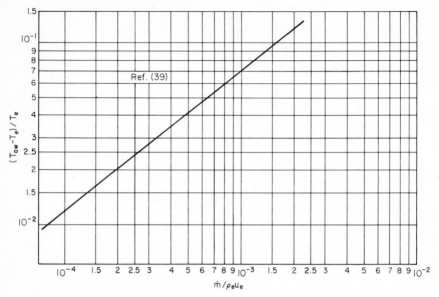

Fig. 76. Dimensionless adiabatic temperature as a function of the mass transfer rate resulting from diffusion thermo for helium injected into a turbulent boundary layer on a flat plate.

a function of the dimensionless mass transfer rate are shown on Fig. 76 for helium injection into a turbulent air boundary layer with a free-stream temperature of approximately 75°F and a free-stream velocity of 100 ft/s. The heat transfer is calculated from the usual relation

$$q = h(T_w - T_{aw}) \tag{9}$$

At high velocities ($Ma \gg 1$) the influence of viscous dissipation becomes important and the measured adiabatic wall temperature, given in terms of the reduced recovery factor, is shown on Fig. 77. The heat transfer is again calculated from Eq. (9).

It is expected that the injection of hydrogen into air boundary layers will also exhibit diffusion thermo effects at low velocities.

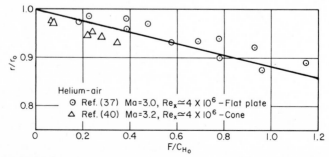

Fig. 77. Normalized recovery factor for helium injected into a turbulent boundary layer on a flat plate and on a cone [40].

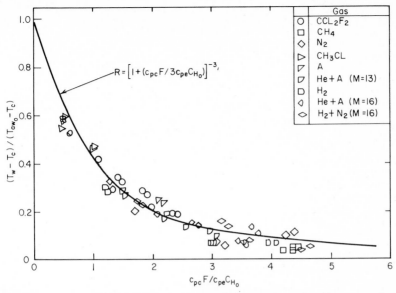

Fig. 78. Dimensionless wall temperature as a function of the dimensionless blowing rate for the injection of foreign gases into a turbulent air boundary layer on a flat plate [41].

e. Binary Mass Transfer Systems—Flat Plate. The most general presentation of transpiration cooling in binary systems is given by Bartle and Leadon [41]. A wide variety of coolant gases were studied and their results have been verified by later investigators [42, 43]. The so-called effectiveness $(T_w - T_c)/(T_{aw_0} - T_c)$ is shown as a function of the modified blowing rate $c_{pc}F/c_{pe}C_{H_0}$ in Fig. 78. The reduced heat transfer q/q_0 (where q_0 is the heat transfer to a flat plate at the same temperature and exposed to the same free-stream condition, calculated by methods of Sec. 8) is calculated from the relation

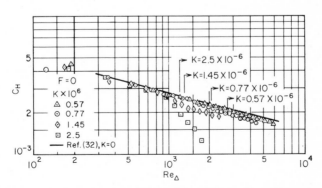

Fig. 79. Stanton number as a function of the enthalpy thickness Reynolds number for various values of the pressure gradient K [44].

$$\frac{T_w - T_c}{T_{aw_0} - T_c} = \frac{1}{1 + (c_{pc}/c_{pe})(F/C_{H_0})(q_0/q)} \tag{37}$$

This method may be applied to the complete Mach number range, provided the wall temperature is not close to the adiabatic wall temperature.

f. Mass Transfer in Turbulent Boundary Layers with Pressure Gradient, Air to Air. Experimental results for air transpiration into a turbulent air-boundary layer indicate that it is possible to use the same flat plate relationship for Stanton number as a function of the enthalpy thickness Reynolds number provided that the pressure gradient is moderate [44]. Specifically results are available for blowing and suction where the initial pressure gradient is zero followed by a pressure gradient described by $K = (\nu/u_e^2)(du_e/dx) = $ constant, then in some cases followed by a zero-pressure gradient. As shown on Fig. 79 for the solid-wall case the zero-pressure gradients shown as a solid line may be used with engineering accuracy for the prediction of the Stanton number for values of K up to 1.45×10^{-6}. When the pressure gradient becomes as large as $K = 2.5 \times 10^{-6}$ there is substantial departure from the zero-pressure gradient $C_H - Re_\Delta$ relation. As shown on Figs. 80 to 83 the influence of blowing reduces the influence of favorable pressure gradient whereas suction increases the effect of the pressure gradient.

4. Application to Liquid Film Cooling and Ablation

If mass transfer cooling is accomplished by the evaporation of a thin film of liquid distributed along the surface to be cooled, there is a unique relationship between the surface temperature and the wall concentration for the given pressure distribution if thermodynamic equilibrium is assumed. The partial pressure of the diffusing vapor at

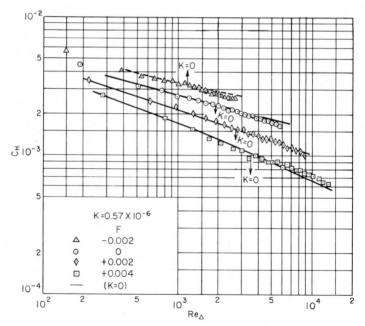

Fig. 80. Dimensionless Stanton number as a function of the enthalpy thickness Reynolds number for various values of the blowing rate and for a fixed pressure gradient $K = 0.57 \times 10^{-6}$ [44].

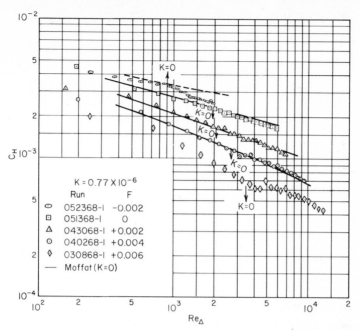

Fig. 81. Dimensionless Stanton number as a function of the enthalpy thickness Reynolds number for various values of the blowing rate and for a fixed pressure gradient $K = 0.77 \times 10^{-6}$ [44].

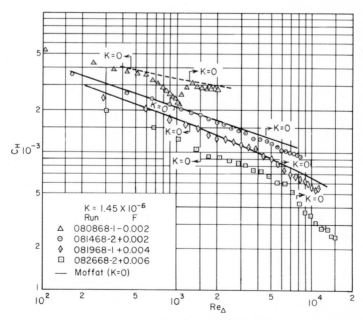

Fig. 82. Dimensionless Stanton number as a function of the enthalpy thickness Reynolds number for various values of the blowing rate and for a fixed pressure gradient $K = 1.45 \times 10^{-6}$ [44].

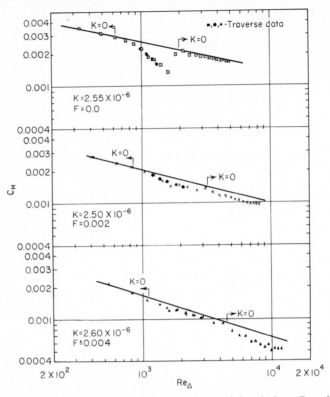

Fig. 83. Dimensionless Stanton number as a function of the enthalpy thickness Reynolds number for various values of the blowing rate and for a fixed pressure gradient $K = 2.55 \times 10^{-6}$ [44].

the surface is approximately equal to the vapor pressure of the liquid at the temperature T_w. Thus when the wall temperature is fixed, the partial pressure is fixed and the wall concentration as wall mass fraction is uniquely determined. Consequently, when the free-stream conditions are known, the wall temperature and the blowing rate are uniquely set by the appropriate combination of the thermodynamics of the system and the conservation equations previously solved. To demonstrate the procedure, if we neglect radiation and conduction along the surface, a local heat balance may be written as

$$m c_{pc}(T_w - T_c) + m h_{fg} = h_x(T_{aw} - T_w) \tag{38}$$

where h_{fg} = latent heat of vaporization. Rearranging

$$\frac{T_w - T_c + (h_{fg}/c_{pc})}{T_{aw} - T_w} = \frac{\mathrm{Nu}_x/\sqrt{\mathrm{Re}_x}}{\mathrm{Pr}[(\dot{m}/\rho_e u_e)\sqrt{\mathrm{Re}_x}](c_{pc}/c_{pe})} \tag{39}$$

or alternatively

$$\frac{T_w - T_c + (h_{fg}/c_{pc})}{T_{aw} - T_c} = \frac{C_H}{(\dot{m}/\rho_e u_e)(c_{pc}/c_{pe})} \tag{39a}$$

These relationships may be easily modified to account for variable properties by the use of the Eckert reference method.

If a value for the surface temperature is assumed, the partial pressure at the wall and accordingly the wall mass fraction are fixed. There is a unique relation between this wall mass fraction and the blowing or mass transfer rate as exemplified by Figs. 12, 13, 29, 30, 39, 40, and 54 through 56. The knowledge of the blowing rate allows the determination of the heat transfer coefficient C_H or $\mathrm{Nu}_x/\sqrt{\mathrm{Re}_x}$ and these values are substituted into the heat balance (Eq. (39)). If the equation is not satisfied, a new wall temperature is assumed and the procedure continued until the equation is satisfied. The extension to sublimation is straightforward [45].

B. GASEOUS FILM COOLING

1. General Remarks

Film cooling involves the introduction of a cooler secondary fluid through holes or slots for the purpose of protecting the surface immediately downstream of the injection parts; alternatively, a hotter gas may be introduced if the purpose is to keep the surface warm (as in prevention of icing on an aircraft wing). The injection geometry may be a slot, a porous section, or a series of holes or louvers. The secondary gas may be the same as the free stream or it may be a foreign gas. The free stream may be subsonic, supersonic, or hypersonic, and the flow over the surface may involve a pressure gradient. With the major exception of film cooling near stagnation regions, the flow is generally turbulent. A comprehensive survey of film cooling in turbulent flow including information on the geometries studied is given by Goldstein [46] and the following sections draw heavily from this reference.

2. Forced-convection Turbulent Flow

a. The Flat Plate with Tangential Injection—Air into a Low-velocity Air Boundary Layer. Considerable experimental data are available for air-to-air film cooling for various slot geometries including the four representative configurations shown on Fig. 84 [47]. Goldstein [46] gives a more complete summary of the slot geometries which have been investigated to date.

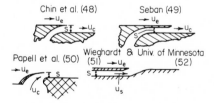

Fig. 84. Four typical slot configurations for gaseous film cooling [47].

In film-cooling applications the prediction of the adiabatic wall temperature distribution is generally the major requirement. It is convenient to present the adiabatic wall temperature in terms of an effectiveness η which is defined as

$$\eta = \frac{T_{aw} - T_e}{T_c - T_e} \tag{40}$$

where T_c is the coolant entrance temperature. There is general agreement that η is a function of the variable ζ, where

$$\zeta = \left(\frac{x}{Ms}\right)\left(\text{Re}_c \frac{\mu_c}{\mu_e}\right)^{-0.25} \tag{41}$$

Here x is measured from the point of injection and Re_c is $\dot{m}s/\mu_c$. The $\eta - \zeta$ relationship is apparently sensitive to variations in geometry, especially to the angle of injection, to the thickness of the slot lip, and to the turbulence level of the free stream. The following formulation of Goldstein and Haji-Sheikh [53] gives a good representation of the available experimental data and is recommended for values of $M \leq 1$

$$\eta = \frac{1.9 \, \text{Pr}^{2/3}}{1 + 0.329 \, (c_{pe}/c_{pc}) \zeta^{0.8}} \tag{42}$$

where

$$\Lambda = 1 + 1.5 \cdot 10^{-4} \, \text{Re}_c \frac{\mu_c}{\mu_e} \frac{m_2}{m_1} \sin \alpha$$

Here α is the angle of injection measured relative to the wall.

The film-cooling effectiveness determined from this equation is shown in Fig. 85 as a function of ζ, where it is compared with data reported by Seban [54, 55].

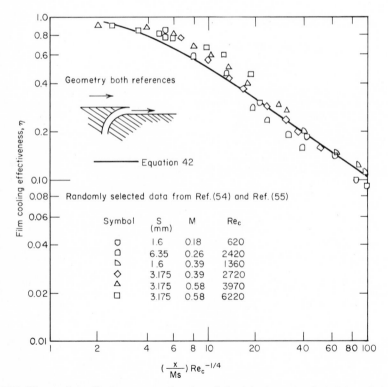

Fig. 85. Film cooling effectiveness as a function of the injection rate of air into a turbulent air boundary layer on a flat plate [46].

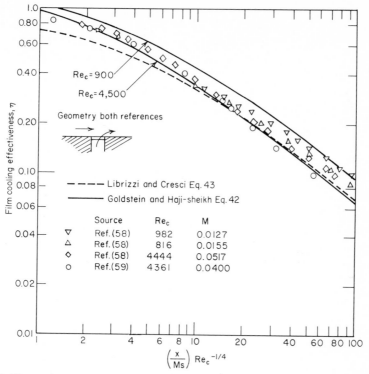

Fig. 86. Film cooling effectiveness resulting from injection of air through a porous section into a turbulent air boundary layer on a flat plate [46].

If the wall is not at the adiabatic wall temperature, there will be a heat transfer to or from the wall. The local heat transfer per unit area from the wall to the free stream is determined from the relation

$$q = h(T_w - T_{aw}) \tag{9}$$

The heat transfer coefficient h is calculated for flow over a solid wall for the same free-stream conditions taking into account the unheated starting length.

b. Normal Injection through a Porous Section on a Flat Plate—Air into a Low-velocity Boundary Layer. The adiabatic wall temperature is determined from Eq. (42) where $\alpha = 90°$. Librizzi and Cresci [56] propose the alternate formulation

$$\eta = \left(1 + 0.329 \frac{c_{pe}}{c_{pc}} \zeta^{0.8}\right)^{-1} \tag{43}$$

Figure 86 reveals that the Goldstein-Haji Sheikh formula gives better agreement with experimental data at low values of ζ. However, both predictions give engineering estimates of the adiabatic wall temperature for film cooling using a porous section.

If the wall is not adiabatic, the heat transfer per unit area from the wall to the main stream may be determined from the relation

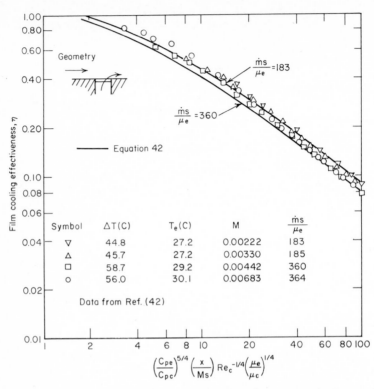

Fig. 87. Film cooling effectiveness for injection of helium through a porous section into a turbulent air boundary layer on a flat plate [46].

$$q = h(T_w - T_{aw})$$

The heat transfer coefficient h is calculated for a solid wall exposed to the same free-stream conditions taking into account the unheated starting length.

c. Foreign Gas Injection into a Low-velocity Air Boundary Layer—Flat Plate. Experimental data are available for helium injection through a tangential slot [57] and through a porous section [58] into a turbulent boundary layer on a flat plate. Freon-12 has also been studied for film cooling [60]. It is proposed that the prediction of Goldstein and Haji Sheikh be used for determining the adiabatic wall temperature in such cases. Figure 87 gives the comparison of the prediction and experiment for helium injection through a porous section. Again the heat transfer coefficient h is calculated for flow over a solid flat plate exposed to the same free-stream conditions, accounting for the unheated starting length.

d. Injection into a Supersonic Air Boundary Layer on a Flat Plate. The adiabatic wall temperature for film cooling of a zero-pressure gradient surface in a supersonic air stream presented in terms of the effectiveness, η:

$$\eta = \frac{T_{aw} - T_{aw_0}}{T'_c - T_{aw_0}} \tag{44}$$

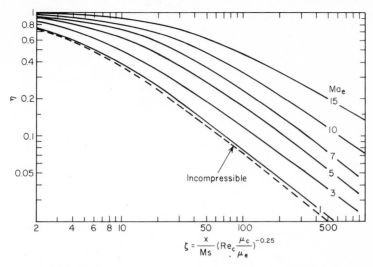

Fig. 88. Influence of the Mach number on film-cooling effectiveness [62].

Here T_{aw_0} is the adiabatic wall temperature in the absence of film cooling and T_c' is the total temperature of the coolant at the slot exit. Goldstein, Eckert, and Wilson [61] and Laganelli [62] use a reference temperature approach in conjunction with the low-speed results. The reference temperature T^* is taken to be

$$T^* = T_e + 0.72 \left(T_{aw_0} - T_e\right)$$

and all properties are evaluated at this temperature. The following formulation is achieved

$$\eta = \frac{1.9 \, (Pr^*)^{2/3}}{1 + 0.33 \, (c_{pe}/c_{pc}) \Lambda \zeta^{*0.8}} \qquad (42a)$$

where

$$\Lambda = 1 + 1.5 \cdot 10^{-4} \, Re_c \frac{\mu_c}{\mu^*} \sin \alpha$$

and

$$\zeta^* = \frac{x}{Ms} \left(Re_c \frac{\mu_c}{\mu^*}\right)^{-0.25} \left(\frac{\rho^*}{\rho_e}\right)$$

Figure 88 gives the effectiveness as a function of the free-stream Mach number as reported by Laganelli [62]. This latter author reports agreement of this correlation with experimental data [61, 63] for foreign gas injection as demonstrated in Fig. 89.

Experimental results are available for the adiabatic wall temperature distribution downstream of a step down slot in a supersonic flow of Mach number 3. For air injected into an air-boundary layer the effectiveness is given on Fig. 90 [64].

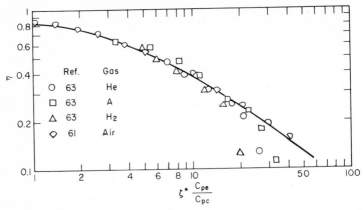

Fig. 89. Film-cooling effectiveness as a function of the dimensionless mass transfer rate for injection of various foreign gases into a turbulent air boundary layer on a flat plate [62].

It is recommended that the heat transfer be calculated in these cases through the use of the reference temperature method, again assuming that the heat transfer coefficient is the same as on a solid flat plate exposed to the same free-stream conditions. The heat transfer is then determined from the relation

$$q = h(T_w - T_{aw})$$

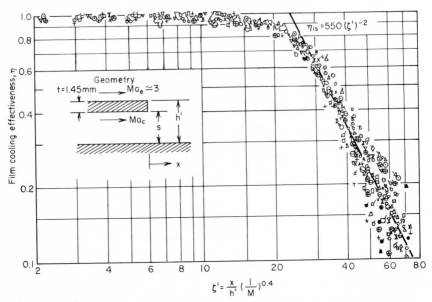

Fig. 90. Film-cooling effectiveness for tangential injection of air into a supersonic air boundary layer on a flat plate [64].

e. Concluding Remarks. Limited data are available for film cooling using distributed holes or louvers. Also some information is available on the influence of pressure gradient on the effectiveness and heat transfer. Reference 46 summarizes these results.

ACKNOWLEDGMENT

This work was supported by the National Science Foundation under its Grant NSF-GK-2972. The author expresses appreciation to the Foundation for its assistance.

REFERENCES

1. J. P. Hartnett and E. R. G. Eckert, Mass Transfer Cooling in a Laminar Boundary Layer with Constant Fluid Properties, *Trans. ASME,* **79**:247 (1957).
2. E. R. G. Eckert, Engineering Relations for Heat Transfer and Friction in High-Velocity Laminar and Turbulent Boundary Layer Flow over Surfaces with Constant Pressure and Temperature, *Trans. ASME,* **78**:1273 (1956).
3. H. A. Simon, C. S. Liu, and J. P. Hartnett, The Eckert Reference Formulation Applied to High Speed Boundary Layers of Nitrogen and Carbon Dioxide, *Int. J. Heat Mass Transfer,* **10**:406 (1967).
4. H. A. Simon, J. P. Hartnett, and C. S. Liu, Transpiration Cooling Correlations for Air and Non-Air Free Streams, in "Progress in Heat and Mass Transfer" (T. F. Irvine and W. Ibele, eds.), vol. 2 (Special Eckert Volume), Pergamon Press, Inc., New York, 1969.
5. J. F. Gross, J. P. Hartnett, D. J. Masson, and Carl Gazley, Jr., A Review of Binary Laminar Boundary Layer Characteristics, *Int. J. Heat Mass Transfer,* **3**:198 (1961).
6. C. S. Liu, J. P. Hartnett, and H. A. Simon, Mass Transfer Cooling of Laminar Boundary Layers with Hydrogen Injected Into Nitrogen and Carbon Dioxide Free Streams, *3d Int. Heat Transfer Conf. Paper 83,* Aug. 1966, Chicago.
7. E. M. Sparrow, W. J. Minkowycz, and E. R. G. Eckert, Mass Transfer Cooling of a Flat Plate with Various Transpiring Gases, *AIAA Journal,* **3**:1341 (1965).
8. E. M. Sparrow and J. B. Starr, The Transpiration-Cooled Flat Plate with Various Thermal and Velocity Boundary Conditions, *Int. J. Heat Mass Transfer,* **9**:508 (1966).
9. V. K. M. Sastri and J. P. Hartnett, Effect of an Unheated Solid Starting Length on Skin Friction and Heat Transfer in a Transpired Laminar Boundary Layer, in "Progress in Heat and Mass Transfer" (T. F. Irvine and W. Ibele, eds.), vol. 2 (Special Eckert Volume), Pergamon Press, Inc., New York, 1969.
10. E. M. Sparrow, W. J. Minkowycz, E. R. G. Eckert, and W. E. Ibele, The Effect of Diffusion Thermo and Thermal Diffusion for Helium Injected into Plane and Axisymmetric Stagnation Flow of Air, *J. Heat Transfer, Trans. ASME,* **E86**:311 (1964).
11. O. E. Tewfik, E. R. G. Eckert, and L. S. Jurewicz, Diffusion Thermo Effects on Heat Transfer From a Cylinder in Cross Flow, *AIAA Journal,* **1**:1537 (1963).
12. E. M. Sparrow, W. J. Minkowycz, and E. R. G. Eckert, Diffusion-Thermo Effects on Stagnation Point Flow of Air with Injection of Gases of Various Molecular Weights into the Boundary Layer, *AIAA Journal,* **2**:652 (1964).
13. E. Reshotko and C. B. Cohen, Heat Transfer at the Forward Stagnation Point of Blunt Bodies, National Advisory Committee, *Aeronaut. Tech. Note 3513,* July 1955.
14. A. F. Gollnick, Jr., Thermal Effects on a Transpiration-Cooled Hemisphere, *J. Aerospace Sci.,* **29**:583 (1962).
15. G. E. Anderson, C. J. Scott, and D. R. Elgin, Mass Transfer Cooling Experiments on a Hemisphere at M = 5, Rosemount Aeronautical Laboratories, University of Minnesota, Res. Rept., Aug. 1959.
16. V. M. Falkner and S. W. Skan, Some Approximate Solutions of the Boundary Layer Equations, *Phil. Mag.,* **12**:865 (1931); *Aeronautical Res. Comm. R & M,* 1314 (1930).
17. W. B. Brown and P. L. Donoughe, Tables of Exact Laminar Boundary Layer Solutions when the Wall is Porous and Fluid Properties are Variable, National Advisory Committee, *Aeronaut. Tech. Note 2479,* Sept. 1951.

18. C. F. Dewey, Jr., and J. F. Gross, Exact Similar Solutions of the Laminar Boundary Layer Equations, in "Advances in Heat Transfer" (T. F. Irvine and J. P. Hartnett, eds.), vol. 4, Academic Press, Inc., New York, 1967.

19. J. P. Hartnett, A Simple Graphical Presentation of the Dewey Gross Boundary Layer Solutions, The Rand Corp., unpublished report.

20. E. M. Sparrow, K. E. Torrance, and L. Y. Hung, Flow and Heat Transfer on Bodies in Cross Flow with Surface Mass Transfer, *Proc. 3d Int. Heat Transfer Conf.*, vol. 3, p. 23, AIChE, 1966.

21. B. V. Johnson and J. P. Hartnett, Heat Transfer in a Cylinder in Cross Flow with Transpiration Cooling, *J. Heat Transfer, Trans. ASME,* **C85**:173 (1963).

22. E. R. G. Eckert and J. N. B. Livingood, Method for Calculation of Laminar Heat Transfer in Air Flow Around Cylinders of Arbitrary Cross Section (Including Large Temperature Differences and Transpiration Cooling), National Advisory Committee, *Aeronaut. Report 1118,* Washington, D.C., 1953.

23. D. B. Spalding and W. M. Pun, A Review of Methods For Predicting Heat-Transfer Coefficients for Laminar Uniform-Property Boundary Layer Flows, *Int. J. Heat Mass Transfer,* **5**:239 (1962).

24. D. J. Wanous and E. M. Sparrow, Longitudinal Flow Over a Circular Cylinder with Surface Mass Transfer, *AIAA Journal,* **3**:147 (1965).

25. D. J. Wanous and E. M. Sparrow, Heat Transfer for Flow Longitudinal to a Cylinder with Surface Mass Transfer, *J. Heat Transfer, Trans. ASME,* **C87**:317 (1965).

26. E. M. Sparrow and J. L. Gregg, Mass Transfer, Flow and Heat Transfer about a Rotating Disk, *J. Heat Transfer, Trans. ASME,* **C82**:294 (1960).

27. C. L. Tien, Mass Transfer in Laminar Flow about a Rotating Cone, *J. Heat Transfer, Trans. ASME,* **C83**:514 (1961).

28. E. M. Sparrow, C. J. Scott, R. J. Forstrom, and W. A. Ebert, Experiments on the Diffusion Thermo Effect in a Binary Boundary Layer with Injection of Various Gases, *J. Heat Transfer, Trans. ASME,* **C87**:321 (1965).

29. O. E. Tewfik and J. W. Yang, The Thermodynamic Coupling Between Heat and Mass Transfer in Free Convection, *Int. J. Heat Mass Transfer,* **6**:915 (1963).

30. E. M. Sparrow, W. J. Minkowycz, and E. R. G. Eckert, Transpiration-Induced Buoyancy and Thermal Diffusion/Diffusion Thermo in a Helium-Air Free Convection Boundary Layer, *J. Heat Transfer, Trans. ASME,* **C86**:508 (1964).

31. E. M. Sparrow, J. W. Yang, and C. J. Scott, Free Convection in an Air-Water Vapor Boundary Layer, *Int. J. Heat Mass Transfer,* **9**:53 (1966).

32. R. J. Moffat and W. M. Kays, The Turbulent Boundary Layer on a Ferrous Plate: Experimental Heat Transfer with Uniform Blowing and Suction, *Int. J. Heat Mass Transfer,* **11**:1547 (1968).

33. D. G. Whitten, "The Turbulent Boundary Layer on a Porous Plate: Experimental Heat Transfer with Variable Suction, Blowing and Surface Temperature," doctoral dissertation, Stanford University, 1967.

34. K. Torii, N. Nishiwaki, and M. Hirata, Heat Transfer and Skin Friction in Turbulent Boundary Layer with Mass Injection, *Proc. 3d Int. Heat Transfer Conf.,* AIChe and ASME, 1966.

35. D. Coles, "Measurements in the Boundary Layer on a Smooth Flat Plate in Supersonic Flow," doctoral dissertation, California Institute of Technology, 1953.

36. V. M. K. Sastri, "Analytical and Experimental Study of the Influence of an Unheated Solid Starting Length in a Transpired Boundary Layer," doctoral dissertation, University of Delaware, June 1968.

37. B. M. Leadon and C. J. Scott, Measurement of Recovery Factors and Heat Transfer Coefficients with Transpiration Cooling in a Turbulent Boundary Layer at M = 3 Using Air and Helium as Coolants, Rosemount Aeronautical Laboratory, University of Minnesota, *Tech. Rept.* 126, Feb. 1956.

38. M. W. Rubesin, The Influence of Surface Injection on Heat Transfer and Skin Friction Associated with the High-Speed Turbulent Boundary Layer, National Advisory Committee, *Aeronaut. Res. Memo. A 55 L 13,* Feb. 1956.

39. O. E. Tewfik, E. R. G. Eckert, and C. J. Shirtliffe, Thermal Diffusion Effects in a Turbulent Boundary Layer with Helium Injection, *Proc. 1962 Heat Transfer and Fluid Mechanics Inst.,* p. 42, Stanford University Press, Stanford, Calif., 1962.

40. B. Rockover and A. F. Gollnick, Jr., Mass Transfer Cooling of a Cone, *M.I.T. Naval Supersonic Lab. TR 424,* 1961.

41. E. Roy Bartle and B. M. Leadon, The Effectiveness as a Universal Measure of Mass Transfer

Cooling for a Turbulent Boundary Layer, *Proc. 1962 Heat Transfer and Fluid Mechanics Inst.,* Stanford University Press, Stanford, Calif., 1962.

42. A. L. Laganelli, "Mass Transfer Cooling on a Porous Flat Plate in Carbon Dioxide and Air Streams," doctoral dissertation, University of Delaware, June 1966.

43. L. W. Woodruff and G. C. Lorenz, Hypersonic Turbulent Transpiration Cooling Including Downstream Effects, *AIAA Journal,* **4**:969 (1966).

44. W. H. Thielbahr, W. M. Kays, and R. J. Moffatt, The Turbulent Boundary Layer: Experimental Heat Transfer with Blowing, Suction and Favorable Pressure Gradient, *Rept. No.* HMT-5, Thermosciences Division, Department of Mechanical Engineering, Stanford University, 1969.

45. J. F. Gross, J. P. Hartnett, D. J. Masson, and Carl Gazley, Jr., A Review of Binary Boundary Layer Characteristics, *R.M.* 2516, The Rand Corp., Santa Monica, Calif., June 1959.

46. R. J. Goldstein, Film Cooling, in "Advances in Heat Transfer" (T. F. Irvine, Jr. and J. P. Hartnett, eds.), vol. 8, Academic Press, Inc., New York, 1971.

47. J. P. Hartnett, A Brief Review of Mass Transfer Cooling, *Institut Francais des Combustibles et de L'Energie, Journees de la Transmission de la Chaleur,* 1961.

48. J. Chin, S. Skirvin, L. Hayes, and A. Silver, Adiabatic Wall Temperature Downstream of a Single Slot, *ASME Paper* 58-A-104, 1958.

49. R. A. Seban, H. W. Chan, and S. Scesa, Heat Transfer to a Turbulent Boundary Layer Downstream of an Injection Slot, *ASME Paper* 57-A-36, 1957.

50. S. Papell and A. M. Trout, Experimental Investigation of Air-Film Cooling Applied to an Adiabatic Wall by Means of an Axially Discharging Jet, *NASA Tech. Note* D-9, 1959.

51. K. Wieghardt, Ueber das Ausblasen von Warmluft fur Enteisen, *Z.B.W. Res. Rept.* 1900, 1943; AAF Translation No1 F-TS-919-RE, Wright Field, 1946.

52. J. P. Hartnett, R. C. Birkebak, and E. R. G. Eckert, Velocity Distributions, Temperature Distributions, Effectiveness and Heat Transfer for Air Injected Through a Tangential Slot into a Turbulent Boundary Layer,

53. R. J. Goldstein and A. Haji-Sheikh, Prediction of Film Cooling Effectiveness, *Semi-international Symposium, Eng.,* Tokyo, 1967.

54. R. A. Seban and L. H. Back, Velocity and Temperature Profiles in Turbulent Boundary Layers with Tangential Injection, *J. Heat Transfer, Trans. ASME,* **84**:45 (1962).

55. R. A. Seban, Heat Transfer and Effectiveness for a Turbulent Boundary Layer with Tangential Fluid Injection, *J. Heat Transfer, Trans. ASME,* **82**:303 (1960).

56. J. Librizzi and R. J. Cresci, Transpiration Cooling of a Turbulent Boundary Layer in an Axisymmetric Nozzle, *AIAA Journal,* **2**:617 (1964).

57. J. E. Hatch and S. S. Papell, Use of a Theoretical Flow Model to Correlate Data for Film Cooling or Heating an Adiabatic Wall by Tangential Injection of Gases of Different Fluid Properties, *NASA TN* D-130, 1959.

58. R. J. Goldstein, R. B. Rask, and E. R. G. Eckert, Film Cooling with Helium Injected into an Incompressible Air Flow, *Int. J. Heat Mass Transfer,* **9**:1341 (1966).

59. R. J. Goldstein, G. Shavit, and T. S. Chen, Film Cooling With Injection Through a Porous Section, *J. Heat Transfer, Trans. ASME,* **C87**:353 (1965).

60. B. R. Pai and J. H. Whitelaw, Paper 29929 HMT 182, Aeronautical Research Council, London, 1967; also, Imperial College Department of Mechanical Engineering EHT/TN/8, London, 1967.

61. R. J. Goldstein, E. R. G. Eckert, and D. J. Wilson, Film Cooling with Normal Injection into a Supersonic Flow, *J. Eng. Ind., Trans. ASME,* **90**:584 (1968).

62. A. L. Laganelli, Downstream Influence of Film Cooling in a Supersonic Turbulent Boundary Layer, *Rept. TFM* 9151-064, Aeromechanics and Materials Laboratory Operation, Re-Entry Systems Department, General Electric Company, Valley Forge, Pa., Aug. 1969.

63. K. Parathasarathy and B. Zakkay, An Experimental Investigation of Turbulent Slot Injector at Mach 6, *Aerospace Res. Lab. Project* 7064, Wright Patterson Air Force Base, Oct. 1968.

64. R. J. Goldstein, E. R. G. Eckert, F. K. Tsou, and A. Haji-Sheikh, Film Cooling with Air and Helium Injection Through a Rear-Falling Slot into a Supersonic Air Flow, *University of Minnesota Heat Transfer Lab. HTL* TR-60, 1965.

NOMENCLATURE

a speed of sound

c_f local skin-friction coefficient $(2\tau_w)/\rho u_e^2$

C_H local Stanton number $(h/\rho c_p u_e)$

c_p specific heat at constant pressure

c_v specific heat at constant volume

D diameter

D_{12} binary diffusion coefficient

F dimensionless blowing rate $\dot{m}/\rho_e u_e$

f_w dimensionless blowing rate $-2(\dot{m}/\rho^* u_e)\sqrt{u_e x/v^*}$

g gravitational acceleration

h local heat transfer coefficient

h_{fg} latent heat of vaporization

H_e total enthalpy of free stream

K pressure gradient parameter $(v/u_e^2)(du_e/dx)$

k_T thermal diffusion ratio

M film cooling parameter $\rho_c v_c/\rho_e u_e = \dot{m}/\rho_e u_e$

Ma Mach number u/a

m_1 molecular weight of injected gas

m_2 molecular weight of free-stream gas

$\dot{m}$ mass flow of injected coolant per unit area per unit time, $\rho_w v_w$ or $\rho_c v_c$

Nu_x local Nusselt number hx/k

Pr Prandtl number $\mu c_p/k$

q heat transferred per unit area per unit time from the wall to the surrounding stream

R effectiveness for turbulent mass transfer $(T_w - T_c)/(T_{aw_0} - T_c)$

r recovery factor defined in Eq. (10)

r_w radial distance from axis of symmetry to surface

Re_c slot Reynolds number $\dot{m}s/\mu_c = \rho_c v_c/\mu_c$

Re_d Reynolds number $U_\infty D/v$

Re_x local Reynolds number $u_e x/v$

Re_Δ enthalpy thickness Reynolds number $u_e \Delta/v$

Re_θ momentum thickness Reynolds number $u_e \theta/v$

s slot thickness

T thermodynamic (absolute) temperature

T_c temperature of coolant gas

T_0 total thermodynamic temperature of free stream

t_w ratio of absolute wall temperature to absolute total temperature of free stream T_w/T_0

u velocity component parallel to surface

v velocity component normal to surface

v_{zw} velocity normal to surface of cone evaluated at the surface (Fig. 61)

U_∞ approach velocity of free-stream

x distance along the surface

x_0 unheated solid starting length (Fig. 17)

y distance normal to surface

Y mass fraction of injected coolant

z distance normal to cone surface (Fig. 61)

α angle

β relates to dimensionless pressure gradient (Eq. (25)) or (Eq. (28)); for low-speed wedge flows is related to wedge angle (Fig. 41)

γ ratio of specific heats c_p/c_v

Δ enthalpy thickness (Eq. (34))

η film cooling effectiveness (Eq. (40))

θ momentum thickness

Λ film cooling parameter (Eq. (42))

μ dynamic viscosity
ν kinematic viscosity
ξ generalized distance parameter (Eq. (26))
$\bar{\xi}$ normalized distance parameter (Eq. (27))
ρ density
σ Eckert number or dissipation parameter (Eq. (29))
τ local shearing stress
ϕ angle measured from forward stagnation point
Ω rotational velocity, radians per unit time
ζ dimensionless temperature ratio defined in Eq. (41)

Subscripts

aw adiabatic wall conditions
c refers to injected coolant gas
e free-stram conditions
w wall conditions
o solid-wall conditions

Superscripts

j equals zero for two-dimensional flow, equals unity for axisymmetric flow
m exponent in generalized flows (Eq. (24))
n exponent in wedge-type flows $u_e = cx^n$
ω viscosity temperature exponent $\mu \simeq T^\omega$
$*$ all properties to be evaluated at reference conditions (Eq. (11))

Section **18**

Heat Exchangers

A. C. MUELLER

Engineering Department, E. I. du Pont de Nemours & Co., Inc.

INTRODUCTION

A heat exchanger is a device to transfer heat from one fluid to another usually separated by walls. The heat transfer coefficient of each fluid is determined by the geometry of the flow passages, fluid flow rates, temperatures, and fluid properties. From these specified conditions the proper method of calculating the film coefficients is selected and an overall coefficient is calculated taking due care to allow for the interaction of individual coefficients through their effect on the wall temperatures. A characteristic of exchanger design is the procedure of specifying a design, then calculating the heat transfer and pressure drops, and checking whether the assumed design satisfies all requirements. If not, the design is modified until a satisfactory solution is obtained. The exchanger may be as simple as a double pipe unit or as complex as an array of several multipass shell-and-tube units operating together as a single exchanger.

This chapter will utilize the methods and equations presented in other chapters together with further methods specific to certain classes of heat exchangers. In all cases, the designer must be aware of the assumptions or experimental limitations of the equations presented in this and other chapters as the actual conditions within a heat exchanger often differ significantly from those on which the equations were derived.

DESIGN METHODS

Heat exchangers are designed by the usual equation $q = UA\Delta T_m$ where the area A is commonly specified as the outside area of the tubes between the tube sheets. The overall coefficient is based on this outside area for which a plain tubular exchanger is

$$\frac{1}{U} = \frac{1}{h_o} + \frac{1}{h_{od}} + \frac{LD_o}{kD_{1m}} + \frac{D_o}{h_{id}D_i} + \frac{D_o}{h_i D_i} \tag{1}$$

where U is the overall coefficient, h is the film coefficient, k is the tube thermal conductivity, D is the tube diameter, L is the wall thickness, and subscripts o is outside, i is inside, d is dirt, lm is log mean.

This overall coefficient is only valid when proper allowance has been made in the calculation of individual films of the effect of the exchanger dimensions, and the interaction that may occur due to film temperature differences such as occur in condensation, boiling, and laminar flow. The mean temperature difference ΔT_m is a function of the inlet and outlet temperatures of both streams, their flow patterns, and sometimes of the definition of the film coefficient, e.g., in laminar flow the film coefficient may be defined in terms of an arithmetic temperature difference. Usually, but not always, the log mean difference is used for turbulent flow and the arithmetic difference for laminar flow.

The logarithmic mean temperature difference

$$\Delta T_{1m} = \frac{(T_1 - t_2) - (T_2 - t_1)}{2.3 \log_{10} [(T_1 - t_2)/(T_2 - t_1)]} \tag{2}$$

is commonly used when designing simple heat exchangers. This equation, however, is based on the following assumptions:

1. Fluid flow is countercurrent.
2. Rate of flow of each fluid is constant.
3. Specific heat of each fluid is constant.
4. Overall heat transfer coefficient U is constant through the exchanger.
5. There is no change of phase or reactions occurring. Only sensible heat effects occur in each fluid.
6. Heat losses are negligible.

The equation for concurrent flow is similar but the terminal temperature differences are $(T_1 - t_1)$ and $(T_2 - t_2)$.

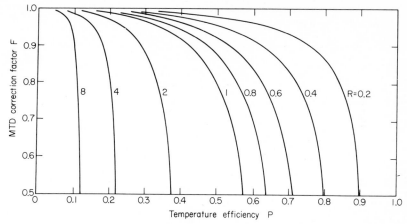

Fig. 1. Correction factor for 1-2 multipass exchangers.

When the design of a heat exchanger results in extreme lengths, it is usual to employ multipass units, several units in series, or various combinations. In a multipass exchanger the flow is both concurrent and countercurrent, hence the correct mean temperature difference lies between the values computed for each type of flow.

There are several methods of presenting curves by which the true mean temperature difference is determined. An early method was the F correction factor which was used to multiply the countercurrent logarithmic temperature difference to determine the true temperature difference. These F factors were plotted as functions of heat capacity ratio $R = (T_1 - T_2)/(t_2 - t_1)$ and thermal effectiveness $P = (t_2 - t_1)/(T_1 - t_1)$. (See Fig. 1.) The advantage of the F factor was that it represented the degree of departure from pure counterflow temperature difference. Limitations of multipass exchangers could be determined from only the terminal temperatures. The disadvantage was that an existing exchanger could not be readily examined for variations in flow or temperature conditions without a trial-and-error solution. To overcome this handicap, the NTU (number of transfer units equals UA/wc) method was developed. The P-NTU method was usually represented in a semilog plot. (See Fig. 2.) Another version was a linear plot (Fig. 57), but here $NTU = UA/(wc)$ min, $R = (wc)$ min/(wc) max, and thermal effectiveness $\epsilon = wc(t_2 - t_1)/(wc)$ min $(T_1 - t_1)$. The NTU method is best suited for examining performance of a known exchanger but is of little use in evaluating the potential design problems in the early stages of design when no configuration has been determined. Another method is presented here which combines all these variables into one plot and is used for all the exchanger flow patterns. Here $\theta = \Delta T_m/(T_1 - t_1) = $ (true mean temperature difference divided by the difference of the inlet temperatures) is plotted against effectiveness P for various ratios of heat capacity R. Thus it is not necessary to calculate the counterflow log mean temperature difference as for the F factor case. Further, for a given exchanger or for variations in performance the slope of a straight line through the origin is equal to wc/UA (reciprocal of number of transfer units). To aid in estimating the departure from counterflow temperature difference, lines for constant correction factors F are drawn. In most of the figures, the lines of constant F were determined by graphical methods and thus might be subject to error in certain regions (e.g., high values of R). However, the F factor is not needed for calculation as $\theta(T_1 - t_1) = F\Delta T_{lm} = \Delta T_m$.

For most shell-and-tube exchangers the following three additional assumptions apply to the six previously mentioned:

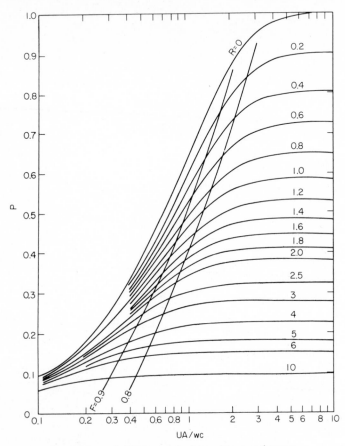

Fig. 2. Thermal effectiveness for 1-2 multipass exchangers.

7. There is equal heat transfer surface in each pass.
8. The temperature of the shell fluid in any shell pass is uniform over the cross section.
9. No fluid or heat leaks between shell passes.

MULTIPASS HEAT EXCHANGERS

With all the above assumptions, plus another of even number of tube passes per shell pass, Fig. 3 is based on the equations below.

For one shell and two tube passes

$$\frac{UA}{wc} = \frac{1}{\sqrt{1 + R^2}} \ln\left[\frac{2 - (1 + R - \sqrt{1 + R^2})P}{2 - (1 + R + \sqrt{1 + R^2})P}\right]$$

While the basic equation was developed for one shell pass and two tube passes, there is little error for any even number of tube passes per shell pass. A recent analysis showed the largest possible error was 4.4 percent for the 1-4 exchanger, rising to 6.8 percent for the 1-12.

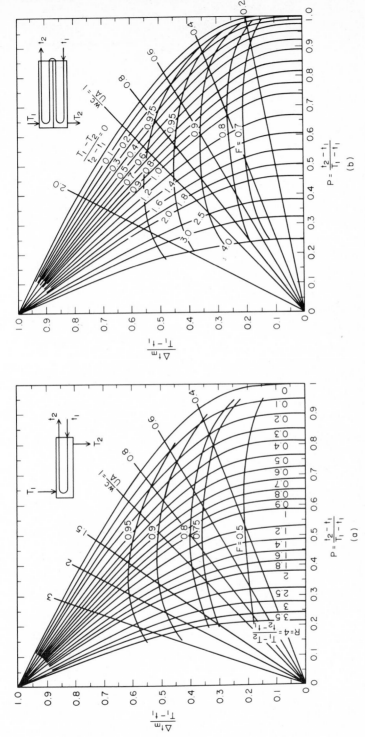

Fig. 3. Mean temperature difference for multipass exchangers with even number of tube passes per shell pass.

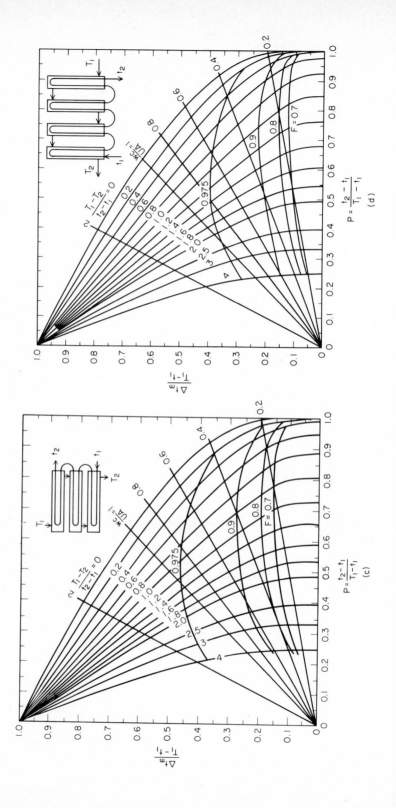

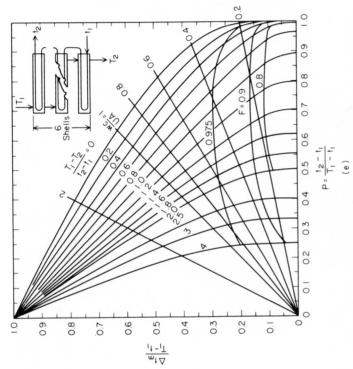

Fig. 3. (*Cont.*). Mean temperature difference for multipass exchangers with even number of tube passes per shell pass.

For the two shell passes and four tube passes the Underwood equation reduces to

$$\frac{UA}{wc} = \frac{2}{\sqrt{1 + R^2}} \ln\left[\frac{(2/P) - 1 - R + (2/P)\sqrt{(1 - P)(1 - PR)} + \sqrt{1 + R^2}}{(2/P) - 1 - R + (2/P)\sqrt{(1 - P)(1 - PR)} - \sqrt{1 + R^2}}\right]$$

For three or more shell passes Bowman [2] has shown

$$p = \frac{[(1 - PR)/(1 - P)]^{1/N} - 1}{[(1 - PR)/(1 - P)]^{1/N} - R}$$

and for $R = 1$

$$p = \frac{P}{N - P(N - 1)}$$

where p is the effectiveness for a 1-2 exchanger, P is the overall effectiveness for N exchangers in series. At this value of p the correction factor F is the same for the 1-2 or N-$2N$ case. The relation for $\theta = \Delta T_m/(T_1 - t_1)$ is

$$\theta_N = \frac{P\theta_1}{pN}$$

where θ_1 is the 1-2 value for p.

The case for odd number of tube passes per shell pass has been solved by Fischer [3]. The best performance is obtained when the shell fluid is flowing countercurrent to more than half the tube passes, but the major benefit is at low values of F where it is generally not desirable to operate. The improvement, however, is not as great as that resulting from the addition of one more pass to the exchanger. Odd number passes are uncommon designs and may result in mechanical and thermal difficulties in fabrication and operation. Fischer's equations require trial and error for solution and Fig. 4 for the 1-3 exchanger is based on his tabulated values. No charts were prepared for the 2-6, etc. exchangers for which Fischer uses the Bowman equation (see above).

Occasionally exchangers may not be arranged in series but in a combination of series and parallel flow because of flow quantities or pressure-drop limitations. Gardner [4] has derived equations which allow calculation of the correction factor for several arrays on the basis of the correction factor for the 1-2 exchanger. The assumption in addition to 2 through 8 is that all exchangers are identical.

For an array of M identical exchangers connected in series on the high-temperature stream and in parallel on the low-temperature stream the equations are:
If neither R nor $R_1 = 1$; $R_1 = R/M$ and

$$P_1 = \frac{M}{R}[1 - (1 - PR)^{1/M}]$$

$$F = \frac{F_1}{M}\left\{\left(\frac{R - M}{R - 1}\right)\frac{\ln[(1 - P)/(1 - PR)]}{\ln[(R - M)/R(1 - PR)^{1/M}] + M/R}\right\} \tag{3}$$

If $R = 1$

$$F = \frac{F_1}{M}\left\{\frac{P(1 - M)}{1 - P}\frac{1}{\ln[(1 - M)/(1 - P)^{1/M} + M]}\right\} \tag{4}$$

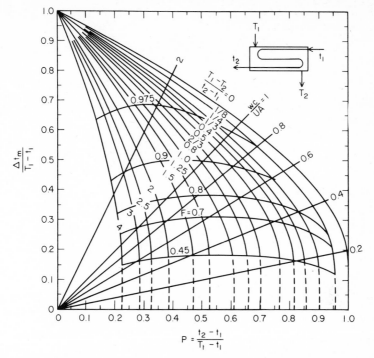

Fig. 4. Mean temperature difference for multipass exchangers with three tube passes (two countercurrent, one concurrent) per shell pass.

If $R_1 = 1$

$$F = F_1 \left\{ \frac{(1 - PM)^{1/M} \ln[(1 - P)/(1 - PM)]}{[1 - (1 - PM)^{1/M}](M - 1)} \right\} \tag{5}$$

When the fluids are interchanged to be series flow on the low-temperature stream and parallel on the high-temperature stream, $R_1 = MR$ and $P_1 = 1 - (1 - P)^{1/M}$, and when neither R nor $R_1 = 1$

$$F = \frac{-F_1}{M} \left\{ \left(\frac{MR - 1}{R - 1} \right) \frac{\ln[(1 - P)/(1 - RP)]}{\ln[(1 - MR)/(1 - P)^{1/M} + MR]} \right\} \tag{6}$$

If $R_1 = 1$

$$F = -F_1 \left\{ \frac{(1 - P)^{1/M}}{(M - 1)[1 - (1 - P)^{1/M}]} \ln \frac{1 - P}{1 - P/M} \right\} \tag{7}$$

When $R = 1$ the solution is the same as given in Eq. (4).

Gardner [4] discusses several other mixed four-exchanger arrays such as one fluid in series on one side and the other fluid in two parallel streams. The original paper [4] should be consulted for these cases. The above equations are useful when each exchanger is a true countercurrent unit and $F_1 = 1$. When each exchanger is a multipass unit and F_1 must be determined, then calculate R_1 and P_1 for the particular flow, and determine F_1 from Fig. 1 and use

$$F = F_1 \left\{ \frac{NP(R_1 - 1)}{MR_1(1 - P)\ln[(1 - P_1)/(1 - P_1R_1)]} \right\} \qquad R = 1,\ R_1 \neq 1$$

$$F = F_1 \left\{ \frac{NR(1 - P_1)\ln[(1 - P)/(1 - PR)]}{MP_1(R - 1)} \right\} \qquad R_1 = 1,\ R \neq 1$$

$$F = F_1 \left[\frac{NP(1 - P_1)}{MP_1(1 - P)} \right] \qquad R = R_1 = 1$$

$$F = F_1 \left\{ \frac{NR}{MR_1} \left(\frac{R_1 - 1}{R - 1} \right) \frac{\ln[(1 - P)/(1 - PR)]}{\ln[(1 - P_1)/(1 - P_1R_1)]} \right\}$$

In these equations, M is the number of exchangers, N is the number of equal streams in which W is divided, and P and R are based on the mixed mean temperature of the parallel streams. R_1, P_1, and F_1 refer to a single unit of the array.

Now let us examine the effect of eliminating some of the assumptions. For the simple counterflow exchanger when assumption 4 is modified to allow U to vary linearly with temperature, Colburn [5] has shown that

$$\frac{q}{A} = (U\,\Delta T)_{\text{lm}} = \frac{U_2(T_1 - t_2) - U_1(T_2 - t_1)}{2.3 \log[U_2(T_1 - t_2)/U_1(T_2 - t_1)]}$$

where U_1 is the value at the T_1 end of the exchanger and U_2 at the T_2 end.

Elimination of assumption 4 for the multipass exchangers has been investigated by Gardner [6] for the case where U varies linearly with the shell fluid temperature T and when the shell fluid is the controlling film. The results are shown in Fig. 5 for the several ratios of U_1/U_2. The values read from these charts are used with the arithmetic average U. These charts may be applied in the useful ranges with only 1 to 2 percent error for any even number of tube passes. Also note that θ does not approach unity as a maximum such as occurs for constant U. Gardner [6] also indicates methods for determining the correction factor if U would vary at rates other than linear. When the tube side coefficient is controlling, then the solution for unbalanced surfaces in passes (Fig. 6) may be used to approximate the solution for a variable heat transfer rate through a single-shell pass–two-tube pass exchanger with equal tube pass surface since $K = A_cU_c/A_pU_p$. Hence, either the surface A or coefficient U may be considered a variable. The subscript c refers to the countercurrent pass and p to the concurrent flow pass. If neither tube nor shell fluid is controlling but the overall coefficient varies through the exchanger, then a judicious interpolation must be made between the values determined from Figs. 5 or 6.

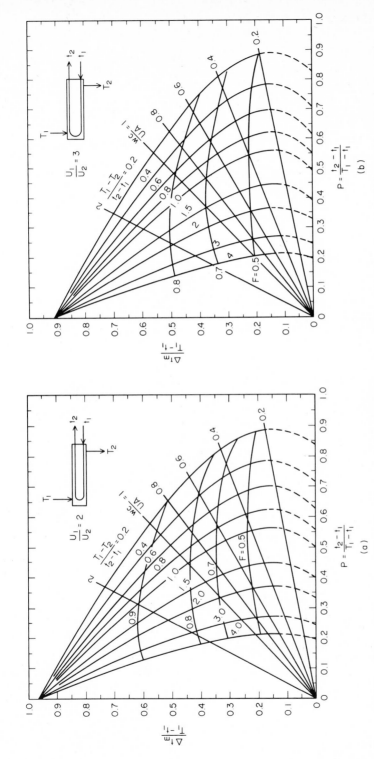

Fig. 5. Mean temperature difference for multipass exchangers with even number of tube passes but with a linearly variable overall coefficient dependent upon shell side fluid.

18-11

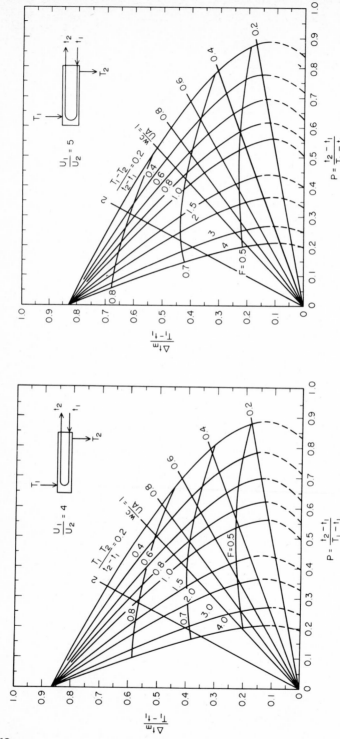

Fig. 5 (*Cont.*). Mean temperature difference for multipass exchangers with even number of tube passes but with a linearly variable overall coefficient dependent upon shell side fluid.

Gardner [7] eliminated assumption 7 and derived equations and curves for exchangers having unequal surface in the counterflow and concurrent flow passes. (See Fig. 6.) As mentioned above, these curves may also be used where the tube fluid is the controlling resistance to heat flow. Here there appears a significant difference in the correction factor for a 1-2 or a 1-4 pass exchanger. However, for the 1-4 exchanger, the results are slightly better than for the balanced ($K = 1$) exchanger. No charts are given for the 1-4 case but, if necessary, refer to the original paper or use the value from Fig. 3a. For more than one shell or shell pass in series, the correction factor for the combination can be determined from the Bowman relations. As the number of passes approaches infinity, the correction factors approach those for cross flow with both fluids completely mixed; any advantage of an unbalanced pass exchanger over balanced passes becomes negligible. Switching the definitions of terms, i.e., shell = hot fluid to shell = cold fluid in the parameters R and P will not result in the same correction factors for cases where U and A are variable as could be done when they are constant. Furthermore, the correction factors would indicate some differences depending upon whether the shell fluid entered at the fixed or floating end. Since higher temperatures generally mean higher coefficients, there will be an advantage of having the shell fluid enter at the fixed end when heating the tube fluid and at the floating end when cooling the tube fluid.

Gardner [8] has also examined the effect of assumption 8, the degree of mixing on the shell side, on the correction factor. Figure 7 shows a cross section of an exchanger with (a) complete mixing where the dividing pass lane is perpendicular to the flow of shell fluid, (b) partial mixing where the lane is parallel to the shell flows, and (c) no mixing as obtained when placing a baffle in the lane of (b). Theoretically, the unmixed case (c) gives the best correction factor; but a practical basis requires that the saving in area can more than offset the increased cost of the necessary baffle. The intermediate case (b) is avoided since the lane parallel to shell fluid flow would permit bypassing of shell fluid causing poorer heat transfer rates and temperature differences within the passes. Case (a) is the practical arrangement used. The unmixed case is the same as a divided shell flow with one tube pass (Fig. 8).

Sometimes, instead of using two exchangers in series, the surface may be placed in one shell having two shell passes. A longitudinal baffle is placed in the shell separating the passes, but heat will leak across the baffle, and fluid may also leak depending upon the baffle construction. Whistler [9] determined the correction factor to be used with the usual log mean temperature difference for the case of only heat conduction, and, in addition to the usual assumptions 2 through 8, the equation is based on only one tube pass per shell pass. Thus, this analysis is a partial removal of assumption 9. The equation is more complicated because other parameters must be added and involve the baffle surface A_b. (The total baffle coefficient is an overall coefficient of the baffle since the fluid and fouling coefficients on each side of the baffle are involved as well as the conductivity through the baffle.) The inlet shell temperature difference is $\Delta t_1 = T_1 - t_2$ for countercurrent flow. The equation for the correction factor which applies to both countercurrent and concurrent flow is

$$F = \frac{Z \ln(\Delta t_1/\Delta t_2)}{[(\Delta t_1 - \Delta t_2)/(\Delta t_1 + \Delta t_2)] \ln[(1 + Z)/(1 - Z)]}$$

$$Z = \sqrt{4\left(\frac{T_1 - T_2}{\Delta t_1 + \Delta t_2}\right)^2 \left(\frac{A_b}{A}\right)\left(\frac{h_b}{U}\right) + \left(\frac{\Delta t_1 - \Delta t_2}{\Delta t_1 + \Delta t_2}\right)^2}$$

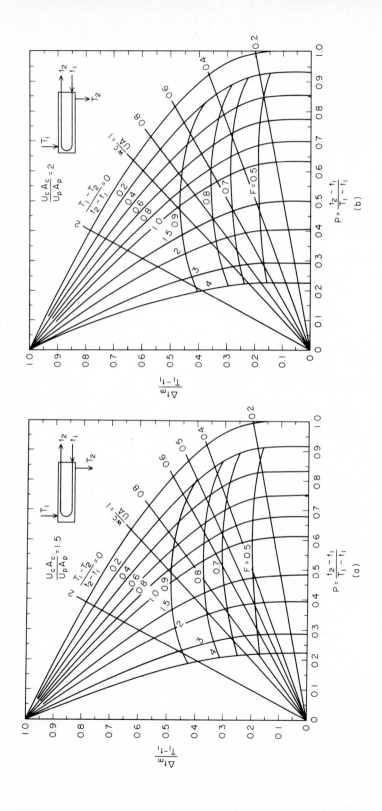

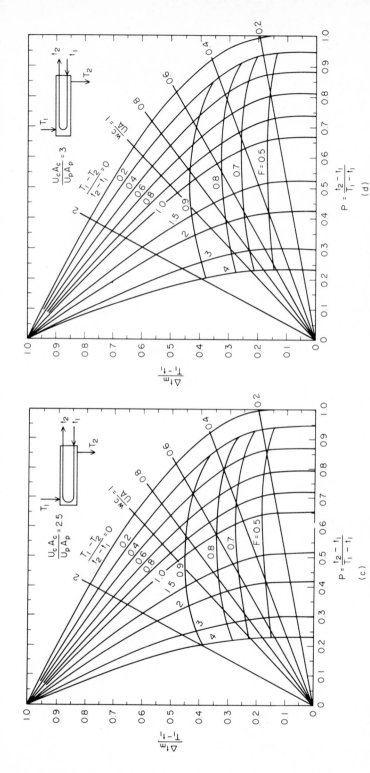

Fig. 6. Mean temperature difference for multipass exchangers with even number of tube passes but with linearly variable overall coefficient dependent upon tube side fluid or unequal areas in the tube passes.

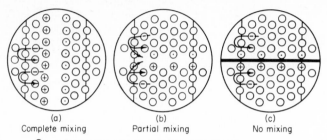

(a)
Complete mixing

(b)
Partial mixing

(c)
No mixing

⊕ Indicates flow of tube fluid into plane of page
⊙ Indicates flow of tube fluid out from plane of page

Fig. 7. Cross-sectional diagrams of heat exchangers showing how mixing may be controlled by baffling.

Divided flow exchangers are used to reduce shell side-pressure drops or where the shell flow is too large for convenient baffle spacing. Divided flow with one-tube pass is the same as the unmixed shell case of a one-shell pass–two-tube pass exchanger which was solved by Gardner [8] and shown in Fig. 8. A divided flow shell for two-tube passes has been considered by several authors, but Fig. 9 is based on the paper by Jaw [10].

Examples. 1. (a) Cool one stream from $T_1 = 300°F$ to $T_2 = 200°F$ with another stream rising from $t_1 = 100°F$ to $t_2 = 200°F$

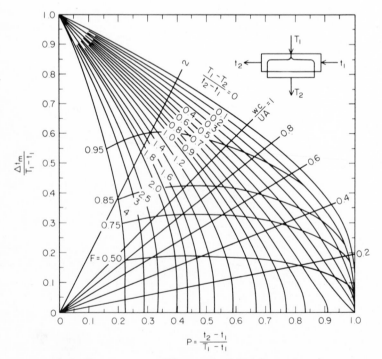

Fig. 8. Mean temperature difference of a divided flow exchanger.

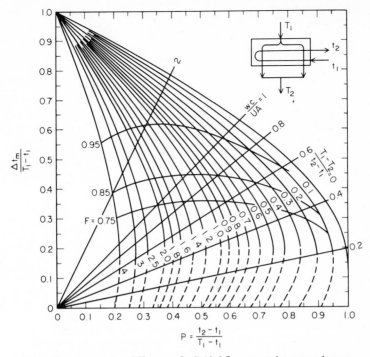

Fig. 9. Mean temperature difference of a divided flow, two-tube pass exchanger.

$$R = \frac{300 - 200}{200 - 100} = 1 \ , \quad P = \frac{200 - 100}{300 - 100} = 0.5$$

Assuming a single multipass exchanger, then from Fig. 3a $\theta = 0.4$ also $F = 0.8$. Hence, $\Delta T_m = 0.4 (300 - 100) = 80°F$. Checking from the log mean temperature difference = 100, then $\Delta T_m = 0.8 \times 100 = 80°F$.

(b) From Fig. 3a $wc/UA = 0.8$, if $w = 75,000$ lb/hr and $c = 0.6$, then required $UA = 75,000 \times 0.6/0.8 = 56,250$. If $U = 150$, then $A = 375$ ft^2.

2. (a) With a flow $w = 75,000$ lb/hr having a sp ht, $c = 0.6$ a 300 ft^2 multipass exchanger is estimated to have a $U = 150$. Fluid is heated from $t_1 = 100°F$ to $t_2 = 200°F$ with another fluid available at $T_1 = 300°F$. $wc/UA = (75,000 \times 0.6)/(150 \times 300) = 1$. $P = (200 - 100)/(300 - 100) = 0.5$. From Fig. 3a, $\theta = 0.5$ and $\Delta T_m = 0.5 (300 - 100) = 100$: $R = 0.74 = (300 - T_2)/(200 - 100)$ or $T_2 = 226°F$. Estimated $F = 0.89$. Log mean temperature difference = 112.5° and $\Delta T_m = 0.89 \times 112.5 = 100°F$. If $C = 1$, then $W = 75,000 \times (0.6/0.74) \times 1 = 60,800$ lb/hr.

(b) Now suppose $t_1 = 80°F$ and the flow remained the same; what is t_2? R is still 0.74 and $wc/UA = 1$. Hence, $P = 0.5 = (t_2 - 80)/(300 - 80)$ or $t_2 = 190°F$.

(c) What is the change in flow rate w to maintain $t_2 = 200°$ in (b)? Then $P = (200 - 80)/(300 - 80) = 0.545$ and from Fig. 3a, assuming U is still 150, hence $wc/UA = 1$. $R = 0.46$ and $\theta = 0.545$. Hence $T_2 = 300 - 0.46 (200 - 80) = 244.9°F$ and $w = 75,000 \times 0.6/0.46 \times 1 = 97,800$ lb/hr. $\Delta T_m = 0.545 (300 - 80) = 120°F$.

Estimate from Fig. 3a F = 0.925. The log mean ΔT = 130° and hence ΔT_m = 0.925 $\times$ 130 = 120°F (close enough check).

CROSS-FLOW HEAT EXCHANGERS

In the cross-flow exchangers the fluids flow at right angles and hence neither the concurrent nor countercurrent form of temperature differences apply. In the following charts (Figs. 10 to 31) the assumptions 2 through 7 apply. In addition, it is important to determine the condition of flow which can be either completely mixed or unmixed on either side and in various combinations for the two streams. The number of combinations mounts as the number of exchangers in series is considered. Fortunately, after three passes or exchangers, the results approach that for counterflow and log mean temperature difference will apply with little error. Most of the charts are based on the calculations by Stevens [11] and his notation will be followed in the equations

$$E_A = \frac{\Delta t_A}{t_1 - T_1} \quad , \quad (NTU_A) = \frac{UA}{C_A} \quad , \quad K = 1 - \exp(-NTU_B) \quad , \quad C_A = (wc)_A$$

$$Z_A = \frac{C_A}{C_B}$$

Single pass:

1. Both fluids mixed (Fig. 10)

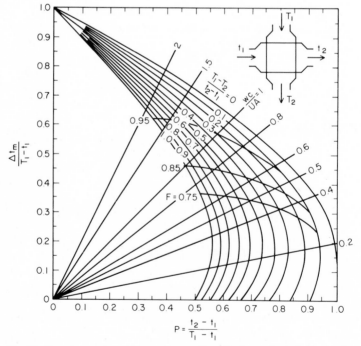

Fig. 10. Mean temperature difference in cross flow. Single pass, both fluids mixed.

$$E_A = \frac{1}{1/[1 - \exp(-NTU_A)] + Z_A/K - 1/(NTU_A)}$$

2. Fluid A mixed, Fluid B unmixed (Fig. 11)

$$E_A = 1 - \exp(-KZ_B)$$

3. Both fluids unmixed (Fig. 12)

$$E_A = \frac{1}{Z_A NTU_A} \sum_{n=0}^{\infty} \left\{ \left[1 - \exp(-NTU_A) \sum_{m=0}^{n} \frac{(NTU_A)^m}{m!} \right] \right.$$

$$\left. \left[1 - \exp(-Z_A NTU_A) \sum_{m=0}^{n} \frac{(Z_A NTU_A)^m}{m!} \right] \right\}$$

Two-pass counter flow:

1. Both fluids mixed between passes, A mixed in each pass, B unmixed in each pass (Fig. 13)

$$E_A = \frac{2E_a - E_a^2(1 + C_A/C_B)}{1 - E_a^2 C_A/C_B}$$

where $E_a = 1 - \exp(-KC_B/C_A)$ using ½ of total NTU.

2. Both fluids mixed between passes, both fluids unmixed in each pass (Fig. 14)

$$E_A = \frac{2E_a - E_a^2(1 + C_A/C_B)}{1 - E_a^2 C_A/C_B}$$

3. Fluid A mixed throughout, Fluid B unmixed throughout, inverted order (Fig. 15)

$$E_A = 1 - \frac{1}{K/2 + (1 - K/2) \exp(2KZ_B)}$$

4. Fluid A mixed throughout, Fluid B unmixed throughout, identical order (Fig. 16)

$$E_A = 1 - \frac{1}{\exp(2K C_B/C_A) - K^2(C_B/C_A) \exp(K C_B/C_A)}$$

5. Fluid A mixed between passes, unmixed in each pass, Fluid B unmixed throughout, inverted order, numerical solution (Fig. 17).

6. Fluid A mixed between passes, unmixed in each pass, Fluid B unmixed throughout, identical order, numerical solution (Fig. 18).

7. Both fluids unmixed throughout, both inverted order, numerical solution (Fig. 19).

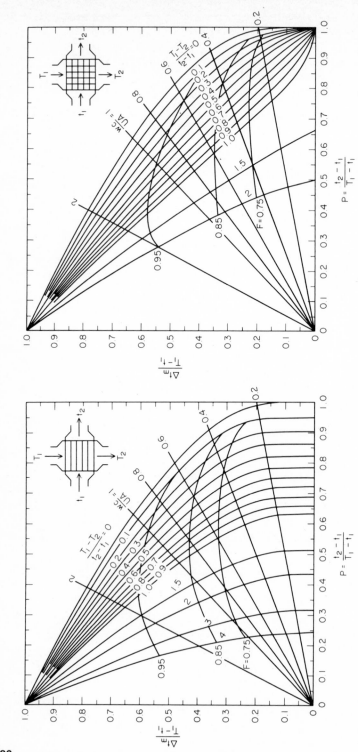

Fig. 11. Mean temperature difference in cross flow. Single pass, one fluid mixed, other unmixed.

Fig. 12. Mean temperature difference in cross flow. Single pass, both fluids unmixed.

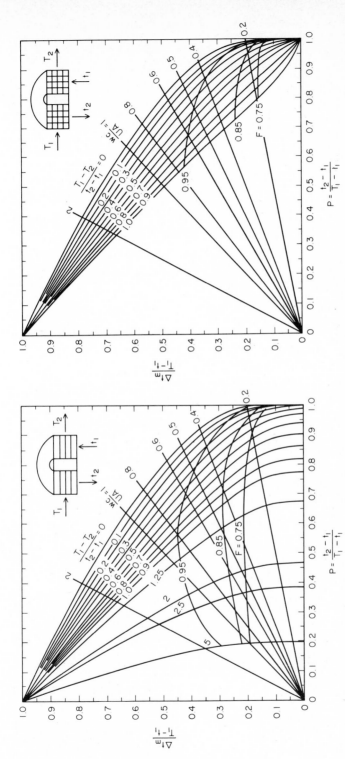

Fig. 13. Mean temperature difference in cross flow. Two pass counter flow. One fluid mixed throughout. Other fluid unmixed in each pass but mixed between passes.

Fig. 14. Mean temperature difference in cross flow. Two pass counterflow. Both fluids unmixed in each pass but mixed between passes.

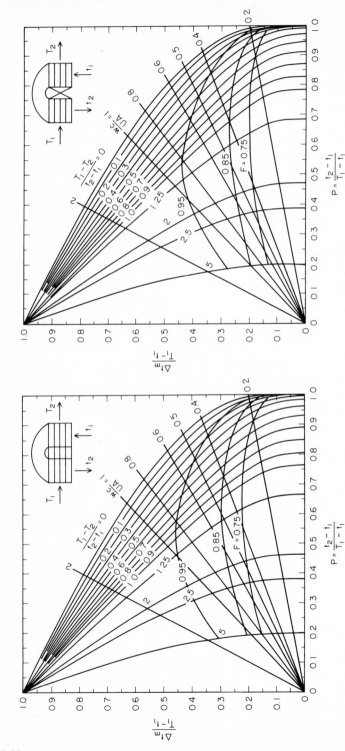

Fig. 15. Mean temperature difference in cross flow. Two pass counterflow. One fluid mixed throughout, other fluid unmixed throughout, inverted order between passes.

Fig. 16. Mean temperature difference in cross flow. Two pass counterflow. One fluid mixed throughout, other fluid unmixed throughout, identical order between passes.

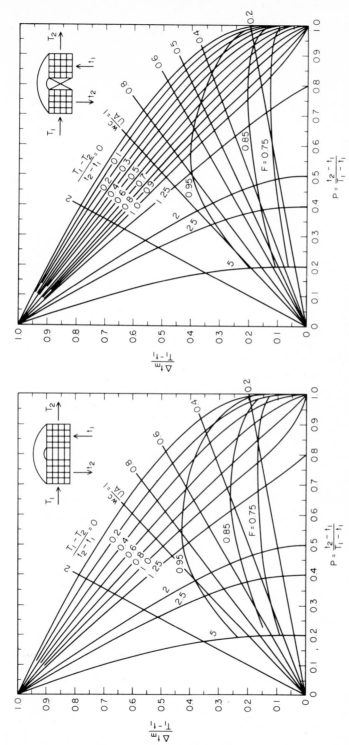

Fig. 17. Mean temperature difference in cross flow. Two pass countercurrent. Both fluids unmixed in passes. One fluid mixed between passes, other fluid unmixed between passes with inverted order.

Fig. 18. Mean temperature difference in cross flow. Two pass counterflow. Both fluids unmixed in passes. One fluid mixed between passes, other fluid unmixed between passes with identical order.

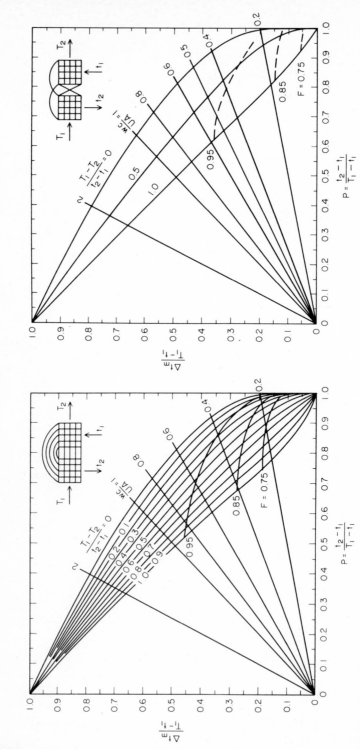

Fig. 19. Mean temperature difference in cross flow. Two pass counterflow. Both fluids unmixed throughout, both in inverted order between passes.

Fig. 20. Mean temperature difference in cross flow. Two pass counterflow. Both fluids unmixed throughout, both in identical order between passes.

8. Both fluids unmixed throughout, both identical order, numerical solution (Fig. 20).

9. Both fluids unmixed throughout, Fluid A inverted order, Fluid B identical order, Fig. 21, numerical solution.

Three-pass counter-cross-flow exchangers:

1. Both fluids mixed throughout (Fig. 22)

$$E_A = \frac{3E_a - 3E_a^2(1 + C_A/C_B) + E_a^3[1 + C_A/C_B + (C_A/C_B)^2]}{1 - (E_a C_A/C_B)(3 - E_a - E_a C_A/C_B)}$$

$$E_a = \frac{1}{1/[1 - \exp(-NTU_A)] + (C_A/C_B)/K - 1/NTU_A} \qquad \text{use of } 1/3 \text{ of total } NTU$$

2. Both fluids mixed between passes; Fluid A mixed in each pass, Fluid B unmixed in each pass (Fig. 23). Same as above equation, but $E_a = 1 - \exp(-KC_B/C_A)$.

3. Both fluids mixed between passes, both fluids unmixed in each pass (Fig. 24).

4. Fluid A mixed throughout, Fluid B unmixed throughout, inverted order (Fig. 25)

$$E_A = 1 - \frac{1}{\left(1 - \frac{K}{2}\right)^2 \exp\left(3K\frac{C_B}{C_A}\right) + \left[K\left(1 - \frac{K}{4}\right) + K^2\frac{C_B}{C_A}\left(1 - \frac{K}{2}\right)\exp\left(K\frac{C_B}{C_A}\right)\right]}$$

5. Fluid A mixed throughout, Fluid B unmixed throughout, identical order (Fig. 26)

$$E_A = 1 - \frac{1}{\left[\exp\left(K\frac{C_B}{C_A}\right) - 2K^2\frac{C_B}{C_A}\right]\exp\left(2K\frac{C_B}{C_A}\right) - \left(1 - K - \frac{K^2}{2}\frac{C_B}{C_A}\right)K^2\frac{C_B}{C_A}\exp\left(K\frac{C_B}{C_A}\right)}$$

6. Fluid A mixed between passes, unmixed in each pass; Fluid B unmixed throughout, inverted order, numerical solution (Fig. 27).

7. Fluid A mixed between passes, unmixed in each pass; Fluid B unmixed throughout, identical order, numerical solution (Fig. 28).

8. Both fluids unmixed throughout; both inverted order, numerical solution (Fig. 29).

9. Both fluids unmixed throughout; both identical order, numerical solution (Fig. 30).

10. Both fluids unmixed throughout; Fluid A inverted order, Fluid B identical order, numerical solution (Fig. 31).

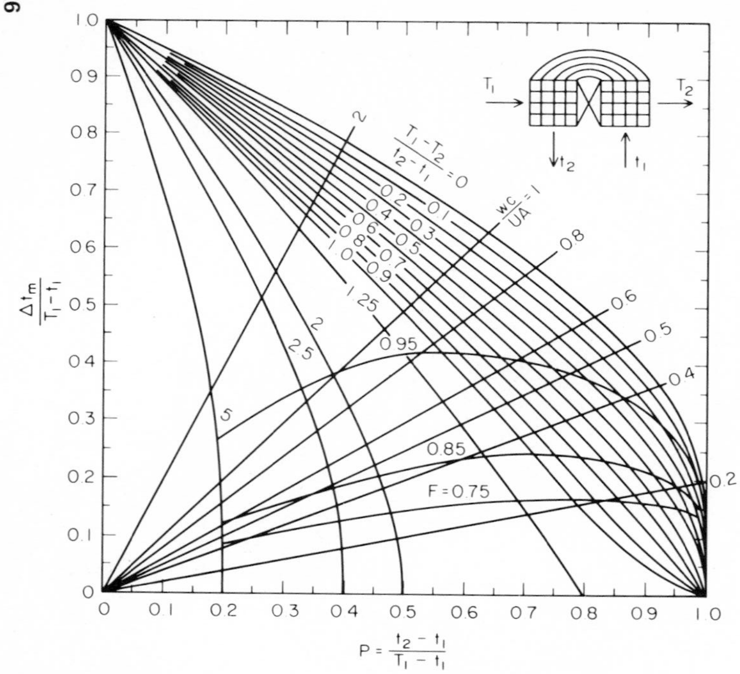

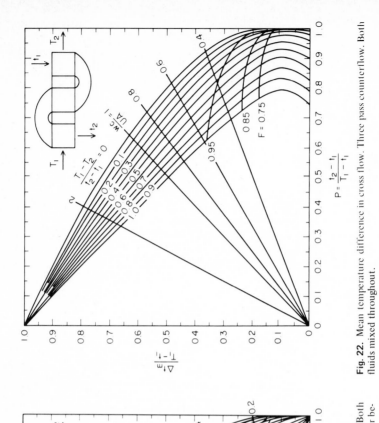

Fig. 21. Mean temperature difference in cross flow. Two pass counterflow. Both fluids unmixed throughout, one fluid identical order, other in inverted order between passes.

Fig. 22. Mean temperature difference in cross flow. Three pass counterflow. Both fluids mixed throughout.

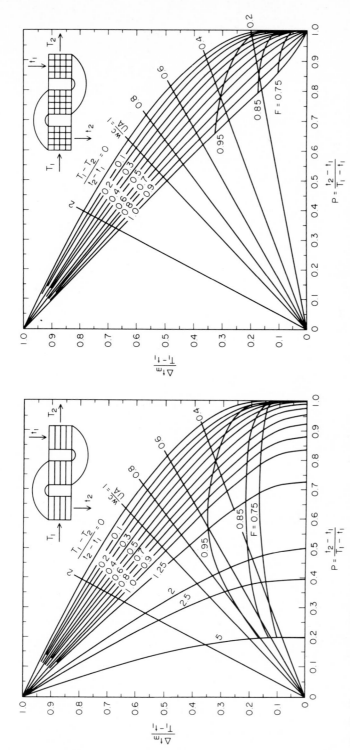

Fig. 24. Mean temperature difference in cross flow. Three pass countercurrent. Both fluids unmixed in each pass and both mixed between passes.

Fig. 23. Mean temperature difference in cross flow. Three pass counterflow. One fluid mixed throughout, other fluid unmixed in passes but mixed between passes.

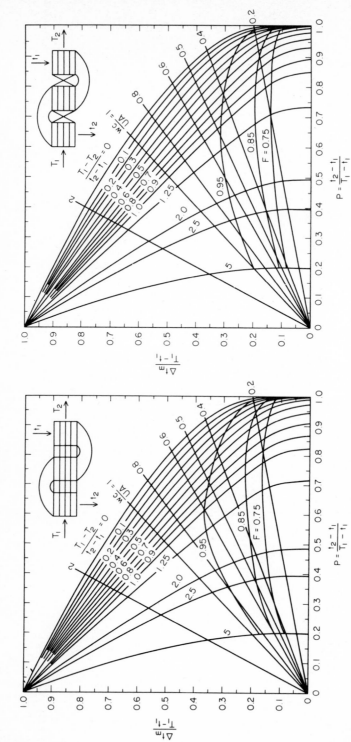

Fig. 25. Mean temperature difference in cross flow. Three pass counterflow. One fluid mixed throughout, other fluid unmixed throughout with inverted order between passes.

Fig. 26. Mean temperature difference in cross flow. Three pass counterflow. One fluid mixed throughout, other fluid unmixed throughout with identical order between passes.

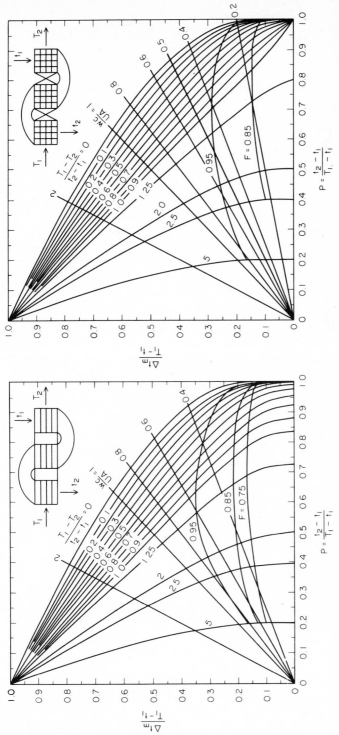

Fig. 27. Mean temperature difference in cross flow. Three pass counterflow. Both fluids unmixed in each pass, one fluid mixed between passes, other fluid unmixed between passes with inverted order.

Fig. 28. Mean temperature difference in cross flow. Three pass counterflow. One fluid unmixed in each pass but mixed between passes, other fluid unmixed throughout with identical order between passes.

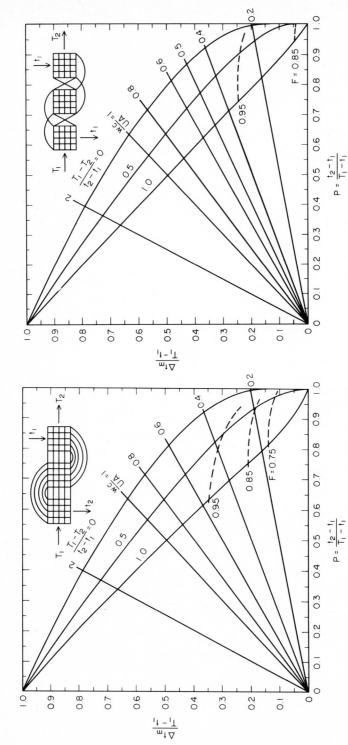

Fig. 29. Mean temperature difference in cross flow. Three pass counterflow. Both fluids unmixed throughout and both in inverted order between passes.

Fig. 30. Mean temperature difference in cross flow. Three pass counterflow. Both fluids unmixed throughout and both in identical order between passes.

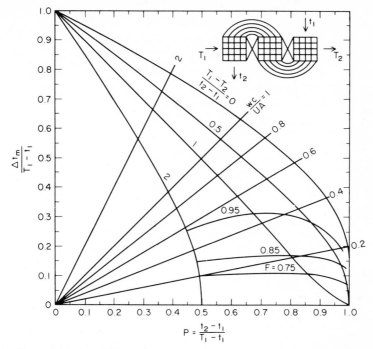

Fig. 31. Mean temperature difference in cross flow. Three pass counterflow. Both fluids unmixed throughout with one fluid in identical order and the other in inverted order between passes.

BATCH HEATING OR COOLING

Many processes operate batchwise and involve heat transfer with a resulting complication in determining the average temperature differences. A number of cases have been solved and are listed below.

Case 1. A simple tank with jacket or coil. The assumptions are:

1. Rate of flow and inlet temperature of heating or cooling medium are constant.
2. Specific heat of each fluid is constant.
3. Only sensible heat is involved.
4. The batch fluid is well agitated and of uniform temperature.
5. Heat losses are negligible.
6. Overall coefficient is constant.

As illustrated in Fig. 32, let MC be the batch heat capacity, T batch temperatures, wc and t the heat capacity and temperature of the heat exchange medium

$$\frac{-UA}{wc} = \ln\left(1 - \frac{MC}{wc\theta} \ln\frac{T_1 - t_1}{T_2 - t_1}\right)$$

Let

$$X = \frac{MC}{wc\theta} \ln\frac{T_1 - t_1}{T_2 - t_1}$$

$$-\frac{UA}{wc} = \ln(1 - X)$$

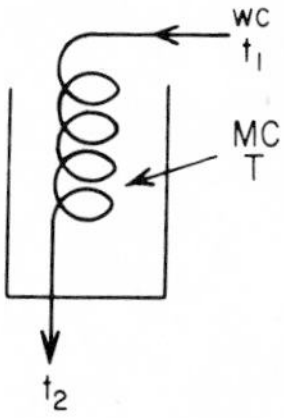

Fig. 32. Batch heating with coil.

Case 1a. When the heat transfer medium is at constant temperature such as condensing steam, then

$$\frac{UA\theta}{MC} = \ln\left(\frac{t_1 - T_1}{t_1 - T_2}\right)$$

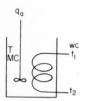

Fig. 33. Batch cooling with a constant heat input.

Case 2. Simple batch cooling with constant heat input by agitation or reaction (Fig. 33). Same as Case 1 except assumption 3 is changed so that a constant rate of heat input q_a is included; e.g., by agitation power

$$\frac{-UA}{wc} = \ln\left[1 - \frac{MC}{wc\theta}\ln\left(\frac{b(T_1 - t_1) - q_a}{b(T_2 - t_1) - q_a}\right)\right]$$

where $b = wc[1 - \exp(-UA/wc)]$

Case 3. A batch tank with an external heat exchanger (Fig. 34) through which the batch is circulated at rate W. Assumptions in addition to Case 1 are:

7. Negligible hold up in exchanger circuit.
8. Counterflow in exchanger.
9. Log mean temperature difference applies.
10. Perfect mixing in tank.

General equation

$$-UA = \frac{WC}{1 - WC/wc}\ln\left[\frac{wc}{Wc}\left(1 - \frac{1 - WC/wc}{1 + (MC/wc\theta)\ln[(T_2 - t_1)/(T_1 - t_1)]}\right)\right]$$

when $WC = wc$

$$\frac{-UA}{WC} = \frac{(MC/WC\theta)\ln[(T_2 - t_1)/(T_1 - t_1)]}{1 + (MC/WC\theta)\ln[(T_2 - t_1)/(T_1 - t_1)]}$$

or when wc has no value as in steam-heated exchangers

$$\frac{-UA}{WC} = \ln\left[1 - \frac{MC}{WC\theta}\ln\left(\frac{t_1 - T_1}{t_1 - T_2}\right)\right]$$

Case 4. Same as 3 but as in 2 a constant rate of heat addition q into tank (Fig. 35)

$$B\theta = \ln\left[\frac{B(T_1 - t_1) - q/MC}{B(T_2 - t_1) - q/MC}\right]$$

where

$$B = \frac{wc}{MC}\left\{1 - \frac{1 - WC/wc}{1 - (WC/wc)\exp[(-UA/wc)(1 - WC/wc)]}\right\}$$

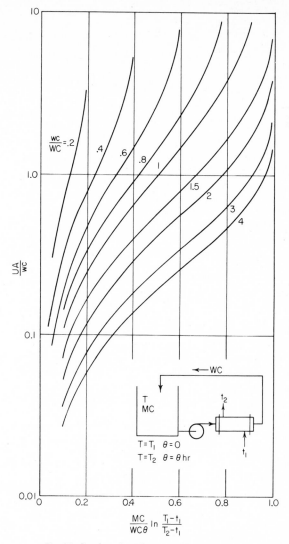

Fig. 34. Batch tank with external heat exchangers.

Case 5. A batch tank with an external steam heated exchanger. Solution with heat capacity wc at temperature T_c is added to contents and is well mixed (Fig. 36)

$$\frac{-UA}{WC + wc} = \ln\left\{1 - \frac{MC}{WC\theta} \ln\left[\frac{t_s - T_c + (WC/wc)(t_s - T_1)}{t_s - T_c + (WC/wc)(t_s - T_2)}\right]\right\}$$

Case 6. A batch tank with coil and continuous flow of solution through the tanks (Fig. 37)

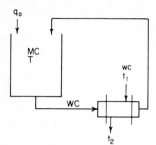

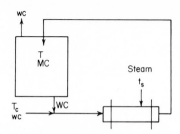

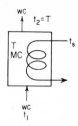

Fig. 35. Batch tank with external heat exchanger and a constant heat input.

Fig. 36. Batch tank with external steam heated exchanger with a constant flow of solution through the tank.

Fig. 37. Batch tank with coil and continuous flow of solution through tank.

$$\ln\left[\frac{UA(t_s - T_1) - wc(T_1 - t_1)}{UA(t_s - T_2) - wc(T_2 - t_1)}\right] = \frac{\theta(UA + wc)}{MC}$$

TYPES OF EXCHANGERS

Shell-and-tube heat exchangers are classified by the methods used for reducing thermal stresses between tubes and the shell. Selection of the type to use depends on such as cost, cleanability, temperatures, pressures, and hazards. Figures 38 to 41 show various types of exchangers with a variety of heads. The double-tube sheets are used when leakage between fluids must be prevented and where doubt exists as to possible leakage at the tube sheet joints. Table 1 lists the nomenclature used for the parts of the exchangers.

A fixed-tube-sheet exchanger is the simplest to fabricate and the cheapest. Both the shell and tube fluids are tightly held. Internal leakage could be due only to leaky tube joints or perforations in the tubes. The primary disadvantages are high thermal stresses between tubes and shell which loosen the tube joints and the impossibility of mechanically cleaning the outside of the tubes. If the shell fluids are clean or if the scale deposits can be removed chemically, then a removable tube bundle is not necessary.

The fixed-tube-sheet exchanger (Fig. 38) with an expansion bellows reduces the stresses between the tubes and shell, but is limited usually to shell pressures below 100 lb/in.^{2}, otherwise the expansion bellows would get too thick to be flexible although some specific design of flexible bellows can be used up to 600 lb/in.^{2}, with several inches allowed for expansion.

The internal floating head (Fig. 39) eliminates the stress between the tubes and shell, the tube bundle is removable for cleaning, and the shell fluid is tightly held preventing external leakage. The principal disadvantage is the internal hidden gasket on the floating end which may fail and cause trouble depending upon the toxicity or reactivity of the fluids.

The U-tube bundle (Fig. 40) is removable, eliminates the shell-to-tube stress, and requires only one shell joint at the fixed end. The disadvantage is that a U-tube is difficult to replace and to clean.

For the outside packed head exchanger (Fig. 41) the gaskets are accessible and the bundle is removable. The packing gland can be designed for 600 lb/in.^{2} pressure, but there always exists the possibility of the shell fluid leaking to the surroundings.

Therefore, a packed head is not recommended when the shell fluid is very toxic, corrosive, expensive, or inflammable.

The choice of the fluid to be on the shell side will influence the selection of the type of exchanger and requires evaluation of the following factors to arrive at a satisfactory compromise.

1. Cleanability. The shell is difficult to clean and requires the cleanest fluid.

2. Corrosion. Corrosion or process cleanliness may dictate the use of expensive alloys; therefore, these fluids are placed inside the tubes in order to save the cost of an alloy shell.

3. Pressure. High pressure shells, because of their diameters, are thick walled and expensive; therefore, high pressure fluids are placed in the tubes.

TABLE 1. Heat Exchanger Nomenclature

1. Shell
2. Shell flange—front head end
3. Shell flange—rear head end
4. Stationary tube sheet
5. Outer stationary tube sheet
6. Shell nozzle
7. Expansion joint
8. Support bracket
9. Support saddle
10. Support flange
11. Bonnet
12. Channel
13. Channel cover
14. Bonnet or channel nozzle
15. Bonnet or channel flange
16. Pass partition
17. Lifting lug
18. Tubes
19. Transverse baffle or support plate
20. Tie rod and spacers
21. Impingement baffle
22. Vent connection
23. Drain connection
24. Instrument connection
25. Floating tube sheet
26. Outer floating tube sheet
27. Packing box flange
28. Packing
29. Packing follower
30. Floating tube sheet skirt
31. Split ring
32. Slip-on backing flange
33. External floating head cover
34. Shell cover
35. Shell cover flange
36. Internal floating head cover
37. Internal floating head flange
38. Internal floating head backing device
39. Internal floating head nozzle
40. Mating nozzle flange
41. Tube bundle support plate
42. Weir
43. Liquid level connection

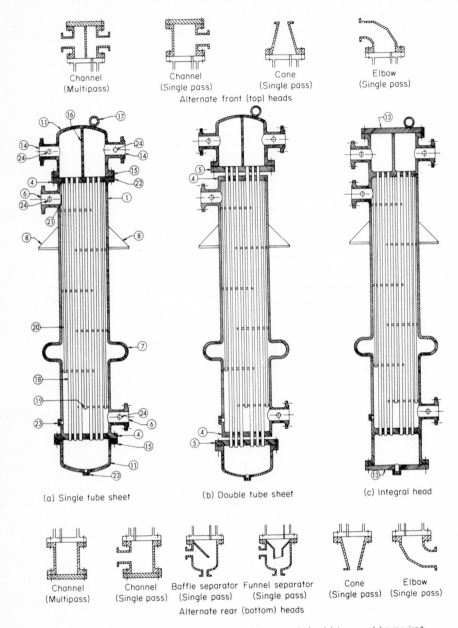

Channel
(Multipass)

Channel
(Single pass)

Cone
(Single pass)

Elbow
(Single pass)

Alternate front (top) heads

(a) Single tube sheet

(b) Double tube sheet

(c) Integral head

Channel
(Multipass)

Channel
(Single pass)

Baffle separator
(Single pass)

Funnel separator
(Single pass)

Cone
(Single pass)

Elbow
(Single pass)

Alternate rear (bottom) heads

Note: Exchangers may be vertical, horizontal or inclined. Shell expansion joint may not be required.
See Table I for Heat Exchanger nomenclature.

Fig. 38. Fixed-tube-sheet exchangers.

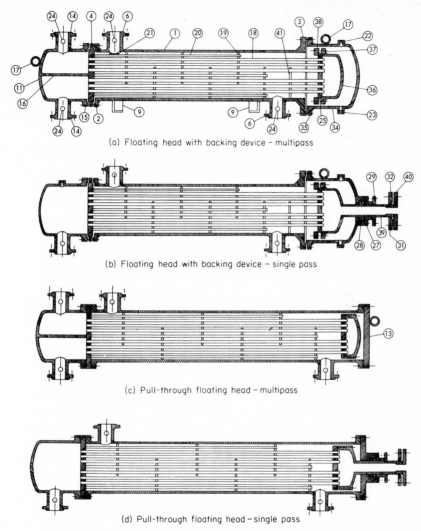

(a) Floating head with backing device – multipass

(b) Floating head with backing device – single pass

(c) Pull-through floating head – multipass

(d) Pull-through floating head – single pass

Note: Exchangers may be vertical, horizontal or inclined.
      See Table I for Heat Exchanger nomenclature.

Fig. 39. Internal floating heat exchanger.

4. Temperatures. The high temperature fluid should be inside the tubes. High temperatures reduce the allowable stresses in materials and the effect is similar to high pressure in determining shell thicknesses. Furthermore, safety of personnel may require the additional cost of insulation if the high temperature fluid is in the shell.

5. Hazardous or expensive fluids. The most hazardous or expensive fluids should be placed on the tightest side of the exchanger which is the tube side of some types of exchangers.

6. Quantity. A better overall design may be obtained when the smaller quantity of fluid is placed in the shell. This effect may be due to avoiding multipass construction

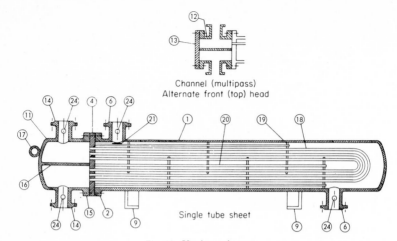

Fig. 40. U-tube exchanger.

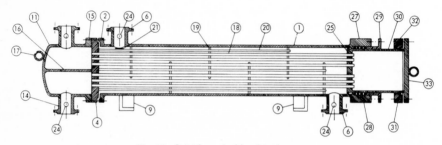

Fig. 41. Outside packed head exchangers.

with consequent loss of temperature drop efficiency or to obtaining turbulent flow in the shell at low Reynolds numbers.

7. Viscosity. The critical Reynolds number for turbulent flow in the shell is about 200; hence when the flow in the tubes is laminar, it may be turbulent if placed in the shell. However, if the flow is still laminar when in the shell, then it is best to place the fluid back inside the tubes as it will be somewhat easier to predict both heat transfer and flow distribution.

8. Pressure Drop. If pressure drop is critical and must be accurately predicted, then place that fluid inside the tubes. Pressure drop inside of tubes can be calculated with less error as the pressure drop in the shell will deviate widely from theoretical values depending upon the shell leakage clearances in the particular exchanger.

TYPES OF BAFFLES

Four types of baffles used in heat exchangers—segmental, disk-and-doughnut, orifice, and strip—are illustrated in Fig. 42.

The segmental baffle is formed by cutting a segment from a disk. The depth of cut as a percentage of the baffle diameter is known as the percent baffle cut. The baffles are arranged in the shell so that alternate baffles have their cut edges rotated 180°. The flow of shell fluid is back and forth across the tubes more or less perpendicularly. One disadvantage of this baffle is the bypassing that occurs because the tube bundle cannot completely fill the shell. A clearance is needed, especially in removable

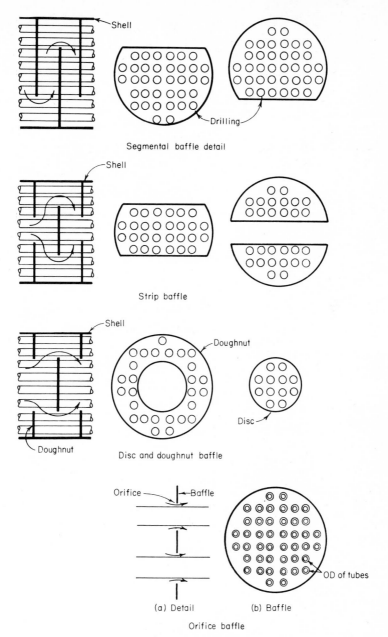

Segmental baffle detail

Strip baffle

Disc and doughnut baffle

Orifice baffle

Fig. 42. Types of baffling used in exchangers.

bundles, to provide retaining edges at the floating end to give the tube sheet strength for the outer row of tubes and to give space for clamping or welding the return heads or channels. This baffle is most common and can be considered as standard. It is efficient and gives good heat transfer rates for the pressure drop and power consumed and all the baffles act as support plates for the tubes.

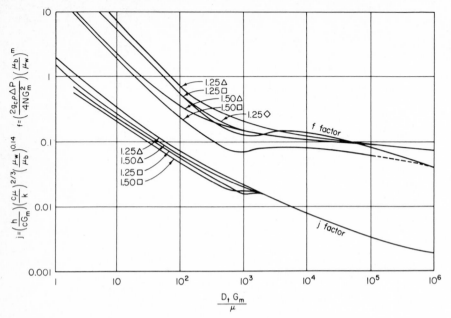

Fig. 43. Heat transfer and friction factors for cross flow in tube bundles.

The strip baffle is much like the segmental baffle but can handle larger flows. Two small segments are placed on opposite sides in the shell and downstream the large center segment is installed. Major advantage is the larger flows that can be handled. It is calculated as though it were a simple segmented baffle.

The disk-and-doughnut baffle consists of alternate disks and doughnut-shaped baffles. Usually the disk is larger than the doughnut hole. This baffle is occasionally used but the test data available for its design are scarce. The bypassing of the tube bundle cannot occur with disk-and-doughnut baffling. The added clearance between bundle and shell only serves to give a larger opening for flow past the disk. The main disadvantages to this design are that all the tie rods lie within the bundle, and the central tubes are supported through the tubes lying in the overlap between the disk and the doughnut.

Orifice baffles consist of disks with oversize holes through which the tubes pass. The fluid flows through the annular orifices and the resulting turbulence influences heat transfer. These baffles are rarely used because they are the least efficient, they cannot be cleaned when plugged with dirt and scale, and they do not provide support for the tubes. In fact, the tubes could rattle and vibrate easily.

SHELL-SIDE HEAT TRANSFER AND PRESSURE DROP

Practical design and fabrication of heat exchangers result in too many variables to allow the use of any simple method to predict either pressure drop or heat transfer even within a ±100 percent error. Tube size and pattern, baffle tolerances, bundle and shell dimensions, pass arrangements, and baffle type, spacing, and cuts all affect the various leakage and bypass streams. These streams are in turn functions of flow rates, physical properties, and tolerances. It is possible by means of complex reiterative programs for high-speed computers to improve prediction methods markedly, but no such program is now in the literature. Therefore, the methods given below can be used for approximation of heat transfer and pressure drop although the reliability is uncertain due to built-in limitations of the method. Generally the predicted results are conservative.

The basis for these methods is the assumption of pure cross flow in tube bundles and correcting for the effect of leakage and bypass. The j and f factor plots for banks of tubes in arrangements used in shell and tube exchangers are given in Fig. 43.

The effect of the various types of leakages and bypass streams on heat transfer and pressure drop are a function of the type of baffling. The leakage between the baffles and the shell is present in all baffle arrangements and seriously affects heat transfer since this stream avoids contact with the heat transfer surfaces. The leakage between the tubes and baffles affects pressure drop more than heat transfer since this stream is leaking past the tubes and some effective heat transfer occurs, as has been demonstrated by the orifice type baffling (e.g., Lee and Knudsen [12]). The bypass stream between the outer tube limit of the bundle and the shell is present only for the segmental baffling. However, the effect of this stream on heat transfer is very important, particularly when cooling viscous fluids. This bypass stream can be a large fraction of the net stream flow through the tube bundle; therefore, special effort must be made to keep this clearance small and to provide, if necessary, auxiliary restrictions such as bypass strips or extra large tie rods between the baffle cuts. A similar bypass stream can exist in multipass exchangers if the gap in the tube bundle for the partition plates in the heads are oriented in the same direction as the shell fluid cross flow.

The method of Bell [13] is reasonably good for standard exchangers built to standard tolerances but can fail in extreme conditions. The sequence of steps in Bell's procedure is:

1. Calculate $\mathrm{Re} = D_t G_m/\mu$, where G_m is computed, assuming that the entire flow is flowing normal to the tubes in the center line cross flow section with no allowance made for the bypass stream.

2. Find j from Fig. 43 for the ideal tube bank. These curves have been based on a ten row bank or $N_c = 10$.

3. Calculate minimum value of F_{BP}

$$F_{BP} = \frac{S_{BP}}{S_m} = \frac{\text{bypass flow area around the tube bundle}}{\text{minimum cross flow area in bundle near exchanger center line}}$$

4. Calculate

$$\xi_h = \exp\left[-\alpha F_{BP}\left(1 - \sqrt[3]{\frac{2N_s}{N_c}}\right)\right] \quad \text{for} \quad N_s \leq \frac{N_c}{2}$$

$$= 1 \quad \text{for} \quad N_s > \frac{N_c}{2}$$

and

$$\alpha = 1.5 \text{ for laminar flow}$$
$$\alpha = 1.35 \text{ for transition or turbulent flow}$$

where N_s = number of sealing strips encountered by the bypass stream during flow across one cross flow section.

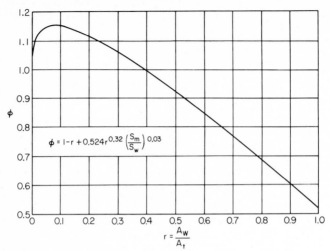

Fig. 44. Plot of ϕ vs. r at $(S_m/S_w)^{0.03} = 1$. (*from "Final Report on Tube Heat Exchangers," by Ball, University of Delaware Engineering Bulletin No. 5*).

N_c = number of constrictions encountered during flow through one cross flow section between baffle tips.

5. Calculate A_t and A_w

A_t = total heat transfer area in shell
A_w = heat transfer area in the window sections of the shell

6. Find ϕ from Fig. 44 or from equation

$$\phi = 1 - \left(\frac{A_w}{A_t}\right) + 0.524 \left(\frac{A_w}{A_t}\right)^{0.32} \left(\frac{S_m}{S_w}\right)^{0.03}$$

7. Calculate $N_c^1 = (N_B + 1) N_c + (N_B + 2) N_w$

where N_B = number of baffles in exchanger.
N_w = effective number of constrictions for cross flow in window area. Tube rows having equal or more than ½ of the number of restrictions that exist in the central tube row.
N_c = number of tube rows between baffle tips in in-line square or triangular pitch and one less for staggered square.

8. Calculate: If Re $\leq$ 100

$$\chi = \left(\frac{N_c^1}{N_c}\right)^{0.18}$$

see steps 4 and 7 for definitions.
If $100 < $ Re < 1000

$$\chi = 1.0$$

If Re $\geq$ 1000

$$X = \frac{(h_m/h_\infty) \text{ ideal tube bank}}{(h_m/h_\infty) \text{ exchanger}}$$

where $[(h_m/h_\infty)$ ideal tube bank] is a function of N_c and $[(h_m/h_\infty)$ exchanger] is a function of N_c^1 determined from the following.

N_c	$(h/h_\infty)^*$	h_m/h_∞
1	0.63	0.63
2	0.76	0.70
3	0.93	0.77
4	0.98	0.83
5	0.99	0.86
6	1.0	0.88
7	1.0	0.90
8	1.0	0.91
9	1.0	0.92
10	1.0	0.93
12	1.0	0.94
15	1.0	0.95
18	1.0	0.96
25	1.0	0.97
35	1.0	0.98
72	1.0	0.99

*Value of h at a given row, h_m is the mean value for n rows.

9. Calculate h_{NL}

$$\frac{h_{NL}}{cG_m} = j\frac{\phi\xi_h}{X}\left(\frac{c\mu}{k}\right)^{-2/3}\left(\frac{\mu_w}{\mu_b}\right)^{-0.14}$$

10. Calculate S_{TB} = tube to baffle leakage flow area per baffle.
S_{SB} = shell to baffle leakage flow area per baffle.
S_L/S_m = total leakage area for baffle/minimum crossflow area at center line.
$(S_{TB} + 2S_{SB})/S_L$

11. Find $\left(1 - h_L/h_{NL}\right)_0$ from Fig. 45 or

$$\left(1 - \frac{h_L}{h_{NL}}\right)_0 = 0.45\left(\frac{S_L}{S_m}\right) + 0.10\left[1 - \exp\left(-30\frac{S_L}{S_m}\right)\right]$$

12. Calculate

$$\left(1 - \frac{h_L}{h_{NL}}\right) \text{exchanger} = \left(1 - \frac{h_L}{h_{NL}}\right)_0\left(\frac{S_{TB} + 2S_{SB}}{S_L}\right)$$

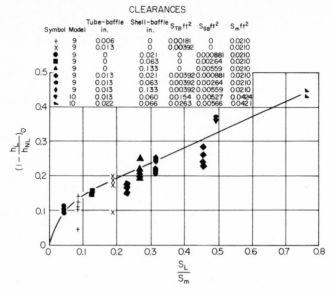

CLEARANCES

Symbol	Model	Tube-baffle in.	Shell-baffle in.	S_{TB} ft²	S_{SB} ft²	S_m ft²
+	9	0.006	0	0.00181	0	0.0210
x	9	0.013	0	0.00392	0	0.0210
●	9	0	0.021	0	0.000881	0.0210
■	9	0	0.063	0	0.00264	0.0210
▲	9	0	0.133	0	0.00559	0.0210
◆	9	0.013	0.021	0.00392	0.000881	0.0210
●	9	0.013	0.063	0.00392	0.00264	0.0210
◆	9	0.013	0.133	0.00392	0.00559	0.0210
▼	10	0.013	0.060	0.0154	0.00527	0.0424
◣	10	0.022	0.066	0.0263	0.00566	0.0421

Fig. 45. Effect of baffle leakage on overall shell side heat transfer coefficient (*from "Final Report on Tube Heat Exchangers" by Ball, University of Delaware Engineering Report No. 5*).

13. Calculate h_L from $(1 - h_L/h_{NL})$ exchanger and h_{NL}.

14. Find f^1 from Fig. 43 for ideal bank.

15. Calculate ΔP_B for one cross-flow section

$$\Delta P_B = 4f^1 N_c \frac{G_m^2}{2g_c\rho}\left(\frac{\mu_w}{\mu_b}\right)^{0.14}$$

16. Calculate ξp. Same equation as in 4 above for ξ except at $R_e < 100$ $\alpha = 5.0$ and at $R_e > 100$ $\alpha = 4.0$. Note it is conservative to take $\xi p = 1$ if N_L is small and stream contractions are significant.

17. Calculate $\Delta P_{BP} = \xi p \Delta P_B$.

18. Calculate V_w and $V_z = \sqrt{V_m V_w}$.

19. Calculate ΔP_w: for $R_e \leq 100$

$$\Delta P_w = 28\left[\frac{V_z\mu}{(p - D_t)g_c}\right]N_w + 26\left(\frac{V_z\mu}{D_V g_c}\right)\left(\frac{L_B}{D_V}\right) + 2\left(\frac{\rho V_z^2}{2g_c}\right)$$

where $D_V = 4$ (net free volume)/(friction surface)
L_B = baffle spacing, ft

for $R_e > 100$

$$\Delta P_w = (2 + 0.6N_w)\frac{\rho V_z^2}{2g_c}$$

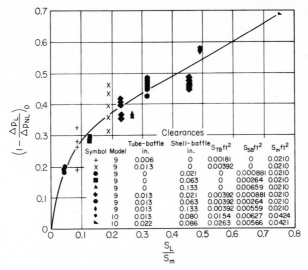

Clearances table (in figure):

Symbol	Model	Tube-baffle in.	Shell-baffle in.	S_{TB} ft²	S_{SB} ft²	S_m ft²
+	9	0.006	0	0.00181	0	0.0210
X	9	0.013	0	0.00392	0	0.0210
●	9	0	0.021	0	0.000881	0.0210
■	9	0	0.063	0	0.00264	0.0210
▲	9	0	0.133	0	0.00659	0.0210
◆	9	0.013	0.021	0.00392	0.000881	0.0210
●	9	0.013	0.063	0.00392	0.00264	0.0210
◆	9	0.013	0.133	0.00392	0.00559	0.0210
▼	10	0.013	0.080	0.0154	0.00627	0.0424
◤	10	0.022	0.086	0.0263	0.00566	0.0421

Fig. 46. Effect of baffle leakage on overall shell side pressure drop (*from "Final Report on Tube Heat Exchangers" by Ball, University of Delaware Engineering Bulletin No. 5*).

20. Find $\left(1 - \Delta P_L / \Delta P_{NL}\right)_0$ from Fig. 46 or from

$$\left(1 - \frac{\Delta P_L}{\Delta P_{NL}}\right)_0 = 0.57\left(\frac{S_L}{S_m}\right) + 0.27\left[1 - \exp\left(-20\,\frac{S_L}{S_m}\right)\right]$$

21. Calculate

$$\left(1 - \frac{\Delta P_L}{\Delta P_{NL}}\right) \text{ exchanger } = \left(1 - \frac{\Delta P_L}{\Delta P_{NL}}\right)_0 \left(\frac{S_{TB} + 2S_{SB}}{S_L}\right)$$

22. Calculate $\Delta P_L / \Delta P_{NL}$ from $(1 - \Delta P_L / \Delta P_{NL})$ exchanger.

23. Calculate

$$\Delta P \text{ total } = 2\,\Delta P_{BP}\left(1 + \frac{N_w}{N_L}\right) + \left[(N_B - 1)\Delta P_{BP} + N_B \Delta P_w\right]\left(\frac{\Delta P_L}{\Delta P_{NL}}\right)$$

Note that first term on right side treats the two end cross-flow sections as nonleakage sections.

The basic method of Tinker [14] is sound but his equations have many so-called constants which are determined by the exchanger tolerances and flow leakage resistance paths. Devore [15] simplified Tinker's equations by using standard tolerances for commercial exchangers plus certain baffle cut ratios. With some further simplifications, the Devore modification of Tinker's method is given below.

1. For specific shell diameter D_1, get C_{14} from Fig. 47 for the particular exchanger design and for the effect of side strip baffling. Then, from Fig. 48 for the proper P/d

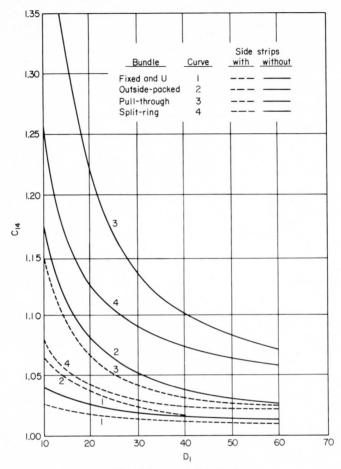

Fig. 47. Plot of C_{14} vs. shell inside diameter.

ratio and tube size and pitch, find N_h using the above C_{14} value. With this N_h calculate the effective area. Find D_3, M, and C_A from Tables 2 and 3.

$$A_e = M C_A l_3 D_3 \left(1 + N_h \sqrt{\frac{D_1}{P}} \right)$$

2. Calculate $G_{\text{eff}} = 144W/A_e$ and $D_t G_{\text{eff}}/\mu$.
3. From Fig. 43 get j and calculate

$$h_{ob} = j c G_{\text{eff}} (\text{Pr})^{2/3} \left(\frac{\mu_b}{\mu_w} \right)^{0.14}$$

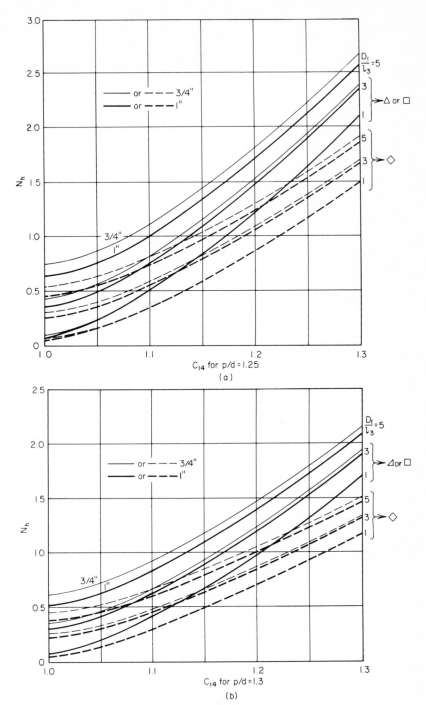

Fig. 48. Plot of N_h vs. C_{14}.

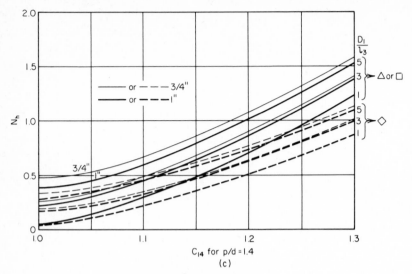

Fig. 48 (*Cont.*). Plot of N_h vs. C_{14}.

TABLE 2. Exchanger Bundle Diameters (Approximate) (D_3)

D_1 Inside dia. in.	Fixed and U tube	Outside packed head	Pull through floating head	Split ring floating head
10.02	9.62	8.52	6.42	8.02
12.09	11.67	10.59	8.49	10.04
13.38	12.95	11.88	9.78	11.30
15.25	14.81	13.75	11.65	13.11
17.25	16.79	15.75	13.65	15.06
19.25	18.78	17.75	15.65	17.00
21.25	20.75	19.75	17.65	18.96
23.00	22.50	21.50	19.40	20.66
27.00	26.46	25.50	23.40	24.56
31.00	30.43	29.50	27.40	28.45
35.00	34.40	33.50	31.30	32.33
39.00	38.37	37.50	35.30	36.25
42.00	41.34	40.50	38.25	39.14
48.00	47.30	46.50	44.20	45.04
51.00	50.27	49.50	47.20	47.93
54.00	53.24	52.50	50.10	50.83
60.00	59.21	58.50	56.00	56.72

4. Where the end baffle zones are much longer than the baffle pitch, Tinker corrects for this effect by $h_o = E_s h_{ob}$ where

$$E_s = \frac{l_2 + (l_1 - l_2)\,[2l_3/(l_1 - l_2)]^{0.6}}{l_1}$$

where l_1 = total tube length, l_2 = baffled tube length.

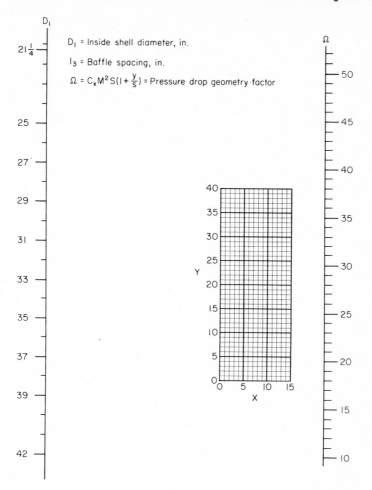

D_1 = Inside shell diameter, in.

l_3 = Baffle spacing, in.

$\Omega = C_x M^2 S(1 + \frac{y}{S})$ = Pressure drop geometry factor

(Pressure Drop Geometry Factor)

d_2 tube O.D., in.	P, pitch, in.	D_1/l_3											
		1.0		1.5		2.0		3.0		4.0		5.0	
		X	Y	X	Y	X	Y	X	Y	X	Y	X	Y
¾"	$^{15}/_{16}$-△	10.0	8.0	6.9	17.5	5.0	24.3	3.5	28.0	2.7	31.0	2.1	31.6
	1-△	10.5	6.7	7.6	15.9	5.7	22.5	4.3	26.0	3.3	28.9	3.1	29.4
	1-□	10.8	6.4	8.1	14.0	6.8	19.5	5.7	22.1	4.8	24.6	4.3	24.3
	1-◇	9.2	11.0	5.9	21.5	3.5	29.3	2.0	34.0	1.0	37.5	0.5	37.9
1"	1¼-△	12.9	0.1	10.0	8.2	8.3	14.1	7.0	17.8	6.5	20.2	6.0	20.6
	1¼-□	12.7	0.0	10.5	6.0	8.9	11.3	8.0	13.9	7.5	15.9	7.1	15.6
	1¼-◇	11.5	3.8	8.6	13.2	6.2	20.3	4.9	24.7	4.0	27.7	3.6	28.1

Fig. 49. Nomograph of factor Ω versus inside shell diameter (*from Devore*, Petroleum Refiner, **40**:*221 (1961)*).

TABLE 3. Values of M*

(D_1/l_3)	$\rightarrow \triangleright$	$\rightarrow \square$	$\rightarrow \lozenge$
1	0.890	0.951	0.853
1.5	0.925	0.967	0.897
2	0.942	0.972	0.922
3	0.960	0.983	0.946
4	0.972	0.987	0.959
5	0.985	0.997	0.975

$$C_A = \frac{0.97\,(P/d) - 1}{P/d} \quad \text{for} \quad \rightarrow \triangleright \quad \rightarrow \square$$

$$C_A = \frac{1.37\,(P/d) - 1}{P/d} \quad \text{for} \quad \rightarrow \lozenge$$

*From Devore, *Petroleum Refiner*, **40**:221 (1961).

TABLE 4. Ratio of Fouled to Clean Pressure Drop*

Fouling resistance	$D_1/l_3 = 1.0$	1.5	2.0	3.0	4.0	5.0
	Viscous-flow region					
0	1.00	1.00	1.00	1.00	1.00	1.00
0.001	1.06	1.11	1.15	1.20	1.24	1.28
0.002	1.13	1.22	1.29	1.39	1.48	1.55
0.003	1.19	1.32	1.44	1.59	1.72	1.83
0.004	1.25	1.43	1.58	1.79	1.96	2.11
> = 0.005	1.32	1.54	1.73	1.99	2.20	2.38
	Transitional and turbulent-flow region					
0	1.00	1.00	1.00	1.00	1.00	1.00
0.001	1.12	1.20	1.28	1.38	1.47	1.55
0.002	1.24	1.42	1.58	1.82	2.03	2.21
0.003	1.37	1.65	1.92	2.31	2.66	2.96
0.004	1.50	1.90	2.28	2.85	3.36	3.82
> = 0.005	1.64	2.17	2.68	3.44	4.14	4.77

*From Devore, *Petroleum Refiner*, **40**:221 (1961).

5. For Re calculated under 2, calculate $\text{Re}_P = M\,\text{Re}$ and find f from Fig. 43 and Ω from Fig. 49.

6. From number of baffles N_B calculate

$$\Delta P_{\text{clean}} = \frac{4.31\,\Omega N_B\, f(G_{\text{eff}}^2/10^4)}{\rho\,(\mu_b/\mu_w)^{0.14}}$$

7. From Table 4 find multiplier for fouled exchanger and calculate ΔP_{fouled}.

The above methods are usually conservative in predicting heat transfer but not on pressure drop where the values are underpredicted. It should be possible to improve the correlation with performance by reevaluating the various flow resistances and tolerances and redetermining the constants used in deriving the above relations.

Devore Method as Example. Two hundred thousand lb/hr MeOH to be cooled $200 \rightarrow 100°F$ with salt water $80° \rightarrow 100°F$. Use ¾-in. tubes on 15/16 $\triangle$ pitch 16 ft long. Take 6¼-in. baffle spacing and a 31-in. split ring floating head unit with bypass strips

$$\frac{D_1}{l_3} = \frac{31}{6.25} = 5$$

1. From Fig. 47 for $D_1 = 31$ in. to SRFH with strips $C_{14} = 1.03$. $P/d = 1.25$; then, from Fig. 48a for $C_{14} = 1.03$ and ¾-in. tubes at $D_1/l_3 = 5$, read $N_h = 0.79$. From Table 3, $M = 0.985$, $C_A = 0.194$. Table 2 gives bundle diameter $D_3 = 28.45$ in.

$$A_e = 0.985 \times 0.194 \times 6.25 \times 28.45 \left(1 + 0.79\sqrt{\frac{31}{1.25}}\right) = 202.5 \text{ in}^2.$$

2. $G = \dfrac{200,000 \times 144}{202.5} = 142,000 \text{ lb/ft}^2\text{-hr}$ or $G_{\text{eff}} = 39.5 \text{ lb/ft}^2\text{-sec}$

$$\frac{dG}{\mu} = \frac{0.75 \times 142,000}{12 \times 2.42 \times 0.32} = 11,500$$

3. From Fig. 43, $j = 0.0078$

$$\text{Pr} = 0.68 \times 2.42 \times \frac{0.32}{0.11} = 4.79 \qquad (\text{Pr})^{2/3} = 2.84$$

$$\left(\frac{\mu_w}{\mu_b}\right)^{0.14} = \left(\frac{0.37}{0.32}\right)^{0.14} = 1.02$$

$$h = \frac{0.0078 \times 0.68 \times 142,000}{1.02 \times 2.84} = 260 \text{ Btu/hr-ft}^2\text{-}^\circ\text{F}$$

4. The end zone correction is needed because of the small baffle spacing. For split ring floating head, the $(l_1 - l_2)$ term can be closely approximated by using three times the nozzle flange diameter. For an 8-in. nozzle $(l_1 - l_2) = 3 \times 13\frac{1}{2} = 40\frac{1}{2}$ in. Allowing 2 in. for tube-sheet thickness, $l_1 = 192 - 4 = 188$ in. l_2 is approximately $188 - 40.5 = 147.5$ in. With ¼-in. baffles, no = $142.5/(6.25 + 25) = 22.8$, say, 22; then $l_2 = 22 \times 6.5 = 149.5$

$$E_s = \frac{149.5 + 40.5\,(2 \times 6.25/40.5)^{0.6}}{188} = 0.895$$

hence $h_o = 0.895 \times 260 = 232 \text{ Btu/hr-ft}^2\text{-}^\circ\text{F}$.

5. Pressure drop

$$\text{Re}_P = M\text{Re} = 0.985 \times 11,500 = 11,300$$

From Fig. 43 $\qquad\qquad\qquad f = 0.122$

From Fig. 49 $\qquad\qquad\qquad \Omega = 34.4$

$$N_B = 23 + 1 = 24$$

6. $\qquad \Delta P_{\text{clean}} = \dfrac{4.31 \times 34.4 \times 24 \times 0.122\,(39.5)^2}{48.5 \times 0.98 \times 10^4} = 1.43 \text{ psi}$

7. To allow for fouling from Table 4

$$\Delta P = 1.55 \times 2.07 = 3.21 \text{ psi}$$

Same Problem with Bell's Method.

1. From Table 2, $D_3 = 28.45$ and rows of tubes across center line = 28.45/(15/16) = 30.3; seal strips are used. Theoretical crossflow area = 6.25 (30.3 − 1) (3/16) = 34.4 in.^{2}

$$G_m = \frac{200,000 \times 144}{34.4} = 838,000 \text{ lb/ft}^2\text{-hr}$$

$$\frac{D_t G_m}{\mu} = \frac{0.75 \times 838,000}{12 \times 2.42 \times 32} = 67,500$$

2. $j = 0.0041$ from Fig. 43.

3. $S_{BP} = (28.83 - 28.45) \, 6.25 = 2.38$ in.^{2} (with seal strips, effective diameter = 28.83 in.)

$$F_{BP} = \frac{2.38}{34.4} = 0.0693$$

4. Turbulent flow $\alpha = 1.35$. Using a sealing strip for each five rows, $N_s/N_c = 0.2$

$$\xi_h = \exp[-(1.35)(0.0693)(1 - \sqrt{2 \times 0.2})] = \exp(-0.0245) = 0.976$$

5.

$$A_t = 822 \times 0.1963 \times 16 = 2585 \text{ ft}^2$$

For $D_1/l_3 = 5H/D_1 = 0.16$ using Tinker's baffle cuts, Baffle cut = 0.16 x 31 = 4.95 in. Segment height for the tube bundle = 4.95 − (31 − 28.45)/2 = 3.68 in. So on bundle $H/D = 3.68/28.45 = 0.129$.

From mathematical tables of segmental areas, area in segment = 0.05933 (28.45)^2 = 48.1 in.^{2} or fraction of bundle area = 48.1/($\sqrt{28.45}^2 \, \pi/4$) = 0.0755. Then $A_w = 0.0755$ x 2585 = 195 ft^2.

6. From Fig. 44, $\phi = 1.15$.

7. From Devore's calculation $N_B = 23$

$$N_c = \frac{31 - 2 \times 4.95}{15/16 \times 0.866} = 26$$

For ½ of central row chord = radius and from tables on segments

$$\frac{H}{D} = 0.0318 \text{ or } H = 0.0318 \times 28.45 = 0.905 \text{ in.}$$

so available rows (3.68 − 0.905)/[(15/16) 0.866] = 3.4 = N_w

$$N_c^1 = (23 + 1) \, 26 + (23 + 2) \, 3.4 = 709$$

8. Re > 1000 $N_c = 26$

$$\frac{h_m}{h_\infty} \text{ for exchanger } (N_c^1 = 709) = 0.99$$

$$\frac{h}{h_\infty} \text{ ideal bank } (N_c = 26) = 0.97$$

$$\chi = \frac{0.97}{0.99} = 0.98$$

9. $\quad h_{mL} = \dfrac{0.0041 \times 0.68 \times 838{,}000 \times 1.15 \times 0.976}{0.98} \times \left(\dfrac{0.68 \times 2.42 \times 0.32}{0.11}\right)^{-2/3} \times$

$$\left(\frac{0.37}{0.32}\right)^{0.14} = 976 \text{ Btu/hr-ft}^2\text{-}^\circ\text{F}$$

10. For TEMA tolerances of 1/32-in. hole diameter over tube size. Tube to baffle leakage area: From 5 above, number of tube holes in baffle = 822 (1 − 0.0755) = 760

$$S_{TB} = 760 \frac{\pi}{4} [(0.7812)^2 - (0.75)^2] = 28.4 \text{ in.}^2.$$

Shell to baffle leakage: TEMA tolerances design, $ID - $ baffle $OD = 0.175$ in. for 31-in. shell. For baffle cut of 4.75, $H/D = 4.95/31 = 0.160$, the angle of the cut from mathematical tables = 94°; so leakage area $S_{SB} = 266/360 \times 0.175/2 \times \pi\ 30.927 = 6.3$ in.^{2}

$$\frac{S_L}{S_m} = \frac{6.3 + 28.4}{34.4} = 1.01$$

$$\frac{S_{TB} + 2S_{SB}}{S_L} = \frac{28.4 + 2 \times 6.3}{34.7} = 1.18$$

11. $\quad \left(1 - \dfrac{h_L}{h_{NL}}\right)_0 = 0.45 \times 1.01 + 0.10\,[1 - \exp(-30.3)] = 0.555$

12. $\quad \left(1 - \dfrac{h_L}{h_{NL}}\right)_{\text{exch}} = 0.555 \times 1.18 = 0.655$

13. $\quad h_L = (1 - 0.655)\,976 = 337 \text{ Btu/hr-ft}^2\text{-}^\circ\text{F}$

This result compares with 260 by Devore's method. No correlation for end baffles is taken by Bell.

14. $f^1 = 0.098$ (Fig. 43).

15. $\quad \Delta P_B = \dfrac{4 \times 0.098 \times 26\,(2325)^2}{2 \times 32.2 \times 48.5} \times 1.02 = 180 \text{ lb/ft}^2$

16. $\quad \alpha = 4$

$$\xi_P = \exp[-4 \times 0.0693\,(1 - \sqrt{0.4})] = \exp(-0.0726) = 0.93$$

17. $\quad \Delta P_{BP} = 0.93 \times 180 = 167.5 \text{ lb/ft}^2$

18. Get S_w as follows. Tubes in window = 822 − 760 = 62, $H/D = 0.16$, Area (from tables) = 0.0811 × 32² = 78 in.², $S_w = 78 - 62 \times (\pi/4)(0.75)^2 = 78 - 27.4 = 50.6$ in.²

$$G_w = \frac{200{,}000 \times 144}{50.6} = 568{,}000 \text{ lb/ft}^2\text{-hr} = 158 \text{ lb/ft}^2\text{-sec}$$

$$V_z = \frac{\sqrt{158 \times 232.5}}{48.5} = 3.95 \text{ ft/sec}$$

19. Re > 100

$$N_w = \frac{0.8H}{\alpha} = \frac{0.8 \times 4.95}{0.866 \times 15/16} = 4.87, \text{ say 5 rows}$$

$$\Delta P_w = (2 + 0.6N_w)\frac{\rho V_z^2}{2g}$$

$$\Delta P_w = (2 + 0.6 \times 5)48.5 \times \frac{3.95^2}{2 \times 32.2} = 58.8 \text{ lb/ft}^2$$

20.
$$\left(1 - \frac{\Delta P_L}{\Delta P_{NL}}\right)_0 = 0.57 \times 1.01 + 0.27[1 - \exp(-20.2)] = 0.845$$

21.
$$\left(1 - \frac{\Delta P_L}{\Delta P_{NL}}\right)_{exch} = 0.845 \times 1.18 = 1.0$$

22.
$$\frac{\Delta P_L}{\Delta P_{NL}} = 0$$

23.
$$\Delta P_{total} = 2 \times 167.4\left(1 + \frac{5}{26}\right) + [(24 - 1) \times 167.5 + 24 \times 58.8] \times 0$$

$$= 400 \text{ lb/ft}^2 = 2.68 \text{ lb/in}^2.$$

The above result implies no pressure drop in the bundle which is due to extrapolating Bell's curves. It is suggested that an upper limit of $\left(1 - \Delta P_L/\Delta P_{NL}\right)_0$ of 0.75 to 0.80 be used. Using 0.75, then

(21)
$$\left(1 - \frac{\Delta P_L}{\Delta P_{NL}}\right)_{exch} = 0.75 \times 1.18 = 0.885$$

or

(22)
$$\frac{\Delta P_L}{\Delta P_{NL}} = 0.115$$

and

(23) $\Delta P_{total} = 400 + (23 \times 167.5 + 24 \times 58.8) \times 0.115 = 1005 \text{ lb/ft}^2$

or 6.97 lb/in.^{2}

This pressure drop is much higher than the above or the one calculated by Devore. If 0.8 had been used instead of 0.75 as a limit, the $\Delta P_{total} = 4.86$ lb/in.^{2} All this really shows is that pressure drop calculations by any method for shell side of heat exchangers still leaves much to be desired.

DISK-AND-DOUGHNUT BAFFLED EXCHANGER

The only data on the disk-and-doughnut baffles are those of Short [16] and are based on his early work. In the absence of other data the only correlation that can be used is

$$\frac{hD_t}{k} = 1.45 \left(\frac{p - D_t}{p} \times D_t\right)^{0.4} \left(\frac{c\mu}{k}\right)^{1/3} \left(\frac{D_t G_{av}}{\mu}\right)^{0.6}$$

where Short defines $G_{av} = (G_A + G_H + 2G_R + 2G_N)/6$. G_A is the mass velocity based on free annular area between the disk and shell, G_H is based on free area of hole, G_N is the radial flow through minimum area region between each pair of baffles, and G_R is the radial flow based on maximum area between each pair of baffles. Care and judgment must be used when applying this equation. The data are based on only one size of exchanger although tube size and pitch and baffle dimensions were varied.

In boilers and exchangers other than the shell and tube, the tube pitch may be greater and the heat transfer and pressure drop results will not follow that given in Fig. 43. For these other bundle arrangements, the results of Huge, Pierson, and Grimison as reported in McAdams [17] are given in Table 5.

The friction factors are shown in Fig. 50. Later work by Fairchild and Welsh [18] shows for in-line arrangements at low Reynolds numbers the Grimison et al. results are conservative. The $j = [(h/cG)(c\mu/k)^{2/3}]$ and $f = \Delta P/N(G^2/2g\mu)$ relations are shown in Figs. 51 and 52. Here G is the mass flow rate through the minimum flow area.

In all tube bank tests where the air flow entering the bundle is not highly turbulent, there is an effect of the number of tube rows on the heat transfer coefficient as indicated in Table 6.

For flow parallel to tube bundles, the results appear to be influenced by the approach conditions. For pure parallel flow, Kattchee and Mackewicz [19] show that the usual Colburn type equation for flow inside tubes applies but the heat transfer is about 15 percent greater. In this approach the Reynolds number was $4w/\mu s$, where w is flow lb/hr, s is wetted perimeter (ft), and μ is viscosity. Sutherland and Kays [20] summarize recent work and provide additional data for $P/d = 1.15$ to 1.25. They find

TABLE 5. Grimison's Values of b_2 and n *

$$h_m D_0/k_f = b_2 (D_0 G_{max}/\mu_f)^n; \quad t_f = t_s - (t_s - t)m/2$$

$x_L = \dfrac{S_L}{D_0}$	$x_r = 1.25$		$x_r = 1.5$		$x_r = 2$		$x_r = 3$	
	b_2	n	b_2	n	b_2	n	b_2	n
Staggered:								
0.600							0.213	0.636
0.900					0.446	0.571	0.401	0.581
1.000			0.497	0.558				
1.125					0.478	0.565	0.518	0.560
1.250	0.518	0.556	0.505	0.554	0.519	0.556	0.522	0.562
1.500	0.451	0.568	0.460	0.562	0.452	0.568	0.488	0.568
2.000	0.404	0.572	0.416	0.568	0.482	0.556	0.449	0.570
3.000	0.310	0.592	0.356	0.580	0.440	0.562	0.421	0.574
In line:								
1.250	0.348	0.592	0.275	0.608	0.100	0.704	0.0633	0.752
1.500	0.367	0.586	0.250	0.620	0.101	0.702	0.0678	0.744
2.000	0.418	0.570	0.299	0.602	0.229	0.632	0.198	0.648
3.000	0.290	0.601	0.357	0.584	0.374	0.581	0.286	0.608

$x_L = S_L/D_0; \quad x_r = S_T/D_0$

*From "Heat Transmission," 3d ed., by W. H. McAdams, McGraw-Hill Book Co., 1954.

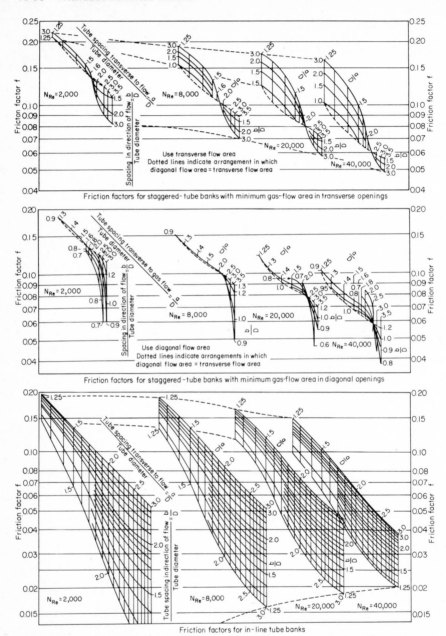

Fig. 50. Friction factors for tube bundles (*from Grimison,* Trans. ASME, **59**:*583 (1937)*).

for uniform heat flux Nu = 0.04 (Re)$^{0.77}$ (Pr)$^{0.6}$. Methods are presented to enable calculation of wall temperatures for nonuniform heating. When $P/d = 1$ the temperature distributions may become critical but heat transfer can be predicted by Nu = 0.01 Re$^{0.8}$ Pr$^{0.43}$. For a cross-flow approach such as through nozzles in a shell and then flow parallel to the tubes, Short [21] shows a different equation wherein

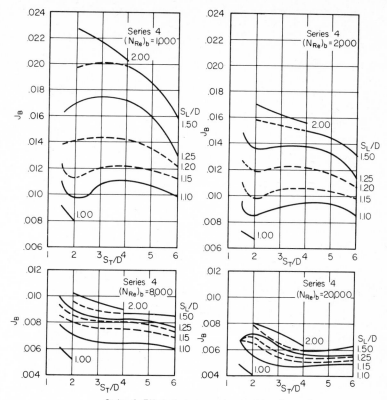

Series 4: Effect of arrangement on heat transfer results

Fig. 51. Effect of tube spacing on heat transfer in tube bundles (*from ASME preprint 61WA250 by Fairchild and Welsh*).

the effect of mass velocity is the 0.6 power compared to 0.8 power for inside tubes. There is also a bypass effect between the bundle and shell that must be allowed. Short gives

$$\frac{hD_t}{h} = 0.16 \left(\frac{D_t G_b}{\mu}\right)^{0.6} \left(\frac{c\mu}{k}\right)^{0.33}$$

where D_t = tube diameter, ft, $G_b = W_b/A_b$

$$W_b = W_s \left[\frac{A_b}{A_b + A_{bx}(D_{bx}/D_b)^{0.715}}\right]$$

$D_b = 4A_b/N_E \pi D_t$
$D_{bx} = 4A_{bx}/\pi D_s$
A_b = net flow area in baffle window (envelope of tubes minus tube cross-sectional area)
A_{bx} = flow area between tubes and shell parallel to tubes
N_t = number of tubes
W_s = total flow through exchanger shell

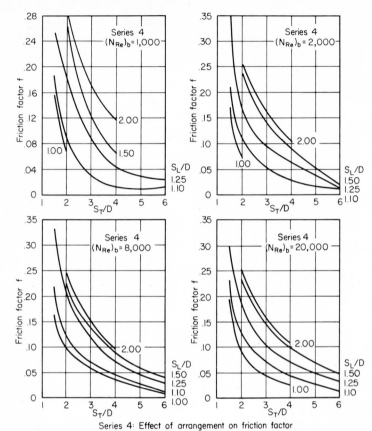

Series 4: Effect of arrangement on friction factor

Fig. 52. Effect of tube spacing on pressure drop in tube bundles (*from* ASME *preprint 61WA250 by Fairchild and Welsh*).

TABLE 6. Ratio of h_m for N Rows Deep to That for 10 Rows Deep*

N	1	2	3	4	5	6	7	8	9	10
Ratio for staggered tubes		0.73	0.82	0.88	0.91	0.94	0.96	0.98	0.99	1.0
Ratio for in-line tubes	0.64	0.80	0.87	0.90	0.92	0.94	0.96	0.98	0.99	1.0

*From "Heat Transmission," 3d ed., by W. H. McAdams, McGraw-Hill Book Co., 1954.

Some other papers for parallel flow offer little concrete results. For instance, some results are reported for closely packed tubes in a paper [22] but insufficient detail of the experimental apparatus makes it difficult to determine the entrance flow condition. Other work of theoretical nature but without supporting data has been done, e.g., Dwyer and Tu [23].

SCALE OR FOULING

Allowance for dirt or scale can be made by either including separate resistances for the scale associated with each fluid or by allowing a percentage reduction of the overall coefficient. The preferred method is the former.

Unfortunately there is no method for calculating a dirt coefficient nor predicting the effect on it of such operating variables as temperature, velocity, fluid properties, or tube materials. A list prepared by the Tubular Exchanger Manufacturers Association (TEMA) and given in Table 7 is widely used. This list is based on experience of the members, but it is doubtful that much test data were available as a basis; therefore, it should be used only as a guide. There are very few quantitative data on fouling rates.

The TEMA tables ignore many important factors which affect fouling. For instance, the tube material is ignored as a factor, but steel tubes do corrode and pit, and consequently collect dirt and scale much easier and faster than more corrosion-resistant and smooth-surfaced materials such as copper and stainless steel. Economics is another very important factor. There are indications that TEMA dirt coefficients may be overly conservative in many instances and in other instances very optimistic. Each exchanger should be considered separately and its operating conditions carefully examined.

As presently defined and used (a scale allowance based on an unspecified time of operation), any value of a dirt coefficient can be specified without theoretically being wrong, but practically be using poor judgment. The proper value to use for a dirt coefficient must be based on economics. The only reason for making an allowance for extra surface by means of the dirt coefficient is to allow a reasonable period between clean-outs. The factors that must balance are the investment, maintenance, repair, tax, and expected return on investment as compared to the losses suffered by cleaning costs and lost production. It should be obvious that the same allowance for fouling under identical conditions cannot be taken when in one case the additional surface costs $1 per square foot and in another case costs $30 per square foot.

A simple economic analysis [23] can be made on fouling coefficients which brings forth the important cost factors and the weakness of present methods of reporting fouling coefficients. Using accounting practices to give profitable designs, the total yearly cost of an exchanger equals repair costs and maintenance exclusive of cleaning plus depreciation plus cleaning costs, which may include allowances for lost production, plus investment cost. Repair costs R and depreciation D are usually given as a fraction of the investment I. Let E be the cost of one cleaning and N be the number of cleanings per year. The investment cost is the gross return Z times the investment where Z is determined from the net return on investment which management specified and adjusted for other expenses and for tax allowances. It is this present-day tax allowance which makes Z so significant.

For the simple case of dirt resistance as a linear function of time ($r = m\theta$)

$$N^2 = \left(\frac{U_c}{h_{1d}}\right)\left[\frac{A_c C_s (R + D + Z)}{E}\right]$$

U_c is the overall coefficient in the *clean* condition. h_{1d} is that dirt coefficient that would be obtained after one year of operation. C_s is the installed cost of the exchanger in dollars per square foot. Note that $A_c C_s (R + D + Z)$ is the minimum annual exchanger cost since it is based on the required surface for a clean exchanger A_c. If the dirt resistance is a function approximated by a parabolic curve ($r = m\theta^2$), then

$$N^3 = 2\left(\frac{U_c}{h_{1d}}\right)\left[\frac{A_c C_s (R + D + Z)}{E}\right]$$

Important conclusions drawn from the equation are that a fouling coefficient is meaningless unless a time interval for building up the specified coefficient is also given as well as information concerning its linearity with time. Also the allowances for

TABLE 7. Fouling Resistances

Typical fouling resistances are referred to the surface on which they occur. In the absence of specific data for setting proper resistances the user may be guided by the values tabulated below.

Water

Temperature of Heating Medium	Up to 240°F		240–400°F	
Temperature of Water	125°F or less		Over 125°F	
Types of Water	Water velocity ft/sec		Water velocity ft/sec	
	3 ft and less	Over 3 ft	3 ft and less	Over 3 ft
Sea water	0.0005	0.0005	0.001	0.001
Brackish water	0.002	0.001	0.003	0.002
Cooling tower and artificial spray pond				
Treated makeup	0.001	0.001	0.002	0.002
Untreated	0.003	0.003	0.005	0.004
City or well water (such as Great Lakes)	0.001	0.001	0.002	0.002
Great Lakes	0.001	0.001	0.002	0.002
River water				
Minimum	0.002	0.001	0.003	0.002
Mississippi	0.003	0.002	0.004	0.003
Delaware, Schuylkill	0.003	0.002	0.004	0.003
East River and New York Bay	0.003	0.002	0.004	0.003
Chicago Sanitary Canal	0.008	0.006	0.010	0.008
Muddy or silty	0.003	0.002	0.004	0.003
Hard (over 15 grains/gal)	0.003	0.003	0.005	0.005
Engine jacket	0.001	0.001	0.001	0.001
Distilled	0.0005	0.0005	0.0005	0.0005
Treated boiler feedwater	0.001	0.0005	0.001	0.001
Boiler blowdown	0.002	0.002	0.002	0.002

*Ratings in columns 3 and 4 are based on a temperature of the heating medium of 240–400°F. If the heating medium temperature is over 400°F and the cooling medium is known to scale, these ratings should be modified accordingly.

Industrial fluids
Oils
Fuel oil 0.005
Transformer oil 0.001
Engine lube oil 0.001
Quench oil 0.004
Gases and vapors
Manufactured gas 0.01
Engine exhaust gas 0.01
Steam (non-oil bearing) 0.0005
Exhaust steam (oil bearing) 0.001
Refrigerant vapors (oil bearing) 0.002
Compressed air 0.002
Industrial organic heat transfer media 0.001
Liquids
Refrigerant liquids 0.001
Hydraulic fluid 0.001
Industrial organic heat transfer media 0.001
Molten heat transfer salts 0.0005
Chemical processing streams
Gases and vapors
Acid gas 0.001
Solvent vapors 0.001
Stable overhead products 0.001

TABLE 7. Fouling Resistances (*Continued*)

Chemical processing streams (*continued*)
Liquids
MEA & DEA solutions 0.002
DEG & TEG solutions 0.002
Stable side draw and bottom product 0.001
Caustic solutions 0.002
Vegetable oils 0.003
Natural gas–gasoline processing streams
Gases and vapors
Natural gas 0.001
Overhead products 0.001
Liquids
Lean oil 0.002
Rich oil 0.001
Natural gasoline & liquefied petroleum gases . 0.001
Oil refinery streams
Crude & vacuum unit gases and vapors
Atmospheric tower overhead vapors 0.001
Light naphthas 0.001
Vacuum overhead vapors 0.002
Crude & vacuum liquids
Crude oil

	0–199°F			200–299°F		
	Velocity, ft/sec			Velocity, ft/sec		
	Under 2 ft	2–4 ft	4 ft and over	Under 2 ft	2–4 ft	4 ft and over
Dry	0.003	0.002	0.002	0.003	0.002	0.002
Salt†	0.003	0.002	0.002	0.005	0.004	0.004

	300–499°F			500°F and over		
	Velocity, ft/sec			Velocity, ft/sec		
	Under 2 ft	2–4 ft	4 ft and over	Under 2 ft	2–4 ft	4 ft and over
Dry	0.004	0.003	0.002	0.005	0.004	0.003
Salt†	0.006	0.005	0.004	0.007	0.006	0.005

†Normally desalted below this temperature range. (Dagger to apply to 200–299°F, 300–499°F, 500°F, and over.)
Gasoline 0.001
Naphtha & light distillates 0.001
Kerosene 0.001
Light gas oil 0.002
Heavy gas oil 0.003
Heavy fuel oils 0.005
Asphalt & residuum 0.010
Cracking & coking unit streams
Overhead vapors 0.002
Light cycle oil 0.002
Heavy cycle oil 0.003
Light coker gas oil 0.003
Heavy coker gas oil 0.004
Bottoms slurry oil (4½ ft/sec minimum) . . . 0.003
Light liquid products 0.002
Catalytic reforming & hydrodesulfurization streams
Reformer charge 0.002
Reformer effluent 0.001

TABLE 7. Fouling Resistances (*Continued*)

Catalytic reforming & hydrodesulfurization streams (*continued*)	
Hydrodesulfurization charge & effluent‡ ...	0.002
Overhead vapors ...	0.001
Liquid product over 50° API ...	0.001
Liquid product 30–50° API ...	0.002
Light ends processing streams	
Overhead vapors & gases ...	0.001
Liquid products ...	0.001
Absorption oils ...	0.002
Alkylation trace acid streams ...	0.002
Reboiler streams ...	0.003
Lube oil processing streams	
Feed stock ...	0.002
Solvent feed mix ...	0.002
Solvent ...	0.001
Extract§ ...	0.003
Raffinate ...	0.001
Asphalt ...	0.005
Wax slurries§ ...	0.003
Refined lube oil ...	0.001

‡Depending on charge characteristics and storage history, charge resistance may be many times this value.
§Precautions must be taken to prevent wax deposition on cold tube walls.
Used by permission of Tubular Exchangers Manufacturing Assoc.

fouling should be smaller when exchanger surface is expensive, net returns and taxes are high, or cleaning costs are low.

Conservatism in the original design, such as maximum load at minimum temperature differences and ample allowances for fouling, may greatly aggravate the rate of fouling due to very low operating velocities, allowing the solids to settle and cake during periods of low production or high temperature differences.

NOISE AND VIBRATIONS IN HEAT EXCHANGERS

The formation and shedding of eddies or vortices when fluids flow past tubes create alternating forces which under proper conditions of the coupling can cause the tubes to vibrate or result in acoustical vibrations. The first is bad because of failure of tubes by fatigue; in the second case the resulting noise is objectionable or may in turn cause problems in duct vibrations. Although some data and theory are available, the problem is far from being solved. Putnam [25] reviews the present status of this subject.

The frequency of tube vibrations is given as

$$\nu_t = a_i \left(\frac{EI}{mL^4} \right)^{1/2}$$

where ν_t = tube frequency
a_i = coefficient specific to mode number and constraints (Table 8)
EI = flexural rigidity
m = mass per unit length of tube
L = unsupported length

TABLE 8. Values of Frequency Coefficient α_i for Tubes

End Condition	Mode				
	1	2	3	4	5
Pin-pin	1.57	6.28	14.1	25.1	39.3
Clamp-free	0.560	3.58	9.82	19.2	31.8
Free-free	3.58	9.82	19.2	31.8	47.5
Clamp-clamp	3.58	9.82	19.2	31.8	47.5
Clamp-pin	2.45	7.96	16.6	28.4	43.3

For a tube $I = \pi(D_0^4 - D_i^4)/64$. From test data it appears to be difficult to obtain a true clamp-clamp condition. Also, for fluid-filled tubes a mass equivalent to two-thirds of the fluid must be added to the tube mass in order to determine frequency.

Acoustical vibrations are of the standing-wave type. In a unit with the length L_e between the open ends, the width L_w, and height (parallel to tubes) L_h the possible frequencies are

$$\nu_n = \frac{C}{2}\left[\left(\frac{n_e}{L_e}\right)^2 + \left(\frac{n_w}{L_w}\right)^2 + \left(\frac{n_h}{L_h}\right)^2\right]^{1/2}$$

where C is the velocity of sound, n is the integers. In all reported cases $n_h = 0$ and L_e is sufficiently greater than L_w that n_e can be assumed zero. Therefore, in most instances

$$\nu_n = \frac{Cn_w}{2L_w}$$

In cylindrical vessels, such as shell-and-tube exchangers, the relation is

$$\nu = \frac{C}{2}\left[\left(\frac{a_{mn}}{R}\right)^2 + \left(\frac{n_h}{L_h}\right)^2\right]^{1/2}$$

where R is the radius of shell and a_{mn} are the values given in Table 9. Again n_h is essentially zero. As neither the radial mode $m = 0$ or the higher transverse modes of $n = 0$ seem probable, the next possible higher mode would appear to be of $m = n = 1$.

TABLE 9. Values of Constant α_{mn}

m	n			
	0	1	2	3
0	0.000	1.220	2.233	3.238
1	0.586	1.697	2.714	3.726
2	0.972	2.135	3.193	4.192
3	1.337	2.551	3.611	4.643

For flow transverse to single cylinders, the Strouhal number $\nu D/V$ has a value of 0.2 for Reynolds number range of 300 to 2×10^5. In the constant Strouhal number range for shapes other than circular

$$\nu \frac{D}{V} = 0.21 \, C_d^{-3/4}$$

where C_d is the drag coefficient.

In-line tube array noise (acoustical) results reported by Putnam are shown in Fig. 53. V_c is the mean velocity through the minimum area. The reference to G_L^{1} is where L/D is replaced by $(G_L/D + 1)$, and G_L replaced by G_L^{1}, the gap between fins.

Staggered tube arrays noise results are best given in Table 10. Here several definitions of Strouhal number are used, but ν_L/V^* seems preferred since it matches closer to the in-line data and correlation.

Tube vibration in heat exchangers results in fatigue failure or wear of the tubes. Although several methods are proposed for predicting the conditions necessary for vibration, they are all based on limited test data plus the complicating facts that flow conditions in a typical shell-and-tube exchanger vary from parallel to cross flow in different sections, the initial stresses in the tubes due to rolling procedures and the exchanger types are unknown, the effect of tube support and fastening on its vibrational characteristics is unknown, and basic data on Strouhal numbers for close pitch tube bundles are missing. A symposium discussing the state of the art was given at the 1970 annual meeting of the American Society of Mechanical Engineers and published in "Flow-Induced Vibration in Heat Exchangers."

Y. N. Chen and M. Weber in the above reference propose the equation

$$\frac{y}{d_h} = \left[1 - \left(\frac{U}{U_c} \right)^2 \right]^{-1} \left(\frac{\beta U}{U_c} \right)^2$$

where y = vibration amplitude
d_h = hydraulic diameter
U = mean flow velocity

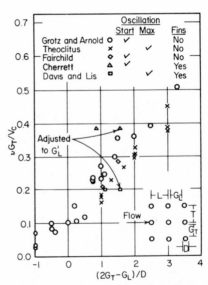

Fig. 53. Correlation of noise data from in-line heat exchangers (*from* ASME *preprint 61WA250 by Fairchild and Welsh*).

TABLE 10. Strouhal Numbers at Maximum Noise Amplitude in Staggered Tube Arrays

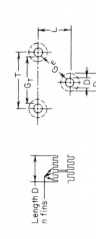

Source	T/D	L/D	G_m/G_T	D_f/D	n	t/D	$\nu D/V$	$\nu D/V^*$	$\nu L/V_{FA}$	$\nu L/V$	$\nu L/V^*$
Hill-Armstrong	2.36	0.95–	0.761				0.249	0.249	0.980		0.565
	2.3	2.3	3.0				0.462	0.608			
Battelle I	2.0	1.0	0.828				0.46	0.556	1.11	0.46	0.556
Battelle II	1.5	1.3	2.0				0.23	0.23	0.92	0.30	0.30
Schenck I	2.48	2.22	2.08	Plate fins	2.96	0.057	0.210	0.210	0.785	0.470	0.470
Schenck II	2.67	2.67	2.37	2.4	3.38	0.024	0.45	0.45	>1.92	1.20	1.20
Godman-Stein	2.6	2.7	2.5	2.3	6.25		0.228	0.228	>1.00	0.616	0.616
Cherrett	2.03	1.25	1.19	1.41	10.66	0.0285	0.507– 0.760	0.507– 0.760	1.42– 2.14	0.634– 0.95	0.634– 0.95

V = average velocity through minimum cross section
V^* = average velocity through minimum cross section between tubes in same row
V_{FA} = average approach velocity

U_c = critical velocity
β = turbulence factor
β = 1 for a well designed entrance to parallel flow bundles
β = 2 for perpendicular entrance to bundle
β = 10 for parallel-cross mixed flow bundles.

Now

$$U_c = \omega_n \frac{1}{n\pi} \left[\frac{\mu}{(\pi/4)\,d^2\rho\,[C_f(1/d)\,+\,1]} \right]^{1/2}$$

where ω_n = the nth order tube natural circular frequency for pinned tubes =

$$\frac{n^2\pi^2}{l^2} \sqrt{\frac{EI}{\mu}}$$

where l = tube length
E = Young's modulus
I = moments of inertia for tube
μ = rod mass per unit length including mass of displaced fluid
d = tube diameter
ρ = fluid density
C_f = coefficient of surface drag along the tube.
These equations were developed basically for axial flow in tube bundles but is also proposed for cross flow with β increased to 10.

H. J. Connors in the above reference proposes another simple equation:

$$\frac{U}{f_0 d} = 9.9 \sqrt{\frac{m_0 \delta_0}{\rho d^2}}$$

where U = velocity through minimum area in./sec
f_0 = natural frequency of tube, H_z
d = diameter, in.
m_0 = mass of tube plus mass of displaced fluid, lb-sec^2/in.^{2}
δ_0 = log decrement of tube in still fluid, dimensionless
ρ = fluid density, lb-sec^2/in.^{4}
This equation defines a stability boundary and when $U/f_0 d$ is less than the critical value the tube array is stable, and large amplitude vibration will develop when $U/f_0 d$ exceeds the critical value.

Lentz [35] proposed for fixed-pinned condition and no axial thrust $V_0 = 7.09\,(d/l)^2(E/\rho)^{1/2}$ but where an axial thrust exists; e.g., fixed tube sheet exchangers, then

$$\left(\frac{P}{P_{cr}}\right) + \left(\frac{V}{V_0}\right)^2 = 1$$

where V = fluid velocity, ft/sec
V_0 = critical velocity = $C\,\dfrac{d}{l^2}\,\dfrac{EI}{\omega}$ $^{1/2}$
P = axial load, lb
P_{cr} = buckling axial load, lb = $\dfrac{\eta\pi^2 EI}{l^2}$

ρ = tube material density, lb/in.^{3}
E = modulus of elasticity, lb/in.^{2}
d = tube diameter, in.
l = tube length, in.
I = moment of inertia, in.^{4}

For pinned-pinned condition η = 1 and C = 12.85; for fixed-pinned η = 2.05 and C = 20.06; and for fixed-fixed η = 4 and C = 29.19. In his derivation, Lentz used a Strouhal number of 0.2 which is for single tubes, but for staggered tube bundles the Strouhal number can be 0.65 and will affect the first equation for V_0; however, it apparently cancels out in the V/V_0 form for fixed tube sheet exchangers.

The above equations demonstrate the variety of approaches to the vibration problem and each has its weaknesses. Much more experimental and theoretical work is needed together with confirming plant experience before reliable methods of predicting vibration are available. There is general agreement, however, that when the driving source frequency, usually the eddy shedding, approaches the natural tube frequency damaging vibrations will result.

REGENERATIVE HEAT EXCHANGERS

In a regenerator the heat is transferred between two fluids, usually gases, by flowing first one gas through the exchanger and then during a second period, the other gas. In this manner heat is removed from the hot gas by heating the regenerator and then this heat is given up by the regenerator when the cold gas flows through. No heat is transferred through walls but instead is alternately transferred into and out of the walls, packing, or filler. The depth of heat penetration is a function of the flow conditions and wall diffusivities.

In order to have continuous operation, either two or more stationary or a single-rotary regenerator may be used. The usual stationary regenerator consists of a chamber packed with refractory of many possible shapes. The gases are alternately switched by valves to flow through the regenerators. In a rotary regenerator the packed rotor alternately passes through the hot- and cold-gas streams in each rotation.

In all regenerators there is the contamination of one gas stream by the other due to the regenerator hold-up volume of gas during a switching operation. If this problem is serious a purge cycle may be used between the heating and cooling cycles. In the rotary regenerators, an additional problem is leakage between the streams through the seals on the rotor.

Numerous analytical and approximate solutions are available but none do a complete solution of the assumptions involved. The problem is very complex due to temperatures varying both in time as well as position within the regenerator. However, the solutions and their limitations are given below.

Approximate Methods. Rummel proposed

$$\frac{q}{A} = \frac{\Delta t_{1m}/(\theta' + \theta'')}{1/h''\theta'' + 1/h'\theta' + 1/2.5 C_s \rho_s r_B + r_B/k(\theta' + \theta'')}$$

where h' and θ' are respectively the coefficient and time cycle for one gas, C_s, ρ_s, and k are solid specific heat, density, and conductivity, and r_B is the solid volume for unit solid surface, Δt_{1m} is the logarithmic mean temperature difference based upon the uniform inlet temperature of each stream and the average outlet temperatures. If the fourth term in the denominator is negligible relative to the other, the resistance to heat flow in the solid is unimportant. If the third denominator term is large relative to the other three, this equation should not be used.

For counterflow regenerators with negligible thermal resistance of the packing, the simplified treatments developed by Hausen and by Hottel can be used as given below.

$$\lambda' = \frac{h'(1 - \epsilon)L}{C_P' G_0' r_B} \quad \text{dimensionless size}$$

$$\lambda'' = \frac{h''(1 - \epsilon)L}{C_P'' G_0'' r_B}$$

where ϵ equals fraction solids and L equals packed length

$$\tau' = \frac{h'\theta'}{C_s \rho_s r_B} \quad \text{dimensionless time period}$$

$$\tau'' = \frac{h''\theta''}{C_s \rho_s r_B}$$

$$\phi_s = \frac{t_2' - t_1'}{t_1'' - t_1'} = \frac{t_1'' - t_2''}{t_1'' - t_1'} \quad \text{temperature efficiency}$$

For a symmetrical cycle $\tau' = \tau'' = \tau_s$ and $\lambda' = \lambda'' = \lambda_s$. Figure 54 is Hottel's chart for symmetrical cycles and negligible thermal resistance in the solid.

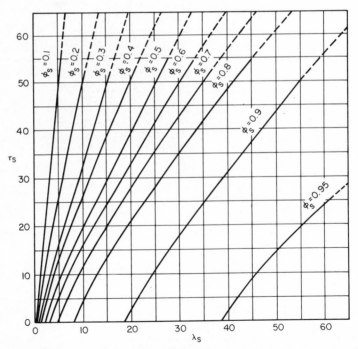

Fig. 54. Hottel's chart for generalized regenerator plot for symmetrical cycles and negligible thermal resistance within the solid (*courtesy of H. C. Hottel from "Heat Transmission," 3d ed., by W. H. McAdams, 1954, used with permission of McGraw-Hill Book Co.*).

For nonsymmetrical cases the rise or fall in temperature of each stream is approximately

$$\frac{t_1'' - t_2''}{\Delta t_{1m}} = \frac{\lambda''/\tau''}{\dfrac{1}{\tau''} + \dfrac{1}{\tau'} + \dfrac{2}{\tau'' + \tau'}\left[\lambda_s\left(\dfrac{1}{\phi_s} - 1\right) - 2\right]}$$

$$\frac{t_2' - t_1'}{\Delta t_{1m}} = \frac{\lambda'/\tau'}{\dfrac{1}{\tau''} + \dfrac{1}{\tau'} + \dfrac{2}{\tau'' + \tau'}\left[\lambda_s\left(\dfrac{1}{\phi_s} - 1\right) - 2\right]}$$

where ϕ_s from Fig. 54 is evaluated at

$$\tau_s = \frac{\tau'' + \tau'}{2} \quad \text{and} \quad \lambda_s = \frac{2}{1/\lambda'' + 1/\lambda'}$$

The maximum fluctuation in outlet gas temperature is for the hot gas

$$\frac{\delta t_2''}{t_2'' - t_1'} = \frac{\tau''}{[(1/\phi_s) - 1](\lambda'' + 1)}$$

and for the cold gas

$$\frac{\delta t_2'}{t_1'' - t_2'} = \frac{\tau'}{[(1/\phi_s) - 1](\lambda' + 1)}$$

Analytical solutions are available for the first time cycle by Nussult, Schumann, and Hausen but the Schumann solutions extended by Furnas are plotted in Figs. 55a and 56a

where T_0 = initial uniform fluid temperature
T_g = fluid temperature at any point at any time
T_s = solid temperature at any point at any time
$$Z = \frac{h_v}{C_{ps}\rho_s(1 - f)}\left(\tau - \frac{X}{fV}\right)$$
$$Y = \frac{h_v V}{C_{pf}\rho_f V}$$
C_{ps}, C_{pf} = specific heat of solid and fluid, respectively, Btu/lb-°F
ρ_s, ρ_f = density of solid and fluid, lb/ft^3
h_v = volumetric heat transfer coefficient, Btu/hr-ft^3-°F
f = fraction voids
X = distance from inlet of regenerator, ft
V = fluid velocity in bed, (ft^3 fluid)/(ft^2 bed cross section)-hr
τ = time, hr

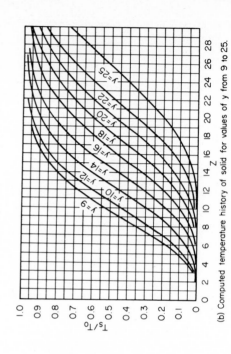

(b) Computed temperature history of solid for values of y from 9 to 25.

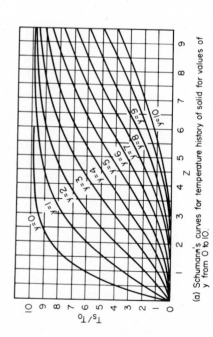

(a) Schumann's curves for temperature history of solid for values of y from 0 to 10.

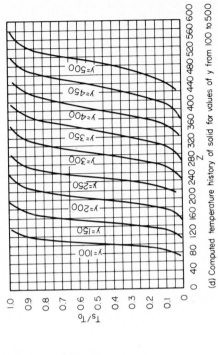

(c) Computed temperature history of solid for values of y from 25 to 100

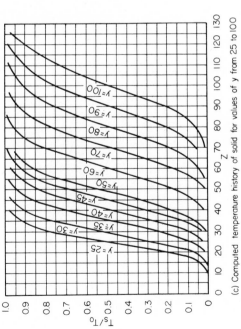

(d) Computed temperature history of solid for values of y from 100 to 500

Fig. 55. Temperature history of solid.

18-71

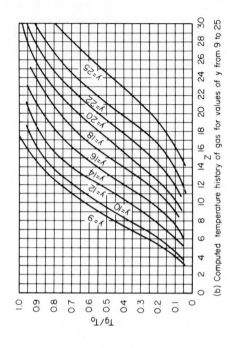

(b) Computed temperature history of gas for values of y from 9 to 25

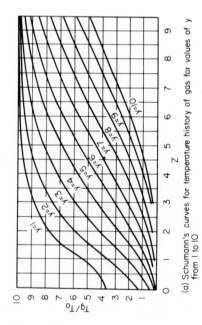

(a) Schumann's curves for temperature history of gas for values of y from 1 to 10

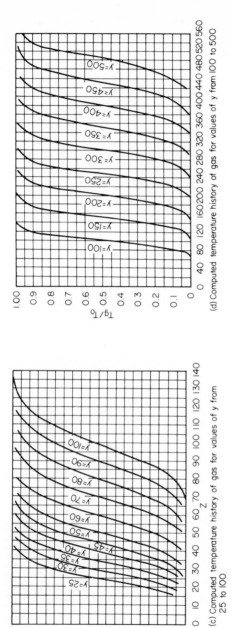

(c) Computed temperature history of gas for values of y from 25 to 100

(d) Computed temperature history of gas for values of y from 100 to 500

Fig. 56. Temperature history of gas.

The regenerator effectiveness has been subjected to much analysis but mostly limited to symmetrical cases. Such limited attempts that have been made for asymmetrical cases fortunately indicate the results are not far different from the symmetrical cases if the degree of asymmetry is not severe.

Kays and London [26] have made some empirical modifications and show curves for various heat capacity flow ratios (Figs. 57 to 60). Here the definitions are

$$NTU_0 = \frac{1}{C_{min}} \left[\frac{1}{(1/hA)_c + (1/hA)_h} \right]$$

C_r the matrix capacity rate = (rev/hr)(Matrix mass)(C Solid). C_{min} = the lower fluid heat capacity rate, C_{max} the larger fluid heat capacity rate, and ϵ = effectiveness =

$$\frac{C_h(t_{h\,in} - t_{h\,out})}{C_{min}(t_{h\,in} - t_{c\,in})} = \frac{C_c(t_{c\,out} - t_{c\,in})}{C_{min}(t_{h\,in} - t_{c\,in})}$$

$$(hA)^* = \frac{(hA)_c}{(hA)_h}$$

Data on various types of packing are given in Figs. 61 to 67.

High effectiveness is desired in regenerators and as developed by A. L. London [27] the irreversibility rate relative to the heat transfer rate is

$$\frac{\text{Irrev}}{q} \approx T_0 \left[\frac{(1 - \epsilon)(T_{H\,in} - T_{c\,in})^2}{(T_{H,lm} \cdot T_{c,lm})} \right] \qquad \text{heat transfer Irrev.}$$

$$+ T_0 \left\{ \frac{Wc_p}{q} \left(\frac{k-1}{k} \right) \left[\left(\frac{\Delta P}{P} \right)_H + \left(\frac{\Delta P}{P} \right)_c \right] \right\} \qquad \text{flow friction Irrev.}$$

$$+ T_0 \left\{ \frac{Wc_p}{q} W_l^* \left[\ln \left(\frac{T_{H\,in}}{T_{c\,out}} \right) + \frac{k-1}{k} \ln \left(\frac{P_{c\,out}}{P_{h\,in}} \right) \right] \right\} \qquad \text{leakage Irrev.}$$

The temperature difference needed for heat transfer is approximately $(1 - \epsilon)(T_{H\,in} - T_{c\,in})$ which is in the numerator of the heat transfer irreversibility. Thus the ineffectiveness $(1 - \epsilon)$ controls this component; e.g., comparing effectiveness of 80 to 90 percent does not seem significant, but $(1 - \epsilon)$ of 11 to 10 percent indicates a 9–10 percent difference. It is also important to note the effect of temperature level in the denominator of this group. $T_{H,lm} \cdot T_{c,lm} = T^2$ level where $T_{level} \approx (T_{H\,in} + T_{c\,in})/2$. Thus it is good practice in gas turbine regenerators to design for an effectiveness of 70 to 90 percent while for cryogenic work effectiveness in excess of 98 percent is standard. For gas turbine cycles $T_{level} \approx 1200\,R$, but for cryogenic levels $T_{level} \approx 300\,R$, a four to one difference. Thus, where an ineffectiveness of 10 to 30 percent is acceptable for gas turbine cycles only one-sixteenth of this magnitude can be tolerated in cryogenic cycles. For the high effectiveness range, it is more useful to use ineffectiveness plots such as shown in Fig. 68.

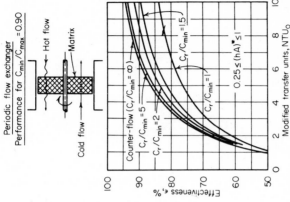

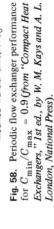

Fig. 57. Periodic flow exchanger performance for $C_{min}/C_{max} = 1$ (from "Compact Heat Exchangers," 1st ed., by W. M. Kays and A. L. London, National Press).

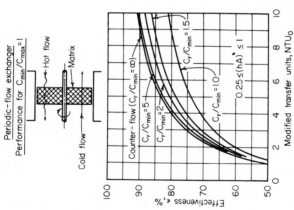

Fig. 58. Periodic flow exchanger performance for $C_{min}/C_{max} = 0.9$ (from "Compact Heat Exchangers," 1st ed., by W. M. Kays and A. L. London, National Press).

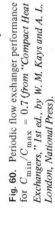

Periodic–flow exchanger
Peformance for $C_{min}/C_{max} = 0.70$

Hot flow

Matrix

Cold flow

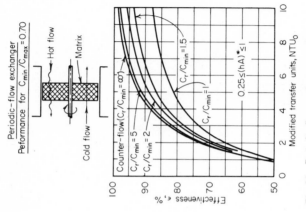

Counter-flow ($C_r/C_{min} = \infty$)

$C_r/C_{min} = 5$

$C_r/C_{min} = 2$

$C_r/C_{min} = 1.5$

$C_r/C_{min} = 1$

$0.25 \leq (hA)^* \leq 1$

Effectiveness ϵ, %

Modified transfer units, NTU_0

Fig. 60. Periodic flow exchanger performance for $C_{min}/C_{max} = 0.7$ (from "Compact Heat Exchangers," 1st ed., by W. M. Kays and A. L. London, National Press).

Periodic–flow exchanger
Performance for $C_{min}/C_{max} = 0.80$

Hot flow

Matrix

Cold flow

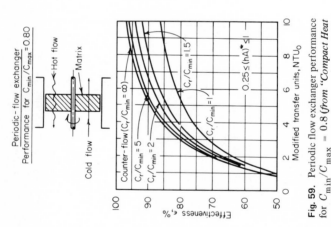

Counter-flow ($C_r/C_{min} = \infty$)

$C_r/C_{min} = 5$

$C_r/C_{min} = 2$

$C_r/C_{min} = 1.5$

$C_r/C_{min} = 1$

$0.25 \leq (hA)^* \leq 1$

Effectiveness ϵ, %

Modified transfer units, NTU_0

Fig. 59. Periodic flow exchanger performance for $C_{min}/C_{max} = 0.8$ (from "Compact Heat Exchangers," 1st ed., by W. M. Kays and A. L. London, National Press).

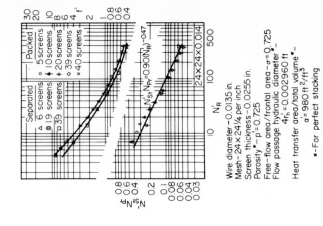

Wire diameter—0.0135 in.
Mesh—24 × 24¼ per inch
Screen thickness—0.0255 in.
Porosity*—p = 0.725
Free-flow area/frontal area—σ = 0.725
Flow passage hydraulic diameter*—
4r'_h = 0.002960 ft
Heat transfer area/total volume*—
α = 980 ft²/ft³
*—For perfect stacking

Fig. 63. Screen matrix surface 24 × 24 × 0.014 (from "Compact Heat Exchangers," 1st ed., by W. M. Kays and A. L. London, National Press).

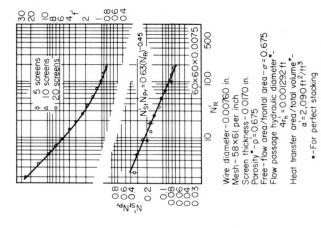

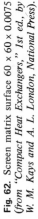

Wire diameter—0.00760 in.
Mesh—58 × 61 per inch
Screen thickness—0.0170 in.
Porosity*—p = 0.675
Free-flow area/frontal area—σ = 0.675
Flow passage hydraulic diameter*—
4r'_h = 0.001292 ft
Heat transfer area/total volume*—
α = 2,090 ft²/ft³
*—For perfect stacking

Fig. 62. Screen matrix surface 60 × 60 × 0.0075 (from "Compact Heat Exchangers," 1st ed., by W. M. Kays and A. L. London, National Press).

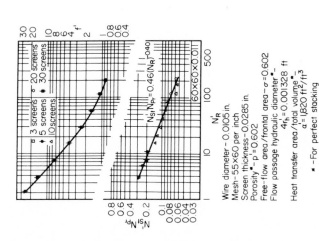

Wire diameter—0.0105 in.
Mesh—55 × 60 per inch
Screen thickness—0.0285 in.
Porosity*—p = 0.602
Free-flow area/frontal area—σ = 0.602
Flow passage hydraulic diameter*—
4r'_h = 0.001328 ft
Heat transfer area/total volume*—
α = 1,820 ft²/ft³
*—For perfect stacking

Fig. 61. Screen matrix surface 60 × 60 × 0.011 (from "Compact Heat Exchangers," 1st ed., by W. M. Kays and A. L. London, National Press).

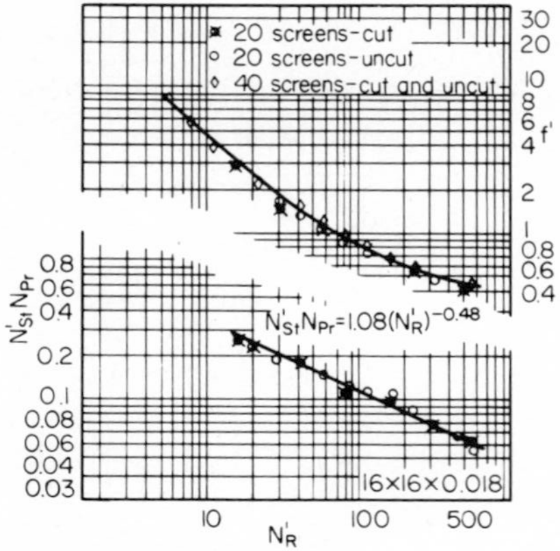

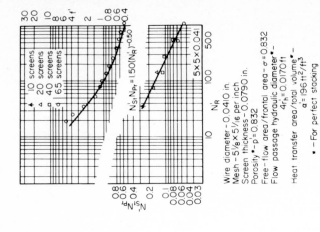

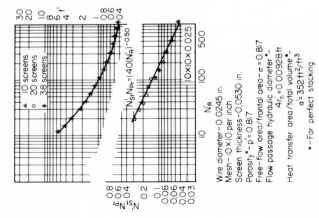

Fig. 64. Screen matrix surface 16 × 16 × 0.018 (*from "Compact Heat Exchangers," 1st ed., by W. M. Kays and A. L. London, National Press*).

Fig. 65. Screen matrix surface 10 × 10 × 0.025 (*from "Compact Heat Exchangers," 1st ed., by W. M. Kays and A. L. London, National Press*).

Fig. 66. Screen matrix surface 5 × 5 × 0.041 (*from "Compact Heat Exchangers," 1st ed., by W. M. Kays and A. L. London, National Press*).

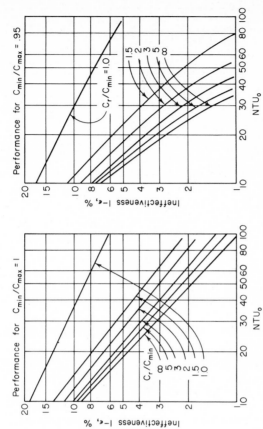

Fig. 68. Design theory results for periodic flow exchanger in the high-effectiveness range (*from "Mechanical Engineering," ASME, May 1964*).

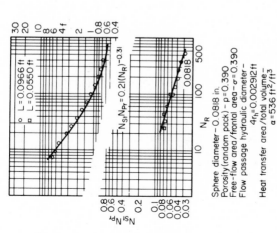

Fig. 67. Sphere matrix surface 0.0818 (*from "Compact Heat Exchangers, 1st ed., by W. M. Kays and A. L. London, National Press*).

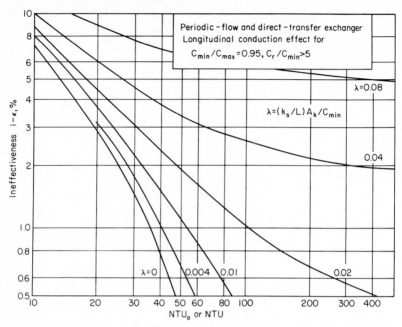

Fig. 69. Computer solution results for longitudinal conduction (*from "Mechanical Engineering,"* ASME, *May 1964*).

For high effectiveness and short regenerators, the effect of longitudinal conduction in the walls can be significant. Bahnke and Howard [28] show if $C_r/C_{min} \geq 5$ the longitudinal conduction effect is virtually the same as for direct transfer counter-current exchanger. Also tables are given for $C_r/C_{min} < 5$ and Fig. 69 shows a means of plotting the effect. The simple assumption that the decrease in effectiveness due to conduction is approximated by $\Delta\epsilon/\epsilon \approx \lambda$ overestimates the conduction effect for $C_r/C_{min} > 2$ and underestimates for $C_r/C_{min} < 2$. For the range $1 \geq C_{min}/C_{max} \geq 0.9$ as seen in Fig. 69, the longitudinal conduction effects become increasingly important for high ϵ and short-flow length exchangers. Also, under reduced load conditions C_{min} becomes smaller and λ increases with a penalty in effectiveness.

Regenerator Example. Determine the effectiveness for a rotary regenerator made of packed screening as in Fig. 63. Thickness of screen pack is 0.84 in. with the face area of 5 ft² for the cold air and 10 ft² for the hot air. Flow rates of air are the same for each stream and are 7,000 lb/hr. Take the Prandtl number equal to 0.68 for each stream and, say, the viscosities are 0.075 and 0.085 lb/ft-hr for the cold and hot streams respectively.

	Cold	Hot
Effective flow area = pA_f	0.725 x 5 = 3.625 ft²	0.725 x 10 = 7.25 ft²
G	7,000/3.625 = 1,930	7,000/7.25 = 965
Reynolds number	0.74 x 10^{-3} x 1,930/0.075 = 190	0.74 x 10^{-3} x 965/0.085 = 84
From figure $N_{ST} N_{Pr}$	0.076	0.11
N_{ST}	0.112	0.162
L/r_h	0.84/12 x 0.74 x 10^{-3} = 94.5	
NTU	0.112 x 94.5 = 106	0.162 x 94.5 = 15.3

$$NTU_0 = \frac{1}{C_{min}}\left[\frac{1}{(1/hA)_c + (1/hA)_h}\right] = \frac{1}{1/NTU_c + (C_c/C_h)/NTU_h}$$

$$C_c = C_h = 7,000 \times 0.25 = 1750$$

or

$$NTU_0 = \frac{1}{1/10.6 + 1/15.3} = 6.25$$

Matrix weight = 366 lb

C_r = (rpm x 60) x mass of matrix x sp ht
  = 3.33 x 60 x 366 x 0.12 = 8750

C_r/C_c = 8750/1750 = 5

From Fig. 57, effectiveness ϵ = 0.855

EXTENDED SURFACE OR COMPACT HEAT EXCHANGER

To reduce size and cost of heat exchangers, various techniques of finning tubes or forming wrinkled or corregated passages between plates separating the streams have been developed. Surfaces are available in a variety of metals and the surfaces held together by soldering, brazing, and welding, or formed by extrusion. There are slight variations in size, thickness, or configuration of these surfaces among all manufacturers such that exact duplication is unusual and, hence, rating or specifying such surfaces almost has to be done on an individual, basis for each manufacturer. The following figures and tables give heat transfer and pressure drop data on a representative number of these designs such that the minor variations can be estimated. In all these extended surface exchangers allowances must be made for the fin efficiency (see Sec. 3.B.2.b).

Finned Tubes. The radial finned tubes are fabricated by wrapping or extruding the fins which are attached to the tube by tension of the fin, mechanical bonding, welding, brazing, or soldering. The fins may be smooth, or ruffled at the base, or slit. The longitudinal finned tube is covered in Sec. 3.B.2.b. Plate fins are also used. Mostly the tubes are round although finned elliptical tubes are commercially available. In addition to fins, studs or wire coils are fastened to tubes or slivers cut from the tube. Selection of a finned tube depends upon evaluating many factors such as cost, reliability of the bond between fin and tube, temperature and material limitations, corrosion especially for bimetal contacts, fouling and cleaning as well as heat transfer and pressure drop. The following correlations for finned tubes are based on papers by Robinson and Briggs [29] and Briggs and Young [30]

$$\frac{h_0 D_r}{k} = 0.134\left(\frac{D_r G_{max}}{\mu}\right)^{0.681}\left(\frac{c\mu}{k}\right)^{1/3}\left(\frac{s}{l}\right)^{0.2}\left(\frac{s}{t}\right)^{0.113}$$

where h_0 is mean heat transfer coefficient Btu/hr-ft^2 external area-°F; D_r is root diameter of tube, ft; k is thermal conductivity; G_{max} is mass rate of flow at minimum cross section, lb/ft^2-hr; s is distance between adjacent fins, in.; l is fin height, in.; t is fin thickness, in.; c is specific heat; μ is viscosity at bulk temperature, lb/ft-hr

$$\frac{\Delta P g_c \rho}{n G_m^2} = f = 18.93\left(\frac{D_r G_m}{\mu}\right)^{-0.316}\left(\frac{P_t}{d_r}\right)^{-0.927}\left(\frac{P_t}{P_l}\right)^{0.515}$$

TABLE 11. Surface Geometry—Plate-fin Surfaces*

Plain fins

Surface designation	Plate spacing b — ft	Plate spacing b — in.	Fins per in.	Hydraulic diameter $4r_h$ — ft	Hydraulic diameter $4r_h$ — in.	Fin thickness δ, in.	Flow length of uninterrupted fin in.	Heat transfer area/ volume between plates 8 ft²/ft³	Fin area/ total area
5.3	0.0392	0.470	5.3	0.02016	0.242	0.006	2.49	188	0.719
6.2	0.0337	0.405	6.2	0.0182	0.218	0.010	1.20	204	0.728
9.03	0.0686	0.823	9.03	0.01522	0.1828	0.008	1.19	244	0.xxx
11.1	0.0208	0.850	11.1	0.01012	0.1213	0.006	2.50	367	0.756
11.11(a)	0.0400	0.480	11.11	0.01153	0.1385	0.008	8.00	312	0.854
14.77	0.0275	0.330	14.77	0.00848	0.1019	0.006	2.51	420	0.844
15.08	0.0348	0.418	15.08	0.00876	0.1052	0.006	6.84	414	0.870
19.86	0.0208	0.250	19.86	0.00615	0.0738	0.006	2.51	561	0.849

Louvered fins

Surface designation	Plate spacing b — ft	Plate spacing b — in.	Fins per in.	Hydraulic diameter $4r_h$ — ft	Hydraulic diameter $4r_h$ — in.	Fin thickness δ, in.	Louver spacing in.	Louver gap in.	Heat transfer area/ volume between plates 8 ft²/ft³	Fin area/ total area
3/8 – 6.06	0.0208	0.250	6.06	0.01460	0.1753	0.006	0.375	0.055	256	0.640
3/8(a)– 6.06	0.0208	0.250	6.06	0.01460	0.1753	0.006	0.375	0.130	256	0.640
1/2 – 6.06	0.0208	0.250	6.06	0.01460	0.1753	0.006	0.500	0.055	256	0.640
1/2(a)– 6.06	0.0208	0.250	6.06	0.01460	0.1753	0.006	0.500	0.130	256	0.640
3/8 – 8.7	0.0208	0.250	8.7	0.01196	0.1437	0.006	0.375	0.055	307	0.705
3/8(a)– 8.7	0.0208	0.250	8.7	0.01196	0.1437	0.006	0.375	0.080	307	0.705
3/16 –11.1	0.0208	0.250	11.1	0.01012	0.1214	0.006	0.1875	0.055	367	0.756
1/4 –11.1	0.0208	0.250	11.1	0.01012	0.1214	0.006	0.250	0.035	367	0.756
1/4(b)–11.1	0.0208	0.250	11.1	0.01012	0.1214	0.006	0.250	0.055	367	0.756
3/8 –11.1	0.0208	0.250	11.1	0.01012	0.1214	0.006	0.375	0.055	367	0.756
3/8(b)–11.1	0.0208	0.250	11.1	0.01012	0.1214	0.006	0.375	0.055	367	0.756
1/2 –11.1	0.0208	0.250	11.1	0.01012	0.1214	0.006	0.500	0.055	367	0.756
3/4 –11.1	0.0208	0.250	11.1	0.01012	0.1214	0.006	0.750	0.040	367	0.756
3/4(b)–11.1	0.0208	0.250	11.1	0.01012	0.1214	0.006	0.750	0.040	367	0.756

Strip fins

Surface designation	Plate spacing b ft	in.	Fins per in.	Hydraulic diameter $4r_h$ ft	in.	Fin thickness δ, in.	Flow length of uninterrupted fin in.	Heat transfer area/volume between plates 8 ft²/ft³	Fin area/total area
1/4(s)–11.1	0.0208	0.250	11.1	0.01012	0.1214	0.006		367	0.756
3/32 –12.2	0.0404	0.485	12.2	0.01120	0.1343	0.004	0.25	340	0.862
1/8 –15.2	0.0346	0.414	15.2	0.00868	0.1042	0.006	0.094	417	0.873
							0.125		

Wavy fins

Surface designation	Plate spacing b ft	in.	Fins per in.	Hydraulic diameter $4r_h$ ft	in.	Fin thickness δ, in.	Wave length in.	Double wave amplitude, in.	Heat transfer area/volume between plates 8 ft²/ft³	Fin area/total area
11.48–3/8W	0.0345	0.413	11.44	0.01060	0.1272	0.006	0.375	0.0775	351	0.847
17.8 –3/8W	0.0345	0.413	17.8	0.00696	0.0836	0.006	0.375	0.0775	514	0.892

Pin fins

Surface designation	Plate spacing b ft	in.	Pin pattern	Hydraulic diameter $4r_h$ ft	in.	Pin diameter in.	Transverse pin spacing, in.	Longitudinal pin spacing, in.	Heat transfer area/volume between plates 8 ft²/ft³	Fin area/total area
AP-1	0.020	0.240	In-line	0.01444	0.1734	0.040	0.125	0.125	188	0.512
AP-2	0.0332	0.398	In-line	0.01172	0.1408	0.040	0.12	0.096	204	0.686
PF-3	0.0625	0.750	In-line	0.00536	0.0644	0.031	0.0602	0.0602	339	0.834
PF-4	0.0418	0.502	Staggered	0.0186	0.223	0.065	0.199	0.125	140	0.704
PF-9	0.0425	0.510	In-line	0.0297	0.356	0.065	0.238	0.196	96.2	0.546

*From "Compact Heat Exchangers," 1st ed., by W. M. Kays and A. L. London, National Press.

TABLE 12. Heat Transfer and Friction Factors*

N_R	Heat transfer ($N_{St}N_{Pr}^{2/3}$)							
	5.3	6.2	9.03	11.1	11.11(a)	14.77	15.08	19.86
300	0.00851							
400	0.00728							0.0113
500	0.00654			0.00840	0.0103	0.00898	0.00930	0.00960
600	0.00606			0.00733	0.00890	0.00791	0.00815	0.00834
800	0.00571	0.00581	0.00692	0.00599	0.00704	0.00663	0.00662	0.00672
1,000	0.00554	0.00496	0.00575	0.00515	0.00586	0.00585	0.00562	0.00567
1,200	0.00535	0.00435	0.00499	0.00471	0.00505	0.00538	0.00491	0.00497
1,500	0.00515	0.00371	0.00421	0.00444	0.00420	0.00495	0.00420	0.00443
2,000	0.00477	0.00312	0.00347	0.00436	0.00375	0.00456	0.00352	0.00410
2,500	0.00448	0.00301	0.00318	0.00424	0.00373	0.00435	0.00322	0.00395
3,000	0.00427	0.00326	0.00310	0.00412	0.00368	0.00417	0.00309	0.00382
4,000	0.00397	0.00333	0.00310	0.00390	0.00353	0.00389	0.00309	0.00363
5,000	0.00373	0.00330	0.00304	0.00372	0.00338	0.00367	0.00310	0.00348
6,000		0.00325	0.00296	0.00356	0.00324	0.00352	0.00308	0.00337
8,000		0.00317	0.00283	0.00333	0.00303	0.00326		0.00320
10,000		0.00310	0.00273	0.00314	0.00288	0.00310		
12,000		0.00303	0.00265					
15,000			0.00255					

Friction factor – f

300								0.0457
400								0.0372
500				0.0350	0.0380	0.0403	0.0405	
600	0.0299			0.0294	0.0319	0.0346	0.0343	0.0314
800	0.0228	0.0211	0.0262	0.0228	0.0243	0.0274	0.0264	0.0242
1,000	0.0189	0.0176	0.0214	0.0190	0.0198	0.0231	0.0215	0.0197
1,200	0.0167	0.0152	0.0182	0.0169	0.0166	0.0202	0.0182	0.0167
1,500	0.0146	0.0127	0.0152	0.0149	0.0137	0.0173	0.0151	0.0142
2,000	0.0127	0.0103	0.0122	0.0139	0.0119	0.0147	0.01205	0.0123
2,500	0.0115	0.00958	0.0106	0.0119	0.0112	0.0133	0.01040	0.0112
3,000	0.0108	0.00923	0.00980	0.0112	0.0105	0.0123	0.00970	0.0104
4,000	0.00978	0.00875	0.00903	0.0103	0.00958	0.0112	0.00925	0.00972
5,000	0.00913	0.00838	0.00870	0.00991	0.00900	0.0106	0.00900	0.00931
6,000	0.00870	0.00807	0.00842	0.00971	0.00862	0.0101	0.00882	0.00900
8,000	0.00806	0.00768	0.00799	0.00923	0.00807	0.00955		0.00851
10,000	0.00764	0.00735	0.00763	0.00878	0.00768	0.00920		
12,000		0.00708	0.00740					
15,000			0.00708					

TABLE 12. Heat Transfer and Friction Factors (*Continued*)

Louvered plate-fin surfaces

$N_{St}N_{Pr}^{2/3}$

N_R	3/8 6.06	3/8 (a) 6.06	1/2 6.06	1/2 (a) 6.06	3/8 8.7	3/8 (a) 8.7	3/16 11.1	1/4 11.1	1/4 (b) 11.1	3/8 11.1	3/8 (b) 11.1	1/2 11.1	3/4 11.1	3/4 (b) 11.1
300														
400														
500	0.0160				0.0169	0.0154	0.0177	0.0168	0.0170	0.0170	0.0161	0.0150	0.0132	0.0122
600	0.0144	0.0149	0.0133	0.0128	0.0149	0.0145	0.0161	0.0155	0.0157	0.0156	0.0148	0.0137	0.0119	0.0112
800	0.0124	0.0140	0.0118	0.0122	0.0126	0.0131	0.0142	0.0137	0.0139	0.0135	0.0130	0.0117	0.0103	0.00981
1,000	0.0112	0.0138	0.0109	0.0118	0.0112	0.0121	0.0130	0.0125	0.0128	0.0121	0.0119	0.0104	0.00928	0.00894
1,200	0.0104	0.0127	0.0102	0.0113	0.0103	0.0113	0.0122	0.0119	0.0120	0.0111	0.0112	0.00950	0.00859	0.00831
1,500	0.00970	0.0119	0.00948	0.0108	0.00951	0.0106	0.0113	0.0112	0.0112	0.0102	0.0105	0.00888	0.00787	0.00762
2,000	0.00900	0.0110	0.00875	0.0100	0.00885	0.00980	0.0103	0.0102	0.0103	0.00930	0.00960	0.00825	0.00711	0.00699
2,500	0.00849	0.0102	0.00829	0.00941	0.00835	0.00950	0.00960	0.00954	0.00972	0.00861	0.00889	0.00777	0.00663	0.00655
3,000	0.00805	0.00970	0.00791	0.00895	0.00794	0.00870	0.00899	0.00901	0.00922	0.00811	0.00835	0.00739	0.00630	0.00621
4,000	0.00738	0.00878	0.00734	0.00809	0.00737	0.00790	0.00802	0.00825	0.00853	0.00793	0.00752	0.00680	0.00576	0.00572
5,000	0.00690	0.00810	0.00690	0.00760	0.00678	0.00730	0.00740	0.00771	0.00800	0.00684	0.00694	0.00640	0.00537	0.00537
6,000	0.00651	0.00760	0.00655	0.00714	0.00640	0.00690	0.00690	0.00728	0.00761	0.00645	0.00650	0.00604	0.00508	0.00510
8,000	0.00593	0.00688	0.00605	0.00645	0.00583	0.00630		0.00666	0.00701	0.00588	0.00590	0.00557	0.00462	0.00469
10,000	0.00551	0.00638	0.00568	0.00598	0.00542					0.00548			0.00432	0.00440
12,000														
15,000														

Friction factor — f

	C1	C2	C3	C4	C5	C6	C7	C8	C9	C10	C11	C12	C13	C14
300														
400														
500	0.0755				0.0793	0.0890	0.0925	0.0850	0.0870	0.0796	0.0741	0.0641	0.0580	0.0565
600	0.0682	0.0962	0.0667	0.0880	0.0700	0.0790	0.0848	0.0772	0.0780	0.0700	0.0659	0.0570	0.0516	0.0500
800	0.0587	0.0860	0.0571	0.0752	0.0585	0.0680	0.0738	0.0670	0.0662	0.0528	0.0550	0.0474	0.0427	0.0416
1,000	0.0532	0.0795	0.0512	0.0680	0.0515	0.0620	0.0662	0.0600	0.0595	0.0513	0.0483	0.0410	0.0370	0.0362
1,200	0.0496	0.0745	0.0474	0.0634	0.0472	0.0580	0.0610	0.0558	0.0550	0.0469	0.0442	0.0368	0.0314	0.0313
1,500	0.0461	0.0696	0.0438	0.0588	0.0430	0.0550	0.0553	0.0512	0.0502	0.0423	0.0406	0.0332	0.0289	0.0288
2,000	0.0426	0.0646	0.0402	0.0540	0.0394	0.0497	0.0491	0.0464	0.0456	0.0375	0.0363	0.0299	0.0244	0.0248
2,500	0.0406	0.0620	0.0381	0.0511	0.0374	0.0470	0.0452	0.0461	0.0430	0.0346	0.0336	0.0283	0.0222	0.0227
3,000	0.0394	0.0596	0.0366	0.0491	0.0359	0.0420	0.0426	0.0408	0.0412	0.0326	0.0319	0.0271	0.0208	0.0213
4,000	0.0375	0.0568	0.0347	0.0463	0.0340	0.0428	0.0390	0.0374	0.0390	0.0300	0.0296	0.0255	0.0190	0.0194
5,000	0.0363	0.0547	0.0332	0.0447	0.0328	0.0410	0.0367	0.0351	0.0375	0.0283	0.0281	0.0242	0.0178	0.0183
6,000	0.0354	0.0531	0.0322	0.0432	0.0319	0.0395	0.0350	0.0333	0.0364	0.0271	0.0271	0.0233	0.0170	0.0175
8,000	0.0340	0.0510	0.0310	0.0413	0.0306	0.0340		0.0309	0.0349	0.0253	0.0257	0.0220	0.0158	0.0168
10,000	0.0331	0.0494	0.0300	0.0400	0.0297					0.0242			0.0151	0.0156
12,000														
15,000														

TABLE 12. Heat Transfer and Friction Factors (Continued)

	Strip-fin plate-fin surfaces $N_{St}N_{Pr}^{2/3}$			Wavy-fin plate-fin surfaces $N_{St}N_{Pr}^{2/3}$	
N_R	1/4 (s) 11.1	3/32 12.22	1/8 15.2	11.44 3/8W	17.8 3/8W
300			0.01810		
400			0.01675		
500	0.0155	0.0205	0.01580	0.0179	
600	0.0139	0.0192	0.01520	0.0175	0.0158
800	0.0122	0.0171	0.01427	0.0165	0.0142
1,000	0.0109	0.0156	0.01373	0.0153	0.0129
1,200	0.0102	0.0146	0.01327	0.0144	0.0120
1,500	0.00940	0.0133	0.01267	0.0132	0.0110
2,000	0.00850	0.0119	0.01177	0.0119	0.00982
2,500	0.00789	0.0108	0.01110	0.0110	0.00900
3,000	0.00740	0.0101	0.01040	0.01025	0.00835
4,000	0.00669	0.00903	0.00959	0.00920	0.00740
5,000	0.00620	0.00828	0.00896	0.00846	0.00675
6,000	0.00580	0.00770	0.00850	0.00794	
8,000	0.00525	0.00688		0.00712	
10,000		0.00629			
12,000					
15,000					

	Friction factor – f			Friction factor – f	
300			0.1390		
400			0.1145		
500	0.0665	0.130	0.1010	0.1045	
600	0.0595	0.113	0.0913	0.0984	0.0738
800	0.0500	0.0942	0.0800	0.0888	0.0643
1,000	0.0438	0.0826	0.0726	0.0819	0.0579
1,200	0.0394	0.0752	0.0676	0.0758	0.0530
1,500	0.0348	0.0680	0.0628	0.0691	0.0478
2,000	0.0298	0.0607	0.0584	0.0615	0.0421
2,500	0.0272	0.0560	0.0558	0.0563	0.0385
3,000	0.0253	0.0530	0.0540	0.0524	0.0358
4,000	0.0231	0.0487	0.0516	0.0469	0.0320
5,000	0.0218	0.0458	0.0498	0.0430	0.0293
6,000	0.0209	0.0440	0.0487	0.0401	
8,000	0.0197	0.0413		0.0359	
10,000		0.0394			
12,000					
15,000					

TABLE 12. Heat Transfer and Friction Factors (*Continued*)

| | Pin-fin plate-fin surfaces | | | | |
| | $N_{St}N_{Pr}^{2/3}$ | | | | |
N_R	AP-1	AP-2	PF-3	PF-4 (F)	PF-9 (F)
200					
250					
300			0.0132		
400			0.0118		
500		0.0222	0.01085		
600		0.0218	0.01020	0.0330	
800	0.0183	0.0209	0.00912		
900					
1,000	0.0175	0.0200	0.00840	0.0301	
1,200	0.0161	0.0190	0.00784	0.0279	
1,500	0.0152	0.0183	0.00720	0.0255	0.0191
2,000	0.0134	0.0168	0.00645	0.0227	0.0173
2,500	0.0123	0.0155		0.0207	0.0161
3,000	0.0116	0.0145		0.0192	0.0152
4,000	0.0105	0.0130		0.0171	0.0137
5,000	0.00970			0.0156	0.0128
6,000	0.00905			0.0144	0.0120
8,000	0.00808			0.0128	0.0108
10,000					0.0101
12,000					0.00951
15,000					0.00880
17,000					0.00842
20,000					0.00800
25,000					0.00740

	Friction factor — f				
200			0.0673		
250			0.0600		
300			0.0551		
400			0.0491		
500		0.180	0.0454		
600	0.0800	0.170	0.0430		
800		0.156	0.0399		
900				0.198	
1,000	0.0755	0.151	0.0383	0.189	
1,200	0.0718	0.150	0.0373	0.182	
1,500	0.0708	0.151		0.175	0.0800
2,000	0.0707	0.158		0.167	0.0760
2,500	0.0725	0.164		0.160	0.0737
3,000	0.0741	0.165		0.156	0.0718
4,000	0.0762	0.166		0.153	0.0692
5,000	0.0780			0.153	
6,000	0.0793			0.153	
8,000	0.0815			0.153	
10,000					
12,000					
15,000					
17,000					
20,000					
25,000					

TABLE 12. Heat Transfer and Friction Factors (*Continued*)

					Flow normal to banks of finned tubes $N_{St}N_{Pr}^{2/3}$				
N_R	CF 7.34	CF 8.72	CF 8.72C	CF 11.46	9.68 0.87	9.1 0.737-S	9.68 0.87-B	9.29 0.737-SR	11.32 0.737-SR
300									
400					0.01150	0.01526	0.01128		0.0163
500		0.0180	0.01915		0.00982		0.01000		0.0142
600	0.01506	0.0165	0.01760		0.00861	0.01377	0.00920	0.01390	0.0127
800	0.01350	0.0143	0.01504		0.00701	0.01175	0.00832	0.01176	0.0108
1,000	0.01210	0.0129	0.01390		0.00600	0.01030	0.00772	0.01045	0.00968
1,200	0.01105	0.0119	0.01285		0.00536	0.00934	0.00727	0.00964	0.00892
1,500	0.00990	0.0108	0.01170		0.00477	0.00831	0.00676	0.00885	0.00816
2,000	0.00860	0.00960	0.01035		0.00427	0.00726	0.00615	0.00802	0.00739
2,500	0.00770	0.00880	0.00945		0.00411	0.00656	0.00573	0.00751	0.00690
3,000	0.00701	0.00820	0.00879		0.00405	0.00608	0.00543	0.00712	0.00655
4,000	0.00609	0.00740	0.00785		0.00385	0.00539	0.00498	0.00660	0.00602
5,000	0.00547	0.00685	0.00727		0.00369	0.00493	0.00468	0.00620	0.00563
6,000	0.00500	0.00645	0.00685		0.00359	0.00460	0.00445	0.00590	0.00533
8,000	0.00465		0.00638		0.00339	0.00417	0.00412	0.00546	0.00468
10,000	0.00388				0.00326	0.00389	0.00390	0.00514	0.00453
12,000									
15,000									

Friction factor — f

N_R									
300			0.0630	0.0522	0.0463	0.0531	0.0550		0.0652
400				0.0481	0.0376		0.0470		0.0565
500									
600	0.0530	0.0572	0.0597	0.0450	0.0321	0.0476	0.0422	0.0569	0.0505
800	0.0490	0.0532	0.0551	0.0408	0.0256	0.0402	0.0366	0.0472	0.0429
1,000	0.0461	0.0483	0.0520	0.0379	0.0219	0.0354	0.0333	0.0416	0.0384
1,200	0.0441	0.0433	0.0495	0.0358	0.0196	0.0320	0.0309	0.0380	0.0353
1,500	0.0420	0.0412	0.0470	0.0337	0.0174	0.0284	0.0284	0.0346	0.0323
2,000	0.0395	0.0388	0.0440	0.0314	0.0155	0.0247	0.0255	0.0315	0.0290
2,500	0.0379	0.0371	0.0418	0.0300	0.0142	0.0223	0.0238	0.0294	0.0269
3,000	0.0365	0.0360	0.0401	0.0290	0.0133	0.0206	0.0228	0.0280	0.0253
4,000	0.0348	0.0344	0.0381	0.0278	0.0122	0.0183	0.0217	0.0261	0.0232
5,000	0.0334	0.0333	0.0368	0.0272	0.0116	0.0167	0.0208	0.0247	0.0218
6,000	0.0326	0.0326	0.0359	0.0267	0.0113	0.0156	0.0201	0.0236	0.0207
8,000	0.0313	0.0314	0.0347	0.0260	0.0108	0.0142	0.0191	0.0219	0.0192
10,000	0.0304	0.0306		0.0255	0.0106	0.0133	0.0183	0.0206	0.0180
12,000									
15,000									

*From "Compact Heat Exchangers," 1st ed., by W. M. Kays and A. L. London, National Press.

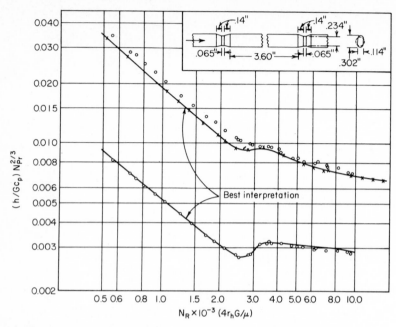

Fig. 70. Flow inside flattened tubes (surface FT-1) *(from "Compact Heat Exchangers," 1st ed., by W. M. Kays and A. L. London, National Press).*

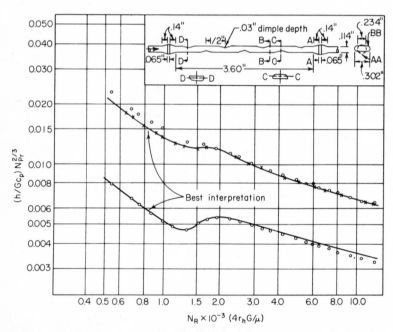

Fig. 71. Flow inside dimpled flattened tubes (surface FTP-1) *(from "Compact Heat Exchangers," 1st ed., by W. M. Kays and A. L. London, National Press).*

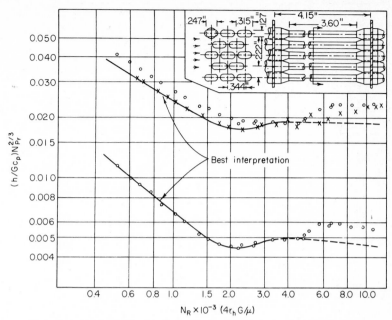

Fig. 72. Flow normal to flattened tubes (*from "Compact Heat Exchangers," 1st ed., by W. M. Kays and A. L. London, National Press*).

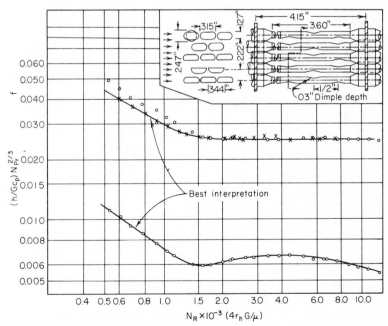

Fig. 73. Flow normal to dimpled flattened tubes (*from "Compact Heat Exchangers," 1st ed., by W. M. Kays and A. L. London, National Press*).

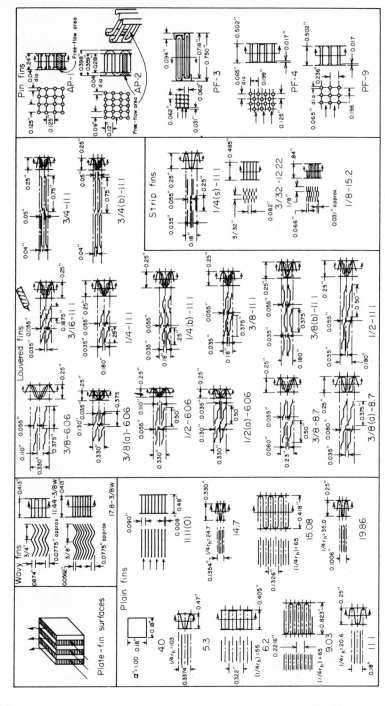

Fig. 74. Diagram of plate-fin surfaces (*from "Compact Heat Exchangers," 1st ed, by W. M. Kays and A. L. London, National Press*).

TABLE 13. Compact Heat Exchangers Surface and Core Geometrical Relations*

$$r_h = L\left(\frac{A_c}{A}\right)$$

$$\sigma_1 = \left(\frac{A_c}{A_{fr}}\right)_1 = \left(\frac{Ar_h}{LA_{fr}}\right)_1 = \frac{(Ar_h)_1}{V} = (\alpha r_h)_1 \qquad \text{(except for matrix surfaces)}$$

$$\sigma_1 = \frac{b_1\beta_1 r_{h1}}{b_1 + b_2 + 2a} \qquad \text{(plate-fin surfaces only)}$$

$$\alpha_1 = \frac{A_1}{V} = \left(\frac{A}{LA_{fr}}\right)_1 = \left(\frac{\sigma}{r_h}\right)_1 \qquad \text{(except for matrix surfaces)}$$

$$\alpha_1 = \frac{b_1\beta_1}{b_1 + b_2 + 2a} \qquad \text{(plate-fin surfaces only)}$$

$$A_{c1} = (\sigma A_{fr})_1 = \left(\frac{Ar_h}{L}\right)_1 = \left(\frac{A\sigma}{L\alpha}\right)_1 \qquad \text{(except for matrix surfaces)}$$

$$\left(\frac{r_h}{L}\right)_1 = \left(\frac{A_c}{A}\right)_1 = \left(\frac{\sigma}{L\alpha}\right)_1 \qquad \text{(except for matrix surfaces)}$$

$$A = \alpha L A_{fr}$$

$$p = \left(\frac{A_c}{A_{fr}}\right) = \alpha r_h \qquad \text{(for matrix surfaces only†)}$$

$$\alpha = \left(\frac{p}{r_h}\right) \qquad \text{(for matrix surfaces only†)}$$

$$A_c = pA_{fr} = \alpha r_h A_{fr} = A\left(\frac{r_h}{L}\right) \qquad \text{(for matrix surfaces only†)}$$

$$A = \left(\frac{\alpha}{p}\right)LA_c \qquad \text{(for matrix surfaces only†)}$$

---

*From "Compact Heat Exchangers," 1st ed., by W. M. Kays and A. L. London, National Press.
†For screen matrices α, r_h, p apply to the condition of perfect stacking, i.e., no separation between screen layers.

where ΔP is static pressure drop lb/ft^2; G_m is mass velocity at minimum cross section through a row of tubes normal to flow, lb/ft^2-hr; n is number of rows in direction of flow; P_t is transverse pitch between adjacent tubes in same row, in.; P_l is longitudinal pitch between adjacent tubes in different rows measured on the diagonal, in.; d_r is root diameter of tube, in.; D_r is root diameter of tube, ft; g_c is 4.18×10^8 ft/hr^2; μ is viscosity, lb/ft-hr.

It should be noted the above equations represent a wide range of tubes and tube arrangements; however, these equations have not been tested against all data and

TABLE 14. Nomenclature*

English symbols

A	Exchanger total heat transfer area on one side, ft^2
A_c	Exchanger minimum free-flow area, ft^2
A_f	Exchanger total fin area on one side, ft^2
A_{fr}	Exchanger total frontal area, ft^2
a	Plate thickness, ft
a	Short side of a rectangular flow passage, ft
b	Plate spacing, ft or in.
b	Long side of a rectangular flow passage, ft
C	Flow-stream capacity rate, (Wc_p), Btu/hr-°F
c_p	Specific heat at constant pressure, Btu/lb-°F
D	Inside diameter of circular tube, ft
D_0	Outside diameter of circular tube, ft
d	Diameter of pins in pin-fin surface, ft
f	Mean friction factor; this is the "small," or "Fanning" friction factor
G	Exchanger flow stream mass velocity, (W/A_c), lb/hr-ft^2
g_c	Proportionality factor in Newton's Second Law, g_c = 32.2 (lb/#)-(ft/sec^2)
h	Unit conductance for thermal convection heat transfer, Btu/hr-ft^2-°F
k	Unit thermal conductivity, Btu/(hr-ft^2-°F/ft)
L	Total heat exchanger flow length, ft
l	Fin length from root to center, $b/2$, ft
l'	Fin dimension in flow direction, ft
p	Porosity for matrix surfaces, dimensionless
p'	Porosity for matrix surfaces based on perfectly stacked screens, dimensionless
q	Heat transfer rate, Btu/hr
r_h	Hydraulic radius, $(A_c L/A)$, ft, $(4r_h$—hydraulic diameter)
r_i	Inner radius of circular fin, ft
r_0	Outer radius of circular fin, ft
t	Temperature, °F
U	Unit overall thermal conductance, Btu/(hr-°F-ft^2 of A)
V	Velocity, ft/sec
V	Volume, ft^3
W	Mass flow rate, lb/hr
x	Distance along the flow passage in direction of flow, ft
X_l	Ratio of longitudinal pitch to tube diameter for flow across banks of circular tubes, dimensionless
X_t	Ratio of transverse pitch to tube diameter for flow across banks of circular tubes, dimensionless

Greek symbols

α	Ratio of total transfer area on one side of the exchanger to total volume of the exchanger, ft^2/ft^3
α^*	Aspect ratio of a rectangular flow passage, b/a, dimensionless
β	Ratio of total heat transfer area on one side of a plate-fin heat exchanger to the volume between the plates on that side, ft^2/ft^3
Δ	Denotes difference
δ	Fin thickness, ft
ϵ	Exchanger effectiveness, dimensionless
η_f	Fin temperature effectiveness, dimensionless
η_0	Total surface temperature effectiveness, dimensionless
σ	Ratio of free-flow area to frontal area, A_c/A_{fr}, dimensionless
μ	Viscosity, lb/hr-ft
ρ	Density, lb/ft^3

*From "Compact Heat Exchangers," 1st ed., by W. M. Kays and A. L. London, National Press.

TABLE 14. Nomenclature (*Continued*)

Dimensionless groupings

N_R	Reynolds number, $(4r_h G/\mu)$, a flow modulus
N_{St}	Stanton number, (h/Gc_p), a heat transfer modulus
N_{Nu}	Nusselt number, $(h4r_h/k)$, a heat transfer modulus
N_{Pr}	Prandtl number, $(\mu c_p/k)$, a fluid properties modulus
$N_{St}N_{Pr}^{2/3}$	Generalized heat transfer grouping; this factor versus N_R defines the heat transfer characteristics of the surface
f	Mean friction factor; this factor versus N_R defines the friction characteristics of the surface
C_{min}/C_{max}	Flow-stream capacity rate ratio, $\left(Wc_p\right)_{min}/\left(Wc_p\right)_{max}$

Subscripts

a	Air side
avg	Average
c	"Cold" side of heat exchanger
h	"Hot" side of heat exchanger
L	Coupling liquid in a liquid-coupled heat exchanger
m	Mean conditions, defined as used
p	Refers to one pass of a multipass heat exchanger
r	Matrix rotor
w	Wall, or water side
1, 2	Indicate different sides of the heat exchanger, or inlet and outlet conditions
max	Maximum
min	Minimum
lma	Log mean average

Force and Mass Units

lb	Denotes pounds mass in distinction to
#	Denoting pounds force

should be used with caution. Sometimes performance data are available in manufacturers' catalogs on specific finned tubes and tube bundles and should be used whenever possible. In absence of such curves, the above equations can be used to approximate designs of finned heat exchangers within about ±20 percent for both radial and plate fins.

Occasionally flattened or dimple tubes are used either alone or with fins. Heat transfer and friction data [26] are given in Figs. 70 to 73 for flow inside and outside of these tubes.

Plate-fin Cores. A wide variety of finned cores are available. These consist of formed fins usually between plates about one-quarter inch apart. The finning arrangement may be different for each of the two fluids and the flow paths may be parallel or perpendicular to one another. Fins can be plain, wavy, cut and louvered, or pins. These are compact-type exchangers where the surface area for heat transfer may range from 100 to over 500 square feet surface per cubic foot of passage volume. With the flow paths paralleled, these surfaces can also be used in regenerators. Kays and London [26] have tested many designs. Dimensions of the units are given in Table 11 and are illustrated in Fig. 74. Tabulated heat transfer and friction factors are given in Table 12. Various ratios used in calculations or specifications of surfaces are given in Table 13 with nomenclature given in Table 14.

AIR-COOLED EXCHANGERS

There are many designs of air-cooled heat exchangers, but they are broadly classified according to the flow of air. In forced draft exchangers the air is blown through the coils by a fan. In induced draft exchangers the air first passes through the coil, then through the fan, and finally a natural draft exchanger where the air is drawn through the coils by means of a chimney on the hot exhaust air. Characteristics of air coolers are shallow tube bundles and large face areas due to the low heat capacity of air and the pressure-drop limitations of the fans. Forced draft results in better heat transfer due to the turbulence imparted by the fans, but offsetting this advantage is possibly poor distribution through the coil and greater chances of air recirculation. With induced designs the air leaves the fan with sufficient velocity to carry the discharge stream high into the atmosphere, thus preventing local air recirculation. Forced draft fans operate cooler and thus more efficiently.

There are numerous designs of tubes and bundles. The tube bundles may be horizontal, vertical, or in V or A frame designs, plus cylindrical shapes. The tubes have extended surfaces of many types: spiked, slit, extruded and wrapped radial fins, and plate fins. In addition to the heat transfer and pressure-drop characteristics of the finned tubes, the other important factors in their selection are cost, material temperature limitation, loosening the bond between tube and fin, and corrosion at this point.

There is no simple means of selecting the best design for a specific application because of the interplay of the numerous factors entering in the selection such as heat transfer, pressure drop, air recirculation, cost, space and location, means of control, corrosion, range of operation and climatic conditions, and mechanical problems with fan drives and controls.

The calculation of heat transfer in air coolers can be made relatively simple by certain techniques. Tube side is covered in other chapters. The air side heat transfer seems to correlate best based on the maximum velocity of air in the bundle (face area minus the blocking effect created by the tubes and the fins). While there are minor differences in heat transfer due to variations in fin designs among the manufacturers, the correlation in Fig. 75 holds for most common commercial designs with ¾- or 1-inch tubes with radial and plate fins with the range of 7 to 11 fins per inch, 1/2- to 5/8-inch fin height. Pressure drops are also shown. The principal problem in calculating heat transfer is that the size of the cooler and hence air flow is not known which normally leads to trial-and-error solutions. However, by specifying the face velocity and the number of tube rows and tube type and fins, a direct solution is available by means of Fig. 76.

Table 15 [31] lists typical ranges of overall coefficients which can be used for preliminary estimating. Actual overall coefficients can be calculated through use of Fig. 75 and suitable equations for inside coefficients. Then Fig. 76 or 77 might be used.

Table 16 gives some typical air velocities for various rows of tubes and values of A' and M for a typical tube. Other values can be readily calculated for any specific design.

Figure 77 is a simple solution if the counterflow temperature difference can be used while Fig. 76 is a similar chart for pure cross flow and Fig. 78 applies to condensation cases. For sensible heat transfer the conditions for $(T_2 - t_1)/(T_1 - t_1)$ are specified. Choosing a specific number of rows and the A' the value M is determined and UM calculated. From these charts value of 1.08 (FV) A_F/WC is read. Since WC and FV are known or specified then face area A_F is determined. Thus, the size of the exchanger is readily specified without a trial-and-error solution. For condensers a simple chart of Fig. 78 is given. Here only $T - t_1$ and UM are needed to obtain $q/1.08$ (FV) A_F and hence A_F. Other types of charts eliminating the trial and error solutions can be devised for specific combinations of FV, U, and A'.

Design Considerations

Field of Use and Economics.   Air coolers can be used in all climates and are not limited to those areas where water supply is low or poor. The economics of using air

TABLE 15. Typical Overall (U_0) Transfer Coefficients for Air
Cooled Exchangers*†

Service	U_0 Btu/hr-ft²-°F (Bare tube surface)
Liquid service	
Jacket water	120-130
50% glycol-water	95-105
Engine lube oil with retarders	20-30
Engine lube oil without retarders	15-20
Light hydrocarbons	75-95
Light naphtha	70-80
Hydroformers and platformers liquids	70
Light gas oil—viscosity, less than 1 cp	60-70
Heavy gas oil—viscosity, 2 to 30 cp	20-25
Heavy lube distillate—viscosity, 10 to 300 cp	8-20
Residuum—viscosity 50 to 1000 cp	10-20
Tar	5-10
Process water	105-120
Fuel oil	20-30
Gas cooling	
Flue gas at 10 psi $\Delta\rho$ = 1 psi	10
Flue gas at 100 psi $\Delta\rho$ = 5 psi	30
Air at 30 to 40 psi	20
Air at 50 to 100 psi	20-30
Air at 100 to 300 psi	30-35
Air at 300 to 600 psi	35-40
Air at 600 to 1000 psi	40-50
Air at 1000 to 3000 psi	50-65
Ammonia reactor stream	80-90
Hydrocarbon gases at 15-50 psi ($\Delta\rho$ = 1 psi)	30-40
Hydrocarbon gases at 50-250 psi ($\Delta\rho$ = 3 psi)	50-60
Hydrocarbon gases at 250-1500 psi ($\Delta\rho$ = 5 psi)	70-90
Hydrocarbon gases at 1500-2500 psi ($\Delta\rho$ = 7 psi)	80-100
Condensing	
Steam (0-20 psig)	130-140
Ammonia	100-120
Amine reactivator	90-100
Light hydrocarbons	80-95
Light gasoline	80
Light naphtha	70-80
"Freon" 12	60-80
Heavy naphtha	60-70
Reactor effluent—platformers, rexformers, hydroformers	60-80
Still overhead—light naphtha, steam, and noncondensable gases	60-70

*From "Air-Cooled Heat Exchangers," by E. C. Smith, *Chemical Engineering,* Nov. 17, 1958.

†Based on Wolverine-type finned tubes, 2-3/8-in. triangular pitch, 1-in. OD, 8 (2.25-in. OD) fins/inch. Values of U_0 largely obtained from Ref. 31.

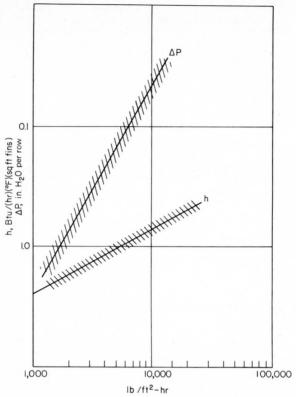

Fig. 75. Heat transfer coefficient and pressure drop for radial finned tube bundles (3/4- to 1-in. ID tubes, 7–11 fin/in., 1/2- to 5/8-in. high fins).

TABLE 16

Rows, deep	3	4	5	6	7
Typical face velocity Std. ft/min (FV)	700	650	600	550	530
Ft² base tube per ft² face area A' *	3.80	5.04	6.32	7.60	8.84
$M = A'/1.08$ (FV)	0.0050	0.0072	0.0097	0.0128	0.0154

Rows, deep	8	9	10	11	12
Typical face velocity Std. ft/min (FV)	510	490	470	450	425
Ft² base tube per ft² face area A' *	10.08	11.36	12.64	13.92	15.20
$M = A'/1.08$ (FV)	0.0183	0.0215	0.025	0.0286	0.0332

*Based on 1-in. finned tube on 2-3/8-in. triangular pitch, 2.25 in. over fins, 8 fins/in.

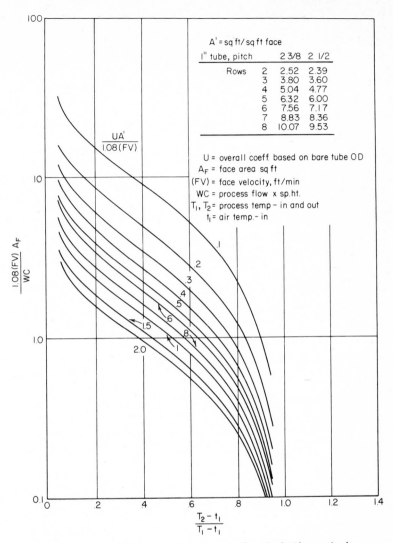

Fig. 76. Heat transfer in air coolers with cross flow—both sides unmixed.

versus water is very complicated due to evaluating all the relevant costs such as space, electrical power, piping, sewers, water costs, maintenance, fouling, etc. In general, air is usually overwhelmingly economic for high-temperature level cooling and can still be economic (depending on local conditions) when temperature levels are lower and the process discharge approaches 30 to 50°F of the ambient air temperature. The surface area is usually larger than with water-cooled units, but the ratio is not large. Air coolers can have air side coefficients of 150–200 Btu/hr-ft^2-°F including fouling while water coefficients may be 400 to 1200 Btu/hr-ft^2-°F, but then must include allowance for fouling making the water plus fouling coefficient ranging 170 to 550

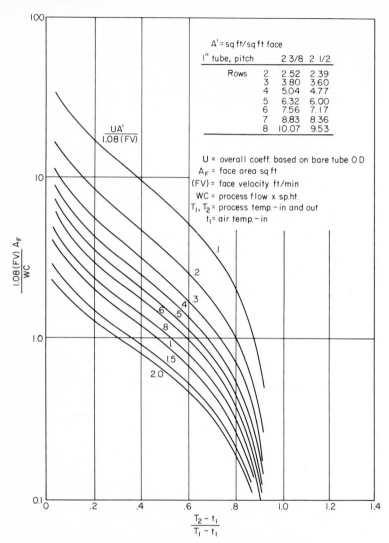

Fig. 77. Heat transfer in air coolers with countercurrent flow.

Btu/hr-ft^2-°F. By the time the process side is included, the net effect may be only an area increase of 10 to 30 percent.

Sizes. The exchangers come in all sizes from small automobile or refrigerator radiators to a large power plant radiator approximately 40 feet high (excluding chimney) and 300 feet diameter. However, most industrial exchangers are built in modules, usually in an 8-foot wide bundle and 24 to 30 feet long. When a number of small exchangers are needed, then several small bundles are mounted together over the fans. The tube bundles usually range from 3 to 6 rows deep but on occasion may extend to 12 rows.

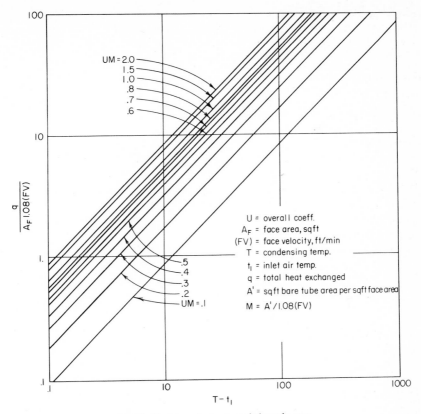

Fig. 78. Heat transfer in air-cooled condensers.

Design Temperature. The ambient air temperature has a large range throughout the year. Usually the maximum air temperature selected for design is that temperature that is exceeded about 2 to 3 percent of the annual hours. Lower design temperatures can be used especially if for peak temperatures (only a few hours per day and few days per year) some good grade of water is sprayed into the inlet air stream cooling it so as to approach the wet bulb temperature.

Fans. Fans are usually 4 to 12 feet in diameter and one or more may be used with a bundle. Larger and smaller sizes have been used. Multiple blades are used and blade pitch is adjustable either manually or automatically. Fan tip speed is usually limited to less than 12,000 ft/min in order to keep the noise level down. These fans handle large volumes of air at relatively low pressure drops of 1 to 2 in water.

Control. Due to rapid ambient temperature changes the control system has to be engineered for the process conditions. Many means of air control are used, e.g., multiple speed fans, turning on or off fans when several are used, pitch control, use of louvers, recirculation, process bypass control. Control can be affected by the design of the cooler. For instance, in a forced draft exchanger, air recirculation can occur particularly if the unit is downwind and next to a larger building. Sun and rain can make a marked effect. Sometimes many of these problems can be avoided by using induced draft units. Winter conditions and possible freezing must be considered. Here auxiliary heating coils or reversing the pitch on some fans to induce recirculation can prevent freezing or air recirculation systems can be designed. In the event of power failure, air coolers are still partially effective due to natural draft created through the coils. In the medium temperature ranges, a cooling capacity of one-third normal is obtainable with the fans turned off. In winter this effect may become a disadvantage.

Drives. Fans are mostly driven by electric motors either direct or through V-belt or gear boxes. Choice depends on operating conditions although gear drives are common for motors of 10 horsepower or larger.

Induced vs. Forced Draft. Some of the advantages and disadvantages of these two types are:

	Induced	Forced
Heat transfer coefficient	Lower: effect of number of rows	High because of fan turbulence
Power	Higher: fan in high temperature stream	Lower
Drive and lubrication	In hot air: limit to 170°F discharge above bundle	In ambient: below bundle
Recirculation due to wind	Very low due to high upward velocity from fan	High possibility due to low bundle discharge velocity
Air distribution	Good	Poor: depends on fan location and design
Rain and sun effects	None: cover and high air velocity prevent rain from entering	Large unless otherwise protected by louvers

Space Problems. Physically air coolers take up more space than an equivalent water-cooled exchanger. Thus more ingenuity is required by the design engineer in laying out his plant. Ground space can be reduced by going to *A*-frame or *V* designs. Space is saved by placing the coolers over pipe bridges, on top of buildings, or placing high enough that space under the coolers can be utilized by other equipment.

Air Cooler Problem

A. Use of Fig. 77. Cool 65,000 lb/hr of a fluid with sp ht equal to 0.8 from 200°F to 120°F in a 4-row–4-pass countercurrent bundle. An overall coefficient of 90 Btu/hr-ft^2-°F is calculated for this flow and with air at 650 ft^2/min face velocity. Air temperature used for design is 90°F. Using the 1-inch OD finned tube (Table 17), $M = 0.0072$

$$UM = 90 \times 0.0072 = 0.647$$

$$\frac{T_2 - t_1}{T_1 - t_1} = \frac{120 - 90}{200 - 90} = 0.273$$

From the chart read

$$\frac{1.08 \, (FV) A_F}{WC} = 2.45$$

$$A_F = \frac{2.45 \times 0.8 \times 650,000}{1.08 \times 650} = 181 \text{ ft}^2$$

A bundle (more or less standard size) of 8 by 24 ft or 192 ft^2 face area will be satisfactory and have a 6 percent safety factor. To check

$$\text{heat load } q = 65{,}000 \times 0.8\,(200 - 120) = 4{,}150{,}000 \text{ Btu/hr}$$

$$\text{air rise for 181 ft}^2 \text{ face} = \frac{4{,}150{,}000}{1.08 \times 181 \times 650} = 32.6°F$$

$$\Delta T_{\text{lm}} = \frac{(200 - 122.6) - (120 - 90)}{\ln 2.58} = 49.2°F$$

$$\text{Required surface area} = \frac{4{,}150{,}000}{49.2 \times 90} = 936 \text{ ft}^2$$

Area of 181 ft^2 face = 181 × 5.04 = 914 ft^2 (close enough check for a graph).

B. Use of Fig. 78. Suppose for some type of surface and air conditions we wished to condense at 150°F a vapor with a total condensing load $q = 2{,}500{,}000$ Btu/hr. How big a bundle is required? Suppose the calculated $U = 100$ Btu/hr-ft^2-°F, then UM = 0.72 and $T - t_1 = 150 - 90 = 60$. From Fig. 78

$$\frac{q}{A_F\, 1.08\,(FV)} = 31$$

$$A_F = \frac{2{,}500{,}000}{31 \times 1.08 \times 650} = 115 \text{ ft}^2$$

or an 8 × 16 foot bundle has 128 square foot face area and an 11 percent safety factor. To check

$$\text{Air rise} = \frac{2{,}500{,}000}{1.08 \times 650 \times 128} = 27.8°F$$

$$\Delta T_{\text{lm}} = \frac{(150 - 90) - (150 - 117.8)}{\ln 1.865} = 44.6°F$$

$$\text{Surface area required} = \frac{2{,}500{,}000}{44.6 \times 100} = 560 \text{ ft}^2$$

Area in 115 ft^2 FA = 115 × 5.04 = 580 ft^2.

Example. Air side coefficient. Calculate air side coefficient for 1-in. OD tube with 8 fin/in. 0.017 in. thick and 5/8 in. high. Tube pitch is 2-3/8 in. and air-face velocity is 500 ft/min at 70°F. ($\rho = 0.075$ lb/ft^3.)

Calculate minimum area for flow. Per foot of tube

$$\text{Gross free area } 12\,(2.375 - 1) = 16.5 \text{ in.}^2$$
$$\text{Fin blockage } 12 \times 8 \times 2 \times 0.017 \times 0.625 = 2.04 \text{ in.}^2$$
$$\text{Net free area} = 14.46 \text{ in.}^2$$

$$\text{Fraction free area/face area} = \frac{14.46}{2.375 \times 12} = 0.508$$

Hence

$$G_m = \frac{500 \times 0.075 \times 60}{0.508} = 4430 \text{ lb/ft}^2\text{-hr}$$

From Fig. 75 h = 9 Btu/hr-ft^2 -$^\circ$F. Fin efficiency: For aluminum fins

$$\sqrt{\frac{2h}{kt}} \times l = \sqrt{\frac{2 \times 9 \times 12}{120 \times 0.017}} \times \frac{0.625}{12} = 0.535$$

$$\frac{D_0}{D_2} = \frac{2.25}{1} = 2.25$$

From fin efficiency curves (Sec. 3.B.2.b) Eff = 86%. Area ratio (approximately): Per foot of tubes, fin area =

$$\frac{12 \times 8 \times 2}{144} \times \frac{\pi}{4}(2.375^2 - 1^2) = 4.85 \text{ ft}^2$$

$$\text{Tube area/ft} = 0.2618 \text{ ft}^2$$

$$\text{Ratio fin area to tube (OD) area} = \frac{4.85}{0.2618} = 18.55$$

Air side coefficient based on tube OD area =

$$9(18.55 \times 0.86 + 1) = 153 \text{ Btu/hr-ft}^2\text{-}^\circ\text{F}$$

DESIGN FACTORS

The usual impulse is to specify for design the worst heat transfer conditions under which the heat exchanger is to operate, e.g., a cooler operating under maximum load at maximum fouling with maximum water temperatures. However, the minimum operating conditions should also be evaluated since it is just as bad to lose control of processes due to excess surface as not to make production with insufficient surface. Both maximum and minimum conditions should be examined to be certain that process control is always possible and there is no possibility of freezing process streams.

The so-called safety factor is really a factor allowing for the uncertainty in the heat transfer equations. The safety factor should be applied to the length of the exchanger if possible and not to the number of tubes since increasing the number reduces the velocities and coefficients, hence the safety factor would not be as large as anticipated. There is no standard in defining the safety or ignorance factor. The factor may be applied to the overall coefficient or to the individual fluid film coefficients. In the latter case the factor usually is not applied to the dirt coefficients, but this procedure then implies that the dirt coefficients are known with greater accuracy than the fluid film coefficients, which is incorrect, or that the dirt coefficient already has a very conservative factor in it, which may be true. However, one should determine the overall coefficient on the best available information and then apply a single factor if the uncertainty on each coefficient is equal or weight the result by applying the variable factor to each film. The normal heat transfer equations are accurate to ±20–25 percent. Of course, if each film has a 25 percent error and if all are of equal magnitude, the probability of all errors occurring in the same direction is small and a lower overall safety factor can be applied. However, if one film is controlling, then the safety factor is the full value. Assuming all films of equal magnitude, then the average safety factor can be determined from the following equation. Average safety factor is

$$\left(\sqrt{e_1^2 + e_2^2 + \cdots e_n^2}\right)\left(\frac{1}{n}\right)$$

where e_1 is the $\pm$ error of one film, e_n of the nth film.

When water is used for cooling, only the inlet temperature may be specified and the designer must select the quantity or temperature rise. In these days of rising costs and with water resources becoming limited, an economic balance must be made. The optimum exit water temperature given in Fig. 79 is the same as given by Colburn [32] but with additional cost factors in the abscissa term to allow for earning a profit on the additional investments involved. Here B is the fraction of a year unit is to be used; C_s is investment for exchanger surface ($\$/ft^2$); C_w is incremented cost of water ($\$/1000$ gal); C_w^1 is investment for water facilities ($\$/gpm$); F is correction factor for log mean temperature difference; U is overall heat transfer coefficient (Btu/hr-ft²-°F); Z is gross return required including allowances for taxes, repair, and maintenance; subscript s for exchanger surface; w for water facilities; T_1 is outlet process temperature; T_2 is inlet process temperature; Δt_1 is temperature difference between process and water at water inlet end; Δt_2 is temperature difference between process and water at water outlet end. There is, however, a practical upper limit of about 70°C (or about 160°F) for water discharge temperatures. At high temperatures, air begins to come out of the water, corrosion rates increase, and scale formation becomes more rapid and serious.

In design of heat exchangers for laminar flow the peculiarities of the system should be kept in mind. Heating of viscous solutions results in a stable system but in cooling the system is unstable when parallel paths of flow are present. If the flow or fluid temperature in one tube should become less than that in the other tubes, then the flow will slow up further because the flow rate is inversely proportional to viscosity. Also, an increase of flow in one tube will result in less

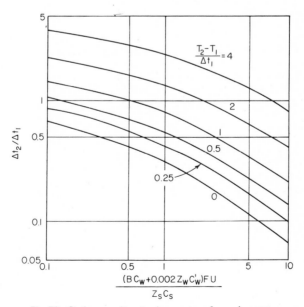

Fig. 79. Optimum exit water temperature for exchangers.

cooling; thus its viscosity is less which, in turn, results in even faster flow rates. Design of coolers, therefore, requires additional surface to take care of the possible maldistribution of flow. The degree of maldistribution is difficult to predict but can be approximated by assuming some tubes run cold and the remaining run hot. Since the total flow is known and the pressure drop is the same for each tube, it is possible to estimate the maldistribution by a trial-and-error calculation. The ratio of cold tubes N_c to the total tubes N_T is

$$\frac{N_c}{N_T} = \frac{1}{1} \frac{-\mu_h/\mu_{av}}{-\mu_h/\mu_c}$$

where the viscosity at bulk temperature for hot tubes is μ_h, for cold tubes μ_c, and for uniform flow μ_{av}. To arrive at the bulk temperatures for the various streams requires a trial-and-error procedure.

Tube sizes, thicknesses, lengths, and pitches all enter into the calculation of heat transfer and must be chosen with care. Sizes used in exchangers range from 1/4-in. OD to 2-in. OD, but the most popular sizes are 5/8-in., 3/4-in., and 1-in. OD. Some new plastic tube exchangers have a tube of 0.1-in. diameter. These sizes are a compromise between the relation that the highest heat transfer occurs in the smallest tubes and results in the smallest exchanger, on the one hand, and the practical viewpoint, on the other, that large tubes are easier to clean and are rugged. In dirty service where tubes must be cleaned, especially by mechanical cleaners, the smallest practical size is 3/4 in., but 1 in. would be preferred. If chemical cleaning can be done, then the smaller tubes can be used, providing that the tubes will never plug. If the tubes plug, then the chemical solutions cannot reach these points, and mechanical rodding is necessary to open the tubes for circulation. Tube lengths are based on even number of feet; and although the TEMA standards are 8, 12, and 16 feet, many exchangers are built with tube lengths ranging from 6 to 20 feet in 2-foot increments. Choice of tube size and length may be further limited by plant standardization. In the new plants emphasis is on standardization to reduce the stock of repair tubes. In order to follow such standards, the safety factor may be greater or less, and the designer must examine the other uncertainties involved in his design to determine the extent of his real gamble in reducing the safety factor.

Tube thickness is governed by many factors such as pressure, corrosion, cost, and rolling the tubes into the tube sheet. The usual thickness used is sufficiently strong to hold pressures of at least 200 lb/in.^{2} Steel tubes are usually several gages heavier than the alloy or nonferrous tubes principally to get more corrosion protection. The 5/8-in. tubes are usually 16 or 18 BWG (Birmingham Wire Gage), 3/4-in. tubes average from 14 for steel to 16 or 18 for other materials and 1-in. tubes range from 10 to 16 BWG with 12 BWG preferred for steel and 16 BWG for other materials. Now the wall thickness is a nominal dimension, and two types of walls are available. The minimum wall tube for ferrous materials has a minimum thickness of the gage and may be 22 percent thicker while the average wall tube has an average thickness of the gage with a ±10 percent tolerance. The tolerances of nonferrous tubes are closer and nominal wall tubes will be used. The choice of type of wall will influence the cost and will influence the pressure drop through the tubes. As an approximation the minimum wall tube is equivalent to an average wall tube of one gage heavier. Wall thickness also affects the rolling operation, as too heavy or too thin walls will give difficulty in getting tight joints. The gages listed above with the tube size will give satisfactory joints for most metals.

Tube pitch is also a compromise. The best heat transfer is obtained with close pitches but such bundles plug easily and are difficult to clean. The tube sheet ligament for close pitches may become too weak for proper rolling of the tubes and result in leaky joints. The current minimum pitch is 1.25 times the tube diameter or, if a cleaning lane is to be provided, 1/4 in. on square pitch. This is the nominal pitch and the actual ligament may be smaller due to drill drift, drill centering, and the need for oversizing the hole for tube insertion. For instance, with a 1-in. tube on a 1¼-in. pitch the TEMA minimum permissible ligament is 0.12 in., which for a 10 BWG tube is too thin and is not recommended. When the exchanger must be tight, it would be best to increase the tube pitch. The two standard types of pitches are the square and the equilateral triangle. The square pitch is mostly used when cleaning is necessary on the bundle. The triangular pitch gives a smaller bundle and shell but cannot be mechanically cleaned. However, it is possible to increase the triangular pitch to obtain cleaning lanes. For instance, 1-in. tubes on 1-3/8-in. triangular pitch is almost equivalent to the 1¼-in. square pitch. The same shell will hold the same number of tubes (within 2–3 percent), the velocities and baffle spacings are the same, the ligaments are 50 percent stronger, and the cleaning lanes are only slightly less (to about 3/16 in.) than the square pitch; furthermore, cleaning can be accomplished from six directions instead of four. The use of wider triangular pitches in place of square pitches offers many advantages.

The usual number of tube passes in a given shell ranges from one to eight. More passes can be obtained but special layouts are required. A fairly standard design is readily adapted to one, two, or four passes. In multipass designs, even numbers of passes are generally used because it is a simpler design. Odd number multipass bundles can be fabricated but are not standard.

Choice of baffle spacing and baffle cut is a variable to be determined. Optimum ratio of baffle cuts and spacing cannot be specified because of insufficient data. Pressure drop on the shell side is usually specified but the proportion of the drop that takes place across the tubes or through the window is determined by the baffling. The practical range of baffle spacing is 1/5 to 1 shell diameter, but the minimum spacing is 2 in. or 1/5 diameter whichever is the larger. The 2-in. specification is based on the ability to clean the bundle and the 1/5 diameter is based on manufacturer's experience in that no further benefit is obtained with closer spacings. The usual range of baffle cuts is from 20 percent to almost 50 percent of the diameter. Large baffle cuts are usually taken with large baffle spacings in order not to get excessive pressure drop across the windows as compared to the bundle. Because of lack of data and the complex leakage factors involved, there is no good rule for determining the relation between baffle cut and baffle spacing. (However, see the section on the Devore-Tinker method of calculating heat transfer for recommended baffle cuts versus baffle spacing.)

There is another maximum baffle spacing which is based on supporting the tubes, but the baffles are then known as support plates and are cut approximately on the center line. Support plates are spaced to give a maximum unsupported tube length of 60 tube diameters; thus the plates are spaced at 30 tube diameters, since each plate supports half the tubes. The support plates are thicker than the baffles and the tube holes have less clearance. The purpose of the support plate is to give the bundle greater stiffness, prevent tubes from vibrating, and reduce the number of leaky joints caused by shipping and handling.

Usually only the space between the nozzles is baffled, not the entire tube length. To baffle the entire length requires special arrangements at the ends.

The heads of the exchanger may be either of the bonnet or channel types. A bonnet head is a one-piece fabricated or cast head with either a side or end entering nozzle. The bonnet head is cheap, but has the disadvantage that inspection or maintenance of the bundle will require removal of the bonnet and breaking of the piping joints. A channel head has a side-entering nozzle and a cover. With this head, only the cover need be removed for inspection or maintenance and piping is not disturbed. Further, the side-entering nozzle gives a better distribution of fluid to the tubes.

The shell nozzles are designed to fit other piping and process equipment, but should not have an excessive velocity that may cause tube vibration or erosion. Velocities range from 3 to 6 ft/sec for liquids. Impingement baffles are used to protect the tubes if erosive conditions are possible, but are not usually used with liquids. As previously mentioned, the presence of the tube bundle partially obstructs the nozzle, and the net velocity from the nozzle entrance should not be excessive and preferably less than the nozzle velocity. The net free area can be considered as the nozzle perimeter times the shell to bundle clearance plus the area between the tubes directly opposite the nozzle.

A number of special designs are available to overcome inherent weaknesses in the exchangers. One weak point is leakage through the tube joints, and double tube sheets are used where intermixing of tube and shell fluid must be prevented. The tubes are rolled into each tube sheet and the space between tube sheets is vented or open. The problems in this design are to satisfactorily roll the tube joint on the inner tube sheet, to be able to locate a leak for rerolling operations, and to take care of additional stresses. These stresses are due to each tube sheet being at different temperatures, thus causing a deflection or shearing action on the tubes, and to the expansion of the tubes causing compression stresses in the tubes if both tube sheets are fastened or held together at their edges. Additional protection to prevent leakage caused by pinholes or corrosion through the tube wall can be secured by using either a double tube with a liquid metal filling the annulus and vented to a space where a change in volume or other detection devices are located [33] or one of the tubes has small longitudinal grooves so when the tubes are drawn together to improve contact and heat transfer the small grooves can still carry any leaking fluid to the space between the double tube sheets for detection before the leak penetrates both tubes [34]. When the tube material is not satisfactory for both the tube and shell fluids then a bimetallic tube may be used. Various combinations of metals are available and tests indicate no contact resistance to heat flow if the tubes have been drawn together to effect a good bond. Use of bimetallic tubes usually also requires double tube sheets or a clad tube sheet. Special techniques are used to strip back the outer material for a portion of the tube sheet thickness and the use of a ferrule as a filler so that the tube fluid is sealed from the jacket metal of the bimetal tube.

REFERENCES

1. A. D. Kraus and D. Q. Kern, Effectiveness of Heat Exchangers with One Shell Pass and Even Number of Tube Passes, Natl. Heat Transfer. Conf., 1965.
2. R. A. Bowman, Mean Temperature Difference Correction in Multipass Exchangers, *Ind. Eng. Chem.*, 28:541 (1936).
3. F. K. Fischer, Mean Temperature Difference Correction in Multipass Exchangers, *Ind. Eng. Chem.*, 30:377 (1938).
4. K. A. Gardner, Mean Temperature Difference in an Array of Identical Exchangers, *Ind. Eng. Chem.*, 34:1083 (1942).
5. A. P. Colburn, *Ind. Eng. Chem.*, 25:873 (1933).
6. K. A. Gardner, Variable Heat Transfer Rate Correction in Multipass Exchangers Shell Side Film Controlling, *Trans. ASME*, 67:31 (1945).
7. K. A. Gardner, Mean Temperature Difference in Unbalanced Pass Exchangers, *Ind. Eng. Chem.*, 33:1215 (1941).
8. K. A. Gardner, Mean Temperature Difference in Multipass Exchangers, *Ind. Eng. Chem.*, 33:1495 (1941).
9. A. M. Whistler, Correction for Heat Conduction through Longitudinal Baffle of Heat Exchanger, *Trans. ASME*, 69:683 (1947).

10. L. Jaw, Temperature Relations in Shell and Tube Exchangers Having One-Pass Split-Flow Shells, *J. Heat Transfer, Trans. ASME,* **C86**:408 (1964).

11. R. A. Stevens, Mean Temperature Difference in Counter-Cross-Flow Heat Exchangers, M.S. thesis, Southern Methodist Univ., 1955.

12. K. S. Lee and J. G. Knudsen, Local Shell-Side Heat Transfer Coefficients and Pressure Drop in a Tubular Heat Exchanger with Orifice Baffles, *AIChE Journal,* **6**:669 (1960).

13. K. J. Bell, Final Report of Cooperative Research Program on Shell and Tube Heat Exchangers, *Univ. of Delaware Eng. Expt. Sta. Bull. 5,* Jan. 1963; also *Petro./Chem. Eng.,* pp. C-26–C-400, Oct. 1960.

14. T. Tinker, Shell Side Characteristics of Shell and Tube Heat Exchangers, *Trans. ASME,* **80**:36 (1958).

15. A. Devore, Try This Simplified Method for Rating Baffled Exchangers, *Pet. Refiner,* **40**:221 (1961).

16. B. E. Short, Heat Transfer and Pressure Drop in Heat Exchangers, Univ. of Texas Bull. 3819, May 1938. See also *Bull.* 4324 (revision), June 1943.

17. H. McAdams, "Heat Transmission," 3d ed., McGraw-Hill Book Company, Inc., New York, 1954.

18. H. N. Fairchild and C. P. Welsh, Convection Heat Transfer and Pressure Drop of Air Flowing Across In-Line Tube Backs at Close Back Spacing, *ASME Paper Cl-WA-250.*

19. N. Kattchee and W. V. Mackewicz, Heat Transfer and Fluid Friction Characteristics of Tube Clusters with Boundary Layer Turbulence Promotions, *ASME Paper 63-HT-1.*

20. W. A. Sutherland and W. M. Kays, Heat Transfer in Parallel Rod Array, *ASME Paper 65-WA/HT-7.*

21. B. E. Short, Flow Geometry and Heat Exchanger Performance, *Chem. Eng. Prog.,* **61**:63 (1965).

22. V. I. Subbotin, P. A. Ushakov, B. N. Gabrianovich, and A. V. Zhukov, Heat Exchanger During Flow of Mercury and Water in a Closely Packed Bundle of Rods, *Reactor Sci. Tech.* (*J. Nuclear Energy*), **17**:455 (1963).

23. D. E. Dwyer and P. S. Tu, Heat Transfer Rates for Parallel Flow of Liquid Metals Through Tube Bundles, *3d Natl. Heat Transfer Conf.,* Paper 119, 1959.

24. A. C. Mueller, Thermal Design of Heat Exchangers, *Purdue Univ. Exp. Sta. Bull.* 121, Sept. 1954.

25. A. A. Putnam, Flow Induced Noise and Vibration in Heat Exchangers, *ASME Paper 64-WA/HT-21.*

26. W. M. Kays and A. L. London, "Compact Heat Exchangers," The National Press, Palo Alto, Calif., 1955.

27. A. L. London, Compact Heat Exchangers, *Mech. Eng.,* **86**:47 (1964).

28. G. D. Bahnke and C. P. Howard, The Effect of Longitudinal Heat Conduction on Periodic Flow Heat Exchanger Performance, *J. Eng. Power, Trans. ASME,* **A86**:105 (1964).

29. K. K. Robinson and D. E. Briggs, Pressure Drop of Air Flowing Across Triangular Pitch Banks of Finned Tubes, *8th Natl. Heat Transfer Conf.,* AIChE Preprint 20, 1965.

30. D. E. Briggs and E. H. Young, Convection Heat Transfer and Pressure Drop of Air Flowing Across Triangular Pitch Banks of Finned Tubes, *Chem. Eng. Prog. Symp. Series No. 41,* 59, 1965.

31. E. C. Smith, *Chem. Eng.,* **63**(23):145, (1958).

32. A. P. Colburn, Heat Transfer by Natural and Forced Convection, *Purdue Univ. Eng. Exp. Sta. Bull.* 84, 1942.

33. T. Trocki and D. B. Nelson, Report on a Liquid Metal Heat Transfer and Steam Generation System for Nuclear Power Plant, *J. Mech. Eng., Trans. ASME,* **A52**:140, 472, 927 (1953).

34. J. T. Cullen, A Leak-Proof Heat Exchanger, *ASME Paper* **A50**:125 (1950); *Mech. Eng.,* **73**:425 (1951) Abstract.

35. G. V. Lentz, Flow Induced Vibration of Heat Exchanger Tubes Near Shell-Side Inlet and Exit Nozzles, *11th National Heat Transfer Conference, 1969,* AIChE Preprint 20.

Section **19**

High-temperature Thermal Protection Systems

H. HURWICZ
Chief Advance Technology Engineer

J. E. ROGAN
Branch Chief
Aero/Thermodynamics and Nuclear Effects
Research and Development, Advance Systems and Technology
McDonnell Douglas Astronautics Company-West

INTRODUCTION

The development of high-temperature thermal protection systems gained impetus with progress in the fields of space exploration and reentry applications. The need for special protection became obvious when conventional structures were unable to withstand the loads in the operating hyperthermal environment. Thus, a new generation of protection has evolved, varying in complexity as the mission requirements increased in severity. These external high-temperature thermal protection systems (referred to in this text as HTTPS) are conveniently classified as either absorptive or radiative, depending on the primary mode used to dissipate the incident energy. The absorptive systems are further subclassified as heat sink, film and transpiration cooling, ablative, and convective. Other possible basic components of the external thermal protection system are insulation materials and internal convective cooling systems.

Most of the design procedures used by practicing engineers involve high-speed computers and two generations of complex computer programs. Wherever possible, reference to typical practice in this area will be made, but detailed discussion of such programs is beyond the scope of this chapter.

It is the intent of this chapter to give an insight into the complexity of thermal protection systems design, to provide some guidelines for their design, and to indicate techniques for "first cut" (as opposed to detailed design procedures) HTTPS weights for certain classes of trajectories and materials. This chapter is not meant to replace previous discussions (e.g., environment, conduction, and ablation) but rather to enhance their utility. As such, the text will emphasize problems associated with definition of requirements and criteria for material selection (rather than the material development process itself) and with system design and test criteria to produce reliable hardware. The environmental factors will be considered only to the extent of their constraints on the HTTPS design.

Finally, the importance of analytical methods and well-designed simulation experiments will be stressed for correlating with sufficient accuracy the material behavior and response, when exposed to its intended environment. It is this last item, as well as the proper statistical evaluation of the experimental information, that are often overlooked in the design of an HTTPS, with ensuing penalties in hardware weight and/or reliability.

Finally, as reuse of spacecraft comes of age, consideration will be given to the design of long-life equipment devoted to space transportation applications.

A. DESIGN INPUTS

The design of a HTTPS is a complex procedure involving coordination between many technical disciplines, requiring inputs from many areas. Although the primary output of the calculations may be the required heat shield thicknesses, additional output parameters (e.g., temperature gradients, backface temperature histories, and mass loss histories) are required by other technical specialists involved in the design cycle. For the purposes of this chapter, the following information will be assumed to be available:

vehicle geometry and dimensions
vehicle trajectory (i.e., time histories for altitude, velocity, and angle of attack)
temperature constraints (design criteria)

In addition, the analyst must have at his disposal accurate data and methods with which to compute the following:

heat transfer rate
duration of heat pulse
total heat input (integrated with respect to time)
local pressure
aerodynamically induced shear

The following sections will provide reference materials from which the reader may extract the data required to make the necessary heating calculations. Many of the areas discussed below have been covered in previous chapters and will not be elaborated upon here.

Atmospheric Properties. A typical source of nominal earth atmospheric properties is Ref. 1. For cases where the atmospheric properties are known to be significantly different from the mean values, the appropriate local properties must be used, e.g., Ref. 2. For Mars and Venus, definitions of the atmospheric properties are contained in Refs. 3 and 4.

Thermodynamic and Transport Properties. For air, the thermodynamic properties in Moeckel and Weston [5] and Hansen [6] can be used for temperatures up to 15,000°K. For gases other than air, the thermodynamic properties may be determined utilizing a computer program such as is described by Horton and Menard [7].

The transport properties of high-temperature air up to temperatures of 15,000°K may be obtained from sources as Ref. 7 while Bennett et al. [8] and Thomas [9] may be used to obtain estimates of the transport properties of gases other than air.

Inviscid Flow Fields. The accurate determination of the inviscid flow field surrounding a vehicle is necessary to provide the boundary conditions for convective heating calculations. For equilibrium flow, it is sufficient to specify two thermodynamic properties, usually static pressure and entropy or enthalpy. For nonequilibrium flow, it is also necessary to specify the gas species mass fractions. An excellent review of the methods available to accomplish this is contained in Nayfeh [10].

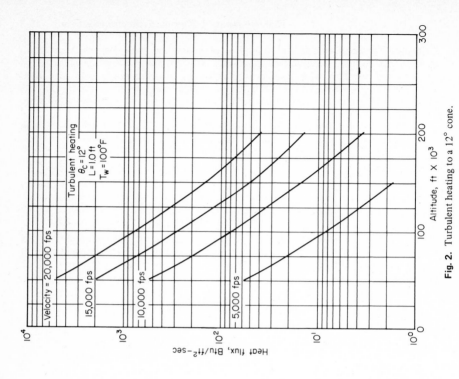

Fig. 2. Turbulent heating to a 12° cone.

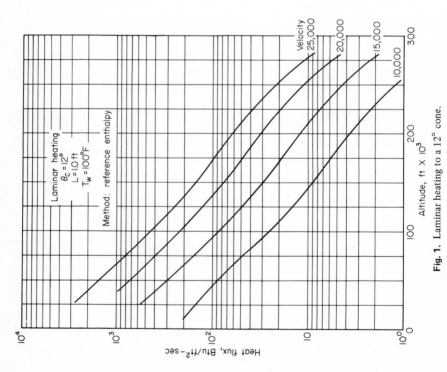

Fig. 1. Laminar heating to a 12° cone.

For simple shapes such as spheres, wedges, and sharp cones, the boundary-layer edge properties may be obtained from Refs. 11 (ideal gas) and 12 (real gas). Another source of useful design information is Ref. 13.

Heat Transfer Techniques. A widely based method for calculating the heating rates on the body of a vehicle is based on Eckert's reference enthalpy approximation. The variation of enthalpy across the boundary layer poses the problem of deciding at which temperature or enthalpy to evaluate the gas properties. Eckert [14] found that the heat transfer could be calculated in a compressible boundary layer using incompressible flow relations if all temperature-dependent properties were evaluated at a reference condition ($h*$). This enthalpy is a weighted mean of local enthalpy, wall enthalpy, and recovery enthalpy and is defined as

$$h^* = h_\delta + 0.5(h_w - h_\delta) + 0.22r(h_s - h_\delta) \tag{1}$$

where r is the recovery factor and is equal to $P_r^{0.5}$ for laminar flow and $P_r^{0.33}$ for turbulent flow.

For a flat plate in laminar flow the heat transfer equation takes the form

$$\dot{q}_L = 0.332 \left(R_{e_x}^*\right)^{1/2} P_r^{-2/3} \frac{\mu^*}{x} (h_r - h_w) \tag{2}$$

Equation (2) can be adapted for use on a cone by applying the Mangler transformation for laminar flow of $\sqrt{3}$.

For turbulent flow the heat transfer can be expressed, independent of pressure gradient and body shape, as

$$\dot{q}_T = 0.029(1 + n)^{1/5} \left(R_{e_x}^*\right)^{0.8} P_r^{-2/3} \frac{\mu^*}{x} (h_r - h_w) \tag{3}$$

where $n = 0$ for 2D flow

$= 1$ for axisymmetric flow

Figures 1 and 2 present some typical heating values for a sharp cone body, with a running length of one foot.

To vary the cone angle

$$\frac{\dot{q}_{(cone)_1}}{\dot{q}_{(cone)_2}} = \left(\frac{\sin\theta_1}{\sin\theta_2}\right)^{1.567} \tag{4}$$

To vary the slant length (or running length)

$$\frac{\dot{q}}{\dot{q}_{L=1\,ft}} = L^{-0.5} \quad \text{(laminar flow)} \tag{5}$$

and

$$\frac{\dot{q}}{\dot{q}_{L=1\,ft}} = L^{-0.2} \quad \text{(turbulent flow)} \tag{6}$$

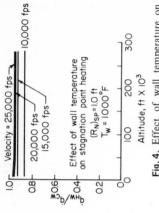

Fig. 4. Effect of wall temperature on stagnation point heating, $T_w = 1000°F$.

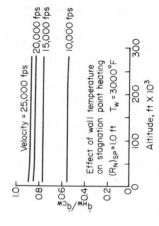

Fig. 5. Effect of wall temperature on stagnation point heating, $T_w = 3000°F$.

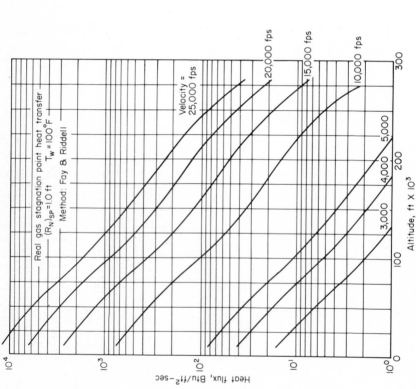

Fig. 3. Stagnation point heat transfer.

Figures 3 to 5 show the stagnation point heating as a function of altitude and velocity and was computed utilizing the theory of Ref. 15 which takes the form

$$\dot{q}_s = 0.537 (n + 1) P_{r_w}^{-0.6} (\rho_w \mu_w)^{0.1} (\rho_s \mu_s)^{0.4} \left[1 + (L_e^{\alpha} - 1) \frac{h_D}{h_s} \right] (h_s - h_w) \left[\left(\frac{du}{dx} \right)_s \right]^{0.5}$$

(7)

where n = 0 for 2D flow
  = 1 for axisymmetric flow
  α = 0.52 for an equilibrium boundary layer
  = 0.63 for a frozen boundary layer

For a spherical nose, utilizing modified Newtonian theory [15], the velocity gradient may be computed from

$$\left(\frac{du}{dx} \right)_s = \frac{1}{R} \left[\frac{2 (P_s - P_\infty)}{\rho_s} \right]^{1/2}$$

(8)

A relationship for this gradient as a function of the density ratio across the shock is given in Ref. 16.

It should be noted that the past decade has produced many references which provide rapid means for estimating the heat transfer to simple vehicle configurations. Some of these are cited [17–19].

1. Design Constraints

A unique design problem [20] for a reentry vehicle is associated with the structural integrity of the heat shield–substructure composite. In a reentry vehicle, it is the purpose of the heat shield to keep the load-carrying structure at a temperature consistent with its load-carrying ability. On the other hand, the strain in the substructure must be compatible with the allowable strains in the heat shield to insure that no heat shield failures can be induced. Further, since the heat shield and the substructure are subject to large temperature changes during reentry, thermally induced strain can cause heat shield failure or even structural failures without consideration of the load-induced strains.

In many heat shield substructure designs, it has been customary to bond the heat shield to the substructure with a high-temperature rigid type of adhesive. For a heat shield design of this type, it is necessary to consider the following possible failure modes:

a. failure during early reentry due to temperature gradients in the heat shield
b. failure in the heat shield due to thermal incompatibilities between the structure and heat shield as the structure heats up
c. failures of the heat shield due to high strains and/or deflections in the substructure due to concentrated loads
d. bond failures between the structure and heat shield due to thermal incompatibilities between the structure and heat shield
e. heat shield or structural failures due to combined thermal induced loading and applied loading

The severity of the heat shield structural design problem is controlled largely by variation with temperature of a number of heat shield properties such as modulus of elasticity, thermal expansion, and strength, as well as the strain to failure in the operating temperature range, and the thermal performance of the heat shield as an ablator and as an insulator. One of the most severe controlling factors is the degree of matching between the thermal expansion of the heat shield and that of the substructure as a function of temperature if a rigid type bond is used. If a flexible bond is used, the problem of substructure heat shield strain compatibility is much reduced, and matching requirements of thermal expansions between the heat shield and the structure are much less stringent. However, availability of flexible bonds over wide temperature ranges is extremely limited. Problems associated with bonding have led to development of integrated composite structures where heat shields also perform the structural function.

The maximum allowable structural temperature can be determined on a weight basis by trading off the increase in structural weight associated with the elevated temperature fall-off in mechanical properties against the decrease in heat shield requirements associated with increased structural temperatures. Such trade-offs—the selection of optimum structural temperatures—are affected to varying degrees by thermal stresses.

Other factors which must be considered in the design of the flight vehicle are [21]:

a. Thermal Limitations. Consideration must be given to the allowable temperature levels and temperature gradients existing during all periods of significant aerodynamic loading. All candidate heat-protection systems must be examined for interrelationships with the substructure, such as matching of thermal-expansion properties; for compatibility with bond materials, if used; for bending, buckling, and expansion requirements; and, if ablative, for char-retention characteristics.

b. Shape Retention. Because of the design difficulties, early attention should be given to areas required to maintain their original shapes or to undergo only a minimal shape change during entry. These areas are usually such small-radius surfaces as nose tips and control surface leading edges that experience the maximum heating rate and total heat input because of their shape and location. Selection of the proper thermostructural system for these locations is limited normally to materials that minimize surface recession (e.g., graphite), an actively cooled structure (transpiration), or some combination of these methods (e.g., an ablative-impregnated porous matrix).

c. Refurbishment or Reuse. If the vehicle is to be reused for a number of missions, the heat shield must be designed for reuse or refurbishment. The anticipated number of missions and the allowable time between missions (turn around time) must be considered in selecting material in the initial design period so that tradeoff studies can be performed between the various heat shield concepts and the different types of segmentation and attachment methods. Possible consequences may be the imposition of machinability, bonding, handling, or other restrictions on the candidate materials.

d. Manufacturability. Regardless of the thermal protection system concept, the materials must be capable of being applied, formed, or machined into the desired shape. The application and curing operations for the bond materials must be considered, along with the effect of these operations on the heat shield and structural properties.

e. Sterilization. If prelaunch sterilization of the vehicle has been stipulated, as may be the case with planetary exploration vehicles, the effects of the heating cycle (required for sterilization) on heat shield and bond-material properties must be determined as some chemical and mechanical degradation may be expected. Heat shield attachment points should be designed to minimize the effects of differential thermal expansion on the thermal protection system as well as on the structure.

f. Operation in a Space Environment. Extended operation in a space environment before atmospheric entry imposes several material-selection and design problems. The vehicle is exposed to solar radiation, possible meteoroid impacts, a cold-temperature soak, and a vacuum environment. To maintain the desired thermal environment within the vehicle, the thermal protection system is required also to regulate and distribute incoming solar energy. This may be accomplished by means of tailored surface coatings, attitude control, heat sinks, and heat exchangers. The space environment can also alter the chemical and mechanical properties of the heat shield materials so that their performance in the subsequent entry-heating environment is substantially degraded. Many effects of space operation on material properties can be effectively determined by means of tests in space-simulation vacuum chambers equipped to program variable radiative heating histories.

g. Ascent Heating. Although ascent-heating loads are nearly always much smaller than those encountered during entry, their effects on the entry thermal protection system must be evaluated. For short-duration ballistic flights, the principal effect is that of increasing the average temperature in the heat shield at the beginning of atmospheric entry. However, certain areas of the vehicle may become sufficiently hot to experience some thermochemical degradation, and this effective loss of material must be included in the design calculations. Alternate design approaches may include the application of some extra thickness of the entry heat shield material, the addition of a thin outer layer of low-conductivity material for ascent flight, or the use of a shroud or housing for protection against ascent heating.

Additional effects that can occur before entry, and that should be accounted for in the thermal protection system design, include ascent vibration loads, separation shocks, and the impingement of separation or control-moter exhausts on the vehicle surface. In cooperation with the other system analysts and designers, the designer of the thermal protection system should account for these problems early in the design process.

h. Prelaunch, Storage, and Aging. The design and performance of the thermal protection system can be affected by events occurring during the prelaunch storage, transportation, testing, and maintenance operations. During storage, the vehicle may be subjected to long-duration compressive loads that would prohibit the use of external materials with high-creep characteristics (e.g., Teflon), or would require special handling methods. During transportation, the vehicle may undergo long periods of vibration loads or, if shipped by air, a cold soak as low as $220°K$, which could permanently degrade the properties of some materials or increase the possibility of damage caused by low-temperature brittleness. The possible importance of such other effects as temperature humidity extremes can be included by examining the known environmental conditions at the various storage locations and at the launch site. The designer should evaluate these ground environments and initiate any precautions necessary to protect the thermal protection system (e.g., protective coverings, special transportation or storage supports, and conditioned storage areas). Finally, storage for prolonged periods prior to use may change the properties of the protection system and should be accounted for in the design.

i. Communications. Severe and unusual communications requirements can also impose restrictions on heat shield design and material selection. For example, the basic design can be affected if antenna windows must be located in a specific area to minimize flow field plasma-attenuation problems or to achieve a particular line-of-sight objective. Or sometimes plasma effects must be minimized by the use of materials of extremely high purity for the ablative heat shield and windows; this can restrict the initial material selection and impose difficult handling and fabrication requirements.

B. THERMAL PROTECTION SYSTEM DESIGN PROCEDURES

1. General Application of Individual Systems

The thermal protection system as defined consists of a material system (shield and/or load carrying member) operating on a given heat dissipation principle. The methods of heat dissipation utilized in various protection systems are generally divided into two basic types: absorptive and radiative, according to the predominant physical mechanism of absorption or rejection of incident energy. The operating principles, advantages, and limitations of such systems (radiation shields, heat sinks, ablation shields, and other internal cooling or transpiration mechanisms) are reviewed subsequently in this section and in the sections on Ablation and Transpiration Cooling. In any given application, it is usually possible to select the most appropriate mode of heat dissipation with relatively little effort and without getting very detailed as regards specific materials.

One of the most important problems in the HTTPS design after the selection of the mode of heat dissipation is the selection of the heat shield material or medium which has the appropriate characteristics for the design environment and which is compatible with the load carrying member. While the difference in aerothermodynamic environment and its transient nature may indicate in a gross manner the choice of an optimum cooling mode, a fairly detailed study is required to stipulate specific thermodynamic measurement criteria for each system and/or application. This requires an understanding of the physical processes involved for prospective materials and their proper mathematical formulation; this in turn permits the determination of the significant material and system parameters useful in performance evaluation.

The steps which are usually followed in the design of a heat shield–substructure composite regardless of the criteria used are generally as follows:

a. Using the predicted flight envelope, calculate the heat transfer from the boundary layer to the vehicle for suspect design trajectories.

b. Assuming several allowable structural temperatures, together with the criteria selected, calculate the temperatures and temperature gradients in the heat shield, bond, and structure at several reentry vehicle positions.

c. Check the heat shield for structural margin using the substructure thicknesses and temperatures selected, using a one-dimensional analysis, for the worse trajectory in the flight corridor. This will allow selection of the possible thermally induced failure modes and a preliminary assessment margin situation. The substructure thicknesses are selected so the thickness is adequate at the temperature selected to carry the load.

d. After Item c is complete, the thickness distribution of the heat shield can be selected. This study will establish the combination of heat shield thickness and substructure thickness which will result in adequate margins and a minimum weight for the heat shield–substructure composite (i.e., the design substructure thickness).

e. The heat shield and substructure are then analyzed for combined thermal induced loadings and applied loads and the margins determined throughout the reentry vehicle. This analysis should include flexural and extensional stresses and will bring in the actual details of the structure.

f. Analyze the complete heat shield–substructure composite for the other loading conditions associated with reentry and launch which could result in strains which could induce heat shield failures.

g. Investigate the effect of cold cycling the reentry vehicle during shipping, handling, and storage. This design condition may in some cases be worse than those during reentry.

h. Investigate the thermal compatibility between the reentry vehicle structure and the payload under temperature gradients associated with reentry and cold cycling.

i. Analyze the interface between the reentry vehicle and spacer since the reentry vehicle is protected by a heavy heat shield while the spacer is subject to greater temperature fluctuations.

In every case, the steps taken for a thermal-structural analysis are the same, i.e., the temperatures and temperature gradients must be determined, the variation of material properties must be known, and a mathematical model for analysis of the TPS must be available.

Before proceeding further, the different heat protection systems must be introduced. Figure 6 illustrates typical thermal protection schemes and the mechanisms involved in the resulting heat balance while Table 1 indicates the general

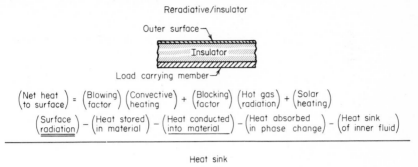

Reradiative/insulator

$$\left(\begin{array}{c}\text{Net heat}\\\text{to surface}\end{array}\right) = \left(\begin{array}{c}\text{Blowing}\\\text{factor}\end{array}\right)\left(\begin{array}{c}\text{Convective}\\\text{heating}\end{array}\right) + \left(\begin{array}{c}\text{Blocking}\\\text{factor}\end{array}\right)\left(\begin{array}{c}\text{Hot gas}\\\text{radiation}\end{array}\right) + \left(\begin{array}{c}\text{Solar}\\\text{heating}\end{array}\right)$$

$$\left(\begin{array}{c}\text{Surface}\\\underline{\text{radiation}}\end{array}\right) - \left(\begin{array}{c}\text{Heat stored}\\\text{in material}\end{array}\right) - \left(\begin{array}{c}\text{Heat conducted}\\\underline{\text{into material}}\end{array}\right) - \left(\begin{array}{c}\text{Heat absorbed}\\\text{in phase change}\end{array}\right) - \left(\begin{array}{c}\text{Heat sink}\\\text{of inner fluid}\end{array}\right)$$

Heat sink

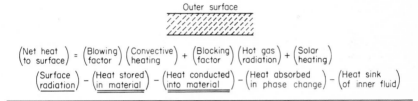

$$\left(\begin{array}{c}\text{Net heat}\\\text{to surface}\end{array}\right) = \left(\begin{array}{c}\text{Blowing}\\\text{factor}\end{array}\right)\left(\begin{array}{c}\text{Convective}\\\text{heating}\end{array}\right) + \left(\begin{array}{c}\text{Blocking}\\\text{factor}\end{array}\right)\left(\begin{array}{c}\text{Hot gas}\\\text{radiation}\end{array}\right) + \left(\begin{array}{c}\text{Solar}\\\text{heating}\end{array}\right)$$

$$\left(\begin{array}{c}\text{Surface}\\\text{radiation}\end{array}\right) - \left(\begin{array}{c}\text{Heat stored}\\\underline{\text{in material}}\end{array}\right) - \left(\begin{array}{c}\text{Heat conducted}\\\underline{\text{into material}}\end{array}\right) - \left(\begin{array}{c}\text{Heat absorbed}\\\text{in phase change}\end{array}\right) - \left(\begin{array}{c}\text{Heat sink}\\\text{of inner fluid}\end{array}\right)$$

Transportation cooling

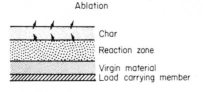

$$\left(\begin{array}{c}\text{Net heat}\\\text{to surface}\end{array}\right) = \left(\begin{array}{c}\text{Blowing}\\\underline{\text{factor}}\end{array}\right)\left(\begin{array}{c}\text{Convective}\\\text{heating}\end{array}\right) + \left(\begin{array}{c}\text{Blocking}\\\text{factor}\end{array}\right)\left(\begin{array}{c}\text{Hot gas}\\\text{radiation}\end{array}\right) + \left(\begin{array}{c}\text{Solar}\\\text{heating}\end{array}\right)$$

$$\left(\begin{array}{c}\text{Surface}\\\underline{\text{radiation}}\end{array}\right) - \left(\begin{array}{c}\text{Heat stored}\\\text{in material}\end{array}\right) - \left(\begin{array}{c}\text{Heat conducted}\\\underline{\text{into material}}\end{array}\right) - \left(\begin{array}{c}\text{Heat absorbed}\\\underline{\text{in phase change}}\end{array}\right) - \left(\begin{array}{c}\text{Heat sink}\\\text{of inner fluid}\end{array}\right)$$

Ablation

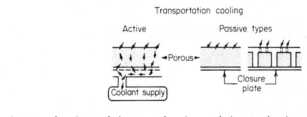

$$\left(\begin{array}{c}\text{Net heat}\\\text{to surface}\end{array}\right) = \left(\begin{array}{c}\text{Blowing}\\\underline{\text{factor}}\end{array}\right)\left(\begin{array}{c}\text{Convective}\\\text{heating}\end{array}\right) + \left(\begin{array}{c}\text{Blocking}\\\text{factor}\end{array}\right)\left(\begin{array}{c}\text{Hot gas}\\\text{radiation}\end{array}\right) + \left(\begin{array}{c}\text{Solar}\\\text{heating}\end{array}\right)$$

$$\left(\begin{array}{c}\text{Surface}\\\underline{\text{radiation}}\end{array}\right) - \left(\begin{array}{c}\text{Heat stored}\\\text{in material}\end{array}\right) - \left(\begin{array}{c}\text{Heat conducted}\\\text{in material}\end{array}\right) - \left(\begin{array}{c}\text{Heat absorbed}\\\underline{\text{in phase change}}\end{array}\right) - \left(\begin{array}{c}\text{Heat sink}\\\text{of inner fluid}\end{array}\right)$$

Fig. 6. Typical thermal protection systems.

TABLE 1. Application of High-temperature Thermal Protection Systems

Thermal protection system	Boost	Application — Entry — Ballistic	Application — Entry — Lifting	Interceptors	Comments
Ablative	Usually low temperature type—typically of class would be cork or thermolag	Ablation initiated at higher temperature—high "ρ", high "k"	Low density, low k; usually has reinforcements to strengthen char	Can utilize less material—usually materials are of a different class	Primary means of thermal protection for severe environments; passive, self regulating
Radiative	Low temperature materials; high ϵ coatings		High temperature materials—high ϵ; many ablatives function in this manner after initial formation of silicious char		Use is limited to low heat flux environments; service limited due to coating developments
Transpiration cooling		Nose tip; impingement areas	Nose tip, impingement areas; less applicable here	Primary utilization would be for shape retention	Utilization has been minimal to date; concept has been seriously considered for high pressure environments
Heat sink		Nose tip and non-survivable afterbody (test vehicles)		Nose tip	Was a primary means of protection in very early designs; still utilized in many test vehicles

applicability of these systems to the various classes of hypersonic flight vehicles. In the text that follows, consideration shall be given to the basic operating principles of each system, the materials generally considered for the system application, and the design procedures utilized to determine the requirement of each. It is this last step-design procedure which causes concern and it is felt necessary to caution the reader. *The procedures as outlined here are not those normally used in the final design of a flight system.* The rigorous techniques as defined in previous chapters are computerized and it is these programs that the analyst uses to analyze a given HTTPS. The scope of the Handbook precludes detailed discussions or instruction in computerized design practice. The calculation procedures presented here do not properly account for many of the surface and subsurface phenomena which occur during the exposure of certain HTTPS to an aerodynamic heating environment. Where possible, the degree of uncertainty associated with the procedures outlined here (relative to the more exact solutions) will be noted.

2. Heat Sinks

A heat sink is the simplest type of absorptive thermal protection system and was used in the design of the early generation of IRBM, ICBM reentry vehicles, and the first two manned Mercury vehicles. Its use is limited to relatively low heating rates and therefore may not be practical for high heat loads encountered in short reentry times.

Heat sinks have the advantages of simplicity, dependability, and for reusable vehicles, ease of refurbishment. Their outstanding disadvantage is their low efficiency. This would cause a heat sink sized to satisfy most current exit or reentry missions to be excessively heavy.

a. Mechanisms and Principles. Heat sinks operate on the basic heat conduction mechanisms described in detail in Section 3 of this Handbook. The net heat flux to a heat sink material is governed by the equation

$$\dot{q} = N_c(h_r - h_w) - \sigma T_w^4 \tag{9}$$

The temperature limits of most of the practical heat sink materials are too low to permit emission of significant amounts of thermal radiation. We can therefore neglect the last term in Eq. (9).

The theoretical maximum heat that could be absorbed by a heat sink is given by

$$Q = \rho b C_p (T_c - T_i) \tag{10}$$

where T_c is either the melting temperature or the temperature at which oxidation begins to dominate the outer surface heat transfer. This theoretical maximum assumes that the entire heat sink is heated uniformly. However, since the heat flux given in Eq. (9) is applied to the outer surface (and, in general, nonuniformly), this maximum is approached only for materials with very high thermal conductivities.

T_c is on the order of 2000 to 2300°F for most heat sink materials as compared to 4000 to 5000°F for most currently used ablators. It is evident from Eq. (9) that this will cause the heat sink to absorb significantly higher heat flux than an ablator under the same conditions. Caution again must be exercised as most heat sink materials are surface-temperature limited for high heating rates where energy cannot be transferred fast enough into the body. Figure 7 illustrates the net heat flux calculated at the stagnation point of a three-stage interceptor for two types of heat protection systems.

b. Materials. Various material design studies have indicated that the most efficient heat sink materials are beryllium, beryllium oxide (beryllia), and copper. Graphite has many desirable heat sink characteristics, but begins to oxidize at temperatures far below those required for best efficiency. The material properties for heat sink materials are given in Section C.

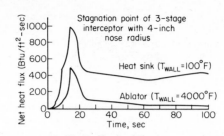

Fig. 7. Heating histories.

The usual configuration for a heat sink system is the amount of material (thickness) required to absorb a given amount of heat within the specified limitations of surface and backface temperatures. Current day heat sinks find their greatest utility in areas where shape retention is required or for relatively clean in-flight measurements (e.g., Ref. 22). Figure 6 illustrates a representative heat sink and the necessary heat balance. Additional information on this mode of TPS may be found in Ref. 23.

c. Design Procedures. The heat sink design is by far the least complex but due to the transient nature of the problem, it will still require some assumptions. For a typical reentry environment, it is the hypothesis that the actual problem, which may involve a variable surface temperature, may be simulated by an equivalent problem involving a constant surface temperature which produces approximately the same backface history as the transient problem.

The problem to be solved is idealized in Fig. 8 and results in one of finding the temperature response in a finite bar, one end of which is held at a constant temperature while the other end is insulated. In addition, it is necessary to determine the rate of heat conducted into the bar, which is

$$\dot{q}_k = \dot{q}_c \left(1 - \frac{h_w}{H_g}\right) - \sigma \epsilon (T_s + 460)^4 \qquad (11)$$

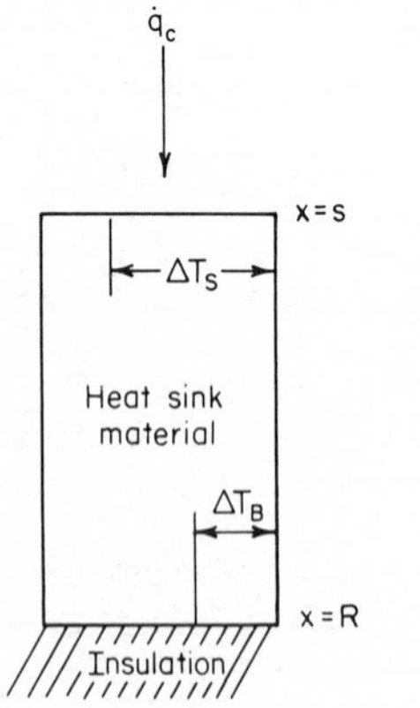

$$\dot{q}_c$$

Heat sink material

$$x = s$$

$$\leftarrow \Delta T_s \rightarrow$$

$$\Delta T_B$$

$$x = R$$

Insulation

At $t = 0$: $T(x,o) = T_0$

At $x = s$: $\dot{q}_c\left(1 - \frac{h_w}{H_g}\right) - \epsilon\sigma T_s^4 = -\left(k \frac{\partial T}{\partial X}\right)_{X=s}$

At $x = R$: $\left(\frac{\partial T}{\partial X}\right)_{X=R} = 0$

Fig. 8. Idealized heat sink.

Neglecting the effects of surface enthalpy and radiation and integrating Eq. (10) over t_i to t_B (t_B is the time at which the backface temperature reaches its design limit T_B), the total heat conducted into the material is found to be

$$Q_c \simeq w c_p \Delta T_m \tag{12}$$

where w = weight of material required to limit the backface temperature rise to the required ΔT_B

ΔT_m = average temperature rise in the bar

At this time some singular value of $k\rho t/c_p W^2$ exists with an associated temperature distribution in the bar. The solution to the latter problem has been solved [24] and is

$$\frac{\Delta T(x, t)}{\Delta T_s} = 1 - \frac{4}{\pi} \sum_{n=0}^{\infty} \frac{(-1)^n}{(2n - 1)} \left\{ \exp\left[-\frac{(2n + 1)^2 \pi^2 \alpha t}{4(R - s)^2} \right] \right\} \cos \frac{(2n + 1)\pi(R - x)}{2(R - s)} \tag{13}$$

Solution of this equation will yield ΔT_m, ΔT_s, or ΔT_B if t_B and one of the ΔT's is known. Equation (13) has been solved for the ratios of $\Delta T_B/\Delta T_s$ and $\Delta T_m/\Delta T_s$ [25] and is presented as a function of the ratio $k\rho t/C_p W^2$ in Fig. 9.

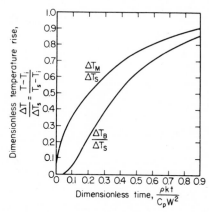

Fig. 9. Backface and mean temperature rise in a finite medium with a constant surface temperature T_s and insulated rear boundary.

Utilizing this information, the heat sink design procedure is:

1. Given $\dot{q}_c$ as a function of time, ρ, k, c_p = $f(T)$, and T_c
2. Compute Q_c from

$$Qc = \int_{t_i}^{t_f} \dot{q}_c \, dt$$

3. Compute ΔT_m
4. Compute $\rho k t/c_p W^2$ utilizing

$$Qc = W c_p \Delta T_m$$

5. Obtain T_s and T_B from Fig. 9

Figure 10 illustrates the weight per unit area of exposed surface for five different heat sink materials with the additional restriction of a rectangular wave heat pulse.

In the actual design of a heat sink TPS, relatively straightforward one- and two-dimensional heat transfer calculations should be used (this is best done on an analog computer or digital computer). The material properties are generally well characterized and it is important that the variation in these properties with temperature be used in all calculations. In addition, a safety factor is usually applied to the critical or melt temperature or the time-dependent heat input. Further information concerning heat sinks may be found in Refs. 26 and 27.

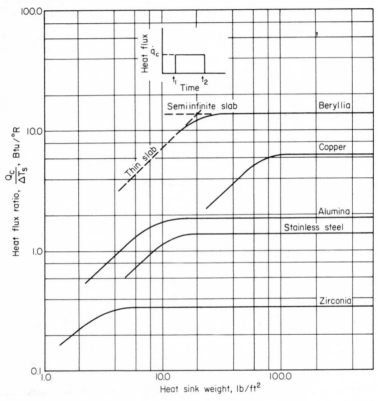

Fig. 10. Heat sink material performance.

3. Radiative Systems

Radiative thermal protection systems offer a conceptually simple and reliable means of thermal protection for a relatively narrow range of operating conditions. The system is normally passive and does not involve significant mass loss. The hot radiating panels are separated from the cooler portion of the vehicle by some type of insulation and may be augmented by active cooling such as a tubular liquid cooling system. Radiative systems were the primary means of thermal protection for the ASSET hypersonic glider and the X-15 aircraft and low thermal flux areas of almost all manned space vehicles. Radiative TPS have advantages and limitations similar to those of heat sinks: simplicity, dependability, and, for reusable vehicles, ease of refurbishment; however, they are like the heat sinks, heating rate and surface

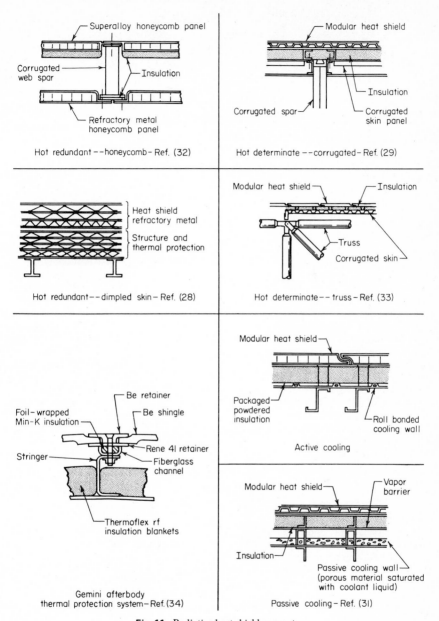

Fig. 11. Radiative heat shield concepts.

oxidation (or melting) limited. It should be noted that for many ablative systems, radiation is the dominant mode of heat rejection. The surface limitations, however, do not apply.

a. Mechanisms and Principles. As the name implies, the radiative system operates on the principle of heat rejection by radiation with the outer surface material being as

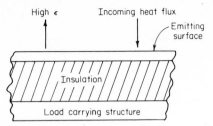

Fig. 12. A typical radiation thermal protection system.

thin as structural requirements allow. As such, energy absorption is small and the surface energy balance can be expressed as

$$\dot{q}_s = \sigma \epsilon T_w^{\,4} \tag{14}$$

where T_w is the radiation equilibrium temperature and is the maximum temperature the surface can experience for a given heat flux. The radiative system is therefore limited to a maximum heat rate (by virtue of its operating temperature limit) rather than a total heat input as an ablator may be or a heat sink which is limited by both. Once an operating temperature limit is defined, the maximum heat flux is correspondingly defined and the system can operate indefinitely at this condition. The penalty associated with increased operating time is the amount of insulation required between the heat shield and inner surface (either the load carrying structure or the payload).

b. Materials. As evidenced by Eq. (14), the effectiveness of a radiative system increases very rapidly with increasing surface temperature. The surface emissivity (ϵ) also exerts an important, although lesser, effect on the amount of energy which can be accommodated. Consequently, the primary development efforts with regard to this type of system have been concerned with the improvement of high emissivity, high temperature coatings, and with increasing the material service temperatures (including that of the internal insulation).

The present temperature limit of unprotected high-strength steels (e.g., [41]) is on the order of 2500°F. This corresponds to a heat flux of around 30 Btu/ft²-sec. The refractory metals (tungsten, tantalum, etc.) retain high strength at considerably higher temperatures and have melting temperatures as high as 5000 to 6000°F, but begin to oxidize rapidly at about the same limit of 200 to 2500°F. Efforts to improve the oxidation resistance by alloying have had only limited success [28] and have usually resulted in a degradation of the mechanical properties. Therefore, other recent development efforts have been in the field of oxidation-resistant coatings [29], primarily the silicide and aluminides. The results, to date, have extended the service temperatures of certain of the refractory metals to around 3000°F and even higher for short-time applications.

The refractory ceramics (oxides, carbides, and borides) are also frequently considered for use as radiative heat shields and, in selected instances, offer some improvement in maximum operating temperatures. Temperature limits of up to 4000°F are claimed for some ceramics, particularly the oxides. Some disadvantages of ceramics are brittleness, poor thermal shock resistance, and the fact that they are difficult to fabricate and attach.

For cases in which some surface recession is tolerable, operating temperatures of 6000°F can be achieved by use of a graphite surface [30, 31]. The use of pyrolytic graphite [31] would, by virtue of its low cross-plane thermal conductivity, also reduce the high temperature requirements imposed on the internal insulation. However, this concept also introduces attachment and refurbishment problems. The use of graphite heat shields is discussed in more detail under the category of ablative materials.

There are three basic concepts, or structural arrangements, for radiative metallic structures. These are (1) the hot redundant structure where the basic load carrying structure is unprotected from the external heating, and acts as the primary radiator; (2) the hot determinate structure, where auxiliary modular heat shields are utilized as the primary radiator and as the aerodynamic surface; and (3) the cool redundant structure, where the modular heat shields are also used as the primary radiator and as the aerodynamic surface. Figure 6 shows various configurations utilizing these concepts.

c. Design Procedures. The technique of utilizing an equivalent problem to obtain a solution to the actual problem will be employed here in much the same manner as in Sec. B.2.c. The problem to be solved is idealized in Figs. 8 and 9, and will follow the procedure outlined in Ref. 23.

The mathematics problem to be solved is

$$\frac{\partial T}{\partial t} = \alpha \frac{\partial^2 T}{\partial t^2} \tag{15}$$

with the following initial and boundary conditions

at
$$t = 0, \quad T(x,0) = T_0 \tag{16}$$

at
$$x = s, \quad \dot{q}_c \left(1 - \frac{h_w}{H_g} \right) - \sigma \epsilon (T_s + 460)^4 = - \left(k \frac{\partial T}{\partial X} \right)_{x = s} \tag{17}$$

at
$$x = R, \quad \left(\frac{\partial T}{\partial X} \right)_{x = R} = 0 \tag{18}$$

To remove the nonlinearities in the boundary conditions, $\dot{q}_c$ will be replaced by an equivalent heating rate with an initial time defined as

$$t_i = t_{\dot{q}(\text{max})} - \frac{\displaystyle\int_{t_0}^{t_{\dot{q}(\text{max})}} \dot{q}_c \, dt}{\dot{q}_{c(\text{max})}} \tag{19}$$

In addition, it is required that the total heat input be the same for both the real and the equivalent problem such that

$$Q_c = \int_{t_i}^{t_f} \dot{q}'_c \, dt = \int_{t_0}^{t_f} \dot{q}_c \, dt \tag{20}$$

The time-dependent gas enthalpy will be replaced with a time averaged H'_g, yielding

$$H'_g = \frac{1}{t_f - t_i} \int_{t_i}^{t_f} H_g \, dt \tag{21}$$

The equivalent heat rate $\dot{q}_c'$ is then chosen such that the rate of heat conducted into the bar is $\dot{q}_k$

$$\dot{q}_c\left(1 - \frac{h_w'}{H_g'}\right) - \sigma\epsilon(T_s + 460)^4 = \dot{q}_k \tag{22}$$

where $\dot{q}_k$ is such that a constant surface temperature results. The hot wall and emissivity corrections are not independent of time and the problem is linearized.

The problem is now that of finding the temperature of a finite bar, with one end held at a constant temperature and the other end perfectly insulated. The solution to this problem was presented in Sec. B.2.c as

$$\frac{\Delta T(x, t)}{\Delta T_s} = 1 - \frac{4}{\pi}\sum_{n=0}^{\infty}\frac{(-1)^n}{(2n + 1)}\left\{\exp\left[-\frac{(2n + 1)^2\pi^2\alpha t}{4(R - s)^2}\right]\right\}\cos\frac{(2n + 1)\pi(R - x)}{2(R - s)} \tag{13}$$

Integration of Eq. (22) from t_i to t_B gives the total heat conducted into the material as

$$Q_c\left(1 - \frac{h_w'}{H_g'}\right) - \epsilon\sigma(T_s + 460)^4(t_B - t_i) = WC_p\Delta T_m \tag{23}$$

where W is the weight required to limit the system to the specified backface temperature rise ΔT_B over the time interval $(t_B - t_i)$, and ΔT_m is the average temperature rise in the bar. At this instant, some singular value of $k\rho t/C_p W^2$ exists with an associated temperature distribution in the bar given by Eq. (13) from which ΔT_m can be calculated for a given t_B and T_s.

For the case where the backface temperature reaches its design limit at some time $t_B < t_B$. At this instant, $T_s = T_B$ and this condition corresponds nearly to a value of unity for $k\rho t/C_p W^2$, yielding

$$W = \sqrt{\frac{\rho k}{C_p}}(t_B - t_i) \tag{24}$$

Solving Eqs. (23) and (24) for $(t_B - t_i)^{1/2}$ gives

$$(t_B - t_i)^{1/2} = \frac{\sqrt{\rho k C_p \Delta T_B^2 + 4\sigma\epsilon(T_B + 460)^4 Q_c(1 - h_w'/H_g')}}{2\sigma\epsilon(T_R + 460)^4} - \frac{\Delta T_B\sqrt{\rho k C_p}}{2\sigma\epsilon(T_B + 460)^4} \tag{25}$$

The required weight may now be found from Eqs. (24) and (25).

For the case where the design limit is reached at the final time, the system reaches a larger value at a later time $t_B > t_f$. An iteration procedure is required for this problem (since Eq. (23) contains two unknowns) and may be accomplished by assuming a reasonable value for T_s to start the iteration. The procedure detailed in Ref. 23 is felt to be reasonable and will be followed here.

Assuming for the first approximation that h_w'/H_g' and $WC_p\Delta T_B$ are both zero, Eq. (23) gives for T_s

$$\left(T_s\right)_1 = \sqrt[4]{\frac{Q_c}{\epsilon\sigma(t_f - t_i)}} \tag{26}$$

The ratio $\Delta T_B / \left(\Delta T_s\right)_1$ may be formed and Eq. (13) utilized to obtain the result. This has been done and is presented in Fig. 9 as a function of the ratio $k\rho t/C_p W^2$. Now let

$$G = \sqrt{\frac{k\rho(t_f - t_i)}{C_p W^2}} \tag{27}$$

or

$$(W)_1 = \sqrt{\frac{k\rho(t_f - t_i)}{C_p(G)_1}} \tag{28}$$

which is the maximum possible value for W. Utilizing $\left(T_s\right)_1$ and the relationship between $\left(h'_w\right)_1$ and $\left(T_s\right)_1$

$$h_w = 33.48[6.8(10^{-3})(T_s + 460) + 3.7(10^{-7})(T_s + 460)^2] \tag{29}$$

a hot wall correction may be applied to Q_c yielding the minimum possible hot-wall heat input. With this and the first approximation for the weight and the average temperature, a second approximation for T_s may be found from Eq. (23)

$$T_{s_2} = \sqrt[4]{\frac{Q_c(1 - h'_w/H'_g) - W_1 C_p\left(\Delta T_m\right)_1}{\epsilon\sigma(t_f - t_i)}} - 460 \tag{30}$$

which is the minimum value of T_s. This procedure is repeated until a suitable convergence is obtained with a final system weight of W_F.

Figure 13 presents a typical boost/reentry trajectory, Fig. 14 the associated heat flux history (for a point on the vehicle centerline), Fig. 15 a typical temperature history, and Fig. 16 illustrates the thermal protection system arrangement for the vehicle being analyzed. The analysis of this vehicle was conducted without the qualifying assumptions presented here.

4. Ablative Systems

Ablation has been defined [35, 36] as a "self-regulating heat and mass transfer process in which incident thermal energy is expended by sacrificial loss of material." In view of this accurate yet simple definition, the effort expended in the development of ablation theories and materials over the past two decades (see Sec. 16) may seem to be inconsistently vast. However, the theories have progressed from the "figure-of-merit" q^* approach to the present-day degree of sophistication described in Sec. 16. The materials as well as the methods used in their development for ablative heat shields have been also significantly improved. The more recent and perhaps the most successful approach is that which utilizes the analytical tools to determine the desirable material properties for a given mission and the subsequent formulation or tailoring of a suitable composite (e.g., [37]). This approach has recently been extended to include the investigation of microscopic effects such as pore size, grain size, or fiber length to assist in the determination of the controlling mechanisms and

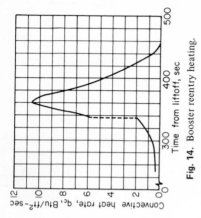

Fig. 14. Booster reentry heating.

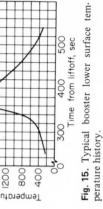
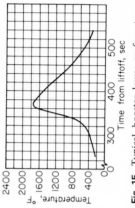

Fig. 15. Typical booster lower surface temperature history.

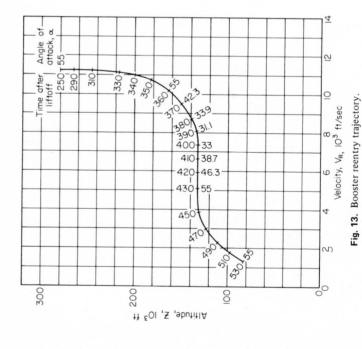

Fig. 13. Booster reentry trajectory.

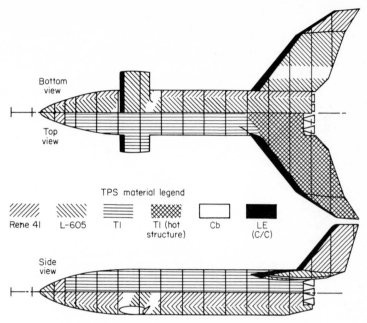

Fig. 16. Typical TPS arrangement for a booster/reentry vehicle.

the subsequent material improvement [38]. Thus, if one considers the various entry applications, man-rating problems of spacecraft, and the actual complexity of the problem, the need for the effort is better understood.

a. Mechanisms and Principles. The ablation process can best be reviewed by examining the reaction of a typical material to a hyperthermal environment. Following the example of Ref. 36, a fiber-reinforced plastic composite will be examined as typical. The incident heat flux to the surface is absorbed and then conducted into the material substrate—at a low rate due to the relatively low thermal conductivity of the material. The rapid rise of the surface temperature is followed by initiation of thermal degradation and the organic constituents of the composite are pyrolyzed into numerous gaseous products. These pyrolytic species are then injected into the boundary layer, reducing the convective heat transfer rate to the surface. The thermal degradation of resin leaves a hard surface residue of porous carbon and the primary region of resin pyrolysis shifts from the surface to a zone beneath the carbonaceous char layer. This char layer is usually refractory in nature and increases the amount of heat reradiated from the surface. The gaseous products formed beneath this layer percolate through the char where they undergo partial vapor phase cracking and deposit pyrolytic carbon on the walls of the porous char.

In a very simple form, the mechanisms described above may be put into an energy balance at the ablating surface as [36]

$$q_0 \;=\; \underbrace{m[C_p(T_w - T_b)}_{\substack{\text{heat capacity}\\\text{of the}\\\text{material}}} + \underbrace{fh_{p,c}}_{\substack{\text{heat ab-}\\\text{sorbed by}\\\text{chemical}\\\text{reaction and}\\\text{phase change}}} + \underbrace{f\beta(Hs - Hw)]}_{\text{heat blockage}} + \underbrace{e_w\sigma T_w^4}_{\substack{\text{heat rerad-}\\\text{iated from}\\\text{the surface}}} \qquad (31)$$

As mentioned previously, the detailed treatment of these mechanisms has been covered in considerable depth in Sec. 16; however, there are times when the student of thermal protection, or the engineer, pressed for quick estimates, must forego analytical accuracy and revert to some of the early methods for qualifying ablative response, namely, "heat of ablation" techniques. This value collectively expresses the ability of an ablative material to absorb, block, and dissipate heat per unit mass expended [36]. In equation form

$$H_{eff} = \frac{\dot{q}_0}{\dot{m}} \tag{32}$$

where H_{eff} = effective heat of ablation

$\dot{q}_0$ = convective heat flux to a nonablating wall at the ablation temperature

$\dot{m}$ = mass ablation rate of the material

More sophisticated techniques for hand calculations and many formulations of the "heat of ablation" expression have been developed; however, for hand calculations, the above technique is felt to provide the best balance between accuracy and computational difficulty.

The other performance parameter utilized in design is the insulation index, defined as that material thickness required to maintain a predetermined temperature, usually at the material backwall. This thickness is analytically determined by computing the material heat content at the end of the heating period and calculating the insulation weight from the equation

$$W_{ins} = \frac{Q_T}{C_p (T_w - T_{bf})} \tag{33}$$

where W_{ins} = insulation weight per unit area

Q_T = heat content of the insulation layer per unit area at the end of the heating period

C_p = specific heat of the material

$T_w - T_{bf}$ = the allowable increase in the backwall temperature

From Eq. (2) it can be seen that the magnitude of the heating rate strongly influences the mass loss rate and the exposure time strongly influences the insulation requirements (Eq. (3)). Together, they determine specific ablative requirements and are thus of prime importance in the selection of an ablative thermal protection system.

The above simplified treatment is intended to give a broad view of the ablation process; it must be reemphasized, however, that the actual process should be studied and applied along the lines of Sec. 16 of this handbook.

b. Materials. As noted before, ablative systems efficiency is particularly sensitive to material performance. Therefore, it is necessary to treat the subject of materials in detail. In the absence of a universally acceptable ablative material, material developers have collaborated with engineers to produce a wide variety of ablative compositions and constructions, usually tailored to satisfy the requirements of a specific vehicle for a specific mission. These thermally protective materials have been arbitrarily categorized by their matrix composition, with typical material given in Table 2 [39]. A discussion of some of the more important classes of materials for reentry vehicles follows and the reader is directed to Schmidt [40] for a discussion of materials for use in a chemical propulsion environment.

Refractories and Ceramics. Graphite thermal protection systems have been studied intensively in recent years. Graphite, which sublimes at temperatures as high as

TABLE 2. Typical Ablative Compositions [39]

Plastic-base	Ceramic-base
Polytetrafluoroethylene	Porous oxide (silica) matrix infiltrated with phenolic resin
Epoxy-polyamide resin with a powdered oxide filler	Porous filament wound composite of oxide fibers and an inorganic adhesive, impregnated with an organic resin
Phenolic resin with an organic (nylon), inorganic (silica), or refractory (carbon) reinforcement	Hot pressed oxide, carbide, or nitride in a metal honeycomb
Precharged epoxy impregnated with a non-charring resin	

Elastomer-base	Metal-base
Silicon rubber filled with microspheres and reinforced with a plastic honeycomb	Porous refractory (tungsten infiltrated with a low melting point metal (silver))
Polybutadiene-acrylonitrile elastomer modified phenolic resin with a subliming powder	Hot pressed refractory metal containing an oxide filler

7000° R, accommodates or rejects the imposed heating through the mechanisms of sensible heat rise, oxidation, the latent heat of sublimation, and surface radiation. The transpiration cooling provided by graphite has little effect on the surface heat balance because of its relatively low ablation rates. Because of its extremely high ablative efficiency, graphite theoretically offers the minimum amount of ablation and shape change in areas subject to high heating rates such as ballistic missile nose tips [41] and small-radius leading edges.

Furthermore, pyrolitic graphite (which is formed by the deposition of a series of layers and is consequently highly anisotropic) has a very low thermal conductivity in the direction normal to the deposition plane so that it is an excellent high-temperature insulator. The thermal conductivity in the plane of the deposited layers is the same as for homogeneous graphite, or almost two orders of magnitude higher than in the normal direction.

The oxidation of graphite in air has been analytically described [38, 39] and verified by experiments [44–46]. It should be noted, however, that these analytical models allow for only one (C_3) or two (C and C_3) carbon species at the surface. The experimental data were obtained at low pressures and thus may not represent high-pressure flight conditions. Some recent analytical models have been constructed on the assumption that several other carbon species may be present and could adversely affect the ablation performance. This assumption, however, has not yet been verified by experiment.

The principal disadvantages of graphite as a thermal protection system are its brittleness and low resistance to thermal stress, which restrict its maximum usable thickness. Further, graphite's high-temperature thermal and structural properties have not been reliably ascertained, and it is still subject to manufacturing inconsistencies. The pyrolytic graphites may experience delamination between the deposited layers at high temperature and high thermal stresses. Although this delamination might be tolerated for certain applications, it is generally undesirable because it is unpredictable and as yet uncontrollable. Pyrolytic graphite is also difficult and expensive to manufacture, particularly in shapes with small radii in relation to their thickness.

Ceramic-base ablators generally have high thermal efficiency, but this capability is difficult to realize because of their susceptibility to thermal stress failure. During thermal shock the material may crack extensively and fail catastrophically. Placing the ceramic in a metal honeycomb tends to alleviate this problem by restricting any cracks to the outer walls of the cell structure [39].

Porous ceramics per se are potential ablative-insulative materials, but by polymer impregnation, their ablation characteristics are greatly enhanced [47–48]. Specifically, the resin component increases the composite strength and thermal shock resistance, decreases the thermal conductivity, and permits high environmental temperatures to be tolerated without exceeding the melting or decomposition temperature of the ceramic. Typical compositions of porous ceramics include silica, zirconia, alumina, magnesia, thoria, carbides, borides, and silicides. They are prepared by such processes as hot pressing, isostatic pressing, slip casting, pyrolysis of organic inclusions, and chemical bonding. Suitable polymeric infiltrants include the phenolics, epoxies, acrylics, polystyrenes, and others. Of the numerous combinations available, phenolic resin impregnated porous silica composite has found the greatest use. The principal limitation of this material (like other internally ablating composites) is a reduced thermal efficiency with exposure time, or loss of molten surface material by the shearing action of the high temperature aerodynamic stream.

Metal-base ablators [50] are another class of thermal protection materials. The most common type of ablator is a porous refractory skeleton containing a lower melting point metallic infiltrant. Tungsten matrices with up to 80 percent porosity are generally employed. They are prepared by fiber felting, or cold pressing powder followed by sintering. The porous matrix is then infiltrated with such metals as copper or silver using high pressures, high vacuum, or a combination of both. The resultant composite has high strength, good thermal shock resistance, and can be easily machined. Its low thermal efficiency (about 1500 Btu/lb), high density, and high thermal conductivity tend to restrict its use to intensely heated areas where the original configuration of the matrix must be maintained. Another source of information on the development of improved metal-ceramic composites is Vogan [51].

Plastics. Plastic-base ablators which employ an organic matrix are the most widely used class of ablative heat protective materials. They respond to a hyperthermal environment in a variety of ways, such as depolymerization-vaporization (polytetra-fluoroethylene), pyrolysis-vaporization (phenolic, epoxy resins), and decomposition-melting-vaporization (nylon fiber reinforced plastic). The principal advantages of plastic-base ablators are their high heat shielding capability and low thermal conductivity. The major limitations are high erosion rates during exposure to very high gasdynamic shear forces, and limited capability to accommodate very high heat loads [39].

The subliming ablators, typified by Teflon, decompose directly from the solid to the gaseous state; that is, the material absorbs sensible energy until the surface reaches the sublimation temperature, which is primarily a function of the local pressure. The temperature of the ablator during sublimation is also dependent on the ablation rate. Energy is absorbed in the phase-change process and the heat flux is reduced by the transpiration effect of the evolving gases. For a low-temperature ablator such as Teflon, the principal mechanisms of heat rejection are the heat of depolymerization and transpiration cooling.

Teflon heat shields have been used on several ballistic missile entry vehicles and on a few current vehicles. Teflon is a low-to-moderate temperature ablator with moderate efficiency, and its thermal performance is reasonably well characterized. Because of its high ablation rates, a Teflon heat shield changes its shape considerably in long-duration heat pulses.

Elastomeric-base low density materials represent a second major class of ablators. They thermally decompose by such processes as depolymerization, pyrolysis, and vaporization. Most of the interest to date has been focused on the silicone polymers, because of their low thermal conductivity, high thermal efficiency at low to moderate heat fluxes, low temperature properties, elongation of several hundred percent at failure, oxidative resistance, low density, and compatibility with other structural substrate materials. Elastomeric materials are generally limited by the amount of structural quality of the char formed during ablation which restricts their use to hyperthermal environments of relatively low mechanical forces [39].

Silicone elastomeric ablators [52] have several advantages over other charring materials for some applications. They form a siliceous melt layer which prevents further oxidation of the char. This results in very high ablative efficiency in the long-duration, low-heat-flux environment characteristic of lifting entry bodies.

The elastomeric materials are frequently fabricated in a fiberglass honeycomb matrix to reinforce the char and to inhibit the flow of the melt layer. Heating rates in excess of approximately 100 Btu/ft^2-sec vaporize the protective melt layer and cause rapid surface recession by diffusion-limited oxidation of the carbonaceous char residue. At heating rates below this level, these materials experience little or no surface recession and char at a very slow rate after the initial surface charring transients. Thus, their primary function for long flight times is to provide high insulation performance.

Many additives, including microballoons and reinforcement materials, are being employed in attempts to improve the insulative and ablative efficiency of the silicone-base materials [53]. Silicone elastomeric heat shields have performed satisfactorily in an entry environment on the Project Gemini and PRIME vehicles [54, 55].

Composites. The thermal protection systems of most interest are the charring-ablator heat shields. These can be made of a homogeneous thermosetting resin, such as the phenolics, epoxies, or silicones, or of the same resin with an organic or refractory fiber, such as glass, asbestos, graphite, or nylon.

As the resin is heated, the temperature increases until the surface reaches a temperature at which the material begins to decompose (pyrolyze) and release gaseous products, leaving a porous, carbonaceous residue. The pyrolysis temperature is a function of the local pressure and ablation rate and is relatively low, from 500 to 1000°F. As the heating continues, the pyrolysis zone proceeds into the material and the decomposition occurs below the surface. The gaseous products diffuse through the porous char to the surface, absorbing energy from the char while undergoing further decomposition (cracking). They finally exit into the boundary layer where they act as a transpirant and may undergo additional chemical reaction with the boundary-layer gas.

The char is primarily carbonaceous and continues to absorb heat until it reaches the temperature at which it oxidizes or sublimes (as previously described for graphite materials), or until it is mechanically removed by external shear forces. For lifting or moderate ballistic entries, oxidation is the dominant thermochemical char-removal mechanism. At surface temperatures below 1500°F, oxidation is limited by reaction-rate kinetics. In this regime, surface recession can be reduced appreciably by incorporating oxidation-resistant additives such as silica. As the surface temperature increases, the oxidation rate increases exponentially until the oxygen at the surface begins to be depleted. At still higher temperatures, the surface recession is limited by the rate at which oxygen can diffuse through the boundary layer. In this regime, the mass rate of char oxidation is virtually independent of material properties. At temperatures of about 6000°F, the char sublimes.

A thick char provides an efficient insulation barrier, radiates a large amount of heat from the surface, and is a very efficient ablator. However, the char formed on a homogeneous plastic is usually weak and brittle. Thus, the material is susceptible to rapid removal by mechanical shear and by spallation (possibly due to thermal stresses and buildup of internal gas pressure). This reduces the insulative efficiency of the char and exposes the cooler internal material to the surface, resulting in less radiative cooling. A discussion of mechanical effects on char layers is presented by Schneider et al. [56].

To improve the char-retention characteristics of the ablative resins, reinforcing fibers are commonly added to the virgin material. Depending on the operating environment, these can be organic or inorganic fibers. The fibers add strength to the char until they reach their own melt or decomposition temperature. The use of reinforcements also increases the complexity of the analytical charring ablation model, since the ablation kinetics of the fibers must be superimposed on that of the resin. However, fiber reinforcements and other additives have provided flexibility in the tailoring of ablative materials to specific applications.

Because the fibers usually possess a higher thermal conductivity than the resin binder, fibers which are normal to the surface will increase the overall conductivity of the composite. When the fibers are placed parallel to the surface, the conductivity approaches that of the resin, but the char shear strength is greatly reduced and the material is subject to delamination. Since any variation between these extremes is possible, the fiber orientation can be selected on the basis of the particular shear stress and heat conduction requirements.

Ablative efficiency is usually proportional to the material density, while the insulation efficiency is inversely proportional to the density. It is therefore useful to reduce the density and thermal conductivity of most ablative materials by adding microballoons—tiny hollow spheres approximately 40 microns in diameter—made of phenolic resin or glass with a wall thickness of 1 or 2 microns [53]. The additives can be graded so that the density varies uniformly through the material and the weight is reduced with a minimal effect on the ablation performance. Some loss of char strength ordinarily accompanies the use of the microballoons.

c. Design Procedures. As mentioned previously, the complexity of the ablation phenomena has been analytically described by equally complex computer programs. These computer programs are the main tools in the final design of any flight vehicle heat shield, and are the recommended design procedure. However, it is often desirable to obtain quick, approximate answers as guidelines in preliminary design or initial material evaluation studies. The method employed for this type of calculation utilizes the previously described heat of ablation [35–37, 58]. It should also be noted that the bulk of the ablative heat shield design calculations yield the amount of material ablated as a fraction of the total amount of material required to maintain a specific structural backface temperature. The calculation procedure presented here outlines the method for obtaining the amount of material ablated. The additional calculation may be obtained using standard conduction techniques [23, 57].

The heating to an ablating surface can be represented by (from Sec. B.4.a)

$$\dot{q}_0 = m[c_p(T_w - T_b) + fh_{p,c} + f\beta(H_s - H_w)] + \epsilon_w \sigma T_w^4 \qquad (34)$$

In the technique presented here, the following assumptions are made

$$f\beta(H_s - H_s) \simeq 1$$

$$fh_{p,c} \gg c_p(T_w - T_b)$$

$$mfh_{p,c} \gg \epsilon_w \sigma T^4 w$$

Further, the material ablated is calculated by assuming that the surface recession is proportional to the net heat flux and utilizing Eq. (32)

$$\dot{s} = \frac{\dot{q}}{H_{eff}} \qquad (35)$$

It is an unfortunate coincidence that each of the assumptions tend to introduce errors in the same direction, i.e., to cause the calculated ablation to exceed the actual. Although more sophisticated hand calculation techniques have been developed, the approach presented here is felt to be the best balance between accuracy and computational difficulty.

The procedure is best illustrated with a sample set of calculations. The vehicle to be analyzed is a typical interceptor missile upper stage (Fig. 17) with a phenolic silica heat shield. The altitude and velocity histories are presented in Fig. 18, and the necessary induced environmental parameters are shown in Fig. 19. The heat transfer and enthalpy histories were calculated for the one-foot station utilizing the methods discussed in Sec. A assuming turbulent flow over the vehicle. The thermochemical heat of ablation for phenolic silica was obtained from Arne [58] and is shown in Fig. 10. Knowing heat flux and recovery enthalpy histories and assuming $h_r \gg h_w$ the recession rates may then be obtained and are shown in Table 3. Also shown is the integrated or total recession for the entire trajectory.

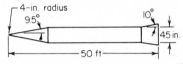

Fig. 17. Interceptor upper stage.

TABLE 3. Recession Rates for a Typical Interceptor Upper Stage

$\rho = 100 \text{ lb/ft}^3$

$h_r \gg h_w$

Turbulent flow

Time	Δh	H_{eff}	$\dot{q}_{cw}$	in.	$\dot{s}$	Δt	$\dot{s}\Delta t$	$\int_0^t \dot{s}\,dt$
0								
	0	3,000	5	0.0017	0.0002	2.0	0.0004	0
2								
	150	3,100	16	0.0052	0.0006		0.0012	0.0004
4								
	390	3,300	40	0.0121	0.0014		0.0028	0.0016
6								
	750	3,500	78	0.0222	0.0027		0.0054	0.0044
8								
	1,150	3,800	147	0.0386	0.0046		0.0092	0.0098
10								
	1,600	4,080	260	0.0636	0.0076		0.0152	0.0190
12								
	2,150	4,450	410	0.0921	0.0110		0.0220	0.0342
14								
	2,700	4,800	600	0.1250	0.0150		0.0300	0.0562
16								
	2,450	4,600	470	0.1020	0.0122		0.0244	0.0862
18								
	2,200	4,470	385	0.0862	0.0103		0.0206	0.1106
20								
	2,000	4,350	320	0.0735	0.0088		0.0176	0.1312
22								
	1,820	4,200	265	0.0632	0.0076		0.0152	0.1488
24								
	1,640	4,100	215	0.0525	0.0063	2.0	0.0126	0.1640 / 0.1766
26								

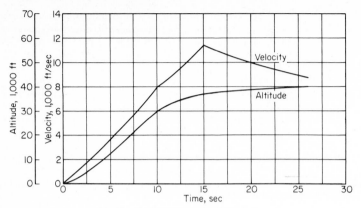

Fig. 18. Typical interceptor trajectory.

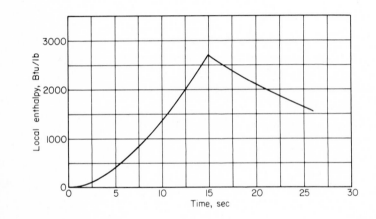

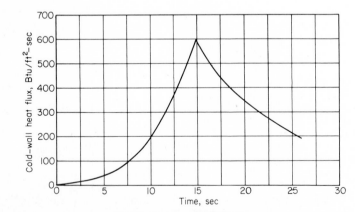

Fig. 19. Environmental histories.

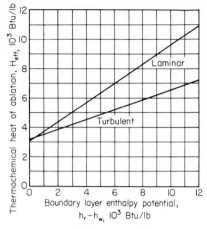

Fig. 20. Heat of ablation of phenolic silica.

5. Transpiration and Film-cooled Systems

The term "transpiration" as used herein includes all forms of active injection of liquid or gaseous material into the boundary layer, such as injection through a porous inert matrix or injection through a series of discrete slots (film cooling). The injectant may or may not be chemically inert in the presence of the boundary-layer gases.

An example of a transpiration-cooled vehicle is shown in Fig. 21. The main elements of a transpiration system for a reentering vehicle include the transpiring structure (in this case a nose tip), a device for regulating the flow of coolant, a coolant storage reservoir, and a high pressure source for expelling the coolant.

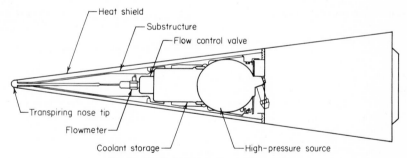

Fig. 21. Typical transpiration cooling system.

In some applications, the shape change caused by the surface recession of an ablating surface is not acceptable for aerodynamic performance reasons. In such cases, if the environment is too severe for radiative or heat sink systems, transpiration cooling may be the only practical solution.

Transpiration cooling makes possible performance in environments that could not otherwise be withstood. However, its mechanical complexity, with the associated reliability problems, and difficulty in the analytical prediction of performance, tend to limit its use.

At present, the development of transpiration cooling systems is in its early stages. The most promising near term applications for transpiration cooling would be in areas subject to extremely high heating rates and where shape changes cannot be tolerated.

Transpiration cooling can also be beneficial where there are stringent communications requirements and thus low signal attenuation may be tolerated. Reference 59 presents the results of a feasibility study of transpiration cooling systems. This system is treated here at more length than others since less useful information is available in the literature.

a. Mechanism and Principle of Operation. From a thermodynamic standpoint, transpiration is the most efficient method for cooling a heated surface. Figure 22 illustrates the principle of transpiration cooling.

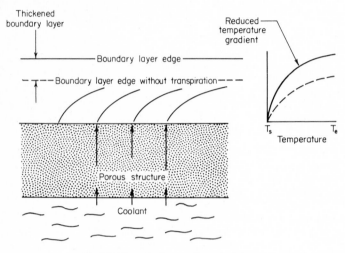

Fig. 22. Transpiration cooling principles.

The structure to be cooled is fabricated with integral coolant passages, permitting the coolant to flow through it and out the heated surface. The structure is maintained cool by two basic mechanisms; heat is conducted to the coolant as it flows through the structure; and as the coolant is ejected out the surface it reduces the surface heat transfer rate by cooling and thickening the boundary layer.

The problem of transpiration cooling may be considered to fall into four categories: The first involves the structural design concept and material selection; the second is the selection of a coolant; the third consists of the analysis of the coolant flow through the structure; and the fourth consists of the effect of the coolant on the external boundary layer.

Porous Flow Analysis. Both the temperature and pressure distribution in the porous structure need to be solved in order to permit an estimate of the coolant flux distribution at the heated surface. It would be desirable to analytically couple the variables, simplifying the solution. However, the boundary conditions are not generally well-behaved analytic functions, and fluid properties are strongly dependent upon temperature and pressure. Therefore it becomes impractical to couple the energy and momentum equations and solutions are found by iterating between the temperature and pressure solutions until convergence is obtained.

The modified Darcy equation is used in evaluating the steady-state coolant flow through the one-dimensional porous structure

$$-g_c \frac{dP}{dx} = \frac{\alpha \mu \dot{m}}{\rho} + \frac{\beta \dot{m}^2}{\rho} \qquad (36)$$

where P = local pressure
   α = material viscous pressure drop constant
   β = material inertial pressure drop constant
   μ = coolant viscosity
   ρ = coolant density
   $\dot{m}$ = local coolant mass flux

If the coolant is in the vapor phase, and assumed to be described by the ideal gas equation, then the modified Darcy equation can be written as

$$-g_c \frac{dP^2}{dx} = \frac{2RT}{M} (\alpha\mu\dot{m} + \beta\dot{m}^2) \tag{37}$$

where R = the universal gas constant
   M = the coolant molecular weight
   T = the coolant temperature

The steady-state temperature distribution in the one-dimensional porous structure is evaluated from the energy equation

$$\frac{d}{dx}\left(\rho\mu H - k_e \frac{dT}{dx}\right) = 0 \tag{38}$$

where H = the coolant local enthalpy
   k_e = the local effective thermal conductivity
   u = coolant velocity

The effective conductivity k_e is determined by

$$k_e = \gamma k_c + \Gamma(1 - \gamma)k_s \tag{39}$$

where K_c = the coolant conductivity
   K_s = the conductivity of the porous structure parent material
   γ = the void fraction of the porous structure
   Γ = the tortuosity factor for the thermal conductivity of the porous structure, that is, $\Gamma = k_{porous}/(1 - \gamma)k_s$

The tortuosity must be evaluated experimentally. Equation (38) is based on the assumption that the coolant and porous structure are in thermal equilibrium, i.e., the coolant locally is at the same temperature as the surrounding structure. This is a very good assumption for porous material that utilizes pores with a small hydraulic diameter ($\lesssim 0.001$ in.) and a dense pore spacing. However, for material with significantly larger pores such that film cooling is approached, the temperature differential between the coolant and surrounding structure must be accounted for, and coupled separate solutions exist for the temperature distribution in the coolant and structure.

As previously stated, transpiration cooling reduces the convective heat transfer to the surface of the structure by cooling and thickening the boundary layer. The surface heat transfer rate is reduced with increased coolant flux, and a high enough flux will cause the boundary layer to separate from the surface. Another significant aspect of transpiration cooling is that injection of coolant into the boundary layer has the tendency to destabilize the boundary layer, causing premature transition and normally the heat transfer rate.

Reduction in Convective Heating.   Numerous studies of transpiration cooling in a laminar boundary layer have been performed (e.g., [60]), and have received some experimental verification. The transpiring turbulent boundary layer has not been solved analytically, but much experimental work has been done [61] and reasonably satisfactory semiempirical heat transfer models have been formulated.

The most common method of presenting experimental data has been to express it in the form of a modified heat transfer coefficient, that is,

$$q_c = N_{c_0} \left(\frac{N_c}{N_{c_0}} \right) (H_r - H_s) \tag{40}$$

where q_c = surface convective heat flux with injection

N_{c_0} = heat transfer coefficient without injection

N_c = apparent heat transfer coefficient with injection

H_r = air recovery enthalpy

H_s = air enthalpy at the surface temperature obtained with injection

In this case, the ratio N_c/N_{c_0} represents the reduction in surface convective heating (or "blocking") due to the alteration in the boundary-layer properties caused by injection of the coolant.

Experimental results have shown that the blocking effect depends on whether the boundary-layer flow is laminar or turbulent, the rate of mass injection, and the molecular weight or specific heat of the coolant.

Empirical correlations that have been derived are: for laminar flow

$$\frac{N_c}{N_{c_d}} = 1 - 0.68 \left(\frac{C_{p\,cool}}{C_{p\,air}} \right)^{0.4} \frac{\dot{m}}{N_{c_0}} + 0.08 \left(\frac{C_{p\,cool}}{C_{p\,air}} \right)^{0.4} \left(\frac{\dot{m}}{N_{c_0}} \right)^2 \tag{41}$$

and for turbulent flow

$$\frac{N_c}{N_{c_0}} = \frac{\left(C_{p\,cool}/C_{p\,air} \right)^{0.8}}{\left[1 + \left(C_{p\,cool}/C_{p\,air} \right)^{0.8} \left(\dot{m}/4N_{c_0} \right) \right]^4 - 1} \tag{42}$$

where $C_{p\,cool}$ = the specific heat of the coolant vapor

$C_{p\,air}$ = air specific heat

$\dot{m}$ = the surface coolant flux

Figures 23 and 24 illustrate the comparison of Eqs. (41) and (42) with laminar and turbulent experimental data.

b. Material Selection. The optimum concept for a transpiration-cooled structure is one that limits the flow of coolant to one direction, i.e., cross flow is not permitted from one coolant passage to another. This simplifies analysis and permits the most efficient distribution of coolant. Coupled with this is the desirability of employing extremely small, densely spaced coolant passages in order to maintain the temperature differential between the coolant and surrounding structure as small as possible. Present fabrication technology has not yet found a solution that satisfactorily meets both objectives. The one-dimensional flow concepts that have been developed and successfully tested utilize coolant passages that are too large to yield true transpiration cooling, and the cooling efficiency that results is more characterized by film cooling. Such structures have included drilled and slotted surfaces as well as structures formed by individual elements that have been bonded together with integral coolant passages between.

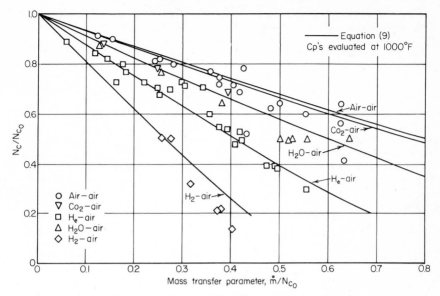

Fig. 23. Correlation of mass transfer effect on laminar heat transfer.

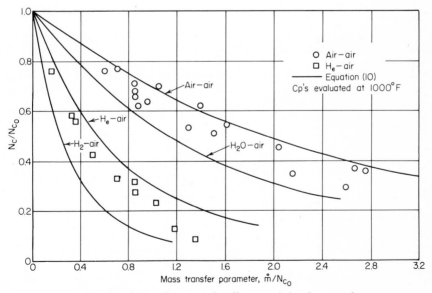

Fig. 24. Correlation of mass transfer effect on turbulent heat transfer.

The only true transpiration structures that have been developed and successfully utilized are those that have been formed from compacted and bonded powders, fibers, or layered mesh type elements. Because of the material uniformity and density required to achieve adequate strength, sintered powders have been generally chosen as the best structure.

The main drawback of sintered powders is that they result in a structure that has truly three-dimensional flow characteristics, and the distribution of coolant becomes strongly influenced by the surface pressure distribution.

For reentry applications, both ceramics and metals have been considered for forming the structure. Generally, aerodynamic loading and thermal stresses have ruled out the use of ceramics, even though they would permit higher surface temperatures and probably lower system weights. The metals that have been utilized include stainless steel, tungsten, and beryllium.

In general, it has been shown that, for a given mass transfer rate, the heat flux reduction is inversely proportional to the molecular weight of the injected gas. Therefore, for cases in which chemical reactions can be neglected, hydrogen would be the most efficient coolant.

However, gaseous coolants can cause storage problems because of the necessity for large volume cryogenic or high-pressure storage containers. Liquid coolants, as well as reducing storage requirements, offer the advantage of providing the latent heat of vaporization for energy disposal. However, design problems can occur if the liquid is permitted to vaporize within the porous structure. The problems arise due to the abrupt increase in pressure gradient as the coolant changes phase. If the structure has three-dimensional flow characteristics, the coolant flow will tend to divert around the location where it is vaporizing, and a hot spot may develop.

Attempts have been made to rank potential coolants on the basis of their thermal efficiency, volume and weight requirements, and compatibility with structural materials. Many fluids have been considered; however, for reentry applications the following ranking probably represents the acceptable coolants:

1. H_2O 3. CF_4
2. NH_3 4. CO_2

c. Design Procedures. Thermal design of a transpiration-cooled reentry system is most conveniently and thoroughly accomplished with the aid of a computer. However, considerable insight into systems requirements, and even a preliminary design can be obtained by performing several hand calculations. The following calculational procedure can be performed utilizing equations that have been established previously in this section.

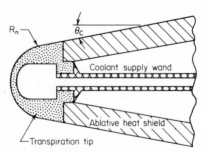

Fig. 25. Typical sphere-cone transpiration cooling system.

A typical sphere/cone geometry is illustrated in Fig. 25. The pertinent geometric variables are the nose cap radius and the cone half angle. It is assumed that these parameters have been fixed on the basis of aerodynamic performance considerations.

Figure 26 presents the altitude and velocity history for a typical ballistic reentry trajectory with severe enough heating in the nose region to require transpiration cooling. It is necessary to utilize the trajectory altitude and velocity history and the nose tip geometry to generate cold wall heating data that is used in the transpiration

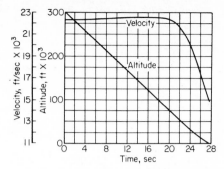

Fig. 26. Typical ballistic vehicle trajectory.

analysis. This heating data can be obtained utilizing the methods outlined in Sec. A. However, the most accurate estimates are normally made utilizing computerized analyses that account for shock curvature, variations in boundary-layer properties, etc.

The aerodynamic heating analysis supplies the local cold wall heat flux, heat transfer coefficient, and recovery enthalpy history that is required to predict the thermally ideal surface cooling flux necessary to maintain the surface of the transpiration-cooled tip at the desired temperature. If the transpiration tip design can accommodate a coolant phase change within the porous structure, a surface temperature near the yield point of the material can be used. This maximizes coolant efficiency by reducing the convective heat flux at the surface and by increasing the amount of heat radiated from the surface. In the event that phase change cannot be tolerated, or a conservative estimate is desired, the surface temperature can be assumed to be at the surface saturation temperature or at some low value like $212°F$.

Thermally Ideal Coolant Requirement. The surface convective heat flux q_c defined in Eq. (40) is equal to the sum of the heat flux q_{cool} required to raise the coolant to the specified surface temperature and the heat flux q_{rad} radiated from the surface at the specified surface temperature. Therefore

$$q_c = q_{cool} + q_{rad} = \left(\frac{N_c}{N_{c_0}}\right) q_{c_0} = \dot{m}\Delta H_{cool} + \epsilon\sigma(T_s^4 - T_{sink}^4) \tag{43}$$

where q_{c_0} is the cold wall heat flux and ΔH_{cool} is the coolant enthalpy change from the storage condition to the vapor phase at the specified surface temperature. q_{c_0}, N_{c_0}, H_r, and H_s are obtained from the aero heating analysis, and ΔH_{cool} and the radiation term can be evaluated from the specified surface temperature. Additionally, the specific heat ratio $C_{p_{cool}}/C_{p_{air}}$ appearing in Eqs. (41) and (42) may be evaluated.

The terms that have not been evaluated are the heat transfer coefficient ratio N_c/N_{c_0} and the thermally ideal coolant flux $\dot{m}$. However, by utilizing Eqs. (41) or (42) in conjunction with Eq. (43) the thermally ideal coolant flux can be determined.

For laminar flow an analytic solution for $\dot{m}$ exists because of the quadratic form of Eq. (41). Equation (43) can be rearranged to yield N_c/N_{c_0}

$$\frac{N_c}{N_{c_0}} = \frac{\dot{m}\Delta H_{cool} + q_{rad}}{q_{c_0}} \tag{44}$$

Equation (44) can be substituted into Eq. (41) and rearranged to yield the following quadratic relationship for the ideal coolant flux

$$\dot{m} = \frac{N_{c_0}}{K}\left[4.25\,K + 0.25\,\Delta H_{cool} - \sqrt{(4.25\,K + 0.25\,\Delta H_{cool})^2 - \frac{0.5K}{N_{c_0}}(q_{c_0} - q_{rad})}\right] \quad (45)$$

where

$$K = \frac{0.04\,q_{c_0}}{N_{c_0}}\left(\frac{C_{p_{cool}}}{C_{p_{air}}}\right)^{0.4}$$

For turbulent flow it is necessary to iterate between Eqs. (42) and (44) until $\dot{m}$ has converged. The recommended iteration procedure is to assume $\dot{m}/N_{c_0} = 1.0$ and thereby evaluate N_c/N_{c_0} from Eq. (42). This ratio can be utilized in Eq. (9) to obtain a value for $\dot{m}$ which can be applied to Eq. (42) for a new evaluation of N_c/N_{c_0}. Five or six iterations of this type are usually sufficient to obtain convergence.

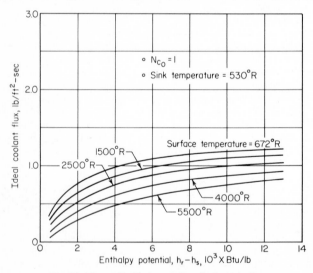

Fig. 27. Thermally ideal water flux for laminar flow.

The ideal water coolant flux is shown graphically in Figs. 27 and 28 for laminar and turbulent flow respectively. The results are based on a value of $N_{c_0} = 1.0$. However, neglecting the surface radiation component q_{rad}, the ratio $\dot{m}/N_{c_0}$ becomes independent of N_{c_0} and the ordinate $\dot{m}$ in Figs. 27 and 28 can be regarded as the ratio $\dot{m}/N_{c_0}$.

Once the ideal coolant flux distribution has been determined over the surface of the transpiration-cooled tip, this distribution can be numerically integrated to give the required coolant flow rate. After evaluating the flow rate at several time points through the trajectory, the flow rate over the trajectory can be numerically integrated with respect to time to yield the total amount of coolant required to cool the tip.

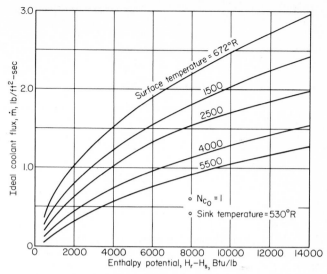

Fig. 28. Thermally ideal water flux for turbulent flow.

Actual Coolant Requirement. Because it is essentially impossible to design a transpiration-cooled nose tip that will yield an actual coolant distribution identical to the thermally ideal, the usual compromise is to design a nose tip that will yield an actual distribution just tangent to the ideal at some point or points on the surface but exceeds it everywhere else. This of course results in some unavoidable coolant wastage required to provide safety margins in operation. In addition, because of uncertainty in the analytical relationships utilized to obtain the ideal coolant flux, current design practice indicates a margin (of about 50 percent) to be applied to the ideal coolant distribution. Thus, the minimum difference between the actual and ideal coolant flux at any point on the transpiration-cooled tip is equal to this margin.

The procedure necessary to obtain the hydraulic design of the nose tip is based on utilizing Eqs. (36) to (39 to predict the coolant pressure and temperature distribution inside the porous structure such that the desired actual coolant distribution is obtained on the nose tip surface.

Generally speaking, if the hydraulic design of the nose tip is one-dimensional, i.e., discrete continuous coolant passages from the interior to the exterior surface, as opposed to three-dimensional, i.e., characterized by sintered metal powders, then more control over the shaping of the actual coolant distribution is possible.

Figures 29 and 30 present typical actual coolant distributions obtainable with one-dimensional and three-dimensional hydraulic designs, respectively. Shown for comparison is the ideal coolant distribution with the margins utilized for calculations.

For making preliminary design estimates or for conducting systems studies it is sometimes desirable to make an estimate of the actual coolant requirement before the tip is designed. This can be done by utilizing multiplying factors that are applied to the ideal coolant requirement. Experience has shown that, for all turbulent flow, one-dimensional designs require about 20 percent more flow than the ideal flow plus margin. For three-dimensional designs, about 50 percent more is required on the spherical cap and about 10 percent more on the conical skirt.

6. Testing

Testing of high-temperature thermal protection systems is required to assist in development and selection of the heat shielding materials or media, verification of

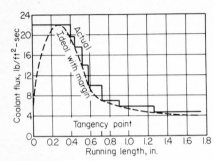

Fig. 29. Typical coolant distribution for a one-dimensional design.

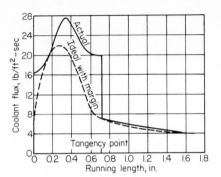

Fig. 30. Typical coolant distribution for a three-dimensional design.

theoretical physio-mathematical models and design procedures, and component assembly qualification. Depending on the system under consideration, various degrees of environment simulation and scale modeling are required to produce the desirable material response simulation. In some cases, nothing short of full scale flight test will satisfy the design requirements. In most cases, it is impossible to simulate the reentry environment in the ground laboratory over the complete flight path history of the reentering body and therefore it is necessary to establish the critical period and degree of entry simulation required to assure proper material selection.

Considering the above difficulties in simulation any test program depends heavily on analytical support. Thus, a mathematical model describing the physical phenomena must be utilized to determine the sensitivity of a given HTTPS to environmental parameters. The result of such parametric sensitivity study will determine the selection of proper test facility, depending on its ability to simulate the critical environment parameter established by the study. The parametric study will also provide the design criterion, while the test results will yield information for verification of the theoretical model to be used for the final design calculations.

a. Heat Sinks. The data presented by Hurwicz, et al. [23] indicated that full environmental simulation was not required for TPS exhibiting no mass transfer. For this class of TPS, if an arc-jet were selected for the test facility, the heat flux levels and enthalpy should be selected to assure that the surface temperature of the material will reach the values expected during flight. The test duration should be sufficient to produce significant temperature response at the rear boundary while the test atmosphere required to observe surface deterioration effects (if any) should also be provided.

During the tests, the surface temperature and emissivity should be measured and the surface behavior observed. The ΔT_b observed from tests of various materials (equal weight samples) will provide a figure of merit for material ranking. The observation of the surface appearance will provide some information on surface resistance to the environment and a more complete temperature traverse will be helpful for internal heat transfer mechanism evaluation.

Once the environmental effects have been established, consideration should be given to quartz-lamp type testing which allows large test panels and preprogramming of a specified temperature profile. This type of testing also allows the determination of the adequacy of other elements of the TPS such as additional insulation, attachment points, and multidimensional effects and falls into the category of component-assembly testing.

b. Radiative Systems. Radiative TPS are normally used to protect against low heating rate environments that can be fairly closely approximated in the laboratory experiment. This is fortunate indeed because the high-redundancy structures used are extremely difficult to analyze on paper with the required degree of accuracy. The main test objectives for radiative TPS are: (1) materials selection and development, and (2) full scale component and assembly qualification. The verification of theoretical models is of less significance here than for ablative system.

1. Simulation Requirements for Materials Selection:
   Exposure time is the critical element in the simulation of the surface reactions and erosion of the radiative system. It is therefore extremely important to approximate the actual (1–15 minutes) mission times during the test. On the other hand for a blunt body the actual Mach number simulation is not critical. It suffices to test the specimen in a supersonic stream that will result in reasonably simulated pressure and velocity gradients on the test specimen front face. A free-stream Mach number of 2 to 3 is often considered satisfactory. It is desirable to simulate the atmospheric conditions encountered in flight. This generally requires a free-stream density corresponding to that found between 100,000 and 300,000 feet altitude. However, since materials performance is relatively independent of environment (except for oxidation), it is possible to determine material performance indices based on material properties and simulation of surface temperature conditions.
2. Simulation Requirements for Component and Assembly Qualification:
   The significant parameters that must be duplicated in component and assembly qualification for radiatively cooled structures are surface temperature and stress. This can be accomplished by heating with quartz lamps and loading concurrently or separately with conventional structural testing equipment.

c. Ablative and Transpiration-cooling Systems. Ablation testing is the method by which the analytical models are verified, many material properties identified, and candidate heat shield material are screened and selected. The simulation requirements for ablative materials and facility performance maps have been presented in Sec. 16. The requirements for transpiration-cooling systems are of the same nature, although additional testing of expulsion system is required. Due to the complex nature of the ablation phenomena coupled with facility limitations, the screening and model verification procedure often requires many test configurations and facilities. Figures 31 through 37 present test models, their usual purpose, as well as the advantages and disadvantages of the depicted type of testing [62].

Test purpose: Low-cost screening tests. Test provides laminar effective heats of ablation at low heat flux, low shear conditions over wide range of enthalpy.

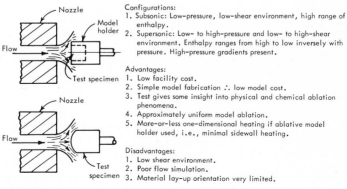

Configurations:
1. Subsonic: Low-pressure, low-shear environment, high range of enthalpy.
2. Supersonic: Low- to high-pressure and low- to high-shear environment. Enthalpy ranges from high to low inversely with pressure. High-pressure gradients present.

Advantages:
1. Low facility cost.
2. Simple model fabrication ∴ low model cost.
3. Test gives some insight into physical and chemical ablation phenomena.
4. Approximately uniform model ablation.
5. More-or-less one-dimensional heating if ablative model holder used, i.e., minimal sidewall heating.

Disadvantages:
1. Low shear environment.
2. Poor flow simulation.
3. Material lay-up orientation very limited.

Fig. 31. Splash configurations.

Test purpose: Screening test for heat shield specimens with realistic material lay-up. Test provides relative performance of materials under high heat flux, high shear conditions in a laminar or turbulent flow environment.

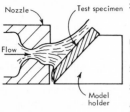

Advantages:
1. Model construction is simple and low cost.
2. High aerodynamic shear on model surface.
3. Test gives insight to physical and chemical ablation phenomena under high shear conditions. Surface recession phenomena can be clearly seen from motion picture observations.

Disadvantages:
1. Significant pressure and heat flux gradients exist on model surface; therefore, flow field simulation is poor.
2. Model erodes in a three-dimensional dish giving rise to complex flow field, further aggravating flow field simulation.
3. Test requires a high-pressure facility; therefore, facility costs are moderately high.
4. Difficult to analyze data due to flow field complexities.
5. Cannot determine time-dependent recession measurements from film observations.
6. Difficult to predict whether flow will be laminar or turbulent.

Fig. 32. Inclined flat plate.

In addition to the pitfalls pointed out in the above figures, care should be exercised in the preparation of test samples to ensure that the particular formulation being tested is identical to that used in the analyses.

C. PROPERTIES OF TYPICAL TPS MATERIALS

Materials utilized in high-temperature thermal protection systems must be capable of withstanding the induced and natural environments without compromising the structural integrity of the aerospace vehicle they are intended to protect. As noted on

Test purpose: Same as for inclined flat plate configuration.

Advantages:
1. Same as for inclined flat plate test.
2. No reversed flow on model surface. All flow over surface has downstream vector component though flow is still three dimensional.
3. Models tend to ablate more two-dimensional than inclined flat plate.
4. Oncoming flow over model is constant because water-cooled leading edge retains its shape.

Disadvantages:
1. Model holder is costly because of water-cooled leading edge feature.
2. As the specimen ablates, a discontinuity is created at the holder specimen interface which tends to perturb the flow field over the specimen.
3. Difficult to analyze data due to flow field complexities.
4. Cannot determine time-dependent recession rates from film observations.
5. Facility costs are moderately high as test required a high-pressure facility.
6. Significant pressure and heat flux gradients exist on model surface; therefore, flow field simulation is poor.
7. Lack of transition criteria results in difficulty to predict laminar or turbulent flow.

Fig. 33. Wedge configuration.

Test purpose: Obtain effective heats of ablation under high shear condition.

Advantages:
1. Effects of lay-up can be investigated. Actual heat shield segments or subscale prototype segments can be used for conic section configuration.
2. Possible to obtain turbulent flow on side of model.
3. Can obtain time-dependent recession data from films since surface is convex.

Disadvantages:
1. Models are relatively costly.
2. Ablation is unsymmetrical.
3. Flow-field is complex.
4. Facility costs are moderately high as a high-pressure jet is required.

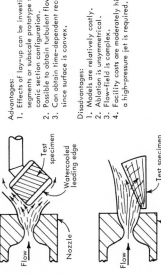

Fig. 34. Cylinder, cone section, and cone at angle of attack.

Test purpose: Laminar and turbulent thermochemical performance data and thermostructural response can be obtained with realistic model pressure distributions.

Advantages:
1. Relatively large models may be tested at high heat flux and pressure allowing use of prototype designs for thermostructural studies.
2. Can test in both laminar and turbulent regimes.
3. By varying gap between model and shroud it is possible to realistically simulate pressure distributions.

Disadvantages:
1. Ablation of model results in a changing flow area between model and shroud which causes boundary conditions to vary during the test.
2. Model fabrication and shroud design are costly.
3. Model surface temperature and photographic observations cannot be made.
4. Test requires rapid stabilization of chamber conditions at test start-up and rapid shutdown capacity since only time average ablation data can be obtained.
5. General test procedure is very complex.

Fig. 35. Shroud configuration.

Test purpose: The test provides thermochemical performance data for laminar and turbulent boundary layer flows over a wide range of test conditions. Overall flow field gradients are small providing a basis for realistic simulation of the heat shield environment. The test may be conducted for subsonic, sonic, and supersonic environments.

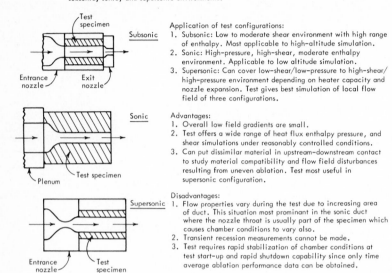

Application of test configurations:

1. Subsonic: Low to moderate shear environment with high range of enthalpy. Most applicable to high-altitude simulation.
2. Sonic: High-pressure, high-shear, moderate enthalpy environment. Applicable to low altitude simulation.
3. Supersonic: Can cover low-shear/low-pressure to high-shear/high-pressure environment depending on heater capacity and nozzle expansion. Test gives best simulation of local flow field of three configurations.

Advantages:

1. Overall low field gradients are small.
2. Test offers a wide range of heat flux enthalpy pressure, and shear simulations under reasonably controlled conditions.
3. Can put dissimilar material in upstream–downstream contact to study material compatibility and flow field disturbances resulting from uneven ablation. Test most useful in supersonic configuration.

Disadvantages:

1. Flow properties vary during the test due to increasing area of duct. This situation most prominent in the sonic duct where the nozzle throat is usually part of the specimen which causes chamber conditions to vary also.
2. Transient recession measurements cannot be made.
3. Test requires rapid stabilization of chamber conditions at test start-up and rapid shutdown capability since only time average ablation performance data can be obtained.

Fig. 36. Pipe configurations.

Test purpose: Test provides thermochemical performance data for laminar and turbulent boundary layer flows over a wide range of test conditions. Overall flow field gradients are small, providing a basis for realistic simulation of the heat shield environment. The test may be conducted for subsonic, sonic, and supersonic environments.

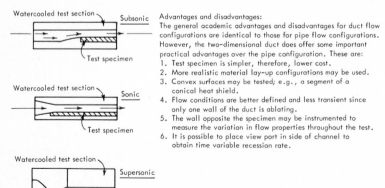

Advantages and disadvantages:

The general academic advantages and disadvantages for duct flow configurations are identical to those for pipe flow configurations. However, the two-dimensional duct does offer some important practical advantages over the pipe configuration. These are:

1. Test specimen is simpler, therefore, lower cost.
2. More realistic material lay-up configurations may be used.
3. Convex surfaces may be tested; e.g., a segment of a conical heat shield.
4. Flow conditions are better defined and less transient since only one wall of the duct is ablating.
5. The wall opposite the specimen may be instrumented to measure the variation in flow properties throughout the test.
6. It is possible to place view port in side of channel to obtain time variable recession rate.

Fig. 37. Rectangular duct.

many previous occasions, the need for and the knowledge of material properties and their variation or degradation, particularly with the change in temperature, is critical in generation of a safe design and thus in achievement of the above goal. Therefore, it was deemed advisable to provide the reader with the information on the type of properties and characteristics of materials he will find useful and necessary. However, the materials selected for description here are only representative of those used in the systems of interest, and their list by no means is all inclusive. Complete property compilation would be beyond the scope of this handbook. Other materials, their properties and characteristics, are described in the vast volumes of references (e.g., [65]), etc. The data on materials of interest are grouped by the protective systems where they have found in the past their greatest utility. They are presented in the following tables and figures:

Systems	Tables	Figures
Heat sinks	4	38–41
Radiation cooled	5	42–46
Ablators*	6–11	46–59
Insulators	12	

*The properties presented for the ablation materials were obtained from Arne [58] and Johnson and Arne [64].

It should be noted that some of the materials described (e.g., graphite) find use and function in more than one mode of HTTPS.

1. Heat Sinks

High thermal diffusivities and/or conductivities and specific heats accompanied by moderate (1500–2300°F) melting temperatures are typical of materials used for heat sink applications. Copper, beryllium, beryllia, and stainless steel are representative of this class and their properties are given in Table 4 and Figs. 38 to 41. Graphites and

TABLE 4. Typical Heat Sink Materials

Property	Material			
	Copper	Beryllium	Beryllia	17-4 pH stainless steel
Density (lb/ft³)	555	116	187	484
Conductivity (Btu/ft-hr-°F)	Fig. 38	Fig. 39	Fig. 40	Fig. 41
Specific heat (Btu/lb)				
Melting temperature (°F)	1950	2300	1450	2250
Emissivity	0.2	0.5	0.45	0.25
Heat of fusion (Btu/lb)	75	470		120
Coefficient of thermal expansion (10⁻⁶ in./in./°F)	14	7	3.3–3.9	8.9
Modulus of elasticity (10⁶ lb/in.²)	18	42	42.8	28
Tensile strength (10³ lb/in.²)	40	20–40	13.8	80

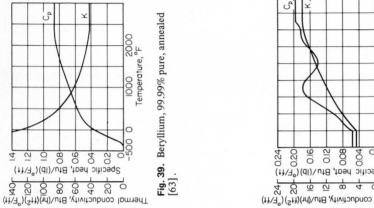

Fig. 39. Beryllium, 99.99% pure, annealed [63].

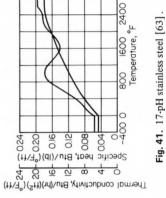

Fig. 41. 17-pH stainless steel [63].

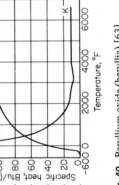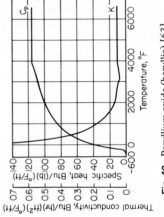

Fig. 38. Copper, 99.999% pure, annealed [63].

Fig. 40. Beryllium oxide (beryllia) [63].

carbonaceous materials have also been used for heat sinks but their use is more typical in ablative systems (Sec. C.3).

2. Radiation-cooled Systems

High surface temperatures (> 2400°F) accompanied by relatively low thermal conductivities are typical of materials used for radiative systems. Molybdenum, tungsten, Al and Zr oxides, and super-alloys fall into this general category and are characterized in Table 5 and Figs. 42 to 46. Again, graphites (Sec. C.3 below) may be used as an ablator.

3. Ablators

The data on ablators are relatively more scarce in the literature, and a more inclusive body of information is given here. It covers such commonly used materials as

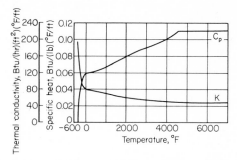

Fig. 42. Molybdenum, 99.99% pure, annealed [63].

TABLE 5. Typical Radiation-cooled System Materials

Property	Material				
	Molybdenum	Tungsten	Alumina	Zirconia	Haynes 25
Density (lb/ft^3)	640	1204	250	363	570
Conductivity (Btu/ft-hr-°F)	Fig. 42	Fig. 43	Fig. 44	Fig. 45	Fig. 46
Specific heat (Btu/lb)					
Melting temperature (°F)	4700 3100*	6100 3400*	3700	4900	2450
Emissivity	0.2 0.75†	0.2 0.75†	0.5	0.4	0.85
Heat of fusion (Btu/lb)	90	70			110
Modulus of elasticity (10^6 lb/in.^{2})	42	50	54	24.8	
Tensile strength (10^3 lb/in.^{2})	147	70–300	37	17.9	100
Coefficient of thermal expansion (10^{-6} in./in./°F)	3–6	2.2–2.9	3.3–3.9	5.0	6.8–9.4

*Oxidation resistant coating limits.
†With oxidation resistant coating.

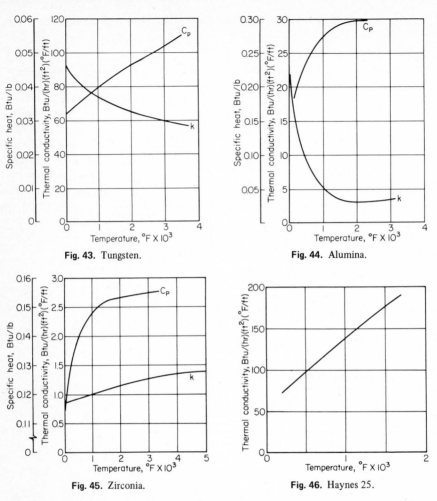

Fig. 43. Tungsten.

Fig. 44. Alumina.

Fig. 45. Zirconia.

Fig. 46. Haynes 25.

phenolic composites of refrasil and carbon, typical graphite, quartz (silica), and porous tungsten.

Phenolic refrasil is a woven, high silica fabric (99.4 percent purity) preimpregnated with 24 percent phenolic resin by weight. The density of the phenolic refrasil is 105 lb/ft^3, with the minimum average values of mechanical properties given in Table 6.

Specific heats of the virgin and charred phenolic refrasil are shown in Fig. 47 and thermal conductivities for end-grain phenolic refrasil are shown in Fig. 48. The

TABLE 6. Phenolic Refrasil

	Temperature	
	77°F	500°F
1. Flexural strength (lb/in.^{2})	18,000	14,000
2. Ultimate tensile strength	10,000	7,000
3. Tensile modulus (lb/in.^{2}, min.)	2.3×10^6	1.3×10^6
4. Edgewise compressive strength (lb/in.^{2})	10,000	12,000

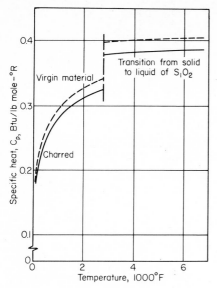

Fig. 47. Specific heat of phenolic refrasil.

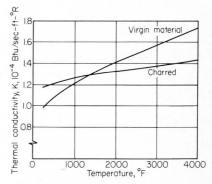

Fig. 48. Thermal conductivity of phenolic refrasil.

specific heats have an expected uncertainty of ±10 percent and the thermal conductivities ±20 percent. The value of the surface emissivity is 0.48.

The ablative characteristics not being "true" properties are not given here. The need for various characteristics or properties of ablators depends on the ablation model or mechanism used (see Sec. 16). Some data of this nature were shown in Ref. 66.

Quartz is a pure silica (SiO_2) ceramic. Typical properties which should be representative are given in Table 7.

TABLE 7. Quartz

Property	
1. Fusing point (°C)	about 1,750
2. Specific gravity	2.0–2.3
3. Hardness, Mohs' scale	4.9
4. Coefficient thermal expansion average (cm/cm-°C)	$0.54 - 0.55 \times 10^{-6}$
5. Thermal conductivity (Btu/sec-ft-°R)	2.22×10^{-4}
6. Compressive strength (lb/in.²)	190,000
7. Tensile strength (lb/in.²)	7,000
8. Young's modulus (lb/in.²)	10.4×10^6

The specific heat of quartz is given as a function of temperature for both crystalline and liquid SiO_2 in Fig. 49 for which the estimated uncertainty is ±10 percent. The thermal conductivity is shown in Fig. 50 and has an estimated uncertainty of ±20 percent. The emissivity of opaque quartz is given by the relationship

$$\epsilon = 0.7\left(1 - \frac{10 - 2}{n\Delta}\right)$$

where $n = 38.5 - 0.003 \, (Tw - 560)$ [16], Tw is in °R, and Δ is the liquid layer

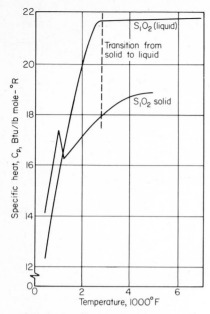

Fig. 49. Specific heat of opaque quartz.

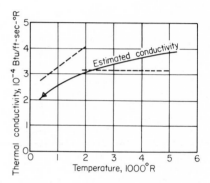

Fig. 50. Thermal conductivity of opaque quartz.

thickness (to be determined by an ablation computer program). The viscosity of quartz is given by

$$\mu = 0.0672 \, \exp\left(\frac{124,000}{T(^\circ R)} - 20\right)$$

Other ablative characteristics of silica may be found [58] and again the reader is cautioned to use properly matched ablation model and ablation characteristics.

Graphite G grade is a high-density, fine-grained, low-anisotropic extruded graphite which is manufactured by the same process used for synthetic graphites and subsequently impregnated and graphitized repeatedly (density equals 119 lb/ft^3). Room-temperature data are given in Table 8. Specific heat versus temperature is shown in Fig. 51 and thermal conductivity versus temperature is shown in Fig. 52. The estimated uncertainty of the specific heat is ±5 percent and of thermal conductivity ±10 percent. No information is available on the emissivity of Graphite G grade, but from emissivity data on ATJ graphite and sandpaper-finished graphite, the emissivity is estimated to be 0.9 above 4000°R surface temperature.

TABLE 8. Graphite G Grade

Typical properties (room temperature)	
1. Tensile strength (lb/in.^{2})	2,900
2. Compressive strength (lb/in.^{2})	11,000
3. Transverse strength (lb/in.^{2})	5,000
4. Modulus of elasticity (lb/in.^{2} × 10^5)	18
5. Thermal expansion ((in./in.-°F) × 10^{-7})	9
6. Thermal conductivity (Btu/ft-hr-°F)	92

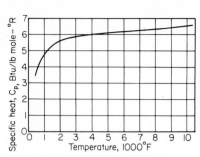

Fig. 51. Specific heat of graphite.

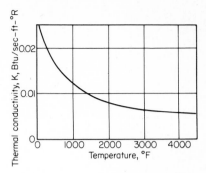

Fig. 52. Thermal conductivity of graphite.

Since various theories concerning graphite ablation exist [38, 42, 43] the reader is again advised to obtain other ablative characteristics as required for the model and the material he is using (e.g., [58]).

Phenolic carbon is composed of a woven carbon cloth impregnated with phenolic resin. Vendor-supplied data* are shown in Table 9. The average density of the material is 82.4 lb/ft³. The specific heats and thermal conductivities of the composite solid (plastic) and char supplied are shown in Figs. 53 and 54, respectively. No information is available on the surface emissivity, but it probably has a value of 0.9 to 0.95 above a surface temperature of 4000°R. Ablative characteristics usage was discussed above in relation to other ablators.

TABLE 9. Phenolic Carbon

Property	Average
1. Tensile strength (lb/in.²) warp	101
2. Tensile strength (lb/in.²) fill	71
3. Oz/yd²	7.54
4. Thickness (in.)	0.0176
5. Thread count, warp	26.7
6. Thread count, fill	23.6
7. Carbon content	99.97%
8. Surface area	0.646 m²/gm

Boron nitride (HBR grade) is a low-moisture absorbing, improved thermal-shock-resistant grade of hot-pressed boron nitride. According to the vendor, unlike most other boron nitrides, this material does not absorb water. In addition, its coefficient of thermal expansion is lower than other boron nitrides. It contains approximately five percent B_2O_3 by weight. The boron nitride supplied by National Carbon Corp. has a density of 133 lb/ft³. The vendor-supplied information is shown in Table 10. Specific heat and thermal conductivity for boron nitride (with estimated uncertainties of ±10 percent) are shown in Figs. 55 and 56 respectively. The surface emissivity of boron nitride with temperature is given in Ref. 64.

Tungsten used here is described as high-purity sintered and forged matrix with a density of 94 percent of theoretical, or approximately 1,200 lb/ft³. The specific heat of tungsten is shown in Fig. 57 and the thermal conductivity of tungsten (95 percent of theoretical density) is shown in Fig. 58. The uncertainties in the specific heat, thermal conductivity, and emissivity are estimated to be ±10 percent.

*As reported by Dr. Adam Sporzynski, Thompson-Ramo Wooldridge Corp.

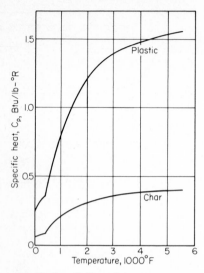

Fig. 53. Specific heat of phenolic carbon.

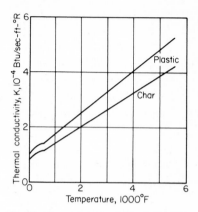

Fig. 54. Thermal conductivity of phenolic carbon.

TABLE 10. Boron Nitride

Physical properties	
Flexural strength, lb/in.^{2} (room temperature)	
Direction A	16,000
Direction B	14,000
Tensile, lb/in.^{2} (room temperature)	
Direction A	6,500
Direction B	8,000
Compressive, lb/in.^{2} (room temperature)	
Direction A	27,500
Direction B	26,500
Young's modulus, lb/in.^{2} (room temperature)	
Direction A	11×10^6
Direction B	9×10^6

Direction A—force applied perpendicular to hot-pressing direction
Direction B—force applied parallel to hot-pressing direction

Coefficient of thermal expansion	0–1200 °C
Perpendicular expansion	$2 \times 10^{-6}/°C$
Parallel expansion	$1 \times 10^{-6}/°C$
Thermal conductivity, Btu/ft-hr-°F	20 (room temperature)
	18 (800–1500°C)
Hardness, Mohs' scale	2

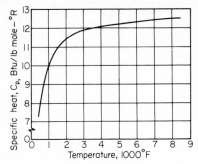

Fig. 55. Specific heat of boron nitride.

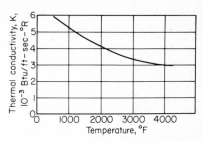

Fig. 56. Thermal conductivity of boron nitride.

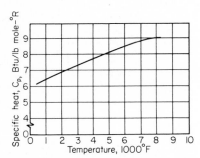

Fig. 57. Specific heat of tungsten.

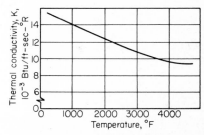

Fig. 58. Thermal conductivity of tungsten.

TABLE 11. Insulators

Material	Density (lb/ft^2)	Property		
		Conductivity (Btu/hr·ft^2·°F)	Specific heat (Btu/lb)	Temperature limit (°F)
MIN-K	20.0	0.015 (0°F) 0.03 (1000°) 0.044 (2000°)	0.18 (0°F) 0.255 (1000°) 0.28 (2000°)	2000
Multilayer foils	0.1	0.001 (1600°F)	0.2	2200
Powders	17.2	0.074 (1400°F)	0.22	2500
Pyrocream foam	14.4	0.125	0.19	2250
Aluminum silicate fibers	6.0	0.25 (1300°F)	0.25	2500
Roamsil	11.0	0.154 (900°F)	0.19	2000
Rokide Z	324.0	0.85 (0°F) 1.02 (1000) 1.15 (2000) 1.26 (3000) 1.35 (4000) 1.39 (4500)	0.115 (0°F) 0.149 (1000) 0.170 (2000) 0.170 (2500)	4500

4. Insulators

Since it is desirable from the point of view of minimizing the weight of HTTPS to operate the backface of the heat shield at highest possible temperatures, but very few structural materials or bonding agents retain sufficient integrity at such temperatures, a high-temperature insulation interface is often provided between the heat shield and structure. The properties of some insulators used for this purpose are given in Table 11.

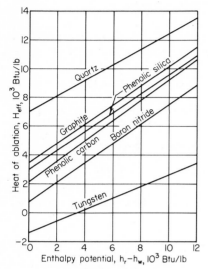

Fig. 59. Laminar heats of ablation.   **Fig. 60.** Turbulent heats of ablation.

ACKNOWLEDGMENTS

The authors wish to gratefully acknowledge the contributions of G. P. Johnson, T. G. Lee, J. R. Schuster, and D. H. Smith, all of McDonnell Douglas Astronautics Co.—West, Huntington Beach, Calif., and Mr. M. M. Sherman of the Space and Reentry Systems Division of Philco-Ford Corp.

REFERENCES

1. U.S. Standard Atmosphere, 1962, U.S. Government Printing Office, Dec. 1962.
2. U.S. Standard Atmosphere Supplements, 1966, U.S. Government Printing Office, 1966.
3. Models of Mars Atmosphere, *NASA* SP-8010, 1967.
4. R. A. Schiffer and A. J. Beck, Models of Venus Atmosphere, *NASA* SP-8011, Dec. 1968.
5. W. E. Moeckel and K. C. Weston, Composition and Thermodynamic Properties of Air in Chemical Equilibrium, *NACA* TN-4265, April 1958.
6. C. F. Hansen, Approximations for the Thermodynamic and Tranport Properties of High Temperature Air, *NASA TR* R-50, 1959.
7. T. E. Horton and W. A. Menard, A Program for Computing Shock-Tube Gasdynamic Properties, *JPL Tech. Rept.* 32-1350, Jan. 1969.
8. S. Bennett, J. M. Yos, C. F. Knopp, J. Morris, and W. L. Bade, Theoretical and Experimental Studies of High-Temperature Gas Transport Properties, Final Report, *AVCO/RAD*-TR-65-7, May 1965.
9. M. Thomas, The High Temperature Transport Properties of Carbon Dioxide, Douglas Aircraft Co., *Rept.* SM-37790, 1960.

10. Ali H. Nayfeh, An Evaluation of Design Analyses Techniques for High Performance Ballistic Vehicle Graphite Nose Tips: Appendix B–Inviscid Flow Fields, *AFML*-TR-69-73, vol. 2, Jan. 1970.
11. Ames Research Staff, Equations, Tables, and Charts for Compressible Flow, *NACA Rept.* 1135, 1953.
12. Saul Feldman, Hypersonic Gas Dynamic Charts for Equilibrium Air, *AVCO RR* 40, Jan. 1957.
13. G. G. Chernyi, "Introduction to Hypersonic Flow," Academic Press, Inc., 1961.
14. E. R. G. Eckert, Engineering Relations for Friction and Heat Transfer to Surfaces in High Velocity Flow, *J. Aeronaut. Sci.,* **22**:585 (1955).
15. J. A. Fay and F. R. Riddell, Theory of Stagnation Point Heat Transfer in Dissociated Air, *J. Aeronaut. Sci.,* **25**:73 (1958).
16. T. Y. Li and R. E. Gieger, Stagnation Point of a Blunt Body in Hypersonic Flow, *J. Aeronaut. Sci.,* **24**:25 (1957).
17. R. H. Weatherford and S. Sayano, Techniques for Rapid Aerodynamic Heat Transfer Calculation, *Douglas Rept.* SM-45932, May 1965.
18. A. C. Thomas, A. Perlbachs, and A. L. Nagel, Advanced Re-Entry Systems Heat Transfer Manual for Hypersonic Flight, *AFFDL*-TR-65-195, Oct. 1966.
19. F. L. Guard, H. D. Schultz, C. G. Forsythe, and Z. Popinski, Compilation of Structural Inflight Heating Data Applicable to Large Launch Vehicles, vol. 1, Theoretical Analysis, *Lockheed Rept.* LR-17562, Feb. 1964.
20. H. Hurwicz, W. C. Broding, and A. J. Hanawalt, An Approach to Re-Entry Thermal Protection System Development, Bumble-Bee Conf., St. Louis, Mo., 1963.
21. M. M. Sherman, NASA Space Vehicle Design Criteria–Entry Thermal Protection, *NASA* SP-8014, Aug. 1968.
22. M. M. Sherman and T. Nakamura, Flight Test Measurements of Boundary Layer Transition on a Non-Ablating 22° Cone, *AIAA Paper* 68-1152, AIAA Entry Vehicle System and Technology Meeting, Dec. 1968.
23. H. Hurwicz et al., Thermal Protection Systems, ML-TDR-64-82, Jan. 1965.
24. H. S. Carslaw and J. C. Jaeger, "Conduction of Heat in Solids," Oxford University Press, New York, 1959.
25. J. D. Brown and F. A. Shukis, An Approximate Method for Design of Thermal Protection Systems, IAS 30th Annual Meeting, New York, N.Y., Jan. 1962.
26. J. V. Beck, A Thermal Study of Composite Heat Sinks, *ASME Paper* 60-SA-22, 1960.
27. J. O. Collins and S. Speil, Ablation, Heat Sink and Radiation: The Materials That Make Them Work, *Materials in Design Engineering,* **53**:3 (1961).
28. R. A. Hirsch, Development of Techniques and Fabrication of a Structural Model for Research on Structures for Hypersonic Aircraft, The Martin Company, NASA Contract NAS 1-3174 (to be published).
29. R. A. Pride, D. M. Royster, and B. F. Helms, Design, Tests, and Analysis of a Hot Structure for Lifting Re-Entry Vehicles, *NASA* TN-D-2186, April 1964.
30. Manufacturing Methods for Insulated and Cooled Double-Wall Structures, *Bell Aerosystems Co.,* ASD-TR-61-7-799, vols. 1 and 2, May 1961.
31. Thermosorb for Spacecraft Thermal Protection, *Vought-Astronautics Rept.* AST/EOR-13192.
32. W. F. Barrett, A. M. Norton, and J. W. McCown, Design and Testing of a Hot Redundant Structures Concept for a Hypersonic Flight Vehicle, AIAA 6th Structures and Materials Conference, April 1965.
33. *Proc. 1962 X-20A (Dyna-Soar) Symposium–Structure and Materials,* ASD-TDR-63-148, vol. 3 (Confidential).
34. T. P. Brooks, Structural Materials Requirements for Manned Space Vehicles, *SAE Paper* 910A, Oct. 1964.
35. M. L. Minges, Ablation Phenomenology–A Review, High Temperature–High Pressures, vol. 1, no. 6, 1969.
36. Eric Baer (ed.), "Engineering Design for Plastics," pp. 815–868, Reinhold Publishing Co., New York, 1964.
37. "Comparative Studies of Conceptual Design and Qualification Procedures for a Mars Probe/Lander, vol. 5, book 2," Space Systems Div., AVCO Corp., AVSSD-006-66-RR, 11 May 1966.
38. K. M. Kratsch et al., Graphite Ablation in High Pressure Environments, *AIAA Paper* 68-1153, Dec. 1968.

39. D. L. Schmidt, Ablative Polymers for Hypersonic Atmospheric Flight, AFML-TR-66-78, May 1966.
40. D. L. Schmidt, Ablative Plastics and Elastomers in Chemical Propulsion Environments, AFML-TR-65-4, April 1965.
41. N. M. Harrington and D. B. Linde, Structural Analysis of High Performance Graphite Nose Tips, J-49, Philco-Ford Corp., Space and Re-Entry Systems Div., Sept. 1967.
42. S. M. Scala and L. M. Gilbert, The Sublimation of Graphite at Hypersonic Speeds, *AIAA Paper* CP-9, pp. 239–258, AIAA Entry Technology Conf., Williamsburg, Va., Oct. 1964.
43. Eugene P. Bartlett, Analytical and Graphical Prediction of Graphite Ablation Rates and Surface Temperatures During Entry at 25,000 to 45,000 Feet Per Second, FTC-TDR-63-40, March 1964.
44. M. R. Dennison and D. A. Dooley, Combustion in the Laminar Boundary Layer of Chemically Active Sublimating Surfaces, *J. Aeronaut. Sci.,* **25**:271 (1958).
45. M. R. Dennison, The Turbulent Boundary Layer on Chemically Active Ablating Surfaces, *J. Aerospace Sci.,* **28**:471 (1961).
46. J. W. Metzger, M. J. Engle, and N. S. Diaconis, Oxidation and Sublimation of Graphite in Simulated Re-Entry Environments, *AIAA Journal,* **5**:451 (1967).
47. E. Strauss, *Amer. Ceramic Soc. Bull.,* **42**:444 (1963).
48. E. Strauss, The Application of Resin-Impregnated Porous Ceramics to Re-Entry Vehicle Heat Shields, in "Aerodynamically Heated Structures" (P. Glaser, ed.), pp. 7–27, Prentice-Hall, Inc., Englewood Cliffs, N.J., 1962.
49. M. Schwartz and T. Greening, Impregnated Foam Ceramic Insulating Materials, "Insulation-Materials and Processes for Aerospace and Hydrospace Applications," sec. 10, Western Periodicals Co., North Hollywood, Calif., 1964.
50. S. Maloof, *Astronautics,* **6**:36 (1961).
51. J. W. Vogan, Metal-Ceramic Structural Composite Materials, *Boeing Document* D2-12057, Sept. 1961.
52. C. M. Dolan, Study for Development of Elastomeric Thermal Shield Materials, *NASA* CR-186, March 1965.
53. Robert T. Swann, William D. Brewer, and Ronald K. Clark, Effect of Composition and Density on the Ablative Performance of Phenolic-Nylon, 8th Natl. Meeting of the Society of Aerospace Materials and Process Engineers, San Francisco, Calif., May 1965.
54. Paul E. Bauer and Donald L. Kummer, Development and Performance of the Gemini Ablative Heat Shield, *J. Spacecraft and Rockets,* **3**:1495 (1966).
55. J. Meltzer, J. Rossoff, J. I. Slaughter, and J. Sterhardt, Structure and Materials Aspects of the PRIME Flight Test Vehicle, SSD-TR-66-54, *Rept.* TDR-669(6108-27)-4, Dec. 1965.
56. P. J. Schneider, T. A. Dolton, and G. W. Reed, Char-Layer Structural Response in High-Performance Ballistic Re-Entry, *AIAA Paper* 66-424, June 1966.
57. C. J. Katsikas, G. K. Castle, and J. S. Higgins, Ablation Handbook—Entry Materials Data and Design, AFML-TR-66-262, Nov. 1966.
58. C. L. Arne, Ablative Materials Subject to Combustion and Thermal Radiation Phenomena, *Douglas Paper* 1851, Jan. 1964.
59. P. Baronti, T. Baurer, H. Fox, S. Linger, and B. Staros, Transpiration Cooled Reentry Vehicle Feasibility Study (U). General Applied Laboratories, Inc., *Final Tech. Rept.* 565, Contract SD-149, ARPA Order No. 396, Project Code 3790, Nov. 1965 (Secret).
60. B. M. Leadon, The Status of Heat Transfer Control by Mass Transfer for Permanent Surface Structures, *Proc. Conf. on Aerodynamically Heated Structures, July 1961* (Peter E. Glaser, ed.), pp. 171–195, Prentice-Hall, Inc., Englewood Cliffs, N.J., 1962.
61. D. B. Spalding, D. M. Auslander, and T. R. Sundaram, The Calculation of Heat and Mass Transfer Through the Turbulent Boundary Layer on a Flat Plate at High Mach Numbers, With and Without Chemical Reaction, in "Supersonic Flow Chemical Processes and Radiative Transfer" (D. B. Olfe and V. Zakkay, eds.), pp. 211–276, Pergamon Press, Inc., New York, 1964.
62. G. P. Johnson, private communication.
63. D. H. Smith, Thermal Properties of Structural and Insulative Materials, *McDonnell Douglas Rept.* DAC-63237, April 1969.
64. G. P. Johnson and C. L. Arne, private communication.
65. Data books of The Thermophysical Properties Research Center, School of Mechanical Engineering, Purdue University, Lafayette, Ind.

66. H. Hurwicz, Aerothermochemistry Studies in Ablation, *Combustion and Propulsion 5th AGARD Colloquium,* Pergamon Press, Inc., New York, 1963.

NOMENCLATURE

b	material thickness (Eq. (10))
c_p	specific heat
G	as defined in Eq. (27)
H_{eff}	effective heat of ablation
h	enthalpy
h_D	dissociation enthalpy
k	conductivity
k_l	local effective conductivity defined by Eq. (39)
$k_c^{\cdot}$	coolant conductivity
k_s	porous structure conductivity
L	running length
Le	Lewis number
$\dot{m}$	mass ablation rate
M	molecular weight
N_c	enthalpy based heat transfer coefficient
N_{c_0}	enthalpy based heat transfer coefficient without injection
P^s	stagnation pressure
P_∞^s	free-stream pressure
Pr	Prandtl number
$\dot{q}$	heat flux
Q	integrated or total heat
R	radius of universal gas constant
Re	Reynolds number
r	recovery factor
$\dot{s}$	linear recession rate (Eq. (35))
t	time
t_i	initial time
t_B	time required to attain a specified backface temperature
t_f	final time
T	temperature
T_c	critical or limit temperature (Eq. (10))
u	coolant velocity
V	velocity
w	weight of material
w_F	final system weight
w_{ins}	insulation weight (Eq. (33))
x	characteristic length
z	altitude
α	material viscous pressure drop constant (Eq. (36)) or angle of attack
β	material inertial pressure drop constant
ϵ	emissivity
Γ	tortuosity factor, $= k_{porous}/(1 - \gamma)k_s$
θ	cone angle (Eq. (4))
ρ	density
σ	Stefan-Boltzmann constant
μ	coolant viscosity
γ	void fraction of porous structure
μ	viscosity
μ^*	viscosity based on reference conditions

Subscripts

B	backface
c	convective
f	final
i	initial
L	laminar
o	without mass addition
r	recovery
s	stagnation
T	turbulent
w	wall
δ	boundary layer edge

Index